MEDICAL BIOCHEMISTRY

FOURTH EDITION

MEDICAL BIOCHEMISTRY

FOURTH EDITION

N. V. BHAGAVAN

Department of Biochemistry and Biophysics
John A. Burns School of Medicine
University of Hawaii

San Diego San Francisco New York Boston London Sydney Tokyo

BS

Sponsoring Editor	Jeremy Hayhurst
Production Managers	Rebecca Orbegoso and Brenda Johnson
Editorial Coordinator	Nora Donaghy
Promotions Manager	Stephanie Stevens
Copyeditor	Janice Stern
Proofreader	Kathy Nida
Preproduction	Supplinc
Composition	TechBooks
Printer	Friesens

Cover photo: © Corbis Corporation/ William Whitehurst, 2001.

This book is printed on acid-free paper. 

Academic Press
A Division of Harcourt, Inc.
525 B Street, Suite 1900, San Diego, California 92101-4495, USA
http://www.academicpress.com

Academic Press
Harcourt Place, 32 Jamestown Road, London NW1 7BY, UK
http://www.academicpress.com

Harcourt/Academic Press
A Division of Harcourt, Inc.
200 Wheeler Road, Burlington, Massachusetts 01803
http://www.harcourt-ap.com

Library of Congress Catalog Card Number: 2001090826

International Standard Book Number: 0-12-095440-0

PRINTED IN CANADA
02 03 04 05 06 FR 9 8 7 6 5 4 3 2

CONTENTS

CHAPTER 23

Nucleic Acid Structure and Properties of DNA

CHAPTER 24

DNA Replication, Repair, and Mutagenesis

CHAPTER 26

Regulation of Gene Expression

CHAPTER 27

Nucleotide Metabolism

CHAPTER 34

Endocrine Metabolism V: Reproductive System

CHAPTER 35

Molecular Immunology

CHAPTER 36

Biochemistry of Hemostasis

CHAPTER 37

Mineral Metabolism

CONTRIBUTORS

Gordon Edlin, Ph.D. (Chapters 23, 24, 25, 26) Department of Biochemistry and Biophysics, John A. Burns School of Medicine, University of Hawaii, Honolulu, Hawaii

Richard J. Guillory, Ph.D. (Chapter 14) Department of Biochemistry and Biophysics, John A. Burns School of Medicine, University of Hawaii, Honolulu, Hawaii

Craig M. Jackson, Ph.D. (Chapters 35, 36) Hemo SAGA Corporation, San Diego, California

Conrad A. Hornick, Ph.D. (Chapters 18, 19, 20) Departments of Physiology and Pathology, Louisiana State University, New Orleans, Louisiana

David A. Lally, Ph.D. (Chapter 21) Department of Physiology, John A. Burns School of Medicine, University of Hawaii, Honolulu, Hawaii

Walter K. Morishige, Ph.D. (Chapters 30, 31, 32, 33, 34) Department of Physiology, John A. Burns School of Medicine, University of Hawaii, Honolulu, Hawaii

James R. Wilson, Ph.D. (Chapter 16) Department of Biology, La-Sierra University, Riverside, California

PREFACE

In keeping with the previous editions, the primary purpose of *Medical Biochemistry, Fourth Edition,* is to present the fundamentals of biochemistry and related materials in a way that is useful to students pursuing medical and other health-related careers. The book was conceived and written with the hope that it would generate interest and enthusiasm among these students, particularly because biochemistry has a crucial role in human health and disease. Since it is assumed that most students in medicine and health-related fields eventually will apply biochemical principles to the art of healing, discussion of the factual information is integrated with frequent use of clinical examples and applications.

A vast number of constituents of the human body—cells, enzymes, hormones, sugars, salts, vitamins, and so forth—vary in normal and abnormal states of human health. Understanding the metabolic and regulatory processes that underlie metabolism is essential to any practice of the healing arts and the relief of human suffering. I have tried to keep this idea foremost.

Progress in biochemistry, molecular biology, cellular biology, endocrinology, and other disciplines has been so rapid and profound in the past 20 years that biochemistry texts obsolesce rapidly. To maximize the usefulness of this book, authors actively involved in research have written several chapters dealing with the most rapidly changing fields.

The overall organization of topics is designed to lead the student logically through the biochemical organization of cells. Emphasis is placed on the structures and functions of the molecular components of cells and on metabolic controls. The text begins with a discussion of water, acids, bases, and buffers, amino acids, proteins, and thermodynamics (Chapters 1–5). This is followed by a detailed discussion of important aspects of enzymology (Chapters 6–8). The broad subject of carbohydrate chemistry (Chapters 9–11, 13, 15, 16) is integrated with chapters that discuss gastrointestinal digestion (Chapter 12), oxidative phosphorylation (Chapter 14), and protein metabolism (Chapter 17). Three chapters on lipids (Chapters 18–20) are integrated with chapters covering muscle systems (Chapter 21) and metabolic homeostasis (Chapter 22).

The principles of molecular biology including nucleic acid chemistry and the regulation of gene expression and protein synthesis are presented in Chapters 23–26. These chapters are followed by ones on nucleotide, hemoglobin, and heme metabolism (Chapters 27–29). The endocrine system and its organs are discussed in Chapters 30–34. Molecular immunology is presented in Chapter 35, and the biochemistry of blood coagulation appears in Chapter 36. The last section discusses mineral and vitamin metabolism and electrolyte balance (Chapters 37–39). Nine appendices contain tables of nutritional values and clinical laboratory measurements that are important in diagnosis.

The importance of human nutrition is emphasized throughout the text and has not been relegated to a single chapter. Likewise, hereditary disorders are discussed throughout, along with other clinical examples that relate the relevant biochemistry to diagnosis and treatment. This book is not small, although conciseness consistent with clarity was a primary goal. Detailed discussions of experiments have, for the most part, been omitted, and discussion of more subtle points has been minimized. References provided at the end of each chapter will lead interested students deeper into particular topics and case studies.

ACKNOWLEDGMENTS

I am grateful to the authors for their contributions and for sharing their knowledge and insights. I also appreciate their cooperation during the lengthy writing and production process. I am indebted to Gordon Edlin, who participated in and supported all aspects of the publication of this book, to Alan P. Goldstein, who reviewed many chapters and helped integrate basic science and clinical medicine, and Chung-eun Ha for spending tireless hours in the preparation of the manuscript during its many developmental stages.

I am especially appreciative to the following individuals, who contributed to this book as reviewers of selected chapters, as writers of subsections, by offering constructive suggestions, or by providing encouragement and support during difficult times: R. A. Dubanoski, J. M. Hardman, K. H. Higa, S. A. A. Honda, K. Harohalli, D. K. Kikuta, W. Lee, A. A. Manoukian, H. F. Mower, D. S. Park, C. E. Petersen, C. N. Rios, R. T. Sakaguchi, C. E. Sugiyama, and L. E. Takenaka.

My thanks also to the following reviewers for their valuable chapter critiques: Dr. James W. Campbell, Department of Biochemistry and Cell Biology, Rice University; Dr. David Daleke, Department of Biochemistry, Indiana University; Dr. Beverly Delidow, School of Medicine, Marshall University; Dr. JoAnne Flynn, School of Medicine, University of Pittsburgh; Dr. Jennifer A. Pietenpol, School of Medicine, Vanderbilt University; Dr. Connie Prosser, Department of Laboratory Medicine, University of Alberta; Dr. Kevin D. Sarge, Department of Biochemistry, University of Kentucky; Dr. Peter B. Smith, School of Medicine, Wake Forest University; Dr. Terry Stoming, Department of Biochemistry and Molecular Biology, Medical College of Georgia; Dr. Francis Vella, International Union of Biochemistry and Molecular Biology; and, Dr. D. Eric Walters, Department of Biochemistry and Molecular Biology, The Chicago Medical School.

Dr. Craig M. Jackson acknowledges the images shown in Chapters 35 and 36 were created using RasMol 2.6, Molecular Graphics Visualization Tool by Roger Sayle, Bio-Molecular Structures Group, Glaxo Research & Development, Greenford, Middlesex, UK. RasMol can be obtained at http://www.umass.edu/microbio/rasmol/. The coordinates were obtained from the Protein Data Bank at Brookhaven National Laboratory, as described in F. C. Bernstein, T. F. Koetzle, G. J. B. Williams, E. F. Meyer, Jr., M. D. Brice, J. R. Rodgers, O. Kennard, T. Shimanouchi, and M. Tasuxni, "The Protein Data Bank: A Computer-Based Archival File for Macromolecular Structures," *J. Mol. Biol.* **112,** 535–542 (1977). Gene information has been obtained from GenBank, as described in C. Burks, M. Cassidy, M. J. Cinkosky, K. E. Cumella, P. Gilna, J. E-D. Hayden, G. M. Keen, T. A. Kelley, M. Kelly, D. Kristofferson, and J. Ryals GenBank. *Nucl. Acids Res.* **19**(Suppl), 2221–2225 (1991).

For the coordinates of the molecules for which structures are shown, see http://www.rcsb.org/.

The author thanks James G. White, M. D., Regents Professor, University of Minnesota, Minneapolis, Minnesota, for the photographs of platelets, and Meir Rigbi,

Professor Emeritus, Hebrew University of Jerusalem, for his many helpful comments on the manuscript.

I also express my thanks to C. Agbayani, J. Gerber, and N. Mead for unfailing assistance in the preparation of the manuscript. Finally, I greatly appreciate the assistance and advice provided to me by the editorial staff of Harcourt/Academic Press, in particular J. R. Hayhurst, N. M. Donaghy, and R. L. Orbegoso.

CHAPTER 1

Water, Acids, Bases, and Buffers

1.1 Properties of Water

Acid and base concentrations in living systems are carefully regulated to maintain conditions compatible with normal life. Biochemical reactions involving acids and bases occur in the body water, whereas buffer systems protect the body from significant variations in the concentrations of acids and bases. This chapter introduces basic concepts of the properties of water, acids, bases, and buffers, and Chapter 39 presents a detailed discussion of both normal and pathological aspects of acid–base metabolism.

Life cannot be sustained without water. Water constitutes 45–73% of total human body weight. It is distributed in intracellular (55%) and extracellular (45%) compartments and provides a continuous solvent phase between body compartments. As the biological solvent, water plays a major role in all aspects of metabolism: absorption, transport, digestion, and excretion of inorganic and organic substances as well as maintenance of body temperature. The unique properties of water are due to its structure.

Hydrogen Bonding

Water (H_2O) is a hydride of oxygen in which the highly electronegative oxygen atom attracts the bonding electrons from two hydrogen atoms. This leads to polar H–O bonds in which the hydrogen atoms have a slightly positive charge (δ^+) and the oxygen atom has a slightly negative charge (δ^-) (Figure 1-1). Water molecules have a relatively high dipole moment because of the angle (104.5°) of the H–O–H bond and the polarity of the bonds. Neighboring liquid water molecules interact with one another to form an extensive lattice-like structure similar to the structure of ice. The intermolecular bonding between water molecules arises from the attraction between the partial negative charge on the oxygen atom and the partial positive charge on the hydrogen atom of adjacent water molecules. This type of attraction involving a hydrogen atom is known as a ***hydrogen bond*** (Figure 1-2).

Hydrogen bonds contain a hydrogen atom between two electronegative atoms (e.g., O and N). One is the formal hydrogen donor; the other is the hydrogen acceptor. The amount of energy required to break a hydrogen bond (bond energy) is estimated to be 2–5 kcal/mol (8.4–20.9 kJ/mol) in the gas phase. Covalent bonds have bond energies of 50–100 kcal/mol (209–418 kJ/mol). The cumulative effect of many hydrogen bonds is equivalent to the stabilizing effect of covalent bonds. In proteins, nucleic acids, and water, hydrogen bonds are essential to stabilize overall structure. In ice, each water molecule forms a hydrogen bond with four other water molecules, giving rise to a rigid tetrahedral arrangement (Figure 1-2). In the liquid state, water maintains a tetrahedrally coordinated structure over short ranges and for short time periods.

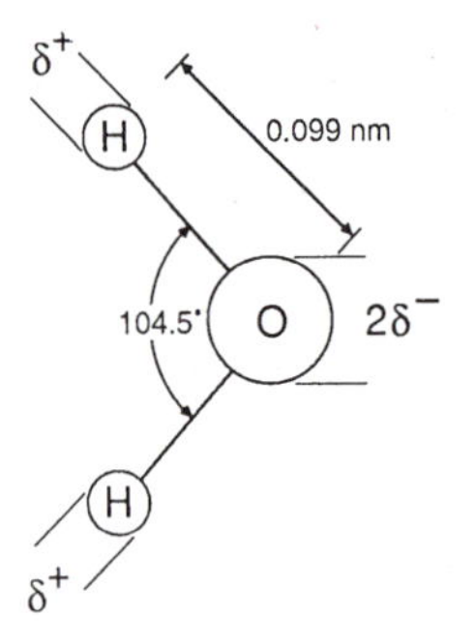

FIGURE 1-1
Structure of the water molecule.

Physical Properties

Other properties of water uniquely suited to biological systems include melting point, boiling point, heat of vaporization (quantity of heat energy required to transform 1 g of liquid to vapor at the boiling point), heat of fusion (quantity of heat energy required to convert 1 g of solid to liquid at the melting point), specific heat (the amount of heat required to raise the temperature of 1 g of substance by 1°C, and surface tension (Table 1-1). All these values for water are much higher than those for other low-molecular-weight substances because of the strong intermolecular hydrogen bonding of water. These properties contribute to maintenance of temperature and to dissipation of heat in living systems. Thus, water plays a major role in thermoregulation in living systems. The optimal body temperature is a balance between heat production and heat dissipation. Impaired thermoregulation causes either *hypothermia* or *hyperthermia* and has serious metabolic consequences; if uncorrected, impaired thermoregulation may lead to death (Chapter 39). Water freezes to form ice at 0°C, but its maximum density is at 4°C. Aquatic organisms survive cold winters because ice floats over and insulates liquid water from sub-zero-degree temperature. If ice were denser than liquid water—which is the case for most liquid-to-solid transformations—the solid form would sink and the entire amount of fluid would solidify rapidly in freezing weather. During hot weather, deep lakes and oceans remain cool because heat generated by sunlight can be dissipated by evaporation of surface water.

FIGURE 1-2
Tetrahedral hydrogen-bonded structure of water molecules in ice. The tetrahedral arrangement is due to the fact that each water molecule has four fractional charges: two negative charges due to the presence of a lone pair of electrons on the oxygen atom and two positive charges, one on each of the two hydrogen atoms. In the liquid phase this tetrahedral array occurs transiently.

TABLE 1-1
Physical Properties of Water*

Density (at 4°C)	1.0 g/mL
Molecular weight	18
Liquid range	0°–100°C
Melting point	0°C
Boiling point	100°C
Heat of fusion	80 cal/g
Heat of vaporization	540 cal/g
Dipole moment	1.86 Debye unit
Dielectric constant (E)	78.4
Solid/liquid density ratio	0.92

*Some of these properties are measured at 1 atm pressure.

Water is transported across cell membranes in one of two ways:

1. by simple diffusion through the phospholipid bilayer and
2. by the action of membrane-spanning transport proteins known as *aquaporins*.

Thus, the concentration of water is in thermodynamic equilibrium across the cell membrane. In the renal collecting duct, water is reabsorbed through a specific aquaporin channel protein (aquaporin 2). This reabsorption of water is regulated by the antidiuretic hormone (also known as *vasopressin*). A defect or lack of functional aquaporin 2, vasopressin, or its receptor leads to enormous loss of water in the urine, causing the disease known as *diabetes insipidus* (Chapter 39). Water plays a significant role in enzyme functions, molecular assembly of macromolecules, and allosteric regulation of proteins. For example, the effect of protein solvation in allosteric regulation is implicated in the transition of deoxyhemoglobin to oxyhemoglobin. During this process about 60 extra water molecules bind to oxyhemoglobin (Chapter 28).

Solutes, Micelles, and Hydrophobic Interactions

Water is an excellent solvent for both ionic compounds (e.g., NaCl) and low-molecular-weight nonionic polar compounds (e.g., sugars and alcohols). Ionic compounds are soluble because water can overcome the electrostatic attraction between ions through solvation of the ions. Nonionic polar compounds are soluble because water molecules can form hydrogen bonds to polar groups (e.g., –OH).

Amphipathic compounds, which contain both large nonpolar hydrocarbon chains (hydrophobic groups) and polar or ionic groups (hydrophilic groups) may associate with each other in submicroscopic aggregations called ***micelles.*** Micelles have hydrophilic (water-liking) groups on their exterior (bonding with solvent water), and hydrophobic (water-disliking) groups clustered in their interior (Figure 1-3). They occur in spherical, cylindrical, or ellipsoidal shapes. Micelle structures are stabilized by hydrogen bonding with water, by van der Waals attractive forces between hydrocarbon groups in the interior, and by energy of hydrophobic reactions. The last is the stabilization energy that would be lost if each hydrocarbon group were transferred from the hydrophobic medium to the polar aqueous solvent. As with hydrogen bonds, each hydrophobic interaction is very weak, but many such interactions result in formation of large, stable structures.

A micelle may contain many hundreds of thousands of amphipathic molecules. The interior molecular organization of micelles has been likened to a "liquid hydrocarbon droplet." However, a recent model departs from this conventional concept and suggests that because of severe constraints in the space-filling requirements of hydrocarbon chains in the interior as well as because of micellar geometry, the chain ends are not uniformly distributed throughout the micelle but tend to be clustered between the center of the micelle and the outer surface, implying that many of the hydrocarbon side chains are bent back upon themselves (Figure 1-3). In this model, there appears to be a progression from ordered (as in crystals) to disordered (as in liquids) structures proceeding from the center of the micelle to the periphery.

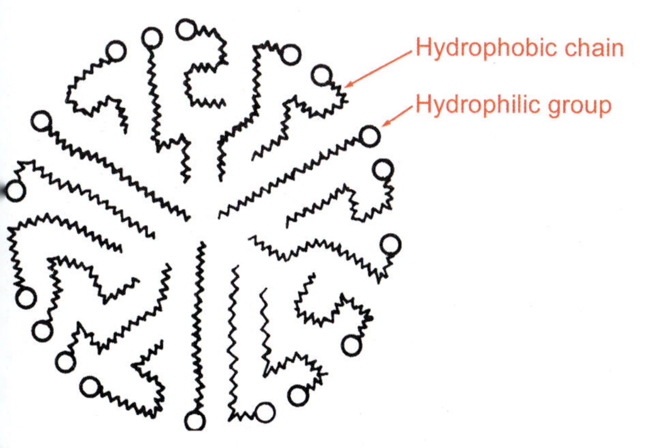

FIGURE 1-3
A geometrical representation of a spherical micelle showing the hydrocarbon chains in the micelle. The hydrophilic groups are attracted to water, and the hydrophobic chains are within the micelle. The ends of the hydrocarbon chains are nonuniformly distributed, and many are located in the middle. The degree of disorder is much higher in the periphery than in the center of the micelle.

Hydrophobic interaction plays a major role in maintaining the structure and function of cell membranes, the activity of proteins, the anesthetic action of nonpolar compounds such as chloroform and nitrous oxide, the absorption of digested fats, and the circulation of hydrophobic molecules in the interior of micelles in blood plasma.

Colligative Properties

The colligative properties of a solvent depend upon the concentration of solute particles. These properties include freezing point depression, vapor pressure depression, osmotic pressure, and boiling point elevation. The freezing point of water is depressed by 1.86°C when 1 mol of nonvolatile solute, which neither dissociates nor associates in solution, is dissolved in 1 kg of water. The same concentration of solute elevates the boiling point by 0.543°C. Osmotic pressure is a measure of the tendency of water molecules to migrate from a dilute to a concentrated solution through a semipermeable membrane. This migration of water molecules is termed ***osmosis.*** A solution containing 1 mol of solute particles in 1 kg of water is a 1-osmolal solution. When 1 mol of a solute (such as NaCl) that dissociates into two ions (Na^+ and Cl^-) is dissolved in 1 kg of water, the solution is 2-osmolal.

Measurement of colligative properties is useful in estimating solute concentrations in biological fluids. For example, in blood plasma, the normal total concentration of solutes is remarkably constant (275–295 milliosmolal). Pathological conditions (e.g., dehydration, renal failure) involving abnormal plasma osmolality are discussed in Chapter 39.

Dissociation of Water and the pH Scale

Water dissociates to yield a hydrogen ion (H^+) and a hydroxyl ion (OH^-).

$$H_2O \rightleftarrows H^+ + OH^- \qquad (1.1)$$

The H^+ bonds to the oxygen atom of an undissociated H_2O molecule to form a hydronium ion (H_3O^+).

$$H_2O + H_2O \rightleftarrows H_3O^+ + OH^-$$

Thus, water functions both as an acid (donor of H^+ or proton) and as a base (acceptor of H^+ or proton). This description of an acid and a base follows from the Brönsted–Lowry theory. According to the Lewis theory, acids are electron pair acceptors and bases are electron pair donors. The equilibrium constant, K, for the dissociation reaction in Equation (1.1) is

$$K = \frac{[H^+][OH^-]}{[H_2O]} \tag{1.2}$$

where the square brackets refer to the molar concentrations of the ions involved. K can be determined by measurement of the electrical conductivity of pure water, which has the value of 1.8×10^{-16} M at 25°C, indicative of a very small ion concentration, where M (molar) is the units of moles per liter. Therefore, the concentration of undissociated water is essentially unchanged by the dissociation reaction.

Since 1 L of water weighs 1000 g and 1 mol of water weighs 18 g, the molar concentration of pure water is 55.5 M. Substitution for K and $[H_2O]$ in Equation (1.2) yields

$$[H^+][OH^-] = (55.5\ M) \times (1.8 \times 10^{-16}\ M)$$
$$[H^+][OH^-] = 1.0 \times 10^{-14}\ M^2 = K_w$$

K_w is known as the *ion product of water*. In pure water, $[H^+]$ and $[OH^-]$ are equal, so that

$$[OH^-] = [H^+] = 1.0 \times 10^{-7}\ M.$$

pH is employed to express these ion concentrations in a convenient form, where the "p" of pH symbolizes "negative logarithm (to the base 10)" of the concentration in question. Thus,

$$pH = -\log_{10}[H^+] = \log \frac{1}{[H^+]}$$

Similarly,

$$pOH = -\log_{10}[OH^-] = \log \frac{1}{[OH^-]}$$

Therefore, for water,

$$\log[H^+] + \log[OH^-] = \log 10^{-14}$$

or

$$pH + pOH = 14.$$

The pH value of 7 for pure water at 25°C is considered to be neutral, and values below 7 are considered acidic and above 7 basic. Table 1-2 illustrates the pH scale extending from −1 to +15. It is important to recognize that as the pH decreases, $[H^+]$ increases. A decrease in one pH unit reflects a 10-fold increase in H^+ concentration. In discussions of acid–base problems in human biochemistry, it is often preferable to express H^+ concentration as nanomoles per liter (nmol/L).

TABLE 1-2
The pH Scale

$[H^+]$, *M*	pH		$[OH^-]$, *M*
10.0	−1	↑	10^{-15}
1.0	0		10^{-14}
0.1	1		10^{-13}
0.01(10^{-2})	2		10^{-12}
10^{-3}	3		10^{-11}
10^{-4}	4	Acidic	10^{-10}
10^{-5}	5		10^{-9}
10^{-6}	6	↓	10^{-8}
10^{-7} ←	7	— Neutral →	10^{-7}
10^{-8}	8	↑	10^{-6}
10^{-9}	9		10^{-5}
10^{-10}	10		10^{-4}
10^{-11}	11	Basic	10^{-3}
10^{-12}	12		0.01
10^{-13}	13		0.1
10^{-14}	14		1
10^{-15}	15	↓	10

1.2 Buffers

Buffers resist change in pH in solutions when acids or bases are added. They are either a mixture of a weak acid (HA) and its conjugate base (A^-) or a mixture of a weak base (B) and its conjugate acid (HB^+).

EXAMPLE 1 Acetic acid (CH_3COOH) and carbonic acid (H_2CO_3) are weak acids. Ammonia (NH_3) is a weak base. CH_3COOH/CH_3COO^-, H_2CO_3/HCO_3^-, and NH_3/NH_4^+ constitute buffer systems.

A buffer solution functions in the following manner to resist changes in acidity or alkalinity. In an acetic acid/sodium acetate buffer system, the species present in solution are CH_3COOH, CH_3COO^-, Na^+, and H_2O. Amounts of H^+ and OH^- are initially assumed to be small.

When acid is added to the buffer, almost all of the H^+ ions react with acetate ions to produce weakly ionized acetic acid ($H^+ + CH_3COO^- \rightleftarrows CH_3COOH$). The H^+ ions are thereby prevented from appreciably changing the pH.

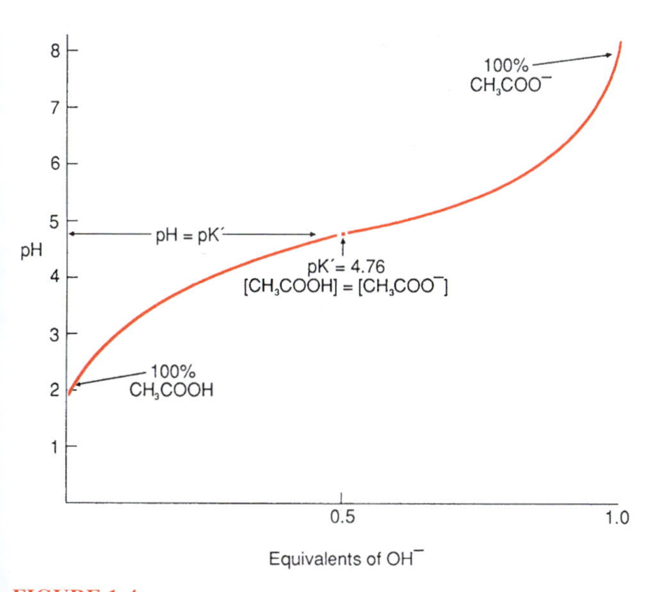

FIGURE 1-4
Titration profile of acetic acid (CH_3COOH) with sodium hydroxide (NaOH). Maximum buffering capacity is at pH = pK′, at which point minimal change in pH occurs upon addition of acid or base.

When OH^- is added, almost all of the hydroxyl radicals react with acetic acid molecules to produce more acetate ions and water ($OH^- + CH_3COOH \rightleftarrows CH_3COO^- + H_2O$). The additional OH^- is thus consumed with little increase in pH.

Adding H^+ or OH^- to a buffer causes only *slight* pH changes provided there is excess salt (CH_3COO^-) or acid (CH_3COOH). If all of the acid is converted to the salt form by the addition of a large amount of OH^-, the solution can no longer behave as a buffer. Adding more OH^- will cause the pH to rise rapidly, as if the solution contained no buffer or only salt. The maximum buffering capacity exists when the molarities of the salt and acid are equal, i.e., when pH = pK′ (or $-\log K'$). The pK′ is at an inflection point on the titration curve and hence is the point of minimum slope or minimum change in pH for a given addition of acid or base (Figure 1-4). In a generalized weak acid buffer reaction,

$$H_2O + HA \rightleftarrows H_3O^+ + A^-$$

a hydronium ion, H_3O^+, is formed by the association of a hydrogen ion with a water molecule. In dilute solutions, the concentration of water changes very little when HA is added; therefore, by convention, the dissociation reaction equation is usually written as

$$HA \rightleftarrows H^+ + A^- \tag{1.3}$$

A weak acid, HA, does not readily dissociate, owing to the high affinity of the conjugate base, A^-, for the hydrogen ion. Similarly in the hydrolysis reaction of a weak base (B) and water, the ions OH^- and HB^+ (the conjugate acid) are produced.

$$B + H_2O \rightleftarrows HB^+ + OH^-$$

The concentration of the conjugate base (or acid) generated from a weak acid (or base) is small, since, by definition, weak acids and bases are only slightly dissociated in aqueous solution. Examples of weak acids are organic acids (e.g., acetic) and of strong acids are mineral acids (e.g., hydrochloric and sulfuric).

Henderson–Hasselbalch Equation

The Henderson–Hasselbalch equation was developed independently by the American biological chemist L. J. Henderson and the Swedish physiologist K. A. Hasselbalch, for relating the pH to the bicarbonate buffer system of the blood (see below). In its general form, the Henderson–Hasselbalch equation is a useful expression for buffer calculations. It can be derived from the equilibrium constant expression for a dissociation reaction of the general weak acid (HA) in Equation (1.3):

$$K = \frac{[H^+][A^-]}{[HA]} \tag{1.4}$$

where K is the equilibrium constant at a given temperature. For a defined set of experimental conditions, this equilibrium constant is designated as K' (K prime) and referred to as an apparent dissociation constant. The higher the value of K', the greater the number of H^+ ions liberated per mole of acid in solution and hence the stronger the acid. K' is thus a measure of the strength of an acid. Rearrangement of Equation (1.4) yields

$$[H^+] = \frac{K'[HA]}{[A^-]} \tag{1.5}$$

Taking logarithms of both sides of Equation (1.5) and multiplying throughout by -1 gives

$$-\log[H^+] = -\log K' - \log[HA] + \log[A^-] \tag{1.6}$$

Substituting pH for $-\log[H^+]$ and pK′ for $-\log K'$ yields

$$\mathrm{pH} = \mathrm{pK}' + \log\frac{[A^-]}{[HA]} \tag{1.7}$$

or

$$\mathrm{pH} = \mathrm{pK}' + \log\frac{[\text{conjugate base}]}{[\text{acid}]} \tag{1.8}$$

This relationship is represented by the Henderson–Hasselbalch equation.

Since a buffer is intended to give only a small change in pH with added H^+ or OH^-, the best buffer for a given pH is the one that gives the smallest change. As may be

TABLE 1-3
*Percent Unprotonated Species and Ratio of Unprotonated Forms Relative to the Difference between the pH and pK′**

%A^-	$[A^-]/[HA]$	Log $[A^-]/[HA]$ (pH − pK′)
"100"	999/1	3.00
99	99/1	2.00
98	98/2	1.69
96	96/4	1.38
94	94/6	1.20
92	92/8	1.06
91	91/9	1.00
90	90/10	0.95
80	80/20	0.60
70	70/30	0.37
60	60/40	0.18
50	50/50	0.00
40	40/60	−0.18
30	30/70	−0.37
20	20/80	−0.60
10	10/90	−0.95
8	8/92	−1.06
6	6/94	−1.20
4	4/96	−1.38
2	2/98	−1.69
1	1/99	−2.00
"0"	1/999	−3.00

*Reproduced, with permission, from J. N. Aronson: The Henderson-Hasselbalch equation revisited. *Biochemical Education* **11**(2), 68 (1981).

seen from the Henderson–Hasselbalch equation, when the pH of the solution equals the pK′ of the buffer, [conjugate base] = [acid], and the buffer can therefore respond equally to both added acid and added base. It also follows from Equation (1.7) that when the pH of the solution is one pH unit above or below the pK′ value, the solution contains approximately 9% unprotonated or protonated species, respectively. Similarly, if the pH of the solution is two units above or below the pK′ value, the solution contains almost entirely (99%) unprotonated or protonated species, respectively. Table 1-3 provides percent unprotonated species and the corresponding unprotonated/protonated ratios for selected (pH–pK′) values.

Buffer Systems of Blood and Exchange of O_2 and CO_2

If the H^+ concentration departs significantly from its normal value in blood, the health and survival of the human body are in jeopardy. H^+ is the smallest ion, and it combines with many negatively charged and neutral functional groups. Changes of $[H^+]$, therefore, affect the charged regions of many molecular structures, such as enzymes, cell membranes, and nucleic acids, and dramatically alter physiological activity. If the plasma pH reaches either 6.8 or 7.8, death may be unavoidable. Despite the fact that large amounts of acidic and basic metabolites are produced and eliminated from the body, buffer systems maintain a fairly constant pH in body fluids (Table 1-4).

TABLE 1-4
pH Values of Human Body Fluids and Secretions

Body Fluid or Secretion	pH
Blood	7.4
Milk	6.6–6.9
Hepatic bile	7.4–8.5
Gall bladder bile	5.4–6.9
Urine (normal)	6.0
Gastric juice (parietal secretion)	0.87
Pancreatic juice	8.0
Intestinal juice	7.7
Cerebrospinal fluid	7.4
Saliva	7.2
Aqueous humor of eye	7.2
Tears	7.4
Urine (range in various disease states)	4.8–7.5
Feces	7.0–7.5
Muscle cell, resting (at 37°C; extracellular pH = 7.4)	6.94–7.06 (intracellular)

The major metabolic product from oxidation of ingested carbon compounds is CO_2. Hydration of CO_2 dissolved in water yields the weak acid H_2CO_3 (carbonic acid). Depending on the type of food ingested and oxidized, 0.7–1.0 mol of CO_2 is produced per mole of O_2 consumed. This results in the metabolic production of about 13 mol of hydrated CO_2 each day in a normal person.

For efficient transport of relatively insoluble CO_2 from the tissues where it is formed to the lungs where it must be exhaled, the buffers of the blood convert CO_2 to the very soluble anionic form HCO_3^- (bicarbonate ion). The principal buffers in blood are bicarbonate-carbonic acid in plasma, hemoglobin in red blood cells, and protein functional groups in both. The normal balance between rates of elimination and production of CO_2 yields a steady-state concentration CO_2 in the body fluids and a relatively constant pH.

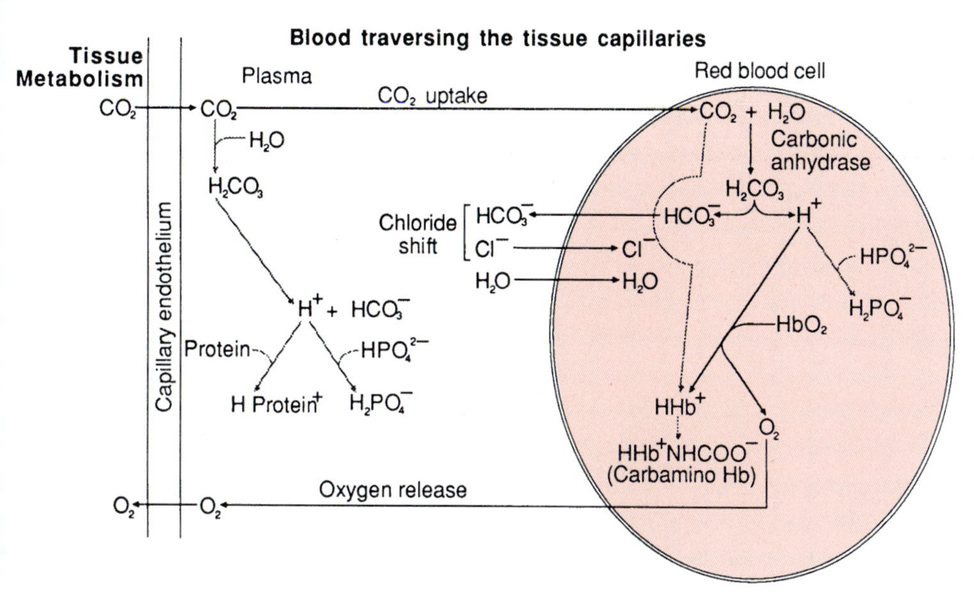

FIGURE 1-5
Schematic representation of the transport of CO_2 from the tissues to the blood. Note that the majority of CO_2 is transported as HCO_3^- in the plasma and that the principal buffer in the red blood cell is hemoglobin. Solid lines refer to major pathways, and broken lines refer to minor pathways. Hb = hemoglobin.

Other acids that are products of metabolism are lactic acid, acetoacetic acid, β-hydroxybutyric acid, phosphoric acid, sulfuric acid, and hydrochloric acid. The organic acids (e.g., lactate, acetoacetate, and β-hydroxybutyrate) are normally oxidized further to CO_2 and H_2O. The hydrogen ions and anions contributed by mineral acids and any unmetabolized organic acids are eliminated via the excretory system of the kidneys. Thus, although body metabolism produces a large amount of acid, a constant pH is maintained by transport of H^+ ions and other acid anions in buffer systems and by elimination of CO_2 through alveolar ventilation in the lungs and through excretion of aqueous acids in the urine.

Metabolic activities continuously release CO_2 to the blood (Figure 1-5), and the lungs continuously eliminate CO_2 (Figure 1-6). As oxygen is consumed in peripheral tissues, CO_2 is formed and its pressure (P_{CO_2}) builds to about 50 mm Hg, whereas the blood entering the tissue capillaries has a P_{CO_2} of about 40 mm Hg. Because of this difference in P_{CO_2} values, CO_2 diffuses through the cell membranes of the capillary endothelium and the blood P_{CO_2} rises to 45–46 mm Hg. Despite this increase in P_{CO_2}, the blood pH value drops by only about 0.03 during the flow from the arterial capillary (pH 7.41) to the venous capillary (pH 7.38) as a consequence of buffering.

About 95% of the CO_2 entering the blood diffuses into the red blood cells. Within the red blood cells, the enzyme carbonic anhydrase catalyzes conversion of most of the CO_2 to H_2CO_3:

$$CO_2 + H_2O \rightleftarrows H_2CO_3$$

H_2CO_3 dissociates to H^+ and HCO_3^-. Although H_2CO_3 is a weak acid, its dissociation is essentially 100% because of removal of H^+ ions by the buffering action of hemoglobin. The presence of CO_2 and the production of H^+ cause a reduction in the affinity of hemoglobin for oxygen. Oxyhemoglobin (HbO_2) consequently dissociates into oxygen and deoxyhemoglobin (Hb). This effect of pH on the binding of O_2 to hemoglobin is known as the ***Bohr effect*** (Chapter 28).

Oxygen diffuses into the tissues because the P_{O_2} in blood is greater than the P_{O_2} in tissue cells and because protonated deoxyhemoglobin (HHb) is a weaker acid than HbO_2 and thereby binds H^+ more strongly than HbO_2. When purified HbO_2 dissociates at pH 7.4 to yield oxygen and Hb, the Hb binds 0.7 mol of H^+ per mole of oxygen released. However, under physiological conditions in whole blood, the Hb combines 0.31 mol of H^+ per mole of oxygen released. This process is reversible. The remainder of the H^+ is buffered by phosphate and proteins other than hemoglobin. The major buffering group involved in the transport of H^+ is an imidazolium group of a histidine residue in hemoglobin. The imidazolium group has a pK′ value of about 6.5 (Figure 1-7 depicts the reactions). The difference in acid–base properties between the two forms of hemoglobin molecules is explained by the conformational change that

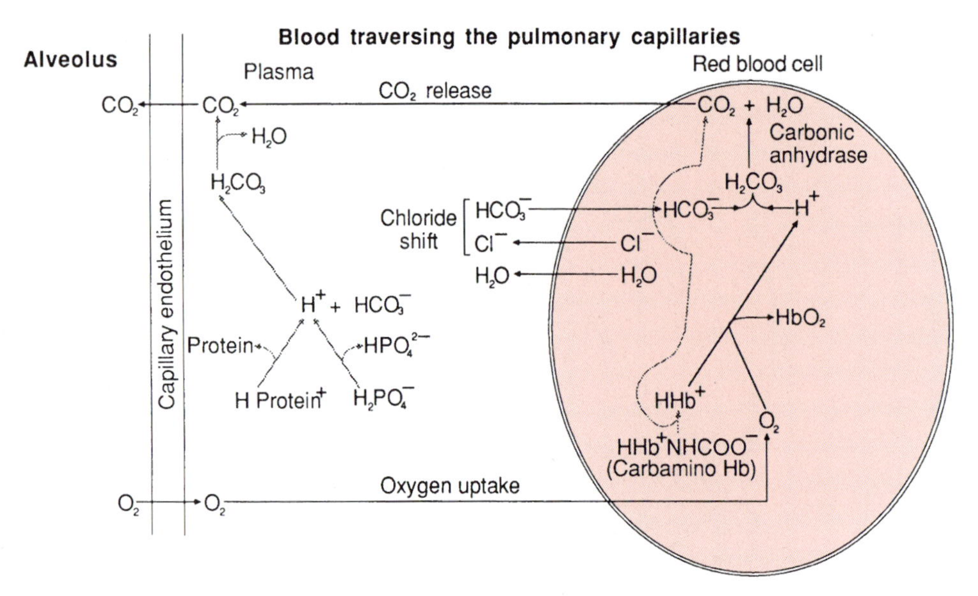

FIGURE 1-6
Schematic representation of the transfer of CO_2 from the alveolus (and its loss in the expired air in the lungs) and oxygenation of hemoglobin. Note that the sequence of events occurring in the pulmonary capillaries is the opposite of the process taking place in the tissue capillaries (Figure 1-5). Solid lines indicate major pathways and broken lines indicate minor pathways. Hb = hemoglobin.

accompanies conversion from HbO_2 to HHb^+ (Chapters 7 and 28).

As the concentration of HCO_3^- (i.e., of metabolic CO_2) in red blood cells increases, an imbalance occurs between the bicarbonate ion concentrations in the red blood cell and plasma. This osmotic imbalance causes a marked efflux of HCO_3^- to plasma and consequent influx of Cl^- from plasma in order to maintain the balance of electrostatic charges. The latter osmotic influx, known as the *chloride shift,* is accompanied by migration of water to red blood cells. Thus, transport of metabolic CO_2 in the blood occurs primarily in the form of plasma bicarbonate formed after CO_2 diffuses into red blood cells.

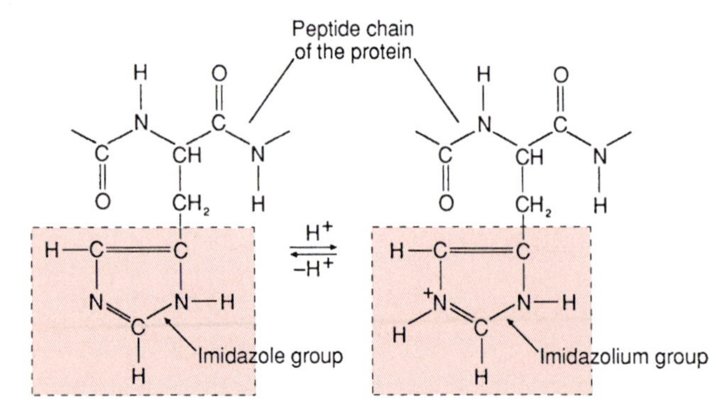

FIGURE 1-7
Buffer function of the imidazole/imidazolium functional groups of histidine residues in protein.

A small percentage of CO_2 entering the red blood cells combines reversibly with an un-ionized amino group ($-NH_2$) of hemoglobin:

$$\text{hemoglobin–NH}_2 + CO_2 \rightleftarrows \text{hemoglobin–NH–COO}^- + H^+$$

Hemoglobin–NH–COO$^-$, commonly known as *carbaminohemoglobin,* is more correctly named hemoglobin carbamate. Formation of this compound causes a lowering of the affinity of hemoglobin for oxygen. Thus, an elevated concentration of CO_2 favors dissociation of oxyhemoglobin to oxygen and deoxyhemoglobin. Conversely, CO_2 binds more tightly to deoxyhemoglobin than to oxyhemoglobin. All of these processes occurring in the red blood cells of peripheral capillaries are functionally reversed in the lungs (Figure 1-6). Since alveolar P_{O_2} is higher than that of the incoming deoxygenated blood, oxygenation of hemoglobin and release of H^+ occur. The H^+ release takes place because HbO_2 is a stronger acid (i.e., has a lower pK′) than deoxyhemoglobin. The released bicarbonate, which is transported to the red blood cells with the corresponding efflux of Cl^-, combines with the released H^+ to form H_2CO_3. Cellular carbonic anhydrase catalyzes dehydration of H_2CO_3 and release of CO_2 from the red blood cells.

Thus, red blood cell carbonic anhydrase which catalyzes the reversible hydration of CO_2, plays a vital role in carbon dioxide transport and elimination. Carbonic anhydrase is a monomeric (M.W. 29,000) zinc metalloenzyme and is present in several different isoenzyme forms. The most prevalent red blood cell isoenzyme of carbonic anhydrase is type I (CAI). Zinc ion, held by coordinate covalent linkage by three imidazole groups of three histidine residues, is involved in the catalytic mechanism of carbonic anhydrase. Water bound to zinc ion reacts with CO_2 bound to the nearby catalytic site of carbonic anhydrase to produce H_2CO_3. The action of carbonic anhydrase is essential for a number of metabolic functions. Some include the formation of H^+ ion in stomach parietal cells (Chapter 12), in bone resorption by the osteoclasts (Chapter 37) and reclamation of HCO_3^- in renal tubule cells (Chapter 39). In osteoclasts and in the renal tubule cells, the isoenzyme CAII catalyzes the hydration reaction of CO_2. A deficiency of CAII caused by an autosomal recessive disorder consists of osteopetrosis (marble bone disease), renal tubular acidosis, and cerebral calcification.

The diffusion of CO_2 from venous blood into the alveoli is facilitated by a pressure gradient of CO_2 between the venous blood (45 mm Hg) and the alveoli (40 mm Hg) and by the high permeability of the pulmonary membrane to CO_2. Blood leaving the lungs has a P_{CO_2} of about 40 mm Hg; thus, essentially complete equilibration occurs between alveolar CO_2 and blood CO_2.

Blood Buffer Calculations

Carbonic acid has the following pK′ values:

$$H_2CO_3 \rightleftarrows H^+ + HCO_3^- \qquad pK'_1 = 3.8 \qquad (1.9)$$

$$HCO_3^- \rightleftarrows CO_3^{2-} + H^+ \qquad pK'_2 = 10.2 \qquad (1.10)$$

It is apparent from the pK′ values that neither equilibrium can serve as a buffer system at the physiological pH of 7.4. However, carbonic acid (the proton donor) is in equilibrium with dissolved CO_2, which in turn is in equilibrium with gaseous CO_2:

$$H_2O + CO_2(\text{aqueous}) \rightleftarrows H_2CO_3 \qquad (1.11)$$

The hydration reaction (1.11), coupled with the first dissociation of carbonic acid (1.9), produces an apparent pK′ of 6.1 for bicarbonate formation. Thus, the summation of Equations (1.9) and (1.11) yields

$$H_2O + CO_2 \rightleftarrows H^+ + HCO_3^- \qquad (1.12)$$

$$pK'\ (\text{apparent}) = \frac{[HCO_3^-][H^+]}{[H_2CO_3]} = 6.1 \qquad (1.13)$$

The ratio of HCO_3^- to H_2CO_3 at a physiological pH of 7.4 can be calculated by using the Henderson–Hasselbalch relationship:

$$7.4 = 6.1 + \log\frac{[HCO_3^-]}{[H_2CO_3]} \qquad (1.14)$$

$$\log\frac{[HCO_3^-]}{[H_2CO_3]} = 1.3$$

Taking antilogarithms,

$$\frac{[HCO_3^-]}{[H_2CO_3]} = \frac{20}{1} = \frac{\text{proton acceptor}}{\text{proton donor}}$$

This ratio is large because the pH is greater than the pK′ (see Table 1-3). At pH 7.4, the bicarbonate system is a good buffer toward acid (i.e., it can neutralize large amounts of acid) but a poor buffer for alkali. However, blood H_2CO_3 is in rapid equilibrium with a relatively large (about 1000 times as much) reservoir of cellular CO_2 and can function as an effective buffer against increases in alkalinity. The HCO_3^-/H_2CO_3 ratio in blood is coupled to the partial pressure of CO_2, i.e., to the metabolic production of CO_2 and to the loss of CO_2 during respiration. In the equilibrium expression for the bicarbonate-carbonic acid buffer system at pH 7.4, the carbonic acid term can be replaced by a pressure term because the carbonic acid concentration is proportional to P_{CO_2} in the blood.

$$pH = 6.1 + \log\frac{[HCO_3^-]}{aP_{CO_2}} \qquad (1.15)$$

where a, a proportionality constant, is defined by the equation

$$[H_2CO_3] = aP_{CO_2} \qquad (1.16)$$

The numerical value of a depends on the solvent and the temperature. For normal plasma at 37°C, $a =$ 0.0301 mmol of dissolved CO_2 per liter of plasma per mm Hg of CO_2 pressure. Assuming that 37°C (310 K) is approximately the same as standard temperature, 25°C (298 K), and that $[H_2CO_3]$ in Equation (1.16) includes *both* carbonic acid and dissolved CO_2, the value of a can be derived from two facts:

1. At 37°C and 760 mm Hg of CO_2 pressure, 521 mL of CO_2 will dissolve per liter of normal plasma;
2. At standard temperature and pressure, 1 mol of dry CO_2 occupies a volume of 22.26 L (not 22.4 L since CO_2 is not an ideal gas).

$$a = \frac{521\ \text{mL } CO_2/\text{L of plasma}}{760\ \text{mm Hg} \times 22.26\ \text{mL/mmol of } CO_2} = 0.0301\,\frac{\text{mmol } CO_2/\text{L of plasma}}{\text{mm Hg of } CO_2 \text{ pressure}}$$

The equation form of the Henderson–Hasselbalch expression [Equation (1.15)] can be further modified by substituting another expression for the bicarbonate term. When excess strong acid is added to plasma, CO_2 is stoichiometrically released from dissolved CO_2, carbonic acid, bicarbonate ions, and carbonate. Carbonate concentration is negligible; thus,

$$\text{Total } [CO_2] = [HCO_3^-] + \text{dissolved}[CO_2] + [H_2CO_3]$$

But

$$\text{Dissolved}[CO_2] + [H_2CO_3] = 0.0301 P_{CO_2}$$

so that

$$[HCO_3^-] = \text{total}[CO_2] - 0.0301 P_{CO_2}$$

Finally,

$$pH = 6.1 + \log\left(\frac{\text{total } [CO_2] - 0.0301 P_{CO_2}}{[0.0301 P_{CO_2}]}\right) \quad (1.17)$$

Equation (1.17) is useful for calculating the pH from the total $[CO_2]$ and P_{CO_2}.

The HCO_3^-/H_2CO_3 buffer system effectively maintains a constant blood pH of 7.4 if bicarbonate and H_2CO_3 concentrations are maintained at a ratio of 20:1. The concentration of HCO_3^- is regulated by its selective excretion and reclamation by the membranes of the renal tubular epithelial cell. P_{CO_2} and $[H_2CO_3]$ in the blood can be altered by changes in the rate and depth of respiration. For examples, *hypoventilation* (slow, shallow breathing) leads to increased blood P_{CO_2}, whereas *hyperventilation* (rapid, deep breathing) has the opposite effect. P_{CO_2} changes mediated by the lungs are more rapid than $[HCO_3^-]$ changes affected through the kidneys (Chapter 39).

Nonbicarbonate Buffers in Blood

Other important nonbicarbonate blood buffers are protein and phosphate. The predominant buffer system in the red blood cells is hemoglobin. Protein amino acid side chains (R-groups) that act as buffers are carboxylate groups of glutamate and aspartate and the weakly basic groups of lysine, arginine, and histidine. To be effective, the pK′ value of a buffer should be close to the pH of the system to be buffered. Except for the R-group of histidine, which is an imidazolium group (Figure 1-7), the pK′ values of the other amino acids mentioned above are not close enough to the physiological pH of blood to be effective buffers. The imidazolium group has a pK′ value of 6.5 but it can vary from 5.3 to 8.3 depending on differences in electrostatic environment either within the same protein molecule or in different proteins. Another potential buffering group in protein is the α-amino group of the amino acid residues at the amino terminus of the protein. This group has a pK′ value ranging from 7.8 to 10.6, with a typical value of about 8 (acid–base properties of amino acids and proteins are discussed in detail in Chapters 2 and 3). In plasma, the protein buffer system has a limited role; the principal plasma buffer is the bicarbonate-carbonic acid system.

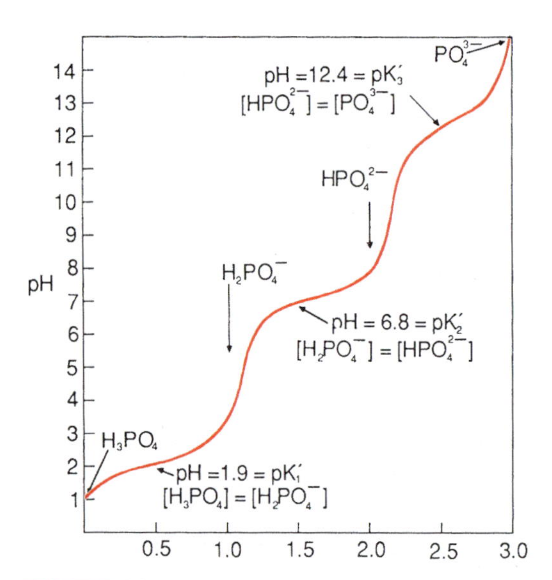

FIGURE 1-8
Titration profile of phosphoric acid (H_3PO_4) with sodium hydroxide (NaOH). The three pK′ values correspond to three buffer regions. The physiological buffering occurs at the pK′ region with $H_2PO_4^-$ (acid) and HPO_4^{2-} (conjugate base) ionic species.

Compared with hemoglobin in the red blood cells and HCO_3^-/H_2CO_3 in plasma, phosphates (both organic and inorganic) play minor roles in physiological buffering. Phosphoric acid (H_3PO_4) has three dissociable protons:

$$H_3PO_4 \rightleftarrows H_2PO_4^- + H^+ \quad pK'_1 = 1.9 \quad (1.18)$$
$$H_2PO_4^- \rightleftarrows HPO_4^{2-} + H^+ \quad pK'_2 = 6.8 \quad (1.19)$$
$$HPO_4^{2-} \rightleftarrows PO_4^{3-} + H^+ \quad pK'_3 = 12.4 \quad (1.20)$$

The titration profile of phosphoric acid with NaOH is shown in Figure 1-8. The principal dissociation expression functioning at a given pH depends on which pK′ is closest to the pH. At a plasma pH of 7.4, the important conjugate pair is $HPO_4^{2-}/H_2PO_4^-$. The Henderson–Hasselbalch equation can be used to obtain the value of the ratio $HPO_4^{2-}/H_2PO_4^-$ at pH 7.4:

$$7.4 = 6.8 + \log\frac{[HPO_4^{2-}]}{[H_2PO_4^-]} \quad (1.21)$$

$$\log\frac{[HPO_4^{2-}]}{[H_2PO_4^-]} = 0.6 \quad \text{and} \quad \frac{[HPO_4^{2-}]}{[H_2PO_4^-]} = 4$$

As was the case for the bicarbonate-carbonic acid system, the conjugate base form (HPO_4^{2-}) of the phosphate buffer is present in large (fourfold) excess compared to the acid form ($H_2PO_4^-$) and provides acid buffering capacity. Since the body metabolism produces more acid than base, this ratio assists in neutralizing acid and in maintaining a constant pH. The $HPO_4^{2-}/H_2PO_4^-$ buffering system plays a minor role in plasma because of the low concentrations of these ions, but it is important in raising the plasma pH through the excretion of $H_2PO_4^-$ via the kidney (Chapter 39). In summary, hemoglobin absorbs a major portion of the hydrogen ions produced by the dissociation of H_2CO_3 generated by the hydration of CO_2—the most important buffer system in the blood. However, since hemoglobin and carbonic anhydrase are present only in red blood cells, the HCO_3^-/H_2CO_3 system in the plasma is an indispensable intermediary in transporting the acid. Thus, the principal method of CO_2 transport is in the form of HCO_3^- in blood plasma.

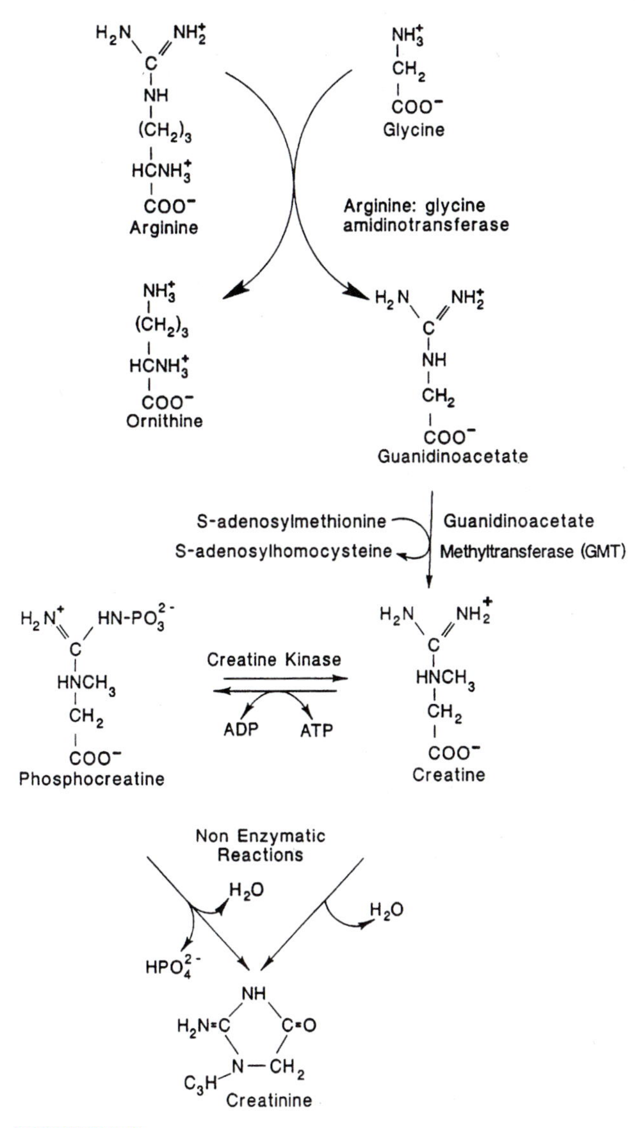

FIGURE 1-9
Pathway of creatine biosynthesis. In GMT deficiency, precursor guanidinoacetate accumulates and the synthesis of creatine and phosphocreatine is severely reduced. Creatinine, a nonmetabolizable end product that is excreted by the renal system, is also diminished.

1.3 Measurement of pH

Blood and urine pH can be measured easily by means of a calibrated glass electrode, whereas pH measurement inside the metabolizing cells is not easily accomplished. Techniques for estimating intracellular pH include glass electrode measurements on homogenates, calorimetric or fluorometric analysis of intracellular distribution of indicator dyes, and microelectrode methods.

Nuclear Magnetic Resonance and Magnetic Resonance Imaging

The noninvasive technique of nuclear magnetic resonance (NMR) spectrometry has been used to measure the concentration of H^+ and other selected ions in isolated cells and tissues. NMR analysis is based on the principle that some atomic nuclei behave like tiny bar magnets because the spinning of charged nuclei generates a magnetic moment along the axis of the spin. If a nucleus with a magnetic dipole (spin) is placed in an external magnetic field, it will acquire an orientation aligned either with the applied field (low-energy state) or against the applied field (high-energy state). The former state is analogous to the way in which a compass needle aligns itself with the earth's magnetic field. Thus, the nuclei, in the presence of an external magnetic field, can remain in either of two *unequal* energy states. If the aligned nuclei are excited with electromagnetic energy of the proper frequency, some of the nuclei in the low-energy state (ground state) will be excited to the high-energy state. Subsequent release of energy by excited nuclei leads to relaxation back to the ground state and completes the resonance cycle between the two energy states.

The NMR spectrum is essentially a measure of the emission of electromagnetic radiation associated with the return of the nuclei from the high-energy state to the low-energy state. Each atomic nucleus has a characteristic spectrum of resonance absorption frequencies that are influenced by the chemical environment surrounding that nucleus and that appear as shifts in the resonance frequency (known as *chemical shifts*). Thus, the chemical shifts (expressed numerically in parts per million [ppm] relative to a standard NMR signal or frequency) can be used to distinguish different chemical compounds containing the same nuclei. NMR spectral features are correlated with spectra of known structures to provide structural information that may permit identification of the molecule

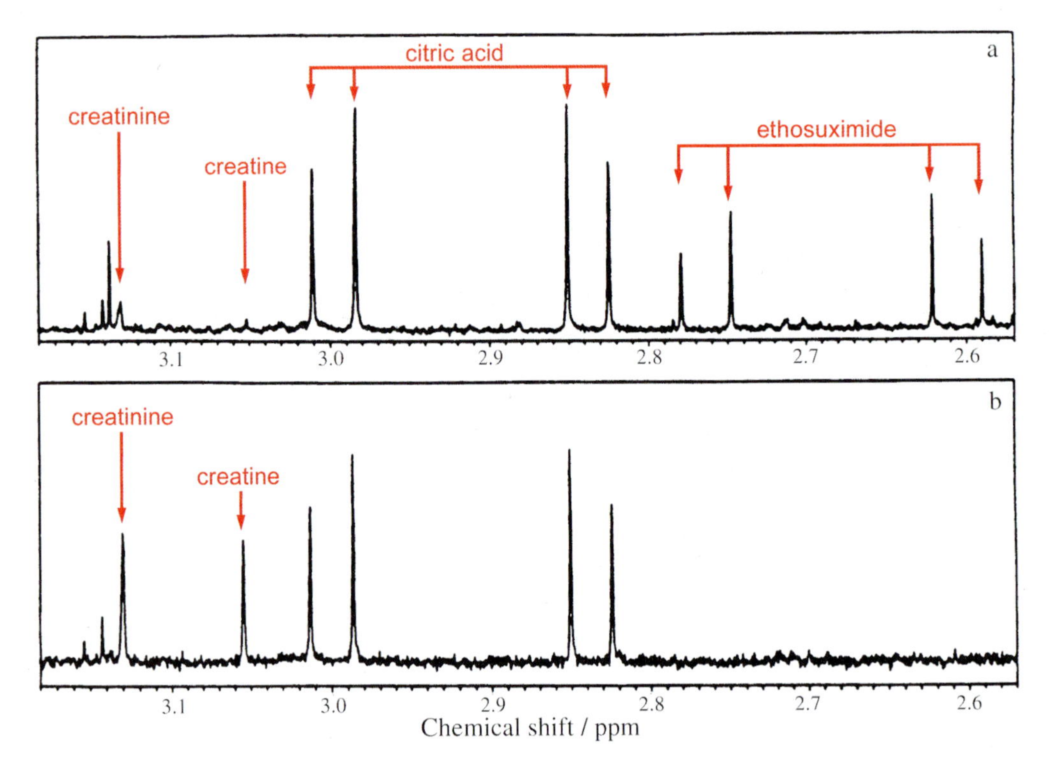

FIGURE 1-10
The ^{1}H-NMR pattern (*in vitro,* 600 MHz) of cerebrospinal fluid (CSF) of a patient with creatine deficiency syndrome (a) compared with normal CSF (b). Note the near absence of creatine and creatinine in the patient's CSF. The ethosuximide observed in the patient's CSF is a drug used in antiepileptic therapy. [Reproduced with permission from A. Schulze et al., Creatine deficiency syndrome caused by guanidinoacetate methyltransferase deficiency*:* diagnostic tools for a new inborn error of metabolism. *J. Pediatr.* **131,** 626 (1997).]

under investigation. Determining the area under the NMR spectrum for a given compound provides a measure of the number of nuclei that are polarized by the magnetic field, and the relative abundance of different nuclei in tissues can be measured.

Atomic nuclei suitable for biological studies include ^{1}H, ^{13}C, ^{15}N, ^{19}F, ^{23}Na, ^{31}P, and ^{39}K. Determination of intracellular pH utilizes ^{31}P-NMR because the resonance frequency of inorganic phosphate (P_i) varies predictably with changes in pH. The exact location of the P_i signal depends on the relative concentrations of $H_2PO_4^-$ and HPO_4^{2-}, which in turn depend on the intracellular pH and pK′ of inorganic phosphate under physiological conditions. Thus, the position of the P_i signal in the NMR spectrum provides a measure of intracellular pH. Apart from measurements of H^+ concentrations in several tissues under varying physiological conditions, NMR analysis of muscle tissues in a patient's forearm has been used to diagnose a hereditary defect in the breakdown of muscle glycogen (*McArdle's syndrome*). In normal individuals during exercise, glycogen in the muscle breaks down to lactic acid, thus decreasing the pH. However, in patients with McArdle's syndrome, the pH does not change with exercise because glycogen is not catabolized to lactic acid (Chapter 15). ^{1}H- and ^{31}P-NMR spectroscopy has been used to measure metabolically significant components in tissues and fluids, including brain and cerebrospinal fluid. In a 4-year-old female patient with a defect in creatine formation due to a deficiency of guanidinoacetate methyltransferase, phosphocreatine synthesis is severely diminished in many tissues including the brain (Figure 1-9). Determination of creatine and creatinine in cerebrospinal fluid by NMR can be used to diagnose *creatine deficiency syndrome* (Figure 1-10). Creatinine is an end product of creatine and phosphocreatine (Chapter 17). The phosphocreatine pool is essential for the storage and transfer of energy via high-energy phosphate compounds (ATP). In this patient who exhibited dystonic-dyskinetic syndrome, seizures, and psychomotor retardation, NMR spectroscopy revealed a depletion of creatine, phosphocreatine, and accumulation of guanidinoacetate (Figure 1-10). Oral administration of creatine resulted in clinical improvement.

High-quality anatomical cross-sectional images can be produced by the use of tomographic methods with NMR instead of x-rays as a probe. Magnetic resonance images

(MRI) are generated by measuring the relaxation time of return to equilibrium for hydrogen nuclei in a constant magnetic field following excitation by a radiofrequency pulse. The time taken by the hydrogen nuclei to return to their original position when the radiofrequency pulse is terminated is known as the T1-relaxation time. The time taken for the hydrogen nuclei to lose the energy that they acquire during the radiofrequency pulse sequence is known as the T2 relaxation time. T2 is always less than T1 and the quality of the MRI depends upon the concentration of hydrogen nuclei (known as proton density or spin density) and the weight given to the T1 and T2 components. In T1-weighted images, lipids have a characteristic short T1 relaxation time and are hyperintense (bright), whereas water has a long T1 relaxation time and is hypointense (dark). Thus, tissues rich in fat appear bright and tissues rich in free water appear dark. In T2 weighted images the opposite is true; lipids appear dark and water bright.

The resolution of anatomical structures is achieved by virtue of different relaxation times of hydrogen nuclei in different tissues. Differences in relaxation time reflect gross chemical characteristics, including fat content, degree of hydration, and presence of paramagnetic substances. Proton density is also an important parameter in determining the intensity of an image, so that soft tissues as well as bone can readily be visualized, making the technique superior to x-ray and other methods of imaging the brain and other soft tissues.

Use of the intravascular contrast agent gadolinium-diethylamine pentaacetic acid (GdDPTA) during the MRI procedure enhances the T1 relaxation time of hydrogen nuclei. This alters the magnetic susceptibility of adjacent tissue and provides information on the integrity of the blood-brain barrier. MRI is the diagnostic procedure of choice in several neurologic diseases. In one autoimmune inflammatory demyelinating disorder of the central nervous system, *multiple sclerosis* (MS), MRI is the preferred imaging procedure both in diagnosis and as a prognostic tool (Figure 1-11). MS is a progressive degenerative disease and exhibits scattered focal lesions of the myelin sheath of the axons. MS usually manifests during the third or fourth decade of life and affects more women (60%) than men (40%). No known risks are associated with MRI, which is another advantage of the procedure.

Gibbs–Donnan Equilibrium

The bicarbonate-carbonic acid buffer system plays a major role in regulating the pH of fluids in tissue spaces outside blood vessels. This fluid, commonly referred to as interstitial fluid and separated from plasma by the membrane barrier known as the capillary endothelium, primarily

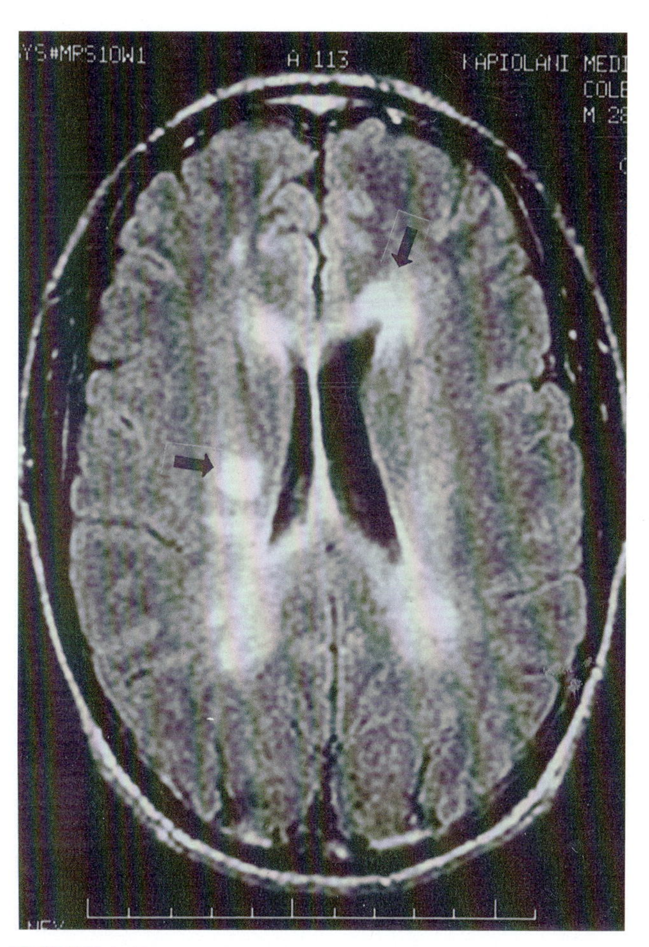

FIGURE 1-11
Magnetic resonance image (T1 weighted) of brain from a patient with multiple sclerosis. The image obtained is a horizontal section at the level of the head of the caudate nucleus showing characteristic marked increase of signal as indicated by arrows. (Courtesy of Robert M. DiMauro and John M. Hardman.)

contains the diffusible ions, Na^+, K^+, Cl^-, and HCO_3^-. Plasma contains proteins in addition to diffusible ions. Membranes (Chapter 10) have a lipid-protein fluid mosaic structure and the membrane proteins may occupy surface positions or extend through the lipid bilayer (Figure 1-12).

Plasma proteins are polyionic at pH 7.4 and cannot diffuse across membranes. The normal difference in concentrations of diffusible ions between the plasma and interstitial compartments is due to the presence of nondiffusible protein in plasma, shown in Table 1-5.

The difference is explained by Gibbs' theory of equilibria and was studied experimentally by Donnan; the overall process is known as the *Gibbs–Donnan equilibrium*.

Gibbs–Donnan equilibria can best be understood in a two-compartment system. Compartment 1 contains the sodium salt of an anionic protein ($Na_n^+P^{n-}$) at an initial concentration C_1, with n representing the number of charges; compartment 2 contains NaCl at an initial

TABLE 1-5
*Concentration of Cations and Anions in Plasma Water and Interstitial Fluid**

Ion	Plasma Water†		Interstitial Fluid‡	
	mEq/L	mmol/L	mEq/L	mmol/L
Cations				
Na^+	153.2	153.2	145.1	145.1
K^+	4.3	4.3	4.1	4.1
Ca^{2+}	3.8	1.9§	3.4	1.7
Mg^{2+}	1.4	0.7ᴷ	1.3	0.65
Total	162.7	160.1	153.9	151.6
Anions				
Cl^-	111.5	111.5	118	118
HCO_3^-	25.7	25.7	27	27
$H_2PO_4^-$–HPO_4^{2-}	2.2	0.66	2.3	0.7
Other	6.3	5.9	6.6	6.2
Protein	17.0	1.5	0	0
Total	162.7	145.3	153.9	151.9
Total mOsm per liter		305.4		303.5

*Reproduced, with permission, from D. M. Woodbury in *Physiology and Biophysics.* T. Ruch, H. Patton, and A. Scher, Eds. (Saunders, 1974).
†Plasma water content assumed to be 93%.
‡Gibbs-Donnan factors used as multipliers are 0.95 for monovalent cations, 0.9 for divalent cations, 1.05 for monovalent anions, and 1.10 for divalent anions.
§Total Ca is 2.7 mmol/L; ionized Ca is about 70% of total Ca.
ᴷTotal Mg is 1 mmol/L; ionized Mg is about 65% of total Mg.

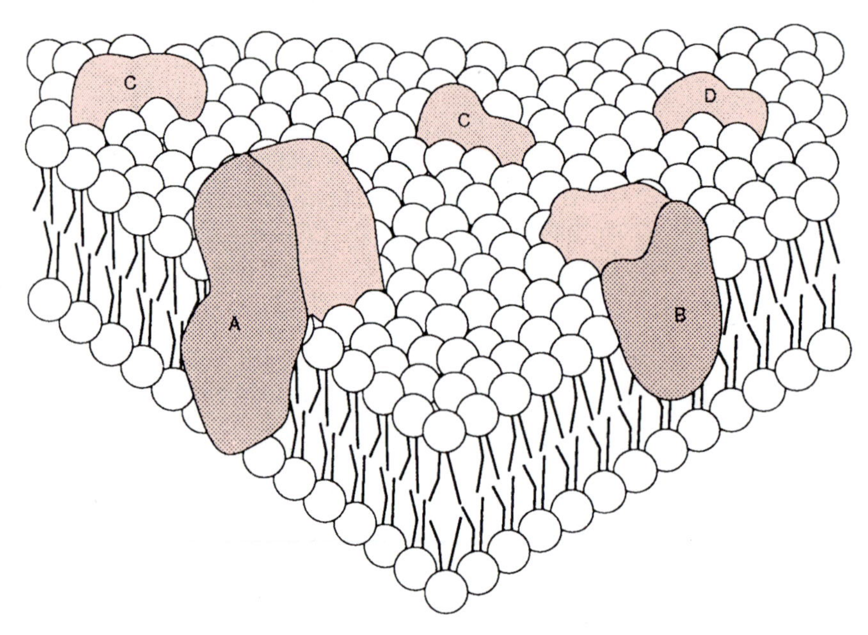

FIGURE 1-12
Fluid mosaic model of a membrane consists of lipids and proteins. Globular proteins are interspersed within the lipid matrix: Protein A traverses the entire bilayer; protein B is embedded in one leaflet; proteins C and D are on the periphery.

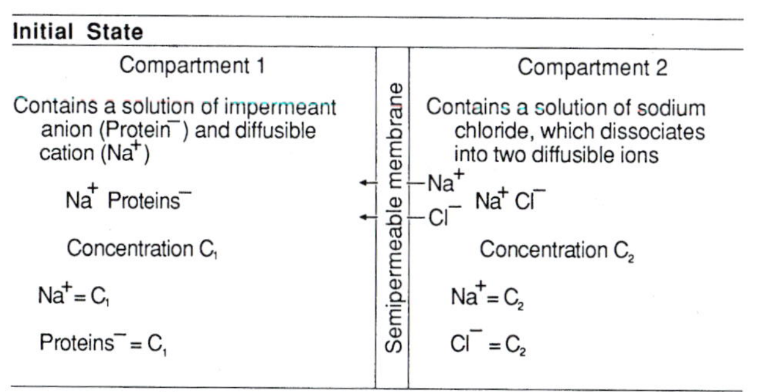

Since there is Cl^- concentration gradient, Cl^- migrates (along with Na^+) from 2 to 1. Let this be x.

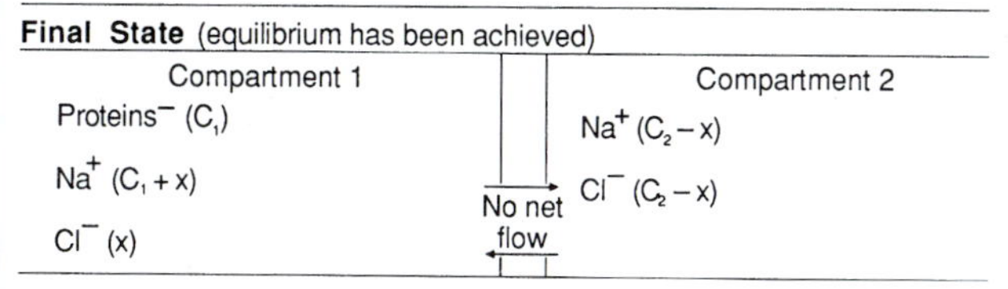

FIGURE 1-13
Gibbs–Donnan equilibrium exists in systems consisting of two fluid compartments separated by a semipermeable membrane, which permits diffusion of some ions (e.g., Na^+, Cl^-) but is impermeable to protein anions. Note that the concentration of diffusible cation in compartment 1 is greater than in compartment 2 and that osmotic differences exist between compartments 1 and 2.

concentration of C_2 (Figure 1-13). For simplicity, assume $n = 1$. The initial concentration of Cl^- is higher in compartment 2 than in compartment 1 so that Cl^- will diffuse into compartment 2. Na^+ also migrates into compartment 2 to maintain electrical neutrality. This net migration occurs until an equilibrium is reached, i.e., when the rate of ion diffusion from 2 into 1 equals that from 1 into 2. If x represents the net concentration of Na^+ or Cl^- transferred to compartment 1, the final equilibrium concentrations differ from the initial values by $\pm x$ (Figure 1-13). The rate of diffusion of NaCl from 2 into 1 is proportional to the product of the concentrations of Na^+ and Cl^- in 2, which is $(C_2 - x)^2$. Similarly, the rate of diffusion of NaCl from 1 into 2 is proportional to the product of the concentrations of Na^+ and Cl^- in 1, which is $(C_1 + x)x$. At equilibrium, the diffusion rates from 2 into 1 and 1 into 2 are equal:

$$(C_1 + x)x = (C_2 - x)^2$$

From which,

$$x = \frac{C_2^2}{C_1 + 2C_2}$$

This unequal equilibrium distribution of solutes depends upon the concentration of both the nondiffusible protein anion C_1 and the value of C_2. This disparity in ion concentrations also causes differences in osmotic pressure.

In the above example, if H^+ replaced Na^+ as the diffusible cation, the Gibbs–Donnan effect would lead to a pH change; compartment 1 would have a decrease in pH, while compartment 2 would have an increase in pH. Consistent with this, relatively protein-rich plasma has a higher $[H^+]$ (i.e., lower pH) than that of protein-poor interstitial fluid. Although Gibbs–Donnan equilibria affect ionic concentrations between many compartments (e.g., blood/interstitial fluid or plasma/red blood cells), major ionic gradients between various compartments are maintained at the expense of energy-requiring transport systems (e.g., ATP-dependent Na^+ and K^+ transport at cell membranes).

1.4 H^+ Concentration and pH

The use of pH to designate $[H^+]$ is due to the fact that a broad range of $[H^+]$ can be compressed within the manageable numerical scale of 0–14. However, in clinical acid–base problems, use of the pH scale has some disadvantages. Since the pH is the logarithm of the reciprocal of $[H^+]$, significant variations of $[H^+]$ in a patient may not be fully appreciated. For example, if the blood pH decreases from 7.4 to 7.1, $[H^+]$ is doubled; or if the pH increases from 7.4 to 7.7, $[H^+]$ is halved (Figure 1-14). In addition, the use of the pH scale masks the relationship between $[H^+]$ and the concentrations of other cations, e.g., Na^+ and K^+. Thus, in clinical situations it is preferable to express

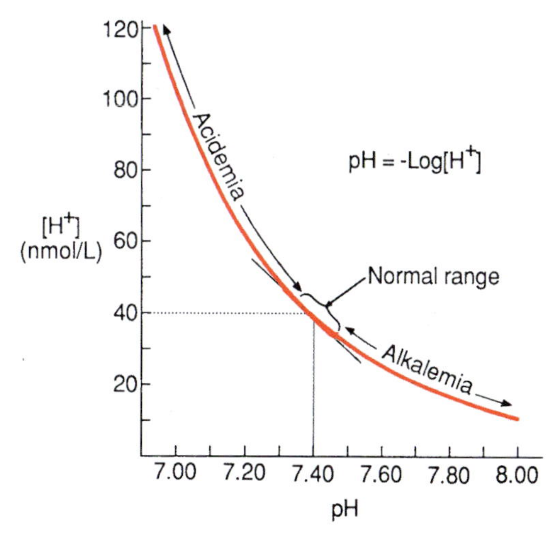

FIGURE 1-14
The relationship of pH to hydrogen ion concentration (in nanomoles per liter). The normal blood pH of 7.40 corresponds to 40 nmol/L of H^+. The solid straight line is drawn to show the linear relationship between the concentration of H^+ and pH, over the pH range of 7.20–7.50. A 0.01-unit change in pH is equivalent to about 1.0 nmol/L change in the opposite direction.

$[H^+]$ directly as nanomoles per liter in order to better evaluate acid–base changes and interpret laboratory tests. A blood pH of 7.40 corresponds to 40 nM $[H^+]$, which is the mean of the normal range (Figure 1-14). The normal range is 7.36–7.44 on the pH scale, or 44–36 nM $[H^+]$. If the pH of blood falls below pH 7.36 ($[H^+] > 44$ nM), the condition is called *acidemia*. Conversely, if the pH rises above pH 7.44 ($[H^+] < 36$ nM), the condition is called *alkalemia*. The suffix *-emia* refers to blood and usually to an abnormal concentration in blood. Over the pH range of 7.20–7.50, for every change of 0.01 pH unit, there is a change of approximately 1 nM $[H^+]$ in the opposite direction.

Since the Henderson–Hasselbalch expression uses pH terms, its utility in clinical situations is less than optimal. Kassirer and Bleich have derived a modified Henderson–Hasselbalch expression that relates $[H^+]$, instead of pH, to P_{CO_2} and HCO_3^-, as follows:

$$H_2CO_3 \rightleftarrows H^+ + HCO_3^-$$

for which

$$K' = \frac{[HCO_3^-][H^+]}{[H_2CO_3]}$$

Substituting $a \times P_{CO_2}$ for $[H_2CO_3]$, as in Equation (1.16), yields

$$K' = \frac{[HCO_3^-][H^+]}{a \times P_{CO_2}}$$

and

$$H^+ = \frac{K'a \times P_{CO_2}}{[HCO_3^-]} \qquad (1.22)$$

In Equation (1.22), K' and a are constants, and the numerical value of $K'a$ is 24 when P_{CO_2} is expressed in mm Hg, $[HCO_3^-]$ in mM, and $[H^+]$ in nM. Therefore,

$$H^+ = \frac{24 \times P_{CO_2}}{[HCO_3^-]} \qquad (1.23)$$

The above formulation expresses the interdependence of three factors; if two of them are known, the third can be readily calculated. For example, at a blood $[HCO_3^-]$ of 24 nM and a P_{CO_2} of 40 mmHg, $[H^+]$ is 40 mM. Clinical applications of Equation (1.23) are discussed in Chapter 39.

Supplemental Readings and References

Properties of Water

M. F. Colombo, D. C. Rau, and V. A. Parsegian: Protein solvation in allosteric regulation: A water effect on hemoglobin. *Science* **256,** 655 (1992).

M. A. Knepper: Molecular physiology of urinary concentrating mechanism: Regulation of aquaporin water channels by vasopressin. *American Journal of Physiology* **272** (Renal Physiology **41**), F3 (1997).

M. A. Knepper, J. G. Verbalis, and S. Nielsen: Role of aquaporins in water balance disorders. *Current Opinion in Nephrology and Hypertension* **6,** 367 (1997).

M. D. Lee, L. S. King, and P. Agre: The aquaporin family of water channel proteins in clinical medicine. *Medicine* **76,** 141 (1997).

R. P. Rand: Raising water to new heights. *Science* **256,** 618 (1992).

P. M. Wiggins: Role of water in some biological processes. *Microbiological Reviews* **54,** 432 (1990).

Acid–Base Chemistry and Respiratory Function of Hemoglobin

H. J. Androgue and N. E. Madias: Management of life-threatening acid–base disorders. *New England Journal of Medicine*; First of two parts **338,** 26 (1998); Second of two parts **338,** 107 (1998).

C. C. W. Hsia: Respiratory function of hemoglobin. *New England Journal of Medicine* **338,** 239 (1998).

Nuclear Magnetic Resonance and Magnetic Resonance Imaging

R. R. Edeman and S. Warach: Medical progress: Magnetic resonance imaging. *New England Journal of Medicine*; First of two parts **328,** 708 (1993); Second of two parts **328,** 785 (1993).

S. Gilman: Medical progress: Imaging of the brain. *New England Journal of Medicine*; First of two parts **338,** 812 (1998); Second of two parts **338,** 889 (1998).

S. J. Knoury and H. L. Weiner: Multiple sclerosis. *Archives of Internal Medicine* **158,** 565 (1998).

L. A. Moulopoulos and M. A. Dimopoulos: Magnetic resonance imaging of the bone marrow in the hematologic malignancies. *Blood* **90,** 2127 (1997).

A. Schulze, T. Hess, R. Wevers, et al.: Creatine deficiency syndrome caused by guanidinoacetate methyltransferase deficiency: Diagnostic tools for a new inborn error of metabolism. *Journal of Pediatrics* **131,** 626 (1997).

S. Stöckler-Ipsiroglu: Creatine deficiency syndromes: A new perspective on metabolic disorders and a diagnostic challenge. *Journal of Pediatrics* **131,** 510 (1997).

CHAPTER 2

Amino Acids

Proteins are the most abundant class of organic compounds in the healthy, lean human body, constituting more than half of its cellular dry weight. Proteins are polymers of amino acids and have molecular weights ranging from approximately 10,000 to more than one million. Biochemical functions of proteins include catalysis, transport, contraction, protection, structure, and metabolic regulation.

Amino acids are the monomeric units, or building blocks, of proteins joined by a specific type of covalent linkage. The properties of proteins depend on the characteristic sequence of component amino acids, each of which has distinctive side chains.

Amino acid polymerization requires elimination of a water molecule as the carboxyl group of one amino acid reacts with the amino group of another amino acid to form a covalent amide bond. The repetition of this process with many amino acids yields a polymer, known as a ***polypeptide.*** The amide bonds linking amino acids to each other are known as ***peptide bonds.*** Each amino acid unit within the polypeptide is referred to as a ***residue.*** The sequence of amino acids in a protein is dictated by the sequence of nucleotides in a segment of the DNA in the chromosomes and the uniqueness of each living organism is due to its endowment of specific proteins.

2.1 L-α-Amino Acids: Structure

Almost all of the naturally occurring amino acids in proteins are L-α-amino acids. The principal 20 amino acids in proteins have an amino group, a carboxyl group, a hydrogen atom, and an R-group attached to the α-carbon (Figure 2-1). *Proline* is an exception because it has a cyclic structure and contains a secondary amine group (called an imino group) instead of a primary amine group (called an amino group). Amino acids are classified according to the chemical properties of the R-group. Except for glycine (R = H), the amino acids have at least one asymmetrical carbon atom (the α-carbon). The absolute configuration of the four groups attached to the α-carbon is conventionally compared to the configuration of L-glyceraldehyde (Figure 2-2). The D and L designations specify absolute configuration and not the dextro (right) or levo (left) direction of rotation of plane-polarized light by the asymmetrical carbon center. In organic chemistry, the assignment of absolute configurations of an asymmetrical center is made by the R and S classification of isomers. This system prioritizes the substituents linked to the asymmetrical carbon atom (e.g., decreasing atomic number or valence density) and the assignment is based upon the clockwise (R) or the counterclockwise (S) positioning of the three higher priority groups.

2.2 Classification

A useful classification of the amino acids is based on the solubility (i.e., ionization and polarity) characteristics of the side chains (R-groups). The R-groups fall into four classes:

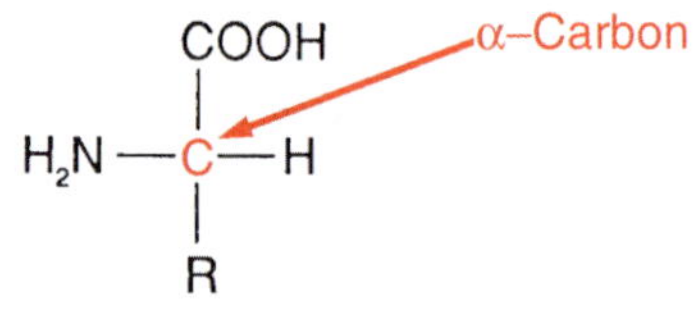

COOH
$H_2N—C—H$
R
α–Carbon

FIGURE 2-1
Basic structure of an α-amino acid.

1. Nonpolar (hydrophobic),
2. Polar, negatively charged (acidic),
3. Polar, positively charged (basic), and
4. Polar, neutral (un-ionized).

Within each class, R-groups differ in size, shape, and other properties. Figure 2-3 shows the structure of each amino acid according to this classification with the R-group outlined. Ionizable structures are drawn as they would exist at pH 7.0. The three-letter and one-letter abbreviations for each amino acid are given in Table 2-1. A *-yl* ending on amino acid residue indicates that the carboxyl group of an amino acid is linked to another functional group (e.g., in a peptide bond).

The eight ***essential amino acids*** (Table 2-1) are those which humans cannot synthesize and which must be supplied in the diet. The remaining amino acids are synthesized in the body by various biochemical pathways (Chapter 17).

D-Glyceraldehyde L-Glyceraldehyde

Fischer perspective formulas

Fischer projection formulas

D-Alanine L-Alanine

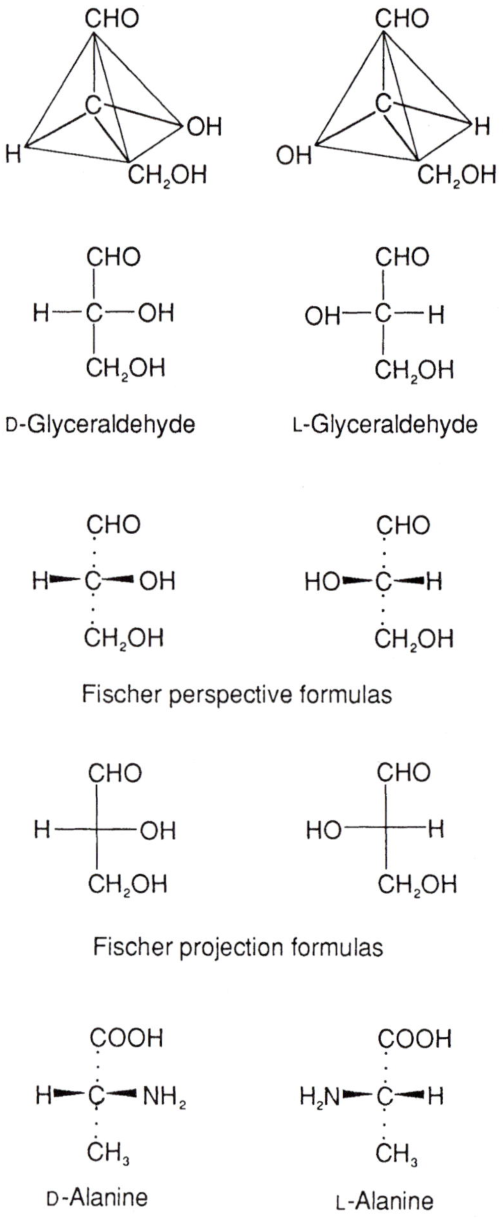

FIGURE 2-2
Different representations of the configurational stereoisomers of glyceraldehyde and of alanine.

Nonpolar Amino Acids

Glycine

Glycine is the smallest amino acid and has an H atom as its R-group. It is the only α-amino acid that is not optically active. The small R-group provides a minimum of steric hindrance to rotation about bonds; therefore, glycine fits into crowded regions of many peptide chains. Collagen, a rotationally restricted fibrous protein, has glycyl residues in about every third position in its polypeptide chains. Glycine is used for the biosynthesis of many nonprotein compounds, such as porphyrins and purines.

Glycine and taurine are conjugated with bile acids, products derived from cholesterol, before they are excreted into the biliary system. Conjugated bile acids are amphipathic and are important in lipid absorption (Chapter 12). Glycine also is a neurotransmitter; it is inhibitory in the spinal cord and excitatory in the cerebral cortex and other regions of the forebrain. *Nonketotic hyperglycinemia* (NKH) is an inborn error of glycine degradation in which a large amount of glycine accumulates throughout the body. NKH causes severe consequences in the central nervous system (CNS) and leads to death (Chapter 17).

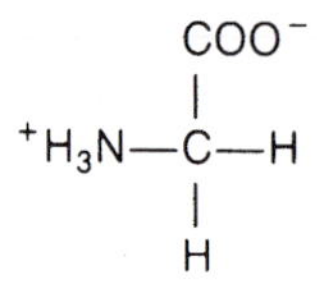

COO^-
$^+H_3N—C—H$
H

Glycine

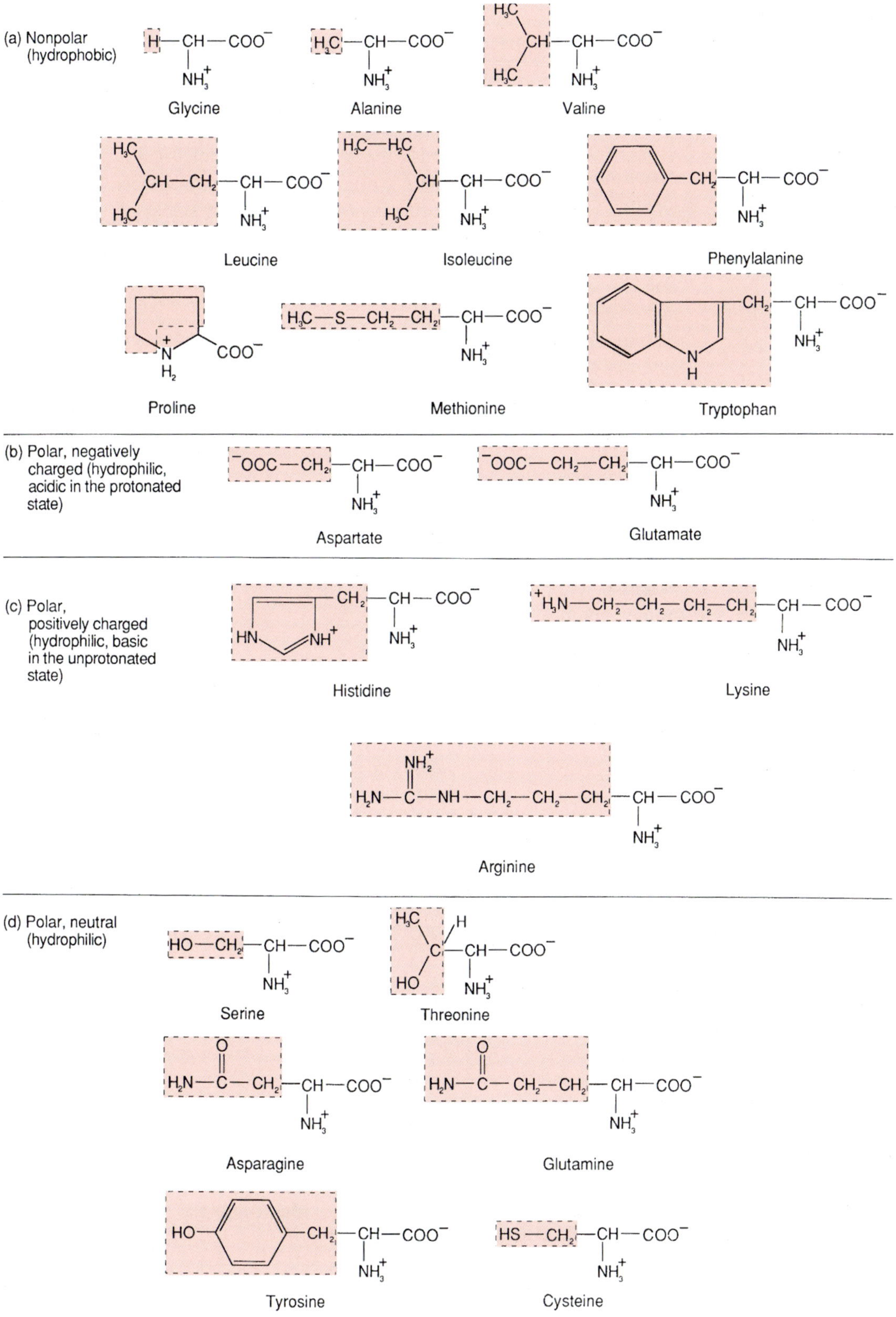

(a) Nonpolar (hydrophobic)

Glycine Alanine Valine

Leucine Isoleucine Phenylalanine

Proline Methionine Tryptophan

(b) Polar, negatively charged (hydrophilic, acidic in the protonated state)

Aspartate Glutamate

(c) Polar, positively charged (hydrophilic, basic in the unprotonated state)

Histidine Lysine

Arginine

(d) Polar, neutral (hydrophilic)

Serine Threonine

Asparagine Glutamine

Tyrosine Cysteine

FIGURE 2-3
Classification of commonly occurring L-α-amino acids based on the polarity of side chains (R-groups) at pH 7.0.

TABLE 2-1
Amino Acid Abbreviations and Nutritional Property

Amino acid	Abbreviation	Designation letter	Nutritional property*
Alanine	Ala	A	non-essential
Arginine	Arg	R	conditionally essential
Asparagine	Asn	N	non-essential
Aspartic acid	Asp	D	non-essential
Cysteine	Cys	C	non-essential
Glutamic acid	Glu	E	non-essental
Glutamine	Gln	Q	conditionally essential
Glycine	Gly	G	non-essential
Histidine	His	H	conditionally essential
Isoleucine	Ile	I	essential
Leucine	Leu	L	essential
Lysine	Lys	K	essential
Methionine	Met	M	essential
Phenylalanine	Phe	F	essential
Proline	Pro	P	non-essential
Serine	Ser	S	non-essential
Threonine	Thr	T	essential
Tryptophan	Trp	W	essential
Tyrosine	Tyr	Y	non-essential
Valine	Val	V	essential

*The eight essential amino acids are not synthesized in the body and have to be supplied in the diet. The conditionally essential amino acids, although synthesized in the body, may require supplementation during certain physiological conditions such as pregnancy. The non-essential amino acids can be synthesized from various metabolites.

Alanine

The side chain of alanine is a hydrophobic methyl group, $-CH_3$. Other amino acids may be considered to be chemical derivatives of alanine, with substituents on the β-carbon. Alanine and glutamate provide links between amino acid and carbohydrate metabolism (Chapter 22).

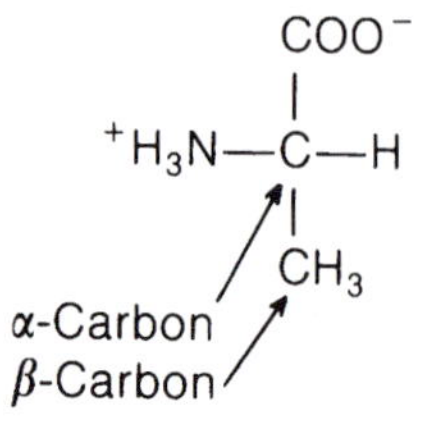

COO^- / $^+H_3N-C-H$ / CH_3

α-Carbon
β-Carbon

Alanine

Valine, Leucine, and Isoleucine

These branched-chain aliphatic amino acids contain bulky nonpolar R-groups and participate in hydrophobic interactions. All three are essential amino acids. A defect in their catabolism leads to *maple syrup urine disease* (Chapter 17). Isoleucine has asymmetrical centers at both the α- and β-carbons and four stereoisomers, only one of which occurs in protein. The bulky side chains tend to associate in the interior of water-soluble globular proteins. Thus, the hydrophobic amino acid residues stabilize the three-dimensional structure of the polymer.

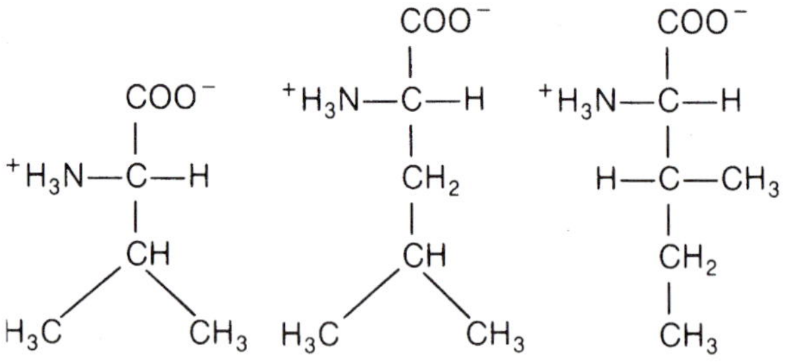

Valine Leucine Isoleucine

Phenylalanine

A planar hydrophobic phenyl ring is part of the bulky R-group of phenylalanine. It is an essential amino acid whose metabolic conversion to tyrosine is defective in *phenylketonuria* (Chapter 17). Phenylalanine, tyrosine, and tryptophan are the only α-amino acids that contain aromatic groups and consequently are the only residues that absorb ultraviolet (UV) light (Figure 2-4). Tryptophan and tyrosine absorb significantly more energy than phenylalanine at 280 nm, the wavelength generally used to measure the concentration of protein in a solution.

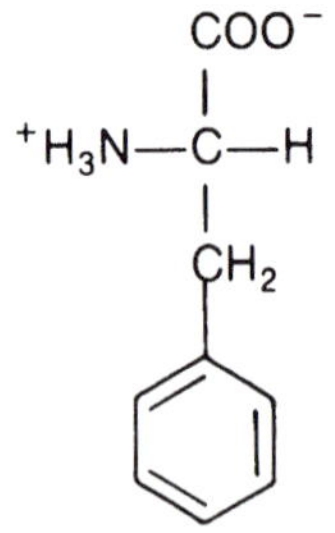

Phenylalanine

Tryptophan

A bicyclic nitrogenous aromatic ring system (known as an indole ring) is attached to the β-carbon of alanine to form the R-group of tryptophan. Tryptophan is a precursor of serotonin, melatonin, nicotinamide, and many

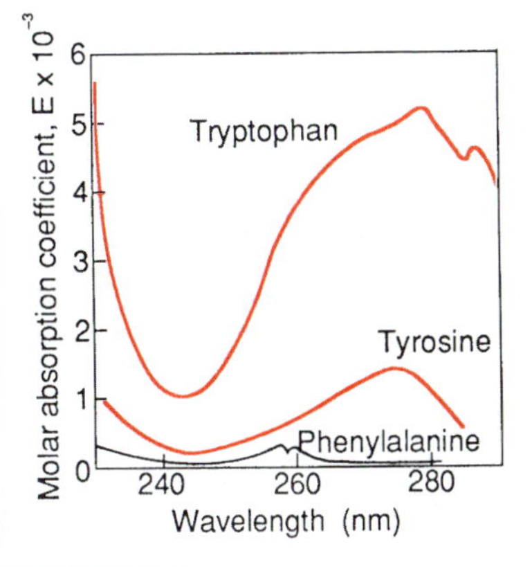

FIGURE 2-4
UV absorption spectra of phenylalanine, tyrosine, and tryptophan.

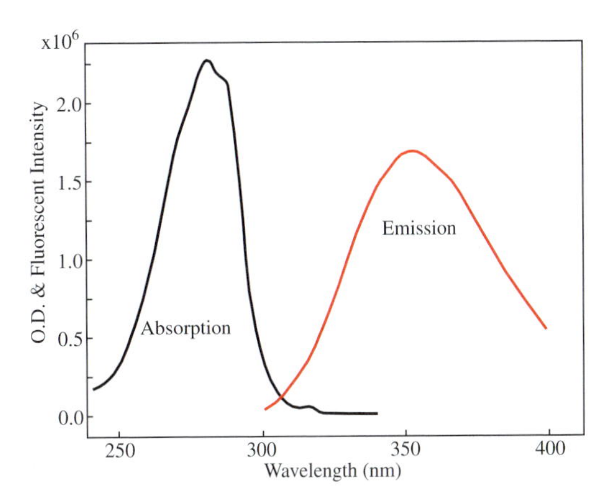

FIGURE 2-5
Tryptophan fluorescence spectrum. The emission spectrum appears at longer wavelengths as compared to the absorption spectrum.

naturally occurring medicinal compounds derived from plants. It is an essential amino acid. The indole group absorbs UV light at 280 nm—a property that is useful for spectrophotometric measurement of protein concentration (Figure 2-4).

Tryptophan and tyrosine both show fluorescence; however, tryptophan absorbs more intensely. When molecules are raised to a higher energy state by absorption of radiant energy, they generally are unstable and return to the ground state. The energy released in this process manifests as heat (radiation energy) or light. The process of light emission is known as ***fluorescence.*** The quantum of energy re-emitted as fluorescence is always less than that of absorbed energy. Thus, the fluorescent light always appears at a longer wavelength (lower energy) than the original absorbed light energy (Figure 2-5). Tryptophan fluorescence studies can provide valuable information regarding protein and protein-ligand conformations due to the effects of surrounding amino acid residues.

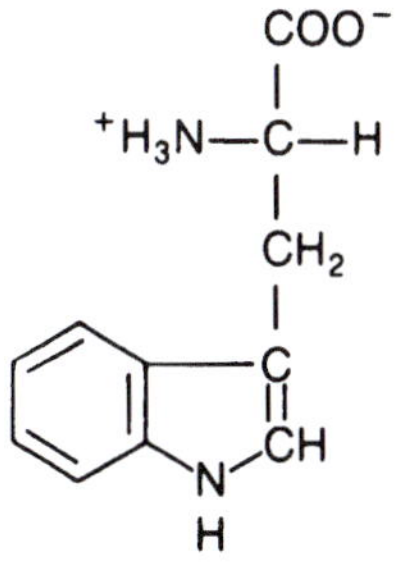

Tryptophan

Methionine

This essential amino acid contains an R-group with a methyl group attached to sulfur. Methionine serves as donor of a methyl group in many transmethylation reactions, e.g., in the synthesis of epinephrine, creatine, and melatonin. Almost all of the sulfur-containing compounds of the body are derived from methionine.

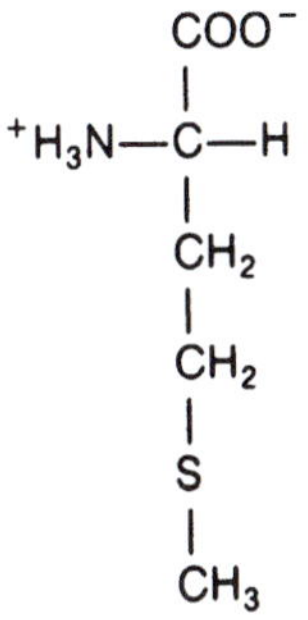

Methionine

Proline

Proline contains a secondary amine group, called an imine, instead of a primary amine group. For this reason, proline is called an imino acid. Since the three-carbon R-group of proline is fused to the α-nitrogen group, this compound has a rotationally constrained rigid-ring structure. As a result, prolyl residues in a polypeptide introduce restrictions on the folding of chains. In collagen, the principal protein of human connective tissue, certain prolyl residues are hydroxylated (Figure 2-6). The hydroxylation occurs during protein synthesis and requires ascorbic acid (vitamin C) as a cofactor. Deficiency of vitamin C causes formation of defective collagen and ***scurvy*** (Chapters 25 and 38).

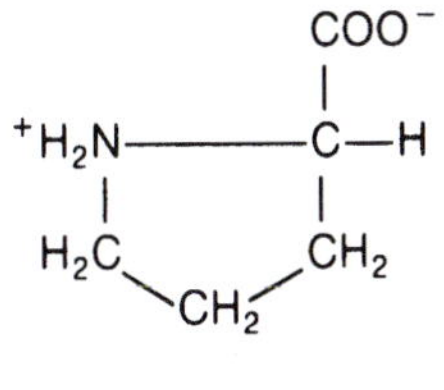

Proline

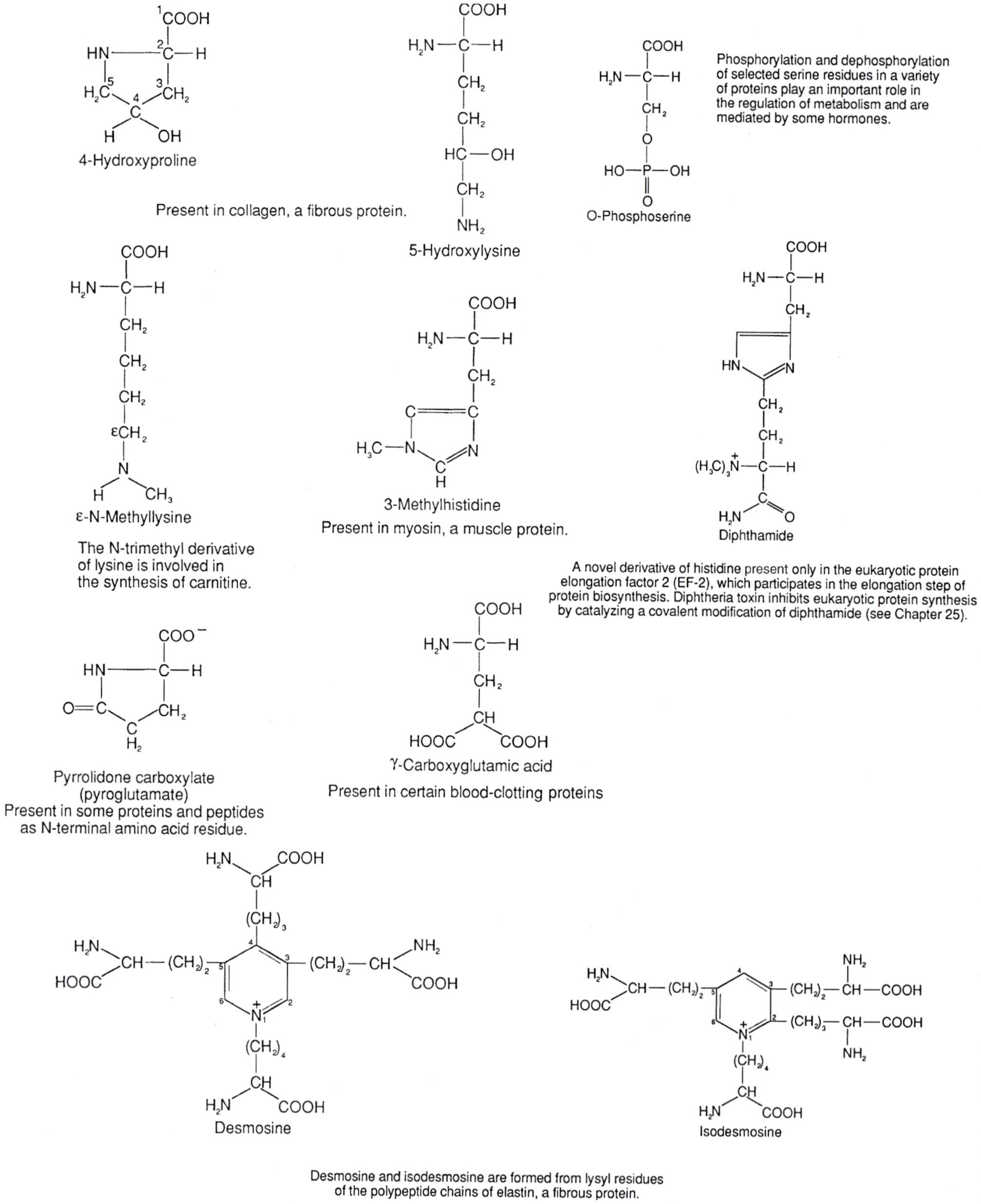

4-Hydroxyproline

5-Hydroxylysine

Present in collagen, a fibrous protein.

O-Phosphoserine

Phosphorylation and dephosphorylation of selected serine residues in a variety of proteins play an important role in the regulation of metabolism and are mediated by some hormones.

ε-N-Methyllysine

The N-trimethyl derivative of lysine is involved in the synthesis of carnitine.

3-Methylhistidine

Present in myosin, a muscle protein.

Diphthamide

A novel derivative of histidine present only in the eukaryotic protein elongation factor 2 (EF-2), which participates in the elongation step of protein biosynthesis. Diphtheria toxin inhibits eukaryotic protein synthesis by catalyzing a covalent modification of diphthamide (see Chapter 25).

Pyrrolidone carboxylate (pyroglutamate)

Present in some proteins and peptides as N-terminal amino acid residue.

γ-Carboxyglutamic acid

Present in certain blood-clotting proteins

Desmosine

Isodesmosine

Desmosine and isodesmosine are formed from lysyl residues of the polypeptide chains of elastin, a fibrous protein.

FIGURE 2-6
Modified derivatives of certain amino acids are found in proteins.

Acidic Amino Acids

Aspartic Acid

The β-carboxylic acid group of aspartic acid has a pK′ of 3.86 and is ionized at pH 7.0 (the anionic form is called *aspartate*). The anionic carboxylate groups tend to occur on the surface of water-soluble proteins, where they interact with water. Such surface interactions stabilize protein folding.

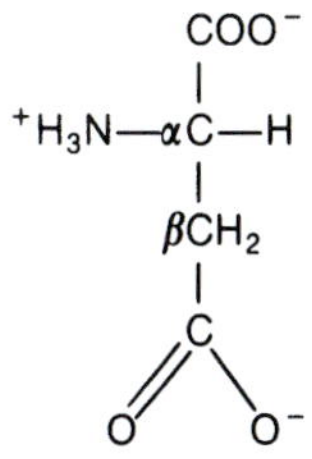

Aspartate

Glutamic Acid

The γ-carboxylic acid group of glutamic acid has a pK′ of 4.25 and is ionized at physiological pH. The anionic groups of glutamate (like those of aspartate) tend to occur on the surfaces of proteins in aqueous environments. Glutamate is the primary excitatory neurotransmitter in the brain. Its levels are regulated by clearance that is mediated by glutamate transfer protein in critical motor control areas in the CNS. In *amyotrophic lateral sclerosis* (ALS) glutamate levels are elevated in serum, spinal fluid, and brain; glutamate excitotoxicity is implicated in the progression of the disease. ALS is a progressive disorder affecting motor neurons in the spinal cord, brain stem, and cortex. The precise molecular basis of the disease is unknown; however, factors involved are glutamate excitotoxicity, genetics, oxidative stress, and diminished neurotrophic factors.

Two drugs that provide neuroprotection against glutamate excitotoxicity are riluzole and gabapentin. Gabapentin is an amino acid structurally related to the neurotransmitter ***γ-aminobutyrate*** (GABA). GABA, an inhibitory neurotransmitter in the CNS, is produced by the decarboxylation of glutamate by glutamate decarboxylase, a pyridoxal phosphate dependent enzyme. GABA, when bound to its receptors, causes an increase in permeability to chloride ions in neuronal cells. A group of tranquilizing drugs known as benzodiazopines enhance the membrane permeability of chloride ions by GABA. In some proteins, the γ-carbon of glutamic acid contains an additional carboxyl group. Residues of ***γ-carboxyglutamic acid*** (Gla) bear two negative charges and can strongly bind calcium ions. γ-Carboxylation of glutamic acid residues is a posttranslational modification and requires ***vitamin K*** as a cofactor. γ-Carboxyglutamate residues are present in a number of blood coagulation proteins (factors II, VII, IX, and X) and anticoagulant proteins C and S (Chapter 36). Osteocalcin, a protein present in the bone, also contains γ-carboxyglutamate residues (Chapter 37).

A cyclic, internal amide derivative of glutamic acid is pyrrolidone carboxylic acid (also known as pyroglutamic acid or 2-oxoproline). Some proteins (e.g., heavy chains of immunoglobulins; Chapter 35) and peptides (e.g., thyrotropin-releasing hormone; Chapter 33) have pyroglutamic acid as their N-terminal amino acid residue.

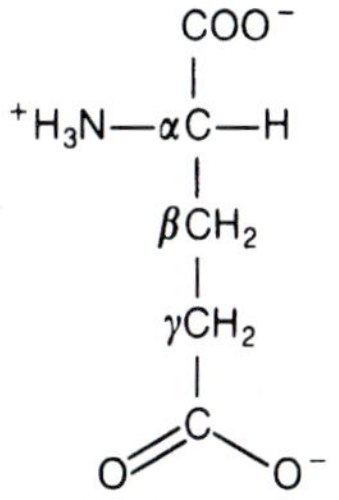

Glutamate

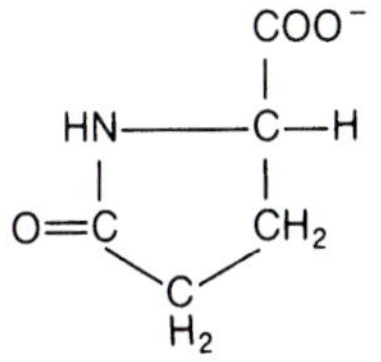

Pyrrolidone carboxylate (pyroglutamate)

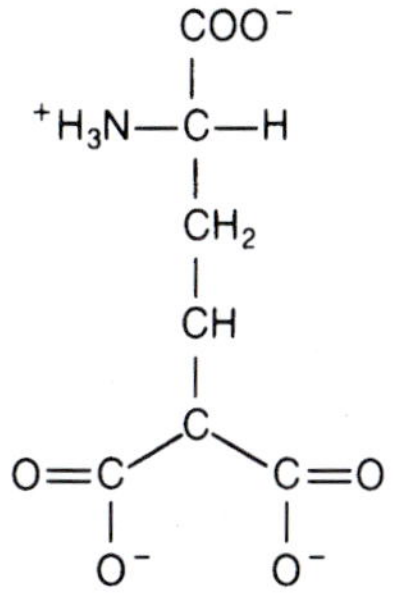

γ–Carboxy glutamate

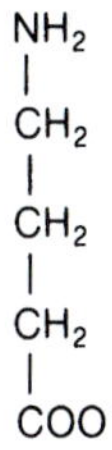

γ–amino butyrate (GABA)

Basic Amino Acids

Lysine

Lysine is an essential amino acid. The long side chain of lysine has a reactive amino group attached to the ε-carbon. The ε-NH_2 (pK′ = 10.53) is protonated at physiological pH. The lysyl side chain forms ionic bonds with negatively charged groups of acidic amino acids. The ε-NH_2 groups of lysyl residues are covalently linked to biotin (a vitamin), lipoic acid, and retinal, a derivative of vitamin A and a constituent of visual pigment.

In collagen and in some glycoproteins, δ-carbons of some lysyl residues are hydroxylated (Figure 2-6), and sugar moieties are attached at these sites. In elastin and collagen, some ε-carbons of lysyl residues are oxidized to reactive aldehyde (–CHO) groups, with elimination of NH_3. These aldehyde groups then react with other ε–NH_2 groups to form covalent cross-links between polypeptides, thereby providing tensile strength and insolubility to protein fibers. Examples of cross-linked amino acid structures are desmosine, isodesmosine (Figure 2-6), dehydrolysinonorleucine, lysinonorleucine, merodesmosine, and dehydromerodesmosine (Chapter 10). Lysyl R-groups participate in a different type of cross-linking in the formation of fibrin, a process essential for the clotting of blood. In this reaction, the ε-NH_2 group of one fibrin polypeptide forms a covalent linkage with the glutamyl residue of another fibrin polypeptide (Chapter 36).

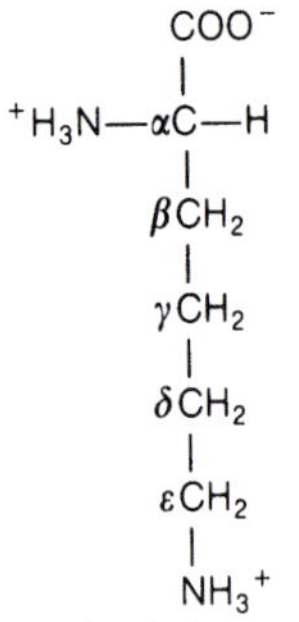

COO^-
^+H_3N—αC—H
βCH_2
γCH_2
δCH_2
εCH_2
NH_3^+
Lysine

Histidine

The imidazole group attached to the β-carbon of histidine has a pK′ value of 6.0. The pK′ value of histidyl residues in protein varies depending on the nature of the neighboring residues. The imidazolium-imidazole buffering pair has a major role in acid–base regulation (e.g., hemoglobin). The imidazole group functions as a nucleophile, or a general base, in the active sites of many enzymes and may bind metal ions. Histidine is nonessential in adults but is essential in the diet of infants and individuals with ***uremia*** (a kidney disorder). Decarboxylation of histidine to yield ***histamine*** occurs in mast cells present in loose connective tissue and around blood vessels, basophils of blood, and ***enterochromaffin-like (ECL) cells*** present in the acid-producing glandular portion (oxyntic cells) of the stomach.

The many specific reactions of histamine are determined by the type of receptor (H_1 or H_2) present in the target cells. The contraction of smooth muscle (e.g., gut and bronchi) is mediated by H_1 receptors and antagonized

COO^-
^+H_3N—C—H
CH_2
C═CH
N NH
C
H
Histidine

H
$H_3\overset{+}{N}$—C—H
CH_2
C═CH
N NH
C
H
Histamine

by diphenhydramine and pyrilamine. H_1-receptor antagonists are used in the treatment of allergic disorders. Secretion of HCl by the stomach (Chapter 12) and an increase in heart rate are mediated by H_2 receptors. Examples of H_2-receptor antagonists of histamine action are cimetidine and ranitidine, agents used in the treatment of gastric ulcers.

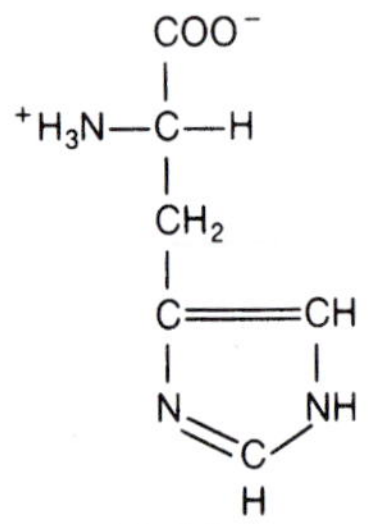

Arginine

The positively charged guanidinium group attached to the δ-carbon of arginine is stabilized by resonance between the two NH_2 groups and has a pK′ value of 12.48. Arginine is utilized in the synthesis of creatine and it participates in the urea cycle (Chapter 17).

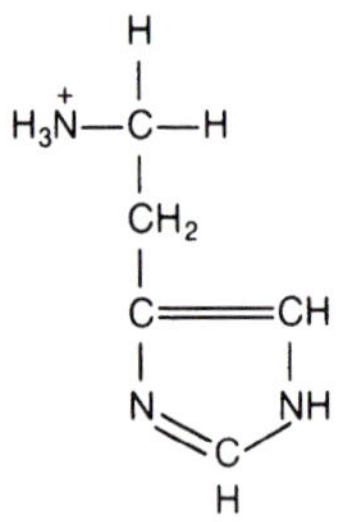

The nitrogen of the guanidino group of arginine is converted to nitric oxide (NO) by nitric oxide synthase. NO is unstable, highly reactive, and has a life span of only a few seconds. However, NO affects many biological activities, including vasodilation, inflammation, and neurotransmission (Chapter 17).

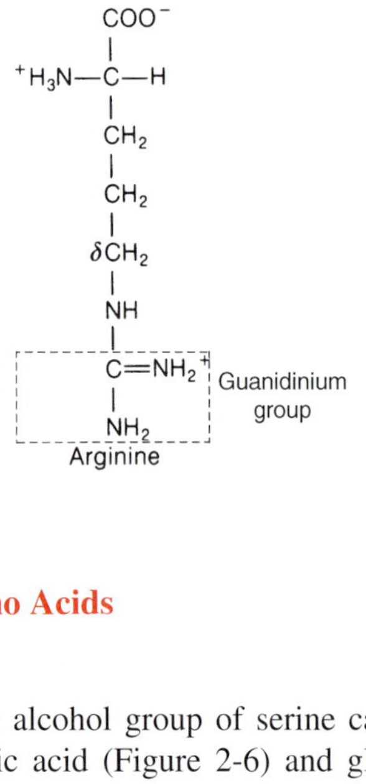

COO^-
^+H_3N—C—H
CH_2
CH_2
δCH_2
NH
C═NH_2^+ Guanidinium group
NH_2
Arginine

Neutral Amino Acids

Serine

The primary alcohol group of serine can form esters with phosphoric acid (Figure 2-6) and glycosides with

COO^-
^+H_3N—C—H
H—C—OH
H
Serine

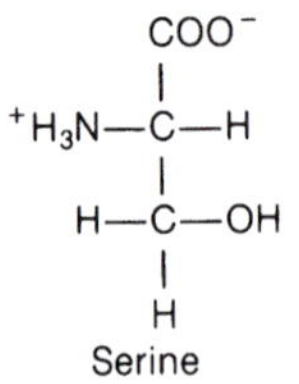

sugars. The phosphorylation and dephosphorylation processes regulate the biochemical activity of many proteins. Active centers of some enzymes contain seryl hydroxyl groups and can be inactivated by irreversible derivatization of these groups. The –OH group of serine has a weakly acidic pK′ of 13.6.

Threonine

This essential amino acid has a second asymmetrical carbon atom in the side chain and therefore can have four isomers, only one of

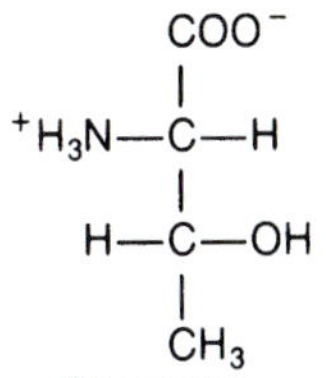

COO^-
^+H_3N—C—H
H—C—OH
CH_3
Threonine

which, L-threonine, occurs in proteins. The hydroxyl group, as in the case of serine, participates in reactions with phosphoric acid and with sugar residues.

Cysteine

The weakly acidic (pK′ = 8.33) sulfhydryl group (–SH) of cysteine is essentially undissociated at physiological pH. Free –SH groups are essential for the function of many enzymes and structural proteins. Heavy metal ions, e.g., Pb^{2+} and Hg^{2+}, inactivate these proteins by combining with their –SH groups. Two cysteinyl –SH groups can be oxidized to form ***cystine.*** A covalent disulfide bond of cystine can join two parts of a single polypeptide chain or two different polypeptide chains through cross-linking of cysteine residues. These –S–S– bonds are essential both for the folding of polypeptide chains and for the association of polypeptides in proteins that have more than one chain, e.g., insulin and immunoglobulins.

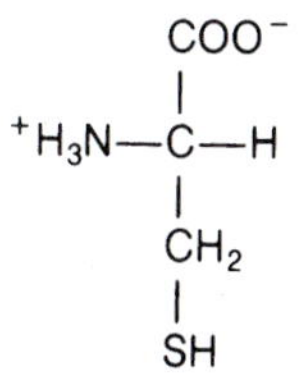

COO^-
^+H_3N—C—H
CH_2
SH
Cysteine

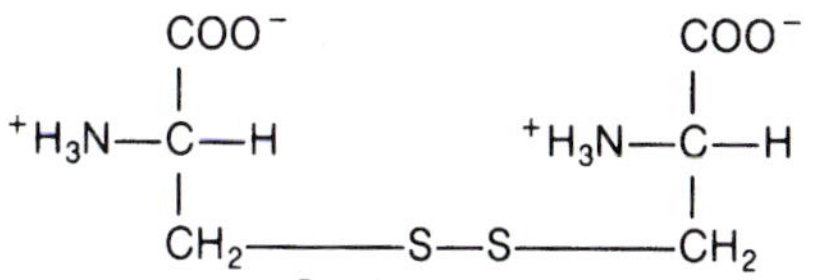

COO^- COO^-
^+H_3N—C—H ^+H_3N—C—H
CH_2—S—S—CH_2
Cystine

Tyrosine

The phenolic hydroxyl group of this aromatic amino acid has a weakly acidic pK′ of about 10 and therefore is un-ionized at physiological pH. In some enzymes, the hydrogen of the phenolic hydroxyl group can participate in hydrogen bond formation with oxygen and nitrogen atoms. The phenolic hydroxyl group of tyrosine residues in protein can be sulfated (e.g., in gastrin and cholecystokinin; see Chapter 12) or phosphorylated by a reaction catalyzed by the tyrosine-specific protein kinase that is a product of some oncogenes (Chapter 26). Tyrosine kinase activity also resides in a family of cell surface receptors that includes receptors for such anabolic polypeptides as insulin, epidermal growth factor, platelet-derived growth factor, and insulin-like growth factor type 1. All of these receptors have a common motif of an external ligand binding domain, a transmembrane segment, and a cytoplasmic tyrosine kinase domain (Chapter 22). Tyrosine accumulates in tissues and blood in ***tyrosinosis*** and ***tyrosinemia,*** which are due to inherited defects in catabolism of this amino acid. Tyrosine is the biosynthetic precursor of thyroxine, catecholamines, and melanin. Tyrosine and its biosynthetic precursor, phenylalanine, both absorb UV light (Figure 2-4).

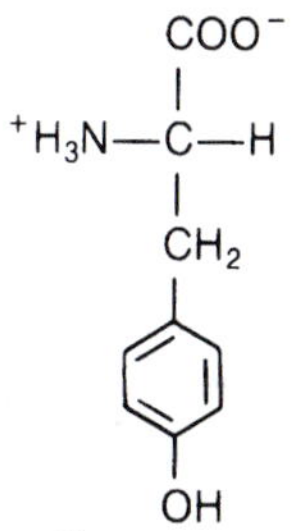

COO^-
^+H_3N—C—H
CH_2
OH
Tyrosine

Asparagine

The R-group of this amide derivative of aspartic acid has no acidic or basic properties but is polar and participates in hydrogen bond formation. It is hydrolyzed to aspartic acid and ammonia by the enzyme asparaginase. In glycoproteins, the carbohydrate side chain is often linked through the amide group of asparagine.

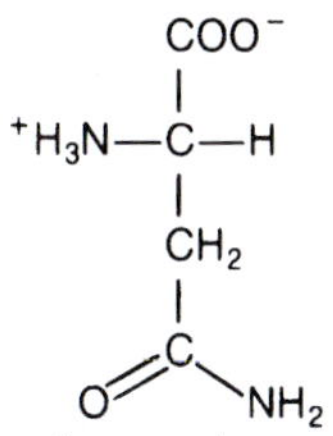

COO^-
^+H_3N—C—H
CH_2
C
O NH_2
Asparagine

Glutamine

This amide of glutamic acid has properties similar to those of asparagine. The γ-amido nitrogen, derived from ammonia, can be used in the synthesis of purine and pyrimidine nucleotides (Chapter 27), converted to urea in the liver (Chapter 17), or released as NH_3 in the kidney tubular epithelial cells. The last reaction, catalyzed by the enzyme glutaminase, functions in acid–base regulation by neutralizing H^+ ions in the urine (Chapter 39).

Glutamine is the most abundant amino acid in the body. It composes more than 60% of the free amino acid pool in skeletal muscle. It is metabolized in both liver and gut tissues. Glutamine, along with alanine, are significant precursors of glucose production during fasting (Chapter 15). It is a nitrogen donor in the synthesis of purines and pyrimidines required for nucleic acid synthesis (Chapter 27). Glutamine is enriched in enteral and parenteral nutrition to promote growth of tissues; it also enhances immune functions in patients recovering from surgical procedures. Thus, glutamine may be classified as a conditionally essential amino acid during severe trauma and illness.

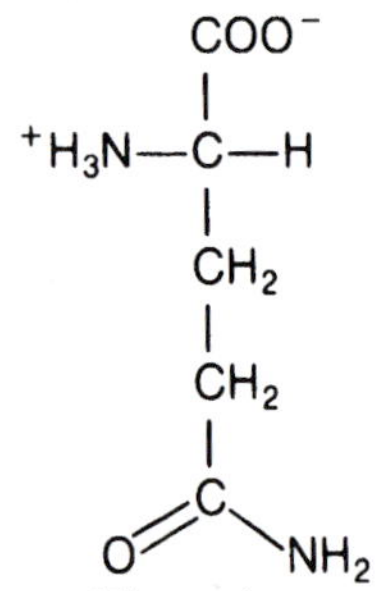

COO^-

$^+H_3N—C—H$

CH_2

CH_2

$O=C—NH_2$

Glutamine

Unusual Amino Acids

Several L-amino acids have physiological functions as free amino acids rather than as constituents of proteins. Examples are as follows:

1. β-Alanine is part of the vitamin pantothenic acid.
2. Homocysteine, homoserine, ornithine, and citrulline are intermediates in the biosynthesis of certain other amino acids.
3. Taurine, which has an amino group in the β-carbon and a sulfonic acid group instead of COOH, is present in the CNS and as a component of certain bile acids participates in digestion and absorption of lipids in the gastrointestinal tract.
4. γ-Aminobutyric acid is an inhibitory neurotransmitter.
5. Hypoglycin A is present in unripe akee fruit and produces severe hypoglycemia when ingested.
6. Some D-amino acids are found in polypeptide antibiotics, such as gramicidins and bacitracins, and in bacterial cell wall peptides.

Amino Acids Used as Drugs

D-Penicillamine (D-β,β-dimethylglycine), a metabolite of penicillin, was first isolated in the urine specimens from patients treated with penicillin with liver disease. It is an effective chelator of metals such as copper, zinc, and lead. It is used in the chelation therapy of ***Wilson's disease,*** which is characterized by excess copper accumulation leading to ***hepatolenticular degeneration*** (Chapter 37).

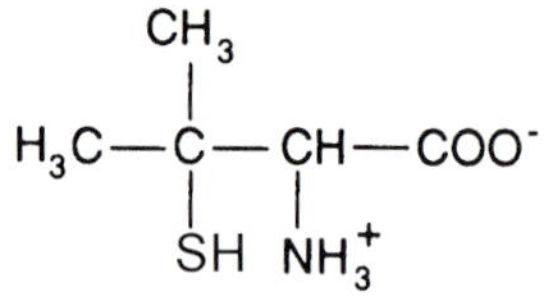

$H_3C—C(CH_3)(SH)—CH(NH_3^+)—COO^-$

D–penicillamine

N-Acetylcysteine is administered in the acetoaminophen toxicity. It replenishes the hepatic stores of glutathione (Chapter 17). *N*-Acetylcysteine is also used in the treatment of pulmonary diseases including cystic fibrosis (Chapter 12). In patients with chronic renal insufficiency, prophylactic oral administration of *N*-Acetylcysteine have been used in the prevention of further renal impairment due to administration of radiographic contrast agents. In this setting presumably *N*-Acetylcysteine functions as an antioxidant and augments the vasodilatory effect of nitric oxide via the formation of S-nitrosothiol (Chapter 17).

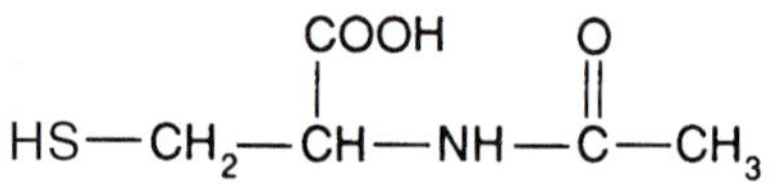

$HS—CH_2—CH(COOH)—NH—C(=O)—CH_3$

N–acetylcysteine

Gabapentin is γ-aminobutyrate covalently linked to cyclohexane to make it lipophilic and to facilitate its transport across the blood-brain barrier. It is used as an anticonvulsant and in amyotrophic lateral sclerosis (ALS).

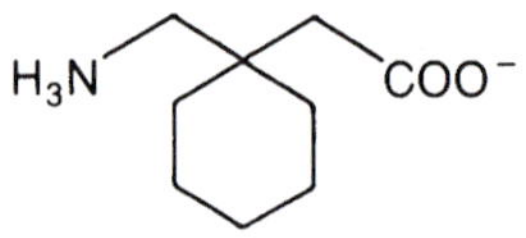

H_3N COO^-

Gabapentin

2.3 Electrolyte and Acid–Base Properties

Amino acids are ampholytes, i.e., they contain both acidic and basic groups. Free amino acids can *never* occur as neutral nonionic molecules:

$$R-CH(NH_2)-C(=O)-OH$$

Instead, they exist as neutral ***zwitterions*** that contain both positively and negatively charged groups:

$$R-CH(NH_3^{\oplus})-CO_2^{\ominus}$$

Nonpolar neutral R-group → R; Ammonium cation → $NH_3^{\oplus}$; Resonance-stabilized carboxylate anion ← $CO_2^{\ominus}$

Zwitterion

Zwitterions are electrically neutral and so do not migrate in an electric field. In acidic solution (below pH 2.0), the predominant species of an amino acid is positively charged and migrates toward the cathode:

$$R-CH(NH_3^{+})-C(=O)-OH$$

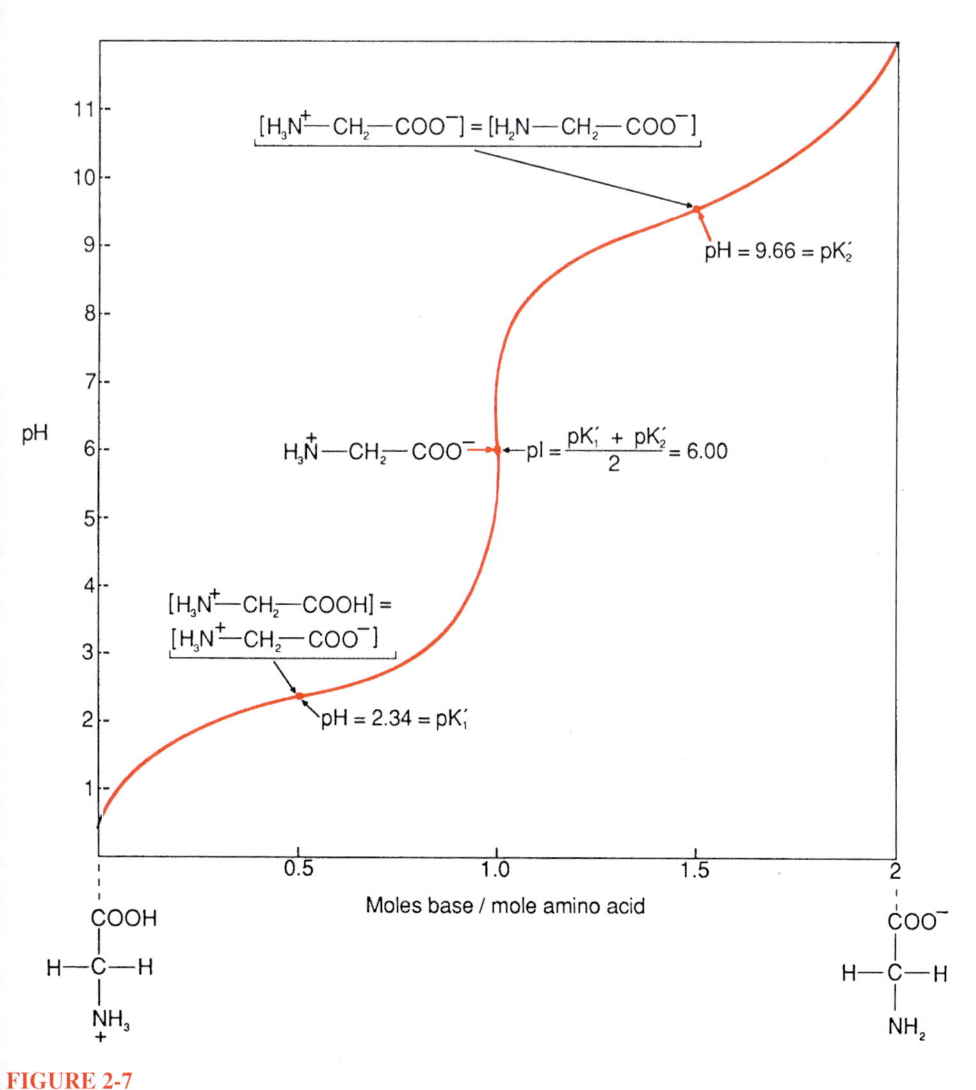

FIGURE 2-7
Titration profile of glycine, a monoaminomonocarboxylic acid.

In basic solution (above pH 9.7), the predominant species is negatively charged and migrates toward the anode:

$$R{-}\underset{\substack{|\\NH_2}}{\overset{\substack{H\\|}}{C}}{-}\overset{\substack{O\\\|}}{C}{-}O^-$$

The ***isoelectric point*** (pI) of an amino acid is the pH at which the molecule has an average net charge of zero and therefore does not migrate in an electric field. The pI is calculated by averaging the pK′ values for the two functional groups that react as the zwitterion becomes alternately a monovalent cation or a monovalent anion.

At physiological pH, monoaminomonocarboxylic amino acids, e.g., glycine and alanine, exist as zwitterions.

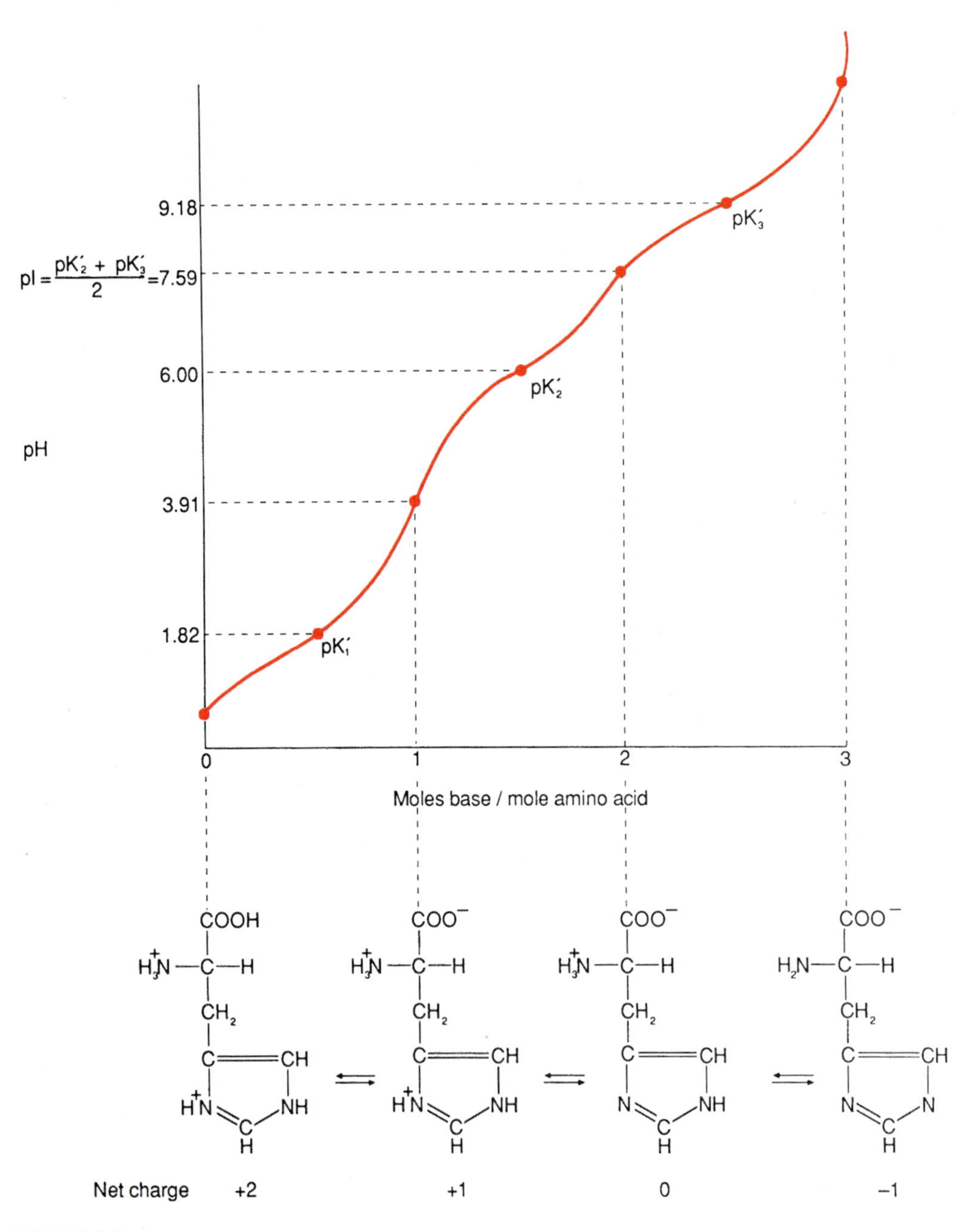

FIGURE 2-8
Titration profile of histidine.

That is, at a pH of 6.9–7.4, the α-carboxyl group (pK′ 2.4) is dissociated to yield a negatively charged carboxylate ion ($-COO^-$), while the α-amino group (pK′ 9.7) is protonated to yield an ammonium group ($-NH_3^+$). The pK′ value of the α-carboxyl group is considerably lower than that of a comparable aliphatic acid, e.g., acetic acid (pK′ 4.6). This stronger acidity is due to electron withdrawal by the positively charged ammonium ion and the consequent increased tendency of a carboxyl hydrogen to dissociate as an H^+. The α-ammonium group is correspondingly a weaker acid than an aliphatic ammonium ion, e. g., ethylamine (pK′ 9.0) because the inductive effect of the negatively charged carboxylate anion tends to prevent dissociation of H^+. The titration profile of glycine hydrochloride (Figure 2-7) is nearly identical to the profiles of all other monoaminomonocarboxylic amino acids with nonionizable R-groups (Ala, Val, Leu, Ile, Phe, Ser, Thr, Gln, Asn, Met, and Pro).

The titration of glycine has the following major features. The titration is initiated with glycine hydrochloride,

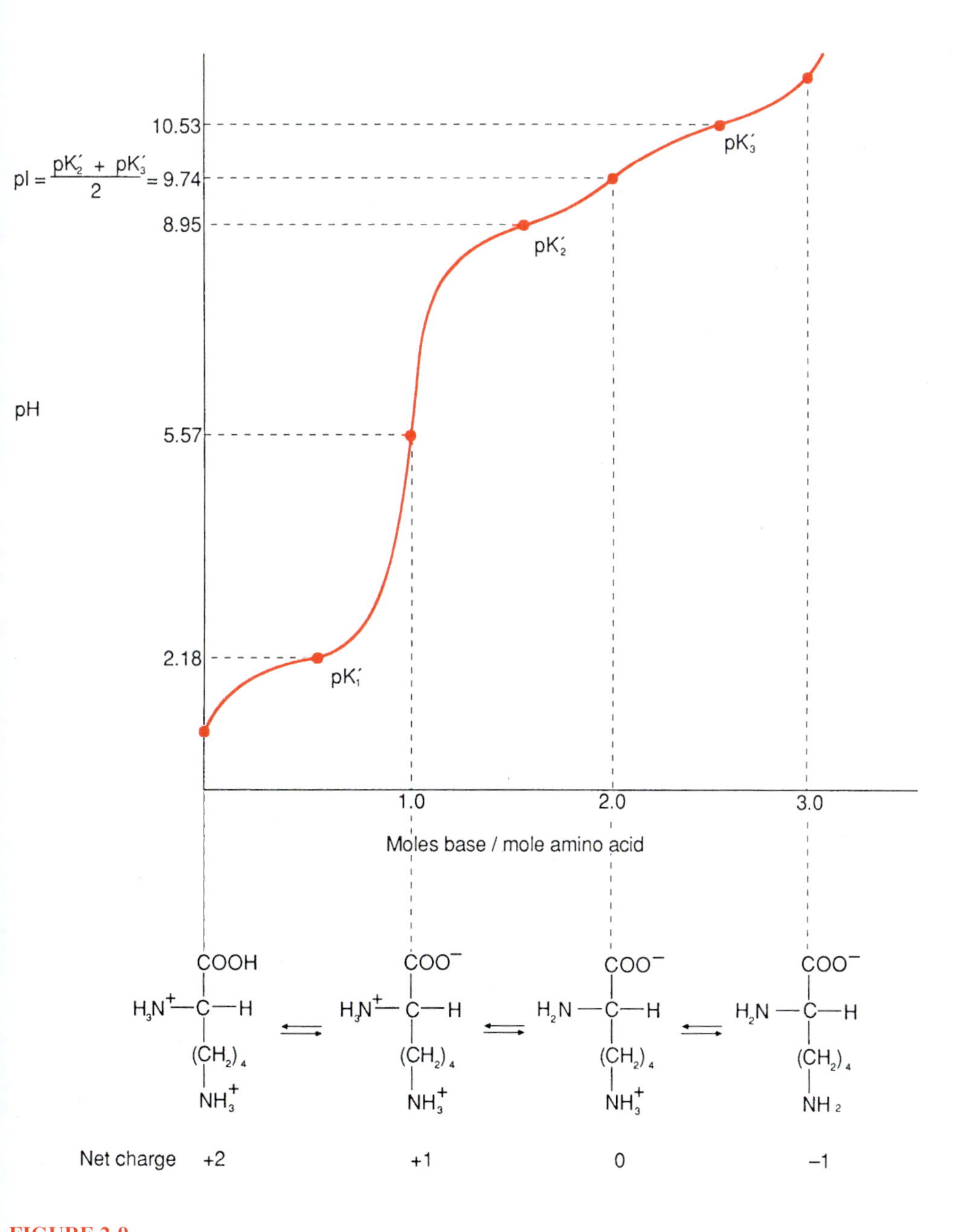

FIGURE 2-9
Titration profile of lysine.

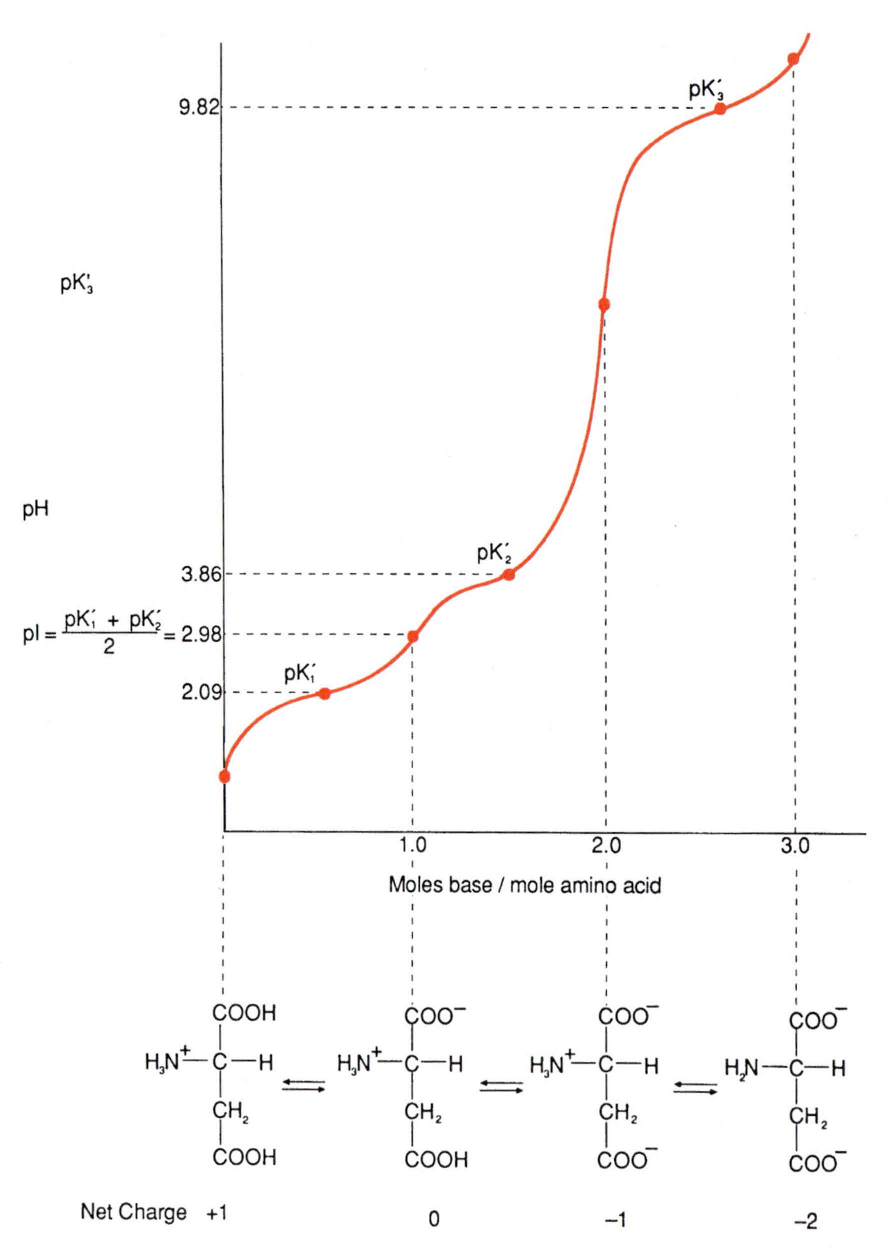

FIGURE 2-10
Titration profile of aspartic acid.

$Cl^-(H_3^+NCH_2COOH)$, which is the fully protonated form of the amino acid. In this form, the molecule contains two acidic functional groups; therefore, two equivalents of base are required to completely titrate 1 mol of glycine hydrochloride. There are two pK′ values: pK_1' due to reaction of the carboxyl group and pK_2' due to reaction of the ammonium group. Addition of 0.5 eq of base to 1 mol of glycine hydrochloride raises the pH 2.34 (pK_1'), whereas addition of 1.5 eq further increases the pH to 9.66 (pK_2'). At low pH values (e.g., 0.4), the molecules are predominantly cations with one positive charge; at pH values of 5–7, most molecules have a net charge of zero; at high pH values (e.g., 11.7), all of the molecules are essentially anions with one negative charge. The midpoint between the two pK′ values [i.e., at pH $= (2.34 + 9.66)/2 = 6.0$] is the pI. Thus, pI is the arithmetic mean of pK_1' and pK_2' values and the inflection point between the two segments of the titration profile.

The buffering capacities of weak acids and weak bases are maximal at their pK values. Thus, monoaminomono-

TABLE 2-2
*pK′ and pI Values of Selected Free Amino Acids at 25°C**

Amino Acid	pK'_1 (α-COOH)	pK'_2	pK'_3	pI
Alanine	2.34	9.69 (α-NH_3^+)		6.00
Aspartic acid	2.09	3.86 (β-COOH)	9.82 (α-NH_3^+)	2.98 $\left(\frac{pK'_1 + pK'_2}{2}\right)$
Glutamic acid	2.19	4.25 (γ-COOH)	9.67 (α-NH_3^+)	3.22 $\left(\frac{pK'_1 + pK'_2}{2}\right)$
Arginine	2.17	9.04 (α-NH_3^+)	12.48 (Guanidinium)	10.76 $\left(\frac{pK'_2 + pK'_3}{2}\right)$
Histidine	1.82	6.00 (Imidazolium)	9.17 (NH_3^+)	7.59 $\left(\frac{pK'_2 + pK'_3}{2}\right)$
Lysine	2.18	8.95 (α-NH_3^+)	10.53 (ε-NH_3^+)	9.74 $\left(\frac{pK'_2 + pK'_3}{2}\right)$
Cysteine	1.71	8.33 (–SH)	10.78 (α-NH_3^+)	5.02 $\left(\frac{pK'_1 + pK'_2}{2}\right)$
Tyrosine	2.20	9.11 (α-NH_3^+)	10.07 (Phenol OH)	5.66 $\left(\frac{pK'_1 + pK'_2}{2}\right)$
Serine	2.21	9.15 (α-NH_3^+)	13.6 (Alcohol OH)	5.68 $\left(\frac{pK'_1 + pK'_2}{2}\right)$

*The pK′ values for functional groups in proteins may vary significantly from the values for free amino acids.

carboxylic acids exhibit their greatest buffering capacities in the two pH ranges near their two pK′ values, namely, pH 2.3 and pH 9.7 (Figure 2-7). Neither these amino acids nor the α-amino or α-carboxyl groups of other amino acids (which have similar pK′ values) have significant buffering capacity in the neutral (physiological) pH range. The only amino acids with R-groups that have buffering capacity in the physiological pH range are histidine (imidazole; pK′ 6.0) and cysteine (sulfhydryl; pK′ 8.3). The pK′ values for R-groups vary with the ionic environment. The titration profile of histidine is shown in Figure 2-8. The pI is the mean of pK'_2 and pK'_3.

Titration profiles of the basic and acidic amino acids lysine and aspartic acid are shown in Figures 2-9 and 2-10. The R-groups are ionized at physiological pH and have anionic and cationic groups, respectively. The pI value for aspartic acid is the arithmetic mean of pK'_1 and pK'_2, whereas for lysine and histidine the pI values are given by the arithmetic mean of pK'_2 and pK'_3. The pK′ and pI values of selected amino acids are listed in Table 2-2.

2.4 Chemical Reactions of Amino Acids

The reactions of amino acids with ninhydrin, carbon dioxide, metal ions, and glucose are described below. The last three are of physiological importance.

Ninhydrin (triketohydrindene hydrate) reacts with α-amino acids to produce CO_2, NH_3, and an aldehyde with one less carbon than the parent amino acid. In most cases, a blue or violet compound (proline and hydroxyproline give a yellow color) is formed owing to reaction of the liberated NH_3 with ninhydrin, as shown in Figure 2-11. Color and CO_2 production provide a basis for the quantitative determination of amino acids.

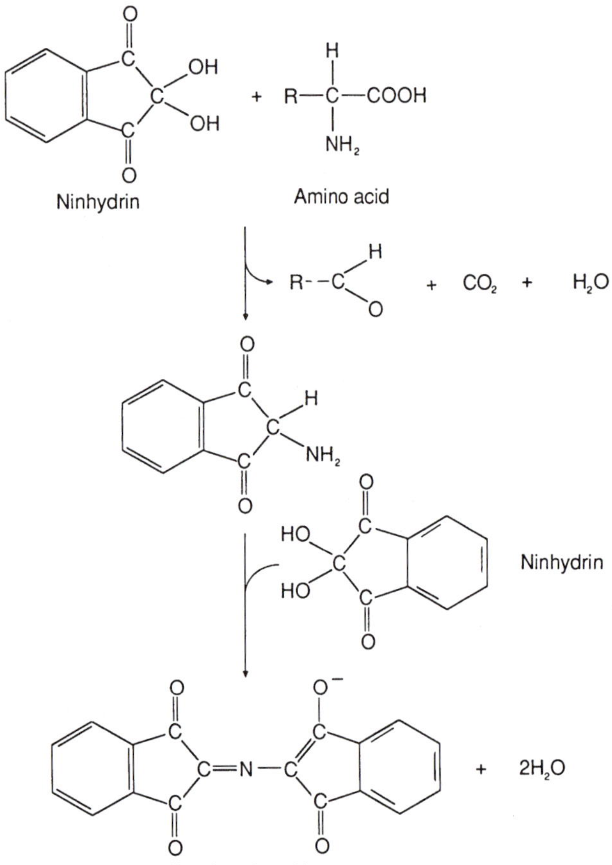

Ninhydrin + Amino acid → $R-CHO$ + CO_2 + H_2O ... + Ninhydrin → Purple-colored ion + $2H_2O$

FIGURE 2-11
Reaction of an α-amino acid with ninhydrin. Two molecules of ninhydrin and the nitrogen atom of the amino acid are involved in the production of the purple product.

Ammonia, some amines, and some proteins and peptides will also yield a colored product but will not generate CO_2. Thus, the colorimetric analysis is not specific for amino acids unless CO_2 release is measured or the amino acids are purified and freed from interfering materials (the usual procedure). The color reaction with ninhydrin is used extensively in manual and automated procedures.

CO_2 adds reversibly to the un-ionized amino group of an amino acid. The product is a carbamate (or carboxyamino) derivative.

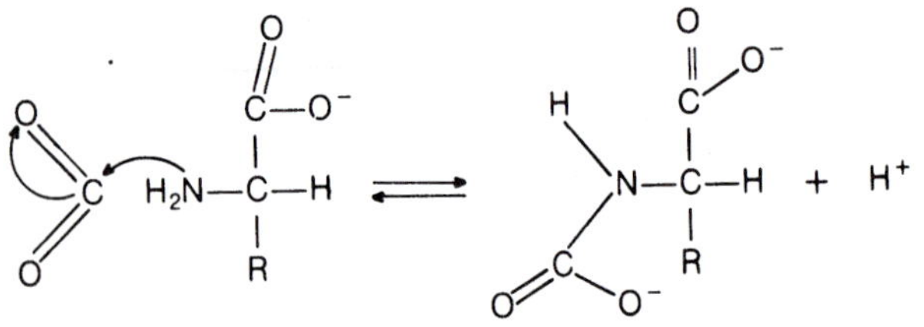

$$O=C=O + H_2N-CH(R)-COO^- \rightleftharpoons {}^-OOC-NH-CH(R)-COO^- + H^+$$

This type of reaction accompanies transport of CO_2 in the blood (Chapter 1). In tissue capillaries, CO_2 combines with free α-amino groups of hemoglobin to form carbaminohemoglobin; in pulmonary capillaries, this reaction is reversed to release CO_2 into the alveoli. This mode of transport is limited to only about 10% of the carbon dioxide transported in the blood.

Metal ions can form complexes with amino acids. Metal ions that function in enzymatic or structural biochemical systems include those of iron, calcium, copper, zinc, magnesium, cobalt, manganese, molybdenum, nickel, and chromium. The functional group that binds a metal ion is called a ***ligand.*** Ligands are electron donors that form noncovalent bonds with the metal ions, usually two, four, or six ligand groups per ion. When four ligand groups bind a metal ion, the complex is either a plane or a tetrahedron; when six ligand groups participate, octahedral geometry results. The term ***chelation*** is applied to a metal-ligand interaction when a single molecule provides two or more ligands (e.g., chelation of iron with four nitrogens in one porphyrin molecule; see Chapter 14).

Metal ions can also react with amino acid functional groups to abolish the biological activity of proteins. Heavy metal ions that form highly insoluble sulfides (e.g., HgS, PbS, CuS, Ag_2S) characteristically react with sulfhydryl groups of cysteinyl residues. If the reactive –SH group is involved in biological activity of the protein, the displacement of the hydrogen and the addition of a large metal atom to the S atom usually cause a major change in protein structure and loss of function. Hence, heavy metals are often poisons.

In contrast, amino acid residues in proteins may undergo nonenzymatic chemical reactions that may or may not alter biological activity. An example of this type of reaction is the formation of glycated proteins. The amino groups of proteins combine with carbonyl groups of sugars (glucose) to form labile aldimines (Schiff bases), which are isomerized (Amadori rearrangement) to yield stable ketoamine (fructosamine) products (Figure 2-12). The degree of glycation achieved in a protein is determined by the concentration of sugar in the environment of the protein. In glycated hemoglobin, a Schiff-base adduct forms between the sugar and the N-terminal group of the β-chains of hemoglobin.

The Amadori sugar-amino acid residue adducts in proteins are produced with prolonged hyperglycemia and undergo progressive nonenzymatic reactions involving dehydration, condensation, and cyclization. These compounds are collectively known as *advanced glycosylation end products* and are involved in the chronic complications of ***diabetes mellitus*** (cataracts and nephropathy) (Chapter 22).

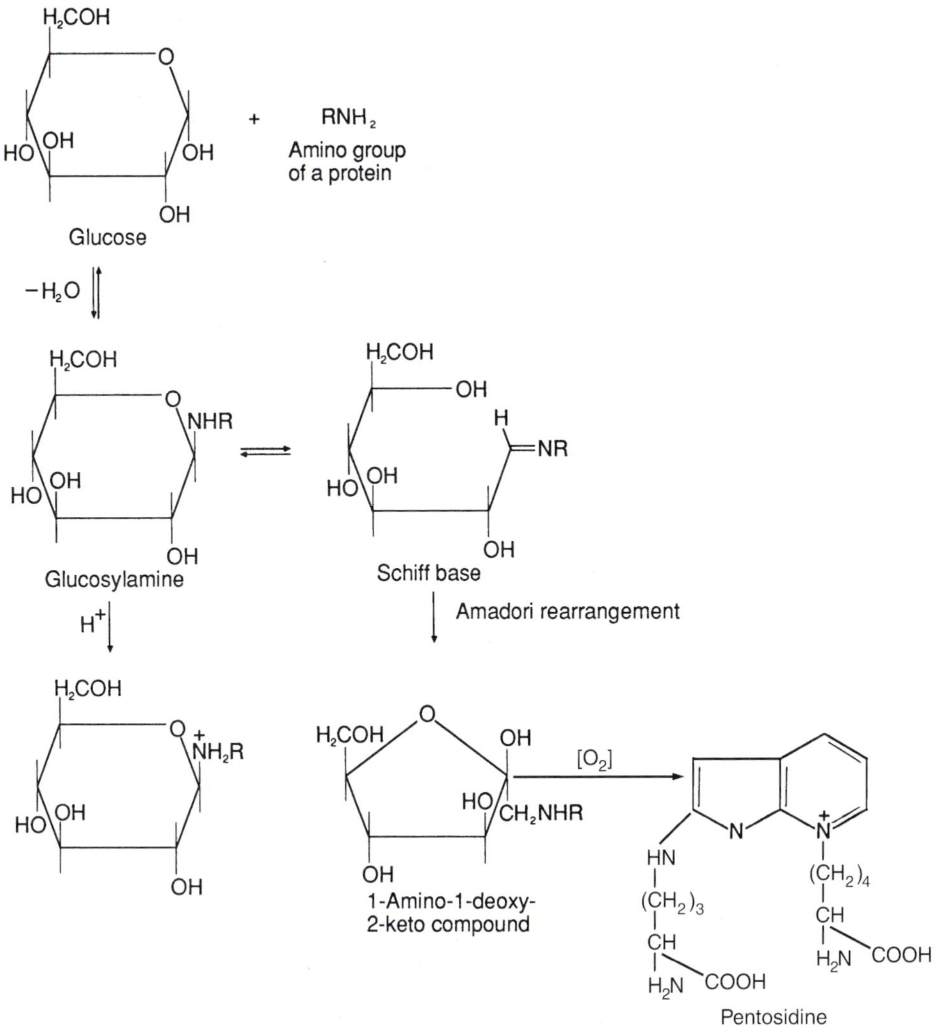

FIGURE 2-12
Nonenzymatic reaction between the glucose and a free amino group of a protein.

Supplemental Readings and References

D. G. Dyer, J. K. Blackledge, S. R. Thorpe, and J. W. Baynes: Formation of pentosidine during nonenzymatic browning of proteins by glucose. *Journal of Biological Chemistry* **266,** 11654 (1991).

M. E. Gurney, F. B. Curring, P. Zhai, and others: Benefit of vitamin E, riluzole, and gabapentin in a transgenic model of familial amyotrophic lateral sclerosis. *Annals of Neurology* **39,** 147 (1996).

R. G. Hankard, M. W. Haymond, and D. Darmaun: Role of glutamine as a precursor in fasting humans. *Diabetes* **46,** 1535 (1997).

S. J. Kuhl and H. Rosen: Nitric oxide and septic shock. *West Journal of Medicine* **168,** 176 (1998).

L. Lacomblez, G. Bensimon, P. N. Leigh, et al.: Dose-ranging study of riluzole in amyotrophic lateral sclerosis. *Lancet* **347,** 1425 (1996).

J. Loscalzo and G. Welch: Nitric oxide and its role in the cardiovascular system. *Progress in Cardiovascular Diseases* **38,** 87 (1995).

R. G. Miller, D. Moore, L. A. Young, et al.: Placebo-controlled trial of gabapentin in patients with amyotrophic lateral sclerosis. *Neurology* **47,** 1383 (1996).

R. Safirstein, L. Andrade, and J. M. Vieira: Acetylcysteine and nephrotoxic effects of radiographic contrast agents—a new use for an old drug. *The New England Journal of Medicine* **343,** 210 (2000).

M. Tepel, M. van der Giet, C. Schwarzeld, and others: Prevention of radiographic-contrast-agent-induced reductions in renal function by acetylcysteine. *The New England Journal of Medicine* **343,** 180 (2000).

H. Vlassara: Recent progress on the biological and clinical significance of advanced glycosylation end products. *J. of Laboratory and Clinical Medicine*, **124,** 19 (1994).

T. R. Ziegler, K. Benfele, R. J. Smith, et al.: Safety and metabolic effects of 1-glutamine administration in humans. *Journal of Parenteral and Enteral Nutrition* **14,** 1375 (1990).

CHAPTER 3

Protein Isolation and Determination of Amino Acid Sequence

To understand the structure and function of a protein, one must know the number and kinds of amino acids present in the protein and their order (called *sequence* or *primary structure*). This information is necessary to understand the effects of mutations, mechanisms of enzyme-catalyzed reactions, and chemical synthesis of species-specific peptides that may eliminate undesirable hypersensitivity reactions. Studies of amino acid sequences in proteins have aided in the understanding of the evolutionary development of living systems. Another application of amino acid sequence determination is in recombinant DNA technology. A desired DNA coding for a given polypeptide can be constructed from knowledge of the precise sequence of that polypeptide (Chapter 23). Through the use of available technologies of sequence determination and peptide synthesis, peptide fragments have been produced that are identical to part of large proteins present on the surface of a virus or other pathogen. In the appropriate host, such peptides elicit antibody production that is effective against the active pathogen and are potentially useful in the development of synthetic vaccines (Chapter 35). Viruses for which immunogenic peptides are being synthesized and clinically applied are hepatitis B, influenza, rabies, mouse leukemia, and hoof-and-mouth disease. Bovine insulin (M.W. 5,700) was the first protein to be completely sequenced (Sanger, 1955). Sanger was also the first to deduce the base sequence of a DNA molecule obtained from phage ϕX 174 (a bacterial virus; see Chapter 23). Bovine insulin comprises two peptide chains of 21 and 30 amino acids each, linked by two interchain disulfide bonds.

In 1960, Hirs, Moore, Stein, and Anfinsen described the first primary structure of the enzyme *ribonuclease* (M.W. 13,700), which has a single peptide chain of 124 amino acid residues and four intrachain disulfide bonds. These investigators established many of the techniques still used in sequence analysis, such as the use of ion exchange resins for separation of peptides and amino acids.

3.1 Quantitative Determination of Proteins

UV absorption at 280 nm is an inaccurate method of protein determination because proteins have different amounts of tyrosine and tryptophan residues, and nucleic acids also absorb at 280 nm.

In the ***biuret test,*** the sample is treated with an alkaline copper sulfate reagent that produces a violet color and requires a peptide with at least two peptide bonds. The violet color is produced through formation of a coordination complex (between peptide nitrogen atoms and cupric ion) that is analogous to the structure of the complex of biuret with cupric ion, as shown below:

$$2\ \mathrm{NH_2{-}C(=O){-}NH{-}C(=O){-}NH_2} + \mathrm{Cu^{2+}} \longrightarrow \text{Violet complex}$$

Biuret   Violet complex

In the ***Folin–Ciocalteu reaction,*** the protein reacts with the phosphomolybdotungstic acid reagent to give a blue color through reaction with tyrosyl residues. The sensitivity of this method depends on the amino acid composition of proteins in each sample.

Protein content, particularly in urine or cerebrospinal fluid, may also be estimated by methods based on precipitation using sulfosalicylic acid (an anionic protein precipitant) or heat. The turbidity, which is a measure of protein concentration, can be quantitated by spectrophotometric absorbance methods or light scattering analysis. Absorbance of a hydrophobic indicator dye that binds to protein and changes color is also used.

Specific proteins in biological fluids can be estimated by more discriminating methods, such as electrophoresis, specific binding techniques, or immunochemical methods.

3.2 Determination of Primary Structure

The determination of the primary structure of a protein consists of the following steps:

1. Obtain a pure protein.
2. Determine the amino acid composition and molecular weight of the pure protein. From amino acid composition and molecular weight data, calculate the number of residues of each amino acid present per protein molecule to the nearest whole number.
3. Reduce disulfide bonds to sulfhydryl groups.
4. Determine amino terminal (N-terminal) and carboxy terminal (C-terminal) amino acids. (A unique residue for each terminus suggests that the native protein contains only one peptide chain.)
5. Fragment aliquots of the polypeptide by enzymatic or chemical hydrolysis and separate the peptide mixtures into individual fragments. (This process will yield overlapping sets of smaller peptides.)
6. Sequence each fragment and, by analyzing the overlapping parts, assemble the sequence of the original protein. The logic for arranging overlapping peptides is as follows:

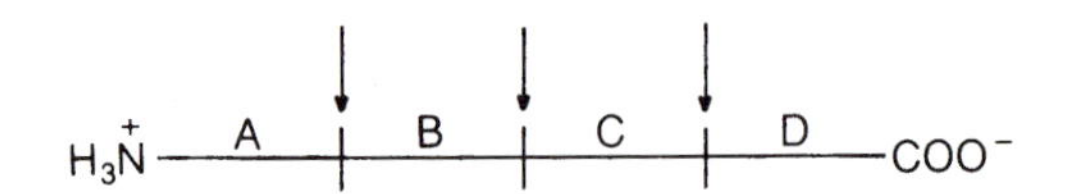

Assume that cleavage of the above polypeptide with a specific cleaving agent yields four peptides, A, B, C, and D. These peptides are separated and sequenced but it is yet not known whether the correct order is ABCD or ACBD. A second hydrolytic procedure gives rise to three peptides: E, F, and G.

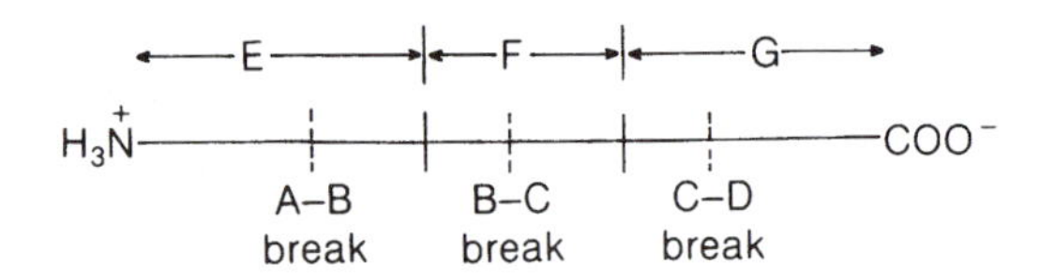

By comparison of the amino acid sequence of fragments E and G, the terminal fragments A and D can be matched with their respective adjacent peptides, B and C. Alternatively, the order of the peptides B and C can be deduced from the overlapping sequence in F.

Purification of a protein is indispensable for establishing its amino acid sequence. Mixtures of proteins will yield mixtures of peptides and ambiguous amino acid positions; a unique amino acid sequence can be obtained only from a pure protein. Protein purification techniques exploit differences in size, shape, charge, solubility, and specific binding affinity of the proteins. The optimal combination of techniques is usually reached by trial and error.

3.3 Separation of Proteins

Proteins are separated on the basis of differences in their size, shape, charge, solubility, and binding affinity for other molecules. Most separations begin with proteins in solution. However, when proteins are an integral part of an organelle's structure, it is first necessary to extract them from membranous elements. Chemicals used to extract proteins from membranous particles include dissociating agents (e.g., urea and mercaptoethanol), chelating agents (e.g., ethylenediaminetetraacetic acid [EDTA]), and organic detergents (e.g., sodium deoxycholate, sodium lauryl sulfate, and Triton X-100).

Since the total quantity (or activity) of a protein in a tissue sample is difficult to determine, the original (100%) quantity of protein at the start of a purification is usually based on measurements made on an aliquot of the initial homogenate. Each step in a purification process should remove extraneous protein and retain most of the protein of interest. A pure protein preparation is operationally defined as one that maintains a high activity per gram of protein following several purification steps, i.e., the optimum specific activity of the protein. Alternatively, a pure protein cannot be further subdivided by the methods described below, e.g., chromatography or electrophoresis.

The initial purification steps usually separate proteins according to general classification, e.g., fibrous (insoluble) or globular (soluble). The fibrous and globular designations are related to shape and solubility. Globular proteins are spherical or ellipsoidal and make up the majority of known proteins. Fibrous proteins contain one or more polypeptide chains. Their molecules are elongated and asymmetrical with lengths that may be many times their diameters. Lateral cross-linking between adjacent polypeptides, by a variety of types of chemical bonding, confers mechanical strength and water insolubility to fibrous proteins; consequently, they are found in connective, elastic, and contractile tissues as well as in hair and skin. An initial aqueous extraction procedure tends to partition globular proteins into the soluble fraction and fibrous proteins into the "insoluble pellet" remaining after centrifugation.

Separation by Molecular Size

Protein can be separated on the basis of their size by ***dialysis, gel filtration,*** and ***membrane filtration.*** Small molecules (e.g., NaCl, amino acids, and sucrose) originally present or added during the separation of organelles can be removed by dialysis through a semipermeable membrane. Dialysis membranes are prepared from cellophane or collodion and contain pores that permit passage of solute molecules whose molecular weight is less than ~5000. Thus, proteins of high molecular weight are retained within a dialysis bag, whereas low-molecular-weight solutes diffuse through the pores into the fluid (dialysate) outside of the bag. Complete removal of low-molecular-weight solutes requires repeated changes of dialysate. Rapid removal of low-molecular-weight solutes and simultaneous concentration of the protein solution can be accomplished by applying pressure to the dialysis solution (or vacuum to the dialysate). Such a process is known as ***ultrafiltration*** (Figure 3-1). The principles of membrane filtration are the same as those of dialysis except that synthetic membranes with specified pore sizes are used.

In gel filtration (or molecular sieving), a column is packed with hydrated, insoluble gel particles with known pore sizes; the protein solution is passed through the column and the effluent solution is collected in fractions that will contain solutes of different sizes (Figure 3-2). The volume of the column is essentially divided into two phases: the gel phase (within the pores) and the solvent phase (outside the gel particles). As a solution migrates in the column, the solute molecules that can penetrate the pores of the gel are distributed both within the gel and outside it. The particles that are larger than the pore size are excluded

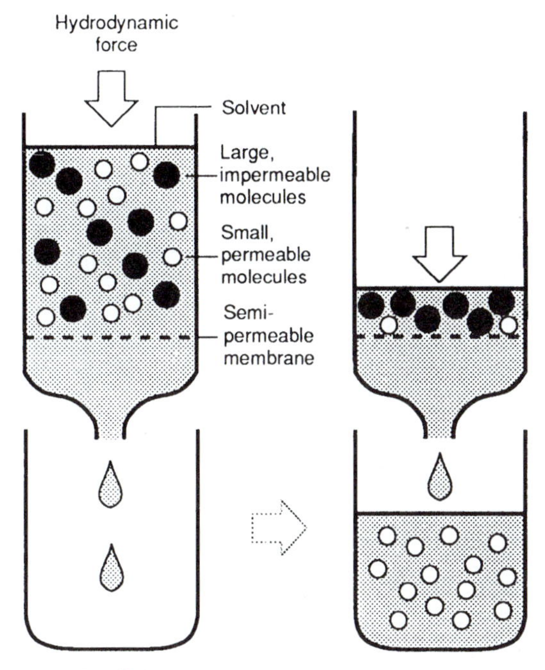

FIGURE 3-1
Separation and concentration of high-molecular-weight solutes from low-molecular-weight solutes by application of hydrodynamic force over the solution above a semipermeable membrane (ultrafiltration).

from the gel phase flow through the column more rapidly and appear first in the effluent. Commonly used gel particles are inert cross-linked dextrans or agarose, which are commercially available with a wide range of exclusion limits. Molecular sieving is effective in the purification of macromolecules and if the column has been calibrated by elution of solutes of known molecular weight, it can also be used in the estimation of molecular weights of proteins or other solutes.

Separation by Chromatography

Chromatographic separations of proteins are based upon the differential partitioning of solute molecules due to their differences in affinity between a moving solvent phase and a fixed or supportive phase. In gas chromatography, the mobile phase is a gas, whereas in liquid chromatography it is a liquid. Gas chromatography is not useful in protein purification because proteins cannot be converted to gases without decomposition. Liquid chromatography of proteins is performed on a variety of mechanically different stationary phases, e.g., paper, finely divided particles coated onto a glass or a plastic surface (thin-layer chromatography), or beads packed in a column.

Many chemically different stationary phases are used in liquid chromatography of proteins. ***Ion exchange chromatography*** uses an ion exchange resin, and the proteins are eluted with buffer solutions differing in ionic strength

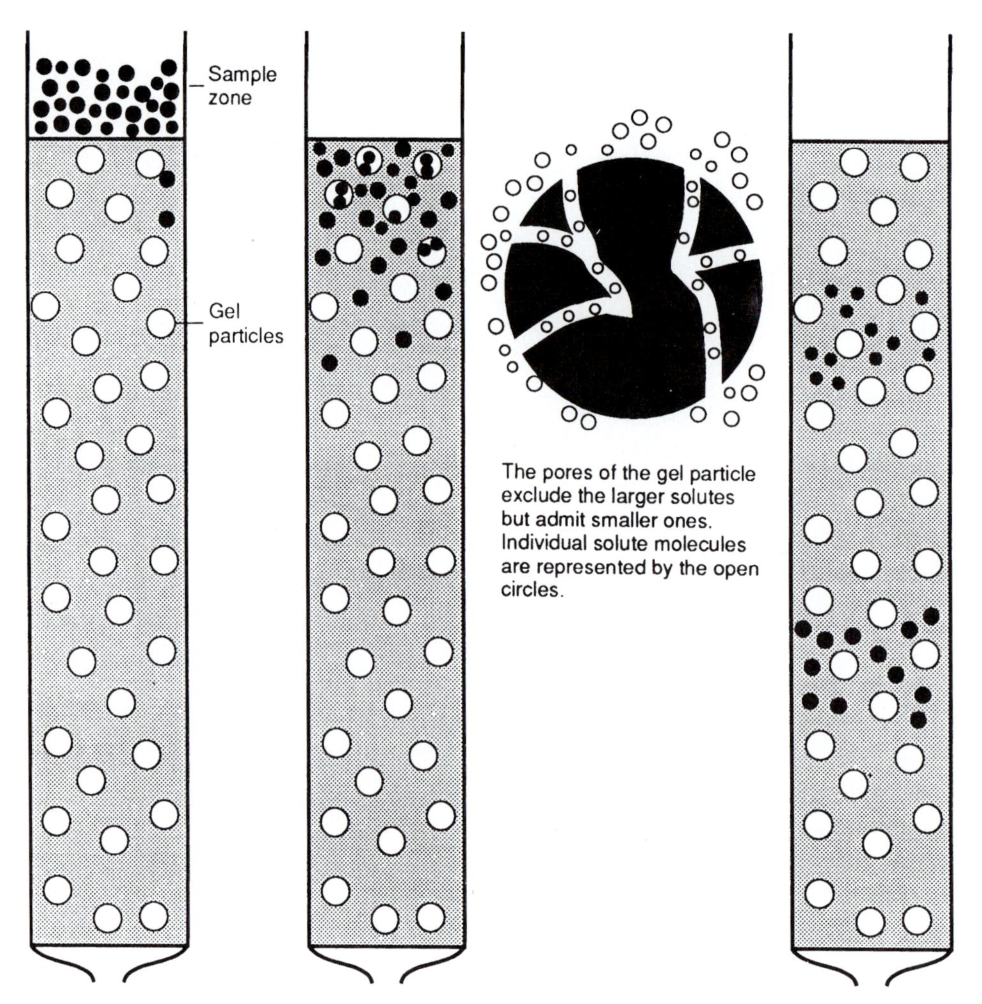
Sample zone
Gel particles
The pores of the gel particle exclude the larger solutes but admit smaller ones. Individual solute molecules are represented by the open circles.

FIGURE 3-2
Gel filtration for separation of solutes by size. Solute molecules smaller than the diameter of the pores of the gel particles enter them and are retained for a longer time. Larger solute molecules cannot penetrate the pores of the gel particles and are eluted off the column first.

and pH. The resins are inert polymers to which ionizable groups have been attached; resins with negative charges are cation exchangers, and those with positive charges are anion exchangers. Two ion exchange resins frequently used in protein purification are diethylaminoethylcellulose (DEAE-cellulose, an anion exchanger) and carboxymethylcellulose (CM-cellulose, a cation exchanger). A protein with a net positive charge at a given pH will combine with the negative groups of a cation exchange resin, and its flow will be retarded; a protein with a net negative charge will migrate through the cation exchange resin unimpeded. Cationic proteins in a mixture going through the column will compete with one another for binding to the negatively charged groups of the resin. The relative migration rates of different molecules depend on three factors: their individual affinities for the charged sites on the resin, the degree of ionization of the functional groups attached to the resin, and the chemical properties and concentrations of competing low-molecular-weight ions, e.g., potassium and sodium. For example, NaCl at the appropriate concentration can displace a cationic protein through competition between sodium ions (Na^+) and the positively charged groups of the protein for the negatively charged sites on the resin (Figure 3-3). Changing the pH of the eluting buffer can also bring about desorption of proteins from the resin through neutralization of the ionizable R-groups of amino acid residues. Decreasing pH will elute proteins in the order of their decreasing isoelectric point values during anion exchange chromatography.

Affinity chromatography takes advantage of specific affinities between protein molecules and analogues of biological molecules that are covalently bound to the column

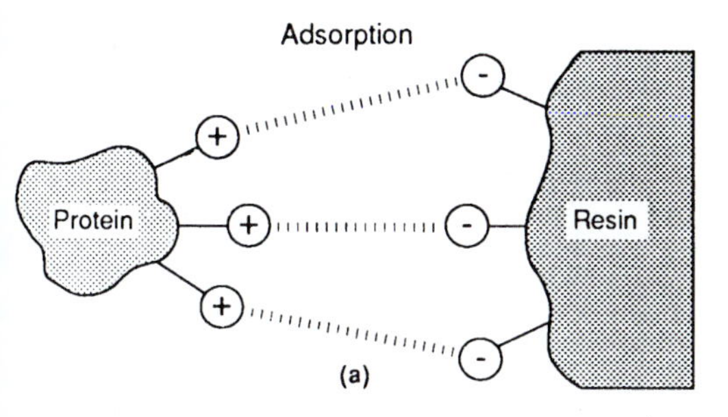

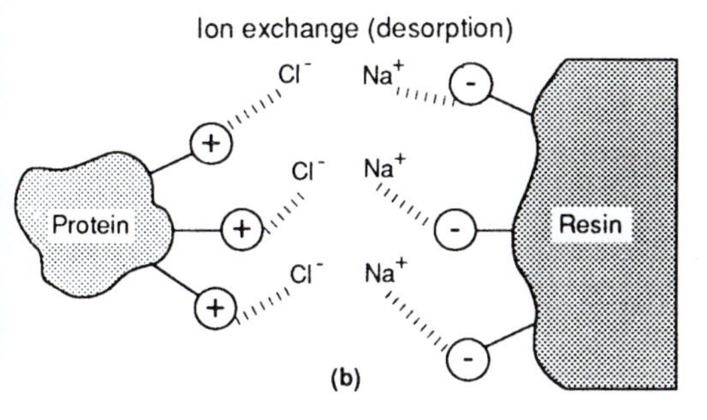

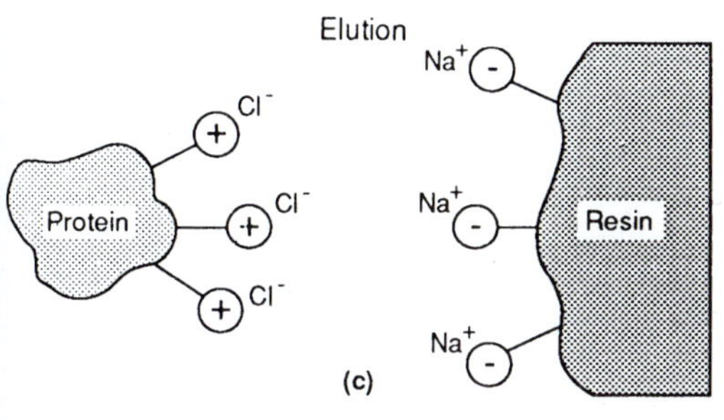

FIGURE 3-3
Steps involved in the ion exchange chromatographic separation of proteins. (a) At an appropriate pH, a protein with positively charged groups is adsorbed to the resin by electrostatic bonding. (b) As the concentration of NaCl is increased in the solvent flowing through the column, Na^+ competes with the positively charged protein for binding with the negatively charged groups or the resin, and desorption of the protein from the resin particles occurs. (c) The released protein is carried away with the flow of the solvent.

matrix (e.g., enzyme–substrate, hormone-receptor, or antigen-antibody interactions). The analogues on the column are usually small molecules resembling enzyme substrates, hormones, or antigens. When a protein solution is applied to the column, only those proteins with a high affinity for the matrix are bound. Proteins that do not specifically bind pass rapidly through the column. The bound proteins can be eluted by altering the pH or ionic strength of the eluent or by adding excess quantities of the ligand, e.g., hormone, antigen, or enzyme substrate or inhibitor. Affinity chromatography is useful in the purification of enzymes, hormones and their receptor sites, immunoglobulins (Chapter 35), and nucleic acids (Chapter 23).

Affinity Tag Chromatography

Affinity tag chromatography permits purification of recombinant proteins from growth media or from cell lysates. New chromatography techniques take advantage of DNA cloning that produces recombinant fusion proteins and allows such proteins to be easily purified. Recombinant proteins can be engineered to contain affinity tag sequences to create a fusion protein. The tag possesses unique affinity characteristics that serve as the basis for subsequent purification. Affinity chromatography is carried out using the immobilized ligand of the tag, which yields a highly purified fusion protein. A variety of affinity tag sequences are used such as ***hexa-histidine*** for metal chelate separation, enzyme tags that allow isolation using immobilized substrate, or epitope sequences for separation by an immobilized monoclonal antibody. An enzyme cleavage site is usually included between the tag and protein for removal of the tag from the fusion protein after purification. Once an effective purification strategy has been established for one fusion protein, it can be used for any protein that is engineered to include the same tag.

One example of this method is the use of glutathione S-transferase (GST) as the affinity tag. The fusion protein is engineered to contain a thrombin cleavage site between GST and the protein N terminus for subsequent removal of the GST (Figure 3-4). The sample is first run through a capture column consisting of the natural substrate for GST, glutathione bound to agarose. The column is then washed, this removing all unbound molecules and cellular debris. A large amount of free glutathione is then added to the column. The free glutathione outcompetes the agarose-bound glutathione for GST causing elution of the fusion protein. The eluted protein is then run through a second column consisting of immobilized thrombin, which will remove the GST affinity tag by cutting at the thrombin cleavage site. An enzyme column is used instead of added enzyme because the bound enzyme maximizes the interaction between substrate and enzyme. Furthermore, it eliminates the need for an enzyme removal step and enables the enzyme column to be reused. The eluant from the thrombin column is passed through a gel filtration column that separates the protein from the affinity tag based on molecular size. Finally, the eluant is fractionated and analyzed for purity.

Separation by Electrophoresis

Electrophoresis separates charged proteins on the basis of their different mobilities in an electric field. When a solution of proteins is subjected to an electrical potential, the

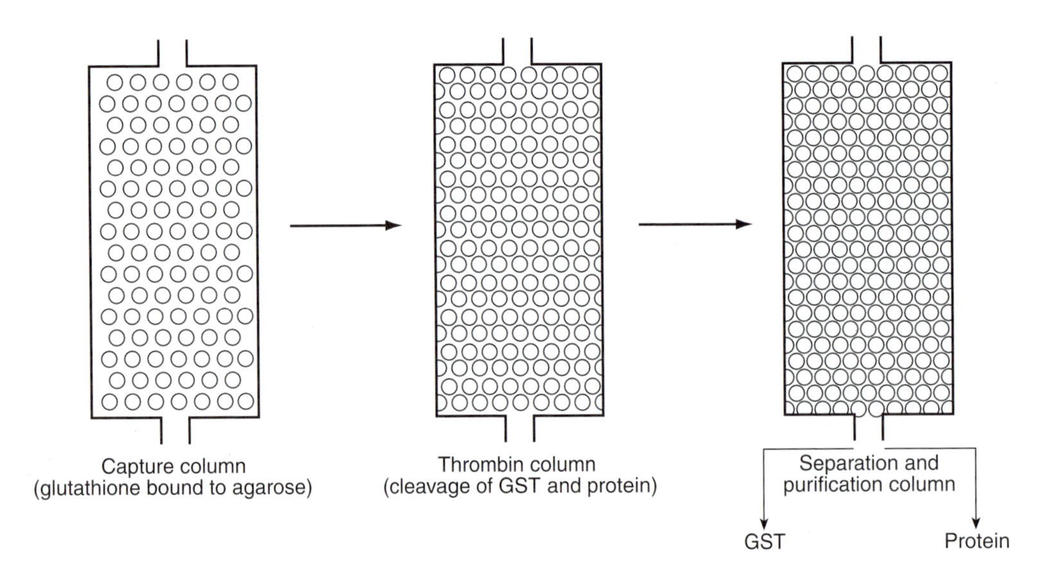

FIGURE 3-4
A schematic diagram of the steps in the purification of a protein fused with glutathione S-transferase. Protein purification using this procedure utilizes three columns and is fully automatic.

charged protein molecules migrate toward either the anode or the cathode. Factors that influence the rate of migration are pH, composition of the medium through which migration occurs, and size and shape of the protein molecule. A protein that does not migrate in an electric field at a given pH has no net charge at that pH. That pH is called the ***isoelectric point*** (pI) of the protein. The pI value is characteristic for each protein (Table 3-1). In a solution at a pH value above its pI, a protein will have a net negative charge; below its pI, a protein will have a net positive charge. The electrophoresis technique known as ***isoelectric focusing*** separates proteins on the basis of differences in their isoelectric points. Proteins whose pI values differ by as little as 0.02 pH unit can be separated by this technique. In isoelectric focusing, a mixture of proteins is placed in a pH gradient; in the presence of an electric field, each protein migrates to a position corresponding to its pI and comes to rest in a narrow band at that pH. The pH gradient is established by placing in an electric field an aqueous mixture of synthetic low-molecular-weight (300–600) ampholytes (oligomers of aliphatic amines, amino acids, and carboxylic acids). On application of an electric potential, the ampholytes migrate and come to rest according to their respective isoelectric points. If proteins are mixed with the ampholytes, the proteins migrate to the positions of their respective isoelectric points within the ampholyte gradient and can thereby be separated and concentrated into narrow bands (Figure 3-5).

Electrophoretic techniques also yield estimates of the molecular weights of proteins and nucleic acids. The detergent sodium dodecyl sulfate (SDS) and proteins form SDS-protein complexes that migrate in polyacrylamide gels according to their molecular weights. SDS dissociates multi subunit proteins into individual polypeptide chains. Each denatured chain has a uniform negative charge per unit mass of protein, since the total negative charge of the sulfonic acid groups of SDS, which are uniformly located along the surface of the protein, far exceeds the

TABLE 3-1
Isoelectric Points of Some Proteins

Protein	pI
Lysozyme	11.0
Cytochrome C	10.6
Pancreatic ribonuclease	9.6
Chymotrypsinogen	9.5
Myoglobin	7.0
Human growth hormone (somatotropin)	6.9
Hemoglobin	6.8
Human serum immunoglobulins	6.4–7.2
Carboxypeptidase	6.0
Catalase	5.6
Fibrinogen	5.5
β-Lactoglobulin	5.2
Urease	5.1
Human serum albumin	4.8
Egg albumin	4.6
Thyroglobulin	4.6
Pepsin	1.0

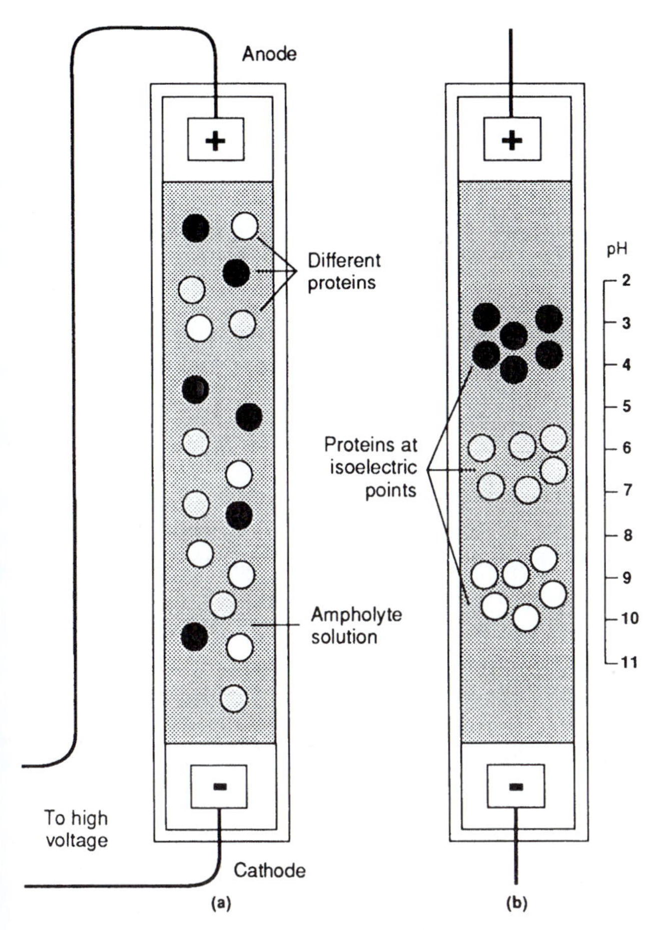

FIGURE 3-5
Separation of proteins by isoelectric focusing. (a) Proteins with different isoelectric points are mixed with an appropriate ampholyte solution. (b) When an electric potential is applied, the ampholytes migrate to their pI values establishing a pH gradient, while the proteins migrate to the positions of their respective isoelectric points.

intrinsic charge of the amino acid residues. SDS-protein complexes, all of which are rod-shaped, migrate toward the anode rapidly if they are small, but slowly if they are large. The polyacrylamide medium retards migration according to the size of the rigid rods. The molecular weight of a protein is established by comparing its rate of migration with those of a number of proteins of known molecular weight.

3.4 Capillary Electrophoresis

Capillary electrophoresis is a technique that can be used to analyze and separate proteins. It has a high resolving power that exceeds other electrophoretic techniques and is capable of distinguishing between proteins that differ only slightly in amino acid composition or glycosylation. Capillary electrophoresis is similar to high-performance liquid chromatography (HPLC) in high level of resolution, speed, on-line detection, and automation. However, it functions like slab gel electrophoresis in which proteins are separated based on mobility differences in an electric field.

In capillary electrophoresis, protein samples are first injected onto a fused silica microcolumn. The diameter of the fused silica microcolumns ranges between 50 and 100 μm. The free silanol groups of the fused silica become ionized at pH values greater than 2, causing the inside surface of the column to be negatively charged; the charge density is pH dependent. The micro-column sustains electric fields that are at least 50-fold more powerful than those used in slab gel electrophoresis. As a result of the high-intensity electric field, separation of proteins is more readily accomplished and the increased sensitivity enables microcharacterization of proteins. Due to the small size of the column, nanoliter quantities can be analyzed and separation occurs within minutes, whereas in slab gels it may take several hours. As the samples come off the column, proteins are measured by a specialized detector.

Separation by Solubility

Proteins may also be purified by selective precipitation, which is usually accomplished by changing the salt concentration (ionic strength) of the solution. Many proteins that are readily soluble in aqueous solutions at low salt concentration exhibit decreased solubility with increased salt concentration. Protein precipitation by increasing salt concentration is known as the ***salting-out*** phenomenon. In this process, the salt ions compete with the protein molecules for interaction with solvent (water) molecules, such that the affinity between protein molecules increases as water molecules are "removed" from the surface functional groups of the proteins. Eventually, the decreased polymer-solvent association results in precipitation of the protein. Since different proteins have different surface functional groups, proteins can be differentially precipitated at different salt concentrations. The related ***salting-in*** phenomenon, which increases the solubility of some proteins by addition of dilute salt solution, is believed to be due to interactions between salt ions and the charged groups on protein molecules, which minimize protein-protein (precipitation) interaction and maximize protein-solvent (dissolution) interaction.

Ammonium sulfate, $(NH_4)_2SO_4$, is the most commonly used compound for salting out of proteins because it is very soluble (706 g/L) and has four ionic charges per molecule. Precipitations are generally performed slowly with cold solutions to minimize protein denaturation due to the heat released on mixing and to allow time for the formation of precipitates.

3.5 Amino Acid Composition

Determination of the amino acid sequence of a purified protein begins with determination of the amino acid composition, i.e., the number of moles of each amino acid residue present per mole of protein, as follows.

The protein is completely hydrolyzed by acid (6 N HCl, 24 hours or longer at 110°C, under vacuum or inert gas) to its constituent amino acids and the resultant hydrolysate is evaporated to dryness. The amino acid composition is determined on protein hydrolysates obtained after 24, 48, and 72 hours of acid treatment. The content of amino acids with bulky aliphatic side chains such as isoleucine, leucine, and valine, which undergo slow hydrolysis, is calculated from an extrapolation of the hydrolysate data to infinite time. The content of hydroxyl-containing amino acids, which are slowly destroyed during hydrolysis, is obtained by a corresponding extrapolation to zero time. Since cysteine, cystine, and methionine residues are somewhat unstable to hydrolysis, these residues are oxidized to cysteic acid and methionine sulfone, respectively, with performic acid before quantitative analysis. Cysteine, or half-cystine, is quantitated as a derivative such as carboxymethyl cysteine after reduction and alkylation, a necessary prerequisite to subsequent sequence analysis. Tryptophan loss due to oxidation during hydrolysis can be greatly reduced by the inclusion of reducing agents—usually low-molecular-weight thiols—or oxygen-halide scavengers in the acid hydrolysis procedure. Tryptophan is substantially preserved if hydrolysis is performed with strong acids that do not contain halogen (e.g., methane sulfonic acid) or with strong alkali (e.g., NaOH). During acid hydrolysis, asparagine (Asn) and glutamine (Gln) are hydrolyzed to aspartic acid (Asp) and glutamic acid (Glu), respectively, and NH_3. Thus, Asn and Gln do not appear in the elution profile, and the Asp and Glu quantities include Asn and Gln, respectively.

The hydrolysate is dissolved in a small volume of an acidic buffer to obtain the protonated form of the amino acid and then chromatographed on a cation exchange resin column, e.g., Dowex 50. The SO_3^- groups of the resin bind all the protonated amino acids at the top of the column. The intensity of binding of each type of amino acid to the resin depends upon the number of positive charges on the molecules, the nature and size of the R-groups, and the pK′ values of the functional groups of the amino acids. Basic amino acids (lysine, histidine and arginine) have more than one positive charge and therefore bind more tightly than the neutral amino acids. Acidic amino acids (glutamic and aspartic acid) bind less tightly.

Amino acids are weakly attracted by hydrophobic and van der Waals interactions through their side chains. The strength of these attractions is determined by the adsorption characteristics of the resin and leads to different

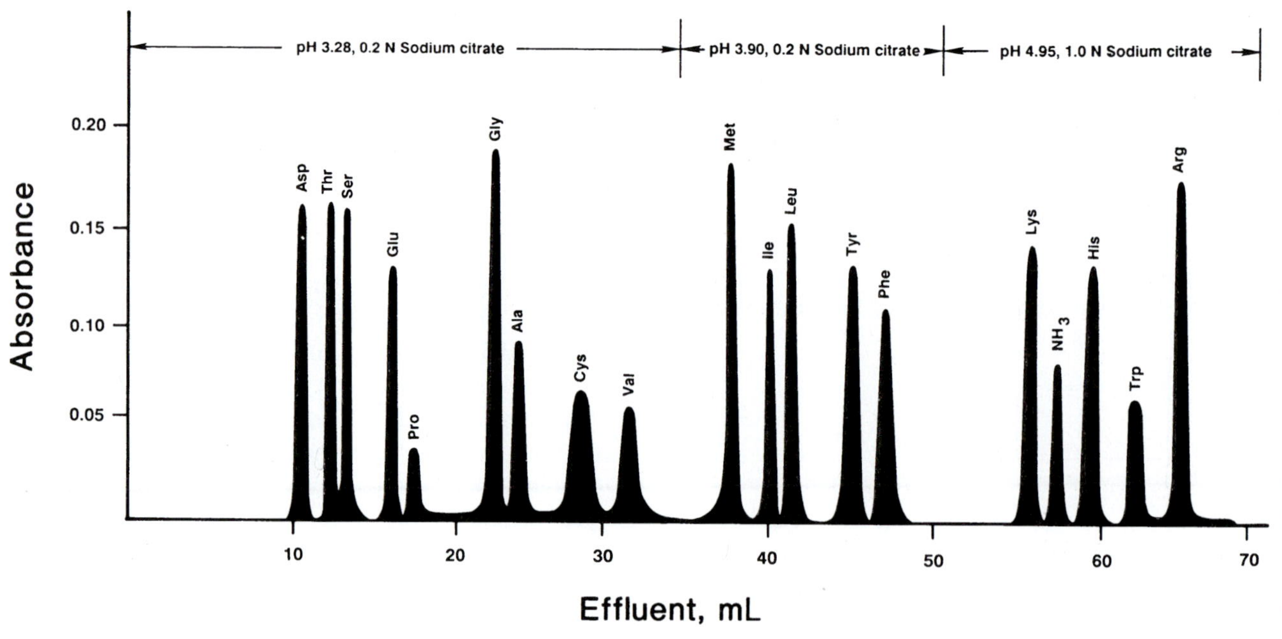

FIGURE 3-6
Separation and quantitation of amino acids by ion exchange chromatography. (Courtesy of Dr. N.S. Reimer.)

elution times for each amino acid. The amino acids are differentially eluted by the use of a stepwise pH gradient varying between pH 3 and pH 5 (Figure 3-6). As each amino acid elutes, it is reacted with ninhydrin at 100°C to produce a deep blue or purple color (yellow for proline and hydroxyproline). The color intensity is measured spectrophotometrically to quantitate the amino acids.

The process of separation and quantitation of amino acids has been automated. In one automated method, a single cation exchange resin column separates all the amino acids in the protein hydrolysate. The analyzer is capable of detecting as little as 1–2 nmol of an amino acid and a complete analysis can be obtained in about 4 hours. In newer procedures, the complete analysis can be performed in about 1hour and permit detection of as little as 1–2 nmol of an amino acid. Picomole amounts of amino acids can be determined when the separated amino acids are coupled to fluorescent reagents such as o-phthalaldehyde. Amino acid separation and quantitation can also be accomplished by reverse-phase high-pressure liquid chromatography of amino acid derivatives—a rapid and sensitive procedure.

3.6 Amino Acid Sequence Determination

Determination of the amino acid sequence of a protein involves the following steps:

1. Identification of the N- and C-terminal amino acid residues,
2. Cleavage of any disulfide bonds present,
3. Limited cleavage of the peptide into overlapping smaller fragments,
4. Purification of the fragments, and
5. Their stepwise cleavage into individual amino acid residues.

Identification of the N-Terminal Residue

Determination of the N-terminal residue is carried out by labeling the free unprotonated α-amino groups. Three alternative labeling reagents are used: 2,4-dinitrofluorobenzene (DNFB; Sanger's reagent), dansyl chloride (1-dimethylaminonaphthalene-5-sulfonyl chloride), and phenylisothiocyanate (PITC; Edman's reagent). DNFB and dansyl chloride react with free amino groups under basic conditions. The labeled peptide is hydrolyzed with acid to yield the labeled N-terminal residue and other free amino acids (Figure 3-7). The 2,4-dinitrophenyl amino acid derivatives (DNP-amino acids) have a yellow color and are separable by chromatographic methods and identifiable by comparison with reference DNP-amino acids. DNFB reacts with the ε-amino groups of lysyl residues to yield ε-DNP-lysine after hydrolysis. N-Terminal lysine produces α,ε-di(DNP)-lysine, whereas an internal lysine produces a derivative with only one dinitrophenyl group (ε-DNP-lysine).

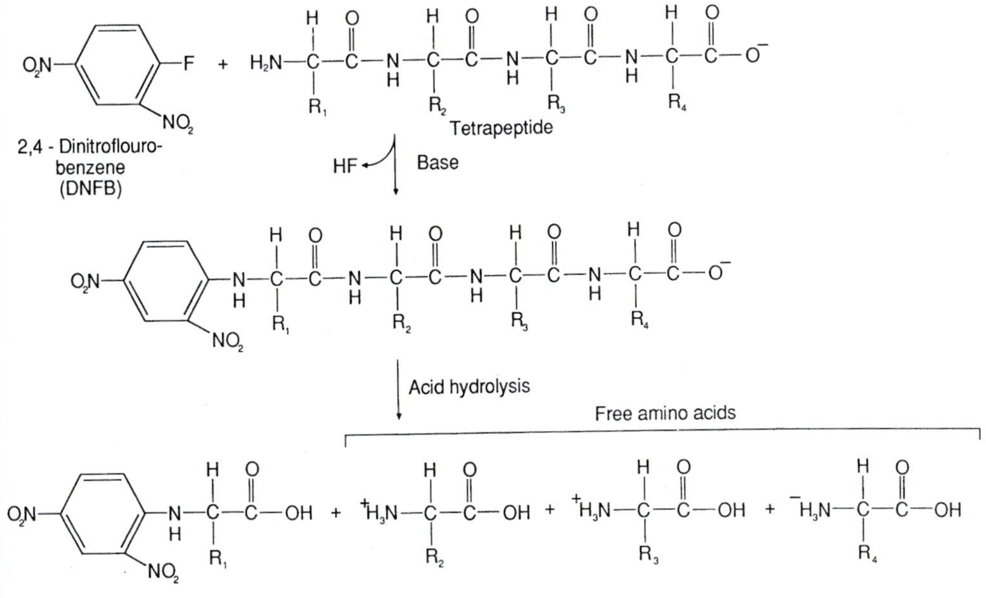

FIGURE 3-7
Determination of N-terminal amino acid residues by use of 2,4-dinitrofluorobenzene (Sanger's reagent).

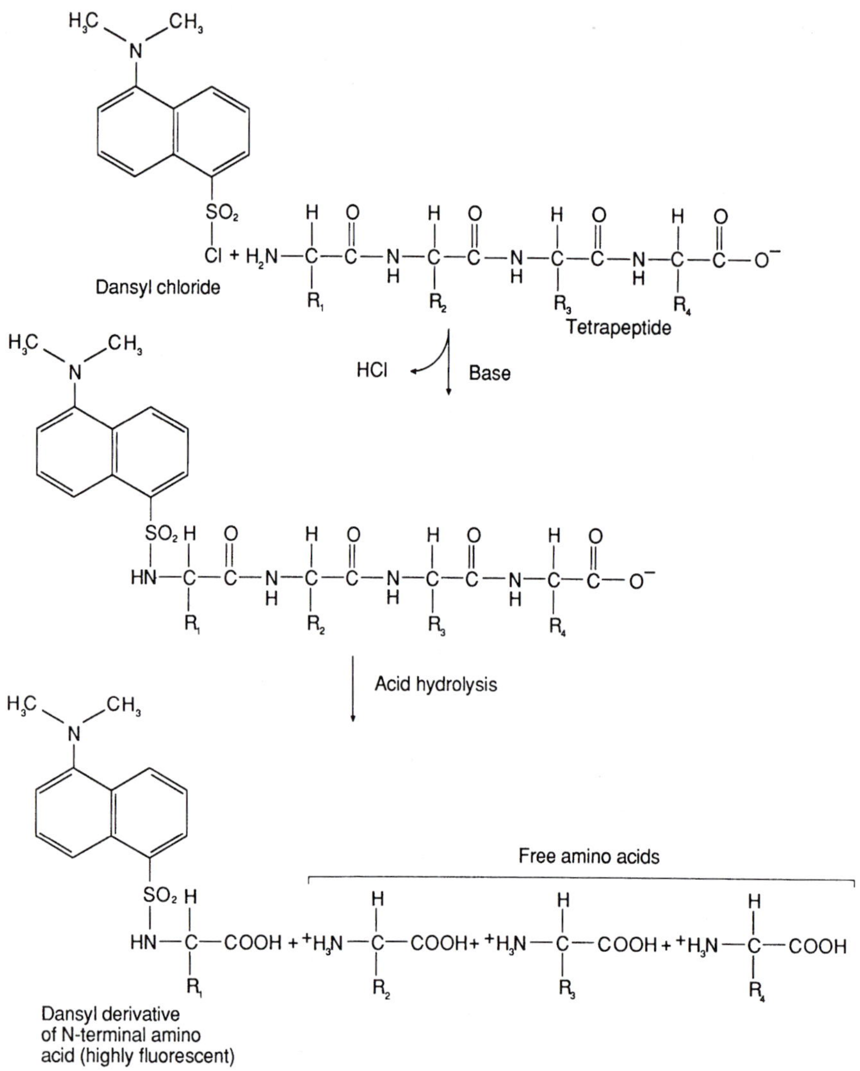

Dansyl chloride

Tetrapeptide

HCl

Base

Acid hydrolysis

Free amino acids

Dansyl derivative
of N-terminal amino
acid (highly fluorescent)

FIGURE 3-8
Determination of N-terminal amino acid residues by use of dansyl chloride.

Treatment of a peptide with dansyl chloride followed by hydrolysis yields a dansyl derivative of the N-terminal amino acid and other unlabeled amino acids (Figure 3-8).

The dansyl amino acid is separated and identified by chromatographic methods. The dansyl procedure is about 100 times more sensitive than the DNFB method because the dansyl amino acids are highly fluorescent and therefore detectable in minute quantities.

In the Edman procedure, PITC reacts under basic conditions with the free α-amino group to form a phenylthiocarbamoyl peptide (Figure 3-9). Treatment with anhydrous acid yields the labeled terminal amino residue plus the remainder of the peptide. In this process, the terminal amino acid is cyclized to the corresponding phenylthiohydantoin derivative (PTH-amino acid), which can be identified by gas chromatography, reverse-phase high-pressure liquid chromatography, thin-layer chromatography, or as the free amino acid after hydrolysis.

A significant advantage of the Edman procedure is that on removal of the N-terminal residue, the remaining peptide is left intact and its N-terminal remaining peptide group is available for another cycle of the procedure. This procedure can thus be used in a stepwise manner to establish the sequence of amino acids in a peptide starting from the N-terminal.

Identification of the C-Terminal Residue

The C-terminal residue is determined by the use of either a chemical reagent or the enzyme carboxypeptidase. The

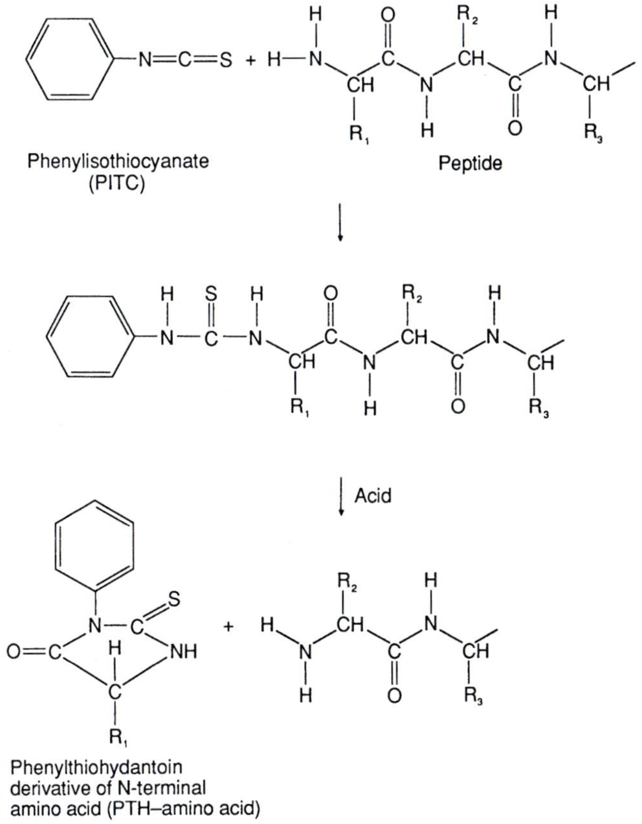

FIGURE 3-9
Determination of the N-terminal residue by the Edman procedure. After removal of the N-terminal amino acid, the remainder of the peptide remains intact and a new N-terminal amino acid is available for removal by the next reaction cycle.

chemical reagent hydrazine forms aminoacyl hydrazides with every residue *except* the C terminus (Figure 3-10). The C terminus is thus readily identified by chromatographic procedures. The disadvantage of hydrazinolysis is that the entire sample is used to determine just one residue.

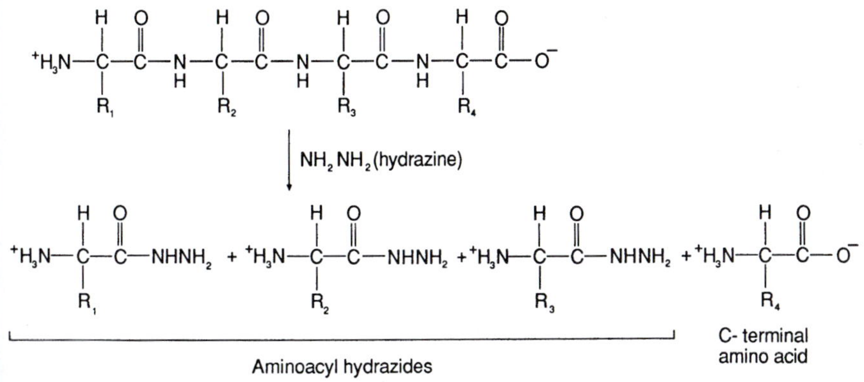

FIGURE 3-10
Determination of C-terminal amino acid residues by use of hydrazine.

Carboxypeptidase is an exopeptidase that specifically hydrolyzes the C-terminal peptide bond and releases the C-terminal amino acid. Two problems are associated with its use: the substrate specificity of the enzyme and the continuous action of the enzyme. The continuous action may yield the second, third, and additional residues from some chains even before the terminal residues on every chain are quantitatively released. Thus, it may be difficult to determine which residue is the C terminus. However, monitoring the sequential release of amino acids can often reveal the sequence of several residues at the C terminus. Concerning specificity, carboxypeptidase A releases all C-terminal residues except Lys, Arg, and Pro; carboxypeptidase B cleaves C-terminal Arg and Lys residues; and carboxypeptidase C hydrolyzes C-terminal Pro residues. Thus, more than one method may be needed to establish the C-terminal amino acid.

Selective Hydrolysis Methods

Cleavage of disulfide bonds occurs before hydrolysis of the protein into peptides. Disulfide bonds may be cleaved oxidatively, or they may be reduced and alkylated. Treatment of the native protein with performic acid, a powerful oxidizing agent, breaks disulfide bonds and converts cystine residues to cysteic acid (Figure 3-11). Reduction of the disulfide linkage by thiols, such as β-mercaptoethanol, yields reactive sulfhydryl groups. These groups may be stabilized by alkylation with iodoacetate or ethyleneimine to yield the carboxymethyl or aminoethyl derivative, respectively.

Hydrolysis of a protein into peptides can be accomplished by group-specific chemical and enzymatic reagents (Table 3-2). N-Bromosuccinimide and cyanogen bromide hydrolyze proteins at tryptophan and methionine (Figure 3-12) residues, respectively. Trypsin hydrolyzes

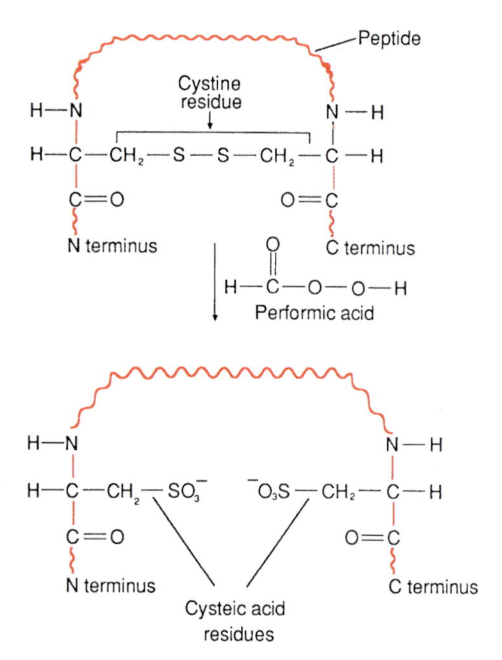

FIGURE 3-11
Cleavage of a disulfide bond (cystine residue) by performic acid.

the peptide linkage on the C-terminal side of lysine and arginine residues. Purification of the hydrolysis products is often the most challenging aspect of sequence determination.

Anion and cation exchange chromatography, paper chromatography, and electrophoresis are useful, and reverse-phase high-pressure liquid chromatography is used increasingly because of its speed and sensitivity. The purified peptides are analyzed for amino acid composition and terminal residues. Small peptides may be sequenced directly, but large peptides must be further hydrolyzed. Proteases such as chymotrypsin, pepsin, and papain, which are much less specific than trypsin, hydrolyze the peptides formed on tryptic digestion. The amino acid sequences of the purified peptides may be determined by the sequential Edman procedure.

Another sequencing technique is indirect analysis following nucleic acid sequencing of a DNA or RNA fragment corresponding to a specific protein. The universal genetic code provides information for translating a nucleic acid sequence into an amino acid sequence. This method will not correctly identify amino acid sequences from proteins that undergo posttranslational modification or proteins derived from eukaryotic genes with intervening sequences that are not translated (Chapter 25). However, it is a rapid means of corroborating sequencing data obtained by the classic slower methods described above.

TABLE 3-2
Hydrolysis of Polypeptides at Specific Sites by Selective Reagents

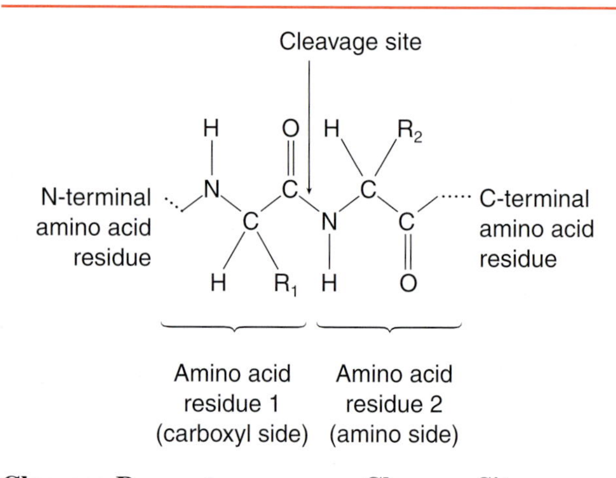

Cleavage Reagent	Cleavage Site
Enzymes	
Trypsin	Lys or Arg at amino acid 1
Chymotrypsin	Phe, Trp, or Tyr at amino acid 1
Thermolysin	Leu, Ile, or Val at amino acid 2
Chemicals	
Cyanogen bromide	Met at aminio acid 1
2-Nitro-5-thiocyanobenzoate	Cys at amino acid 2
N-Bromosuccinimide	Tryptophan or tyrosine at amino acid 1
Hydroxylamine	R_1=Asn; R_2=Gly
Mild acid hydrolysis (70% HCOOH or 0.1 *N* HCl)	R_1=Asp; R_2=Pro

Peptide Sequence Confirmation

Once the sequence has been determined, the proper arrangement of individual peptides in the protein can be established by identifying the overlapping sequences between peptides obtained by different cleavage procedures (Figure 3-13). The ultimate confirmation of sequence determination is protein synthesis. Chemical synthesis of peptides and proteins of known amino acid sequence can be accomplished by an elegant, automated, solid-phase procedure developed by Merrifield et al. Synthesis begins with the C-terminal amino acid, with each successive residue added in a stepwise manner. The C-terminal amino acid is covalently bound to a solid phase by a reaction between the carboxyl group and a chloromethyl group linked to a phenyl group of the resin polystyrene. Since

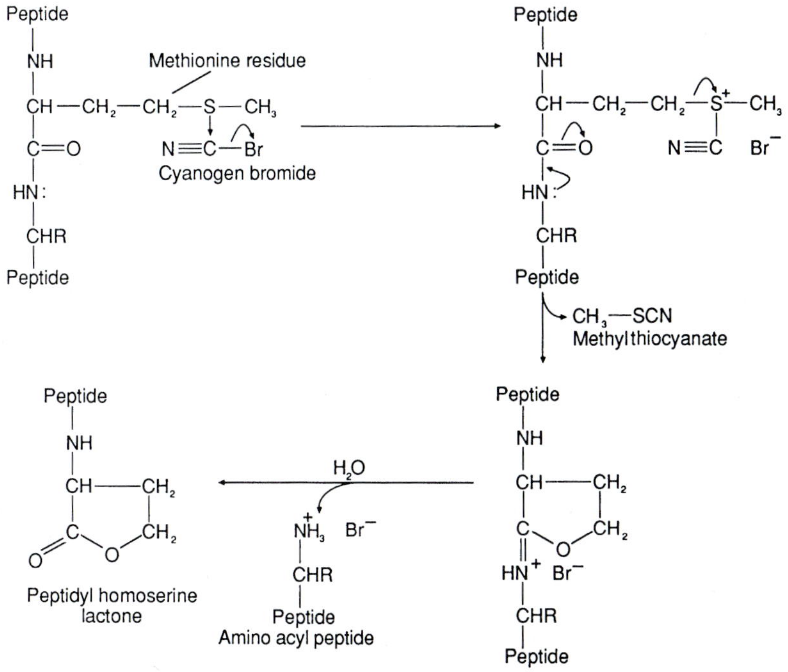

FIGURE 3-12
Cleavage of a peptide chain at methionine residue by cyanogen bromide. The methionine residue is modified to a carboxy-terminal homoserine lactone residue.

the amino group of an amino acid is reactive, it must be protected by a blocking group (tertiary butyloxycarbonyl; t-BOC), so that it does not react with the chloromethyl groups. The t-BOC is later removed by acid, enabling the amino group to participate in peptide bond formation. The degradation products of t-BOC, isobutylene and carbon dioxide, are removed as gases. The carboxyl group of a second amino acid, added as a t-BOC derivative, reacts with the amino group of the anchored amino acid in the presence of the condensing agent, dicyclohexylcarbodiimide (DCC). DCC, which removes H_2O from the two functional groups forming a peptide bond, is converted to dicyclohexylurea. In the next cycle, the t-BOC of a second amino acid of the solid-phase dipeptide is similarly removed, and a third t-BOC-amino acid is added along with DCC. The stepwise process of peptide synthesis is continued until the desired peptide is formed (Figure 3-14). The finished peptide can be easily cleaved from the polystyrene resin without affecting the peptide linkages.

Advantages of the solid-phase method are amenability to automation, the almost 100% yield of product for each reaction, the ease of removal of excess reagents and waste products by washing and filtration of resin particles, the lack of need for purification of intermediates, and speed. Peptides or proteins that have been synthesized by the solid-phase method include ribonuclease, bradykinin, oxytocin, vasopressin, somatostatin, insulin, and the β-chain of hemoglobin. The sequence analyses for these substances were confirmed by demonstrating that the synthetic products, constructed on the basis of sequence data, had the same biological activities as those of the corresponding natural substances.

Peptides obtained from trypsin cleavage:
Ala-Gly-Glu-Lys
Gly-Ala-Met-Arg
Ile-Val-Phe
Peptides obtained from cyanogen cleavage:
Ala-Gly-Glu-Lys-Gly-Ala-homoserine lactone*
Arg-Ile-Val-Phe
*This residue is derived from methionyl residue.
The complete sequence deduced from the above overlapping peptides:
Ala-Gly-Glu-Lys-Gly-Ala-Met-Arg-Ile-Val-Phe

FIGURE 3-13
Deduction of peptide sequence from analysis of overlapping sequences of component peptides.

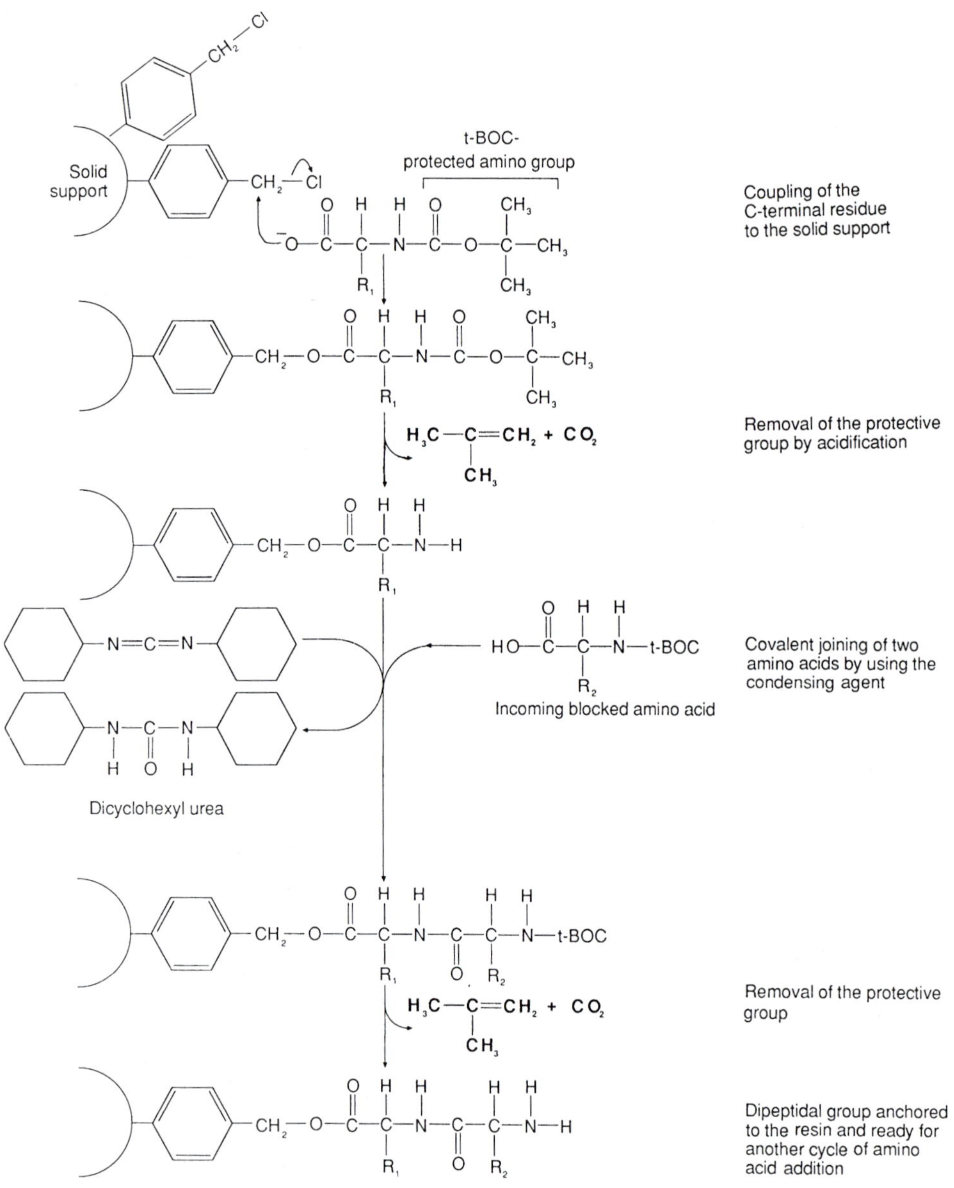

FIGURE 3-14
Steps in the chemical synthesis of a polypeptide by the solid-phase technique.

Fmoc Solid-Phase Peptide Synthesis

Solid-phase polypeptide synthesis that uses the t-BOC group to protect the N^{α}-amino group is a major procedure used by protein chemists. However, another solid-phase procedure is frequently used for the chemical synthesis of peptides and proteins of interest; this is the 9-fluorenylmethoxycarbonyl (Fmoc) procedure (Figure 3-15). The Fmoc method for solid-phase peptide synthesis uses the Fmoc group for protecting the N^{α}-amino group. In contrast to the t-BOC group method which requires acid for removal of the N^{α}-amino protective group, the Fmoc group can be removed by a mild base. A piperidine solution in N,N-dimethylformamide or N-methylpyrrolidone is generally used for the removal of the Fmoc group; anhydrous trifluoroacetic acid is used to remove the t-BOC group. This Fmoc procedure is technically simpler and chemically less complex than the t-BOC procedure. Because the Fmoc protecting group can be removed by base, the linkage of the peptide to the resin support does not have to be stable under acidic conditions as in the t-BOC procedure.

The Fmoc procedure offers more flexible reaction conditions and more reagent options. Because of the milder conditions of peptide synthesis, the Fmoc procedure is widely used for the synthesis of modified polypeptides that are phosphorylated, sulfated, or glycosylated.

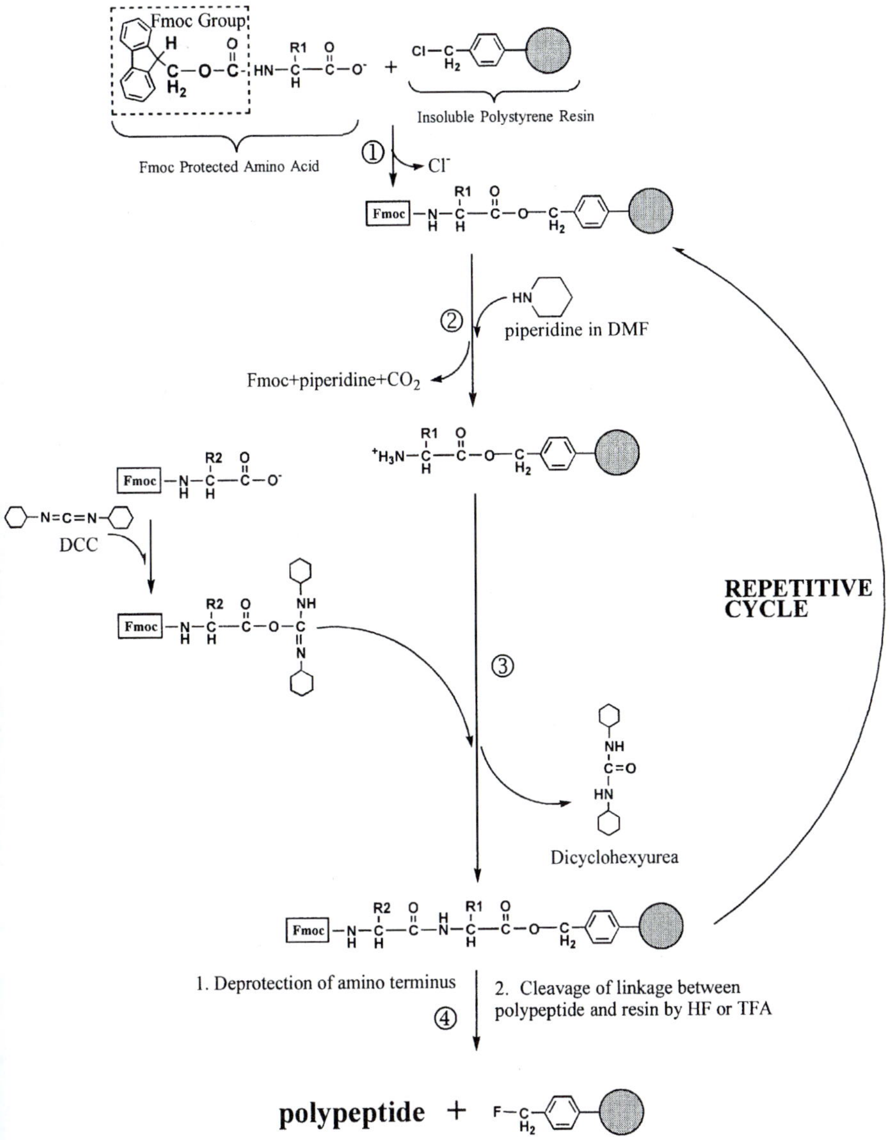

FIGURE 3-15
Schematic representation of 9-fluorenylmethoxycarbonyl (Fmoc) solidphase polypeptide synthesis. (1) Amino-acid containing Fmoc as N^{α}-amino protecting group is attached to the insoluble polystyrene resin. (2). The Fmoc protecting group is removed by piperidine in a N,N-dimethylformamide (DMF) solution. (3). N^{α}-amino group of amino acid attacks dicyclohexylcardodiimide (DCC)-activated carboxyl group of amino acid 2 to form a peptide bond. (4). Deprotection of N^{α}-amino group of the newly integrated amino acid by piperidine in the DMF solution the synthesized polypeptide is ultimately released from likage to the resin by treatment with hydrogen fluoride (HF) or trifluoroacetic acid (TFA). Steps (2) and (3) are repeated until the specific polypeptide chain is synthesized.

Since the Fmoc procedure uses a base for protection of the N-amino group, acid-labile compounds are used to protect the side chains. Side chain protecting compounds generally use a t-butyl moiety such as t-butyl ethers for Ser, Thr, and Tyr; t-butyl esters for Asp and Glu; and the t-BOC group for His and Lys, respectively. Also, the trityl group is used to protect Cys, Asn, and Gln; the 4-methoxy 2,3,6,-trimethylbenzenesulphonyl or the 2,2,5,7,8,-pentamethylChroman-6-sulfonyl groups have been used to protect the Arg guanidino group.

Supplemental Readings and References

P. Alewood, D. Alewood, L. Miranda, et al.: Rapid in situ neutralization protocols for t-Boc and Fmoc solid-phase chemistries. *Methods in Enzymology* **289,** 14 (1997).

J. A. Borgia and G. B. Fields: Chemical synthesis of proteins. *Trends in Biotechnology* **18,** 243 (2000).

K. A. Denton and S. A. Tate: Capillary electrophoresis of recombinant proteins. *Journal of Chromatography* B **697,** 111 (1997).

C. Jones, A. Patel, S. Griffin, J. Martin, et al.: Current trends in molecular recognition and bioseparation. *Journal of Chromatography* A **707,** 3 (1995).

B. L. Karger, Y-H. Chu, and F. Foret: Capillary electrophoresis of proteins and nucleic acids. Annual Review of Biophysical Biomolecular Structure **24,** 579 (1995).

M. Lynch, P. Lynch, N. Gordon, D. Whitney, et al.: Strategies for rapid purification of fusion proteins: Protein targets for drug discovery. American Biotech. Lab. **16,** 8 (1998).

K. H. Mayo: Recent advances in the design and construction of synthetic peptides: for the love of basics or just for the technology of it. *Trends in Biotechnology* **18,** 212 (2000).

D. A. Wellings and E. Atherton: Standard Fmoc protocols. *Methods in Enzymology* **289,** 44 (1997).

CHAPTER 4

Three-Dimensional Structure of Proteins

Proteins that consist of a single polypeptide chain are generally considered at three levels of organization: ***primary, secondary,*** and ***tertiary structure.*** For proteins that contain two or more polypeptide chains, each chain is a subunit and there is a ***quaternary*** level of structure. The primary structure is the unique sequence of amino acids that make up a particular polypeptide; primary structure is maintained by covalent bonds; secondary, tertiary, and quaternary structures are maintained principally by noncovalent bonds; disulfide bridges may also be considered at the secondary and tertiary levels. Secondary structure arises from repeated hydrogen bonding within a chain, as in the ***α-helix, β-pleated sheet,*** and ***β-turns*** (discussed later). Tertiary structure describes the three-dimensional stereochemical relationships of all of the amino acid residues in a single protein chain. *Folding* of a polypeptide is an orderly sequential process by which the polypeptide attains the lowest possible state of energy. The folding of the polypeptide into its secondary structure is determined primarily by the primary structure. Once the secondary structures are in place, a tertiary structure is formed and stabilized by interactions among amino acids which may be far from each other in the primary sequence but which are close to each other in the three-dimensional structure.

In a discussion of protein structure, it is necessary to differentiate the terms "configuration" and "conformation." *Configuration* refers to the absolute arrangement of atoms or substituent groups in space around a given atom. Configurational isomers cannot be interconverted without breaking one or more covalent bonds. For example, D and L-amino acids (Chapter 2), which have different amino acid configurations around the asymmetrical carbon atom, are not interconvertible without the breaking and remaking of one or more covalent bonds. *Conformation* refers to a three-dimensional arrangement of groups of atoms that can be altered without breaking any covalent bonds. For example, rotation around single bonds allows molecules to undergo transitions between conformational isomers (conformers), as in the eclipsed and staggered conformers of ethane (Figure 4-1). Since rotation is relatively unrestricted around the $H_3C–CH_3$ bond, the two conformers rapidly interconvert.

Proteins contain many single bonds capable of free rotation. Theoretically, therefore, proteins can assume an infinite number of possible conformations but under normal biological conditions, they assume only one or a very small number of "most stable" conformations. Proteins depend upon these stable conformations for their specific biological functions. A functional protein is said to be in its *native* form, usually the most stable one. The three-dimensional conformation of a polypeptide chain is ultimately determined by its amino acid sequence (primary structure). Changes in that sequence, as they arise from mutations in DNA, may yield conformationally altered (and often less stable, less active, or inactive) proteins. Since the biological function of a protein depends on a

Eclipsed conformation

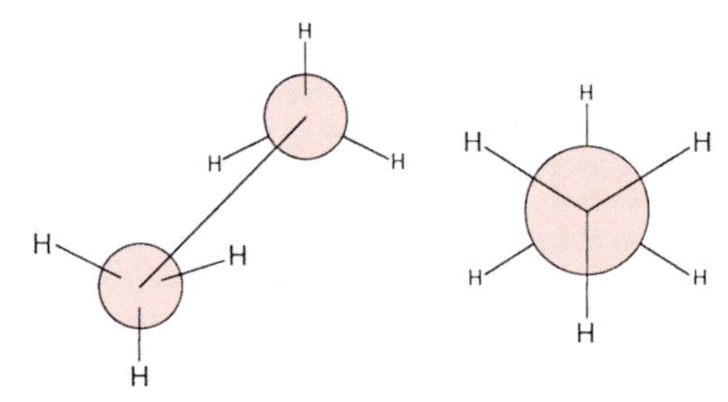

Staggered conformation

FIGURE 4-1
Conformational isomers of ethane ($H_3C–CH_3$). Eclipsed and staggered conformations for ethane are possible by virtue of the unrestricted rotation about the carbon-carbon single bond. There is a potential energy difference between the two forms, the staggered form being at the minimum and the eclipsed form at the maximum.

particular conformation, changes such as ***denaturation*** (protein unfolding) can lead to loss of biological activity.

4.1 Attractive and Repulsive Forces in Proteins

Both attractive and repulsive interactions occur among different regions of polypeptide chains and are responsible for most secondary and tertiary structure.

Attractive Forces

Covalent bonds involve the equal sharing of an electron pair by two atoms. Examples of important covalent bonds are ***peptide*** (amide) and ***disulfide bonds*** between amino acids, and C–C, C–O, and C–N bonds within amino acids.

Coordinate covalent bonds involve the unequal sharing of an electron pair by two atoms, with both electrons (originally) coming from the same atom. The electron pair donor is the ligand, or Lewis base, whereas the acceptor is the central atom (because it frequently can accept more than one pair of electrons), or Lewis acid. These bonds are important in all interactions between transition metals and organic ligands (e.g., Fe^{2+} in hemoglobin and the cytochromes).

Ionic interactions arise from electrostatic attraction between two groups of opposite charge. These bonds are formed between positively charged (α-ammonium, ε-ammonium, guanidinium, and imidazolium) side chains and negatively charged (ionized forms of α-carboxyl, β-carboxyl, γ-carboxyl, phosphate, and sulfate) groups.

Hydrogen bonds involve the sharing of a hydrogen atom between two electronegative atoms that have unbonded electrons. These bonds, although weak compared to the bonds discussed above, are important in water-water interactions and their existence explains many of the unusual properties of water and ice (Chapter 1). In proteins, groups possessing a hydrogen atom that can be shared include N–H (peptide nitrogen, imidazole, and indole), –SH (cysteine), –OH (serine, threonine, tyrosine, and hydroxyproline), $-NH_2$ and $-NH_3^+$ (arginine, lysine, and α-amino), and $-CONH_2$ (carbamino, asparagine, and glutamine). Groups capable of sharing a hydrogen atom include $-COO^-$ (aspartate, glutamate, and α-carboxylate), $-S-CH_3$ (methionine), –S–S– (disulfide), and $>C{=}O$ (in peptide and ester linkages).

Van der Waals attractive forces are due to a fixed dipole in one molecule that induces rapidly oscillating dipoles in another molecule through distortion of the electron cloud. The positive end of a fixed dipole will pull an electron cloud toward it; the negative end will push it away. The strength of these interactions is strongly dependent on distance, varying as $1/r^6$ where r is the interatomic separation. The Van der Waals forces are particularly important in the nonpolar interior structure of proteins, where they provide attractive forces between nonpolar side chains.

Hydrophobic interactions cause nonpolar side chains (aromatic rings and hydrocarbon groups) to cling together in polar solvents, especially water. These interactions do not produce true "bonds," since there is no sharing of electrons between the groups involved. The groups are pushed together by their "expulsion" from the polar medium. Such forces are also important in lipid-lipid interactions in membranes.

Repulsive Forces

Electrostatic repulsion occurs between charged groups of the same charge and is the opposite of ionic (attractive) forces. This kind of repulsion acts according to Coulomb's law: q_1q_2/r^2, where q_1 and q_2 are the charges and r is the interatomic separation.

Van der Waals repulsive forces operate between atoms at very short distances from each other and result from the dipoles induced by the mutual repulsion of electron clouds. Since there is no involvement of a fixed dipole (in contrast to van der Waals attractive forces), the dependence on distance in this case is even greater ($1/r^{12}$). These repulsive forces operate when atoms not bonded to each other approach more closely than the sum of their atomic radii and are the underlying forces in steric hindrance between atoms.

4.2 Primary Structure

Peptide Bond

Peptide bonds have a planar trans configuration and undergo very little rotation or twisting around the amide bond that links the α-amino nitrogen of one amino acid to the carbonyl carbon of the next (Figure 4-2). This effect is due to amido-imido tautomerization. The partial double-bond character of the N–C bond in the transition state probably best represents what exists in nature. Electrons are shared by the nitrogen and oxygen atoms, and the N–C and C–O bonds are both (roughly) "one-and-one-half" bonds (intermediate between single and double). The short carbonyl carbon-nitrogen bond length, 0.132 nm (the usual carbon-nitrogen single bond length is 0.147 nm), is consistent with the partial double-bond character of the peptide linkage. The planarity and rigidity of the peptide bond are accounted for by the fact that free rotation cannot occur around double bonds.

Whereas most peptide bonds exist in the trans configuration to keep the side chains (R-groups) as far apart as possible, the peptide bond that involves the –NH group of the rigid pyrrolidone ring of proline can occur in both trans and cis arrangements (Figure 4-3). However, x-ray data suggest that the trans form occurs more frequently in proteins than does the cis form. It has been further postulated that some proline residues (known as "permissive" proline residues) can exist in either the cis or trans configuration. For example, of the four proline residues of ribonuclease A, at least two are thought to be in the trans configuration in order to form a native structure, whereas one or both of the other residues may be accommodated in either the cis or trans configuration.

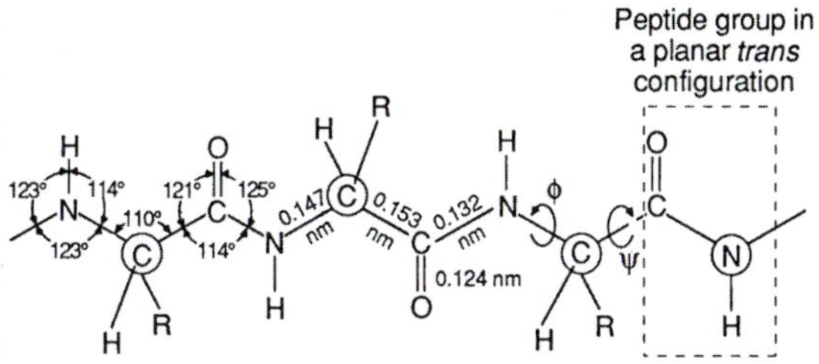

The bonds on either side of the α-carbon (i.e., between the α-carbon and the nitrogen, and between the α-carbon and the carbonyl carbon) are strictly single bonds. Rotation is possible around these single bonds. The designation for the angle of rotation of the α-carbon-nitrogen bond is ϕ, whereas that of the α-carbon-carbonyl carbon bond is ψ. Although theoretically an infinite number of ϕ and ψ angles are possible around single bonds, only a limited number of ϕ and ψ angles are actually possible in proteins. A polypeptide has specific ϕ and ψ values for each residue that determines its conformation.

FIGURE 4-2
Geometry of a peptide (amide) linkage. For the peptide bond, bond angles and bond lengths indicate that carbon-nitrogen bonds have a significant amount of double-bond character and that the C, O, N, and H atoms all lie in the same plane. The ϕ and ψ refer to rotations about the single bonds connecting the α-carbon with the α-nitrogen and the α-carbonyl carbon, respectively.

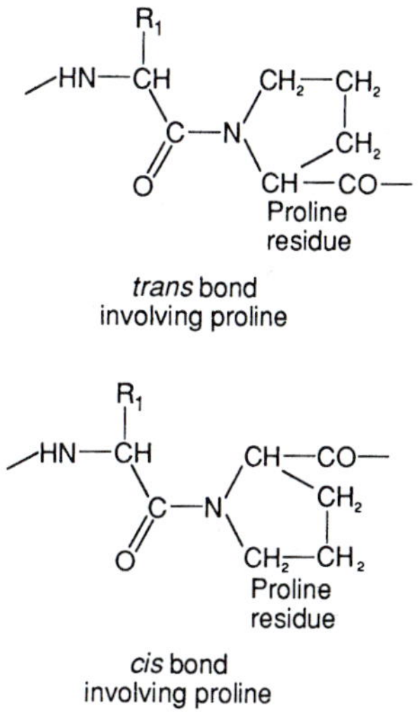

FIGURE 4-3
The trans and cis configurations of peptide bonds involving proline.

4.3 Secondary Structure

The folding of polypeptide chains into ordered structures maintained by repetitive hydrogen bonding is called *secondary structure*. The chemical nature and structures of proteins were first described by Linus Pauling and Robert Corey who used both fundamental chemical principles and experimental observations to elucidate the secondary structures. The most common types of secondary structure are the right-handed ***α-helix,*** parallel and antiparallel ***β-pleated sheets,*** and ***β-turns.*** The absence of repetitive hydrogen-bonded regions (sometimes erroneously called "random coil") may also be part of secondary structure. A protein may possess predominantly one kind of secondary structure (α-keratin of hair and fibroin of silk contain mostly α-helix and β-pleated sheet, respectively), or a protein may have more than one kind (hemoglobin has both α-helical and non-hydrogen-bonded regions). ***Globular proteins*** usually have mixed and fibrous proteins have predominantly one kind of secondary structure.

α-Helix

The rod-shaped right-handed α-helix, one of the most common secondary structures found in naturally occurring proteins, consists of L-α-amino acids (Figure 4-4). In the right-handed α-helix the helix turns counterclockwise (C-terminal to N-terminal) and in the left-handed it turns clockwise. The left-handed α-helix is less stable than a right-handed α-helix because its carbonyl groups and the R-groups are sterically hindered. The helical structure is stabilized by intrachain hydrogen bonds involving each –NH and –CO group of every peptide bond. These hydrogen bonds are parallel to the axis of the helix and form between the amido proton of the first residue and the carbonyl oxygen of the fourth residue, and so on, producing 3.6 amino acid residues per turn of the helix. The rise per residue is 0.15 nm and the length of one turn is 0.54 nm (Figure 4-5).

In some proteins, α-helices contribute significantly to the secondary structure (e.g., α-keratin, myoglobin, and hemoglobin), whereas in others, their contribution may be small (e.g., chymotrypsin and cytochrome c) or absent (e.g., collagen and elastin). Whether a polypeptide segment forms an α-helix depends on the particular R-groups of the amino acid residues. Destabilization of an α-helix may occur for a variety of reasons: electrostatic repulsion between similarly charged R-groups (Asp, Glu, His, Lys, Arg); steric interactions due to bulky substitutions on the β-carbons of neighboring residues (Ile, Thr); and formation of side-chain hydrogen or ionic bonds. Glycine residues can be arranged in an α-helix; however, the preferred and more stable conformation for a glycine-rich polypeptide is the β-pleated sheet because the R-group of glycine (–H) is small and gives rise to a large degree of rotational freedom around the α-carbon of this amino acid. Prolyl and hydroxyprolyl residues usually create a bend in an α-helix because their α-nitrogen atoms are located in rigid ring structures that cannot accommodate the helical bonding angles. Moreover, they do not have an amido hydrogen and therefore can form neither the necessary hydrogen bond nor the usual planar peptide bond. However, some proteins such as rhodopsin do contain proline residues embedded in α-helical segments.

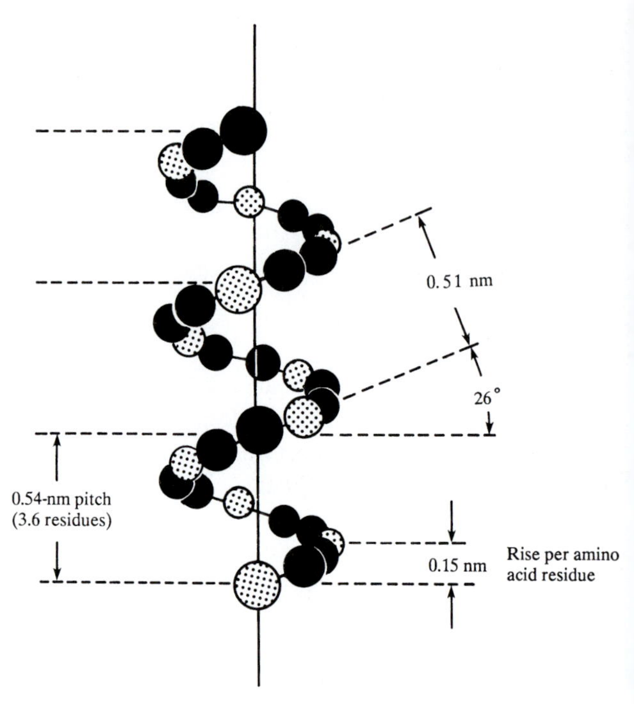

FIGURE 4-5
Average dimensions of an α-helix. Only the atoms of the α-carbon, the carbonyl carbon and the nitrogen of the peptide bonds are shown. The rise per residue and the length of one turn are 0.15 and 0.54 nm corresponding to minor and major periodicity, respectively.

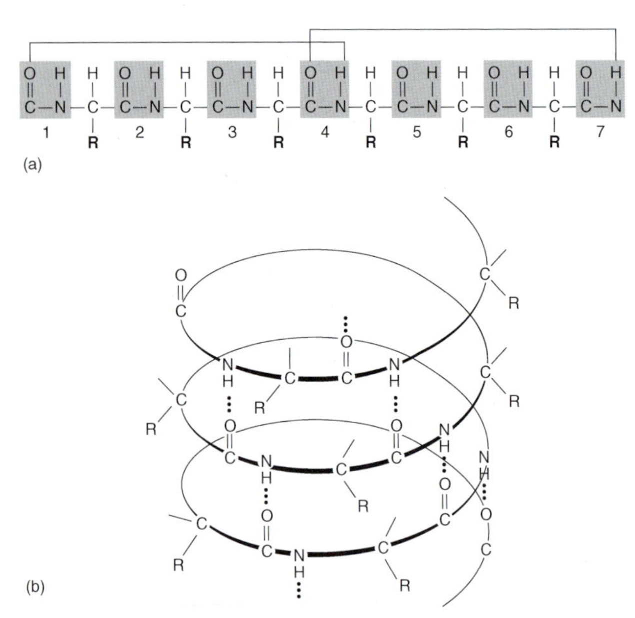

FIGURE 4-4
Hydrogen bonds in the α-helix. (a) Each peptide group forms a hydrogen bond with the fourth peptide group in each direction along the amino acid chain. (b) Coiling of an amino acid chain brings peptide groups into juxtaposition so that the hydrogen bonds shown in (a) can form. The multiple hydrogen bonds (indicated by the three dots) stabilize the helical configuration.

In some proteins, the α-helices twist around each other to form rope-like structures (coiled coils) to give rise to a ***supersecondary structure.*** Examples of such proteins are the α-keratins, which are major protein components of

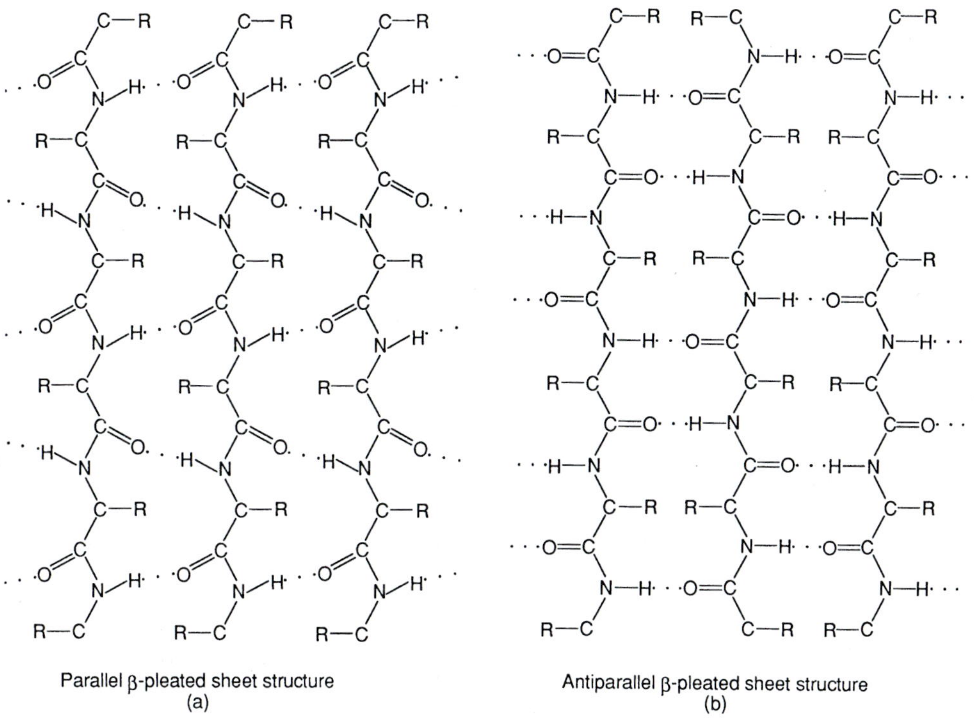

Parallel β-pleated sheet structure
(a)

Antiparallel β-pleated sheet structure
(b)

FIGURE 4-6
Hydrogen-bonding pattern of parallel (a) and antiparallel (b) β-pleated sheet structures.

hair, skin, and nails. These proteins are rich in amino acid residues that favor the formation of an α-helix. In addition, consistent with their properties of water insolubility and cohesive strength, α-keratins are rich in hydrophobic amino acid residues and disulfide cross-links. The α-helices are arranged parallel to their length with all the N-terminal residues present at the same end. Three α-helical polypeptides are intertwined to form a left-handed supercoil, called a *protofibril* (the α-helix itself is right handed). Eleven protofibrils form a microfibril. The polypeptides within the supercoil are held together by disulfide linkages and are also stabilized by van der Waals interactions between the nonpolar side chains. The number of disulfide cross-linkages in α-keratins varies from one source to another. Skin is stretchable because of fewer cross-links, whereas nails are inflexible and tough because of many more cross-links.

β-Pleated Sheet

The β-structure has the amino acids in an extended confirmation with a distance between adjacent residues of 0.35 nm (in the α-helix, the distance along the axis is 0.15 nm). The structure is stabilized by intermolecular hydrogen bonds between the –NH and –CO groups of *adjacent* polypeptide chains. The β-structure can occur between separate peptide chains (e.g., silk fibroin) or between segments of the same peptide chain, where it folds back upon itself (e.g., lysozyme). Two types of β-pleated sheets exist: ***parallel*** and ***antiparallel.*** In the parallel sheet structure, adjacent chains are aligned in the same direction with respect to N-terminal and C-terminal residues, whereas in the antiparallel sheet structure, the alignments are in the opposite directions (Figure 4-6). Some amino acid residues promote the formation of β-pleated sheets. For examples, in silk fibroin, which consists almost entirely of antiparallel β-structures, every other amino acid is glycine and alanine predominates in the remaining positions. Thus, one side of the peptide has only H in the R-position, whereas the other side has predominantly methyl groups. These small R-groups allow the formation of stacked β-pleated sheet structures. The methyls of one sheet fit into the pleat between the hydrogens of the overlying sheet. Thus, layers of silk proteins are associated but not covalently bonded. Bulky or similarly charged R-groups that cannot form pleated sheets create regions of flexibility in the otherwise uniform silk structure. Disrupting the hydrogen bonds of silk with heat does not appreciably change the length of the fiber because the β-structure is fully extended, but it does break interchain associations, whereas heat disruption of the intrachain hydrogen bonds in α-keratin facilitates conversion of the fiber from the helical to the extended β-structure (nonsheet).

The β-pleated sheet occurs as a principal secondary structure in proteins found in persons with ***amyloidosis.*** The generic name ***β-fibrilloses*** has been suggested for this group of disorders. The proteins that accumulate are called amyloid and are aggregates of twisted β-pleated sheet fibrils. They derive from endogenous proteins (e.g., immunoglobulins) on selective proteolysis and other chemical modifications. The fibrillar proteins are insoluble and relatively inert to proteolysis. Their accumulation in tissues and organs can severely disrupt normal physiological processes. The amyloid deposit, which occurs in several different tissues, is produced in certain chronic inflammatory diseases, in some cancers, and in the brain with some disorders, e.g., ***Alzheimer's disease.*** Partial or total disappearance of amyloid deposits in mice has been noted on administration of dimethyl sulfoxide, which disrupts hydrogen bonds.

β-Turns

β-Turns which are stabilized by a hydrogen bond, cause polypeptide chains to be compact molecules (e.g., globular proteins of spherical or ellipsoidal shape). The four amino acid residues of a β-turn form a hairpin structure in a polypeptide chain, thus providing an energetically economical and space-saving method of turning a corner. Two tetrapeptide conformations can accomplish a β-turn that is stabilized by a hydrogen bond (Figure 4-7).

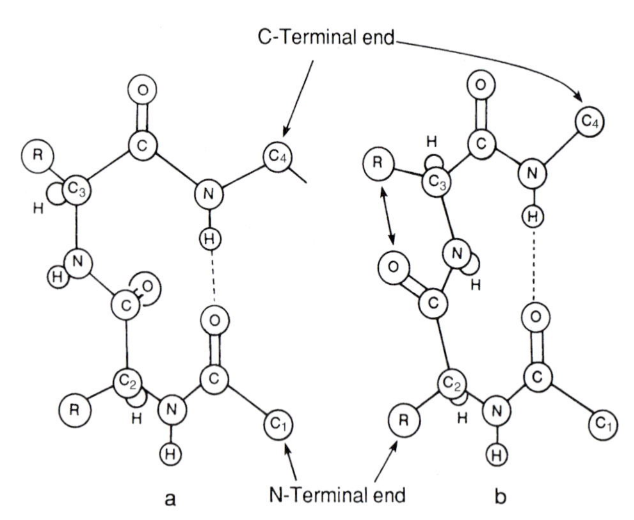

FIGURE 4-7
Two forms of β-turns. Each is a tetrapeptide and accomplishes a hairpin turn. The amino acid residues are identified by numbering the α-carbons 1–4. The CO group of residue 1 is hydrogen-bonded to the NH group of residue 4. Structure (b) is stable only if a glycine (R = H) residue is present as the third residue because of steric hindrance between the R-group and the carbonyl oxygen (double-headed arrow).

Random Coil

Certain regions of peptides may not possess any definable repeat pattern in which each residue of the peptide chain interacts with other residues, as in an α-helix. However, a given amino acid sequence has only one conformation, or possibly a few, into which it coils itself. This conformation has minimal energy. Since energy is required to bring about change in protein conformation, the molecule may remain trapped in a conformation corresponding to minimal energy, even though it is not at absolute minimum internal energy. This concept of a molecule seeking a preferred, low-energy state is the basis for the tenet that the primary amino acid sequence of proteins determines the secondary, tertiary, and quaternary structures.

Determination of Secondary Structure by Using Circular Dichroism (CD) Spectroscopy

Circular dichroism (CD) spectroscopy is widely used to determine the amount of α-helix, β-pleated sheet, and random coil structures in a protein molecule. The principle of CD is based on the fact that asymmetrical structures absorb light in an asymmetrical manner. Natural light vibrates in all planes perpendicular to its direction of travel; but its plane of polarization can be fixed to possess either left or right orientation. However, in circular polarization the direction of polarization rotates with the frequency of the light. If the rotation is *clockwise,* it is called *right* circularly polarized light and if *counterclockwise* it is called *left* circularly polarized light.

Absorption of light by a given molecule is equal to $E \times C$, where E is the molar extinction coefficient and C is the molar concentration. Asymmetrical molecules absorb right or left circularly polarized light differently. This difference in absorption $\varepsilon_L - \varepsilon_R$ is called *molar ellipticity,* which is dependent upon the wavelength of the incident light. A plot of $\varepsilon_L - \varepsilon_R$ versus the wavelength provides a CD spectrum characteristic of different secondary structures (Figure 4-8).

Other Types of Secondary Structure

Other distinct types of protein secondary structure include the type present in collagen, a fibrous connective tissue protein and the most abundant of all human proteins. Collagen peptide chains are twisted together into a three-stranded helix. The resultant "three-stranded rope" is then twisted into a ***superhelix*** (Chapter 10).

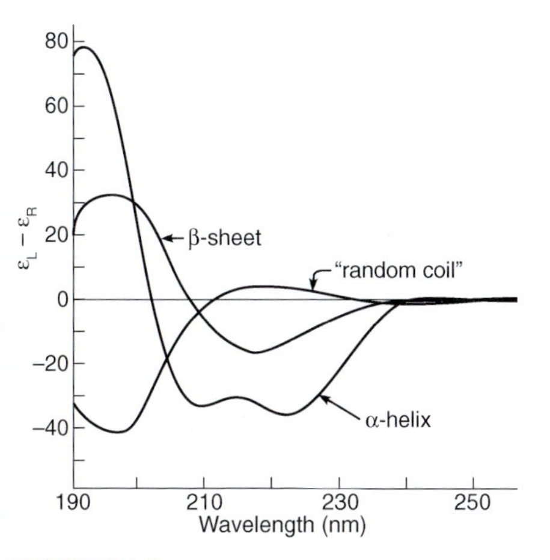

FIGURE 4-8
Circular dichroism spectra for polypeptides with different conformations. Note that for α-helix conformations, a characteristic "dip" is observed in the region of 210 nm.

4.4 Tertiary Structure

Three-dimensional tertiary structure in proteins is maintained by ionic bonds, hydrogen bonds, –S–S– bridges, van der Waals forces, and hydrophobic interactions.

The first protein whose tertiary structure was determined is ***myoglobin,*** an oxygen-binding protein consisting of 153 amino acid residues; its structure was deduced from x-ray studies by Kendrew, Perutz, et al. after a 19-year analysis (Figure 4-9). Their studies provided not only the first three-dimensional representation of a globular protein but also insight into the important bonding modes in tertiary structures. The major features of myoglobin are described below.

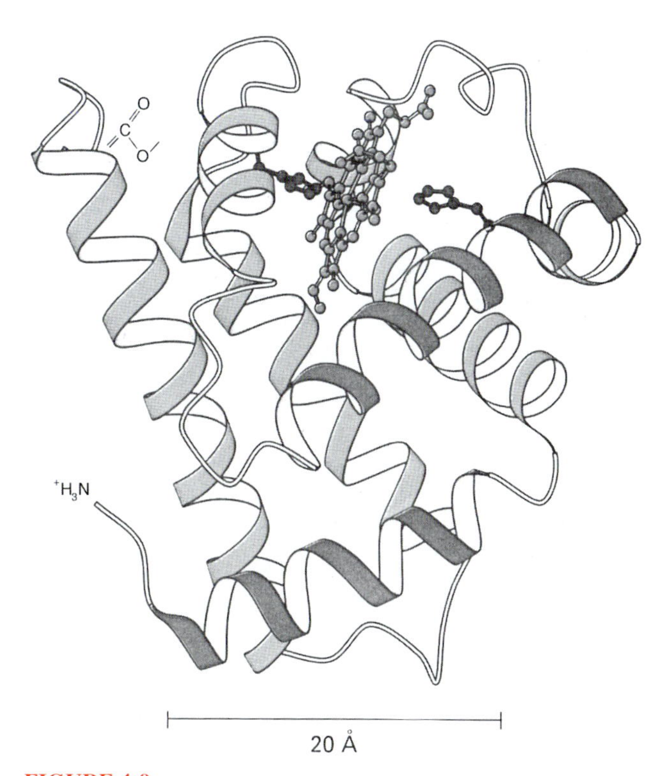

FIGURE 4-9
A high-resolution, three-dimensional structure of myoglobin.

1. Myoglobin is an extremely compact molecule with very little empty space, accommodating only a small number of water molecules within the overall molecular dimensions of 4.5 × 3.5 × 2.5 nm. All of the peptide bonds are planar with the carbonyl and amide groups in trans configurations to each other.
2. Eight right-handed α-helical segments involve approximately 75% of the chain. Five nonhelical regions separate the helical segments. There are two nonhelical regions, one at the N terminus and another at the C terminus.
3. Eight terminations of the α-helices occur in the molecule, four at the four prolyl residues and the rest at residues of isoleucine and serine.
4. Except for two histidyl residues, the interior of the molecule contains almost all nonpolar amino acids. The two histidyl residues interact in a specific manner with the iron of the heme group required for myoglobin activity (Chapter 21). The exterior of the molecule contains polar residues that are hydrated and nonpolar amino acid residues. The amino acid residues whose R-groups contain both polar and nonpolar portions (e.g., threonine, tyrosine, and tryptophan) are oriented so that the nonpolar group faces inward and the polar group faces outward, allowing only the polar portion to come in contact with water.
5. Myoglobin has no S–S bridges that generally help stabilize the conformation formed by amino acid interactions.

Three-dimensional analyses of proteins other than myoglobin reveal the incorporation of every type of secondary structure into tertiary globular arrays. For example, triose phosphate isomerase, a glycolytic enzyme (Chapter 13), contains a core of eight β-pleated sheets surrounded by eight α-helices arranged in a symmetrical cylindrical barrel structure known as $\alpha\beta$-barrels (Figure 4-10). This type of structure has also been noted among other glycolytic as well as nonglycolytic enzymes (e.g., catalase, peroxidase). A few generalizations can be made. Larger globular proteins tend to have a higher percentage of hydrophobic amino acids than do smaller proteins.

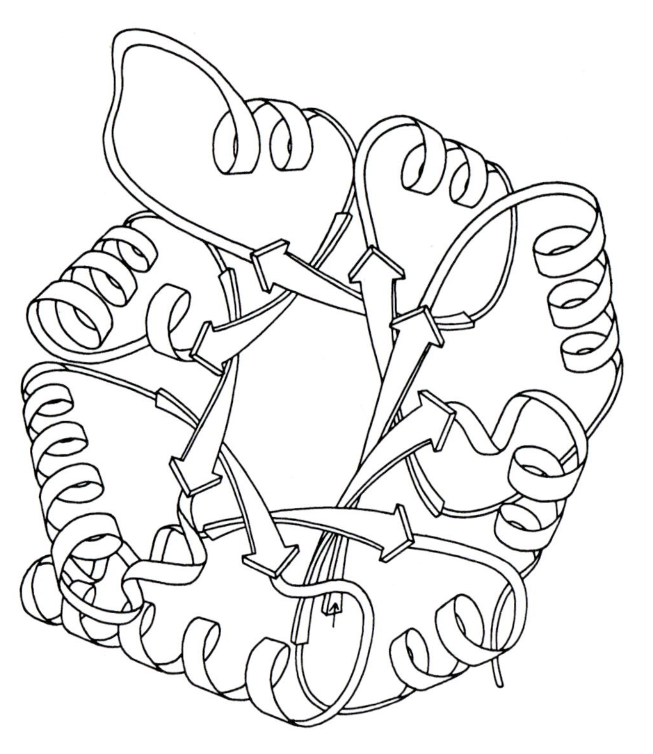

FIGURE 4-10
The $\alpha\beta$-barrel structure seen in triose phosphate isomerase. This structure consists of eight β-pleated sheets (represented by arrows) arranged in a cylindrical manner and surrounded by eight α-helices. (Reproduced with permission from J.S. Richardson.)

Membrane proteins tend to be rich in hydrophobic amino acids that facilitate their noncovalent interaction with membrane lipids. In both large globular and membrane proteins, the ratio of aqueous solvent–exposed surface to volume is small; hence, the need for polar, ionic, and hydrogen bonding groups is minimized, whereas the need for hydrophobic interaction is maximized. Small globular proteins tend to have more S–S bonds, which contribute significantly to tertiary structure stability, than do large globular proteins. These S–S bonds may be critical for proteins that lack sufficient stabilization from hydrophobic interactions.

Hydrophobic interactions are considered to be a major determinant in the maintenance of protein conformation. Hydrophobic interactions are due to the tendency of nonpolar side chains to interact with other nonpolar side chains rather than with water. Such interactions may involve the side chains of different adjacent molecules, or they may occur between side chains on the same protein molecule.

Nonpolar side chains are not attracted to water molecules either ionically or through hydrogen bonds (van der Waals attractions do occur). When hydrophobic molecules are put into an aqueous solution, they disturb the stable hydrogen-bonded water structure. To minimize this disturbance (and to return to the lowest possible free-energy state), the hydrophobic groups clump together as much as possible, so that the surface-to-volume ratio of the hydrophobic material is minimal. Once the chains are brought close together, van der Waals attractive forces can assist in holding them there. The van der Waals forces are weak, however, and most of the "hydrophobic bond energy" probably is from the free energy made available when water is able to self-associate in a stable structure instead of solvating all the hydrophobic groups in an unstable array. The driving force for the reaggregation of water in the presence of hydrophobic material can be envisioned in two ways:

1. The water surrounding a hydrophobic group is ice-like (highly ordered) because of limitations imposed on its movement by the presence of the hydrophobic material. A decrease in hydrophobic surface area causes a decrease in order and hence an increase in entropy that helps lower the free energy of the solution.
2. Hydrophobic regions in an aqueous solution break up hydrogen bond networks. A decrease in the area of the hydrophobic surface permits more hydrogen bonds to form, thus lowering the free energy of the solution.

4.5 Quaternary Structure

Quaternary structure exists in proteins consisting of two or more identical or different polypeptide chains (***subunits***). These proteins are called ***oligomers*** because they have two or more subunits. The quaternary structure describes the manner in which subunits are arranged in the native protein. Subunits are held together by noncovalent forces; as a result, oligomeric proteins can undergo rapid conformational changes that affect biological activity. Oligomeric proteins include the hemoglobins (Chapter 28), allosteric enzymes (Chapter 7) responsible for the regulation of metabolism, and contractile proteins such as actin and tubulin (Chapter 21).

4.6 Denaturation

Denaturation of a native protein may be described as a change in its physical, chemical, or biological properties. Mild denaturation may disrupt tertiary or quaternary structures, whereas harsher conditions may fragment the chain. Mild denaturation normally is a reversible process. Some of the changes in properties that may be caused by denaturation are as follows:

1. decreased solubility (often but not invariably);
2. alteration in the internal structure and arrangement of peptide chains that does not involve breaking the peptide bonds (e.g., separation of subunits of oligomeric proteins);
3. disrupted secondary structure (e.g., loss of helical structure);
4. increased chemical reactivity of functional groups of amino acids, particularly ionizable and sulfhydryl groups (e.g., shift of pK values);
5. increased susceptibility to hydrolysis by proteolytic enzymes;
6. decrease or total loss of the original biological activity; and
7. loss of crystallizability.

Studies by Anfinsen of the reversible denaturation of the pancreatic enzyme ribonuclease prompted the hypothesis that secondary and tertiary structures are derived inclusively from the primary structure of a protein (Figures 4-11 and 4-12). RNase A, which consists of a single polypeptide chain of 124 amino acid residues, has four disulfide bonds. Treatment of the enzyme with 8 M urea, which disrupts noncovalent bonds, and β-mercaptoethanol, which reduces disulfide linkages to cysteinyl residues, yields a random coil conformation.

However, if both reagents are removed and the cysteinyl residues are allowed to oxidize and re-form disulfide bonds, of the 105 different possible intramolecular combinations of disulfide linkages, only the four correct bonds form, and the denatured enzyme returns to its original, biologically active structure (Figure 4-12). These experiments are taken as proof that the primary structure (which is genetically controlled) determines the unique three-dimensional structure of a protein. However, as described below, it is now known that the folding of some proteins is assisted by other proteins.

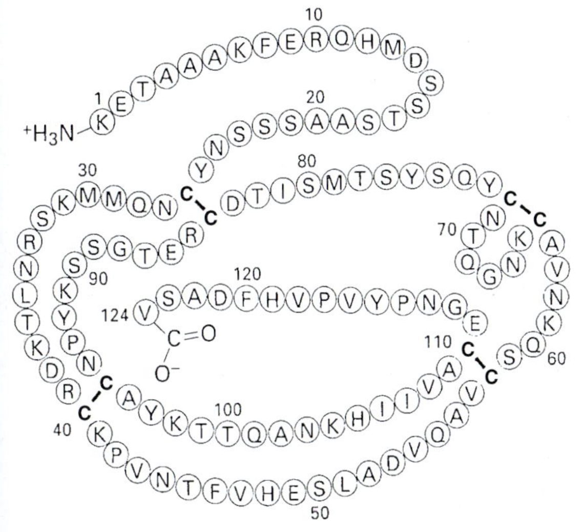

FIGURE 4-11
Amino acid sequence of bovine ribonuclease A. The molecule contains four disulfide bridges.

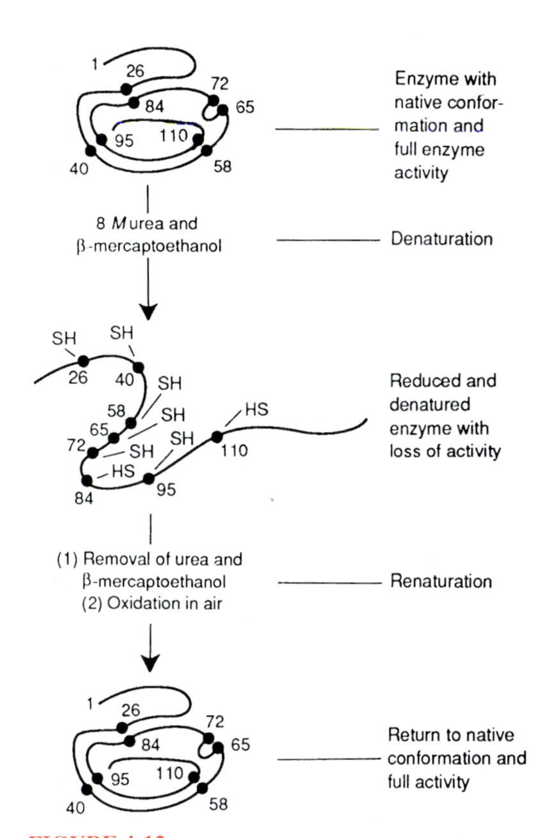

FIGURE 4-12
Denaturation and renaturation of ribonuclease A.

4.7 Protein Folding and Associated Diseases

Proteins are synthesized on ribosomes as nascent polypeptides in the lumen of the endoplasmic reticulum (ER). The amino acid sequence of proteins that determines the secondary and tertiary structures is dictated by the nucleotide sequence of mRNA. In turn, mRNA sequences are determined by DNA sequences (Chapters 23–25). As discussed earlier, the classic experiments of Pauling and Anfinsen led to the concepts that certain key amino acids at the proper positions are essential for the folding of proteins into a three-dimensional, functional, unique conformation. It is amazing that, of hundreds of millions of conformational possibilities, only a single conformational form is

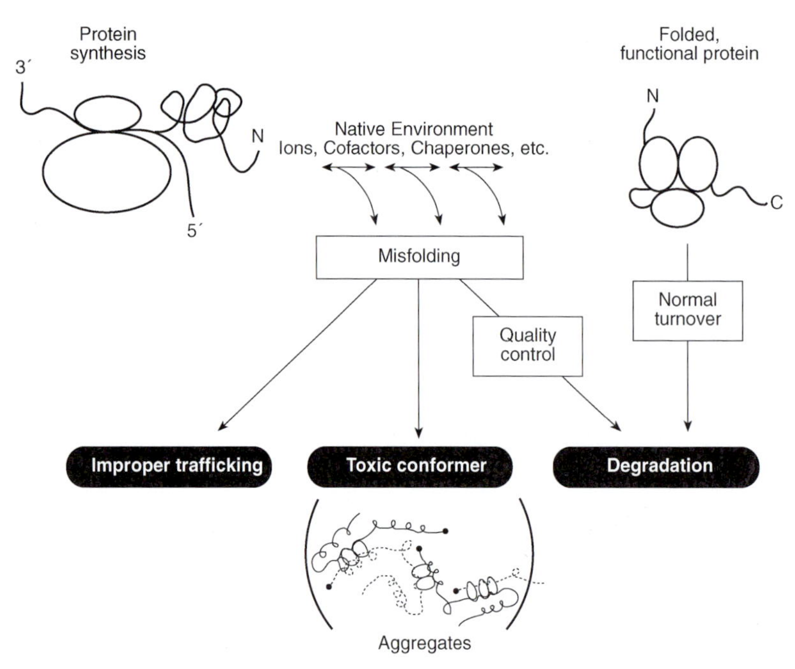

FIGURE 4-13
Pathway of protein folding. Normal folding occurs with the help of chaperones and other factors. Misfolding of polypeptides can lead to targeting to inappropriate cellular locations, or to degradation as a part of the quality control process, or to aggregation. The aggregated product is often resistant to proteolysis and forms aggregates, such as amyloid plaques.

associated with a functional protein. The process of directing and targeting the folding of intermediate polypeptides to the fully folded structures is aided, in some instances, by proteins known as ***molecular chaperones*** (also called *chaperonins*) (Figure 4-13).

Chaperones bind reversibly to unfolded polypeptide segments and prevent their misfolding and premature aggregation. This process involves energy expenditure provided by the hydrolysis of ATP. A major class of chaperones is ***heat-shock proteins*** (hsp), which are synthesized in both prokaryotic and eukaryotic cells in response to heat shock or other stresses (e.g., free-radical exposure). There are many classes of heat-shock chaperones (HSP-60, HSP-70, and HSP-90) that are present in various organelles of the cell. HSP-70 chaperones contain two domains: an ATPase domain and a peptide binding domain. They stabilize nascent polypeptides and also are able to reconform denatured forms of polypeptides. The HSP-70 family of chaperones shows a high degree of sequence homology among various species (e.g., *E. coli* and human HSP-70 proteins show 50% sequence homology).

Another chaperone, ***calnexin,*** is a 90 kDa Ca^{2+}-binding protein and is an integral membrane phosphoprotein of ER. Calnexin monitors the export of newly synthesized glycoproteins by complexing with misfolded glycoproteins that have undergone glycosylation (Chapter 16). If a protein cannot be folded into its proper conformation, chaperones assist in destruction. The process of folding is also facilitated by the ionic environment, cofactors, and enzymes. For example, folding is affected by protein disulfide isomerase, which catalyzes the formation of correct disulfide linkages, and by peptidyl prolyl isomerases, which catalyze the cis-trans isomerization of specific amino acid–proline peptide bonds.

Several disorders of protein folding are known that have the characteristic pathological hallmark of protein aggregation and deposits in and around the cells. The protein deposits are called amyloid and the disease is known as *amyloidosis*. Protein folding diseases, also known as conformational diseases, have many different etiologies, such as changes in the primary structure of proteins, defects in chaperones, and the inappropriate presence or influence of other proteins. A list of protein folding disorders is given in Table 4-1; some are discussed below and others in subsequent chapters. A common though not invariable aspect of conformational protein diseases is that the aggregation of polypeptides is made up of ***β-structures.*** This is primarily due to a transition from α-helical structure to β-structure. Another feature is that the aggregates are resistant to normal proteolysis.

A dementia syndrome characterized by an insidious progressive decline in memory, cognition, behavioral

TABLE 4-1
*Examples of Protein Folding Diseases**

Disease	Mutant protein/protein involved	Molecular phenotype
Inability to fold		
Cystic fibrosis	CFTR	Misfolding/altered Hsp70 and calnexin interactions
Marfan syndrome	Fibrillin	Misfolding
Amyotrophic lateral sclerosis	Superoxide dismutase	Misfolding
Scurvy	Collagen	Misfolding
Maple syrup urine disease	α-Ketoacid dehydrogenase complex	Misassembly/misfolding
Cancer	p53	Misfolding/altered Hsp70 interaction
Osteogenesis imperfecta	Type I procollagen pro α	Misassembly/altered BiP expression
Toxic folds		
Scrapie/Creutzfeldt-Jakob/ familial insomnia	Prion protein	Aggregation
Alzheimer's disease	β-Amyloid	Aggregation
Familial amyloidosis	Transthyretin/lysozyme	Aggregation
Cataracts	Crystallins	Aggregation
Mislocalization owing to misfolding		
Familial hypercholesterolemia	LDL receptor	Improper trafficking
α_1- Antitrypsin deficiency	α_1- Antitrypsin	Improper trafficking
Tay–Sachs disease	β-Hexosaminidase	Improper trafficking
Retinitis pigmentosa	Rhodopsin	Improper trafficking
Leprechaunism	Insulin receptor	Improper trafficking

*Reproduced with permission from P. J. Thomas, B-H. Qu and P. L. Pedersen. *Trends in Biochemicals Sciences* **20,** 456 (1995).

stability, and independent function was described by ***Alois Alzheimer*** and is known as ***Alzheimer's disease*** (AD). Age is an important risk factor for AD; it affects 10% of persons over 65 years of age and about 40% of those over age 85. The characteristic neuropathological changes include formation of extracellular neuritic plaques and intraneuronal tangles with associated neuronal loss in hippocampus and neocortex (Figure 4-14).

The major constituent of the extracellular plaques is ***amyloid β-protein*** (Aβ), which aggregates into 8 nm filaments. Aβ is a peptide of 40 or 42 amino acid residues and is proteolytically derived from a transmembrane glycoprotein known as ***β-amyloid precursor protein*** (βAPP). The enzymes that cleave βAPP to Aβ are known as secretases. βAPP is widely expressed, particularly in brain, and its gene has been localized to chromosome 21q. Two major observations have aided in understanding the role of Aβ peptides in the pathology of Alzheimer's disease. The first is that patients with ***Down syndrome*** have trisomy 21 (i.e., three chromosome 21's instead of two), exhibit Aβ deposits, and develop classical features of Alzheimer's disease at age 40 years or earlier. Second, several missense mutations in βAPP have been identified in cases of autosomal dominant Alzheimer's disease. These dominant mutations in βAPP adversely affect the action of secretases either by increasing the absolute rate of Aβ excretion (N-terminal mutations) or by increasing the ratio Aβ_{42} to Aβ_{40} (C-terminal mutations).

Inherited disorders of Alzheimer's disease represent less than 1% of all cases. The Aβ peptides aggregate to form β-structures leading to fibrils. The Aβ_{42} peptides are more neurotoxic and produce toxic effects by many interrelated mechanisms. These may involve oxidative injury, changes in intracellular Ca^{2+} homeostasis, cytoskeletal reorganization, and actions by cytokines.

The intraneuronal tangles are bundles of long paired helical filaments that consist of the microtubule-associated protein tau. The normal function of tau protein is

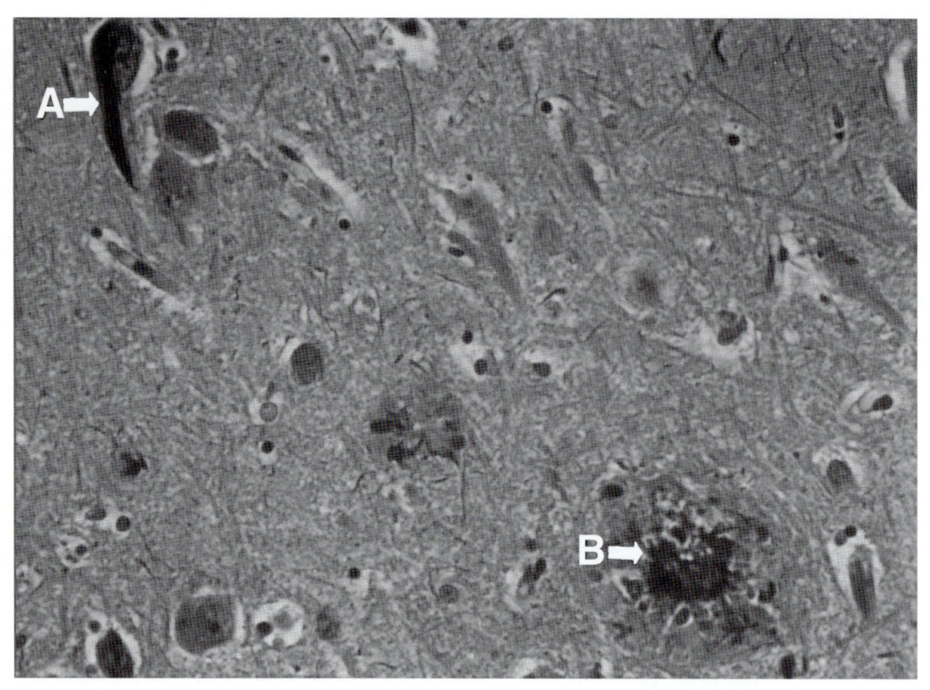

FIGURE 4-14
Section of cerebral cortex from a patient with Alzheimer's disease containing neurofibrillary tangle (A) and neuritic plaque (B). The section was processed with Bielschowsky's stain. (Courtesy of John M. Hardman.)

to stabilize microtubules in neurons by enhancing polymerization of *tubulin*. Normally, tau protein is soluble; however, when it is excessively phosphorylated, it turns into an insoluble filamentous polymer. The dysregulation of phosphorylation/dephosphorylation events has been attributed to an enhanced activity of certain kinases and a diminished activity of certain phosphatases. Whereas plaques are pathognomonic for AD, tangles are found in etiologically different neurological diseases. Disorders of abnormal hyperphosphorylation and aberrant aggregation of tau protein into fibrillar polymers are known as ***taupathies.*** Examples of taupathies in addition to AD are *progressive supranuclear palsy, Pick's disease, corticobasal degeneration,* and *frontotemporal dementias*.

Two other genes in addition to βAPP have been implicated in the early onset of autosomal dominant forms of Alzheimer's disease. The other two causative genes are located on chromosomes 14 and 1 and code for transmembrane proteins ***presenilin 1*** (consisting of 467 amino acid residues) and ***presenilin 2*** (consisting of 448 amino acid residues). These proteins are synthesized in neurons but their functions are not known. However, mutations in the presenilin genes lead to excessive production of $A\beta_{42}$ peptides.

Sporadic forms of Alzheimer's disease, responsible for 90% of all cases, are complex diseases and may represent the combined action of both environmental factors and genetic traits that manifest over long time spans. Various forms of a polymorphic gene for ***apolipoprotein E*** (apo E) which is on chromosome 19 have been found to occur in higher frequency in persons with Alzheimer's disease. There are three alleles of the apo E gene with six combinations: $\varepsilon2/\varepsilon2$, $\varepsilon3/\varepsilon3$, $\varepsilon4/\varepsilon4$, $\varepsilon2/\varepsilon3$, $\varepsilon2/\varepsilon4$, and $\varepsilon3/\varepsilon4$. Apo E is a lipid carrier protein that is primarily synthesized in the liver; however it is also synthesized in astrocytes and neurons. The function of apo E proteins in lipoprotein metabolism and their relation to premature atherosclerosis are discussed in Chapter 20.

Of the several genotypes for apo E, the acquisition of two apo E $\varepsilon4$ alleles may increase the risk for Alzheimer's disease up to eightfold. Each copy of the apo E gene increases the risk and shifts the onset to lower ages. The biochemical mechanism by which apo E $\varepsilon4$ protein participates in formation of tangles and plaques is unclear. Several mechanisms have been suggested, namely, interaction with tau protein and generation, and clearance of $A\beta$ peptides.

Pharmacological therapy for Alzheimer's disease consists of correcting the ***cholinergic deficit*** by administering ***acetylcholinesterase inhibitors*** (e.g., tacrine, donepezil). ***Estrogen therapy*** in women with Alzheimer's disease has been associated with improved cognitive performance. Estrogen's beneficial effect may be due to cholinergic and neurotrophic actions. Other therapeutic strategies are directed at inhibiting or decreasing the formation of neurotoxic peptides. In addition, drugs that selectively digest the aggregated peptides may prove useful. An experimental vaccine which contains AP peptide administered to plaque-producing mice leads to less plaque formation in younger mice and the disappearance of plaques in the older mice. The alterations in the plaque formation in mice were associated with preservation of memory and learning ability. The vaccination did not trigger an autoimmune response or toxic reaction in the experimental animals. Thus, these studies have provided impetus for the development of a human vaccine.

In evaluating a patient for Alzheimer's disease, it is essential that other treatable causes of dementia be excluded by determining critical biochemical and clinical parameters. Some of the treatable, relatively common abnormalities that produce dementia include drug abuse, electrolyte imbalance, thyroid abnormalities, and vitamin B_{12} deficiency; less common abnormalities are tumor, stroke, and ***Wernicke's encephalopathy.***

Transthyretin amyloidosis (also called familial amyloid polyneuropathy) is an autosomal dominant syndrome characterized by peripheral neuropathy. This disease results from one of five mutations identified thus far in the gene for transthyretin. Transthyretin is also called *prealbumin* (although it has no structural relationship to albumin) because it migrates ahead of albumin in standard electrophoresis at pH 8.6. Transthyretin is synthesized in the liver and is a normal plasma protein with a concentration of 20–40 mg/dL. It transports thyroxine and retinol binding protein (Chapter 38). The concentration of transthyretin is significantly decreased in malnutrition and plasma levels are diagnostic of disorders of malnutrition (Chapter 17).

The gene for transthyretin resides on chromosome 18 and it is expressed in a constitutive manner. The primary structure of transthyretin consists of 127 amino acid residues and eight β-sheet structures arranged in an antiparallel conformation on parallel planes (Figure 4-15).

Protein folding disorders of an unusual nature may account for a group of *transmissible spongiform encephalopathies* (TSE) involving proteins called *prions* (PrP). These disorders, known as *prion diseases,* are all characterized by amyloid deposition in the brain of animals and humans. The clinical features include neurological symptoms with loss of motor control, dementia, paralysis, and wasting. Incubation periods for prion diseases are months in animals and years in humans. No treatments are available for any of these diseases. TSEs occur in several species of animals and humans, and animal models have been essential in deciphering the molecular basis of these diseases. Examples of prion diseases occuring in animals and humans are:

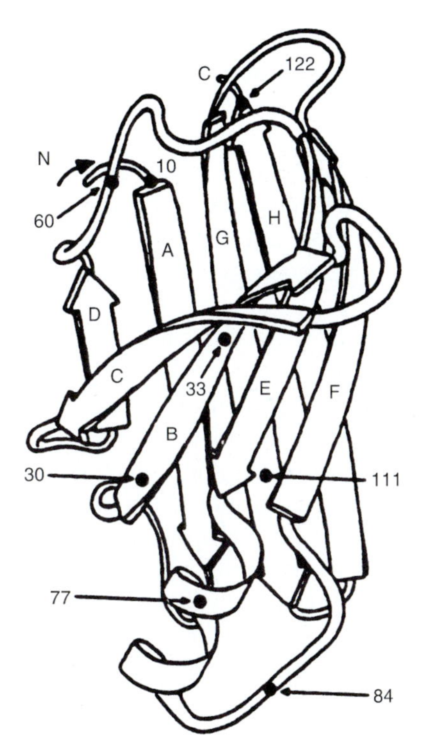

FIGURE 4-15
The structure of transthyretin. The molecule contains eight antiparallel β-strands (A–H) arranged in two parallel planes. The circulating form of transthyretin is a tetramer. Some mutations in the transthyretin gene are associated with amyloidosis and eight of the amino acid alterations causing this disease are indicated. In plasma, transthyretin is a tetramer composed of identical monomers. It appears that mutations cause the monomeric unfolded intermediate of transthyretin to aggregate into an insoluble β-amyloid fibril formation.

Cats	: *transimissible feline encephalopathy*
Cows	: *bovine spongiform encephalopathy* (BSE)
Mink	: *transmissible mink encephalopathy*
Mule deer and elk:	*chronic wasting disease*
Sheep	: *scrapie*
Humans	: *Creutzfeld-Jakob disease* (CJD), *Kuru, fatal-familial insomnia syndrome,* and *Gerstmann-Straussler-Schenker disease*

TSEs can exhibit inherited, infectious and sporadic presentations. Additionally, the inherited disease can also be infectious. CJD occurs both as an inherited autosomal dominant disorder and in a transmissible form. In the "protein only" hypothesis, the abnormal prion protein, either introduced from external sources or produced by the mutated prion protein gene, affects normal protein folding and shifts the prion protein folding towards the formation of abnormal prion protein. The conversion of the normal prion protein, whose function is unknown, to an aberrant form involves a conformational change rather than a covalent modification. The abnormal prion protein functions as a seed that induces the normal cellular prion protein towards the abnormal amyloidogenic rich, β-structure proteins which can be propagated and transmitted to other cells. The aggregated form of prion protein forming amyloid is resistant to proteolysis.

The conversion of naturally occurring protease-sensitive prion protein to a protease-resistanct form occurs in vitro by mixing the two proteins. However, these protease-resistant prion proteins are not infectious. Thus, in the "protein-only" hypothesis of prion infection, the acquisition of an abherrant conformation is not sufficient for the propagation of infectivity. However, in the yeast (*Saccharomyces cerevisiae*) system, the abnormal prion form of the yeast protein, introduced by liposome fusion, is able to seed a self-propagating conformational change of the normal proteins, which accumulate as aggregates. The aggregates are transmissible to daughter yeast cells along with the propagation of abnormal phenotype.

Recently a serious public health problem has arisen by showing that a prion disease in cattle can cross species barriers and infect humans. This occurred when cattle were fed meal made from sheep infected with scrapie. The cattle developed BSE (commonly called "mad cow disease"). Subsequently, when people consumed prion-contaminated beef, a small number, primarily in Great Britain, developed a variant of CJD (vCJD) approximately five years afterward. The variant form of CJD is a unique form of prion disease occurring in a much younger population than would be expected from inherited or sporadic CJD. Both BSE and vCJD share many similar pathologic characteristics suggesting an etiologic link between human vCJD and cattle BSE.

The tumor suppressor protein *p*53 provides yet another example of protein misfolding that can lead to pathological effects, in this case cancers (*p* is for protein and 53 is for its approximate molecular weight of 53,000). The gene for *p*53 is located on the short arm of chromosome 17 (17*p*) and codes for a 393-amino-acid phosphoprotein. In many cancers the *p*53 gene is mutated and the lack of normal *p*53 protein has been linked to the development of as many as 40% of human cancers.

Normal *p*53 functions as a tumor suppressor and is a transcription factor that normally participates in the regulation of several genes required to control ***cell growth, DNA repair,*** and ***apoptosis*** (programmed cell death). Normal *p*53 is a tetramer and it binds to DNA in a sequence-specific manner. One of the *p*53-regulated genes produces a protein known as *p*21, which interferes with the cell cycle by binding to cyclin kinases. Other genes regulated by *p*53 are MDM2 and BAX. The former gene codes for a protein that inhibits the action of *p*53 by functioning as a part of a regulatory feedback mechanism. The protein made by the BAX gene is thought to play a role in *p*53-induced apoptosis.

Most mutations of *p*53 genes are somatic missense mutations involving amino acid substitutions in the DNA binding domain. The mutant forms of *p*53 are misfolded proteins with abnormal conformations and the inability to bind to DNA, or they are less stable. Individuals with the rare disorder ***Li–Fraumeni syndrome,*** (an autosomal dominant trait) have one mutated *p*53 gene and one normal *p*53 gene. These individuals have increased susceptibility to many cancers, such as leukemia, breast carcinomas, soft-tissue sarcomas, brain tumors, and osteosarcomas.

Clinical trials are underway to investigate whether the introduction of normal *p*53 gene into tumor cells by means of gene therapy (Chapter 23) has beneficial effects in the treatment of cancer. Early results with *p*53 gene therapy indicate that it may shrink the tumor by triggering apoptosis.

Supplemental Readings and References

Protein Folding and Its Defects

R. Aurora, T. P. Creamer, R. Srinivasan and G. D. Rose: Local interactions in protein folding: Lessons from the α-helix. *Journal of Biological Chemistry* **272,** 1413 (1997).

J. R. Beasley and M. H. Hecht: Protein design: The choice of de novo sequences. *Journal of Biological Chemistry* **272,** 2031 (1997).

M. Blaber, X.-J. Zhang, and B. W. Mathews: Structural basis of amino acid α-helix propensity. *Science* **260,** 1637 (1993).

R. W. Carrell and D. A. Lomas: Conformational disease. *Lancet* **350,** 134 (1997).

W. D. Kohn, C. T. Mant, and R. S. Hodges: α-helical protein assembly motifs. *Journal of Biological Chemistry* **272,** 2583 (1997).

R. W. Ruddon and E. Bedows: Assisted protein folding. *Journal of Biological Chemistry* **272,** 3125 (1997).

P. J. Thomas, B.-H. Qu, and P. L. Pedersen: Defective protein folding as a basis of human disease. *Trends in Biochemical Sciences* **20,** 456 (1995).

Alzheimer's Disease, *p*53, and Prions

J. Avila: Tau aggregation into fibrillar polymers: taupathies. *FEBS Letters* **476,** 89 (2000).

A. Bossers, R. de Vries, M. A. Smits: Susceptibility of sheep for scrapie as assessed by in vitro conversion of nine naturally occurring variants of PrP. *Journal of Virology* **74,** 1407 (2000).

F. E. Cohen: Prion, peptides and protein misfolding. *Molecular Medicine Today* **6,** 292 (2000).

T. M. Gottlieb and M. Oren: *p*53 in growth control and neoplasms. *Biochimica Biophysica Acta* **1287,** 77 (1996).

A. M. Haywood: Transmissible spongiform encephalopathies. *New England Journal of Medicine* **337,** 1821 (1997).

L. Helmuth: Further progress on a β-amyloid vaccine. *Science* **289,** 375 (2000).

A. F. Hill, M. Antoniou, and J. Collinge: Protease-resistant prion protein produced in vitro lacks detectable infectivity. *Journal of General Virology* **80,** 11 (1999).

M. Horiuchi, S. A. Priola, J. Chabry, et al.: Interactions between heterologous fortms of prion protein: binding, inhibition of conversion, and species barriers. *Proceedings of the Society of National Academy of Sciences* (USA) **97,** 5836 (2000).

D. A. Kocisko, J. H. Come, S. A. Priola, et al.: Cell-free formation of protease-resistant prion protein. *Nature* **370,** 471 (1994).

D. Malkin: *p*53 and the Li-Fraumeni syndrome. *Biochimica Biophysica Acta* **1198,** 197 (1994).

C. L. Masters and K. Beyreuther: Alzheimer's disease. *British Medical Journal* **316,** 446 (1998).

S. B. Pruisner: Molecular biology and pathogenesis of prion diseases. *Trends in Biochemical Sciences* **21,** 482 (1996).

A. D. Roses: Alzheimer's disease: The genetics of risk. *Hospital Practice* **32(7),** 51 (July 15, 1997).

D. J. Selkoe: Amyloid β-protein and genetics of Alzheimer's disease. *Journal of Biological Chemistry* **271,** 18295 (1996).

D. J. Selkoe: Alzheimer's disease: Making progress through a multidisciplinary approach. Proceedings of symposium. *American Journal of Medicine* **104(4A),** I S-32S (1998).

H. E. Sparrer, A. Santoso, F. C. Szoka Jr., et al.: Evidence for the prion hypothesis: induction of the yeast [PSI^+] factor by in vitro converted sup35 protein. *Science* **289,** 595 (2000).

C. W. Stephen and D. P. Lane: Mutant conformation of *p*53. Precise epitope mapping using filamentous phage epitope library. *Journal of Molecular Biology* **225,** 577 (1992).

M. F. Tuite: Sowing the protein seeds of prion propagation. *Science* **289,** 556 (2000).

E. Yonish-Rouach: The *p*53 tumor suppressor gene: A mediator of a Gi growth arrest and of apoptosis. *Experentia* **52,** 1001 (1996).

CHAPTER 5

Thermodynamics, Chemical Kinetics, and Energy Metabolism

Nearly all chemical changes involving breakage and formation of covalent bonds in a living organism (absorption, digestion, metabolism, locomotion, putrefaction, etc.) are catalyzed by enzymes. Without these catalysts, the reactions would proceed too slowly for biological systems to function at any significant rate.

5.1 Methods of Altering the Rate of Reactions

All reaction rates can be enhanced by increasing the concentration of reactants; however, this is neither practical nor desirable for many intracellular chemical substances whose concentration cannot easily be increased, would be toxic if increased, or would require physiologically unacceptably large volumes of cellular material to maintain suitable osmotic conditions.

Reaction rates can be accelerated by raising the temperature because this increases thermal motion and energy. However, the human body functions at a constant temperature, and so temperature can play no significant role in altering the rates of chemical processes within the body. Conceivably, subtle local differences in body temperature may affect chemical reaction rates (e.g., hormone-receptor interactions or conformational changes of macromolecules) that involve weak binding forces. Organisms such as reptiles have a minimal capacity to regulate their body temperature; as a result, their body temperature fluctuates depending on environmental temperature. In such organisms, chemical reactions, which are also enzyme-catalyzed, largely depend on body temperature. In the human body, under abnormal conditions leading to either increased temperature (fever) or decreased temperature (hypothermia), metabolism is increased or decreased frequently with deleterious consequences. In certain clinical conditions, however, ***hypothermia*** can be useful. Organs maintained at low temperatures (0–4°C) by perfusion with a cold perfusate (usually a balanced electrolyte solution) remain viable 10 times longer because of reduction in the rate of chemical reactions and in metabolic requirements. Hypothermic storage of cadaver kidneys is extensively used prior to transplantation. Skin, cornea, and blood are preserved at low temperatures and hypothermia is used in open heart surgery to prevent organ damage that occurs during interruption of the circulation. A third method of increasing the rate of a reaction is to add a ***catalyst*** (an enzyme), which speeds up the reaction by reducing the free energy of activation (see below) by providing alternative reaction pathways. Catalysts accelerate reactions without being consumed and therefore need not appear in a balanced chemical equation. In an enzyme-catalyzed reaction, $A + B \rightleftharpoons C + D$, the enzyme influences the reaction velocity of both forward and backward reactions to the same extent as long as both reactions are thermodynamically feasible. Therefore, in a closed system in which neither reactants nor products are removed, the enzyme *cannot affect the equilibrium of the reaction;* it only affects the rate at which equilibrium is attained. If products are removed as they are formed (as in a biochemical pathway), the enzyme will continue to accelerate the reaction in the forward direction; such a state of affairs occurs frequently in biological systems.

Chemical reactions in living systems involve orderly release, storage, or utilization of energy. Knowledge of thermodynamics and kinetics is essential to appreciate how this occurs. Thermodynamics deals with the changes in energy content between reactants and products, whereas kinetics is concerned with the reaction rates.

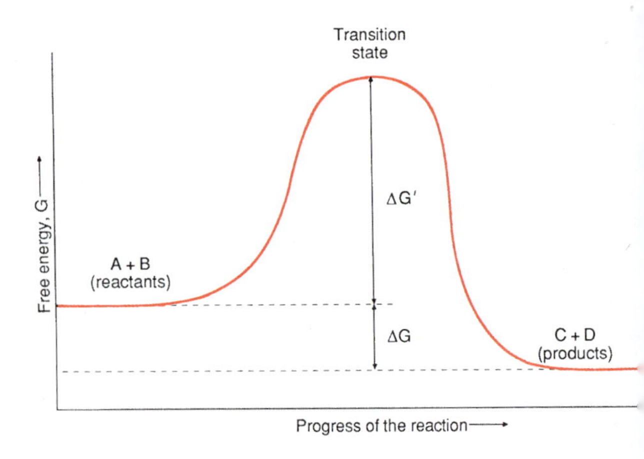

FIGURE 5-1
Energy diagram for a thermodynamically favorable forward chemical reaction, $A + B \rightarrow C + D$. $\Delta G'$ = free energy of activation and ΔG = free energy difference between reactants and products that has a negative value ($G_{C+D} - G_{A+B} < 0$).

5.2 Thermodynamics

The energy changes that occur in cellular reactions obey the two laws of thermodynamics. The first law states that energy can be neither created nor destroyed (i.e., the energy of the universe is constant) but can be converted from one form into another. The second law states that in all processes involving energy changes under a given set of conditions of temperature and pressure, the entropy of the system and the surroundings (i.e., the universe) increases and attains a maximum value at equilibrium. *Entropy* (S) may be considered a measure of disorder or randomness, or that amount of energy unavailable for useful work. In applying the laws of thermodynamics to human metabolism, the primary concern is how the concentrations and energy levels of reactants and products affect the direction of a reaction. Less emphasis is placed upon factors such as temperature and pressure, since they are maintained at nearly constant levels.

In considering thermodynamic parameters, e.g., heat and work, we do not need to know the exact chemical pathway taken by the reactants in conversion to products. Using thermodynamics, we can obtain information about reactions that cannot be studied directly in living systems. Thermodynamics predicts, on the basis of the known energy levels of reactants and products, whether a reaction can be expected to occur spontaneously or how much energy must be supplied to drive the reaction in one direction or another. Such information is crucial in establishing reaction routes in metabolic pathways. Thermodynamics explains how equilibrium constants are related to changes in temperature. Thermodynamics also explains the basis for enzyme catalysis.

A typical energy profile of a chemical reaction, $A + B \rightleftharpoons C + D$, is shown in Figure 5-1. During the conversion of reactant molecules ($A + B$) to product molecules ($C + D$), there is a change in the potential energy, called the ***Gibbs free energy*** (G), of the molecules involved which is shown by plotting the change in free energy of the participating molecules as a function of the reaction progress. The phrase "progress of the reaction" on the abscissa of the graph indicates generality and includes all changes in bond lengths and angles (i.e., in the shapes and atomic interactions) of the reactant molecules as they are interchangeably converted to product molecules. "Interchangeability" implies reversibility: if both reactants and products are present in a reaction mixture, molecules of reactant will be converted to product and "product" molecules to "reactant" molecules at any given moment. Thus, the equilibrium state is dynamic, not static. In the chemical reaction shown in Figure 5-1, the free energy of reactants has to be raised to that of the transition state before the products are formed. The transition state is an activated state in which the probability is high that appropriate chemical bonds are broken or re-formed in the conversion of reactants to products. Thus, the rate of reaction is proportional to the concentration of the activated species that exist at the transition state. The energy required to raise all of the reactant molecules to the transition state species is known as the ***free energy of activation,*** $\Delta G'$. Prime designates the activated state and Δ the difference between two energy states. The free-energy difference between reactants and products is indicated by ΔG. Thus, a negative value for ΔG for a chemical reaction (as in Figure 5-1) indicates that energy is released in converting reactants to products; such a reaction is termed ***exergonic.*** For reactions with $\Delta G < 0$, we can predict that they are *thermodynamical*

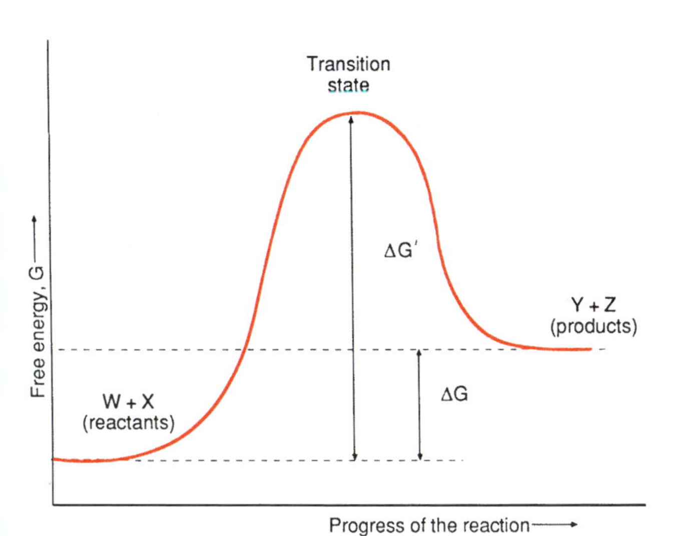

FIGURE 5-2
Energy diagram for a thermodynamically unfavorable forward chemical reaction, $W + X \rightarrow Y + Z$. $\Delta G'$ = free-energy of activation, and ΔG = free energy difference between reactants and products that has a positive value ($G_{Y+Z} - G_{W+X} > 0$).

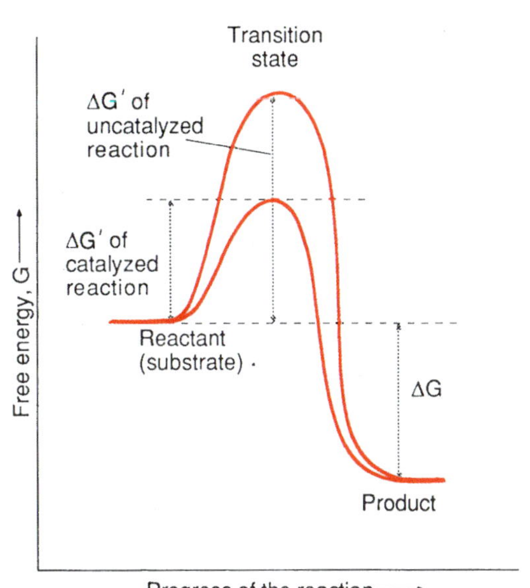

FIGURE 5-3
Energy diagram showing the reduction in free energy of activation ($\Delta G'$) in an enzyme-catalyzed reaction. The free-energy difference between reactant and product (ΔG) is *unchanged*.

favorable and that *the reaction can occur spontaneously*. If a chemical reaction has a positive ΔG value (as in Figure 5-2), the reaction is *thermodynamically unfavorable* and can occur only with the assistance of an external source of energy (e.g., coupling of an unfavorable reaction with a favorable one); such a reaction is termed ***endergonic.***

An enzyme accelerates a chemical reaction by *decreasing* $\Delta G'$, which is usually accomplished by providing an alternate pathway for the reaction (Figure 5-3). While $\Delta G'$ is decreased by the enzyme, the ΔG value (i.e., free-energy difference between reactant and product) is *unchanged.* In enzyme catalysis, the reactant is termed ***substrate.*** The ability of an enzyme to decrease the $\Delta G'$ value is illustrated by hydrogen peroxide decomposition ($2H_2O_2 \rightarrow 2H_2O + O_2$). In the absence of catalase, $\Delta G'$ for the reaction is 18 kcal · mol^{-1} (79.3 kJ · mol^{-1}), whereas in the presence of catalase, $\Delta G'$ is 7 kcal · mol^{-1} (29.2 kJ · mol^{-1}). A calorie is the amount of heat required to raise the temperature of 1 g of water by 1°C (from 14.5°C to 15.5°C). One kilocalorie (kcal) is 1000 calories. A joule (J) is the amount of energy required to apply a force of 1 N (newton) over a distance of 1 m. One kilojoule (kJ) is 1000 J (1 kcal = 4.184 kJ).

For a reversible chemical reaction, the dependence of ΔG on the temperature and concentration of reactants and products is given by the expression

$$\Delta G = \Delta G^\circ + \mathrm{RT}\ \ln \frac{[\text{products}]}{[\text{reactants}]} \tag{5.1}$$

where ΔG° is the standard free-energy change in calories per mole of reactant consumed; R is the gas constant (1.987×10^{-3} kcal · mol^{-1} · deg^{-1}); T is the absolute temperature in degrees Kelvin (= °C + 273); ln is the natural logarithm (base e); and [products] and [reactants] are the corresponding molar concentrations (for solvents and solutes) or pressures in atmospheres (for gases).

Since at equilibrium $\Delta G = 0$, this equation can be written

$$\Delta G^\circ = -RT \ln K_{eq} \text{ where } K_{eq} = \frac{[\text{products}]}{[\text{reactants}]}.$$

Note that the standard free-energy change is proportional to the absolute temperature and the equilibrium constant, K_{eq}.

The equilibrium constant depends only on the temperature of the system. This equation provides a way of determining ΔG° from equilibrium concentration data. Another commonly used form is

$$\Delta G^\circ = -RT\,(\ln 10)\,(\log K_{eq}) = -2.303RT \log K_{eq} \tag{5.2}$$

where log means logarithm (base 10).

The standard state for ΔG° is one in which the temperature is 25°C, all gases are present at pressures of 1 atm, all liquids and solids are pure (i.e., not mixtures), and all solutes have unit activity. Activity is similar to concentration but takes into account interactions of solute molecules with each other and with the solvent. Activities are difficult to measure, however, and molar concentrations are commonly used in their place. Since concentration and activity are almost equal in dilute solutions and many biological reactions occur at low concentrations, this approximation is valid.

A physical interpretation of ΔG° can be obtained by noting that if [products]/[reactants] = 1, then $\Delta G = \Delta G^\circ$. ΔG° is the amount of useful work that can be obtained by conversion of 1 mol of each reactant in its standard state to 1 mol of each product in its standard state.

The standard free-energy change for a reaction is also useful for predicting whether a reaction will occur under standard conditions. If $\Delta G^\circ < 0$, the reaction will occur spontaneously provided standard conditions prevail; if $\Delta G^\circ > 0$, the reaction will not occur by itself. It is ΔG, however, that determines whether or not a reaction will occur under conditions different from the standard state, such as those existing within a cell, for two reasons.

1. Most solutes are present at fairly low concentrations (< 0.1 M) and water is present at high concentration (55.6 M for pure water); the water concentration is thus assumed to be constant. If water enters into a reaction, it is incorporated into K_{eq}; if K_{eq} is determined at pH 7.0, it is indicated by K'_{eq}.
2. If H^+ or OH^- participates in a reaction, its concentration influences ΔG°. Since most biological reactions occur in systems buffered to pH $\sim$ 7, it is useful to define $\Delta G^{\circ\prime}$ as the value of ΔG° measured at pH = 7.0, i.e., where $[H^+] = 10^{-7}$ M. $\Delta G^{\circ\prime}$ values can be compared to each other but not to values of ΔG°.

   Therefore, for biochemical systems, Equation (5.2) can be written as

$$\Delta G^{\circ\prime} = -2.303 RT \log K'_{eq} \quad (5.3)$$

Substituting for $R = 1.987 \times 10^{-3}$ kcal · mol^{-1} · deg^{-1} (8.314×10^{-3} kJ · mol^{-1} · deg^{-1}) and $T = 298$ K (i.e., 273 + 25°C),

$$[\text{in kcal/mol}]\Delta G^{\circ\prime} = -1.36 \log K'_{eq};$$
$$\text{or } K'_{eq} = 10^{-\Delta G^{\circ\prime}/1.36}$$
$$[\text{in kcal/mol}]\Delta G^{\circ\prime} = -5.69 \log K'_{eq};$$
$$\text{or } K'_{eq} = 10^{-\Delta G^{\circ\prime}/5.69} \quad (5.4)$$

From Equation (5.4), at 25°C, each −1.36 kcal/mol value of $\Delta G^{\circ\prime}$ corresponds to a factor of 10 in the equilibrium constant in favor of product formation (Table 5-1). As K'_{eq} increases, the value for $\Delta G^{\circ\prime}$ decreases and the tendency for reactions to occur increases spontaneously. Thus, we can calculate $\Delta G^{\circ\prime}$ by knowing K'_{eq} and the temperature at which K'_{eq} is determined. In living systems, chemical reactions rarely reach complete equilibrium states but do attain near-equilibrium and nonequilibrium conditions. These conditions are maintained by utilizing energy from the external source. In a metabolic pathway, one of the nonequilibrium reactions usually becomes the rate-determining step of that pathway and its regulation provides directionality to the pathway. In a reaction sequence

$$A \rightleftharpoons B \rightleftharpoons C \rightleftharpoons \cdots$$

the reaction A → B can be prevented from reaching near-equilibrium by removing B (converting it to C) as fast as it can be made from A. However, the conversion of B to C may attain near-equilibrium. The near-equilibrium reactions are reversible and allow for reversal of the biochemical pathway. All reactions in the body are interrelated and the system as a whole is in a ***steady-state condition,*** with individual reactions operating in either near- or nonequilibrium conditions. A change in concentration of any component (product of one reaction used as reactant in another reaction) shifts the concentration of all other components linked to it by means of a sequence of chemical reactions, resulting in the attainment of a new steady state.

The value of ΔG for a nonequilibrium system can be determined from $\Delta G^{\circ\prime}$, and the actual concentrations of

TABLE 5-1
Relationship between Standard Free Energy Change ($\Delta G^{0\prime}$) and Equilibrium Constant (K'_{eq}) at pH 7.0 and 25°C*

	$\Delta G^{0\prime}$	
K'_{eq}	kcal/mol	kJ/mol
10^{-6}	8.18	34.22
10^{-5}	6.82	28.53
10^{-4}	5.46	22.84
10^{-3}	4.09	17.11
10^{-2}	2.73	11.42
10^{-1}	1.36	5.69
1	0	0
10	−1.36	−5.69
10^2	−2.73	−11.42
10^3	−4.09	−17.11
10^4	−5.46	−22.84
10^5	−6.82	−28.53
10^6	−8.18	−34.22

*$K'_{eq} = 10^{-\Delta G^{0\prime}\,1.36}$ when $\Delta G^{0\prime}$ is expressed in kcal/mol at 25°C.
$K'_{eq} = 10^{-\Delta G^{0\prime}\,5.69}$ when $\Delta G^{0\prime}$ is expressed in kJ/mol at 25°C.

products and reactants in the cell can be determined by use of the common logarithm analogue of Equation (5.1):

$$\Delta G = \Delta G^{\circ\prime} + 2.303RT \log \frac{[\text{products}]}{[\text{reactants}]}.$$

If ΔG is negative, then *under cellular conditions* the reaction proceeds to form products. The $\Delta G^{\circ\prime}$ value for a given chemical reaction may be positive suggesting immediately that the reaction cannot occur (without external assistance), yet the ΔG value may be negative indicating that the reaction can occur under the prevailing concentrations of reactant and products. The reaction of glycolysis, catalyzed by the enzyme aldolase (Chapter 13), furnishes a good illustration.

$$\text{Fructose 1,6-bisphosphate} \rightleftharpoons \text{glyceraldehyde 3-phosphate} + \text{dihydroxyacetone phosphate}$$

The K'_{eq} for this reaction at 25°C is 6.7×10^{-5} M. We can calculate $\Delta G^{\circ\prime}$ from Equation (5.3):

$$\begin{aligned}\Delta G^{\circ\prime} &= -2.303RT \log K'_{eq} \\ &= -2.303 \times 1.98 \times 10^{-3} \times 298 \times \log\left(6.7 \times 10^{-5}\right) \\ &= +5.67 \text{ kcal/mol (or} + 23.72 \text{ kJ/mol).}\end{aligned}$$

The positive value for $\Delta G^{\circ\prime}$ indicates that the reaction is not favored in the forward direction but can proceed in the reverse direction. Now we calculate the ΔG value using the concentration of 50 μmol/L (or 50×10^{-6} M) for both reactant and product, which is close to their physiological concentrations:

$$\begin{aligned}\Delta G &= \Delta G^{\circ\prime} + 2.303RT \\ &\quad \times \log \frac{\begin{bmatrix}\text{glyceraldehyde} \\ \text{3-phosphate}\end{bmatrix}\begin{bmatrix}\text{dihydroxyacetone} \\ \text{phosphate}\end{bmatrix}}{[\text{Fructose 1,6-bisphosphate}]} \\ &= +5.67 \text{ kcal/mol} + 2.303 \times 1.98 \times 10^{-3} \times 298 \\ &\quad \times \log \frac{[(50 \times 10^{-6}) \times (50 \times 10^{-6})]}{(50 \times 10^{-6})} \\ &= 5.67 \text{ kcal/mol} - 5.85 \text{ kcal/mol} \\ &= -0.18 \text{ kcal/mol(or} - 0.75 \text{ kJ/mol).}\end{aligned}$$

The negative value of ΔG suggests that under physiological conditions the forward reaction is favored, despite the fact that $\Delta G^{\circ\prime}$ is positive. Thus, under physiological conditions ΔG values predict better than $\Delta G^{\circ\prime}$ values whether a reaction can occur spontaneously. Similarly, whereas under certain conditions of reactants and products the conversion of reactant to product is favored (e.g., the conversion of fructose 1,6-bisphosphate to glyceraldehyde 3-phosphate and dihydroxyacetone phosphate during glycolysis), the direction of this reaction can be reversed when appropriate changes occur in the concentrations of reactants and products (formation of fructose 1,6-bisphosphate from glyceraldehyde 3-phosphate and dihydroxyacetone phosphate during gluconeogenesis). The same enzyme catalyzes both the forward and the reverse reactions.

The concept of ΔG is further clarified by an example. Consider a self-operating heat engine that functions by taking in an agent (e.g., steam) at temperature T_1 and releasing it at T_2 ($T_1 > T_2$). The heat extracted (Q) is converted as efficiently as possible into useful work. If W represents the maximum useful work available,

$$W = Q\frac{(T_1 - T_2)}{T_1} = Q - Q\frac{T_2}{T_1}$$

where T_1 and T_2 are absolute (Kelvin) temperatures. Unless $T_2 = 0$ or $T_1 = \infty$, the useful work is always less than the total energy supplied by a factor of $Q(T_2/T_1) = (Q/T_1)T_2$. The ratio Q/T_1 is the entropy (S) of the system, and the amount of energy unavailable for useful work (T_2S) is that which is lost in the process of energy transfer; it can be thought of as the amount of randomness or disorder introduced into the system during the transfer.

For chemical systems, G can be defined by the equation $G = H - TS$. Here, G (free energy) is analogous to W, the maximum useful work available; H (enthalpy) is analogous to Q, the heat content of the system at constant pressure; S (entropy) is analogous to (Q/T_1), the wasted heat energy; and T is again the absolute temperature. This relationship $G = H - TS$ is more commonly used to describe ***changes*** in these quantities. If a system goes from state I (with $G = G_I$, $H = H_I$, $S = S_I$) to state II ($G = G_{II}$, $H = H_{II}$, $S = S_{II}$) at constant temperature,

$$G_{II} - G_I = (H_{II} - H_I) - T\,(S_{II} - S_I)$$

or, simply,

$$\Delta G = \Delta H - T\,\Delta S.$$

Since entropy (S) is a measure of disorder (randomness), it is the energy that is unavailable for useful work. Entropy values of denatured molecules are high relative to those of native structures (as in protein denaturation). In a living system, entropy is kept at a minimum by utilization of free energy (G) from outside and by increase in entropy of the surroundings.

Enthalpy (H) is related to the internal energy of a system by the following equation, where E is the internal energy, P the external pressure on the system, and V the volume of the system. In terms of changes between states, the equation becomes

$$\Delta H = \Delta E + \Delta(PV) = \Delta E + P\Delta V + V\Delta P.$$

At constant pressure, $\Delta P = 0$ and $\Delta H = \Delta E + P\Delta V$. If, in addition, the volume is constant, then $\Delta V = 0$ and $\Delta H = \Delta E$. In many biological reactions, $\Delta V = \Delta P = 0$. Consequently, for a change of state, the maximum amount of energy that can be released as heat (ΔH) is equal to ΔE, the change in the internal energy of the system.

ΔH can be measured by burning the particular compound in a bomb calorimeter at constant pressure. The heat released is measured as the rise in temperature of a large water bath surrounding the combustion chamber. The values so obtained indicate the total energy available in a compound when it is completely oxidized. Factors that usually contribute to ΔH under these conditions are the heat of fusion (melting), the heat of vaporization (boiling), and the heat of combustion (bond making and breaking). The first two factors are important, for example, if at the standard temperature (25°C) a solid is combusted to a combination of liquids and gases, as in the examples below.

Just as with ΔG, ΔH for a process is equal to $-\Delta H$ for the reverse process:

$$\Delta H\ (\text{liquid} \rightarrow \text{gas}) = -\Delta H\ (\text{gas} \rightarrow \text{liquid})$$
$$\Delta H\ (\text{liquid} \rightarrow \text{solid}) = -\Delta H\ (\text{solid} \rightarrow \text{liquid})$$
$$\Delta H\ (\text{making a bond}) = -\Delta H\ (\text{breaking the same bond}).$$

ΔH values for various food substances and their application to human energy metabolism are discussed later in this chapter.

A number of processes in living systems result in the transfer of electrons (oxidation and reduction) and can be understood in thermodynamic terms. ***Oxidation*** is the loss of electrons or hydride (H^-) ions (but not hydrogen [H^+] ions) by a molecule, atom, or ion. ***Reduction*** is the gain of electrons or hydride (H^-) ions by a molecule, atom, or ion. Transfer of one hydride ion results in the transfer of *two* electrons.

The amount of work required to add or remove electrons is called the ***electromotive potential or force*** (***emf***) and is designated E. It is measured in volts (joules per coulomb, where a coulomb is a unit of electric charge or a quantity of electrons).

The standard emf, E° (or, at pH 7, $E^{\circ\prime}$), is the emf measured when the temperature is 25°C and the materials being oxidized or reduced are present at concentrations of 1.0 M. In biological systems, $E^{\circ\prime}$ is most commonly used.[1] The emf measured for an oxidation-reduction (***redox***) reaction under nonstandard conditions, E_h, is mathematically related to the emf measured for the same reaction under standard conditions, $E^{\circ\prime}$, through the Nernst equation:

$$E_h = E^{\circ\prime} + \frac{2.303RT}{nF}\log\frac{(\text{electron acceptor})}{(\text{electron donor})}$$

where E_h = observed potential, $E^{\circ\prime}$ = standard redox potential, R = gas constant (8.31 J deg^{-1}· mol), T = temperature in Kelvin, n = number of electrons being transferred, and F = the *Faraday constant* (96,487 J/V). It is evident from the Nernst equation that $E_h = E^{\circ\prime}$ when [electron acceptor] = [electron donor]. Thus, the $E^{\circ\prime}$ value represents the midpoint in a redox reaction, which is analogous to the measurement of the pK′ value of a weak acid, HA, when $[A^-] = [HA]$ (see discussion of the Henderson–Hasselbalch equation in Chapter 1). In this manner, $E^{\circ\prime}$ values for many redox reactions of biochemical importance have been determined (Chapter 14).

A ***half-reaction*** is a reaction in which electrons or hydride ions are written explicitly. Values of E° and $E^{\circ\prime}$ are tabulated with the half-reactions for which they are measured. For example,

$$\text{NADH} \rightarrow \text{NAD}^+ + \text{H}^- \qquad E^{\circ\prime} = -0.32\ \text{V}$$

and

$$\text{H}_2\text{O} \rightarrow \frac{1}{2}\text{O}_2 + 2\text{H}^+ + 2e^- \qquad E^{\circ\prime} = -0.816\ \text{V}$$

are half-reactions because they show electrons (e^-) or hydride ions (H^-). NADH and NAD^+ are abbreviations for reduced and oxidized forms of nicotinamide adenine dinucleotide (whose structure is given in Chapter 6 and whose metabolic functions are discussed elsewhere).

Alternatively, these half-reactions can be written as

$$\text{NAD}^+ + \text{H}^- \rightarrow \text{NADH} \qquad E^{\circ\prime} = +0.32\ \text{V}$$
$$2e^- + 2\text{H}^+ + \frac{1}{2}\text{O}_2 \rightarrow \text{H}_2\text{O} \qquad E^{\circ\prime} = +0.816\ \text{V}.$$

Reversing the direction of a reaction changes the sign of the potential charge (just as with ΔG and ΔH).

The oxidation potential is a measure of the ease of removal of electrons from a material compared to the ease of removal of electrons from hydrogen in the half-reaction:

$$\text{H}_2 \rightarrow 2\text{H}^+ + 2e^- \qquad E^\circ = 0.00\ \text{V}.$$

E° in this reaction is defined as zero when H_2 gas is at 1.0 atm, [H^+] is 1.0 M (i.e., pH 0), and the temperature is 25°C, thus fixing the scale of E° value for other reactions. If it is easier to remove electrons from a material, $E^{\circ\prime}$ will be negative; if it is harder, $E^{\circ\prime}$ will be positive. E° for the hydrogen half-reaction is used as the zero point even when reference is made to $E^{\circ\prime}$ values. Thus,

$$\text{H}_2 \rightarrow 2\text{H}^+ + 2e^- \qquad E^{\circ\prime} = -0.42\ \text{V}.$$

It is easier to remove electrons from hydrogen (and produce H^+) when $[H^+] = 10^{-7}$ M than when $[H^+] = 1.0$ M.

[1] Some texts use the notation E_0' for the standard oxidation-reduction potential.

The choice of a negative sign for the standard emf to signify "easier to remove electrons" is arbitrary.[2]

Since free electrons combinc rapidly with whatever is at hand, half-reactions never occur by themselves. Something must accept electrons as fast as they are released. The substance releasing electrons (or H^-) is the ***reductant,*** or ***reducing agent*** (because it is oxidized), and the substance accepting electrons is the ***oxidant,*** or ***oxidizing agent*** (because it is reduced). Two half-reactions when combined give a redox reaction. When balanced, such reactions never show free (uncombined) electrons. For example,

$$H^+ + NADH + \frac{1}{2}O_2 \rightarrow NAD^+ + H_2O$$

$$\Delta E^{\circ\prime} = -0.32 - (+0.816) = -1.136 \text{ V}.$$

$\Delta E^{\circ\prime}$ for this reaction is calculated according to

$$\Delta E^{\circ\prime} = \Delta E^{\circ\prime} \text{ (reductant)} - \Delta E^{\circ\prime} \text{ (oxidant)}.$$

Under standard conditions (and pH = 7.0), the reaction will occur as written if $\Delta E^{\circ\prime} < 0$. Otherwise the reaction will proceed from right to left (the reverse of the way it is written).

When $\Delta E^{\circ\prime}$ in volts is converted to calories per mole, the result is $\Delta G^{\circ\prime}$ for the redox reaction being considered. This conversion is accomplished by means of the equation

$$\Delta G^{\circ\prime} = \frac{nF \cdot \Delta E^{\circ\prime}}{4.184} = (23.061)\, n \cdot \Delta E^{\circ\prime}$$

where, as before, F is Faraday's constant = 96,487 J/V; n is the number of electrons transferred per mole of material oxidized or reduced; $\Delta E^{\circ\prime}$ is $E^{\circ\prime}$ (reductant) − $E^{\circ\prime}$ (oxidant) in volts; and $\Delta G^{\circ\prime}$ is the standard free energy change in calories per mole of material oxidized or reduced (since there are 4.184 J/cal).

In the case of the reaction just considered

$$H^+ + NADH + \frac{1}{2}O_2 \rightarrow NAD^+ + H_2O$$

$$\Delta E^{\circ\prime} = -1.136 \text{ V},$$

n equals 4 gram-equivalents of electrons per mole of O_2 or 2 gram-equivalents of electrons per mole of H_2O, NAD^+, or NADH. Then,

$$\begin{aligned}\Delta G^{\circ\prime} &= 23.061n \times (-1.136) = -26.197n \\ &= -26.197 \times 2 = -52.394 \\ &\quad \text{calories/mol of } H_2O \text{ or } NAD^+ \text{ formed} \\ &\quad \text{or NADH converted} \\ &= -26.197 \times 4 \\ &= -104.788 \text{ calories/mol of } O_2 \text{ transformed}\end{aligned}$$

[2]Although the International Union of Pure and Applied Chemistry recommends the opposite convention, the older convention is retained here because it is commonly encountered in medically and biologically related publications.

As in all reactions, the actual value of $\Delta G^{\circ\prime}$ depends on the reactant or product for which it is calculated.

The significance of free-energy changes and redox reactions lies in the fact that life on earth depends on the redox reaction in which CO_2 is reduced by H_2O to yield (CH_2O), using sunlight as the energy source:

$$CO_2 + H_2O + \text{energy} \underset{\text{Oxidation (respiration)}}{\overset{\text{Reduction (photosynthesis)}}{\rightleftharpoons}} O_2 + (CH_2O)$$

where (CH_2O) represents carbohydrates. The $\Delta E^{\circ\prime}$ for the forward reaction is about +1.24 V, or $\Delta G^{\circ\prime}$ is about 114.3 kcal/(CH_2O). Thus, for glucose formation, $\Delta G^{\circ\prime}$ is +685.8 kcal.

In plants, energy from the sun is absorbed by the chloroplasts of green plants, causing water to be oxidized and carbon dioxide reduced, producing oxygen and carbohydrate [represented by $(CH_2O)_m$]. In oxidation reactions, carbohydrate (or lipids) and oxygen are consumed, and energy, water, and carbon dioxide are released. The energy is captured primarily as adenosine triphosphate (ATP), which is used as an immediate source of energy in most cellular endergonic processes.

5.3 Standard Free Energy of Hydrolysis of ATP

In the living organism, ATP functions as the most important energy intermediate, linking exergonic with endergonic processes. Exergonic processes (e.g., oxidation of glucose, glycogen, and lipids) are coupled to the formation of ATP, and endergonic processes are coupled to the expenditure of ATP (e.g., biosynthesis, muscle contraction, active transport across membranes). ATP is called a "high-energy compound" because of its large negative free energy of hydrolysis

$$ATP^{4-} + H_2O \rightarrow ADP^{3-} + HPO_4^{2-} \text{ (or } P_i) + H^+$$

$$\Delta G^{\circ\prime} = -7.3 \text{ kcal/mol } (-30.5 \text{ kJ/mol}).$$

The reason for the large release of free energy associated with hydrolysis of ATP is that the products of the reaction, ADP and P_i, are much more stable than ATP. Several factors contribute to their increased stability: relief of electrostatic repulsion, resonance stabilization, and ionization.

At pH 7.0, ATP has four closely spaced negative charges with a strong repulsion between these charges (Figure 5-4). Upon hydrolysis, the strain in the molecule due to electrostatic repulsion is relieved by formation of less negatively charged products (ADP^{3-} and P_i^{2-}). Furthermore, since these two products are also negatively charged, their

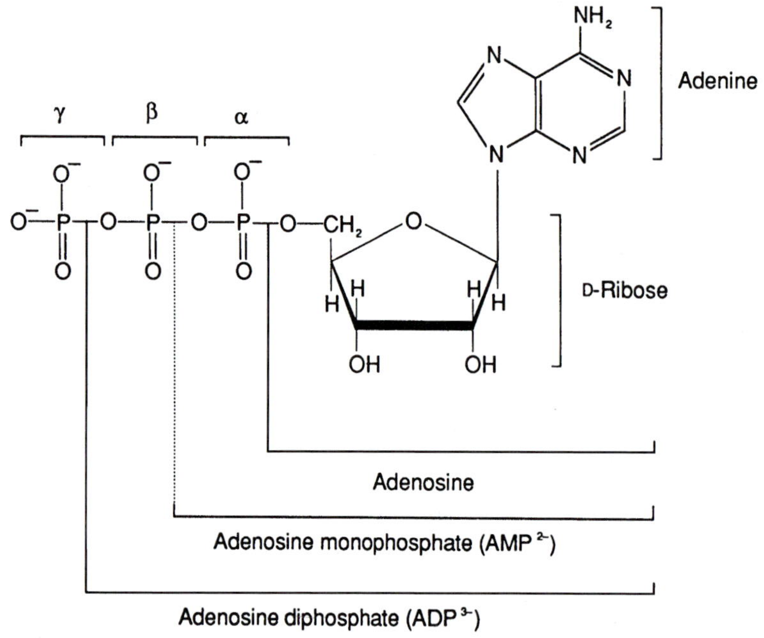

FIGURE 5-4
Structural formula of adenosine triphosphate (ATP) at pH 7.0. The three phosphate groups are identified by Greek letters α, β, and γ. The γ- and β-phosphate groups are linked through phosphoanhydride bonds and their hydrolysis yields a large negative $\Delta G^{\circ\prime}$, whereas the α-phosphate linked by a phosphate ester bond has a much lower negative $\Delta G^{\circ\prime}$. *In vivo* most ATP is chelated to magnesium ions via two of the anionic oxygens ($Mg \cdot ATP^{2-}$).

tendency to approach each other to re-form ATP is minimized. The number of resonance forms of ADP and P_i exceeds those present in ATP so that the products are stabilized by a larger resonance energy. Ionization of $H_2PO_4^- \rightleftharpoons HPO_4^{2-} + H^+$ has a large negative $\Delta G^{\circ\prime}$ contributing the $\Delta G^{\circ\prime}$ for ATP hydrolysis.

The free energy of hydrolysis of ATP is also affected by Mg^{2+} and the physiological substrate is $Mg^{2+} \cdot ATP^{4-}$ (or $Mg \cdot ATP^{2-}$). The presence of magnesium ions favors ATP hydrolysis because both ADP and P_i have higher affinity (about six times) for Mg^{2+} than ATP. In vivo, the ΔG value for ATP hydrolysis is probably much higher than $\Delta G^{\circ\prime}$ owing to the prevailing intracellular concentration of the reactants and products (in erythrocytes, ~ -12.4 kcal/mol).

In some energy-consuming reactions, ATP is hydrolyzed to AMP and pyrophosphate, with a value of free energy of hydrolysis comparable to that of ATP to ADP and P_i

$$ATP^{4-} + H_2O \rightarrow AMP^{2-} + HP_2O_7^{3-} + H^+$$
$$\Delta G^{\circ\prime} = -7.7 \text{ kcal/mol } (-32.2 \text{ kJ/mol})$$

In vivo, pyrophosphate is usually hydrolyzed to inorganic phosphate, which also has a large negative $\Delta G^{\circ\prime}$,

$$HP_2O_7^{3-} + H_2O \rightarrow 2HPO_4^{2-} + H^+$$
$$\Delta G^{\circ\prime} = -7.17 \text{ kcal/mol } (-30 \text{ kJ/mol})$$

The pyrophosphate hydrolysis, although not coupled to any particular endergonic reaction, still ensures completion of the forward reaction or process (e.g., activation of fatty acids and amino acids, synthesis of nucleotides and polynucleotides).

The hydrolysis of AMP to adenosine and P_i does not yield a high negative $\Delta G^{\circ\prime}$ because the phosphate is linked in a normal ester bond, as opposed to the other two linkages (β and γ), which are phosphoanhydride bonds (Figure 5-4). The $\Delta G^{\circ\prime}$ values of hydrolysis of other nucleoside triphosphates (e.g., UTP, CTP, GTP) are similar to that of ATP and they are utilized in the biosynthesis of carbohydrates, lipids, and proteins (discussed elsewhere). In addition to undergoing hydrolysis, ATP may act as a donor of phosphoryl (e.g., in the formation of glucose 6-phosphate; see Chapter 13), pyrophosphoryl (e.g., in the formation of phosphoribosylpyrophosphate; see Chapter 27), adenylyl (e.g., in the adenylylation of glutamine synthetase; see Chapter 17), or adenosyl groups (in the formation of *S*-adenosylmethionine; see Chapter 17).

Since ATP synthesis from ADP and P_i is energy-dependent, it requires compounds or processes that yield larger negative $\Delta G^{\circ\prime}$ values than ATP hydrolysis does. Some of these compounds are organic phosphates (Table 5-2). Other high-energy compounds are thioesters,

TABLE 5-2
Standard Free Energy of Hydrolysis ($\Delta G^{0'}$) of Some Organophosphates

Compound/Hydrolyzed Product	$\Delta G^{0'}$ kcal/mol (kJ/mol)
Phosphoenolpyruvate → Pyruvate	−14.8 (−61.9)
1,3-Bisphosphoglycerate → 3-Phosphoglycerate	−11.8 (−49.3)
Phosphocreatine → Creatine	−10.3 (−43.1)
ATP → ADP	−7.3 (−30.5)
ATP → AMP	−7.7 (−32.2)
ADP → AMP	−6.6 (−27.6)
Glucose 1-Phosphate → Glucose	−5.0 (−20.9)
Fructose 6-Phosphate → Fructose	−3.8 (−15.9)
AMP → Adenosine	−3.4 (−14.2)
Glucose 6-Phosphate → Glucose	−3.3 (−13.8)
Glycerol 3-Phosphate → Glycerol	−2.2 (−9.2)

aminoacyl esters, sulfonium derivatives, and sugar nucleoside diphosphates.

5.4 Chemical Kinetics

The study of chemical reaction rates is called *chemical kinetics*. Whereas thermodynamics deals with the relative energy states of reactants and products, kinetics deals with how fast a reaction occurs and with the chemical pathway (***mechanism***) it follows.

The conversion $A \xrightarrow{k} B$ (where k is the ***rate constant***) can be represented diagrammatically by the increase in concentration of product, [B], with respect to time (Figure 5-5). The average velocity, $\bar{v}$, for the period t_1 to t_2 is the slope,

$$\bar{v} = \frac{\Delta[B]}{\Delta t} = \frac{[B]_2 - [B]_1}{t_2 - t_1}$$

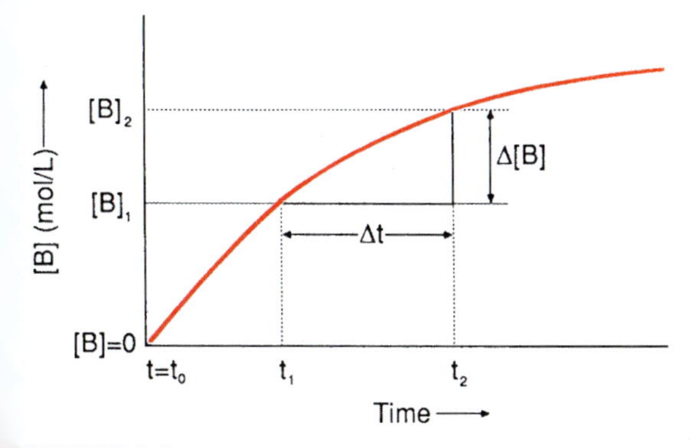

FIGURE 5-5
Time course of a chemical reaction.

The instantaneous velocity, $\bar{v}$, at some time t' ($\neq t_0$) is expressed as follows:

$$v = \frac{d[B(t)]}{dt} \text{ evaluated at } t'.$$

The equation $d[B(t)]/dt$ is the first derivative of $[B(t)]$ with respect to t. It is the limit (lim) of the slope $\Delta[B]/\Delta t$ as $\Delta t \to 0$, or

$$v = \lim_{\Delta t \to 0} \bar{v} = \lim_{\Delta t \to 0} \frac{\Delta[B]}{\Delta t} = \frac{\Delta[B(t)]}{dt}.$$

Alternatively, one can consider the change in concentration of the reactants. Thus,

$$v = \frac{d[B(t)]}{dt} = -\frac{d[A(t)]}{dt}$$

since the reactant concentration, [A], decreases at the same rate at which the product concentration, [B], increases. The expressions $[B(t)]$ and $[A(t)]$ imply a detailed knowledge of the way in which the concentrations of A and B vary with time and the way in which the reaction progresses. Since the preceding is generally not the case, [B] and [A] will be used instead of $[B(t)]$ and $[A(t)]$. The rate of formation of [B] should increase if [A] increases. For the reaction $nA \xrightarrow{k} B$, it is appropriate to write

$$\frac{\Delta[B]}{\Delta t} = k[A]^n$$

where k is again the rate constant, and $n = 0, 1, 2, \ldots$ is the order of the reaction with respect to A. (The order is said to be zero, first, second, etc., with respect to A.)

The rate constant expresses the proportionality between the rate of formation of B and the molar concentration of A and is characteristic of a particular reaction. The units of k depend on the order of the reaction. For *zero order,* they are moles liter^{-1} time^{-1} (time is frequently given in seconds). For first order, they are time^{-1}, and for second order, liters moles^{-1} time^{-1}, etc. The units of k are whatever is needed for $d[B]/dt$ to have the unit of moles liters^{-1} time^{-1}. If $n = 0$, $v = k$; this means that in a *zero order reaction,* [B] changes at a constant rate independent of the concentration of reactants, which is especially important in enzyme kinetics. A plot of [B] versus t for such a reaction is a straight line. A somewhat more complicated example is

$$A + B \xrightarrow{k} C + D$$

$$v = \frac{d[C]}{dt} = \frac{d[D]}{dt} = \frac{d[\text{products}]}{dt} = k[A][B]$$

This reaction is first order with respect to A, first order with respect to B, and second order overall. Most reactions are *zero, first,* or *second* order.

The concept of reaction order can be related to the number of molecules that must collide simultaneously for the reaction to occur. In a *first-order reaction,* no collisions are required (since only one reactant molecule is involved). Every molecule with sufficient free energy to surmount the activation barrier may spontaneously convert to products. In a second-order reaction, not only must two molecules have enough free energy but they also must collide with each other for the products to form. A third-order reaction requires the simultaneous meeting of three molecules, a less likely event. Reactions of orders higher than second are seldom encountered in simple chemical conversions. Higher apparent orders are encountered, however, in some cases in which an overall rate equation is written for a process that proceeds in several steps via one or more intermediates. In such situations, the individual steps seldom involve a third- or higher order process. Therefore, for a chemical reaction consisting of multiple steps, the order of the reaction provides information about the overall rate equation but does not provide information regarding the number of atoms or molecules that must actually collide to form products during each step. This information is obtained by knowing the molecularity of the reaction, which is defined as the number of molecules that come together in a reaction. In a single-step reaction, the molecularity (which is the same as the order of reaction) determines the exponents in the rate equation. In a multistep reaction sequence, each step has unique molecularity; knowing this provides information about the reaction mechanism.

Zero-order reactions can be explained in two ways. In a process that is truly zero order with respect to all reactants (and catalysts, in the case of enzymes), either the activation energy is zero or every molecule has sufficient energy to overcome the activation barrier. This kind of reaction is rare in homogeneous reactions in gases or solutions.

Alternatively, a reaction can be zero order with respect to one or more (but not all) of the reactants. This kind of reaction is important in enzyme kinetics and assays of enzymes of clinical importance (Chapter 8). If one of the reactants (or a catalyst, such as an enzyme) is limited, then increasing the availability (concentrations) of the other reactants can result in no increase in the velocity of the reaction beyond that dictated by the limiting reagent. Hence, the rate is independent of the concentrations of the nonlimiting materials and zero-order with respect to those materials.

Theoretically all reactions are reversible. The correct way to write $A \rightarrow B$, then, is

$$A \underset{k_{-1}}{\overset{k_1}{\rightleftarrows}} B$$

where k_1 is the rate constant for the conversion of A to B, and k_{-1} is the rate constant for the reverse reaction, i.e., conversion of B to A. The reaction is kinetically reversible if $k_1 \cong k_{-1}$; it is irreversible for all practical purposes (i.e., the reverse reaction is so slow that it can be ignored) if $k_{-1} \ll k_1$. The rate of a reversible reaction is written

$$v = \frac{d[B]}{dt} = k_1[A] - k_{-1}[B].$$

For the reversible reaction,

$$A + B \underset{k_{-1}}{\overset{k_1}{\rightleftarrows}} C + D$$

$$v = \frac{d[\text{products}]}{dt} = \frac{d[C]}{dt} = \frac{d[D]}{dt}$$
$$= k_1[A][B] - k_{-1}[C][D].$$

Kinetic schemes can become complicated with many linked reactions and intermediates. Kinetics are important in biological systems, however because enzymes make reversible many otherwise kinetically irreversible reactions.

The rate at which a reaction will proceed (measured by k) is directly related to the amount of energy that must be supplied before reactants and products can be interconverted. This ***activation energy*** (E_a) comes from the kinetic energy of the reactants. This energy may be translational and rotational (needed if two molecules must collide to react) or vibrational and electronic (useful when one molecule rearranges itself or eliminates some atoms to form the product). The larger the activation energy, the slower the reaction rate and the smaller the rate constant.

ΔE_a is a free energy, so it contains both ΔH and ΔS terms. Not only must the reactants have enough energy (ΔH) to make and break the requisite bonds, but they must also be properly oriented (ΔS) for the products to form.

The Arrhenius equation expresses the relation between rate constants and activation energies:

$$k = A \cdot e^{-E_s/RT}$$

where E_s is the Arrhenius activation energy in calories per mole; R is the gas constant (1.987 calories mole^{-1} degrees K^{-1}); T is the absolute temperature (K); A is the Arrhenius pre-exponential factor, a constant; k is the rate constant; and e is 2.71828, the base of the natural logarithms. This equation is empirical, whereas E_a, the free energy of activation, is a theoretical concept, derived from the kinetic theory of reaction rates. From the Arrhenius equation, it is apparent that the larger the value of E_s (or E_a), the smaller the value of k. Hence, the basis for kinetic irreversibility is the same as that for thermodynamic irreversibility; the more energy a process takes (the more positive the value of ΔG or E_a), the less likely the reaction is to occur.

Although values for A are not well understood, the Arrhenius equation can, on rearrangement, be used to calculate E_s:

$$-R \ln k = \frac{E_s}{T} - R \ln \mathrm{A}.$$

The slope of a plot of $-R \ln k$ versus $(1/T)$ equals E_s, and $\Delta H_a = E_s - RT$.

5.5 Energy Metabolism

In order to carry out the body's essential functions (e.g., growth and repair, pregnancy, lactation, physical activity, maintenance of body temperature), food must be consumed and utilized and body constituents synthesized. The term "metabolism" encompasses the numerous chemical transformations that occur within the human body. Metabolism comprises ***anabolism*** and ***catabolism.*** Anabolism is concerned with the synthesis of new molecules, usually larger than the reactants, and is an energy-requiring process. Catabolism is concerned with degradation processes, usually the breaking down of large molecules into smaller ones, and is an energy-yielding process. ***Intermediary metabolism*** refers to all changes that occur in a food substance beginning with absorption and ending with excretion.

Two main sources of energy for metabolism are carbohydrates and fats (lipids). Proteins have less importance as an energy supplier. Knowledge of the total energy (ΔH) content of the major body fuels is necessary to understand how energy requirements are met by different fuels.

The reactions for representative food fuels, glucose and palmitic acid, will be used to indicate differences in ΔH values. The subscripts s, ℓ, and g (solid, liquid, and gas, respectively) indicate the phase of the material under 1 atm, the pressure at which the reaction is carried out, and at the given temperature. The ΔH^0 values given are ***standard heats of formation,*** since all reactants are in their standard (natural) state for the temperature given.

Oxidation of 1 mol of glucose (a carbohydrate) to carbon dioxide and water:

$$C_6H_{12}O_{6(s)} + 6O_{2(g)} \rightarrow 6CO_{2(g)} + 6H_2O_{(\ell)}$$

At 20°C, $\Delta H^\circ = -673$ kcal/mol or -2816 kJ/mol.

Oxidation of 1 mol of palmitic acid (a fatty acid) to carbon dioxide and water:

$$C_{16}H_{32}O_{2(s)} + 23O_{2(g)} \rightarrow 16CO_{2(g)} + 16H_2O_{(l)}$$

At 20°C, $\Delta H^\circ = 2380$ kcal/mol or -9.96 MJ/mol (1 MJ = 1 megajoule = 1000 kJ).

The average ΔH values in kilocalories per gram are as follows: carbohydrate, 4.1; lipid, 9.4; and protein, 5.6. These values vary within each class of food depending on chemical structure. For example, the value for starch is 4.1 kcal/g, whereas for glucose it is 3.8. Furthermore, in humans the end product of protein metabolism is urea instead of CO_2 and nitrogen, which are the end products obtained after complete oxidation in the bomb calorimeter. Therefore, a realistic estimate of energy derived from protein in the body, taking into account incomplete oxidation and specific dynamic action, is 4.1 kcal/g. In human nutrition, the commonly used values for energy yield in kilocalories (or kilojoules) per gram are as follows: carbohydrate, 4 (16.7); lipid, 9 (37.7); and protein, 4 (16.7). During the metabolism of food substances in the body, oxygen is consumed and CO_2 is produced. The molar ratio of CO_2 produced to oxygen consumed is known as the ***respiratory quotient*** (RQ) and is characteristic of a given substrate. For example, in the complete oxidation of glucose presented above,

$$\mathrm{RQ} = \frac{\text{volume of } CO_2 \text{ produced}}{\text{volume of } O_2 \text{ consumed}} = \frac{6CO_2}{6O_2} = 1.0.$$

For the corresponding oxidation of palmitic acid,

$$\mathrm{RQ} = \frac{16CO_2}{23O_2} = 0.7.$$

The RQ for protein is difficult to measure because it is not oxidized completely in the body and part of its carbon, hydrogen, and oxygen is lost in the urine and feces. However, the RQ is usually taken to be 0.8. The measurement of RQ thus provides a means for assessing the type of food that is being metabolized. An overall RQ of 0.85 is obtained in a normal adult who consumes a mixed diet. This value is enhanced by carbohydrate feeding and decreased by lipid feeding. Heat production can be calculated from oxygen consumption and RQ as follows. For glucose oxidation:

$$\underset{180\text{ g}}{C_6H_{12}O_6} + \underset{\substack{192\text{ g}\\(134\text{L})}}{6O_2} = 6CO_2 + 6H_2O$$

Since the energy yield of 1 g of glucose is 3.8 kcal, 180 g of glucose yields 180 × 3.8 or 684 kcal. The energy yield per liter of oxygen consumed equals (180 × 3.8)/134 = 5.1 kcal (21.3 kJ). Again, for palmitate oxidation,

$$\underset{256\text{ g}}{C_{16}H_{32}O_2} + \underset{\substack{736\text{ g}\\(513.7\text{ L})}}{23O_2} \rightarrow 16CO_2 + 16H_2O.$$

Since the energy yield of 1 g of fat is 9 kcal, 256 g of fat yields 256 × 9, or 2304 kcal. The energy yield per liter of oxygen consumed equals (256 × 9)/513.7 = 4.5 kcal (18.8 kJ).

In the body, the energy derived from food is released as body heat and also used in the synthesis of ATP. The energy captured in ATP is then transformed into other forms, i.e., chemical (synthesis of new compounds), mechanical (muscle contraction), electrical (nerve activity), electrochemical (various ion pumps), thermal (maintenance of body temperature), and informational (base sequences in nucleic acids, amino acids in proteins). In general, the energy of food provides for the specific dynamic action of food, the maintenance of the body's basal metabolism, and the energy expenditure associated with various types of activity.

The term ***specific dynamic action*** (SDA) describes the thermogenic effect of food, which consists of the rise in the metabolic rate during the assimilation of food above the metabolic rate during fasting. The mechanism of the thermogenic effect is not understood, but it may be due to the work performed in the assimilation of food and in its preparation for energy storage or energy yield. These processes require energy, which is derived from the body's energy stores. The values for SDA vary with the type of food substance ingested. The SDA value for protein is about 30% of its energy value because protein must undergo many changes before the carbon skeletons of its constituent amino acids can be used appropriately. SDA values for carbohydrates and lipids are, respectively, 5% and 4% of their energy value. The energy utilized as a result of SDA is wasted and appears as heat. It is maximal with protein intake. Thermodynamically, all energy derived from the metabolism of substrates to CO_2 and H_2O must of course eventually appear as heat.

Basal metabolism (determined by the ***basal metabolic rate*** [BMR]) is an expression of the body's vital energy needs during *physical, emotional, and digestive rests.* It represents the energy required for the maintenance of body temperature, muscle tone, the circulation of blood, the movement of respiration, and the glandular and cellular activities of a person who is awake and not involved in physical, digestive, or emotional activities. The BMR is measured in an individual at rest after a 12-hour fast and in comfortably warm, pleasant, quiet surroundings. Although the BMR could be obtained directly by measurement of the amount of heat produced, an indirect method that measures the volume of oxygen consumed and carbon dioxide evolved per unit time is less cumbersome and provides acceptable values. Oxygen consumption is measured in a respirometer. For every liter of oxygen consumed, the energy equivalent is 4.83 kcal (20.2 kJ). This value is based on the average value for the oxidation of carbohydrates and lipids (see above). BMR values are standardized and usually expressed as kilocalories per square meter of body surface per hour. The metabolic rate that occurs during sleep is approximately equal to the BMR. Less stringent conditions often used for hospitalized patients provide the resting metabolic rate, which is about 3% less than the BMR. The normal values for the resting rate of energy expenditure are shown in Table 5-3. It is clear that the resting metabolic energy requirements are closely related to lean body mass (as the body fat increases, the energy requirement decreases). Muscle, endocrine glands, and organs such as the liver are metabolically more active (i.e., consume a larger amount of oxygen per unit weight) than adipose tissue and bone.

The BMR for women (who have a greater proportion of fat per total body weight than men) is lower than that for men. In both sexes, the development of muscle tissue as a result of increased exercise increases the metabolic rate and the basal metabolic needs. Body fat is rapidly determined with calipers by skinfold measurement at several different sites. In normal young men and women, the average body fat constitutes about 12% and 26% of total body weight, respectively; if these values are exceeded by more than 20% and 30% of the average values respectively, they indicate the presence of obesity. The BMR changes with age. It is high at birth, increases up to 2 years of age, then gradually decreases except for a rise at puberty. This general decline in the BMR (approximately 2% per decade after age 21 years) is proportional to the reduction in muscle tissue (or lean body mass) and the accompanying increase in body fat. BMR is also affected by menstruation, pregnancy, and lactation.

The BMR is altered in a number of pathological states. The BMR is increased in ***hyperthyroidism,*** fever (approximately 12% elevation for each degree Celsius above normal body temperature), *Cushing's syndrome*, tumors of the adrenal gland, anemia, leukemia, polycythemia, cardiac insufficiency, and injury. BMR is decreased in *hypothyroidism*, starvation, malnutrition, hypopituitarism, hypoadrenalism (e.g., ***Addison's disease***), and anorexia nervosa.

To calculate energy requirements in an individual, it is necessary to take into account BMR, physical activity (muscular work), age, sex, height, and weight (Appendix I). The energy requirements of muscular work can be measured. Table 5-4 lists the energy expenditure for various physical activities.

Maintenance of desirable body weight at any age depends on the balance between energy intake and energy output or requirement. During growth, pregnancy, or recovery from illness, energy demands are greater; hence,

TABLE 5-3
*Normal Values for the Resting Rate of Metabolic Energy Expenditure of Adults Expressed in kcal/min (kJ/min)**

Men	Women	Percent Body Fat	Weight in kg (lbs)							
			45(99)	50(110)	55(121)	60(132)	65(143)	70(154)	75(165)	80(176)
Thin		5	—	0.98 (4.10)	1.05 (4.39)	1.12 (4.69)	1.20 (5.02)	1.27 (5.31)	1.32 (5.52)	1.40 (5.86)
Average		10	—	0.94 (3.93)	1.00 (4.18)	1.08 (4.52)	1.15 (4.81)	1.22 (5.10)	1.27 (5.31)	1.34 (5.61)
Plump	Thin	15	0.82 (3.42)	0.89 (3.72)	0.96 (4.02)	1.04 (4.35)	1.11 (4.64)	1.18 (4.94)	1.22 (5.10)	1.30 (5.44)
Fat	Average	20	0.79 (3.30)	0.84 (3.51)	0.90 (3.77)	0.98 (4.10)	1.05 (4.39)	1.12 (4.69)	1.18 (4.94)	1.24 (5.19)
	Plump	25	—	0.79 (3.31)	0.86 (3.60)	0.94 (3.93)	1.00 (4.18)	1.08 (4.52)	1.12 (4.69)	1.20 (5.02)
	Fat	30	—	—	0.82 (3.43)	0.89 (3.72)	0.96 (4.02)	1.04 (4.35)	1.08 (4.52)	1.15 (4.81)

*Data from S. Davidson, R. Passmore, J. F. Brock, and A. S. Truswell: *Human Nutrition and Dietetics,* 7th ed. (Churchill Livingstone, 1979).

intake should be higher than output. ***Obesity*** and ***cachexia*** (a disorder characterized by general physical wasting and malnutrition) are extreme examples of problems affecting the energy stores of the body. The appetite and satiety centers in the central nervous system normally are sensitive regulators that adjust energy consumption to energy requirements. Obesity is rarely caused by damage to those centers. Obesity results from an excess of energy intake over physiological needs. Factors predisposing or contributing to the development of obesity are heredity,

TABLE 5-4
*Examples of Rate of Energy Expenditure during Various Types of Physical Activities of Adults**

Type of Activity	Rate of Energy Expenditure, kcal/min (kJ/min)	
	Man, 70 kg	Woman, 58 kg
Very light: Sitting, standing, painting, driving, working in a laboratory, typing, playing musical instrument, sewing, ironing.	up to 2.5 (up to 10.5)	up to 2.0 (up to 8.3)
Light: Walking on level (2.5–3 mph), tailoring, pressing, working on automobiles, washing, shopping, golfing, sailing, playing table tennis or volley ball.	2.5–4.9 (10.5–20.5)	2.0–3.9 (8.37–16.3)
Moderate: Walking (3.5–4 mph), plastering, weeding and hoeing, loading and stacking bales, scrubbing floors, shopping with heavy load, cycling, skiing, playing tennis, dancing.	5.0–7.4 (20.9–31.0)	4.0–5.9 (16.7–24.7)
Heavy: Walking uphill with load, tree felling, working with pick and shovel, basketball, swimming, climbing, playing football.	7.5–12.0 (31.4–50.2)	6.0–10.0 (25.1–41.8)

*Adapted from *Recommended Dietary Allowances,* 9th ed. (Food and Nutrition Board, National Research Council—National Academy of Sciences, Washington, D.C., 1980).

endocrine disorders, behavior, lifestyle and physical activity, eating habits, and culture. Obesity is the most common nutritional disorder in developed countries.

A sedentary lifestyle is a significant contributing factor in insidious weight gain. Sedentary lifestyles may not only predispose individuals to obesity but also lead to degenerative arterial diseases. A sound exercise program coupled with good nutrition can help maintain health and quality of life. Obese individuals who wish to lose weight by dietary restriction and increased physical activity should be under medical supervision. The human body needs a wide variety of nutrients to function optimally. *Recommended Dietary Allowance* (*RDA*), published by the Food and Nutrition Board of the National Academy of Sciences National Research Council (USA), provides a broad guideline of the amounts of some of the well-established nutrients (protein, vitamins, and minerals) that should be consumed *daily* to maintain normal health (RDA values are given in Appendix IV). To use the RDA properly, keep in mind the following:

1. RDA values are overestimates of nutrient requirements; minimum quantities of nutrients required to maintain normal function and health in humans are either unknown or incompletely known.
2. RDA values relate primarily to populations of developed countries and are intended to meet the needs of virtually all healthy individuals. Thus, the allowances are far higher than demanded by physiological needs, and for many populations of the world, the RDA estimates may be inappropriately high. If the intake of a given nutrient is below the RDA value, nutritional inadequacy does not necessarily result. However, if a particular deficiency is accompanied by biochemical and clinical abnormalities, then corrective action is needed. Clinical conditions include trauma and stress, prematurity, inherited metabolic disorders, disease states and rehabilitation, and use of certain drugs.
3. Since the RDA applies to healthy people, it does not provide requirements in disease states or in drug-nutrient interactions.
4. RDA does not address the fact that some nutrients possess pharmacological activities unrelated to nutrient function. Excessive intake of such nutrients can have toxic effects.
5. RDA does not provide values for all essential nutrients, and therefore the daily requirement should be met by selection from a wide variety of foods. No single food source is nutritionally complete. Foods are categorized into four groups: 1) milk and milk products, 2) meat and meat alternatives (e.g., legumes, nuts, and milk products), 3) fruits and vegetables, and 4) breads and cereals. Table 5-5 lists the key nutrients in some representative foods in each group. Not only is consumption of adequate amounts of nutrients essential but so is consumption of appropriate quantities of foods that provide amounts of energy commensurate with physical activity and the physiological state. Knowledge of a given nutrient *concentration* (also known as nutrient *density*) per unit of energy (e.g., 1000 kcal) in a specific food is useful in adjusting the diet to individual need. Dietary fiber, although not nutritionally essential, is important in maintaining good health. Dietary fiber consists of polymers of sugars and other indigestible substances. Fiber affects nutrient absorption and makes the contents of the gut bulky and soft. In terms of the four food groups (Table 5-4), fiber can be obtained from the second group by selection of dry beans, peas, and nuts in place of (or in addition to) meat, poultry, and fish

TABLE 5-5
*Food Groups**

Group	Major Nutrients Provided
1. Milk and milk products (e.g., milk, yogurt, cheese, ice cream)	Calcium, riboflavin, protein
2. Meat (e.g., beef, pork, poultry, fish) and meat alternatives (e.g., legumes, nuts, milk products)	Protein, niacin, iron, thiamine (vitamin B1),
3. Fruits and vegetables (e.g., green leafy, deep yellow, or orange vegetables, citrus fruits)	Vitamins A and C
4. Breads, cereals, and grains	Carbohydrate, thiamine, iron, niacin

*A well-balanced diet includes a variety of foods on a daily basis, selected from each of the four food groups. With proper planning a vegetarian diet can be nutritionally adequate. Special foods consumed by ethnic groups should be considered in diet planning. Basic principles of a good diet include consumption of a variety of foods; consumption to maintain ideal body weight; moderation in intake of foods or beverages rich in saturated fats, cholesterol, sugar, and sodium salts; and consumption of adequate quantities of fruits, vegetables, and whole grains.

TABLE 5-6
BMI Values Corresponding to Different Height–Weight Combinations*

BMI→		19		20		21		22		23		24		25	
Height*		Weight**													
ft.in.	m	lb	kg	lb	kg	lb	kg	lb	kg	lb	kg	lb	kg	lb	kg
4'10"	1.473	91	41.28	96	43.55	100	45.36	105	47.63	110	49.90	115	52.16	119	53.98
4'11"	1.499	94	42.64	99	44.91	104	47.17	109	49.44	114	51.71	119	53.98	124	56.25
5'0"	1.524	97	44.00	102	46.27	107	48.54	112	50.80	118	53.52	123	55.79	128	58.06
5'1"	1.549	100	45.36	106	48.08	111	50.35	116	52.62	122	55.34	127	57.61	132	59.88
5'2"	1.575	104	47.17	109	49.44	115	52.16	120	54.43	126	57.15	131	59.42	136	61.69
5'3"	1.600	107	48.54	113	51.26	118	53.52	124	56.25	130	58.97	135	61.24	141	63.96
5'4"	1.626	110	49.90	116	52.62	122	55.34	128	58.06	134	60.78	140	63.50	145	65.77
5'5"	1.651	114	51.71	120	54.43	126	57.15	132	59.88	138	62.60	144	65.32	150	68.04
5'6"	1.676	118	53.52	124	56.25	130	58.97	136	61.69	142	64.41	148	67.13	155	70.31
5'7"	1.702	121	54.89	127	57.61	134	60.78	140	63.50	146	66.23	153	69.40	159	72.12
5'8"	1.727	125	56.70	131	59.42	138	62.60	144	65.32	151	68.49	158	71.67	164	74.39
5'9"	1.753	128	58.06	135	61.24	142	64.41	149	67.59	160	72.58	162	73.48	169	76.66
5'10"	1.778	132	59.88	139	63.05	146	66.23	153	69.40	162	73.48	167	75.75	174	78.93
5'11"	1.803	136	61.69	143	64.86	150	68.04	157	71.22	165	74.84	172	78.02	179	81.19
6'0"	1.829	140	63.50	147	66.68	154	69.85	162	73.48	169	76.66	177	80.29	184	83.46
6'1"	1.854	144	65.32	151	68.49	159	72.12	166	75.30	174	78.93	182	82.56	189	85.73
6'2"	1.880	148	67.13	155	70.31	163	73.94	171	77.57	179	81.19	186	84.37	194	88.00
6'3"	1.905	152	68.95	160	72.58	168	76.20	176	79.83	184	83.46	192	87.09	200	90.72
6'4"	1.930	156	70.76	164	74.39	172	78.02	180	81.65	189	85.73	197	89.36	205	92.99

BMI→		26		27		28		29		30		35		40	
Height*		Weight**													
ft.in.	m	lb	kg	lb	kg	lb	kg	lb	kg	lb	kg	lb	kg	lb	kg
4'10"	1.473	124	56.25	129	58.51	134	60.78	138	62.60	143	64.86	167	75.75	191	86.64
4'11"	1.499	128	58.06	133	60.33	138	62.60	143	64.86	148	67.13	173	78.47	198	89.81
5'0"	1.524	133	60.33	138	62.60	143	64.86	148	67.13	153	69.40	179	81.19	204	92.53
5'1"	1.549	137	62.14	143	64.86	148	67.13	153	69.40	158	71.67	185	83.92	211	95.71
5'2"	1.575	142	64.41	147	66.68	153	69.40	158	71.67	163	73.94	191	86.64	218	98.88
5'3"	1.600	146	66.23	152	68.95	158	71.67	163	73.94	169	76.66	197	89.36	225	102.06
5'4"	1.626	151	68.49	157	71.22	163	73.94	169	76.66	174	78.93	204	92.53	232	105.24
5'5"	1.651	156	70.76	162	73.48	168	76.20	174	78.93	180	81.65	210	95.26	240	108.86
5'6"	1.676	161	73.03	167	75.75	173	78.47	180	81.65	186	84.37	216	97.98	247	112.04
5'7"	1.702	166	75.30	172	78.02	178	80.74	185	83.92	191	86.64	223	101.15	255	115.67
5'8"	1.727	171	77.57	177	80.29	184	83.46	190	86.18	197	89.36	230	104.33	262	118.84
5'9"	1.753	176	79.83	182	82.56	189	85.73	196	88.91	203	92.08	236	107.05	270	122.47
5'10"	1.778	181	82.10	188	85.28	195	88.45	202	91.63	209	94.80	243	110.22	278	126.10
5'11"	1.803	186	84.37	193	87.54	200	90.72	208	94.35	215	97.52	250	113.40	286	129.73
6'0"	1.829	191	86.64	199	90.27	206	93.44	213	96.62	221	100.25	258	117.03	294	133.36
6'1"	1.854	197	89.36	204	92.53	212	96.16	219	99.34	227	102.97	265	120.20	302	136.99
6'2"	1.880	202	91.63	210	95.26	218	98.88	225	102.06	233	105.69	272	123.38	311	141.07
6'3"	1.905	208	94.35	216	97.98	224	101.61	232	105.24	240	108.86	279	126.55	319	144.70
6'4"	1.930	213	96.62	221	100.25	230	104.33	238	107.96	246	111.59	287	130.18	328	148.78

*Without shoes **Without clothes

from the third group; and from the fourth group by selection of whole-grain products (See also Chapter 9).

5.6 Obesity

The first law of thermodynamics states that the amount of stored energy equals the difference between energy intake and energy expenditure. The principal storage of energy is that of triglycerides in adipose tissue. Energy stores are essential for survival during times of energy deprivation (Chapters 18 and 22).

Energy stores in an adult are maintained at a relatively constant level throughout life. However, even a small imbalance in energy intake over long periods of time will have a significant effect on energy storage. For example, suppose a non obese adult's energy intake exceeds expenditure by about 1% daily for one year. This amounts to an excess of 9,000 kcal and corresponds to a weight gain of about 2.5 lb (1.15 kg) per year. 3,500 kcal of chemical energy is equivalent to 1 lb (0.45 kg) of adipose tissue. Weight gain in most people is attributable to overconsumption of palatable, energy-dense foods (e.g., lipids) and a sedentary lifestyle. Childhood obesity is a risk factor for obesity in adulthood.

Obesity is a consequence of a positive energy balance, i.e., input is greater than output. Body mass index (BMI) is the most useful parameter in assessing the magnitude of obesity (Table 5-6). A BMI of 19–25 is considered healthy, 26–29.9 is moderately overweight, and 30 or greater is obese. Obesity is the most common health problem in the developed world, and it is estimated that obesity affects about one third of the U.S. adult population. In developed nations, the prevalence of obesity is higher among economically deprived people, whereas in developing nations, the relatively affluent have a higher risk.

A BMI greater than 28 is an independent risk factor (3–4 times higher than the general population) for cardiovascular diseases (Chapter 20), diabetes mellitus type 2 (Chapter 22), and stroke. The prevalence of obesity-associated morbidity depends on the location of fat distribution in the body. Intra-abdominal or visceral fat deposits are associated with higher health risks than gluteofemoral adipose tissue fat accumulation.

In experimental animals and possibly in humans, energy intake influences the aging process as well as the onset of aging-associated diseases. Energy restriction in animals, without altering their optimal nutritional status, increases the average life span but not the maximal life span. This beneficial biological response to energy restriction has been attributed to the attenuation of oxidative damage to proteins, lipids, and DNA (Chapter 14). During aging, despite increased body fat, there is a linear decrease in food intake and metabolism over the life span.

Biochemical Mediators of Obesity

The regulation of energy intake and expenditure is achieved by coordinating the effects of endocrine mediators and neural signals that arise from adipose tissue, endocrine glands, neurological and gastrointestinal systems. All of the information finally is integrated by the central nervous system (Figure 5-6).

One of the most significant mediators of the energy store in the adipose tissue is ***leptin*** (from the Greek *leptos,* meaning "thin"). Leptin is a protein of 167 amino acid residues that is synthesized in adipocytes. Its synthesis is increased by insulin, glucocorticoids, and estrogens and is decreased by β-adrenergic agonists. The role of leptin in obesity comes from studies on rodents. In genetically obese mice (Ob/Ob), the observed gross obesity is due to absence of leptin production in the adipocytes. Leptin's action on energy metabolism is mediated by receptors in many cells and it binds specifically to a receptor in the hypothalamus.

The action of leptin involves at least two pathways. During starvation and weight loss, adipose tissue is decreased with consequent low levels of leptin. The low level of leptin leads to production of neuropeptide Y, which is synthesized in the arcuate nucleus of the hypothalamus and is transported axonally to the paraventricular nucleus. Neuropeptide Y binds to its receptor and functions as a potent appetite stimulant. The overall effect is increased appetite, decreased energy expenditure and temperature, decreased reproductive function (infertility), and increased parasympathetic activity. An opposite set of events occurs when the leptin levels rise, except that the effects are mediated by melanocyte-stimulating hormone (MSH) that binds to the melanocortin 4 receptor (MC4-R). The MSH binding to MC4-R initiates several biological responses, including decreased appetite, increased energy expenditure, and increased sympathetic activity (Figure 5-7).

In humans, obesity is a complex disease because of the redundancy of systems that regulate energy storage. There are inherited disorders of hyperphagia leading to obesity with associated clinical features such as hypogonadism and mental retardation. One hereditary disorder is ***Prader–Willi syndrome*** (Chapter 26), which is the most prevalent form of dysmorphic genetic obesity (1 in 10,000–20,000

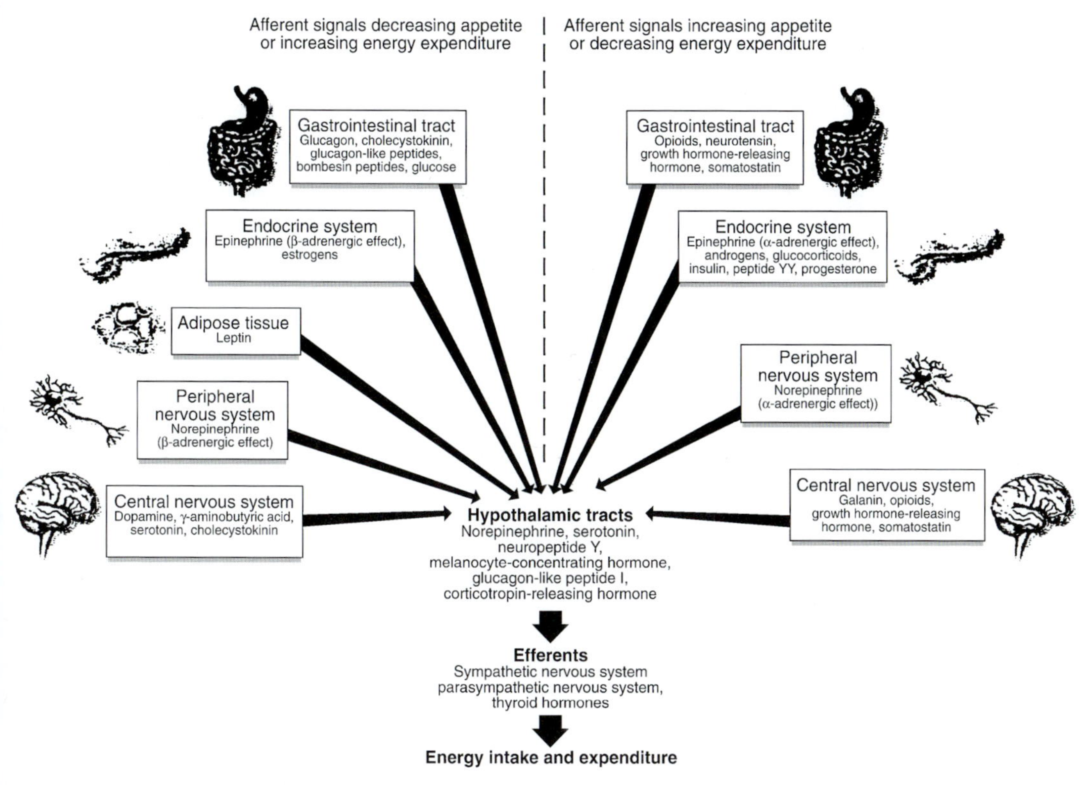

FIGURE 5-6
Afferent and efferent pathways of regulation of energy intake and expenditure. [Reproduced with permission from M. Rosenbaum, R. L. Leibel, and J. Hirsch. Obesity. *N. Engl. J. of Med.* **337**:396 (1997).]

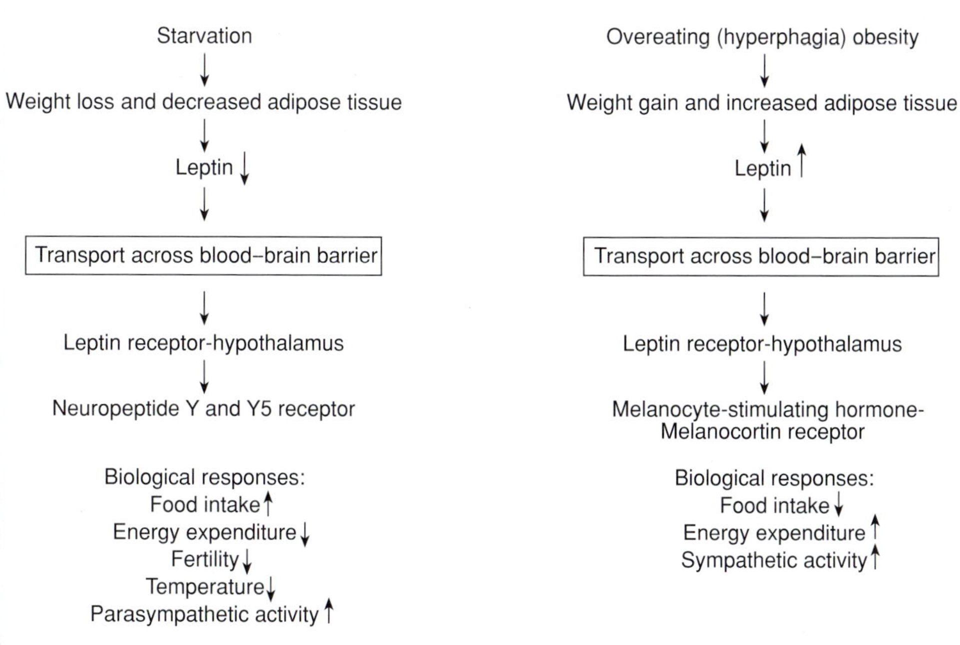

FIGURE 5-7
An overview of the action of leptin on food intake, energy expenditure, and biological responses (↑ increased and ↓ decreased).

births). Deletions in chromosome 15 account for a majority of cases of Prader–Willi syndrome. In other cases, a novel genetic mechanism of uniparental disomy is the cause of this syndrome. Congenital human leptin deficiency is associated with early-onset obesity. Homozygous mutations in the human leptin receptor gene result in a truncated leptin receptor lacking both transmembrane and intracellular domains; such mutations are associated with early-onset obesity, absence of pubertal development, and pituitary dysfunction.

In most obese humans, the plasma levels of leptin are high due to excess adipose tissue. In spite of abundant leptin, there is continued overeating. This puzzling observation may ultimately be explained by as yet-to-be identified defects in leptin metabolism.

Supplemental Readings and References

P. Björntorp: Obesity. *Lancet* **350,** 423 (1997).

J. F. Caro, J. W. Kolaczynski, M. R. Nyce, et al.: Decreased cerebrospinal-fluid/serum leptin ratio in obesity: A possible mechanism for leptin resistance. *Lancet* **348,** 159 (1996).

K. Clement, C. Vaisse, N. Lahlou, et al.: A mutation in the human leptin receptor gene causes obesity and pituitary dysfunction. *Nature* **392,** 398 (1998).

J. M. Friedman and J. L. Halaas: Leptin and the regulation of body weight in mammals. *Nature* **395,** 763 (1998).

C. T. Montague, I. S. Farooqi, J. P. Whitehead, et al.: Congenital leptin deficiency is associated with severe early-onset obesity in humans. *Nature* **387,** 903 (1997).

J. E. Morley: Anorexia of aging: Physiologic and pathologic. *American Journal of Clinical Nutrition* **66,** 760 (1997).

Recommended Dietary Allowances, 10th ed. Food and Nutrition Board, National Research Council–National Academy of Sciences, Washington, DC, (1989).

M. Rosenbaum, R. L. Liebel, and J. Hirsch: Obesity. *New England Journal of Medicine* **337,** 396 (1997).

M. W. Schwartz, E. Peskind, M. Raskind, et al.: Cerebrospinal fluid leptin levels: Relationship to plasma levels and to adiposity in humans. *Nature Medicine* **2,** 589 (1996).

J. Stevens, J. Cai, E. R. Parnuk, et al.: The effect of age on the association between body-mass index and mortality. *New England Journal of Medicine* **338,** 1 (1998).

L. A. Tartaglia, M. Dembroski, X. Weng, et al.: Identification and expression of cloning of a leptin receptor, OB-R. *Cell* **83,** 1263 (1995).

R. Weindruch and R. S. Sohal: Caloric intake and aging. *New Journal of Medicine* **337,** 986 (1997).

CHAPTER 6

Enzymes I: General Properties, Kinetics, and Inhibition

This chapter deals with enzyme catalysis, kinetics, inhibition, and mechanisms. Enzymes are catalysts and are functional units of cellular metabolism. Enzymes are usually proteins but may include RNA molecules as well. This chapter and the following two chapters discuss protein enzymes. RNA molecules are ribonucleases (ribozymes) that recognize specific nucleotide sequences in the target RNA and hydrolyze phosphodiester bonds. Ribosomal RNA functions as a peptidyl transferase in protein biosynthesis (Chapter 25). Some RNA molecules also undergo a self-splicing process. One example is the transformation of pre-mRNA to mature, functional mRNA by the splicing out of intervening sequences (***introns***) and ligation of the coding sequences (***exons***). The splicing–ligation process requires ***small nuclear ribonucleoprotein particles*** (snRNPs) and other proteins. Enzymatic RNA also can act on other RNA molecules. An example is ribonuclease P (RNase P), which is involved in the conversion of precursor tRNA to functional tRNA by generating 5′-phosphate and 3′-hydroxyl termini. RNase P contains both a catalytic RNA moiety and an associated protein (Chapter 25). Many enzymes are bound either covalently or noncovalently to nonprotein components essential for enzymatic activity. The term ***prosthetic group*** is generally reserved for those nonprotein components that are bound tightly, whereas the term ***coenzyme*** applies to less tightly bound nonprotein components. The coenzymes are complex organic compounds, frequently derived from vitamins. The term ***cofactor*** is used for metal ions and simple organic compounds that partipate in enzyme catalysis. The protein portion of an enzyme is the ***apoenzyme,*** and the fully functional enzyme with its attached nonprotein component is the ***holoenzyme*** (i.e., apoenzyme + coenzyme = holoenzyme).

6.1 Nomenclature

Enzymes are generally named for the ***substrate*** or chemical group on which they act, and the name takes the suffix *-ase*. Thus, the enzyme that hydrolyzes urea is named urease. Examples of exceptions to this terminology are trypsin, pepsin, and papain, which are trivial names. Systematic nomenclature for the enzymes has been developed by the Enzyme Commission of the International Union of Biochemistry. This system provides a rational and practical basis for identification of all enzymes currently known as well as for new enzymes. The systematic name describes the substrate, the nature of the reaction catalyzed, and other characteristics. A unique numerical code consisting of four numbers separated by periods (e.g., EC. 4.2.1.1) is designated. The prefix "EC" denotes "Enzyme Commission." The first number in this designation specifies the class to which the enzyme belongs. All enzymes are assigned to one of six classes on this basis of the type of reaction they catalyze:

1. oxidoreductases (oxidation–reduction),
2. transferases (transfer of groups),
3. hydrolases (hydrolysis),
4. lyases (nonhydrolytic and nonoxidative cleavage of groups),
5. isomerases (isomerization), and
6. ligases or synthetases (joining of two molecules with the breaking of a pyrophosphate bond).

The next two numbers in the code indicate the subgroup and the sub-subgroup; the last number is the special serial number given to each enzyme in its sub-subgroup.

Consider the systematic nomenclature for the enzyme with the trivial name carbonic anhydrase, which facilitates the transport of CO_2 by catalyzing the following reaction (Chapter 1):

$$H_2CO_3\ (\text{or } H^+ + HCO_3^-) \rightleftharpoons H_2O + CO_2$$

or

$$H_2CO_3(\text{or } H^+ + HCO_3{}^-) \rightleftarrows H_2O + CO_2$$

or

$$H^+ + HO{-}\overset{\overset{\displaystyle O}{\|}}{C}{-}O^- \rightleftarrows H_2O + O{=}C{=}O.$$

The systematic name for carbonic anhydrase is carbonate hydro-lyase, and its numerical code is EC. 4.2.1.1. The first number identifies it as a lyase; the second as an enzyme that catalyzes the breakage of a carbon–oxygen bond, leading to unsaturated products; and the third as a hydro-lyase, participating in a reaction involving the elimination of water. The last number is the specific serial number assigned to this enzyme. In this text, the trivial names are used. Names of a selected list of clinically useful enzymes with their EC codes, systematic names, other common names, and abbreviations are given in Appendix V.

6.2 Catalysis

Specificity of Enzyme Catalysis

Enzymes are highly specific and usually catalyze only one type of reaction. Some enzymes show absolute specificity. For example, pyruvate kinase catalyzes the transfer of a phosphate group only from phosphoenolpyruvate to adenosine diphosphate during glycolysis (Chapter 13). Examples of enzymes that show less specificity are:

- Hexokinase, which transfers the phosphate group from adenosine triphosphate to several hexoses (D-glucose, 2-deoxy-D-glucose, D-fructose, and D-mannose) at almost equivalent rates.
- Phosphatase, which hydrolyzes phosphate groups from a large variety of organic phosphate esters.
- Esterase, which hydrolyzes esters to alcohols and carboxylic acids, with considerable variation in chain length in both the alkyl and acyl portions of the ester.
- Proteinase, which hydrolyzes peptide bonds irrespective of the chemical nature of the substrate.

Many enzymes show stereoisomeric specificities. For example, human α-amylase catalyzes the hydrolysis of glucose from the linear portion of starch but not from cellulose. Starch and cellulose are both polymers of glucose, but in the former the sugar residues are connected by $\alpha(1 \rightarrow 4)$ linkages, whereas in the latter they are connected by $\beta(1 \rightarrow 4)$ linkages (Chapter 9).

Active Site and Enzyme–Substrate Complex

An enzyme-catalyzed reaction is initiated when the enzyme binds to its substrate to form an enzyme–substrate complex. In general, enzyme molecules are considerably larger than the substrate molecules. Exceptions are proteinases, nucleases, and amylases that act on macromolecular substrates. Irrespective of the size of the substrate, binding to the enzyme occurs at a specific and specialized region known as the ***active site,*** a cleft or pocket in the surface of the enzyme that constitutes only a small portion of the enzyme molecule. Catalytic function is accomplished at this site because various chemical groups important in substrate binding are brought together in a spatial arrangement that confers specificity on the enzyme. Thus, the unique catalytic property of an enzyme is based on its three-dimensional structure and on an active site whose chemical groups may be brought into close proximity from different regions of the polypeptide chain. The stereospecificity of an enzyme for a substrate has been compared to a lock-and-key relationship. This analogy implies that the enzyme has an active site that fits the exact dimensions of the substrate (Figure 6-1). However, after attachment of substrate, the enzyme may undergo conformational changes that provide a more perfect fit between it and the substrate. This process has been described as ***induced fit*** (Figure 6-2).

Factors Governing the Rate of Enzyme-Catalyzed Reactions

The overall reaction involving conversion of substrate (S) to product (P) with formation of the enzyme–substrate

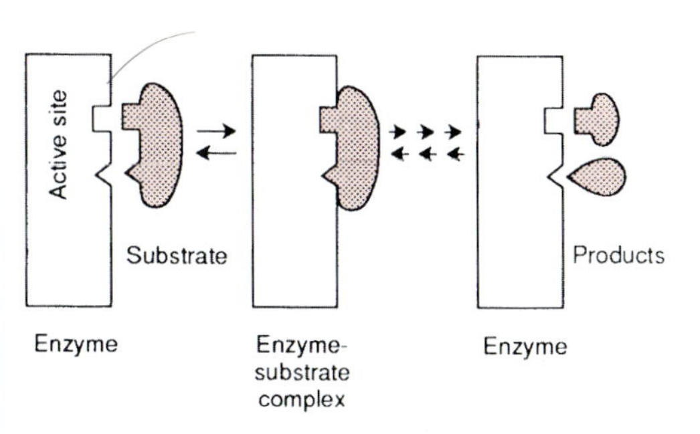

FIGURE 6-1
Schematic diagram of a lock and key relationship between an enzyme and its substrate. The shape of the active site is complementary to that of the substrate molecule.

complex, ES can be written:

$$E + S \underset{k_{-1}}{\overset{k_1}{\rightleftarrows}} ES \underset{k_{-2}}{\overset{k_2}{\rightleftarrows}} E + P,$$

where k_1, k_{-1}, k_2, and k_{-2} are rate constants for the designated steps; the positive subscripts indicate forward reactions, and the negative subscripts indicate backward reactions. Because an enzyme increases the rate of a particular reaction by decreasing the free energy of activation without itself being consumed or permanently altered, a small number of enzyme molecules can convert an extremely large number of substrate molecules to products very rapidly. Enzymes do not alter the equilibrium constant of a chemical reaction because they catalyze the forward and backward reactions

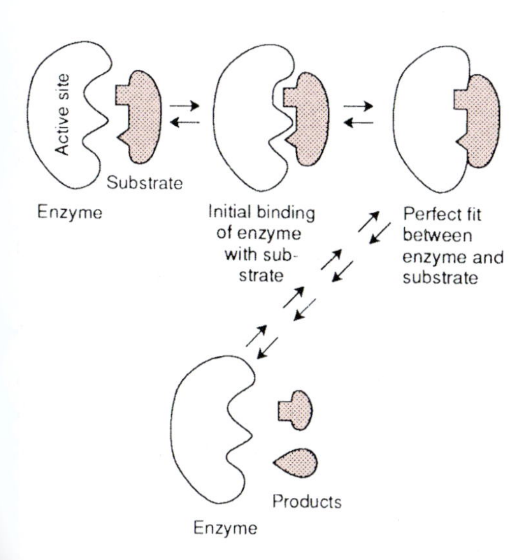

FIGURE 6-2
Schematic diagram of the induced-fit model for the relationship between an enzyme and its substrate. The shape of the active site of the enzyme conforms to that of the substrate *only* after substrate binding induces appropriate conformational changes in the enzyme.

to the same extent. Thus, enzymes affect only the *rate* at which equilibrium is established between reactants and products. However, under steady-state conditions, which is the normal state of affairs in the human body, the net effect of the enzyme is to convert substrates to products as rapidly as the products are removed (Chapter 5).

Effect of Temperature

The rates of almost all chemical reactions increase with a rise in temperature, which causes both the average kinetic energy and the average velocity of the molecules to increase, resulting in a higher probability of effective reaction collisions. Enzymes, however, undergo denaturation and are inactivated at high temperatures. Below the denaturation temperature, the reaction rate will approximately double for every rise of 10°C. The ratio by which the rate changes for a 10°C increase in temperature is known as Q_{10}; this ratio varies from 1.7 to 2.5. The optimal temperature for most enzymes is close to the normal temperature of the organism at which catalysis occurs at the maximum rate. In humans, most enzymes have an optimal temperature of 37°C.

Effect of pH

The activity of most enzymes depends on pH. The pH–enzyme activity profile of most enzymes delineates a bell-shaped curve (Figure 6-3), exhibiting an optimal pH at which activity is maximal. This pH is usually the same as the pH of the fluid in which the enzyme functions. Thus, most enzymes have their highest activity between pH 6 and pH 8 (the pH of human blood is about 7.4). However, pepsin, which must function at the low pH of gastric juice, has maximal activity at about pH 2.

The pH dependence of enzyme activity is the result of several effects. Ionizable groups in the active site of the enzyme (or elsewhere), in the substrate, or in the enzyme–substrate complex can affect catalysis depending on whether the protons on the reactive groups are dissociated or undissociated. Ionization of these groups depends on their pK values, the chemical properties of surrounding groups, and the pH of the reaction medium. Changes in pH affect the binding of the substrate at the active site of the enzyme and also the rate of breakdown of the enzyme–substrate complex. Thus, it may be possible to infer the identity of an ionizable group that participates at the active site from the pH–activity profile for a given enzyme.

The enzymes in living systems function at nearly constant pH because they are in an environment that contains buffers (Chapter 1).

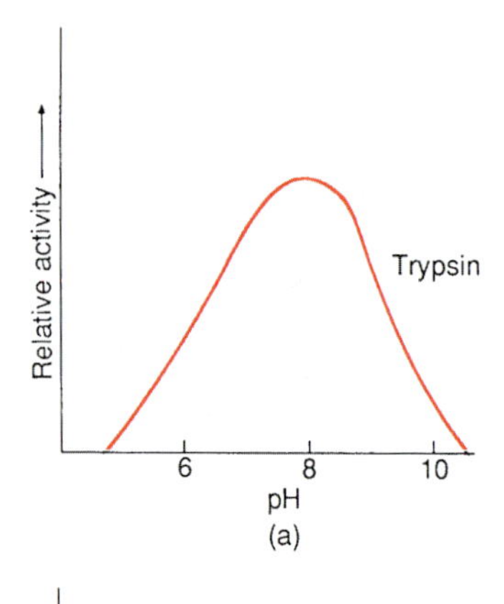

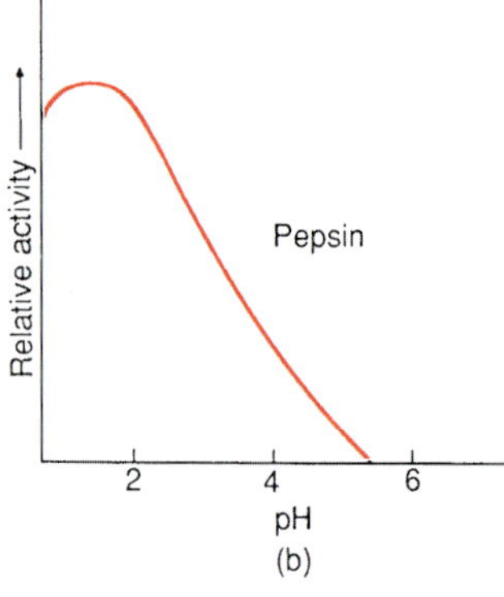

FIGURE 6-3
pH dependence of the rates of enzyme-catalyzed reactions. (a) Optimal activity at alkaline pH. (b) Optimal activity of acidic pH.

Effect of Concentration of Enzyme and Substrate

The reaction rate is directly proportional to the concentration of the enzyme if an excess of free substrate molecules is present. Thus, enzyme–substrate interactions obey the mass-action law. For a given enzyme concentration, the reaction velocity increases initially with increasing substrate concentration. Eventually, a maximum is reached, and further addition of substrate has no effect on reaction velocity (v) (Figure 6-4). The shape of a plot of v versus [S] is a rectangular hyperbola and is characteristic of all nonallosteric enzymes (Chapter 7). At low substrate concentrations, the reaction rate is proportional to substrate concentration, with the reaction following first-order kinetics in terms of substrate concentration.

Michaelis–Menten Treatment of the Kinetic Properties of an Enzyme

A model for enzyme kinetics that has found wide applicability was proposed by Michaelis and Menten in 1913 and later modified by Briggs and Haldane. The Michaelis–Menten equation relates the initial rate of an enzyme-catalyzcd reaction to the substrate concentration and to a ratio of rate constants. This equation is a ***rate equation,***

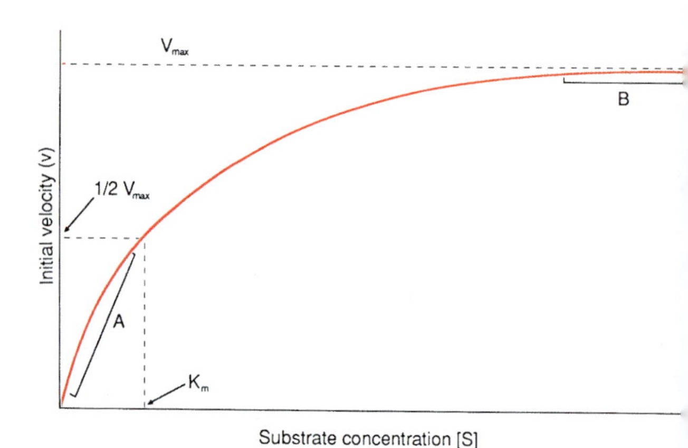

FIGURE 6-4
Plot of substrate concentration versus initial velocity of an enzyme-catalyzed reaction. Segment A: At low substrate concentration, the reaction follows first-order kinetics with respect to substrate concentration; i.e., $v = k[S]$, where k is a reaction rate constant. Segment B: At high substrate concentration, maximum velocity (V_{max}) is attained (saturation kinetics), and any further increase in substrate concentration does not affect the reaction rate; the reaction is then zero-order with respect to substrate but first-order with respect to enzyme. K_m is the value of [S] corresponding to a velocity of $\frac{1}{2}V_{max}$.

derived for a single substrate–enzyme–catalyzed reaction. In the reaction

$$E + S \underset{k_{-1}}{\overset{k_1}{\rightleftharpoons}} ES \overset{k_2}{\rightarrow} E + P$$

an enzyme E combines with substrate S to form an ES complex with a rate constant k_1. The ES complex formed can dissociate back to E and S with a rate constant k_{-1}, or it can give rise to a product P and regenerated enzyme with a rate constant k_2. In the latter process, the formation of product is essentially irreversible; the reverse reaction ($E + P \overset{k_{-2}}{\rightarrow} ES$) does not occur to any appreciable extent. Thus, k_{-2} is much less than k_2, which corresponds to a large, negative ΔG in going from ES to E + P.

In the Michaelis–Menten expression, $[E_T]$ represents the total concentration (in moles per liter) of enzyme, equal to [E] + [ES]. Therefore, the concentration of free enzyme [E] is equal to $[E_T] - [ES]$. The following assumptions are made:

1. In the reaction mixture, [S] is greater than [E]. However, [S] is *not* so large that all enzyme molecules are in the ES form, but [S] must be sufficiently large that [S] does not rapidly become so small that [S] < [E]. Excess concentration of substrate is the most common condition in enzymatic studies in which the catalyst is present at very low concentration. Under these conditions, [ES] rapidly becomes constant, so that the rate of formation of ES

equals its rate of breakdown, and steady-state kinetics then applies.

2. Only initial velocities are measured. Hence, during the period of measurement, [S] is not significantly depleted as [P] increases.
3. [ES] is constant during the period of measurement, which means that (a) enough time has elapsed since mixing E and S for [ES] to build up; (b) not enough time has passed for the rate of formation of ES to decrease owing to substrate depletion; and (c) $k_2 < k_{-1}$, so that [ES] can build up. If $k_2 > k_{-1}$, ES breaks down as rapidly as it is formed, and a steady-state can never be established.

$$\text{Rate of formation of ES} = \frac{d[\text{ES}]}{dt} = k_1\,[\text{E}]\,[\text{S}] = k_1\,([\text{E}_\text{T}] - [\text{ES}])\,[\text{S}]\,.$$

$$\text{Rate of breakdown of ES} = \frac{d[\text{ES}]}{dt} = k_{-1}\,[\text{ES}] + k_2\,[\text{ES}] = (k_{-1} + k_2)\,[\text{ES}]\,.$$

Under steady-state conditions, the rate of formation of ES equals its rate of breakdown, so that

$$k_1\,([\text{E}_\text{T}] - [\text{ES}])\,[\text{S}] = (k_{-1} + k_2)\,[\text{ES}]\,. \qquad (6.1)$$

Rearranging the above Equation yields

$$\frac{([\text{E}_\text{T}] - [\text{ES}])\,[\text{S}]}{[\text{ES}]} = \frac{k_{-1} + k_2}{k_1} = K_\text{m}. \qquad (6.2)$$

Note that K_m (the *Michaelis constant*) is *not* an equilibrium constant but a ratio of rate constants.

Equation (6.2) can be rearranged in the following ways:

$$[\text{S}][\text{E}_\text{T}] - [\text{S}][\text{ES}] = K_\text{m}[\text{ES}]$$

or

$$[\text{S}][\text{E}_\text{T}] = [\text{ES}]\,(K_\text{m} + [\text{S}])$$

or

$$[\text{S}] = \frac{[\text{ES}]}{[\text{E}_\text{T}]}\,(K_\text{m} + [\text{S}])\,. \qquad (6.3)$$

It is usually difficult to measure [ES] in a reaction mixture. Consequently, Equation (6.3) is not useful experimentally. On the other hand, the velocity (v) and the maximum velocity (V_{max}) are readily determined by a variety of methods.

As indicated in Figure 6-4, V_{max} is the limiting value that v approaches as $[S] \to \infty$. In that situation, all enzyme molecules have substrate bound to them, and so $[E\,] = 0$ and $[E_T] = [ES]$. Thus, $v = k_2[ES]$, and $V_{max} = k_2[E_T]$. These expressions can both be solved for k_2:

$$k_2 = \frac{v}{[\text{ES}]} \text{ and } k_2 = \frac{V_\text{max}}{[\text{E}_\text{T}]}.$$

Therefore,

$$\frac{v}{[\text{ES}]} = \frac{V_\text{max}}{[\text{E}_\text{T}]} \text{ or } \frac{v}{V_\text{max}} = \frac{[\text{ES}]}{[\text{E}_\text{T}]}.$$

Substituting the latter equation for $[ES]/[E_T]$ in Equation (6.3), we obtain

$$[\text{S}] = \frac{v}{V_\text{max}}\,(K_\text{m} + [\text{S}])\,.$$

Rearranging and solving for v,

$$v = \frac{[\text{S}]V_\text{max}}{K_\text{m} + [\text{S}]}. \qquad (6.4)$$

Equation (6.4) is known as the ***Michaelis–Menten equation,*** and the following points should be noted.

1. K_m is a constant for a particular enzyme and substrate and is independent of enzyme and substrate concentrations.
2. V_{max} depends on enzyme concentration, and at saturating substrate concentration, it is independent of substrate concentration.
3. K_m and V_{max} may be influenced by pH, temperature, and other factors.
4. A plot of v versus [S] fits a rectangular hyperbolic function (Figure 6-4).
5. If an enzyme binds more than one substrate, the K_m values for the various substrates can be used as a relative measure of the affinity of the enzyme for each substrate (the smaller the value of K_m, the higher the affinity of the enzyme for that substrate).
6. In a metabolic pathway, K_m values for enzymes that catalyze the sequential reactions may indicate the rate-limiting step for the pathway (the highest K_m corresponds roughly to the slowest step).
7. When $k_{-1} \gg k_2$, [ES] is assumed to be at equilibrium with [E] and [S], ES is dissociating more often to yield E and S than to yield product. *Under this condition,* the dissociation constant of the enzyme–substrate complex K_S for this equilibrium ($ES \rightleftharpoons E + S$) is

$$K_\text{S} = \frac{[\text{E}]\,[\text{S}]}{[\text{ES}]} = \frac{k_{-1}}{k_1}$$

But

$$K_\text{m} = \frac{k_{-1} + k_2}{k_1}$$

so that

$$K_m = \frac{k_{-1}}{k_1} = K_S.$$

8. When $k_2 \gg k_{-1}$, the rate of dissociation of ES to E and S is low, so that products are usually formed. For practical purposes, the reaction sequence can be thought of as consisting of two irreversible reactions:

$$E + S \xrightarrow{k_1} ES \xrightarrow{k_2} E + P$$

The overall reaction rate is always determined by [ES]. At low values of [S], ES is formed by a second-order reaction whose rate is proportional to [E][S]. At high values of [S], [ES] is constant and the rate is determined by the first-order breakdown of ES to product.

9. When $[S] \gg K_m$, the characteristic property of the ***turnover number*** for an enzyme can be invoked. This number provides information on how many times the enzyme performs its catalytic function per unit time, or how many times it forms the ES complex and is regenerated (turned over) by yielding product. The turnover number can be determined from V_{max} and $[E_T]$. We know that

$$\frac{d\,(P)}{dt} = V_{max} = k_2[ES] = k_2[E_T]$$

under saturation conditions. Therefore, if $[E_T]$ is increased n-fold while $[S] > [E_T]$ is maintained, then V_{max} will also increase n-fold. The rate is proportional to (or first order with respect to) $[E_T]$. The kinetic constant k_2, denoted as k_{cat} under saturation conditions, provides the value for the turnover number. For example, if a reaction mixture contains an enzyme at a concentration of x molar, and the product is formed at a rate such that its concentration increases by y molar per second when the enzyme is saturated,

$$k_{cat} = \frac{y}{x} \text{ per second.}$$

This means that y/x substrate molecules are converted to product molecules every second. Correspondingly, the time required for a single conversion is x/y seconds. Turnover numbers for most enzymes usually range from 1 to 10^4 per second. A few enzymes have turnover numbers above 10^5 (Table 6-1). The ability of a cell to produce a given amount of product by an enzymatic reaction during its life span is proportional to the turnover number and the number of molecules of that enzyme in the cell. Because turnover numbers can be measured only for pure enzymes, the activity of an enzyme is expressed as specific activity, defined as micromoles (μmol) of substrate converted to product per minute per milligram (mg) of enzyme protein. The International Union of Biochemistry recommends use of the unit known as ***katal*** (kat); one katal is the amount of enzyme that converts one mole of substrate to product per second. In clinical disorders, the activity of a variety of enzymes is measured in biological fluids. For this purpose, the activity is usually defined as that quantity of enzyme which catalyzes the conversion of 1 μmol of substrate to product per minute under a defined set of optimal conditions. This unit, referred to as the ***International Unit*** (U), is expressed in terms of U/mL of biological specimen (e.g., serum), or U/L.

TABLE 6-1
Turnover Numbers of Some Enzymes

Enzyme	Turnover Number (per second)
Catalase	5×10^6
Carbonic anhydrase	6×10^5
Acetylcholinesterase	2.5×10^5
α-Amylase	1.9×10^4
Lactate dehydrogenase	1×10^3
β-Galactosidase	2×10^2
Chymotrypsin	10^2
Phosphoglucomutase	21
Tryptophan synthase	2
Lysozyme	0.5

10. When $[S] \ll K_m$, the rate of product formation increases linearly with increase of [S]. In other words, the reaction is first order with respect to [S]. The rate is also proportional to $[E_T]$ (since doubling $[E_T]$ should double [ES]); thus, the reaction velocity at low values of [S] becomes

$$v = k'[E][S].$$

It can also be shown that

$$k' = \frac{k_{cat}}{K_m},$$

a fact that must be included in a description of any model of enzyme activity.

In summary, the rate equation for an enzyme reaction must satisfy a second-order dependence on $[S][E_T]$ when [S] is small as well as a first-order dependence on $[E_T]$ alone when [S] is large. Furthermore, the rate constants are k_{cat} for the first-order reaction and k_{cat}/K_m for the second-order reaction in which K_m is the value of [S] when $v = 1/2V_{max}$. Since k_{cat} is the rate constant for the conversion of ES to product and enzyme, it is a measure of how active the enzyme is in carrying out the reaction *after the enzyme has bound to the substrate*. The constant k_{cat}/K_m, however, measures the total activity of the enzyme, which includes the ability of the enzyme to bind to a particular substrate.

11. K_m (the ***Michaelis constant***) equals [S] when the substrate concentrations results in the velocity that is half-maximal ($V_{max}/2$). To demonstrate that $K_m = [S]$ when $v = V_{max}/2$, we substitute for v in Equation (6.4):

$$\frac{V_{max}}{2} = \frac{V_{max}[S]}{K_m + [S]}$$
$$(K_m + [S])(V_{max}) = 2V_{max}[S]$$
$$K_m + [S] = 2[S]$$

whence, as indicated above,

$$K_m = [S] \text{ when } v = \frac{V_{max}}{2}.$$

From Equation (6.4), it can be seen that v increases with increase in V_{max} at constant [S] and K_m and that v decreases with increase in K_m at constant [S] and V_{max}.

Linear Plots for Michaelis–Menten Expression

Because straight-line plots are easier to evaluate than curves, it is convenient to reformulate Equation (6.4) to yield straight-line plots. Two such reformulations can be performed. The ***Lineweaver–Burk plot*** is a double-reciprocal plot, obtained by taking reciprocals of both sides of Equation (6.4) and rearranging:

$$\frac{1}{v} = \frac{K_m + [S]}{[S]V_{max}} \quad (6.5)$$

$$\frac{1}{v} = \frac{K_m}{V_{max}}\frac{1}{[S]} + \frac{1}{V_{max}}. \quad (6.6)$$

According to Equation (6.6), if the system obeys Michaelis–Menten kinetics, a plot of $1/v$ versus $1/[S]$ is a straight line having a slope of K_m/V_{max} and an intercept of $1/V_{max}$ on the $1/v$ axis. Furthermore, the intercept on the $1/[S]$ axis (occurring when $1/v = 0$) is $-1/K_m$. This is seen by setting $1/v$ equal to zero. If

$$0 = \frac{K_m}{V_{max}}\frac{1}{[S]} + \frac{1}{V_{max}},$$

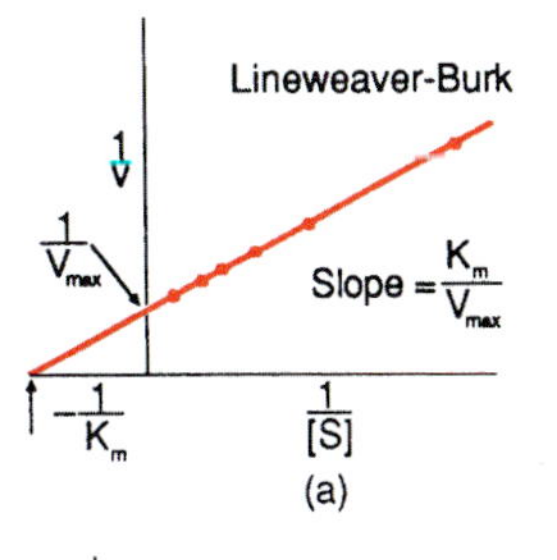

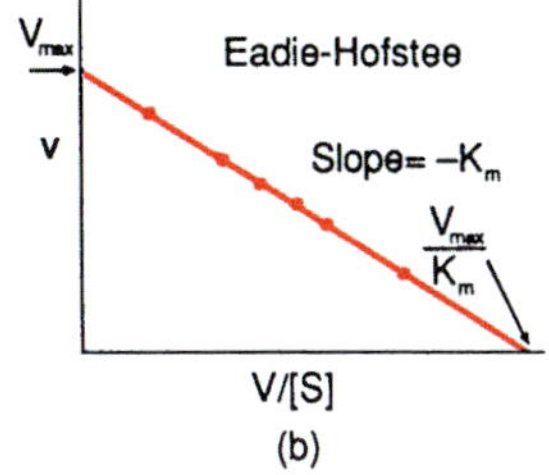

FIGURE 6-5
Two linear plots for the same data obtained for an enzyme-catalyzed reaction that obeys Michaelis–Menten kinetics.

then

$$\frac{1}{[S]} = \frac{-(1/V_{max})}{(K_m/V_{max})}$$
$$\frac{1}{[S]} = \frac{1}{K_m}.$$

Figure 6-5a shows a Lineweaver–Burk plot. The disadvantage of this plot is that it depends on less precisely determined points obtained at low values of [S], whereas the more accurate points obtained at high values of [S] cluster and so are less valuable in establishing the linear plot.

The ***Eadie–Hofstee plot*** represents an improvement over the Lineweaver–Burk plot in that the experimental points are usually more equally spaced. We obtain the equation for the Eadie-Hofstee plot by rearranging Equation (6.4):

$$v\left(\frac{K_m + [S]}{[S]}\right) = V_{max}$$
$$v\left(\frac{K_m}{[S]}\right) + v = V_{max}$$
$$v = -K_m\left(\frac{v}{[S]}\right) + V_{max} \quad (6.7)$$

From Equation (6.7), a plot of v versus $v/[S]$ has a slope of $-K_m$, an intercept of V_{max} on the v axis, and an intercept of V_{max}/K_m on the $V/[S]$ axis (Figure 6-5b).

6.3 Kinetics of Enzymes Catalyzing Two-Substrate Reactions

Most enzymatic reactions involving two-substrate reactions show more complex kinetics than do one-substrate reactions. Examples are catalyzed by dehydrogenases and aminotransferases. Hydrolytic reactions are bisubstrate reactions in which water is one of the substrates. The change in water concentration is negligible and has no effect on the rate of reaction. A two-substrate reaction can be written as

$$\text{Substrate A} + \text{Substrate B} \underset{}{\overset{\text{Enzyme}}{\rightleftharpoons}} \text{Product C} + \text{Product D}$$

The enzyme–substrate interactions can proceed by either ***single-displacement*** or ***double-displacement reactions*** (commonly known as "ping-pong" reactions). A substrate reaction proceeding by way of a single-displacement reaction can be shown by

$$\text{E} + \text{A} \rightleftharpoons \text{EA} \quad \text{or} \quad \text{E} + \text{B} \rightleftharpoons \text{EB}$$
$$\text{EA} + \text{B} \rightleftharpoons \text{EAB} \quad \text{or} \quad \text{EB} + \text{A} \rightleftharpoons \text{EAB}$$
$$\text{EAB} \rightleftharpoons \text{C} + \text{D}$$

Note that the ternary complex EAB can be formed in two different ways. If the formation of EAB can occur with either substrate binding first, the reaction is known as a ***random*** single-displacement reaction. Many reactions catalyzed by phosphotransferases are of this type. If a particular substrate must bind first with the enzyme before the second substrate can bind, the reaction is known as an ***ordered*** single-displacement reaction. Many reactions catalyzed by dehydrogenases are of this type. The values for K_m and V_{max} for each substrate can be obtained from experiments in which the concentration of one substance is held constant at saturating levels while the concentration of the second substrate is varied. Kinetic analyses can distinguish between these types of reactions.

In a double-displacement reaction, at first only one substrate is bound; the release of one product and covalently modified enzyme (E^*) follows. E^* then combines with the second substrate to form a second product. Reactions catalyzed by aminotransferases are of this type. In this mechanism, no ternary complex EAB is formed. The double displacement reaction sequence is shown below:

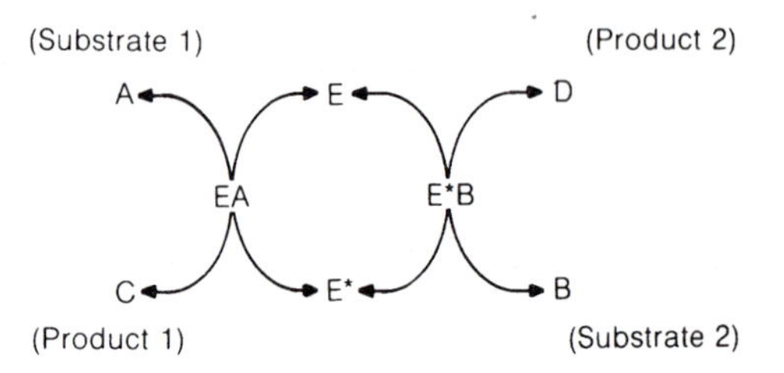

6.4 Inhibition

Enzyme inhibition is one of the ways in which enzyme activity is regulated experimentally or naturally. Most therapeutic drugs function by inhibition of a specific enzyme. Inhibitor studies have contributed much of the available information about enzyme kinetics and mechanisms. In the body, some of the processes controlled by enzyme inhibition are blood coagulation (hemostasis), blood clot dissolution (fibrinolysis), complement activation, connective tissue turnover, and inflammatory reactions.

Enzyme inhibitors are ***reversible*** or ***irreversible.***

Reversible Inhibition

In reversible inhibition, which is further subdivided into competitive, noncompetitive, and uncompetitive types, the activity of the enzyme is fully restored when the inhibitor is removed from the system (by dialysis, gel filtration, or other separation techniques) in which the enzyme functions. In reversible inhibition, equilibrium exists between the inhibitor, I, and the enzyme, E:

$$\text{E} + \text{I} \rightleftharpoons \text{EI}.$$

The equilibrium constant for the ***dissociation*** of the enzyme–inhibitor complex, known as the inhibitor constant K_i, is given by the Equation

$$K_i = \frac{[\text{E}][\text{I}]}{[\text{EI}]}$$

Thus, K_i is a measure of the affinity of the inhibitor for the enzyme, somewhat similar to the way in which K_m reflects the affinity of the substrate for the enzyme.

In ***competitive inhibition,*** the inhibitor is a structural analogue that competes with the substrate for binding at the active site. Thus, two equilibria are possible:

$$\text{E} + \text{S} \rightleftharpoons \text{ES} \rightarrow \text{E} + \text{P}$$

and

$$\text{E} + \text{I} \rightleftharpoons \text{EI}.$$

A modified Michaelis–Menten equation that relates the velocity of the reaction in the presence of inhibitor to the concentrations of substrate and inhibitor can be derived:

$$v = \frac{V_{max}[\text{S}]}{[\text{S}] + K_m\left(1 + \frac{[\text{I}]}{K_i}\right)}$$

In this relationship, K_m is modified by a term that includes the inhibitor concentration, [I], and the inhibitor constant

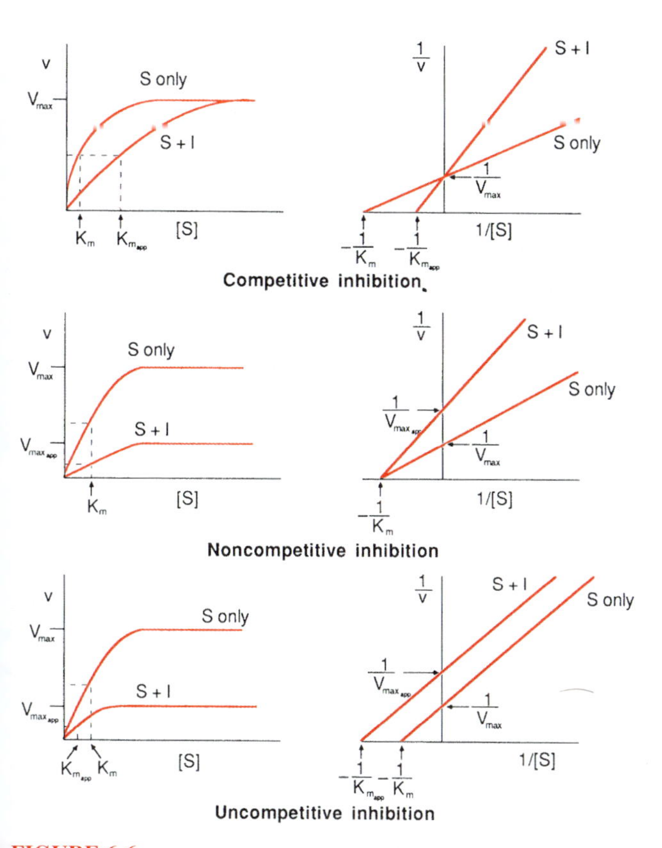

FIGURE 6-6
Effects of three types of reversible inhibitors on the velocity-substrate plots and on the Lineweaver–Burk double-reciprocal plots.

K_i, but V_{max} is unchanged. Thus, the profiles of plots of v against [S] show the same V_{max} in the presence and the absence of inhibitor but with an apparent change in K_m, to a larger value $(K_{m_{app}})$, because S and I compete for binding at the same site and a higher concentration of S is required to achieve half-maximal velocity. Therefore, the value of K_m is increased in the presence of inhibitor and depends on the concentration of I. The Lineweaver–Burk plot for competitive inhibition shows that the lines for [S] and ([S] + [I]) intersect at the same point on the ordinate (because V_{max} is unaltered) but have different slopes and intercepts on the abscissa (because of differences in K_m). Figure 6-6 illustrates these plots, and Table 6-2 shows the values for apparent K_m.

Competitive inhibition occurs in four different circumstances.

(1) Structural analogues compete with the substrate for binding at the active site of the enzyme. Several specific examples are described below.

Inhibition of Succinate Dehydrogenase A classic example of competitive inhibition occurs with the succinate dehydrogenase reaction:

$$\underset{\text{Succinate}}{^{-}OOC{-}CH_2{-}CH_2{-}COO^{-}} + FAD \xrightarrow{\text{Succinate dehydrogenase}} \underset{\text{Fumarate}}{^{-}OOC{-}CH{=}CH{-}COO^{-}} + FADH_2$$

In this reaction, flavin adenine dinucleotide (FAD), a coenzyme, serves as a hydrogen acceptor. This enzyme is competitively inhibited by malonate, oxalate, or oxaloacetate, all structural analogues of succinate.

$COO^{-}{-}CH_2{-}COO^{-}$	$COO^{-}{-}COO^{-}$	$COO^{-}{-}C{=}O{-}CH_2{-}COO^{-}$
Malonate	Oxalate	Oxaloacetate

Inhibition of Folate Synthesis Competitive inhibition of a biosynthetic step in folate synthesis accounts for the antimicrobial action of sulfonamides, which are structural analogues of p-aminobenzoic acid (PABA):

$H_2N{-}C_6H_4{-}C(=O){-}OH$	$H_2N{-}C_6H_4{-}S(=O)_2{-}NH_2$
p-Aminobenzoic acid	Sulfanilamide

PABA is used by bacteria in the synthesis of folic acid (pteroylglutamic acid), which functions as a coenzyme in one-carbon transfer reactions that are important in amino acid metabolism, in the synthesis of RNA and DNA, and thus in cell growth and division. Sulfonamides inhibit the bacterial enzyme responsible for incorporation of PABA into 7,8-dihydropteroic acid (Figure 6-7) and lead to the inhibition of growth (bacteriostasis) of a wide range of gram-positive and gram-negative microorganisms. Microorganisms susceptible to sulfonamides are those which synthesize their own folic acid or which cannot absorb folic acid derived from the host. Sulfonamides, however, have no effect on host cells (or other mammalian cells) that require preformed folic acid.

Inhibition of Dihydrofolate Reductase Folate-dependent reactions in the body are inhibited by folate analogues (or antagonists, e.g., methotrexate). Before it can function as a coenzyme in one-carbon transfer reactions, folate (F) must be reduced by dihydrofolate reductase to tetrahydrofolate (FH_4). Dihydrofolate

TABLE 6-2

Effects of Competitive Inhibitors on K_m and V_{max} (Michaelis-Menten Kinetic Parameters)

Inhibitor Type	V_{max}	K_m
Competitive	None ($V_{max_{app}} = V_{max}$)	$K_{m_{app}} = K_m\left(1 + \frac{1}{K_i}\right)$
Noncompetitive	$V_{max_{app}} = \dfrac{V_{max}}{\left(1 + \frac{1}{K_i}\right)}$	None ($K_{m_{app}} = K_m$)
Uncompetitive	$V_{max_{app}} = \dfrac{V_{max}}{\left(1 + \frac{1}{K_i}\right)}$	$K_{m_{app}} = \dfrac{K_m}{\left(1 + \frac{1}{K_i}\right)}$

Note: $K_i = \frac{[E][I]}{[EI]}$ or $\frac{[ES][I]}{[ESI]}$. In uncompetitive inhibition, the inhibitor binds equally well to E and ES.

reductase is competitively inhibited by methotrexate. Since FH_4 is needed for the synthesis of DNA precursors, a deficiency causes most harm to those cells which synthesize DNA rapidly. Certain types of cancers (e.g., the leukemias) exhibit an extremely high rate of cell division and are particularly susceptible to folate antagonists. The reactions of one-carbon metabolism are discussed in Chapter 27.

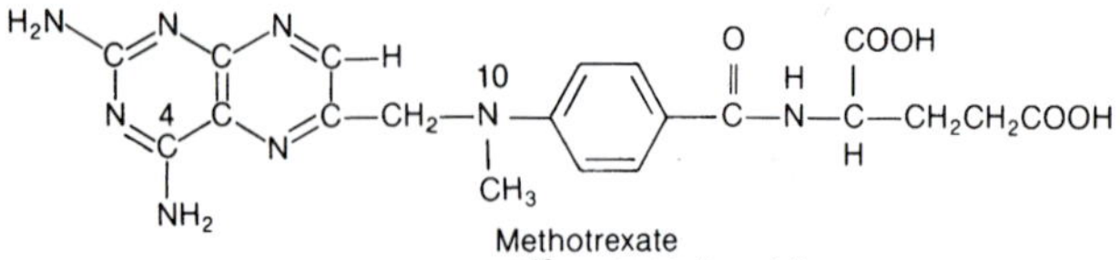

Methotrexate
(4-amino-N^{10}-methyl folic acid)

Inhibition of Xanthine Oxidase Uric acid, the end product of purine catabolism in humans, is formed by the serial oxidation of hypoxanthine and of xanthine, catalyzed by xanthine oxidase.

$$\text{Hypoxanthine} \xrightarrow[\text{oxidase}]{\text{xanthine}} \text{xanthine} \xrightarrow[\text{oxidase}]{\text{xanthine}} \text{uric acid}$$

Allopurinol, a structural analogue of hypoxanthine, is a competitive inhibitor as well as a substrate for xanthine oxidase:

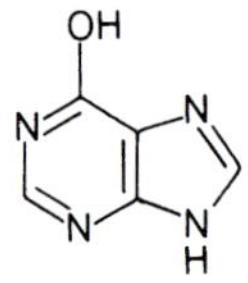

Hypoxanthine 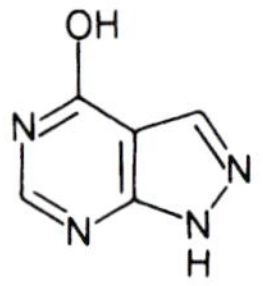   Allopurinol

Allopurinol inhibits the formation of xanthine and of uric acid, catalyzed by xanthine oxidase, and is itself transformed into alloxanthine by the enzyme. Alloxanthine bears the same structural relationship to xanthine that allopurinol does to hypoxanthine:

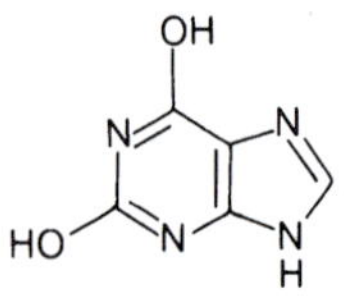

Xanthine 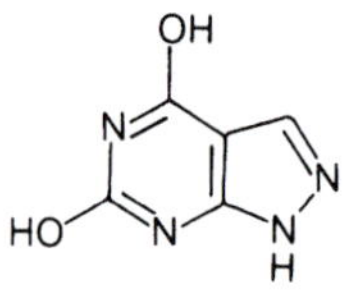   Alloxanthine

Alloxanthine, however, remains tightly bound to the active site of the enzyme by chelation with Mo^{4+}. Since xanthine oxidase is a molybdenum-dependent enzyme whose catalytic cycle requires the reversible oxidation and reduction of Mo^{4+} to Mo^{6+} (Chapter 27), the reoxidation of Mo^{4+} to Mo^{6+} in the presence of alloxanthine is very

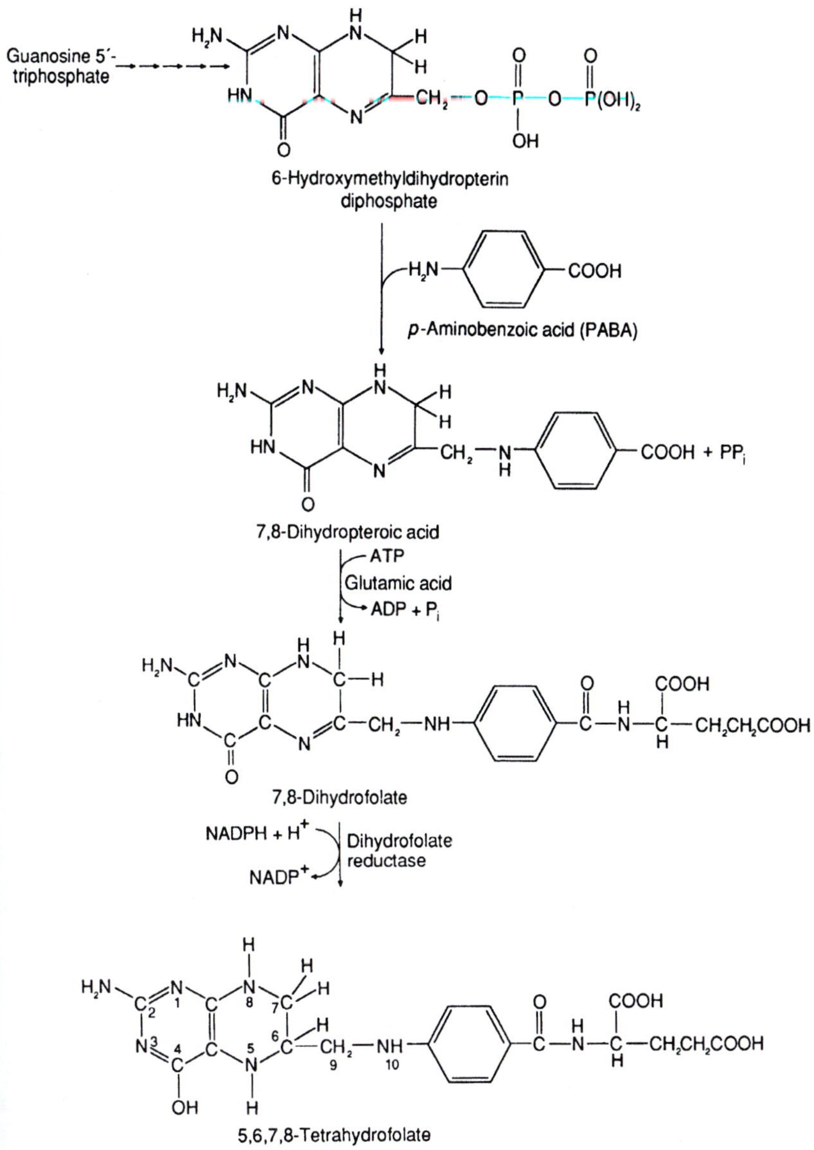

FIGURE 6-7
Folate biosynthetic pathways in bacteria. The incorporation of *p*-aminobenzoic acid into 7,8-dihydropteroic acid is competitively inhibited by sulfanilamide.

slow, thereby affecting the overall catalytic process. Thus, in this type of inhibition, an inhibitor bearing a particular structural similarity to the substrate binds to the active site of the enzyme and, through the catalytic action of the enzyme, is converted to a reactive compound that can form a covalent (or coordinate covalent) bond with a functional group at the active site. This type of inhibition, known as ***mechanism-based enzyme inactivation,*** depends on both the structural similarity of the inhibitor to the substrate and the mechanism of action of the target enzyme. The substrate analogue is called a ***suicide substrate*** because the enzyme is inactivated in one of the steps of the catalytic cycle. The suicide substrate, by virtue of its high selectivity, provides possibilities for many *in vivo* applications (e.g., development of rational drug design). Suicide substrates with potential clinical applications for several enzymes (e.g., penicillinase, prostaglandin cyclooxygenase) have been synthesized. Examples of suicide substrates harmful to the body are given in Chapter 27.

Allopurinol, which affects both the penultimate and ultimate steps in the production of uric acid, is used to lower plasma uric acid levels in conditions associated with excessive urate production (e.g., gout, hematologic disorders, and antineoplastic therapy). Sodium urate has a low solubility in biological fluids and tends to crystallize in derangements of purine metabolism that result in hyperuricemia. The crystalline deposits of sodium urate are responsible for recurrent attacks of acute arthritis or of renal colic (pain in kidney due to either stone formation or acute inflammation; see also discussion of purine catabolism in Chapter 27).

(2) In two-substrate enzyme-catalyzed reactions with a double-displacement reaction sequence, high concentrations of the second substrate may compete with the first substrate for binding. For example, in the reaction catalyzed by aspartate aminotransferase,

$$\text{L-Aspartate} + \alpha\text{-ketoglularate} \rightleftharpoons \text{L-glutamate} + \text{oxaloacetate},$$

this enzyme is inhibited by excess concentrations of α-ketoglutarate. The inhibition is competitive with respect to L-aspartate.

(3) Competitive inhibition can occur in freely reversible reactions owing to accumulation of products. Even in reactions that are not readily reversible, the product can function as an inhibitor. In the alkaline phosphatase reaction, in which hydrolysis of a wide variety of organic monophosphate esters into the corresponding alcohols (or phenols) and inorganic phosphates occurs, the inorganic phosphate acts as a competitive inhibitor. Both the inhibitor and the substrate have similar enzyme binding affinities (i.e., K_m and K_i are of the same order of magnitude).

(4) In reactions that require metal ions as cofactors, similar metal ions can compete for the same binding site on the enzyme. For example, Ca^{2+} inhibits some enzymes that require Mg^{2+} for catalytic function. Pyruvate kinase catalyzes the reaction

$$\text{Phosphoenolpyruvate} + \text{ADP} \rightarrow \text{ATP} + \text{pyruvate}$$

for which K^+ is an obligatory activator, whereas Na^+ and Li^+ are potent competitive inhibitors.

Competitive Substrates in Treatment of Some Intoxications

Competitive inhibition is the basis for the treatment of some intoxications (e.g., methyl alcohol, ethylene glycol). Methanol, which is widely used industrially as a solvent, is added to ethanol (ethyl alcohol) to make it unsuitable for human consumption. Such adulterated alcohol is commonly known as denatured alcohol. Methanol is metabolized primarily in the liver and kidney by oxidation to formaldehyde and formic acid:

$$\underset{\text{Methanol}}{CH_3OH} \xrightarrow[\uparrow \text{ Inhibited by ethanol}]{\text{Alcohol dehydrogenase}} \underset{\text{Formaldehyde}}{HCHO} \rightarrow \underset{\text{Formic acid}}{HCOOH}$$

Major toxic effects are caused by formaldehyde and formic acid. The former is responsible for damage to retinal cells that may cause blindness, while the latter produces severe acidosis that may eventually lead to death. A minor effect of methanol is depression of the central nervous system (CNS). Retardation of the first step in the oxidation of methanol is accomplished by administration of ethanol, the oxidation products of which are not as toxic as those of methanol. Other therapeutic modalities include removal of methanol by gastric lavage (to prevent further absorption), hemodialysis (to remove absorbed methanol), and administration of exogenous bicarbonate (for treatment of severe acidosis).

Ethylene glycol, which is widely used as an antifreeze for automobile radiators, upon ingestion causes depression of the CNS, metabolic acidosis, and severe renal damage. Its oxidation in the body requires the action of alcohol dehydrogenase:

Ethylene glycol ($HOCH_2-CH_2OH$)

↓ Alcohol dehydrogenase ($NAD^+ \rightarrow NADH + H^+$); Inhibited by ethanol or 4–methylpyrazole

Glycoaldehyde ($HOCH_2-CHO$)

↓

Glycolic acid ($HOCH_2-COOH$)

↓

Glyoxylic acid ($OHC-COOH$) → Formic acid HCOOH → CO_2

↓

Oxalic acid ($HOOC-COOH$)

Kidney damage results from precipitation of oxalate crystals in the convoluted tubules. The elevated anion-gap metabolic acidosis is caused by glycolic acid and lactic acid. The latter is formed from pyruvate due to a shift in the redox potential favoring the production of lactate. The treatment is the same as that for methanol intoxication.

4-Methylpyrazole (fomepizole), a more effective inhibitor of alcohol dehydrogenase with none of the adverse effects of ethanol, has been used in the treatment of ethylene glycol toxicity.

Isopropanol, present in rubbing alcohol, hand lotions, and antifreeze preparations can be ingested accidentally by young children and intentionally by alcoholics or suicidal adults. Isopropanol is oxidized to acetone, a toxic, nonmetabolizable product, by alcohol dehydrogenase. It should be noted that acetone is also present in diabetic ketoacidosis (Chapter 22). However, diabetic ketoacidosis is accompanied by glucosuria, whereas isopropanol ingestion shows the presence of acetone without glucose. Toxicity of isopropanol includes profound CNS depression, with coma as the most common presentation, gastritis, vomiting, and hemorrhage are also common. Acetone is also a CNS depressant and can be detected in the urine. The treatment of isopropanol toxicity is supportive and is removed by hemodialysis.

In evaluating individuals for toxicity to substances such as methanol, ethylene glycol, and isopropanol, the following serum components are measured: Na^+, K^+, Cl^-, HCO_3^-, glucose, urea nitrogen, blood osmolality, blood gases, and urine ketones. The severity of the osmolality gap, anion gap, and metabolic acidosis (Chapter 39), along with pertinent clinical history, will aid in the identification of the offending substance. In particular, the assessment of osmolality gap, which represents the difference between the measured serum osmolality and the calculated osmolality is useful in the diagnosis of low-molecular-weight toxic substances. Elevated serum levels of low-molecular-weight toxic chemicals contribute to serum osmolality significantly.

$$\begin{aligned}\text{Serum osmolal gap} &= \text{measured osmolality} \\ &\quad - \text{calculated osmolality} \\ \text{Calculated osmolality} &= 2 \times Na^+ \text{ (mM/L)} \\ &\quad + \text{glucose (mM/L)} \\ &\quad + \text{urea nitrogen (mM/L)}.\end{aligned}$$

In ***noncompetitive inhibition,*** the inhibitor does not usually bear any structural resemblance to the substrate, and it binds to the enzyme at a site distinct from the substrate binding site. No competition exists between the inhibitor and the substrate, and the inhibition cannot be overcome by increase of substrate concentration. An inhibitor may bind either to a free enzyme or to an enzyme–substrate complex; in both cases, the complex is catalytically inactive:

$$\begin{aligned}E + I &\rightleftharpoons EI \text{ (inactive)} \\ ES + I &\rightleftharpoons ESI \text{ (inactive)}\end{aligned}$$

The value of V_{max} is reduced by the inhibitor, but K_m is unaffected because the affinity of S for E is unchanged. Thus, comparison of the double-reciprocal plots, with and without the presence of inhibitor (Figure 6-6), shows that both the slope and the intercept on the y axis $(1/V_{max})$ are altered, whereas the x intercept $(-1/K_m)$ is unchanged.

Examples of noncompetitive inhibition are

1. Enzymes with sulfhydryl groups (–SH) that participate in maintenance of the three-dimensional conformation of the molecule are noncompetitively inhibited by heavy metal ions (e.g., Ag^+, Pb^{2+}, and Hg^{2+}):

   $$E\text{–}SH + Hg^{2+} \rightleftharpoons E\text{–}S\text{–}Hg^+ + H^+$$

   Lead poisoning causes anemia (low levels of hemoglobin) owing to inhibition of heme synthesis at two sites at least: porphobilinogen synthase and ferrochelatase, both of which contain sulfhydryl groups (Chapter 29). Heavy metal ions react with S-containing, O-containing, and N-containing ligands, present as –OH, $–COO^-$, $–OPO_3H^-$, –C=O, $–NH_2$, and –NH groups.
2. Enzymes that are dependent on divalent metal ion (e.g., Mg^{2+} and Ca^{2+}) for activity are inhibited by chelating agents (e.g., ethylenediamine tetraacetate) that remove the metal ion from the enzyme.
3. Enolase catalyzes a step in the metabolism of glucose (see the discussion of glycolysis in Chapter 13),

   $$\text{2-Phospho-D-glycerate} \rightleftharpoons \text{phosphoenolpyruvate} + H_2O$$

   a reaction that has an absolute requirement for a divalent metal ions (e.g., Mg^{2+} or Mn^{2+}) complexed to the enzyme before the substrate is bound. This reaction is inhibited by fluoride ion (F^-) in a process involving the formation of a complex with phosphate giving rise to a phosphofluoridate ion that binds magnesium ions. Thus, addition of fluoride ions inhibits the breakdown of glucose in the glycolytic pathway. For this reason, F^- is used as a preservative in clinical specimens (e.g., blood) in which glucose determinations are to be made.

In ***uncompetitive inhibition,*** inhibitor, I, combines with ES to form an enzyme–substrate-inhibitor complex:

$$ES + I \rightleftharpoons ESI$$

The double-reciprocal plot with the inhibitor yields parallel lines (i.e., the slope remains constant), but the intercepts on both the x and y axes are altered by the presence of the inhibitor (Figure 6-6). The apparent V_{max} and the apparent K_m are both divided by a factor of $(1 + [I]/K_i)$ (Table 6-2). Uncompetitive inhibition is rarely observed in single-substrate reactions. A noteworthy example in clinical enzymology is the inhibition of intestinal alkaline phosphatase by L-phenylalanine. Uncompetitive inhibition is more common in two-substrate reactions with a double-displacement reaction mechanism.

Irreversible Inhibition

Irreversible inhibition occurs when the inhibitor reacts at or near the active site of the enzyme with covalent modification of the active site or when the inhibitor binds so tightly that, for practical purposes, there is no dissociation of enzyme and inhibitor. The latter situation occurs in the case of proteinase inhibitors (see below). Thus, physical separative processes are ineffective in removing the irreversible inhibitor from the enzyme. Irreversible inhibitor reaction is written

$$E + I \rightarrow EI \text{ (inactive enzyme)}$$

Examples of irreversible inhibitors of enzymes are:

1. Enzymes that contain free sulfhydryl groups at the active site (e.g., glyceraldehyde-3-phosphate dehydrogenase; see Chapter 13) react with an alkylating reagent, iodoacetic acid, resulting in inactivation of the enzyme.

$$\text{Enzyme–SH} + \underset{\text{Iodoacetic acid}}{ICH_2COOH} \rightarrow \underset{\text{Inactive covalent derivative of enzyme}}{\text{enzyme–S} - CH_2COOH} + HI$$

   The imidazole ring of histidine also undergoes alkylation on reaction with iodoacetate. In ribonuclease, two residues (His 12 and His 119) are alkylated with loss of activity when the enzyme is treated with iodoacetate at pH 5.5.

2. Enzymes with seryl hydroxyl groups at the active sites can be inactivated by organophosphorous compounds. Thus, diisopropylphosphofluoridate (DPF) inactivates serine hydrolases by phosphorylation at the active site:

Enzyme–CH_2–OH (Active seryl residue) + F–P(=O)(O–CH(CH_3)$_2$)$_2$ (Diisopropylphosphofluoridate (DPF)) → (– HF) Enzyme–CH_2–O–P(=O)(O–CH(CH_3)$_2$)$_2$

   A specific example is inactivation of acetylcholinesterase (Table 6-1), which catalyzes hydrolysis of acetylcholine to acetate and choline. Acetylcholine is a ***neurotransmitter,*** a chemical mediator of a nerve impulse at a junction—known as a ***synapse***—between two neurons or between a neuron and a muscle fiber. On arrival of a nerve impulse at the ending of the neuron, acetylcholine (which is stored in the vesicles of the presynaptic nerve terminal) is released. The released acetylcholine acts on the postsynaptic membrane to increase the permeability of Na^+ entry across the membrane. Depolarization results in the inside of the membrane becoming more positive than the outside; normally, the inside of the membrane is more negative than the outside. This process may propagate an action potential along a nerve fiber, or it may lead to contraction of a muscle (Chapter 21). Acetylcholine is quickly destroyed by acetylcholinesterase present in the basal lamina of the neuromuscular junction (Figure 6-8). If, however, acetylcholine is not destroyed, as in the case of inactivation of acetylcholinesterase by DPF, its continued presence causes extended transmission of impulses. In muscle fibers, continuous depolarization leads to paralysis. The cause of death in DPF intoxication is respiratory failure due to paralysis of the respiratory muscles (including the diaphragm and abdominal muscles). Several organophosphorous compounds are used as agricultural insecticides, improper exposure to which can result in toxic manifestations and death.

Knowledge of the mechanisms of action of acetylcholinesterase and of the reaction of organophosphorous compounds with esterases led to the development of drugs useful in the treatment of this kind of intoxication. The active site of acetylcholinesterase consists of two subsites:

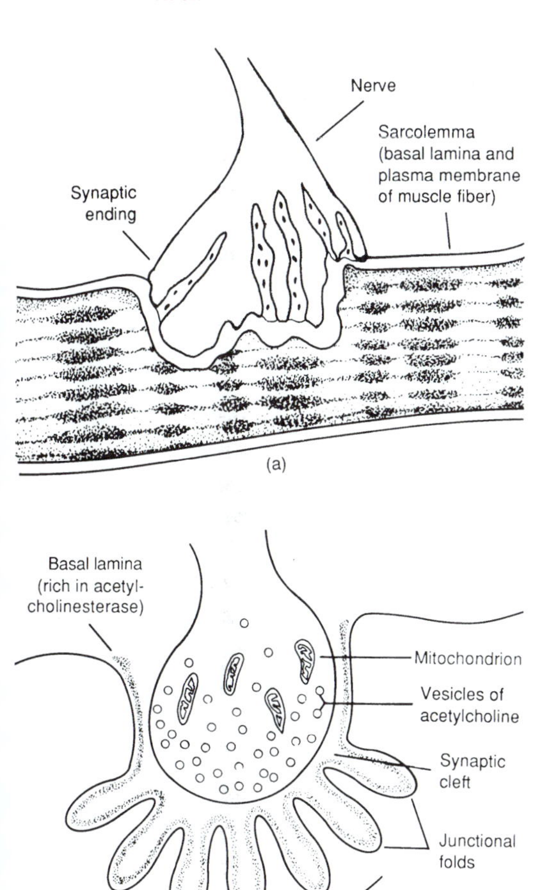

FIGURE 6-8
The neuromuscular junction. (a) Connection between a nerve and a muscle fiber. (b) Invagination of a nerve terminal into the muscle fiber. The acetylcholine receptor-rich domains are located on the crest of the fold.

an esteratic site and a site consisting of a negative charge (Figure 6-9).

The esteratic site contains a seryl hydroxyl group whose nucleophilicity toward the carbonyl carbon of the substrate is enhanced by an appropriately located imidazole group of histidine that functions as a general base catalyst. A side-chain carboxyl group that is suitably located apparently functions to hold the imidazole and imidazolium ion in place. The substrate is positioned on the enzyme so that its positively charged nitrogen atom is attracted to the negatively charged active site of the enzyme by both coulombic and hydrophobic forces. The acyl carbon of the substrate is subjected to nucleophilic attack by the oxygen atom of the serine hydroxyl group. This catalytic mechanism is similar to that of other serine hydrolases (esterases and the proteinases, i.e., chymotrypsin, trypsin, elastase, and thrombin). Although they have different functions, these enzymes have a common catalytic mechanism, which supports the view that they evolved from a common ancestor.

The inactive phosphorylated acetylcholinesterase undergoes hydrolysis to yield free enzyme, but the reaction is extremely slow. However, a nucleophilic reagent (e.g., hydroxylamine, hydroxamic acids, and oximes) can reactivate the enzyme much more rapidly than spontaneous hydrolysis. Wilson and co-workers accomplished the reactivation by use of the active-site-specific nucleophile pralidoxime. This compound, oriented by its quaternary nitrogen atom, brings about a nucleophilic attack on the phosphorus atom, leading to the formation of an oxime–phosphonate–enzyme complex that dissociates into oxime–phosphonate and free enzyme (Figure 6-10). Pralidoxime has found use in the treatment of organophosphorous poisoning and is most active in relieving the inhibition of skeletal muscle acetylcholinesterase. The phosphorylated enzyme can also lose an isopropoxy residue; such an enzyme–inhibitor complex has been named an "aged" enzyme. The aged enzyme is resistant to regeneration by pralidoxime because the phosphorus atom is no longer an effective center for nucleophilic attack. The severity and duration of toxicity of organophosphorous compounds depend on their lipid solubility, the stability of the bond linking the phosphorus atom to the oxygen of the serine hydroxyl group, and the ease with which the aged enzyme complex is formed. Some organophosphorous compounds are so toxic that they can be fatal within a few seconds of exposure. Other manifestations (e.g., excessive mucus secretion and bronchoconstriction) are reversed by administration of atropine, which does not relieve enzyme inactivation but, by binding to some of the acetylcholine receptor sites, renders ineffective the accumulated acetylcholine.

Organophosphorous compounds have been used in the identification of functional groups essential for catalysis. Another approach to identification of functional amino acid residues is ***affinity labeling.*** The labeling is produced by a synthetic substrate–like reagent designed to form a covalent linkage with some amino acid residue at or near the active site of the enzyme. After the labeling, the enzyme is subjected to classic techniques of degradative protein chemistry (Chapter 3) to determine the amino acid sequence of the structurally altered portion of the enzyme, which is inferred to be the active site.

Inactivation and Reactivation of Cytochrome Oxidase

Cyanides are among the most rapidly acting toxic substances. Cyanide (CN^-) inhibits cellular respiration to

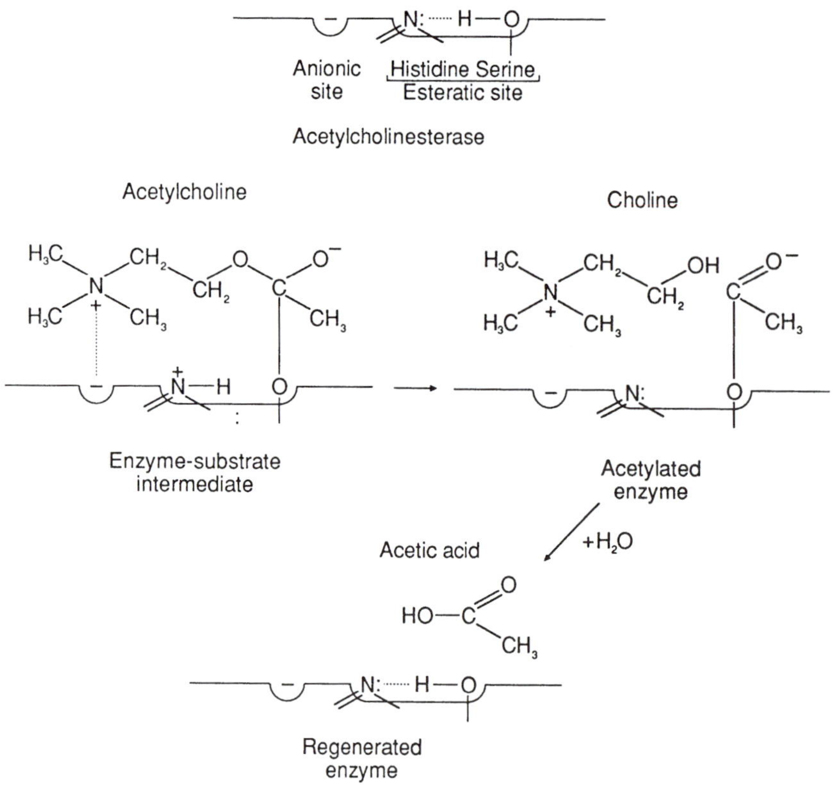

FIGURE 6-9
Hydrolysis of acetylcholine by acetylcholinesterase.

cause tissue hypoxia by binding to the trivalent iron of cytochrome oxidase, a terminal component of the mitochondrial electron transport chain. This chain consists of electron-transferring proteins and other carriers arranged sequentially in the inner mitochondrial membrane. The reducing equivalents, obtained from a variety of substrates, are passed through the electron transport system to molecular oxygen with the formation of water and energy (Chapter 14). Thus, cyanide severely impairs the normal energy-generating functions of mitochondria by inhibition of mitochondrial respiration, leading to cell death, particularly affecting the central nervous system. Death in acute cyanide poisoning is due to respiratory failure.

Cytochrome oxidase is a multienzyme complex that contains oxidation–reduction centers of iron–porphyrin prosthetic groups as well as centers of copper ion. Cyanide has a higher affinity for the oxidized form of cytochrome oxidase than for the reduced form. CN^- probably forms a loose complex with Fe^{2+} of porphyrin. When Fe^{2+} is oxidized to Fe^{3+}, the latter forms a stable complex with CN^-. This complex cannot be reduced, thus preventing electron flow and uptake of O_2.

Cyanide and cyanide precursors occur widely in nature. Foods that contain moderate to high levels of cyanogenic glycosides include cassava (a dietary staple in several regions of Africa), kernels of some fruits (peach, cherry, plum, and apricot), lima beans, sorghum, linseed, sweet potato, maize, millet, and bamboo shoots. ***Amygdalin*** (Figure 6-11) is one of the principal cyanogenic glycosides of dietary origin. Hydrogen cyanide is released when amygdalin undergoes enzymatic hydrolysis in the gastrointestinal tract:

$$\underset{\text{Amygdalin}}{C_{20}H_{27}NO_{11}} + 2H_2 \rightarrow \underset{\text{Glucose}}{2C_6H_{12}O_6} + \underset{\text{Benzaldehyde}}{C_6H_5CHO} + \underset{\text{Hydrogen cyanide}}{HCN}$$

The toxicity of amygdalin is directly related to the release of hydrogen cyanide. The enzymes that catalyze the hydrolysis of amygdalin are supplied by the microbial flora of the intestine, which probably explains why amygdalin is many times more toxic when taken by mouth than when given intravenously. Amygdalin and the related compound ***laetrile*** have been at the center of a controversy regarding their efficacy as anticancer agents. Their anticancer activity is claimed to depend on selective hydrolysis at the tumor site by β-glucuronidase or β-glucosidase with local release of HCN to cause cell death. Presumably

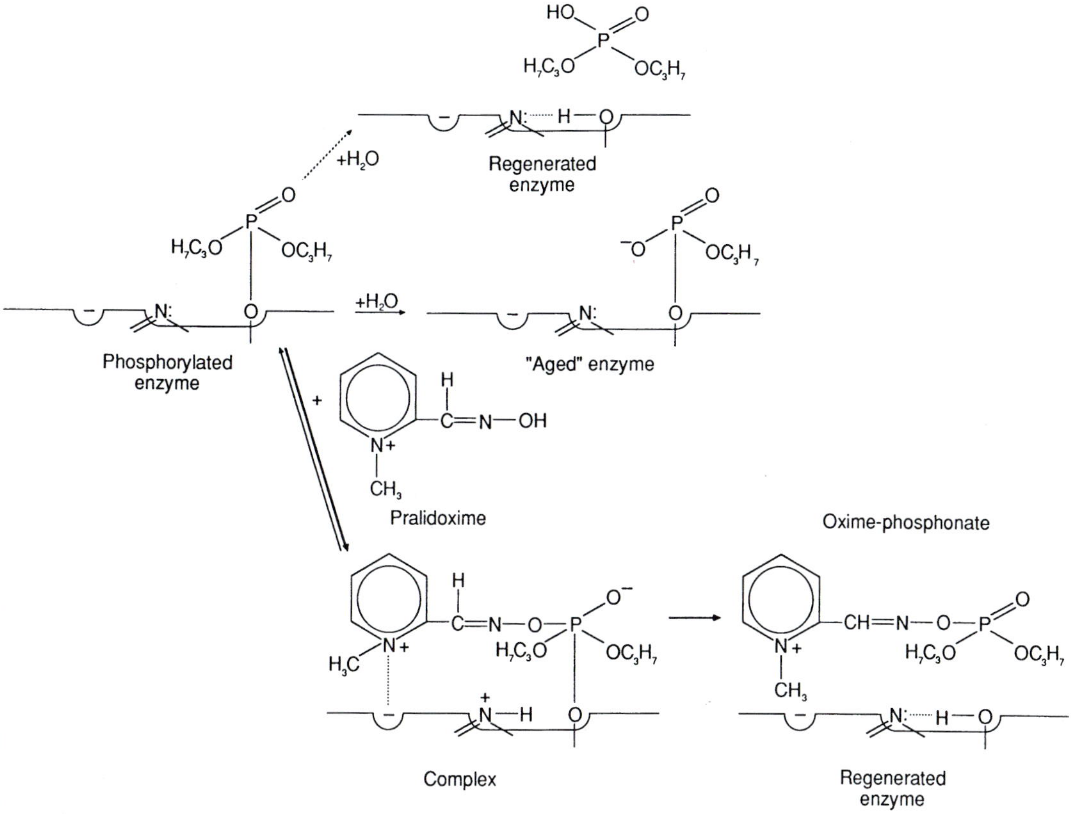

FIGURE 6-10
Reactivation of alkylphosphorylated acetylcholinesterase. Spontaneous reactivation (dashed arrow) by water occurs at an insignificant rate; however, loss of one isopropoxy group occurs at a much more rapid rate, yielding the "aged" enzyme, which is resistant to reactivation. The regeneration of the enzyme by pralidoxime is shown at bottom.

normal cells are not affected by HCN because they contain CN^--inactivating enzyme (rhodanese; see below). No objective evidence demonstrates any therapeutic value for these compounds as anticancer agents. Their ingestion has led to severe toxicity in some reported cases.

Nondietary sources of cyanide include sodium nitroprusside (a hypotensive agent), succinonitrile (an antidepressant agent), acrylonitrile (used in the plastic industry and as a fumigant to kill dry-wood termites), and tobacco smoke. Chronic exposure to cyanogenic compounds leads to toxic manifestations such as demyelination, lesions of the optic nerves, ataxia (failure of muscle coordination), and depressed thyroid functions. This last effect arises from accumulation of thiocyanate, the detoxified product of cyanide in the body (see below). Thiocyanate inhibits the active uptake of iodide by the thyroid gland and, therefore, the formation of thyroid hormones (Chapter 33).

Acute cyanide poisoning requires prompt and rapid treatment. The biochemical basis for an effective mode of treatment consists of creating a relatively nontoxic porphyrin–ferric complex that can compete effectively with cytochrome oxidase for binding the cyanide ion. This is accomplished by the administration of nitrites ($NaNO_2$ solution intravenously, and amylnitrite by inhalation), which convert a portion of the normal oxygen-carrying hemoglobin with divalent iron to oxidized hemoglobin (i.e., methemoglobin with trivalent iron, which does not transport oxygen). Methemoglobin binds cyanide to form cyanomethemoglobin, whose formation is favored because of an excess of methemoglobin relative to cytochrome oxidase. This process may lead to the restoration of normal cytochrome oxidase activity (Figure 6-12). Cyanomethemoglobin is no more toxic than methemoglobin and can be removed by normal processes that degrade erythrocytes and by the reaction catalyzed by rhodanese. Rhodanese (thiosulfatecyanide sulfurtransferase) catalyzes the reaction involving cyanide and thiosulfate to form thiocyanate:

$$\underset{\text{Thiosulfate}}{SSO_3^{3-}} + \text{enzyme} \rightleftharpoons \underset{\text{Sulfite}}{SO_3^{2-}} + \underset{\text{Sulfur-substituted enzyme}}{\text{enzyme–S}}$$

$$\text{Enzyme–S} + \underset{\text{(Cyanide)}}{CN^-} \rightarrow \underset{\text{Thiocyanate}}{SCN^-} + \text{enzyme}$$

Amygdalin (D-mandelonitrile-β-D-glucosido-6-β-D-glucoside)

Laetrile (1-mandelonitrile-β-glucuronic acid)

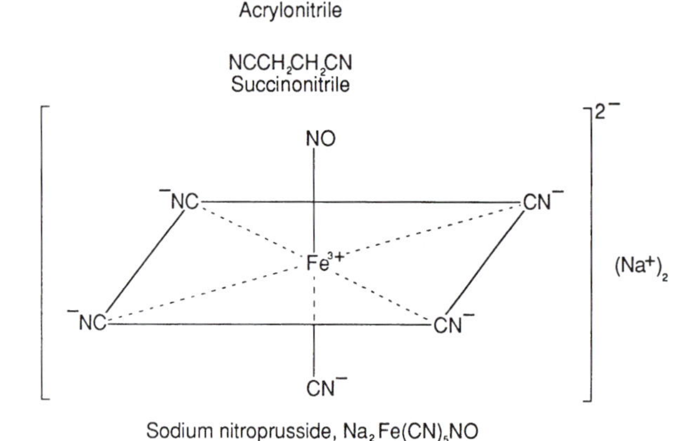

Sodium nitroprusside, $Na_2Fe(CN)_5NO$

FIGURE 6-11
Structures of some cyanogenic compounds.

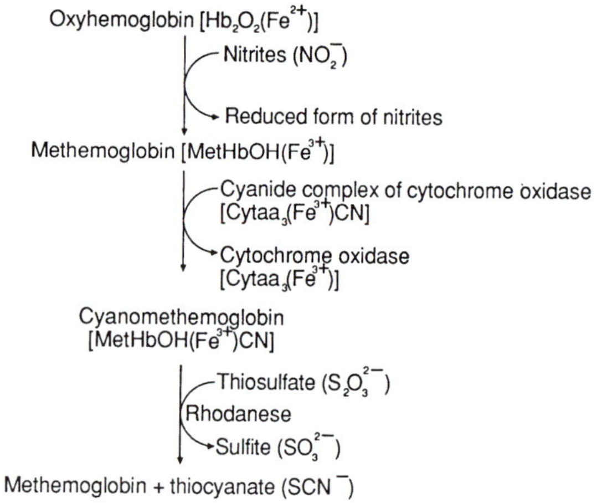

Rhodanese is present in the mitochondria, particularly of liver and kidney cells. A double-displacement mechanism has been proposed for its biochemical action. The steps are as follows: The free enzyme reacts with a sulfane sulfur-containing compound (a sulfane sulfur is one that is divalent and covalently bonded only to other sulfurs), cleaving the S–S bond of the donor substrate (e.g., SSO_3^{2-}) to form the sulfur-substituted enzyme. The latter reacts with the cyanide (a thiophilic acceptor) to form thiocyanate in an essentially irreversible reaction.

Nitrite administration has been augmented by thiosulfate administration (intravenously) in the treatment of cyanide poisoning. Cobalt-containing compounds (e.g., cobalt chloride and cobalt ethylenediaminetetraacetate) have also been used to form complexes with cyanide, in

Oxyhemoglobin [$Hb_2O_2(Fe^{2+})$]
Nitrites (NO_2^-)
Reduced form of nitrites
Methemoglobin [$MetHbOH(Fe^{3+})$]
Cyanide complex of cytochrome oxidase [$Cytaa_3(Fe^{3+})CN$]
Cytochrome oxidase [$Cytaa_3(Fe^{3+})$]
Cyanomethemoglobin [$MetHbOH(Fe^{3+})CN$]
Thiosulfate ($S_2O_3^{2-}$)
Rhodanese
Sulfite (SO_3^{2-})
Methemoglobin + thiocyanate (SCN^-)

FIGURE 6-12
Reactivation of cytochrome oxidase and inactivation of cyanide.

order to decrease the amount of cyanide available for binding with cytochrome oxidase.

Proteinase Inhibitors and Their Clinical Significance

In the body, enzymes are compartmentalized and function under highly restricted conditions. Some enzymes (e.g., proteinases) are not substrate-specific. When present in active form in an inappropriate part of the body, they act indiscriminately and cause considerable damage to the tissue. Inhibitors inactivate these enzymes at sites where their action is not desired. Proteinase inhibitors, which are themselves proteins, are widely distributed in intracellular and extracellular fluids. Protein inhibitors of enzymes other than proteinases are relatively rare. Such inhibitors are available for α-amylases, deoxyribonuclease I, phospholipase A, and protein kinases.

Many proteinase inhibitors are present in blood plasma and participate in the control of blood coagulation (Chapter 36), dissolution of blood clots (Chapter 36), activation of the complement cascade (Chapter 35), formation and destruction of some peptide hormones, and inactivation of proteinases released from phagocytic cell (Table 6-3). The proteinase inhibitors rapidly combine with their target enzymes to form stable complexes that are practically nondissociable. Inhibition occurs through binding of the reactive site (a substrate-like region) of the inhibitor to the active site (the substrate binding region) of the proteinase. At or near the center of the reactive site of the inhibitor, a specific amino acid residue recognizes the primary binding site of the target proteinase. Some of the

TABLE 6-3
*Principal Proteinase Inhibitors in Human Blood Plasma**

Name	Concentration (mg/100 mL)	Mol. Wt.	Number of Polypeptide Chains	Heads[†]
α_1-Proteinase inhibitor	290±45	52,000	1	1
α_1-Antichymotrypsin	49±7	69,000	1	1
Inter-α-trypsin inhibitor	50	160,000	1	2
α_2-Antiplasmin	7±1	70,000	1	1
Antithrombin III	24±2	65,000	1	1
C_1-inactivator	24±3	70,000	1	1
α_2-Macroglobulin	260±70	720,000	4	1–2 or more

*Reproduced, with permission, from M. Lakowski, Jr., and I. Kato: Protein inhibitors of proteinases. *Annu. Rev. Biochem.*, **49,** 593 (1980). © 1980 by Annual Reviews, Inc.
[†]Maximal number of enzyme molecules simultaneously inhibited by one inhibitor molecule.

proteinase inhibitors share significant homology at their reactive site regions.

Neutrophils and macrophages function protectively against foreign organisms in an inflammatory process. In this process, a number of proteinases escape to surrounding tissues. One such enzyme is elastase, which catalyzes the hydrolysis of elastin, a protein of connective tissue (Chapter 10). Normally, the activity of elastase is inhibited by the α_1-***proteinase inhibitor*** (also known as α_1-***antitrypsin***), which inhibits a broad spectrum of proteinases containing serine in their active sites (e.g., elastase, trypsin, and chymotrypsin). Genetic deficiency of an α_1-proteinase inhibitor strongly predisposes to pulmonary emphysema and liver disease. Emphysema (which means swelling or inflation) results from the breakdown of alveolar walls due to coalescence of alveoli resulting in formation of air sacs, which leads to a gradual decrease in the effectiveness of CO_2 elimination and oxygenation of hemoglobin.

A collection of high-molecular-weight proteinase inhibitors with broad specificity, known as α_2-***macroglobulins,*** are present in the plasma of all mammals and exhibit an interesting inhibition pattern. When α_2-macroglobulins combine with a wide variety of proteinases, they inhibit only the proteolytic activity of the enzyme toward large protein substances without significantly affecting the catalysis of low-molecular-weight substrates.

Proteinases and their inhibitors play a major role in metastasis of cancer. Metastasis of tumors requires remodeling of extracellular matrix (ECM). Remodeling is a balance between proteolysis and respective proteinase inhibitors. This process is aided by proteolytic enzymes synthesized by the tumor cells. Examples of proteolytic enzymes include serine proteinases, cathepsins, and matrix metalloproteinases (Table 6-4). Some of the ***cathepsins*** are cysteine proteinases and their inhibitors belong to the cystatin superfamily. The ***cystatin family*** consists of three subfamilies: *stefins*, *cystatins*, and *rinogens*. Progressive loss of expression of the proteinase inhibitors may be responsible for metastasis. The loss of control over proteinase expression and their respective inhibitors is influenced by a variety of biological response modifiers such as growth factors, cytokines, tumor promoters, and suppressor genes.

Viral proteinases (also called proteases) offer unique targets for antiviral drugs. An example is in the treatment of human immunodeficiency virus (HIV) infections by HIV protease inhibitors (Figure 6-13). HIV is a retrovirus and its RNA undergoes reverse transcription in cells to produce double stranded DNA (Chapter 26). This step is inhibited by nucleoside analogues such as zidovuline, didanosine, zalcitabine, stavudine, and lambivudine. The viral DNA, upon integration into the host genome, produces a protease as well as polypeptides. The HIV protease is an aspartyl protease and the enzymatic activity resides in the homodimers. The HIV protease cleaves polypeptides at phenylalanine–proline or tyrosine–proline bonds, which are unusual sites of cleavage for mammalian proteases. That cleavage results in the production of three large proteins (p24, p17, and p7) and three smaller proteins (p6, p2, and p1). The three large proteins are essential for RNA packaging and in the structure of the virion. Viral protease is essential in the production of functionally mature and infectious particles and thus is a prime target for inhibition by drugs.

TABLE 6-4
Proteinases and Their Putative Extracellular Matrix Targets

Proteinases	Functions/Targets of Degradation
Serine proteinases	
Urokinase plasminogen activator (uPA)	Activation of plasminogen
Tissue-type plasminogen activator (tPA)	Activation of plasminogen
Plasminogen	
Plasmin	Laminin, type IV collagen
Thrombin	
Elastase	
Cathepsins	Collagens, laminin
Cathepsin D (aspartic proteinase)	
Cathepsin B, L (cysteine proteinases)	
Cathepsin G (serine proteinase)	
Cathepsin O2 (similar to cathepsins S and L)	
Integral membrane proteinases	Localized ECM degradation
Matrix metalloproteinases (MMP)	
MMP-1 interstitial collagenase	Degradation of interstitial collagen types I, II, III, VII
Neutrophil collagenase	Collagen I
MMP-2 (type IV collagenase, gelatinase) (two major forms)	Types I, II, III, IV, V, VII collagens, fibronectin
MMP-9 (type IV collagenase)	
MMP-3 (Stromelysin-1)	Types IV, IX collagens, laminin, fibronectin
MMP-10 (Stromelysin-2, transin)	Types III, IV, V collagens, fibronectin
MMP-11 (Stromelysin-3)	
MMP-4	α_1-chain of type I collagen
MMP-5	Native $^3/_4$ collagen fragments
MMP-6 Acid metalloproteinase	Cartilage proteoglycan
MMP-7 (pump-1)	Gelatin of types I, III, IV, V collagens, fibronectin

Reproduced with permission from *The Genetics of Cancer,* G. V. Sherbet and M. S. Lakshmi. Academic Press, 1997.

6.5 Kinetics of Ligand–Receptor Interaction

The kinetics of enzyme–substrate interactions can be applied to ligand–receptor relationships. A ligand (e.g., a hormone or a drug) can saturate the receptor sites as well as dissociate from them, similar to enzyme–substrate interactions. Nonspecific ligand binding, on the other hand, shows a relatively low affinity to the receptor and nonsaturation kinetics. In ligand–receptor binding, no product is formed, but effects responsible for the ligand's biological action are produced (Chapter 30). Cells that contain a receptor protein specific for a particular hormone are targets for the action of that hormone. The activity of a ligand on a target cell depends on the concentration of ligand, the binding affinity of the ligand to the receptor sites, the number of receptors bound, and the duration of binding. Ligands that compete for binding to a receptor at a specific site and produce the same biological response are known as ***agonists.*** A ligand that competes with an agonist for binding to a receptor and causes a blockage or inhibition of the biological response of the cell is known as an ***antagonist.*** Antagonists are similar to competitive inhibitors of enzymes. Consider a ligand–receptor interaction,

$$\mathrm{L} + \mathrm{R} \rightleftharpoons \mathrm{LR} \tag{6.8}$$

where L is the ligand, R is the free receptor site, and LR is the ligand bound to the receptor site. We apply the law of mass action to the above equilibrium,

$$K_a = \frac{[\mathrm{LR}]}{[\mathrm{L}]\,[\mathrm{R}]} \tag{6.9}$$

where K_a is the association constant. The reciprocal of K_a represents the dissociation constant known as K_d. Thus,

$$K_\mathrm{a} = \frac{1}{K_\mathrm{d}} \text{ or } K_\mathrm{d} = \frac{1}{K_\mathrm{a}}.$$

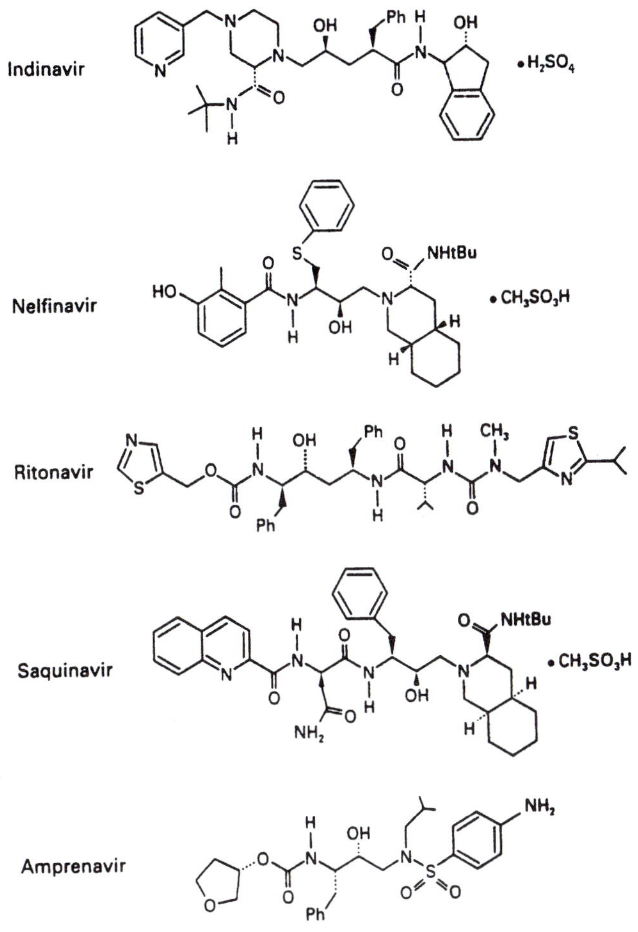

Indinavir

Nelfinavir

Ritonavir

Saquinavir

Amprenavir

FIGURE 6-13
Structures of HIV protease inhibitors. NHtBu denotes an amino-tertiary butyl and Ph denotes a phenyl group.

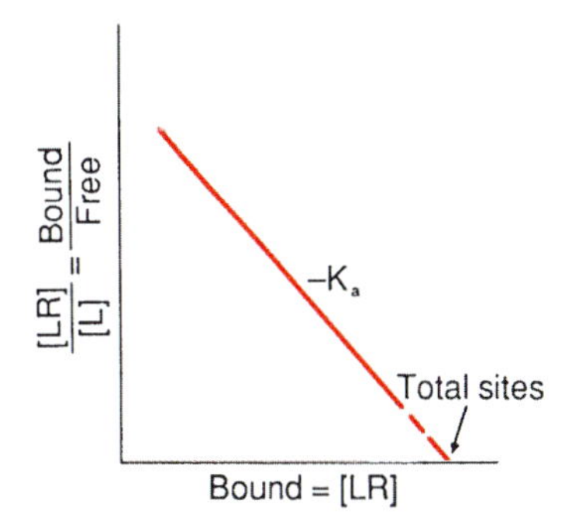

FIGURE 6-14
Scatchard plot of ligand–receptor interaction.

The total number of receptor binding sites, [TR], is equal to the sum of the unbound receptor sites, [R], and the bound receptor sites, [LR]. Therefore,

$$[R] = [TR] - [LR] \tag{6.10}$$

Substituting for [R] in Equation (6.8) yields

$$K_a = \frac{[LR]}{[L]([TR] - [LR])} \tag{6.11}$$

which on rearrangement gives

$$\frac{[LR]}{[L]} = K_a([TR] - [LR]) \tag{6.12}$$

where

$$\frac{[LR]}{[L]} = \frac{\text{concentration of bound ligand}}{\text{concentration of unbound ligand}}$$

A plot of [LR]/[L] versus [LR] yields a straight line and is known as a ***Scatchard plot*** (Figure 6-14). The slope of the line gives the value $(-K_a)$, while the intersection of the line with the abscissa yields the value for the total number of sites [TR]. Comparing this relationship to the Michaelis–Menten analysis of enzyme kinetics shows that $1/K_a$ (where 50% of the sites are occupied) is equivalent to K_m and the maximum concentration (or number) of bound ligand sites is equivalent to V_{max}. The Scatchard plot is similar to the Eadie–Hofstee plot (Figure 6-5), but the axes are reversed. These principles have been applied in the assay of ligands in human biological fluids by the use of either specific receptor proteins (radioreceptor assay) or specific binding proteins, namely, antibodies (radioimmunoassay (RIA)). In these assays, the unlabeled ligand (the quantity of which is to be determined) competes with a predetermined amount of added radioactive ligand to bind a limited amount of specific receptor protein. This step is followed by separation of protein-bound ligand from the unbound (free) ligand and measurement of radioactivity of each fraction. Separation of the protein-bound ligand from the free ligand is accomplished by protein precipitation, adsorption of the free form, chromatography, or electrophoresis. Under standard conditions of time and temperature, the amount of radioactivity found in the protein-bound ligand fraction is inversely proportional to the concentration of unlabeled ligand and the amount of radioactivity found in the unbound (free) ligand fraction is directly proportional to unlabeled ligand concentrations. This procedure has been widely used for the assay of hormones, vitamins, drugs, and other compounds. Since the concentrations of hormones in body fluids are very low (10^{-8} to 10^{-12} M), assay sensitivity is increased by ligands with high specific radioactivity.

6.6 Mechanisms of Enzyme Action

The mechanism of a reaction catalyzed by an enzyme is a detailed description of the chemical interactions occurring among the substrates, enzymes, and cofactors.

For example, although the overall reaction catalyzed by acetylcholinesterase consists of hydrolysis of acetylcholine to choline and acetic acid, the detailed mechanism is a two-step displacement reaction in which an alcohol (choline) is produced first, followed by an acid (acetic acid). The same is true for the rhodanese-catalyzed reaction. The types of experiments that are carried out to arrive at the description of an enzyme mechanism are as follows.

Isotope Exchange If a particular atom of the substrate or product is labeled with an isotope other than the natural one, its location at various intermediate reaction steps can be followed, providing information on what is happening to that part of the molecule.

Irreversible Inhibition Studies If an enzyme is inactivated by a reagent that reacts with a specific amino acid residue, then one (or more) of the reagent-specific amino acid residues is involved in catalysis by that enzyme.

pH Dependence Usually only one form (protonated or unprotonated) of an acidic or basic residue is involved in a catalysis.

Structural studies, particularly those that elucidate the three-dimensional relationships of the amino acid residues in the enzyme, permit judgments about whether the involvement of one or the other side chain is physically and sterically possible. Such techniques include nuclear magnetic resonance, electron spin resonance, single-crystal x-ray diffraction, and cross-linking studies. The important question to be answered by mechanistic studies is how an enzyme catalyzes a particular reaction so rapidly. Two general factors seem to be involved.

1. Because all the reactants are bound to a small region on the enzyme, their effective concentrations are greatly increased; i.e., the adjacent binding makes it easier for an effective collision to occur between reactants.
2. As a reaction proceeds, the conformation of the reacting molecules must change from that of the substrates to that of the products. Alterations in bonds, bond lengths, and bond angles are involved. The activation energy represents, at least in part, the work needed to bring about this transformation. The conformation that lies between those of the substrates and products and that has the highest energy is termed the ***transition state.*** Additionally, the binding energy of an enzyme distorts the substrates, so that they approach or attain the conformation of the transition state. Detailed structures and catalytic mechanisms are known for many enzymes, some of which are discussed elsewhere (e.g., lysozyme in Chapter 11). Serine proteinases are among the most thoroughly studied enzymes, and they share similar catalytic mechanisms (see above). However, they act on different substrates because of differences in their substrate binding sites. Chymotrypsin, trypsin, and elastase have the same amino acid sequence around the active serine residue (*),

   –Gly–Asp–Ser*–Gly–Gly–Pro–

   and their tertiary structures, as revealed by x-ray studies, are similar. Their amino acid sequences are 40% identical, leading to the hypothesis that these three enzymes may have evolved from a common ancestral gene. Duplication of this gene probably occurred several times during evolution and, with mutation, eventually yielded genes that now code for these three proteinases, each with its own substrate specificity. This type of evolution is known as ***divergent evolution.*** Chymotrypsin and subtilisin provide an example of ***independent convergent evolution.*** Subtilisin is a serine proteinase of bacterial origin whose mechanism of action is similar to that of chymotrypsin, but it differs from chymotrypsin in the amino acid sequence around the active serine residue:

   –Gly–Thr–Ser*–Met–Ala–Ser–
   Subtilisin
   –Gly–Asp–Ser*–Gly–Gly–Pro–
   Chymotrypsin

   Chymotrypsin and subtilisin also differ in their amino acid sequences, number of disulfide bridges (chymotrypsin has five, whereas subtilisin has none), and overall three-dimensional structures. The striking difference in structure and common catalytic mechanism are taken as evidence of an independent but convergent evolutionary process.

Coenzymes, Prosthetic Groups, and Cofactors

Many enzymes require the presence of low-molecular-weight nonprotein molecules. These small molecules may be bound to the enzyme by a covalent, tight, or noncovalent linkage. Prosthetic groups are usually bound by covalent or tight linkages and coenzymes by noncovalent linkages. Cofactors usually are metal ions or organic compounds (some are vitamins, e.g., ascorbic acid).

Prosthetic groups and coenzymes are complex organic compounds, many of which are derived from vitamins. These compounds are recycled and are needed only in catalytic amounts to convert a large amount of reactants to products. Coenzymes function as substrates in two-substrate reactions, being bound only momentarily to the enzyme during catalysis. They are chemically altered during the reaction and are reconverted to their original forms by the same or another enzyme (e.g., NAD^+- or $NADP^+$-dependent dehydrogenase reactions). Prosthetic groups form part of the active center and undergo cyclic chemical changes during the reaction (e.g., pyridoxal phosphate in aminotransferases). Principal coenzymes, prosthetic groups, and cofactors are listed below. Their metabolic roles are discussed elsewhere, e.g., with the reactions they catalyze and the vitamins from which they are derived.

1. Compounds involved in H-transfer reactions and derived from vitamin precursors.
   a. ***Nicotinamide adenine dinucleotide*** (NAD^+), also known as diphosphopyridine nucleotide (DPN^+). The reduced form is NADH (DPNH).
   b. ***Nicotinamide adenine dinucleotide phosphate*** ($NADP^+$), also known as triphosphopyridine nucleotide (TPN^+). The reduced form is NADPH (TPNH).
   These compounds are derived from the vitamin niacin (nicotinic acid, nicotinamide) and require it for their synthesis. Small amounts of niacin are derived from the essential amino acid tryptophan.
   c. ***Flavin mononucleotide*** (FMN). The reduced form is $FMNH_2$.
   d. ***Flavin adenine dinucleotide*** (FAD). The reduced form is $FADH_2$.
   These compounds are derived from the vitamin riboflavin. They are not true nucleotides, because the flavin ring system is attached to ribitol, a sugar alcohol, rather than ribose.
2. Compounds involved in H-transfer reactions and derived from nonvitamin precursors.
   a. ***Lipoic acid*** apparently can be synthesized in sufficient quantity by humans. It is also involved in acyl group transfer during oxidative decarboxylation of α-keto acids.
   b. ***Biopterin*** (a pteridine-containing compound) participates in certain hydroxylase reactions (e.g., phenylalanine hydroxylase).
   c. ***Coenzyme Q*** (CoQ, ubiquinone) is a group of closely related compounds differing only in the length of the side chain. They can be synthesized in humans from farnesyl pyrophosphate, an intermediate in cholesterol biosynthesis. Although the structure of coenzyme Q is similar to that of vitamin K_1, no interrelationship has been established.
3. Compounds involved in group transfer reactions and derived from vitamin precursors.
   a. ***Coenzyme A*** (CoA, CoASH) takes part in acetyl and other acyl group transferase and requires the vitamin pantothenic acid for its synthesis.
   b. ***Thiamine pyrophosphate*** (TPP, cocarboxylase) is used for oxidative decarboxylation of α-keto acids and in the transketolase-catalyzed steps of the pentose phosphate pathway. It is derived from thiamine (vitamin B_1).
   c. ***Pyridoxal phosphate*** is a cofactor for a wide variety of types of reactions on amino acids, namely, racemization, decarboxylation, transamination, elimination of water or hydrogen sulfide. It is derived from pyridoxine, pyridoxal, and pyridoxamine (collectively called vitamin B_6).
   d. ***Tetrahydrofolic acid*** (FH_4) is a carrier of one-carbon fragments such as formyl, methylene, methyl, and formimino groups. It is derived from folic acid (folacin).
   e. ***Biotin*** is vitamin tightly bound to the apoenzyme in an amide linkage to the ε-amino group of a lysyl residue and involved in carboxylation reactions.
   f. ***Cobamide coenzyme*** (5,6-dimethylbenzimidazole cobamide 5′-dehydroadenosyl cobalamine) contains cobalt bound in a porphyrin-like ring system and is unusual in that it contains a cobalt–carbon (organometallic) bond. It is involved in methyl-transfer reactions and participates in the 1,2 shift of a hydrogen atom accompanied by a 2,1-shift of an alkyl, carboxyl, hydroxyl, or amino group. Very small amounts, derived from cyanocobalamin (vitamin B_{12}), are required by the body.
   g. ***Vitamin K*** participates in carboxylation reactions of glutamic acid residues of some proteins involved in blood coagulation and its regulation to yield γ-carboxyglutamic acid residues required for Ca^{2+} binding. The vitamins K are derived from naphthoquinones with isoprenoid side chains of varying lengths.
   h. ***Vitamin C*** participates as a cofactor in the hydroxylation reactions of proline and lysine of collagen.
4. Compounds involved in group transfer reactions and derived from nonvitamin precursors.
   a. ***Adenosine triphosphate*** (ATP) can be a donor of phosphate, adenosine, and adenosine monophosphate (AMP) for various purposes. The exergonic hydrolysis reaction of ATP provides the energy for driving thermodynamically unfavorable reactions.

b. ***Cytidine diphosphate*** (CDP) is a carrier of phosphorylcholine, diacylglycerols, and other molecules during phospholipid synthesis.
c. ***Uridine diphosphate*** (UDP) is a carrier of monosaccharides and their derivatives in a variety of reactions (see bilirubin, lactose, galactose and mannose metabolism, glycogen synthesis, and other pathways).
d. ***Phosphoadenosine phosphosulfate*** (PAPS, "active sulfate") is a sulfate donor in the synthesis of sulfur-containing mucopolysaccharides as well as in the detoxification of sterols, steroids, and other compounds (see metabolism of the sulfur-containing amino acids in Chapter 17). PAPS is derived from ATP and inorganic sulfate.
e. ***S-Adenosylmethionine*** (SAM, "active methionine") is a methyl group donor in biosynthetic reactions. It is formed from ATP and the essential amino acid methionine.
f. ***Heme*** proteins, containing the iron–protoporphyrin group, participate in oxygen transport (hemoglobin), oxygen storage (myoglobin), electron transport (cytochromes), hydrogen peroxide inactivation (catalase, peroxidase), hydroxylases, oxygenases, and other processes.

5. Metal cofactors.
   a. Mg^{2+} is required by most enzymes that use ATP. The active form of ATP is a Mg^{2+}–ATP^{4-} complex.
   b. Ca^{2+} is involved in a wide variety of processes, notably muscle contraction, blood clotting, nerve impulse transmission, and cAMP-mediated processes.
   c. Fe^{2+}/Fe^{3+} (***ferrous/ferric iron***) is required as heme (see above) or nonheme iron in the functioning of many proteins.
   d. Cu^{+}/Cu^{2+} (***cuprous/cupric copper***). Cytochrome oxidase in mitochondrial electron transport contains Fe and Cu. Tyrosinase and lysyl oxidase also require Cu. Superoxide dismutase contains both Cu and Zn.
   e. Zn^{2+} (***zinc ion***) is used by lactic acid dehydrogenases, alcohol dehydrogenases, carbonic anhydrase, carboxypeptidase A, DNA polymerase, superoxide dismutase, and others.
   f. Mo^{6+} (***molybdenum ion***). Xanthine oxidase contains Mo and Fe.
   g. Mn^{2+} (***manganous ion***) is required by acetyl-CoA carboxylase, deoxyribonuclease (Mg^{2+} can replace Mn^{2+} in this case), arginase, and other enzymes.
   h. ***Se*** (***selenium***) is required for the functioning of glutathione peroxidase.

Metal cofactors in enzymes may be bound reversibly or firmly. Reversible binding occurs in metal-activated enzymes (e.g., many phosphotransferases); firm (or tight) binding occurs in metalloenzymes (e.g., carboxypeptidase A). Metals participate in enzyme catalysis in a number of different ways. An inherent catalytic property of a metal ion may be augmented by the enzyme protein, or metal ions may form complexes with the substrate and the active center of the enzyme and promote catalysis, or metal ions may function in electron transport reactions between substrates and enzymes.

Supplemental Readings and References

J. Brent, K. McMartin, S. Phillips, et al.: Fomepizole for the treatment of ethylene glycol poisoning. *New England Journal of Medicine* **340,** 832, (1999).

K. K. Burkhart and K. W. Kulig: The other alcohols. Methanol, ethylene glycol, and isopropanol. *Emergency Medicine Clinics of North America* **8,** 913 (1990).

W. D. N. Chin, R. Barnett, and G. Sheehan: Methanol poisoning. *Intensive Care Medicine* **18,** 391 (1992).

C. Flexner: HIV-protease inhibitors, *New England Journal of Medicine* **338,** 1281 (1998).

HIV-1 protease inhibitors: A review for clinicians. *Journal of American Medical Association* **277,** 145 (1997).

G. V. Sherbet and M. S. Lakshmi: *The Genetics of Cancer*, Academic Press, San Diego (1997).

P. Houeto, J. R. Hofman, M. Imbert, P. Levillain, and F. J. Baud: Relation of blood cyanide to plasma cyanocobalamin concentration after a fixed dose of hydroxycolalamin in cyanide poisoning. *Lancet* **346,** 605 (1995).

D. Jacobsen: The treatment for ethylene glycol poisoning. *New England Journal of Medicine*. **340,** 879 (1999).

H. A. James and I. Gibson: The therapeutic potential of ribozymes, *Blood* **91,** 371 (1998).

J. A. Kruse and P. Cadnapapnornchai. The serum osmole gap. *Journal of Clinical Care* **9,** 185 (1994).

CHAPTER 7

Enzymes II: Regulation

The metabolic rate of key substances, which can proceed in multiple pathways, is regulated and integrated. A close interrelationship exists among products formed by different metabolic pathways from a common metabolite. For example, glucose can be consumed either by oxidation to CO_2 or by conversion to glycogen, lipid, nonessential amino acids, or other sugar molecules. The glucose supply of the body can be derived either from the diet or from the breakdown of glycogen, a polymer of glucose (primarily from the liver and the kidney), or it can be synthesized from some amino acids or lactate (predominantly in the liver). These processes of glucose utilization and synthesis are under tight regulation. In fact, the plasma glucose level is maintained at the level at which tissues (e.g., brain, erythrocytes, kidney medulla, and the lens and cornea of the eye) that require glucose as a primary substrate are not deprived of this essential fuel. Each of these metabolic pathways is mediated by enzymes that are unique for a given pathway and that are under control. Metabolic pathways do not usually compete with each other for utilization of a substrate, and they operate only to serve a particular physiological need or function.

7.1 Types of Regulation

A metabolic pathway involves many enzymes functioning in a sequential manner or in some unique arrangement to carry out a particular metabolic process. Control of a pathway is accomplished through modulation of the activity of only one or a few key enzymes. These ***regulatory enzymes*** usually catalyze the first or an early reaction in a metabolic sequence. A regulatory enzyme catalyzes a ***rate-limiting*** (or rate-determining) chemical reaction that controls the overall pathway. It may also catalyze a chemical reaction unique to that pathway, which is known as a ***committed step.*** In the metabolic pathway for the formation of E from A,

$$A \overset{E_1}{\rightarrow} B \overset{E_2}{\rightarrow} C \overset{E_3}{\rightarrow} D \overset{E_4}{\rightarrow} E$$

the conversion of A to B, catalyzed by the enzyme E_1, is the rate-limiting step and also a committed step. The rate-limiting step need not be the same as the committed step. In the branched metabolic pathway,

$$A \xrightarrow{E_1} B \xrightarrow{E_2} C \xrightarrow{E_3} D \xrightarrow{E_4} E$$
$$B \xrightarrow{E_5} L \xrightarrow{E_6} M \xrightarrow{E_7} N$$

if the conversion of A to B is the rate-limiting step, the committed step in the pathway for the formation of N is the conversion of B to L ($B \rightarrow L$), catalyzed by the enzyme E_5. Those enzymes which catalyze the rate-limiting step or the committed step of a pathway are under regulation. When the end product exceeds the steady-state level concentration, it inhibits the regulatory enzyme in an attempt to normalize the overall process. This type of control, known as ***feedback inhibition*** (see below), ensures a high degree of efficiency in the utilization of materials and of energy in living systems.

Regulation may be achieved in other ways. The absolute amount of a regulatory enzyme may be altered through mechanisms that control gene expression (Chapter 26). This regulation at the genetic level occurs during various phases of reproduction, growth, and development, with different metabolic pathways being turned on or off in accordance with the special requirements of each phase. In eukaryotic cells, regulation at the genetic level is relatively a long-term process. Several short-term regulatory mechanisms control metabolic activity rapidly (see below). Both substrates and some hormones play significant roles in regulating the concentration of key enzymes at this level. Many drugs or other chemicals can increase levels of enzymes that affect their own metabolism. Thus, phenobarbital and polycyclic hydrocarbons increase the levels of microsomal enzyme systems involved in their metabolism.

Regulation of metabolic processes can be accomplished by other methods. One is the use of a ***multienzyme complex*** (e.g., pyruvate dehydrogenase complex or fatty acid synthase complex) in which various enzymes are organized such that the product of one becomes the substrate for an adjacent enzyme. A single polypeptide chain may contain multiple catalytic centers that carry out a sequence of transformations (e.g., the mammalian fatty acid synthase; see Chapter 18). Such multifunctional polypeptides increase catalytic efficiency by abolishing the accumulation of free intermediates and by maintaining a stoichiometry of 1:1 between catalytic centers.

Another type of regulation is accomplished by a series of ***proenzymes*** in which activation of the initial proenzyme by a biological signal activates the second proenzyme which, in turn, activates the third, and so on. Such an enzyme cascade process provides great amplification in terms of the amount of final product formed. Examples are blood coagulation, the dissolution of blood clots, complement activation, and glycogen breakdown.

Regulation is also accomplished by compartmentalization of enzyme systems involved in anabolic and catabolic pathways into different cell organelles. For example, fatty acid synthesis occurs in the soluble fraction of the cytoplasm, whereas fatty acid oxidation takes place in mitochondria. Heme synthesis begins and is completed in mitochondria, but some of the intermediate reactions take place in the cytosol. Heme catabolism is initiated in the smooth endoplasmic reticulum. Transport of key metabolites across an organelle membrane system is also a form of regulation.

Many enzymes occur in several molecular forms called ***isoenzymes*** (or isozymes), and those which are genetically determined may be called primary isoenzymes. The different primary isoenzymes catalyze the same chemical reaction but may differ in their primary structure and kinetic properties. The tissue distribution of isoenzymes imparts distinctive properties and specific patterns of metabolism to organs of the body. The presence of isoenzymes may reveal differences not only between organs but also between cells that make up an organ or between organelles of a cell. During different stages of differentiation and development from embryonic life to adulthood, the isoenzyme distribution in an organ undergoes characteristic changes. When an adult organ reverts to the embryonic or fetal state (e.g., in cancer), the isoenzyme distributions change to those characteristic of that developmental state. The existence of isoenzymes in human tissues has important implications in the study of human disease.

Zymogen (e.g., trypsinogen and chymotrypsinogen) synthesis, secretion, transport, and activation and the rate of inactivation of the active enzyme by inhibitors may all be considered means of enzyme regulation.

Enzyme activity can be regulated by covalent modification or by noncovalent (allosteric) modification. A few enzymes can undergo both forms of modification (e.g., glycogen phosphorylase and glutamine synthetase). Some covalent chemical modifications are phosphorylation and dephosphorylation, acetylation and deacetylation, adenylylation and deadenylylation, uridylylation and deuridylylation, and methylation and demethylation. In mammalian systems, phosphorylation and dephosphorylation are most commonly used as means of metabolic control. Phosphorylation is catalyzed by protein kinases and occurs at specific seryl (or threonyl) residues and occasionally at tyrosyl residues; these amino acid residues are not usually part of the catalytic site of the enzyme. Dephosphorylation is accomplished by phosphoprotein phosphatases:

$$\text{Enzyme–Ser–OH} \; \underset{\text{Phosphoprotein phosphatase} \;(H_2O \rightarrow HO\text{–}P(=O)(O^-)\text{–}O^-)}{\overset{\text{ATP} \rightarrow \text{ADP},\ \text{Protein kinase}}{\underset{Mg^{2+}}{\rightleftharpoons}}} \; \text{Enzyme–Ser–O–}\overset{\overset{\text{O}}{\|}}{\underset{\underset{\text{O}^-}{|}}{\text{P}}}\text{–O}^-$$

The overall process of phosphorylation and dephosphorylation is a cascade of reactions and consists of an extracellular signal, commonly referred to as ***first messenger*** (e.g. hormones, such as glucagon); a specific receptor on the cell membrane of the target cell; a transducer coupled to the receptor that produces an intracellular signal (i.e., a ***second messenger***); one or more modifying enzymes whose activity is affected by the intracellular signal; a target enzyme, which is the substrate for the modifying enzyme and subjected to covalent modification with consequent metabolic alteration; and an enzyme that reverses the modification of the target enzyme. The transducer of the system is adenylate cyclase, and it catalyzes the cyclization of ATP to cyclic AMP (cAMP), the second messenger. The modifying enzyme is a protein kinase activated by cAMP. Other protein kinases are activated by non-cAMP-dependent or Ca^{2+}-dependent systems (Chapter 30). Phosphorylation of a target enzyme may be either stimulatory or inhibitory. For example, phosphorylation converts glycogen phosphorylase to an activated form and glycogen synthase to an inactivated form, thus preventing the simultaneous occurrence of glycogen breakdown and synthesis (Chapter 15).

7.2 Allosteric Enzyme Regulation

Those enzymes in metabolic pathways whose activities can be regulated by noncovalent interactions of certain compounds at sites other than the catalytic are known as ***allosteric enzymes.*** They are usually rate-determining enzymes and play a critical role in the control and integration of metabolic processes. The term "allosteric" is of Greek origin, the root word "allos" meaning "other." Thus, an ***allosteric site*** is a unique region of an enzyme *other* than the substrate binding site that leads to catalysis. At the allosteric site, the enzyme is regulated by noncovalent interaction with specific ligands known as ***effectors, modulators,*** or ***modifiers.***

The properties of allosteric enzymes differ significantly from those of nonregulatory enzymes. Ligands (in some instances even the substrate) can bind at such sites by a cooperative binding process. Cooperativity describes the process by which binding of a ligand to a regulatory site affects binding of the same or of another ligand to the enzyme. Allosteric enzymes have a more complex structure than nonallosteric enzymes and do not follow Michaelis–Menten kinetics. An allosteric site is specific for its ligand, just as the active site is specific for its substrate. Binding of an allosteric modulator causes a change in the conformation of the enzyme (see below) that leads to a change in the binding affinity of the enzyme for the substrate. The effect of a modulator may be positive (activatory) or negative (inhibitory). The former leads to increased affinity of the enzyme for its substrate, whereas the reverse is true for the latter. Activatory sites and inhibitory sites are separate and specific for their respective modulators. Thus, if an end product of a metabolic pathway accumulates in excess of its steady-state level, it can slow down or turn off the metabolic pathway by binding to the inhibitory site of the regulatory enzyme of the pathway. As the concentration of the end product (inhibitor) decreases below the steady-state level, the number of enzymes having bound inhibitor decreases and they revert to their active form. In this instance, the substrate and the negative modulator bear no structural resemblance. An allosteric enzyme may be positively modulated by the substrate itself or by a metabolite of another pathway that depends on production of the end product in question for its eventual utilization (e.g., pathways of synthesis of purine and pyrimidine nucleotides in the formation of nucleic acids; see Chapter 27).

Most allosteric enzymes are ***oligomers*** (i.e., they consist of two or more polypeptide chains or subunits). The subunits are known as ***protomers.*** Two types of interaction occur in allosteric enzymes: ***homotropic*** and ***heterotropic.*** In a homotropic interaction, the same ligand influences positively the cooperativity between different modulator sites. An example is a regulatory enzyme modulated by its own substrate. Thus, this class of enzyme has at least two substrate binding sites which respond to situations that lead to substrate excess by increasing its rate of removal. Heterotropic interaction refers to the effect of one ligand on the binding of a *different* ligand. For example, a regulatory enzyme modulated by a ligand other than its substrate constitutes a heterotropic system, in which the cooperativity can be positive or negative. Some allosteric enzymes exhibit mixed homotropic and heterotropic interactions.

Kinetics of Allosteric Proteins

The kinetic properties of allosteric enzymes vary significantly from those of nonallosteric enzymes, exhibiting cooperative interactions between the substrate, the activator, and the inhibitor sites. These properties are responsible for deviations from the classic Michaelis–Menten kinetics that apply to nonallosteric enzymes. Nonallosteric enzymes yield a rectangular hyperbolic curve when the initial velocity (v) is plotted against the substrate concentration [S]. For allosteric enzymes, a plot of v versus [S] yields curves of different shapes, including sigmoid-shaped curves in some cases. (A sigmoidal curve can result from other mechanisms.)

The v versus [S] plot for a homotropic enzyme is shown in Figure 7-1. The following features should be noted:

1. The substrate functions as a positive modulator; i.e., there is positive cooperativity between the substrate binding sites so that binding of the substrate at one

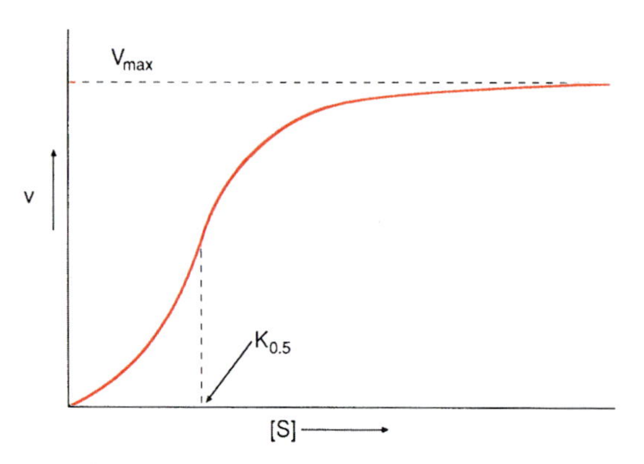

FIGURE 7-1
Relationship between the initial velocity (v) and the substrate concentration [S] for an allosteric enzyme that shows a homotropic effect. The substrate functions as a positive modulator. The profile is sigmoidal, and during the steep part of the profile, small changes in [S] can cause large changes in v. $K_{0.5}$ represents the substrate concentration corresponding to half-maximal velocity.

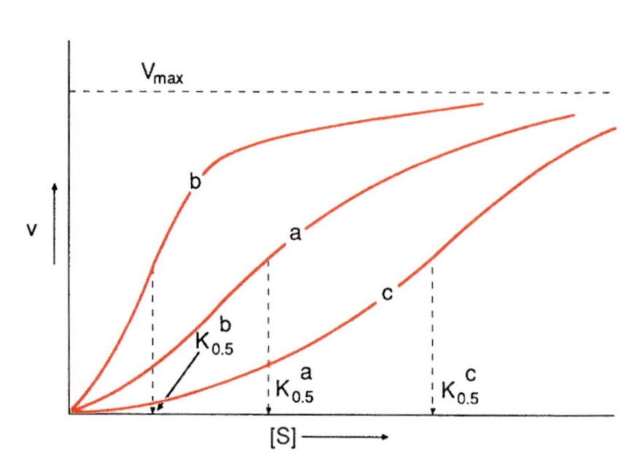

FIGURE 7-2
Relationship between the initial velocity (v) and the substrate concentration [S] for an allosteric enzyme that shows a heterotropic effect with constant V_{max} but with varying $K_{0.5}$. Curve a is obtained in the absence of any modulators, curve b in the presence of a positive modulator, and curve c in the presence of a negative modulator. Regulation is achieved by modulation of $K_{0.5}$ without change in V_{max}.

binding site greatly enhances binding of the substrate at the other sites. As the substrate concentration increases, there is a large increase in the velocity of the reaction.

2. Owing to the above effect, the shape of the curve is sigmoidal.
3. The value of the substrate concentration corresponding to half-maximal velocity is designated as $K_{0.5}$ and not K_m since the allosteric kinetics do not follow the hyperbolic Michaelis–Menten relationship.
4. Maximum velocity (V_{max}) is attainable at a rather high substrate concentration, implying saturation of the catalytic site of the enzyme.

The v versus [S] plot for heterotropic enzymes is more complex. The kinetic profiles can be divided into two classes, depending upon whether the allosteric effector alters $K_{0.5}$ and maintains a constant V_{max} or alters V_{max} and maintains a nearly constant $K_{0.5}$. The v versus [S] profile of an allosteric enzyme that follows the former set of properties is shown in Figure 7-2. In the absence of any modulators, the profile is sigmoidal (curve a). In the presence of a positive modulator (curve b), the value for $K_{0.5}$ is decreased; i.e., a lower substrate concentration is required to attain half-maximal velocity. Curve b is more hyperbolic than sigmoidal. Curve c obtained with a negative modulator is more sigmoidal than curve a, and the $K_{0.5}$ value is increased, reflecting a decreased affinity for the substrate, i.e., a higher substrate concentration is required to attain half-maximal velocity. Regulation of the enzyme is achieved through positive and negative modulators. Thus, at a given substrate concentration (e.g., steady-state level), the activity of the enzyme can be turned on or off with appropriate modulators. Figure 7-3 shows v versus [S] plots for allosteric enzymes modulated by changes in V_{max} but retaining an essentially constant $K_{0.5}$. This type of modulation is less common than the two previous cases considered. The positive modulator increases V_{max} (curve b), whereas the negative modulator decreases V_{max} (curve c).

Examples of Allosteric Proteins

We will consider two specific examples of allosteric proteins—one an enzyme and the other an oxygen transport protein.

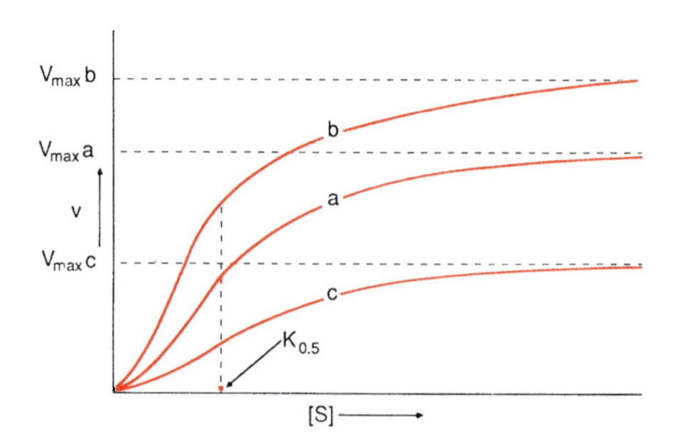

FIGURE 7-3
Relationship between the initial velocity (v) and the substrate concentration [S] for an allosteric enzyme that shows a heterotropic effect with constant $K_{0.5}$ but with varying V_{max}. Curve a is obtained in the absence of any modulator, curve b in the presence of a positive modulator, and curve c in the presence of a negative modulator. Regulation is achieved by modulation of V_{max} without change in $K_{0.5}$.

1. ***Aspartate transcarbamoylase*** (ATCase) is an allosteric enzyme of the bacterium *Escherichia coli,* which has been extensively studied. This enzyme catalyzes the transfer of the carbamoyl group from carbamoyl phosphate to the α-amino group of aspartate:

$$H_2N{-}\overset{O}{\overset{\|}{C}}{-}O{-}\overset{O}{\overset{\|}{P}}(O^-){-}O^- + {}^+H_3N{-}\overset{H}{C}(CH_2COO^-){-}COO^-$$

Carbamoyl phosphate Aspartate

$$\longrightarrow \; HO{-}\overset{O}{\overset{\|}{P}}(O^-){-}O^- \text{ (Inorganic phosphate } (P_i)) \; + \; H_2N{-}\overset{O}{\overset{\|}{C}}{-}NH{-}\overset{H}{C}(CH_2COO^-){-}COO^- + H^+$$

N-Carbamoyl aspartate

This committed reaction, catalyzed by a regulatory enzyme, is the second step (Figure 7-4) in the biosynthesis of pyrimidines. N-carbamoylaspartate ultimately is converted to the pyrimidine nucleotide cytidine triphosphate (CTP), which is the negative modulator of the enzyme. That is, as the CTP concentration increases, it inhibits the ATCase by decreasing its affinity for substrates. The preceding is an example of an allosteric feedback inhibition. ATCase has a molecular weight of 310,000 and consists of 12 polypeptide chains (subunits). The catalytic and the regulatory sites reside on two different sets of subunits: six larger subunits (M.W. 33,000) have catalytic sites, and six smaller subunits (M.W. 17,000) have regulatory sites. X-ray crystallographic studies of ATCase by Lipscomb showed that the molecule is roughly triangular. The six catalytic subunits (C) are present as two trimers (C_3), one above the other (but not exactly parallel), with the dimeric regulatory subunits (R_2) forming an equatorial belt around the edges of the trimers (Figure 7-5). In the center of the molecule, an aqueous cavity is accessible through several openings. ATCase can be dissociated into two unequal clusters with different subunit arrangements by treatment with mercurials (e.g., *p*-hydroxymercuribenzoate). The larger cluster containing three catalytic subunits (C_3)

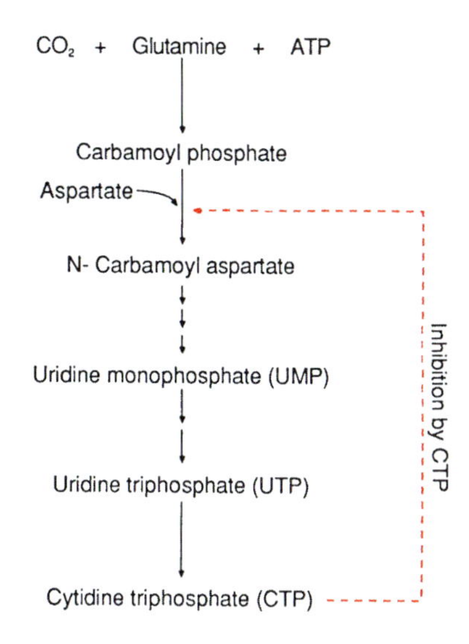

FIGURE 7-4
Regulation of pyrimidine nucleotide biosynthesis in *E. coli*.

is catalytically active. The v versus [S] plot exhibits a hyperbolic profile and is insensitive to CTP. The smaller cluster is a dimer of regulatory subunits (R_2) and, as expected, shows no catalytic activity but does bind to CTP. The native enzyme can be reconstituted with return of the allosteric activity by removal of the mercurial compound and mixing of both catalytic and regulatory clusters:

$$\underset{\text{Catalytic cluster}}{2C_3} + \underset{\text{Regulatory cluster}}{3R_2} \rightarrow \underset{\text{Native enzyme}}{R_6C_6}$$

Reconstitution also requires zinc ions (Zn^{2+}), since the native enzyme contains six atoms of Zn^{2+}. C_3 and R_2 clusters can be dissociated further into their respective, inactive monomeric subunits by strong denaturing agents (Figure 7-5).

ATCase can assume different conformations depending on whether it is active or inactive. In the presence of the substrate or positive modulator, it is in a catalytically more active conformation known as the ***relaxed state*** or ***R-state.*** In the presence of the negative modulator, the enzyme is in less active conformation known as the ***taut state*** or ***T-state.*** The allosteric kinetic effects of ATCase are shown in Figure 7-6. The interaction of substrates with the enzyme is cooperative (an example of homotropic cooperativity), as indicated by the sigmoidal shapes of the v versus [S] plots, CTP being an inhibitor and ATP an activator. These modulators compete for the same regulatory site and modulate the affinity of the enzyme for its

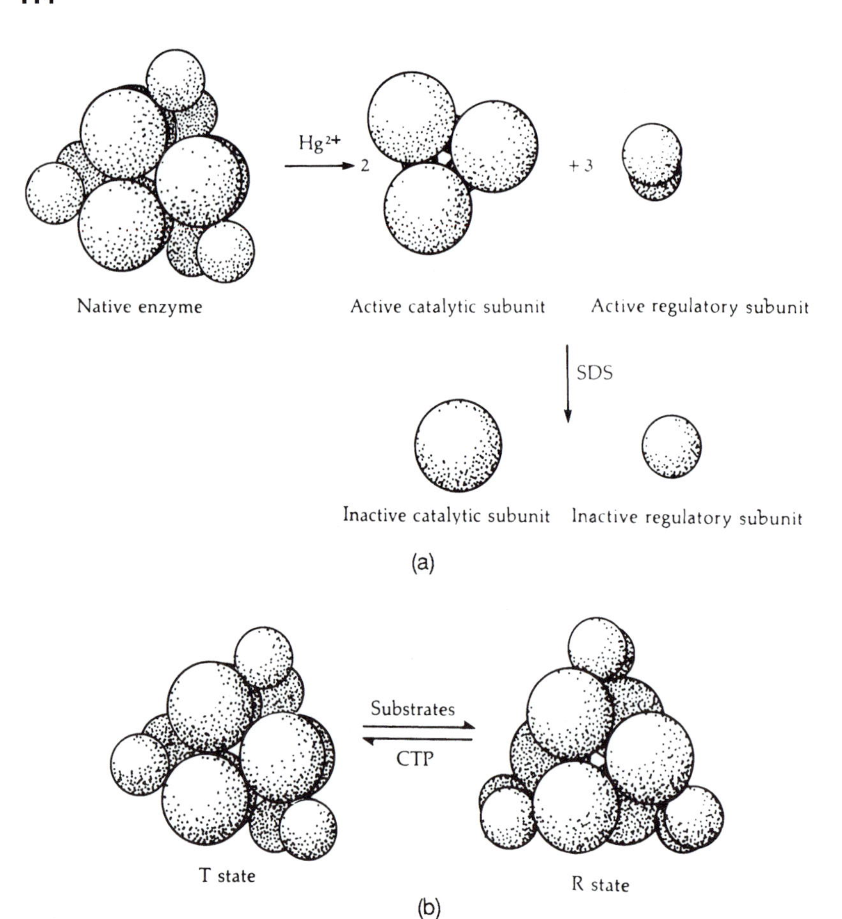

FIGURE 7-5
Schematic representation of the subunit structure of aspartate transcarbamoylase and its dissociation into catalytic and regulatory subunits by mercurials, which can be further converted to inactive monomeric subunits by strong denaturing agents (e.g., sodium dodecyl sulfate). The native enzyme consists of two catalytic trimers placed one above the other, along with three dimeric regulatory subunits surrounding the catalytic trimers in an equatorial plane (a). Substrate maintains the enzyme in the catalytically more active relaxed (R) conformation, while cytidine triphosphate maintains it in the catalytically less active taut (T) conformation (b). [Reproduced, with permission from E. L. Smith, R. L. Hill, I. R. Lehman, et al: *Principles of Biochemistry: General Aspects,* 7th ed. McGraw-Hill, New York, 1983].

substrates, without altering V_{max}. The biological significance of the activation by the purine nucleotide ATP can be appreciated, since both ATP and CTP are eventually utilized in the biosynthesis of double-helical DNA, which contains equal amounts of purines and pyrimidines. Thus, modulation of ATCase activity equalizes the rates of formation of purine and pyrimidine nucleotides. Other regulatory and metabolic aspects of purines and pyrimidines are discussed in Chapter 27.

2. Hemoglobin binding of oxygen is a classic example of the homotropic effect. Hemoglobin also shows heterotropic effects with specific molecules in its environment. These effects are intimately related to the function of hemoglobin as a carrier not only of oxygen but of H^+ and CO_2 (Chapters 1 and 28). The heterotropic modulators are H^+, CO_2, and

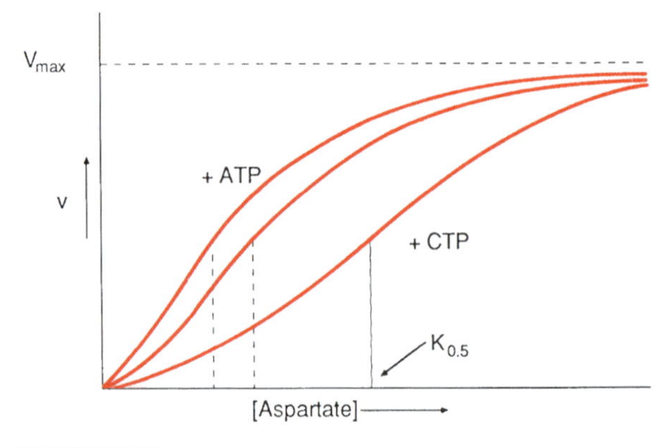

FIGURE 7-6
Relationship between the initial velocity (v) of aspartate transcarbamoylase and the substrate concentration. Note that ATP is a positive allosteric modulator, which causes decreased $K_{0.5}$, whereas CTP is a negative allosteric modulator, which causes increased $K_{0.5}$.

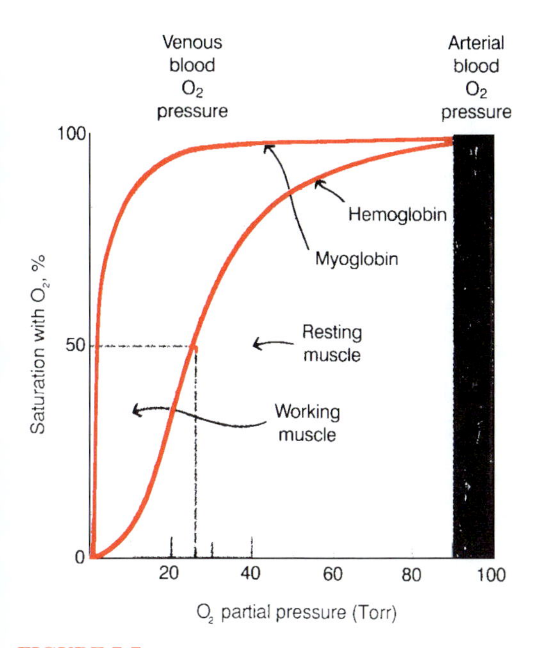

FIGURE 7-7
Profiles of oxygenation of myoglobin and hemoglobin as a function of partial pressure of oxygen. Myoglobin shows a typical Michaelis–Menten type of rectangular hyperbolic saturation curve, whereas hemoglobin shows a sigmoidal saturation curve, consistent with its allosteric properties. Myoglobin at any partial pressure of oxygen has much higher affinity for oxygen than does hemoglobin. [Reproduced with permission from A. Lehninger, *Principles of Biochemistry.* Worth, New York, 1982.]

2,3-bisphosphoglycerate (BPG or DPG). The cooperativity of oxygen binding to hemoglobin and the alterations of hemoglobin by various ligands provide the most extensively investigated molecular regulation of a biological process.

Hemoglobin carries oxygen from the lungs to the tissues and carries CO_2 and H^+ back from the tissues to the lungs (Chapter 1), whereas myoglobin functions as an oxygen store in muscle. Consistent with its function, myoglobin has a higher affinity for oxygen at any partial pressure of oxygen than does hemoglobin (Figure 7-7). Thus, oxygen can be transferred easily from hemoglobin to myoglobin. Hemoglobin is a ***tetramer*** consisting of two different subunit types (e.g., α and β in hemoglobin A). Each polypeptide contains one heme group (an iron–porphyrin prosthetic group) that binds to one oxygen molecule by a cooperative process. Myoglobin, a monomeric protein with one heme group, remains monomeric under a wide range of concentrations and does not show cooperative binding with oxygen. The polypeptides of myoglobin and hemoglobin exhibit many differences with respect to their primary structures. For example, the many amino acid residues present on the surface of myoglobin are polar, whereas many of those in the individual hemoglobin polypeptide chains are capable of participating in nonpolar interactions with other subunits. Despite these differences, critical regions are conserved in the polypeptide chains of myoglobin and hemoglobin, namely, the proximal and distal histidyl residues that interact with the heme iron, the hydrophobic amino acid residues that surround the heme group, and certain prolyl residues that interrupt the helical regions to allow the chain to fold back upon itself. The region of the polypeptide chain in contact with the heme group is known as the heme pocket. The amino acid residues in this pocket maintain the heme iron in the divalent state, which is the functional oxidation state of iron in both myoglobin and hemoglobin. Thus, the single polypeptide chain of myoglobin and the two different chains of hemoglobin are remarkably similar in secondary and tertiary structures (Figure 7-8). These similarities support the hypothesis that myoglobin and hemoglobin evolved by gene duplication and subsequent mutation from a common ancestral oxygen-binding heme protein.

The binding of oxygen to myoglobin is not cooperative, but the binding of oxygen to hemoglobin is cooperative. This difference can be accounted for kinetically by considering the equilibrium for dissociation of oxymyoglobin (MbO_2) to deoxymyoglobin (Mb) and oxygen (O_2):

$$MbO_2 \rightleftharpoons Mb + O_2 \tag{7.1}$$

The equilibrium constant,

$$K_d = \frac{[Mb][O_2]}{[MbO_2]} \tag{7.2}$$

is expressed in moles per liter, and its value depends on pH, ionic strength, and temperature. Since myoglobin has a single oxygen binding site, a single equilibrium defines the dissociation of oxymyoglobin.

So that we can deal with measurable parameters, Equation (7.2) needs to be modified by the introduction of two terms, Y and P_{50}. Y is defined as the fractional saturation of myoglobin, e.g., when $Y = 0.3$, 30% of the available sites on the myoglobin are occupied by oxygen. Thus,

$$Y = \frac{\text{number of binding sites occupied by } O_2}{\text{total number of binding sites available for binding } O_2}$$

or

$$Y = \frac{[MbO_2]}{[Mb] + [MbO_2]} \tag{7.3}$$

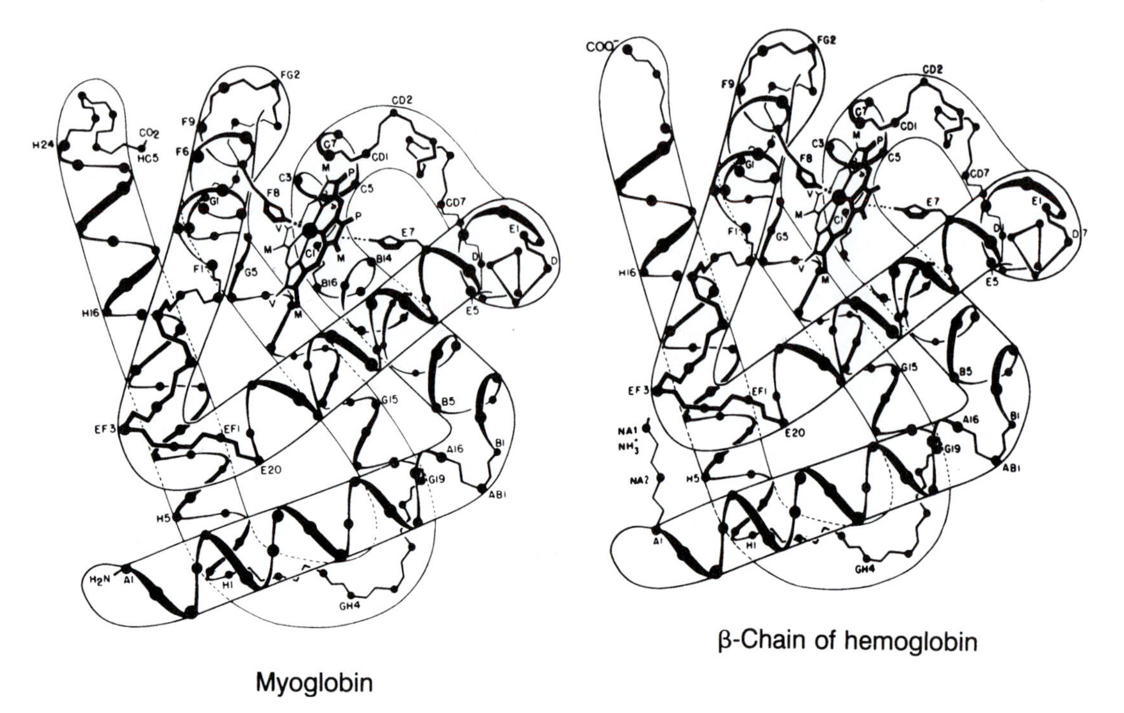

FIGURE 7-8
The remarkable similarity in the conformations of myoglobin and of the β-chain of hemoglobin. [Reproduced with permission from A. Fersht, *Enzyme Structure and Mechanism.* (W. H. Freeman, New York, 1977.]

Substituting Equation (7.2) into Equation (7.3), we obtain

$$Y = \frac{[O_2]}{[O_2] + K_d} \tag{7.4}$$

Because O_2 is a gas, it is convenient to express $[O_2]$ in terms of its partial pressure in units of Torr (or mm Hg; multiply by 0.133 to obtain kilopascals). Therefore,

$$Y = \frac{P_{O_2}}{P_{O_2} + K_d} \tag{7.5}$$

Now we can substitute for K_d the term P_{50}, which is defined as the partial pressure of oxygen at which 50% of the sites are occupied (i.e., when $Y = 0.5$), because this situation is analogous to the Michaelis–Menten treatment of enzyme kinetics.

$$Y = \frac{P_{O_2}}{P_{O_2} + P_{50}} \tag{7.6}$$

A plot of Y versus P_{O_2} yields an oxygen saturation profile that is a rectangular hyperbola (Figure 7-9), indicating that the binding of oxygen to myoglobin is noncooperative. Equation (7.6) can be rearranged to yield a linear plot as follows:

$$\frac{Y}{1 - Y} = \frac{P_{O_2}}{P_{50}} \tag{7.7}$$

Taking the logarithms of both sides of Equation (7.7) yields

$$\log\left(\frac{Y}{1 - Y}\right) = \log P_{O_2} - \log P_{50} \tag{7.8}$$

Equation (7.8) is known as the ***Hill equation.*** A plot of $\log(Y/1 - Y)$ versus $\log P_{O_2}$ yields a straight line with a slope of 1 (the Hill coefficient) (Figure 7-10). Thus, a value

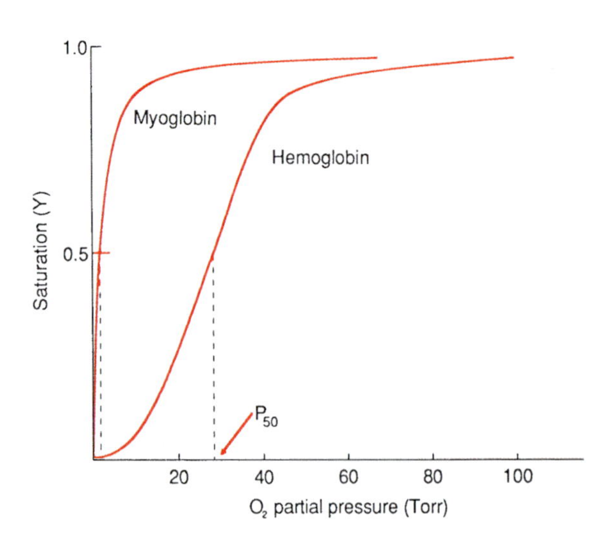

FIGURE 7-9
Profiles of fractional saturation of myoglobin and of hemoglobin with oxygen as a function of partial pressure of oxygen. Under physiological conditions, P_{50} for myoglobin is only 1 or 2 torr, whereas for hemoglobin it is 26 torr, indicating that oxygen is bound much more tightly to myoglobin than to hemoglobin. The loading and unloading to oxygen are cooperative in the case of hemoglobin but not cooperative in the case of myoglobin.

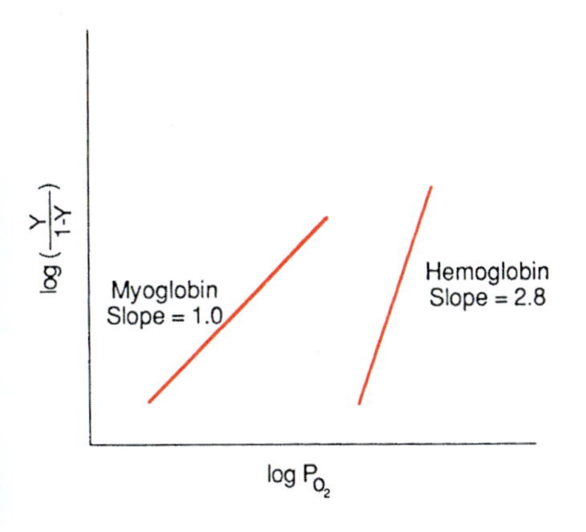

FIGURE 7-10
Hill plots for myoglobin and hemoglobin. A slope of 1.0 for myoglobin is consistent with noncooperative oxygen binding, whereas a slope of 2.8 for hemoglobin is indicative of cooperative oxygen binding.

of 1 for the Hill coefficient is indicative of noncooperativity; as we will see with hemoglobin, values greater than 1 indicate positive cooperativity, and values less than 1 indicate negative cooperativity.

A molecule of hemoglobin (Hb) can bind four molecules of oxygen, and therefore four equilibrium expressions can be written to describe the dissociations of HbO_2, $Hb(O_2)_2$, $Hb(O_2)_3$, and $Hb(O_2)_4$. For simplicity, we consider an equilibrium in which $Hb(O_2)_4$ dissociates into deoxyhemoglobin and four molecules of oxygen:

$$Hb(O_2)_4 \rightleftharpoons Hb + 4O_2 \quad (7.9)$$

The equilibrium constant for Reaction (7.9) is:

$$K = \frac{[Hb]\,[O_2]^4}{[Hb\,(O_2)_4]} \quad (7.10)$$

Based upon considerations similar to those for myoglobin dissociation, the value for Y is

$$Y = \frac{(P_{O_2})^4}{(P_{O_2})^4 + (P_{50})^4} \quad (7.11)$$

A plot of Y versus P_{O_2} yields a sigmoidal curve, indicating cooperative binding of oxygen to hemoglobin. A general expression for the dissociation of oxyhemoglobin to deoxyhemoglobin and oxygen may be written as

$$Hb(O_2)_n \rightleftharpoons Hb + nO_2 \quad (7.12)$$

where n is the number of molecules of oxygen. The equilibrium constant for reaction (7.12) is

$$K = \frac{[Hb]\,[O_2]^n}{[Hb\,(O_2)_n]} \quad (7.13)$$

and

$$Y = \frac{(P_{O_2})^n}{(P_{O_2})^n + (P_{50})^n} \quad (7.14)$$

Rearrangement of Equation (7.14) yields

$$\frac{Y}{1-Y} = \left(\frac{P_{O_2}}{P_{50}}\right)^n$$

The Hill equation is obtained by taking the logarithm of both sides:

$$\log\left(\frac{Y}{1-Y}\right) = n \log P_{O_2} - n \log P_{50}$$

A plot of $\log(Y/1-Y)$ versus $\log P_{O_2}$ yields a straight line with a slope of n, the Hill coefficient. For hemoglobin, $n = 2.8$ (Figure 7-10), which signifies that the binding of oxygen to hemoglobin exhibits positive cooperativity.

From a physiological point of view, the cooperative oxygen binding characteristics of hemoglobin are eminently suited for the transport of oxygen from the lungs to the tissues. In the alveolar spaces of the lungs, the partial pressure of oxygen is about 100 Torr, and about 97% of the hemoglobin is combined with oxygen (i.e., 97% saturation with O_2). As the oxygenated blood passes through the tissue capillaries, where the partial pressure of oxygen often falls below 40 Torr (in actively exercising muscle, the P_{O_2} is about 20 Torr), about 30% of the oxygen is unloaded from oxyhemoglobin to tissue cells. This process is cooperative, so that as O_2 is released from oxygen-saturated hemoglobin because of a drop in P_{O_2}, the loss of a single O_2 molecule causes rapid release of the remaining ones. Similarly in the lungs, the affinity of hemoglobin for binding the first O_2 molecule is low; however, once this molecule is bound, the affinity increases. Myoglobin, in conformity with its storage function, has higher affinity for oxygen than does hemoglobin at any partial pressure of oxygen (Figure 7-9).

X-ray studies have shown that oxyhemoglobin (R form) and deoxyhemoglobin (T form) have different three-dimensional conformations. However, no changes in the tertiary structure of the individual subunits have been observed. The molecular mechanisms of cooperative binding of oxygen are known, and the details of this process along with other ligand interactions with hemoglobin are discussed in Chapter 28.

Theoretical Models for Allosteric Effect

Two theoretical models for allosteric effects have been proposed to explain the mechanism for ligand-protein cooperative interactions: the concerted (or symmetry) model of Monod, Wyman, and Changeux and the sequentially induced-fit model of Koshland. The nomenclature associated with allosterism and cooperativity originated from the concerted model. Both models assume that

1. Each subunit of an oligomeric protein exists in two forms, T and R, which bind the ligand with low and high affinity, respectively; and
2. The T $\rightleftharpoons$ R transformations involve noncovalent bonds and result in changes in the quaternary structure of the enzyme.

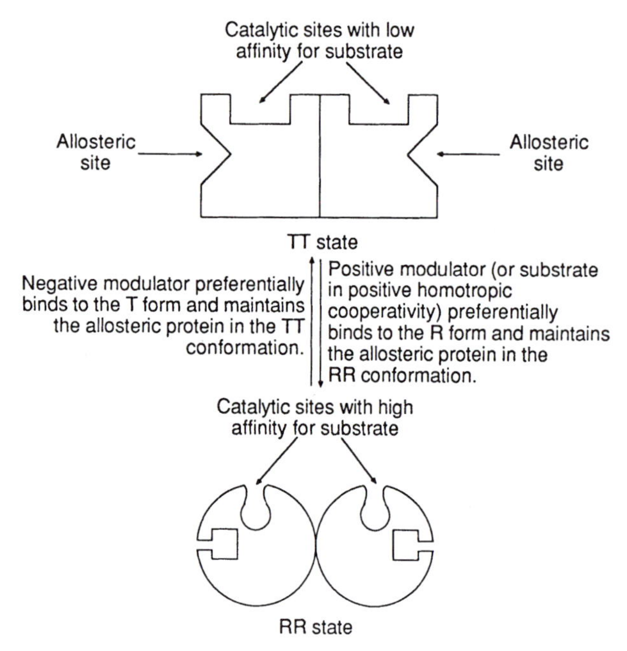

FIGURE 7-11
Schematic diagram of conformational changes of concerted model for a dimeric allosteric enzyme. All subunits are either in the T-form with low affinity for the substrate or in the R-form with high affinity for the substrate.

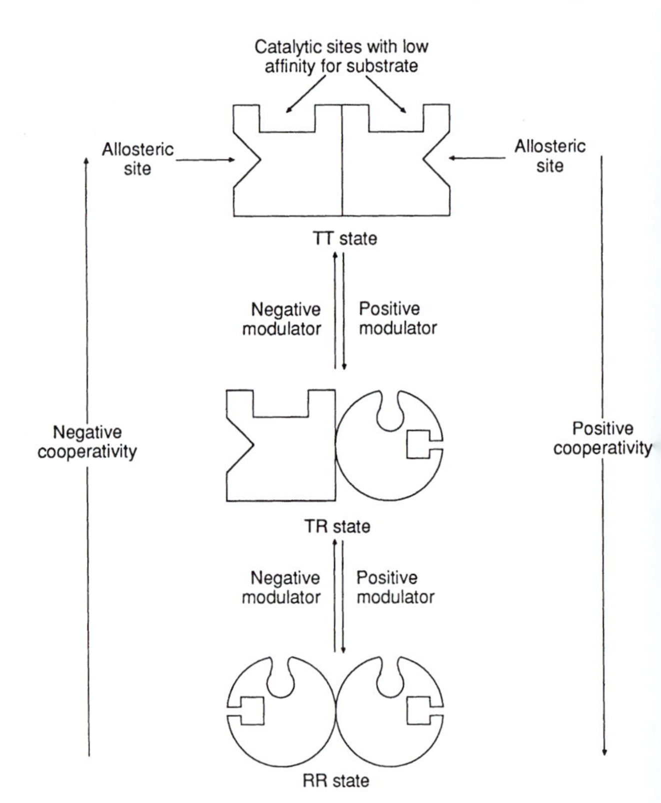

FIGURE 7-12
Schematic diagram of conformational changes of sequentially induced-fit model for a dimeric allosteric enzyme. The TT conformation is progressively converted to the RR conformation via the intermediate TR conformation through cooperative interaction in the presence of the positive modulator. In the presence of the negative modulator, the opposite conformational changes occur. In this model, the notion of symmetry is discarded and the concept of induced fit is emphasized.

The concerted model has the following features. The allosteric protein exists only in two states: T or R. If, for example, an allosteric protein contains two subunits, the sole permissible conformation states are TT and RR (RT is not allowed because R and T cannot form a stable pair). Thus, in this model (also called the "all or none" model), symmetry of the allosteric protein is preserved (Figure 7-11). It is assumed that the T and R forms are in equilibrium, that significant binding of ligand occurs *only* to the R form, and that this binding shifts the equilibrium strongly in favor of formation of the R form. Thus, a conformational change in one subunit that occurs after it binds to a ligand causes a corresponding change in all of the subunits without formation of hybrid species. The concerted model accounts for the kinetic behavior of many allosteric proteins, but it cannot account for negative cooperativity; i.e., the decrease in affinity for a ligand by the allosteric protein as the sites become occupied.

The sequential model proposes that as the ligand binds to a subunit a conformational change is induced that stabilizes the ligand-allosteric protein complex such that a second ligand molecule is bound more readily than the first (Figure 7-12). This effect is of positive cooperativity. However, if the initial binding of the ligand results in a ligand-protein complex with decreased stability, the binding of additional ligand molecules, or ligand-induced substrate binding, becomes increasingly difficult. This effect is negative cooperativity. In this model, the notion of symmetry is discarded, since a conformational change induced by a ligand that binds to a subunit also induces a conformational change in an adjacent subunit. Figure 7-13 shows sequential transmission of conformational changes through contiguous subunits upon ligand binding. The behavior of some allosteric proteins is best explained by the concerted model, whereas for others the sequential model is more appropriate. For many proteins, neither model is satisfactory and a more complex model may be required.

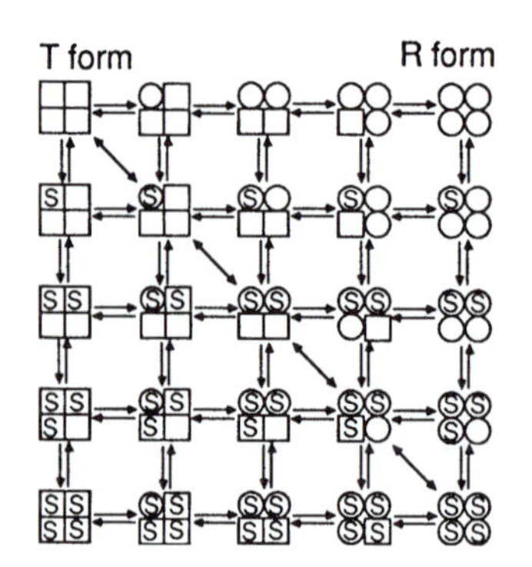

FIGURE 7-13
Sequential transmission of conformational changes through contiguous subunits induced by a ligand (S). In idealized form, the conformational changes proceed diagonally as the protein progressively binds ligand.

Supplemental Readings and References

Enzyme Regulation (General)

Z. Dische: The discovery of feedback inhibition; *Trends in Biochemical Sciences* **1,** 269 (1976).

D. E. Koshland, G. Nemethy, and D. Filmer: Comparison of experimental binding data and theoretical models in proteins containing subunits; *Biochemistry* **5,** 365 (1966).

J. Monod, J.-P. Changeux, and F. Jacob: Allosteric proteins and cellular control systems. *Journal of Molecular Biology* **6,** 306 (1963).

J. Monod, J. Wyman, and J.-P. Changeux: On the nature of allosteric transitions: A plausible model. *Journal of Molecular Biology* **12,** 88 (1965).

H. L. Segal: Enzymatic interconversions of active and inactive forms of enzymes. *Science* **180,** 25 (1973).

G. Weber, (Ed): *Advances in Enzyme Regulation,* Vols. 1–6. Pergamon Press, London, 1963–1979.

Allosteric Properties of Aspartate Transcarbamoylase and Hemoglobin Aspartate Transcarbamoylase

P. Hensley and H. K. Schachman: *Communication between dissimilar subunits in aspartate transcarbamoylase: Effect of inhibitor and activator on the conformation of the catalytic polypeptide chains.* Proceedings of the National Academy of Sciences USA. **76,** 3732 (1979).

E. V. Kantrowitz, S. C. Pastra-Landis, and W. N. Lipscomb: *E. coli aspartate transcarbamoylase. Part 2. Structure and allosteric interactions.* Trends in Biochemical Sciences **5,** 150 (1980).

I. L. Monoco, J. L. Crawford, and W. N. Lipscomb: *Three dimensional structures of aspartate transcarbamoylase from* Escherichia coli *and its complex with cytidine triphosphate.* Proceedings of the National Academy of Sciences USA. **75,** 5276 (1978).

Hemoglobin

J. Baldwin and C. Chothia: Hemoglobin: The structural changes related to ligand binding and its allosteric mechanism. *Journal of Molecular Biology* **129,** 175 (1979).

M. F. Perutz: Stereochemistry of cooperative effects of hemoglobin. *Nature* **228,** 726 (1970).

M. F. Perutz: Hemoglobin structure and respiratory transport. *Scientific American* **239**(6), 92 (1978).

M. F. Perutz: Regulation of oxygen affinity of hemoglobin: Influence of structure of the globin on the heme iron. *Annual Review of Biochemistry* **48,** 327 (1979).

CHAPTER 8

Enzymes III: Clinical Applications

Enzymes are used clinically in three principal ways:

1. As indicators of enzyme activity or enzyme concentration in body fluids (e.g., serum and urine) in the diagnosis and prognosis of various diseases;
2. As analytical reagents in the measurement of activity of other enzymes or nonenzyme substances (e.g., substrates, proteins, and drugs) in body fluids; and
3. As therapeutic agents.

8.1 Diagnosis and Prognosis of Disease

Enzyme assays employed in the diagnosis of diseases are one of the most frequently used clinical laboratory procedures. The most commonly used body fluid for this purpose is serum, the fluid that appears after the blood has clotted. The liquid portion of unclotted blood is called *plasma*. Serum is used for many enzyme assays because the preparation of plasma requires addition of anticoagulants (e.g., chelating agents) that interfere with some assays. Enzymes in circulating plasma are either ***plasma-specific*** or ***non-plasma-specific.*** Plasma-specific enzymes are normally present in plasma, perform their primary function in blood, and have levels of activity that are usually higher in plasma than in tissue cells. Examples are those enzymes involved in blood clotting (e.g., thrombin), fibrinolysis (e.g., plasmin), and complement activation, as well as cholinesterase (also called pseudocholinesterase or choline esterase II to distinguish it from the acetylcholinesterase of nervous tissue; see Chapter 6) and ceruloplasmin. These enzymes are synthesized mainly in the liver and released into the circulation at a rate that maintains optimal steady-state concentrations. Hereditary enzyme defects or impaired liver function can cause suboptimal levels. For example, deficiency of thrombin or plasmin can cause defective blood coagulation or fibrinolysis, respectively. Cholinesterase deficiency can cause prolonged muscle paralysis when succinylcholine is administered as an adjunct to anesthesia to produce skeletal muscle relaxation. Succinylcholine, a structural analogue of acetylcholine, competes with acetylcholine receptors at the motor end-plate of the neuromuscular junction and brings about depolarization that leads to muscle relaxation. The effect of succinylcholine is terminated through the enzymatic action of plasma cholinesterase but is not significantly altered by acetylcholinesterase at the neuromuscular junction. Measurement of plasma cholinesterase before surgery may prevent this adverse effect.

Non-plasma-specific enzymes are intracellular enzymes normally present in plasma at minimal levels or at concentrations well below those in tissue cells. Their presence in plasma is normally due to turnover of tissue cells, but they are released into the body fluids in excessive concentrations as a result of cellular damage or impairment of membrane function. Tissue injury and impairment of membrane function can be caused by diminished oxygen supply (e.g., myocardial infarction), infection (e.g., hepatitis), and toxic chemicals. Proliferation of cells, with consequent increased turnover, can also raise levels in plasma of enzymes characteristic of those cells (e.g., elevation of serum acid phosphatase in prostatic carcinoma). Table 8-1 lists clinically useful enzymes and their tissue of origin. Intracellular enzymes are essentially confined to their cells of origin. A few are secretory enzymes that are secreted by some selected tissue (salivary gland, gastric mucosa, or pancreas) into the gastrointestinal tract, where they participate in digestion of food constituents (Chapter 12). Plasma levels of secretory enzymes increase when their cells of origin undergo damage or membrane impairment or when the usual pathways of enzyme secretion are obstructed. For example, large amounts of pancreatic amylase and triacylglycerol lipase (commonly known as *lipase*) enter the blood circulation in patients suffering from pancreatitis. These enzymes can digest the pancreas itself and surrounding adipose tissue in a process known as *enzymatic necrosis* (death of tissue cells).

The diagnosis of organ disease is aided by measurement of a number of enzymes characteristic of that tissue or organ. Most tissues have characteristic enzyme patterns (Table 8-2) that may be reflected in the relative serum concentrations of the respective enzymes in disease. The diseased tissue can be further identified by determination of the isoenzyme pattern of one of these enzymes (e.g., lactate dehydrogenase, creatine kinase) in the serum, since many tissues have characteristic isoenzyme distribution patterns for a given enzyme. For example, creatine kinase (CK) is a dimer composed of two subunits, M (for muscle) and B (for brain), that occur in three isoenzyme forms, BB(CK_1), MB(CK_2) and MM(CK_3), which catalyze the reversible phosphorylation of creatine with adenosine triphosphate (ATP) as the phosphate donor:

$$\text{Creatine} + \text{ATP} \overset{\text{CK}}{\rightleftharpoons} \text{phosphocreatine} + \underset{\text{(adenosine diphosphate)}}{\text{ADP}}$$

This reaction provides ATP for muscle contraction (Chapter 21). Skeletal muscle contains predominantly CK_3, whereas heart muscle (myocardium) contains CK_3 and CK_2. Serum normally contains a small amount of CK_3 derived predominantly from skeletal muscle. Detection of CK_2 in serum (in an appropriate clinical setting) is strongly suggestive of myocardial damage. Since an abnormal isoenzyme level may occur with apparently normal total activity of the enzyme, determination of the isoenzyme concentration is essential in the diagnostic enzymology.

TABLE 8-1
Tissues of Origin of Some Diagnostically Important Serum Enzymes

Enzyme	Principal Tissue Source
Acid phosphatase	Prostate
Alanine aminotransferase (glutamate pyruvate transaminase)	Liver
Alcohol dehydrogenase	Liver
Alkaline phosphatase	Bone, intestinal mucosa, hepatobiliary system, placenta, kidney
Amylase*	Pancreas, salivary glands
Arginase	Liver
Aspartate aminotransferase (glutamate oxaloacetate transaminase)	Heart and skeletal muscle, liver kidney, brain
Ceruloplasmin	Liver
Cholinesterase	Liver
Chymotrypsin(-ogen)*	Pancreas
Creatine kinase	Skeletal and heart muscle, brain
Fructose-bisphosphate aldolase	Skeletal and heart muscle
γ-Glutamyltransferase	Kidney, hepatobiliary system, prostate, pancreas
Glutamate dehydrogenase	Liver
Isocitrate dehydrogenase	Liver
Lactate dehydrogenase	Skeletal and heart muscle, liver, kidney, erythrocytes, pancreas, lungs
Leucine aminopeptidase	Hepatobiliary system, intestine, pancreas, kidney
Ornithine carbamoyl-transferase	Liver
Pepsin(-ogen)*	Gastric mucosa
Prostatic specific antigen (a serine proteinase)	Prostate
Sorbitol dehydrogenase	Liver
Triacylglycerol lipase* (lipase)	Pancreas
Trypsin(-ogen)*	Pancreas

*Secretory enzymes.

Factors Affecting Presence and Removal of Intracellular Enzymes from Plasma

Many factors are taken into consideration in the clinical interpretation of serum enzyme levels. Membrane permeability changes and cell destruction affect the release of

TABLE 8-2
Properties of Cardiac Markers Used in the Evaluation of Acute Myocardial Infarction (AMI)

Marker	Molecular Weight (Daltons)	Elevation after AMI (Hours)	Peak Elevation (Hours)	Return to Normal Levels (Hours or Days)	Clinical Usefulness
Myoglobin	17,200	1–3	4–6	20–24 hrs	Earliest marker; not cardiac specific.
Total CK	86,000	4–8	12–24	36–48 hrs	A relatively early marker; not cardiac specific.
CKMB	86,000	4–6	12–24	24–48 hrs	A relatively early marker; has higher cardiac specificity over total CK.
Cardiac Troponin T	39,700	4–6	18–36	5–14 days	A relatively early marker; cardiac specific; however elevated in diseases of regeneration of skeletal muscle and chronic renal disease.
Cardiac Troponin I	24,000	4–6	12–24	5–9 days	A relatively early marker; cardiac specific.
LDH	135,000	8–12	48–72	10–15 days	Late marker; not cardiac specific; LD1/LD2 determination increases cardiac specificity.

intracellular enzymes. These changes can result from a decrease in intracellular ATP concentration due to any of the following conditions: deficiency of one or more of the enzymes needed in ATP synthesis (e.g., pyruvate kinase in red blood cells); glucose deprivation; localized hypoxia; and high extracellular K^+ (ATP depletion results from increased activity of the Na^+, K^+-ATPase in the cell plasma membrane required to maintain the proper K^+/Na^+ ratio between the intracellular and extracellular environments). Localized hypoxia can result from poor blood flow, the result of obstruction of blood vessels responsible for the territorial distribution of the blood (a condition known as ***ischemia***). Ischemia may result from narrowing of the lumen of the blood vessels (e.g., deposition of lipids in the vessel wall, which leads to atherosclerosis; see Chapter 20) or from formation of blood clots within the vessels. Ischemic necrosis leads to formation of an infarct (the process is called *infarction*). When cells of the myocardium die as a result of severe ischemia, the lesion is known as a *myocardial infarct*.

The amount of enzymes released depends on the degree of cellular damage, the intracellular concentrations of the enzymes, and the mass of affected tissue. The nature of the insult (viral infection, hypoxia, or surgical, chemical, or mechanical trauma) has no bearing on the enzymes released into the circulation.

The nature of the enzymes released also reflects the severity of the damage. Mild inflammatory conditions are likely to release cytoplasmic enzymes, whereas necrotic conditions yield mitochondrial enzymes as well. Thus, in severe liver damage (e.g., hepatitis), the serum aspartate aminotransferase (AST) level is extremely high (much greater than that of alanine aminotransferase) because the mitochondrial isoenzyme of AST is released in addition to the corresponding cytoplasmic isoenzyme.

The amount of enzyme released into the plasma from an injured tissue is usually much greater than can be accounted for on the basis of tissue enzyme concentration and magnitude of injury. Loss of enzymes from cells may stimulate further synthesis of enzymes. In experimental animals, biliary occlusion leads to increased synthesis of hepatic alkaline phosphatase. Many drugs cause an increase in drug-metabolizing enzymes as a result of an increase in *de novo* synthesis (i.e., enzyme induction). These drug-metabolizing enzymes, located in the smooth endoplasmic reticulum (microsomal fraction; see Chapter 2) of liver and other tissues, catalyze the following chemical reactions: hydroxylation, demethylation, deethylation, acetylation, epoxidation, deamination, glucuronidation, and dehalogenation. Thus, serum levels of some of these enzymes may be elevated following exposure to enzyme-inducing agents, such as ethanol, barbiturates, phenytoin, and polycyclic hydrocarbons. While plasma levels of enzymes can become elevated because of tissue injury, the levels may drop (in spite of continued progress of the injury) to normal (or below normal) levels when the blood circulation is compromised or when the functional part of the tissue is replaced by repair or nonfunctional tissue (e.g., connective tissue, as in extensive fibrosis of the liver in the disease known as *cirrhosis*).

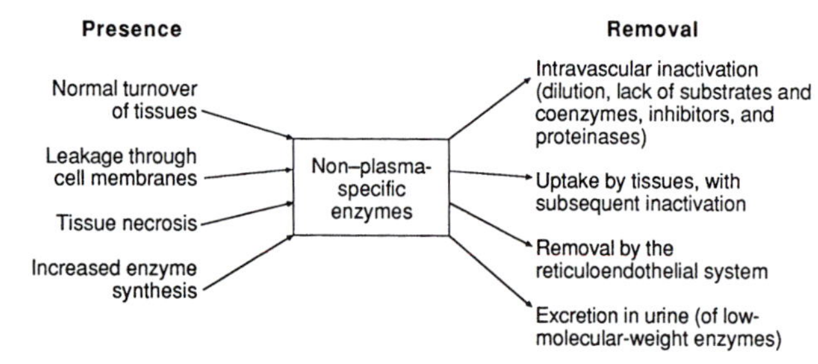

FIGURE 8-1
Factors affecting the presence and removal of non-plasma-specific enzymes from plasma.

Inactivation or removal of plasma enzymes may be accomplished by several processes: denaturation of the enzyme due to dilution in plasma or separation from its natural substrate or coenzyme; presence of enzyme inhibitors (e.g., falsely decreased activity of amylase in acute pancreatitis with hyperlipemia); removal by the reticuloendothelial system; digestion by circulating proteinases; uptake by tissues and subsequent degradation by tissue proteinases; and clearance by the kidneys of enzymes of low molecular mass (amylase and lysozyme).

A schematic representation of the causes for the appearance and disappearance of non-plasma-specific enzymes is shown in Figure 8-1. Since enzymes differ in the rates of their disappearance from plasma, it is important to know when the blood specimen was obtained relative to the time of injury. It is also important to know how soon after the occurrence of injury various enzyme levels begin to rise. The biological half-lives for enzymes and their various isoenzymes are different. We can illustrate this by examining the disappearance rates of the lactate dehydrogenase (LDH) isoenzymes. LDH consists of four subunits of two different types: H (heart) and M (muscle). The subunits combine to yield a series of five tetramers: $LDH_1(H_4)$, $LDH_2(H_3M)$, $LDH_3(H_2M_2)$, $LDH_4(HM_3)$, and $LDH_5(M_4)$. The overall rate of disappearance of LDH activity from plasma is biphasic because of the more rapid disappearance of LDH_5 (the predominant isoenzyme of liver and skeletal muscle), followed by the relatively slower removal of LDH_1 (the predominant isoenzyme of heart, kidney cortex, and erythrocytes). The remaining LDH isoenzymes have intermediate rates of disappearance. In chronic disease states, the plasma enzyme levels continue to be elevated.

The use of appropriate normal ranges is important in evaluating abnormal levels of plasma enzymes. However, an abnormal isoenzyme pattern may occur despite normal total activity (see above). The standard unit for enzyme activity was discussed in Chapter 6. The normal range is affected by a variety of factors: age, sex, race, degree of obesity, pregnancy, alcohol or other drug consumption, and malnutrition. Drugs can alter enzyme level *in vivo* and interfere with their measurement *in vitro*.

Enzyme activities may also be measured in urine, cerebrospinal fluid, bone marrow cells or fluid, amniotic cells or fluid, red blood cells, leukocytes, and tissue cells. Cytochemical localization is possible in leukocytes and biopsy specimens (e.g., from liver and muscle). Under ideal conditions, both the concentration of the enzyme and its activity would be measured. Radioimmunoassay (RIA) and its alternative modes such as fluorescence immunoassay (FIA), fluorescence polarization immunoassay (FPIA), and chemiluminescence immunoassay (CLIA) (discussed later), can be used to measure enzyme concentration as well as other clinically important parameters.

Measurement of Enzyme Activity

In the assessment of enzyme levels in the clinical laboratory, the most frequently used procedure consists of measuring the rate of the enzyme-catalyzed chemical reaction. The initial rate of an enzymatic reaction is directly proportional to the amount of enzyme present when the substrate concentrations are maintained at saturating levels (i.e., zero-order kinetics with respect to substrate concentration) and other factors (e.g., pH, temperature, and cofactors) are maintained at optimal and constant levels (see Chapter 6). Under these conditions, the rate of substrate removal or product formation in a given time interval is directly related to enzyme activity. The reaction rate is determined by measuring product formation (or substrate removal) between two fixed times or by continuously monitoring the reaction with time. For accuracy, it is preferable to measure product formation rather than substrate removal because the former involves measurement of a change in concentration from an initial zero or low level to higher levels, which is analytically more reliable than the reverse. Both methods, fixed-time and continuous monitoring, are kinetic procedures. A schematic diagram of an enzymatic reaction is shown in Figure 8-2. The true rate of enzyme activity is obtained only when the reaction rate is measured in the linear region (a period of constant reaction rate and of maximum velocity). Inaccurate results are obtained if the rate is measured during nonlinear or nonmaximum phases. In fixed-time procedures, in which a single given interval of time is chosen to measure enzyme activity, it is essential to select reaction conditions such as concentration of substrate, pH, temperature, and cofactors, that will provide linearity with maximum slope during the time period in question. False negative values for enzyme rates may arise from substrate depletion, inhibition of the enzyme by product, increase in the reverse reaction due to product accumulation, denaturation of the

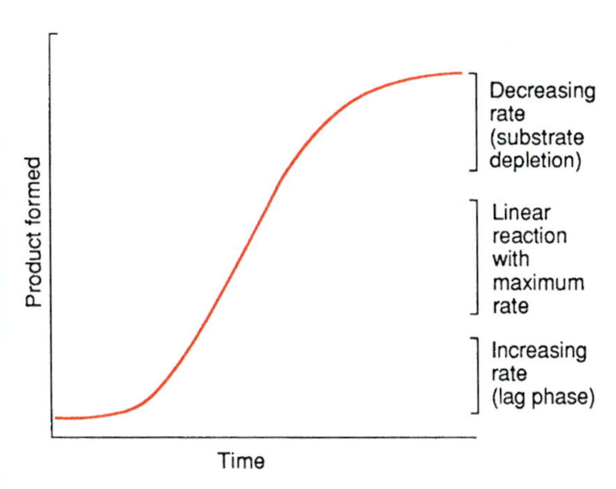

FIGURE 8-2
Diagrammatic representation of the measurement of the activity of an enzyme, showing product formation as a function of time. The true activity of the enzyme is calculated from data obtained when the reaction rate is linear with maximum velocity. The reaction rate is directly proportional to the amount of active enzyme only when the substrate concentrations are maintained at saturating levels and when other variables (e.g., pH, and cofactors) are held constant at optimal conditions.

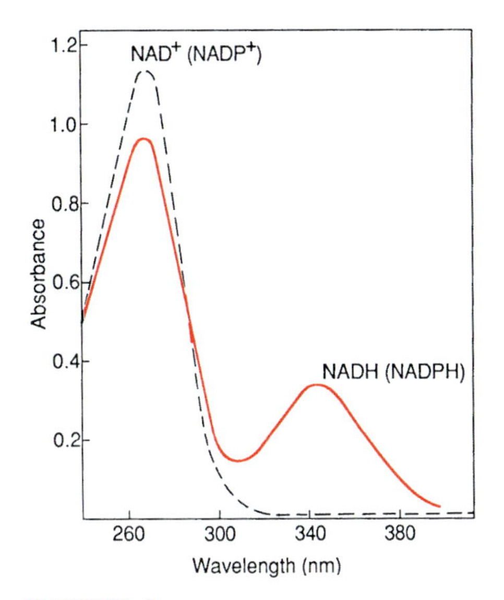

FIGURE 8-3
Absorption spectra of NAD^+ ($NADP^+$) and NADH (NADPH). At 340 nm, the reduced coenzymes (NADH or NADPH) show significant absorbance, whereas the oxidized forms (NAD^+ or $NADP^+$) show negligible absorbance. Thus, many enzymatic reactions can be monitored at 340 nm if the reaction is directly or indirectly dependent upon a dehydrogenase reaction involving a nicotinamide adenine dinucleotide as a coenzyme.

enzyme, presence of certain endogenous metabolites, and drugs (or their metabolites).

In the design of an enzyme assay, a key element is selection of a suitable physical or chemical technique for following the appearance of a product or the disappearance of a substrate. The rate of enzymatic reactions can be monitored continuously by photometric procedures (such as light absorption, fluorescence, or optical rotation) if the reaction is accompanied by a change in optical properties. For example, serum alkaline phosphatase, used in the diagnosis of hepatobiliary and bone diseases, is measured by employing its ability to catalyze the hydrolysis at alkaline pH of the colorless synthetic substrate *p*-nitrophenylphosphate to the yellow-colored product *p*-nitrophenol. The rate of reaction can be measured from changes in absorption at 405 nm. Many enzymatic reactions can be followed by a change in absorbance at 340 nm if they are linked (either directly or indirectly) to a dehydrogenase reaction requiring a nicotinamide adenine-dinucleotide (NAD^+ or $NADP^+$). The coenzyme absorbs ultraviolet light at 340 nm in the reduced state but not in the oxidized state (Figure 8-3). Hence, changes in reaction rate in either direction can be followed by measurement of the absorbance at 340 nm. An example of one such enzyme assay is given below.

Red cell glucose-6-phosphate dehydrogenase (Chapter 15) can be specifically assayed in a red cell hemolysate. The enzyme catalyzes the reaction

$$\text{D-Glucose 6-phosphate} + NADP^+ \rightleftharpoons \text{D-glucono-}\delta\text{-lactone 6-phosphate} + NADPH + H^+$$

Since NADPH absorbs light at 340 nm and $NADP^+$ does not (Figure 8-3), this reaction can be followed by measurement of the change in absorbance at this wavelength with time. A similar assay (with lactate and NAD^+) can be used for serum lactate dehydrogenase. The difference in absorbance is measured so readily that a number of assays make use of it directly or indirectly. In the latter case, the reaction to be measured is linked by one or more steps to a reaction in which a nicotinamide adenine dinucleotide is oxidized or reduced. Thus, in the measurement of aminotransferases (Chapter 17), one of the products of the primary reaction serves as a substrate for a secondary reaction in which NADH is required. For example, in the assay of aspartate aminotransferase,

$$\text{L-Aspartate} + \underset{\text{(also known as 2-oxoglutarate)}}{\alpha\text{-ketoglutarate}} \rightleftharpoons \text{oxaloacetate} + \text{L-glutamate (primary reaction)}$$

the oxaloacetate formed in the primary reaction is reduced to L-malate by NADH in the presence of malate dehydrogenase:

$$\text{Oxaloacetate} + NADH + H^+ \xrightarrow[\text{dehydrogenase}]{\text{malate}} \text{L-malate} + NAD^+ \text{ (indicator reaction)}$$

This secondary reaction is an indicator reaction, i.e., the one whose rate is followed. The primary, not the secondary, reaction must be the rate-limiting step. The indicator reaction should be capable of instantaneous removal of the product formed in the primary reaction to prevent or minimize reversal of the primary reaction. Maintenance of a large excess of the second enzyme (malate dehydrogenase) in the reaction mixture fulfills this requirement. The indicator reaction must function at the concentration of the substrate (oxaloacetate) of the indicator enzyme, so that the rate of the indicator reaction is directly proportional to substrate concentration. Under these conditions, the rate of reaction catalyzed by the indicator enzyme is directly proportional to the rate of product formation in the primary reaction. Thus, during the initial phase of the reaction, the indicator reaction must wait for a period of time known as the ***lag phase*** for the primary reaction to generate adequate quantities of oxaloacetate to permit the maximum rate of conversion of NADH to NAD^+.

In nicotinamide adenine dinucleotide–linked assays, endogenous metabolites and enzymes in the serum may cause oxidation or reduction of the coenzyme, thereby introducing errors. For example, in the AST assay, the NADH of the indicator reaction may be oxidized by the presence of pyruvate and of LDH in the serum:

$$\text{Pyruvate} + \text{NADH} + \text{H}^+ \rightleftarrows \text{L-lactate} + \text{NAD}^+$$

This reaction falsely raises the measured activity of AST. A way to prevent this is to include a preincubation step in the assay procedure. This step consists of a time period during which all components of the reaction mixture except one of the substrates of the primary reaction (in this case, α-ketoglutarate) are allowed to incubate. This period permits all undesirable reactions to reach completion before the primary reaction is initiated by addition of α-ketoglutarate. Because of the need for rapid and accurate measurement, automated procedures have been introduced.

8.2 Serum Markers in the Diagnosis of Tissue Damage

Myocardium

Coronary artery occlusion causes heart tissue damage due to ischemia and can lead to *myocardial infarction* (MI). The immediate and common cause of artery obstruction is formation of a thrombus. Antithrombolytic therapy (Chapter 36), with streptokinase or recombinant tissue plasminogen activator, protects the myocardium from permanent damage by restoring of blood flow. An early diagnosis of acute MI (AMI) is crucial for proper management. A patient's history, presence of chest pain, and electrocardiograms are problematic in the diagnosis of AMI. Therefore, measurements of circulatory proteins (enzymes and nonenzyme proteins) released from the necrotic myocardial tissue are useful in the diagnosis of AMI.

Characteristics of an ideal myocardial injury marker are cardiac specificity, rapid appearance in the serum, substantial elevation for a clinically useful period of time and ease and rapidity of the analytical essay. At present, no serum marker fulfills all of these criteria. Some cardiac markers that appear in plasma are myoglobin, LDH, CK, and troponins. Measurements are made at the appropriate time intervals based upon the appearance of the marker in the plasma (Table 8-1). Frequently utilized markers are CKMB and cardiac troponin I (cTn I). Troponins consist of three different proteins I, C, and T, and are expressed in both cardiac and skeletal muscle. The tripartite troponin complex regulates the calcium-dependent interaction of myosin with actin (Chapter 21). Troponins are encoded by different genes. Cardiac I and T isoforms have unique structural differences from their skeletal muscle counterparts. However, cTn T like CKMB undergoes ontogenic recapitulation and is reexpressed in regenerating skeletal muscle and in patients with chronic renal failure. CKMB is also present in the skeletal muscle, albeit in small concentration. Compared to the myocardium (about 360– 400 g), skeletal muscle consists of a much larger mass (about 40% of total body mass). In rhabdomyolysis (disintegration or breakdown of muscle) CKMB and myoglobin appear in plasma in significant quantities. Myoglobin, although an early marker (Table 8-1), lacks specificity, and CKMB is elevated in many circumstances other than cardiac injury, e.g., rhabdomyolysis, chronic renal disease, and degenerative diseases of skeletal muscle. CKMB has a subform in the plasma due to the fact the M subunit undergoes cleavage of a lysine residue from the carboxy terminus by the plasma enzyme carboxypeptidase N. Tissue and the plasma subforms of CKMB have been designated as CKMB2 and CKMB1, respectively. The ratio of the serum levels, CKMB2/CKMB1 can also yield information useful in the early diagnosis of MI. Measurement of LDH isoenzymes (LDH1 and LDH2) also lacks specificity; they appear significantly later after the myocardial injury and show ontogenic recapitulation. Cardiac troponin I, a highly specific marker for AMI, does not suffer from these disadvantages and appears in the plasma as early as CKMB and persists as long as LDH isozymes. For these reasons, serial serum cTn I measurements may be superior in the detection of MI even in patients with chronic renal disease, rhabdomyolysis, or diseases of skeletal muscle regeneration (e.g., muscle dystrophies).

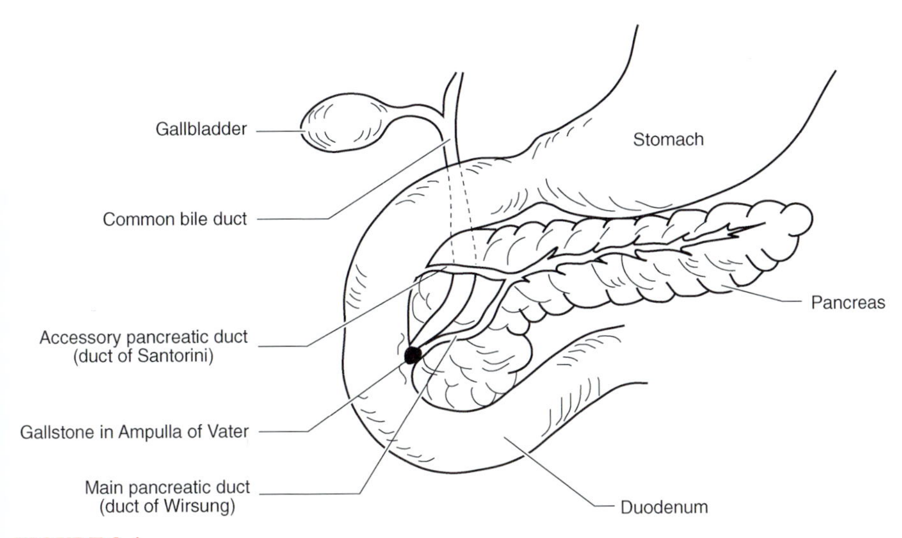

FIGURE 8-4
Pancreatic duct obstruction by a gallstone at the ampulla of Vater. Obstruction can lead to pancreatitis, from induction of bile reflux, which eventually damages acinar cells of the pancreas.

Pancreas

The pancreas is both an exocrine and an endocrine gland. The exocrine function is digestion of food substances (Chapter 12); the endocrine function involves glucose homeostasis (Chapter 22). *Acute pancreatitis* is characterized by epigastric pain; it is an inflammatory process and potentially fatal. Obstruction of the pancreatic duct (Figure 8-4), which delivers pancreatic juice to the small intestine, by gallstones (Chapter 12) or alcohol abuse is the most common cause of acute pancreatitis, representing 80% of all causes. The pathophysiology is due to inappropriate release of pancreatic enzymes and their premature activation. The principal pancreatic enzyme is trypsinogen, which after activation to trypsin converts many other enzymes to their active forms. Some of these are kallikrein, phospholipase A2, elastase, enzymes of blood coagulation and fibrinolysis, and complement. Effects of these abnormal processes are autodigestion of the pancreas, vasodilation, increased capillary permeability, and disseminated intravascular coagulation. These can result in circulatory collapse, renal insufficiency, and respiratory failure.

Laboratory diagnosis of acute pancreatitis involves the measurement of the pancreatic digestive enzymes: amylase and lipase (Chapter 12). Elevated serum amylase level is a sensitive diagnostic indicator in the assessment of acute pancreatitis, but it has low specificity because there are many nonpancreatic causes of *hyperamylasemia*. Furthermore, amylase (M.W. 55,000) is rapidly cleared by the kidneys and returns to normal levels by the third or fourth day after onset of the abdominal pain. Amylase activity in the serum appears within 2–12 hours after the onset of pain. Serum lipase also is used to assess pancreatic disorders and has a higher specificity than serum amylase. It appears in the plasma within 4–8 hours, peaks at about 24 hours and remains elevated for 8–14 days. Measurements of amylase and lipase provide 90–95% accuracy in the diagnosis of acute pancreatitis with abdominal pain. Pancreatic isoamylase level has not proven useful in comparison with measurements of both amylase and lipase. The catalytic activity of lipase requires the presence of both bile salts and colipase (Chapter 12) and must be incorporated in assays of serum lipase activity.

Since exocrine cells of the pancreas contain many enzymes, attempts have been made to use markers other than amylase and lipase to diagnose acute pancreatitis. One such enzyme is *trypsinogen*. It is a 25,000-Da protein that exists in two isoenzyme forms: trypsinogen-1 (cationic) and trypsinogen-2 (anionic). Both forms are readily filtered through the kidney glomeruli. However, the tubular reabsorption of trypsinogen-2 is less than for trypsinogen-1; a dipstick method has been developed to detect trypsinogen-2 in the urine of patients suspected of having acute pancreatitis. The test strip contains monoclonal antibodies specific for trypsinogen-2.

Liver

The liver is the largest glandular organ and its parenchymal cells are called hepatocytes. Liver has numerous functions, including metabolism, detoxification, formation and excretion of bile, storage, and synthesis. Liver diseases include alcohol abuse, medication, chronic hepatitis B and C infections, steatosis and steatohepatitis, autoimmune hepatitis, hemochromatosis, Wilson's disease, α_1-antitrypsin deficiency, malignancy, and poisons and infectious agents. These disorders require specific laboratory testing procedures and are discussed at the appropriate places in the text. The serum enzymes used in assessment of liver function are divided into two categories: (1) markers used in hepatocellular necrosis and (2) markers that reflect cholestasis. Serum enzymes used as markers of cholestasis include alkaline phosphatase, 5′-nucleotidase, and γ-glutamyl transferase. Alanine aminotransferase and aspartate aminotransferases are markers for hepatocellular necrosis. Other tests used in the assessment of liver disorders include measurement of bilirubin, albumin, and α-fetoprotein.

8.3 Enzymes as Analytical Reagents

The use of enzymes as analytical reagents in the clinical laboratory has found widespread application in the measurement of substrates, drugs, and activity of other enzymes. Above we discussed the use of enzymes as reagents in the coupled assay of aspartate aminotransferase. Enzyme-dependent procedures, used in the assay of substrates such as glucose, urea, uric acid, and cholesterol, provide many advantages over classical chemical procedures. These advantages include specificity of the substrate that is being measured and direct measurement of the substrate in a complex mixture that avoids preliminary separation and purification steps such as serum. A reagent enzyme can be used in substrate measurements by two methods: end-point and rate assays. In an end-point assay, the substrate is completely converted to product before it is measured; in the rate assay, a change in substrate concentration produced during a fixed-time interval is measured. This second method, also referred to as the two-point kinetic method, when carried out under constant conditions of pH, temperature, amount of enzyme, and time interval, yields very accurate values for substrate concentration. This procedure is calibrated with standard solutions of the substrate. Of course, the first-order reaction condition with respect to substrate must be maintained by keeping the reagent enzyme concentration high. Under these conditions, when average velocities during the chosen time interval are plotted against various standard substrate concentrations, the profile obtained should exhibit a straight line from which unknown values of substrate can be calculated.

Optical properties can be used in monitoring assay procedures as they are in measuring enzyme activity. If the primary substrate or product does not have a suitable optical property (e.g., absorption in visible or ultraviolet light), then a coupled assay is constructed in which one or more auxiliary enzymes are employed to form an ultimate product that possesses an easily measured optical property. For example, glucose in a biological specimen can be measured by the following reactions under optimal conditions:

$$\text{D-Glucose} + \text{ATP} \xrightarrow{\text{hexokinase}} \text{D-glucose 6-phosphate} + \text{ADP}$$

$$\text{D-Glucose 6-phosphate} + \text{NADP}^+ \underset{}{\overset{\text{glucose-6-phosphatase dehydrogenase}}{\rightleftharpoons}} \text{D-glucono-}\delta\text{-lactone 6-phosphate} + \text{NADPH} + \text{H}^+$$

The second reaction (in which NADPH is formed) functions as the indicator reaction. Although the hexokinase reaction is relatively nonspecific and acts on hexoses other than glucose (Chapter 13) to form 6-phosphate esters, the indicator reaction is specific for glucose 6-phosphate making the overall assay procedure highly specific for glucose.

Enzymes immobilized on an insoluble surface can be used repeatedly in substrate assays. This type of assay is feasible if the reaction product can be measured directly. Immobilization can be accomplished through chemical linkages involving diazo, triazine, and azide groups with many types of insoluble matrices, such as diethylaminoethylcellulose, carboxymethylcellulose, agarose, microcrystalline cellulose, and inner walls of plastic tubings. Enzymes immobilized on the inner surface of plastic tubings are particularly useful in continuous-flow analyzers. Reagent enzymes that have been immobilized include glucose oxidase, urease, α-amylase, trypsin, and leucine aminopeptidase.

Reagent enzymes and antibodies formed against specific molecules can be combined to determine the concentration of a variety of molecules to which antibodies (Chapter 35) can be formed. Such analytical procedures are known as ***enzyme immunoassays*** (EIAs). The reagent enzyme can be linked to antibodies or antigens such that the complexes possess immunological or enzymatic activity. Antibodies can be raised in vertebrate animals when injected with specific proteins (antigens) foreign to them. Macromolecules other than proteins can also be antigenic. Low-molecular-weight compounds (e.g., substrates, drugs) by themselves do not elicit antibodies but do so if covalently linked to a carrier protein (e.g., albumin) before injection. The term ***hapten*** designates the low-molecular-weight substance that can combine with the antibody produced against the carrier protein complex. EIAs are either ***heterogeneous*** or ***homogeneous.*** A heterogeneous assay consists of at least one separation step in which the bound enzyme-labeled

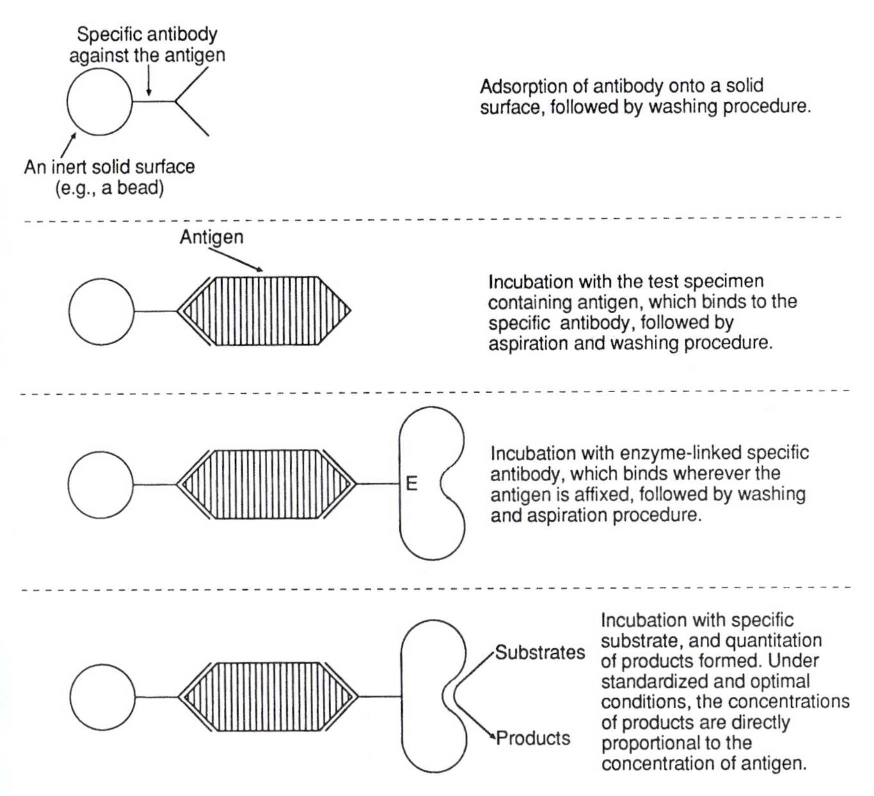
Specific antibody against the antigen
An inert solid surface (e.g., a bead)
Adsorption of antibody onto a solid surface, followed by washing procedure.
Antigen
Incubation with the test specimen containing antigen, which binds to the specific antibody, followed by aspiration and washing procedure.
E
Incubation with enzyme-linked specific antibody, which binds wherever the antigen is affixed, followed by washing and aspiration procedure.
Substrates
Products
Incubation with specific substrate, and quantitation of products formed. Under standardized and optimal conditions, the concentrations of products are directly proportional to the concentration of antigen.

FIGURE 8-5
Diagrammatic representation of an ELISA (solid-phase, sandwich technique) for an antigen.

reagent is separated from the unbound enzyme, enabling the measurement of either bound or free activity. A procedure based upon this principle is the enzyme-linked immunosorbent assay (ELISA), which is analogous to an RIA.

An example of a heterogeneous immunoassay that uses a solid-phase system is shown in Figure 8-5. This method consists of coating or adsorbing antibodies generated in an animal (e.g., guinea pig) against a human antigen, whose concentration is to be determined in a test serum sample, onto a solid surface (e.g., beads or inner surface of a test tube) and incubating with the serum specimen. During the first incubation step, the antigen combines with its specific antibody and is fixed to a solid surface. Subsequently, the unreacted serum components are removed by aspiration of the fluid, and the solid surface (containing antibody-antigen complexes) is washed with suitable buffers. Incubation with antibodies covalently linked to enzyme molecules follows. During this step, the enzyme-labeled antibody (E-Ab) will combine wherever the antigens are affixed; a washing procedure then removes the unbound E-Ab conjugates. Finally, incubation with the substrate specific for the enzyme yields a colored product that is quantitated by suitable spectrophotometric methods. The time interval for the substrate incubation step and other reaction conditions must be kept constant for both test sera and the standards. The enzyme reaction is stopped by altering the pH of the reaction mixture by adding an acid solution (e.g., 1 N hydrochloric acid). The change in absorbance of color is proportional to the concentration of antigen in the test sera. A standard curve is obtained by plotting known antigen concentrations against their absorbances.

Various modifications of this technique are in use. For example, if the concentration of an antibody is to be determined, the process can begin with immobilization of the antigen onto the solid phase.

The enzymes used as labels should have a high turnover number, act on substrates that are stable and soluble, yield easily measurable products, be obtainable in a highly purified form, and be relatively inexpensive. Enzymes that fit these criteria are horseradish peroxidase, glucose oxidase, *E. coli* or calf-intestine alkaline phosphatase, catalase, and β-D-galactosidase. The covalent linkage of the enzyme to the protein (antigen or antibody) through free amino groups can be accomplished with the bifunctional reagent glutaraldehyde. The periodate oxidation method (Chapter 9), involving the carbohydrate side-chains of proteins, also yields reactive groups useful in the preparation of enzyme conjugates. The ELISA exhibits high levels of both sensitivity and specificity. The sensitivity, which is due to the amplification factor introduced by way of the catalytic activity of the labeled enzyme, can equal (with comparable precision) that obtainable with RIA. The specificity is attributable to specific molecular recognition properties of antigens and antibodies. The ELISA may be more advantageous than RIA because ELISA reagents have longer shelf lives than do many radioisotopes (e.g., ^{125}I-labeled reagents) and present no health hazards. The ELISA also has been adapted to identify antigens or antibodies in tissue slices by the use of protein-peroxidase conjugates.

Homogeneous EIA procedures have been especially useful for rapid determination of low-molecular-weight substances (haptens). Thus, these procedures have found application in therapeutic drug monitoring and toxicology. A commercial reagent for the assay of many drugs is available under the trade name of EMIT (Enzyme-Multiplied Immunoassay Technique). Figure 8-6 diagrams the steps involved in this procedure. The principle is based on competition between the enzyme-labeled hapten and free hapten in the test serum for binding with a limited amount of a specific antibody. This assay requires a hapten covalently linked to an enzyme (e.g., glucose-6-phosphate dehydrogenase, lysozyme, or malate dehydrogenase), with retention of enzymatic activity, and antibodies raised against the hapten (through injection into an animal of the hapten covalently linked to a carrier protein) that inhibit enzyme activity when they combine with the hapten-linked enzyme reagents. The inhibition by the antibody of catalytic activity of the enzyme-linked hapten complex may be due to steric hindrance or to conformational constraints affecting access of substrate of the active center. This reaction system consists of the test specimen, the enzyme-labeled hapten, a limited amount of the antibody specific to the hapten, and the substrate. Any haptens present in the test serum compete with the hapten linked to the enzyme for binding with the antibody. The unbound hapten-linked enzyme acts on the substrate to convert it to an easily measurable product. Thus, the enzyme activity is directly correlated with the free hapten concentration in the test specimen. The method is calibrated with known standard concentrations of the hapten.

The ELISA and EIA receptor systems measure substances at concentrations as low as a few nanograms (10^{-9} g). This sensitivity is not sufficient for detecting many substances and alternate methods have been devised. One is chemiluminescence immunoassay (CLIA), which can measure concentrations in femtogram (10^{-15} g) quantities. CLIA depends on the detection of emitted light associated with the dissipation of energy from a substance

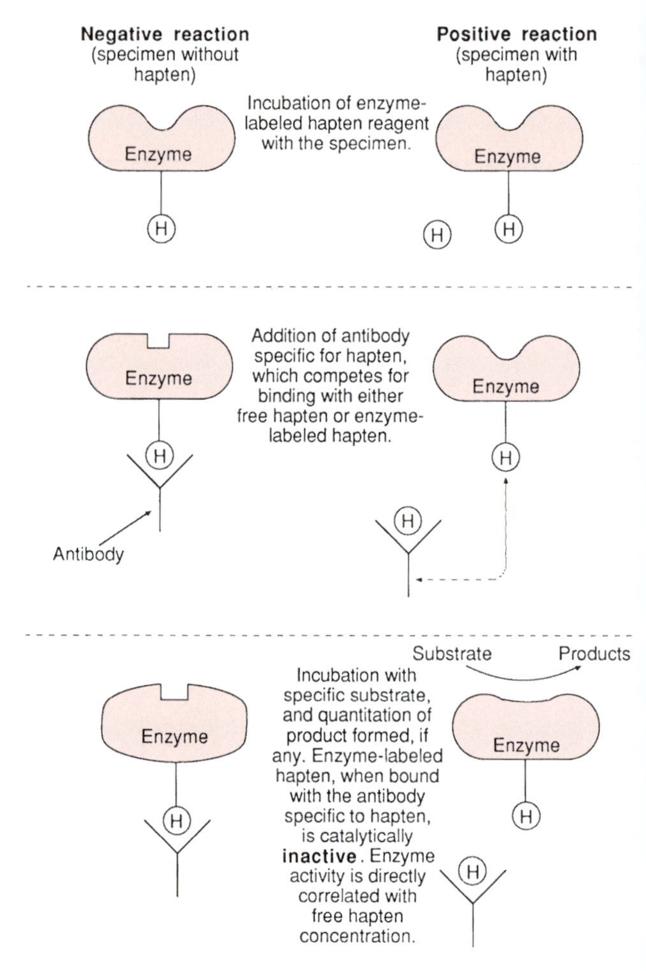

FIGURE 8-6
Diagrammatic representation of a homogeneous enzyme immunoassay procedure for the determination of low-molecular-weight compounds (haptens).

excited electronically as a result of a chemical reaction. An example of a chemiluminescent tracer is acridinium ester conjugated to a desired protein, polypeptide, or other organic molecule (Figure 8-7).

CLIA is similar to EIA and ELISA techniques except that the final receptor enzyme assay is replaced with a chemiluminescent tracer followed by measurement of light released as a result of the chemical reaction. The principles of a chemiluminescence competitive binding assay are shown in Figure 8-8.

8.4 Enzymes as Therapeutic Agents

Enzymes have found a few applications as therapeutic agents. Some examples are transfusion of fresh blood or its active components in bleeding disorders, oral administration of digestive enzymes in digestive diseases

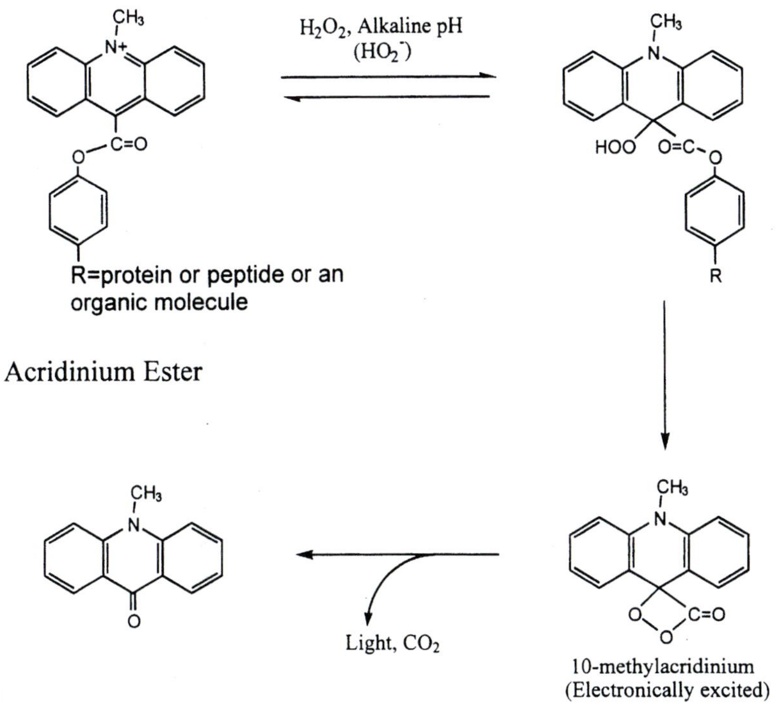
CH3
N+
C=O
O
R=protein or peptide or an organic molecule
Acridinium Ester
H2O2, Alkaline pH
(HO2-)
CH3
N
HOO
O=C
O
R
CH3
N
O
Light, CO2
CH3
N
O
C=O
O
10-methylacridinium
(Electronically excited)

FIGURE 8-7
Acridinium ester used as a chemiluminescent tracer. Labeled acridinium ester in the presence of alkaline hydrogen peroxide undergoes chemical changes producing light that is detected at 430 nm by a photomultiplier detector.

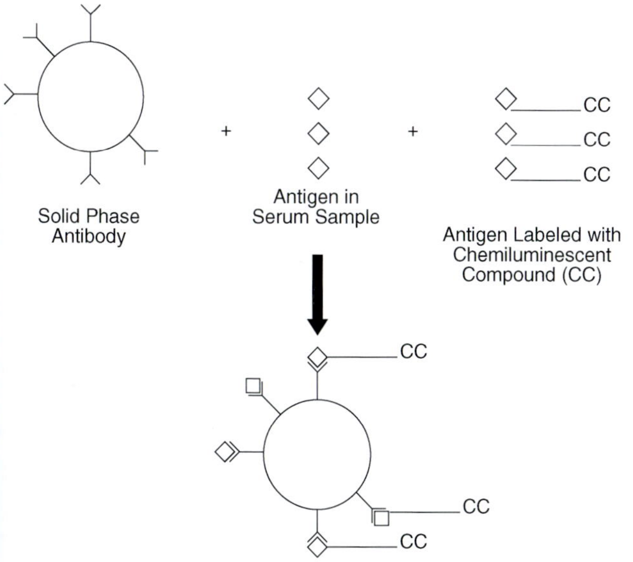
Solid Phase Antibody
+
Antigen in Serum Sample
+
CC
CC
CC
Antigen Labeled with Chemiluminescent Compound (CC)
CC
CC
CC

FIGURE 8-8
Competitive binding chemiluminescence assay. The chemiluminescent competitve binding assay is similar to a RIA. The acridinium ester–labeled substance and the endogenous substance compete with the specific antibody that is available in limited concentration. The concentration is determined by the quantity of tracer that binds to the antibody. The use of solid-phase antibody simplifies separation of free from bound tracer.

(e.g., cystic fibrosis),administration of fibrinolytic enzymes (e.g., streptokinase) to recanalize blood vessels occluded by blood clots (thrombi) in thromboembolic disorders (e.g., pulmonary embolism, acute myocardial infarction), treatment of selected disorders of inborn errors of metabolism (e.g., *Gaucher's disease*), and cancer therapy (e.g., L-asparaginase in acute lymphocytic leukemia). For enzymes to be therapeutically useful, they should be derived from human sources to prevent immunological problems. Although enzymes derived from human blood are readily obtainable, enzymes derived from tissues, which would be particularly useful in the treatment of inborn errors of metabolism, are difficult to obtain in adequate quantities. Transport of specific enzymes to target tissues is also a problem, but some recent advances and commercial applications (e.g., propagation of human tissue culture cell lines, isolation and cloning of specific genes) have the potential of overcoming these difficulties. Such techniques have been used in the production of peptide hormones such as somatostatin and insulin, interferon, and tissue plasminogen activator. In the treatment with enzymes or proteins, covalent attachment of an inert polymer polyethylene glycol (PEG) provides many therapeutic benefits. These include slowing the clearance, diminished immunogeneity, prevention of degradation, and binding to antibody. PEG-enzyme therapy is used in the treatment of immunodeficiency disease caused by adenosine deaminase deficiency (Chapter 27) and PEG-interferon alfa complex (peginterferon alfa-2a) is used in treatment of chronic hepatitis C infection. Ultimately, however, when a gene is cloned, techniques will be developed to incorporate it into the genome of persons lacking the gene or having a mutated gene.

Supplemental Readings and References

N. V. Bhagavan, A. P. Goldstein, S. A. A. Honda, et al.: Role of cardiac troponin I in the evaluation of myocardial injury. *Journal of Clinical Laboratory Analysis* **12,** 276 (1998).

L. Coudrey: The troponins. *Archives of Internal Medicine* **158,** 1173 (1998).

R. F. Dudley: Chemiluminescence immunoassay: An alternative to RIA. *Laboratory Medicine* **21,** 216 (1990).

A. H. Gershlick and R. S. More: Treatment of myocardial infarction. *British Medical Journal* **316,** 280 (1998).

E. A. Kemppainen, J. I. Hedotrom, P. A. Puolakkainen, et al.: Rapid Measurement of Urinary Trypsinogen-2 as a screening test for acute pancreatitis *New England Journal of Medicine* **336,** 1788 (1997).

K. Mergener and J. Baillie: Acute Pancreatitis. *British Medical Journal* **316** 44 (1998).

D. S. Pratt and M. M. Kaplan: Evaluation of abnormal liver-enzyme results in asymptomatic patients. *New England Journal of Medicine* **342,** 1266 (2000).

J. J. Reichling and M. M. Kaplan: Clinical use of serum enzymes in liver disease. *Digestive Diseases and Sciences* **33,** 1601 (1988).

W. Steinberg and S. Tenner: Acute pancreatitis. *New England Journal of Medicine* **330,** 1198 (1994).

S. Zeuzem, S. V. Feinman, J. Rasenack, et al: Peginterferon alfa-2a in patients with chronic hepatitis C. *New England Journal of Medicine* **343,** 1666 (2000).

CHAPTER 9

Simple Carbohydrates

Carbohydrates are major functional constituents of living systems. The primary source of energy in animal cells, carbohydrates are synthesized in green plants from carbon dioxide, water, and solar energy. They provide the skeletal framework for tissues and organs of the human body and serve as lubricants and support elements of connective tissue. Major energy requirements of the human body are met by dietary carbohydrates. They confer biological specificity and provide recognition elements on cell membranes. In addition, they are components of nucleic acids and are covalently linked with lipids and proteins.

9.1 Classification

Carbohydrates consist of polyhydroxyketone or polyhydroxyaldehyde compounds and their condensation products. The term "carbohydrate" literally means hydrate of carbon, a compound with an empirical formula $(CH_2O)_n$. This formula applies to many carbohydrates, such as glucose, which is $C_6H_{12}O_6$, or $(CH_2O)_6$. However, a large number of compounds are classified as carbohydrates even though they do not have this empirical formula; these compounds are derivatives of simple sugars (e.g., deoxyribose, $C_5H_{10}O_4$). Carbohydrates may be classified as monosaccharides, oligosaccharides, and polysaccharides; the term ***saccharide*** is derived from the Greek word for sugar. Monosaccharides are single polyhydroxyaldehyde (e.g., glucose) or polyhydroxyketone units (e.g., fructose), whereas oligosaccharides consist of two to ten monosaccharide units joined together by glycosidic linkages. Sucrose and lactose are disaccharides, since they are each made up of two monosaccharide units. Names of common monosaccharides and disaccharides take the suffix *-ose* (e.g., glucose, sucrose). Polysaccharides, also known as ***glycans,*** are polymers that may contain many hundreds of monosaccharide units. They are further divided into homopolysaccharides and heteropolysaccharides. The former contain only a single type of polysaccharide unit (e.g., starch and cellulose, both of which are polymers of glucose), whereas the latter contain two or more different monosaccharide units.

Monosaccharides

Monosaccharides are identified by their carbonyl functional group (aldehyde or ketone) and by the number of carbon atoms they contain. The simplest monosaccharides are the two trioses: glyceraldehyde (an aldotriose); and dihydroxyacetone (a ketotriose). Four-, five-, six-, and seven-carbon-containing monosaccharides are called tetroses, pentoses, hexoses, and heptoses, respectively. Structures of some monosaccharides are shown in Figure 9-1. All monosaccharides, with the exception of dihydroxyacetone, contain at least one asymmetrical or chiral carbon atom, and therefore two or more stereoisomers are

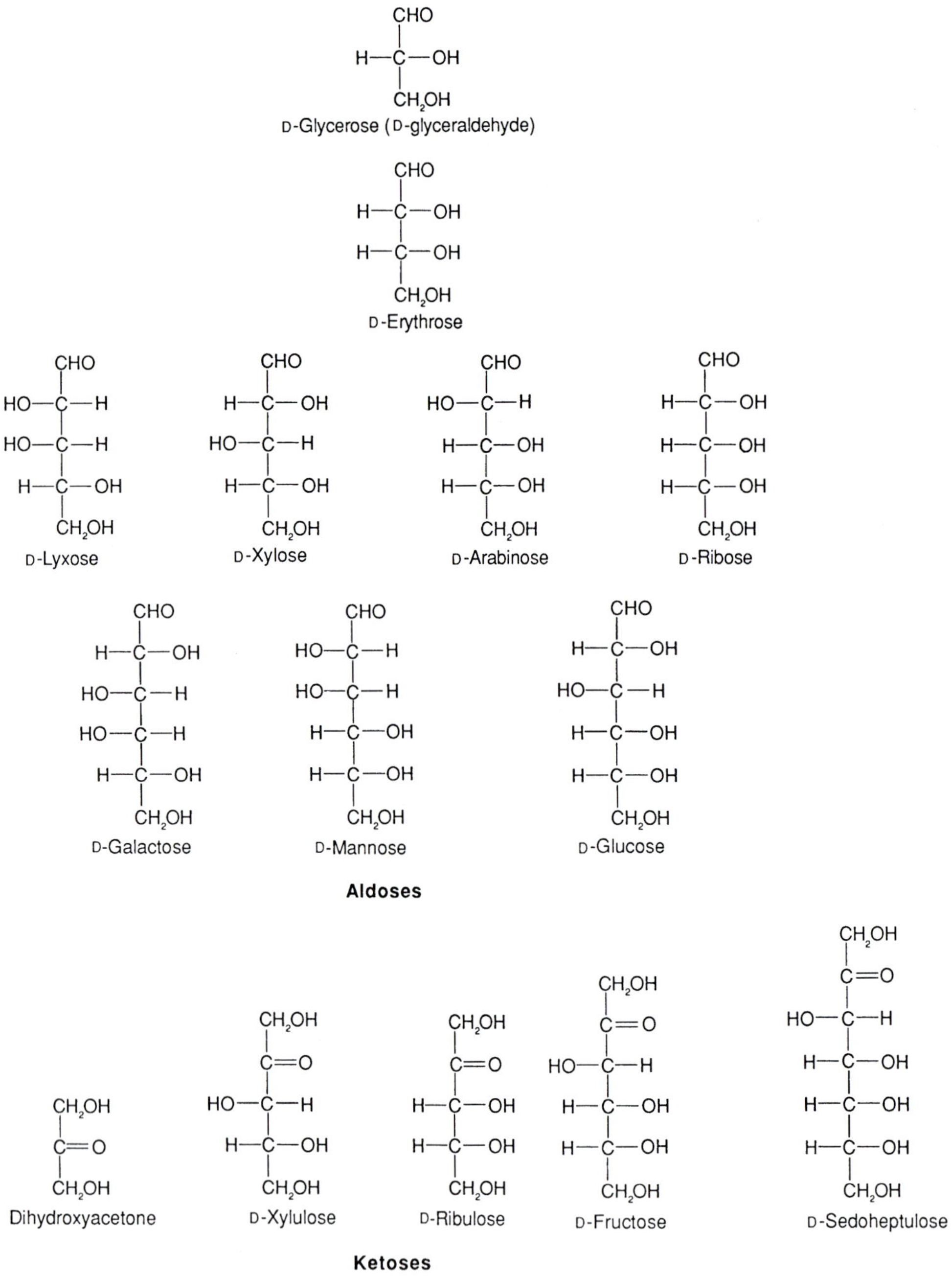

D-Glycerose (D-glyceraldehyde)

D-Erythrose

D-Lyxose D-Xylose D-Arabinose D-Ribose

D-Galactose D-Mannose D-Glucose

Aldoses

Dihydroxyacetone D-Xylulose D-Ribulose D-Fructose D-Sedoheptulose

Ketoses

FIGURE 9-1
Structures of some monosaccharides.

possible for each monosaccharide depending on the number of asymmetrical centers it contains. Glyceraldehyde, with one asymmetrical center, has two possible stereoisomers, designated D and L forms (Figure 9-2). A method for representing the D and L forms is the *Fischer projection formula*. The D and L representations are used for all monosaccharides. The designation of D or L, given to a monosaccharide with two or more asymmetrical centers, is based on the configuration of the asymmet rical carbon atom located farthest from the carbonyl func tional group. Thus, if the configuration at that carbon i the same as that of D-glyceraldehyde (with the hydroxy group on the right-hand side), it belongs to the D series A similar relationship exists between L-glyceraldehyd (with the hydroxyl group on the left-hand side) and th L series of monosaccharides. The optical rotation of

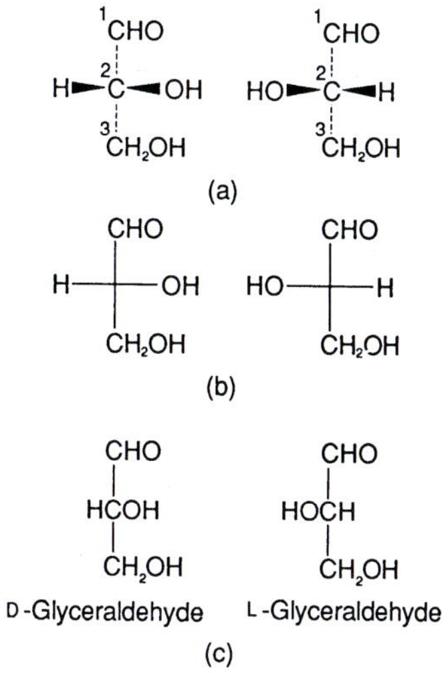

(a)

(b)

CHO HCOH CH_2OH D-Glyceraldehyde

CHO HOCH CH_2OH L-Glyceraldehyde

(c)

FIGURE 9-2
Stereoisomers of glyceraldehyde. (a) Perspective formulas showing tetrahedral arrangement of the chiral carbon 2 with four different substituents. (b) Projection formulas in which the horizontal substituents project forward and the vertical substituents project backward from the plane of the page. (c) A common method of representation.

monosaccharide with multiple asymmetrical centers is the net result of contributions from the rotations of each optically active center. Thus, the prefix D or L provides no information with regard to optical rotation; it indicates only the configuration around the asymmetrical carbon atom located farthest from the carbonyl carbon.

The numbering system for monosaccharides depends on the location of the carbonyl carbon (or potential carbonyl carbon), which is assigned the lowest possible number. For glucose (an aldohexose), carbon C_1 bears the carbonyl group and the farthest asymmetrical carbon atom is C_5 (the penultimate carbon), the configuration around which determines the D and L series. For fructose (a ketohexose), C_2, bears the carbonyl group and C_5 is the highest numbered asymmetrical carbon atom. Glucose and fructose have identical configurations around C_3 to C_6.

The structure of glucose, written in straight-chain form (Figure 9-1), shows four asymmetrical centers. In general, the total number of possible isomers with a compound of n asymmetrical centers is 2^n. Thus, for aldohexoses having four asymmetrical centers, 16 isomers are possible, 8 of which are mirror images of the other 8 (***enantiomers***). These two groups constitute members of the D and L series of aldohexoses. Most of the physiologically important isomers belong to the D series, although a few L-isomers are also found. In later discussions, the designation of D and L is omitted, and it is assumed that a monosaccharide belongs to the D series unless it is specifically designated an L-isomer. Of the D-series of aldohexoses, three are physiologically important: D-glucose, D-galactose, and D-mannose. Structurally, D-glucose and D-galactose differ only in the configuration around C_4; D-glucose and D-mannose differ only in the configuration around C_2. Pairs of sugars (e.g., glucose and galactose, glucose and mannose), which differ only in the configuration around a single carbon, are known as ***epimers.*** D-Galactose and D-mannose are *not* epimers, since they differ in configurations around both C_2 and C_4 (Figure 9-1). D-Fructose, one of eight 2-ketohexoses, is the physiologically important ketohexose. Monosaccharides with five or more carbons occur predominantly in cyclic (ring) forms owing to a reaction between the carbonyl group (aldehyde or ketone) and an alcohol group:

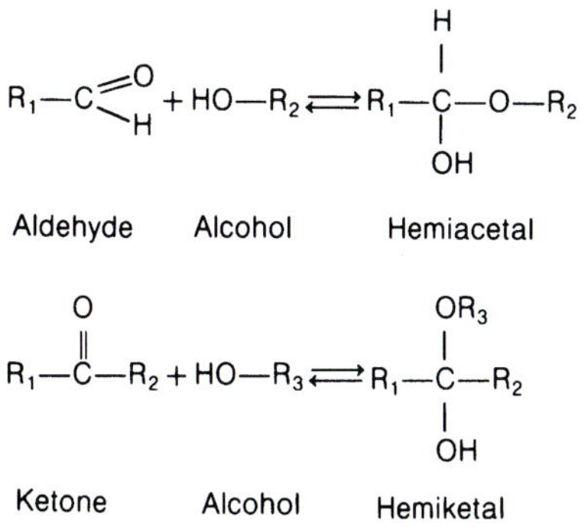

$$R_1{-}CHO + HO{-}R_2 \rightleftharpoons R_1{-}CH(OH){-}O{-}R_2$$

Aldehyde Alcohol Hemiacetal

$$R_1{-}C(=O){-}R_2 + HO{-}R_3 \rightleftharpoons R_1{-}C(OH)(OR_3){-}R_2$$

Ketone Alcohol Hemiketal

Formation of the cyclic forms of monosaccharides is favored because these structures have lower energies than the straight-chain forms. Cyclic forms of D-glucose are formed by the hemiacetal linkage between the C_1 aldehyde group and the C_4 or C_5 alcohol group. If the ring structure is formed between C_1 and C_4, the resulting five-membered ring structure is named D-glucofuranose because it resembles the compound furan:

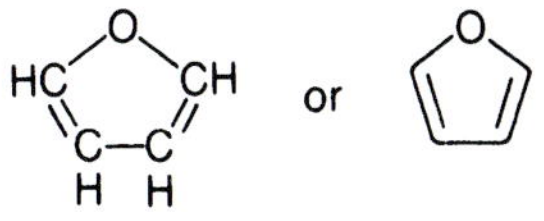

HC CH O C C H H or

If the ring structure is formed between C_1 and C_5, the resulting six-membered ring is named D-glucopyranose because it resembles the compound pyran:

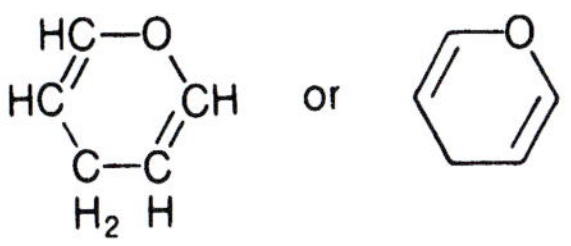

HC—O HC CH C—C H_2 H or

Aldohexoses exist in solutions mainly in six-membered pyranose ring forms, since these forms are thermodynamically more stable than furanose ring forms.

Cyclization of a monosaccharide results in the formation of an additional asymmetrical center, known as

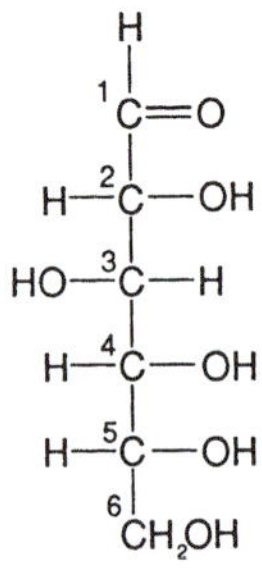

D-Glucose
A linear representation of D-glucose

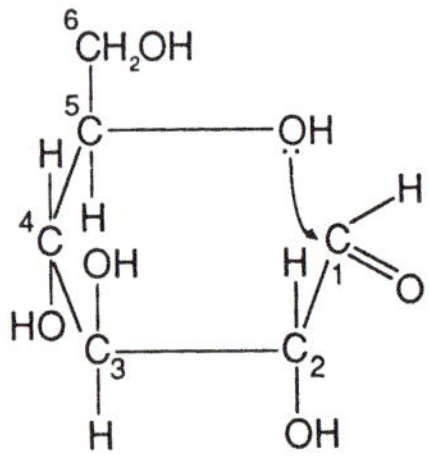

A modified representation of the D-glucose molecule, showing the formation of the hemiacetal linkage

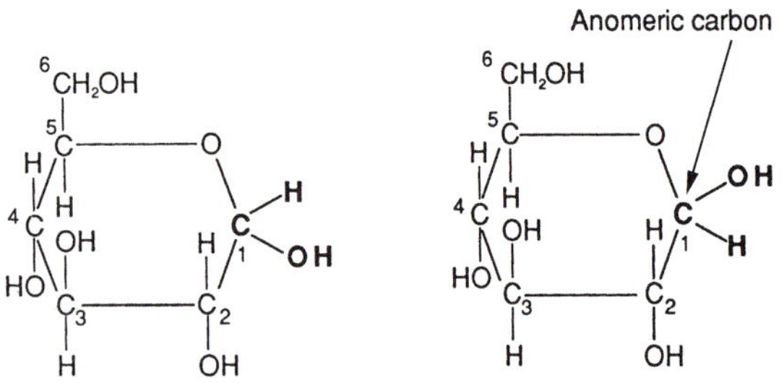

Anomeric carbon

α-D-Glucopyranose β-D-Glucopyranose

α- and β-Anomers of D-glucopyranose

FIGURE 9-3
Formation of α- and β-anomers from D-glucose.

α-D-Glucopyranose

β-D-Glucopyranose

FIGURE 9-4
Haworth projection formulas of anomers of D-glucopyranose. The thick line of the structure projects out toward the observer, and the upper edge (thin line) projects behind the plane of the paper.

the ***anomeric carbon,*** when the carbon of the carbonyl group reacts with the C_5 hydroxyl group. The two possible stereoisomers resulting from the cyclization are called α- and β-anomers (Figure 9-3). Aldohexoses in their cyclic forms have five asymmetrical centers and therefore 32 stereoisomers. In other words, each of the 16 isomers that belong to the D or L series has two anomeric forms. The systematic names for these two anomers are α-D-glucopyranose and β-D-glucopyranose. Three-dimensional representations of ring structures are frequently shown as Haworth projection formulas (Figure 9-4), in which the lower edge of the ring is presented as a thick line, to indicate that this part of the structure projects out toward the observer, and the upper edge as a thin line that projects behind the plane of the paper. Carbon atoms of the ring are not explicitly shown but occur at junctions of lines representing bonds. Sometime the hydrogen atoms are also omitted and are presume to exist wherever a bond line ends without a specifie group.

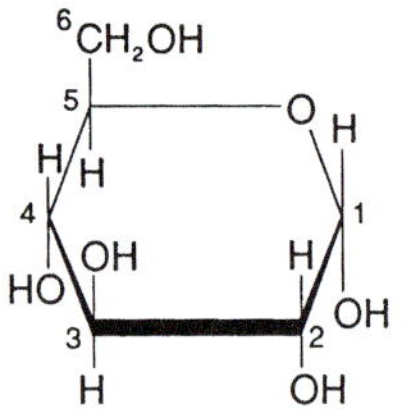

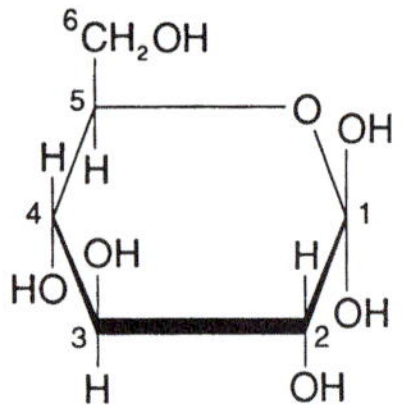

The pyranose ring is not planar, being similar to that o cyclohexane. The bond angles in cyclohexane are simila to those between the bonds of a tetrahedral carbon (i.e. 109°). In the pyranose ring, all bond angles are similar t those of cyclohexane, including the hemiacetal C–O–C bond angle, which is 111°. Most pyranoses occur in th chair conformation, in which most of the substituents ca assume equatorial positions (i.e., lie approximately in th same plane as the ring) instead of axial positions (i.e. lie approximately vertically above or below the plane o the ring). In the equatorial positions, the bulky substituen groups (–OH, –CH_2OH) can more easily be accommo dated than in the axial positions, and the preferred con formation is usually the chair conformation (Figure 9-5)

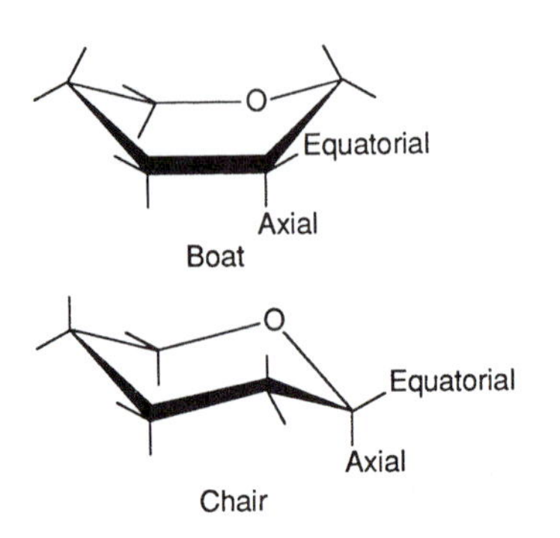

FIGURE 9-5
Conformational formulas for the boat and chair forms of pyranose.

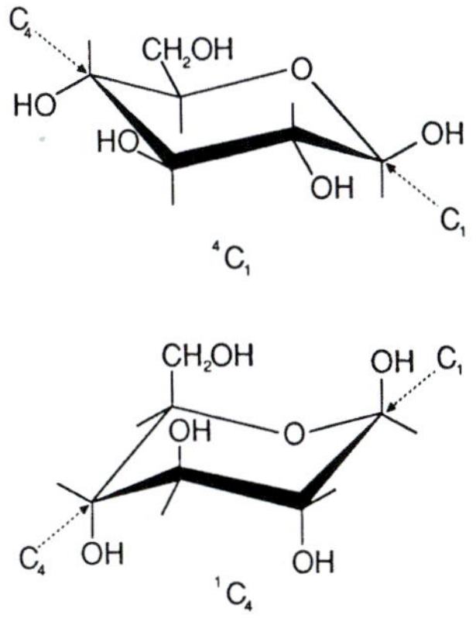

FIGURE 9-6
The two chair conformational formulas for β-D-glucopyranose.

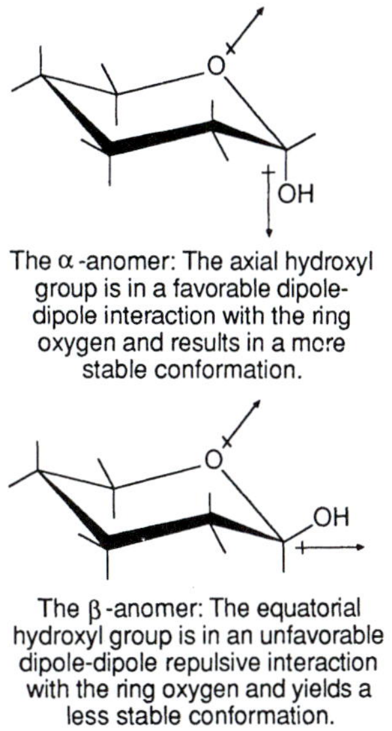

The α-anomer: The axial hydroxyl group is in a favorable dipole-dipole interaction with the ring oxygen and results in a more stable conformation.

The β-anomer: The equatorial hydroxyl group is in an unfavorable dipole-dipole repulsive interaction with the ring oxygen and yields a less stable conformation.

FIGURE 9-8
Dipole-dipole interaction between the C_1-hydroxyl group and the ring oxygen in aldohexoses. The arrows indicate the direction of the dipole moment.

However, two chair forms can be drawn for each of the two anomers of D-glucopyranose, as shown in Figure 9-6 for β-D-glucopyranose. The structure on top is designated the ${}^4C_1$ form because it contains carbon 4 and carbon 1 above and below the plane of the molecule, respectively. The designation of ${}^1C_4$ for the structure on the bottom conveys the opposite sense. In the ${}^4C_1$ conformation, all the large substituent groups are in equatorial positions, whereas in the ${}^1C_4$ conformation, they are in axial positions. ${}^4C_1$ is the more stable and therefore the preferred chair conformation.

The conformational formulas for α- and β-D-glucopyranose are shown in Figure 9-7. In aqueous solutions, the β-anomer of glucose is better solvated, more stable, and therefore the predominant form. However, for other aldohexoses (e.g., mannose and galactose), the α-anomer is more stable than the β-anomer because of the dipole effect. The dipole effect involves the C_1-hydroxyl group and the ring oxygen. In the α-anomer, the dipoles of the axial C_1-hydroxyl group and the ring oxygen are nearly antiparallel, whereas in the β-anomer, the dipoles of the equatorial C_1-hydroxyl group and the ring oxygen are nearly parallel (Figure 9-8). The dipoles impart stability to a structure when they are antiparallel. Although the dipole effect is also applicable to α- and β-anomers of glucose, the solvation effect overrides the dipole effect.

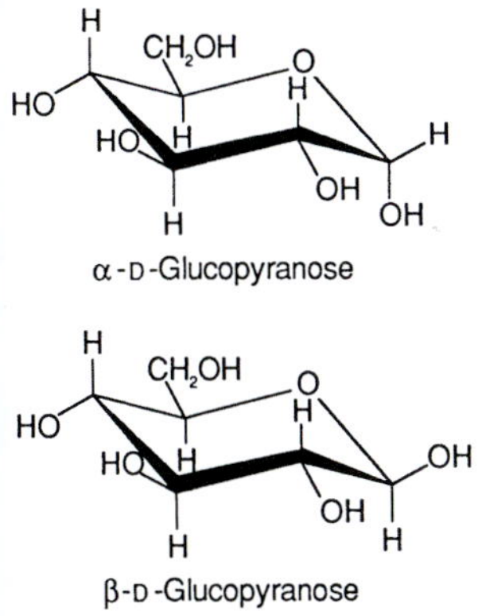

α-D-Glucopyranose

β-D-Glucopyranose

FIGURE 9-7
Conformational formulas for α- and β-D-glucopyranose.

Interconversion of α- and β-forms can be followed in a polarimeter by measuring the optical rotation. Crystallization of D-glucose from water yields α-D-glucopyranose, a form which is least soluble and has a specific rotation of $+112.2^\circ$. Ordinary crystalline glucose is in the α-form, but if the crystallization of D-glucose takes place in pyridine, β-D-glucopyranose, which has a specific rotation of $+18.7^\circ$, is obtained. Freshly prepared solutions of the α- and β-forms show specific rotations of $+112.2^\circ$ and $+18.7^\circ$, respectively. However, over a period of a few hours at room temperature, the specific rotation of both forms in aqueous solution changes and attains a stable value of $+52.7^\circ$. This change in optical rotation, ***mutarotation,*** is characteristic of sugars that form cyclic structures. It represents the interconversion of the α- and β-forms, yielding an equilibrium mixture consisting of about two thirds β-form, one third α-form, and a very small amount of the noncyclic form (Figure 9-9). Thus, the change in structure can occur in solution and attain equilibrium, which favors the formation of more stable (lowest energy) forms. However, the less stable forms may occur under certain conditions, such as during the formation of

α-D-Glucopyranose ⇌ D-Glucose ⇌ β-D-Glucopyranose

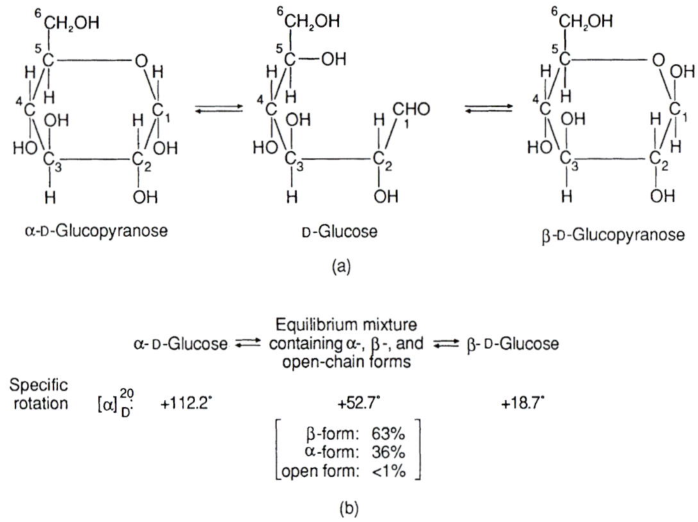

(a)

α-D-Glucose ⇌ Equilibrium mixture containing α-, β-, and open-chain forms ⇌ β-D-Glucose

Specific rotation $[\alpha]_D^{20}$: +112.2° | +52.7° | +18.7°

β-form: 63%
α-form: 36%
open form: <1%

(b)

FIGURE 9-9
(a) Interconversion of the anomeric forms of D-glucopyranose via the intermediate open-chain form. (b) Mutarotation of D-glucose.

enzyme–substrate complexes (e.g., during the hydrolysis of polysaccharides, catalyzed by the enzyme lysozyme; see Chapter 11).

D-Fructose, a ketohexose, can potentially form either a five-membered (furanose) or a six-membered (pyranose) ring involving formation of an internal hemiketal linkage between C_2 (the anomeric carbon atom) and the C_5 or C_6 hydroxyl group, respectively. The hemiketal linkage introduces a new asymmetrical center at the C_2 position. Thus, two anomeric forms of each of the fructofuranose and fructopyranose ring structures are possible (Figure 9-10). In aqueous solution at equilibrium, fructose is present predominantly in the β-fructopyranose form. However, when fructose is linked with itself or with other sugars or when it is phosphorylated, it assumes the furanose form. Fructose 1,6-bisphosphate is present in the β-fructofuranose form, with a 4:1 ratio of β- to α-anomeric forms.

Fructose is a major constituent (38%) of honey; the other constituents are glucose (31%), water (17%), maltose (a glucose disaccharide, 7%), sucrose (a glucose-fructose disaccharide, 1%), and polysaccharide (1%). The variability of these sugars in honey from different sources is quite large.

Sugars that contain a carbonyl functional group are reducing agents. Therefore, an oxidizing agent (e.g.,

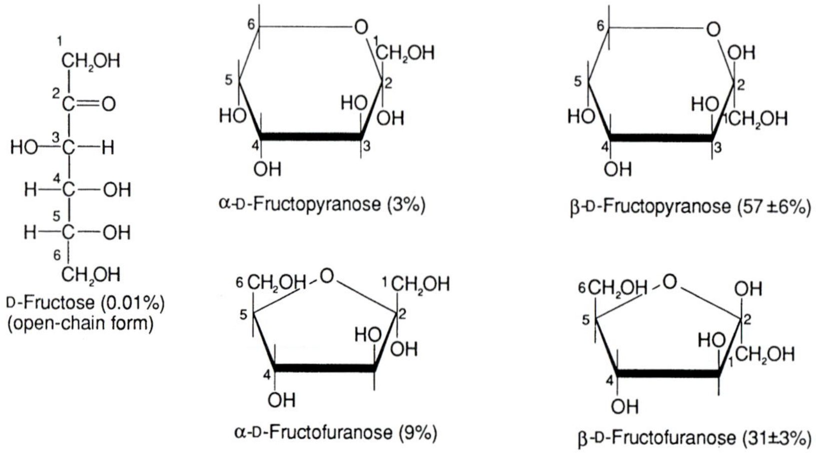

D-Fructose (0.01%) (open-chain form)

α-D-Fructopyranose (3%)

β-D-Fructopyranose (57±6%)

α-D-Fructofuranose (9%)

β-D-Fructofuranose (31±3%)

FIGURE 9-10
Structures of open-chain and Haworth projections of D-fructose and the approximate percentage of each at equilibrium in aqueous solution.

ferricyanide, cupric ion, or hydrogen peroxide) becomes reduced with the simultaneous oxidation of the carbonyl group of the sugar. This property is exploited in the estimation of reducing sugars. The individual monosaccharides are better quantitated by specific procedures, such as an enzymatic procedure (e.g., for glucose, the glucose oxidase method).

A reaction frequently used in the determination of carbohydrate structure and for its identification in tissue preparations is the ***periodate reaction.*** Periodate (sodium periodate, $NaIO_4$) oxidatively cleaves carbon–carbon bonds bearing adjacent oxidizable functional groups, e.g., vicinal glycols, α-hydroxyaldehydes, and ketones. The vicinal glycols, upon periodate oxidation, yield a dialdehyde; this reaction is quantitative. If the periodate oxidation is carried out with a compound containing three consecutive carbon atoms, each of which bears hydroxyl groups, the products are a dialdehyde and formic acid. The latter is derived from the oxidation of the central carbon atom. This observation is used to determine whether a given sugar molecule exists in the furanose or pyranose ring conformation. For example, β-D-methylfructofuranoside consumes only 1 mol of periodate with no formic acid production, whereas β-D-methylglucopyranoside yields 1 mol of formic acid with the consumption of 2 mol of periodate.

Periodate cleavage of glycogen (a polyglucose; see below) yields a polyaldehyde that can be coupled to a visible dye reaction. This is referred to as *periodic acid–Schiff* (PAS) *staining*. A tissue slice is treated with periodate, followed by staining with Schiff's reagent (basic fuchsin bleached with sulfurous acid). The polyaldehyde, if present, will combine with the NH_2 groups of the bleached dye to produce a magenta or purple complex. The specificity of the reaction can be confirmed by first treating another tissue slice with α-amylase (which breaks down the glycogen to smaller fragments that are removed by washing) and then staining it with PAS. Absence of color confirms that the material is glycogen, whereas the appearance of color suggests the presence of a nonglycogen carbohydrate, such as a glycoprotein.

Some Physiologically Important Monosaccharide Derivatives

Sugar Alcohols

Sugar alcohols are polyhydric alcohols (or polyols) formed when the carbonyl group of the monosaccharide is reduced to a hydroxyl group (Figure 9-11). For example, reduction of glyceraldehyde or dihydroxyacetone yields glycerol, a component of triacylglycerols (fatty acid esters of glycerol) and of phospholipids. D-Sorbitol (also known as D-glucitol) is formed when D-glucose is reduced. Keto sugars of more than three carbons can yield more than one sugar alcohol. For example, the chemical reduction of D-fructose yields a mixture of D-sorbitol and D-mannitol (sugar alcohol of D-mannose) because of the creation of a new asymmetrical center at C_2 when the >C=O group is converted to a H–C–OH group.

Sorbitol, with about 35–60% of the sweetness of sucrose, is used as a sweetener and flavoring agent. It is hygroscopic and therefore is used as a humectant. Sorbitol (and other sugar alcohols) can accumulate in tissues such as the lens, sciatic nerve, and renal papillae in certain disorders (e.g., *diabetes mellitus* and *galactosemia*) and can lead to pathological changes. In these instances, the intracellular accumulation of sugar alcohols is due to the

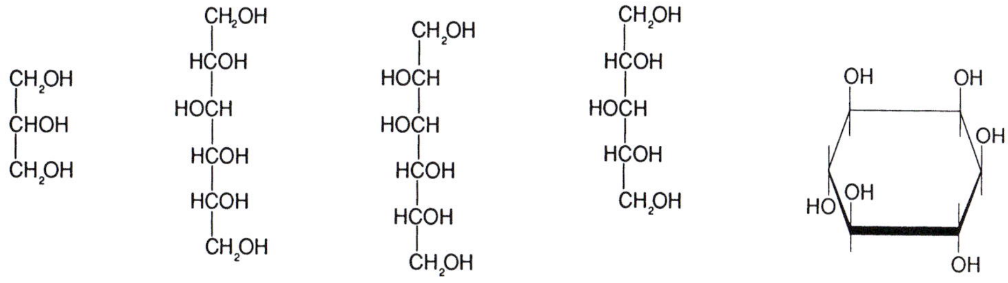

Glycerol: CH_2OH–CHOH–CH_2OH. Obtained from the reduction of either D-glyceraldehyde or dihydroxyacetone.

D-Sorbitol: CH_2OH–HCOH–HOCH–HCOH–HCOH–CH_2OH. Obtained from the reduction of either the C_1 carbonyl group of glucose or the C_2 carbonyl group of fructose.

D-Mannitol: CH_2OH–HOCH–HOCH–HCOH–HCOH–CH_2OH. Obtained from the reduction of either the C_2 carbonyl group of D-fructose or the C_1 carbonyl group of D-mannose.

Xylitol: CH_2OH–HCOH–HOCH–HCOH–CH_2OH. Obtained from the reduction of either the C_1 carbonyl group of D-xylose or the C_2 carbonyl group of D-xylulose.

Myo-inositol (OH, OH, OH, HO, OH, OH)

FIGURE 9-11
Structures of selected sugar alcohols and *myo*-inositol.

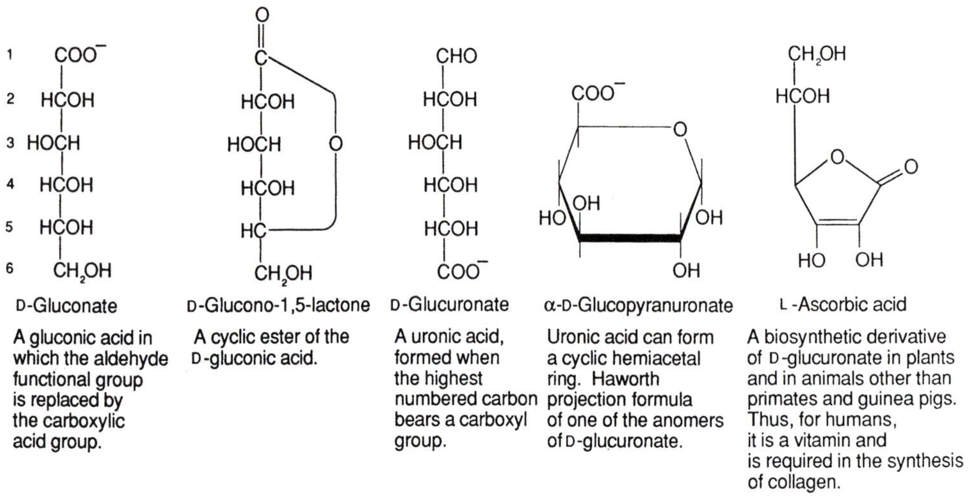

D-Gluconate: A gluconic acid in which the aldehyde functional group is replaced by the carboxylic acid group.

D-Glucono-1,5-lactone: A cyclic ester of the D-gluconic acid.

D-Glucuronate: A uronic acid, formed when the highest numbered carbon bears a carboxyl group.

α-D-Glucopyranuronate: Uronic acid can form a cyclic hemiacetal ring. Haworth projection formula of one of the anomers of D-glucuronate.

L-Ascorbic acid: A biosynthetic derivative of D-glucuronate in plants and in animals other than primates and guinea pigs. Thus, for humans, it is a vitamin and is required in the synthesis of collagen.

FIGURE 9-12
Sugar acids and their derivatives.

high levels of precursor sugars in the plasma, which are converted enzymatically to the respective sugar alcohols in the cytoplasm. Sugar alcohols are not metabolized as rapidly as their precursors, and reconversion to the precursor is also slow.

Mannitol is widely distributed in nature and occurs in the exudates of many plants. It has about half the sweetness of sucrose. Clinically, mannitol is administered intravenously as an osmotic diuretic in patients with acute renal failure. It is not metabolized appreciably, is filtered by the glomerulus, and is not reabsorbed by the tubules; hence, it is excreted in urine (Chapter 39). The nonreabsorbable solute holds water, limits back-diffusion, and thus maintains urine volume in the presence of decreased glomerular function. Intravenous mannitol is also used to relieve an increase in pressure and in volume of cerebrospinal fluid.

Xylitol, a five-carbon polyol, is widely distributed in the plant kingdom. It has about twice the sweetness of sucrose. A potential benefit of xylitol and other sugar alcohols used as sucrose substitutes may be the prevention of *dental caries*. The beneficial effect may be due to the inability of the oral microorganisms to utilize the sugar alcohols. Sugar alcohols, though derived from reducing monosaccharides, are not reducing agents and exist only in the straight-chain form, since they cannot form cyclic hemiacetal or hemiketal linkages. However, cyclic compounds related to sugar alcohols do exist, e.g., *myo*-inositol, a component of some phospholipids (Figure 9-11).

Sugar Acids

The sugar acids are obtained when a carbonyl group or a hydroxyl group is oxidized to a carboxylic acid group (Figure 9-12). The physiologically important sugar acids are ***aldonic*** and ***uronic acids.*** An aldonic acid is obtained when the aldehyde group in an aldo sugar is oxidized; thus, oxidation of D-glucose at C_1 yields D-gluconic acid.

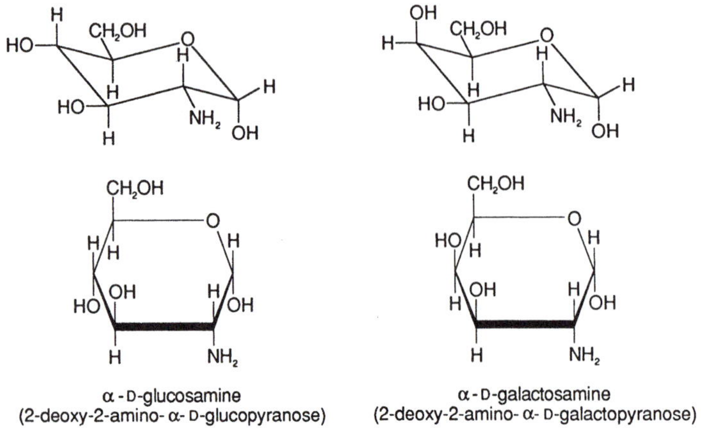

α-D-glucosamine (2-deoxy-2-amino-α-D-glucopyranose)

α-D-galactosamine (2-deoxy-2-amino-α-D-galactopyranose)

FIGURE 9-13
Structures of two amino sugars.

Aldonic acids cannot exist in hemiacetal ring forms but can form cyclic structures by forming an ester linkage between the carboxylic group and one of the hydroxyl groups of the same molecule. This type of cyclic ester is known as a ***lactone*** (e.g., D-glucono-1,5-lactone, Figure 9-12).

Uronic acids, in which the highest numbered carbon atom bears the carboxyl group, can exist in cyclic hemiacetal ring structures because they have an accessible carbonyl group. They can form glycosidic bonds (see below) with other uronic acids or monosaccharides to yield polysaccharides. At physiological pH, uronic acids exist in the ionized form (e.g., D-glucuronate). L-Ascorbate (vitamin C) in plants and in animals other than primates and guinea pigs is a biosynthetic derivative of D-glucuronate (Figure 9-12).

Amino Sugars

Amino sugars are obtained by replacing a hydroxyl group of a monosaccharide by an amino group (Figure 9-13). The most common amino sugars are the 2-aminoaldohexoses, namely, D-glucosamine and D-galactosamine. The amino groups usually occur as the N-acetyl derivatives. Amino sugars are components of structural polysaccharides and of glycosphingolipids of membranes (Chapter 10). N-Acetylmuramic acid (Figure 9-14), a constituent of a bacterial cell wall polysaccharide, has a lactyl side chain linked to C_3 of glucosamine through an ether linkage. The polysaccharide is a polymer of alternating N-acetylmuramic acid and N-acetylglucosamine residues linked in a $\beta(1 \rightarrow 4)$ glycosidic bond. *Lysozyme* (also called *muramidase*) catalyzes the hydrolysis of these bonds. The polysaccharide chains are cross-linked through short peptide chains, and the overall structure is known as a peptidoglycan (Chapter 11). N-Acetylneuraminic acid (Figure 9-15), known as *sialic*

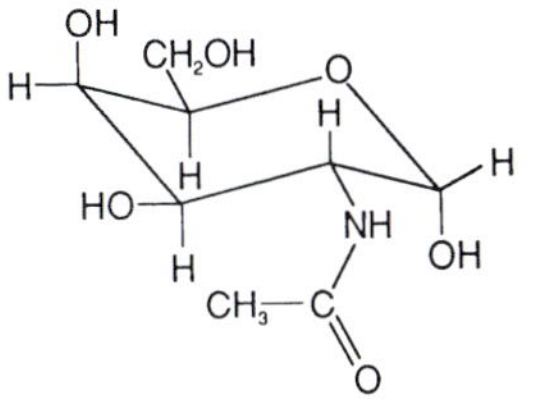

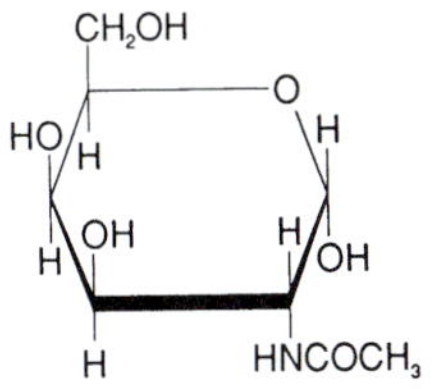

N-Acetyl-α-D-galactosamine (2-acetamido-2-deoxy-α-D-galactosamine)

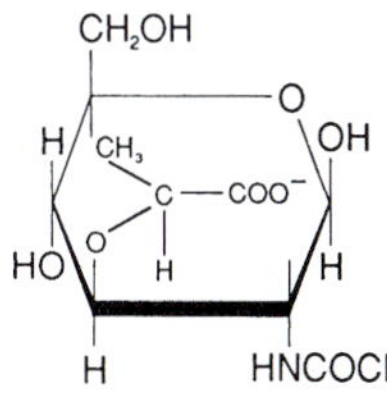

β-N-Acetylmuramic acid (3-O-lactyl-N-acetyl-β-D-glucosamine)

FIGURE 9-14
Structures of N-acetylated amino sugars.

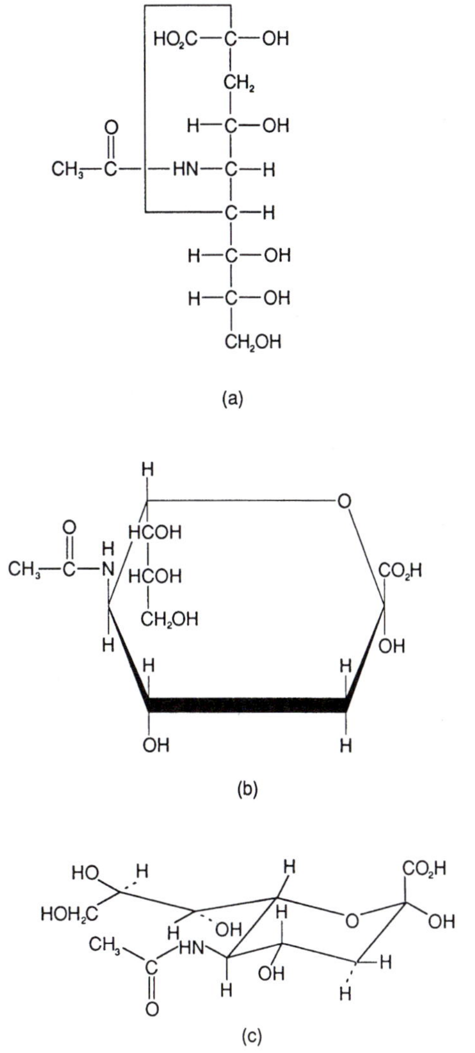

(a) (b) (c)

FIGURE 9-15
Structure of N-acetyl-D-neuraminic acid in three different representations: (a) Fischer projection; (b) Haworth projection, and (c) conformational formula.

acid, is a derivative of a ketonanose. It occurs in the oligosaccharide side chains of some glycoproteins and in gangliosides, both of which are constituents of cell membranes (Chapter 10).

Sugar Phosphates

Sugar phosphates, which are phosphoric acid esters of monosaccharides, occur as intermediates in carbohydrate metabolism. Two of these compounds, namely, ribose phosphate and deoxyribose phosphate, are constituents of nucleotides and nucleic acids. Glucose can be phosphorylated either at the C_6 primary hydroxyl group to yield glucose 6-phosphate or at the C_1 anomeric hydroxyl group to yield glucose 1-phosphate (Figure 9-16). In glucose 1-phosphate, the phosphate group can exist in either the α- or β-position. These two forms are not interconverted in solution because the substitution of the anomeric hydroxyl group by any group prevents the ring opening responsible for the equilibration of anomers. The reducing property is also lost.

Another class of sugar phosphates consists of nucleoside diphosphate sugars, in which a monosaccharide is attached through the anomeric hydroxyl group to a nucleoside diphosphate. A nucleoside contains D-ribose, an aldopentose, attached to a purine or a pyrimidine base, as in uridine diphosphate glucose (Figure 9-16). Such compounds are important in the synthesis of polysaccharides, the interconversion of sugars, and the synthesis of glycosides.

Deoxy Sugars

In deoxy sugars, one or more hydroxyl groups of the pyranose or furanose ring is substituted by hydrogen. A well-known example is 2-deoxyribose, which is a component of deoxyribonucleotides, the repeating units of deoxyribonucleic acids (DNA). Another example is L-fucose

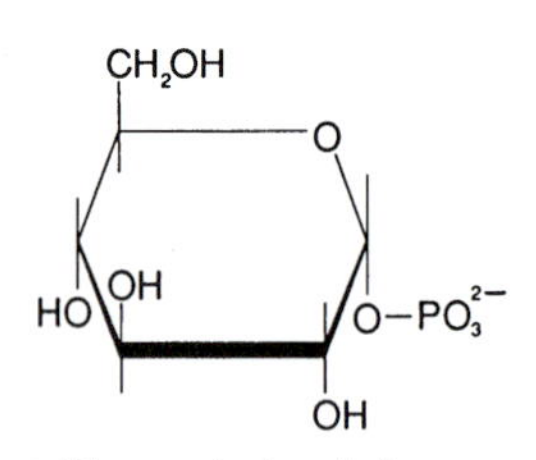

2-Deoxy-D-ribose-5-phosphate
(2-deoxy-D-ribofuranosyl-5-phosphate)

D-Glucose-1-phosphate
(α-D-glucopyranosyl phosphate)

D-Glucose-6-phosphate
(6-phospho-D-glucopyranose)

Uridine diphosphate glucose

FIGURE 9-16
Sugar phosphates and a nucleoside diphosphate sugar. When the anomeric position of a sugar is not substituted, the configuration at this position is not specified because the sugar can exist in either the α- or β-anomeric form.

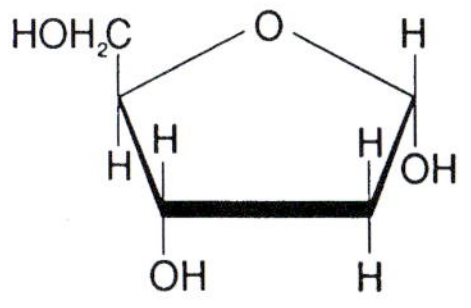

2-Deoxy-α-D-ribose

The phosphorylated derivative of deoxyribose (Fig 9–16) is a component of DNA.

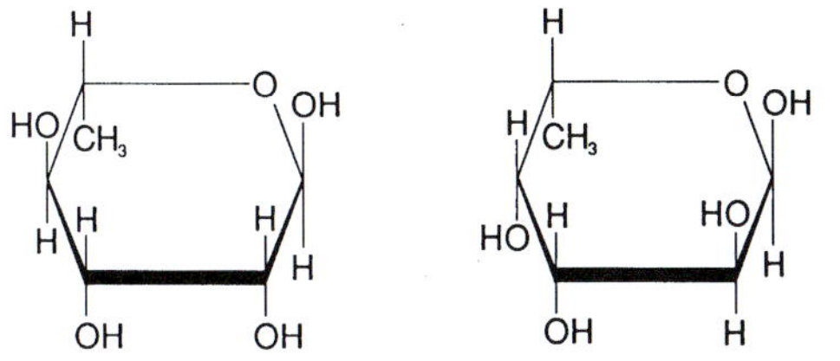

α-L-Rhamnose (6-deoxy-L-mannose)

α-L-Fucose (6-deoxy-L-galactose)

These two deoxy sugars are the rare monosaccharides that are in the L-configuration. L-Rhamnose is a constituent of lipopolysaccharides of the outer membrane layer of gram-negative bacteria. L-Fucose is a constituent of glycoproteins in cell membranes.

FIGURE 9-17
Deoxy sugars.

(6-deoxyl-L-galactose), a constituent of cell membrane glycoproteins and glycolipids and one of the few monosaccharides that exist in the L-configuration (Figure 9-17).

L-Fucose occurs either at terminal or preterminal positions of many cell surface oligosaccharide ligands. These fucosylated oligosaccharides mediate cell-cell recognition and adhesion-signaling pathways. Some of these processes include early embryologic development, blood group recognition (Chapter 10), and pathological processes such as inflammation and neoplasia. The cell-cell adhesion mediated by fucosylated oligosaccharide involves cell surface calcium-dependent binding proteins known as ***selectins.***

Glycosides

Glycosides are formed when the anomeric (hemiacetal or hemiketal) hydroxyl group of a monosaccharide undergoes condensation with the hydroxyl group of a second molecule, with the elimination of water. Formation of glycosides is an example of ***acetal*** formation, which is a reaction between a hemiacetal group and another hydroxyl group. The linkage resulting from such a reaction is known as a ***glycosidic bond.*** Glycosides are named for the sugar that provides the hemiacetal group. Thus, if glucose provides the hemiacetal group, the resultant molecule is a ***glucoside;*** if galactose provides the hemiacetal group, the result is a ***galactoside.*** An example of the formation of a glycoside is shown in Figure 9-18. When glucose reacts with methanol at an elevated temperature in the presence of an acid catalyst, a mixture of α- and β-methyl glucopyranosides is obtained. Once the anomeric hydroxyl group is substituted, properties associated with the anomeric carbon atom, namely, mutarotation, reduction, and ring size (pyranose versus furanose), are permanently lost.

The noncarbohydrate moiety of a glycoside is known as the ***aglycone.*** Methanol, glycerol, sterols, and phenols may serve as aglycones. Glycosides, which stimulate cardiac muscle contractions and are used therapeutically,

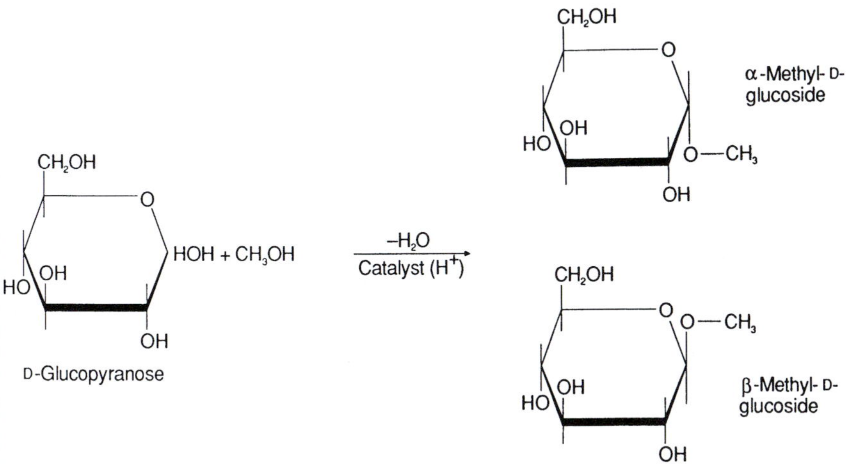

D-Glucopyranose + CH_3OH $\xrightarrow[\text{Catalyst } (H^+)]{-H_2O}$ α-Methyl-D-glucoside; β-Methyl-D-glucoside

FIGURE 9-18
Formation of glycosides.

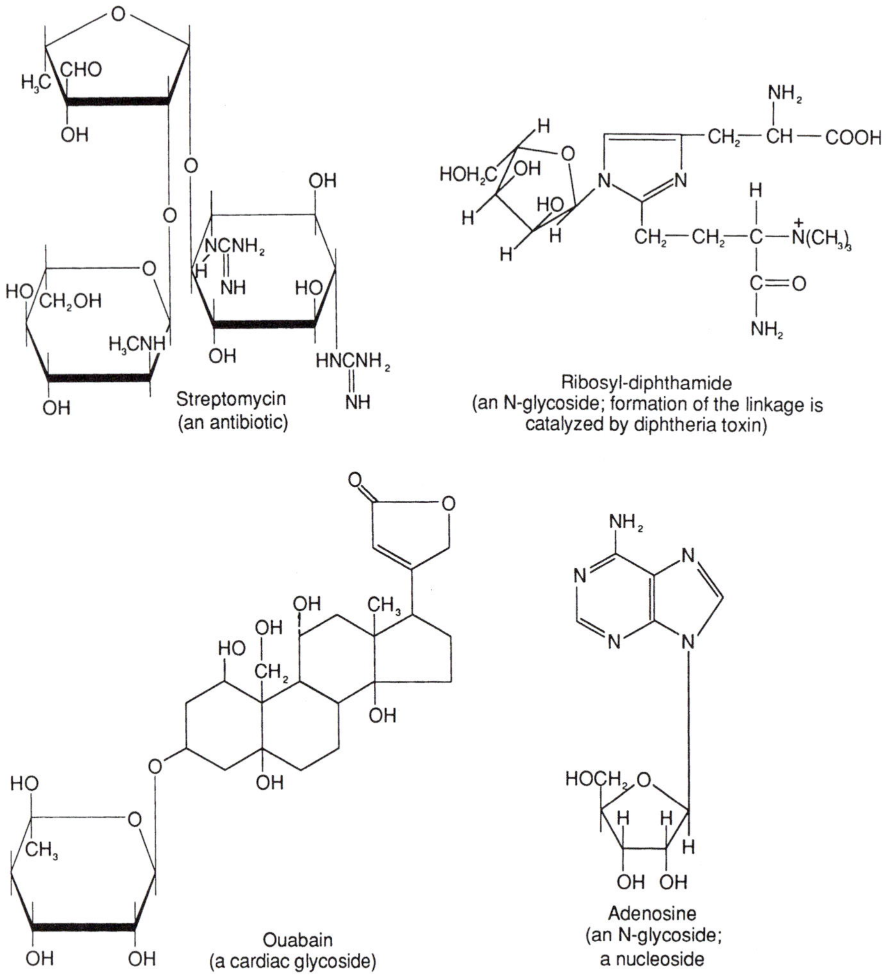

Streptomycin
(an antibiotic)

Ribosyl-diphthamide
(an N-glycoside; formation of the linkage is catalyzed by diphtheria toxin)

Ouabain
(a cardiac glycoside)

Adenosine
(an N-glycoside;
a nucleoside

FIGURE 9-19
Structures of glycosides.

contain a steroid as the aglycone component (Figure 9-19). Adenosine, a major constituent of nucleotides and nucleic acids, is an N-glycoside.

If the glycosidic linkage occurs between two monosaccharides, the compound is a ***disaccharide.*** Since the second monosaccharide contains several hydroxyl groups, various linkages are possible, thus giving rise to a number of isomers in which the anomeric carbon of at least one of the monosaccharides forms part of the glycosidic linkage.

Disaccharides

Structures of some disaccharides are illustrated in Figure 9-20.

Maltose is composed of two glucose residues joined by an α-glycosidic linkage between C_1 of one residue and C_4 of the other residue [designated $\alpha(1 \rightarrow 4)$]. In maltose, the second sugar residue has an unsubstituted anomeric carbon atom and therefore can function as a reducing agent as well as exhibit mutarotation. In ***trehalose,*** two glucose residues are joined by an α-linkage through both anomeric carbon atoms; therefore, the disaccharide is not a reducing sugar, nor does it exhibit mutarotation. ***Lactose,*** synthesized only by secretory cells of the mammary gland during lactation, is a disaccharide consisting of galactose and glucose. The glycosidic bond is a β-linkage between C_1 of galactose and C_4 of glucose (Figure 9-20). Lactose is a reducing sugar and exhibits mutarotation by virtue of the anomeric C_1 of the glucose residue. ***Lactulose*** is a disaccharide consisting of galactose and fructose linked through a β-linkage between C_1 of galactose and C_4 of fructose. It is used in the treatment of some forms of chronic liver disease (such as ***hepatic encephalopathy***) in which the ammonia content in the blood is elevated (***hyperammonemia***). Normally, ammonia produced in the gastrointestinal

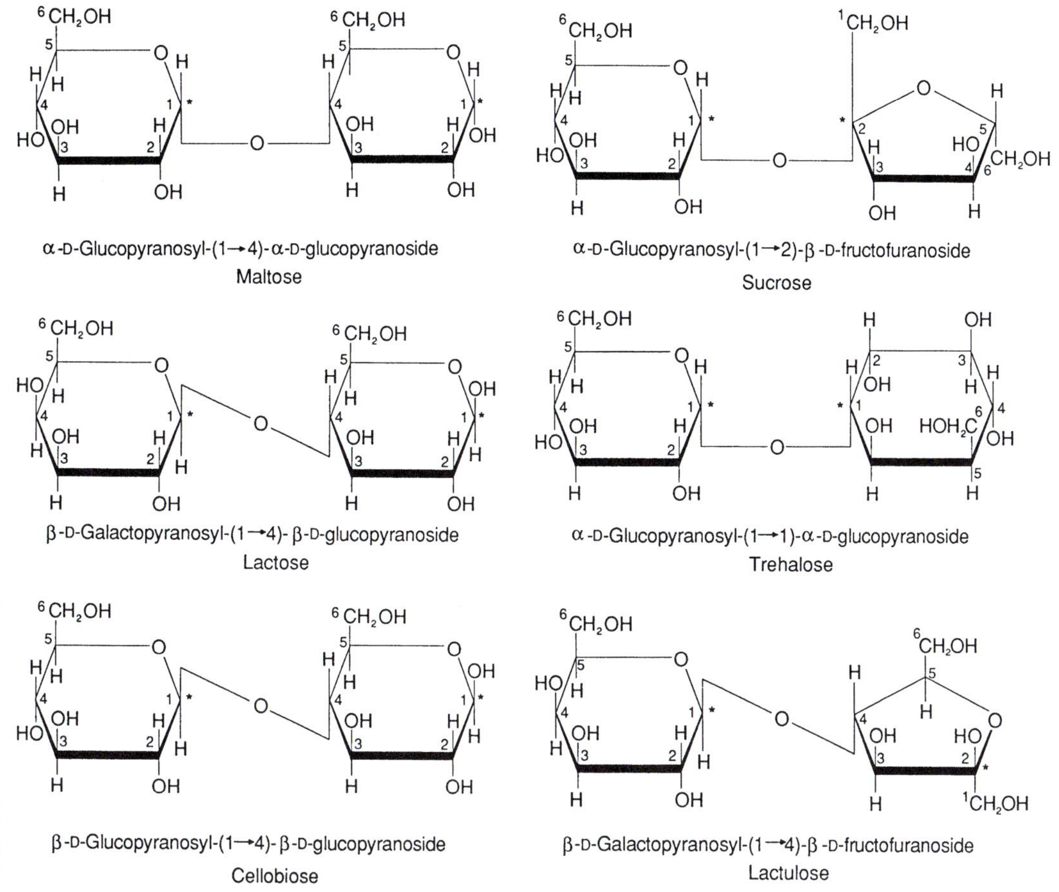

α-D-Glucopyranosyl-(1→4)-α-D-glucopyranoside
Maltose

α-D-Glucopyranosyl-(1→2)-β-D-fructofuranoside
Sucrose

β-D-Galactopyranosyl-(1→4)-β-D-glucopyranoside
Lactose

α-D-Glucopyranosyl-(1→1)-α-D-glucopyranoside
Trehalose

β-D-Glucopyranosyl-(1→4)-β-D-glucopyranoside
Cellobiose

β-D-Galactopyranosyl-(1→4)-β-D-fructofuranoside
Lactulose

FIGURE 9-20
Structures of some disaccharides.*Anomeric carbon atoms. All structures are Haworth projections.

tract, principally in the colon by microbial action, is transported to the liver via the portal circulation and inactivated by conversion to urea (Chapter 17). Oral administration of lactulose relieves hyperammonemia by microfloral conversion in the colon to a variety of organic acids (e.g., lactate) that acidify the colonic contents. Lactulose is neither broken down nor absorbed in the small intestine. Reduction of the colonic luminal pH favors conversion of ammonia (NH_3) to ammonium ion (NH_4^+), which is not easily absorbed, and thus its absorption is decreased. Reduction of luminal pH may additionally promote a microflora that causes a decrease in the production of ammonia as well as an increase in its utilization. The osmotic activity of the disaccharide and its metabolites causes an osmotic diarrhea, which is useful in eliminating toxic waste products.

Another nonabsorbable disaccharide used in the treatment of hepatic encephalopathy is lactitol (β-galactosidosorbitol). Compared to lactose, lactitol has the advantage of higher palatability and fewer side effects (e.g., flatulence). Ammonia production in the colonic lumen by urease-producing bacteria can be reduced by administering antibiotics such as neomycin or metronidazole. The therapeutic effect of the combined use of a nonabsorbable disaccharide and an antibiotic may result from the metabolism of the disaccharide by antibiotic-resistant bacteria.

Sucrose, a widely occurring disaccharide found in many plants (cane sugar and beet sugar), consists of glucose and fructose moieties linked together through C_1 of glucose and C_2 of fructose. Sucrose is not a reducing sugar and does not mutarotate.

Because of its sweet taste sucrose is consumed in large amounts. The perception of sweetness is mediated by taste buds submerged in the tongue and oral mucous membranes. The taste bud, a pear-like organ, consists of sensory cells (taste cells) interwoven with a branching network of nerve fibers. The taste bud contains two additional cell types: basal and supporting cells. Sensory cells have a short life span of about 10 days, and new cells are derived from basal cells that continually undergo mitosis. Sensory cells contain microvilli (thin hair-like projections on the surface of the cells). The microvilli protrude through the pores of the taste buds to provide a receptor surface for the perception of taste. Highly soluble and diffusible substances, such as salt (NaCl) and sugars, enter the taste pores and produce taste sensations. The chemical stimuli received by the sensory cells are transduced into electrical impulses. These impulses, in turn, are passed on to the nerve fibers through neurotransmitters. Less soluble and diffusible compounds, such as starch and protein, produce correspondingly less taste sensation.

The four primary taste sensations are sweet, salty, bitter, and sour. Each taste bud possesses different degrees of sensitivity for all four qualities, but it usually has greatest sensitivity to one or two. The integration of taste perception occurs in the cerebral cortex, which receives nerve signals arising from the taste buds that pass through the medulla and the thalamus. An important function of taste perception is to provide reflex stimuli that regulate the output of saliva. A pleasant taste perception increases saliva production, whereas an unpleasant taste reduces output. Taste also affects the overall digestive process by affecting gastric contractions, pancreatic flow, and intestinal motility.

Choice of food and dietary habits are influenced by taste and smell, which are interrelated. The sense of smell resides in receptors of specialized bipolar neuronal cells (olfactory cells) located on each side of the upper region of the nasal cavity. Like taste sensory cells, the receptors of the olfactory cells with cilia protruding into the mucus covering the epithelium, also undergo continuous renewal but with a longer turnover rate of about 30 days. The receptors of olfactory cells are stimulated by volatile airborne compounds. Since perceptions of taste and smell are triggered by chemicals, they are called *chemosensory perceptions*. At the molecular level, they are mediated by ionotropic channels and G protein coupled receptors (Chapter 30). Chemosensory perceptions are affected by a number of factors. Normal aging leads to perceptual as well as anatomical losses in chemosensory processes. Increased thresholds for both taste and smell accompany aging. For example, aged persons need two to three times as much sugar or salt as young persons to produce the same degree of taste perception. The reduction in chemosensory acuity may contribute to weight loss and malnutrition in elderly persons. Other causes of chemosensory disorders include aberrations of nutrition and hormones, infectious diseases, treatment with drugs, radiation, or surgery. Sugars exhibit different degrees of sweetness (Table 9-1). Sucrose is sweeter than the other common disaccharides, maltose and lactose. D-Fructose is sweeter than either D-glucose or sucrose. D-Fructose is manufactured commercially starting with hydrolysis of cornstarch to yield D-glucose, which is subsequently converted to D-fructose by the plant enzyme glucose isomerase.

TABLE 9-1
The Relative Sweetness of Sugars, Sugar Alcohols, and Noncarbohydrate Sweeteners

Type of Compounds	Percent Sweetness Relative to Sucrose
Disaccharides	
Sucrose	100
Lactose	20
Maltose	30
Monosaccharides	
Glucose	50–70
Fructose	130–170
Galactose	30
Sugar alcohols	
Sorbitol	35–60
Mannitol	45–60
Xylitol	200–250
Noncarbohydrate sweeteners	
Saccharin	40,000
Aspartame	16,000

Synthetic noncarbohydrate compounds can also produce a sweet taste. Saccharin, a synthetic compound, tastes 400 times as sweet as sucrose and has the following structure:

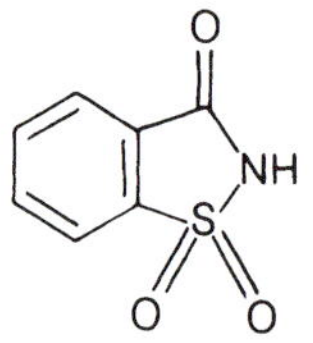

Saccharin

Another synthetic sweetener is aspartame (L-aspartyl-L-phenylalanine methyl ester), a dipeptide. Aspartame is 160 times as sweet as sucrose and, unlike saccharin, is said to have no aftertaste. Artificial sweeteners, because of their high degree of sweetness on a weight-for-weight basis compared to sucrose, contribute very little energy in human nutrition. They are useful in the management of obesity and diabetes mellitus. However, use of aspartame during pregnancy, particularly by individuals heterozygous or homozygous for phenylketonuria (Chapter 17), may be hazardous to the fetus.

The perception of sweet taste can be elicited by a wide range of chemical compounds. Two naturally occurring sweet proteins, *thaumatin* and *monellin,* are derived from the fruits of two African plants called katemfe and serendipity berries, respectively. These two proteins are intensely sweet and produce a perception of sweetness at a concentration as low as 10^{-8} mol/L. Despite the similarity in sweetness, thaumatin and monellin bear no significant structural similarities with respect to amino acid sequence or crystalline structure. However, they do exhibit immunological cross-reactivity suggesting a common chemical and structural feature. Sweet substances may act as short-term antidepressants, presumably by raising serotonin (a metabolite of tryptophan) levels in the central nervous system. This property of sweet-tasting carbohydrates may unwittingly contribute to the development of obesity in susceptible individuals.

Polysaccharides

Polysaccharides, also known as *glycans,* contain many monosaccharide units joined together by glycosidic linkages. They may be homopolysaccharides (e.g., glycogen, starch, and cellulose), which contain only one type of monomeric residue, or heteropolysaccharides, which consist of two or more different types of monosaccharide units glycosidically joined in different ways. The heteropolysaccharides have complex structures, and they may also be found covalently linked with proteins and lipids (e.g., proteoglycans and glycosphingolipids).

Starch and glycogen are energy storage forms of carbohydrate and are thus known as ***storage carbohydrates.*** When the supply of carbohydrate exceeds the needs of the cell, the excess is converted to storage forms. When the situation is reversed, the storage forms are converted to usable forms of carbohydrate. Therefore, a storage carbohydrate should be capable of rapid synthesis as well as breakdown in response to the energy requirements of the cell. As monosaccharide accumulates in the cell, its rapid conversion to insoluble, high-molecular-weight polysaccharide prevents an osmotic imbalance and also maintains a favorable concentration gradient between the intra- and extracellular compartments, which facilitates sugar transport. Starch, the storage polysaccharide of most plants and particularly of tubers (e.g., potatoes) and seeds (corn and rice), consists of a mixture of ***amylose*** and ***amylopectin.*** Amylose is an unbranched polymer of glucose in which the glucosyl residues are linked in $\alpha(1 \rightarrow 4)$ glycosidic linkages (Figure 9-21). The conformation of amylose has been elucidated by the use of stable amylose complexes prepared by reacting amylose with iodine. X-ray diffraction studies of such complexes have revealed a helical conformation with six glucose residues per turn of the helix. The amylose-iodine complex has an intense blue color, which provides the basis for the iodine test for starch.

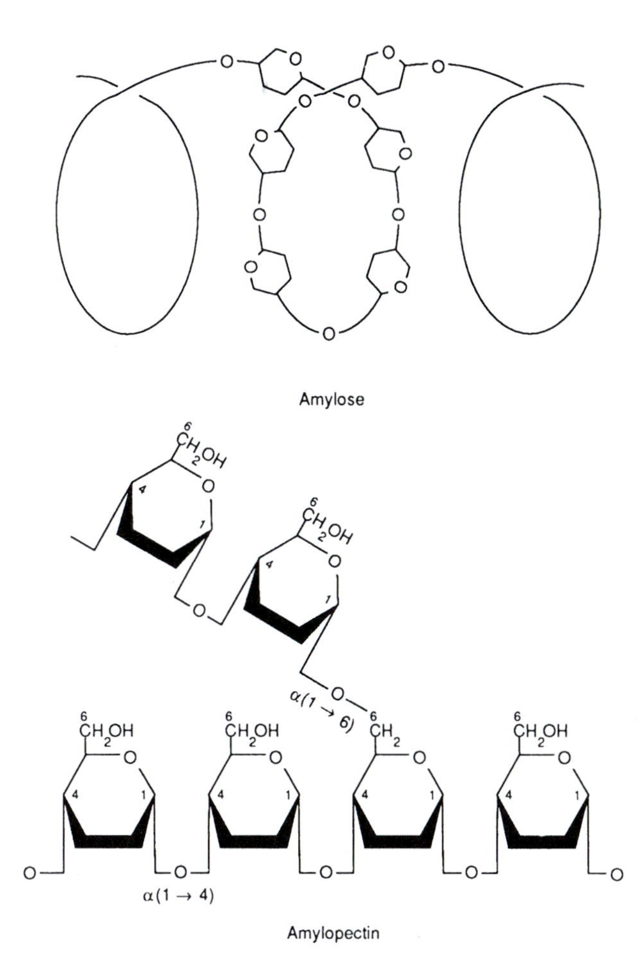

FIGURE 9-21
Structures of the starch polysaccharides amylose and amylopectin.

Amylopectin contains glucosyl units joined together in both $\alpha(1 \rightarrow 4)$ and $\alpha(1 \rightarrow 6)$ linkages, the latter linkages being responsible for branch points (Figure 9-21). Unlike amylose, amylopectin is unable to assume a stable helical conformation because of the branching. Amylopectin complexes with iodine to a much lesser extent than amylose; therefore, the amylopectin-iodine complex has a red–violet color that is much less intense than the blue of the amylose-iodine complex. Starch from different sources contains different amounts of amylose and amylopectin. In most plants, amylopectin is the more abundant form (about 75–80%). Virtually no amylose is found in starch obtained from some waxy varieties of maize (corn) and rice. Starch is easily digested by humans (Chapter 12).

Glycogen is the animal equivalent of starch in plants and functions as the main storage polysaccharide in humans. It is a branched polysaccharide of D-glucose and, like amylopectin, contains both $\alpha(1 \rightarrow 4)$ and $\alpha(1 \rightarrow 6)$ linkages, the latter forming branch points (Figures 9-22 and 9-23). Each molecule of glycogen contains one reducing glucose

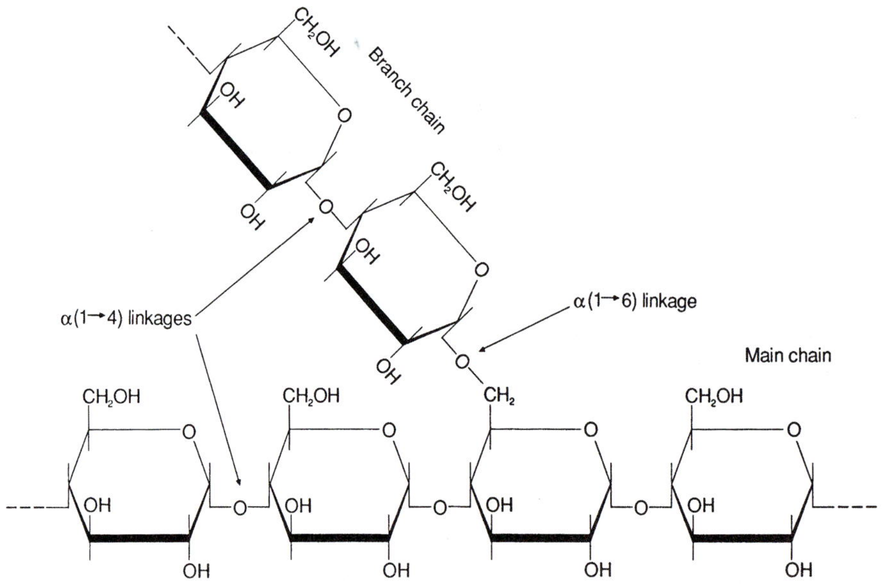

FIGURE 9-22
Linkages of glucose residues in glycogen. Glycogen is structurally similar to amylopectin but more highly branched.

residue (i.e., unsubstituted C_1 hydroxyl group), which is the terminal unit on one of the chains. At each of the other termini, the glucose residue has a free hydroxyl group at C_4, while the C_1 hydroxyl group participates in the glycosidic linkage. Synthesis and breakdown take place at these termini.

A glycogen molecule contains about 10^5 glucose units. It has no discrete molecular weight, since its size varies considerably depending on the tissue of origin and its physiological state. In the human body, most glycogen is found in liver and muscle. The functional roles of glycogen in these two tissues are quite different: in muscle, glycogen serves as an energy reserve mostly for contraction, whereas liver glycogen supplies glucose to other tissues via the blood circulatory system.

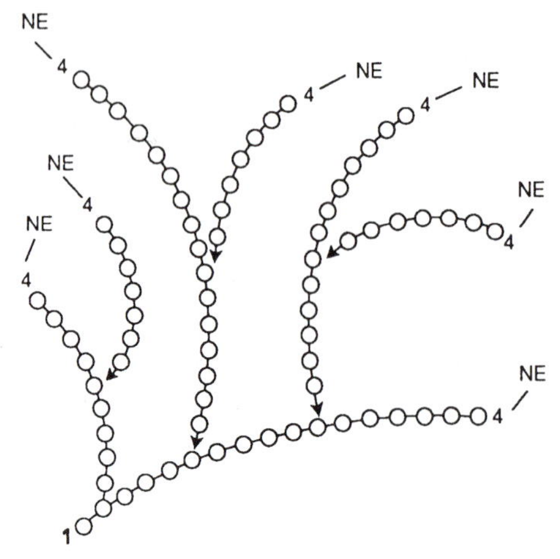

FIGURE 9-23
A diagrammatic representation of a portion of a glycogen molecule. The glucose residues are shown by circles; ○–○ indicates $\alpha(1 \rightarrow 4)$ linkages, and ○ → ○ indicates $\alpha(1 \rightarrow 6)$ linkages that are branch points. The ratio of 1,4 to 1,6 linkages varies between 12:1 and 18:1. Note that each molecule of glycogen has many nonreducing (NE) termini; further addition or cleavage of glucose units occurs at the nonreducing ends. Numbers 1 and 4 refer to C_1 (reducing end) and C_4 (nonreducing end), respectively.

Cellulose, the most abundant carbohydrate on earth, is an unbranched polymer with glucosyl residues joined together in $\beta(1 \rightarrow 4)$ linkages (Figure 9-24). Cellulose from different sources varies in molecular weight, and the number of glucose units lies in the range of 2,500–14,000. Cellulose is the structural polysaccharide in plants. Cellulose microfibrils are tightly packed aggregates of cellulose molecules, which are chemically inert and insoluble and possess significant mechanical strength.

Because of its β linkages, the preferred conformation of cellulose is one in which the ring oxygen of one residue forms a hydrogen bond with the C_3 hydroxyl group of the next residue (Figure 9-24). The 40 or more individual cellulose molecules that aggregate to form microfibrils are held together by intermolecular hydrogen bonds.

In many fungal cell walls and invertebrates (shells of crustaceans and exoskeletons of insects), the main structural polysaccharide is ***chitin,*** which is a polymer of N-acetylglucosamine linked in $\beta(1 \rightarrow 4)$ glycosidic linkage. In humans, cellulose is not digested in the small intestine but is digested in the large intestine to varying degrees by the microflora to yield short-chain fatty acids,

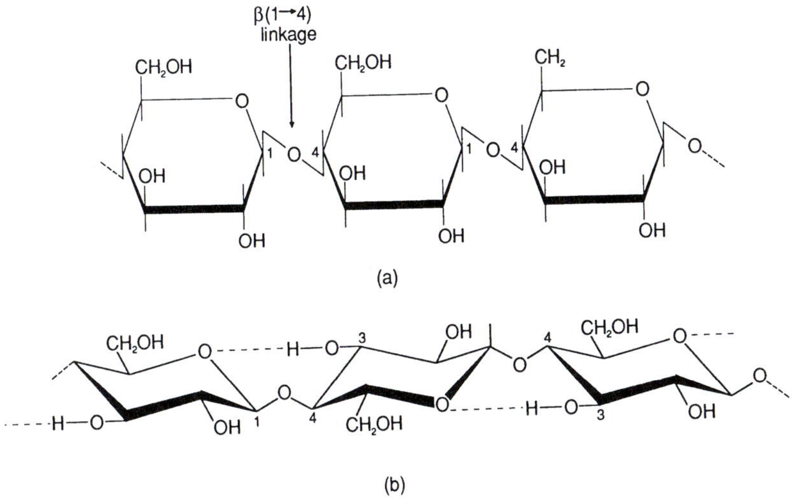

FIGURE 9-24
Structure of cellulose. (a) $\beta(1 \rightarrow 4)$ linkage of glucose residues in cellulose. (b) Conformational formula showing hydrogen bonds (-----) between the ring oxygens and the C_3 hydroxyl groups.

hydrogen, carbon dioxide, and methane. Undigested cellulose forms a part of the indigestible component of the diet, known as ***dietary fiber.*** Ruminants and termites are able to digest cellulose, which is a primary energy source for them, because they harbor microorganisms in their intestinal tract that elaborate cellulase, which catalyzes hydrolysis of the $\beta(1 \rightarrow 4)$ glucosidic linkages.

Dietary fiber consists of cellulose and other polysaccharides (hemicelluloses, pectins, gums, and alginates) and a nonpolysaccharide component, lignin. These different types of dietary fiber are derived from the cell wall and the sap components of plant cells, and their properties are listed in Table 9-2. The exact role of dietary fiber in human nutrition is not clearly understood. Each component has a different chemical composition and exhibits different actions in the gastrointestinal tract. Dietary fiber has effects on the rate of absorption of nutrients and on the fecal mass. Depending on the amount and type of the indigestible component, gastric emptying is decreased, thus delaying of entry of digestible components into the small intestine, where the nutrients are absorbed. Thus, the absorption of nutrients can be modulated by the presence of fiber, particularly pectins or guar gums. Pectins, guar gums, and lignins may reduce plasma cholesterol levels. Thus, fiber may ameliorate *diabetes mellitus* (Chapter 22) and *atherosclerosis* (Chapter 20). Since the degree of satiety is related to the bulk of food that is consumed, the amount of fiber in the diet regulates the total energy consumed. Fiber also affects nutrient density, so that diets high in fiber are normally low in energy content. These aspects are important in the development and management of obesity. The relationship of dietary fiber to colonic cancer has been the subject of epidemiological studies showing an inverse relationship between fiber consumption and risk of colonic cancer. The postulated mechanism is that the high-fiber diet results in rapid intestinal transit, so that potential carcinogens or cocarcinogens have less time to interact with the mucosa. Various components of fiber may also bind carcinogens or dilute their concentrations by increasing the intraluminal bulk.

Although dietary fiber is resistant to human digestive enzymes, some fiber is digested by microorganisms that reside in the colon, yielding organic anions such as acetate, propionate, and butyrate. Fiber and the osmotically active anions increase the size and wetness of the stools because they attract water and form gels. The overall effect of a high-fiber diet is a larger fecal mass, decrease in intraluminal pressure, reduction in transit time through the colon, and greater ease and frequency of defecation. These effects are useful in the treatment of some gastrointestinal disorders. For example, a high-fiber diet has been recommended in the treatment of ***diverticulosis,*** in which outpouchings of colonic mucosa occur through the muscularis layer at points of weakness in the muscle wall. The diverticuli can become infected (*diverticulitis*) and produce bleeding, altered bowel function, and pain. Cross-cultural epidemiological studies link low-fiber diets with diverticulosis, cancer of the colon and rectum, coronary heart disease, diabetes mellitus, appendicitis, varicose veins (dilated tortuous veins), and hemorrhoids (enlarged veins of the lower rectum and anus). Despite gaps in our knowledge about the role of dietary fiber in human nutrition, adequate amounts probably are beneficial in the prevention of some chronic diseases. Consumption of carbohydrates in their

TABLE 9-2

Composition and Properties of Dietary Fiber with Respect to Human Nutrition

Type of Dietary Fiber and Food Sources	Chemical Composition	Properties
Cellulose Whole-wheat flour, Bran, Cabbage family, Dried peas/beans, Apples, Root vegetables	Polymer of glucose with $\beta(1\rightarrow4)$ linkages.	Not digested in the small intestine; a significant amount is digested by the bacterial flora of the large intestine.
Hemicellulose Bran, Cereals, Whole grains	Heterogeneous group of polysaccharides; the major group consists of pentose polymers (pentosans), xylans, and arabinosylans; the second group consists of hexose polymers, galactans, and mannans; and the third group consists of uronic acid polymers containing either galacturonic or glucuronic acid.	Associated with cellulose in plant tissues not digested in the small intestine but digested by the bacterial flora of the large intestine. Can be extracted from cell walls by alkaline solutions.
Pectins Apples, Citrus fruits, Strawberries	Mixture of galactouronan [a polymer of galacturonic acid linked in $\beta(1\rightarrow4)$ linkages] and galactan and arabinan in varying proportions depending on the source. Carboxylate groups of the uronic acids are either free or esterified with methyl groups.	Function as intercellular cementing material in plant tissues. When solubilized, the product has a characteristic gel texture (viscous and sticky properties). Not digested in the small intestine but digested to a small extent by the bacterial flora of the large intestine. Pectins are extracted by acidic solutions or solutions containing chelating agents.
Gums and Alginates Oatmeal, Dried beans, Other legumes	Heterogeneous group of polysaccharides. Some gums are galactomannans (e.g., guar gum). Alginates are polymannuronic acids.	Gums form viscous solutions; alginates are used as thickening agents and stabilizers in food. Not digested in the small intestine but digested to varying extents by the bacterial flora of the large intestine.
Lignins Mature vegetables, Wheat	Nonpolysaccharide polymer found in woody plant tissues. The nature of the monomers is not completely known, but appears to involve coniferyl and sinapyl or related alcohols.	Totally indigestible even by ruminants.

ative form rather than in a processed state may provide an dequate amount of fiber. Fiber should be derived from a ariety of plant foods such as high- and low-fiber breads nd cereals, fruits, vegetables, legumes, nuts, and seeds, nd eaten as part of normal meals. The recommended di-tary fiber intake for adults is 20–25 g/day or 10–13 g per 000 kcal (4184 kJ) of energy intake. A suggested intake f dietary fiber for children older than 2 years is the age of he child plus 5–10 g/day. The consumption of dietary fiber n childhood is also associated with health benefits, such s normal laxation, and may reduce the future risk of car-iovascular disease, some cancers, and diabetes mellitus ype 2.

Supplemental Readings and References

L. B. Ferzoco, V. Raptopouoos, and W. S. Len: Acute diverticulitis. *New England Journal of Medicine* **338,** 1521 (1998).

J. S. Hampl, N. M. Betts, and B. A. Benes: The "age +5" rule: Comparison of dietary fiber intake among 4- to 10-years old children. *Journal of the American Dietetic Association* **98,** 1418 (1998).

J. L. Listinsky, G. P. Siegal, and C. L. Listinsky: α-L-Fucose: A potentially critical molecule in pathologic processes including neoplasia. *American Journal of Clinical Pathology* **110,** 425 (1998).

Position of the American Dietetic Association: Health implications of dietary fiber. *Journal of the American Dietetic Association* **97,** 1157 (1997).

S. M. Riordan and R. Williams: Treatment of hepatic encephalopathy. *New England Journal of Medicine* **337,** 473 (1997).

C. L. Williams (Ed.): The role of dietary fiber in childhood. *Pediatrics* **96,** 985 (1995).

CHAPTER 10

Heteropolysaccharides I: Glycoproteins and Glycolipids

10.1 Glycoproteins

Glycoproteins, a wide range of compounds of diverse structure and function, are components of cell membranes, intercellular matrices, and extracellular fluids, such as plasma. They occur in soluble and membrane-bound forms (Table 10-1).

The proportion of carbohydrate varies considerably in glycoproteins derived from different tissues or from various sources. Collagen, for example, contains an amount of carbohydrate that varies with the source: skin tissue collagen, about 0.5%; cartilage collagen, about 4%; and basement membrane collagen, more than 10%. Glycoproteins containing high amounts of carbohydrate include glycophorin, a membrane constituent of human erythrocytes, about 60%, and soluble blood group substances, as much as 85%.

In glycoproteins, the protein and the carbohydrate residues are bound in covalent linkage. There are five common types of carbohydrate, four of which are common to human glycoproteins (Figure 10-1). All the linkages are either N- or O-glycosidic bonds. The amino acid residue that participates in the N-glycosidic linkage is asparagine, and the amino acid residues that participate in O-glycosidic linkage are serine, threonine, hydroxylysine, and hydroxyproline. The glycoproteins exhibit microheterogeneity, i.e., glycoproteins with an identical polypeptide sequence may vary in the structure of theire oligosaccharide chains. Microheterogeneity arises from incomplete synthesis or partial degradation and poses problems in the purification and characterization of glycoproteins. For example, human serum α_1-acid glycoprotein has five linkage sites for carbohydrates and occurs in at least 19 different forms as a result of differences in oligosaccharide structures.

The oligosaccharide side chains of glycoproteins consist of only a limited number (about 11) of different monosaccharides. These monosaccharides are hexoses and their derivatives (N-acetylhexosamine, uronic acid, and deoxyhexose), pentoses, and sialic acids derived from neuraminic acid, a nine-carbon sugar. The most common of the many different types of sialic acids is N-acetylneuraminic acid. (All of these sugars are in the D-configuration, unless indicated otherwise.) The sugar residues of the oligosaccharide chains are not present in serial repeat units (unlike the proteoglycans, Chapter 11), but they do exhibit common structural features. For example, the oligosaccharide chains bound in N-glycosidic linkages may be envisaged as consisting of two domains. The inner domain, common to all glycoproteins, is attached to the protein via an N-β-glycosidic linkage between an asparaginyl residue and N-acetylglucosamine. The inner domain consists of a branched polysaccharide, which is trimannosyl-di-N-acetylglucosamine (Figure 10-2). The peripheral mannose residue of the inner domain is linked

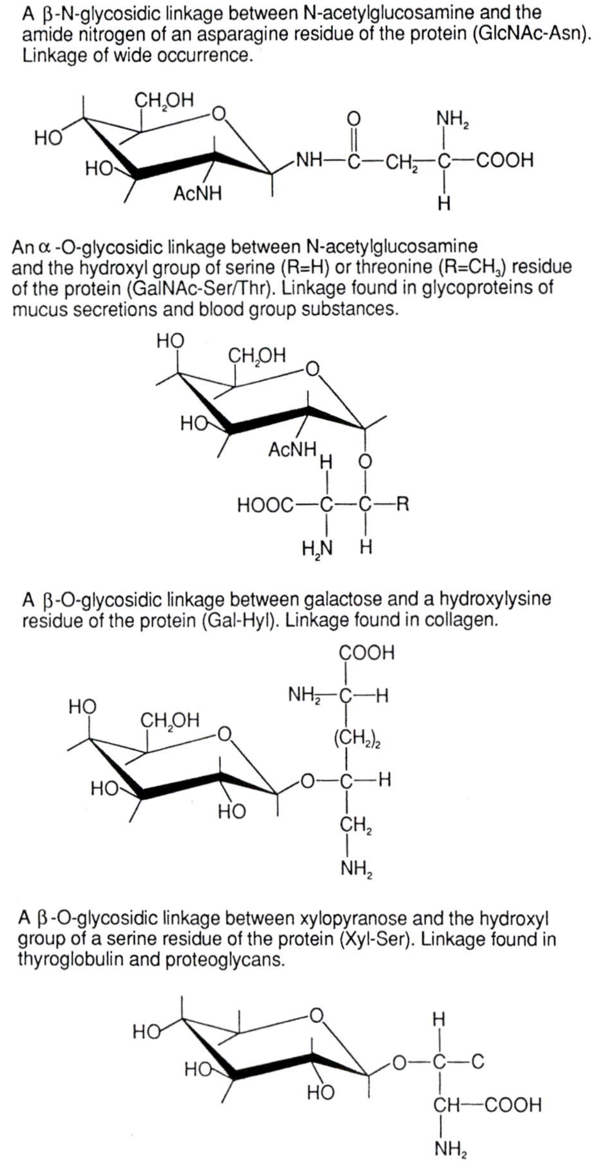

A β-N-glycosidic linkage between N-acetylglucosamine and the amide nitrogen of an asparagine residue of the protein (GlcNAc-Asn). Linkage of wide occurrence.

An α-O-glycosidic linkage between N-acetylglucosamine and the hydroxyl group of serine (R=H) or threonine (R=CH_3) residue of the protein (GalNAc-Ser/Thr). Linkage found in glycoproteins of mucus secretions and blood group substances.

A β-O-glycosidic linkage between galactose and a hydroxylysine residue of the protein (Gal-Hyl). Linkage found in collagen.

A β-O-glycosidic linkage between xylopyranose and the hydroxyl group of a serine residue of the protein (Xyl-Ser). Linkage found in thyroglobulin and proteoglycans.

FIGURE 10-1
Carbohydrate-peptide linkages in glycoproteins.

to an oligosaccharide chain, known as the outer domain. The outer domain is made up of either oligosaccharides consisting of mannose residues (oligomannosidic types) or N-acetyllactosamine units (complex types, Figure 10-3). The latter are disaccharides consisting of galactosyl-β-(1 $\rightarrow$ 4)-N-acetylglucosamine units.

Glycoproteins with O-glycosidic linkages (Figure 10-4) do not show the common features of glycoproteins with N-glycosidic linkages. The number of sugar residues may vary from one (e.g., collagen) to many (e.g., blood group substances). Many glycoproteins, however, do show the presence of a common disaccharide constituent, namely, galactosyl-β-(1 $\rightarrow$ 3)-N-acetylgalactosamine, which is

TABLE 10-1
Some Functions of Glycoproteins in the Body

Function	Examples
Structure	Collagen
Lubrication and protection	Epithelial mucins, synovial fluid glycoproteins
Transport	Ceruloplasmin (copper carrier), transferrin (iron carrier)
Endocrine regulation	Thyrotropin, chorionic gonadotropin, erythropoietin
Catalysis	Proteases, nucleases, glycosidases, hydrolases
Defense against infection	Immunoglobulins, complement proteins, interferon
Membrane receptors	Receptors for hormones (e.g., insulin), acetylcholine, cholera toxin, electromagnetic radiation (e.g., rhodopsin)
Antigens	Blood group substances
Cell–cell recognition and adhesion	Fibronectin, laminin, chondronectin
Miscellaneous	Glycophorin (an intrinsic red blood cell membrane constituent), intrinsic factor (essential for absorption of dietary vitamin B_{12}), clotting factors (e.g., fibrinogen)

linked to either serine or threonine. In collagens, the O-glycosidic linkages occur via hydroxyproline or hydroxylysine residues, which are unique to collagens. A given glycoprotein may contain oligosaccharide chains of both N- and O-glycosidic types. If more than one N-glycosidic carbohydrate linkage site are present in a glycoprotein, they are usually separated by several amino acid residues. In contrast, O-glycosidic carbohydrate linkages

Man α(1→6)
Man α(1→3) Man β(1→4) GlcNAc β(1→4) GlcNAc β→Asn

or schematically,

FIGURE 10-2
Structure of the common branched inner domain of oligosaccharides linked in N-glycosidic linkages with asparagine. Man = Mannose ▼, GlcNAc = N-acetylglucosamine ●, Asn = Asparagine. [Adapted with permission from N. Sharon and H. Lis, *Chem. Eng. News*, p. 28 (March 30 1981) ©1981 by the American Chemical Society.]

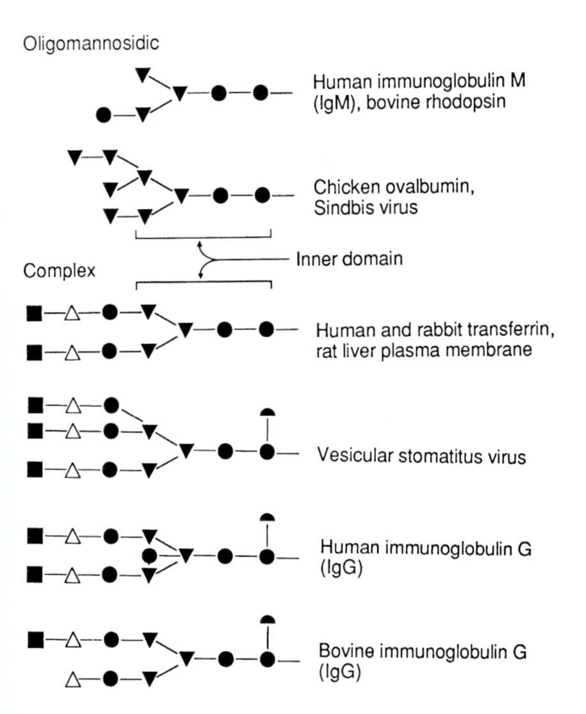

FIGURE 10-3
N-Glycosidically linked oligosaccharides. The oligomannosidic type is rich in mannose residues, whereas the complex type contains carbohydrate residues in the outer domain. Both types have a common core. Δ = galactose, ▼ = mannose, ▲ = L-fucose, ● = N-acetylglucosamine, ■ = N-acetylneuramine acid. [Adapted, with permission, from N. Sharon and H. Lis, Special Report, *Chem. Eng. News*, p. 28 (March 30, 1981) © 1981 by the American Chemical Society.]

may be found in adjacent hydroxyamino acid residues, or they may occur in close proximity, e.g., in glycophorin, human chorionic gonadotropin, and antifreeze glycoprotein. This last is found in the blood of Arctic and Antarctic fish species and other species on the eastern coast of North America. It contains a very high amount of carbohydrate, since every threonine residue of the glycoprotein is linked with a galactosyl-β-(1 $\rightarrow$ 3)-N-acetylgalactosamine unit. The protein consists of the repeating tripeptide sequence of alanyl-alanyl-threonine. The presence of antifreeze glycoprotein, along with high concentrations of NaCl in the blood, prevents water from freezing in blood vessels and permits survival at the low temperatures of polar seawater. This freezing-point depression by antifreeze glycoproteins has been attributed to their highly hydrated and expanded structure, which interferes with the formation of ice crystals.

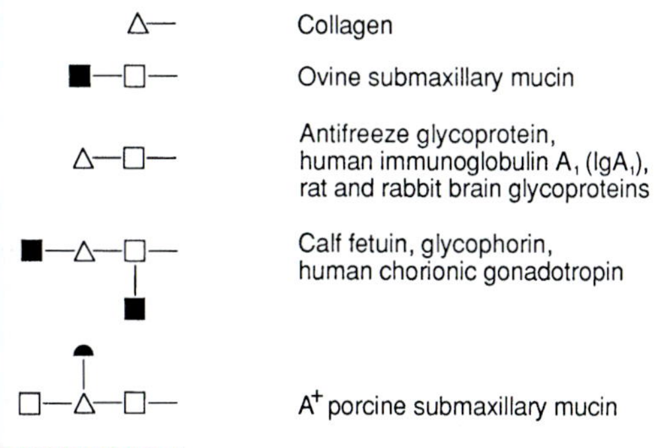

FIGURE 10-4
O-Glycosidically linked oligosaccharides. The sugar residues vary from one to many and are not arranged in any particular pattern. Δ = Galactose; ▲ = L-Fucose; □ = GalNAc; ■ = N-acetylneuraminic acid. [Adapted with permission from N. Sharon and H. Lis, Special report, *Chem. Eng. News*, p. 28 (March 30, 1981). ©1981 by the American Chemical Society.]

In all glycoproteins, the polypeptide component is synthesized first on the membrane-bound ribosomes of the rough endoplasmic reticulum; carbohydrate side chains are added during passage through the endoplasmic reticulum and Golgi apparatus. The carbohydrate additions involve specific glycosyltransferases and their substrates (uridine diphosphate sugars) and, in some glycoproteins, an oligosaccharide carrier known as dolichol (a lipid). (See also Chapters 16 and 25.)

Glycoproteins can also be formed by addition of carbohydrate residues without any of the complex enzymatic pathways of carbohydrate addition. This process, known as nonenzymatic glycation, proceeds by the condensation of a monosaccharide, usually glucose, with certain reactive amino groups on the protein. The initial, labile Schiff base adduct slowly rearranges to the stable ketoamine or fructosamine form (Chapter 2). For example, a small fraction of hemoglobin A, the major hemoglobin of adult humans, is present in the red blood cells as glycated hemoglobin (HbA_{1C}). The glycation of hemoglobin is a continuous process occurring throughout the 120-day life span of the red cell. In HbA_{1C}, glucose is incorporated via an N-glycosidic linkage into the N-terminal amino group of valine of each β-chain. Enhanced levels of HbA_{1C} occur in individuals with diabetes mellitus (Chapter 22), and measurement of glycated hemoglobin has been useful in monitoring the effects of therapy. Human serum albumin, which has a half-life of 19 days, is also subjected to nonenzymatic glycation, producing a stable condensation product known as fructosamine. Fructosamine is a generic term applied to the stable condensation product of glucose with serum proteins, of which albumin is quantitatively the largest fraction. Measurement of fructosamine concentration provides a means by which short-term (1–3 weeks) plasma glucose levels can be estimated, whereas measurement of HbA_{1C} concentration reflects integrated plasma glucose levels over a longer period (2–3 months). Human lens proteins, α-, β-, and γ-crystallins, which have much longer life spans than other proteins in the body, also undergo age-dependent, nonenzymatic glycation at the ε-amino groups of their lysine residues. In diabetics, this process occurs twice as often as in normal individuals of comparable age. Lens cells, like red blood cells, do not require insulin for the inward transport of glucose. However, the extent of nonenzymatic glycation of crystallins is much lower than that of hemoglobin (about 2.4% versus 7.5% at age 50) because

1. The lens cells contain about one sixth the glucose of red cells.
2. The content of lysine is lower in crystallin compared to hemoglobin
3. The lysine residues are inaccessible in crystallins owing to the high content of β-pleated sheet structures packed in a structured array oriented orthogonally to the lens optic axis, restricting rotational and translational movement. Crystallins contain almost no α-helix, whereas hemoglobin, a globular protein, possesses a high content of α-helical structure and can freely rotate in the fluid media, exposing the lysine residues.

Crystallins constitute 90% of the soluble proteins of the lens cells (also called *fiber cells*). The human lens, a transparent, biconvex, elliptical, semisolid, avascular structure, is responsible for focusing the visual image onto the retina. The lens grows throughout life at a slowly decreasing rate, building layer upon layer of fiber cells around a central core and never shedding the cells. Crystallin turnover is very slow or nonexistent. In addition to nonenzymatic glycation, crystallins undergo other age-dependent, post-translational modifications *in vivo:* formation of disulfide bonds and other covalent cross-links, accumulation of high-molecular-weight aggregates, deamidation of asparagine and glutamine residues, partial proteolysis at characteristic sites, racemization of aspartic acid residues, and photo-oxidation of tryptophan. Some of these processes contribute to the increasing amount of insoluble crystallins during aging. Nonenzymatic incorporation of carbohydrates into proteins *in vivo* can be extensive and may contribute to the pathophysiology of diabetes mellitus and galactosemia.

10.2 Cell Membrane Constituents

Various aspects of the cell membrane are discussed throughout this text, and a brief introduction is presented here. The living system's ability to segregate from and protect itself against—and interact with and against—changes in the external environment is accomplished by membranes. In the body, membranes function at the level of tissues, cells, and intracellular domains. They function as protective barriers and as transducers of extracellular messages carried by the chemical agents because they have recognition sites that interact with metabolites, ions, hormones, antibodies, or other cells in a specific manner. This characteristic selectivity of membranes to interact with specific molecules confers unique properties on a given cell type. Within the cell, the membranes of organelles are highly differentiated and have properties consistent with metabolic function. Examples include electron transport and energy conservation systems in the mitochondrial membrane, protein biosynthesis in the rough endoplasmic reticulum, modification and packaging of proteins for export in the membranes of the Golgi complex, drug detoxification in the smooth endoplasmic reticulum, and light reception and transduction in the disk membranes or retinal cells.

The membrane constituents are lipids (phospholipids, glycosphingolipids, and cholesterol; Figure 10-5), carbohydrates, and proteins. The ratio of protein : lipid : carbohydrate on a weight basis varies considerably from membrane to membrane. For example, the human erythrocyte membrane has a ratio of about 49:43:8, whereas myelin has a ratio of 18:79:3. The composition of the normal human erythrocyte membrane is shown in Table 10-2. All membrane lipids are ***amphipathic*** (i.e., polar lipids). The polar heads of the phospholipids may be neutral, anionic, or dipolar. The surface of the membrane bears a net negative charge. The distribution of lipid constituents in the bilayer is asymmetrical. For example, in the erythrocyte membrane, phosphatidylethanolamine and phosphatidylserine are located primarily in the internal monolayer, whereas phosphatidylcholine and sphingomyelin are located in the external monolayer.

Lipids are organized in bilayers that account for most of the membrane barrier properties. Membrane proteins may be ***peripheral*** (extrinsic) or ***integral*** (intrinsic). Peripheral proteins are located on either side of the bilayer and are easily removed by ionic solutions, whereas integral proteins are embedded in the bilayer to varying degrees (Figure 10-6). Some integral proteins penetrate the bilayer and are exposed to both external and internal environments. By spanning both external and internal environments of the cell, these proteins may provide a means of communication across the bilayer that may be useful in the transport of metabolites, ions, and water, or in the transmission of signals in response to external stimuli provided by hormones, antibodies, or other cells. Because of hydrophobic interactions, isolation of integral proteins requires harsh methods, such as physical disruption of the bilayer and chemical extraction procedures using

TABLE 10-2

*Composition of Normal Human Red Blood Cell Membranes (Ghosts)**

Component	WT%	Grams/Ghost ($\times 10^{13}$)	Approximate Number of Molecules/Ghost ($\times 10^{6}$)	% in Outer Half of Bilayer	% in Inner Half of Bilayer
Proteins and glycoproteins	55	5.7	3.7		
Lipids					
Phospholipids	28	3.0	250		
Sphingomyelin	7.1	0.76	65	80	20
Phosphatidylcholine	7.9	0.85	70	75	25
Phosphatidylethanolamine	7.8	0.84	70	20	80
Phosphatidylserine	3.8	0.41	35	0	100
Phosphatidylinositols	0.4	0.04	3	20	80
Phosphatidic acid	0.4	0.04	3	Unknown	Unknown
Other	0.6	0.06	5	Unknown	Unknown
Cholesterol	13	1.3	195	~50	~50
Glycolipids	3	0.3	10	100	0
Free fatty acids	1	0.1	20	Unknown	Unknown
	100	10.4	480		

*Reproduced with permission from J. B. Stanbury, J. W. Wyngaarden, D. S. Fredrickson, et al., Eds.: *The Metabolic Basis of Inherited Disease,* 5th ed. (McGraw-Hill, 1983).

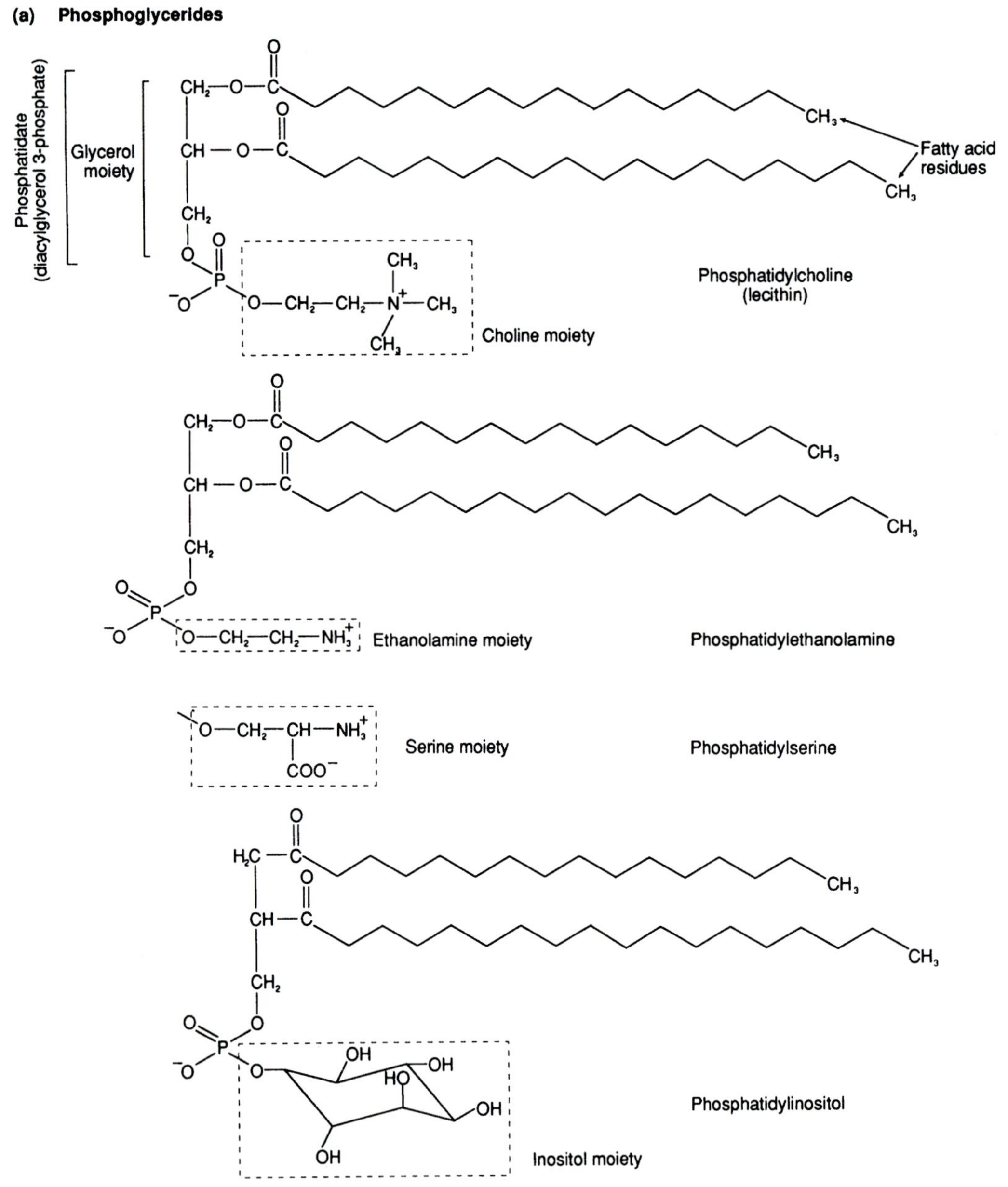

(a) Phosphoglycerides

Phosphatidate (diacylglycerol 3-phosphate)
Glycerol moiety
Fatty acid residues
Choline moiety
Phosphatidylcholine (lecithin)
Ethanolamine moiety
Phosphatidylethanolamine
Serine moiety
Phosphatidylserine
Inositol moiety
Phosphatidylinositol

FIGURE 10-5
Structures of some membrane lipids.

synthetic detergents or bile salts to disrupt the lipid-protein interactions.

Carbohydrate residues are covalently linked (exclusively on the external side of the bilayer) to proteins or lipids to form glycoproteins or glycolipids, respectively, both of which are asymmetrically distributed in the lipid bilayer (Figure 10-7). Fluidity of the membrane structure is determined by the degree of unsaturation of the hydrocarbon chains of the phospholipids and by the amount of cholesterol in the membrane. Hydrocarbon chains with cis-double bonds produce kinks and allow a greater degree of freedom of movement for the neighboring alkyl side chains; hence, these unsaturated chains give rise to more fluidity than do saturated alkyl chains, which associate in ordered arrays. Cholesterol, an inflexible polycyclic molecule, is packed between fatty alkyl chains, the ring bearing the polar hydroxyl group interacting with the polar groups of phospho- and glycolipids. The presence of cholesterol disrupts the orderly stacking of alkyl side chains, restricts their mobility, and causes increased

(b) Phosphosphingolipids

These lipids contain a long unsaturated hydrocarbon chain amino alchohol known as **sphingosine:**

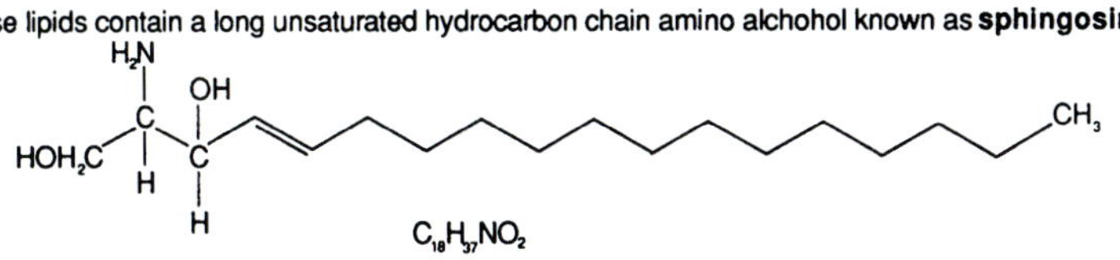

$C_{18}H_{37}NO_2$

A derivative of sphingosine, in which a fatty acid is linked by an amide linkage, is **ceramide:**

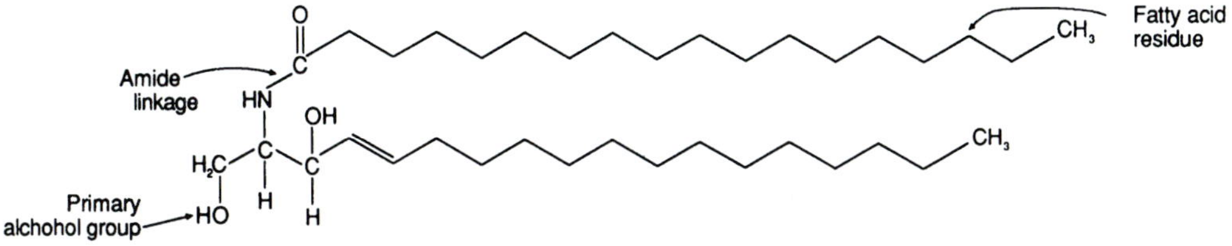

The product obtained when the alchohol hydroxyl group of sphingosine is esterified with phosphorylcholine is **sphingomyelin**. The conformations of phosphatidylcholine and sphingomyelin are similiar.

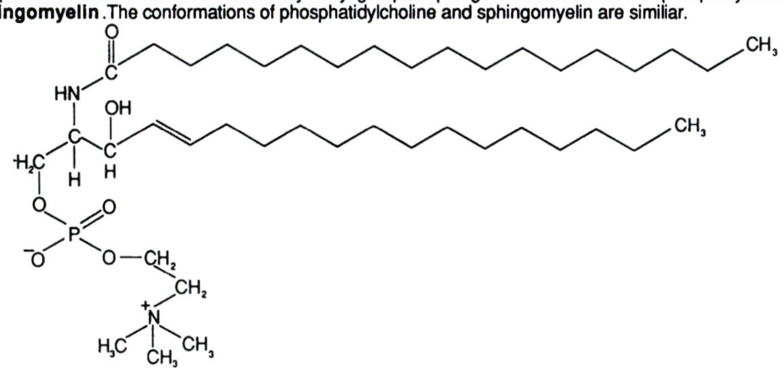

(c) Glycosphingolipids (cerebrosides)

A derivative of ceramide that contains a monosaccharide unit (either glucose or galactose) linked in a β-glycosidic linkage is a cerebroside. These neutral lipids occur most abundantly in the brain and myelin sheath of nerves. The specific galactocerebroside, phrenosine, contains a 2-hydroxy-24-carbon fatty acid residue. Following is a structure of a glucocerebroside.

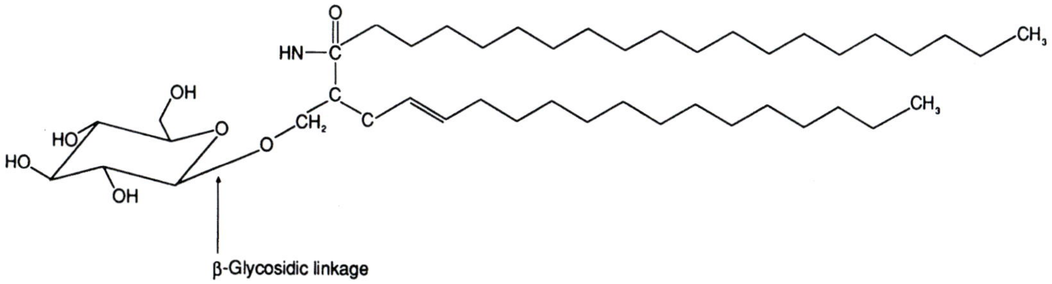

FIGURE 10-5 (*Continued*)

membrane viscosity. Thus, the lipid composition of the membrane at physiological temperatures can have significant effects on fluidity and permeability. Some correlation appears to exist between the concentrations of sphingomyelin and cholesterol in different membranes. Plasma membranes (e.g., red blood cells and myelin sheaths) are rich in both; the inner membrane of mitochondria contains neither.

Membrane proteins show considerable mobility in the plane of the bilayer (lateral motion). There is no evidence that proteins migrate from one side of the bilayer to the other. The frequency of reorientation of lipid components (flip-flop migration) is extremely slow or nonexistent, for thermodynamic reasons.

Artificial membrane systems (***liposomes***) have increased our understanding of the dynamic nature of lipid

(d) Cholesterol

Cholesterol (3-hydroxy-5,6-cholestene) belongs to a family of compounds derived from a fused, reduced, nonlinear four-ring system of cyclopenta[α]-phenanthrene. Bile acids, steroid hormones, and vitamin D metabolites are derived from cholesterol.

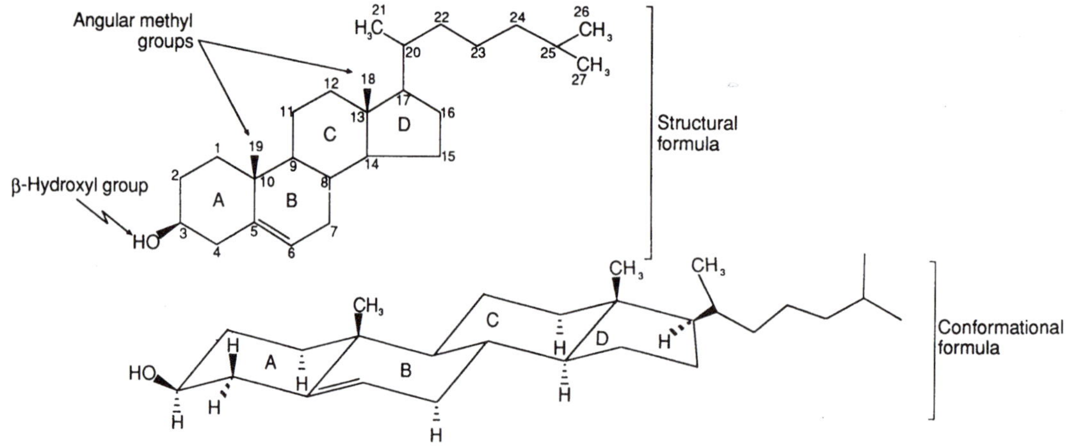

(e) Glycosphingolipids (gangliosides)

Ceramide oligosaccharides consisting of at least one residue of sialic acid are known as gangliosides. These are abbreviated by the letter G with a subscript M (mono), D (di), or T (tri) to indicate the number of sialyl residues and a number (or letter) to distinguish different members of a group from each other. Gangliosides are particularly rich in nervous tissues. Following is the structure of ganglioside G_{M1}.

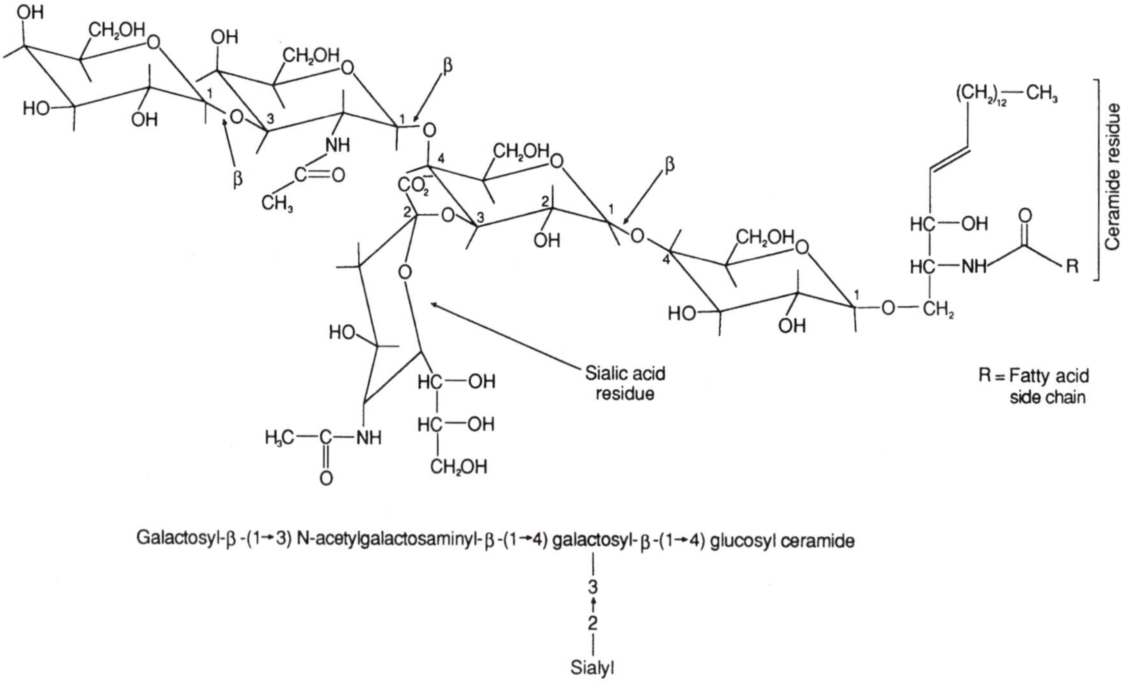

Galactosyl-β-(1→3) N-acetylgalactosaminyl-β-(1→4) galactosyl-β-(1→4) glucosyl ceramide

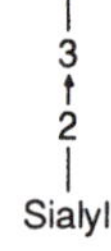

FIGURE 10-5 (*Continued*)

components of the bilayers and their role in natural membranes. Liposomes are formed when phospholipids are shaken vigorously in an aqueous medium. Electron microscopic studies of liposome vesicle reveal "sealed" concentric bilayers with the aqueous phase trapped on the inside (Figure 10-8). Known substances can be entrapped in liposomes, and permeability properties can be studied under experimental conditions. Liposomes have potential applications in medicine. Drugs and macromolecules (e.g., enzymes and nucleic acids) encapsulated in liposome systems can be targeted to a particular cell population or organ system.

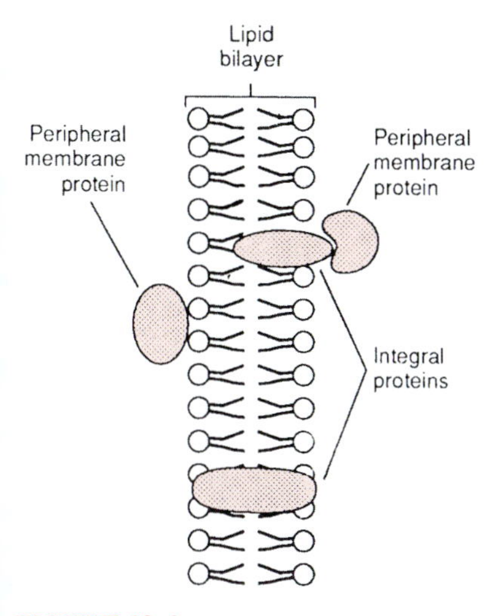

FIGURE 10-6
A lipid bilayer that contains both integral and peripheral proteins.

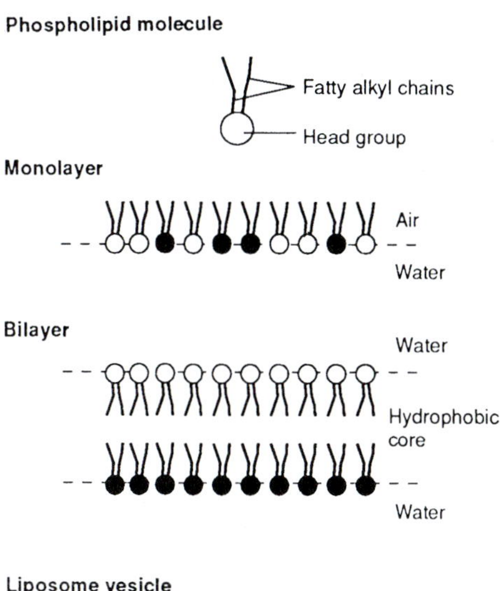

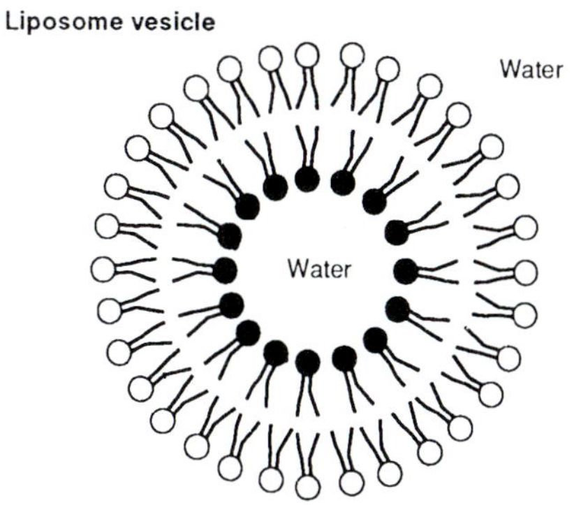

FIGURE 10-8
Interactions of phospholipids in an aqueous medium and the formation of liposome vesicle. Phospholipids spontaneously form lipid bilayers in which the polar head groups interact with water, whereas the hydrophobic tails interact among themselves to form an environment that excludes water. The lipid bilayers are stabilized by noncovalent interactions.

10.3 Cell-Surface Glycoproteins

Glycoproteins play major roles in antigen-antibody reactions, hormone function, enzyme catalysis, and cell-cell interactions. Membrane glycoproteins have domains of hydrophilic and hydrophobic sequences and are amphipathic molecules. The carbohydrate moieties of glycoproteins are distributed asymmetrically in cell membranes, cluster near one end of the protein molecule (Figure 10-7), and constitute a hydrophilic domain of amino acid residues (Chapter 21) as well as carbohydrates. The hydrophobic domain of the molecule interacts with the lipid bilayer.

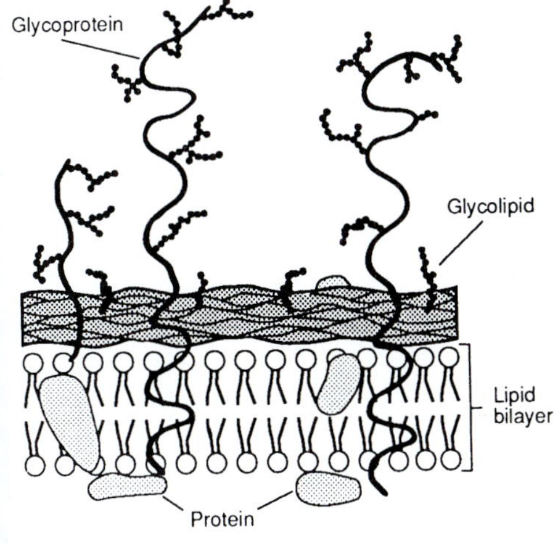

FIGURE 10-7
Membrane asymmetry with respect to location of glycoproteins and glycolipids. These carbohydrate-containing molecules are exclusively present on the external membrane.

The role of glycoproteins in cell-cell interaction is coordination and regulation of adhesion, growth, differentiation of cells, and cell size. Disruption of these processes may lead to loss of control of cell division and growth, a property characteristic of cancer cells. Normal cells grown in tissue culture show ***contact inhibition.*** When cells are allowed to grow in a tissue culture medium under optimal conditions on a surface, such as that of a Petri dish, they grow and divide until the surface is covered with a monolayer and further growth is inhibited. However, when these cells are transformed by treatment with certain viruses or carcinogenic chemicals, they lose contact inhibition and continue to grow beyond the monolayer to form multilayered masses of cells. Cancer cells (malignant neoplastic cells), like transformed cells, show continued growth and invasiveness (spreading) in tissue culture. In both transformed and cancer cells, cell-cell interaction has

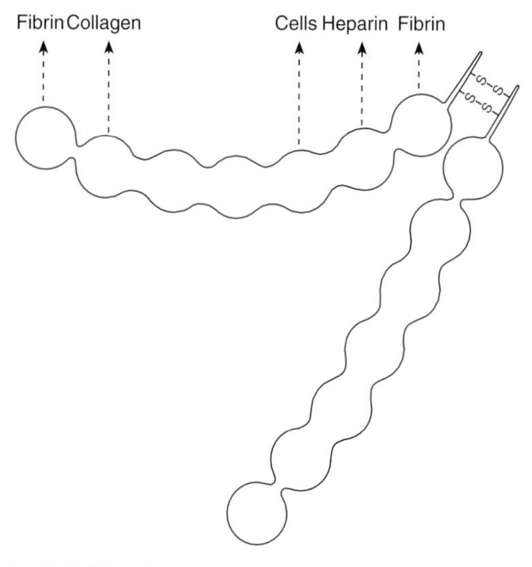

FIGURE 10-9
Schematic representation of a fibronectin molecule. It is a dimer of similar subunits (M.W. of each ~ 250,000) joined by a pair of disulfide linkages. The various functional domains by which fibronectin can interact with other protein and membrane components are indicated.

been altered. Reduced adhesion of cancer cells to a matrix has been related to reduced synthesis of fibronectin and collagen.

Fibronectin, a glycoprotein abundant on the cell surface of normal cells, promotes that attachment and subsequent spreading of many cell types. Known also as a cell surface protein, fibronectin is a large, external, transformation-sensitive protein that binds to a number of substances (e.g., collagen and fibrin). The name derives from Latin *fibra* ("fiber") and Latin *nectere* ("to tie"). Two types of fibronectin are recognized: cell surface fibronectin, which occurs as dimers and multimers, and plasma fibronectin, which occurs primarily as dimers. Fibronectin is a multifunctional molecule (Figure 10-9) containing regions that recognize glycoconjugates (e.g., glycolipids) of the cell surface, proteoglycans, and collagen fibers. A schematic model of an adhesion site is shown in Figure 10-10. Fibronectin contains about 5% carbohydrate by weight. Plasma fibronectin plays several roles in wound repair: in the formation of a fibrin clot as cross-linked fibronectin, in some reactions of platelets, in the enhancement of the opsonic activity of macrophages (important for removal of foreign material and necrotic tissue), and in attracting fibroblasts (which participate in the production of repair components such as collagen and proteoglycans in the extracellular matrix). A unique form of fibronectin, known as *fetal fibronectin,* is found in the extracellular matrix surrounding the extravillous trophoblast at the uteroplacental junction. The presence of fetal fibronectin in cervicovaginal secretions may be used as a marker in assessing the risk for preterm delivery. It is measured by enzyme immunoassay procedures (Chapter 8).

In some cell types, fibronectin is not involved in cell adhesion. For example, the extracellular matrix adjacent to epithelial cells and chondrocytes does not contain fibronectin but rather two other glycoproteins, ***laminin*** and ***chondronectin.*** Laminin mediates in adhesion of epithelial cells, whereas chondronectin mediates

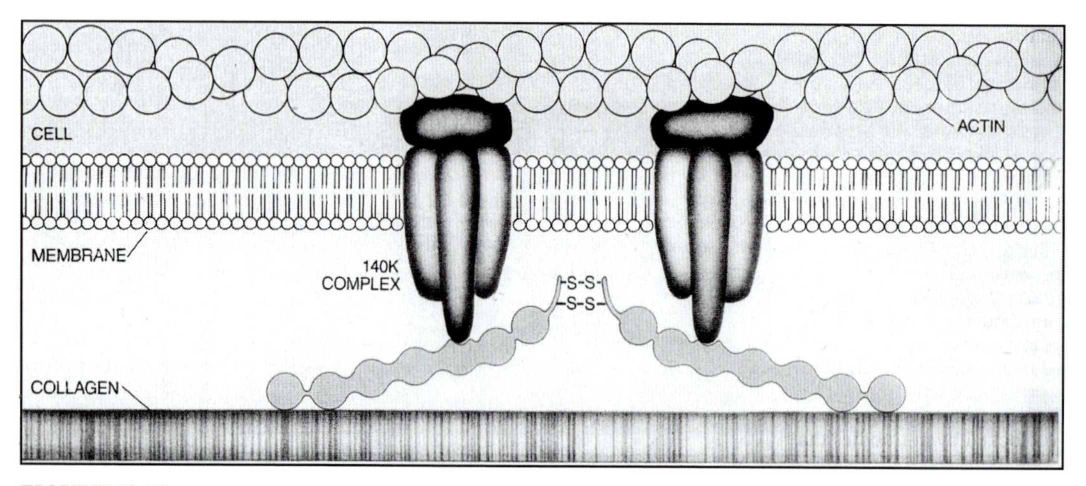

FIGURE 10-10
Fibronectin's role in adhesion of a cell with its extracellular collagenous matrix. Fibronectin, through one of its sites (see Figure 10-9), binds to specific regions of collagen fibers. At another site, fibronectin binds to a cell surface receptor protein complex (140K complex) that interacts with the actin filaments inside the cell. [Reproduced with permission from R. O. Hynes: Fibronectins. *Sci. Am.* **254,** 42 June 1986. ©1986 by Scientific American, Inc., All rights reserved.]

in the attachment of chondrocytes to collagen. A family of cell surface adhesion receptor proteins known as *integrins* binds with fibronectin. Integrins are a family of proteins containing $\alpha\beta$ heterodimers and possess receptors not only for fibronectin but for collagens, laminin, fibrinogen, vitronectin, and integral membrane protein of the immunoglobulin superfamily (Chapter 35). A sequence of 3 amino acids (RGD) in fibronectin is a recognition site for binding with integrin. Other amino acid sequences that are present in proteins for integrin recognition include KQAGDV, DGEA, EILDV, and GPRP.

Red Blood Cell Membrane and Membrane Skeleton Proteins

A major integral membrane glycoprotein of human red blood cells, ***glycophorin A,*** has been characterized. It has a molecular weight of 31,000. Each membrane contains about 400,000 molecules of glycophorin, accounting for about 1.5% of the weight of the membrane, while the carbohydrate portion constitutes about 40% of the weight. The polypeptide consists of 131 amino acid residues (Figure 10-11). The transmembrane protein has three distinct domains: a hydrophilic amino terminal end that extends outside the membrane and contains all of the oligosaccharide side chains; a hydrophobic middle region buried in the lipid bilayer; and a hydrophilic region rich in charged residues that protrudes into the cytosol. Glycophorin A has 16 oligosaccharide units, of which 15 are linked by O-glycosidic bonds to serine or threonine residues and one is linked by an N-glycosidic bond. All of the O-linked oligosaccharides have the following disialotetrasaccharide:

$$\left.\begin{array}{r}\text{Sia}-\alpha(2\rightarrow 6)-\text{Gal}-\beta(1\rightarrow 3)\\ \text{Sia}-\alpha(2\rightarrow 6)\end{array}\right\rangle \text{GalNAc}-\text{O}-\text{Ser/Thr}$$

where Sia = sialic acid, Gal = galactose, GalNAc = N-acetylgalactose, and Ser/Thr = serine/threonine.

The N-linked oligosaccharide contains mannose. All of the oligosaccharide chains appear in the 50 residues of the amino terminal domain, in which amino acid residues 2–4 and 10–15 all carry an oligosaccharide unit in O-glycosidic linkage. The middle hydrophobic domain is believed to have an α-helical structure. Glycophorin contains antigenic determinants for blood groups M and N (see below). The M and N antigenic determinants are present on glycophorin A and glycophorin B, respectively. The former differs from the latter at positions 1 and 5 of the amino terminal end:

$$\overset{1}{\text{H}_2\text{N}-\text{Ser}}-\text{Ser}-\text{Thr}-\text{Thr}-\overset{5}{\text{Gly}}-$$
Glycophorin A

$$\overset{1}{\text{H}_2\text{N}-\text{Leu}}-\text{Ser}-\text{Thr}-\text{Thr}-\overset{5}{\text{Glu}}-$$
Glycophorin B

Glycophorin A is present in type M, glycophorin B in type N, and both glycophorins in type MN individuals. Antibodies directed against M and N do not cross-react and can distinguish sequence differences between the two glycophorins. Although some antibody preparations may

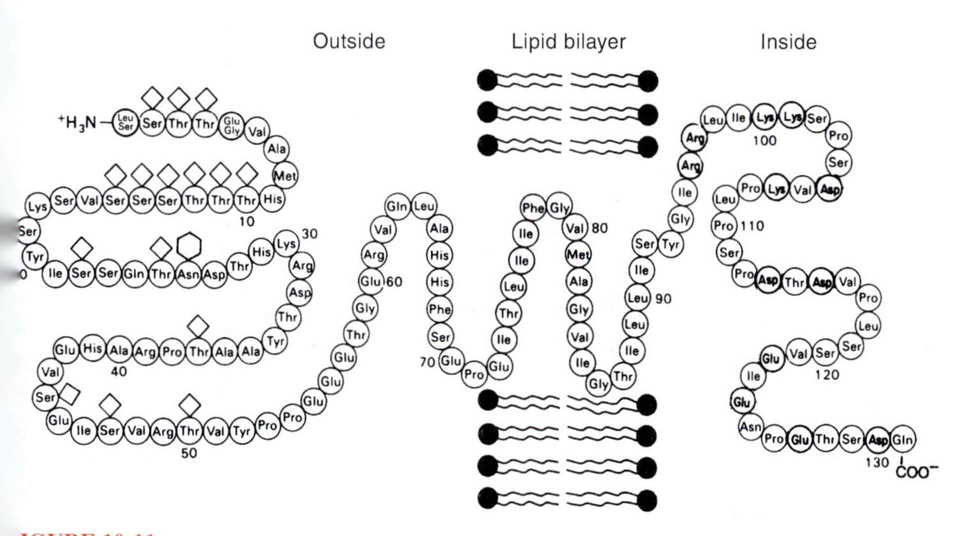

FIGURE 10-11
Primary structure of glycophorin A. There are 15 O-linked and 1 N-linked carbohydrates attached to the protein. [Adapted from L. W. Stryer, *Biochemistry*, 3rd ed. W. H. Freeman, New York, 1995.]

not react with desialylated glycophorins, the M and N blood group specificities reside in the sequences at the amino terminus of glycophorins N and B. The red blood cell membrane contains a number of other proteins, which have been separated and characterized into 10–12 major components by polyacrylamide gel electrophoresis in sodium dodecyl sulfate (Chapter 2). A schematic illustration of electrophoretic patterns of the major red blood cell membrane proteins and of membrane skeletal proteins is shown in Figure 10-12; their properties are given in Table 10-3.

Normal red blood cells are deformable biconcave disks. Their shape is determined by the external environment of the cell, the metabolic activity of the cell, the nature of hemoglobin, the membrane skeleton (see below), and the age of the cell. A normal human red blood cell has a life span of about 120 days and travels a distance of about 175 miles. Much of this travel occurs in capillary channels of the microcirculation, where flow rates are very slow. Here, particularly at branch points, the shape of the cell undergoes striking deformations and can squeeze through openings as small as one-twentieth the cell diameter. Thus, the primary determinant of blood flow and viscosity is

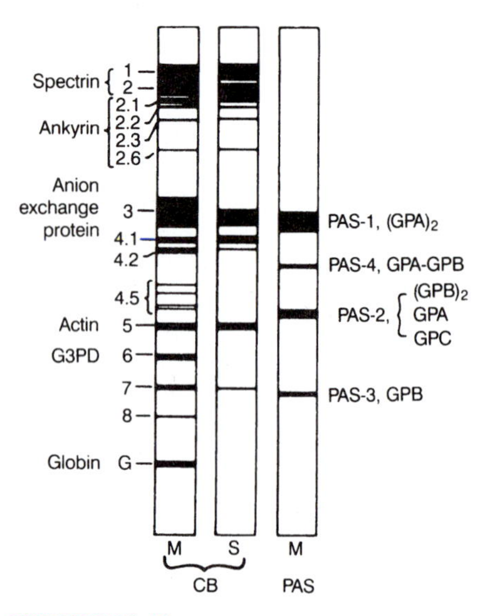

FIGURE 10-12
Schematic representations of sodium dodecyl sulfate polyacrylamide gel electrophoretic patterns of red blood cell membranes (M) and membrane skeletons (S), based on work by Fairbanks and Steck. Proteins are stained with Coomassie blue (CB) and sialoglycoproteins with periodic acid–Schiff (PAS). GPA, GPB, and GPC are glycophorin A, B, and C, respectively; G3PD is glyceraldehyde-3-phosphate dehydrogenase. $(GPA)_2$ and $(GPB)_2$ are dimers, and GPA-GPB is a heterodimer. [Reproduced with permission from J. B. Stanbury, J. B. Wyngaarden, D. S. Fredrickson, et al. (Eds.), *The Metabolic Basis of Inherited Disease*, 5th ed. McGraw-Hill, New York, 1983.]

red blood cell membrane deformability. Abnormalities of vascular channels (e.g., presence of platelet clumps and fibrin strands) and intrinsic defects in red blood cells (e.g., presence of abnormal hemoglobins; Chapter 28) can alter the dynamics of the red cell shape, affect the oxygen supply to the tissue being traversed, and reduce their own life span.

The red blood cell's durability and flexibility are due to its submembranous protein network, the ***membrane skeleton.*** The membrane skeleton consists predominantly of four proteins: ***spectrin, actin, protein 4.1,*** and ***ankyrin*** (also known as ***syndein***) (Figure 10-13). Spectrin, an extrinsic protein located on the cytoplasmic surface of the membrane, was so named because of its extraction from red blood cell ghosts (specters). It constitutes about 25% by weight of the membrane proteins. Spectrin is a long and unusually flexible molecule consisting of two structurally and functionally distinct polypeptide chains (α and β) aligned side by side to form a heterodimer. It assumes a variety of conformations that may be essential for pliancy and deformability. The lateral connections of spectrin are established by head-to-head association of heterodimers to form heterotetramers or perhaps higher order oligomers. Spectrin can also bind at one end to short filaments of actin consisting of 10–20 monomers. The monomer is known as G-actin and the polymer as F-actin. Actin is present in all eukaryotic cells, and its role in muscle contraction is well established (Chapter 21). The spectrin-actin interaction is cooperatively strengthened by protein 4.1, a globular protein that binds to spectrin at the tail end of the molecule in close proximity to the actin binding site. Spectrin is attached to the inner membrane surface by means of two proteins: *ankyrin* (from the Greek word for *"anchor"*), or *syndein* (from the Greek word for *"binding together"*) and anion exchange protein (protein 3). Spectrin is bound to ankyrin, a large protein of pyramidal shape, which tethers the membrane skeleton via its connection to the anion exchange protein. The latter is an integral membrane glycoprotein that spans the lipid bilayer and functions in exchange of Cl^- for HCO_3^- (Chapter 1). Protein 4.1 may also bind the actin-spectrin complex to the transmembrane glycoprotein glycophorin.

The intricate interactions of the spectrin-protein 4.1-actin complex may be of central importance in maintaining the structural integrity of the red cell membrane. Two genetic disorders affecting the red cell membrane skeleton are ***hereditary spherocytosis*** and ***hereditary elliptocytosis.*** The former, the most common congenital form of hemolytic anemia in persons of northern European descent, exhibits an autosomal dominant inheritance pattern. The red blood cells are spherical, osmotically fragile, and considerably reduced in life span. They underg

TABLE 10-3
Major Erythrocyte Membrane Proteins*

SDS Gel Band*	Protein	Molecular Mass (kDa)		Monomer Molecules/ Cell, Thousands	Oligomeric State	Approximate Proportion, %	Peripheral or Integral	Chromosome Location	Associated Diseases**
		Gel	Calc						
1	α Spectrin	240	281	242±20	Heterodimer/ tetramer/ oligomer	14	P	1q22-q25	HE,HPP,HS
2	β Spectrin	220	246	242±20		13	P	14q23-q24.2	HE,HPP,HS
2.1	Ankyrin′	210	206	124±11	Monomer	6′	P	8p11.2	HS
2.9	α Adducin	103	81	~30	Tetramer	<1	P	4p16.3	
	β Adducin	97	80	~30		<1	P	—	—
3	AE1	90–100	102	~1200	Dimer or tetramer	29	I	17q12-q21	HS,SAO,HAc
4.1	Protein 4.1	80+78	66	~200	Monomer	5	P	1p33-p34.2	HE
4.2	Pallidin	72	77	~250	? Dimer or trimer	5	P	15q15-q21	HS
4.9	Dematin′	48+52	43+46	~140	Trimer′	1	P	—	—
	p55	55	53	~80	? Dimer		P	Xq28	—
5	β Actin	43	42	~500	Oligomer (~14)	6	P	7pter-q22	—
	Tropomodulin	43	41	~30	Monomer		P	9q22	—
6	G3PD	35	36	~500	Tetramer	5	P	12p13	—
7	Stomatin	31	32	—	—	4	I	9q34.1	?HSt
	Tropomyosin	27+29	28	~70	Heterodimer′		P	1q31	—
8	Protein 8	23	—	~200	—	1–2	P	—	—
PAS-1	Glycophorin A	36	14	~1000	Dimer	1.6	I	4q31	None
PAS-2	Glycophorin C	32	14	~200′	—	0.1	I	2q14-q21	HE
PAS-3	Glycophorin B	20	8	~200	?Dimer	0.2	I	4q31	None
	Glycophorin D	23	11	~200	—	0.02	I	2q14-q21	HE
	Glycophorin E	—	6	—	—	—	I	4q31	—

*Reproduced with permission from P. S. Becker and S. E. Lux: Hereditary Spherocytesis and Hereditary Elliptocytosis. In *The Metabolic Basis of Inherited Disease,* 7th Ed., C. R. Scriver, A. L. Beaudet, W. S. Sly, et al., Eds. McGraw-Hill, p. 3513, 1995.
**HE = Hereditary elliptocytosis; HPP = hereditary pyropykilocytosis; HS = hereditary spherocytosis; SAO = Southeast Asian ovalocytosis; HSt = hereditary stomatocytosis; HAc = hereditary acanthocytosis.

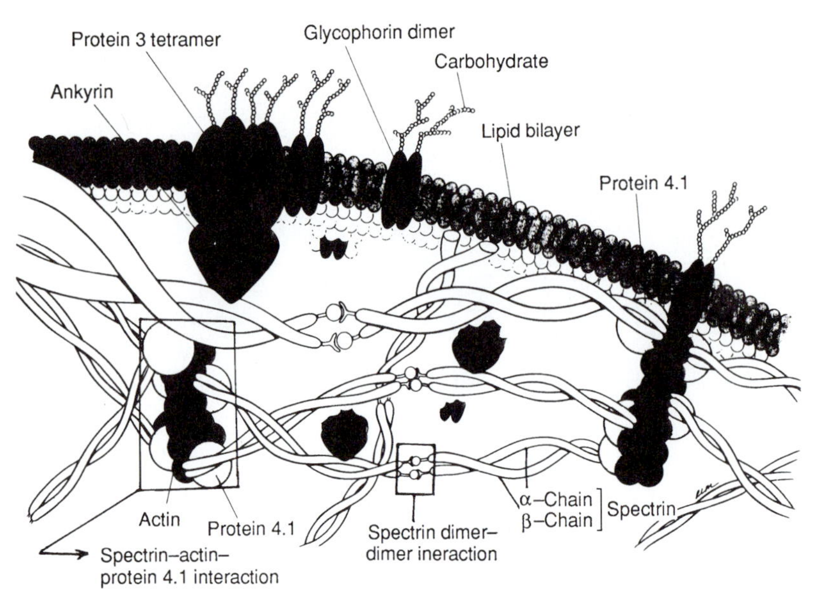

FIGURE 10-13
Model of the organization of the red cell membrane skeleton. [Reproduced, with permission, from S. Lux and S. B. Shohet, The erythrocyte membrane skeleton: Biochemistry. *Hosp. Prac.* **19(10),** 82 (1984). R. Margulies, illustrator.]

fragmentation and loss of membrane as a result of circulatory and metabolic stresses and are selectively trapped and removed by the spleen. In some kindreds with this disorder, a qualitative defect in spectrin leads to poor binding with protein 4.1 and to a weakened spectrin–protein 4.1–actin complex. A spherocytic hemolytic anemia, having an autosomal recessive inheritance pattern associated with about 40% deficiency of the α- and β-chains of spectrin, has also been described. In hereditary elliptocytosis, the red blood cells are elliptical. Molecular lesions identified in a number of kindreds include deficiency of protein 4.1 and diminished spectrin-spectrin or ankyrin-protein 3 interactions. Both diseases are biochemically heterogeneous. In ***hereditary pyropoikilocytosis,*** the isolated spectrin seems incapable of forming higher oligomers (dimer-dimer association) *in vitro,* possibly as a result of a structural change in the α-subunit of spectrin. The red cells show abnormal and bizarre shapes (poikilocytosis). The cells hemolyze at a temperature 2–3° below that (50°C) required to hemolyze normal cells. Normal spectrin melts (converts from ordered to disordered structure) at 69°C.

Blood Group Antigens

Antigenic determinants (Chapter 35) known as *blood group substances* are present on the surface of human red blood cells. The blood group substances are inherited according to Mendel's law. Knowledge of blood group substances is essential for blood transfusion and valuable in forensic medicine and anthropological studies. The antigens are recognized by specific antigen-antibody interactions that produce agglutination. For example, when red blood cells containing a specific antigenic determinant are mixed with plasma containing specific antibodies to that antigen, cells will agglutinate through formation of a network of antigen-antibody linkages. More than 100 different blood group antigens have been categorized on the basis of their structural relationships into 15 independent blood group systems. The blood group substances are found on red blood cell membranes and in a variety of tissue cells. Soluble blood group substances are found as glycoproteins in saliva, gastric juice, milk, seminal fluid, urine, fluids produced in ovarian cysts, and amniotic fluid.

The most widely investigated antigenic determinant are the ABO and the Lewis blood group systems. The antigenic specificities are established by carbohydrate residues occurring at the nonreducing ends of oligosaccharides of the glycoproteins and glycolipids (Table 10-4) of red blood cell membranes.

The ABO blood group system is associated with three antigens: A, B, and H. The specificity of these antigens resides in difference of the terminal carbohydrate residue. Addition of an N-acetylgalactosamine residue to an H antigen yields an A antigen, whereas addition of a galactose residue yields a B antigen (Table 10-4). The H antigen itself is synthesized by addition of a fucosyl residue to

TABLE 10-4
The Carbohydrate Antigenic Determinants of Selected Blood Group Substances

Blood Group System	Sequence at Nonreducing Terminal of Oligosaccharide
ABO Locus	
A_1 antigen	Fucα1↘ $^{2}_{3}$Galβ1→ 3GlcNAcβ1→ 3Galβ1→ 4Glc→ GalNAcα1↗
	(In type A_2, the linkage between Gal and GlcNAc is β1→ 4.)
B_1 antigen	Fucα1↘ $^{2}_{3}$Galβ1→ 3GlcNAcβ→ 3Galβ1→ 4Glc→ Galα1↗
H or O antigen	Fucα1→ 2Galβ1→ 3GlcNAcβ1→ 3Galβ1→ 4Glc→
The difference between the above antigens resides in the absence or presence of either N-acetylgalactosamine or galactose linked to the penultimate galactose residue in α(1→ 3) linkage. The oligosaccharide determinants occur in glycoproteins and in glycosphingolipids.	
Lewis (L) Locus	
Le^a antigen	Galβ1↘ $^{3}_{4}$GalNAcβ1→ 3Galβ1→ 4Glc-Cer Fucα1↗
Le^b antigen	Fucα1→ 2Galβ1↘ $^{3}_{4}$GalNAcβ1→ 3Galβ1→ 4Glc-Cer Fucα1↗
The Lewis antigens are not components of developng red blood cells. They are acquired as glycosphingolipids during circulation by adsorption from plasma low- and high-density lipoproteins (Chapter 20).	
P Locus	
P antigen	GalNAcβ1→ 3Galα1→ 4Galβ1→ 4Glcβ1-Cer
P^K antigen	Galα1→ 4Galβ1→ 4Glcβ1-Cer
I Locus	
I antigen	Galβ1→ 4GlcNAcβ1↘ $^{6}_{3}$Galβ1→ 4-X* Galβ1→ 4GlcNAcβ1↗
i antigen	Galβ1→ 4GlcNAcβ1→ 3Galβ1→ 4-X*
	X*=GlcNAcβ1→ 3Galβ1→ 4Glcβ1-Cer

Gal=galactose, GalNAc=N-acetylgalactosamine, Glu=glucose, GluNAc=N-acetylglucosamine, Fuc=fucose (6-deoxy-L-galactose), Cer=ceramide (a sphingolipid).

precursor oligosaccharide. The classification of individuals into ABO types is based upon whether A or B antigens are present or not; type A individuals have the A antigen, type B, the B antigen, and type O, the H antigen, respectively. Normally, an antigen and its specific antibody are not present simultaneously. For example, individuals having type A red blood cells possess anti-B antibodies; those of type B possess anti-A antibodies; those of type AB have neither of these antibodies; and those of type O have both anti-A and anti-B antibodies (Table 10-5).

The antigenic variation of blood group substances is due to specific glycosyltransferases (the primary gene products) responsible for synthesis of the oligosaccharide determinants (the secondary gene products).

Under appropriate experimental conditions, the blood type of red blood cells can be changed by addition or removal of specific carbohydrate residues. For example, when type O red blood cells are incubated with the galactose donor substrate (uridine diphosphate galactose) and the specific α-galactosyltransferase enzyme, they are

TABLE 10-5
*ABO Blood Group System and the Agglutination Reaction**

Blood Type	Type of Oligosaccharide Antigens Present on Red Blood Cells	Antibody in Serum	Types of Serum That Cause Agglutination When Mixed with Red Blood Cells
O	H	Anti-A, anti-B	None
A	A	Anti-B	O, B
B	B	Anti-A	O, A
AB	A and B	None	O, A, B

*Red blood cells from type O individuals can be donated to individuals with any other cells from type within the ABO system without causing agglutination. Individuals with type AB can accept red blood cells from all other types but cannot donate to individuals with other types except AB. Thus, type O individuals are known as universal donors and AB as universal acceptors.

converted to type B red blood cells. The reverse conversion can be accomplished by incubation of cells in the presence of α-galactosidase, which removes the terminal galactose residue. Glycophorins A and B possess antigenic determinants for the M and N blood group antigens, respectively.

Rh blood group antigens are clinically important due to their involvement in hemolytic disease of the newborn, transfusion medicine, and autoimmune hemolytic anemia. The designation of Rh stands for *Rhesus* because the antibody specificity was identical to that of antibodies generated in rabbits injected with red blood cells from Rhesus monkeys. The Rh blood group system consists of more than 50 nonglycosylated protein antigens; however, only five are commonly identified; these are encoded by two genes termed RHD and RHCED. Marked racial differences are observed for Rh alleles in different populations. Typically, 85% of all Caucasians are Rho(D)-positive. Anti-Rho(D) immunoglobulin G (IgG) prepared from human plasma is administered intramuscularly as means of passive immunization to Rho(D)-negative individuals exposed to Rho(D)-positive cells.

Rh proteins have a molecular weight of about 32,000 and span the red blood cell membrane. Rh proteins are incorporated into the erythrocyte membrane after palmitolyation via thioester linkages involving free sulfhydryl groups of cysteine residues. The exact function of Rh proteins is not known, however, individuals lacking all Rh protein expression exhibit multiple red blood cell abnormalities including abnormal morphology and survival. This condition is known as ***Rh_{null} syndrome.*** A number of protein blood group antigens possessing a variety of functions are shown in Table 10-6.

Blood group antigens including both oligosaccharides and proteins are expressed on other cell surface structures as well as on erythrocytes. These antigens function as receptors for a variety of infectious agents. Examples of oligosaccharide antigens include P antigen, which is the receptor for B19 parvovirus; individuals who do not express this antigen show no serological evidence of past infections with this common childhood pathogen. Le^b antigen is the epithelial receptor for ***Helicobacter pylori,*** which is the causative agent of peptic ulcer disease (Chapter 12). Lewis antigens function as receptors for a wide variety of pathogens including *S. aureus, N. meningitidis, H. influenzae, N. gonorrhoeae,* and *C. albicans.*

Blood group antigen-bearing proteins also serve as receptors for infection agents (Table 10-7). A notable example is the presence of receptors in the Duffy blood group antigens to *Plasmodium vivax* and *Plasmodium knowlesi.* Individuals who do not express Duffy antigens are resistant to infection by these malarial parasites.

10.4 Serum Glycoproteins

Almost all serum proteins, with the notable exception of albumin, are glycoproteins. The sugar residues found most commonly in the outer domain of the oligosaccharides of these glycoproteins are galactose, N acetylhexosamine, and sialic acid. L-Fucose is a minor constituent. The structure of a typical oligosaccharide chain of a serum glycoprotein is shown in Figure 10-14. The liver plays a major role in the synthesis and catabolism of these proteins. Glycoproteins lose their terminal sialic acid residues through the action of neuraminidase (sialidase) during circulation in the blood, which exposes the galactose residues (Figure 10-14). The resulting galactose-terminated glycoproteins, known as

TABLE 10-6
Functions of Blood Group Protein Antigens of Erythrocytes

Blood Group (abbreviation)	Locus Name	Protein
a. Proteins that contribute to the structural integrity		
Rhesus (Rh)	RH	Rh acylproteins
Gerbich (Ge)	GYPC	Glycophorins C and D
Diego (Di), Wright (Wr)	AE1	Anion channel protein
XK	Kx	Kx protein
b. Proteins with complement-related functions		
Cromer (Cr)	DAF	Decay accelerating factor (CD55)
c. Proteins with receptor functions		
Duffy (Fy)	Fy	Chemokine receptor
d. Proteins with transport functions		
Diego (Di), Wright (Wr)	AE1	Anion channel protein
Kidd (Jk)	JK	Urea transporter
Colton (Co)	AQP1	Aquaporin 1 water channel protein
e. Proteins with enzymatic activity		
Kell (K,k,Kp,Js)	KEL	93kD protein, metalloproteinase (?)
Cartwright (Yt)	YT	Acetylcholinesterase

asialoglycoproteins, are taken up after binding to receptors on hepatocytes. The bound asialoglycoprotein is internalized by a process known as ***receptor-mediated endocytosis*** (Chapter 11) and subjected to lysosomal degradation. In liver disease, the plasma levels of asialoglycoproteins are elevated.

TABLE 10-7
Blood Group Protein Antigens as Microbial Receptors*

Oligosaccharides	
P	*Escherichia coli*, Parvovirus B19
Lewis (Leb)	*Helicobacter pylori*
H, A	*Candida albicans*
Proteins	
Glycophorin A (MN)	Plasmodium falciparum
Decay accelerating factor (Cromer)	Echovirus, *E. coli*
CD44 (Indian)	Poliovirus
Duffy	*Plasmodium vivax*, *Plasmodium knewlesi*
AnWj	*H. influenzae*

*Modified and reproduced with permission from M. J. Telen. Erythrocyte Blood Group Antigens: Polymorphisms of functionally Important Molecules. Seminars in Hematology **33**, 302 (1996).

The internalized glycoproteins are catabolized to their monomeric units by the lysosomal enzymes. The oligosaccharides are degraded sequentially by specified hydrolases, starting from the nonreducing termini. Hereditary deficiency of some of these enzymes has been reported: α-D-mannosidase in mannosidosis, α-L-fucosidase in fucosidosis, glycoprotein-specific α-neuraminidase in sialidosis, and aspartylglycosaminidase in aspartylglycosaminuria. In these disorders, undigested or partially digested oligosaccharides derived from glycoproteins accumulate within the lysosomes. Keratin sulfate (a glycosaminoglycan, Chapter 11) also accumulates, presumably because the above enzymes are required for its degradation. These disorders are inherited as autosomal recessive traits and can be diagnosed prenatally. Clinically, they resemble mild forms of mucopolysaccharidosis (Chapter 11). There is no definitive treatment.

Removal of senescent red blood cells from the circulation has been attributed to desialylation of the membrane glycoproteins. *In vitro* removal of sialic acid from human red blood cells and introduction of the modified cells into the circulatory system result in drastic shortening of their life span. However, aging and removal of red blood

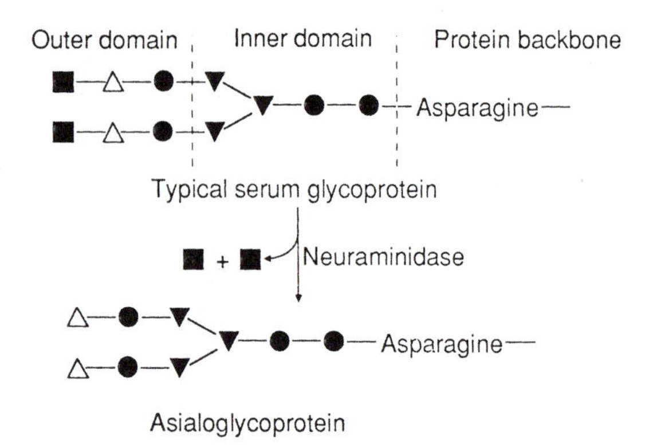

FIGURE 10-14
Oligosaccharide side chain of a serum glycoprotein. The removal of the terminal sialic acid residue by neuraminidase exposes the penultimate galactose residue. The resulting asialoglycoprotein is cleared from the circulation by a galactose receptor–mediated process in the hepatocytes. ■ = Sialic acid, ● = N-acetylglucosamine, △ = galactose, ▼ = mannose.

cells depend on many other factors (e.g., inactivation of enzymes and oxidation of sulfhydryl proteins). If desialylation occurs normally as part of the physiological mechanism in clearing glycoproteins and cells of the circulatory system, the anatomical location of this process is not known. Other clearance systems for glycoproteins have been identified, e.g., one in the reticuloendothelial system that terminates with mannose and N-acetylglucosamine. Information of this type has potential application in targeting biologically active molecules to a specific tissue site. For example, mannose-terminated catalytically active lysosomal enzymes can be targeted to reticuloendothelial cells. Such targeting of lysosomal enzymes could be useful in the treatment of lysosomal enzyme-deficiency diseases. The same principle can be applied to targeting drugs to specific tissue sites by coupling the drugs to glycoproteins with the appropriate terminal sugar residues.

10.5 Molecular Mimicry of Oligosaccharides and Host Susceptibility

In the previous section we discussed the presence of oligosaccharide and protein blood group substances that cause susceptibility to certain pathogens. In this section we discuss an infectious agent that contains a common antigenic epitope with the host and that elicits antibodies that cause a disease. ***Guillain–Barré syndrome*** (GBS) is one such disease. GBS is an acute inflammatory neuropathy with progressive weakness in both arms and legs leading to paralysis. It is a self-limited autoimmune disease. Most GBS cases have resulted from an antecedent, acute infection of bacterial or viral origin; in children GBS has also been identified following vaccination.

One major cause of bacterial gastroenteritis is *Campylobacter jejuni* and this infection also may be followed by GBS. Some serotypes of *C. jejuni* possess liposaccharides that contain terminal tetrasaccharides identical to ganglioside GM1. Antibodies made in response to the bacterial infection also attack GM1 which is widely present in the nervous system. The antibodies directed against the GM1 epitope cause an immune-mediated destruction of nerve fibers. Thus, the sharing of homologous epitopes between bacterial liposaccharides and gangliosides is an example of molecular mimicry, which causes disease. Host factors may also influence a person's susceptibility to GBS. Plasma exchange and intravenous immunoglobulin administration are used for immunomodulation in the therapy of GBS.

Supplemental Readings and References

Extracellular Matrix

M. H. Ascarelli and J. C. Morrison: Use of fetal fibronectin in clinical practice. *Obstetrical and Gynecological Survey* **52**(Suppl.), S1 (1997).
R. O. Hynes: Integrins: Versatility, Modulation and Signalling in Cell Adhesion. *Cell* **69,** 11 (1992).
K. A. Piez: History of extracellular matrix: A personal review. *Matrix Biology* **16,** 85 (1997).
J. Labat-Robert, M. Bihari-Varga, and L. Robert: Extracellular Matrix. *FEBS Letters* **268,** 386 (1990).
W. K. Stadelmann, A. G. Digens, and G. R. Tobin: Physiology and healing dynamics of chronic cutaneous wounds. *American Journal of Surgery* **176**(Suppl. 2A), 265 (1998).
W. K. Stadelmann, A. G. Digens, and G. R. Tobin: Impediments to wound healing. *American Journal of Surgery* **176**(Suppl. 2A), 395 (1998).

Blood Group Antigens

P. Agre and J-P. Cartron: Molecular biology of Rh antigens. *Blood* **78,** 551 (1991).
C-H. Huang, P. Z. Liu, and J. G. Cheng: Molecular biology and genetics of the Rh blood group system. *Seminars in Hematology* **37,** 150 (2000).
A. O. Pogo and A. Chaudhuri: The Duffy protein: a malarial and chemokine receptor. *Seminars in Hematology* **37,** 122 (2000).
M. Rios and C. Bianco: The role of blood group antigens in infectious disease. *Seminars in Hematology* **37,** 177 (2000).
M. J. Telen: Erythrocyte blood group antigens: Polymorphisms of functionally important molecules. *Seminars in Hematology* **33,** 302 (1996).
M. J. Telen: Red blood cell surface adhesion molecules: Their possible role in normal human physiology and disease. *Seminars in Hematology* **37,** 130 (2000).

Molecular Mimicry

T. E. Feasby and R. A. C. Hughes: Campylobacter jejuni, antiganglioside antibodies, and Guillain–Barré syndrome. *Neurology* **51,** 340 (1998).

A. F. Hahn: Guillain–Barré syndrome. *The Lancet* **352,** 635 (1998).

T. Lasky, G. J. Terraccians, L. Magder, et al.: The Guillain–Barré syndromes and the 1992–1993 and 1993–1994 influenza vaccines. *New England Journal of Medicine* **339,** 1797 (1998).

J. J. Ma, M. Nishimura, H. Mine, et al.: HLA and T-cell receptor gene polymorphisms in Guillain–Barré syndrome. *Neurology* **51,** 379 (1998).

A. H. Ropper and M. Victer: Influenza vaccination and Guillain–Barré syndrome. *New England Journal of Medicine* **339,** 1845 (1998).

K. A. Sheikn, I. Nachamkin, T. W. Ho, et al.: Campylobacter jejuni lipopolysaccharides in Guillain–Barré syndrome. *Neurology* **51,** 371 (1998).

Disorders of Red Blood Cell Membrane Skeleton

P. S. Becker and S. E. Lux: Hereditary Spherocytosis and Hereditary Elliptocytosis. *In the Metabolic Bases of Inherited Disease,* 7th ed., C. R. Scriver, A. L. Beaudet, W. S. Sly, D. Valle., (Eds.) McGraw-Hill, New York, 1995, p. 3513.

S. C. Liu and L. H. Derick: Molecular anatomy of the red cell membrane skeleton: Structure-function relationships. *Seminars in Hematology* **29,** 231 (1992).

CHAPTER 11

Heteropolysaccharides II: Proteoglycans and Peptidoglycans

11.1 Protein Fibers and Proteoglycans

Connective tissues are composed of insoluble protein fibers (the glycoprotein ***collagen*** and the nonglycoprotein ***elastin***) embedded in a matrix of proteoglycans (ground substance). The connective tissues bind tissues together and provide support for the organs and other structures of the body. Their properties depend on the proportion of different components present. A tissue of very high tensile strength, the Achilles tendon, is composed of about 32% collagen and 2.6% elastin, whereas an elastic tissue, the ligamentum nuchae, is composed of about 32% elastin and 7% collagen. The proteins and proteoglycans are synthesized by connective tissue cells: ***fibroblasts*** (generalized connective tissue), ***chondroblasts*** (cartilage), and ***osteoblasts*** (bone). Connective tissue also contains blood and lymphs vessels and various transient cells including macrophages and mast cells. Adipose tissue is a specialized form of connective tissue consisting of a collection of ***adipocytes*** (stores of triacylglycerol) that cluster between the protein fibers.

Collagen

Collagens are extracellular proteins of connective tissue and they make up about one third of all body protein. They are a family of related glycoproteins in which hydroxylysyl residues provide the sites for attachment of glucose, galactose, or an $\alpha(1 \rightarrow 2)$ glucosylgalactose residue via a β-O-glycosidic linkage (Figure 11-1). In brief, the synthesis of collagen can be considered to occur in two stages: intracellular and extracellular. The intracellular stage consists of the production of procollagen from precursor polypeptide chains that undergo, in sequence, hydroxylation, glycosylation, formation of a triple helix, and secretion. The extracellular stage consists of the conversion of procollagen to tropocollagen by limited proteolysis from the amino and carboxyl termini, self-assembly of tropocollagen molecules into fibrils, and finally cross-linking of the fibrils to form collagen fibers (Chapter 25).

Collagen exists predominantly as fibrous protein; however, in the basement membrane of many tissues, including kidney glomeruli and the lens capsule, it is present in a nonfibrous form. The unique property of each connective tissue (e.g., the flexibility of skin, rigidity of bone, elasticity of large arteries, and strength of tendons) depends on the composition and organization of collagen and other matrix components.

Collagen Types

More than 16 different types of collagen have been reported. They constitute the most abundant family of proteins in the human body. The collagens are encoded by 28 genes dispersed in at least 12 different chromosomes.

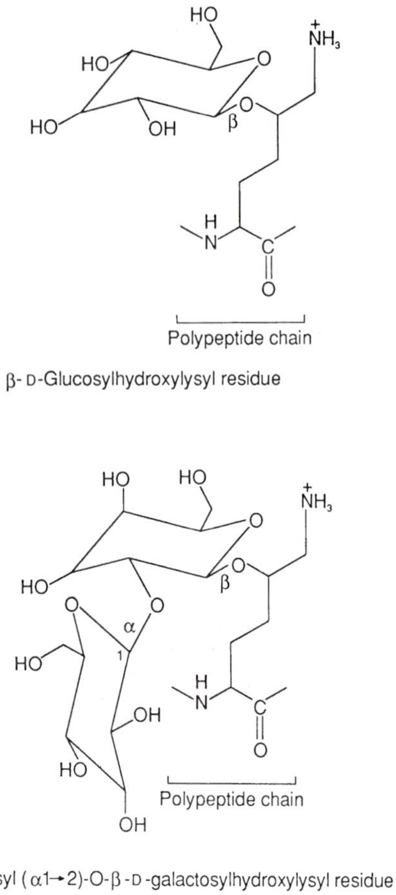

β-D-Glucosylhydroxylysyl residue

D-Glucosyl (α1→2)-O-β-D-galactosylhydroxylysyl residue

FIGURE 11-1
Glycosides of the collagen polypeptide chain.

The individual genes are identified by both collagen type and the polypeptide chain. For example, COL1A2 gene codes for type I collagen (COL1) and $\alpha 2$ (A2) polypeptide chain. The tissue distribution of collagens also varies (Table 11-1).

Type I collagen consists of two identical chains of $\alpha 1$(I) and one chain of $\alpha 2$; it is the major connective tissue protein of skin, bone, tendon, dentin, and some other tissues. Type II collagen consists of three identical chains of $\alpha 1$(II); it is found in cartilage, cornea, vitreous humor, and neural retinal tissue. Type III collagen consists of three identical chains of $\alpha 1$(III); it is present, along with type I collagen, in skin, arteries, and uterine tissue. Type IV collagens are composed of $\alpha 1$(IV) to $\alpha 5$(IV) and are found in the basement membranes of various tissues. Unlike collagen types I, II, III, V, and XI which are fibrillar, type IV collagen is not fibrillar in structure.

The synthesis of collagens in cultured cells has aided the understanding of collagen biochemistry. The tissue specificity of various types of collagen is also reflected in cultured cells obtained from appropriate tissues. For example, human fibroblasts and smooth muscle cells synthesize both types I and III collagens; epithelial and endothelial cells synthesize type IV collagen; and chondroblasts synthesize type II collagen.

Structure and Function

Each collagen molecule contains three polypeptide chains coiled around each other in a triple helix. These chains are called α-chains and are designated by Roman numerals according to the chronological order of their discovery. The three polypeptide chains may be identical or may consist of two identical chains and one dissimilar chain.

Each α-polypeptide chain consists of about 1000 amino acid residues, of which every third following a glycine is also a glycine. Thus, the molecular formula of an α-chain may be written as $(\text{Gly–X–Y})_{333}$, where X and Y represent amino acid residues other than glycine. In mammalian collagens, about 100 of the X-positions are occupied by proline residues, and 100 of the Y-positions are occupied by 4-hydroxyproline residues. At a few X-positions, 3-hydroxyproline residues are present; however, they only occur next to 4-hydroxyproline. Most hydroxyproline residues are present as the trans isomers. Although all collagen polypeptides have the general structure $(\text{Gly–X–Y})_n$, differences between the various collagen types are associated with the particular sequences of amino acid residues in the X and Y positions. Hydroxyproline residues are not common in proteins; other than the collagens, hydroxyproline residues are found in elastin, acetylcholinesterase, and the C1q subcomponent of the complement system (Chapter 35).

Another unique amino acid residue found in collagen is hydroxylysine, which occurs at the Y-position. The number of hydroxylysine residues per polypeptide chain lies in the range of 5 to 50. The hydroxylysine residues provide the sites for β-O-glycosidic linkage with galactose or glucose or $\alpha(1 \rightarrow 2)$ glucosylgalactose. The collagens differ in their ratio of monosaccharide to disaccharide residues, as well as in their total carbohydrate content. For example, the carbohydrate contents of collagen from skin, cartilage, and basement membrane are about 0.5%, 4%, and more than 10%, respectively. Because collagen biosynthesis involves intracellular and extracellular post-translational modifications (e.g., hydroxylation, glycosylation, fibril formation, and formation of cross-links), a given genetically determined collagen may show a great deal of heterogeneity. This is particularly true for type I collagen.

Collagen also contains alanine residues in relatively high quantities. The only amino acid not found in collagen

TABLE 11-1

Types of Collagen, Their Distributions in Tissues and Properties

Type*	Composition	Distribution in Tissues	Examples of Known Disorders
I	$[\alpha 1(I)]_2\alpha 2$ $\alpha 1(I)_3$	Bone, Tendon, Skin, Dentin, Fascia, Arteries	Osteogensis Imperfacta (OI) and Ehlers-Danlos Syndrome (EDS). Both syndromes are clinically heterogeneous due to genetic defects that affect the biosynthesis, assembly, postranslational modification, secretion, fibrillogenesis, or other extracellular matrix components.
II	$[\alpha 1(II)]_3$	Cartilage, Vitreous humor	Chondrodysplasia: Spondyloepiphysial dysplasia, a chondrogenesis and Stickler Syndrome
III	$[\alpha 1(III)]_3$	Skin, Blood vessels, Uterus	EDS Type IV: Mutation affects synthesis, structure or secretion
IV	$[\alpha 1(IV)]_2\alpha_2(IV)$ other forms are also known	Basement membrane	Alport Syndrome: characterized by nephritis and sensorineural deafness
V	$[\alpha 1(V)]_3$ $[(\alpha 1(V)_2)]_2\alpha_2(V)$ $\alpha 1(V)\alpha_2(V)\alpha_3(V)$	Skin, Placenta, Blood vessels, Chorion uterus	
VII	$[\alpha(VII)]_3$	Anchoring fibrils	Epidermolysis bullosa
VIII	Not yet known	Cornea, Blood vessels, Network-forming collagen	
IX	$[\alpha 1(IX)\alpha_2(IX)\alpha_2(IX)$	Cartilage, Fibril associated collagen	
X	$[\alpha 1(X)]_3$	Cartilage	
XI	$\alpha 1(XI)\alpha 2(XI)\alpha 1(II)$	Cartilage	
XII	$[\alpha 1(XII)]_3$	Soft tissues	

*Collagen Types I, II, III, V, and XI are fibrillar.

is tryptophan (an essential amino acid). Collagen consists essentially of four amino acids in abundant quantities and negligible amounts of almost all other amino acids. For this reason, collagen is an inferior protein nutritionally. It is very insoluble in water and, for the most part, indigestible. However, collagen can be converted to soluble and digestible products by hydrolysis of some covalent bonds (by the use of heat or by certain plant proteinases) to yield gelatin. In the human body, collagen is resistant to most proteinases. Polymerized fibrillar collagen is a stable component of the extracellular matrix and its turnover rate is insignificant except in areas where tissue remodeling and repair occur. However, nonfibrillar collagens (e.g., type IV) are susceptible to proteolytic attack. Specific proteases (collagenases) cleave collagen at specific sites.

The amino acid sequence of collagen produces its unique secondary and tertiary structures. A ***tropocollagen*** molecule contains three polypeptide chains, each coiled into a left-handed helix having about three amino acid residues per turn (Figure 11-2). This type of helix is unique to collagen and differs significantly from the α-helix in its periodicity and dimensions (Chapter 4). The three helical polypeptides twist tightly around each other, except for the two short nonhelical regions at the C and N termini to form a right handed triple-stranded superhelix. Tropocollagen has a molecular weight of about 300,000, a length of 300 nm, and a diameter of 1.5 nm. The glycine residue, which occurs at every third position, has the smallest R-group and is thus able to fit into the restricted space where the three polypeptide chains are closest to each other. Since the proline and hydroxyproline residues hinder free rotation around the N–C bond (Chapter 4), the polypeptide chain has a rigid and kinked conformation. These stereochemical properties are responsible for the superhelix. Hydrogen bond formation between the NH group of a glycyl residue in one chain and the CO group of a prolyl or other amino acid residue in the X-position of an adjacent chain stabilizes the triple helix. The hydrogen bond between the glycyl residue of one chain and the prolyl residue of another chain is direct. If the prolyl residue is replaced by any other amino acid, the interchain hydrogen bond occurs through a water-bridged structure. These bonds are further stabilized by hydrogen bonding with the hydroxyl group of the *trans*-4-hydroxyprolyl residue, which occurs at the Y-position. Additionally, the chains are held together by covalent linkages involving lysine residues (Figure 11-3).

Polypeptide chains consisting only of glycine, proline, and hydroxyproline residues (in that order) form an extremely stable triple helix. Furthermore, the stability (thermal) of the triple helix decreases in the following order for repeating sequences of the chain: Gly–Pro–Hyp > Gly–Pro–Y > Gly–X–Pro > Gly–X–Y. In a given polypeptide chain of native tropocollagen, about one third of the molecule contains the Gly-Pro-Hyp sequence and two thirds involve Gly–X–Y, which decreases the stability of the triple helix. Amino acid residues other than proline and hydroxyproline that occupy the X- and Y-positions decrease helix stability but are essential for the next level of organization of collagen—the formation of microfibrils.

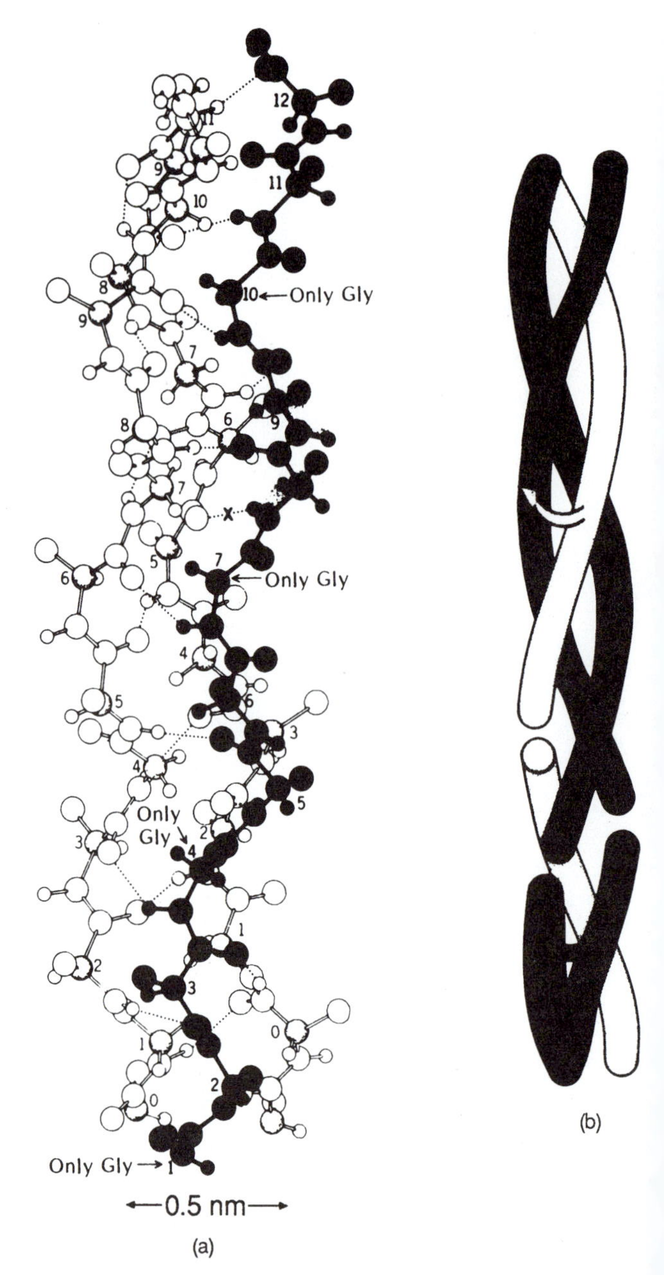

FIGURE 11-2
Structure of tropocollagen. (a) The coiling of three left-handed helices of collagen polypeptides. The dotted lines indicate hydrogen bonds. (b) The right-handed triple-stranded superhelix. The curved arrow indicates the covalent linkage between two chains. [Reproduced with permission from I. Geis and R. E. Dickerson.]

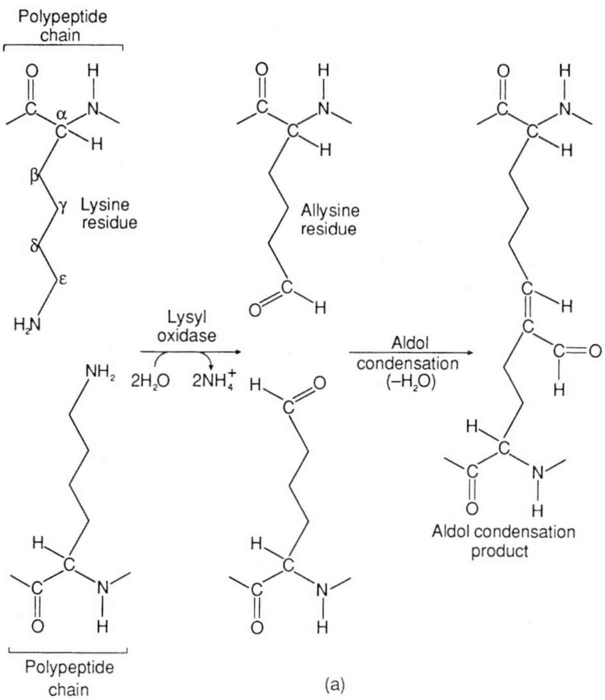

(a)

Formation of aldol condensation product—linkage involving nonhelical regions of the collagen molecule.

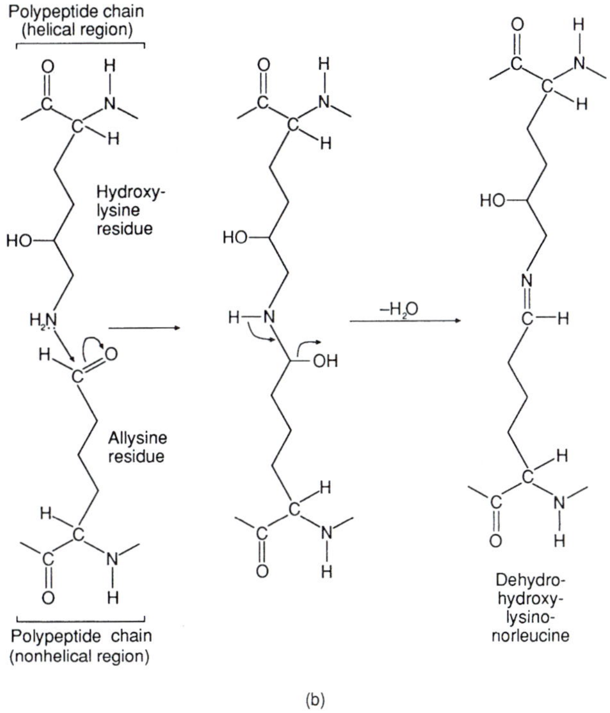

(b)

Formation of dehydrohydroxylysinonorleucine linkage.

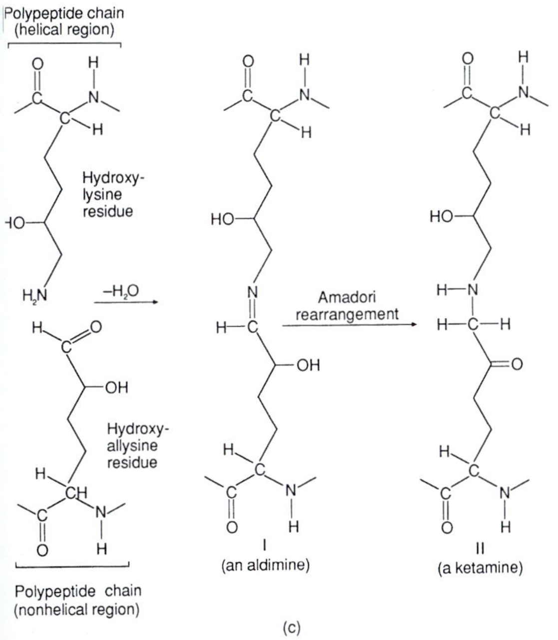

(c)

Formation of dehydrohydroxylysinohydroxynorleucine(I), an aldimine, and hydroxylysino-5-keto-norleucine(II), a ketamine. The linkage in II is more stable.

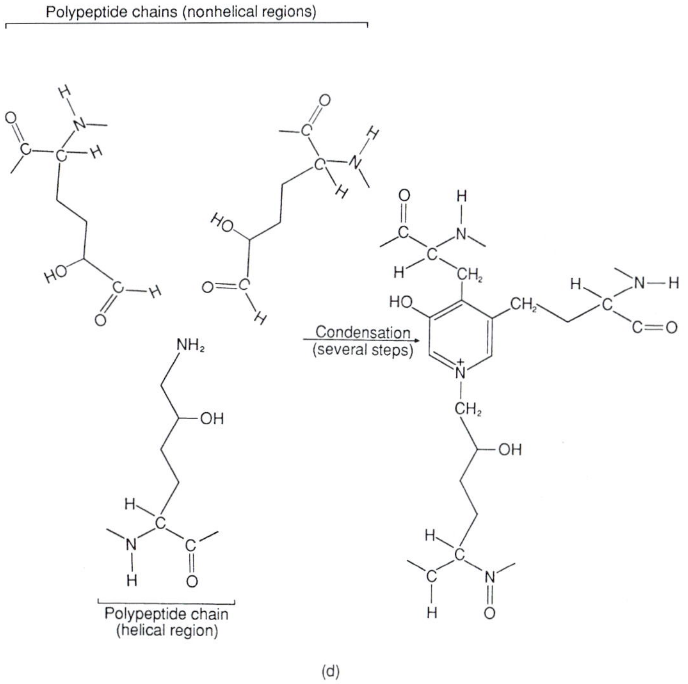

(d)

Formation of pyridinoline cross-linkage, a polyfunctional cross-linkage.

FIGURE 11-3

Formation of cross-links in collagen. All cross-links are derived from lysyl and hydroxylysyl amino acid residues. The initial step is the oxidative deamination of the ε-NH_2 group of two amino acid residues located at certain strategic positions (e.g., short nonhelical segments at both ends of the collagen molecule), with the formation of corresponding aldehyde-containing residues, allysine and hydroxyallysine. This reaction occurs extracellularly and is catalyzed by lysyl oxidase, a copper-dependent enzyme. The aldehyde groups react spontaneously with other aldehyde groups located in the adjacent chain of the same molecule or adjacent molecule. The structures do not show carbohydrate moieties for the sake of clarity.

Hydroxylysine glycosides occur at the Y-position and may play a role in determining fibril diameter. The side chains of these amino acids project outward from the center of the triple helix, permitting hydrophobic and ionic interactions between tropocollagen molecules. These interactions determine the manner in which individual tropocollagen molecules aggregate to form microfibrils initially, then larger fibrils, and eventually fibers.

The microfibril, about 4 nm wide, consists of four to eight tropocollagen molecules that aggregate in a highly ordered and specific manner owing to interactions of amino acid residues at the X- and Y-positions. In this ordered arrangement, each molecule is displaced longitudinally by about one-fourth of its length from its nearest neighbors. The longitudinally displaced tropocollagen molecules are not linked, and there is a gap of about 40 nm between the end of one triple helix and the beginning of the next (Figure 11-4). These holes may provide sites for deposition of hydroxyapatite [$Ca_{10}(PO_4)_6(OH)_2$] crystals in the formation of bone (Chapter 37). Electron microscopic studies of negatively stained collagen fibrils reveal alternating light and dark regions. The light region, where the stain does not deposit, represents the overlapping of tropocollagen molecules; the dark region corresponds to a hole where the stain is deposited.

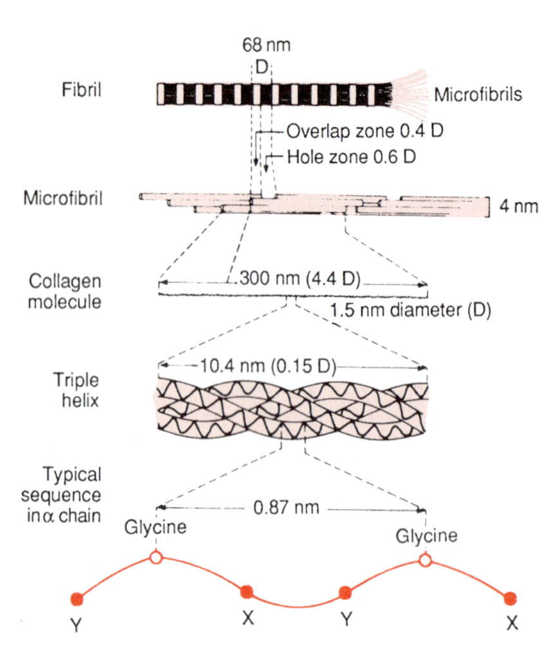

FIGURE 11-4
Packing of collagen polypeptides into a fibril. The individual tropocollagen molecules are quarter-staggered with respect to their nearest neighbors by a distance of 68 nm (D). A gap of about 40 nm (0.6 D) known as the hole region comes between the end of one tropocollagen molecule and the beginning of the next. The overlap zones are about 27 nm (0.4 D) in length and do not contain any hole zones. [Modified and reproduced, with permission, from A. Cohen, Ed.: *Rheumatology and Immunology* (W.B. Saunders, 1979).]

The tensile strength of collagen fibrils is determined by covalent cross-links involving lysyl and hydroxylysyl side chains. The packing arrangement of tropocollagen molecules provides tensile strength and also prevents sliding of molecules over one another. This organization does not allow for stretching, which does occur in elastin. The nature and extent of crosslinking depend on the physiological function and age of the tissue. With age, the density of cross-linkages increases, rendering connective tissue rigid and brittle. The arrangement of fiber bundles varies with the tissue; it is random in bone and skin, sheet-like in blood vessels, crossed in the cornea, and parallel in tendons.

Turnover of Collagen and Tissue Repair

The catabolism of collagen in the connective tissue matrix is carried out by enzymes known as ***collagenases.*** Part of the degradation may also involve neutral proteinases. A metalloenzyme specific for collagen catalyzes the cleavage of the triple helix at a single peptide bond, located at a distance from its N terminus corresponding to about three fourths the length of the tropocollagen molecule. The cleavage sites in type I collagen are the peptide bonds of Gly–Ile of the α1-chain and of Gly–Leu of the α2-chain. The role and regulation of collagenase activity in vivo are unclear. The activity of the enzyme may be regulated via the formation of an enzyme–inhibitor complex or by the activation of a proenzyme, or both.

A substantial amount (20–40%) of newly synthesized polypeptide chains of collagen undergo ***intracellular degradation.*** This degradation, which appears to occur in lysosomes, may be important in regulating the amount of collagen synthesized and in removing any defective or abnormal polypeptide chains that may be synthesized. Turnover of collagen in humans has been estimated by measurement of urinary hydroxyproline, which occurs mostly in the form of a peptide. Hydroxyproline makes up about 9–13% of collagen residues and is not reutilized. Two aspects of hydroxyproline metabolism affect its use in the assessment of the true rate of collagen turnover:

1. Hydroxyproline is rapidly metabolized in a pathway initiated by the enzyme hydroxyproline oxidase (Chapter 17).
2. Hydroxyproline is present in C1q, which is synthesized at a rate estimated to be about 4.5 mg/kg per day.

The first aspect has been overcome by measurement of urinary hydroxyproline in individuals with an inherited deficiency of hydroxyproline oxidase. Such persons excrete

0.3 g of hydroxyproline per day in urine, which corresponds to about 2.25 g of hydroxyproline-containing proteins, most of which is collagen. Total protein catabolism in a normal well-fed adult is about 300 g; thus, collagen catabolism constitutes only about 0.8% of total protein catabolism.

Collagen turnover has important clinical implications. The location, amount, type, and form of collagen depend on the coordinated control of its synthesis and degradation. In tissue injury (physical, chemical, infectious, or radiation), repair process comprises regeneration and fibrous connective tissue formation. Regeneration is the most desirable form of repair: a cut surface of an epidermis is replaced with new epidermis; scattered dead liver cells are replaced with new liver cells.

Degradation products of type I collagen, namely N-telopeptides, C-telopeptides, hydroxyproline, and the collagen cross-links pyridinolone and deoxypyridinolone, have been used as markers of osteoclast activity in bone resorption. Collagen metabolite markers used for bone formation reflecting osteoblast activity are procollagen type I carboxyterminal propeptide and procollagen type I N-terminal propeptide. Both of these propeptides are cleaved byproducts generated from the conversion of procollagen to tropocollagen. Osteocalcin and bone-specific alkaline phosphatase are also used as bone formation markers. Biochemical markers of bone turnover are used for assessing and monitoring therapy in a number of diseases of bone including osteoporosis (Chapter 37).

The cells of the body fall into three groups according to regenerative capacity: ***labile, stable,*** or ***permanent.*** The labile cells multiply throughout life and are maintained at an optimal level by continual proliferation of reserve cells. Labile cells are found in all epithelial surfaces, the spleen, and lymphoid and hematopoietic tissues. Stable cells have the potential to regenerate but do not normally undergo replication. However, under appropriate stimuli, they can proliferate rapidly. Stable cells include parenchymal cells of all glandular organs of the body and mesenchymal cells (e.g., fibroblasts, smooth muscle cells, osteoblasts, chondroblasts, and vascular endothelial cells). Permanent cells do not undergo significant replication postnatally. Nerve cells, skeletal muscle, and cardiac muscle cells are permanent cells.

In fibrous tissue scar formation (a tough mass of collagen), normal cells are permanently lost. Many tissue injuries are repaired partly by regeneration and partly by scar formation (e.g., healing of dermal cut, improperly united bone fracture, and damaged liver). In tissues with permanent cells, injury and repair result in the formation of scar tissue only (e.g., myocardium). Although the liver normally undergoes repair by regeneration, chronic liver disease resulting from ethanol abuse, some forms of malnutrition, viral infections, or exposure to chemicals can lead to hepatic fibrosis (cirrhosis) with distortion of the liver architecture by newly synthesized collagen fibers. In hepatic fibrosis, decreased collagenolytic activity and increased collagen synthesis may determine the net collagen deposition. A defect in the turnover of collagen may be responsible for the pathogenesis of *idiopathic pulmonary fibrosis.* This usually fatal disorder of the lungs is characterized by chronic inflammation of alveolar structures (alveolitis) and progressive interstitial fibrosis. An active collagenase in the lower respiratory tract may be responsible for sustained collagen lysis followed by disordered resynthesis.

In rheumatoid arthritis, collagenous tissues of the joint (including articular cartilage) are eroded by collagenases and neutral proteinases derived from proliferating cells of the synovial membrane and leukocytes. Therapeutic approaches may depend on regulation of collagen synthesis, breakdown, or both.

In a large family with osteoarthritis, which is a disorder of progressive degeneration of joint cartilage, in all affected members the α(II) collagen contained a single-base mutation converting the codon for arginine at position 519 to a codon of cysteine, an amino acid not found in normal α(II). In unaffected members of the family, this mutation does not occur.

Other disorders of collagen metabolism, heritable or acquired, are discussed in Chapter 25.

Elastin

Structure and Function

Elastin is a fibrous, insoluble protein that is not a glycoprotein but is present with collagen in the connective tissues. Connective tissues rich in elastic fibers exhibit a characteristic yellow color. Elastic fibers are highly branched structures responsible for physiological elasticity. They are capable of stretching in two dimensions and are found most notably in tissues subjected to continual high-pressure differentials, tension, or physical deformation. Elastin imparts to these tissues the properties of stretchability and subsequent recoil that depend only on the application of some physical force. Tissues rich in elastic fibers include the aorta and other vascular connective tissues, various ligaments (e.g., ligamentum nuchae), and the lungs. Microscopically, elastic fibers are thinner than collagen fibers and lack longitudinal striations.

Elastin fibers can be separated into amorphous and fibrillar components. The amorphous component consists of elastin, which is characterized by having 95% nonpolar amino acid residues and two unique lysine-derived amino acid residues, desmosine and isodesmosine.

A major fibrillar component is a glycoprotein called ***fibrillin.*** The structure of fibrillin consists of several cysteine-rich motifs and exhibits a multidomain organization similar to epidermal growth factor. The structure is stabilized by disulfide linkages. Fibrillin monomers undergo aggregation extracellularly into supramolecular structures. The fibrillar structures have a diameter of 10–12 nm and surround the amorphous component elastin. During fetal development, the fibrillar component appears first in the extracellular matrix. With increasing fetal age and with maturation of the fibers the amorphous component is deposited within the framework of the microfibrillar bundles producing mature, elastic microfibrils.

Fibrillin is encoded by a gene located in the long arm of chromosome 15. Mutations in the fibrillin gene lead to an autosomal dominant trait known as *Marfan's syndrome.* The incidence of this disorder is 1:10,000, and 15–30% of cases are caused by new mutations in the fibrillin gene. Consistent with the function of fibrillin in the elastic connective tissues, the clinical manifestations present as disorders of cardiovascular, musculoskeletal, and opthalmic systems. For example, dissecting aneurysm of the aorta, preceded by a dilatation, is a potentially fatal cardiac manifestation.

Mature elastin is a linear polypeptide, ***tropoelastin,*** which has a molecular weight of about 72,000 and contains about 850 amino acid residues. Although glycine accounts for one third of the residues, the repeat sequence Gly–X–Y characteristic of collagen is not present in elastin. Instead, glycine residues are present in the repeat units Gly–Gly–Val–Pro, Pro–Gly–Val–Gly–Val, and Pro–Gly–Val–Gly–Val–Ala. Elastin is relatively rich in the nonpolar amino acids; alanine, valine, and proline. In contrast to collagen, only a few hydroxyproline residues are present in elastin. Elastin contains no hydroxylysine or sugar residues.

A feature of mature elastin is the presence of covalent cross-links that connect elastin polypeptide chains into a fiber network. The major cross-linkages involve desmosine and isodesmosine, both of which are derived from lysine residues. Several regions rich in lysine residues can provide cross-links. Two such regions that contain peptide sequences that are repeated several times in tropoelastin have the primary structure –Lys–Ala–Ala–Ala–Lys– and

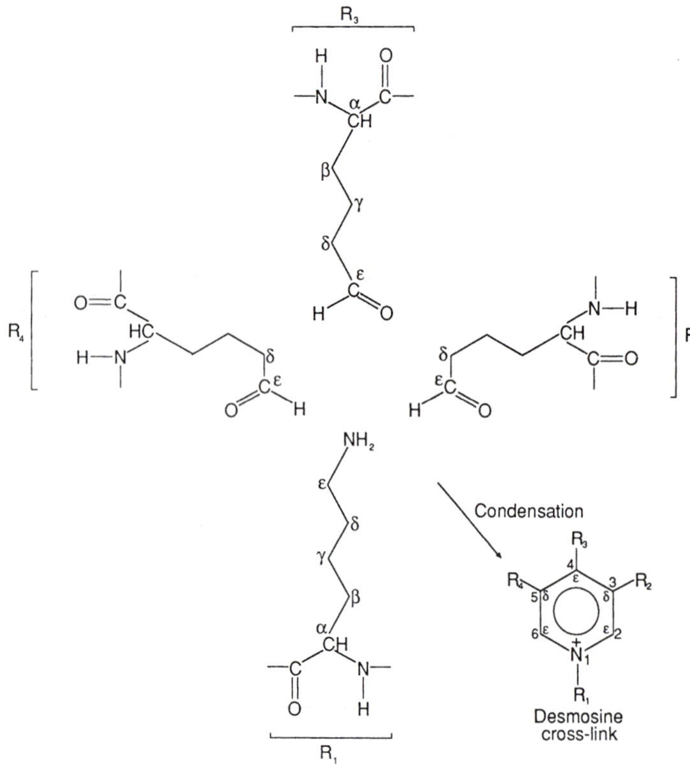

FIGURE 11-5

Formation of desmosine and isodesmosine covalent cross-links in elastin. Three allysine residues (R_2, R_3, and R_4) and one lysyl residue (R_1) condense to give a desmosine cross-link. The allysine residues (ε-aldehydes) are derived from the oxidative deamination of lysyl residues. The isodesmosine cross-link is formed similarly, except that it contains a substitution at position 2 rather than at position 4, along with substitutions at 1, 3, and 5 on the pyridinium ring.

–Lys–Ala–Ala–Lys–. The clustering of lysine residues with alanine residues provides the appropriate geometry for the formation of cross-links. Cross-linking elastin occurs extracellularly. The polypeptides are synthesized intracellularly by connective tissue cells (such as smooth muscle cells), according to the same principles for the formation of other export proteins (e.g., collagen; see Chapter 25). After transport into the extracellular space, the polypeptides undergo crosslinking. A key step is the oxidative deamination of certain lysyl residues, catalyzed by lysyl oxidase (a Cu^{2+} protein). These lysyl residues are converted to very reactive aldehydes (α-aminoadipic acid δ-semialdehyde) known as allysine residues. Through an unknown mechanism, three allysine residues and one unmodified lysine residue react to form a pyridinium ring, alkylated in four positions, which is the basis for the cross-links (Figure 11-5). The cross-links may occur within the same polypeptide or involve two to four different polypeptide chains. Current models of elastin structure suggest that only two polypeptide chains are required to form the desmosine cross-link, one chain donating a lysine and an allysine residue, the other contributing two allysine residues. Elastin polypeptides are also cross-linked by condensation of a lysine residue with an allysine residue, followed by reduction of the aldimine to yield a lysinonorleucine residue (Figure 11-6). These linkages are present to a much lesser extent than in collagen.

Formation of cross-linkages in elastin can be prevented by inhibition of lysyl oxidase. Animals maintained on a copper-deficient diet and those administered lathyrogens such as β-aminopropionitrile or α-aminoacetonitrile develop connective tissue abnormalities. Lathyrogens are so-called because of their presence in certain peas of the genus *Lathyrus*. They are noncompetitive inhibitors of lysyl oxidase, both *in vitro* and *in vivo*. Consumption of certain types of sweet pea (e.g., *Lathyrus odoratus*) can lead to lathyrism. In osteolathyrism, the abnormalities involve bone and other connective tissues, the responsible compounds being the above-mentioned nitrites. The toxic agent responsible for neurolathyrism, β-N-oxalyl-L-α,β-diaminopropionic acid, has been isolated from *Lathyrus sativus* but its mode of action is not known. Thus, copper deficiency or administration of lathyrogens in animals prevents the formation of the highly insoluble cross-linked elastin but results in the accumulation of soluble elastin. Soluble elastin obtained in this manner is particularly useful in structural studies. Purification of the insoluble elastin without disruption of the integrity of the polypeptide chains is a formidable task.

Several models of the macromolecular structure of elastin have been suggested to account for its elasticity: cross-linked globular elastin subunits, cross-linked random elastin chains, the oiled-coil model, and the fibrillar model. The fibrillar model incorporates a unique conformation due to the presence of a glycine residue followed by an amino acid residue with a bulky R-group (e.g., Pro–Gly–Gly–Val). This type of sequence gives rise to near right angle turns (β-turns; Chapter 4) in the polypeptide backbone. The oiled-coil model assumes that each polypeptide chain is fibrillar and consists of alternating segments of cross-linked regions and "oiled coils," each chain being linked to many others to form a three-dimensional network similar to a mattress spring. It is postulated that these coils can stretch up to 2–2.5 times their original length. The driving force for recoil of elastic fibers is due to a change in entropy.

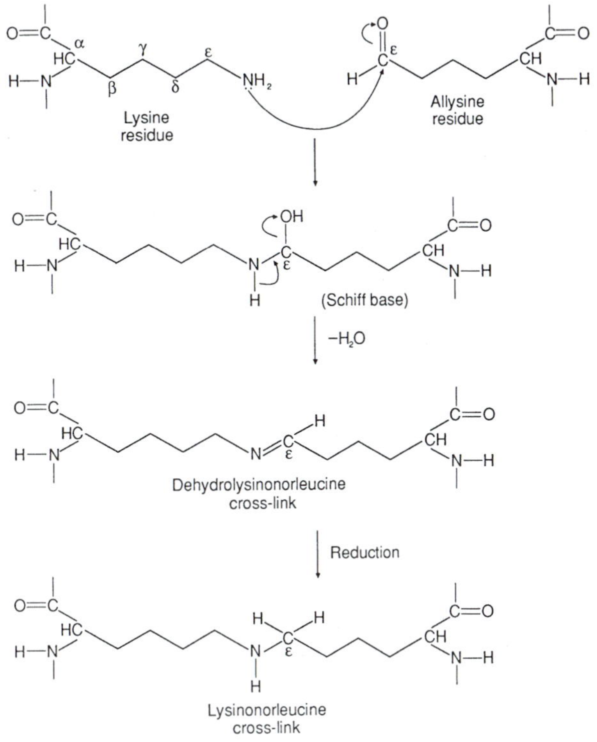

FIGURE 11-6
Formation of the lysinonorleucine cross-link between an allysine and a lysine residue in elastin.

Turnover of Elastin

The turnover rate of mature elastin in healthy persons is relatively low. Insoluble elastin in healthy elastic tissue is usually stable and subjected to minimal proteolytic degradation. In several clinical conditions (e.g., emphysema, advanced atherosclerosis, pancreatitis), increased degradation of fragmentation of elastic fibers may play a significant role. The interaction between insoluble elastin and soluble elastolytic enzymes, and the regulation of these enzymes, may shed light on certain cardiovascular diseases, in view of the role of elastin in arterial dynamics.

Elastolytic proteinases (elastases) are found in pancreatic tissue, polymorphonuclear leukocytes, macrophages, and platelets. These enzymes exhibit broad specificity. They catalyze preferential cleavage of peptide bonds adjacent to aliphatic amino acids, namely, glycine, alanine, and valine, which are present in high amounts in elastin. Elastases degrade elastin at neutral or slightly alkaline pH. The pancreas secretes elastase as an inactive precursor (zymogen) known as *proelastase,* which is converted to its active form by trypsin in the duodenum; elastase takes part in the digestion of dietary proteins (Chapter 12). This enzyme has a structure homologous to that of other pancreatic serine proteinases, including the amino acid sequence at the active site (Chapter 6). Polymorphonuclear leukocyte elastase is also a serine proteinase. Elastases are inactivated by serum α_1-proteinase inhibitor and α_2-macroglobulin; thus, their proteolytic action may be checked to prevent indiscrimination digestion of elastin-containing tissues (Chapter 6). An enzyme that degrades only soluble forms of elastin and exhibits trypsin-link activity has been described. Its biological significance is unknown. Several heritable and acquired disorders of elastin are known (see Chapter 25).

Proteoglycans

Proteoglycans are high-molecular-weight, complex molecules with diverse structures and functions. They are polyanionic substances containing a core protein to which at least one ***glycosaminoglycan*** (also known as mucopolysaccharide) chain is covalently attached. Proteoglycans are major components of connective tissue and participate with other structural protein constituents, namely, collagen and elastin, in the organization of the extracellular matrix.

Types, Structures, and Functions of Glycosaminoglycans

Six classes of glycosaminoglycans have been described. All are heteropolysaccharides and contain repeating disaccharide units. The compositions and structures of the repeating disaccharide units are shown in Table 11-2 and Figure 11-7. In five glycosaminoglycans, the

TABLE 11-2
Composition, Properties, and Distribution of Glycosaminoglycans

Glycosaminoglycan (and range of molecular weight)	Amino Sugar	Uronic Acid	Type of Sulfate Linkage (extent of sulfation is variable)	Tissue Distribution
Hyaluronate (4 to 80×10^6)	D-Glucosamine	D-Glucuronate	None	Connective tissues, cartilage, synovial uid, vitreous humor, umbilical cord
Chondroitin sulfate (5,000 to 50,000)	D-Galactosamine	D-Glucuronate	4-0- and/or 6-0-sulfate on galactosamine	Cartilage, bone, skin, cornea, blood vessel walls
Dermatan sulfate (15,000 to 40,000)	D-Galactosamine	L-Iduronate, D-Glucuronate	4-0-sulfate on galactosamine; 2-0-sulfate on iduronate	Skin, heart valve, tendon, blood vessel walls
Keratan sulfate (4,000 to 19,000)	D-Glucosamine	None (but contains D-Galactose)	6-0-sulfate on both carbohydrate residues	Cartilage, cornea, intervertebral disks
Heparan sulfate (10^4 to 10^5)	D-Glucosamine	D-Glucuronate (major), L-Iduronate (minor)	6-0-sulfate and N-sulfate (or N-acetyl) on glucosamine; 2-0-sulfate on iduronate	Lung, blood vessel walls, many cell surfaces
Heparin (10^3 to 10^6)	D-Glucosamine	L-Iduronate (major) D-Glucuronate (minor)	2-0-sulfate on iduronate; 6-0-sulfate and N-sulfate (or N-acetyl) on glucosamine	Lung, liver, skin, intestinal mucosa (mast cells)

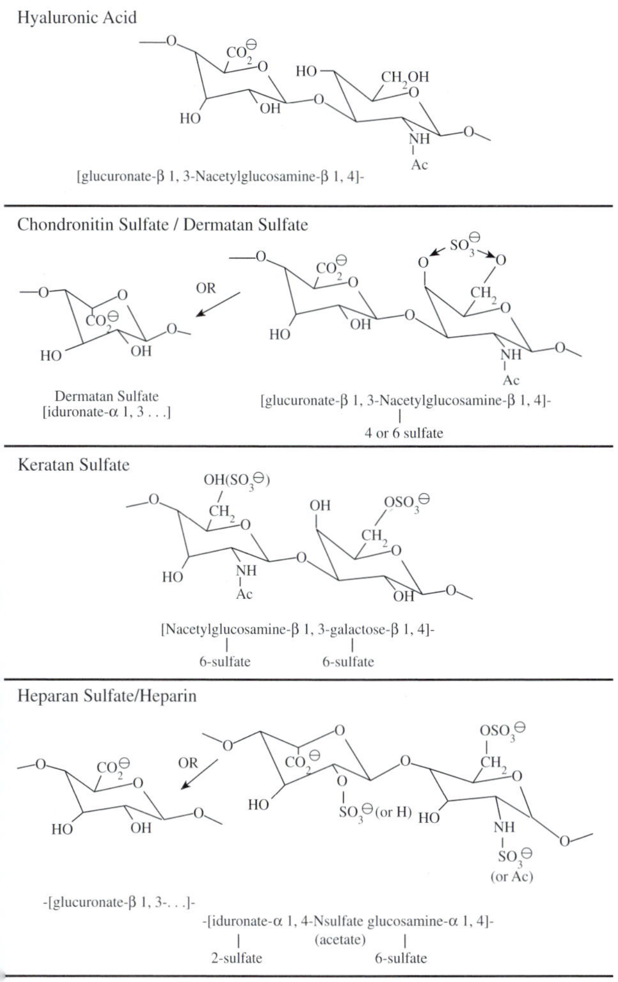

Hyaluronic Acid

[glucuronate-β 1, 3-Nacetylglucosamine-β 1, 4]-

Chondronitin Sulfate / Dermatan Sulfate

Dermatan Sulfate
[iduronate-α 1, 3 . . .]

[glucuronate-β 1, 3-Nacetylglucosamine-β 1, 4]-
4 or 6 sulfate

Keratan Sulfate

[Nacetylglucosamine-β 1, 3-galactose-β 1, 4]-
6-sulfate 6-sulfate

Heparan Sulfate/Heparin

-[glucuronate-β 1, 3-. . .]-

-[iduronate-α 1, 4-Nsulfate glucosamine-α 1, 4]-
(acetate)
2-sulfate 6-sulfate

FIGURE 11-7
Basic repeating disaccharide units of six classes of glycosaminoglycans. Chondroitin sulfate and dermatan sulfate differ in type of principal uronic acid residue; the former contains glucuronate and the latter iduronate. Heparin and heparan sulfate differ in that heparin contains more iduronate 2-sulfate and N-sulfated glucosamine residues than heparan sulfate. In the sulfate-containing glycosaminoglycans, the extent of sulfation is variable. All of these molecules have a high negative charge density and are extremely hydrophilic. [Reproduced with permission from V. C. Hascall and J. H. Kimura; Proteoglycans: Isolation and characterization. In *Meth. Enzymol.* **82**(A), L. W. Cunningham and D. W. Frederiksen. Eds. Academic Press, San Diego (1982).]

lisaccharide units consist of amino sugars (usually)-galactosamine) alternating with uronic acids. In one ype, keratan sulfate, the uronic acid is replaced by galacose. The amino sugars are usually present as N-acetyl lerivatives. In heparin and heparan sulfate, however, most mino sugars are found as N-sulfates in sulfamide linkge, with a small number of glucosamine residues as N-acetyl derivatives. With the exception of hyaluronate, ll glycosaminoglycans contain sulfate groups in ester inkages with the hydroxyl groups of the amino sugar esidues.

The presence of carboxylate and sulfate groups provides the glycosaminoglycans with a high negative charge density and makes them extremely hydrophilic. Almost all water in the intercellular matrix is bound to glycosaminoglycans, which assume a random conformation in solution, occupying as much solvent space as is available by entrapping the surrounding solvent molecules. In solution, glycosaminoglycans also undergo aggregation and impart a high viscosity. In order to maintain electrical neutrality, negatively charged (anionic) groups of glycosaminoglycans, fixed to the matrix of the connective tissue, are neutralized by positively charged (cationic) groups (e.g., Na^+). Osmotic pressure within the matrix is thereby elevated, contributing turgor to the tissue. Thus, proteoglycans, because they contain glycosaminoglycans, are polyanionic and usually occupy large hydrodynamic volumes relative to glycoproteins or globular proteins of equivalent molecular mass. Usually, they have high buoyant densities, and this property has been utilized in their isolation and characterization. The physical properties of connective tissue also depend on collagen or elastin fibers. Collagen fibers resist stretching and provide shape and tensile strength; the hydrated network of proteoglycans resists compression and provides both swelling and pressure to maintain volume (Figure 11-8). Thus, the properties of these different macromolecules are complementary, and their relative compositions vary according to tissue function. For example, sulfated glycosaminoglycans tend to provide a solid consistency, whereas hyaluronate provides a softer, more fluid consistency. The vitreous space behind the lens of the eye is filled with a viscous, gelatinous solution of hyaluronate free of fibrous proteins and transparent to light. Hyaluronate is not sulfated, and there is no evidence that it is linked to a protein molecule, as are the other glycosaminoglycans. Hyaluronate is also present in synovial fluid in joint cavities that unite long bones, bursae, and tendon sheaths. The viscous and elastic properties of hyaluronate contribute to the functioning of synovial fluid as a lubricant and shock absorber.

Hyaluronate is depolymerized by hyaluronidase, which hydrolyzes the $\beta(1 \rightarrow 4)$ glycosidic linkages between N-acetylglucosamine and glucuronate. Some pathogenic bacteria secrete hyaluronidase, which breaks down the protective barrier of hyaluronate and renders the tissue more susceptible to infection. Hyaluronidase, found in spermatozoa, may facilitate fertilization by hydrolyzing the outer mucopolysaccharide layer of the ovum, thereby enabling the sperm to penetrate it. Hyaluronidase (prepared from mammalian testes) is used therapeutically to enhance dispersion of drugs administrated in various parts of the body.

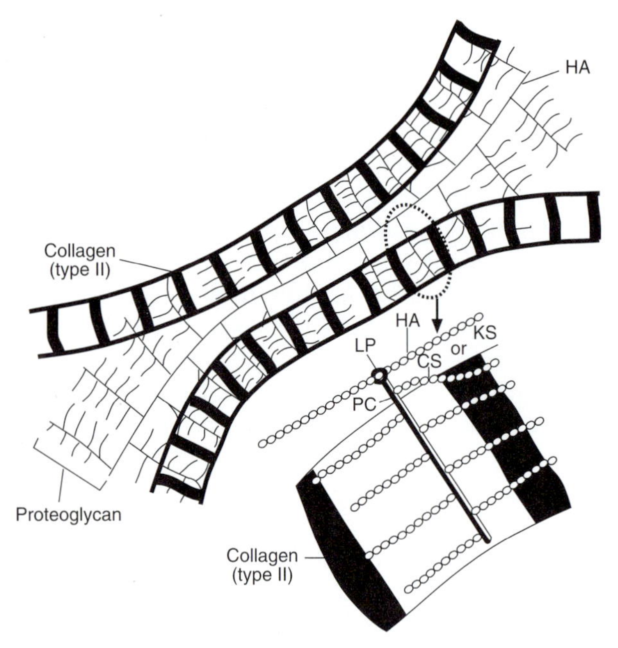

FIGURE 11-8
Schematic representation of the molecular organization of structural elements in cartilage matrix. Collagen fibrils provide tensile forces, and proteoglycans, because of their large solvent domains, accommodate reversible compressible forces. LP = link protein; HA = hyaluronate; KS = keratan sulfate; CS = chondroitin sulfate; PC = core protein. [Reproduced with permission from L. C. Junqueira, J. Carneiro, and J. A. Long, *Basic Histology*, 5th ed. Appleton & Lange, Norwalk, CT, 1986. ©1986 Appleton-Century-Crofts.]

Heparin is a heterogeneous glycosaminoglycan found in tissues that contain mast cells (e.g., lungs and perivascular connective tissue). Its primary physiological function is unknown, but it is a powerful inhibitor of blood clotting and is used therapeutically for that purpose. The therapeutic anticoagulant action of heparin is due to its ability to produce conformational changes in the proteinase inhibitor antithrombin. Both activated factor X (factor Xa) and thrombin are inactivated by a heparin-antithrombin complex (Chapter 36). The antithrombin is homologous in structure with the α_1-antitrypsin family of proteinase inhibitors and is a suicide substrate for factor Xa and thrombin. The proteinase binds to a specific Arg–Ser peptide bond present at the reactive site of antithrombin. This 1:1 antithrombin-thrombin is a stable inactive complex. The function of heparin is catalytic. After the formation of the heparin-mediated antithrombin-thrombin complex, heparin is released to initiate another cycle. Factor Xa is inhibited by a specific heparin pentasaccharide bound to antithrombin. However, the inhibition of thrombin requires an antithrombin bound to the heparin consisting of the specific pentasaccharide that consists of chains of at least 18 monosaccharide units.

Heparin therapy has a serious risk. In about 3% of patients on heparin therapy, thrombocytopenia (low platelet levels) develops. This can be life threatening if it leads to thromboembolic complications such as disseminated intravascular coagulation. Thrombocytopenia is caused by a heparin-induced immune-mediated process. The antibody is directed against a cryptic epitope present on a platelet factor (PF4) that appears only after the protein binds to heparin. The multimolecular complex consisting of antibody–heparin–PF4 causes platelet activation by binding to specific receptors (FcϒRIIA) on the platelet membrane. A genetic predisposition to this disorder is due to a specific mutation in the FcϒRIIA gene changes arginine to histidine at position 131. This amino acid change causes a strong affinity for the Fc region of immunoglobulin IgG_1 and IgG_2. Thus, identification of patients with homozygosity for the Arg 131 $\rightarrow$ His 131 mutation in FcϒRIIA and who are on heparin therapy has important therapeutic implications. Heparin-induced thrombocytopenia occurs less frequently with the use of low-molecular-weight heparin (M.W. 5,000) compared to unfractionated heparin (M.W. 3,000–30,000). The treatment of life threatening complications of heparin-induced thrombocytopenia requires cessation of heparin therapy. Heparin's anticoagulant action can be reversed rapidly by intravenous administration of prolamine sulfate. Protamines are basic proteins, isolated from fish sperm and bind tightly to heparin.

One of the laboratory tests used in the identification of patients with heparin-induced thrombocytopenia, employs a micro-ELISA plate coated with purified PF4-heparin complex. Upon incubation of patient's plasma, if antibody is present, it binds to the PF4-heparin complex. The bound antibody is detected with goat antihuman antibody coupled to peroxidase, followed by incubation with substrates *o*-phenylenediamine/H_2O_2 (Chapter 8).

Tendons have a high content of collagen and sulfated glycosaminoglycans (chondroitin and dermatan sulfates). Tendons are fibrous cords that fuse with skeletal muscle at each end and penetrate bones at the two sides of a joint Thus, tendons are aligned along their long axis, providing flexible strength in the direction of the muscle pull.

Remarkable progress has been achieved in understanding the organization of the macromolecular components of cartilage. Cartilage is one type of dense connective tissue; bone is the other (Chapter 37). Cartilage is less resistant to pressure than is bone, and its weightbearing capacity is exceeded by that of bone. It has a smooth resilient surface. During growth and development of an embryo, cartilage provides the temporary framework for the formation and development of bone. In postnatal life it permits the long bones of the extremities to grow until

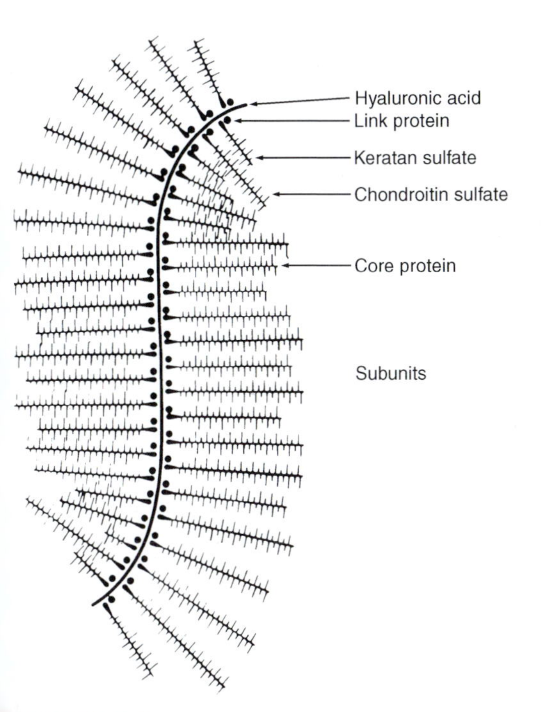

FIGURE 11-9
Schematic representation of cartilage proteoglycan. The monomers, consisting of glycosaminoglycan chains linked to a core protein by covalent linkage, extend laterally at intervals from opposite sides of a very long filament of hyaluronate. The interaction between the core protein and hyaluronate is noncovalent and is aided by the link protein. The entire structure is highly hydrated and occupies a large volume. [Reproduced with permission from W. J. Lennarz, *The Biochemistry of Glycoproteins and Proteoglycans,* Plenum Press, New York, 1980].

their longitudinal growth ceases. Its smooth surface provides for freely movable joints (e.g., knee and elbow). The properties of cartilage depend on its content of collagen (or collagen and elastin) and proteoglycans. Cells responsible for the synthesis and maintenance of cartilage are known as ***chondrocytes.***

The proteoglycan monomer of cartilage consists of glycosaminoglycan chains of chondroitin sulfate and keratan sulfate covalently linked to a core protein (Figure 11-9) about 300 nm long with molecular weight of 250,000. Each molecule of core protein contains about 30 chondroitin sulfate and 100 keratan sulfate chains. The linkage between the oligosaccharides and the protein can be O-glycosidic between xylose and serine residues, O-glycosidic between N-acetylgalactosamine and serine or threonine residues, or N-glycosidic between N-acetylglucosamine and the amide nitrogen of asparagine. The glycosaminoglycan chains with their negatively charged groups extend out of the core protein to occupy a large volume and to entrap surrounding solvent. One end of the core protein has relatively few or no attached glycosaminoglycan chains and resembles a glycoprotein; at this end, the core protein possesses an active site that binds to a very long filament of hyaluronate by a noncovalent interaction at intervals of 30 nm. The interaction between the core protein and the hyaluronate is aided by a protein (M.W. 50,000) known as ***link protein.*** The macromolecular complex is negatively charged and highly hydrated; it interacts electrostatically with basic charges on collagen fibrils, which define tissue space and provide tensile strength. The interspersed proteoglycan aggregates provide a hydrated, viscous gel for resistance to compressive loads. Proteoglycans originating in different tissues and exhibiting differences in molecular structure are presumed to be products of different genes. Synthesis and secretion of proteoglycans consist of several different phases of the internal membrane systems of the cell. Synthesis of the protein component occurs on ribosomes attached to the rough endoplasmic reticulum, with translocation of the newly synthesized polypeptide chains to the cisternal side of the membrane. The polypeptide chains travel through the cisternae of the endoplasmic reticulum to the Golgi apparatus, where the oligosaccharide side chains are synthesized by the addition of one sugar residue at a time. Further modification of the sugar residues (the conversion of selected residues of D-glucuronate to L-iduronate, and sulfation) takes place after completion of the oligosaccharide (Chapter 16). The fully synthesized proteoglycans are then transferred from the Golgi apparatus, in the form of vesicles, to the cell membrane for the purpose of secretion into the extracellular spaces.

The proteoglycan content in animal tissues appears to change with aging. Tissues rich in keratan sulfate show a continual increase in amount of this proteoglycan throughout life. The amounts of chondroitin sulfate in cartilage and in nucleus pulposus (of intervertebral disk) and of hyaluronic acid in skin decrease with age. Administration of growth hormone to an older animal appears to reverse the pattern of proteoglycan synthesis, making it similar to that of a younger animal. Some of the effects of growth hormone are mediated indirectly by a group of peptides known as somatomedins, which are secreted by the liver (and possibly by the kidney) under the influence of growth hormone (Chapter 31). Somatomedin A (formerly known as sulfation factor) exhibits cartilage-stimulating activity, in part by affecting the incorporation of sulfate into proteoglycans. Testosterone, a steroid hormone (Chapter 34), increases the rate of hyaluronate synthesis. Insulin deficiency causes diminished synthesis of proteoglycans in rats that are rendered diabetic experimentally; synthesis is restored to normal following insulin administration. Some chronic complications of diabetes mellitus, namely, accelerated vascular degeneration, poor wound healing, and increased susceptibility to infection, may be partly attributable to altered proteoglycan metabolism.

Turnover of Proteoglycans and Role of Lysosomes

Proteoglycans undergo continuous turnover at rates dependent upon the nature of the proteoglycan and tissue location. Their half-life is between one and several days. Degradation of proteoglycans is initiated by proteolytic enzymes that release glycosaminoglycans; the latter are subsequently degraded by ***lysosomal enzymes.*** Some of the products of hydrolysis (e.g., dermatan sulfate and heparan sulfate) are excreted in the urine.

Lysosomes are subcellular organelles in which a wide range of catabolic enzymes are stored in a closed, protective membrane system. They are major sites of intracellular digestion of complex macromolecules derived from both intracellular (autophagic) and extracellular (heterophagic) sources. Lysosomal enzymes show optimal activity at acidic pH. The pattern of enzymes in lysosomes may depend upon the tissue of origin, as well as upon the physiological or developmental state of the cells.

Lysosomal enzymes are synthesized on the ribosomes of the rough endoplasmic reticulum, passed through the Golgi apparatus, and packaged into vesicles. The hydrolyases are glycoproteins, some of which contain mannose 6-phosphate markers necessary for the normal uptake of glycoproteins into lysosomes. Thus, carbohydrates may also serve as determinants of recognition in the *intracellular* localization of the glycoproteins following their synthesis. In two biochemically related disorders of lysosomal function, I-cell disease (mucolipidosis II) and pseudo-Hurler's polydystrophy (mucolipidosis III), the lesion is in the posttranslational modification step for acid hydrolases destined to be packaged into lysosomes. In normal cells, these acid hydrolyases are glycoproteins carrying mannose 6-phosphate markers that direct them to lysosomes through a receptor-mediated process. In other words, the presence of phosphomannose residues on the newly synthesized acid hydrolases and of phosphomannose receptors on selected membranes leads to the segregation of these enzymes in the Golgi apparatus, with subsequent translocation into lysosomes. In mucolipidosis II, the acid hydrolases are not phosphorylated; in mucolipidosis III, the enzymes either are not phosphorylated or have significantly diminished phosphate content. The lack of mannose 6-phosphate leads to defective localization of acid hydrolases that, instead of being packaged in lysosomes, are exported outside the cell; thus, the enzyme activity in plasma reaches high levels. These enzymes could cause indiscriminate damage. At least eight acid hydrolases (glycosidases, sulfatases, and cathepsins) appear to be affected in this manner. However, not all lysosomal enzymes and tissue cells are affected. For example, lysosomal acid phosphatase and β-glucosidase, and hepatocytes and neurons, appear to be spared from this defect. In mucolipidosis II and III, large inclusions of undigested glycosaminoglycans and glycolipids occupy almost all of the cytoplasmic space in cultured skin fibroblasts. In addition to the phosphorylation defect, the acid hydrolases are much larger than their normal counterparts, presumably owing to lack of the limited proteolysis of the hydrolases that occurs in normal lysosomes. Both disorders are inherited as autosomal recessive traits, affect primarily connective tissue, and are characterized by psychomotor retardation, skeletal deformities, and early death.

The segregation of lysosomal enzymes into lysosomes requires carbohydrate recognition markers (phosphomannose in some) and also the formation of coated vesicles into which the enzymes are sequestered. Coated vesicles shuttle macromolecules between organelles and may be responsible for selectivity in intercompartmental transport (e.g., from endoplasmic reticulum to Golgi, from Golgi to lysosomes). The major component of coated vesicles is ***clathrin,*** a nonglycosylated protein (M.W. 180,000). It is located on the outer surface (cytoplasmic side) of the coated vesicles. Clathrin and its tightly bound light chains (M.W. 33,000 and 36,000) form flexible lattices that function as structural scaffolds surrounding the vesicles.

In addition to their role in intracellular transport, coated vesicles are involved in ***receptor-mediated endocytosis.*** This process accomplishes internalization of macromolecules (ligands) by binding them to receptors on the cell membranes located in specialized regions of clathrin-containing coated pits, which invaginate into the cell to form coated vesicles. Inside the cell, the vesicles lose their clathrin coat (which may be reutilized) and fuse with one another to form endosomes, whose contents are acidified by proton pumps driven by free energy of hydrolysis derived from ATP. In the endosome, the ligand and receptor undergo dissociation. The endosome fuses with the primary lysosome to form a secondary lysosome. Receptors may also be recycled or degraded. Thus, there are several pathways of receptor-mediated endocytosis. In some cells the receptors migrate continuously to coated pits and undergo internalization whether or not ligands are bound to them (e.g., receptors for low-density lipoproteins, transferrin, and asialoglycoproteins). In other cells, the receptors are diffusely distributed and do not migrate to coated pits unless they are bound with ligands (e.g., epidermal growth factor).

Following internalization, the receptor-ligand complex may be disposed of by one of four routes:

1. Ligand degradation and receptor reutilization (e.g., low-density lipoproteins, Chapter 20; asialoglycoproteins, Chapter 10; transcobalamin II, Chapter 38; some peptide hormones;
2. Both ligand and receptor reutilization (e.g., transferrin, Chapter 29; class I major histocompatibility complex molecules on T cells and class II molecules on macrophages, Chapter 35);
3. Both receptor and ligand degradation (e.g., epidermal growth factor, immune complexes);
4. Both ligand and receptor transportation out of the cell (e.g., maternal IgG, secretory IgA Chapter 35).

Receptors for different ligands have been characterized and show multiple functional domains (see low-density lipoproteins, Chapter 20; insulin, Chapter 22).

Phagocytosis is a transport system that does not involve receptor-mediated endocytosis coupled to the formation of coated vesicles. This type of internalization consists of forming phagosomes, which are membrane-bound vesicles containing ingested substances (e.g., bacteria or their products, extracellular material). Secondary lysosomes are formed when primary lysosomes fuse with phagosomes; they can also be formed when primary lysosomes fuse with autophagosomes, which contain internally derived defective organelles (e.g., mitochondria or rough endoplasmic reticulum). Within the secondary lysosomes, the ingested macromolecules are broken down by a wide variety of hydrolases into their respective monomeric units (e.g., sugars, amino acids, purines, pyrimidines, fatty acids, and cholesterol). Eventually these molecules appear in the cytoplasm and are usable for metabolic purposes. Undigested material is retained within the lysosomes and forms a residual body. A schematic representation of the formation and some functions of lysosomes is shown in Figure 11-10.

In some circumstances, lysosomal enzymes are extruded into the extracellular milieu and cause destruction of the surrounding matrix. This extracellular catabolic process releases materials that are taken up in phagosomes for further breakdown by the cells. This process occurs in the remodeling of bone and cartilage. In the thyroid gland, lysosomes participate in the regulation of hormone release from the cells to the blood circulation. The thyroid hormones, thyroxine and triiodothyronine, are synthesized on a polypeptide, thyroglobulin, and stored within the thyroid follicles. Under appropriate stimuli, the thyroglobulin is internalized (endocytosed) and hydrolyzed by lysosomal hydrolases to release the free hormones, which are then transferred to the blood (Chapter 33). *In vitro,* estradiol, testosterone, and vitamins A and D have disruptive effects on the lysosomal membrane, causing the release of lysosomal enzymes. The anti-inflammatory steroid hormone *cortisol* has an opposite effect. The destructive potential of leukocytic lysosomal proteinases is checked by α_1-proteinase inhibitor and α_2-macroglobulin in plasma and synovial fluids. Enhanced catabolism of proteoglycans of articular cartilage occurs in rheumatoid arthritis as a result of the action of lysosomal enzymes on articular cartilage.

Many pathological conditions occur as a result of a deficiency of lysosomal enzymes (e.g., storage disorders of glycoproteins and glycosphingolipids). At least 41 distinct *lysosomal storage diseases* are known. Most of these diseases are due to deficiency of a particular enzyme and a few are due to defects in the nonlysosomal proteins that are necessary for lysosomal biogenesis. Lysosomal storage disorders are predominantly autosomal recessive, with the exception of *Fabry's disease* and *mucopolysaccharidosis type II,* which are X-linked recessive disorders. Overall, lysosomal storage disorders are relatively common with a prevalence of 1 per 7700 live births, although some disorders are very rare.

Mucopolysaccharidoses

Mucopolysaccharidosis (MPS) encompasses disorders in which undegraded or partly degraded glycosaminoglycans accumulate in the lysosomes of many tissues owing to a deficiency of specific lysosomal enzymes. Table 11-3 lists the missing enzymes and gives relevant clinical and laboratory findings. These disorders, with the exception of Hunter's syndrome, which is an X-linked trait, are inherited in the autosomal recessive manner. All of the deficient enzymes are acid hydrolases except for acetyltransferase in Sanfilippo's syndrome type C. In mucopolysaccharidoses the catabolism of heparan sulfate, dermatan sulfate, and keratan sulfate is affected. Their degradation proceeds from the nonreducing end of the carbohydrate chain by the sequential actions of lysosomal exoglycosidases, exosulfatases, and an acetyltransferase (Figures 11-11 through 11-13).

These disorders are rare; collectively, they may occur in 1 in 20,000 live births. Since proteoglycans are widely distributed in human tissues, the syndromes can affect a wide variety of tissues; thus, the clinical features vary considerably. All types are characterized by reduced life expectancy, with the exception of Scheie's syndrome. All types are characterized by skeletal abnormalities, which are particularly severe in types IV and VI. Types IH, IS, IV, VI, and VII usually exhibit clouding of the

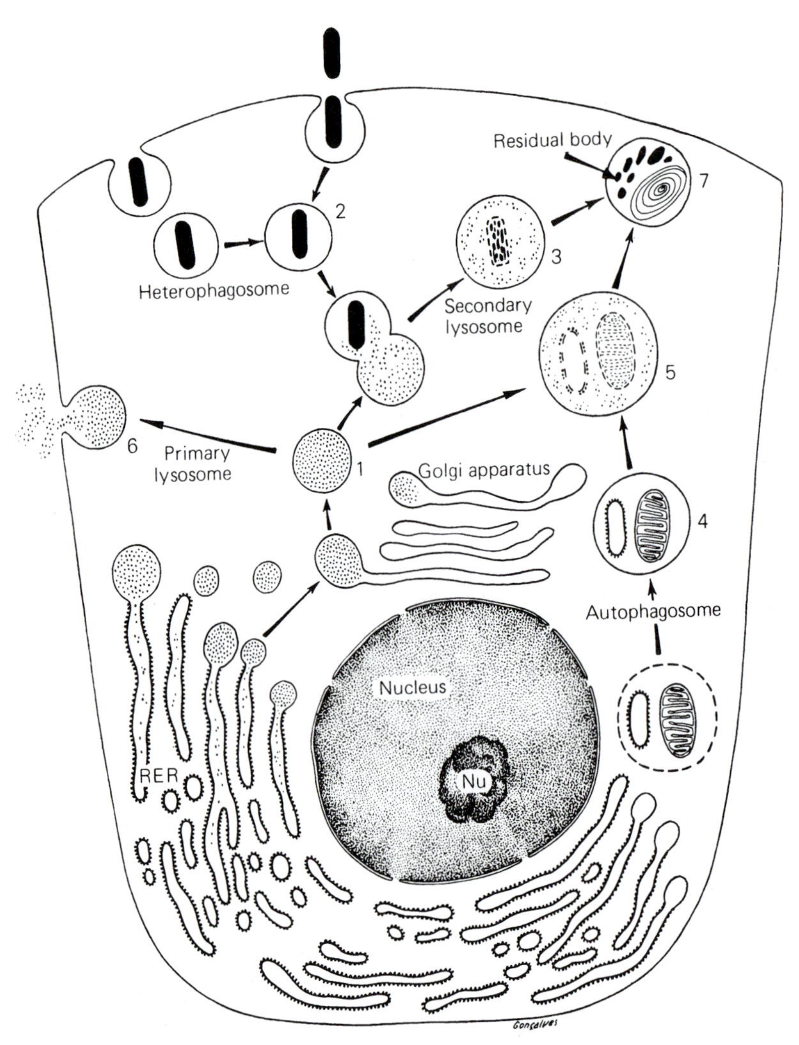

FIGURE 11-10

Formation and function of lysosomes. Lysosomal enzymes are synthesized in the rough endoplasmic reticulum. Those with appropriate recognition markers are processed and packaged into primary lysosomes in the complex network of the Golgi apparatus. 1, 2, and 3 represent digestion of extracellular substance segregated into a heterophagosome; 1, 4, and 5, digestion of segregated cytoplasmic material (autophagosome); 1 and 6, extrusion of lysosomal enzymes that will act extracellularly (e.g., collagenase); and 7, a residual body. [Reproduced with permission from L. C. Junqueira, J. Carneiro, and J. L. Long, *Basic Histology,* 5th ed. Appleton & Lange, Norwalk, CT, 1986. ©1986 Appleton-Century-Crofts.]

cornea. Types IH, II, III, and VIII are characterized by severe mental retardation, while types IS, IV, and VI exhibit normal intelligence. Patients with different enzyme deficiencies may exhibit phenotypic similarities (pleiotropism), and they may also exhibit variation in clinical severity with the same enzyme deficiency (allelic variants). The enzyme deficiency can be established by assays on peripheral lymphocytes or cultured fibroblasts. Prenatal diagnosis is possible but requires successful culture of amniotic fluid cells and assay of specific enzymes. No specific therapy is available. Management focuses on providing prognostic information and counseling.

11.2 Peptidoglycans

Peptidoglycans are components of bacterial cell wall and consist of heteropolysaccharide chains cross-linke by short peptide chains. These cell walls bear the anti genic determinants; when exposed to them, humans (an other mammalian species) develop specific antibodies t defend against bacteria. Bacterial virulence is also relate to substances associated with the cell wall. Cell wall syn thesis is the target for the action of the penicillins an cephalosporins.

Bacterial cell walls are rigid and complex, enable th cells to withstand severe osmotic shock, and survive in

TABLE 11-3
*Classification of the Mucopolysaccharidoses (MPS)**

Number	Eponym	Clinical Manifestations	Enzyme Deficiency	Glycosaminoglycan Affected
MPS I H	Hurler	Corneal clouding, dysostosis multiplex, organomegaly, heart disease, mental retardation, death in childhood	α-L-Iduronidase	Dermatan sulfate, heparan sulfate
MPS I S	Scheie	Corneal clouding, stiff joints, normal intelligence and life span	α-L-Iduronidase	Dermatan sulfate, heparan sulfate
MPS I H/S	Hurler/Scheie	Phenotype intermediate between I H and I S	α-L-Iduronidase	Dermatan sulfate, heparan sulfate
MPS II (severe)	Hunter (severe)	Dysostosis multiplex, organomegaly, no corneal clouding, mental retardation, death before 15 years	Iduronate sulfatase	Dermatan sulfate, heparan sulfate
MPS II (mild)	Hunter (mild)	Normal intelligence, short stature, survival to 20s to 60s	Iduronate sulfatase	Dermatan sulfate, heparan sulfate
MPS III A	Sanfilippo A	Profound mental deterioration, hyperactivity, relatively mild somatic manifestations	Heparan *N*-sulfatase	Heparan sulfate
MPS III B	Sanfilippo B	Phenotype similar to III A	α-*N*-Acetylglucosaminidase	Heparan sulfate
MPS III C	Sanfilippo C	Phenotype similar to III A	Acetyl CoA:α-glucosaminide acetyltransferase	Heparan sulfate
MPS III D	Sanfilippo D	Phenotype similar to III A	*N*-Acetylglucosamine 6-sulfatase	Heparan sulfate
MPS IV A	Morquio A	Distinctive skeletal abnormalities, corneal clouding, odontoid hypoplasia; milder forms known to exist	Galactose 6-sulfatase	Keratan sulfate, chondroitin 6-sulfate
MPS IV B	Morquio B	Spectrum of severity as in IV A	β-Galactosidase	Keratan sulfate
MPS V	No longer used	—	—	—
MPS VI	Maroteaux-Lamy	Dysostosis multiplex, corneal clouding, normal intelligence; survival to teens in severe form; milder forms known to exist	*N*-Acetylgalactosamine 4-sulfatase (arylsulfatase B)	Dermatan sulfate
MPS VII	Sly	Dysostosis multiplex, hepatosplenomegaly; wide spectrum of severity, including hydrops fetalis and neonatal form	β-Glucuronidase	Dermatan sulfate, heparan sulfate, chondroitin 4-, 6-sulfates
MPS VIII	No longer used	—	—	—

*Reproduced with permission from E. F. Neufeld and J. Muenzer: In Metabolic Basis of Inherited Disease 7th ed., C. R. Scriver, A. L. Beaudet, W. S. Sly and D. Valle, Eds. (McGraw-Hill 1995) page 2466.

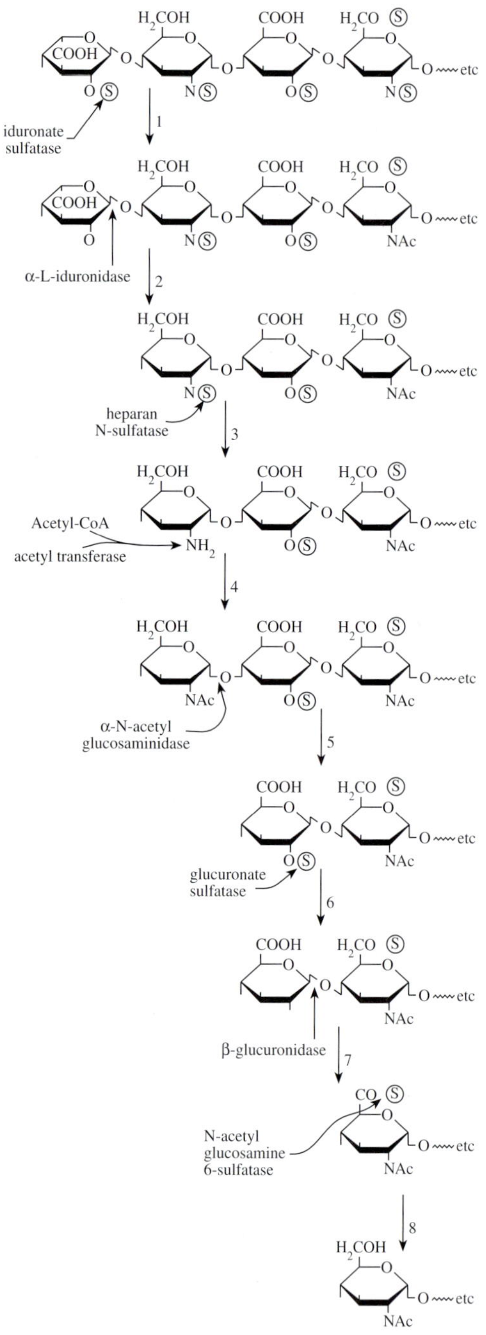

FIGURE 11-11
Stepwise degradation of heparan sulfate. The deficiency diseases corresponding to the numbered reactions are: 1 = mucopolysaccharidosis (MPS) II, Hunter's syndrome; 2 = MPS I, Hurler's, Scheie's, and Hurler–Scheie's syndromes; 3 = MPS III A, Sanfilippo's syndrome type A; 4 = MPS III C, Sanfilippo's syndrome type C; 5 = MPS III B, Sanfilippo's syndrome type B; 6 = no deficiency disease yet known; 7 = MPS VII, Sly's syndrome; 8 = MPS III D, Sanfilippo's syndrome type D. The schematic drawing depicts all structures known to occur within heparan sulfate and does not imply that they occur stoichiometrically. Very few of the glucuronic acid residues are sulfated. [Reproduced with permission from E. F. Neufeld and J. Muenzer. In: *Metabolic Basis of Inherited Disease,* 7th ed., C. R. Scriver, A. L. Beaudet, W. S. Sly, and D. Valle (Eds). McGraw-Hill, New York, 1995, p. 2468.]

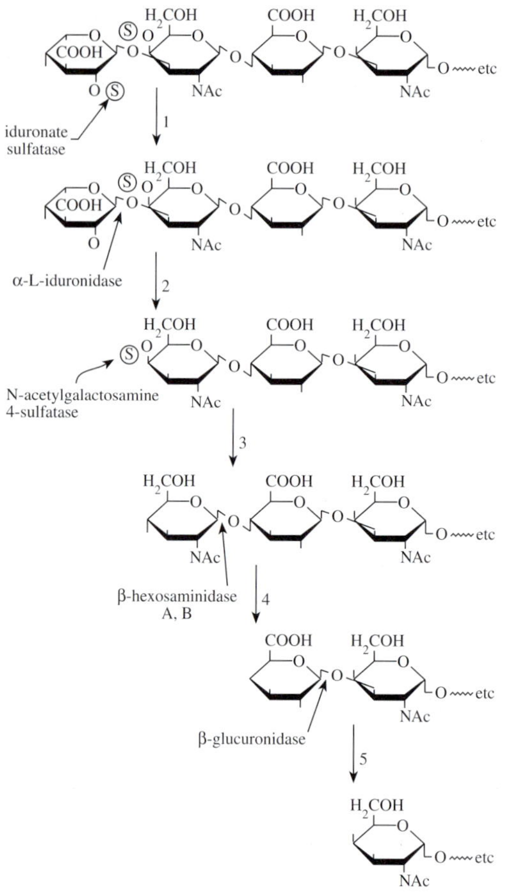

FIGURE 11-12
Stepwise degradation of dermatan sulfate. The deficiency diseases corresponding to the numbered reactions are: 1 = MPS II, Hunter's syndrome; 2 = MPS I, Hurler's, Scheie's, and Hurler–Scheie's syndromes; 3 = MPS VI, Maroteaux–Lamy syndrome; 4 = Sandhoff's disease; and 5 = MPS VII, Sly's syndrome. This schematic drawing depicts all structures known to occur within dermatan sulfate and does not imply that they occur in equal proportion. For instance, only a few of the L-iduronic acid residues are sulfated, and L-iduronic acid occurs much more frequently than glucuronic acid. [Reproduced with permission from E. F. Neufeld and J. Muenzer. In: *Metabolic Basis of Inherited Disease,* 7th ed., C. R. Scriver, A. L. Beaudet, W. S. Sly, and D. Valle (Eds). McGraw-Hill, New York, 1995, p. 2467.]

hypotonic environment. The contents of a bacterium can exert an osmotic pressure as high as 20 atm. At cell division, the walls rupture and reseal rapidly.

Bacteria are classified into two groups on the basis of a staining reaction discovered by Gram in 1884. In this reaction, the cells do or do not retain a crystal violet-iodine dye complex after an alcohol wash. Cells that retain the stain are gram-positive; those that do not are gram-negative. This empirical classification divides bacteria into two classes that differ in cell wall structure.

Gram-positive bacteria (e.g., *Staphylococcus aureus*) are surrounded by a cytoplasmic membrane with a bilayer

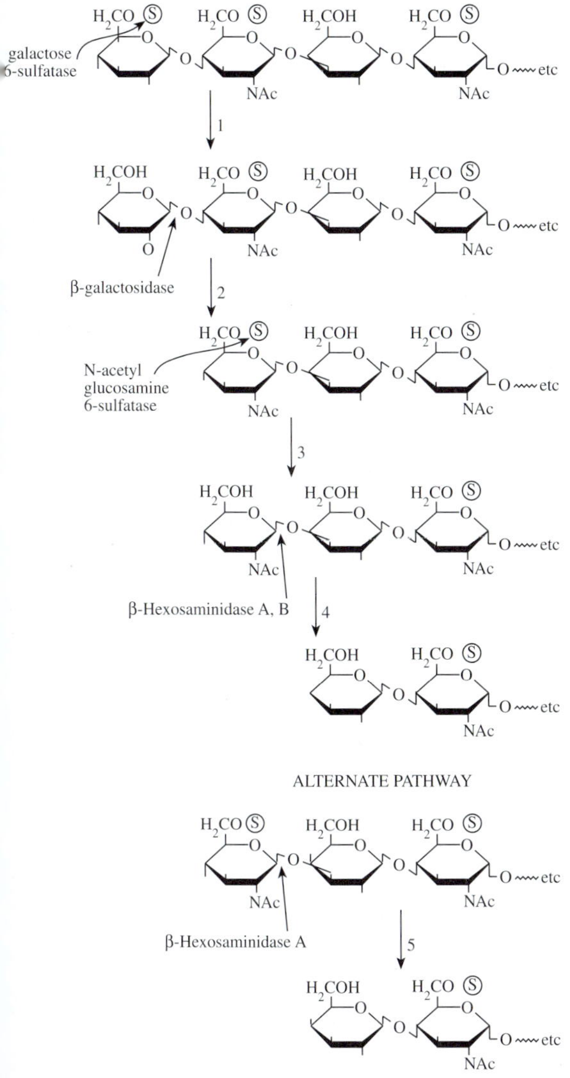

FIGURE 11-13

Stepwise degradation of keratan sulfate. The deficiency diseases corresponding to the numbered reactions are: 1 = MPS IV A, Morquio syndrome type A; 2 = MPS IVB, Morquio syndrome type B; 3 = MPS III D, Sanfilippo's syndrome type D; 4 = Sandhoff's disease; and 5 = Tay-Sachs and Sandhoff's disease. The alternate pathway releases intact N-acetylglucosamine-6-sulfate, a departure from the usual stepwise cleavage of sulfate and sugar residues. [Reproduced with permission from E.F. Neufeld and J. Muenzer: In *Metabolic Basis of Inherited Disease,* 7th ed., C. R. Scriver, A. L. Beaudet, W. S. Sly, and D. Valle (Eds). McGraw-Hill, New York, 1995, p. 2468.]

structure similar to that of eukaryotes and a thick, bag-shaped cell wall. The cell wall, about 25 nm wide, consists of peptidoglycans (Figure 11-14) and polyol phosphate polymers known as teichoic acids. Gram-negative bacteria (e.g., *Escherichia coli* and *Salmonella typhimurium*) do not contain teichoic acids but have an outer membrane system external to the plasma membrane and the peptidoglycan. The constituents of this second outer membrane system are phospholipids, lipopolysaccharides, and proteins. The peptidoglycan layer is connected to the outer layer via lipoprotein. The region between the inner membrane and the peptidoglycan layer is known as the periplasmic space. This space contains enzymes and other proteins that digest impermeable nutrients and transport low-molecular-weight nutrients, such as amino acids, sugars, and specific ions.

Lipopolysaccharides are complex amphipathic molecules located on the outer leaflet of the outer membrane. Their structure in Enterobacteriaceae comprises three regions: a phospholipid (A) (the hydrophobic portion of the molecule), a phosphorylated core oligosaccharide, and a polysaccharide chain known as the 0 side chain or 0 antigen. The core oligosaccharide and the 0 side chain regions contain several unusual sugars and form the hydrophilic portion of the molecule that projects outward. Lipopolysaccharides serve as a barrier against invading organisms, are highly toxic and are also known as endotoxins. The fever that occurs in gram-negative bacterial infections is caused by endotoxins.

The outer membrane also contains a protein known as *porin*. The trimeric forms of porin form channels for the passage of small ionic molecules through the outer membrane. Other transport systems exist for higher molecular weight ionic substances.

Both gram-negative and gram-positive bacteria contain peptidoglycan as the main structural component of their cell walls. However, gram-negative bacteria have only a single layer of peptidoglycan, whereas gram-positive cells have several layers with cross-linkages between them. The polymeric structure of peptidoglycan surrounds the entire inner membrane and the cytoplasm of the cell. Consisting of a glycan composed of amino sugars in one dimension, a cross-linked peptide moiety in the second dimension, and an interpeptide bridge in the third dimension, peptidoglycan projects in different planes (Figure 11-14). This three dimensional, covalently bound network offers considerable resistance to outward pressure and confers mechanical stability on the cell wall.

The glycan is called ***murein*** (from the Latin word for "wall") and is a polymer of alternating units of N-acetylglucosamine (GlcNAc) and N-acetylmuramic acid (MurNAc). The sugar residues are linked by $\beta(1 \rightarrow 4)$ glycosidic bonds (Figure 11-15). N-Acetylmuramic acid is the 3-O-D-lactic acid ether of N-acetylglucosamine, the carboxyl group of its lactic acid side chain being condensed with the cross-linking peptide. This peptide usually consists of four amino acid residues that are alternately L- or D-amino acids. The bacterial cell wall is one of the few structures containing a significant amount of D-amino acid residues. In *S. aureus,*

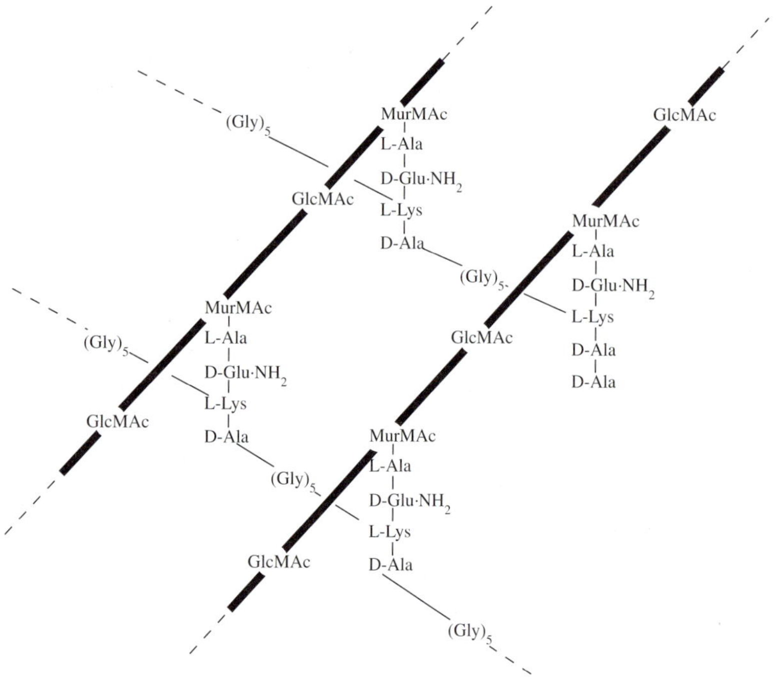

FIGURE 11-14
Structure of peptidoglycan in the cell wall of the gram-positive bacterium *Staphylococcus aureus*. The structure consists of an alternating polymer of N-acetylglucosamine (GlcNAc) and N-acetylmuramic acid (MurNAc) linked by $\beta(1 \rightarrow 4)$ glycosidic bonds in one dimension, a cross-linking peptide that contains alternating L- and D-amino acid residues in the second dimension, and a pentaglycine moiety that cross-links the peptide chains in the final dimension. This three-dimensional network is continuous and surrounds the entire cell. [Reproduced with permission from M. J. Osborn, Structure and biosynthesis of the bacterial cell wall. *Annu. Rev. Biochem.* **38,** 301 (1969). ©1969 by Annual Reviews Inc.]

the tetrapeptide sequence is L-alanyl-D-γ-isoglutaminyl-L-lysyl-D-alanine. The interpeptide chain is a pentaglycine chain that connects the terminal D-alanyl residue of one tetrapeptide to an L-lysyl residue in the third amino acid position in another tetrapeptide (Figure 11-14). The chemical linkages are as follows: the amino-terminal group of pentaglycine is joined with the carboxyl group of D-alanine, and the carboxyl terminal of pentaglycine is linked with the ε-NH_2 of L-lysine. Both of the sequence variations of the tetrapeptide and the cross-linking pattern are characteristic of the shape and species of the bacterium.

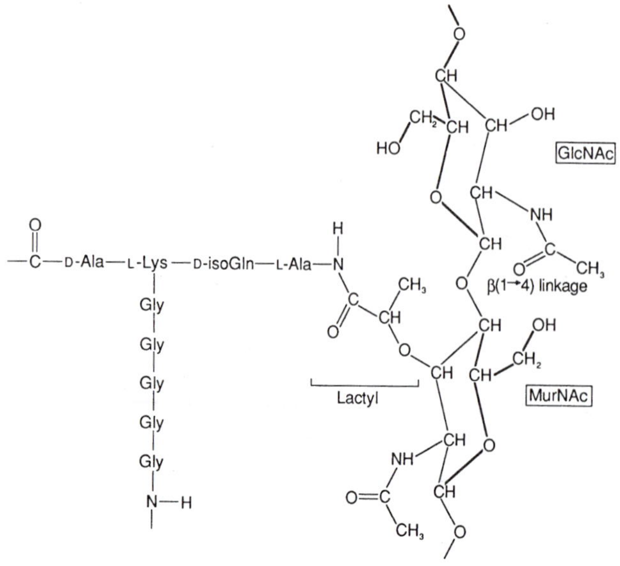

In some bacteria, in addition to the cell wall, substances form capsules or slime layers external to the cell wall. These layers are not essential for growth and multiplication but may be important for survival of the organism in harsh environments, e.g., in preventing desiccation and serving as a barrier against phage (bacterial virus) attack. The capsular layer also provides a charged surface. *Streptococcus mutans,* which plays a significant role in dental caries and plaque formation, produces an extracellular 1,3-glucan that enables the bacteria to adhere to teeth. In the formation of dental caries, possibly the most widespread human pathological process, dietary carbohydrates (sugar in general and sucrose in particular) promote bacterial growth and produce noxious metabolites.

Penicillins prevent cell wall synthesis in susceptible organisms by inhibiting a late step in the enzymatic synthesis of peptidoglycan (Chapter 16).

FIGURE 11-15
Repeating unit of the peptidoglycan of *Staphylococcus aureus*.

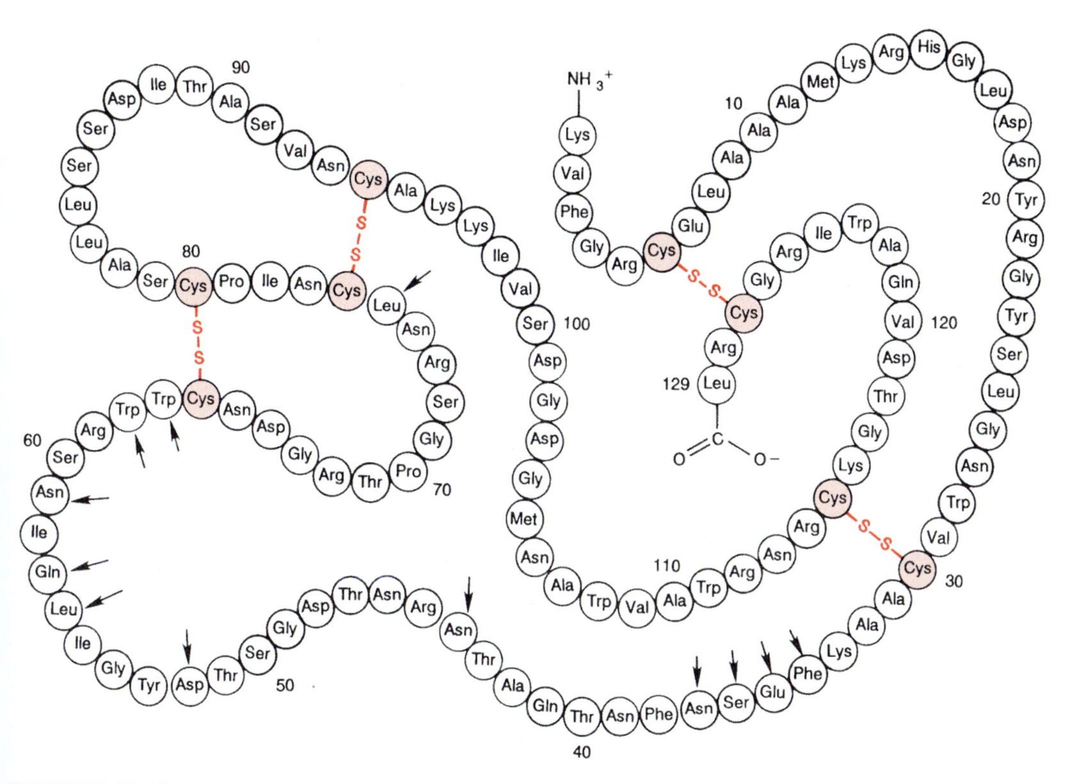

FIGURE 11-16
Primary sequence of the hen egg-white lysozyme. Amino acid residues thought to be part of the active site are identified by the arrows.

Lysis of Peptidoglycans by Lysozymes

Lysozyme breaks down the peptidoglycans by hydrolysis of the $\beta(1 \rightarrow 4)$ glycosidic bond between N-acetylglucosamine and N-acetylmuramic acid. Lysozyme occurs in tears, nasal and bronchial secretions, gastric secretions, milk, and tissues and may have a protective effect against air- and food-borne bacterial infections.

The structure and catalytic mechanism of lysozyme have been extensively studied. Its catalytic mechanism provides an example of steric distortion induced in the substrate by the enzyme before products are formed. Hen egg-white lysozyme is abundant and easy to purify. It is a roughly ellipsoidal, compact molecule ($4.5 \times 3.0 \times 3.0$ nm) of molecular weight 14,600 consisting of a single chain of 129 amino acid residues with four disulfide bridges (Figure 11-16). It is also the first enzyme whose structure was elucidated by x-ray crystallography, that contains no prosthetic group or metal ions, and that consists of regions of antiparallel β-pleated sheet, α-helix (small amount), and nondescriptive ("random") structure (large amount). As in hemoglobin and myoglobin (Chapter 4), its interior contains almost entirely nonpolar amino acid residues. Lysozymes and α-lactalbumin show striking similarity in sequence and structure, which suggests a common evolutionary ancestry. However, α-lactalbumin functions as an enzyme modifier in the synthesis of lactose in the mammary gland after parturition (Chapter 15), so this may be an example of divergent evolution. The mechanism of lysozyme action has been elucidated by the use of synthetic substrates and inhibitors. GlcNAc–GlcNAc–GlcNAc is an inhibitor that forms a stable enzyme–inhibitor complex. The active site of the enzyme has been established by x-ray crystallography with and without inhibitor and also by model building. The enzyme contains a central crevice running horizontally across the molecule. The hexasaccharide

$$\underset{A}{\text{GlcNAc}}\text{–}\underset{B}{\text{MurNAc}}\text{–}\underset{C}{\text{GlcNAc}}\text{–}\underset{D}{\text{MurNAc}}\text{–}\underset{E}{\text{GlcNAc}}\text{–}\underset{F}{\text{MurNAc}}$$

binds in the cleft noncovalently, with residue C making the largest contribution. This binding distorts the normal chair conformation of sugar residue D. Since the bond between residues D and E is hydrolyzed, this binding region contains the active site of the enzyme. The rate of catalysis is not significant unless the substrate is a pentamer. The main features of a proposed mechanism for hydrolysis of the bond between D and E are shown in Figure 11-17. Ring D is distorted resulting in formation of a transition state intermediate in which the conformation lies between those of substrate and products, and which has the maximum energy. With the strained conformation (half-chair) of ring D, the susceptible bond between D and E is drawn in close proximity to Glu-35 and Asp-52 residues of the

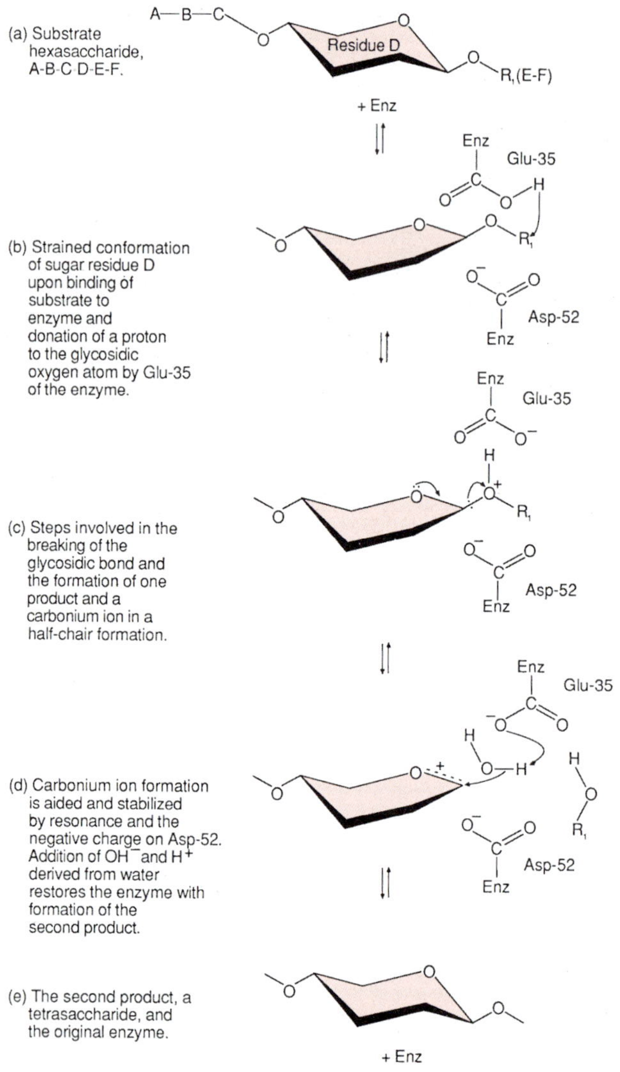

(a) Substrate hexasaccharide, A-B-C-D-E-F.

(b) Strained conformation of sugar residue D upon binding of substrate to enzyme and donation of a proton to the glycosidic oxygen atom by Glu-35 of the enzyme.

(c) Steps involved in the breaking of the glycosidic bond and the formation of one product and a carbonium ion in a half-chair formation.

(d) Carbonium ion formation is aided and stabilized by resonance and the negative charge on Asp-52. Addition of OH^- and H^+ derived from water restores the enzyme with formation of the second product.

(e) The second product, a tetrasaccharide, and the original enzyme.

FIGURE 11-17
Proposed mechanism for the catalytic hydrolysis of a hexasaccharide lysosome.

enzyme. Glu-35 functions as a general acid catalyst donating a proton to the glycosidic oxygen between sugar residues D and E, which results in cleavage of the bond and release of the disaccharide E–F. The remainder of the substrate (a tetrasaccharide), however, is still bound to the enzyme with D acquiring a positive charge (a carbonium ion) and continuing to be in the half-chair conformation. The negatively charged Asp-52 now stabilizes the positively charged carbonium ion intermediate. A hydroxyl ion and a proton from water then react with the carbonium ion and the enzyme to generate a tetrasaccharide and the original enzyme.

In this mechanism, Asp-52 is ionized and Glu-35 is un-ionized in the enzyme, consistent with their functions. The optimum pH of the enzyme is 5.0 and the pH-rate profile is bell-shaped (activity rapidly falling off on either side of the optimum pH). At pH 5.0, Asp-52 (pK′ ~3.8) is ionized and Glu-35 (pK′ ~6.7) is un-ionized. The pK′ of Glu-35 is significantly different from that of the free amino acid (4.25) because Glu-35 is in a nonpolar region of the molecule. Asp-52, on the other hand, is in a polar environment, where its ionization is facilitated. The importance of acidic amino acid residues in catalysis is further substantiated by the fact that lysozyme remains active when all carboxyl groups, except those of Glu-35 and Asp-52, are chemically modified (e.g., by esterification).

11.3 Lectins

The lectins are a group of proteins, originally discovered in plant seeds (now known to occur more widely), that bind carbohydrates and agglutinate animal cells. They have two or more stereospecific sites that bind noncovalently with the terminal (and often penultimate) residue at the nonreducing end of an oligosaccharide chain. A number of plant lectins have been purified and their binding properties investigated. Wheat germ agglutinin binds to N-acetylglucosamine and its glycosides; concanavalin A from jack beans binds to mannose, glucose, and glycosides of mannose and glucose; peanut agglutinin binds to galactose and galactosides; and red kidney bean lectin binds to N-acetylglucosamine. Since lectins have high affinity for specific sugar residues, they have been used to identify specific carbohydrate groups and used in the purification of carbohydrate-containing compounds. Lectins may be involved in carbohydrate transport, specific cellular recognition, embryonic development, cohesion, or binding of carbohydrates. As noted in Chapter 10, hepatocytes bind to serum glycoproteins with exposed galactose residues. This binding is thought to be a lectin-mediated clearance of partially degraded glycoproteins. Hepatocytes also contain a Ca^{2+}-dependent fucose-binding lectin. A lectin that binds to N-acetylglucosamine- and mannose-terminated glycoproteins in reticuloendothelial cells has been identified. Many lectins are glycoproteins.

Some lectins cause agglutination of red blood cells and can be used in typing of blood groups. Soybean lectin, a galactose-binding protein, binds selectively to T lymphocytes and causes their agglutination. Thus, it has been used in selective removal of mature T lymphocytes from bone marrow preparations. In treatment of disorders such as immunodeficiencies, blood cancers, and some hemoglobinopathies, bone marrow grafts are made after ensuring that donor and recipient are histocompatible. If they are not, the donor cells destroy the recipient's cells, and graft-versus-host disease (GVHD) ensues. However, GVHD does not occur if the T lymphocytes are removed from the donor marrow, even without histocompatibility. A highly selective method of removal of T lymphocytes uses monoclonal antibodies (Chapter 35).

Supplemental Readings and References

D. D. Bikle: Biochemical markers of bone metabolism: An overview, *Clinical Biochemistry* **30,** 573 (1997).

P. H. Byers: Disorders of collagen biosynthesis and structure. In: *Metabolic Basis of Inherited Disease,* 7th ed., C. R. Scriver, A. L. Beaudet, W. S. Sly, and D. Valle (Eds). McGraw-Hill, New York, 1995, p. 4029.

B. H. Chong and M. Eisbacher: Pathophysiology and laboratory testing of heparin-induced thrombocytopenia. *Seminars in Hematology* **35**(No. 4, Suppl. 5), 3 (1998).

S. Elgavish and B. Shaanan: Lectin-carbohydrate interactions: Different folds common recognition principles. *Trends in Biochemical Sciences* **22,** 462 (1997).

E. Holtzman: *Lysosomes,* Plenum Press, New York, 1992.

P. J. Meikle, J. J. Hopwood, A. E. Clague, and W. F. Carey: Prevalence of lysosomal storage disorders, *Journal of the American Medical Association* **281,** 249 (1999).

E. F. Neufeld and J. Muenzer: *The mucopolysaccharidosis* in: *Metabolic Basis of Inherited Disease,* 7th ed., C. R. Scriver, A. L. Beaudet, W. S. Sly, and D. Valle (Eds). McGraw-Hill, New York, 1995, p. 2465.

F. Ramirez, M. Godfrey, B. Lee, and P. Tsipouras: Marfan syndrome and related disorders In: *Metabolic Basis of Inherited Disease,* 7th ed., C. R. Scriver, A. L. Beaudet, W. S. Sly, and D. Valle (Eds). McGraw-Hill, New York, 1995, p. 4079.

J. Rosenbloom, W. R. Abrams, and R. Mecham: Extracellular matrix 4: The elastic fiber. *FASEB Journal* **7,** 1208 (1993).

J. I. Weitz: Low-molecular-weight heparins, *New England Journal of Medicine* **337,** 688 (1997).

CHAPTER 12

Gastrointestinal Digestion and Absorption

The gastrointestinal (or alimentary) system (or tract) is esponsible for digestion and absorption of nutrients and luids. The tract extends from the mouth to the anus and in n adult human is about 10 m long. ***Digestion*** is the hydroysis of complex food substances into simpler units such as nonosaccharides, amino acids, fatty acids, and glycerol. ***bsorption*** is the transport of the products of digestion nd of vitamins, minerals, and water across the intestinal pithelium to the lymphatic or blood circulatory systems.)igestion and absorption involve not only the gastroinestinal (GI) tract but also secretions from salivary glands, iver (and gallbladder), and pancreas. These processes in he GI system are interrelated. The system provides highly elective, efficient, and elaborate absorptive surfaces that ontain enzymes and secretions of enzyme-containing fluds, electrolytes, and other substances required for digesion and absorption. Secretory and absorptive activities are egulated by hormonal and neural mechanisms.

2.1 Anatomy and Physiology of the GI Tract

'he GI tract consists of mouth and esophagus, stomach, mall intestine (duodenum, jejunum, and ileum), and large itestine (colon and rectum).

Mouth and Esophagus

In the mouth, food is mixed with saliva, chewed to break up large particles, and propelled into the esophagus by swallowing. Saliva is secreted by three pairs of glands (parotid, submaxillary, and sublingual) and by numerous small buccal glands and is under autonomic nervous system control. Parasympathetic nervous stimulation causes profuse secretion of saliva. Thus, atropine and other anticholinergic agents diminish salivary secretion and make the mouth dry. The presence of food in the mouth and the act of chewing stimulate secretion of saliva by reflex nervous stimulation. Salivary secretion can also be a conditioned response, so that the sight, smell, and even thought of food can elicit salivary secretion; however, in humans, this type of conditioned response is weak.

The composition, pH, and volume of salivary excretion for different salivary glands vary. Usually 1–2 L of saliva are secreted per day. Saliva is a hypotonic solution with a pH of about 7.0; principal cations are Na^+ and K^+, and anions are Cl^- and HCO_3^-. Aldosterone (a steroid hormone of the adrenal cortex) modulates Na^+ and K^+ levels in saliva as it does in the kidney (Chapter 32). Aldosterone increases reabsorption of Na^+ and secretion of K^+ by direct action on the salivary gland ducts. In ***hypoaldosteronism*** (e.g., *Addison's disease*), the salivary Na^+/K^+ ratio is high, and in adrenocortical hyperfunction (e.g., *aldosteronism*), this ratio is low. The chief salivary organic constituent is ***mucin,*** which is a mixture of glycoproteins containing 60–85% oligosaccharide by weight. The oligosaccharide units are linked by O-glycosidic linkage to either serine or threonine residues (Chapter 9). Mucin-rich saliva is secreted by submaxillary and sublingual glands, and because of the high carbohydrate content, mucins form viscous solutions responsible for much of the lubricating action of saliva that aids swallowing. ***Amylase*** is a salivary enzyme that catalyzes the hydrolysis of $\alpha(1 \rightarrow 4)$ glucosidic linkages of starch and glycogen. Its role is minimal because of the limited time spent by the food in the mouth and the inactivation of amylase in the stomach by the acid pH. Other enzymes detected in saliva include carbonic anhydrase, a lipase (lingual lipase), a phosphatase, and a protease called kallikrein. Whether these enzymes function in saliva is not known. Lingual lipase initiates hydrolysis of dietary fat in the stomach and facilitates the duodenal-jejunal hydrolysis of triacylglycerols.

Even when no food is present in the mouth, saliva is secreted to keep the mouth moist and facilitates speech. Saliva also has some antibacterial action. Patients with ***xerostomia*** (dry mouth) exhibit higher than normal incidence of dental caries. Saliva maintains the oral cavity at a pH of about 7.0. The teeth do not lose calcium to oral fluids because of the high concentration of calcium in the saliva at pH 7.0.

Like the thyroid gland (Chapter 33), salivary glands (and the gastric mucosa) can take up iodide ions against a concentration gradient.

The mouth is the normal point of entry of food and drink. There, solid food is reduced in size by mastication, blended with saliva, and temperature-moderated before being swallowed. Conditions that interfere with any of these processes (e.g., tooth loss) can affect food choice and hence the nutritional status and health of the individual. The esophagus is a muscular tube through which masticated food is transported from the mouth to the stomach. It consists of both striated muscle (upper one third) and smooth muscle (lower two thirds). These muscles undergo periodic contractions in the form of peristaltic waves that push the swallowed boluses toward the stomach. The esophagus has no digestive function but secretes mucus to protect the esophageal mucosa from excoriation.

Stomach

The stomach stores food temporarily, retards its entry into the small intestine, and secretes pepsin to begin the digestion of protein. Hydrogen ions in the stomach activate pepsinogen to form pepsin and aid in maintaining the sterility of the upper GI tract. The stomach also secretes intrinsic factor (a glycoprotein), which is required for vitamin B_{12} absorption in the ileum (Chapter 38), helps prepare some of the essential minerals (e.g., iron) for absorption by the small intestine (Chapter 29), and provides mucus for protective and lubricative functions.

The stomach wall is made up of four concentric layers: the mucosa, the submucosa, the muscularis external, and the serosa. Functionally, the gastric mucosa is divided into the oxyntic gland area (which includes the body and fundus of the stomach) and the pyloric gland area. The oxyntic gland area contains tightly packed parallel glands and is covered with a layer of tall, columnar surface epithelial cells that secrete mucous and line the gastric pits. Undifferentiated cells in the region joining the gastric gland to the pit (the neck region) undergo division with some daughter cells migrating upward and differentiating into surface epithelial cells, with other daughter cells migrating downward and differentiating into neck mucus cells, and with ***chief*** and ***parietal*** cells. Thus, the mucosa renews itself rapidly, approximately every 3–5 days in humans. Replacement of the mucosa after injury occurs even more rapidly. Pepsinogen is synthesized and secreted by chief cells and HCl by parietal cells.

A proteolytic enzyme secreted by gastric mucosa of infants is ***chymosin*** (rennin), which functions to clot milk and promote its digestion by preventing rapid passage from the stomach. Chymosin hydrolyzes casein, a mixture of several related milk proteins, to paracasein, which reacts with Ca^{2+} to yield the insoluble curd. Pepsin performs the same functions as chymosin. Chymosin is found in the fourth stomach of ruminants. Calf stomach is a source of this enzyme, which is used in the manufacture of cheese. Chymosin has been synthesized by recombinant DNA techniques and successfully used in the production of cheese.

The pyloric mucosa contains surface epithelial cells that secrete mucus, which increases during digestion of food. Inadequate secretion of mucus can lead to stomach ulcers. The cells of the pyloric glands secrete a fluid similar in composition to an ultrafiltrate of plasma. These cells also secrete a pepsinogen that differs structurally from the pepsinogen secreted by the oxyntic gland mucosa. Secretion from all gastric glands amounts to 2–2.3 L/day and is regulated by neural and humoral systems. Neural control is mediated by the vagus nerves. The sight, smell, and thought of pleasant-tasting food can elicit gastric secretion during the ***cephalic phase.*** Neural control is also exerted by local autonomic reflexes mediated by cholinergic neurons. Anger and hostility can cause gastric secretion through nerve impulses down the vagi. Secretory signals can arises from the cerebral cortex.

Food and its digestion products stimulate receptors in the stomach that increase gastric secretion during the *gastric phase,* mediated primarily by secretion of gastrin. ***Gastrin*** is a gastrointestinal hormone synthesized and stored in G cells of the pyloric glands and mucosa of the proximal part of the small intestine. When food moves into the small intestine, the intestinal phase of gastric secretion involves complex interactions of stimulatory and inhibitory factors. Thus, ***chyme*** (the semifluid, homogeneous material produced by gastric digestion) in the upper jejunum inhibits gastric secretion, whereas amino acids in the middle and lower jejunum stimulate gastric secretion. These activities are primarily mediated through GI hormones (see below). The three phases of gastric secretion (cephalic, gastric, and intestinal) are interdependent and proceed more or less simultaneously.

The rate of passage of food into the duodenum depends on the type of food and on the osmotic pressure it exerts in the duodenum. Food rich in fat moves most slowly, food rich in carbohydrate moves most rapidly, and food rich in protein moves at an intermediate rate. Hyperosmolality decreases gastric emptying via duodenal osmoreceptors. Some GI hormones also inhibit gastric motility and secretion.

Small Intestine

Most digestion and absorption of food constituents occurs in the small intestine. The small intestine is also the major site for absorption of water and electrolytes. It is a convoluted tube about 6 m long, beginning at the pylorus and ending at the ileocecal valve. It is divided into a duodenum (a C-shaped, short, fixed segment), the mobile jejunum, and ileum. Its wall consists of the same four concentric layers as in other parts of the GI tract: ***mucosa, submucosa, muscularis externa,*** and ***serosa.*** Finger-like projections known as ***villi*** that arise from the luminal side of the mucosal membrane provide a large surface area in the small intestine. Each villus consists of a single layer of tall, columnar epithelial absorptive cells (***enterocytes***) that contain ***microvilli*** which further increase the absorptive surface and give it a brush-like appearance (brush border). The surface membrane of the microvillus contains digestive enzymes and transport systems. The core of the microvillus consists of a bundle of filamentous structures that contain actin and associated proteins (e.g., myosin, tropomyosin, villin, fimbrin; see Chapter 21), which may have a structural role. Enterocytes are abundantly supplied with mitochondria, endoplasmic reticulum, and other organelles. The outermost surface of the microvillus is the carbohydrate-rich ***glycocalyx,*** which is anchored to the underlying membrane. The glycocalyx is rich in neutral and amino sugars and may protect cells against digestive enzymes. Villi also contain ***goblet cells,*** which secrete mucus. The absorptive cells are held together by junctional complexes that include "tight junctions." However, the tight junctions contain pores that permit the transport of water and small solutes into the intercellular spaces according to the osmotic pressure gradient.

Endocrine cells are distributed throughout the small intestine and other sections of the GI tract. About 20 different cell types have been identified. Some, easily identifiable by their staining characteristics, are argentaffin or enterochromaffin cells containing 5-hydroxytryptophan, the precursor of serotonin (Chapter 17). Neoplastic transformation of these cells leads to excessive production and secretion of serotonin, which causes diarrhea, flushing, and bronchoconstriction (*carcinoid syndrome*). Other types of cells lack 5-hydroxytryptophan but take up amine precursors (e.g., amino acids) to synthesize and store biologically active peptides (e.g., hormones).

The mucosal cells of the small intestine and stomach are among the most rapidly replaced cells of the human body, being renewed every 3–5 days. Millions of cells are sloughed off (or exfoliated) from the tips of villi every minute; about 17 billion such cells are shed per day. These mucosal cells amount to about 20–30 g of protein, most of which is reclaimed as amino acids after hydrolysis in the lumen. These cells are replaced through division of undifferentiated cells in the crypts of the villi (crypts of Lieberkühn)and subsequent migration to the top of the villus, followed by maturation. Mitotic poisons (e.g., antineoplastic agents), injury, and infection can adversely affect mucosal cell proliferation, producing a flat mucosa devoid of villi and leading to severe malnutrition.

Digestion in the small intestine requires the biliary and pancreatic secretions that are emptied into the duodenum (usually at a common site).

Formation, Secretion, and Composition of Bile

Bile is formed and secreted continuously by polygonally shaped liver parenchymal cells called ***hepatocytes.*** An aqueous buffer component (e.g., HCO_3^-) is added to the bile by the hepatic bile duct cells that carry the secretion toward the common bile duct. The membrane of the hepatocytes in contact with the blood has microvilli that facilitate the exchange of substances between plasma and the cells. Hepatocytes are rich in mitochondria and endoplasmic reticulum. Hepatic bile flows into the gallbladder, where it is concentrated, stored, and emptied into the duodenum when the partially digested contents of the stomach enter

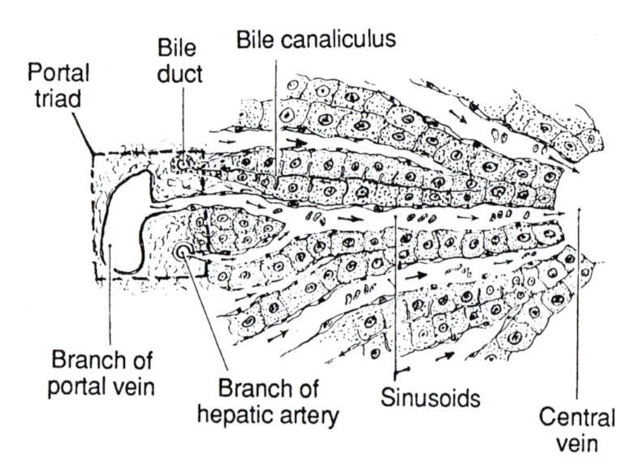

FIGURE 12-1
Schematic representation of the radial architecture of the plates formed by the liver cells. Blood from the portal vein and the hepatic artery flows into sinusoids and eventually enters the central vein. The bile canaliculi are located between the liver cells. Bile flows in the opposite direction and empties into the bile duct in portal triads. [Reproduced with permission from A. W. Ham *Histology,* 8th ed., J. B. Lippincott, Philadelphia, 1979.]

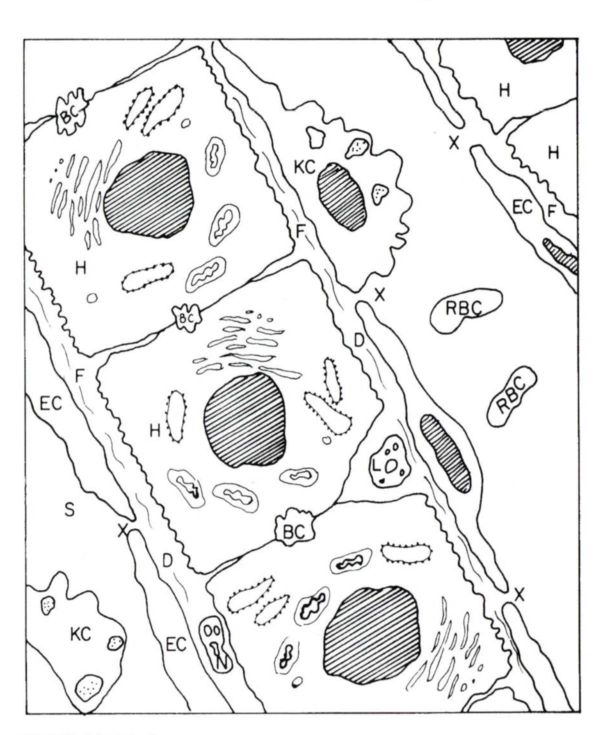

FIGURE 12-2
Schematic representation of structures within the hepatic lobule. H = hepatocyte, BC = bile canaliculus, KC = Kupffer cell, EC = endothelial cell, N = nerve fiber, F = reticulin fibers, S = sinusoid, D = space of Disse, X = gap between sinusoid lining cells, and RBC = red blood cell.

the duodenum. This movement is accomplished by contraction of the gallbladder mediated by cholecystokinin, a GI hormone.

The ***liver*** is the largest gland and second largest organ in the body and weighs about 1.5 kg in an adult. It functions in synthesis, storage, secretion, excretion, and specific modification of endogenously and exogenously derived substances. The liver takes up material absorbed from the small intestine, since it is mainly supplied by venous blood arriving from the GI tract. It is uniquely susceptible to changes in alimentation and to toxic substances. The liver participates in numerous metabolic functions.

The liver is organized into ***lobules,*** functional aggregates of hepatocytes arranged in radiating cords or plates surrounding a central vein (Figure 12-1). In adults, the branching plates are usually one-cell thick, and the vascular channels between them are known as ***sinusoids.*** Blood in the sinusoids passes slowly between the rows of cells, facilitating the exchange of substances between cells and the plasma. The blood supply of the sinusoids is furnished by the portal vein (75%) and by the right and left hepatic arteries (25%). Blood flows from the periphery of the lobules into the central vein. Sinusoids anastomose to form central veins that join to form larger veins which feed through the hepatic vein into the inferior vena cava. Sinusoids are larger than capillaries and are lined by reticuloendothelial cells, including the phagocytic ***Kupffer cells.*** This layer contains numerous, small fenestrations that provide access of sinusoidal plasma to the surface microvilli of hepatocytes via the intervening space, called the ***space of Disse.*** Fine collagen fibers (reticulin) within this space provide the support for the liver cell plates (Figure 12-2).

Bile produced by hepatocytes is secreted into the bile canaliculi between adjacent hepatic cells. The wall of the canaliculus is formed by the plasma membrane of the hepatocytes, which are held together by tight junctions. Canaliculi arise near central veins and extend to the periphery of the lobules. The direction of bile flow in the canaliculi is centrifugal, whereas that of the blood flow is centripetal. Canaliculi coalesce to form ducts, which are lined by epithelium, and the ducts coalesce to form the right and left hepatic ducts. Outside the liver these ducts form the common hepatic duct.

The volume of hepatic bile varies in a normal human adult from 250 to 1100 mL, depending on the rate a which bile acids recirculate in the enterohepatic cycle (see below). Some GI hormones increase the volume and bicarbonate content of bile through stimulation of the duct epithelium. Hepatic bile has a pH of 7.0–7.8 and is isotonic with plasma. During the interdigestive period it is diverted into the gallbladder, since the ***sphincter of Oddi*** is closed. Bile in the gall bladder is concentrated 12- to 20-fold by absorption of electrolytes and water Nonetheless, gallbladder bile is isotonic with plasma because cations (mainly Na^+) associate with the osmotically inactive bile acid micellar aggregates. Gallbladder capacity is 50–60 mL. Emptying requires a coordinated contraction of the gallbladder and relaxation of the sphincter of Oddi, both mediated by cholecystokinin.

Bile contains bile acids, bile pigments, cholesterol, phosphatidylcholine (lecithin), and electrolytes. Bile pigments are glucuronide conjugates of bilirubin and biliverdin derived from the degradation of heme, a prosthetic group of many proteins. Most of the bilirubin formed in the body comes from hemoglobin catabolized in reticuloendothelial cells (liver is one of the sites where these cells are formed). The bilirubin is bound to plasma albumin and transported to hepatocytes, where it is converted to glucuronide conjugates and secreted into the bile, giving it a golden yellow color. If bilirubin remains in the gallbladder, it is oxidized to biliverdin, which is greenish brown. Bile pigments are toxic metabolites (Chapter 29).

The bile acids are 24-carbon steroid derivatives. The two primary bile acids, cholic acid and chenodeoxycholic acid, are synthesized in the hepatocytes from cholesterol by hydroxylation, reduction, and side chain oxidation. They are conjugated by amide linkage to glycine or taurine before they are secreted into the bile (see cholesterol metabolism, Chapter 19). The mechanism of secretion of bile acids across the canalicular membrane is poorly understood. Bile acids are present as anions at the pH of the bile, and above a certain concentration (critical micellar concentration) they form polyanionic molecular aggregates, or micelles (Chapter 11). The critical micellar concentration for each bile acid and the size of the aggregates are affected by the concentration of Na^+ and other electrolytes and of cholesterol and lecithin. Thus, bile consists of mixed micelles of conjugated bile acids, cholesterol, and lecithin. While the excretion of osmotically active bile acids is a primary determinant of water and solute transport across the canalicular membrane, in the canaliculi they contribute relatively little to osmotic activity because their anions aggregate to form micelles.

Bile is also a vehicle for excretion of other endogenous (e.g., bilirubin and steroids) and exogenous (e.g., drugs and dyes) compounds. The dye sulfobromophthalein is removed from the bloodstream by the hepatocytes, excreted in the bile, and eliminated through the GI tract. Its rate of removal from the circulation depends on the functional ability of the hepatocytes and of the hepatic blood flow, and so it is used to assess hepatic function.

Cholesterol is virtually insoluble in water but is solubilized through the formation of micelles, which also contain phosphatidylcholine and bile acid anions in specific proportions. Cholesterol can be precipitated to form stones (known as ***gallstones***) when the critical concentration of micellar components is altered, i.e., when the concentration ratio of cholesterol to bile acids is increased. Gallstones may also form because of precipitation of bilirubin, e.g., with increased production of bile pigments due to hemolysis, associated with infection and bile stagnation. Presumably, under these conditions the bilirubin diglucuronide is converted to free bilirubin by β-glucuronidase. Bilirubin is insoluble and forms calcium-bilirubinate stones (called pigment stones). Gallstones occur in 10–20% of the adult population in Western countries. Cholesterol stones account for 80% (see Chapter 19).

More than 90% of the bile acids are reabsorbed, mostly in the ileum. About 5–10% undergoes deconjugation and dehydroxylation catalyzed by bacterial enzymes in the colon to form deoxycholic acid and lithocholic acid. These acids are known as secondary bile acids. Deoxycholic acid is reabsorbed, but lithocholic acid, which is relatively insoluble, is poorly absorbed. The reabsorbed bile acids enter the portal blood circulation, are reconjugated in the liver, and are secreted into bile. This cycling is known as the ***enterohepatic circulation*** and occurs about six to eight times daily. The rate of cycling increases after meals. The lithocholic acid is conjugated with either taurine or glycine, sulfated at the ring hydroxyl group, and again secreted into the bile. Sulfated lithocholate is not reabsorbed. Free lithocholic acid is highly toxic to tissues and causes necrosis.

The bile acid pool normally consists of about 2–4 g of conjugated and unconjugated primary and secondary bile acids. Daily loss of bile acids in feces, mostly as lithocholate, is about 0.2–0.4 g. Hepatic synthesis of bile acids equals this amount, so that the size of the bile acid pool is maintained at a constant level.

Exocrine Pancreatic Secretion

The pancreatic gland is 10–15 cm long and resides in the retroperitoneal space on the posterior wall of the abdomen, with its broad and flat end ("head") fitting snugly into the loop formed by the duodenum. The endocrine function of the pancreas consists of synthesis, storage, and secretion of the hormones insulin, glucagon, and somatostatin into the venous circulation that drains the pancreas. These hormones are essential in the regulation of metabolism of carbohydrates and other substrates (Chapter 22). The endocrine cells are located in the ***islets of Langerhans,*** distributed throughout the pancreas but separated from the surrounding exocrine-acinar tissue by collagen fibers. The exocrine function of the pancreas consists of synthesis, storage, and secretion of digestive enzymes and bicarbonate-rich fluids.

The acinar cells of the pancreas are arranged in ellipsoidal structures known as *acini*. Each acinus contains about a dozen cells arranged around a lumen, which forms the beginning of the pancreatic duct system, and centroacinar cells, which secrete a bicarbonate solution. The acinar cells are joined to one another by tight junctions, desmosomes, and gap junctions. The tight junctions probably prevent leakage of pancreatic secretions into the extracellular spaces. Acinar cells are pyramidal in shape with the nucleus located away from the lumen and secretory granules located toward the lumen. Centroacinar and acinar cells have microvilli on their free borders.

Enzymes are synthesized in the acinar cells by ribosomes situated on the rough endoplasmic reticulum and sequestered in membrane-bound granules (Chapter 25) stored in the apex of the cell. Upon stimulation by the vagus nerve or by cholecystokinin, the contents of the granules are released into the lumen.

The centroacinar and duct cells secrete bicarbonate ions under the control of gastrointestinal hormones. The bicarbonate ions are formed by the action of carbonic anhydrase:

$$CO_2 + H_2O \xrightleftharpoons{\text{carbonic anhydrase}} H_2CO_3 \rightleftharpoons H^+ + HCO_3^-$$

The exact mechanism of HCO_3^- secretion into ductules is not known.

Composition of Pancreatic Juice

Pancreatic juice is alkaline because of its high content of HCO_3^-. It neutralizes the acidic chyme from the stomach, aided by the bile and small intestinal juices. In the jejunum, chyme is maintained at a neutral pH as required for the optimal functioning of the pancreatic digestive enzymes. A normal pancreas secretes about 1.5–2 L of juice per day.

Pancreatic juice contains the proenzymes trypsinogen, chymotrypsinogen, procarboxypeptidases, and proelastase. All are activated by trypsin in the intestinal lumen. Enteropeptidase located in the brush border of the jejunal mucosa converts trypsinogen to trypsin. A trypsin inhibitor in pancreatic juice protects against indiscriminate autodigestion from intraductal activation of trypsinogen. Other enzymes of pancreatic juice and their substrates are listed below.

Enzyme	Substrate
Prophospholipase, activated to phospholipase	Phospholipids
Cholesteryl ester hydrolase	Cholesteryl esters
Lipase and procolipase, converted to colipase	Triacylglycerols
Amylase	Starch and glycogen
Ribonuclease	RNA
Deoxyribonuclease	DNA

Large Intestine

The large intestine extends from the ileocecal valve to the anus. It is wider than the small intestine except for the descending colon, which when empty may have the same diameter as the small intestine. Major functions of the colon are absorption of water, Na^+, and other electrolytes, as well as temporary storage of excreta followed by their elimination. The colon harbors large numbers of mostly anaerobic bacteria that can cause disease if they invade tissues. These bacteria metabolize carbohydrates to lactate, short-chain fatty acids (acetate, propionate, and butyrate), and gases (CO_2, CH_4, and H_2). Ammonia, a toxic waste product, is produced from urea and other nitrogenous compounds. Other toxic substances are also produced in the colon. Ammonia and amines (aromatic or aliphatic) are absorbed and transported to the liver via the portal blood, where the former is converted to urea (Chapter 17) and the latter is detoxified. The liver thus protects the rest of the body from toxic substances produced in the colon. Colonic bacteria can also be a source of certain vitamins (e.g., vitamin K, Chapter 36).

12.2 Gastrointestinal Hormones

Gastrointestinal hormones are chemical messengers that regulate intestinal functions (e.g., motility and secretion), which are also regulated by the autonomic nervous system. The endocrine cells of the GI tract are interspersed with mucosal cells and do not form discrete glands. The GI tract has been described as the largest endocrine organ in the body. In addition, the GI tract is noted for its special significance in the development of endocrinology. The first hormone discovered and characterized as a chemical messenger by Bayliss and Starling in 1902 was the GI hormone secretin.

The GI hormones are a heterogeneous group of peptides that are released into the bloodstream in response to specific stimuli, bind to specific receptors on target cells to produce biological responses, and are subject to feedback regulation. Some may affect neighboring cells by being transported through intercellular spaces (paracrine secretion), and others serve as neurotransmitters in peptidergic neurons (Chapter 30). Peptidergic neurons are present in the GI tract and in the peripheral and central nervous systems. These peptides are synthesized in the GI tract and in the brain, although in the GI tract some may originate from vagal nerve endings. This dual localization of "brain-GI tract" peptides is an example of biological conservation.

Of the many regulatory gastrointestinal peptides, four are well understood and are considered to be authentic

TABLE 12-1
*Selected Peptide Candidate Hormones of the Gut**

Name	Source	Mechanism of Release and Putative Function
Bombesin	Intestinal mucosa	Mechanism of release unknown; stimulates gastric and pancreatic secretions directly and by release of gastrin and CCK, respectively; inhibits absorption of water, Na^+, and Cl^- from the jejunum.
Bulbogastrone	Mucosa of duodenal bulb	Released by acid; inhibits gastric secretion.
Enteroglucagon	Duodenal mucosa	Functions like glucagon (Chapter 22).
Motilin	Duodenum and jejunum	Released by vagal stimulation and passage of nutrients; modulates interdigestive motility, clearing the intestine between meals and preparing it for another nutritional bolus.
Somatostatin	Intestinal mucosa, pancreas and nerves; hypothalamus	Mechanism of release not clear; multiple inhibitory actions.
Substance P	GI tract (highest concentrations in small intestine and colon)	Released after food ingestion, contraction of intestinal smooth muscle and gall bladder, reduction of bile flow, and increase in pancreatic juice output.
Vasoactive intestinal peptide	GI tract (all layers and nerve fibers)	Mechanism of release not clear; inhibits gastric acid and pepsin secretion, potent stimulator of intestinal secretion, vasodilator, possibly neurotransmitter.

*A number of these peptides are found in the gut (in both enteroendocrine cells and enteric neurons) and brain.

hormones: ***secretin, cholecystokinin*** (CCK), ***gastrin,*** and ***gastric inhibitory peptide.*** The hormonal status of other peptides (sometimes referred to as candidate hormones, Table 12-1) is not yet established. On the basis of structural similarity, these hormones fall into two families:

1. The gastrin family, which consists of *gastrin, CCK, cerulein, enkephalin,* and *motilin,* and
2. the secretin family, which consists of *secretin, glucagon, vasoactive intestinal peptide* (VIP), *gastric inhibitory peptide* (GIP), and *bombesin.*

Other candidate hormones (e.g., neurotensin, substance P, and somatostatin) belong to neither family. Somatostatin, considered a candidate hormone in the GI tract, is an authentic hypothalamic hormone (Chapter 31).

The amino acid sequences of some members of the gastrin and secretin families are shown in Tables 12-2 and 12-3, respectively. The hormones exhibit both macro- and microheterogeneity. For example, gastrin exists in many forms, probably including the principal hormone, biosynthetic precursors, and metabolites. The sulfate group on tyrosine at position 12 of gastrin may be missing without significant change in biological activity. The biological activity of gastrin and CCK resides in the carboxyterminal tetrapeptide and heptapeptide, respectively.

Gastrin

Gastrin is synthesized and stored in G cells in the antral mucosa of the stomach; it is also found, to a lesser extent, in the mucosa of the duodenum and jejunum. During fetal life, gastrin is also found in the pancreatic islet cells. Gastrin-secreting tumors, known as ***gastrinomas*** (e.g., Zollinger–Ellison syndrome), may develop in the pancreas. The main form of gastrin obtained from G cells is a heptadecapeptide. At the N-terminal end of the heptadecapeptide., there is a pyroglutamyl residue, and at the C-terminal end, there is a phenylalaninamide residue, both of which protect the peptide from amino and carboxy-peptidase activity, respectively. Two forms of the heptadecapeptide exist: one in which the tyrosyl residue at position 12 is free (gastrin 1) and the other in which it is sulfated (gastrin 2). Though larger and smaller forms of gastrin have been identified, the biological activity of gastrin resides in the C-terminal tetrapeptide. A synthetic pentapeptide is available (pentagastrin, Table 12-2).

Gastrin is released in response to chemical, mechanical, or neural stimuli on the G cell. Hypoglycemia (plasma glucose concentration $\leqslant$40 mg/dL), acting through the hypothalamus, causes release of gastrin mediated through the vagus nerve. A rise in calcium ion concentration in plasma causes the release of gastrin. The main physiological actions of gastrin are stimulation of acid secretion by

TABLE 12-2
*Primary Structure of Gastrin Family**

Hormone	Molecular Weight	Structure
Gastrin†		
Big gastrin (G-34)	3839	Pyr‡-Leu-Gly-Pro-Gln-Gly-His-Pro-Ser-Leu-Val-Ala-Asp-Pro-Ser-Lys-Lys-Gln-Gly-Pro-Trp-Leu-Glu-Glu-Glu-Glu-Glu-Ala-Tyr-Gly-Trp-Met-Asp-Phe-NH_2
Little gastrin (G-17)	2098	Pyr‡-Gly-Pro-Trp-Leu-Glu-Glu-Glu-Glu-Glu-Ala-Tyr-Gly-Trp-Met-Asp-Phe-NH_2
Minigastrin (G-14)	1833	Trp-Leu-Glu-Glu-Glu-Glu-Glu-Ala-Tyr-Gly-Trp-Met-Asp-Phe-NH_2
Pentagastrin	768	N-t-butyloxycarbonyl-β-Ala-Trp-Met-Asp-Phe-NH_2
Cholecystokinin§		
CCK-39	4678	Tyr-Ile-Gln-Gln-Ala-Arg-Lys-Ala-Pro-Ser-Gly-Arg-Val-Ser-Met-Ile-Lys-Asn-Leu-Gln-Ser-Leu-Asp-Pro-Ser-His-Arg-Ile-Ser-Asp-Arg-Asp-Tyr-Met-Gly-Trp-Met-Asp-Phe-NH_2
CCK-33	3918	Lys-Ala-Pro-Ser-Gly-Arg-Val-Ser-Met-Ile-Lys-Asn-Leu-Gln-Ser-Leu-Asp-Pro-Ser-His-Arg-Ile-Ser-Asp-Arg-Asp-Tyr-Met-Gly-Trp-Met-Asp-Phe-NH_2
CCK-8	1143	Asp-Tyr-Met-Gly-Trp-Met-Asp-Phe-NH_2
CCK-4	597	Trp-Met-Asp-Phe-NH_2
Caerulein	1352	Asp-Arg-Asp-Tyr-Met-Gly-Trp-Met-Asp-Phe-NH_2

*Reproduced, with permission, from J. C. Thompson and M. Marx: Gastrointestinal hormones. *Curr. Probl. Surg.*, **21(6),** 19 (1984). © 1984 by Year Book Medical Publishers.
†Human.
‡Pyr, glutamine in pyriform.
§Porcine.

parietal cells, secretion of pepsin by chief cells, increase in gastric mucosal blood flow, stimulation of gastric motility, and promotion of the growth of oxyntic mucosa and exocrine pancreatic tissue. Release of gastrin is suppressed by acidification of the antral mucosa. In disorders in which H^+ is not excreted owing to destruction or absence of functioning parietal cells (e.g., ***pernicious anemia, atrophic gastritis***), gastrin plasma levels are highly elevated. Plasma levels of gastrin are also elevated in the ***Zollinger–Ellison syndrome,*** in which hypersecretion of acid, peptic ulcer disease, and hyperplasia of the gastric mucosa occur. Gastrin release is suppressed by all members of the secretin family (Table 12-3).

The mechanism of acid secretion by parietal cells is complex and not completely understood. These cells have many receptors and are subjected to a variety of stimuli that can act independently or modulate one another's action. If intracellular pH is assumed to be 7 and luminal pH 1, parietal cells secrete H^+ at a concentration a million-fold higher than that inside the cell. Parietal cells contain the largest number of mitochondria found in eukaryotic cells. They are polar: at the apical membrane, HCl is secreted into the gastric lumen, and at the basolateral membrane, HCO_3^- is secreted. The basolateral membrane contains many different receptors, ion channels, and transport pathways. The resting cell is packed with membrane-bound vesicles called tubulovesicles the membranes of which contain H^+, K^+-ATPase (a proton pump), the enzyme responsible for acid secretion. When the cell is stimulated, these tubulovesicles interact by means of cytoskeletal elements to form a secretory canaliculi. A schematic representation of receptor

TABLE 12-3
Primary Structure of Secretin Family*

Hormone	Molecular Weight	Structure
Secretin	3055 (27 amino acids)	His-Ser-Asp-Gly-Thr-Phe-Thr-Ser-Glu-Leu-Ser-Arg-Leu-Arg-Asp-Ser-Ala-Arg-Leu-Gln-Arg-Leu-Leu-Gln-Gly-Leu-Val-NH_2
Glucagon	3484 (29 amino acids)	His-Ser-Gln-Gly-Thr-Phe-Thr-Ser-Asp-Tyr-Ser-Lys-Tyr-Leu-Asp-Ser-Arg-Arg-Ala-Gln-Asp-Phe-Val-Gln-Trp-Leu-Met-Asp-Thr
VIP	3326 (28 amino acids)	His-Ser-Asp-Ala-Val-Phe-Thr-Asp-Asn-Tyr-Thr-Arg-Leu-Arg-Lys-Gln-Met-Ala-Val-Lys-Lys-Tyr-Leu-Asn-Ser-Ile-Leu-Asn-NH_2
GIP	5104 (43 amino acids)	Tyr-Ala-Glu-Gly-Thr-Phe-Ile-Ser-Asp-Tyr-Ser-Ile-Ala-Met-Asp-Lys-Ile-Arg-Gln-Gln-Asp-Phe-Val-Asn-Trp-Leu-Leu-Ala-Gln-Gln-Lys-Gly-Lys-Lys-Ser-Asp-Trp-Lys-His-Asn-Ile-Thr-Gln

*Reproduced, with permission, from J. C. Thompson and M. Marx: Gastrointestinal hormones. *Curr. Probl. Surg.*, **21(6),** 19 (1984). © 1984 by Year Book Medical Publishers.
†Human.
‡Pyr, glutamine in pyriform.
§Porcine.

systems and ion pathways is shown in Figure 12-3. The mediators that stimulate acid secretion are acetylcholine, gastrin, and histamine. Bombesin, a gastrin-releasing peptide, is also released from enteric neurons by vagal stimulation. The action of acetylcholine can be inhibited by anticholinergic agents (e.g., atropine). The action of histamine (the decarboxylated product of histidine, Chapter 17) is mediated by H_2 receptors. H_1 receptors mediate the action of histamine on smooth muscle, causing bronchoconstriction, as in acute anaphylaxis and allergic asthma. The actions of histamine can be selectively inhibited by specific H_1- and H_2-receptor antagonists. The introduction of H_2-receptor antagonists (e.g., cimetidine and ranitidine; see Figure 12-4) helps to heal duodenal and gastric ulcers. Secretion of gastric acid can also be suppressed by prostaglandin E derivatives (Chapter 18) and substituted benzimidazoles that inhibit H^+, K^+-ATPase (e.g., omeprazole, pantoprazole). Omeprazole consists of a sulfinyl group which bridges two ring systems consisting of benzimidazole and pyridine. The active drug is formed by protonation in the stomach and rearranges to form sulfenic acid and sulfenamide. H^+, K^+-ATPase contains critical sulfhydryl groups located on the luminal side. Sulfenamide is covalently attached to H^+, K^+-ATPase, causing an irreversible inhibition (Figure 12-5). Complete inhibition occurs when two molecules of sulfenamide bind to the enzyme. The degree of proton pump inhibition is dose-dependent and the drug is particularly useful in patients with peptic ulcer disease not well controlled by H_2-receptor antagonists.

Although most of the action of gastrin is mediated via the H_2 receptor, residual stimulation due to gastrin alone suggests the presence of specific receptors for gastrin.

The precise intracellular events that lead to acid secretion are not clear, but the second messengers in this process appear to be cAMP and Ca^{2+}. The H_2 receptor is coupled to the adenylate cyclase system (Chapter 30), and its activation results in the intracellular elevation of cAMP concentration. Stimulation of cholinergic receptor systems is coupled to increased Ca^{2+} permeability.

The H^+ ions in the stimulated parietal cell are produced by the action of carbonic anhydrase:

$$H_2O + CO_2 \xrightarrow{\text{carbonic anhydrase}} H_2CO_3 \rightleftharpoons H^+ + HCO_3^-$$

and the H^+ ions formed are secreted with the aid of H^+, K^+-ATPase, which secretes H^+ in exchange for K^+ through the free-energy hydrolysis of ATP. K^+ is transported into the lumen via a specific ion channel. In the lumen, the accompanying anion for H^+ is Cl^- secreted from the parietal cell. The chloride ions are derived from the plasma and transported into the parietal cell via the electroneutral Cl^-/HCO_3^- exchange. Thus, for every H^+ secreted into the gastric lumen, there is a

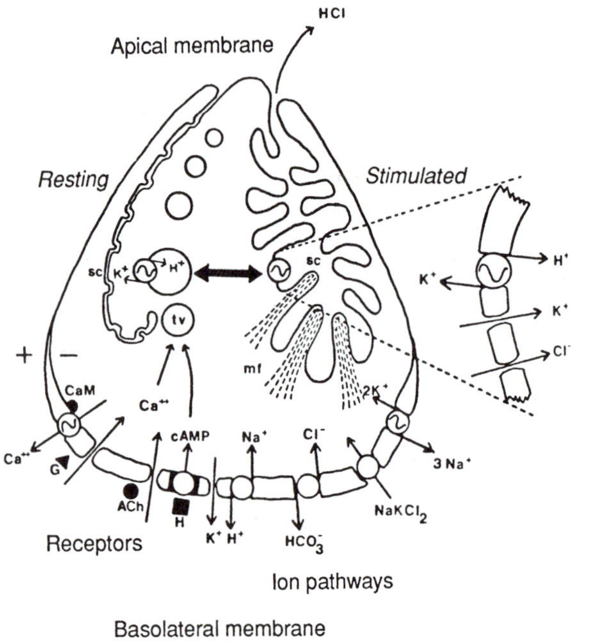

FIGURE 12-3
Schematic representation of the resting (left side) and stimulated (right side) state of the parietal cell. Basolateral membrane contains three major receptor classes: gastrin (G), acetylcholine (ACh), and histamine (H). Their actions are mediated by cAMP responses, Ca^{2+} changes, or both. In addition, there are a number of ion transport pathways. In the stimulated state, the apical membrane acquires H^+,K^+-ATPase contained in the tubulovesicles (tv) as well as the property of K^+ and Cl^- conductance, both of which are essential in the secretion of HCl. A change in cytoskeletal arrangement is also associated with stimulation. CaM = calmodulin; SC = secretory canaliculus; mf = microfilaments. [Reproduced with permission from D. H. Malinowska and G. Sachs, Cellular mechanisms of acid secretion, *Clin. Gastroenterol.* **13,** 322 (1984).]

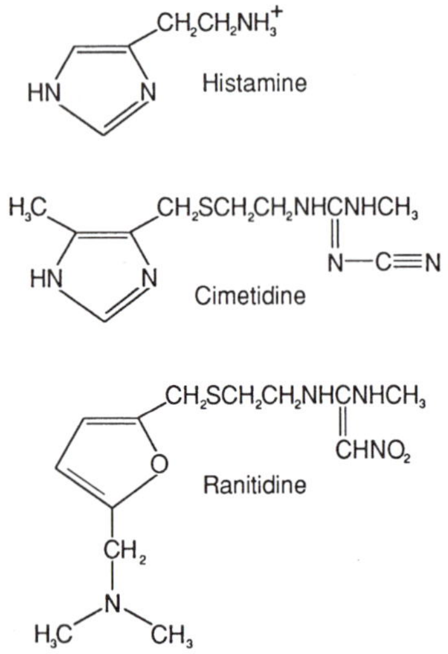

FIGURE 12-4
Comparison of the structures of histamine with cimetidine and ranitidine. The latter two are H_2-receptor antagonists and act on the gastric parietal cells to inhibit gastric acid production. Ranitidine, which has a furan rather than an imidazole structure, is a more potent competitive inhibitor than cimetidine.

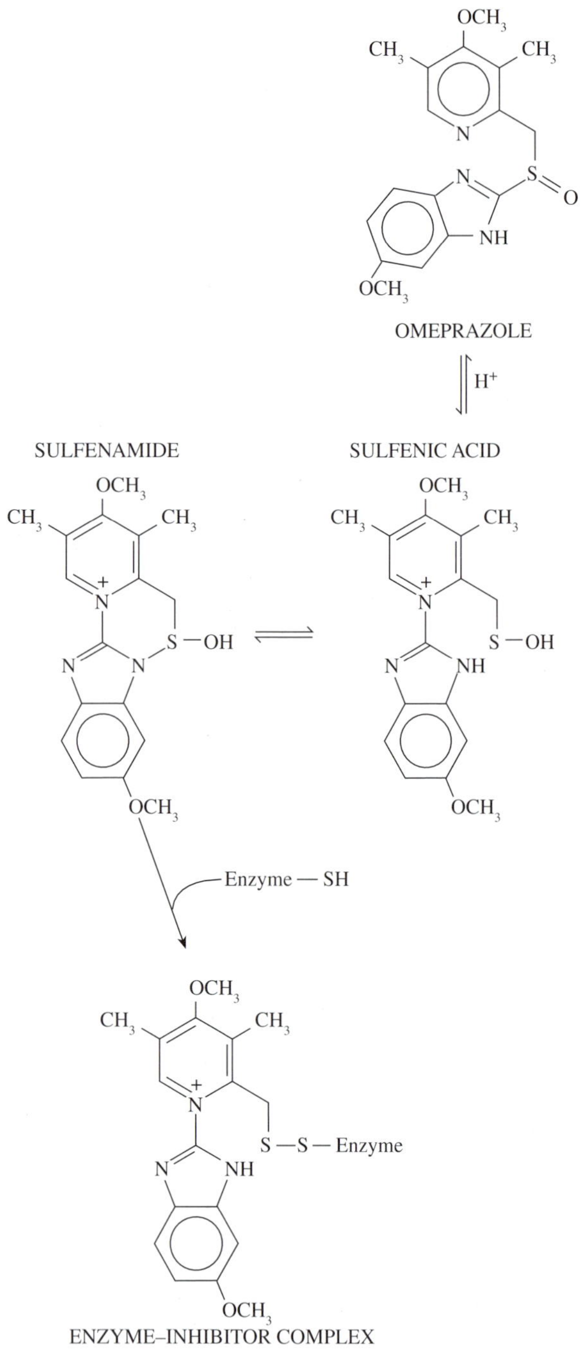

FIGURE 12-5
Inhibition of H^+,K^+-ATPase by omeprazole. In the acidic pH of the stomach, omeprazole is converted to sulfenamide, which forms an enzyme–inhibitor complex by disulfide linkage.

transfer of an HCO_3^- into the plasma in exchange for Cl^-, which is ultimately secreted into the gastric lumen. In instances of excessive loss of gastric fluids (e.g., persistent vomiting or nasogastric suction), metabolic alkalosis results (Chapter 39).

Peptic Ulcer Disease

Peptic ulcers are caused by an imbalance of acid secretion by the parietal cells and lack of mucosal protective barriers. There are two major causes of peptic ulcer disease. One is drug induced and the other is caused by a bacterial infection. The drugs that are related to ***gastropathy*** (and also possibly renal insufficiency) belong to a group of compounds known as ***nonsteroidal anti-inflammatory drugs*** (NSAIDs). NSAIDs encompass many different classes of chemical compounds including aspirin (acetylsalicylic acid). NSAIDs exert anti-inflammatory action and are used in the treatment of many rheumatic conditions. The mechanism of action of NSAIDs involves inhibition of the cyclooxygenase group of enzymes which are responsible for the synthesis of prostaglandins (Chapter 18). The inhibition of those enzymes by aspirin is irreversible and is caused by covalent modification involving acetylation of a key serine residue of cyclooxygenase.

At least two isoforms of cyclooxygenase (COX) are known, COX1 and COX2. COX1 is a constitutive enzyme and is responsible for the synthesis of prostaglandins that are essential for normal function. Gastric prostaglandins maintain mucosal integrity by modulating parietal cell acid production, stimulate mucus and bicarbonate production in the mucous gel layer, and regulate mucosal blood flow. COX2 is induced during inflammation by cytokines and inflammatory mediators. NSAIDs inhibit both COX1 and COX2. While they produce a desirable therapeutic effect as anti-inflammatory agents, NSAIDs undesirably inhibit the glandular prostaglandin production necessary for normal function. Thus, drugs that selectively inhibit COX2 enzymes are better anti-inflammatory agents; one such NSAID is celecoxib (Chapter 18). A stable analogue of prostaglandin E_1, misoprostol has cytoprotective effects in the treatment of peptic ulcer disease. Aspirin has other pharmacological effects such as inhibition of platelet aggregation (Chapter 36).

Peptic ulcer disease is associated with *Helicobacter pylori* infection in 90% of patients with gastric and duodenal ulceration. Elimination of *H. pylori* infection with antibiotics heals the peptic ulcer and the associated symptoms. Combination therapy with antibiotics, anti-secretory agents, namely H_2-receptor antagonists or proton pump inhibitors, and bismuth salts has significantly improved the clinical outcome of peptic ulcer disease. Not all strains of *H. pylori* cause peptic ulcer disease, and other factors are necessary for *H. pylori* colonization and disease to occur. Flagellated motile bacteria resist peristalsis and adhere to gastric epithelium in a highly specific manner.

Several adhesins and their ligands have been identified on the host cells. Lewis blood group antigen has been identified as an epithelial cell receptor for *H. pylori* binding (Chapter 10). Enhanced acid production due to chronic *H. pylori* infection has been attributed to the production of inflammatory mediators (e.g., cytokines, Chapter 35) and pH alterations. Urease produced by the bacteria converts urea to ammonia and carbon dioxide. Ammonia increases the pH and is essential for the survival of bacteria at acidic pH. A change in pH toward alkalinity may increase the levels of gastrin, thus causing increased acid production and a vicious cycle. Urease activity is conserved among all *H. pylori* species as is the primary structure of the enzyme.

H. pylori organisms that produce vacuolating cytotoxin (coded by the gene *vacA*) and a high-molecular-weight protein known as CagA (coded by the gene *cagA*) are implicated in ulcerogenesis and gastritis. *H. pylori* strains are genetically diverse, due to their ability to mutate and the ease with which they exchange genes. *H. pylori* infection may also be a risk factor in adenocarcinoma of the antrum and the body of the stomach and ***non-Hodgkin's lymphoma*** of the stomach.

Infection by *H. pylori* is detected by serological markers produced by host immune responses (e.g., antibodies to antigens of *H. pylori*) and a breath test. The latter, known as the urea breath test, consists of oral administrations of radioactively labeled urea. This is metabolized to labeled CO_2 and ammonia by the urease of *H. pylori* present in the gastric mucosa. The presence of labeled CO_2 measured in the exhaled air confirms infection.

H_2-receptor antagonists and proton pump inhibitors are used to reduce gastric acidity and are useful in the treatment of ***gastroesophageal reflux disease*** (GERD). In GERD, the contents of the stomach reflux into the esophagus and is the most common malady of the esophagus. Therapy also involves increasing of lower esophageal sphincter tone. Surgical therapy involves strengthening esophageal sphincters by wrapping part of the stomach around lower esophagus and placing pressure on the sphincter to assist its closure. This procedure, known as ***fundoplication,*** can be performed by abdominal surgery or, more commonly, by laparoscopy. GERD complications involve structures contiguous to the esophagus namely laryngeal, pharyngeal, pulmonary, sinusal and dental disorders. Lifestyle changes that can alleviate GERD include cessation of smoking and consumption of alcohol; avoidance of spicy food, coffee, and bedtime meals; correction of obesity; and elevation of the head by at least 15 cm (about 6 inches) during sleep. Therapeutic agents that contain theophylline, anticholinergic agents, and progesterone should be avoided because they delay gastric emptying and decrease lower esophageal sphincter tone. One of the changes of chronic GERD that may occur in some patients is that the healing epithelium of the esophagus is replaced not with the normal squamous cells but with specialized columnar cells. Columnar epithelial cells normally are found in the small intestine. Hence, this process is known as intestinal metaplasia or ***Barrett's esophagus*** and is regarded as a premalignant condition.

Cholecystokinin

Cholecystokinin (CCK) is found throughout the small intestine but is located predominantly in the mucosal I cells of the duodenum and jejunum. In the ileum and colon, it is localized in nerve endings, and it is widely distributed throughout the peripheral and central nervous systems. CCK consists of 33 amino acid residues (CCK-33, Table 12-2), and shows macro- and microheterogeneity. Several other forms are known: CCK-58, CCK-39, and CCK-8. Naturally occurring CCK has a sulfated tyrosyl residue at position 27, and removal of sulfate changes the biological activity to that of gastrin. The C-terminal tetrapeptide is identical with that of gastrin. A synthetic C-terminal octapeptide is about three times more potent than CCK-33. Cerulein, a decapeptide present in the skin and GI tract of certain amphibians, has a C-terminal octapeptide sequence identical to that of CCK (Table 12-2). These two peptides have similar biological properties, and cerulein is clinically useful for the stimulation of gallbladder contraction. CCK is secreted as a result of stimuli caused by the products of digestion of proteins and lipids. The secretion of CCK is terminated when these digestion products are absorbed or migrate into the lower portions of the GI tract. The principal physiological actions of CCK are to stimulate gallbladder contraction, to relax the sphincter of Oddi, and to stimulate secretion of pancreatic juice rich in digestive enzymes. Other functions are stimulation of bicarbonate-rich fluid secretion, insulin secretion, and intestinal motility. CCK can induce satiety in laboratory animals and humans, the gastric vagal fibers being necessary for this effect.

Secretin

Secretin is synthesized by the S cells of the duodenum and jejunum and is also present in the brain. Its amino acid sequence is similar to that of glucagon, VIP, and GIP (Table 12-3). All 27 amino acid residues are required for biological activity. Chyme (pH < 4.5) in the duodenum stimulates release of secretin. Secretin stimulates the secretion of pancreatic juice rich in bicarbonate, which neutralizes the acid chyme and inhibits further secretion of the hormone. This action appears to be mediated by membrane-bound adenylate cyclase and increased concentrations of intracellular cAMP.

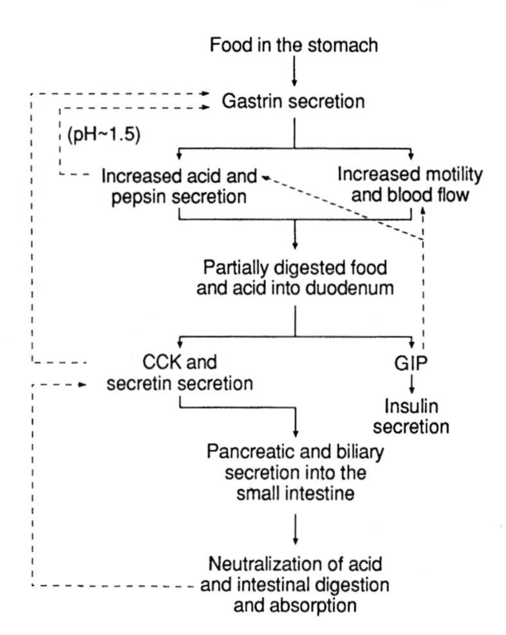

FIGURE 12-6
Integrated function of gastrointestinal hormones in regulation of digestion and absorption of food. Dashed arrows indicate inhibition.

Gastric Inhibitory Peptide

Gastric inhibitory peptide (GIP) is found in the K cells of the duodenum and jejunum: It consists of 43 amino acid residues and occurs in multiple molecular forms. Its secretion is stimulated by the presence of glucose and lipids in the duodenum. It has two main functions:

1. To stimulate insulin secretion that prepares the appropriate tissues for the transport and metabolism of nutrients obtained from the GI tract (Chapter 22), and
2. To inhibit gastric secretion and motility.

The physiological functions of the GI hormones discussed above are integrated and coordinated to facilitate the digestion and absorption of food (Figure 12-6).

12.3 Digestion and Absorption of Major Food Substances

Carbohydrates

In most diets, carbohydrates are a major source of the body's energy requirements. The predominant digestible carbohydrates are starches (amylose and amylopectin). Depending upon the diet, the digestible carbohydrates may

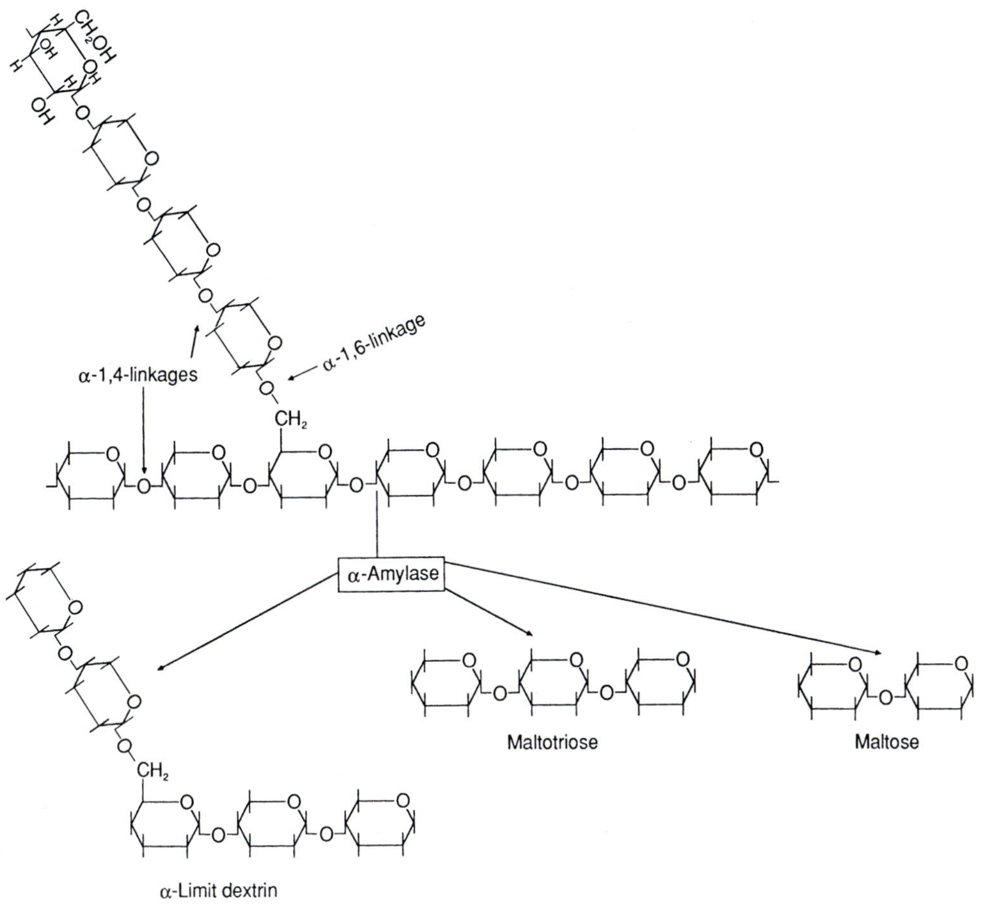

FIGURE 12-7
Hydrolysis of starch by α-amylase.

include glycogen and the disaccharides sucrose and lactose. The disaccharide present in mushrooms, trehalose, is also digestible. The specific patterns of carbohydrate intake are influenced by cultural and economic factors. Many plant carbohydrates are not digestible, and they constitute the dietary fiber, which includes cellulose, hemicelluloses, pectins, gums, and alginate (Chapter 9). The digestion of starch (and glycogen) begins in the mouth during mastication and the mixing of food with salivary α-amylase, which hydrolyzes starch to some extent. The digestive action of salivary α-amylase on starch is terminated in the acidic environment of the stomach. Starch digestion is resumed in the duodenum by another α-amylase secreted by the pancreas. Salivary and pancreatic α-amylases share some properties and exist in several isoenzyme forms that are separable by electrophoresis. In pancreatic disease, the α-amylase level in the serum increases, and its measurement and isoenzyme pattern are useful in diagnosis. Carbohydrate digestion and absorption take place in a well-defined sequence:

1. Intraluminal hydrolysis of starch and glycogen by α-amylase to oligosaccharides of variable length and structure;
2. Brush-border surface hydrolysis of oligosaccharides and disaccharides (e.g., sucrose, lactose, and trehalose) to their monomers by specific oligosaccharidases that are integral to the cell membrane of the enterocyte; and
3. Transport of monosaccharides (e.g., glucose, galactose, and fructose) into enterocytes.

Digestion of Starch

α-Amylase hydrolyzes starch into ***α-limit dextrins*** (branched oligosaccharides of five to nine glucose residues), maltotriose, and maltose (Figure 12-7). The α-amylase is an endoglycosidase, and its action does not yield free glucose. It cannot catalyze the hydrolysis of $\alpha(1 \rightarrow 6)$ linkages (the branch points). α-Amylase has optimal activity at pH 7.1, an absolute requirement for the presence of Cl^-, and is stabilized by Ca^{2+}.

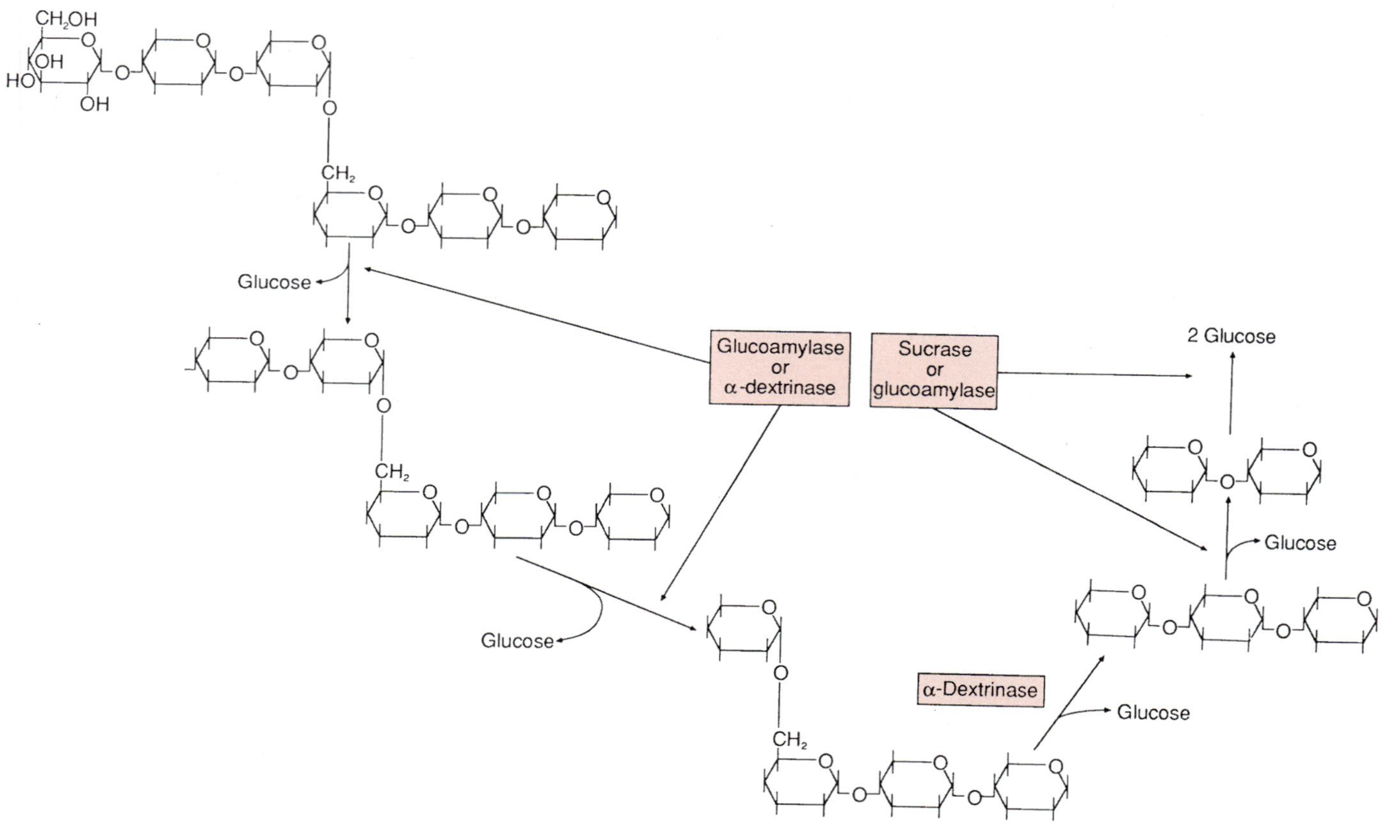

Glucoamylase or α-dextrinase

Sucrase or glucoamylase

α-Dextrinase

Glucose

Glucose

Glucose

Glucose

2 Glucose

CH_2OH OH HO OH O CH_2

FIGURE 12-8
Hydrolysis of α-limit dextrin to glucose by brush border membrane enzymes.

Brush-Border Surface Hydrolysis

Products of α-amylase digestion of starch and ingested disaccharides are hydrolyzed by oligosaccharidases on enterocyte cell membranes to yield monosaccharides that are transferred across the brush-border bilayer. The oligosaccharidases are large glycoproteins (M.W. > 200,000) that are integral constituents of the cell membrane (Chapter 10). Their active sites project toward the luminal side. They have pH optima at about 6 and K_m values for substrates in the range of 3–20 mmol/L. Some oligosaccharidases are the following:

1. ***Exo-1,4-α-D-glucosidase,*** also called glucoamylase or maltase, catalyzes sequential hydrolysis of terminal glucosyl units linked in $\alpha(1 \rightarrow 4)$ linkages from the nonreducing ends of malto-oligosaccharides or maltose.
2. ***Sucrose α-D-glucohydrolase,*** also called sucrose-α-dextrinase or sucrase, catalyzes hydrolysis of the $\alpha(1 \rightarrow 2)$ linkage of sucrose to release glucose and fructose; of the $\alpha(1 \rightarrow 4)$ linkage of maltose to release two glucose units; and of the $\alpha(1 \rightarrow 6)$ linkage in α-limit dextrins and isomaltose. It is a single gene product that is posttranslationally cleaved to form two distinct polypeptide chains, one chain catalyzing the hydrolysis of an $\alpha(1 \rightarrow 4)$ linkage and the other of an $\alpha(1 \rightarrow 6)$ linkage.
3. ***β-D-Galactoside galactohydrolase,*** also called *lactase*, catalyzes hydrolysis of the $\beta(1 \rightarrow 4)$ linkage of lactose (or terminal nonreducing ***β-D-galactose*** units in ***β-D-galactosides***) into galactose and glucose.
4. ***α,α-Trehalose glucohydrolase,*** also called *trehalase,* catalyzes the hydrolysis of trehalose into two glucose units.

Hydrolysis of branched-chain oligosaccharides (α-limit dextrins) to glucose thus requires sequential action of three enzyme activities (Figure 12-8). Hydrolysis of oligosaccharides is rapid and is not the rate-limiting step in their absorption. However, accumulation of monosaccharides in the lumen is limited by end-product inhibition of the oligosaccharidases. The rate-determining step in absorption is the monosaccharide transport system, with the following exception. Mucosal lactase activity is the lowest oligosaccharidase activity, and so hydrolysis rather than absorption is rate limiting. Oligosaccharide digestion is virtually complete by midjejunum.

Transport of Monosaccharides into the Enterocyte

Glucose and galactose compete for a common transport system. This system is an active transport system; i.e., the monosaccharides are absorbed against a concentration gradient, it is saturable and obeys Michaelis–Menten

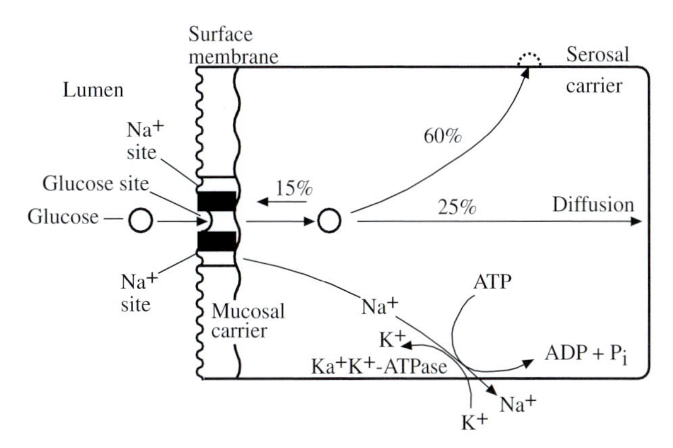

FIGURE 12-9
Schematic representation of glucose (or galactose) transport by the enterocyte. Glucose binds to the receptor, facilitated by the simultaneous binding of two Na^+ at separate sites. The glucose and Na^+ are released in the cytosol as the receptor affinity for them decreases. The Na^+ are actively extruded at the basolateral surface into the intercellular space by Na^+,K^+-ATPase, which provides the energy for the overall transport. Glucose is transported out of the cell into the intercellular space and thence to portal capillaries, both by a serosal carrier and by diffusion. (Reproduced with permission from G. M. Gray, *Carbohydrate Absorption and Malabsorption in Gastrointestinal Physiology*. Raven Press, New York, 1981.]

kinetics, and it is carrier-mediated and Na^+-dependent. Translocation of glucose (or galactose) is shown in Figure 12-9. On the luminal side, one molecule of glucose and two sodium ions bind to the membrane carrier (presumably, Na^+ binding to the carrier molecule increases the affinity for glucose because of a conformational change). The carrier-bound Na^+ and glucose are internalized along the electrochemical gradient that results from a low intracellular Na^+ concentration. Inside the cell, the sodium ions are released from the carrier, and the diminished affinity of the carrier for glucose releases the glucose. The sodium ions that enter in this manner are transported into the lateral intercellular spaces against a concentration gradient by the free energy of ATP hydrolysis catalyzed by a Na^+,K^+-ATPase. Thus, glucose and Na^+ are transported by a common carrier, and energy is provided by the transport of Na^+ down the concentration and electrical gradient. The low cytoplasmic Na^+ concentration is maintained by the active transport of Na^+ out of the cell coupled to K^+ transport into the cell by the Na^+,K^+-ATPase. Since the sugar transport does not use ATP directly, it can be considered as secondary active transport. Although this mode of glucose transport is the most significant, passive diffusion along a concentration gradient may also operate if the luminal concentration of glucose exceeds the intracellular concentration.

The intracellular glucose is transferred to the portal capillary blood by passive diffusion and by a carrier-mediated system. Intracellular glucose can be converted to lactate

(Chapter 13), which is transported via the portal blood system to the liver, where it is reconverted to glucose (gluconeogenesis, Chapter 15). The quantitative significance of this mode of glucose transport is probably minimal.

Fructose transport is distinct from glucose-galactose transport and requires a specific saturable membrane carrier (facilitated diffusion).

Na^+,K^+-ATPase

Na^+,K^+-ATPase, in addition to participating in Na^+-driven uptake of glucose and amino acids (see below), is responsible for maintaining high intracellular concentrations of K^+ and low concentrations of Na^+ (the reverse of the relative concentrations of these ions in the extracellular fluid). The Na^+ and K^+ gradient across the cell membrane is involved in the maintenance of osmotic equilibrium, propagation of nerve impulses, reabsorption of solutes by the kidney, and other processes that require the electrochemical energy of the ion gradients. Thus, Na^+,K^+-ATPase plays a critical role in many important functions of the body.

Na^+,K^+-ATPase is a transmembrane protein found almost exclusively in the plasma membrane. It has two major subunits: α (M.W. ~ 95,000) and β (M.W. ~ 55,000). The latter is a glycoprotein and is exposed to the exterior of the cell. The former spans the entire membrane (Figure 12-10). Minimum subunit stoichiometry required for the activity of the enzyme is $\alpha\beta$, and the most probable native enzyme structure is $(\alpha\beta)_2$. The enzyme is asymmetrically oriented and drives active transport only in one direction. The ATP binding site is located on the cytoplasmic aspect of the α-subunit. *Ouabain*, a cardiac glycoside (similar to digitalis glycosides), which inhibits the enzyme, also binds to the α-subunit but at a site that projects to the exterior of the cell. The inhibition of Na^+, K^+-ATPase activity indirectly leads to an increase in intracellular Ca^{2+} concentration, which stimulates contraction in muscle cells (Chapter 21), thus accounting for the therapeutic effect of cardiac glycosides on the heart. Na^+,K^+-ATPase requires the presence of Na^+, K^+, Mg^{2+}, and ATP. Each cycle of enzyme activity results in the extrusion of three Na^+ coupled to the transport of two K^+ into the cell, with the hydrolysis of one molecule of ATP. Thus, the enzyme utilizes the energy derived from ATP hydrolysis to transport K^+ into the cell and Na^+ out of the cell, against concentration gradients. Since unequal numbers of monovalent cations are transferred across the plasma membrane, a transmembrane electric current is generated.

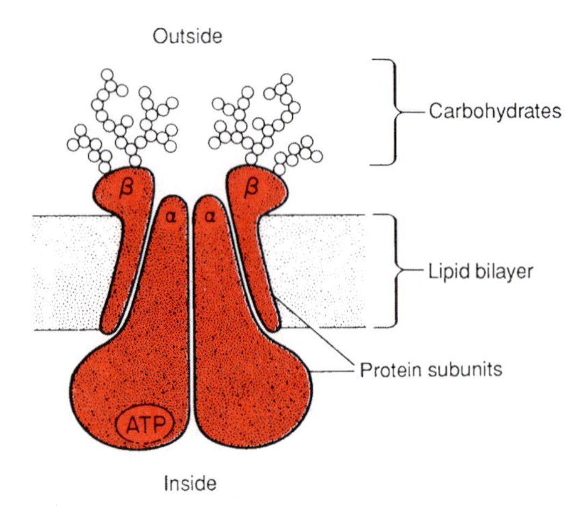

FIGURE 12-10
Schematic representation of Na^+, K^+-ATPase. The minimum subunit stoichiometry of the active enzyme is $\alpha\beta$, and the native enzyme most likely has $(\alpha\beta)_2$ structure. The enzyme spans the plasma membrane, with the smaller β glycoprotein subunits projecting outside the cell. Each functional unit has binding sites on both sides of the membrane; the outer surface has K^+ and cardiac glycoside binding sites, and the inner surface has Na^+ and ATP binding sites. [Reproduced with permission from K. J. Sweadner and S. M. Goldin, Active transport of sodium and potassium ions. *N. Engl. J. Med.* **302,** 777 (1980).]

A model for the mechanism of action of the enzyme is shown in Figure 12-11. It proposes that Na^+,K^+-ATPase can exist in two (or more) conformational states: one binding Na^+ or ATP (or both) and the other binding K^+ or phosphate (or both). On the cytoplasmic side, Na^+ binding initiates transient phosphorylation of an aspartate residue at the active site, resulting in a cyclic process with translocation of Na^+ from inside to outside and of K^+ from outside to inside. The vectorial equation for the transport is

$$3Na_i^+ + 2K_o^+ + ATP^{4-} + H_2O \rightarrow 3\,Na_o^+ + 2K_i^+ + ADP^{3-} + HPO_4^{2+} + H^+$$

where i = inside and o = outside. Thyroid hormone increases Na^+,K^+-ATPase activity (Chapter 33). Other ATPases participate in the transport of other ions (e.g., K^+,H^+-ATPase, above; Ca^{2+}-ATPase, Chapter 21).

Disorders of Carbohydrate Digestion and Absorption

Carbohydrate malabsorption can occur in a number of diseases that cause mucosal damage or dysfunction (e.g., gastroenteritis, protein deficiency, gluten-sensitive enteropathy). Disorders due to deficiencies of specific oligosaccharidases are discussed below.

Lactose Intolerance (Milk Intolerance) Lactose intolerance is the most common disorder of carbohydrate absorption. Lactase deficiency occurs in the majority of human adults throughout the world and appears to be genetically determined. The prevalence is high in persons of African and Asian ancestry ($\geqslant$ 65%) and low in persons of Northern European ancestry. Lactase deficiency in which mucosal lactase levels are low or absent at birth

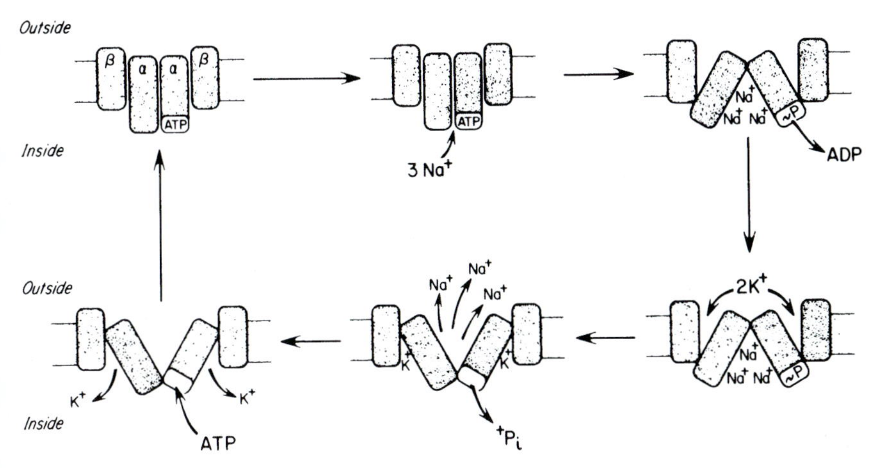

FIGURE 12-11
Model for transport of Na^+ and K^+ by Na^+,K^+-ATPase. Inside the cell, Na^+ initiates the one-way ion exchange cycle by the phosphorylation of an aspartate residue at the active site by ATP, which eventually leads to the translocation of Na^+ and K^+. Conformational changes of the enzyme occur during the exchange of ions. [Reproduced, with permission, from K. J. Sweadner and S. M. Goldin, Active transport of sodium and potassium ions. *New Engl. J. Med.* **302,** 777 (1980).]

is rare and is transmitted as an autosomal recessive trait. Prompt diagnosis and a lactose-free diet assure the infant's normal growth. Since hydrolysis of lactose by lactase is rate-limiting, any mucosal damage will cause lactose intolerance.

In full-term human infants, lactase activity attains peak values at birth and remains high throughout infancy. As milk intake decreases, lactase levels drop and lactose intolerance may develop. The extent of the decrease of lactase activity distinguishes lactose-tolerant from intolerant populations.

Ingestion of milk (or lactose) by individuals who have lactose intolerance leads to a variety of symptoms (bloating, cramps, flatulence, and loose stools). Severity of the symptoms depends on the amount of lactose consumed and the enzyme activity. Symptoms disappear with elimination of lactose from the diet. Eight ounces of milk contains about 12 g of lactose. Some individuals show symptoms with consumption of as little as 3 ounces of milk. Some lactose-intolerant individuals can drink milk if it is consumed along with meals and show no reaction to milk added to cereal or coffee. Lactose-depleted milk or fermented milk products with negligible amounts of lactose are good substitutes for milk. The intestinal problems are due primarily to osmotic effects of lactose and its metabolites in the colon. The lactose not absorbed in the small intestine increases the osmolarity and causes water to be retained in the lumen. In the colon, it is metabolized by bacterial enzymes to a number of short-chain acids, further increasing osmolarity and aggravating fluid reabsorption. The bacterial fermentation also yields gaseous products (H_2, CO_2, and CH_4), hence the bloating, flatulence, and sometimes frothy diarrhea.

A lactose tolerance test is performed by oral administration of a load of lactose (100 g in an adult), measurement of blood glucose at specific time intervals, and occurrence of characteristic symptoms. The values for blood glucose are compared with those from an oral glucose tolerance test performed the previous day on the same individual. The hydrogen breath test can also be used in the assessment of lactose malabsorption. Hydrogen is a product of bacterial fermentation of undigested sugar in the colon and is not a product of human metabolism. Some of this hydrogen is absorbed and excreted by the lungs. Measurement of the amount of hydrogen in the expired air by gas chromatographic studies provides a measure of the unabsorbed carbohydrate in the colon. Both tests can be adapted for the diagnosis of malabsorption disorders for other carbohydrates.

Intolerance of Other Carbohydrates Intolerance to sucrose and α-limit dextrins may be due to deficiency of sucrase-α-dextrinase or to a defect in glucose-galactose transport. These disorders are rare autosomal recessive traits; clinical problems can be corrected by removing the offending sugars from the diet. Lactulose, a synthetic disaccharide consisting of galactose and fructose with a $\beta(1 \rightarrow 4)$ linkage, is hydrolyzed not in the small intestine but in the colon and is converted to products similar to those derived from lactose fermentation. It has been used in the treatment of patients with liver disease. Normally, ammonia (NH_3) produced in the GI tract is converted in the liver to urea (Chapter 17), hence, in patients with severe liver disease, blood ammonia levels increase. Absorption of ammonia can be decreased by administration of lactulose, which acidifies the colonic constituents so that NH_3 is trapped as NH_4^+ ions (Chapter 9).

Proteins

Protein is an essential nutrient for human growth, development, and homeostasis. The nutritive value of dietary proteins depends on its amino acid composition and digestibility. Dietary proteins supply ***essential amino acids,*** which are not synthesized in the body. Nonessential amino acids can be synthesized from appropriate precursor substances (Chapter 17). In human adults, essential amino acids are valine, leucine, isoleucine, lysine, methionine, phenylalanine, tryptophan, and threonine. Histidine (and possibly arginine) appears to also be required for support of normal growth in children. In the absence from the diet of an essential amino acid, cellular protein synthesis does not occur. The diet must contain these amino acids in the proper proportions. Thus, quality and quantity of dietary protein consumption and adequate intake of energy (carbohydrates and lipids) are essential. Protein constitutes about 10–15% of the average total energy intake.

Animal proteins, with the exception of collagen (which lacks tryptophan), provide all of the essential amino acids. Vegetable proteins differ in their content of essential amino acids (Table 12-4), but a mixture of these proteins will satisfy the essential amino acid requirement. For example, the lysine lacking in grains can be provided by legumes. This combination also corrects for the methionine (which is supplied in corn) deficiency of legumes. Such combinations (e.g., lentils and rice, chick-peas and sesame seeds, spaghetti and beans, corn and beans) are widely used in different cultures to provide for an optimal protein requirement. The recommended allowance for mixed proteins in an adult in the U.S.A. is 0.85 g per kilogram of body weight per day (Chapter 5). The allowances are increased during childhood, pregnancy, and lactation (Appendix IV).

TABLE 12-4
Limiting Essential Amino Acids in Plant Proteins

Source of Plant Proteins	Limiting Amino Acids
Grain	
Maize or corn	Lysine, tryptophan
Millet, oats, wheat, or rice	Lysine, threonine
Legumes	
Beans, immature	Methionine, isoleucine
Beans, mature	Methionine, valine
Peas	Methionine, tryptophan
Peanuts	Lysine, threonine
Nuts and oil seeds	Lysine, threonine
Coconut	Lysine, threonine
Vegetables	Methionine, isoleucine

Besides dietary protein, a large amount of endogenous protein undergoes digestion and absorption. Endogenous protein comes from three sources:

1. Enzymes, glycoproteins, and mucins secreted from the salivary glands, stomach, intestine, biliary tract, and pancreas, which together constitute about 20–30 g/day;
2. Rapid turnover of the gastrointestinal epithelium, which contributes about 30 g/day; and
3. Plasma proteins that normally diffuse into the intestinal tract at a rate of 1–2 g/day.

In several disorders of the GI tract (e.g., protein-losing gastroenteropathy), loss of plasma proteins is considerable and leads to hypoproteinemia.

Digestion

Protein digestion begins in the stomach, where protein is denatured by the low pH and is exposed to the action of pepsin. The low pH also provides the optimal H^+ concentration for pepsin activity. The zymogen precursor pepsinogen (M.W. 40,000) is secreted by the chief cells and is converted to pepsin (M.W. 32,700) in the acid medium by removal of a peptide consisting of 44 amino acid residues. This endopeptidase hydrolyzes peptide bonds that involve the carboxyl group of aromatic amino acid residues, leucine, methionine, and acidic residues (Table 12-5). The products consist of a mixture of oligopeptides.

Chyme contains potent secretagogues for various endocrine cells in the intestinal mucosa. CCK and secretin cause release of an alkaline pancreatic juice containing trypsinogen, chymotrypsinogen, proelastase, and procarboxypeptidases A and B. Activation begins with that of trypsinogen to trypsin by enteropeptidase (previously called enterokinase) present in the brush-border membranes of the duodenum.

Enteropeptidase cleaves between Lys-6 and Ile-7 to release a hexapeptide from the N-terminus. Trypsin autocatalytically activates trypsinogen and activates the other zymogens. The importance of the initial activation of trypsinogen to trypsin by enteropeptidase is manifested by children with congenital enteropeptidase deficiency who exhibit hypoproteinemia, anemia, failure to thrive, vomiting, and diarrhea.

Trypsin, elastase, and chymotrypsin are ***endopeptidases.*** Carboxypeptidases are ***exopeptidases*** (Table 12-5). The combined action of these enzymes produces oligopeptides having two to six amino acid residues and free

TABLE 12-5
Specificities of the Proteolytic Enzymes Involved in Gastrointestinal Protein Digestion

Peptide bond cleaved

H_2N-----NH–CH(R_1)–C(=O)–↓–NH–CH(R_2)–C(=O)-----C(=O)OH

Enzyme*	Preferentially cleaves peptide bonds in which
Trypsin	R_1 = Arg or Lys; R_2 = any amino acid residue
Chymotrypsin	R_1 = aromatic amino acid (Phe, Tyr, Trp); R_2 = any amino acid residue
Carboxypeptidase A	R_1 = any amino acid residue; R2 = any *C-Terminal residue* except Arg, Lys, or Pro
Carboxypeptidase B	R_1 = any amino acid residue; R2 = Arg or Lys *at the C-terminus* of a polypeptide
Elastase	R_1 = Neutral (uncharged) residues; R_2=any amino acid residue
Aminopeptidase	R_1 = Most *N-terminal residues* of a polypeptide chain; R_2 = any residue except Pro
Pepsin	R_1 = Trp, Phe, Tyr, Met, Leu; R_2 = any amino acid residue
Dipeptidases	H_2N–CH(R_1)–C(=O)–↑–NH–CH(R_2)–C(=O)OH Peptide bond cleaved
Tripeptidases	H_2N–CH(R_1)–C(=O)–↑–NH–CH(R_2)–C(=O)–↑–NH–CH(R_3)–C(=O)OH
R_1, R_2, R_3 = any amino acid residue	Choice of bond cleaved depends on specific enzyme. In a given molecule, only one bond is cleaved by a tripeptidase.

*Trypsin, chymotrypsin, elastase, and pepsin are endopeptidases (trypsin is the most specific and pepsin is the least specific): carboxypeptidase A and B and aminopeptidase are exopeptidases.

amino acids. Hydrolysis of oligopeptides by the brush-border aminopeptidases releases amino acids. Leucine aminopeptidase, a Zn^{2+}-containing enzyme, is an integral transmembrane glycoprotein with a carbohydrate-rich hydrophilic portion and active site protruding into the luminal side. It cannot degrade proline-containing dipeptides, which are largely hydrolyzed intracellularly. Dipeptidases and tripeptidases are associated with the brush-border membranes, but their functions are not clearly understood. The major products of intraluminal digestion of protein are mixtures of amino acids (30–40%) and small peptides (60–70%). Chymotryptic activity can be measured by oral administration of N-benzoyl-L-tyrosyl-*p*-amino acid (Figure 12-12). On hydrolysis, *p*-aminobenzoic acid (PABA) and N-benzoyl-L-tyrosine are released; PABA is absorbed, conjugated to glycine by the liver, and excreted in the urine to give an index of exocrine pancreatic function.

Absorption of Amino Acids and Oligopeptides

Dipeptides and tripeptides that escape brush border membrane peptidases are actively transported against a concentration gradient by Na^+-dependent mechanisms. Inside the cell they are hydrolyzed.

Free amino acids are transported into enterocytes by four active, carrier-mediated, Na^+-dependent transport systems remarkably similar to the system for glucose. These systems transport, respectively, neutral amino acids; basic amino acids (Lys, Arg, His) and cystine; aspartic and glutamic acids; and glycine and imino acids. Some amino acids (e.g., glycine) have affinities for more

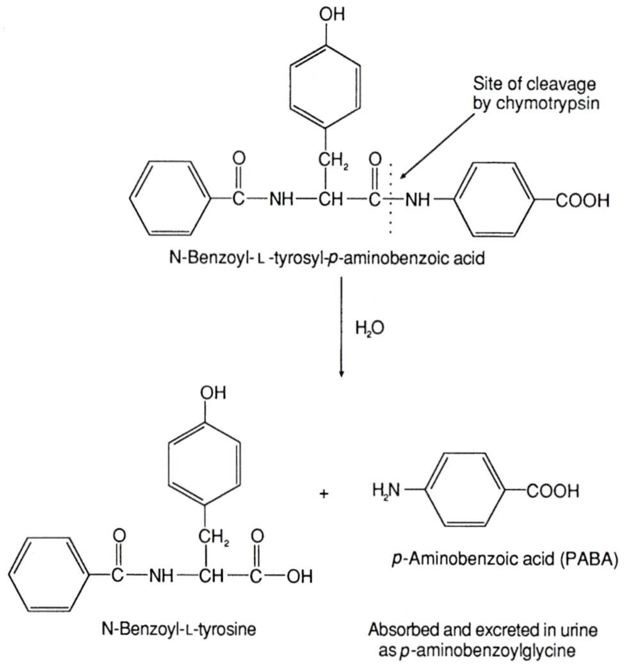

FIGURE 12-12
Hydrolysis of a synthetic tripeptide by chymotrypsin in the small intestine. The hydrolysis yields PABA, which is absorbed and eventually excreted in the urine. The concentration of PABA is a measure of exocrine pancreatic function.

than one transport system; within each group, amino acids compete with each other for transport. These systems are specific for L-Amino acids. D-Amino acids are transported by a passive diffusion process. Amino acid transport in the renal cells may use similar systems (Chapter 39). Entry of amino acids into cell compartments elsewhere in the body may require different transport systems (e.g., the γ-glutamyl cycle, Chapter 17). In the enterocyte, amino acids may be metabolized or transported to the liver. Glutamate, glutamine, aspartate, and asparagine are metabolized in the enterocyte (Chapter 17). Small amounts of protein (e.g., dietary, bacterial, and viral) may be absorbed intact by nonselective pinocytosis. This absorption may be more common in neonatal life. Absorption of food proteins (or antigenic peptides from them) can cause allergic manifestations, whereas bacterial and viral antigens stimulate immunity by production and secretion of secretory IgA (Chapter 35). Since thyrotropin-releasing hormone (pGlu-His-Pro-NH_2) is resistant to hydrolysis, it is effective if taken orally (Chapter 33).

Disorders of Protein Digestion and Absorption

The principal causes of protein maldigestion and malabsorption are diseases of the exocrine pancreas and small intestine. Primary isolated deficiency of pepsinogen or pepsin, affecting protein assimilation, has not been described. Deficiencies of trypsinogen and enteropeptidase are rare. Defects in neutral amino acid transport (*Hartnup disease*), in basic amino acids and cystine (*cystinuria*), dicarboxylic aminoaciduria, and aminoglycinuria have been reported. The clinical severity of these disorders is usually minimal and relates to the loss of amino acids or relative insolubility of certain amino acids in the urine. In cystinuria, for example, cystine can precipitate in acidic urine to form stones. In Hartnup disease, severe nutritional deficiencies are uncommon, since the essential amino acids are absorbed as dipeptides or oligopeptides. Tryptophan, a precursor of nicotinamide (a vitamin; see Chapter 38), and NAD^+ and $NADP^+$ (Chapter 17) are lost in this disease; skin and neuropsychiatric manifestations characteristic of nicotinamide deficiency respond to oral nicotinamide supplementation.

Lipids

Dietary fat provides energy in a highly concentrated form and accounts for 40–45% of the total daily energy intake (100 g/day in the average Western diet). Lipids contain more than twice the energy per unit mass than carbohydrates and proteins (Chapter 5). The efficiency of fat absorption is very high; under normal conditions, almost all ingested fat is absorbed, with less than 5% appearing in the feces. The predominant dietary lipid is triacylglycerol (previously called triglyceride), which contains three long-chain (16-carbon or longer) fatty acids (Chapter 18). The dietary lipids include essential fatty acids (Chapter 18) and the lipid-soluble vitamins A, D, E, and K (Chapters 36–38).

Digestion and absorption of lipids involves the coordinated function of several organs but can be divided into three phases: luminal, intracellular, and secretory.

Intraluminal Phase

Digestion of lipid in the mouth and stomach is minimal. However, lipases secreted by lingual glands at the base of the tongue are active at acid pH and initiate the hydrolysis of triacylglycerol without a requirement for bile acids. The fatty acids released stimulate the release of CCK and a flow of bile and pancreatic juice. The free fatty acids also stabilize the surface of triacylglycerol particles and promote binding of pancreatic colipase. This phase aids in the optimal action of pancreatic lipase and is particularly important in disorders of pancreatic function or secretion (e.g., in prematurity, cystic fibrosis, congenital deficiency of pancreatic lipase).

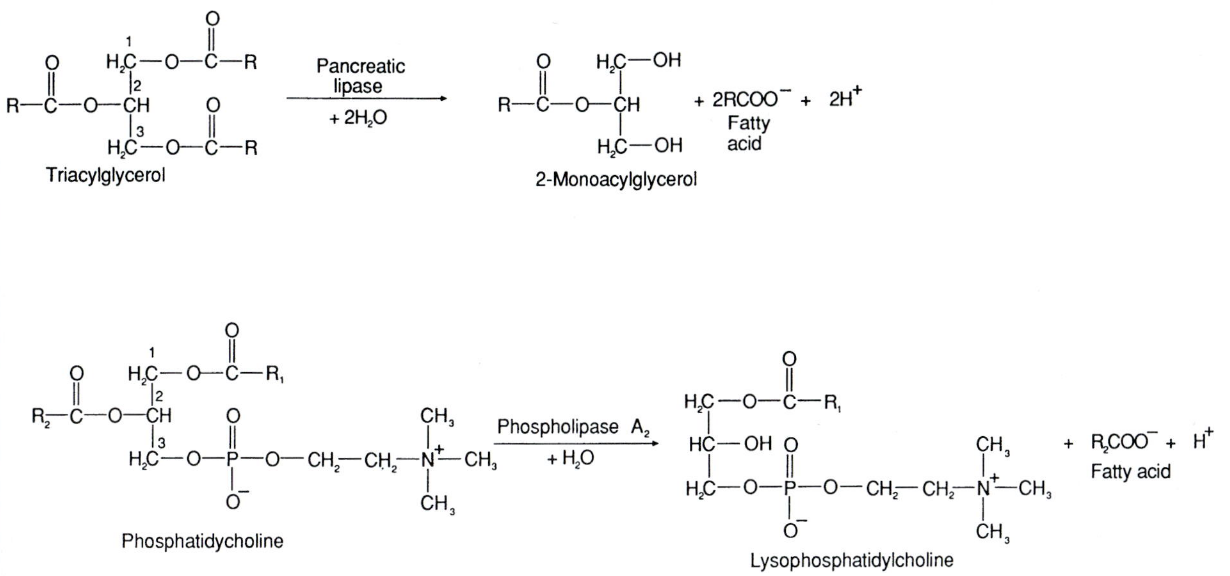

FIGURE 12-13
Hydrolysis of triacylglycerol and phosphatidylcholine by pancreatic lipase and phopholipase A_2, respectively.

The major functions of the stomach in fat digestion are to store a fatty meal, to contribute to emulsification by the shearing actions of the pylorus, and to gradually transfer the partially digested emulsified fat to the duodenum by controlling the rate of delivery. The hydrolysis of triacylglycerol in the duodenum and jejunum requires bile and pancreatic juice. Bile acids are powerful detergents that, together with monoacylglycerol and phosphatidylcholine, promote the emulsification of lipids. Emulsification is also aided by the churning action of the GI tract which greatly increases the area of the lipid-water interface that promotes the action of pancreatic lipase. This triacylglycerol lipase hydrolyzes ester linkages at the 1- and 3-positions to yield 2-monoacylglycerol and fatty acids (Figure 12-13). The products of digestion are relatively insoluble in water but are solubilized in micelles. Micelles also contain lipid-soluble vitamins, cholesterol, and phosphatidylcholine. Pancreatic lipase functions at the lipid-water interface, its activity being facilitated by colipase (M.W. $\sim$ 10,000), also secreted by the pancreas as procolipase activated by tryptic hydrolysis of an Arg-Gly bond in the N-terminal region. Colipase anchors the lipase to the triacylglycerol emulsion in the presence of bile salts by forming a 1:1 complex with lipase and protects lipase against denaturation. Colipase deficiency (with normal lipase) is accompanied by significant lipid malabsorption, as is pancreatic lipase deficiency. Pancreatic juice contains esterases that act on short-chain triacylglycerols and do not require bile salts, as well as cholesteryl esterase. Phosphatidylcholine in the diet (4–8 g/day) and in bile secretions (17–22 g/day) is hydrolyzed to lysophosphatidylcholine and fatty acid by phospholipase A_2, a pancreatic enzyme with an absolute requirement for Ca^{2+} ions and for bile acids. The secreted form, prophospholipase A_2, is activated by tryptic hydrolysis of an Arg-Ala bond in the N-terminal region. Phospholipase A_2 also hydrolyzes fatty acids esterified at the 2-position of phosphatidylethanolamine, phosphatidylglycerol, phosphatidylserine, phosphatidylinositol, and cardiolipin but has no effect on sphingolipids.

Lipid absorption in the duodenum and jejunum appears to be a passive diffusion process. Lipid-laden micelles migrate to the microvilli, and the fatty acids, monoacylglycerols, and lysophosphoglycerols are transferred across the membrane according to their solubility within the lipid bilayer of the cell surfaces. Bile acids are not absorbed into the enterocyte but migrate to the ileum, where they are actively absorbed and transferred to the liver via the portal venous system. The bile acid pool is recycled several times daily (enterohepatic circulation) to meet the demands of lipid digestion, and disorders that interfere with this process lead to malabsorption of lipids. A cytoplasmic fatty acid–binding protein with high affinity for long-chain fatty acids transports fatty acids to the smooth endoplasmic reticulum for resynthesis of triacylglycerol.

Digestion and absorption of triacylglycerols with medium-chain fatty acids ($\leqslant$12 carbons) proceed by a different pathway. Medium-chain triacylglycerols are partly water-soluble, are rapidly hydrolyzed by lingual and pancreatic lipases, and do not require the participation of bile acids. Some are absorbed intact and hydrolyzed inside the absorptive cell. Medium-chain fatty acids enter the portal vein. Thus, medium-chain triacylglycerols can be digested and absorbed in the presence of minimal amounts of pancreatic lipase and in the absence of bile salts. For this reason, they are used to supplement energy intake in patients with malabsorption syndromes. Coconut oil is rich in trioctanoyl-glycerol (8-carbon) and tridecanoylglycerol (10-carbon).

Intracellular (Mucosal) Phase

Fatty acids (long-chain) are activated and monoacylglycerols are converted to triacylglycerols at the smooth endoplasmic reticulum. The steps involved are as follows:

1. Conversion of fatty acids to acyl CoA derivatives by acyl-CoA synthetase.

$$RCOO^- + HSCoA + ATP^{4-} \xrightleftharpoons{Mg^{2+},\ K^+} RCOSCoA + AMP^{2-} + PP_i^{3-}$$

2. Esterification of monoacylglycerol to diacylglycerol and triacylglycerol catalyzed by monoacylglycerol transacylase and diacylglycerol transacylase, respectively.

H_2COH / R_2COOCH / H_2COH (Monoacylglycerol) + RCOSCoA (Acyl-CoA) ⇄ H_2COOCR_1 / R_2COOCH / H_2COH (Diacylglycerol) + CoASH

H_2COOCR_1 / R_2COOCH / H_2COH (Diacylglycerol) + $R_3COSCoA$ (Acyl-CoA) ⇄ H_2COOCR_1 / R_2COOCH / H_2COOCR_3 (Triacylglycerol) + CoASH

In a minor alternative pathway, triacylglycerol is synthesized from glycerol-3-phosphate and acyl-CoA by esterification at the 1,2-positions of glycerol, removal of the phosphate group, and esterification at C_3 (Chapter 19).

The triacylglycerols are incorporated into a heterogeneous population of spherical lipoprotein particles known as ***chylomicrons*** (diameter, 75–600 nm) that contain about 89% triacylglycerol, 8% phospholipid, 2% cholesterol, and 1% protein. Phospholipids of the chylomicron arise by *de novo* synthesis (Chapter 19) or from reacylation of absorbed lysolecithin. Cholesterol is supplied by ***de novo*** synthesis (Chapter 19) or is absorbed. The protein apolipoprotein B-48 (apo B-48) forms a characteristic protein complement of chylomicrons and is synthesized in the enterocyte. Synthesis of apo B-48 is an obligatory step in chylomicron formation. Absence of apo B-48 synthesis, as in the rare hereditary disease *abetalipoproteinemia,* leads to fat malabsorption. Enterocytes are involved in the synthesis of other lipoproteins (Chapter 20).

Secretion

Vesicles that contain chylomicrons synthesized within the endoplasmic reticulum and the saccules of the Golgi apparatus migrate toward the laterobasal membrane, fuse with it, and extrude the chylomicrons into the interstitial fluid, where they enter the lymphatic vessels through fenestrations. Medium-chain triacylglycerols are absorbed and transported by portal blood capillaries without formation of micelles or chylomicrons. Chylomicrons enter the bloodstream at the left subclavian vein via the thoracic duct. In the bloodstream, they are progressively hydrolyzed by endothelial lipoprotein lipase activated by apolipoprotein C-II. Fatty acids so released are taken up by the tissues (e.g., muscle and adipose) as blood passes through them. The chylomicron remnants consist primarily of lipid-soluble vitamins and cholesterol (and its esters) and are metabolized in the liver. Chylomicrons normally begin to appear in the plasma within 1 hour after ingestion of fat and are completely removed within 5–8 hours (see Chapters 18–20).

Disorders of Lipid Digestion and Absorption

Normally more than 95% of ingested lipid is absorbed. When a large fraction is excreted in the feces, it is called ***steatorrhea.*** Measurement of fecal lipid with adequate lipid intake is a sensitive indicator of lipid malabsorption. Malabsorption can result from impairment in lipolysis (Table 12-6), micelle formation (Table 12-7), absorption, chylomicron formation, or transport of chylomicrons via the lymph to blood.

General Malabsorptive Problems

A malabsorptive disorder caused by proteins found in wheat, rye, and barley produces chronic sensitivity that damages the small intestine in susceptible individuals. This disorder is known as ***gluten-sensitive enteropathy*** or ***celiac disease.*** The toxicity of the cereals are associated with a group of proteins known as ***gliadins*** that leads to production of antibodies to endomysium of smooth muscle (Chapter 21). The damage to the small bowel consists of conversion of normal columnar mucosal cells to cuboidal cells, villous flattening, crypt hyperplasia, and infiltration of lymphocytes and plasma cells into the lamina propria. Celiac disease has characteristics of an autoimmune disorder associated with a genetic predisposition. High risk populations for celiac disease include patients with Down's syndrome, insulin-dependent diabetes, those with

TABLE 12-6
Abnormalities of Lipid Digestion Due to Impaired Lipolysis

Type of Defect	Biochemical Disturbance	Examples of Disease States
Rapid gastric emptying	Reduction in the efficiency of lipid interaction with bile and pancreatic secretions	Gastrectomy, as in treatment of ulcer or in neoplasms of stomach
Acidic duodenal pH	Inactivation of pancreatic lipase and decreased ionization of bile acids	Zollinger-Ellison syndrome
Decreased CCK release	Deficiency of bile and pancreatic secretions	Disorders associated with mucosal destruction; regional enteritis, gluten enteropathy
Congenital lipase or colipase deficiency	Defective lipolysis	
Pancreatic insufficiency	Defective lipolysis	Chronic pancreatitis, pancreatic duct obstruction (e.g., cystic fibrosis)
Absent or decreased bile salts	Decreased lipolysis due to impaired micelle formation	(See Table 12–7)

other autoimmune disorders such as lupus and rheumatoid arthritis, and relatives of patients with celiac disease.

The diagnosis of celiac disease includes characteristic intestinal biopsy findings (discussed above) and serological testing for antigliadin and antiendomysial antibodies. The treatment of patients with celiac disease consists of lifelong complete abstinence of gluten-containing foods, acceptable foods are rice, potato, and maize.

Cystic fibrosis (CF) is a disease of multiple exocrinopathy and generalized malabsorption due to lack of delivery of pancreatic digestive enzymes to the small intestine. CF is an autosomal recessive disease and is due to lethal homozygous mutations. Among Caucasians the incidence is 1 in 2000–3000 births. The prevalence of heterozygous individuals is about 1 in 20. Abnormalities in affected persons are found in airways, lungs, pancreas, liver, intestine, vas deferens, and sweat glands. The primary abnormality in CF is in electrolyte transport, specifically in Cl^- secretion from the apical membrane of the epithelial cells of sweat glands, airways, pancreas, and intestine. Defective Cl^- secretion causes hyperactivity of Na^+ absorption and these two processes cause the secreted mucus to become viscous and sticky. The most life-threatening clinical feature is related to pulmonary disease. The sticky, viscous mucus clogs the airway and compromises the normal beating of the cilia that cover the apical surface of the airway epithelium. These conditions encourage lung infections by bacteria such as *Pseudomonas aeruginosa* and *Staphylococcus aureus*.

The clinical abnormalities related to the gastrointestinal tract are not life-threatening and can be treated. In newborns with CF, intestinal obstruction (meconium ileus) can occur in 10–20% of cases due to failure of digestion of intraluminal contents due to lack of pancreatic enzymes *in utero*. Exocrine pancreatic enzyme deficiency is present from birth affecting both lipid and protein digestion. In general, carbohydrate digestion is not severely impaired.

CF is associated with a plethora of clinical conditions such as diseases of the heptobiliary tract and genitourinary tract (e.g., male infertility). Elevation of chloride concentration in the sweat is the most consistent functional

TABLE 12-7
Abnormalities of Lipid Digestion Due to Impaired Micelle Formation Leading to Decreased Lipolysis

Type of Defect	Examples of Disease States
Decreased hepatic synthesis of bile salts	Severe parenchymal liver disease
Decreased delivery of bile salts to the intestinal lumen	Biliary obstruction due to stone, tumor, or primary biliary cirrhosis
Decreased effective concentration of conjugated bile acids	Zollinger-Ellison syndrome (causes hyperacidity); bacterial overgrowth and stasis; administration of drugs, neomycin, and cholestyramine
Increased intestinal loss of bile salts	Ileal disease or resection

abnormality in CF and the determination of the sweat chloride concentration is used as the standard diagnostic test. The cause of elevated Cl^- concentration in the sweat is due to failure of reabsorption in the reabsorptive portion of the sweat gland. Normally sweat is a hypotonic fluid as it emerges at the surface of the skin. This occurs because the secretory and absorptive activities of the sweat gland are located in two different regions.

The CF gene encodes a protein designated as the cystic fibrosis transmembrane conductance regulator (CFTR) and is on chromosome 7q31.2. CFTR is a glycosylated protein containing 1480 amino acid residues. It is a 170- to 180-kDa protein and the variations in molecular weight are due to differences in glycosylation. CFTR belongs to a family of channel proteins known as ATP-binding cassette (ABC) transporters which are essential to virtually all cells. Abnormalities in channel proteins have both inherited and noninherited causes, and associated disorders have been called ***channelopathies.*** Other ABC transport defects are multidrug resistance (mdr) transporter (known as P-glycoprotein) and sulfonylurea receptors (SUR1 and SUR2). P-Glycoprotein gene is up-reregulated in its expression in response to certain chemotherapeutic drugs (e.g., vinca alkaloids). This causes the cells to become multidrug resistant, because the drugs are exported out of the target cells in an ATP-dependent process. The normal physiological role of P-glycoprotein may reside in its participation in the transport of phosphatidylcholine and other phospholipids (Chapter 19). It is also thought that the P-glycoprotein functions in the detoxification process of pumping toxins that are xenobiotic out of cells. The sulfonylurea-receptor transporter is an ATP-sensitive K^+ channel found in the β cells of the pancreatic islets of Langerhans and in cardiac and skeletal muscle. Regulation of insulin secretion from β cells is governed by sulfonylurea receptor proteins (Chapter 22). TAP1 and TAP2, subunits of the major histocompatibility complex, are ABC transporters involved in peptide transport into the endoplasmic reticulum of antigen-presenting cells. Recurrent respiratory infections and bronchiectasis occur in patients with TAP2 defects. ABC transporters are also involved in many other biochemical processes. These include peroxisome biogenesis (*X-linked adrenoleukodystrophy, Zelweger's syndrome,* Chapter 18) cholesterol efflux, *Tangier disease,* Chapter 20), and all-*trans* retinaldehyde efflux from the inside of the disk to the cytoplasm in retinal rod cells (*early-onset macular degeneration,* Chapter 28).

The CFTR protein consists of five domains: two membrane spanning domains (MSDs), two nucleotide-binding domains (NBDs), and a regulatory domain (Figure 12-14). Both the amino terminus and carboxy terminus are located in the cytoplasm and each of the two membrane spanning domains contains six transmembrane

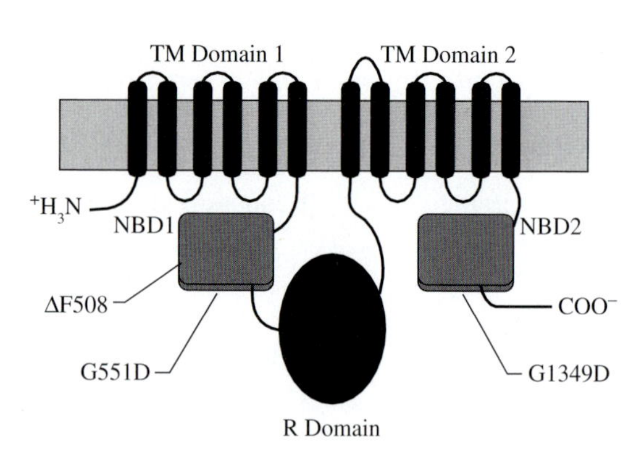

FIGURE 12-14

Diagrammatic representation of cystic fibrosis transmembrane regulatory protein (CFTR). CFTR consists of two transmembrane (TM) domains, two nucleotide binding domains (NBD) and one regulatory R-domain. The opening of the chloride channel requires ATP binding to at least one of the NBDs and phosphorylation of R-domain by cAMP-dependant protein kinase. There are more than 700 mutations and polymorphisms of CFTR. NBDs are hot spots for mutation, and three severe CF causing mutations ΔF508, G551D, and G1349D are shown. The ΔF508 mutation is the most common mutation, occurring in about 70% of CF patients.

segments. The two nucleotide binding domains interact with ATP. The unique regulatory domain contains several consensus phosphorylation sites. The phosphorylation of the R-protein occurs both by cAMP dependent protein kinase and protein kinase C. CFTR protein is located in the apical cell membrane of most epithelial cells; however, it is also found in the basolateral cell membranes and presumably participates in Cl^- reabsorption. The basolateral cell membrane sites are located in sweat duct epithelial cells and the proximal convoluted tubule and the thick ascending limbs of Henle's loop in the kidney.

The understanding of CFTR function in chloride transport has been aided by insertion of CFTR into artificial lipid membranes and by recording currents flowing through single membrane channels using patch-clamp technology. Activation of the channel for Cl^- secretion requires phosphorylation of sites located in the R-domain and binding and/or hydrolysis of ATP at the nucleotide binding domain. Phosphorylation by protein kinases and dephosphorylation by phosphatases at the R-subunit controls Cl^- secretion. Thus, if CFTR is defective, Cl^- secretion does not occur. If CFTR is kept continually active by increasing intracellular cAMP levels, the Cl^- transport is enormously increased with accompanying fluid loss (Figure 12-15).

More than 700 mutations located throughout the gene have been identified in the CFTR gene in CF patients. The mutations include missense, nonsense, and frameshift mutations. In terms of functional defects, CFTR mutations have been grouped into four categories:

1. Protein production,
2. Protein processing,

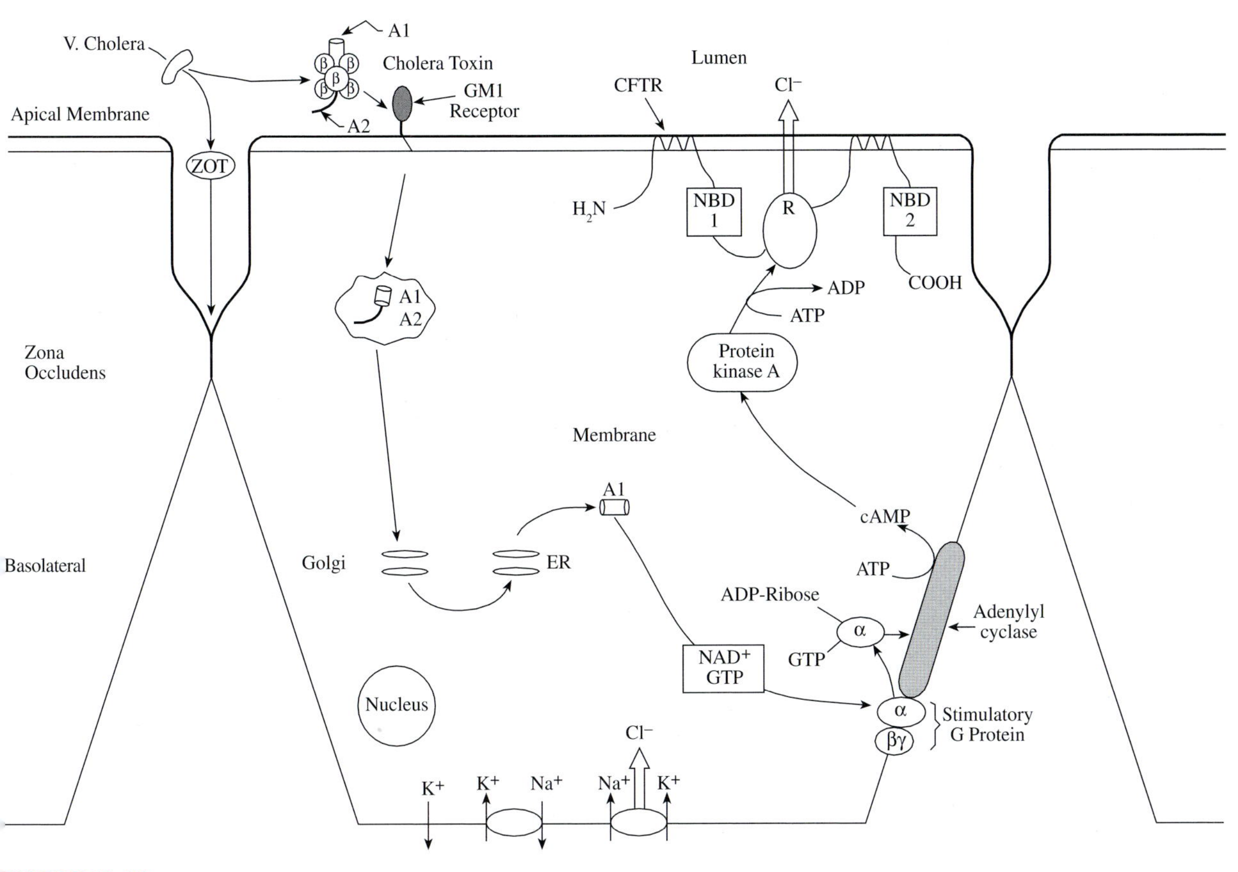

FIGURE 12-15
Mechanism of action of cholera toxin. The steps consist of cholera toxin (CT) binding to GM_1 receptors anchored by its ceramide moiety; internalization of CT in endosomes; the release of A1-subunit of CT from the trans-Golgi network and endoplasmic reticulum (ER); ADP ribosylation of the α-subunit of stimulatory G-protein by A1; activation of adenylyl cyclase and production of cAMP; activation of protein kinase A; phosphorylation of the regulatory (R) subunit of CFTR; and finally opening of the chloride channel. *Vibrio cholerae* also produces a zona occludens toxin (ZOT) which increases the ionic permeability of the zona occludens.

3. Regulation of the channel, and
4. Conduction.

The most common mutation causing CF is a 3-base pair deletion resulting in loss of a phenylalanine (F) residue at position 508. This deletion (ΔF508), is found in about 70% of CF patients worldwide and is associated with severe disease. The ΔF508 mutation belongs to the category 2 defect and involves the folding of the first nucleotide binding domain. The phenylalanine-deficient CFTR protein gets trapped in the endoplasmic reticulum and is degraded. The G1349D mutation occurs in the second nucleotide binding domain. In general, genotype-phenotype correlations with clinical manifestations are difficult to interpret in CF patients.

As discussed earlier, the Cl^- transport defect occurs in association with hyperabsorption Na^+ ions. The mechanism of regulatory interaction between CFTR and the Na^+ channel is not clear. However, two possible mechanisms may play a role:

1. A direct protein-protein interaction or,
2. A physical coupling of channel proteins mediated by the cytoskeleton.

It is thought that CFTR participates in the regulation of Cl^- transport within intracellular membranes as being a regulator of other ion conductances.

The diagnosis of CF includes the clinical features of chronic obstructive lung disease, persistent pulmonary infection, meconium ileus, and pancreatic insufficiency with failure-to-thrive syndrome. Family history is also very helpful. The diagnosis can be confirmed by a positive sweat chloride test in which a concentration of chloride ions is greater than 60 meq/l. The genotype is confirmed by DNA analysis, and carrier detection in CF families is useful in genetic counseling. Newborn screening, although not universally accepted, consists of quantitation immunoreactive trypsin in association with mutational analysis using dried blood. Treatment of CF patients involves a comprehensive approach. Use of antibiotics and pancreatic enzyme replacement therapy has been helpful in treating pulmonary infection and the maldigestion of food substances, respectively. Recombinant DNase (pulmozyme) and N-acetylcysteine have been used in clearing the airways in patients. A potentially useful agent in clearing airways is *recombinant gelsolin* which degrades actin present in cell lysates and decreases viscosity. Aerosolized drugs such as amiloride, an inhibitor of Na^+ channel, and nonhydrolyzable forms of UTP (UTPϒS), an activator of non-CFTR chloride channels, are being tested clinically.

Other experimental molecular strategies for managing CF include manipulation of endogenous chaperones that interact with CFTR protein, enhancement of gene transcription by phenylbutyrate and other analogs (used in Sickle cell anemia, see Chapter 28) and agents that read-through stop codons in the CFTR messenger RNA (e.g., aminoglycosides), protein replacement therapy and gene therapy. Two major vehicles to introduce normal CF genes *in vivo* are cationic liposomes and replication-deficient adenoviruses.

Chronic alcoholism is frequently associated with generalized malabsorption of major foods and vitamins because of liver and pancreatic involvement and mucosal dysfunction.

12.4 Absorption of Water and Electrolytes

Fluid turnover of about 9 L occurs daily in the GI tract. Ingested water contributes about 2 L, and the remainder arises from secretions of the GI tract mucosa and associated glands. Nearly all of this water is reabsorbed, and about 200 mL (2%) or less is excreted (Table 12-8). If the amount of water excreted in feces exceeds 500 mL, diarrhea results. Similarly, only 2% of Na^+ and 10% of K^+ in gastrointestinal fluids appear in the feces. Most absorption of fluid and electrolytes occurs in the small intestine.

The gastric mucosa is relatively impermeable to water, but the small intestine is highly permeable, and water transport occurs in both directions depending on osmotic gradients. The osmolality of the duodenal contents can be low or high, depending on the food ingested. However, in the jejunum, the luminal fluid becomes isotonic and remains so throughout the rest of the small intestine. Water absorption is a passive process and is coupled to the transport of organic solutes and electrolytes. The simultaneous presence of glucose and Na^+ facilitates the absorption of all three. In the treatment of diarrhea, oral administration

TABLE 12-8
Daily Turnover of Water in the GI Tract

Source of Fluids	Quantity (L/d)	Composition
Input		
Diet	2.0	Variable
Saliva	1.5	Hypotonic, alkalin
Gastric juice	2.5	Isotonic, acid
Bile	0.5	Isotonic, alkaline
Pancreatic juice	1.5	Isotonic, alkaline
Intestinal juice	1.0	Isotonic, neutral
Total	9.0	
Reabsorption		
Jejunum	5.5	
Ileum	2.0	
Colon	1.3	
Total	8.8	
Lost in the stool (9.0–8.8) =	0.2	

of solutions containing glucose and NaCl replaces fluid and electrolyte.

Na^+ is also absorbed separately from organic solute through two coupled transport systems, of which one absorbs Na^+ in exchange for H^+, while the other absorbs Cl^- in exchange for HCO_3^-. In the lumen, H^+ and HCO_3^- combine to give rise to CO_2 and H_2O, which enter the mucosa cell or pass through to the plasma. The Na^+ absorbed i pumped out by Na^+, K^+-ATPase in the basolateral membrane. Cl^- follows Na^+ passively, with transfer of NaC to plasma.

In the colon Na^+, Cl^-, and water are efficiently absorbed. Na^+ absorption is regulated by aldosterone (Chapter 32). K^+ is secreted into the lumen as a component of mucus but is reabsorbed by passive diffusion The amount of K^+ in the feces is usually far below th daily intake; however, in chronic diarrhea the loss of ilea and colonic fluids can cause negative K^+ balance.

Disorders of Fluid and Electrolyte Absorption

If the intestinal contents become hyperosmolar, water enters the lumen to produce iso-osmolarity, and fluid and electrolyte loss occur. This condition is seen in lactose intolerance, ingestion of nonabsorbable laxatives such a magnesium salts, and ingestion of sugars such as lactulose. Bile acids inhibit the absorption of Na^+ in the colon Under normal circumstances, bile acids do not reach th colon because they are actively reabsorbed in the ileum. When the ileum is diseased or resected, sufficient bile salts enter the colon to inhibit absorption of Na^+ and water and cause diarrhea. Hydroxylated fatty acids (e.g., ricinoleic acid, the active ingredient of castor oil) also inhibit salt and water absorption. In disorders with significant mucosal abnormalities (e.g., gluten-sensitive enteropathy), fluid and electrolyte absorption is impaired.

Loss of fluids and electrolytes in ***cholera*** results from stimulation of a secretory process. The toxin secreted by *vibrio cholerae* causes a diarrhea of up to 20 L/day, resulting in dehydration and electrolyte imbalance, which may lead to death. Bacteria in contaminated food attach chiefly to the ileal mucosa and secrete enterotoxin consisting of one fraction that binds to specific sites on the cell membrane and another responsible for the characteristic biochemical activity. The binding moiety consists of five identical polypeptide subunits (B) (M.W. 11,300) that surround the active moiety (A). The A-subunit consists of two unequally sized polypeptides, A_1 (M.W. 23,500) and A_2 (M.W. 5,500), linked by a disulfide bridge. The A_2 polypeptide appears to connect the A_1 polypeptide to the B-subunit. The B-subunit binds rapidly to monosialogangliosides in the membrane (G_{M1}, Chapters 10 and 19). The A_1 polypeptide then migrates through the membrane and catalyzes transfer of the ADP-ribose group from NAD^+ to the stimulatory guanine nucleotide-binding protein (G_s) that regulates adenylate cyclase activity. The adenylate cyclase, which catalyzes the conversion of ATP to cAMP (Chapter 30) is stimulated or inhibited by the active forms of the guanine nucleotide-binding protein, G_s and G_i, respectively. The G_s protein is activated by the binding of GTP and inactivated when GTP is converted to GDP by a GTPase intrinsic to G_s protein. G proteins contain α-, β-, and γ-subunits. The ADP ribosylation of the α-subunit decreases GTP hydrolysis and thus leads to sustained activation of adenylate cyclase activity, increased intracellular levels of cAMP, and secretion of isotonic fluid throughout the entire small intestine. Cholera toxin does not cause fluid secretion in the stomach, has minimal effects on the colon, and does not affect Na^+-dependent absorption of glucose and amino acids. Its effects can be readily reversed by oral or intravenous administration of replacement fluids.

The mechanism by which cholera toxin causes secretory diarrhea is through continuous stimulation of the CFTR-regulated Cl^- channel (Figure 12-15). In cystic fibrosis, CFTR defects cause the abolition of intestinal chloride secretion without affecting the absorptive capacity. In homozygous CF patients, the disease is eventually lethal (discussed earlier). In cholera infections, however, CFTR is overactivated with fluid and electrolyte losses that lead to intravascular volume depletion, severe metabolic acidosis, and profound hypokalemia. These metabolic changes can result in cardiac and renal failure with fatal consequences. It has been postulated that persons who are heterozygous for a CF mutation may have selective advantage during cholera epidemics. This speculation has been tested in a mouse model; indeed, homozygous CF mice treated with cholera toxin did not show intestinal secretion of fluids despite an increase in intracellular cAMP levels. In heterozygous CF mice the intestinal secretion is intermediate compared to that of controls.

A human pathogen known as *Escherichia coli* O157:H7 may cause nonbloody diarrhea, *hemorrhagic colitis, hemolytic uremic syndrome,* and death. The interval between exposure and illness averages only 3 days. The designation of O157:H7 derives from the fact that the bacterium expresses the 157th somatic (O) antigen and the 7th flagellar (H) antigen. The pathogen is transmitted by contaminated food (e.g., ground beef, fruit and vegetables, and water) from one person to another and occasionally through occupational exposure. The pathogenicity of *E. coli* O157:H7 is due to its ability to produce a molecule composed of an enzymatic subunit (A_1) and a multimer of five receptor binding (B) subunits. The genome for the synthesis of the toxin resides on a bacteriophage inserted into *E. coli* O157 DNA. The A_1-subunit is linked to a carboxy terminal A_2 fragment by a single disulfide bond.

Shiga toxin produced by *Shigella dysenteriae* has similar structural features. The toxin binds to a glycolipid (Gb3), undergoes endocytosis, and the enzymatic A_1 fragment, which is a specific N-glycosidase, removes adenine from one particular adenosine residue in the 28S RNA of the 60S ribosomal subunit. Removal of the adenine inactivates the 60S ribosome, blocking protein synthesis. *Ricin*, *abrin*, and a number of related plant proteins inhibit eukaryotic protein synthesis in a similar manner (Chapter 25).

Several *E. coli* strains also elaborate heat-labile enterotoxins that cause diarrheal disease ("traveller's" diarrhea) by similar mechanisms. In *V. cholerae*, the same enterotoxin is produced by all pathogenic strains and is chromosomally determined, whereas in *E. coli*, different enterotoxins are produced and the toxin genes are carried on plasmids.

The actions of diphtheria and pertussis toxins are also mediated by ADP-ribosylation. Diphtheria toxin inhibits eukaryotic protein synthesis by ADP ribosylation of elongation factor II (Chapter 23). Pertussis toxin inactivates G_i by ADP ribosylation of its A-subunit and causes an increase in cAMP production. Unlike cholera toxin, pertussis and diphtheria toxins gain access to many tissues to produce diverse biological effects. Severe watery diarrhea occurs in a pancreatic islet-cell tumor (pancreatic cholera) that produces large quantities of VIP (Table 12-1). VIP activates adenylate cyclase by a different mechanism from that of cholera toxin.

12.5 Thermic Effect of Food

Energy balance depends on energy intake, energy expenditure, and existing energy stores (Chapter 5). Energy is expended in digestion, absorption, transport, metabolism, and storage of food. The level varies with type of food ingested and its metabolic rate. Energy costs for the processing of lipids, carbohydrates, and proteins are 4%, 5%, and 30% of their energy content, respectively. Lipogenesis from carbohydrate also has a high metabolic cost. Part of this energy appears as heat energy and is variously referred to as the *thermic effect of food, diet-induced thermogenesis,* or *specific dynamic action* of food. The magnitude of this thermic effect depends upon the food, nutritional state, and antecedent diet. The thermic effect of food accounts for about 10% of the daily energy expenditure and exhibits interindividual variation. Activation of the sympathetic nervous system and secretion of thyroid hormone (Chapter 33) contribute significantly to diet-induced thermogenesis. Some forms of obesity may result from decreased thermic effect of food.

In laboratory rodents, diet-induced thermogenesis by sympathetic nervous stimulation in brown adipose tissue plays a significant role in energy expenditures. In most mammals, brown adipose tissue is present to a much smaller extent and is located in the interscapular, subscapular, and axillary regions, the nape of the neck, along the length of the great vessels in the thorax and abdomen, and in small patches between the ribs. Hibernating animals and those adapted to living in a cold environment have an abundant supply of brown adipose tissue. Brown adipose tissue cells are metabolically more active than adipocytes and contain a larger supply of mitochondria and usually several small lipid droplets.

Norepinephrine secreted at sympathetic nerve endings binds to β-adrenergic receptors on brown adipose tissue cells and initiates cAMP-dependent triacylglycerol lipase activity (Chapter 22). Accelerated lipolysis of triacylglycerol and oxidation in mitochondria of fatty acids ensue. These mitochondria have a specific regulator mechanism for proton conductance. Fatty acids provide substrates and promote proton conductance, so that a proton gradient generated by electron transport and required to drive ATP synthesis is not established (Chapter 14). Uncoupling of mitochondria leads to increased heat production. In laboratory-inbred obese mice, obesity is associated with defective thermogenesis in brown adipose tissue and decreased energy expenditure. No direct evidence links obesity in humans with defective thermogenesis in brown adipose tissue.

Supplemental Readings and References

Q. Aziz and D. G. Thompson: Brain-gut axis in health and disease. *Gastroenterology* **114,** 559 (1998).

M. J. Blaser: *Helicobacter pylori* and gastric diseases. *British Medical Journal* **316,** 1507 (1998).

D. Branski and R. Troncone: Celiac disease: A reappraisal. *Journal of Pediatrics* **133,** 181 (1998).

Gastrointestinal hormones in medicine. *Endocrinology and Metabolism Clinics of North America* **22**(4), 709 (1993). This reference includes many articles on gastrointestinal hormones.

First Multi-Disciplinary International Symposium on Supraesophageal Complications of Reflux Disease. R. Shaker, Guest Editor. *American Journal of Medicine* **103(5A),** 1997 (The whole issue deals with gastrointestinal esophageal reflux disease).

L. B. Lovat; Age-related changes in gut physiology and nutritional status. *Gut* **38,** 306 (1996).

L. D. McBean and G. D. Miller: Allaying fears and fallacies about lactose intolerance. *Journal of American Dietetic Association* **98,** 671 (1998).

P. S. Mead and P. M. Griffin: Escherichia coli 0157:H7. *Lancet* **352,** 1207 (1998).

A. D. O'Brien, V. L. Tesh, A. Donohue-Rolf, et al.: Shiga toxin: Biochemistry, genetics, mode of action, and role in pathogenesis. *Current Topics in Microbiology and Immunology* **180,** 65 (1992).

R. M. Peek and M. J. Blaser: Pathophysiology of *Helicobacter pylori*-induced gastritis and peptic ulcer disease. *American Journal of Medicine* **102,** 200 (1997).

J. P. Raufman: Cholera. *American Journal of Medicine* **104,** 386 (1997)

S. M. Schweibert, D. J. Benos, and C. M. Fuller: Cystic Fibrosis: A multiple exocrinopathy caused by dysfunctions in a multifunctional transport protein. *American Journal of Medicine* **104,** 576 (1998).

R. C. Stern: The diagnosis of cystic fibrosis. *New England Journal of Medicine* **336,** 487 (1997).

CHAPTER 13

Carbohydrate Metabolism I: Glycolysis and TCA Cycle

Carbohydrates are metabolized by several metabolic pathways, each with different functions. Although these pathways usually begin with glucose, other sugars may enter a pathway via appropriate intermediates. Glucose can be stored as glycogen (glycogenesis), which, in turn, can be broken down to glucose (glycogenolysis), synthesized from noncarbohydrate sources (gluconeogenesis), converted to nonessential amino acids, used in the formation of other carbohydrates or their derivatives (e.g., pentoses, hexoses, uronic acids) and other noncarbohydrate metabolites, converted to fatty acids (the reverse process does not occur in humans) and stored as triacylglycerols, used in the biosynthesis of glycoconjugates (e.g., glycoproteins, glycolipids, proteoglycans), or catabolized to provide energy for cellular function (glycolysis, tricarboxylic acid cycle, and electron transport and oxidative phosphorylation).

13.1 Glycolysis

Glycolysis is common to most organisms and in humans occurs in virtually all tissues. Ten reactions culminate in formation of two pyruvate molecules from each glucose molecule. All 10 reactions occur in the cytoplasm and are anaerobic. In cells that lack mitochondria (e.g., erythrocytes) and in cells that contain mitochondria but under limiting conditions of oxygen (e.g., skeletal muscle during heavy exercise), the end product is lactate. Under aerobic conditions in cells that contain mitochondria (i.e., most cells of the body), pyruvate enters the mitochondria, where it is oxidized to acetyl-coezyme A (acetyl-CoA), which is then oxidized through the tricarboxylic acid (TCA) cycle. Thus, the pathway is the same in the presence or absence of oxygen, except for the end product formed.

Source and Entry of Glucose into Cells

Glucose can be derived from exogenous sources by assimilation and transport of dietary glucose (Chapter 12) or from endogenous sources by glycogenolysis or gluconeogenesis (Chapter 13). The blood circulation transports glucose between tissues (e.g., from intestine to liver, from liver to muscle). Control and integration of this transport are discussed in Chapter 22. Glucose transport across cell membranes can occur by carrier-mediated active transport or by a concentration gradient-dependent facilitated transport that requires a specific carrier. The latter type can be either insulin-independent or insulin-dependent. The active transport system is Na^+-dependent and occurs in intestinal epithelial cells (Chapter 12) and epithelial cells of the renal tubule (Chapter 39).

The properties of glucose transporter proteins (GLUT) consist of tissue specificity and differences in functional properties reflected in their glucose metabolism. Five transporter proteins, Glut 1–5, have been identified

TABLE 13-1
Properties of Human Glucose Transporters (GLUT)

Transporters	Major Tissue Distribution	Properties
GLUT 1	Brain, microvessels, red blood cells, placenta, kidney, and many other cells	Low K_m (about 1mM), ubiquitous basal transporter
GLUT 2	Liver, pancreatic β–cell, small intestine	High K_m (15–20mM)
GLUT 3	Brain, placenta, fetal muscle	Low K_m, provide glucose for tissue cells metabolically dependent on glucose
GLUT 4	Skeletal and heart muscle, fat tissue (adipocytes)	K_m (5mM), insulin responsive transporter
GLUT 5	Small intestine, testes	Exhibits high affinity for fructose

(Table 13-1). The tranporters combine with glucose and facilitate its transport across the intervening membrane for entry into the cells. The GLUT has features characteristic of Michaelis–Menten kinetics, namely, bidirectionality and competitive inhibition. GLUT proteins belong to a family of homologous proteins coded by multiple genes. They vary from 492–524 amino acids and share 39–65% identity among primary sequences. All GLUT proteins are single-polypeptide chains and contain 12 transmembrane α-helical domains with both amino and carboxytermini extending into the cytoplasm.

Insulin-stimulated glucose uptake in muscle and adipose tissue cells is mediated by GLUT4. Insulin's role in recruiting GLUT4 proteins from intracellular vessels to the plasma membrane consists of trafficking through multiple intracellular membrane compartments (Chapter 22). Defects in GLUT4 can result in insulin resistance. GLUT5 is located both at the luminal and basolateral sides of the intestinal epithelial cells. At the luminal side, it functions in tandem with Na^+ glucose symporter and at the basolateral site it is involved in the transport of glucose from the absorptive epithelial cells into portal blood circulation. ***GLUT2*** is located in the liver and pancreatic β-cell membranes. It has a high K_m for glucose and, therefore, the entry of glucose is proportional to blood glucose levels. In the liver, glucose can be stored as glycogen or converted to lipids when the plasma glucose levels are high (***hyperglycemia***), and during low levels of plasma glucose (***hypoglycemia***) the liver becomes a provider of glucose to extrahepatic tissues by glycogenolysis, gluconeogenesis, or both (Chapter 15). In pancreatic β cells, plasma membrane GLUT2 participates in insulin secretion (Chapter 22). GLUT1 and 3 are present in many cell membranes and are basal transporters of glucose at a constant rate into tissues which are metabolically dependent on glucose (e.g., brain and red blood cells). These transporters have lower K_m for glucose than GLUT2 and, therefore, transport glucose preferentially.

The importance or GLUT1 in brain metabolism is illustrated in a report of two infants with a syndrome of poorly controlled seizures and delayed development and who have a genetic defect in GLUT1 protein. Glucose is an essential fuel for the brain and is transported by GLUT1 across the plasma membranes of the brain endothelial cells of the blood-brain barrier system. In these patients despite normal blood glucose levels, low levels of glucose in cerebrospinal fluid (CSF) (***hypoglycorrhachia***) as well as low levels of CSF lactate were observed. Since GLUT1 is present in both red blood cells and brain endothelial cells, the more accessible GLUT1 in red blood cells is used in clinical studies of disorders of brain glucose transport. Two patients with a primary defect of glucose transport into the brain were treated with a ketogenic diet. The metabolism of ketogenic substrates does not depend on the glucose transporter and thus can provide a large fraction of the brain's energy requirement.

Reactions of Glycolysis

The overall pathway is shown in Figure 13-1, and some properties of these reactions and the enzymes involved are listed in Table 13-2. Glycolytic enzymes can be classified into six groups according to the type of reaction catalyzed: kinase, mutase, dehydrogenase, cleaving enzyme, isomerase, and enolase.

Phosphorylation of Glucose

Glucose is phosphorylated by hexokinase (in extrahepatic tissues) or glucokinase (in the liver).

$$\alpha\text{-D-Glucose} + ATP^{4-} \xrightarrow{Mg^{2+}} \alpha\text{-D-glucose} - 6\text{-phosphate}^{2-} + ADP^{3-} + H^+$$

The Mg^{2+} is complexed with ATP^{4-} and is present as $MgATP^{2-}$. This reaction, essentially an irreversible

IGURE 13-1
athway of glycolysis.

nonequilibrium) reaction, is accompanied by a substantial oss of free energy as heat. This phosphorylation initiates lycolysis and leads to intracellular trapping of intermeiates (because the plasma membrane is not permeable to hosphate esters).

Four isoenzymes of hexokinases (types I–IV) contitute a family of enzymes that probably arose from common ancestral gene by gene duplication and fuion events. Hexokinases I–III (M.W. ~100,000) have K_m of about 0.1 mM and are allosterically inhibited y glucose-6-phosphate. Due to their low K_m, hexokases I–III are saturated with the substrate glucose at concentrations found in plasma. Thus, the overall control of glucose-6-phosphate formation resides in the rate of glucose transport across the plasma membrane. Hexokinase I is expressed in many tissues and is considered a "housekeeping" enzyme. Hexokinase II is found primarily in insulin-sensitive tissues, namely, heart, skeletal muscle, and adipose tissue. In these tissues, the glucose transporter GLUT4 and hexokinase II function in concert in glucose utilization. Abnormalities in insulin, GLUT4, or hexokinase II production may be associated with disorders of insulin resistance, obesity, and diabetes mellitus (Chapter 22).

TABLE 13-2
Properties of Enzymes of Glycolysis

Enzyme	Coenzymes and Cofactors	Allosteric Modulators Positive	Allosteric Modulators Negative	Equilibrium Constant at pH 7.0 (K'_{eq})	$\Delta G^{\circ\prime}$ kcal/mol (kJ/mole)
Hexokinase	Mg^{2+}	ATP, P_i	Glucose 6-phosphate	650	−4.0 (−16.7) (nonequilibrium)*
Glucokinase	Mg^{2+}	—	—	—	(nonequilibrium)
Glucose-phosphate isomerase	Mg^{2+}	—	—	0.5	+0.4 (+1.7) (near-equilibrium)†
6-Phosphofructokinase	Mg^{2+}	Fructose 2,6-bisphosphate, ADP, AMP, P_i, K^+, NH_4^+	ATP, citrate	220	−3.4 (−14.2) (nonequilibrium)*
Fructose-bisphosphate aldolase		—	—	0.001	+5.7 (+23.8) (near-equilibrium)‡
Triose-phosphate isomerase	Mg^{2+}	—	—	0.075	+1.8 (+7.5) (near-equilibrium)
Glyceraldehyde-3-phosphate dehydrogenase	NAD	—	—	0.08	+1.5 (+6.3) (near-equilibrium)
Phosphoglycerate kinase	Mg^{2+}	—	—	1,500	−4.5 (−18.8) (near-equilibrium)
Phosphoglyceromutase	Mg^{2+}, 2,3-bisphosphoglycerate	—	—	0.02	+1.1 (+4.6) (near-equilibrium)
Enolase	Mg^{2+}	—	—	0.5	+0.4 (+1.7) (near-equilibrium)
Pyruvate kinase§	Mg^{2+}, K^+	Fructose 1,6 bisphosphate	ATP, alanine, acetyl-CoA	200,000	−7.5 (−31.4) (nonequilibrium)*
Lactate dehydrogenase	NAD	—	—	16,000	−6.0 (−25.1) (near-equilibrium)

*Physiologically irreversible reactions.
†Physiologically reversible reactions.
‡This reaction, despite a high positive $\Delta G^{0\prime}$ value, is reversible under *in vivo* conditions.
§Pyruvate kinase is also regulated by cAMP-dependent phosphorylation. The dephosphorylated form is more active, and the phosphorylated form is less active.

Hexokinase IV, known as *glucokinase,* differs functionally from hexokinases I–III. Glucokinase, a monomeric protein (M.W. 50,000), has a higher K_m for glucose (5 mM) compared to hexokinases I–III, is found only in hepatocytes (liver parenchymal cells) and pancreatic (islet) β cells, and is not inhibited by glucose-6-phosphate. The function of glucokinase in hepatocytes and β cells are different and this difference is consistent with their metabolic function. During postprandial and hyperglycemic periods, hepatocyte glucose uptake is increased due to increased levels of glucokinase. Thus, the hepatocyte glucokinase initiates the metabolism of glucose and maintains a gradient of inward flow with glucose being converted to other metabolic products. During periods of hypoglycemia (e.g., starvation), the hepatocyte glucokinase levels are reduced and the liver becomes provider of glucose.

In β cells, glucokinase functions as a glucose sensc and modulates insulin secretion. As the plasma levels c glucose increase with the resultant increase in cytosol glucose concentration in β cells, glucokinase determine the rate of glucose phosphorylation, glycolytic flux, an hence, the ATP/ADP ratio. As ATP levels increase, it binc to the ATP-binding K^+ cassette transporter located in tl plasma membrane (Chapter 12) and blocks the K^+ efflu from the cell. This ATP-dependent K^+ channel is know as the ***sulfonylurea receptor*** because sulfonylurea dru used in the treatment of diabetes mellitus type 2 bind a block the channel. Inhibition of K^+ efflux and the resu ing cell depolarization opens the voltage-sensitive Ca

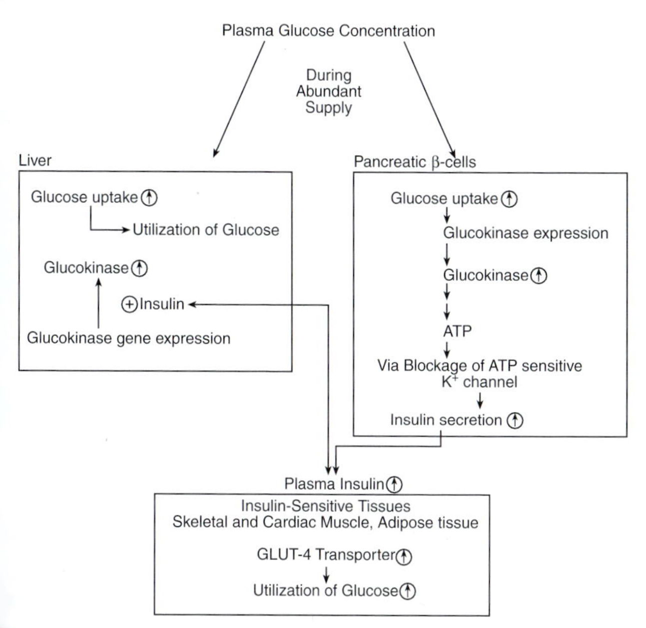

FIGURE 13-2
The role of tissue-specific glucokinases of liver and pancreatic β cells. During the periods of abundant supply of glucose (e.g., postprandial conditions), the plasma levels of glucose are increased. This induces glucokinase expression in pancreatic islet β cells with eventual insulin secretion. In the hepatocytes, insulin induces the glucokinase expression with accompanying increased metabolism. Insulin also enhances glucose utilization by recruiting glucose transporters, GLUT-4 in insulin-sensitive tissues. $\uparrow$ = Increased; $\oplus$ = positive effect.

channels. The increased cytosolic Ca^{2+} triggers the secretion of insulin (Chapter 22).

Tissue-specific glucokinases expressed in liver and pancreatic islet β cells are differentially regulated. The glucokinase gene consists of two different transcription control regions. Other examples of genes that contain different transcription control regions (promoters) yet produce mRNAs that code for identical proteins, are α-amylase and α_1-antitrypsin genes. In the hepatocytes, insulin promotes glucokinase expression, whereas in β cells insulin has no effect but glucose promotes enzyme expression. In the hepatocytes, cAMP (and thus glucagon) and insulin have opposing effects on the expression of glucokinase (Chapter 22). Tissue-specific glucose transporters and glucokinases play critical roles in glucose homeostasis (Figure 13-2). Mutations that alter these proteins may lead to inherited forms of diabetes mellitus (Chapter 22).

Isomerization of Glucose-6-Phosphate to Fructose-6-Phosphate

This freely reversible reaction requires Mg^{2+} and is specific for glucose-6-phosphate and fructose-6-phosphate. It is catalyzed by glucose–phosphate isomerase.

$$\alpha\text{-D-glucose-6-phosphate} \rightleftharpoons \alpha\text{-D-fructose-6-phosphate}$$

Phosphorylation of Fructose-6-Phosphate to Fructose-1,6-Bisphosphate

This second phosphorylation reaction is catalyzed by 6-phosphofructokinase (PFK-1).

$$\text{D-Fructose-6-phosphate}^{2-} + \text{ATP}^{4-} \xrightarrow{Mg^{2+}} \text{D-fructose-1,6-bisphosphate}^{4-} + \text{ADP}^{3+} + \text{H}^+$$

The reaction is essentially irreversible under physiological conditions and is a major regulatory step of glycolysis. PFK-1 is an inducible, highly regulated, allosteric enzyme. In its active form, muscle PFK-1 is a homotetramer (M.W. 320,000) that requires K^+ or NH_4^+, the latter of which lowers K_m for both substrates. When adenosine triphosphate (ATP) levels are low during very active muscle contraction, PFK activity is modulated positively despite low concentration of fructose-6-phosphate. Allosteric activators of muscle PFK-1 include adenosine monophosphate (AMP), adenosine diphosphate (ADP), fructose-6-phosphate, and inorganic phosphate (P_i); inactivators are citrate, fatty acids, and ATP.

The most potent regulator of liver PFK-1 is fructose-2,6-bisphosphate, which relieves the inhibition of PFK-1 by ATP and lowers the K_m for fructose-6-phosphate. Fructose-2,6-bisphosphate is a potent inhibitor of fructose-1,6-bisphosphatase (which is important in gluconeogenesis) and thus ensures the continuation of glycolysis. Metabolism of fructose-2,6-bisphosphate, its role as activator of PFK-1 and inhibitor of fructose-1,6-bisphosphatase, and its allosteric and hormonal regulation are discussed in Chapter 15.

Cleavage of Fructose-1,6-Bisphosphate into Two Triose Phosphates

In this reversible reaction, aldolase (or fructose-1,6-bisphosphate aldolase) cleaves fructose-1,6-bisphosphate into two isomers:

$$\text{D-Fructose-1,6-bisphosphate}^{4-} \rightleftharpoons \text{D-Glyceraldehyde 3-phosphate}^{2-} + \text{dihydroxyacetone phosphate}^{2-}$$

Isomerization of Dihydroxyacetone Phosphate to Glyceraldehyde 3-Phosphate

In this reversible reaction, triose-phosphate isomerase converts dihydroxyacetone phosphate to D-glyceraldehyde 3-phosphate, which is the substrate for the next reaction. Thus, the net effect of the aldolase reaction is to yield

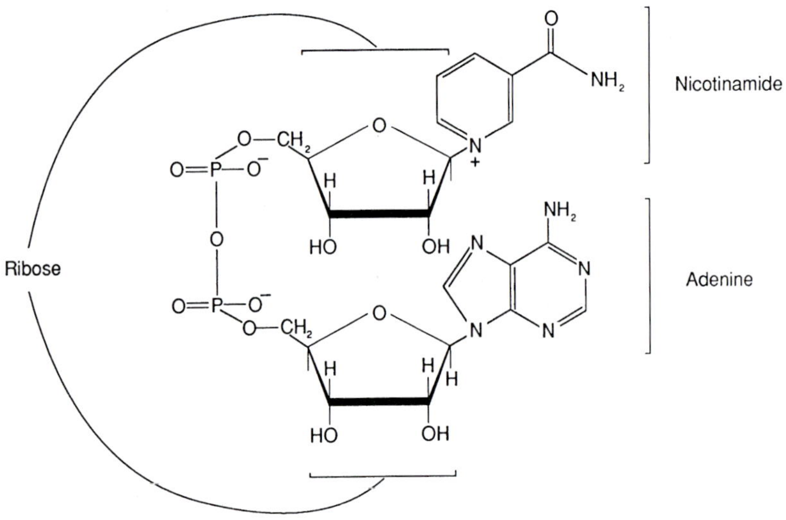

FIGURE 13-3
Nicotinamide adenine dinucleotide (NAD^+).

two molecules of D-glyceraldehyde 3-phosphate from one molecule of fructose-1,6-bisphosphate. This reaction concludes the first phase of glycolysis, during which one molecule of glucose yields two triose phosphates and two ATP molecules are consumed. The second phase of glycolysis consists of energy conservation and begins with the oxidation of glyceraldehyde 3-phosphate.

Dehydrogenation of Glyceraldehyde 3-Phosphate

In a reversible reaction, glyceraldehyde 3-phosphate is converted by glyceraldehyde phosphate dehydrogenase to an energy-rich intermediate, 1,3-bisglycerophosphate (or 3-phosphoglyceroyl-phosphate).

$$\text{D-Glyceraldehyde 3-phosphate}^{2-} + NAD^+ + P_i^{2-} \rightleftharpoons \text{1,3-bisphosphoglycerate}^{4-} + NADH + H^+$$

The rabbit skeletal enzyme is a homotetramer (M.W. 146,000). Each subunit contains a binding region for glyceraldehyde 3-phosphate and another for nicotinamide adenine dinucleotide (NAD^+). NAD^+, a cosubstrate of this reaction, participates in many hydrogen transfer reactions. NAD^+ (Figure 13-3) contains the vitamin nicotinamide (Chapter 38). Nicotinamide can also be obtained from the amino acid tryptophan. An –SH group present at the active site plays a prominent role in the catalysis, and iodoacetate (ICH_2COO^-) and heavy metals (e.g., Hg^{2+}, Pb^{2+}), which react with sulfhydryl groups covalently, inactivate the enzyme (E):

$$E\text{–}SH + ICH_2COO^- \rightarrow E\text{–}S\text{–}CH_2COO^- + HI$$

The reaction mechanism of glyceraldehyde-3-phosphate dehydrogenase has several steps (Figure 13-4). First, the enzyme–SH group attacks the carbonyl group of the substrate to form a thiohemiacetal, which is then oxidized to a thioester by transfer of a hydride ion (a hydrogen with two electrons, H^-) to an enzyme-bound NAD^+, with concurrent release of a proton (H^+). Thus, in effect, two hydrogen atoms are removed from the substrate and the overall NAD^+-dependent reaction can be written as

$$\text{Reduced substrate} + NAD^+ \rightleftharpoons \text{oxidized substrate} + NADH + H^+$$

Once NADH is formed, its affinity for the enzyme decreases, so that a free NAD^+ displaces this NADH. The thioester is an energy-rich intermediate, and by phosphorolysis the high-energy 1,3-bisphosphoglycerate is generated with the release of the free enzyme. Thus, the substrate aldehyde group is oxidized to a carboxylic acid group, with conservation of most of the energy of oxidation in formation of the anhydride bond between carboxylic and phosphoric acids. If arsenate is used in place of phosphate, the intermediate formed is 1-arseno-3-phosphoglycerate, which spontaneously hydrolyzes to arsenate and 3-phosphoglycerate with release of heat. Thus, arsenate uncouples this phosphorylation coupled to substrate oxidation. This reaction is a model for similar reactions in which energy released by oxidation of a substrate is conserved in the terminal phosphoanhydride bond of ATP via the formation of a high-energy intermediate (the next reaction of glycolysis).

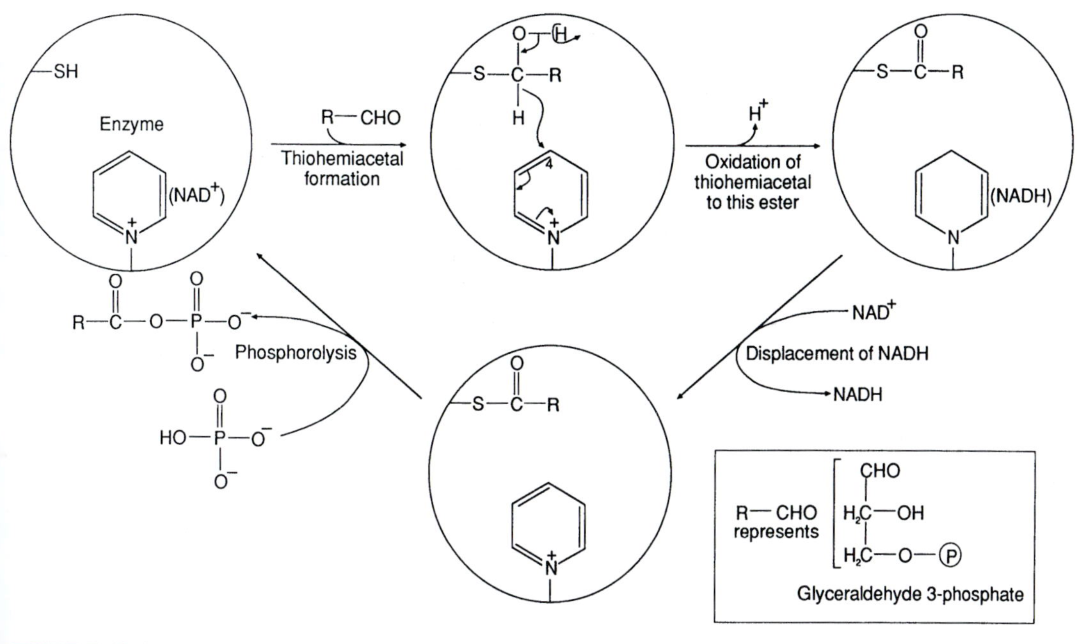

FIGURE 13-4
Glyceraldehyde 3-phosphate is oxidized by NAD^+, and inorganic phosphate (P_i) is incorporated into the product to form an acyl phosphate, 1,3-bisphosphoglycerate. NAD^+ is reduced by transfer of a hydride ion (H^-) from thiohemiacetal to the fourth position on the nicotinamide ring of NAD^+.

Phosphorylation of ADP from 1,3-Bisphosphoglycerate

In a reversible reaction, the phosphoryl group of the acid anhydride of glyceroyl phosphate is transferred to ADP by phosphoglycerate kinase.

$$\text{1,3-Bisphosphoglycerate}^{4-} + \text{ADP}^{3-} \overset{\text{Mg}^{2+}}{\rightleftharpoons} \text{3-phosphoglycerate}^{3-} + \text{ATP}^{4-}$$

The net result of this reaction is the formation of two molecules of ATP per molecule of glucose. This formation of ATP is known as ***substrate-level phosphorylation*** to distinguish it from the oxidative phosphorylation coupled to electron transport that occurs in mitochondria (Chapter 14). In erythrocytes, 1,3-bisphosphoglycerate is converted to 2,3-bisphosphoglycerate (also called 2,3-diphosphoglycerate or 2,3-DPG) by 1,3-bisphosphoglycerate mutase, which possesses 2,3-bisphosphoglycerate phosphatase activity and is responsible for the formation of 3-phosphoglycerate (which reenters glycolysis) and inorganic phosphate. In this bypass pathway, the free energy associated with the anhydride bond of 1,3-bisphosphoglycerate is dissipated as heat. Erythrocytes contain a high level of 2,3-DPG, which functions as an allosteric modulator of hemoglobin and decreases affinity for oxygen (Chapter 28). 2,3-DPG is present in trace amounts in most cells, functions as a cofactor in the next reaction, and is synthesized by 3-phosphoglycerate kinase in the reaction

$$\text{3-Phosphoglycerate}^{2-} + \text{ATP}^{4-} \rightarrow \text{2,3-bisphosphoglycerate}^{4-} + \text{ADP}^{3-} + \text{H}^{+}$$

In the next set of reactions of glycolysis, the remaining phosphate group of glycerate is elevated to a high-energy state, at which it can phosphorylate ADP in another substrate-level phosphorylation reaction. The formation of the high-energy intermediate occurs by way of the following two reactions.

Isomerization of 3-Phosphoglycerate to 2-Phosphoglycerate

In this reversible reaction, the phosphate group is transferred to the 2-position by phosphoglycerate mutase:

$$\text{3-Phosphoglycerate}^{2-} \overset{\text{Mg}^{2+}}{\rightleftharpoons} \text{2-phosphoglycerate}^{2-}$$

The reaction is similar to the interconversion catalyzed by phosphoglucomutase (Chapter 15). The reaction is primed by phosphorylation of a serine residue of the enzyme by 2,3-bisphosphoglycerate and occurs in the following steps shown below.

$$\begin{array}{l} COO^- \\ | \\ CHOH \\ | \\ CH_2OPO_3^{2-} \end{array} + E{-}OPO_3^{2-} \quad \begin{array}{l} COO^- \\ | \\ CHOPO_3^{2-} \\ | \\ CH_2OPO_3^{2-} \end{array} + E{-}OH$$

3-Phosphoglycerate — 2,3-Biphosphoglycerate

$$\downarrow$$

$$E{-}OPO_3^{2-} + \begin{array}{l} COO^- \\ | \\ HCOPO_3^{2-} \\ | \\ CH_2OH \end{array}$$

2-Phosphoglycerate

Dehydration of 2-Phosphoglycerate to Phosphoenolpyruvate

In this reversible reaction, dehydration of the substrate causes a redistribution of energy to form the high-energy compound phosphoenolpyruvate.

$$\text{2-Phosphoglycerate} \xrightleftharpoons{Mg^{2+}} \text{phosphoenolpyruvate} + H_2O$$

Enolase (2-phospho-D-glycerate hydrolyase) is a homodimer (M.W. 88,000) that is inhibited by fluoride, with formation of the magnesium fluorophosphate complex at the active site. This property of fluoride is used to inhibit glycolysis in blood specimens obtained for measurement of glucose concentration. In the absence of fluoride (or any other antiglycolytic agent), the blood glucose concentration decreases at about 10 mg/dL (0.56 mM/L) per hour at 25°C. The rate of decrease is more rapid in blood from newborn infants owing to the increased metabolic activity of the erythrocytes and in leukemia patients because of the larger numbers of leukocytes.

Neuron-specific and non-neuron-specific enolase isoenzymes have been used as markers to distinguish neurons from nonneuronal cells (e.g., glial cells that are physically and metabolically supportive cells of neurons) by immunocytochemical techniques. Neuron-specific enolase is extremely stable and resistant to a number of *in vitro* treatments (e.g., high temperature, urea, chloride) that inactivate other enolases. The functional significance of these isoenzymes is not known.

Phosphorylation of ADP from Phosphoenolpyruvate

In this physiologically irreversible (nonequilibrium) reaction, the high-energy group of phosphoenolpyruvate is transferred to ADP by pyruvate kinase, thereby generating ATP (i.e., two molecules of ATP per molecule of glucose). This reaction is the second substrate-level phosphorylation reaction of glycolysis.

$$\text{Phosphoenolpyruvate}^{3-} + ADP^{3-} + H^+ \rightarrow \text{pyruvate}^- + ATP^{4-}$$

The pyruvate kinase reaction has a large equilibrium constant because the initial product of pyruvate, the enol form, rearranges nonenzymatically to the favored keto form:

$$\begin{array}{l} COO^- \\ | \\ COH \\ \| \\ CH_2 \end{array} \longrightarrow \begin{array}{l} COO^- \\ | \\ C{=}O \\ | \\ CH_3 \end{array}$$

Enol-pyruvate — Keto-pyruvate

Several isoenzyme forms of pyruvate kinase are known (M.W. 190,000–250,000, depending on the source). Each is a homotetramer exhibiting catalytic properties consistent with the function of the tissue in which it occurs. Enzyme activity is dependent on K^+ (which increases the affinity for phosphoenolpyruvate) and Mg^{2+}.

Pyruvate kinase is an allosteric enzyme regulated by several modifiers. The liver isoenzyme (L-type) shows sigmoidal kinetics with phosphoenolpyruvate. Fructose-1,6-bisphosphate is a positive modulator and decreases K_m for phosphoenolpyruvate; ATP and alanine are negative modulators and increase K_m for phosphoenolpyruvate. The former is an example of positive feed-forward regulation, and the latter are examples of negative feedback regulation. Alanine, a gluconeogenic precursor, is obtained by proteolysis or from pyruvate by amino transfer (Chapter 17). The modulation of pyruvate kinase is consistent with the function of the liver; when glucose abounds, its oxidation is promoted, and when glucose is deficient, its formation is favored by gluconeogenesis (Chapter 15).

The L-type is also regulated by diet and hormones. Fasting or starvation decreases activity, whereas a carbohydrate-rich diet increases it. Insulin increases activity, whereas glucagon decreases it. These hormones also have reciprocal effects on gluconeogenesis, which insulin inhibits and glucagon promotes. Glucagon action is dependent on the cAMP-mediated cascade process of reversible phosphorylation and dephosphorylation of the enzyme (Chapter 30). The cAMP cascade begins with stimulation of membrane-bound adenylate cycles by glucagon to form cAMP and is followed by activation of cAMP-dependent protein kinase, which phosphorylates pyruvate kinase. The phospho enzyme is less active than the dephospho form, has a higher K_m for phosphoenolpyruvate

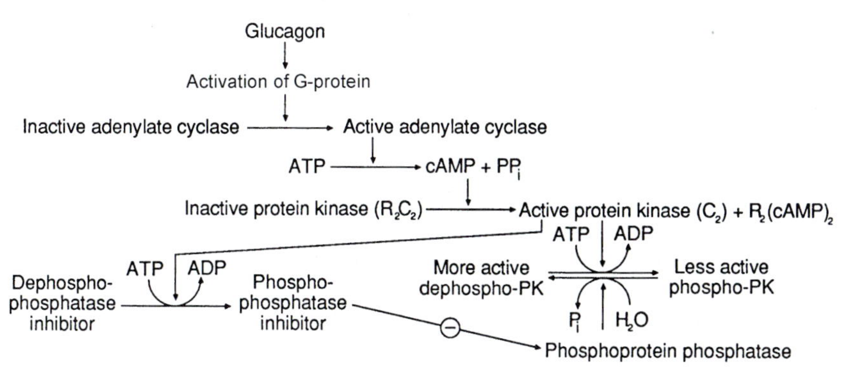

FIGURE 13-5
Action of glucagon on the liver pyruvate kinase (PK) via the cAMP-cascade system (see also Chapter 30). Glucagon, by combining at specific receptor sites on the hepatocyte plasma membrane, activates adenylate cyclase, which converts ATP to cAMP. The latter activates (cAMP-dependent) protein kinase, which phosphorylates PK (phospho-PK), converting it to a less active form and making it more susceptible to allosteric inhibition by alanine and ATP. Phospho-PK is converted to the more active form by removal of phospho groups by a phosphoprotein phosphatase, whose activity is also regulated by cAMP-dependent protein kinase via the phosphorylation of phosphatase inhibitor. The overall effect of glucagon is to diminish glycolysis, whereas the effect of insulin is to promote glycolysis. PP_i = pyrophosphate; ⊖ = Inhibition.

has a higher susceptibility to inhibition by the negative modulators alanine and ATP, and has virtually no activity in the absence of fructose-1,6-bisphosphate. The phospho enzyme is converted to the dephospho form by a phosphoprotein phosphatase, the activity of which is also regulated by cAMP-dependent protein kinase (Figure 13-5).

The muscle and brain isoenzyme (M-type) shows hyperbolic kinetics with phosphoenolpyruvate and is inhibited by ATP, with an increase in K_m for phosphoenolpyruvate and development of sigmoidal kinetics. Fructose-1,6-bisphosphate and alanine have no effect on this isoenzyme.

The overall reaction for glycolysis in cells that possess mitochondria and are under aerobic conditions is:

$$\text{Glucose} + 2\text{NAD}^+ + 2\text{ADP}^{3-} + 2\text{HPO}_4^{2-} \rightarrow 2\text{pyruvate}^- + 2\text{NADH} + 2\text{H}^+ + 2\text{ATP}^{4-} + 2\text{H}_2\text{O}$$

Pyruvate is transported into mitochondria, where it is oxidized in the TCA cycle (see below). Although NADH is not transported into mitochondria, its reducing equivalent is transported into mitochondria (Chapter 15), where it is oxidized in the electron transport system coupled to oxidative phosphorylation (Chapter 14).

Cells that lack mitochondria (e.g., erythrocytes) or contain mitochondria but under hypoxic conditions (e.g., vigorously and repeatedly contracting muscle) reduce pyruvate to lactate by lactate dehydrogenase (LDH), which uses NADH as a reductant. This reduction regenerates NAD^+, which is required for continued oxidation of glyceraldehyde 3-phosphate.

Reduction of Pyruvate to Lactase

This reversible reaction is the final step of glycolysis and is catalyzed by (LDH):

$$\text{Pyruvate}^- + \text{NADH} + \text{H}^+ \rightleftharpoons \text{L-lactate}^- + \text{NAD}^+$$

The overall reaction of glycolysis becomes

$$\text{Glucose} + 2\text{ADP}^{3-} + 2\text{P}_i^{2-} \rightarrow 2\text{lactate}^- + 2\text{ATP}^{4-} + 2\text{H}_2\text{O}$$

LDH (M.W. 134,000) occurs as five tetrameric isoenzymes composed of two different types of subunits. Subunits M (for muscle) and H (for heart) are encoded by loci in chromosomes 11 and 12, respectively. Two subunits used in the formation of a tetramer yield five combinations: H_4(LDH-1), H_3M(LDH-2), H_2M_2(LDH-3), HM_3(LDH-4), and M_4(LDH-5). The tissue distributiont of LDH isoenzymes is variable. For example, LDH-1 and LDH-2 are the principal isoenzymes in heart, kidney, brain, and erythrocytes; LDH-3 and LDH-4 predominate in endocrine glands (e.g., thyroid, adrenal, pancreas), lymph nodes, thymus, spleen, leukocytes, platelets, and nongravid uterine muscle; and LDH-4 and LDH-5 preponderate in liver and skeletal muscle. In tissue injury or insult, the appropriate tissue isoenzymes appear in plasma (Chapter 8); thus, determination of LDH isoenzyme composition has diagnostic value.

Serum LDH isoenzymes can be separated by electrophoresis on agarose gel or cellulose acetate membrane, usually at pH 8.6. After separation, their location is determined by incubation of the support medium in a solution containing lactate and NAD^+. NADH generated at the LDH zones is detected either by its fluorescence when excited by long-wave ultraviolet light (365 nm) or by its reduction of a tetrazolium salt (nitroblue tetrazolium, NBT) to form an insoluble colored complex via an intermediate redox carrier (e.g., phenazine methosulfate, PMS):

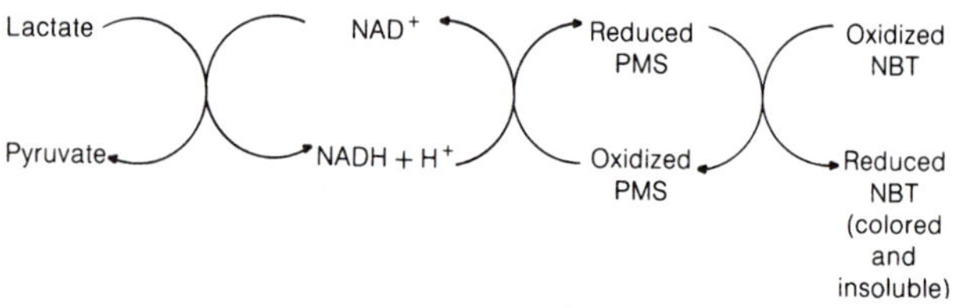

Lactate → Pyruvate; NAD^+ → $NADH + H^+$; Reduced PMS → Oxidized PMS; Oxidized NBT → Reduced NBT (colored and insoluble)

As long as the total activity of LDH is not very high so that the substrates are not limiting, the intensity of the color is approximately proportional to the various isoenzyme activities. Thus, densitometric scanning of an electropherogram provides an estimate of the activities of the individual enzymes as well as quantitation of their relative intensities. LDH-1 is the most negatively charged and fastest moving isoenzyme (mobility comparable to that of the α_1-globulin region in serum protein electrophoresis). LDH-5 is the least negatively charged and slowest migrating isoenzyme (mobility comparable to that of γ-globulin); the other isoenzymes possess intermediate mobilities. The normal relative intensity of LDH isoenzyme fractions in serum is

LDH-1 < LDH-2 > LDH-3 > LDH-4 <=> LDH-5

Since myocardium is rich in LDH-1, injury to that tissue results in the elevation of LDH-1 and the ratio of LDH-1 to LDH-2; similarly, since liver is rich in LDH-4 and LDH-5, elevations of these isoenzymes suggests liver disease. Because a particular isoenzyme pattern may result from injury to more than one tissue (Table 13-3), further diagnostic tests are required. For example, measurement of creatine kinase and its isoenzyme determination and troponin I in the serum is helpful in the evaluation of acute injury to the myocardium. Because of differences in half-lives in serum, serial determination of total enzyme activity and relative isoenzyme composition aids in the diagnosis and assessment of the magnitude of the injury of the tissue in question (Chapter 8). In patients with acute myocardial infarction, serial analysis every 8–12 hours during the first 48 hours after the onset of symptoms is useful. *In vitro,* LDH-1(H_4) has a low K_m for pyruvate and is strongly inhibited by high concentration of pyruvate. This form may favor the oxidation of lactate to pyruvate in aerobic tissues such as the heart. LDH-5(M_4) exhibits a higher K_m value for pyruvate and is less susceptible to inhibition by high concentrations of pyruvate than is LDH-1.

TABLE 13-3
Serum LDH Isoenzyme Patterns in Various Disorders*

Isoenzyme Pattern	Disorder
Elevation of LDH-1 and LDH-2, frequently LDH-1 > LDH-2	Myocardial infarction Renal cortical infarction Pernicious anemia Hemolysis Muscular dystrophy (later stages)
Elevation of LDH-5	Liver disease Skeletal muscle damage Some cancers
Elevation of LDH-3, frequently LDH-3 > LDH-2	Some neoplastic diseases Lymphoproliferative disorders Platelet-related disorders
Elevation of LDH-2 and LDH-3	Pulmonary infarction
All isoenzymes elevated	Widespread tissue injury

*Normal distribution: LDH-1 < LDH-2 > LDH-3 > LDH-4 <=> LDH-5

This form may occur in anaerobic tissues, in which lactate is the end product of glycolysis. However, *in vitro* differences in kinetic properties of the isoenzyme may be inappropriate to explain actual physiological actions for several reasons: differences in kinetic properties between LDH-1 and LDH-5 are less marked at 37°C (body temperature) than at 25°C, high intracellular concentrations of the enzyme are present, and differences in the actual ratio of ketopyruvate to enolpyruvate exist (the enol form may be the more potent inhibitor). Furthermore, the occurrence of similar isoenzyme patterns in widely different tissues with divergent metabolic goals (e.g., LDH-5 in liver and muscle; LDH-1 in heart and erythrocytes) points out our lack of understanding of their precise role.

Alternative Substrates of Glycolysis

The glycolytic pathway is also utilized by fructose, galactose, mannose, glycogen, and glycerol. The metabolism of the monosaccharides and glycogen is discussed in Chapter 15. Glycogenolysis catalyzed by phosphorylase in a nonequilibrium reaction yields glucose-1-phosphate which is then converted to glucose-6-phosphate (Chapter 15), bypassing the initial phosphorylation reaction of glycolysis. Therefore, the conversion of a glucosyl unit of glycogen to two lactate molecules yields three ATP, which is 50% more than the yield from a glucose molecule. Glycerol released by hydrolysis of triacylglycerol enters the glycolytic pathway by way of dihydroxyacetone phosphate, as follows:

$$\text{Glycerol} + \text{ATP}^{4-} \xrightarrow{\text{glycerol kinase, Mg}^{2+}} \text{glycerol 3-phosphate}^{2-} + \text{ADP}^{3-} + \text{H}^+$$

$$\text{Glycerol 3-phosphate}^{2-} + \text{NAD}^+ \xrightleftharpoons{\text{glycerol-3-phosphate dehydrogenase}} \text{dihydroxyacetone phosphate}^{2-} + \text{NADH} + \text{H}^+$$

Role of Anaerobic Glycolysis in Various Tissues and Cells

In tissues that lack mitochondria or function under limiting conditions of oxygen (e.g., lens and erythrocytes), glycolysis is the predominant pathway providing ATP. On the basis of speed of contraction and metabolic properties, skeletal muscle fibers may be classified into three types: type I (slow-twitch, oxidative), type IIA (fast-twitch, oxidative-glycolytic), and type IIB (fast-twitch, glycolytic). Short-term, sudden energy output is derived from glycolytic fibers. The glycolytic fibers (particularly type IIB) have few mitochondria, with relative enrichment of myofibrils and poor capillary blood supply. Thus, these fibers can perform a large amount of work in a short period of time, in contrast to slow-twitch fibers, which are rich in mitochondria, have good capillary blood supply, and power sustained efforts (Chapter 21).

A high rate of glycolysis also occurs in lymphocytes, kidney medulla, skin, and fetal and neonatal tissues. The anaerobic capacity of lymphocytes is increased for growth and cell division. Kidney medulla (Chapter 39), which contains the loops of Henle and collecting tubules, receives much less blood than the cortex and is rich in glycogen. In contrast, kidney cortex, which contains the glomeruli, proximal tubules, and parts of the distal tubules, receives a large amount of blood through the renal arterioles. The renal cortex derives its energy from oxidation of glucose, fatty acids, ketone bodies, or glutamine to CO_2 and H_2O.

Some rapidly growing tumor cells exhibit a high rate of glycolysis. Inadequate oxygen supply (hypoxia) causes tumor cells to grow more rapidly than the formation of blood vessels that supply oxygen. Thus, in hypoxic tumor cells, glycolysis is promoted by increased expression of glucose transporters and most glycolytic enzymes by hypoxia-inducible transcription factor (HIF-1).

Glycolytic Enzyme Deficiencies in Erythrocytes

The only pathway that provides ATP in mature erythrocytes is glycolysis. Because these cells lack mitochondria, a nucleus, and other organelles required for protein synthesis, deficiency of glycolytic enzymes may reduce their normal 120-day life span. For example, deficiency of hexokinase, glucose-phosphate isomerase, 6-phosphofructokinase, aldolase, triose-phosphate isomerase, or phosphoglycerate kinase is associated with hemolytic anemia, whereas lactate dehydrogenase deficiency is not. The most common deficiency is that of pyruvate kinase, (PK), which is inherited as an autosomal recessive trait. The PK deficiency occurs worldwide, however, it is most commonly found in kindreds of Northern European ancestry. Erythrocyts of PK individuals have elevated levels of 2.3-DPG and decreased ATP level. Several mutations of the PK gene have been identified. Individuals with PK deficiency suffer from lifelong chronic hemolysis to a varying degree. Splenectomy ameliorates hemolytic process in severe cases.

Some enzymopathies of erythrocytes may be associated with multisystem disease (e.g., aldolase deficiency with mental and growth retardation). Individuals with 6-phosphofructokinase deficiency exhibit hemolysis and myopathy and have increased deposition of muscle glycogen (a glycogen storage disease; see Chapter 15). The myopathy is usually characterized by muscle weakness and exercise intolerance. (See also Chapters 10, 15, and 28.)

13.2 Pyruvate Metabolism

Pyruvate has several metabolic fates. It can be reduced to lactate, converted to oxaloacetate in a reaction important in gluconeogenesis (Chapter 15) and in an anaplerotic reaction of the TCA cycle (see below), transminated to alanine (Chapter 17), or converted to acetyl-CoA and CO_2. Acetyl-CoA is utilized in fatty acid synthesis, cholesterol (and steroid) synthesis, acetylcholine synthesis, and the TCA cycle (Figure 13-6).

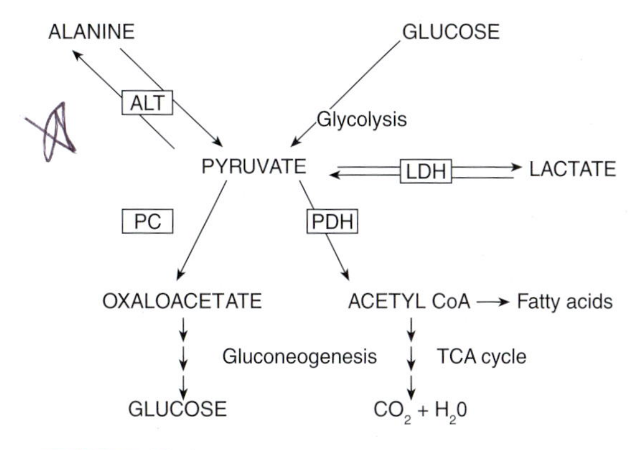

FIGURE 13-6
Major pathways of pyruvate metabolism. Pyruvate is metabolized through four major enzyme pathways: Lactate dehydrogenase (LDH), pyruvate dehydrogenase complex (PDH), pyruvate carboxylase (PC), and alanine aminotransferase (ALT). Arrows indicate multiple steps.

Conversion of pyruvate to ethanol by certain yeast strains occurs in two steps. It is first decarboxylated to acetaldehyde by pyruvate decarboxylase, which utilizes thiamine pyrophosphate (TPP) as coenzyme.

$$CH_3COCOOH \xrightarrow[\text{TPP}]{\text{mg}^{2+}} CH_3CHO + CO_2$$

In the second step, acetaldehyde is reduced to ethanol by alcohol dehydrogenase, an NAD-dependent enzyme.

$$CH_3CHO + NADH + H^+ \rightleftarrows CH_3CH_2OH + NAD^+$$

The NADH is derived from glyceraldehyde 3-phosphate dehydrogenation. Thus, yeast fermentation yields ethanol rather than lactate as an end product of glycolysis. Small amounts of ethanol are produced by the microbial flora of the gastrointestinal tract. Other types of fermentation using similar reactions occur in microorganisms and yield a variety of products (e.g., acetate, acetone, butanol, butyrate, isopropanol, hydrogen gas).

Lactic Acidemia and Lactic Acidosis

As discussed earlier, some tissues produce lactate as an end product of metabolism. The lactate produced is L-lactate and is commonly referred to simply as lactate. As we will see in the following discussion, D-lactate is also produced under certain pathological conditions, which presents a unique clinical problem.

Under normal conditions, lactate is metabolized in the liver and the blood lactate level is between 1 and 2 mM. Lactate accumulation in body fluids can be due to increased production and/or decreased utilization. Blood lactate-to-pyruvate ratio below 25 suggests defects in a gluconeogenic enzyme (Chapter 15) or pyruvate dehydrogenase (discussed later). A common cause of *lactic acidosis* is tissue hypoxia caused by shock, cardiopulmonary arrest, and hypoperfusion. Inadequate blood flow leads to deprivation of oxygen and other nutrients to the tissue cells as well as to the removal of waste products. Oxygen deprivation leads to decreased ATP production and accumulation of NADH, which promotes conversion of pyruvate to lactate.

A major cause of acidosis that occurs during inadequate cellular oxygen delivery is continual hydrolysis of the available supply of ATP that releases protons:

$$ATP^{4-} + H_2O \rightarrow ADP^{3-} + HPO_4^{2-} + H^+$$

Laboratory assessment includes measurements of blood lactate, pyruvate, β-hydroxybutyrate, and acetoacetate (Chapter 39). The primary treatment of lactic acidosis involves correcting the underlying cause such as reversal of circulatory failure.

D-Lactic Acidosis

This unusual form of lactic acidosis is due to increased production and accumulation of D-lactate in circulation. The normal isomer synthesized in the human body is L-lactate but the D-lactate isomer can occur in patients with jejunoileal bypass, small bowel resection, or other types of short bowel syndrome. In these patients, ingested starch and glucose bypass the normal metabolism in the small intestine and lead to increased delivery of nutrients to the colon where gram-positive, anaerobic bacteria (e.g., *Lactobacilli*) ferment glucose to D-lactate. The D-lactate is absorbed via the portal circulation.

A limited quantity of D-lactate is converted to pyruvate by a mitochondrial flavoprotein enzyme D-2-hydroxy acid dehydrogenase. Thus, the development of D-lactate acidosis requires excessive production of D-lactate and an impairment in its metabolism. The clinical manifestations of D-lactic acidosis are characterized by episodes of encephalopathy after ingestion of foods containing carbohydrates.

The diagnosis of D-lactic acidosis is suspected in patients with disorders of the small intestine causing malabsorption and when the serum anion gap (Chapter 39) is elevated in the presence of normal serum levels of L-lactate and other organic acids. Measurement of serum D-lactate requires special enzymatic procedures utilizing D-lactate dehydrogenase and NADH. As D-lactate is converted to pyruvate, NADH is oxidized to NAD^+ which is detected spectrophotometrically (Chapter 8).

The treatment of D-lactic acidosis consists of oral administration of antibiotics, limitations of oral carbohydrate intake, and recolonization of the colon by bacterial flora which do not produce D-lactate.

Oxidation of Pyruvate to Acetyl-CoA

Pyruvate must first be transported into mitochondria by a specific carrier that cotransports a proton to maintain electrical neutrality. Inside mitochondria pyruvate undergoes oxidative decarboxylation by three enzymes that function sequentially and are present as a complex known as the pyruvate dehydrogenase complex. The overall reaction is physiologically irreversible, has a high negative $\Delta G^{0'}$ (−8.0 kcal/mol, or −33.5 kJ/mol), and commits pyruvate to the formation of acetyl-CoA:

$$\text{Pyruvate} + NAD^+ + CoASH \rightarrow \text{acetyl-CoA} + NADH + H^+ + CO_2$$

In the above reaction CoASH stands for coenzyme A (Figure 13-7), and it contains the vitamin pantothenic acid (Chapter 38). Coenzyme A functions as a carrier of

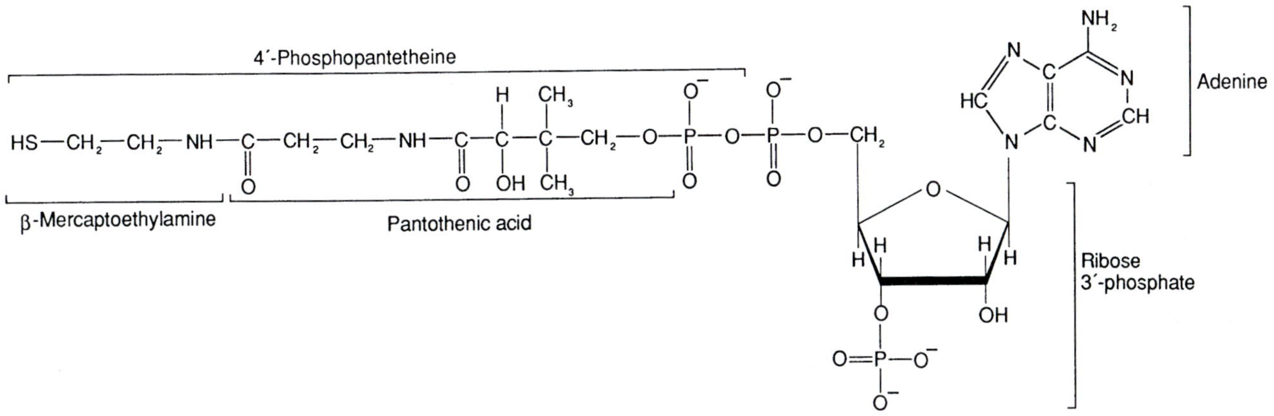

4′-Phosphopantetheine

$HS-CH_2-CH_2-NH-CO-CH_2-CH_2-NH-CO-CH(OH)-C(CH_3)_2-CH_2-O-PO_2^--O-PO_2^--O-CH_2$-(ribose 3′-phosphate)-adenine

β-Mercaptoethylamine | Pantothenic acid | Adenine | Ribose 3′-phosphate

FIGURE 13-7
Coenzyme A. The terminal sulfhydryl group is the reactive group of the molecule.

acetyl (and acyl groups) by formation of thioesters, and the acetyl–sulfur bond has a $\Delta G^{0'}$ comparable to that of the phosphoanhydride bond of ATP.

Dehydrogenation and decarboxylation of pyruvate occur in five reactions requiring three enzymes and six coenzymes, prosthetic groups, and cofactors (Mg^{2+}, TPP, lipoic acid, CoASH, flavin adeninedinucleotide [FAD], and NAD^+). The mammalian complex (M.W. $7–8.5 \times 10^6$) consists of 20 or 30 units of pyruvate dehydrogenase (depending on the source), 60 units of dihydrolipoyl transacetylase (also called dihydrolipoamide acetyltransferase), and 5 or 6 units of dihydrolipoyl dehydrogenase (also called dihydrolipoamide reductase). A kinase and a phosphatase are tightly bound to the pyruvate dehydrogenase subunit and participate in the regulation of the activity.

In the first reaction, pyruvate is decarboxylated by reaction with TPP bound to pyruvate dehydrogenase (E_1), with formation of a hydroxyethyl group attached to the thiazole ring of TPP:

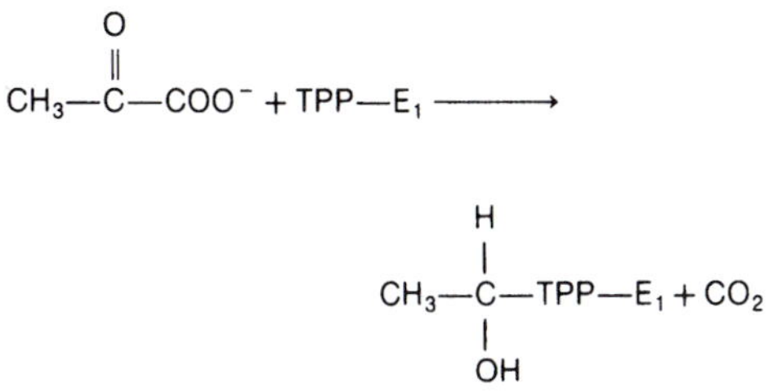

$$CH_3-CO-COO^- + TPP-E_1 \longrightarrow CH_3-CH(OH)-TPP-E_1 + CO_2$$

In the second step, also catalyzed by pyruvate dehydrogenase, the hydroxyethyl group is transferred to the oxidized form of the lipoyl-lysyl prosthetic group of dihydrolipoyl transacetylase (E_2):

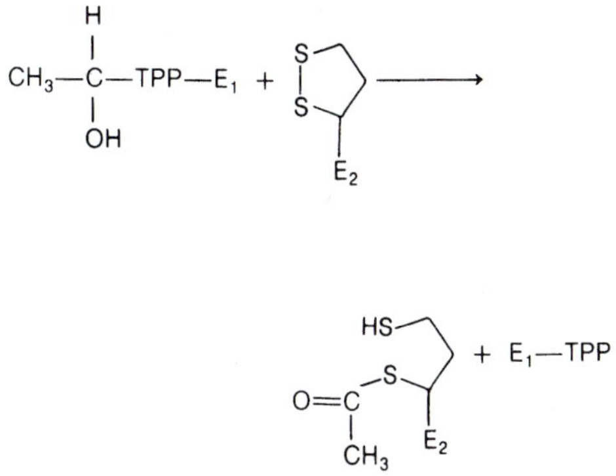

$$CH_3-CH(OH)-TPP-E_1 + \text{(oxidized lipoyl)}-E_2 \longrightarrow \text{(HS)(}CH_3CO-S\text{)lipoyl}-E_2 + E_1-TPP$$

A detailed mechanism of the first two reactions is shown in Figure 13-8. In the third step, catalyzed by dihydrolipoyl transacetylase, the acetyl group attached to E_2 is transferred to CoASH, with formation of the dithiol form of the lipoyl group and acetyl-CoA.

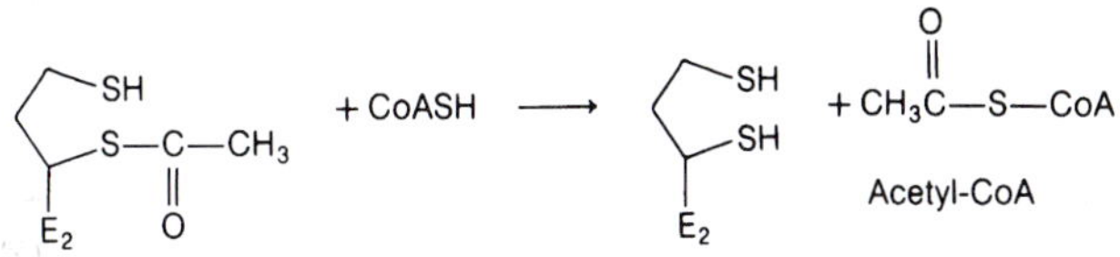

$$\text{(HS)(}CH_3CO-S\text{)lipoyl}-E_2 + CoASH \longrightarrow \text{(HS)}_2\text{lipoyl}-E_2 + CH_3CO-S-CoA$$

Acetyl-CoA

In the fourth step, catalyzed by the same enzyme, the hydrogen atoms of the dithiol group of E_2 are transferred to the FAD prosthetic group of dihydrolipoyl dehydrogenase (E_3).

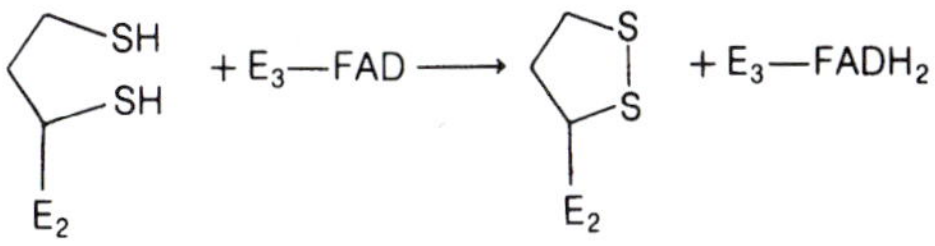

$$\text{(HS)}_2\text{lipoyl}-E_2 + E_3-FAD \longrightarrow \text{(S–S)lipoyl}-E_2 + E_3-FADH_2$$

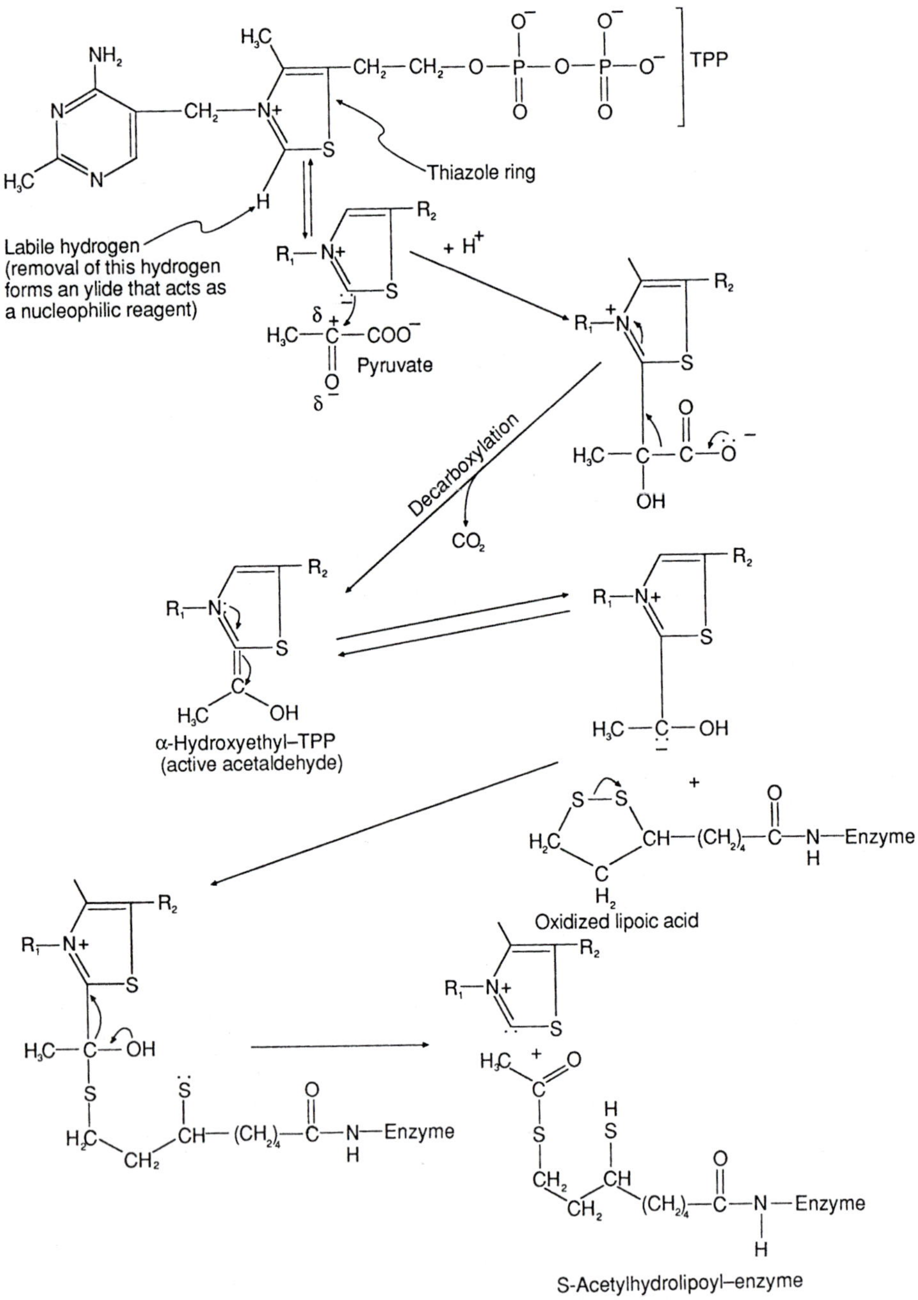

TPP
Thiazole ring
Labile hydrogen (removal of this hydrogen forms an ylide that acts as a nucleophilic reagent)
Pyruvate
Decarboxylation
CO_2
α-Hydroxyethyl–TPP (active acetaldehyde)
Oxidized lipoic acid
S-Acetylhydrolipoyl–enzyme

FIGURE 13-8
Steps involved in the decarboxylation and formation of the S-acetylhydrolipoyl enzyme of the dehydrogenase complex. TPP = Thiamine pyrophosphate.

The structure of FAD that contains the vitamin riboflavin is given in Figure 13-9. In the last step, catalyzed by dihydrolipoyl dehydrogenase, the hydrogens (or reducing equivalents) are transferred to NAD^+, with the formation of NADH and the oxidized flavoprotein.

$$E_3\text{–}FADH_2 + NAD^+ \rightarrow E_3\text{–}FAD + NADH + H^+$$

Thus, the flow of reducing equivalents in the pyruvate dehydrogenase complex is from pyruvate to lipoyl to FAD to NAD^+. Conversion of pyruvate to acetyl-CoA requires four vitamins: thiamine, pantothenic acid, riboflavin, and niacin. In contrast, in glycolysis, niacin is the only vitamin used. NADH generated in this reaction is oxidized to NAD^+ in the electron transport

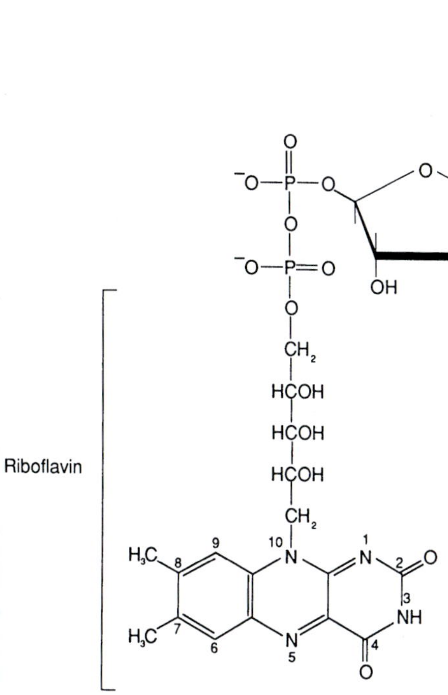

FIGURE 13-9
Flavin adenine dinucleotide (FAD).

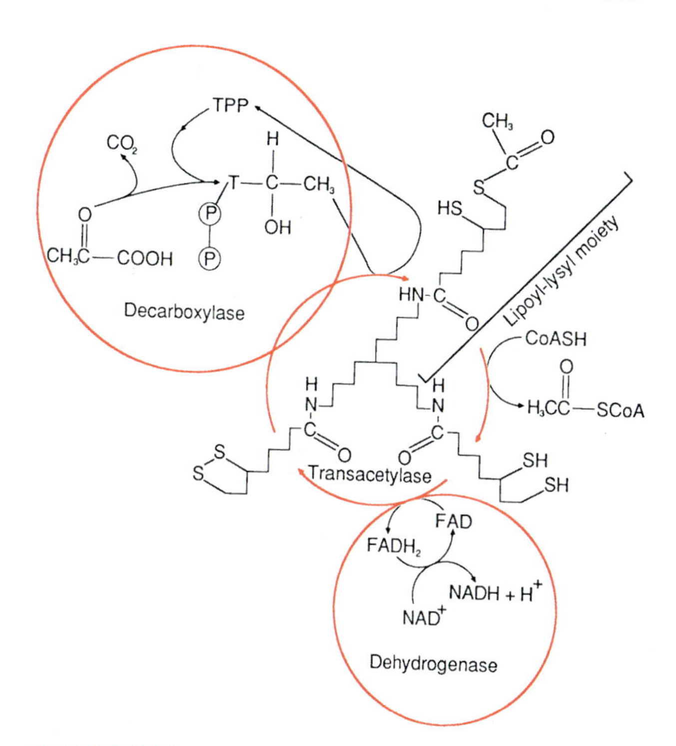

FIGURE 13-10
Schematic representation of the relationship between the three enzymes of the pyruvate dehydrogenase complex. The lipoyl-lysyl moiety of the transacetylase delivers the acetyl group to CoA and the reducing equivalents to FAD. TPP = Thiamine pyrophosphate.

chain, with the production of ATP (Chapter 14). Dihydrolipoyl transacetylase plays a central role in transferring hydrogen atoms and acetyl groups from one enzyme to the next in the pyruvate dehydrogenase complex. This is made possible by the lipoyl-lysyl swinging arm of the transacetylase, which is about 1.4 nm long (Figure 13-10).

In a similar way, other α-keto acids, e.g., α-ketoglutarate (in the TCA cycle; see below) and branched-chain α-keto acids derived by transamination from the branched-chain amino acids valine, leucine, and isoleucine (Chapter 17), undergo decarboxylation and dehydrogenation catalyzed by enzyme complexes. These enzyme complexes differ in specificity of E_1 and E_2, but all contain the same E_3 (the dihydrolipoyl dehydrogenase).

Regulation of Pyruvate Dehydrogenase Activity

The pyruvate dehydrogenase complex catalyzes an irreversible reaction that is the entry point of pyruvate into the TCA cycle (see below) and is under complex regulation by allosteric and covalent modification of the pyruvate dehydrogenase component of the complex. The end products of the overall reaction (NADH and acetyl-CoA) are potent allosteric inhibitors of the pyruvate dehydrogenase component of the complex. They also function as effectors in a non-cAMP-dependent reversible phosphorylation and dephosphorylation cycle of the dehydrogenase. Phosphorylation occurs by an ATP-specific pyruvate dehydrogenase kinase at three serine residues of the α-subunit of the enzyme and leads to inactivation. The kinase is activated by elevated [acetyl-CoA]/[CoA], [NADH]/[NAD^+], and [ATP]/[ADP] and inhibited by increases in [pyruvate], [Ca^{2+}], and [K^+]. The phospho enzyme is converted to the active dephospho enzyme by a pyruvate dehydrogenase phosphatase, an Mg^{2+}/Ca^{2+}-stimulated enzyme that is also stimulated by insulin in adipocytes. The kinase and phosphatase are associated with pyruvate dehydrogenase. The regulation of pyruvate dehydrogenase is shown in Figure 13-11. In general, when the levels of ATP, NADH, or acetyl-CoA are high, the oxidation of pyruvate to acetyl-CoA is markedly decreased. For example, fatty acid oxidation (Chapter 18) provides all three metabolites and thus decreases the need for pyruvate oxidation.

An analogue of pyruvate, dichloroacetate ($CHCl_2COO^-$), inhibits pyruvate dehydrogenase kinase and maintains the pyruvate dehydrogenase complex

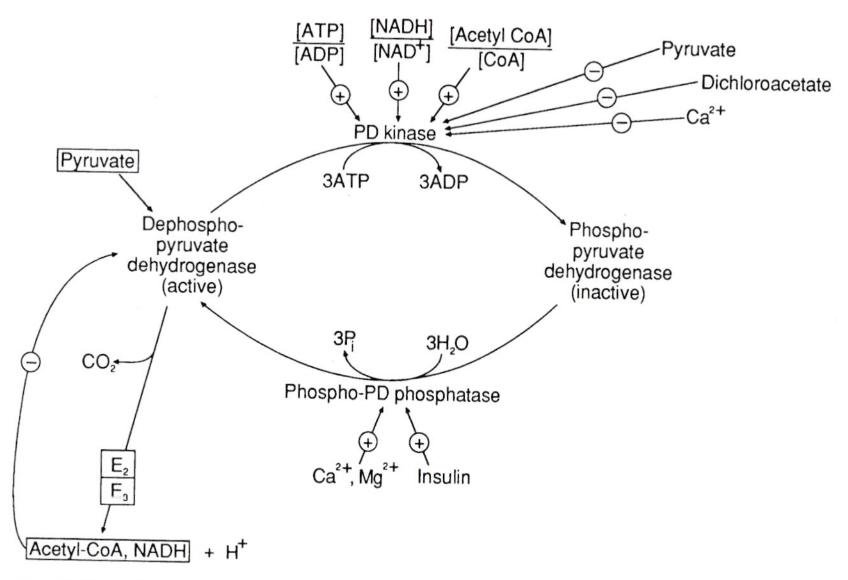

FIGURE 13-11
Regulation of pyruvate dehydrogenase (PD) by inactivation and reactivation by a non-cAMP-dependent phosphorylation-dephosphorylation cycle. Although PD kinase phosphorylates three specific seryl residues in the α-subunit of PD, phosphorylation at any of these sites inactivates PD. The kinase and the phosphatase are under the influence of several regulators, and the dephospho-active PD is also regulated by end products. ⊕ = Activation; ⊖ = inhibition; E_2 = dihydrolipoyl transacetylase; E_3 = dihydrolipoyl dehydrogenase.

in the active state. In experimental animals and in humans with lactic acidemia due to a variety of causes, dichloroacetate administration lowers the concentration of lactate through promotion of pyruvate oxidation. Compounds like dichloroacetate have potential applications in disorders associated with lactic acidemia (e.g., diabetes mellitus).

Abnormalities of Pyruvate Dehydrogenase Complex

Thiamine deficiency causes decreased pyruvate oxidation, leading to accumulation of pyruvate and lactate, particularly in the blood and brain, and is accompanied by impairment of the cardiovascular, nervous, and gastrointestinal systems (Chapter 38). Inherited deficiency of pyruvate dehydrogenase complex is accompanied by lactic acidemia and abnormalities of the nervous system (e.g., ataxia and psychomotor retardation). Pyruvate carboxylase deficiency causes similar abnormalities (Chapter 15). Both inherited disorders of pyruvate utilization are autosomal recessive.

Nervous system abnormalities may be attributed in part to diminished synthesis of neurotransmitters rather than to inadequate synthesis of ATP. In pyruvate dehydrogenase complex deficiency, diminished levels of acetyl-CoA cause decreased production of acetylcholine; in pyruvate carboxylase deficiency, decreased production of oxaloacetate may lead to deficiency of amino acid neurotransmitters (e.g., γ-aminobutyrate, glutamate, aspartate). The interrelationships of these amino acids are discussed in Chapter 17. Although no particular therapeutic method is well established in treatment of the inherited disorders of pyruvate metabolism, ketogenic diets have been beneficial in pyruvate dehydrogenase complex deficiency, since they provide the product of the deficient reaction (acetyl-CoA). Administration of large doses of thiamine may be of benefit because mutations of pyruvate dehydrogenase complex may give rise to decreased affinity for thiamine pyrophosphate. In one patient, administration of dichloroacetate was beneficial even in the absence of a ketogenic diet.

In pyruvate carboxylase deficiency, administration of diets supplemented with aspartate and glutamate demonstrated sustained improvement in neurological symptoms. These two amino acids cross the blood-brain barrier after amidation to asparagine and glutamine in nonneural tissues. The toxicity of organic arsenicals and arsenite (AsO_3^{3-})is due to their abiltiy to bind functional sulfhydryl groups of enzymes. One important target is the dithiol form of the lipoyl group of pyruvate dehydrogenase and α-ketoglutarate dehydrogenase complexes. In the earlier times arsenicals were used in the treatment of syphilis however due to their toxicity it has been replaced with better drugs such as penicillin.

13.3 Tricarboxylic Acid (TCA) Cycle

The tricarboxylic acid (TCA) cycle (also called the ***citric acid cycle*** or ***Krebs cycle***) consists of eight enzymes that oxidize acetyl-CoA with the formation of carbon dioxide, reducing equivalents in the form of NADH and $FADH_2$, and guanosine triphiosphate (GTP):

$$CH_3COSCoA + 3NAD^+ + FAD + GDP^{3-} + HPO_4^{2-} + 2H_2O \rightarrow CoASH + 2CO_2 + 3NADH + FADH_2 + GTP^{4-} + 2H^+$$

The reducing equivalents are transferred to the electron transport chain, ultimately to reduce oxygen and generate ATP through oxidative phosphorylation (Chapter 14). Thus, oxidation of acetyl-CoA yields a large supply of ATP.

Since the intermediates in the cycle are not formed or destroyed in its net operation, they may be considered to play catalytic roles. However, several intermediates are biosynthetic precursors of other metabolites (amphibolic role) and hence may become depleted. They are replenished (anaplerotic process) by other reactions to optimal concentrations.

The TCA cycle is the major final common pathway of oxidation of carbohydrates, lipids, and proteins, since their oxidation yields acetyl-CoA. Acetyl-CoA also serves as precursor in the synthesis of fatty acids, cholesterol, and ketone bodies. All enzymes of the cycle are located in the mitochondrial matrix except for succinate dehydrogenase, which is embedded in the inner mitochondrial membrane. Thus, the reducing equivalents generated in the cycle have easy access to the electron transport chain. TCA cycle enzymes, with the exception of α-ketoglutarate dehydrogenase complex and succinate dehydrogenase, are also present outside the mitochondria. The overall TCA cycle is shown in Figure 13-12.

Reactions of TCA Cycle

Condensation of Acetyl-CoA with Oxaloacetate to Form Citrate

The first reaction of the TCA cycle is catalyzed by citrate synthase and involves a carbanion formed at the methyl group of acetyl-CoA that undergoes aldol condensation with the carbonyl carbon atom of the oxaloacetate:

SCoA–*C=O–*CH_3 (Acetyl-CoA) + O=C(–COO^-)–CH_2–COO^- (Oxaloacetate) $\xrightarrow{H_2O}$ *CH_2–*COO^- / HO–C–COO^- / CH_2–COO^- (Citrate) + $CoASH + H^+$

The condensation reaction is thought to yield a transient enzyme-bound intermediate, citroyl-CoA, which undergoes hydrolysis to citrate and CoASH with loss of free energy. This reaction is practically irreversible and has a $\Delta G^{0'}$ of −7.7 kcal/mol (−32.2 kJ/mol). Formation of citrate is the committed step of the cycle and is regulated by allosteric effectors. Depending on the cell type, succinyl–CoA (a later intermediate of the cycle), NADH, ATP, or long-chain fatty acyl-CoA functions as the negative allosteric modulator of citrate synthase. Citrate formation is also regulated by availability of substrates, and citrate is an allosteric inhibitor.

Citrate provides the precursors (acetyl-CoA, NADPH) for fatty acid synthesis and is a positive allosteric modulator of acetyl-CoA carboxylase, which is involved in the initiation of long-chain fatty acid synthesis (Chapter 18). It regulates glycolysis by negative modulation of 6-phosphofructokinase activity (see above). All of the above reactions occur in the cytoplasm, and citrate exits from mitochondria via the tricarboxylate carrier.

Isomerization of Citrate to Isocitrate

In this reaction, the tertiary alcoholic group of citrate is converted to a secondary alcoholic group that is more readily oxidized. The isomerization catalyzed by aconitate dehydratase (or aconitase) occurs by removal and addition of water with the formation of an intermediate, *cis*-aconitate:

CH_2–COO^- / HO–C–COO^- / CH_2–COO^- (Citrate) $\xrightarrow{-H_2O}$ [CH_2–COO^- / C–COO^- ‖ CH–COO^-] (*cis*-Aconitate) $\xrightarrow{H_2O}$ CH_2–COO^- / CH–COO^- / HO–CH–COO^- (Isocitrate)

The *cis*-aconitate may not be an obligatory intermediate, since the enzyme presumably can isomerize

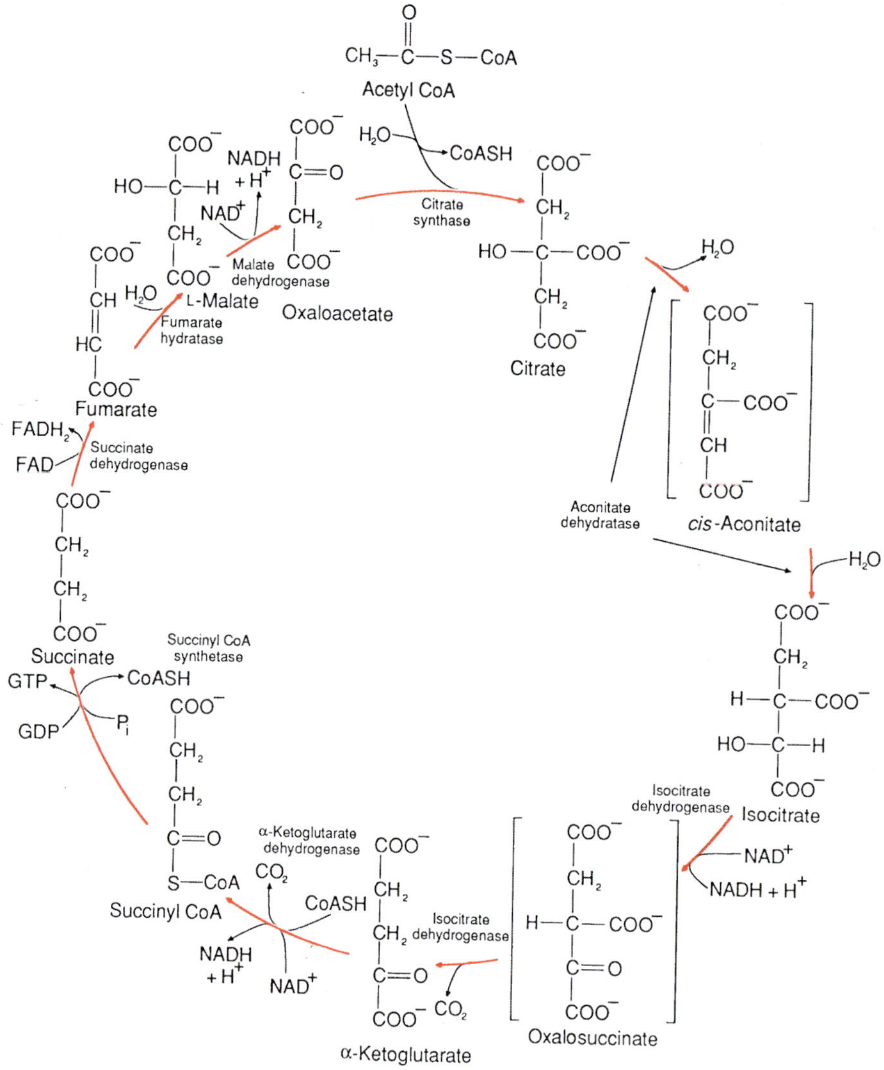

FIGURE 13-12
Reaction of the tricarboxylic acid (TCA) cycle.

citrate to isocitrate directly. Aconitate dehydratase has an iron-sulfur center, the exact function of which is not known. The reaction is considered to be near-equilibrium, even though under physiological conditions the ratio of citrate to isocitrate is about 9:1 owing to rapid conversion of isocitrate to subsequent products of the cycle.

Although citrate is a symmetrical molecule, its two $-CH_2COOH$ groups are not identical with regard to the orientation of the –OH and –COOH groups. Thus, citrate reacts in an asymmetrical manner with the enzyme, so that only the $-CH_2COOH$ group derived from oxaloacetate is modified. A three-point attachment of the enzyme to the substrate accounts for the asymmetrical nature of the reaction.

A powerful competitive inhibitor (highly toxic) of aconitate dehydratase is fluorocitrate, an analogue of citrate:

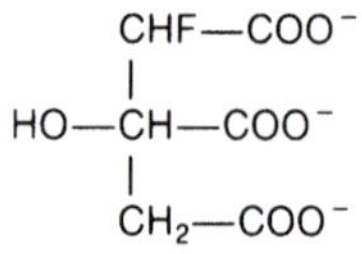

$$\begin{array}{c} CHF{-}COO^- \\ | \\ HO{-}CH{-}COO^- \\ | \\ CH_2{-}COO^- \end{array}$$

In vivo, fluorocitrate can be formed by condensation of fluoroacetyl-CoA with oxaloacetate by action of citrate synthase. Fluoroacetyl-CoA itself is synthesized from fluoroacetate by acetyl-CoA synthase.

Oxidative Decarboxylation of Isocitrate to α-Ketoglutarate

Isocitrate dehydrogenase catalyzes the first of two decarboxylations and dehydrogenations in the cycle. Three different isocitrate dehydrogenases are present: one specific for NAD^+ and found only in mitochondria, the other two specific for $NADP^+$ and found in mitochondria and cytoplasm. The NAD^+-specific enzyme is the primary enzyme with regard to TCA cycle operation. All three require Mg^{2+} or Mn^{2+}. The reaction yields α-ketoglutarate (2-oxoglutarate), NAD(P)H, and CO_2 and involves enzyme-bound oxalosuccinate as an intermediate.

$$\begin{array}{l} CH_2-COO^- \\ | \\ CH-COO^- \\ | \\ HO-CH-COO^- \end{array} \xrightarrow{NAD^+ \;\; NADH + H^+}$$

Isocitrate

$$\begin{array}{l} CH_2-COO^- \\ | \\ CH-COO^- \\ | \\ O{=}C-COO^- \end{array} \xrightarrow{CO_2} \begin{array}{l} H_2C-COO^- \\ | \\ CH_2 \\ | \\ O{=}C-COO^- \end{array}$$

Oxalosuccinate (enzyme-bound) — α-Ketoglutarate (2-oxoglutarate)

The reaction is nonequilibrium in type and has a $\Delta G^{0\prime}$ of −5.0 kcal/mol (−20.9 kJ/mol). Although in mitochondria both NAD^+- and $NADP^+$-linked enzymes are involved in the cycle, the NAD^+-linked enzyme, which is also under allosteric regulation, is the more predominant. Positive effectors are ADP and Ca^{2+}, and negative effectors are ATP and NADH. Thus, under conditions of abundance of energy the enzyme is inhibited, and under conditions of low energy the enzyme is stimulated.

Oxidative Decarboxylation of α-Ketoglutarate to Succinyl-CoA

The α-ketoglutarate dehydrogenase complex is analogous to the pyruvate dehydrogenase complex and consists of three enzymes: α-ketoglutarate dehydrogenase, dihydrolipoyl transsuccinylase, and dihydrolipoyl dehydrogenase (the latter dehydrogenase is identical in both complexes). The cofactor, prosthetic groups, and coenzyme requirements are identical to those of pyruvate oxidation.The overall reaction is

$$\begin{array}{l} CH_2-COO^- \\ | \\ CH_2 \\ | \\ O{=}C-COO^- \end{array} + CoASH + NAD^+ \xrightarrow{CO_2}$$

α-Ketoglutarate

$$\begin{array}{l} CH_2-COO^- \\ | \\ CH_2CO-SCoA \end{array} + NADH + H^+$$

The reaction is of the nonequilibrium type, with a $\Delta G^{0\prime}$ of −8 kcal/mol (−33.5 kJ/mol). Unlike the pyruvate dehydrogenase complex, the α-ketoglutarate dehydrogenase complex does not possess a complex regulatory mechanism involving a kinase and a phosphatase. However, activity is inhibited by high ratios of [ATP]/[ADP], [succinyl-CoA]/[CoASH], and [NADH]/[NAD^+] and stimulated by Ca^{2+}. α-Ketoglutarate is reversibly transaminated to glutamate and is utilized in the hydroxylation of prolyl or lysyl residues of collagen (Chapter 25).

Conversion of Succinyl-CoA to Succinate Coupled to Formation of GTP

In this complex reaction catalyzed by succinyl-CoA synthase (succinate thiokinase), the energy-rich thioester linkage of succinyl-CoA is hydrolyzed with release of free energy that is conserved in the substrate phosphorylation of GDP with phosphate to form GTP:

$$\begin{array}{l} CH_2COO^- \\ | \\ CH_2CO\text{-}SCoA \end{array} + GDP^{3-} + P_i^{2-} \rightleftarrows$$

Succinyl-CoA

$$\begin{array}{l} CH_2COO^- \\ | \\ CH_2COO^- \end{array} + GTP^{4-} + CoASH$$

Succinate

This reaction is readily reversible (a near-equilibrium reaction) and has a $\Delta G^{0\prime}$ of −0.7 kcal/mol (−2.9 kJ/mol). It proceeds with the formation of intermediates of the enzyme with succinyl phosphate and with phosphate (P_i), the latter being linked to histidyl residue of the enzyme (E):

$$E + \text{succinyl–CoA} + P_i^{2+} \rightleftharpoons \text{E–succinyl phosphate} + CoASH$$

$$\text{E–succinyl phosphate} \rightleftharpoons \text{E–phosphate} + \text{succinate}$$

$$\text{E–phosphate} + GDP \rightleftharpoons E + GTP$$

GTP is converted to ATP by nucleoside-diphosphate kinase:

$$GTP + ADP \rightleftharpoons GDP + ATP$$

Succinyl–CoA can also be synthesized from propionyl–CoA by way of methylmalonyl–CoA, which is formed in the oxidation of branched-chain amino acids (e.g., valine, isoleucine) and in the terminal stage of oxidation of odd-chain-length fatty acids (Chapter 18). Succinyl–CoA is utilized in the activation of acetoacetate (Chapter 18) and the formation of δ-aminolevulinate, a precursor of prophyrin (Chapter 29).

Dehydrogenation of Succinate to Fumarate

Succinate is dehydrogenated to the trans-unsaturated compound fumarate by succinate dehydrogenase, an FAD enzyme:

$$\underset{\text{Succinate}}{\begin{matrix} CH_2COO^- \\ | \\ CH_2COO^- \end{matrix}} + E\text{—}FAD \longrightarrow \underset{\text{Fumarate}}{\begin{matrix} HCCOO^- \\ \| \\ {}^-OOCCH \end{matrix}} + E\text{—}FADH_2$$

This reaction is the only dehydrogenation reaction of the TCA cycle in which NAD^+ does not mediate the transport of reducing equivalents. The FAD is covalently linked to a histidyl residue of the enzyme, which is an integral protein of the inner mitochondrial membrane and contains iron-sulfur centers that undergo redox changes ($Fe^{3+} + e^- \rightleftharpoons Fe^{2+}$). Reducing equivalents from $FADH_2$ enter the electron transport chain at the coenzyme Q level, bypassing one of the sites of oxidative phosphorylation (Chapter 14). The enzyme is stereospecific for the trans hydrogen atoms of the methylene carbons of the substrate, so that only the trans isomer is produced (the cis isomer is maleate).

Succinate dehydrogenase is competitively inhibited by malonate, the next lower homologue of succinate (Chapter 6).

Hydration of Fumarate to Malate

In this reaction, water is added across the double bond of fumarate by fumarate hydratase (fumarase) to yield the hydroxy compound L-malate:

$$\underset{\text{Fumarate}}{\begin{matrix} HC\text{—}COO^- \\ \| \\ {}^-OOC\text{—}CH \end{matrix}} + H_2O \rightleftharpoons \underset{\text{L-Malate}}{\begin{matrix} HO\text{—}CH\text{—}COO^- \\ | \\ CH_2\text{—}COOH \end{matrix}}$$

The enzyme has absolute specificity for the double bond of the *trans*-unsaturated acid and for the formation of L-hydroxy acid. The reaction is freely reversible (near-equilibrium).

Dehydrogenation of Malate to Oxaloacetate

In the last reaction of the cycle, L-malate is oxidized to oxaloacetate by malate dehydrogenase, an NAD^+-linked enzyme:

$$\underset{\text{L-Malate}}{\begin{matrix} HO\text{—}CHCOO^- \\ | \\ CH_2\text{—}COO^- \end{matrix}} + NAD^+ \rightleftharpoons \underset{\text{Oxaloacetate}}{\begin{matrix} O \\ \| \\ C\text{—}COO^- \\ | \\ CH_2\text{—}COO^- \end{matrix}} + NADH + H^+$$

Although the equilibrium of this reaction favors malate formation, *in vivo* the reaction proceeds toward the formation of oxaloacetate, since the latter is rapidly removed by the citrate synthase reaction to initiate the next round of the cycle.

Stereochemical Aspects of the TCA Cycle

Aconitate dehydratase, succinate dehydrogenase, and fumarase yield stereospecific products. Labeling experiments with ^{14}C in methyl and carboxyl carbons of acetyl-CoA or in all of the carbons of oxaloacetate yield the following results in terms of the intermediates or product formed.

When only labeled acetyl-CoA is used, the label does not appear in CO_2 during the first revolution of the cycle, and upon completion of the first revolution, two of the four carbon atoms of oxaloacetate are labeled. Thus, during the first revolution, neither carbon atom of acetyl-CoA is oxidized to CO_2 and the carbon atoms of the CO_2 produced are derived from oxaloacetate. This finding is due to discrimination of the two $–CH_2COO^-$ groups of citrate by aconitate dehydratase, so that citrate is treated as a chiral molecule at the asymmetrical active center of the enzyme. However, during the succeeding revolutions of the cycle, the label is randomized because succinate, a symmetrical compound, is treated as such by the enzyme without discriminating between its two carboxyl groups. If only labeled oxaloacetate is used, half of the label is retained in the oxaloacetate at the end of the first revolution of the cycle.

Amphibolic Aspects of the TCA Cycle

At each turn of the TCA cycle, oxaloacetate is regenerated and can combine with another acetyl-CoA molecule. The TCA cycle is amphibolic; i.e., it serves as a catabolic and an anabolic pathway. Reactions that utilize intermediates of the cycle as precursors for the biosynthesis of other molecules are listed below. Some of these reactions occur outside the mitochondria.

1. Citrate + ATP + CoA $\rightarrow$ acetyl–CoA + oxaloacetate + ADP + P_i. This reaction takes place in the cytoplasm and is a source of acetyl-CoA for fatty acid biosynthesis.
2. α-Ketoglutarate + alanine $\rightleftharpoons$ glutamate + pyruvate.
3. α-Ketoglutarate $\rightarrow$ succinate + CO_2. This reaction is involved in the hydroxylation of prolyl and lysyl residues of protocollagen, a step in the synthesis of collagen.
4. Succinyl-CoA + glycine $\rightarrow$ δ-aminolevulnic acid (ALA). ALA is then utilized for the synthesis of heme.
5. Succinyl-CoA + acetoacetate $\rightarrow$ acetoacetyl-CoA + succinate. This reaction is important in the activation of acetoacetate (a ketone body) and hence for its utilization in extrahepatic tissues.
6. Oxaloacetate + alanine $\rightleftharpoons$ aspartate + pyruvate.
7. Oxaloacetate + GTP $\xrightarrow{Mg^{2+}}$ phosphoenolpyruvate + GDP + CO_2.

Utilization of intermediates in these reactions leads to their depletion and hence to a slowdown of the cycle unless they are replenished by the replacement or anaplerotic reactions listed below:

1. Pyruvate + CO_2 + ATP $\xrightarrow{\text{biotin, } Mg^{2+}}$ oxaloacetate + ADP + P_i. This reaction, catalyzed by pyruvate carboxylase, is the most important anaplerotic reaction in animal tissues and occurs in mitochondria. Pyruvate carboxylase is an allosteric enzyme that requires acetyl-CoA for activity (see gluconeogenesis; Chapter 15).
2. Pyruvate + CO_2 + NADPH + H^+ $\rightleftharpoons$ L-malate + $NADP^+$.
3. Oxaloacetate and α-ketoglutarate may also be obtained from aspartate and glutamate, respectively, by transaminase reactions.
4. Oxaloacetate obtained in the reverse of reaction (7), above.

Regulation of the TCA Cycle

TCA cycle substrates oxaloacetate and acetyl-CoA and the product NADH are the critical regulators. The availability of acetyl-CoA is regulated by pyruvate dehydrogenase complex. The TCA cycle enzymes citrate synthase, isocitrate dehydrogenase and the α-ketoglutarate dehydrogenase complex are under regulation by many metabolites (discussed above) to maintain optimal cellular energy needs. Overall fuel homeostasis is discussed in Chapter 22.

Energetics of the TCA Cycle

Oxidation of one molecule of acetyl-CoA in the cycle yields three NADH, one $FADH_2$, and one GTP. Transport of reducing equivalents in the electron transport chain from one NADH molecule yields three ATP and from $FADH_2$, two ATP (see Chapter 14). Thus, oxidation of one molecule of acetyl-CoA yields 12 ATP. Since one glucose molecule yields two acetyl-CoA, the yield of ATP is 24. In addition, in the oxidation of glucose to acetyl-CoA, two other steps yield four NADH (i.e., glyceraldehyde-3-phosphate dehydrogenase and pyruvate dehydrogenase reactions), since one glucose molecule yields two triose phosphate molecules, which accounts for 12 more ATP. Oxidation of one molecule of glucose to two of pyruvate yields two ATP (see under glycolysis) by substrate-level phosphorylation. Thus, complete oxidation of one molecule of glucose can yield 38 ATP. In cells dependent only on anaerobic glycolysis, glucose consumption has to be considerably greater to derive an amount of energy equal to that obtainable from aerobic oxidation (two ATP versus 38 ATP per glucose). However, in cells capable of both aerobic and anaerobic metabolism, glycolysis, the TCA cycle, and oxidation in the electron transport chain are integrated and regulated so that only just enough substrate is oxidized to satisfy the energy needs of a particular cell. In the presence of oxygen, a reduction in glucose utilization and lactate production takes place, a phenomenon known as the ***Pasteur effect.*** The depression of rate of flux through glycolysis can be explained in part by the accumulation of allosteric inhibitors (e.g., ATP) of 6-phosphofructokinase and pyruvate kinase, together with the supply of cofactors and coenzymes of certain key enzymes (e.g., glyceraldehyde-3-phosphate dehydrogenase).

Supplemental Readings and References

D. C. De Vivo, R. R. Trifiletti, R. I. Jacobson, et al.: Defective glucose transport across the blood-brain barrier as a cause of persistent hypoglycorrhachia, seizures and developmental delay. *New England Journal of Medicine* **325,** 703 (1991).

O. N. Elpeleg, W. Ruitenbeek, C. Jakobs, et al.: Congenital lacticacidemia caused by lipoamide dehydrogenase deficiency with favorable outcome. *Journal of Pediatrics* **126,** 72 (1995).

R. A. Fishman: The glucose-transporter protein and glucopenic brain injury. *New England Journal of Medicine* **325,** 731 (1991).

R. S. Hotchkiss and I. E. Karl: Reevaluation of the role of cellular hypoxia and bioenergetic failure in sepsis. *Journal of the American Medical Association* **267,** 1503 (1992).

S. J. Hunter and T. Garvey: Insulin action and insulin resistance: Diseases involving defects in insulin receptors, signal transduction, glucose transport effector system. *American Journal of Medicine* **105,** 331 (1998).

M. A. Magnuson: Glucokinase gene structure. *Diabetes* **39,** 523 (1990).

A. Mattevi, M. Bolognesi, and G. Valentini: The allosteric regulation of pyruvate kinase. *FEBS Letters* **389,** 15 (1996).

M. A. Permutt, K. C. Chiu, and Y. Tanizawa: Glucokinase and NIDDM. *Diabetes* **41,** 1367 (1992).

S. J. Pilkis, I. T. Weber, R. W. Harrison, and G. I. Bell: Glucokinase: Structural analysis of a protein involved in susceptibility to diabetes. *Journal of Biological Chemistry* **269,** 2192 (1994).

B. H. Robinson: Lactic acidemia. In: *The Metabolic and Molecular Bases of Inherited Disease,* 7th ed. C. R. Scriver, A. L. Bendet, and S. Sly (Eds). McGraw-Hill, New York, 1995, p. 1479.

K. R. Tanaka and D. E. Paglia: Pyruvate kinase and other enzymopathies of the erythrocyte. In: *The Metabolic and Molecular Bases of Inherited Disease,* 7th ed. C. R. Scriver, A. L. Bendet, and S. Sly (Eds). McGraw-Hill, New York, 1995, p. 3485.

J. Uribarri, M. S. Oh, and H. J. Carroll: D-Lactic acidosis. *Medicine* **77,** 73 (1998).

CHAPTER 14

Electron Transport and Oxidative Phosphorylation

The energy requirements of aerobic cells are met by the energy released in the oxidation of carbohydrates, fatty acids, and amino acids by molecular oxygen. In these oxidation processes, the reducing equivalents from substrates are transferred to NAD^+, flavin mononucleotide (FMN), or flavin adenine dinucleotide (FAD). The hydrogens, electrons, or hydride (H^-) ions are removed catalytically from NADH and reduced flavin nucleotides and transferred through a series of coupled reduction-oxidation (redox) reactions to oxygen, which serves as the ultimate electron acceptor. The reduced oxygen is then converted to water. The entire process is known as ***cell respiration.*** The numerous coupled redox reactions are tightly linked and take place in the inner mitochondrial membrane. This reaction sequence is called the respiratory chain ***electron transport.*** Electrons in substrate or cofactor begin with a high potential energy and end at oxygen with a lower potential energy. During this electron flow, a portion of the free energy liberated is conserved by an energy-transducing system (by which electrical energy is changed to chemical energy). Since the energy is conserved in the terminal phosphoanhydride bond of ATP through phosphorylation of ADP to ATP, the overall coupled process is known as ***oxidative phosphorylation.***

The following reactions yield NADH or $FADH_2$ during the breakdown of glucose by glycolysis and in the TCA cycle (Chapter 13). The NADH-producing reactions are

Glyceraldehyde 3-phosphate → 1,3-bisphosphoglycerate
Pyruvate → acetyl-CoA
Isocitrate → α-ketoglutarate
α-Ketoglutarate → succinyl-CoA
Malate → oxaloacetate

The $FADH_2$-producing reaction is

Succinate → fumarate

Except for the conversion of glyceraldehyde 3-phosphate to 1,3-bisphosphoglycerate occurring in the cytoplasm, all of the above reactions take place in the mitochondria. Since mitochondria are not permeable to NADH, two shuttle pathways transport the reducing equivalents of cytoplasmic NADH into the mitochondria.

The TCA cycle functions for both the oxidation and the production of reducing equivalents from a variety of metabolites such as carbohydrates, amino acids, and fatty acids. In the mitochondria fatty acids undergo successive removal of two carbon units in the form of acetyl-CoA, which is converted in the TCA cycle to CO_2 and reducing equivalents found in NADH and $FADH_2$. The latter two substances are then fed into the electron transport chain. The formation of each acetyl-CoA molecule requires the removal of four H atoms and is catalyzed by two dehydrogenases (Chapter 18). One dehydrogenase yields $FADH_2$; the other yields NADH.

The term "one reducing equivalent" means that 1 mol of electrons is present in the form or one gram-equivalent of a reduced electron carrier. One mole of NAD^+, when reduced to 1 mol of NADH, utilizes 2 mol of electrons from 1 mol of substrate (Chapter 13):

$$NAD^+ + SH_2 \text{ (reduced substrate)} \rightleftharpoons NADH + H^+ + S \text{ (oxidized substrate)}$$

In this reaction, the 2 mol of electrons is transferred in the form of 1 mol of hydride ion $(H:)^-$.

14.1 Mitochondrial Structure and Properties

Mitochondria are present in the cytoplasm of aerobic eukaryotic cells. They are frequently found in close proximity to the fuel sources and to the structures that require ATP for maintenance and functional activity (e.g., the contractile mechanisms, energy-dependent transport systems, and secretory processes). The number of mitochondria in a single cell varies from one type of cell to another; a rat liver cell contains about 1000, while one giant amoeba has about 10,000. In a given cell, the number of mitochondria may also depend on the cell's stage of development or functional activity.

The size and shape of mitochondria vary considerably from one cell type to another. Even within the same cell, mitochondria can undergo changes in volume and shape depending on the metabolic state of the cell. In general, they are 0.5–1.0 μm wide and 2–3 μm long and are known to aggregate end to end, forming long filamentous structures.

Mitochondria consist of two membranes, one encircling the other, creating two spatial regions: the intermembrane space and the central space, called the matrix. The outer membrane is 6–7 nm thick, smooth, unfolded (Figure 14-1), and freely permeable to molecules with molecular weights below 10,000. It contains a heterogeneous group of enzymes that catalyze certain reactions of lipid metabolism as well as hydroxylation reactions (Table 14-1). The intermembrane space (5–10 nm) contains the enzymes that catalyze interconversion of adenine nucleotides.

The inner membrane (6–8 nm thick) has many folds directed toward the matrix. These invaginations, known as cristae, increase the surface area of the inner membrane. The lipid component, almost all of which is phospholipid, constitutes 30–35% by weight of the inner membrane.

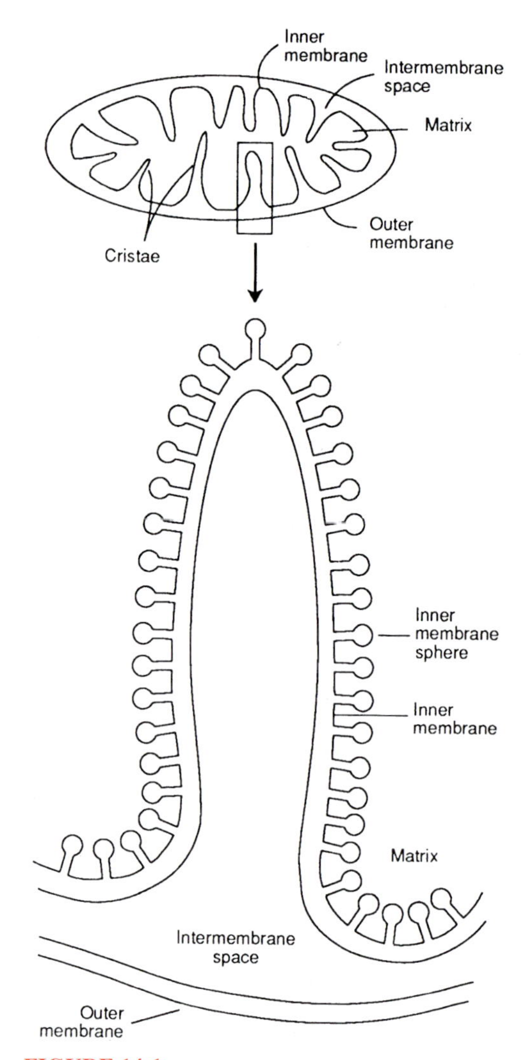

FIGURE 14-1
Morphology of a mitochondrion (transverse section).

Phospholipids are asymmetrically distributed in the lipid bilayer, with phosphatidylethanolamine predominating on the matrix side and phosphatidylcholine on the cytoplasmic side. Seventy-five percent of the cardiolipin is present on the matrix side of the membrane. The fatty acid composition of the phospholipids depends on the species, tissue, and diet. In all cases, sufficient unsaturated fatty acids are contained in the phospholipids to provide a highly fluid membrane at physiological temperatures.

The inner membrane is studded with spheres, each 8–10 nm in diameter, that are attached via stalks 4–5 nm in length. These inner membrane spheres are present on the matrix side (M-side) but absent from the cytoplasmic side (C-side). The components of the inner membrane include respiratory chain proteins, a variety of transport molecules and a part of the ATP-synthesizing apparatus (the base piece of ATP synthase). The ATP–ADP translocase and

TABLE 14-1
Distribution of Enzymes in Mitochondrial Compartments

Outer Membrane	Intermembrane Space	Inner Membrane	Matrix
NADH cytochrome b_5 reductase	Adenylate kinase (myokinase)	NADH-coenzyme Q reductase	Pyruvate dehydrogenase complex
Cytochrome b_5	Nucleoside diphosphokinase	Succinate-coenzyme Q reductase	α-Ketoglutarate dehydrogenase complex
Monoamine oxidase	Nucleoside monophosphokinase	Coenzyme QH_2-cytochrome c reductase	Citrate synthase
Kynurenine hydroxylase	Sulfite oxidase	Cytochrome oxidase	Aconitase
Glycerolphosphate acyltransferase		Oligomycin-sensitive ATPase	Malate dehydrogenase
Lysophosphatidyl transferase		β-Hydroxybutyrate dehydrogenase	Isocitrate dehydrogenase [NAD^+]
Phosphatidate phosphatase		Pyridine nucleotide transhydrogenase	Isocitrate dehydrogenase [$NADP^+$]
Phospholipase A		Carnitine palmitoyltransferase	Fumarase
Nucleoside diphosphokinase		Ferrochelatase (heme synthase)	Glutamate dehydrogenase
Fatty acid elongation enzyme system		Adenine nucleotide carrier and other carrier proteins	Pyruvate carboxylase
		Carbamoyl phosphate synthetase I	Aspartate aminotransferase
		Dihydroorotate dehydrogenase (C-side)	Ornithine carbamoyltransferase
		Glycine cleavage enzyme complex	Fatty acyl-CoA synthetase
			Fatty acyl-CoQ dehydrogenase
			Enoyl hydrase
			β-Hydroxyacyl-CoA dehydrogenase
			β-Ketoacyl-CoA thiolase
			Amino acid activating enzymes
			RNA polymerase
			DNA polymerase
			δ-Aminolevulinic acid synthase
			HMG-CoA synthase
			HMG-CoA lyase
			Acetyl-CoA acetyltransferase
			β-Ketoacid CoA-transferase
		Rhodanese (detected in both inner membrane and matrix)	
		*Coproporphyrinogen oxidase	
		*Protoporphyrinogen oxidase	

*The exact location is not yet known.

he ATP synthase together make up at least two thirds of all protein within the inner mitochondrial membrane. Small uncharged molecules (e.g., water, oxygen, carbon dioxide, ammonia, and ethanol) can diffuse through the inner membrane, but all other molecules that pass through require specific transport systems. The mitochondrial outer and inner membranes are vastly different in their constituents and function:

1. The outer membrane contains two to three times more phospholipids per unit of protein;
2. Cardiolipin is localized in the inner membrane; and
3. Cholesterol is found predominantly in the outer membrane.

The permeability of the outer membrane to charged or uncharged substances up to a molecular weight of 10,000

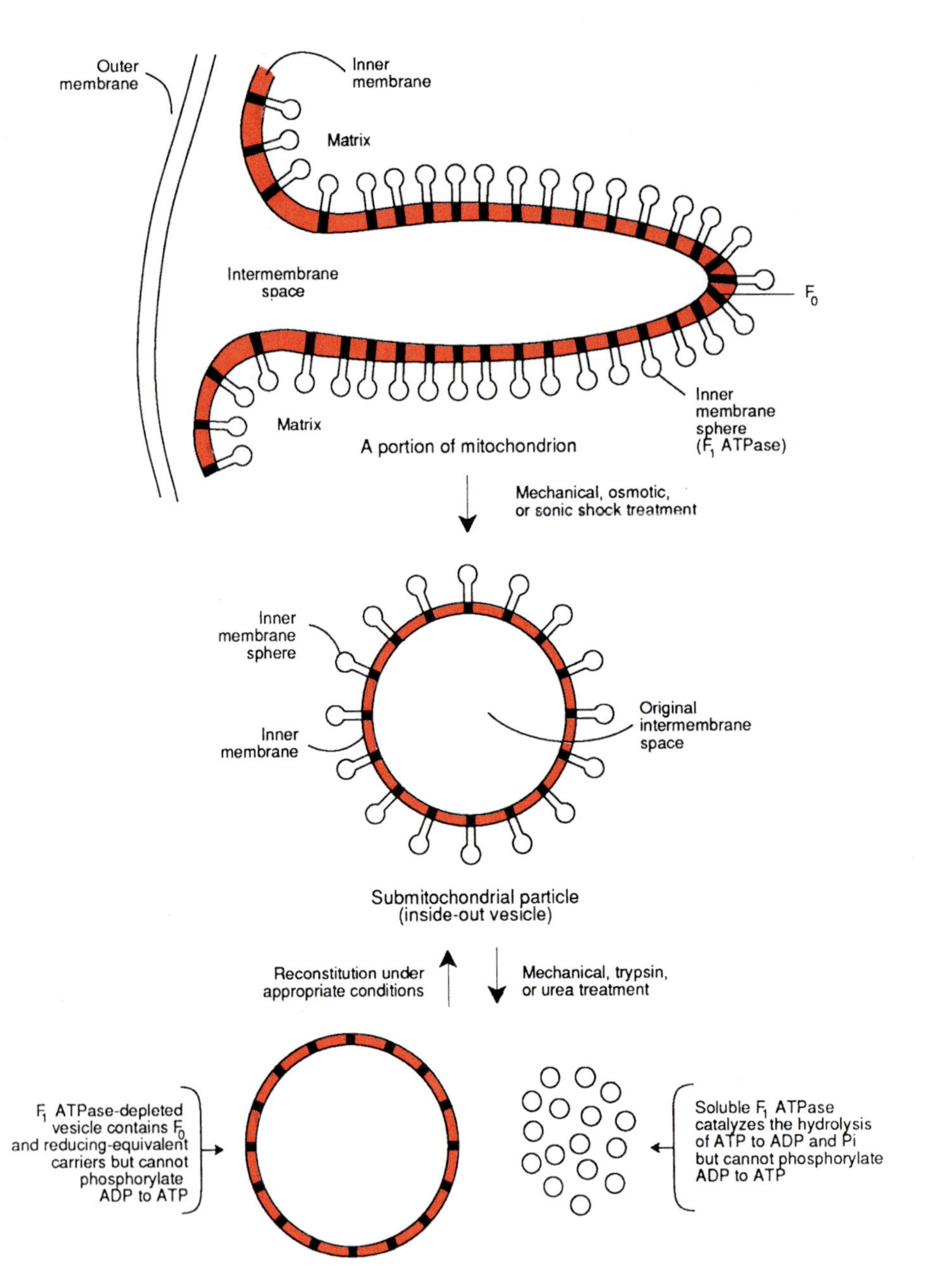

FIGURE 14-2
Formation of submitochondrial particles. These resealed inside-out inner membrane vesicles are formed when mitochondria are subjected to a variety of disruptive forces; they can support both reducing-equivalent transport and phosphorylation of ADP. Removal of the inner membrane spheres from the vesicles leads to the loss of ability to phosphorylate ADP.

is due to pore-like structures that consist of a protein (M.W. 30,000) embedded in the phospholipid bilayer. *In vitro*, addition of the purified pore protein renders phospholipid bilayers permeable to a variety of substances. An analogous pore protein, known as porin, has been identified in the outer membrane of gram-negative bacteria (Chapter 11).

The matrix is viscous and contains all TCA cycle enzymes except succinate dehydrogenase, which is a component of electron transport complex II and is located within the inner membrane. Enzymes of the matrix or the inner mitochondrial membrane mediate reactions of fatty acid oxidation, ketone body formation and oxidation, and biosynthesis of urea, heme, pyrimidines, DNA, RNA, and protein. Mitochondria are involved in programmed cell death apoptosis, (see Chapter 26).

The structure, mechanism of replication of mitochondrial DNA, and processes of transcription and translation are unique in several respects (discussed later).

Submitochondrial Particles

Submitochondrial particles (SMPs) are produced by disruption of the mitochondria by mechanical, osmotic, or sonic shock treatment. The fragmentation results in the release of water-soluble components and inner membrane fragments that re-form into vesicles (Figure 14-2). The components are then separated by differential centrifugation. The membranes of SMPs have the characteristic inner membrane spheres on their outside. SMPs are capable of electron transport and oxidative phosphorylation (i.e., the synthesis of ATP from ADP and phosphate). Removal of the inner membrane spheres by further mechanical treatment, urea, or trypsin results in the dissociation of the electron transport assembly from ATP synthesis. The ability to synthesize ATP resides in the overall structure, which includes the spheres (called F_1), the stalks, and a membrane protein subunit (called F_o). The F_o-subunit spans the inner membrane and thus is retained in the vesicles.

The spheres removed from SMPs do not support ATP synthesis but do hydrolyze ATP to ADP and phosphate. Thus, ATP synthesis is carried out by F_o/F_1–ATPase (ATP synthase). The subscript "o" in F_o indicates that it contains the site at which a potent antibiotic inhibitor, oligomycin, binds and inhibits oxidative phosphorylation. Oligomycin does not bind F_1–ATPase and does not inhibit ATP hydrolysis to ADP and phosphate.

$$\mathrm{ADP} + \mathrm{P_i} \underset{\substack{F_1\text{–ATPase} \\ \text{(Not inhibited by oligomycin)}}}{\overset{\substack{F_oF_1\text{–ATPase} \\ \text{(ATP synthase;} \\ \text{inhibited by oligomycin)}}}{\rightleftharpoons}} \mathrm{ATP} + \mathrm{H_2O}$$

The fully active SMPs can be reconstituted by adding F_1–ATPase to depleted vesicles under appropriate conditions.

Components of the Electron Transport Chain

The electron transport system can be reconstituted into discrete enzyme complexes that catalyze the following four reactions:

1. $\mathrm{NADH} + \mathrm{H^+} + \text{coenzyme Q (Q)} \xrightarrow{\text{NADH-coQ reductase}} \mathrm{NAD^+} + \mathrm{QH_2}$
2. $\text{Succinate} + \mathrm{Q} \xrightarrow{\text{Succinate-CoQ reductase}} \text{fumarate} + \mathrm{QH_2}$
3. $\mathrm{QH_2} + 2\text{ferricytochrome c} \xrightarrow{\text{coQ cytochrome c reductase}} \mathrm{Q} + 2\text{ ferrocytochrome c} + 2\mathrm{H^+}$
4. $2\text{ Ferrocytochrome c} + 2\mathrm{H^+} + 1/2\mathrm{O_2} \xrightarrow{\text{cytochrome c oxidase}} 2\text{ ferricytochrome c} + \mathrm{H_2O}$

Each complex can be considered as a functional unit composed of a fixed number of electron carriers. The individual components of the four complexes are firmly bonded together and are not dissociated by mild fractionation procedures, whereas the bonds holding unlike complexes are relatively weak and can be dissociated. The functional organization of the four complexes in the inner mitochondrial membrane is shown in Figure 14-3. In addition to these complexes, the F_0/F_1–ATPase, which is required for ATP synthesis, may be considered as complex V. The relative ratios of complexes I, II, III, IV, and V have been estimated to be 1:2:3:6:6. Complexes I, II, III, and IV can be combined in the presence of cytochrome c (which separates during fractionation) to form a single unit with all of the enzymatic properties of the intact electron transport system except coupled phosphorylation.

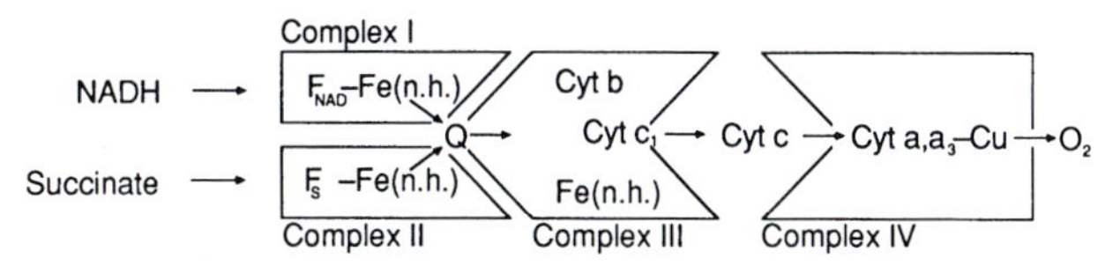

FIGURE 14-3
Diagram of the functional complexes of the electron transport system within the respiratory chain. F_{NAD} = NADH dehydrogenase flavoprotein; F_s = succinate dehydrogenase flavoprotein; Fe(n.h.) = nonheme iron.

The individual electron carriers of the four complexes of the respiratory chain, shown in Figure 14-4, are arranged in accordance with their redox potentials, with the transfer of electrons from NADH to oxygen associated with a potential drop of 1.12 V, and that of succinate to oxygen of 0.8 V. In the electron transport system, the electrons can be transferred as hydride ions $(\mathrm{H{:}})^-$ or as electrons (e.g., in the cytochromes).

Electron Transport Complexes

Complex I

Complex I catalyzes an NADH-CoQ reductase activity, and it contains the NADH dehydrogenase flavoprotein. It has two types of electron-carrying structures: FMN and several iron-sulfur centers. FMN is a tightly bound prosthetic group of the dehydrogenase enzyme, and it is reduced to $\mathrm{FMNH_2}$ by the two reducing equivalents derived from NADH:

$$\mathrm{NADH} + \mathrm{H^+} + \mathrm{E\text{–}FMN} \rightleftharpoons \mathrm{NAD^+} + \mathrm{E\text{–}FMNH_2}$$

The electrons from $\mathrm{FMNH_2}$ are transferred to the next electron carrier, coenzyme Q, via the iron-sulfur centers of the NADH-CoQ reductase. The iron–sulfur centers consist of iron atoms paired with an equal number of acid-labile sulfur atoms. The respiratory chain iron–sulfur clusters are of the $\mathrm{Fe_2S_2}$ or $\mathrm{Fe_4S_4}$ type. The iron atom, present as nonheme iron, undergoes oxidation-reduction cycles $(\mathrm{Fe^{2+}} \rightleftharpoons \mathrm{Fe^{3+}} + \mathrm{e^-})$. In the $\mathrm{Fe_4S_4}$ complexes, the centers

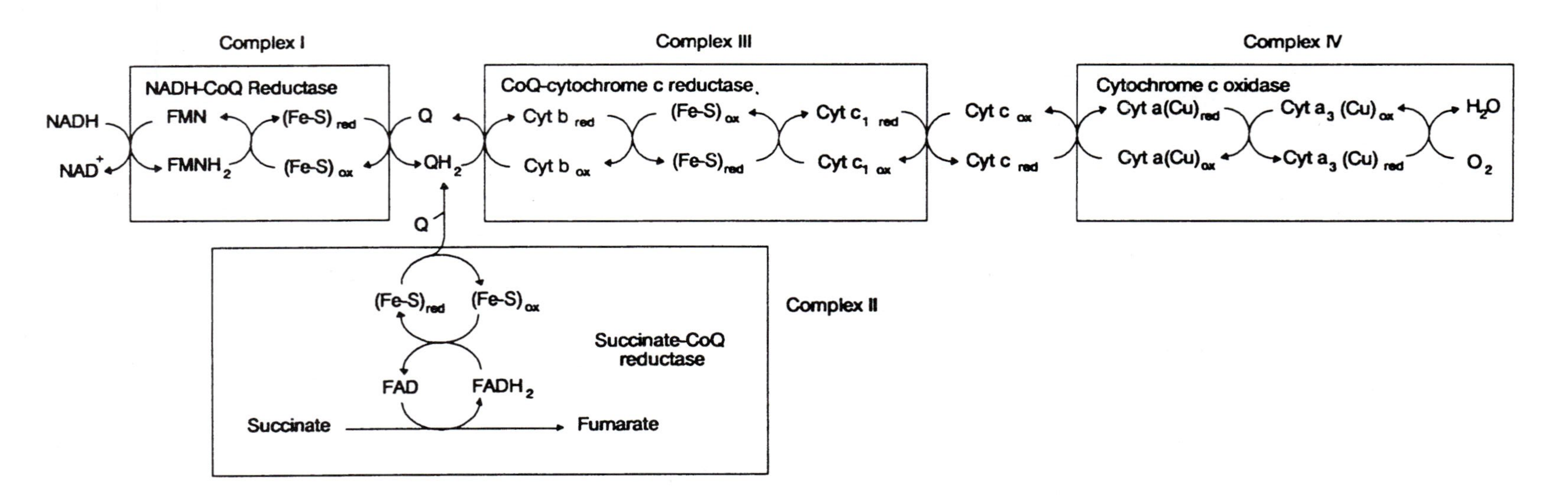

FIGURE 14-4
Mitochondrial electron transport system.

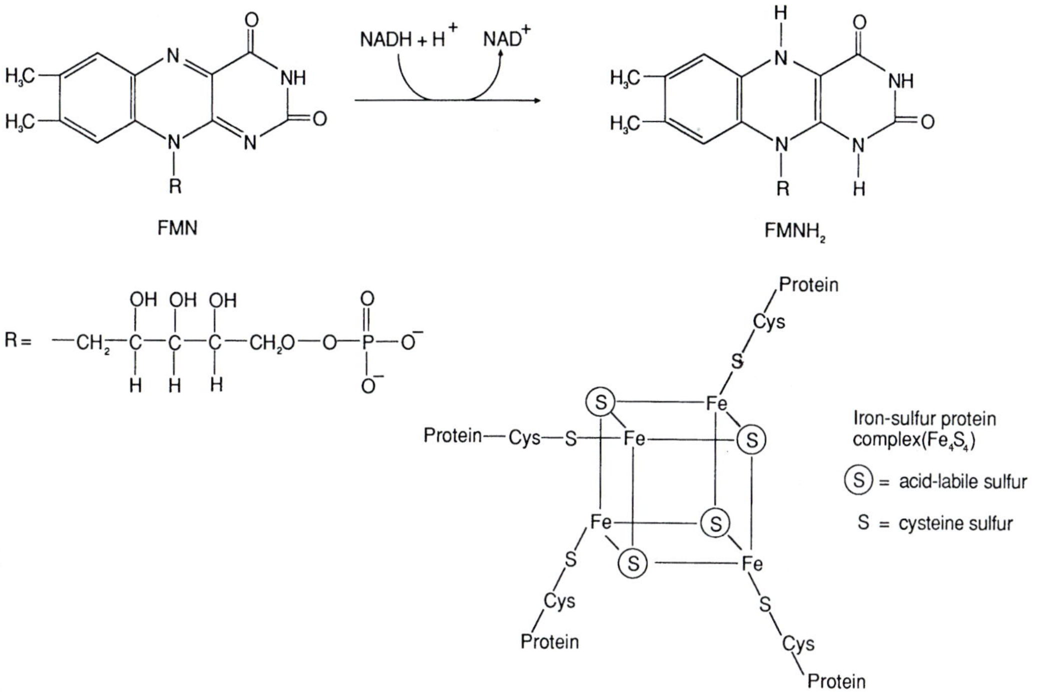

FIGURE 14-5
Transport of reducing equivalents from NADH to FMN and structure of the iron-sulfur protein complex that mediates electron transport from $FMNH_2$ to CoQ. Both FMN and the iron-sulfur centers are components of NADH-CoQ reductase.

are organized such that iron and sulfur atoms occupy alternate corners of a cube. The four iron atoms are covalently linked via the cysteinyl sulfhydryl groups of the protein (Figure 14-5). Complex I is inhibited by rotenone (a natural toxic plant product), amobarbital (a barbiturate), and piericidin A (an antibiotic) (Figure 14-6). all of which act at specific points and are useful in the study of electron transport.

Complex II

Complex II contains succinate dehydrogenase and its iron-sulfur centers. The complex has also been reported to contain a specific cytochrome, cytochrome b_{558}. ***Cytochromes*** are heme proteins that undergo oxidation-reduction reactions and are differentiated on the basis of their apoprotein structure, heme structure, and optical absorption in the visible spectrum. The mitochondrial electron transport chain contains at least six different cytochromes classified into three groups (a, b, and c). It is usual to indicate the absorption maximum of the α-band of a particular cytochrome (e.g., cytochrome b_{558}). Succinate dehydrogenase, an FAD-containing enzyme, is part of the TCA cycle and catalyzes the trans elimination of two hydrogens from succinate to form fumarate (Chapter 13).

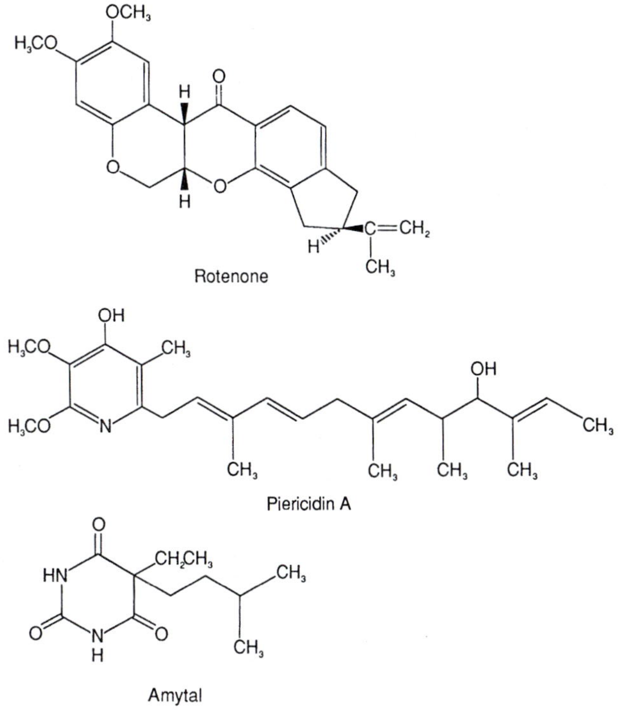

FIGURE 14-6
Inhibitors of NADH–CoQ reductase: rotenone (a toxic plant product), piericidin A (an antibiotic), and amytal (a barbiturate).

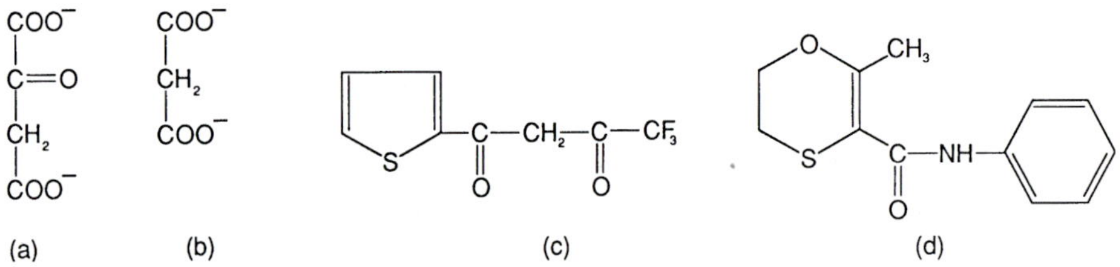

(a) (b) (c) (d)

FIGURE 14-7
Inhibitors of complex II. (a) Oxaloacetate, (b) malonate, (c) thenoyltrifluoroacetone, and (d) carboxin.

The hydrogens are accepted by FAD, which is covalently bound to the apoprotein via a histidine residue. In many flavoproteins, the flavin nucleotide is bound to the apoprotein not covalently but rather via ionic linkages with the phosphate group. The reducing equivalents of $FADH_2$ are passed on to coenzyme Q (CoQ or Q) via the iron-sulfur centers. Thus, the overall reaction catalyzed by complex II is

Succinate + FAD—(Fe—S)$_n$—E
↓
Fumarate + $FADH_2$—(Fe—S)$_n$—E
↓ ← Q
FAD—(Fe—S)$_n$—E + QH_2

During the terminal stages of electron transfer in complex II, cytochrome b_{558} is involved; however, its specific function is not understood. Oxaloacetate and malonate are competitive inhibitors of succinate dehydrogenase and compete with the substrate for binding at the active site (Chapters 6 and 13). Carboxin and thenoyltrifluoroacetone (Figure 14-7) inhibit electron transfer from $FADH_2$ to CoQ.

Complex III

Complex III contains cytochromes b_{562} and b_{566}, (collectively called cytochrome b), cytochrome c_1, and an iron-sulfur protein. Complex III catalyzes the transport of reducing equivalents from coQ to cytochrome c:

$$QH_2 + 2\text{Cyt c } (Fe^{3+}) \rightarrow Q + 2\text{Cyt c } (Fe^{2+}) + 2H^+$$

Coenzyme Q, also called *ubiquinone* because of its ubiquitous occurrence in microorganisms, plants, and animals, is lipid-soluble and not tightly or covalently linked to a protein, although it carries out its electron transport function together with specific CoQ-binding peptides. It plays a central role in the electron transport chain because it collects reducing equivalents from NADH- and $FADH_2$-linked dehydrogenases and passes them on to the terminal cytochrome system. CoQ is a substituted 1,4-benzoquinone containing a polyisoprenoid side chain at C_6 (Figure 14-8). In bacteria, CoQ usually contains six isoprenoid units (Q_6), whereas in most mammalian mitochondria it has ten (Q_{10}). The reduction of Q to QH_2 (a hydroquinone) requires two electrons and two protons and probably occurs via a one-electron intermediate:

$$Q \underset{1e^-, 1H^+}{\overset{1e^-, 1H^+}{\rightleftharpoons}} QH^{\cdot} \underset{1e^-, 1H^+}{\overset{1e^-, 1H^+}{\rightleftharpoons}} QH_2$$

where the dot associated with QH represents an unpaired electron (a free radical). Antimycin A (a *Streptomyces* antibiotic) inhibits the transfer of electrons from QH_2 to cytochrome c (Figure 14-9).

Cytochrome c transfers electrons from complex III to complex IV, and cytochromes a and a_3 transfer electrons

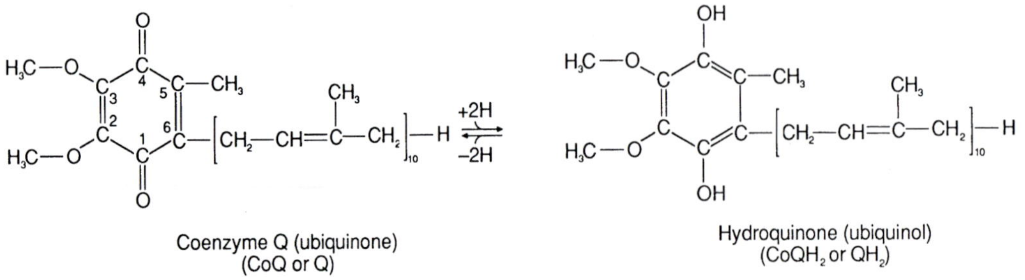

Coenzyme Q (ubiquinone) (CoQ or Q) $\underset{-2H}{\overset{+2H}{\rightleftharpoons}}$ Hydroquinone (ubiquinol) ($CoQH_2$ or QH_2)

FIGURE 14-8
Structure and redox reaction of coenzyme Q. In most mammalian tissues, CoQ has 10 isoprenoid units. CoQ collects reducing equivalents from NADH dehydrogenase and from other flavin-linked dehydrogenases.

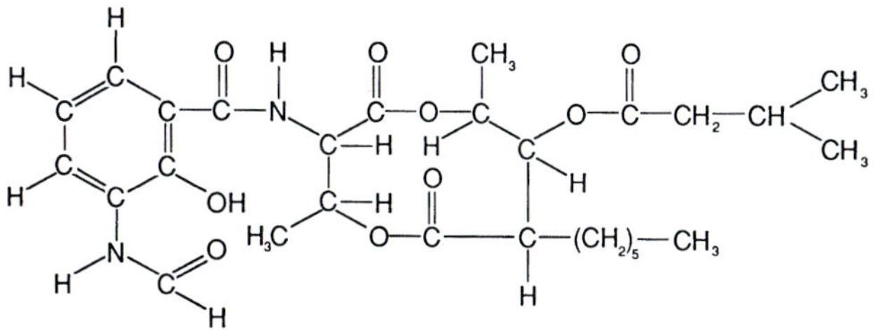

FIGURE 14-9
Structure of antimycin A, an antibiotic that inhibits electron transport from CoQ to cytochrome c.

to oxygen in complex IV. The structure of the heme prosthetic group (iron-protoporphyrin IX) in cytochromes b, c, and c_1 is the same as that present in hemoglobin and myoglobin but differs from the heme group (heme A) of cytochromes a and a_3 (Figure 14-10). The heme groups in cytochromes c and c_1 are covalently linked to the apoprotein by thioether bonds between sulfhydryl groups of two cysteine residues and the vinyl groups of the heme. The heme of cytochromes b and a_3 is bound by strong hydrophobic interactions between the heme and the apoprotein.

Cytochromes b, c_1, a, and a_3 are integral membrane proteins, whereas cytochrome c is a peripheral protein located on the C side of the membrane and is easily isolated from mitochondria. The amino acid sequence of cytochrome c has been established for a wide variety of species and consists of 100–113 amino acid residues (M.W. 13,000). Cytochrome c has played an important role in our understanding the evolutionary relationships among species. Twenty-eight residues are invariant among 67 species sequenced, presumably because a hydrophobic environment around the heme appears to be essential. Amino acids in other positions vary from one species to another and reflect the times of divergence of the different species. For example, only two amino acid residues differ in duck and chicken, but 48 differences are found between horse and yeast cytochrome c, consistent with their long, divergent evolutionary history. Cytochrome c's are identical in chicken and turkey, and in pig, cow, and sheep.

Complex IV

Complex IV, also called cytochrome c oxidase, is the terminal component of the respiratory chain. It consists of four redox centers: cytochrome a, cytochrome a_3, and two Cu ions. Like the iron atoms, the copper ions function as one-electron carriers:

$$Cu^{2+} + e^- \rightleftarrows Cu^+$$

Cytochrome c transfers electrons from cytochrome c_1, the terminal component of complex III, to the four redox centers of the cytochrome oxidase complex. The transfer of four electrons from each of the four redox centers of the cytochrome oxidase complex to an oxygen molecule

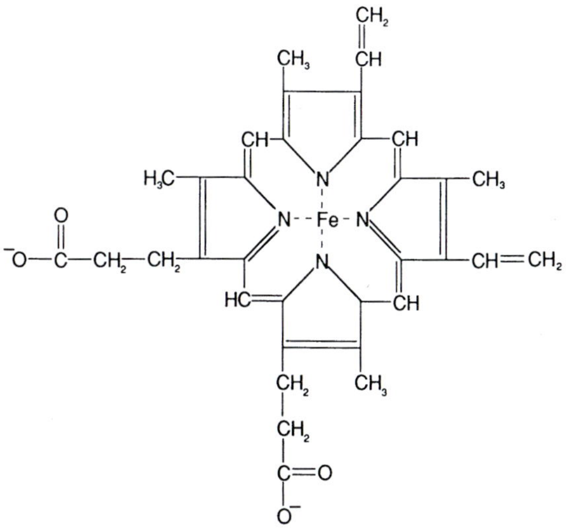

Heme
(iron-protoporphyrin IX)

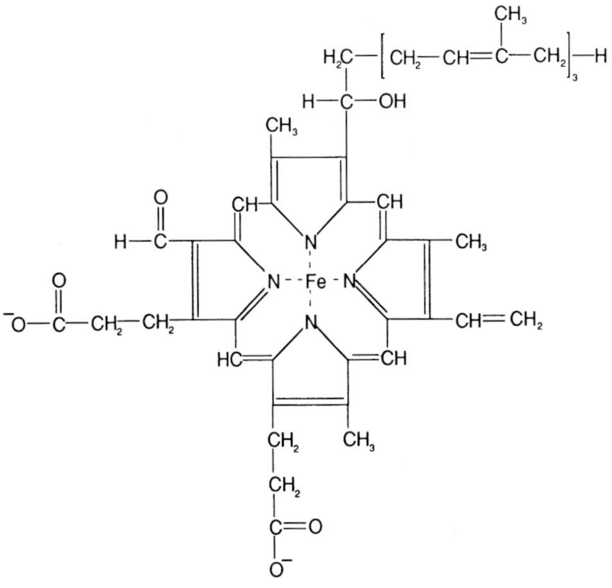

Heme A

FIGURE 14-10
Structure of heme (present in cytochromes b, c, and c_1) and of heme A (present in cytochromes a and a_3).

occurs in a concerted manner to yield two molecules of water:

$$4e^- + O_2 + 4H^+ \rightarrow 2H_2O$$

More than 90% of metabolic oxygen is consumed in the cytochrome oxidase reaction. Oxygen contains an unconventional distribution of its two valence electrons. These two electrons occupy different orbitals and are not spin paired; thus oxygen is a diradical. The reduction of an oxygen molecule with less than four electrons results in the formation of an active oxygen species. One electron transfer yields Superoxide radical (O^{2-}) and the two electron transfer yields hydrogen peroxide (H_2O_2). Hydroxyl free radical ($HO^\bullet$) formation can take place from hydrogen peroxide in the presence of ferrous iron or cuprous chelates. Both O_2^- and $HO^\bullet$ free radicals are cytotoxic oxidants. In the mitochondrial electron transport system, leakage of electrons at any one of the redox-centers due to aging or pathological conditions results in the formation of superoxide. Antioxidant enzymes, namely superoxide dismutases (SOD), catalase, and glutathione peroxidase participate in the elimination of toxic oxygen metabolites. Three different SODs are present in human cells; they are located in mitochondria, cytosol and extracellular fluid. The importance of mitochondrial SOD (labeled as SOD2), which is a manganese containing enzyme, is exemplified in the homozygous SOD2 knockout mice. These SOD2 knockout mice have low birth weights and they die shortly after their birth from dilated cardiomyopathy.

Redox reactions are a required part of normal metabolism. In neutrophils, for example, the killing of the invading microorganisms requires reactive oxygen metabolites (discussed later and also see Chapter 16).

Oxidants also are involved in gene expression (e.g., the variety of protein kinases) and in the regulation of redox homeostasis. Perturbation of redox homeostasis causes oxidative stress and may contribute to chronic inflammatory diseases and malignancy. Cytochrome oxidase in inhibited by cyanide (CN^-; see Chapter 6), carbon monoxide (CO), and azide (N_3^-).

Organization of the Electron Transport Chain

The arrangement of components of the electron transport chain was deduced experimentally. Since electrons pass only from electronegative systems to electropositive systems, the carriers react according to their standard redox potential (Table 14-2). Specific inhibitors and spectroscopic analysis of respiratory chain components are used to identify the reduced and oxidized forms and also aid in the determination of the sequence of carriers.

TABLE 14-2

Standard Oxidation-Reduction Potential ($E^{0'}$) of Components of the Electron-Transport Chain

Redox Component	$E^{0'}$ (in volts)
$NADH/NAD^+$	−0.32
$FMNH_2$/FMN (of NADH dehydrogenase)	−0.11
$CoQH_2$/CoQ	+0.10
Cyt b (red)/Cyt b (ox)	+0.12
Cyt c_1 (red)/Cyt c_1 (ox)	+0.23
Cyt c (red)/Cyt c (ox)	+0.25
Cyt aa_3 (red)/Cyt aa_3 (ox)	+0.29
$\frac{1}{2}O_2^{2-}/\frac{1}{2}O_2$ (or $H_2O/\frac{1}{2}O_2$)	+0.82

Complexes I–IV of the respiratory chain are organized asymmetrically in the inner membrane (Figure 14-11) as follows:

1. The flavin prosthetic groups of NADH dehydrogenase and succinate dehydrogenase face the *M* side of the membrane;
2. CoQ and cytochrome b of complex III are probably inaccessible from either side of the membrane;
3. Cytochrome c interacts with cytochrome c_1 and cytochrome a, all located on the C side;

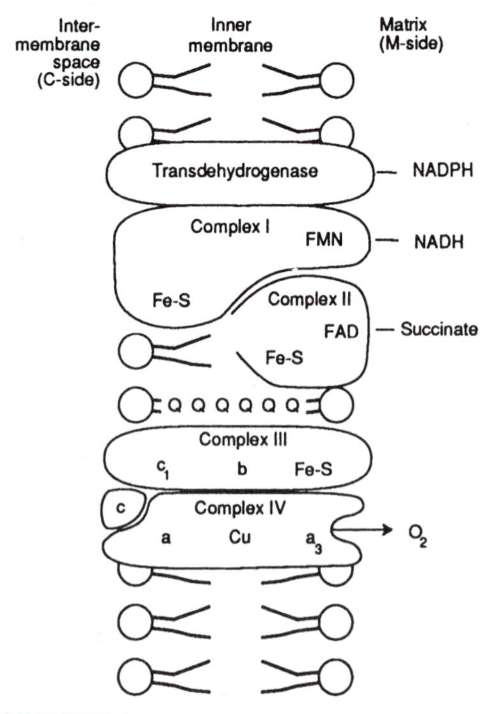

FIGURE 14-11

Orientation of the components of the electron transport complexes within the inner mitochondrial membrane. Fe-S = Iron-sulfur center; *b*, *c*, c_1, *a*, and a_3 = cytochromes; Cu = copper ion.

4. Complex IV spans the membrane, with cytochrome a oriented toward the C side; copper ions and cytochrome a_3 are oriented toward the M side.
5. Nicotinamide nucleotide transhydrogenase, which catalyzes the reaction $NADPH + NAD^+ \rightleftharpoons NADH + NADP^+$, spans the membrane, but its catalytic site faces the M side.

The anisotropic organization of electron carriers across the membrane accounts for the vectorial transport of protons from the inside to the outside of the membrane, which occurs with the passage of electrons. The coupling of this proton gradient to a proton-translocating ATP synthase (also known as ATP synthetase) accounts for the chemiosmotic coupling in oxidative phosphorylation.

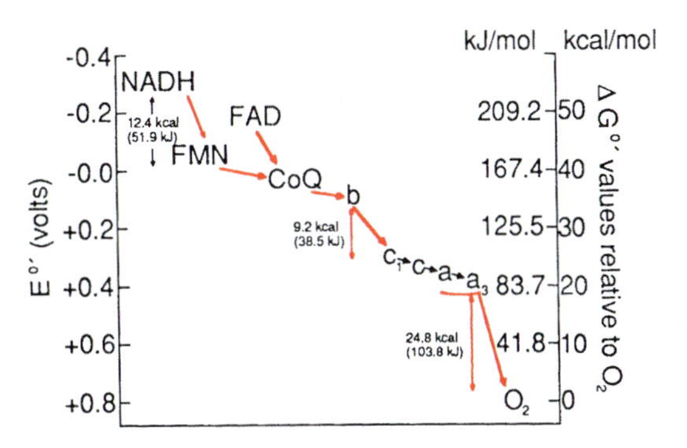

FIGURE 14-12
Flow of reducing equivalents in the respiratory chain and their relationship to energy availability to drive ATP synthesis. The largest free energy changes occur between NADH and FMN, between cytochromes b and c_1, and between cytochromes (a + a_3) and molecular oxygen; ATP formation is coupled to these three sites.

14.2 Oxidative Phosphorylation

ATP is synthesized from ADP and phosphate during electron transport in the respiratory chain. This type of phosphorylation is distinguished from substrate-level phosphorylation, which occurs as an integral part of specific reactions in glycolysis and the TCA cycle. The free energy available for the synthesis of ATP during electron transfer from NADH to oxygen can be calculated from the difference in the value of the standard potential of the electron donor system and that of the electron acceptor system. The standard potential of the $NADH/NAD^+$ redox component is −0.32 V and that of $H_2O/\frac{1}{2}O_2$ is +0.82 V; therefore, the standard potential difference between them is

$$\begin{aligned} \Delta E^{o'} &= (\text{electron acceptor} - \text{electron donor}) \\ &= 0.82 - (-0.32) = +1.14\ \text{V} \end{aligned}$$

The standard free energy is calculated from the expression

$$\Delta G^{0'} = -n\text{F}\,\Delta E^{0'}$$

where n is the number of electrons transferred, and F is the *Faraday constant,* which is equal to 23,062 cal/eV. Thus, the standard free energy for a two-electron transfer from NADH to molecular oxygen is

$$\begin{aligned} \Delta G^{o'} &= -n\text{F}\,\Delta E^{o'} = -2 \times 23,062 \times 1.14 \\ &= -52.6\ \text{kcal/mol}\ (-220\ \text{kJ/mol}) \end{aligned}$$

Although 52.6 kcal of free energy is available from the reaction, only 21.9 kcal is conserved in the formation of phosphoanhydride bonds at ATP. Formation of each phosphoanhydride bond requires 7.3 kcal (30.5 kJ), and 21.9 kcal accounts for three ATPs synthesized. The remainder of the energy is assumed to be dissipated as heat.

In the mitochondria of brown adipose tissue, very little ATP is synthesized, and most of the energy liberated in the electron transport system is converted to heat.

The conservation of energy in the electron transport chain occurs at three sites where there is a large decrease in free energy (Figure 14-12): between NADH and FMN, between cytochromes b and c_1, and between cytochromes (a + a_3) and molecular oxygen. Because electrons from succinate bypass complex I and enter at the CoQ level, only two moles of ATP are synthesized per mole of succinate. The number of ATP molecules synthesized depends on where the reducing equivalents enter the respiratory chain and is indicated by the ratio (or quotient) of phosphate esterified (or ATP produced) to oxygen consumed (P/O or ATP/O) per two-electron transport. The NAD^+-linked substrates (e.g., malate, pyruvate, α-ketoglutarate, isocitrate, β-hydroxybutyrate, and β-hydroxyacyl-CoA) have P/O values of 3, and some flavin-linked substrates (e.g., succinate, glycerol phosphate, and fatty acyl-CoA) have P/O ratios of 2. The complete oxidation of 1 mol of glucose yields either 36 or 38 mol of ATP, depending on which shuttle pathway is used in the transport of cytoplasmic NADH to mitochondria. A list of the energy-yielding reactions in glucose oxidation is shown in Table 14-3.

Mechanisms of Oxidative Phosphorylation

Three hypotheses have been proposed to explain how the mechanism of energy conservation is coupled to electron transport (the energy transduction system): the chemical, conformational, and chemiosmotic hypotheses.

The ***chemical hypothesis*** proposes that the energy is conserved by the formation of high-energy intermediates as reducing equivalents pass from one carrier to the

TABLE 14-3
Energy-Yielding Reactions in the Complete Oxidation of Glucose

Reaction	Net Moles of ATP Generated per Mole of Glucose
Glycolysis (phosphoglycerate kinase, pyruvate kinase; two ATPs are expended)	2
NADH shuttle glycerol-phosphate shuttle (or malate-aspartate shuttle)	4(6)
Pyruvate dehydrogenase (NADH)	6
Succinyl CoA synthetase (GTP is equivalent to ATP)	2
Succinate dehydrogenase (succinate$\rightarrow$fumarate + $FADH_2$)	4
Other TCA cycle reactions (isocitrate$\rightarrow\alpha$-ketoglutarate, α-ketoglutarate$\rightarrow$succinyl CoA, malate$\rightarrow$oxaloacetate; total of 3 NADH generated)	18
Total	36(38)

next. The energy is ultimately stored as a phosphoric acid anhydride bond in ATP. The formulation, in general terms, is

$$\begin{array}{l} AH_2 + B + C \rightleftharpoons A \sim C + BH_2 \\ A \sim C + ADP + P_i \rightleftharpoons A + C + ATP \\ \hline \text{Sum: } AH_2 + B + ADP + P_i \rightleftharpoons A + BH_2 + ATP \end{array}$$

where A and B represent the known redox pair, C is a hypothetical ligand, and A $\sim$ C is a hypothetical high-energy intermediate. The above mechanism can be modified to include other phosphorylated intermediates.

A model reaction that supports the above mechanism is the glycolytic substrate-linked phosphorylation, which proceeds via a thiol ester prior to the formation of the phosphorylated intermediate (Chapter 13). Although the chemical hypothesis is consistent with the substrate-linked phosphorylation mechanism, it is deficient in explaining the oxidative phosphorylation in mitochondria for two reasons:

1. The postulated high-energy chemical intermediates, either phosphorylated or nonphosphorylated, have never been identified despite many attempts to find them, and
2. The chemical-coupling mechanism does not explain why the inner mitochondrial membrane must be present as a completely closed vesicle for oxidative phosphorylation to occur.

The ***conformational hypothesis*** proposes that the energy-yielding steps generate protein conformational changes that are used in ATP synthesis. The conformational changes that occur in the redox catalysts are transmitted to the energy-transducing units via protein-protein interactions, the formation of covalent intermediates, or the proton-motive force. Current opinion holds that the conformational changes are linked with a proton-motive force (see below).

Morphological changes do occur in the inner membranes of the mitochondria when active respiration is stimulated by ADP. Fluorescent probes, such as 1-aminonaphthalene-8-sulfonate (ANS) and the antibiotic aurovertin, bind either to the inner membrane (ANS) or directly to ATP synthase (aurovertin). The binding enhances or diminishes fluorescence in response to changes in conformation or hydrophobicity of the inner membrane. Results support the hypothesis that ATP synthase undergoes conformational changes during respiration and oxidative phosphorylation (discussed later).

According to the ***chemiosmotic hypothesis,*** developed by Peter Mitchell, an electrochemical gradient (pH gradient), generated across the inner mitochondrial membrane by the passage of reducing equivalents along the respiratory chain provides the driving force for the synthesis of ATP. There are three prerequisites for achieving oxidative phosphorylation according to this hypothesis:

1. An anisotropic (direction-oriented) proton-translocating respiratory chain capable of vectorial transport of protons across the membrane;
2. A coupling membrane impermeable to ions except via specific transport systems; and
3. An anisotropic ATP synthase whose catalytic activity is driven by an electrochemical potential.

The transport of reducing equivalents in the respiratory chain generates a proton gradient across the membrane by virtue of the specific vectorial arrangement of the redox components within the inner mitochondrial membrane. The proton gradient is generated by ejection of protons from the matrix into the intermembrane space during proton-absorbing reactions, which occur on the M side of the inner membrane, and the proton-yielding reactions, which occur on the C side, to form redox loops (Figure 14-13).

According to the chemiosmotic hypothesis, ejection of two or more protons occurs at each of three sites in complexes I, III, and IV. Thus, in the transfer of two reducing equivalents from NADH to oxygen, at least six protons

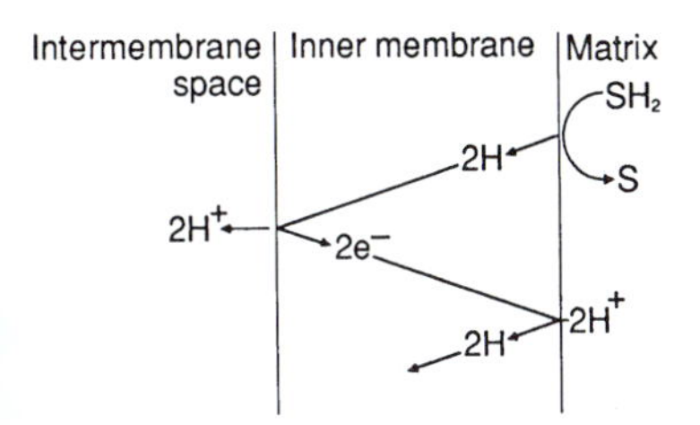

FIGURE 14-13
Schematic representation of a transmembrane redox loop, in which $2H^+$ are ejected into the intermembrane space as the substrate is oxidized on the matrix side.

(2×3) are ejected. Development of an electrochemical potential for protons is the mechanism by which conservation of free energy occurs during transport of reducing equivalents in the respiratory chain.

The electrochemical potential is proportional to the proton-motive force (pmf, designated as Δp or $\Delta\tilde{\mu}_{H^+}$), given by the following equation:

$$\Delta\tilde{\mu}_{H^+} = \Delta\psi - Z\,\Delta\text{pH}$$

where $\Delta\psi$ is the membrane potential resulting from charge separation, the matrix compartment being negative; ΔpH is the pH difference across the membrane, due to proton translocation; and Z is equal to -2.303 RT/F, the factor used to convert pH units into millivolts, the unit of the other two terms of the expression; Z equals -59 at 25°C.

The anisotropic arrangement of the respiratory chain and the vectorial transport of protons are supported by experimental observations. The distribution of the redox carriers required by the chemiosmotic hypothesis is remarkably similar to that derived from studies on enzyme topology of the inner membrane. In respiring mitochondria, the intramembrane space is more acidic than the matrix space by about 1.4 pH units, and the transmembrane potential is about 0.180–0.220 V. Thus, the basic premise of the chemiosmotic hypothesis is

$$\text{Transport of reducing equivalents} \rightarrow \Delta\tilde{\mu}_{H^+} \rightarrow \text{ATP synthesis}$$

The proton-motive force drives ATP synthesis via the reentry of protons. The ATP synthase (F_o/F_1 complex) is driven by this vectorial transfer of protons from the intramembrane space into the matrix. ATP synthesis occurs in the inner membrane spheres (the F_1 component of the synthase).

ATP synthase consists of three parts (Figure 14-14):

1. The catalytic part, F_1, composed of five firmly associated subunits with a subunit composition of $\alpha_3\beta_3\gamma\delta\varepsilon$;
2. A hydrophobic complex of three or four proteins (also called proteolipids), or F_0 complex, located in the inner membrane, which is thought to contain a proton-translocating channel; and
3. A stalk consisting of protein components that connect F_1 with F_0.

When the ATP synthase forms a part of the intact membrane system, it catalyzes proton translocation and ATP synthesis. However, when F_1 is dissociated from F_0, the F_1 no longer catalyzes ATP synthesis but rather ATP hydrolysis. Thus, F_1 becomes an ATPase rather than an ATP synthase. Submitochondrial vesicles devoid of F_1 can transport reducing equivalents, since they contain the redox carriers, but are unable to support ATP synthesis. Careful reconstitution of membrane vesicles by addition of F_1 allows the complex to regain its capacity for ATP synthesis.

The antibiotics oligomycin and aurovertin and the reagents N,N′-dicyclohexylcarbodiimide (DCCD) and 4-chloro-7-nitrobenzofurazan (Nbf-Cl) inhibit ATP synthesis by binding to different sites of ATP synthase. Oligomycin and DCCD bind to proteolipid (F_0) components and block translocation of protons, aurovertin binds to a specific β-subunit of F_1, and Nbf-Cl binds with a specific tyrosine residue of the β-subunit of F_1.

The mechanism of ATP formation by the ATP synthase driven by the passage of protons is attributed to conformational changes that occur in the $\alpha\beta$ dimers of the F_1 complex (Figure 14-15). The F_1 complex has three interacting nucleotide-binding and conformationally distinct $\alpha\beta$ domains:

1. Open conformation (O): has very low affinities for substrates and products and is catalytically inactive.
2. Loose conformation (L): binds loosely to nucleotides and is catalytically inactive.
3. Tight conformation (T): binds nucleotides tightly and is catalytically active.

In this model, originally proposed by Paul Boyer, the conformational changes of $\alpha\beta$ assemblies are brought about by the energy dependent rotation of-the γ subunit. The γ subunit rotation occurs in discrete steps of 120° and is fuelled by proton passage through F_0 channels. In step 1 rotation (Figure 14-15), the conformation change at the T site causes ATP release at the O site. In step 2 rotation, ADP and P_i bound at the L site, now bind at the T site and undergo conversion to ATP. Thus, in the conformational cycle, ATP is synthesized only in the T site and is released only in the O site. The energy dependent process is the ATP release step which is accomplished by energy dependent rotation of the γ subunit.

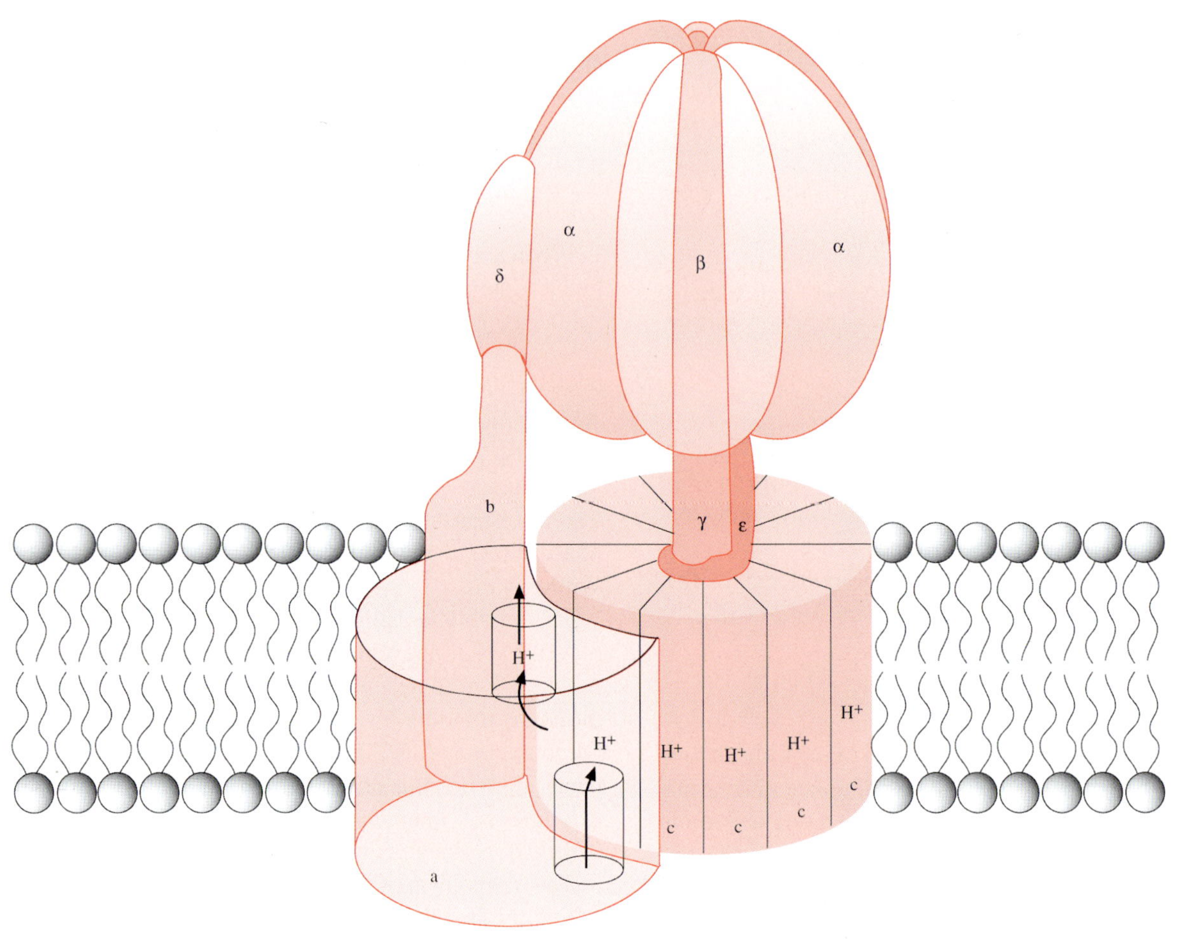

FIGURE 14-14
Structure of ATP synthase. The F_1 complex situated above the membrane consists of three $\alpha\beta$ dimers and single γ, δ, ε subunits. The membrane segment F_0, also known as a stalk, consists of H^+ channels. Protons are conducted via the C-subunit channels. The rotation of C-subunits relative to a subunit of the stalk drives the rotation of the γ-subunit. [Reproduced with permission from Y. Zhou, T. M. Duncan and R. L. Crass: Subunit rotation in *Escherichia coli* F_0F_1–ATP synthase during oxidative phosphorylation. *Proc. Natl. Acad. Sci.* **94,** 10583 (1997). [©1997 by the National Academy of Sciences.]

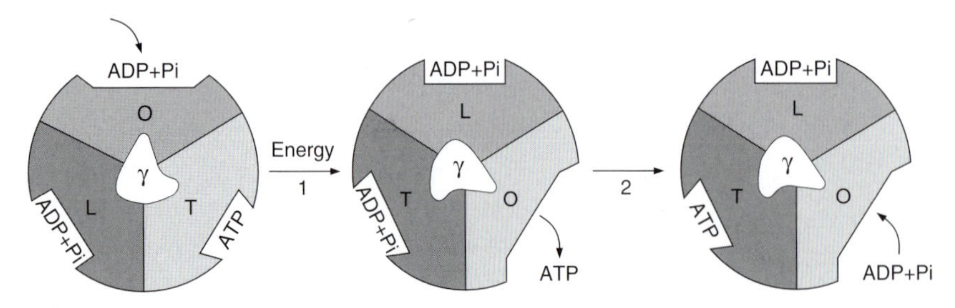

FIGURE 14-15
Nucleotide binding conformational changes of F_1 ATP synthase complex. See text for details. [Reproduced with permission from Y. Zhou, T. M. Duncan and R. L. Crass: Subunit rotation in *Escherichia coli* F_0F_1-ATP synthase during oxidative phosphorylation. *Proc. Natl. Acad. Sci.* **94,** 10583 (1997). ©1997 by the National Academy of Sciences.]

The central idea of chemiosmosis, that ATP synthesis is driven by the proton-motive force generated during respiration, is well accepted and is supported by several lines of experimental evidence. However, some aspects of the chemiosmotic theory are still to be resolved. For example, the precise location of proton-generating sites, the pathways of proton transport, the mechanism of proton flow, and how many protons are ejected per site are not well understood. It was originally proposed that two protons were ejected at each site, but the number may be more.

Uncoupling Agents of Oxidative Phosphorylation

Uncoupling agents dissociate ATP synthesis and other energy-dependent membrane functions from the transport of reducing equivalents in the respiratory chain. Normally, these processes are tightly coupled. These uncoupling agents cause a several-fold stimulation of respiration with unimpeded utilization of substrate and dissipation of energy as heat. A classic example of an uncoupling agent is 2,4-dinitrophenol (Figure 14-16);

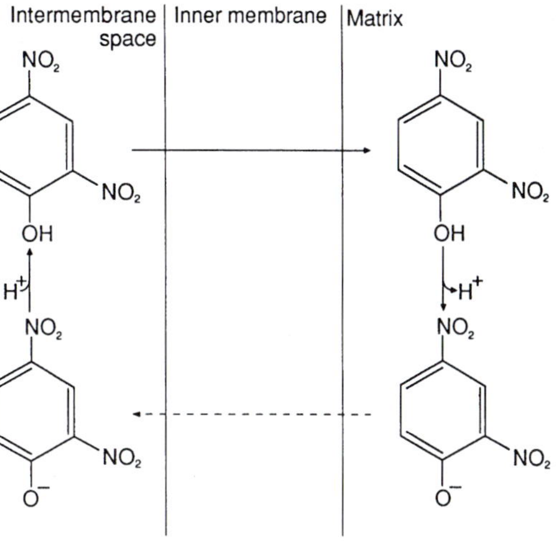

A group of antibiotics (e.g., valinomycin, nigericin, and gramicidin A) transport cations across the cell membrane. Such agents, known as ***ionophores,*** are widely used to probe membrane structure and function. Ionophores uncouple oxidative phosphorylation. Valinomycin, a cyclic peptide (Figure 14-17), forms a lipid-soluble complex with K^+ that readily passes through the inner membrane, whereas K^+ by itself does not. In the valinomycin-K^+ complex, hydrophobic groups, present on the outside, facilitate transport of the complex in the lipid environment; K^+, located on the inside, interacts with the hydrophilic groups. The matrix side of the inner membrane has a negative potential, so the positively charged valinomycin-K^+ complex is drawn inward. The complex uncouples oxidative phosphorylation by decreasing the membrane potential. Unlike valinomycin, it exchanges H^+ for K^+ across the membrane. Its

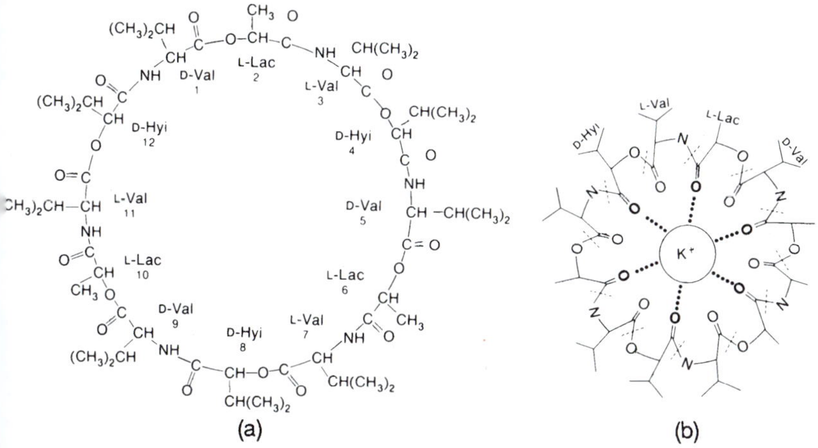

FIGURE 14-16
Uncoupling of oxidative phosphorylation by 2,4-dinitrophenol (2,4-DNP). The anionic form of 2,4-DNP is protonated in the intermembrane space, is lipid soluble, and crosses the inner membrane readily. In the matrix, the protonated form dissociates, abolishing the proton gradient established by substrate oxidation. The ionized form of 2,4-DNP is poorly soluble in the membrane lipids and therefore is not easily transported across the membrane (dashed arrow). It is lipophilic and capable of transporting protons from one side of the membrane to the other (a ***protonophore***), thus abolishing the proton gradient.

IGURE 14-17
tructure of valinomycin (a) and its complex with K^+ (b). Valinomycin, which consists of three identical fragments of -valyl-L-lactyl-L-valyl-D-α-hydroxyisovaleric acid (D-Val-L-Lac-L-Val-D-Hyi), is a mobile or channel-forming nophore and an uncoupler of oxidative phosphorylation. Note the hydrophobic exterior and the hydrophilic interior of e complex. [Structure (b) is reproduced, with permission, from B. C. Pressman: Biological application of ionophores. *nnu. Rev. Biochem.* **45,** 501 (1976). © 1976 by Annual Reviews Inc.]

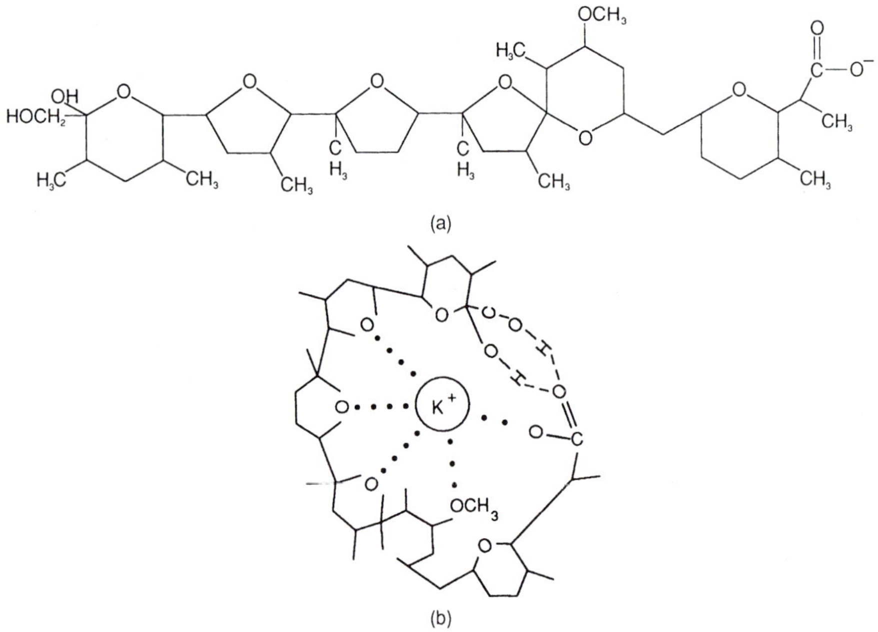

(a)

(b)

FIGURE 14-18
Structure of nigericin (a) and its complex with K^+ (b). Nigericin is a mobile or channel-forming ionophore; at one side of the membrane, the anionic form complexes with K^+ and then migrates to the other side, where K^+ is discharged and the anionic form is protonated. The protonated form migrates to the opposite side and discharges H^+, and the anionic species formed initiates the next cycle. Thus, nigericin exchanges H^+ for K^+ across the membrane and uncouples oxidative phosphorylation. Nigericin, a polycyclic ether carboxylate like valinomycin, forms a complex with K^+ that diffuses through the membrane.

carboxylate group interacts with the cation to form an electroneutral complex. In the exchange of H^+ for K^+, although the membrane potential is not altered, the pH differential is abolished, thereby uncoupling oxidative phosphorylation.

Gramicidin A (Figure 14-19), a linear peptide, forms a head-to-head dimer that creates an aqueous helical channel, which allows transport of a variety of monovalent cations (e.g., Na^+, K^+, and H^+). The channel is a helix formed by the coiling of two β-pleated sheets and is lined with

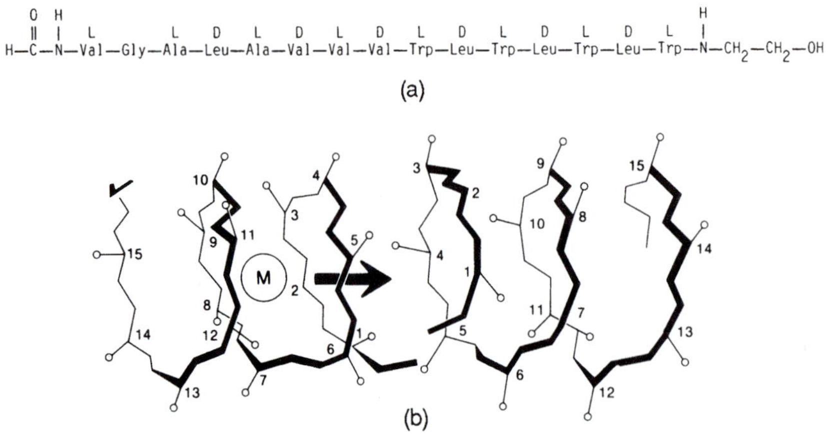

H—C—N—Val—Gly—Ala—Leu—Ala—Val—Val—Val—Trp—Leu—Trp—Leu—Trp—Leu—Trp—N—CH_2—CH_2—OH

(a)

(b)

FIGURE 14-19
Structure of gramicidin A (a) and the formation of a helical pore through a lipid bilayer by assembly of two gramicidin A molecules via their formyl groups at the N termini (b). A variety of monovalent cations are transported through the static pore. [Structure (b) is reproduced with permission from Y.A. Ovchinnikov; Physico-chemical basis of ion transport through biological membranes: Ionophores and ion channels. *Eur. J. Biochem.* **94,** 321 (1979).]

TABLE 14-4
*Energy States of Mitochondria**

	State 1	State 2	State 3	State 4	State 5
Characteristics	Aerobic	Aerobic	Aerobic	Aerobic	Anaerobic
ADP level	Low	High	High	Low	High
Substrate level	Low-endogenous	Approaching 0	High	High	High
Respiration rate	Slow	Slow	Fast	Slow	
Rate-limiting component	Phosphate acceptor (ADP)	Substrate	Respiratory chain	Phosphate acceptor (ADP)	Oxygen

*Reproduced, with permission, from B. Chance and G. R. Williams: Respiratory enzymes in oxidative phosphorylation: the steady state. *J. Biol. Chem.* **217,** 409 (1955).

carbonyl groups of the peptide bonds. Unlike valinomycin and nigericin, gramicidin A forms a static pore in the membrane.

14.3 Mitochondrial Energy States

Normal mitochondria can exist in one of five energy states (Table 14-4). In the presence of nonlimiting amounts of respiratory chain components, the rate of oxygen consumption is affected by changes in the $NADH/NAD^+$ ratio, the phosphorylation ratio (or phosphate potential, $[ATP]/[ADP][P_i]$), and pO_2. In a resting (unstimulated) mitochondrion, in the presence of nonlimiting concentrations of substrate (reductant), oxygen, and phosphate, respiratory activity is controlled by the availability of ADP (state 4, Table 14-4). In this state, the rate of endogenous energy dissipation is low and is regulated by ADP. On addition of ADP, the respiratory rate increases dramatically (state 3) until the ADP supply is depleted. Repeated ADP-induced respiratory cycles between states 4 and 3 cause depletion of substrate (state 2), phosphate, or oxygen. If oxygen is unavailable, respiration ceases (state 5). Dinitrophenol abolishes respiratory control because it uncouples the transport of reducing equivalents in the respiratory chain from the phosphorylation of ADP.

In the oxidation of carbohydrates, energy is conserved at all three stages: glycolysis, TCA cycle, and oxidative phosphorylation. Each of these stages is regulated at specific sites, and all three are coordinated in such a manner that ATP is synthesized only to the extent that it is needed in the cell. For example, when the [ATP]/[ADP] ratio is high, glycolysis, TCA cycle, and oxidative phosphorylation are inhibited. The key regulatory enzyme, 6-phosphofructokinase, is activated by ADP and inhibited by ATP (see also Chapters 13, 15, and 22.)

Energy-Linked Functions of Mitochondria Other Than ATP Synthesis

The energy derived from the respiratory chain, although primarily coupled to the formation of ATP, is also utilized for purposes such as heat production, ion transport, and the transhydrogenase reaction.

Brown adipose tissue is responsible for nonshivering thermogenesis, which is important in the arousal of hibernating animals and for maintaining body temperature in hairless neonates of mammals, including humans. Diet-induced thermogenesis also occurs in brown adipose tissue (Chapter 12). Brown adipose tissue, in contrast to white adipose tissue, consists of small cells that are rich in mitochondria and lipid dispersed in the cytoplasm as distinct droplets. Mitochondria of brown adipose tissue are naturally uncoupled, so that oxidation of substrates generates heat rather than a proton-motive force. The inner mitochondrial membrane contains, instead of the normal proton-translocating channel in F_0, an H^+ uniport that causes the dissipation of energy as heat without the concomitant synthesis of ATP. When the requirement for thermogenesis ceases, the H^+ uniport is blocked by a protein called ***thermogenin*** (M.W. 32,000). Thermogenin becomes functional after binding to purine nucleotides, of which GDP is the most effective and ADP and ATP are less effective. This protein recouples phosphorylation with the energy released in the respiratory chain. It is located at the entrance to the H^+ channel on the C side of the inner membrane. Stimulation of thermogenesis in brown adipose tissue is initiated by stimulation of the sympathetic nervous system, which releases norepinephrine at nerve endings. Norepinephrine combines with the β-adrenergic receptors of the brown adipose tissue cells and initiates cAMP-dependent activation of triacylglycerol lipase (Chapter 22), leading to the elevated intracellular concentrations of fatty acids that are oxidized in the mitochondria.

TABLE 14-5
*Mitochondrial Metabolite Translocators**

Metabolites	Species Translocated (In⇌out)
Dicarboxylates	$Malate^{2-}$⇌$phosphate^{2-}$
Tricarboxylates	$Citrate^{3-}+H^{+}$⇌$malate^{2-}$
α-Ketoglutarate	α-$Ketoglutarate^{2-}$⇌$malate^{2-}$
Phosphate	$Phosphate^{-}$⇌OH^{-}
Pyruvate	$Pyruvate^{-}$⇌OH^{-}
Glutamate	$Glutamate^{-}$⇌OH^{-}
Glutamine	Glutamine⇌$glutamate^{-}+H^{-}$
Ornithine	$Ornithine^{+}$⇌H^{+}
Neutral amino acids	Neutral amino acids
Acyl carnitine, carnitine	Acyl carnitine⇌carnitine
Glutamate, aspartate	$Glutamate^{-}+H^{+}$⇌$aspartate^{-}$
ADP, ATP	ADP^{3-}⇌ATP^{4-}

*Reproduced, with permission, from L. Ernster and G. Schatz: Mitochondria: A historical review. *J. Cell Biol.* **91**,227s (1981). ©1981 by the Rockefeller University Press.

A transhydrogenase catalyzes the transfer of reducing equivalents from NADH to $NADP^{+}$ to form NADPH by utilizing energy captured by the energy conservation mechanisms of the respiratory chain without the participation of the phosphorylation system. Similarly, ion translocation occurs at the expense of energy derived from substrate oxidation in the respiratory chain. For example, Ca^{2+} is transported from the cytoplasm to the mitochondrial matrix, through the inner mitochondrial membrane, at the expense of proton-motive force. The Ca^{2+} efflux from mitochondria is regulated so that levels of cytoplasmic Ca^{2+} that are optimal for function are achieved. Increased cytoplasmic Ca^{2+} levels initiate or promote muscle contraction (Chapter 21), glycogen breakdown (Chapter 15), and oxidation of pyruvate (Chapter 13). Decreased levels of Ca^{2+} have the opposite effect.

Translocation systems of the inner mitochondrial membrane are listed in Table 14-5. Anion translocators are responsible for electroneutral movement of dicarboxylates, tricarboxylates, α-ketoglutarate, glutamate, pyruvate, and inorganic phosphate. Specific electrogenic translocator systems exchange ATP for ADP, and glutamate for aspartate, across the membrane. The metabolic function of the translocators is to provide appropriate substrates (e.g., pyruvate and fatty acids) for mitochondrial oxidation that is coupled to ATP synthesis from ADP and P_i.

The ATP/ADP translocator, also called adenine nucleotide translocase, plays a vital role in the metabolism of aerobic cells because mitochondrial ATP is primarily consumed outside the mitochondria to support biosynthetic reactions. The translocator is the most abundant protein in the mitochondrion; two molecules per unit of respiratory chain are present. The translocator consists of two identical hydrophobic peptides. The export of ATP^{4-} is necessarily linked to the uptake of ADP^{3-}. This transport system is electrogenic owing to the transport of nucleotides of unequal charge, the driving force being the membrane potential. Although made up of two identical subunits, the translocator is asymmetrical in its orientation; on the C side, it binds with ADP, and on the M side, it binds with ATP. This asymmetry of nucleotide transport has been demonstrated by use of the inhibitors atractyloside and bongkrekic acid (Figure 14-20). Atractyloside is a glycoside found in the rhizomes of a Mediterranean thistle; bongkrekic acid is a branched-chain unsaturated fatty acid synthesized by a fungus found in decaying coconut meat. The former inhibitor binds to the C side of the translocator at the ADP site, whereas the latter binds to the M side of the translocator at the ATP site.

Transport of Cytoplasmic NADH to Mitochondria

NADH generated in the cytoplasm during glycolysis must be transported into the mitochondria if it is to be oxidized in the respiratory chain. However, the inner mitochondrial membrane not only is impermeable to NADH and NAD^{+} but contains no transport systems for these substances. Thus, the $[NAD^{+}]/[NADH]$ ratio is many times higher in the cytoplasm (about 1000) than in mitochondria (about 8). The high value for this ratio in the cytoplasm favors the glyceraldehyde-phosphate dehydrogenase reaction (which is essential for glycolysis) and contributes to a negative ΔG; the low value for the ratio in mitochondria favors the oxidation of NADH in the respiratory chain. Impermeability to NADH is overcome by indirect transfer of the reducing equivalents through shuttling substrates that undergo oxidation-reduction reactions. This process consists of the following steps: a reaction in which NADH reduces a substrate in the cytoplasm; transport of the reduced substrate into mitochondria; and oxidation of the reduced substrate in the respiratory chain. Two pathways that transport reducing equivalents from NADH into mitochondria have been characterized and are known as the ***glycerol-phosphate shuttle*** and the ***malate-aspartate shuttle.***

The glycerol-phosphate shuttle is mainly associated with wing muscle mitochondria in insects and is not quantitatively significant in mitochondria of mammalian muscle. However, its activity is higher in mammalian brain and liver cells than in mammalian muscle. The shuttle (Figure 14-21) involves two glycerol-3-phosphate dehydrogenases, one NAD^{+}-linked and present in the cytoplasm, the other FAD-linked and present on the C side of the inner mitochondrial membrane. The process begins in the cytoplasm with the reduction of dihydroxyacetone

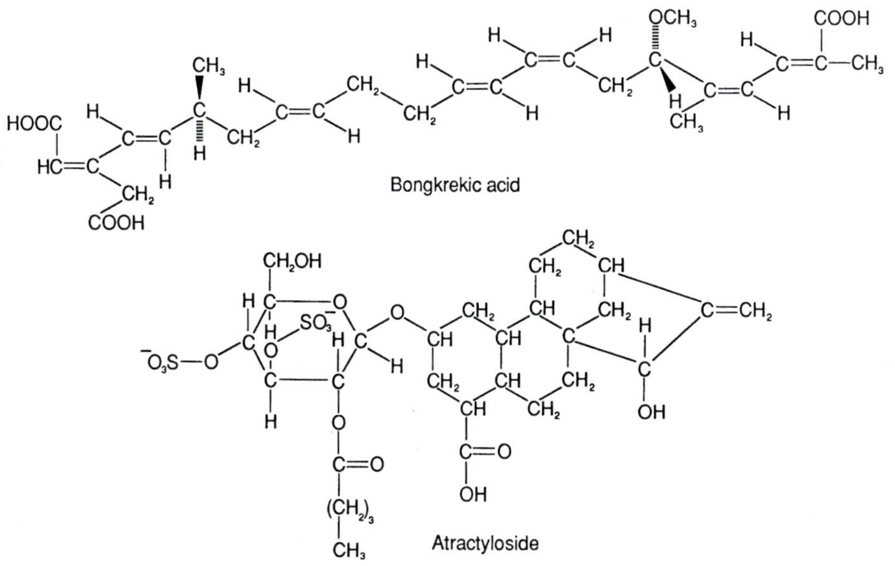

Bongkrekic acid

Atractyloside

FIGURE 14-20
Inhibitors of adenine nucleotide (ATP/ADP) translocation. Bongkrekic acid binds to the ATP binding site, whereas atractyloside binds to the ADP binding site of the translocator.

phosphate to glycerol 3-phosphate by NAD^+-dependent glycerol-3-phosphate dehydrogenase:

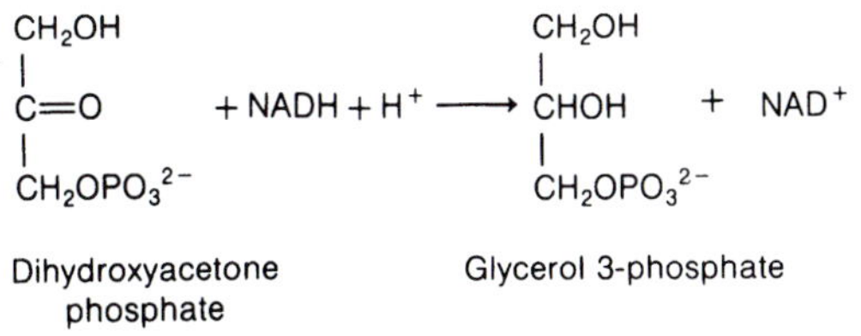

$$\begin{array}{l} CH_2OH \\ | \\ C{=}O \\ | \\ CH_2OPO_3^{2-} \end{array} + NADH + H^+ \longrightarrow \begin{array}{l} CH_2OH \\ | \\ CHOH \\ | \\ CH_2OPO_3^{2-} \end{array} + NAD^+$$

Dihydroxyacetone phosphate — Glycerol 3-phosphate

Glycerol 3-phosphate in the intramembrane space is oxidized to dihydroxyacetone phosphate by the FAD-linked dehydrogenase. The $FADH_2$ thus generated is then oxidized by CoQ of the respiratory chain. This shuttle differs from the malate-aspartate shuttle in three important respects:

1. The substrate glycerol 3-phosphate, which carries the reducing equivalents, does not penetrate the inner mitochondrial membrane, since it is oxidized on the C side of the membrane.
2. The oxidation of one molecule of cytoplasmic NADH yields only two molecules of ATP, since the reducing equivalents enter the respiratory chain at the CoQ level.
3. The shuttle is bidirectional.

The malate-aspartate shuttle is quantitatively more significant in all vertebrate tissues. It is unidirectional, requiring cytoplasmic and mitochondrial malate dehydrogenases and aspartate aminotransferases as well as two membrane-bound carrier systems (Figure 14-22). In this process, the reducing equivalents of NADH are

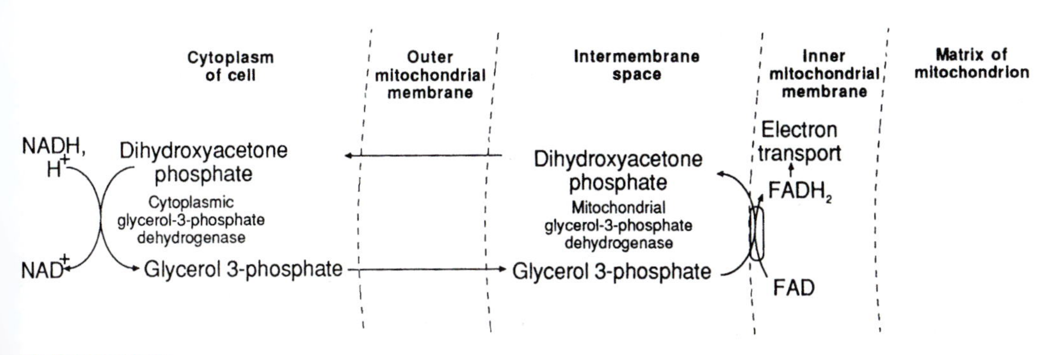

FIGURE 14-21
Glycerol 3-phosphate shuttle for the transport of cytoplasmic reducing equivalents to the inner mitochondrial membrane. Glycerol 3-phosphate, which carries the reducing equivalents, is oxidized to dihydroxyacetone phosphate by an FAD-linked dehydrogenase located on the C side of the inner membrane.

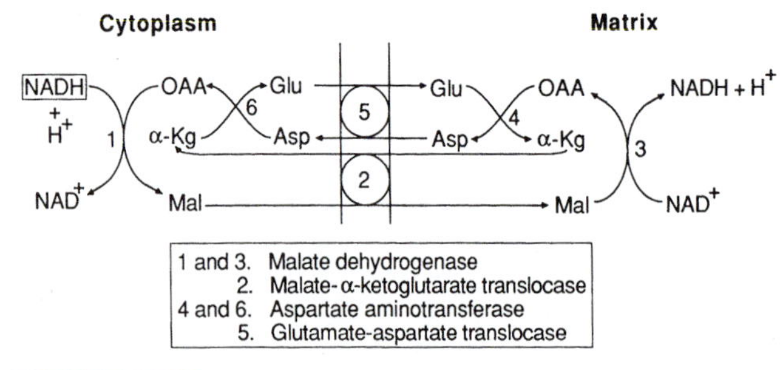

FIGURE 14-22
Malate-aspartate shuttle for the transport of cytoplasmic reducing equivalents across the inner membrane of mitochondria. Malate, which carries the reducing equivalents, is oxidized to oxaloacetate with the generation of NADH in the matrix. To complete the unidirectional cycle, oxaloacetate is transported out of the matrix as aspartate. Mal = malate; OAA = oxaloacetate; α-KG = α-ketoglutarate; Glu = glutamate; Asp = aspartate.

transferred in the cytoplasm to oxaloacetate, which is converted to malate. Malate is transported, via the malate-α-ketoglutarate translocase, to the mitochondrial matrix, where it is oxidized to oxaloacetate, accompanied by the formation of NADH, which, by respiratory chain oxidation, produces three molecules of ATP. To complete the shuttle, oxaloacetate must be transported from the matrix to the cytoplasm but there is no such transport system. Instead, oxaloacetate is first converted to aspartate by aspartate aminotransferase (Chapter 17) and then transported out of the mitochondria via the glutamate-aspartate translocase.

14.4 The Mitochondrial Genome

The human mitochondrion contains its own genome; a circular double-stranded DNA molecule of 16,569 base pairs. Mitochondrial DNA (mtDNA) is a highly compact molecule and one of its strands has greater density (heavy-, H-strand) than the other (light-, L-strand). The only non-coding region, which consists of about 1 kb of mtDNA, is known as a ***displacement loop*** (D-loop). The D-loop contains the replication origin for H-strand replication and promoters for transcription from both strands. The mtDNA consists of 37 genes that encode 13 polypeptides that are components of the respiratory chain complexes I, III, IV, and V. Complex I (NADH: ubiquinone oxidoreductase) contains seven mtDNA gene products. Complex III (ubiquinol: cytochrome c oxidoreductase) incorporates one (cytochrome b); complex IV (cytochrome c oxidoreductase) incorporates three; and complex V incorporates two (Table 14-6). The remainder of the mtDNA genes encode 12S and 16S rRNAs and 22 tRNAs.

TABLE 14-6
The Mitochondrial Genes Contributing to the Complexes of the Mitochondrial Oxidative Phosphorylation Machinery

Complex	Total Subunits	Encoded by mtDNA
Complex I, NADH: CoQ reductase	40	7
Complex II, succinate dehydrogenase	4	0
Complex III, ubiquinone: cytochrome c reductase	10	1
Complex IV, cytochrome oxidase	13	3
Complex V, ATP synthase	13	2

Nuclear DNA (nDNA) provides more than 65 oxidative phosphorylation gene products. Overall, oxidative phosphosphorylation requires at least 100 gene products; proteins encoded in nuclear genes are synthesized on cytoplasmic ribosomes and imported and assembled within the mitochondria.

A typical human cell contains many mitochondria with each mitochondrion having more than one mtDNA. Cytoplasmic location and multiple mtDNAs give mitochondria special characteristics:

- The mtDNA is transmitted through the oocyte cytoplasm and is maternally inherited.
- The occurrence of a mtDNA mutation creates a mixed intracellular population of mutant and original molecules (heteroplasmy). As the heteroplasmic cells divide during mitosis or meiosis, the mutant and original mtDNAs are randomly distributed to the daughter cells. This results in drifting of the mitochondrial genotype. The replicative segregation ultimately results in cells with either mutant or normal mtDNAs (homoplasmy).
- Severe mtDNA defects reduce cellular energy outputs until they decline below the minimum energy level (energetic threshold) for normal tissue function. Such energetic thresholds differ among tissues, with the brain, heart, muscle, kidney and endocrine organs being most reliant on mitochondrial energy.
- The mtDNA has a very high mutation rate, some 10–17 times higher than nDNA, which affects both germ line and somatic tissue mtDNAs. Germ line mutations result in maternally transmitted diseases or predispose individuals to late-onset degenerative

diseases. Somatic mutations accumulate in somatic tissues and exacerbate inherited oxidative phosphosphorylation defects. The high mtDNA mutation rate may be due to the high concentration of oxygen radicals at the mitochondrial inner membrane, the lack of efficient mtDNA repair mechanisms and/or the absence of histones.

- Cellular oxygen utilization decreases as a function of age and the ATP generating capacity of the cell is a function of age. The attenuation of ATP synthesis is associated with an age-related increase in somatic mtDNA damage in postmitotic tissue. The "normal" decline in ATP generating capability may facilitate disease occurrence when it is associated with an inherited oxidative phosphorylation mutation.

Mitochondrial Biogenesis

Mitochondrial biogenesis incorporates two distinct processes:

1. The *synthesis* of mitochondrial membranes in which the synthesis of both outer and inner compartments is linked closely to the cell cycle; and
2. *Differentiation* of the organelle for respiratory chain phosphorylation, requiring coordinated control of both nuclear and mitochondrial genes. This is independent of cell growth and division.

The supply of lipids for mitochondrial membrane biosynthesis depends largely on lipids synthesized elsewhere in the cell, especially the endoplasmic reticulum. However, one major lipid component of the inner mitochondrial membrane, ***cardiolipin*** (diphosphatidylglycerol), is synthesized within the mitochondria.

The formation of functional enzyme complexes within the inner mitochondrial membrane proceeds through the sequential assembly of subunits. In assembly of complex I, nuclear encoded subunits form the inner core of the complex followed by attachment of mtDNA encoded subunits. Mitochondrial proteins synthesized in the cytosol contain information that guide them to the mitochondria (intracellular targeting) and that direct their incorporation to a specific site. This information resides in a signal sequence since mitochondrial proteins lacking all or a part of their sequences are not transported into mitochondria. The signal sequences share the property of having a high content of basic and hydroxylated amino acids, no acidic amino acids, and stretches of hydrophobic sequences.

One disorder of mitochondrial transport in humans is due to a failure to transport the methylmalonyl-CoA-mutase precursor protein. The mutation responsible for this disorder affects the signal sequence region. Another disorder has been ascribed to the transport of alanine glyoxylate aminotransferase, an enzyme normally located in peroxisomes of human liver, to mitochondria. The transport of proteins into mitochondria is aided by an outer membrane protein, the general insertion protein (GIP). Proteins that reside in the matrix space, inner membrane, or intermembrane space are routed from the GIP, across the contact site to the matrix space. This transport requires an electrical potential across the inner membrane. After crossing the mitochondrial membrane, precursor polypeptides are converted to mature proteins by the action of proteolytic enzymes that remove signal sequences.

Expression of mtDNA

In addition to limited initiation sites for transcription and the presence of overlapping reading frames, mtDNA has other distinctive features, including a codon assignments that differ from those of nDNA.

1. UGA is used as a tryptophan codon (instead of a stop codon).
2. During translation in mitochondria, unusual codon recognition is a "two out of three" base interaction between codon and anticodon.
3. There are no AGA or AGG (arginine) codons in mtDNA genes. Also, no tRNAs are made in mitochondria for these codons.
4. A single $tRNA^{met}$ specifies both methionine tRNA and N-formylmethionine tRNA by secondary modification of the primary transcript.

Thus, mitochondria require only 22 tRNA molecules to read the genetic code as compared to 31 required for the cytosolic system.

14.5 Nuclear Control of Respiratory Chain Expression

Nuclear genes contribute the majority of respiratory subunits and all of the proteins required for mtDNA transcription, translation, and replication. It is estimated that more than 80% of the genes encoding the subunits of the respiratory chain are located in the cell nucleus. Despite the predominant role for nuclear gene products in the biogenesis and function of the mitochondrial respiratory chain, few nuclear genes have been implicated in mitochondrial genetic disorders. Certain abnormalities of mtDNA are transmitted as Mendelian traits, and it is thought that such characteristics are caused by mutation in identified nuclear genes.

TABLE 14-7
*Correlations Found between Mitochondrial DNA Mutations and Human Diseases**

A. Nucleotide Substitutions	Clinical Features
1. Mildly deleterious base substitutions	Familial deafness, Alzheimer's disease, Parkinson's disease
2. Moderately deleterious nucleotide substitutions	Leber's Hereditary Optic Neuropathy (LHON), Myoclonic Epilepsy and Ragged-Red Fiber disease (MERRF)
3. Severe nucleotide substitutions	Leigh's Syndrome dystonia
B. (mt)DNA Rearrangements	
1. Milder rearrangements (duplications)	Maternally inherited adult-onset diabetes and deafness
2. Severe rearrangements (deletions)	Adult-onset, Chronic Progressive External Ophthalmoplegia (CPEO), Kearns-Sayre Syndrome (KSS), Lethal Childhood Disorders, Pearsons Marrow/Pancreas Syndrome

*Modified from D. C. Wallace *J. of Bioenergetics and Biomembranes* **26** (1994) 241–250.

A number of nuclear-derived transcription factors has been implicated in respiratory chain expression and may play an important role in control of mitochondrial functions. One nuclear respiratory factor (NRF-1) binds to genes encoding respiratory proteins, to the rate-limiting enzyme in heme biosynthesis, and to components of mtDNA transcription and replication. A second respiratory factor (NRF-2) is required for maximal production of cytochrome c oxidase subunit IV (COXIV) and Vb (COXVb). The two regulatory proteins act on a subset of overlapping nuclear genes required for mitochondrial respiratory activity and integrate the expression of nuclear and mitochondrial genes. Complexes III, IV, V, and cytochrome c have at least one subunit whose expression is under NRF control. Another regulatory protein (mtTFA) stimulates transcription initiated by mitochondrial RNA polymerase from heavy (H) and light (L) strand promoters within the mtDNA D-loop. The essential role of mtTFA in the transcription and maintenance of mtDNA makes it a prime candidate for a regulatory function in nuclear-mitochondrial interactions.

Another nDNA regulatory protein functions in the termination of mitochondrial transcription. The mitochondrial termination factor TERF promotes termination *in vitro*. The factor binds at the junction between 16S rRNA and leucyl-tRNA genes, a bidirectional termination site. The binding activity is associated with a 34-kDa protein. A point mutation associated with mitochondrial myopathy, encephalomyopathy, lactic acidosis and stroke-like episodes diminishes the termination of 16S rRNA transcription *in vitro* and reduces the binding affinity of the 34-kDa protein. These findings suggest a potential link between human disease and a nDNA regulatory factor.

14.6 Mitochondrial Diseases

The bioenergetic defects resulting from mtDNA mutations may be a common cause of degenerative diseases (Table 14-7). Defects in nuclear-cytoplasmic interaction are generally the resultant of autosomal dominant mutations; complex disease states result from depletion of mtDNAs from tissues. Mitochondrial DNA mutations are associated with a broad spectrum of chronic degenerative diseases with a variety of clinical presentations. Identical mutations are associated with very different phenotypes and the same phenotype can be associated with different mutations.

Base Substitution Mutations

Diseases involving mtDNA nucleotide substitution represent a broad array of clinical phenotypes typically featuring nervous system involvement. LHON (***Leber's hereditary optic neuropathy***), the best studied mtDNA nucleotide substitution disease, is characterized by a rapid, painless, bilateral loss of central vision due to optic nerve atrophy. The disease is maternally inherited, and its occurrence within families is variable, with young adult males most commonly affected. Modulating factors also influence disease expression. LHON is a genetically heterogeneous disease associated with at least 16 different missense mutations in mtDNA. Three of these—ND1 gene, ND4 gene, and ND6 gene—are the primary causes of LHON. The latter mutation is believed to cause not only LHON but also childhood dystonia associated with bilateral basal ganglia degeneration (LHON + dystonia).

Two heteroplasmic missense mutations (ATPase G gene) result in highly variable disease phenotypes within

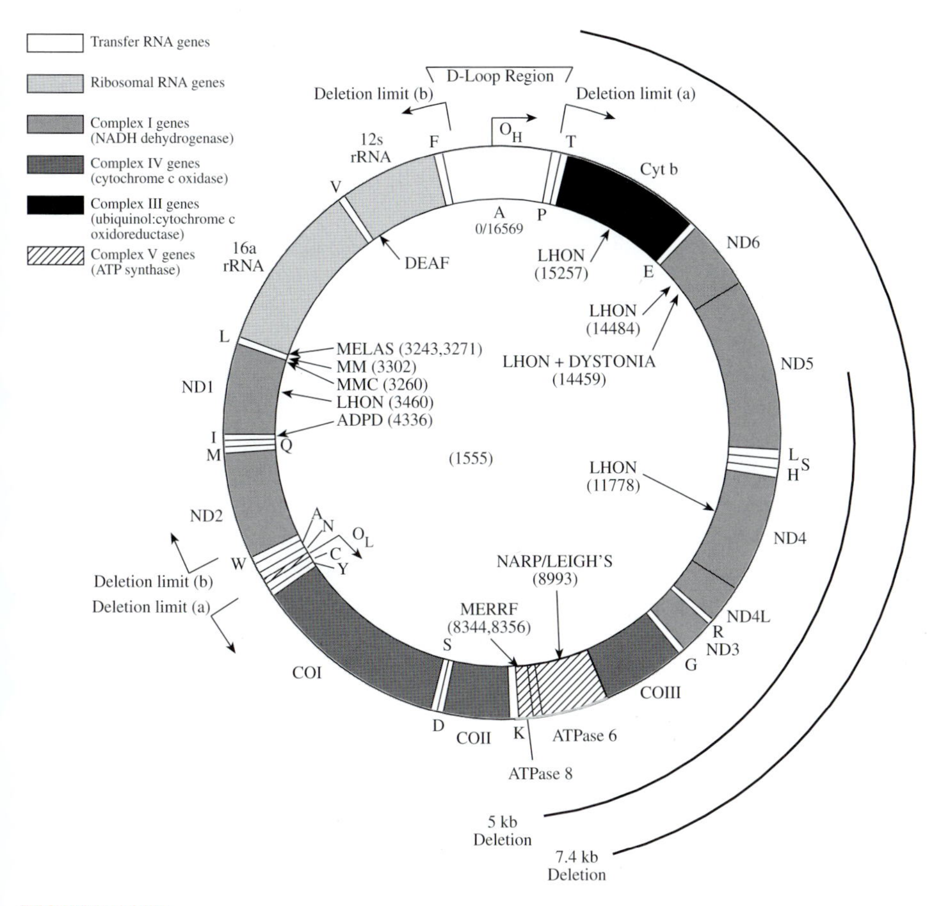

FIGURE 14-23
Human mitochondrial genome showing locations of disease-causing mutations. [Reproduced with permission from M.D. Brown and D.C. Wallace: Molecular basis of mitochondrial DNA disease, *J. Bioenerg. Biomembranes* **26,** 273 (1994).]

families. In one, the mutation leads to neurogenic muscle weakness, ataxia, and retinitis pigmentosa (NAPP) while in most individuals the presentation is described as ***Leigh's disease.*** These mutations cause replacement of a highly conserved leucine residue resulting in loss of ATP production. The differences in the clinical phenotypes of the various mtDNA substitution mutations (LHON, LHON and dystonia, and NARP/Leigh's disease) illustrate the clinical variability of mtDNA diseases.

The LHON group of diseases demonstrates that mildly deleterious mtDNA point mutations can accumulate on a mitochondrial chromosome during evolution, thus creating a "predisposing" mtDNA genotype. Such mutations are relatively benign and require other factors such as age-related decline in oxidative phosphorylation to reduce energy output sufficiently to cause disease. Consequently, these mtDNA mutations are generally associated with late-onset disease and also may predispose individuals to Alzheimer's and Parkinson's disease. Figure 14-23 presents a map of the human mitochondrial genome showing positions of point mutations and deletions.

Transfer RNA (tRNA) Mutations

Fourteen mtDNA tRNA mutations have been associated with maternally inherited disease. Such mutations are typically associated with severe mitochondrial myopathies, characterized by "ragged red" skeletal muscle fibers upon Gomori trichrome staining and the accumulation of structurally abnormal mitochondria in muscle. Mutations in tRNAs exemplify the threshold effect whereby (due to replicative segragation) individuals may not exhibit clinical signs until the proportion of mutant mtDNA exceeds 80–90%. *Myoclonic epilepsy* and *ragged red fiber* (MERRF) *disease, mitochondrial encephalomyopathy lactic acidosis (MELAS),* as well as maternally inherited myopathy and cardiomyopathy (MMC) are well-characterized mitochondrial diseases.

Because oxidative phosphorylation capacity declines with age, individuals with identical mtDNA genotypes may express different clinical signs if they are of different ages. Individuals under the age of 20 with 80% mutant mtDNAs can be asymptomatic; individuals with the same mtDNA mutation over the age of 60 may suffer severe multisystem neurological disease. In general, older individuals have a lower threshold for oxidative phosphorylation dysfunction without clinical consequences.

Two $tRNA^{Leu(UUR)}$ point mutations account for the majority of MELAS patients. An A-to-G transition at 3243 has been found in approximately 80% of MELAS patients. The 3243 mutation alters a highly conserved nucleotide within the dihydrouridine loop of the leucine tRNA. Unlike the MERRF variants, the 3243 mutation can result in a number of different clinical syndromes in addition to MELAS. These include diabetes mellitus and deafness, cardiomyopathy and ocular myopathy. Other mutations to $tRNA^{Leu}$ result in mitochondrial myopathy as well as late-onset hypertrophic cardiomyopathy and myopathy. Overall, the mtDNA tRNA mutations affect both CNS and skeletal/cardiac muscle tissue while missense mutations primarily affect nervous tissue. Only one mitochondrial rRNA point mutation has been associated with disease. The homoplasmic A-to-G transition at position 1555 of the 12S rRNA gene is associated with maternally inherited deafness and aminoglycoside-induced deafness.

mtDNA Deletions and Duplications

Large mtDNA deletions account for most cases of *ocular myopathy* and *Pearson's marrow/pancreas syndrome*. Ocular myopathy patients can exhibit a variety of clinical symptoms, from mild chronic progressive external ophthalmoplegia (CPEO) to *Kearns–Sayre Syndrome (KSS)*. These diseases are characterized by an early onset of ophthalmoplegia, atypical retinitis pigmentosa, mitochondrial myopathy, and usually cerebellar syndrome and cardiac conduction abnormalities. More than 120 different mtDNA deletions have been identified from patients' tissues. Partial duplications of mtDNA have been detected in *ocular myopathy* and *Pearson's syndrome*; however, duplications are much rarer than spontaneous deletions in patients with these conditions. Exactly how partial mtDNA duplications arise is unknown.

Patients with mitochondrial disorders may present at any age and show variation in both the severity and kind of symptoms associated with a single genetic abnormality. For example, the $tRNA^{Leu(UUR)}$ mutation is predominantly associated with the neurological syndrome MELAS but may also be manifested as CPEO, myopathy, diabetes, and deafness.

Mitochondriopathies associated with severe limitation of aerobic metabolism in such organs as liver, kidney, heart, and brain are probably incompatible with survival. Milder mitochondrial disorders may become clinically observable when cellular energy demand is not satisfied (e.g., exercise intolerance). Despite major advances in our understanding of mitochondrial disease, treatment options are limited. Pharmacological therapies have been reported to be of some benefit in isolated cases. In two sisters with NADH-CoQ reductase deficiency who exhibited exercise intolerance, the ^{31}P-NMR study of forearm muscles showed an abnormally rapid decrease in phosphocreatine levels during light exercise and a very slow recovery to normal phosphocreatine and pH levels in the postexercise period. ^{31}P-NMR spectroscopy has also been used to evaluate the bioenergetic capacity of skeletal muscle in a patient with a history of progressive muscle weakness and lactic acidosis. The defect resides between Q and cytochrome c (i.e., in complex III); in particular, it is related to reduced levels of cytochrome b and to a deficiency of at least four additional polypeptides of complex III. Administration of appropriate redox mediators, menadione (vitamin K_3), and ascorbate (vitamin C) bypasses the deficient complex III and allows for the transport of electrons from Q to cytochrome c:

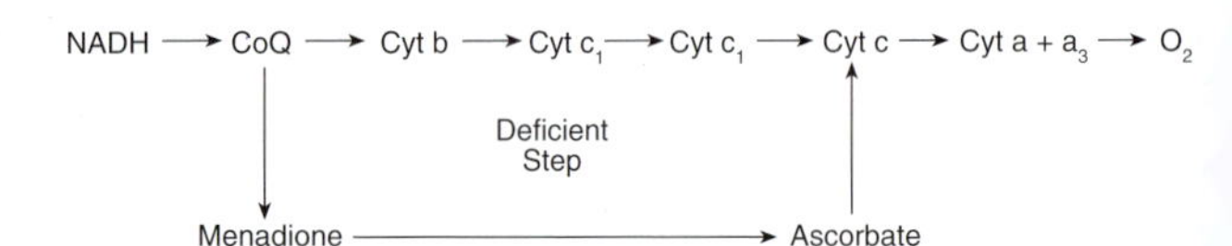

Although the bypass pathway for electron transport theoretically decreases ATP production by 33% of normal mitochondria, in this individual it improves phosphorylation and functional activity.

14.7 Other Reducing-Equivalent Transport and Oxygen-Consuming Systems

In most cells, more than 90% of the oxygen utilized is consumed in the respiratory chain that is coupled to the production of ATP. However, electron transport and oxygen utilization occur in a variety of other reactions, including those catalyzed by oxidases or oxygenases. Xanthine oxidase, an enzyme involved in purine catabolism (Chapter 27), catalyzes the oxidation of hypoxanthine to xanthine, and of xanthine to uric acid. In these reactions, reducing equivalents are transferred via FAD, and Fe^{3+} and Mo^{6+}, while the oxygen is converted to superoxide anion (O_2^-):

Hypoxanthine (H_2O, O_2 → H^+, O_2^-) →

Xanthine (H_2O, O_2 → H^+, O_2^-) → Uric acid

D-Amino acid oxidase is a flavoprotein located in peroxisomes. D-Amino oxidases catalyze the oxidation of a D-amino acid to the corresponding keto acid:

$$\underset{\text{D-Amino acid}}{H{-}\overset{\overset{\displaystyle COO^-}{|}}{\underset{\underset{\displaystyle R}{|}}{C}}{-}NH_3^+} + H_2O \xrightarrow{O_2 \;\; \curvearrowright \;\; H_2O_2} \underset{\text{Keto acid}}{R{-}\overset{\overset{\displaystyle O}{\|}}{C}{-}COO^-} + NH_4^+$$

The O_2^- and H_2O_2 produced in the above two reactions are potentially toxic. Superoxide anion is also produced by oxygen-reducing enzymes of phagocytes (neutrophils, eosinophils, and mononuclear phagocytes) that defend the host against invading organisms by producing reactive oxidants from oxygen. The reducing equivalents are provided by NADPH derived from the hexose monophosphate shunt (Chapter 15).

Neutrophils are the most active participants in ***phagocytosis.*** Unstimulated neutrophils circulate in the bloodstream with a life span of a few days. Upon bacterial infection, neutrophils actively migrate to the infected site, where they kill bacteria by the process of phagocytosis. During phagocytosis a large amount of oxygen is consumed by the neutrophils in a reaction termed the *respiratory burst*. The burst was initially thought to represent an energy requirement for phagocytosis; however, it was subsequently found to be insensitive to inhibitors of the mitochondrial respiratory chain (cyanide, antimycin) and associated with an increased turnover of the hexose monophosphate shunt (Chapter 15).

The oxygen consumed during phagocytosis is utilized by a unique enzyme system termed the *respiratory burst oxidase* or NADPH oxidase. The oxidase generates superoxide anion (O_2^-), a one-electron reduced species, driven by intracellular NADPH,

$$NADPH + 2O_2 \rightarrow NADP^+ + 2O_2^- + H^+$$

Superoxide anion is also the source of a number of other microbicidal oxidants. It functions as an electron donor and is converted to hydrogen peroxide by superoxide dismutase:

$$2O_2^- + 2H^+ \rightarrow H_2O_2 + O_2$$

A number of highly reactive oxidizing agents are produced from H_2O_2 and O_2^-, such as the conjugate acid of superoxide, the hydroperoxy radical ($HO_2^\bullet$), hydroxyl radical ($OH^\bullet$), and hypochlorite (OCl^-). These reactive agents are powerful microbicides:

$$O_2^- + H_2O_2 \rightarrow OH^\bullet + OH^- + O_2$$

and

$$Cl^- + H_2O_2 \rightarrow OCl^- + H_2O$$

The former reaction is promoted by metal ions (e.g., Fe^{2+}), and the latter is catalyzed by myeloperoxidase present in neutrophil granules. By acquiring a single electron, oxygen can give rise to a variety of toxic products. For example, tissue destruction is enhanced when x-ray treatment is used in conjunction with hyperbaric oxygen.

The NADPH oxidase consists of five components. In the unstimulated cell, two of the components are membrane-bound ($p22^{phox}$ and $gp91^{phox}$) and three components are present in cytosol ($p47^{phox}$, $p67^{phox}$, and $p40^{phox}$). The designations gp, phox, and p represent glycoprotein, phagocytic oxidase, and protein, respectively. The membrane-bound components occur as a heterodimeric flavohemoprotein known as cytochrome b_{558}.

The NADPH oxidase complex is dormant in resting phagocytes and becomes assembled and activated for superoxide formation upon bacterial invasion. The respiratory burst is stimulated *in vitro* as well as *in vivo* by a variety of reagents, among which are phorbol esters (PMA, phorbol 12-myristate 13-acetate), heat-aggregated IgG, unsaturated fatty acids and analogues of bacterial peptides (FMLP, formylmethyonyl-leucyl phenylalanine).

The importance of NADPH oxidase is highlighted by ***chronic granulomatous disease*** (CGD), a rare inherited disease in which affected individuals are incapable of mounting a sustained *respiratory burst*. As a consequence, persons with CGD are incapable of generating the reactive oxygen compounds necessary for the intracellular killing of phagocytized microorganisms. The CGD patient suffers severe and recurrent bacterial and fungal infections. Studies using CGD patients' neutrophils as well as a cell-free system have helped identify the cellular components responsible for O_2^- generation.

Four different mutations cause CGD and these cause four defects in different components of the NADPH oxidase. These are the two membrane bound components of cytochrome b_{558} and two cytosolic factors $p47^{phox}$ and $p67^{phox}$. The loci for the four proteins have been mapped in chromosomes X, 16, 7, and 1, respectively. CGD patients are therefore classified as being either of an X-linked type or autosomal recessive. All reported cases of CGD may be explained by abnormalities in the genes coding for one or more of these four components of the oxidase complex. Recent studies have identified an additional cytosolic factor, $p40^{phox}$ and a small G protein (rac) as being involved in the activation of the O_2^- generating system.

Unsaturated fatty acid residues of membrane lipids form α-hydroperoxyalkenes by reacting with oxygen:

$$>C{=}C{-}\overset{|}{\underset{|}{C}}{-} + O_2 \rightarrow >C{=}C{-}\overset{OOH}{\underset{|}{C}}{-}$$

These hydroperoxides undergo cleavage to give rise to short-chain aldehydes. Peroxidation can also cause damage to DNA and proteins. In the latter, sulfhydryl groups give rise to disulfide linkages. Vitamin E functions as an antioxidant and can prevent oxygen toxicity (Chapter 38). Glutathione peroxidase, a selenium containing enzyme, inactivates peroxides:

$$2\,GSH + \begin{matrix} | \\ HC{-}O \\ | \quad | \\ HC{-}O \\ | \end{matrix} \longrightarrow GSSG + \begin{matrix} | \\ HC \\ \| \\ HC \\ | \end{matrix} + H_2O_2$$

Glutathione   Oxidized glutathione

Glutathione, which protects sulfhydryl groups of proteins, is regenerated by glutathione reductase, which uses reducing equivalents from NADPH.

$$GSSG + NADPH + H^+ \rightarrow 2GSH + NADP^+$$

The hydrogen peroxide formed above is decomposed by catalase:

$$2H_2O_2 \rightarrow 2H_2O + O_2$$

Aerobic organisms are protected from oxygen toxicity by three enzymes: glutathione peroxidase, catalase, and superoxide dismutase. Superoxide dismutases are metalloenzymes (the cytoplasmic enzyme contains Cu^{2+} and Zn^{2+}, whereas the mitochondrial enzyme contains Mn^{2+}) widely distributed in aerobic cells. The role of superoxide dismutases in preventing oxygen toxicity is still controversial. For example, some aerobic cells (e.g., adipocytes and some bacteria) lack superoxide dismutase, whereas some strict anaerobes possess this enzyme.

Human superoxide dismutase has been prepared in large quantities by recombinant DNA methods in *Escherichia coli*. It has potential clinical use in preventing oxygen toxicity, for example, during the reestablishment of blood flow through dissolution of a blood clot by thrombolytic agents after a myocardial infarction, during reperfusion after kidney transplant, and in the lungs of premature infants. Allopurinol, a xanthine oxidase inhibitor (Chapter 27), is potentially useful in preventing the tissue injury brought about by ischemia followed by reperfusion. During ischemia, hypoxanthine production increases by enhanced adenine nucleotide catabolism (ATP → ADP → AMP → adenosine → inosine → hypoxanthine; see Chapter 27), providing a larger supply of substrate for xanthine oxidase and thus increased formation of superoxide anions. Complete reduction of O_2 by single-electron transfer reactions (the univalent pathway) yields superoxide anion, hydrogen peroxide, and hydroxyl radical as intermediates:

$$O_2 \xrightarrow{e^-} O_2^- \xrightarrow{e^-,2H^+} H_2O_2 \xrightarrow[H_2O]{e^-,H^+} \cdot OH \xrightarrow{e^-,H^+} H_2O$$

In contrast, most O_2 reduction in aerobic cells occurs by cytochrome c oxidase, which prevents the release of toxic oxygen intermediates:

$$O_2 + 4H^+ + 4\,e^- \rightarrow 2H_2O$$

Mitochondrial superoxide dismutase maintains intramitochondrial superoxide anion at very low steady-state concentrations.

The overall process of reducing oxygen to water can be depicted as follows:

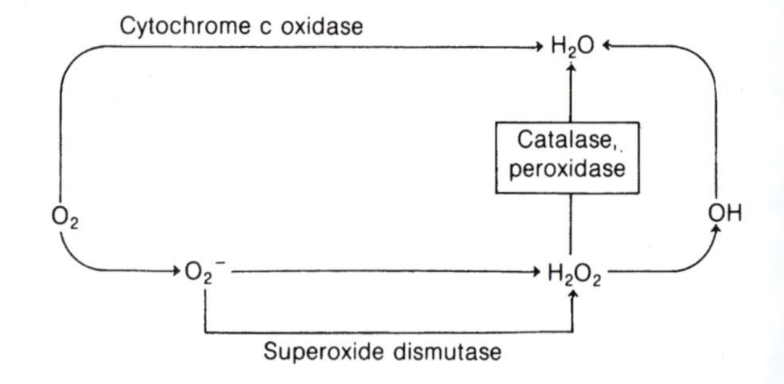

Paraquat, a herbicide that is highly toxic to humans, causes respiratory distress that can lead to death. Damage to membranes of the epithelial cells lining the bronchioles and alveoli that occurs with paraquat poisoning has been ascribed to excessive production of superoxide anion. Paraquat readily accepts electrons from reduced substrates of high negative potential, while the reduced paraquat reacts with molecular oxygen to form superoxide anion and an oxidized paraquat molecule:

Oxidized electron-donor   Reduced electron-donor

$H_3C{-}N$(pyridyl)–(pyridyl)$\overset{+}{N}{-}CH_3$ ⇌ $H_3C{-}\overset{+}{N}$(pyridyl)–(pyridyl)$\overset{+}{N}{-}CH_3$

O_2 O_2^-

Reduced paraquat   Oxidized paraquat

Oxygenases catalyze reactions in which an oxygen atom or molecule is incorporated into organic substrates. They may therefore be ***monooxygenases*** or ***dioxygenases.*** An example of a dioxygenase is tryptophan-2,3-dioxygenase (a heme enzyme), which participates in the catabolism of tryptophan (Chapter 17):

$$\text{Tryptophan} + O_2 \longrightarrow \text{Formylkynurenine}$$

Monooxygenases, also called hydroxylases or mixed-function oxidases, catalyze the incorporation of one atom of an oxygen molecule into the substrate, while the other atom is reduced to water. A second reductant is required (e.g., NADH, NADPH or tetrahydrobiopterin); the overall reaction is

$$RH + AH_2 + O_2 \rightarrow ROH + A + H_2O$$

where AH_2 is the second reductant.

There are many classes of hydroxylases, depending upon the nature of the second reductant. One group, which acts on a variety of substrates including foreign compounds (xenobiotics), contains ***cytochrome P-450.*** The latter derives its name from the absorption maximum at 450 nm of its carbon monoxide adduct, the P standing for pigment. The cytochrome P-450 monooxygenase system is widely distributed in nature. In humans, it is present in most organs, but the highest concentrations are found in the liver. It is found in membranes of the endoplasmic reticulum (microsomal fraction of the cell), mitochondria, and nucleus.

The hepatic microsomal cytochrome P-450 system has been extensively investigated. It consists of a flavin

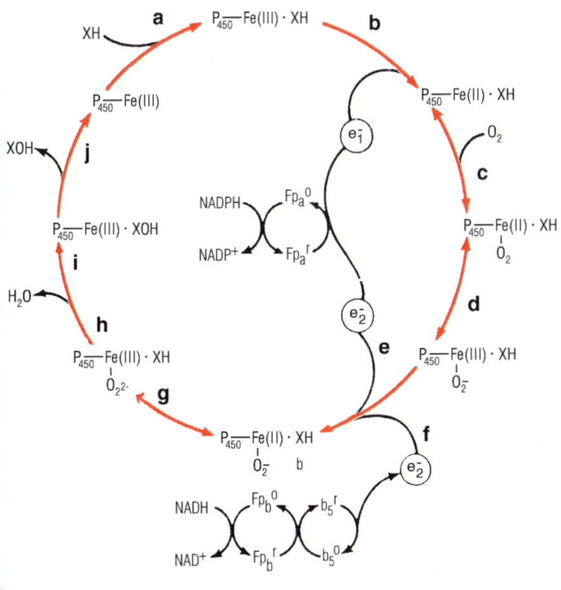

FIGURE 14-24
Cytochrome P-450 monooxygenase system in the hepatic endoplasmic reticulum. P-450–Fe(III) = ferricytochrome P-450; P-450–Fe(II) = ferrocytochrome P-450; b_5 = cytochrome b_5; Fp_a = NADPH-cytochrome P-450 reductase; Fp_b = NADH-cytochrome b_5 reductase; XH = substrate. [Reproduced with permission from G. Mannering, K. W. Renton, R. el Azhary, and L. B. Delor, Effects of interferon-inducing agents on the hepa cytochrome P-450 drug metabolizing systems. *Ann. N.Y. Acad. Sci.* **350,** 314 (1980).]

TABLE 14-8
Cytochrome P-450–Dependent Oxidations

Oxidation of carbon atoms

Aliphatic hydroxylation

$$R{-}CH_3 \longrightarrow RCH_2OH$$

$$R{-}CH_2{-}CH_3 \longrightarrow R{-}CH(OH){-}CH_3$$

Epoxidation

$$R_1{-}CH{=}CH{-}R_2 \longrightarrow R_1{-}CH{-}O{-}CH{-}R_2 \ (\text{epoxide})$$

Aromatic hydroxylation

$$R{-}C_6H_5 \longrightarrow R{-}C_6H_4{-}OH \quad (\text{via } [R{-}\text{arene oxide}])$$

O-, N-, S-dealkylation

$$R{-}O{-}CH_3 \longrightarrow ROH + HCHO$$

$$R{-}NH{-}CH_3 \longrightarrow RNH_2 + HCHO$$

$$R{-}S{-}CH_3 \longrightarrow RSH + HCHO$$

Oxidation of nitrogen atom

N-Hydroxylation

$$R_1{-}NH{-}C(=O){-}R_2 \longrightarrow R_1{-}N(OH){-}C(=O){-}R_2$$

reductase (NADPH-cytochrome P-450 reductase) and one of at least six molecular species of cytochrome P-450. The overall reaction is initiated by combining the substrate (XH) with the ferric form of P-450 to produce the ferrous form by accepting electrons from NADPH-cytochrome P-450 reductase (steps a and b, Figure 14-24). The reductase contains one molecule each of FMN and FAD. The reduced P-450-substrate complex binds molecular oxygen, which becomes activated upon acceptance of an electron from the heme iron (steps c and d). The P-450-substrate-oxygen complex accepts a second electron from either NADPH-cytochrome-P-450 reductase (step e) or cytochrome b_5 (step f). In the final steps (g–j), one oxygen atom receives two protons to form a water molecule, and the other oxygen forms the hydroxyl group of the substrate. The regenerated ferric form of P-450 initiates the next cycle. Supply of the second electron pair is via the cytochrome b_5 electron transport system, which is a minor pathway used principally for the desaturation of fatty acids. Cytochrome b_5, a microsomal membrane-heme protein, receives electrons from NADH via the flavoprotein NADH-cytochrome b_5 reductase. Superoxide anion can also be formed as a byproduct in the cytochrome P-450 system.

The hepatic cytochrome P-450 system exhibits a broad substrate specificity, and many lipophilic compounds including drugs, chemicals, and endogenous metabolites are oxidized by it (Table 14-8). The hydroxylated compounds are converted to more polar metabolites by conjugation with glucuronate, sulfate, amino acids, or acetate that is catalyzed by appropriate transferases. The polar metabolites are excreted by either the biliary-intestinal or the renal system.

Occasionally, the cytochrome P-450 system converts some chemicals to reactive species with carcinogenic potential (e.g., polycyclic hydrocarbons). The hepatic microsomal cytochrome P-450 system is inducible by many of its substrates. The cytochrome P-450 of adrenal cortical mitochondria is involved in steroid hydroxylase reactions, and this system contains iron-sulfur (Fe_2S_2) proteins.

Supplemental Readings and References

T. E. Andreoli: Free radicals and oxidative stress. *American Journal of Medicine* **108,** 650 (2000).

B. M. Babior: NADPH oxidase: An update. *Blood* **93,** 1464 (1999).

B. M. Babior: Phagocytes and oxidative stress. *American Journal of Medicine* **109,** 33 (2000).

H. Beinert, R. H. Holon, and E. Munck. Iron-sulfur clusters: Nature's modulator, multipurpose structures. *Science* **277,** 653 (1997).

P. D. Boyer: The ATP synthase—a splendid molecular machine. *Annual Review of Biochemistry* **66,** 717 (1997).

S. Iwata, J. W. Lee, K. Okada, et al.: Complete structure of the 11 subunit bovine mitochondrial cytochrome bc_1 complex. *Science* **281,** 64 (1998).

N. G. Larsson and D. A. Clayton: Molecular genetic aspects of human mitochondrial disorders. *Annual Review of Genetics* **29,** 151 (1995).

G. S. Shadel and D. A. Clayton: Mitochondrial DNA maintenance in vertebrates. *Annual Review of Biochemistry* **66,** 409 (1997).

B. L. Trumpower: The proton motive Q cycle. *Journal of Biological Chemistry* **265,** 1409 (1990).

L. H. Underhill: Mitochondrial DNA and Disease. *New England Journal of Medicine* **333,** 638 (1995)

D. Xi, C. A. Chang, H. Kim, et al.: Crystal structure of the cytochrome bc_1 complex from bovine heart mitochondria. *Science* **277,** 60 (1997).

M. Zeviani, V. Tiranti, and C. Piantadosi: Mitochondrial disorders. *Medicine* **77,** 59 (1998).

Y. Zhou, T. M. Duncan, and R. L. Cross: Subunit rotation in *Escherichia Coli* F_0F_1–ATP synthase during oxidative phosphorylaton. *Proc. Natl. Acad. Sci.* **94,** 10583 (1997).

CHAPTER 15

Carbohydrate Metabolism II: Gluconeogenesis, Glycogen Synthesis and Breakdown, and Alternative Pathways

In this chapter, carbohydrate metabolism is discussed in terms of nondietary sources of glucose and nonglycolytic pathways. Gluconeogenesis and glycogen synthesis and breakdown make up the first category, while the pentose phosphate pathway (also called hexose monophosphate shunt) and the glucuronic acid pathway make up the second. Metabolic pathways of fructose, galactose, and some other sugars belong in both categories.

15.1 Gluconeogenesis

Metabolic Role

Gluconeogenesis (literally, "formation of new sugar") is the metabolic process by which glucose is formed from noncarbohydrate sources, such as lactate, amino acids, and glycerol. Gluconeogenesis provides glucose when dietary intake is insufficient to supply the requirements of the brain and nervous system, erythrocytes, renal medulla, testes, and embryonic tissues, all of which use glucose as a major source of fuel. Gluconeogenesis has three additional functions.

1. ***Control of acid–base balance.*** Production of lactate in excess of its clearance causes metabolic acidosis, and resynthesis of glucose from lactate is a major route of lactate disposal. Since glycolysis is almost totally anaerobic in erythrocytes, renal medulla, and some other tissues, even under normal conditions lactate is continually released. Other tissues, particularly muscle during vigorous exercise, can produce large amounts of lactate, which must be removed or lactic acidosis will result (Chapter 21). The continuous conversion of lactate to glucose in the liver and of glucose to lactate by anaerobic glycolysis, particularly in muscle, forms a cyclical flow of carbon called the ***Cori cycle*** (Chapter 22). Deamination of amino acids prior to gluconeogenesis in the kidney also provides a supply of NH_3 to neutralize acids excreted in the urine (Chapter 39).
2. ***Maintenance of amino acid balance.*** Metabolic pathways for the degradation of most amino acids and for the synthesis of nonessential amino acids involve some steps of the gluconeogenic pathway. Imbalances of most amino acids, whether due to diet or to an altered metabolic state, are usually corrected in the liver by degradation of the excess amino acids or by synthesis of the deficient amino acids through gluconeogenic intermediates.
3. ***Provision of biosynthetic precursors.*** In the absence of adequate dietary carbohydrate intake, gluconeogenesis supplies precursors for the synthesis of glycoproteins, glycolipids, and structural carbohydrates.

Gluconeogenic Pathway

Gluconeogenesis from pyruvate is essentially the reverse of glycolysis, with the exception of three nonequilibrium reactions (Figure 15-1). These reactions are

$$\text{Glucose} + \text{ATP}^{4-} \xrightarrow{\text{hexokinase or glucokinase}} \text{glucose-6-phosphate}^{2-} + \text{ADP}^{3-} + \text{H}^+ \quad (15.1)$$

$$\text{Fructose-6-phosphate}^{2-} + \text{ATP}^{4-} \xrightarrow{\text{6-phosphofructokinase}} \text{fructose-1,6-bisphosphate}^{4-} + \text{ADP}^{3-} + \text{H}^+ \quad (15.2)$$

$$\text{Phosphoenolpyruvate}^{3-} + \text{ADP}^{3-} + \text{H}^+ \xrightarrow{\text{pyruvatekinase}} \text{pyruvate}^- + \text{ATP}^{4-} \quad (15.3)$$

In gluconeogenesis, these reactions are bypassed by alternate steps also involving changes in free energy and also physiologically irreversible.

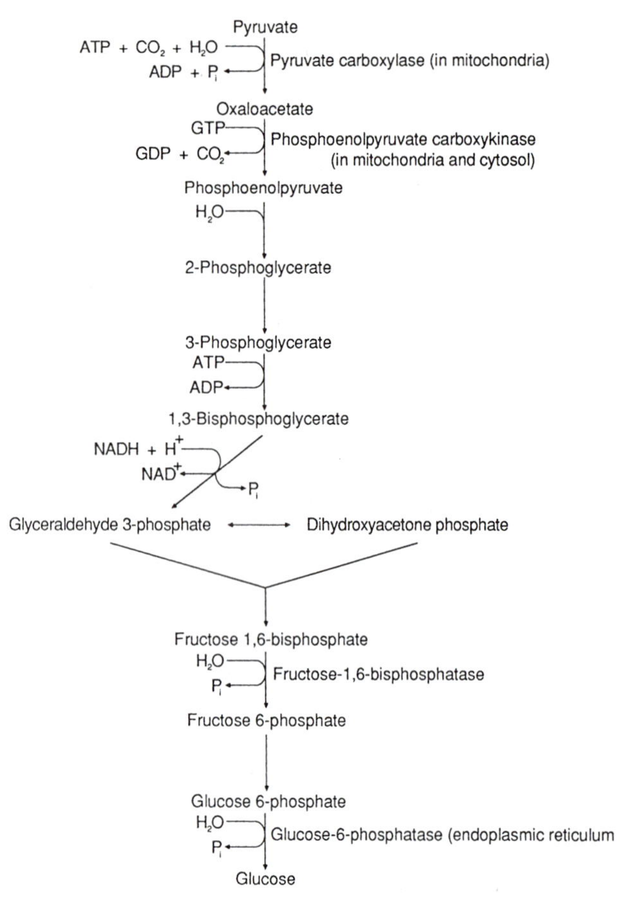

FIGURE 15-1
Pathway of gluconeogenesis from pyruvate to glucose. Only enzymes required for gluconeogenesis are indicated; others are from glycolysis. The overall reaction for the synthesis of one molecule of glucose from two molecules of pyruvate is $2\text{Pyruvate} + 4\text{ATP}^{4-} + 2\text{GTP}^{4-} + 2\text{NADH} + 2\text{H}^+ + 6\text{H}_2\text{O} \rightarrow \text{Glucose} + 2\text{NAD}^+ + 4\,\text{ADP}^{3-} + 2\text{GDP}^{3-} + 6\text{P}_i^{2-} + 4\text{H}^+$

Conversion of pyruvate to phosphoenolpyruvate involves two enzymes and the transport of substrates and reactants into and out of the mitochondrion. In glycolysis, conversion of phosphoenolpyruvate to pyruvate results in the formation of one high-energy phosphate bond. In gluconeogenesis, two high-energy phosphate bonds are consumed ($\text{ATP} \rightarrow \text{ADP} + \text{P}_i$; $\text{GTP} \rightarrow \text{GDP} + \text{P}_i$) in reversing the reaction. Gluconeogenesis begins when pyruvate, generated in the cytosol, is transported into the mitochondrion—through the action of a specific carrier—and converted to oxaloacetate:

$$\text{Pyruvate}^- + \text{HCO}_3^- + \text{ATP}^{4-} \xrightarrow[\text{pyruvate carboxylase}]{\text{acetyl-CoA, Mg}^{2+}} \text{oxaloacetate}^{2-} + \text{ADP}^{3-} + \text{P}_i^{2-} + \text{H}^+$$

Like many CO_2-fixing enzymes, pyruvate carboxylase contains ***biotin*** bound through the ε-NH_2 of a lysyl residue (Chapter 18).

The second reaction is the conversion of oxaloacetate to phosphoenolpyruvate:

$$\text{Oxaloacetate}^{2-} + \text{GTP}^{4-}(\text{or ITP}^{4-}) \xrightarrow{\text{phosphoenolpyruvatecarboxylinase (PEPCK)}} \text{phosphoenolpyruvate}^{3-} + \text{CO}_2 + \text{GDP}^{3-}(\text{or IDP}^{3-})$$

In this reaction, inosine triphosphate (ITP) can substitute for guanosine triphosphate (GTP), and the CO_2 lost is the one fixed in the carboxylase reaction. The net result of these reactions is

$$\text{Pyruvate} + \text{ATP} + \text{GTP (or ITP)} \rightarrow \text{phosphoenolpyruvate} + \text{ADP} + \text{P}_i + \text{GDP (or IDP)}$$

Pyruvate carboxylase is a mitochondrial enzyme in animal cells, whereas PEPCK is almost exclusively mitochondrial in some species (e.g., pigeons) and cytosolic in others (e.g., rats and mice). In humans (and guinea pigs), PEPCK occurs in both mitochondria and cytosol. An interesting consequence of this species differences is that quinolinate, an inhibitor of cytoplasmic PEPCK, causes hypoglycemia in rats but is less active in humans and guinea pigs.

In humans, oxaloacetate must be transported out of the mitochondrion to supply the cytosolic PEPCK. Because there is no mitochondrial carrier for oxaloacetate and its diffusion across the mitochondrial membrane is slow, it is transported as malate or asparate (Figure 15-2). The malate shuttle carries oxaloacetate and reducing equivalents, whereas the aspartate shuttle, which does not require a preliminary reduction step, depends on the availability of glutamate and α-ketoglutarate in excess of tricarboxylic acid (TCA) cycle requirements.

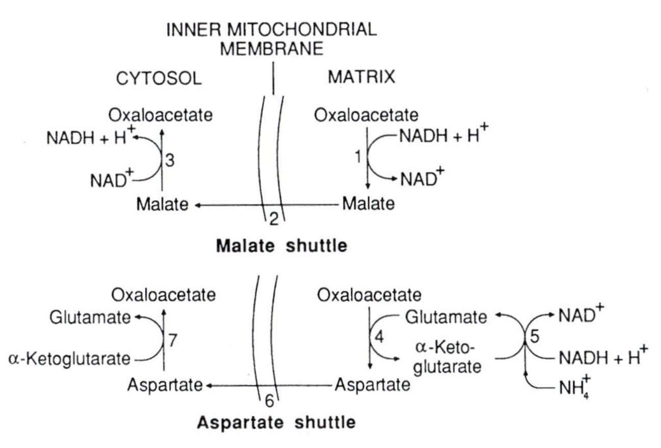

FIGURE 15-2
Shuttle pathways for transporting oxaloacetate from mitochondria into the cytosol. The shuttles are named for the molecule that actually moves across the mitochondrial membrane. 1 and 3 = malate dehydrogenase; 2 = malate translocase; 4 and 7 = aspartate aminotransferase; 5 = glutamate dehydrogenase; 6 = aspartate translocase.

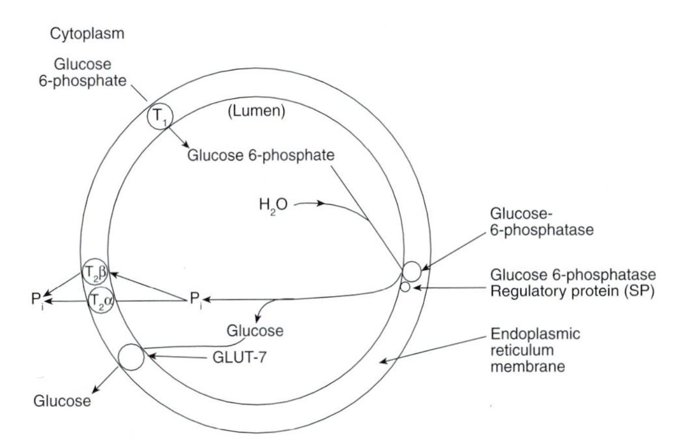

FIGURE 15-3
Schematic representation of the hepatic microsomal glucose-6-phosphatase system. The hydrolysis of glucose phosphate to glucose and inorganic phosphate (P_i) consists of (1) glucose-6-phosphate transport protein (T_1), (2) glucose-6-phosphatase with its catalytic site located on the luminal side of the membrane, (3) glucose-6-phosphatase Ca^{2+}-binding stabilizing protein (SP) also known as regulatory protein, (4) phosphate transport protein ($T_2\beta$), (5) phosphate, pyrophosphate, and carbamoyl phosphate transport protein ($T_2\beta$), and (6) glucose transport protein (GLUT7).

The proportion of oxaloacetate carried by each shuttle probably depends on the redox state of the cytosol. If most of the pyruvate is derived from lactate, the NADH / NAD^+ ratio in the cytosol is elevated. In this situation, there is no need to transport reducing equivalents out of the mitochondria, and the asparate shuttle predominates. However, if alanine is the principal source of pyruvate, no cytosolic reduction occurs, and the glyceraldehyde-phosphate dehydrogenase reaction (which requires NADH) requires transport of reducing equivalents via the malate shuttle. In species in which oxaloacetate is converted in mitochondria to phosphoenolpyruvate (which is readily transported to the cytosol, perhaps via its own carrier system), no transport of oxaloacetate or reducing equivalents is required. In pigeons, which have virtually no cytosolic PEPCK, the rate of gluconeogenesis from pyruvate is much slower than it is from lactate, since there is no source of cytosolic reducing equivalents.

Phosphoenolpyruvate is converted to fructose-1,6-bisphosphate by reversal of glycolysis in the cytosol via reactions that are at near-equilibrium and whose direction is dictated by substrate concentration. Conversion of fructose-1,6-bisphosphate to fructose-6-phosphate is a nonequilibrium step, catalyzed by fructose-1,6-bisphosphatase:

$$\text{Fructose-1,6-bisphosphate}^{4-} + H_2O \xrightarrow{Mg_2^+} \text{fructose-6-phosphate}^{2-} + P_i^{2-}$$

Fructose-6-phosphate is then converted to glucose-6-phosphate by reversal of another near-equilibrium reaction of glycolysis. In the last reaction in gluconeogenesis, glucose-6-phosphate is converted to glucose by glucose-6-phosphatase:

$$\text{Glucose-6-phosphate}^{2-} + H_2O \rightarrow \text{glucose} + P_i^{2-}$$

Glucose-6-phosphatase is part of a multi-component integral membrane protein system (Figure 15-3) located in the endoplasmic reticulum of liver, kidney, and intestine but not of muscle or adipose tissue.

This system consists of six different proteins. These are

1. Glucose-6-phosphate transport protein (T1), which transports glucose-6-phosphate into the lumen of the endoplasmic reticulum;
2. Catalytic subunit of glucose-6-phosphatase (M.W. 36,500), which hydrolyzes glucose-6-phosphate into glucose and phosphate at the luminal surface;
3. Glucose-6-phosphatase regulating protein (M.W. 21,000), which stabilizes the activity of glucose-6-phosphatase;
4. Microsomal phosphate transport protein ($T_2\alpha$) which mediates the efflux of P_i, an inhibitor of the glucose-6-phosphatase, from the lumen of the endoplasmic reticulum to the cytosol;
5. Microsomal phosphate/pyrophosphate transport protein ($T_2\beta$, M.W. 37,000), which transports phosphate, pyrophosphate, and carbamoyl phosphate, which are substrates for glucose-6-phosphatase; and
6. Microsomal glucose transport protein (GLUT7), which is a member of the family of facilitative glucose transport proteins (Chapter 13) and which transports glucose into the cytosol.

These actions of all four transport proteins are reversible. Phosphate transport protein $T_2\beta$ can transport pyrophosphate and carbamoyl phosphate as well as P_i, providing substrates and removing products for several other reactions (of unknown metabolic significance), which are also catalyzed by glucose-6-phosphatase. The importance of each member of the glucose-6-phosphate hydrolysis system is reflected in the occurrence of glycogen storage diseases. Deficiency of any of these five proteins leads to glycogen storage diseases (discussed later).

Thus, gluconeogenesis requires the participation of enzymes of the cytosol, mitochondrion, and smooth endoplasmic reticulum, as well as of several transport systems, and it may involve more than one tissue. The complete gluconeogenic pathway, culminating in the release of glucose into the circulation, is present only in liver and kidney. Most tissues contain only some of the necessary enzymes. These "partial pathways" are probably used in glycerogenesis and in replenishing tricarboxylic acid (TCA) intermediates. Muscle can also convert lactate to glycogen, but this probably takes place only in one type of muscle fiber and only when glycogen stores are severely depleted and lactate concentrations are high, such as after heavy exercise.

Under normal conditions, the liver provides 80% or more of the glucose produced in the body. During prolonged starvation, however, this proportion decreases, while that synthesized in the kidney increases to nearly half of the total, possibly in response to a need for NH_3 to neutralize the metabolic acids eliminated in the urine in increased amounts (Chapter 22).

Gluconeogenesis is a costly metabolic process. Conversion of two molecules of pyruvate to one of glucose consumes six high-energy phosphate bonds ($4ATP + 2GTP \rightarrow 4ADP + 2GDP + 6P_i$) and results in the oxidation of two NADH molecules (Figure 15-1). In contrast, glycolytic metabolism of one molecule of glucose to two of pyruvate produces two high-energy phosphate bonds ($2ADP + 2P_i \rightarrow 2ATP$) and reduces two molecules of NAD^+. For gluconeogenesis to operate, the precursor supply and the energy state of the tissue must be greatly increased. Using some gluconeogenic precursors to provide energy (via glycolysis and the TCA cycle) to convert the remainder of the precursors to glucose would be inefficient, even under aerobic conditions. Usually, the catabolic signals (catecholamines, cortisol, and increase in glucagon/insulin ratio) that increase the supply of gluconeogenic precursors also favor lipolysis, which provides fatty acids to supply the necessary ATP.

When amino acid carbons are the principal gluconeogenic precursors, the metabolic and physiological debts are particularly large compared to those incurred when lactate or glycerol is used. Amino acids are derived by breakdown of muscle protein, which is accompanied by a loss of electrolytes and tissue water. One kilogram of muscle contains about 150 g of protein, which can be used to form about 75 g of glucose. However, 1 kg of muscle also contains 1,200 mM of K^+, 27 mM of phosphate, and 8 mM of Mg^{2+}, all of which must be excreted by the kidney. The ammonia released from catabolism of the amino acids is converted to urea in an energy-requiring process, and the urea further increases the osmotic load on the kidney. Renal excretion of these solutes necessitates mobilization of at least 2 L of water from other tissues in addition to the 750 mL released from the muscle, and osmotic diuresis may result from catabolism of even a relatively small amount of muscle. Finally, if fat metabolism is stimulated, as in starvation and diabetes mellitus, increased plasma levels of fatty acids and ketone bodies causes acidemia. To prevent severe acidosis, renal proton excretion is increased by increasing secretion of ammonia into the renal tubules. Excretion of nitrogen as ammonia avoids energy consumption due to urea synthesis, but it doubles the osmotic load per nitrogen excreted, causing even greater diuresis. Without a compensatory increase in water intake, profound depletion of blood volume can occur (Chapter 22).

Gluconeogenic Precursors

Gluconeogenic precursors include lactate, alanine, and several other amino acids, glycerol, and propionate (Chapter 22).

Lactate, the end product of anaerobic glucose metabolism, is produced by most tissues of the body, particularly skin, muscle, erythrocytes, brain, and intestinal mucosa. In a normal adult, under basal conditions, these tissues produce 1,300 mM of lactate per day, and the normal serum lactate concentration is less than 1.2 mM/L. During vigorous exercise, the production of lactate can be increased several fold. Lactate is normally removed from the circulation by liver and kidney. Because of its great capacity to use lactate, liver plays an important role in the pathogenesis of lactic acidosis, which may be thought of as an imbalance between the relative rates of production and utilization of lactate (Chapter 39).

Alanine, derived from muscle protein and also synthesized in the small intestine, is quantitatively the most important amino acid substrate for gluconeogenesis. It is converted to pyruvate by alanine aminotransferase (Chapter 17):

$$\text{Alanine} + \alpha\text{-ketoglutarate}^{2-} \xrightleftharpoons{B_6\text{-}PO_4} \text{pyruvate}^- + \text{glutamate}^-$$

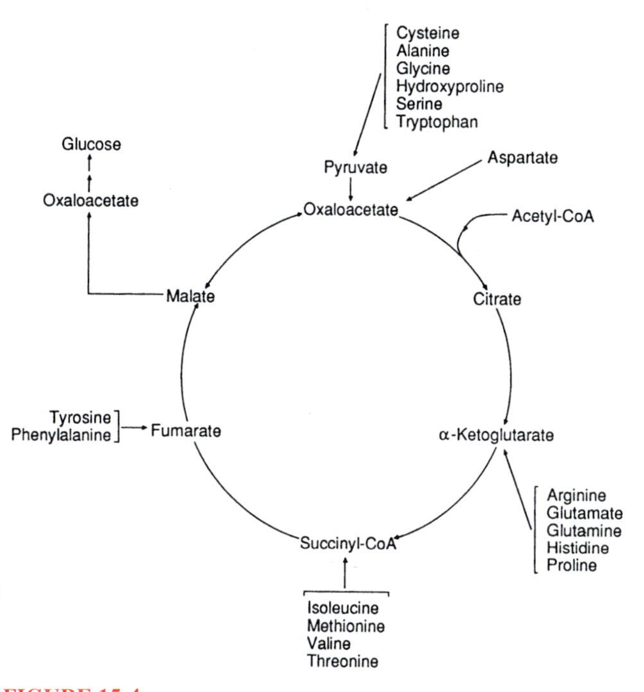

FIGURE 15-4
Points of entry of amino acids into the pathway of gluconeogenesis.

where B_6–PO_4 = pyridoxal phosphate. An amino acid is classified as ketogenic, glucogenic, or glucogenic/ketogenic depending on whether feeding it to a starved animal increases plasma concentrations of ketone bodies (Chapter 18) or of glucose. Leucine and lysine are purely ketogenic because they are catabolized to acetyl-CoA, which cannot be used to synthesize glucose (Chapter 17). Isoleucine, phenylalanine, tyrosine, and tryptophan are both glucogenic and ketogenic, and the remaining amino acids (including alanine) are glucogenic. Entry points of the amino acids into the gluconeogenic pathway are shown in Figure 15-4.

Glycerol is more reduced than the other gluconeogenic precursors, and it results primarily from triacylglycerol hydrolysis in adipose tissue. In liver and kidney, glycerol is converted to glycerol 3-phosphate by glycerol kinase:

$$\underset{\text{Glycerol}}{\begin{matrix} CH_2OH \\ | \\ CHOH \\ | \\ CH_2OH \end{matrix}} + ATP^{4-} \longrightarrow \underset{\text{Glycerol 3-phosphate}}{\begin{matrix} CH_2OH \\ | \\ CHOH \\ | \\ CH_2OPO_3^{2-} \end{matrix}} + ADP^{3-} + H^+$$

Glycerol 3-phosphate is oxidized to dihydroxyacetone phosphate by glycerol-3-phosphate dehydrogenase:

$$\underset{\text{Glycerol 3-phosphate}}{\begin{matrix} CH_2OH \\ | \\ CHOH \\ | \\ CH_2OPO_3^{2-} \end{matrix}} + NAD^+ \longrightarrow \underset{\text{Dihydroxyacetone phosphate}}{\begin{matrix} CH_2OH \\ | \\ C{=}O \\ | \\ CH_2OPO_3^{2-} \end{matrix}} + NADH + H^+$$

Dihydroxyacetone phosphate is the entry point of glycerol into gluconeogenesis. Glycerol cannot be metabolized in adipose tissue, which lacks glycerol kinase, and the glycerol 3-phosphate required for triacylglycerol synthesis in this tissue is derived from glucose (Chapter 22). During fasting in a resting adult, about 210 mM of glycerol per day is released, most of which is converted to glucose in the liver. During stress or exercise, glycerol release is markedly increased (Chapter 22).

Propionate is not a quantitatively significant gluconeogenic precursor in humans, but it is a major source of glucose in ruminants. It is derived from the catabolism of isolecucine, valine, methionine, and threonine; from β-oxidation of odd-chain fatty acids; and from the degradation of the side chain of cholesterol. Propionate enters gluconeogenesis via the TCA cycle after conversion to succinyl-CoA (Chapter 18).

Regulation of Gluconeogenesis

Gluconeogenesis is regulated by the production and release of precursors and by the activation and inactivation of key enzymes. Glucagon and glucocorticoids stimulate gluconeogenesis, whereas insulin suppresses it. See Chapter 22 for a discussion of the overall metabolic control and physiological implications of the several seemingly competing pathways.

Carboxylation of Pyruvate to Oxaloacetate

The pyruvate carboxylase reaction is activated by Mg^{2+} and, through mass action, by an increase in either the [ATP]/[ADP] or the [pyruvate]/[oxaloacetate] ratio. It is virtually inactive in the absence of acetyl-CoA, an allosteric activator. The enzyme is allosterically inhibited by glutamate, since oxaloacetate formed in excess would flood the TCA cycle and result in a buildup of α-ketoglutarate and glutamate. Pyruvate carboxylase activity is indirectly stimulated by the catabolic hormones glucagon (via cAMP) and epinephrine (independent of cAMP), in four ways:

1. By increasing substrate availability through stimulation of mitochondrial respiration. This decreases the intramitochondrial concentration of H^+ and increases the rate of pyruvate transport. The [ATP]/[ADP] ratio also rises.
2. By increasing the rate of fatty acid oxidation, which results in an increase in the mitochondrial concentration of acetyl-CoA, an allosteric activator of pyruvate carboxylase.
3. By decreasing the mitochondrial concentration of glutamate, an inhibitor of pyruvate carboxylase, through stimulation of the TCA cycle (secondary to the increase in mitochondrial acetyl-CoA) and the aspartate shuttle (secondary to the increase in cytosolic PEPCK induced by glucagon).
4. By decreasing the activity of glycolytic enzymes competing for the same substrate; in this case, by inactivating pyruvate kinase in the cytosol and pyruvate dehydrogenase in the mitochondria (via cAMP stimulation of protein kinase).

Conversion of Oxaloacetate to Phosphoenolpyruvate

Short-term regulation of this reaction is accomplished by changes in the relative proportions of substrates and products. Increased concentrations of oxaloacetate and GTP (or ITP) increase the rate, and accumulation of phosphoenolpyruvate and GDP (or IDP) decreases it. The cytosolic PEPCK is also under long-term regulation by hormones. Its synthesis is increased by corticosteroids. Starvation and diabetes mellitus increase the synthesis of cytosolic PEPCK, whereas refeeding and insulin have the opposite effect.

The net effect of increasing the ratio of cytosolic PEPCK to mitochondrial PEPCK is to maintain gluconeogenesis even during periods of increased fatty acid metabolism. When gluconeogenesis is strongly stimulated, lipolysis and intramitochondrial fatty acid oxidation increase with the activity of cytosolic PEPCK. The increase in fatty acid oxidation elevates the [NADH]/[NAD$^+$] ratio within the mitochondria, inhibiting phosphoenolpyruvate formation by channeling oxaloacetate into formation of malate or aspartate. At higher [NADH]/[NAD$^+$] ratios, aspartate synthesis is favored because of increased formation of glutamate by mitochondrial glutamate hydrogenase:

$$\alpha\text{-Ketoglutarate} + NH_4^+ + NADH \rightleftharpoons \text{glutamate} + NAD^+ + H_2O$$

The increase in glutamate favors transamination of oxaloacetate and limits oxaloacetate availability for phosphoenolpyruvate synthesis. When the [NADH]/[NAD$^+$] ratio is low, malate formation occurs more readily. The cytosolic PEPCK is relatively unaffected by the mitochondrial [NADH]/[NAD$^+$] ratio. Once malate and aspartate are transported to the cytosol and they are reconverted to oxaloacetate, cytosolic PEPCK can convert it to phosphoenolpyruvate.

Conversion of Fructose-1,6-Bisphosphate to Fructose-6-Phosphate

This reaction is catalyzed by fructose-1,6-bisphosphatase (FBPase-1). This enzyme is a tetramer (M.W. 164,000) of identical subunits. It is inhibited by AMP, inorganic phosphate, and fructose-2,6-bisphosphate, all of which are allosteric activators of 6-phosphofructo-1-kinase (PFK-1), the competing glycolytic enzyme (Figure 15-5). This inhibition prevents the simultaneous activation of the two enzymes and helps prevent a futile ATP cycle. However, this futile cycle apparently does operate at a low rate in mammalian muscle, and in the bumblebee it is essential for maintaining the flight muscle at 30°C.

Activation of the futile cycle occurs after exposure to certain anesthetic agents such as suxamethonium and/or volatile halogenated compounds. In susceptible persons, ***malignant hyperthermia*** is characterized by a hypermetabolic state and is accompanied by a catastrophic rise in body temperature (to 42°C or higher), massive increase in oxygen consumption, rhabdomyolysis, (disintegration of muscle, associated with excretion of myoglobin in the urine) and metabolic acidosis. ***Rhabdomyolysis*** is assessed by measurement of serum creatine kinase levels (Chapter 8). In susceptible persons, the hypermetabolic events can also be triggered by excessive exercise under hot conditions, infections, and neuroleptic drugs.

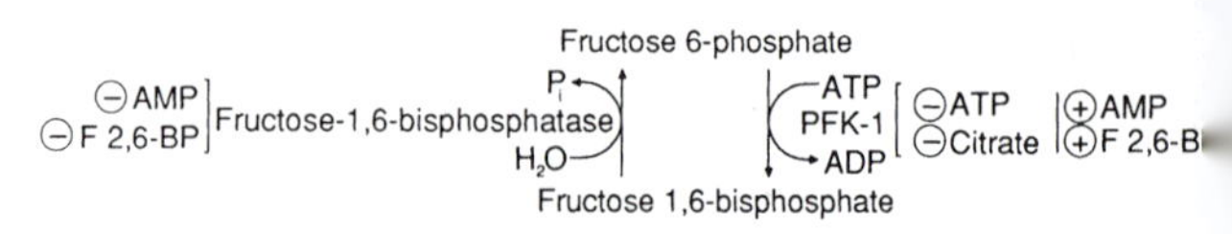

FIGURE 15-5
Regulation of liver 6-phosphofructokinase and fructose-1,6-bisphosphatase. These multimodulated enzymes catalyze nonequilibrium reactions, the former in glycolysis and the latter in gluconeogenesis. Note the dual action of fructose-2,6-bisphosphate (F-2,6-BP), which activates phosphofructokinase (PFK-1) and inactivates fructose-1,6-bisphosphatase. The activity of F-2,6-BP is under hormonal and substrate regulation (Figure 15-6). ⊕ = positive effectors; ⊖ = negative effectors.

The primary biochemical abnormality of malignant hyperthermia is a membrane defect that leads to a rapid and sustained rise in Ca^{2+} levels in the cytosol of skeletal muscle cells. Sarcoplasmic reticulum is the main storage site of intracellular Ca^{2+} and it is released into the myoplasm by excitation-contraction coupling (Chapter 21). Dantrolene, a drug useful in the treatment of malignant hyperthermia, inhibits excitation–contraction coupling. However, dantrolene does not affect neuromuscular transmission or the electrical properties of skeletal muscle. In the animal model, pigs susceptible for malignant hyperthermia exhibit a genetic defect in a Ca^{2+} release channel receptor located in the sarcoplasmic reticulum. This receptor protein is named ***ryanodine receptor (RYR)*** because it binds to a plant alkaloid ryanodine and upon binding causes Ca^{2+} release. The mutation that alters the receptor protein causes a substitution of cysteine for arginine at position 615.

The concentration of fructose-2,6-bisphosphate is controlled by two competing enzyme activities in a single multifunctional protein (M.W. 110,000) with two identical subunits (Figure 15-6). Phosphorylation by cAMP-dependent protein kinase produces one phosphoserine per subunit, inhibiting 6-phosphofructo-2-kinase (PFK-2) activity and stimulating fructose-2,6-bisphosphatase (FBPase-2) activity. The phosphotransferase inactivation results from an increased K_m for fructose-6-phosphate and a small decrease in V_{max}. An increase in V_{max} and a decrease in K_m for fructose-2,6-bisphosphate increases the phosphohydrolase activity. The protein phosphatase that reverses the cAMP-dependent phosphorylation has not yet been characterized. The nomenclature of these enzyme activities (PFK-1, PFK-2, FBPase-1, and FBPase-2) reflects both the order of their discovery and the carbon atom of fructose that is phosphorylated or dephosphorylated.

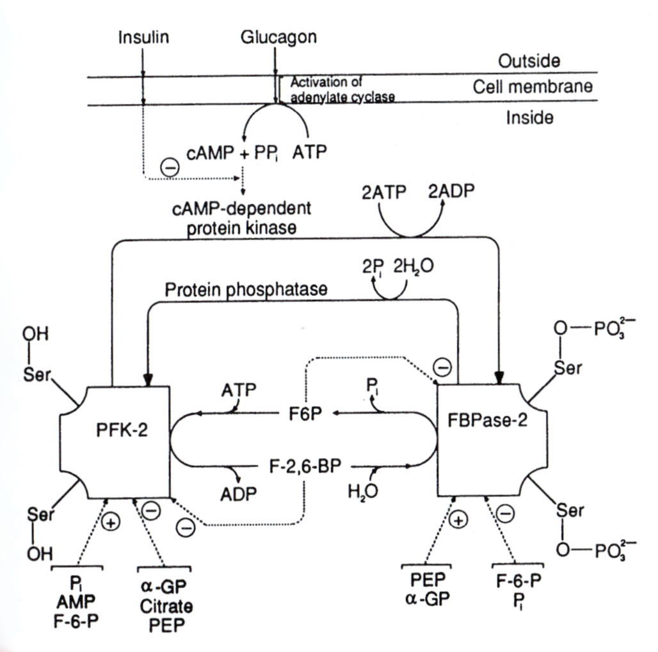

FIGURE 15-6
Regulation of fructose-2,6-bisphosphate (F-2,6-BP) concentration in liver. F-2,6-BP is a major factor controlling the relative activities of fructose-1,6-bisphosphatase (FBPase-1) and 6-phosphofructo-1-kinase (PFK-1), key enzymes in gluconeogenesis and glycolysis, respectively. 6-Phosphofructo-2-kinase (PFK-2) and fructose-2,6-bisphosphatase (FBPase-2) are separate enzymes expressed by a single multifunctional protein. Insulin can antagonize the effect of glucagon on cAMP, but the mechanism is not known. Dotted lines = regulatory factors; ⊖ = inhibitory; ⊕ = stimulatory. F6P = fructose-6-phosphate; PEP = phosphoenolpyruvate; α-GP = α-glycerol phosphate; P_i = inorganic phosphate; F-6-P = fructose-6-phosphate.

Phosphorylation is regulated by insulin and glucagon. Diabetes mellitus (in which the ratio of glucagon to insulin is increased) and glucagon therapy reduce the hepatic activity of PFK-2 and increase that of FBPase-2. The concentration of fructose-2,6-bisphosphate is thus reduced, thereby stimulating gluconeogenesis.

The concentration of fructose-2,6-bisphosphate is also influenced by nonhormonal factors. PFK-2 is allosterically stimulated by inorganic phosphate, AMP, and fructose-6-phosphate and inhibited by citrate, fructose-2,6-bisphosphate, phosphoenolpyruvate, and α-glycerol phosphate. FBPase-2, on the other hand, is stimulated by phosphoenolpyruvate and α-glycerol phosphate and inhibited by phosphate and fructose-6-phosphate. Glucose increases the concentration of fructose-2,6-bisphosphate *in vivo,* probably by increasing the availability of fructose-6-phosphate, thereby stimulating PFK-2, the kinase for which this is a substrate and inhibiting the phosphatase, FBPase-2. The effect is to increase glycolysis and inhibit gluconeogenesis. The concentration of α-glycerol phosphate *in vivo* can be altered by ischemia consumption of ethanol, and feeding of various sugars; it may be an important regulatory molecule in determining the levels of fructose-2,6-bisphosphate. A rise in α-glycerol phosphate decreases fructose-2,6-bisphosphate concentration, decreasing glycolysis and increasing gluconeogenesis.

Conversion of Glucose-6-Phosphate to Glucose

Glucose-6-phosphatase and the corresponding glycolytic enzyme, glucokinase, are not controlled by the metabolites that affect PFK-1 and FBPase-1. Glucose-6-phosphatase seems to be regulated only by the concentration of glucose-6-phosphate, which is increased by glucocorticoids, thyroxine, and glucagon. The concentration of glucose-6-phosphate, which is also elevated in diabetes mellitus and following ethanol administration, is increased 300-fold after a 48-hour fast. Consumption of a low-protein diet reduces the concentration of glucose-6-phosphate. A futile cycle involving glucokinase and glucose-6-phosphatase does not occur because glucose-6-phosphatase is active only when the concentration of glucose-6-phosphate is high, and glucokinase is active only when glucose-6-phosphate concentration is low.

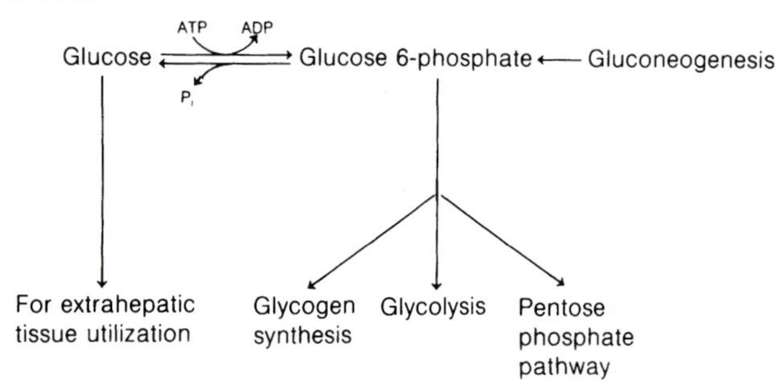

Abnormalities of Gluconeogenesis

Glucose is the predominant fuel for cells that depend largely on anaerobic metabolism, cells that lack mitochondria, and tissues such as brain that normally cannot use other metabolic fuels. An adult brain represents only 2% of total body weight, but it oxidizes about 100 g of glucose per day, accounting for 25% of the basal metabolism. Brain cannot utilize fatty acids because they are bound to serum albumin and so do not cross the blood-brain barrier. Ketone bodies are alternative fuels, but their concentration in blood is negligible, except in prolonged fasting or diabetic ketoacidosis (Chapters 18 and 22). Brain and liver glycogen stores are small relative to the needs of the brain, and gluconeogenesis is essential for survival. Decreased gluconeogenesis leads to hypoglycemia and may cause irreversible damage to the brain. Blood glucose levels below 40 mg/dL (2.2 mM/L) in adults or below 30 mg/dL (1.7 mM/L) in neonates represent severe hypoglycemia. The low levels may result from increased glucose utilization, decreased glucose production, or both. Increased glucose utilization can occur in hyperinsulinemia secondary to an insulinoma. In a diabetic pregnancy, the mother's hyperglycemia leads to fetal hyperglycemia and causes fetal and neonatal hyperinsulinemia. Insufficiency of glucocorticoids, glucagon, or growth hormone and severe liver disease can produce hypoglycemia by depressing gluconeogenesis (Chapters 22 and 32). Ethanol consumption can cause hypoglycemia, since a major fraction of ethanol is oxidized in the liver by ***cytosolic alcohol dehydrogenase,*** and the NADH and acetaldehyde generated inhibit gluconeogenesis:

$$CH_3CH_2OH + NAD^+ \rightarrow CH_3CHO + NADH + H^+$$

Excessive NADH decreases the cytosolic $[NAD^+]/[NADH]$ ratio, increasing lactate dehydrogenase activity and thereby increasing conversion of pyruvate to lactate. The decrease in pyruvate concentration inhibits pyruvate carboxylase. Acetaldehyde inhibits oxidative phosphorylation, increasing the [ADP]/[ATP] ratio, promoting glycolysis, and inhibiting gluconeogenesis. Hypoglycemia and lactic acidosis are common findings in chronic alcoholics.

Pyruvate carboxylase deficiency can cause intermittent hypoglycemia, ketosis, severe psychomotor retardation, and lactic acidosis. The deficiency of either cytosolic or mitochondrial isoenzyme form of phosphoenolpyruvatecarboxykinase (PEPCK), is characterized by a failure of gluconeogenesis. Both these disorders are rare and the mitochondrial PEPCK deficiency is inherited as an autosomal recessive trait. The major clinical manifestations include hypoglycemia, lactic acidosis, hypotonia, hepatomegaly and failure to thrive. The treatment is supportive and is based upon symptoms. Fructose-1,6-bisphosphatase deficiency, inherited as an autosomal recessive trait, severely impairs gluconeogenesis, causing hypoglycemia, ketosis, and lactic acidosis. Glucose-6-phosphatase deficiency, also an autosomal recessive trait, causes a similar condition but with excessive deposition of glycogen in liver and kidney (discussed later).

Hypoglycin A (2-methylenecyclopropylalanine), the principal toxin of the unripe akee fruit, produces severe hypoglycemia when ingested. A less toxic compound, hypoglycin B, is a γ-glutamyl conjugate of hypoglycin A. The akee tree is indigenous to western Africa and grows in Central America and the Caribbean. Hypoglycin causes vomiting ("vomiting sickness") and central nervous system depression. In Jamaica, symptoms apparently occur in epidemic proportions during colder months when the akee fruit is unripe and other food sources are limited. The edible portion of the ripe akee fruit, which contains only small amounts of hypoglycin, is a main dietary staple in Jamaica.

Hypoglycin A causes hypoglycemia by inhibiting gluconeogenesis. In the liver, hypoglycin A forms nonmetabolizable esters with CoA (shown below) and carnitine depleting the CoA and carnitine pools, thereby inhibiting fatty acid oxidation (Chapter 18). Since the principal source of ATP for gluconeogenesis is mitochondrial oxidation of long-chain fatty acids, gluconeogenesis is stopped Intravenous administration of glucose relieves the hypoglycemia.

$H_2C{=}C$(—CH_2—, cyclopropane)—CH—CH_2—$CH(NH_2)$—$COOH$ Hypoglycin A

↓ α-Ketoglutarate → Glutamate (Transmination)

$H_2C{=}CH$(—CH_2—, cyclopropane)—CH—CH_2—$C(=O)$—$COOH$

Methylenecyclopropyl α-ketopropionic acid

↓ NAD^+, CoASH → $NADH + H^+$, CO_2 (Oxidative decarboxylation)

$H_2C{=}C$(—CH_2—, cyclopropane)—$CHCH_2C(=O)$—S—CoA

Methylenecyclopropyl α-ketoproprionyl-CoA

Overactivity of gluconeogenesis due to increased secretion of catecholamines, cortisol, or growth hormone or an increase in the glucagon/insulin ratio (Chapter 22) leads to hyperglycemia and causes many metabolic problems.

15.2 Glycogen Metabolism

Animals have developed a method of storing glucose that reduces the need to catabolize protein as a source of gluconeogenic precursors between meals. The principal storage form of glucose in mammals is glycogen. Storage of glucose as a polymer reduces the intracellular osmotic load, thereby decreasing the amount of water of hydration and increasing the energy density of the stored glucose.

In muscle, glycogen is stored in the cytosol and endoplasmic reticulum as granules, called ***β-particles,*** each of which is an individual glycogen molecule. In the liver, the β-particles aggregate, forming larger, rosette-shaped ***α-particles*** that can be seen with the electron microscope. The average molecular weight of glycogen is several million (10,000–50,000 glucose residues per molecule). The storage granules also contain the enzymes needed for glycogen synthesis and degradation and for regulation of these two pathways.

Glycogen is present in virtually every cell in the body, but it is especially abundant in liver and skeletal muscle. The amount stored in tissues varies greatly in response to metabolic and physiological demands, but in a resting individual after a meal, liver usually contains roughly 4–7% of its wet weight as glycogen and muscle about 1%. Since the body contains 10 times more muscle than hepatic tissue, the total amount of glycogen stored in muscle is greater than that in liver.

Muscle needs a rapidly available supply of glucose to provide fuel for anaerobic glycolysis when, as during bursts of muscle contraction, blood may provide inadequate supplies of oxygen and fuel. The liver, under most conditions, oxidizes fatty acids for this purpose; however, liver is responsible for maintaining blood glucose levels during short fasts, and it integrates the supply of available fuels with the metabolic requirements of other tissues in different physiological states. Liver glycogen content changes primarily in response to the availability of glucose and gluconeogenic precursors. Muscle glycogen stores vary less in response to dietary signals, but they depend on the rate of muscular contraction and of oxidative metabolism (tissue respiration).

Glycogenesis (glycogen synthesis and storage) and glycogenolysis (glycogen breakdown) are separate metabolic pathways having only one enzyme in common, namely, phosphoglucomutase. Glycogen synthesis and breakdown are often reciprocally regulated, so that stimulation of one inhibits the other. The control mechanisms are more closely interrelated than are those for glycolysis and gluconeogenesis, perhaps because the glycogen pathways ensure the availability of only one substrate, whereas intermediates of several other metabolic pathways are produced and metabolized in glycolysis and gluconeogenesis (Chapter 13).

Glycogen Synthesis

Glycogenesis begins with the phosphorylation of glucose by glucokinase in liver and by hexokinase in muscle and other tissues (Chapter 13):

$$\text{Glucose} + ATP^{4-} \xrightarrow{Mg^{2+}} \text{glucose-6-phosphate}^{2-} + ADP^{3-} + H^+$$

The second step in glycogenesis is conversion of glucose-6-phosphate to glucose-1-phosphate by phosphoglucomutase in a reaction similar to that catalyzed by phosphoglyceromutase. The phosphoryl group of the enzyme participates in this reversible reaction in which glucose-1,6-bisphosphate serves as an intermediate:

$$\text{E-P} + \text{glucose-6-phosphate} \rightleftharpoons \text{E} + \text{glucose-1,6-bisphosphate} \rightleftharpoons \text{E-P} + \text{glucose-1-phosphate}$$

In the third step, glucose-1-phosphate is converted to uridine diphosphate (UDP) glucose, the immediate precursor of glycogen synthesis, by reaction with uridine triphosphate (UTP). This reaction is catalyzed by glucose-1-phosphate uridylyltransferase (or UDP-glucose pyrophosphorylase):

$$\text{UTP}^{4-} + \text{Glucose 1-phosphate}^{2-} \rightleftharpoons \text{UDP-glucose} + \text{PPi}^{4-}$$

The reaction is freely reversible, but it is rendered practically irreversible by hydrolysis of pyrophosphate (PP_i) through the action of pyrophosphatase.

$$PP_i + H_2O \rightarrow 2P_i$$

Thus, two high-energy phosphate bonds are hydrolyzed in the formation of UDP-glucose. Nucleoside diphosphate sugars are commonly used as intermediates in carbohydrate condensation reactions. ADP-glucose is the precursor for the synthesis of starch in plants and of other storage polysaccharides in bacteria. In each case, the high-energy bond between the sugar and the nucleoside diphosphate provides the energy needed to link the new sugar to the nonreducing component of another sugar or polysaccharide.

The next step in the biogenesis of glycogen is addition of a glucosyl residue at the C-1 position of UDP-glucose to a tyrosyl group located at position 194 of the enzyme glycogenin, which is a Mg^{2+}-dependent autocatalytic reaction. Glycogenin (M.W. 37,000) extends the glucan chain, again by autocatalysis, by six to seven glucosyl units in $\alpha(1 \rightarrow 4)$ glycosidic linkages using UDP-glucose as substrate (Figure 15-7). The glucan primer of glycogenin is elongated by glycogen synthase using UDP-glucose. Initially, the primer glycogenin and glycogen synthase are firmly bound in a 1:1 complex. As the glucan chain grows, glycogen synthase dissociates from glycogenin.

The branching of the glycogen is accomplished by transferring a minimum of six $\alpha(1 \rightarrow 4)$ glucan units from the elongated external chain into the same or a neighboring chain by $\alpha(1 \rightarrow 6)$ linkage. This reaction of $\alpha(1 \rightarrow 4)$ to $\alpha(1 \rightarrow 6)$ transglucosylation creates new non-reducing ends and is catalyzed by a branching enzyme. A mature glycogen particle is spherical, containing one molecule of

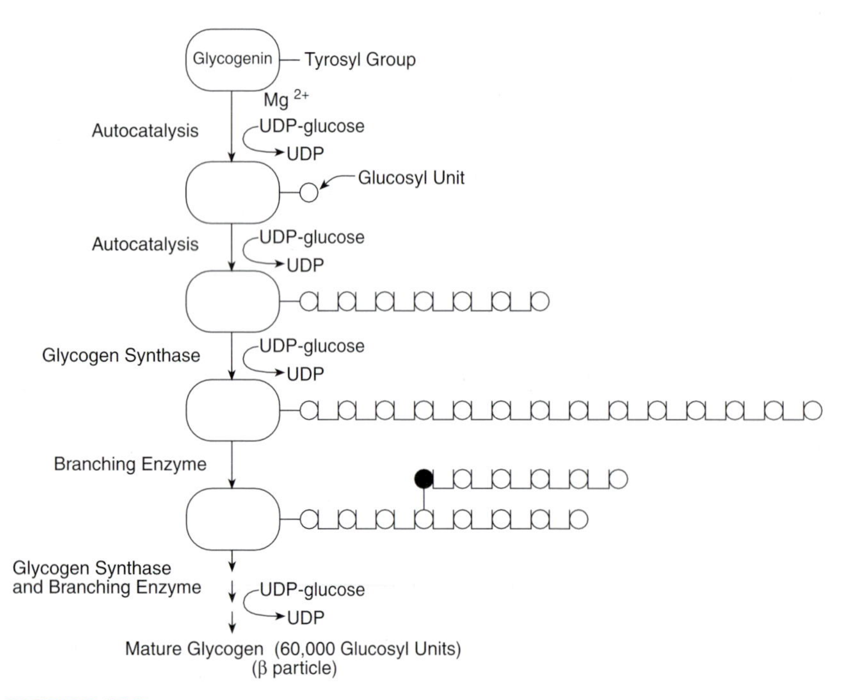

FIGURE 15-7

Schematic representation of glycogen biosynthesis. ●, Glucosyl units, $(1 \rightarrow 6)$ linkage.

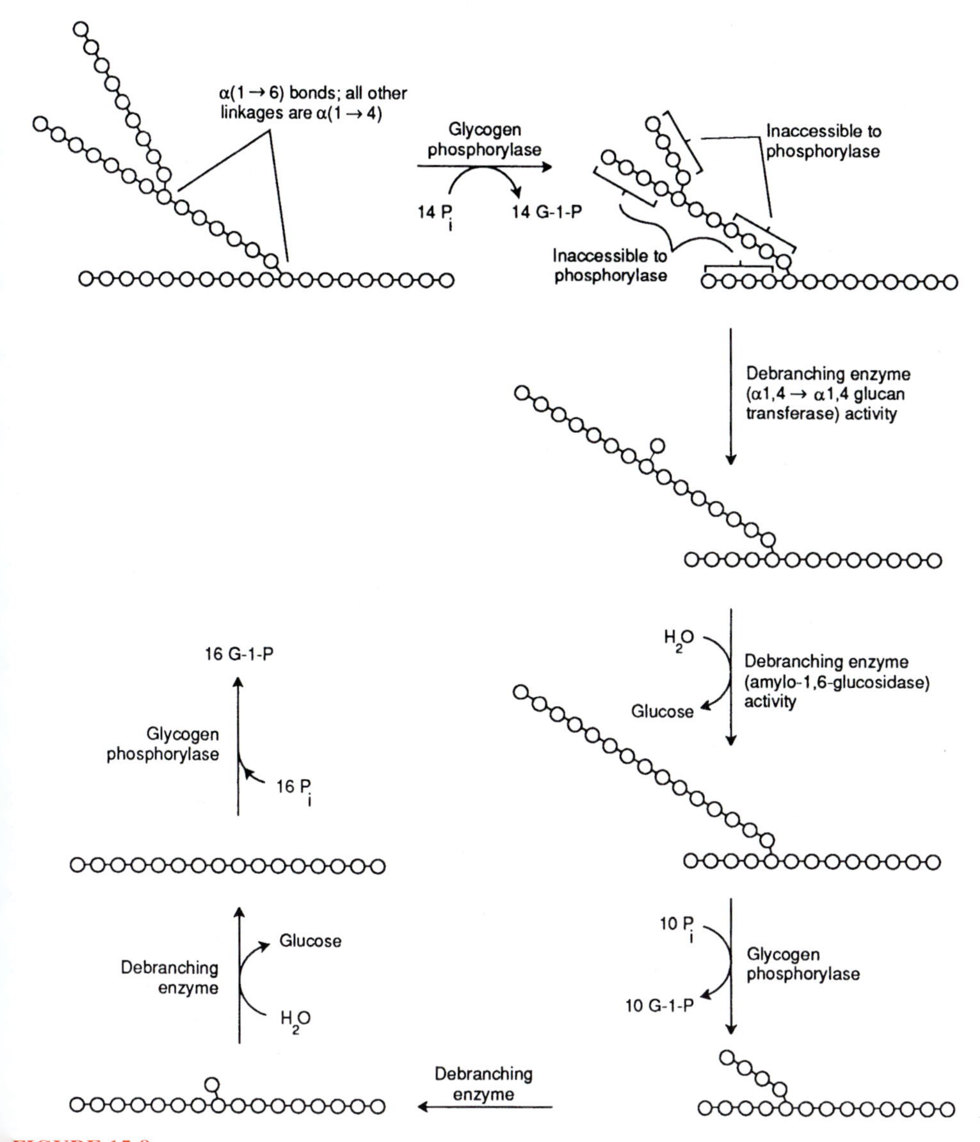

FIGURE 15-8
Catabolism of a small glycogen molecule by glycogen phosphorylase and debranching enzyme. P_i = inorganic phosphate; G-1-P = glucose-1-phosphate.

glycogenin and up to 60,000 glucosyl units (β-particles). In the liver, 20–40 β-particles are aggregated into rosettes, known as (α-particles).

Glycogen Breakdown

Glycogenolysis is catalyzed by two enzymes unique to the pathway: glycogen phosphorylase and debranching enzyme. The former normally regulates the rate of glucose release from glycogen. The progressive degradation of glycogen is illustrated in Figure 15-8. Glycogen phosphorylase catalyzes the release of glucose-1-phosphate from the terminal residue of a nonreducing end of a glycogen branch, by means of phosphorolysis. A molecule of inorganic phosphate attacks the C_1 side of an $\alpha(1 \rightarrow 4)$ glycosidic bond, leaving a hydroxyl group on C_4 that remains in the glycogen polymer. The reaction is analogous to hydrolysis, in which water attacks and cleaves bonds. The phosphorolysis reaction is summarized below.

$$(\text{Glucose})_n + P_i \rightarrow (\text{glucose})_{n-1} + \text{glucose-1-Phosphate}$$

The energy stored in the $\alpha(1 \rightarrow 4)$ glycosidic bond during the condensation reaction in glycogen synthesis is sufficient to permit the formation of a glucose-phosphate bond without using ATP.

Glucose-1-phosphate is next converted by phosphoglucomutase to glucose-6-phosphate. The latter may then enter the glycolytic pathway, but if glucose-6-phosphatase is present, free glucose can be formed.

Glycogen phosphorylase sequentially removes the glucosyl residues from a glycogen branch until further action is sterically hindered by a branch point. This occurs when the branch is four residues long from the branch point. Debranching enzyme, a multifunctional protein, first removes the trisaccharide "stump" on the branch and then removes the branch point itself. The $\alpha(1 \rightarrow 4)$ glycosidic bond linking the trisaccharide to the branch point residue is first cleaved, and the trisaccharide is transferred to the nonreducing end of an adjacent branch. This elongated branch can now be cleaved, one residue at a time, by glycogen phosphorylase. The glucose residue that remains, linked by an $\alpha(1 \rightarrow 6)$ glycosidic bond, is then cleaved by hydrolysis to yield free glucose. Thus, one molecule of glucose is released for each branch point removed, even in muscle, which lacks glucose-6-phosphatase. Roughly 7% of the glycosidic bonds in glycogen are $\alpha(1 \rightarrow 6)$ linkages.

Under normal conditions, glycogen phosphorylase and debranching enzyme act simultaneously at different regions of the glycogen molecule. Deficiency of either enzyme prevents complete glycogen degradation. Glycogen phosphorylase deficiency leaves the original glycogen molecule untouched. Deficiency of debranching enzyme results in a glycogen molecule smaller than the original, with very short chains on the outer branches but with the inner core unchanged (a limit dextrin).

Regulation of Glycogen Metabolism

Metabolism of glycogen in muscle and liver is regulated primarily through control of glycogen synthase and glycogen phosphorylase. The activities of these enzymes vary according to the metabolic needs of the tissue (as in muscle) or of other tissues that use glucose as a fuel (as in liver). Proximal control is exerted on synthase and phosphorylase by phosphorylation/dephosphorylation and by allosteric effectors such as glucose, glucose-6-phosphate, and several nucleotides (ATP, ADP, AMP, and UDP).

The concept that allosteric effectors are of primary importance in regulating synthase and phosphorylase activities was based mostly on *in vitro* experiments. For example, inactive (phosphorylated) glycogen synthase can be activated, without dephosphorylation, by glucose-6-phosphate. For this reason, the active and inactive forms of this enzyme were formerly called glycogen synthase I (glucose-6-phosphate *i*ndependent) and glycogen synthase D (glucose-6-phosphate *d*ependent). It now appears that the concentration of glucose-6-phosphate may not vary widely enough, particularly in muscle, to change synthase activity significantly, although glucose-6-phosphate binding may help determine the basal activity of the enzyme. The two forms are now known as glycogen synthase a (dephosphorylated) and b (phosphorylated). The same type of nomenclature is used for glycogen phosphorylase, except that phosphorylase a, the active form, is phosphorylated, while phosphorylase b is dephosphorylated. The regulation of glycogen metabolism in liver and in muscle differs in several ways. Although the control mechanisms are not completely understood, particularly in liver, the differences probably are due to the receptors in each tissue and to the presence of glucose-6-phosphatase in liver rather than to differences in the intrinsic regulatory properties of the enzymes involved. This aspect of tissue differences in glycogen metabolism between muscle and liver probably also applies to that in brain, myocardium, and other tissues. In all tissues, the rate of glycogen synthesis must be inversely proportional to the rate of glycogenolysis to avoid futile cycling.

Muscle

In muscle, glycogen is used as a fuel for anaerobic metabolism during brief periods of high-energy output (e.g., sprinting). Glycogenolysis is initiated and glycogenesis inhibited by the onset of muscle contraction and by factors such as epinephrine that signal a need for muscular activity. Since muscle glycogen is not a source of glucose for other tissues, it is not sensitive to blood glucose levels.

Control of Glycogen Synthase

Muscle glycogen synthase (M.W. ~340,000) is a tetramer of identical subunits that exists in several forms, which differ in catalytic activity and degree of covalent modification. Glycogen synthase a, an active dephosphorylated form, can interconvert with several less active phosphorylated forms, collectively called glycogen synthase b. The enzyme contains at least nine serine residues located near the extremities of the molecule, which can be phosphorylated by protein kinases (Figure 15-9). For the most part, the sites can be phosphorylated in any order. An exception is site C42, which undergoes phosphorylation by glycogen synthase kinase-3 only after phosphorylation at site C46 by casein kinase-2. Once C46 and C42 are phosphorylated in that order, glycogen synthase kinase-3 phosphorylates at sites C38, C34, and C30. In general, phosphorylation reduces synthase activity, and an increase in the number of phosphorylated sites additively decreases the activity. Reduced synthase activity may be manifested as increased K_m for UDP-glucose, increased

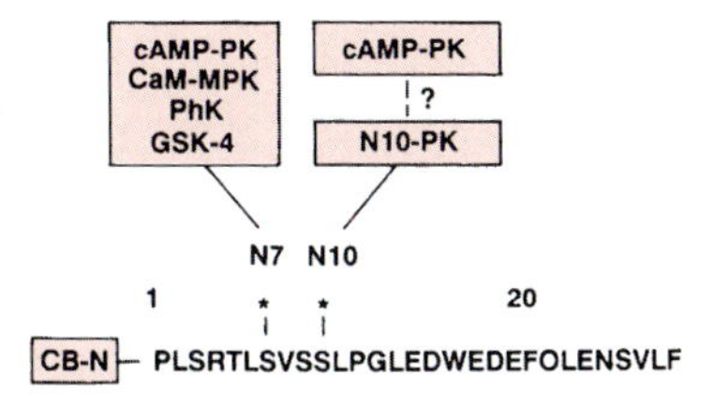

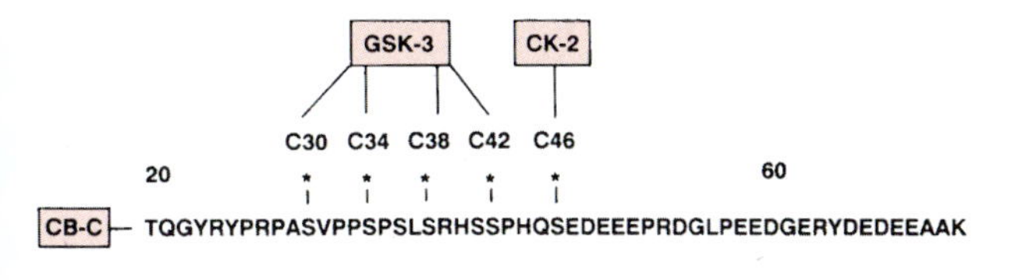

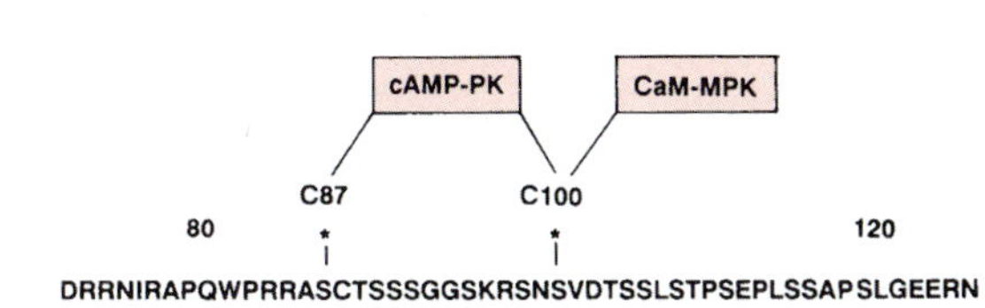

FIGURE 15-9
Regulation of glycogen synthase by multisite phosphorylation. The location of phosphorylation sites (*) and the protein kinases that phosphorylate at these sites (boxes) are shown. Phosphorylations occur only at N- and C-terminal regions of the enzymes, as indicated by CB-N and CB-C, respectively. The single-letter abbreviations for amino acids are used (see Chapter 2). cAMP-PK = cyclic AMP-dependent protein kinase; CAM-MPK = Ca^{2+}/calmodulin-dependent multiprotein kinase; PhK = phosphorylase kinase; GSK = glycogen synthase kinase; CK = casein kinase; N10-PK = A novel protein kinase. [Reproduced with permission from P. Cohen, Protein phosphorylation and hormone action. *Proc. R. Soc. Lond. (Biol.)* **234,** 115 (1988).]

K_a for glucose-6-phosphate, or decreased K_i for ADP, inorganic phosphate, or other inhibitors, or a change in V_{max}.

At least seven different protein kinases can phosphorylate muscle glycogen synthase *in vitro,* but whether all are involved *in vivo* is not established. Some are moderately specific for glycogen synthase, while others phosphorylate a variety of proteins. They are listed below, together with the sites that they phosphorylate (Figure 15-9).

1. cAMP-dependent protein kinase (sites N7, C87, and C100, broad specificity; mediates effects of many hormones).
2. Phosphorylase kinase (site N7; main function *in vivo* probably is activation of glycogen phosphorylase).
3. Ca^{2+}/calmodulin-dependent multiprotein kinase (sites N7 and C100; relatively specific for glycogen synthase).
4. Glycogen synthase kinase-3 (GSK-3; sites C30, C34, C38 and C42; also phosphorylates inhibitor-2 (see below), thereby activating protein phosphatase-1).
5. Glycogen synthase kinase-4 (GSK-4; site N7; relatively specific for glycogen synthase).
6. Casein kinase-2 (site C46; broadly specific; physiologically important substrates not yet known).
7. A novel protein kinase that may be activated by cAMP-dependent protein kinase (site N10).

In muscle, cAMP-dependent protein kinase (Chapter 30) mediates the inhibitory effect of epinephrine on glycogenesis by phosphorylating protein phosphatase inhibitor-1 and phosphorylating sites N7, C87, and C100. The inhibitory effect of cytosolic Ca^{2+} on glycogen synthesis may be mediated by calmodulin-dependent multiprotein kinase. GSK-3 may be a key control enzyme; it phosphorylates sites on glycogen synthase that have the greatest effect on synthase activity. GSK-3 also activates protein phosphatase-1 by phosphorylating inhibitor-2, however, and these seemingly opposing effects have not been fully explained. Stimulation by insulin of glycogen synthesis in muscle may be mediated by GSK-3 or phosphatases that remove the phosphate group at the GSK-3 sites.

Reactivation of glycogen synthase requires removal of the phosphate groups by a phosphatase. Four enzymes that dephosphorylate glycogen synthase or one or more kinases *in vitro* have been isolated from skeletal muscle.

1. Protein phosphatase-1 (Mg^{2+}/ATP-dependent phosphatase; multisubstrate protein phosphatase; M.W. of catalytic subunit is 35,000; major enzyme in regulation of glycogen metabolism in skeletal muscle; dephosphorylates glycogen phosphorylase, β-subunit of phosphorylase kinase, and at least three sites of glycogen synthase; regulated by inhibitor-1, inhibitor-2, and GSK-3 + Mg^{2+}· ATP).
2. Protein phosphatase-2A (polycation-dependent phosphatase, M.W. ~200,000; M.W. of catalytic subunit ~38,000; function unclear).
3. Protein phosphatase-2B (calcineurin; dimer; catalytic subunit of M.W. 60,000 has Ca^{2+} binding site; Ca^{2+}-binding subunit of M.W. ~18,000, partially homologous with calmodulin and troponin C; N-terminal glycine blocked by amide linkage with myristic acid; absolute requirement for Ca^{2+}; regulated by Ca^{2+}/calmodulin; dephosphorylates α-subunit of phosphorylase kinase and inhibitory but not glycogen phosphorylase or synthase or β-subunit of phosphorylase kinase).
4. Protein phosphatase-2C (monomer; M.W. 43,000; represents a small fraction of glycogen synthase phosphatase in skeletal muscle; may primarily regulate other metabolic pathway *in vivo*).

Other proteins involved in glycogen synthase control are calmodulin, inhibitor-1, and inhibitor-2. Troponin C, when complexed with Ca^{2+}, can activate phosphorylase kinase, but phosphorylase kinase may have little or no role in the inactivation of glycogen synthase *in vivo*. Inhibitor-1 (M.W. 18,600), when phosphorylated on threonine 35 by cAMP-dependent protein kinase, binds to and inhibits protein phosphatase-1. The concentration of protein phosphatase-1 is lower than that of inhibitors, and its activity is sensitive to the latter, suggesting that inhibitor-1 phosphorylation may be an important regulatory mechanism for this phosphatase. Inhibitor-2 (M.W. 30,500) appears to be an integral part of protein phosphatase-1, forming an inactive 1:1 complex, called Mg^{2+}/ATP-dependent phosphatase, with the catalytic subunit of this enzyme. Phosphorylation of inhibitor-2 (also at a threonine residue) activates protein phosphatase-1. This phosphorylation, which requires Mg^{2+}-ATP, is catalyzed by an "activating factor" (F_A), GSK-3.

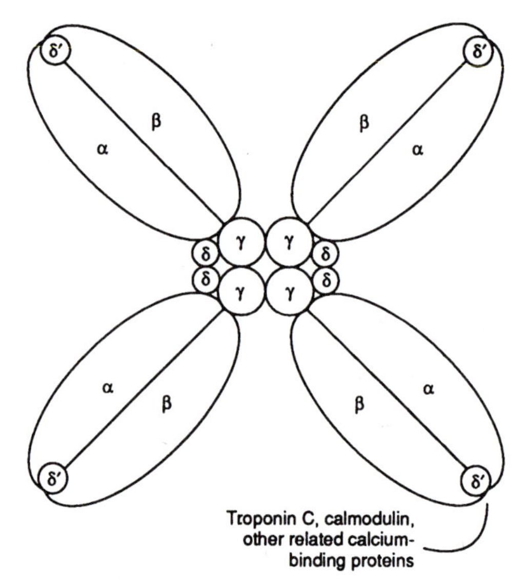

FIGURE 15-10
Subunit structure of muscle phosphorylase kinase. The α- and β-subunits are regulatory proteins containing the sites at which phosphorylation can activate the kinase. The γ-subunit contains the catalytic site. The δ-subunit is a Ca^{2+}-binding protein, essentially identical with calmodulin, Ca^{2+}. Calcium binding to the δ-subunit causes a change in the quaternary structure of the enzyme, allowing binding of the δ'-subunit, thereby increasing kinase activity even in the absence of phosphorylation of the β-subunit. [Modified and reproduced with permission from A. K. Campbell: *Intercellular Calcium: Its Universal Role as Regulator*, Wiley, New York, 1983. © 1983 by John Wiley & Sons, Inc.]

Control of Glycogen Phosphorylase

Regulation of glycogen phosphorylase in muscle is accomplished by many of the same enzymes that control glycogen synthesis. Phosphorylase kinase converts the dimeric phosphorylase from the inactive to the active form by Mg^{2+} and ATP-dependent phosphorylation of two identical serine residues. The principal enzyme that removes this phosphate may be protein phosphatase-1 (phosphorylase phosphatase).

Phosphorylase kinase is a hexadecamer having four different subunits: α (M.W. 145,000) or α' (M.W. 133,000), β (M.W. 128,000), γ (M.W. 44,700), and δ (M.W. 16,700), with the stoichiometry α_4(or α'_4)$\beta_4\gamma_4\delta_4$ (Figure 15-10). Whether the α- or α'-subunit is present depends on the tissue. The subunits apparently differ in Ca^{2+} sensitivity, since the δ'-subunit, discussed later, will not bind to the α'-isozyme.

A catalytic site on the γ-subunit has considerable homology with the catalytic subunit of cAMP-dependent protein kinase (Chapter 30). Evidence for active sites on the α- and β-subunits is weak.

Phosphorylase kinase activity has an absolute requirement for Ca^{2+}, which binds to the δ-subunit. The amino acid sequence of this subunit is nearly identical to that of calmodulin, with four calcium binding sites, but unlike calmodulin, the δ-subunit is an integral part of the enzyme and does not dissociate from it in the absence of Ca^{2+}. In the presence of Ca^{2+}, kinase activity is further increased by phosphorylation of the α- and β-subunits, catalyzed by cAMP-dependent protein kinase and several other kinases. Phosphorylation may activate the enzyme by disinhibiting it, since limited proteolysis (which degrades the α- and β-subunits) and dissociation of the holoenzyme into partial complexes ($\beta\gamma\delta$ and $\gamma\delta$) also cause activation. Activation by covalent modification increases the affinity of the kinase for phosphorylase b. The kinase also is activated by the binding of a δ'-subunit (Ca^{2+}/calmodulin or Ca^{2+}/troponin C) to a site formed by the α- and β-subunits. There are sites for four δ'-subunits on the holoenzyme. Activation by Ca^{2+}/troponin C is 20–30 times greater than by Ca^{2+}/calmodulin. This route may provide a means of coordinating muscle contraction, initiated by binding of Ca^{2+} to troponin C (Chapter 21), with glycogenolysis. Finally, phosphorylase kinase is allosterically activated by the binding of up to 8 mol of ADP.

Only 20–30% of isolated phosphorylase kinase is associated with glycogen particles, whereas all glycogen phosphorylase activity is associated with glycogen particles. This fact, together with the subcellular localization of phosphorylase kinase in intact cells and its biochemical properties, suggests that phosphorylase kinase may control other metabolic pathways as well. Suggested phosphorylase kinase targets include the Na^+, K^+- and Ca^{2+}-ATPases in the sacrolemma and sarcoplasmic reticulum, respectively.

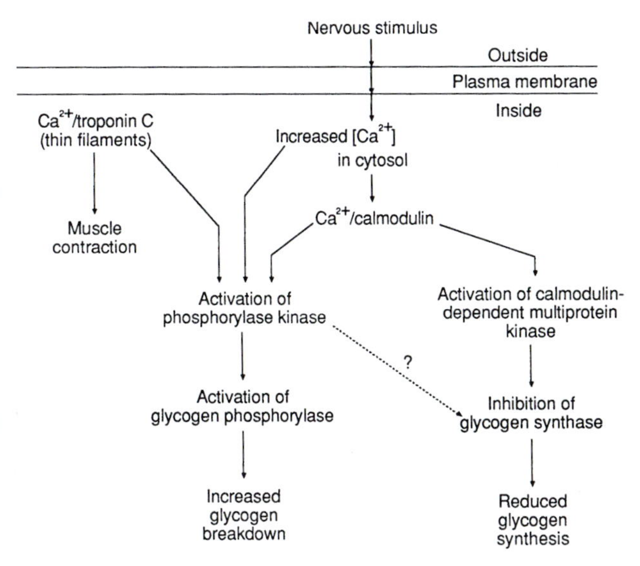

FIGURE 15-11
Possible mechanism for regulation of glycogen metabolism in skeletal muscle by changes in cytosolic calcium. Increased glycogen breakdown may be coordinated with muscle contractions, as indicated here. The actual control scheme is probably more complicated, since phosphoprotein phosphatases are also involved. Interactions with cAMP-activated reactions, which also may complicate regulation, are not included. Whether glycogen synthase is a substrate for phosphorylase kinase *in vivo* is unclear. [Modified and reproduced with permission from P. Cohen, Protein phosphorylation and the control of glycogen metabolism in skeletal muscle. *Philos. Trans. R. Soc. Lond. (Biol.)* **302,** 13 (1983).]

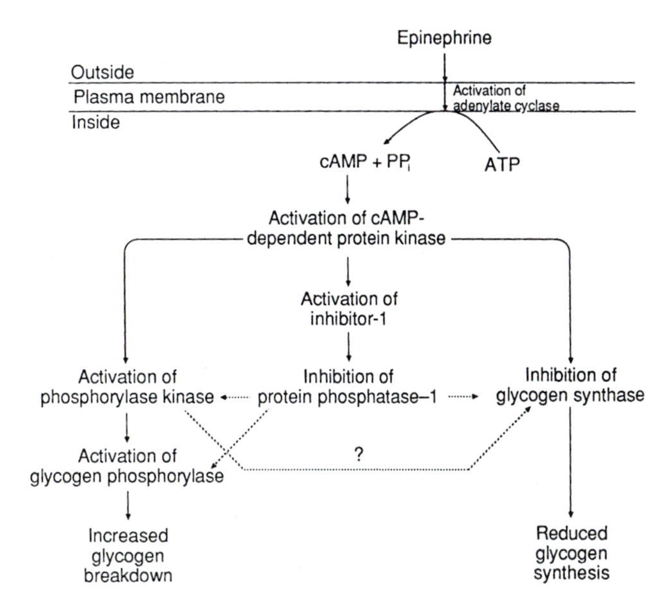

FIGURE 15-12
Possible mechanism for regulation of glycogen metabolism in skeletal muscle by changes in cyclic AMP. The actual control scheme is probably more complicated, since changes in cytoplasmic calcium concentration are likely also to be important. Whether glycogen synthase is a substrate for phosphorylase kinase *in vivo* is unclear (indicated by ?). Dashed arrows from protein phosphatase-1 to glycogen synthase and phosphorylase and phosphorylase kinase indicate that inhibition of dephosphorylation may be less important than activation of phosphorylation in changing the activities of these enzymes. [Modified and reproduced with permission from P. Cohen, Protein phosphorylation and the control of glycogen metabolism in skeletal muscle. *Philos. Trans. R. Soc. Lond. (Biol.)* **302** (1983).]

Integrated Regulation of Muscle Glycogen Metabolism

Figures 15-11 and 15-12 provide a model for the regulation of glycogen metabolism in muscle. This model is consistent with many of the known facts, but alternative interpretations are equally plausible. Only two of the external stimuli that affect muscle glycogen metabolism are considered in these figures.

1. Initiation of muscle contraction by action potential from an α motor neuron depolarizes the muscle-cell membrane and causes an increase in the myoplasmic $[Ca^{2+}]$ (Chapter 21). The increase in calcium activates phosphorylase kinase and Ca^{2+}/calmodulin-dependent kinase, which inactivate glycogen synthase and active glycogen phosphorylase (Figure 15-11). This step coordinates muscle contraction with glycogenolysis. These steps are reversed by one or more phosphatases at the end of contraction.
2. Binding of epinephrine to an adrenergic receptor away from a motor end plate on the muscle membrane causes glycogenolysis to increase and glycogenesis to decrease. This activates adenylate cyclase, increasing the cytosolic concentration of cAMP and activating the cAMP-dependent protein kinase (Figure 15-12). The net effect *in vivo* is an increase in phosphorylation of glycogen synthase and phosphorylase. This may occur in several ways, and the relative importance of increased phosphorylation compared to decreased dephosphorylation remains unclear.

Insulin also regulates skeletal muscle glycogen metabolism. Skeletal muscle is the major site of insulin-stimulated glucose uptake from the bloodstream, most of which is converted to glycogen. The insulin-induced glucose uptake is mediated by the glucose transporter GLUT4 (Chapter 13). Binding of insulin to its cell surface receptor initiates a cascade of phosphorylation reactions followed by dephosphorylation reactions (Chapter 22). Activation of glycogen synthase occurs by dephosphorylation at sites phosphorylated by cAMP-dependent kinase and GSK-3. Protein phosphatase-1 is a key component of the insulin signaling pathway and it activates glycogen synthase; it simultaneously inactivates phosphorylase a and phosphorylase kinase, promoting glycogen synthesis. Inactivation of GSK-3 via the protein kinase B pathway has also been implicated in insulin-mediated activation of glycogen synthase.

Liver

In liver, glycogen serves as the immediate glucose reserve for other tissues, and regulation of its metabolism is closely linked to maintenance of blood glucose concentration. An important feature of glucose transport in hepatocytes is its bidirectional flux across the plasma membrane that is mediated by glucose transporter GLUT2. Insulin does not regulate GLUT2 and its function is not rate limiting. Therefore, the concentrations of glucose in the blood and hepatocytes are equal. During energy deprivation for short periods of time such as overnight fasting, moderate exercise, and stress, which cause hypoglycemia (e.g., hypovolemic shock), glycogenolysis is stimulated. During sustained periods of starvation, gluconeogenesis maintains blood glucose homeostasis. During refeeding, hepatocytes replenish their stores of glycogen by using glucose-6-phosphate obtained from either plasma glucose or from the gluconeogenic pathway using amino acids, glycerol, and propionate as precursors. Regulatory factors include glucagon, vasopressin, angiotensin II, and α-adrenergic agonists (including epinephrine). Glycogen synthesis is stimulated by a rise in insulin or blood glucose concentration.

Control of Glycogen Synthase

Hepatic glycogen synthase is similar to the muscle enzyme, although it is encoded by different genes. It is inactivated by phosphorylation and activated by dephosphorylation and may contain 12 mol of alkali-labile phosphate per mole of subunit. The phosphorylation sites have not been mapped, and the specificities of hepatic glycogen synthase kinases are not known.

Several enzymes that phosphorylate glycogen synthase (muscle or liver) *in vitro* have been identified in liver. cAMP-dependent protein kinase, thought to mediate the inhibition of glycogen synthesis by glucagon, is similar to the muscle enzyme. A Ca^{2+}/calmodulin-dependent protein kinase may mediate Ca^{2+}-dependent inactivation of the synthase by vasopressin, angiotensin II, and α-adrenergic agonists. Casein kinase I and GSK-3 have also been found in liver, and there are undoubtedly other hepatic glycogen synthase kinases. As in muscle, hepatic phosphorylase kinase is probably not important as a glycogen synthase kinase *in vivo,* despite its ability to readily phosphorylate the synthase *in vitro.* Muscle and liver thus have major regulatory pathways in common but differ in more subtle aspects of control.

Protein phosphatases-1, 2A, 2B, and 2C occur in mammalian liver and, as in skeletal muscle, possess essentially all of the phosphatase activity toward enzymes and regulatory proteins of glycogen metabolism. In liver, however, the ratios of the activities of phosphatase-2A and 2C to that of phosphatase-1 are seven-fold higher than in muscle. Although protein phosphatase-I sediments with glycogen particles in both tissues, a much smaller fraction is glycogen-associated in liver than in muscle. The specific activity of phosphatase-2B is lower in liver than in muscle. Protein phosphatase inhibitors-1 and 2 have been identified in liver, where they appear to function as they do in muscle. A disinhibitor protein (M.W. ~9,000) of liver can block the effects of inhibitors-1 and 2 on phosphatase-1.

Control of Glycogen Phosphorylase

Liver glycogen phosphorylase exists in an inactive, dephosphorylated form and in at least one active, phosphorylated form. Conversion of phosphorylase b to phosphorylase a is catalyzed by phosphorylase kinase, which is activated by vasopressin, angiotensin II, and α-adrenergic agonists (mediated by Ca^{2+}) and by glucagon (which elevates cAMP). Glucagon activation of phosphorylase is in some way antagonized by insulin.

Dephosphorylation of glycogen phosphorylase a is probably catalyzed by either protein phosphatase-1 or 2A. Binding of free glucose to phosphorylase a makes it a better substrate for the phosphatase, increasing the rate of inactivation of glycogenolysis. Phosphorylase a is a competitive inhibitor of the reaction between glycogen synthase b and phosphatase, suggesting that the same enzyme dephosphorylates the synthase and kinase, activating one and inhibiting the other.

Integrated Regulation of Liver Glycogen Metabolism

Glycogen metabolism in liver is regulated by phosphorylation and dephosphorylation of regulatory and metabolic enzymes. Control of the phosphorylation state is mediated by Ca^{2+}, cAMP, cytosolic glucose concentration, and perhaps, in the case of insulin, by another mechanism.

When the glucose concentration rises in the hepatocyte, the rate of conversion of glycogen phosphorylase a to phosphorylase b increases. This decreases the rate of glycogenolysis and, initially, causes no change in glycogen synthesis. As the concentration of phosphorylase a falls, however, its ability to competitively inhibit glycogen synthase b dephosphorylation decreases. The rate of activation of glycogen synthase then increases, as does the rate of synthesis of glycogen. Glucose can be an effective regulator because its concentration in the hepatocyte varies with its concentration in the blood. The postprandial rise in blood glucose contributes to the inhibition of hepatic glycogenolysis and to the activation of glycogen synthesis.

Following a meal, the blood level of insulin also rises, while that of glucagon falls. As glucagon concentration decreases, activity of the hepatocyte adenylate cyclase decreases and cytosolic cAMP concentration falls owing to degradation by phosphodiesterase. The result is a decreased phosphorylation of glycogen synthase and phosphorylase, activation of the synthase, and inhibition of the phosphorylase. Insulin, which may accelerate this shift, can also directly antagonize the cAMP-mediated effects of glucagon. The mechanisms of insulin action may include activation of a low-K_m cAMP-phosphodiesterase, reduction in basal activity of cAMP-dependent protein kinase, or increase in basal activity of glycogen synthase phosphatase. Since the effects on the kinase and phosphatase are on basal activities, insulin may affect the metabolism of glycogen that has not been previously stimulated by glucagon. The net effect of blood glucose, insulin, and glucagon, then, is to coordinate the relative rates of hepatic glycogen synthesis and breakdown (see Chapter 22).

In shock, inadequate perfusion of the heart, brain, and other organs is due to a variety of causes. The body's responses include release of epinephrine from the adrenal medulla, release of vasopressin from the neurohypophysis, and activation of the renin-angiotensin system with elevation of angiotensin II. In the liver, epinephrine (acting as an α-adrenergic agonist), vasopressin, and angiotensin II increase the intrahepatocytic [Ca^{2+}], stimulating glycogenolysis. The consequent rise in blood glucose provides a rapid energy source in times of stress. Since shock is often caused by an absolute (hemorrhage) or relative (vasodilation) decrease in intravascular volume, release of the large amount of water of hydration stored with glycogen may contribute to restoration of blood volume.

Glycogen Storage Diseases

This group of diseases is characterized by accumulation of normal or abnormal glycogen due to deficiency of one of the enzymes of glycogen metabolism. Although all are rare (overall incidence of ~1:25,000 births), they have contributed greatly to the understanding of glycogen metabolism. They are summarized in Table 15-1 and Figure 15-13.

Infants with type I disease (subtypes Ia, Ib, Ic, and Id) develop hypoglycemia even after feeding because of inability to convert glucose-6-phosphate to glucose. Lactate is produced at a high rate in extrahepatic tissues and is transported to the liver for gluconeogenesis. The low insulin and high glucagon levels caused by hypoglycemia promote glycogenolysis, leading to an unusual metabolic substrate cycle in these infants (Figure 15-14). The small amount of glucose released by the action of debranching enzyme is the only endogenous source of glucose available to these patients.

Many characteristics of type I disease are attributed to the attendant hypoglycemia, and patients have been treated with frequent daytime feedings and continuous nocturnal intragastric feeding with a high-glucose formula. This regimen produces substantial improvement in growth, reduction in hepatomegaly, and normalization of other biochemical parameters. Feeding uncooked cornstarch every 6 hours resulted in normoglycemia, resumption of normal growth, and reduction in substrate cycling and liver size. The success of this simple nutritional therapy is thought to depend on slow hydrolysis of uncooked starch in the small intestine by pancreatic amylase, with continuous release and absorption of glucose. (Cornstarch that is cooked or altered in other ways is ineffective, presumably because of too rapid hydrolysis.) The therapy is ineffective in patients with low pancreatic amylase activity due, for example, to prematurity.

15.3 Alternative Pathways of Glucose Metabolism and Hexose Interconversions

Glucuronic Acid Pathway

The ***glucuronic acid pathway*** (Figure 15-15) is a quantitatively minor route of glucose metabolism. Like the pentose phosphate pathway, it provides biosynthetic precursors and interconverts some less common sugars to ones that can be metabolized.

The first steps are identical to those of glycogen synthesis, i.e., formation of glucose-6-phosphate, its isomerization to glucose-1-phosphate, and activation of glucose-1-phosphate to form UDP-glucose. UDP-glucose is then oxidized to UDP-glucuronic acid by NAD^+ and UDP-glucose dehydrogenase. An aldehyde intermediate is formed that remains bound to an amino group on the enzyme as a Schiff base. Since this is a four-electron oxidation, two molecules of NAD^+ are used for each molecule of UDP-glucuronic acid formed. UDP-glucuronic acid is bound to the enzyme as a thioester and is released by hydrolysis.

UDP-glucuronic acid is utilized in biosynthetic reactions that involve condensation of glucuronic acid with a variety of molecules to form an ether (glycoside), an ester, or an amide, depending on the nature of the acceptor molecule. As in condensation reactions that use nucleotide sugars as substrates, the high-energy bond between UDP and glucuronic acid provides the energy to form the new

TABLE 15-1
Glycogen Storage Diseases

Type	Eponym	Deficiency	Glycogen Structure	Comments
Ia	von Gierke's disease (hepatorenal glycogenosis)	Glucose-6-phosphatase in liver, intestine, and kidney	Normal	Hypoglycemia; lack of glycogenolysis by epinephrine or glucagon; ketosis, hyperlipidemia, hyperuricemia; hepatomegaly; autosomal recessive. Galactose and fructose not converted to glucose.
Ib	. . .	Glucose-6-phosphate transporter in hepatocyte microsomal membrane	Normal	Clinically identical to type Ia.
Ic	. . .	Phosphate transporter in hepatocyte microsomal membrane	Normal	Clinically identical to type Ia.
Id	. . .	Glucose transporter GLUT-7	Normal	Clinically identical to type Ia.
II	Pompe's disease (generalized glycogenesis)	Lysosomal α-1,4-glucosidase (acid maltase)	Normal	How this deficiency leads to glycogen storage is not well understood. In some cases, the heart is the main organ involved; in others, the nervous system is severely affected; autosomal recessive.
III	Forbes' disease, Cori's disease (limit dextrinosis)	Amylo-1,6-glucosidase (debranching enzyme)	Abnormal; outer chains missing or very short; increased number of branched points	Hypoglycemia; diminished hyperglycemic response to epinephrine or glucagon; normal hyperglycemic response to fructose or galactose; autosomal recessive. Six subtypes have been defined based on relative effects on liver and muscle and on properties of the enzyme.
IV	Andersen's disease (branching deficiency; amylopectinosis)	$(1,4 \rightarrow 1,6)$-transglucosylase (branching enzyme)	Abnormal; very long inner and outer unbranched chain	Rare, or difficult to recognize; cirrhosis and storage of abnormal glycogen; diminished hyperglycemic response to epinephrine; abnormal liver function; autosomal recessive.

Type	Name	Enzyme deficiency	Glycogen structure	Clinical features
V	McArdle's disease	Muscle glycogen phosphorylase (myophosphorylase)	Normal	High muscle glycogen content (2.5–4.1% versus 0.2–0.9% normal); fall in blood lactate and pyruvate after exercise (normal is sharp rise) with no postexercise drop in pH; normal hyperglycemic response to epinephrine (thus normal hepatic enzyme); myoglobinuria after strenuous exercise; autosomal recessive.
VI	Hers' disease	Liver glycogen phosphorylase (hepatophosphorylase)	Normal	Not as serious as glucose-6-phosphatase deficiency; liver cannot make glucose from glycogen but can make it from pyruvate; mild hypoglycemia and ketosis; hepatomegaly due to glycogen accumulation; probably more than one disease; must be distinguished from defects in the glycogen phosphorylase-activating system.
VII	Tarui's disease	Muscle phosphofructokinase	Normal	Shows properties similar to type V; autosomal recessive; it is not completely clear why this defect results in increased glycogen storage.
VIII	. . .	Reduced activation of phosphorylase in hepatocytes and leukocytes	Normal	Hepatomegaly; increased hepatic glycogen stores; probably X-linked, but there may be more than one type, with some autosomally inherited; must be distinguished from glycogen phosphorylase deficiency.

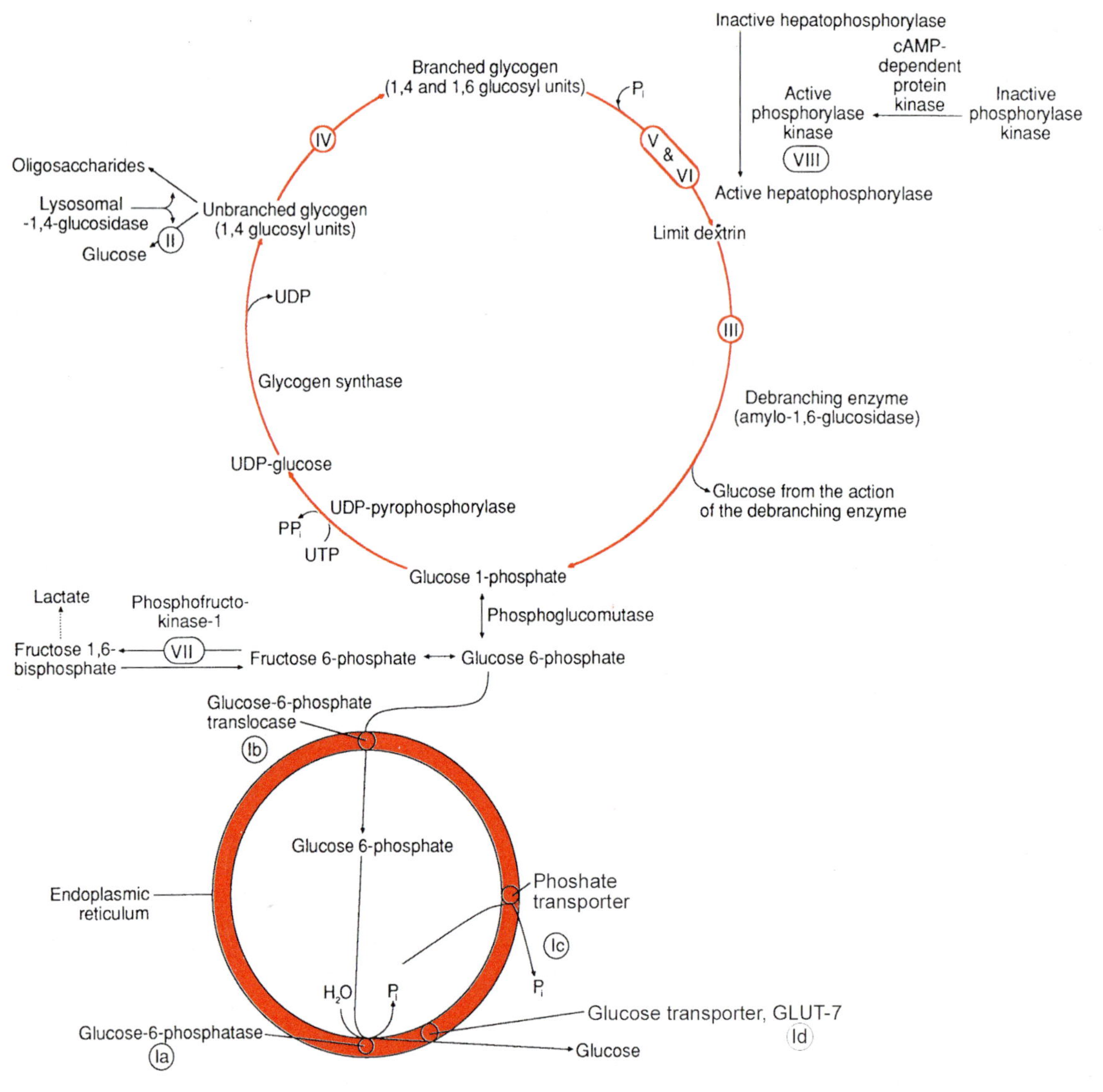

FIGURE 15-13
Locations of the glycogen storage disease enzyme defects in the overall scheme of glycogenesis and glycogenolysis. The numbers correspond to the disease types listed in Table 15-1.

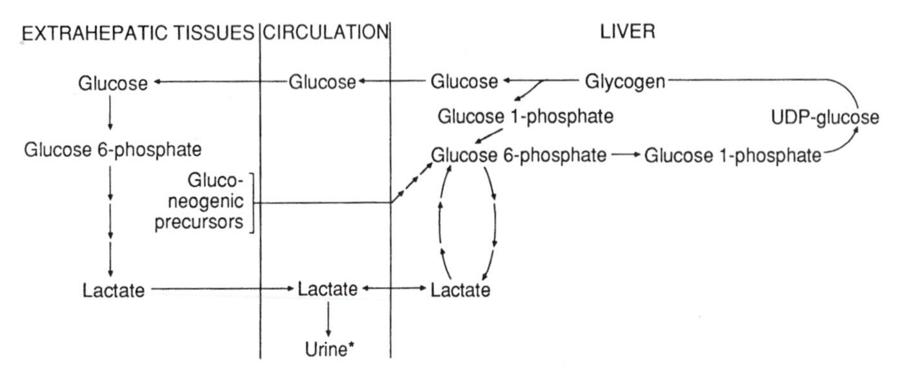

FIGURE 15-14
Lactic acidosis and substrate cycling in glucose-6-phosphatase deficiency.
*Urinary lactate output is increased 10- to 300-fold in patients with type I glycogen storage disease.

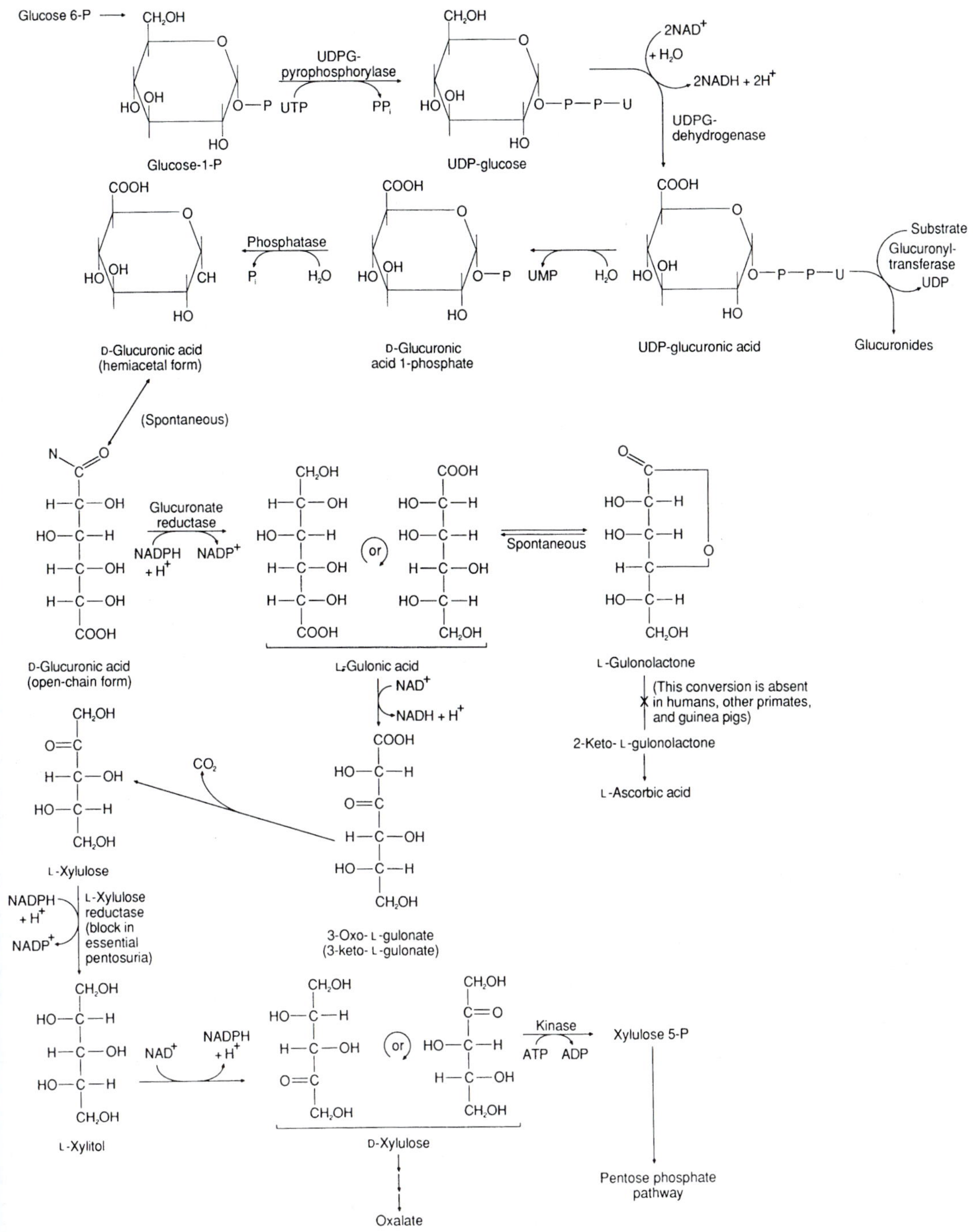

FIGURE 15-15
Glucuronic acid pathway. The first part consists of synthesis of UDP-glucuronic acid and release of free D-glucuronic acid. The second part is the metabolism of D-glucuronic acid. D-Glucuronic acid is written as both the cyclic hemiacetal and the open-chain aldohexose and two orientations of L-gulonic acid and D-xylulose are shown. P = phosphate.

bond in the product. UDP is hydrolyzed to UMP (uridine monophosphate) and inorganic phosphate, further ensuring the irreversibility of the coupling reaction.

Because glucuronic acid is highly polar, its conjugation with less polar compounds, such as steroids, bilirubin, and some drugs, can reduce their activity and make them more water-soluble, thus facilitating renal excretion. Glucuronic acid is a component of the structural polysaccharides called glycosaminoglycans (hyaluronic acid and other connective tissue polysaccharides; see

Chapters 11 and 16). Glucuronic acid is usually not a component of glycoproteins or glycolipids.

Other metabolic routes available to UDP-glucuronic acid are shown in Figure 15-15. Hydrolysis to free glucuronic acid and NADPH-dependent reduction together lead to L-gulonic acid, which spontaneously cyclizes to L-gulonolactone. In many animals and higher plants, this can be converted to 2-ketogulonolactone, a precursor of ascorbic acid (vitamin C), by gulonolactone oxidase. In humans and other primates and in guinea pigs, absence of this enzyme makes ascorbic acid an essential dietary ingredient (Chapter 38).

Gulonic acid is oxidized and decarboxylated to L-xylulose. Entry of this sugar into the pentose phosphate pathway requires isomerization to D-xylulose. This is accomplished by reduction (NADPH) to xylitol via ***L-xylulose reductase*** and by oxidation to D-xylulose with reduction of NAD^+. D-Xylulose, after conversion of xylulose 5-phosphate, is metabolized in the pentose phosphate pathway or converted to oxalate via formation of xylulose-1-phosphate, glycoaldehyde, and glycolate. In *essential pentosuria*, a clinically benign inborn error of metabolism, L-xylulose reductase (also known as ***NADP-linked xylitol dehydrogenase***) is abnormal or absent, and large amounts (1–4 g/day) of L-xylulose appear in the urine. Because L-xylulose is a reducing sugar, tests for urinary reducing substances that are intended to detect glucose may result in an erroneous diagnosis of diabetes mellitus. A positive test for urinary reducing substances, especially in the absence of clinical symptoms of diabetes mellitus, should be verified by a method that is specific for glucose.

Alimentary pentosuria may follow ingestion of large quantities of fruit, with L-arabinose and L-xylose occurring in high concentrations in urine. Ribosuria may occur in some muscular dystrophy patients. In both conditions, urine is positive for reducing substances.

Fructose and Sorbitol Metabolism

Fructose is a ketohexose found in honey and a wide variety of fruits and vegetables. Combined with glucose in an $\alpha(1 \rightarrow 2)\beta$ linkage, it forms sucrose (Chapter 9). It makes up one sixth to one third of the total carbohydrate intake of most individuals in industrialized nations. Inulin, found in some plants, is a polymer containing about 40 fructose residues connected in $\beta(2 \rightarrow 1)$ linkages with some $\beta(2 \rightarrow 6)$ branch points. It cannot be hydrolyzed by any human enzyme and has been used for measurement of renal clearance.

Sorbitol, a sugar alcohol, is a minor dietary constituent. It can be synthesized in the body from glucose by NADPH-dependent aldose reductase (Figure 15-16). It is clinically

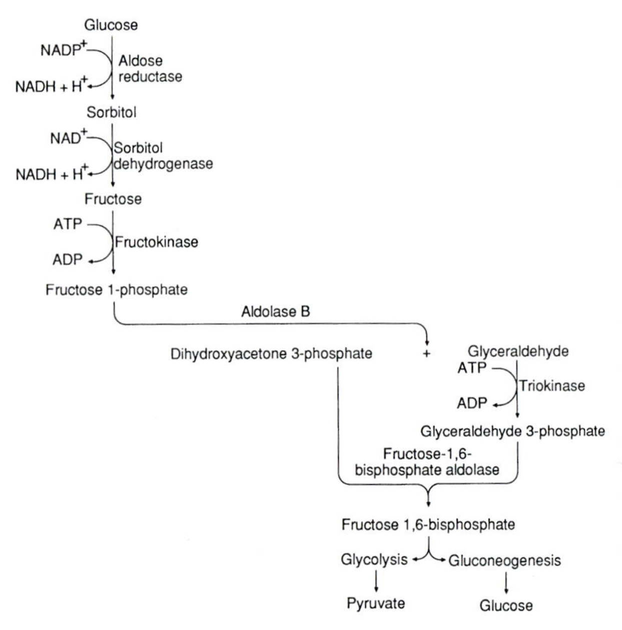

FIGURE 15-16
Metabolic pathway for sorbitol and fructose. Sorbitol dehydrogenase is sometimes known as iditol dehydrogenase. Aldolase B is also called fructose-1-phosphate aldolase, in contrast to fructose-1,6-bisphosphate aldolase.

important because of its relationship to cataract formation in diabetic patients. Fructose and sorbitol are catabolized by a common pathway (Figure 15-16). Fructose transport and metabolism are insulin-independent; only a few tissues (e.g., liver, kidney, intestinal mucosa, and adipose tissue, but not brain) can metabolize it. Fructose metabolism, which is much less tightly regulated, is more rapid than glucose metabolism. The renal threshold for fructose is very low, and fructose is more readily excreted in urine than is glucose. Despite these differences, the metabolic fates of glucose and fructose are closely related because most fructose is ultimately converted to glucose. Fructose metabolism (Figure 15-16) starts with phosphorylation and formation of fructose-1-phosphate, and this reaction is the rate-limiting step. The K_m of hexokinase for fructose is several orders of magnitude higher than that for glucose, and at the concentrations of glucose found in most tissues, fructose phosphorylation by hexokinase is competitively inhibited. In tissues that contain fructokinase, such as liver, the rate of fructose phosphorylation depends primarily on fructose concentration, and an increase in fructose concentration depletes intracellular ATP. ***Essential fructosuria*** is caused by absence of hepatic fructokinase activity. The condition is asymptomatic but, as with pentosuria, may be misdiagnosed as diabetes mellitus.

The next step in fructose metabolism is catalyzed by aldolase B (fructose-1-phosphate aldolase), which cleaves fructose-1-phosphate to the trioses dihydroxyacetone

3-phosphate and glyceraldehyde. Glyceraldehyde is phosphorylated by triokinase, and glyceraldehyde 3-phosphate and dihydroxyacetone 3-phosphate can either enter the glycolytic pathway or be combined to form fructose-1,6-bisphosphate by the action of fructose-1,6-bisphosphate aldolase. Thus, fructose metabolism bypasses phosphofructokinase, the major regulatory site of glycolysis.

Most dietary fructose is converted to glucose by way of gluconeogenesis, through condensation of the triose phosphates to fructose-1,6-bisphosphate. However, administration of large doses of fructose (i.e., by intravenous feeding) may lead to hypoglycemia and lactic acidosis because of saturation of aldolase B, causing accumulation of fructose-1-phosphate, and depletion of intracellular ATP and inorganic phosphate. This situation may be likened to ***hereditary fructose intolerance,*** which is caused by inadequate amounts of aldolase B activity. Although normally asymptomatic, individuals with this condition exhibit hypoglycemia, metabolic acidosis, vomiting, convulsions, coma, and signs of liver failure following ingestion of fructose or sucrose. The fructose-induced hypoglycemia arises from accumulation of fructose-1-phosphate, reduction in the [ATP]/[ADP] ratio, and depletion of inorganic phosphate. Fructose-1-phosphate depresses gluconeogenesis and promotes glycolysis, while inorganic phosphate depletion inhibits ATP synthesis (Chapter 14). Glycogenolysis is inhibited by fructose-1-phosphate at the level of phosphorylase. If sucrose, fructose, and sorbitol are eliminated from the diet, complete recovery occurs. High levels of fructose also increase purine turnover owing to enhanced ATP utilization, which leads to increased production of purine degradation products—inosine, hypoxanthine, xanthine, and uric acid (Chapter 27).

Fructose and sorbitol have been recommended as substitutes for sucrose in diabetic diets because they are much sweeter than sucrose. Foods can be made more palatable even when the total carbohydrate content is reduced by replacing sucrose with fructose or sorbitol. Their insulin-independent metabolism has also been cited as a reason for substituting them for glucose or sucrose in diabetic diets. In severe physical injury, hyperglycemia and insulin resistance are common, and inclusion of fructose and sorbitol in the parenteral nutrition formulations for these patients has been suggested. In small amounts, fructose and sorbitol could benefit both groups of patients. In large amounts, however, they can severely damage the liver by depleting ATP stores, and they can cause hypoglycemia and lactic acidosis.

When sorbitol and fructose are taken by mouth, the increase in blood fructose is modulated by their rates of absorption from the intestine, preventing the serious metabolic problems caused by high concentrations of these sugars. Most normal diets do not result in adverse effects, except in patients with hereditary fructose intolerance. However, when sorbitol and fructose are given parenterally, this important modulating action is lost, and serious side effects can occur. Sorbitol and fructose as intravenous nutrients are not widely employed in the United States.

Sorbitol has been implicated in cataract formation in diabetics. In general, cataracts are formed when the lens of the eye becomes cloudy, probably because of a change in solubility of lens proteins. Because entry of glucose into the lens does not require insulin, intravenous and intralenticular glucose concentrations increase in parallel. In diabetics, quite high concentrations can be achieved. In the lens, glucose is converted to sorbitol by aldose reductase, an NADPH-dependent enzyme present in many tissues. Unlike glucose, sorbitol does not diffuse readily across the lenticular membrane but accumulates within the lens, increasing osmolarity and causing water retention. A reduction in lens glutathione concentration parallels the increase in water content and may reflect the depletion of NADPH by the aldose reductase reaction and lead to oxidation and denaturation of lens proteins. The lens is particularly susceptible to oxidative damage because it is exposed to a high oxygen concentration.

Galactose Metabolism

Most galactose ingested by humans is in the form of lactose (Chapter 9), the principal sugar in human and bovine milk. Milk sugar other than lactose is found in the sea lion and marsupials, whose first pouch milk contains a trisaccharide of galactose. Lactose is hydrolyzed to galactose and glucose by lactase, located on the microvillar membrane of the small intestine (Chapter 12). Following absorption, galactose is transported to the liver, where it is converted to glucose (Figure 15-17). The enzymes required are found in many tissues, but the liver is the quantitatively most important site for this epimerization.

Galactose is a poor substrate for hexokinase; it is phosphorylated by galactokinase. Galactose-1-phosphate is converted to UDP-galactose by galactose-1-phosphate uridylyltransferase. This enzyme may be regulated by substrate availability, since the normal hepatic concentration of galactose-1-phosphate is close to the K_m for this enzyme. The transferase is inhibited by UDP, UTP, glucose-1-phosphate, and high concentrations of UDP-glucose. Deficiency of galactokinase or transferase can cause galactosemia.

Galactose is isomerized to glucose by UDP-galactose-4-epimerase in what may be the rate-limiting step in galactose metabolism. The reaction, which is freely

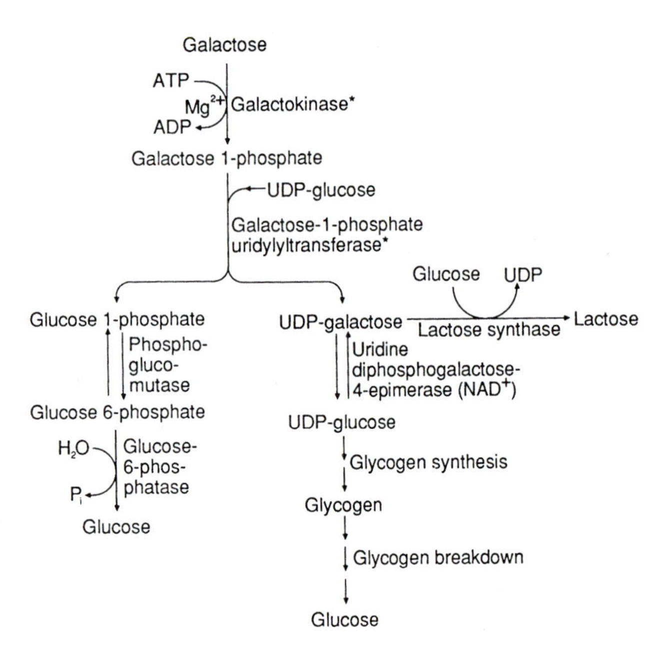

FIGURE 15-17
Metabolic pathway of galactose. *Inherited defects that lead to galactosemia.

reversible, converts UDP-glucose to UDP-galactose for the synthesis of lactose, glycoproteins, and glycolipids. The epimerase requires NAD^+ and is inhibited by NADH. It may also be regulated by the concentrations of UDP-glucose and other uridine nucleotides. Because of this epimerase, preformed galactose is not normally required in the diet. However, infants with deficiency of epimerase in hepatic and extrahepatic tissues require small quantities of dietary galactose for normal development and growth. An isolated deficiency of epimerase in erythrocytes, which is clinically benign, has been described.

Genetically determined deficiencies of galactokinase and galactose 1-phosphate uridylyltransferase cause clinically significant galactosemia. Galactokinase deficiency is a rare, autosomal recessive trait in which high concentrations of galactose are found in the blood, particularly after a meal that includes lactose-rich foods, such as milk and nonfermented milk products. Patients frequently develop cataracts before 1 year of age because of accumulation of galactitol in the lens. Galactose diffuses freely into the lens, where it is reduced by aldose reductase, the enzyme that converts glucose to sorbitol. Galactitol may cause lens opacity by the same mechanisms as does sorbitol. Galactose in the urine is detected by nonspecific tests for reducing substances, as are fructose and glucose.

Deficiency of the transferase causes a more severe form of galactosemia. Symptoms include cataracts, vomiting, diarrhea, jaundice, hepatosplenomegaly, failure to thrive, and mental retardation. If galactosuria is severe, nephrotoxicity with albuminuria and aminoaciduria may occur. The more severe clinical course may be due to accumulation of galactose-1-phosphate in cells. Measurement of transferase activity in red blood cells is the definitive test for this disorder.

In about 70% of transferase alleles in the white population, the DNA in cells of transferase-deficient patients possesses an A-to-G transition that leads to the Q186R mutation.

Both forms of galactosemia are treated by rigorous exclusion of galactose from the diet. In pregnancies for which the family history suggests that the infant may be affected, the best outcomes have been reported when the mother was maintained on a galactose-free diet during gestation.

Condensation of glucose with UDP-galactose, catalyzed by lactose synthase (Figure 15-17), forms lactose, the only disaccharide made in large quantities by mammals. This enzyme is a complex of galactosyltransferase, a membrane-bound enzyme that participates in glycoprotein synthesis (Chapter 16), and α-lactalbumin, a soluble protein secreted by the lactating mammary gland. Binding of α-lactalbumin to galactosyltransferase changes the K_m of the transferase for glucose from 1–2 mol/L to 10^{-3} mol/L. Synthesis of α-lactalbumin by the mammary gland is initiated late in pregnancy or at parturition by the sudden decrease in progesterone level that occurs at that time. Prolactin promotes the rate of synthesis of galactosyltransferase and α-lactalbumin.

Metabolism of Amino Sugars

In amino sugars, one hydroxyl group, usually at C_2, is replaced by an amino group. Amino sugars are important constituents of many complex polysaccharides, including glycoproteins and glycolipids (Chapters 10 and 16), and of glycosaminoglycans (Chapters 11 and 16).

De novo synthesis of amino sugars starts from glucose-6-phosphate, but salvage pathways can also operate. Some reactions involved are shown in Figure 15-18. Incorporation of amino sugars into biological macromolecules is described in Chapter 16.

Pentose Phosphate Pathway

The series of cytoplasmic reactions known as the pentose phosphate pathway is also called the hexose monophosphate (HMP) shunt (or cycle) or the phosphogluconate pathway. The qualitative interconversions that take place

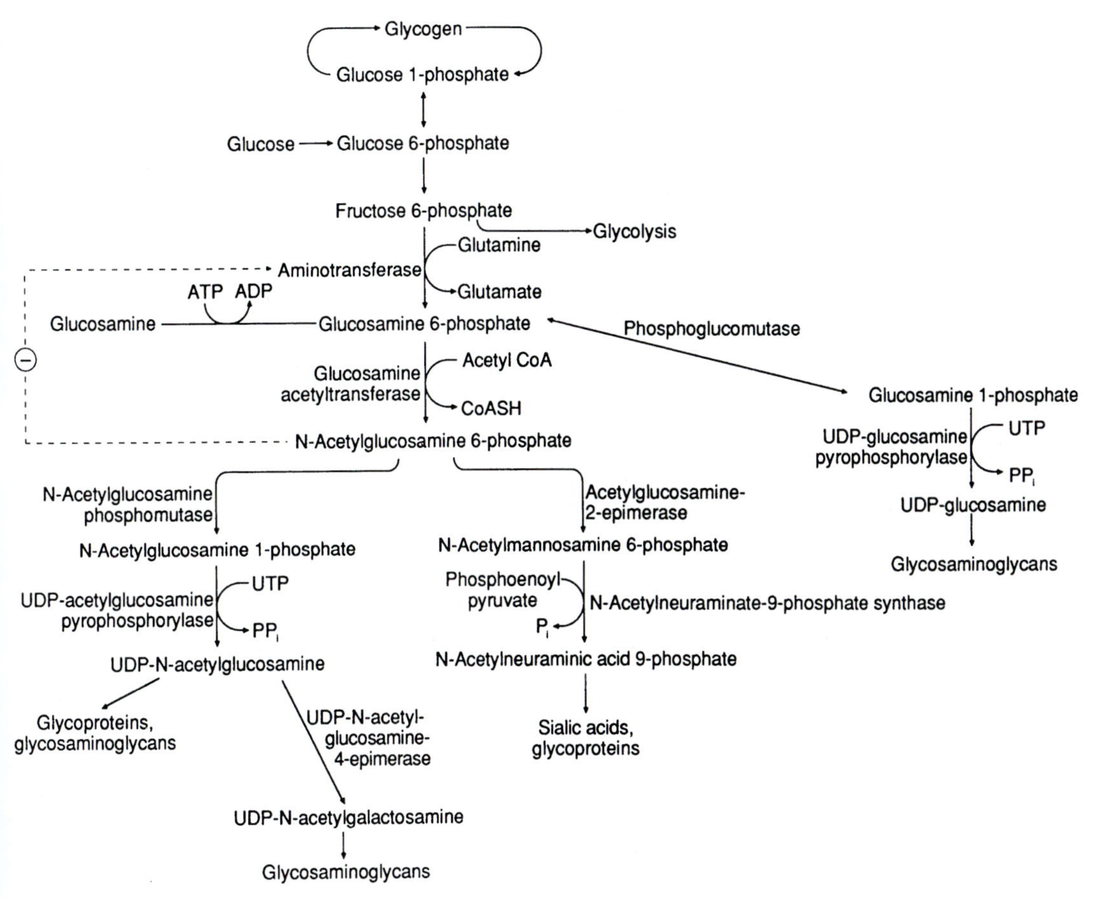

FIGURE 15-18
Summary of some interconversions and synthetic reactions in which amino sugars participate. Substrates for the pathway can be derived from glucose, glycogen, and gluconeogenesis.

are summarized in Figure 15-19, in which stoichiometry is ignored.

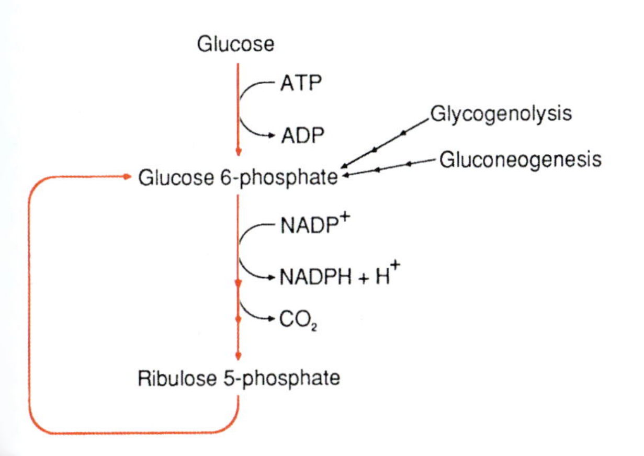

FIGURE 15-19
Summary of the pentose phosphate pathway. This diagram is intended to show the two major parts of the pathway: oxidation and decarboxylation of glucose-6-phosphate to ribulose 5-phosphate, and resynthesis of the former from the latter. The stoichiometry of the pathway is ignored.

The pentose phosphate pathway can be thought of as two separate pathways:

1. The oxidative conversion of glucose-6-phosphate to ribulose 5-phosphate and CO_2 and
2. Resynthesis of glucose-6-phosphate from ribulose 5-phosphate. Because the pathway begins and ends with glucose-6-phosphate, it is a cycle, or shunt. For every six molecules of glucose-6-phosphate that enter the pathway, five molecules of glucose-6-phosphate $+ 6CO_2$ are produced. Thus, there is a net loss of one carbon from each glucose-6-phosphate that enters the cycle.

Interpreting the pathway as a means of oxidizing glucose to ribulose 5-phosphate and recycling the ribulose 5-phosphate back to glucose-6-phosphate is too narrow. The pathway performs a variety of functions, the least important of which seems to be that of an alternative pathway for glucose metabolism. Production of ribose 5-phosphate for nucleotide synthesis by *de novo* and salvage pathways (Chapter 27), generation of NADPH for biosynthetic reactions in the cytosol, and interconversion of pentoses and hexoses are more valuable processes. Thus, the pathway provides a means for generating glucose from ribose and other pentoses that can be converted to ribose 5-phosphate.

The pentose phosphate pathway is active in a wide variety of cell types, particularly those which have a high rate of nucleotide synthesis or which utilize NADPH in large amounts. Nucleotide synthesis is greatest in rapidly dividing tissues, such as bone marrow, skin, and gastric mucosa. NADPH is needed in the biosynthesis of fatty acids (liver, adipose tissue, lactating mammary gland), cholesterol (liver, adrenal cortex, skin, intestine, gonads), steroid hormones (adrenal cortex, gonads), and catecholamines (nervous system, adrenal medulla) and in other reactions that involve tetrahydrobiopterin. NADPH is also needed for maintenance of a reducing atmosphere in cells exposed to high concentrations of oxygen radicals including erythrocytes, lens and cornea of the eye, and phagocytic cells, which generate peroxide and superoxide anions during the process of killing bacteria.

Oxidative Phase

The first reaction in this phase is the oxidation of glucose-6-phosphate to 6-phosphoglucono-δ-lactone and reduction of $NADP^+$ to NADPH, catalyzed by glucose-6-phosphate dehydrogenase (G6PD):

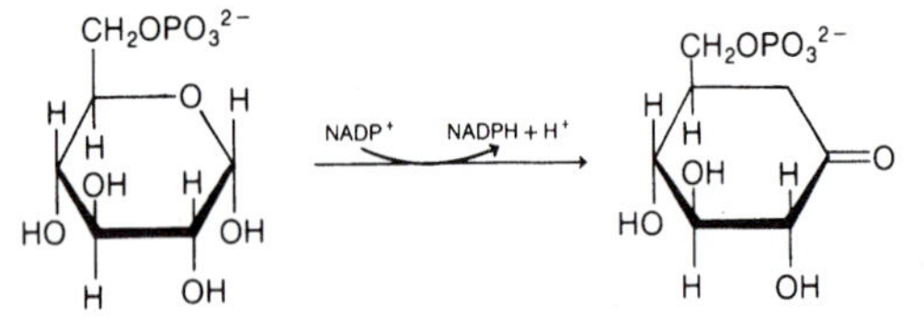

Glucose 6-phosphate $\xrightarrow{NADP^+ \;\; NADPH + H^+}$ 6-Phosphoglucono-δ-lactone

The equilibrium of the reaction lies far to the right.

Under physiological conditions, G6PD is a dimer of identical subunits of M.W. 55,000. The active enzyme contains a molecule of $NADP^+$, removal of which causes dissociation into inactive monomers. Enzymes from rat, cow, and human are very similar, and active interspecies hybrid enzymes can be formed *in vitro*. The enzyme is inhibited by NADPH at the concentration normally found in hepatocytes and erythrocytes, by ATP competing with glucose-6-phosphate for binding, and by cyanate. A large reserve capacity of G6PD activity exists in the red blood cell and, presumably, in other tissues.

In the second step, 6-phosphoglucono-δ-lactone is hydrolyzed to 6-phosphogluconate by 6-phosphogluconolactonase:

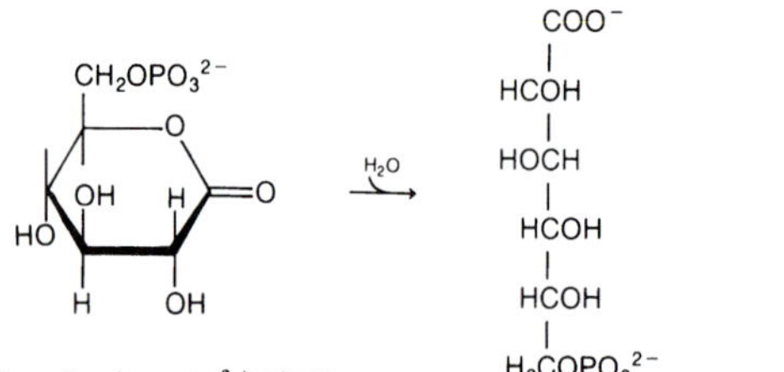

6-Phosphoglucono-δ-lactone $\xrightarrow{H_2O}$ 6-Phosphogluconate

The equilibrium of this reaction lies far to the right.

In the third step, 6-phosphogluconate is oxidatively decarboxylated to ribulose 5-phosphate in the presence of $NADP^+$, catalyzed by 6-phosphogluconate dehydrogenase:

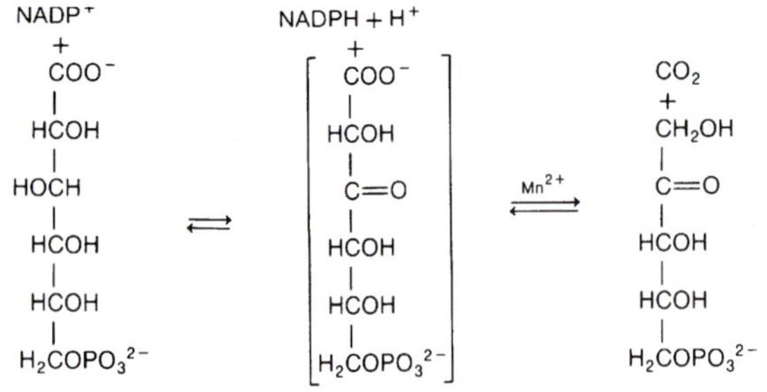

6-Phosphogluconate + $NADP^+$ $\rightleftharpoons$ [3-Keto-6-phosphogluconate] + $NADPH + H^+$ $\xrightleftharpoons{Mn^{2+}}$ D-Ribulose-5-phosphate + CO_2

3-Keto-6-phosphogluconate is a probable intermediate. The reaction is similar to those catalyzed by malic enzyme (in gluconeogenesis) and by isocitrate dehydrogenase (in the TCA cycle).

Nonoxidative Phase

In this phase, ribulose 3-phosphate is converted to glucose-6-phosphate. Stoichiometrically, this requires the rearrangement of six molecules of ketopentose phosphate to five molecules of aldohexose phosphate (Figure 15-20). It has been claimed that the pathway shown in Figure 15-20 occurs primarily in fat tissue and that a modified pathway involving arabinose 5-phosphate and octulose 8-phosphate occurs in liver cells. The overall scheme of the nonoxidative phase should be considered tentative.

There are only four types of reactions in this part of the pathway:

1. Interconversion of a keto sugar (ribulose 5-phosphate or fructose-6-phosphate) and an aldo sugar (ribose 5-phosphate or glucose-6-phosphate) by an isomerase.
2. Inversion of the optical configuration of an optically active carbon atom, as in a conversion of ribulose 5-phosphate to xylulose 5-phosphate by an epimerase.

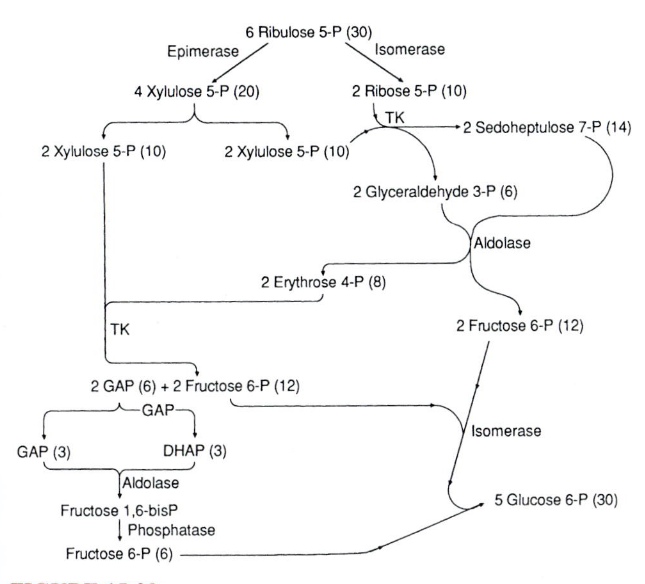

FIGURE 15-20
Nonoxidative phase of the pentose phosphate pathway. Numbers in parentheses show the distribution of carbon atoms between the various branches of each reaction. TK = transketolase; GAP = glyceraldehyde 3-phosphate; DHAP = dihydroxyacetone phosphate; P = phosphate.

3. Transfer of a two-carbon unit from a 2-keto sugar to the carbonyl carbon (C_1) of an aldose by a transketolase, which requires thiamine pyrophosphate and magnesium as cofactors. A covalent enzyme-substrate intermediate is formed similar to the one that occurs during the pyruvate dehydrogenase reaction (Chapter 13).
4. Transfer of carbons 1–3 from a ketose phosphate to the carbonyl carbon (C_1) of an aldose phosphate by a transaldolase. The reaction mechanism is similar to that of aldolase, with formation of a Schiff base intermediate involving the ε-amino group of a lysyl side chain on the enzyme.

The flexibility of this phase of the pathway allows for interconversion of a number of sugars and glycolytic intermediates, in part because of the ready reversibility of the reactions and regulation of the enzymes by substrate availability. Thus, the pathway adjusts the concentrations of a number of sugars rather than act as a unidirectional anabolic or catabolic route for carbohydrates.

Pentose Phosphate Pathway in Red Blood Cells

In the erythrocyte, glycolysis, the pentose phosphate pathway, and the metabolism of 2,3-bisphosphoglycerate (Chapter 28) are the predominant pathways of carbohydrate metabolism. Glycolysis supplies ATP for membrane ion pumps and NADH for reoxidation of methemoglobin. The pentose phosphate pathway supplies NADPH to

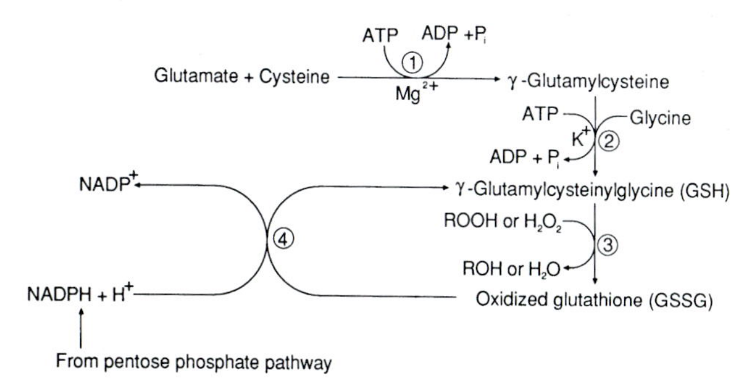

FIGURE 15-21
Metabolism of glutathione. (1) γ-Glutamylcysteine synthase; (2) GSH synthase; (3) glutathione peroxidase (Se-containing enzyme); 4. glutathione reductase (an FAD enzyme).

reduce glutathione for protection against oxidant injury. Glutathione (γ-glutamylcysteinylglycine; GSH) is synthesized by γ-glutamylcysteine synthase and GSH synthase (Figure 15-21). In the steady state, 99.8% of the glutathione is in the reduced form (GSH), and only 0.2% is in the oxidized form (GSSG), because of theNADPH-dependent reduction of oxidized glutathione, catalyzed by glutathione reductase:

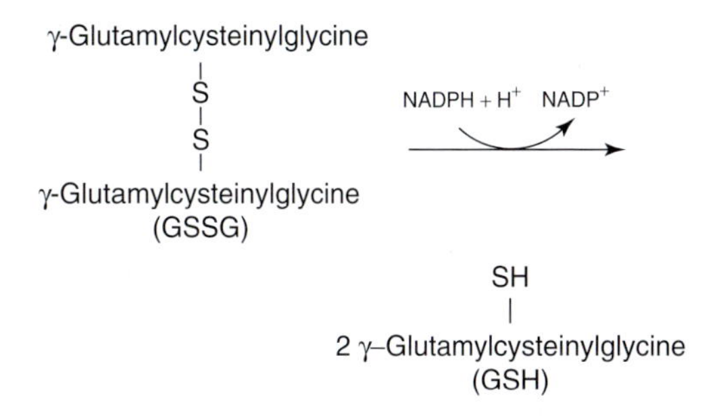

Glutathione reductase also catalyzes reduction of mixed disulfides of glutathione and proteins (Pr):

$$\text{Pr–S–SG} + \text{NADPH} + \text{H}^+ \rightarrow \text{Pr–SH} + \text{GSH} + \text{NADP}^+$$

Thus, the active sulfhydryl groups of hexokinase, glyceraldehyde-phosphate dehydrogenase, glutathione reductase, and hemoglobin (–SH group at the β-93 position) are maintained in the reduced form. Glutathione reductase is an FAD enzyme composed of two identical subunits encoded by a single locus on the short arm of human chromosome 8.

GSH is used in the inactivation of potentially damaging organic peroxides (e.g., a peroxidized unsaturated fatty acid) and of hydrogen peroxide. Hydrogen peroxide is formed through the action of superoxide dismutase (Chapter 14):

$$2\text{O}_2^- + 2\text{H}^+ \rightarrow \text{H}_2\text{O}_2 + \text{O}_2$$

Peroxide inactivation is catalyzed by glutathione peroxidase, a selenium-containing enzyme, in the following reactions:

$$2GSH + ROOH \rightarrow GSSG + H_2O + ROH$$

and

$$2GSH + H_2O_2 \rightarrow GSSG + 2H_2O$$

The concentrations of substrates for glutathione reductase and peroxidase determine the rate at which the pentose phosphate pathway operates in erythrocytes.

Genetically determined deficiencies of γ-glutamylcysteine synthase and glutathione synthase can cause hemolytic anemia (Chapter 17). Abnormalities of glutathione metabolism can also result from nutritional deficiencies of riboflavin or selenium. Glutathione reductase is an FAD enzyme that requires riboflavin for activity, and riboflavin deficiency can cause hemolytic anemia. An inherited lack of glutathione reductase apoenzyme has been described. Selenium deficiency diminishes activity of glutathione peroxidase and can lead to peroxidative damage. This can be partially ameliorated by vitamin E, an antioxidant (Chapters 37 and 38).

Glucose-6-Phosphate Dehydrogenase Deficiency

G6PD deficiency is the most common inherited enzyme deficiency known to cause human disease, occurring in about 100 million people. Most clinical manifestations are related to hemolysis, which results from impaired ability to produce cytosolic NADPH. Over 150 variants of the G6PD structural gene are known, many of which show either abnormal kinetics or instability of the enzyme. The erythrocytes are most severely affected because of their long half-lives and inability to carry out protein synthesis. Since persons with G6PD deficiency can usually make an adequate supply of NADPH under normal conditions, the defect may not become apparent until the patient takes a drug, such as primaquine, that greatly increases the demand for NADPH. The severity of the reaction depends, in part, on the particular inherited mutation.

In most persons with G6PD deficiency, hemolysis is observed as an acute phenomenon only after severe oxidative stress, leading to loss of perhaps 30–50% of the circulating red cells. The urine may turn dark, even black, from the high concentration of hemoglobin, and a high urine flow must be maintained to prevent damage to the renal tubules by the high-protein load. Because G6PD activity is highest in reticulocytes and decreases as the cell ages, the older erythrocytes (more than 70 days old) are destroyed. For this reason, measurement of red cell G6PD activity following a hemolytic crisis can lead to a spuriously high value, even within the normal range. If the patient survives the initial crisis, with or without transfusions, recovery usually occurs as the reticulocyte count increases.

TABLE 15-2
Partial List of Drugs and Chemicals That Cause Acute Hemolytic Anemia in Persons with G6PD Deficiency

8-Aminoquinoline derivatives (pamaquine, pentaquine, primaquine)	Phenylhydrazine
Methylene blue	Sulfonamides and sulfones (dapsone, sulfacetamide, sulfamethoxazole, sulfanilamide, sulfapyridine, salicylazosulfapyridine, thiazolesulfone)
Nalidixic acid	
Naphthalene	
Neoarsphenamine	
Niridazole	Toluidine blue
Nitrofurantoin	Trinitrotoluene

The characteristics of the disorder were elucidated during investigations into the hemolytic crises observed in some patients following administration of 8-aminoquinoline derivatives, such as primaquine and pamaquine, used for prophylaxis and treatment of malaria. The relatively high frequency of such crises in some geographic areas is due to the extensive overlap in the distributions of endemic malaria and G6PD deficiency. The geographic distribution of G6PD variants, like that of sickle cell trait (Chapter 28), suggests that heterozygosity for G6PD deficiency may confer some protection against ***falciparum malaria.*** A number of other drugs (Table 15-2) also cause hemolytic crises.

Oxidation of glutathione by these drugs beyond the capacity of the cell to generate NADPH for GSSG reduction causes the acute crisis. Since these drugs do not cause hemolysis *in vitro,* additional steps must take place *in vivo.* Following administration of a drug known to promote hemolysis, Heinz bodies are seen in erythrocytes in the peripheral blood. These also occur in some ***thalassemias*** and consist of oxidized and denatured forms of hemoglobin known as ***hemichromes*** (Chapter 28). Heinz bodies impair movement of the cells through the splenic pulp and probably are excised there, presumably together with the adjacent piece of plasma membrane, leaving a red cell that is more susceptible to destruction by the reticuloendothelial system. Infection and diabetic ketoacidosis can also cause hemolysis in persons with G6PD deficiency, possibly through depletion of NADPH.

Favism is characterized by acute hemolysis following ingestion of fava beans (*Vicia fava*) by persons with G6PD deficiency. The fava bean is a vegetable staple of the Mediterranean region, an area in which G6PD deficiency is endemic. Infants are especially susceptible to favism. The disorder is frequently fatal unless a large amount of blood is transfused rapidly. Not everyone with G6PD deficiency reacts to fava beans, and ingestion of fava beans does not invariably destroy ^{51}Cr-labeled G6PD-deficient erythrocytes *in vivo*. Thus, there may be a fundamental difference between favism and the hemolytic response to other agents. It has been suggested that another mutation must also be present for the expression of favism, but the nature of this hypothetical mutation is unknown.

Studies of the genetics of human G6PD variants have contributed to the understanding of G6PD deficiency and of more general aspects of human genetics. G6PD deficiency is inherited as an X-linked trait, as are hemophilia (Chapter 36) and color blindness (Chapter 38). If the X chromosome carrying an abnormal G6PD allele is designated X*, then the three possible genotypes containing X are

1. *X**Y—hemizygous male, with full phenotypic expression of the abnormal allele;
2. *XX**—heterozygous female, with a clinically normal phenotype in spite of the abnormal allele expressed in about half her cells; and
3. *X***X**—homozygous female, with full phenotypic expression of the abnormal allele. Sons of affected males are usually normal (because they receive their X chromosome from their mothers), and daughters of affected males are usually heterozygotes (because they receive one X chromosome from their fathers). The rarest genotype is that of the homozygous female, since it requires that both parents have at least one abnormal X chromosome.

Females heterozygous for a G6PD variant are phenotypic mosaics. They have two erythrocyte populations, one containing normal G6PD, the other the variant. In fact, in heterozygotes, every tissue has some cells expressing the normal, and some the abnormal, G6PD gene. Random X-chromosome inactivation early in embryonic development causes only one of the two X chromosomes to be active (***Lyon's hypothesis***).

Severity of the disorder in homozygotes and hemizygotes depends on a number of factors. G6PD variants are classified into five groups (class I being the most severe) depending on the presence or absence of chronic anemia and on the amount of enzyme activity present in the erythrocytes. The most common normal activity (class IV) allele is G6PD B. Another common electrophoretic variant with normal activity is G6PD A. Among North American blacks, the gene for the absence of G6PD A (G6PD A$^-$; class III) has an incidence of about 11%. Hemizygous males have only 5–15% of normal erythrocyte G6PD activity and exhibit a mild hemolytic anemia following an oxidative insult, such as primaquine administration. Hemolysis may cease even with continued administration of the drug because the reticulocytes, which increase in proportion following hemolysis, have adequate G6PD activity. In persons of Mediterranean and Middle Eastern ancestry, the most common abnormal allele is G6PD M (G6PD Mediterranean; class II) associated with severe hemolysis following administration of an appropriate drug (Table 15-2). The average incidence of this allele is approximately 5–10%, but a subpopulation of Kurdish Jews is reported to have an incidence of 50%. Enzyme activity in erythrocytes from these patients is often less than 1% of normal, and transfusion is usually required following a hemolytic crisis. The difference in the clinical severity of the diseases associated with G6PD A$^-$ and G6PD M can be explained by the characteristics of the two variants. G6PD A$^-$ has a normal K_m for glucose-6-phosphate (50–70 μmol/L) and for NADP$^+$ (2.9–4 μmol/L), but it exhibits an abnormal pH activity curve. The deficiency is due to an accelerated rate of inactivation of G6PD A$^-$ protein. Bone marrow cells and reticulocytes have normal amounts of G6PD activity, while activity in older red cells is very low. However, the residual enzyme seems to be more resistant to inhibition by NADPH. In contrast, G6PD M molecules have an intrinsically lower catalytic activity, and both reticulocytes and older red cells have decreased amounts of G6PD activity. Consequently, when one of the drugs in Table 15-2 is given, more cells are susceptible, and hemolysis is greater than with G6PDA$^-$. The low K_m of G6PD M for G6P and NADP$^+$ may account for the near-normal survival of erythrocytes in the absence of oxidative stress.

A number of other enzymopathic substances (e.g., pyruvate kinase, Chapter 13; and pyrimidine-5′-nucleotidase, Chapter 27), abnormal hemoglobins (Chapter 28), and abnormalities of the erythrocyte cytoskeleton (Chapter 10) may cause hemolytic anemia. Because many enzymes in the red cell are identical to those in other tissues, defects in these enzymes may have pleiotropic effects. Thus, in addition to hemolytic anemia, triose phosphate isomerase deficiency causes severe neuromuscular disease, and phosphofructokinase deficiency causes a muscle glycogen storage disease (Chapter 13). Mutations that result in decreased enzyme stability are usually most strongly expressed in erythrocytes because of their inability to synthesize proteins.

Phagocytosis and the Pentose Phosphate Pathway

The pentose phosphate pathway is crucial to the survival of erythrocytes because of its ability to provide NADPH for reduction of toxic, spontaneously produced oxidants. In phagocytic cells, the pentose phosphate pathway generates oxidizing agents that participate in the killing of bacteria and abnormal cells engulfed by the phagocytes.

In humans, the two principal types of phagocytic cells are polymorphonuclear leukocytes (PMN leukocytes, neutrophilic granulocytes, or neutrophils) and those of the mononuclear phagocyte system (MPS). The MPS, also called the reticuloendothelial system (RES), includes peripheral blood monocytes, tissue macrophages (inflammatory macrophages, Kupffer cells, histiocytes, alveolar macrophages, and others), and their precursor cells in bone marrow. These cells are normally active against bacteria, foreign cells, and particulate matter such as asbestos fibers, silicon dioxide particles, and coal dust. Neutrophils are particularly important in protecting against acute bacterial infections.

Phagocytosis is the process whereby phagocytes move toward, engulf, and digest foreign material. Attraction of phagocytes is mediated by chemotactic factors such as C5a (derived from complement; Chapter 35), bacterial endotoxins (Chapter 16), kallikrein, fibrinopeptide B (Chapter 36), and leukotrienes (Chapter 18). The contractile systems, which enable the phagocytes to extend pseudopods for movement and engulfment, are discussed in Chapter 21.

Following invagination of the phagocyte plasma membrane to surround a particle or microorganism, the edges of the vesicle fuse to form a cytoplasmic vacuole or phagosome. Lysosomes or, in the case of neutrophils, primary (azurophilic) and secondary (specific) cytoplasmic granules then fuse with the phagosome to form a phagolysosome. The contents of these organelles are emptied into the phagolysosome and contribute to the killing and digestion of the phagocytized material. Most of the cytocidal activity of phagocytes is due to the generation of highly reactive forms of oxygen from NADPH and molecular oxygen. In the absence of stimulation, the activity of the pentose phosphate pathway in phagocytes is low. Within 15–60 seconds after a phagocyte is alerted to the presence of foreign material, there is a 20- to 200-fold increase in oxygen consumption, the ***respiratory burst,*** and a 10- to 20-fold increase in pentose phosphate pathway activity. These changes are initiated by interactions between receptors on the plasma membranes of the phagocytes and a wide variety of stimuli, including immune complexes, chemotactic peptides, and opsonized bacteria. Phosphatidyl inositol, prostaglandins, cAMP, and changes in cytosolic $[Ca^{2+}]$ appear to be essential for this process. The magnitude of the response is related to the type of inducing signal and to the cell's ability to respond to that signal.

Most of the oxygen consumed in the respiratory burst is converted to superoxide anion (O_2^-). This highly reactive oxygen radical anion is one of the substances responsible for oxidative damage to red cells. In erythrocytes, superoxide anion occurs spontaneously, as a byproduct of heme-oxygen interactions. In the phagocyte, it is formed by the NADPH oxidase on the outer surface of the plasma membrane. The reaction involves transfer of a single electron from NADPH to oxygen:

$$\text{NADPH} + 2O_2 \xrightarrow[\text{oxidase}]{\text{NADPH}} 2O_2^- + \text{NADP}^+ + \text{H}^+$$

NADPH oxidase is thought to be an electron transport chain that includes a flavoprotein, and cytochrome b_{558} (Chapter 14).

The NADPH is provided by the pentose phosphate pathway. Glucose-6-phosphate dehydrogenase and 6-phosphogluconate dehydrogenase are inhibited by NADPH at the concentrations found in the unstimulated phagocyte. As NADPH is consumed by NADPH oxidase, inhibition of the pentose phosphate pathway is reduced, increasing the rate of formation of NADPH.

Phagosomes contain some of the activated NADPH oxidase from the cell surface, and superoxide is produced within the phagolysosome. Although superoxide is only bactericidal against organisms that lack superoxide dismutase, it is the substrate for formation of a number of cytotoxic compounds.

Formation of phagolysosomes is accompanied by a decrease in the pH of their interior, which causes spontaneous conversion of superoxide anion to oxygen and hydrogen peroxide:

$$2O_2^- + 2H^+ \rightarrow H_2O_2 + O_2$$

Hydrogen peroxide is a substrate for myeloperoxidase, a multisubunit heme protein of M.W. ~150,000, present in primary neutrophilic granules. The active prosthetic groups are two hemes covalently attached to the apoenzyme. This enzyme catalyzes many kinds of oxidation reactions, but oxidation of halide ions to hypohalite ions appears to be the most important. Hypochlorite ion is the principal compound formed, although Br^-, I^-, and SCN^- (a pseudohalide) can also serve as substrates. The reaction catalyzed is

$$Cl^- + H_2O_2 \rightarrow ClO^- + H_2O$$

The hypochlorite ion has an antimicrobial potency 50-fold greater than that of hydrogen peroxide. It oxidizes a variety of biological molecules and inactivates α_1-antitrypsin and other protease inhibitors.

In the presence of metals, such as iron, hydrogen peroxide reacts with superoxide anion to produce hydroxyl radicals ($OH^•$) and singlet oxygen ($^1O_2$) by the reaction shown below:

$$O_2^- + H_2O_2 \rightarrow {}^1O_2 + OH^- + OH^•$$

Hydroxyl radicals are highly reactive, oxidizing a variety of biological molecules. Singlet oxygen, which has an excited (high-energy) electronic configuration, can decay spontaneously to O_2 or interact with and oxidize some other molecule. In either case, part of the energy of the singlet state is emitted as light. This chemiluminescence, like the respiratory burst, is characteristic of actively phagocytic cells capable of killing microorganisms; it has been used for evaluating the functional capacity of these cells.

Persons with ***chronic granulomatous disease*** (CGD) suffer from recurrent bacterial and fungal infections. About 300 cases have been reported, of which about 80% were inherited as an X-linked, and 20% as an autosomal, recessive trait. CGD is actually a group of diseases caused by lack of activation of NADPH oxidase due to an abnormality in the enzyme itself or in the activating system. Phagocytes from CGD patients fail to show either a respiratory burst or chemiluminescence. Chemotaxis, recognition, engulfment, degranulation, and the presence of enzymes of the pentose phosphate pathway are normal in the phagocytes from CGD patients. Many infections are caused by bacteria that make catalase. The small amounts of hydrogen peroxide made in these organisms are destroyed by this endogenous catalase. Bacteria that lack catalase release small amounts of H_2O_2 into the phagolysosome, where myeloperoxidase, the presence of which is normal, can use it as a substrate to generate bactericidal products. Therapy of CGD involves prevention of infections (e.g., by antibiotic prophylaxis) and treatment of infections and their complications. A promising new prophylactic agent is ***γ-interferon,*** which is synthesized by recombinant DNA technology. This agent, a cytokine, alters the development of very early neutrophil precursors and eventually leads to improvement in neutrophil function.

Severe G6PD deficiency (<5% of normal; class I) may cause a disease with symptoms similar to those of CGD, together with nonspherocytic hemolytic anemia. This disorder occurs only rarely, however, because leukocytes have a large excess of G6PD activity. In common G6PD deficiency (G6PD A^-) the enzyme is produced in normal amounts and is functionally normal. Although unstable, the enzyme retains adequate activity during the lifetime of the phagocyte to provide NADPH to support bactericidal activity.

The oxidative intermediates produced in phagocytes (O_2^-, H_2O_2, $^1O_2$, and $OH^•$) are bactericidal *in vitro,* reacting readily with many biological molecules, including DNA and membrane lipids. However, hypohalide ions and hence the myeloperoxidase pathway are thought to be the major source of bactericidal activity in phagocytes. All of these highly reactive substances are potentially cytotoxic to phagocytes themselves and to surrounding tissue, if leakage occurs. Protection against such an eventuality is provided by glutathione, glutathione peroxidase, and superoxide dismutase. Cytoplasmic (Cu/Zn) superoxide dismutase (SOD-1) and catalase can together convert superoxide ion to oxygen and increase the bactericidal capacity of phagocytes under conditions in which oxygen supply is limited. Disorders of cellular adhesion, chemotaxis, ingestion, and degranulation of phagocytes are also known.

Supplemental Readings and References

M. Bollen, S. Keppens, and W. Stalmans: Specific features of glycogen metabolism in the liver. *Biochemical Journal* **336,** 19 (1998).

X. Chen, N. Iqbal, and G. Boden: The effects of free fatty acids on gluconeogenesis and glycogenolysis in normal subjects. *The Journal of Clinical Investigation* **103,** 365 (1999).

M. Denborough: Malignant hyperthermia. *Lancet* **352,** 1131 (1998).

L. J. Elsas, S. Langley, E. Steele, et al.: Galactosemia: A strategy to identify new biochemical phenotypes and molecular genotypes. *American Journal of Human Genetics* **56,** 630 (1995).

R. Halse, J. J. Rochford, J. G. McCormack, et al.: Control of glycogen synthesis in cultured human muscle cells. *Journal of Biological Chemistry* **274,** 776 (1999).

M. B. Hampton, A. J. Keftle, and C. C. Winterbourn: Inside the Neutrophil phagosome: Oxidants, myeloperoxidase and bacterial killing. *Blood* **92,** 3007 (1998).

J. C. Lawrence, Jr. and P. J. Roach: New insights into the role and mechanism of Glycogen Synthase activation by insulin. *Diabetes* **46,** 541 (1997).

C. G. Proud and R. M. Denton: Molecular Mechanisms for the control of translation of insulin. *Biochemical Journal* **328,** 329 (1997).

L. Ragolia and N. Begum: Protein phosphatase-1 and insulin action. *Molecular and Cellular Biochemistry* **182,** 49 (1998).

A. K. Srivastava and S. K. Pandey: Potential mechanism(s) involved in the regulation of glycogen synthesis by insulin. *Molecular and Cellular Biochemistry* **182,** 135 (1998).

C. Villar-Palasi and J. J. Guinovart: The role of glucose-6-phosphate in the control of glycogen synthase. *FASEB Journal* **11,** 544 (1997).

CHAPTER 16

Carbohydrate Metabolism III: Glycoproteins, Glycolipids, GPI Anchors, Proteoglycans, and Peptidoglycans

Glycoconjugates are composed of linear and branching arrays of sugars attached to polypeptide or lipid backbones. They include glycoproteins, glycosphingolipids, glycosylphosphatidyl inositol (GPI), glycosaminoglycans (together with proteoglycans), and peptidoglycans.

Most of the biosynthetic pathways for oligosaccharide chains have been determined within the last decade and genes for a number of the enzymes in these pathways have been isolated and sequenced. Better methods for substrate and product identification have also improved understanding of the details in specific reactions. It is important to understand that many factors and pathways come into play within specific membrane compartments for proper glycosylation to occur. Because so many functions, including proper folding, intracellular routing of glycoproteins, and cellular recognition, are ascribed to oligosaccharides, mistakes in synthesis may result in various syndromes. Conversely, altered oligosaccharide biosynthesis may occur as a result of disease states that hamper appropriate glycosylation. This may lead to further complications such as a change in blood type, which is based on the presence of specific terminal sugars on red blood cells (RBCs) and other cells.

The elaboration of oligosaccharide chains occurs at specific sites in all cells. In prokaryotes, the enzymes are located on the plasma membrane, whereas in eukaryotes, they occur predominantly on membranes of the endoplasmic reticulum and Golgi apparatus. The reactions require energy in the form of a nucleotide-sugar donor molecule. This use of nucleotides as carriers is similar to that seen in the synthesis of phospholipids (Chapter 19) and glycogen (Chapter 15). Transfer of the sugar portions of the donors to acceptor molecules is catalyzed by glycosyltransferases. The general reaction is summarized below.

$$\text{XDP-sugar (A)} + \text{sugar (B)} \cdots \text{(C)} \rightarrow \text{(A)–O–(B)} \cdots \text{(C)} + \text{XDP}$$

In this reaction, XDP-sugar (A) is the nucleoside diphosphate sugar donor, and sugar (B)· · ·(C) is the acceptor, usually an oligosaccharide attached to a protein or lipid (C). In the product (A)–O–(B)· · ·(C), the sugar from the donor (A) is linked to the terminal sugar (B) on the acceptor by the glycosidic bond. Energy to form the glycosidic bond is provided by the phosphate ester linkage between the sugar and the nucleoside diphosphate. Dolichol phosphate and dolichol pyrophosphate are intermediate sugar carriers in the synthesis of asparagine-linked glycoproteins. The topography of glycosylation reactions is important, with some transferase reactions, such as dolichol-oligosaccharide synthesis, occurring on the cytoplasmic membrane face and the majority of other reactions taking place within the lumen of either the endoplasmic reticulum or Golgi apparatus.

Assembly of an oligosaccharide having 10 or more sugars may require almost the same number of glycosyltransferases, since each enzyme is usually specific for the formation of one glycosidic bond between a particular donor sugar and a particular acceptor molecule. This specificity has led to the hypothesis of a unique glycosyltransferase for every linkage between two monosaccharides in an oligosaccharide. Although not strictly true, the concept of "one linkage-one enzyme" is useful in understanding glycoconjugate synthesis. Many glycosyltransferases are highly specific, using only one particular nucleotide sugar as donor and requiring as acceptor an oligosaccharide with the correct terminal and penultimate sugars in the proper linkage. Other enzymes are less specific. For example, β-1,4-galactosyltransferase attaches galactose to different acceptors, provided the terminal sugar is N-acetylglucosamine.

Synthesis of glycoconjugates, in contrast to that of DNA, RNA, and protein, is not directed by a template but depends on the specificity of the enzymes that catalyze addition of each sugar residue. Thus, oligosaccharide chain biosynthesis may be less predictable than DNA, RNA, or protein synthesis and more often results in incomplete oligosaccharide chains. Furthermore, because there is no template from which to deduce the oligosaccharide size or sequence in a given glycoconjugate, the synthetic reactions can best be described by sequencing the product. The oligosaccharide reflects its biosynthetic history, and understanding of the biosynthetic pathways has hinged on structural analysis of oligosaccharides.

The recent advances in the determination of oligosaccharide structure using high-performance liquid chromatography (HPLC), gas chromatography–mass spectrometry (GC-MS), and nuclear magnetic resonance (NMR) have increased the capacity to separate and sequence sugars and their linkages in oligosaccharides. Analytical procedures must overcome several unique sequencing problems that include:

1. Lack of unique characteristics (such as catalytic activity) that can be used to follow purification;
2. Presence of more than one type of oligosaccharide side chain in a particular glycoprotein;
3. Presence of more than one type of glycosidic bond within an oligosaccharide, in contrast to the peptide bond that universally links amino acids to proteins;
4. Chemical similarity of one sugar to another, which leads to poor separation in a number of analytical systems and makes identification of a particular sugar more difficult. Except for the purification steps, determination of oligosaccharide structure is now relatively routine.

Carbohydrate structures, whether attached to N-glycans, O-glycans, or glycosphingolipids, generally have a core structure with arms containing terminal sugar sequences (Figure 16-2). In some cases the arms may consist of repeating Gal and GlcNAc referred to as polylactosamine. Figure 16-1 shows the structures of several oligosaccharides found in different types of glycoconjugates, together with the abbreviations used in writing these structures (see also Chapters 10 and 11).

The high degree of specificity of the glycosyltransferases necessitates a detailed scheme for their nomenclature. Criteria used to classify glycosyltransferases are the nucleotide-sugar, the transferred sugar, the monosaccharide at the acceptor site, and the linkage. In addition, transferases can show specificity for the location of the acceptor site, i.e., the protein or lipid to which the acceptor is attached or the exact location of the acceptor on the macromolecule.

This specificity can be indicated in one of two ways:

1. by specifying the sugar acceptor, linkage, and sugar donor (e.g., GlcNAc-β1,4-galactosyltransferase) or
2. when there are a number of glycosyltransferases with similar acceptor and linkages, a Roman numeral is placed after the name derived from the above criteria (e.g., a family of UDP-N-acetylglucosaminyltransferases are numbered I–VI based on the particular acceptor sites to which they transfer GlcNAc (Figure 16-3).

Uridine diphosphate (UDP) is the most common nucleotide carrier for sugars, although cytidine monophosphate (CMP)) and guanosine diphosphate (GDP) are also used. The most abundant nucleotide-sugar (and the first discovered) is UDP-glucose. Its structure and a list of other nucleotide-sugar types are given in Figure 16-4. See Chapter 15 for a discussion of the synthesis of UDP-glucose and its role in glycogen synthesis. Figure 16-5 summarizes biosynthetic pathways for UDP-glucose and other nucleotide-sugars important for glycoconjugate synthesis. The sugar donors can be synthesized from glucose, provided that the requisite enzymes are present, and the interconversion of fucose to other sugars does not occur. Similarly, N-acetylmannosamine *in vivo* seems to be entirely metabolized to N-acetylneuraminic acid (sialic acid), despite the existence of alternative reaction pathways, as shown in Figure 16-5). As a result, radioactively labeled fucose and N-acetylmannosamine, when administered experimentally in animals, can mark specific glycosylation sites within cells.

Several sugar-nucleotide pool defects have been identified: ***Type I carbohydrate-deficient glycoprotein syndrome*** (CDGs type I), a defect in phosphomannomutase, one of the enzymes responsible for converting glucose to GDP-mannose, results in an absence of

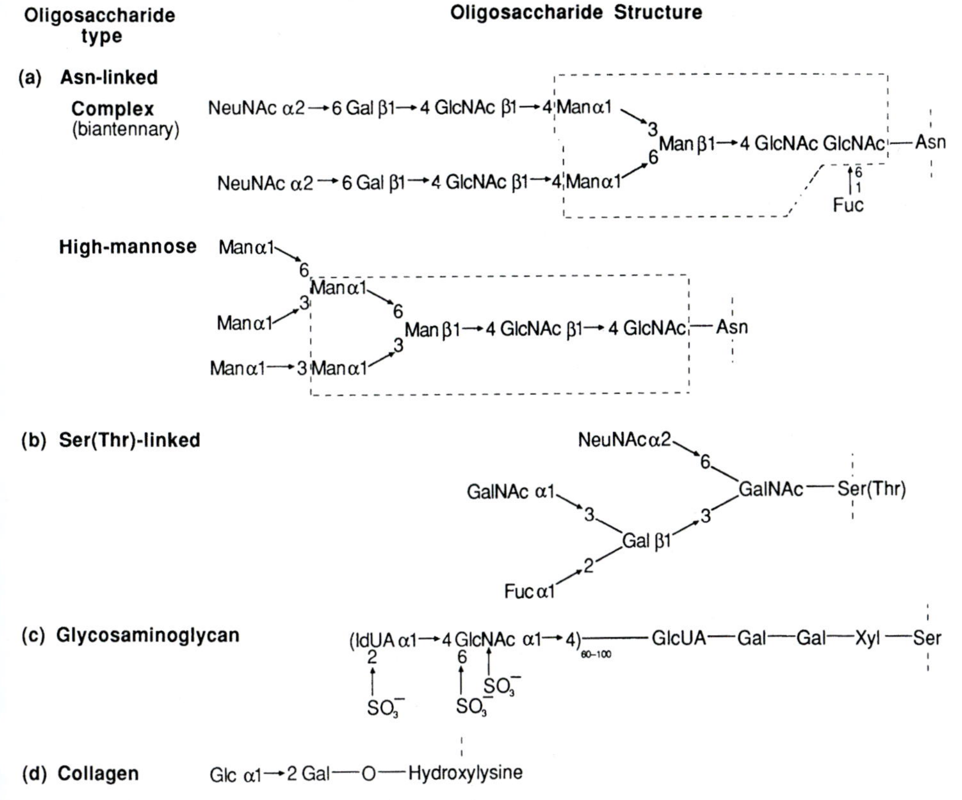

FIGURE 16-1

Structures of representative oligosaccharide groups that occur as parts of glycoconjugates. $-SO_3^-$ = sulfate.

a) Asn-linked: The core structure is shown in the dashed box. The sugar-protein linkage is always an N-glycosidic bond. In addition to the two types shown here, an intermediate, or mixed, structure has features of both.

b) Ser(Thr)-linked: The structure shown is for one of the oligosaccharides found in porcine submaxillary mucin. The sugar-protein linkage is an O-glycosidic bond to serine or threonine.

c) Glycosaminoglycan: The example shown is the predominant structure of heparin. The presence of a link oligosaccharide and the variability in the length of the chain of repeating disaccharide units are features common to many proteoglycans.

d) Collagen: The carbohydrate groups are attached as a late posttranslational step via O-glycosidic linkages to hydroxylysyl residues generated in an earlier posttranslational reaction.

The following abbreviations for sugars and related compounds are used throughout this chapter: Gal = D-galactose, Man = D-mannose, Fuc = L-fucose (6-deoxy-L-galactose), IdUA = L-iduronic acid, GalNAc = N-acetyl-D-galactosamine, NeuNAc = N-acetyl-D-neuraminic acid, NeuNGly N-glycolyl-D-neuraminic acid, Glc = D-glucose, Xyl = D-xylose, GlcUA = D-glucuronic acid, GlcNAc = N-acetyl-D-glucosamine.

GDP-mannose that is essential for glycoprotein synthesis. This defect is autosomal recessive and has been mapped to human chromosome 16p13.3–p13.12. All infants with CDGs have neurological abnormalities, failure to thrive, developmental delay, and dysmorphic features. Female patients with CDGs type I have hypergonadotropic hypogonadism and do not attain secondary sexual characteristics.

Gal-1-P uridyltransferase deficiency is autosomal recessive and results in galactosemia. Removal of all galactose from the diet resolves acute symptoms. However, long-term outcome may still be poor with frequent developmental delay, characteristic speech abnormalities, and ovarian failure. The hypergonadotropic ovarian failure associated with this deficiency may be related to improper glycosylation of follicle-stimulating hormone (FSH) that impairs binding to receptor.

Gal-4-epimerase deficiency causes severe developmental delay with poor growth. Acute symptoms can be resolved with a galactose-free diet; however, small quantities of galactose are supplemented to allow for synthesis of UDP-Gal in the absence of dietary galactose.

Leucocyte adhesion deficiency type II (LADII) patients lack products of fucosyltransferases; and therefore, they express the rare Bombay blood group phenotype and are negative for the Lewis blood group. They show severe mental retardation, are short in stature, and have dysmorphic features (low hairline, hypertelorism, broad

Blood Groups	Side Chain Structure	
A	GalNAcα1–3Galβ- |α1–2 Fuc	
B	Galα1–3Galβ- |α1–2 Fuc	
H or O	Galβ- |α1–2 Fuc	
T	Galβ1–3GalNAcα-	
Sialyl-T	SAα2–3Galβ1–3GalNAcα-	
Tn	GalNAcα-	
Sialyl-Tn (SA-Tn)	SAα2–6GalNAcα-	
Forssman	GalNAcα1–3GalNAcβ1–3Gal-	
S[d](Cad)	GalNAcβ1–4(SAα2–3)Galβ-	
i	Gal1–4GlcNAcβ1–3Galβ-	
I	Galβ1–4GlcNAc |β1–6 Galβ1–4GlcNAcβ1–3Galβ-	
	Type 1 chain	**Type 2 chain**
Le[a]	Galβ1–3GlcNAcβ1–3Gal- |α1–4 Fuc	
Le[b]	Galβ1–3GlcNAcβ1–3Gal- |α1–2 |α1–4 Fuc Fuc	
Le[x] (SSEA-1)		Galβ1–4GlcNAcβ1–3Gal- |α1–3 Fuc
Le[y]		Galβ1–4GlcNAcβ1–3Gal- |α1–2 |α1–3 Fuc Fuc
Sialy-Le[a] (SA-Le[a])		SAα2–3Galβ1–3GlcNAcβ1–3Gal- |α1–4 Fuc
Sialy-Le[x] (SA-Le[x])		SAα2–3Galβ1–4GlcNAcβ1–3Gal- |α1–3 Fuc
Sialyl-dimeric Le[x] (SA-dimeric Le[x])		SAα2–3Galβ1–4GlcNAcβ1–3Galβ1–4GlcNAc- |α1–3 |α1–3 Fuc Fuc

FIGURE 16-2

Carbohydrate blood group and tissue antigens. [Reproduced with permission from I. Brockhausen, Clinical aspects of glycoprotein biosynthesis. *Crit. Rev. Clin. Lab. Sci.* **30**:70 (1993).]

GlcNAc-transferase		Branches of multiantennary oligosaccharides: bi	tri	tri*	tetra	penta	bisected
V	GlcNAcβ6 (→ Manα6)			+	+	+	
VI	GlcNAcβ4 — Manα6					+	
II	GlcNAcβ2 (→ Manα6)	+	+	+	+	+	
III	GlcNAcβ4Manβ4GlcNAcβ4GlcNAcβ-R						+
IV	GlcNAcβ4 (→ Manα3)		+		+	+	
I	GlcNAcβ2 — Manα3	+	+	+	+	+	

FIGURE 16-3

Bisected penta-antennary oligosaccharide found in hen ovomucoid. The GlcNAc-transferases (named with Roman numerals) are responsible for the addition of the GlcNAc residues that initiate antennae. The branches present in various multiantennary and bisected N-glycans are indicated. [Reproduced with permission from I. Brockhausen, Clinical aspects of glycoprotein biosynthesis. *Crit. Rev. Clin. Lab. Sci.* **30**:68 (1993).]

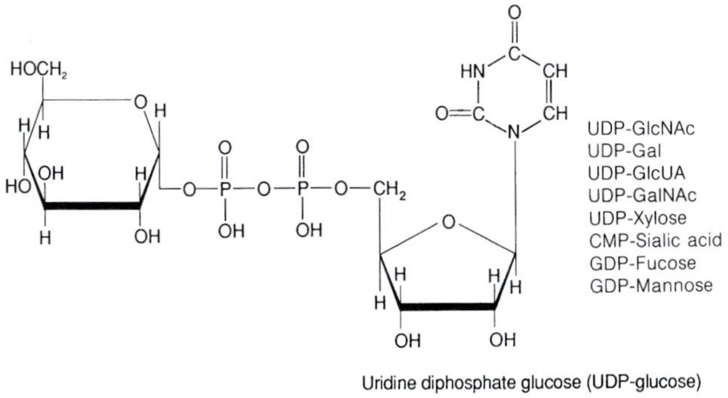

UDP-GlcNAc
UDP-Gal
UDP-GlcUA
UDP-GalNAc
UDP-Xylose
CMP-Sialic acid
GDP-Fucose
GDP-Mannose

Uridine diphosphate glucose (UDP-glucose)

FIGURE 16-4
Examples of nucleotide sugars involved in glycoconjugate biosynthesis. The structures of these compounds are the same as that of UDP-glucose but with different nucleotides and sugars.

nasal bridge, protruding tongue, prominent mandible, and a narrow or high-arched palate). The reduced fucosylation is most likely due either to a failure of fucose transport into the Golgi apparatus or in the synthetic pathway of fucose. Loss of fucosylation results in poor adhesion of leucocyte selectin receptors to endothelial cell oligosaccharides.

16.1 Biosynthesis of Glycoproteins

Glycoproteins can be classified into two major types with differing biosynthetic pathways. The first is the asparagine-linked glycoproteins (***N-glycans***), which contain an N-glycosidic linkage between N-acetylglucosamine and the β-amide nitrogen of an asparagine residue in the protein (abbreviated GlcNAc-N-Asn). This type of linkage predominates in serum glycoproteins. The second is the serine (threonine)-linked glycoprotein (***O-glycans***), such as salivary mucin and a number of other proteins and proteoglycans, which contain an O-glycosidic linkage between the hydroxyl groups of serine or threonine and N-acetylgalactosamine (abbreviated GalNAc-O-Ser/Thr). Details of monosaccharide units and linkages are usually indicated by structures of the kind shown in Figure 16-1.

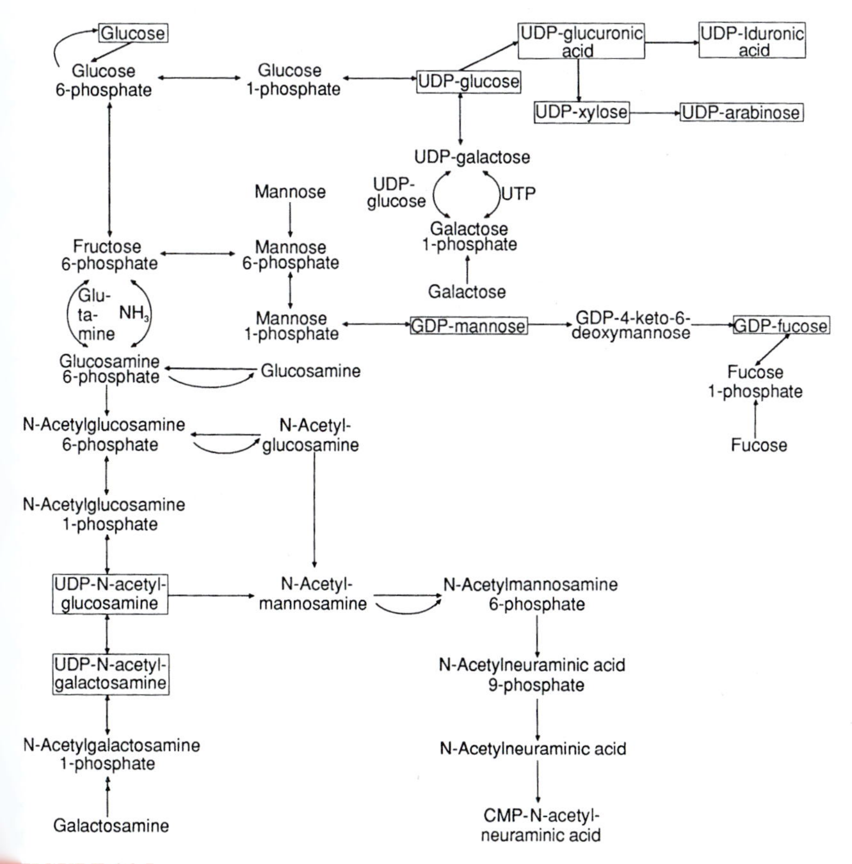

FIGURE 16-5
Biosynthetic pathways for nucleotide sugars. [Reproduced with permission from H. Schachter, Glycoprotein biosynthesis. In: *The Glycoconjugates,* Vol. 2, M.I. Horowitz and W. Pigman (Eds). Academic Press, San Diego, 1977.]

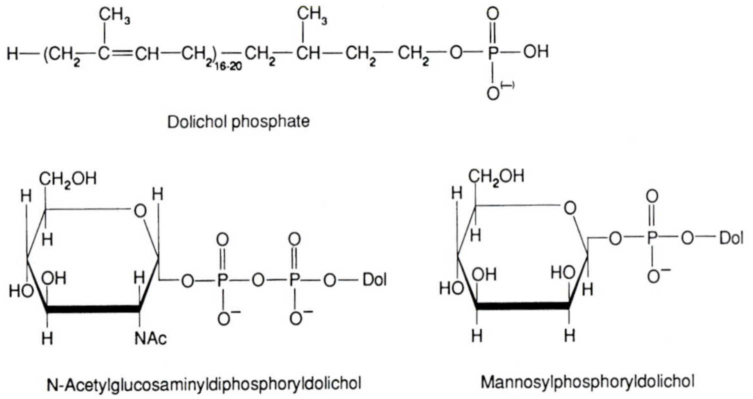

Dolichol phosphate

N-Acetylglucosaminyldiphosphoryldolichol

Mannosylphosphoryldolichol

FIGURE 16-6
Structures of dolichol phosphate, GlcNAc-P-P-Dol, and Man-P-Dol. GlcNAc-P-P-Dol is the first compound formed in the biosynthesis of the lipid-linked oligosaccharide shown in Figure 16-5. Replacement of GlcNAc by –OH in GlcNAc-P-P-Dol yields dolichol diphosphate (dolichol pyrophosphate). Replacement of Man by Glc in Man-P-Dol yields Glc-P-Dol.

N-Glycan Asn-Linked Glycoproteins

The discovery that dolichol phosphates and pyrophosphates (diphosphates) are carriers for oligosaccharides in eukaryotic cells initiated the modern era of glycoprotein biochemistry. These lipid-linked oligosaccharides are precursors for the carbohydrate side chains of the asparagine-linked glycoproteins. Because of this common precursor, all of the side chains of these glycoproteins share the same carbohydrate core, shown within the dashed box in Figure 16-1a.

Dolichols are a family of long-chain polyisoprenols that occur only in eukaryotes. They are related in structure and function to undecaprenyl phosphate, a prokaryotic polyisoprenol used in the synthesis of peptidoglycan (see below). Dolichols contain 65–110 carbons arranged as 13–22 isoprene units, two in the trans configuration, the remainder in the cis configuration. The initial isoprene unit is saturated and carries a hydroxyl group, which is esterified with orthophosphoric (monophosphoric) or pyrophosphoric (diphosphoric) acid in dolichol phosphate and pyrophosphate, respectively (Figure 16-6).

The pathway for *de novo* synthesis of dolichol phosphate is the same as for cholesterol until production of farnesyl pyrophosphate (Chapter 19). Two salvage pathways provide for the reutilization of dolichol and dolichol pyrophosphate once transfer of oligosaccharide from the dolichol to glycoprotein has occurred. A significant portion of the dolichol in liver is present as the free alcohol or as a fatty acid ester, and dolichol absorbed from the diet may enter this pool. Phosphorylation of dolichol is catalyzed by a dolichol kinase via cytidine triphosphate (CTP). Removal of one phosphate also permits recycling of the dolichol pyrophosphate released by transfer of the oligosaccharide to a protein acceptor.

Biosynthesis of an asparagine-linked oligosaccharide chain proceeds in three steps. A lipid-linked oligosaccharide precursor is first synthesized and then transferred *en bloc* to an available asparagine residue, in a nascent or newly completed polypeptide chain. Synthesis and transfer together make up the initiation step. After transfer, several sugar residues are removed from the oligosaccharide (the processing step), following which the peripheral sugars of the mature glycoprotein are added (the elongation step). If all steps proceed normally, the mature side chain is of the "complex" type. Glycoproteins with partially processed side chains may also be released. If they contain none of the peripheral sugars, they are termed "high-mannose" oligosaccharides. "Hybrid" oligosaccharides have features of both high-mannose and complex structures.

The above steps, not including synthesis of the lipid-linked oligosaccharide, are outlined in Figure 16-7. Structures of the side chains of mature glycoproteins depend on the specificities of the processing and elongation enzymes (glycosidases and transferases, respectively) and on the order in which they function. Regulatory steps may occur in this pathway, but their function is poorly understood.

(1) ***Initiation*** includes synthesis of the oligosaccharide-lipid precursor and transfer of the oligosaccharide to a protein. Evidence for a possible role for retinol (vitamin A) as a carrier in this step is discussed in Chapter 38.

The lipid-oligosaccharide is built by sequential attachment of mannose, N-acetylglucosamine, and glucose units

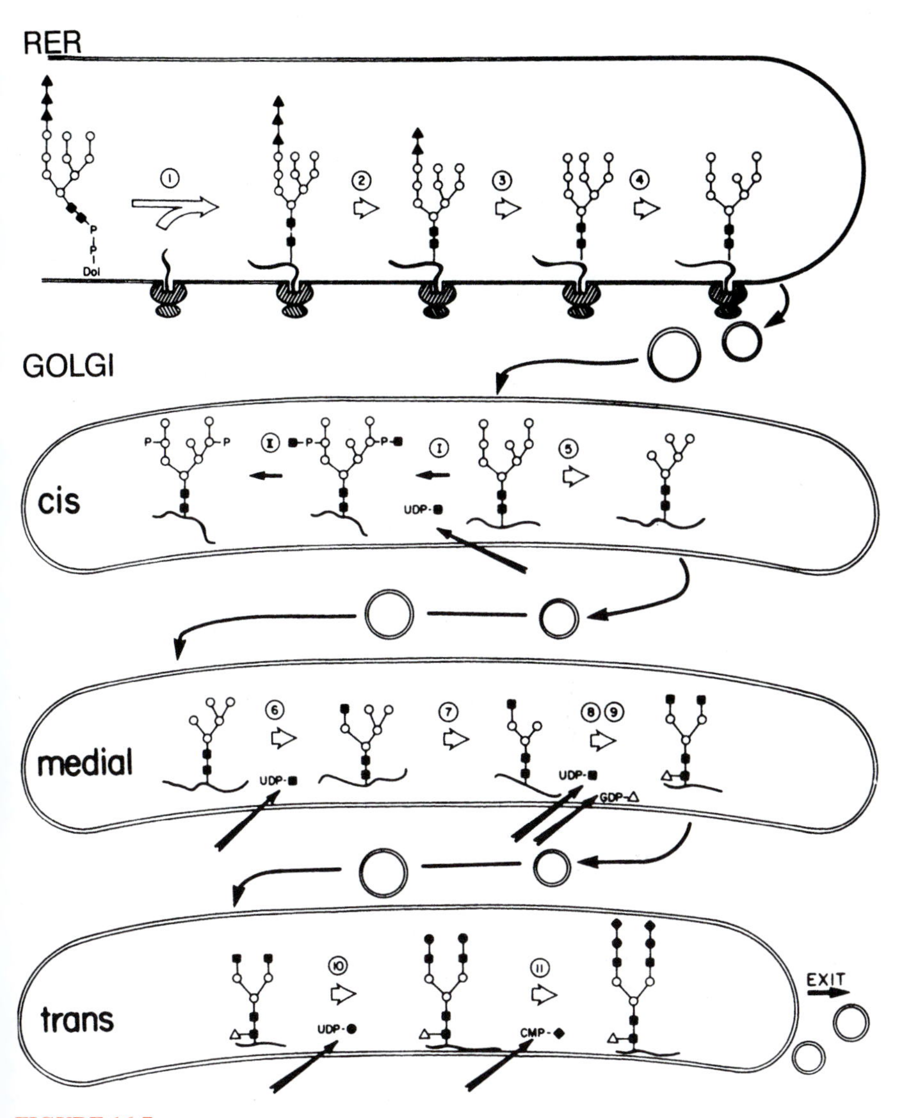

FIGURE 16-7
Proposed pathway for the biosynthesis of Asn-linked glycoproteins, including phosphorylation of high-mannose structures on glycoproteins destined for inclusion in lysosomes. Glycosylation begins during protein synthesis in the rough endoplasmic reticulum. Transfer of glycoproteins between subcellular compartments is thought to be accomplished by vesicle transport. Nucleotide sugars must enter the Golgi cisternae from their site of synthesis in the cytosol. Completed glycoproteins leave the trans Golgi for their ultimate intra- or extracellular destinations in membrane vesicles. The enzymes catalyzing the numbered reactions are (1) oligosaccharyltransferase, (2) α-glucosidase I; (3) α-glucosidase II, (4) endoplasmic reticulum α-1,2-mannosidase, (5) Golgi α-mannosidase I, (6) N-acetylglucoaminyltransferase I, (7) Golgi α-mannosidase II, (8) N-acetylglucosaminyltransferase II, (9) fucosyltransferase, (10) galactosyltransferase, (11) sialyltransferase, (I) N-acetylglucosaminylphosphotransferase, (II) N-acetylglucosamine-1-phosphodiester α-N-acetylglucosaminidase. [Reproduced with permission from R. Kornfeld and S. Kornfeld, Assembly of asparagine-linked oligosaccharides. *Annu. Rev. Biochem.* **54**:631 (1985). © 1985 by Annual Reviews Inc.]

to dolichol phosphate. The final chain contains nine mannose, two GlcNAc, and three glucose residues attached to a dolichol chain through a diphosphate linkage. A scheme for the addition of these sugars is shown in Figure 16-8. The dolichol phosphate is located in the rough endoplasmic reticulum, with the phosphate probably oriented toward the cytosol. The first two GlcNAc and five mannose residues that are attached to dolichol are derived from UDP-GlcNAc and GDP-mannose, respectively. Their addition occurs on the cytoplasmic face of the rough endoplasmic reticulum. The oligosaccharide-dolichol is then translocated from the cytoplasmic to the luminal side of the membrane. Further attachment of four mannose and three glucose residues occurs within the

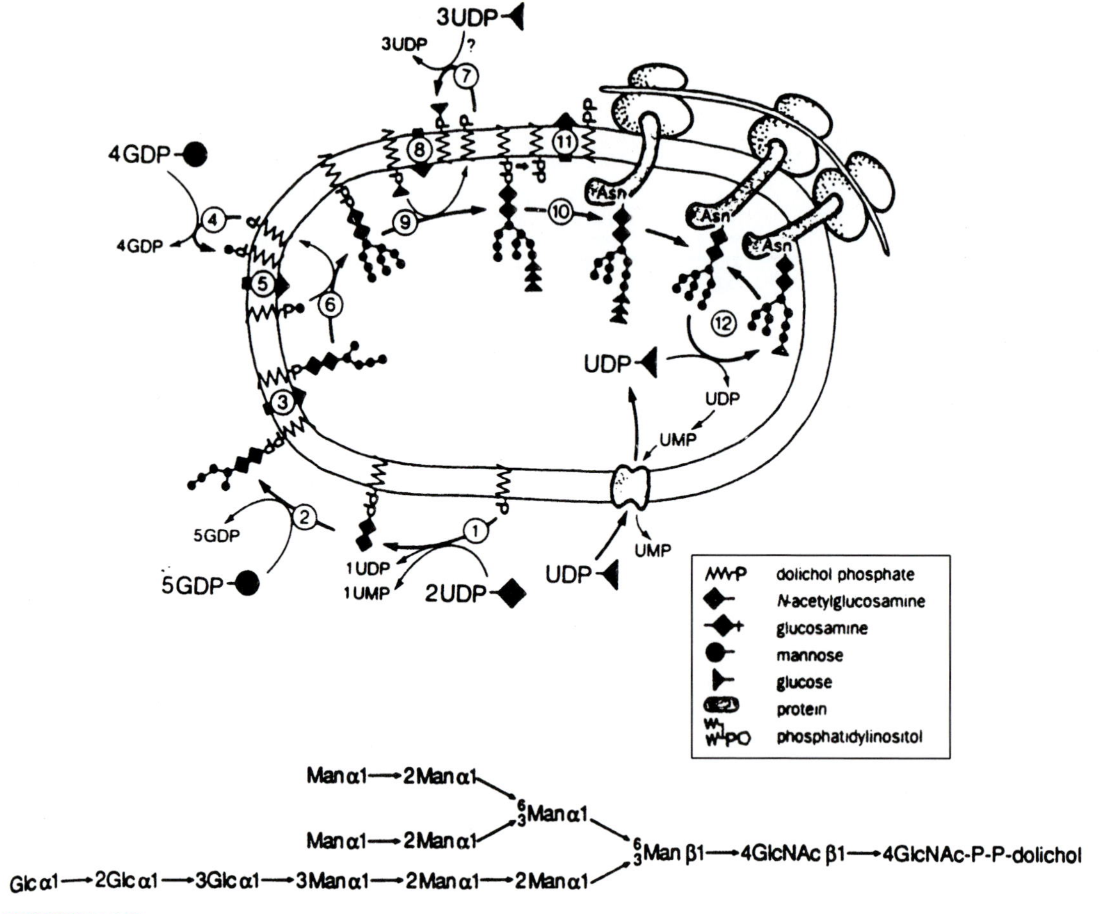

FIGURE 16-8
Topography of glycosylation in the rough endoplasmic reticulum. Initiation of N-glycosylation and transient reglucosylation. [Reproduced with permission from C. Abeijon and C. B. Hirschberg, Topography of glycosylation reactions in the endoplasmic reticulum. *Trends Biol. Sci.* **17**:34 (1992).]

lumen from dolichol-P-mannose and possibly dolichol-P-glucose donor substrates that were synthesized on the cytoplasmic side of the membrane and then translocated to the lumen. The dolichol-P-mannose transfer to the membrane lumen is essential since GDP-mannose cannot be transported across the endoplasmic reticulum. Both UDP-GlcNAc and UDP-Glc can be transported across the membrane. The final lipid-linked oligosaccharide is then ready for transfer to nacsent polypeptides that extend into the cisternae from the cytosolic surface to the rough endoplasmic reticulum.

Transfer of the oligosaccharide to an asparagine residue on a polypeptide occurs in the cisternae of the rough endoplasmic reticulum (Figure 16-8), during or soon after the synthesis of the protein. The oligosaccharyltransferase is a large heteroligomeric membrane protein complex that shows significant conservation of sequence between vertebrate, invertebrate, plant, and fungal subunits. The transferase is specific for the sequence N–X–T/S [Asn–X–Ser(Thr)], in which X can be any residue except prolyl or aspartyl. Proline at the +3 position also prevents transfer. The N-linked carbohydrates have also been shown to be attached to N-X-C sequences for human protein C and von Willebrand factor. It has been proposed that the seryl (or threonyl) hydroxyl in the triplet participates in catalysis by activation of the amide nitrogen of asparagine for reaction with the donor substrate, oligosaccharide-dolichol. Statistical analysis of glycosylated consensus sites in glycoproteins showed that N-X-T sites were glycosylated about twice as often as N-X-S sequons. Experiments using peptides showed a 40-fold preference for N-X-T over sequences containing serine. In addition, glycosylation sites located toward the COOH-terminal end of proteins were less frequently glycosylated. Clearly, additional mechanisms must play a role in N-glycosylation.

The secondary or tertiary structure can prevent glycosylation of sites within those regions, and disruption of tertiary structure allows glycosylation to occur at new sites in certain glycoproteins. Availability of active transferase and of lipid-linked oligosaccharide also affects the rate of glycosylation.

(2) ***Processing*** rapidly follows transfer of the oligosaccharide to the polypeptide. Stepwise removal of the glucose residues and of some of the $\alpha(1 \rightarrow 2)$-linked mannose residues drastically alters the structure. The glycosidases are located in the rough endoplasmic reticulum and in the cis and medial portions of the Golgi apparatus (Figure 16-7). Processing begins in the rough endoplasmic reticulum with removal of three glucose residues and one mannose residue. The partially processed, high-mannose glycoprotein is then transferred to the cis Golgi region closest to the endoplasmic reticulum, probably by budding of vesicles from the rough endoplasmic reticulum that move to and fuse with the Golgi membrane. In the cis Golgi cisternae, high-mannose glycoproteins destined for inclusion in lysosomes are phosphorylated, while other glycoproteins are further processed by removal of three more mannose residues and transferred to the medial Golgi region. This transfer is probably mediated also by budding of vesicles from the cis Golgi membrane.

(3) ***Elongation*** begins in the medial Golgi by addition of GlcNAc to the 3′-core mannose, followed by removal of the two mannose residues on the 6′-mannose branch. Further elongation to form a complex-type oligosaccharide proceeds in the medial and trans Golgi regions with addition of GlcNAc, Gal, Fuc, and sialic acid. Transfer from medial to trans Golgi and transport of completed glycoproteins to elsewhere in the cell probably, occur by vesicular transport. Many glycosyltransferases that catalyze elongation reactions require an acceptor having the correct terminal sugar and correct sugars at specific locations in the oligosaccharide. For example, the enzyme that transfers fucose from GDP-fucose to the innermost GlcNAc of the core oligosaccharide cannot function unless a GlcNAc has first been attached to the 3′-mannose at the outermost end of the structure.

The final oligosaccharide product of this elongation process is significantly influenced by the types and amounts of glycosyltransferases present in the medial and trans Golgi. Activation of *Ras* genes, associated with the metastatic spread of tumors, have also been shown to increase the expression of GlcNAcTransferase V. This enzyme may play a role in cell motility and therefore its correlation with tumor metastasis is significant, since motility is an important factor in the initiation of metastasis.

Phosphorylation of Oligosaccharide Chains on Lysosomal Enzymes

Some Asn-linked glycoproteins (e.g., the acid hydrolases destined for incorporation into lysosomes) undergo phosphorylation of oligosaccharide chains. The phosphate groups are attached to two penultimate mannose residues on incompletely processed high-mannose side chains. The first step is transfer of GlcNAc-l-phosphate to the mannose acceptor sites, catalyzed by UDP-GlcNAc:lysosomal enzyme GlcNAc-l-phosphotransferase, to form a phosphodiester linkage. This phosphotransferase is specific for the Man $\alpha(1 \rightarrow 2)$ Man linkage characteristic of high-mannose structure and for an amino acid sequence in the lysosomal protein. The GlcNAc is then removed by an α-GlcNAc-phosphodiesterase, leaving a phosphomannose residue (Figure 16-9). Several potential phosphorylation sites exist on a high-mannose oligosaccharide and multiple isomers can occur.

Phosphate group transfer is important because the phosphomannosyl residues on these enzymes act as recognition markers for their proper routing to lysosomes. Two human lysosomal storage disorders (Chapter 11) are caused by a genetically determined reduction in the activity of GlcNAc-phosphotransferase. In inclusion cell disease (***I-cell disease,*** mucolipidosis II), the activity is completely absent, whereas in pseudo-Hurler's polydystrophy (mucolipidosis III), the activity is severely reduced. Both diseases are characterized by high concentrations of certain lysosomal enzymes in the plasma and by greatly reduced activities of these and other acid hydrolases in lysosomes.

The plasma lysosomal enzymes in these disorders probably represent a particularly stable subset of misdirected enzymes. Some have a higher molecular weight and more negative charge than the corresponding enzymes in normal lysosomes. The increased size may occur because the enzymes have not undergone the limited proteolysis that normally occurs within the lysosomes. The increased negative charge is due to the presence of sialic acid residues added during further processing and elongation of the nonphosphorylated proteins. Acid phosphatase and β-glucosidase occur in normal amounts in the lysosomes, suggesting that a different mechanism directs them to the lysosomes.

Inhibitors of Glycoprotein Biosynthesis

The oligosaccharide chains on some glycoproteins may be needed for receptor recognition, antigenicity, intracellular transport and secretion, protection against proteolytic digestion, and stability at extremes of temperature and pH. In other glycoproteins, however, absence of the carbohydrate

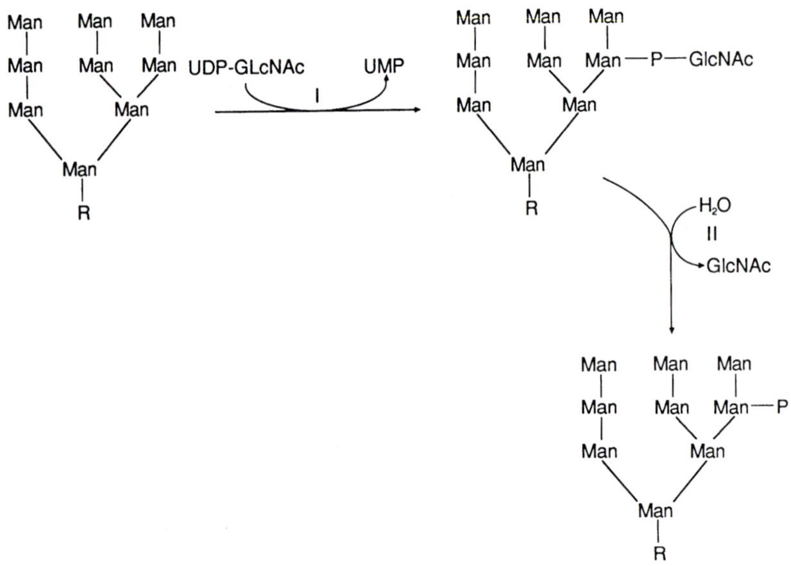

FIGURE 16-9
Proposed mechanism for the phosphorylation of high-mannose oligosaccharides on glycoproteins (primarily acid hydrolases) being directed to lysosomes. The enzymes catalyzing the reactions are I, UDP-N-acetylglucosamine: lysosomal enzyme N-acetylglucosaminyl-1-phosphotransferase and II, α-N-acetylglucosaminylphosphodiesterase. R = GlcNAcβ1 → 4GlcNAc → Asn. *Additional mannose residues that are phosphorylated in other isomers.

has no discernible effect on biological function. Information regarding the function of carbohydrate side chains has been obtained either after their removal from glycoproteins by glycosidases or by inhibition of some step in the glycosylation pathway.

Most inhibitors affect Asn-linked glycoprotein synthesis. Some have antiviral activity (tunicamycin, deoxy sugars) or inhibit peptidoglycan synthesis in bacteria (tunicamycin, streptovirudin); others are isolated from plants that are toxic to humans or animals (swainsonine, castanospermine). The range of substances includes inhibitors of dolichol phosphate synthesis, of initiation or polymerization of the lipid-linked oligosaccharide, of processing of the oligosaccharide after transfer to glycoprotein, and of elongation of the processed side chain.

Inhibitors of dolichol phosphate formation may interfere with the synthesis of polyprenyl diphosphate or prevent recycling or phosphorylation of polyisoprenols. Dolichols are synthesized from precursors of cholesterol synthesis (Chapter 19). A regulatory enzyme in the precursor synthesis is β-hydroxy-β-methylglutaryl-CoA (HMG-CoA) reductase. Inhibition of HMG-CoA reductase by 25-hydroxycholesterol and drugs such as mevinolin prevents synthesis of dolichol and inhibits glycosylation in some systems. Cholesterol synthesis is also inhibited, which is disruptive to cell function. Bacitracin inhibits *in vitro* dephosphorylation of dolichol diphosphate, preventing recycling of dolichol phosphate. Glycosylation stops in the absence of dolichol phosphate. *In vivo,* bacitracin does not affect glycoprotein biosynthesis, presumably because it does not enter cells.

The transfer of GlcNAc-P from UDP-GlcNAc to dolichol phosphate is inhibited by tunicamycin, streptovirudin, and amphomycin. Tunicamycin may be a competitive inhibitor of GlcNAc-1-phosphate transferases and inhibits the initial step in peptidoglycan synthesis and in the synthesis of the oligosaccharide units linking teichoic acids to peptidoglycans. Initiation is also inhibited by the antibiotics diumycin and (at high concentrations) showdomycin. Diumycin has a greater inhibitory effect on the synthesis of Man-P-Dol, whereas showdomycin blocks formation of Glc-P-Dol. Both prevent polymerization of the lipid-linked oligosaccharide.

Sugar analogues and deoxy sugar inhibitors include 2-deoxy-D-glucose, 2-deoxy-2-fluoro-D-glucose, 2-deoxy-2-amino-D-glucose, and 2-deoxy-2-fluoro-D-mannose, no two of which have the same effect on glycosylation. Many are metabolized to UDP- or GDP-sugars and may be incorporated into oligosaccharides, making addition of further sugars impossible, thereby causing synthesis of truncated oligosaccharides. Others may inhibit one or more glycosyltransferase, and more than one step is often affected.

Another class of inhibitors that affect processing or "trimming" of the glucosylated, high-mannose side chains have in common the ability to inhibit one or more of the specific processing glycosidases. Deoxynojirimycin and castanospermine inhibit glucosidase I, which catalyzes removal of the outermost glucose residue. Deoxynojirimycin is also a competitive inhibitor of sucrase in the intestinal brush border, while castanospermine causes gastrointestinal irritation when consumed as a constituent of the Australian legume *Castanospermum australe*. Bromoconduritol, a derivate of myo-inositol, is an irreversible inhibitor of glucosidase II, which acts after glucosidase I to remove both inner glucose residues. However, it interferes with removal of only the innermost sugar, the second step catalyzed by this enzyme. Swainsonine, a mannosidase inhibitor, is a toxic constituent of locoweed; it inhibits lysosomal α-mannosidase, causing accumulation within lysosomes of mannose-rich oligosaccharides. In oligosaccharide processing, swainsonine inhibits α-mannosidase II, which acts late in the trimming process and leads to side chains with hybrid (partly high-mannose, partly complex) structures.

Finally, drugs that disrupt the integrity of the Golgi apparatus (monensin, brefeldin A, bafilomycin, okadaic acid, nocodazole) will alter glycoconjugate synthesis. The molecular target of some of these drugs remains unknown; however, some of these drugs disrupt vesicle transport (brefeldin A) and others mimic the dispersion of Golgi that takes place during mitosis (okadaic acid).

O-Glycan Ser(Thr)-Linked Oligosaccharides

Oligosaccharide chains linked by O-glycosidic bonds to seryl or threonyl residues show more structural variability than do the asparagine-linked oligosaccharides; the serine (threonine)-linked oligosaccharides range in size from 1 to 20 or more monosaccharide residues. The residues are added one at a time, directly on the protein, rather than preassembled on a lipid carrier. Assembly is not random, however, and the glycosyltransferases involved have acceptor specificities that render certain structures preferable.

The O-linked oligosaccharides are associated with mucins (Chapter 10), glycoproteins in which carbohydrate accounts for at least half the molecular weight and which are produced by mucus-secreting organs and glands (e.g., submaxillary glands and the respiratory, urogenital, and intestinal tracts). Antifreeze glycoproteins (Chapter 10), which are O-linked glycoproteins, in the circulation of certain Arctic fish allow the fish to survive freezing temperatures. Elongation depends on the glycosyltransferases available and on their acceptor specificities.

Polypeptide α-GalNAc transferase initiates O-glycan synthesis by attachment of GalNAc to Ser/Thr resides of proteins. There does not appear to be a specific amino acid sequence signal for GalNAc transfer; however, Pro is often

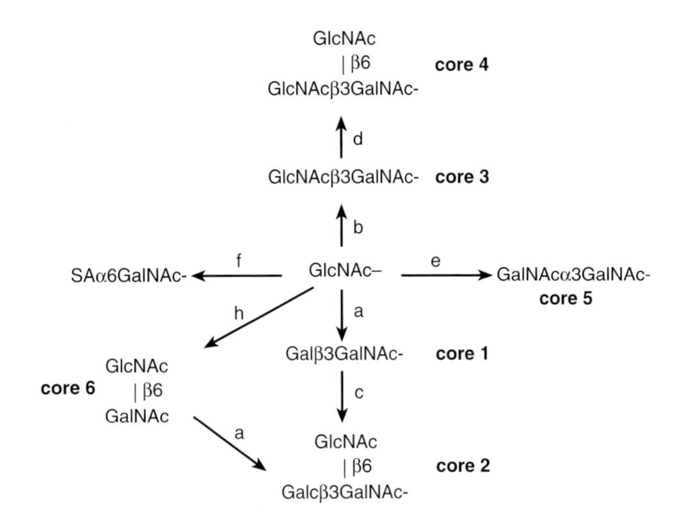

FIGURE 16-10
Synthesis of O-glycan cores: (a) core 1 β3-Gal-T; (b) core 3 β3-GlcNAc-T; (c) core 2 β6-GlcNAc-T; (d) core 4 β6-GlcNAc-T; (e) core 5 α3-GalNAc-T; (f) α6-SA-T I blocks further O-glycan synthesis; (h) possibly a human-specific pathway. [Reproduced with permission from I. Brockhausen, Clinical aspects of glycoprotein biosynthesis. *Crit. Rev. Clin. Lab. Sci.* **30**:81 (1993).]

found near glycosylated Ser/Thr and may provide a three-dimensional structure that favorably exposes Ser/Thr for glycosylation. Initiation and most O-glycosylation takes place in the cis-Golgi. Six O-glycan core types have been described and their synthetic pathways are shown in Figure 16-10. The elongation of core O-glycans depends upon the types, amounts, and localization of glycosyltransferases and proceeds in a fashion similar to that of N-glycans. The elongation reactions of core 1 O-glycan are shown in Figure 16-11; because of the specificity of transferases for their substrates only certain pathways are possible. Sialylation and sometimes fucosylation of substrates may act as a STOP signal and terminate further branch elongation of oligosaccharide chains. The products of α2-fucosyltransferase (g in Figure 6-11) can be glycosylated either by A-dependent α3-GalNAc-tranferase or B-dependent α3-Gal-transferase, or by a combination of both to give rise to the blood group antigens A, B, or AB, respectively.

The A and B transferases have been mapped to human chromosome 9q34 and only differ by four amino acids. In the absence of both transferases, the O (or H) blood group will contain the α2 fucose. Individuals missing the blood group H α2-fucosyltransferase are said to contain the "Bombay" type blood group and are seriously at risk from transfusion reactions from any of the common blood types. Terminal sequences containing blood group antigens also occur on other glycoprotein and glycolipid oligosaccharides. Sulfate may also occur

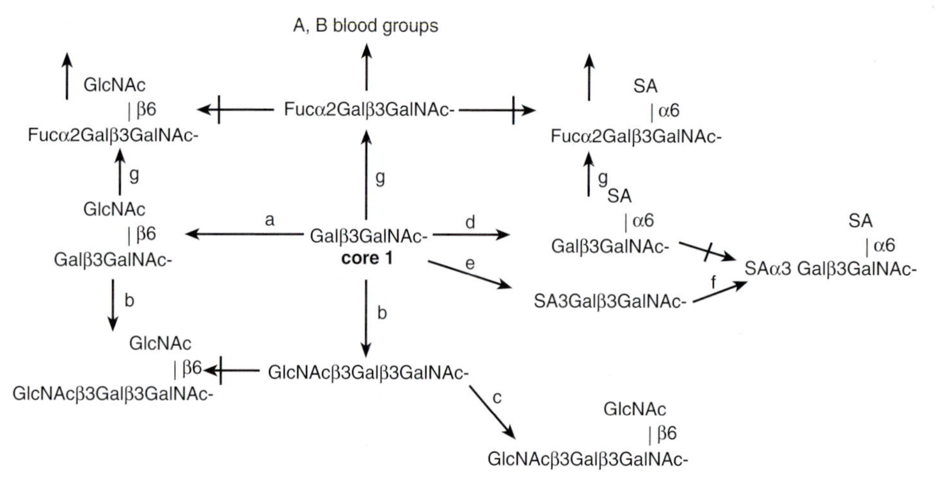

FIGURE 16-11

Processing reactions of O-glycan core 1: (a) core 2 β6-GlcNAc-T; (b) elongation β3-GlcNAc-T; (c) I β6-GlcNAc-T; (d) α6-SA-T I; (e) α3-SA-T; (f) α6-SA-T II; (g) α2-Fuc-T. [Reproduced with permission from I. Brockhausen, Clinical aspects of glycoprotein biosynthesis. *Crit. Rev. Clin. Lab. Sci.* **30**:80 (1993).]

on O-glycans on Gal, GlcNAc, or GalNAc at various positions.

Another interesting O-glycosylation type which may be related to cellular regulation is the attachment of single O-GlcNAc to specific Ser or Thr on many cytoplasmic and nuclear proteins. The O-glycosylating GlcNAcT is a unique, highly conserved 110 kDA catalytic subunit that is trimerized when active. In contrast to other glycosyltransferases involved in synthesis of O-glycans, this GlcNAcT has cytosolic and nuclear localization. All O-GlcNAc modified proteins to date also occur as phosphorylated proteins, and protein phosphorylation and O-GlcNAc modification appear to be reciprocal and their interrelationship complex. There are important implications that O-GlcNAc glycosylation may play a role in disease processes. The alpha toxin of *Clostridium novyi* (the gangrene-causing bacteria) is an O-GlcNAcT that is toxic due to its glycosylation of Rho subfamily proteins. In Alzheimer's disease, Tau microtubule-associated protein is hyperphosphorylated, whereas in unaffected brain, Tau is repeatedly O-GlcNAc modified.

Biosynthesis of GPI-Anchored Proteins

The synthesis of glycosylphosphatidylinositol (GPI) anchors of proteins, including Thy-1 of neurons and lymphocytes, is primarily carried out within the lumen of the rough endoplasmic reticulum. The anchor consists of a phosphatidylinositol-glucosamine-mannose$_3$-ethanolamine phosphate core that is linked to the α-carboxyl of the carboxyterminal amino acid of a protein (Figure 16-12). This addition takes place within minutes of

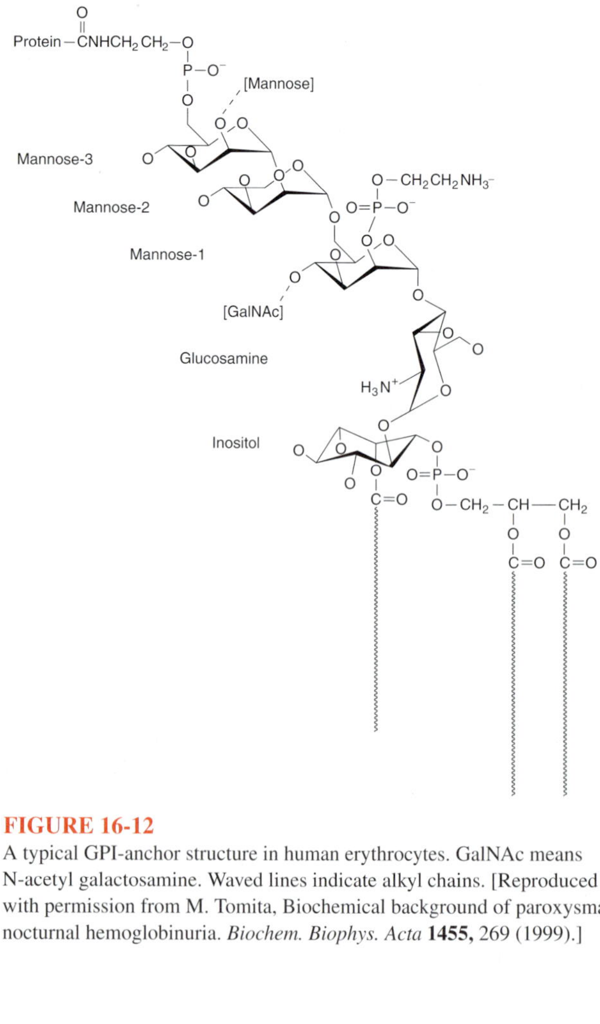

FIGURE 16-12

A typical GPI-anchor structure in human erythrocytes. GalNAc means N-acetyl galactosamine. Waved lines indicate alkyl chains. [Reproduced with permission from M. Tomita, Biochemical background of paroxysmal nocturnal hemoglobinuria. *Biochem. Biophys. Acta* **1455,** 269 (1999).]

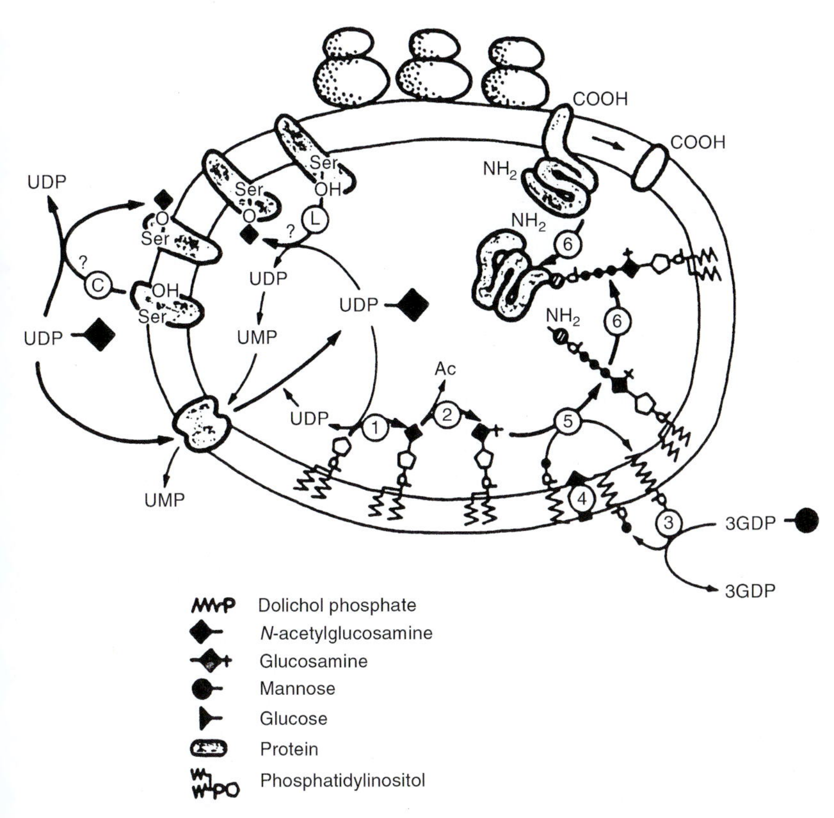

FIGURE 16-13
Topography of glycosylation in the rough endoplasmic reticulum. Addition of O-linked N-acetylglucosamine to protein and phosphatidylinositol anchoring. Ac, Acetyl; Ser, serine; L, lumenal; C, cytosolic. [Reproduced with permission from C. Abeijon and C. B. Hirschberg, Topography of glycosylation reactions in the endoplasmic reticulum. *Trends Biol. Sci.* **17**:34 (1992).]

peptide chain termination and occurs just after or concomitant with cleavage of a C-terminal signal anchor sequence from the protein. Studies using yeast and Thy-1-negative T lymphocytes as well as intact ER vesicles suggest the following pathway: synthesis of the anchor oligosaccharide portion with GlcNAc transfer to inositol, deacetylation of GlcNAc, mannose transfer via dolichol-P-mannose after GDP-mannose transfer to dolichol on the cytoplasmic side of the ER, and translocation of the dolichol-P-mannose to the lumen (Figure 16-13). Finally, the terminal mannose of the core is phosphorylated with ethanolamine phosphate which serves as the linker to the protein. The core GPI may later be modified by attachment of fatty acids, mannose, N-acetylgalactosamine, and additional mannose in the ER or galactose in the Golgi. GPI-anchored proteins are absent on both red and white cells in the human disease ***paroxysmal nocturnal hemoglobinuria*** (PNH). In some cells the anchor is resistant to PI-phospholipase C; however, cells with a deacylase activity that removes an acyl chain from inositol are sensitive to release of GPI-anchored proteins (Figure 16-12). A mutation in the gene coding for the first enzyme in the pathway for GPI anchor synthesis results in PNH, which is an anemia caused by complement-mediated hemolysis. Two complement regulatory proteins that protect host cells from complement do not get properly GPI-anchored on the surface of PNH erythrocytes causing them to be abnormally sensitive to complement lysis. Patients with PNH, in addition to having hemolytic anemia and the resulting hemoglobinuria, also have a reduced ability to combat infections and a tendency for thrombosis. The excessive accumulation of iron by overloaded kidneys due to the hemolysis has caused some patients to set off airport metal detectors. The gene, PIG-A, the cause of PNH, is located on the X chromosome. Thus, a single PIG-A mutation can account for the loss of GPI anchor synthesis even in females, depending upon which X chromosome is inactivated. While many cells can survive without their GPI-anchored proteins, it is the specific loss of specialized, complement-protective proteins by erythrocytes that is the problem in PNH. While only a small percentage of proteins in humans is anchored to membranes using GPI anchors, in some species, such as protozoa, many or all of the membrane-bound proteins utilize GPI anchors. An understanding of the differences in the biochemistry of GPI anchors may lead to novel therapeutic agents to fight infections and possibly other diseases.

Biosynthesis of Glycosphingolipids

The glycosphingolipids (GSLs, glycolipids) include neutral glycosphingolipids, gangliosides, and sulfatoglycosphingolipids. The terms "cerebroside" and "ganglioside" reflect their abundance in gray and white matter of the central nervous system. Glycolipids are widely distributed in the body. Gangliosides G_{M1} and GT_{1b} have been found to enhance the effect of nerve growth factor (NGF) on neurite outgrowth. Such studies may lead to effective treatments after spinal cord injury. Gangliosides also are known to modulate protein kinase and NGF receptor function. All glycosphingolipids contain an oligosaccharide chain of one or more residues of galactose, glucose, N-acetylglucosamine, N-acetylgalactosamine, or fucose. Gangliosides also have one or more residues of sialic acid, whereas the sulfatoglycosphingolipids contain sulfate groups in ester linkage on some residues. Structures of the glycosphingolipids are presented in Chapter 10.

GSLs are classified by the sequence and type of the first four sugars attached to ceramide (Table 16-1). The prefixes used to indicate the core structure are listed in Table 16-1, along with the tissue in which the different types are predominantly found. This system reflects the synthetic history of a structure and is useful for understanding biosynthetic pathways. Structures of the three most abundant glycolipids and the tissues in which they occur are shown in Table 16-1.

GSLs of the ganglio- series are designated by the capital letter G and the sialic acid content is designated by a subscripted capital letter (A = asialo; M = monosialo; D = disialo; etc.). The length of the neutral sugar chain is designated by a subscripted number using the formula $(5 - n)$ where n equals the number of neutral sugars in the ganglioside. Further designations of α or β relate to the positioning of the sialic acid. Using this nomenclature, G_{M1} is a ganglioside containing one sialic acid and four neutral sugars (Figure 16-14). The initial glucosylation of ceramide occurs on the cytoplasmic side of the Golgi while subsequent stepwise sugar transfers occur within the Golgi lumen and may result in 300–400 possible GSL types in different cells. Subcellular fractionation and drug studies that disrupt Golgi function have suggested that the distribution of glycosyltransferases synthesizing GSLs may differ from enzymes involved in N-linked glycoprotein synthesis (Figure 16-15).

During the progression of cancer, many changes in glycosyltransferases result in alteration of oligosaccharides

TABLE 16-1
Names, Structures, and Major Tissue Distributions of Oligosaccharides Found in Glycolipids

Name/Prefix*	Structure†	Distribution‡
Lacto	Galβ1→3GlcNAcβ1→3Galβ1→4Glcβ1→Cer	Central nervous system
Lactoneo	Galβ1→4GlcNAcβ1→3Galβ1→4Glcβ1→Cer	Central nervous system, blood group antigens
Muco	Galβ1→3Galβ1→4Galβ1→4Glcβ1→Cer	
Gala (GL_{2b})	(GalNAc?1→3Gal?1)→4Galα1→4Galβ1→Cer	Central nervous system, kidney
Globo (GL_{4a})	GalNAc?1→3Galα1→4Galβ1→4Glcβ1→Cer	Kidney, heart, erythrocytes
Ganglio (G_{M1})	Galβ1→3GalNAcβ1→4Galβ1→4Glcβ1→Cer	Central nervous system
Lactosyl ceramide (GL_{2a})	Galβ1→4Glcβ1→Cer	Major glycolipid in most non-neural tissue
Glucosyl ceramide (GL_{1a})	Glcβ1→Cer	Major neutral glycolipid in central nervous system
Galactosyl ceramide (GL_{1b})	Galβ1→Cer	Major neutral glycolipid in plasma

*The names on the first six lines are the prefixes used to designate the core sugars when naming glycolipids. For all the structures, the names in parentheses are the old designations for the corresponding glycosylceramide. The only exception is GL_{2b}, which stands for Gal-Gal-Cer, not for the Gala tetrasaccharide.

†Although the Gala structure is shown as a tetrasaccharide, no Gala-type glycolipids having structures more complex than Gal-Gal-Cer (galabiosyl ceramide) have been purified from natural sources. Gal = Galactose, Glc = glucose, GalNAc = N-acetylgalactosamine, Cer = ceramide.

‡The tissues listed in this column are those in which relatively large amounts of glycolipid having the indicated structure are found. Many of these compounds are also found in other tissues at lower concentrations.

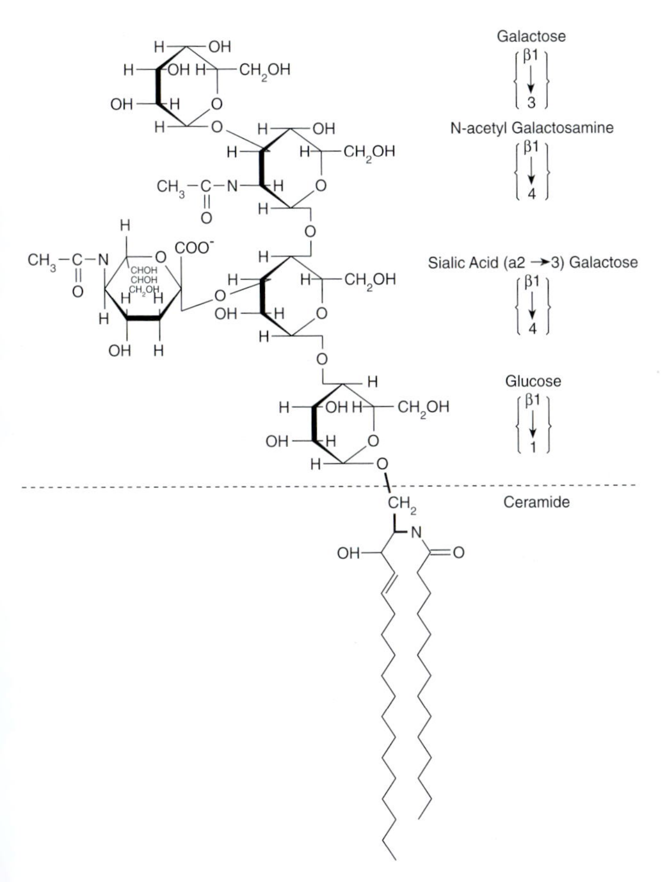

FIGURE 16-14
Structure of ganglioside G_{M1}. Ceramide portion of a representative ganglioside is shown beneath dashed line. Ceramide is synthesized by a condensation of serine (boldface) with palmitoyl-CoA followed by an N-acylation. Ceramide is then glycosylated stepwise to produce a glycosphingolipid, such as the ganglioside G_{M1} shown here. [Reproduced with permission from C. B. Zeller and R. B. Marchase, Gangliosides as modulators of cell function. *Am. J. Physiol.* **262**:C1342 (1992).]

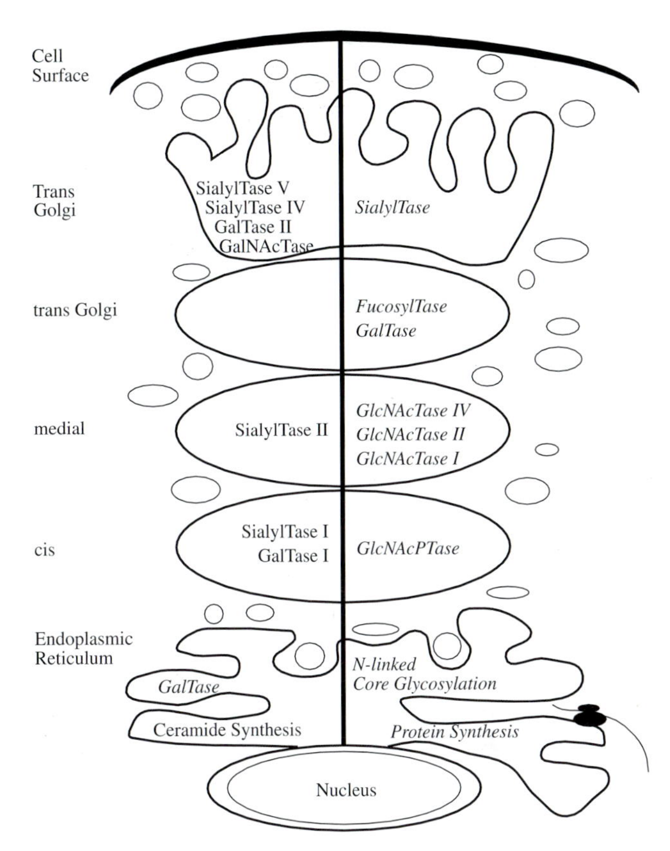

FIGURE 16-15
Comparison of topography of glycosphingolipid and N-linked glycoprotein glycosylation reactions. Approximate topographies of major glycosylation reactions important to ganglioside synthesis are depicted at left and those pertaining to N-linked glycosylation of glycoproteins at right. Distributions of glycosphingolipid glycosyltransferases are much less well defined than are their glycoprotein counterparts. [Reproduced with permission from C. B. Zeller and R. B. Marchase, Gangliosides as modulators of cell function. *Am. J. Physiol.* **262**:C1344 (1992).]

on glycolipids and glycoproteins. In particular, some cancer cells end up with oligosaccharide structures identical to those used by normal leucocytes to attach to blood vessel receptors (selectins) for invasion of tissues; unfortunately, in cancer cells this change promotes metastasis. The expression of new terminal carbohydrate structures by cancer cells can be prognostic as to disease outcome. For example, patients with either *colorectal (Dukes C grade)* or *breast carcinoma* that did not express sialosyl-GalNAc-(sialosyl-Tn antigen) terminal disaccharides had a much better survival rate than patients positive for this antigen. Many of these carbohydrate structures can occur as terminal sequences on either glycolipids or glycoproteins. It is interesting that drugs (cyclosporin A, tamoxifen, and verapamil) that reverse multidrug resistance inhibit ceramide glucosylation and, thereby, GSL synthesis.

Glyceride glycolipids have been reported. Monogalactosyldiacylglycerol (MGD) represents approximately 0.5% of the total lipid of adult nervous tissue and occurs in kidneys but not elsewhere. MGD and digalactosyldiacylglycerol (DGD) increase in concentration during myelination. Sulfated forms of diacylglycero- and alkylacylglycerogalactolipid make up about 2–7% of rat brain sulfolipids. Sulfoalkylacylglycerogalactolipid ("seminolipid") is the major sulfolipid of mammalian sperm and testes. An elevated concentration of seminolipid occurs in the primary spermatocyte stage of sperm development and is correlated with a rise in galactolipid sulfotransferase activity.

Breakdown of GSL occurs in the lysosomes, and inherited deficiencies of lysosomal hydrolases or of sphingolipid activator proteins (SAPs, saposins) can give rise to sphingolipid storage diseases. The SAPs are necessary to elevate the shorter oligosaccharide chains out of the membrane to provide access to exoglycosidase digestion. The accumulation of undegraded GSLs in cells, especially brain cells, can cause characteristic neurological symptoms. ***Tay-Sachs, Gaucher's*** and ***Hurler's diseases*** are all human lysosomal storage diseases. Studies using the drug N-butyldeoxynojirimycin to block GSL biosynthesis have demonstrated that the lysosomal storage disease, Tay-Sachs, can be prevented in a mouse model of the disease.

GSLs, because they are shed into the CSF, may be potential markers for brain pathology. In addition, shed GSLs or bacterial GSL structural mimics may also act as antigens in peripheral neuropathies like ***Guillain-Barré syndrome*** and ***multifocal motor neuropathy.***

Biosynthesis of Glycosaminoglycans

Glycosaminoglycans (GAGs; mucopolysaccharides) are long, highly negatively charged, unbranched polymers of repeating disaccharide units. They are classified according to the hexosamine units that alternate with uronic acid in the polymer; heparin, heparin sulfate (HS), and hyaluronic acid (HA) are glucosaminoglycans, whereas chondroitin sulfate (CS) and dermatan sulfate (DS) are galactosaminoglycans (Figure 16-16). The N-sulfation of glucosamine and O-sulfation at various positions on each sugar unit coupled with the hexuronic acid gives these polymers some of the highest negative charge density of any biological molecules. GAGs are present in most cells and tissues but are most often associated with connective tissue and

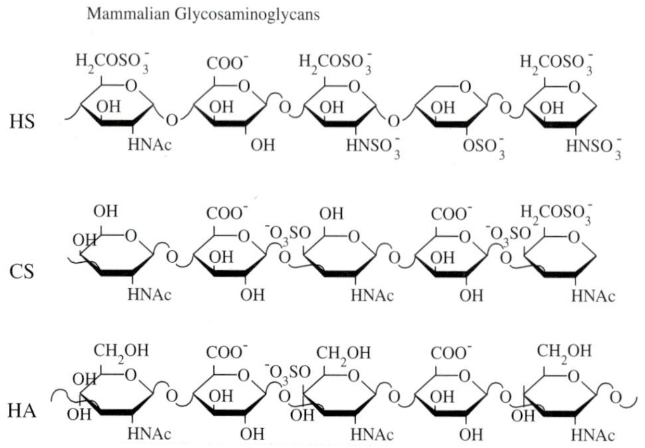

Mammalian Glycosaminoglycans

FIGURE 16-16
Structures of mammalian glycosaminoglycans: the anti-thrombin-binding pentasaccharide in heparin/HS, a CS polysaccharide, and a mammalian HA. [Reproduced with permission from K. Lidholt, Biosynthesis of glycosaminoglycans in mammalian cells and in bacteria. *Biochem. Soc. Trans.* **25**:866 (1997).]

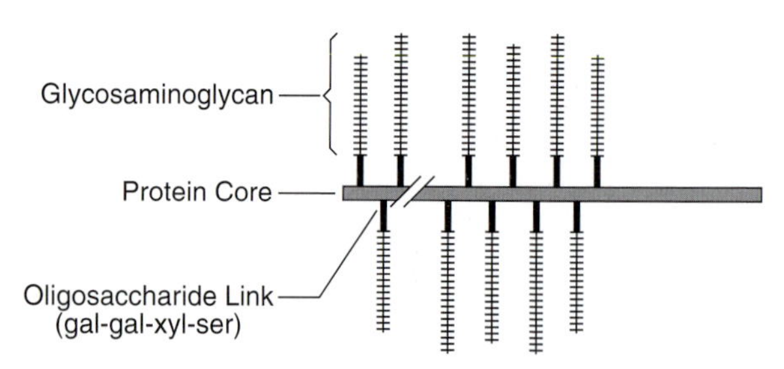

FIGURE 16-17
Structure of a proteoglycan. [Reproduced with permission from J. E. Silbert and G. Sugumaran, Intracellular membranes in the synthesis, transport, and metabolism of proteoglycans. *Biochim. Biophys. Acta* **1241**:372 (1995).]

extracellular matrix. Most GAGS are a part of a larger molecule called a proteoglycan. GAGs are attached to the core protein of proteoglycans through an oligosaccharide link region of gal-gal-xyl-ser (Figure 16-17). Many of these proteoglycans and glycosaminoglycans are found in the extracellular matrix including hyaluronic acid, proteochondroitin/dermatan sulfate (aggrecan, versican, decorin, biglycan), and proteoheparan sulfate. The proteoheparan sulfate perlecan is a major constituent of basement membranes. Proteoglycans attached to the cell surface may have heparan sulfate GAGs (glypican, cerebroglycan) or both chondroitin sulfate or dermatan sulfate and heparan sulfate GAGs on the same core protein (syndecan, β-glycan). Cell surface proteoglycans can either be integral membrane proteins or attached via a GPI anchor to the plasma membrane.

In addition to the role of proteoglycan and GAGs as highly solvated, structural molecules, they have now been identified as important mediators of receptor function and may also have a role in normal morphology and malignancies. In particular, a family of transmembrane cell surface heparan sulfate proteoglycans, the syndecans, have been noted for their role in cell surface interactions. The four mammalian syndecans have different tissue distribution and their core protein expression is developmentally regulated.

Syndecans can interact with a number of extracellular molecules (fibronectin, collagens I, III and V, tenascin, thrombospondin; and antithrombin III) and can act as coreceptors for high-affinity receptors of fibroblast growth factor type 2 (FGF2). All four syndecans have a cleavage site within their peptide sequence near the exoplasmic site of their insertion into the plasma membrane. This provides another means to rapidly down-regulate surface expression of GAGs. Mutant cells that lack proteoheparan sulfate do not respond to FGF2. Studies also show that bacterial gonococcal attachment to human mucosal cells is directed at cell surface heparan sulfate syndecan-like proteoglycans. Chemical modification or enzymatic removal of proteoglycan from the cell surface prevented gonococcal attachment and subsequent invasion. An understanding of the GAG fine structure involved in such mechanisms may be useful medically to prevent such infections.

The sugar residues in these polymers are often modified by N-deacetylation and N- or O-sulfation. The sulfotransferases use phosphoadenosine-phosphosulfate (PAPS; "active sulfate"; see Chapter 17) as the source of sulfate groups. A modification unique to glycosaminoglycans is epimerization of glucuronic acid residues to iduronic acid, which occurs before polysaccharide polymerization. Because modification of the glycosaminoglycan chains is often incomplete, extensive microheterogeneity in the oligosaccharides is not related to the glycosyltransferase reactions.

The chondroitin sulfates, heparin, and heparan sulfate are composed of

1. A core protein;
2. A carbohydrate-core protein link region that contains xylose, galactose, and glucuronic acid;
3. A repeating disaccharide that undergoes extensive postsynthetic modification.

Based on the "one enzyme–one linkage" concept, at least six glycosyltransferases and four or more modifying enzymes are necessary for their biosynthesis. Except for keratin sulfate, which has both N- and O-linked oligosaccharides, the biosynthesis of most proteoglycans begins with synthesis of the core protein followed by xylosylation of specific Ser amino acids in the polypeptide. The consensus sequence for xylose attachment is Ser-Gly in regions predominantly containing acidic amino acids and devoid of basic amino acids (using the example of syndecan synthesis). Xylosylation probably begins in the endoplasmic reticulum and continues in the Golgi. The oligosaccharide linker is completed by the sequential attachment of two Gal and one GlcUA residues. At this point the attachment of either GalNAc or GlcNAc to this linkage oligosaccharide will direct the synthesis toward either CS/DS or heparin/HS type GAGs, respectively. The hexosaminyltransferases that attach GalNAc and GlcNAc to the link oligosaccharide are not the same transferases that carry out the subsequent polymerization reaction. The polymerization reaction alternates the addition of more hexosamine residues and GlcUA or Gal to form large heteropolymer GAG chains. These chains are then sequentially modified by variable N-deacetylation/N-sulfation, and/or O-sulfation, and possible epimerization of GlcUA to IdUA. Sulfation is catalyzed by sulfotransferases that use PAPS to modify sugars at 2′-, 3′-, 4′- or 6′-positions depending upon the sugars. These modifications result in the fine structural differences in the GAGs that are important for their function. Heparin and heparan sulfate use the enzymes UDP-GlcNAc: GlcUA oligosaccharide GlcNActransferase and UDP-GlcUA:GlcNAc oligosaccharide glucuronyltransferase, in that order. For chondroitin sulfates, the GlcNAc transferase is replaced by GalNAc transferase. Epimerization of glucuronic acid and O-sulfation of C_4 and C_6 of galactosamine and C_2 iduronic acid complete the synthesis of chondroitin sulfates. There is usually one sulfate per disaccharide unit, but the particular uronic acid present and the position of the sulfate are variable and lead to microheterogeneity. Regulation of the degree and position of sulfation may be important for function. Indeed, two diseases of proteoglycan synthesis, the ***diastrophic dystrophy type of osteochondrodysplasia*** and ***macular corneal dystrophy type I,*** both result from the absence of sulfate groups on proteoglycans.

In heparan sulfate and heparin, deacetylation of the GlcNAc residues in the repeating disaccharides is a prerequisite for further modification, and the activity of deacetylase may regulate the extent of modification. For example, a specific pentasaccharide sequence in heparin binds to and activates antithrombin III (Chapter 36). This sequence, containing at least four sulfate groups, is required for binding. Since N-deacetylation must precede addition of these groups, the deacetylase partly regulates the anticoagulant activity of heparin. Heparin also has some sulfate groups at C_3 of the glucosamine residues within the antithrombin III-binding sequence.

Other modification reactions include N-sulfation of the glucosamine residue, formation of iduronate by epimerization of glucuronate residues, and O-sulfation of C_6 on glucosamine and C_2 on iduronic acid. Epimerization may be linked to O-sulfation of iduronic acid, since experimental elimination of PAPS blocks the epimerization reaction. O-sulfation may stabilize the iduronic configuration around C_5, favoring its formation.

Heparin, but not heparan sulfate, is a good anticoagulant, probably because of higher charge density. Heparan sulfate is found in virtually all tissues, but heparin appears to be produced only by mast cells. Their different distributions may reflect differences in activity of the deacetylase.

Hyaluronic acid (HA) is not sulfated and is the only glycosaminoglycan that is not part of a proteoglycan. Consequently, chain initiation does not require a core protein. HA may be polymerized at the plasma membrane by sequential addition of GlcNAc and GlcUA from the corresponding UDP-sugar to the reducing ends of the growing polysaccharide. HA is a major constituent of the vitreous humor and synovial fluid of the eye.

Keratan sulfate occurs in proteoglycans as N- and O-linked polysaccharides. In the peptidoglycan of corneal stroma, one to three chains are attached to the core protein by N-glycosidic bonds between N-acetylglucosaminyl and asparaginyl residues. The carbohydrate links between the keratan sulfate and the protein may resemble the complex type of oligosaccharide in N-linked glycoproteins synthetically and structurally. In cartilage, keratan sulfate chains are attached to the core by O-glycosidic bonds between N-acetylgalactosaminyl and seryl or threonyl residues. Biosynthesis of the carbohydrate linkages in these molecules and assembly of the disaccharide polymers of keratan sulfate are probably similar to synthesis of other glycosaminoglycans.

The turnover of glycosaminoglycans in the lysosomes follows the same pathways as other glycoconjugates, however, N-sulfate must be removed and the deacetylated glycosamine must be reacetylated for digestion by lysosomal β-hexosaminidase. Patients with the lysosomal storage disease, ***Sanfillipo C mucopolysaccharidosis,*** are unable to reacetylate glucosamine and a buildup of undigested glycosaminoglycan occurs.

16.2 Biosynthesis of Peptidoglycans

An understanding of the biochemistry of peptidoglycan (PG; murein) that comprises bacterial cell walls is very important medically since blockage of its synthesis was the first, and continues to be a primary, point of attack in the control of bacterial infection. In addition to inhibition of cell wall synthesis, antimicrobial drug's main mechanisms are interference with nucleic acid synthesis, inhibition of folate metabolism, and binding to ribosomes to disrupt protein synthesis (Table 16-2).

PGs are considered to be primarily responsible for the protective and shape-maintaining properties of walls. They are a biologically unique class of macromolecules because they consist of netlike polymers that are linked together by three different chemical bonds (glycosidic, amide, and peptide). The exact chemical structure of a PG may vary depending on environmental factors. PG, along with endotoxin, is also a primary target for the CD14 molecule, which is expressed on different types of immune cells, especially on monocytes/macrophages. The interaction between these PGs and CD14 leads to production and release of cytokines and other factors that cause local and generalized inflammatory reactions.

PG biosynthesis involves about 30 enzymes and occurs in three stages:

TABLE 16-2
Mechanisms of Action of Antibacterial Agents

Mechanisms	Antibacterial Agents
Inhibition of cell-wall synthesis	β-Lactams Vancomycin
Ribosomal binding	Tetracyclines Macrolides Chloramphenicol Clindamycin Aminoglycosides
Interference with nucleic acid synthesis	Quinolones Rifampin Metronidazole
Inhibition of folic acid pathway	Trimethoprim-sulfamethoxazole

1. N-Acetylmuramic acid (MurNAc) peptide chain synthesis;
2. Synthesis of the GlcNAc-MurNAc repeating units; and
3. Cross-linking of the polymer side chains.

These stages occur in the cytoplasm, at the bacterial cell membrane (probably on the cytoplasmic surface), and outside the membrane in the cell wall, respectively. The wall grows by an inside-to-outside mechanism, and new material compensates for loss of outer wall and provides for expansion for cell growth.

The first stage (Figure 16-18) starts with modification of UDP-GlcNAc to UDP-MurNAc. Phosphoenolpyruvate is enzymatically transferred to UDP-GlcNAc, yielding UDP-GlcNAc-3-enolpyruvate ether. An NADPH-linked reductase then reduces the pyruvyl group to lactyl, forming UDP-MurNAc. The pentapeptide side chain is built up by the sequential transfer of amino acids by specific enzymes in ATP-dependent reactions. The last two amino acids are added as a dipeptide (D-Ala-D-Ala). Both racemization of L-Ala to D-Ala and synthesis of this dipeptide are competitively inhibited by the antibiotic D-cycloserine, a structural analogue of D-alanine. The final product of this stage is UDP-MurNAc-pentapeptide ("Park nucleotide").

The second stage begins with the transfer of MurNAc-phosphate-pentapeptide to membrane-bound undecaprenyl phosphate, accompanied by the release of UMP. Undecaprenyl phosphate ("bactoprenol phosphate") is a 55-carbon isoprenol similar to dolichol phosphate. It is also a carrier in the synthesis of the O-specific antigens of gram-negative, and of the teichoic acids of gram-positive bacterial cell walls. The undecaprenylpyrophosphate-MurNAc-peptide then undergoes modification of the sugar

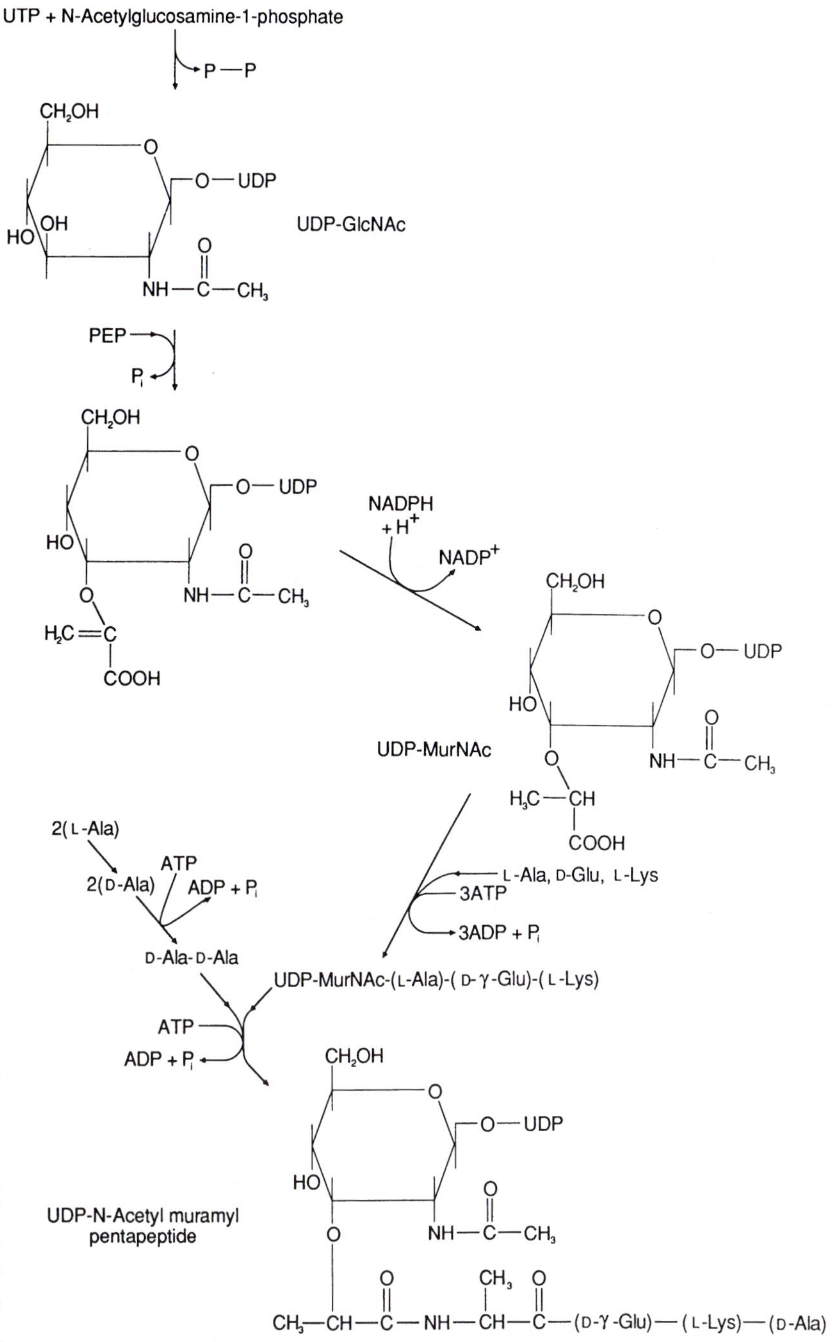

FIGURE 16-18
Formation of UDP-N-acetylmuramic acid pentapeptide: stage 1 of peptidoglycan biosynthesis. The steps in this pathway are discussed in the text. MurNAc = N-acetylmuramic acid, PEP = phosphoenolpyruvate.

nd peptide. N-Acetylglucosamine is enzymatically transferred to MurNAc from UDP-GlcNAc, forming the repeating disaccharide unit found in peptidoglycan. This bond is susceptible to hydrolysis by lysozyme, an enzyme found in many bodily secretions, including tears (Chapter 11). In most bacterial species, the peptide is modified by ATP-dependent amidation of glutamic acid residues and by sequential addition of five glycine residues linked to the α-amino group of lysine. The resulting lipid-linked disaccharide-peptide units are then polymerized to form linear peptidoglycan molecules (Figure 16-19). The high-energy phosphate ester linkage provides the energy for the transfer. Undecaprenyl pyrophosphate is liberated and hydrolyzed to undecaprenyl phosphate for

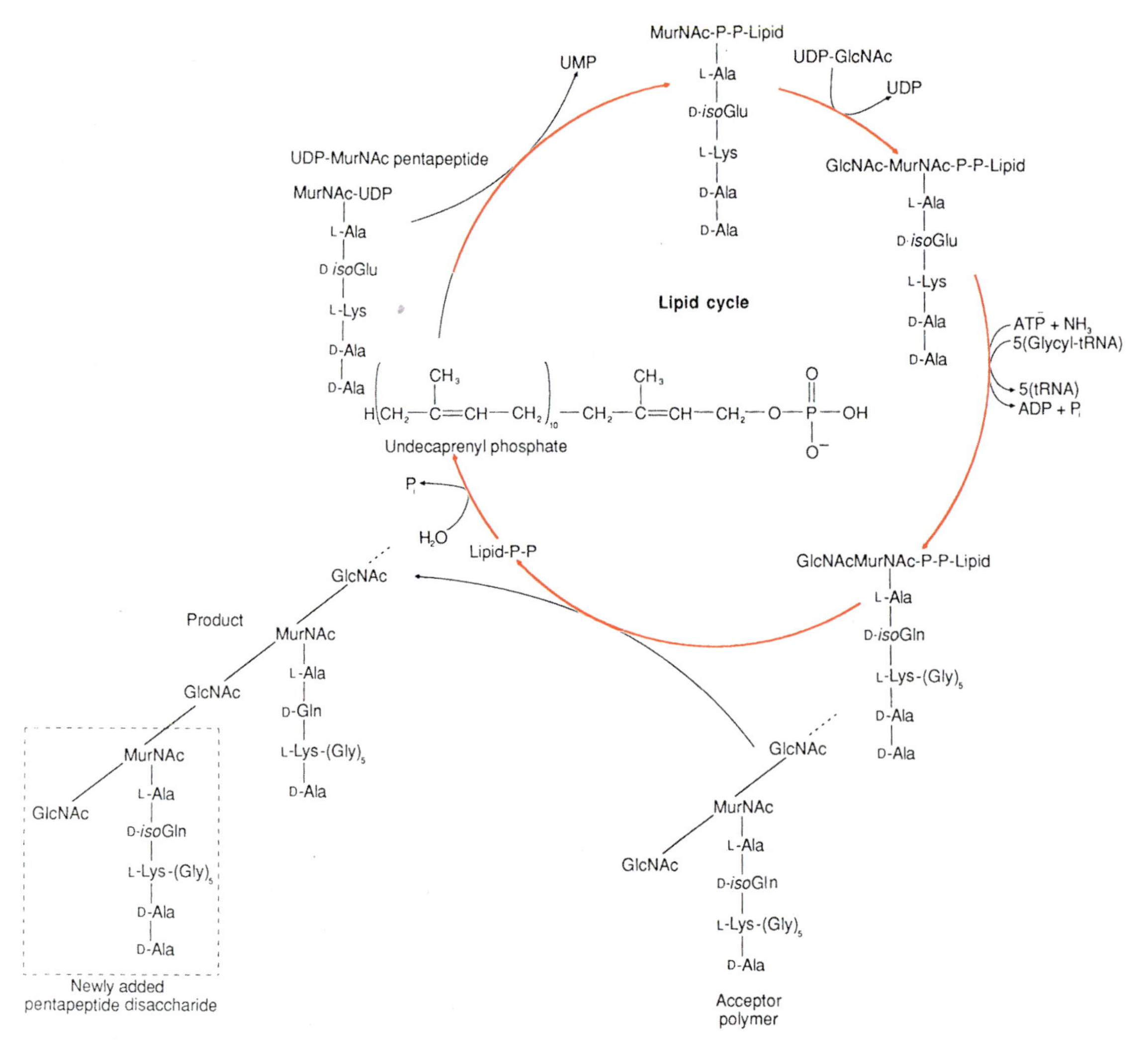

FIGURE 16-19
Polymerization of MurNAc-pentapeptide units to form linear peptidoglycans: stage 2 of peptidoglycan biosynthesis. The lipid in these structures is undecaprenol; lipid-P-P is undecaprenyl pyrophosphate. The use of tRNA as a carrier and activator of glycine is unusual, since ribosomes and mRNA are not involved in the reaction.

reuse. Bacitracin may exert its bactericidal activity partly by inhibiting this hydrolysis. During assembly, the linear peptidoglycan ends up outside the cell membrane at the site of cell wall assembly. The carrier lipid may participate in translocation of the disaccharide-peptide unit across the membrane. Polysaccharide cleavage from the lipid carrier, which must occur before final cell wall assembly (third stage), is inhibited by the antibiotic vancomycin.

The third stage is insertion and attachment by cross-linking of the linear peptidoglycan into the growing three-dimensional peptidoglycan shell. The cross-linking involves a transpeptidation in which the terminal glycine residue of a peptidoglycan displaces the carboxy-terminal D-alanine on an adjacent peptidoglycan, releasing a D-alanine and forming a pentaglycine cross-linkage. Cross-linking ranges from 20–50% of the available side chain in gram-positive and gram-negative bacilli to over 90% in some cocci. In staphylococci, cross-linking of the peptide moiety of peptidoglycan is mediated via an additional spacer, the interpeptide bridge, consisting of five glycine residues. Transpeptidation is the key reaction inhibited by penicillins and cephalosporins.

The energy for cross-link formation is supplied by hydrolysis of the peptide bond between the terminal D-alanine residues of the pentapeptide (Figure 16-20).

D-Alanine carboxypeptidases (CPases) catalyze removal of the terminal D-alanyl residue from uncross-linked pentapeptides, and the resultant tetrapeptide cannot serve as substrate for transpeptidation. In this way CPases may regulate the degree of cross-linking. Some CPases catalyze transpeptidation reactions between sid

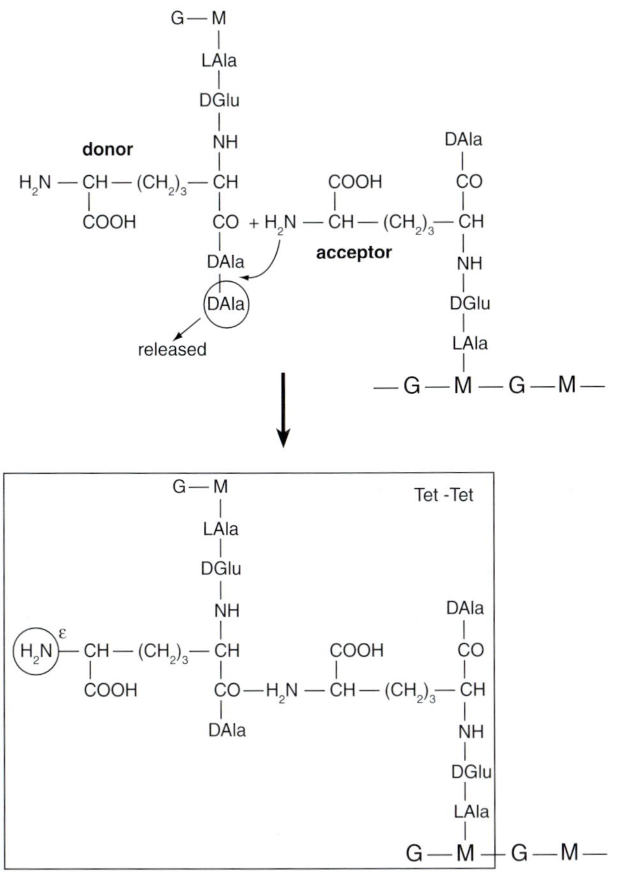

FIGURE 16-20
Cross-link formation by a transpeptidase reaction of the disaccharide pentapeptide (donor) and a disaccharide tetrapeptide (acceptor) which forms part of existing murein. The terminal D-Ala of the pentapeptide is released during this process and the ε-amino group of donor dipeptidylaminopeptidase remains free. The free amino group can be labeled by dansylation. The box indicates the dominant bisdisaccharide tetratetra compound (Tet–Tet). M,N-acetylmuramic acid; G,N-acetylglucosamine. [Reproduced with permission from N. Nanninga: Cell division and peptidoglycan assembly in *Escherichia coli*. *Mol. Microbiol.* **5**:792 (1991).]

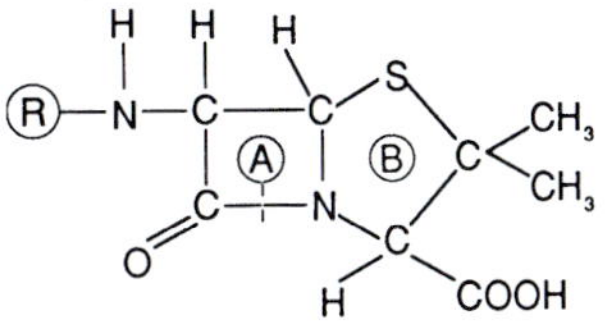

6-Aminopenicillanic acid

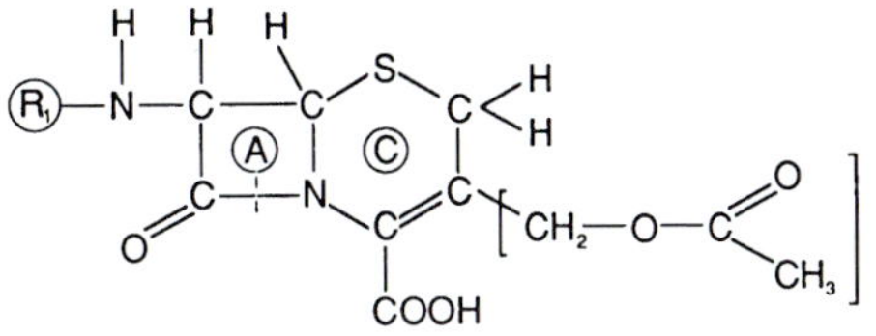

7-Aminocephalosporanic acid

FIGURE 16-21
Structures of parent compounds for the β-lactam antibiotics. In the first structure, if R = H, the compound is 6-aminopenicillanic acid; in the penicillin antibiotics, R is an acyl group. In the second structure, if R_1 = H, the compound is 7-aminocephalosporanic acid; in the cephalosporins, R_1 is an acyl group. The group in parentheses attached to ring C may also vary in the cephalosporins. Ring A is a β-lactam ring; ring B is a thiazolidine ring; ring C is a dihydrothiazine ring. The dotted line through the C–N bond in the β-lactam rings indicates the site at which penicillins and cephalosporins are inactivated by β-lactamase (penicillinase), a bacterial enzyme.

chains on linear proteoglycans that have already been attached to the cell wall. Since the degree of cross-linking is directly related to the extent to which this secondary transpeptidation occurs, CPases may also influence the extent of cross-linking in this manner.

Penicillins and Cephalosporins

The penicillins contain a bicyclic ring system formed by fusion of a thiazolidone ring with a β-lactam ring. Substitution in this ring system produces 6-aminopenicillanic acid, the parent compound for the penicillins (Figure 16-21). The cephalosporins, discovered in 1948, are derivatives of 7-aminocephalosporanic acid (Figure 16-21), which contains a dihydrothiazine-β-lactam ring system closely related to the penicillin ring system. Bactericidal activity of the penicillins and the cephalosporins is based primarily on their ability to inhibit bacterial cell wall synthesis.

Penicillins inhibit the transpeptidation reaction that cross-links linear peptidoglycans to form the saccular cell wall. Eubacteria have three to eight distinct ***penicillin-binding proteins*** (PBPs) in the cytoplasmic cell membrane (the inner cell membrane in gram-negative bacteria). All are integral membrane proteins, requiring detergents for solubilization, and have molecular weights ranging from 40,000 to 140,000. There are approximately 1000 to 10,000 molecules of PBP per cell (about 1% of the cell membrane protein). Individual PBPs make up 1–90% of the total penicillin-binding activity of a particular species. The PBPs involved in lateral wall synthesis and those (including PBP 3) involved in septum formation during cell division are different. The β-lactam mecillinam is a specific inhibitor of lateral wall elongation, while piperacillin is a specific inhibitor of septum formation.

PBPs can be classified into two groups, large and small PBPs. Large PBPs, i.e., those having molecular weights ranging from about 60,000 to 140,000, are sensitive to both penicillins and cephalosporins. They are quantitatively minor components of the cell membrane

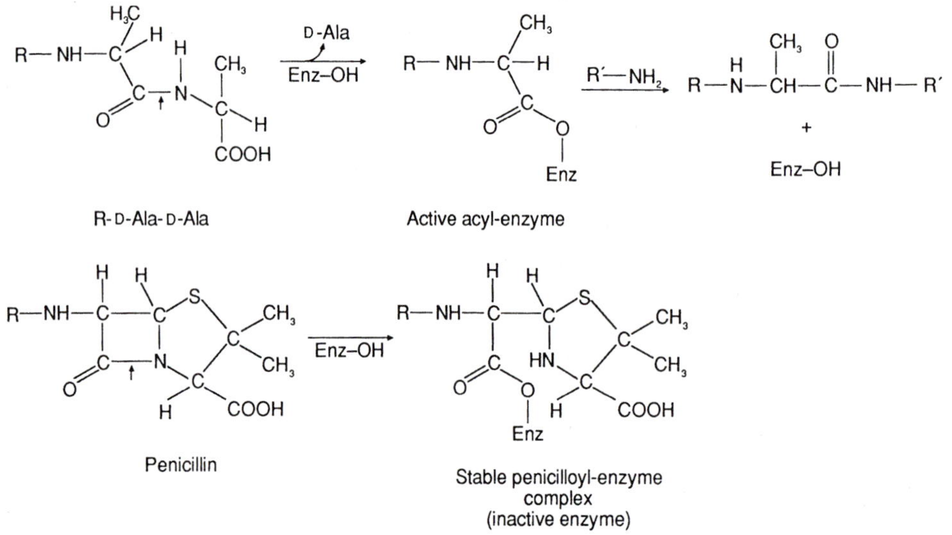

R-D-Ala-D-Ala

Active acyl-enzyme

Penicillin

Stable penicilloyl-enzyme complex (inactive enzyme)

FIGURE 16-22
Mechanism of action of penicillin. The conformation of the D-alanyl-D-alanine portion of the peptidoglycan of bacterial cell walls is similar to that of the lactam ring of penicillin, so that the enzyme cleaves the lactam ring (at the site marked ↑) to form a stable penicilloyl-enzyme complex.

and act neither as transpeptidases nor as D-alanine carboxypeptidases *in vitro* using model substrates. This does not necessarily mean that they lack these activities *in vivo,* but rather that they are denatured, that some factor essential for activity is lost during purification, or that they may require larger, more complex, or more specific molecules as substrates. Large PBPs are frequently necessary for cell viability. In contrast, small PBPs, i.e., those having molecular weights from about 40,000 to 50,000, are generally less sensitive to penicillins and are insensitive to most cephalosporins. Small PBPs are relatively abundant, are enzymatically active *in vitro,* and do not appear to be necessary for cell viability.

Interest has centered on large PBPs as lethal targets for β-lactams because of the relative insensitivity of small PBPs to these antibiotics and because of the importance of large PBPs for cell viability.

The interactions of β-lactams with PBPs indicate that these compounds are structural analogues of R-D-alanyl-D-alanine, the natural substrate of peptidoglycan transpeptidases and D-alanine carboxypeptidases, where R is the remainder of the pentapeptide. The mechanisms of the transpeptidase and carboxypeptidase reactions are thought to involve formation of an acyl-enzyme intermediate that can react with either a primary amine (e.g., an α-amino group) to form a peptide bond, or with water to form a carboxylic acid. In both reactions D-alanine is released before the acyl enzyme is formed. When a β-lactam antibiotic enters the binding site, the β-lactam bond is hydrolyzed, and the resulting acyl group reacts with the active-site serine hydroxyl group to form a stable acyl enzyme (Figure 16-22). Thus, penicillin functions as a suicide substrate (Chapter 6). The β-lactam in antibiotics is highly reactive because of the strain inherent in a four-membered ring. In addition, the other bonds prevent the release of a group equivalent to the D-alanine that is released during reaction with the usual substrates. This reaction by which penicillin forms a stable enzyme complex blocks the active site of the enzyme. Some specific interaction between the thiazolidine or dihydrothiazine ring and the enzyme may stabilize the enzyme-inhibitor complex. β-Lactam antibiotics bind to and acylate the catalytic site of the enzyme.

Glycopeptide antibiotics, including vancomycin and teicoplanin, are large, rigid molecules that inhibit a late stage in bacterial cell wall peptidoglycan synthesis. The three-dimensional structure contains a cleft into which peptides of highly specific configuration can fit (L-aa-D-aa-D-aa). Such sequences are found only in bacterial cell walls making these glycopeptides selectively toxic. As a result of binding to L-aa-D-Ala-D-Ala- groups in wall intermediates, glycopeptides inhibit, apparently by steric hinderance, the formation of backbone glycan chains (catalyzed by peptidoglycan polymerase) from the simple wall subunits as they are extruded through the cytoplasmic membrane. As a result, subsequent transpeptidation reactions that impart rigidity to the cell wall are also inhibited.

Bacteria susceptible to penicillin can mutate to resistant forms by developing an enzyme, lactamase, that hydrolyzes the lactam ring of penicillin and inactivates its therapeutic effect. In addition, highly β-lactam-resistant enterococcal bacteria strains may overexpress penicillin-binding proteins and/or reduce their affinity to β-lactam antibiotics. Resistance to vancomycin may be more difficult to achieve because it involves binding of the bulky inhibitor to substrate outside the membrane so that the active sites of two enzymes cannot align themselves correctly. It has been found that enterococci can express an inducible resistance to high levels of the glycopeptides vancomycin and teicoplanin. This resistance is due mainly to synthesis of altered peptide sequences that reduce vancomycin affinity. Because β-lactam antibiotics, such as penicillin, also trigger the production of autolytic enzymes that degrade the peptidoglycan, another option is to treat with combinations of drugs; the β-lactam antibiotics can weaken the cell wall and provide "holes" for the entry of other antibiotics.

Supplemental Readings and References

General

E. Dabelsteen: Cell surface carbohydrates as prognostic markers in human carcinomas. *Journal of Pathology* **179**:358 (1996).

H. Schachter: Molecular basis of glyoconjugate disease (Editorial). *Biochim. Biophys. Acta* **1473**:61 (1999).

A. Varki: Biological roles of oligosaccharides: All of the theories are correct. *Glycobiology* **3**:97 (1993).

D. Vestweber: Ligand-specificity of the selectins. *Journal of Cellular Biochemistry* **61**:585 (1996).

Glycoproteins

C. Abeijon, C. B. Hirschberg: Topography of glycosylation reactions in the endoplasmic reticulum. *Trends in Biological Science* **17**:32 (1992).

I. Brockhausen: Clinical aspects of glycoprotein biosynthesis. *Critical Reviews in Clinical Laboratory Science* **30**:651 (1993).

I. Brockhausen: Biosynthesis and functions of O-glycans and regulation of mucin antigen expression in cancer. *Biochemical Society Transactions* **25**:871 (1997).

I. Brockhausen: Pathways of O-glycan biosynthesis in cancer cells. *Biochimica Biophysica Acta* **1473**:67 (1999).

T. Burda, M. Aebi: The dolichol pathway of N-linked glycosylation. *Biochimica Biophysica Acta* **1426**:239 (1999).

F. I. Conler and G. W. Hart: O-glycosylation of nuclear and cytosolic proteins: Dynamic interplay between O-GlcNAc and O-Phosphate. *Journal of Biological Chemistry*

J. W. Dennis, M. Granovsky, and C. F. Warren: Glycoprotein glycosylation and cancer progression. *Biochimica Biophysica Acta* **1473**:21–34 (1999).

A. Dinter and E. G. Berger: Golgi-disturbing agents. *Histochemistry Cell Biology* **109**:571 (1998).

E. F. Hounsell, M. Young, and M. J. Davies: Glycoprotein changes in tumors: A renaissance in clinical applications [editorial]. *Clinical Science (Colch.)* **93**:287 (1997).

G. McDowell and W. A. Gahl: Inherited disorders of glycoprotein synthesis: cell biological insights. *Proceedings of the Society for Experimental Medicine and Biology* **215**:145 (1997).

P. Sears and C.-H. Wong: Enzyme action in glycoprotein synthesis. *Cellular and Molecular Life Science* **54**:223 (1998).

S. Silberstein and R. Gilmore: Biochemistry, molecular biology, and genetics of the oligosaccharyltransferase. *FASEB Journal* **10**:849 (1996).

J. Thyberg and S. Moskalewski: Role of microtubules in the organization of the Golgi complex. *Experimental Cell Research* **246**:263 (1999).

GPI-Anchored Proteins

M. A. J. Ferguson, J. S. Brimacombe, I. R. Brown, et al.: The GPI biosynthetic pathway as a therapeutic target for African sleeping sickness. *Biochimica Biophysica Acta* **1455**:327–340 (1999).

T. Kinoshita, N. Inoue, and J. Takeda: Role of phosphatidylinositol-linked proteins in paroxysmal nocturnal hemoglobinuria pathogenesis. *Annual Review of Medicine* **47**:1 (1996).

T. Kinoshita, K. Ohishi, and J. Takeda: GPI-anchor synthesis in mammalian cells: genes, their products, and a deficiency. *Journal of Biochemistry (Tokyo)* **122**:251 (1997).

A. K. Menon, N. A. Baumann, W. van't Hof, and J. Vidugiriene: Glycosylphosphatidylinositols: Biosynthesis and intracellular transport. *Biochemical Society Transactions* **25**:861 (1997).

O. Nosjean, A. Briolay, and B. Roux: Mammalian GPI proteins: Sorting, membrane residence and functions. *Biochimica Biophysica Acta* **1331**:153 (1997).

J. Takeda and T. Kinoshita: GPI-anchor biosynthesis. *Trends in Biological Science* **20**:367 (1995).

A. M. Tartakoff: Biological functions and biosynthesis of glycolipid-anchored membrane proteins. *Subcellular Biochemistry* **21**:81 (1993).

M. Tomita: Biochemical background of paroxysmal nocturnal hemoglobinuria. *Biochimica Biophysica Acta* **1455**:269–286 (1999).

A. J. Turner: PIG-tailed membrane proteins. *Essays in Biochemistry* **28**:113 (1994).

Glycosphingolipids

P. Fredman and A. Lekman: Glycosphingolipids as potential diagnostic markers and/or antigens in neurological disorders. *Neurochemical Research* **22**:1071 (1997).

K. Furukawa: Recent progress in the analysis of gangliosides biosynthesis. *Nagoya Journal of Medical Science* **61**:27 (1998).

S.-I. Hakomori and Y. Igarashi: Functional role of glycosphingolipids in cell recognition and signaling. *Journal Biochemistry (Tokyo)* **118**:1091 (1995).

S.-I. Hakomori: Tumor malignancy defined by aberrant glycosylation and sphingo(glyco)lipid metabolism. *Cancer Research* **56**:5309 (1996).

S.-I. Hakomori and Y. Zhang: Glycosphingolipid antigens and cancer therapy. *Chemistry and Biology* **4**:97104 (1997).

K. Sandhoff, T. Kelter, and G. Van Echten-Deckert: Sphingolipid metabolism. Sphingoid analogs, sphingolipid activator proteins, and the pathology of the cell. *Annals of the New York Academy of Sciences* **845**:139 (1998).

C. B. Zeller and R. B. Marchase: Gangliosides as modulators of cell function. *Americal Journal of Physiology* **262**:C1341 (1992).

Glycosaminoglycans and Proteoglycans

D. J. Carey: Biological functions of proteoglycans: Use of specific inhibitors of proteoglycan synthesis. *Molecular and Cellular Biochemistry* **104**:21 (1991).

J. R. Couchman and A. Woods: Syndecans, signaling, and cell adhesion. *Journal of Cellular Biochemistry* **61**:578 (1996).

P. Inki and M. Jalkanen: The role of syndecan-1 in malignancies. *American Journal of Medicine* **28**:63 (1996).

K. Lidholt: Biosynthesis of glycosaminoglycans in mammalian cells and in bacteria. *Biochemical Society Transactions* **25**:866 (1997).

K. Prydz and K. T. Dalen: Synthesis and sorting of proteoglycans. *J. Cell Science* **113**:193 (2000).

A. C. Rapraeger: The coordinated regulation of heparan sulfate, syndecans and cell behavior. *Current Opinion in Cell Biology* **5**:844 (1993).

J. E. Silbert and G. Sugumaran: Intracellular membranes in the synthesis, transport, and metabolism of proteoglycans. *Biochimica Biophysica Acta* **1241**:371 (1995).

S. E. Stringer and J. T. Gallagher: Heparan Sulphate. *International Journal of Biological Chemistry* **29**:709 (1997).

Peptidoglycans, Antibiotics, and Resistance

D. Bramhill: Bacterial cell division. *Annual Review of Cell and Developmental Biology* **13**:395 (1997).

I. C. Hancock: Bacterial cell surface carbohydrates: Structure and assembly. *Biochemical Society Transactions* **25**:1 83–7 (1997).

Y. Lavie, H. Cao, A. Volner, et al.: Agents that reverse multidrug resistance, tamoxifen, verapamil, and cyclosporin q block glycosphingolipid metabolism by inhibiting ceramide glycosylation in human cancer cells. *Journal of Biological Chemistry* **272**:1682 (1997).

M. C. McManus: Mechanisms of bacterial resistance to antimicrobial agents [see comments]. *American Journal of Health System Pharmacy* **54**:1420; quiz 1444 (1997).

N. Nanninga: Cell division and peptidoglycan assembly in *Escherichia coli. Molecular Microbiology* **5**:791 (1991).

CHAPTER 17

Protein and Amino Acid Metabolism

In contrast to the case of lipids and carbohydrates, no special storage forms of either the nitrogen or the amino acid components of proteins exist. Dietary protein in excess of the requirement is catabolized to provide energy and ammonia, a toxic metabolite that is converted to urea in the liver and excreted by the kidneys. All body proteins serve a specific function (e.g., structural, catalytic, transport, regulatory) and are potential sources of carbon for energy production.

Proteins constantly undergo breakdown and synthesis. During growth, even though there is net deposition of protein, the rates of synthesis and breakdown are increased. Total protein turnover in a well-fed, adult human is estimated at about 300 g/day, of which approximately 100 g is myofibrillar protein, 30 g is digestive enzymes, 20 g is small intestinal cell protein, and 15 g is hemoglobin. The remainder is accounted for by turnover of cellular proteins of various other cells (e.g., hepatocytes, leukocytes, platelets) and oxidation of amino acids, and a small amount is lost as free amino acids in urine. Protein turnover rates vary from tissue to tissue, and the relative tissue contribution to total protein turnover is altered by aging, disease, and changes in dietary protein intake. Several proteins (e.g., many hepatic enzymes) have short turnover times (less than 1 hour), whereas others have much longer turnover times (e.g., collagen >1000 days). Turnover of myofibrillar protein can be estimated by measurement of 3-methylhistidine in the urine. Histidyl residues of actin and myosin (Chapter 21) are released during catabolism of these proteins and are excreted in the urine.

Protein turnover is not completely efficient in the reutilization of amino acids. Some are lost by oxidative catabolism, while others are used in synthesis of nonprotein metabolites. For this reason, a dietary source of protein is needed to maintain adequate synthesis of protein. During periods of growth, pregnancy, lactation, or recovery from illness, supplemental dietary protein is required. These processes are affected by energy supply and hormonal factors. An overview of amino acid metabolism is presented in Figure 17-1.

17.1 Essential and Nonessential Amino Acids

Plants and some bacteria synthesize all 20 amino acids (see also Chapter 2). Humans (and other animals) can synthesize about half of them (the nonessential amino acids) but require the other half to be supplied by the diet (the essential amino acids). Diet must also provide a digestible source of nitrogen for synthesis of the nonessential amino acids. The eight ***essential amino acids*** are isoleucine, leucine lysine, methionine, phenylalanine, threonine, tryptophan, and valine. In infants, histidine (and possibly arginine) is required for optimal development and growth and is thus essential. In adults, histidine is nonessential, except in uremia. Under certain conditions,

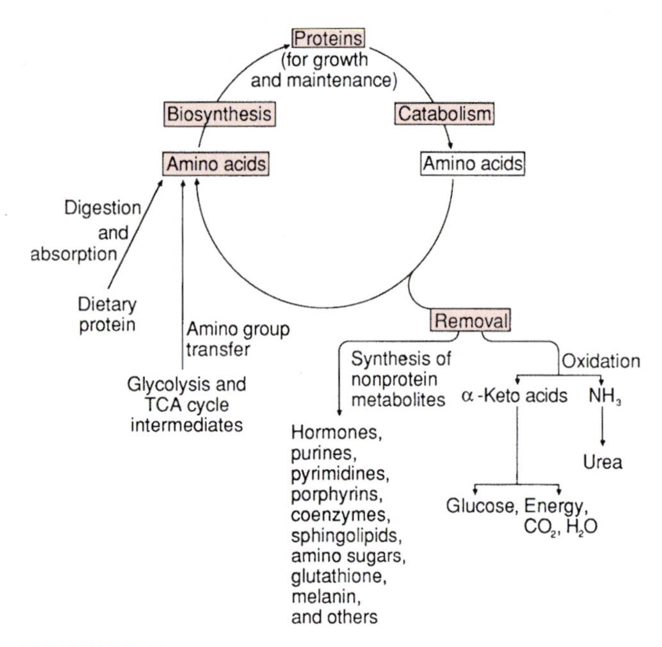

FIGURE 17-1
Overall metabolism of proteins and amino acids. Body protein is maintained by the balance between the rates of protein synthesis and breakdown. These processes are influenced by hormones and energy supply.

some nonessential amino acids may become essential. For example, when liver function is compromised by disease or premature birth, cysteine and tyrosine become essential because they cannot be formed from their usual precursors (methionine and phenylalanine).

Glutamine, a nitrogen donor in the synthesis of purines and pyrimidines required for nucleic acid synthesis (Chapter 27), aids in growth, repair of tissues, and promotion of immune function. Enrichment of glutamine in enteral and parenteral nutrition augments recovery of seriously ill patients. Arginine may be considered as a semiessential amino acid. It participates in a number of metabolic pathways, namely, formation of urea and ornithine, creatine and creatinine, spermine, agmatine and citrulline, and nitric oxide (NO). The endothelial cells lining the blood vessels produce NO from arginine, which has a major role in vasodilator function (discussed later). Dietary arginine supplementation improves coronary blood flow, reduces episodes of angina, and helps in patients with walking pain due to claudication.

Nitrogen Balance

For protein synthesis to occur, all 20 amino acids must be present in sufficient quantities. Absence of any one essential amino acid leads to cessation of protein synthesis, catabolism of unused amino acids, increased loss of nitrogen in urine, and reduced growth. Negative nitrogen balance exists when the amount of nitrogen lost from the body (as nitrogen metabolites excreted in urine and feces) exceeds that taken in. This state continues until the essential amino acid deficiency is corrected. Negative nitrogen balance also occurs in malabsorption syndromes, fever, trauma, cancer, and excessive production of catabolic hormones (e.g., hypercortisolism; see Chapter 32). When the dietary nitrogen intake equals nitrogen losses, the body is in nitrogen balance. In normal adults, anabolism equals catabolism. When nitrogen intake exceeds nitrogen losses, there is a positive nitrogen balance, with anabolism exceeding catabolism. The body retains nitrogen as tissue protein, which is a characteristic of active growth and tissue repair (e.g., growth in children, pregnancy, recovery from an emaciating illness).

Quality and Quantity of Dietary Protein Requirement

Dietary protein provides organic nitrogen and the essential amino acids. The quantitative estimation of protein requirement must take into account the quality of protein, as determined by its essential amino acid composition. Dietary protein should provide all of the essential amino acids in the appropriate amounts. If the concentration of one amino acid is significantly greater or less than that of the others (in a protein or amino acid mixture), utilization of the others may be depressed and will be reflected in growth failure.

A procedure for assessment of protein quality consists of feeding growing rats various levels of the test protein and assessing the slope of regression lines relating growth rate and protein intake. Wheat protein is deficient in lysine when compared with lactalbumin, which contains all of the essential amino acids in desirable concentrations. Wheat protein is therefore assessed to be 20% as effective as an equivalent amount of lactalbumin. Similarly, proteins from corn, which are also deficient in lysine, do not support optimal growth. However, genetic selection and breeding programs have yielded strains of corn with higher lysine content. Proteins of animal origin—namely, meats, eggs, milk, cheese, poultry, and fish—are of good quality since they provide all of the essential amino acids. ***Gelatin,*** the protein derived from collagen, lacks tryptophan and is of poor quality. In general, plant proteins are of poor quality because they lack one or more essential amino acids. The best quality plant proteins are found in legumes and nuts. Therefore, the diet of a pure vegetarian requires careful planning to achieve a combination of proteins that provide necessary amounts of all essential amino acids. Combinations of complementary vegetable proteins include rice and black-eyed peas; whole wheat or parched crushed wheat (bulgur) with soybeans and sesame seeds cornmeal and kidney beans; and soybeans, peanuts, brown rice, and bulgur (Chapter 12). Deficiency of vitamin B_{12} (Chapter 38) may occur in persons on a pure vegetarian diet. Although plant proteins used singly do not provide all of the essential amino acids, their inclusion in the diet provides nonessential amino acids that would otherwise have to be synthesized at the expense of the nitrogen of the essential amino acids. The following estimates of daily protein needs are for persons on a Western diet: adults, 0.8 g/kg; newborns, 2.2 g/kg; infants (0.5–1 year), 2 g/kg. During pregnancy and lactation protein intake above the normal adult level is recommended (Appendix IV). These protein requirements are valid only when energy needs are adequately met from nonprotein sources. If intake of carbohydrates and lipids is insufficient to meet the energy expenditure, dietary protein is utilized to meet the energy deficit and results in negative nitrogen balance.

Protein Energy Malnutrition

Two disorders of protein energy nutrition that are widespread among children in economically depressed areas are kwashiorkor (in Ghana, "the disease the first child gets when the second is on the way") and marasmus (from the Greek "to waste away").

Kwashiorkor occurs after weaning and is due to inadequate intake of good-quality protein and a diet consisting primarily of high-starch foods (e.g., yams, potatoes, bananas, maize, cassava) and deficient in other essential nutrients. Victims have decreased mass and function of heart and kidneys; decreased blood volume, hematocrit, and serum albumin concentration; atrophy of pancreas and intestines; decreased immunological resistance; slow wound healing; and abnormal temperature regulation. Characteristic clinical signs include edema, ascites, growth failure, apathy, skin rash, desquamation, pigment changes and ulceration, loss of hair, liver enlargement, anorexia, and diarrhea.

Marasmus results from deficiency of protein and energy intake, as in starvation, and results in generalized wasting (atrophy of muscles and subcutaneous tissues, emaciation, loss of adipose tissue) Edema occurs in kwashiorkor but not in marasmus; however, the distinction between these disorders is not always clear. The treatment of marasmus requires supplementation of protein and energy intake.

Protein energy malnutrition occurs with high frequency (30%–50%) in hospitalized patients as well as in populations in chronic care facilities as either an acute or a chronic problem. These individuals suffer from inadequate nutrition due to a disease or depression, and are susceptible to infections due to impaired immune function. Surgical patients with protein energy malnutrition exhibit delayed wound healing with increased length of stay in the hospital. Thus, protein energy malnutrition can cause morbidity, mortality and also has economic consequences. Acute stressful physiological conditions such as trauma, burn, or sepsis can also precipitate protein energy malnutrition due to hypermetabolism caused by the neuroendocrine system. Prompt diagnosis and appropriate nutritional intervention is required in the management of patients with protein energy malnutrition.

Measurements of the levels of serum proteins such as albumin, transthyretin (also known as prealbumin), transferrin and retinol-binding protein are used as biochemical parameters in the assessment of protein energy malnutrition (Table 17-1). An ideal protein marker should have rapid turnover and present in sufficiently high concentrations in serum to be measured accurately. Transthyretin has these properties; it is a sensitive indicator of protein deficiency and is effective in assessing improvement with refeeding.

Transport of Amino Acids into Cells

Intracellular metabolism of amino acids requires their transport across the cell membrane. Transport of L-amino acids occurs against a concentration gradient and is an active process usually coupled to Na^+-dependent carrier systems as for transport of glucose across the intestinal mucosa (Chapter 12). At least five transport systems for amino acids (with overlapping specificities) have been identified in kidney and intestine. They transport neutral amino acids, acidic amino acids, basic amino acids, ornithine and cystine, and glycine and proline, respectively. Within a given carrier system, amino acids may compete for transport (e.g., phenylalanine with tryptophan). Na^+-independent transport carriers for neutral and lipophilic amino acids have also been described. D-Amino acids are transported by simple diffusion favored by a concentration gradient.

Inherited defects in amino acid transport affect epithelial cells of the gastrointestinal tract and renal tubules. Some affect transport of neutral amino acids (***Hartnup disease***), others that of basic amino acids and ornithine and cystine (***cystinuria***), or of glycine and proline (Chapter 12). ***Cystinosis*** is an intracellular transport defect characterized by high intralysosomal content of free cystine in the reticuloendothelial system, bone marrow, kidney, and eye. After degradation of endocytosed protein to amino acids within lysosomes, the amino acids normally are transported to the cytosol. The defect in cystinosis may reside in the ATP-dependent efflux system for cystine transport, and particularly in the carrier protein.

A different mechanism for translocation of amino acids in some cells is employed in the ***γ-glutamyl cycle*** (Figure 17.2). Its operation requires six enzymes (one membrane-bound, the rest cytosolic), glutathione (GSH; γ-glutamylcysteinylglycine present in all tissue cells),

TABLE 17-1

Properties of Serum/Plasma Biochemical Markers Used in the Assessment of Protein Energy Malnutrition

Biochemical Marker	Site of Synthesis and Molecular Weight	Half-life in the Blood	Serum/Plasma Concentration during Normal(N), Mild, Moderate, and Severely Depleted States	Properties and Clinical Utility
Albumin	Liver M.W.: 66,000	15–19 days	N: 3.5–5.0 g/dl Mild: 2.8–3.5 g/dl Moderate: 2.1–2.7 g/dl Severe: <2.1 g/dl	Transports many endogenous and exogenous ligands, major determinant of plasma oncotic pressure. Large body pool and long half-life makes it a poor index of acute malnutrition
Transthyretin (also known as prealbumin)	Liver plasma circulating form is a tetramer composed of four identical monomers. M.W.: 55,000	1–2 days	N: 20–40 mg/dl Mild: 10–15 mg/dl Moderate: 5–10 mg/dl Severe: <5 mg/dl	Circulates in plasma in a 1:1 complex with retinol-binding protein, transports thyroxine, has a small body pool, and has a short half-life. Sensitive indicator of protein deficiency and in the improvement with protein refeeding.
Retinol-binding protein (RBP)	Liver M.W.: 21,000	10–12 hours	N: 3.5–9.0 mg/dl	Circulates in plasma in 1:1 complex with transthyretin, transports retinol and thyroxine, plasma levels influenced by glomerular filtration rate, retinol and zinc status, considered to be too sensitive and therefore has limited value
Transferrin	Liver M.W.: 77,000	8.8 days	N: 225–400 mg/dl Mild: 150–200 mg/dl Moderate: 100–150 mg/dl Severe: <100 mg/dl	Transports iron in the ferric state; serum levels rise proportional to iron deficiency, a better marker than albumin however, not suitable in the short-term hospital setting.

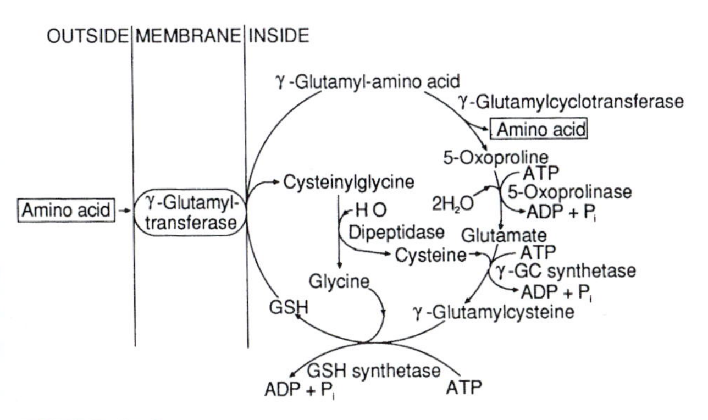

FIGURE 17-2
The γ-glutamyl cycle for the transport of amino acids. GSH = Glutathione; γ-GC = γ-glutamylcysteine.

and ATP (three ATP molecules are required for each amino acid translocation). In this cycle, there is no net consumption of GSH, but an amino acid is transported at the expense of the energy of peptide bonds of GSH, which has to be resynthesized. Translocation is initiated by membrane-bound γ-glutamyltransferase (γ-GT, γ-glutamyl transpeptidase), which catalyzes formation of a γ-glutamyl amino acid and cysteinylglycine. The latter is hydrolyzed by dipeptidase to cysteine and glycine, which are utilized in resynthesis of GSH. The former is cleaved to the amino acid and 5-oxoproline (pyroglutamic acid) by γ-glutamylcyclotransferase. The cycle is completed by conversion of 5-oxoproline to glutamate and by resynthesis of GSH by two ATP-dependent enzymes, γ-glutamylcysteine synthetase and glutathione synthetase, respectively. GSH synthesis appears to be regulated by nonallosteric inhibition of γ-glutamylcysteine synthetase.

GSH has several well-established functions: it provides reducing equivalents to maintain –SH groups in other molecules (e.g., hemoglobin, membrane proteins; Chapter 15); it participates in inactivation of hydrogen peroxide, other peroxides, and free radicals (Chapter 14); it participates in other metabolic pathways (e.g., leukotrienes; Chapter 18); and it functions in inactivation of a variety of foreign compounds by conjugation through its sulfur atom. The conjugation reaction is catalyzed by specific glutathione S-transferases and the product is eventually converted to mercapturic acids and excreted.

Inherited deficiency of GSH synthetase, γ-glutamylcysteine synthetase γ-glutamyltransferase, and 5-oxoprolinase have been reported. Red blood cells, the central nervous system, and muscle may be affected. In GSH synthetase deficiency, γ-glutamylcysteine accumulates from lack of inhibition of γ-glutamylcysteine synthetase by glutathione, and it is converted to 5-oxoproline and cysteine by γ-glutamylcyclotransferase. 5-Oxoproline is excreted so that GSH synthetase deficiency causes 5-oxoprolinuria (or ***pyroglutamic aciduria***).

Measurement of serum γ-GT activity has clinical significance. The enzyme is present in all tissues, but the highest level is in the kidney; however, the serum enzyme originates primarily from the hepatobiliary system. Elevated levels of ***serum γ-GT*** are found in the following disorders: intra- and posthepatic biliary obstruction (elevated serum γ-GT indicates cholestasis, as do leucine aminopeptidase, 5′-nucleotidase, and alkaline phosphatase); primary or disseminated neoplasms; some pancreatic cancers, especially when associated with hepatobiliary obstruction; alcohol-induced liver disease (serum γ-GT may be exquisitely sensitive to alcohol-induced liver injury); and some prostatic carcinomas (serum from normal males has 50% higher activity than that of females). Increased activity is also found in patients receiving phenobarbital or phenytoin, possibly due to induction of γ-GT in liver cells by these drugs.

General Reactions of Amino Acids

Some general reactions that involve degradation or interconversion of amino acids provide for the synthesis of nonessential amino acids from α-keto acid precursors derived from carbohydrate intermediates.

Deamination

Removal of the α-amino group is the first step in catabolism of amino acids. It may be accomplished oxidatively or nonoxidatively.

Oxidative deamination is stereospecific and is catalyzed by L- or D-amino acid oxidase. The initial step is removal of two hydrogen atoms by the flavin coenzyme, with formation of an unstable α-amino acid intermediate. This intermediate undergoes decomposition by addition of water and forms ammonium ion and the corresponding α-keto acid: L-Amino acid oxidase occurs in the liver

$$R{-}CH(NH_3^+){-}COO^- \xrightarrow[\text{Flavin} \rightarrow \text{Flavin-}H_2]{\text{L-Amino acid oxidase}}$$

α-Amino acid

Flavin-H_2 + O_2 → Flavin + H_2O_2

and kidney only. It is a flavoprotein that contains flavin mononucleotide (FMN) as a prosthetic group and does not attack glycine, dicarboxylic, or β-hydroxy amino acids. Its activity is very low.

High levels of D-amino acid oxidases are found in the liver and kidney. The enzyme contains flavin adenine dinucleotide (FAD) and deaminates many D-amino acids and glycine. The reaction for glycine is analogous to that for D-amino acids.

$$H_3N\text{—}CH_2\text{—}COO^- \text{ (Glycine)} \xrightarrow[\text{FAD} \rightarrow \text{FADH}_2;\ H_2O_2 \leftarrow O_2]{\text{D-Amino acid oxidase}} {}^+H_2N\text{=}CH\text{—}COO^- \text{ (}\alpha\text{-Imino acid)} \xrightarrow[-NH_4^+]{H_2O} O\text{=}CH\text{—}COO^- \text{ (Glyoxylate)}$$

D-Amino acid oxidase occurs in peroxisomes containing other enzymes that produce H_2O_2 (e.g., L-α-hydroxy acid oxidase, citrate dehydrogenase, and L-amino acid oxidase) and catalase and peroxidase, which destroy H_2O_2. In leukocytes, killing of bacteria involves hydrolases of lysosomes and production of H_2O_2 by NADPH oxidase (Chapter 15). Conversion of D-amino acids to the corresponding α-keto acids removes the asymmetry at the α-carbon atom. The keto acids may be aminated to L-amino acids. By this conversion from D- to L-amino acids, the body utilizes D-amino acids derived from the diet:

$$H\text{—}C(R)(COO^-)\text{—}NH_3^+ \xrightarrow{\text{D-Amino acid oxidase}} R\text{—}C(\text{=}O)\text{—}COO^- \xrightarrow{\text{Reamination (transaminase)}} {}^+H_3N\text{—}C(R)(COO^-)\text{—}H$$

D-Amino acid (not useful for protein synthesis) → α-Keto acid (has no asymmetric α-carbon) → L-Amino acid (metabolically useful)

Nonoxidative deamination is accomplished by several specific enzymes.

Amino acid dehydratases deaminate hydroxyamino acids:

$$\left.\begin{array}{r}\text{Serine}\\ \text{Threonine}\\ \text{Homoserine}\end{array}\right\} \xrightarrow[\text{(Pyridoxal phosphate)}]{\text{Specific dehydratase},\ -H_2O} \alpha\text{-Imino acid} \xrightarrow[H_2O]{-NH_3} \alpha\text{-Keto acid}$$

For serine dehydratase the reaction is

$$HO\text{—}CH_2\text{—}CH(NH_3^+)\text{—}COO^- \text{ (Serine)} \xrightarrow{-H_2O} H_2C\text{=}C(NH_3^+)\text{—}COO^- \rightarrow H_3C\text{—}C(\text{=}NH_2^+)\text{—}COO^- \xrightarrow[-NH_3]{H_2O} H_3C\text{—}C(\text{=}O)\text{—}COO^- \text{ (Pyruvate)}$$

Dehydrogenation of L-Glutamate

Glutamate dehydrogenase plays a major role in amino acid metabolism. It is a zinc protein, requires NAD^+ or $NADP^+$ as coenzyme, and is present in high concentrations in mitochondria of liver, heart, muscle, and kidney. It catalyzes the (reversible) oxidative deamination of L-glutamate to α-ketoglutarate and NH_3. The initial step probably involves formation of α-iminoglutarate by dehydrogenation. This step is followed by hydrolysis of the imino acid to a keto acid and NH_3:

$$^-OOC\text{—}CH_2\text{—}CH_2\text{—}CH(NH_3^+)\text{—}COO^- + NAD(P)^+ \rightleftharpoons$$

L-Glutamate

$$^-OOC\text{—}CH_2\text{—}CH_2\text{—}C(\text{=}NH_2^+)\text{—}COO^- + NAD(P)H + H^+$$

α-Iminoglutaric acid

$$\Big\Updownarrow\ H_2O$$

$$^-OOC\text{—}CH_2\text{—}CH_2\text{—}C(\text{=}O)\text{—}COO^- + NH_4^+$$

α-Ketoglutarate

Glutamate dehydrogenase is an allosteric protein modulated positively by ADP, GDP, and some amino acids and negatively by ATP, GTP, and NADH. Its activity is affected by thyroxine and some steroid hormones *in vitro*. Glutamate dehydrogenase is the only amino acid dehydrogenase present in most cells. It participates with appropriate transaminases (aminotransferases) in the deamination of other amino acids.

R—CH—COO⁻ (NH_3^+) (L-Amino acid) + α-Ketoglutarate ⇌ R—C(=O)—COO⁻ (α-Keto acid) + L-Glutamate [Transaminase (aminotransferase) pyridoxal phosphate enzyme]

L-Glutamate + NAD(P)⁺ + H_2O ⇌ α-Ketoglutarate + NAD(P)H + NH_4^+ [Glutamate dehydrogenase]

These reactions are at near-equilibrium, so their overall effect depends upon concentrations of the substrates and products. Aminotransferases occur in cytosol and mitochondria, but their activity is much higher in cytosol. Since glutamate dehydrogenase is restricted to mitochondria, transport of glutamate (generated by various transaminases) into mitochondria by a specific carrier becomes of central importance in amino acid metabolism. The NH_3 produced in deamination reactions must be detoxified by conversion to glutamine and asparagine or to urea, which is excreted in urine. The NADH generated is ultimately oxidized by the electron transport chain.

Transamination

Transamination reactions combine reversible amination and deamination, and they mediate redistribution of amino groups among amino acids. Transaminases (aminotransferases) are widely distributed in human tissues and are particularly active in heart muscle, liver, skeletal muscle, and kidney. The general reaction of transamination is

$$\underset{\text{Amino acid (1)}}{\mathrm{R_1{-}CH(NH_3^+){-}COO^-}} + \underset{\text{Keto acid (2)}}{\mathrm{R_2{-}C(=O){-}COO^-}} \underset{\text{(pyridoxal phosphate)}}{\overset{\text{Transaminase}}{\rightleftharpoons}} \underset{\text{Keto acid (1)}}{\mathrm{R_1{-}C(=O){-}COO^-}} + \underset{\text{Amino acid (2)}}{\mathrm{R_2{-}CH(NH_3^+){-}COO^-}}$$

The α-ketoglutarate/L-glutamate couple serves as an amino group acceptor/donor pair in transaminase reactions. The specificity of a particular transaminase is for the amino group other than the glutamate. Two transaminases whose activities in serum are indices of liver damage catalyze the following reactions:

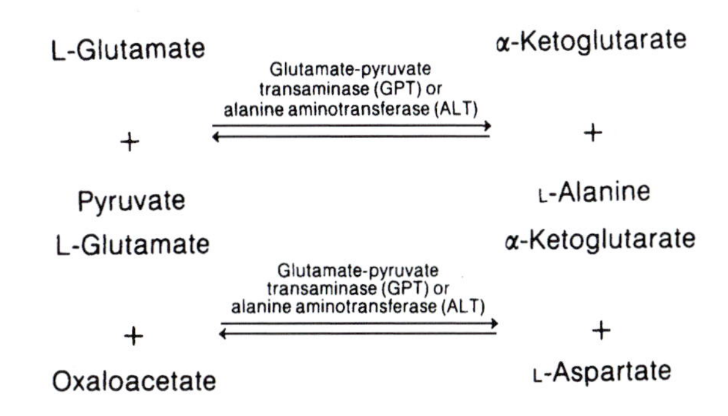

All of the amino acids except lysine, threonine, proline, and hydroxyproline participate in transamination reactions. Transaminases exist for histidine, serine, phenylalanine, and methionine, but the major pathways of their metabolism do not involve transamination. Transamination of an amino group not at the α-position can also occur. Thus, transfer of δ-amino group of ornithine to α-ketoglutarate converts ornithine to glutamate-γ-semialdehyde.

All transaminase reactions have the same mechanism and use pyridoxal phosphate (a derivative of vitamin B_6; Chapter 38). Pyridoxal phosphate is linked to the enzyme by formation of a Schiff base between its aldehyde group and the ε-amino group of a specific lysyl residue at the active site and held noncovalently through its positively charged nitrogen atom and the negatively charged phosphate group (Figure 17-3). During catalysis, the amino acid substrate displaces the lysyl ε-amino group of the enzyme in the Schiff base. An electron pair is removed from the α-carbon of the substrate and transferred to the positively charged pyridine ring but is subsequently returned

FIGURE 17-3
Binding of pyridoxal phosphate to its apoenzyme. The carbonyl carbon reacts with the ε-amino group of the lysyl residue near the active site to yield a Schiff base. Ionic interactions involve its positively charged pyridinium ion and negatively charged phosphate group.

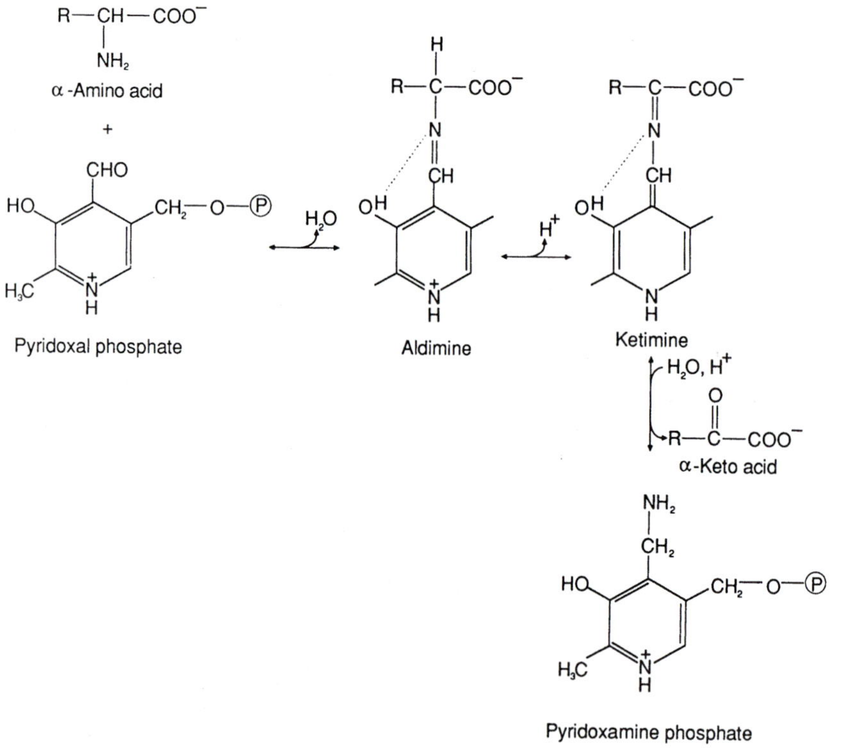

R—CH—COO⁻ | NH₂ α-Amino acid + Pyridoxal phosphate ⇌ (H_2O) Aldimine ⇌ (H^+) Ketimine → (H_2O, H^+) α-Keto acid + Pyridoxamine phosphate

FIGURE 17-4
Mechanism of the first phase of transamination. The –NH_2 group from the amino acid is transferred to pyridoxal phosphate, with formation of the corresponding α-keto acid. The second phase occurs by the reversal of the first phase reactions and is initiated by formation of a Schiff base with the α-keto acid substrate and pyridoxamine phosphate. The transamination cycle is completed with formation of the corresponding α-amino acid and pyridoxal phosphate.

to the second substrate, the α-keto acid. Thus, pyridoxal phosphate functions as a carrier of amino groups and as an electron sink by facilitating dissociation of the α-hydrogen of the amino acid (Figure 17-4). In the overall reaction, the amino acid transfers its amino group to pyridoxal phosphate and then to the keto acid through formation of pyridoxamine phosphate as intermediate.

Pyridoxal phosphate is also the prosthetic group of amino acid decarboxylases, dehydratases, desulfhydrases, racemases, and aldolases, in which it participates through its ability to render labile various bonds of an amino acid molecule (Figure 17-5). Several drugs (Figure 17-6) inhibit pyridoxal phosphate–dependent enzymes. Isonicotinic hydrazide (used in the treatment of tuberculosis) and hydralazine (a hypertensive agent) react with the aldehyde group of pyridoxal (free or bound) to form pyridoxal hydrazones, which are eliminated in the urine. Isonicotinic acid hydrazide is normally inactivated in the liver by acetylation; some individuals are "slow acetylators" (an inherited trait) and may be susceptible to pyridoxal deficiency from accumulation of the drug. Cycloserine (an amino acid analogue and broad-spectrum antibiotic) also combines with pyridoxal phosphate.

Role of Specific Tissues in Amino Acid Metabolism

The specific roles of various tissues and organs and their interdependence on amino acid metabolism are discussed here. An overview of this topic is given in Chapter 22. In the post absorptive state, maintenance of steady-state concentrations of plasma amino acids depends on release

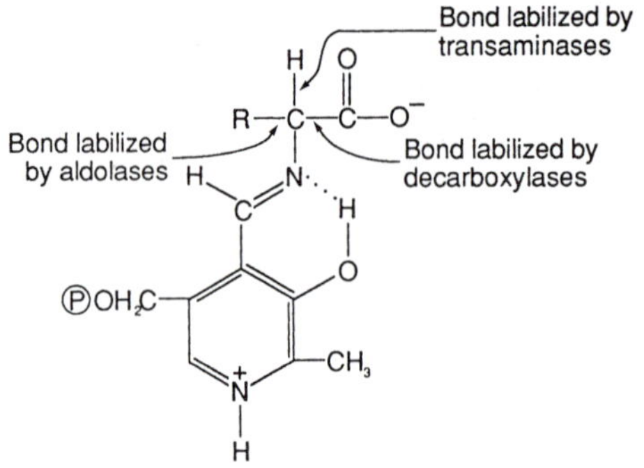

Bond labilized by transaminases
Bond labilized by aldolases
Bond labilized by decarboxylases

FIGURE 17-5
Labilization of bonds of an amino acid bound to pyridoxal phosphate–containing enzymes. Given the appropriate apoenzyme, any atom or group on the carbon atom proximal to the Schiff base can be cleaved.

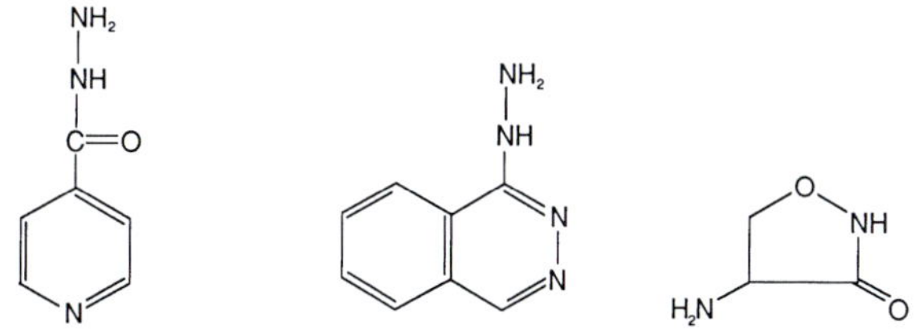

Isonicotinic acid hydrazide (antituberculosis drug)

Hydralazine (hypertensive agent)

Cycloserine (antibiotic)

FIGURE 17-6
Structures of compounds that inhibit pyridoxal phosphate–containing enzymes.

of amino acids from tissue protein. After a meal, dietary amino acids enter the plasma and replenish the tissues that supply amino acid during fasting.

Liver plays a major role, since it can oxidize all amino acids except leucine, isoleucine, and valine (see Chapter 22). It also produces the nonessential amino acids from the appropriate carbon precursors. Ammonia formed in the gastrointestinal tract or from various deaminations in the liver is converted to urea and excreted in urine (discussed later).

Skeletal muscle tissue constitutes a large portion of the body weight and accounts for a significant portion of nonhepatic amino acid metabolism. It takes up the amino acids required to meet its needs for protein synthesis, and metabolizes alanine, aspartate, glutamate, and the branched-chain amino acids. Amino acids are released from muscle during the postabsorptive state (i.e., in fasting or starvation). Alanine and glutamine constitute more than 50% of the α-amino acid nitrogen released. During starvation, the total amino acid pool increases from catabolism of contractile proteins. However, the amino acid composition of these proteins does not account for the large amount of alanine and glutamine released. Amino acids that give rise to pyruvate can be transaminated to alanine. For example, aspartate can be converted to alanine as follows:

$$\text{Aspartate} \xrightarrow{\text{transaminase}} \text{Oxaloacetate} \xrightarrow{\text{PEPCK}} \text{Phosphoenolpyruvate} \xrightarrow{\text{pyruvate kinase}} \text{Pyruvate} \xrightarrow{\text{transaminase}} \text{Alanine}$$

Similarly, amino acids that produce tricarboxylic acid (TCA) cycle intermediates (Chapter 15) produce alanine by conversion to oxaloacetate. During starvation or intake of a carbohydrate-poor diet, conversion of pyruvate to alanine is preferred because pyruvate dehydrogenase is inactivated by oxidation of fatty acids and ketone bodies (Chapters 13 and 18).

Glutamine is synthesized from glutamate and ammonia by glutamine synthase:

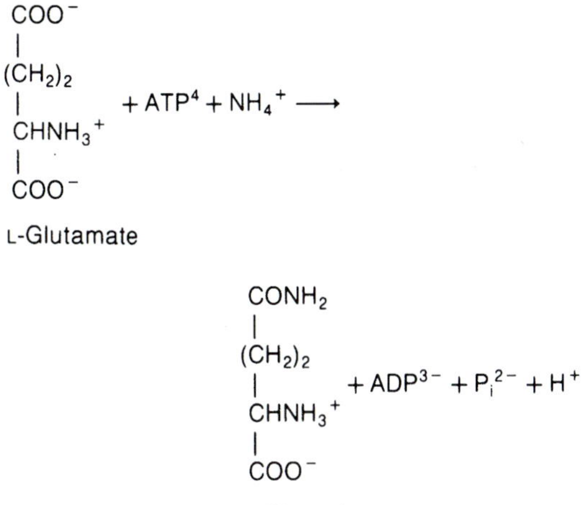

$$\underset{\text{L-Glutamate}}{^{-}OOC\text{-}CH(NH_3^+)\text{-}(CH_2)_2\text{-}COO^-} + ATP^4 + NH_4^+ \longrightarrow \underset{\text{L-Glutamine}}{^{-}OOC\text{-}CH(NH_3^+)\text{-}(CH_2)_2\text{-}CONH_2} + ADP^{3-} + P_i^{2-} + H^+$$

Glutamate is derived by transamination of α-ketoglutarate produced in the TCA cycle from citrate via oxaloacetate and acetyl-CoA (Chapter 13). All of the amino acids can produce acetyl-CoA. All except leucine and lysine (which are oxidized solely to acetyl-CoA) can be used in net synthesis of α-ketoglutarate to enhance glutamate synthesis. Ammonia is generated in glutamate dehydrogenase and AMP deaminase reactions (Chapter 21).

The mucosa of the ***small intestine*** metabolizes dietary glutamine, glutamate, asparagine, and aspartate by oxidation to CO_2 and H_2O, or by conversion to lactate, alanine, citrulline, and NH_3. These intermediates and the unmetabolized dietary amino acids are transferred to the portal blood and then to the liver for further metabolism.

In the fasting state, the intestinal mucosa depends on other tissues for metabolites to provide energy and precursors for protein and nucleotide synthesis to maintain the rapid cell division characteristic of that tissue. Glutamine, released from liver and muscle, is utilized for purine nucleotide synthesis (Chapter 27), is oxidized to provide energy, and can be converted to aspartate for pyrimidine nucleotide synthesis (Chapter 27). Thus, glutamine is important in cells undergoing rapid division. Intestine can also oxidize glucose, fatty acids, and ketone bodies to provide energy.

Kidney releases serine and small (but significant) quantities of alanine into the blood, and takes up glutamine, proline, and glycine. Amino acids filtered in the glomeruli are reabsorbed by renal tubule cells. Glutamine plays an important role in acid-base regulation by providing ammonia, which forms the NH_4^+ ion eliminated in urine (Chapter 39). It can provide two ammonia molecules, by glutaminase and glutamate dehydrogenase, respectively, in renal tubular mitochondria. Its carbon skeleton can be oxidized or converted to glucose since renal tissue is capable of gluconeogenesis (Chapter 15).

Brain takes up significant quantities of valine and may be a major (if not primary) site of utilization of branched-chain amino acids. Glutamate, aspartate, and glycine are ***neurotransmitters.*** Glutamate is a precursor of γ-aminobutyrate; tyrosine of dopamine, norepinephrine, and epinephrine; and tryptophan of serotonin, all of which are neurotransmitters. Inactivation of neurotransmitters involves deamination with production of ammonia, which is removed by formation of glutamine. N-acetylaspartate occurs in high levels in the brain but its function is not known. It is synthesized from acetyl-CoA and aspartic acid catalyzed by acetyl-CoA aspartate N-acetyl transferase. Aspartoacylase catalyzes the hydrolysis of N-acetylaspartate to acetate and aspartic acid. The deficiency of aspartoacylase, which is inherited as an autosomal recessive trait, is associated with degenerative brain changes. Patients of this disorder, also known as ***Canavan dystrophy,*** are usually of Eastern European Jewish heritage.

17.2 Metabolism of Ammonia

Ammonia (at physiological pH, 98.5% exists as NH_4^+), the highly toxic product of protein catabolism, is rapidly inactivated by a variety of reactions. Some products of these reactions are utilized for other purposes (thus salvaging a portion of the amino nitrogen), while others are excreted. The excreted form varies quite widely among vertebrate and invertebrate animals. The development of a pathway for nitrogen disposal in a species appears to depend chiefly on the availability of water. Thus, urea is excreted in terrestrial vertebrates (***ureotelic organisms***); ammonia in aquatic animals (***ammonotelic organisms***); and uric acid (in semisolid form) in birds and land-dwelling reptiles (***uricotelic organisms***). During their aquatic phase of development amphibia excrete ammonia but the adult frog excretes urea; during metamorphosis the liver produces the enzymes required for their synthesis. In humans, ammonia is excreted mostly as urea, which is highly water-soluble, is distributed throughout extracellular and intracellular body water, is nontoxic and metabolically inert, has a high nitrogen content (47%), and is excreted via the kidneys.

Ammonia is produced by deamination of glutamine, glutamate, other amino acids, and adenylate. A considerable quantity is derived from intestinal bacterial enzymes acting on urea and other nitrogenous compounds. The urea comes from body fluids that diffuse into the intestine, and the other nitrogenous products are derived from intestinal metabolism (e.g., glutamine) and ingested protein. The ammonia diffuses across the intestinal mucosa to the portal blood and is converted to urea in the liver.

Ammonia is particularly toxic to brain but not to other tissues, even though levels in those tissues may increase under normal physiological conditions (e.g., in muscle during heavy exercise in kidney during metabolic acidosis). Several hypotheses have been suggested to explain the mechanism of neurotoxicity.

In brain mitochondria, excess ammonia may drive the reductive amination of α-ketoglutarate by glutamate dehydrogenase. This step may deplete a key intermediate of the TCA cycle and lead to its impairment, with severe inhibition of respiration and considerable stimulation of glycolysis. Since the $[NAD^+]/[NADH]$ ratio will be high in mitochondria, there will be a decrease in the rate of production of ATP. This hypothesis does not explain why the same result does not occur in tissues that are not affected by ammonia. A more plausible hypothesis is depletion of glutamate which is an excitatory neurotransmitter. Glutamine, synthesized and stored in glial cells, is the most likely precursor of glutamate. It is transported into the neurons and hydrolyzed by glutaminase. Ammonia inhibits glutaminase and depletes the glutamate concentration. A third hypothesis invokes neuronal membrane dysfunction, since elevated levels of ammonia produce increased permeability to K^+ and Cl^- ions, while glycolysis increases H^+ ion concentration (NH_4^+ stimulates 6-phosphofructokinase; Chapter 13). Encephalopathy of hyperammonemia is characterized by brain edema and astrocyte swelling. Edema and swelling have been attributed to intracellular accumulation of glutamine which causes osmotic shifts of water into the cell.

Behavioral disorders such as ***anorexia,*** sleep disturbances, and ***pain insensitivity*** associated with hyperammonemia have been attributed to increased tryptophan transport across the blood-brain barrier and the accumulation of its metabolites. Two of the tryptophan-derived metabolites are serotonin and quinolinic acid (discussed later). The latter is an excitotoxin at the N-methyl-D-aspartate (NMDA) glutamate receptors. Thus, the mechanism of the ammonium-induced neurological abnormalities is multifactorial. Normally only small amounts of NH_3 (i.e., NH_4^+) are present in plasma, since NH_3 is rapidly removed by reactions in tissues of glutamate dehydrogenase, glutamine synthase, and urea formation.

Urea Synthesis

Ammonia contained in the blood flowing through the hepatic lobule is removed by the hepatocytes and converted into urea. Periportal hepatocytes are the predominant sites of urea formation. Any ammonia that is not converted to urea may be incorporated into glutamine catalyzed

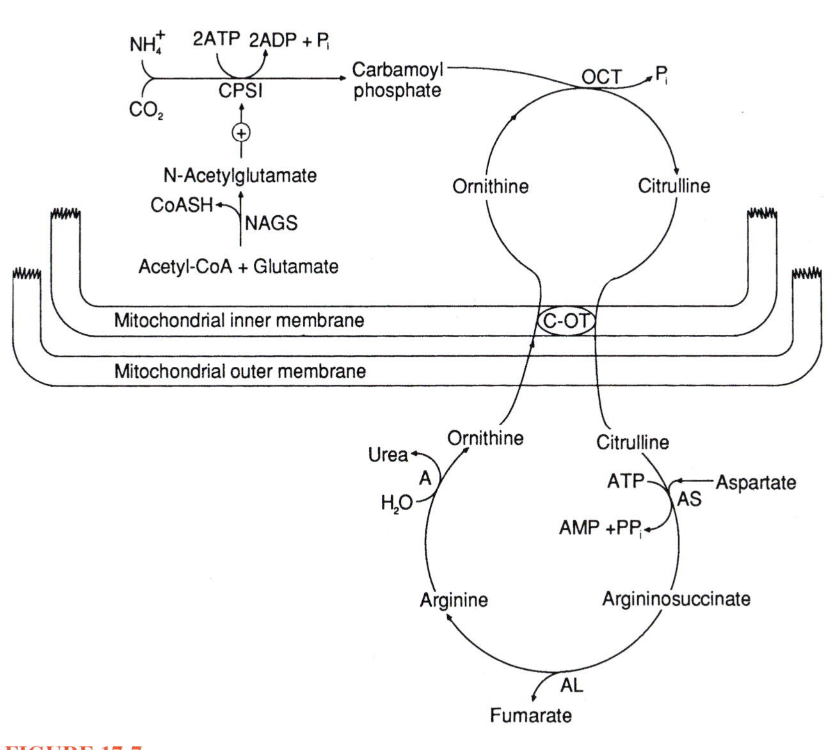

FIGURE 17-7
Formation of urea in hepatocytes. NAGS = N-acetylglutamate synthase; CPSI = carbamoylphosphate synthase I; OCT = ornithine carbamoyltransferase; C-OT = citrulline-ornithine translocase; AS = argininosuccinate synthase; AL = argininosuccinate lyase; A = arginase. —⊕→ indicates the absolute requirement of N-acetylglutamate for CPSI activity.

by glutamine synthase located in pericentral hepatocytes. Formation of urea requires the combined action of two enzymes to produce carbamoyl phosphate and of four enzymes that function in a cyclic manner in the urea cycle (Figure 17-7). Although some of these enzymes occur in extrahepatic tissues and urea formation has been shown to occur in several cell lines in tissue culture, the most important physiological site of urea formation is the liver. In hepatocytes the first three enzymes are mitochondrial and the others are cytosolic. A citrulline-ornithine antiport is located in the inner mitochondrial membrane.

Formation of Carbamoyl Phosphate

Carbamoyl phosphate synthesis requires amino acid acetyltransferase (N-acetylglutamate synthase, mitochondrial) and carbamoyl-phosphate synthase I (CPSI). N-Acetylglutamate (NAG) is an obligatory positive effector of CPSI. NAG synthase is under positive allosteric modulation by arginine and product inhibition by NAG. Depletion of CoA-SH decreases NAG synthesis and ureagenesis. This situation can occur in ***organic acidemias*** (e.g., propionic acidemia; Chapter 18), in which organic acids produced in excess compete for CoA-SH for formation of CoA derivatives that are also competitive inhibitors of N-acetylglutamate synthase and inhibitors of CPSI. ***Hyperammonemia*** often accompanies organic acidemias.

$$\underset{\text{Acetyl-CoA}}{CH_3\overset{O}{\overset{\|}{C}}\text{—SCoA}} + \underset{\text{Glutamate}}{\overset{COO^-\,|\,(CH_2)_2\,|}{CHNH_3^+}\underset{|\,COO^-}{}} \xrightarrow[\ominus\ \text{N-Acetylglutamate}]{\text{Arg}\ \oplus} \text{CoASH} + \overset{COO^-\,|\,(CH_2)_2\,|}{CHNHCOCH_3}\underset{|\,COO^-}{} + H^+$$

CPSI catalyzes the reaction

$$NH_4^+ + HCO_3^- + 2ATP^{4-} \xrightarrow{\text{NAG},K^+,Mg^{2+}} H_2N\text{—}\overset{O}{\overset{\|}{C}}\text{—}OPO_3^{2-} + 2ADP^{3-} + P_i^{2-} + 2H^+$$

NAG binding changes the conformation and subunit structure of CPSI, with preponderance of the monomers. Carbamoylglutamate is also an activator of CPSI. Glutamate and α-ketoglutarate compete with NAG for binding. CPSI is subject to product inhibition by Mg-ADP. It possesses two binding sites for ATP. One ATP is utilized in activation of bicarbonate by forming an enzyme-bound carboxyphosphate that reacts with ammonium ion to form an enzyme-bound carbamate, with elimination of inorganic phosphate. Carbamoyl phosphate is generated when the second ATP reacts with the enzyme-bound carbamate, with release of ADP and free enzyme.

In humans, there are two immunologically distinct carbamoyl phosphate synthases, one mitochondrial (CPSI) and the other cytosolic (CPSII). CPSI is involved in ureagenesis, uses NH_3 exclusively as the nitrogen donor, and requires binding of NAG for activity. CPSII uses glutamine as substrate, is not dependent on NAG for activity, and is required for synthesis of pyrimidine (Chapter 27). Normally, the mitochondrial membrane is not permeable to carbamoyl phosphate, but when the concentration increases, carbamoyl phosphate spills into the cytosol and promotes synthesis of orotic acid and uridine 5′-phosphate.

Formation of Citrulline

Ornithine carbamoyltransferase (ornithine transcarbamoylase) catalyzes the condensation between carbamoyl phosphate and ornithine to yield citrulline in mitochondria:

$$H_2N{-}\overset{\overset{O}{\|}}{C}{-}O{-}PO_3^{2-} + NH_3^+{-}CH_2{-}CH_2{-}CH_2{-}CHNH_3^+{-}COO^- \rightleftharpoons NH_2{-}C(=O){-}NH{-}CH_2{-}CH_2{-}CH_2{-}CHNH_3^+{-}COO^- + P_i^{2-} + H^+$$

Carbamoyl phosphate, L-Ornithine, L-Citrulline ($NH_2{-}C{=}O$: Transferred carbamoyl group)

Although the equilibrium constant strongly favors citrulline formation, the reaction is reversible. Citrulline is transported out of the mitochondria by the citrulline-ornithine antiporter.

Formation of Argininosuccinate

The condensation of citrulline and aspartate to argininosuccinate is catalyzed by argininosuccinate synthase in the cytosol and occurs in two steps. In the initial step, the ureido group is activated by ATP to form the enzyme-bound intermediate adenylylcitrulline. In the second step, nucleophilic attack of the amino group of aspartate displaces AMP and yields argininosuccinate. The overall reaction is shown below:

$$NH_2{-}C(=O){-}NH{-}CH_2{-}CH_2{-}CH_2{-}CHNH_3^+{-}COO^- + {}^-OOC{-}CHNH_3^+{-}CH_2{-}COO^- + ATP^{4-} \rightleftharpoons NH{-}C(=NH^+{-}CH(COO^-){-}CH_2{-}COO^-){-}NH{-}CH_2{-}CH_2{-}CH_2{-}CHNH_3^+{-}COO^- + AMP^{2-} + PP_i^{3-} + H^+$$

L-Citrulline, L-Aspartate, L-Argininosuccinate

The reaction is driven forward by hydrolysis of pyrophosphate to inorganic phosphate. Argininosuccinate formation is considered as the rate-limiting step for urea synthesis. This reaction incorporates the second nitrogen atom of the urea molecule donated by aspartate.

Formation of Arginine and Fumarate

Argininosuccinate lyase in cytosol catalyzes cleavage of argininosuccinate to arginine and fumarate:

$$NH_2{-}C(=NH^+{-}CH(CO^-)(=O){-}CH_2{-}COO^-){-}NH{-}CH_2{-}CH_2{-}CH_2{-}CHNH_3^+{-}COO^- \rightleftharpoons NH_2{-}C(=NH_2^+){-}NH{-}CH_2{-}CH_2{-}CH_2{-}CHNH_3^+{-}COO^- + {}^-OOC{-}CH{=}HC{-}COO^-$$

L-Argininosuccinic acid, L-Arginine, Fumarate

This is the pathway for synthesis of arginine, a nonessential amino acid; however, in the event of physiological deficiency, as in premature infants, or a defect in any of the enzymes discussed above, an exogenous supply of arginine is required.

Formation of Urea and Ornithine

This irreversible reaction is catalyzed by arginase in the cytosol:

$$
\begin{array}{c}
NH_2 \\ | \\ C{=}NH_2^+ \\ | \\ NH \\ | \\ CH_2 \\ | \\ CH_2 \\ | \\ CH_2 \\ | \\ CHNH_3^+ \\ | \\ COO^- \\ \text{L-Arginine}
\end{array}
\; + \; H_2O \longrightarrow
\begin{array}{c}
NH_2 \\ | \\ C{=}O \\ | \\ NH_2 \\ \text{Urea}
\end{array}
\; + \;
\begin{array}{c}
NH_3^+ \\ | \\ CH_2 \\ | \\ CH_2 \\ | \\ CH_2 \\ | \\ CHNH_3^+ \\ | \\ COO^- \\ \text{L-Ornithine}
\end{array}
$$

The urea so formed is distributed throughout the body water and excreted. The renal clearance of urea is less than the glomerular filtration rate because of passive tubular back-diffusion. Diffusion of urea in the intestine leads to formation of ammonia, which enters the portal blood and is converted to urea in liver. Reentry of ornithine into mitochondria initiates the next revolution of the urea cycle. Ornithine can be converted to glutamate-γ-semialdehyde (which is in equilibrium with its cyclic form Δ'-pyrroline-5-carboxylate) by ornithine aminotransferase and decarboxylated to putrescine by ornithine decarboxylase. Ornithine is also produced in the arginine-glycine transamidinase reaction.

The availability of substrates (ammonia and amino acids) in the liver determines the amount of urea synthesized. Urea excretion increases with increased protein intake and decreases with decreased protein intake.

Energetics of Ureagenesis

The overall reaction of ureagenesis is

$$NH_3 + HCO_3^- + \text{aspartate} + 3ATP \rightarrow \text{urea} + \text{fumarate} + 2ADP + 4P_i + AMP$$

Hydrolysis of four high-energy phosphate groups is required for the formation of one molecule of urea. If fumarate is converted to aspartate (by way of malate and oxaloacetate), one NADH molecule is generated that can give rise to three ATP molecules through the electron transport chain, so that the energy expenditure becomes one ATP molecule per each molecule of urea.

Hyperammonemias

Hyperammonemias are caused by inborn errors of ureagenesis and organic acidemias, liver immaturity (transient hyperammonemia of the newborn), and liver failure (hepatic encephalopathy). Neonatal hyperammonemias are characterized by vomiting, lethargy, lack of appetite, seizures, and coma. The underlying defects can be identified by appropriate laboratory measurements (e.g., assessment of metabolic acidosis if present and characterization of organic acids, urea cycle intermediates, and glycine).

Inborn errors of the six enzymes of ureagenesis and NAG synthase have been described. The inheritance pattern of the last is not known, but five of the urea cycle defects are autosomal recessive and ornithine carbamoyltransferase (OCT) deficiency is X-linked.

Carriers of OCT deficiency (estimated to be several thousand women in the U.S.A.) can be identified by administration of a single oral dose of allopurinol, a purine analogue, followed by measurement of urinary orotidine excretion. The underlying principle of this assay is that when the intramitochondrial carbamoyl phosphate accumulates in OCT heterozygotes, it diffuses into the cytoplasm stimulating the biosynthesis of pyrimidines. One of the intermediates in this pathway—***orotidine***—accumulates, leading to ***orotidinuria*** (Figure 17-8).

The sensitivity of this test is increased by increasing the flux in the pyrimidine biosynthetic pathway. The enhanced flux is accomplished by allopurinol, which by way of oxypurinol ribonucleotide inhibits the formation of final product uridine 5′-phosphate (UMP) in the pyrimidine biosynthesis (Chapter 27).

Antenatal diagnosis for fetuses at risk for the urea cycle enzyme disorders can be made by appropriate enzyme assays and DNA analysis in the cultured amniocytes.

Acute neonatal hyperammonemia, irrespective of cause, is a medical emergency and requires immediate and rapid lowering of ammonia levels to prevent serious effects on the brain. Useful measures include hemodialysis, exchange transfusion, peritoneal dialysis, and administration of arginine hydrochloride. The general goals of management are to

1. Decrease nitrogen intake so as to minimize the requirement for nitrogen disposal,
2. Supplement arginine intake, and
3. Promote nitrogen excretion in forms other than urea.

The first can be accomplished by restriction of dietary protein and administration of α-keto analogues of essential amino acids. Arginine supplementation as a precursor of ornithine is essential to the urea cycle. The diversion

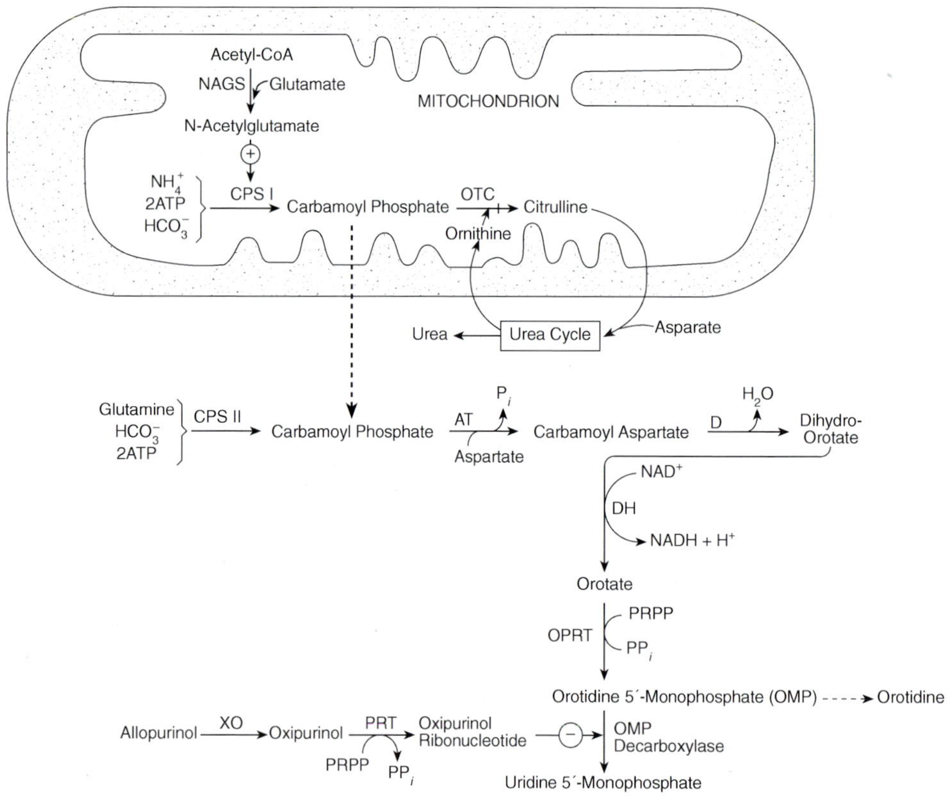

FIGURE 17-8

The metabolic interrelationship between mitochondrial carbamoyl phosphate synthesis to urea formation and to cytosolic carbamoyl phosphate channeled into pyrimidine biosynthesis. In ***ornithine transcarbamoylase (OTC) deficiency,*** mitochondrial carbamoyl phosphate diffuses into the cytosol and stimulates pyrimidine biosynthesis, leading to *orotidinuria.* Administration of allopurinol augments orotidinuria by increasing the flux in the pyrimidine biosynthetic pathway. CPS = Carbamoyl phosphate synthase, AT = aspartate transcarbamoylase, D = dihydroorotase, DH = dihydroorotate dehydrogenase, OPRT = orotate phosphoribosyltransferase, XO = xanthine oxidase, PRT = phosphoribosyltransferase, PRPP = 5-phosphoribosyl-1-pyrophosphate.

of nitrogen to products other than urea is achieved by administration of sodium benzoate or sodium (or calcium) phenylacetate. Administration of benzoate leads to elimination of hippurate (benzoylglycine):

$$C_6H_5{-}C(=O){-}O^- + ATP^{4-} + CoASH \xrightarrow{\text{Activating enzyme}} C_6H_5{-}C(=O){-}S{-}CoA + AMP^{2-} + PP_i^{3-}$$

Benzoate → Benzoyl-CoA

Hippurate is rapidly secreted since its clearance is five times greater than its glomerular filtration rate. The glycine nitrogen is derived from ammonia in a complex reaction that uses CO_2, NADH, N^5,N^{10}-methylenetetrahydrofolate (a source of a single carbon unit; Chapter 27) and pyridoxal phosphate, catalyzed by mitochondrial glycine synthase (glycine cleavage enzyme).

Phenylacetate or phenylbutyrate administration increases excretion of phenylacetylglutamine:

$$C_6H_5{-}CH_2{-}COO^- + ATP^{4-} + CoASH \xrightarrow{\text{Activating enzyme}} C_6H_5{-}CH_2{-}C(=O){-}SCoA + AMP^{2-} + PP_i^{3-}$$

Phenylacetyl-CoA

$$C_6H_5{-}CH_2{-}C(=O){-}SCoA + H_3^+N{-}CH(COO^-){-}(CH_2)_2{-}CONH_2 \xrightarrow{\text{Conjugating enzyme}} C_6H_5{-}CH_2{-}C(=O){-}NH{-}CH(COO^-){-}(CH_2)_2{-}CONH_2 + CoASH + H^+$$

Glutamine → Phenylacetylglutamine

The excretion of phenylacetylglutamine produces loss of two nitrogen atoms.

NAG synthase deficiency cannot be treated by administration of NAG, since NAG undergoes cytosolic inactivation by deacylation and is not readily permeable across the inner mitochondrial membrane. An analogue of NAG,

N-carbamoylglutamate, activates CPSI, does not share the undesirable properties of NAG, and has been effective in management of this deficiency.

The most common cause of hyperammonemia in adults is disease of the liver (e.g., due to ethanol abuse, infection, or cancer). The ability to detoxify ammonia is decreased in proportion to the severity of the damage. In advanced disease (e.g., cirrhosis), hyperammonemia is augmented by shunting of portal blood that carries ammonia from the intestinal tract and other splanchnic organs to the systemic blood circulation (bypassing the liver) and leads to portal-systemic encephalopathy. In addition to dietary protein restriction, colonic growth of bacteria must be suppressed by antibiotics (e.g., neomycin) and administration of lactulose (Chapter 9), a nonassimilable disaccharide. Enteric bacteria catabolize lactulose to organic acids that convert NH_3 to NH_4^+, thereby decreasing absorption of NH_3 into the portal circulation. Catabolism of lactulose also leads to formation of osmotically active particles that draw water into the colon, produce loose, acid stools, and permit loss of ammonia as ammonium ions.

17.3 Metabolism of Some Individual Amino Acids

Mammalian tissues synthesize the nonessential amino acids from carbon skeletons derived from lipid and carbohydrate sources or from transformations that involve essential amino acids. The nitrogen is obtained from NH_4^+ or from that of other amino acids. Nonessential amino acids (and their precursors) are glutamic acid (α-ketoglutaric acid), aspartic acid (oxaloacetic acid), serine (3-phosphoglyceric acid), glycine (serine), tyrosine (phenylalanine), proline (glutamic acid), alanine (pyruvic acid), cysteine (methionine and serine), arginine (glutamate-γ-semialdehyde), glutamine (glutamic acid), and asparagine (aspartic acid).

Amino acids may be classified as ketogenic, glucogenic, or glucogenic and ketogenic, depending on whether feeding of a single amino acid to starved animals or animals with experimentally induced diabetes increases plasma or urine levels of glucose or ketone bodies (Chapter 18). Leucine and lysine are ketogenic; isoleucine, phenylalanine, tyrosine, and tryptophan are glucogenic and ketogenic; and the remaining amino acids are glucogenic. Points of entry of amino acids into the gluconeogenic pathway are discussed in Chapter 15.

Arginine

Arginine participates in a number of metabolic pathways depending on the cell type. It is synthesized as an intermediate in the urea cycle pathway and is also obtained from dietary proteins. A number of key metabolites such as nitric oxide, phosphocreatine, spermine and ornithine are derived from arginine. During normal growth and development, under certain pathological conditions (e.g., endothelial dysfunction) and if the endogenous production of arginine is insufficient, a dietary supplement of arginine may be required. Thus, arginine is considered a semiessential amino acid.

Metabolism and Synthesis of Nitric Oxide

Nitric oxide (NO) is a reactive diatomic gaseous molecule with an unpaired electron (a free radical). It is lipophilic and can diffuse rapidly across biological membranes. NO mediates a variety of physiological functions such as endothelial derived relaxation of vascular smooth muscle, inhibition of platelet aggregation, neurotransmission, and cytotoxicity. The pathophysiology of NO is a double-edged sword. Insufficient production of NO has been implicated in the development of hypertension, impotence, susceptibility to infection, and atherogenesis. Excessive NO production is linked to septic shock, inflammatory diseases, transplant rejection, stroke, and carcinogenesis.

NO is synthesized from one of the terminal nitrogen atoms or the guanidino group of arginine with the concomitant production of citrulline. Molecular oxygen and NADPH are cosubstrates and the reaction is catalyzed by nitric oxide synthase (NOS). NOS consists of several isoforms and is a complex enzyme containing bound FMN, FAD, tetrahydrobiopterin, heme complex, and nonheme iron. A calmodulin binding site is also present. NO formation from arginine is a two-step process requiring five-electron oxidations. The first step is the formation of N^G-hydroxylarginine (N^G denotes guanidinium nitrogen atom):

$$\text{Arginine} + O_2 + \text{NADPH} + H^+ \rightarrow \text{HO–N}^G\text{–Arg} + \text{NADP}^+ + H_2O$$

This step is a mixed-function oxidation reaction similar to the one catalyzed by cytochrome P-450 reductase and there is considerable homology between NOS and cytochrome P-450 reductase. In the second step, further oxidation of N^G-hydroxyl arginine yields NO and citrulline:

$$\text{HO–N}^G\text{–Arg} + O_2 + \frac{1}{2}(\text{NADPH} + H^+) \rightarrow \text{citrulline} + \text{NO} + H_2O + \frac{1}{2}\text{NADP}^+$$

The overall reaction is:

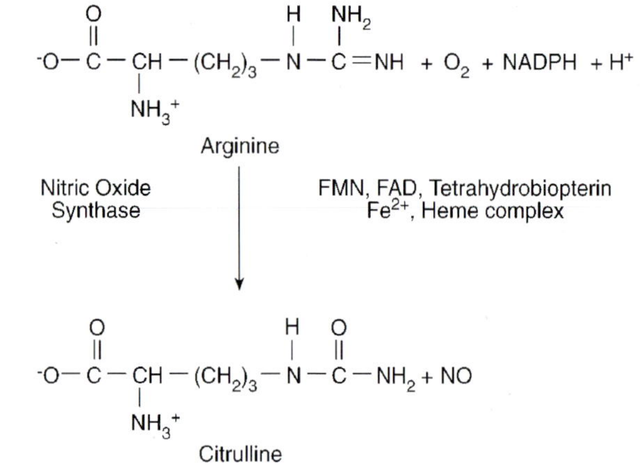

$$^{-}O-\overset{O}{\overset{\|}{C}}-\underset{NH_3^+}{\underset{|}{CH}}-(CH_2)_3-\overset{H}{\overset{|}{N}}-\overset{NH_2}{\overset{|}{C}}=NH + O_2 + NADPH + H^+$$

Arginine

Nitric Oxide Synthase — FMN, FAD, Tetrahydrobiopterin Fe^{2+}, Heme complex

$$^{-}O-\overset{O}{\overset{\|}{C}}-\underset{NH_3^+}{\underset{|}{CH}}-(CH_2)_3-\overset{H}{\overset{|}{N}}-\overset{O}{\overset{\|}{C}}-NH_2 + NO$$

Citrulline

The NOS activity is inhibited by N^G-substituted analogues of arginine, such as N^G-nitroarginine and N^G-monomethyl-L-arginine.

Isoforms (Also Known as Isozymes) of Nitric Oxide Synthase

There are three major isoforms of ***Nitric Oxide Synthase (NOS)*** ranging in molecular size from 130 to 160 kDa. Amino acid similarity between any two isoforms is about 50–60%. Isoforms of NOS exhibit differences in tissue distribution, transcriptional regulation, and activation by intracellular Ca^{2+}. Two of the three isoforms of NOS are constitutive enzymes (cNOS) and the third isoforin is an inducible enzyme (iNOS). The cNOS isoforms are found in the vascular endothelium (eNOS), neuronal cells (nNOS), and many other cells, and are regulated by Ca^{2+} and calmodulin. In the vascular endothelium, agonists such as acetylcholine and bradykinin activate eNOS by enhancing intracellular Ca^{2+} concentrations via the production of inositol 1,4,5-trisphosphate, which activates the phosphoinositide second-messenger system (Chapter 30). The NO produced in the vascular endothelium maintains basal vascular tone by vasodilation which is mediated by vascular smooth muscle cells. Organic nitrates used in the management of ischemic heart disease act by denitration with the subsequent formation of NO. Sodium nitroprusside, an antihypertensive drug, is an NO donor. Thus, organic nitrates and sodium nitroprusside are prodrugs, and the exact mechanism by which these prodrugs yield NO is not yet understood. Inhaled NO can produce pulmonary vasodilation. This property of NO has been used in the management of hypoxic respiratory failure associated with primary pulmonary hypertension in neonates. NO produced by cNOS in neuronal tissue functions as a neurotransmitter.

The inducible class of NOS (iNOS) is found in macrophages and neutrophils and is Ca^{2+}-independent. Bacterial endotoxins, cytokines (e.g., interleukin-1, interferon-γ), or bacterial lipopolysaccharides can induce and cause expression of NOS in many cell types. Glucocorticoids inhibit the induction of iNOS. In stimulated macrophages and neurophils, NO and superoxide radical (O_2^-) react to generate peroxynitrite, a powerful oxidant, and hydroxyl radicals. These reactive intermediates are involved in the killing of phagocytized bacteria (Chapter 14). Excessive production of NO due to endotoxinemia produces hypotension and vascular hyporeactivity to vasoconstrictor agents, and leads to septic shock. NOS inhibitors have potential therapeutic application in the treatment of hypotensive crisis.

Signal Transduction of NO

NO is lipophilic and diffuses readily across cell membranes. It interacts with molecules in the target cells producing various biological effects. One mechanism of action of NO is stimulation of guanylate cyclase, which catalyzes the formation of cyclic guanosine monophosphate (cGMP) from GTP, resulting in increased intracellular cGMP levels (Figure 17-9). NO activates guanylate cyclase by binding to heme iron. The elevation of cGMP levels may activate cGMP-dependent protein kinases.

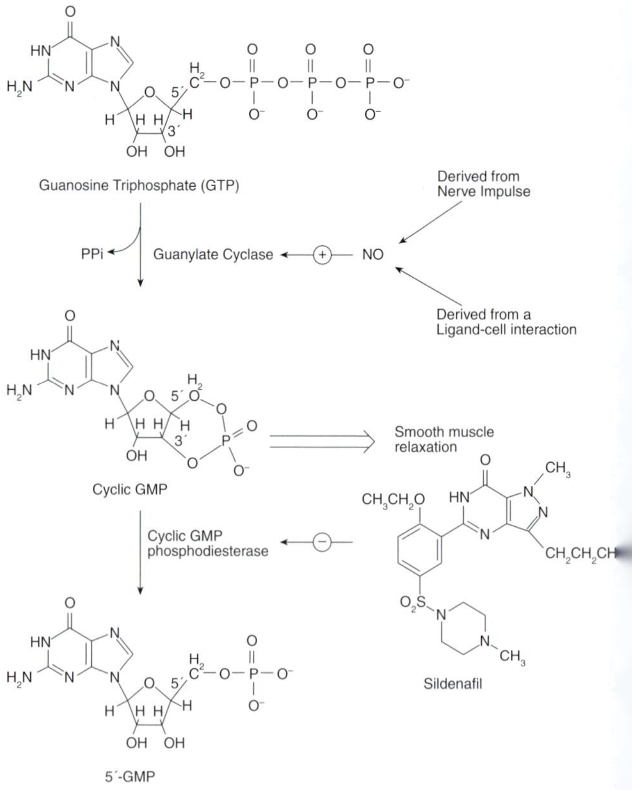

Guanosine Triphosphate (GTP)

PPi — Guanylate Cyclase ← (+) — NO

Derived from Nerve Impulse

Derived from a Ligand-cell interaction

Cyclic GMP

Smooth muscle relaxation

Cyclic GMP phosphodiesterase ← (−) — Sildenafil

5′-GMP

FIGURE 17-9
NO-mediated synthesis of cGMP from GTP in the corpus cavernosum that leads to smooth muscle relaxation. Sildenafil potentiates the effects of NO by inhibiting cGMP phosphodiesterase.

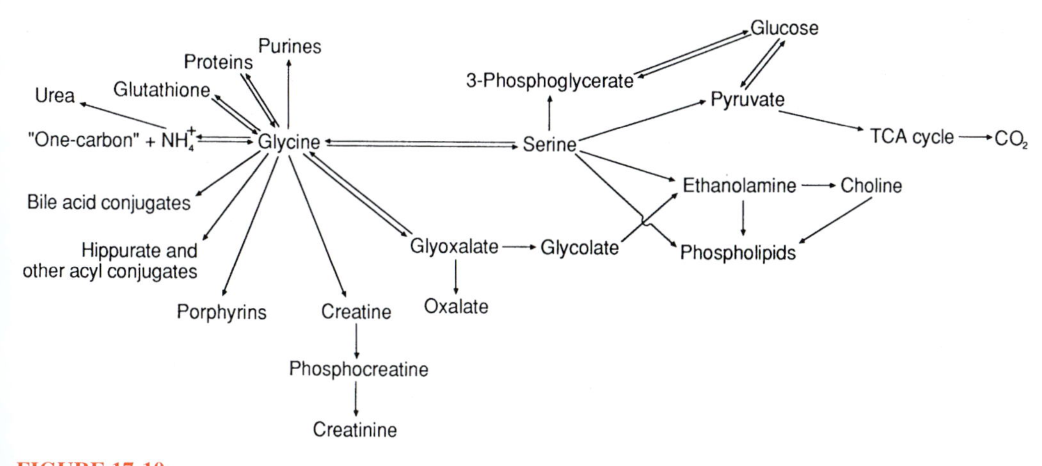

FIGURE 17-10
Overview of glycine and serine metabolism.

These kinases phosphorylate specific proteins that may be involved in removal or sequestration of Ca^{2+} or other ions, resulting in physiological stimuli. The physiological actions of cGMP are terminated by its conversion to 5′-GMP by cGMP-phosphodiesterase. Inhibitors of cGMP phosphodiesterase promote the actions of NO.

Sildenafil is a selective inhibitor of a specific cGMP phosphodiesterase (type 5) present in the corpus cavernosum. This compound (structure shown in Figure 17-9) is used orally in the therapy of some types of ***erectile dysfunction.*** NO is the principal transmitter involved in the relaxation of penile smooth muscle. During central or reflex sexual arousal, NO production is enhanced leading to increased production of cGMP. Smooth muscle relaxation permits the corpus cavernosum to fill with blood. Since the therapeutic effect of sildenafil potentiates the action of cGMP, the drug is ineffective in the absence of sexual arousal. The relaxation of cavernosal smooth muscle caused by cGMP involves inhibition of Ca^{2+} uptake. Prostaglandin E_1 (alprostadil) inhibits the uptake of Ca^{2+} smooth muscle by a separate mechanism and causes erections in the absence of sexual arousal. Blood flow through corpus cavernosum may also be increased by α-adrenergic blocking agents (e.g., phentolamine mesylate). Coadministration of NO donor drugs with the NO potentiation drug sildenafil may have severe consequences on the cardiovascular system.

Signal transduction of NO by cGMP-independent mechanisms include ADP-ribosylation of glyceraldehyde-3-phosphate dehydrogenase (GADPH), an enzyme of the glycolytic pathway (Chapter 13), and interactions with many heme-containing and nonheme iron-sulfur containing proteins. NO activates ADP-ribosyltransferase which catalyzes the transfer of ADP-ribose from NAD^+ to GADPH. This results in the inactivation of GADPH causing inhibition of glycolysis and decreased ATP production. The antiaggregability of platelets and the neurotoxicity of NO have been attributed to inhibition of glycolysis by NO.

Glycine

Glycine participates in a number of synthetic pathways and is oxidized to provide energy (Figure 17-10). The interconversion of glycine and serine by serine hydroxymethyltransferase is shown below:

$$\underset{\text{glycine}}{NH_3^+\text{–}CH_2\text{–}COO^-} + N^5,N^{10}\text{-methylene-}FH_4 + H_2O \xrightleftharpoons{\text{pyridoxal phosphate}} \underset{\text{serine}}{HOH_2C\text{–}CHNH_3^+\text{–}COO^-} + FH_4$$

The one-carbon carrier N^5,N^{10}-methylenetetrahydrofolate is derived from reactions of the one-carbon pool (Chapter 27). [The term ***one-carbon pool*** refers to all single-carbon-containing metabolites (e.g., $-CH_3$, $-CHO$, $NH{=}C-$, etc.) that can be utilized in biosynthetic reactions such as formation of purine and pyrimidine.] These reactions include oxidation of glycine by glycine cleavage enzyme complex (glycine synthase):

$$\underset{\text{glycine}}{NH_3^+\text{–}CH_2\text{–}COO^-} + FH_4 + NAD^+ \xrightleftharpoons{\text{pyridoxal phosphate}} NH_4^+ + CO_2 + NADH + N^5,N^{10}\text{-methylene–}FH_4$$

This reaction favors glycine degradation, but the formation of glycine may also occur. The enzyme complex is mitochondrial and contains a pyridoxal phosphate–dependent glycine decarboxylase, a lipoic acid–containing protein that is a carrier of an aminomethyl moiety, a tetrahydrofolate-requiring enzyme, and lipoamide dehydrogenase. The reactions of glycine cleavage resemble those of oxidative decarboxylation of pyruvate (Chapter 13).

Glycine is also oxidized by D-amino acid oxidase, an FAD protein:

$$\underset{\text{glycine}}{NH_3^+\text{–}CH_2\text{–}COO^-} + O_2 + H_2O \rightarrow \underset{\text{glyoxalate}}{CHO\text{–}COO^-} + NH_4^+ + H_2O_2$$

Glyoxalate can be transaminated to glycine, reduced to glycolate, converted to α-hydroxy-β-ketoadipate by reaction with α-ketoglutarate, or oxidized to oxalate and excreted in urine. The first three reactions require pyridoxal phosphate, NADH, and thiamine pyrophosphate, respectively. In humans, ascorbic acid (vitamin C) is a precursor of urinary oxalate (Chapter 38). Since calcium oxalate is poorly soluble in water, it can cause nephrolithiasis and nephrocalcinosis due to ***hyperoxaluria.***

Disorders of Glycine Catabolism

Nonketotic hyperglycinemia is an inborn error due to a defect in the glycine cleavage enzyme complex in which glycine accumulates in body fluids and especially in cerebrospinal fluid. It is characterized by mental retardation and seizures. Glycine is an inhibitory neurotransmitter in the central nervous system, including the spinal cord. Strychnine, which produces convulsions by competitive inhibition of glycine binding to its receptors, gives modest results in treatment but not very effective. Sodium benzoate administration reduces plasma glycine levels but does not appreciably alter the course of the disease. Exchange transfusion may be useful. ***Ketotic hyperglycinemia*** also occurs in propionic acidemia but the mechanism has not been established.

Primary hyperoxaluria type I is due to a deficiency of cytosolic α-ketoglutarate-glyoxylate carboligase, which catalyzes the following reaction:

$$\underset{\text{Glyoxalate}}{CHO\text{—}COO^-} + \underset{\alpha\text{-Ketoglutarate}}{^-OOC\text{—}CH_2\text{—}CH_2\text{—}C(=O)\text{—}COO^-}$$

Carboligase | Thiamine pyrophosphate ↓

$$CO_2 + \underset{\alpha\text{-Hydroxy-}\beta\text{-ketoadipate}}{^-OOC\text{—}CH_2\text{—}CH_2\text{—}C(=O)\text{—}CH(OH)\text{—}COO^-}$$

The glyoxylate that accumulates is converted to oxalate.

Creatine and Related Compounds

Phosphocreatine serves as a high-energy phosphate donor for ATP formation (e.g., in muscle contraction; see Chapter 21). Synthesis of ***creatine*** (methyl guanidinoacetate) requires transamidination, i.e., transfer of a guanidine group from arginine to glycine, to form guanidinoacetate (glycocyamine) by mitochondrial arginine-glycine amidinotransferase

$$\underset{\text{Arginine}}{NH_2\text{—}C(=NH_2^+)\text{—}NH\text{—}(CH_2)_3\text{—}CHNH_3^+\text{—}COO^-} + \underset{\text{Glycine}}{NH_3^+\text{—}CH_2\text{—}COO^-} \rightleftharpoons \underset{\text{Guanidinoacetate (glycocyamine)}}{NH_2\text{—}C(=NH_2^+)\text{—}NH\text{—}CH_2\text{—}COO^-} + \underset{\text{Ornithine}}{NH_3^+\text{—}(CH_2)_3\text{—}CHNH_3\text{—}COO^-}$$

In the next step guanidinoacetate is methylated by S-adenosylmethionine by cytosolic S-adenosylmethionine guanidinoacetate-N-methyltransferase to form creatine.

$$\underset{\text{S-Adenosyl-methionine}}{CH_3\text{—}S^+\text{-Adenosyl}\text{—}(CH_2)_2\text{—}CHNH_3^+\text{—}COO^-} + \underset{\text{Guanidino-acetate}}{NH_2\text{—}C(=NH_2^+)\text{—}NH\text{—}CH_2\text{—}COO^-} \longrightarrow \underset{\text{Creatine}}{NH_2\text{—}C(=NH_2^+)\text{—}N(CH_3)\text{—}CH_2\text{—}COO^-} + \underset{\text{S-Adenosyl-homocysteine}}{S\text{-Adenosyl}\text{—}(CH_2)_2\text{—}CHNH_3^+\text{—}COO^-} + H^+$$

These reactions occur in liver, kidney, and pancreas, from which creatine is transported to organs such as muscle and brain. Creatine synthesis is subject to negative modulation of amidinotransferase by creatine. Phosphocreatine production is catalyzed by creatine kinase:

$$\underset{\text{Creatine}}{NH_2\text{—}C(=NH_2^+)\text{—}N(CH_3)\text{—}CH_2\text{—}COO^-} + ATP^{4-} \rightleftharpoons \underset{\text{Phosphocreatine}}{NH\text{—}PO_3^{2-}\text{—}C(=NH_2^+)\text{—}N(CH_3)\text{—}CH_2\text{—}COO^-} + ADP^{3-} + H^+$$

This kinase is a dimer of M and B (M = muscle, B = brain) subunits produced by different structural genes. Three isozymes are possible: BB (CK-1), MB (CK-2), and MM (CK-3). Another isozyme differs immunologically and electrophoretically and is located in the intermembrane space of mitochondria. Tissues rich in CK-

are brain, prostate, gut, lung, bladder, uterus, placenta, and thyroid; those rich in CK-3 are skeletal and cardiac muscle. Cardiac muscle contains significant amounts of CK-2 (25–46% of total CK activity, as opposed to less than 5% in skeletal muscle), so that in myocardial infarction the rise in serum total CK activity is accompanied by a parallel rise in that of CK-2 (Chapter 8).

Phosphocreatine undergoes a slow and nonenzymatic cyclization to creatinine.

$$\text{Phosphocreatine } (NH{-}PO_3^{2-},\ C{=}NH_2^+,\ N{-}CH_3,\ CH_2,\ COO^-) + H^+ \longrightarrow \text{Creatinine } (H_2N{=}C,\ NH,\ C{=}O,\ N{-}CH_2,\ H_3C) + P_i^{2-} + H_2O$$

Phosphocreatine   Creatinine

Creatinine has no useful function and is eliminated by renal glomerular filtration and to a small extent by renal tubular secretion. Creatinine clearance approximately parallels the ***glomerular filtration rate*** (GFR) and is used as a kidney function test. It is calculated as follows:

$$\text{Creatinine clearance} = \frac{\text{urine creatinine (mg/L)}}{\text{plasma creatinine (mg/L)}} \times \text{urine volume per unit time}$$

Creatinine concentrations are measured from a precisely timed urine specimen (e.g., 4-hour, 24-hour) and a plasma specimen drawn during the urine collection period. Excretion of creatinine depends on skeletal muscle mass and varies with age and sex. However, day-to-day variation in a healthy individual is not significant. ***Creatinuria,*** the excessive excretion of creatine in urine, may occur during growth, fever, starvation, diabetes mellitus, extensive tissue destruction, muscular dystrophy, and hyperthyroidism.

Use of Creatine as a Dietary Supplement

The creatine pool in the human body comes from both endogenous synthesis and the diet which provides 1–2 g. Red meat provides large amounts of dietary creatine and vegetables a limited amount. Using glycine, arginine, and methionine, creatine is synthesized in the liver, pancreas, and kidney. Creatine transported in blood crosses muscle and nerve cell membranes by means of a specific creatine transporter system against a concentration gradient of 200:1. Intracellularly, creatine is converted to phosphocreatine by ATP, a reaction catalyzed by creatine kinase. Phosphocreatine, with its high phosphoryl transfer potential, regenerates ATP from ADP, thereby maintaining a high level of ATP required during intense exercise. A large pool of phosphocreatine resides in the skeletal muscle. It has been theorized that in order to maximize phosphocreatine stores in the skeletal muscle to replenish ATP during rapid muscle contractions, an exogenous source of creatine may be beneficial.

Double-blind placebo-controlled studies of oral supplementation of creatine in human subjects have shown increased performance during short duration, strenuous, high-intensity exercise. Such activities require that ATP be replenished rapidly from phosphocreatine stores during anaerobic metabolism. These studies usually consisted of ingestion of 20 g of creatine per day for 5 days followed by a maintenance dose of 5–10 g/day. Studies on creatine as an ergogenic aid have not been uniformly positive; some have shown no beneficial effect and still others have been equivocal and indicated that creatine supplementation did not enhance athletic activities. The safety issues of long-term creatine supplementation on kidney, liver, nerve, muscle, and other tissues are not known.

Serine

Synthesis of ***serine*** from 3-phosphoglycerate, an intermediate of glycolysis (Chapter 13), requires oxidation of 3-phosphoglycerate to 3-phosphohydroxypyruvate, transamination of 3-phosphohydroxypyruvate by glutamate, and hydrolysis of 3-phosphoserine to serine (Figure 17-11). This cytosolic pathway is regulated by inhibition of phosphoserine phosphatase by serine. Serine is converted to pyruvate by cytosolic serine dehydratase. More importantly, it is converted in mitochondria to 2-phosphoglycerate by way of hydroxypyruvate and D-glycerate; and the enzymes involved are a transaminase, a dehydrogenase, and a kinase. Serine is interconvertible with glycine (Figure 17-10) and is involved in phospholipid (Chapter 19) and in cysteine synthesis.

Proline

Proline arises from and gives rise to glutamate. Synthesis is by reduction of glutamate to glutamate-γ-semialdehyde by way of an enzyme-bound γ-glutamyl phosphate. The γ-semialdehyde spontaneously cyclizes to Δ'-pyrroline-5-carboxylate, which is then reduced by NAD(P)H to proline (Figure 17-12). Proline is converted to Δ'-pyrroline-5-carboxylate by proline oxidase, which is tightly bound to the inner mitochondrial membrane in liver, kidney, heart, and brain. Δ'-Pyrroline-5-carboxylate is in equilibrium with glutamate-γ-semialdehyde, which can be transaminated to ornithine or reduced to glutamate (Figure 17-12).

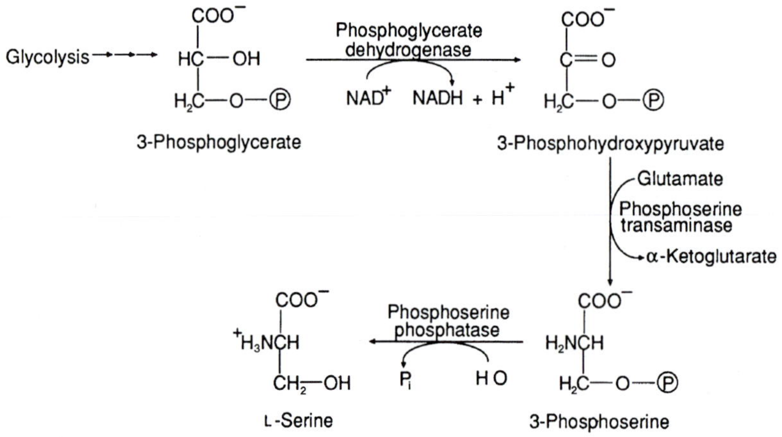

FIGURE 17-11
Synthesis of serine from 3-phosphoglycerate. $P = PO_3^{2-}$; $P_i = HPO_4^{2-}$.

Decarboxylation of ornithine to putrescine by ornithine decarboxylase serves as a source of the polyamines spermidine and spermine. ***Ornithinemia*** results from deficiency of ornithine aminotransferase or ornithine decarboxylase. Ornithine aminotransferase deficiency is associated with gyrate atrophy of the choroid and retina (Chapter 38). Proline and hydroxyproline (produced by posttranslational modification) are major constituents

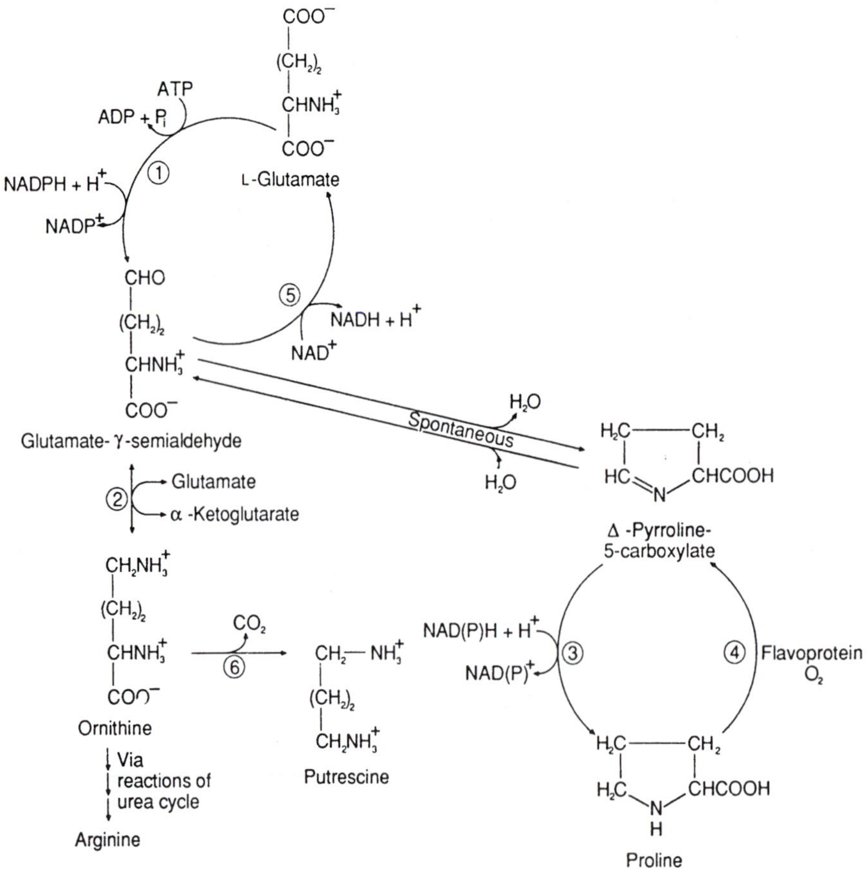

FIGURE 17-12
Metabolism of proline (1) Δ′-pyrroline-5-carboxylate (P5C) synthase; (2) ornithine aminotransferase; (3) P5C reductase; (4) proline oxidase; (5) P5C dehydrogenase; (6) ornithine decarboxylase.

of collagen (Chapter 25). Hydroxyproline released by collagen turnover undergoes degradation similar to that of proline. Hydroxyproline cleavage initiated by hydroxyproline oxidase eventually yields glyoxylate and pyruvate. In ***hyperprolinemia*** type I, proline oxidase is deficient, and in type II, Δ′-pyrroline-5-carboxylate dehydrogenase is deficient. ***Hydroxyprolinemia*** results from hydroxyproline oxidase deficiency. All are clinically harmless autosomal recessive traits.

Histidine

Histidine is not essential for adults except in persons with uremia. It is essential for growth in children. Histidine is synthesized from 5-phosphoribosyl-1-pyrophosphate and ATP, forming N′-1′-phosphoribosyl-ATP, catalyzed by the allosteric enzyme ATP phosphoribosyltransferase. This reaction is analogous to the initial reaction of purine nucleotide biosynthesis (Chapter 27). Histamine breakdown produces a one-carbon unit (N^5-formiminotetrahydrofolate) and glutamate (Figure 17-13) by a nonoxidative deamination to urocanate, cleavage of the imidazole ring to N-formiminoglutamate (Figlu), and transfer of the formimino group (–CH=NH) to tetrahydrofolate (Chapter 27).

Folate deficiency leads to accumulation of Figlu, which is excreted in urine. The excretion is very pronounced after a loading dose of histidine, a test used to detect folate deficiency. More sensitive radioisotopic assays use folate binders to the vitamins. High urinary levels of ***Figlu*** may coexist with elevated levels of serum folate. Thus, in vitamin B_{12} (***cobalamin***) deficiency, since cobalamin participates in the following reaction, FH_4 is trapped as N^5-methyltetrahydrofolate and is unavailable as a carrier for the formimino group of Figlu.

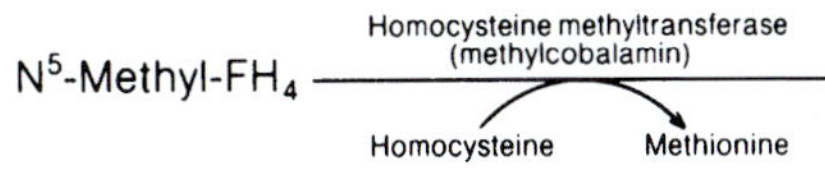

$$N^5\text{-Methyl-}FH_4 \xrightarrow[\text{Homocysteine} \;\to\; \text{Methionine}]{\text{Homocysteine methyltransferase (methylcobalamin)}} FH_4$$

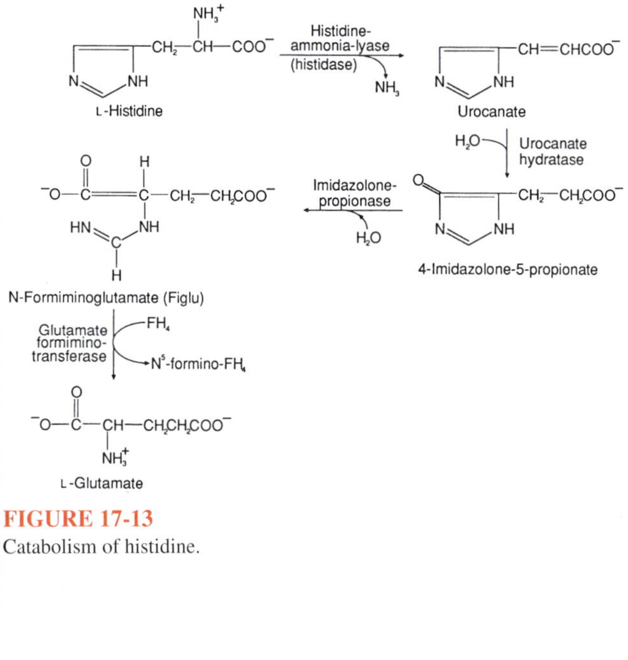

FIGURE 17-13
Catabolism of histidine.

Similarly, a deficiency of glutamate formiminotransferase leads to accumulation of Figlu and high levels of serum folate.

Histidinemia results from deficiency of histidine ammonia-lyase. With a normal diet, histidine (and the products imidazole-pyruvate, imidazole-lactate, imidazole-acetate) accumulates in plasma, cerebrospinal fluid, and urine. This rare autosomal recessive disease may be benign or may manifest with mental retardation and speech defects.

Histidine and β-alanine yield the dipeptide carnosine (present in muscle), and histidine and γ-aminobutyrate yield homocarnosine (found in brain). Methylhistidyl residues are found in some proteins (e.g., actin; Chapter 21) as a result of posttranslational modification. ***Histamine*** is decarboxylated histidine.

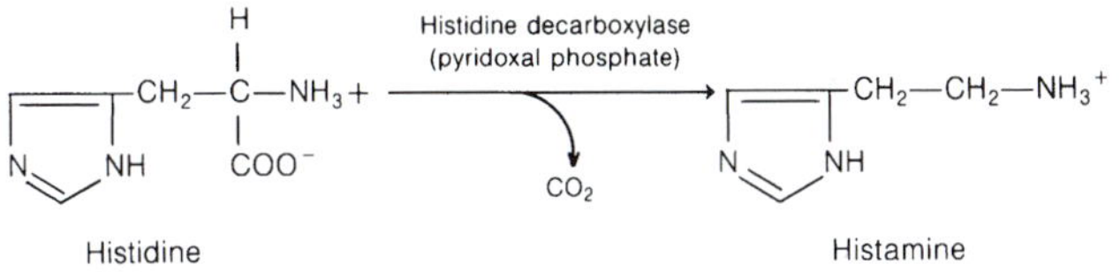

$$\text{Histidine} \xrightarrow[-CO_2]{\text{Histidine decarboxylase (pyridoxal phosphate)}} \text{Histamine}$$

Histamine occurs in blood basophils, tissue mast cells, and certain cells of the gastric mucosa and other parts of the body (e.g., anterior and posterior lobes of the pituitary, some areas of the brain). Histamine is a neurotransmitter in certain nerves ("histaminergic") in the brain. In mast cells found in loose connective tissue and capsules, especially around blood vessels, and in basophils, histamine is stored in granules bound by ionic interactions to a heparin-protein complex and is released (by degranulation, vacuolization, and depletion) in immediate hypersensitivity reactions, trauma, and nonspecific injuries (infection, burns). Degranulation is affected by oxygen, temperature, and metabolic inhibitors. Release of histamine from gastric mucosal cells is mediated by acetylcholine (released by parasympathetic nerve stimulation) and gastrin and stimulates secretion of hydrochloric acid (Chapter 12). Histamine causes contraction of smooth muscle in various organs (gut, bronchi) by binding to H_1 receptors. The conventional antihistaminic drugs (e.g., diphenhydramine and pyrilamine) are H_1-receptor antagonists and are useful in the management of various allergic manifestations. However, in acute anaphylaxis, bronchiolar constriction is rapidly relieved by epinephrine (a physiological antagonist of histamine). Its effect on secretion of hydrochloric acid is mediated by H_2 receptors. H_2-receptor antagonists are cimetidine and ranitidine (Chapter 12), which are useful in treatment of gastric ulcers. Histamine is rapidly inactivated by methylation from S-adenosylmethionine of one of the nitrogen atoms of the imidazole ring (catalyzed by N-methyltransferase) or of the terminal amine group (catalyzed by methyltransferase). Ring-methylated histamine is deaminated by monoamine oxidase to methyl imidazole acetic acid, which is readily excreted. Inactivation also results from deamination of histamine by diamine oxidase. The imidazole acetic acid formed is then excreted as 1-ribosylimidazole-4-acetic acid. This reaction is the only known reaction in which ribose is used for conjugation.

Branched-Chain Amino Acids

Leucine, isoleucine, and valine are essential amino acids but can be derived from their respective α-keto acids. A single enzyme may catalyze transamination of all three. The α-keto acids, by oxidative decarboxylation, yield the acyl-CoA thioesters, which, by α,β-dehydrogenation, yield the corresponding α,β-unsaturated acyl-CoA thioesters. The catabolism of these thioesters then diverges. Catabolism of leucine yields acetoacetate and acetyl-CoA via β-hydroxy-β-methylglutaryl-coenzyme A (HMG-CoA)—also an intermediate in the biosynthesis of cholesterol and other isoprenoids (Chapter 19). Catabolism of isoleucine yields propionyl-CoA (a glucogenic precursor) and acetyl-CoA. Catabolism of valine yields succinyl-CoA (Figure 17-14). Thus, leucine is ketogenic and isoleucine and valine are ketogenic and glucogenic.

Oxidative decarboxylation of the α-keto acids is catalyzed by a branched-chain keto acid dehydrogenase (BCKADH) complex analogous to that of the pyruvate dehydrogenase and α-ketoglutarate dehydrogenases complexes. BCKADH is widely distributed in mammalian tissue mitochondria (especially in liver and kidney). It requires Mg^{2+}, thiamine pyrophosphate, CoA-SH, lipoamide, FAD, and NAD^+ and contains activities of α-keto acid decarboxylase, dihydrolipoyl transacylase, and dihydrolipoyl dehydrogenase. Like the pyruvate dehydrogenase complex, BCKADH is regulated by product inhibition and by phosphorylation (which inactivates) and dephosphorylation (which activates).

The α,β-dehydrogenation is catalyzed by an FAD protein and is analogous to the dehydrogenation of straight-chain acyl-CoA thioesters in β-oxidation of fatty acids (Chapter 18). Methylenecyclopropylacetyl-CoA derived from the plant toxin hypoglycin (Chapters 15 and 18), which inhibits this step in β-oxidation, also inhibits it in the catabolism of branched-chain amino acids.

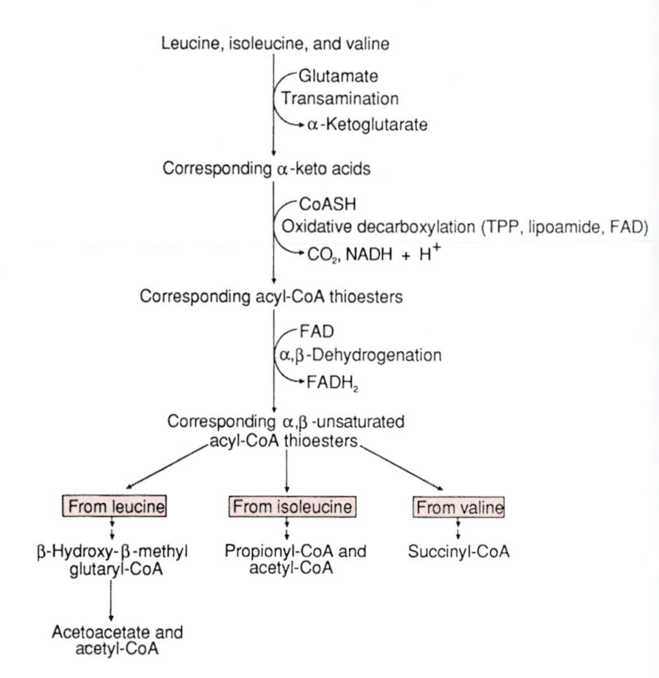

FIGURE 17-14
Overview of the catabolism of branched-chain amino acids. TPP = thiamin pyrophosphate.

Hypoglycin produces hypoglycemia and metabolic acidosis, which frequently are fatal.

Branched-chain ketoaciduria (*maple syrup urine disease*), an autosomal recessive disorder characterized by ketoacidosis starts early in infancy and is due to a defect in the oxidative decarboxylation step of branched-chain amino acid metabolism. The name derives from the characteristic odor (reminiscent of maple syrup) of the urine of these patients. Five different variants (classic, intermittent, intermediate, thiamine-responsive, and dihydrolipoyl dehydrogenase deficiency) are known, of which the first, which is due to deficiency of branched-chain α-keto acid decarboxylase, is the most severe. The incidence of *maple syrup urine disease* in the U.S. population is 1 in 250,000–400,000 live births. In Mennonite populations the incidence is extremely high (1 in 760). Neonatal screening programs consist of measuring leucine levels in dried blood spots using a bacterial inhibition assay. Neonatal screening programs usually include testing for a number of other treatable metabolic disorders such as ***hypothyroidism, phenylketonuria, galactosemia,*** and others. If the screening test is positive for a given metabolic disease, a confirmatory test is performed. For maple syrup urine disease, the confirmation requires quantitation of the serum levels of branched-chain amino acids and urine levels of both the branched-chain amino acids and their ketoacids. Long-term management includes dietary restriction of the branched-chain amino acids. Frequent measurement of plasma concentrations of these amino acids is necessary to monitor the degree of dietary restriction and patient compliance.

Many aminoacidurias and their metabolites give rise to abnormal odors, maple syrup urine disease is one example. Some of the others are ***phenylketonuria*** (musty odor), ***tyrosinemia type I*** (boiled cabbage), ***glutaric aciduria*** (sweaty feet), ***3-methylcrotonyl glycinuria*** (cat's urine), and ***trimethylaminuria*** (fish). In patients with trimethylaminuria the compound responsible for the fish odor is trimethylamine which is a byproduct of protein catabolism by the large intestinal bacterial flora. Normally, trimethylamine is inactivated by hepatic flavin monooxygenases. Several different mutations in the gene for flavin monooxygenases have been identified in trimethylaminuric patients. An inhibitor of flavin monooxygenases is indole-3-carbinol found in dark green vegetables (e.g., broccoli). The amelioration of symptoms of bad odor in trimethylaminuria may be achieved by limiting intake of dark green vegetables and protein, and by administering low doses of antibiotics to reduce intestinal bacterial flora.

Sulfur-Containing Amino Acids

Methionine and cysteine are the principal sources of organic sulfur in humans. Methionine is essential (unless adequate homocysteine and a source of methyl groups are available), but cysteine is not, since it can be synthesized from methionine.

Methionine

Methionine is utilized primarily in protein synthesis, providing sulfur for cysteine synthesis, and is the body's principal methyl donor. In methylation reactions, S-adenosylmethionine (SAM) is the methyl group donor. SAM is a sulfonium compound whose adenosyl moiety is derived from ATP as follows:

$$\underset{\text{Methionine}}{CH_3{-}S{-}(CH_2)_2{-}\overset{NH_3^+}{\overset{|}{C}H}{-}COO} \xrightarrow[\text{Methionine adenosyltransferase}]{ATP, H_2O \quad PP_i + P_i}$$

$${}^-OOC{-}\overset{NH_3^+}{\overset{|}{C}H}{-}(CH_2)_2{-}\overset{+}{\underset{CH_3}{\underset{|}{S}}}{-}CH_2{-}\text{(ribose, HO OH)}{-}\text{adenine}(NH_2)$$

S-Adenosylmethionine
(active methionine)

$$\underset{\text{S-Adenosylmethionine}}{Adenosine{-}\overset{+}{\underset{CH_3}{\underset{|}{S}}}{-}(CH_2)_2{-}\overset{NH_2}{\overset{|}{C}H}{-}COOH} \xrightarrow[\text{Acceptor (examples)} \quad \text{Acceptor—}CH_3 \text{ (examples)}]{\text{Specific methyltransferase}} \underset{\text{S-Adenosylhomocysteine}}{Adenosine{-}S{-}(CH_2)_2{-}\overset{NH_2}{\overset{|}{C}H}{-}COOH}$$

Acceptor (examples)	Acceptor — CH_3 (examples)
1. Guanidinoacetic acid	1. Creatine
2. Nicotinamide	2. N-Methylnicotinamide
3. Norepinephrine	3. Epinephrine
4. Phosphatidylethanolamine	4. Phosphatidylcholine (three cycles of methylation)
5. N-Acetyl-serotonin	5. Melatonin

FIGURE 17-15
Selected methyl transfer reactions involving S-adenosylmethionine.

The methyl group is transferred to appropriate acceptors by specific methyltransferases with production of S-adenosylhomocysteine (Figure 17-15), which is hydrolyzed to homocysteine and adenosine by adenosylhomocysteinase:

$$\underset{\text{S-Adenosylhomocysteine}}{Adenosyl{-}S{-}(CH_2)_2{-}\overset{NH_3^+}{\overset{|}{C}H}{-}COO^-} \xrightarrow{H_2O \quad Adenosine} \underset{\text{Homocysteine}}{HS{-}(CH_2)_2{-}\overset{NH_3^+}{\overset{|}{C}H}{-}COO^-}$$

Homocysteine can be recycled back to methionine either by transfer of a methyl group from betaine catalyzed by betaine-homocysteine methyltransferase, or from N^5-methyltetrahydrofolate (N^5-methyl-FH_4) catalyzed by N^5-methyl-FH_4-methyltransferase, which requires methyl cobalamin:

(1)

$$\underset{\text{Homocysteine}}{HS{-}(CH_2)_2{-}\overset{NH_3^+}{\overset{|}{C}H}{-}COO^-} + \underset{\text{Betaine}}{(CH_3)_3{}^+N - CH_2{-}COO^-} \xrightarrow{\text{Betaine-homocysteine methyltransferase}}$$

$$\underset{\text{Methionine}}{H_3C{-}S{-}(Ch_2)_2{-}\overset{NH_3^+}{\overset{|}{C}H}{-}COO^-} + \underset{\text{Dimethylglycine}}{(CH_3)_2N{-}CH_2{-}COO^-}$$

(2)

$$HS{-}(CH_2)_2{-}\overset{NH_3^+}{\overset{|}{C}H}{-}COO^- + N^5{-}Methyl{-}FH_4 \xrightarrow{N^5\text{-Methyl-}FH_4\text{-homocysteine methyltransferase (methylcobalamin)}}$$

$$H_3C{-}S{-}(CH_2)_2{-}\overset{NH_3^+}{\overset{|}{C}H}{-}COO^- + FH_4$$

Betaine (an acid) is obtained from oxidation of choline (an alcohol) in two steps:

$$\underset{\text{Choline}}{(CH_3)_3{}^+N{-}CH_2{-}CH_2OH} \xrightarrow[\text{Choline oxidase}]{FAD \quad FADH_2} \underset{\text{Betaine aldehyde}}{(CH_3)_3{}^+N{-}CH_2{-}CHO}$$

$$\xrightarrow[\text{Betaine aldehyde dehydrogenase}]{NAD^+ \quad NADH + H^+} \underset{\text{Betaine}}{(CH_3)_3N{-}CH_2{-}COO^-}$$

Cysteine

In the biosynthesis of cysteine, the sulfur comes from methionine by transsulfuration, and the carbon skeleton and the amino group are provided by serine (Figure 17-16). Cysteine regulates its own formation by functioning as an allosteric inhibitor of cystathionine γ-lyase. α-Ketobutyrate is metabolized to succinyl-CoA by way of propionyl-CoA and methylmalonyl-CoA.

Cysteine is required for the biosynthesis of glutathione and of CoA-SH. A synthetic derivative, N-acetylcysteine, is used to replenish hepatic levels of glutathione and prevent hepatotoxicity due to overdosage with acetaminophen. When high concentrations of acetoaminophen are present in the liver, the drug undergoes N-hydroxylation to form N-acetyl-benzoquinoneimine, which is highly reactive with sulfhydryl groups of proteins and glutathione and causes hepatic necrosis. N-Acetylcysteine is used as a mucolytic agent (e.g., in cystic fibrosis) because it cleaves disulfide linkages of mucoproteins. Cysteine and cystine are interconverted by NAD-dependent cystine reductase and nonenzymatically by an appropriate redox agent (e.g., GSH).

The major end products of cysteine catabolism in humans are inorganic sulfate, taurine, and pyruvate. Taurine is a β-amino acid that has a sulfonic acid instead of a carboxylic acid group. Taurine is conjugated with bile acids in the liver (Chapter 19) and is readily excreted by the kidney. It is a major free amino acid of the central nervous system (where it may be an excitatory neurotransmitter) and the most abundant in the retina; it also occurs in other tissues (e.g., muscle, lung).

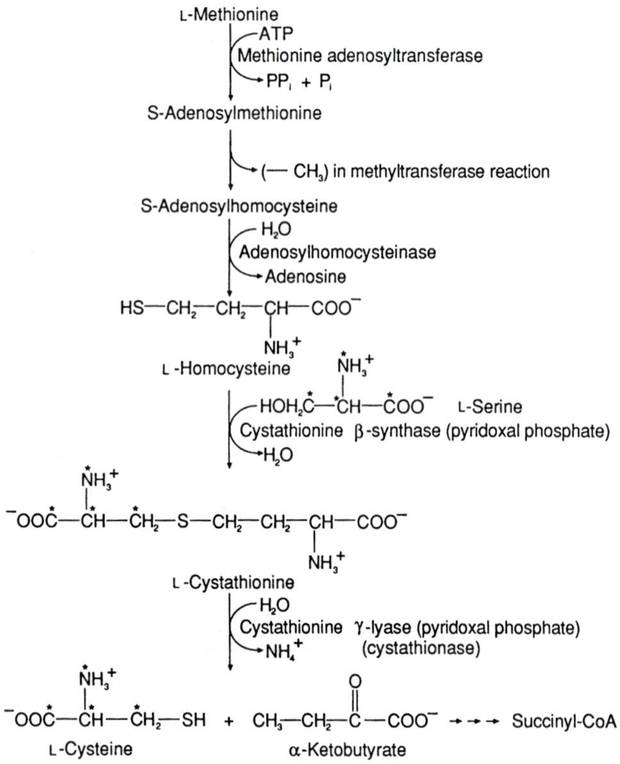

Sulfate can be converted to the sulfate donor compound 3′-phosphoadenosine-5′-phosphosulfate (PAPS) in a two-step reaction (Figure 17-17). PAPS participates in the sulfate esterification of alcoholic and phenolic functional groups (e.g., in synthesis of sulfolipids and glycosaminoglycans).

L-Methionine
ATP
Methionine adenosyltransferase
$PP_i + P_i$
S-Adenosylmethionine
($-CH_3$) in methyltransferase reaction
S-Adenosylhomocysteine
H_2O
Adenosylhomocysteinase
Adenosine
$HS-CH_2-CH_2-CH(NH_3^+)-COO^-$
L-Homocysteine
$HOH_2C-CH(NH_3^+)-COO^-$ L-Serine
Cystathionine β-synthase (pyridoxal phosphate)
H_2O
$^-OOC-CH(NH_3^+)-CH_2-S-CH_2-CH_2-CH(NH_3^+)-COO^-$
L-Cystathionine
H_2O
Cystathionine γ-lyase (pyridoxal phosphate) (cystathionase)
NH_4^+
$^-OOC-CH(NH_3^+)-CH_2-SH$ + $CH_3-CH_2-C(=O)-COO^-$ → → → Succinyl-CoA
L-Cysteine α-Ketobutyrate

FIGURE 17-16
Biosynthesis of cysteine. The sulfur is derived from methionine, and the carbon skeleton and amino group are derived from serine.

Abnormalities Involving Sulfur-Containing Amino Acids

Deficiencies of methionine adenosyltransferase, cystathionine β-synthase, and cystathionine γ-lyase have been described. The first leads to ***hypermethioninemia*** but no other clinical abnormality. The second leads to ***hypermethioninemia, hyperhomocysteinemia,*** and ***homocystinuria.*** The disorder is transmitted as an autosomal recessive trait. Its clinical manifestations may include skeletal abnormalities, mental retardation, ectopia lentis (lens dislocation), malar flush, and susceptibility to arterial and venous thromboembolism. Some patients show reduction in plasma methionine and homocysteine concentrations and in urinary homocysteine excretion after large doses of pyridoxine. Homocystinuria can also result from a deficiency of ***cobalamin*** (vitamin B_{12}) or folate metabolism. The third, an autosomal recessive trait, leads to cystathioninuria and no other characteristic clinical abnormality.

Hereditary sulfite oxidase deficiency can occur alone or with xanthine oxidase deficiency. Both enzymes contain molybdenum (Chapter 27). Patients with sulfite oxidase deficiency exhibit mental retardation, major motor seizures, cerebral atrophy, and lens dislocation. Dietary deficiency of molybdenum (Chapter 37) can cause deficient activity of sulfite and xanthine oxidases.

Cystinuria is a disorder of renal and gastrointestinal tract amino acid transport that also affects lysine, ornithine, and arginine. The four amino acids share a common transport mechanism (discussed above). Clinically, it presents as urinary stone disease because of the insolubility of cystine. In cystinosis, cystine crystals are deposited in tissues because of a transport defect in ATP-dependent cystine efflux from lysosomes (discussed above).

Homocysteine

Homocysteine is an amino acid not found in proteins. Its metabolism involves two pathways; one is

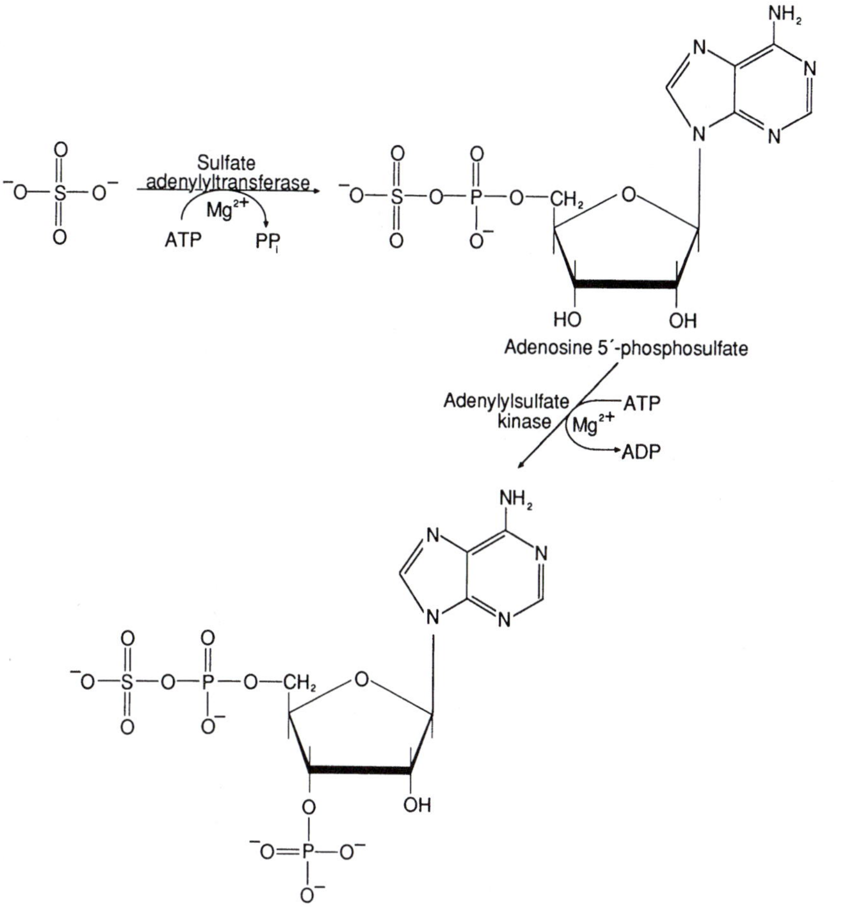

FIGURE 17-17
Formation of 3′-phosphoadenosine-5′-phosphosulfate (PAPS).

the methylation of homocysteine to methionine using N^5-methyltetrahydrofolate (N^5-methyl-FH_4) catalyzed by a vitamin B_{12}-dependent enzyme. The second is the transsulfuration pathway where homocysteine condenses with serine to form cystathionine; this is catalyzed by cystathionine β-synthase (CBS) which is a pyridoxal-5′-phosphate enzyme. End products of the transsulfuration pathway are cysteine, taurine, and sulfate (Figure 17-18). The methyl donor N^5-methyl-FH_4 is synthesized from N^5,N^{10}-methylene-FH_4 and the reaction is catalyzed by N^5,N^{10}-methylenetetrahydrofolate reductase (MTHFR). MTHFR is a FAD-dependent enzyme. Thus, the metabolism of homocysteine involves four water soluble vitamins, folate, vitamin B_{12}, pyridoxine, and riboflavin. Any deficiencies or impairment in the conversion of the four vitamins to their active coenzyme forms will affect homocysteine levels. Severe cases of ***hyperhomocysteinemia*** occur due to deficiencies of enzymes in the homocysteine remethylation or transsulfuration pathways. Individuals with a homozygous defect in cystathionine β-synthase have severe hyperhomocysteinemia (plasma concentrations >50 μM/L) and their clinical manifestations are premature atherosclerosis, thromboembolic complications, skeletal abnormalities, ectopia lentis and mental retardation.

In plasma, homocysteine is present as both free ($<1\%$) and oxidized forms ($>99\%$). The oxidized forms include protein (primarily albumin)–bound homocysteine mixed disulfide (80–90%), homocysteine-cysteine mixed disulfide (5–10%), and homocystine (5–10%). Several studies have shown the relationship between homocysteine and altered endothelial cell function leading to thrombosis. Thus, hyperhomocysteinemia appears to be an independent risk factor for occlusive vascular disease. Five to ten percent of the general population have mild hyperhomocysteinemia.

It has been shown that a thermolabile form of MTHFR is a major cause of mildly elevated plasma homocysteine levels, which have been associated with coronary heart disease. The thermolabile MTHFR gene has a

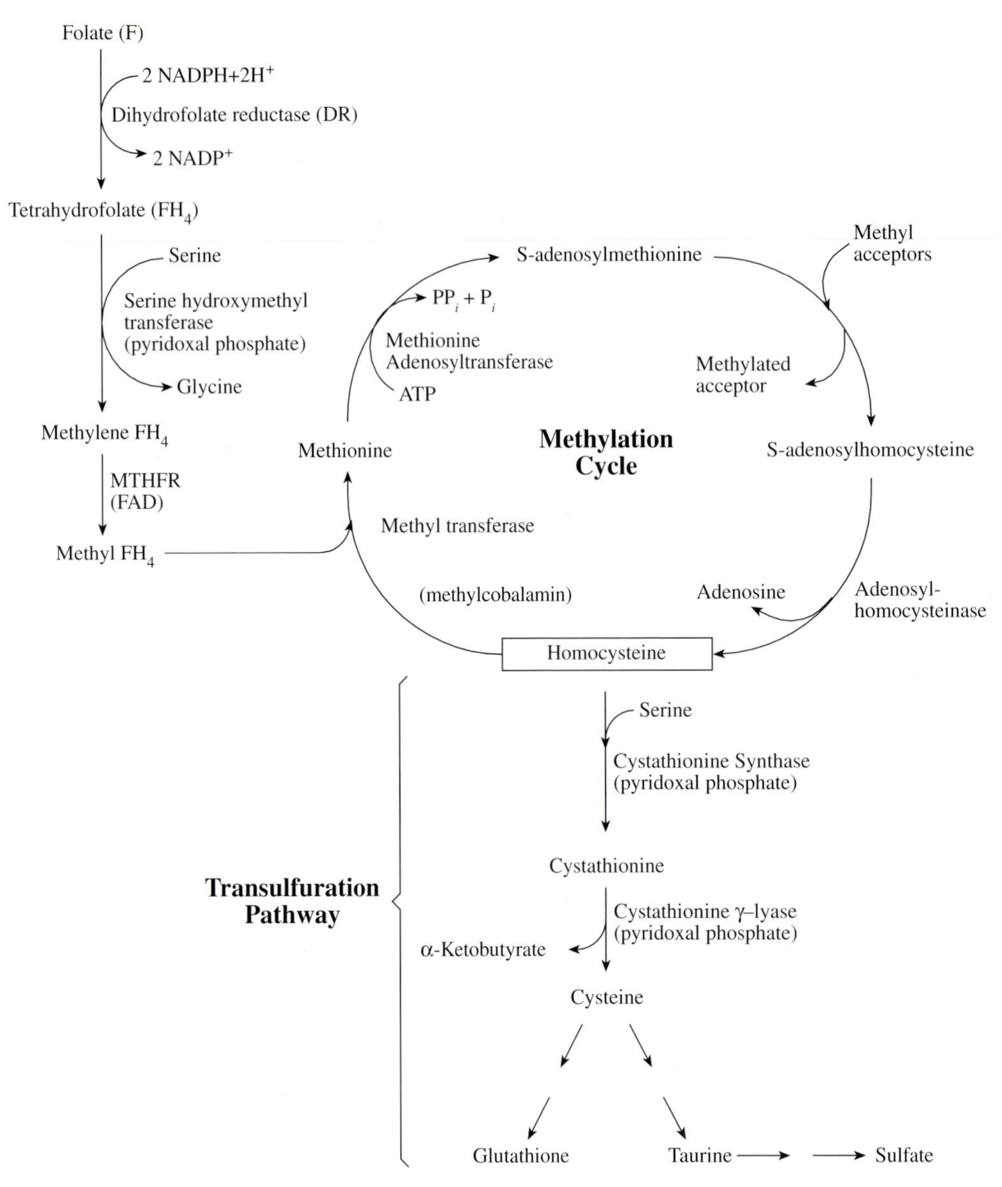

FIGURE 17-18
Homocysteine metabolism.

mutation of C to T at nucleotide position 677 which causes an alanine-to-valine amino acid substitution in the protein. The mechanism by which homocysteine mediates vascular pathology remains to be understood. The targets for homocysteine damage are connective tissue, endothelial cells, smooth muscle cells, coagulation factors, nitric oxide metabolism, plasma lipids and their oxidized forms (Chapter 20). Vitamin supplementation with B_{12}, folate, and B_6 has reduced total plasma homocysteine levels. Vitamin supplementation may decrease the morbidity and mortality from atherosclerotic vascular disease due to hyperhomocysteinemia. However, further studies are required to assess the utility of vitamin supplementation.

Phenylalanine and Tyrosine

Phenylalanine is an essential amino acid. Tyrosine is synthesized by hydroxylation of phenylalanine and therefore is not essential. However, if the hydroxylase system is deficient or absent, the tyrosine requirement must be met from the diet. These amino acids are involved in synthesis of a variety of important compounds, including thyroxine, melanin, norepinephrine, and epinephrine

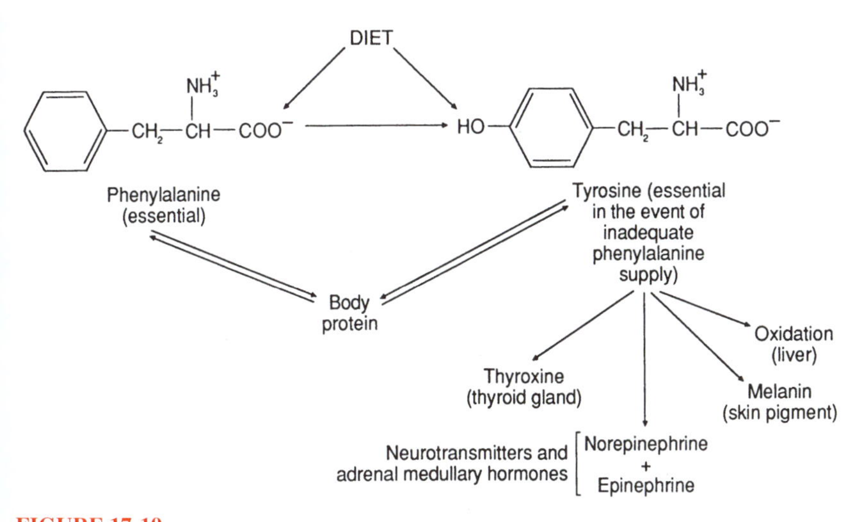

FIGURE 17-19
Overview of the metabolism of phenylalanine and tyrosine.

(Figure 17-19). The conversion of phenylalanine to tyrosine and its degradation to acetoacetate and fumarate are shown in Figure 17-20.

The phenylalanine hydroxylase reaction is complex, occuring principally in liver but also in kidney. The hydroxylating system is present in hepatocyte cytosol and contains phenylalanine hydroxylase, dihydropteridine reductase, and tetrahydrobiopterin as coenzyme. The hydroxylation is physiologically irreversible and consists of a coupled oxidation of phenylalanine to tyrosine and of tetrahydrobiopterin to a quinonoid dihydroderivative with molecular oxygen as the electron acceptor:

$$\text{Phenylalanine} + O_2 + \text{tetrahydrobiopterin} \xrightarrow{\text{phenylalanine hydroxylase}} \text{tyrosine} + H_2O + \text{quinonoid-dihydrobiopterin}$$

The tetrahydrobiopterin is regenerated by reduction of the quinonoid dihydrobiopterin in the presence of NAD(P)H by dihydropteridine reductase:

$$\text{Quinonoid dihydrobiopterin} + \text{NAD(P)H} + H^+ \xrightarrow{\text{dihydropteridine reductase}} \text{NAD(P)}^+ + \text{tetrahydrobiopterin}$$

NADH exhibits a lower K_m and higher V_{max} for the reductase than NADPH. Thus, the pterin coenzyme functions stoichiometrically (in the hydroxylase reaction) and catalytically (in the reductase reaction). Deficiency of dihydropteridine reductase causes a substantial decrease in the rate of phenylalanine hydroxylation. Dihydropteridine reductase and tetrahydrobiopterin are involved in hydroxylation of tyrosine and of tryptophan to yield neurotransmitters and hormones (dopamine, norepinephrine, epinephrine, and serotonin). Unlike phenylalanine hydroxylase, dihydropteridine reductase is distributed widely in tissues (e.g., brain, adrenal medulla).

Human liver phenylalanine hydroxylase is a multimeric homopolymer whose catalytic activity is enhanced by phenylalanine and has a feed-forward metabolic effect. Phosphorylation of phenylalanine hydroxylase by cAMP-dependant kinase leads to increased enzyme activity and dephosphorylation has an opposite effect. Thus, glucagon and insulin have opposing effects on the catalytic activity of phenylalanine hydroxylase.

Quinonoid dihydrobiopterin is an extremely unstable compound that can rapidly rearrange (by tautomerization) to 7,8-dihydrobiopterin (Figure 17-21) and be reduced to the tetrahydro form by dihydrofolate reductase:

$$\text{7,8-Dihydrobiopterin} + \text{NADPH} + H^+ \xrightarrow{\text{dihydrofolate reductase}} \text{NADP}^+ + \text{tetrahydrobiopterin}$$

This enzyme also catalyzes conversion of dihydrofolate (FH_2) to tetrahydrofolate (FH_4), and folic acid contains a pteridine ring system (see the discussion of one-carbon metabolism in Chapter 27). However, regeneration of tetrahydrobiopterin by the dihydrofolate reductase reaction, however, is too slow to support normal rates of phenylalanine hydroxylation.

Tetrahydrobiopterin is synthesized starting from GTP and requires at least three enzymes. The first committed step is GTP-cyclohydrolase, which converts GTP to dihydroneopterin triphosphate. 6-Pyruvoyltetrahydrobiopterin synthase transforms dihydroneopterin triphosphate into 6-pyruvoyltetrahydrobiopterin. The latter is reduced to tetrahydrobiopterin by NADPH-dependent sepiapterin reductase. Deficiency of GTP-cyclohydrolase and

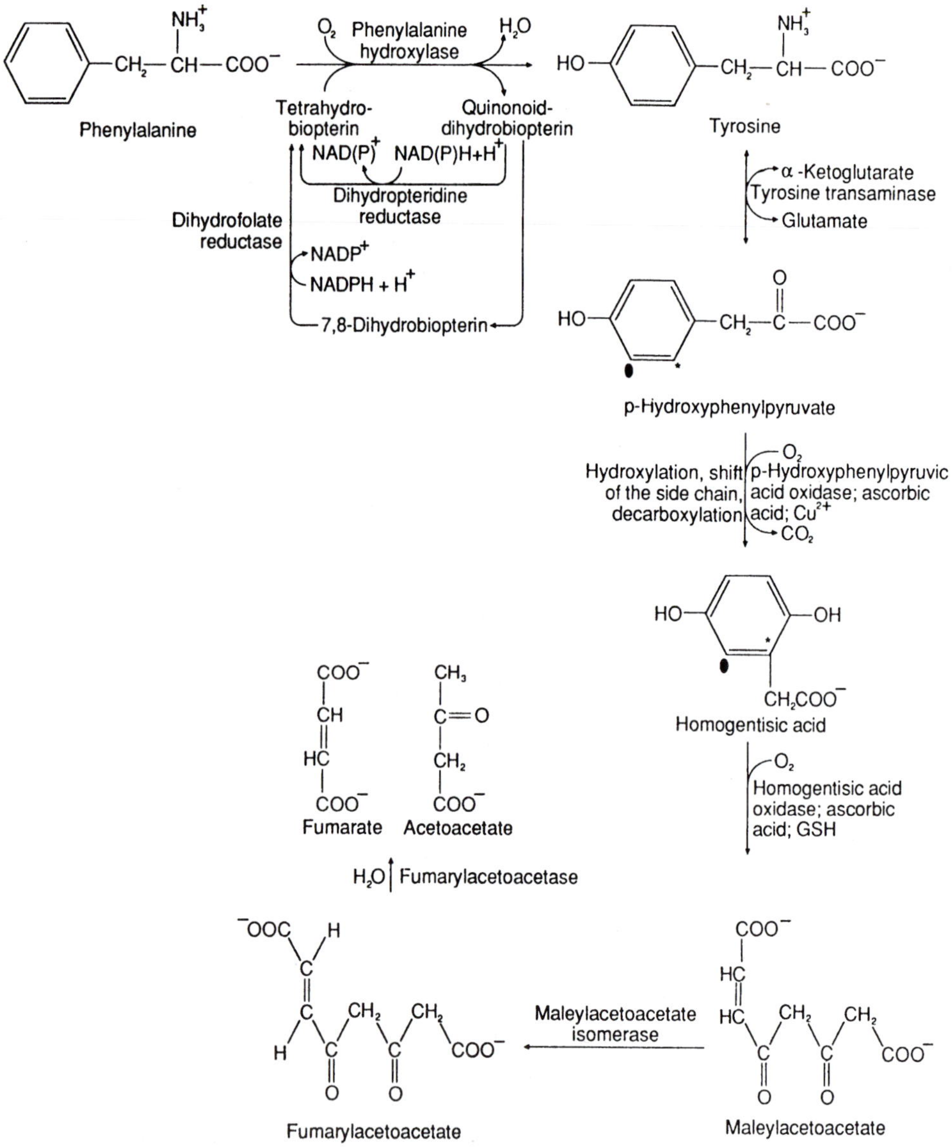

FIGURE 17-20
Conversion of phenylalanine to tyrosine and the oxidative pathway of tyrosine.

6-pyruvoyl tetrahydrobiopterin synthase leads to ***hyperphenylalaninemia.***

Phenylketonuria (PKU)

Deficiency of phenylalanine hydroxylase, tetrahydrobiopterin, or dihydropteridine reductase results in phenylketonuria (PKU), an autosomal recessive trait. Because phenylalanine accumulates in tissues and plasma (hyperphenylalaninemia), it is metabolized by alternative pathways and abnormal amounts of phenylpyruvate appear in urine (Figure 17-22). Phenylalanine hydroxylase deficiency may be complete (classic PKU, type I) or partial (types II and III). Many mutations of the phenylalanine hydroxylase gene have been identified (missense, nonsense, insertions, deletions, and duplications) leading to PKU or non-PKU hyperphenylalaninemia.

The incidence of classic PKU is about 1 in 10,000–20,000 live births and exhibits considerable geographi variation (the incidence in Ireland is 1 in 4000, wherea the condition is rare among blacks and Asians). Abou 2% of hyperphenylalaninemic infants have a deficienc of biopterin or biopterin reductase. The most importar clinical presentation is mental impairment. Diagnosis ca be made early in the neonatal period by measureme of phenylalanine concentration in blood collected fro

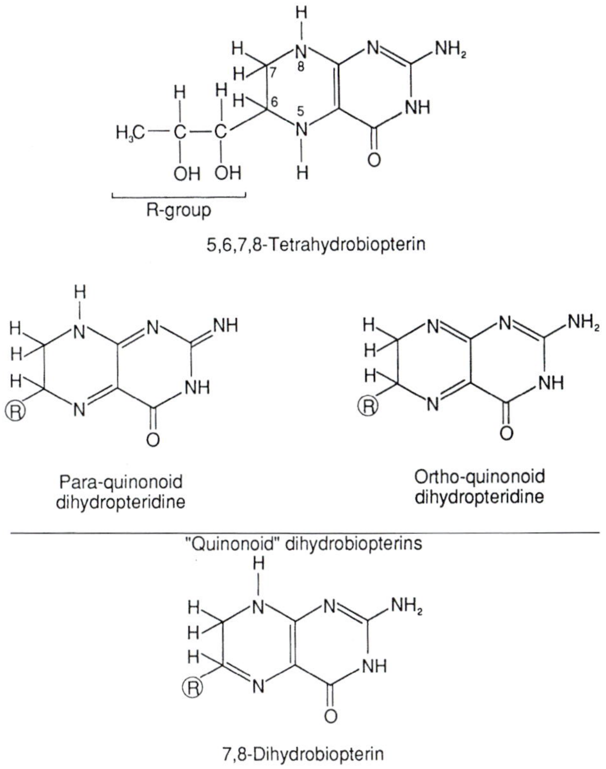

5,6,7,8-Tetrahydrobiopterin

Para-quinonoid dihydropteridine

Ortho-quinonoid dihydropteridine

"Quinonoid" dihydrobiopterins

7,8-Dihydrobiopterin

FIGURE 17-21
Structures of biopterin derivatives.

a heel prick onto filter paper. Treatment of phenylalanine hydroxylase deficiency consists of a diet low in phenylalanine but which maintains normal nutrition. This diet is effective in preventing mental retardation, and its continuation throughout the first decade, or for life, may be necessary.

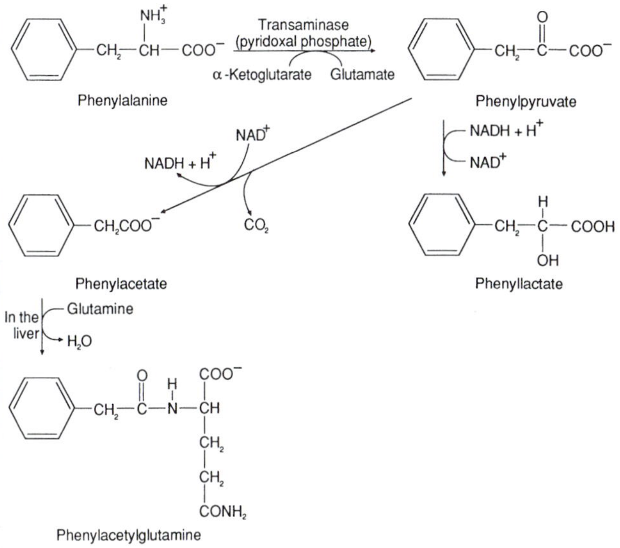

Treatment of biopterin and biopterin reductase deficiency consists not only of regulating the blood levels of phenylalanine but of supplying the missing form of coenzyme and the precursors of neurotransmitters, namely, dihydroxyphenylalanine and 5-hydroxytryptophan, along with a compound that inhibits peripheral aromatic decarboxylation. This compound is necessary because the amine products do not cross the blood-brain barrier.

Successfully treated females who have reached reproductive age may expose their offspring (who are obligate heterozygotes) to abnormal embryonic and fetal development. These effects include spontaneous abortion, microcephaly, congenital heart disease, and intrauterine growth retardation, and they correlate with the plasma level of phenylalanine of the pregnant mother. Thus, reinstitution of a low-phenylalanine diet during pre- and postconception periods may be necessary. The diet should also restrict intake of phenylalanine-containing substances, such as the synthetic sweetener aspartame (L-aspartyl-L-phenylalanyl methyl ester). Because defective myelination occurs in the brain in PKU, there is an increased incidence of epileptic seizures and abnormal electroencephalograms are common. The biochemical basis for the severe mental impairment is not understood. One factor may be inhibition of glutamate decarboxylase by phenylpyruvate and phenylacetate:

$$^{-}OOC{-}(CH_2)_2{-}CH(NH_3^+){-}COO^{-} \xrightarrow[CO_2]{\text{Glutamate decarboxylase}} {^{-}OOC{-}(CH_2)_3{-}NH_3^+}$$

L-Glutamate → γ-Aminobutyrate

The substrate is an excitatory neurotransmitter and the product an inhibitory one in the central nervous system.

Abnormal indole derivatives in the urine and low levels of serotonin (a product of tryptophan metabolism) in blood and brain point to a defect in tryptophan metabolism in PKU. 5-Hydroxytryptophan decarboxylase, which catalyzes the conversion of 5-hydroxytryptophan to serotonin, is inhibited *in vitro* by some of the metabolites of phenylalanine. Phenylalanine hydroxylase is similar to the enzyme that catalyzes the hydroxylation of tryptophan to 5-hydroxytryptophan, a precursor of serotonin. *In vitro,* phenylalanine is also found to inhibit the hydroxylation of tryptophan. The mental defects associated with PKU may be caused by decreased production of serotonin. High phenylalanine levels may disturb the transport of amino

Phenylalanine — Transaminase (pyridoxal phosphate); α-Ketoglutarate → Glutamate — Phenylpyruvate

Phenylpyruvate — NADH + H^+ → NAD^+ — Phenyllactate

Phenylpyruvate — NAD^+ → NADH + H^+; CO_2 — Phenylacetate

Phenylacetate — In the liver: Glutamine → H_2O — Phenylacetylglutamine

FIGURE 17-22
Formation of metabolites of phenylalanine that accumulate in abnormal amounts and are excreted in phenylketonuria.

acids into cells. Variations in the clinical manifestations of PKU may reflect differences in this disturbance. Defects in pigmentation of skin and hair (light skin and hair) may be caused by interference of melanin formation by phenylalanine and its metabolites and also by lack of tyrosine.

Melanin

Melanin is an insoluble, high-molecular-weight polymer of 5,6-dihydroxyindole, which is synthesized from tyrosine (Figure 17-23). It is produced by pigment cells (melanocytes) in cytoplasmic organelles (melanosomes). In the epidermis, melanocytes are associated with keratinocytes, which contain melanosomes supplied by melanocytes via dendritic processes. Color variation in human skin reflects the amount of melanin synthesized in melanosomes. Melanin synthesis is apparently under hormonal and neural regulation.

The first two steps in the synthesis of melanin are catalyzed by tyrosinase, a copper-containing oxidase, which converts tyrosine to dopaquinone. All subsequent reactions presumably occur through nonenzymatic auto-oxidation, in the presence of zinc, with formation of the black to brown pigment eumelanin. The yellow to reddish brown, high-molecular-weight polymer known as pheomelanin and the low-molecular-weight trichromes result from addition of cysteine to dopaquinone and further modification of the products. Pheomelanins and trichromes are primarily present in hair and feathers.

Abnormalities of Tyrosine Metabolism

Hepatic cytosolic tyrosine aminotransferase (tyrosine transaminase) deficiency produces ***tyrosinemia type II,*** an autosomal recessive trait marked by hypertyrosinemia and tyrosinuria. Clinical manifestations may include corneal erosions and plaques, inflammation (from intracellular crystallization of tyrosine), and mental retardation. Low-tyrosine and low-phenylalanine diets are beneficial.

Tyrosinosis is presumably due to fumarylacetoacetate hydrolase deficiency and has a high prevalence in the French–Canadian population of Québec. It is associated with abnormal liver function, renal tubular dysfunction, anemia, and vitamin D-resistant rickets. Transient ***tyrosinemia*** of the newborn, particularly in premature infants, is the most common form of tyrosinemia in infancy.

Alcaptonuria is a rare metabolic hereditary disease in which homogentisic acid is eliminated in urine, which darkens upon exposure to air owing to oxidation of homogentisic acid. The biochemical lesion is homogentisic acid oxidase deficiency. Clinical features include pigmentation of cartilage and other connective tissues (ochronosis) later in life from deposition of oxidized homogentisic acid. Patients nearly always develop arthritis in later years, but the relationship between pigment deposition and arthritis is not understood.

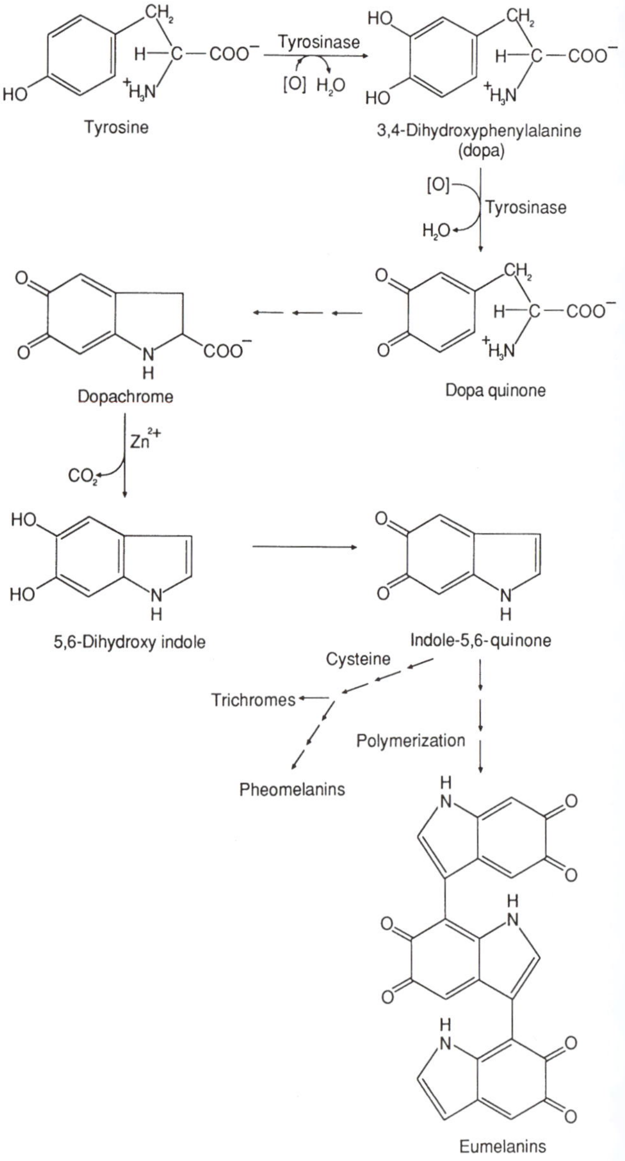

Lack of melanin production (***hypomelanosis***) gives rise to several hereditary disorders collectively known as ***albinism.*** Some forms result from deficiency of tyrosinase The inheritance pattern of albinism varies with type. Affected individuals have increased susceptibility to various

FIGURE 17-23
Biosynthesis of melanins.

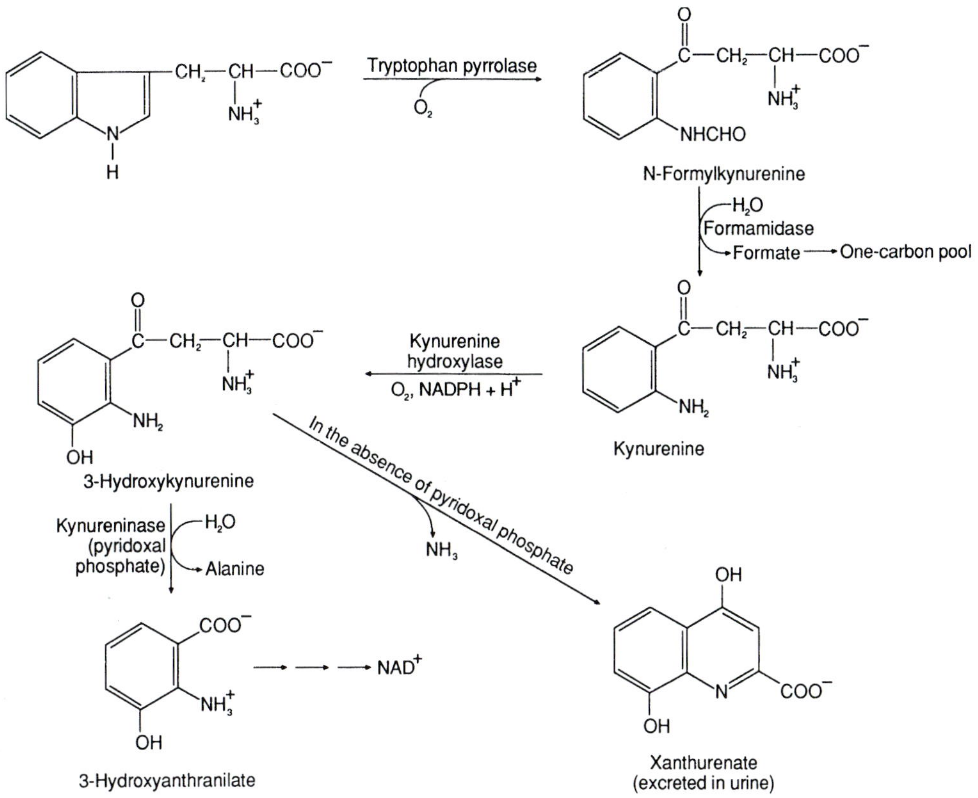

FIGURE 17-24
Pathway for the synthesis of NAD from tryptophan.

types of carcinoma (from the effect of solar radiation on DNA). When the eyes are involved, photophobia, subnormal visual acuity, strabismus, and nystagmus may be present.

Tryptophan

Tryptophan is an essential amino acid involved in synthesis of several important compounds. Nicotinic acid (amide), a vitamin required in the synthesis of NAD^+ and $NADP^+$, can be synthesized from tryptophan (Figure 17-24). About 60 mg of tryptophan can give rise to 1 mg of nicotinamide. The synthesis begins with conversion of tryptophan to N-formylkynurenine by tryptophan pyrrolase, an inducible iron-porphyrin enzyme of liver. N-Formylkynurenine is converted to kynurenine by removal of formate, which enters the one-carbon pool. Kynurenine is hydroxylated to 3-hydroxykynurenine, which is converted to 3-hydroxyanthranilate, catalyzed by kynureninase, a pyridoxal phosphate–dependent enzyme. 3-Hydroxyanthranilate is then converted by a series of reactions to nicotinamide ribotide, the immediate precursor of NAD. In deficiency of pyridoxal phosphate, 3-hydroxykynurenine accumulates and is converted to xanthurenate, which is excreted in urine. Thus, vitamin B_6 deficiency can be diagnosed by measurement of urinary xanthurenate after administration of a standard dose of tryptophan (tryptophan load test).

5-Hydroxytryptamine (serotonin) is found in enterochromaffin cells, brain, and platelets. In the former two, it is produced from tryptophan, whereas in platelets, serotonin is taken up from plasma. Synthesis involves hydroxylation of tryptophan by tryptophan 5-hydroxylase and decarboxylation by aromatic L-amino acid decarboxylase (Figure 17-25). Hydroxylation is the rate-limiting reaction, is analogous to that of phenylalanine, and requires molecular oxygen and tetrahydrobiopterin. Serotonin is a powerful vasoconstrictor and stimulator of smooth muscle contraction. In the brain it is a neurotransmitter, and in the pineal gland it serves as a precursor of melatonin. Synthesis of melatonin requires N-acetylation of serotonin, followed by methylation (Figure 17-26). The role of melatonin in humans is not understood; in frogs, it lightens the color of skin melanocytes and blocks the action of melanocyte-stimulating hormone (MSH) and

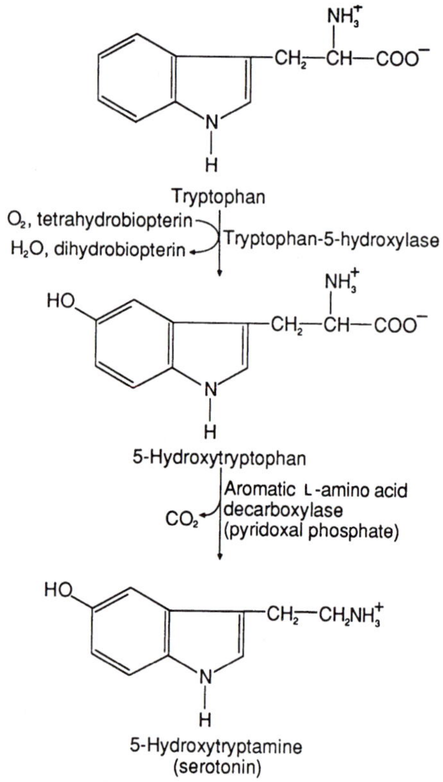

Tryptophan
O_2, tetrahydrobiopterin
H_2O, dihydrobiopterin
Tryptophan-5-hydroxylase
5-Hydroxytryptophan
Aromatic L-amino acid decarboxylase (pyridoxal phosphate)
CO_2
5-Hydroxytryptamine (serotonin)

FIGURE 17-25
Biosynthesis of serotonin from tryptophan.

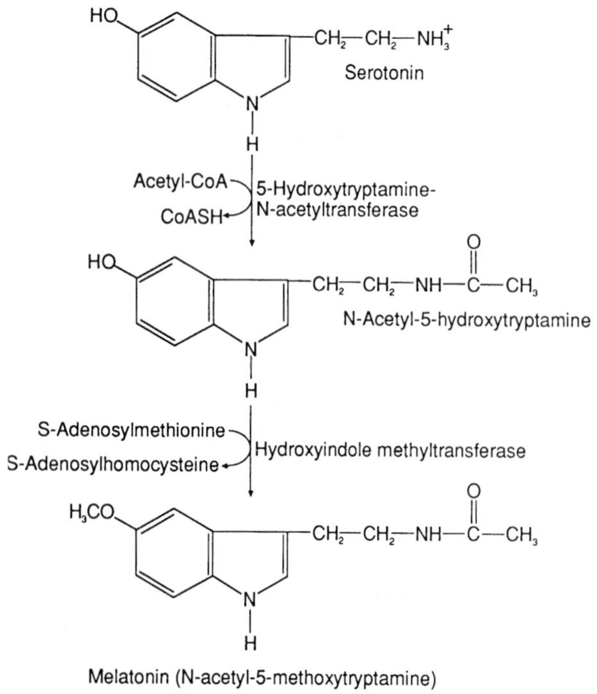

Serotonin
Acetyl-CoA
CoASH
5-Hydroxytryptamine-N-acetyltransferase
N-Acetyl-5-hydroxytryptamine
S-Adenosylmethionine
S-Adenosylhomocysteine
Hydroxyindole methyltransferase
Melatonin (N-acetyl-5-methoxytryptamine)

FIGURE 17-26
Biosynthesis of melatonin from serotonin.

adrenocorticotropic hormone (ACTH). In rats, melatonin regulates the breeding cycle, and its secretion is increased by exposure to light via adrenergic stimulation of the pineal gland.

N-Acetyltransferase is activated by increased concentrations of cytosolic cyclic AMP and Ca^{2+} that is mediated by the activation of adrenergic receptors of the pineal gland. In humans, melatonin synthesis and its release follows a circadian rhythm which is stimulated by darkness and inhibited by light. The blood levels of melatonin increase by passive diffusion from the central nervous system after the onset of darkness, reaching a peak value during the middle of the night and declining during the second half of the night. Melatonin secretion is also regulated endogenously by signals from the suprachiasmatic nucleus. Since melatonin's peak concentration in plasma coincides with sleep, exogenous administration of the hormone can affect the circadian rhythm. Melatonin supplementation has been used to ameliorate subjective and objective symptoms of jet lag caused by travel across time zones. At high levels, melatonin promotes sleep. Short-term and long-term biological effects of melatonin supplementation have yet to be determined.

Serotonin is degraded to 5-hydroxyindoleacetic acid (5-HIAA) by monoamine oxidase and aldehyde dehydrogenase acting in sequence. 5-HIAA is excreted in the urine. Its excretion is markedly increased in subjects with carcinoid tumor (found most frequently in the gastrointestinal tract and lungs). ***Carcinoid tumors*** are frequently indolent and asymptomatic; however, a significant number of these tumors can manifest as metabolic problems. Although increased production of serotonin is a characteristic feature of the carcinoid tumor, cells also synthesize other substances. These include kinins, prostaglandins, substance P, gastrin, somatostatin, corticotropin, and neuron-specific enolase. Carcinoid tumor cells are known as enterochromaffin cells because they stain with potassium dichromate and also are known as argentaffin cells because they take up and reduce silver salts. The symptoms of the carcinoid tumor are due to synergistic biochemical interactions between serotonin and the above-mentioned active metabolites. Characteristics of the carcinoid syndrome are flushing, diarrhea, wheezing, heart valve dysfunction, and pellagra. Lifestyle conditions that precipitate symptoms include intake of alcohol or spicy foods, and strenuous exercise. The treatment options for the carcinoid syndrome are multidisciplinary, including lifestyle changes, inhibitors of serotonin release and serotonin receptor antagonists, somatostatin analogues, hepatic artery embolization, chemotherapy, and surgery.

Hartnup disease is a disorder of renal tubular and intestinal absorption of tryptophan and other neutral amino acids.

Supplemental Readings and References

J. Abrains: The role of nitrates in coronary heart disease. *Archives of Internal Medicine* **155,** 357 (1995).

M. L. Batshaw: Inborn errors of urea synthesis. *Annals of Neurology* **35,** 133 (1994).

D. Bohn: Nitric oxide in acute hypoxic respiratory failure: From the bench to the bedside and back again. *Journal of Pediatrics* **134,** 387 (1999).

A. Brezezinski: Melatonin in Humans. *The New England Journal of Medicine* **331,** 186 (1997).

M. E. Caplin, J. R. Buscombe, A. J. Hilson, et al.: Carcinoid tumour. *Lancet* **352,** 799 (1998).

K. J. Carpenter: Protein requirements of adults from an evolutionary perspective. *American.Journal of Clinical Nutrition* **55,** 913 (1992).

D. Christensen: What's that smell? *Science News* **155,** 316 (1999).

D. T. Chuang: Maple syrup urine disease: It has come a long way. *Journal of Pediatrics* **132,** S17 (1998).

J. P. Cooke and P. S. Tasao: Arginine: A new therapy for artherosclerosis? *Circulation* **97,** 311 (1997).

B. R. Crane, A. S. Arvai, R. Gachhui, et al.: The structure of nitric oxide synthase oxygenase domain and inhibitor complexes. *Science* **278,** 425 (1997).

E. L. Dobyns, D. N. Cornfield, N. G. Anas, et al.: Multicenter randomized control trial of the effects of inhaled nitric oxide therapy on gas exchange in children with acute hypoxic respiratory failure. *Journal of Pediatrics* **134,** 406 (1999).

E. B. Feldman: Creatine: A Dietary supplement and ergogenic aid. *Nutrition Reviews* **57,** 45 (1999).

I. Goldstein, T. F. Lue, H. Padma-Nathan, et al.: Oral sildenafil in the treatment of erectile dysfunction. *The New England Journal of Medicine* **338,** 1397 (1998).

R. Jalan and P. C. Hayes: Hepatic encephalopathy and ascites. *Lancet* **350,** 1309 (1997).

S. G. Korenman: New insights into erectile dysfunction. *American Journal of Medicine* **105,** 135 (1998).

J. Loscalzo and G. Welch: Nitric oxide and its role in the cardiovascular system. *Progress in cardiovascular diseases* **38,** 87 (1995).

N. E. Maestri, D. Clisold, and S. W. Brusilow: Neonatal onset ornithine transcarbamylase deficiency: A retrospective analysis. *Journal of Pediatrics* **134,** 268 (1999).

M. M. Malinow, P. B. Duell, D. L. Hess, et al: Reduction of plasma homocysteine levels by breakfast cereal fortified with folic acid in patients with coronary heart disease. *The New England Journal of Medicine* **338,** 1009 (1998).

A. J. Maxwell and J. P. Cooke: Cardiovascular effects of L-arginine. *Current Opinion in Nephrology and Hypertension* **7,** 63 (1998).

D. J. Millward: Aging, protein requirements, and protein turnover. *American Journal of Clinical Nutrition* **66,** 774 (1997).

J. D. Parker and J. O. Parker: Nitrate therapy for stable angina pectoris. *The New England Journal of Medicine* **338,** 520 (1998).

S. M. Riordan and R. Williams: Treatment of hepatic encephalopathy. *The New England Journal of Medicine* **337,** 473 (1997).

A. M. Spiekerman: Proteins used in nutritional assessment. *Clinical Laboratory Medicine* **13,** 353 (1993).

J. H. Stein and P. E. McBride: Hyperhomocysteinemia and atherosclerotic vascular disease. *Archives of Internal Medicine* **158,** 1301 (1998).

J. G. Theone: Treatment of urea cycle disorders. *Journal of Pediatrics* **134,** 255 (1999).

R. D. Utiger: A pill for impotence. *The New England Journal of Medicine* **338,** 1458 (1998).

G. Wagner and I. S. de Tejada: Update on male erectile dysfunction. *British Medical Journal* **376,** 678 (1998).

N. J. Wald, H. C. Watt, M. R. Law, et al.: Homocysteine and ischemic heart disease. *Archives of Internal Medicine* **158,** 862 (1998).

G. N. Welch and J. Loscalzo: Homocysteine and atherothrombosis. *The New England Journal of Medicine* **338,** 1042 (1998).

M. Wyss, R. Kaddurah-Daouk: Creatine and creatine metabolism. Physiological Reviews **80,** 1107 (2000).

CHAPTER **18**

Lipids I: Fatty Acids and Eicosanoids

Lipids (or fats) are a heterogeneous group of organic compounds defined by their solubility in nonpolar solvents such as chloroform, ether, and benzene and by their poor solubility in water.

Lipids may be polar or nonpolar (amphipathic). Polar lipids have limited solubility in water because they are amphipathic, i.e., they possess hydrophilic and hydrophobic regions in the same molecule. Major polar lipids include ***fatty acids, cholesterol, glycerophosphatides,*** and ***glycosphingolipids.*** Very short chain fatty acids and ketone bodies are readily soluble in water. Nonpolar lipids serve principally as storage and transport forms of lipid and include ***triacylglycerols*** (also called triglycerides) and ***cholesteryl esters.***

Lipids have numerous functions including the following: thermal insulation, energy storage (as triacylglycerol), metabolic fuels, membrane components (phospholipids and cholesterol; Chapter 10), hormones (steroids and vitamin D metabolites; Chapters 32 and 37, respectively), precursors of ***prostanoids*** (discussed on p. 391) and ***leukotrienes,*** (vitamins A, C, D, E, and K; Chapters 36–38), emulsifying agents in the digestion and absorption of lipids (bile acids; Chapters 12 and 19), and surfactants in the alveolar membrane (phosphatidylcholine; Chapter 19). The metabolism of fatty acids (saturated and unsaturated) is discussed in this chapter. The metabolism of phospholipids, glycosphingolipids, and cholesterol is considered in Chapter 19.

Fatty acids that contain no carbon-carbon double bonds are known as saturated and those with carbon-carbon double bonds as unsaturated. Fatty acids that contain an even number of carbon atoms and are acyclic, unbranched, nonhydroxylated, and monocarboxylic make up the largest group. The most abundant saturated fatty acids in animals are palmitic and stearic acids (Table 18-1). The melting point of fatty acids rises with increase in chain length, the even-numbered saturated fatty acids having higher melting points than the odd-numbered. Among the even-numbered, the presence of cis double bonds lowers the melting point significantly. Free fatty acids at physiological pH are ionized (pK ~4.85) and exist only in small quantities; in plasma, they typically are bound to albumin. They are usually present as esters or amides.

Digestion and absorption of lipids are discussed in Chapter 12. The Western diet contains about 40% fat, mostly as triacylglycerol (100–150 g/day). Triacylglycerols packaged as chylomicrons in the intestinal epithelial cell are delivered to the blood circulation via the lymphatic system and are hydrolyzed to glycerol and fatty acids by endothelial lipoprotein lipase. Fatty acids are taken up by the cells of the tissue where the hydrolysis occurs, whereas glycerol is metabolized in the liver and kidney (Chapter 15). Another means of triacylglycerol transport is very-low-density lipoprotein (VLDL), which is synthesized in the liver. Its triacylglycerol is also hydrolyzed by endothelial lipoprotein lipase. The metabolism of plasma

TABLE 18-1
Naturally Occurring Saturated Fatty Acids

Common Name	Systematic Name*	Molecular Formula	Structural Formula	Melting Point (°C)
Capric	*n*-Decanoic	$C_{10}H_{20}O_2$	$CH_3[CH_2]_8COOH$	31
Lauric	*n*-Dodecanoic	$C_{12}H_{24}O_2$	$CH_3[CH_2]_{10}COOH$	44
Myristic	*n*-Tetradecanoic	$C_{14}H_{28}O_2$	$CH_3[CH_2]_{12}COOH$	58
Palmitic†	*n*-Hexadecanoic	$C_{16}H_{32}O_2$	$CH_3[CH_2]_{14}COOH$	63
Stearic†	*n*-Octadecanoic	$C_{18}H_{36}O_2$	$CH_3[CH_2]_{16}COOH$	70
Arachidic	*n*-Eicosanoic	$C_{20}H_{40}O_2$	$CH_3[CH_2]_{18}COOH$	76
Behenic	*n*-Docosanoic	$C_{22}H_{44}O_2$	$CH_3[CH_2]_{20}COOH$	80
Lignoceric	*n*-Tetracosanoic	$C_{24}H_{48}O_2$	$CH_3[CH_2]_{22}COOH$	84
Cerotic	*n*-Hexacosanoic	$C_{26}H_{52}O_2$	$CH_3[CH_2]_{24}COOH$	88
Montanic	*n*-Octacosanoic	$C_{28}H_{56}O_2$	$CH_3[CH_2]_{26}COOH$	92

*Systematic name is based on replacing the final letter "e" of the parent hydrocarbon with "oic."
†Most abundant fatty acids present in animal lipids.

lipoproteins is discussed in Chapter 20. Fatty acids are released by hydrolysis of triacylglycerol in adipocytes by hormone-sensitive lipase, particularly during starvation, stress, and prolonged exercise (Chapter 22).

Interrelationships of tissues in lipid metabolism are discussed in Chapter 22.

18.1 Oxidation of Fatty Acids

The overall fatty acid oxidation process in mitochondria consists of uptake of fatty acids, their activation to acyl-CoA, then to thioesters, and finally translocation into mitochondria which involves a carnitine transesterification shuttle and β-oxidation. Fatty acids released from chylomicrons and VLDLs are transferred across cell membranes by passive diffusion, which depends on the concentration gradient. Fatty acids are also obtained from the hydrolysis of triacylglycerol stored in adipose tissue which are bound to albumin and transported in blood. Fatty acids serve as substrates for energy production in liver, skeletal and cardiac muscle during periods of fasting. Although the brain does not utilize fatty acids for generating energy directly, brain cells can utilize ketone bodies synthesized from acetyl-CoA and acetoacetyl-CoA. The latter two are obtained from β-oxidation of fatty acids in the liver. All of the enzymes involved in mitochondrial fatty acid β-oxidation are encoded by nuclear genes. After their synthesis in the cytosolic endoplasmic reticulum, the enzymes are transported to mitochondria. The transport of the enzymes into mitochondria, in many instances, requires the presence of N-terminal extensions to guide the protein across the mitochondrial membrane, receptor-mediated ATP-dependent uptake, and proteolytic processing to form fully assembled, mature enzymes. During β-oxidation of acyl-CoA, the chain length of the substrate is shortened by two carbon atoms (acetyl-CoA) each cycle. Thus, β-oxidation requires a group of enzymes with chain length specificity.

Activation of Fatty Acids

At least three acyl-CoA synthases, each specific for a particular size of fatty acid, exist: acetyl-CoA synthase acts on acetate and other low-molecular-weight carboxylic acids, medium-chain acyl-CoA synthase on fatty acids with 4–11 carbon atoms, and acyl-CoA synthase on fatty acids with 6–20 carbon atoms. The activity of acetyl-CoA synthase in muscle is restricted to the mitochondrial matrix. Medium-chain acyl-CoA synthase occurs only in liver mitochondria, where medium-chain fatty acids obtained from digestion of dietary triacylglycerols and transported by the portal blood are metabolized. Acyl-CoA synthase, the major activating enzyme, occurs on the outer mitochondrial membrane surface and in endoplasmic reticulum. The overall reaction of activation is as follows:

$$RCOO^- + ATP^{4-} + CoASH \rightleftharpoons \underset{\text{Acyl-CoA (a thioester)}}{RC(=O)\text{—}SCoA} + AMP^{2-} + PP_i^{3-}$$

The reaction favors the formation of fatty acyl-CoA, since the pyrophosphate formed is hydrolyzed by pyrophosphatase: $PP_i + H_2O \rightarrow 2P_i$. Thus, activation of a fatty acid molecule requires expenditure of two high-energy phosphate bonds. The reaction occurs in two steps (E = enzyme):

$$RCOO^- + ATP + E \xrightarrow{\text{Step 1}} E\text{—}AMP\text{—}\overset{\overset{\large O}{\|}}{C}\text{—}R + PP_i$$

$$E\text{—}AMP\text{—}\overset{\overset{\large O}{\|}}{C}\text{—}R \xrightarrow[\text{Step 2}]{CoASH} E + R\text{—}\overset{\overset{\large O}{\|}}{C}\text{—}SCoA + AMP$$

A mitochondrial acyl-CoA synthase, which utilizes GTP, has also been identified:

$$R \cdot COO^- + CoASH + GTP^{4-} \rightarrow RCO \cdot SCoA + GDP^{3-} + P_i^{2-}$$

Its role is not known.

Transport of Acyl-CoA to Mitochondrial Matrix

This transport is accomplished by carnitine (L-β-hydroxy-γ-trimethylammonium butyrate), which is required in catalytic amounts for the oxidation of fatty acids (Figure 18-1). Carnitine also participates in the transport of acetyl-CoA for cytosolic fatty acid synthesis. Two carnitine acyltransferases are involved in acyl-CoA transport: carnitine palmitoyltransferase I (CPTI), located on the outer surface of the inner mitochondrial membrane, and carnitine palmitoyltransferase II (CPTII), located on the inner surface. The overall translocation reaction is as follows:

$$\underset{\text{Acyl-CoA}}{R\text{—}\overset{\overset{\large O}{\|}}{C}\text{—}SCoA} + \underset{\text{Carnitine}}{H_3C\text{—}\overset{\overset{\large CH_3}{|}}{\underset{\underset{\large CH_3}{|}}{N^+}}\text{—}CH_2\text{—}\underset{\underset{\large OH}{|}}{CH}\text{—}CH_2\text{—}COOH}$$

$$\Big\Updownarrow \text{ Carnitine palmitoyltransferase}$$

$$H_3C\text{—}\overset{\overset{\large CH_3}{|}}{\underset{\underset{\large CH_3}{|}}{N^+}}\text{—}CH_2\text{—}\overset{\overset{\large H}{|}}{\underset{\underset{\underset{\underset{\large C=O\ \text{(Acyl group)}}{|}}{O}}{|}}{C}}\text{—}CH_2\text{—}COOH + CoASH \quad (\text{acyl group: } C{=}O\text{—}R)$$

The standard free-energy change of this reaction is about zero, and therefore the O-ester bond of acylcarnitine may be considered as a high-energy linkage. Malonyl-CoA, a precursor in the synthesis of fatty acids, is an allosteric inhibitor of CPTI in liver and thus prevents a futile cycle of simultaneous fatty acid oxidation and synthesis.

Carnitine is synthesized from two essential amino acids, lysine and methionine. S-Adenosylmethionine donates three methyl groups to a lysyl residue of a protein with the formation of a protein-bound trimethyllysyl. Proteolysis yields trimethyllysine, which is converted to carnitine (Figure 18-2). In humans, liver and kidney are major sites of carnitine production; from there it is transported to skeletal and cardiac muscle, where it cannot be synthesized.

Four inherited defects of carnitine metabolism lead to impaired utilization of long-chain fatty acids for energy production. These include defects of plasma

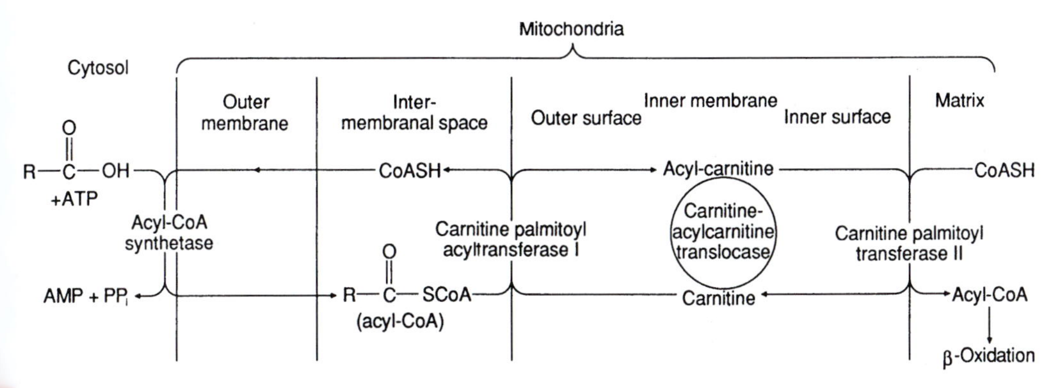

FIGURE 18-1
Role of carnitine in transport of fatty acids into the mitochondrial matrix. Entry of acyl-carnitine is linked to the exit of carnitine, both mediated by a translocase.

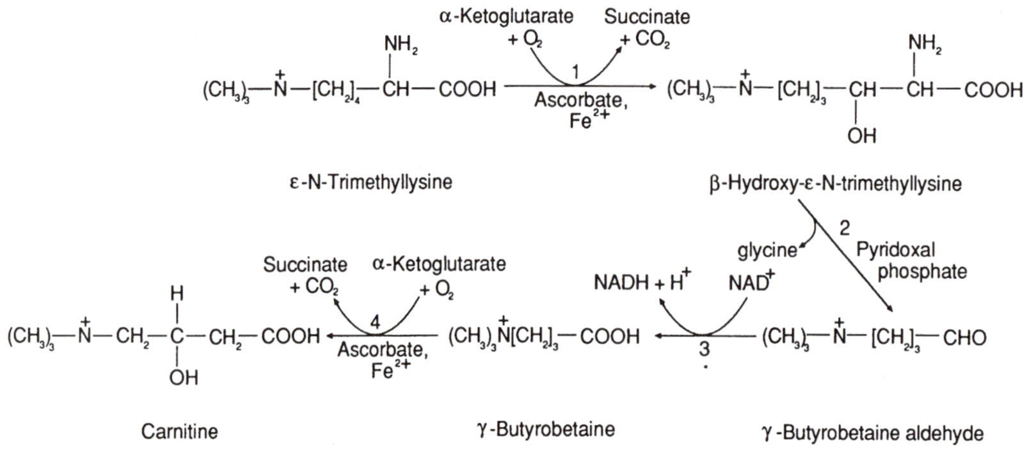

ε-N-Trimethyllysine → (1; α-Ketoglutarate + O_2 → Succinate + CO_2; Ascorbate, Fe^{2+}) → β-Hydroxy-ε-N-trimethyllysine → (2; Pyridoxal phosphate; glycine) → γ-Butyrobetaine aldehyde → (3; NAD^+ → NADH + H^+) → γ-Butyrobetaine → (4; α-Ketoglutarate + O_2 → Succinate + CO_2; Ascorbate, Fe^{2+}) → Carnitine

FIGURE 18-2
Carnitine biosynthesis in humans. A lysyl residue is trimethylated by S-adenosylmethionine, with subsequent proteolytic release of trimethyllysine, the starting material. The reactions are catalyzed by (1) trimethyllysine β-hydroxylase, (2) β-hydroxy-trimethyllysine aldolase (pyridoxal phosphate), (3) γ-trimethylaminobutyraldehyde dehydrogenase, and (4) γ-butyrobetaine hydroxylase.

membrane carnitine transport, carnitine palmitoyltransferase I (CPTI), carnitine palmitoyltransferase II (CPTII), and carnitine-acylcarnitine translocase. Clinical manifestations of disorders of carnitine metabolism and fatty acid oxidation disorders (discussed later) span a wide spectrum and can be affected by the severity and site of the defect (e.g., muscle, liver, and kidney). The disorders may be characterized by hypoketotic hypoglycemia, hyperammonemia, liver disease, skeletal muscle weakness, and cardiomyopathy. In some instances, dietary intervention brings about marked improvement in the clinical manifestations; for example, patients with carnitine transport defect respond well to carnitine therapy.

β-Oxidation

The major pathway for fatty acid oxidation, β-oxidation (Figure 18-3), involves oxidation of acyl-CoA at the β-carbon and removal of two carbon fragments as acetyl-CoA and takes place entirely in the mitochondrial matrix.

Oxidation of a saturated acyl-CoA with an even number of carbon atoms to acetyl-CoA requires repeated sequential action of four enzymes.

1. Acyl-CoA dehydrogenase dehydrogenates acyl-CoA at the α- and β-carbon atoms to yield the α,β-unsaturated acyl-CoA (or Δ^2-unsaturated acyl-CoA). Each one of four distinct dehydrogenases is specific for a given range of fatty acid chain length. All four are flavoproteins and contain a tightly bound molecule of flavin adenine dinucleotide (FAD). The electrons from the acyl-CoA dehydrogenase are transferred to the main respiratory chain (Chapter 14) through mitochondrial electron transfer flavoprotein (ETF) and ETF-ubiquinone oxidoreductase (ETF-QO) (Figure 18-4). Both ETF and ETF-ubiquinone oxidoreductase are nuclear encoded proteins. They also mediate transfer of electrons from dimethylglycine dehydrogenase and sarcosine dehydrogenase. Inherited defects in ETF and ETF-QO cause accumulation of organic acids (acidemia) and their excretion in the urine (acidurias) Examples of these disorders are glutaric acidemia type I and type II which are inherited as autosomal recessive traits. Glutaric acid is an intermediate in the metabolism of lysine, hydroxy lysine, and tryptophan. *Glutaric acidemia type I* is caused by deficiency of glutaryl-CoA dehydrogenase which catalyzes the conversion of glutaryl-CoA to crotonyl-CoA. *Glutaric acidemia type II* is caused by defects in the ETF/ETF-QO proteins. The clinical manifestations of these disorders are similar to medium-chain acyl-CoA dehydrogenase deficiency (discussed later). The Δ^2 double bond formed by the acyl-CoA dehydrogenase has a trans configuration. The double bonds in naturally occurring fatty acids are generally in the cis configuration. The oxidation of unsaturated *cis*-fatty acids requires two auxiliary enzymes, enoyl-CoA isomerase and 2,4-dienoyl-CoA reductase.

Acyl-CoA dehydrogenase (especially butyryl-CoA dehydrogenase) is irreversibly inactivated by methylene cyclopropylacetyl-CoA through the formation of covalent adduct with the FAD of the enzyme. The inhibitor is derived by transamination

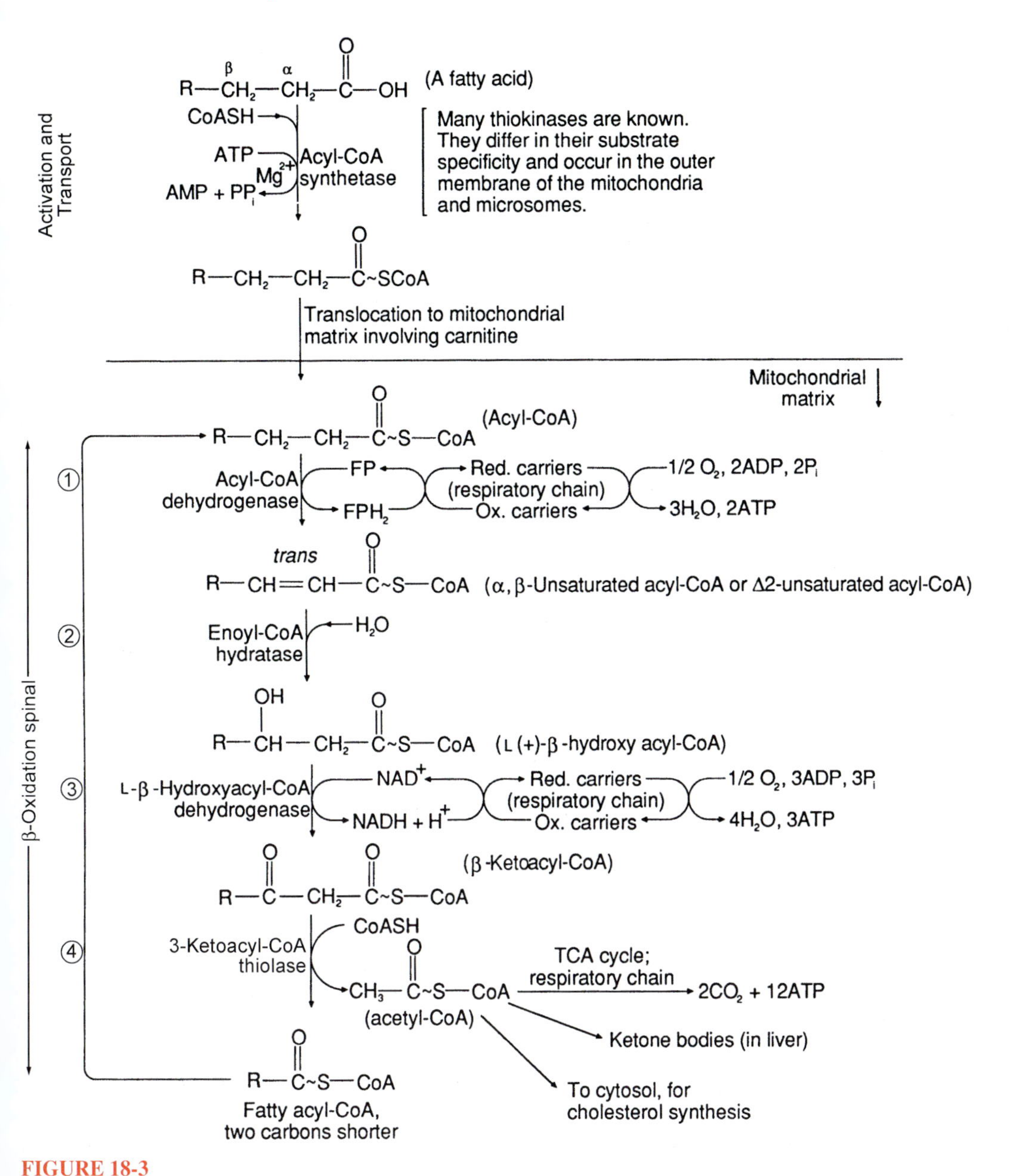

FIGURE 18-3
Fatty acid activation, transport, and β-oxidation. The shortened fatty acyl-CoA from one cycle is further oxidized in successive passes until it is entirely converted to acetyl-CoA. Odd-chain fatty acids produce one molecule of propionyl-CoA. Ox. = Oxidized; Red. = reduced; respiratory chain = oxidative phosphorylation and electron transport; $\sim$ = high-energy bond; FP = flavoprotein.

and oxidative decarboxylation of the amino acid hypoglycin (Chapter 15). Ingestion of hypoglycin causes severe hypoglycemia due to the inhibition of β-oxidation and corresponding decrease in ATP synthesis. Gluconeogenesis, which is important in maintaining fasting glucose levels, is dependent on adequate supplies of ATP. The action of hypoglycin thus serves to emphasize the importance of β-oxidation in gluconeogenesis under normal circumstances.

Among the fatty acid oxidation disorders, *medium-chain acyl-CoA dehydrogenase deficiency (MCAD)* is the most common and its frequency is similar to that of phenylketonuria. The disorder can be identified by mutant alleles and some key abnormal metabolites. An A $\rightarrow$ G transition mutation occurs at position 985 of MCAD-cDNA in about 90% of cases. This mutation leads to replacement of lysine with glutamate at position 329 (K329E) of the polypeptide.

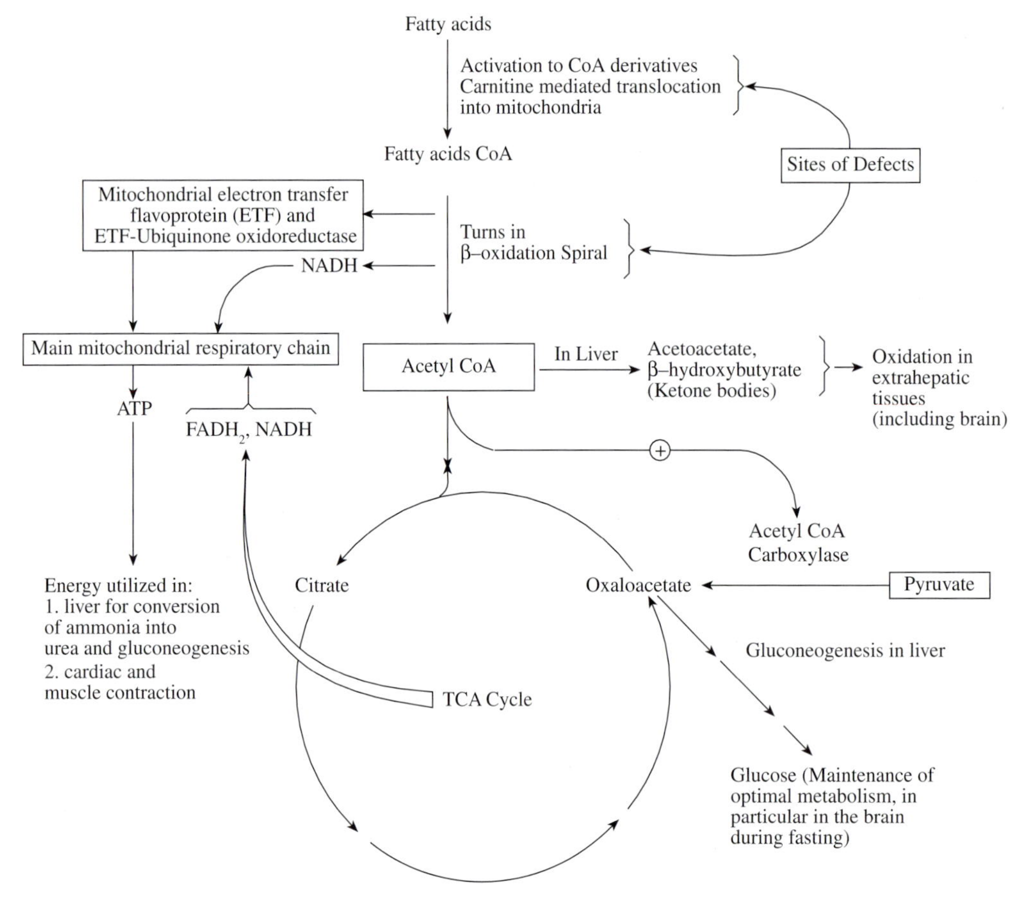

FIGURE 18-4
Spectrum of consequences of defects in fatty acid oxidation. The primary effect is inadequate production of acetyl-CoA, which leads to decreased flux through the TCA cycle and lack of ketone body synthesis in the liver. Both of these events cause energy deficits and changes in metabolic regulatory processes. Alterations in hepatic metabolism lead to hypoglycemia and hyperammonemia. Abnormalities also occur in skeletal and cardiac muscle and in the central nervous system.

The MCAD deficiency primarily affects hepatic fatty acid oxidation and the most common clinical presentation is episodic hypoketotic hypoglycemia initiated by fasting. The major metabolic derangement in MCAD is an inadequate supply of acetyl-CoA (Figure 18-4). The deficiency of acetyl-CoA leads to a decreased flux through the tricarboxylic acid (TCA) cycle causing diminished ATP production, decreased ketone body formation (the ketone bodies are metabolites used by the extrahepatic tissues), decreased citrate synthesis, and decreased oxaloacetate synthesis from pyruvate catalyzed by pyruvate carboxylase for which acetyl-CoA is the primary activator. The decreased flux through the TCA cycle during deficiency of acetyl-CoA causes a diminished synthesis of citrate from oxaloacetate and acetyl-CoA, as well as inhibition of α-ketoglutarate dehydrogenase due to an elevated ratio of fatty acyl-CoA to CoA. The formation of oxaloacetate is crucial for gluconeogenesis (Chapter 15). Accumulation of octanoate, which occurs in MCAD, may be responsible for encephalopathy and cerebral edema. In *Reye's syndrome,* octanoate also is elevated and may be responsible for its phenotypical similarity with MCAD. MCAD is managed with avoidance of fasting, stress, and treatment with intravenous glucose during acute episodes.

2. Enoyl-CoA hydratase catalyses the hydration of Δ^2 unsaturated acyl-CoA. This enzyme has broad specificity and can act on α-, β- (or Δ^2-) unsaturated CoA in trans or cis configuration. The products formed are

$$\text{2-}trans\text{-Enoyl-CoA} \rightarrow \text{L}(+)\text{-}\beta\text{-hydroxyacyl-CoA} \quad [\text{or L}(+)\text{-3-hydroxyacyl-CoA}]$$

$$\text{2-}cis\text{-Enoyl-CoA} \rightarrow \text{D}(-)\text{-}\beta\text{-hydroxyacyl-CoA} \quad [\text{or D}(-)\text{-3-hydroxyacyl-CoA}]$$

The latter reaction occurs in the oxidation of natural unsaturated fatty acids, and an epimerase converts the product to the L-isomer, which is the substrate of the next enzyme.

3. β-Hydroxyacyl-CoA dehydrogenase oxidizes β-hydroxyacyl-CoA by an NAD^+-linked reaction that is absolutely specific for the L-stereoisomer. The electrons from the NADH generated are passed on to NADH dehydrogenase of the respiratory chain.
4. 3-Ketoacyl-CoA thiolase (β-ketothiolase) catalyzes a thiolytic cleavage, has broad specificity, and yields acetyl-CoA and acyl-CoA shortened by two carbon atoms. The reaction is highly exergonic ($\Delta G^{0\prime} = -6.7$ kcal/mol) and favors thiolysis. The enzyme has a reactive —SH group on a cysteinyl residue, which participates as follows (E = enzyme):

$$\underset{\beta\text{-Ketoacyl-CoA}}{R{-}\overset{O}{\overset{\|}{C}}{-}CH_2{-}\overset{O}{\overset{\|}{C}}{-}SCoA} + HS{-}E \rightleftharpoons R\overset{O}{\overset{\|}{C}}{-}S{-}E + \underset{\text{Acetyl-CoA}}{CH_3{-}\overset{O}{\overset{\|}{C}}{-}SCoA}$$

$$R{-}\overset{O}{\overset{\|}{C}}{-}S{-}E + HS{-}CoA \rightleftharpoons \underset{\text{Acyl-CoA (with two less carbon atoms)}}{R{-}\overset{O}{\overset{\|}{C}}{-}SCoA} + \underset{\text{(Regenerated enzyme)}}{HS{-}E}$$

Three enzyme activities—long-chain enoyl-CoA hydratase, β-hydroxyacyl-CoA dehydrogenase, and long-chain β-ketoacyl-CoA thiolase (reactions 2–4, Figure 18-3)—are associated with a trifunctional protein complex consisting of four α- and four β-subunits bound to inner mitochondrial membrane. Each of the four α-subunits possesses hydratase and dehydrogenase enzyme activities at the N-terminal and C-terminal domains, respectively. The active site for the thiolase activity resides in the four β-subunits of the protein complex. Deficiencies of the dehydrogenase activity or all of the three enzyme activities for oxidation of long-chain fatty acids have been described. These deficiencies can cause nonketotic hypoglycemia during fasting, hepatic encephalopathy, and cardiac and skeletal myopathy. In some instances, women carrying fetuses with a deficiency of long-chain β-hydroxyacyl-CoA dehydrogenase may themselves develop acute liver disease, hemolysis, and a low platelet count. This clinical disorder is associated with a high risk of maternal and neonatal morbidity and mortality, and is known as HELLP (hemolysis, elevated liver enzyme levels and low platelet count). A change of Glu 474 to Gln (E474Q) in the α-subunits of the trifunctional protein has been identified in three unrelated children whose mothers had an acute fatty liver episode or HELLP syndrome during pregnancy. The fetal-maternal interaction that leads to toxic effects in women during pregnancy may be due to transport of long-chain β-hydroxyacyl metabolites produced by the fetus and placenta to the maternal liver.

Other inborn errors of fatty acid oxidation include defects in short-chain β-hydroxy acyl-CoA dehydrogenase and medium-chain β-ketoacyl-CoA thiolase. The spectrum of clinical findings in these and other fatty acid oxidation defects are variable; typical symptoms include fasting intolerance, cardiomyopathy, and sudden death. Children with long-chain fatty acid oxidation disorders are treated with frequent feeding of a low-fat diet consisting of medium-chain triglycerides. This dietary regimen can prevent hypoketotic hypoglycemic liver dysfunction.

Energetics of β-Oxidation

Palmitoyl-CoA yields 8 acetyl-CoA molecules and 14 pairs of hydrogen atoms, by seven cycles through the β-oxidation system. Acetyl-CoA can be oxidized in the TCA cycle, used for the synthesis of fatty acid or cholesterol, or used for the formation of ketone bodies in liver. β-Oxidation of an acyl-CoA with an uneven number of carbon atoms also yields a propionyl-CoA during the acetyl-CoA acyltransferase reaction of the last cycle.

Two high-energy bonds are consumed in the activation of a fatty acid molecule. Every mole of fatty acyl-CoA that cycles through reactions 1–4 produces 1 mol of $FADH_2$, 1 mol of NADH, and 1 mol of acetyl-CoA. On the last pass of an even-chain-length fatty acid, 2 mol of acetyl-CoA are formed; and the final pass of an odd-chain-length molecule releases 1 mol of propionyl-CoA. The amount of ATP formed from complete oxidation of a hexanoic acid is calculated as shown in Table 18-2.

Fatty acid oxidation produces more moles of ATP per mole of CO_2 formed than does carbohydrate oxidation. In this case, oxidation of 1 mol of hexose produces at most (assuming malate shuttle operation exclusively) 38 mol of ATP.

Complete oxidation of one molecule of palmitic acid yields 129 ATP molecules:

$$C_{15}H_{31}COOH + 8CoASH + ATP + 7FAD + 7NAD^+ + 7H_2O \rightarrow 8CH_3COSCoA + AMP + PP_i + 7FADH_2 + 7NADH + 7H^+$$

Each molecule of acetyl-CoA yields 12 ATP ($12 \times 8 = 96$); $FADH_2$ yields 2 ATP ($7 \times 2 = 14$); NADH yields 3ATP ($7 \times 3 = 21$); and two high-energy bonds are consumed (-2; ATP $\rightarrow$ AMP + PP_i). Thus, net ATP production is 129. The energy yield from total combustion of palmitic acid in a bomb calorimeter (Chapter 5) is

TABLE 18-2

Reaction	Direct Consequences of the Reaction	Moles of ATP Gained or Lost per Mole of Hexanoic Acid
Activation reaction	Hexanoic acid → Hexanoyl-CoA	−2
First dehydrogenation	Dehydrogenation of acyl-CoA; 2(FAD → $FADH_2$)	+4
Hydration	Hydration of α, β-unsaturated fatty acyl-CoA	0
Second dehydrogenation	Dehydrogenation of β-hydroxy-acyl-CoA; 2(NAD^+ → NADH + H^+)	+6
Oxidation of acetyl-CoA	Formation of 3 mol of acetyl-CoA, and their oxidation in the TCA cycle and electron transport system	+(3 × 12) = +36
		Total ATP = +44

−2380 kcal/mol:

$$C_{16}H_{32}O_2 + 23O_2 \rightarrow 16CO_2 + 16H_2O \text{ at } 20^\circ C,$$
$$\Delta H^\circ = -2380 \text{ kcal/mol}$$

In biological oxidation, the energy conserved as ATP is about 942 kcal/mol (129 × 7.3). Thus, the percentage of standard free energy of oxidation of palmitic acid conserved as high-energy phosphate is about 40% (942/2380 × 100).

Regulation of Fatty Acid Oxidation

Regulation of fatty acid oxidation involves diet, cofactors, competing substrates and hormones of fatty acid mobilization. Adipose tissue triacylglycerol lipolysis is one of the major sites of regulation. The other site is CPTI. The latter is inhibited by malonyl-CoA which is involved in fatty acid biosynthesis (discussed later). Thus, fatty acid oxidation and synthesis do not occur simultaneously. Insulin inhibits fatty acid oxidation by blocking lipolysis in adipose tissue, and it stimulates lipogenesis and synthesis of malonyl-CoA. Glucagon stimulates fatty acid oxidation by inhibiting synthesis of acetyl-CoA carboxylase which leads to decreased synthesis of malonyl-CoA. This causes enhanced activity of CPTI, and promotion of fatty acid oxidation. In the fed state, the glucagon/insulin ratio is low, and fatty acid synthesis is promoted in the liver. In the fasting state, the glucagon/insulin ratio is high and mobilization of free fatty acids from adipose tissue and mitochondrial fatty acid oxidation are augmented.

Peroxisomal Fatty Acid Oxidation

Peroxisomes have a single membrane and contain a fairly homogeneous, moderately electron-dense matrix. They are present in many mammalian cells and are particularly abundant in liver and kidney. A normal hepatocyte contains about a thousand peroxisomes, whose proliferation is inducible by hypolipidemic drugs such as clofibrate (Chapter 20). Peroxisomes contain H_2O_2-producing oxidases and also H_2O_2-inactivating catalase. Oxidation of very long-chain, saturated, unbranched fatty acids (C_{24}–C_{26}) appears to take place mainly, if not exclusively, in peroxisomes after the acyl-CoA derivatives are transported across the membrane without involvement of carnitine. Oxidation is mediated by flavoprotein dehydrogenases that yield H_2O_2 and acetyl-CoA and terminates with octanoyl-CoA. Octanoyl- and acetyl-CoA are transferred to mitochondria for further oxidation. Phytinic acid monooxygenase, which initiates the catabolism of phytinic acid (a 20-carbon branched chain fatty acid of dietary origin), is probably a peroxisomal enzyme. Peroxisomal oxidation does not yield ATP. All the energy produced appears as heat.

Three genetic disorders (*Zellweger's syndrome, neonatal adrenoleukodystrophy,* and *childhood adrenoleukodystrophy*) exhibit defective formation of peroxisomes (in Zellweger's syndrome no morphologically detectable peroxisomes are present) or deficiency of one or more constituent enzymes. All three disorders are characterized by a marked accumulation of very long chain, saturated, unbranched fatty acids (tetracosanoic and hexacosanoic acids) in liver and central nervous system tissues, severe neurological symptoms, and early death.

Peroxisomes contain dihydroxyacetone phosphate acyltransferase and alkyldihydroxyacetone phosphate synthase, which are involved in synthesis of the plasmalogens (Chapter 19). Peroxisomes may also participate in the biosynthesis of bile acids. The conversion of trihydroxycholestanoic acid to cholic acid (Chapter 19) has been localized to peroxisomes.

Other Pathways of Fatty Acid Oxidation

Propionyl-CoA Oxidation

β-Oxidation of fatty acids with an odd number of carbon atoms yields propionyl-CoA. Since the concentration of such fatty acids in the diet is small, little propionyl-CoA is produced. Important sources of propionyl-CoA are the catabolism of isoleucine, valine, methionine, and threonine (Chapter 17). Cholesterol side chain oxidation also yields propionyl-CoA. Thus, propionyl-CoA is derived from the catabolism of lipids and proteins. In ruminants, propionate is largely derived from bacterial fermentation in the rumen.

Propionyl-CoA is converted to succinyl-CoA, which is oxidized or converted to glucose by way of oxaloacetate and pyruvate (gluconeogenesis; Chapter 15). Succinyl-CoA may also form δ-aminolevulinate, a precursor of porphyrin biosynthesis (Chapter 29). Formation of succinyl-CoA from propionyl-CoA requires three mitochondrial enzymes and two vitamins (Figure 18-5).

1. Propionyl-CoA carboxylase is a tetramer of nonidentical subunits, α and β. The native enzyme (M.W. ~540,000) appears to have the structure $(\alpha\beta)_4$. Biotin is bound through an amide linkage to an ε-amino group of a lysyl residue in the α-subunit. Carboxylation is a two-step reaction similar to that of acetyl-CoA carboxylase (see below). The first step requires ATP and Mg^{2+} and fixes CO_2 with the formation of an apoenzyme-biotin-CO_2 complex. In the second step, the carboxyl group from the biotinyl complex is transferred to propionyl-CoA to form D-methylmalonyl-CoA.
2. Methylmalonyl-CoA racemase converts D-methylmalonyl-CoA to the L-isomer by labilization of an α-hydrogen atom, followed by uptake of a proton from the medium.
3. Methylmalonyl-CoA mutase utilizes 5′-deoxyadenosylcobalamin (Chapter 38) to catalyze intramolecular isomerization by the migration of the –COSCoA group. The only other cobalamin-dependent reaction in the mammalian system is methylation of homocysteine to methionine (Chapters 17, 27, and 38).

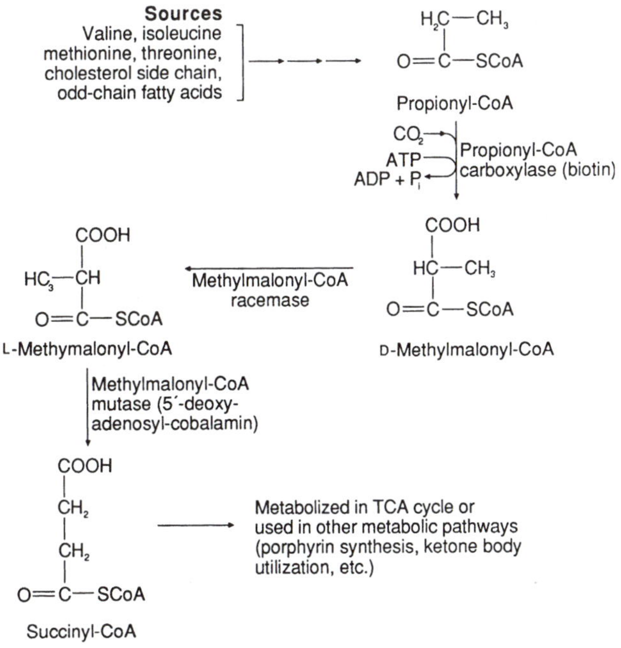

FIGURE 18-5
Metabolism of propionyl-CoA.

Inborn errors of metabolism may be due to propionyl-CoA carboxylase deficiency, defects in biotin transport or metabolism, methylmalonyl-CoA mutase deficiency, or defects in adenosylcobalamin synthesis. The former two defects result in propionic acidemia, the latter two in methylmalonic acidemia. All cause metabolic acidosis and developmental retardation. Organic acidemias often exhibit hyperammonemia, mimicking ureagenesis disorders, because they inhibit the formation of N-acetylglutamate, an obligatory cofactor for carbamoyl phosphate synthase (Chapter 17). Some of these disorders can be partly corrected by administration of pharmacological doses of the vitamin involved (Chapter 38). Dietary protein restriction is therapeutically useful (since propionate is primarily derived from amino acids). Propionic and methylmalonyl acidemia (and aciduria) results from vitamin B_{12} deficiency (e.g., pernicious anemia; Chapter 38).

α-Oxidation

α-Oxidation is important in the catabolism of branched-chain fatty acids. The general reaction, catalyzed by a monooxygenase, requires O_2, Fe^{2+}, and either ascorbate or tetrahydropteridine. It has been demonstrated in plants and in microsomes from brain and other tissues.

$$R—CH_2—CH_2—COOH + \text{Reduced cofactor} + O_2 \xrightarrow{\text{Monooxygenase}}$$

$$R—CH_2—CH(OH)—COOH + \text{Oxidized cofactor} + H_2O$$

α-Hydroxy fatty acid

This reaction is also a route for the synthesis of hydroxy fatty acids. The α-hydroxy fatty acid can be further oxidized and decarboxylated to a fatty acid one carbon shorter than the original. Thus, if an odd-chain-length compound is used initially, an even-chain-length acid is produced that can be further oxidized by β-oxidation.

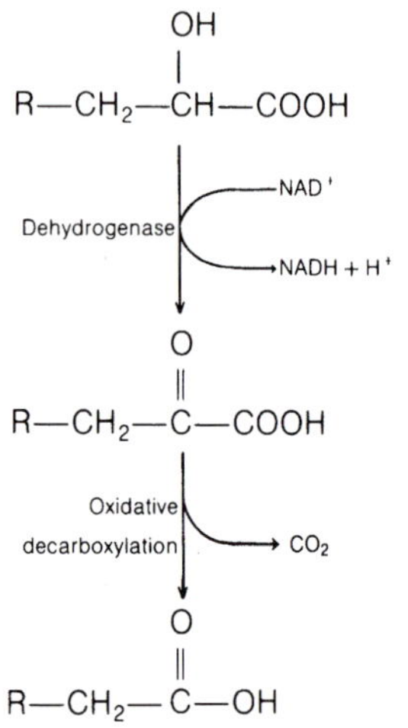

$$R{-}CH_2{-}CH(OH){-}COOH \xrightarrow[\text{Dehydrogenase}]{NAD^+ \rightarrow NADH + H^+} R{-}CH_2{-}C(=O){-}COOH \xrightarrow[\text{Oxidative decarboxylation}]{\rightarrow CO_2} R{-}CH_2{-}C(=O){-}OH$$

In Refsum's disease, an autosomal recessive disorder, the defect is probably in the α-hydroxylation of phytanic acid. Phytanic acid is a 20-carbon, branched-chain fatty acid derived from the plant alcohol phytol, which is present as an ester in chlorophyll. Thus, its origin in the body is from dietary sources. The oxidation of phytanic acid is shown in Figure 18-6. The clinical characteristics of Refsum's disease include peripheral neuropathy and ataxia, retinitis pigmentosa, and abnormalities of skin and bones. Significant improvement has been observed when patients are kept on low–phytanic acid diets for prolonged periods (e.g., diets that exclude dairy and ruminant fat).

ω-*Oxidation*

ω-Oxidation is oxidation of the carbon atom most remote from the carboxyl group in a fatty acid. The basic reaction, catalyzed by a monooxygenase that requires NADPH, O_2, and cytochrome P-450, is shown below. It has been observed in liver microsomes and some bacteria.

$$H_3C{-}(CH_2)_n{-}COOH + O_2 + NADPH + H^+ \xrightarrow{\omega\text{-oxidation}} HO{-}CH_2{-}(CH_2)_n{-}COOH + NADP^+ + H_2O$$

Further oxidation of the ω-hydroxy acids produces dicarboxylic acids, which can be β-oxidized from either end.

Oxidation of Mono- and Polyunsaturated Fatty Acids

Oxidation of unsaturated fatty acids requires Δ^3-*cis*-,Δ^2-*trans*-enoyl-CoA isomerase and NADPH-dependent 2,4-dienoyl-CoA reductase, in addition to the enzymes of β-oxidation. The enoyl-CoA isomerase produces the substrate for the hydration step. The reductase catalyzes the reduction of Δ^2-*trans*-,Δ^4-*cis*-decadienoyl-CoA to Δ^3-*trans*-decenoyl-CoA. The latter is isomerized to Δ^2-*trans*-decenoyl isomerase, which is a normal β-oxidation intermediate. These reactions are illustrated for oxidation of oleic and linoleic acids in Figures 18-7 and 18-8.

18.2 Metabolism of Ketone Bodies

Ketone bodies consist of acetoacetate, D-β-hydroxybutyrate (D-3-hydroxybutyrate), and acetone. They are

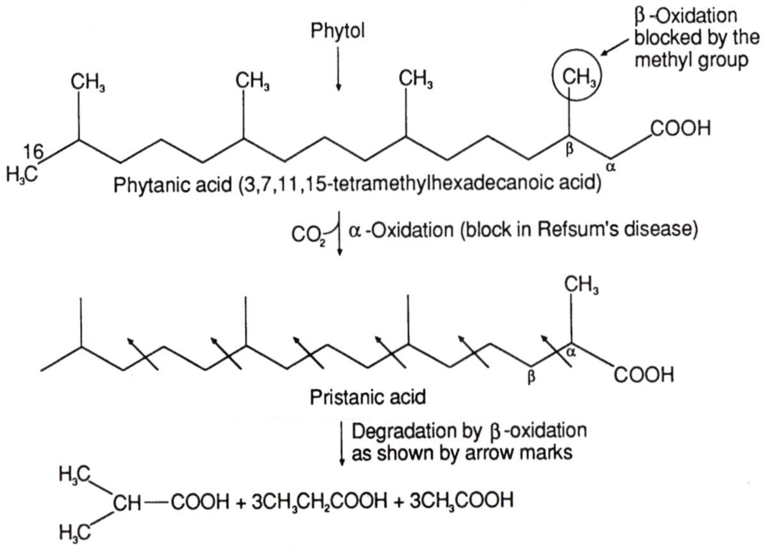

Phytol
β-Oxidation blocked by the methyl group
CH_3 CH_3 CH_3 CH_3 COOH 16 H_3C β α
Phytanic acid (3,7,11,15-tetramethylhexadecanoic acid)
CO_2 α-Oxidation (block in Refsum's disease)
CH_3 α β COOH
Pristanic acid
Degradation by β-oxidation as shown by arrow marks

$$(H_3C)_2CH{-}COOH + 3CH_3CH_2COOH + 3CH_3COOH$$

FIGURE 18-6
Oxidation of phytol and phytanic acid.

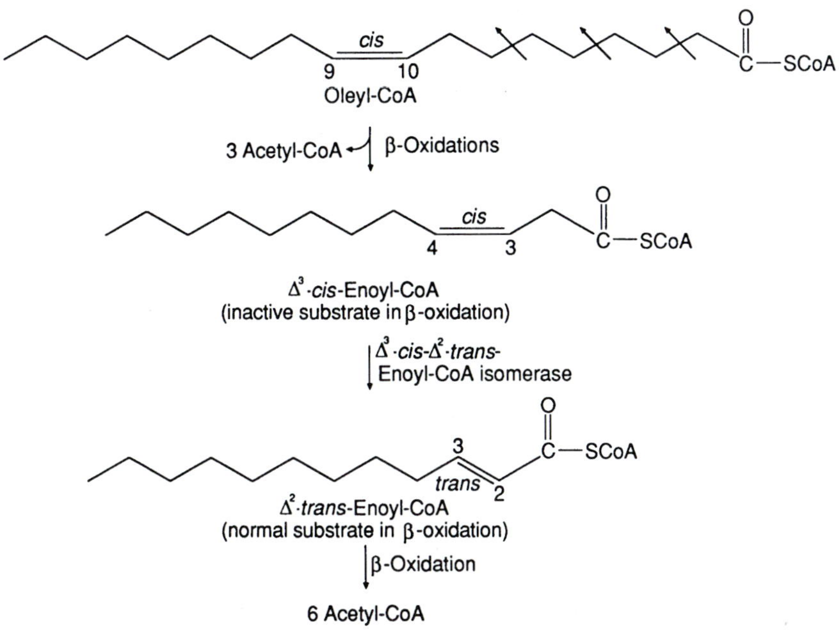

cis
9 10
Oleyl-CoA
C—SCoA

3 Acetyl-CoA ← β-Oxidations

cis
4 3
C—SCoA

Δ^3-*cis*-Enoyl-CoA
(inactive substrate in β-oxidation)

Δ^3-*cis*-Δ^2-*trans*-
Enoyl-CoA isomerase

3
C—SCoA
trans 2

Δ^2-*trans*-Enoyl-CoA
(normal substrate in β-oxidation)

β-Oxidation

6 Acetyl-CoA

FIGURE 18-7
Oxidation of oleic acid.

synthesized in liver mitochondria. The overall steps involved in the formation of ketone bodies include the mobilization of fatty acids by lipolysis from adipose tissue triacylglycerol by hormone-sensitive triacylglycerol lipase, plasma fatty acid transport, fatty acid activation, fatty acid transport into mitochondria (with acylcarnitine as an intermediate), and β-oxidation. The regulatory reactions are those of lipolysis and of acyl-CoA transport across the mitochondrial membrane (CPTI). Synthesis of ketone bodies from acetyl-CoA consists of three steps: formation of acetoacetyl-CoA, formation of acetoacetate, and reduction of acetoacetate to β-hydroxybutyrate. Nonenzymatic decarboxylation of acetoacetate yields acetone, which is eliminated via the lungs.

The pathways of formation of ketone bodies are shown in Figure 18-9. The major pathway of production of acetoacetate is from β-hydroxy-β-methylglutaryl-CoA (HMG-CoA). Hydrolysis of acetoacetyl-CoA to acetoacetate by acetoacetyl-CoA hydrolase is of minor importance because the enzyme has a high K_m for acetoacetyl-CoA. HMG-CoA is also produced in the cytosol, where it is essential for the synthesis of several isoprenoid compounds and cholesterol (Chapter 19). The reduction of acetoacetyl-CoA to β-hydroxybutyrate depends on the mitochondrial $[NAD^+]/[NADH]$ ratio.

Ketone bodies are oxidized primarily in extrahepatic tissues (e.g., skeletal muscle, heart, kidney, intestines, brain) within mitochondria. β-Hydroxybutyrate is oxidized to acetoacetate by NAD^+-dependent β-hydroxybutyrate dehydrogenase by reversal of the reaction that occurred during ketogenesis:

$$\underset{\text{D-}\beta\text{-Hydroxybutyrate}}{CH_3CH(OH)CH_2CO^-} + NAD^+ \rightarrow \underset{\text{acetoacetate}}{CH_3COCH_2COO^-} + NADH + H^+$$

Activation of acetoacetate requires transfer of coenzyme A from succinyl-CoA, derived from the TCA cycle, by succinyl-CoA-acetoacetate-CoA transferase (thiophorase):

$$\underset{\text{Succinyl-CoA}}{\begin{array}{c}CH_2COSCoA\\ |\\ CH_2COO^-\end{array}} + \underset{\text{Acetoacetate}}{\begin{array}{c}CH_2COO^-\\ |\\ COCH_3\end{array}} \longrightarrow \underset{\text{Succinate}}{\begin{array}{c}CH_2COO^-\\ |\\ CH_2COO^-\end{array}} + \underset{\text{Acetoacetyl-CoA}}{\begin{array}{c}CH_2COSCoA\\ |\\ COCH_3\end{array}}$$

The activation occurs at the expense of conversion of succinyl-CoA to succinate in the TCA cycle and formation of GTP (Chapter 13). Acetoacetyl-CoA is cleaved to two molecules of acetyl-CoA by acetyl-CoA acetyltransferase, the same enzyme involved in the synthesis of acetoacetyl-CoA (Figure 18-9). Acetyl-CoA is oxidized in the TCA cycle. Thus, formation of ketone bodies in the liver and their oxidation in extrahepatic tissues are dictated by the ratio [substrates]/[products].

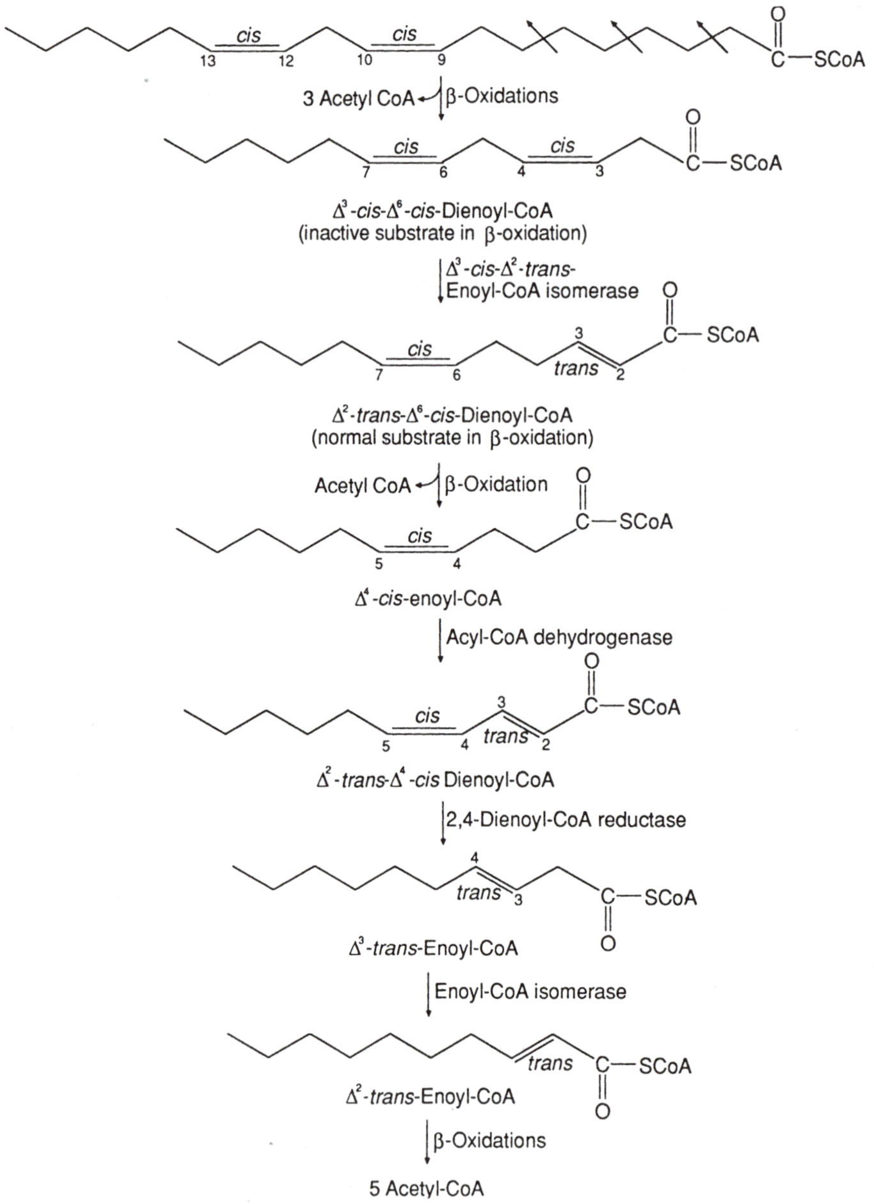

3 Acetyl CoA ← β-Oxidations

Δ^3-*cis*-Δ^6-*cis*-Dienoyl-CoA
(inactive substrate in β-oxidation)

Δ^3-*cis*-Δ^2-*trans*-Enoyl-CoA isomerase

Δ^2-*trans*-Δ^6-*cis*-Dienoyl-CoA
(normal substrate in β-oxidation)

Acetyl CoA ← β-Oxidation

Δ^4-*cis*-enoyl-CoA

Acyl-CoA dehydrogenase

Δ^2-*trans*-Δ^4-*cis* Dienoyl-CoA

2,4-Dienoyl-CoA reductase

Δ^3-*trans*-Enoyl-CoA

Enoyl-CoA isomerase

Δ^2-*trans*-Enoyl-CoA

β-Oxidations

5 Acetyl-CoA

FIGURE 18-8
Oxidation of linoleic acid.

Physiological and Pathological Aspects of Metabolism of Ketone Bodies

Acetoacetate and β-hydroxybutyrate are products of normal metabolism of fatty acid oxidation and serve as metabolic fuels in extrahepatic tissues. Their level in blood depends on the rates of production and utilization. Oxidation increases as their plasma level increases. Some extrahepatic tissues (e.g., muscle) oxidize them in preference to glucose and fatty acid. Normally, the serum concentration of ketone bodies is less than 0.3 mM/L.

The rate of formation of ketone bodies depends on the concentration of fatty acids derived from hydrolysis of adipose tissue triacylglycerol by hormone-sensitive lipase. Insulin depresses lipolysis and promotes triacylglycerol synthesis and storage, while glucagon has the opposite effects. Thus, insulin is antiketogenic and glucagon is ketogenic (Chapter 22). Uncontrolled insulin-dependent diabetes may result in fatal ketoacidosis (Chapter 39). Although ketonemia and ketonuria are generally assumed to be due to increased production of ketone bodies in the liver, studies with depancreatized rats have shown that ketosis may also arise from their diminished oxidation.

Ketosis can occur in starvation, in ethanol abuse, and following exercise, the last because of a switch in blood

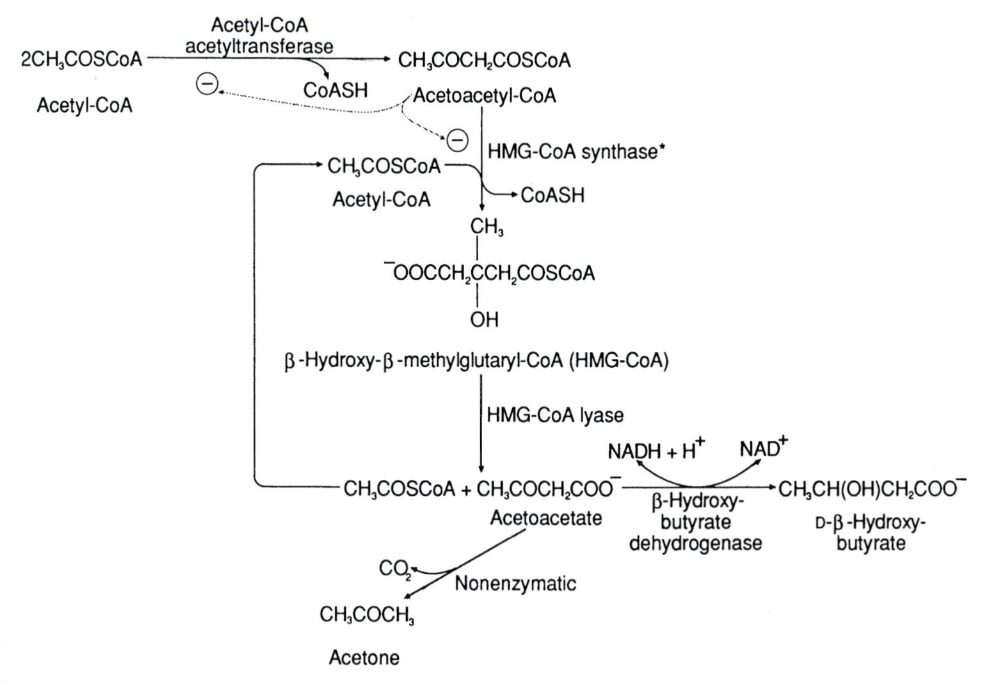

FIGURE 18-9
Ketogenesis in the liver. All reactions occur in mitochondria; the rate-controlling reactions (not shown) are release of fatty acids from adipose tissue and uptake of acyl-CoA into mitochondria, in particular, the CPTI reaction (see Figure 18-1). Acetoacetyl-CoA may regulate ketogenesis by inhibiting the transferase and the synthase.
*This enzyme is similar to citrate synthase (Chapter 13) which catalyzes an analogous reaction.

flow. During sustained exercise, the blood flow to the liver, intestines, and kidneys is substantially decreased, with a corresponding increase in blood flow to working muscles, so that more fatty acids mobilized from adipose tissue are delivered to the muscle. Thus, the formation of ketone bodies is severely curtailed. But during the postexercise period, with the resumption of normal blood flow to liver, ketone bodies are generated as a result of increased mobilization of fatty acids. Reduced ketone body utilization in the extrahepatic tissues can occur due to deficiency of either succinyl-CoA-acetoacetate-CoA transferase or acetyl-CoA acetyltransferase. These patients are susceptible to attacks of ketoacidosis and the presence of persistent ketone bodies in the urine.

18.3 Metabolism of Ethanol

Ethanol is consumed widely. Microbial fermentation in the large intestine of humans can produce about 3 g of ethanol per day. Ethanol is rapidly absorbed throughout the gastrointestinal tract or, when inhaled, through the lungs. It is metabolized in the liver by a process having zero-order kinetics; i.e., the rate of oxidation is constant with time. The amount metabolized per unit time depends on liver size (or body weight); the average rate in an adult is about 30 mL in 3 hours. The energy content of alcohol is about 7 kcal/g.

Ethanol oxidation begins with conversion to acetaldehyde by alcohol dehydrogenase (M.W. ~85,000), a zinc-containing, NAD^+-dependent enzyme that is a relatively nonspecific cytoplasmic enzyme with a K_m of about 1 mM/L:

$$CH_3CH_2OH + NAD^+ \rightarrow CH_3CHO + NADH + H^+$$

Acetaldehyde is rapidly converted to acetate by NAD^+-dependent aldehyde dehydrogenase:

$$CH_3CHO + NAD^+ + H_2O \rightarrow CH_3COOH + NADH + H^+$$

Ethanol is also oxidized by the mixed-function oxidase of smooth endoplasmic reticulum, which requires NADPH, oxygen, and a cytochrome P-450 electron transport system (Chapter 14):

$$CH_3CH_2OH + NADPH + H^+ + 2O_2 \rightarrow CH_3CHO + 2H_2O_2 + NADP^+$$

Many drugs are metabolized by this enzyme, hence the competition between ethanol and other drugs (e.g., barbiturates). Peroxisomal catalase catalyzes the reaction:

$$CH_3CH_2OH + H_2O_2 \rightarrow CH_3CHO + 2H_2O$$

The K_m for this catalase and for the mixed-function oxidase is about 10 mM/L. The extent to which these two enzymes metabolize ethanol is not known.

Acetaldehyde is converted to acetate in the liver by NAD^+-linked aldehyde dehydrogenases, one in the cytosol ($K_m = 1$ mM/L) and another in mitochondria ($K_m = 0.01$ mM/L):

$$CH_3CHO + NAD^+ + H_2O \rightarrow CH_3COO^- + NADH + 2H^+$$

Disulfiram (tetraethylthiuram disulfide)

$$(H_5C_2)_2N{-}C({=}S){-}S{-}S{-}C({=}S){-}N(C_2H_5)_2$$

causes irreversible inactivation of these aldehyde dehydrogenases by reacting with sulfhydryl groups, with a buildup of acetaldehyde that produces the acetaldehyde syndrome (vasodilation, intense throbbing, pulsating headache, respiratory difficulties, copious vomiting, sweating, thirst, hypotension, vertigo, blurred vision, and confusion). Disulfiram by itself is relatively nontoxic. It is used in the treatment of chronic alcoholism but does not cure it. Disulfiram provides a willing patient with a deterrent to consumption of alcohol. A shorter acting reversible inhibitor of aldehyde dehydrogenase is calcium carbimide, which causes accumulation of acetaldehyde and unpleasant symptoms. Thus, calcium carbimide can also be used as a deterrent to alcohol consumption.

Symptoms similar to the disulfiram-ethanol reaction occur in high proportion in certain ethnic groups (e.g., Asians and Native Americans) who are extremely sensitive to ethanol consumption. The ethanol sensitivity in these populations is accompanied by a higher acetaldehyde steady-state concentration in the blood, which may be due to a rapid rate of formation of acetaldehyde by alcohol dehydrogenase or to a decreased rate of its removal by aldehyde dehydrogenase. Both of these dehydrogenases are present in several isozyme forms and exhibit extensive polymorphism among racial groups. An alcohol dehydrogenase variant found in the ethanol-sensitive populations has a relatively higher rate of activity at physiological pH and may account for more rapid oxidation of ethanol to acetaldehyde. However, a more important cause of acetaldehyde accumulation appears to be deficiency of an isozyme of aldehyde dehydrogenase, which has a low K_m for acetaldehyde. Thus, the cause of ethanol sensitivity may be impaired rate of removal of acetaldehyde rather than its excessive formation. Individuals who are predisposed to ethanol sensitivity should avoid ethanol intake in any form.

Acetate produced from ethanol is converted to acetyl-CoA by acetyl-CoA synthase in hepatic and extrahepatic tissues.

$$CH_3COO^- + ATP^{4-} + CoASH \rightarrow CH_3COSCoA + AMP^{2-} + PP_i^{3-}$$

Acetyl-CoA is oxidized in the TCA cycle and is used in liver and adipose tissue for biosynthesis of fatty acids and triacylglycerol.

Alcoholism affects about 10% of the drinking population and alcohol (ethanol) abuse has been implicated in at least 20% of admissions to general hospitals. This chronic disease exhibits high mortality due to a wide variety of factors. Ethanol produces effects in virtually every organ system. The biochemical effects of ethanol are due to increased production of NADH that decreases the $[NAD^+]/[NADH]$ ratio in the cytoplasm of liver cells at least tenfold from the normal value of about 1000. Increased production of lactate and inhibition of gluconeogenesis (Chapter 15) result. The hyperuricemia associated with ethanol consumption has been attributed to accelerated turnover of adenine nucleotides and their catabolism to uric acid (Chapter 27). Alcohol increases hepatic fatty acid and triacylglycerol synthesis and mobilization of fat from adipose tissue, which can lead to fatty liver, hepatitis, and cirrhosis. These effects are complicated by a deficiency of B vitamins and protein.

Alcohol increases the plasma level of VLDL and of HDL cholesterol (Chapter 20).

Many actions of ethanol may be attributed to a membrane-disordering effect. Changes in membrane fluidity can affect membrane-bound enzymes (e.g., Na^+, K^+-ATPase, adenylate cyclase) and phospholipid architecture. Alcohol also affects several neurotransmitter systems in the brain. These include dopamine (mediates pleasurable effects), γ-aminobutyric acid (GABA), glutamate, serotonin, adenosine, norepinephrine, and opioid peptides. Potential drug therapy for alcohol dependence consists of the use of antagonists and agonists of alcohol affected neurotransmitter systems. For example, naltrexone, a μ-opioid antagonist, inhibits alcohol-induced dopamine release, thus minimizing the pleasurable effect of alcohol and reducing the desire to consume alcohol. Another drug, acamprosate, reduces the craving for alcohol presumably by an agonist activity at GABA receptors and an inhibitory activity at N-methyl-D-aspartate receptors. A selective antagonist of serotonin receptor 5-HT_3, ondansetron, reduces alcohol consumption in patients with early onset alcoholism. The 5-HT_3 receptors are densely distributed in mesocorticolimbic neuronal terminals and regulate dopamine release. Attenuation of dopamine release reduces alcohol consumption.

In chronic alcoholics, heavy drinking and decreased food intake lead to ketoacidosis. Accelerated lipolysis arising from reduced insulin and increased glucagon secretion caused by hypoglycemia leads to ketosis with a high [β-hydroxybutyrate]/[acetoacetate] ratio. Treatment requires normalization of fluid and electrolyte balance (Chapter 39) and of glucose level. Administration of glucose provokes insulin release and depresses glucagon release, thus suppressing the stimuli for ketogenesis. The distinction between diabetic ketoacidosis and alcoholic ketoacidosis may be difficult to determine, and in some patients plasma glucose levels may not discriminate between the two entities (although in diabetic ketoacidosis plasma glucose levels are usually high, whereas in alcoholic ketoacidosis these levels may be low, normal, or marginally elevated). Fluid and electrolyte replacement and glucose administration in ketoacidosis are essential regardless of etiology.

Ethanol is a teratogen partly because it inhibits embryonic cellular proliferation. Maternal alcoholism causes fetal alcohol syndrome, which is characterized by abnormal function of the central nervous system, microcephaly, cleft palate, and micrognathia.

18.4 Synthesis of Long-Chain Saturated Fatty Acids

The reactions of *de novo* fatty acid biosynthesis are shown in Figure 18-10. They are carried out by two multienzyme systems functioning in sequence. The first is acetyl-CoA carboxylase, which converts acetyl-CoA to malonyl-CoA. The second is fatty acid synthase, which sequentially joins two-carbon units of malonyl-CoA, eventually producing palmitic acid. Both complexes consist of multifunctional subunits. The various catalytic functions can be readily separated in plant cells and prokaryotes, but in yeasts, birds, and mammals, attempts to subdivide catalytic functions lead to loss of activity. Important features of this system are as follows:

1. *De novo* synthesis takes place in the cytosol (whereas fatty acid oxidation occurs in mitochondria).
2. All carbon atoms are derived from acetyl-CoA (obtained from carbohydrates or amino acids), and palmitate (C_{16}) is the predominant fatty acid produced. Fatty acids longer than 16 carbons, those that are unsaturated, and hydroxy fatty acids are obtained by separate processes of chain elongation, desaturation, or α-hydroxylation, respectively.
3. The committed (rate-controlling) step is the biotin-dependent carboxylation of acetyl-CoA by acetyl-CoA carboxylase. Important allosteric effectors are citrate (positive) and long-chain acyl-CoA derivatives (negative).
4. Although the initial step requires CO_2 fixation, CO_2 is not incorporated into fatty acids. The labeled carbon in $^{14}CO_2$ (as $H^{14}CO_3^-$) is not incorporated into the carbons of fatty acids synthesized.
5. Synthesis is initiated by a molecule of acetyl-CoA that functions as a primer. Its two carbons become C_{15} and C_{16} of palmitate. The acetyl group is extended by successive addition of the two carbons of malonate originally derived from acetyl-CoA, the unesterified carboxylic acid group being removed as CO_2. In mammalian liver and mammary gland, butyryl-CoA is a more active primer than acetyl-CoA. Odd-chain-length fatty acids found in some organisms are synthesized by priming the reaction with propionyl-CoA instead of acetyl-CoA.
6. Release of the finished fatty acid occurs when the chain length reaches C_{16} by action of thioester hydrolase, which is specific for long-chain acyl-CoA derivatives. A thioester hydrolase of mammary gland is specific for acyl residues of C_8, C_{10}, or C_{12}.

The overall reaction for palmitate synthesis from acetyl-CoA is

$$8\text{Acetyl-CoA} + 14\text{NADPH} + 14\text{H}^+ + 7\text{ATP} + \text{H}_2\text{O} \rightarrow \text{palmitate} + 8\text{CoASH} + 14\text{NADP}^+ + 7\text{ADP} + 7\text{P}_i$$

The reducing equivalents of NADPH are derived largely from the pentose phosphate pathway.

Acetyl-CoA carboxylase is a biotin-dependent enzyme. It has been purified from microorganisms, yeast, plants, and animals. In animal cells, it exists as an inactive protomer (M.W. ~400,000) and as an active polymer (M.W. 4–8 million). The protomer contains the activity of biotin carboxylase, biotin carboxyl carrier protein (BCCP), transcarboxylase, and a regulatory allosteric site. Each protomer contains a biotinyl group bound in amide linkage to the ε-amino group of a lysyl residue.

Citrate shifts the equilibrium from inactive protomer to active polymer. The polymeric form appears as long filaments in electron micrographs.

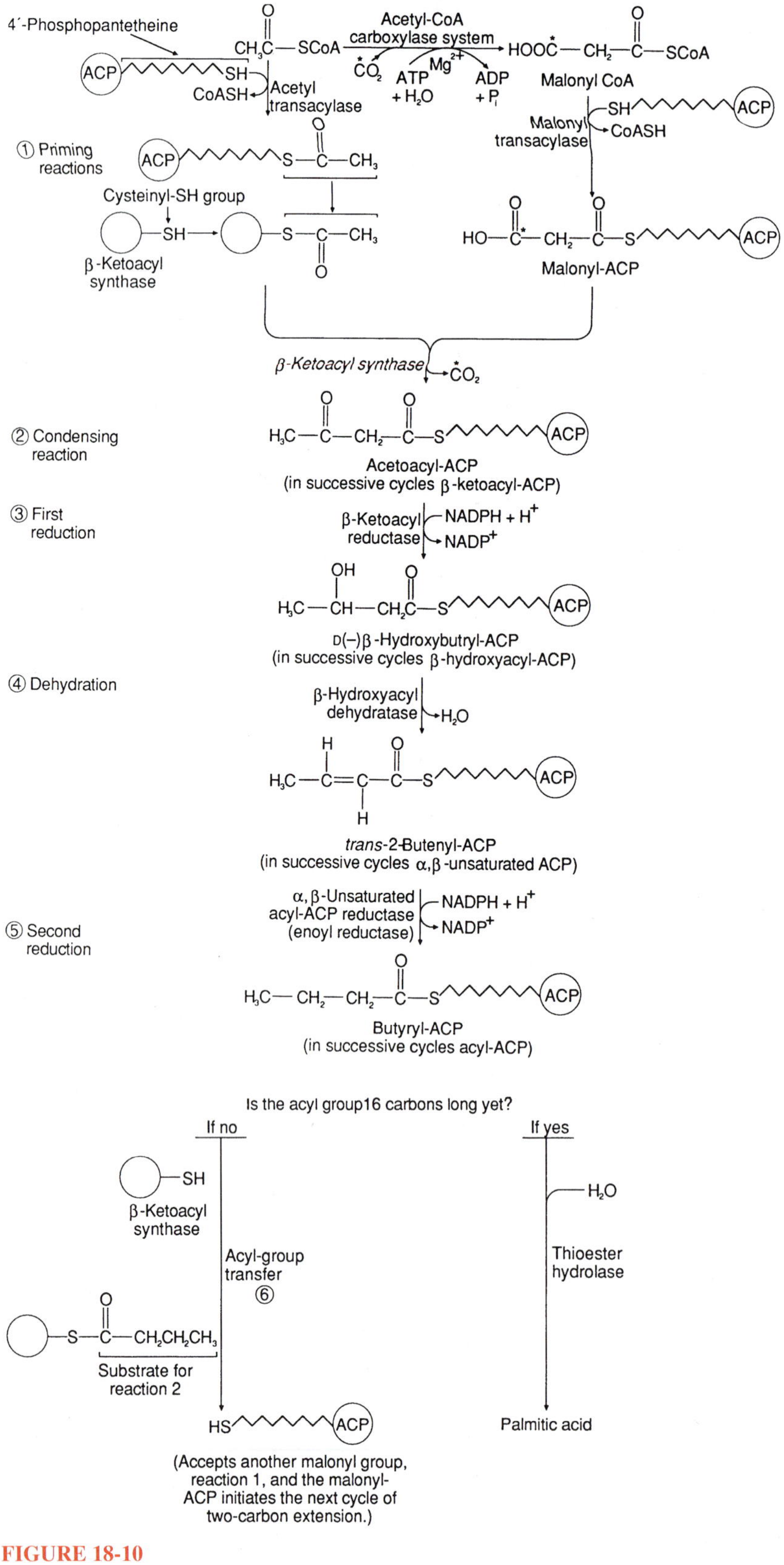

4′-Phosphopantetheine

$CH_3\overset{O}{\overset{\|}{C}}—SCoA$ → (Acetyl-CoA carboxylase system; $\overset{*}{C}O_2$, ATP + H_2O, Mg^{2+}, ADP + P_i) → $HOO\overset{*}{C}—CH_2—\overset{O}{\overset{\|}{C}}—SCoA$ Malonyl CoA

ACP~SH; CoASH; Acetyl transacylase

Malonyl transacylase; SH~ACP; CoASH

① Priming reactions

$ACP\text{~}S—\overset{O}{\overset{\|}{C}}—CH_3$

Cysteinyl-SH group

○—SH → ○—$S—\underset{O}{\underset{\|}{C}}—CH_3$

β-Ketoacyl synthase

$HO—\overset{O}{\overset{\|}{\overset{*}{C}}}—CH_2—\overset{O}{\overset{\|}{C}}—S\text{~}ACP$

Malonyl-ACP

β-Ketoacyl synthase → $\overset{*}{C}O_2$

② Condensing reaction

$H_3C—\overset{O}{\overset{\|}{C}}—CH_2—\overset{O}{\overset{\|}{C}}—S\text{~}ACP$

Acetoacyl-ACP (in successive cycles β-ketoacyl-ACP)

③ First reduction

β-Ketoacyl reductase; NADPH + H^+ → $NADP^+$

$H_3C—\overset{OH}{\overset{|}{C}H}—CH_2\overset{O}{\overset{\|}{C}}—S\text{~}ACP$

D(−)β-Hydroxybutryl-ACP (in successive cycles β-hydroxyacyl-ACP)

④ Dehydration

β-Hydroxyacyl dehydratase → H_2O

$H_3C—\overset{H}{\overset{|}{C}}=\underset{H}{\underset{|}{C}}—\overset{O}{\overset{\|}{C}}—S\text{~}ACP$

trans-2-Butenyl-ACP (in successive cycles α,β-unsaturated ACP)

⑤ Second reduction

α,β-Unsaturated acyl-ACP reductase (enoyl reductase); NADPH + H^+ → $NADP^+$

$H_3C—CH_2—CH_2—\overset{O}{\overset{\|}{C}}—S\text{~}ACP$

Butyryl-ACP (in successive cycles acyl-ACP)

Is the acyl group16 carbons long yet?

If no

○—SH β-Ketoacyl synthase

Acyl-group transfer ⑥

○—$S—\overset{O}{\overset{\|}{C}}—CH_2CH_2CH_3$

Substrate for reaction 2

HS~ACP

(Accepts another malonyl group, reaction 1, and the malonyl-ACP initiates the next cycle of two-carbon extension.)

If yes

H_2O

Thioester hydrolase

Palmitic acid

FIGURE 18-10

Synthesis of fatty acid. ACP = Functional unit of acyl-carrier-protein segment of fatty acid synthase. The cysteinyl-SH group of β-ketoacyl synthase accepts the acetyl group or the acyl group, and its catalytic site, which is adjacent to the –SH group, catalyzes the condensation reaction.

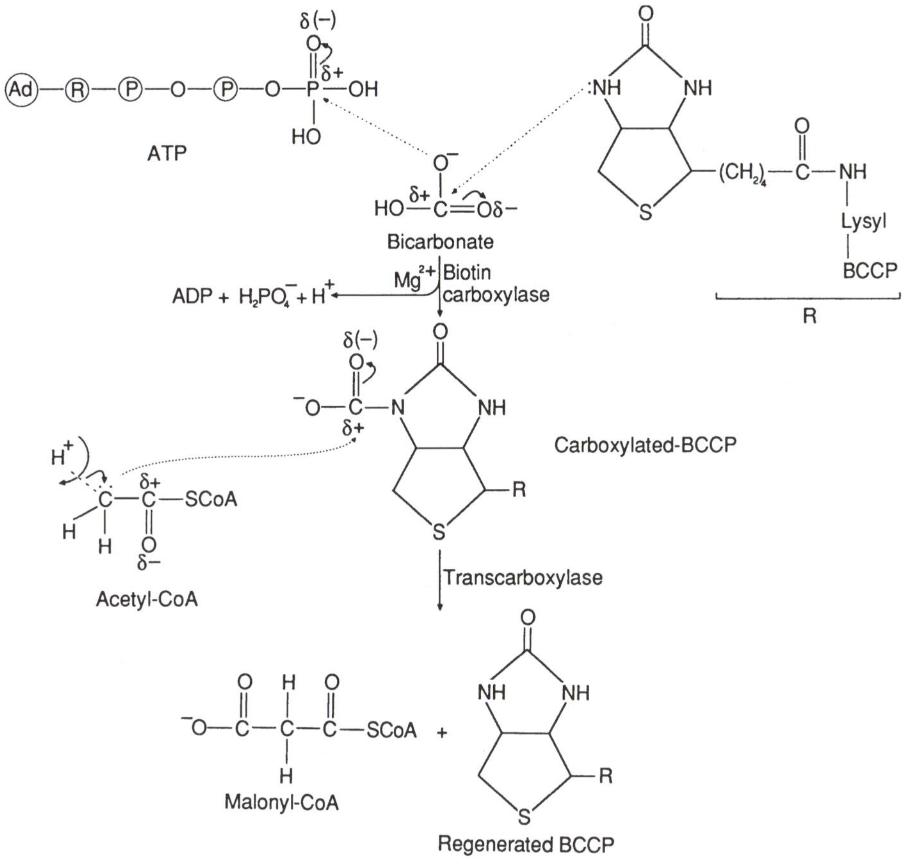

FIGURE 18-11
Mechanism of carboxylation of acetyl-CoA. BCCP = Biotin carboxyl carrier protein; Ad = adenosine.

The mechanism of the carboxylation reaction consists of two half-reactions:

$$\text{ATP} + \text{HCO}_3^- + \text{BCCP} \xrightleftharpoons[\text{Mg}^{2+}]{\text{biotin carboxylase}} \text{ADP} + \text{P}_i + \text{BCCP–COO}^-$$

and

$$\text{BCCP–COO}^- + \text{acetyl-CoA} \rightleftarrows \text{BCCP} + \text{malonyl-CoA}$$

The overall reaction is

$$\text{ATP} + \text{HCO}_3^- + \text{acetyl-CoA} \rightleftarrows \text{malonyl-CoA} + \text{ADP} + \text{P}_i$$

The presumed reaction mechanisms are shown in Figure 18-11.

Other biotin-dependent enzymes include propionyl-CoA carboxylase and pyruvate carboxylase (Chapter 15). The latter, like acetyl-CoA carboxylase, is subject to allosteric regulation. Pyruvate carboxylase, a mitochondrial enzyme, is activated by acetyl-CoA and converts pyruvate to oxaloacetate which, in turn, is converted to glucose via the gluconeogenic pathway or combines with acetyl-CoA to form citrate. Some of the citrate is transported to the cytosol, where it activates the first step of fatty acid synthesis and provides acetyl-CoA as substrate (see below). Other carboxylation reactions use bicarbonate but are dependent on vitamin K, the acceptor being glutamyl residues of glycoprotein clotting factors II, VI, IX, and X and anticlotting factors Protein C and Protein S (Chapter 36).

Acetyl-CoA carboxylase is under short- and long-term control. Allosteric modulation functions as a short-term regulator. Positive modulators are citrate and isocitrate; negative modulators are long-chain acyl-CoA derivatives. The binding of citrate increases the activity by polymerization of the protomers, whereas negative modulators favor dissociation of active polymers to inactive monomers. Acetyl-CoA carboxylase is also regulated by covalent modification by phosphorylation, which inhibits activity, and by dephosphorylation, which restores activity. Phosphorylation can occur by action of cAMP-dependent protein kinase through β-adrenergic agonists and glucagon or by action of calcium-dependent protein kinase through α-adrenergic agonists. It is not known whether the kinases act at different sites. Insulin suppresses cAMP levels and promotes activity of acetyl-CoA carboxylase. Insulin may also increase the activity of acetyl-CoA carboxylase phosphatase, which is complexed with the carboxylase. This phosphatase also dephosphorylates glycogen synthase, phosphorylase a, and HMG-CoA reductase. Thus, common mediators (e.g., insulin, glucagon, and catecholamines) regulate fatty acid synthesis and carbohydrate metabolism.

Long-term regulation of acetyl-CoA carboxylase involves nutritional, hormonal (e.g., insulin, thyroxine), and other factors. In animals on high-carbohydrate diets, fat-free diets, choline deprivation, or vitamin B_{12} deprivation, the activity is enhanced. However, fasting, high intake of fat or of polyunsaturated fatty acids, and prolonged biotin deficiency leads to decreased activity. In diabetes, the enzyme activity is low, but insulin administration raises it to normal levels.

Fatty acid synthesis is also carried out by a multienzyme complex and leads from acetyl-CoA, malonyl-CoA, and NADPH to palmitic acid. The overall process is as follows:

$$\underset{\text{Acetyl-CoA}}{\underset{*}{C}H_3\underset{*}{\overset{\overset{\displaystyle O}{\|}}{C}}\text{—SCoA}} + \underset{\text{Malonyl-CoA}}{7HOOC\text{—}CH_2\overset{\overset{\displaystyle O}{\|}}{C}\text{—SCoA}} + 14NADPH + 14H^+ \longrightarrow$$

$$\underset{\text{Palmitic acid}}{\overset{*}{C}H_3\overset{*}{C}H_2(CH_2CH_2)_6CH_2COOH} + 7CO_2 + 14NADP^+ + 8CoASH + 6H_2O$$

The fatty acid synthesis complex contains seven catalytic sites.

Acetyl transacylase catalyzes the reaction

$$CH_3\overset{\overset{\displaystyle O}{\|}}{C}\text{—SCoA} + ACP\text{—SH} \longrightarrow CH_3\overset{\overset{\displaystyle O}{\|}}{C}\text{—S—ACP} + CoASH$$

and malonyl transacylase catalyzes the reaction

$$HO\text{—}\overset{\overset{\displaystyle O}{\|}}{C}\text{—}CH_2\text{—}\overset{\overset{\displaystyle O}{\|}}{C}\text{—SCoA} + ACP - SH \rightarrow HO\text{—}\overset{\overset{\displaystyle O}{\|}}{C}\text{—}CH_2\text{—}\overset{\overset{\displaystyle O}{\|}}{C}\text{—S—ACP}$$

(ACP–SH is the 4′-phosphopantetheine–SH of the acyl carrier domain.)

These two priming reactions load a cysteinyl-SH of one subunit with acetyl-CoA, and the –SH of the 4′-phosphopantetheine arm of the acyl carrier site on the other subunit with malonyl-CoA.

β-Ketoacyl-ACP synthase (condensing enzyme) has a active cysteine–SH group that forms an acetyl-S (or acyl S) intermediate (E = enzyme).

$$CH_3\overset{\overset{\displaystyle O}{\|}}{C}\text{—S—ACP} + E\text{—SH} \rightleftharpoons CH_3\overset{\overset{\displaystyle O}{\|}}{C}\text{—S—E} + ACP\text{—SH}$$

The condensation occurs with release of CO_2:

$$CH_3\overset{\overset{\displaystyle O}{\|}}{C}\text{—S—E} + HO\text{—}\overset{\overset{\displaystyle O}{\|}}{C}\text{—}CH_2\text{—}\overset{\overset{\displaystyle O}{\|}}{C}\text{—S—ACP} \longrightarrow CH_3\overset{\overset{\displaystyle O}{\|}}{C}CH_2\overset{\overset{\displaystyle O}{\|}}{C}\text{—S—ACP} + CO_2 + E\text{—SH}$$

β-Ketoacyl-ACP reductase catalyzes the first reductio reaction:

$$CH_3\overset{\overset{\displaystyle O}{\|}}{C}\text{—}CH_2\overset{\overset{\displaystyle O}{\|}}{C}\text{—S—ACP} + NADPH^+ + H^+ \rightleftharpoons CH_3\text{—}\overset{\overset{\displaystyle OH}{|}}{C}HCH_2\overset{\overset{\displaystyle O}{\|}}{C}\text{—S—ACP} + NADP^+$$

It is stereospecific, and the product formed is D(−)-β hydroxybutyryl-ACP (or D(−)-β-hydroxyacyl-ACP, i subsequent reactions).

β-Hydroxyacyl-ACP dehydratase catalyzes the reac tion:

$$CH_3\overset{\overset{\displaystyle OH}{|}}{C}HCH_2\overset{\overset{\displaystyle O}{\|}}{C}\text{—S—ACP} \rightleftharpoons \textit{trans}\text{-}CH_3CH{=}CH\overset{\overset{\displaystyle O}{\|}}{C}\text{—S—ACP} + H_2O$$

In this stereospecific reaction, the D(−) isomer is converte to a *trans*-α,β-unsaturated acyl-ACP derivative.

Enoyl-ACP reductase catalyzes the second reductio with NADPH.

$$CH_3CH{=}CH\overset{\overset{\displaystyle O}{\|}}{C}\text{—S—ACP} + NADPH + H^+ \rightleftharpoons CH_3CH_2CH_2\overset{\overset{\displaystyle O}{\|}}{C}\text{—S—ACP} + NADP^+$$

The product formed is a saturated acyl-thioester of ACP.

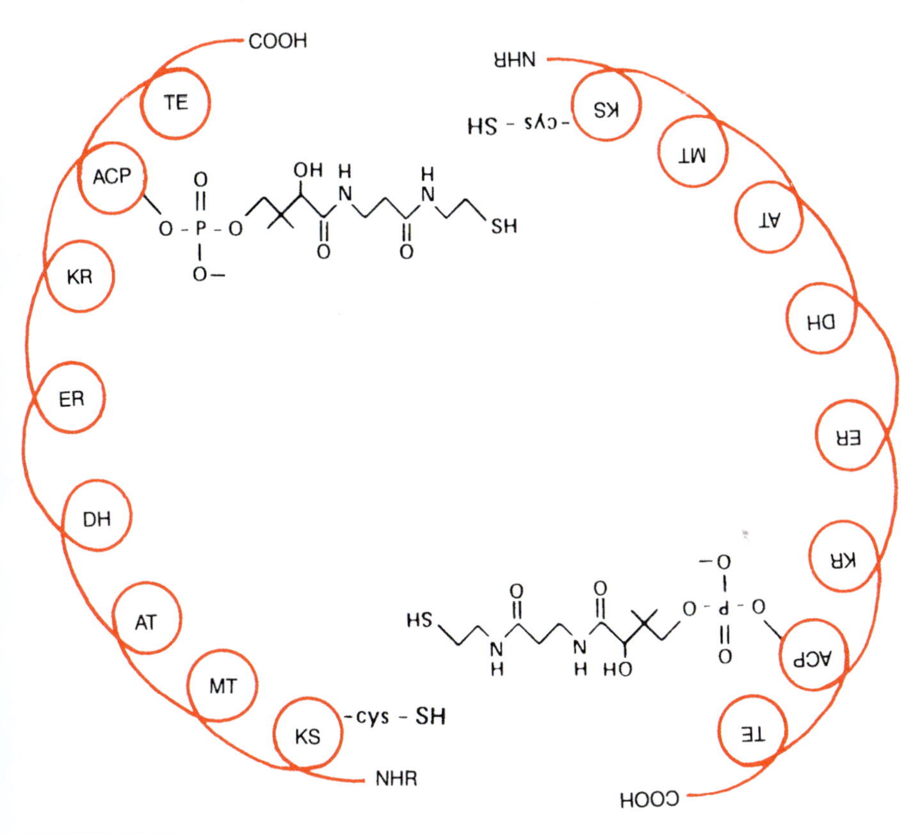

FIGURE 18-12
Diagram of a fatty acid synthase dimer with its head-to-tail association of the two multifunctional polypeptides. KS = β-Ketoacyl synthase; MT = malonyl transacylase; AT = acetyl transacylase; DH = dehydratase; ER = enoyl reductase; KR = β-ketoacyl reductase; ACP = acyl carrier site; TE = thioesterase. [Reproduced with permission from S. J. Wakil, J. K. Stoops, and V. C. Joshi, Fatty acid synthesis and its regulation. *Annu. Rev. Biochem.* **52,** 537 (1983). © 1983 by Annual Reviews Inc.]

Thioester hydrolase catalyzes the removal of palmitate from the 4′-phosphopantetheine arm of the acyl carrier site.

$$CH_3(CH_2)_{14}\overset{\overset{\displaystyle O}{\|}}{C}—S—ACP + H_2O \longrightarrow$$

$$CH_3(CH_2)_{14}COOH + ACP—SH$$

The enzyme has an active serine residue and is specific for long-chain acyl derivatives.

Functional Organization of Fatty Acid Synthase

In animal cells, fatty acid synthase (FAS) consists of two subunits, each having a molecular weight of 250,000. Evidence from negative-stain electron microscopy indicates that FAS is a linear polypeptide with a series of globular domains representing areas of catalytic activity. Fatty acid synthase mRNA is large enough to code for the ~2300 amino acids required. Animal FAS is the largest known multifunctional protein. Although each subunit contains all of the catalytic activities required to synthesize palmitate, the monomer lacks β-ketoacyl synthase activity. The two subunits must be juxtaposed head to tail to bring the cysteine-SH of the β-ketoacyl synthase of one close to the 4′-phosphopantetheine-SH of the acyl carrier site of the other to obtain a fully functional dimer.

Figure 18-12 presents schematically the active FAS homodimer complex. Synthesis may proceed from either end of the active complex.

The assembly of rat liver FAS involves three stages: synthesis of the multifunctional polypeptide chains, formation of the dimer, and attachment of a 4′-phosphopantetheine group by an enzyme-catalyzed reaction. This assembly process is influenced by changes in developmental, hormonal, and nutritional states. The FAS complex provides considerable catalytic efficiency, since free intermediates do not accumulate and the individual activities are present in equal amounts.

The central role of the acyl carrier domain is to carry acyl groups from one catalytic site to the next. The

4′-phosphopantetheine (20 nm long) derived from coenzyme A is bound as a phosphodiester through the hydroxyl group of a specific seryl residue. The acyl intermediates are in thioester linkage with the –SH of the prosthetic group, which serves as a swinging arm to carry acyl groups from one catalytic site to the next.

The structure of 4′-phosphopantetheine attached to the serine residue of ACP is

$$\text{Seryl residue}-O-\overset{O}{\overset{\|}{\underset{OH}{\underset{|}{P}}}}-O-CH_2-\overset{CH_3}{\overset{|}{\underset{CH_3}{\underset{|}{C}}}}-\underset{OH}{\underset{|}{CH}}-\overset{O}{\overset{\|}{C}}-\underset{H}{N}-CH_2-CH_2-\overset{O}{\overset{\|}{C}}-\underset{H}{N}-CH_2-CH_2-SH$$

Sources of NADPH for Fatty Acid Synthesis

The reducing agent for fatty acid synthesis is NADPH. It is mostly supplied by the pentose phosphate pathway (Chapter 15) in the reactions

$$\text{Glucose 6-phosphate} \xrightarrow[\text{NADP}^+ \;\rightarrow\; \text{NADPH} + \text{H}^+]{\text{Glucose-6-phosphate dehydrogenase}} \text{6-Phosphate-gluconolactone}$$

and

$$\text{6-Phosphogluconate} \xrightarrow[\text{NADP}^+ \;\rightarrow\; CO_2 + \text{NADPH} + \text{H}^+]{\text{6-Phosphogluconate dehydrogenase}} \text{Ribulose 5-phosphate}$$

Oxidation of malate also provides NADPH:

$$\text{Malate} \xrightarrow[\text{NADP}^+ \;\rightarrow\; CO_2 + \text{NADPH} + \text{H}^+]{\text{Malic enzyme}} \text{Pyruvate}$$

These three enzymes, like the fatty acid synthase complex, are located in the cytosol. Active lipogenesis occurs in liver, adipose tissue, and lactating mammary glands, which contain a correspondingly high activity of the pentose phosphate pathway. Thus, lipogenesis is closely linked to carbohydrate oxidation. The rate of lipogenesis is high in humans on carbohydrate-rich diets. Restricted energy intake, a high-fat diet, and insulin deficiency decrease fatty acid synthesis.

Source and Transport of Acetyl-CoA

Acetyl-CoA is synthesized in mitochondria by a number of reactions: oxidative decarboxylation of pyruvate, catabolism of some amino acids (e.g., phenylalanine, tyrosine, leucine, lysine, and tryptophan; see Chapter 17), and β-oxidation of fatty acids (see above). Since acetyl-CoA cannot be transported directly across the inner mitochondrial membrane to the cytosol, its carbon atoms are transferred by two transport mechanisms.

1. Transport dependent upon carnitine: Carnitine participates in the transport of long-chain acyl-CoA into the mitochondria and plays a similar role in the transport of acetyl-CoA out of mitochondria. However, carnitine acetyl transferases have a minor role in acetyl-CoA transport.
2. Cytosolic generation of acetyl-CoA ("citrate shuttle"): This pathway is shown in Figure 18-13. Citrate synthesized from oxaloacetate and acetyl-CoA is transported to the cytosol via the tricarboxylate anion carrier system and cleaved to yield acetyl-CoA and oxaloacetate.

$$\text{Citrate}^{3-} + \text{ATP}^{4-} + \text{CoA} \xrightarrow[\text{(or citrate cleavage enzyme)}]{\text{ATP citrate-lyase}} \text{acetyl-CoA} + \text{oxaloacetate}^{2-} + \text{ADP}^{3-} + \text{P}_i^{2-}$$

Thus, citrate not only modulates the rate of fatty acid synthesis but also provides carbon atoms for the synthesis. The oxaloacetate formed from pyruvate may eventually be converted (via malate) to glucose by the gluconeogenic pathway. The glucose oxidized via the pentose phosphate pathway augments fatty acid synthesis by providing NADPH. Pyruvate generated from oxaloacetate can enter mitochondria and be converted to oxaloacetate, which is required for the formation of citrate.

Regulation of Fatty Acid Synthase

Like acetyl-CoA carboxylase, FAS is under short- and long-term control. The former is due to negative or positive allosteric modulation or to changes in the concentrations of substrate, cofactor, and product. The latter usually consists of changes in enzyme content as a result of protein synthesis or decreased protein degradation. Variation in levels of hormones (e.g., insulin, glucagon, epinephrine, thyroid hormone, and prolactin) and in the nutritional state affect fatty acid synthesis through short- and long-term mechanisms. In the diabetic state, hepatic fatty acid synthesis is severely impaired but is corrected by administration of insulin. The impairment may be due to defects in glucose metabolism that lead to a reduced level of an inducer or increased level of a repressor of transcription of the FAS gene, or both. Glucagon and epinephrine raise intracellular levels of cAMP, and their inhibitory effect on fatty acid synthesis may be due to phosphorylation or dephosphorylation of acetyl-CoA carboxylase. They also stimulate the action of hormone-sensitive triacylglycerol lipase and raise intracellular levels of long-chain acyl-CoA.

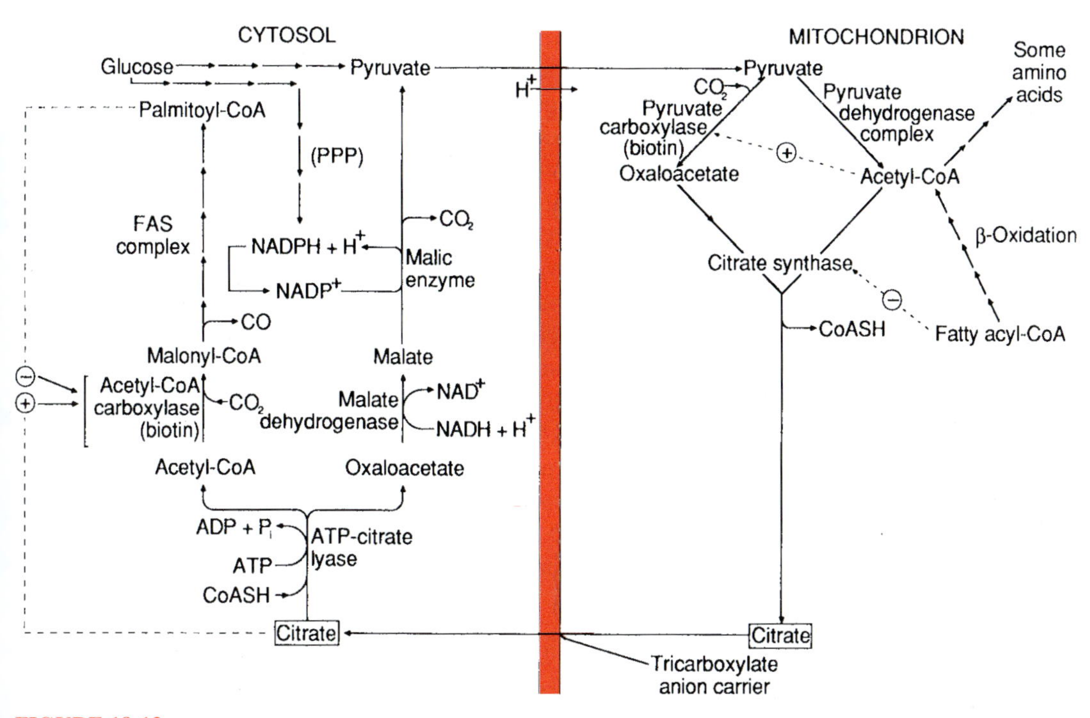

FIGURE 18-13
Cytoplasmic generation of acetyl-CoA via citrate transport and related reactions. PPP = Pentose phosphate pathway; FAS = fatty acid synthase. —⊖→ Negative allosteric modifier. —⊕→ Positive allosteric modifier.

As a result, acetyl-CoA carboxylase and citrate synthase are inhibited. The stimulatory effect of prolactin is confined to the mammary gland and may involve synthesis of the enzyme.

Fatty Acid Elongation

Cytoplasmic fatty acid synthase yields palmitate. Human triacylglycerol contains fatty acids with 18, 20, 22, and 24 carbon atoms, which are synthesized by elongation of palmitate in endoplasmic reticulum or mitochondria. Elongation in the endoplasmic reticulum occurs mainly in liver and involves C_{10-16}–saturated and C_{18}–unsaturated fatty acids by successive addition of two-carbon groups derived from malonyl-CoA (Figures 18-14 and 18-15). The reductant is NADPH. The intermediates, however, are CoA thioesters.

The mitochondrial system uses acetyl-CoA, not malonyl-CoA, by a slightly modified reversal of β-oxidation. The substrates are saturated and unsaturated C_{12}, C_{14}, and C_{16} fatty acids, and the products are C_{18}, C_{20}, C_{22}, and C_{24} fatty acids. The first reduction step utilizes NADH, and the enzyme is β-hydroxyacyl dehydrogenase (a β-oxidation enzyme); the second reduction step utilizes NADPH, and the enzyme is enoyl reductase. Mitochondrial fatty acid elongation occurs primarily when the [NADH]/[NAD$^+$] ratio is high (e.g., anaerobiosis, excessive ethanol oxidation).

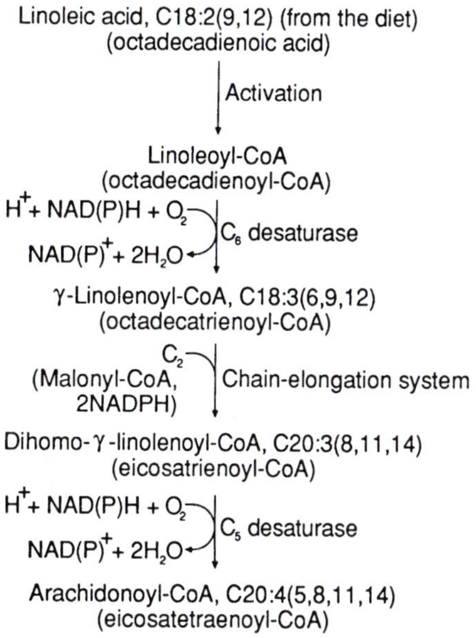

Linoleic acid, C18:2(9,12) (from the diet)
(octadecadienoic acid)

↓ Activation

Linoleoyl-CoA
(octadecadienoyl-CoA)

H^+ + NAD(P)H + O_2 → NAD(P)$^+$ + 2H_2O ↓ C_6 desaturase

γ-Linolenoyl-CoA, C18:3(6,9,12)
(octadecatrienoyl-CoA)

C_2 (Malonyl-CoA, 2NADPH) ↓ Chain-elongation system

Dihomo-γ-linolenoyl-CoA, C20:3(8,11,14)
(eicosatrienoyl-CoA)

H^+ + NAD(P)H + O_2 → NAD(P)$^+$ + 2H_2O ↓ C_5 desaturase

Arachidonoyl-CoA, C20:4(5,8,11,14)
(eicosatetraenoyl-CoA)

FIGURE 18-14
Synthesis of arachidonic acid from linoleic acid. The desaturation and chain elongation occur in microsomes.

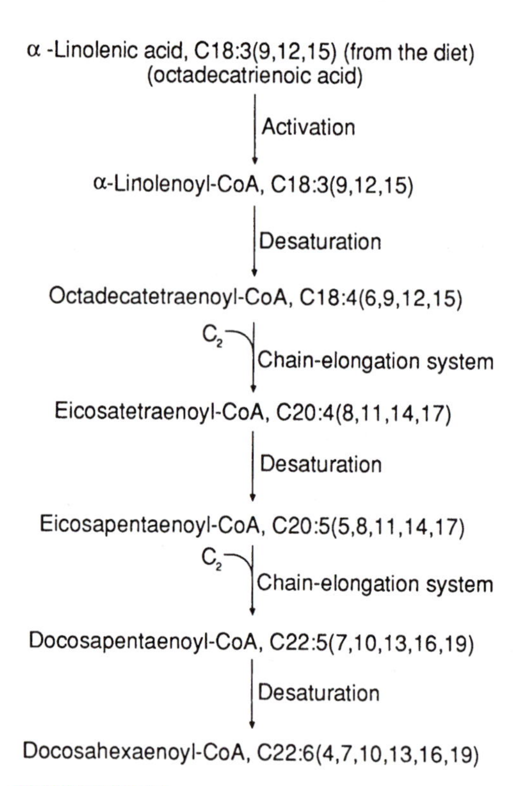

FIGURE 18-15
Synthesis of docosahexaenoic acid from α-linolenic acid. Desaturation and chain elongation are similar to the use described in Figure 18-13.

18.5 Metabolism of Unsaturated Fatty Acids

Structure and Nomenclature of Unsaturated Fatty Acids

Unsaturated fatty acids contain one or more double bonds. A common method for designating fatty acids gives the carbon chain length, number of double bonds, and double-bond positions (in parentheses). Thus, palmitoleic acid is designated C16:1(9) and linoleic acid is C18:2(9,12). The location of the double bond is sometimes indicated by Δ; for example, Δ^9 signifies that the double bond is between carbon 9 and carbon 10. In both methods, the carboxyl carbon is carbon 1. The double-bond position can also be related to the ω-end of the fatty acid molecule (i.e., the methyl carbon farthest from the carboxyl end); oleic acid is an ω-9 acid; linoleic acid has double bonds at ω-6 and ω-9 carbons. The structures and names of some naturally occurring unsaturated fatty acids are given in Table 18-3.

The presence of a double bond in the hydrocarbon chain gives rise to geometrical isomerism, which is due to restricted rotation around carbon–carbon double bonds and is exemplified by fumaric and maleic acids.

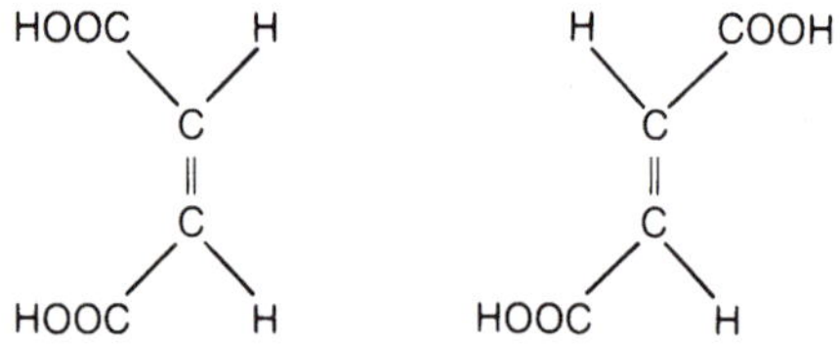

Maleic acid (*cis*-form)

Fumaric acid (*trans*-form)

Almost all naturally occurring, unsaturated, long-chain fatty acids exist as the cis isomers, which are less stable than the trans isomers. The cis configuration introduces a bend (of about 30°) in the molecule, whereas the trans isomer resembles the extended form of the saturated chain (Figure 18-16). Arachidonic acid with four cis double bonds is a U-shaped molecule. Some cis isomers are biologically active as essential fatty acids. The trans isomers cannot substitute for them but are metabolized like the saturated fatty acids.

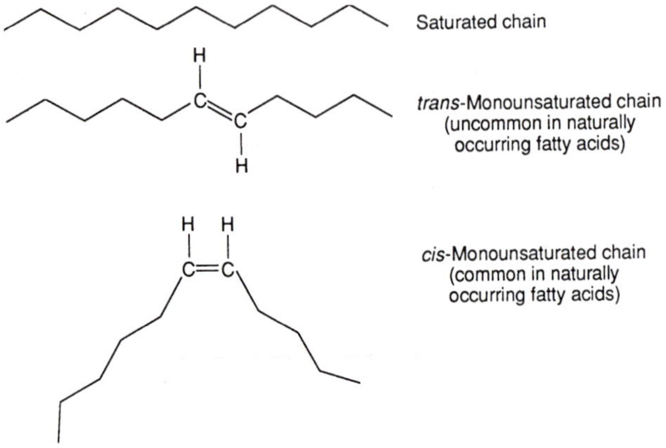

Saturated chain

trans-Monounsaturated chain (uncommon in naturally occurring fatty acids)

cis-Monounsaturated chain (common in naturally occurring fatty acids)

FIGURE 18-16
Geometry of saturated, trans monounsaturated, and cis monounsaturated chains.

Functions of Unsaturated Fatty Acids

The cis unsaturated fatty acids provide fluidity of triacylglycerol reserves and phospholipid membranes and many serve as precursors of eicosanoids (prostaglandins, prostacyclins, thromboxanes, and leukotrienes). The importance of membrane fluidity and its relationship to the membrane constituent phospholipids are discussed in Chapter 10. Eicosanoids have numerous functions (see below).

18.6 Nonessential Fatty Acids

Palmitoleic and oleic acids, the two most abundant monounsaturated fatty acids of animal lipids, can be

TABLE 18-3

Naturally Occurring Unsaturated Fatty Acids

Common Name	Systematic Name*	Molecular Formula	Structural Formula	ω-Series†	Melting Point (°C)
Palmitoleic‡	9-Hexadecenoic	$C_{16}H_{30}O_2$	$CH_3(CH_2)_5CH{=}CH(CH_2)_7COOH$	ω-7	0.5
Oleic‡	9-Octadecenoic	$C_{18}H_{34}O_2$	$CH_3(CH_2)_7CH{=}CH(CH_2)_7COOH$	ω-9	13
Vaccenic	*trans*-11-Octadecenoic	$C_{18}H_{34}O_2$	$CH_3(CH_2)_5CH{=}CH(CH_2)_9COOH$	ω-7	43
Linoleic‡	9,12-Octadecadienoic	$C_{18}H_{32}O_2$	$CH_3(CH_2)_4CH{=}CHCH_2CH{=}CH(CH_2)_7COOH$	ω-6	–5
α-Linolenic	9,12,15-Octadecatrienoic	$C_{18}H_{30}O_2$	$CH_3CH_2CH{=}CHCH_2CH{=}CHCH_2CH{=}CH(CH_2)_7COOH$	ω-3	–11
γ-Linolenic	6,9,12-Octadecatrienoic	$C_{18}H_{30}O_2$	$CH_3(CH_2)_4CH{=}CHCH_2CH{=}CHCH_2CH{=}CH(CH_2)_4COOH$	ω-6	–11
Arachidonic‡	5,8,11,14-Eicosatetraenoic	$C_{20}H_{32}O_2$	$CH_3(CH_2)_4{-}(CH{=}CH{-}CH_2)_4{-}(CH_2)_2{-}COOH$	ω-6	–50
Nervonic	15-Tetracosenoic	$C_{24}H_{46}O_2$	$CH_3(CH_2)_7CH{=}CH(CH_2)_{13}COOH$	ω-9	39

*All double bonds are in the *cis* geometric configuration, except where indicated.

†This series is based on the number of carbon atoms present between the terminal methyl group and the nearest double bond; ω-3 and ω-6 are essential fatty acids.

‡Most abundant unsaturated fatty acids in animal lipids.

synthesized from the respective saturated fatty acid coenzyme-A esters. Desaturase is a monooxygenase system present in endoplasmic reticulum of liver and adipose tissue. The overall reaction for palmitoleic acid synthesis is

$$\text{Palmitoyl-CoA} + \text{NAD(P)H} + \text{H}^+ + \text{O}_2 \rightarrow \text{palmitoleyl-CoA} + \text{NAD(P)}^+ + 2\text{H}_2\text{O}$$

One molecule of oxygen accepts two pairs of electrons, one from palmitoyl-CoA and the other from NADPH or NADH. The electrons NAD(P)H are transported via cytochrome-b_5 reductase to cytochrome b_5 (microsomal electron transport; Chapter 14). An enzyme-bound superoxide radical is responsible for the oxidation of acyl-CoA. Four desaturases specific for introducing cis double bonds at C_9, C_6, C_5, and C_4, respectively, are known. If the substrate is saturated, the first double bond introduced is C_9. With an unsaturated substrate, other double bonds are introduced between the carboxyl group and the double bond nearest the carboxyl group. Desaturation yields a divinylmethane arrangement of double bonds ($-CH{=}CH{-}CH_2{-}CH{=}CH-$). Usually desaturation alternates with chain elongation. Desaturation is inhibited by fasting and diabetes. The oxidation of unsaturated fatty acids occurs in mitochondria.

18.7 *trans*-Fatty Acids

trans-Fatty acid metabolism is similar to that of saturated fatty acids. During the partial dehydrogenation of vegetable oils (e.g., in the manufacture of margarine), the cis fatty acids are isomerized to *trans*-fatty acid forms. The "hydrogenated" margarines contain 15–40% of *trans*-fatty acids.

The hypercholesterolemic effect of *trans*-fatty acids may be due to impairment of the first step in the formation of bile acids from cholesterol. Since the steady-state level of cholesterol depends on its conversion to bile acid (Chapter 19), any perturbation in this process affects cholesterol levels. Both metabolic and epidemiological studies have shown that the consumption of *trans*-fatty acids increases the risk of coronary heart disease. This risk appears to be even higher when compared on a per-gram basis with saturated fatty acids. The adverse effects of *trans*-fatty acids are attributed to the elevation of atherogenic low-density lipoprotein (LDL) cholesterol and a decrease in the antiatherogenic (or cardioprotective) high-density lipoprotein (HDL) cholesterol level. Thus, the ratio of LDL cholesterol to HDL cholesterol is significantly higher with the *trans*-fatty acid diet compared to a saturated-fat diet. A diet rich in oleic acid has a lower ratio of LDL cholesterol to HDL cholesterol compared to either of the other diets. The metabolism of lipoproteins and their role in atherosclerosis are discussed in Chapter 20.

18.8 Essential Fatty Acids

Polyunsaturated fatty acids not synthesized in the body but required for normal metabolism are essential fatty acids (EFAs). EFAs are linoleic acid, linolenic acids (α and γ), and arachidonic acid. All contain at least one double bond located beyond C-9 or within the terminal seven carbon atoms (Table 18-3).

ω-Carbon

$$\overset{1}{H_3C}-\overset{2}{CH_2}-\overset{3}{CH_2}-\overset{4}{CH_2}-\overset{5}{CH_2}-\overset{6}{CH_2}-\overset{7}{CH_2}-\overset{8}{CH_2}-(CH_2)_n-COOH$$

Introduction of double bonds does not occur in humans (carbons 1–7)

Double bonds can be introduced in this region (from carbon 8)

A double bond within the terminal seven carbon atoms can be present at ω-3 or ω-6. γ-Linolenic acid is an ω-6 EFA and α-linolenic acid an ω-3 EFA. Other ω-3 EFA are eicosapentaenoic acid (EPA) and docosahexaenoic acid (DCHA), both abundant in edible fish tissues. Vegetable oils are rich in ω-6 EFA (Table 18-4). Plants contain α-linolenic acid, which can be converted in the body to EPA and DCHA, but it is found within chloroplast membranes and not in seed oils; hence, it may not be available in significant quantities in the diet. The ω-3 and ω-6 EFA have different metabolic effects (see below). Particularly rich sources of EPA are fishes (e.g., salmon, mackerel, blue fish, herring, menhaden) that live in deep, cold waters. These fishes have fat in their muscles and their skin. In contrast, codfish, which have a similar habitat, store fat in liver rather than muscle. Thus, cod liver oil is a good source of EPA, but it also contains high amounts of vitamins A and D, which can be toxic in large quantities (Chapters 38 and 37, respectively). Shellfish also contain EPA. Plankton are the ultimate source of EPA.

Linoleic acid can be converted in mammalian liver to γ-linolenic acid and arachidonic acid by the microsomal desaturation and chain elongation process (Figure 18-14). Thus, the requirement for arachidonic acid may be dispensed with when the diet contains adequate amounts of linoleic acid. Similarly, α-linolenic acid is converted by desaturation and chain elongation to EPA and DCHA (Figure 18-15).

Deficiency of Essential Fatty Acids

The clinical manifestations of EFA deficiency in humans closely resemble those seen in animals. They include dry, scaly skin, usually erythematous eruptions (generalized or localized and affecting the trunk, legs, and intertriginous areas), diffuse hair loss (seen frequently in infants), poor wound healing, failure of growth, and increased metabolic rate. Abnormalities in ECG patterns may be due to membrane alterations, which may also account for structural and functional abnormalities observed in mitochondria. Surgical patients maintained on glucose-amino acid solutions for prolonged periods develop EFA deficiency, manifested as anemia, thrombocytopenia, hair loss and sparse hair growth, increased capillary permeability, dry scaly skin, desquamating dermatitis, and a shift in the oxygen-dissociation curve of hemoglobin to the left. Oral or intravenous administration of linoleic acid is necessary to correct these problems. Fat emulsions containing linoleic acid are commercially available for intravenous use. An adult requires 10 g of linoleic acid per day. The recommended dietary allowance for EFA is 1–2% of the total energy intake.

EFA deficiency can also occur in infants with highly restricted diets (e.g., primarily skim milk intake), in patients receiving total parenteral hyperalimentation without supplements of unsaturated lipids, and in those with severe malabsorptive defects.

In EFA deficiency, oleic acid can be dehydrogenated to yield polyunsaturated fatty acids (PUFAs) that are nonessential and do not substitute for the essential fatty acids. One such PUFA is 5,8,11-eicosatrienoic acid, which occurs in significant amounts in heart, liver, adipose tissue, and erythrocytes of animals fed diets deficient in EFAs but decreases after supplementation with linoleic or linolenic acids. Its appearance in tissues and plasma has been used in the assessment of EFA deficiency.

Most vegetable oils are relatively rich in EFAs (coconut oil is an exception), low in saturated fatty acids, and lack cholesterol. Animal fats (except those in fish), on the other hand, are generally low in EFAs, high in saturated fats, and contain cholesterol. The EFA content of body and milk fat of ruminants can be increased by the feeding of EFA encased in formalin-treated casein. The EFA is released at the site of absorption by dissolution of the capsules. These dietary manipulations in ruminants are accompanied by an increase in carcass EFA, a decrease in saturated fatty acids, and no change or an increase in cholesterol content. Table 18-4 summarizes the fatty acid composition of some fats of animal and plant origin. The recommended daily diet does not exceed 30–35% of the total energy intake as fat (current average consumption in North America is 40–45%), with equal amounts of saturated, monounsaturated, and polyunsaturated fats, and a cholesterol intake of no more than 300 mg/day (current average consumption in North America is about 600 mg/day).

Substitution of ω-6 polyunsaturated for saturated fats in the diet lowers plasma cholesterol levels through reduction in levels of VLDL and LDL. Diets rich in polyunsaturated fats lead to higher biliary excretion of sterols, although this effect may not be directly related to reduced levels of plasma lipoproteins. Diets low in EFA (linoleic acid) have been associated with high rates of coronary heart disease. A significantly lower proportion of EFA in the adipose tissue of people dying from coronary heart disease has been reported, and an inverse relationship has been found between the percentage composition of EFA in serum cholesteryl esters and mortality rates from coronary heat disease. Consumption of ω-3 polyunsaturated fatty acids markedly decreases plasma triacylglycerol and, to a lesser extent, cholesterol levels in some hyperlipoproteinemic patients (Chapter 20). Consumption of fish-oil fatty acids decreases the biosynthesis of fatty acids and of VLDL by the liver and also decrease the platelet and monocyte function. These effects of ω-3 fatty acids appear to prevent or delay atherogenesis. Low death rates from coronary heart disease are found among populations with high intake of fish (e.g., Greenland Eskimos, people of fishing villages of Japan, people of Okinawa). Metabolic and functional differences exist between ω-3 and ω-6 fatty acids. They have opposing physiological effects and their balance in the diet is important for homeostasis and normal development.

18.9 Metabolism of Eicosanoids

The eicosanoids—***prostaglandins*** (PGs), ***thromboxanes*** (TXs), ***prostacyclins*** (PGIs), and ***leukotrienes*** (LTs)—are derived from essential fatty acids and act similarly to hormones (Chapter 30). However, they are synthesized in almost all tissues (unlike hormones, which are synthesized in selected tissues) and are not stored to any significant extent; their physiological effects on tissues occur near sites of synthesis rather than at a distance. They function as paracrine messengers and are sometimes referred to as autacoids.

The four groups of eicosanoids are derived, respectively, from a 20-carbon fatty acid with three, four, or five double bonds: 8,11,14-eicosatrienoic acid (dihomo-γ-linolenic acid), 5,8,11,14-eicosatetraenoic acid (arachidonic acid), and 5,8,11,14,17-eicosapentaenoic acid (Figure 18-17). In humans, the most abundant precursor is arachidonic acid. Secretion of eicosanoids in response

TABLE 18-4
Fatty Acid Composition of Some Fats of Animal and Plant Origin

	Saturated* C_{14}, C_{16}, C_{18} (predominantly C_{16} and C_{18})	Monounsaturated* C16:1; C18:1 (predominantly C18:1)	Polyunsaturated*, mostly as linoleic acid (C18:2)
Animal Fats			
Butter	59	37	4
Beef	54	44	2
Chicken	40	38	22
Pork	40	46	14
Fish (salmon and tuna)	28	29	23+20†
Vegetable Oils			
Safflower	11	11	78
Corn	14	26	60
Sesame	14	43	43
Soybean	15	27	58
Peanut	20	45	35
Cottonseed	29	19	52
Coconut	92	6	2
Palm	53	37	10
Olive	16	69	15
Sunflower	12	18	70

*All values expressed as weight percentages of total fatty acids.
†Other polyenoic acids, e.g., eicosapentaenoic acid, docosahexaenoic acid.

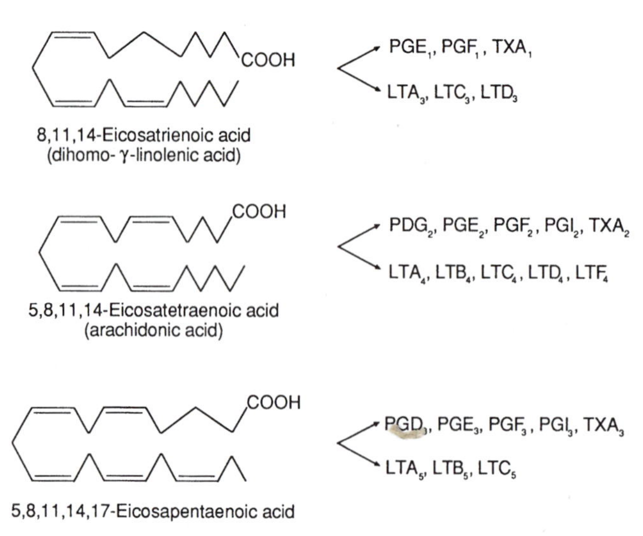

FIGURE 18-17
Precursor and product relationships of eicosanoids: prostaglandins (PGs), prostacyclins (PGIs), thromboxanes (TXs), and leukotrienes (LTs). Arrows arising from each fatty acid indicate two different synthetic pathways: one for prostanoids (PG, PGI, TX) and the other for leukotrienes. The numerical subscript of an eicosanoid indicates the total number of double bonds in the molecule and thus the series to which it belongs. The prostanoids contain two fewer double bonds than does the precursor fatty acid.

to stimuli requires mobilization of precursor fatty acids bound by ester, ether, or amide linkages. This utilization is generally considered the rate-limiting step. Membrane phosphoglycerides contain essential fatty acids in the 2-position. In response to stimuli and after the activation of the appropriate phospholipase, the essential fatty acid is released.

Known phospholipases and their hydrolytic specificities are shown in Figure 18-18. The primary source of

FIGURE 18-18
Hydrolysis of phosphoglycerides by phospholipases, whose cleavage sites are shown by vertical arrows.

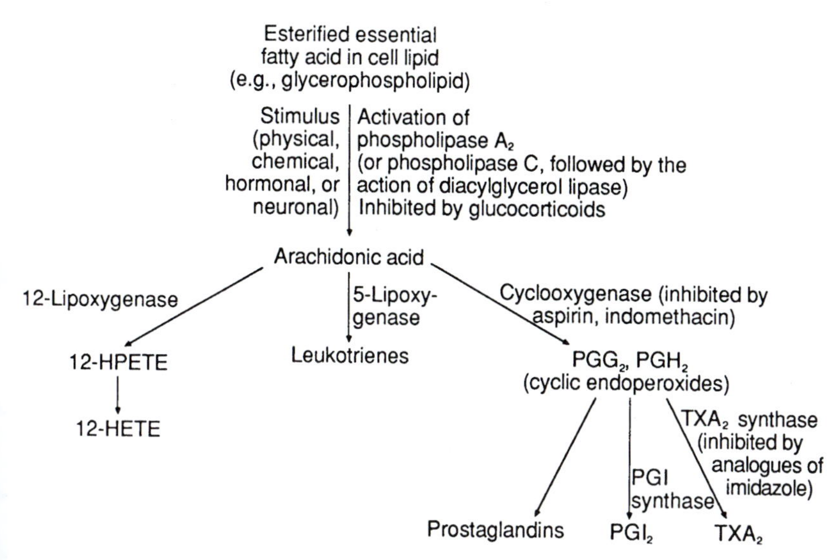

FIGURE 18-19
Pathways of arachidonate metabolism in eicosanoid synthesis and their inhibitors. HPETE = Hydroperoxyeicosatetraenoic acid; HETE = hydroxyeicosatetraenoic acid; PG = prostaglandin; PGI = prostacyclin; TX = thromboxane. Conversions of arachidonic acid by various enzymes can be inhibited by analogues of the natural fatty acid, e.g., the acetylenic analogue 5,8,11,14-eicosatetraynoic acid.

arachidonic acid is through action of phospholipase A_2. It may also be derived through action of phospholipase C, which liberates diacylglycerol; the latter is then acted on by diacylglycerol lipase. Stimuli that increase the biosynthesis of eicosanoids cause increased mobilization of intracellular calcium, which with calmodulin is thought to activate membrane-bound phospholipases A_2 and C (phagocytosis, Chapter 15; and mechanism of hormone action, Chapter 30).

Glucocorticoids (e.g., cortisol) inhibit phospholipase A_2 activity by induction of synthesis of a phospholipase inhibitor protein, which partly explains their anti-inflammatory effects.

The major metabolites of arachidonic acid (Figure 18-19) arise from the 12-lipoxygenase, the 5-lipoxygenase, and the fatty acid cyclooxygenase pathway. The 5-lipoxygenase pathway yields leukotrienes, and the cyclooxygenase pathway yields cyclic endoperoxides, which are converted to PGs, TXs, and PGIs.

Prostaglandins (PG) were discovered in human semen more than 50 years ago. Their name derives from the prostate gland, but they are produced in many tissues. In fact, the high concentrations found in semen arise in the seminal vesicles rather than the prostate. The chemical parent compound is a 20 carbon unnatural fatty acid known as *prostanoic acid* that contains a five-membered (cyclopentane) ring. Derivatives that contain this structure (PGs, TXs, and PGIs) are known collectively as *prostanoids*.

Prostanoic acid

Differences among various PGs are attributable to differences in substituents and in their positions on the five-membered ring (Figure 18-20). PGs are identified by a letter (e.g., PGE, PGF), characteristic for ring substituents, and by a numerical subscript (e.g., PGE_1, PGF_2), which indicates the number of double bonds (Figure 18-17). The location and type of double bonds are as follows: PG_1, *trans*-C_{13}; PG_2, *trans*-C_{13} *cis*-C_5; PG_3, *trans*-C_{13} *cis*-C_5C_{17}. All PGs have a hydroxy group at C_{15} except PGG, which has a hydroperoxy group (–OOH). The hydroxy group at C_{15} is in the S-configuration in the naturally occurring prostaglandins. The α and β notations (e.g., $PGF_{2\alpha}$) designate the configuration of the substituent at C_9 on the cyclopentane moiety, as used in steroid chemistry (α for below and β for above the plane of the projection of the cyclopentane ring). The natural compounds are α-derivatives.

PGs are synthesized in a stepwise manner by microsomal enzymes. The metabolic pathways discussed here use arachidonic acid as an example. Similar pathways are applicable to other polyenoic fatty acids. PG synthesis is started by microsomal prostaglandin endoperoxide

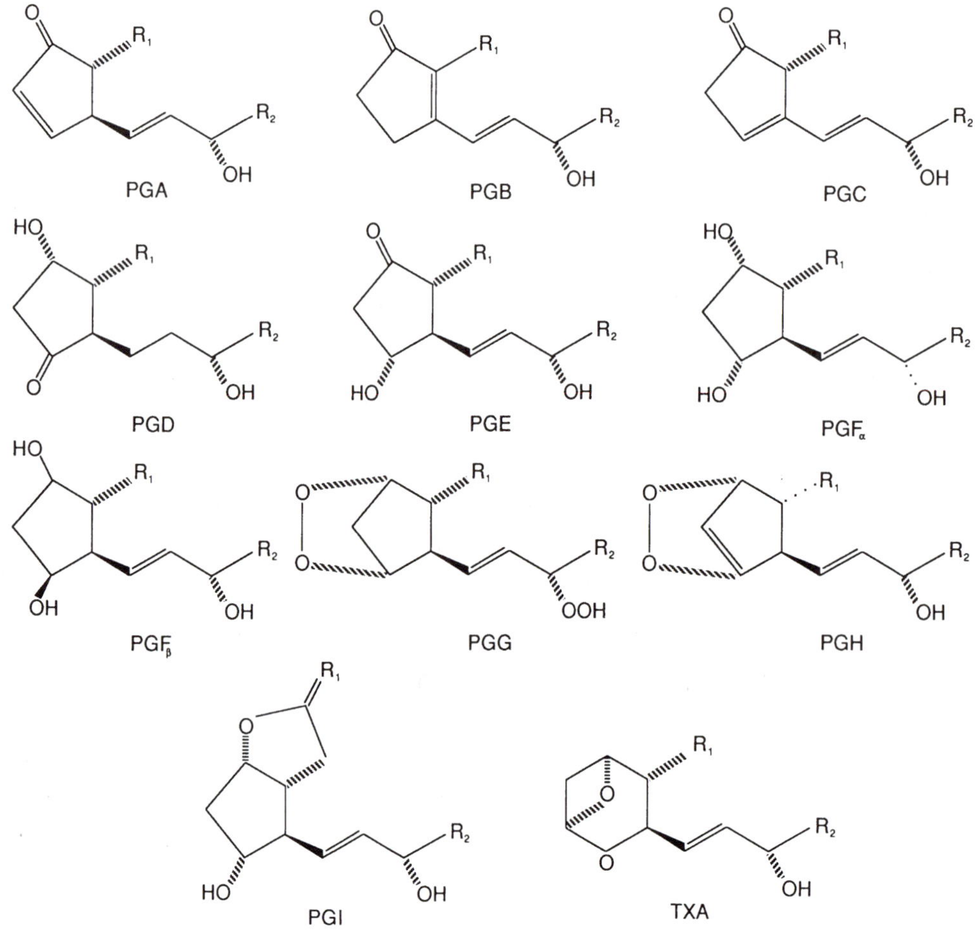

FIGURE 18-20
Ring structures of prostaglandins (PG), prostacyclin (PGI), and thromboxane (TXs). Groups that lie behind the plane of the ring are shown by ||| and those that lie above the plane by ◄.

synthase, which is a cyclooxygenase and a peroxidase. Cyclooxygenase activity (the rate-limiting reaction) results in 15-hydroperoxy-9,11-endoperoxide (PGG_2), which is converted to a 9,11-endoperoxide (PGH_2) by the peroxidase activity. There are two isoforms of cyclooxygenase (COX), which have been designated as COX1 and COX2. Both forms are membrane-associated enzymes. COX1 is constitutively expressed in many tissues, where arachidonic acid metabolites play a role in protective "housekeeping" homeostatic functions. Some of the COX1 mediated normal physiological functions include gastric cytoprotection and limiting acid secretion (Chapter 12), maintenance of renal blood flow, vascular homeostasis, and hemostasis (e.g., antiplatelet effects, Chapter 36). COX2 activity, on the other hand, is normally undetectable in most tissues and it is principally an inducible enzyme. In cells such as monocytes, macrophages, synoviocytes, endothelial cells, and chondrocytes, COX2 is expressed at high levels after induction by inflammatory mediators (e.g., interleukin-1 and tumor necrosis factor) and growth factors. COX2 enzymatic activity initiates the synthesis of arachidonic acid metabolites that mediate pain, inflammation, cellular differentiation and mitogenesis. For example, PGE_2 is chemotactic for neutrophils and PGI_2 causes changes in vascular permeability facilitating extravasation of leukocytes. Although COX2 is generally an inducible enzyme, it is constitutively expressed in developing kidney and brain and, therefore, may be involved in their normal development and proper maturation.

The two unique isoforms, COX1 and COX2, are structurally similar but they are encoded by separate genes differing in their tissue distribution and expression. The COX1 gene contains a promoter region without a TATA sequence and is constitutively expressed. In contrast, the COX2 gene contains DNA segments that allow for rapid up regulation in response to appropriate stimuli. The anti-inflammatory action of glucocorticoids have no effect on the regulation of the COX1 gene.

The amino acid sequence homology of COX1 and COX2 is about 60%. However, in the region of the active site the amino acid homology is about 90% and both isoforms contain a long narrow largely hydrophobic channel with a hairpin bend at the end to accommodate the substrate arachidonic acid. A unique single-amino-acid difference in the wall of the hydrophobic channel (position 523) of COX1 and COX2 has been used to develop specific COX2 inhibitors. At position 523, COX1 has an isoleucine residue whereas COX2 has a valine residue which is smaller by a single CH_2 group. The presence of the less bulky valine residue in the COX2 hydrophobic channel provides access for COX2 selective inhibitors. In COX1 the bulkier isoleucine residue prevents the entry of COX2 selective inhibitors.

In the treatment and management of pain and inflammation produced by arachidonic acid metabolites, COX inhibitors are widely used. These agents arc known as ***non-steroidal anti-inflammatory drugs*** (NSAIDs). Acetylsalicylate (aspirin) is the classic anti-inflammatory and analgesic drug. Aspirin is an irreversible inhibitor of both COX1 and COX2 and it inhibits by acetylation of the hydroxyl group of the serine residue located at the active site of the enzymes. There are nonaspirin NSAIDs, the majority of which are organic acids (e.g., indomethacin, ibuprofen), that are reversible inhibitors of both COX1 and COX2. These inhibitors form a hydrogen bond with an arginine residue at position 120 of both COX1 and COX2 in the channel and block the entry of arachidonic acid. Because of their non-selectivity, aspirin and nonaspirin NSAIDs cause undesirable side effects due to inhibition of the "housekeeping" COX1 enzyme. The side effects include gastrointestinal disorders, renal dysfunction and bleeding tendency. Thus, a COX2 selective or preferential inhibitor that spares COX1 activity is valuable in the treatment of pain and inflammation. Based on the biochemical differences between COX1 and COX2, drugs have been designed with COX2 inhibitor activity, which are associated with a markedly lower incidence of gastrointestinal injury. These drugs often possess sulfonyl, sulfone, or sulfonamide functional groups that bind with the COX2 side pocket in the hydrophobic channel. Examples of COX2 inhibitors are celecoxib, which is a 1,5-diarylpyrazole sulfonamide, and rofecoxib, which is a methylsulfonylphenyl derivative (Figure 18-21). Since nitric oxide (NO) protects gastric mucosa (Chapter 17), a NO moiety linked to conventional NSAIDs may negate the gastric toxic effects due to prostaglandin deficiency. Such drugs of NO-NSAIDs are currently being tested.

Other potential uses for COX inhibitors (in particular for COX2 inhibitors) may include the treatment of Alzheimer's disease and colon cancer. In Alzheimer's disease it is thought that an inflammatory component may lead to deposition of β-amyloid protein in neuritic plaques in the hippocampus and cortex (Chapter 4). The potential use of COX2 inhibitors in colon cancer arises from studies with experimental animals in which COX2 activity is related to the promotion and survival of intestinal adenomas and colon tumors. The cyclooxygenase reaction is also inhibited by arachidonic acid analogues such as

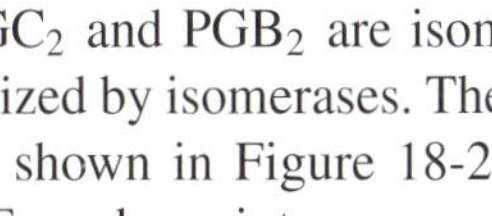

(a)

(b)

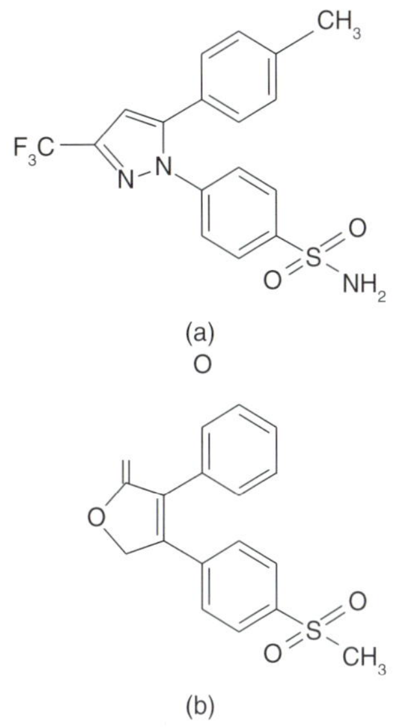

FIGURE 18-21
Structures of cyclooxygenase-2 (COX2) selective inhibitors. (A) Celecoxib and (B) Rofecoxib.

5,8,11,14-Eicosatetraynoic acid

PGH_2 is converted to PGD_2, PGE_2, $PGF_{2\alpha}$, prostacyclin (PGI_2), and thromboxane A_2 (TXA_2) by specific enzymes (Figure 18-22). PGA_2 is obtained from PGE_2 by dehydration. Since PGC_2 and PGB_2 are isomers of PGA_2, they can be synthesized by isomerases. The formation of these compounds is shown in Figure 18-23. In some tissues, PGE_2 and PGF_2 undergo interconversion:

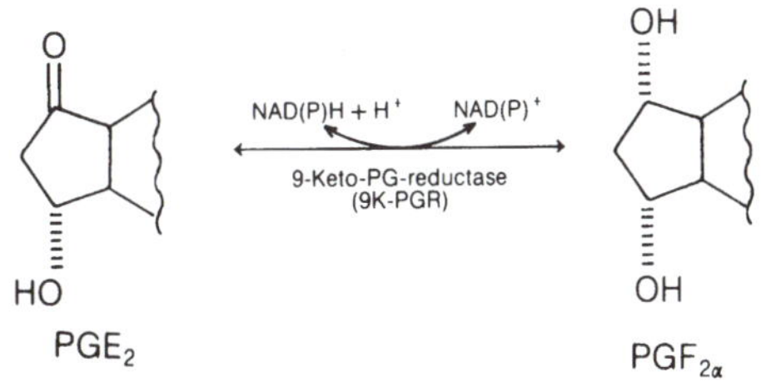

PGE_2 ⇌ $PGF_{2\alpha}$ ($NAD(P)H + H^+$ / $NAD(P)^+$; 9-Keto-PG-reductase (9K-PGR))

The $NAD(P)^+$ inhibits the conversion of PGE_2 to $PGF_{2\alpha}$, while reducing agents favor the formation of $PGF_{2\alpha}$.

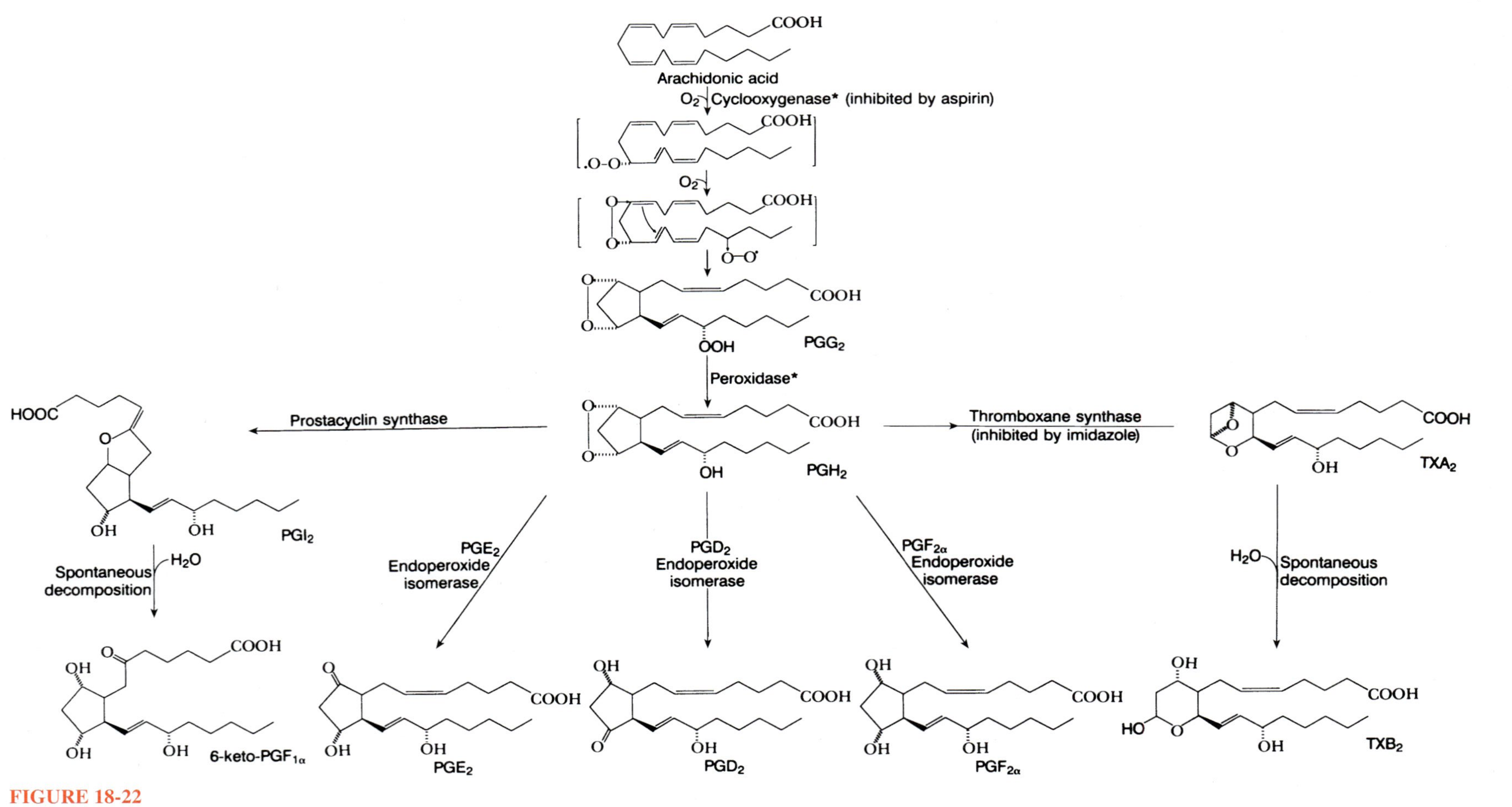

FIGURE 18-22
Synthesis of prostanoids from arachidonic acid. *Both activities reside in one enzyme.

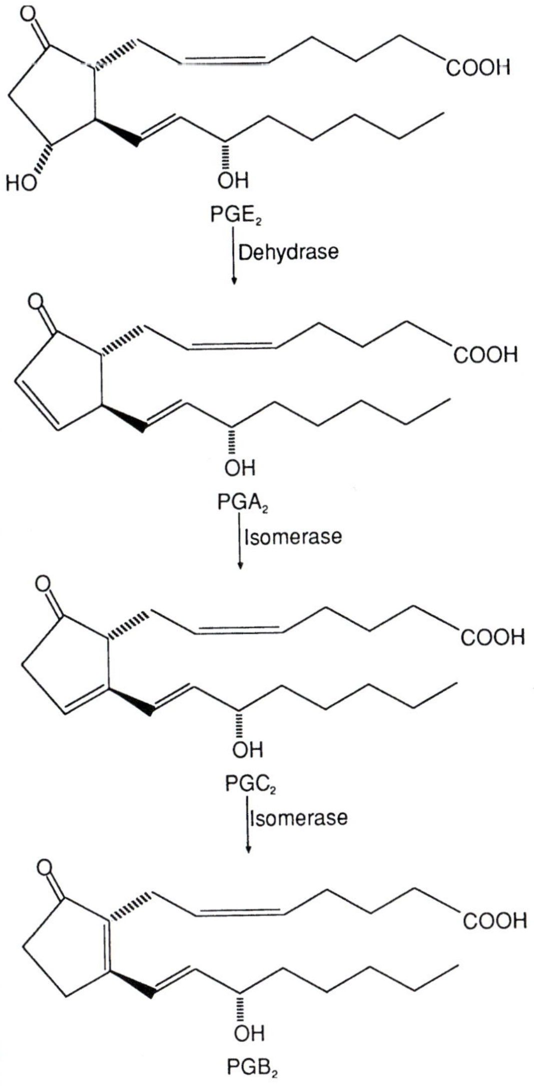

FIGURE 18-23
Conversion of PGE_2 to PGA_2, PGC_2, and PGB_2.

However, PGE_2 formation is favored by glutathione. In the initial catabolic reaction of both compounds by 15-hydroxy-PG-dehydrogenase (15-PGDH), the reduced NAD(P) formed in that reaction inhibits the first step. Thus, the ratio of reduced to oxidized NAD(P) may control the interconversion of PGE_2 and $PGF_{2\alpha}$ and also the first step in their catabolism. This finding is important because in many tissues PGE and PGF have opposing effects. PG biosynthesis can also be regulated by activation of latent forms of cyclooxygenase, promoted by catecholamines and serotonin. The PGs synthesized differ from tissue to tissue; within the same tissue, different cells may yield products with antagonistic actions. For example, the lung parenchymal cells may produce TXA_2, while the lung vascular endothelial cells may produce PGI_2.

Catabolism of prostanoids occurs throughout the body, but the lungs can remove most of the plasma PGs during a single circulatory cycle. Despite this rapid removal, the PGs have adequate access to target organs. Catabolism starts with the reactions of 15-PGDH (oxidation of allylic –OH group at C_{15}) and of PG reductase (reduction of the Δ^{13} double bond). 15-PGDH is found in the cytoplasm (lungs), requires NAD^+, and is specific for the C_{15}(S) alcohol group. These reactions are followed by β-oxidation, ω-oxidation of the alkyl side chains, and elimination of the products. The catabolism of PGE_2 and $PGF_{2\alpha}$ is shown in Figure 18-24. The ***thromboxanes*** (TX), first isolated from human and equine thrombocytes (platelets), contain an oxane ring. TXA_2 is synthesized from PGH_2 by microsomal thromboxane synthase. Thromboxane synthase is inhibited by imidazole derivatives. TXA_2 has a very short half-life ($t_{1/2} = 30$ seconds at 37°C and pH 7.5) and undergoes rapid, nonenzymatic hydrolysis to the inactive TXB_2 (Figure 18-22).

Prostacyclin (PGI_2) is an active and unstable metabolite ($t_{1/2} = 3$ minutes at 37°C and pH 7.5) formed from PGH_2 by prostacyclin synthase. PGI_2 has a double-ring structure and is converted by nonenzymatic hydrolysis to 6-keto-$PGF_{1\alpha}$ (Figure 18-22).

Biological Properties of Prostanoids

Many effects of prostanoids are mediated through adenylate cyclase or mobilization of Ca^{2+} from intracellular stores. PGs increase cAMP in adenohypophysis, corpus luteum, fetal bone, lung platelets, and thyroid but decrease it in adipose tissue. Thromboxanes block the production of cAMP by PGs and mobilize intracellular Ca^{2+}. Thus, many endocrine glands (e.g., adrenal cortex, ovary, pancreatic islets, parathyroids) secrete hormones in response to PGs. Some of these effects are stimulation of steroid hormone production in the adrenal cortex, insulin release, thyroid hormone production, and progesterone secretion from the corpus luteum. *In vitro* PGEs, notably PGE_1, inhibit adipocyte lipolysis—the basal rate as well as that stimulated by catecholamine and other lipolytic hormones. Low doses of PGE_1 in humans tend to stimulate lipolysis through stimulation of release of catecholamines. PGEs stimulate the activity of osteoclasts with the mobilization of Ca^{2+} from bone, an effect independent of that of parathyroid hormone (Chapter 37).

Problems in delineating the primary actions of PGs arise from their frequently opposing effects and from the difficulty of distinguishing between physiological and pharmacological actions. In general, PGE_2 and $PGF_{2\alpha}$ have opposing effects on smooth muscle tone, release of mediators of immediate hypersensitivity, and cyclic nucleotide

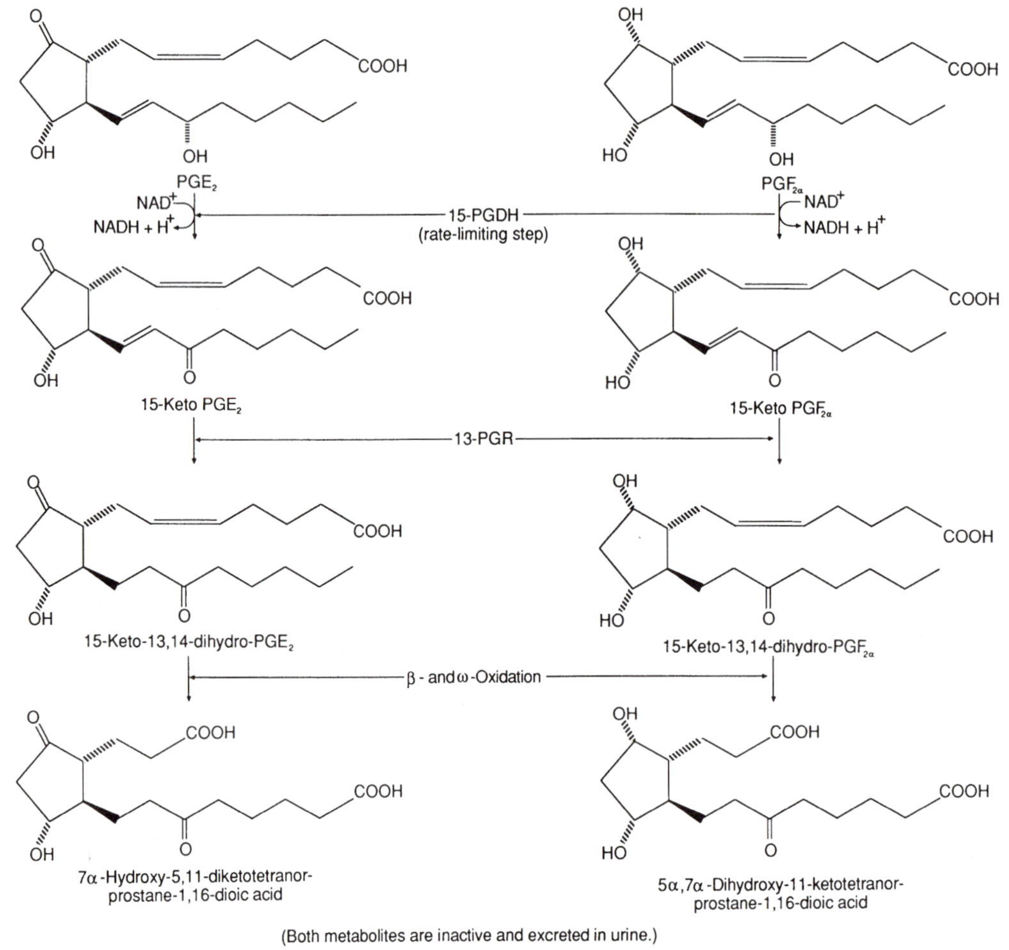

PGE_2

$PGF_{2\alpha}$

NAD^+ / $NADH + H^+$ — 15-PGDH (rate-limiting step) — NAD^+ / $NADH + H^+$

15-Keto PGE_2

15-Keto $PGF_{2\alpha}$

13-PGR

15-Keto-13,14-dihydro-PGE_2

15-Keto-13,14-dihydro-$PGF_{2\alpha}$

β- and ω-Oxidation

7α-Hydroxy-5,11-diketotetranor-prostane-1,16-dioic acid

5α,7α-Dihydroxy-11-ketotetranor-prostane-1,16-dioic acid

(Both metabolites are inactive and excreted in urine.)

FIGURE 18-24
Catabolism of PGE_2 and $PGF_{2\alpha}$. 15-PGDH = 15-Hydroxy-PG-dehydrogenase; 13-PGR = 13-PG-reductase.

levels. Thus, the ratio between E and F compounds (due to changes in the $[NAD^+]/[NADH]$ ratio) may be a crucial factor in control of a given physiological response. The relative proportions of TXs and leukotrienes, as opposed to PGI, also appear to exert an important influence on physiological response.

In most animal species, PGI_2, PGEs, and PGAs are vasodilators, while TXA_2 is a vasoconstrictor. $PGF_{2\alpha}$ and 15-methyl $PGF_{2\alpha}$ are used for induction of mid-trimester abortions because they stimulate uterine muscle. Several PGs suppress gastric HCl production, which has therapeutic potential in the treatment of gastric ulceration and may explain the effect of aspirin to increase HCl secretion by inhibition of PG synthesis. PGE_2 elevation causes fever by increasing firing rates of neurons that control thermoregulation in the hypothalmus.

The effect of prostanoids on platelets has received considerable interest. TXA_2 synthesized in platelets induces platelet aggregation, whereas PGI_2 generated in the vessel wall inhibits platelet aggregation. PGI_3 (Figure 18-17), a product of eicosapentaenoic acid (an ω-3 fatty acid), inhibits TXA_2 synthesis by inhibiting release of arachidonate from phospholipids and by competing for thromboxane synthase. TXA_3 is a much weaker aggregator of platelets than is TXA_2, while PGI_3 is a stronger antiaggregator than is PGI_2. The net effect is an antiplatelet effect, which may be beneficial in patients with thrombotic complications (e.g., myocardial infarction). The low incidence of coronary thrombosis in Greenland Eskimos, whose diet is almost completely derived from marine sources rich in ω-3 fatty acids, has been attributed to antiplatelet effects. This diet is also associated with lower levels of serum cholesterol and triacylglycerol than typical Western diets.

Leukotrienes

Leukotrienes (LTs) are most commonly found in leukocytes, mast cells, platelets, and vascular tissues of the lung and heart. They are formed chiefly from arachidonic acid

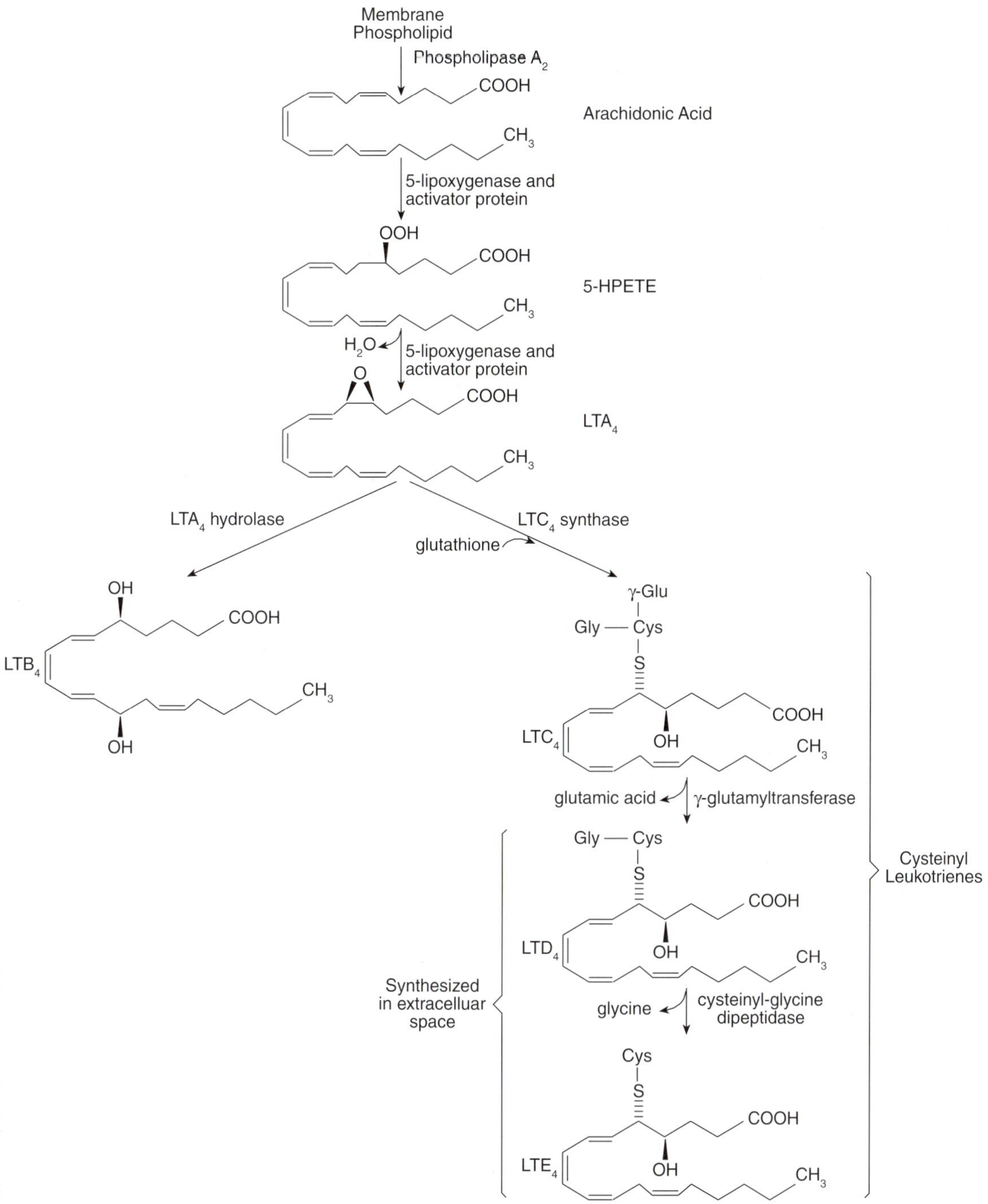

FIGURE 18-25
Pathway for the biosynthesis of leukotrienes.

but they may be derived from eicosatrienoic and eicosapentaenoic acids. The name "leukotrienes" derives from their discovery in leukocytes and from the conjugated triene structure they contain. In the most active LTs, the conjugated triene is in a trans, trans, cis arrangement. They are distinguished by letters A–E and by a subscript that indicates the number of double bonds present.

LTs are produced in the 5-lipoxygenase pathway (Figure 18-25). Their synthesis begins with arachidonic acid obtained from cleavage of the membrane phospholipid pool due to the action of phospholipase A_2. Arachidonic acid is converted in a catalytic sequence by 5-lipoxygenase complex and its activating protein to 5-hydroperoxyeicosatetraenoic acid (5-HPETE) and then to leukotriene A_4 (5,6-oxido-7,9-*trans*-11,14-*cis*-eicosatetraenoic acid). Leukotriene A_4 (LTA_4) is transformed by LTA_4 hydrolase into 5,12-dihyroxy-eicosatetraenoic acid (leukotriene B_4, LTB_4) or into a glutathione adduct with the formation of a thioether linkage at C_6, (leukotriene C_4, LTC_4) by leukotriene C_4 synthase (also known as glutathione S-transferase). Leukotriene D_4 (LTD_4) and LTE_4 are synthesized in the extracellular space from LTC_4. A specific transmembrane transporter exports LTC_4 to the extracellular space. In the extracellular space, removal of the glutamyl residue from LTC_4 by γ-glutamyltransferase yields LTD_4 and the removal of the glycyl residue from LTD_4 by a variety of dipeptidases results in the formation LTE_4 (Figure 18-25).

The three cysteinyl linked leukotrienes, namely LTC_4, LTD_4, and LTE_4 are known collectively as cysteinyl leukotrienes. All three cysteinyl leukotrienes are potent mediators of inflammation and cause microvascular permeability, chemotaxis (particularly eosinophils), mucus hypersecretion, and neuronal stimulation. The potential role of LTC_4 as a neuromessenger or modulator has been implicated in an infant with *LTC_4 synthase deficiency.* The clinical features include muscular hypotonia, psychomotor retardation, failure to thrive, microcephaly, and a fatal outcome. In lung tissue mast cells, eosinophils and alveolar macrophages possess the enzyme activities to synthesize cysteinyl leukotrienes and cause, in addition to above mentioned biological actions, bronchial smooth muscle constriction and proliferation. Thus, cysteinyl leukotrienes are important mediators of immune-mediated inflammatory reactions of anaphylaxis and are constituents of substances originally called "slow reacting substances of anaphylaxis" (SRS-A). They are several times more potent than histamine in constricting airways and promoting tissue edema formation. The proinflammatory effect of LTE_4 is less than that of LTC_4 and LTD_4; it is excreted in the urine and is used as a marker of leukotriene production.

Antileukotriene agents, which can be used in treatment of allergen and exercise-induced asthma and allergic rhinitis, inhibit 5-lipoxygenase or the binding of the activator protein with 5-lipoxygenase or antagonists of leukotriene receptors at the target cell (e.g., airway epithelial cell). The traditional drugs used for treatment of asthma include inhaled corticosteroids, β_2-agonists, and theophyllines. Leukotriene receptor antagonists are orally active and are a new class of antiasthmatic therapeutic agents (Figure 18-26).

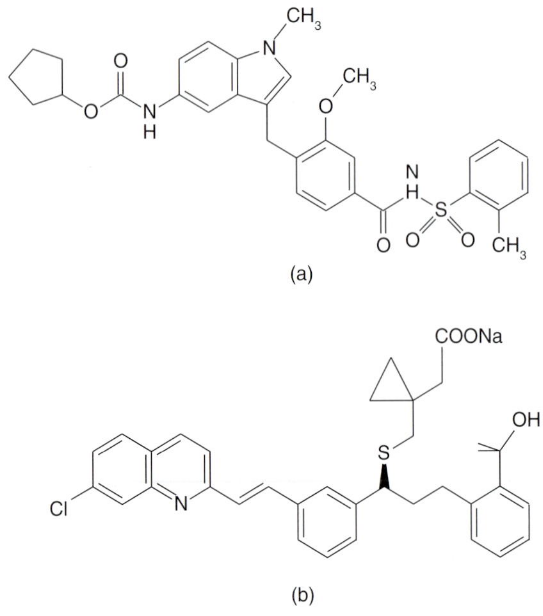

(a)

(b)

FIGURE 18-26
Structures of leukotriene receptor antagonists. (a) Zafirlukast and (b) montelukast.

Supplemental Readings and References

A. Ascherio, M. B. Katan, P. L. Zook, et al.: Trans fatty acids and coronary heart disease. *New England Journal of Medicine* **340,** 1994 (1999).

R. G. Boles, E. A. Buck, M. G. Blitzer, et al.: Retrospective biochemical screening of fatty acid oxidation disorders in postmortem livers of 418 cases of sudden death in the first year of life. *Journal of Pediatrics* **132,** 924 (1998).

J. M. Drazen, E. Israel, and P. M. O'Byrne: Treatment of asthma with drugs modifying the leukotriene pathway. *New England Journal of Medicine* **340,** 197 (1999).

S. Eaton, K. Bartlett, and M. Pourfarzam: Mammalian mitochondrial β-oxidation. *Biochemical Journal* **320,** 345 (1996).

C. J. Hawkey: COX2 inhibitors. *Lancet* **353,** 307 (1999).

J. A. Ibdah, M. J. Bennett, P. Rinaldo, et al.: A fetal fatty-acid oxidation disorder as a cause of liver disease in pregnant women. *New England Journal of Medicine* **340,** 1723 (1999).

S. M. Innis, H. Sprecher, D. Hachey, et al.: Neonatal polyunsaturated fatty acid metabolism. *Lipids* **34,** 139 (1999).

B. A. Johnson, J. D. Roache, M. A. Javors, et al.: Ondansetron for reduction of drinking among biologically predisposed alcoholic patients. *Journal of American Medical Association* **284,** 963 (2000).

H. R. Kranzler: Medications for alcohol dependence—New vistas. *Journal of American Medical Association* **284,** 1016 (2000).

R. G. Kurumbail, A. M. Stevens, J. K. Gierse, et al.: Structural basis for selective inhibition of cyclooxygenase-2 by anti-inflammatory agents. *Nature* **384,** 644 (1996).

D. R. Lichtenstein, M. M. Wolfe: COX2 Selective NSAIDs. New and improved? *Journal of American Medical Association* **284,** 1297 (2000).

B. J. Lipworth: Leukotriene-receptor antagonists. *Lancet* **353,** 57 (1999).

E. Mayatepek and B. Flock: Leukotriene C_4-synthesis deficiency: a new inborn error of metabolism linked to a fatal developmental syndrome. *Lancet* **352,** 1514 (1998).

A. A. M. Morris, S. I. Olpin, M. Brivet, et al.: A patient with carnitine-acylcarnitine translocase deficiency with a mild phenotype. *Journal of Pediatrics* **132,** 514 (1998).

P. M. O'Byrne, F. Israel, and J. M. Drazen: Antileukotrienes in the treatment of asthma. *Annals of Internal Medicine* **127,** 472 (1999).

M. R. Pierce, G. Pridpan, S. Morrison, and A. S. Pickoff: Fatal carnitine palmitoyltransferase II deficiency in a newborn: New phenotypic features. *Clinical Pediatrics* **38,** 13 (1999).

P. Rinaldo, K. Raymond, A. Al-Odaib, and M. J. Bennett: Clinical and biochemical features of fatty acid oxidation disorders. *Current Opinion in Pediatrics* **10,** 615 (1998).

D. R. Silverstein, G. Faich, J. L. Goldstein, et al.: Gastrointestinal toxicity with celecoxib vs. nonsteroidal anti-inflammatory drugs for osteoarthritis and rheumatoid arthritis. *Journal of American Medical Association* **284,** 1247 (2000).

R. M. Swift: Drug therapy for alcohol dependence. *New England Journal of Medicine* **340,** 1482 (1999).

T. L. Wall, C. M. Petersen, K. P. Peterson, et al.: Alcohol metabolism in Asian-American men with genetic polymorphisms of aldehyde dehydrogenase. *Annals of Internal Medicine* **127,** 376 (1997).

G. N. Wilson: Structure-function relationships in the peroxisome: Implications for human disease. *Biochemical Medicine and Metabolic Biology* **46,** 288 (1991).

M. M. Wolfe, D. R. Lichtenstein, and G. Singh: Gastrointestinal toxicity of nonsteroidal antiinflammatory drugs. *New England Journal of Medicine* **340,** 1888 (1999).

CHAPTER 19

Lipids II: Phospholipids, Glycosphingolipids, and Cholesterol

Phospholipids and glycosphingolipids are amphipathic lipid constituents of membranes (Chapter 10). They play an essential role in the synthesis of plasma lipoproteins (Chapter 20) and eicosanoids (Chapter 18). They function in transduction of messages from cell surface receptors to second messengers that control cellular processes (Chapter 30) and as surfactants. Cholesterol is mainly of animal origin and is an essential constituent of biomembranes (Chapter 10). In plasma, cholesterol is associated with lipoproteins (Chapter 20). Cholesterol is a precursor of bile acids formed in the liver; of steroid hormones secreted by adrenals, gonads, and placenta; and 7-dehydrocholesterol of vitamin D formed in the skin. In tissues, cholesterol exists primarily in the unesterified form (e.g., brain and erythrocytes), although appreciable quantities are esterified with fatty acids in liver, skin, adrenal cortex, and plasma lipoproteins.

19.1 Phospholipids

Phospholipids can be ***glycerolipids*** or ***sphingolipids.*** Examples of glycerolipids are phosphatidylcholine, phosphatidylethanolamine, phosphatidylserine, phosphatidylinositol, and phosphatidylglycerol (Figure 19-1). To distinguish between the two primary alcohol-carbon atoms of asymmetrically substituted glycerol derivatives, the glycerol carbon atoms are numbered 1 through 3 from top to bottom and the C_2 hydroxyl group is written to the left. This system is known as the stereochemical numbering (*sn*) convention. Thus, the structural formula for *sn*-1, 2-diacylglycerol is

```
                    O
          1         ‖
         H2C—O—C—R′
           |
R″—C—O—C²—H
           |
         H2C³OH
```

Phosphatidylcholines

Phosphatidylcholines, or ***lecithins,*** are zwitterionic over a wide pH range because of the presence of a quaternary ammonium group and a phosphate moiety. Phosphatidylcholines are the most abundant phospholipids in animal tissues and typically contain palmitic, stearic, oleic, linoleic, or arachidonic acid, usually with saturated fatty acids in the *sn*-1 position and unsaturated fatty acids at *sn*-2.

The *de novo* pathways for phospholipid synthesis use cytidine triphosphate (CTP) for activation of intermediate species (analogous to the role of UTP in glycogen biosynthesis; Chapter 15). The principal pathway of

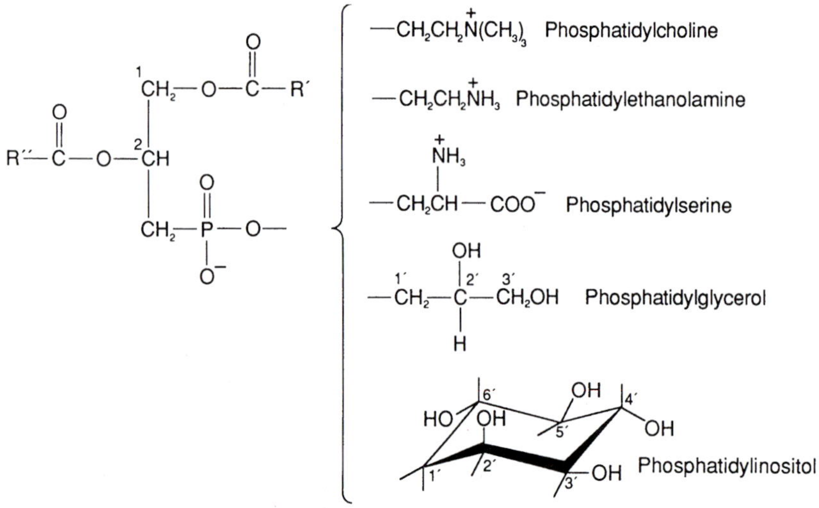

FIGURE 19-1
Structure of some glycerophospholipids.

phosphatidylcholine biosynthesis uses cytidine diphosphate (CDP) choline (Figure 19-2). Many reactions of phospholipid synthesis occur in the endoplasmic reticulum. Choline is first phosphorylated by ATP to phosphocholine, which reacts with CTP to form CDP-choline, from which phosphocholine is transferred to *sn*-1,2-diacylglycerol. The rate-limiting step in this pathway appears to be that catalyzed by CTP:phosphocholine cytidylyltransferase, which is activated by fatty acids.

Phosphatidylcholine can also be synthesized by the methylation pathway that converts phosphatidylethanolamine to phosphatidylcholine, principally in the liver. The methyl donor is S-adenosylmethionine (Chapter 17). Phosphatidylethanolamine-N-methyltransferase transfers three methyl groups in sequence to produce phosphatidylcholine. The fatty acid components of phosphatidylcholine can then be altered by deacylation-reacylation reactions.

Phosphatidylcholine is degraded by phospholipases that cleave preferentially at specific bonds (Chapter 18). Choline released is phosphorylated by choline kinase and reutilized in phosphatidylcholine synthesis. However, in liver mitochondria, choline is also oxidized to betaine (N-trimethylglycine):

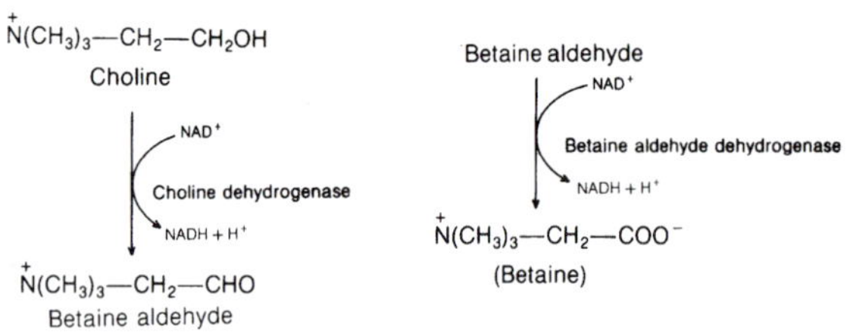

$\overset{+}{N}(CH_3)_3-CH_2-CH_2OH$ (Choline) $\xrightarrow[\text{Choline dehydrogenase}]{NAD^+ \rightarrow NADH + H^+}$ $\overset{+}{N}(CH_3)_3-CH_2-CHO$ (Betaine aldehyde)

Betaine aldehyde $\xrightarrow[\text{Betaine aldehyde dehydrogenase}]{NAD^+ \rightarrow NADH + H^+}$ $\overset{+}{N}(CH_3)_3-CH_2-COO^-$ (Betaine)

Betaine functions as a methyl donor (e.g., in methionine biosynthesis from homocysteine; Chapter 17), and it can also be converted to glycine.

Other Glycerophospholipids

Phosphatidylethanolamines, or cephalins (so-called because they were first obtained from brain tissue), can be synthesized by reactions analogous to those of *de novo* synthesis of phosphatidylcholine. Ethanolamine is first phosphorylated by ATP and ethanolamine kinase to phosphoethanolamine, which then reacts with CTP to form CDP-ethanolamine. CTP:phosphoethanolamine cytidylyltransferase is not located on the endoplasmic reticulum, nor do fatty acids activate it as they do the analogous enzyme of phosphatidylcholine synthesis. Finally, 1,2-diacylglycerol phosphoethanolamine transferase catalyses the reaction of diacylglycerol with CDP-ethanolamine to form phosphatidylethanolamine.

Ethanolamine $\xrightarrow{ATP \rightarrow ADP}$ Phosphoethanolamine $\xrightarrow{CTP \rightarrow PP_i}$ CDP-ethanolamine $\xrightarrow{\text{Diacylglycerol} \rightarrow CMP}$ Phosphatidylethanolamine

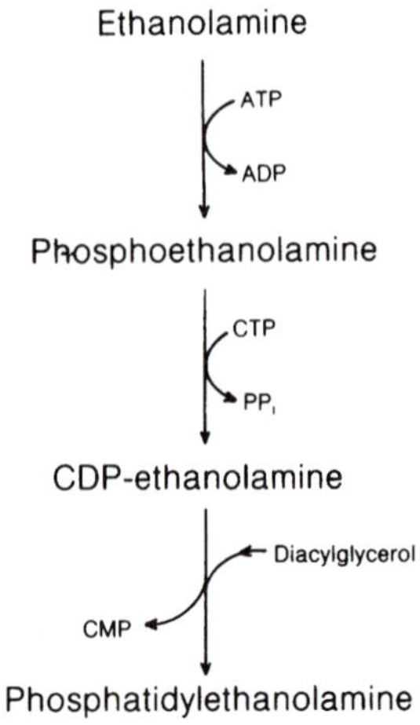

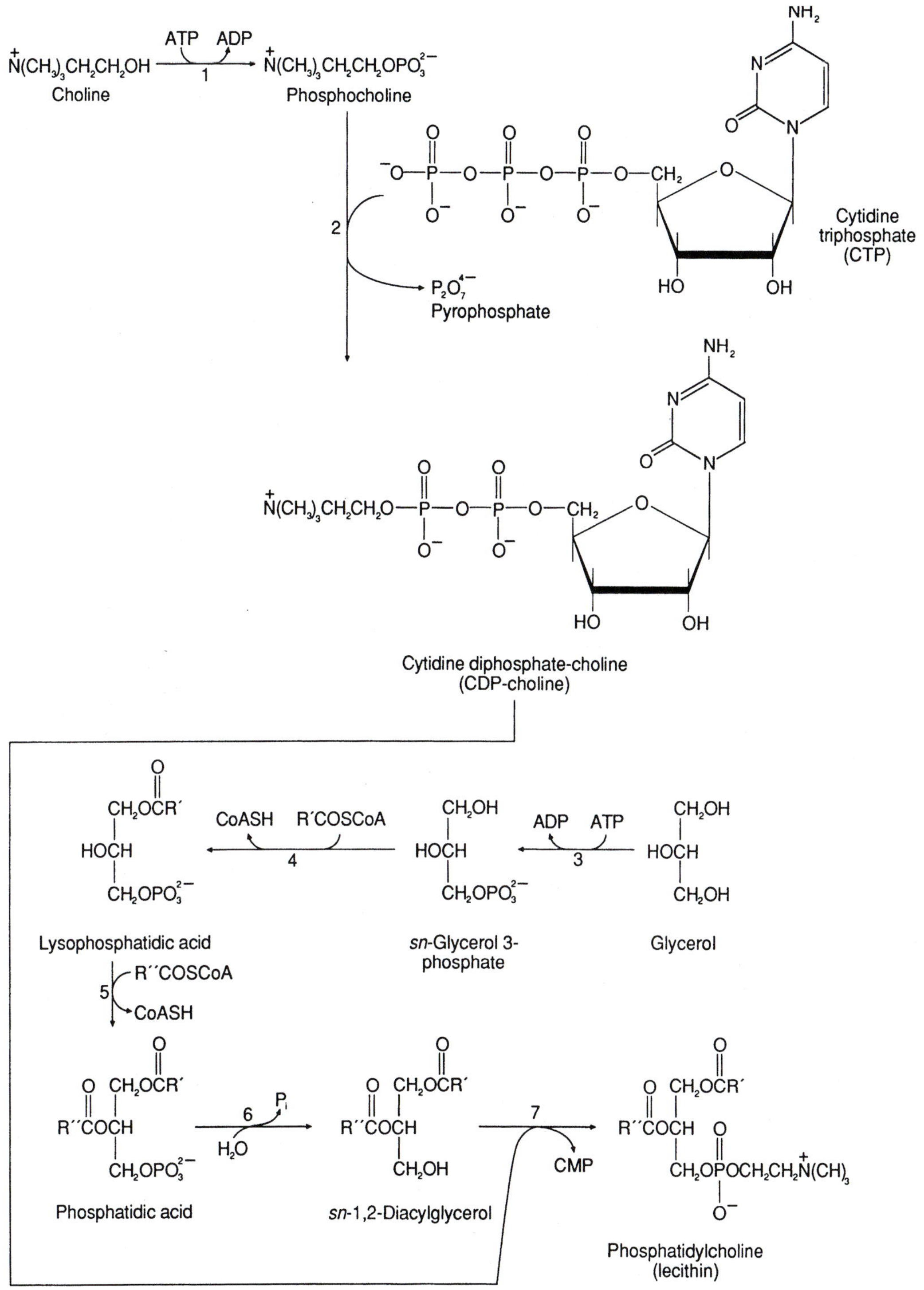

FIGURE 19-2

Synthesis of phosphatidylcholine. The rate-limiting reaction is that catalyzed by cytidylyltransferase (reaction 2) which appears to be active only when attached to the endoplasmic reticulum, although it is also found free in the cytosol. Cytidylyltransferase is inactivated by a cAMP-dependent protein kinase and activated by a phosphatase. Translocation to the endoplasmic reticulum can be stimulated by substrates such as fatty acyl Coenzyme A (CoA). Choline deficiency can result in deposition of triacylglycerol in liver and reduced phospholipid synthesis. Enzymes: (1) choline kinase; (2) CTP:phosphocholine cytidylyltransferase; (3) glycerol kinase; (4) acyl-CoA: glycerol-3-phosphate acyltransferase; (5) acyl-CoA: acyl glycerol-3-phosphate acyltransferase; (6) phosphatidic acid phosphatase; (7) CDP-choline: diacylglycerol phosphocholine transferase.

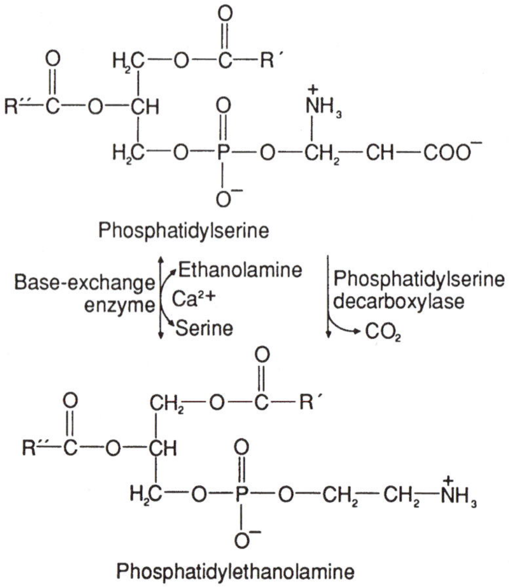

Phosphatidylethanolamines can also be synthesized by decarboxylation of phosphatidylserine and in mammals principally through action of the Ca^{2+}-mediated base exchange enzyme (Figure 19-3). Phosphatidylserine production in liver occurs at the cytosolic face of the endoplasmic reticulum. In brain tissue, this phospholipid accounts for up to 15% of the total phospholipid content.

Phosphatidylinositols and phosphatidylglycerols are synthesized via CDP-diacylglycerol (Figure 19-4). Phosphatidylinositols are enriched with arachidonic acids by deacylation and reacylation. Two other phosphoinositides are synthesized and degraded in the phosphoinositol cycle (Chapter 30) by the enzymes on the plasma membrane, endoplasmic reticulum, and cytosol of brain and liver cells. These anionic phospholipids contain *myo*-inositol, which can be derived from food or produced via cyclization of glucose-6-phosphate. Phosphatidylinositol 4-phosphate and phosphatidylinositol 4,5-bisphosphate are derived from phosphatidylinositol by ATP-dependent phosphorylation of the 4- and 5-positions of the inositol ring. Dephosphorylation reactions are carried out by phosphomonoesterases. The exact proportion of the three forms in cell membranes is unknown, although the

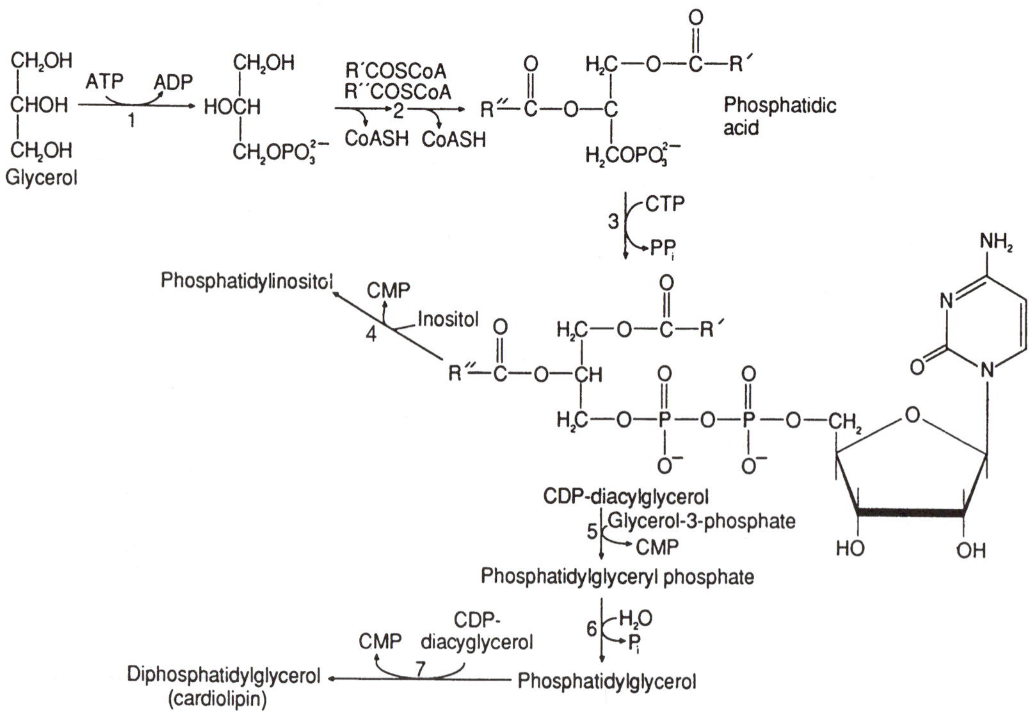

Phosphatidylserine

Base-exchange enzyme
Ethanolamine
Ca^{2+}
Serine

Phosphatidylserine decarboxylase
CO_2

Phosphatidylethanolamine

FIGURE 19-3
Biosynthesis of phosphatidylethanolamine from phosphatidylserine. The base-exchange enzyme on the cytosolic face of the endoplasmic reticulum can interconvert these phospholipids in the presence of Ca^{2+} and the alternate head group, serine or ethanolamine. The decarboxylase is localized in the inner membrane of mitochondria and catalyzes the nonequilibrium conversion of phosphatidylserine to phosphatidylethanolamine.

Glycerol
ATP ADP
1
R′COSCoA
R″COSCoA
2
CoASH CoASH
Phosphatidic acid
3
CTP
PP_i
Phosphatidylinositol
CMP
Inositol
4
CDP-diacylglycerol
5
Glycerol-3-phosphate
CMP
Phosphatidylglyceryl phosphate
6
H_2O
P_i
CMP
CDP-diacyglycerol
7
Diphosphatidylglycerol (cardiolipin)
Phosphatidylglycerol

FIGURE 19-4
Biosynthesis of phosphatidylinositols and phosphatidylglycerols. Enzymes: (1) glycerol kinase; (2) acyltransferases; (3) phosphatidate cytidyltransferase; (4) CDP-diacylglycerol: inositol phosphatidate transferase; (5) CDP-diacylglycerol: *sn*-glycerol-3-phosphate phosphatidate transferase; (6) phosphatase; (7) phosphatidylglycerol: CDP-diacylglycerol phosphatidate transferase.

phosphatidylinositol content greatly exceeds that of the other two. The activity of a number of hormones, growth factors, and neurotransmitters depends upon hydrolysis of phosphatidylinositol 4,5-bisphosphate to inositol triphosphate and diacylglycerol, which serve to initiate parallel metabolic cascades that can mobilize intracellular calcium stores, activate protein kinase C and release arachidonic acid. Each of these events in turn can regulate a number of cellular processes (Chapter 30).

The transfer of phosphatidic acid from CDP-diacylglycerol to phosphatidylglycerol yields diphosphatidylglycerol, or cardiolipin (Figure 19-4), which is found in highest concentration in the inner membranes of mitochondria of cardiac muscle. Cardiolipin isolated from beef heart is used as an antigen in serological flocculation and precipitation tests for syphilis, a sexually transmitted disease caused by *Treponema pallidum.* Anticardiolipin antibodies are formed in response to lipoidal material released from damaged host cells early in the infection and that present on the cell surface of the treponeme. This test is nonspecific. In a specific test for syphilis, antigen derived from *T. pallidum* itself is used to detect the presence of antibodies specific to it.

In the synthesis of phosphatidic acid, the starting material may be glycerol, or dihydroxyacetone phosphate, a product of aldolase action on fructose-1,6-bisphosphate (Chapter 13). Analogous reactions are used in the synthesis of plasmalogens, which, like phosphoglycerides, have the common glycerol backbone. However, at the C_1 position, an α,β-unsaturated fatty ether is present rather than a fatty acid ester:

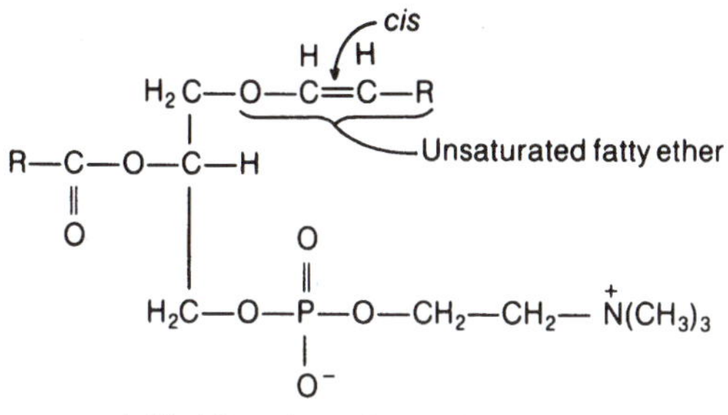

cis

Unsaturated fatty ether

1-Alkyl-2-acyl-*sn*-glycero-3-phosphorylcholine (phosphatidalcholine, a plasmalogen)

The α,β-unsaturated fatty ether is an aldehydogenic group because its hydrolysis releases an α,β-unsaturated primary alcohol that readily tautomerizes to an aldehyde. Choline, ethanolamine, and serine plasmalogens are found in cardiac and skeletal muscle, brain, and liver. The biosynthesis of phosphatidylethanolamine is shown in Figure 19-5.

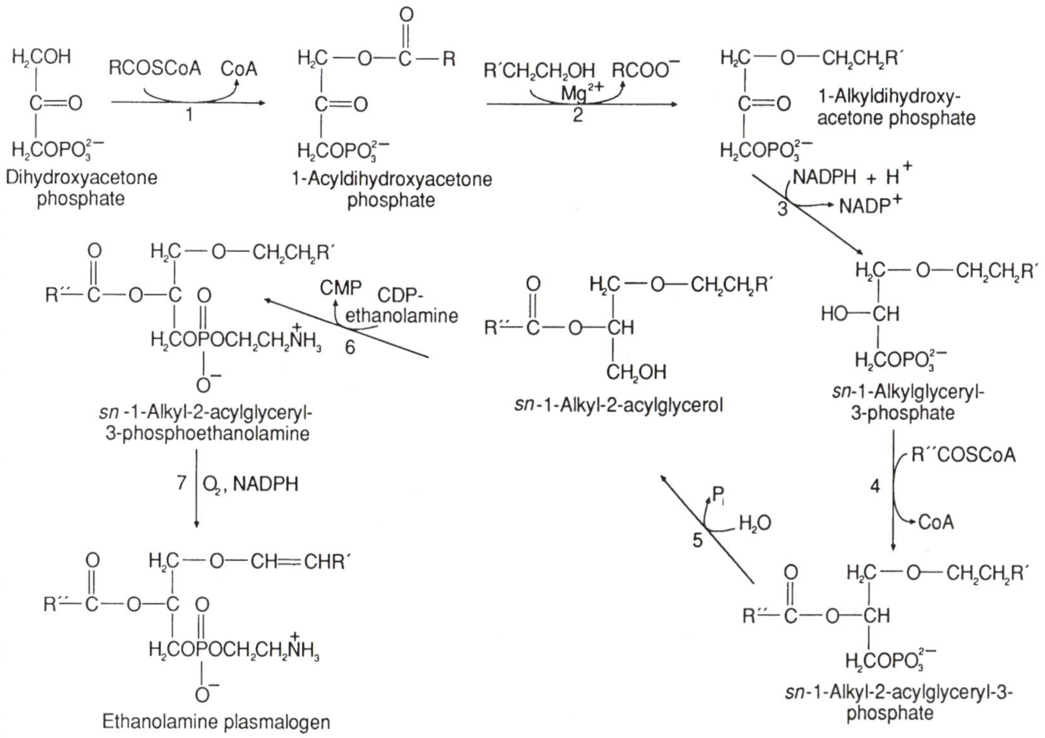

Dihydroxyacetone phosphate

1-Acyldihydroxyacetone phosphate

1-Alkyldihydroxyacetone phosphate

sn-1-Alkylglyceryl-3-phosphate

sn-1-Alkyl-2-acylglyceryl-3-phosphate

sn-1-Alkyl-2-acylglycerol

sn-1-Alkyl-2-acylglyceryl-3-phosphoethanolamine

Ethanolamine plasmalogen

FIGURE 19-5

Biosynthesis of ethanolamine plasmalogen. Enzymes: (1) acyltransferase; (2) synthase; (3) oxidoreductase; (4) acyltransferase; (5) phosphatase; (6) transferase; (7) Δ^1-alkyl desaturase.

Synthesis of phosphatidylcholine can occur by base-exchange reaction, by methylation of phosphatidalethanolamine, or by the coupled action of phospholipase and CDP-choline:cholinephosphotransferase on phosphatidalethanolamine. Although the function of most ether-containing phosphoglycerides is unknown, a "platelet-activating factor" (PAF) that promotes platelet aggregation has the following structure:

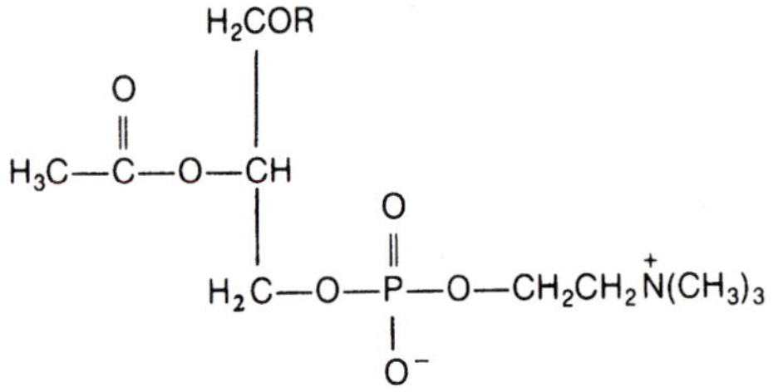

1-Alkyl-2-acetyl-*sn*-glyceryl-3-phosphocholine

PAF released from I_gE-sensitized basophilic leukocytes (and probably other mononuclear cells, such as mast cells) in response to antigen stimulation causes aggregation of platelets and release of their granular constituents (e.g., serotonin). This action of PAF is potent and rivals that of thromboxane A_2 (Chapter 18). PAF also is a potent antihypertensive agent when given intravenously to hypertensive rats. It is inactivated by deacetylation by a specific acetylhydrolase.

Phosphosphingolipids

The ***sphingomyelins*** are structurally similar to phosphatidylcholine but contain N-acylsphingosine (ceramide) instead of *sn*-1,2-diacylglycerol (Figure 19-6). They occur in high concentration in myelin and in the brain and are a nearly ubiquitous constituent of membranes.

Sphingolipid biosynthesis is catalyzed by membrane-bound enzymes of the endoplasmic reticulum. Sphingosine, an acylaminoalcohol, is synthesized from palmitoyl-CoA and serine in a reaction that requires pyridoxal phosphate, NADPH, and Mn^{2+} (Figure 19-7). The exact pathway of ceramide synthesis is not known. The acyl group may be added to the 2-amino group of sphinganine, either as a fatty acid (the reverse of hydrolysis of ceramide) or from an acyl-CoA by an acyltransferase. In such a pathway, the double bond in the aliphatic chain would be inserted after the acylation step. Direct transfer of an acyl group to sphingosine is an alternative pathway of ceramide synthesis. The acyl groups of sphingolipids may be those of long-chain fatty acids synthesized by the fatty acid synthase complex (e.g., palmitic or stearic), very-long-chain fatty acids synthesized by the microsomal chain elongation system (e.g., behenic, lignoceric), monoenoic fatty acids (e.g., oleic, nervonic), or α-hydroxy very-long-chain fatty acids. The presence of transferases specific for different chain lengths determines the acyl composition of sphingolipids in a given tissue.

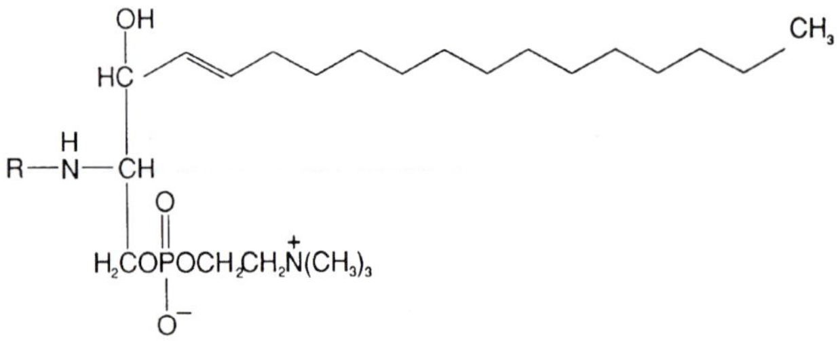

FIGURE 19-6
Structure of sphingomyelin.

Sphingomyelin is probably synthesized by an exchange reaction in which the phosphorylcholine moiety of phosphatidylcholine is transferred to ceramide:

$$\text{Ceramide} + \text{phosphatidylcholine} \rightarrow \text{sphingomyelin} + \text{diacylglycerol}$$

Synthesis of glycosphingolipids and sulfoglycosphingolipids involves the addition of sugar and sulfate residues to ceramide from UDP-sugar derivatives or the activated sulfate donor 3′-phosphoadenosine-5′-phosphosulfate (Chapter 17), and appropriate transferases. These pathways are discussed in Chapter 16. Catabolism of sphingolipids is by specific lysosomal hydrolases. Several inherited disorders associated with the deficiencies of these enzymes are discussed below.

19.2 Phospholipids and Glycosphingolipids in Clinical Medicine

Pulmonary Surfactant Metabolism and Respiratory Distress Syndrome

Pulmonary surfactant is a complex of lipids and proteins with unique surface active properties that is synthesized exclusively in alveolar type II cells. The composition of surfactant is 90% lipids and 5–10% surfactant-specific proteins. The lipid component is made up of dipalmitoylphosphatidylcholine (also called lecithin, 70–80%) and another major phospholipid, phosphatidylglycerol (PG, 10%). The remainder of the phospholipids of surfactant are phosphatidylinositol (PI), phosphatidylethanolamine (PE), and phosphatidylserine (PS). Immature surfactant contains higher amounts of PI compared to PG. Thus, a low ratio of PG to PI indicates lung immaturity. Cholesterol, a neutral lipid, is also a constituent of the lipid component of surfactant.

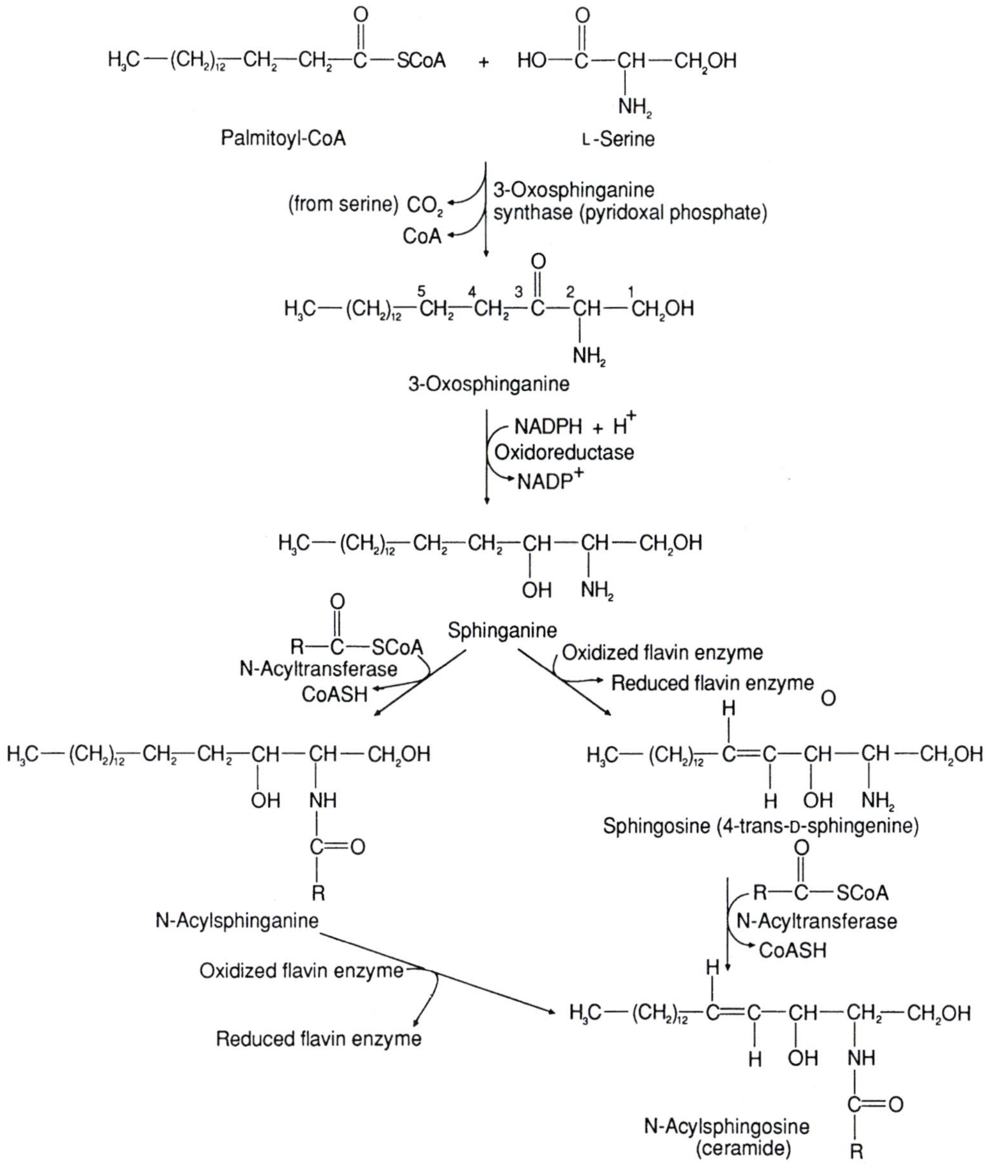

$H_3C-(CH_2)_{12}-CH_2-CH_2-C(=O)-SCoA$ + $HO-C(=O)-CH(NH_2)-CH_2OH$

Palmitoyl-CoA L-Serine

(from serine) CO_2 ← ; CoA ← ; 3-Oxosphinganine synthase (pyridoxal phosphate)

$H_3C-(CH_2)_{12}-\overset{5}{CH_2}-\overset{4}{CH_2}-\overset{3}{C}(=O)-\overset{2}{CH}(NH_2)-\overset{1}{CH_2}OH$

3-Oxosphinganine

NADPH + H^+ → Oxidoreductase → $NADP^+$

$H_3C-(CH_2)_{12}-CH_2-CH_2-CH(OH)-CH(NH_2)-CH_2OH$

Sphinganine

$R-C(=O)-SCoA$ → N-Acyltransferase → CoASH

Oxidized flavin enzyme → Reduced flavin enzyme

$H_3C-(CH_2)_{12}-CH_2-CH_2-CH(OH)-CH(NH-C(=O)-R)-CH_2OH$

N-Acylsphinganine

$H_3C-(CH_2)_{12}-CH=CH-CH(OH)-CH(NH_2)-CH_2OH$

Sphingosine (4-trans-D-sphingenine)

$R-C(=O)-SCoA$ → N-Acyltransferase → CoASH

Oxidized flavin enzyme → Reduced flavin enzyme

$H_3C-(CH_2)_{12}-CH=CH-CH(OH)-CH_2(NH-C(=O)-R)-CH_2OH$

N-Acylsphingosine (ceramide)

FIGURE 19-7
Biosynthesis of sphingosine and ceramide.

After synthesis in the various compartments of endoplasmic reticulum of alveolar type II cells, surfactant components are assembled in the cytosol into lamellar bodies. In the process of formation of lamellar bodies, the transfer of phospholipids between membranes is facilitated by phospholipid transfer proteins, which are nonenzymatic proteins found in all eukaryotic cells and which play an important role in lipid metabolism. There are three well-characterized phospholipid transfer proteins:

1. PC-specific transfer protein.
2. PI- and PC-specific transfer protein.
3. Phospholipids and cholesterol-nonspecific lipid transfer protein (also known as sterol carrier protein).

All three of these proteins are present in the lung. The lamellar bodies are secreted into alveolar lumen where they are transformed into an extracellular form of surfactant that has a quadratic lattice structure called *tubular myelin*. The three-dimensional tubulin-myelin structures spread in a monolayer at the air-liquid interface. This spreading decreases the surface tension, prevents alveolar collapse at the end of expiration, and confers mechanical stability to the alveoli. The surfactant system is in a continuous state of flux, and surfactant is recycled by uptake and endocytosis in alveolar type II cells. Thus, the type II cells are involved in both the synthesis and the recycling of surfactant.

The phospholipids are mainly synthesized starting from glycerol 3-phosphate which is derived from glucose. The CDP-choline pathway is utilized in the synthesis of phosphatidylcholine or lecithin (Figure 19-2). The protein component of surfactant is lung-specific and consists of four proteins designated SP-A, SP-B, SP-C, and SP-D. These surfactant proteins perform important functions that lead to a reduction in alveolar surface tension during respiration. These include structural transformation of lamellar body to tubular myelin (SP-A and SP-B in the presence of Ca^{2+}), enhancement of surface-tension lowering properties and promotion of adsorption of surfactant phospholipids at the air-liquid interface (SP-B and SP-C), reuptake by endocytosis of surfactant by type II cells, and the activation of alveolar macrophages to facilitate surfactant clearance. Both SP-A and SP-D possess antimicrobial properties. SP-A is chemotactic for macrophages and promotes bacterial phagocytosis.

The most abundant surfactant protein is a water-soluble asialoglycoprotein that is a multimer consisting of six triple helical structures. The primary structure of SP-A is highly conserved among several species. It has two domains; the N terminus is collagen-like with Gly-X-Y repeats (where Y is frequently a prolyl residue), and the C terminus has lectin-like properties. SP-B and SP-C are highly hydrophobic proteins. SP-D is a glycoprotein and has a structure similar to SP-A. The importance of one surfactant protein is demonstrated in neonates who are born with an inherited deficiency of SP-B. Infants with a SP-B deficiency require ventilatory support and extracorporeal membrane oxygenation. However, almost all die during the first year of life due to progressive respiratory failure. Lung transplantation is the only therapy by which the SP-B-deficient infants can be saved from death.

Surfactant biosynthesis is developmentally regulated. The capacity for the fetal lung to synthesize surfactant occurs relatively late in gestation. Although the type II cells are identifiable at 20–22 weeks of gestation, the secretion of surfactant into the amniotic fluid occurs during 30–32 weeks of gestation. Thus, for the infant a consequence of prematurity is *respiratory distress syndrome* (RDS), which is a leading cause of neonatal morbidity and mortality in developed countries. The synthesis of surfactants is regulated by factors that include glucocorticoids, thyroid hormones, prolactin, estrogens, androgens, catecholamine (functioning through β-adrenergic receptors and cAMP), growth factors, and cytokines. Glucocorticoids stimulate lung maturation, thus glucocorticoid therapy in women in preterm labor prior to 34 weeks of gestation can significantly decrease the incidence of RDS in the premature neonates.

Thyroid hormones also accelerate fetal lung maturation. Fetal thyroid hormone levels may be increased by antenatal administration of thyrotropin-releasing hormone (TRH), a tripeptide that crosses the placental barrier, stimulates fetal pituitary production of thyroid stimulating hormone (TSH), and which, in turn, increases fetal thyroid hormone production (Chapter 33). This indirect method of enhancement of fetal thyroid hormone production is utilized because thyroid hormones do not readily cross the placental barrier. Insulin delays surfactant synthesis and so fetal hyperinsulinemia in diabetic mothers may increase the incidence of RDS even in the full-term infant. Androgen synthesized in the fetal testis is the probable cause of a slower onset of surfactant production in male fetuses. Prophylactic, or after onset of RDS, administration of synthetic or natural pulmonary surfactants intratracheally to preterm infants improves oxygenation and decreases pulmonary morbidity.

In adults, a severe form of lung injury can develop in association with sepsis, pneumonia, and injury to the lungs due to trauma or surgery. This catastrophic disorder is known as *acute respiratory distress syndrome* (ARDS) and has a mortality rate of more than 40%. In ARDS, one of the major problems is a massive influx of activated neurophils which damage both vascular endothelium and alveolar epithelium and result in massive pulmonary edema and impairment of surfactant function. Neutrophil proteinases (e.g., elastase) break down surfactant proteins. A potential therapeutic strategy in ARDS involves administration of both surfactant and antiproteinases (e.g., recombinant α_1-antitrypsin).

Biochemical Determinants of Fetal Lung Maturity

The need for surfactant production does not become essential until birth because no air liquid interface exists in the alveoli *in utero,* and fetal oxygen needs are met by maternal circulation. The pulmonary system, including surfactant production, is among the last of the fetal organ systems to attain functional maturity. Since preterm birth is associated with significant neonatal morbidity and mortality in inadequate oxygen supply to an immature pulmonary lung system, the assessment of antenatal fetal lung maturity is necessary to develop a therapeutic strategy in the management of a preterm infant. The biochemical determinants are measured primarily in the amniotic fluid obtained by amniocentesis.

In a normal pregnancy, the lung is adequately developed by about the 36th or 37th week. Biochemical changes occurring during this period of gestation can be used to evaluate fetal lung maturity when early delivery is planned. One such measurement is that of lecithin to sphingomyelin (L/S) ratio in amniotic fluid. In a normal pregnancy, the L/S ratio is less than 1 before the 31st week, rises to about 2 by the 34th week, to about 4 at the 36th week, and to about 8 at term (39 weeks). The change is due to an increase in lecithin synthesis rather than a decrease in synthesis of sphingomyelin. These values vary in normal gestations and in abnormal pregnancies (due to maternal, fetal, or placental disorders), the ratio may be elevated or reduced without regard to gestational age. A low L/S ratio is not inevitably associated with RDS. While an L/S ratio greater than 2 is associated with the absence of serious RDS, one lower than 2 is not uniformly predictive of the development of RDS.

Pulmonary surfactant in amniotic fluid can also be measured by its ability to generate stable foam in the presence of ethanol. This *foam stability test* (FST), or shake test, correlates well with the L/S ratio and with fetal lung maturity. In some instances, in the presence of an L/S ratio of less than 2, the FST has indicated lung maturity (without subsequent respiratory distress). This discrepancy may be due to the presence of surfactants other than lecithin that stabilize the neonatal alveoli, namely, phosphatidylglycerol (PG) and phosphatidylinositol (PI). These acidic phospholipids are synthesized in stepwise fashion during the last trimester of normal pregnancy.

The test for PG employs thin-layer chromatographic separation or a slide agglutination test using an antisera specific for PG. The advantage of PG measurement is that its value is not altered by the blood, meconium, vaginal secretions, or other contaminants, whereas the L/S is altered by the same contaminants. However, a disadvantage of PG determination in the assessment of fetal lung maturity is its late appearance (after 35 weeks of gestation) during pregnancy. Other amniotic fluid tests of lung maturity include measurement of lamellar bodies, either by measuring optical absorbance at 650 nm or by actual counting by procedures using standard hematological counters. An optical density of 0.15 or greater and a lamellar body count of 30,000–50,000/μL indicate pulmonary maturity. A fluorescent polarization technique which consists of competitive binding of a fluorescent probe to albumin and surfactant is employed in the assessment of fetal lung maturity. The net polarization for the albumin bound probe yields a high value, whereas a surfactant-bound probe yields a low value. A value of 55 mg of surfactant or greater per gram of albumin indicates maturity. In unanticipated premature births, the risk of RDS can be assessed by measurement of surfactant in gastric aspirates from the newborn, since the newborn swallows amniotic fluid *in utero*.

A number of factors (such as hypoxia and acidosis) depress phospholipid synthesis, and administration of glucocorticoids to mothers accelerates the rate of fetal lung maturation. The fetal lung undergoes an abrupt transition from a P_{O_2} of about 20 mm Hg to a P_{O_2} of 100 mm Hg. This change from a hypoxic to a relatively hyperoxic condition may lead to increased production of potentially cytotoxic O_2 metabolites such as superoxide radical ($O_2\cdot$), hydrogen peroxide (H_2O_2), hydroxyl radical (OH·), singlet oxygen ($^1O_2$), and peroxide radical (ROO·). The antioxidant enzyme system consists of superoxide dismutase, glutathione peroxidase, and catalase (see Chapter 14). In addition to these enzymes, other potential antioxidants are vitamin E, ascorbate, β-carotene, and thiol compounds (e.g., glutathione, cysteine). Infants born immaturely are particularly susceptible to deficiency of both surfactant and antioxidant defense. Administration of surfactant and the antioxidant enzymes using liposome technology has potential application in the management of RDS. Administration of surfactant to the lungs of very premature infants through an endotracheal tube has reduced morbidity and mortality from RDS.

Catabolism and Storage Disorders of Sphingolipids

There are four groups of glycosphingolipids: cerebrosides, sulfatides, globosides, and gangliosides. Cerebrosides contain a single sugar residue linked to ceramide, which is an N-acylsphingosine. Sulfatides contain a sulfate group attached to sugar residue. Globosides contain two or more sugar residues and an N-acetylgalactosamine group linked to ceramide. Gangliosides (G) contain oligosaccharide chains that contain sialic acid residues. They are classified based upon the number of sialic acid (N-acetylneuraminic acid, NANA) residues they contain and the sequence of sugar residues. G_M, G_D, G_T and G_Q contain gangliosides with one, two, three, and four sialic acid residues, respectively. The number associated with M, D, T, and Q signifies the sequence of sugar residues:

1. Represents Gal–NacGal–Gal–Glc–ceramide;
2. Represents NacGal–Gal–Glc–ceramide;
3. Represents Gal–Glc–ceramide. Thus, the structure of ganglioside G_{M1}, is:

```
Gal — NacGal — Gal — Glc — Ceramide
                |
              NANA
          (sialic acid)
```

Sphingolipids are in a continuous state of turnover. They are catabolized by lysosomal enzymes by stepwise removal of sugar residues beginning at the nonreducing end

TABLE 19-1

Characteristics of Glycosphingolipid Storage Disorders

Disorders	Major Lipids Accumulated	Other Compounds Affected	Enzyme or Activator Protein Lacking*	Remarks
G_{M2} gangliosidosis, type II (Tay-Sachs variant; Sandhoff's disease)	Globoside and G_{M2} ganglioside		Hexosaminidases A and B (① and ⑥) G_{M2}-activator	Same clinical picture as Tay-Sachs desease but progresses more rapidly; no racial predilection; hepatosplenomegaly, cardiomyopathy.
Fabry's disease (glycosphingolipid lipidosis)	Gal-(4←1α)-Lac-Cer	Gal-(4←1α)-Gal-Cer accumulates.	α-Galactosidase (②)	X-linked recessive; hemizygous males have a characteristic skin lesion usually lacking in heterozygous females; pain in the extremities; death usually in the fourth decade results from renal failure or cerebral or cardiovascular disease.
Ceramide lactoside lipidosis	Gal-(4←1β)-Glc-Cer		Ceramide lactoside β-galactosidase (③)	Liver and spleen enlargement; slowly progressive brain damage neurological impairment.
Gaucher's disease (glucosyl ceramide lipidosis; three types; see text)	Glc-Cer	G_{M3} ganglioside accumulates most frequently; other compounds occasionally.	β-Glucocerebrosidase (glucosylceramidase;④) Activator protein sap-C	Hepatosplenomegaly; frequently fatal; no known treatment; occurrence of Gaucher's cells (reticuloendothelial cells that contain accumulations of erythrocyte-derived glucocerebroside).
G_{M1} gangliosidosis (two types; see text)	G_{M1}- and desialo-G_{M1}-gangliosides	Keratan sulfate–related polysaccharide accumulates.	G_{M1}-β-galactosidase (⑤)	Mental and motor deterioration; accumulation of mucopolysaccharides is as significant as accumulation of gangliosides; invariably fatal; autosomal recessive; blindness, cherry red macula (30%); hepatosplenomegaly; vacuolated lymphocytes; startle response to sound, dysostosis multiplex.

G_{M2} gangliosidosis, type I (Tay-Sachs disease; see text)	G_{M2}- and desialo-G_{M2}-gangliosides	Other desialo hexosyl ceramides accumulate; other compunds occasionally.	Hexosaminidase A (⑥)	Red spot in retina; mental retardation; severe psychomotor retardation; seizures; blindness; startle response to sound; invariably fatal; autosomal recessive; panracial but especially prevalent among Northern European Jews.
Metachromatic leukodystrophy (MLD; sulfatide lipidoses; at least three types; see text)	3-sulfate-galacto-sylcerebroside	Cerebrosides other than sulfatides are decreased; ceramide dihexoside sulfate accumulates.	Sulfatidases (⑧); arylsulfatases) Activator protein sap-B	Demyelination; progressive paralysis and dementia; death usually occurs within the first decade; autosomal recessive inheritance.
Krabbe's disease (globoid cell leukodystrophy; galactosyl ceramide lipidosis)	Galactocerebroside	Salfatides are also greatly decreased, probably as a secondary feature.	Galactocerebroside-β-galactosidase (⑨)	Mental retardation; demyelination; psychomotor retardation; failure to thrive; progressive spasticity; globoid cells in brain white matter; invariably fatal; autosomal recessive inheritance.

*The circled numbers refer to reactions in Figure 19-8. The abbreviations are the same as in that figure.

of the molecule. Each sugar residue removed involves a specific exoglycosidase. Sulfatases are required for the removal of sulfate groups from sulfolipids. The degradation of sphingolipids, in addition to their requirement for specific hydrolases, is also dependent on nonenzymatic glycoproteins, known as ***sphingolipid activator proteins*** (SAPs). SAP-stimulated degradation of sphingolipids is thought to involve the binding of the activator protein with the sphingolipids so that the water-soluble hydrolases can access the specific sites of hydrolysis. Genes for SAPs are located in chromosomes 5 and 10. The SAP gene that resides in chromosome 5 codes for the activator of hexoseaminidase A, which hydrolyzes G_{M2}. The gene on chromosome 10 codes for a precursor which, after synthesis in the endoplasmic reticulum, is exported to the cell surface followed by its importation into the lysosomal compartment. In the lysosomes, the precursor protein is processed to yield four mature activator proteins: sap-A, sap-B, sap-C, and sap-D. The activator function of these proteins are as follows: sap-A stimulates glucosylceramidase and galactosylceramidase in the presence of detergents; sap-B is a nonspecific activator that stimulates hydrolysis of about 20 glycolipids as well as hydrolysis of sulfatide by arylsulfatidase A; sap-C is essential for the action of glucosylceramidase; and function of sap-D is unknown.

The importance of SAPs is exemplified in disorders where these activator proteins are not made as a result of mutations. A defect in the synthesis of the enzyme or its activator protein can both result in the same phenotype. Examples are hexoseaminidase A deficiency or its activator protein (Ganglioside G_{M2} activator) resulting in *Tay-Sachs disease*; arylsulfatase A deficiency or its activator protein sap-B resulting in *juvenile metachromatic leukodystrophy*; and glucosylceramidase deficiency or its activator protein sap-C resulting in *Gaucher's disease.* All of these disorders are accompanied by pronounced accumulation of the respective precursor lipids in the reticuloendothelial system. Sphingomyelin is hydrolyzed to ceramide and phosphorylcholine by sphingomyelinase:

$$\text{Sphingomyelin} + H_2O \rightarrow \text{phophorylcholine} + \text{ceramide}$$

Deficiency of sphingomyelinase leads to *Nieman-Pick disease* A and B in which sphingomyelin accumulates in reticuloendothelial cells, peripheral tissues, and central nervous system and affects all of these tissues and organs. *Nieman-Pick disease* C (and D) has normal tissue sphingomyelinase levels but exhibits defects in intracellular trafficking of exogenous cholesterol leading to lysosomal unesterified cholesterol accumulation. The C variant is characterized by hepatic damage and neurologic disease. Each of these disorders is inherited as an autosomal recessive trait.

Ceramide is hydrolyzed to sphingosine and fatty acid by ceramidase:

$$\text{Ceramide} + H_2O \rightarrow \text{sphingosine} + \text{fatty acid}$$

A nonlysosomal ceramidase in some tissues functions optimally at neutral or alkaline pH and participates in the synthesis and breakdown of ceramide. Deficiency of lysosomal (acid) ceramidase in *Farber's disease* (lipogranulomatosis) causes accumulation of ceramide. The disease is inherited as an autosomal recessive trait and is characterized by granulomatous lesions in the skin, joints, and larynx and moderate nervous system dysfunction; it may also involve heart, lungs, and lymph nodes. It is usually fatal during the first few years of life.

Sphingosine is catabolized to *trans*-2-hexadecanal and phosphoethanolamine by way of sphingosine phosphate and its cleavage by a lyase. Catabolism of glycosphingolipids involves removal of successive glycosyl residues from their nonreducing end until ceramide is released.

Abnormalities usually involve specific exoglycosidases and their activator proteins (discussed earlier) that hydrolyze the glycosidic bonds, except in metachromatic leukodystrophy, in which there is deficiency of a sulfatidase.

Catabolic pathways for the glycosphingolipids are given in Figure 19-8 and associated disorders are summarized in Table 19-1. Some comments are warranted:

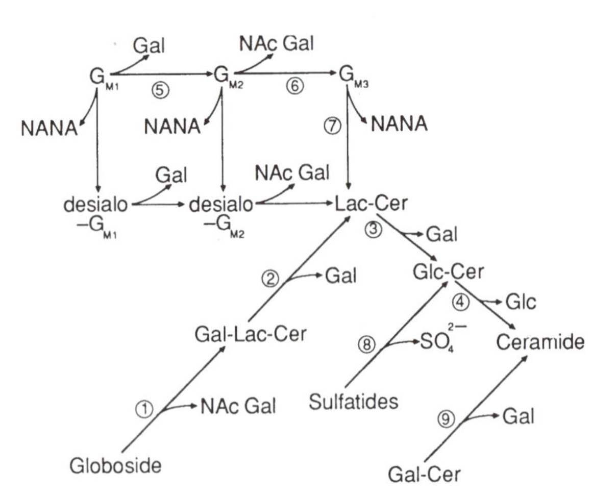

FIGURE 19-8

Degradative pathways for glycosphingolipids. $G_{M1,2,3}$ are gangliosides. Their structures and those of globosides and sulfatides should be inferred from their respective catabolic routes. The circled numbers correspond to the metabolic lesions listed in Table 19-1. Gal = Galactose; Glc = glucose; NAcGal = N-acetyl-galactose-2-amine; NANA = N-acetyl neuraminic acid (a sialic acid); Lac = lactose [galactosyl (β-1 $\rightarrow$ 4) glucose]; Cer = ceramide (N-acylsphingosine,); desialo = without a sialic acid (NANA) residue.

1. Accumulation of a specific lipid in these disorders is frequently accompanied by deposition of one or more polysaccharides structurally related to the lipid.
2. Treatment is generally palliative or nonexistent. Enzyme replacement therapy has proved useful in some of these disorders. Because the exogenous enzymes are unable to cross the blood-brain barrier, their efficacy in the glycosphingolipidoses that have neurological involvement is doubtful. Attempts to modify the enzymes to overcome this difficulty offer some hope.
3. Considerable progress has been made in the identification of carriers and in prenatal diagnosis of homozygotes. Thus, laboratory assays of enzyme activity in leukocytes or cultured skin cells using chromogenic or fluorogenic synthetic substrates have dramatically reduced the incidence of *Tay-Sachs disease.*

Gaucher's disease is the most common lysosomal storage disorder and also the most common inherited disease among Ashkenazi Jews, with a carrier frequency of about 1 in 14. Four mutations in the gene encoding for β-glucocerebrosidase account for at least 90% of the symptomatic patients. Gaucher's disease has three forms in which genetic effects appear to be due to errors in the same or related genetic loci. *Type I, chronic nonneuronopathic* (adult), is the most common variety. It comprises a heterogeneous group of patients characterized by the presence of hematological abnormalities (anemia, thrombocytopenia) and erosion of the cortices of long bones. *Type II, acute neuronopathic,* usually appears before 6 months of age and is fatal by 2 years. Mental damage is a primary characteristic, and the disease progresses rapidly from its onset. *Type III, subacute neuronopathic* (juvenile), comprises a heterogeneous group in which death occurs between infancy and about 30 years of age. The cerebral abnormalities usually appear at least 2 years postnatally.

All three types share common features: hepatosplenomegaly, Gauchers cells in the bone marrow (accumulation of glucocerebroside in reticuloendothelial cells in liver, spleen, and bone marrow), and autosomal recessive inheritance. Some studies have shown a correlation between the residual β-glucocerebrosidase activity and clinical severity, but the molecular basis for the genetic heterogeneity is not known. The Gaucher cells obtained from bone marrow aspirates exhibit a characteristic appearance of the cytoplasm owing to rod-shaped striated inclusion bodies composed primarily of glucocerebroside. Patients with Gaucher's disease have elevated levels of acid phosphatase activity in serum and spleen, increased iron stores, increased angiotensin converting enzyme activity, and a relative deficiency of clotting factor IX (Chapter 36).

Enzyme replacement therapy with purified macrophage-targeted human β-glucocerebrosidase in type I Gaucher's disease causes breakdown of stored glucocerebrosides. This results in a reduction in the size of the liver and spleen, in improvement in hematological abnormalities (anemia and thrombocytopenia), increased bone mineralization, and decreased bone pain. Two sources of human β-glucocerebrosidase are available; one of them is derived from human placenta (aglucerase) and the other is synthesized by recombinant DNA technology (imiglucerase). Both enzymes are modified in their oligosaccharide side chains to expose terminal mannose residues. Macrophages, through their mannose receptors, internalize the modified enzyme.

Studies of cases of *GM_1-gangliosidosis* have revealed two distinct types. In *generalized gangliosidosis,* GM_1 and desialo-GM_1-gangliosides accumulate in brain and viscera. The three β-galactosidase activities isolated from normal human liver all are absent. The disease begins at or near birth, progresses rapidly, and ends fatally, usually by 2 years of age. In *juvenile GM_1-gangliosidosis,* psychomotor abnormalities usually begin at about 1 year, and death ensues at 3–10 years. Two liver β-galactosidase activities are absent, possibly accounting for the lack of lipid accumulation in this organ. This enzymatic finding supports the genetic separation into two forms.

GM_2-gangliosidosis is of two types: Tay-Sachs disease, due to β-hexoseaminidase A (Hex-A) deficiency, and Sandhoff's disease, due to deficiency of β-hexoseaminidase A and B (Hex-A, Hex-B). The relationship between these diseases is based on the subunit composition of the two affected enzymes. Hex-A, a heteropolymer, consists of two α-chains (coded for on chromosome 15), a β-chain (coded for on chromosome 5), and an activator protein. Hex-B is a tetramer of β-chains. Mutations at the α-locus give rise to Tay-Sachs disease. A variant form can arise from mutation at the activator protein locus; however, it shows normal *in vitro* Hex-A activity with chromogenic substrates. Mutations at the β-locus yield *Sandhoff's disease* and affect Hex-A and Hex-B, both of which contain the β-subunit.

Treatment of sphingolipidoses is primarily symptomatic and supportive. For example, in patients with anemia due to Gaucher's disease, thrombocytopenia associated with hypersplenism is relieved by splenectomy. Infusion of appropriate purified human placental tissue enzymes in patients with Gaucher's disease and Fabry's disease reduced the accumulated glycolipids in the circulation and liver. Recent advances in the cloning and amplification of human DNA segments in bacterial plasmids and subsequent isolation of the gene product has yielded enough enzyme required for treatment (Chapter 23). Use of the recipient's erythrocytes in which the enzyme is entrapped is under investigation to minimize immunological complications. Exposing erythrocytes to hypotonic conditions in the presence of the enzyme causes formation of pores in the membrane that allow rapid exchange of the enzyme with the cellular contents. Restoration to isotonicity reseals the membrane and entraps some of the enzyme. Other enzyme carriers are liposomes, concentric lipid bilayers prepared from cholesterol, lecithin, and phosphatidic acid. The ideal treatment for these disorders would be addition, or replacement, of genetic material coding for the missing gene product. Replacement therapy with a polyethylene glycol–modified form of the missing enzyme, which has an extended half-life and reduced immunogenicity, may provide a promising approach to treatment (see *Adenosine deaminase deficiency,* Chapter 27).

Alterations in Cell Surface Glycosphingolipids

Changes in cell surface glycosphingolipids occur during fetal development and are reflected in several properties of the cell, namely, receptor specificity, antigenic specificity, adhesion, and possibly cell growth regulation. Glycosphingolipid metabolism is affected during oncogenic transformation in cultured cells and may be responsible for some properties of tumor cells (e.g., lack of contact inhibition of growth).

19.3 Cholesterol

Cholesterol (3-hydroxy-5,6-cholestene) is a steroid and contains the carbon skeleton of cyclopentanoperhydrophenanthrene, which consists of three six-membered rings and a five-membered ring. It is also a monohydroxyalcohol and contains a double bond between C_5 and C_6:

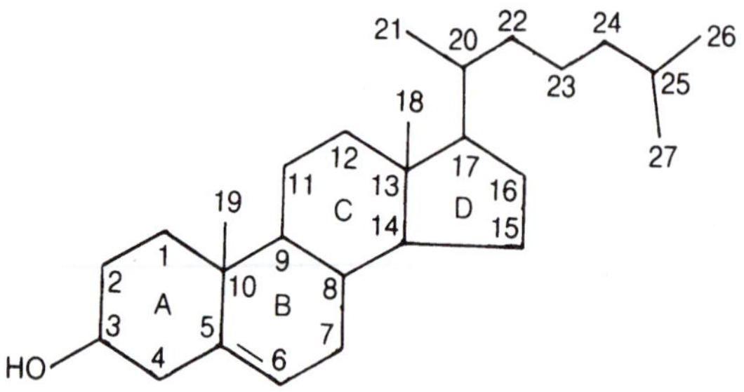

Cholestanol (dihydrocholesterol) has the following conformation:

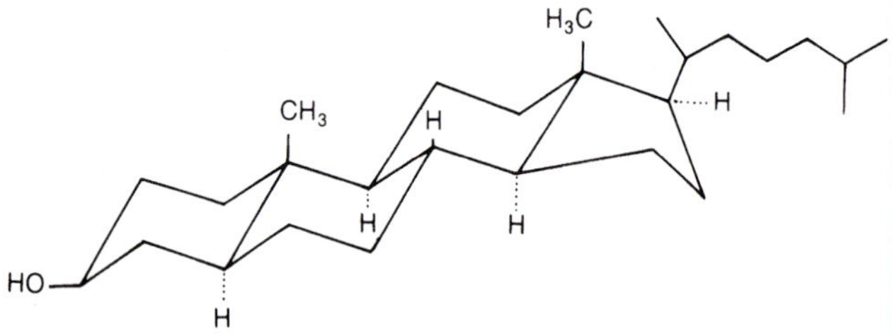

In this structure, all of the ring fusions (between A and B, B and C, and C and D) are trans, the hydrogen atoms or methyl groups attached to the bridgehead carbons project to opposite sides of the rings, and the rings are in the more stable chair conformation. In cholesterol, the double bond between C_5 and C_6 distorts the conformation of the rings A and B and leads to the conformation shown below:

The angular methyl groups at C_{18} and C_{19}, the 3-hydroxyl group, and the side chain at C_{17} all project toward the same side of the ring system. These substituents are indicated by solid lines and designated as β. A substituent group situated below the plane of the ring is in the α-orientation and is indicated with a dotted line. In general, the angular methyl groups are β-oriented, but the 3-hydroxyl group may be present in either the α- and β-orientation. In cholesterol, the 3-OH is in the β-orientation.

In some naturally occurring compounds (e.g., β-coprostanol), the junction between rings A and B is cis. This compound occurs in large quantities in feces, where it is produced from cholesterol by action of the intestinal flora.

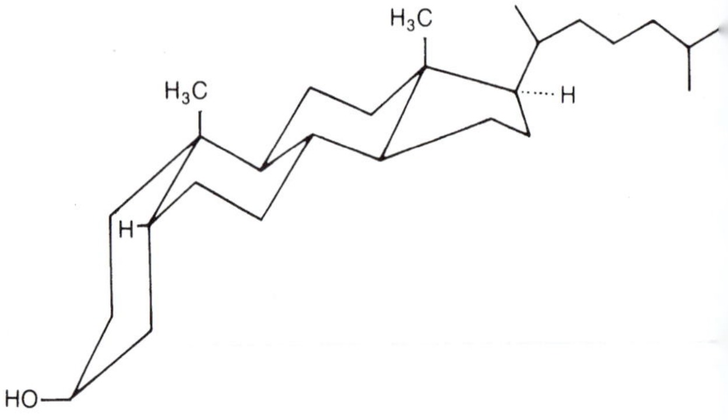

Adults normally synthesize approximately 1 g of cholesterol and consume about 0.3 g/day. Dietary cholesterol is primarily derived from foods of animal origin such as eggs and meat. Plants, yeasts, and fungi contain sterols that are structurally similar to cholesterol—sitosterols and ergosterols—but are poorly absorbed by the human intestinal tract. A rare inherited autosomal sterol storage disorder (*sitosterolemia*) is due to defects in the ATP-binding cassette-family of transporters that mediate cholesterol efflux. Treatment consists of diets low in plant sterol content and with cholestyramine to enhance sterol excretion (Chapter 20). In intestinal mucosal cells, most of the absorbed cholesterol is esterified with fatty acids and incorporated into chylomicrons that enter the blood through the lymph. After chylomicrons unload most of their triacylglycerol content at the peripheral tissues, chylomicron remnants are rapidly taken up by the liver (Chapter 20). The routing of nearly all of the cholesterol derived from dietary sources to the liver facilitates the balance of the steroid content in the organism, since the liver is the principal site of cholesterol production. Although the intestinal tract, adrenal cortex, testes, skin, and other tissues can also synthesize cholesterol, their contribution is a minor one.

Cholesterol biosynthesis proceeds via the isoprenoids in a multistep pathway. The end product, cholesterol, and the intermediates of the pathway participate in diverse cellular functions. The isoprenoid units give rise to dolichol, CoQ, heme A, isopentenyl-tRNA, farnesylated proteins, and vitamin D (in the presence of sunlight and 7-dehydrocholesterol). Dolichol is used in the synthesis of glycoproteins, CoQ in the mitochondrial electron transport chain, farnesylation and geranylgeranylation by posttranslational lipid modification that is required for membrane association and function of proteins such as $p21^{ras}$ and G-protein subunits.

Cholesterol has several functions including involvement in membrane structure, by modulation of membrane fluidity and permeability, serving as a precursor for steroid hormone and bile acid synthesis, in the covalent modification of proteins, and formation of the central nervous system in embryonic development. The latter role of cholesterol was discovered through mutations and pharmacological agents that block cholesterol biosynthesis that occurs in six steps:

1. Conversion of acetyl-CoA to 3-hydroxy-3-methylglutaryl coenzyme-A (HMG-CoA);
2. Conversion of HMG-CoA to mevalonate, the rate-limiting step in cholesterol biosynthesis;
3. Conversion of mevalonate to isoprenyl pyrophosphates with loss of CO_2;
4. Conversion of isoprenyl pyrophosphates to squalene;
5. Conversion of squalene to lanosterol; and
6. Conversion of lanosterol to cholesterol.

The biosynthetic reactions involve a series of condensation processes and are distributed between cytosol and microsomes. All of the carbons of cholesterol are derived from acetyl-CoA, 15 from the "methyl" and 12 from the "carboxyl" carbon atoms. Acetyl-CoA is derived from mitochondrial oxidation of metabolic fuels (e.g., fatty acids) and transported to cytosol as citrate (Chapter 18) or by activation of acetate (e.g., derived from ethanol oxidation) by cytosolic acetyl-CoA synthase (Chapter 18). All of the reducing equivalents are provided by NADPH.

Conversion of Acetyl-CoA to HMG-CoA

In the cytosol, three molecules of acetyl-CoA are condensed to HMG-CoA through successive action of thiolase and HMG-CoA synthase, respectively (Figure 19-9). HMG-CoA synthase is under transcriptional regulation by the sterol end products.

HMG-CoA is also synthesized in mitochondria by the same sequence of reactions but yields the ketone bodies acetoacetate, D(−)-β-hydroxybutyrate, and acetone (Figure 19-10). Mitochondrial HMG-CoA also arises from oxidation of leucine (Chapter 17), which is ketogenic. Although HMG-CoA derived from leucine is not utilized in mevalonate synthesis, the carbon of leucine can be incorporated into cholesterol by way of acetyl-CoA. Thus, two distinct pools of HMG-CoA exist: one

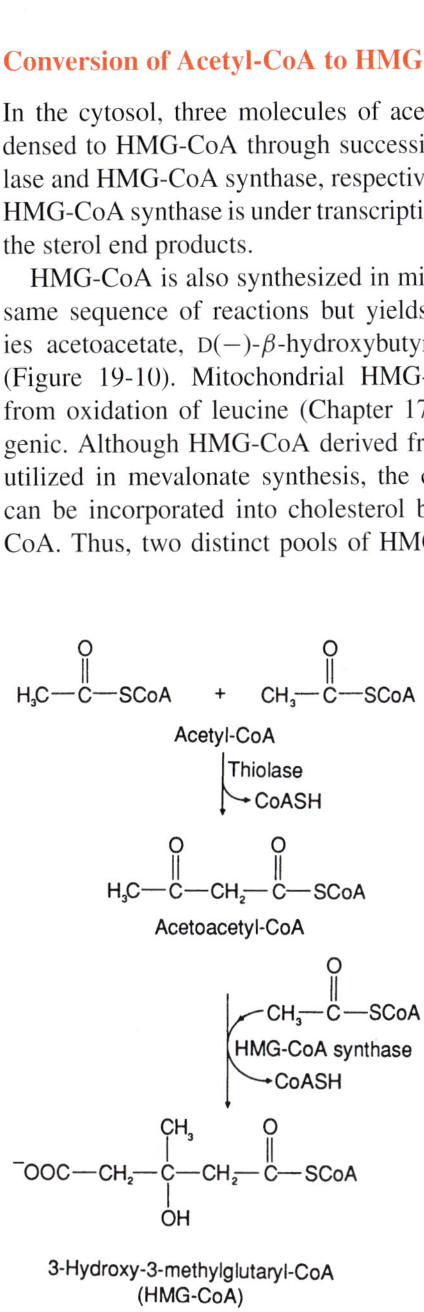

$H_3C-C(=O)-SCoA + CH_3-C(=O)-SCoA$

Acetyl-CoA

Thiolase → CoASH

$H_3C-C(=O)-CH_2-C(=O)-SCoA$

Acetoacetyl-CoA

$CH_3-C(=O)-SCoA$

HMG-CoA synthase → CoASH

$^{-}OOC-CH_2-C(CH_3)(OH)-CH_2-C(=O)-SCoA$

3-Hydroxy-3-methylglutaryl-CoA (HMG-CoA)

FIGURE 19-9
Biosynthesis of HMG-CoA.

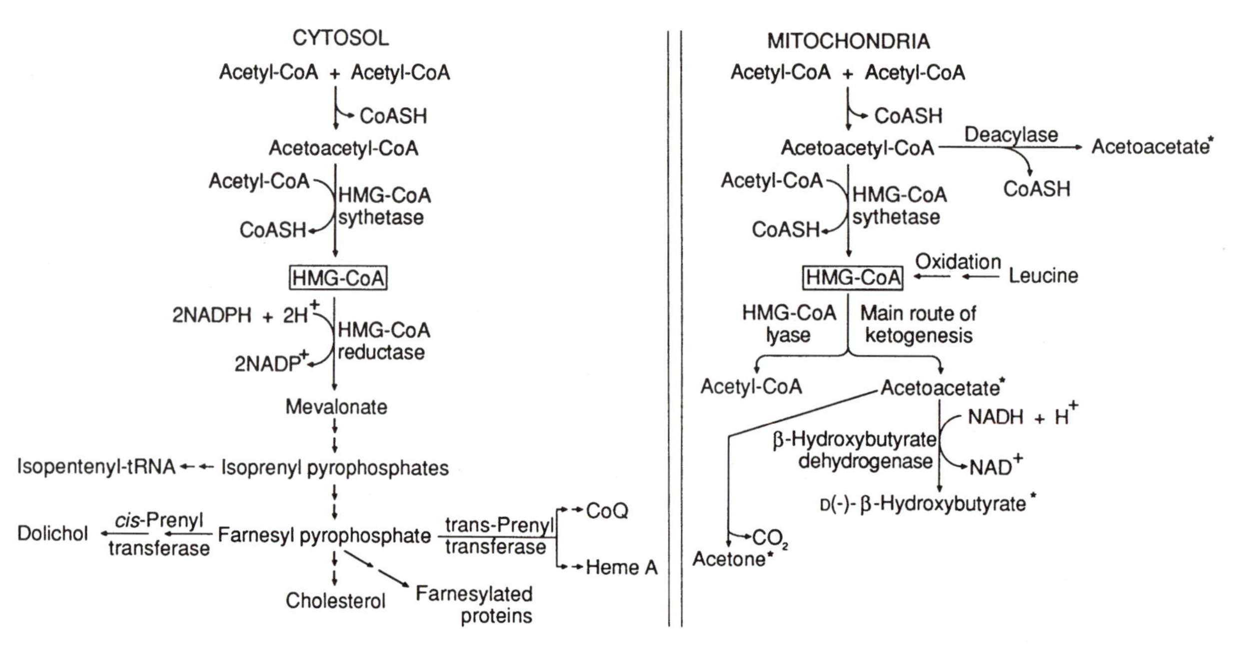

FIGURE 19-10
Mitochondrial and cytosolic biosynthesis and utilization of HMG-CoA in the liver. The molecules indicated by an asterisk are the ketone bodies. Acetoacetate and β-hydroxybutyrate (after conversion to acetoacetate) are metabolized in extrahepatic tissues. Acetone is excreted in the lungs. Note the cytosolic multifunctional isoprenoid pathway for cholesterol biosynthesis. The double arrow indicates a multistep pathway.

mitochondrial and concerned with formation of ketone bodies, the other extramitochondrial and involved with synthesis of isoprenoid units.

Conversion of HMG-CoA to Mevalonate

This two-step reduction reaction is the rate-limiting step in cholesterogenesis. Cytosolic HMG-CoA is reduced by NADPH to mevalonate by HMG-CoA reductase through the production of an enzyme-bound aldehyde intermediate:

$$^{-}O-\overset{\overset{O}{\|}}{C}-CH_2-\underset{\underset{OH}{|}}{\overset{\overset{CH_3}{|}}{C}}-CH_2\overset{\overset{O}{\|}}{C}-SCoA + 2NADPH + 2H^{+}$$

HMG—CoA

$$\downarrow \text{HMG-CoA reductase}$$

$$^{-}O-\overset{\overset{O}{\|}}{C}-CH_2-\underset{\underset{OH}{|}}{\overset{\overset{CH_3}{|}}{C}}-CH_2-CH_2OH + CoASH + 2NADP^{+}$$

Mevalonate

HMG-CoA reductase is an integral protein of the endoplasmic reticulum and the primary site of regulation of synthesis of cholesterol and nonsterol isoprenoid derivatives. Its activity has a well-defined diurnal rhythm in rats and mice, coinciding with that of the enzyme's synthesis and of the mRNA concentration. Activity is highest at about the middle of the dark period and lowest at about the middle of the light period. Its mechanism may be related to food consumption. Rats are nocturnal animals and consume food in the dark; the increased bile production and excretion depletes liver cholesterol and may stimulate the increased synthesis of HMG-CoA reductase as a compensatory mechanism.

HMG-CoA reductase is regulated via synthesis-degradation and phosphorylation-dephosphorylation. Phosphorylation decreases activity, whereas dephosphorylation increases it (Figure 19-11). The reductase kinase phosphorylation is not cAMP-dependent. However, cAMP dependence arises by way of activation of a protein kinase, which phosphorylates a protein inhibitor of phosphatase. The two phosphatases are identical. Thus, increase in cAMP concentration inhibits phosphatase activity, resulting in marked decrease in HMG-CoA reductase activity. Elevation of the plasma glucagon level (e.g., during fasting) activates cAMP production and reduces cholesterol production. Activity is also inhibited by oxygenated sterols (e.g., 27-hydroxycholesterol) but

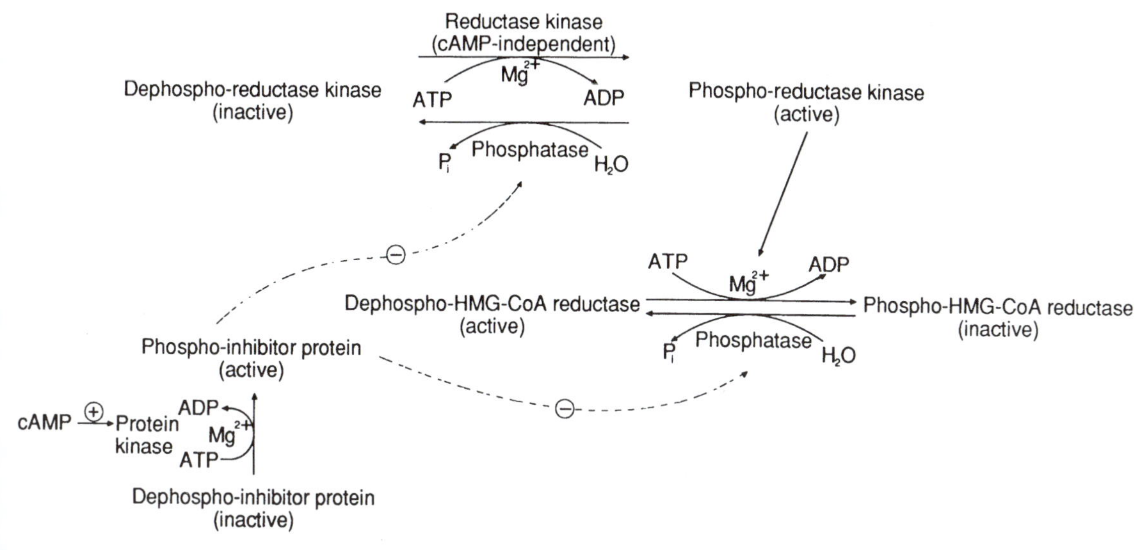

FIGURE 19-11
Regulation of HMG-CoA reductase by phosphorylation-dephosphorylation; phosphorylation leads to loss of catalytic activity. The two phosphatases are identical and are inhibited by an inhibitor protein that is more active when phosphorylated by a cAMP-dependent protein kinase. Phosphorylation of reductase kinase is not cAMP-dependent.

not in enucleated cells, indicating involvement of the nucleus. The oxygenated sterols are synthesized in mitochondria and may repress the HMG-CoA reductase gene or activate genes for enzymes that degrade the reductase. A rare familial sterol storage disease, *cerebrotendinous xanthomatosis,* is characterized by accumulation of cholesterol (and its reduced product cholestanol) in every tissue, especially in brain, tendons, and aorta, causing progressive neurological dysfunction, tendon xanthomas, premature atherosclerosis, and myocardial infarction.

Patients also develop cholesterol gallstones from a defect in bile acid synthesis. The defect is in the mitochondrial C_{27}-steroid 27-hydroxylase. In these patients, the reduced formation of normal bile acids, particularly chenodeoxycholic acid, leads to the up-regulation of the rate limiting enzyme 7α-hydroxylase of the bile acid synthetic pathway (discussed later). This leads to accumulation of 7α-hydroxylated bile acid intermediates that are not normally utilized.

The inhibition of cholesterol synthesis by oxygenated sterols involves the following steps. After synthesis in the mitochondria, oxygenated sterols in the cytoplasm inhibit the activation of *sterol regulatory element binding proteins* (SREBPs), eventually leading to suppression of cholesterol biosynthesis. In cholesterol depleted states, activation of SREBPs requires participation of the SREBP-cleavage activating protein and two proteases. The mature SREBPs are translocated to the nucleus. In the nucleus, SREBPs function as transcription factors and activate, along with other factors, several genes by interacting at promoter sites consisting of a 10-base pair *cis* element known as sterol regulatory element-1 (SRE-1). Examples of activated genes include HMG-CoA synthase, HMG-CoA reductase, and low-density lipoprotein (LDL) receptors. The latter internalizes LDL to provide cholesterol to cells.

HMG-CoA reductase is inhibited competitively by structural analogues. These compounds are commonly known as "statins" and are used pharmacologically in cholesterol reduction which can reduce the risk for *coronary artery disease* and stroke (Chapter 20). Statins inhibit HMG-CoA reductase at a much lower concentration (1 μM) compared to the K_m for HMG-CoA (10 μM). The structures of clinically effective statins are shown in Figure 19-12. Lovastatin, simvastatin, and pravastatin are derivatives of naturally occurring fungal products and fluvastatin, atorvastatin and cerivastatin are entirely synthetic compounds. Lovastatin and simvastatin are inactive lactones that are activated by the liver; others are active hydroxy-acids. Naturally occurring statins are found in a dietary supplement known as cholestin, which is obtained from rice fermented in red yeast. In China red yeast rice has been used as a coloring and flavoring agent. Cholestin's safety and effectiveness as hypocholesterolemic agent awaits long-term clinical studies.

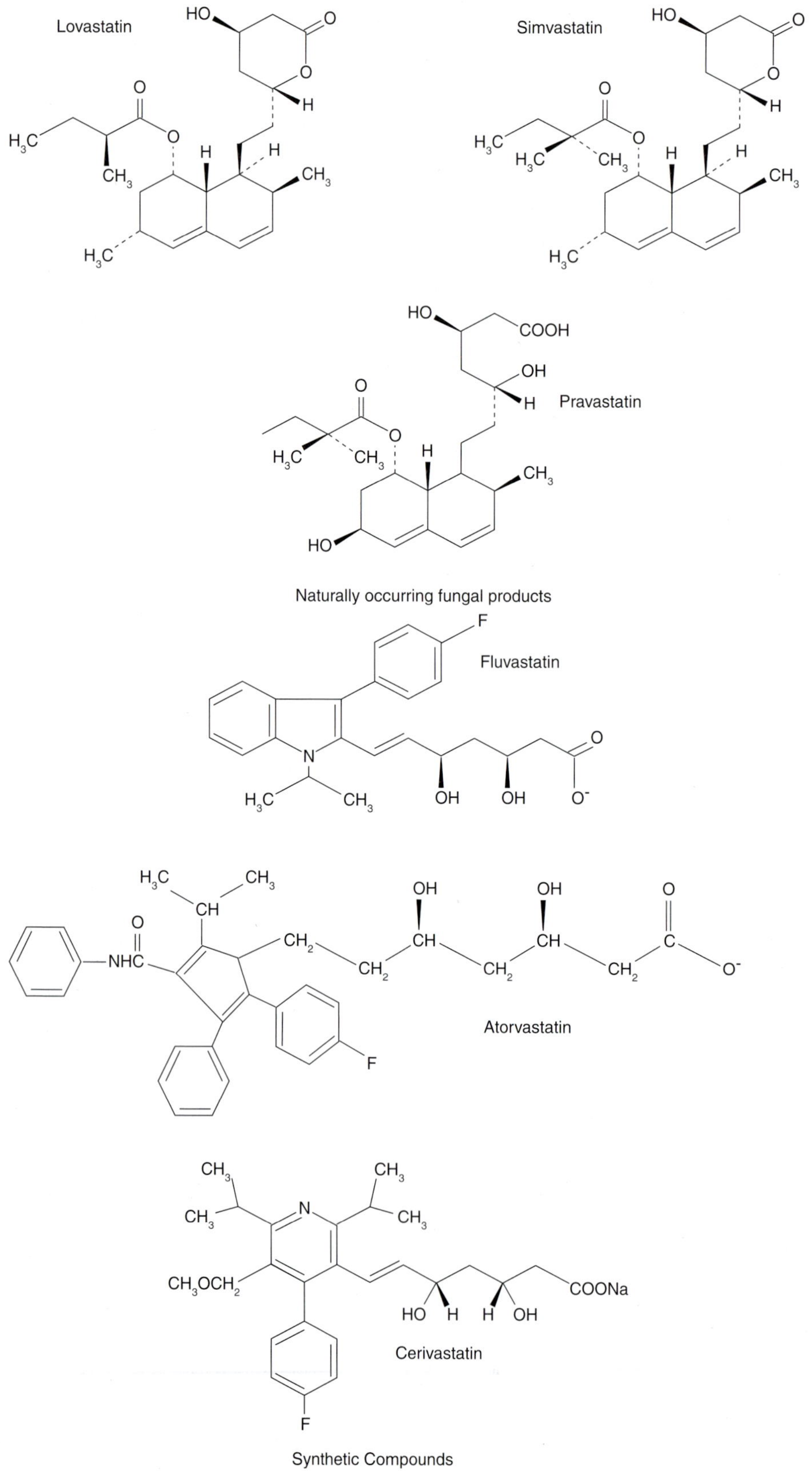

Lovastatin

Simvastatin

Pravastatin

Naturally occurring fungal products

Fluvastatin

Atorvastatin

Cerivastatin

Synthetic Compounds

FIGURE 19-12
Structures of HMG-CoA reductase inhibitors (statins).

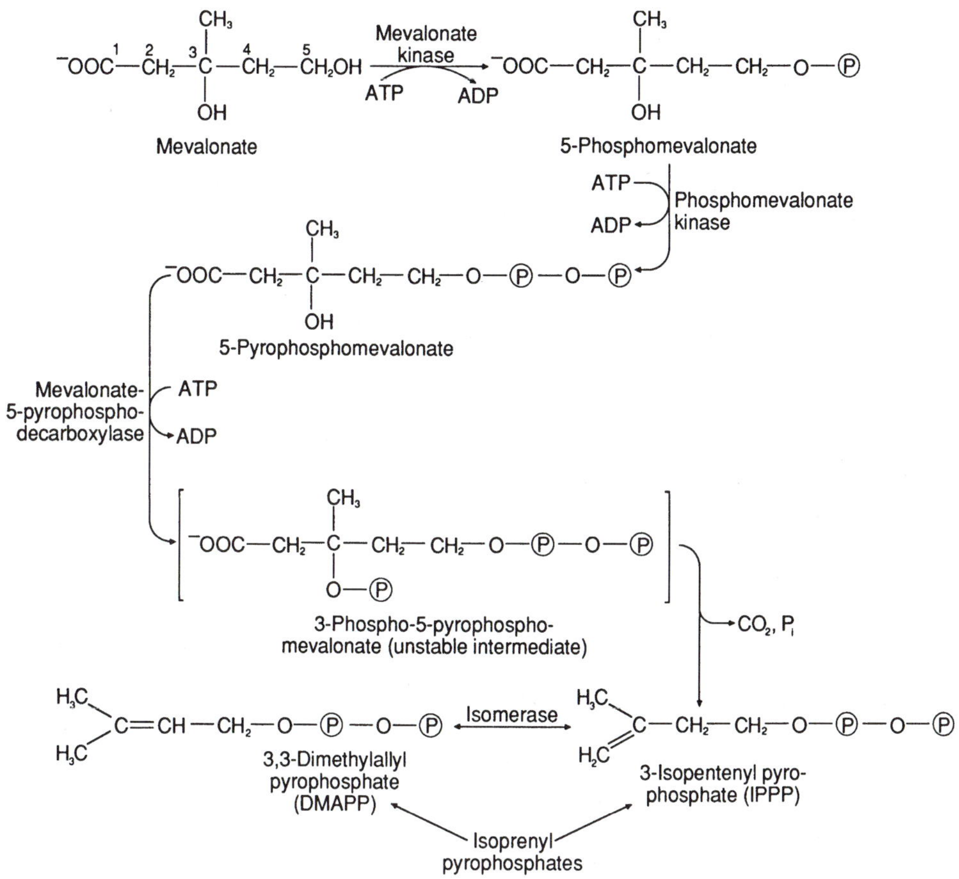

Mevalonate kinase; ATP → ADP

Mevalonate → 5-Phosphomevalonate

Phosphomevalonate kinase; ATP → ADP

5-Pyrophosphomevalonate

Mevalonate-5-pyrophosphodecarboxylase; ATP → ADP

3-Phospho-5-pyrophosphomevalonate (unstable intermediate)

CO_2, P_i

3,3-Dimethylallyl pyrophosphate (DMAPP) ⇌ Isomerase ⇌ 3-Isopentenyl pyrophosphate (IPPP)

Isoprenyl pyrophosphates

FIGURE 19-13
Synthesis of isoprenoid units from mevalonate.

Despite inhibition of HMG-CoA reductase by statins, ells compensate by increasing enzyme expression several old. However, the total body cholesterol is reduced by 0–40% due to increased expression of LDL-receptors fter statin administration; this enhances LDL (the maor cholesterol carrying lipoprotein) clearance from serum vith a net reduction of serum cholesterol (Chapter 20). ndividuals who lack functional LDL-receptors (homozyous familial hypercholesterolemia, Chapter 20) do not enefit from statin therapy. However, statin therapy is seful in the treatment of heterozygous familial hyercholesterolemia. Since HMG-CoA reductase plays pivotal role in the synthesis of many products vial for cellular metabolism, inhibitors of the enzyme nay have toxic effects. Monitoring of liver and musle function may be necessary to detect any toxicity f statin drug therapy. A decreased risk of bone fracures with statin therapy has been observed in subjects ge 50 years or older, who are being treated for hyperholesterolemia. The mechanism of action of statins in one metabolism may involve inhibition of prenylation of signaling proteins of osteoclast cell membrane (Chapter 37).

Conversion of Mevalonate to Isoprenyl Pyrophosphate

Isoprenyl pyrophosphates are synthesized by successive phosphorylation of mevalonate with ATP to yield the 5-monophosphate, 5-pyrophosphate, and 5-pyrophospho-3-monophospho derivatives. This last compound is very unstable and loses the 3-phosphate and the carboxyl group to yield isopentenyl pyrophosphate (IPPP), which is isomerized to 3,3-dimethylallyl pyrophosphate (DMAPP). These reactions, catalyzed by cytosolic enzymes, are shown in Figure 19-13.

Patients with severe forms of inherited mevalonate kinase deficiencies exhibit mevalonic aciduria, failure to thrive, developmental delay, anemia, hepatosplenomegaly, gastroenteropathy, and dysmorphic features during neonatal development. These clinical manifestations underscore the importance of the formation of isoprenyl pyrophosphates not only for sterol synthesis but also for the nonsterol isoprene compounds dolichol, CoQ, heme A, isopentenyl-tRNA and farnesylated proteins.

Condensation of Isoprenyl Pyrophosphate to Form Squalene

IPPP, a nucleophile (by virtue of its terminal vinyl group), and DMAPP, an electrophile, undergo condensation with elimination of pyrophosphate to yield geranyl pyrophosphate (an electrophile), which condenses with a molecule of IPPP to yield a farnesyl pyrophosphate and pyrophosphate. These reactions are probably catalyzed by the same cytosolic enzyme complex. Two molecules of farnesyl pyrophosphate and then condense head-to-head to form squalene by action of microsomal squalene synthase (Figure 19-14).

The farnesyl pyrophosphate generated in this pathway is also used in the farnesylation of proteins. The farnesyl group is attached to a protein via the thioether linkage involving a cysteine residue found in the C terminus. Several proteins that are modified by farnesyl groups have been identified, e.g., growth-regulating *ras* proteins (Chapter 26) and nuclear envelope proteins. Proteins attached to a geranyl-geranyl group (a 20-C isoprene unit) have also been identified. The modification of proteins by these lipid moieties increases their hydrophobicity and may be required for these proteins to interact with other hydrophobic proteins and for proper anchoring in the cell membrane. The importance of farnesylation of proteins is exemplified by blockage of cell growth when mevalonate synthesis is inhibited.

Conversion of Squalene to Lanosterol

This step comprises cyclization of squalene to lanosterol (the first sterol to be formed) and conversion of lanosterol to cholesterol. The cyclization begins with conversion of squalene to squalene-2,3-epoxide by a microsomal mixed-function oxidase that requires O_2, NADPH, and FAD (Figure 19-15). Cyclization of squalene-2,3-epoxide to lanosterol occurs by a series of concerted 1,2-methyl group

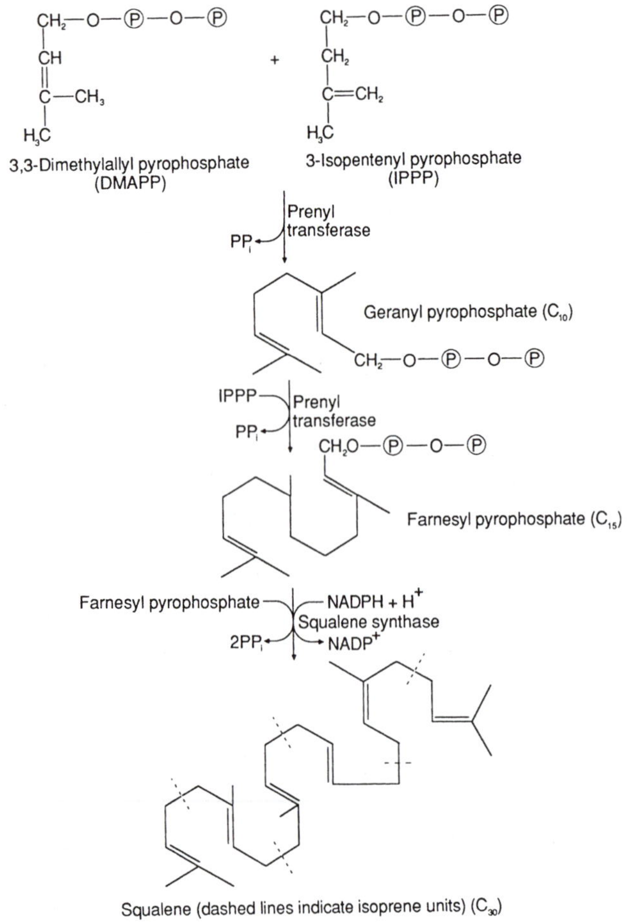

FIGURE 19-14
Synthesis of squalene from isomeric pentenyl pyrophosphates.

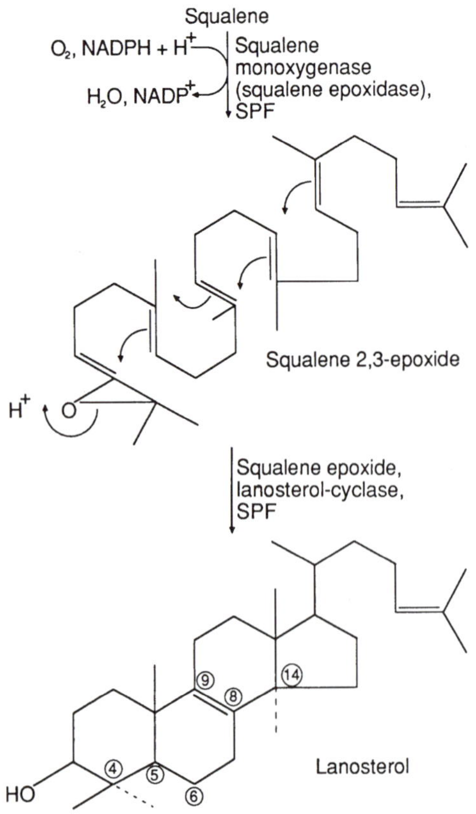

FIGURE 19-15
Cyclization of squalene to lanosterol. Supernatant protein factor (SPF), a cytosolic protein, promotes both stages of the cyclization.

and hydride shifts along the squalene chain. In both stages, the reactants are bound to supernatant protein factor (SPF), a cytosolic carrier that promotes conversion of squalene to lanosterol.

Conversion of Lanosterol to Cholesterol

Transformation of lanosterol to cholesterol (Figure 19-16) is a complex, multistep process catalyzed by enzymes of the endoplasmic reticulum (microsomes). A cytosolic sterol carrier protein is also required and presumably functions as a carrier of steroid intermediates from one catalytic site to the next but may also affect activity of the enzymes. The reactions consist of removal of the three methyl groups attached to C_4 and C_{14}, migration of the double bond from the 8,9- to the 5,6-position, and saturation of the double bond in the side chain. Conversion of lanosterol to cholesterol occurs principally via 7-dehydrocholesterol and to a minor extent via desmosterol.

The importance of cholesterol biosynthesis in embryonic development and formation of the central nervous system is reflected in patients with disorders in the pathway for the conversion of lanosterol to cholesterol. Three enzyme deficiencies have been identified (Figure 19-16):

1. 3β-Hydroxysteroid-Δ^{24}-reductase (also known as sterol-Δ^{24}-reductase);
2. 3β-Hydroxysteroid-Δ^{8},Δ^{7}-isomerase (commonly known as sterol-Δ^{8}-isomerase);
3. 3β-Hydroxysteroid-Δ^{7}-reductase (also known as 7-dehydrocholesterol reductase).

Sterol-Δ^{8}-isomerase deficiency, known as *Conradi–Hünermann syndrome* ($CDPX_2$), is an X-linked dominant disorder. Clinical manifestations of this disorder include skeletal abnormalities, chondrodysplasia punctata, craniofacial anomalies, cataracts, and skin abnormalities. The 7-dehydrocholesterol reductase deficiency, known as *Smith–Lemli–Opitz syndrome (SLO)* is an autosomal recessive disorder occurring in about 1 in 20,000 births. Clinical manifestations of affected individuals include craniofacial abnormalities, microcephaly, congenital heart disease, malformation of the limbs, psychomotor retardation, cerebral maldevelopment, and urogenital anomalies. Measurement of 7-dehydrocholesterol in amniotic fluid during second trimester or in neonatal blood specimen has been useful in the identification of the disorder. The sterol-Δ^{24}-reductase deficiency causes a developmental phenotype similar to SLO syndrome and is associated with accumulation of desmosterol. The inability of *de novo* fetal synthesis of cholesterol combined with its inadequate transport from the mother to the fetus appears to be involved in the multiple abnormalities of morphogenesis. SLO infants treated with rich sources of dietary cholesterol after birth have shown fewer growth abnormalities. However, it is not known whether long-term dietary cholesterol supplement can improve cognitive development, particularly since cholesterol is not transported across the blood-brain barrier.

An appreciation of the relationship between cellular cholesterol metabolism and a family of signaling molecules that participate in embryonic development is emerging. These signaling molecules are known as *hedgehog proteins* which were initially identified in *Drosophila.* The vertebrate counterparts of hedgehog proteins, participate in embryonic processing, including the neural tube and its derivatives, the axial skeleton, and the appendages. The hedgehog protein is a self-splicing protein that undergoes an autocatalytic proteolytic processing giving rise to an N-terminal and a C-terminal product. In *Drosophila,* hedgehog protein cleavage occurs between Gly-257 and Cys-258. Cholesterol is covalently attached to the carboxy terminal end of the N-terminal cleavage product. Both the autocatalytic proteolysis and intramolecular cholesterol transferase activities are located in the C-terminal portion of the hedgehog protein. The covalent modification of the N-terminal segment of the hedgehog protein is necessary for proper localization on the cell membrane at target sites to initiate downstream events (e.g., transcription of target genes). Thus, perturbations of cholesterol biosynthesis due to mutations or pharmacological agents can lead to defects in embryonic development.

Utilization of Cholesterol

Cholesterol is utilized in formation of membranes (Chapter 10), steroid hormones (Chapters 30, 32, and 34), and bile acids. 7-Dehydrocholesterol is required for production of vitamin D (Chapter 37). Under steady-state conditions, the cholesterol content of the body is maintained relatively constant by balancing synthesis and dietary intake with utilization. The major consumer of cholesterol is formation of bile acids, of which about 0.8–1 g/day are produced in the liver and lost in the feces. However, secretion of bile acids by the liver is many times greater (15–20 g/day) than the rate of synthesis because of their enterohepatic circulation (Chapter 12). Cholesterol is also secreted into bile, and some is lost in feces as cholesterol and as coprostanol, a bacterial reduction product (about 0.4–0.5 g/day). Conversion of cholesterol to steroid hormones and of 7-dehydrocholesterol to vitamin D and elimination of their inactive metabolites, are of minor significance in the disposition of cholesterol, amounting to approximately 50 mg/day. A small amount of cholesterol

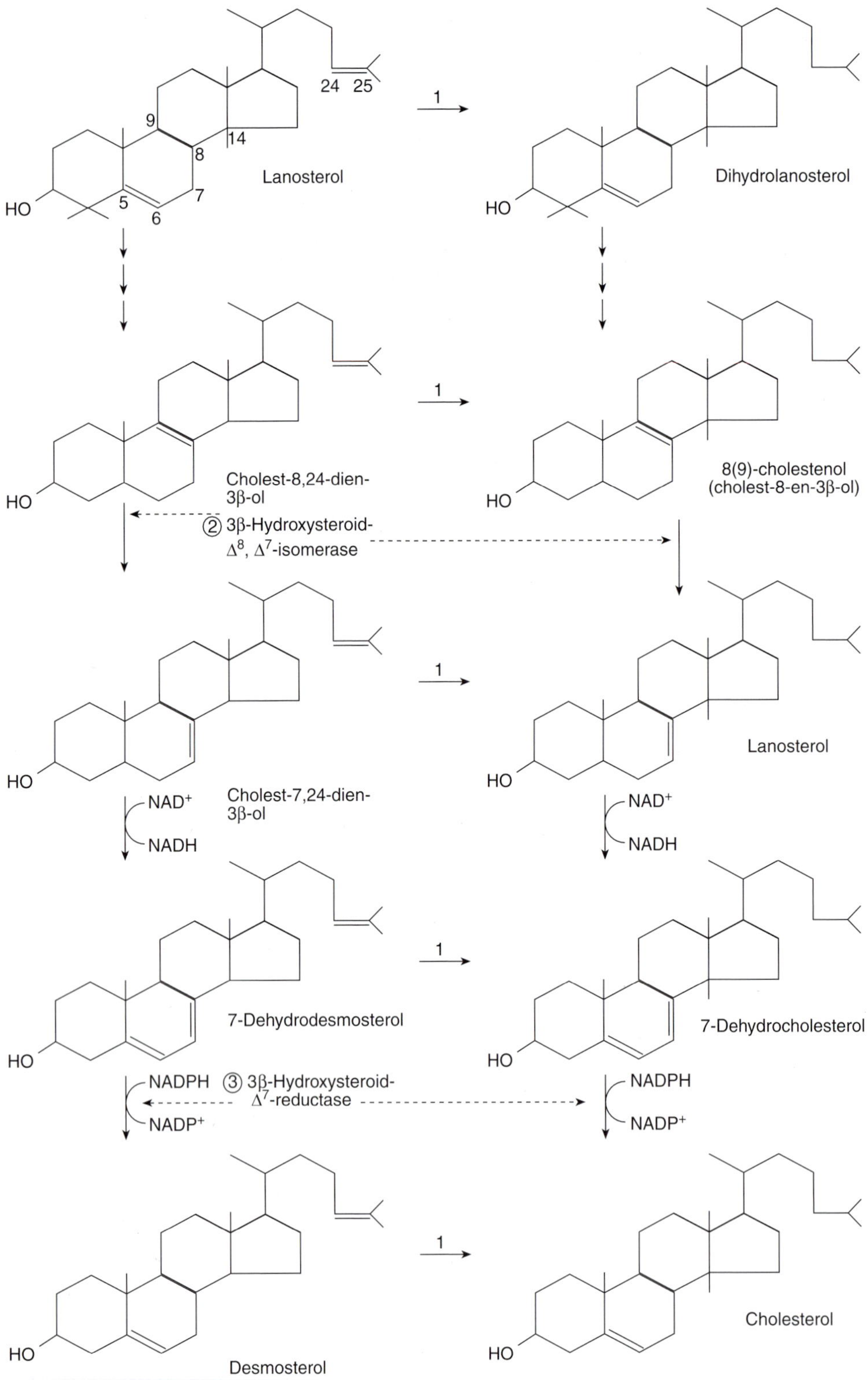

FIGURE 19-16
A partial pathway for the conversion of lanosterol to cholesterol. The complete process consists of 19 steps. The $C_{24} = C_{25}$ double bond can be reduced by 3β-hydroxylsteroid-Δ^{24}-reductase at several steps along the pathway (indicated by 1), and deficiency of this enzyme leads to accumulation of desmosterol. Deficiency of enzyme 2 and enzyme 3 results in CDPX2 and SLO syndromes, respectively (see text).

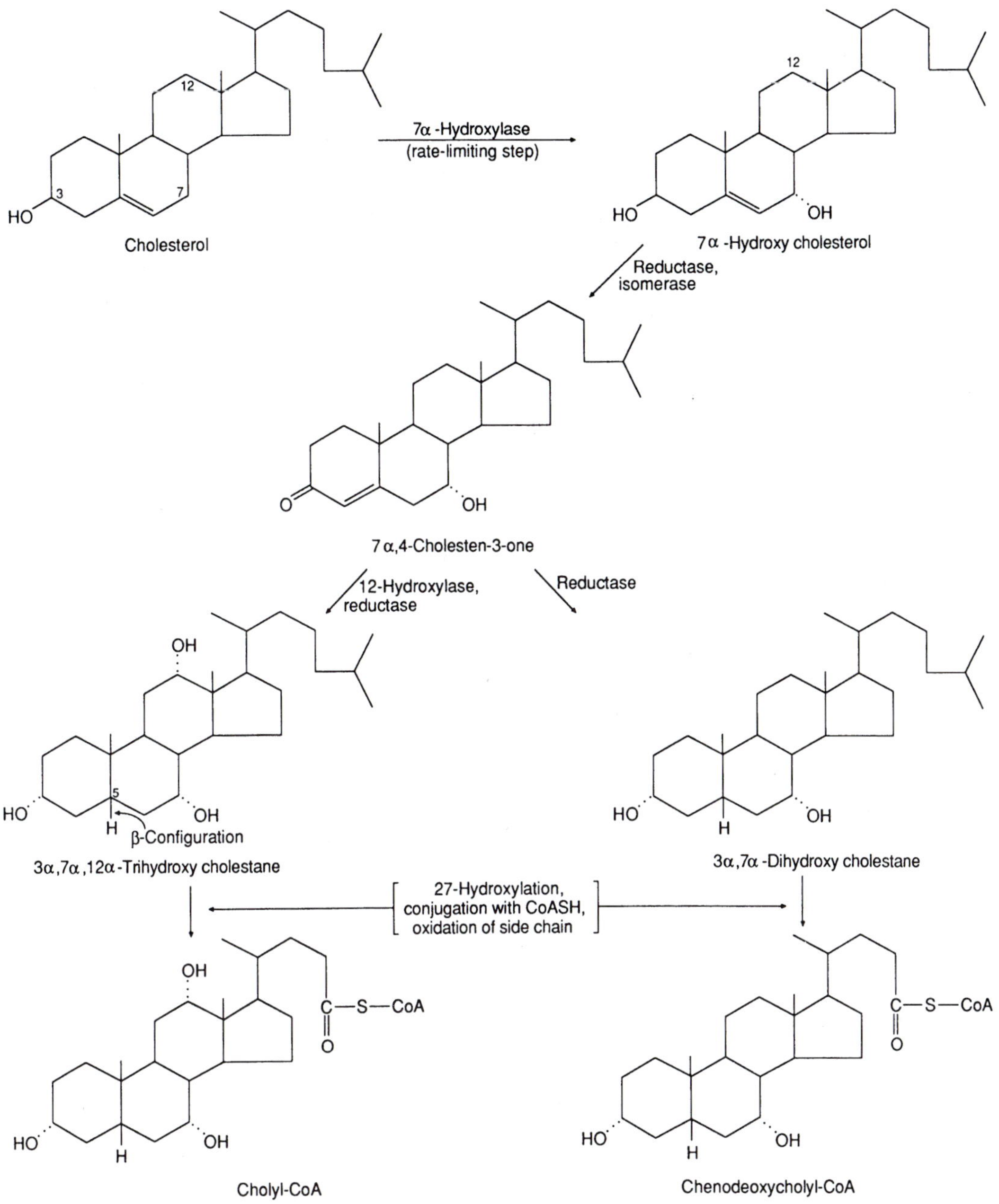

FIGURE 19-17
Formation of bile acids.

(about 80 mg/day) is also lost through shedding of the outer layers of the skin.

19.4 Bile Acids

Bile acids are 24-carbon steroid compounds. Primary bile acids (cholic and chenodeoxycholic) are synthesized in the liver from cholesterol (Figure 19-17). In human bile, about 45% is chenodeoxycholic acid, 31% cholic acid and 24% deoxycholic acid (a secondary bile acid formed in the intestine). Early studies in rodents showed the preferred substrate to be newly synthesized cholesterol. However, whole-body turnover studies in humans using radioactive markers indicate that approximately two thirds of bile acid is derived from HDL cholesterol delivered to the liver. Formation is initiated by 7α-hydroxylation, the committed and rate-limiting step catalyzed by microsomal

7α-hydroxylase; the reaction requires NADPH, O_2, cytochrome P-450, and NADPH: cytochrome P-450 reductase. Reactions that follow are oxidation of the 3β-hydroxyl group to a 3-keto group, isomerization of the Δ^5 double bond to the Δ^4-position, conversion of the 3-keto group to a 3α-hydroxyl group, reduction of the Δ^4 double bond, 12α-hydroxylation in the case of cholic acid synthesis, and oxidation of the side chain. 12α-Hydroxylase, like 7α-hydroxylase, is associated with microsomes and requires NADPH, molecular oxygen, and cytochrome P-450. Unlike 7α-hydroxylase, its activity does not exhibit diurnal variation. Its activity determines the amount of cholic acid synthesized. Side chain oxidation starts with 27-hydroxylation and is followed by oxidative steps similar to those of β-oxidation of fatty acids (Chapter 18). The 27-hydroxylation catalyzed by a mixed-function hydroxylase probably occurs in mitochondria and requires NADPH, O_2, and cytochrome P-450. Bile acid deficiency in cerebrotendinous xanthomatosis (see above) is due to a deficiency of 27-hydroxylase. Since the substrates for bile acid formation are water insoluble, they require sterol carrier proteins for synthesis and metabolism.

Bile acids are conjugated with glycine or taurine (Figure 19-18) before being secreted into bile, where the ratio of glycine- to taurine-conjugated acids is about 3:1. Sulfate esters of bile acids are also formed to a small extent. At the alkaline pH of bile and in the presence of alkaline cations (Na^+, K^+), the acids and their conjugates are present as salts (ionized forms), although the terms *bile acids* and *bile salts* are used interchangeably.

Regulation of Bile Acid Synthesis

Regulation of bile acid formation from cholesterol occurs at the 7α-hydroxylation step and is mediated by the concentration of bile acids in the enterohepatic circulation. 7α-Hydroxylase is modulated by a phosphorylation-dephosphorylation cascade similar to that of HMG-CoA reductase (Figure 19-11) except that the phosphorylated form of 7α-hydroxylase is more active.

As noted earlier, the major rate-limiting step of cholesterol biosynthesis is reduced synthesis of HMG-CoA. 7α-hydroxycholesterol, the first intermediate of bile acid formation, inhibits HMG-CoA reductase. The activities of 7α-cholesterol hydroxylase and HMG-CoA reductase undergo parallel changes under the influence of bile acid levels. In the rat, they show similar patterns of diurnal variation, with highest activities during the dark period. Bile acids and intermediates do not appear to function as allosteric modifiers. In the intestines, bile acids may regulate cholesterol biosynthesis, in addition to their role in cholesterol absorption. In humans, the presence of excess cholesterol does not increase bile acid production proportionally, although it suppresses endogenous cholesterol synthesis and increases excretion of fecal neutral steroids.

Hepatic bile acid synthesis amounts to about 0.8–1 g/day. When loss of bile occurs owing to drainage

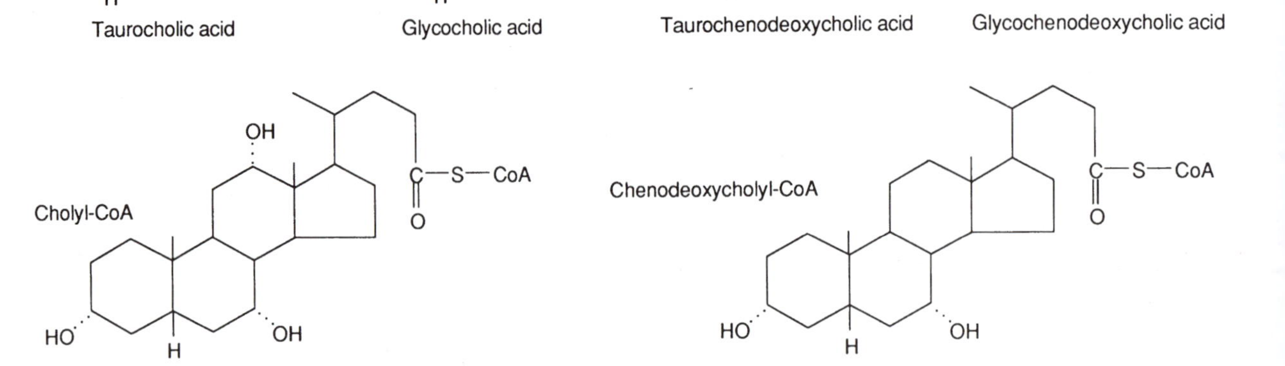

Cholyl-CoA ($R-CO-SCoA$) + Taurine ($HO_3S-CH_2-CH_2-NH_2$) → CoASH + Taurocholic acid ($R-CO-NH-CH_2-CH_2-SO_3H$)

Cholyl-CoA + Glycine (NH_2-CH_2-COOH) → CoASH + Glycocholic acid ($R-CO-NH-CH_2-COOH$)

Chenodeoxycholyl-CoA ($R'-CO-SCoA$) + Taurine ($HO_3S-CH_2-CH_2-NH_2$) → CoASH + Taurochenodeoxycholic acid ($R-CO-NH-CH_2-CH_2-SO_3H$)

Chenodeoxycholyl-CoA + Glycine (NH_2-CH_2-COOH) → CoASH + Glycochenodeoxycholic acid ($R'-CO-NH-CH_2-COOH$)

Cholyl-CoA

Chenodeoxycholyl-CoA

FIGURE 19-18
Conjugation of bile acids with taurine and glycine.

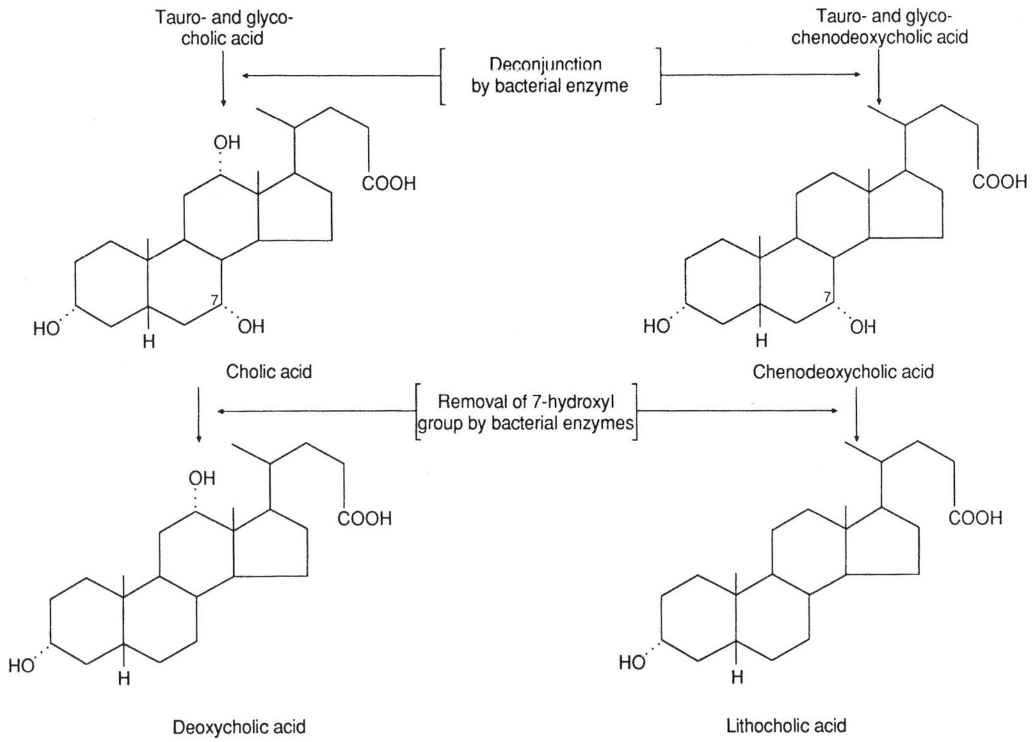

FIGURE 19-19
Conversion of primary to secondary bile acids by microbial enzymes.

through a biliary fistula, to administration of bile acid–complexing resins (e.g., cholestyramine), or to ileal exclusion, activity of 7α-hydroxylase increases several fold, with consequent increase in bile acid formation. The latter two methods are used to reduce cholesterol levels in hypercholesterolemic patients (Chapter 20).

Disposition of Bile Acids in the Intestines and Their Enterohepatic Circulation

Bile is stored and concentrated in the gallbladder, a saccular, elongated, pear-shaped organ attached to the hepatic duct. Bile contains bile acids, bile pigments (i.e., bilirubin glucuronides; see Chapter 29), cholesterol, and lecithin. The pH of gallbladder bile is 6.9–7.7. Cholesterol is solubilized in bile by the formation of micelles with bile acids and lecithin. Cholesterol gallstones can form as a result of excessive secretion of cholesterol or of insufficient amounts of bile acids and lecithin relative to cholesterol in bile. Inadequate amounts of bile acids result from decreased hepatic synthesis, decreased uptake from the portal blood by hepatocytes, or increased loss from the gastrointestinal tract.

With ingestion of food, cholecystokinin (Chapter 12) is released into the blood and causes contraction of the gallbladder, whose contents are rapidly emptied into the duodenum by way of the common bile duct. In the duodenal wall, the bile duct fuses with the pancreatic duct at the ampulla of Vater. Bile functions include absorption of lipids and the lipid-soluble vitamins A, D, E, and K by the emulsifying action of bile salts (Chapter 12), neutralization of acid chyme, and excretion of toxic metabolites (e.g., bile pigments, some drugs and toxins) in the feces.

The secondary bile acids, deoxycholic and lithocholic acids, are derived by 7-dehydroxylation from the deconjugated primary bile acids, cholic and chenodeoxycholic acids, respectively (Figure 19-19), through action of bacterial enzymes primarily in the large intestine. The major portion ($>90\%$) of bile acids in the intestines is reabsorbed by an active transport system into the portal circulation at the distal ileum and transported bound to albumin. They are taken up by the liver, promptly reconjugated with taurine and glycine, and resecreted into bile. Both ileal absorption and hepatic uptake of bile acids may be mediated by Na^+-dependent (carrier) transport mechanisms. This cyclic transport of bile acids from intestine to liver and back to the intestine is known as the enterohepatic circulation (Figure 19-20). During a single passage of portal blood through the liver, about 90% of the bile acids are extracted. The bile acid pool size in the enterohepatic circulation is 2–4 g and circulates about twice during digestion

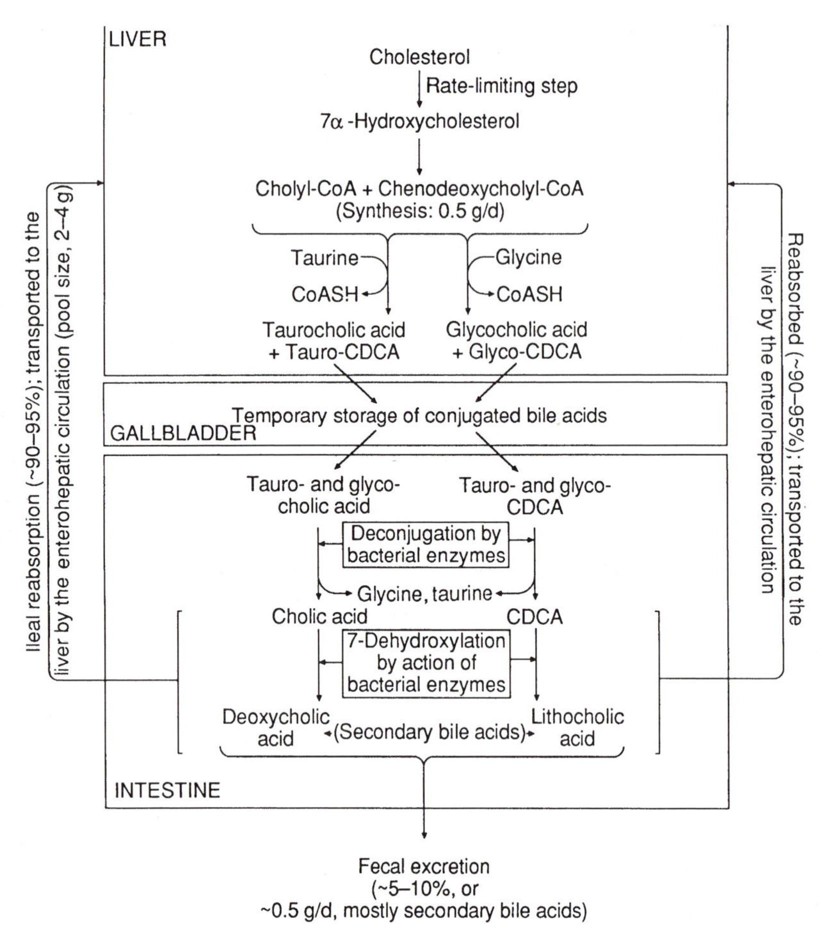

FIGURE 19-20
Formation, enterohepatic circulation, and disposition of the bile acids. CDCA = Chenodeoxycholic acid.

of each meal. The amount of bile acids lost in feces is about 0.8–1 g/day and consists mostly of secondary bile acids (particularly lithocholic acid, the least soluble of the bile acids). The loss is made up by synthesis of an equal amount in the liver.

Bile Acid Metabolism and Clinical Medicine

In liver disorders, serum levels of bile acids are elevated, and their measurement is a sensitive indicator of liver disease. Bile acids are not normally found in urine owing to efficient uptake by the liver and excretion into the intestines. In hepatocellular disease and obstructive jaundice, however, their urinary excretion increases.

Lithocholic acid is toxic and can cause hemolysis and fever. Effects associated with hyperbile acidemia include pruritus, steatorrhea, hemolytic anemia, and further liver injury.

The main cause of ***cholelithiasis*** (presence or formation of gallstones) is precipitation of cholesterol in bile. Elevated biliary concentrations of bile pigments (bilirubin glucuronides) can also lead to formation of concretions known as pigment stones (Chapter 29). Since biliary cholesterol is solubilized by bile acids and lecithin, an excess of cholesterol along with decreased amounts of bile acids and lecithin, cause bile to become supersaturated with cholesterol with the risk of forming cholesterol stones. The limits of solubility of cholesterol in the presence of bile salts and lecithin has been established by using a ternary phase diagram. If the mole ratio of bile salts and phospholipids to cholesterol is less than 10:1, the bile is considered to be lithogenic (stone forming), but this ratio is not absolute. The pattern of food intake in Western nations, which consists of excess fat, cholesterol, and an interval of 12–14 hours between the evening meal and breakfast, leads to a fasting gallbladder bile that is saturated or supersaturated with cholesterol. In individuals who have gallstones, the problems associated with production of lithogenic bile appear to reside in the liver and not in the gallbladder. When the activities of HMG-CoA reductase and 7α-hydroxylase were measured in the liver of patients who had gallstones and compared with those of controls, cholesterogenesis was found to be increased in the patients with gallstones, and bile acid synthesis was reduced. Under these conditions, the ratio of cholesterol to bile acids secreted by the liver is increased. The gallbladder provides an environment where concentration of the lithogenic bile triggers formation of gallstones. A genetic predisposition to formation of cholesterol gallstones is common in certain groups (e.g., 70% of Native American women older than 30 years have cholesterol gallstones). In general, the incidence of gallstones in women is three times higher than in men, and this proportion increases with age. Obesity and possibly multiparity are also associated with formation of gallstones. Any disorder of the ileum (e.g., *Crohn's disease*) or its resection can result in gallstones owing to impaired absorption of bile acids and depletion of the bile acid pool. Agents that increase biliary secretion of cholesterol (e.g., estrogen, oral contraceptives, and clofibrate) and those which prevent bile acid reabsorption in the intestines (e.g., cholestyramine) are also predisposing factors.

Cholelithiasis is frequently treated by surgical removal of the gallbladder (cholecystectomy). Oral treatment with ursodeoxycholic (ursodiol) and chenodeoxycholic (chenodiol) acids has effectively solubilized gallstones in a number of patients. These bile acids apparently reduce HMG-CoA reductase activity, thereby lowering cholesterol levels while enriching the bile acid pool. The increased bile acid to cholesterol ratio in the bile apparently aids in solubilizing the stones already present. However, the effect of chenodiol is transient, and therapy may have to be long-term. In addition, the treatment appears to be promising only in patients who have radiolucent gallstones and functioning gallbladders; and because it raises serum transaminase levels, the significance of which is not clear, the measurement of other liver function tests may be necessary. However, ursodiol has few side effects and is effective at a lower dosage. Ursodiol is not a human metabolite and differs from chenodiol only in the orientation of a hydroxyl group at C_7 (Figure 19-17). Chenodiol is in the α-orientation, whereas ursodiol is in the β-orientation. Two other nonsurgical treatments undergoing clinical evaluation are extracorporeal shock-wave lithotripsy to fragment gallstones and direct infusion into the gallbladder with the solvent methyl tert-butyl ether to dissolve cholesterol stones. Oral treatment with ursodiol in conjunction with extracorporeal shock-wave lithotripsy is more effective than lithotripsy alone.

Supplemental Readings and References

S-H. Bae and Y-K. Paik: Cholesterol biosynthesis from lanosterol: Development of a novel assay method and characterization of rat liver microsomal lanosterol Δ^{24}-reductase. *Biochemical Journal* **326,** 609 (1997).

C. S. Baker, T. W. Evans, B. J. Randle, and P. L. Haslam: Damage to surfactant-specific protein in acute respiratory distress syndrome. *Lancet* **353,** 1232 (1999).

N. Braverman, P. Lin, F. F. Moebius, et al.: Mutations in the gene encoding 3β-hydroxysteroid-Δ^8,Δ^7-isomerase cause X-linked dominant Conradi–Hünermann syndrome. *Nature Genetics* **22,** 291 (1999).

J. Charrow, J. A. Esplin, T. J. Gribble, et al.: Gaucher disease: Recommendations on diagnosis, evaluation, and monitoring. *Archives of Internal Medicine* **158,** 1754 (1998).

S. Y. Cho, J-H. Kim, and Y-K. Paik: Cholesterol biosynthesis from lanosterol: Differential inhibition of sterol Δ^8-isomerase and other lanosterol-converting enzymes by tamoxifen. *Molecules and Cells* **8,** 233 (1998).

C. Cunniff, L. E. Kratz, A. Mosher, et al.: Clinical and biochemical spectrum of patients with RSH/Smith–Lemli–Opitz syndrome and abnormal cholesterol metabolism. *American Journal of Medical Genetics* **68,** 263 (1997).

J. M. J. Derry, E. Gormally, G. D. Means, et al.: Mutations in a Δ^8–Δ^7 sterol isomerase in the tattered mouse and X-linked dominant chondrodysplasia punctata. *Nature Genetics* **22,** 286 (1999).

S. B. Dubin: The laboratory assessment of fetal lung maturity. *American Journal of Clinical Pathology* **97,** 836 (1992).

R. V. Farese, Jr. and J. Herz: Cholesterol metabolism and embryogenesis. *Trends in Genetics* **14,** 115 (1998).

J. L. Goldstein and M. S. Brown: Regulation of the mevalonate pathway. *Nature* **343,** 425 (1990).

H. P. Haagsman and L. M. G. van Golde: Synthesis and assembly of lung surfactant. *Annual Review of Physiology* **53,** 441 (1991).

C. E. M. Hollack, E. P. M. Corssmit, J. M. F. G. Aerts, et al.: Differential effects of enzyme supplementation therapy on manifestations of type I Gaucher disease. *American Journal of Medicine* **103,** 185 (1997).

R. I. Kelley: RSH/Smith–Lemli–Opitz syndrome: Mutations and metabolic morphogenesis. *American Journal of Human Genetics* **63,** 322 (1998).

J. M. Klein, M. W. Thompson, J. M. Snyder, et al.: Transient surfactant protein B deficiency in a term infant with severe respiratory failure. *Journal of Pediatrics* **132,** 244 (1998).

M. C. Maberry: Methods to diagnose fetal lung maturity. *Seminars in Perinatology* **17,** 241 (1993).

G. Paumgartner and T. Sauerbruch: Gallstones: Pathogenesis. *Lancet* **338,** 1117 (1991).

J. A. Porter, K. E. Young, and P. A. Beachy: Cholesterol modification of hedgehog signaling proteins in animal development. *Science* **274,** 255 (1996).

M. Trauner, P. J. Meier, and J. L. Boyer: Molecular pathogenesis of cholestasis. *Mechanisms of Disease* **339,** 1217 (1998).

K. W. A. Wirtz: Phospholipid transfer proteins revisited. *Biochemical Journal* **324,** 353 (1997).

CHAPTER 20

Lipids III: Plasma Lipoproteins

Lipids, by virtue of their immiscibility with aqueous solutions, depend on protein carriers for transport in the bloodstream and extracellular fluids. Fat-soluble vitamins and free fatty acids are transported as noncovalent complexes. Vitamin A is carried by retinol-binding protein and free fatty acids on plasma albumin. However, the bulk of the body's lipid transport occurs in elaborate molecular complexes called *lipoproteins*.

20.1 Structure and Composition

Lipoproteins are often called pseudomicellar because their outer shell is in part composed of amphipathic phospholipid molecules. Unlike simple micelles, lipoproteins contain ***apolipoproteins,*** or ***apoproteins,*** in their outer shell and a hydrophobic core of ***triacylglycerol*** and ***cholesteryl esters.*** Unesterified, or free, cholesterol, which contains a polar group, can be found as a surface component and in the region between the core and surface (Figure 20-1). Most lipoproteins are spherical. However, newly secreted high-density lipoproteins (HDLs) from the liver or intestine are discoidal and require the action of lecithin-cholesterol acyltransferase (LCAT) in plasma to expand their core of neutral lipid and become spherical. The hydrophobic core of the low-density lipoprotein (LDL) molecule may contain two concentric layers: one of triacylglycerol and another of cholesteryl ester.

The apoproteins are distinct physically, chemically, and immunochemically and have important roles in lipid transport and metabolism (Table 20-1). In keeping with their individual metabolic functions, they have specific structural domains. Amino acid substitutions or deletions in critical domains result in functional abnormalities. The apoproteins share a common structure in the form of an amphipathic helix, in which the amino acid residues have hydrophobic side chains on one face of the helix and hydrophilic polar residues on the other. The hydrophilic face is believed to interact with the polar head groups of the phospholipids, while the hydrophobic residues interact with their fatty acid portions.

The laws of mass action govern the interactions of lipids and most apoproteins in lipoproteins, so that as the affinities between surface components change during lipoprotein metabolism, apoproteins may dissociate from one particle and bind to another. In fact, all of the apoproteins, with the possible exception of apoprotein B (apo B), can change their lipoprotein associations. The reason for the unique behavior of apo B remains a mystery. On the basis of their principal transport function, lipoproteins may be divided into two classes according to the composition of their major core lipids. The principal triacylglycerol carriers are chylomicrons and very-low-density lipoproteins (VLDLs), whereas most cholesterol transport occurs via LDLs and HDLs.

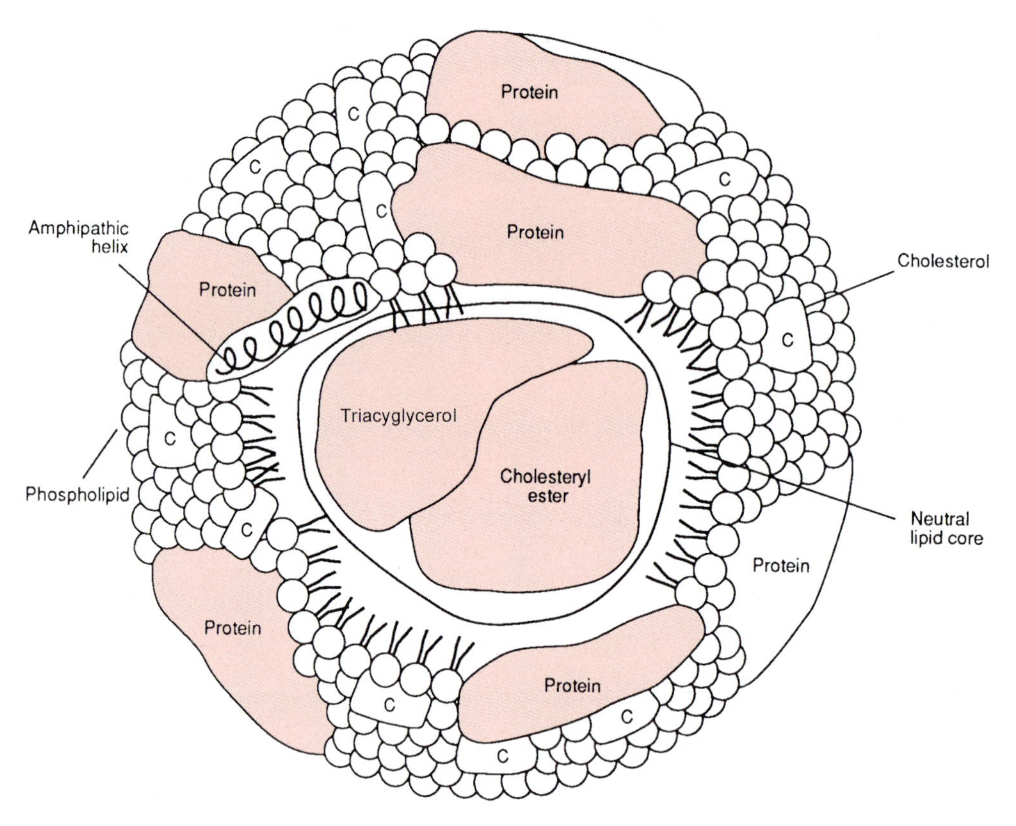

FIGURE 20-1
Generalized structure of a lipoprotein molecule showing the distribution of polar components in an outer shell composed of free cholesterol, phospholipids, and amphipathic proteins and in an inner core composed of neutral triacylglycerols and cholesteryl esters. Phospholipids are oriented with polar head groups toward the aqueous environment and hydrophobic tails toward the neutral core, analogous to their positioning in the outer leaflet of the typical cell membrane.

These four major groups of plasma lipoproteins can be separated and characterized by electrophoresis and ultracentrifugation (Table 20-2). Each group is heterogeneous and can be subdivided on the basis of variation in apoprotein and lipid compositions. The groups share several apoprotein components, e.g., apo B occurs in chylomicrons, LDLs, and VLDLs. Some apoproteins belong to families of polypeptides, one or another of which may predominate in a particular lipoprotein group. For example, apo B, when formed in chylomicrons synthesized in intestinal mucosa, is apo B-48, while that in VLDL from the liver is B-100. The designations B-48 and B-100 reflect the relative molecular masses of these proteins, B-48 being 48% of the mass of B-100. Intestinal apo B contains 2152 of the 4563 amino acids that make up the hepatic form. Oligonucleotide hybridization studies have shown the intestinal and hepatic genes to be identical. The shortened intestinal form of apo B is produced by a single nucleotide substitution of uracil for cytosine at position 6457 in the nucleotide sequence of the apo B mRNA. This changes the codon CAA (glutamine) to the termination codon UAA. The tissue-specific apo B mRNA editing, which consists of site-specific deamination of cytosine to uracil, is mediated by a multicomponent cytidine deaminase. This unique form of mRNA editing eliminates the carboxy terminal portion of apo B to form the B-48 of chylomicrons. Since the carboxy terminal portion of the apo B sequence contains the apo B-binding domain, this deletion ensures a distinct metabolic routing of the chylomicron particle.

In a tube of plasma or serum collected from a nonfasting individual and allowed to stand overnight, chylomicrons collect at the top surface in a milky layer because of their low density ($d < 0.95$). If a pure sample of chylomicron is required, it is better to obtain it from lymph or to separate VLDLs from chylomicrons in plasma by chromatography. Sequential ultracentrifugation is used to obtain various lipoprotein fractions from a single sample. Ultracentrifugation of fasting serum or plasma at its own density (1.006) for 18 hours at $100{,}000 \times g$ will float the VLDL fraction ($d < 1.006$). After removal of the VLDL, the density of the remaining serum is raised to 1.063 and

TABLE 20-1
*Properties of Human Apolipoproteins**

Apolipoprotein	Plasma Concentration (mg/dL) (approximate)	Molecular Weight	Major Density Class	Biological Function in Addition to Structural and Transport Role
A-I	100–150	29,016	HDL	Activates LCAT, cell-receptor binding (?)
A-II	30–50	17,414	HDL	Inhibits LCAT (?) enhances hepatic lipase activity.
A-IV	15	44,465	Chylomicron	Unknown.
B-100	80–100	512,723	VLDL, LDL	Necessary for triacylgycerol secretion; binding to cell receptors.
B-48	<5	241,000	Chylomicron	. . .
C-I	7	6630	Chylomicron, VLDL	Cofactor with LCAT.
C-II	4	8900	Chylomicron, VLDL	Activates LPL.
C-III	12	8800	Chylomicron, VLDL	Regulates LPL reaction; inhibits uptake of remnants via apo E receptor.
D	6	19,000	HDL	LCAT reaction, cholesteryl ester transfer.
E_{2-4}	3–7	34,145	VLDL, IDL, HDL	Binding to specific receptors
(a)	10	300,000–800,000	LDL, HDL	Competes with plasminogen for its receptors and may inhibit thrombolysis.

*LCAT = Lecithin-cholesterol acyltranferase; LPL = lipoprotein lipase; VLDL = very-low-density lipoprotein; LDL = low-density lipoprotein; HDL = high-density lipotrotein; IDL = intermediate density lipoprotein.

centrifuged at 100,000 × *g* overnight, and the LDL fraction is removed from the top. The density of the remaining material is raised to 1.21. After centrifugation for 2 days at 100,000 × *g*, the HDL fraction is obtained from the top.

In electrophoresis of plasma or serum at alkaline pH, with either paper or agarose as a support medium, the lipoproteins migrate to different positions. The lipoprotein bands are identified by staining with appropriate lipid

TABLE 20-2
Physical Data for the Major Types of Lipoprotein

Lipoprotein	Hydrated Density (g/ml)	S_f*	Position of Migration in Paper or Agarose Electrophoresis in Comparison with Serum Globulins
Chylomicrons	<0.95	>400	Origin
VLDL	0.95–1.006	20–400	α_2-(migrates ahead of β-globulin to pre-β position)
LDL	1.006–1.063	0–20	β (with β-globulin)
HDL_2	1.063–1.125	—	α_1 (with α_1-globulin)
HDL_3	1.125–1.210	—	
Lp(a)	1.040–1.090	—	Slow pre-β-LP

*The rate at which lipoprotein floats up through a solution of NaCl of specific gravity 1.063 is expressed in Svedberg flotation units or S_f; one S_f unit = 10^{-13} cm/sec/dyne/g at 26°C. S_f can be thought of as a negative sedimentation constant.

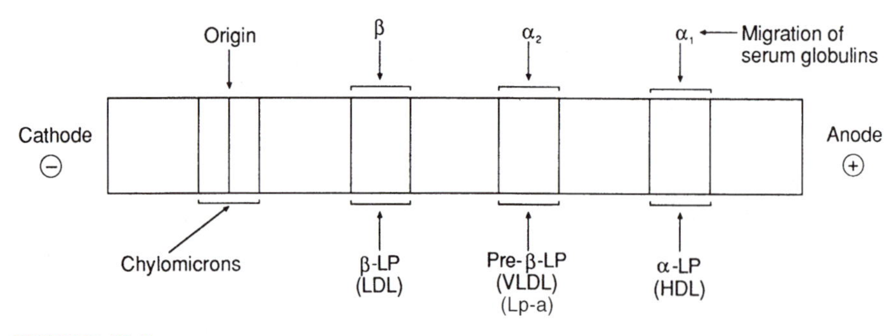

FIGURE 20-2
Migration of plasma lipoproteins on paper electrophoresis at pH 8.6 in barbital buffer. LP = Lipoprotein; other abbreviations as in Table 20-1.

stains (e.g., oil red O, fast red 7B, Sudan black B). Chylomicrons appear at the origin (point of application of the sample), VLDLs at the α_2-region (the pre-β region; hence, "pre-β lipoproteins"), LDLs at the β-globulin region ("β-lipoproteins"), and HDLs at the α_1-globulin region ("α-lipoproteins"). The electrophoretic pattern of the lipoproteins separated on paper (or on thin agarose films) is shown in Figure 20-2, and a correlation of the classification of lipoproteins by electrophoresis and hydrated density is shown in Table 20-2.

As knowledge of the lipoproteins and of their physiological roles has increased, the major species have been subdivided. In the density class 1.006–1.063 g/mL, a relatively pure LDL fraction may be isolated between the densities of 1.019 and 1.063 g/mL. Lipoproteins isolated in the $d = 1.006$–1.019 range are known as *intermediate-density lipoproteins* (*IDLs*) and are thought to represent VLDLs being processed to LDLs. The HDL fraction is commonly subdivided into HDL_2 and HDL_3. HDL_2 occurs in the density range 1.063–1.125 g/mL and HDL_3 at 1.125–1.21 g/mL. The composition of the major and minor subclasses of lipoproteins is given in Table 20-3.

Other lipoproteins are lipoprotein (a) and lipoprotein-X. Lipoprotein (a) [Lp(a)], a variant of LDL, occurs in the plasma of normal subjects in the density range 1.040–1.090 g/mL. Lp(a) is similar to LDL in lipid composition and also contains apo B-100, which is bound to apolipoprotein (a) by a disulfide linkage. Apo(a) is a large glycoprotein that exhibits size heterogeneity among individuals in a range of M.W. 300,000–800,000 on sodium dodecyl sulfate (SDS–PAGE). On agarose electrophoresis, Lp(a) migrates with VLDL (pre-β electrophoretic mobility). For this reason, it is also known as the "sinking pre-β-lipoprotein." The anodal electrophoretic mobility of Lp(a) relative to LDL may be due to its high content of sialic acid.

The exact physiological function of Lp(a) is not known. However, apo(a) and plasminogen share extensive sequence homology. Both Lp(a) and plasminogen possess multiple tandem repeats of triple loop structural motifs, known as ***kringles*** (Chapter 36). Type 2 kringle IV is found in apo(a) and varies from 11 to greater than 50 tandem repeats. The exact number of repeats is an inherited property and defines different isoforms of Lp(a). The apo(a) gene is close to the plasminogen gene (6q2.6–q2.7). Plasma Lp(a) primarily arises in the liver.

TABLE 20-3
Composition of Lipoproteins (percent of mass)

Protein	Major Apoproteins	Phospholipid	Cholesterol	Cholesteryl Ester	Triacylglycerol	Diameter (nm)
Chylomicrons	2 (A-1, A-II, A-IV, B-48)	3	2	3	90	75–1200
VLDL	6 (B-100, C, E)	14	4	16	60	30–80
IDL	18 (B-100, E)	22	8	32	20	25–35
LDL	21 (B-100)	22	8	42	7	18–28
LP(a)	21 [B-100, apo(a)]	20	8	45	6	25–30
HDL_2	37 (A-I, A-II)	32	5	20	6	9–12
HDL_3	54 (A-I, A-II)	26	2	15	3	5–9

*Represents the sum of both free and esterified cholesterol.

Due to the structural similarity between apo(a) and plasminogen, it is thought that Lp(a) interferes with fibrinolysis and thus promotes thrombosis. Normally, formation of a blood clot (thrombus) and its dissolution by fibrinolysis is carefully regulated and the two processes are in a delicate balance (Chapter 36). The clot is cleaved and solubilized by the serine protease plasmin which is derived from its precursor plasminogen by proteolysis. Lp(a) exerts its antifibrinolytic activity by competitively inhibiting the binding of plasminogen to endothelial cells and monocytes, and to fibrin which is required for formation of plasmin (Chapter 36). Elevated homocysteine levels (homocysteinemia) promote fibrin binding of Lp(a), thus further preventing the formation of plasmin and subsequent impaired thrombolysis.

Several prospective studies have shown that an excess plasma level of Lp(a) is a risk factor and an independent predictor of coronary heart disease. Survivors of myocardial infarcts have higher Lp(a) levels than do controls. In individuals with Lp(a) levels greater than 30 mg/dL (0.78 mM/L), the relative risk for coronary heart disease is 2- to 3.5-fold higher than that of controls. Population studies have shown that plasma Lp(a) levels in Africans are several-fold higher compared with Asian and Caucasian populations. Lp(a) levels are elevated in end-stage renal disease. In postmenopausal women, Lp(a) is increased but estrogen replacement therapy can lower the Lp(a) levels. Most lipid-lowering drugs (except nicotinic acid) do not affect plasma Lp(a) concentration.

Lipoprotein-X (Lp-X), an abnormal lipoprotein, occurs in patients with obstructive liver disease or LCAT deficiency. Lp-X floats in the density range of LDL and has the same electrophoretic mobility as LDL. It can be separated from LDL, however, by hydroxyapatite chromatography or by zonal centrifugation. The composition of Lp-X differs from that of LDL, and it does not react with antisera to LDL. The major apoproteins of Lp-X isolated from patients with LCAT deficiency are albumin, apo C, and apo A. Lp-X also contains small amounts of apo D and apo E. Lp-X from patients with obstructive liver disease has been reported to lack apo A-I, a powerful activator of LCAT. The lipid constituents of Lp-X are cholesterol (almost entirely unesterified) and phospholipids. In electron microscopy, negatively stained Lp-X preparations appear as stacks of disk-like structures (*rouleaux*).

20.2 Metabolism

Plasma lipoproteins are in a dynamic state. Their continuous synthesis and degradation are accompanied by rapid exchanges of lipid and protein components between the different lipoprotein classes. Major sites of plasma lipoprotein synthesis are the intestine and liver. Synthesis takes place in rough and smooth endoplasmic reticulum. The necessary components are triacylglycerols, cholesterol (and cholesteryl esters), phospholipids, and apoproteins. At various points in the metabolic cycle four enzymes play an important role in the delivery, storage, and mobilization of lipoprotein lipids.

1. ***Lipoprotein lipase*** (LPL) is a glycoprotein that belongs to a family of serine esterases that include hepatic lipase and pancreatic lipase. LPL is synthesized by many types of parenchymal cells in the body but is concentrated mainly in muscle and adipose tissue. After its synthesis in the parenchymal cells, LPL undergoes secretion and translocation to the lumenal surface of endothelial cells lining the vascular beds where it is bound to heparan sulfate. Purified LPL is a dimer in the active state and loses activity when it dissociates into monomers. Each subunit of LPL contains binding sites for glycosaminoglycans and apo C-II, both of which promote dimerization of LPL. Apo C-II is required for the activation of LPL *in vivo*. Liver is the site of apo C-II synthesis and, in the plasma, it recycles between HDL, triacylglycerol-rich lipoproteins, chylomicrons, and VLDLs.

   LPL is the major enzyme involved in the processing of chylomicrons and VLDL by hydrolysis of fatty acids from triglycerides. Phospholipids may also serve as substrates for LPL. Apo C-III inhibits the activation of LPL by apo C-II *in vitro,* albeit at high levels, but the physiological importance of inhibition is not understood.

   In the postprandial state, elevated serum insulin increases LPL activity in adipose tissue (but not in muscle) and promotes fuel storage as triacylglycerols. In the postabsorptive state, serum insulin levels decrease causing a decrease in LPL activity in adipose tissue; however, LPL activity in muscle remains high or increases releasing fatty acids from VLDL particles for use as fuel. In general, LPL has different functions in different tissues. In cardiac and skeletal muscle it provides energy; in white adipose tissue it stimulates triacylglycerol storage; in brown adipose tissue it is thermogenesis; and in lactating breast it is triacylglycerol synthesis for milk production. LPL also is involved in surfactant synthesis in the lungs and in phospholipid and glycosphingolipid synthesis in the brain.
2. ***Hepatic lipase,*** like LPL, is synthesized in the parenchymal cells of liver and is localized on the

membrane surfaces of the hepatic endothelial cells of blood vessels bound to heparan sulfate. It also occurs on the endothelial cells of blood vessels of adrenals and gonads. The role of hepatic lipase is not completely understood, however, it continues the lipolysis of VLDL and IDL in their stepwise conversion to LDL. Hepatic lipase also hydrolyzes phospholipids and HDL-triacylglycerol. It performs these functions both on the endothelial cell surface and within endosomes as lipoproteins are endocytosed into cells via receptor-mediated endocytosis. Hepatic lipase activity responds positively to androgens and negatively to increasing levels of estrogen.

3. ***Lecithin-cholesterol acyltransferase*** (LCAT) is a glycoprotein synthesized in the liver. LCAT circulates in the plasma with HDL, LDL, apo D, and cholesteryl ester transfer protein (CETP). It is activated by apo A-I. LCAT catalyzes the transfer of long-chain fatty acids from phospholipids to cholesterol, forming cholesteryl esters and permitting the storage and transport of cholesteryl esters in the lipoprotein core. Cholesteryl esters are exchanged between lipoproteins, a process mediated by CETP. Apo D bound to HDL may also play a role in the formation of cholesteryl esters by providing a binding site for the cholesterol. There is a net transfer of cholesteryl esters from HDL to LDL and VLDL with an exchange of triacylglycerol. The VLDL is transformed into LDL in the circulation (discussed later). The cholesterol content of LDL is increased thus promoting its atherogenicity.

   HDL is antiatherogenic and removes cholesterol from peripheral cells and tissues for eventual transport to hepatocytes and excretion in the bile directly or after conversion into bile acids. The efflux of cholesterol from peripheral cells is mediated by the ATP-binding cassette (ABC) transporter protein (discussed later). The flux of cholesterol transport from extrahepatic tissues (e.g., blood vessel wall) toward liver for excretion is known as the reverse cholesterol transport pathway. In contrast, the forward cholesterol pathway involves the transport of cholesterol from liver to the peripheral cells and tissues via the VLDL → IDL → LDL pathway. It should be noted, however, that the liver plays a major role in the removal of these lipoproteins. Thus, the system of reverse cholesterol transport consisting of LCAT, CETP, apo D, and their carrier lipoproteins is critical for maintaining cellular cholesterol homeostasis. The role of CETP is exemplified in clinical studies involving patients with polymorphic forms of CETP that promote cholesteryl ester transfer from HDL to LDL or that increase plasma CETP levels. These patients exhibit an increased risk of coronary heart disease. Homozygous CETP deficiency found in some Japanese people results in increased HDL cholesterol levels; there is anecdotal evidence of longevity in this group of individuals.

4. ***Acyl-CoA:cholesterol acyltransferase*** (ACAT) esterifies free cholesterol by linking it to a fatty acid. ACAT is an intracellular enzyme that prepares cholesterol for storage as in liver parenchymal cells.
   The roles of these enzymes in lipoprotein metabolism will become more apparent in the discussion of the origin and fate of each class of lipoprotein.

Chylomicrons

After partial hydrolysis in the gut, dietary fatty acids, monoacylglycerols, phospholipids, and cholesterol are absorbed into the mucosal enterocytes lining the small intestine (Chapter 12). Once within the cell, the lipids are reesterified and form a lipid droplet within the lumen of the smooth endoplasmic reticulum. These droplets consists of triacylglycerol and small amounts of cholesteryl esters and are stabilized by a surface film of phospholipid. At the junction of the smooth and the rough endoplasmic reticulum, the droplet acquires apoproteins B-48, A-I, A-II, and A-IV, which are produced in the lumen of the rough endoplasmic reticulum in the same way as other proteins bound for export. The lipoprotein particle is then transported to the Golgi stacks where further processing yields chylomicrons, which are secreted into the lymph and then enter the blood circulation at the thoracic duct.

Synthesis and secretion of chylomicrons are directly linked to the rate of dietary fat absorption. When fat is absent from the diet, small chylomicrons with a diameter of about 50 nm are secreted at a rate of approximately 4 g of triacylglycerol per day. On a high-fat diet, the mass of lymphatic triacylglycerol transport may increase 75-fold, owing partly to greater production of chylomicrons but primarily to a dramatic increase in size of the particles, which may have diameters of 1200 nm, and a 16-fold increase in the amount of triacylglycerol within their core.

In the circulation, chylomicrons undergo a number of changes (Figure 20-3). First, they acquire apo C and apo E from plasma HDL in exchange for phospholipids. Next, hydrolysis of triacylglycerols by endothelial LPL (e.g., in adipose tissue and skeletal muscle) begins. Progressive hydrolysis of triacylglycerol through diacylglycerol and

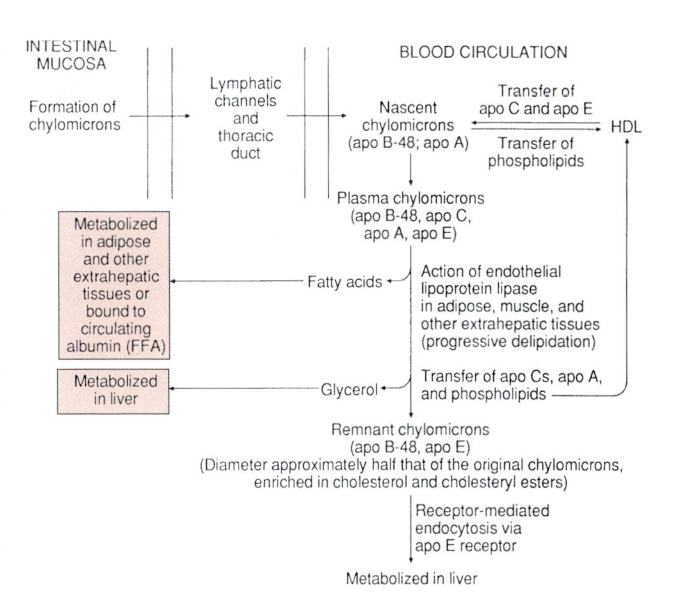

FIGURE 20-3
Steps involved in the metabolism of chylomicrons (the exogenous pathway of lipid transport).

monoacylglycerol to fatty acids and glycerol ensues. The released fatty acids cross the endothelium and enter the underlying tissue cells, where they undergo reesterification to form triacylglycerol for storage (adipocytes) or are oxidized to provide energy (e.g., muscle). The earlier acquisition of apo C-II by chylomicrons is important because this apoprotein is an essential activator of LPL. Chylomicron triacylglycerol has a half-life of about 5 minutes in the circulation. LPL is most active after meals, when it is stimulated by elevated levels of plasma insulin. A low-affinity form (high K_m) of LPL, found principally in adipose tissue, is most active when triacylglycerol levels in plama are high; it promotes lipid storage after meals. A high-affinity (low K_m) form of LPL predominates in heart (and striated muscle tissues) and is active when triacylglycerol levels are low, as in the postabsorptive state. The high-affinity enzyme thus hydrolyzes triacylglycerols into fatty acids at sites where they will be required for energy production. Mammary gland LPL is a high-affinity type that facilitates uptake of fatty acids to promote milk-fat synthesis during lactation.

As the core triacylglycerols of a chylomicron are depleted, often reducing its diameter by a factor of 2 or more, the surface components are also modified. A substantial portion of the phospholipids and of apo A and C is transferred to HDL. The C apoproteins thus cycle repeatedly between newly produced chylomicrons and HDL. The ***chylomicron remnant*** is consequently rich in cholesteryl esters and apo B-48 and E.

Catabolism of chylomicron remnants may be viewed as the second step in the processing of chylomicrons. After the loss of apo C-II and other C and A apoproteins, LPL no longer acts upon the remnants, and they leave the capillary surface. Chylomicron remnants are rapidly removed by uptake into liver parenchymal cells via receptor-mediated endocytosis. Apo E is important in this uptake process. The chytomicron receptors in liver are distinct from the B-E receptor that mediates uptake of LDL. The hepatic receptor for chylomicrons binds with apo E, but not apo B-48. Another receptor, known as the LDL receptor–related protein (LRP), may also function in chylomicron uptake. Chylomicron remnants are transported into the lysosomal compartment where acid lipases and proteases complete their degradation. In the liver, fatty acids so released are oxidized or are reconverted to triacylglycerol, which is stored or secreted as VLDL. The cholesterol may be used in membrane synthesis, stored as cholesteryl ester, or excreted in the bile unchanged or as bile acids.

Very-Low-Density Lipoproteins

VLDLs are produced by the parenchymal cells of the liver from lipid and apoprotein constituents in a way similar to that of chylomicron formation in enterocytes. However, while the triacylglycerol core of chylomicrons is derived exclusively from absorbed dietary fatty acids and monoacylglycerols, VLDL triacylglycerols derive from

1. Stored fat released from adipose tissue as fatty acids,
2. Conversion of carbohydrates to fatty acids in the liver, and
3. Hydrolysis of lipoprotein triacylglycerols on capillary endothelia and in the liver.

Although synthesis of apoprotein B-100 is necessary for VLDL secretion, addition of carbohydrate moieties in the Golgi apparatus does not appear to be required, since blockage of this function by tunicamycin does not reduce VLDL secretion.

The amount of VLDL secreted by the liver is extremely variable and can be affected in a number of ways. A primary determinant of VLDL output is the flux of free fatty acids entering the liver. The liver responds to an increase in free fatty acids by synthesis of more and larger VLDLs. If saturated fatty acids predominate in the formation of triacylglycerol, the VLDL particles will be more numerous but smaller than if polyunsaturated fatty acids predominate. This finding may be related to the reduction in plasma cholesterol levels that results from elevating the proportion of polyunsaturated fats in the diet. The surface-to-volume ratio is smaller in the larger VLDLs. Since cholesterol

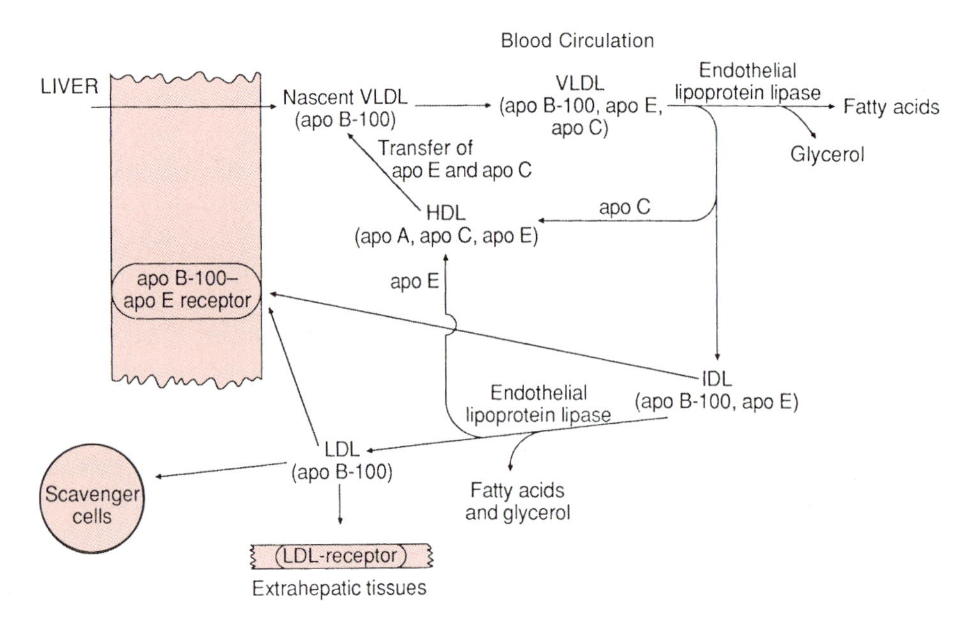

FIGURE 20-4
Conversion of very-low-density lipoproteins (VLDLs) to low-density lipoproteins (LDLs), via intermediate-density lipoproteins (IDLs) (the endogenous pathway of lipid transport). Ox-LDL = Oxidized LDL.

is primarily a surface component in these particles, its secretion would be reduced. If the fatty acid flux to the liver is large, the process of secretion of triacylglycerols in VLDLs can become saturated. The resulting triacylglycerol deposition in cytoplasmic droplets is seen in "fatty liver."

A high-carbohydrate diet results in substantial elevation of plasma VLDL concentration. A high-cholesterol diet alters the composition of VLDL, with cholesterol esters substituting for triacylglycerol as core components, and leads to a marked increase in apo E synthesis.

Like nascent chylomicrons, newly secreted VLDL undergoes changes in the plasma (Figure 20-4). Nascent VLDL acquires apo C and E from HDL. In chylomicrons, the apo B is B-48, whereas in VLDL only B-100 is found. LDL contains exclusively apo B-100, indicating that VLDL rather than chylomicrons is the principal precursor of LDL. In some species (e.g., the rat), most VLDL remnants, like chylomicron remnants, are rapidly taken up by the liver. In humans, as the core triacylglycerols are removed and the C apoproteins are lost, approximately half of the VLDL is rapidly removed by the liver via the apo B-100-E receptor pathway. The rest remains in circulation as VLDL remnants. Since some of these remnants have a density between 1.006 and 1.019, they are called IDLs and are analogous to chylomicron remnants. Thus, the liver plays a major role in clearing remnant lipoproteins. The remaining IDL are subjected to further catabolism by hepatic lipase to produce LDL, a cholesterol-rich particle containing almost exclusively apo B-100. In cholesterol-fed animals, including humans, a lipoprotein in the VLDLs density range is formed that shows pre-β mobility on electrophoresis. These cholesterol-rich VLDL contain high concentrations of apo E and apo B-48, indicating their intestinal origin. They are removed from the circulation by a specific β-VLDL receptor found on macrophages and endothelial cells. By promoting cholesterol storage in macrophages, β-VLDL may play an important role in the formation of macrophage-derived foam cells found in atherosclerotic plaques.

Since LDL is the principal plasma-cholesterol carrier and its concentration in plasma correlates positively with the incidence of coronary heart disease, LDL is the most intensively studied plasma lipoprotein. Production in humans, via the pathway VLDL $\rightarrow$ IDL $\rightarrow$ LDL, accounts for all of the LDL normally present. However, in familial hypercholesterolemia or on a high-cholesterol diet, VLDL is produced that is higher in cholesterol content, smaller in size, and within the LDL density range (1.019–1.063 g/mL).

As IDL loses apo E and is converted to LDL with apo B-100 as its sole apoprotein, the residence of LDL in plasma increases from several hours to 2.5 days. This long-lived, cholesterol-rich LDL serves as a source of cholesterol for most tissues of the body; although most cells can synthesize cholesterol under normal conditions, most endogenous production occurs in the liver and intestine, from which it is distributed to peripheral tissues by LDL. This arrangement provides an efficient balance between endogenous production and dietary intake of cholesterol.

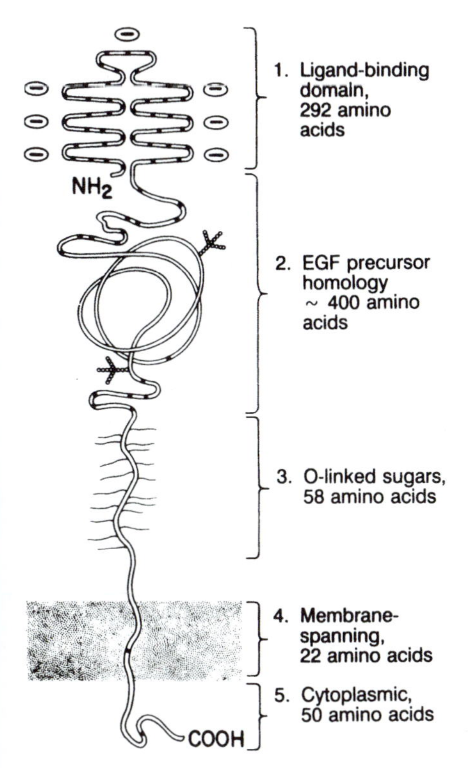

FIGURE 20-5
Schematic representation of human LDL receptor showing five domains. The receptor mediates internalization of LDL and IDL by specific interaction of their B-100 and E apoprotein moieties with two cysteine-rich negatively charged areas of domain 1. Domain 2 is homologous to epidermal growth factor (EGF) precursor, suggesting a possible evolutionary or functional relationship. [Reproduced with permission from M. S. Brown and J. L. Goldstein, A receptor-mediated pathway for cholesterol homeostasis. *Science* **232,** 34 (1986).]

Low-Density Lipoproteins

Distribution and delivery of cholesterol to peripheral tissues are mediated by binding of LDL to specific receptors on the plasma membrane of target cells. The receptor is a glycoprotein (whose 5.3-kilobase cDNA has been cloned) of 839 amino acid residues distributed into five distinct domains (Figure 20-5). Chemical modification of apo B and apo E, the ligands for the LDL receptor, reveals that positively charged lysine and arginine residues are essential for receptor-ligand binding. Binding is probably mediated by electrostatic interaction between the lysine and arginine residues of apo B and apo E and by the clustered negatively charged residues of the cysteine-rich region of the LDL receptor. The gene for the LDL receptor protein (LRP) is more than 40 kilobases in length. It contains 18 exons, a number of which encode protein sequences homologous to those in other proteins, namely, a precursor for epidermal growth factor (EGF), C_9 component of the complement, and three clotting and anticlotting proteins (factor IX and X and protein C).

The number of LDL receptors on the cell membrane depends on the degree of accumulation of intracellular cholesterol, which down-regulates the transcription of the LDL receptor gene. The population of LDL receptors may be reduced 10-fold by this mechanism. The LDL receptor migrates to areas of the cell membrane specialized for endocytosis called *coated pit* because of the protein coat on the cytoplasmic side that consists predominantly of clathrin (M.W. 180,000). *In vitro,* clathrin spontaneously self-associates into pentagonal and hexagonal structures. *In vivo,* these structures form a basket-like scaffolding that mediates invagination of the receptor-rich coated pit. Once inside the cell, the clathrin dissociates from the endosomal vesicle. ATP-dependent proton pumps in the vesicular membrane lower the internal pH of these structures, causing LDL to dissociate from its receptor. The receptor is recycled to the cell surface. The endosome fuses with a primary lysosome whose hydrolases (e.g., cathepsins, acid lipase) degrade LDL to its monomeric constituents (e.g., amino acids, cholesterol), which are then released into the cytoplasm (Figure 20-6).

Cholesterol released inside cells is incorporated into an intracellular pool, which is used for membrane synthesis and for reactions that require a sterol nucleus (e.g., formation of steroid hormones or bile acids). Cholesterol (or its metabolite 26-hydroxycholesterol) suppresses its own synthesis by inhibiting two sequential enzymes of mevalonate synthesis—3-hydroxy-3-methylglutamyl coenzyme A (HMG-CoA) synthase and HMG-CoA reductase, the rate-limiting enzyme of cholesterol biosynthesis (Chapter 19). This cholesterol also regulates the number of receptor sites on the cell membrane and, therefore, cellular uptake of LDL. Genes for cholesterol synthesis undergo coordinate induction or repression with regard to formation of the respective mRNAs (Chapter 19). Some of the cholesterol is esterified (with oleate and palmitoleate) into cholesteryl esters by microsomal ACAT, whose activity is stimulated by cholesterol. Cholesteryl esters of LDL contain linoleate, whereas those produced by ACAT contain predominantly oleate and palmitoleate. In hepatocytes, cholesterol can lead to the increased activity of cholesterol 7α-hydroxylase, the rate-limiting enzyme in synthesis of bile acids (Chapter 19).

Thus, the receptor-mediated, LDL-derived cholesterol meets cellular requirements for cholesterol and prevents its overaccumulation by inhibiting *de novo* cholesterol synthesis, suppressing further entry of LDL, and storing unused cholesterol as cholesteryl esters or exporting it from the liver as bile acids or other sterol-derived products. About 75% of high-affinity LDL uptake occurs in the liver.

Despite this elaborate regulatory system, cells can accumulate excessive amounts of cholesteryl esters when the

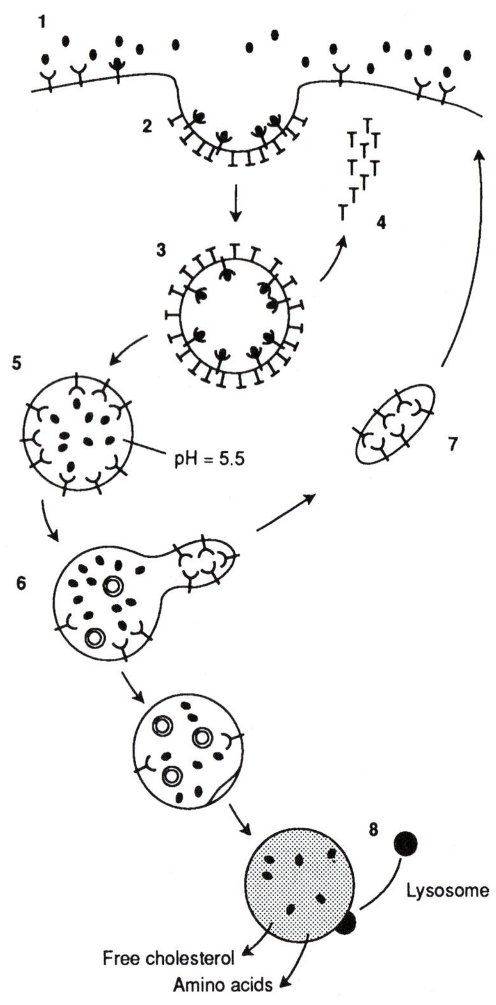

FIGURE 20-6
Cellular uptake and metabolism of LDL. (1) LDL in extracellular fluid. (2) LDL is bound to receptors that cluster in clathrin-coated pits. (3) The pits are endocytosed to form coated vesicles. (4) Clathrin is recycled to the surface, leaving uncoated endosomes. (5) ATP-dependent proton pumps in the endosomal membrane lower the intravesicular pH, resulting in the separation of ligand and receptor. (6) Multivesicular bodies form, as receptors are segregated into finger-like projections. (7) Most receptors are recycled to the cell surface. (8) Multivesicular bodies fuse with primary lysosomes to form secondary lysosomes, which digest LDL to free cholesterol and amino acids. • = LDL; Y = LDL receptor; T = clathrin coat.

plasma LDL saturates the high-affinity, receptor-mediated LDL uptake process. Under these conditions, LDL enters cells by a nonspecific endocytic process known as "bulk-phase pinocytosis." This mechanism seems to play no role in regulation of *de novo* synthesis of cholesterol and leads to its excessive accumulation, with pathological consequences (e.g., atherosclerosis). Abnormally high plasma LDL levels cause scavenger cells (e.g., macrophages) to take up lipids, which results in xanthoma in tendons and skin.

High-Density Lipoproteins

HDLs are secreted in nascent form by hepatocytes and enterocytes (Figure 20-7). Loss of surface components, including phospholipids, free cholesterol, and protein from chylomicrons and VLDL as they are acted on by lipoprotein lipase, may also contribute to formation of HDL in plasma. Discoidal, nascent HDL is converted to spherical, mature HDL by acquiring free cholesterol from cell membranes or other lipoproteins. This function of HDL in peripheral cholesterol removal may underlie the strong inverse relationship between plasma HDL levels and incidence of coronary heart disease. After esterification of HDL surface cholesterol by LCAT, which is activated by apo A-I, HDL sequesters the cholesteryl ester in its hydrophobic core. This action increases the gradient of free cholesterol between the cellular plasma membrane and HDL particles. Cholesteryl esters are also transferred from HDL to VLDL and LDL via apo D, the cholesteryl ester transfer protein (Figure 20-8).

Removal of cholesterol from cells requires an active transport system involving an ATP-binding cassette (ABC1) transporter. ABC1-transporter is a member of a superfamily of proteins involved in energy-dependent transport of several substances across cell membranes. It is activated by protein kinases via phosphorylation. The role of ABC1-transporter in cholesterol efflux is exemplified by an autosomal recessive disease known as *Tangier disease* in which disorder mutations in the gene encoding the ABC1-transporter lead to accumulation of cholesterol esters in the tissues with almost complete absence of HDL cholesterol (discussed later). Other inherited diseases caused by mutations in ABC-proteins are *cystic fibrosis* (Chapter 12), *early onset macular degeneration* (Chapter 38), *sitosterolemia* (Chapter 19), *adrenoleukodystrophy, Zellweger syndrome,* and *progressive familial intrahepatic cholestasis.*

Under normal physiological conditions, HDL exists in two forms: HDL_2 ($d = 1.063$–1.125 g/mL) and HDL_3 ($d = 1.125$–1.210 g/mL). Fluctuations in plasma HDL levels have been principally associated with changes in HDL_2. This fraction is often found in much higher concentration in females and may be associated with their reduced risk for atherosclerotic disease. Clinically, the cholesterol fraction of total HDL ($d = 1.063$–1.210) is commonly measured, and low values are frequently associated with increased risk of coronary heart disease.

The primary determinant of HDL cholesterol levels in human plasma appears to be the cholesterol efflux mediated by ABC1-transporter (also called cholesterol efflux regulatory protein). A defect in this protein causes *Tangier*

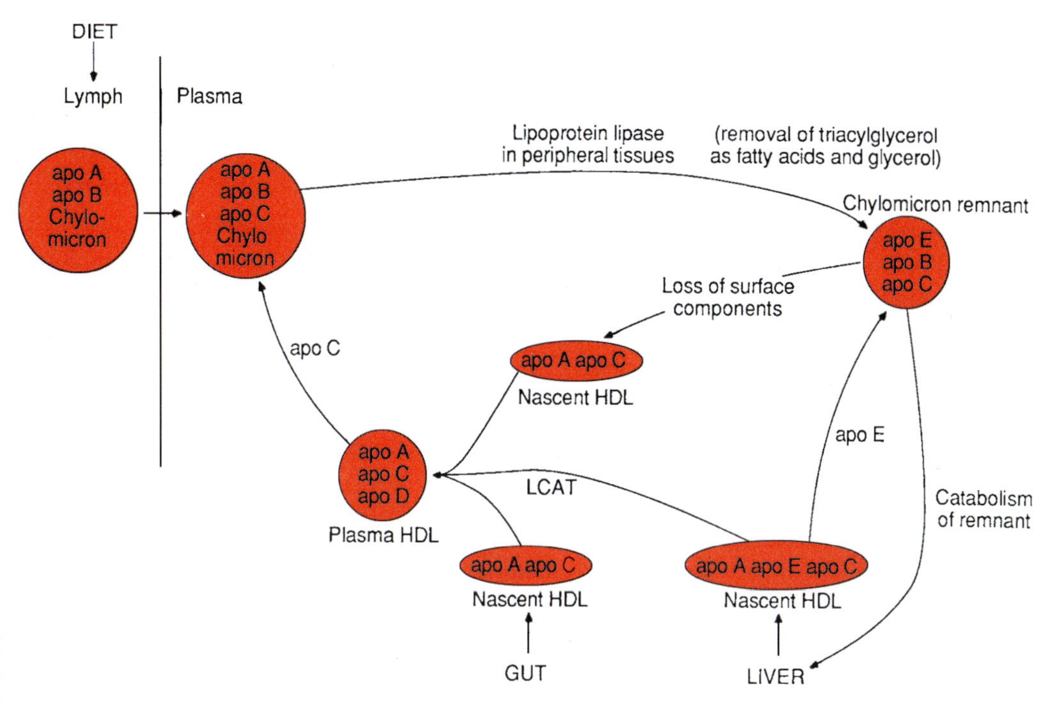

FIGURE 20-7
Sources of HDL. HDL is derived from *de novo* production in the intestinal mucosa and liver, as well as from breakdown of chylomicrons and possibly VLDL. Nascent HDL is discoidal when first formed and becomes spherical in plasma as the formation and storage of cholesteryl ester in its cores (via LCAT) lowers the surface-to-volume ratio. Nascent chylomicrons pick up apo C from plasma HDL, which serves to hinder chylomicron uptake by the liver. After losing a substantial portion of their triacylglycerols to peripheral tissues via lipoprotein lipase, chylomicrons recycle most of their apo C back to HDL, and then gain apo E, which mediates their hepatic uptake as chylomicron remnants.

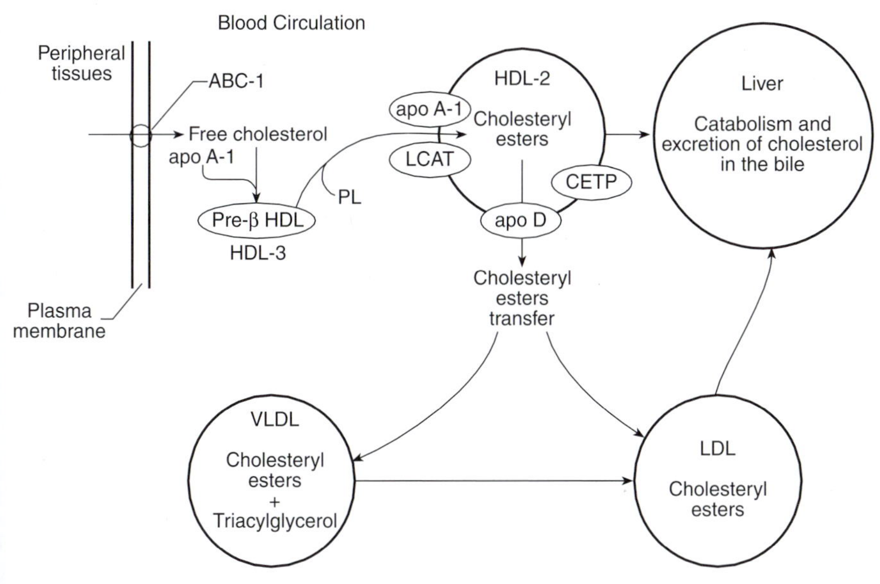

FIGURE 20-8
Role of HDL in peripheral cholesterol removal. Free cholesterol in plasma membranes of peripheral tissues is transferred to apo A-I containing pre-β-HDL (nascent HDL), via an ATP-binding cassette transporter 1 (ABC1). Cholesterol is esterified by LCAT and stored in the core of the HDL particle. In the presence of suitable acceptor lipoproteins (VLDL or LDL), cholesteryl esters are transferred from HDL via apo D and CETP to the lower density lipoproteins, thereby shortening the half-life of plasma cholesterol since VLDL and LDL have a much faster turnover time than HDL. PL = Phospholipid.

disease, which is rare, and the more common *familial HDL deficiency*. Polymorphisms in the ABC1-transporter protein may lead to variations in HDL cholesterol levels increasing the risk for premature coronary artery disease. The rate of apo A-I catabolism is regulated by the size and composition of the particle on which it resides. Smaller lipid-depleted particles have a higher rate of catabolism. In the absence of cholesterol efflux from cells, nascent HDL is not formed and apo A-I is rapidly cleared from the plasma. Modifications in HDL size are the result of the action of enzymes which either esterify cholesterol (LCAT), transfer cholesteryl esters (CETP), or hydrolyze HDL lipids (hepatic lipase, lipoprotein lipase). Apo A-II, which is present on larger HDL particles, inhibits hydrolysis of HDL-triacylglycerol and phospholipids by hepatic lipase, thereby retarding catabolism of larger HDL particles.

HDL binding sites have been reported in several tissues. The interaction appears to be mediated by apo A-I. However, a lipid-lipid interaction may mediate HDL-to-cell binding in fibroblasts and smooth muscle cells. In macrophages, binding occurs via apo A-I. HDL in the macrophage then gains apo E and cholesterol from intracellular lipid droplets. The apo E-enriched and cholesterol-enriched HDL are then secreted (retroendocytosis) into the plasma. An increase in the apo E concentration increases HDL uptake by the liver. The liver and kidneys are believed to be the principal organs for HDL catabolism.

20.3 Lipoprotein-Associated Disorders

Individuals in a given population who have plasma cholesterol and triacylglycerol values that fall in the upper 5–10% of the normal range are arbitrarily considered abnormal. These values are age and sex-dependent. However, "normal" values for one population may be associated with a high risk of heart disease, whereas for another population they may be quite benign. This difference between the statistical "normal" for a population and the levels generally associated with good health should be recognized. Values in Table 20-4 represent the 95th percentile for the North American population.

A practical classification of disorders of lipid transport is based on concentrations of the major classes of lipoproteins in plasma (Table 20-5). Lipoprotein phenotyping is carried out by measuring cholesterol and triacylglycerol levels and by analyzing lipoprotein electrophoretograms. However, the phenotypes so described do not adequately represent current understanding of the pathophysiology underlying these disorders. The disorders can be characterized on the basis of the underlying defects in apolipoproteins, enzymes, or cellular receptors.

TABLE 20-4
Generalized Cutoff Values for Hyperlipidemia (95th percentile)*

Age (years)	Cholesterol (mg/dL)	Triacylglycerol (mg/dL)
1–9	200	120
10–19	205	140
20–29	210	140
30–39	240	150
40–49	265	160
>50	265	190

*Values were obtained from subjects who had fasted for 12–14 hours. Hyperlipidemia occurs when levels are above these values.

Hyperlipoproteinemias can be primary (genetic) or secondary. Some causes of secondary hyperlipoproteinemias include diabetes mellitus, hypothyroidism, nephrotic syndrome, uremia, ethanol abuse, primary biliary cirrhosis, and intake of oral contraceptives. Primary disorders can be due to a single-gene defect or to a combination of genetic defects. The latter type, known as multifactorial or polygenic hyperlipoproteinemias, are affected by secondary disorders and environmental insults, such as a diet high in saturated fat and cholesterol, obesity, and ethanol abuse. In general, the polygenic hyperlipoproteinemias are more common and exhibit lower plasma lipid levels than do the single-gene defects.

Hyperlipidemias

Hypertriacylglycerolemias

Lipoprotein lipase (LPL) deficiency is inherited as an autosomal recessive disorder. *Hyperchylomicronemia* is present from birth. Upon fat ingestion, triacylglycerol levels may rise to 5000–10,000 mg/dL. Chylomicron levels are greatly elevated but not the VLDL levels (*type I hyperlipoproteinemia*). Type I hyperlipoproteinemia can also be

TABLE 20-5
Phenotyping Based on Hyperlipoproteinemia

Phenotype	Lipoprotein Present in Excess
I	Chylomicrons
IIa	LDL
IIb	LDL + VLDL
III	β-VLDL
IV	VLDL
V	Chylomicrons + VLDL

due to deficiency of apo C-II, an essential cofactor for LPL, or to the presence of an inhibitor of LPL in the plasma. Hyperchylomicronemia may be accompanied by an increase in VLDL (type V hyperlipoproteinemia) in which LPL is not absent but may be defective. Occasionally, this disorder is linked to low levels of apo C-II.

The severe hypertriacylglycerolemia of types I and V may result in "eruptive xanthomas," elevated papules in the skin containing lipid-laden phagocytic cells. Phagocytosis of lipoproteins by macrophages in liver and spleen may produce hepatosplenomegaly, abdominal pain, and occasionally acute pancreatitis. Since insulin may be required for LPL production, LPL deficiency may be secondary to diabetes mellitus. Hyperchylomicronemia that occurs with LPL deficiency does not usually predispose to atherosclerosis. However, in some patients with hyperchylomicronemia due to mutations in the LPL gene, premature atherosclerosis does occur. In most patients morbidity and mortality is caused by pancreatitis. Dietary fat restriction reduces the hyperchylomicronemia and patients lead fairly normal lives.

Reduced hepatic uptake of chylomicron remnants and reduction in conversion of plasma VLDL to LDL lead to accumulation of two populations of remnant particles in plasma, a condition known as *dysbetalipoproteinemia* (type III hyperlipoproteinemia). On paper electrophoresis, the abnormal VLDL produces a band in the β to the pre-β region ("broad β" pattern). Relative to their concentration of apo B, both populations of remnants are depleted in apo C and enriched in mutant apo E. Since hepatic uptake of remnant particles depends on apo E, the increased residence in plasma of these remnants suggests a structural abnormality in apo E. The six phenotypes comprise three homozygous (E-4:E-4, E-3:E-3, and E-2:E-2) and three heterozygous forms (E-4:E-3, E-3:E-2, and E-4:E-2). The apo-E isoforms differ by single amino acid substitutions at two sites. Apo E-2 has cysteine at residues 112 and 158, apo E-3 has cysteine at 112 and arginine at 158, and apo E-4 has arginine at 112 and 158. Patients with *dysbetalipoproteinemia* exhibit phenotype E-2:E-2; apo E-2, which is different from the predominant wild-type apo E-3, is known as mutant E-2. Apo E-2 has a much lower affinity (1%) for the hepatic receptors compared with other phenotypes. *Type III hyperlipoproteinemia* is a multifactorial disorder. The expression of the disorder not only involves the inheritance of two alleles for apo E-2:apo E-2, but also requires other genetic, hormonal, and environmental factors. For example, diabetes mellitus and hypothyroidism are frequently associated with type III hyperlipoproteinemia. Age, sex, nutritional status, and alcohol consumption are all factors that affect this disorder. There are other rare variants of apo E (e.g., Arg-142 → Cys, Arg-145 → Cys, Lys-146 → Glu) which can give rise to type III hyperlipoproteinemia, independent of other factors. The association of apo E-4 and Alzheimer's disease has been confirmed in many populations, but its significance in the pathology of the disorder awaits explanation (Chapter 4).

Patients with type III hyperlipoproteinemia exhibit increased plasma cholesterol and triacylglycerol and the presence of β-VLDL. Dysbetalipoproteinemics are prone to premature vascular disease, eruptive xanthomas on elbows and knees, and planar xanthomas in the palmar and digital creases. These patients respond well to therapy. Dietary therapy is preferred, but drug therapy (see below) may also be necessary.

Primary elevation of VLDL, or *hyperprebetalipoproteinemia* (type IV hyperlipoproteinemia), occurs in familial hypertriacylglycerolemia or familial combined hyperlipidemia. The former is characterized by overproduction of VLDL triacylglycerols but normal synthesis of apo B-100. A familial form is characterized by overproduction of triacylglycerols and apo B-100. These disorders differ from familial combined hyperlipidemia in not showing increased levels of LDL. A third form of primary hypertriacylglycerolemia has been identified in which there is decreased catabolism of VLDL rather than overproduction. Mild hyperchylomicronemia may appear to complicate these disorders based on defective VLDL metabolism; however, it is usually due to obesity or excess dietary fat intake and should not be confused with the severe hyperchylomicronemias mentioned earlier. Patients with either familial combined hyperlipidemia or familial hypertriglyceridemia are at risk for mortality from cardiovascular disease. Expression of the apolipoprotein (APO) gene (*APOAV*) that is located proximal to *APOAI/CIII/AIV* gene cluster, decreases plasma triacylglycerol levels. Single nucleotide polymorphisms in the *APOAV* gene locus are associated with high levels of plasma triacylglycerol. Prospective studies have shown that plasma triacylglycerol levels greater than 150mg/dl are an independent risk factor for cardiovascular disease and mortality.

Hypercholesterolemias

Familial hypercholesterolemia (FH) is an inborn error of metabolism due to a defective LDL-receptor protein. Five classes of mutations have been identified consisting of more than 150 different alleles. The defects in the receptor function can be grouped into five types:

1. The receptor may not be synthesized at all;
2. The receptor may not be transported to the surface;
3. The receptor may fail to bind LDL;
4. The receptor may fail to cluster in coated pits; or
5. The receptor may fail to release LDL in the endosome.

Recombinant DNA technology has begun to link specific mutations to altered biochemical expression. For example, the failure to cluster in coated pits, the internalization-defective phenotype, arises from a mutation that results in a loss of the membrane-spanning and cytoplasmic domains of the LDL-receptor protein (Figure 20-5). The abnormal protein is secreted from the cell with about 30% remaining on the cell surface; having no transmembrane portion, it is unable to migrate to coated pits. The autosomal dominant gene of FH is expressed in heterozygous form in approximately 1 in 500 individuals. Thus, 1 in 250,000 marriages will pair FH heterozygotes; and one fourth of their offspring, or 1 of 1,000,000 individuals, will be homozygous for FH. FH heterozygotes have normal triacylglycerol and HDL levels, but their LDL cholesterol levels are typically between 320 and 500 mg/dL. LDL residence time in plasma is increased to about 2.5 times normal. About half of affected individuals have palpable tendon xanthomas and experience onset of coronary artery disease in their third or fourth decade of life. In homozygous subjects, cholesterol levels between 600 and 1200 mg/dL are common, tendon xanthomas and corneal opacification are usually present, and onset of coronary disease before 10 years of age is typical. Since the appearance of accelerated atherogenesis is often accompanied by no other risk factors, it contributes important evidence linking elevated plasma cholesterol levels with atherosclerosis.

Defects in the LDL receptor result in decreased uptake of LDL and also in increased production caused by an increase in the fraction of IDL that is converted to LDL (Figure 20-9). Members of kindreds with FH may show primary elevation of only LDL (type IIa) or combined elevations of LDL and VLDL (type IIb hyperlipoproteinemia). These patterns may also be seen in familial combined hyperlipidemia, for which the molecular defects remain unknown, although the basic abnormality appears to be overproduction of lipoproteins containing apo B. In many individuals with this disorder, the secretion rates of VLDL are elevated and result in *hypertriacylglycerolemia* (*type IV hyperlipoproteinemia*), whereas other members of the same kindred may show hypercholesterolemia, presumably due to excessive conversion of VLDL to LDL or to defective LDL clearance. Some individuals may show elevations of VLDL and LDL levels, so that diagnosis of familial combined hyperlipidernia can be made only by family screening. As opposed to FH patients, individuals with familial combined hyperlipidemia rarely exhibit hyperlipidemia before adulthood. Hypertriacylglycerolemia in these patients may be exacerbated by obesity or diabetes. In many patients with primary hypercholesterolemia in which LDL is grossly elevated, a monogenetic inheritance cannot be demonstrated. In this case, alteration of multiple gene products may be responsible for the elevation of LDL, and the defect is classified as polygenic hypercholesterolemia. Secondary hypercholesterolemia may occur in hypothyroidism from delayed clearance of LDL, the result of down-regulation of LDL receptors. In the nephrotic syndrome, overproduction of VLDL results either in elevated VLDL itself or in elevated LDL due to an increase in conversion. Elevated LDL cholesterol has also been reported in porphyria, anorexia nervosa, and Cushing's syndrome.

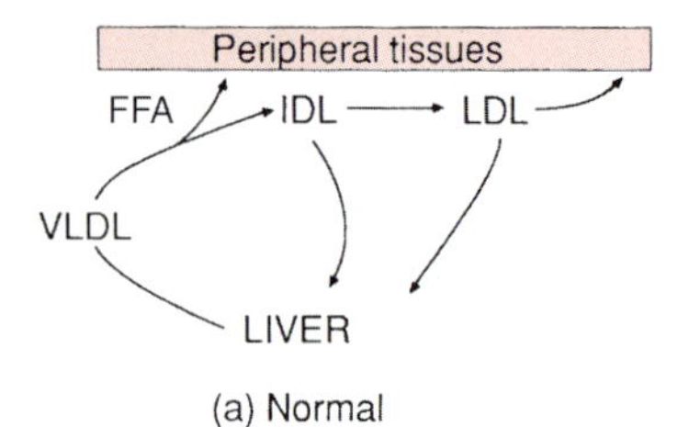

(a) Normal

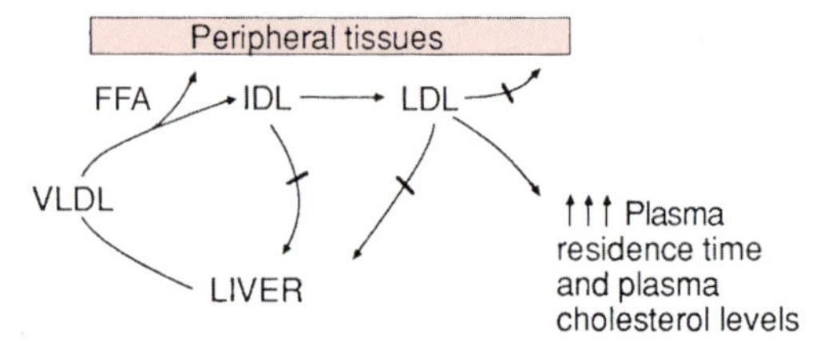

(b) Down-regulated, or defective receptor

FIGURE 20-9
LDL-receptor deficiency. In the normal condition (a), VLDL produced by the liver loses triacylglycerol as free fatty acids (FFA) via lipoprotein lipase to peripheral tissues and then proceeds down the metabolic cascade to IDL and LDL. A major portion of these two lipoprotein species is taken up by the liver or peripheral tissues via the LDL (apo B, E) receptor. In individuals with down-regulated or genetically defective LDL receptors (b), the residence time in the plasma of IDL is increased, a greater proportion being converted to LDL. LDL production and turnover time are increased, and total plasma cholesterol levels become grossly abnormal.

Hypolipidemias

In *abetalipoproteinemia* (acanthocytosis; *Bassen–Kornzweig syndrome*), lipoproteins containing apo B (LDL, VLDL, and chylomicrons) are absent. This experiment of nature indicates that apo B is absolutely essential for the formation of chylomicrons and VLDL (and hence LDL). Clinical findings include retinitis pigmentosa, malabsorption of fat, and ataxic neuropathic disease. Erythrocytes are distorted and have a "thorny" appearance due to protoplasmic projections of varying sizes and shapes (hence the term *acanthocytosis,* from Greek *akantha* "thorn"). Serum triacylglycerol levels are low. The genetic defect may completely prevent apo B synthesis, although other possibilities exist. Since it is transmitted as an autosomal recessive trait, heterozygotes cannot be detected without an affected offspring.

In *hypobetalipoproteinenia* the plasma LDL level is decreased (10–20% of normal), but that of HDL is normal, and that of VLDL is mildly lowered. Of 23 affected individuals from the four known affected families, one had central nervous system dysfunction and fat malabsorption. The others had mild or no pathological changes. The disease is inherited as an autosomal dominant trait. The benign nature of this condition is in sharp contrast with the seriousness of hyperbetalipoproteinemia. In the latter, LDL cholesterol concentrations are two to six times normal, and patients are predisposed to premature atherosclerosis. In another form of hypobetalipoproteinemia, the patient synthesized apo B-48 and secreted chylomicrons but did not produce apo B-100 or secrete VLDL.

Familial apolipoprotein A-I and C-III deficiency is characterized by mild corneal opacification, coronary artery disease, marked HDL deficiency, and a normal ratio of free cholesterol to esterified cholesterol in plasma in all probands, also yellow orange, lipid-laden plaques on the trunk, eyelids, neck, chest, anus, and backs of several affected individuals. The LDL is often found to be triacylglycerol-rich. The mode of inheritance is autosomal codominant.

HDL deficiency with planar xanthomas was identified in a 48-year-old Swedish woman who had a history of yellow discoloration of skin around the eyes and groin. She also exhibited hypothyroidism, angina pectoris, facial neuralgia, and corneal opacification. Both eyelids were thickened and infiltrated with small firm nodules. Moderate hepatomegaly was noted. Plasma triacylglycerol and VLDL cholesterol levels were elevated, LDL cholesterol was normal, and HDL cholesterol was very low. LCAT activity and the ratio of free cholesterol to total cholesterol was normal. Apo A-I_{Milano} is the result of a substitution of cysteine for arginine at residue 173 of apo A-I. This mutation apparently results in enhanced uptake and hepatic catabolism of HDL. No evidence of coronary artery disease, corneal opacification, xanthomas, or hepatosplenomegaly has been reported. Plasma concentrations of HDL and apo A-I are greatly reduced. Triacylglycerol is elevated, and both LDL and HDL are triacylglycerol-rich in affected kindred. LCAT activity is normal. The mode of inheritance is autosomal codominant.

Familial hypoalphalipoproteinemia has been identified in several kindreds with a history of accelerated coronary disease. Plasma lipid and lipoprotein values were all normal except for reduced levels of HDL cholesterol (50% of normal). Slight reductions in apo A-II were reported (78% of normal), but apo A-I was normal, as were LCAT and hepatic lipase activities.

Tangier disease is an autosomal recessive disorder. In the homozygous state it is characterized by the absence of plasma HDL cholesterol and deposition of cholesterol esters in the reticuloendothelial system with hepatosplenomegaly, enlarged tonsils with yellowish orange color, enlarged lymph nodes, and peripheral neuropathy. Patients with Tangier disease, despite low levels of HDL cholesterol, are not uniformly at risk for coronary artery disease. The molecular basis for Tangier disease is a defect of cholesterol efflux from cells. The defect has been localized to mutations in a specific ABC1-transporter protein. Defective export of cholesterol from cells, particularly from tissue macrophages causes accumulation of cholesteryl esters. Plasma apo A-I levels in Tangier disease undergoes rapid clearance since nascent HDL is not made due to lack of cholesterol efflux from the cells.

LCAT deficiency is due either to absence of the enzyme or to synthesis of defective enzyme. LCAT catalyzes the following reaction:

Phosphatidylcholine (lecithin) + cholesterol →
cholesteryl ester + 1-acylglycerolphosphocholine (lysolecithin)

This reaction is responsible for formation of most of the cholesteryl ester in plasma. The preferred substrate is phosphatidylcholine, which contains an unsaturated fatty acid residue on the 2-carbon of the glycerol moiety. HDL and LDL are the major sources of the phosphatidylcholine and cholesterol. Apo A-I, which is a part of HDL, is a powerful activator of LCAT. Apo C-I has also been implicated as an activator of this enzyme; however, activation may depend on the nature of the phospholipid substrate. LCAT is synthesized in the liver. The plasma level of LCAT is higher in males than in females. The enzyme converts excess free cholesterol to cholesteryl ester with the simultaneous conversion of lecithin to lysolecithin. The products are subsequently removed from circulation. Thus, LCAT plays a significant role in the removal of cholesterol and lecithin from the circulation, similar to the role of lipoprotein lipase in the removal of triacylglycerol contained in chylomicrons and VLDL. Since LCAT regulates the levels of free cholesterol, cholesteryl esters, and phosphatidylcholine in plasma, it may play an important role in maintaining normal membrane structure and fluidity in peripheral tissue cells.

Primary deficiency of LCAT, initially found in three Norwegian sisters, is rare and is inherited as an autosomal recessive trait. Manifestations include corneal opacity, normocytic anemia due to decreased erythropoiesis and to increased erythrocyte destruction (membrane erythrocytes contain increased amounts of cholesterol and phosphatidylcholine), proteinuria, and hematuria. The plasma lipoproteins contain abnormally high amounts of unesterified cholesterol and phospholipids, an abnormal lipoprotein (Lp-X; see above), VLDL with β-electrophoretic mobility, and two types of abnormal HDL particle.

LCAT deficiency can be selective with regard to the rate of plasma cholesterol esterification that occurs in VLDL, LDL, and HDL. In *fisheye disease,* the LCAT activity directed toward VLDL and LDL is normal, whereas LCAT activity directed toward HDL is absent. Fisheye disease is characterized by severe corneal opacification beginning in the teenage years, with eventual bilateral visual impairment. This loss of vision has not been reported in other instances of corneal opacification. Analysis of the cornea after transplantation revealed vacuolization of the stroma and Bowman's layer and a greatly increased cholesterol content. Patients with fisheye disease do not exhibit hematological and renal abnormalities, which are characteristic features of familial LCAT deficiency. Hypertriacylglycerolemia with elevated VLDL and triacylglycerol-rich LDL accompany a marked reduction in HDL levels.

Atherosclerosis and Coronary Heart Disease

Atherosclerosis (athero = fatty and sclerosis = scarring or hardening) of the coronary and peripheral vasculature is the leading cause of morbidity and mortality worldwide. Lesions (called *plaque*) are initiated by an injury to endothelium and thicken the intima of arteries, occlude the lumen, and compromise delivery of nutrients and oxygen to tissue (*ischemia*). Atherosclerotic lesions primarily occur in large and medium-sized elastic and muscular arteries and progress over decades of life. These lesions cause ischemia, which can result in infarction of the heart (myocardial infarction) or brain (stroke), as well as abnormalities of extremities. The proximate cause of occlusion in these pathological conditions is thrombus formation.

The normal artery wall is composed of three layers: intima, media, and adventita (Figure 20-10). On the luminal side, the intima contains a single layer of endothelial cells. These cells permit passage of water and other substances from blood into tissue cells. On the peripheral side, the intimal layer is surrounded by a fenestrated sheet of elastic fibers (the internal elastic lamina). The middle portion of the intimal layer contains various extracellular components of connective tissue matrix and fibers and occasional smooth cells, depending on the type of artery, and the age and sex of the subject.

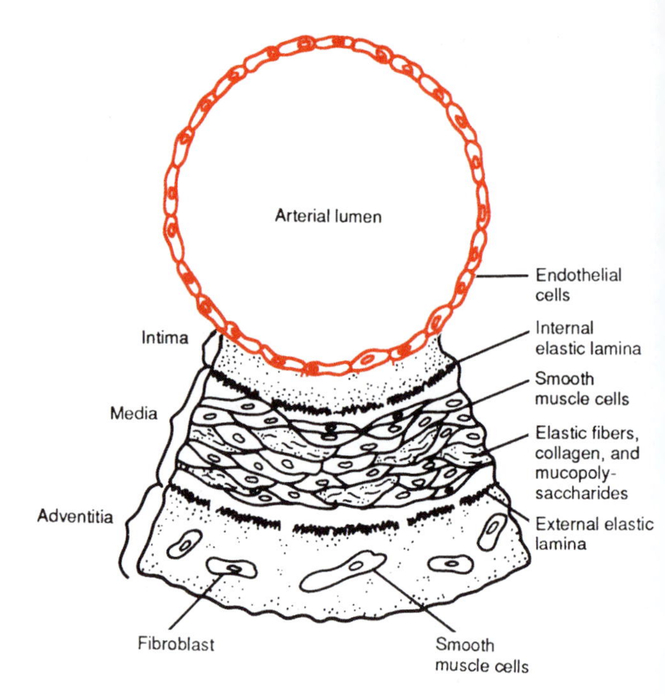

FIGURE 20-10
Cross-sectional view of an artery.

The media is composed of diagonally oriented smooth muscle cells. They are surrounded by collagen, small elastic fibers, and glycosaminoglycans (proteoglycans). Most cells are attached to one another by specific junctional complex, which are arranged as spirals between the elastic fibers and support the arterial wall.

The adventitia is separated from the media by a discontinuous sheet of elastic tissue (the external elastic lamina) and consists of smooth muscle cells, fibroblasts, collagen fibers, and glycosaminoglycans. This layer is supplied with blood vessels to provide nutrients.

Atherosclerosis affects mainly the intimal layer. Lesions occur because of

1. Proliferation of smooth muscle cells,
2. Lipid accumulation in smooth muscle cells and scavenger cells (resident macrophages and migrant plasma monocytes), and
3. Connective tissue formation.

Perhaps the most satisfying hypothesis for the formation of atherosclerotic lesions is that of "response to injury" in which lesions are precipitated by some form of injury to endothelial cells. The injury may be caused by elevated plasma levels of LDL and modified LDL (oxidized LDL), free radicals (e.g., caused by cigarette smoking), diabetes mellitus, hypertension-induced shear stress, and other factors that lead to focal desquamation of endothelial cells such as elevated plasma homocysteine levels, genetic factors, infectious microorganisms (e.g., herpes viruses, *Chlamydia pneumoniae*), or combinations of these factors. The endothelial response to injury which eventually leads to atherosclerosis involves platelets, leukocytes, and a variety of cytokines, e.g., vasoactive molecules, adhesion molecules, and growth factors (Chapter 35). Once the subendothelial region is exposed, circulating platelets and monocytes become attached and release factors with chemotactic and mitogenic properties. Platelet-derived growth factor (PDGF), for example, binds with high affinity to specific binding sites on the surface of smooth muscle cells. PDGF induces DNA synthesis, thereby initiating the proliferation of smooth muscle cells. It also increases binding and degradation of LDL. Monocytes enter the arterial wall at the site of injury and produce a mitogenic factor that appears to be as potent as PDGF.

In the pathogenesis of atherothrombosis inflammation may play a significant role. An acute-phase systemic inflammatory marker, C-reactive protein (CRP) is elevated in patients with acute ischemia or myocardial infarction. In a prospective study consisting of apparently healthy men, individuals with high baseline plasma CRP levels (1.51 vs. 1.31 mg/L) had a higher risk of myocardial infarction or ischemic stroke than men in the lower quartile of CRP values. The use of aspirin, an anti-inflammatory and antiplatelet drug, reduced the risk of myocardial infarction among men in the highest quartile of plasma CRP levels. These data suggest that the plasma CRP level in apparently healthy men can predict myocardial infarction and ischemic stroke, and is an independent marker not altered by plasma lipid levels, body mass index, smoking, or blood pressure.

Oxidative modification of LDL (e.g., lipid peroxidation) occurs primarily in the arterial intima and is a prerequisite for LDL uptake by macrophages. This uptake by macrophages is not regulated like LDL uptake by receptor-mediated endocytosis, and the uptake is done by a scavenger receptor. This unregulated uptake leads to accumulation of LDL so that macrophages become foam cells. The foam cells form yellow patches on the arterial wall that are called *fatty streaks* (discussed later). Since oxidized LDL is an important and possibly an essential component in the pathogenesis of atherosclerosis, the inhibition of LDL oxidation by variety of antioxidants has been considered in treating coronary artery disease. There are two lipophilic antioxidants, α-tocopherol (vitamin E) and β-carotene (provitamin A), and the water-soluble antioxidant ascorbate (vitamin C). The antioxidants also help maintain normal endothelial function (e.g., preserving endothelial-derived nitric oxide production). Epidemiological studies have shown an inverse relationship between intake of antioxidant vitamins and coronary artery disease. However, a cause-and-effect association between intake of antioxidant vitamins and prevention of coronary artery disease has not yet been determined. Probucol, in addition to its hypocholesterolemic action (discussed later), also has antioxidant properties.

Prostaglandins and thromboxanes play major roles in platelet adhesion and aggregation to damaged intima (Chapter 18). Normal endothelium synthesizes prostacyclin (PGI_2), which inhibits platelet aggregation. In damaged endothelium, platelets aggregate and form thromboxanes; thromboxanes and endoperoxide (which has increased owing to lack of utilization in the synthesis of PGI_2) accelerate the platelet aggregation and lead to the formation of thrombus. Aspirin is an inhibitor of cyclooxygenase which initiates the synthesis of prostaglandins and thromboxanes (Chapter 18). Aspirin therapy is beneficial in the prevention of heart attacks and strokes by interfering with the function of platelets. However, aspirin may increase the risk of hemorrhagic stroke. Thus, the benefits and risks of aspirin therapy in patients with known coronary artery disease or history of stroke have to be assessed before the initation of aspirin therapy.

Three types of atherosclerotic lesion exist:

1. The fatty streak, which is found at a young age ubiquitously among the world's population, consists of smooth muscle cell aggregates surrounded by cholesteryl ester deposits; it is a sessile lesion that does not obstruct and presents no clinical symptoms.
2. The fibrous plaque, which protrudes into the lumen, is composed chiefly of cholesteryl ester-loaded smooth muscle cells proliferating in the intimal layer, surrounded by collagen, elastic fibers, and proteoglycans; it is enclosed in a fibrous cap.
3. The complicated lesion is associated with occlusive disease; the contents of the fibrous plaque become calcified and are altered as a result of hemorrhage, cell necrosis, and mural thrombosis.

Although only the complicated plaque results in significant (>75%) coronary lumenal narrowing, evidence indicates that complicated lesions originate during the first two decades of life. Such lesions were found on autopsy in more than a third of teenaged military casualties in Vietnam and Korea. Although the final event that results in blockage of nutrient flow and myocardial infarction may be episodic rather than gradual, the degenerative buildup of material appears to begin in childhood. It is therefore not surprising that the serum cholesterol level in children in those countries where coronary disease is rare is significantly lower than that of "normal" children in the U.S.A. For example, children between the ages of 10 and 14 years in Mexico have a mean serum cholesterol value of 100 mg/dL, whereas children of comparable age in the U.S.A. have a value of 185 mg/dL.

Epidemiological studies have established that reduction in plasma total cholesterol levels reduces the risk of coronary heart disease (CHD). For example, when the bile acid sequestrant cholestyramine was used to reduce plasma cholesterol levels in a randomized, double-blind study, a 19% reduction in CHD risk was associated with each decrement of 8% in total plasma cholesterol levels. The incidence of CHD in men who experienced a fall of 25% in total plasma cholesterol was half that of men who remained at pretreatment levels. Several studies show that therapy with cholesterol-reducing HMG-CoA reductase inhibitors (called statins) in patients with coronary artery disease and elevated cholesterol levels reduces both fatal and nonfatal heart attacks. Cardiovascular benefits from statin therapy have been observed even among patients with "normal" cholesterol levels. These data and the significant decreases in CHD mortality in industrialized nations in recent years, which were accompanied by declines in total plasma cholesterol levels, suggest that although a large fraction of an individual's plasma cholesterol level may be genetically determined, sociocultural determinants can strongly influence susceptibility to heart disease.

The relationship between the development of premature CHD and LDL cholesterol is well established. Recent analyses of epidemiological data have suggested an inverse correlation between HDL cholesterol and premature CHD. At any given level of LDL cholesterol, the probability of CHD increases as the level of HDL cholesterol decreases. These observations suggest that a lowered HDL cholesterol level (or HDL deficiency) is an independent risk factor for premature CHD. In one study of men aged 50–69, the prevalence (number of cases per 1000 persons) of CHD at plasma HDL cholesterol levels of < 25, 25–34, 35–44, and 45–54 mg/dL was 180, 123.6, 94.6, and 77.9, respectively. The cardioprotective effect of HDL cholesterol may involve removal of cholesterol from the peripheral tissues and its transport to the liver for removal. As discussed earlier, the efflux of cholesterol from peripheral cells is mediated by ABC1-transporter 1. Defects and polymorphisms of this protein are determining factors in the plasma HDL cholesterol levels. In animal studies it has been shown that regulation and expression of the ABC1 protein play a significant role in cholesterol efflux and reverse cholesterol transport. Furthermore the ABC1 protein present in the intestinal cell membrane mediates the cholesterol efflux into the lumen, thus preventing cholesterol absorption. Potential mechanisms of the antiatherogenic effects of HDL cholesterol include:

1. Inhibition of conversion of LDL to oxidized LDL which is preferentially taken up by the tissue macrophages,
2. The prevention of adhesion of monocytes to the endothelium, and
3. The prolongation of the half-life of prostacyclin produced by endothelial cells to promote vasodilatory effects.

The inhibition of LDL oxidation by HDL has been attributed to both an HDL-associated, calcium-dependent enzyme known as paraoxonase, and to a phospholipid fraction. HDL cholesterol and triacylglycerol levels are inversely correlated across various populations. Decreased HDL levels are associated with male sex, obesity, physical inactivity, cigarette smoking, and hypertriacylglycerolemia. Increased HDL levels are associated with female sex and vigorous exercise undertaken regularly (e.g., running). The incidence of atherosclerosis in premenopausal women is significantly lower than in men. This difference has been attributed to estrogen. In fact, in postmenopausal women with a decline in estrogen levels, the risk of athersclerosis rises and this risk can be reduced with estrogen therapy. The atheroprotective effect of estrogen involves both rapid effects (nongenomic) and long-term effects (genomic). The direct effects of estrogen on endothelial cells and smooth muscle cells of blood vessels include vasodilation and inhibition of response to vascular injury. The vasodilation has been attributed to the production of nitric oxide and prostacyclin in the blood vessel. The long-term effects of estrogen involve changes in gene expression mediated by estrogen receptors that are ligand-activated transcription factors (Chapter 34).

Several parameters are affected by estrogen. Some of these include changes in the serum lipid concentration, coagulation and fibrinolytic systems, and antioxidant systems. Estrogen in women lowers total serum cholesterol, LDL cholesterol levels, and Lp(a) concentrations and elevates HDL cholesterol. These effects of estrogen on serum lipoproteins are largely due to estrogen-mediated effects of the expression of apolipoprotein genes in the liver. Progesterone can blunt estrogen's effect on serum lipid levels. In addition to environmental factors, genetic factors influence HDL levels. The familial aggregation of CHD is well recognized, and children of CHD patients have lower HDL concentations. The approach taken in the management of individuals with low HDL cholesterol level (Table 20-6) is similar to that of patients in whom LDL cholesterol levels are high (discussed later). First, it is essential to exclude secondary causes of lipid abnormalities, which include acute illness, chronic diseases, and some commonly prescribed drugs (Table 20-7). Second, it is

TABLE 20-6
Plasma Levels of Cholesterol and Risk of Atherosclerosis in Adults

Cholesterol Level	Desirable	Moderate Risk	High Risk
Total	<200 mg/dl (5.20 mmol/L)	200–239 mg/dl (5.2–6.20 mmol/L)	>240 mg/dl (>6.20 mmol/L)
LDL <100	<100 mg/dl* (2.60 mmol/L)	130–159 mg/dl (3.35–4.10 mmol/L)	>160 mg/dl (>4.15 mmol/L)
HDL	40–59 mg/dl (1.03–1.53 mmol/L) Cardioprotective: >60 mg/dl (1.55 mmol/L)	—	<40 mg/dl (1.03 mmol/L)

*LDL level 100–129 mg/dl considered near optimal/above optimal.

The Total Plasma and LDL-Cholesterol Levels in Children and Adolescents from Families with Hypercholesterolemia or Premature Cardiovascular Disease

Cholesterol Level	Acceptable	Borderline	High
Total	<170 mg/dl (<4.42 mmol/L)	170–199 mg/dl (4.42–5.17 mmol/L)	>200 mg/dl (52 mmol/L)
LDL	<100 mg/dl (<2.6 mmol/L)	110–129 mg/dl (2.86–3.35 mmol/L)	>130 mg/dl (>3.38 mmol/L)

TABLE 20-7
Examples of Secondary Causes That Can Alter Plasma Lipoprotein Levels

Acute Illnesses
Surgical procedures
Viral illness
Burns
Myocardial infarction
Acute inflammatory disorders
Chronic Conditions
Diabetes mellitus
Thyroid disease
Uremia
Nephrotic syndrome
Liver disease
Commonly Prescribed Drugs
Diuretics
Progestins
Androgens
β-Adrenergic blocking agents

important to initiate lifestyle changes that include weight loss in obese individuals, dietary modifications (e.g., reduced saturated fat intake including *trans*-fatty acids), exercise, avoidance of tobacco, and moderate alcohol consumption. Individuals who are homozygous for a slow oxidizing alcohol dehydrogenase gene (ADH3) have higher HDL-cholesterol levels and a lower risk of myocardial infarction. In the absence of positive results with lifestyle therapy, the pharmacological agents for increasing HDL cholesterol levels includes niacin and estrogen in postmenopausal women. Modest increases in HDL cholesterol levels are also observed with HMG-CoA reductase inhibitors and fibric acids.

Another factor that regulates HDL cholesterol levels is the plasma level of cholesteryl ester transfer protein (CETP). CETP, a hydrophobic glycoprotein (M.W. 741,000), facilitates the transfer of cholesteryl esters in HDL and triacylglycerols in LDL and VLDL (see above). In CETP deficiency due to a point mutation (G $\rightarrow$ A) in a splice donor site that prevents normal processing of mRNA, the plasma HDL cholesterol levels of affected individuals are markedly high, with decreased LDL cholesterol. In the affected families, there was no evidence of premature atherosclerosis and, in fact, there was a trend toward longevity. These observations support the role of CETP and the antiatherogenic property of HDL. However, not all factors that elevate HDL levels may be protective against atherosclerosis. Because hepatic lipase deficiency causes an increase in triglyceride-enriched HDL (and IDL), it is associated with premature atherosclerosis. Examples of low plasma HDL cholesterol which are not associated with coronary artery disease are also known. In both Tangier disease and fisheye disease (discussed earlier), despite low plasma or HDL cholesterol levels, coronary artery disease is an inconsistent finding. In individuals with apo A-I_{Milano}, despite low HDL cholesterol levels, longevity has been observed in the affected individuals.

Elevated Lp(a) is a major independent risk factor for atherosclerosis in patients with familial hypercholesterolemia. This risk is independent of levels of LDL, HDL, age, sex, and smoking habits, and is primarily dependent on genetic factors. The exact mechanism of Lp(a) acceleration of atherosclerosis is not understood but may be attributable to its potential inhibition of blood clot dissolution caused by its structural similarity with plasminogen (see above and Chapter 36).

In evaluating the plasma cholesterol level with respect to CHD, the measurement of LDL and HDL cholesterol is therefore useful. These two levels can be determined by measurement of plasma triacylglycerol and cholesterol after a 12- to 14-hour fast. In the absence of chylomicrons and remnant lipoproteins, VLDL cholesterol is equal to triacylglycerol divided by 5. HDL cholesterol is estimated in the supernatant obtained from precipitation of non-HDL lipoproteins by heparin-Mn^{2+}, dextran sulfate 500, or phosphotungstate-Mg^{2+}. LDL cholesterol is calculated from the following formula (all values are expressed as mg/dL):

$$\text{LDL cholesterol} = \text{total cholesterol} - (\text{VLDL cholesterol} + \text{HDL cholesterol})$$

or

$$\text{LDL cholesterol} = \text{total cholesterol} - \left(\frac{\text{triacylglycerol}}{5} + \text{HDL cholesterol}\right)$$

Direct measurements of serum LDL cholesterol and HDL cholesterol are currently used. For example, a direct LDL cholesterol measurement is obtained by immunoseparation. In this procedure, addition of antibodies against apo A-I and apo E removes the HDL and VLDL fractions, respectively, allowing LDL cholesterol to be measured directly in the filtrate. Measurement of apo A-I, apo B, and apo E may prove to be equally useful, if not better, for the assessment of CHD. For example, the predictive values for CHD of LDL cholesterol and apo B-100 values are similar, as are those of HDL cholesterol and apo A-I.

Lipid-Lowering Methods

Since hypercholesterolemia (in particular, LDL cholesterol) increases the risk of CHD, it seems reasonable to lower cholesterol levels in patients whose levels put them at risk. Before treatment, other risk factors such as hypertension, cigarette smoking, obesity, and glucose intolerance need to be evaluated and corrected. Disorders that exacerbate hyperlipoproteinemia (e.g., chronic ethanol abuse, hypothyroidism, diabetes mellitus) need to be treated before lipid-lowering measures are taken (discussed earlier, Table 20-7).

In uncomplicated hyperlipoproteinemia, the first step is to reduce dietary intake of cholesterol and saturated fats and to establish an ideal body weight. Daily dietary cholesterol consumption should be limited to 300 mg or less and fat intake to 30% of total energy intake, with a polyunsaturated fat to saturated fat ratio of 1:1. Polyunsaturated fats include the ω-3 family (found mostly in fish oils) and the ω-6 family of essential fatty acids (found mostly in vegetable oils). The lipid-lowering effects of ω-3 fatty acids (particularly of decreases in fatty acid and VLDL synthesis in liver) are much greater than those of ω-6 fatty acids. The former also decrease platelet aggregation. The lipid-lowering effect and decreased platelet aggregation are beneficial in lowering the risk of CHD. The low death rate from CHD among Greenland Eskimos is probably due to daily consumption of 5–10 g of ω-3 essential fatty acids that enter their diet by way of marine phytoplankton eaten by fish, which in turn are eaten by seals, walruses, and whales.

Hypocholesterolemia can also be brought about by diets rich in water-soluble fibers. Sources include pectins, gums, certain hemicellulosees, and storage carbohydrates (Chapter 9). Water-soluble fibers are found in fruits, oats, barley, and legumes; oat bran and dried beans are particularly rich sources. The cholesterol-lowering effect is not shared by water-insoluble fibers (e.g., cellulose, found in wheat bran). The specific cholesterol-reducing mechanisms of water-soluble fiber may be related to its ability to bind bile acids in the intestine. This decreases bile acid reabsorption and results in less cholesterol being available for lipoprotein synthesis in the liver. In addition to their lipid-lowering effect, water-soluble fiber increases intestinal transit time, delays gastric emptying, and slows glucose absorption. The last effect may be beneficial in the treatment of diabetes mellitus.

If dietary therapy is unsuccessful, drug therapy should be employed. Five classes of drugs are available for treatment of hyperlipoproteinemias; their effects are due to decreased production or enhanced removal of lipoprotein from plasma.

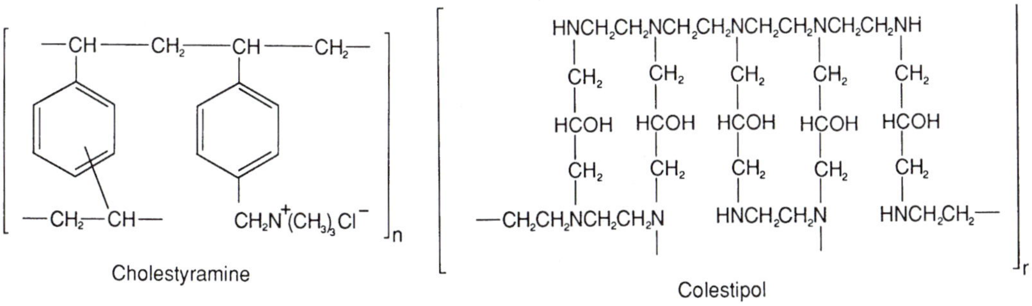

Cholestyramine

Colestipol

FIGURE 20-11
Structures of bile acid sequestrants. Cholestyramine and colestipol are hydrophilic yet water-insoluble, nondigestible, and nonabsorbable synthetic resins. They bind bile acids in the intestine to increase their loss in feces and thereby decrease plasma cholesterol levels.

1. The steroid ring of cholesterol is not degradable within the body. The principal avenue of cholesterol secretion is via bile acid excretion in the feces. Cholestyramine and colestipol (Figure 20-11) are nonabsorbable resins that bind bile acids in the intestine, causing their excretion in feces and interruption of their enterohepatic circulation. The decreased return of bile acids to the liver increases conversion of cholesterol to bile acids and increases the concentration of LDL receptors on hepatocytes. This action mediates increased removal of plasma LDL and decreased cholesterol levels. Bile acid sequestrants are most effective in patients with heterozygous FH or polygenic hypercholesterolemia. It is not effective in patients with homozygous FH who are completely deficient in LDL receptors.
2. Nicotinic acid reduces the levels of VLDL and LDL by inhibiting hepatic secretion of VLDL and by suppressing mobilization of fatty acid from adipose tissues. It is used in the treatment of types II, III, IV, and V hyperlipoproteinemia because of its ability to bring about large reductions in cholesterol and triacylglycerol levels, but many patients experience side effects such as flushing, hyperpigmentation, gastrointestinal upset, diarrhea, liver function abnormalities, hyperuricemia, and glucose intolerance. The effect of nicotinic acid on lipoprotein metabolism is not shared by nicotinamide and is not related to its vitamin function (Chapter 38).
3. The fibric acid derivatives *clofibrate* and *gemfibrozil* (Figure 20-12) promote rapid turnover of VLDL by activating lipoprotein lipase. Gemfibrozil also inhibits VLDL secretion. Use of fibric acids may result in a modest elevation of plasma HDL cholesterol in hypertriacylglycerolemic subjects. Occasional side effects include abdominal discomfort and cholesterol gallstones.
4. *Probucol* (Figure 20-12) significantly reduces plasma cholesterol levels but has no effect on triacylglycerols. It may act via blockage of intestinal cholesterol transport. HDL cholesterol levels are reduced by this drug. No consistent side effects have been reported.
5. HMG-CoA reductase inhibitors (statins) inhibit the regulatory step in the biosynthesis of cholesterol. They lower serum cholesterol and LDL cholesterol by inhibition of hepatic cholesterol synthesis and, more importantly, by up-regulating LDLreceptor activity.

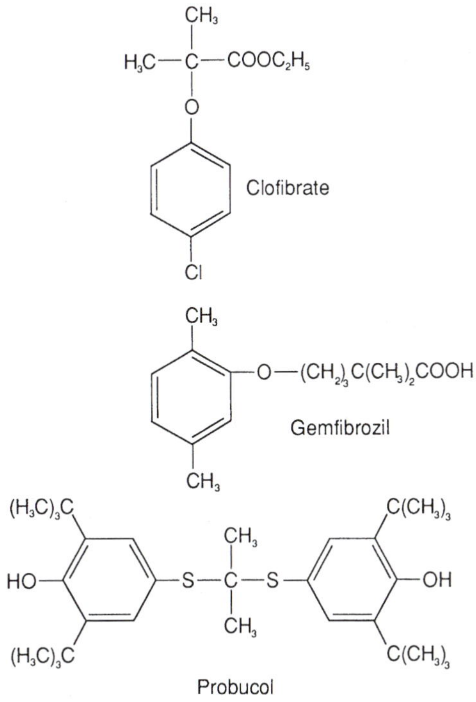

Clofibrate

Gemfibrozil

Probucol

FIGURE 20-12
Structures of plasma lipid-lowering drugs.

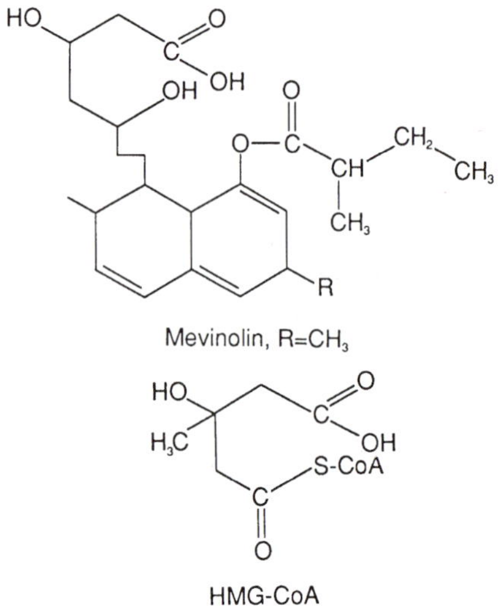

Mevinolin, R=CH_3

HMG-CoA

FIGURE 20-13
Competitive inhibitors of HMG-CoA reductase, the rate-controlling enzyme of cholesterol biosynthesis. Note the similarity in structures.

Statins are fungal derivatives or chemically synthesized therapeutic agents (Chapter 19). The structure of a statin is compared to that of HMG-CoA in Figure 20-13. Other properties of statins include activation of endothelial nitric oxide synthase, decreased fibrinogen levels and viscosity, diminished uptake of aggregated LDL by vascular smooth cells, a modest increase in HDL cholesterol levels, and an enhanced immune tolerance after transplantation. Because of various biological effects, statin therapy can reduce the effects of both coronary and cerebrovascular diseases and can decrease the need for coronary artery bypass surgery and angioplasty. Since statins inhibit a key step in a multifunctional cholesterol biosynthetic pathway, their long-term toxicity is not known (Chapter 19).

A combination of bile acid sequestrants with nicotinic acid or probucol or an HMG-CoA reductase inhibitor can be used to produce synergistic effects in lowering plasma lipoprotein levels, particularly LDL. The efficacy of drug treatment was shown in a recent study in which lovastatin and colestipol were used to reduce cholesterol levels in men with CHD. The rate of progression of coronary lesions was decreased and that of regression increased. These changes also were associated with reduced cardiovascular abnormalities.

Treatment of homozygous FH individuals with diet or drug therapy is not effective. Other modes of therapy for this condition include frequent plasma exchange and creation of a portacaval shunt. In one patient, liver transplantation was highly successful in reducing plasma LDL levels, underscoring the important role of the liver in cholesterol homeostasis.

In patients with blocked coronary blood vessels, coronary artery bypass grafting or angioplasty may be required. A potential new therapy involves agents that promote the formation of new blood vessels (angiogenesis). However, drugs that interfere with the formation of new blood vessels (angiogenesis inhibitors) may have a promising future in the treatment of cancer since they can cut off blood supply to tumors.

Supplemental Readings and References

M. A. Austin, B. McKnight, K. L. Edwards, et al.: Cardiovascular disease mortality in familial forms of hypertriglyceridemia: A 20-year prospective study. *Circulation* **101,** 2777 (2000).

M. Bodzioch, E. Orso, J. Klucken, et al.: The gene encoding ATP-binding cassette transporter I is mutated in Tangier disease. *Nature Genetics* **22,** 347 (1999).

A. G. Bostom, H. Silbershatz, I. H. Rosenberg, et al.: Nonfasting plasma total homocysteine levels and all-cause and cardiovascular disease mortality in elderly Framingham men and women. *Archives of Internal Medicine* **159,** 1077 (1999).

M. N. Diaz, B. Frei, J. A. Vita, et al.: Antioxidants and atherosclerotic heart disease. *New England Journal of Medicine* **337,** 408 (1997).

J. J. Genest Jr., S. S. Martin-Munley, J. R. McNamara, et al.: Familial lipoprotein disorders in patients with premature coronary artery disease. *Circulation* **85,** 2025 (1992).

A. M. GoHo: Prognostic and therapeutic significance of low-levels of high-density lipoprotein cholesterol. *Archives of Internal Medicine* **159,** 1038 (1999).

D. J. Gordon and B. M. Rifkind: High-density lipoprotein—The clinical implications of recent studies. *New England Journal of Medicine* **321,** 1311 (1989).

A. Graham, D. G. Hassall, S. Rafique, and J. S. Owen: Evidence for a paraoxonase-independent inhibition of low-density lipoprotein oxidation by high-density lipoprotein. *Atherosclerosis* **135,** 193 (1997).

S. M. Grundy: Cholesterol and coronary heart disease. *Archives of Internal Medicine* **157,** 1177 (1997).

C. R. Harper and T. A. Jacobson: New perspectives on the management of low levels of high-density lipoprotein cholesterol. *Archives of Internal Medicine* **159,** 1049 (1999).

R. H. Knopp: Drug treatment of lipid disorders. *New England Journal of Medicine* **341,** 498 (1999).

S. Kochl, F. Fresser, E. Lobentanz, et al.: Novel interaction of apolipoprotein(a) with β-2 glycoprotein I mediated by the kringle IV domain. *Blood* **90,** 1482 (1997).

J. A. Kuivenhoven, J. W. Jukema, A. H. Zwinderman, et al.: The role of a common variant of the cholesteryl ester transfer protein gene in the progression of coronary atherosclerosis. *New England Journal of Medicine* **338,** 86 (1998).

A. Maitra, S. V. Hirany, and I. Jialal: Comparison of two assays for measuring LDL cholesterol. *Clinical Chemistry* **43,** 1040 (1997).

M. E. Mendelsohn and R. H. Karas: The protective effects of estrogen on the cardiovascular system. *New England Journal of Medicine* **340,** 1801 (1999).

C. N. B. Merz, A. Rozanski, and J. S. Forrester: The Secondary Prevention of Coronary Artery Disease. *The American Journal of Medicine* **102,** 572 (1997).

J. J. Repa, S. D. Turley, J-M.A. Lobaccaro, et al.: Regulation of absorption and ABC1-mediated efflux of cholesterol by RXR heterodimers. *Science* **289,** 1524 (2000).

P. M. Ridker, M. Cushman, M. J. Stampfer, et al.: Inflammation, aspirin and the risk of cardiovascular disease in apparently healthy men. *New England Journal of Medicine* **336,** 973 (1997).

R. Ross: Atherosclerosis—An inflammatory disease. *New England Journal of Medicine* **340,** 115 (1999).

H. B. Rubins, S. J. Robins, D. Collins, et al.: Gemfibrozil for the secondary prevention of coronary heart disease in men with low levels of high-density lipoprotein cholesterol. *New England Journal of Medicine* **341,** 410 (1999).

C. von Schakey, P. Angerer, W. Kotheny, et al.: The effect of dietary ω-3 fatty acids on coronary atherosclerosis: A randomized, double-blind, placebo-controlled trial. *Annals of Internal Medicine* **130,** 54 (1999).

J. H. Stein and R. S. Rosenson: Lipoprotein Lp(a) excess and coronary heart disease. *Archives of Internal Medicine* **157,** 1170 (1997).

N. G. Stephens, A. Parsons, P. M. Schofield, et al.: Randomised controlled trial of vitamin E in patients with coronary disease: Cambridge Heart Antioxidant Study (CHAOS). *Lancet* **347,** 781 (1996).

T. Syrovets, J. Thellet, M. J. Chapman, and T. Simmet: Lipoprotein(a) is a potent chemoattractant for human peripheral monocytes. *Blood* **90,** 2027 (1997).

A. Tybjaerg-Hansen, R. Steffensen, H. Meinertz, et al.: Association of mutations in the apolipoprotein B gene with hypercholesterolemia and the risk of ischemic heart disease. *New England Journal of Medicine* **338,**1577 (1998).

J. L. Witztum: The oxidation hypothesis of atherosclerosis. *Lancet* **344,** 793 (1994).

CHAPTER 21

Muscle and Nonmuscle Contractile Systems

All cells produce movement internally, and many are capable of motility or of changing shape. In some, movement is related to the function of individual cells, as in the migratory and engulfing movements of phagocytic cells or the swimming movements of sperm cells. In other cases, cells generate movement as one aspect of tissue function, as in the ciliary transport of mucus by the bronchial epithelium. Cells specialized for changing the dimensions or shape of anatomical structures or for movement of body parts with respect to each other are called muscle cells.

The ability of a cell to hold or to change shape and to move organelles within it depends on the existence of a cytoskeleton, comprising ***actin filaments*** (microfilaments), ***microtubules,*** and ***intermediate filaments*** (IFs) capable of transmitting force. Actin filaments and microtubules contain predominantly actin and tubulin, respectively. (Table 21-1). The diameter of intermediate filaments (10 nm) is between that of actin filaments (6–7 nm) and microtubules (25 nm). They are structural proteins not directly involved in motion.

Five classes of intermediate filament proteins have been described, generally referred to as types I through V. Types I and II are the acidic and basic polypeptides, respectively, which comprise the keratins and cytokeratins, a family of heteropolymers that are abundant in epithelia. Type III consist of vimentin, desmin, glial fibrillary acidic protein, and peripherin. Of these, desmin has specific importance in muscle cells. Desmin links adjacent Z-disks to one another and is involved in linking Z-disks to cell membrane integrins at the costameres. It may also help bind filaments into myofibrils. Type IV includes the neurofilament proteins NF-L, NF-M, and NF-H plus another protein from a similarly organized gene called nestin.

In neurons, microtubules are responsible for axonal transport and longitudinal axon growth, while neurofilaments are related to radial growth, so that axonal diameter (and thus conduction velocity) are roughly proportional to neurofilament content. Type V IFs are the lamins associated with the nuclear membrane. Actin, tubulin, and several of the intermediate filament proteins are found in all cells.

Controlled transformation of the chemical energy of nucleoside triphosphates into mechanical energy is called ***chemomechanical transduction.*** In addition to the actin filaments and microtubules, the "motor" proteins ***myosin*** and ***dynein*** or ***kinesin*** are needed for chemomechanical transduction. Several other proteins are associated with these, including regulatory proteins that control contractile activity and enzymes involved in maintaining the supply of high-energy phosphate.

This chapter discusses muscle, an actin filament system, and cilia, a microtubule system. These do not typify all contractile systems. In some actin filament systems, movement occurs in the absence of myosin, being driven

TABLE 21-1
Comparison of Actin-Based and Tubulin-Based Motility Systems

	Actin/ Myosin	Tubulin/ Dynein or Kinesin
Structural arrangement	Actin (thin) filaments and myosin (thick) filaments; myosin heads form bridges during contraction; filaments are stable in sarcomeres and stereocilia, but often labile in cytoplasm.	Microtubules with dynein arms or cytoskeletal microtubules to which cytosolic dynein or kinesin may attach; dynein forms intertubule bridges during movement; tubules are stable in cilia and flagella, but often labile in cytoplasm.
Static element/dynamic element	Actin filament/myosin filament.	Microtubule/dynein or kinesin.
Type of motion	Sliding of actin and myosin filaments past each other.	Sliding of microtubules restrained at one end produce bending in cilia and flagella; dynien or kinesin translate organelles or other cargo along cytoskeletal microtubules.
Energy transducer	Myosin is an actin-activated ATPase.	Dynein and kinesin are tubulin-activated ATPases.
Properties of static protein	Actin is globular, MW 42,000; monomer binds MgATP; ATP and calcium promote polymerization; polymerizes as a two-stranded helix 6–7 nm across; each subunit interacts with four adjacent ones; monomer can be added or lost at either end: encoded by multigene family.	Tubulin is heterodimer of globular proteins, MW 50,000 each; dimer binds two GTP, calcium may inhibit polymerization; polymerizes in protofilaments which bond in parallel to form a cylinder of 13 filaments; each dimer interacts with 4 or 6 adjacent dimers; dimer can be added or lost at either end; encoded by a multigene family.
Properties of dynamic protein	Myosin: MW 475,000; globular and helical regions; heterohaxamer; ATPase activity; functional unit is bipolar aggregate of 10 to 500 myosin monomers; some types are activated by phosphorylation; encoded by multigene family.	Dynein: MW 1,500,000+; functional unit is heteromultimer; ATPase activity; encoded by a multigene family. Kinesin: MW 380,000; heteromultimer of 2 heavy and 2 intermediate chains; ATPase activity; encoded by a multigene family.
Regulatory factors	Cytoplasmic calcium concentration; phosphorylation of myosin light chains and (less important) other proteins; interactions with actin-binding proteins.	Cytoplasmic calcium concentration; phosphorylation of several proteins; interaction with MAPs, tau and other proteins.

by growth or shortening of bundles of actin filaments; and in several microtubule systems, the motor protein is kinesin rather than dynein. However, muscle and cilia have many features in common with most other actin and tubulin systems, and are vital to normal functioning of the organism.

21.1 Muscle Systems

Mammals have four types of cells specialized for contraction: skeletal muscle, cardiac muscle, smooth muscle, and myoepithelial cells (Table 21-2). Skeletal and cardiac muscles contract with more force and much more speed, shorten less, and consume much more ATP during maintenance of tension than smooth muscle or myoepithelium. Characteristics of these cell types reflect the functional roles of the tissues in which they occur; however, the fundamental mechanism of contraction is the same in all tissues.

Structure and Development of Skeletal Muscle

The skeletal muscles of the torso and limbs arise from the mesoderm of the somites, while those of the head arise from the mesoderm of the somitomeres which contribute to the branchial (pharyngeal) arches. They form by the fusion and elongation of numerous precursor cells called

TABLE 21-2
Comparison of Four Types of Mammalian Cells Specialized for Contraction

Cell Type	Structure	Contractile Properties	Function
Skeletal muscle	Long syncytial, multinucleated cells; orderly arrangement of myosin and actin filaments gives striated appearance, each fiber is directly innervated by a motor neuron.	Rapid, powerful contractions; can shorten to 60–80% of resting length; contraction is initiated by the central nervous system under voluntary control.	Movement of the bony parts across joints.
Cardiac muscle	Similar to skeletal muscle but extrinsic innervation is only at the specialized nodal pacemakers; the action potential is conducted from cell to cell via gap junctions (nexuses).	Similar to skeletal muscle but contraction is initiated by automatic firing of pacemaker cells; contraction is slower and more prolonged than in skeletal muscle.	Movement of blood by repetitive rhythmic contraction; beats about 3 billion times during a normal lifetime.
Smooth muscle	Elongated, tapering cells; mononuclear; no striations; occurs singly, in small clusters, or in sheets enclosing organs; innervated by local plexuses and extrinsically by autonomic nerves.	Slow contractions under involuntary control; can shorten to 25% of resting length.	Control of shape and size of hollow organs such as the digestive, respiratory, genital, and urinary tracts and the vascular system.
Myoepithelia	Basket-shaped, mononuclear cells surrounding the acini of exocrine glands; have cytoplasmic fibrils resembling smooth muscle; derived from ectoderm rather than mesoderm.	Contraction stimulated by hormones (e.g., oxytocin) and presumably by autonomic nerves; may have noncontractile functions such as pressure transduction in the renal cortex.	Contraction to expel contents of exocrine glands (salivary, sweat, mammary); form the dilator muscle of the iris; may be the pressure transducers in juxtaglomerular cells.

myoblasts. Some stem cell precursors of myoblasts remain in an adult animal; they are located between the sarcolemma and basement membrane of mature muscle cells, and are called satellite cells. Since each myoblast contributes its nucleus to the muscle cell, skeletal muscle fibers are all multinucleated, the longest having 200 or more nuclei.

In the past decade, several growth and differentiation factors have been identified which play a role in causing embryonic stem cells to become committed to the muscle cell lineage, and influence the rate and extent of their proliferation and differentiation. These include four "myogenic regulatory factors" (or MRFs) of the helix-loop-helix (HLH) family of DNA-binding transcription factors, called MyoD, Myf5, Mrf4 and myogenin. There are also inhibitory factors such as myostatin, the absence of which causes substantially greater than normal muscle mass in animals with the corresponding gene deletion. The MRFs interact with the promoter regions of many muscle-specific genes, and with other transcription factors (especially MEF2). A summary of major myogenic regulators and their actions is given in Table 21-3.

Figure 21-1 schematically illustrates the structure of skeletal muscle. Individual muscle cells, or fibers, are elongated, roughly cylindrical, and usually unbranched, with a mean diameter of 10–100 μm. The plasma membrane of muscle fibers is called the ***sarcolemma,*** and fibers are surrounded by structural filaments of the extracellular matrix which are often described as forming a basement membrane.

Within each fiber is a longitudinal network of tubules called the sarcoplasmic reticulum (SR), analogous to the endoplasmic reticulum of other cells (Figure 21-2). Release of Ca^{2+} from the SR is a the key step in coupling

TABLE 21-3
Factors Known to Influence Muscle Development

Regulatory Factor	Role in Development	Comments
***myo*D**	Can initiate the differentiation of stem cells to myoblasts, may stimulate proliferation of myoblasts	Either ***myo*D** or ***myf*5** must be present: deletion of both prevents myoblast formation
***myf*5**	Same as ***myo*D**	
myogenin	Required for terminal differentiation of myoblasts, myoblast fusion, and development of myotubes	Deletion of myogenin gene results in accumulation of myoblasts instead of formation of muscle cells
***mrf*4**	Required for development of normal amount of myofibrils; may be needed for maintenance of mature muscle cells	
myostatin	Inhibits proliferation of myoblasts and formation of myotubes; plays major role in terminating the myogenic program in small animals, lesser role in large animals	Defect in myostatin gene in cattle explains the unusual degree of muscularity in Belgian Blue and Piedmontese breeds

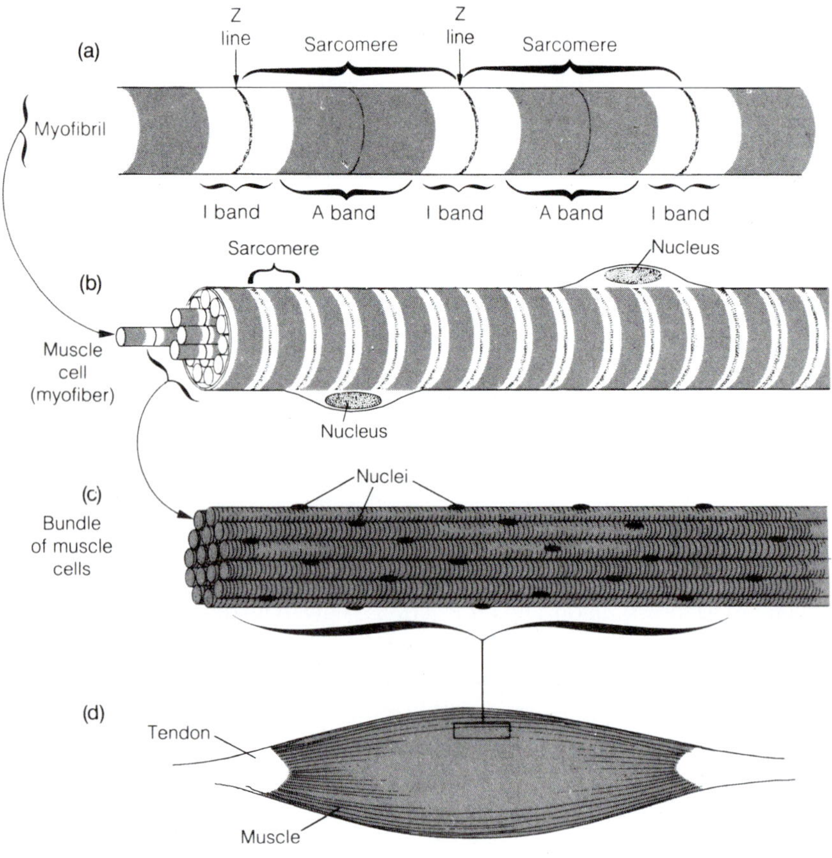

FIGURE 21-1
Structure of skeletal muscle. (a) A myofibril is a long cylinder consisting of tandemly arranged sarcomeres. A sarcomere consists of a darkly staining central region, the A-band, and two adjacent lightly staining regions, the I-bands. Very darkly staining Z-lines mark the borders of successive sarcomeres. (b) Many myofibrils are packed inside one muscle cell (myofiber). The myofibrils are in parallel register and impart the sarcomere banding pattern to the muscle cell. (c) A bundle of muscle cells. Muscle cells are multinucleated. (d) Bundles of muscle cells are in turn arranged into larger bundles within a muscle.

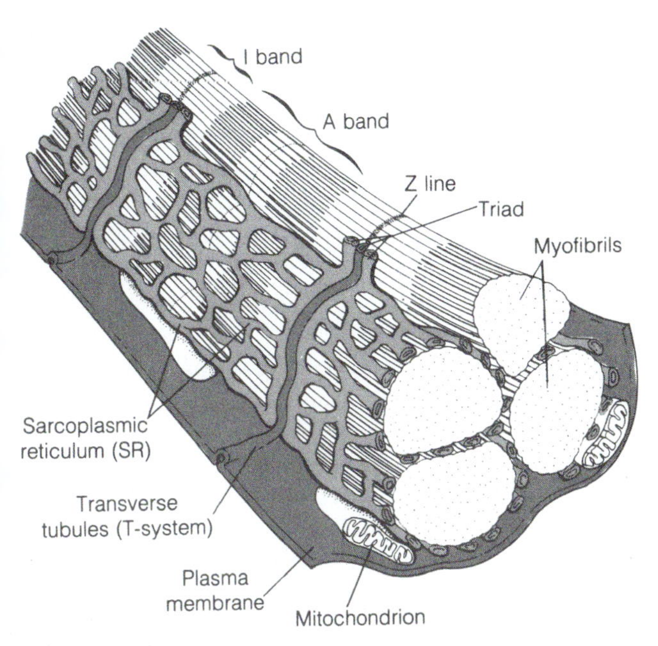

FIGURE 21-2
Schematic representation of the internal organization of a muscle fiber. Note the parallel arrangement of myofibrils and the location of the sarcoplasmic reticulum and T-tubules with respect to them. This arrangement is typical of a mammalian fast-twitch fiber, with the T-tubules and terminal cisternae located at or near the Z-lines.

the sarcolemma action potential to activation of contraction. SR membranes bear large numbers of Ca^{2+} pump proteins whose role is to pump Ca^{2+} into the lumen of the SR where it is sequestered until the fiber is stimulated. Crossing the SR perpendicular to the cell axis are transverse tubules (T-tubules), which are invaginations of the sarcolemma. They conduct the action potential into the interior of the cell which triggers the release of Ca^{2+} from SR.

Muscle cells contain many mitochondria which are often present as reticulum-like structures extending longitudinally in the fiber near the sarcolemma, rather than as discrete ellipsoidal organelles found in many other cells. These provide much of the high-energy phosphate needed to power contraction and to operate the Ca^{2+} pumps that control the cytosolic calcium concentration. Different types of skeletal muscle fibers differ considerably in the extent and organization of both their SR and mitochondria.

Muscle fibers contain long cylindrical myofibrils (typically 1–2 μm in diameter) aligned longitudinally and consisting of interpenetrating arrays of thin myofilaments (6–7 nm in diameter) and thick filaments (15–16 nm in diameter). These structures are the contractile apparatus of the fiber (Figure 21-3).

Myofibrils

Skcletal and cardiac muscle exhibit alternating light and dark bands (striae) visible by light microscopy and are called striated muscle. This banded pattern is due to regions of overlap alternating with nonoverlap of the thick and thin filaments (Figure 21-3). In cross-sections, these filaments are arranged in interpenetrating hexagonal arrays. Between the thick and thin filaments, cross-bridges may form.

The sarcolemma has more or less regularly spaced electron-dense patches or rings of large membrane-spanning multiunit proteins. These are analogous to desmosomes, and the proteins therein are structural proteins linking the meshwork of cytoskeletal filaments that tie the myofibrils together to the extracellular matrix. The complex of cytoskeletal actin and IFs and the membrane proteins is called a ***costamere.*** One end of each thin filament is anchored to a dense fibrous structure (***Z-band***) whose central region contains actin, α-actinin, and other proteins. The Z-disk periphery contains predominantly actin and desmin and lesser amounts of filamin, vimentin, and synemin.

In vitro, α-actinin and filamin bind actin, the major protein of thin filaments. One α-actinin molecule binds to two actin filaments, one from each side of the Z-disk. Four α-actinin molecules bind to each actin filament at 90° angles, so that the actin filaments are bound into the Z-disks in a square array, although in most of the length of the sarcomere their arrangement is hexagonal. ***Desmin,*** an intermediate filament protein, forms a network from one Z-disk to the next across the myofibril. Such links, aided by attachment of desmin and dystrophin to the sarcolemma, help hold the sarcomere structure in

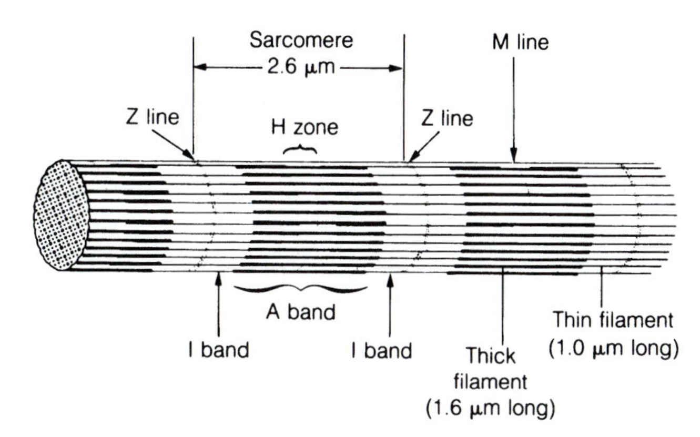

FIGURE 21-3
Portion of a myofibril. The Z-lines mark the ends of sarcomeres. An A-band consists of thick filaments. An I-band consists of thin filaments only. The thin filaments extend part way into the A-band leaving a central region, the H-zone, free of thin filaments. A faintly staining M-line is present in the center of the H-zone.

register, and play an important role in transmitting force produced in the myofibrils to adjacent myofibrils and, via the costamere structure, to the extracellular matrix. It appears that this mechanism of force transmission, rather than transmission across the tapered ends of the cells, is the major means of force transmission from myofibril to tendon.

Between the Z-disks, other regularly recurring features are seen (Figure 21-3). Each complete filament assembly, from one Z-disk to the next, is called a sarcomere, and is the basic structural and functional unit of the fibril. The A-band corresponds to the bundle of thick filaments centered between Z-disks, while the I-band is the region on either side of a Z-disk that contains only thin filaments. Thick and thin filaments partially overlap, making the A-band darker at its ends and leaving a light area in the middle where there is no overlap. A clue to the mechanism of contraction was the finding that during contraction, the H- and I-bands shorten, while the A-bands do not.

In the center of the A-band, the thick filaments are linked by a network of proteins forming the M-line. The pattern of cross-linking of these proteins, such as myomesin, and their ability to bind the thick filament protein myosin probably explains the hexagonal arrangement of the thick filaments. Cytosolic creatine kinase activity also localizes extensively to the M-line region and may actually be a property of one of the myosin cross-linking M-line proteins.

Another set of filaments, sometimes called the "third filament system," helps maintain sarcomere structure. A very long protein (***titin***) holds the thick filament array centered in the sarcomere. Titin is an enormous protein consisting of more than 27,000 amino acids with a M.W. of about 3 million Da. One end binds to one or more M-line proteins and the other binds to one or more proteins in the Z-disk. Most of the molecule comprises repeats of large domains resembling fibronectin 3 (FN3) and immunoglobulin. There is a super-repeat of these motifs occurring 11 times within the A-band region of the molecule, which suggests a functional relationship with a group of thick filament accessory proteins that occur in 11 bands in each half of the A-band. Thus, there may be multiple attachments of titin to thick filaments.

There are also globular regions of the molecule located between the thick filaments and the Z-disk, which apparently unravel one by one as increasing tension is applied to the molecule; these fold back into globules as tension is released. This region of titin is like a long spring capable of considerable elongation before reaching its elastic limit. In cardiac muscle, which has a low compliance, this region of titin contains only 163 amino acids, while in soleus muscle (whose compliance is 10 times that of cardiac muscle), it is over 2000 amino acids long. It appears that passive tension in muscle is accounted for by the myofibrillar structure inside the fiber, not the connective network outside it, and titin is probably the main structural protein involved.

Another very long protein (***nebulin***) is associated with the thin filaments. Nebulin has a molecular weight of about 700,000. An abundant protein in skeletal muscle, nebulin extends from either side of the Z-disks along the entire length of the thin filaments. It may serve as a template for thin-filament assembly, and may interact with tropomyosin, and also may have a regulatory role.

Thin Myofilaments

Thin myofilaments are polymers of globular (G) actin molecules (M.W. 42,000). Although the actin gene has apparently undergone considerable duplication, only six actin genes are expressed in mammals; three are α-actins, which are the actins found in sarcomeric thin filaments, each in a different type of muscle. These actins are 375 amino acids long; the ones found in skeletal and cardiac muscle have valine at position 17, unlike all others. There are also two γ-actins and one β-actin. The β-actin and one of the γ-actins are cytoskeletal actins found in almost all nonmuscle cells. They lack the N-terminal aspartate of skeletal muscle actin and have 374 amino acids. The N-terminal residue of G-actin is always acetylated, and typically some of these are sequentially removed as part of the posttranslational modifications common in the N-terminal region. The other γ-actin seems to be expressed only in gut smooth muscle. For all six, essential amino acids (especially leucine, isoleucine, and threonine) account for about 41% of the molecule. Skeletal G-actin contains N^3-methylhistidine as does myosin. Other proteins contain methylhistidine (e.g., acetylcholinesterase), but in quantitatively insignificant amounts compared to actin and myosin. Consequently, urinary methylhistidine excretion has been used as a marker for contractile protein turnover and as an indirect indicator of muscle mass.

G-actin has four globular domains (arranged in a U-shape) designated domains I through IV with a cleft between domains II and IV (Figure 21-4). This structure is stabilized by the binding of Mg-ATP in the cleft (called an ATPase fold) by ionic and hydrogen bonds between the Mg-ATP and adjacent amino acid side chains. Several other ATP-binding proteins (e.g., hexokinase) have been found to have similar ATPase folds in which the bottom of the cleft acts as a hinge region allowing the cleft to be bent open. Binding of Mg-ATP (or Mg-ADP) holds the cleft closed and stabilizes the molecule. Without bound nucleotide, actin denatures readily.

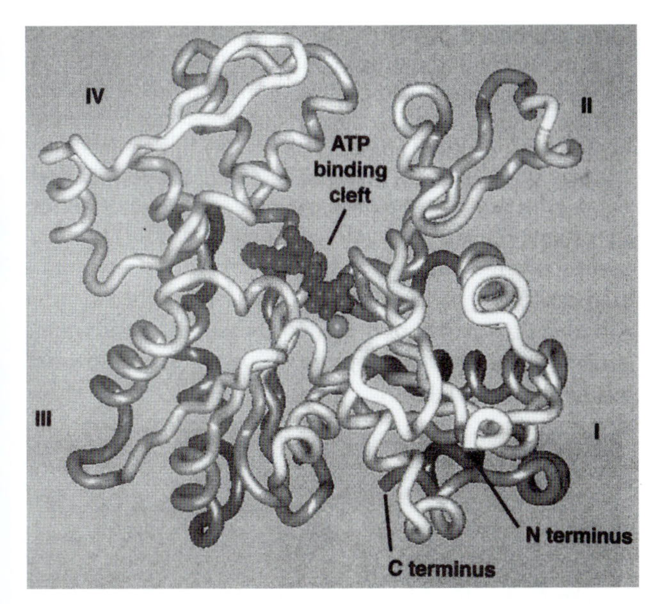

FIGURE 21-4
Structure of β-actin from a nonmuscle cell. As described in the text, α- and β-actins are very similar. The shape is like a rectangular pillow, with dimensions of roughly $5.5 \times 5.5 \times 3.5$ nm, with a cleft in it. The U-shape formed by this arrangement of the four major domains is stabilized by the binding of Mg-ATP in the cleft, as shown. In the absence of bound nucleotide, actin monomers denature readily. [From *Molecular Cell Biology*, 3rd ed., by Lodish et al. (Eds). W.H. Freeman, New York, 1995, p. 995.]

At concentrations exceeding 0.1 mM, ATP-G-actin spontaneously polymerizes into filaments of F-actin in solutions of physiological ionic strength and depolymerizes at low ionic strength. Other structural or catalytic proteins are not required. The bound ATP hydrolyzes during polymerization but this is not an energy-requiring process; Mg-ADP and hydrolysis-resistant ATP analogues also permit polymerization, albeit at a slower rate. Each of the four G-actin domains binds noncovalently with an adjacent G-actin, such that each G-actin is thought to interact with four other G-actins in the growing filament. The four interactin sites are specific, so that all of the G-actins assembling into a filament have the same polarity, i.e., each cleft points toward the same end of the filament. Thus, one end of the filament has exposed I and III domains and another has exposed II and IV domains, so that the whole filament has a polarity. The affinities of the two ends for new G-actin differ, so that in the normal range of cytoplasmic G-actin concentration (1 mM), actin polymerizes 5–10 times faster at one end of the filament [called the (+) end] than at the other [(−) end]. Since the critical concentration for polymerization is lower at the (+) end (0.1 vs. 0.8 mM), it is possible for a filament to grow at one end and shorten at the other. In most cases, capping proteins prevent growth at one end so that the filament grows only at the free end. In muscle, the (+) end is ultimately capped by a protein (CapZ) which also binds to α-actinin or other Z-disk proteins, and probably plays some role in anchoring thin filaments to the Z-disk. The (−) end is capped by an unrelated protein (tropomodulin) whose capping effectiveness is enhanced by tropomyosin. Thus, the tropomodulin-capped (−) end extends toward the center of the sarcomere on either side of the Z-disk.

G-actin is very highly conserved, both across actin genes within a species and across species. Apparently, the need for so many functional binding sites in a molecule of that size leaves few options for nonlethal mutations. Among the actins sequenced from 30 widely divergent species, there were only 32 amino acid substitutions. One implication of this is that when differences in contractile properties are observed between various types of muscle, those differences must be due to the motor protein (myosin) or to the various regulatory proteins.

Thin filaments contain nebulin and two regulatory proteins called *tropomyosin* and *troponin* (Figure 21-5). Tropomyosin (M.W. 68,000) is a coiled-coil α-helical

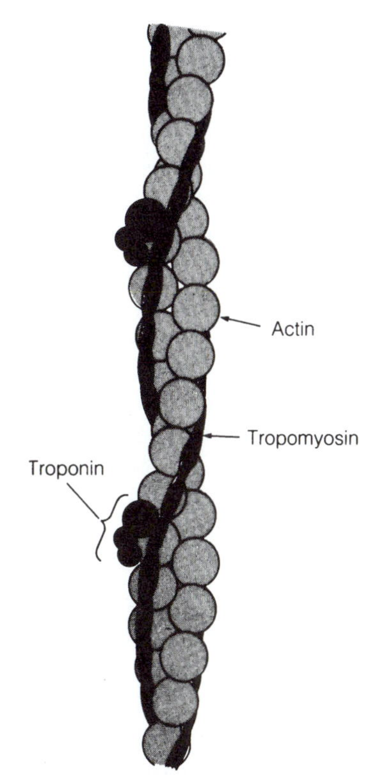

FIGURE 21-5
Model of a thin filament. Two tropomyosin filaments, each composed of two subfilaments, wind around the actin chain. They block the binding of the globular heads of myosin molecules (in thick filaments) to the actin molecules. Troponin consists of three polypeptides and binds to both actin and tropomyosin.

heterodimer. The proportions of α- and β-chains present, and which α- and β-chain genes are expressed, varies with the type of muscle fiber. In large mammals, cardiac tropomyosin is predominantly a homodimer of the same α-tropomyosin found in slow-twitch skeletal muscle. Fast-twitch fibers have primarily the α–β heterodimer. Tropomyosin is phosphorylated at specific sites in some fiber types. The molecule is rigid and insoluble, and binds to F-actin along the grooves formed by the spiraling of the filament. The tropomyosins are arranged end-to-end, are about 40 nm long, and each one spans seven G-actin units. Tropomyosin normally inhibits interaction between actin and myosin by interfering with a projection on the myosin head distal to the actin binding site.

Near one end of each tropomyosin is a molecule of troponin (M.W. ~76,000), a trimer of noncovalently bound subunits called troponins C, I, and T (Tn C, Tn I, and Tn T). Tn T binds to the C-terminal region of tropomyosin and to Tn C, Tn I, and actin. Tn I binds to Tn C in a Ca^{2+}-dependent manner, and to actin, and is thought to determine the position of the tropomyosin-troponin complex on the actin filament. Tn C is structurally and functionally similar to calmodulin (Chapters 15 and 30). Each Tn C molecule can cooperatively bind up to four Ca^{2+} ions (three in cardiac Tn C). Under physiological resting conditions, typically two of these sites would be occupied by Mg^{2+} and two would be unoccupied. Binding of Ca^{2+} to Tn C (during the calcium spike induced by depolarization of the cell) reverses the tropomyosin inhibition of actin-myosin binding. Tn I is a 179 to 216 amino acid polypeptide (M.W. ~23,000), with a preponderance of basic amino acids in the actin binding region. It is so named because Tn I binding to actin can inhibit contraction even in the absence of tropomyosin. Tn T is an elongated 259 amino acid polypeptide (M.W. ~37,000). *In vivo,* about half of these amino acids are charged (with negative charges near the N-terminus and positive charges near the C-terminus).

Thick Myofilaments

Thick myofilaments are made primarily of myosin. Myofibrillar myosin is one member of the multi-gene myosin II sub-family in the myosin super-family. Of the 15 types of myosin that have been described, at least 7 occur in mammalian species. Of these seven, three are fairly well characterized. Myosin I has a single heavy chain and one to three (usually one) light chain, which is calmodulin. Myosin V has a heavy chain dimer, and carries six light chains, which are usually calmodulin. Both types I and V are motors that carry organelles and vesicles along the actin filaments of the cytoskeleton and probably also perform other functions. Myosin II is a large hexameric protein (M.W. ~475,000) consisting of two myosin heavy chains (MHCs; M.W. ~200,000) and four myosin light chains (MLCs; M.W. ~16,000 to 20,000). These are non-covalently linked. The size and properties of the light chains vary by tissue.

About half the amino acids of the MHCs form α-helices which twist together in a coiled-coil to form a rigid, insoluble tail. The N-terminal halves of the chains form discrete globular heads consisting of 839 to 850 amino acids. To each head are bound two different MLCs, as discussed below. The head of the myosin molecule has a site which binds to a complementary site on actin and, about 3.5 nm away, an ATP binding site which has a high ATPase activity only when myosin is bound to actin. The affinity of the binding to actin is regulated by events at the ATP binding site.

Papain readily cleaves myosin at site 2 (Figure 21-6), releasing the heads, called S_1, fragments and an almost complete tail. Brief exposure to trypsin or chymotrypsin cleaves myosin at site 1, yielding two large fragments called *heavy meromyosin* (HMM; M.W. ~350,000) and *light meromyosin* (LMM; M.W. ~125,000). LMM forms the core of thick filaments and is responsible for the low solubility of myosin under physiological conditions. HMM comprises the heads attached by short neck regions to an arm region. Digestion of HMM by papain releases the S_1 fragments and an S_2 fragment that corresponds to the arm. There is probably a relatively flexible region of the molecule at site 1, and the S_2 fragments normally have little charge and thus little interaction with LMM. These features allow the heads to project outward from the filament to a distance where actin-myosin interactions would not be influenced by the properties of LMM. Sequence information and imaging techniques have led to a detailed model of this complex structure in relation to F-actin (Figure 21-7). The light chains bind to the neck region, presumably stiffening it and permitting it to act as a lever arm amplifying movement that occurs in the head during the binding of myosin to actin and the subsequent binding and hydrolysis of ATP.

Several nomenclatures have been used for the light chains. Each head is said to have one "essential" and one "regulatory" light chain. The term "essential" persists despite the fact that myosin devoid of light chains retains actin-activated ATPase activity. Phosphorylation of, or Ca^{2+} binding to, the regulatory chains can alter the behavior of myosin greatly. The MLCs have also been named according to method of preparation. Light chains released by exposure to alkali are called alkali light chains and those released by exposure to the solvent DTNB [5,5′-dithiobis(2-nitrobenzoate)] are called DTNB light chains.

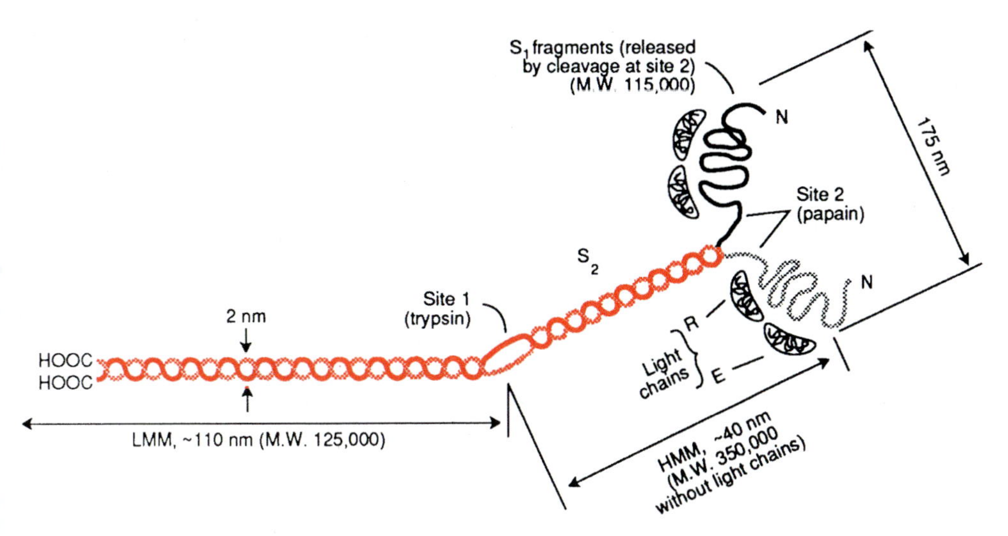

FIGURE 21-6
Major structural features of the myosin molecule. The myosin rod, extending from the C terminus to cleavage site 2, consists of two α-helical polypeptides coiled around each other. Cleavage by trypsin at site 1, which is located in the hinge region, yields light meromyosin (LMM) and heavy meromyosin (HMM). Digestion of HMM with papain cleaves HMM at site 2 (swivel region), releasing two S_1 fragments (myosin heads) and one S_2 fragment (myosin arm) for each molecule of HMM. The locations of the light chains (R = regulatory; E = essential) are indicated roughly in this figure. The actual locations are shown in Figure 21-7.

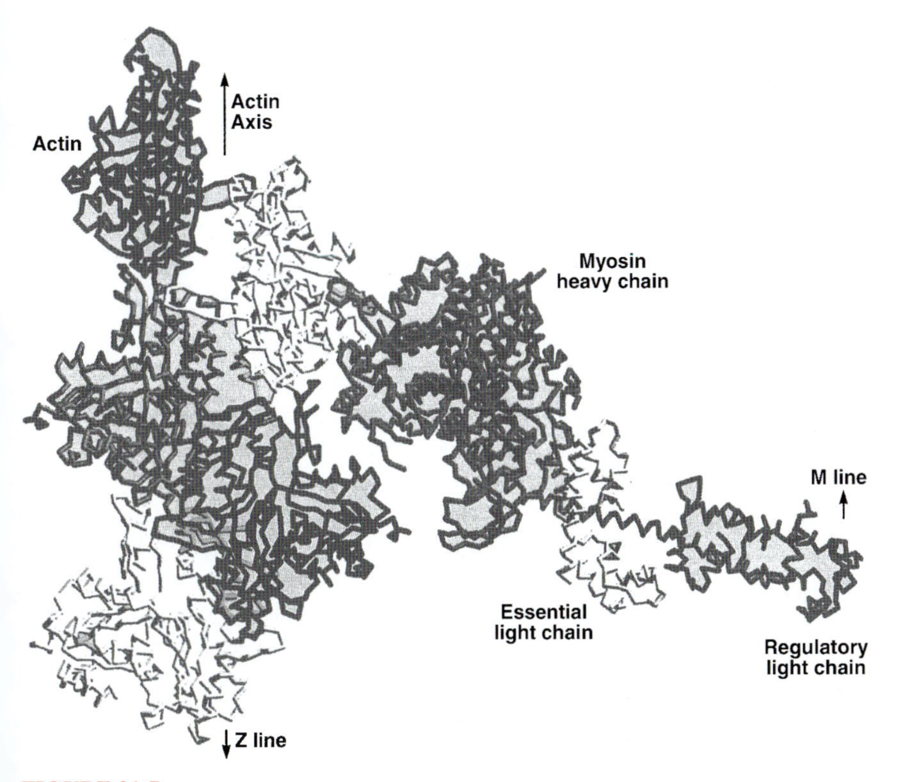

FIGURE 21-7
The head domain of myosin shown in its relation to the actin filament. The NH_2-terminal end of the myosin heavy chain is in the globular head. The light chains bind to the neck region of the MHC. In this figure, the orientation of the myosin to the actin is that of the rigor bond, i.e., at the end of the power stroke. [From M. Irving and G. Piazzesi, Motions of myosin heads that drive muscle contraction. *News Physiol. Sci.* **12**(6), 249–254 (December 1997).]

TABLE 21-4
Equivalent Light Chain Nomenclature

Alkaline light chains	Essential light chains	A_1	LC_{1f}
		A_2	LC_{3f}
		A	LC_{1s}
DTNB light chains	Regulatory light chains	DTNB	LC_{2f}
			LC_{2s}

There is also a designation based on the source of the MLC, with the subscripts "f" and "s" indicating the MLCs are from fast or slow muscle fibers, respectively. The relationship of one nomenclature to another is shown in Table 21-4. As explained under control of contraction, the regulatory light chain (rather than troponin) is the major regulatory protein under some circumstances in smooth muscle.

The physiologically active form of myosin is the thick myofilament (Figure 21-8), which is typically about 1.6 μm long and is formed by aggregation of myosin tails to create a bundle with the heads sticking outward. Head pairs are spaced linearly 14.3 nm apart axially, and 60° apart angularly around the filament, which corresponds to the spacing of the thin filaments in a hexagonal pattern around the thick filaments. The axial repeat distance of the thick filaments is 43 nm. Apparently, myosin molecules initially bind tail to tail, and thereafter assemble head to tail, so that there is a "central bare zone" where there are only tails and no heads. Spontaneous polymerization is an intrinsic property of actin, myosin, and tubulin. *In vitro,* myosin can self-assemble into filaments similar to those found *in vivo*. About 500 myosin molecules aggregate to form one skeletal muscle thick filament.

Filament assembly depends on characteristics of the myosin tail, especially the LMM, in which much of the amino acid sequence exhibits a heptapeptide repeat. If the repeat is represented as ABCDEFG, residues A and D are hydrophobic and lie at points where the α-helices in the coiled-coil touch. E, which is usually acidic, and G, which is usually basic, probably create ionic interactions that help bind the myosins into a filament.

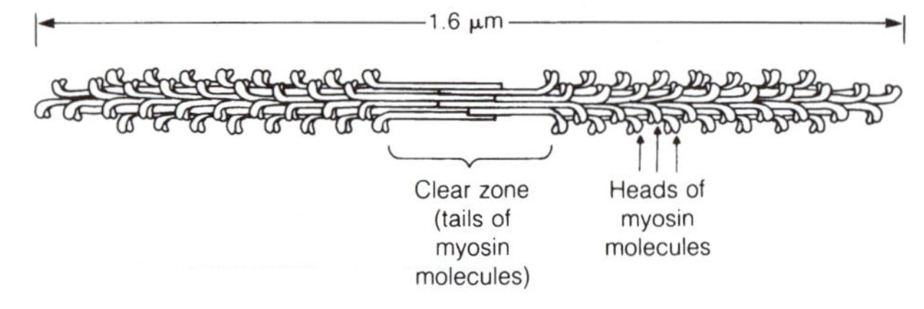

FIGURE 21-8
Organization of myosin in striated-muscle thick filaments. Filament formation begins with tail-to-tail (antiparallel) binding of myosin molecules, with subsequent parallel binding of myosin molecules to the ends of the initial nucleus, leaving the central clear zone. There are approximately 500 myosin molecules per striated thick filament.

Organization and Properties of Muscle Fibers

With few exceptions, a vertebrate skeletal muscle fiber is innervated at one region along its length by a branch of an axon from an α-motor neuron (α-MN). Each α-MN may innervate many fibers, from 10 or fewer in small muscles used for finely graded movements, to 2000 or more in proximal limb muscles. The group of fibers innervated by a single α-MN is called a muscle unit, and the muscle unit and the α-MN driving it form a motor unit. The muscle unit is the basic functional unit of skeletal muscle, since normally all fibers in a muscle unit are stimulated when their α-MN is active. In contrast, cardiac muscle contracts in response to action potentials generated spontaneously in the heart and transmitted from cell to cell through gap junctions. Contraction of the heart is modified, but not initiated, by its innervation. Organization of smooth muscle varies considerably by tissue.

Skeletal muscle fibers have different mechanical and metabolic properties, and fiber classification schemes are based on these differences. All fibers within a muscle unit are similar (but not identical) with respect to twitch characteristics, as these are largely determined by the innervation, and all fibers in a muscle unit are innervated by the same α-MN. However, even within a muscle unit, there may be appreciable interfiber differences in metabolic profile. Consequently, the following classifications should be viewed as useful simplified categories rather than a literal description of populations of fibers.

Contractile Properties

Most fibers can be classified as slow-twitch (ST) or fast-twitch (FT). The latter have a shorter time to peak tension, shorten faster, and have a shorter relaxation time than the former, and are used for movements requiring high speed or power. ST fibers are used for sustained activity and control of posture. The speed of shortening is related to the ATPase activity of the myosin, while the twitch duration parameters are related to the extent of the SR, and so these two are not directly connected. However, fiber histology and myosin gene expression have some controlling influences in common, so that speed of contraction and twitch duration vary inversely.

pH Dependence of Myosin ATPase Activity

The myosins in muscle differ in their inhibition by high or low pH, and can be classified on this basis. Serial thin sections of muscle specimens can be incubated in solutions at two different pH's with ATP and Ca^{2+} and stained for phosphate liberation. In the Brooke and Kaiser classification scheme, fibers that are light at pH >10 are type I; those that are dark are called type II. This staining pattern is reversed at pH <4.4. Fibers that are light at pH 4.4 but moderately dark at pH 4.6 are called type IIB; those that are light both at pH <4.4 and at pH 4.6 are type IIA. An additional type seen in small numbers in human cells stains dark at both acid pH's and at the alkaline pH, and is called IIC. Since the speed of shortening depends on which myosin is present, the relationship between the Brooke and Kaiser fiber typing and the FT vs. ST classification is quite consistent: type I fibers are slow, IIB are the fastest, and IIA are intermediate in speed. The I, IIA, IIB scheme has proven useful clinically and in analysis of muscle performance, but does not convey the near-continuum of myosin ATPase activity actually present across fibers.

Metabolic Profile

Histochemical staining techniques enable the semi-quantitative assessment of activity of enzymes in the energy pathways of cells, such as succinate dehydrogenase, glyceraldehyde-3-phosphate dehydrogenase, and adenylate kinase (myokinase). By staining serial sections for myosin ATPase typing and enzyme profiling, it has been found that generally type I fibers are high in oxidative enzyme activity and low in glycolytic activity, while type IIB fibers are the reverse. Type IIA fibers are moderately high in both oxidative and glycolytic enzymes. Hence, type I fibers are often called slow oxidative (SO), while type IIB are called fast glycolytic (FG) fibers. Type IIA are called FOG, or fast oxidative-glycolytic. However, the Brooke and Kaiser typing and energy pathway profiles reflect quite different specializations of the cell, and there is heterogeneity of enzyme profile within a muscle unit.

The proportions of fibers within a muscle are determined by a combination of genetic and developmental factors and by the pattern of recruitment of the muscle unit. Due to this latter influence, the twitch and enzyme characteristics of muscle fibers are somewhat malleable, being influenced by training (especially endurance training) or by detraining (as in bed rest).

Multigene Families Encode Muscle Proteins

So far as we know, multiple isoforms of all of the myofibrillar proteins exist. These are encoded by families of genes in probably all mammalian species. Expression of these genes tends to be tissue-specific or fiber type-specific, and for many there are fetal, adult, and (for some) neonatal isoforms.

The human genome contains 20 or more actin genes and *pseudogenes* (or large segments thereof), distributed on several chromosomes. Apparently, actin genes were frequently duplicated in the course of evolution. As mentioned earlier, six actin genes are expressed in significant quantity in mammalian species in a tissue-specific manner. Skeletal and myocardial actin genes are located on chromosomes 1 and 15, respectively.

The type I muscle MHC gene is located on chromosome 14 in both humans and mouse. This gene also codes for cardiac βMHC, although they are not the identical protein. The skeletal muscle type II MHCs are on chromosome 17 in humans (11 in mice). There is also a cardiac α MHC and embryonic and neonatal cardiac and skeletal MHCs.

Similarly, there are multiple genes for the regulatory proteins. In the case of troponin, for example, there are two skeletal muscle genes for Tn I, one expressed in fast and the other in slow fibers, and a cardiac Tn I specific to myocardium. Fetal heart expresses this Tn I together with slow skeletal Tn I. Cardiac Tn I is 30–32 amino acids longer than either skeletal Tn I and thus, easily distinguished from them. In myocardial infarct (MI) patients, Tn I appears in plasma about 4 hours post-MI and remains elevated for approximately seven days. For Tn C, there are also two skeletal genes and one cardiac gene. Tn T also has two skeletal muscle forms, one fast and one slow. There are also two isoforms of adult cardiac Tn T, called Tn T_1 and Tn T_2, and two fetal cardiac Tn T isoforms. At each age, the two forms are thought to result from alternative RNA splicing. The predominant adult isoform is Tn T_2, and it has been found that serum Tn T_2 rises about four hours post-MI and remains detectable for approximately 14 days. Although it has been reported that both Tn T and Tn I are 90% (or more) sensitive and specific for MI, Tn I has been found not to undergo ontogenic recapitulation in tissue injury, which gives it an advantage over CKMB (or Tn T) in suspected MI. Either Tn T or Tn I can be used clinically, and the troponin assay has become part of the standard of care in cases of suspected MI. In persons with unstable angina, those with elevated cardiac troponins (especially Tn I and Tn T) are much more likely than others to have a cardiac event in the coming months, so that the troponin assay has predictive as well as diagnostic value.

Similarly, multiple genes, alternative RNA splicing, and posttranslational modifications result in multiple essential and regulatory light chains, tropomyosins, titins, and other myofibrillar proteins. Energy pathway enzymes are differentially expressed in various skeletal fiber types, in cardiac and smooth muscle, and at different stages of development. This also applies to Ca^{2+} regulatory proteins such as the

SR Ca^{2+}-ATPase, where there is one gene expressed in FT fibers, called SERCA1, and another in ST and cardiac fibers, called SERCA2.

Nucleotide sequences of multiple copies of genes tend to diverge over time. However, the primary structure of actin must be conserved, as mentioned earlier, due to the large number of specific binding sites in relation to the number of amino acids. The sequence of actin from the slime mold *Physarum polycephalum* differs only 8% from that of mammalian skeletal muscle actin. The various myosin families vary much more, especially in their tail and neck regions, and some have only one heavy chain. The number of light chains can vary from one to six. However, there is much less variation in myosin II. For example, the repeat structure of the tail regions in the muscle MHCs is highly conserved, and there are several segments of the actin binding and ATPase sites in the heads that are conserved not only within the myosin II subfamily, but across myosin families. There are also striking structural similarities between the heads of myosin and the microtubule motor kinesin despite a general lack of sequence similarity, and there are some key sequences in the nucleotide binding regions that are highly conserved. *G-proteins* are another class of proteins that, like myosin heads, link two other structures in a nucleotide-dependent manner, and G-proteins also strongly resemble myosin heads structurally and have striking sequence homology in the region of the nucleotide binding site. Thus, these molecular motors and the G-proteins share structural features, and it is likely that insight into one will increase understanding of the others.

21.2 Mechanism of Muscle Contraction: Overview

Our current understanding of the mechanism of contraction is reflected in the ***sliding filament model*** and the ***cross-bridge hypothesis.*** The sliding filament model holds that the shortening of sarcomeres characteristic of contraction in muscle is due to the sliding of the thick and thin filaments past one another due to interactions between them, such that the thin filaments are pulled toward the center of the thick filament array. This movement between filaments is believed to be driven by interaction between the myosin heads and the thin filaments in which the binding of myosin to actin, or cross-bridge formation, triggers changes in the head structure that create mechanical tension. This is the *cross-bridge hypothesis*. One prediction of this model of contraction is that the tension elicited by stimulation of a muscle will depend on the degree of overlap between thick and thin filaments, since this would determine the number of cross-bridges that can be formed. Thus, there would be a characteristic relation between length and isometric tension, which is, in fact, observed. For significant shortening to occur by this mechanism, cross-bridges must repeatedly form, create tension, be broken, and re-form. This is called cross-bridge cycling. The speed of sarcomere shortening depends on the mean cross-bridge cycling rate. Since the early 1960s, methods of studying the contraction mechanism have become increasingly sophisticated; recent developments such as optical traps or laser tweezers allow measurement of force (usually 5–10 piconewtons) and movement (from 5–11 nm) produced by a single myosin head acting on an actin filament.

Mechanism of Contraction: Excitation/Contraction Coupling

The primary intracellular event in the activation of contraction is an increase in $[Ca^{2+}]_i$ in the sarcoplasm from a resting level of about 0.05 μmol/L to about 5 μmol/L or more during repetitive stimulation. This Ca^{2+} surge can be elicited by α-MN stimulation or by direct electrical stimulation of the fiber. The details of how this occurs differ greatly between skeletal, cardiac, and smooth muscle. In skeletal muscle, an action potential from an α-MN in the ventral horn of the spinal cord triggers acetylcholine (Ach) release from the presynaptic membrane. At the postsynaptic membrane, binding of Ach to its receptors depolarizes the membrane to threshold, initiating an action potential that propagates all along the sarcolemma and down the *t*-tubules. This depolarization results in Ca^{2+} release from the sarcoplasmic reticulum. The abundant SR Ca-ATPase creates a concentration ratio from SR to sarcoplasm of 10^4 to 10^5, so that the release of Ca^{2+} requires only that the SR conductance to Ca^{2+} be increased. The connection between depolarization and Ca^{2+} release is called ***excitation-contraction (or E-C) coupling.*** A component of botulinum toxin (Botox) cleaves synaptobrevin, a protein required for docking of transmitter vesicles to the presynaptic membrane. Botox blocks Ach release, causing paralysis. It is used to treat spasticity, tremor, hypertonia, and other muscle conditions of localized hyperactivity.

E-C coupling involves a large protein in the t-tubular membrane called the DHP receptor (DHPR), and another protein in the SR membrane of the terminal cistern called the ryanodine receptor (RyR). The DHPR has been shown to be a voltage-gated Ca^{2+} channel. It has an α_1 subunit similar to the voltage-dependent fast Na^+ channel, and four other proteins, with a total molecular weight of ~415,000. The RyR has four large subunits, each with a molecular weight of about 565,000. It has been shown to be a ligand-gated Ca^{2+} channel, for which an operative ligand is moderate $[Ca^{2+}]_i$. This led to the hypothesis of ***calcium-induced calcium release,*** or ***CICR,*** as the explanation for E-C coupling: Ca^{2+} entering through the DHPR would trigger opening of the RyR, thus serving as "trigger calcium." However, since skeletal muscle contracts in Ca^{2+}-free media, entry of external Ca^{2+} through DHPR is not necessary, and the CICR hypothesis requires elaborate modification in skeletal muscle. It is now known that the skeletal muscle DHPR is a good voltage sensor, but a poor calcium channel, with a low conductance and slow kinetics.

It is now widely believed that in skeletal muscle a depolarization-induced change in the structure of the DHPR α_1 subunit directly influences the RyR in such a way as to markedly increase its conductance for Ca^{2+}. Since the resulting increase in $[Ca^{2+}]_i$ can open other RyR channels, this produces a surge of Ca^{2+} release. In contrast, cardiac muscle expresses a different $\alpha 1$ DHPR subunit than skeletal muscle. The cardiac subunit has much higher channel conductance and faster kinetics than the skeletal muscle type, and so admits much more extracellular calcium during the action potential. Thus, CICR does play an important role in cardiac muscle. In both cases, high $[Ca^{2+}]_i$ reduces RyR conductance, by direct binding of Ca^{2+} (or Ca^{2+}-calmodulin) to RyR, by activation of a kinase that phosphorylates RyR, and probably both. At the same time, Ca-calmodulin (Ca-CaM) activates a protein kinase that phosphorylates the SR Ca-ATPase, which increases its activity 10- to 100-fold. These two mechanisms combine to terminate the Ca^{2+} spike.

As mentioned above, DHPR are named for their binding by dihydropyridines, which include the drugs nifedipine and nimodipine, which block the DHPR (and other closely related) Ca^{2+} channels. These channels are also blocked by phenylalkylamines (e.g., verapamil) and benzothiazipines (e.g., diltiazem). Since Ca^{2+} entry through these channels is not required for E-C coupling in skeletal muscle, these drugs have little effect in this tissue. Vascular smooth muscle, however, is almost completely dependent upon entry of extracellular Ca^{2+} for contraction, and in this tissue these drugs significantly reduce tension. Myocardium, which depends partially on CICR for E-C coupling, is intermediate in sensitivity to these drugs.

Three RyR genes are expressed in humans. RyR1 predominates in skeletal muscle and in Purkinje cells of the cerebellum. RyR2 is very predominant in heart and is the most abundant form in brain. About 2–5% of the RyR in skeletal muscle is RyR3. Smooth muscle also has RyR3 in small amounts, with more abundant RyR1 and RyR2. The sequence homology between any two of these is about 70%. Ryanodine at low concentration (<10 mM) binds to RyR in the open state, and holds these channels open, producing sustained contracture, but at high concentration ryanodine blocks the channels. Dantrolene sodium is a drug that blocks RyR1 channels in skeletal muscle and is a direct-acting muscle relaxant, while producing little effect on the RyR2 channels in cardiac muscle. Neomycin and other aminoglycosides also inhibit skeletal muscle SR Ca^{2+} release by binding to the same part of RyR1 as ryanodine.

SR contains two other types of protein important to normal function: calcium-binding proteins and structural proteins. ***Calsequestrin*** is a Ca^{2+}-binding protein (M.W. 44,000), especially abundant at the terminal cisterns. It is highly acidic, with glutamate and aspartate accounting for about 37% of the amino acids. One calsequestrin binds up to 43 Ca^{2+} ions, with a mean K_m of about 1 mM. There is evidence indicating that calsequestrin is required for normal RyR activity. This protein has an FT (chromosome 1q21) and an ST/cardiac form (1p11–p13.3). Another protein, called ***high-affinity Ca^{2+}-binding protein,*** has a lower capacity but a higher affinity for Ca^{2+} than calsequestrin. Both of these buffer the $[Ca^{2+}]$ in the SR lumen, increasing the SR Ca^{2+} capacity and limiting the $[Ca^{2+}]$ gradient against which the ATPase has to work. ***Triadin*** (M.W. 94,000) is a transmembrane glycoprotein that exists as a disulfide-linked multimer of uncertain structure with most of its mass inside the SR, where it is believed to bind to RyR and probably calsequestrin. The extracellular portion links to the α_1-subunit of the DHPR. Thus, triadin may functionally link the DHPR to the RyR, or it may simply serve to colocalize the DHPR, RyR, and calsequestrin. ***Junctin*** (M.W. 26,000) is another protein that binds calsequestrin. Its 210 amino acids include a short cytoplasmic domain and a large luminal domain that binds calsequestrin, anchoring it to the T-tubule/SR junctional region of the SR membrane.

Methyl xanthines, especially caffeine, enhance Ca^{2+} release from SR, increasing the size and duration of a twitch. If the caffeine concentration is high enough, Ca^{2+} release can occur without depolarization and cause calcium contractures. This is the mechanism of muscle spasm in severe caffeine toxicity. Electrically silent contractures also occur in a rare disorder called ***malignant hyperthermia,*** in which exposure to halogenated anesthetics and/or muscle relaxants such as succinylcholine causes a sustained increase in sarcoplasmic $[Ca^{2+}]$, with resulting increases in SR Ca^{2+}-ATPase activity, muscle contracture, and both aerobic and anaerobic metabolism. The ensuing increase in body temperature, hyperkalemia, and acidosis can be lethal. In about 50% of humans so affected, there is a mutation in RyR1 that results in a reduced threshold $[Ca^{2+}]$ for CICR, among other abnormalities.

Twitch duration is largely a function of how extensive the SR is. In FT fibers, SR accounts for roughly 15% of the cell volume, whereas it is only about 3–5% in ST fibers. The total RyR Ca^{2+} conductance and the total SR Ca^{2+}-ATPase activity reflects this, so that the rates of Ca^{2+} release in the E-C coupling phase, the peak sarcoplasmic [Ca^{2+}] achieved in a twitch, and the reuptake rate of Ca^{2+} are all 2–4 times greater in FT fibers than in ST fibers. These differences result in the shorter time to peak tension, higher twitch tension, and shorter half-relaxation time seen in FT fibers compared to ST.

The ATP-dependent Ca^{2+} pump of SR (sarcoplasmic/endoplasmic reticulum Ca^{2+}-ATPase, or SERCA) belongs to a group of ion pumps called the P class, which typically have four transmembrane subunits, two large α-subunits and two β-subunits. The α-subunit is phosphorylated during the transport process, and the transported ion is believed to move through this protein. This class includes the ubiquitous Na^+, K^+-ATPase and the plasma membrane Ca^{2+}-ATPase. In SR, this pump apparently occurs as a single α-subunit with a molecular weight of 100,000, coupled to a smaller protein called ***phospholamban.*** These are densely packed on regions of the SR and account for 60–80% of the transmembrane protein in SR. SERCA transports 2 Ca^{2+} for each ATP hydrolyzed and has very high affinity for Ca^{2+} on its sarcoplasmic face (K_m about 0.1 μM); this allows the protein to pump the sarcoplasmic [Ca^{2+}] to the 0.1 μM range and below. Phospholamban regulates the activity of the pump, probably by regulating the affinity of the α-subunit for Ca^{2+}. Phosphorylation of phospholamban increases the Ca^{2+}-ATPase activity 10- to 100-fold.

Three SERCA genes are known. The SERCA1 gene produces two isoforms via alternative splicing (SERCA1a and SERCA1b) which are found in adult FT fibers and neonatal fibers, respectively. SERCA2 also produces two proteins by alternative splicing, SERCA2a and SERCA2b, which are found in ST/cardiac fibers and smooth muscle, respectively. SERCA3 is found in the ER of most nonmuscle tissue.

Mechanism of Contraction: Activation of Contraction

As described above, depolarization leads to transient Ca^{2+} release from SR. As sarcoplasmic [Ca^{2+}] rises, so does Ca^{2+} binding to Tn C. As Tn C saturates with Ca^{2+}, it reverses the tropomyosin inhibition of myosin binding to actin. Our current understanding is that this occurs by a small movement of the tropomyosin induced by the dimensional change that occurs in Tn C upon Ca^{2+} binding, thus relieving a steric block. However, a mechanism based on more subtle interactions between tropomyosin and actin cannot be ruled out (Figure 21-9). The result is that myosin heads are able to contact actin with formation of active cross-bridges and generation of tension. The initiation of cross-bridge formation by calcium is called *activation of contraction*. Activation becomes detectable at a $[Ca^{2+}]_i$ of about 0.1 μM, is half-maximal at 1 μM, and plateaus at 10 μM. SR reuptake of calcium poststimulus lowers sarcoplasmic [Ca^{2+}], desaturating the Tn C and allowing the tropomyosin inhibition of cross-bridge formation to be reasserted. It has recently been shown that nebulin (see subsection **Myofibrils,** p. 457) can inhibit actomyosin ATPase activity, preventing actin from being physically translated by myosin, and that this inhibition can be reversed by Ca-CaM. This suggests a possible thin filament-linked regulation in striated muscle.

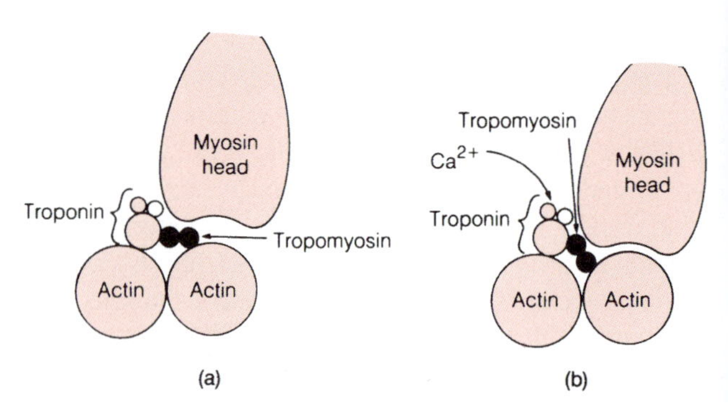

FIGURE 21-9
Schematic the interaction of myosin and actin. (a) In the absence of Ca^{2+}, tropomyosin prevents the binding of myosin heads to actin. (b) When the [Ca^{2+}] rises, Ca^{2+} binds to a subunit of troponin, which causes the tropomyosin to shift slightly into the groove of the actin thin filament. The shift in position of tropomyosin allows the myosin heads to bind to actin. Lowering of the [Ca^{2+}] results in reversal of these events.

Mechanism of Contraction: Cross-Bridge Cycling

The nature of the S2 region allows the myosin heads to be in very close proximity to the actin. When the tropomyosin-mediated inhibition of contraction is reversed, the myosin heads interact strongly with actin. Although there is still debate over the details of how this interaction produces mechanical work, there is a consensus on the general outline. Myosin bonds initially via relatively flexible loop regions near the catalytic site, followed by progressively more extensive hydrogen bonding. In this process, a large surface area on each protein (totaling 15–20 nm^2) is removed from interaction with cell water during formation of the bond between them. In cases where proteins bond without changes in conformation, burying this much area would be associated with an energy change of 60–80 kJ/mol of bonds, a number roughly twice as great

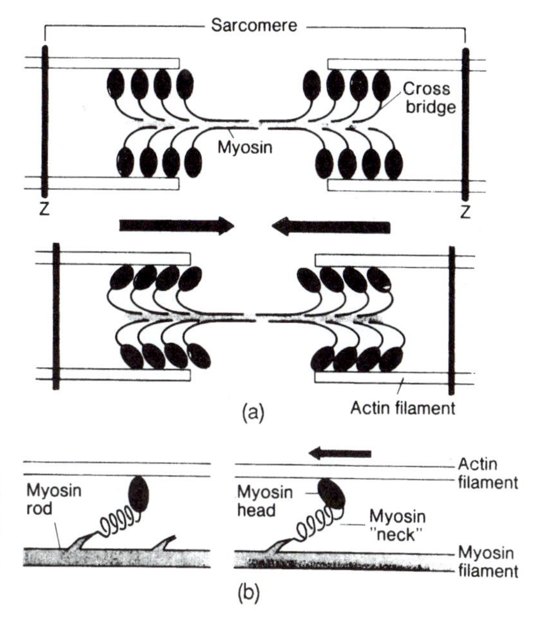

FIGURE 21-10
Schematic illustration of the sliding filament-rotating head mechanism of force generation in muscle. Cross-bridges form approximately at right angles to the thin filaments (a). This angle changes to about 45 degrees at the end of the cross-bridge cycle when the bridge is released. Recent measurements indicate that the initial and final angles in intact sliding filaments are more nearly 80 and 50°, respectively. In the attached bridge, a conformational change occurs, putting tension on the neck region. This may be due to an abrupt change in the angle of the head (b). Movement of the thick and thin filaments relative to each other relieves the stress on the neck. The product of the tension (force) and the distance moved is the work done per stroke by a cross-bridge.

as that actually observed for binding of myosin to actin. This suggests the capture of energy internally in the myosin head, actin, or both. Apparently, as the myosin moves into this tightly bound configuration, energy is captured in the form of deformation within the myosin head, which applies force (5–10 piconewtons) to the head's attachment to the thick filament (the neck region). Much, probably most, of this energy is available to do mechanical work as the filaments slide past one another (the power stroke illustrated schematically in Figure 21-10), with myosin ending up very firmly bound to actin. Subsequent binding of Mg-ATP to myosin, which is also a very high-affinity binding, provides the free-energy input to the system to alter myosin's actin binding site to a low-affinity state, permitting detachment. ATP binding lowers the affinity of myosin for actin by a factor of about 10^4. Hydrolysis of the ATP then occurs, with little change in the free energy of the system. Subsequent events release P_i and ADP from the ATPase site.

Thus, in the cross-bridge cycle, myosin is bound with high affinity alternately to actin and to ATP. Since the energy changes associated with myosin binding to actin and MgATP are internal to the system, the only free energy changes externally observable are the free-energy change from ATP in solution to ADP and P_i in solution, which equals the sum of the mechanical work performed plus the heat released. Thus the overall result is the conversion of energy of hydrolysis of ATP (about 50 kJ/mol under physiological conditions) to work (and heat), a process called ***chemomechanical transduction.*** The efficiency of this process in mammalian skeletal muscle is 60–70%.

There is uncertainty over the timing of ATP hydrolysis and release of ADP and P_i with respect to the actomyosin binding states and the power stroke. It is currently thought that ATP hydrolysis occurs after transition of the actomyosin-ATP (A-M-ATP) complex to a weakly-bound state and may sometimes occur after release of myosin from actin. The complex A-M-ADP-P_i may remain weakly bound until dislodged by movement of the filaments. The released M-ADP-P_i has moderate affinity for actin, and upon reattachment, forming A-M-ADP-P_i, the phosphate release step occurs. This creates a state called A-M*-ADP which is the high-affinity state associated with initiating the power stroke. As the structural changes produced by increasingly tight binding to actin produce strain in the myosin head and therefor force and movement, the affinity of the ATPase site for ADP also changes, releasing the ADP. Thus, at the end of the power stroke, cross-bridges are typically in the A-M state (called the rigor state), their most tightly bound rigid state, in which they will remain unless ATP is available to bind to the ATPase site and alter the affinity of the actin binding site (Figure 21-11). In normal circumstances it is almost impossible to deplete ATP to the point that a large proportion of myosin heads form rigor bonds, but it does happen in severely ischemic muscle and post-mortem (rigor mortis). When Mg-ATP is available, binding occurs and alters the actin binding site to a low-affinity configuration, and hydrolysis follows, so that the cross-bridge will probably be in the weakly-bound A-M-ADP-P_i state until once again pulled free. So long as $[Ca^{2+}]_i$ remains high, this cycle will continue, provided that adequate ATP concentration and other appropriate conditions of the internal environment are maintained. The rate-limiting step is the P_i release step, and all of the steps following P_i release up through ATP hydrolysis happen quickly, so that the M-ADP-P_i and A-M-ADP-P_i states predominate.

In a general way, transport ATPases are similar to myosin. The initial binding of the transported substance to the transport protein corresponds to tight binding of myosin to actin. Reorientation of the binding site toward the opposite face of the membrane is analogous to the force-producing conformational change in the myosin head, and conversion of the substrate binding site to a low-affinity state is accomplished by binding of ATP. ATP

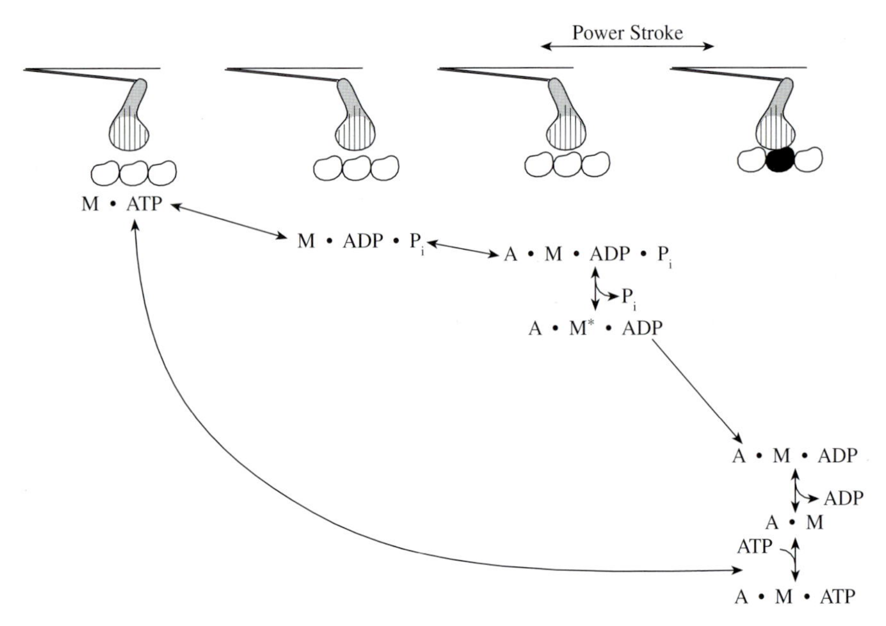

FIGURE 21-11
A simplified model of the interactions thought to occur during the cross-bridge cycle. Four major structural states are depicted, but the actual number of structurally and energetically distinct states through which this system transitions is uncertain. The vertical drop in the figure crudely represents the free-energy change in the transitions shown. The functional change in myosin structure is shown as a change in the relative orientation of the neck (cross-hatched) to the catalytic head region (hatched). A is actin, M is myosin, and P_i is phosphate. M · ATP represents myosin complexed with ATP immediately after detachment from actin. Here, the neck is shown in the same configuration as at the end of the power stroke. Hydrolysis of ATP leads to M · ADP · P_i, with weak bonding to actin both before and after hydrolysis. ATP hydrolysis is energetically favored, but due to the products remaining bound, the free-energy change is not large. Hydrolysis is shown here as eliciting a change in orientation of the neck, producing the conformation thought to exist at the initiation of the power stroke, but the relationship between the hydrolysis events and the mechanical events is still speculative. Release of P_i from A · M · ADP · P_i following binding to actin is associated with conformational changes that produce the strained, or force-producing, state A · M* · ADP. Transition from A · M* · ADP to A · M coincides with performance of mechanical work and formation of a very strong A · M bond, and a reorientation of the neck region is thought to occur during this transition. In striated muscle, a large part of the free energy driving the cross-bridge cycle is lost between these two states. A · M · ADP and A · M are thought to differ little in configuration, and the energy change associated with ADP release is small. Subsequent binding of ATP to A · M dramatically lowers the affinity of myosin for actin, probably producing release of M · ATP from A without altering the neck orientation, returning the system to the state shown at the upper left. [Adapted from: R. Cooke, Actomyosin interaction in striated muscle. *Physiol. Rev.* **77**(3): 671–697, 1997.]

hydrolysis is associated with return of the substrate binding site to its original orientation, and ADP and P_i are released into solution.

It is important to note that this model predicts the hydrolysis of one ATP for every cross-bridge cycle of every myosin head: there is evidence that, in high-speed contractions at least, there may be multiple attachments and detachments per hydrolysis. Thus, our understanding is obviously not complete. Nevertheless, the ATP consumption associated with contraction can be enormous. The *metabolic scope* (ratio of maximal to resting energy consumption) of skeletal muscle can reach 100:1, and there must necessarily be metabolic specializations to meet this peak demand and to do so quickly. Energy metabolism is discussed in Chapters 13–15 and 18.

21.3 Energy Supply in Muscle

During muscle activation, ATP consumed by the myofibrils must be replaced by aerobic and/or anaerobic resynthesis. The magnitude of the task is increased by the augmentation of the SR Ca^{2+}-ATPase activity that must also take place. SERCA activity is not very great at rest, being only about 7% of the resting energy consumption, but it increases more than ten-fold during contraction. Na^+, K^+-ATPase activity also increases due to the ion fluxes accompanying repeated action potentials. The metabolic specializations for meeting these demands vary by fiber type (see **Organization and Properties of Muscle Fibers,** p. 462). Type I fibers are generally oxidative. Their ATP resynthesis is largely dependent on supplying reducing equivalents to the cytochromes via β-oxidation of fatty acids and via glycolysis and the tricarboxylic acid (TCA) cycle. Type I fibers usually have only modest activity of glycolytic enzymes and myokinase, and normally do not produce pyruvate at rates much in excess of the rate at which they can oxidize it. Type IIb fibers are metabolically the mirror image of type I, and type IIa is a mixed fiber type metabolically.

Ordinarily, slow-twitch oxidative fibers are the first recruited, and if the force and speed thus produced are adequate for the task, they are the only fibers recruited. The relatively low actomyosin ATPase activity of these fibers enables production of tension with less ATP expenditure than in fast fibers, and so, at the low speeds characteristic of prolonged work, they are more efficient than faster fibers. Aerobic energy production in these fibers offers many important advantages over the low-oxidative high-glycolytic strategy of IIb fibers. First, complete oxidation of glucose yields 18 times as much ATP per glucose as glycolysis only. Second, the metabolites are CO_2, which is easily eliminated via the lung, and H_2O, which is needed during work. This is in contrast to the lactic acid produced in glycolysis, which lowers pH considerably until it can be metabolized aerobically. Low intramuscular pH inhibits SR Ca^{2+} release, impairs activation, and slows cross-bridge cycling. Third, glucose is the only available substrate for glycolysis, while oxidative metabolism can also consume lipid and amino acids in addition to glucose. This latter feature enables a vastly greater volume of work to be performed aerobically than anaerobically, even when the anaerobic work is intermittent to allow recovery from the lactic acidosis. A typical value for glucose stored as glycogen in muscle is about 430 g, and about 70 g in liver, equivalent to 2100 kcal total (or roughly the energy needed to run 20 miles, assuming that all of it could be metabolized). Lipid stores in adipose tissue, muscle, and elsewhere are typically 5.5 kg in an active young adult male, equivalent to almost 52,000 kcal (or enough energy to run more than 500 miles, again assuming that all of it could be metabolized), and this is in a relatively lean person. The derivation of tricarboxylic acid (TCA) cycle intermediates from some amino acids and the conversion of glycogenic amino acids to glucose enables protein to also serve as an energy source. For these reasons, the energy substrate available to the slow oxidative fibers is almost inexhaustible, and, due to their low actomyosin ATPase activity, replenishment of ATP by oxidative phosphorylation can keep up with ATP hydrolysis. Thus, SO fibers are fatigue-resistant, while fast glycolytic fibers which use only glucose (and inefficiently at that) are quite fatiguable. The intermediate FOG fibers are much more fatigue-resistant than FG fibers, but less so than SO fibers.

An additional consideration is that each gram of glycogen stored is associated with 3–4 grams of water, so that the energy per gram of wet weight (about 4 kJ/g) is much lower than for lipid (39 kJ/g). Carrying stored energy around as lipid is therefor less expensive than carrying it as carbohydrate, but only the aerobic fiber types draw significantly upon this store. When malnourished persons are fed, there is rapid synthesis of glycogen, with a corresponding movement of water into the cells (primarily in muscle and liver, where glycogen synthase activity is high). Potassium is the predominant intracellular cation, so there is also a movement of K^+ into cells. Since muscle may be half the body mass, and since malnourished persons are often hypovolemic, these movements are proportionately large. The resulting *hypokalemia* and further *hypovolemia* may be fatal unless adequate hydration and K^+ supplementation are provided.

Aerobic fiber types have high activities of the enzymes for β-oxidation of lipid, and their mitochondria also have abundant acylcarnitine palmitoyltransferases and associated substances (Chapter 18). However, the low solubility of fatty acids limits their rate of diffusion through aqueous media, and the numerous steps involved in mobilizing fatty acids from triacylglycerols, activating them to acyl-CoA and transporting them into mitochondria, and splitting off two-carbon pieces as acetyl-CoA, also slows flux through β-oxidation. The net result is that the rate of ATP production attainable by this pathway is about one half the maximum rate achievable by aerobic glycolysis.

When the volume of work is such that glycogen is functionally depleted, muscle power output must decrease to the level that can be supported by fatty acid oxidation. During prolonged moderate-intensity exercise, most of the fatty acid oxidized is derived from lipolysis in adipose tissue during the exercise. Regulation of lipolysis varies regionally in humans: the exercise-induced lipolysis is 50% greater in abdominal than in gluteal adipocytes, but the underlying biochemical differences have not been explained. Adipocyte lipolysis depends at least in part on activation of lipoprotein lipase by catecholaminergic stimulation and other hormonal changes accompanying exercise, and takes many minutes (20–40) to fully accelerate. In the shorter term, and in higher intensity work where the metabolic demand exceeds the cardiovascular delivery of lipid, lipolysis of triglyceride droplets stored in the muscle fibers becomes the most important source of fatty acids. Type I, and to a somewhat lesser extent type IIa, fibers store appreciable triglyceride, presumably by esterifying fatty acids from the blood since activity of enzymes involved in lipid synthesis is low in human muscle. Lipid storage in muscle increases with endurance training. Utilization of lipid previously present in the plasma as protein-bound fatty acids or as triglyceride can begin promptly, and is not dependent on hormonal mechanisms. Activation of capillary lipoprotein lipase (e.g., by heparin) measurably increases muscle lipid oxidation in endurance exercise.

Rate of lipid oxidation as a function of exercise intensity is an inverted U shaped curve, increasing to a maximum of roughly 40 $\mu M \cdot kg^{-1} \cdot min^{-1}$ at 60–70% of maximum aerobic power, and decreasing to about three fourths of this rate at 85% of maximum aerobic power. As the power requirement approaches the limits of β-oxidation, the flux through glycolysis increases rapidly. In the rat, this has been shown to cause increased citrate release from mitochondria with a consequent increase in malonyl-CoA synthesis via acetyl-CoA carboxylase (ACC). Malonyl-CoA strongly inhibits carnitine palmitoyltransferase–1, thereby inhibiting lipid oxidation (Chapter 18). However, in humans, muscle ACC activity is low, as is malonyl-CoA, and malonyl-CoA has not been found to be negatively correlated with rate of β-oxidation. It seems unlikely that the phenomenon is explained merely by competition for CoA or for NAD, and so the mechanism is unclear. The significance is that as power output exceeds 70% of maximum, aerobic glycolysis not only supplements β-oxidation, it progressively replaces it.

Type IIb fibers rely on glycolytic rather than oxidative metabolism for energy. In these fibers, glucose derived from muscle glycogen and from glycogenolysis and gluconeogenesis in liver is split to pyruvate at rates far greater than the rate at which pyruvate can be oxidized. Though producing only 2 ATP per glucose molecule (3 per glucosyl unit from glycogen), these fibers can generate such a high flux through glycolysis that the peak ATP production rate is twice the maximum rate of oxidative ATP production in type I fibers. Aerobic and anaerobic glycolysis are discussed in Chapter 13.

In glycolysis, conversion of glyceraldehyde 3-phosphate to 1,3-bisphosphoglycerate by glyceraldehyde-3-phosphate dehydrogenase (G3PD) requires reduction of NAD to NADH. The NAD pool is small, and without rapid oxidation of NADH back to NAD, glycolysis would immediately stop at the G3PD step. Oxidation of NADH can be accomplished by conversion of pyruvate to lactate via lactate dehydrogenase (LDH). Maximum values of lactate production in glycolytic fibers in humans range from 0.5 to 0.9 $mM \cdot g^{-1} \cdot s^{-1}$, while in oxidative fibers it is only 0.25 $mM \cdot g^{-1} \cdot s^{-1}$. Since the maximal reported rates of appearance of lactate in blood are only one tenth the FG fiber production rate, it is clear that the maximal production rate far exceeds the efflux capacity and that high-intensity exercise must necessarily produce high intramuscular lactate concentration. Intramuscular lactate concentrations as high as 45–50 mM/kg of cell water have been reported in humans. Since the pK of lactic acid is 3.9, most (>95%) of this acid will be dissociated in the physiological range of pH, imposing an almost equimolar H^+ load on the cell's buffer systems. Intracellular pH in muscle fibers is about 7.0, and decreases 0.4–0.8 pH unit during intense exercise.

Lactate efflux from muscle occurs by simple diffusion of undissociated lactic acid and by carrier-mediated lactate-proton cotransport. The latter probably accounts for 50–90% of the lactate efflux, depending on fiber type and pH. It was thought that fast glycolytic fibers had an abundant lactate transporter with a low K_m, providing rapid efflux to support continued glycolysis, while slow oxidative fibers had a less abundant transporter or one with higher K_m, or both, so that lactate would be retained as a redox buffer. Retained lactate maintains an NAD/NADH ratio that stimulates oxidation, and ensures subsequent oxidation or gluconeogenesis in the muscle. There are three or more muscle lactate transporters, one of which has been sequenced in rat muscle and found to have 494 amino acids (M.W. 53,000) and 86% sequence identity with the corresponding protein from human erythrocytes. However, K_m for lactate has not been found to differ significantly between fiber types, being around 30 mM. Moreover, the transport capacity is almost twice as great in SO fibers as in FG. The high K_m implies that most of the lactate produced will be retained during exercise regardless of fiber type, and the relative amounts of carrier imply that lactate efflux from FG fibers is greater than from SO fibers only because the rate of production is usually much higher, creating greater gradient for efflux. It follows that lactate efflux from FG fibers is not adequate to prevent eventual inhibition at the G3PD step and pronounced acidification of the fibers. The greater lactate transport capacity of oxidative fibers also enhances lactate entry into these fibers when blood lactate rises during moderate-intensity work in which the SO fibers produce little lactate.

The fate of the lactate produced is primarily oxidative (55–70%). Lactate released from muscle enters exclusively oxidative tissues in which lactate concentration is low (like heart, diaphragm and brain), is converted back to pyruvate, and is oxidized. Retained lactate is oxidized to generate ATP needed to replenish phosphocreatine stores following exercise and to restore normal distribution of Ca^{2+}, Na^+, and K^+. About 15% is used in gluconeogenesis and subsequent glycogenesis. The balance is converted to alanine, glutamate, or other substances.

Utilizing amino acids as fuel requires eventual elimination of an equimolar amount of ammonia, which may also require eliminating more water. It also requires a longer recovery time, since replacement of the hydrolyzed proteins may be slow, and requires higher dietary protein intake. In most circumstances, protein is the macronutrient in least supply. For all of these reasons, protein metabolism is not a preferred pathway for energy supply. Nevertheless, protein metabolism contributes 2–3% of the energy requirement in exercise of a few minutes duration and rises to as much as 12% after several hours of physical work. Replenishment of TCA cycle intermediates such as α-ketoglutarate (derived from glutamate) or oxaloacetate (from aspartate or asparagine) is probably important to offset the loss of TCA cycle intermediates from mitochondria over time. This role of protein catabolism in supporting glucose and lipid oxidation, called ***anaplerosis,*** may be more important than its direct contribution to energy supply.

Nitrogen transport from muscle to liver seems to be accounted for mainly by release from muscle of alanine synthesized *de novo* by amination of pyruvate. Liver production of urea, however, is thought to be driven mainly by plasma glutamine and ammonia concentrations. Muscle release of glutamine does not increase during exercise (although ammonia efflux increases), nor does the plasma glutamine concentration, and there is no significant increase in urea production during exercise of moderate intensity and duration. Most alanine reaching the liver in this circumstance is deaminated and used in gluconeogenesis, but the immediate disposition of the amino nitrogen is still unclear.

Special significance is often attributed to branched-chain amino acids (BCAAs: leucine, isoleucine, and valine) in the context of muscular performance. Increased BCAA uptake and oxidation by muscle does occur during exercise, and it has been claimed that intramuscular BCAA concentration exerts a regulatory influence on rate of protein synthesis. It is also claimed that, since BCAA and tryptophan (TRP) compete for transport across the blood-brain barrier, decreasing plasma BCAA concentration leads to increased brain TRP uptake and serotonin synthesis and a host of putative effects thereof. Although these hypotheses are attractive and a segment of the supplement industry relies on them, available studies show that BCAA supplementation in humans neither offers any performance benefit over other energy supplements nor alters mood or perception during or after exercise.

The ***purine nucleotide cycle*** also is involved in muscle energy production. During intense stimulation, or when O_2 supply is limited, the high-energy bond of ADP is used to synthesize ATP via the myokinase reaction (Figure 21-12). The resulting AMP can dephosphorylate to adenosine, which diffuses out of the cell. Conversion of AMP to IMP via adenylate deaminase and then to adenylosuccinate helps sustain the myokinase reaction, especially in FG fibers, by reducing accumulation of AMP. It may also reduce the loss of adenosine from the cell, since nucleosides permeate cell membranes while nucleotides do not.

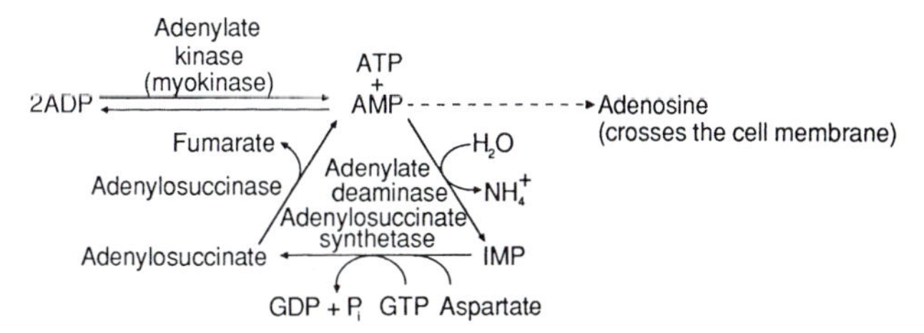

FIGURE 21-12
Role of purine nucleotides in muscle energy metabolism. The conversion of AMP to IMP prevents loss of adenosine from the cell.

Lack of muscle adenylate deaminase has been found in some muscle disorders in which other abnormalities could not be identified. Failure to produce ammonia during intense effort may be diagnostic. In the heart, AMP accumulation is typically due to ischemia, and the adenosine released as a result is a potent coronary vasodilator. Accordingly, myocardium has less adenylate deaminase than skeletal muscle. However, large losses of adenosine from myocardium are dangerous because they lead to a decrease in ATP concentration that does not respond to dilation, thrombolytic, or oxygen therapy. Lack of adenosine deaminase in lymphoid tissue causes a severe immunodeficiency (Chapter 27).

Phosphocreatine Shuttle

Energy transfer between mitochondria and the myofibrillar ATPases is mediated by phosphocreatine (Chapter 17). The phosphocreatine shuttle is illustrated schematically in Figure 21-13. Synthesis of ATP in mitochondria is closely coupled to that of phosphocreatine. Since the reaction

$$\text{ATP} + \text{C} \rightarrow \text{ADP} + \text{P} \sim \text{C}$$

is energetically favored when the ATP/ADP ratio is high, phosphocreatine is rapidly formed from ATP releasing ADP, which stimulates mitochondrial respiration. Phosphocreatine diffuses from the mitochondria to the various sites of energy utilization, where CK reverses this reaction to form ATP and creatine. Creatine diffuses back to the mitochondria for rephosphorylation. Since creatine elicits formation of ADP in mitochondria, the increase in creatine helps stimulate respiration when muscle activity increases.

Cellular adenine nucleotides are compartmentalized by their very low diffusibility (due to their size and charge) with pools in the mitochondria, at the myofibrils, SR, and other sites of energy utilization. CK is located at those sites. Phosphocreatine is much smaller and less charged, and therefore much more mobile in cells than ATP. ATP produced by substrate-level phosphorylation in glycolysis may be used to rephosphorylate creatine in the sarcoplasm;

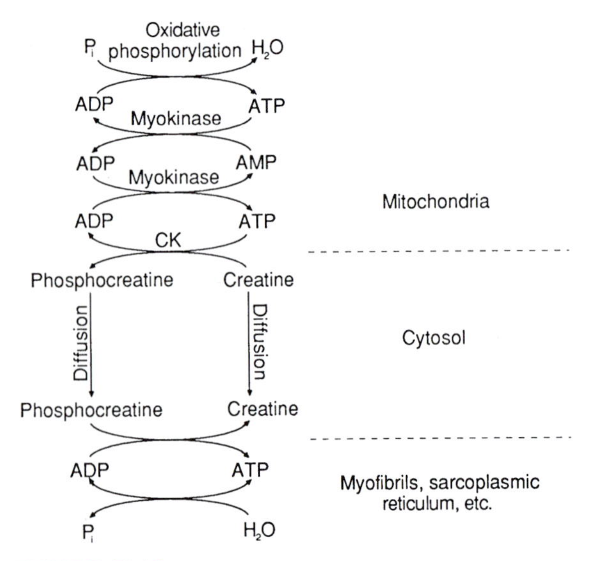

FIGURE 21-13
Phosphocreatine shuttle. A myokinase (adenylate kinase) cascade between oxidative phosphorylation and creatine kinase (CK) has been postulated. A similar myofibrillary cascade may exist at the myofibrillar ATPase site.

however, CK and glycolytic enzymes are not colocalized. Phosphocreatine can be replenished rapidly only by oxidative metabolism.

In addition to its energy transport function, phosphocreatine functions as an intracellular energy store. ATP inhibits mitochondrial respiration, limiting the maximum achievable concentration of ATP, but phosphocreatine can accumulate without inhibiting respiration and ATP synthesis. Phosphocreatine provides a reserve of immediately available energy that can be used for brief bursts of activity as in throwing or jumping, and which is able to cover the energy needs for a few seconds at the beginning of sprint-type activity while glycolysis is accelerating.

Supplementation with creatine has not been found to enhance aerobic work performance despite measurable increases in muscle creatine and phosphocreatine, suggesting that energy transfer in muscle cells is not normally limited by the total creatine concentration. However, ability to perform intermittent high-intensity work is enhanced, indicating that the energy storage function of PC is increased by creatine supplementation.

Creatine kinase (also called creatine phosphokinase, or CPK) is a dimer of subunit molecular weight 40,000. The brain isozyme is a dimer of B subunits. In skeletal muscle, the principal form is a homodimer of M subunits. In cardiac muscle, 80–85% of the CK is MM, the balance is MB. These isozymes are electrophoretically distinct, as is mitochondrial CK. Depending on fiber type, 10–30% of CK activity is on the outer side of the inner mitochondrial membrane, 3–4% is at the M-lines, and the remainder is in the cytoplasm or bound to SR, sarcolemma, or other sites of ATP utilization.

Measurement of serum concentration of total CK and CK-2 (i.e., MB) have long been used to assess the extent of myocardial damage in suspected MI (Chapter 8). Measurement of LDH isozymes has been used similarly. In recent years, however, reliance on cardiac troponin assays has increased and use of CK and LDH assays will probably decrease.

Regulation of Smooth and Cardiac Muscle

In all the actin-based motility systems, and especially skeletal, cardiac and smooth muscle, contraction is initiated primarily by an increase in cytoplasmic [Ca^{2+}]. However, the major differences in histology and function of these muscle types are associated with great variety in how contraction is controlled. Smooth muscle especially differs from the model presented for skeletal muscle. Skeletal muscle is activated by Ca^{2+} released from SR in response to sarcolemmal action potentials. Ca^{2+} exerts its effect by binding to troponin on the thin filaments, which reverses the tropomyosin inhibition of cross-bridge formation.

Smooth muscle is much more dependent on entry of extracellular calcium. Smooth muscle cells are much smaller, having lengths about equal to the diameter of small skeletal muscle fibers, and so have much bigger surface-to-volume ratios than skeletal muscle cells. Therefore, Ca^{2+} entry from the extracellular space can increase cytoplasmic [Ca^{2+}] much more readily in smooth muscle than in skeletal muscle. Most smooth muscle accordingly need not depend on Ca^{2+} release from SR, may not have much SR, and does not need action potentials to trigger SR Ca^{2+} release.

Smooth muscle exhibits very diverse behaviors depending on which control mechanisms are present. Vascular smooth muscle, for example, lacks fast voltage-dependent Na^+ or Ca^{2+} channels and so does not have action potentials or Ca^{2+} spikes. It has slow voltage-dependent Ca^{2+} channels that admit calcium in a graded fashion in response to fluctuations in membrane potential induced by humoral or transmitter effects on membrane ion conductances, and it has several membrane receptor–initiated second-messenger cascades that control Ca^{2+} entry and Ca^{2+} release from its limited SR, and which moderate the effectiveness of Ca^{2+}. Vascular smooth muscle contraction is thus tonic rather than phasic, and is very dependent on extracellular Ca^{2+}; therefore Ca^{2+} channel blockers effectively inhibit contraction. In contrast, gut smooth muscle does have fast voltage-dependent channels sufficient to produce action potentials and more SR than vascular smooth muscle, and also has gap junctions through which ion fluxes can occur. It also has receptor-mediated Ca^{2+}

control similar to vascular smooth muscle. Therefore, gut smooth muscle can contract tonically or phasically, and depolarization can be directly transmitted from one cell to the next.

As the name implies, smooth muscle lacks the highly ordered sarcomere structure of striated muscle, having thick and thin filaments in less orderly arrays with relatively less myosin (one fifth as much) than in striated muscle. Smooth muscle thin filaments have tropomyosin but generally lack troponin. Myosin in smooth muscle is found in monomeric form as well as small thick filaments, and phosphorylation is almost essential for condensation of monomeric myosin into filaments. Thus, the amount of myosin available to cross-bridge with actin may be physiologically adjustable. Like other myosin II types, smooth muscle myosin is a hexamer, and several isoforms of the heavy chains and both light chains are known. The SM-1 isoform (M.W. 204,000) has an unusually long COOH–terminal tail not found in SM-2 (M.W. 200,000). Some smooth muscle myosins have an extra 7 amino-acid segment in the head near the aminoterminal which is associated with higher ATPase activity and velocity of shortening. The light chains are called LC_{17} and LC_{20}, based on their molecular weight, for each of which two forms have been described. The LC_{17} isoform particularly seems to influence speed of shortening and tension development, the LC_{17a} form being associated with higher speed.

In skeletal muscle, disinhibition of actin is necessary for contraction to occur, and control of contraction is said to be actin-linked. In smooth muscle, phosphorylation of myosin light chains (MLCs) is required for contraction. Several mechanisms alter MLC phosphorylation, and so in smooth muscle, control of contraction is primarily myosin-linked. Three control proteins have been identified in smooth muscle: *myosin light chain kinase* (MLCK); *caldesmon* (CaD); and *calponin* (CaP). Figure 21-14

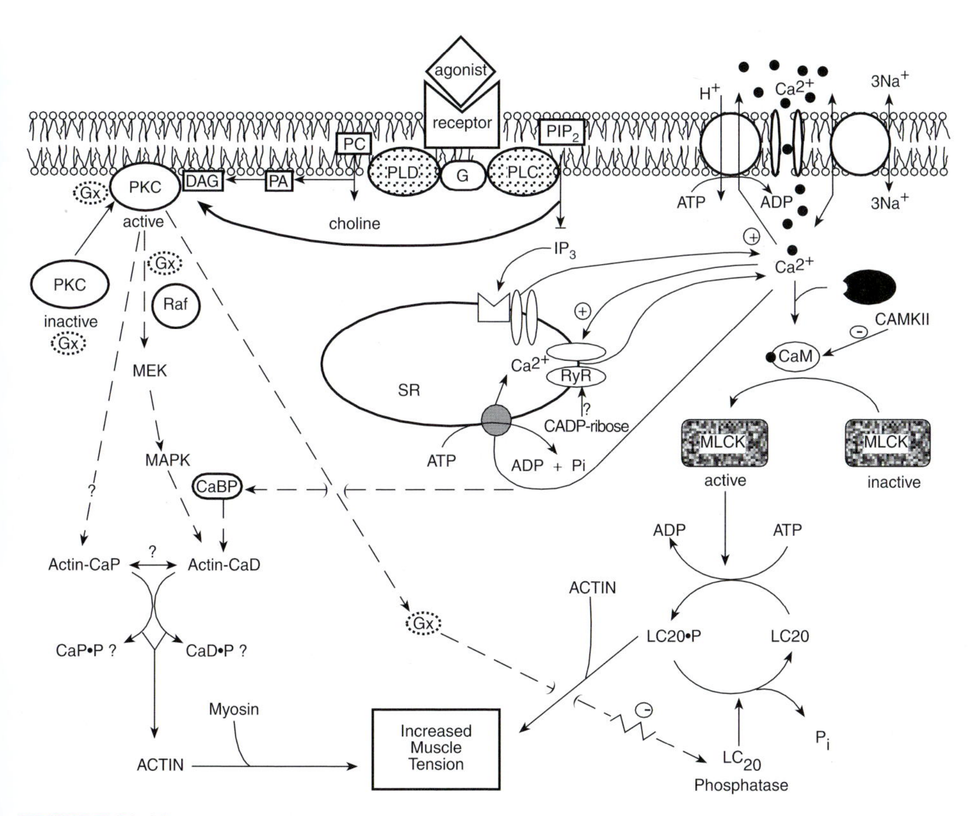

FIGURE 21-14
Pathways activating contraction in vascular smooth muscle. As in other muscle types, increased cytosolic Ca^{2+} is the most important signal, but control of cytosolic $[Ca^{2+}]$ and the mode of Ca^{2+} action is more complicated in smooth muscle, and other mechanisms may contribute. Key: G, GTP-binding protein; PLC, phospholipase C; PLD, phospholipase D; PIP_2, phosphatidylcholine 4,5-bisphosphate; PC, phosphatidylcholine; IP_3, inositol 1,4,5-triphosphate; PA, phosphatidic acid; DAG, diacylglycerol; SR, sarcoplasmic reticulum; CaK calmodulin; PKC, protein kinase C; MLCK, myosin light chain kinase; LC_{20}, 20-kDa myosin light chain; CaD, caldesmon; CaP, calponin; CaMKII, Ca/calmodulin protein kinase II; MAPK, mitogen-activated protein kinase; MEK, MAP/ERK kinase; G_x, small GTP-binding protein; CaBP, calcium-binding protein; RyR, ryanodine receptor. Dashed lines indicate pathways not fully defined or that are speculative. The role of membrane potential is not indicated in this figure. [Reproduced with permission from A. Horowitz, et al., Mechanisms of smooth muscle contraction. *Physiol. Rev.* **76**(4), 967–1003 (1996).]

summarizes the control mechanisms in smooth muscle that are discussed below.

MLCK is a very specific kinase whose only known physiological substrate is MLC in myosin II. The only phosphate donor is Mg-ATP. Smooth muscle MLCK (actually found in almost all cells) phosphorylates LC_{20} at Ser^{19}. There is also a skeletal muscle-specific MLCK and an embryonic MLCK. The various MLCKs are about 950–1000 amino acids long. MLCK has functional domains for: calmodulin binding, actin binding, substrate (ATP and myosin) binding, a catalytic domain, several target sites for various protein kinases, and a putative autoinhibitory site. The sequence and structure of skeletal and smooth muscle MLCKs are generally similar at these functional sites, but are quite divergent otherwise. MLCK is inactive in the absence of Ca-CaM, and its activity increases with increasing $[Ca^{2+}]_i$ such that the phosphorylation of LC_{20} becomes half-maximal at about 0.3 μM Ca^{2+}. Phosphorylation of a serine in the CaM binding domain decreases the affinity of MLCK for CaM and causes a decrease in the sensitivity of MLCK activity to increases in $[Ca^{2+}]_i$. This site appears to be phosphorylated *in vivo* by Ca-CaM-dependent protein kinase II (CaMKII), but not by protein kinase C (PKC) or cAMP-dependent kinases. However, some studies indicate that this site is phosphorylated by protein kinase A (PKA), which might play a role in relaxation induced by dilators that increase cAMP, but the evidence is conflicting. Also, it appears that the $[Ca^{2+}]_i$ required to activate CaMKII is about three times higher than that required to activate MLCK. This implies that Ca-CaM would bind to MLCK and shield the target serine before CaMKII could act on it, raising doubts about the significance of CaMKII in control of smooth muscle. It is well established only that Ca-CaM activates MLCK.

LC_{20} phosphorylation produces two pronounced effects: it promotes the ability of myosin monomers to assemble into filaments, and it markedly increases (100-fold) the ATPase activity of myosin compared to unphosphorylated filamentous myosin. How unphosphorylated myosin filaments can remain stable is not certain, but may be related to binding to myosin of independently expressed MLCK fragments containing the myosin binding domain of MLCK. How LC_{20} phosphorylation increases the ATPase activity so dramatically is still unknown.

LC_{20} phosphatase dephosphorylates LC_{20}. This is a trimeric type I protein phosphatase (PP-I) comprising a catalytic subunit (M.W. 37,000), a myosin-binding subunit (M.W. 130,000), and a 20,000-M.W. subunit that also binds myosin. This PP-I may be inhibited by high arachidonic acid, and inhibited or stimulated by phosphorylation by PKC or cAMP-dependent kinases, respectively. However, since the phosphatase is bound to myosin, and the ratio of phosphatase to myosin is only about 1:50, there is some doubt that LC_{20} phosphorylation is regulated via the phosphatase. MLCs may spontaneously dephosphorylate at a significant rate.

In some smooth muscle (especially vascular smooth muscle), tension elicited by agonists or K^+-induced depolarization is often tonically maintained after LC_{20} phosphorylation and sometimes $[Ca^{2+}]_i$ has fallen to near basal levels. These tonic contractions are characterized by very low cross-bridge cycling rates and ATP turnover, and are referred to as the ***latch phenomenon.*** It is currently believed that dephosphorylation of attached cross-bridges converts them to a form in which ADP is stably bound, called *latch bridges,* thereby dramatically slowing the events of the cross-bridge cycle subsequent to the force production step. It also appears that in the complete absence of Ca^{2+}-induced LC_{20} phosphorylation, the latch state cannot be sustained.

Protein kinase C (PKC) may play a role in tonic tension. PKC refers to a family of related serine/threonine kinases, five of which are found in smooth muscle. PKC is activated by diacylglycerol, or DAG. DAG (and also IP_3) is liberated from membranes by the action of phospholipase C (PLC) on phosphatidylinositol 4,5-bisphosphate, or by the action of phospholipase D (PLD) on phosphatidylcholine. A number of receptor-mediated events are transduced by activation of these lipases. Some agents that elicit tonic contraction (e.g., angiotensin II) activate PLD, thus producing DAG (but not IP_3), which activates PKC with no effect on $[Ca^{2+}]_i$. There are at least three sites on LC_{20} that can be phosphorylated by PKC, but it is not known which one, if any, of these is involved in the induction of contraction and latch.

Caldesmon, or CaD, is a thin filament-linked regulator in smooth muscle. CaD is a single long peptide (M.W. ~90,000). It binds to actin, tropomyosin, myosin, and Ca-CaM. The affinity of CaD binding to thin filaments is much higher in the presence of tropomyosin than in its absence, and *in vivo,* CaD is probably bound only to tropomyosin-containing filaments. CaD inhibits actin-activated myosin ATPase by competing with myosin for a binding site on actin, and substantially reduces ATPase activity at a CaD/actin ratio of 1:10. CaD displaced by myosin remains attached to actin, however, via a second binding site for which myosin does not compete. It has been found that CaD can also bind to the S_2 region of the myosin head, which might play a role in stabilizing thick filaments that have dephosphorylated.

Reversal of CaD-induced inhibition can occur by several means. High concentrations of Ca-CaM inhibit CaD binding to actin/tropomyosin. In some species, another calcium-binding protein called caltropin (M.W. ~11,000) has also been shown to reverse the CaD inhibition of actin-myosin interaction in a Ca^{2+}-dependent manner. Phosphorylation is another such mechanism. CaD can be phosphorylated by CaMKII, PKC, PKA, casein kinase II, cdc2 kinase, and mitogen-activated protein kinase (MAP kinase). Mammalian CaD is phosphorylated at two sites by MAP kinase, which is probably the physiologically relevant kinase in mammals. The extent to which actin-myosin interaction is normally inhibited by CaD is not clear, and so the importance of these mechanisms regulating CaD in the control of smooth muscle contraction is debated. It has been argued that CaD phosphorylation is required to permit formation of latch bridges.

Calponin is another polypeptide monomer (M.W. ~32,000) that can inhibit actin-activated myosin ATPase activity. In contrast to CaD, CaP exerts its effect in the absence of tropomyosin and completely inhibits motility in a 2/3 ratio with actin. CaP inhibits myosin binding to actin, but does so by reducing the affinity of actin for myosin rather than competing for the same site. CaP can be phosphorylated by PKC and CaMKII, both of which reverse CaP's inhibitory activity. As with caldesmon, many questions remain. The ratio of CaP to actin actually observed in smooth muscle is in the range 1:10 to 1:16, far from the 2/3 ratio found to produce near-complete inhibition of motility. Therefore, the importance of CaP and its regulation by phosphorylation is still debatable.

Relaxation of smooth muscle (especially airway smooth muscle) by β_2 agonists remains a puzzle. It had been thought that G-protein-mediated cAMP production stimulated cAMP-dependent kinases such as PKA, eliciting phosphorylation of MLCK, which prevented Ca-CaM binding and subsequently inhibited MLCK. This would result in decreasing LC_{20} phosphorylation and loss of tension. It is clear that β_2 stimulation does indeed increase [cAMP] and activity of cAMP-dependent kinases in smooth muscle. However, β-agonists that relax tracheal smooth muscle can reduce the sensitivity of LC_{20} phosphorylation to $[Ca^{2+}]_i$ without reducing the Ca^{2+} sensitivity of MLCK activity, which is not consistent with a mechanism based on inhibition of Ca-CaM binding to MLCK. β_2-agonists generally elicit hyperpolarization by increasing K^+ conductance, particularly in a K^+ channel population called large-conductance calcium- and voltage-activated K^+ (KCa) channels, also called maxi-K channels. The probability of these channels being open is increased by phosphorylation by cAMP-dependent kinases, and a subpopulation may also be directly opened by β_2-receptor activation via a G-protein interaction. Hyperpolarization reduces the open probability of voltage-dependent Ca^{2+} channels and reduces $[Ca^{2+}]_i$, which reduces MLCK activity. There are several other mechanisms by which β stimulation can alter $[Ca^{2+}]_i$, and there are circumstances in which β-adrenergic relaxation occurs without hyperpolarization. It has also been claimed that the light chain phosphatase, PP-I, is stimulated by β-adrenergic pathways. Given the diversity of smooth muscle, it is quite possible that the relative importance of various activating and relaxing mechanisms varies with the specific smooth muscle type and, perhaps, with the recent electrical and chemical history of the cell.

Control of cardiac muscle is somewhat similar to gut smooth muscle while its mechanical properties are similar to skeletal muscle. Myocardial cells are stimulated to contract by action potentials generated by a combination of fast Na^+ and Ca^{2+} voltage-dependent channels, and which propagate from cell to cell via gap junctions, as in gut smooth muscle. As in smooth muscle, various transmitters and hormones can alter extracellular Ca^{2+} entry, but in myocardium these effects are largely confined to the plateau phase of the action potential. Myocardial cells have large T-tubules and moderately extensive SR, with DHPR and RyR directly apposed to one another as in skeletal muscle. Myocardial DHPR is a better Ca^{2+} channel than skeletal DHPR, and calcium-induced calcium release triggered by Ca^{2+} entry during the plateau phase plays a significant role in E-C coupling in myocardium, but a fully activating $[Ca^{2+}]_i$ is not normally attained in the absence of sympathetic stimulation. In the absence of action potentials, no normally occurring chemical stimulus induces sufficient Ca^{2+} entry to initiate contraction, and so myocardium is entirely phasic. Phosphorylation of myosin light chains is not required for contraction in myocardium but significantly increases cross-bridge cycling rate. Sympathetic (norepinephrine) stimulation of the heart increases both Ca^{2+} entry and light chain phosphorylation, thus increasing both force and speed of contraction. Ca^{2+} channel blockers reduce force of contraction and oxygen demand in myocardium, although the reflex increase in sympathetic drive to the heart (in response to the decrease in blood pressure induced by these agents) may largely offset these effects.

It has recently become possible to alter the relation between force and $[Ca^{2+}]_i$ in myocardium, a useful trick in congestive heart failure. ***Pimobendan*** modestly increases the affinity of cardiac Tn C (cTn) for Ca^{2+}, thereby increasing the activation of contraction at any given $[Ca^{2+}]_i$. The more potent ***levosimendan*** binds to the N-terminal region of cTn C in a Ca^{2+}-dependent manner, and amplifies the effect of Ca^{2+}, perhaps by increasing the stability of the Ca^{2+}-induced conformational change in cTn C or by enhancing cooperativity in the thin filament.

21.4 Inherited Diseases of Muscle

Many disorders of skeletal and cardiac muscle due to genetic defects have been described. Taken as a group, the *hypertrophic cardiomyopathies* are the most common, with a combined prevalence of about 1 in 500. By comparison, the most common muscular dystrophy, ***Duchenne's muscular dystrophy*** (DMD), has a prevalence of roughly 1 in 3500 male births. It is not known whether genetic defects specific to myocardium are actually more common than those affecting skeletal muscle, or whether cardiac defects are simply more serious due to the incessant and vital nature of cardiac work. Most of these genetic defects do not affect both skeletal and cardiac muscle because there are cardiac-specific forms of almost all sarcomeric proteins and some enzymes. These disorders fall into the following broad categories: cardiomyopathies, muscular dystrophies, channelopathies (including the myotonias), metabolic diseases, and mitochondrial gene defects (see Chapter 14). The most prevalent examples of each are summarized in Table 21-5.

Hypertrophic cardiomyopathy (HCM) is a syndrome characterized by dyspnea and chest pain associated with decreased diastolic compliance and outflow obstruction and left ventricular hypertrophy, usually without dilation. More than 100 different mutations have been found in HCM patients in genes coding for β-MHC, essential and regulatory MLC, Tn T, Tn I, α-tropomyosin, and myosin-binding protein C. Most of these are transmitted as autosomal dominant traits with variable penetrance. Those cases due to defects in β-MHC and Tn T are the most severe and have the worst prognosis, and heavy chain mutations are the most common type. About 15% of HCM is due to mutations in chromosome 11p11.2, which codes for myosin-binding protein C, a structural protein that binds myosin and titin, and may also play a role in mediating the contractility response to adrenergic stimulation. Protein C is 1274 amino acids long, and is coded for by a complex gene comprising 24,000 base pairs in at least 37 exons.

Duchenne's muscular dystrophy (DMD), is a muscle degenerative disease due to a recessive mutation in the X chromosome. The mutation involves large defects in or complete deletion of the gene coding for ***dystrophin.*** The involved gene was identified by producing probes to the DNA surrounding the site of a large deletion in the X chromosome of a DMD patient. Applying these probes to the DNA of normal individuals, it was possible to isolate DND cDNA. Subsequent determination of the DNA sequence made it possible to identify dystrophin (M.W. 426,000); the gene coding dystrophin has over 2 million bases.

Dystrophin has some similarities to *spectrin* and *laminin,* which led to the realization that dystrophin is involved in attaching the cytoskeleton to the extracellular matrix (ECM). That is, dystrophin binds to webs of actin filaments and to transmembrane glycoproteins which, in turn, bind to ECM proteins such as laminin and agrin. Dystrophin is required to transmit force from the myofibrils through the costamere structure of the fiber to the ECM, and ultimately, to the tendons of the muscle (see subsection on **Myofibrils,** p. 457). Similarly, dystrophin allows forces applied to the sarcolemma via the ECM to be transmitted to the myofibrils and be borne by them rather than the sarcolemma, which cannot support tension by itself. Lack of dystrophin results in mechanical stresses in muscle tearing holes in the sarcolemma, which causes sustained high $[Ca^{2+}]_i$ and activation of Ca-dependent proteases such as *calpains.* This leads to focal destruction in the fiber, migration of polymorphonuclear leukocytes (PMNs) into the site, and either remodeling or total destruction of the fiber. ***Becker's muscular dystrophy*** (BMD) is allelic to DMD. Both DMD and BMD patients often have defects or deletions of genes flanking the dystrophin gene, and so present a varied clinical picture in which the muscular dystrophy is usually the most prominent feature.

Channelopathies are a varied group of rare hereditary disorders due to defects, usually point mutations, in genes for ion channel proteins. Most of these affect voltage-gated channels. Two that do not are ***Thomsen's disease*** (autosomal dominant myotonia congenita) and ***Becker's disease*** (autosomal recessive generalized myotonia), which are both due to mutations in a gene coding for a skeletal muscle Cl^- channel called ClC-1. Since Cl^- conductance stabilizes the membrane potential by allowing movement of Cl^- in response to depolarization, loss of 75% or more of the Cl^- conductance makes membranes unusually susceptible to the generation of action potentials by random depolarizing stimuli and delays repolarization, resulting in the unwanted twitches and contractions and difficulty in releasing voluntary contractions seen in these disorders. The channel is an oligomer, and some mutations in one ClC-1 gene result in a protein that interacts negatively with the normal (wild-type) peptide from the nonmutated gene, suppressing its ion conductance. In such mutations, association of a mutant channel with its normal allele produces a dysfunctional channel, so that only one mutated gene is required to produce symptoms (Thomsen's disease). Mutations that do not affect the properties of the normal protein produce symptoms only in homozygous individuals (Becker's disease).

Most channelopathies have some features in common. There are paroxysmal attacks of myotonia or paralysis, migraine, or ataxia precipitated by physiological stressors. They are often suppressed by membrane-stabilizing agents

TABLE 21-5
Inherited Diseases of Muscle

Description	Inheritance*
Hypertrophic cardiomyopathy	
Dyspnea, chest pain or syncope, usually features ventricular hypertrophy with impingement on LV volume and often LV outflow obstruction. Most cases hereditary, but some are new mutations. Defective genes for β-MHC, either MLC, TnT, TnI, α-tropomyosin, myosin binding protein C.	Most AD with variable penetrance
Muscular dystrophy	
Duchene (DMD) and **Becker's** (BMD) muscular dystrophy. Progressive number and severity of lesions in muscle fibers leading to inflammation and rapid (DMD) or slow (BMD) loss of fibers. DMD due to deletion of most or all of the gene for dystrophin, BMD due to defective dystrophin. See text.	X-linked recessive, or new mutations
Limb girdle (Erb's disease) Either pelvic or shoulder girdle initially,with spread to the other. Variable progression.	Usually AR, maybe dominant
Facioscapulohumeral (Lou Gehrig's disease) Face and shoulder girdle initially, then pelvic girdle and legs. Progression usually slow.	AD
Distal Onset in extremities, progresses proximally. Usually slow progression.	AD
Ocular External ocular muscles, some involvement of face, neck, arms.	Most cases AD
Oculopharyngeal As in ocular, but with dysphagia.	AD
Myotonic dystrophy (Steinert's disease) Muscle stiffness and impaired relaxation after contraction, wasting of muscles of face, neck, distal limbs, many other serious symptoms. Defect in gene coding for a serine-threonine kinase, resulting in impaired Na-K ATPase activity and perhaps other problems.	AD
Metabolic diseases: Glycogen storage diseases (GSD)	
GSD type II, Pompe's disease Deficiency of acid maltase causes excess glycogen accumulation in lysosomes. See text.	AR
GSD type III, Cori's disease Deficiency of debranching enzyme activity causes accumulation of limit dextrins. See text.	AR
GSD type V, McArdle's disease Myophosphorylase deficiency, inability to utilize glycogen. See text.	AR
GSD type VII, Tarui 's disease Muscle phosphofructokinase deficiency causes pronounced decrease in exercise tolerance.	AR
Other metabolic diseases	
Carnitine deficiency Impaired β-oxidation with lipid accumulation, see text.	AR
Adenylate deaminase deficiency Impaired anaerobic tolerance.	AR
Channelopathies	
Myotonias (muscle stiffness, impaired relaxation, fasciculation) due to various defects in skeletal muscle chloride channel 1 (CLCN1 or C1C-1):	
Becker's generalized myotonia (with moderate weakness and atrophy)	AR
Myotonia levior	AD
Thomsen's myotonia congenita (without weakness, may hypertrophy)	AD
Central core storage disease Abnormal SR with myofibrillar degeneration near fiber axis. Due to RyR1 gene defect.	?
Hypokalemic periodic paralysis Due to defective gene for one type of DHPR.	AD
Hyperkalemic periodic paralysis Due to defects in a skeletal muscle sodium channel, SCN4A.	AD
Paramyotonia congenita and **"pure" myotonias** Are due to varied (and different) mutations in the gene for SCN4A.	AD
Mitochondrial myopathy Due to defects in mitochondrial genes.	See text
Mitochondrial myopathy Fiber degeneration, weakness, "ragged red fibers" histological appearance.	most cases tRNA gene defects
CPEO (chronic progressive external opthalmoplegia) Paralysis of eye muscles and mitochondrial myopathy.	
KSS (Kearns-Sayre syndrome) CPEO plus retinal degeneration, cardiomyopathy, hearing loss, diabetes, renal failure.	
MELAS (mitochondrial encephalopathy, lactic acidosis, and strokelike episodes) Cognitive symptoms, seizures, transient paralysis, and mitochondrial myopathy.	

*AD=Autosomal dominant, AR=autosomal recessive

such as mexilitine, and by acetazolamide. Several channelopathies are listed in Table 21-5.

Metabolic disorders of muscle include those of glycogen storage, substrate transport and utilization, and electron transport chain and ATP metabolism. Some produce dynamic syndromes with symptoms occurring primarily during exertion, some cause degenerative syndromes, and some produce both. A few are discussed below.

Degenerative Syndromes

Acid maltase is a lysosomal enzyme that is not in the energy pathway of the cell, so that its deficiency does not produce dynamic symptoms in muscle. Also called α-1,4-glucosidase, acid maltase hydrolyzes the α-1,4 bonds in glycogen. Since lysosomes degrade glycogen along with other macromolecules in the normal process of cellular turnover, this deficiency causes marked accumulation of glycogen in lysosomes. This leads to a vacuolar degeneration of muscle fibers. Heart and other tissues are also affected. Deficiency of debranching enzyme (amylo-1,6-glycosidase) is another disorder of glycogen metabolism. Since phosphorylase cannot act at or near branch points (Chapter 15), lack of debranching enzyme results in great accumulation of limit dextrins in muscle, liver, heart, and leukocytes, with swelling and functional impairment. Since this enzyme is in the energy pathway, its absence causes dynamic symptoms and, more importantly, vacuolar degeneration. Carnitine deficiency causes a disorder of lipid metabolism. Carnitine is derived both from the diet and from ε-N-trimethyllysine produced by catabolism of methylated proteins including myosin, and is required for the transport of fatty acids across the mitochondrial membranes (Chapter 18). If any of the enzymes or cofactors required for carnitine synthesis are deficient or defective, carnitine deficiency may develop. This limits the energy supply available from β-oxidation, and causes a lipid storage myopathy.

Dynamic Syndromes

Myophosphorylase deficiency is the classic example of a carbohydrate-related dynamic syndrome. Affected persons are unable to mobilize glycogen; therefore they cannot perform high-intensity work and must rely extensively on lipid metabolism. Several other defects of glycolysis produce similar symptoms. All are characterized by inability to do anaerobic work and to produce lactate during ischemic exercise, which is the basis for the customary screening test for these disorders. Patients are asked to perform maximal hand grip contractions at the rate of one per second for 60 seconds with the forearm circulation occluded by inflation of a cuff on the upper arm. Following release of the cuff pressure, venous effluent from the exercised arm is sampled and analyzed for lactate. In addition to reduced anaerobic work capacity, the low flux through glycolysis reduces maximal muscle power output and maximal aerobic power as well. In phosphorylase deficiency, muscular performance can often be improved by glucose infusion, while patients with other defects of glycolysis are dependent on lipid metabolism and show little or no improvement with glucose infusion.

21.5 Nonmuscle Systems

Actin

Actin is present in all eukaryotic cells where it has structural and mobility functions. Most movement associated with microfilaments requires myosin. The myosin-to-actin ratio is much lower in nonmuscle cells, and myosin bundles are much smaller (10–20 molecules rather than about 500), but the interaction between myosin and actin in nonmuscle cells is generally similar to that in muscle. As in smooth muscle, myosin aggregation and activation of the actin-myosin interaction are regulated primarily by light chain phosphorylation. Myosins involved in transporting organelles along actin filaments are often activated by Ca-CaM.

Actin filaments are relatively permanent structures in muscle, whereas in nonmuscle cells microfilaments may be transitory, forming and dissociating in response to changing requirements. The contractile ring that forms during cell division to separate the daughter cells and the pseudopodia formed by migrating phagocytes comprise transient actin filaments. Belt desmosomes in epithelial cells and microvilli on intestinal epithelial cells comprise relatively permanent filaments.

The rate-limiting step in actin polymerization appears to be nucleation, the formation of an actin cluster large enough (typically three or four G-actins) for the rate of monomer association to exceed the rate of dissociation. Once filaments of this size form, they continue to grow, and the concentration of G-actin monomers decreases until it is in equilibrium with F-actin. The concentration of monomeric actin at equilibrium is called the critical concentration, C_c. *In vitro,* C_c is 0.1 mM. The value *in vivo* is variable, depending in part on the concentration of ATP. In the presence of ADP, instead of ATP, both ends of the filament grow at the characteristic slow rate. ATP speeds up the rate of polymerization and lowers the effective C_c.

If nascent actin filaments anchored to the cytoskeleton (by binding proteins such as those listed in Table 21-6)

TABLE 21-6

*Actin–Cross-Linking Proteins***

Protein*	MW	Domain Organization†	Location
Group I			
30 Kd	33,000	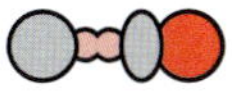	Filopodia, lamellipodia, stress fibers
EF-1a	50,000		Pseudopodia
Fascin	55,000	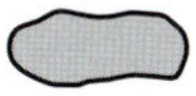	Filopodia, lamellipodia, stress fibers, microvilli, acrosomal process
Scruin	102,000	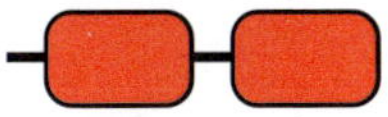	Acrosomal process
Group II			
Villin	92,000		Intestinal and kidney brush border microvilli
Dematin	48,000		Erythrocyte cortical network
Group III			
Fimbrin	68,000	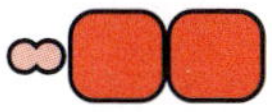	Microvilli, stereocilia, adhesion plaques, yeast actin cables
α-Actinin	102,000	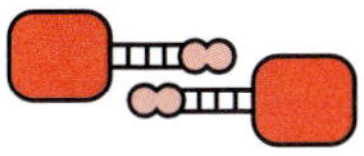	Filopodia, lamellipodia, stress fibers, adhesion plaques
Spectrin	α: 280,000 β: 246,000–275,000	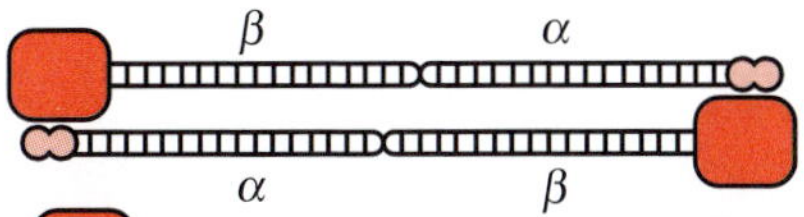 β α / α β	Cortical networks
Dystrophin	427,000		Muscle cortical networks
ABP 120	92,000		Pseudopodia
Filamin	280,000	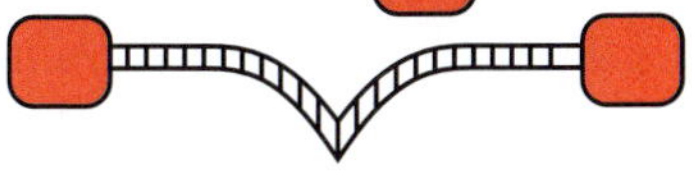	Filopodia, pseudopodia, stress fibers

*Cross-linking proteins are placed into three groups. Group I proteins have unique actin-binding domains; Group II have a 7,000-MW actin-binding domain; and Group III have pairs of a 26,000-MW actin-binding domain.
†Calmodulin-like calcium-binding domains (light blue), actin-binding domains (dark blue).

*Actin-Capping and Actin-Severing Proteins***

Protein	MW	Domain Organization*	Activity
gCAP39	40,000		Capping
Severin (fragmin)†	40,000		Capping, severing
Gelsolin	87,000		Capping, severing
Villin	92,000		Capping, severing, cross-linking

*Actin monomer binding domains are white; F-actin binding domains are shaded.
†Severin and fragmin are synonyms for the same protein.
**Reproduced with permission from: H. Lodish *et al. Molecular Cell Biology,* 3rd edition, W. H. Freeman and Co., New York 1995. Page 997 and page 1009.

grow toward the cell membrane and continue to grow at the membrane, they will create a projection of membrane and cytoplasm in the direction of growth, especially if the filaments are connected by short linking proteins into rigid bundles. In this way, microvilli extend from the surfaces of many cell types, including highly specialized structures such as stereocilia in sensory cells of the ear. Projection of the acrosomal process through the zona pellucida is driven in this way, as are the extension of filopodia and lamellipodia at the leading edges of migrating cells.

The structure and properties of actin filaments can be regulated by controlling the transformation of G-actin to F-actin or the length of the F-actin filaments, and by modulating the aggregation of actin filaments into bundles or three-dimensional arrays. Proteins that bind actin monomers reduce polymerization. Gelsoin, villin, and other proteins affect actin polymerization and actin filament elongation by capping the growing filament and blocking elongation. Some accelerate nucleation, perhaps by binding to and stabilizing dimers and trimers. Many capping proteins can also sever actin filaments without depolymerizing them, markedly reducing the viscosity of F-actin gels. The severing and capping activities of gelsolin require Ca^{2+}. Although nucleation and severing increase the number of free ends available for growth, the net effect of these proteins is a greater number of short actin filaments and an increased concentration of monomeric actin.

Cytochalasins, drugs that inhibit cellular processes that require actin polymerization and depolymerization (e.g., phagocytosis, cytokinesis, clot retraction, etc.), also act by severing and capping actin filaments. Actin filaments can be stabilized by ***phalloidin,*** derived from the poisonous mushroom *Amanita phalloides*. Assembly of actin filaments into bundles (as in microvilli) and three-dimensional networks is accomplished by two groups of cross-linking proteins (Table 21-6).

Cilia

Tubulin and microtubules occur in all plant, animal, and prokaryotic cells and participate in a number of essential processes. As is the case for actin filaments, microtubules occur in highly organized, relatively permanent forms such as cilia and flagella, and as transient cytoplasmic structures. Other similarities between actin filaments and microtubules are listed in Table 21-1.

Cilia are hair-like cell surface projections, typically 0.2–0.3 μm in diameter and 10 μm long, found densely packed on many types of cells. In eukaryotes, they move fluid past cells by a characteristic whip-like motion (Figure 21-15). Eukaryotic ***flagella*** are basically elongated cilia, one or two per cell, that propel the cell through a fluid medium. Flagellar movement is wavelike and distinct from ciliary beating, although the microtubular arrangement and mechanism of movement are quite similar in the two structures. Bacterial flagella are structurally and functionally different from other flagella.

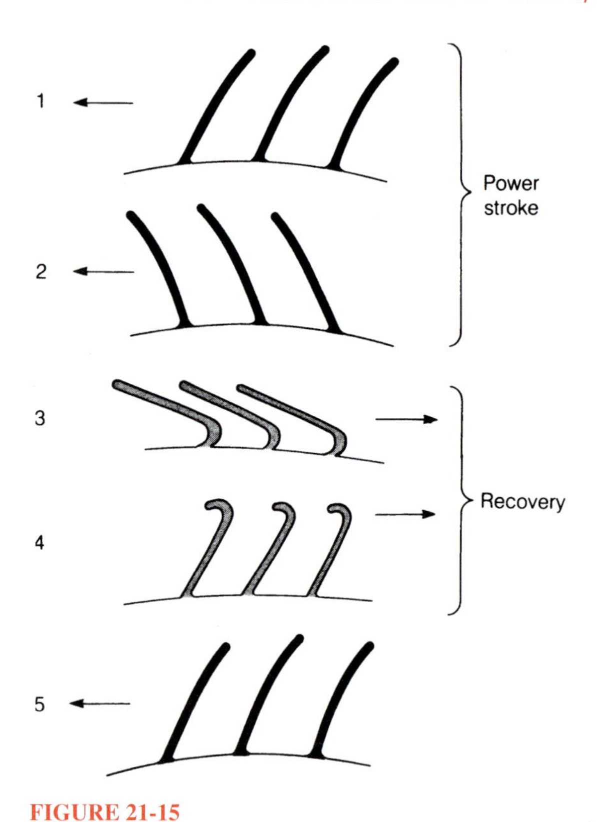

FIGURE 21-15
Cycle of ciliary stroke shown for three cilia beating in synchrony. (1) Beginning of the power stroke. The cilia are straight and stiff. (2) Half completion of the power stroke. (3) Start of recovery stroke. The cilia are flexible, and by bending reduce frictional drag. (4) Near completion of the recovery stroke. (5) Start of the next power stroke.

Figure 21-16 shows a cross-section of a cilium, showing the "9 + 2" arrangement of microtubules found in the axoneme (core) of cilia and flagella. Nine asymmetrical doublet microtubules are arranged in the periphery, and a symmetrical pair of singlet tubules is in the center. Attached to the A microtubules of each doublet are dynein arms extending toward the B microtubule of the adjacent doublet. The protein tektin, a highly helical protein about 48 nm long, runs along the outside of each doublet where the A and B microtubules join. It is regarded as a structural rather than a regulatory protein despite the similarity to tropomyosin. Three types of links preserve the axoneme structure. The central singlets are linked by structures, called inner bridges, like rungs on a ladder, and are wrapped in a fibrous structure called the inner sheath. Adjacent outer doublets are joined by links of nexin (M.W. 150,000–165,000) spaced every 86 nm. Outer doublets are

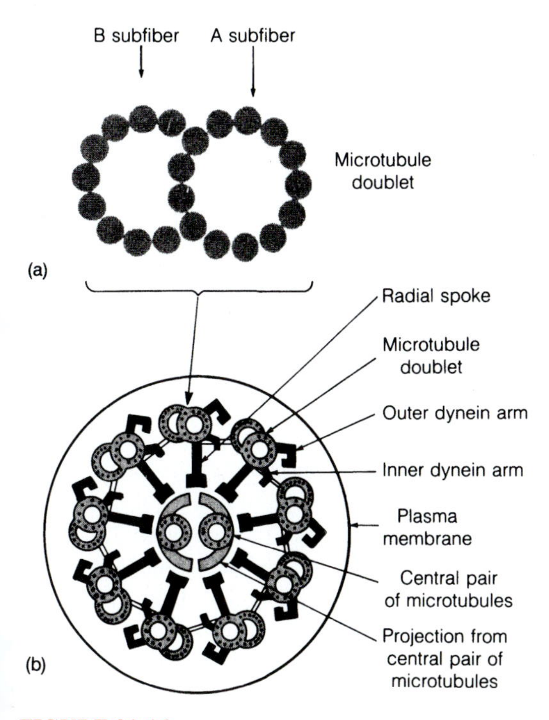

FIGURE 21-16
(a) A microtubule doublet consisting of an A subfiber with 13 protofilaments of tubulin and a B subfiber with 11 protofilaments. (b) A cilium contains nine microtubule doublets and a central pair of microtubules. The dynein arms provide for sliding of one doublet along another during the beating of the cilium. The functions of radial spokes and the projections from the central pair of microtubules have not been as well defined.

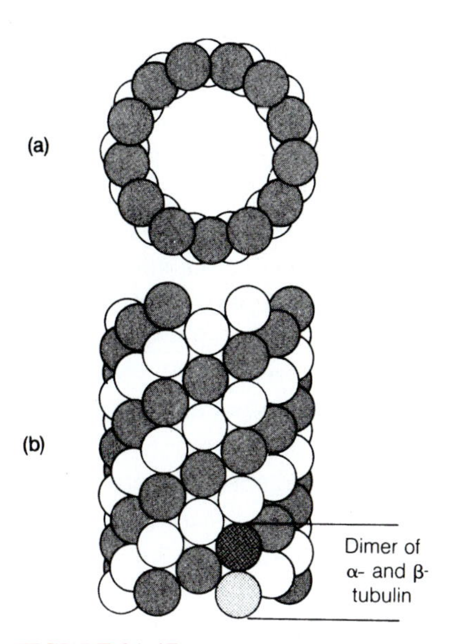

FIGURE 21-17
Drawing of a microtubule. (a) Cross-sectional view of the 13 protofilaments. (b) Longitudinal view.

joined to central singlets by radial spokes, which have a relatively globular end, or head, that interacts with the central singlet. The radial spokes are complex structures, which seem to have 17 or more constituent proteins, at least 6 of which are in the head. They are arranged in pairs every 96 nm along the microtubules. The inner sheath comprises primarily thin protein arms extending from the central microtubules at roughly 14-nm intervals, and the spoke heads may interact with these.

Cilia grow from ***basal bodies,*** one of several types of microtubule organizing centers in cells. Each basal body contains nine fused triplets of microtubules that act as nucleation centers for the growth of microtubules down the axoneme. Each triplet contains one complete A microtubule, that is continuous with the A microtubule of the axonemal doublet, and an incomplete B microtubule that is continuous with the B microtubule of the doublet. A second incomplete microtubule, the C microtubule, is fused to the B but does not extend beyond the basal body. The basal body does not have central singlets. There are numerous proteins beside the α- and β-tubulin of the microtubules in the basal bodies, which are thought to be involved in controlling polymerization, stabilizing the axoneme structure, and securing the basal body to the cytoskeleton. These include a third type of tubulin, γ-tubulin, which is believed to form a ring that serves as the nucleus from which the tubules grow.

Microtubules are constructed of protofilaments (Figure 21-17), the axis of each being parallel to the axis of the microtubule. In cilia, the A microtubule and the central singlets have 13 protofilaments, while the B microtubule has 10 protofilaments of its own and shares 4 with the A microtubule (Figure 21-16a). The protofilaments are chains of heterodimers of α- and β-tubulin arranged end-to-end. There are two models of protofilament arrangement. In one, the heterodimers in adjacent protofilaments are only slightly staggered, forming spiraling rows of α- and β-tubulin in the microtubule wall as in Figure 21-17. In the other, the α- and β-tubulin units are staggered a half-unit apart, yielding a checkerboard pattern. The strongest bonds in the structure are clearly along the axis of the filament. In depolymerizing conditions, the dissociating ends of the microtubules have a frayed appearance due to the separation of protomers from one another.

The ***tubulins*** are globular proteins with a mean diameter of 4 nm. They are about 450 amino acids long, (M.W. 50,000). The human genome contains 15–20 genes for both the α- and β-forms. Of these, probably five or six of each are expressed in a tissue, developmental, or cell cycle-dependent manner, and the rest are pseudogenes. Tubulin synthesis is regulated at the level of transcription and translation. Monomeric tubulin binds to ribosomes during tubulin synthesis and causes degradation of tubulin mRNA, thus providing a negative feedback on tubulin synthesis. There is typically about 90% homology between the various tubulins for both the α- and β-forms, and the tubulins are functionally quite alike, but some additional diversity can be created by posttranslational modification. Removal of the C-terminal tyrosine from α-tubulin by a specific carboxypeptidase (detyrosinase) affects primarily α-tubulin that is incorporated into a microtubule. Tyrosine can be reattached by a second enzyme, tubulintyrosine ligase, which acts preferentially on monomeric α-tubulin in a reaction that requires hydrolysis of ATP. Tyrosinated α-tubulin is found in all microtubules, while detyrosinated α-tubulin is found preferentially in the more stable tubular structures such as cilia. In some organisms, acetylation of a lysine in α-tubulin occurs during or just after incorporation into the microtubule. Acetylation has no known effect on the behavior of microtubules in which it occurs. Some serine residues in β-tubulin can be phosphorylated, but again the significance is unknown. Oxidation and blocking of sulfhydryl groups can prevent tubule assembly and may promote tubule disassembly.

Tubulin is similar to actin and myosin in its ability to spontaneously polymerize *in vitro*. Like actin, there is a critical concentration, C_c, of $\alpha\beta$ dimers at which equilibrium between association and dissociation occurs, and the microtubule, like actin, has a polarity and undergoes polymerization and depolymerization more rapidly at one end [(+) end] than at the other [(−) end]. The heterodimers polymerize α to β all along the protomer, so that the protomer has a polarity, α at one end, β at the other, and all of the protomers associate with the same polarity. So one end of a microtubule is ringed with α-tubulin and the other with β-tubulin. The $\alpha\beta$ heterodimers each bind two molecules of GTP. The α-tubulin site binds GTP irreversibly and does not hydrolyze it because β-tubulin shields the site from water, but the β-tubulin site binds GTP reversibly and hydrolyzes it to GDP after incorporation into the protomer, and especially after additional dimers have been added. This site is called the exchangeable site because GTP can displace GDP from it. Microtubules, then, tend to be capped with GTP dimers, but may be capped with GDP dimers if the microtubule has shortened, exposing GDP-tubulin, or when the microtubule has not grown for such a long time that most GTP has hydrolyzed. When the terminal dimers at the (+) end of the microtubule have GTP at the exchangeable site, the structure grows at dimer concentrations above C_c and slowly depolymerizes below C_c. When the (+) end dimers have GDP, however, depolymerization is rapid below C_c. This raises the surrounding dimer concentration and accelerates the growth of those microtubules that are elongating. Thus, microtubules are inherently unstable unless restrained in a structure like cilia by numerous accessory proteins.

Formation and stabilization of microtubules other than in cilia and flagella usually involves a group of proteins called microtubule-associated proteins (MAPs). One class, called assembly MAPs, cross-link microtubules in the cytosol. These have two large domains, a basic microtubule binding domain and an acidic projection domain, so named because it is a filamentous structure projecting from the microtubule. The projection domain binds to membrane proteins, intermediate filaments, or other microtubules. In the latter case its length determines the spacing between microtubules. MAPs are divided into two groups. Type I MAPs—MAP1A and MAP1B—are large proteins found mainly in neurons but also in other cell types. Each is derived from a larger precursor which is proteolytically processed to yield one light chain and one heavy chain. Type II MAPs include MAP2, MAP4, and tau protein. These are smaller than type 1, and are characterized by three or four repeats of an 18-amino-acid sequence in the microtubule binding region. MAP2 proteins are found only in dendrites where they connect microtubules to each other and to cytoskeletal IFs. The tau proteins are a group of four or more proteins derived from the same gene by alternative mRNA splicing. They accelerate polymerization and cross-link microtubules. They may play a role in stabilizing the long and quite permanent microtubules found in axons.

MAPs generally inhibit depolymerization, and some enhance polymerization. Several MAPs can be phosphorylated on their projection domains by mitogen-activated protein kinase, or MAP kinase, which prevents them from binding microtubules. There is some evidence that MAPs can also be phosphorylated by Ca-CaMKII and that they may be directly influenced by Ca-CaM. Thus, MAPs may be modulated by signal pathways using Ca^{2+} or MAP kinase, although specific roles for such control have not been clearly demonstrated.

Dynein is the motor protein in cilia and flagella. Dynein arms are arranged in two groups called inner and outer dynein arms, spaced at 24-nm intervals along the A microtubule. Dynein is a large family of microtubule-based motor proteins. Axonemal dyneins are multimers of (very) heavy chains, light chains, and intermediate chains. Each heavy chain has a large globular region with two stalks extending from it. One stalk connects to a cluster of proteins forming a "base" that attaches to the A microtubule. The other stalk projects 10 nm toward the B microtubule and forms a small head that binds to it. The heavy chain is quite large, about 4500 amino acids long with a molecular weight of more than 540,000. It has ATPase activity, especially when bound to the B microtubule, which is believed to be localized to the head, near the tubulin binding site. Most of the smaller chains are clustered at the base

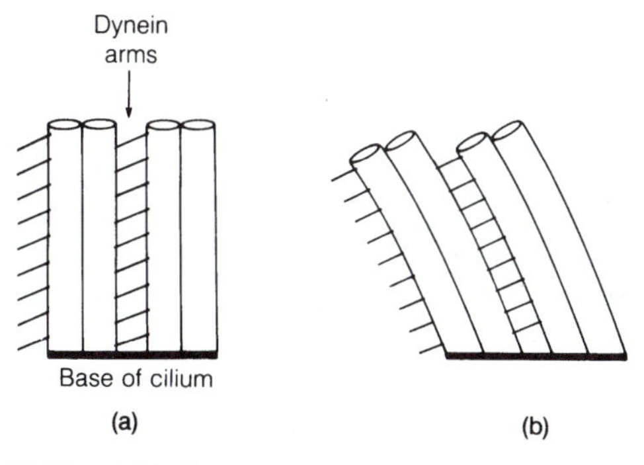

FIGURE 21-18
Sliding of one microtubule along another in an intact cilium causes the cilium to bend, creating a power stroke.

and are believed to form a fixed attachment of the dynein complex to the A microtubule, and they probably play a role in regulating the activity of the dynein, although not much is known about the regulatory mechanisms. As with myosin, there are multiple dynein genes, and the axoneme contains eight or more different heavy chains. The outer dynein arms have two or three heavy chains; the inner arms have two.

The scheme of chemomechanical transduction by dynein-tubule systems is similar in a general way to that in myosin–actin systems, but the structure of dynein is very different from that of myosin. Force generation begins by formation of a dynein cross-bridge to the adjacent B microtubule. This is followed by a conformational change in the dynein that pulls the two microtubules past each other. Dynein is a (−)-directed motor, i.e., it tries to pull its base toward the end of the adjacent microtubule, which is toward the basal body in cilia and flagella. Free energy of hydrolysis of ATP is required to release the dynein head and allow the dynein bridge to recycle. As in the case of actomyosin ATPase, it is thought that the coupling of ATP hydrolysis to mechanical movement is effected at the product release step and that this step is rate limiting.

Since the microtubules are all fixed at one end, the only way they can slide past one another is to bend (Figure 21-18). Activation of the dynein seems normally to occur from the base to the outer tip.

21.6 Drugs Affecting Microtubules

Microtubule systems are used in the cell for many other functions, such as transport of organelles and vesicles, and separating genetic material on the mitotic spindle and other motile events of the cell cycle. Substances that interfere with microtubule growth or turnover, or with microtubule interaction with motor proteins, will disrupt these functions. The classic example of such a drug is *colchicine,* an alkaloid derived from the autumn crocus (*Colchicum autumnale*). Colchicine in high concentration causes cytosolic microtubules to depolymerize. In low concentrations, it does not produce this effect, but binds to tubulin dimers. The tubulin-colchicine complex, even at quite low concentration, can add on to the end of a growing (or at least stable) microtubule and block further reactions at that end. Only one or two colchicine-tubulin units at the end of a microtubule prevents any further addition or removal of dimers at that end. In cells that are replicating, this freezes the cell at metaphase. Drugs that produce such an effect are useful as ***antineoplastic agents.*** The well-known effect of colchicine in gouty arthritis is probably due to its inhibiting the migration of granulocytes into the inflammatory area by interfering with a microtubule-based component of their motility.

Vinca alkaloids (vincristine, vinblastine, vinorelbine) are derived from the periwinkle plant (*Vinca rosea*). These agents work by binding to tubulin at a site different than colchicine or paclitaxel. They block polymerization, which prevents the formation of the mitotic spindle, and are used as antineoplastic agents. Taxanes produce a stabilization of microtubules similar to colchicine, but by a different mechanism, and also halt cells in metaphase. Paclitaxel (taxol) is the taxane used clinically. It is derived from the bark of the pacific yew. Taxol disrupts several microtubule-based functions as completely as inhibitors of polymerization, emphasizing the importance of assembly/disassembly balance in microtubule function. Recently, it has been found that paclitaxel also binds to and inhibits the function of a protein called bcl-2, an inhibitor of one or more pathways involved in mediating apoptosis. Paclitaxel's interference with this function promotes apoptosis in addition to its microtubule-related inhibition of cell division.

Immotile Cilia Syndrome

Defects in proteins needed for normal assembly and functioning of microtubules can cause cellular dysfunction. Several inherited disorders of this type have been identified that cause dyskinetic or completely immotile cilia and flagella. Kindreds have been identified in whom dynein arms, radial spokes, central sheath, or one or both central singlets are missing or defective. These disorders may result from a mutation in a gene needed for one of the axonemic structures, or in a regulatory gene controlling assembly of the microtubule system in cilia and flagella.

Affected individuals manifest bronchiectasis and chronic sinusitis. Because the cilia of the respiratory epithelium are defective, mucociliary clearance is reduced or absent, leading to frequent pneumonia, colds, and ear infections. These defects were originally described in conjunction with situs inversus (lateral transposition of the thoracic and abdominal viscera). Some embryonic cells have a single flagellum that is presumed to be important for movement to their proper location. The flagellar defect may make the cells unable to migrate properly, giving rise to situs inversus. Only about half of persons with immotile cilia have situs inversus, suggesting that chirality (handedness) of embryonic organization is random in the absence of flagellar function. Association of the three abnormalities (bronchiectasis, sinusitis, situs inversus) is known as *Kartagener's syndrome*. These defects also cause infertility in males because the sperm are immotile. Affected females have nearly normal fertility despite the probable lack of ciliary activity in the oviducts. The function of microtubules other than in cilia and flagella is presumed to be normal; otherwise cell division could not occur.

Kinesins

Kinesins are another large family of motor proteins. Thirteen have been described altogether, 11 in mouse brain. How many occur in humans is uncertain. Cytosolic kinesin is a tetramer of two heavy chains of unit M.W. 124,000 and two light chains of unit M.W. 64,000. The heavy chains have a head region and a tail, similar to myosin except that the tail is largely globular rather than a rod. There are structural similarities between myosin and kinesin heads, and in key functional groups, such as the helices flanking the ATPase site, the sequence homology is high. The light chains are associated with the tail. Typically, kinesin is a (+)-directed motor, i.e., it tries to pull whatever it is attached to toward the (+) end of the microtubule. In an axon, that would be away from the cell body. Cytosolic dynein, which is smaller and simpler than axonemal dynein, is a (−)-directed motor. In most kinesins, the motor region is at the N-terminal half of the molecule, but in some it is in the C-terminal region. Most of these latter kinesins are, like dynein, (−)-directed.

Kinesin has in common with cytosolic dynein and transport myosins (e.g., myosin I) a property called ***processivity.*** This is the term used to refer to the fact that a single motor molecule, when dragging a vesicle or other cargo along a filament or a microtubule, cannot let go of that filament or microtubule lest the motor and its cargo drift away or continually reattach to the same site. Instead the motor must remain bound to one G-actin or tubulin while "reaching for" the next one. A similar argument applies to other motor proteins, such as DNA helicase which crawls along DNA unzipping the strands, and the ribosomal motors which pull the RNA and the nascent polypeptide through the ribosome. In contrast, myosins whose functional form is filaments, such as myosin II, need not exhibit this property: any given myosin molecule need not be attached to the actin filament because attachment and orientation of the myosin filament to the actin filament will be maintained so long as one or two myosins anywhere in the filament remain attached. Which features of these motor molecules determine the presence or absence of processivity remains a mystery.

Supplemental Readings and References

M. J. Ackerman and D. E Clapham: Ion channels—basic science and clinical disease. *New England Journal of Medicine* **336,** 1575 (1997).

L. A. Amos and R. A. Cross: Structure and dynamics of molecular motors. *Current Opinion in Structural Biology* **7,** 239 (1997).

G. Bonne, L. Carrier, P. Richard, B. Hainque, and K. Schwartz: Familial hypertrophic cardiomyopathy: from mutations to functional defects. *Circulation Research* **83,** 580 (1998).

R. Cooke: Actomyosin interaction in striated muscle. *Physiological Reviews* **77,** 671 (1997).

M. A. Geeves and K. C. Holmes: Structural mechanism of muscle contraction. *Annual Review of Biochemistry* **68,** 687 (1999).

A. M. Gordon, E. Honsher, and M. Regnier: Regulation of contraction in striated muscle. *Physiological Reviews* **80,** 853 (2000).

N. Hirokawa: Kinesin and dynein superfamily proteins and the mechanism of organelle transport. *Science* **279,** 519 (1998).

A. Horowitz, C. B. Menice, R. Laporte, and K. G. Morgan: Mechanisms of smooth muscle contraction. *Physiological Reviews* **76,** 967 (1996).

H. Lodish, D. Baltimore, A. Berk, et al.: Microfilaments: Cell motility and control of cell shape. In: *Molecular Cell Biology.* H. Lodish *et al.* (Eds)., W. H. Freeman, New York, 1995, p. 991.

G. J. Lutz and R. L. Lieber: Skeletal muscle myosin II structure and function. *Exercise Sport Science Review* **27,** 63 (1999).

L. A. Sabourin and M. A. Rudnicki: The molecular regulation of myogenesis. *Clinical Genetics* **57,** 16 (2000).

S. Sach, F. J. Kull, and E. Mandelkow: Motor proteins of the kinesin family. Structures, variations, and nucleotide binding sites. *European Journal of Biochemistry* **262,** 1 (1999).

S. Schiaffino and C. Reggiani: Molecular diversity of myofibrillar proteins: Gene regulation and functional significance. *Physiology Reviews* **76,** 371 (1996).

C. A. Sewry: Immunocytochemical analysis of human muscular dystrophy. *Microsc. Res. Tech.* **48,** 142 (2000).

J. M. Squire and E. P. Morris: A new look at thin filament regulation in vertebrate skeletal muscle. *FASEB Journal* **12,** 761 (1998).

M. J. Tanasijevic, C. P. Canon, and E. M. Antman: The role of cardiac troponin-I (CTnI) in risk stratification of patients with unstable coronary artery disease. *Clinical Cardiology* **22,** 11 (1999).

R. H. Wade and A. A. Hyman: Microtubule structure and dynamics. *Current Opinion in Cell Biology* **9,** 12 (1997).

D. C. Wallace: Mitochondrial DNA in aging and disease. *Scientific American* **227,** 40 (1997).

A. Weiss and L. A. Leinwand: The mammalian myosin heavy chain gene family. *Annual Review of Cell Devopmental Biology* **12,** 417 (1996).

CHAPTER 22

Metabolic Homeostasis

22.1 Metabolic Homeostasis

The functions of cells and tissues in all organisms require energy that normally is obtained from ingested food containing proteins, lipids, and carbohydrates. Human beings follow an intermittent food/fast schedule; however, energy must be supplied to cells and tissues continuously. Thus, excess chemical substances capable of supplying energy are stored and released as needed to maintain ***homeostasis,*** the tendency for biological systems to maintain relatively constant chemical conditions in their internal environments. Some organs, tissues, and cells are metabolic providers; others are users. Each cell membrane acts as a barrier to hold its cellular components and, by expending energy, is able to regulate the gradient of ions and metabolites moving into and out of the cell. This chapter discusses how energy is stored and released by tissues and how homeostasis is maintained.

Organs serve as reservoirs or sites for synthesis of the metabolites required for homeostasis; at all levels homeostasis is regulated by the endocrine system. Fluxes of metabolites through biochemical pathways are regulated by stoichiometric need, allosteric control of enzyme activity, modification of regulatory enzymes, and changes in their rate of enzyme synthesis or degradation. Hormones often influence these processes and, in turn, metabolites modulate the secretion of hormones. Thus, many interconnected feedback loops provide the positive and negative regulation that maintains homeostasis.

Human beings possess considerable metabolic versatility for utilization of major fuel classes such as carbohydrate, lipid, and protein to maintain energy requirements. Metabolism of these substances is organized to:

1. Maintain the blood glucose level within narrow limits;
2. Maintain an optimal supply of glycogen;
3. Maintain an optimal supply of protein.

If carbohydrates or proteins are ingested in excess of the amounts necessary to maintain optimal supplies of glycogen in tissues or protein, the excess is converted to fat. Conversion of excess carbohydrate and/or protein into fat is biochemically an irreversible process. As a result, the body conserves the compounds that it can interconvert, uses them to supply energy when needed, or converts them to fatty acids when it is more efficient to store the carbon in the form of fat. The body tends to conserve its protein reserves and to draw on fat reserves preferentially in time of energy demand.

The tricarboxylic acid (TCA) cycle and β-oxidation (Chapter 14) are tightly coupled to electron transport via the nicotinamide and flavin nucleotides and by ADP and ATP. The TCA cycle in cells functions only when ADP is present (i.e., when ATP is being utilized). Thus, oxidation of primary foodstuffs is determined by energy need and cannot occur simply to convert excess foodstuffs to products that can be readily eliminated. As a consequence, regulation of body weight is determined by three factors:

1. Food intake,
2. Heat loss, and
3. Energy expenditure (i.e., exercise).

The **metabolic activity** of each organ in the body depends on its specific functions and the energy required to perform those functions. The amount of energy needed can be determined from the amount of ATP generated since most of the oxygen is used by mitochondria to produce ATP. At rest, the adult brain (~2% of body weight) uses 20% of the oxygen consumed; in contrast, skeletal muscle (~40% of body weight) consumes only 30% of the oxygen. During heavy muscular work, oxygen consumption by the body may increase 5–10 fold with a corresponding 10- to 25-fold increase in oxygen consumption by the skeletal muscle. During heavy muscular work, however, oxygen consumption by the brain does not change. These differences between tissues illustrate the metabolic flexibility that can be achieved. The large change in oxygen consumption by muscle tissue is accompanied by large changes in fuel utilization and in production of metabolites while overall homeostasis is maintained.

Skeletal muscle has the highest glycolytic capacity reflecting the need to generate ATP rapidly for muscle contraction. Skeletal muscle relies on glycolysis for short bursts of high-intensity work under conditions where oxygen is in short supply (Chapter 21). Other tissues that also possess a high glycolytic capacity are the brain and the heart. However, the heart, which has a much greater capacity than skeletal muscle to derive energy oxidatively from sources other than glucose, rarely relies on glycolysis for energy. The brain, on the other hand, preferentially uses glucose for energy, first converting it to pyruvate and then to CO_2.

The liver and kidney, which do not metabolize glucose to a significant degree, rely primarily on fatty acid

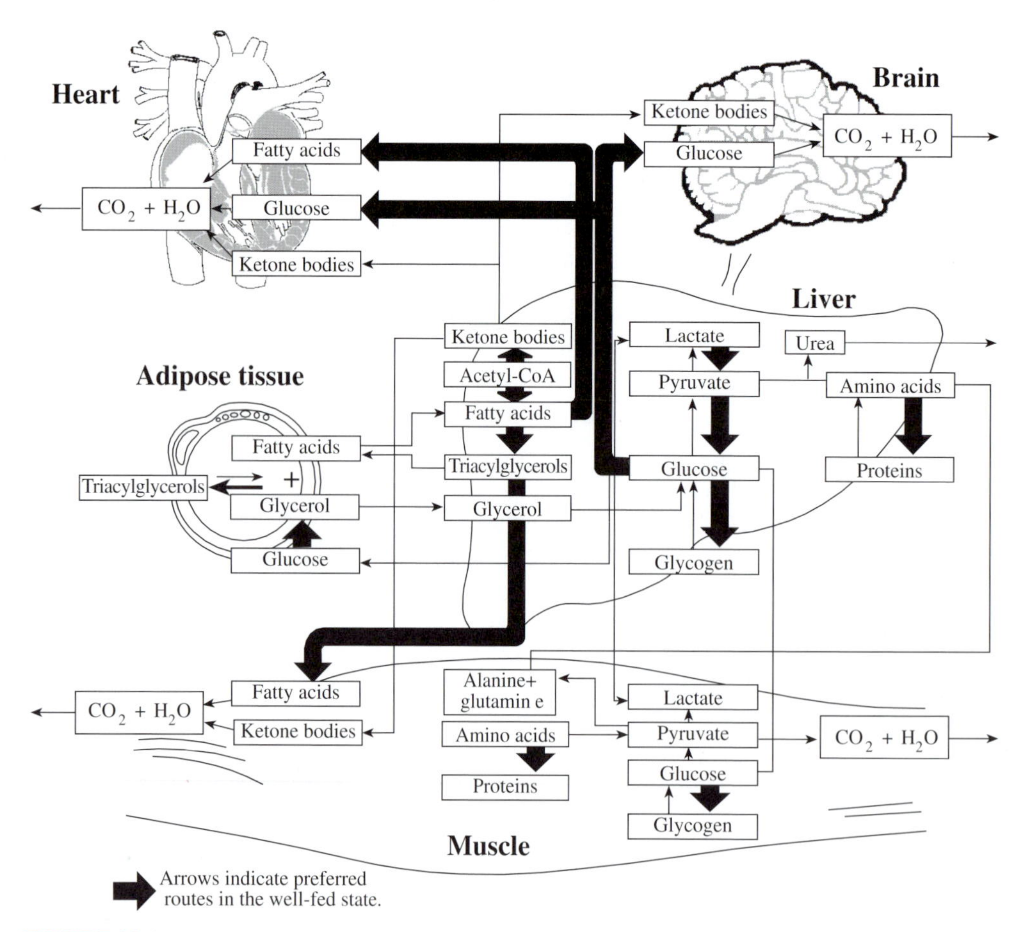

FIGURE 22-1
Interrelationships of organs to maintain homeostasis. [Reproduced with permission from R. H. Garrett and C. M. Grisham, *Biochemistry,* 2nd ed. Harcourt College Publishers, 1995, p. 940.]

oxidation for energy. The rate-limiting enzymes of gluconeogenesis (pyruvate carboxylase, phosphoenolpyruvate carboxykinase, and fructose bisphosphatase) show high activities in the liver and kidney, but very low glycolytic activities reflect the fact that the liver and kidney are glucose producers whereas the other tissues are glucose users. The overall metabolic patterns in the various tissues and organs of the body are illustrated in Figure 22-1.

22.2 Metabolic Roles of Organs

Liver

The liver (see also Chapter 12) in a normal, 70-kg human adult weighs approximately 1.5 kg, or about 2% of body weight. (Metabolic fluxes quoted throughout this chapter refer to the typical 70-kg human.) Its major cell type, the hepatocyte, is of epithelial origin. Interposed between the intestinal tract and the general circulation, the liver is uniquely related to the endocrine pancreas; insulin and glucagon, released from the pancreas, play a prominent role in metabolic homeostasis. The liver receives a large volume of blood from the intestinal tract via the portal vein and a small volume of blood via the hepatic artery, and it drains via the hepatic veins into the inferior vena cava. It is the first organ to meet nutrients delivered from the intestine, with the exception of lipid, and to meet the secreted insulin and glucagon. By delivering bile (which contains bile acids and cholesterol) into the intestine, the liver predominates in cholesterol homeostasis. It has the greatest metabolic flexibility of any organ and shows tremendous adaptability to fluxes of metabolites and nutrients. It rapidly undergoes changes in size and glycogen and protein content In the fed state, 5–10% of its wet weight consists of glycogen but after a 24-hour fast the glycogen disappears almost entirely. After a day or two on a high-protein diet, the liver shows a large increase in activity of enzymes involved in amino acid metabolism and in gluconeogenesis. A high–carbohydrate diet produces the opposite effect. The liver is the primary site of glycogen deposition and blood glucose maintenance. It also plays a central role in lipid, protein, and nitrogen homeostasis. In the typical adult, the liver exports daily 180 g of glucose, 100 g of triacylglycerol, and 14 g of albumin. Its metabolic energy is derived primarily from fatty acid oxidation.

Adipose Tissue

Cosmetically, adipose tissue is viewed as an enemy; however its importance in energy homeostasis is second only

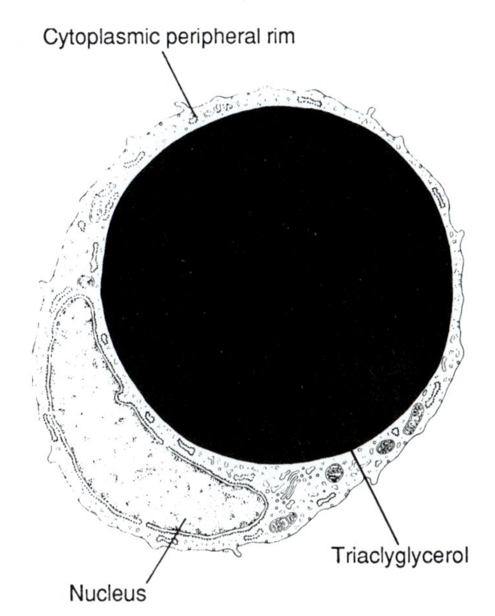

FIGURE 22-2
Schematic representation of a white adipocyte.

to that of the liver. The typical adult has 13 kg of adipose tissue located extensively under the skin, in the abdominal cavity, around internal blood vessels, in skeletal muscle, and in mammary glands. **Obesity** results when this amount increases. White and brown adipocytes, the two types of fat cell, are schematically depicted in Figures 22-2 and 22-3. White adipocytes have a characteristic spherical shape with a large central droplet of triacylglycerol surrounded by a thin rim of cytoplasm, and nucleus, mitochondria, and endoplasmic reticulum situated peripherally. Brown adipocytes have multiple lipid droplets and numerous mitochondria, which impart a

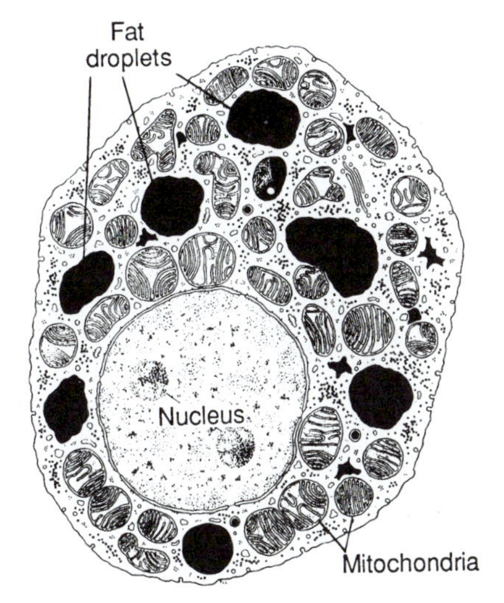

FIGURE 22-3
Schematic representation of a brown adipocyte.

red-brown color. White adipocytes store triacylglycerol for export as fatty acid to serve as fuel for other tissues. In brown adipocytes, the lipid is oxidized within the tissue to CO_2 and leads to the production of heat. Brown adipocytes play a principal role in thermogenesis (Chapter 14).

Human adipose tissue is mostly of the white type. It responds rapidly to the supply of and demand for fuel in the organism. Storage of fuel as triacylglycerol is advantageous over storage as glycogen, because triacylglycerols are more highly reduced and release more energy per gram when oxidized; in addition, lipids are stored anhydrously and therefore very compactly, whereas glycogen must have an aqueous environment. As stored, triacylglycerols yield four to five times more energy than do equivalent amounts of glycogen. If the adipose tissue in the typical adult were replaced by an energy-equivalent amount of glycogen, body weight would increase by 60% (from 70 to 115 kg). Adipocytes are highly active metabolically. They synthesize triacylglycerols from glucose or from fatty acids delivered by chylomicrons or very-low-density lipoproteins (VLDLs). Glucose supplies acetyl-CoA for the biosynthesis of fatty acids, the reducing power in the form of NADPH produced by the pentose phosphate shunt, and glycerol phosphate for esterification of fatty acids. Glycerol phosphate is derived from intermediates of glycolysis because adipocytes lack glycerol kinase. Adipocytes hydrolyze the triacylglycerol and release fatty acids. Energy for adipocyte function is derived primarily from fatty acid oxidation and TCA cycle activity.

Skeletal Muscle

The typical adult has about 35 kg of skeletal muscle (see also Chapter 21), which contains 5–6 kg of contractile protein. Thus, skeletal muscle contains the major portion of the body's nonlipid fuels. In severe wasting diseases or starvation, it can be depleted by about 40%. In the well-fed state, it contains about 1% of its wet weight as glycogen. Because of its mass, muscle contains almost four times as much glycogen as does the liver. Muscle glycogen is not directly available as a source of blood glucose because muscle lacks glucose-6-phosphatase, but when mobilized to support muscular activity, glycogen becomes available, via lactate for conversion into blood glucose in the liver (discussed later). During starvation, muscle protein provides amino acids, which become the primary carbon source for maintenance of glucose homeostasis. Since there is no "storage" form of protein, the protein that is degraded in muscle is the contractile protein. Resting muscle maintains itself mainly on energy

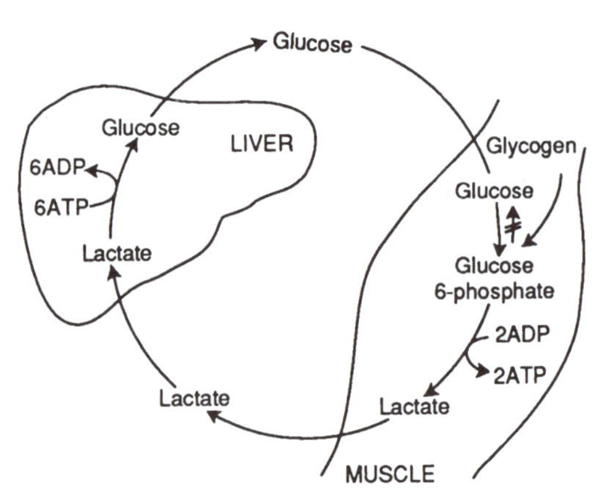

FIGURE 22-4
Cori cycle. Skeletal muscle anaerobic glycolysis yields lactate, which in the liver is converted to glucose by gluconeogenesis, which provides more glucose for support of muscle energy needs. The 2 mol of ATP generated in muscle occurs at the expense of 6 mol of ATP in liver. Hepatic metabolism is maintained by fatty acid oxidation. Muscle glycogen cannot serve directly as a source of blood glucose because of the absence of glucose-6-phosphatase.

provided by fatty acid oxidation. Muscle contraction can be supported by anaerobic or aerobic glucose utilization and by oxidation of fatty acids and amino acids, depending on the intensity of the exercise and the availability of glucose and glycogen and of oxygen. The density of mitochondria is lower in skeletal muscle than in heart or liver, and severe exercise can result in the oxidative capacity's becoming insufficient. During anaerobic glycolysis, 2 mol of ATP is formed per mole of glucose whereas aerobic oxidation produces 34–36 mol. However, anaerobic glycolysis permits muscle to function quickly at high intensity in the absence of oxygen. The lactate produced can be removed by the circulation and oxidized by the heart or used by the liver for glucose homeostasis. Lactate can be oxidized in muscle during rest following intense exercise. The cyclic process of muscular anaerobic glycolysis maintained by hepatic gluconeogenesis is termed the ***Cori cycle*** (Figure 22-4). This cycle is energetically expensive, since gluconeogenesis from lactate requires the expenditure of 6 mol of ATP per mole of glucose formed (Chapter 15). At rest, muscle can be a major contributor of amino acid carbon for gluconeogenesis. During exercise it becomes the predominant user of metabolic fuels.

Brain

The brain is a constant user of metabolic fuel, provided by other tissues. It uses ~20% of the total oxygen consumed by an adult in the resting state. Perhaps two-thirds of this demand is used to maintain the transmembrane potential by action of membrane enzymes such as Na^+,K^+-ATPase. In the well-fed human, the brain uses glucose exclusively for energy (with little or no formation of lactate) and depends on its continual supply, since it stores little glycogen. A decrease in the blood glucose level below ~50 mg/L can cause symptoms of dizziness and lightheadedness. More marked drops lead to coma and death if uncorrected.

Heart

Although the molecular design of the contractile apparatus of heart muscle is very similar to that of skeletal muscle, the two tissues differ metabolically. Under essentially all circumstances, metabolism in heart muscle is aerobic and uses fatty acids or ketone bodies, lactate, and pyruvate. The heart stores some glycogen but appears to utilize it only when occlusion of a coronary artery prohibits aerobic metabolism. In starvation, cardiac muscle increases its endogenous glycogen stores approximately twofold. The heart can function well on glucose, oxidizing it anaerobically or aerobically, but if presented with glucose and fatty acids, it preferentially uses the fatty acids for its energy needs. Cardiac muscle does not contribute significantly to fuel homeostasis, but its ability to utilize any metabolic fuel makes it a "scavenger."

Kidneys

The kidneys (see also Chapter 39) eliminate noxious material while preserving important metabolites; normally, very little glucose, ketone bodies, or amino acids is wasted. Renal tissue is important in amino acid homeostasis (see below) and has the same gluconeogenic capacity per gram of tissue as does liver, although the liver provides a larger amount because of its larger size. About 80% of the total energy produced by the kidneys is utilized in the active transport processes involved in urine formation. It readily oxidizes fatty acids, ketone bodies, glucose, and amino acids.

Gastrointestinal System

The principal nutrients of the body are of the same type as those that it uses as stores, namely, complex carbohydrates, triacylglycerol, and protein. These are hydrolyzed within the intestinal tract to produce primarily mono- and disaccharides, amino acids and small peptides, glycerol, and free fatty acids (Chapter 12).

Blood and Other Body Fluids

Blood distributes metabolic fuels among tissues. Approximately 40–50% of the body fluid of an adult is intracellular and separated from other fluid compartments by the cell membranes whose properties and transport systems determine which metabolites pass across them. Figure 22-5 shows the relationships and volumes of the fluid compartments of the body. If homeostasis is well maintained, the plasma concentration of a metabolite does not change because its utilization by one organ is matched by its release from another; thus, a low plasma concentration does not indicate a low flux.

Simple monosaccharides (principally glucose) are the means for transporting carbohydrates. Glycogen and phosphorylated sugars do not occur to any significant extent in plasma. All 20 amino acids used in protein synthesis are present in plasma but at ratios that do not reflect those found in proteins. Particularly high are the levels of alanine and glutamine, which are important carriers of carbon and nitrogen between muscle, liver, and other tissues (Chapter 17). Organic acids present as anions include lactate and pyruvate (from anaerobic glycolysis in muscle and red blood cells), acetoacetate, β-hydroxybutyrate (produced by oxidation of fatty acids), and several intermediates of the TCA cycle, although little importance has been attached to their interorgan transport. Other constituents of plasma reflect the metabolism, function, and metabolic processes of particular organs. Thus, urea, synthesized in liver, is the major vehicle for eliminating excess nitrogen; bilirubin is a reticuloendothelial cell product of porphyrin degradation; creatinine and its metabolite creatine, which is the primary elimination product, are derived almost entirely from muscle, where phosphocreatine serves as an immediate source of energy for contraction (Chapter 21). These specific plasma constituents can be assayed to evaluate disease processes.

Macromolecules found in plasma include simple proteins, metalloproteins, glycoproteins, and lipoproteins. Albumin and the lipoproteins play a special role in metabolic

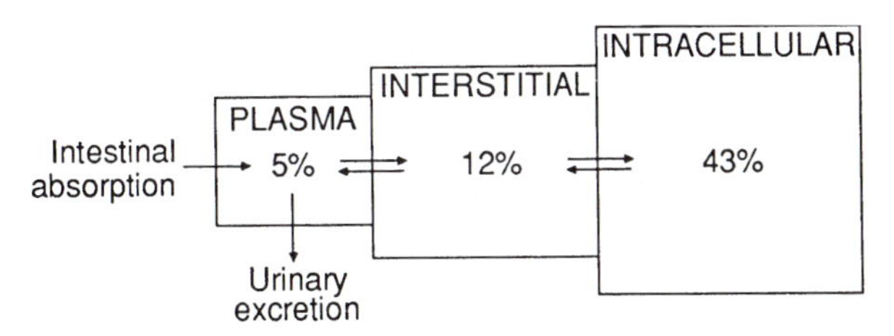

FIGURE 22-5
Blood–fluid relationships. Volumes are expressed as a percentage of body weight.

TABLE 22-1
Constituents of Endocrine Pancreas

Cell Type	Hormone	Structure	Structurally Analogous Hormones
α	Glucagon	29 amino acids	Secretin Vasoactive intestinal peptide Gastric inhibitory polypeptide Glicentin
β	Insulin	Two chains: A=21 amino acids B=30 amino acids	IGF-I (somatomedin C) IGF-II Relaxin Nerve growth factor
δ	Somatostatin	Two forms: S-14=14 amino acids S-28=28 amino acids	None known
F	Pancreatic polypeptide	36 amino acids	None known

homeostasis. The protein constituents of plasma do not serve as significant sources of carbon and nitrogen for cells.

Albumin

One of the smallest and the most abundant plasma proteins, **albumin** plays a significant role in osmotic regulation and transport of free fatty acids. Albumin is synthesized in the liver at a rate of approximately 14 g/d, or 10% of the total protein synthesis of the body. Deviations from the normal concentration of albumin in plasma can indicate the state of hepatic function. Albumin is also present in interstitial fluid.

Lipoproteins

The plasma lipoproteins play a central role in metabolic homeostasis. As discussed in Chapter 20, a **lipoprotein** is a complex composed of triacylglycerol, protein, cholesterol (free or esterified), and phospholipids. The phospholipid and protein form an external coat to render the particles soluble in plasma for transport around the body. The roles of lipoproteins in metabolic homeostasis are presented in Chapter 20.

22.3 Endocrine Pancreas and Pancreatic Hormones

The endocrine pancreas is pivotal in metabolic homeostasis and an integral component of metabolic regulation (see also Chapter 30). It is composed of 1–2 million islets of Langerhans scattered throughout the organ and forming 2–3% of its total weight. The islets contain at least four cell types—α, β, δ, and F—that secrete glucagon, insulin, somatostatin, and pancreatic polypeptide, respectively (Table 22-1). The α and β cells make up 20% and 75% of the total, respectively. Most islets have a characteristic distribution of cells; α and δ cells are located at the periphery and β cells the center, but close intercellular communication exists between different cell types. The islet appears to function as an integrated unit rather than as four independent types of cell. Islets in the posterior portion of the head of the pancreas may contain up to 70% of F cells; those in the other regions of the pancreas may have fewer or none.

Insulin

Structure and Synthesis

The structures of insulin and proinsulin are given in Figure 22-6. Proinsulin is a single polypeptide chain of 86 amino acids that permits correct alignment of three pairs of disulfide bonds. Insulin is composed of an A chain of 21 amino acids and a B chain of 30 amino acids, the chains being held together by two disulfide bonds. A third disulfide bond is present within the A chain. Human insulin differs from porcine insulin by a single amino acid and from beef insulin by three amino acids. These substitutions do not significantly affect activity; hence the widespread use of bovine and porcine insulin in clinical therapy. Human insulin and its analogues have been synthesized by recombinant technology for clinical use. In one analogue, two amino acid residues, namely prolyl and lysyl residues at positions 28 and 29 on the B chain, respectively, are switched. This analogue is called *lispro*

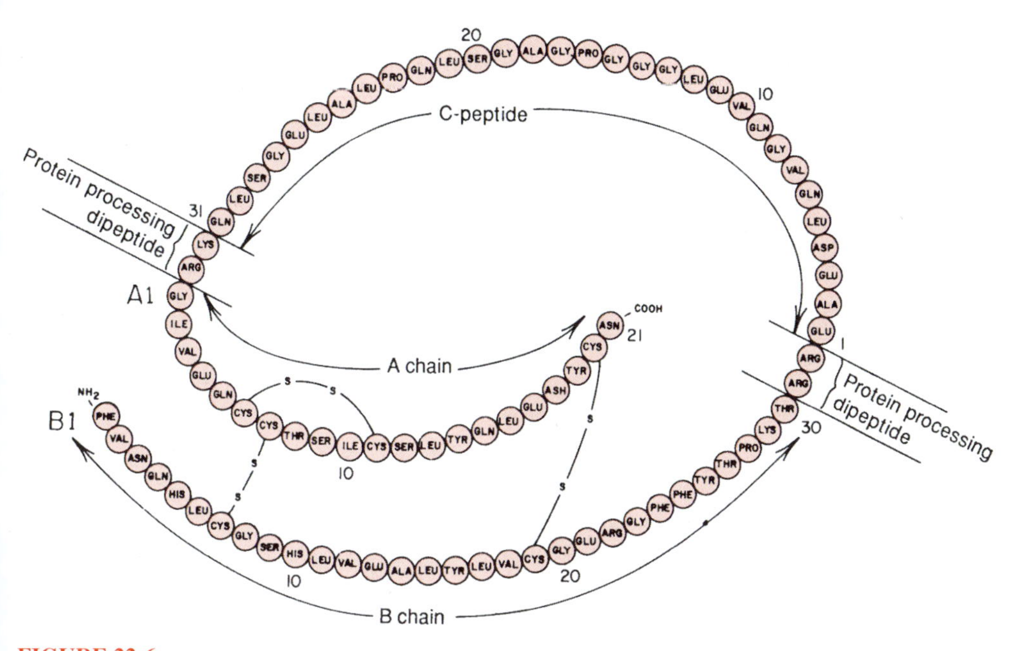

FIGURE 22-6
Structures of human proinsulin and insulin. Insulin is derived from proinsulin by cleavage at the dipeptides Arg-Arg and Lys-Arg to give A and B chains held together by disulfide bonds. In the pig, B30 is Ala. In the cow, A8 is Ala, A10 is Val, and B30 is Ala. Bovine and porcine insulins are used extensively in clinical practice.

insulin and has the same biological activity as normal human insulin. However, unlike normal human insulin, lispro insulin has a reduced capacity for self-association and thus begins its biological activity within 15 minutes of subcutaneous administration; it has become the insulin of choice in many patients with diabetes (discussed later).

Insulin biosynthesis resembles that of other export proteins: gene transcription, processing and maturation of precursor mRNA, translation of mature RNA, translocation of preproinsulin to cisternae of the endoplasmic reticulum with removal of the N-terminal leader or signal sequence of 23 amino acids, folding of proinsulin and formation of the proper disulfide bridges, packaging of proinsulin into secretory granules, and conversion of proinsulin to equimolar amounts of insulin and connecting peptide (C-peptide) by site-specific enzymatic cleavages. The human insulin gene is located on the short arm of chromosome 11 and consists of a 5′ untranslated region split by an intron; an exon that codes for the leader peptide; a region coding for the B chain; a region coding for C-peptide that is split by an intron; a region coding for the A chain; and a 3′ untranslated region.

Conversion of proinsulin to insulin and C-peptide in secretory granules involves site-specific cleavages at the Arg-Arg and Lys-Arg sequences (Figure 22-6); these serve as signals for proteolytic processing of many other proteins. Cleavage occurs at the C-terminal end of each pair by trypsin-like enzymes and is followed by removal of the basic residues by a carboxypeptidase B-like enzyme.

Insulin monomers undergo noncovalent dimerization by formation of antiparallel β-pleated sheet associations between monomers involving the C-terminal portion of the B chain. As discussed earlier, lispro insulin, in which the B28 and B29 is reversed from the normal prolyl and lysyl sequence, does not dimerize. Three insulin dimers subsequently self-associate to form hexamers in the presence of Zn^{2+}. The Zn^{2+} hexameric array of insulin probably gives the β-cell granule its unusual morphologic characteristics.

Zn^{2+} is released when insulin is secreted. Conversion of proinsulin to insulin in the secretory granule is not complete and some proinsulin is also released upon secretion of insulin. Proinsulin has less than 5% of the biological activity of insulin. The C-peptide has no physiological function but assay of C-peptide helps distinguish between endogenous and exogenous sources of insulin.

Like genes for other proteins, the insulin gene may undergo mutation and produce an abnormal product. This process may be suspected in individuals who exhibit ***hyperinsulinemia*** without hypoglycemia or evidence of insulin resistance. Three abnormal insulins have been documented: one in which Ser replaces Phe at B24, another in which Leu replaces Phe at B25, and a third in which the Lys-Arg basic amino acid pair is replaced by Lys-X (X = nonbasic amino acid). The former two mutations occur in the nucleotide sequence that codes for an invariant

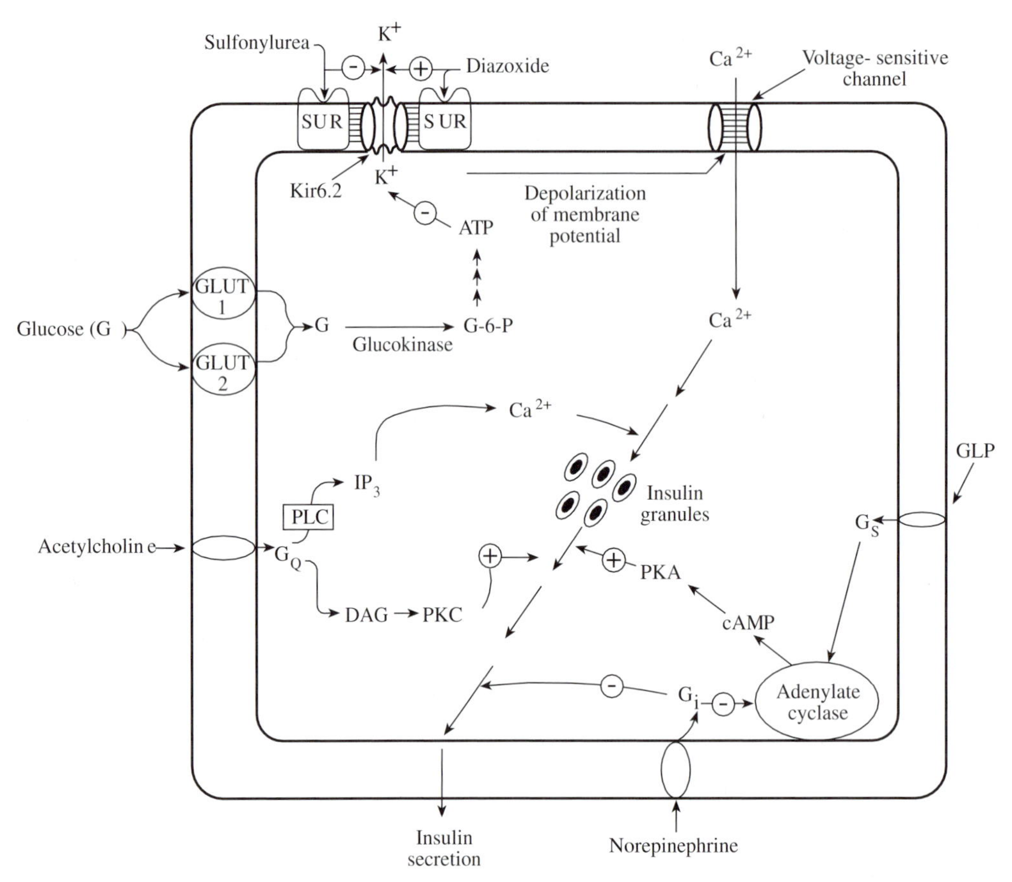

FIGURE 22-7
Diagrammatic representation of insulin secretion from pancreatic β cells. The sequence of events of insulin secretion coupled to glucose entry into β cells consists of glucokinase action, ATP production, inhibition of the ATP-sensitive K^+ channel, membrane depolarization, Ca^{2+} influx, and insulin release. Neurotransmitters acetylcholine and norepinephrine stimulate and inhibit insulin secretion via trimeric G-proteins G_Q and G_i, respectively. Glucagon-like peptide (GLP) promotes insulin release via the G-protein G_s. Sulfonamides and diazoxide have direct effects on sulfonylurea receptors (SURs); the former promotes insulin release and the latter inhibits insulin release. +, Stimulation; −, inhibition. Other abbreviations are given in the text.

tetrapeptide sequence (B23–B26), and hence the abnormal insulins have greatly diminished biological activity. The third mutation yields an abnormal form of proinsulin in which the C-peptide remains attached to the A chain after processing. It is expressed as an autosomal dominant trait (***familial hyperproinsulinemia***). Affected individuals show no signs of insulin resistance and may or may not exhibit mild hyperglycemia.

Secretion

Multiple factors regulate insulin secretion from the pancreatic β cells. Glucose, amino acids, glucagon, acetylcholine and β-adrenergic agents stimulate insulin secretion, whereas somatostatin and α-adrenergic agents exert inhibitory influences. Most notable of the insulin secretagogue is glucose. The sequence of events that leads to insulin secretion by the β cells as the threshold of blood glucose increases is shown in Figure 22-7.

The normal fasting plasma glucose is maintained between 70 and 105 mg/dL (3.89–5.83 mmol/L). Glucose enters the β cells by means of **glucose transporters** GLUT1 and GLUT2 (Chapter 13). GLUT1 is a constitutive glucose transporter and GLUT2 is a low-affinity glucose transporter capable of sensing the glucose concentration in β cells. Glucose transport is not a rate-limiting step in the β-cell glucose metabolism because the transporters are present in greater abundance relative to physiological rates of glucose entry. After glucose enters the β cells, it is converted to glucose-6-phosphate by glucokinase, which is an isoenzyme of hexokinase. Glucokinase has a low affinity for glucose, is the rate-limiting step for glucose metabolism and is not affected by feedback inhibition. The glucose-sensing device that allows rapid and precise quantitation of the ambient glucose level appears to be glucokinase, which is ultimately responsible for glucose-regulated insulin secretion.

The oxidation of glucose-6-phosphate via glycolysis, the TCA cycle, and the electron transport system coupled to oxidative phosphorylation leads to increased ATP levels (or ATP/ADP ratio) in the β cell. The elevated ATP levels cause a closure of the ATP-sensitive K^+ channel, leading to inhibition of K^+ efflux and depolarization of the β-cell membrane potential. The ATP-sensitive K^+ channel consists of two different types of protein subunits, namely, the sulfonylurea receptor (SUR) and a potassium channel (Kir6.2). SUR is the regulatory subunit and belongs to the family of ATP-binding cassette proteins. Kir6.2 subunits participate in the actual conduction of K^+. It has been suggested that the overall functional K^+ channel is an octameric complex composed of equal numbers of SUR and Kir6.2 subunits. The depolarization of β-cell membrane that occurs with the closure of the ATP-sensitive K^+ channel activates the voltage-sensitive Ca^{2+} channel causing Ca^{2+} influx, which ultimately leads to release of insulin from stored insulin granules. The exocytosis of insulin involves both phosphorylation and exocytosis-related ATPases. Along with insulin and C-peptide, a 37-amino-acid peptide known as amylin is also secreted during exocytosis. Amylin is obtained by proteolytic processing of an 89-amino-acid precursor molecule. The precise physiological role of amylin remains to be understood.

Mechanisms that alter the level of cytosolic Ca^{2+} in β cells, other than glucose metabolism, also affect insulin release. For example, a regulatory protein associated with the ATP-sensitive K^+-channel, when occupied with sulfonylurea, inhibits K^+ efflux causing insulin secretion. Sulfonylureas are drugs used in the management of type 2 diabetes mellitus (discussed later). Diazoxide has an opposite effect to that of sulfonylureas. It either prevents the closing or prolongs the open time of the ATP-sensitive K^+ channel resulting in the inhibition of insulin secretion and thus hyperglycemia. Somatostatin inhibits Ca^{2+} influx and causes diminished insulin secretion. Acetylcholine causes elevation of cytosolic Ca^{2+} followed by insulin secretion caused by activation of G_Q-protein, phospholipase C-inositol trisphosphate-Ca^{2+}, and protein kinase C (Chapter 30). Norepinephrine and epinephrine depress insulin secretion by binding at the α-adrenergic receptor sites and by inhibiting adenylate cyclase mediated by the activation of the inhibitory G-protein (G_i). This leads to the inhibition of cAMP production and results in decreased activity of protein kinase A. Decreased protein kinase A levels determine exocytosis-related phosphorylation required for insulin secretion. Release of epinephrine during stress signals the need for catabolic rather than anabolic activity. Depression of insulin secretion during exercise or trauma also is associated with epinephrine (catecholamine) secretion. A gastrointestinal hormone known as glucagon-like peptide (GLP-1) promotes insulin secretion via G-protein activation of the adenylate cyclase-cAMP-protein kinase A system. Gastrointestinal hormones that regulate insulin secretion may act in a feed-forward manner to signal gastrointestinal activity and metabolic fuel intake. Pancreatic glucagon, simply known as glucagon, stimulates secretion of insulin while somatostatin depresses it. Thus, various regulatory inputs to the β cells are integrated to maintain secretion of optimal quantities of insulin and to maintain glucose homeostasis. The coordinated activity of three pancreatic hormones (insulin, glucagon, and somatostatin) is essential for fuel homeostasis. Furthermore, the action of insulin is opposed by glucagon and by other counterregulatory hormones, namely, epinephrine, cortisol, and growth hormone. All of these hormones correct hypoglycemia by maintaining adequate levels of glucose in tissues such as brain, which is primarily dependent on glucose as a fuel source.

Some amino acids also function as secretagogues of insulin secretion. An example is leucine, which stimulates the release of insulin by allosteric activation of glutamate dehydrogenase. Glutamate dehydrogenase, a mitochondrial enzyme, converts glutamate to α-ketoglutarate by oxidative deamination (Chapter 17). Glutamate dehydrogenase is positively modulated by ADP and negatively modulated by GTP. The α-ketoglutarate is subsequently oxidized to provide ATP which blocks the ATP-sensitive K^+ channel, eventually causing insulin release.

The overall glucokinase–glucose sensor mechanism as the primary regulator of glucose-controlled insulin secretion of β cells has been substantiated by identifying mutations that affect human glucokinase. Both gain-in-function and loss-of-function mutants of glucokinase are known. The activating glucokinase mutation (Val455Met) with increased affinity for glucose results in hyperinsulinism with fasting hypoglycemia. Other mutations have been identified that impair glucokinase activity. These defects result in hyperglycemia and diabetes mellitus, known as *maturity-onset diabetes of the young* (MODY), which also occurs as a result of mutations in genes that encode hepatocyte nuclear factors 1α, 4α, 1β, and insulin promoter factor 1. Hyperinsulinemic hypoglycemia can be caused by mutations in the SUR/Kir6.2 components of the K^+ channel. Abnormalities can also result from defects in glutamate dehydrogenase function. Conversion of glutamate to α-ketoglutarate by glutamate dehydrogenase provides substrates for ATP production and the enzyme is inhibited by GTP at an allosteric site. The importance of the sensitivity of glutamate dehydrogenase to inhibition by

GTP is illustrated in patients with hyperinsulinism with episodes of hypoglycemia. All of these patients carry mutations affecting the allosteric domain of glutamate dehydrogenase and result in insensitivity to GTP inhibition. This caused increased α-ketoglutarate production followed by ATP formation and consequent insulin secretion. In addition to hyperinsulinemia, these patients also have high levels of plasma ammonia (*hyperammonemia*) that result from hepatic glutamate dehydrogenase activity. In liver mitochondria glutamate is converted to N-acetylglutamate, which is the required positive allosteric effector of carbamoyl-phosphate synthetase. Carbamoyl-phosphate synthetase catalyzes the first step in the conversion of ammonia to urea (see Chapter 17 for ammonia metabolism). Increased activity of glutamate dehydrogenase causes depletion of glutamate required for the synthesis of N-acetylglutamate. The decreased level of N-acetylglutamate impairs ureagenesis in the liver and is accompanied by hyperammonemia.

Biological Actions of Insulin

Insulin affects virtually every tissue. Insulin is an anabolic signal and promotes fuel storage in the form of glycogen and triacylglycerols while inhibiting the breakdown of these two fuel stores. Insulin also promotes protein synthesis while inhibiting its breakdown. Regulation of expression of several genes, either positive or negative, is mediated by insulin. The genes involved in the expression of enzymes that participate in fuel storage (e.g., hepatic glucokinase) are induced; those that encode catabolic enzymes (e.g., hepatic phosphoenolpyruvate carboxykinase) are inhibited. The principal effect of insulin on blood glucose is its uptake into muscle and adipose tissue via the recruitment and translocation of glucose transporter 4 (GLUT4).

The mechanism of action of insulin is complex and can be divided into three parts. The first part is the binding of insulin to its receptor on the cell membrane; the second consists of postreceptor events; and the third consists of biological responses.

Insulin Receptor

The insulin receptor is derived from a single polypeptide chain that is the product of a gene located on the short arm of chromosome 19. The proreceptor undergoes extensive posttranslational processing consisting of glycosylation and proteolysis. The proteolytic cleavage yields α (M.W. 135,000) and β (M.W. 95,000) subunits that are assembled into a heterotetrameric ($\alpha_2\beta_2$) complex. The subunits are held together by both disulfide linkages and noncovalent interactions. The heterotetrameric receptor is incorporated into the cell membrane with the α-subunits projecting into the extracellular space and the β-subunit with its transmembrane subunit projecting into cytosolic space (Figure 22-8). The α-subunits contain the insulin binding domain, but it is not clear whether insulin binds to both α-subunits. Insulin binding to one site induces negative cooperativity for the second insulin binding site. The intracellular portion of the β-subunit has a tyrosine-specific protein kinase (tyrosine kinase) activity. Tyrosine kinase activity initiated by insulin binding to α-subunits results in autophosphorylation on at least six tyrosine residues in β-subunits. Other protein substrates also are targets of a series of phosphorylations catalyzed by the activated β-subunit tyrosine kinase. One of the key proteins that is phosphorylated is insulin receptor substrate-1 (IRS-1), which has many potential tyrosine phosphorylation sites as well as serine and threonine residues. Phosphorylated IRS-1 interacts and initiates a cascade of reactions involving other proteins. Phosphotyrosine motifs of IRS-1 have binding sites for SH2 domains (*src homology domain 2*) contained in signal transducing proteins such as phosphatidylinositol-3-kinase

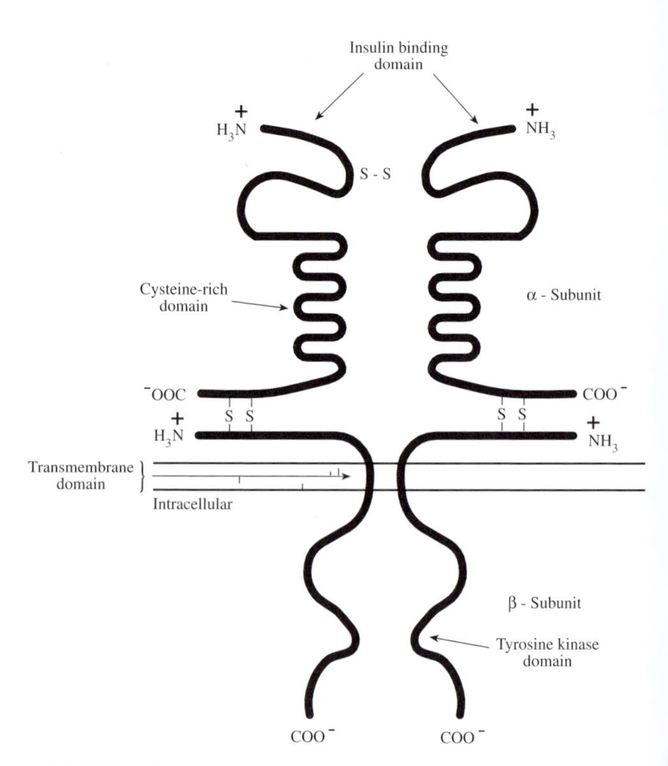

FIGURE 22-8
Model of the insulin receptor. The receptor contains two α- and two β-subunits, held together by disulfide linkages. The α-subunits are entirely extracellular and contain the insulin binding site. The β-subunits have transmembrane and intracellular domains. Both autophosphorylation and tyrosine kinase activity reside in the β-subunit and are markedly enhanced upon insulin binding.

(PI-3-kinase). SH2 domains are conserved sequences of about 100 amino acids but the role of PI-3-kinase is not understood.

Insulin-mediated IRS-1 phosphorylation also leads to activation of ras proteins. The ras proteins are proto-oncogene products that mediate signaling pathways from cell membrane receptors, the regulation of cellular proliferation, differentiation, or apoptosis. They are posttranslationally modified by farnesylation at a conserved cysteine residue in the C-terminal CAAX motif; the tripeptide AAX is removed by proteolysis and the newly exposed C terminus is methylated. The ras protein is functionally active when GTP is bound and inactive when GDP is bound. The ras-GTP/GDP cycle is regulated by GTPase-activating proteins and guanine nucleotide exchange factor proteins. The pathway of IRS-1 to activation of ras protein involves a number of proteins such as growth factor receptor binding protein-2 (GRB-2). Activated ras protein stimulates a cascade of serine/threonine kinases affecting transcriptional activation of specific genes. These pathways also involve proteins with SH2 domains. In summary, responses due to insulin binding to its receptor involve a cascade of phosphorylation and dephosphorylation steps that lead to gene regulation.

One of the major effects of insulin is to promote glucose transport from the blood into muscle and adipose tissue cells. This is accomplished by recruiting and localizing GLUT4 receptor molecules in the plasma membrane. In the absence of insulin GLUT4 remains in the intracellular vesicles. Insulin-mediated GLUT4 trafficking to the cell membrane is constitutive, multicompartmental, and involves the participation of several proteins. Impairment in any of the steps of GLUT4 transport can cause insulin resistance and diabetes mellitus.

The importance of the biological actions of insulin on target cells is underscored by defects in any of the five steps involved in receptor function. These steps are analogous to the scheme described for LDL receptor gene defects (Chapter 20). Mutations can lead to:

1. Impaired receptor biosynthesis,
2. Impaired transport of receptors to the cell surface,
3. Decreased affinity of insulin binding,
4. Impaired tyrosine kinase activity, and
5. Accelerated receptor degradation.

Examples of disorders caused by mutations in the insulin receptor gene are *leprechaunism* and *type A insulin resistance.* A severe form of leprechaunism is due to mutations in both alleles of the insulin receptor gene. These patients exhibit insulin resistance, intrauterine growth retardation, and many other metabolic abnormalities. Patients with type A insulin resistance exhibit insulin resistance, acanthosis nigricans, and hyperandrogenism. The latter two have been ascribed to toxic effects of insulin on the skin and ovaries. Insulin resistance is also associated with *hyperandrogenism* and *polycystic ovary disease syndrome* (*diabetes mellitus* is discussed later in this chapter).

Insulin is catabolized (inactivated) primarily in the liver and kidney (and placenta in pregnancy). Liver degrades about 50% of insulin during its first passage through this organ. An insulin-specific protease and glutathione-insulin transdehydrogenase are involved. The latter reduces the disulfide bonds with separation of A and B chains, which are subjected to rapid proteolysis.

Glucagon

Glucagon is a single-chain polypeptide of 29 amino acids that has a structural homology with secretin, vasoactive intestinal polypeptide (VIP), and gastric inhibitory polypeptide (GIP). The glucagon structure is also contained within the sequence of glicentin ("gut glucagon"). The synthesis of glucagon in the pancreatic α cells probably involves a higher molecular weight precursor. Glucagon secretion is inhibited by glucose and stimulated by arginine and alanine. Depressed plasma glucagon levels result in depressed hepatic glucose output at times when glucose is available by intestinal absorption. Amino acid-stimulated glucagon secretion counteracts the effects of the coincidental secretion of insulin, which otherwise would provoke hypoglycemia. Secretion of glucagon from the α cells and somatostatin from the δ cells is regulated by the other pancreatic hormones. Figure 22-9 illustrates the coordination of islet hormone secretion. This coordination may occur by the product of one cell type regulating secretion by the others and by direct intercommunication between cells

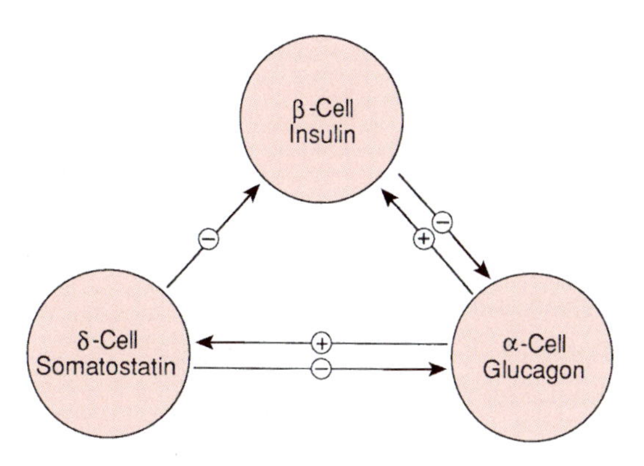

FIGURE 22-9
Paracrine regulation of islet cell hormonal secretion. ⊕, Stimulation; ⊖, inhibition. [Drawn after R. H. Unger, R. E. Dobbs, and L. Orci, Insulin, glucagon, and somatostatin secretion in the regulation of metabolism. *Annu. Rev. Physiol.* **40,** 307 (1978). © 1978 by Annual Reviews Inc.]

(of the same or different types). Glucagon secretion is also stimulated by ***gastrin*** and cholecystokinin, presumably as an anticipatory response to intestinal fuel absorption. Stress and catecholamines provoke glucagon secretion, which increases output of hepatic glucose and adipocyte fatty acids to provide the extra energy requirements.

The primary targets for glucagon are the liver and adipose tissue. It stimulates hepatic glycogenolysis, gluconeogenesis, and ketogenesis and inhibits glycogen synthesis. Glucagon stimulates adipocyte lipolysis to provide fatty acids to tissues for which glucose is not the obligatory fuel. In both liver and adipose tissues, glucagon activates adenylate cyclase, increases cAMP level, activates cAMP-dependent protein kinase, and increases phosphorylation of the controlling enzymes in the various metabolic pathways (Chapter 15).The initial step in the glucagon signaling pathway involves its binding to the receptor on the cell membrane and activation of G-protein complex (Chapter 30).

The major function of insulin and glucagon is to maintain fuel homeostasis. Since glucagon opposes the actions of insulin, the [insulin]/[glucagon] ratio determines the metabolic fate of fuels due to the induction or repression of appropriate enzymes. Enzymes induced by a high [insulin]/[glucagon] ratio and repressed by a low ratio are glucokinase, citrate cleavage enzyme, acetyl-CoA carboxylase, β-hydroxy-β-methylglutaryl-CoA (HMG-CoA) reductase, pyruvate kinase, 6-phosphofructo-1-kinase, 6-phosphofructo-2-kinase, and fructose-2,6-bisphosphatase. Enzymes induced by a low [insulin]/[glucagon] ratio and repressed by a high ratio are glucose-6-phosphatase, phosphoenolpyruvate carboxykinase, and fructose-1,6-bisphosphatase. Metabolic implications of these actions are discussed later.

Somatostatin

Two forms of somatostatin exist: S14 and S28, which are single-chain polypeptides of 14 and 28 amino acids, respectively. Somatostatin is synthesized in the δ cells of the islets, the gut, the hypothalamus, and several other areas of the brain. Its actions appear to be primarily paracrine in nature (i.e., on other cell types in the immediate vicinity of its secretion). In the islets it blocks insulin and glucagon secretion; in the pituitary gland it inhibits growth hormone and thyroid-stimulating hormone (TSH) release; in the gut it blocks secretion of gastrin and motilin, inhibits gastric acid and pepsin secretion, and suppresses gallbladder contraction. The action of somatostatin on the gastrointestinal tract leads to decreased delivery of nutrients to the circulation.

Pancreatic Polypeptide

Pancreatic polypeptide, a peptide of 36 amino acids, is secreted in response to fuel ingestion and potentially affects pancreatic exocrine secretion of bicarbonate and protein.

22.4 Stored Fuels

Table 22-2 lists the amounts of fuels stored in the human body. The amount of glycogen present would not suffice for even 1 day of normal energy expenditure. Triacylglycerol and protein are the principal energy stores, with lipid being in highest abundance. Although glucose, amino acids, fatty acids, and triacylglycerols are present in body fluids, their amounts would not support life for 1 hour. Glycogen is rapidly mobilized in the first 24–36 hours of fasting; little is mobilized after the fourth day of fast. Protein degradation starts off at a

TABLE 22-2
*Metabolic Fuels Stored in a Normal 70-kg Human**

Tissues	Wt(kg)	Available Energy kcal(kJ)
Triacylglycerol (adipose tissue)	15	141,000 (589,944)
Glycogen (liver and muscle)	0.4	1,500 (6,276)
Protein (primarily muscle)	6	24,000 (100,416)
Extracellular Fluids		
Glucose	0.02	80 (335)
Lipid (fatty acid and triacylglycerol)†	0.05	400 (1,674)
Amino acids	0.03	120 (502)

*Data from G. F. Cahill: Starvation in man. *N. Engl. J. Med.* **282,** 668 (1970).
†Total lipid varies from this level down to 0.005 kg between periods of eating and fasting.

low level, increases to a maximum by about the fourth, and then decreases to a point below the initial level. Protein is not mobilized rapidly and even at maximum utilization accounts for only ~25% of fuel expenditure.

Protein degradation is limited mainly to the maintenance of blood glucose. Starting postprandially, oxidation of lipid accounts for the bulk of fuel utilization for metabolic homeostasis. As fasting progresses, the total energy utilized slowly decreases, reflecting the decreasing energy expenditure, but lipid continues to be the major energy source. Few people in developed countries undergo direct prolonged starvation, but disorders such as alcoholism and anorexia nervosa reflect a similar pattern of fuel utilization. The starvation pattern is very helpful for understanding metabolic control and will be used throughout this chapter. The metabolic controls that function in a long-term fast are similar to those that occur in the normal day-to-day pattern of fasting and feeding.

Appetite, Hunger, and Control of Food Intake

The perception that the extent of food intake by humans is poorly controlled is false because most people, once they have matured physiologically, maintain a stable body weight. There appears to be a set point for weight (as there is for body temperature) in each individual. During fasting, body weight decreases. Upon refeeding, an initially greater intake of food returns the body to the original weight. Conversely, if a mammal is force-fed body weight increases, but when force-feeding is stopped, the animal voluntarily decreases food intake until the body has returned to the original weight. Even in most cases of obesity, a set body weight is manifest, albeit at a higher than normal level. A 70-kg adult typically consumes 5–10 tons of food between the ages of 20 and 40. Since body weight will most frequently be maintained at its set point or close to it, food intake is finely controlled to within ±1% over time. A 1% increase in food intake over this 20-year period would result in a 100-pound gain in weight. Both short-term regulation of starting and stopping of eating and long-term regulation of total food intake occur. Both composition and content are regulated. Different foods have different energy contents. Thus, the amount of food eaten decreases if it has a higher energy content per unit of weight; the result is a constant energy intake.

After ingestion, food may either be stored or utilized, with release of energy. Metabolic homeostasis requires a balance between energy input and energy output. Energy output occurs as heat produced, work produced, or maintenance of cellular integrity. These factors are balanced by the signals for hunger, appetite, and satiety. Eating starts and stops voluntarily in a manner that is not related simply to body weight (short-term regulation) but also to the need for energy and specific constituents (long-term regulation).

Regulators of food intake, fuel stores, and energy balance are many and involve a complex set of physiological signals. Some of the regulators are blood glucose level, which is maintained by insulin and other counterregulatory hormones, the mass of adipose tissue and the adipocyte-derived hormone leptin, cerebral cortex and hypothalamic neurons, hypothalamic-pituitary-adrenal axis, distention of the gastrointestinal tract, and release of gastrointestinal and gonadal hormones. The prevalence of overweight and obesity are global health problems. The risk of death from all causes is increased in both conditions and is discussed later.

22.5 Carbohydrate Homeostasis

Most tissues can use fatty acids as their primary, if not sole, source of metabolic energy. Brain and other nervous tissues, except in long-term fasting use glucose as the sole energy source; even in long-term fasting they require significant amounts of glucose. Red blood cells can obtain energy only by anaerobic glycolysis. Skeletal muscle at rest uses predominantly fatty acids, but in heavy exercise it also draws on muscle glycogen and blood glucose. Because brain and red blood cells depend on glucose for energy, glucose must always be available.

Glucose occurs in plasma and interstitial fluid at a concentration of approximately 80 mg/dL; this value corresponds to about 20 g in the extracellular compartment of a typical person. Approximately 180 g of glucose is oxidized per day. The body must therefore replenish the total blood glucose concentration about nine times a day; nevertheless, the concentration in blood remains remarkably constant. The glucose level following an overnight fast is approximately 80 mg/dL. Following a meal, such as breakfast, the level rapidly rises by 30–50 mg/dL, but within 2 hours it returns to the previous level where it remains until the next meal when the pattern is repeated. The remarkable stability of the blood glucose level is an indication of the balance between supply and utilization. Maintenance of this balance is discussed below.

Carbohydrate as a Food

Carbohydrate is essential for the survival of some tissues and as a structural constituent of nucleic acids, glycoproteins, proteoglycans, and glycolipids. The normal adult can synthesize all the needed carbohydrate from noncarbohydrate sources, namely, amino acids and glycerol. Thus, humans can exist with little or no dietary carbohydrate intake. In Western cultures, the diet generally consists of approximately 45% carbohydrate, 43% lipid, and 10% protein. Of the carbohydrate, about 60% is starch, 30% sucrose, and most of the remainder lactose. In less affluent cultures, the typical diet contains more grains and vegetables, with correspondingly higher levels of carbohydrate. Dietary carbohydrate is digested to glucose (80%), fructose (15%), and galactose (5%). Glucose and galactose are actively transported directly into the blood by the intestinal cells. Fructose is absorbed by a specific system that is distinct from glucose and galactose absorption (Chapter 12). Part of the fructose is metabolized by the intestinal cells, and the rest enters the portal blood. Liver and kidney are the other sites of fructose utilization (Chapter 13).

Disposition of High Glucose Intake

A temporary rise in plasma glucose concentration immediately follows a meal, but within 2–3 hours the level returns to the preprandial level. Several factors influence the blood glucose profile.

1. Digestion and absorption of carbohydrate are very rapid. The rate of appearance of glucose in plasma is determined by gastric motility and emptying and by intestinal absorption. The glucose peak is characteristically only 30–50% above the fasting level. If the total amount of glucose in a typical meal (50–100 g) were to enter the blood and not be used immediately, total blood glucose would go up almost fourfold. This rise does not occur because the glucose entering the plasma is rapidly taken up by the tissues. The influx of glucose following a meal is accompanied by a rapid decrease in hepatic secretion of glucose, while utilization by tissues dependent on glucose remains unchanged. Thus the rise in blood glucose level after a meal reflects only a portion of the glucose entering the system from digestion and absorption. During fasting blood glucose remains at a constant level because utilization is matched by formation.
2. The rise in blood glucose level is quickly followed by a rise in the blood insulin level which increases three- to tenfold. An elevated blood glucose level directly stimulates pancreatic insulin release, as do leucine and arginine derived from digestion of protein. Release of gastrin, cholecystokinin, gastric inhibitory peptide, and glicentin appears to stimulate insulin release in an anticipatory manner in addition to regulating other digestive responses.
3. Concomitant with insulin release are changes in glucagon release, whose magnitude and direction depend on dietary composition. If the diet contains only carbohydrate, glucagon secretion falls precipitously because of direct inhibition of the α cells by glucose and of the insulin released from β cells. On a diet high in protein, glucagon secretion is stimulated as a consequence of amino acid influx. After a meal that contains both carbohydrate and protein, plasma glucagon levels may not change perceptibly.
4. The liver is freely permeable to glucose and typically extracts about half of the carbohydrate load. This glucose is phosphorylated to glucose-6-phosphate by glucokinase and hexokinase. After a glucose load, glucokinase is more important because it has a high K_m, is inducible, and is sensitive in a wide range of glucose concentrations. After overnight fasting, most carbohydrate taken up by the liver is converted to glycogen. This regulation is illustrated in Figure 22-10.

   Insulin promotes dephosphorylation of the inactive glycogen synthase D to the active glycogen synthase I form. Depressed levels of glucagon and elevated levels of insulin decrease hepatic cAMP levels and the cAMP-dependent protein kinase inactivation of glycogen synthase. High levels of glucose promote high levels of glucose-6-phosphate, a feed-forward, allosteric, positive modifier of glycogen synthase, and result in hepatic glucose storage. Glycogen synthesis is affected by the level of glycogen itself, which inhibits glycogen synthase phosphatase. All of these events occur rapidly. Insulin also has a long-term effect on the liver by inducing the synthesis of glucokinase (see also Chapter 15).

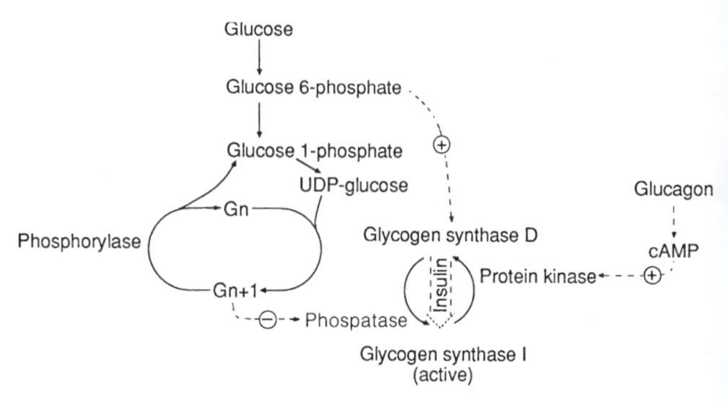

FIGURE 22-10
Hepatic glycogen synthesis after a meal. The permeability of the liver to glucose provides the substrate for hepatic glycogen synthesis. This synthesis is controlled by (1) activation of glycogen synthase by insulin, (2) allosteric activation of glycogen synthase by glucose-6-phosphate, and (3) absence of cAMP, blocking the cAMP-dependent protein kinase inactivation of glycogen synthase. Synthesis ends when glycogen stores are high as a result of inhibition of glycogen synthase phosphatase by glycogen.

5. The liver plays a major role in converting excess glucose into triacylglycerols, packaging them into VLDLs, and exporting them to nonhepatic tissues. Thus, glucose in excess of that needed to restore glycogen levels ends up stored as triacylglycerol in adipocytes.
6. Glucose that is not sequestered in the liver is eventually distributed within the extracellular space and accounts for the increased plasma glucose levels. This glucose is rapidly metabolized by the other tissues.
7. Insulin stimulates uptake of glucose into muscle by recruiting more glucose transporters to the cell membrane. The glucose that is taken up will be used to replenish muscle stores of glycogen through reaction mechanisms similar to those depicted in Figure 22-10 (except that glucagon does not regulate muscle metabolism). Muscle glycogen levels will be restored, while any protein degraded for gluconeogenesis during fasting will be replenished. The signal for increased protein synthesis is insulin.
8. Much of the glucose not taken up by liver and muscle is taken up by adipocytes under the influence of insulin-stimulated transport of glucose into the cell. Esterification of fatty acids for storage as triacylglycerols is dependent on the availability of α-glycerophosphate derived *in situ* from glucose. The glucose can also be converted into fatty acids from acetyl-CoA generated via glycolysis and the pyruvate dehydrogenase reaction and from NADPH obtained via the pentose phosphate pathway.
9. With the exception of brain, liver, and blood cells, insulin directly stimulates the entrance of glucose that is used for anabolic processes in most cells of the body.

The disposition of a high glucose intake is presented in Figure 22-11.

As much glucose as possible is stored in liver and muscle, and the remainder is used for biosynthetic purposes or converted to fatty acids and stored as triacylglycerol. Fatty acid synthesis in liver leads to secretion of triacylglycerols as VLDLs and transport to adipose tissue. Plasma glucose levels do not exceed the renal threshold of 200 mg/dL for glucose. Insulin stimulates a variety of anabolic processes in addition to those of glycogen, protein, and triacylglycerol synthesis by mechanisms that may involve rapid, covalent modification of key regulatory enzymes or regulation of their synthesis. Insulin stimulates glycolysis for provision of anabolic metabolites. Low levels of glucagon, glucocorticoids, and catecholamines, which oppose the action of insulin, also contribute to rapid removal of glucose.

Glucose Tolerance

Intake of carbohydrate leads to a characteristic change in blood glucose level that may be affected by many factors, such as abnormalities in insulin secretion or effectiveness. Normal and abnormal glucose tolerance responses are shown in Figure 22-12. The test is performed on a patient who has fasted overnight and is at rest. The patient ingests 75 g of glucose, and the blood glucose level is measured over a 2- to 4-hour period. In a normal individual, the fasting level is constant before and shortly after glucose ingestion. Only values outside the range of normal fasting values (denoted by the shaded bar) are helpful diagnostically. Glucose intolerance is manifested by a slower return to the fasting level. Many factors affect the shape of the curve. After an overnight fast, hepatic glycogen stores will be low, but if the fast is shorter they may be much higher and the plasma glucose levels would remain elevated longer. Tissue sensitivity to insulin is modulated by past dietary composition, obesity, stress, age, and exercise. After a diet rich in carbohydrate, for example, insulin levels would be increased by down-regulation of insulin receptors, tissue insulin insensitivity, and glucose intolerance. After fasting or on a low-carbohydrate diet, insulin responsiveness and glucose tolerance would be greater. Exercise enhances insulin sensitivity. The pattern shown in Figure 22-12 is typical of glucose intolerance observed at the onset of insulin-independent diabetes. The fasting level of glucose is frequently in the normal range in mild diabetes, but after glucose ingestion the level stays higher for a longer period of time.

As the plasma glucose level decreases, it falls slightly below the fasting level before returning to normal, since restoration of insulin to the normal fasting level occurs slightly after that of the blood glucose level. This depression can be heightened in "reactive hypoglycemia." If, after a fast of 18–24 hours a small amount of glucose in a highly digestible form is consumed (such as the sucrose in a candy bar), the blood glucose level will be temporarily elevated followed by rapid elevation of insulin level, rapid entry of glucose into tissues, and depression of plasma glucose level below normal.

Glucose Homeostasis during Fasting

During most of the day the plasma glucose level remains constant, even after a 24-hour fast. In very prolonged fasts, the plasma glucose level decreases very slightly. During these periods, glucose is being actively metabolized by tissues such as brain and red blood cells, and is replenished. During a 24-hour fast, the body uses approximately 180 g

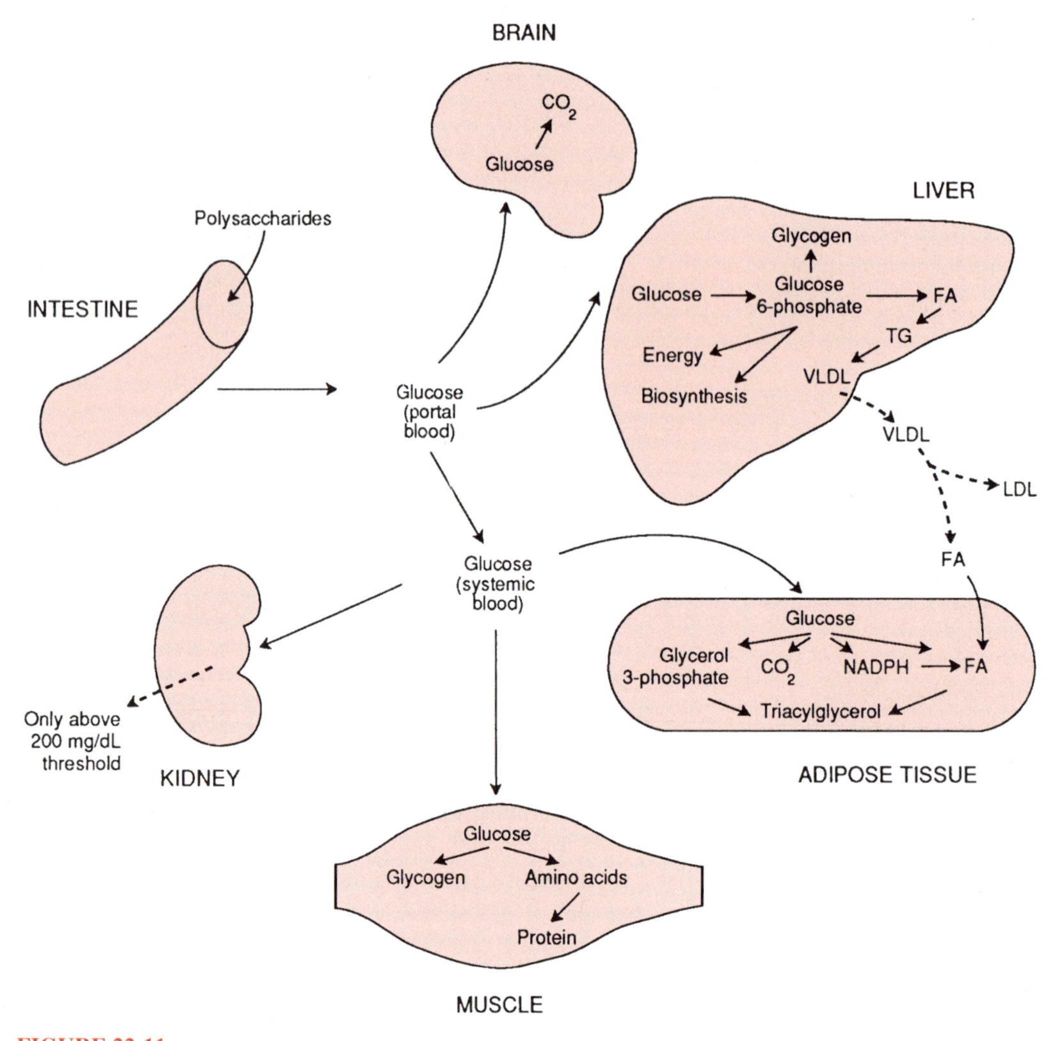

FIGURE 22-11
Disposition of high glucose intake. FA, Fatty acid; TG, triacylglycerol; VLDL, very-low-density lipoprotein; LDL, low-density lipoprotein.

of glucose, initially derived from hepatic glycogenolysis but with an ever-increasing contribution from gluconeogenesis.

Utilization of Hepatic Glycogen

The signal for glycogen breakdown is glucagon secreted in response to hypoglycemia. Glucagon stimulates hepatic glycogenolysis by the adenylate cyclase cascade (Figure 22-13). The hepatic plasma membrane is markedly sensitive to glucagon, which stimulates adenylate cyclase with formation of cAMP, activation of phosphorylase kinase and phosphorylase, and degradation of glycogen to glucose-1-phosphate, then to glucose-6-phosphate, and finally to blood glucose. In hypoglycemia, insulin levels are low, and no hepatic glycogen synthesis occurs. Thus,

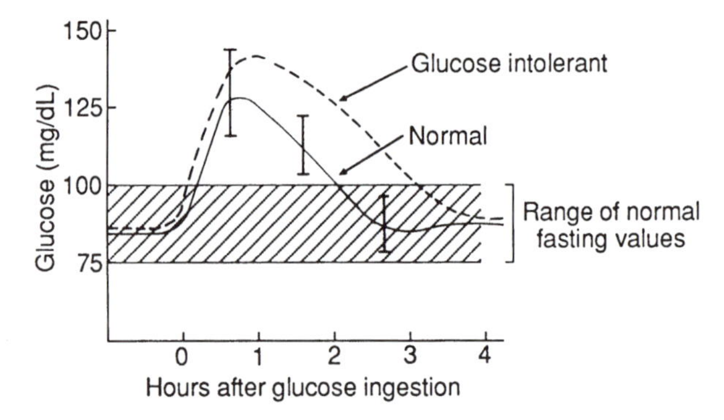

FIGURE 22-12
Typical glucose tolerance response for a normal individual and one with mild glucose intolerance. The shaded area shows the range for plasma glucose levels in normal fasting individuals. At the peak value, the level for a glucose-intolerant individual often is not significantly different from that of a normal individual; thus, intolerance is best detected at the later time points when glucose values remain higher.

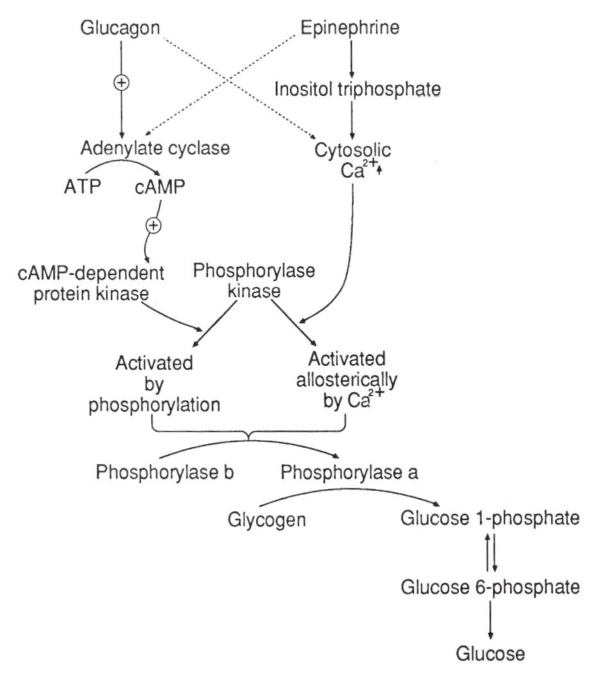

FIGURE 22-13
Regulation of hepatic glycogenolysis. Glucagon and epinephrine can activate hepatic glycogenolysis and lead to glucose release. Glucagon acts primarily by increasing cytosolic levels of cAMP, whereas epinephrine acts predominantly by increasing the release of calcium from endoplasmic reticulum into cytosol. In either circumstance, phosphorylase kinase is activated. cAMP mediates the phosphorylation of phosphorylase kinase, while calcium activates phosphorylase kinase allosterically. The result is the phosphorylation and activation of phosphorylase b. Some overlap between the two mechanisms occurs.

glucagon and insulin function to regulate glycogen synthesis and degradation that directly reflects the blood glucose level. In the initial phases of starvation, this is the major glucose-producing mechanism. Hepatic glycogenolysis is also regulated by catecholamines. Catecholamine release is less sensitive to hypoglycemia than glucagon release but plays a significant role in stress and high-intensity exercise. Catecholamines act via stimulation of calcium release from endoplasmic reticulum along with allosteric activation of phosphorylase kinase that results in activation of phosphorylase, degradation of glycogen, and hepatic glucose output.

Utilization of Skeletal Muscle Glycogen

The muscle of a 70-kg human contains about 300 g of glycogen, but this glycogen is not readily available to maintain blood glucose levels because muscle lacks glucose-6-phosphatase. However, the Cori cycle provides a means for muscle to function anaerobically during intense exercise. Thus, muscle glycogen contributes to plasma glucose homeostasis, although its conversion to lactate is regulated by the metabolic demands of muscle contraction. The ATP/ADP cycle links muscle contraction and conversion of muscle glycogen to lactate. Since there are only limited amounts of adenine nucleotides, these processes are tightly coupled. The coordination of muscle glycogen utilization and contraction is presented in more detail in Figure 22-14.

Regulation occurs because

1. ADP is a required substrate for glycolysis (stoichiometric regulation),
2. Phosphofructokinase catalyzes an irreversible step in glycolysis and is subject to allosteric inhibition by

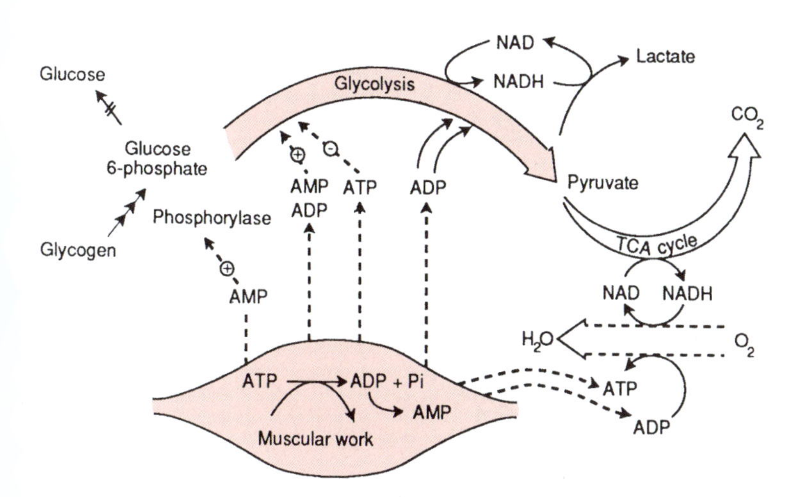

FIGURE 22-14
Coordinated regulation of skeletal muscle metabolism by nucleotides.

ATP and activation by inorganic phosphate, 5′-AMP, and ADP (Chapter 13), and

3. Skeletal muscle phosphorylase b is allosterically activated by 5′-AMP. Conversion of phosphorylase b to phosphorylase a is affected by epinephrine, through increased levels of cAMP, in a sequence of reactions similar to that indicated in Figure 22-13. Following depolarization of the muscle cell membrane, calcium is released into the sarcoplasm from the sarcoplasmic reticulum, resulting in the calcium-dependent activation of phosphorylase kinase and conversion of phosphorylase to the "a" form.

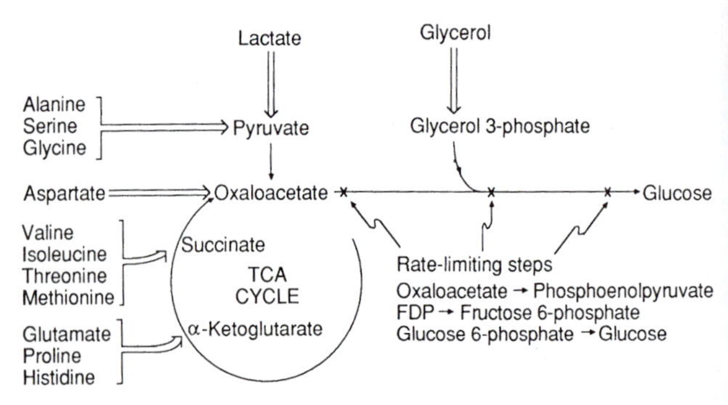

FIGURE 22-15
Entry of precursors into the pathway of gluconeogenesis.

Gluconeogenesis

As hepatic glycogen stores become depleted, gluconeogenesis becomes increasingly important. Although the kidney is a source of glucose during protracted starvation, during brief fasting at least 90% of total gluconeogenesis occurs in the liver.

The three sources for this glucose synthesis are the following.

1. ***Glycerol:*** The hydrolysis of triacylglycerol to fatty acids by adipocytes releases glycerol, which is released into the plasma and converted to glucose in the liver. It makes only a minor contribution to the total glucose formed, but it helps to conserve protein.
2. ***Lactate:*** In sedentary humans 10–30% of plasma glucose is recycled from lactate that originates from glycolysis in blood cells. During exercise, anaerobic glycolysis in muscle increases leading to enhanced gluconeogenesis.
3. ***Amino acids:*** Amino acids are derived from proteolysis in muscle. All except leucine and lysine are potentially gluconeogenic in mammals; however, alanine is the primary amino acid released from muscle. This alanine is derived from muscle protein and from the metabolism of many other amino acids.

Figure 22-15 shows the points of entry of precursors into hepatic gluconeogenesis. Lactate and alanine enter the gluconeogenic pathway at the pyruvate carboxylase reaction. Glycerol is phosphorylated and then oxidized to dihydroxyacetone phosphate. A number of amino acids are converted to intermediates of the TCA cycle. Conversion of gluconeogenic substrates to glucose is not a direct reversal of glycolysis, since the pathway differs at three rate-limiting steps (Chapter 15).

Regulation of Gluconeogenesis

Tissues involved in glucose conservation are liver, skeletal muscle, and adipose. Signals that attune the body to the status of gluconeogenesis include glucagon (an acute modulator), glucocorticoid (a chronic modulator), and absence of insulin. The general effects of these hormones are shown in Figure 22-16.

Glucagon is directed toward glycogen degradation, gluconeogenesis, triacylglycerol hydrolysis, and fatty acid oxidation. In the liver, it stimulates glycogen degradation (Figure 22-13) and gluconeogenesis, possibly by inactivation of pyruvate kinase. Glucagon also modulates the levels of fructose-2,6-bisphosphate (Chapter 15). Insulin opposes these effects by stimulating glycogen synthesis, glycolysis, and fatty acid synthesis. The stimulation of glycolysis by insulin is not meant to increase energy production, since glycolysis is amphibolic and provides metabolites for biosynthesis, e.g., acetyl-CoA for fatty

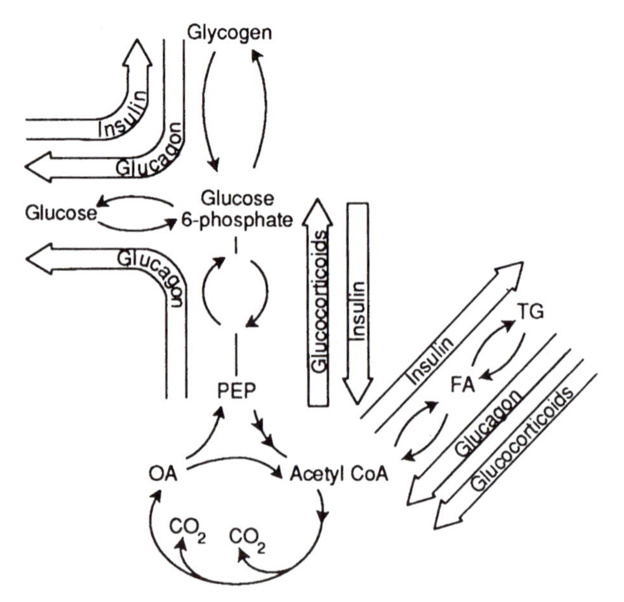

FIGURE 22-16
Effects of glucagon, glucocorticoids, and insulin on the pathways of carbohydrate and lipid metabolism. G-6-P, Glucose-6-phosphate; PEP, phosphoenolpyruvate; OA, oxaloacetate; FA, fatty acid; TG, triacylglycerol.

acid synthesis. Glucocorticoids stimulate fatty acid breakdown and gluconeogenesis and increase the rate of glycogen synthesis. This action appears counterproductive, but glucocorticoids are a major signal for the degradation of muscle proteins and thus for gluconeogenesis. Glucocorticoid effects are long term (whereas those of glucagon are short term), but they are synergistic and act through induction of enzymes of gluconeogenesis. Regulation by glucocorticoids may involve increases in the amounts of regulatory proteins that are involved in the actions of glucagon. Many allosteric regulators in addition to hormones fine-tune the gluconeogenic pathway at the steps catalyzed by pyruvate carboxylase, phosphoenolpyruvate carboxykinase, fructose-bisphosphatase, and glucose-6-phosphatase. A summary of glucose homeostasis during fasting is shown in Figure 22-17.

The sources of carbon for hepatic glucose formation are hepatic glycogen, lactate from red blood cells or skeletal muscle, amino acids from muscle protein, and glycerol from adipocytes. The kidney reabsorbs the glucose in the glomerular filtrate and is a site of gluconeogenesis. In mammals, there is no net conversion of fatty acids to glucose because during oxidation of acetyl-CoA to CO_2 via the TCA cycle there is no net production of oxaloacetate. However, fatty acids, play an important role in glucose homeostasis (Figure 22-17). Since brain and red blood cells use glucose, the heart functions well on fatty acids without depleting glucose levels. When glucose levels are low, only tissues that have a requirement for glucose and are insulin-independent use glucose because, with low insulin levels, there is essentially no glucose uptake by insulin-dependent tissues. After fasting for a 24-hour period, 180 g of glucose is produced from glycogen or gluconeogenesis, of which 75% is used by the brain and the remainder by red and white blood cells, etc. From the latter, about 36 g of lactate return to the liver for gluconeogenesis,

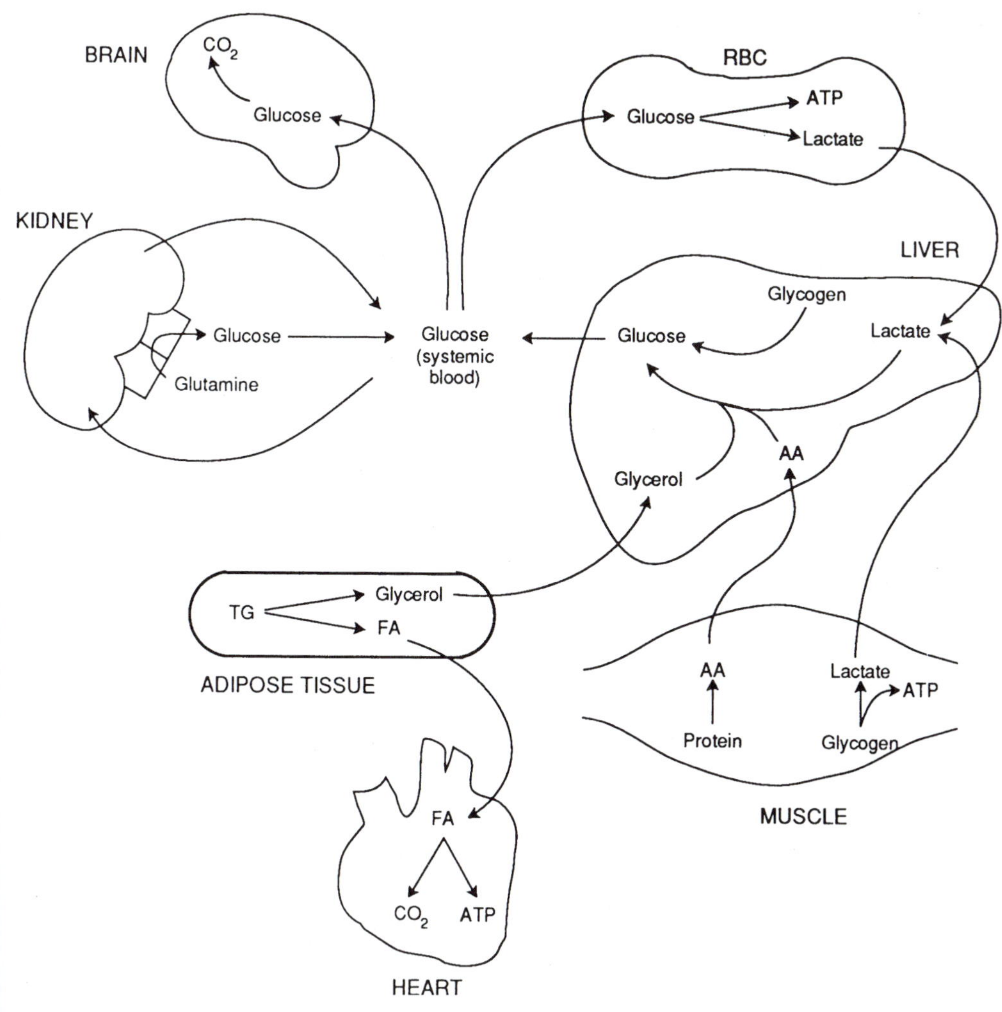

FIGURE 22-17
Glucose homeostasis in the fasting state. TG, Triacylglycerol; RBC, red blood cells; FA, fatty acid; AA, amino acid.

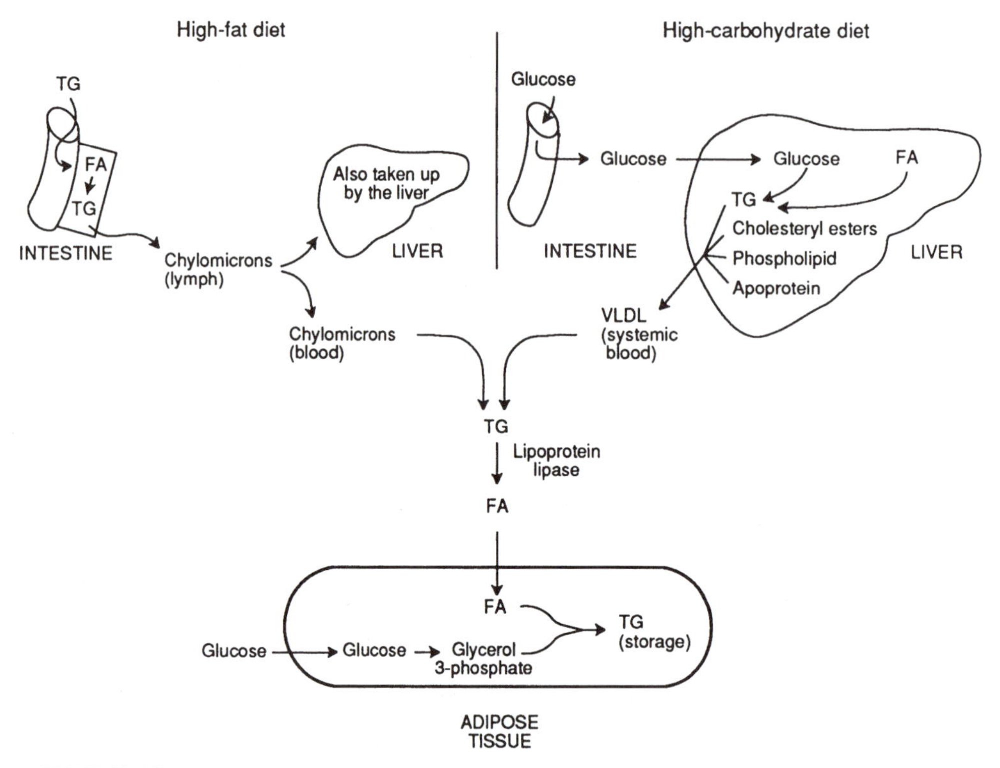

FIGURE 22-18
Triacylglycerol storage in high-fat and high-carbohydrate diets. FA, Fatty acid; TG, triacylglycerol.

supported by amino acids from 75 g of protein and 16 g of glycerol.

22.6 Lipid Homeostasis

The two principal types of lipid in fuel homeostasis are triacylglycerols and fatty acids. Other lipids serve roles in cellular membrane structure and other functions. Sufficient triacylglycerol normally is stored in humans to support many weeks of fasting (Table 22-2). Triacylglycerols constitute a very compact form of energy and their complete oxidation yields a high amount of energy. They are transported between organs as components of lipoproteins. Free fatty acids are amphiphilic and act as a detergent. For interorgan transport, fatty acids are bound to albumin.

Lipid Digestion and Absorption

Dietary lipid consists mostly of triacylglycerol from plant and animal sources. It supplies about 45% of the energy in a typical Western diet. Except for the essential polyene fatty acids (Chapter 18), the dietary requirement for lipid can be met by carbohydrate or protein. However, the Eskimo has a satisfactory diet consisting of 80–90% lipid. Digestion of triacylglycerol is discussed in Chapter 12. A small proportion of the free fatty acids, notably the short-chain acids in milk and milk products, is absorbed directly into the portal blood and transported to the liver. The bulk of the fatty acids are esterified in intestinal cells into triacylglycerol and packaged within the Golgi apparatus into chylomicrons that contain triacylglycerol, phospholipids, cholesterol and cholesterol esters, and apoproteins. Chylomicrons are secreted into lymphatic capillaries that surround the intestinal cells and fuse into larger vessels to form the thoracic duct, which opens into the systemic circulation by a one-way valve at the junction of the left jugular and subclavian veins. Chylomicrons appear in the plasma shortly after a fatty meal and reach a maximum concentration within 4–5 hours. Their large size results in light scattering and gives plasma a turbid appearance termed *postalimentary lipemia.*

Disposition of Absorbed Triacylglycerol

In various tissues of the body, chylomicrons are acted upon by lipoprotein lipase that hydrolyzes the associated triacylglycerol. This lipase is secreted by adipocytes, hepatocytes, and cardiac and mammary tissue, and it becomes associated with the external surface of endothelial cells of the vasculature in response to the need of these tissues for fatty acids. Its half-life is a matter of hours. Its synthesis and export are stimulated by insulin and glucose and inhibited by catecholamines. Mammary gland lipoprotein lipase activity is elevated maximally at parturition when lipid is needed for milk formation. Cardiac lipoprotein lipase is elevated during starvation. The lipoproteins transport triacylglycerol and cholesterol (Chapter 20).

Production of Triacylglycerol from Carbohydrate

Processes involved in triacylglycerol storage are presented in Figure 22-18. When dietary carbohydrate exceeds requirements, it is converted into triacylglycerol within adipocytes or hepatocytes. The liver daily synthesizes 40–100 g of triacylglycerols, partly from glucose and partly from free fatty acids present in excess of its energy needs. Their transport to adipocytes occurs via VLDLs, whose synthesis, secretion, and fate are analogous to those of chylomicrons.

Release of Lipid from Adipose Tissue Stores

During fasting or exercise and at times of stress, lipid is used as a source of energy. Hormone-sensitive lipase catalyzes the sequential hydrolysis of triacylglycerol to yield ultimately 3 mol of fatty acids and 1 mol of glycerol. Hormone-sensitive lipase is regulated by phosphorylation by cAMP-dependent protein kinase. Listed in Table 22-3 are hormones that affect lipase. Epinephrine, glucagon, and the other rapid stimulators release free fatty acids within minutes. The actions of growth hormone and of glucocorticoids require hours, but their sites of action are unknown. Insulin and prostaglandin E_1 suppress lipolysis by depressing cAMP levels.

TABLE 22-3
Hormonal Regulation of Adipocyte Lipolysis

Rapid Stimulators	Slow Stimulators	Inhibitors
Epinephrine	Glucocorticoids	Insulin
Norepinephrine	Growth hormone	Prostaglandin E_1
Glucagon		
ACTH		
Secretin		
Vasopressin		

The major activities of adipocytes are summarized in Figure 22-19. Glucagon, an indicator of the need for

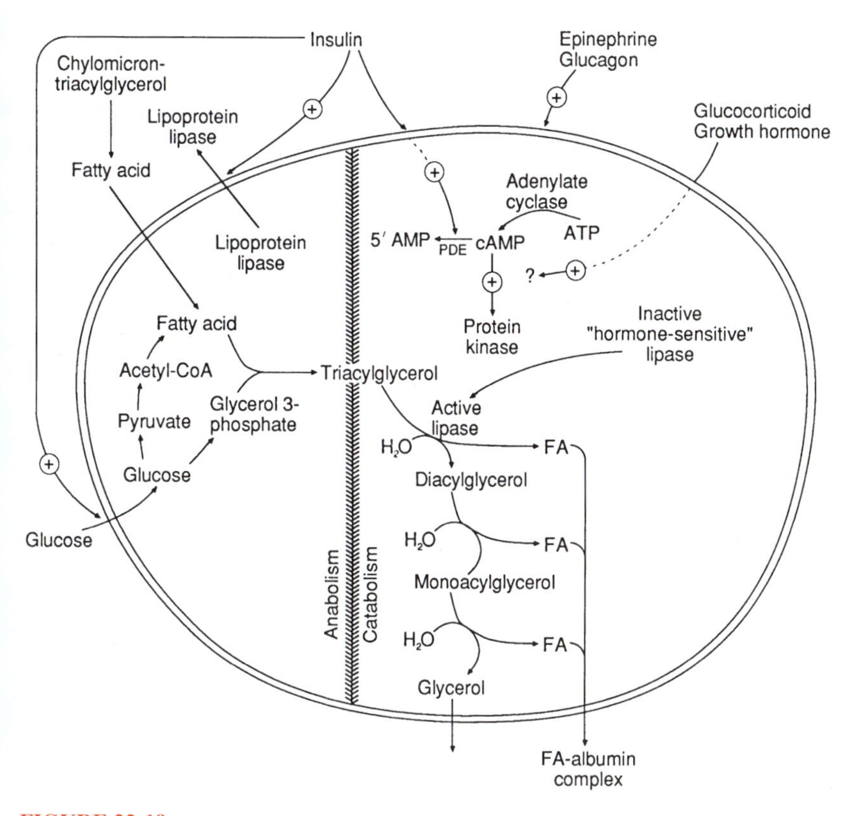

FIGURE 22-19
Regulation of adipocyte triacylglycerol synthesis and degradation. FA, Fatty acid; PDE, phosphodiesterase.

carbohydrate, increases hepatic glycogenolysis, gluconeogenesis, and ketogenesis at the expense of fatty acid oxidation and enhances adipocyte lipolysis. Epinephrine and glucocorticoids are, respectively, acute and chronic indicators of stress. They increase hepatic glucose output and peripheral tissue fatty acid utilization. Growth hormone regulates many processes associated with growth of the organism. The energy for much of this growth derives from fatty acid oxidation.

As fatty acids leave the adipocyte, they form complexes with albumin. The availability of separate pathways for fat deposition and release in these cells permits appropriate regulation under anabolic or catabolic circumstances. The absence of glycerol kinase in the adipocyte decreases the potential for resynthesis of the triacylglycerol and promotes hepatic gluconeogenesis. Triacylglycerols are deposited in the adipocyte when glucose and insulin are readily available.

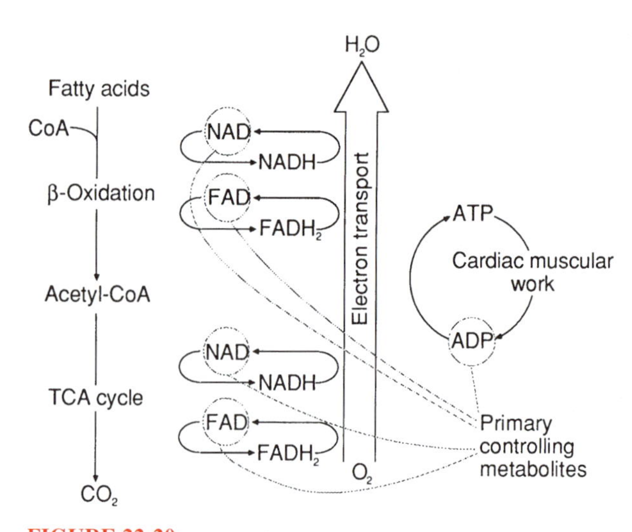

FIGURE 22-20
Regulation of fatty acid oxidation in cardiac muscle.

Tissue Utilization of Fatty Acids

The level of plasma albumin-bound fatty acids increases when lipolysis rates are high. The tissue uptake of fatty acids is proportional to their concentration in plasma and is therefore largely dependent on blood flow. During intense exercise, the flow of blood through the splanchnic bed is reduced and more fatty acids are available to skeletal muscle. With the exception of nerve tissue and blood cells, tissues can use fatty acids by β-oxidation and by the TCA cycle. Fatty acid uptake is not regulated by hormones or intracellular effectors. Free fatty acids readily diffuse across the plasma membrane of cells where they are used strictly in response to supply and demand. This is illustrated for cardiac muscle in Figure 22-20.

Fatty acid oxidation requires adequate amounts of NAD and FAD. Oxidation of NADH and $FADH_2$ occurs in mitochondria via electron transport and the use of molecular oxygen. However, the electron transport chain can only function if ADP is present, so that fatty acid breakdown occurs at the rate necessary to maintain cellular levels of ATP which decrease if metabolic or other work is performed. Tight coupling of electron transport to oxidative phosphorylation (Chapter 14) is of considerable importance in fuel economy. Without it, fatty acids would be utilized in an unproductive manner. In the liver, fatty acids in excess of requirements are converted to triacylglycerols and secreted as VLDLs.

Many tissues preferentially utilize fatty acids as an energy source. The sparing effect on glucose is illustrated in Figure 22-21 and reflects the tight coupling of oxidation of fatty acids to ATP production. When sufficient fatty acids are available ATP concentrations will be high and ADP concentrations low. ADP is required in two reactions in glycolysis; ATP is an allosteric inhibitor, and AMP is an allosteric activator of phosphofructokinase.

Ketone Body Production and Utilization

Fatty acids released from adipose tissue are the source of ketone bodies (Figure 22-22). Fasting, high levels of glucagon and catecholamine, and a low level of insulin result in rapid lipolysis and ready availability of fatty acids. Fatty acids, after being converted to CoA thioesters, are oxidized in the mitochondria. The rate-limiting step in the oxidation process is the transport of fatty acyl-CoA

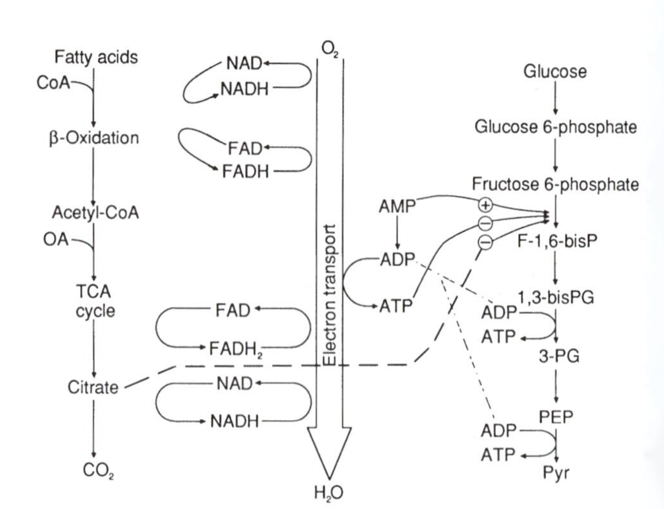

FIGURE 22-21
Glucose-sparing effect of fatty acid oxidation. Note the inhibition of glucose utilization by high levels of ATP and citrate. OA, Oxaloacetate; F-1, 6-bisP, fructose-1,6-bisphosphate; 1,3-bisPG, 1,3-bisphosphoglycerate; 3-PG, 3-phosphoglycerate; PEP, phosphoenolpyruvate; Pyr, pyruvate.

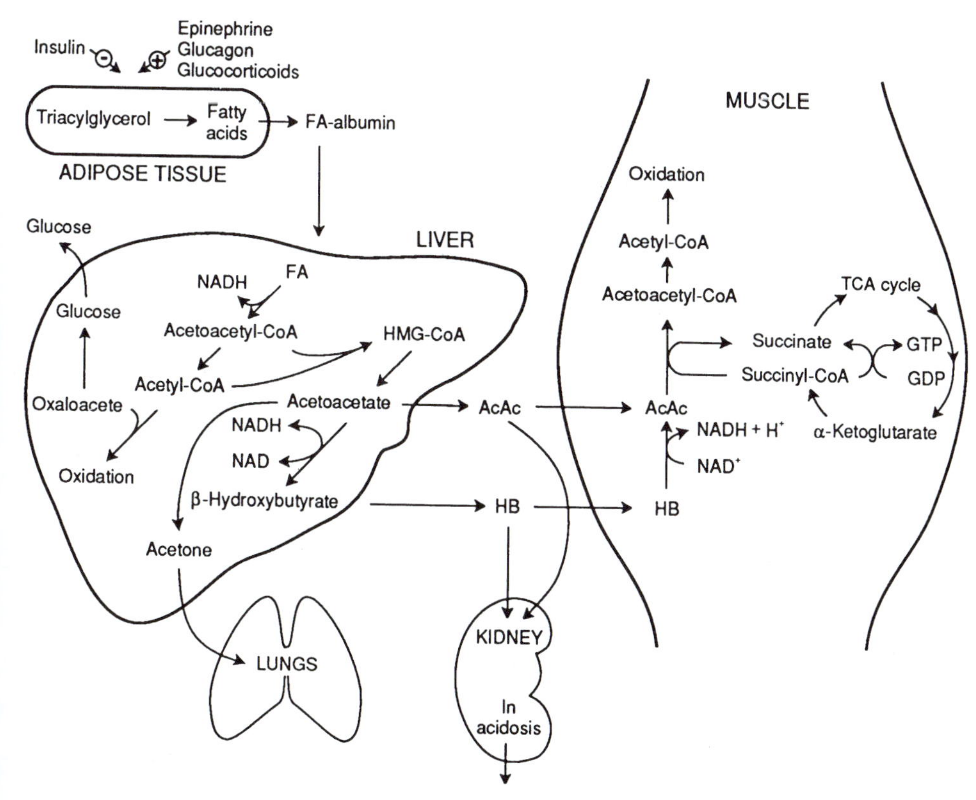

FIGURE 22-22

Ketone body production and utilization. Ketone bodies are produced in the liver from fatty acids derived from adipocyte lipolysis. They are released and used as fuel in peripheral tissues. The initial step in acetoacetate metabolism is activation to acetoacetyl-CoA by succinyl-CoA. HMG-CoA, β-hydroxy-β-methylglutaryl-CoA; HB, β-hydroxybutyrate.

across the inner mitochondrial membrane. The transport is a transesterification process and involves carnitine and two enzymes, carnitine palmitoyltransferase (CPT) types I and II. CPTI is inhibited by malonyl-CoA, primary substrate for fatty acid synthesis that occurs in the cytosol. In the presence of a low insulin:glucagon ratio, malonyl-CoA formation is inhibited, leading to active CPTI and promoting fatty acid oxidation and consequent ketone body production. The opposite situation occurs at a high insulin: glucagon ratio.

Hepatic β-oxidation, without oxidation of acetyl-CoA through the TCA cycle, produces a substantial amount of energy. At such a time, liver is actively engaged in gluconeogenesis so that mitochondrial oxaloacetate is depleted, TCA cycle activity is depressed, and acetyl-CoA levels rise. The last reaction in β-oxidation is conversion of acetoacetyl-CoA to acetyl-CoA, with an equilibrium in favor of high levels of acetoacetyl-CoA. Thus, acetyl-CoA and acetoacetyl-CoA accumulate and form HMG-CoA; cleavage of this last compound yields acetoacetate, which is reduced to β-hydroxybutyrate. Acetone results from nonenzymatic decarboxylation of acetoacetate. Ketone body formation occurs exclusively in liver (see Chapter 18) and is prominent in starvation and diabetes owing to the low insulin:glucagon ratio in these conditions. The ketone bodies readily leave the liver and enter the bloodstream. Acetone is eliminated in the urine and exhaled by the lungs. Because of its odor, it has long been diagnostic in urine and in breath in clinical situations. Because production of ketone bodies was associated with starvation and disease, these compounds were not considered to be metabolically useful. During periods of starvation, acetoacetate and β-hydroxybutyrate make a considerable contribution to fuel homeostasis. The ratio of their plasma concentrations in starvation and diabetes is about 1:5, reflecting high hepatic levels of NADH, which favors the reduction of acetoacetate to β-hydroxybutyrate. In utilization of ketone bodies in skeletal muscle, β-hydroxybutyrate is converted to acetoacetate and then to acetoacetyl-CoA, which is cleaved, and the acetyl-CoA formed is oxidized through the TCA cycle. Activation of acetoacetate to acetoacetyl-CoA requires conversion of succinyl-CoA to succinate.

Under normal conditions, the brain cannot use ketone bodies because it lacks the enzyme needed to activate acetoacetate. However, this enzyme is induced in brain after about 4 days of starvation, permitting the brain to obtain 40–70% of its energy from ketone body oxidation while needing less glucose from the liver. Two ketone bodies are acids and give rise to acidosis which is decreased by their elimination in urine. In a 24-hour fast, a 70-kg person hydrolyzes approximately 160 g of triacylglycerol, with release of 160 g of fatty acid and 16 g of glycerol. The glycerol is used for gluconeogenesis. Of the fatty acids, approximately 40 g is metabolized to ketone bodies in liver; the remainder and these ketone bodies provide for the energy needs of all other tissues.

22.7 Protein Synthesis and Nitrogen Homeostasis

Protein Synthesis and Proteins as Energy Source

While carbohydrates and lipids are essential for structural and functional needs of cells and organs, the principal changes in their fluxes reflect fuel homeostasis and energy requirements. Protein homeostasis is different because no protein serves purely as an energy storage form. All proteins—in addition to being potential sources of carbon for energy—serve some structural, catalytic, transport, or regulatory function. Many fluctuations observed in protein metabolism reflect changes in the need for the function of that protein. Thus, in starvation the total rate of skeletal muscle protein synthesis decreases but that of protein degradation increases because the amounts of proteases and enzymes involved in the catabolic routes for amino acids increase. Thus, starvation stimulates the specific synthesis of some proteins.

Protein anabolism and catabolism are governed by many forces. Some rules apply:

1. Which specific proteins are synthesized in a cell is governed entirely by the needs or functions of the cell at that time.
2. Synthesis of specific proteins is triggered by a specific signal.
3. Synthesis of a specific protein may not occur if the general "anabolic state" of the cell is low.
4. The anabolic state is reflected by metabolite availability and the presence of anabolic or catabolic hormones.
5. Synthesis occurs if the strength of the specific trigger overrides that of catabolic messages or other hormones or metabolites.
6. Degradation of a protein may occur if
   a) the protein has no function in the cell,
   b) a specific trigger to initiate degradation is present, or
   c) the state of the cell is sufficiently catabolic as a result of the presence or absence of specific metabolites or hormones.

Many hormones modulate the amounts and types of proteins synthesized and their rates of degradation. One group functions on very few target cells, e.g., TSH (thyroid-stimulating hormone) on the thyroid gland, FSH (follicle-stimulating hormone) and LH (luteinizing hormone) on the gonads, or ACTH (adrenocorticotropic hormone) on the adrenals. Another group has widespread action on protein metabolism, e.g., insulin, glucocorticoids, growth hormone, and thyroid hormone. Many sites of action of insulin and glucocorticoids remain to be elucidated. Glucocorticoids act at the gene level and promote peripheral tissue protein degradation, delivery of amino acids for gluconeogenesis, and adipocyte lipolysis.

Insulin stimulates peripheral tissue protein synthesis by stimulating amino acid uptake and protein synthesis at the level of translation and by inhibiting protein degradation. At low insulin levels, muscle proteolysis occurs. As the levels increase, proteolysis decreases and protein synthesis is favored. Exercise decreases proteolysis and increases protein synthesis, whereas disuse results in muscle wasting and depressed protein synthesis. Exercise increases sensitivity to insulin, whereas disuse makes the tissue insulin-resistant. Obesity, pregnancy, and glucocorticoids also increase insulin resistance.

Nitrogen Balance

Unlike most carbohydrate and lipid, protein contains nitrogen required for incorporation into a wide range of compounds (Table 22-4). Although the synthesis of proteins fixes nitrogen into those proteins, they are synthesized in direct response to specific needs. Dietary protein nitrogen is used for the synthesis of nitrogen-containing compounds in response to specific hormones suchas insulin.

TABLE 22-4
Compounds for Which Amino Acids Serve as Precursors

Amino Acid Precursor	Product
Glycine, aspartate, glutamine	Purines
Aspartate, glutamine	Pyrimidines
Glycine	Porphyrins
Arginine, methionine	Polyamines
Arginine, glycine	Creatine
Tryptophan	Serotonin, melatonin
Tyrosine	Catecholamines, dopamine, melanin
Histidine	Histamine
Arginine	Nitric oxide

Excess nitrogen is excreted primarily as urea (Chapter 17). Since nitrogen is essential but cannot be stored, the body attempts to maintain "nitrogen balance." In positive nitrogen balance, more nitrogen is ingested than is eliminated. This occurs during growth, pregnancy, lactation, and recovery from trauma, such as major surgery. If less nitrogen is ingested than is excreted, the body is in negative nitrogen balance. In times when dietary nitrogen is not available, negative nitrogen balance occurs because some nitrogen is always excreted as urea and ammonia.

Of the 20 amino acids in proteins, the body can readily synthesize eight if an appropriate nitrogen source is available. Two others can be synthesized from other amino acids of the diet: tyrosine from phenylalanine and cysteine from methionine. The rest must be provided in the diet (Chapter 17), since the body can synthesize none or an insufficient amount. The dietary requirement depends on several factors. Beside essential amino acids, the diet should provide the nitrogen required for synthesis of the nonessential amino acids.

Ammonia Toxicity

Ammonia, a metabolite formed from many nitrogen-containing compounds, is used for biosynthesis. High ammonia levels are toxic. The energy requirements of the brain, which is sensitive to high levels of ammonia, are met almost exclusively by glucose oxidation, for which a source of oxaloacetate is essential. This condition is satisfied by the carboxylation of pyruvate, even though the pyruvate carboxylase is somewhat limiting. For efficient TCA cycle activity, considerable recycling of oxaloacetate is necessary, since *de novo* synthesis is limited. With high levels of ammonia, the equilibrium for glutamate dehydrogenase favors formation of glutamate and glutamine, which diminishes the ammonia level but decreases TCA cycle activity. The brain is thus deprived of its source of ATP generation. In addition, glutamate and aspartate have neurotransmitter functions.

Within the liver, elimination of ammonia occurs via urea synthesis (Chapter 17). Since urea is uncharged, it does not disturb the acid-base balance. Many interorgan relationships in protein and nitrogen homeostasis arose because of the role that the liver plays in excess nitrogen excretion.

Nitrogen Transfer between Compounds and Tissues

Nitrogen is eliminated from the body as urea and ammonia. Urea synthesized in the liver is excreted by the kidneys. Urinary ammonia is produced in the kidney. The nitrogen in other tissues is transported to the liver and kidney in the form of amino acids. The general reactions involved in these processes are described below.

Methods for Directly Transferring Nitrogen

Nitrogen for biosynthetic processes is derived from the α-amino group of amino acids and from the amido group of glutamine and asparagine by transamination and transamidation, respectively (Chapter 17). Because of the wide variety of transaminases, transamination can provide the right balance of all nonessential amino acids and nitrogen-containing compounds derived from them. Transamidation is less ubiquitous. Other reactions utilizing the nitrogen of amino acids include the incorporation of glycine into purines and its partial incorporation into porphyrins.

Reactions in Which Ammonia Is Released

Ammonia is produced by oxidative and nonoxidative deaminations catalyzed by glutaminase and glutamate dehydrogenase (Chapter 17). Ammonia is also released in the purine nucleotide cycle. This cycle is prominent in skeletal muscle and kidney. Aspartate formed via transamination donates its α-amino group in the formation of AMP; the amino group is released as ammonia by the formation of IMP.

Reactions That "Fix" Ammonia

In muscle, ammonia is used to synthesize other nitrogen-containing compounds or is eliminated. In either case, glutamine and glutamate are formed by glutamine synthetase and glutamate dehydrogenase, respectively. The glutamate dehydrogenase reaction is readily reversible, but that of glutamine synthetase is not because it is driven by the hydrolysis of ATP. Glutamine synthesis is the major mechanism for nonhepatic tissues to eliminate ammonia. In liver, ammonia is fixed by the formation of urea (Chapter 17).

Disposition of Dietary Intake of Protein

Ingested protein is digested in a stepwise fashion in the stomach, small intestinal lumen, and small intestinal mucosal cells (Chapter 12). Peptides formed in the intestinal lumen are absorbed into the mucosal cells and degraded to free amino acids. The outflow of amino acids to the portal vein does not reflect the amino acid composition of the ingested protein. Thus, alanine levels increase two- to fourfold, and glutamine, glutamate, and aspartate are absent. These changes arise from amino acid interconversions within the intestinal cell.

With the exception of the branched-chain amino acids (valine, leucine, and isoleucine), most amino acids released from the intestine are metabolized by the liver. The liver carries out

1. Formation of other molecules from amino acid precursors, e.g., nonessential amino acids, purines, pyrimidines, porphyrins, and plasma proteins (such as albumin, VLDL, and transferrin), for its needs and those of other tissues;
2. Utilization of the amino acid carbon skeletons in gluconeogenesis or glycogen or lipid synthesis; and
3. Formation of urea.

After a protein meal, the plasma amino acids are those delivered from the intestine and those synthesized by the liver. The levels of branched-chain amino acids are greatly increased because they bypass the liver, reaching a peak at about 4 hours and returning to normal by 8–10 hours. They account for up to 60% of the total amino acids entering the systemic circulation. After uptake by skeletal muscle and transamination, followed by oxidation of the α-keto acid derivatives by the pathways described in Chapter 17, they can meet most energy requirements of muscle. The branched-chain amino acids have been called *amino fats*. Their α-amino nitrogen is converted to alanine, which is transported to liver for elimination as urea. Figure 22-23 summarizes the various pathways for amino acids following protein ingestion. Alanine is formed in intestine and muscle from many of the amino acids.

Following a protein meal, insulin secretion increases, which stimulates peripheral tissue uptake of amino acids and net protein synthesis. On a diet high in carbohydrate and low in protein, insulin levels are elevated and glucagon levels are depressed. On a diet high in protein and low in carbohydrate, insulin and glucagon levels are elevated. While amino acids and glucose stimulate insulin release, amino acids stimulate and glucose inhibits glucagon release. Thus, on a high-carbohydrate diet, insulin promotes hepatic and peripheral tissue utilization of glucose without the need for hepatic gluconeogenesis. Because of the low level of glucagon, hepatic glucose formation is depressed. On a high-protein, low-carbohydrate diet, insulin secretion is stimulated by amino acids and results in their uptake by peripheral tissue and utilization for glucose uptake. However, glucose homeostasis is maintained because glucagon stimulates hepatic glucose release.

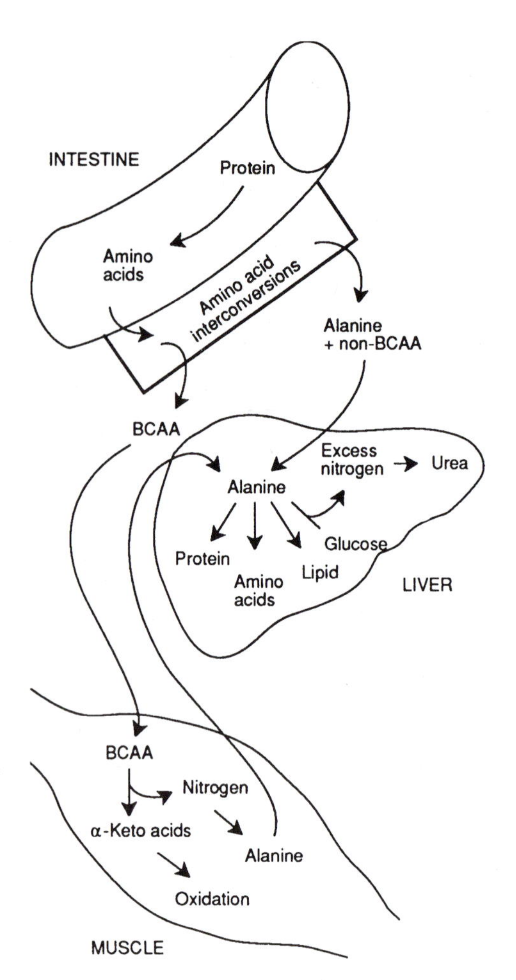

FIGURE 22-23
Amino acid metabolism following dietary protein intake. Digestion of protein produces amino acids. Within the intestinal cell, amino acid interconversions form alanine, which is delivered by the portal blood to liver, where it serves as the source of α-amino nitrogen and pyruvate, which is converted to lipid and glucose. Excess nitrogen is converted to urea. Branched-chain amino acids (BCAAs) are not taken up by the liver but enter peripheral tissues such as muscle where they serve as an important fuel source. Their α-amino nitrogen is transported to liver in the form of alanine.

Protein Catabolism during Starvation

During starvation, following depletion of hepatic glycogen, amino acids become the major source for glucose homeostasis because of the decrease in plasma insulin level and the rise in glucocorticoid level. The pattern of amino acids released by skeletal muscle during starvation does not reflect the composition of muscle protein. Alanine and glutamine account for over half of the amino acids released. Alanine is taken up by liver, its carbon chain converted to glucose, and the nitrogen to urea. In early starvation, the principal site of glutamine metabolism is the gut. One product is alanine. The special role of glutamine in gut may be due to the high demand for glutamine in purine synthesis because of the active shedding of intestinal cells. In long-term starvation, a major site of glutamine metabolism is the kidney, since the excretion of ketone bodies requires NH_4^+ as a counterion formed from ammonia produced by glutaminase. The resulting glutamate is

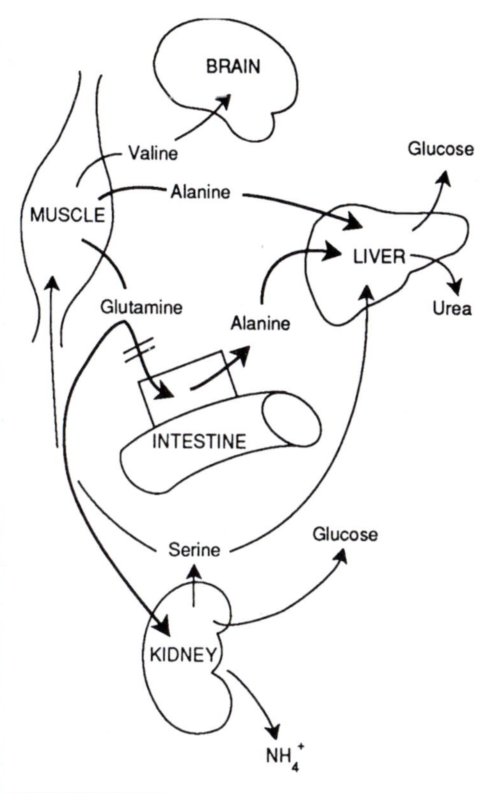

FIGURE 22-24
Postabsorptive metabolism of amino acids and the acid-base metabolic switch. In the early postabsorptive phase, the liver is the primary source of plasma glucose formed from alanine, delivered from muscle or intestine. The primary nitrogen product is urea. As starvation progresses, the need for acid-base balance results in decreased glutamine uptake by the splanchnic bed and increased uptake by the kidney. In the kidney, glutamine produces ammonium ion as counterion for β-hydroxybutyrate and acetoacetate. The carbon chain of glutamine is used in gluconeogenesis. If starvation progresses, the need for acid-base balance results in decreased glutamine uptake by the splanchnic bed and increased uptake by the kidney. The branched-chain amino acids serve as major substrates for muscle metabolism. Later in starvation, some of the branched-chain amino acids, notably valine, are released from muscle and can be used by the brain as a source of energy.

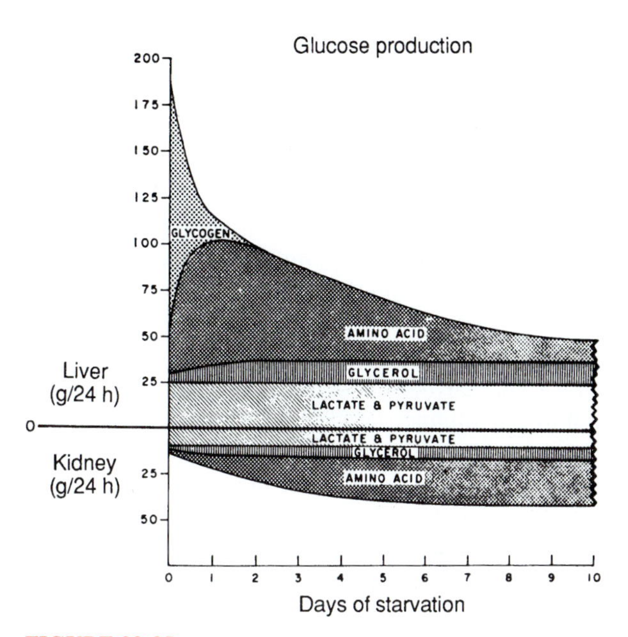

FIGURE 22-25
Sources of glucose production during starvation. [Reproduced with permission from F. J. Cahill, Physiology of insulin in man: *Diabetes* **20,** 783 (1971).]

utilized for renal gluconeogenesis, the activity of which is quantitatively equivalent to that of liver. The change in location of glutamine metabolism with long-term starvation has been termed the *acid-base metabolic switch.* Figure 22-24 summarizes the organ interrelationships in amino acid fluxes postprandially and during extended starvation. Figure 22-25 shows the sources and sites of glucose production during starvation. In early starvation, hepatic glycogen becomes depleted as gluconeogenesis increases. As starvation progresses, gluconeogenesis diminishes in the liver but increases in the kidney as the need for ammonia excretion increases. This switch is reflected by the nitrogen excretion products. In the fed state, urea predominates. In fasting, the total nitrogen excreted decreases, and as starvation progresses, urea excretion decreases while ammonia excretion increases, coinciding with the increased rate of renal gluconeogenesis.

22.8 Abnormalities of Metabolic Homeostasis

In this section two major health problems are discussed—diabetes mellitus and obesity.

Diabetes Mellitus

Diabetes mellitus is a syndrome consisting of a group of metabolic diseases all of which are characterized by hyperglycemia. The development of diabetes comprises several pathogenic processes involving autoimmune destruction of pancreatic β cells with consequent insulin deficiency and resistance to insulin action. The abnormalities of insulin deficiency or its action on target tissues lead to derangements in carbohydrate, lipid, and protein metabolism. The symptoms of hyperglycemia consist of polyuria, polydipsia, polyphagia, weight loss, blurred vision, and susceptibility to infections. The acute life-threatening complications of diabetes are ketoacidosis in the absence of insulin therapy or the nonketotic hyperosmolar coma (Chapter 39). Long-term complications of diabetes are development of retinopathy with potential loss of vision, nephropathy with end-stage renal disease, neuropathy with a potential for foot infections and amputation, and macrovascular disease such as atherosclerosis.

Etiologic Classification of Diabetes Mellitus

In the new system of classification the arabic numerals 1 and 2 are used in place of roman numerals I and II. The previously used terms, i.e., insulin-dependent diabetes (IDDM) and non-insulin-dependent diabetes (NIDDM), are no longer used.

Type 1 Diabetes Mellitus This disorder is primarily caused by autoimmune destruction of pancreatic β cells. However, in some patients evidence of autoimmunity may not be present and the cause of the destruction of the β cells may be undetermined (idiopathic). Type 1 diabetics are insulin-dependent and prone to ketoacidosis. The anti-β-cell antibodies may be directed against many different antigens but the major ones are insulin and glutamic acid decarboxylase (GAD). The antibodies appear in the plasma during the preclinical stage; in high-risk individuals the presence of the autoantibodies, e.g., antibodies to GAD can have predictive value in predicting development of diabetes mellitus. The inheritance of type 1 diabetes mellitus does not follow simple Mendelian inheritance and appears to be a polygenic disease with environmental factors playing a role in its initiation. It has been proposed that some infectious agents (e.g., coxsackie virus) may trigger the autoimmune destruction of β cells by molecular mimicry.

Type 2 Diabetes Mellitus This is the most prevalent form of diabetes and is characterized by both an insulin secretion defect and insulin resistance. Maturity-onset diabetes of the young (MODY), attributable to mutations of the glucose kinase gene (discussed earlier), may also be classified as type 2 diabetes mellitus. Obesity is a contributory factor and may predispose to insulin resistance with eventual development of type 2 diabetes mellitus. The precise mechanism by which obesity leads to insulin resistance in the target tissues is not understood. However, in several animal models (e.g., *ob/ob* mouse, *db/db* mouse) mutations have been identified that cause both obesity and diabetes mellitus. Unlike type 1 diabetes mellitus, type 2 is not an autoimmune disease. Studies with monozygotic twins have revealed a 90% concordance rate for type 2 diabetes mellitus, suggesting the involvement of genetic factors in the development of the disease.

Gestational Diabetes Mellitus (GDM) Glucose intolerance that is first recognized during pregnancy is classified as GDM. The prevalence of GDM varies among ethnic groups. Studies of predominately white women have shown that the prevalence of GDM is between 4% and 7%. GDM is associated with an increase in maternal and fetal neonatal morbidity. The clinical risks may include macrosomia (excessive fetal growth) associated with complications of labor and delivery, jaundice, respiratory distress syndrome, hypocalcemia, and polycythemia. Thus, proper diagnosis (discussed later) with appropriate intervention and treatment is required during the antepartum period of GDM. The causes of macrosomia in GDM is attributed to multiple disturbances that occur in both maternal and fetal metabolism. Fetal hyperinsulinemia due to increased delivery of maternal glucose may be one of the contributory factors. Two additional risks associated with GDM have been identified in epidemiological studies:

1. Persons exposed to GDM *in utero* have an increased risk of obesity, and
2. GDM increases the risk of development of type 2 diabetes mellitus between 5 and 16 years after the index pregnancy.

Specific Mutations That Cause Diabetes Mellitus This group of disorders is relatively uncommon and include mutations in genes that affect various aspects of β-cell function including mitochondrial DNA and the insulin receptor gene; these disorders have been discussed earlier.

Endocrinopathies Excessive production of insulin counterregulatory hormones can cause diabetes mellitus. Examples include excessive production of growth hormone (*acromegaly*), cortisol (*Cushing's syndrome*), epinephrine (*pheochromocytoma*), and glucagon (*glucagonoma*).

Miscellaneous Causes of Diabetes Mellitus These include diseases of the exocrine pancreas (e.g., *cystic fibrosis, hemochromatosis, pancreatitis*) and drug-induced causes (e.g., thiazides, glucocorticoids).

Diagnosis of Diabetes Mellitus Since diabetes mellitus is a heterogeneous group of metabolic disorders characterized by hyperglycemia, the diagnosis involves first excluding other causes of hyperglycemia. For example, all of the endocrinopathies that have other pathognomonic features, such as acromegaly, Cushing's syndrome, glucagonoma, and pheochromocytoma, should be identified with appropriate diagnostic procedures. After excluding all such causes of hyperglycemia, the diagnosis of type 1 or type 2 diabetes mellitus should be considered. The diagnostic criteria for diabetes mellitus are given

TABLE 22-5
Criteria for the Diagnosis of Diabetes Mellitus

1. Symptoms of diabetes plus casual plasma glucose concentration ≥200 mg/dl (11.1 mmol/l). Casual is defined as any time of day without regard to time since last meal. The classic symptoms of diabetes include polyuria, polydipsia, and unexplained weight loss.

or

2. FPG ≥126 mg/dl (7.0 mmol/l). Fasting is defined as no caloric intake for at least 8 h.

or

3. 2hPG ≥200 mg/dl during an OGTT. The test should be performed as described by WHO using a glucose load containing the equivalent of 75-g anhydrous glucose dissolved in water.

In the absence of unequivocal hyperglycemia with acute metabolic decompensation, these criteria should be confirmed by repeat testing on a different day. The third measure (OGTT) is not recommended for routine clinical use.

FPG = Fasting plasma glucose; OGTT = Oral glucose tolerance test; 2hPG = 2 hour post load glucose. The diagnostic criteria are based upon the report by the Expert committee on the diagnosis and classification of Diabetes mellitus. *Diabetes Care* **24**(suppl. 1), 55 (2001).

in Table 22-5. Subjects who have fasting plasma glucose levels between ≥110 mg/dL and <126 mg/dL are said to have impaired glucose tolerance and may require additional as well as periodic testing.

The diagnosis of gestational diabetes mellitus consists of two parts. A screening glucose challenge test is performed between 24 and 28 weeks of gestation after the oral administration of 50-g of glucose. This test is performed at any time of the day and irrespective of food intake. If the plasma glucose after the 50-g challenge is ≥140 mg/dL (7.8 mmol/L), the test is considered positive and requires a diagnostic test consisting of oral ingestion of 100 g of glucose after a fasting for 8 hours, followed by 3-hour oral glucose tolerance test (Table 22-6). In the 50-g challenge test, if the threshold for the abnormality is decreased to ≥130 mg/dL (7.2 mmol/L), the sensitivity for identifying patients with gestational diabetes mellitus increases from 80% to 90%.

Since diabetes mellitus is an insidious disorder, testing of asymptomatic patients may be desirable under certain conditions, including age 45 years or older; obesity; first-degree relatives of diabetics; members of high-risk ethnic population (e.g., Native American, Hispanic, African-American); women who have delivered an infant weighing more than 9 lb (4.08 kg) or have had gestational diabetes mellitus; hypertension; abnormal lipid studies; recurring skin, genital, or urinary tract infections; and previous impaired glucose intolerance.

TABLE 22-6
Screening and Diagnosis Scheme for Gestational Diabetes Mellitus (GDM)

Plasma Glucose	50-g Screening Test	100-g Diagnostic Test
Fasting	—	95 mg/dl
1-h	140 mg/dl	180 mg/dl
2-h	—	155 mg/dl
3-h	—	140 mg/dl

Screening for GDM may not be necessary in pregnant women who meet all of the following criteria: <25 years of age, normal body weight, no first degree relative with diabetes, and not Hispanic, Native-American, Asian-, or African-American. The 100-g diagnostic test is performed on patients who have a positive screening test. The diagnosis of GDM requires any two of the four plasma glucose values obtained during the test to meet or exceed the values shown above. To convert values for glucose to mmol/L, multiply by 0.05551. The diagnostic criteria is based upon the report by the American Diabetes Association on Gestational Diabetes Mellitus. *Diabetes Care* **23**(suppl. 1), S77 (2000).

After the diagnosis of diabetes mellitus has been made and with the initiation of appropriate therapy (discussed later), assessment of other biochemical parameters is necessary in the management phase of the disorder to maintain the fasting blood glucose level as close to normal as possible and to prevent long-term complications. These biochemical tests include measurement of a stable form of glycosylated hemoglobin (hemoglobin A_{1c}), determination of urine albumin excretion rate, measurement of serum fructosamine levels, and self-monitoring of blood glucose levels. Hemoglobin A_{1c} levels are used to assess average glucose control over a 2 to 3-month period, since the red blood cell's life span is about 120 days. Entry of glucose into red blood cells depends only on the prevailing plasma glucose concentration. (Formation of hemoglobin A_{1c}, which is nonenzymatic, is discussed in Chapters 2 and 10). The normal hemoglobin A_{1c} concentration is about 4–6%; spurious values for hemoglobin A_{1c} levels can occur in uremic states, hemoglobinopathies (Chapter 28), hemolytic anemia and blood transfusion. Fructosamine is a generic term applied to the stable condensation product of glucose with serum proteins (albumin, with a circulating half-life of about 20 days, is the major contributor). Thus, serum fructosamine levels reflect glucose control over a period of 2–3 weeks.

One of the chronic complications of diabetes mellitus is diabetic nephropathy, which leads to end-stage renal disease. An initial biochemical parameter of diabetic nephropathy in the asymptomatic state is a persistent increase in urine albumin excretion rate between 20 and 200 μg/min (or 30–300 mg/d). This degree of albumin loss in the urine is called microalbuminuria and is a harbinger of renal failure and other complications of diabetes. It is important to identify individuals with microalbuminuria because with appropriate therapeutic intervention attenuation of loss of renal function can be accomplished. Therapeutic interventions include blood glucose control, treatment for high blood pressure if present, and the inhibition of formation of angiotensin II by use of angiotensin-converting enzyme (ACE) inhibitors. The formation of angiotensin II involves the following steps: release of the protease renin by justaglomerular cells in response to decreased renal perfusion pressure, release of angiotensin I from angiotensinogen by the action of renin, and conversion of angiotensin I to angiotensin II by ACE. Physiologic functions of angiotensin II in the kidney include restoration of normal renal blood flow and glomerular filtration and stimulation of aldosterone secretion, which affects Na^+ reabsorption (Chapter 32). In diabetes mellitus, renal hemodynamics are altered due to glomerular hyperfiltration. Angiotensin II has been postulated to contribute to the progression of renal disease. Some of the deleterious effects of angiotensin II may involve its action on vasomotor functions and induction of cytokines (e.g., TGF-β, see Chapter 35) that lead to hypertrophy of mesangial cells. Thus, ACE inhibitors as well as angiotensin II receptor antagonists can prevent or slow the rate of progression of renal disease in diabetes mellitus. Clinical studies have shown that even in normotensive diabetic patients, ACE inhibitors are beneficial in the management of renal disease.

Chronic Complications of Diabetes Mellitus These stem from elevated plasma glucose levels *per se* and involve tissues that do not require insulin (e.g., lens, retina, peripheral nerve) for the uptake and metabolism of glucose. In these tissues, the intracellular level of glucose parallels that in plasma. The chronic complications, which cause considerable morbidity and mortality, are atherosclerosis, microangiopathy, retinopathy, nephropathy, neuropathy, and cataract. The biochemical basis of these abnormalities may be attributed to increased tissue ambient glucose concentration and may involve the following mechanisms: nonenzymatic protein glycation (Chapter 2), increased production of sorbitol, and decreased levels of *myo*-inositol. The ramifications of glycation of proteins are not clear. Sorbitol is produced from glucose by NADPH-dependent reduction catalyzed by aldose reductase. Sorbitol may also be converted to fructose by NAD^+-dependent oxidation. Sorbitol (and fructose) diffuses poorly across cell membranes and accumulates inside the cell causing osmosis-induced disturbances (e.g., cataract formation). The *myo*-inositol depletion may be due in part to the competition of glucose with *myo*-inositol for its intracellular transport; glucose and *myo*-inositol are strikingly similar in structure (Chapter 9). *Myo*-inositol can also be synthesized from glucose-6-phosphate. Decreased *myo*-inositol may lead to decreased phosphoinositide turnover. The latter yields at least two active catabolites (inositol polyphosphates and diacylglycerol) that function as second messengers (Chapter 30). The diminished activity of Na^+,K^+-ATPase found in nerve fibers is thought to be related to the altered phosphoinositide turnover. Potential therapeutic use of *myo*-inositol supplementation and aldose reductase inhibitor administration is being explored.

Management of Diabetes Mellitus The primary goal in the management of all types of diabetes mellitus (type 1, type 2, and GDM) is to maintain near-normal plasma glucose levels in order to relieve symptoms (polydipsia, polyuria, polyphagia) and to prevent both acute and chronic complications. The glycemic control is assessed by monitoring of glucose (by self and in clinical settings), hemoglobin A_{1c}, fructosamine, and microalbuminuria. In type 1 diabetes mellitus due to β-cell destruction, administration of insulin is required throughout the person's lifetime. There are many insulin preparations that differ in duration of action (ultrashort, short, intermediate, and long acting) and in their origin (human, porcine, and bovine). Human insulin analogues *lispro* and *aspart* (B28) do not undergo polymerization and are rapid acting insulins. The side effects of insulin therapy include hypoglycemia and weight gain.

The management of type 2 diabetes mellitus is based on both behavioral changes with lifestyle modifications and pharmacological measures. Diet, exercise, and weight loss (in the obese) are the cornerstone of treatment in maintaining euglycemia. Physical activity stimulates insulin sensitivity, whereas physical inactivity leads to insulin resistance. In the absence or failure of lifestyle and behavioral measures in the management of type 2 diabetes mellitus, pharmacological therapy is employed. Pharmacological agents used in the treatment of type 2 diabetes mellitus include insulins, sulfonylureas, α-glucosidase inhibitors, biguanidines, and thiazolidinediones (Figure 22-26). Sulfonylureas bind to specific β-cell plasma membrane receptors and cause closure of the ATP-dependent K^+ channel with accompanying depolarization. The depolarization leads to influx of Ca^{2+} followed by insulin secretion. A benzoic acid derivative (repaglinide) is structurally similar to sulfonylurea without the sulfur

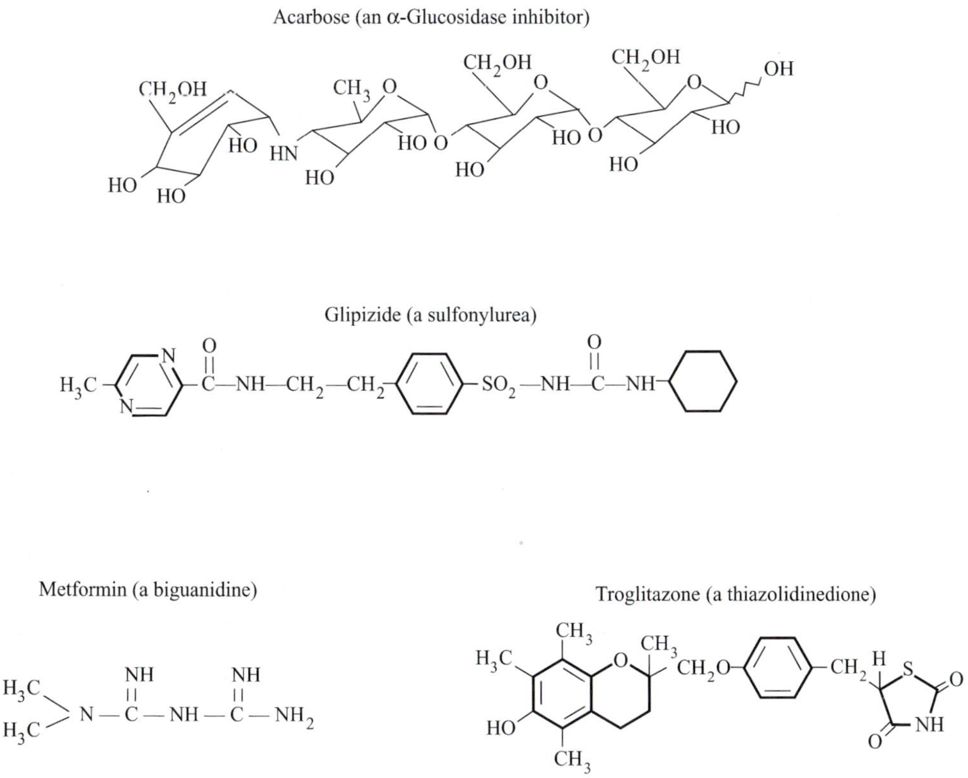

Acarbose (an α-Glucosidase inhibitor)

Glipizide (a sulfonylurea)

Metformin (a biguanidine)

Troglitazone (a thiazolidinedione)

FIGURE 22-26
Structures of four classes of hypoglycemic agents.

group and belongs to the class meglitinides. This drug inhibits a different K^+ channel than SUR in promoting secretion and has rapid-onset and short-acting actions. An example of an α-glucosidase inhibitor is acarbose, a nitrogen-containing pseudotetrasaccharide. Acarbose inhibits the breakdown of carbohydrates into glucose by inhibiting intestinal brush-border α-glucosidase and pancreatic amylase (Chapter 12) and brings about a reduction in postprandial hyperglycemia. Metformin belongs to the class of biguanidines and has many effects on carbohydrates and lipid metabolism. Its mechanism of action may include inhibition of hepatic gluconeogenesis and glycogenolysis and, in muscle, increased insulin receptor tyrosine kinase activity and enhanced GLUT4 transporter system. An adverse effect of metformin is lactic acidosis.

Troglitazone is an example of the thiozolidinedione class of antidiabetic agents. It enhances insulin sensitivity in liver, muscle, and adipose tissue. Troglitazone promotes the conversion of non-lipid-storing preadipocytes to mature adipocytes (discussed later) with increased insulin sensitivity. The cellular target for the action of troglitazone, which functions as a transcription factor, is the nuclear receptor known as peroxisome proliferation-activated receptor-γ (PPAR-γ). PPAR-γ upon binding with troglitazone is activated and forms a heterodimer with retinoid X receptor (RXR). The PPAR-γ/RXR heterodimer regulates the transcription of genes encoding proteins that control metabolic pathways (e.g., acetyl CoA-synthetase, lipoprotein lipase, GLUT4, mitochondrial uncoupling protein). An adverse effect of troglitazone is hepatic toxicity; periodic monitoring of liver function by measurement of serum alanine aminotransferase is recommended. When a diabetic patient becomes refractory to monotherapy, combination therapy (e.g., sulfonylurea and insulin) is used to achieve glycemic control.

Obesity

Overweight and obesity are defined based on the body mass index (BMI), which is calculated as weight in kilograms divided by the square of height in meters ($BMI = kg/m^2$). A BMI between 25 and 29.9 is considered as overweight and ≥ 30 is considered obese (Chapter 6). Body weight is determined by a balance between energy intake and energy expenditure. The energy expenditure is required to maintain basal metabolic functions, absorption and digestion of foods (thermic effect of food), physical activity and, in children, linear growth and development. If there is a net excess energy intake, BMI increases, eventually leading to overweight and ultimately to obesity and morbid obesity. In a population with stable genetic factors, increase in obesity is primarily attributable to consumption of excess foods with high fat content and decreased physical activity. Obesity is a worldwide health problem. It is a risk factor for development of diabetes mellitus, hypertension, and heart disease, all of which cause decreased quality of life and life expectancy.

Feeding behavior and energy balance are regulated by a complex set of short-term and long-term physiological signals. Mechanisms that lead to obesity involve interactions between genetic, environmental, and neuroendocrine factors. The short-term regulators of hunger and satiety may include plasma levels of glucose and amino acids, cholecystokinin and other hormones, and body temperature. One of the long-term regulators of food intake and energy expenditure is an adipocyte-derived hormone leptin (Chapter 5). Leptin is a polypeptide of 167-amino acid residues that functions in the afferent signal pathway of a negative feedback loop in regulating the size of adipose tissue and energy balance. The physiological role of leptin was established by using *ob/ob* and *db/db* mice that have identical phenotypes of obesity and diabetes mellitus. In the *ob/ob* mice, leptin levels are low due to mutations in the leptin gene, and in the *db/db* mice the leptin receptors are defective due to inactivating mutations. The role of leptin and its receptor was investigated by joining the circulatory systems (parabiosis) of *ob/ob* and *db/db* mice. In *ob/ob* mice, obesity, diabetes, and sterility are corrected by leptin administration. These animal studies have led to understanding of several aspects of obesity in humans.

In humans, leptin is secreted in a pulsatile manner, in a nyctohemeral rhythm, and in proportion to size of the adipose tissue. The pathway of leptin's action involves the neuroendocrine system of the hypothalamus (Figure 22-27). When leptin levels are low, appetite increases via the activation of the neuropeptide Y

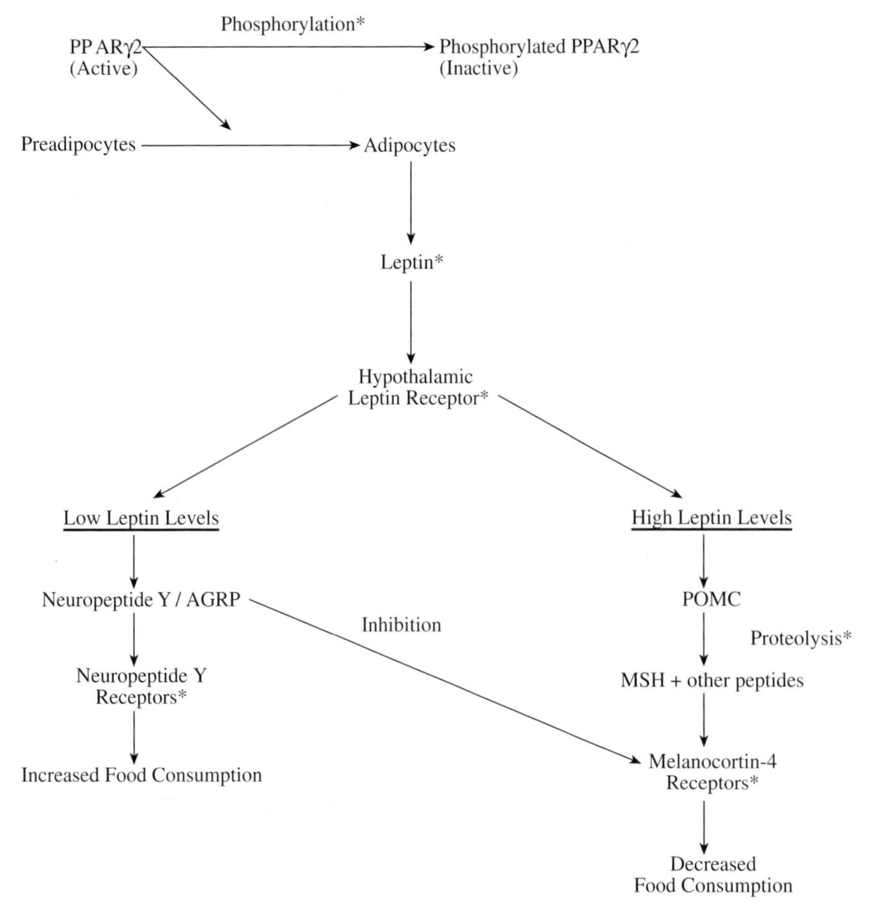

FIGURE 22-27
A model for the regulation of human food consumption. Asterisks (*) indicate obesity-causing mutations. Abbreviations are given in the text.

pathway, eventually leading to stimulation of food intake. Neuropeptide Y contains 36 amino acid residues and has tyrosine at both the N and C termini (hence the designation Y). The actions of neuropeptide Y result in the stimulation of food intake followed by the restoration of depleted adipose stores and inhibition of sympathetic nervous system outflow with a consequent decrease in energy expenditure. When leptin levels are high, food intake decreases; this is mediated by the pro-opiomelanocortin (POMC)-α-melanocyte stimulating hormone and melanocortin-4 receptor pathway. Inhibition of the melanocortin-4 receptor signal transduction system by the agouti protein leads to obesity in mice. Agouti protein (AGRP) is normally synthesized in hair follicles and antagonizes the action of α-melanocyte-stimulating hormone in eumelanin synthesis (Chapter 17) by binding to melanocortin-4 receptor. In autosomal-dominant mice that overexpress agouti protein in skin and hypothalamus, the phenotype is obese, yellow mice.

In humans, the pathways of regulation of appetite and food consumption have been verified by identifying mutations in genes that participate in regulation of body weight (Figure 22-27). The regulators of conversion of preadipocytes to adipocytes also determine the adipose tissue mass. One of the regulators is a transcription factor known as peroxisome proliferator-activated receptor-γ2 (PPAR-γ2). Troglitazone, a hypoglycemic agent (previously discussed) that increases insulin sensitivity, is a synthetic ligand for PPAR-γ2. However, contrary to studies in rodents, administration of troglitazone in humans does not cause significant weight gain as compared to other oral antidiabetic drugs. Phosphorylation of PPAR-γ2 at ser114 makes the receptor less active in converting preadipocytes to adipocytes. A mutation converting proline to glutamine at position 115 blocks phosphorylation at the serine site, maintains the protein in the active state, and leads to obesity.

Leptin therapy has corrected obesity in a child with congenital leptin deficiency. In obese individuals, the presence of circulating high levels of leptin has been attributed to resistance or some other defect in the leptin receptors. This apparent paradox of high leptin levels associated with obesity is analogous to insulin resistance seen in type 2 diabetes mellitus. In general, in the vast majority of obese patients, the molecular defects remain unknown. Diet and exercise are the mainstays in the management of obesity.

22.9 Metabolic Homeostasis during Exercise

At rest, skeletal muscle utilizes the catabolism of fatty acids and branched-chain amino acids to maintain cellular integrity. In exercise, oxygen utilization by muscle may increase 30-fold, and additional energy can arise from anaerobic glycolysis. The substrates used depend on the intensity of exercise and on training. High-intensity exercise can start almost instantaneously (e.g., the 100-m dash or the clean-and-jerk in weight lifting), but increase of oxygen utilization is limited by the rates of blood flow, respiration, and oxygen delivery to the tissue. After the onset of high-intensity exercise, the body and muscle are initially in oxygen deficit until a new steady-state, increased rate of oxygen uptake can be achieved. When the exercise ends, this oxygen debt is repaid during the recovery phase. The availability of oxygen determines the metabolic fuels utilized. Skeletal muscle has fewer mitochondria per unit weight than heart or liver. A major adaptation induced by endurance exercise training is an increase in the potential for delivery and utilization of oxygen: the heart hypertrophies; total blood volume increases; muscle capillarity increases; and skeletal muscle mitochondrial size, number, and capacity increase (see Chapter 21).

Exercise Generating Maximum Power

The substrates for muscular activity are intracellular ATP, phosphocreatine, glycogen, plasma glucose, fatty acids, ketone bodies, triacylglycerols, and dietary branched-chain amino acids. In exercise that generates maximum power, the primary sources are ATP and phosphocreatine. The total amount of ATP, and of ATP generated by the adenylate kinase reaction (2ADP $\rightarrow$ ATP + AMP), is sufficient for only a fraction of a second. The immediate reserve is ATP produced by the creatine kinase reaction (phosphocreatine + ADP $\rightarrow$ ATP + creatine). Maximum-power-output exercise rapidly depletes phosphocreatine, can be sustained for short periods (15–20 seconds) and ends as phosphocreatine is depleted. Some anaerobic glycolysis from glycogen may occur during such a period but would reach a maximum rate only after the work had stopped. Maximum-power-output work exceeds the capacity of muscle to generate energy oxidatively even if this could be initiated rapidly.

High-Intensity Endurance Exercise

This exercise is typified by an individual working in the range of 60–80% of maximum capacity of oxygen uptake (e.g., running distances of 5 km up to a marathon). The exercise is terminated by exhaustion. Phosphocreatine is the initial energy source; glycogenolysis ensues, anaerobically at first, with lactate production, but with increasing aerobic oxidation as oxygen availability increases. Glycogen utilization is the major fuel for the first 15 minutes, but it is superseded by fatty acid and plasma glucose utilization. Most of the plasma glucose used is oxidized entirely but 10–20% is converted to lactate. During the exercise and depending on the past dietary history, branched-chain amino acid oxidation contributes some energy, and alanine is formed to eliminate the α-amino nitrogen. Lactate and alanine are used in the liver for gluconeogenesis and can provide 5–15% of the glucose required by skeletal muscle. The rest of the glucose is derived from glycogenolysis. Thus, in endurance exercise, hepatic glycogen serves as a major fuel for muscle, and muscle glycogen is the initial fuel. As plasma glucose and fatty acids are used, glycogenolysis in muscle diminishes; despite its smaller contribution, muscle glycogen plays a critical role since exhaustion occurs with its total depletion in spite of the availability of fatty acids and plasma glucose. Thus, "glycogen loading" is used to extend endurance. Exercise-depleted glycogen stores can be increased to above-normal levels for about 24 hours by high-carbohydrate refeeding. With elevated levels of tissue glycogen, an individual's endurance ability is increased. Several other phenomena contribute to the pattern of metabolite utilization in high-intensity endurance exercise.

1. Although fatty acids are avidly utilized by skeletal muscle during exercise, their plasma concentrations do not differ significantly from those at rest. Their utilization is being balanced by production from adipocyte lipolysis, probably as a result of an elevated catecholamine level.
2. Glucose uptake by skeletal muscle increases with intensity of exercise even though levels of circulating insulin decrease. Either this glucose transport is insulin-independent or exercise enhances insulin sensitivity.
3. The increase in glucose uptake with increased intensity of exercise is matched by hepatic glucose production to the extent that arterial glucose concentrations are elevated by very intense exercise. Hepatic glycogenolysis probably is enhanced by a decrease in the level of insulin and by an increase in levels of glucagon and epinephrine.
4. Exercise may lower insulin levels even though plasma glucose concentrations are increased. This effect may result from α-adrenergic inhibition of β-cell secretion.

Low-Level Nonfatiguing Exercise

This exercise is characteristic of an individual in normal occupational tasks that could reasonably be continued for 8–12 hours. The principal substrates used are similar to those in long endurance exercise but without depletion of phosphocreatine and with minimal muscle glycogen utilization. The main source of energy is the aerobic oxidation of fatty acids, glucose, and branched-chain amino acids.

Supplemental Readings and References

P. H. Bennett, S. Haffner, B. Kasiske, et al.: Diabetic renal disease recommendations. *American Journal of Kidney Disease* **25,** 107 (1995).

E. E. Calle, M. J. Thun, J. M. Petrelli, et al.: Body-mass index and mortality in a prospective cohort of U.S. adults. *New England Journal of Medicine* **341,** 1097 (1999).

G. W. Cline, K. F. Petersen, M. Krssak, et al.: Impaired glucose transport as a cause of decreased insulin-stimulated muscle glycogen synthesis in type 2 diabetes. *New England Journal of Medicine* **341,** 240 (1999).

S. Dagogo-Jack and J. V. Santiago: Pathophysiology of type 2 diabetes and modes of action of therapeutic interventions. *Archives of Internal Medicine* **157,** 1802 (1997).

R. DeFronzo: Pharmacologic therapy for type 2 diabetes mellitus. *Annals of Internal Medicine* **131,** 281 (1999).

Expert Committee on the Diagnosis and Classification of Diabetes Mellitus: Report of the expert committee on the diagnosis and classification of diabetes mellitus. *Diabetes Care* **20,** 1183 (1997).

I. S. Farooqi, S. A. Jebb, G. Langmack, et al.: Effects of recombinant leptin therapy in a child with congenital leptin deficiency. *New England Journal of Medicine* **341,** 879 (1999).

F. L. Ferris III, M. D. Davis, and L. W. Aiello: Treatment of diabetic retinopathy. *New England Journal of Medicine* **341,** 667 (1999).

J. M. Friedman and J. L. Halaas: Leptin and the regulation of body weight in mammals. *Nature* **395,** 763 (1998).

B. Glaser, P. Kesavan, M. Heyman, et al.: Familial hyperinsulinism caused by an activating glucokinase mutation. *New England Journal of Medicine* **338,** 226 (1998).

S. B. Heymsfield, A. S. Greenberg, K. Fujioka, et al.: Recombinant leptin for weight loss in obese and lean adults. *Journal of American Medical Association* **282,** 1568 (1997).

F. Holleman and J. B. L. Hoekstra: Insulin lispro. *New England Journal of Medicine* **337,** 176 (1997).

S. J. Hunter and W. T. Garvey: Insulin action and insulin resistance: diseases involving defects in insulin receptors signal transduction, and the glucose transport effector system. *American Journal of Medicine* **105,** 331 (1998).

S. L. Kjos and T. A. Buchanan: Gestational diabetes mellitus. *New England Journal of Medicine* **341,** 1749 (1999).

A. Must, J. Sadano, E. H. Coakley, et al.: The disease burden associated with overweight and obesity. *Journal of American Medical Association* **282,** 1523 (1999).

O. E. Owen, K. J. Smalley, D. A. D'Alessio, et al.: Protein, fat, and carbohydrate requirements during starvation: anaplerosis and cataplerosis *American Journal of Clinical Nutrition* **68,** 12 (1998).

A. Rebollo and C. Martinez-A: *Ras* protein: recent advances and new functions. *Blood* **94,** 2971 (1999).

M. Ristow, D. Miller-Wieland, A. Pfeiffer, et al.: Obesity associated with a mutation in a genetic regulator of adipocyte differentiation. *New England Journal of Medicine* **339,** 953 (1998).

E. Ritz and S. R. Orth: Nephropathy in patients with type 2 diabetes mellitus. *New England Journal of Medicine* **341,** 1127 (1999).

J. N. Roemmich and A. D. Rogol: Role of leptin during childhood growth and development. *Endocrinology and Metabolism Clinics of North America* **28,** 749 (1999).

M. W. Schwartz and R. J. Seeley: Neuroendocrine responses to starvation and weight loss. *New England Journal of Medicine* **336,** 1802 (1997).

P. R. Shepherd and B. B. Kahn: Glucose transporters and insulin action. *New England Journal of Medicine* **341,** 248 (1999).

C. A. Stanley, Y. K. Lieu, B. Y. L. Hsu, et al.: Hyperinsulinism and hyperammonemia in infants with regulatory mutations of the glutamate dehydrogenase gene. *New England Journal of Medicine* **338,** 1352 (1998).

J. A. Tamada, S. Garg, I. Jovanich, et al.: Noninvasive glucose monitoring. *Journal of the American Medical Association* **282,** 1839 (1999).

A. H. Xiang, R. K. Peters, E. Trigo, et al.: Multiple metabolic defects during late pregnancy in women at high risk for type 2 diabetes. *Diabetes* **48,** 848 (1999).

J. A. Yanovski and S. Z. Yanovski: Recent advances in basic obesity research. *Journal of American Medical Association* **282,** 1504 (1999).

CHAPTER 23

Nucleic Acid Structure and Properties of DNA

The genetic information in all living cells is carried in molecules of deoxyribonucleic acid (DNA), which is primarily found in chromosomes. However, DNA is also present in cellular organelles such as mitochondria and chloroplasts. Viruses carry genetic information in either DNA or RNA (ribonucleic acid) molecules. When RNA viruses infect cells, the genetic information in RNA is converted to DNA prior to the replication and synthesis of new viral particles. In the case of certain RNA viruses (retroviruses) such as human immunodeficiency virus (HIV), the DNA that is copied from the infecting RNA is permanently integrated into the host chromosomes and the viral genome becomes an integral part of the cells' genetic information.

The structure of DNA was elucidated by James Watson and Francis Crick in 1953. The structure they proposed made it apparent for the first time how genetic information in chemically stored in cells and how it is replicated and transmitted from one generation to the next. The proposed DNA structure also provided insight into the chemical nature of mutations and how they might occur during replication. Scarcely fifty years has passed since that monumental discovery and now we are deciphering the complete sequence of a human genome. This will lead to a chemical understanding of thousands of genetic disorders and the ability to diagnose, prevent, and treat many inherited diseases and cancers.

In this chapter we discuss the structure and properties of DNA. Subsequent chapters examine the functions of DNA such as replication, mutation, recombination, repair, and the expression of genes.

23.1 Components of Nucleic Acids

A nucleic acid is a ***polynucleotide***—a linear polymer of nucleotides that is defined by its three components (Figure 23-1):

1. A nitrogenous heterocyclic base (either a purine or a pyrimidine) attached to the 1′-carbon atom of the sugar by an N-glycosidic bond. In DNA the purine bases are adenine (A) and guanine (G) and the pyrimidine bases are cytosine (C) and thymine (T). The bases in RNA are the same except that uracil (U), a pyrimidine, replaces thymine.
2. A cyclic five-carbon sugar which is ribose for RNA and deoxyribose for DNA. The difference in structure between ribose and 2′-deoxyribose is shown in Figure 23-1. The carbon atoms of the sugar are numbered with a prime to distinguish them from the carbon atoms in the base in the same nucleotide structure.
3. A phosphate group attached to the 5′-carbon atom of the sugar by a phosphodiester linkage. This phosphate group is responsible for the strong negative charge of both nucleotides and nucleic acids.

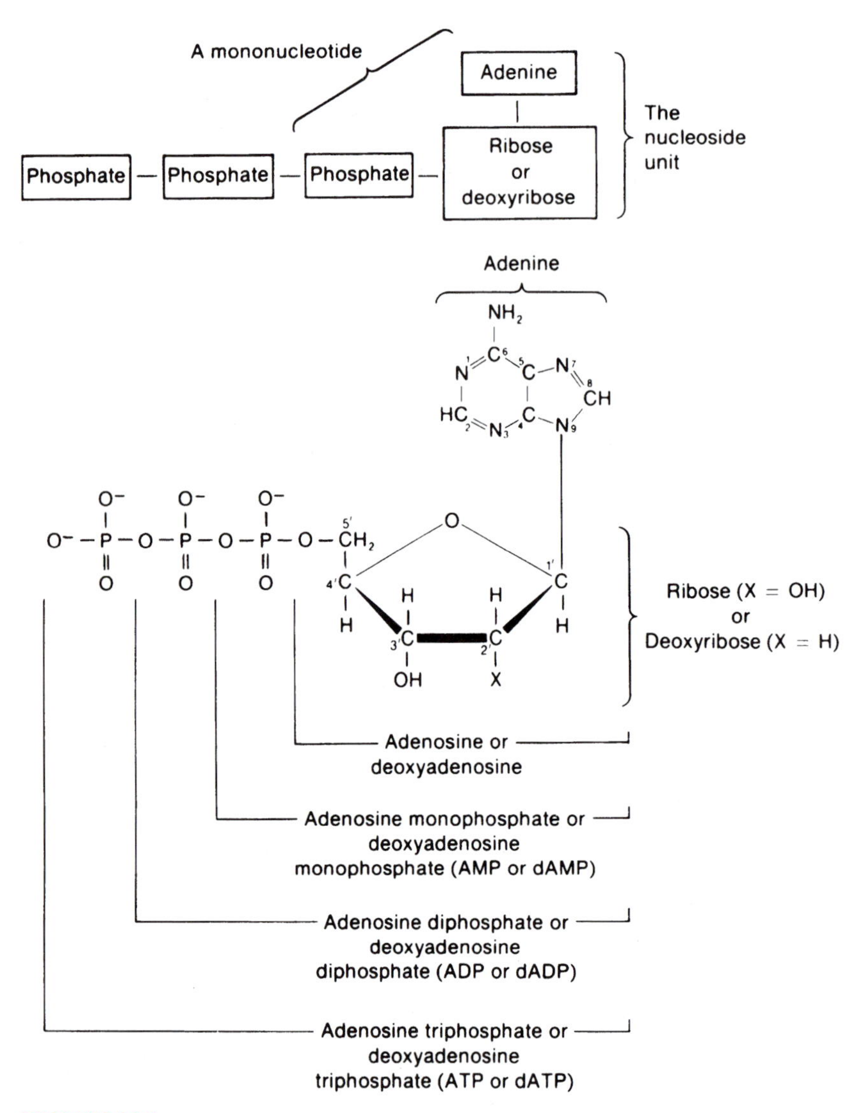

FIGURE 23-1
Structure of the nucleotide in DNA and RNA. The base adenine is used as an example. See Table 27-1 for a complete nomenclature of nucleosides and nucleotides.

A base linked to a sugar is called a ***nucleoside;*** a base linked to a sugar linked to a phosphate is called a ***nucleotide*** or a nucleoside phosphate. The nucleotides in nucleic acids are joined to one another by a second phosphodiester bond that joins the 5′-phosphate of one nucleotide to the 3′-OH group of the adjacent nucleotide. Such a doubly esterified phosphate is called a ***phosphodiester group*** (Figure 23-2).

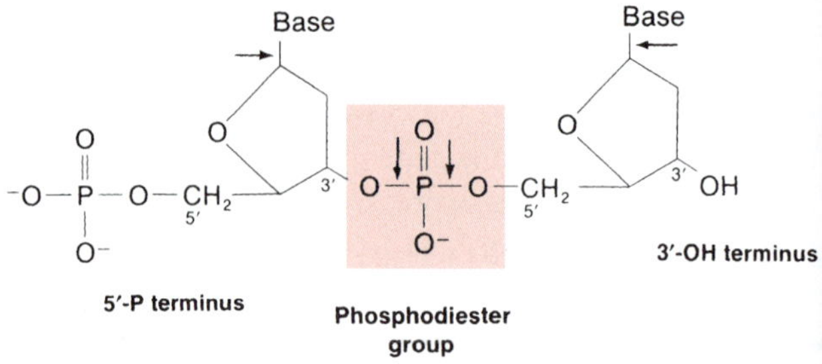

FIGURE 23-2
Structure of a phosphodiester group. The vertical arrows show the bonds in the phosphodiester group that are free to rotate. The horizontal arrows show the N-glycosidic bond about which the base can rotate freely. A polynucleotide consists of many nucleotides linked together by phosphodiester groups.

Base Pairing and Base Composition

The molar content of the four bases in DNA (called the base composition) always satisfies the equalities [A] = [T] and [G] = [C] where [ ] denotes molar concentration. These equalities arise because of the base pairing rules in

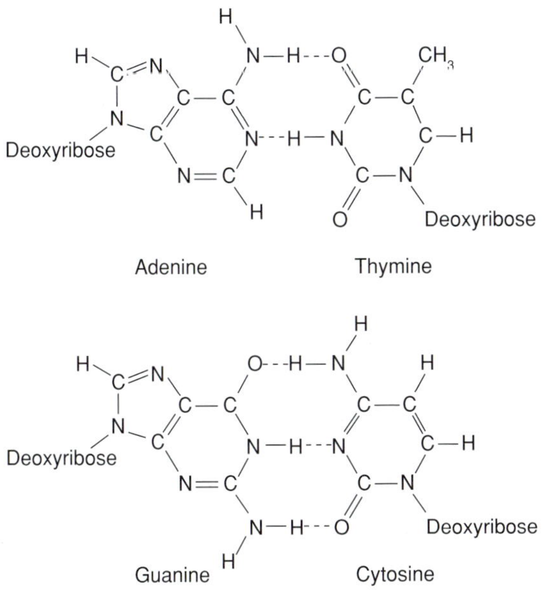

FIGURE 23-3
The normal base pairs in DNA. Adenine in one polynucleotide chain pairs with thymine in the complementary chain; guanine pairs with cytosine. A–T base pairs are joined by two hydrogen bonds; G–C base pairs are joined by three hydrogen bonds.

DNA which require that A pair with T and G pair with C in the double-stranded DNA molecule (Figure 23-3). Because of the base pairing rules it follows that [purines] = [pyrimidines] in all DNA molecules.

The overall base composition of DNA varies considerably among organisms. ***Base composition*** is expressed as the fraction of all bases in DNA that are GC pairs divided by the total number of base pairs, such as ([G] + [C])/[all bases]. This fraction is termed the GC content or percent GC. For human beings and other primates, the value of the GC content is approximately 0.5. For lower organisms the value can vary widely; the most extreme variation is found in bacteria where the GC content varies from 0.27 to 0.76 from one genus to another; *E. coli* DNA has a GC value of 0.5, which may reflect its close association and evolutionary history with human beings. The higher the GC content of DNA, the more stable is the double-stranded helical molecule. This is because GC base pairs share three hydrogen bonds whereas AT base pairs share only two (Figure 23-3).

Tautomerization of Bases

Although the bases are chemically quite stable, certain hydrogen atoms bound to the bases are able to undergo ***tautomerization*** in which they change their locations on the bases (Figure 23-4). The preferred tautomeric forms of adenine and cytosine are the ***amino*** configurations, however, with low probability each can assume the ***imino*** configuration. The preferred tautomeric forms of guanine and cytosine are the ***keto*** configuration. However, again with low probability, each can assume the ***enol*** configuration.

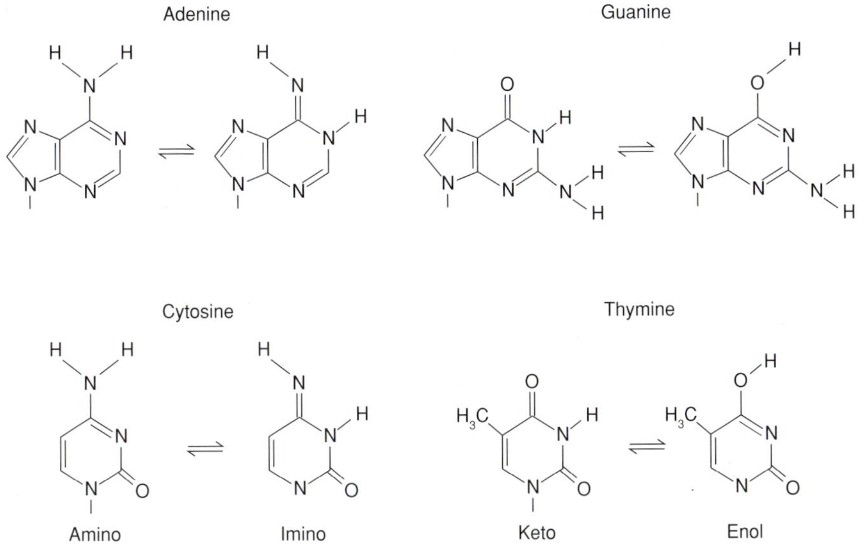

FIGURE 23-4
Tautomerization of bases in DNA. The most stable forms of adenine and cytosine are the amino conformations. With low probability these bases can tautomerize into the imino form; if this occurs during replication, an incorrect base pair (a point mutation) may result. The stable forms of guanine and thymine are the keto conformations; the enol conformations also can result in mistakes in base pairing during replication.

Tautomerization of the bases can occur as the free base and as polynucleotides. If tautomerization of a base should occur at the moment of replication of that region of DNA, an incorrect base may be inserted. For example, the tautomeric form of adenine can pair with cytosine or the tautomeric form of thymine can pair with guanine. During subsequent rounds of DNA replication these mismatched base pairs can result in point mutations in half of the DNA molecules in subsequent cycles of DNA replication and cell divisions. During replication, tautomerization of bases in DNA and ensuing mistakes in base pairing are extremely rare, but experiments show that point mutations do arise in this manner and contribute to genetic variation. However, most mistakes in base pairing that occur during DNA replication are corrected by enzymatic editing or repair functions.

Methylation of Bases

After the four bases have been incorporated into DNA, they can be modified by ***methylation,*** the addition of methyl groups at various positions (Figure 23-5). The bases that are most frequently methylated are guanine and cytosine. Methylation of cytosine residues influences gene regulation in higher organisms, and about 70% of GC base pairs in mammalian cells are methylated. The pattern of methylation of cytosine residues is inherited and is specific for each species. However, methylation of DNA is not universal; for example, the DNA in the fruitfly *Drosophilia* is completely unmethylated.

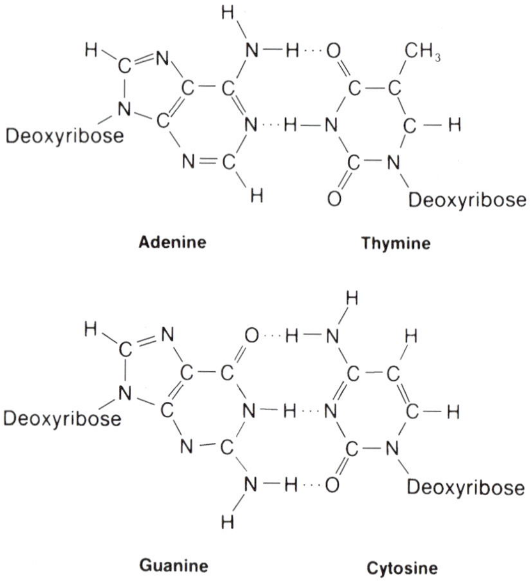

Methylation of DNA is primarily the result of two enzymes, ***Dam methylase*** and ***Dem methylase.*** Dam methylase transfers a methyl group from S-adenosylmethionine to an adenine contained in any GATC sequence in DNA. Dem methylase acts in a similar fashion on cytosine residues in the sequence CCAGG. One or both cytosines in opposite strands of DNA are methylated. Methylated segments of DNA are recognized by proteins that interact with DNA in such processes as replication, recombination, and gene expression. Methylation of DNA also serves an important function in bacteria. Methylation of specific sequences in bacterial DNA protects the bacterial DNA from cleavage by endogenously synthesized restriction endonucleases.

Many human genes are methylated differently in maternal and paternal chromosomes at CpG nucleotides, a mechanism that is referred to as ***imprinting.*** As a result of genomic imprinting, the expression of genes on maternal and paternal chromosomes differs; the loss of imprinting through mutation or some other mechanism can result in overexpression of critical genes and severe disease. Two inherited diseases that are the result of faulty imprinting are *Prader-Willi/Angelmann's syndrome (PWS)* and *Beckwith-Wiedemann syndrome (BWS)*. The symptoms of PWS include neonatal hypotonia, hypogonadism, obesity, and short stature; the symptoms of BWS are omphalocele (abdominal wall defect), thickening of long bones, and renal abnormalities.

The altered gene in BWS codes for insulin-like growth factor-2 (*IGF2*) and is located on the short arm of chromosome 15. Usually *IGF2* is silent in the maternal chromosome and active in the paternal chromosome. In some cases of BWS, the child receives two copies of chromosome 15 from the father, a condition knows as ***disomy.*** These individuals have twice the normal level of IGF2 and suffer from "overgrowth." Low levels of IGF2 may also result from disomy and lead to abnormal "undergrowth." Other inherited diseases and cancers may result from mutations that affect imprinting and normal transmission of chromosomes (Chapter 26).

Deoxyribose

Adenine

Thymine

Deoxyribose

Guanine

Cytosine

FIGURE 23-5
Methyl groups can be added to bases in DNA. Except for some yeasts and insects, all DNA contains methylated bases. The attachment of methyl groups to bases at specific sequences in DNA acts to regulate the functions of DNA. One function of site-specific methylation in prokaryotes is to protect endogenous bacterial DNA from digestion by restriction endonucleases produced by the bacteria.

23.2 Physical and Chemical Structure of DNA

The Watson-Crick DNA Structure

In DNA, two polydeoxynucleotide strands are coiled about one another in a double-helical structure as originally proposed by the Watson-Crick (W-C) model (Figure 23-6). The important features of the W-C model are as follows:

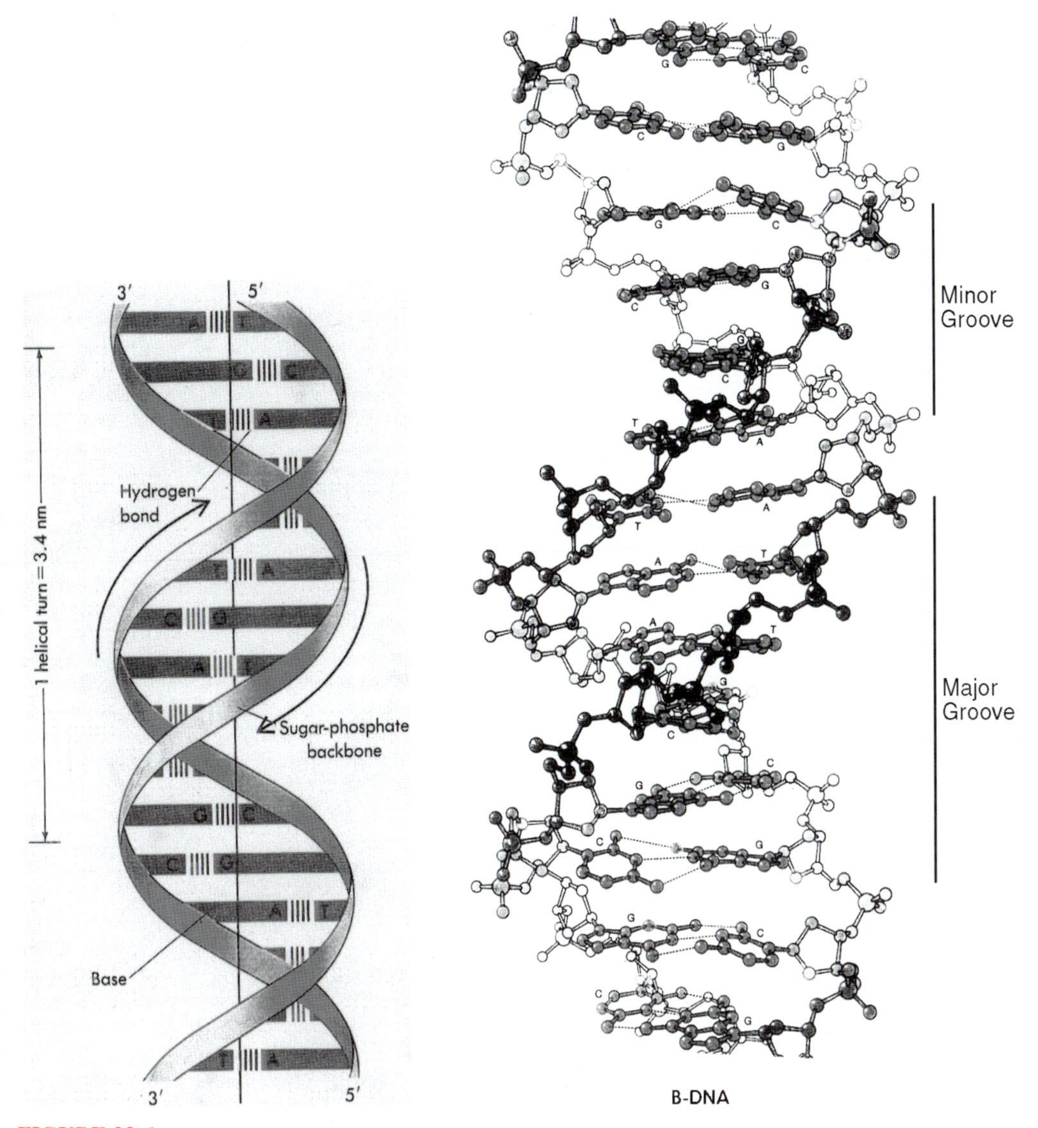

FIGURE 23-6
(a) Schematic diagram indicating the dimensions and helical structure of double-stranded DNA. (b) The structure of B-DNA showing the orientation of the atoms and the major and minor grooves in DNA. [From R. E. Dickerson, *Sci. Am.* December 1983, pp. 100–104; Copyright ©1983 by Scientific American, Inc. All rights reserved. Illustration C. Irving Geis.]

1. The sugar-phosphate backbones of the double helix follow helical paths at the outer edge of the molecule. The sugar-phosphate strands are oriented in antiparallel directions such that the 5′-phosphate end of one strand is opposite the 3′-OH end of its partner. As a result of the antiparallel orientation, if one reads a sequence of bases in a 5′ to 3′ direction on one strand, one is reading the complementary bases on the other strand in a 3′ to 5′ direction (Figure 23-6a). The overall width of the double helix is 2.0 nm.
2. The two strands in the double helix are ***complementary*** because of the base pairing rules that dictate A = T and G = C. Because of the base pairing rules, the information in DNA is redundant; knowing the sequence of bases in one strand dictates the sequence of bases in the opposite strand.
3. Each base pair in the double helix lies in a plane that is perpendicular to the axis of the helix (Figure 23-6b). Adjacent pairs of bases in DNA are separated by 0.34 nm and rotated with respect to one another so that 10 base pairs occupy each turn of the helix, which repeats every 3.4 nm.
4. Space filling models of DNA reveal two grooves that run the length of the molecule (Figure 23-6b). The ***major groove*** is wide and deep and the ***minor groove*** is narrow and shallow. These grooves in DNA provide

space for other strands of nucleic acids and also for binding of regulatory proteins.

Alternative DNA Structures

The W-C DNA structure is a right-handed helix as one looks down either axis of the molecule and the helix rotates in a clockwise direction. This conformation of DNA is called ***B-DNA*** and is the form found in solution and inside cells. The double helix is stabilized by a number of forces, including hydrophobic interactions and van der Waals forces, which also help stabilize single polynucleotide chains. Collectively, these two forces are known as ***stacking interactions*** because of their contribution to the stacked arrangement of the bases in DNA.

Other stabilizing forces in DNA are the hydrogen bonds between the AT and GC base pairs. AT base pairs share two hydrogen bonds and GC base pairs share three. These hydrogen bonds are of sufficient strength to discriminate between insertion of a correct and an incorrect base during DNA replication.

Although B-DNA is the physiologically significant form of DNA, two other conformations, ***A-DNA*** and ***Z-DNA,*** have been observed. When water is removed from solutions of DNA (such as in DNA prepared for x-ray analysis), a structure referred to as A-DNA is observed. While A-DNA is still a right-handed helix, the helix is wider and more condensed than in B-DNA. Also, the bases are tilted with respect to the helix axis and the minor groove in the molecule almost disappears. A-DNA is not believed to have any biological significance.

A polynucleotide that consists of repeating CGCGCGCGCG base pairs forms a structure in solution that is a left-handed double helix with only one major groove. Also, because the repeating units are dinucleotide pairs, the phosphate groups in the backbone are rearranged to form a zig-zag configuration; this feature accounts for the term Z-DNA. While it is not known what biological role, if any, Z-DNA plays in cells, the fact that long CG tracts exist in DNA *in vivo* suggests that segments of DNA may assume different structural configurations and that DNA exists in a dynamic rather than a static state in chromosomes.

Plasmid DNA

Circular DNA molecules that are of great significance in nature as well as in numerous biotechnology applications are ***plasmids,*** which are widespread among bacteria. Plasmids are small circular DNA molecules consisting of just a few genes to more than a hundred. Among the important genes carried in plasmids are those coding for resistance to a wide range of antibiotics and genes that allow plasmids to be transferred to other bacteria, even to other species of bacteria. The process of transfer of plasmid and chromosomal DNA from one bacterium to another is called ***conjugation.***

Transfer of antibiotic resistance genes among bacteria in nature has created serious problems in the treatment of many infectious diseases such as tuberculosis, gonorrhea, pneumonia, staphylococcus, and others. The widespread use of antibiotics in agriculture and in hospitals has resulted in the selection and evolution of pathogenic microorganisms that are resistant to many, sometimes all, of the antibiotics normally used to treat infections by these microorganisms.

Plasmids also have been genetically engineered in a multitude of ways so that they can carry and express foreign genes in bacteria. For example, the genes coding for human insulin and human growth hormone have been inserted into plasmids which are then reintroduced into bacteria such as *E. coli*. The genetically engineered bacteria are then used as biological factories to produce the desired drugs.

Circular and Supercoiled DNA

Most DNA molecules isolated from prokaryotes and from some virus particles are circular. A circular molecule may be covalently closed, consisting of two unbroken complementary single strands, or nicked, i.e., having one or more interruptions (nicks) in one or both strands. With few exceptions, covalently closed circles assume the form of ***supercoils*** (Figure 23-7).

The two ends of a linear DNA helix can be brought together and joined in such a way that each strand is continuous. Consider a molecule in which one of the ends is rotated 360 degrees with respect to the other in the unwinding direction, and then the ends are joined. If the hydrogen bonds re-form, the resulting covalent circle will twist in the opposite sense to form a twisted circle, in order to relieve the strain of underwinding. Such a molecule will resemble a figure 8 and will have one crossover point. If it is underwound by 720 degrees before the ends are joined, the resulting superhelical molecule will have two crossover points. In the case of a 720-degree unwinding of the helix, 20 base pairs must be broken (because the linear molecule has 10 base pairs per turn of the helix). However, such a DNA molecule tries to maintain a right-handed (positive) helical structure with 10 base pairs per turn; it will deform itself in such a way that the underwinding is compensated for by negative (left-handed) twisting of the circle.

All naturally occurring, superhelical DNA molecules are initially underwound and, hence, form negative

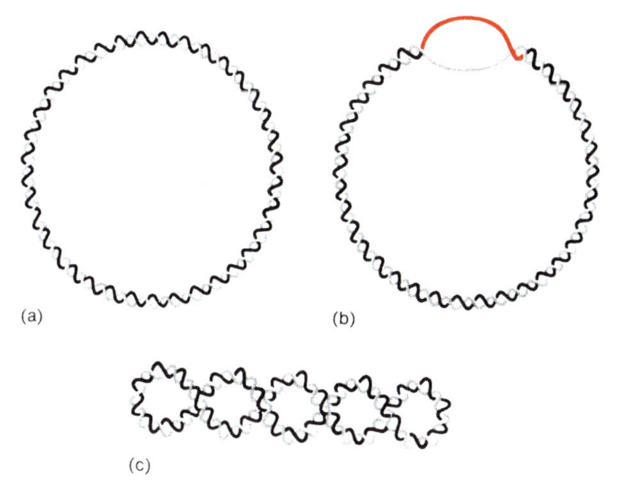

FIGURE 23-7
Different states of a covalent DNA circle. (a) A nonsupercoiled covalent DNA having 36 turns of the helix. (b) An underwound covalent circle having only 32 turns of the helix. (c) The molecule in (b), but with four superhelical turns to eliminate the underwinding. In solution, (b) and (c) would be in equilibrium; the equilibrium would shift toward (b) with increasing temperature.

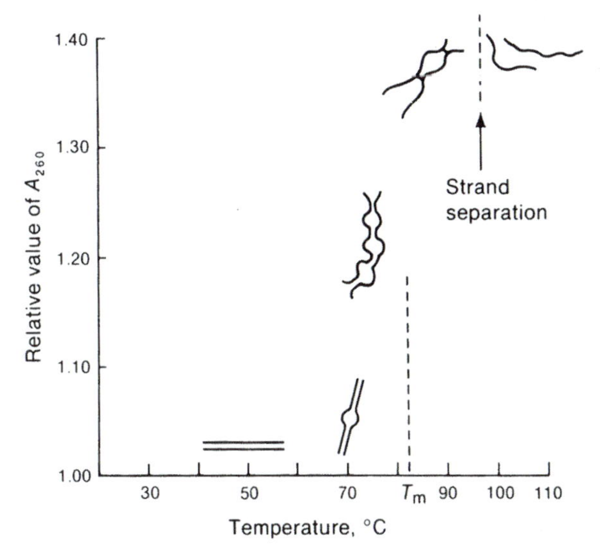

FIGURE 23-8
Melting curve of DNA showing the melting temperature (T_m) and possible molecular conformations for various degrees of melting.

superhelices. In bacteria, the underwinding of superhelical DNA is not a result of unwinding before end joining but is introduced into preexisting circles by an enzyme called ***DNA gyrase,*** which is one of a class of enzymes called ***topoisomerases*** (Chapter 24). In eukaryotes, the underwinding is a result of the structure of chromatin, a DNA-protein complex of which chromosomes are composed. In chromatin, DNA is wound about specific protein molecules, in a direction that introduces underwinding.

23.3 Denaturation of DNA

The three-dimensional structures of DNA, RNA, and proteins are determined by weak noncovalent interactions, principally hydrogen bonds and hydrophobic interactions. The free energies of these interactions are not much greater than the energy of thermal motion at room temperature, so that at elevated temperatures the structures of these molecules are disrupted. A macromolecule in a ***disrupted state*** is said to be denatured; the ***ordered state,*** which is presumably that originally present in nature, is called the native state. A transition from the native to the denatured state is called ***denaturation.*** When double-stranded (native) DNA is heated, the bonding forces between the strands are disrupted and the two DNA strands separate; thus, completely denatured DNA is single stranded.

Much information about the structure and stabilizing interactions has been obtained by studying denaturation. Some property of DNA that changes as denaturation proceeds is measured, e.g., the absorption of ultraviolet light. A change in the ultraviolet absorbance (or some other property) as a function of temperature is called a ***melting curve*** (Figure 23-8). Many reagents either break hydrogen bonds or weaken hydrophobic interactions, and are powerful denaturants. Thus, denaturation is also studied by varying the concentration of a denaturant at a constant temperature. For DNA, the simplest way to detect denaturation is to monitor the ability of DNA in a solution to absorb ultraviolet light at a wavelength of 260 nm. The absorbance of DNA at 260 nm, A_{260}, is not only proportional to its concentration (as is the case for most light-absorbing molecules) but also depends on the structure of the molecule; the more ordered the structure, the less light is absorbed. Therefore, free nucleotides absorb more light than a single-stranded polymer of DNA (or RNA), which in turn absorbs more light than a double-stranded DNA molecule. For example, solutions of double-stranded DNA, single-stranded DNA, or free bases, each at 50 μg/mL, have the following A_{260} values:

$$\text{Double-stranded DNA: } A_{260} = 1.00$$
$$\text{Single-stranded DNA: } A_{260} = 1.37$$
$$\text{Free bases: } A_{260} = 1.60$$

If a DNA solution is heated slowly and the A_{260} is measured at various temperatures, a melting curve such as that shown in Figure 23-8 is obtained. The following features of this curve should be noted:

1. The A_{260} remains constant up to temperatures well above those encountered by living cells in nature.

2. The rise in A_{260} occurs over a range of 6–8°C.
3. The maximum A_{260} is about 37% higher than the starting value.

The hypothetical state of a DNA molecule in different regions of the melting curve is also shown in Figure 23-8. Before the rise begins, the molecule is fully double stranded. In the region of rapid denaturation, base pairs in various segments of the molecule are broken; the number of broken base pairs increases with temperature. A convenient parameter to characterize a melting transition is the temperature at which the rise in A_{260} is half complete. This temperature is called the ***melting temperature,*** T_m. The value of T_m varies both with base composition and experimental conditions. In particular. T_m increases with increasing percent G + C, which is a result of the hydrogen bonds in a GC pair (three) versus an AT pair (two). A higher temperature is required to disrupt a GC pair than an AT pair. Reagents such as urea and formamide, which can hydrogen-bond with the DNA bases, reduce T_m. These denaturing agents maintain the unpaired state at a temperature at which a broken base pair would normally pair again, so that permanent melting of a section of paired bases requires less thermal energy.

Other reagents either enhance the interaction of weakly soluble substances (such as the nucleic acid bases) with water or disrupt the water shell; such substances should weaken hydrophobic interactions. An example of the former type of substance is methanol, which increases the solubility of the bases. Sodium trifluoracetate is an example of the second type. The addition of both these reagents greatly reduces T_m because hydrophobic interactions are also important in stabilizing the DNA structure. In fact, the three-dimensional structure of DNA is one that minimizes contact between bases and water and maximizes the contact of the highly soluble phosphate group with water. Minimization of base-water contact is accomplished by stacking of the bases, which occurs even in single-stranded DNA. The bases of double-stranded DNA are more stacked than those in single-stranded DNA because of the hydrogen bonds between the two strands.

Both hydrogen bonds and hydrophobic interactions are weak and easily disrupted by thermal motion. Maximum hydrogen bonding is achieved when all bases are oriented in the right direction. Similarly, stacking is enhanced if the bases are unable to tilt or swing out from a stacked array. Clearly, stacked bases are more easily hydrogen-bonded, and correspondingly, hydrogen-bonded bases, which are oriented by the bonding, stack more easily. Thus, the two interactions act cooperatively to form a very stable structure. If one interaction is eliminated, the other is weakened, which explains why T_m drops so markedly following addition of a reagent that destroys either type of interaction. When hydrogen bonds and hydrophobic interactions are eliminated, the helical structure of DNA is disrupted and the molecule loses its rigidity. This collapse of the ordered structure is accompanied by complete disentanglement of the two strands. At high pH, the charge of several groups engaged in hydrogen bonding is changed and base pairing is reduced. At a pH greater than 11.3, all hydrogen bonds are eliminated and DNA is completely denatured.

When a DNA solution is heated above 90°C, the value of A_{260} increases by 37% and the solution consists entirely of single strands whose bases are unstacked. If the solution is then rapidly cooled to room temperature and the salt concentration is greater than 0.05 mol/L, the value of A_{260} drops significantly because random intrastrand hydrogen bonds re-form between distant short tracts of bases whose sequences are complementary (or nearly so). After cooling, about two-thirds of the bases are either hydrogen-bonded or in such close proximity that stacking is restored and the molecule is very compact. In contrast, if the salt concentration is 0.01 mol/L or less, the electrostatic repulsion due to unneutralized phosphate groups keeps the single strands sufficiently extended that the bases cannot approach one another. Thus, after cooling, no hydrogen bonds are re-formed. At a sufficiently high DNA concentration and in a high salt solution, interstrand hydrogen bonding competes with the intrastrand bonding just mentioned. This effect can be used to re-form native DNA from denatured DNA.

23.4 Renaturation of DNA

If a DNA solution is heated to a temperature at which most (but not all) hydrogen bonds are broken and then cooled slowly to room temperature, A_{260} drops immediately to the initial, undenatured value and the native structure is restored. Thus, if strand separation is not complete and denaturing conditions are eliminated, the helix rewinds. A related observation is that if two separated strands come in contact and form even a single base pair at the correct position in the molecule, the native DNA molecule will re-form. This phenomenon is called ***renaturation,*** or ***reannealing.***

Two requirements are necessary for renaturation to occur.

1. The salt concentration must be high enough that electrostatic repulsion between the phosphates in the two strands is eliminated—usually 0.15–0.50 M NaCl is used.

2. The temperature must be high enough to disrupt the random, intrastrand hydrogen bonds described above. However, if the temperature is too high, stable interstrand base pairing will not occur or be maintained. The optimal temperature for renaturation is 20–25°C below the value of T_m.

Renaturation is slow compared with denaturation. The rate-limiting step is not the rewinding of the helix but the precise collision between complementary strands such that base pairs are formed at the correct positions. Since two molecules participate in the rate-limiting step, renaturation is a concentration-dependent process requiring several hours under typical laboratory conditions. In particular, the kinetics of association follows a simple second-order rate law with the association rate increasing with DNA concentration. The kinetics are conveniently described by an equation that relates the fraction, f, of single-stranded (dissociated) DNA and the time elapsed after exposing a DNA sample to renaturing conditions (high salt concentration, elevated temperature):

$$f = 1/(1 + kC_0t)$$

in which C_0 is the initial DNA concentration, t is the elapsed time in seconds, and k is a rate constant. A plot of f versus the logarithm of the product C_0t yields a sigmoid curve commonly called a C_0t ("cot") curve. A set of such curves for several DNA samples is shown in Figure 23-9. A notable feature of these curves is that the renaturation rate is related to the molecular weight of the DNA. A useful index for characterizing these curves is $C_0/t_{1/2}$, with the value of C_0t corresponding to renaturation of half of the DNA ($f = 1/2$). Comparison of the $C_0/t_{1/2}$ values for *E. coli* DNA (M.W. $= 2.7 \times 10^9$) and T_4 DNA (M.W. $= 1.1 \times 10^8$) shows that T_4 DNA renatures roughly 50 times faster than *E. coli* DNA. The reason

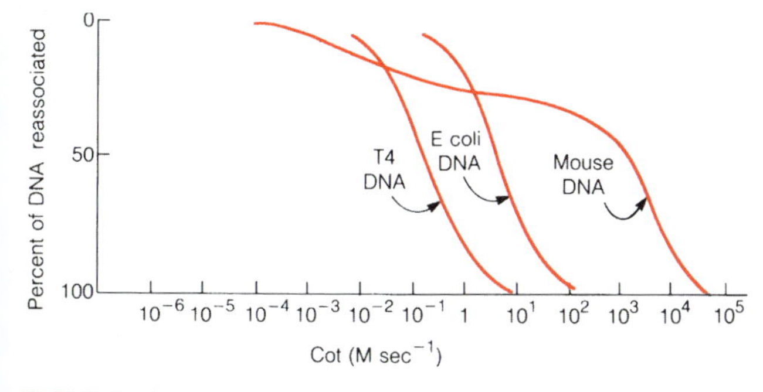

FIGURE 23-9
Reassociation "C_0t" curves for bacteriophage T_4 DNA, *E. coli* DNA, and mouse DNA. [Reproduced with permission from D. M. Prescott: *Cells: Principles of Molecular Structure and Function.* Jones and Bartlett, London, 1988.]

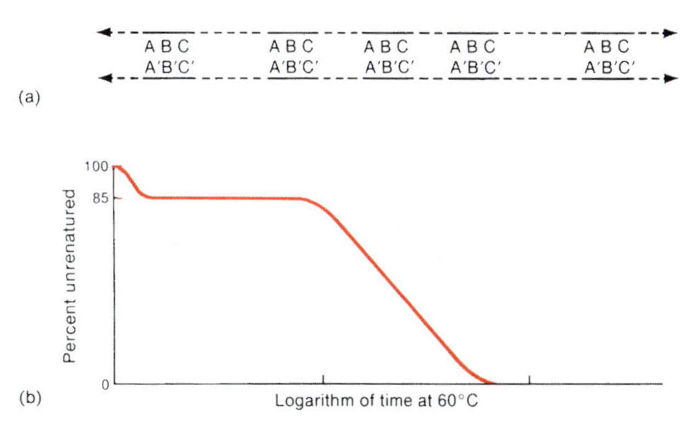

FIGURE 23-10
(a) Hypothetical DNA molecule having a base sequence that is 3% of the total length of the DNA and is repeated five times. The dashed lines represent the nonrepetitive sequences: they account for 85% of the total length. (b) Renaturation curve for the DNA in (a). Time is logarithmic to keep the curve on the page.

is that if the two DNA samples have equal molar concentrations of nucleotides, the T_4 sample will contain many more DNA molecules than the *E. coli* sample.

C_0t analysis is not usually done with intact DNA molecules but rather with fragments having lower molecular weights. This breakage does not affect the relative values of $C_0/t_{1/2}$ since the number of different kinds of fragments of T_4 DNA is smaller and hence their concentration larger than the corresponding number of fragments in the *E. coli* sample. Studies of a variety of prokaryotic DNAs show that the value of $C_0/t_{1/2}$ is directly related to the total size of the DNA of the organism (the genome size). However, this relation does not apply to the DNA of eukaryotes because of the presence of highly repetitive sequences.

The fragmentation of the DNA molecules allows a new feature of base sequences to be seen. Figure 23-10(a) shows a long hypothetical DNA molecule having a repeating base sequence. If the unbroken denatured DNA were allowed to renature, the C_0t curve would be a smooth curve in which $C_0/t_{1/2}$ would be proportional to the size of the DNA. However, if the DNA were broken into small pieces, several fragments from each molecule would contain the repeated sequence, and these fragments would renature more rapidly than the bulk of the sequences. For example, if the repeated sequence contains 3% of the total number of bases of the DNA and there are five copies of the repeated sequence in the DNA, the C_0t curve would be that shown in Figure 23-10b, in which the more rapidly renaturing component accounts for 15% of the transition; hence, the conclusion that 15% of the DNA (on a weight basis) contains a repeating sequence. Approximately 30% of human DNA contains sequences of bases that are repeated 20 times or more.

23.5 Repetitive DNA Sequences

C_0t curves from various eukaryotic organisms show the presence of repetitive sequences to varying degrees. Unique sequences represent genes, approximately 100,000 in the human genome, but constitute only 1–2% of the total genetic information in the human genome. Some genes may be slightly repetitive if they are related to ***pseudogenes,*** sequences of DNA that resemble genes but that are no longer expressed.

About one-third of the human genome consists of sequences of bases that are repeated at least 20 times or more. *Moderately repetitive* refers to sequences that usually contain between 10 and several hundred copies; *highly repetitive* refers to sequences that are repeated thousands or millions of times throughout the genome. The distinction among various classes of repetitive sequences is frequently blurred and additional categories are sometimes defined.

Repetitive sequences are also defined by the number of base pairs in each repeated segment. Sequences that have from 100 to 500 bp are referred to as SINES (short interspersed repeated sequences) and sequences that have several thousand base pairs are referred to as LINES (long interspersed repeated sequences.) Thus, repetitive DNA sequences can be described both by the length of the segment and the degree to which it is repeated.

Some examples of moderately repetitive sequences are transposable elements including retroviruses, histone genes that are repeated 30–40 times in the human genome, and genes coding for ribosomal RNAs and transfer RNAs. Most of the moderately repetitive genes do not code for proteins but serve other functions in the cell. Histones are involved in condensation of DNA in chromosomes and both ribosomal RNAs and transfer RNA are involved in protein synthesis.

One special class of a highly repetitive sequence is ***satellite DNA*** which consists of 6–8 bp (such as ATAAACT) and may account for 10–20% of the total DNA. The term *satellite DNA* derives from the observation that if fragmented DNA is centrifuged in a density gradient solution that separates DNA by weight, a small band of DNA is observed that is separate from the bulk of the DNA. The base composition of satellite DNA is AT-rich, which explains why it separates from the rest of the DNA.

Satellite DNA is thought to play a structural role in chromosomes. Certain satellite DNA sequences are concentrated near the centromeres of chromosomes, the site where spindle fibers attach when sister chromatids are separated. Other satellite DNA sequences are located in telomeres, structures at the ends of chromosomes that play a critical role in DNA replication (Chapter 24).

Another class of highly repetitive sequences is the ***Alu family,*** which consist of about 300 bp that are repeated millions of times throughout the genome. Any Alu sequence is at least 85% homologous in base sequence to any other Alu sequence; hence, the family of genes has been highly conserved. Each Alu sequence contains a restriction site that is recognized by the AluI restriction enzyme, from which the name of the family of sequences is derived.

23.6 Degradation of DNA

Nucleases

A variety of enzymes break phosphodiester bonds in nucleic acids; ***deoxyribonucleases (DNases)*** cleave DNA and ***ribonucleases (RNases)*** cleave RNA. DNases usually are specific for single- or double-stranded DNA although some DNases can cleave both. DNases can act as ***exonucleases*** in which they remove one nucleotide at a time from either the 3′ or 5′ end of the strand. Other DNases function as ***endonucleases*** and are specific for cleaving between particular pairs of bases.

Restriction Enzymes

An important class of DNA endonucleases are ***restriction enzymes*** (restriction endonucleases) that recognize specific sequences of bases in DNA (a restriction site) and make two cuts, one in each strand that generates fragments of double-stranded DNA. Microorganisms use their restriction enzymes to degrade any foreign DNA that may enter the cell in the form of viruses, plasmids, or naked DNA. Hundreds of different restriction enzymes have been isolated from microorganisms; these enzymes generally recognize sequences of four, six, eight, or rarely more bases that have an axis of symmetry (Table 23-1). Restriction sites that have an axis of symmetry are called ***palindromes,*** which means that the restriction site can be rotated 180 degrees and the sequence of bases will remain the same.

Restriction enzymes can cut DNA in either of two ways. Each strand can be cleaved along the axis of symmetry generating fragments of double-stranded DNA that have blunt (flush) ends. Symmetrical cuts that are staggered around the axis of symmetry generate fragments of double-stranded DNA that have single-stranded cohesive ends (Figure 23-11). Fragments of DNA with cohesive ends

TABLE 23-1
Some Restriction Endonucleases and Their Cleavage Sites*

Microorganism	Name of Enzyme	Target Sequence and Cleavage Sites
Generates cohesive ends		**Axis of symmetry**
Escherichia coli RY 13	EcoRI	G↓A A T T C C T T A A↑G
Bacillus amyloliquefaciens H	BamHI	G↓G A T C C C C T A G↑G
Bacillus globigii	BglII	A↓G A T C T T C T A G↑A
Haemophilus aegyptius	HaeII	Pu G C G C↓Py Py↑C G C G Pu
Haemophilus influenzae R_d	HindIII	A↓A G C T T T T C G A↑A
Providencia stuartii	PstI	C T G C A↓G G↑A C G T C
Streptomyces albus G	SalI	G↓T C G A C C↓A G C T↑G
Xanthomonas badrii	XbaI	T C T A G A A G↓A T C↑T
Thermus aquaticus	TaqI	T C G A A G C↑T
Nocardia otitdis	Not I	G C↓G G C C G C C G C C G G↑C G
Streptomyces sp.	Sgf I	G C G A T↓C G C C G C↑T A G C G
Generates flush ends		
Brevibacterium albidum	BalI	T G G ↓ C C A A C C ↑ G G T
Haemophilus aegyptius	HaeI	(A/T) G G ↓ C C (T/A) (T/A) C C ↑ G G (A/T)
Serratia marcescens	SmaI	C C C ↓ G G G G G G ↑ C C C

*The vertical dashed line indicates the axis of dyad symmetry in each sequence. Arrows indicate the sites of cutting. The enzyme TaqI yields cohesive ends consisting of two nucleotides, whereas the cohesive ends produced by the other enzymes contain four nucleotides. The enzyme HaeI recognizes the sequence GGCC whether the adjacent pair is A·T or T·A, as long as dyad symmetry is retained. Pu and Py refer to any purine and pyrimidine, respectively.

are useful for constructing novel DNA molecules, which is the goal of ***recombinant DNA (rDNA)*** technology.

Since a restriction enzyme recognizes a unique sequence, the number of cuts and the number of DNA fragments depends on the size of the molecule. In general, restriction sites consisting of four bases will occur more frequently by chance than sites consisting of six or eight bases. Thus, four-cutters will generate more fragments than six-cutters which, in turn, will generate more fragments than eight-cutters.

A particular restriction enzyme generates a unique family of fragments for a particular DNA molecule.

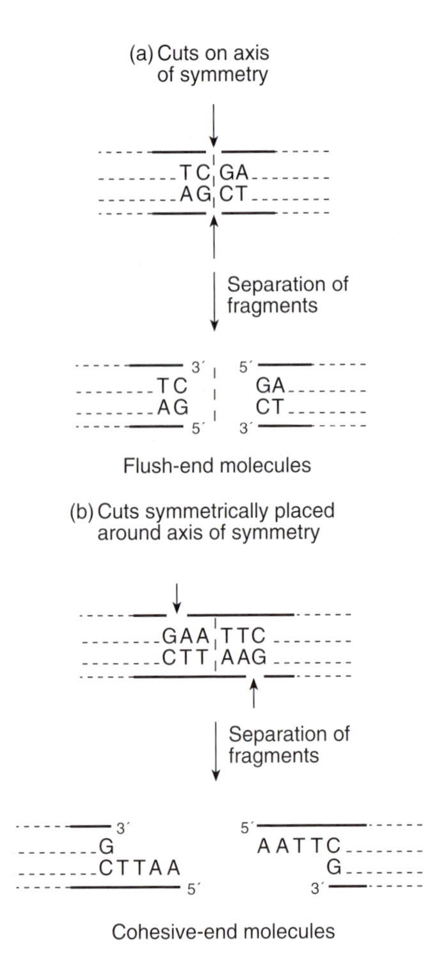

FIGURE 23-11
Two types of cuts made by restriction enzymes. The arrows indicate the cleavage sites. The dashed line is the axis of symmetry of the sequence.

Figure 23-12a shows the positions of restriction sites in bacteriophage lambda (λ) DNA for the restriction enzymes EcoRI and BamHI. This arrangement of restriction sites is known as a ***restriction map.*** The family of fragments generated by one or more restriction enzymes is detected by agarose gel electrophoresis that separates DNA molecules according to their molecular weights (Figure 23-12b).

23.7 Diagnostic and Clinical Applications of DNA

DNA Probes

The ability to isolate specific fragments of DNA containing known sequences of genes gives rise to ***DNA probes*** that can be used for a variety of diagnostic, forensic, and therapeutic purposes. DNA probes can be labeled with either radioactive or nonradioactive markers. A DNA probe has a strong interaction with (ideally) a specific DNA target and can be detected after the interaction. DNA probes consisting of 20 bases or fewer usually will have a unique target even in a large set of DNA molecules. The probability that any base will follow any other base in DNA is one in four or 0.25. Therefore, the probability of a specific sequence of 20 bases occurring in a DNA molecule by chance is 0.25^{20}, a vanishingly small number.

The use of DNA probes in various aspects of medical diagnostics is increasing rapidly. DNA probes can be used to identify infectious agents if sequences specific to different pathogens are known. Identification of a pathogen by DNA probes can be done in hours as compared to days or weeks by conventional culturing of microorganisms.

DNA probes are now used routinely to detect the presence of mutant alleles in fetal cells obtained by amniocentesis, as well as in cells removed from affected adults or carriers. Many inherited disorders, such as sickle cell disease, cystic fibrosis, Huntington's disease, Duchenne's muscular dystrophy, and dozens of other Mendelian (single-gene) disorders, can now be diagnosed in fetuses and adults. In addition to inherited disorders, DNA probes are used to detect the presence of active oncogenes or inactive tumor suppressor genes in cancerous tissues removed from patients (Chapter 26).

Southern Blot Analysis

The ***Southern blot*** (named for its inventor, E. M. Southern) is a method for hybridizing one or more labeled DNA probes to a large number of DNA fragments and discriminating among them. The procedure depends on the ability of denatured DNA single strands to bind tightly to nitrocellulose under certain conditions.

The Southern blot procedure is outlined schematically in Figure 23-13. The DNA to be investigated is digested with several restriction enzymes to generate DNA fragments of varying sizes which are then separated by agarose gel electrophoresis. After the fragments are separated, the gel is immersed in a denaturing solution that converts all DNA to single strands. Then the gel, which is in the form of a broad, flat slab, is placed on top of filter paper supported by a glass plate. A nitrocellulose filter is placed over the gel and covered with another sheet of filter paper. Finally, paper towels and weights are placed on top of the gel.

The glass plate supporting the filter papers and gel is adjusted so that the ends of the filter paper are suspended in buffer and act as wicks. The buffer moves upward through the gel and into the wad of paper towels that act to absorb the buffer. The DNA migrates from the gel to the nitrocellulose filter that binds the single-stranded DNA fragments.

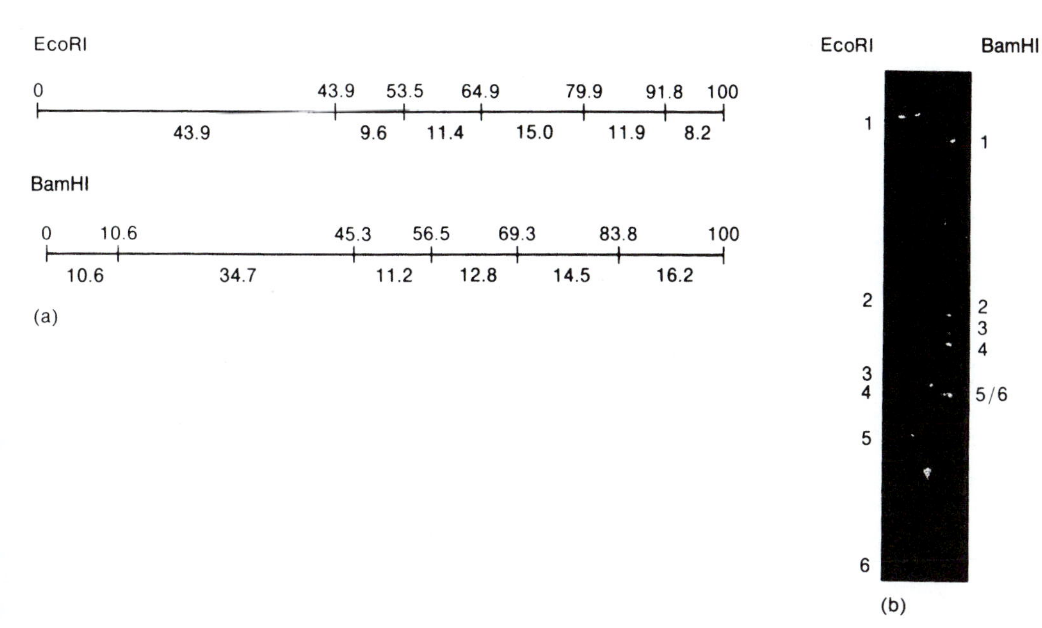

FIGURE 23-12
(a) Restriction maps of λ DNA for EcoRI and BamHI nucleases. The vertical bars indicate the sites of cutting. Thc top numbers indicate the percentage of the total length of DNA measured from the gene A end of the molecule. The bottom numbers are the lengths of each fragment, again expressed as percentage of total length. (b) Gel electrophoretogram of EcoRI and BamHI restriction enzyme digests of λ DNA. The bands labeled "cohered ends" contain molecules consisting of two terminal fragments, joined by the normal cohesive ends of λ DNA. Numbers indicate fragments in order from largest 1 to smallest 6. Bands 5 and 6 of the BamHI digest are not resolved.

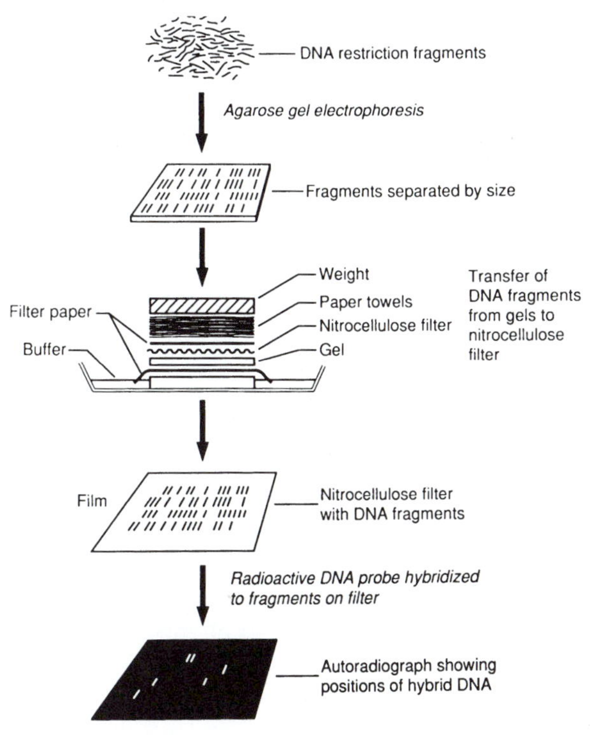

FIGURE 23-13
Procedure for a Southern blot. This analysis allows detection of a specific fragment of DNA by hybridization of a radioactive probe with a complementary sequence.

The nitrocellulose filter is dried and exposed to high temperature to keep the DNA in a denatured state. A labeled DNA probe containing a known sequence or gene is then incubated with the nitrocellulose filter under a variety of hybridization conditions to identify a particular fragment of DNA from the sample. Strands of the DNA probe bind to fragments of the separated DNA that contain similar or identical sequences and hybridize together to form double-stranded DNA. After the hybridization reaction is complete, the solution is cooled and the itrocellulose filter is washed to remove unbound probe DNA and dried. The labeled DNA fragments can be visualized by autoradiography (if the probe was radioactive) or by fluorescence. The DNA fragments of interest also can be eluted from the nitrocellulose filter for further analysis.

Polymorphisms

Approximately 25% of all human genes occur among individuals in a population in multiple allelic forms referred to as ***polymorphisms.*** For example, the ABO locus, an important factor in blood transfusions, consists of three different alleles (alternative states of a gene) called A, B, and O. Individuals can be homozygous or heterozygous for any pair of A, B, and O alleles. The HLA family of genes that play a vital role in the rejection of tissue transplants consists of hundreds of different alleles (Chapter 35). No two individuals, except for identical twins, share exactly the same set of HLA alleles. For a gene to be polymorphic, alternative alleles must be present in at least 1% of individuals in a population. The relative frequencies of the different alleles at a polymorphic locus usually vary from one human population to another.

The remaining 75% of human genes consist of a single allelic form and are said to be ***monomorphic.*** For example, all human beings have the same α and β genes for hemoglobin, presumably because the protein structure has been optimized by millions of years of human evolution. All persons, irrespective of race or ethnicity, have hemoglobin genes that are identical in sequence. The only exceptions to the monomorphic state of the hemoglobin genes are the rare individuals who inherit or acquire a mutation in the α or β genes that gives rise to the several hundred characterized human hemoglobinopathies (Chapter 28).

Restriction enzyme sites in DNA also are highly polymorphic in human populations. The pattern of DNA fragments produced by digesting DNA from different individuals will usually show a difference if a sufficient number of fragment sizes are examined; again, only identical twins have identical restriction enzyme sites. The different fragments produced by digesting DNA with one or more restriction enzymes are called ***restriction fragment length polymorphisms (RFLPs).***

RFLPs have been extremely useful in constructing a restriction map of the human genome (which can be correlated with the genetic and physical maps), in screening human populations for the presence of mutant alleles, and for diagnosing hereditary disorders in fetuses and prospective parents. An early success in mapping the human genome was determining the location of more than 10,000 RFLPs on the 23 human chromosomes.

Most mutant genes that cause hereditary disorders are linked to specific RFLP differences among affected and unaffected persons that can be detected by using DNA probes carrying the gene in question. For example, the single-base change found in all cases of sickle cell anemia also changes a restriction site located within the β-globin gene that is recognized by the restriction enzyme MstII (Figure 23-14). When DNA from unaffected and affected individuals is digested with MstII and a Southern blot analysis is performed with the appropriate probe, affected individuals show a different pattern of DNA fragments as compared with unaffected individuals.

Huntington's disease (HD) is an autosomal dominant disorder characterized by progressive chorea, dementia,

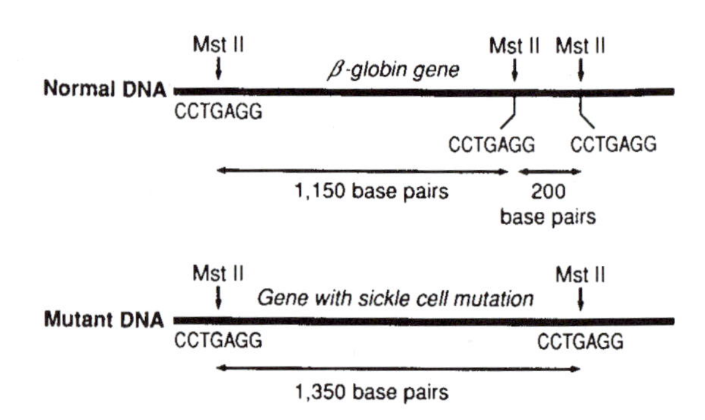

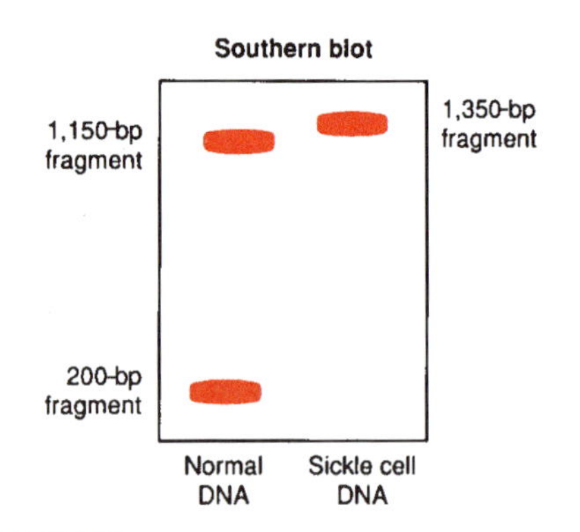

FIGURE 23-14
Detection of sickle cell anemia by RFLP analysis and Southern blot.

and ultimately death. It is a late-onset hereditary disease that usually manifests after age 40. Any son or daughter who had a parent with Huntington's disease has a 50% probability of inheriting the mutant allele from the affected parent. A HindIII restriction site is closely linked to the *HD* gene in asymptomatic family members who carry the mutant allele but is absent in family members who have not inherited the mutant allele. Southern blot analysis can determine which family members are destined to die of the disease later in life. (Not all families that carry a *HD* gene mutation have this particular mutation that was characteristic of the original large family investigated.)

Genetic testing for susceptibility to Huntington's disease (and many others) raises profound ethical questions for society, for families, and for individuals who may or may not have inherited disease-causing mutant alleles, and who may or may not want to know their disease status. Persons diagnosed with a gene for an inherited disorder may face discrimination in obtaining health insurance, life insurance, job opportunities, and so path. Hundreds of Mendelian (single-gene) disorders can now be detected with appropriate DNA probes (Table 23-2). However, genetic screening generally is not advised, particularly if no medical treatment is available for the particular disease.

TABLE 23-2
A Partial List of Diseases for Which Single Genes Have Been Isolated and Characterized*

Familial amyloid polyneuropathies	Niemann-Pick disease
Phenylketonuria	X-linked Charcot-Marie-Tooth disease
Familial hypercholesterolemia	Familial colon cancer
Christmas factor deficiency	Hereditary retinal degenerative diseases
Retinoblastoma, Wilson's disease	Ototoxic deafness
Type IV collagen deficiencies	XSCID
Hydatidiform moles and choriocarcinomas	Compulsive disorders
Tay-Sachs disease	Myotonic muscular dystrophy
Alzheimer's disease	Familial dysalbuminemic hyperthuroxinemia
Gaucher's disease	Long QT syndrome
Charcot-Marie-Tooth 1A disease	Cerebral autosomal dominant arteriopathy
Williams syndrome	Cancer-associated mar binding protein
Von Hippel-Lindau disease	Breast or ovarian cancer
Breast cancer (BRCA1)	Idiopathic dilated cardiomyopathy
Prostate cancer	Hereditary neuropathy with liability

*Each of these mutant alleles contributes to a greater or lesser degree to the pathology of the disease or inherited disorder. As of December 1997, patents had been issued to universities or companies for the commercial use of these genes.

Forensic DNA Analysis

RFLPs are also widely used in forensic pathology, criminology, and cases of contested paternity. A particular set of DNA probes is specific for hypervariable sequences in the human genome that are inherited in a Mendelian pattern identical to the inheritance of genes. ***Hypervariable sequences*** are highly polymorphic minisatellite loci that are unique to each individual just as each individual has a unique set of genes. These hypervariable sequences can be detected by a special set of DNA probes called ***Jeffreys probe*** after their discoverer. A Southern blot analysis of fragments of DNA from an individual using the Jeffreys probe constitutes a unique "genetic fingerprint." Only monozygous twins have the same pattern of hypervariable sequences and identical genetic fingerprints (Figure 23-15).

DNA obtained from a blood stain, a human hair cell, or even a few sperm provides enough material to match with the DNA obtained from a suspect. DNA testing of suspects in rape, murder, and other crimes in which a sample of DNA was obtained at the scene of the crime is now a routine procedure.

DNA sequences can be amplified by the ***polymerase chain reaction (PCR)*** technique. Only an infinitesimal amount of DNA is needed, e.g., a sample of DNA from a single hair follicle or sperm. The DNA to be amplified is mixed with three other components in an automated PCR procedure:

1. A heat-stable DNA polymerase that is isolated from a thermophilic bacterium,
2. An excess of two short primer DNAs that are complementary to opposite strands of the DNA fragment that is to be amplified, and
3. An excess of deoxyribonucleotide triphosphates.

PCR consists of repeated cycling of three reactions: denaturation of the DNA by heating, reannealing of the primers with the target DNA by cooling, and synthesis of new DNA strands. The sequence of reactions is automatically repeated at defined intervals to yield an exponential increase in the amount of DNA. Twenty cycles of PCR amplify DNA by about a factor of 10^6 and 30 cycles by about 10^9.

PCR amplification of DNA is one of the most widely used techniques in medical research and diagnostics. PCR is used in forensic pathology (to identify human remains), in rapid identification of infectious microorganisms, in diagnosis of inherited diseases, and in archaeology and anthropology where small DNA samples can be recovered.

Sequencing DNA

Until the development of automated DNA sequencing machines in the 1990s, two techniques were used to sequence the bases in a segment of DNA. Each of the techniques involves the isolation of a restriction fragment containing a few hundred or a few thousand base pairs. The DNA is denatured and each strand is sequenced separately so

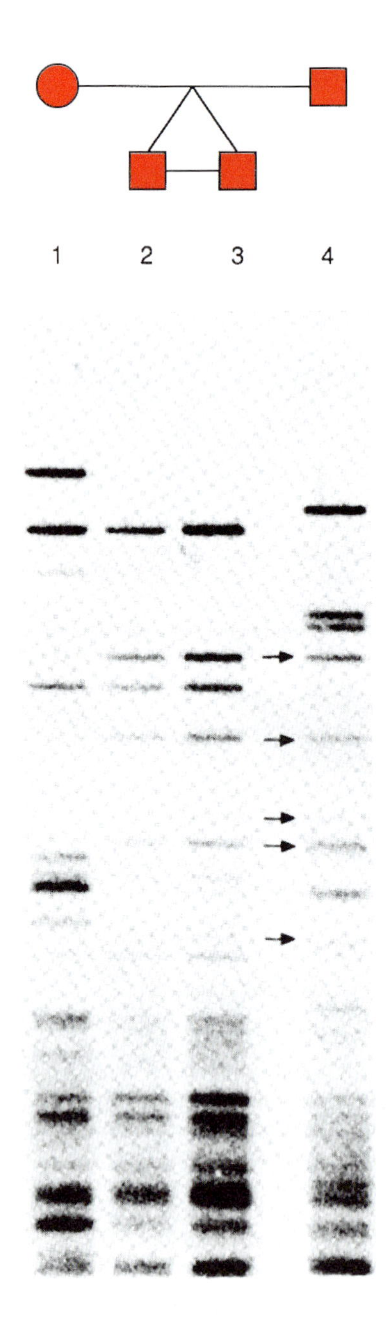

FIGURE 23-15
Genetic fingerprint of monozygotic twins (columns 2 and 3) and their parents (columns 1 and 4). The arrows indicate hypervariable sequences inherited from the father. Similar hypervariable sequences can be identified from the mother. [Reproduced with permission from A. J. Jeffreys, V. Wilson, and S. L. Thein: Individual-specific "fingerprints" of human DNA. *Nature* **316,** 76 (1985). © 1985 by Macmillan Magazines Ltd.]

that the sequences can be compared in order to eliminate errors.

One sequencing technique is the ***Sanger method*** (developed by ***Fred Sanger***) which uses dideoxynucleotides that stop chain elongation at the site of their incorporation. The four dideoxynucleotide substrates are labeled with radioactivity and used in four separate reactions *in vitro* in which a single-stranded DNA template is copied. A series of radioactive DNA fragments are separated according to length by agarose gel electrophoresis (fragments differing by only one nucleotide are separated), and the DNA sequence can be read directly from the pattern of radioactive bands in the gel.

Another sequencing technique is the ***Maxam-Gilbert method*** (developed by Alan Maxam and Walter Gilbert). In this method, a single DNA strand is labeled at the 5′ end with radioactive phosphorus (P^{32}). The radioactive DNA is divided into four portions and each one is exposed to different chemical reactions. Each reaction causes a 5′ cleavage adjacent to either

1. A or T,
2. G alone,
3. C or T, or
4. C alone.

The reactions are carried out for a short time so that, on average, only one cleavage occurs in each DNA molecule.

This produces a set of DNA fragments (one set for each reaction), whose length identifies that position of a particular base. For example, a fragment containing 19 nucleotides in the G-only reaction mixture identifies G at position 20 from the 5′ end. Similarly, a fragment containing 27 nucleotides present in the C or T reaction but not in the C-only reaction indicates that T is at position 28. The lengths of DNA fragments are determined by ***polyacrylamide gel electrophoresis,*** a technique that can separate DNA fragments differing in length by only one nucleotide.

Gene Therapy

One expected advance in medical treatment is the use of gene therapy to ameliorate or cure a variety of inherited disorders as well as diseases caused by somatic mutations such as cancer. The goal of gene therapy is the replacement of a defective, disease-causing gene in an individual by normal, functioning copies. In essence, gene therapy is a novel form of drug therapy; it uses the biochemical capacity of a patient's cells to synthesize the therapeutic agent from the introduced gene.

The first attempts at gene therapy were directed at trying to correct ***severe combined immunodeficiency (SCID)*** which is caused, in some cases, by a deficiency of adenosine deaminase (ADA), that is expressed in all tissues. ADA deaminates both adenosine and deoxyadenosine and, in the absence of ADA, deoxyadenosine accumulates in cells. Deoxyadenosine can be phosphorylated by the enzyme deoxcytidine kinase to produce deoxyadenosine triphosphate, a toxic substance that kills dividing cells. As a result, individuals with an ADA deficiency are defective in both humoral and cell-mediated immunity.

In the mid-1980s, the first clinical gene therapy trials were attempted to correct ADA deficiency in two children with SCID. The ADA gene was cloned into a viral vector and inserted into peripheral T cells removed from the affected children. After a period of growth of the genetically modified cells *in vitro,* they were reintroduced into the patients. Synthesis of ADA could be detected for a time but the activity disappeared as the introduced cells died. Several other trials of gene therapy for ADA deficiency showed promise, but it was still necessary to maintain the children with SCID on enzyme replacement therapy.

Finally, in 2000, successful treatment of an X-linked form of SCID (SCID-X1) by gene therapy was reported by a French medical team. Almost a year after a normal gene was introduced into their cells, two children were still synthesizing the enzyme that they lacked. This was the first success for gene therapy in curing a disease after years of effort.

Since the 1980s, hundreds of clinical trials of gene therapy for cystic fibrosis, osteogenesis imperfecta, Gaucher's disease, Fanconi's anemia, and several forms of cancer have been attempted. The problems that need to be overcome in developing a successful strategy for gene therapy are:

1. Development of safe and effective vectors for cloning the gene and inserting it into palients cells,
2. Regulation of expression of the desired gene product so that it is produced at the right time and in the correct amounts in the appropriate tissues, and
3. Stability of the inserted genetic construct in cells so that the product will continue to be produced.

Many different viral vectors have been developed to deliver genes to various organs in the body; these include retroviruses that can insert themselves and the genes they carry into the chromosomes of cells; adenovirus, a DNA virus that causes respiratory infections, which has been inactivated by removal of many viral genes; and lentiviruses, slow-growing retroviruses. All of the viral vectors carry with them some risks. In 1999, a gene therapy patient died of complications from the use of adenovirus that was being tested as a therapy for an inherited liver disease caused by a deficiency in the enzyme ornithine transcarbamylase (OTC). Despite the problems that have beset gene therapy trials, it is expected that development of safer vectors and new techniques for delivering genes to cells will eventually make gene therapy a vital part of medical treatment.

23.8 DNA Vaccines

An extension of gene therapy, and one that may turn out to be of worldwide importance, is the use of naked DNA as a vaccine to prevent viral diseases. For example, plasmid DNA can be injected into tissues; upon entry, the DNA expresses any cloned gene, such as a viral antigen, that is carried by the plasmid. DNA vaccines have the advantage that the viral protein that is expressed in the cells stimulates both humoral and cell-mediated immunity. Fragments of the synthesized viral protein are carried to the cells' surface where they stimulate $CD8^+$ cytotoxic T cells and, thereby, cell mediated-immunity (Figure 23-16).

In experiments with mice, ***DNA vaccines*** have proven to be very effective. The gene coding for the core protein of the influenza virus was cloned into a plasmid vector and injected into mice. The mice developed immunity not only to the strain of influenza from which the gene was derived but from other strains of influenza virus as well. Inducing an immune response with a viral core antigen is thought to be more effective than capsid antigens because viral core proteins from related viral strains do not differ much in structure or antigenicity. Viral capsid proteins, on the other hand, evolve rapidly; such changes, for example, account for the different strains of influenza virus that arise each year.

Before DNA vaccines can replace conventional vaccines that use inactivated or attenuated viruses, the safety of the plasmid vectors must be rigorously proved. The plasmids might occasionally integrate into the host genome or they might stimulate an immune response to tissues containing the plasmid DNA. Either event might dictate against the widespread use of DNA vaccines.

Antibodies to DNA

Both single- and double-stranded DNA are antigenic; antibodies to DNA are normally found in the circulation, but in some individuals overproduction of DNA antibodies causes disease. In particular, patients with ***systemic lupus erythematosus (SLE)*** show abnormal levels of antibodies to double-stranded DNA. Antibodies to single-stranded DNA also bind to bases, nucleosides, nucleotides, oligonucleotides, and the ribose-phosphate backbone of RNA. Antibodies to double-stranded DNA also bind to base pairs, chromatin, nucleosomes, type IV collagen, and the deoxyribose-phosphate backbone of DNA.

Antibodies to DNA consist of both IgM and IgG classes. Healthy individuals usually have low-affinity IgM antibodies to DNA; however, if these undergo an isotype switch to IgG, they may become pathogenic. Tests for DNA antibodies help establish a diagnosis of SLE,

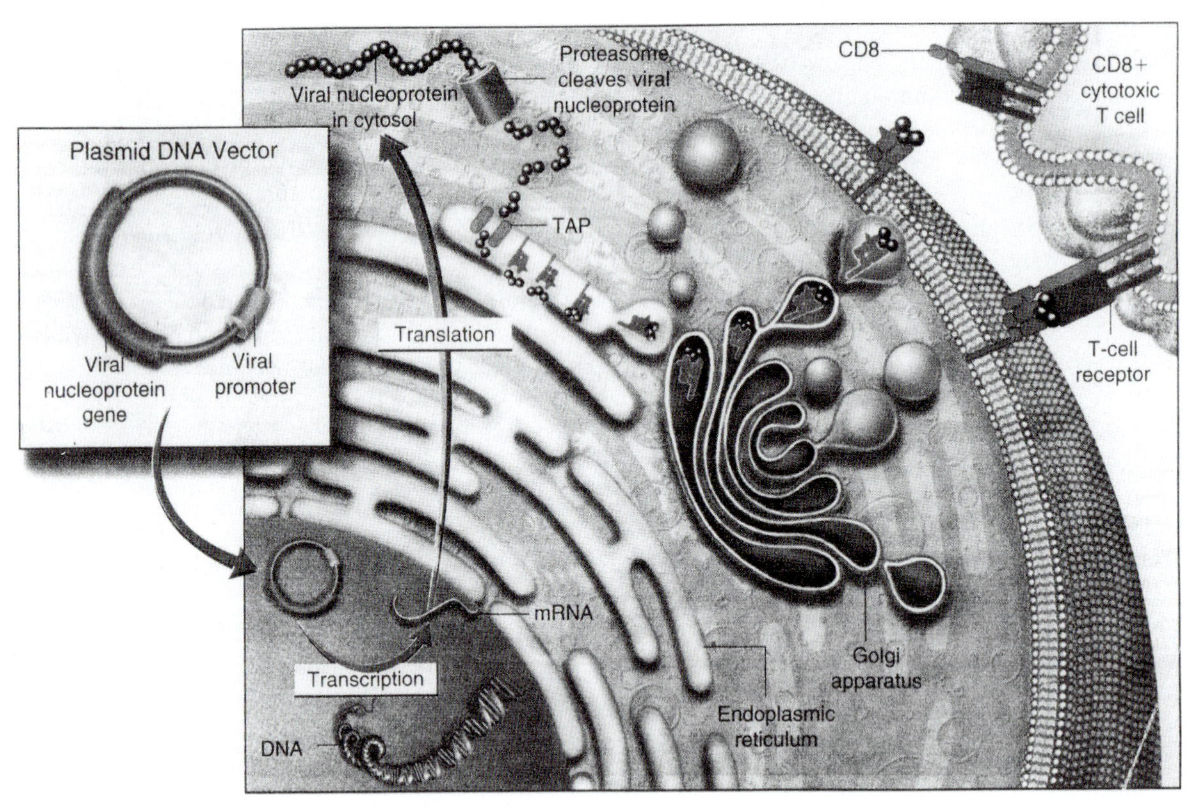

FIGURE 23-16
How a DNA vaccine might work. A viral gene coding for a core or capsid protein is inserted into a plasmid. The plasmids are injected into cells where the plasmid migrates to the nucleus. Viral proteins are expressed in the cells producing both a cell-mediated and humoral immune response. [Reproduced with permission from W. M. McDonnel and F. K. Ashari, DNA vaccines, *N. Engl. J. Med.* **314,** 44 (1996)].

although the level of such antibodies is not always predictive of the severity of the disease. Genetic susceptibility also seems to play a role in the pathogenicity of DNA antibodies in some individuals.

Antibodies to double-stranded DNA may cause glomerulonephritis by forming antibody-DNA complexes that become trapped in the glomeruli. In some cases, DNA antibodies can be recovered from damaged tissues in patients with SLE or glomerulonephritis, indicating that the antibodies are involved in tissue damage. The production of DNA antibodies is reduced by immunosuppressive drugs, and immunosuppressive therapy is the accepted treatment for SLE and related diseases caused by antibodies directed against DNA.

23.9 The Human Genome Project

A complete human genome, the total genetic information in a haploid set of chromosomes, contains approximately 3.1 billion bases and 50,000–1000,000 genes that encode proteins. The **Human Genome Project** was established with federal funds to obtain the complete sequence of a human genome or, more accurately, the sequence of pooled DNA extracted from cells donated by several anonymous individuals. Initially a public research effort, the sequencing task later was undertaken by a private company, Celera Genomics (Rockville, MD). The ensuing competition led to a joint announcement by the U.S. Government and Celera Genomics on June 26, 2000 that the human genome had essentially been sequenced although there were still gaps and errors in the sequence. The two groups agreed to continue cooperating to complete the sequence of a generic human genome as soon as possible.

Any two human genomes are approximately 99.9% identical in sequence. Of course, the 0.1% represents a difference of 3 million bases between any two individuals, and these base differences may have profound effects on disease susceptibility, behavior, intelligence, personality, and other traits. Approximately 75% of all human genes have the same DNA sequence in all individuals except for those with rare mutations.

The type of genetic markers employed in the dissection of the human genome include

1. **Restriction sites:** sequences of DNA digested by restriction enzymes;
2. ***Variable-number tandem repeats (VNTRs):*** sequences of bases that are repeated in tandem almost without change up to 40 copies;
3. ***Simple sequence repeats (SSRs):*** sequences of bases, most commonly CA, that are repeated dozens of times and occur at thousands of sites in the genome; and
4. ***Single-nucleotide polymorphisins (SNPs):*** sites in human DNA that differ by a single base pair.

Because SSRs and SNPs occur so frequently in the genome, they can be used to distinguish one chromosome from another in an individual. An SSR may occur 19 times at one location in a particular chromosome and 21 times in a homologous chromosome. Since an SSR is inherited like a gene, the variability can be followed from person to person in a pedigree. The length of a particular SSR can be determined by gel electrophoresis without any knowledge of the exact sequence. SNP differences occur about once in every thousand bases in DNA. That is, at a position on a chromosome where an A–T base pair occurs, a G–C pair might occur on the homologous chromosome.

Because only a small percentage of the total number of nucleotides in the human genome is actually used to encode information for the synthesis of different proteins, techniques have been developed to search for and map genes that are expressed. Knowing the location of these protein-encoding genes is extremely useful as mutated forms of these expressed genes are likely implicated in many diseases.

Two techniques have contributed to the identification of many active genes in the human genome. ***Expressed sequence tags (ESTs)*** for as many as 50,000 human genes have been generated by isolating messenger RNAs and copying them into DNA. These DNA probes can then be used to identify the location of the expressed genes on the various chromosomes by hybridization. Specific chromosomal sites also can be identified by using PCR primers to amplify a small segment of a chromosomal locus. The sites identified by using PCR amplification are called ***sequence tagged sites (STSs).***

Completing the sequencing of the human genome is only the first step in the revolution in medicine that many are predicting will result in the 21st century. However, knowing the base sequence, chromosomal position, and even the function of the protein that is encoded in the gene is only the first step in using the information to benefit patients with inherited defects (Figure 23-17). The sequence information does not specify how and when a gene is turned or or off, in what cells the gene is active or inactive, or in what stage of development expression of the gene is needed. The expression of most genes is modulated in complex ways and coordinated by numerous intracellular and extracellular factors. How a particular protein affects the biochemistry, physiology, and phenotype of an individual must be understood before therapeutic messures can be developed. Often inherited defects are manifested during fetal development and the damage is irreversible by the time the child is born. Some of the anticipated medical, social, and ethical consequences deriving from the human genome project are described below.

Genetic testing: Hundreds of genetic tests are available that can be used to determine if a person carries a defective gene and to estimate the risks of developing a disease as well as the risk of passing the gene on to progeny. In many instances, determining the risk of passing on genes for Hungtington's disease, cystic fibrosis, hemophilia, Duchenne's muscular dystrophy, or sickle cell anemia provides prospective parents with information useful in making reproductive choices. However, an asymptomatic person with a family history of Huntington's disease faces a difficult choice. Finding out while young that you carry a gene that eventually will produce a lethal, neurological disease creates enormous emotional and psychological stress since there is no treatment or cure that can be offered. Faced with the choice of genetic testing for Huntington's disease, many possible carries of the gene opt not to be tested.

Genetic tests also can determine if a person carries genes that increase the chances of developing breast cancer, colon cancer, and other polygenic-multifactorial diseases. Again, in most cases, only marginally helpful medical help can be offered to individuals who test positive for cancer-causing susceptibility genes. Some young women who are homozygous for the *BRCA1* gene have elected to undergo prophylactic bilateral mastectomy to reduce the risk of breast cancer later in life. If one tests positive for colon cancer susceptibility genes, the only recourse is frequent examination of the colon and immediate removal of any polps that appear. Clearly, the consequences of many of the genetic tests currently available create serious problems for those who discover that they carry disease-causing genes of one kind or another. As hundreds of additional genetic tests become available, patients and physicians will have to make difficult choices regarding the risks and benefits of genetic testing.

Genetic discrimination: Profound social, ethical, and financial questions are raised by wide spread use of genetic testing and identification of persons at risk for diseases. Employers and insurance companies would like to have access to such information since it would be in their financial interest to know who is likely to become sick or disabled.

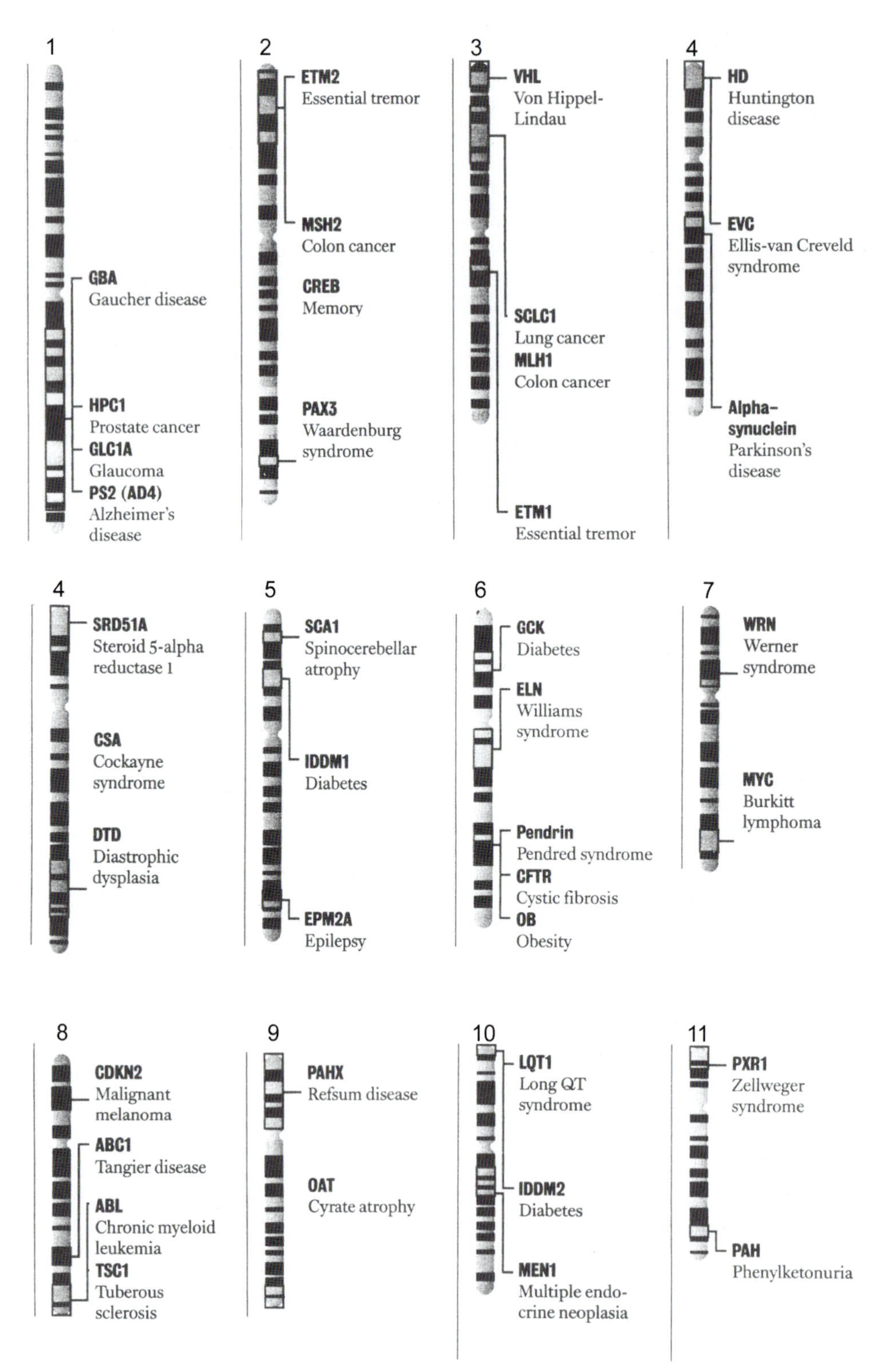

FIGURE 23-17
A sampling of single-gene defects (Mendelian disorders) that have been mapped to specific chromosomal loci. More than 8000 inherited disorders and defects have been identified as stemming from mutations in single genes, and many of the altered base changes have been mapped to specific chromosomal loci. As the analysis of the DNA sequence of the human genome proceeds, it is expected that eventually thousands more mutated alleles that cause inherited disorders and defects will be identified and mapped.

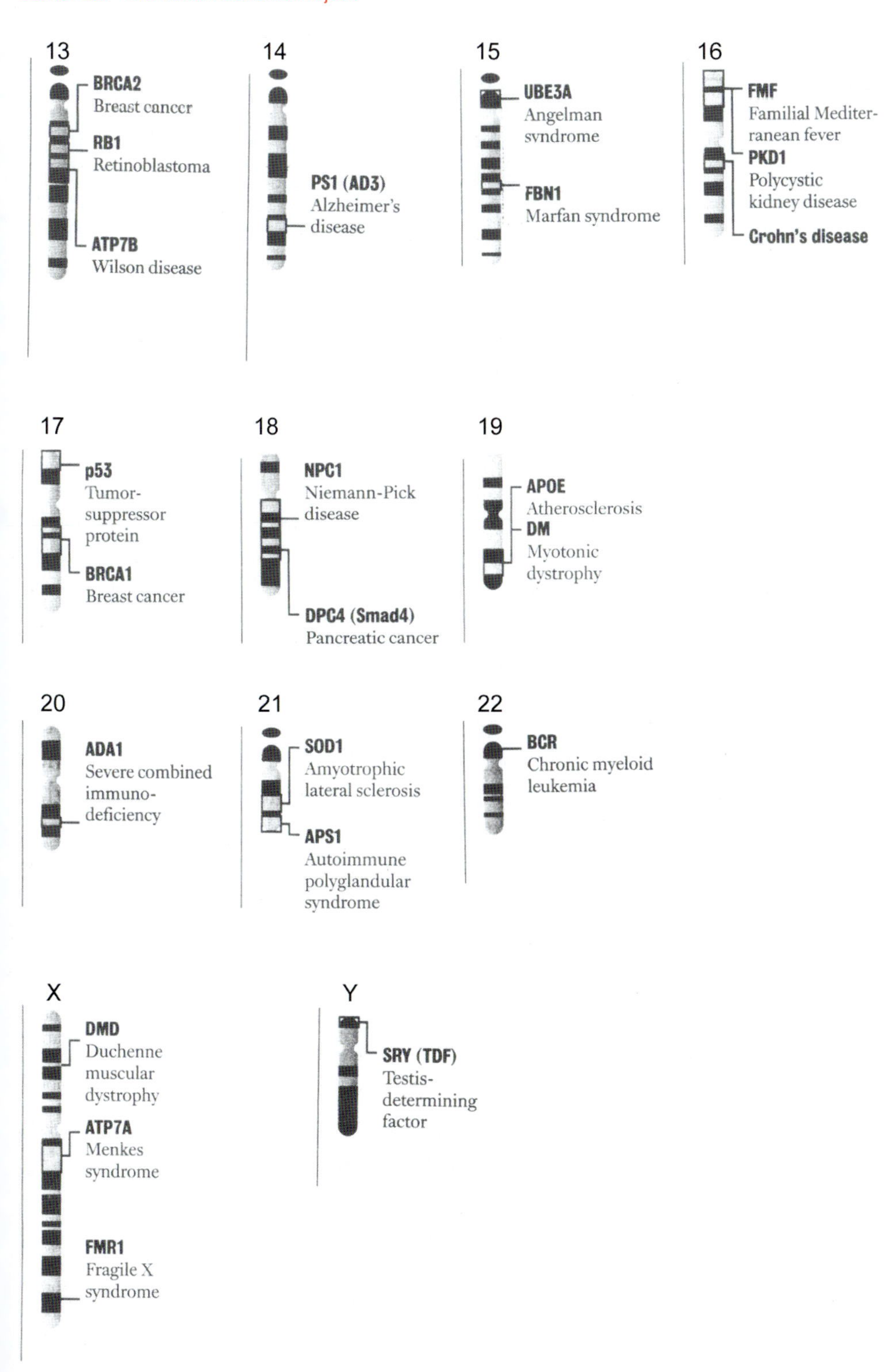
13
BRCA2
Breast cancer
RB1
Retinoblastoma
ATP7B
Wilson disease
14
PS1 (AD3)
Alzheimer's disease
15
UBE3A
Angelman syndrome
FBN1
Marfan syndrome
16
FMF
Familial Mediterranean fever
PKD1
Polycystic kidney disease
Crohn's disease
17
p53
Tumor-suppressor protein
BRCA1
Breast cancer
18
NPC1
Niemann-Pick disease
DPC4 (Smad4)
Pancreatic cancer
19
APOE
Atherosclerosis
DM
Myotonic dystrophy
20
ADA1
Severe combined immuno-deficiency
21
SOD1
Amyotrophic lateral sclerosis
APS1
Autoimmune polyglandular syndrome
22
BCR
Chronic myeloid leukemia
X
DMD
Duchenne muscular dystrophy
ATP7A
Menkes syndrome
FMR1
Fragile X syndrome
Y
SRY (TDF)
Testis-determining factor

FIGURE 23-17 (*Continued*)

On the other hand, such use of personal medical information would generate a society that accepts the concept of genetic discrimination. Since our society has made great efforts to eliminate sex, racial, and age-related discrimination, the overwhelming sentiment is to not allow genetic discrimination. The question for society is how to guarantee that genetic discrimination will not be practiced either legally or illegally as more and more DNA information becomes part of peoples' medical records. Laws against genetic discrimination have been introduced in many states but there is no federal law (as of 2000) that guarantees a person's right to genetic privacy.

Drug design: One of the anticipated benefits of the identification of human genes involved in diseases is the development of highly specific drugs that can be targeted to correct the biochemical and physiological consequences of a defective gene and its abnormal protein. It is expected that drugs eventually can be developed that will be targeted to correct the effects of specific mutations that individuals carry. Drug design and drug therapy may evolve to the point where each person's genotype determines the drug that will be used. Pharmacogenomics is the study of the relation between human genetic variability in the activity, toxicity, and metabolism of drugs. Examples include deficiencies of catabolic enzymes dihydropyrimidine dehydrogenase and thiopurine S-methyl transferase which can cause profound toxicity to fluoropyrimidine and mercaptopurine, respectively. These are used in cancer chemotherapy (Chapter 27).

23.10 Genomics and Proteomics

Enormous progress has been made in recent years in deciphering the genetic basis of different human phenotypes and clinical diseases that result from inherited mutations. Technological breakthrough have made the identification of genes and their sequencing a relatively straightforward procedure. ***Genomics*** is the study of DNA and the genes that it contains. While genomics has resulted in the characterization of many human genes that are responsible for abnormal phenotypes and genetic disorders, identification of the protein corresponding to the gene is often difficult and, in many cases, is still unknown long after the gene has been identified and sequenced. Often the activity of the gene does not correlate with the amount of the protein or its activity in the cell. Posttranslational modifications cannot be determined by genomics, which only allows determination of the primary amino acid sequence as deduced from the nucleotide sequence of the gene.

Proteomics is the direct study of proteins produced by both healthy and diseased cells. For example, two-dimensional gel electrophoresis can be used to display and quantify thousands of proteins within a cell. A change in the level of a particular protein, either higher or lower, can be detected by differential protein abundance analysis. When a cell receives a chemical signal, such as the binding of a growth factor or cytokine, the immediate response is at the level of proteins that are unrelated to the target gene. Cell surface receptor proteins are modified in response to the external signal. Also, transmission of information from the cell surface to the nucleus is mediated by the physical movement of proteins. Proteomic technologies have been developed that can detect and analyze such protein changes. In addition, the level of a particular protein in the cell may change dramatically in response to a change in transcriptional regulation of one or more genes, and this can be monitored (Chapter 26).

Proteomics is a valuable new technology used in the development of novel and more effective drugs following the identification and characterization of disease-causing proteins. Proteins also provide useful molecular markers for the diagnosis of specific diseases. In essence, proteomics provides the link between genes, proteins, and diseases and complements the information obtained by genomics.

Supplemental Readings and References

Structure of DNA

B. Alberts, A. B. Bray, J. Lewis, et al.: *Molecular Biology of the Cell.* New York: Garland (1994).

S. Henikoff, E. A. Greene, S. Pietrokovski, et al.: Gene families: The taxonomy of protein paralogs and chimeras, *Science* **278,** 609 (1997).

D. W. Ross: The human genome: information content and structure. *Hospital Practice,* June 15, 49 (1999).

Sequencing DNA

L. Alphey: *DNA Sequencing.* New York: Springer-Verlag (1997).

F. R. Blattner, G. Plunkott, C. A. Bloah, et al.: The complete genome sequence of *Escherichia coli* K-12. *Science* **277,** 1453 (1997).

F. S. Collins: Sequencing the human genome. *Hospital Practice,* January 15, 35 (1997).

F. S. Collins: Medical and societal consequences of the human genome project, *New England Journal of Medicine* **341,** 28 (1999).

N. Rosenthal: Fine structure of a gene—DNA sequencing. *New England Journal of Medicine* **332,** 589 (1995).

L. Rowen, G. Mahairas, and L. Hood: Sequencing the human genome. *Science* **278,** 669 (1997).

Properties of DNA

B. H. Hahn: Antibodies to DNA. *New England Journal of Medicine* **338,** 1359 (1998).

D. E. Housman: DNA on trial—The molecular basis of DNA fingerprinting. *New England Journal of Medicine* **332,** 534 (1995).

S. P. Naber: Molecular pathology—diagnosis of infectious DNA. *New England Journal of Medicine* **331,** 1212 (1994).

W. Reik and M. A. Surami, eds.: *Frontiers in Molecular Biology*. New York: Oxford University Press (1997).

C. W. Schmid, N. Deka, and G. Matera: Repetitive human DNA: The shape of things to come. In *Chromosomes: Eukaryotic, Prokaryotic, and Viral.* Volume 1. K. W. Adolph, ed., Boca Raton, FL: CRC Press (1990).

R. A. Seder and S. Gurunathan: DNA vaccines—Designer vaccines for the 21st century. *New England Journal of Medicine* **341,** 277 (1999).

Gene Therapy

H. M. Blau and M. I. Springer: Muscle-mediated gene therapy. *New England Journal of Medicine* **333,** 1554 (1995).

M. Cavazzana-Calvo, S. Hacein-Ray, G. de Saint Basile, et al.: Gene therapy of human severe combined immunodeficiency (SCID)-X1 disease. *Science* **288,** 669 (2000).

S. J. Eck, ed.: Gene therapy. *Hematology/Oncology Clinics of North America* **12** (1998).

W. M. McDonnell and F. K. Askari: DNA vaccines. *New England Journal of Medicine* **334,** 42 (1996).

L-C. Tsui and P. Durie: Genotype and phenotype in cystic fibrosis. *Hospital Practice*, June 15, 15 (1997).

Human Genome Project

G. J. Annas: Rules for research on human genetic variation—Lessons from Iceland. *New England Journal of Medicine* **342,** 1830 (2000).

E. Green: The Human Genome Project and its impact on the study of human disease. In *Metabolic and Molecular Bases of Inherited Disease,* C. R. Scriver, ed., 8th edition. New York, McGraw-Hill (2000).

V. Hatzimanikatis, L. H. Choe, and K. H. Lee: Proteomics: Theoretical and experimental considerations. *Biotech. Prod.* **15,** 312 (1999).

N. A. Holtzman and T. M. Marteau: Will genetics revolutionize medicine? *New England Journal of Medicine* **343,** 141 (2000).

G. J. B. van Ommen, E. Bakker, and J. T. den Dunnen: The human genome project and the future of diagnostics, treatment, and prevention. *The Lancet* **354,** SI8, (Molecular Medicine) (1999).

D. B. Searls: Using bioinformatics in gene and drug discovery. *Drug Discovery Today* **5,** 135 (2000).

CHAPTER 24

DNA Replication, Repair, and Mutagenesis

DNA is the permanent repository for all genetic information in every cell and, as such, must maintain the fidelity of that information from one cell generation to the next as well as from one human generation to the next. This means, in the case of human beings, accurately copying the approximately 6×10^9 base pairs in each diploid cell prior to its undergoing mitosis (somatic cells) or meiosis (germ cells).

Over billions of years of biological history, cells have developed intricate mechanisms to ensure faithful replication of DNA molecules and for the detection and repair of errors in the sequence of bases when they do occur. However, errors *must* occur; the evolution of new organisms and new species depends on a supply of new genomes, the source of genetic diversity on which natural selection acts and which fuels biological evolution. This is the paradox of DNA replication. On the one hand, DNA must replicate itself with the utmost accuracy; on the other hand, it must allow a low frequency of genetic change to occur during each round of replication, particularly during meiosis, which produces the cells (sperm and ova) that pass DNA to succeeding generations.

This apparent paradox is resolved by the mechanisms of mutation and recombination that provide for a small but essential amount of genetic change to occur in each generation. Mutations occur spontaneously during DNA replication because not all of the replication errors or damage to templates are detected and repaired by the enzymes responsible for proofreading and repairing DNA. The rate of mutations can be markedly increased by ionizing radiation (x-rays), mutagenic chemicals (cigarette smoke), or retroviruses that integrate in DNA (HIV). Thus, mutations are essential to the species because they allow for adaptive changes but often are harmful to the individual by giving rise to cells that cause cancer, congenital defects, and other diseases.

Recombination occurs primarily in germ cells of eukaryotes and during cell division in partially diploid prokaryotes. In general, recombination does not occur in mitosis of eukaryotic cells. However, recombination is an integral part of meiosis in which recombinant sperm and ova provide new genotypes and phenotypes in each generation which may be favored by natural selection.

In this chapter, the basic mechanisms of DNA replication, recombination, repair, and mutation are discussed. Much of our understanding of these mechanisms derived from the study of bacteria. However, as further evidence for the unity of biology, the processes that replicate and modify DNA in bacteria also apply to human DNA with some modifications.

24.1 DNA Replication

Problems of Replication

The replication process requires that each double-helical molecule of DNA produce two identical molecules of DNA. This means that wherever a G-C or A-T base pair occurs in the parental molecule, the identical base pair must occur in the progeny molecules. However, many factors interfere with accurate replication of DNA. If an A should pair with C or G with T as a result of tautomerization (Chapter 23), a ***point mutation*** (a change in one base pair) will result. Occasionally, a segment of DNA will be replicated more than once (***duplication***) or a segment may fail to be replicated (***deletion***). These and other aberrations in DNA replication do occur, but the mechanism of replication has evolved to minimize such mistakes.

Semiconservative Replication

The information in each strand of the double helix serves as the template for the construction of a new double-helical DNA molecule; this is called ***semiconservative replication*** since one old strand of DNA is paired with one new strand to produce the daughter DNA molecule. All DNA molecules in all organisms replicate semiconservatively. This basic fact was first proved by Matthew Meselson and Frank Stahl who labeled DNA in *E. coli* by growing cells for several generations in medium containing a heavy isotope of nitrogen (^{15}N). Cells were then transferred to medium containing ^{14}N and continued to grow exponentially. After one generation of growth in ^{14}N medium, the DNA was extracted from a sample of cells and analyzed by cesium chloride density centrifugation. All DNA was of intermediate (hybrid) density; one strand contained ^{15}N and the newly synthesized strand contained ^{14}N. After two generations of growth in ^{14}N medium, 50% of the DNA contained only ^{14}N (light DNA) and 50% is hybrid. This important experiment confirmed a prediction of the Watson-Crick model for DNA. Subsequent experiments also showed that human DNA undergoes semiconservative replication.

Origin and Direction of DNA Replication

Another unifying principle of DNA replication is that each molecule of DNA has one or more specific ***origins of replication*** where DNA synthesis begins. In bacteria that carry their genetic information in a circular DNA molecule, only one origin of replication exists. Plasmid DNA also contains a unique origin of replication. Origins of replication can be moved from one location in DNA to another, or from one DNA molecule to another just as genes can be moved by genetic engineering techniques.

In both bacterial and plasmid DNA, replication is initiated at a unique origin and proceeds in both directions along the DNA molecule to a ***termination site*** that is located approximately 180 degrees from the origin. Thus, replication of DNA is, in most cases, ***bidirectional replication.*** The ability of DNA to replicate from many origins and in both directions means that cells can replicate all of their DNA in a fairly short period of time. The shortest generation time for the bacterium *E. coli* is approximately 30 minutes in rich medium in the laboratory whereas the shortest division time for a human cell is approximately 24 hours (Table 24-1).

TABLE 24-1

Comparison of Parameters of DNA Replication in E. coli and Human Cells

DNA	*E. coli*	Human
Amount of DNA	~3.9×10^6 bp	~3×10^9 bp
Rate of replication at each replication fork	~850 bases/sec	~6–90 bases/sec
Number of origins of replication	one	10^3 to 10^4
Time for cell division	~30 min	~24 hours

Unwinding a double helix during replication presents a serious mechanical problem. Either the two daughter DNA molecules at the ***replication fork*** (the Y-shaped fork) must rotate around one another, or the unreplicated segment of DNA must rotate (Figure 24-1). This necessity for DNA rotation during replication creates topological problems for covalently closed circular DNA in bacteria and for the enormously long condensed and folded DNA in human chromosomes. The unwinding of strands of the DNA double helix is accomplished by a special group of enzymes called ***helicases;*** positive and negative coiling of DNA to maintain the topology necessary for replication is the function of another group of enzymes called ***topoisomerases*** (discussed in a later section).

Replicons

Much of the basic understanding of DNA replication was gleaned from studying the replication of ***plasmids*** (Chapter 23): small, circular, extrachromosomal DNA elements in bacteria. Some plasmids, such as the F (fertility) factor in *E. coli,* have an origin of replication but replicate only unidirectionally from that single origin by a rolling-circle mechanism of replication (Figure 24-2). If an F factor

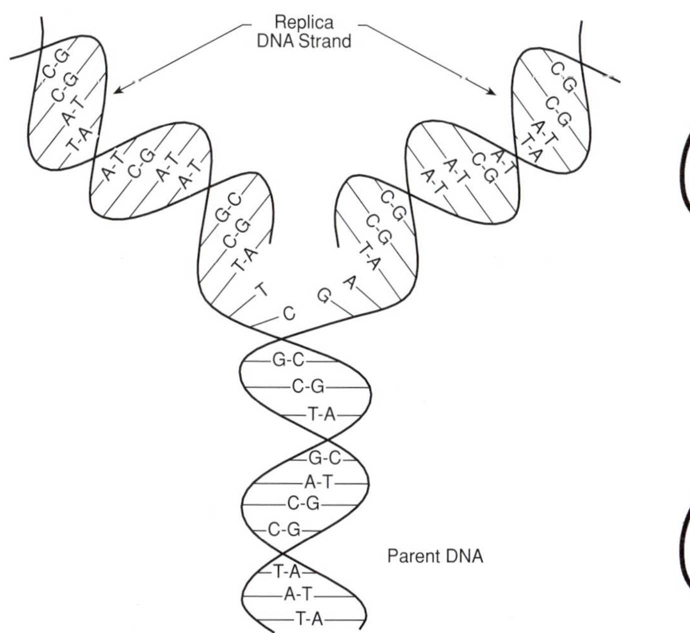

FIGURE 24-1
Semiconservative replication of DNA. The two replicas consist of one parental strand plus one daughter strand. Each base in a daughter strand is selected by the requirement that it form a base pair with the parental base.

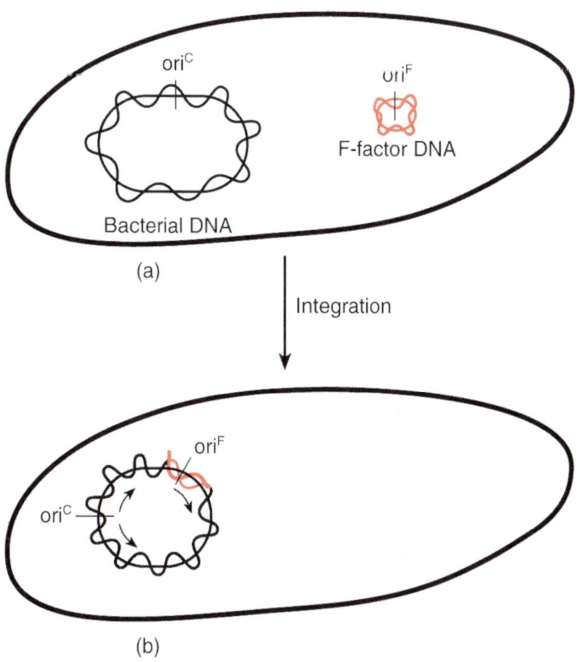

FIGURE 24-3
Replication of bacterial DNA with an integrated F factor. There are two origins of DNA replication, ori^C and ori^F. Replication is bidirectional from ori^C and unidirectional from ori^F. Somehow the cell solves this topological dilemma when cell division occurs.

integrates into the bacterial chromosome, it can replicate its own DNA and bacterial DNA as well during ***conjugation,*** a process in which bacteria physically join and transfer DNA from the donor bacterium to the recipient bacterium. In this situation, bacteria replicate DNA simultaneously in a bidirectional manner from their own origin of replication (ori^C) and unidirectionally from the origin of the integrated F factor (ori^F). Such replication clearly raises topological problems during DNA replication (Figure 24-3). Studies of bacterial and plasmid DNA replication led to the concept of a ***replicon,*** which is:

1. A site on DNA consisting of a sequence of nucleotides that defines an origin of replication, and
2. Structural genes coding for proteins that recognize and bind to the origin to initiate DNA replication.

These genes and the segment of DNA that is replicated from the origin are collectively defined as a replicon. Thus, a bacterial chromosome or a plasmid is a single replicon, whereas a single human chromosome may contain thousands of replicons.

Discontinuous DNA Replication

In the diagram of replication shown in Figure 24-1, both daughter molecules are shown as having continuous strands. In truth, DNA cannot replicate by copying both strands continuously because DNA polymerases, the enzymes that synthesize new DNA, can *only* add nucleotides to a 3′-OH group. Because of the antiparallel nature of the

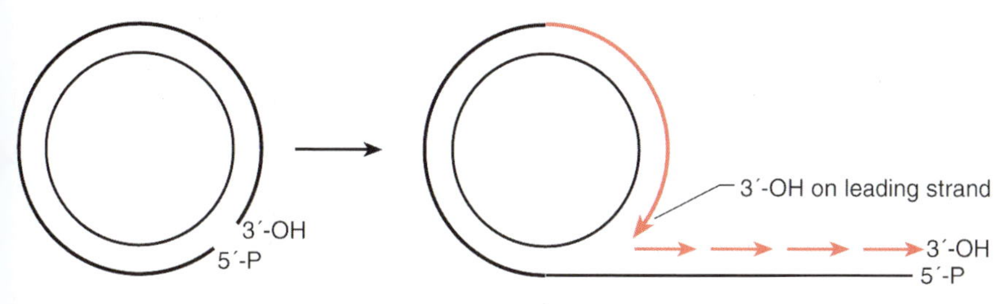

FIGURE 24-2
Rolling-circle or σ replication. Newly synthesized DNA is shown in color.

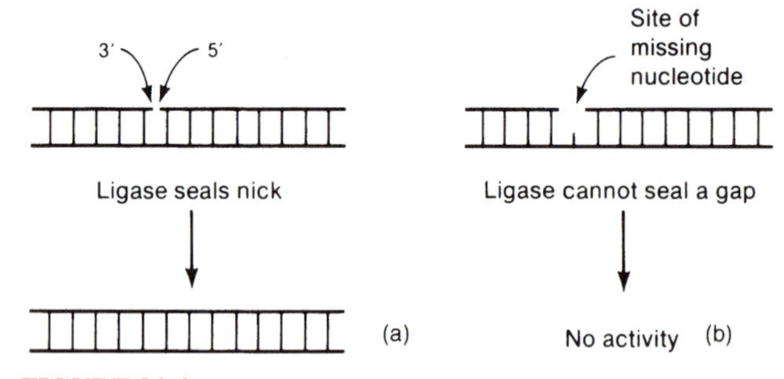

FIGURE 24-4
Action of DNA ligase. (a) A nick having a 3′-OH and a 5′-P terminus sealed (left panel). (b) If one or more nucleotides is absent, the gap cannot be sealed.

DNA strands, synthesis of one new strand of DNA can be continuous but synthesis along the other strand must occur by ***discontinuous replication.***

At each replication fork one DNA strand has a free 3′-OH group; the other has a free 5′-PO_4^{2-} group. Since DNA can only elongate in a 5′ to 3′ direction, synthesis of one strand (the ***leading strand***) is continuous; synthesis of the opposite strand (the ***lagging strand***) is discontinuous.

Short fragments of DNA are synthesized in the replication fork on the lagging strand in a 5′to 3′ direction; these are called ***Okazaki fragments*** after the scientist (Reiji Okazaki) who first demonstrated their existence during replication. These Okazaki fragments are joined in the replication fork by the enzyme DNA ligase that can form a phosphodiester bond at a single-strand break in DNA. DNA ligase joins a 3′-OH at the end of one DNA fragment to the 5′-monophosphate of the adjacent fragment (Figure 24-4). However, if even one nucleotide is missing in the DNA strand, ligase cannot seal the sugar-phosphate backbone.

24.2 Enzymology of DNA Replication

The fundamental enzymology of DNA replication derives from both *in vivo* and *in vitro* studies with cells and extracts derived from *E. coli*. Many of the enzymes involved in DNA replication were identified by isolation of conditional lethal mutants of the bacterium, e.g., mutants that are unable to replicate DNA (and unable to grow) at high temperatures (42°C) but that replicate and grow normally at low temperatures (30°C).

The synthesis of DNA is a complex process because of the need for faithful replication, enzyme specificity, and topological constraints. Approximately 20 different enzymes are utilized in bacteria to replicate DNA. In addition to polymerization reactions, DNA replication requires accurate initiation, termination, and proofreading to eliminate errors.

DNA Polymerases

Three DNA polymerases have been characterized in *E. coli* and are designated polymerase I, polymerase II, and polymerase III (Table 24-2). Although present in very low concentration in the cell, polymerase III, also called ***replicase,*** is the polymerase that elongates both strands of the bacterial DNA in the replication fork. Polymerase I is primarily a DNA repair enzyme and is responsible for excision of the short RNA primer that is required to initiate DNA synthesis on both the leading and lagging strands of DNA during replication. It also can remove mismatched base pairs during replication and fill in gaps in single-stranded DNA that is joined in a double helix. The function of polymerase II is not clear but it probably also has repair functions.

All DNA polymerases select the nucleotide that is to be added to the 3′-OH end of the growing chain

TABLE 24-2
Properties of E. coli DNA polymerases

Property	Polymerase		
	I	II	III
Molecular weight	105,000	90,000	130,000
Molecules/cell	~400	~100	~10
Nucleotides/sec	~20	~5	~1,000
3′ exonuclease activity	yes	yes	no
5′ exonuclease activity	yes	no	no
Biological activity	RNA primer excision, DNA repair	SOS DNA repair?	Replicase

and catalyze formation of the phosphodiester bond. The substrates for DNA polymerases are the four deoxynucleoside-5′-triphosphates (dATP, dCTP, dGTP, and dTTP) and a single-stranded template DNA. The overall chemical reaction catalyzed by all DNA polymerases is

$$\text{Poly(nucleotide}_n\text{)-3}'\text{-OH} + \text{dNTP} \rightleftarrows \text{poly(nucleotide}_{n+1}\text{)-3}'\text{-OH} + \text{PP}_\text{i}$$

in which PP_i represents pyrophosphate cleaved from the dNTP. That is, a reaction occurs between a 3′-OH group at a terminus of a DNA strand and the phosphoryl group (the one linked to the sugar) of an incoming nucleoside triphosphate.

Even though the hydrolysis of the nucleoside triphosphates has a large negative ΔG, the polymerization reaction as written still has a positive ΔG at concentrations present in a cell and in laboratory reactions ($+0.5$ kcal/mol $= 2.1$ kJ/mol). Thus, in the absence of any other reaction DNA polymerases would catalyze depolymerization rather than polymerization. Indeed, if excess pyrophosphate and a polymerase are added to a solution containing a partially replicating DNA molecule, the polymerase acts like a nuclease. In order to drive the reaction to the right, pyrophosphate must be removed, and this is accomplished by a potent pyrophosphatase, a widely distributed enzyme that breaks down pyrophosphate to inorganic phosphate. Hydrolysis of pyrophosphate has a large negative free energy, so essentially all of the pyrophosphate is rapidly removed.

No DNA polymerase can catalyze the reaction between two free nucleotides, even if one has a 3′-OH group and the other a 5′-triphosphate. Polymerization can occur only if the nucleotide with the 3′-OH group is hydrogen-bonded to the template strand. Such a nucleotide is called a ***primer*** (Figure 24-5). The primer can either be a single nucleotide or the terminus of a hydrogen-bonded oligonucleotide. When an incoming nucleotide is joined to a primer it supplies another free 3′-OH group, so that the growing strand itself is a primer. Since polymerization occurs only at the 3′-OH end, strand growth is said to proceed in the $5' \rightarrow 3'$ direction. All known polymerases (both DNA and RNA) are capable of chain growth only in the $5' \rightarrow 3'$ direction. This unidirectional feature of polymerases complicates the simultaneous replication of both strands of DNA.

Polymerization is not confined to addition of a nucleotide to a growing strand in a replication fork. For example, pol I can also add nucleotides to the 3′-OH group at a single-strand (a nick) in a double helix. This activity results from the ability of pol I both to recognize a 3′-OH group anywhere in the helix and to displace the base-paired strand ahead of the available 3′-OH group.

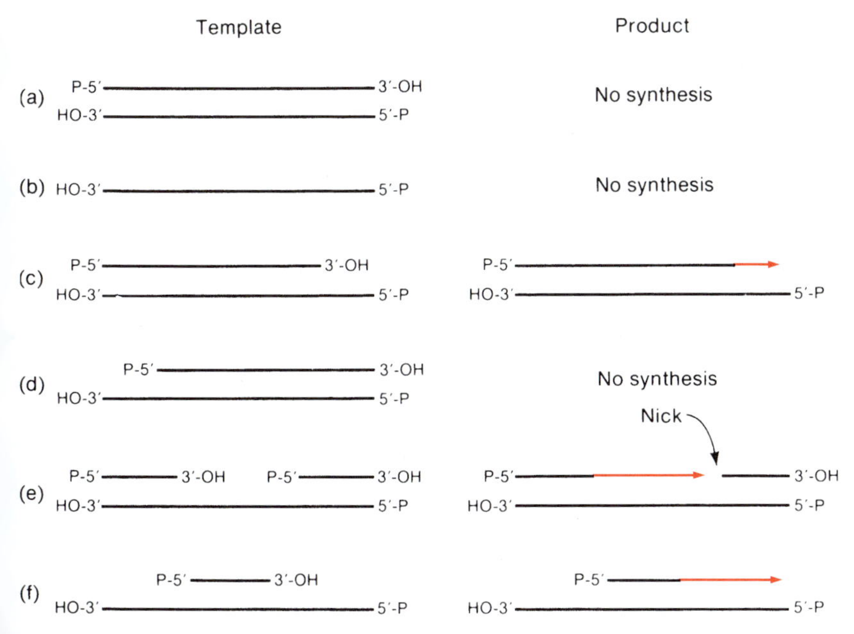

FIGURE 24-5
Effect of various templates used in DNA polymerization reactions. A free 3′-OH on a hydrogen-bonded nucleotide at the strand terminus and a non-hydrogen-bonded nucleotide at the adjacent position on the template strand are needed for strand growth. Newly synthesized DNA is shown in color.

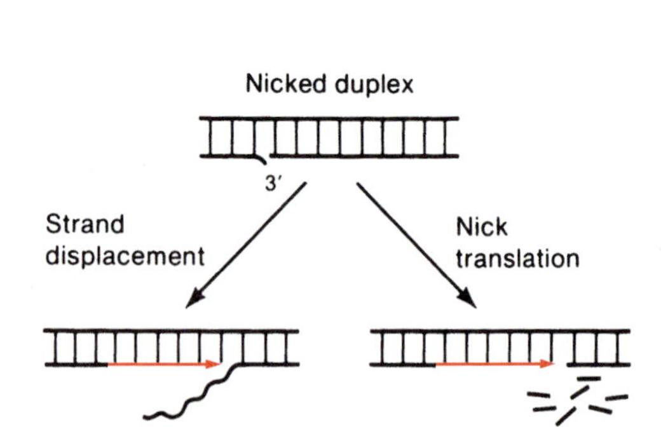

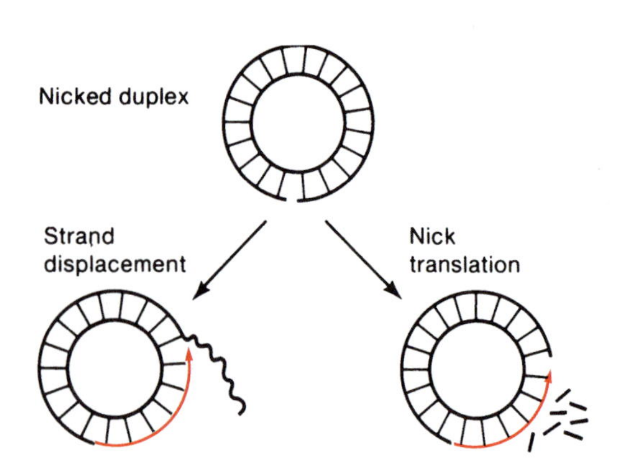

FIGURE 24-6
Strand displacement and nick translation on linear and circular molecules. In nick translation, a nucleotide is removed by an exonuclease activity for each nucleotide added. The growing strand is shown in color.

This reaction is called ***strand displacement*** (Figure 24-6). Not all DNA polymerases can carry out strand displacement.

Pol I has several enzymatic activities other than its ability to polymerize; one of these is important in maintaining continuity of the daughter strand, and the other improves the fidelity of replication. These two functions are a $5' \rightarrow 3'$ exonuclease activity and a $3' \rightarrow 5'$ exonuclease activity.

Exonuclease Activities of Polymerase I

Occasionally DNA polymerase, in error, adds a nucleotide to the 3′-OH terminus that cannot hydrogen-bond to the corresponding base in the template strand. Such a nucleotide would clearly alter the information content of the daughter DNA molecule, and mechanisms exist for correcting such incorporation errors.

Once having added an incorrect nucleotide and moved on to the next position, pol I cannot add another nucleotide because the enzyme requires a primer that is correctly hydrogen-bonded. Where such an impasse is encountered, a $3' \rightarrow 5'$ exonuclease activity, which may be thought of simply as pol I running backward or in the $3' \rightarrow 5'$ direction, is stimulated and the unpaired base is removed. After removal of this base, the exonuclease activity stops, polymerizing activity is restored, and chain growth resumes. This exonuclease activity is called the ***proofreading*** or ***editing function*** of polymerase I.

Pol I also has a potent $5' \rightarrow 3'$ exonucleolytic activity (Figure 24-7). This activity is directed against a base-paired strand and consists of stepwise removal of nucleotides one by one from the 5′-P terminus. Furthermore, the nucleotide removed can be either of the deoxyribo or the ribo type. The $5' \rightarrow 3'$ exonucleolytic activity also can be coupled to the polymerizing activity. Recall that pol I can add nucleotides to a 3′-OH group at a nick and displace the downstream strand. Under certain conditions, the displacement reaction does not occur and instead the $5' \rightarrow 3'$ exonuclease acts on the strand that would otherwise be displaced, removing one downstream nucleotide for each nucleotide added to the 3′ side of the nick. Thus, the position of the nick moves along the strand; this reaction is called ***nick translation.*** It is an important laboratory procedure for producing labeled DNA that can be used as probes; simply by carrying out the reaction in the presence of radioactive or chemically labeled nucleotides, an unlabeled DNA molecule with nicks in both strands can be converted to a labeled molecule. DNA probes are used for many diagnostic and forensic purposes (Chapter 23). The main function of the $5' \rightarrow 3'$ exonuclease activity is to remove ribonucleotide primers that are used in DNA replication.

Pol I also has a $5' \rightarrow 3'$ endonuclease activity. An endonucleolytic cut is made between two base pairs that follow a 5′-P-terminated segment of unpaired bases, as shown below.

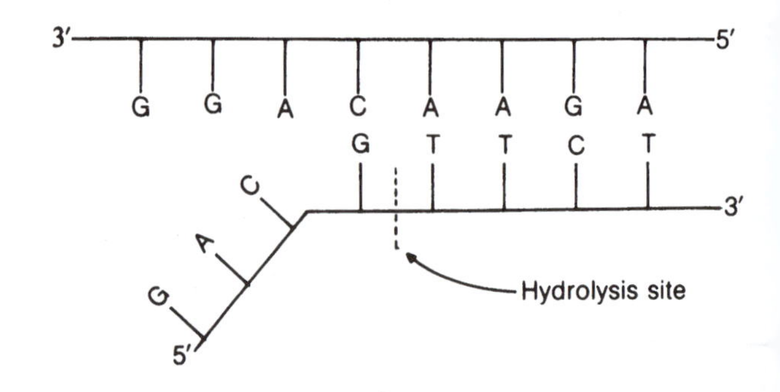

This type of endonucleolytic activity is unimportant in normal replication but is important in excision repair.

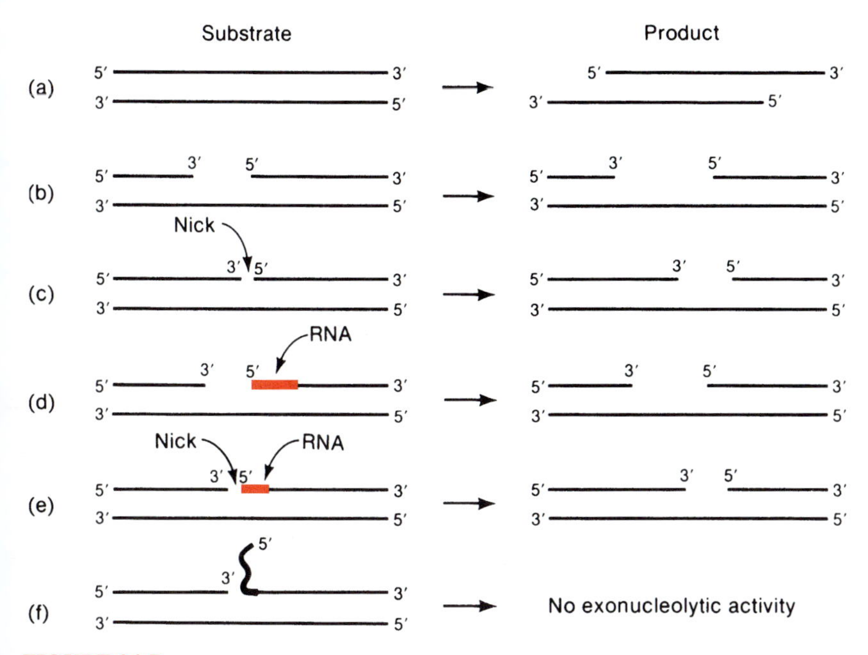

FIGURE 24-7
Several substrates and products alter activity of the polymerase I 5′ → 3′ exonuclease. The 5′ terminus attacked by the enzyme is shown by the small straight arrow. (a) 5′-Terminal nucleotides are removed from each end of the molecule. (b) A gap is enlarged by cleavage at a 5′ terminus. (c) A gap is formed by removal of a 5′-terminal nucleotide at a nick. The gap is then enlarged as in (b). (d) RNA (colored line) at the boundary of a gap is removed by cleavage at the 5′ terminus of the RNA. (e) A gap is formed from a nick at a DNA-RNA boundary by removal of 5′-P-terminal ribonucleotides. (f) A non-hydrogen-bonded 5′ terminus is resistant to the exonuclease; there is an endonucleolytic activity.

Polymerase III

Polymerase I plays an essential role in the replication process in *E. coli,* but it is not responsible for the overall polymerization of the replicating strands. The enzyme that accomplishes this is a less abundant enzyme, polymerase III (pol III). (A DNA polymerase II has also been isolated from *E. coli,* but it probably plays no role in DNA synthesis.) Pol III catalyzes the same polymerization reaction as pot I but has certain distinguishing features. It is a very complex enzyme and is associated with eight other proteins to form the pol III holoenzyme. (The term ***holoenzyme*** refers to an enzyme that contains several different subunits and retains some activity even when one or more subunits is missing.) Pol III is similar to pol I in that it has a requirement for a template and a primer but its substrate specificity is much more limited. For a template pol III cannot act at a nick nor can it unwind a helix and carry out strand displacement. The latter deficiency means that an auxiliary system is needed to unwind the helix ahead of a replication fork. Pol III, like pol I, possesses a 3′ → 5′ exonuclease activity, which performs the major editing function in DNA replication. Polymerase III also has a 3′ exonuclease activity, but this activity does not seem to play a role in replication.

Pol I and pol III holoenzyme are both essential for *E. coli* replication. The need for two polymerases seems to be characteristic of all cellular organisms but not all viruses, e.g., *E. coli.* Phage T_4 synthesizes its own DNA polymerase, which is capable of carrying out all functions necessary for synthesizing phage DNA.

In the usual polymerization reaction, the activation energy for phosphodiester bond formation comes from cleaving of the triphosphate. Since DNA ligase can use a monophosphate, another source of energy is needed. This energy is obtained by hydrolyzing either ATP or NAD; the energy source depends on the organism from which the DNA ligase is obtained.

Ligases have two major functions: the sealing of single-strand breaks produced randomly in DNA molecules by nucleases and the joining of fragments during a particular stage of replication. ***DNA ligases*** are enzymes that can form a phosphodiester bond at a single-strand break in DNA, a reaction between a 3′-OH group and a 5′-monophosphate. These groups must be termini of adjacent base-paired deoxynucleotides (Figure 24-4).

Bacteria usually contain a single species of ligase. Mammalian cells possess two DNA ligases (I and II) present in very small amounts compared with bacteria. Both

TABLE 24-3
Properties of Eukaryotic DNA Polymerases

Property	Polymerase				
	α	β	γ	δ	ε
Location in cell	**Nucleus**	**Nucleus**	**Mitochondria**	**Nucleus**	**Nucleus**
Associated primase	Yes	No	No	No	No
$3'$ exonuclease	No	No	Yes	Yes	Yes
Sensitivity to aphidicolin	High	Low	Low	High	Low
Biological activity	Replication (lagging strand)	DNA repair	Replication	Replication (leading strand)	Replication

eukaryotic ligases are located in the nucleus. Ligase I is predominant in proliferating cells and presumably plays a role in DNA replication; ligase II predominates in resting cells.

Eukaryotic Polymerases

Five polymerizing enzymes have been isolated from many mammalian cells (Table 24-3). Three—pol α, pol β, and pol γ—function in replication. Pol α is the major polymerase of mammalian cells; it is found in the nuclei and is analogous to *E. coli* pol III. It is a multisubunit enzyme with a core (4–5 subunits) responsible for polymerization and a holoenzyme form possessing additional subunits and activities. It lacks the $3' \rightarrow 5'$ exonuclease editing function. An intriguing protein subunit in the holoenzyme enables it to bind ApppppA (diadenosine tetraphosphate), a small molecule that stimulates replication of resting mammalian cells and is hypothesized to be a growth signal. The pol α holoenzyme of rat liver also possesses a DNA primase activity (see below), a feature not found in prokaryotic enzymes. Pol β is a nuclear polymerase, probably analogous in function to *E. coli* pol I. Pol γ is found in mitochondria and is responsible for replication of mitochondrial DNA. It functions in the same way as pol III but is a single polypeptide.

24.3 The Replication Fork

DNA replication requires not only an enzymatic mechanism for adding nucleotides to the growing chains but also a means of unwinding the parental double helix. These are distinct processes, and the unwinding of the helix is closely related to the initiation of synthesis of precursor fragments.

The pol III holoenzyme cannot unwind the helix. In order for unwinding to occur, hydrogen bonds and hydrophobic interactions must be eliminated, which requires energy. Pol I utilizes the free energy of hydrolysis of the triphosphate for unwinding as it synthesizes a DNA strand in a way that other polymerases cannot; instead, helix unwinding is accomplished by enzymes called ***helicases.*** These enzymes hydrolyze ATP and utilize the free energy of hydrolysis for unwinding.

Unwinding of the helix by a helicase is not sufficient in itself for advance of the replication fork. Accessory proteins called ***single-stranded DNA-binding proteins (SSB proteins)*** are usually needed. As a helicase advances, it leaves in its wake two single-stranded regions: a longer one that is copied discontinuously and a shorter one just ahead of the leading strand. In order to prevent the single-stranded regions from reannealing or from forming intrastrand hydrogen bonds, the single-stranded DNA is protected with SSB proteins (Figure 24-8). SSB proteins bind tightly to both single-stranded DNA and to one another and hence are able to cover extended regions. As the polymerase advances, it must displace the SSB proteins so that base pairing of the added nucleotide can occur.

Some phage replication systems utilize a single protein that functions as both a helicase and an SSB protein; the gene-32 protein of phage T_4 is the prototype. It binds very tightly to single-stranded DNA and exceedingly tightly to itself, and its binding energy is great enough to unwind the helix.

As a replication fork moves along a circular helix, rotation of the daughter molecules around one another causes the individual polynucleotide strands of the unreplicated portion of the molecule to become wound more tightly, i.e., overwound. (This may be difficult to visualize but can be seen by taking two interwound circular strings and pulling them apart at any point.) Thus, advance of the replication forks causes positive supercoiling (Chapter 23) of the unreplicated portion. This supercoiling obviously cannot increase indefinitely because soon the unreplicated portion becomes coiled so tightly that further advance of the

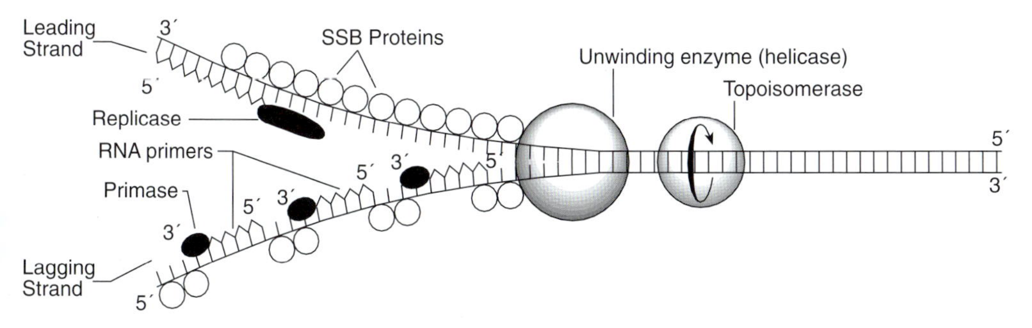

FIGURE 24-8
Simplified overview of a bacterial DNA replication fork. The leading strand is initiated with an RNA primer and then elongated by a DNA polymerase, pol III (***replicase***). The lagging strand is synthesized in a series of fragments, each initiated by ***primase*** which inserts short sequences of RNA primers that will be elongated by a replicase. The RNA fragments are removed by another DNA polymerase, pol I, and the breaks in the sugar-phosphate backbone are sealed by ***DNA ligase.***

replication fork is impossible. This additional coiling can be removed by a ***topoisomerase*** that can introduce negative superhelicity. In *E. coli,* the enzyme ***DNA gyrase*** produces ***negative superhelicity*** in nonsupercoiled covalent circles and is also responsible for removing the positive superhelicity generated during replication. The evidence for this comes from *in vivo* experiments using drugs (discussed later) that inhibit DNA gyrase; addition of any of these drugs to a growing bacterial culture or raising the temperature of cells with a temperature-sensitive gyrase protein inhibits DNA synthesis. Furthermore, *in vitro* replication of circular DNA can proceed only if DNA gyrase or a similar topoisomerase is present in the reaction mixture.

Inhibitors of DNA Replication

Inhibitors of DNA synthesis are used in the laboratory and in treatment of bacterial, viral, and neoplastic diseases. Successful treatment of these conditions requires careful attention to dosage and the fine difference between drug effectiveness and toxicity.

Inhibitors of DNA synthesis can be divided into three main classes:

1. Those that prevent or reduce the synthesis of precursors (bases, nucleotides),
2. Those that affect either the template or the priming ability of the growing strand, and
3. Those that act directly on polymerases or other enzymes needed for replication.

A variety of inhibitors bind to DNA and thereby eliminate its template activity. Notable among these are the ***intercalating agents,*** which slip in between base pairs (e.g., the acridines, phenanthridium derivatives), and the anthracyclines (daunorubicin, doxorubicin, and plicamycin). Other agents bind to DNA covalently and cause chain breakage (bleomycin, zinostatin) and interstrand cross-links (alkyl sulfonates, anthramycin, mitomycin, nitrogen mustards). Several of these compounds are also useful antitumor agents. A variety of platinum coordination compounds bind to DNA and inhibit its template activity, apparently by binding to guanine. Substances that prevent extension of the growing chain (2′,3′-dideoxyribonucleosides, cordycepin) are incorporated into the growing chain, but since they lack a 3′-OH group, further extension is not possible.

Only a few substances act directly on DNA polymerases; they often are effective only on one or a small number of polymerases. For example, ***acyclovir*** inhibits the DNA polymerase of herpes simplex. Aphidicolin inhibits pol α, pol δ (but not pol β or pol γ), many viral polymerases, and pol I and pol II of yeast; this compound is extremely valuable in laboratory research on DNA replication. Other components of the replication complex can also be inhibited; for example, 2′-dideoxyazidocytidine is an inhibitor of bacterial primase, and cournermycin, novobiocin, oxolinic acid, and nalidixic acid are effective inhibitors of DNA gyrase in bacteria.

Topoisomerase I Inhibitors

A variety of antibiotics and antineoplastic drugs exert their therapeutic effects by interaction with topoisomerase I and disruption of DNA synthesis during phase of dividing cells. Topoisomerase I is essential for DNA replication and cell growth. The enzyme relieves torsional stress in DNA by inducing reversible single-strand breaks. The interaction of topoisomerase I and certain drugs produces double-strand breaks in DNA that are irreversible and can lead to cell death.

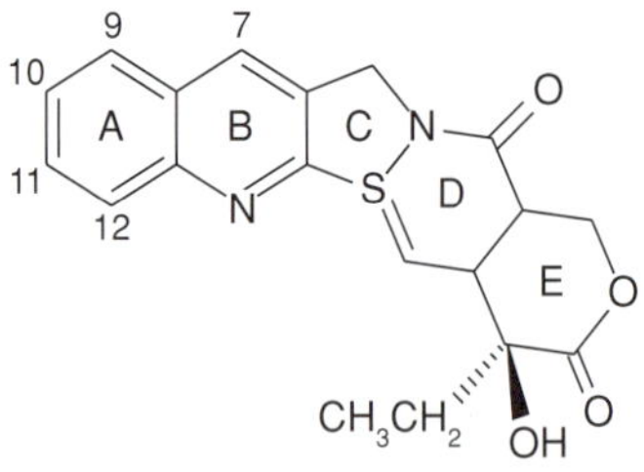

FIGURE 24-9
The basic structure of camptothecin. Analogues of camptothecin that are used to treat various neoplastic diseases involve substitutions at positions C-7, C-9, and C-11 of the basic molecule. All analogues as well as the parent molecule bind to topoisomerase I and interfere with its functions during DNA replication.

A variety of antibiotics and antineoplastic drugs function as topoisomerase I inhibitors; these include the quinolone antibiotics, anthracyclines (doxorubicin), epipodophylotoxins (etoposide), and the ***camptothecins,*** which are active in treating lung, ovarian, and colorectal cancers. The camptothecins also are used in the treatment of myelomonocytic syndromes, ***chronic myelomonocytic leukemia (CMML),*** acute leukemia, and ***multiple myeloma.*** A variety of synthetic analogues of natural camptothecins are being tested clinically for efficacy and safety in the treatment of these aggressive cancers (Figure 24-9).

The camptothecins were discovered in extracts from the Chinese tree *Camptotheca acuminata.* Initial studies showed that camptothecins had antitumor activity, but clinical trials demonstrated severe side effects and toxicity. Numerous camptothecin analogues have been synthesized; several have received FDA approval (irinotecan and topotecan), whereas others are still in clinical trials.

24.4 Chromosome Replication

DNA in mammalian cells is organized in complex structures called chromosomes (prokaryotes do not have a nucleus, do not divide by mitosis, and do not, strictly speaking, have chromosomes). The DNA in the chromosomes of human and other eukaryotic cells is intimately associated with two classes of proteins called ***histones*** and ***nonhistones.*** Collectively, DNA, histones, and nonhistones constitute ***chromatin,*** from which the name chromosome is derived. The DNA in a chromosome is an extremely long, linear molecule that must be condensed and organized to fit into the chromosomes in the nucleus. (The DNA in the 46 human chromosomes would be about 1 m long if fully extended.) Histones are responsible for the structural organization of DNA in chromosomes; the nonhistone proteins

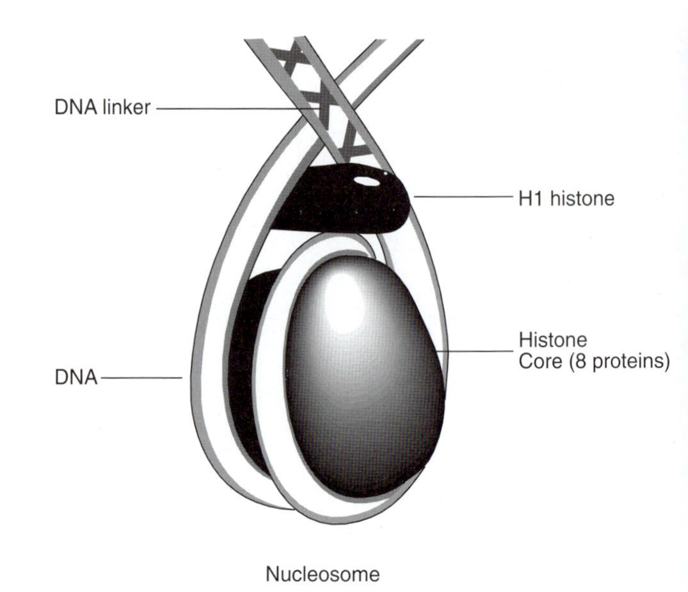

FIGURE 24-10
Structure of a nucleosome. DNA is looped around a core of eight histone proteins (pairs of four different histone proteins) and connected to adjacent nucleosomes by linker DNA and another histone (H1).

regulate the functions of DNA including replication and gene expression. The positive charge of histones, due to the presence of numerous lysine and arginine residues, is a major feature of the molecules, enabling them to bind to the negatively charged phosphate groups in DNA. The electrostatic attraction is an important stabilizing force in chromatin. If chromosomes are placed in solutions of high salt that break down electrostatic interactions, chromatin dissociates into free histones and DNA. Chromatin also can be reconstituted by mixing purified histones and DNA in concentrated salt solutions and gradually removing the salt by dialysis.

Histones share a similar primary structure among eukaryotic species. However, they undergo various posttranslational modifications such as phosphorylation, acetylation, methylation, and ADP ribosylation.The chemical modifications of histones can alter their net charge, shape, and other properties affecting DNA binding.

Pairs of four different histones (H2A, H2B, H3, and H4) combine to form an eight-protein bead around which DNA is wound; this bead-like structure is called a ***nucleosome*** (Figure 24-10). A nucleosome has a diameter of 10 nm and contains approximately 200 base pairs. Each nucleosome is linked to an adjacent one by a short segment of DNA (linker) and another histone (H1). The DNA in nucleosomes is further condensed by the formation of thicker structures called ***chromatin fibers,*** and ultimately DNA must be condensed to fit into the metaphase chromosome that is observed at mitosis (Figure 24-11).

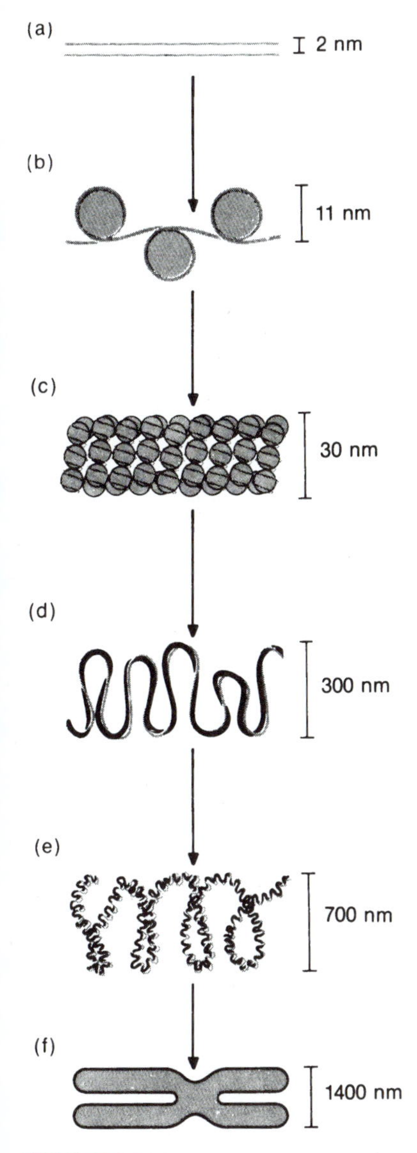

FIGURE 24-11
Hypothetical stages in the condensation of DNA with chromatin to form a chromosome. (a) Double-stranded DNA. (b)–(e) Formation of nucleosome beads and fibers consisting of histones and condensed DNA to form (f) a metaphase chromosome. These proposed intermediates are derived from dissociation reactions with intact chromosomes.

Despite the dense packing of DNA in chromosomes, it must be accessible to regulatory proteins during replication and gene expression. At a higher level of organization, chromosomes are divided into regions called ***euchromatin*** and ***heterochromatin.*** Transcription of genes seems to be confined mainly to euchromatic regions while DNA in heterochromatic regions is genetically inactive. When DNA is replicated during the S phase of the cell cycle, the histone and nonhistone proteins also are duplicated and combine with the daughter DNA molecules.

Replication of DNA in Chromosomes

The rate of movement of a replication fork in *E. coli* is about 10^5 nucleotides per minute; in eukaryotes the DNA polymerases move only about one-tenth as fast. To replicate all of the DNA in a human cell in a few hours means that replication must be initiated at thousands of origins of replication in each chromosome and move bidirectionally. To accomplish this, mammalian cells have thousands of times more DNA polymerase available than is found in bacteria. As replication proceeds through each nucleosome, the histones must dissociate to allow the replication fork to proceed; after replication, the nucleosome re-forms. The separation of parental and daughter DNA also requires the synthesis of new nucleosomes. Recent studies suggest that DNA is replicated in "replication factories," i.e., fixed sites within the nucleus consisting of the numerous proteins needed for replication. The DNA is replicated by being drawn through the replication factories rather than having the replication proteins move along the DNA.

Replication at the Ends of Chromosomes

Since DNA in chromosomes is a linear molecule, problems arise when replication comes to the ends of the DNA. Synthesis of the lagging strand at each end of the DNA requires a primer so that replication can proceed in a 5′ to 3′ direction. This becomes impossible at the ends of the DNA and 50–100 bp is lost each time a chromosome replicates. Thus, at each mitosis of a somatic cell, the DNA in chromosomes becomes shorter and shorter. Ultimately, after a limited number of divisions, a cell enters a nondividing state, called ***replicative senescence,*** which may play an important role in biological aging.

To prevent the loss of essential genetic information during replication, the ends of DNA in chromosomes contain special structures called ***telomeres*** that are synthesized by a specific enzyme called ***telomerase.*** Intact telomerase consists of an RNA primer and associated proteins so telomerase is actually a reverse transcriptase. The activity of telomerase in replenishing telomeres is regulated by a number of telomere-specific DNA-binding proteins, TRF1 and TRF2. TRF1 regulates the length of telomeres and TRF2 protects the ends. Overexpression of TRFI results in progressive shortening of telomeres; underexpression results in lengthening. Another telomere regulatory protein is ***tankrase*** (*T*RF1-interacting *ank*yrin-*r*elated ADP-ribose polymer*ase*), which alters the activity of TRF1.

Telomeres in human chromosomes consist of tandem repeats of the sequence TTAGGG. In most adult somatic cells, telomerase activity is very low or absent. Even in hematopoietic stem cells that do have residual telomerase activity, telomere shortening is observed at the level of granulocyte and mononuclear cell fractions. Accelerated telomere shortening has been observed in cells from patients with aplastic anemia, suggesting that abnormal telomere shortening is associated with disease and aging.

A characteristic of malignant tumor cells is that they can replicate indefinitely. The immortality of tumor cells appears to result, at least in part, from enhanced levels of telomerase that allow them to repair and elongate telomeres at the ends of DNA. This hypothesis is supported by the observation that ectopic expression of the catalytic subunit of telomerase (a product of the *hTERT* gene) enabled human retinal pigmented epithelial cells and fibroblasts to avoid senescence and to maintain their differentiated state when grown *in vitro*. Current research is focused on drugs that can promote or inhibit the action of telomerase or telomere-associated regulatory proteins. It is hoped that increasing telomerase activity in cells approaching senescence will retard aging or that decreasing telomerase activity in tumor cells will result in arrested tumor growth.

24.5 DNA Repair

DNA can be damaged by external agents and by replication errors. Since maintenance of the correct base sequence of DNA and of daughter DNA molecules is essential for hereditary fidelity, repair systems have evolved that restore the correct base sequence.

Mismatch Repair and Methylation of DNA

DNA polymerases occasionally catalyze incorporation of an "incorrect" base that cannot form a hydrogen bond with the template base in the parental strand; such errors usually are corrected by the editing function of these enzymes. The editing process occasionally fails, so a second system, called ***mismatch repair,*** exists for correcting the errors that are not edited out. In mismatch repair, a pair of non-hydrogen-bonded bases (e.g., G · · · T; Figure 24-12a) within a helix are recognized as aberrant and a polynucleotide segment of the daughter strand is excised, thereby removing one member of the unmatched pair. The resulting gap is filled in by pol I, which presumably uses this "second chance" to form correct base pairs; then the final seal is made by DNA ligase.

If it is only to correct and not create errors, the mismatch repair system must be capable of distinguishing the correct base in the parental strand from the incorrect base in the daughter strand. Rare methylated bases (methyl-A and methyl-C) provide the basis for this distinction. In *E. coli,*

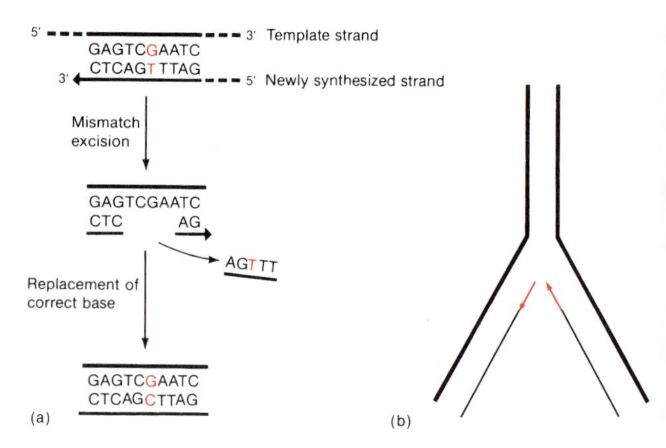

FIGURE 24-12
Mismatch repair. (a) Excision of a short segment of a newly synthesized strand and repair synthesis. (b) Methylated bases in the template strand direct the excision mechanism to the newly synthesized strand containing the incorrect nucleotide. The regions in which methylation is complete are black lines; the regions in which methylation may not be complete are shown in color.

an enzyme, ***DNA methylase,*** methylates adenine in the sequence GATC. Methylation occurs soon (but not immediately) after the replication fork has synthesized such a sequence in the daughter strand. Thus, the nucleotides in the daughter strand are usually not methylated near the fork, whereas those in the parental strand are always completely methylated (Figure 24-12b). The mismatch repair system recognizes the degree of methylation of a strand and preferentially excises nucleotides from the undermethylated strand. The daughter strand is always the undermethylated strand, so that parental information is retained.

Glycosylases

Occasionally, uracil or other incorrect bases may become incorporated into new DNA strands. These bases are usually removed by a pathway that starts with the cleavage of the N-glycosidic bond by an enzyme called a glycosylase (Figure 24-13). Many glycosylases are known, and each is base-specific (e.g., uracil N-glycosylase). This enzyme cleaves the N-glycosidic bond and leaves the deoxyribose in the backbone. A second enzyme (AP endonuclease) makes a single cut, freeing one end of the deoxyribose. (AP stands for apurinic acid, a polynucleotide from which purines have been removed by hydrolysis of the N-glycosidic bonds.) This step is followed by removal of the deoxyribose and several adjacent nucleotides (probably by a second enzyme that acts at the other side of the apurinic site), after which pol I fills the gap with correct nucleotides. This sequence, endonuclease-enlargement-polymerase, is an example of a general repair mechanism called ***excision repair*** (see below).

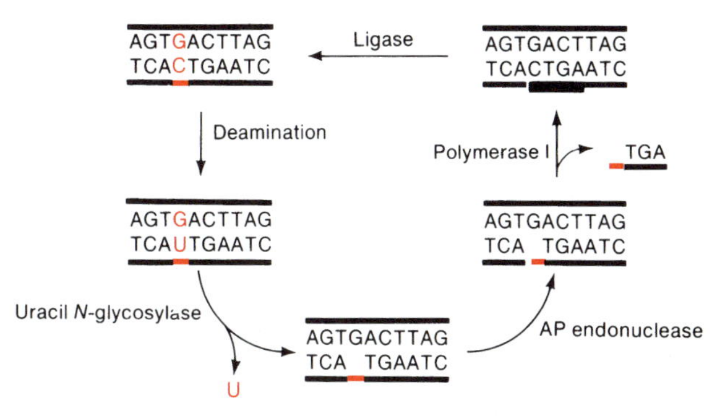

FIGURE 24-13
Scheme for repair of cytosine deamination. The same mechanism could remove a uracil that is accidentally incorporated.

Alterations of DNA Molecules

Several agents can break phosphodiester bonds. Among the more common are peroxides and various metal ions (e.g., Fe^{2+}, Cu^{2+}). Ionizing radiation also efficiently produces strand breaks. DNases present in cells probably also occasionally break phosphodiester bonds. Double-strand breakage, i.e., two single-strand breaks opposite one another, results from exposure to all forms of ionizing radiation. Single-strand breaks can be repaired by DNA ligases, although sometimes additional enzymes are required. Double-strand breaks rarely are repaired.

Bases can be changed into different compounds by a variety of chemical and physical agents. For instance, ionizing radiation can break purine and pyrimidine rings and can cause several types of chemical substitutions, the most common being in guanine and thymine. The best studied altered base is the dimer formed by covalent linkage of two adjacent pyrimidine rings; these are produced by ultraviolet (UV) irradiation. The most prevalent of these dimers is the ***thymine dimer.*** The significant effects of the presence of thymine dimers are the following:

1. The DNA helix becomes distorted as the thymines, which are in the same strand, are pulled toward one another (Figure 24-14).
2. As a result of the distortion, hydrogen bonding to adenines in the opposite strand is significantly weakened, causing inhibition of advance of the replication fork.

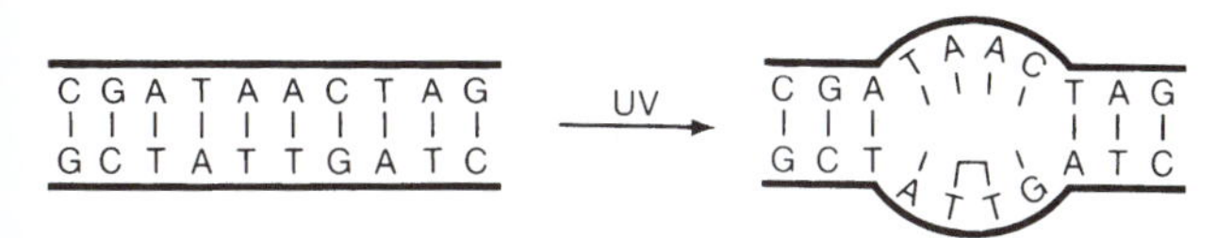

FIGURE 24-14
Ultraviolet radiation damage of two adjacent thymines of DNA. Distortion of the DNA helix caused by two thymines moving closer together when joined in a dimer. The dimer is shown as two joined lines.

General Mechanisms for Repair of DNA

Repair of damaged bases was first observed and is best understood in bacteria. It is a widespread and probably universal phenomenon in both prokaryotes and eukaryotes. Some systems that repair dimers repair other types of DNA damage also.

Four major pathways for DNA repair exist that can be subdivided into two classes: light-induced repair (photoreactivation) and light-independent repair (dark repair). The latter can be accomplished by three distinct mechanisms:

1. Excision of the damaged nucleotides (excision repair),
2. Reconstruction of a functional DNA molecule from undamaged fragments (recombinational repair), and
3. Disregard of the damage (SOS repair).

Photoreactivation

Photoreactivation is a light-induced (300–600 nm) enzymatic cleavage of a thymine dimer to yield two thymine monomers. It is accomplished by ***photolyase,*** an enzyme that acts on dimers contained in single- and double-stranded DNA.

The enzyme-DNA complex absorbs light and uses the photon energy to cleave specific C-C bonds of the cyclobutylthymidine dimer. Photolyase is also active against cytosine dimers and cytosinethymine dimers, which are also formed by UV irradiation but much less frequently.

Excision Repair

Excision repair is a multistep enzymatic process. Several mechanisms are known, but only two will be described (Figure 24-15). All require an early incision step, in which a nuclease recognizes the distortion produced by a thymine dimer and makes a cut in the sugar-phosphate backbone. Following this, a DNA polymerase mediates a strand displacement step. In *E. coli,* two cleavages occur; the first is 12 nucleotides from the 5′ side of the dimer and the second is 4–5 nucleotides from the 3′ side. Each cut produces a 3′-OH and a 5′-P group. The 3′-OH group of the first cut is recognized by pol I, which then synthesizes a new strand, displacing the dimer-containing DNA strand. When the second cut is reached, the displaced fragment falls away and DNA ligase joins the 3′-OH and 5′-P groups. In ***Micrococcus luteus,*** the first step is breakage of the N-glycosidic bond of the thymine at the 5′ end of the dimer (by a dimer-specific glycosylase), leaving a free deoxyribose which is removed, leaving a free 3′-OH group. As

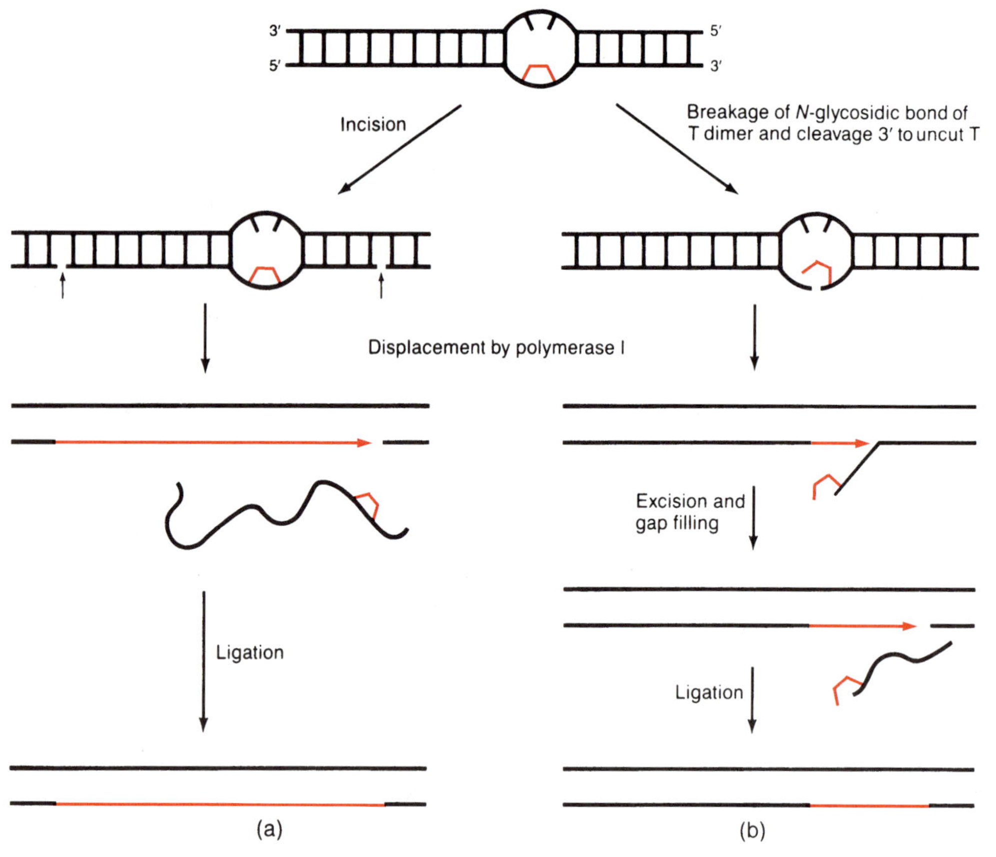

FIGURE 24-15
Two modes of excision repair. (a) ***Escherichia coli*** mechanism. Two incision steps are followed by gap-filling and displacement by polymerase I. (b) ***Micrococcus luteus*** mechanism. A pyrimidine dimer glycosylase breaks an N-glycosidic bond and makes a single incision. Pol I displaces the strand, which is removed by an exonucleolytic event. In both mechanisms, the final step is ligation.

in *E. coli,* pol I polymerizes from this end and displaces the dimer-containing strand. After the dimer has been displaced, a second cut is made, the displaced strand falls away, and ligase forms the final phosphodiester bond. Repair of dimers in mammalian cells is more complicated and poorly understood, and requires a larger collection of enzymes.

Recombination Repair

Recombination repair is a mechanism for generating a functional DNA molecule from two damaged molecules. It is an essential repair process for dividing cells because a replication fork may arrive at a damaged site, such as a thymine dimer, before the excision repair system has eliminated damage.

When pol III reaches a thymine dimer, an adenine is added to the growing strand. However, the distortion of the helix caused by the dimer weakens the hydrogen bond and activates the polymerase editing function, and the adenine is removed. The cycle begins again—an adenine is added and then removed—the net result of which is that the replication fork fails to advance. A cell in which DNA synthesis is permanently stalled cannot complete a round of replication and does not divide. However, in a way that is not understood, after a pause of ~5 seconds per dimer, chain growth begins again beyond the thymine dimer block. The result of this process is that the daughter strands have large gaps, one for each unexcised thymine dimer. Viable daughter cells cannot be produced by continued replication alone because the strands having the thymine dimer will continue to turn out defective daughter strands and the first set of daughter strands would be fragmented when the growing fork enters a gap. However, by a recombination mechanism called ***sister-strand exchange*** proper double-stranded molecules can be made.

The essence of sister-strand exchange is that a single-stranded segment free of any defects is excised from an undamaged strand on the homologous DNA segment at the replication fork and somehow inserted into the gap created by the thymine dimer. The combined action of polymerase I and DNA ligase joins this inserted piece to adjacent regions, thus filling in the gap. The gap formed in the donor molecule is also repaired. If this exchange and gap filling are done for each thymine dimer, two complete single daughter strands can be formed, and each can serve in the next round of replication as a template for synthesis of normal DNA molecules. The system fails if two dimers in opposite strands are very near one another because no undamaged segments are available. Since recombinational repair occurs after DNA replication, in contrast with excision repair, it is often called ***postreplicational repair.***

SOS Repair

SOS repair includes a bypass system that allows DNA chain growth across damaged segments at the cost of fidelity of replication. It is an error-prone process; even though intact DNA strands are formed, the strands are often altered.

As described above, activation of the editing system stalls replication at a thymine dimer. In SOS repair, the editing system is relaxed to allow polymerization to proceed across a dimer. Relaxation of the editing system means a loss of the ability to remove "incorrect" bases added to the growing strand. Most of the time, pol III inserts two adenines at a dimer site. However, the distortion increases the error frequency, allowing other nucleotides to be added to the chain. This error-prone repair is the major cause of UV-induced mutagenesis.

An important nuclear protein, conserved from yeast to mammals, is the Ku **heterodimer.** This protein binds to DNA and repairs double strand breaks caused by x-rays and other agents. The Ku heterodimer is essential in maintaining chromosome integrity.

Human Diseases and DNA Repair Deficiency

Human disease may result from inability to carry out certain stages of DNA repair. The best studied disease, ***xeroderma pigmentosum,*** is a result of mutations in genes that encode the UV excision system. Cells cultured from tissue obtained from affected individuals are killed by much smaller doses of UV light than are normal cells. Furthermore, the removal of thymine dimers in DNA from these cells is very inefficient. People with this disease develop skin lesions when exposed to sunlight and commonly develop one of several kinds of skin cancer.

Ataxia telangiectasia is characterized by severe abnormalities in various organ systems and a high incidence of lymphoreticular cancer. Defective DNA repair was suspected when patients developed an unexpected severe or fatal reaction while undergoing radiotherapy for cancer. As predicted, nontumor cells cultured from these patients are hypersensitive to x-rays.

Fanconi's syndrome, a lethal aplastic anemia, is also due to defective DNA repair. Cells from affected persons cannot repair interstrand cross-links or damage induced by x-rays. Two premature aging disorders (***Hutchinson-Gilford syndrome*** and ***Bloom's syndrome***) and several other disorders (***Cockayne's syndrome*** and ***retinoblastoma***) are also associated with defects in DNA repair. Cells from patients with some chromosome abnormalities (e.g., Down syndrome) may also show aberrant DNA repair.

Several human DNA mismatch repair genes are associated with ***hereditary nonpolyposis colon cancer (HNPCC).*** One of these mismatch repair genes (*hMSH2*) is located on the short arm of chromosome 2; others are located on chromosome 3. Defects in any of these DNA repair genes predispose individuals to colon cancer as well as to other cancers.

24.6 DNA Mutation

Mutation refers to any change in the base sequence of DNA. The most common change is a substitution, addition, rearrangement, or deletion of one or more bases. A mutation need not give rise to a mutant phenotype.

A ***mutagen*** is a physical agent or chemical reagent that causes mutations. For example, nitrous acid reacts with some DNA bases, changing their chemistry and hydrogen bonding properties, and is a mutagen.

Mutagenesis is the process of producing a mutation. If it occurs in nature without the action of a known mutagen, it is called *spontaneous mutagenesis* and the resulting mutations are spontaneous mutations. If a mutagen is used, the process is called *induced mutagenesis.*

Types of Mutations

Mutations can be categorized in several ways. One system is based on the nature of the change, specifically on the number of bases changed. Thus, we distinguish a ***point mutation,*** in which a single base pair is changed from a multiple mutation, in which two or more base pairs differ from the wild-type sequence. A point mutation may be a ***base substitution,*** a ***base insertion,*** or a ***base deletion,*** but the term most frequently refers to a ***base substitution.***

A second system is based on the consequence of the change in terms of the amino acid sequence that is affected. For example, if there is an amino acid substitution, the mutation is a ***missense mutation.*** If the substitution produces a protein that is active at one temperature (typically 30°C) and inactive at a higher temperature (usually

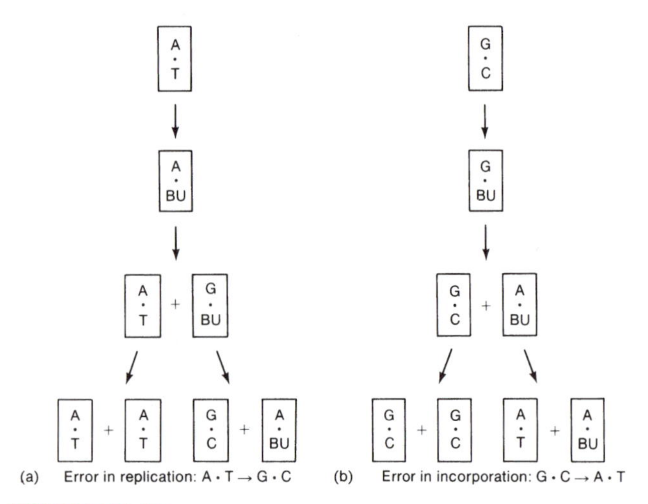

FIGURE 24-16
Two mechanisms of 5–bromouracil (BU)-induced mutagenesis. (a) During replication, BU in its usual keto form substitutes for T, and the replica of an initial A-T pair becomes an A-BU pair. In the first mutagenic round of replication, BU in its rare enol form pairs with G. In the next round of replication, the G pairs with a C, completing the transition from an A-T pair to a G-C pair. (b) During replication of a G-C pair, a BU in its rare enol form pairs with a G. In the next round of replication, the BU is again in the common keto form and it pairs with A, so the initial G-C pair becomes all A-T pair. The replica of the A-BU pair produced in the next round of replication is another A-T pair.

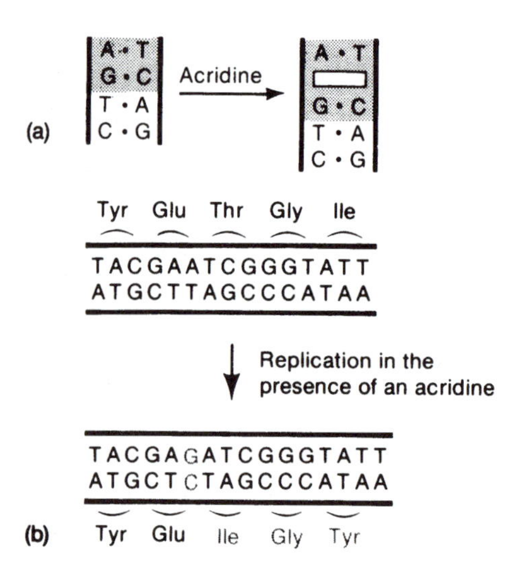

FIGURE 24-17
Mutagenesis by intercalating substances (e.g., acridine). (a) Separation of two base pairs (shown as a box) by an intercalating agent. (b) A base addition resulting from replication in the presence of an acridine. The change in amino acid sequence read from the upper strand in groups of three bases is also shown.

40–42°C), the mutation is called a ***temperature-sensitive (Ts) mutation.*** Sometimes no amino acid corresponds to a new base sequence (Chapter 25); in that case, termination of synthesis of the protein occurs at that point, and the mutation is called a ***chain termination mutation*** or ***nonsense mutation.*** These mutations generate one of the three nonsense codons: UAA, UAG, or UGA (Chapter 25).

Replication errors can be introduced by base analogues that can be added to a replicating DNA molecule. Such a base analogue must be able to pair with the base on the complementary strand being copied, or the $3' \rightarrow 5'$ editing function would remove it. However, if it can tautomerize (Chapter 23), or if it has two modes of hydrogen bonding, it will be mutagenic. The substituted base 5-bromouracil (BU) is an analogue of thymine inasmuch as the bromine atom has about the same van der Waals radius as the methyl group of thymine. BU forms a nucleoside triphosphate, and DNA polymerases will add BU to a DNA strand opposite an adenine. However, BU tautomerizes at high frequency to a form that can pair with guanine. Thus, if BU replaces a thymine, in subsequent rounds of replication it occasionally generates a guanine in the complementary strand, which in turn specifies cytosine, resulting in formation of a GC pair (Figure 24-16).

A chemical mutagen is a substance that can alter a base that has already been incorporated into DNA and thereby change its hydrogen bonding specificity. ***Nitrous acid,*** a powerful mutagen, converts amino groups to keto groups by oxidative deamination. Thus, cytosine, adenine, and guanine are converted to uracil (U), hypoxanthine (H), and xanthine (X), respectively, which can form the base pairs UA, HC, and XC, respectively. As a consequence, in a later round of replication, GC-to-AT and AT-to-GC mutations can occur. The chemical mutagen ***hydroxylamine*** reacts specifically with cytosine and converts it to a modified base that pairs only with adenine; thus, a GC pair ultimately becomes an AT pair.

An interesting class of mutagen are intercalating agents, e.g., the acridines. These planar molecules insert between base pairs. When replication occurs in the region of an intercalated molecule, one or both daughter strands are synthesized that either lack one or more nucleotides or have additional ones. These changes alter the reading frame of the base sequence of a gene and hence are called ***frameshift mutations*** (Figure 24-17).

Triplet Repeats and Fragile Sites

Inherited and acquired (sporadic) mutations account for an enormous diversity of human diseases and disorders. At least 300 genetic disorders include mental retardation as one of the symptoms, implying that many genes can affect brain development and mental functions. In the 1970s, a defect at the tip of the long arm of X chromosomes, called a ***fragile site,*** was discovered in a large number of patients with mental retardation. It is now known that a specific mutation is responsible for the fragile site on

TABLE 24-4
Inherited Diseases Involving Repeated Trinucleotides That Are Present in an Excessive Number of Copies*

Disease	Trinucleotide repeat	Symptoms
Fragile X syndrome	CGG 5–54 unaffected 60–230 carriers 230–4000 affected	Mental retardation
Huntington's disease	CAG 11–30 unaffected 36–121 affected	Chorea, progressive dementia, death
Myotonic dystrophy	CTG	Muscle weakness, myotonia
Spinocerebellar ataxia Type 1	CAG 19–36 unaffected 40–81 affected	Ataxia, dysarthria, rigidity, abnormal eye movements
Fredreich's ataxia	GAA 7–22 unaffected 200–900 affected	Ataxia, dysarthria, hypertropic cardiomyopathy, diabetes
Machado-Joseph disease	CAG 12–37 unaffected 61–84 affected	Spasticity, dystonia

*The extent of the repeated triplet nucleotide usually determines the severity of the disease and the age of onset. These diseases may be caused by slippage of the DNA polymerase during replication that increases the length of the repeated triplet from generation to generation. The patterns of inheritance are: Fragile X syndrome (X-linked); Huntington's disease, myotonic dystrophy, Spinocerebellar ataxia Type I, Machado-Joseph disease (autosomal dominant); Fredreich's ataxia (autosomal recessive).

the X chromosome and the associated inherited disorder ***fragile X syndrome.*** This particular mutation accounts for about 20% of all mental retardation; about 1 in 1500 boys and 1 in 2500 girls are born with fragile X syndrome.

Although fragile X syndrome is classified a Mendelian dominant sex-linked disorder, it does not behave as a classically dominant gene inherited disease. Only about 80% of males and 35% of females who carry the mutation and a fragile site suffer from mental retardation. For many years it remained a mystery how both men and women could carry a fragile X chromosome and be unaffected.

In the 1990s, the molecular explanation for fragile X syndrome was revealed. The gene responsible for fragile X syndrome (*FMR1*) was found to contain a repeat of the trinucleotide CGG. In unaffected individuals, the number of CGG repeats ranges from about 5 to 54 copies. In mentally retarded individuals, the number of CGG repeats is dramatically increased to hundreds or thousands of copies. When the CGG repeat exceeds 230, the DNA becomes abnormally methylated in that region and the gene becomes nonfunctional. Presumably the amplification of the CGG repeat occurs during DNA replication in germ cells prior to meiosis perhaps due to "stuttering" of the DNA polymerase as it reads the repeating trinucleotides. Why amplification of the CGG repeats occurs in some X chromosomes but not others and in some individuals and not others is unresolved.

Triplet repeats account for a growing list of muscular-neurologic diseases including Huntington's disease, myotonic dystrophy, Friedreich's ataxia, and others (Table 24-4). Triplet repeats represent an entirely new class of mutations. They are not caused by any external mutagenic agent but are generated during normal DNA replication.

Fragile sites are chromosomal regions that are prone to breakage during chromosome movement or recombination; fragile sites can be visualized by a variety of cytological techniques and dyes that bind to metaphase chromosomes. Fragile sites are relatively common in chromosomes and there is a growing suspicion that they may play a greater role in disease, especially cancer, than previously thought. The chromosomes in many tumor cells exhibit fragile sites that are not present in normal cells. What is not yet clear is whether a fragile site initiates the conversion of a normal cell to a cancer cell or is merely the consequence of the altered growth properties of the tumor cell. One possibility is that fragile sites are more susceptible to carcinogenic chemicals and that mutations in fragile sites contribute to the transformation of a normal cell to a tumor cell.

Supplemental Readings and References

S. G. Arbuck and C. H. Takimoto: An overview of topoisomerase I-targeting agents. *Seminar in Hematogy* **35**(Suppl. 4), 3 (1998).

S. E. Ball, F. M. Gibson, S. Rizzo, et al.: Progressive telomere shortening in aplastic anemia. *Blood* **91,** 3582 (1998).

V. A. Bohr: Gene specific DNA repair. *Carcinogenesis* **12,** 1983 (1991).

D. Bootsma: The genetic defect in DNA-repair deficiency syndromes. *European Journal of Cancer* **29,** 1482 (1993).

C. E. Bronner, S. M. Baker, P. T. Morrison, et al.: Mutation in the DNA mismatch repair gene homologue hMLH1 is associated with hereditary non-polyposis colon cancer. *Nature* **368,** 258 (1994).

C. H. C. M. Buys: Telomeres, telomerase, and cancer. *New England Journal of Medicine* **342** (2000).

J. E. Cleaver: Stopping DNA replication in its tracks. *Science* **285,** 212 (1999).

A. Durr, M. Cossee, Y. Agid, et al.: Clinical and genetic abnormalities in patients with Friedreich's ataxia. *New England Journal of Medicine* **335,** 1169 (1996).

M. Engelhardt, R. Kumar, J. Albanell, et al.: Telomerase regulation, cell cycle, and telomere stability in primitive hematopoietic cells. *Blood* **90,** 183 (1997).

M. Fessel: Telomerase and the aging cell. *Journal of the American Medical Association* **279,** 1732 (1998).

E. C. Friedberg, G. C. Walker, and W. Siede: *DNA Repair and Mutagenesis.* Washington, DC: ASM Press (1995).

M. Grunstein: Histone acetylation in chromatin structure and transcription. *Nature* **389,** 349 (1007).

J. H. J. Hoeijmakers: Nucleotide excision repair II: from yeast to mammals. *Trends in Genetics* **9,** 211 (1993).

B. Kremer, P. Goldberg, S. E. Andrew, et al.: A worldwide study of the Huntington disease mutation. *New England Journal of Medicine* **330,** 1401 (1994).

C. Lengauer, K. W. Kinzler, and B. Vogelstein: Genetic instabilities in human cancers. *Nature* **396,** 543 (1998).

T. Lindahl and R. D. Wood: Quuality conntrol by DNA rpair. *Science* **286,** 1897 (1999).

J.-L. Mandel: Breaking the rule of three. *Nature* **386,** 767 (1997).

D. W. Ross: Cancer: The emerging molecular biology. *Hospital Practice* **35,** 63 (2000).

S. Smith, I. Giriat, A. Schmitt, et al.: Tankyrase, a poly (ADPribose) polymerase at human telomeres. *Science* **282,** 1484 (1998).

T. A. Steitz: DNA polymerises: Structural diversity and common mechanisms. *Journal of Biological Chemistry* **274,** 17395 (1999).

S. T. Warren: Trinucleotide repetition and fragile X syndrome. *Hospital Practices* **32,** 73 (1997).

CHAPTER 25

RNA and Protein Synthesis

Most proteins are enzymes that catalyze the myriad of chemical reactions in cells that are necessary for life; other proteins form structural functions as in bone and muscle. The information for making proteins resides in the sequence of bases in DNA in chromosomes and in organelles such as mitochondria. Converting the information contained in genes into proteins involves two complex processes. ***Transcription*** is the first step in which the sequence of bases in a gene is converted into a complementary sequence of bases in a molecule of RNA. Three chemically identical but functionally quite different molecules of RNA are transcribed from DNA: ***messenger RNA (mRNA)*** carries the genetic information contained in a gene; ***transfer RNA (tRNA)*** and ***ribosomal RNA (rRNA)*** are also transcribed from genes but are used to convert the information in the sequence of bases in mRNA into the corresponding sequence of amino acids in a protein.

Translation is the process by which an mRNA is "read" by tRNAs, ***ribosomes*** (complex structures consisting of rRNAs and ribosomal proteins), and numerous other enzymes. Each type of cell is programmed to synthesize only those proteins necessary for its particular cellular functions (Chapter 26). The difference between a neuron and a liver cell is the kind of proteins that are synthesized even though both cells contain exactly the same genetic information. Cellular differentiation is due to differential gene expression; a tumor cell invariably is a cell that has lost the ability to regulate and express its genetic information correctly and generally grows in an unregulated manner, as opposed to normal cells whose growth is regulated.

The flow of information in all cells is from DNA to RNA to protein, which is known as the ***central dogma*** of molecular biology; it was formulated by Francis Crick shortly after the discovery of the structure of DNA. Information also can flow from DNA to DNA in both cells and among viruses that infect cells. Information also flows from RNA to RNA during the replication of RNA viruses such as the polio virus. The final permitted information transfer is from RNA to DNA, which only occurs in the case of retroviruses such as *human immunodeficiency virus (HIV)*. The only information transfer that is prohibited by the central dogma is from protein to RNA or to DNA. The permitted information transfers in cells (infected or uninfected) is summarized below.

$$\circlearrowleft \text{DNA} \rightleftarrows \circlearrowleft \text{RNA} \longrightarrow \text{Protein}$$

The synthesis of a protein in a human cell can be broadly outlined as follows. A mRNA molecule is transcribed from a single strand of DNA (the "sense" strand) in the nucleus. The mRNA is processed by splicing out nontranslatable segments of nucleotides (***introns***) and rejoining the translatable segments (***exons***). Additional chemical modifications are made to both ends of the mRNA molecule and it is transported into the cytoplasm. The mRNA is translated in the cytoplasm where groups of three bases in the mRNA (***codons***) are recognized by ***anticodons*** in specific tRNAs that carry a particular amino acid.

The ***genetic code*** consists of 64 different codons that specify all 20 amino acids as well as codons that function to initiate and terminate translation. More than one codon may specify the same amino acid, which is called degeneracy of the genetic code. Finally, every organism from bacteria to human uses the same codons to specify the same amino acids; this is why the genetic code is said to be universal.

Amino acids are joined together in a specific order determined by tRNAs, ribosomes, and associated enzymes that translate the mRNA. Each amino acid is joined to its neighbor by a peptide bond. The specific amino acid sequence of a protein specifies its three-dimensional structure. Some proteins require the help of ***chaperonins*** to fold into a functional configuration. When synthesis of a polypeptide chain is completed on a ribosome, it is released from the ribosome and may join with one or more similar or different polypeptides to constitute a functional protein.

25.1 Structure of RNA

RNA is a single-stranded polynucleotide containing the nucleosides adenosine, guanosine, cytosine, and uridine. Roughly one-third to one-half of the nucleotides are engaged in intrastrand hydrogen bonds, with single-stranded segments interspersed between double-stranded regions that may contain up to about 30 base pairs. The base pairing produces conformations that are important to the function of the particular RNA molecules.

Ribosomal RNA (rRNA)

Ribosomes contain several different RNA molecules, three in prokaryotic ribosomes and four in eukaryotic ribosomes. For historical reasons, each class is characterized by its sedimentation coefficient, which represents a typical size. For prokaryotes, the three ***Escherichia coli*** rRNA molecules are used as size standards; they have sedimentation coefficients of 5S, 16S, and 23S. The *E. coli* rRNA molecules have been sequenced and contain 120, 1541, and 2904 nucleotides, respectively. The sizes of the prokaryotic rRNA molecules vary very little from one species of bacterium to another.

Eukaryotic rRNA molecules are generally larger and there are four eukaryotic rRNA molecules. Rat liver rRNA molecules are used as standards; the S values and the average number of nucleotides are 5S (120), 5.8S (150), 18S (2100), and 28S (5050), respectively. The eukaryotic 5.8S species corresponds functionally to the prokaryotic 5S species; no prokaryotic rRNA molecule corresponds to the eukaryotic 5S rRNA.

Transfer RNA (tRNA)

Transfer RNA molecules range in size from 73 to 93 nucleotides. Since they function as amino acid carriers, they are named by adding a superscript that designates the amino acid carried, e.g., $tRNA^{Ala}$ for alanine tRNA. All tRNA molecules studied contain extensive double-stranded regions and form a cloverleaf structure in which open loops are connected by double-stranded stems. By careful comparison of the sequences of more than 200 different tRNA molecules, common features have been found and a "consensus" tRNA molecule consisting of 76 nucleotides arranged in a cloverleaf form has been defined (Figure 25-1). By convention, the nucleotides are numbered 1 through 76 starting from the 5′-P terminus. The standard tRNA molecule has the following features:

1. The 5′-P terminus always is base-paired, which probably contributes to the stability of tRNA.
2. The 3′-OH terminus always is a four-base single-stranded region having the base sequence XCCA-3′-OH, in which X can be any base. This is called the CCA or acceptor stem. The adenine in the CCA sequence is the site of attachment of the amino acid by the cognate synthetase.
3. tRNA has many "modified" bases. A few of these, dihydrouridine (DHU), ribosylthymine (rT), ***pseudouridine*** (ψ), and inosine (I), occur in particular regions.
4. tRNA has three large single-stranded loops. The anticodon loop contains seven bases. The loop containing bases 14–21 is called the DHU loop; it is not constant in size in different tRNA molecules. The loop containing bases 54–60 almost always contains the sequence TψC and is called the TψC loop.
5. Four double-stranded regions called stems (or arms) often contain GU base pairs. The names of the stems match the corresponding loop.
6. Another loop, containing bases 44–48, is also present. In the smallest tRNAs it contains four bases, whereas in the largest tRNA molecule it contains 21 bases. This highly variable loop is called the *extra arm.*

25.2 Messenger RNA

Messenger RNA molecules in prokaryotic and eukaryotic cells are similar in some structural aspects but also differ significantly. All messenger RNAs contain the same four nucleotides, A, C, G, and U, and utilize the codon AUG to

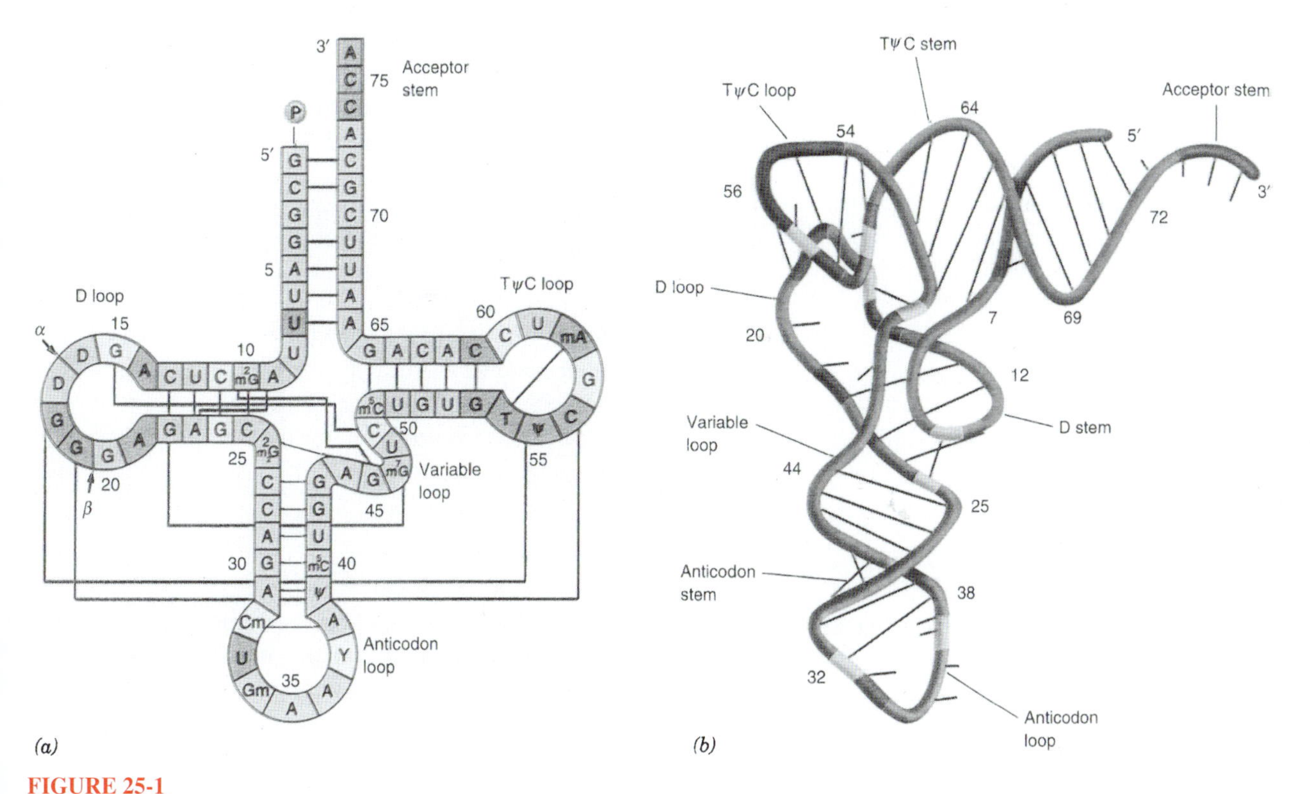

FIGURE 25-1
(a) The currently accepted "standard" tRNA cloverleaf with its bases numbered. A few bases present in almost all tRNA molecules are indicated. (b) Schematic diagram of the three-dimensional structure of yeast $tRNA^{Phe}$. (Courtesy of Dr. Sung-Hou Kim.)

initiate translation of a polypeptide and the codons UAG, UGG, and UAA to terminate translation. Prokaryotic mRNAs are polycistronic (polygenic) and usually carry information for the synthesis of several polypeptides from a single mRNA. The triplet codons in prokaryotic mRNA are transcribed from the sense strand of DNA and subsequently are translated continuously from the 5′-PO_4 end of the mRNA to the 3′-OH end. Since prokaryotic DNA is not separated from the cytoplasm by a nuclear membrane, translation begins on mRNA molecules before transcription is completed. Thus, transcription and translation are coupled in prokaryotes. Synthesis of each polypeptide chain in a polycistronic mRNA is determined by an AUG initiation codon and one or more nonsense codons that release the finished polypeptide from the ribosome.

Eukaryotic mRNAs differ from prokaryotic mRNAs in several respects (Figure 25-2). Eukaryotic genes invariably contain information for only a single polypeptide but each gene may consist of millions of nucleotides because eukaryotic genes contain introns and exons. The mRNA that is transcribed (primary transcript) is processed in several ways:

1. The ***introns*** (intervening sequences) are spliced out of the primary transcript and the ***exons*** (expressed sequences) are joined together. The splicing reactions and removal of introns from the primary transcript are carried out by ***small nuclear ribonucleoproteins*** (snRNPs).
2. While transcription is in process, the 5′ end of the mRNA is capped with a methyl guanine nucleotide (m^7 Gppp).
3. After the primary transcript is complete, a polyA tail ($-AAA_n\ A_{OH}$) is added to the 3′ terminus.

Prokaryotic mRNA

5′ AUG UAG AUG Gene 1 UAG AUG Gene 2 UAG 3′ Gene 3

Eukaryotic mRNA (Primary transcript)

5′-m²G Cap AUG I E I E I E UAG AAAn A-3′ poly A tail

One Gene

I = intron
E = exon

FIGURE 25-2
Structures of prokaryotic and eukaryotic primary transcripts (mRNAs). Prokaryotic mRNAs are polygenic, do not contain introns or exons, and are short lived in the cell. Eukaryotic mRNAs are monogenic, contain introns and exons, and usually are long lived in the cell.

4. Other modifications of the primary transcript are possible such as ***alternative splicing*** which produces mRNAs with different sets of exons and ***RNA editing*** in which bases are modified or changed in the original transcript.
5. The functional mRNA is transported to the cytoplasm where translation occurs on ribosomes bound to the endoplasmic reticulum of the cell.

25.3 Enzymatic Synthesis of RNA

The basic chemical features of the synthesis of RNA are the following (Figure 25-3):

1. The precursors of RNA synthesis are the four ribonucleoside 5′-triphosphates (rNTPs): ATP, GTP, CTP, and UTP. The ribose portion of each NTP has an -OH group on both the 2′ and the 3′ carbon atoms.
2. In the polymerization reaction, a 3′-OH group of one nucleotide reacts with the 5′-triphosphate of a second nucleotide; a pyrophosphate is removed and a phosphodiester bond is formed. This same reaction occurs in the polymerization of DNA.
3. The sequence of bases in an RNA molecule is determined by the base sequence of the DNA template strand. Each base added to the growing end of an RNA chain is chosen by base pairing with the appropriate base in the template strand; thus, the bases C, T, G, and A in a DNA strand cause incorporation of G, A, C, and U, respectively, in the newly synthesized RNA molecule. The RNA is complementary to the template DNA strand, which is called the ***coding (sense) strand*** or template strand.
4. The RNA chain grows in the 5′ → 3′ direction, which is the same as the direction of chain growth in DNA synthesis. The RNA strand and the DNA template strand are also antiparallel.
5. RNA polymerases, in contrast with DNA polymerases, can initiate RNA synthesis, i.e., no primer is needed.
6. Only ribonucleoside 5′-triphosphates participate in RNA synthesis, and the first base to be laid down in the initiation event is a triphosphate. Its 3′-OH group is the point of attachment of the subsequent nucleotide. Thus, the 5′ end of a growing RNA molecule terminates with a triphosphate. In tRNAs and rRNAs, and in eukaryotic mRNAs, the triphosphate group is removed.

E. coli RNA polymerase consists of five subunits—two identical α subunits and one each of β, β', and σ—having a total molecular weight of 465,000; it is one the largest enzymes known.

The α subunit is easily dissociated from the enzyme and, in fact, does so shortly after polymerization is initiated. The term ***core enzyme*** describes the σ-free unit ($\alpha_2\beta\beta'$); the complete enzyme is called the ***holoenzyme*** ($\alpha_2\beta\beta'\sigma$). In this chapter, the name RNA polymerase is used when the holoenzyme is meant. Several different RNA polymerases exist in eukaryotes and are described below.

An *E. coli* cell contains 3000–6000 RNA polymerase molecules; the number is greater when cells are growing rapidly. In eukaryotes, the number of RNA polymerase molecules varies significantly with cell type and is greatest in cells that actively make large quantities of protein, e.g., secretory cells.

25.4 Prokaryotic Transcription

The first step in prokaryotic transcription is the binding of RNA polymerase to DNA at a particular region called

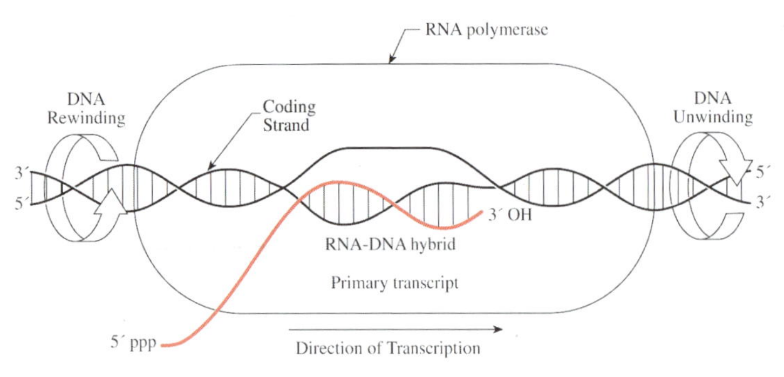

FIGURE 25-3
Model of RNA synthesis by RNA polymerase from the sense strand of DNA. See text for details.

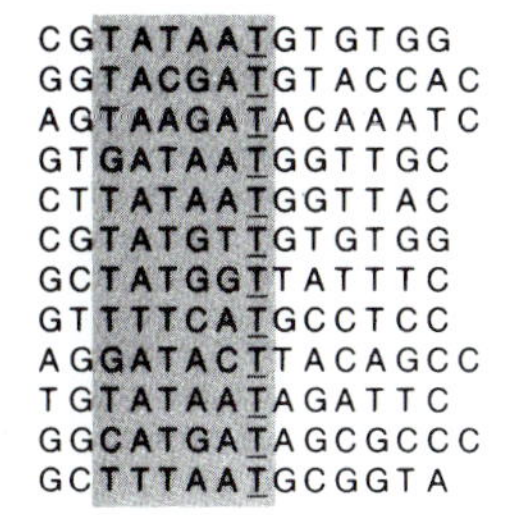

FIGURE 25-4
Segments of the coding strand of conserved regions from various genes showing the common sequence of six bases. The start point for mRNA synthesis is indicated by the heavy letters. The "conserved T" is underlined.

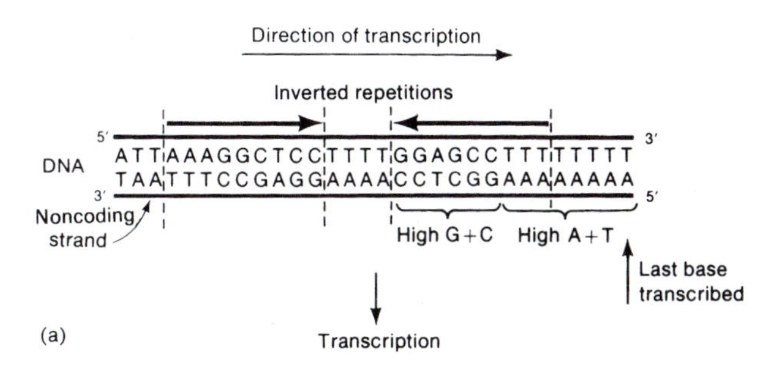

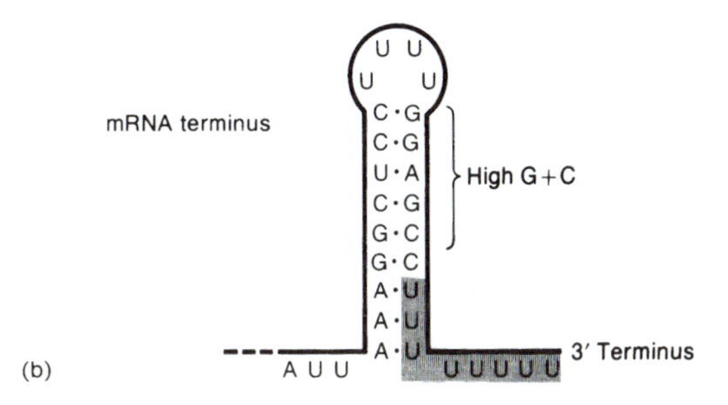

FIGURE 25-5
Base sequence of (a) the DNA of the *E. coli* trp operon at which transcription termination occurs and of (b) the 3′ terminus of the mRNA molecule. The inverted-repeat sequence is indicated by reversed arrows. The mRNA molecule is folded to form a stem-and-loop structure.

a ***promoter***—a specific nucleotide sequence, which for different RNA classes range in length from 20 to 200 nucleotides. In bacteria, a promoter is divided into subregions called the –35 sequence (8–10 bases long) and the –10 or ***Pribnow box*** (6 bases). The bases in the –35 sequence are quite variable but the Pribnow boxes of all promoters are similar (Figure 25-4). Both the –35 sequence and the Pribnow box interact with RNA polymerase to initiate RNA synthesis.

Following RNA polymerase binding to a promoter, a conformational change occurs such that a segment of the DNA is unwound and RNA polymerase is positioned at the polymerization start site. Transcription begins as soon as the RNA polymerase-promoter complex forms and an appropriate nucleotide binds to the enzyme. RNA polymerase contains two nucleotide binding sites called the ***initiation site*** and the ***elongation site.*** The initiation site binds only purine triphosphates (ATP and GTP), and one of these (usually ATP) becomes the first nucleotide in the RNA chain. Thus, the first DNA base that is copied from the DNA is usually thymine. The initiating nucleoside triphosphate binds to the enzyme and forms a hydrogen bond with the complementary DNA base. The elongation site is then filled with a nucleoside triphosphate that is selected strictly by its ability to form a hydrogen bond with the next base in the DNA strand. The two nucleotides are then joined together, the first base is released from the initiation site, and initiation is complete. The dinucleotide remains hydrogen-bonded to the DNA. The elongation phase begins when the polymerase releases the base and then moves along the DNA chain.

The drug ***rifampin*** binds to bacterial RNA polymerases and is a useful experimental inhibitor of initiation of transcription. It binds to the β subunit of RNA polymerase, blocking the transition from the chain initiation phase to the elongation phase; it is an inhibitor of chain initiation but not of elongation. ***Actinomycin D*** also inhibits initiation but does so by binding to DNA. These drugs have limited clinical use because of their toxicity.

After several nucleotides have been added to the growing chain, RNA polymerase holoenzyme changes its structure and loses the σ subunit. Thus, most elongation is carried out by the core enzyme, which moves along the DNA, binding a nucleoside triphosphate that can pair with the next DNA base and opening the DNA helix as it moves. The open region extends over about 30 base pairs. Chain elongation does not occur at a constant rate but slows down or stops at various points along the DNA molecule and, in some cases, may have a regulatory function.

Termination of RNA synthesis occurs at specific base sequences within the DNA molecule. Many prokaryotic termination sequences have been determined and most have the following three characteristics (Figure 25-5):

1. An inverted-repeat base sequence containing a central nonrepeating segment; the sequence in one DNA strand would read ABCDEF-XYZ-F′E′D′C′B′A′, in which A and A′, B and B′, and so on, are complementary bases. The RNA transcribed from this segment is capable of intrastrand base pairing, forming a stem and loop.
2. A sequence having a high G + C content.
3. A sequence of AT pairs in DNA (which may begin in the stem) that results in a sequence of 6–8 U's in the RNA.

Termination of transcription includes the following steps:

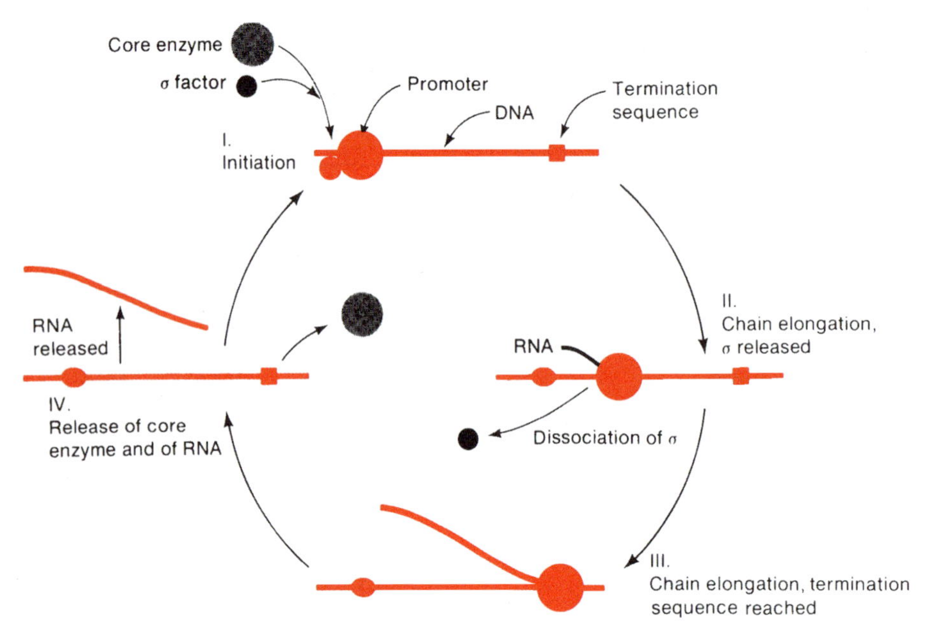

FIGURE 25-6
Transcription cycle of *E. coli* RNA polymerase showing dissociation of the σ subunit shortly after chain elongation begins, dissociation of the core enzyme during termination, and re-formation of the holoenzyme from the core enzyme and the σ subunit. A previously joined core enzyme and a subunit will rarely become rejoined; instead, reassociation occurs at random.

1. Cessation of RNA elongation,
2. Release of newly formed RNA, and
3. Release of the RNA polymerase from the DNA.

There are two kinds of termination events in prokaryotes: those that are self-terminating (dependent on the DNA base sequence only) and those that require the presence of a termination protein called ***Rho.*** Both types of events occur at specific but distinct base sequences. Rho, an oligomeric protein, does not bind to the core polymerase or to the holoenzyme and binds only very weakly to DNA. The action of Rho is poorly understood. Some microorganisms regulate transcription of certain genes by inhibiting Rho thereby allowing transcription to continue into adjacent genes, a process called ***antitermination.*** The transcription cycle of *E. coli* RNA polymerase is shown in Figure 25-6.

Various drugs inhibit chain elongation. Cordycepin is converted to a 5′-triphosphate form and then acts as a substrate analogue, blocking chain elongation.

Lifetime of Prokaryotic mRNA

All mRNA molecules are subject to attack by RNases, and this degradation is an essential aspect of the regulation of gene expression. Proteins are not made when they are not needed, and the rate of protein synthesis is determined by a balance between the rates of RNA synthesis and RNA degradation. The half-life of a typical prokaryotic mRNA molecule is only a few minutes, so constant production of a bacterial protein requires continued transcription. In contrast, eukaryotic mRNA molecules have a lifetime of hours to days. Presumably, the reason for the difference is that bacteria must adapt to rapidly changing environments, whereas eukaryotic cells receive a constant supply of nutrients that maintain a uniform environment.

25.5 Transcription in Eukaryotes

The chemistry of transcription in eukaryotes is the same as in prokaryotes. However, the promoter structure and the mechanism for initiation are strikingly different.

Eukaryotic RNA Polymerases

Eukaryotic cells contain three classes of RNA polymerases, denoted I, II, and III, which are distinguished by their requirements for particular ions and by their sensitivity to various toxins. All are found in the nucleus. Minor RNA polymerases are found in mitochondria and chloroplasts. Polymerase I molecules are located in the nucleolus and are responsible for synthesis of 5.8S, 18S, and 28S rRNA molecules. Polymerase II synthesizes all

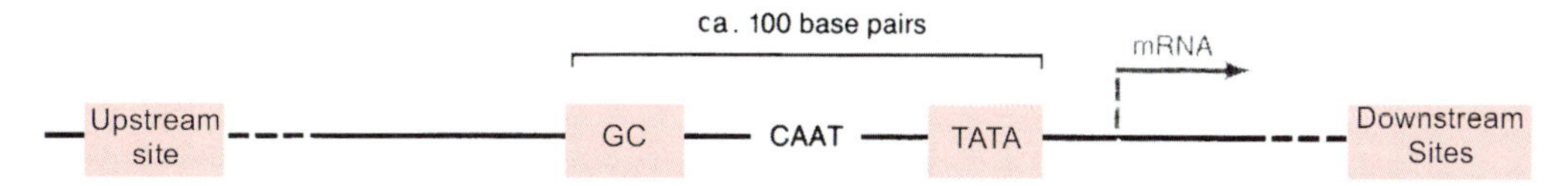

FIGURE 25-7
Sequences found in and near some RNA polymerase II promoters. Only the TATA box is represented in many promoters (some promotors lack this sequence). The CAAT box occurs much less frequently, and the GC box has only occasionally been observed. Upstream sites are very common but are not considered to be part of the promoter.

RNA molecules destined to become mRNA. Polymerase III synthesizes 5S rRNA and the tRNAs.

Polymerases II and III are inhibited by α-amanitin, the toxic product of *Amanita phalloides* mushrooms, and are identified by their sensitivity to this substance. Rifampin, a powerful inhibitor of bacterial RNA polymerase, is inactive against the eukaryotic nuclear polymerases but inhibits mitochondrial RNA polymerases, although at higher concentrations than needed to inhibit bacterial polymerases. Thus, rifampin is used with other drugs (e.g., isoniazid) in treating tuberculosis. Actinomycin D is a general inhibitor of eukaryotic transcription by virtue of its binding to DNA. It has been useful in the treatment of childhood neoplasms (e.g., Wilms' tumor) and choriocarcinoma. However, it inhibits rapidly proliferating cells of both normal and neoplastic origin and hence produces toxic side effects.

RNA Polymerase II Promoters

The structure of eukaryotic promoters is more complex than that of prokaryotic promoters. DNA sequences, hundreds of base pairs (bp) upstream from the transcription start site, control the rate of initiation. Furthermore, initiation requires numerous specific proteins (transcription factors) that bind to particular DNA sequences. Without the transcription factors, RNA polymerase II cannot bind to a promoter. However, KNH polymerase II itself is not a transcription factor. The complexity of initiation may derive, in part, from the fact that eukaryotic DNA is in the form of chromatin, which is inaccessible to RNA polymerases. Many RNA polymerase II promoters have the following features:

1. A sequence, TATAAAT, about 25 bp upstream from the transcription start site, known as the ***TATA*** or ***Hogness box.*** (Note its similarity to the Pribnow box.) The TATA box probably determines the base that is first transcribed.
2. A common sequence in the –75 region, GG(T/C)CAATCT, in which T and C appear with equal frequency at the third position. This sequence is called the CAAT box. A third element, recognized ina few promoters, is the GC box, GGGCGG (Figure 25-7).

Note: The term "upstream" refers to regions in the DNA that are to the left (or 5′) of the start of transcription of a gene; the term "downstream" refers to regions to the right (or 3′) of the gene. Regulatory regions affecting gene expression may be located close to the transcriptional start site, thousands of bases or more upstream or downstream from the gene, or in introns.

RNA Polymerase III Promoters

RNA polymerase III promoters differ significantly from RNA polymerase II promoters in that they are located downstream from the transcription start site and within the transcribed segment of the DNA. For example, in the 5S RNA gene of the South African toad (*Xenopus laevis*) the promoter is between 45 and 95 nucleotides downstream from the start point. Thus, the binding sites on RNA polymerase III are reversed with respect to the transcription direction, as compared with RNA polymerase II. That is, RNA polymerase II reaches forward to find the start point, and RNA polymerase III reaches backward. In fact, RNA polymerases can slide in either direction along a DNA template; however, they can only synthesize RNA molecules in a $5' \rightarrow 3'$ direction.

Eukaryotic mRNA Synthesis

Eukaryotic mRNA used in protein synthesis is usually about one-tenth the size of the primary transcript. This size reduction results from excision of noncoding sequences called ***intervening sequences*** or ***introns.*** After excision, the coding fragments are rejoined by RNA splicing enzymes (Figure 25-8).

Capping and Polyadenylylation

Capping occurs shortly after initiation of synthesis of the mRNA and precedes other modifications that protect the mRNA from degradation by nucleases. The poly(A)

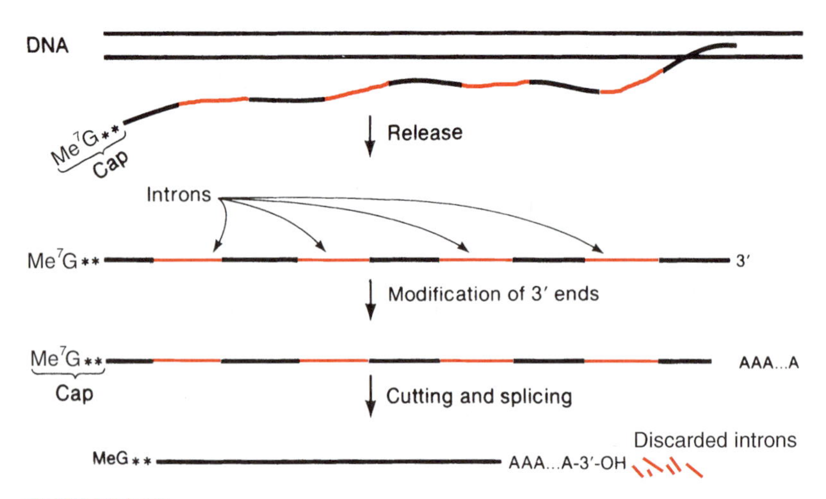

FIGURE 25-8
Schematic drawing showing production of eukaryotic mRNA. The primary transcript is capped before it is released. Then its 3′-OH end is modified, and finally the intervening regions are excised. MeG, 7-methylguanosine. **Two nucleotides whose riboses may be methylated.

terminus is synthesized by a nuclear enzyme, poly(A) polymerase. However, adenylate residues are not added directly to the 3′ terminus of the primary transcript. Instead, RNA polymerase transcribes past the recognition site for addition of poly(A). After synthesis of the complete RNA, endonucleolytic cleavage occurs at the poly(A) recognition site in the RNA, and poly(A) is added. A sequence, AAUAAA, located 10–25 bases upstream from the poly(A) addition site, is also required for enzyme recognition of the site of polyadenylation.

Splicing of RNA in Eukaryotes

A distinguishing feature of most primary transcripts of higher eukaryotes is the presence of untranslated intervening sequences (***introns***) that interrupt the coding sequence and are excised from the primary RNA transcript. In the processing of RNA in higher eukaryotes, the amount of discarded RNA ranges from 30% to nearly 90% of the primary transcript. The remaining coding segments (***exons***) are joined together by splicing enzymes to form translatable mRNA molecules. The excision of the introns and the formation of the final mRNA molecule by joining of the exons is called RNA splicing. The 5′ segment (the cap) of the primary transcript is never discarded and hence is always present in the completely processed mRNA molecule; the 3′ segment is also usually retained. Thus, the number of exons is usually one more than the number of introns. The number of introns per gene varies considerably (Table 25-1). Furthermore, within different genes the introns are distributed differently and have many sizes (Figure 25-9), and introns are usually longer than exons.

The splicing reaction is remarkably precise: cuts are made at unique positions in transcripts that contain thousands of bases. The fidelity of the excision and splicing reaction is extraordinary, for if an error of even one base were made, the correct reading frame would be destroyed. Such fidelity is achieved by recognition of particular base sequences by splicing enzymes.

Base sequence studies of the regions adjacent to several hundred different introns indicate that common sequences can be found at each end of an intron. The sites at which cutting occurs are always 5′ to GU and 3′ to AG. The rule is that the base sequence of an intron begins with GU and ends with AG.

Introns are excised one by one, and ligation occurs before the next intron is excised; thus, the number of different nuclear RNA molecules present at any instant is huge. Translation does not occur until processing is complete.

TABLE 25-1
Translated Eukaryotic Genes in Which Introns Have Been Demonstrated

Gene	Number of Introns
α-Globulin	2
Immunoglobulin L chain	2
Immunoglobulin H chain	4
Yeast mitochondrial cytochrome *b*	6
Ovomucoid	6
Ovalbumin	7
Ovotransferrin	16
Conalbumin	17
α-Collagen	52

*At present the histone and interferon genes are the only known translated genes in the higher organisms that do not contain introns.

FIGURE 25-9
Diagram of the conalbumin primary transcript and the processed mRNA. The 16 introns, which are excised from the primary transcript, are shown in color.

Before RNA splicing was discovered, the nucleus was observed to contain a significant amount of seemingly untranslated RNA. The collection of RNA molecules of widely varied sizes was given the name ***heterogeneous nuclear RNA (hnRNA),*** a term that is still sometimes used for nuclear RNA.

Splicing and Ribozymes

Different snRNPs are found in eukaryotic cells which function in removing introns from primary RNA transcripts. The association of small RNAs, nuclear proteins, and the introns that they attach to is referred to as a ***spliceosome.*** Small nuclear RNAs (designated U1 through U6) provide specificity to different spliceosomes so that they recognize different classes of introns. Introns are distinguished according to their three-dimensional structures and each class of introns is spliced out by a different mechanism.

Class I introns were originally discovered in ciliated protozoa and subsequently were found in fungi, bacteriophages, and other organisms. The RNA itself in a class I intron has catalytic activity and class I introns remove themselves from primary RNA transcripts by a self-splicing reaction. Class I introns are not true enzymes in that they function only once. The nucleotides in the intron that is spliced out are recycled in the cell.

Class II introns are removed from RNA by a self-splicing reaction that proceeds through an intermediate structure called a lariat. The removal of class II introns from RNAs also results in the splicing together of exons on either side of the intron. The ability of class II introns to specifically bind to a 5′ exon has led to their use as reagents to construct novel RNA molecules. Chemical derivatives of class II introns have been constructed that can carry out the reverse of the splicing reaction. When these introns insert themselves into RNA, they can be used to shuffle sequences or to link one RNA molecule to another.

Class II introns, which are found in bacteria, plant organelles, yeast, and fungi, also are capable of reintegrating into DNA after being excised from an RNA molecule. Class II introns encode a multifunctional intron-encoded protein (IEP) that has reverse transcriptase activity, RNA splicing activity, and DNA endonuclease activity. These three enzymatic activities allow the intron to be excised from RNA, copied into an RNA-DNA heteroduplex, and inserted into DNA at specific target sites recognized by the IEP-intron complex. Recently, class II introns have been modified so that they can be targeted to specific genes in any DNA molecule. This novel molecular mechanism raises the possibility that suitably engineered class II introns can be used therapeutically in gene therapy and possibly to inactivate viruses such as HIV by integration and disruption of essential genes.

Ribonuclease P (RNase P) consists of both protein and an RNA component that has catalytic activity. RNase P functions in eukaryotic cells to process the 5′ end of precursor tRNA molecules. RNase P also can be directed to cleave any RNA molecule when the target is complexed with a short complementary oligonucleotide called an external guide.

Ribozymes refer to catalytic RNA molecules that recognize specific target sequences in other RNA molecules. This activity and specificity makes ribozymes potentially useful therapeutic agents. For example, a ribozyme could be directed to silence the expression of a deleterious gene by destroying the mRNA transcript before it can be expressed. Or ribozymes might be constructed that would inactivate the mRNA or oncogenes or the expression of genes in RNA viruses such as HIV. Laboratory studies have supported the feasibility of ribozyme therapy but its application in clinical practice is still far in the future.

25.6 Genetic Code

"Universal" Genetic Code

Production of proteins from mRNA requires translation of the base sequence into an amino acid sequence. The collection of base sequences (codons) that correspond to each amino acid and to signals for termination of translation is the genetic code. The code consists of 64 triplets of bases (Table 25-2). The codons are written with the 5′ terminus

TABLE 25-2
*"Universal" Genetic Code**

First Position (5′ end)	Second Position				Third Position (3′ end)
	U	C	A	G	
U	Phe	Ser	Tyr	Cys	U
	Phe	Ser	Tyr	Cys	C
	Leu	Ser	Stop	Stop	A
	Leu	Ser	Stop	Trp	G
C	Leu	Pro	His	Arg	U
	Leu	Pro	His	Arg	C
	Leu	Pro	Gln	Arg	A
	Leu	Pro	Gln	Arg	G
A	Ile	Thr	Asn	Ser	U
	Ile	Thr	Asn	Ser	C
	Ile	Thr	Lys	Arg	A
	[Met]	Thr	Lys	Arg	G
G	Val	Ala	Asp	Gly	U
	Val	Ala	Asp	Gly	C
	Val	Ala	Glu	Gly	A
	[Val]	Ala	Glu	Gly	G

*The boxed codons are used for initiation. GUG is very rare.

of the codon at the left. The following features of the code should be noted:

1. Sixty-one codons correspond to 20 different amino acids.
2. The codon AUG has two functions. It corresponds to the amino acid methionine when AUG occurs within a coding sequence in the mRNA, i.e., within a polypeptide chain. It also serves as a signal to initiate polypeptide synthesis—with methionine for eukaryotic cells but with N-formylmethionine for prokaryotic cells. How the protein-synthesizing system distinguishes an initiating AUG from an internal AUG is discussed below. The codon GUG also has both functions, but it is only rarely used in initiation. Once initiation has occurred at an AUG codon, the reading frame is established and the subsequent codons are translated in order.
3. Three codons—UAA, UAG, and UGA—do not represent any amino acid but serve as signals to terminate the growing polypeptide chain.
4. Except for methionine (AUG) and tryptophan (UGG), most amino acids are represented by more than one codon. The assignment of codons is not random; with the exception of serine, leucine, and arginine, all synonyms (codons corresponding to the same amino acid) are in the same box and differ by only the third base. For example, GGU, GGC, GGA, and GGG all code for glycine.

The fidelity of translation is determined by two features of the system:

1. Attachment of the correct amino acid to a particular tRNA molecule by the corresponding aminoacyl synthetase, and
2. Correct codon-anticodon pairing. The former results from the specificity of interaction of the enzyme, the amino acid, and the tRNA molecule. The latter is assured by base pairing.

A striking aspect of the code is that, with few exceptions, the third base in a codon appears to be unimportant, i.e., XYA, XYB, XYC, and XYD are usually synonymous. This finding has been explained by the base pairing properties of the anticodon-codon interaction. In DNA, no base pairs other than GC and AT are possible because the regular helical structure of double-stranded DNA imposes steric constraints. However, since the anticodon is located within a single-stranded RNA loop, the codon-anticodon interaction does not require formation of a structure with the usual dimensions of a double helix. Model building indicates that the steric requirements are less stringent at the third position of the codon, a feature called *wobble*. The wobble hypothesis allows for inosine (I), a nucleoside in which the base is hypoxanthine and which is often found in anticodons, to base-pair with A, U, or C. In addition, the wobble hypothesis allows for U to base-pair with G. This explains how a tRNA molecule carrying a particular amino acid can respond to several different codons. For example, two major species of yeast $tRNA^{Ala}$ are known; one responds to the codons GCU, GCC, and GCA and has the anticodon IGC. Recall that the convention for naming the codon and the anticodon always has the 5′ end at the left. Thus, the codon 5′-GCU-3′ is matched by the anticodon 5′-IGC-3′. The other $tRNA^{Ala}$ has the anticodon CGC and responds only to GCG.

The function of three stop codons derives from the fact that no tRNA exists that has an anticodon that can pair with the stop codons.

Genetic Code of Mitochondria

The genetic code of mitochondria is not the same as the "universal" code, which has implications for the evolution of mitochondria. Human mitochondria contain a set of tRNA molecules that are not found elsewhere in the

TABLE 25-3
Genetic Code of Human Mitochondria

First Position (5′ end)	Second Position				Third Position (3′ end)
	U	C	A	G	
U	Phe	Ser	Tyr	Cys	U
	Phe	Ser	Tyr	Cys	C
	Leu	Ser	Stop	Trp	A
	Leu	Ser	Stop	Trp*	G
C	Leu	Pro	His	Arg	U
	Leu	Pro	His	Arg	C
	Leu	Pro	Gln	Arg	A
	Leu	Pro	Gln	Arg	G
A	[Ile]	Thr	Asn	Ser	U
	Ile	Thr	Asn	Ser	C
	[Met]*	Thr	Lys	Stop*	A
	[Met]	Thr	Lys	Stop*	G
G	Val	Ala	Asp	Gly	U
	Val	Ala	Asp	Gly	C
	Val	Ala	Glu	Gly	A
	Val	Ala	Glu	Gly	G

*These entries are found in mitochondria but not in the univeral code. Boxed codons are used as start codons. The mitochondrial codes of other organisms exhibit further differences.

cell and a circular DNA molecule containing 16,569 base pairs. This DNA molecule encodes some mitochondrial enzymes (Chapter 14) and is the template for synthesis of all mitochondrial tRNA and rRNA molecules. Human mitochondrial DNA sequences contain the genes for 12S and 16S ribosomal RNA, 22 different tRNA molecules, three subunits of the enzyme cytochrome oxidase (whose amino acid sequence is known), cytochrome b, and several other enzymes. The human mitochondrial code is shown in Table 25-3; entries shown with an asterisk differ from the universal code (cf. Table 25-2). The differences are striking in that most are in the initiation and termination codons. That is, in mammalian mitochondria,

1. UGA codes for tryptophan and not for termination.
2. AGA and AGG are termination codons rather than codons for arginine.
3. AUA and AUU are initiation codons, as is AUG. Both AUA and AUG also code for methionine. AUU also codes for isoleucine, as in the universal code.
4. AUA codes for methionine (and initiation, as shown in item 3) instead of isoleucine.

Maize mitochondria use CGG for tryptophan rather than for arginine, and CGU, CGC, and CGA for arginine. Yeast mitochondria use CUX, where X is any base, for threonine rather than for leucine. Both maize and yeast use AGA and AGG for arginine. Evidently, various mitochondrial codes can differ from each other as well as from the universal code.

The number of mammalian mitochondrial tRNA molecules is 22, which is less than the minimum number (32) needed to translate the universal code. This is possible because in each of the fourfold redundant sets—e.g., the four alanine codons GCU, GCC, GCA, and GCG—only one tRNA molecule (rather than two, as explained above) is used. In each set of four tRNA molecules, the base in the wobble position of the anticodon is U or a modified U (not I). It is not yet known whether this U is base-paired in the codon-anticodon interaction or manages to pair weakly with each of the four possible bases. For those codon sets that are doubly redundant—e.g., the two histidine codons CAU and CAC—the wobble base always forms, a G·U pair, as in the universal code. The structure of the human mitrochondrial tRNA molecule is also different from that of the standard tRNA molecule (except for mitochondrial $tRNA^{Leu}UUX$). (X = any nucleotide.) The most notable differences are the following:

1. The universal sequence CTψCXA is lacking in mitochondrial tRNA.
2. The "constant" 7-bp sequence of the TψC loop varies from three to nine bases.
3. The invariant bases U8, A14, G15, G18, G19, and U48 of the standard tRNA molecule are not invariant in mitochondrial tRNA.

In standard tRNA molecules, each of these bases participates in bonds that produce the folded L-shaped molecule. Thus, the mitochondrial tRNA molecule seems to be stabilized by fewer interactions. The three-dimensional configurations of these molecules are not known with certainty; possibly they differ from the standard L-shape, and mitochondrial tRNA engages in a different type of interaction with the ribosome than standard tRNA molecules do.

Most DNA molecules contain long noncoding segments (***spacers***) between genes. In mitochondrial DNA, there are either a few bases or none. In each case, there is a start codon (AUG, AUA, or AUU) at the 5′ end of the mRNA molecule or within a few bases of it. This arrangement differs from the usual arrangement for eukaryotes in that ordinarily eukaryotic mRNA molecules commence with a short leader sequence thought to be responsible for binding to the ribosome. Such a leader is not present in the mitochondrial mRNA molecules, which suggests that the 5′ terminus itself of a mitochondrial mRNA molecule binds to the ribosome. The plethora of diseases and clinical abnormalities caused by mitochondrial mutations discussed in Chapter 14.

25.7 Attachment of Amino Acid to tRNA Molecule

When an amino acid is covalently linked to a tRNA molecule, the tRNA is said to be ***aminoacylated*** or ***charged.*** The notation for a tRNA charged with serine is seryl tRNA. The term ***uncharged tRNA*** refers to a tRNA molecule lacking an amino acid.

Acylation is accomplished in two steps, both of which are catalyzed by an ***aminoacyl tRNA synthetase.*** In the first, or activation, step an aminoacyl AMP is generated in a reaction between an amino acid and ATP:

$$H_3N^+-\underset{R}{\overset{H}{\overset{|}{\underset{|}{C}}}}-\overset{O}{\overset{\|}{C}}-O^- + ATP \rightleftharpoons H_3N^+-\underset{R}{\overset{H}{\overset{|}{\underset{|}{C}}}}-\overset{O}{\overset{\|}{C}}-O-\underset{O^-}{\overset{O}{\overset{\|}{\underset{|}{P}}}}-\text{Ribose}-\text{Adenine} + PP$$

In the second, or transfer, step, the aminoacyl AMP reacts with tRNA to form an aminoacylated tRNA and AMP:

$$\text{Aminoacyl AMP} + \text{tRNA} \rightleftharpoons \text{aminoacyl tRNA} + \text{AMP}$$

As in DNA synthesis, the reaction is driven to the right by hydrolysis of pyrophosphate, so the overall reaction is

$$\text{Amino acid} + \text{ATP} + \text{tRNA} + H_2O \rightarrow \text{aminoacyl tRNA} + \text{AMP} + 2P_i$$

Usually only one aminoacyl synthetase exists for each amino acid. However, for a few amino acids specified by more than one codon, more than one synthetase does exist.

25.8 Initiator tRNA Molecules and Selection of Initiation Codon

The initiator tRNA molecule in prokaryotes—$\text{tRNA}^{\text{fMet}}$—has several properties that distinguish it from all other tRNA molecules. One feature is that the tRNA is first acylated with methionine, and then the methionine is modified. Acylation is by methionyl tRNA synthetase, which also charges tRNA^{Met}. However, the methionine of charged $\text{tRNA}^{\text{fMet}}$ is immediately recognized by another enzyme, tRNA methionyl transformylase, which transfers a formyl group from N^{10}-formyltetrahydrofolate (fTHF) to the amino group of the methionine to form

$$\text{tRNA}^{\text{fMet}} + \text{Met} \xrightarrow{\text{Methionyl-tRNA synthetase}} H_3N-\underset{\underset{\underset{CH_3}{|}}{\underset{S}{|}}}{\underset{(CH_2)_2}{|}}{\overset{H}{\overset{|}{C}}}-\overset{O}{\overset{\|}{C}}-O-\text{tRNA}^{\text{fMet}}$$

$$\xrightarrow[\text{THF} \quad \text{fTHF}]{\text{Transformylase}} \overset{O}{\overset{\|}{\underset{H}{C}}}-\overset{H}{\overset{|}{N}}-\underset{\underset{\underset{CH_3}{|}}{\underset{S}{|}}}{\underset{(CH_2)_2}{|}}{\overset{H}{\overset{|}{C}}}-\overset{O}{\overset{\|}{C}}-O-\text{tRNA}^{\text{fMet}}$$

N-formylmethionine (fMet): Transformylase does not formylate methionyl tRNA^{Met} because tRNA^{Met} and $\text{tRNA}^{\text{fMet}}$ are structurally different. A second feature is its ability to initiate polypeptide synthesis.

Eukaryotic initiator tRNA molecules differ from the prokaryotic initiator molecule in several ways. The most striking difference is that whereas eukaryotic organisms produce both a normal tRNA^{Met} and an initiator tRNA, which is also charged with methionine, the methionine does not undergo formylation. In eukaryotes, the first amino acid in a growing polypeptide chain is Met and not fMet. The codon for both kinds of tRNA molecules in eukaryotes is AUG, just as for prokaryotes.

The sequence AUG serves as both an initiation codon and the codon for methionine. Since methionine occurs within protein chains, some signal in the base sequence of the mRNA must identify particular AUG triplets as start codons. In eukaryotes, initiation usually begins at the AUG triplet nearest to the 5′ terminus of the mRNA molecule, i.e., no particular signal is used (although sequences do play a role in the efficiency of initiation). Presumably, the ribosome and the mRNA interact at the 5′ terminus and slide with respect to one another until an AUG is encountered. This AUG determines the reading frame. When a stop codon is encountered, the ribosome and the mRNA dissociate. Only a unique polypeptide is translated from a particular eukaryotic mRNA molecule.

In prokaryotes, the situation is quite different. At a fixed distance upstream from each AUG codon used for initiation is the sequence AGGAGGU (or a single-base variant). This sequence, known as the ***Shine-Dalgarno sequence,*** is complementary to a portion of the 3′-terminal sequence of one of the RNA molecules in the ribosome (in particular, the 16S rRNA in the 30S particle; see below). Base pairing between these complementary sequences orients the initiating AUG codon on the ribosome. Thus, initiation in prokaryotes is restricted to an AUG on the 3′ side of the Shine-Dalgarno sequence.

Many prokaryotic mRNA molecules are polycistronic and contain coding sequences for several polypeptides. Thus, a polycistronic mRNA molecule must possess a series of start and stop codons for each polypeptide. The mechanism for initiating synthesis of the first protein molecule in a polycistronic mRNA is the same as that in a monocistronic mRNA. However, if a second protein in a polycistronic mRNA is to be made, protein synthesis must reinitiate after termination of the first protein. This is usually accomplished by the start codon of the second protein being so near the preceding stop codon that reinitiation occurs before the ribosome and the mRNA dissociate. Otherwise, a second initiation signal is needed.

25.9 Ribosomes

A ribosome is a multicomponent structure that serves to bring together a single mRNA molecule and charged tRNA molecules so that the base sequence of the mRNA molecule is translated into an amino acid sequence. The ribosome also contains several enzymatic activities needed for protein synthesis. The protein composition of the prokaryotic ribosome is well known, but less is known about the composition of eukaryotic ribosomes. However, both share the same general structure and organization. The proper ties of the bacterial ribosomes are constant over a wide range of species, and the *E. coli* ribosome has been analyzed in great detail and serves as a model for all ribosomes.

Chemical Composition of Prokaryotic Ribosomes

A prokaryotic ribosome consists of two subunits. The intact particle is called a 70S ribosome because of its S (sedimentation) value of 70 Svedberg units. The subunits, which are unequal in size and composition, are termed 30S and 50S subunits. The 70S ribosome is the form active in protein synthesis. At low concentrations of Mg^{2+}, ribosomes dissociate into ribosomal subunits. At even lower concentrations, the subunits dissociate, releasing indi vidual RNA and protein molecules.

The composition of each subunit is as follows:

30S subunit : one 16S rRNA molecule + 21 different proteins

50S subunit : one 5S rRNA molecule + one 23S rRNA molecule + 34 different proteins

The proteins of the 30S subunit are termed S1, S2, . . ., S21 (in this context, S is for "small" subunit); one copy of each is contained in the 30S subunit. The proteins of the 50S subunit are denoted by L (for "large" subunit). Each 50S subunit contains one copy of each protein molecule, except for proteins L7 and L12, of which there are four copies, and L26, which is not considered to be a true component of the 50S subunit. Most ribosomal proteins are very basic proteins, containing up to 34% basic amino acids. This basicity probably accounts in part for their strong association with the RNA, which is acidic. The amino acid sequences of most of the *E. coli* ribosomal proteins are known. A summary of the structure of the *E. coli* 70S ribosome is shown in Figure 25-10.

Ribosomes Are Ribozymes

Ever since the discovery of the role of ribosomes in protein synthesis there has been uncertainty as to the functional roles of the rRNAs and the many ribosomal proteins. Conventional wisdom held that the rRNAs provided the scafolding (structural role) that oriented the ribosomal proteins so that one or more could carry out the enzymatic

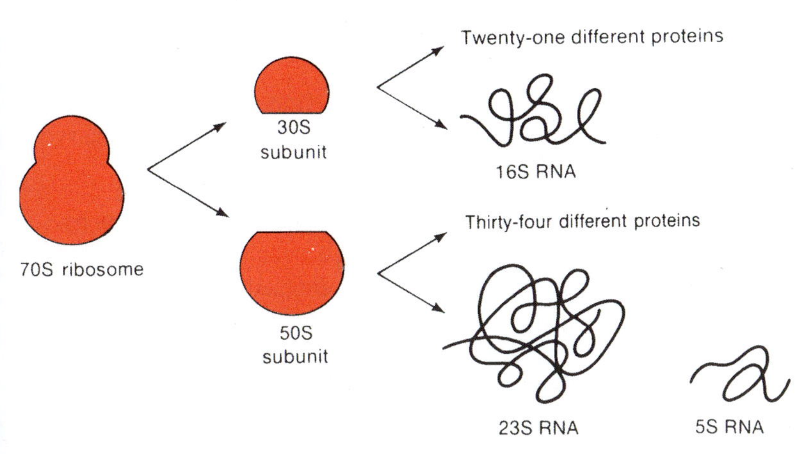

FIGURE 25-10
Dissociation of a prokaryotic ribosome. The configuration of two overlapping circles is used throughout this chapter, for the sake of simplicity.

reactions necessary for protein synthesis. It was not until crystallization of the 50S subunit provided an atomic-level view of its structure that it was finally proved that the opposite is actually correct. The ribosomal proteins provide the structural framework for the 23S which actually carries out the peptidyltransferase reaction. Thus, the 23S rRNA is an enzyme and the ribosome is a ***ribozyme.*** Although the activity of the 23S rRNA is dependent on the ribosomal proteins, the 50S subunit that is substantially deproteinized still can perform the peptidyltransferase reaction.

A crucial feature of the peptidyltransferase reaction is a particular adenine that is conserved in rRNAs extracted from the large ribosomal subunits of thousands of different organisms from all three kingdoms. Sequence analysis shows that this adenine base ($_{2451}$A) always is present in the active site of the ribosome and acts as a general acid-base catalyst by deprotonating the nucleophilic amine as shown below.

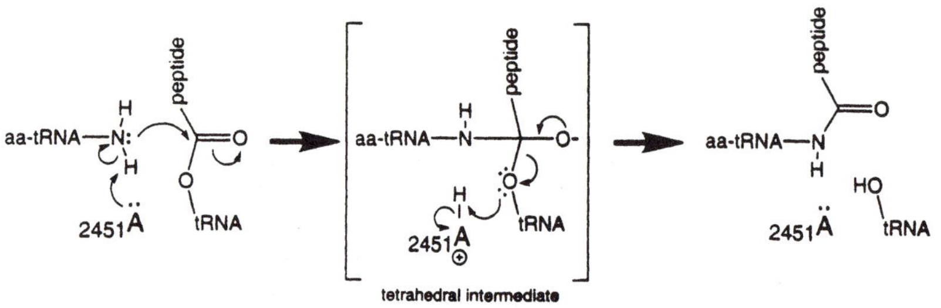

tetrahedral intermediate

The positive charge on $_{2451}$A in the tetrahedral intermediate is probably transferred to adjacent tRNA nucleotides that undergo tautomeric shifts. The evidence that ribosomal RNAs perform enzymatic functions critical to protein synthesis lends weight to the evolutionary significance of RNA in primordial chemical reactions that eventually led to the development of cells.

Chemical Composition of Eukaryotic Ribosomes

The basic features of eukaryotic ribosomes are similar to those of bacterial ribosomes, except for their larger size. They contain a greater number of proteins (about 80) and an additional RNA molecule.

A typical eukaryotic ribosome has a sedimentation coefficient of about 80 and consists of two subunits: 40S and 60S. These sizes may vary by as much as 10% from one organism to another in contrast with the homogeneity of bacteria ribosomes. The components of the subunits of eukaryotic ribosomes are as follows:

40S subunit : one 18S rRNA molecule + 30–35 proteins

60S subunit : one 5S, one 5.8S, and one 28S rRNA molecule + 45–50 proteins

25.10 Protein Synthesis

Protein synthesis has three stages:

1. Polypeptide chain initiation,
2. Chain elongation, and
3. Chain termination.

The main features of the initiation step are binding of mRNA to the ribosome, selection of the initiation codon, and binding of the charged tRNA bearing the first amino acid. In the elongation stage, there are two steps: joining of adjacent amino acids by peptide bond formation, and moving the mRNA and the ribosome with respect to one another so that codons are translated successively (***translocation***). In the termination stage, the completed polypeptide dissociate from the ribosomes, which are released to begin another cycle of synthesis.

Every polypeptide has an amino terminus and a carboxyl terminus. In both prokaryotes and eukaryotes, synthesis begins at the amino terminus. For a protein having the sequence H_2N–Met–Trp–Asp . . . Pro–Val–COOH, the f Met (or Met) is the initiating amino acid and Val is the last amino acid added to the chain. Translation of mRNA molecules occurs in the $5' \rightarrow 3'$ direction.

Stages of Protein Synthesis

The mechanismsof protein synthesis in prokaryotes and eukaryotes differ slightly in detail, but the prokaryotic mechanism is used as a general model:

1. *Initiation.* Protein synthesis in bacteria begins by the association of one 30S subunit (not the 70S ribosome), an mRNA, a charged $tRNA^{fMet}$, three protein initiation factors, and guanosine 5′-triphosphate (GTP). These molecules make up the 30S preinitiation complex. Association occurs at an initiator AUG codon, whose selection was described above. A 50S subunit joins to the 30S subunit to form a 70S initiation complex (Figure 25-11). This joining process requires hydrolysis of the GTP contained in the 30S preinitiation complex. There are two tRNA

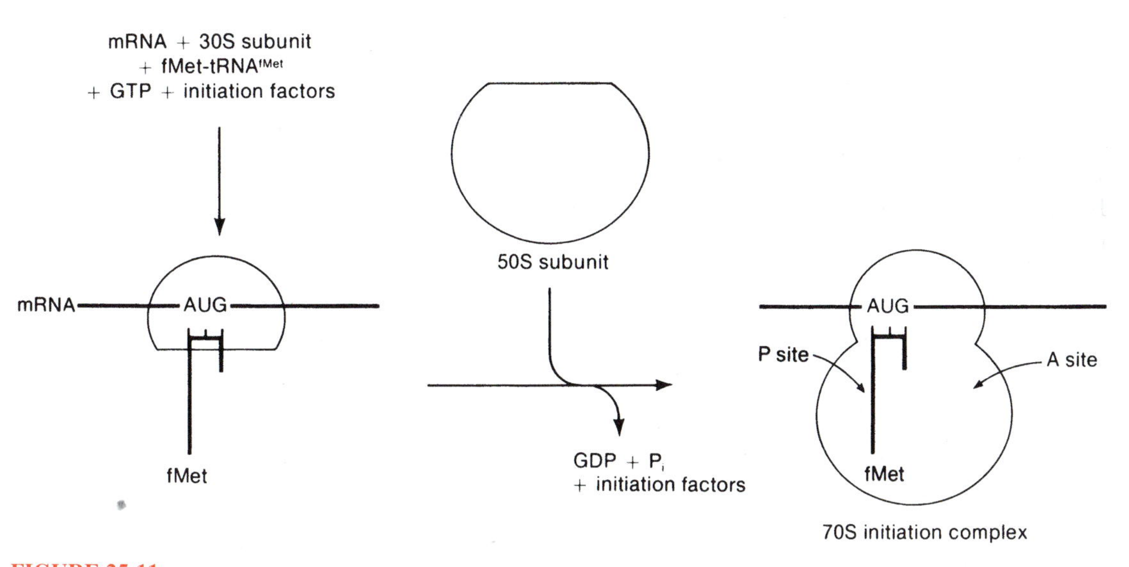

FIGURE 25-11
Early steps in protein synthesis: in prokaryotes: formation of the 30S preinitiation complex and 70S initiation complex.

binding sites on the 50S subunit: the ***aminoacyl(A) site*** and the ***peptidyl(P) site;*** each site consists of several L proteins and 23S rRNA. The 50S subunit is positioned in the 70S initiation complex such that the $tRNA^{fMet}$, which was previously bound to the 30S preinitiation complex, occupies the P site of the 50S subunit. Positioning $tRNA^{fMet}$ in the P site fixes the position of the anticodon of $tRNA^{fMet}$ so that it can pair with the initiator codon in the mRNA. in this way, the reading frame is unambiguously defined.

2. *Binding of a second tRNA to the A site.* The A site of the 70S initiation complex is available to any tRNA molecule whose anticodon can pair with the codon adjacent to the AUG complex initiation codon. However, entry to the A site by the tRNA requires the activity of a protein called ***elongation factor*** EF-Tu (Figure 25-12). EF-Tu binds GTP and an aminoacyl tRAA, forming a ternary complex, aminoacyl tRNA [EF-Tu-GTP]. This ternary complex enters the A site for the codon-dependent placement of the aminoacyl tRNA at the A site. During this process of placement, GTP is hydrolyzed to GDP and P_i, and EF-Tu is released. EF-Tu-GTP is regenerated from EF-Tu-GDP through the inter action of another protein, EF-Ts, and GTP.
3. *Chain elongation (formation of the first peptide bond).* After a charged tRNA is positioned in the A site, a peptide bond between f Met and the adjacent amino acid forms. The peptide bond is formed by an enzyme complex called peptidyltransferase, whose active site resides in 23SrRNA of 50S subunit *as previously discussed* (Figure 25-13). As the peptide bond is formed, the Met is cleaved from the $tRNA^{fMet}$ in the P site by another ribosomal protein, tRNA deacylase.
4. *Translocation of the ribosome.* After the peptide bond forms, an uncharged tRNA occupies the P site and a dipeptidyl tRNA is in the A site. At this point three events, which together make up the translocation step, occur: the deacylated $tRNA^{fMet}$ leaves the P site, the peptidyl tRNA moves from the A site to the P site, and the ribosome moves a distance of three bases on the mRNA to position the next codon at the A site. The translocation step requires the activity of another elongation protein, EF-G, and hydrolysis of GTP to provide the energy to move the ribosome. Thus, the total amount of energy expended in the synthesis of one peptide bond comes from the hydrolysis of four high-energy phosphate bonds (equal to about $30 \text{ kcal} = 7.5 \times 4$). Recall that the synthesis of each molecule of ammoacyl tRNA consumes two high-energy phosphate bonds ($ATP \rightarrow AMP + PP_i$; $PP_i + 2P_i$). (In these calculations, the one GTP molecule expended in the formation of the initiation complex is not included because of its small contribution to the overall synthesis of a polypeptide.)
5. *Refilling of the A site.* After translocation has occurred, the A site is again available for a charged tRNA molecule with a correct anticodon. When this occurs, the series of elongation reactions is repeated.
6. *Termination.* When a chain termination codon is reached, no aminoacyl tRNA is available that can fill the A site, so chain elongation stops. Since the polypeptide chain is still attached to the tRNA occupying the P site, release of the protein is

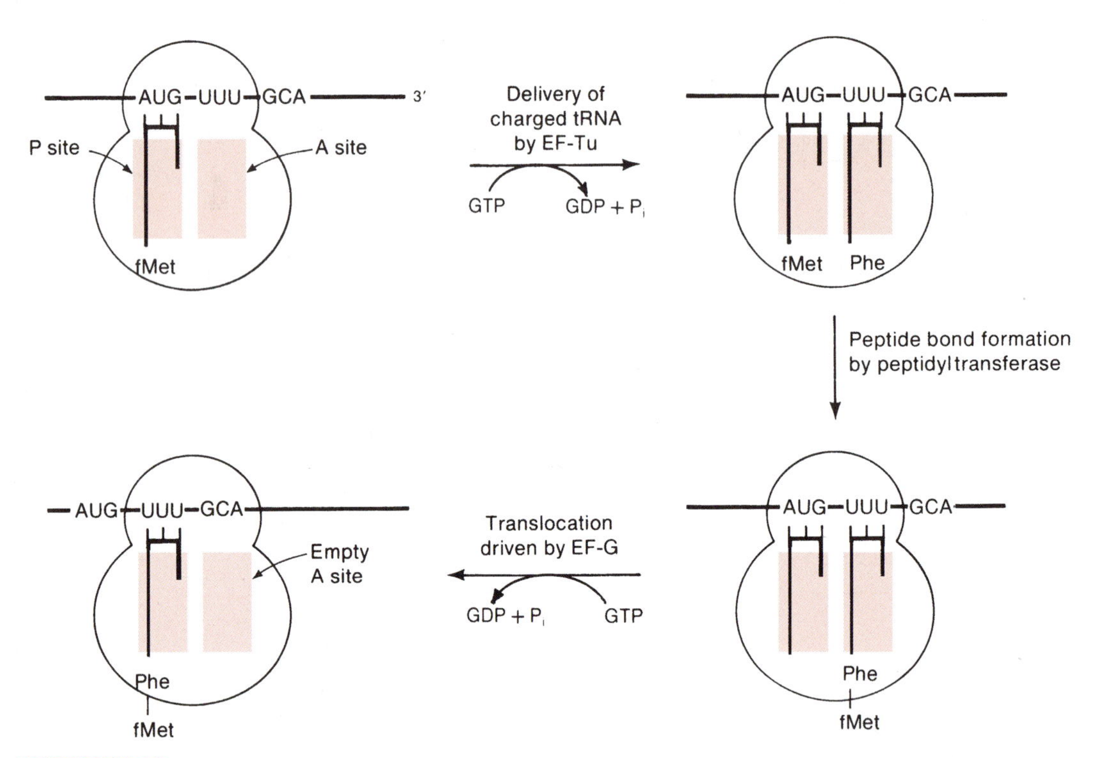

FIGURE 25-12
Elongation phase of protein synthesis: binding of charged tRNA, peptide bond formation, and translocation. Note that this phase consumes 2 GTP during each cycle of chain elongation.

accomplished by release factors (RFs), which are proteins that, in part, respond to chain termination codons. In the presence of release factors, peptidyltransferase catalyzes the reaction of the bound peptidyl moiety with water rather than with the free aminoacyl tRNA (Figure 25-14). Thus, the polypeptide chain, which has been held in the ribosome solely by the interaction with the tRNA in the P site, is released from the ribosome.

7. *Dissociation of the 70S ribosome.* The 70S ribosome dissociates into 30S and 50S subunits, which may start synthesis of another polypeptide chain.

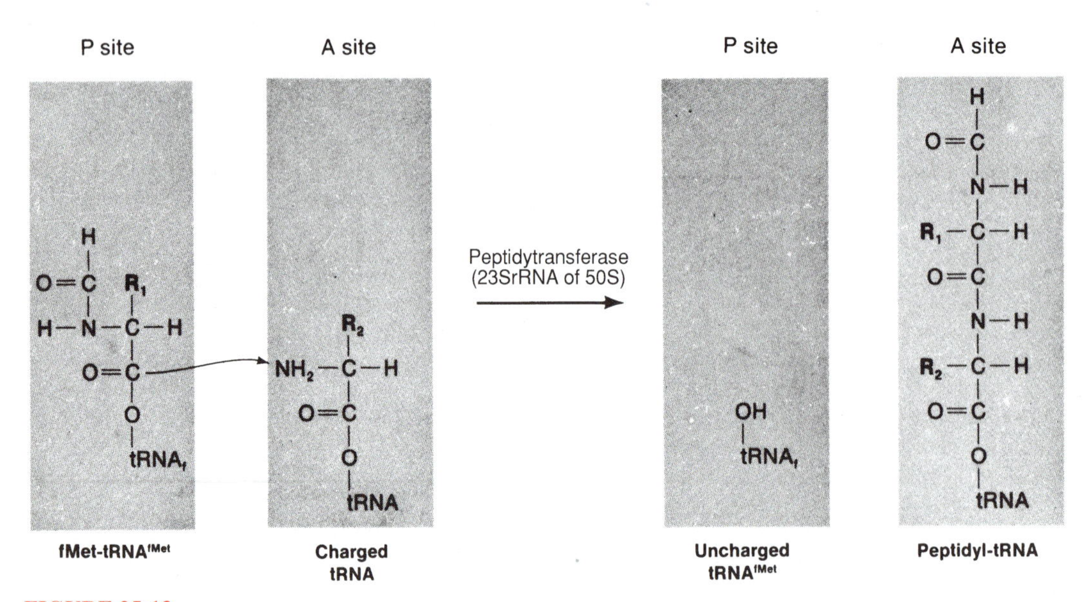

FIGURE 25-13
Peptide bond formation in prokaryotes. In eukaryotes, the chemistry is the same, although the initiating tRNA is not $tRNA^{fMet}$ but $tRNA^{Met}_{init}$.

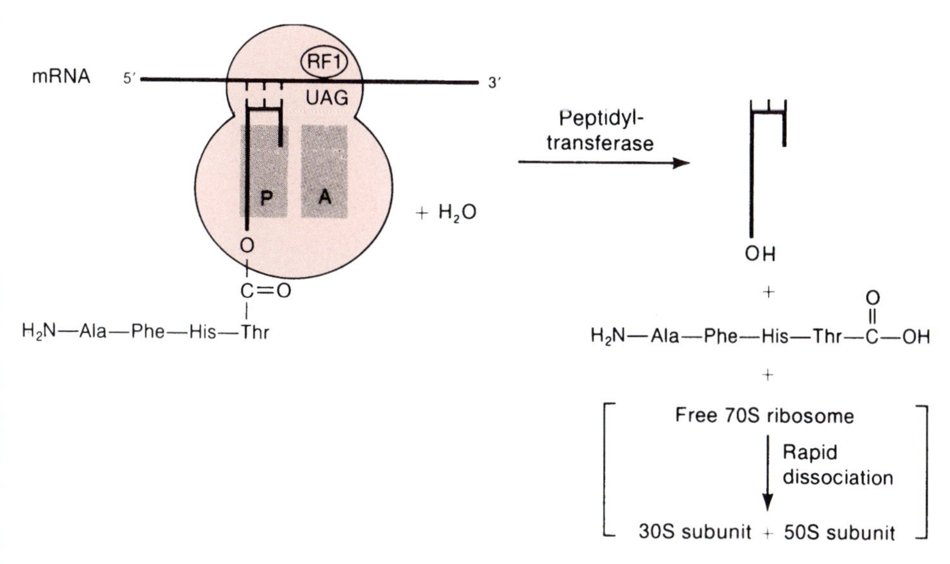

FIGURE 25-14
The steps of termination of protein synthesis.

Role of GTP

GTP, like ATP, is an energy-rich molecule. Generally, when such molecules are hydrolyzed, the free energy of hydrolysis is used to drive reactions that otherwise are energetically unfavorable. This does not seem to be the case in protein synthesis. The reaction sequence indicates that GTP facilitates binding of protein factors either to tRNA or to the ribosome. Furthermore, hydrolysis of GTP to GDP always precedes dissociation of the bound factor. Comparison of the structure of the free factor and the factor-GTP complex indicates that the factor undergoes a slight change in conformation when GTP is bound. The function of GTP is to induce a conformational change in a macromolecule by binding to it. Since it is easily hydrolyzed by various GTPases, the use of GTP as a controlling element allows cyclic variation in macromolecular shape. When GTP is bound, the macromolecule has an active conformation, and when the GTP is hydrolyzed or removed, the molecule resumes its inactive form. GTP plays a similar role in hormone activation systems (Chapter 30).

Posttranslational Modification of Proteins

The protein molecule ultimately needed by a cell often differs from the polypeptide chain synthesized. Modification of the synthesized chain occurs in several ways:

1. In prokaryotes, fMet is never retained as the NH_2, terminal amino acid. In roughly half of all proteins, the formyl group is removed by the enzyme deformylase, leaving methionine as the NH_2 terminal amino acid. In both prokaryotes and eukaryotes, the fMet, methionine, and possibly a few more amino acids, are often removed; their removal is catalyzed by a hydrolytic enzyme called aminopeptidase. This hydrolysis may occur as the chain is being synthesized or after the chain is released from the ribosome. The choice of deformylation versus removal of fMet usually depends on the identity of the adjacent amino acids. Deformylation predominates if the second amino acid is arginine, asparagine, aspartic acid, glutamic acid, isoleucine, or lysine, whereas f Met is usually removed if the adjacent amino acid is alanine, glycine, proline, threonine, or valine.
2. Newly created NH_2 terminal amino acids are sometimes acetylated, and amino acid side chains may also be modified. For example, in collagen a large fraction of the prolines and lysines are hydroxylated (see below). Phosphorylation of serine, tyrosine, and threonine occurs in many organisms. Various sugars may be attached to the free hydroxyl group of serine or threonine to form glycoproteins. Finally, a variety of prosthetic groups such as heme and biotin are covalently attached to some proteins.
3. Two distant sulfhydryl groups in two cysteines may be oxidized to form a disulfide bond.
4. Polypeptide chains may be cleaved at specific sites. For instance, chymotrypsinogen is converted to the digestive enzyme chymotrypsin by removal of four amino acids from two different sites. In some cases, the uncleaved chain represents a storage form of the protein that can be cleaved to generate the active protein when needed. This is true of many mammalian digestive enzymes, e.g., pepsin is formed by cleavage of pepsinogen. An interesting precursor is a huge protein synthesized in animal cells infected

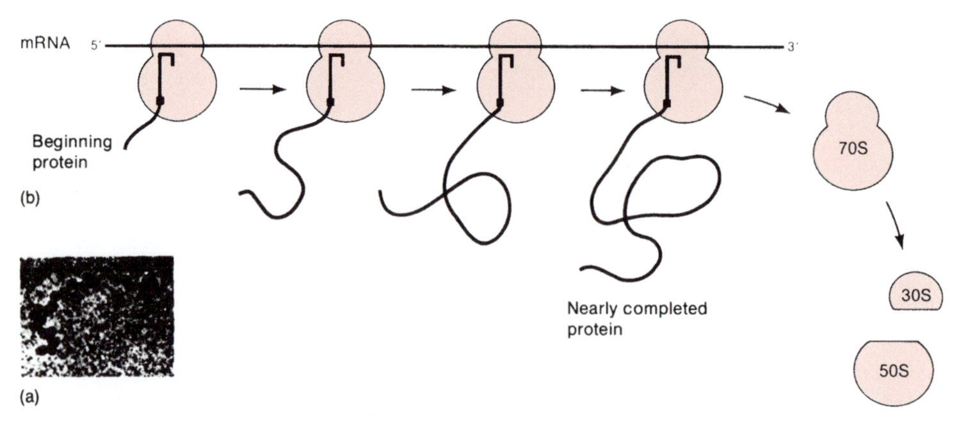

FIGURE 25-15
Polysomes. (a) Electron micrograph of an *E. coli* polysome. (Courtesy of Barbara Hamkalo.) (b) Diagram showing relative movement of the 70S ribosome and the mRNA and growth of the protein chain.

with some viruses; this molecule is cleaved at several sites to yield different active proteins and hence is called a ***polyprotein.*** Some cellular enzymatic systems are also formed by cleavage of polyproteins.

Coupled Transcription and Translation

After about 25 amino acids have been joined together in a polypeptide chain in prokaryotes, the AUG initiation site of the mRNA molecule becomes exposed and a second initiation complex then forms. The overall configuration is of two 70S ribosomes moving along the mRNA at the same rate. When the second ribosome has moved a distance similar to that traversed by the first, a third ribosome is able to attach. This process—movement and reinitiation—continues until the mRNA is covered with ribosomes separated by about 80 nucleotides. This large translation unit is called a ***polyribosome*** or ***polysome*** (Figure 25-15). Polysomes greatly increase the overall rate of protein synthesis, since 10 ribosomes traversing a single segment of mRNA clearly can make 10 times as many polypeptides per unit of time as can a single ribosome. In prokaryotes, transcription and translation are coupled; this can not occur in eukaryotes since transcription occurs in the nucleus and translation in the cytoplasm.

Endoplasmic Reticulum

In most eukaryotic cells two major classes of ribosomes exist: attached ribosomes and free ribosomes. The attached ribosomes are bound to an extensive cytoplasmic network of lipoprotein membranes called the endoplasmic reticulum. The ***rough endoplasmic reticulum*** consists of bound ribosomes; the ***smooth endoplasmic reticulum*** is devoid of ribosomes. There is no structural difference between a free and an attached ribosome, and attachment to the membrane occurs after synthesis of particular proteins begins.

Most endoplasmic reticulum membranes enclose large, irregularly shaped, discrete regions of the cell called *cisternae.* In this sense, the membrane system has an inside and an outside, and ribosomes are bound only to the outside. Cells responsible for secreting large amounts of a particular protein (e.g., hormone-secreting cells) have an extensive endoplasmic reticulum. Most proteins destined to be secreted by the cell or to be stored in intracellular vesicles such as lysosomes (which contain degradative enzymes) and peroxisomes (which contain enzymes for eliminating hydrogen peroxide) are synthesized by attached ribosomes. These proteins are primarily found in the cisternae of the endoplasmic reticulum. In contrast, most proteins destined to be free in the cytoplasm are made on free ribosomes.

The formation of the rough endoplasmic reticulum and the secretion of newly synthesized proteins through membrane is explained by the ***signal hypothesis*** (Figure 25-16). The basic idea is that the signal for attachment of the ribosome to the membrane is a sequence of very hydrophobic amino acids near the amino terminus of the growing polypeptide chain. When protein synthesis begins, the ribosome is free. Then the amino terminal hydrophobic amino acids interact with lipophilic membrane components and somehow direct the association of the large ribosomal subunit to a ribosomal receptor protein on the membrane surface. As protein synthesis continues, the protein moves through the membrane to the cisternal side of the endoplasmic reticulum. A specific protease termed the ***signal peptidase*** cleaves the amino terminal signal sequence.

Intracellular compartmentation of the newly synthesized proteins is a complex process, and disorders in this process lead to severe abnormalities (see below).

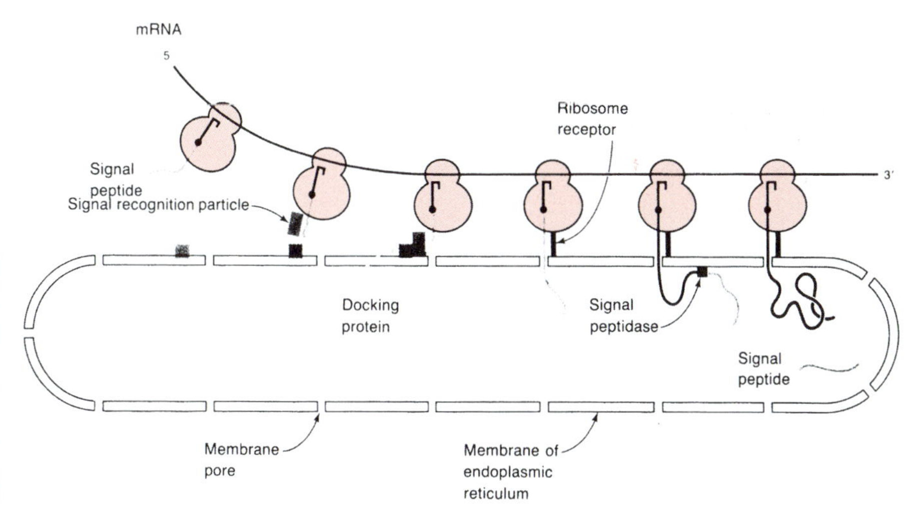

FIGURE 25-16
Signal hypothesis for the synthesis of secretory and membrane proteins. Shortly after initiation of protein synthesis, the amino-terminal sequence of the polypeptide chain binds a signal recognition protein, which then binds to a docking protein. The signal peptide is released from the signal recognition protein as the ribosome binds to a ribosome receptor, which is adjacent to a pore. Translation continues with the signal peptide passing through the pore. Once through the endoplasmic reticulum, the signal sequence is excised by the signal peptidase within the vesicle. When protein synthesis is completed, the protein remains within the vesicle and the ribosome is released.

In summary, the following events take place. When the predominant portion of the protein is synthesized and migrates in the endoplasmic reticulum, the signal sequence is removed by proteolytic cleavage. The finished protein, which is sequestered on the cisternal side of the endoplasmic reticulum, undergoes many posttranslational modifications, one of which is the addition of a series of carbohydrate residues to form glycoproteins (Chapter 16). Glycoproteins are transported to the Golgi complex, where further modification of the carbohydrate residues occurs. The finished glycoproteins are then packaged into lysosomes, peroxisomes, or secretory vesicles. The last fuses with the plasma membrane, discharging its contents into the extracellular fluid. Defects in compartmentation processes may lead to mislocation of the proteins and cause deleterious effects. Many proteins that are destined for various organelles (e.g., mitochondria) are not routed through endoplasmic reticulum and the Golgi complex. These proteins are made on cytosolic polysomes. They possess presequences at amino terminal ends that target them to receptors on appropriate organelles. Import of proteins into cell organelles is aided by ***chaperone proteins*** (Chapter 4).

Compartment Disorders

Two defects of compartmentation are known. One, ***mucolipidosis*** or ***I-cell disease*** (Chapter 16), is characterized by mislocalization of lysosomal enzymes (acid hydrolases). These glycoproteins lack mannose-6-phosphate residues, and consequently they are secreted into extracellular fluids instead of being sequestered into lysosomal vesicles. The mannose-6-phosphate serves as a marker that allows the enzymes to bind with mannose-6-phosphate receptors that direct the enzymes into the vesicles to be packaged into lysosomes. The lysosomal enzymes in the extracellular fluid bring about indiscriminate destruction of tissues.

The second example of the compartmentation defect is α_1-***antitrypsin deficiency*** (also known as α_1-proteinase inhibitor deficiency). The defect in this disorder is almost exactly the opposite of that in I-cell disease, i.e., I-cell disease is a disorder of a defect in (intracellular) retention of proteins and α_1-antitrypsin deficiency is a defect in the secretion of a protein. α_1-Antitrypsin (α_1-AT) consists of a single polypeptide chain of 394 amino acid residues with three oligosaccharide side chains (Figure 25-17), all of which are attached to asparagine residues. It contains three β sheets and eight α helices. It has an overall molecular weight of 51,000. It is a polar molecule and readily migrates into tissue fluids. It is synthesized in hepatocytes and secreted into the plasma, where it has a half-life of 6 days. Its normal serum concentration is 150–200 mg/dL. Although the presence of α_1-AT has been noted in other cell types, the primary source in plasma is the hepatocyte. α_1-AT does not possess a propeptide but does contain a 24-residue signal peptide, which is eliminated during its passage through the endoplasmic reticulum. α_1-AT is a defense protein, and its synthesis and release into circulation

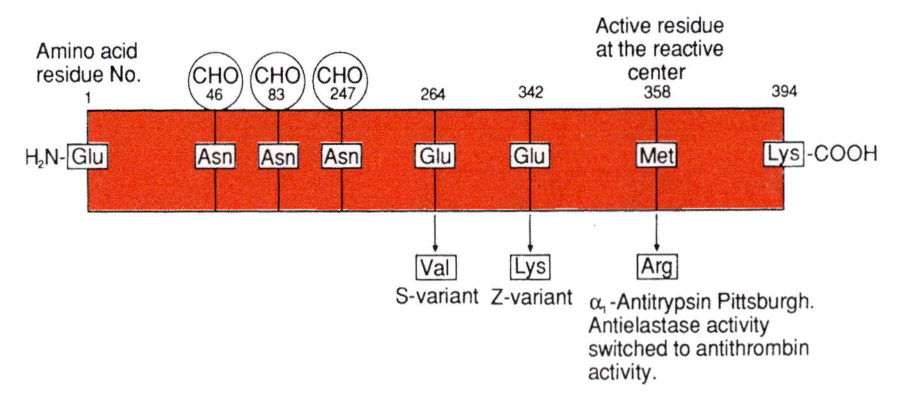

FIGURE 25-17
Schematic representation of the amino acid sequence of α_1-antitrypsin (PiMM) and some of its variants. The amino acid residues shown are not drawn to scale; CHO represents the oligosaccharide units.

increase in response to trauma and inflammatory stimulus. Thus, it is known as an acute-phase reactant. α_1-AT functions by forming a tight complex with the active site of the target enzyme that inhibits its proteolytic activity. The enzyme-inhibitor complexes are rapidly cleared by the reticuloendothelial cells. Thus, α_1-AT is a suicide protein. Many genetic variants of α_1-AT have been identified by isoelectric focusing. The α_1-AT gene is on chromosome 14, is 10.2 kb long, and contains four introns. The phenotype is based on a codominant pattern with both alleles expressed equally. Variants are classified by a system known as ***Pi (protease inhibitor).*** The most common form is PiMM (carried by 95% of the U.S. population). Most polymorphism observed in α_1-AT are due to single-amino-acid substitutions.

α_1-AT is a broad-spectrum serine proteinase inhibitor (Chapter 6). Its principal action is to inhibit leukocyte (neutrophil) elastase. This function appears to be most important in maintaining the integrity of the elastic fibers for the elastic recoil of normal lung tissue. Recall that elastic fibers contain an amorphous core of elastin surrounded by an envelope of microfibrils (Chapter 16). The target substrate for neutrophil elastases is elastin. The turnover rate of mature elastin is extremely low, and if it is destroyed, replacement is severely limited. Thus, the risk of development of a lung disease (***emphysema***) characterized by dilatation of air spaces distal to the terminal bronchiole and destruction of bronchiole walls is high in α_1-AT deficiency.

A human phenotype in which a severe deficiency of α_1-AT (10–15% of normal serum concentration) has been observed is designated PiZ. The risk of development of emphysema in PiZZ (homozygotic ZZ individuals) is 20 times that in PiMM (a normal genotype). In addition, cigarette smoking by PiZ individuals accelerates development of destructive lung disease for several reasons.

1. The levels of α_1-AT are very low, and smoking causes an increase in elastolytic load owing to change in the phagocytic population.
2. A reduction of the proteinase inhibitory activity is due to oxidation of the reactive methionine residue of the active center by the cigarette smoke as well as by oxygen radicals produced by leukocytes and macrophages. Emphysema is a common disease, and its most common cause is cigarette smoking. Only 1–2% of cases are due to genetic deficiency of α_1-AT.

The substitution of a lysine for a glutamic acid at position 342 leads to the deficiency of α_1-AT in PiZ individuals (Figure 25-17). This substitution of a basic for an acidic amino acid is consistent with a base change of cytosine to thymine in DNA. The amino acid change prevents normal processing of oligosaccharide side chains of the protein and therefore its secretion. As a result, Z-α_1-AT accumulates in hepatocytes. An association of hepatic disease with severe α_1-AT deficiency has been observed; however, the relationship is not understood. The sequestered α_1-AT may cause damage to hepatocytes or render them more susceptible to injury, toxins, or viruses. Liver transplantation has been attempted with limited success in patients with hepatocellular failure. The therapy for destructive lung disease is administration of an anabolic steroid (danazol), which increases the serum levels of α_1-AT by about 40%, or direct replacement of α_1-AT by transfusion.

Another structural variant of α_1-AT that has been studied in detail is S-α_1-AT (PiSS). This protein differs from M-α_1-AT at position 264, where a glutamic acid residue is substituted by valine (Figure 25-17). The S-α_1-AT undergoes normal addition of oligosaccharides, and the protein is secreted into the blood. Affected individuals have sufficiently high levels of S-α_1-AT to protect against neutrophil elastase and so are not at higher risk for developing

a

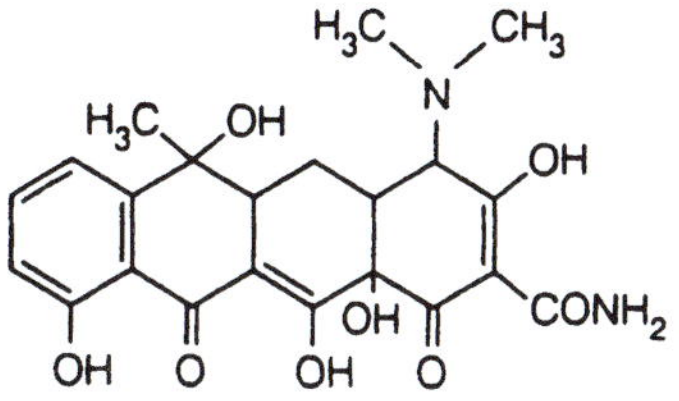

Tetracycline: Inhibits the binding of aminoacyl tRNAs to the ribosome in bacteria.

b

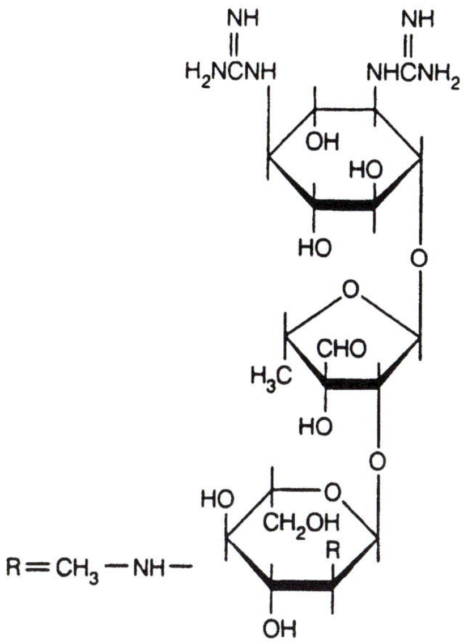

Streptomycin: Interferes with pairing between aminoacyl tRNAs and RNA codons, causing misreading.

c

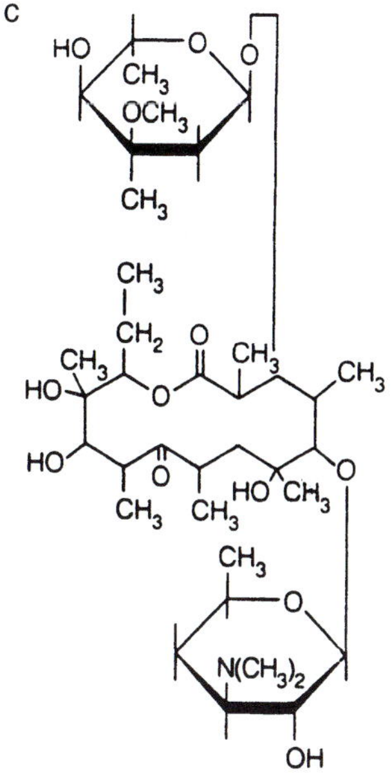

Erythromycin: Binds to a specific site on the 23S RNA and blocks elongation by interfering with the translocation step.

d

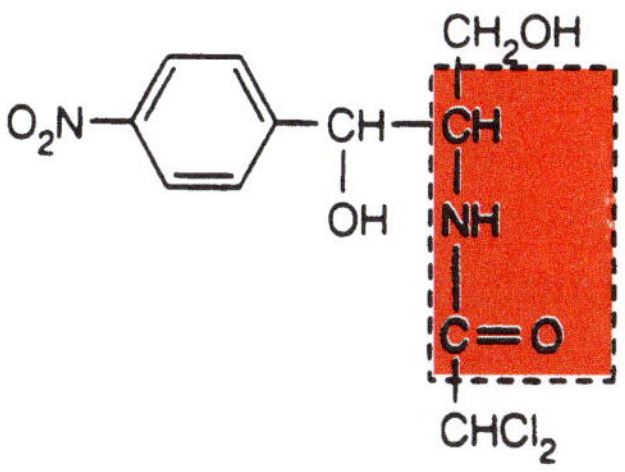

Chloramphenicol: Blocks elongation, apparently by acting as a competitive inhibitor for the peptidyltransferase complex. The amide link (red) resembles a peptide bond.

e

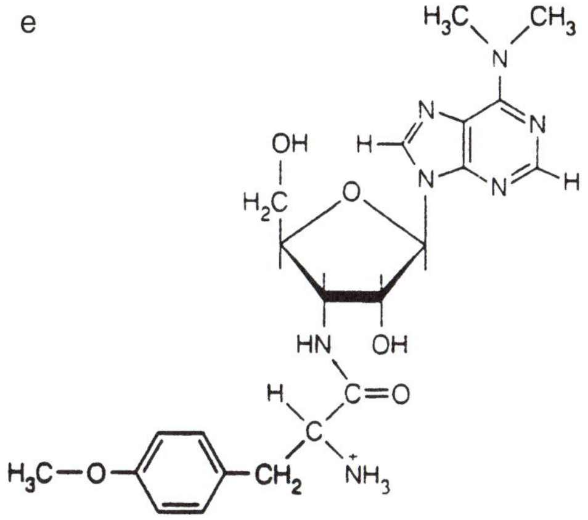

Puromycin: Causes premature chain termination. The molecule resembles the 3' end of the amino-acylated tRNA and will enter the A site. It transfers to the growing chain causing premature termination.

FIGURE 25-18
Antibiotic inhibitors of protein synthesis.

emphysema. Prenatal diagnosis of α_1-AT can be done by the use of either synthetic oligonucleotide probes or restriction fragment length polymorphism analysis.

The reactive site of α_1-AT is at the methionine residue, position 358. The sequence around the reactive center of several proteinase inhibitors, including α_1-AT, antithrombin III, and several unrelated plant proteinase inhibitors, is strikingly similar. In fact, α_1-AT, antithrombin III, and ovalbumin (a protein without known inhibitory function) share extensive homologies and may all have evolved from a common ancestral protease inhibitor more than 500 million years ago. A fatal bleeding disorder due to substitution of a methionine for an arginine residue at position 358 of α_1-AT has been reported. This particular alteration brought a change in the specificity of the inhibitor. This altered protein (known as α_1-antitrypsin Pittsburgh) exhibits antithrombin activity while losing antielastase activity. Antithrombin III, which has an important regulatory role in hemostasis (Chapter 36), has a reactive arginine at its active center. Since the α_1-AT and antithrombin III are structurally similar, the substitution of methionine of α_1-AT for arginine at the active center alters its specificity. This bleeding disorder is further complicated by the fact that the concentration of normal antithrombin III does not increase in response to inflammatory stress as does the level of α_1-AT. Therefore, with each hemolytic episode, the abnormal α_1-AT increases, creating a cycle resulting in uncontrolled bleeding.

Inhibitors of Protein Synthesis and Related Disorders

Many antibacterial agents (antibiotics) have been isolated from fungi. Antibiotics are used both clinically and as reagents for unraveling the details of protein, RNA, and DNA synthesis.

Antibiotics that have no effect on eukaryotic translation either fail to penetrate the cell membrane (which is quite common) or do not bind to eukaryotic ribosomes. The differences in effectiveness of antibiotics on the two classes of cells *in vivo* is the basis of their usefulness as therapeutic agents for bacterial infections (Figure 25-18).

Some antibiotics are active against both bacterial and mammalian cells. One example is chloramphenicol, which inhibits peptidyltransferase in both bacterial and mitochondrial ribosomes, although eukaryotic cytoplasmic ribosomes are unaffected. Such a drug may be clinically useful if a concentration range can be maintained in the patient in which the antibacterial action is substantial but toxic effects on host cells are minimal. However, because of the potential for toxicity, such antibiotics are used only in serious infections when other drugs fail.

Three antibiotic inhibitors that act only on prokaryotes are streptomycin, tetracycline, and erythromycin. Other antibiotics act primarily on eukaryotic cells (Table 25-4). Eukaryotic protein synthesis can also be inhibited by toxins of bacterial origin. An example is the toxin produced by *Corynebacterium diphtheriae* bacteria carrying a lysogenic bacteriophage. Uninfected bacteria do not elaborate the toxin, and therefore acquisition of the phage is essential for the toxic effect. ***Diphtheria*** is an acute infectious disease usually localized in the pharynx, larynx, and nostrils and occasionally in the skin. Initially, the cytotoxic effects are restricted to those tissues immediately adjacent to the bacterial growth. With increased bacterial growth, production of toxin increases and is disseminated to the blood, lymphatics, and other organ systems (e.g., cardiovascular and nervous systems), where it brings about destructive changes.

The pathogenesis of diphtheria is due entirely to the toxin's inhibition of protein synthesis in the host cells. Diphtheria toxin consists of a single polypeptide chain (M.W. 63,000) with two intramolecular disulfide linkages. Toxin binds to the outer surface of susceptible cells at specific sites and enters by receptor-mediated endocytosis. The toxin is cleaved proteolytically into a small fragment, A (M.W. 21,000), and a larger fragment, B (M.W. 42,000), during its internalization. Fragment A is the biologically active part of the molecule and presumably penetrates the cell membrane with the aid of fragment B.

Upon entry into the cytoplasm, fragment A catalyzes the ADP ribosylation of the transfer factor, EF-2, leading to its inactivation and the interruption of protein synthesis. The ADP-ribose group is donated by NAD^+. The ADP ribosylation reaction, catalyzed by the toxin, is specific for EF-2 of eukaryotic cells; other proteins of eukaryotic and bacterial cells are not substrates. This specificity is due to an unusual amino acid residue in EF-2, diphthamide, which is the acceptor of the ADP ribosyl group. Diphthamide derives from the posttranslational modification of histidine. The acute symptoms are treated with antitoxin. The bacteria, which are gram-positive, succumb to a variety of antibiotics, including penicillin. Diphtheria is effectively prevented by immunization with toxoid (inactivated toxin) preparations.

ADP ribosylation is also involved in the action of ***cholera*** toxin and in certain pathogenic strains of *E. coli* (Chapter 12). Cholera toxin catalyzes the ADP ribosylation of the guanidinium. group of a specific arginine residue in the guanine nucleotide-binding protein of the adenylate cyclase. This activates adenylate cyclase, which catalyzes the formation of cAMP from ATP. The cAMP formed stimulates secretion of water and electrolytes from intestinal epithelial cells. Thus, patients infected with ***Vibrio cholerae*** secrete enormous quantities (up to 20 L/d) of water and electrolytes. Without adequate and prompt replacement, death can ensue owing to dehydration and electrolyte imbalance. Interestingly, in the case of diphtheria toxin, ADP-ribosylation leads to inactivation of a protein, whereas with cholera toxin, ADP ribosylation causes activation of the protein.

A group of plant ***lectins,*** such as abrin, ricin, and modeccin, are highly toxic to eukaryotic cells. Their mode of action consists of inhibition of protein synthesis by enzymatically inactivating the EF-2 binding region of the 60S ribosomal subunit, whereas the diphtheria toxin inactivates the EF-2 protein itself. Ricin is isolated from castor beans and has a molecular weight of 66,000. Like most plant and bacterial toxic proteins, ricin contains two

TABLE 25-4
Inhibitors of Protein Synthesis in Eukaryotes

Inhibitor	Action
Abrin, ricin	Inhibits binding of aminoacyl tRNA
Diphtheria toxin	Catalyzes a reaction between NAD* and EF2 to yield an inactive factor. Inhibits translocation.
*Chloramphenicol	Inhibits peptydyltransferase of mitochondrial ribosomes. Is inactive against cytoplasmic ribosomes.
*Puromycin	Causes premature chain termination by acting as an analogue of charged tRNA.
*Fusidic acid	Inhibits translocation by altering an elongation factor.
Cycloheximide	Inhibits peptidyltransferase.
Pactamycin	Inhibits positioning of $tRNA_f^{Met}$ on the 40S ribosome.
Showdomycin	Inhibits formation of the eIF2-$tRNA_f^{Met}$-GTP complex.
Sparsomycin	Inhibits translocation.

*Also active on prokaryotic ribosomes.

polypeptide chains with two different but complementary functions. The A chain possesses enzymatic activity and is responsible for toxicity, and the B chain, which is a lectin, binds to galactose-containing glycoproteins or glycolipids on the cell surface. The A and B chains are linked by a labile disulfide linkage. Upon binding of the ricin molecule to the carbohydrate receptors of the cell surface via the B chain, the A chain enters the cytoplasm, presumably by receptor-mediated endocytosis, where it inhibits protein synthesis by irreversible inactivation of the 60S ribosomal subunit. Toxins have been used to develop highly selective cytotoxic agents targeted against specific cells. For example, ricin A has been coupled to agents that selectively bind to cell surface membrane components. The selective agents may be monoclonal antibodies (Chapter 35), hormones, or other cell surface ligands. These conjugates act as selective cytotoxic agents and may have potential therapeutic applications (e.g., in the treatment of cancer). Potential clinical application for diphtheria toxin may be impossible because of the prevalence of the diphtheria antitoxin in human populations.

25.11 Collagen Biosynthesis and Its Disorders

Collagen occurs in several genetically distinct forms and is the most abundant body protein; most of the body scaffolding is composed of collagen (Chapter 11). Its structure is uniquely suited for this structural role. It is a fibrillar protein but also exists in a nonfibrillar form in the basement membrane. The basic structural unit of collagen, ***tropocollagen,*** consists of three polypeptide chains. Each polypeptide chain has the general formula $(\text{–Gly–X–Y–})_{333}$. Some of the amino acid residues at the X and Y positions are proline, hydroxyproline, lysine, hydroxylysine, and alanine. Collagen is a glycoprotein and contains only two types of carbohydrate residues, namely, glucose and galactose linked in O-glycosidic bonds to hydroxylysyl residues.

In collagen, each polypeptide chain is coiled into a special type of a rigid, kinked, left-handed helix, with about three amino acid residues per turn. The three helical polypeptides in turn are wrapped around each other to form a right-handed triple-stranded superhelix that is stabilized by hydrogen bonding. The collagen molecules are aggregated in an ordered quarter-staggered array to give rise to microfibrils, which in turn combine to give fibrils. Covalent cross-linkages occur at various levels of collagen fiber organization and provide great mechanical strength. Collagen biosynthesis is unusual in that it consists of many posttranslational modifications. The unique posttranslational reactions of collagen biosynthesis are

1. Hydroxylation of selected prolyl and lysyl residues.
2. Glycosylation of certain hydroxylysyl residues.
3. Folding of procollagen polypeptides into a triple helix.
4. Conversion of procollagen to collagen.
5. Self-assembly into fibrils.
6. Oxidative deamination of ε-amino groups of strategically located lysyl and hydroxylysyl residues to provide reactive aldehydes. These form cross-linkages between polypeptide chains of the same molecule as well as between the adjacent molecules that give strength and stability to the fibrils.

The first three processes take place inside the cell, whereas the last three are extracellular modifications.

Collagen disorders can result from primary defects in the structure of procollagen or collagen or from secondary changes that affect collagen metabolism. Collagen disorders are both acquired and inherited. ***Ehlers-Danlos syndrome*** and ***osteogenesis imperfecta*** are examples of inherited primary collagen diseases; ***scurvy*** and various fibrotic processes (e.g., pulmonary fibrosis and cirrhosis) are examples of secondary collagen diseases. The mechanisms that lead to collagen diseases are:

1. Aberration in the control mechanisms that alter the balance between synthesis and degradation. This imbalance can lead either to excessive deposition or depletion of collagen.
2. Synthesis of structurally altered collagen due to defects at the level of DNA transcription, RNA processing, translation, or posttranslational modifications.
3. Imbalance in the relative rates of synthesis of genetically distinct collagens.
4. Abnormalities in the packing of collagen molecules into a fiber or in the interaction of collagen fibers with other extracellular components of the connective tissue.

Transcription and Translation of Collagen Polypeptides

In general, the transcription and translation of collagen polypeptides resemble other proteins. The primary RNA transcript is capped and polyadenylated, introns are removed, and the functional mRNA is transported to the cytoplasm. The procollagen mRNA, in addition to being one of the largest mRNAs in eukaryotic cells, contains unusually high amounts of G and C, because these bases are overrepresented in the codons for glycine and proline. As is the case for other secretory proteins, the procollagen mRNA encodes a signal sequence coding for hydrophobic amino acids at the amino terminal ends. The signal sequences of proα1 chain are different from the proα2 chain and are unusually long (100 amino acid residues) in comparison with other secreted proteins (15–30 residues). Because of their hydrophobicity, these signal sequences are thought to facilitate binding and transport of the nascent polypeptide into cisternae of the endoplasmic reticulum. The signal sequences are removed by proteolysis within the membrane during the translation. The synthetic rate of proα chain is rather slow and is probably related to the unfolding of the secondary structure of procollagen mRNA, which is rich in GC content.

Molecular defects at the levels of transcription and translation of collagen polypeptides have not yet been clearly established but may cause one type of ***Ehlers-Danlos syndrome (type IV).*** This disorder is characterized by decreased synthesis of type III collagen in the aorta, intestinal tract, skin, and probably other tissues. Among the Ehlers-Danlos disorders, type IV is the most severe because of the threat of arterial rupture or gastrointestinal perforation. Ehlers-Danlos syndrome is a group of inherited disorders that share similar clinical symptoms; many have been attributed to specific errors in the structure or metabolism of collagen (Table 25-5). The major phenotypic features of Ehlers-Danlos syndrome include hyperextensible skin, hypermobile joints, easy bruisability, and friability of tissues. The clinical severity of these disorders is highly variable. Defects in splicing, transcription, or translation may also cause ***osteogenesis imperfecta.*** In this disease, fibroblasts or tissue samples have a markedly diminished capacity for type I collagen

TABLE 25-5
Ehlers-Danlos Syndrome

Type	Inheritance	Biochemical Defect	Collagen Fibril Diameter
I	AD	Unknown	Increased
II	AD	Unknown	Increased
III	AD	Unknown	Increased
IV	AD or AR	Decreased type III collagen synthesis and its intracellular accumulation	Heterogeneous
V	XL	Unknown	Heterogeneous
	XL	Unknown	Increased
VI	AR	Lysyl hydroxylase deficiency	Decreased
VII	AD	Deletion of exons from the genes of proα_1(I) and pro α_2(I) that encode the amino-terminal cleavage site; required for the action of procollagen aminoprotease	Heterogeneous
VII	AD	Unknown	Not determined
IX	SL	Lysyl oxidase deficiency	Increased
X	AR	Fibronectin deficiency	Not determined

AD = Autosomal dominant; AR = autosomal recessive; XL = X-linked.

TABLE 25-6
Osteogenesis Imperfecta

Type	Inheritance	Biochemical Defect
I	AD	Decreased synthesis of proα_1(I), leading to less production of type I collagen because of several defects in the gene.
II	AR	Markedly diminished synthesis of type I collagen; presence of hydroxylysine-enriched collagens associated with delayed helix formation.
III	AR	Delayed secretion of type I procollagen associated with excess mannose residues in the carboxyterminal propeptide region; absence of α_2(I) synthesis.
IV	AD	Point mutations in α_2(I) gene, resulting in substitution for glycine residues.

AD=Autosomal dominant; AR=autosomal recessive.

synthesis. Osteogenesis imperfecta is a genetically and clinically heterogeneous disorder (Table 25-6). It is one of the most common inherited connective tissue disorders and has an incidence of 1:15,000. It is characterized clinically by defects in bone, tendon, dentin, ligament, and skin.

Intracellular Posttranslational Modifications

Hydroxylations of Selected Prolyl and Lysyl Residues

The hydroxylation reactions are catalyzed by three enzymes: prolyl 4-hydroxylase (usually known as prolyl hydroxylase), prolyl 3-hydroxylase, and lysyl hydroxylase. These enzymes are located within the cisternae or rough endoplasmic reticulum; as the procollagen chains enter this compartment, the hydroxylations begin.

All three enzymes have the same cofactor requirements: ferrous ion, α-ketoglutarate, molecular oxygen, and ascorbate (vitamin C). The reducing equivalents required for the hydroxylation reaction are provided by the decarboxylation of equimolar amounts of α-ketoglutarate to succinate and carbon dioxide. One atom of the O_2 molecule is incorporated into succinate while the other is incorporated into the hydroxyl group. The data on enzyme kinetics and mechanism of reaction are consistent with the ordered binding of Fe^{2+}, α-ketoglutarate, O_2, and the peptide substrate and an ordered release of hydroxylated peptide, CO_2, succinate, and Fe^{2+}. The Fe^{2+} does not necessarily dissociate from the enzyme during each catalytic cycle. The activation of oxygen is required and may involve the formation of superoxide (O_2^-). The requirement for ascorbate is specific with purified enzyme preparations. It is not consumed stoichiometrically during the hydroxylation reaction. The exact role of ascorbate is not known, but presumably it reduces either the enzyme iron complex or the free enzyme.

Humans, other primates, and guinea pigs lack the enzyme required for the conversion of gluconate to ascorbate (Chapter 15). However, most mammals do possess this enzyme and are able to synthesize ascorbate from glucose by way of glucuronic acid. Other reducing agents, such as dithiothreitol, L-cysteine, and some reduced pteridines, in high concentrations, can partially replace ascorbate in *in vitro* assays. The generalized hydroxylation reaction is shown below.

$$\text{Substrate–H} + O_2 + \underset{\alpha\text{-Ketoglutarate}}{\begin{array}{c}COO^-\\ |\\ CH_2\\ |\\ CH_2\\ |\\ C{=}O\\ |\\ COO^-\end{array}} \xrightarrow[\text{Fe}^{2+}\ \text{Ascorbate}]{\text{Enzyme}} \text{Substrate–OH} + \underset{\text{Succinate}}{\begin{array}{c}COO^-\\ |\\ CH_2\\ |\\ CH_2\\ |\\ COO^-\end{array}} + CO_2$$

The substrates for all three hydroxylases are highly specific. Free proline and lysine are not substrates. The residue to be hydroxylated must be in a peptide linkage (the minimum is a tripeptide) and in the correct X or Y position in the chain. The proline hydroxylase and the lysyl hydroxylase catalyze the hydroxylation of only prolyl or lysyl residues in the Y positions of peptides with the sequence –X–Y–Gly–, whereas the prolyl 3-hydroxylase catalyzes the hydroxylation of prolyl residues at the X position only if Y is 4-hydroxyproline. The conformation of the substrate is also important for catalytic activity. The substrates have to be nonhelical; triple-helical polypeptides do not function as substrates. Thus, the hydroxylation reactions must occur before triple-helix formation. Many important functions are associated with the hydroxylation of prolyl and lysyl residues. The 4-hydroxyprolyl residues participate in interchain hydrogen bonding that aids in the maintenance of triple-helical structures. For example, nonhydroxylated polypeptide chains cannot form stable triple-helical structures at body temperature. The biological role of 3-hydroxyprolyl residues is not known. The hydroxyl groups of hydroxylysyl residues are sites of attachment of carbohydrate and also participate in the intermolecular collagen cross-links.

The importance of lysyl hydroxylation is seen in patients with the type VI variant of Ehlers-Danlos syndrome (Table 25-5). The collagen in these individuals has a decreased fibril diameter and profound changes in mechanical properties. Skin fibroblasts show virtually no lysyl hydroxylase activity. Furthermore, hydroxylysine formation can be severely affected in some tissues, mildly affected in others, and unaffected in still others (e.g., cartilage). These observations suggest the presence of tissue-specific lysyl hydroxylases.

Glycosylation of Hydroxylysyl Residues

The glycosylation occurs as the N-terminal ends of the polypeptide chains move into cisternae at specific hydroxylysyl residues. The carbohydrates found are galactose and the disaccharide, glucosylgalactose. These reactions are catalyzed by two specific enzymes: hydroxylysyl galactosyltransferase and galactosylhydroxylysyl glucosyltransferase. The first enzyme catalyzes the transfer of galactose from UDP-galactose to hydroxylylsyl residues, and the second enzyme transfers a glucose from UDP-glucose to galactosylhydroxylysyl residues. Both enzymes require a bivalent cation, preferably manganese. The substrate has to be in the nonhelical conformation, and the glycosylation ceases when the collagen propeptides fold into a triple helix. Thus, both hydroxylations and glycosylation must occur before triple-helix formation, which is an intracellular process.

The exact function of the carbohydrate unit in the collagen molecule is not clear, but it may have some role in the organization of fibrils. For example, an inverse relationship between carbohydrate content and fibril diameter has been observed. Deficiency of galactosylhydroxylysyl glucosyltransferase has been seen in the members of kindred with a dominantly inherited disease known as ***epidermolysis buflosa simplex.*** This disease belongs to a group of inherited disorders characterized by blister formation in response to minor skin trauma. Further investigation of this disorder may provide information about the role of glycosylation of collagen.

Collagen propeptides contain oligosaccharide units typical of glycoproteins in the proregions of the polypeptide chain (Figure 25-19). These oligosaccharide units may be present in either the N- or C-terminal proregions of the peptide or both. In type I procollagen, the oligosaccharide units are at the C-terminal, whereas in type II procollagen, the units are found at both ends, These oligosaccharicles do not appear in the finished collagen molecule because the proregions are excised. The oligosaccharide units probably are synthesized through

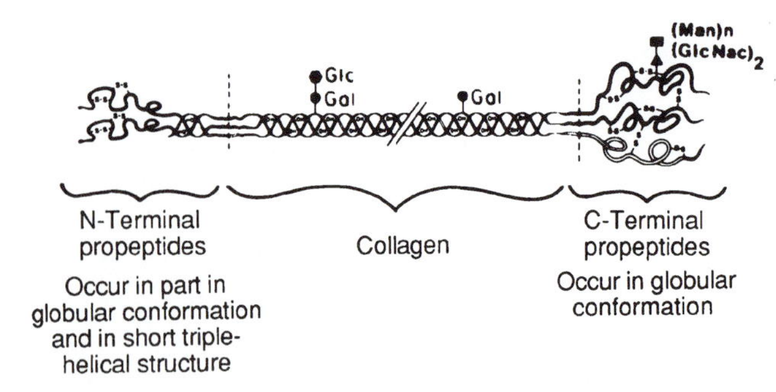

FIGURE 25-19
Structure of type I procollagen molecule, which consists of two proα_1(I) and one proα_2(I) chains. Note the presence of oligosaccharide units at the C-terminal propeptide region. In other collagen types, the oligosaccharide units may be present in either one of the regions or both. The disulfide linkages are also present in these regions. These linkages may be either intra- or interchain linkages and may be involved in the triple-helical formation. In the finished collagen molecule, the N- and C-propeptide regions are absent because they are excised (shown by vertical dotted lines). Glc, Glucose; Gal, galactose; Man, mannose; GlcNAc, N-acetylglucosamine. [Reproduced with permission from D. J. Prockop and K. I. Kivirikko, Heritable diseases of collagen. *N Engl J Med* **311**:376 (1984).]

dolichol intermediates (Chapter 16) and linked to the nascent peptide chains within the rough endoplasmic reticulum. The function of these asparaginyl-linked oligosaccharide units is not known.

Since ascorbate is required for hydroxylation reactions, its deficiency causes accumulation of the collagen polypeptides in the endoplasmic reticulum and their eventual secretion. However, these proteins lack the proper modifications and therefore cannot be used for assembly of collagen fibrils. Thus, the role of ascorbate in the hydroxylation of collagen is one of its major physiological roles (see also Chapter 38). ***Scurvy,*** a connective tissue disorder, is due to vitamin C deficiency in the diet. Most pathological changes of scurvy can be attributed to failure to synthesize collagen. Scurvy patients have defective blood vessels and poor intravascular support, leading to frequent hemorrhages, defective formation of bone and teeth, and poor wound healing. All of these manifestations can be corrected by administration of vitamin C.

Formation of Disulfide Linkages and Assembly of Procollagen Polypeptides into a Triple Helix

The propeptide regions of procollagen polypeptides contain cysteine residues that can form both intra- and interchain disulfide linkages (Figure 25-19). As is the case for other proteins that contain disulfide linkages, it is not known whether synthesis of these linkages requires an enzyme or whether it occurs spontaneously. The assembly of procollagen polypeptides into a triple helix appears to have two requirements:

1. The 4-hydroxylation of at least 100 prolyl residues per proα chain;
2. The formation of *interchain* disulfide linkages in the C-terminal propeptide. The latter cannot be formed until translation is nearly finished. The rate of disulfide bond and triple-helix formation varies greatly from one cell type to another—only minutes in tendon cells that synthesize type I collagen but an hour in cells that synthesize basement membrane collagen. These differences in synthesis time may account for the variations in hydroxylation and glycosylation. Thus, the extent of posttranslational modifications depends not only on levels of enzyme and cofactors but also on the time available.

Translocation and Secretion of Procollagen

After the procollagen polypeptides are assembled into a triple helix, they are secreted by the classical route. They pass through the smooth endoplasmic reticulurn and the Golgi complex, where they are packaged into membranous vesicles and secreted into the extracellular space by exocytosis. This process requires ATP and may involve microtubules and microfilaments. The conformation of procollagen markedly affects the rate of secretion. Prevention of the formation of a triple helix (e.g., lack of 4-hydroxylation due to vitamin C deficiency) leads to the accumulation of nonhelical propolypeptides within the cisternae of the rough endoplasmic reticulum and a delayed rate in its secretion.

Extracellular Posttranslational Modification

Conversion of procollagen to collagen requires at least two proteases: a procollagen aminoprotease and a procollagen carboxyprotease. The former catalyzes removal of the N-terminal propeptide and the latter removal of the C-terminal propeptide. The two enzymes are endopeptidases, function at neutral pH, require a bivalent cation such as Ca^{2+}, and show a preference for the helical conformation. There appears to be no preferential order for the cleavage of the propeptides.

The conversion of procollagen to collagen by removal of propeptides seems to be essential for the formation of collagen fibrils. This supposition is supported by studies of two heritable diseases, one found in humans and the other in cattle, sheep, and cats. In both, the defect lies in the removal of N-terminal propeptides and results in impaired fibril formation. The human disorder is the type VII variant of Ehlers-Danlos syndrome (Table 25-5). Affected individuals exhibit marked joint hypermobility, dislocation of joints, short stature, and minor changes in skin elasticity. Their skin fibroblasts show normal N-terminal procollagen protease activity; however, the defect resides in the deletions of the exon that encodes the protease cleavage sites, thus preventing normal cleavage of the N-terminal propeptide. In animals, however, the defect is the N-terminal procollagen protease deficiency, which produces skin that is easily torn; thus, the disease is known as ***dermatesparaxis.***

Formation of Collagen Fibrils from Ordered Aggregation of Collagen Molecules

The collagen molecules formed by removal of the propeptides spontaneously assemble into fibrils. At this stage, the fibrils are still immature and lack tensile strength, which is acquired by cross-linking. The initial step in cross-link formation is the oxidative deamination of α-amino groups in certain lysyl and hydroxylysyl residues catalyzed by lysyl oxidase. The enzyme is a copper-dependent (probably cupric) protein, and the reaction requires molecular oxygen and pyridoxal phosphate for full activity. Only native collagen fibrils function as substrates.

$$\underset{\text{Lysyl residue}}{\text{HN}\backslash\,\text{HC}(\text{C=O})\text{—}(\text{CH}_2)_4\text{—NH}_3^+} + \text{O}_2 \xrightarrow{\;-\text{NH}_3\;} \underset{\text{Allysine residue}}{\text{HN}\backslash\,\text{HC}(\text{C=O})\text{—}(\text{CH}_2)_3\text{—}\overset{\text{H}}{\text{C}}\text{=O}} + \text{H}_2\text{O}$$

In a similar reaction, the hydroxyallysine residue is formed from hydroxylysine residues. The aldehyde groups react spontaneously with each other or with amino groups, with formation of intra- and intermolecular linkages. Lysyl oxidase is also involved in the formation of elastin cross-linkages.

Several collagen disorders result from defects in the formation of cross-links (Chapter 11). The cross-linking disorders can be due to a hereditary cleficiency of lysyl oxidase, inhibition of lysyl oxidase, deficiency of copper, defects in the formation of cross-links, or defects in their stabilization. The genetic deficiency of lysyl oxidase is characterized by ***Ehlers-Danlos syndrome type IX*** (Table 25-5) and some forms of ***cutis laxa.*** Type IX Ehlers-Danlos syndrome patients exhibit extreme extensibility of the skin, "cigarette-paper" scarring, and easy bruisability. In cutis laxa, the skin is loose and inelastic and appears to be too large for the surface it covers. Some affected individuals exhibit deficiency of lysyl oxidase, presumably limited to skin. The animal counterpart of lysyl oxidase deficiency is seen in several variants of aneurysm-prone mice. All of these disorders are X-linked.

Lysyl oxidase undergoes irreversible inactivation when exposed to certain nitriles. β-Aminopropionitrile ($H_2N–CH_2–CH_2–CN$), which is found in certain peas (e.g., ***Lathyrus odoratus***), inhibits the lysyl oxidase presumably by binding covalently to the enzyme. Ingestion of these peas results in ***lathyrism,*** which is characterized by multiple defects in collagen- and elastin-containing tissues. In experimental animals, lathyrism is produced by administering β-aminopropionitrile during their active growth period. Lysyl oxidase is also irreversibly inhibited by carbonyl reagents (e.g., hydroxylamine) and copper-chelating agents.

Impaired cross-linking can also occur as a result of copper deficiency, since lysyl oxidase is a copper-dependent enzyme. In experimental animals, copper deprivation causes skeletal and cardiovascular abnormalities. These abnormalities are similar to a human disorder known as ***Marfan's syndrome,*** an autosomal dominant trait whose molecular defect is unknown. A genetic defect affecting the gastrointestinal absorption of copper, known as ***Menkes' (kinky-hair) syndrome,*** exhibits neurological, connective tissue, pigmentary, and hair abnormalities. These defects can be explained by deficient activities of various copper-requiring enzymes (Chapter 37). The connective tissue changes are attributed to the deficiency of lysyl oxidase.

Cross-linking can be inhibited by agents that can react with the aldehyde groups of allysyl and hydroxyllysine residues. Penicillamine reacts with the aldehyde groups, forming a thiazolidine complex and rendering the aldehyde groups unavailable for cross-link formation.

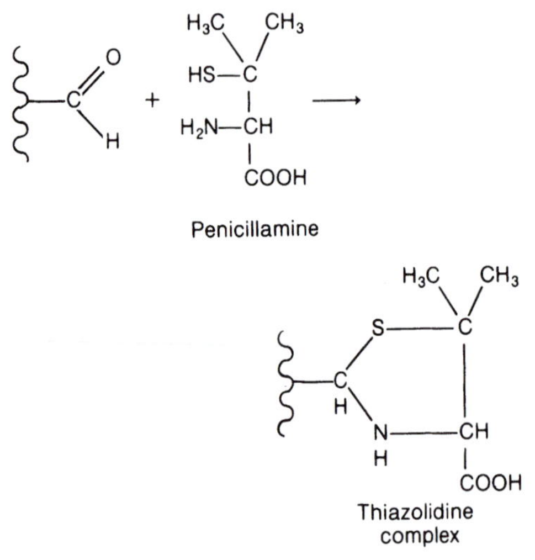

Penicillamine

Thiazolidine complex

Penicillamine is an effective chelator of copper, thereby inhibiting lysyl oxidase. Penicillamine is used therapeutically in the treatment of disorders involving copper accumulation (***Wilson's disease***) or in lead and mercury poisoning.

Several inherited disorders of methionine metabolism (Chapter 17) give rise to excessive production of homocysteine, $HS–CH_2–CH_2CH(NH_3^+)COO^-$, and its excretion in urine. The most common form of ***homocystinuria*** is due to a deficiency of cystathionine synthase (Chapter 17). A major clinical manifestation of homocystinuria is connective tissue abnormalities that are probably due to the accumulation of homocysteine, which either inactivates the reactive aldehyde groups or impedes the formation of polyfunctional cross-links.

The structure or the metabolism of collagen is abnormal in a variety of other disorders. For example, in ***diabetes mellitus*** (Chapter 22), poor wound healing, atrophy of the skin, and thickening of the basement membranes have been attributed to changes in collagen metabolism. Fibrotic tissue and scars consist primarily of collagen fibrils. The fibrosis is a normal process of wound healing in response to trauma and injury. However, excessive fibrosis in parenchymal tissues (e.g., liver and lungs) can severely limit their function. Thus, attempts are being made to develop specific agents that would inhibit collagen synthesis or increase its catabolism. Both growth and aging involve changes in collagen metabolism. These changes occur in the type, quantity, and quality of collagen. Growth of connective tissues requires the appropriate type and amounts of collagen. It is not known to what extent changes in collagen metabolism are responsible for aging.

Supplemental Readings and References

N. Ban, P. Nissen, J. Hansen, et al.: The complete atomic structure of the large ribosomal subunit at 2.4 Å Resolution. *Science* **289,** 905 (2000).

P. Cramer, D. A. Bushnell, J. Fu, et al.: Architecture of RNA polymerase II and implications for the transcription mechanism. *Science* **288,** 640 (2000).

D. E. Draper: Protein-RNA recognition. *Annual Reviews in Biochemistry* **64,** 593 (1995).

L. Ellgaard, M. Molinari, and A. Helenius: Setting the standards: quality control in the secretory pathway. *Science* **286,** 1882 (1999).

R. Green and H. Noller: Ribosomes and translation. *Annual Reviews in Biochemistry* **66,** 40 (1997).

H. Guo, M. Karberg, M. Long, et al.: Group II introns designed to insert into therapeutically relevant DNA target sites in human cells. *Science* **289,** 452 (2000).

V. Hatzimanikatis, L. H. Choe, and K. H. Lee: Proteomics: theoretical and experimental considerations. *Biotechnology Progress* **15,** 312 (1999).

J. W. B. Hershey, M. B. Mathews, and N. Sonnenberg, eds.: *Translational Control,* Cold Spring Harbor Laboratory Press, 1996.

P. H. von Hippel: An integrated model of the transcription complex in elongation, termination, and editing. *Science* **281,** 660 (1998).

M. Ibba and D. Söll: Quality control mechanisms during translation. *Science* **286,** 1893 (1999).

H. A. James and I. Gibson: The therapeutic potential of ribozymes. *Blood* **91,** 371 1998.

G. Kuznetsov and S. K. Nigam: Folding of secretory and membrane proteins. *New England Journal of Medicine* **339,** 1688 (1998).

R. Lund and J. E. Dahlberg: Proofreading and aminoacylation of tRNAs before export from the nucleus. *Science* **282,** 2082 (1998).

G. W. Muth, L. O-Donnelly, and S. A. Strobel: A single adenosine with a neutral pK_a in the ribosomal peptidyl transferase center. *Science* **289,** 947 (2000).

W. Neuport: Protein transport into mitochondria. *Annual Reviews in Biochemistry* **66,** 863 (1997).

V. M. Pain: Initiation of protein synthesis in eukaryotic cells. *European Journal of Biochemistry* **236,** 747 (1996).

T. V. Pestova, S. I. Borukhov, and C. U. T. Hellen: Eukaryotic ribosomes require initiation factors I and IA to locate initiation codons. *Nature* **394,** 854 (1998).

A. M. Pyle: Ribozymes: a distinct class of metalloenzymes. *Science* **261,** 709 (1993).

H. Riezman: The ins and outs of protein translocation. *Science* **278,** 1728 (1997).

C. H. Schilling, J. S. Edwards, and B. O. Palsson: Toward metabolic phenomics: analysis of genomic data using flux balances. *Biotechnology Progress* **15,** 288 (1999).

P. Schimmel and E. Schmidt: Making connections: RNA-dependent amino acid recognition. *Trends in Biomedical Science* **20,** 1 (1995).

H. Stark, E. V. Orlova, J. Rinke-Appel, et al.: Arrangement of tRNAs in pre- and posttranslational ribosomes revealed by electron cryomicroscopy. *Cell* **88,** 19 (1997).

M. D. Wang, M. J Schnitzer, H. Yin, et al.: Force and velocity measured for single molecules of RNA polymerase, *Science* **282,** 902 (1998).

S. Wikner, M. Maurizi, and S. Gottesman: Posttranslational quality control: folding, refolding, and degrading proteins. *Science* **286,** 1888 (1999).

C. L. Will, C. Schnaider, R. Read, et al.: Identification of both shared and distinct proteins in the major and minor spliceosomes. *Science* **284,** 2003 (1999).

CHAPTER 26

Regulation of Gene Expression

The number of molecules of a protein produced per unit time from a particular gene differs from gene to gene, sometimes greatly. This variation can occur in several ways. For some genes, the strength of a promoter or a ribosome binding site is sufficient to regulate the required concentration of a particular protein. However, other gene products are needed only occasionally, e.g., when a particular nutrient is present in the surrounding medium; still other gene products are required only during particular stages of differentiation or in specific cell types. For such genes, the products are usually not present in significant amounts except when gene expression is activated, and the level of gene expression is determined by an on-off switch of some sort. In this chapter, numerous mechanisms are described by which differences in gene expression are achieved.

26.1 Regulation of mRNA

Regulation of gene expression is achieved by

1. Regulation of the number of mRNA molecules produced per unit time,
2. Regulation of the translation of mRNA,
3. Regulation of the number of copies of a gene, and
4. Posttranslational modification.

In prokaryotes, mRNA synthesis can be controlled simply by regulating initiation of transcription. In eukaryotes, mRNA is formed from a primary transcript followed by a series of processing events (e.g., intron excision, polyadenylylation). Eukaryotes regulate not only transcription initiation but the various stages of processing as well.

An important aspect of ***mRNA regulation*** is determined by the turnover of mRNA molecules, i.e., translation can occur only as long as the mRNA remains intact. In bacteria, mRNA molecules have a lifetime of only a few minutes, and continued synthesis of mRNA molecules is needed to maintain synthesis of the proteins encoded in the mRNAs. In eukaryotes, the lifetime of mRNA is generally quite long (hours or days), thereby enabling a small number of transcription initiation events to produce proteins over a long period of time.

Metabolic pathways normally consist of a large number of enzymes; in some cases, the individual enzymes are used in a particular pathway and nowhere else. In these cases, it is efficient to regulate expression of all or none of the enzymes. In bacterial systems, all enzymes of the pathway are encoded in a single polycistronic mRNA molecule, and synthesis of the mRNA produces all the enzymes. In eukaryotes, common signals for transcription of different genes may be used or the primary transcript can be differentially processed to yield a set of mRNA molecules each of which encodes one protein. Regulation of synthesis of the primary transcripts simultaneously regulates synthesis of all the gene products.

In prokaryotes, gene expression fluctuates because bacteria must be able to respond rapidly to a changing environment. However, in the differentiation of cells of the higher eukaryotes, changes in gene expression are usually irreversible, as in the differentiation of a muscle cell from a precursor cell. By changing the number of copies of a gene during differentiation, eukaryotic cells can regulate levels of gene expression.

26.2 Gene Regulation in Prokaryotes

Several common patterns of transcription regulation have been observed in bacteria. The particular pattern depends on the type of metabolic activity of the system being regulated. For example, in a catabolic (degradative) pathway, the concentration of the substrate for the first enzyme in the pathway often determines whether all the enzymes in the pathway are synthesized. In contrast, the final product is often the regulatory substance in a biosynthetic pathway. In the simplest mode, absence of an end product stimulates transcription and presence of an end product inhibits it. Even for a gene in which a single type of protein is synthesized from monocistronic mRNAs, the protein may "autoregulate" itself in the sense that the transcriptional activity of the promoter is determined directly by the concentration of the protein. The molecular mechanisms for each regulatory system vary considerably but usually fall in one of two major categories—negative regulation and positive regulation.

In ***negative regulation,*** an inhibitor, which keeps transcription turned off, is present in the cell and an anti-inhibitor—an inducer—is needed to turn the system on. In ***positive regulation,*** an effector molecule (which may be a protein, small molecule, or molecular complex) activates a promoter. Negative and positive regulation are not mutually exclusive, and some degradative pathways, such as the utilization of lactose in *E. coli,* are both positively and negatively regulated.

Lactose (lac) Operon

The prototype for negative regulation is the system in *E. coli* for metabolizing lactose (lac) (Figure 26-1). The key chemical reaction carried out by the lac system is a cleavage of lactose to galactose and glucose. This reaction enables bacteria to utilize lactose as a carbon source when glucose is not available. The regulatory mechanism of the lac system, known as the ***operon model,*** was the

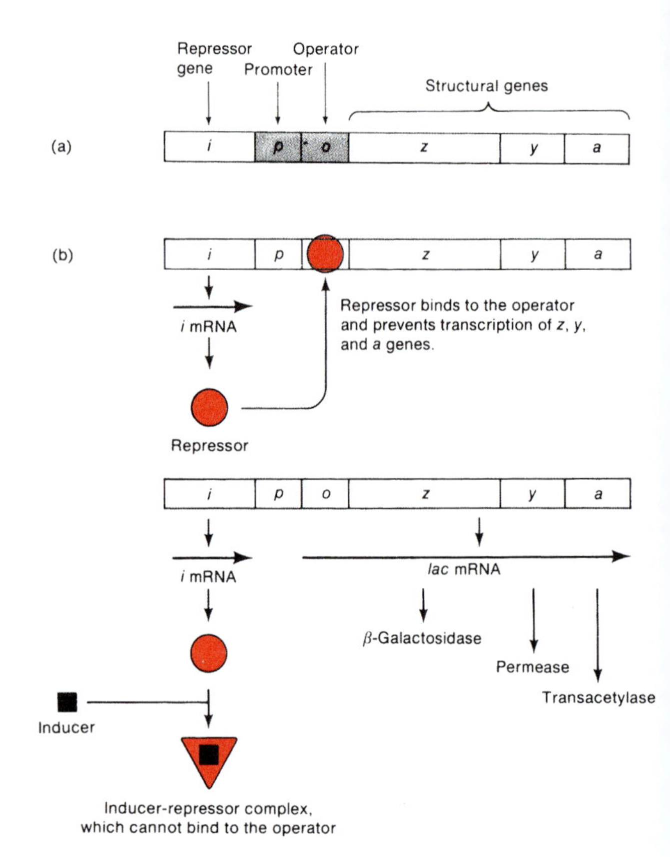

FIGURE 26-1
(a) Genetic map of the lac operon, not drawn to scale; the p and o sites are actually much smaller than the genes. (b) Diagram of the lac operon in (I) repressed and (II) induced states. The inducer alters the shape of the repressor, so the repressor can no longer bind to the operator.

first system described in detail, and it defines most of the terminology and concepts in current use. The principal regulatory features of the lac operon are as follows:

1. The products of two genes are required for lactose utilization. The *lacz* gene encodes an enzyme, *β*-galactosidase, that degrades lactose; the *lacy* gene encodes a protein, lac permease, needed to transport lactose and concentrate it within the cell. A third gene, *laca*, encodes an enzyme that participates in lactose metabolism only in certain circumstances; some lactose-fermenting bacteria do not have the *laca* gene, and it is not essential for lactose utilization in bacteria.
2. Lac enzymes are encoded in a single polycistronic mRNA molecule (lac mRNA), and enzyme levels are regulated by controlling synthesis of that mRNA.
3. Synthesis of lac mRNA is determined by the activity of a specific regulatory protein—the lac repressor. This protein is encoded in the *laci* gene, which is transcribed in a monocistronic mRNA molecule distinct from lac mRNA. When the repressor is active,

lac mRNA is not made. This is the negative regulation of the lac operon. Mutants have been isolated that produce inactive repressor proteins; these cells make lac mRNA continuously (constitutive synthesis), even in the absence of lactose. Actually, some lac mRNA is always made at a level of about one or two transcription events per generation because repression is never complete. This basal level synthesis is responsible for a very small amount of the proteins.

4. A sequence of bases (in the DNA)—the operator lac o—is adjacent to the lac mRNA promoter and binds repressor. When the repressor protein is bound to the operator, attachment of RNA polymerase to the promoter is prevented by steric interference, and initiation of transcription of lac mRNA does not occur. A mutation in the operator, which eliminates repressor binding, also leads to constitutive synthesis.
5. An inducer (a small effector molecule) is needed to initiate transcription of lac mRNA. The inducer of the lac operon, which is allolactose, [β-D-galactopyranosyl-(1 $\rightarrow$ 6)-β-D-glucopyranose] a structural isomer of lactose formed by basal synthesis of β-galactosidase, binds to the repressor and alters its three-dimensional structure such that it is unable to bind to the operator. Thus, in the presence of lactose or other inducers the operator is unoccupied and the promoter is available for initiation of mRNA synthesis. It is common to refer to inactivation of the repressor by an inducer as derepression.

Thus, the overall regulatory pattern is as follows: A bacterium growing in a medium without lactose does not make either the lac z or lac y product because the repressor that is made is bound to the operator and prevents synthesis of lac mRNA. The cell grows by utilizing whatever other carbon source is available. If lactose is added (and if glucose is absent; see below), basal level lac y and lac z proteins bring the lactose into the cell and convert it to allolactose. As a result, the repressor is inactivated, RNA polymerase binds to the promoter, and synthesis of lac mRNA begins. Lac mRNA synthesis continues until the lactose is exhausted, in which case the inactive repressor would be reactivated and repression reestablished.

The lac operon is also positively regulated, presumably because of the role of glucose in general metabolism. The function of β-galactosidase is to generate glucose by cleaving lactose, so if both glucose and lactose are available, the cell can use glucose and there is no reason for the lac operon to be induced. (The other cleavage product, galactose, is also converted to glucose by enzymes of the galactose operon.) Indeed, when glucose is present in the medium, little or no lac mRNA is made because glucose indirectly prevents RNA polymerase from binding to the lac promoter. This positive regulation is accomplished by ***cyclic AMP,*** a regulatory molecule in both prokaryotes and eukaryotes. Cyclic AMP (cAMP) is synthesized from ATP by the enzyme adenylate cyclase, and its concentration is regulated by glucose metabolism. In a bacterial culture that is starved of glucose, the intracellular concentration of cAMP is very high. If a culture is growing in a medium containing glucose, the cAMP concentration is very low. The observation that in a medium containing a carbon source that cannot enter the glycolytic pathway the cAMP concentration is high (Table 26-1) suggests that some glucose metabolite or derivative is an inhibitor of adenylate cyclase.

TABLE 26-1
Concentration of Cyclic AMP in Cells Growing in Media Having the Indicated Carbon Sources

Carbon Source	cAMP Concentration
Glucose	Low
Glycerol	High
Lactose	High
Lactose+glucose	Low
Lactose+glycerol	High

Cyclic AMP does not act directly as a regulator but is bound to a protein called the ***cAMP receptor protein (CRP),*** which forms a complex with cAMP (cAMP-CRP). This complex is active in the lac system and in many other operons involved in catabolic pathways. The cAMP-CRP complex binds to a base sequence in the DNA in the lac promoter region and stimulates transcription of lac mRNA; thus, cAMP-CRP functions as a positive regulator. The cAMP-CRP complex is not needed for binding of RNA polymerase to the lac promoter, for a free promoter binds the enzyme. However, an open-promoter complex does not form unless cAMP-CRP is also bound to the appropriate DNA sequence. Thus, when glucose is absent, the cAMP concentration is high and there is enough cAMP-CRP to allow an active transcription complex to form; if glucose is present, cAMP levels are low, no cAMP-CRP is available, and transcription cannot begin. Thus, the lac operon is independently regulated both positively and negatively, a feature common to many carbohydrate utilization operons (Figure 26-2).

Tryptophan (trp) Operon

The tryptophan operon is responsible for the production of the amino acid tryptophan, whose synthesis occurs in

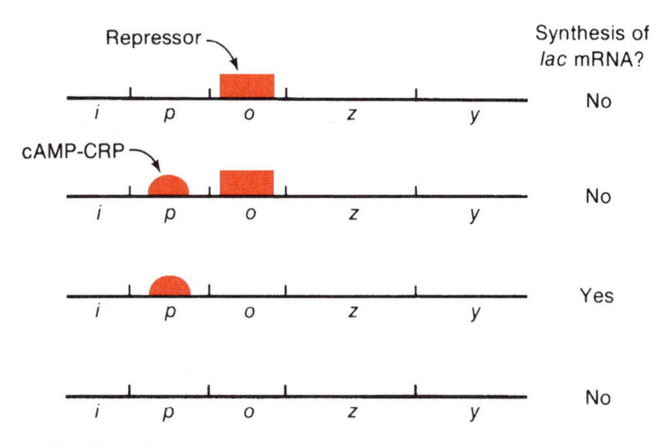

FIGURE 26-2
Three states of the lac operon showing that lac mRNA is made only if cAMP-CRP is present and repressor is absent.

five steps, each requiring a particular enzyme. In *E. coli*, these enzymes are translated from a single polycistronic mRNA. Adjacent to the enzyme coding sequences in the DNA are a promoter, an operator, and two regions called the ***leader*** and the ***attenuator*** (Figure 26-3). The leader and attenuator sequences are transcribed. Another gene (*trpR*) encoding a repressor is located some distance from this gene cluster.

Regulation of the trp operon is determined by the concentration of tryptophan: when adequate tryptophan is present in the growth medium, there is no need for tryptophan biosynthesis. Transcription is turned off when a high concentration of tryptophan is present and is turned on when tryptophan is absent. The regulatory signal is the concentration of tryptophan itself and, in contrast with lactose, tryptophan is active in repression rather than induction.

The trp operon has two levels of regulation—an on-off mechanism and a modulation system. The protein product of the *trpR* gene—the trp aporepressor—cannot bind to the operator in contrast with the lac repressor. However, if tryptophan is present, the aporepressor and the tryptophan molecule join together to form an active repressor complex that binds to the operator. When the external supply of tryptophan is depleted (or reduced substantially), the operator becomes exposed, and transcription begins. This type of on-off mechanism—activation of an aporepressor by the product of the biosynthetic pathway—has been observed in other biosynthetic pathways.

When the trp operon is derepressed, which is usually the case unless the concentration of tryptophan in the medium is very high, the optimal concentration of tryptophan is maintained by a modulating system in which the enzyme concentration varies with the concentration of tryptophan. This modulation is effected by

1. Premature termination of transcription before the first structural gene is reached, and
2. Regulation of the frequency of premature termination by the concentration of tryptophan.

Located between the 5′ end of the trp mRNA molecule and the start codon of the *trpE* gene is a 162-base segment called the ***leader*** (a general term for such regions). Within the leader is a sequence of bases (bases 123 through 150) having regulatory activity. After initiation of mRNA synthesis, most mRNA molecules are terminated in this region (except in the complete absence of tryptophan), yielding a short RNA molecule consisting of only 40 nucleotides and terminating before the structural genes of the operon. This region in which termination occurs is a regulatory region called the ***attenuator.*** The base sequence around which termination occurs (Figure 26-4) has the usual features of a transcription termination site—namely, a possible stem-and-loop configuration in the mRNA followed by a sequence of eight AU pairs.

The leader sequence has an AUG codon that is in-phase with a UGA stop codon; together these start-stop signals encode a polypeptide of 14 amino acids. The leader sequence has an interesting feature—at positions 10 and 11 are two adjacent tryptophan codons.

Premature termination of mRNA synthesis is mediated through translation of the leader peptide. The two tryptophan codons make translation of the leader polypeptide

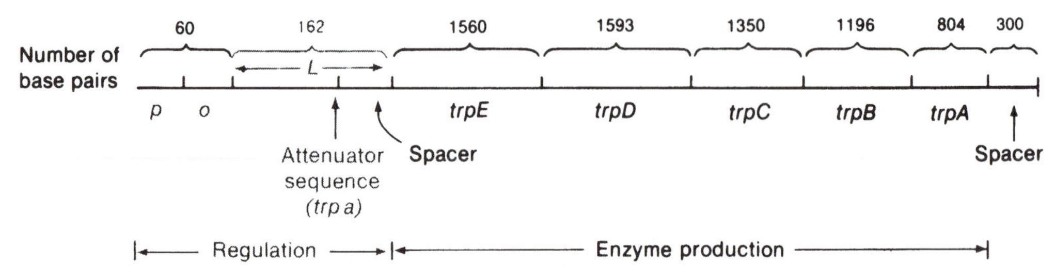

FIGURE 26-3
Escherichia coli trp operon. For clarity, the regulatory region is enlarged with respect to the coding region. The proper size of each region is indicated by the number of base pairs. L is the leader.

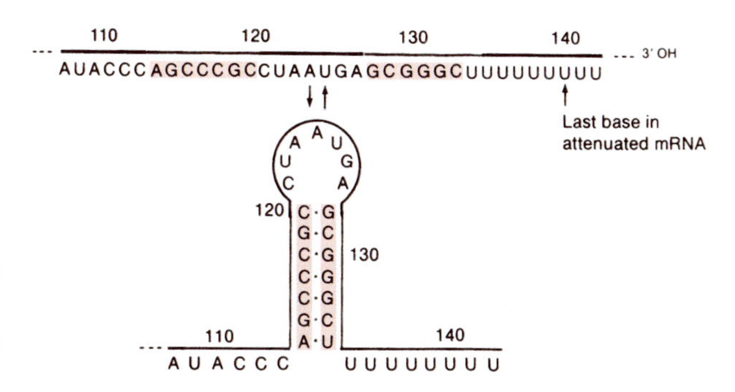
110
120
130
140
3′ OH
AUACCCAGCCCGCCUAAUGAGCGGGCUUUUUUUU
Last base in
attenuated mRNA
130
140
AUACCC
UUUUUUUU

FIGURE 26-4
Terminal region of the trp leader mRNA (right end of L in Figure 26-3). The base sequence given is extended past the termination site at position 140 to show the long stretch of U's. The bases (colored lines) form an inverted repeat sequence that could lead to the stem-and-loop configuration shown (segment 3–4 in Figure 26-5).

sequence quite sensitive to the concentration of charged tRNA^{Trp}. If tryptophan is limiting, there will be insufficient charged tRNA^{Trp} and translation will pause at the tryptophan codons. Thus, regulation depends on two characteristics of gene regulation in bacteria.

1. Transcription and translation are coupled.
2. Base-pair formation in mRNA is eliminated in any segment of the mRNA that is in contact with the ribosome.

Figure 26-5 shows that the end of the trp leader peptide is in segment 1 and a ribosom is in contact with about 10 bases in the mRNA past the codons being translated. When the final codons are being translated, segments 1 and 2 are not paired. In a coupled transcription-translation system, the leading ribosome is not far behind the RNA polymerase molecule. Thus, if the ribosome is in contact with segment 2, when synthesis of segment 4 is being completed, then segments 3 and 4 are free to form the duplex region 3–4 without segment 2 competing for segment 3. The presence of the 3–4 stem-and-loop configuration allows termination to occur when the terminating sequence of seven U's is reached.

If exogenous tryptophan is not present or is present in very small amounts, the concentration of charged tRNA^{Trp} will be limiting, and occasionally a translating ribosome will be stalled for an instant at the tryptophan codons. These codons are located 16 bases before the beginning of segment 2. Thus, segment 2 will be free before segment 4 has been translated and the 2–3 duplex will form. In the absence of the 3–4 stem and loop, termination will not occur and the complete mRNA molecule will be made, including the coding sequences for the *trp* genes. Thus, if tryptophan is present in excess, termination occurs and little enzyme is synthesized; if tryptophan is absent, there is no termination and the enzymes are made. At intermediate concentrations, the frequency of ribosome pausing will be such as to maintain the optimal concentrations of enzymes. This tryptophan regulatory mechanism is called ***attenuation*** and has been observed for several amino acid biosynthetic operons, e.g., histidine and phenylalanine.

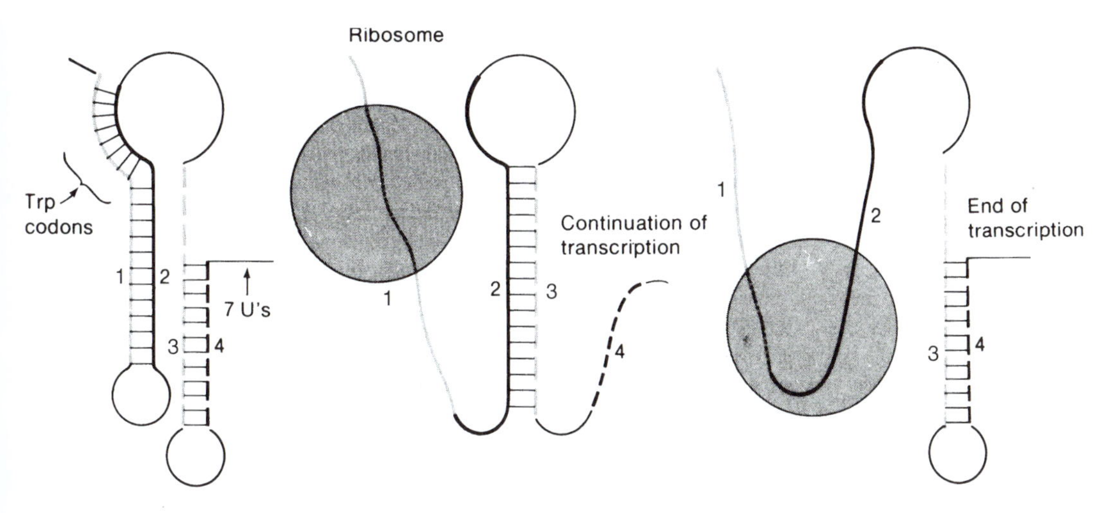
Ribosome
Trp codons
7 U's
Continuation of transcription
End of transcription
(a) Free mRNA. Base pairs between 1 and 2 and between 3 and 4.
(b) Low concentration of tryptophan. Ribosome stalled in region 1 permits formation of 2–3 before transcription of region 4 is completed.
(c) High concentration of tryptophan. Ribosome reaches region 2 before region 4 is completed, and permits formation of 3–4.

FIGURE 26-5
Model for the mechanism of attenuation in the *E. coli* trp operon.

Some bacterial operons are regulated solely by attenuation without repressor-operator interactions.

Temporal mRNA Regulation in Phage Systems

The lac and trp operons are regulated by molecules that turn gene expression on and off in response to the concentration of nutrients. However, some organisms require regulation of genes in time. For example, in the production of phage (bacteriophage or bacterial viruses) in an infected cell, various biosynthetic activities must occur on schedule. When a bacterium is infected by a phage, the enzymes responsible for breaking open the cell and releasing progeny phage must act late in the life cycle after phage DNA replication and packaging has occurred; otherwise, the cell could burst before any phage formed. Almost all phage systems accomplish this temporal regulation by selective and sequential transcription of particular sets of genes. This transcription gives rise to various classes of mRNA which are usually termed ***early*** and ***late mRNA.*** Three *E. coli* phage systems, which accomplish temporal regulation in different ways, are described briefly below.

Phage T7 contains several promoters, but only one is recognized by *E. coli* RNA polymerase. Transcription of T7 DNA begins at this promoter. Two important proteins are encoded there by transcript (early mRNA). One phosphorylates and inactivates the *E. coli* RNA polymerase, thus preventing *E. coli* from making any bacterial mRNA. The second protein is a new RNA polymerase that does not recognize any *E. coli* promoters but is active at the remaining phage promoters. In the absence of *E. coli* RNA polymerase, the early mRNA is no longer made, but the new T7 RNA polymerase makes the second transcript, which encodes the DNA replication enzymes and other proteins needed early in the life cycle. Transcription at the third promoter is delayed because injection of the phage DNA into the bacterium occurs very slowly. Roughly halfway through the life cycle of the phage this promoter enters the cell, transcription occurs by T7 RNA polymerase, and the structural proteins and lysis enzymes are synthesized. Lysis does not occur until phage particles have been assembled.

Phage T4 has numerous promoters only a few of which can be recognized by *E. coli* RNA polymerase. However, unlike T7, the late promoters are made available by successive modification of the *E. coli* enzyme. These modifications are of two types: addition of phage-encoded protein subunits and chemical modification of preexisting subunits. Temporal regulation occurs because the gene responsible for the first modification is encoded in the first set of mRNAs, that for the second modification in the second set, and so on. To ensure that the late mRNA, which encodes the structural proteins and the lysis enzyme, is not synthesized until adequate DNA has been made by the replication system, the template for this late mRNA cannot be parental T4 DNA but must be a replica.

E. coli ***phage λ*** also uses *E. coli* RNA polymerase throughout its life cycle. With this phage, regulation of transcription is accomplished in two ways: modification of the positive regulatory elements, such as the cAMP-CRP complex, needed for recognition of certain promoters; specific repressors are also used to turn off expression of certain genes. Figure 26-6 shows a portion of the genetic map of λ, which includes four regulatory genes (*cro*, *N*, *Q*, and *cII*), three promoters (pL, pR, and pR2), and five termination sites (tLl, tRl, tR2, tR3, and tR4). Seven mRNA molecules are also shown; the L and R series are transcribed leftward and rightward respectively from complementary DNA strands. The genes *O* and *P* encode proteins required for λ DNA replication, and the late genes encode the structural proteins of the phage and the lysis enzyme. The basic transcription sequence is: Two early mRNAs, LI and RI, are made that encode the regulatory proteins N and Cro, respectively. N is an antitermination

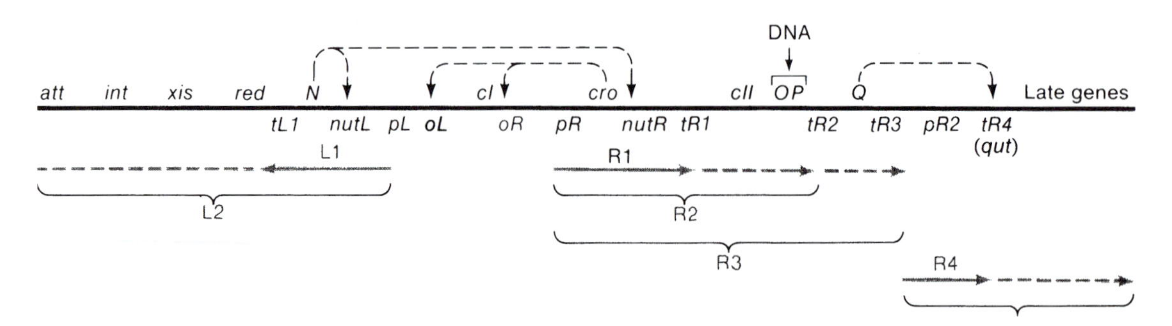

FIGURE 26-6
Genetic map of the regulatory genes of phage λ. Genes are listed above the line; sites are below the line. The mRNA molecules are heavy lines. The dashed black arrows with thin lines indicate the sites of action of the N, Cro, and Q proteins.

protein and enables *E. coli* RNA polymerase to ignore certain transcription termination sites and thereby extend synthesis of these mRNAs. It acts together with a bacterial protein, NusA, by binding to a site called nutR in the λ DNA. When RNA polymerase, which has initiated transcription at pR, reaches this site, it picks up the N-NusA complex and is thereby modified such that it is able to ignore the tR1 and tR2 terminators. A similar site, *nut*L, is present downstream from the pL promoter. Because of this antitermination effect, in time RI is extended until the DNA replication proteins and another regulatory protein, Q, are made. Q is also an antitermination protein. A small constitutively synthesized RNA, R4, is made from the outset of the infection. The Q protein binds to a site (*qut*) downstream from the promoter for R4, causing RNA polymerase to antiterminate, and R4 is extended to form an mRNA that encodes the head, tail, and lysis proteins. From the extended R1 mRNA, the gene-Cro protein is made. The concentration of Cro ultimately reaches a value at which the protein dimerizes, producing the active form. The Cro dimer acts as a repressor at both pL and pR; this activity turns off synthesis of early proteins that are no longer needed and prevents excess synthesis of DNA replication proteins.

Lambda, like many phages (but not T4 or T7), can engage in two alternative life cycles. In the ***lytic cycle,*** progeny phage particles are produced and the cell ultimately lyses, releasing the phage to the surrounding medium. In the ***lysogenic cycle,*** injected phage DNA is repressed and becomes inserted into the bacterial chromosome. At a later time, if the bacterial DNA is damaged sufficiently, the phage DNA is excised and a lytic cycle ensues. DNA damage, in a complicated way, leads indirectly to inactivation of the cell repressor and subsequent excision of the phage DNA and production of progeny phage. The *int* gene (Figure 26-6) encodes the enzyme that causes insertion of the phage DNA into the bacterial chromosome. Neither the cI repressor nor the Int protein is needed in the lytic cycle; however, and both are needed in the lysogenic cycle and are coordinately regulated. The two products, CI and Int, are encoded in different mRNA molecules but synthesis of the mRNA is initiated by a common signal. The product of the gene *cII* is a positive regulatory element. Like the cAMP-CRP complex in the lac operon, the *cII* product must be bound to the DNA adjacent to the promoters for the mRNAs encoding the cI and Int proteins. In this way, the choice of the lytic versus the lysogenic cycle depends on the concentration of the cII protein. If the concentration is low, neither cI repressor nor Int is made and the lytic cycle is followed; if the concentration is high, both cI repressor and Int are made and the lysogenic cycle occurs. The concentration of cII protein is regulated in response to environmental influences.

Regulons

One fairly well-studied regulon is the ***heat-shock*** or ***high-temperature protection regulon (htp)*** which synthesizes a variety of proteins when bacteria are exposed to temperatures above 40°C. In *E. coli*, a single protein, the product of the *htpR* gene, is a positive effector of all mRNAs of the regulon. The C-terminal end of the HtpR protein is homologous to the RNA polymerase σ subunit, which suggests that the heat-shock response involves a reprogramming of RNA polymerase by this σ-like protein, enabling RNA polymerase to initiate transcription at a class of promoters that is not otherwise recognized.

The inducible SOS repair system was described in Chapter 24. This system allows the frequency of replication errors to increase when repair is necessary and is regulated in order to keep the normal error frequency low. Since the repair system is needed only following certain types of DNA damage, some feature of the damage may act as the inducer. Several genes representing functions not directly related to repair are also components of this system, which is called the ***SOS regulon.*** Common to all mRNAs of the SOS regulon is an operator region to which a repressor, ***LexA,*** binds. When DNA is damaged, a protein called RecA binds to the damaged segment. Binding causes a conformational change in the protein and converts it to a specific protease that is active against only a small number of repressor proteins, LexA among them. Cleavage of LexA prevents it from binding to the SOS operator, so transcription of all mRNAs of the SOS regulon occurs, yielding Uvr enzymes, RecA, and SOS repair proteins. LexA normally represses its own synthesis (it is autoregulated). However, the large amount of RecA protein made after the LexA protein has been cleaved continues to be activated for proteolysis and cleaves LexA; thus, all proteins of the SOS regulon continue to be made. Once DNA repair is completed, RecA loses its proteolysis activity, and LexA is no longer cleaved. Without RecA protease activity, newly made LexA rapidly accumulates, binds to the SOS operators, and turns off the SOS regulon; thus, the state of the cell existing before DNA damage occurred is reestablished.

26.3 Gene Regulation in Eukaryotes

Regulationof gene expression in eukaryotes proceeds primarily by control of transcription as in prokaryotes. Some systems are also regulated at the translational level.

However, synthesis of mRNA in eukaryotes is not a simple matter of initiation at a promoter, as it is in prokaryotes, but includes several steps in which the primary transcript is converted to mRNA (Chapter 25). Control of these processing steps is also used to regulate gene expression in eukaryotes.

A variety of differences between the regulatory mechanisms of prokaryotes and eukaryotes are evident. First, transcription of related genes is initiated in response to a single signal. Second, a single polygenic primary transcript may be differentially processed to yield a set of distinct mRNA molecules each encoding one protein. Third, large proteins can be processed into small, active polypeptides. The second and third mechanisms are unique to eukaryotes.

Housekeeping Genes and RNases

Many proteins in eukaryotic cells such as those involved in glycolysis (Chapter 13) are needed continuously; the genes encoding such proteins are called ***housekeeping genes.*** A variety of regulatory mechanisms have evolved to ensure a constant supply of these gene products. In some instances the amount of each protein may be regulated by the strength of the promoter and ribosomal binding site. However, in housekeeping genes with strong promoters, the gene product functions as a repressor and binds to a site adjacent to the gene, thus regulating the level of transcription. This mechanism is called ***autoregulation.*** If the gene product is in short supply, transcription is activated; as the concentration of the product increases, the level of transcription is reduced.

In bacteria, mRNAs are rapidly degraded, which is essential for an organism that must adapt to rapidly changing environments. The RNase E protein (product of the *rne* gene) is essential in controlling the stability of mRNAs in *E. coli*. The RNase E mRNA, in turn, is autoregulated; an amino terminal fragment of RNase E acts as a repressor of the *rne* gene. In human cells an analogous RNase activity has been purified using antibodies against RNase E and both RNases recognize the same sequence (AUUUA). At least four RNase families can be identified in eukaryotes that have significant homology to pancreatic RNases. Overall, it is estimated that there should be as least 100 different eukaryotic RNases. Because of the crucial roles RNases and their structural homologues play in regulation of gene expression, splicing of primary transcripts, organogenesis, and other cellular activities, RNases are often referred to as housekeeping enzymes.

RNase activity in serum and cell extracts is elevated in a variety of cancers and infectious diseases. The level of RNases is regulated by both activators and inhibitors. The cellular content of RNase depends on both regulation of endogenous synthesis as well as the uptake of enzymes synthesized in the pancreas. Because of the multiple regulatory roles of RNases in cells, elevated levels are associated with a wide spectrum of diseases (Table 26-2) and serve as useful markers in diagnosis. RNases also are being investigated for antifungal, antiviral, and antitumor activity, and drugs are being developed that are based on specific RNase activities.

TABLE 26-2
Diseases Associated with Elevated Levels of RNase Activity in Body Fluids or Cell Extracts

Disease	RNase
Colorectal adenocaricnoma	RNase L
CML monocytes	RNase L
Chronic fatigue syndrome	RNase L
Liver diseases	poly-C degrading RNase
Chronic myeloid leukemia (CML)	>200 folds increase in endoribonuclease activity
Pancreatic necrosis	pancreatic RNase
Neurological infectious diseases	poly-C degrading RNase

Gene Families

Eukaryotic genes are not arranged in operons, since each mRNA contains only one gene. However, many systems are grouped into gene families either because of their location or, more commonly, because of their related function. For example, the set of tRNA genes, whose transcription is correlated by some unknown mechanism, is an example of a gene family. Another example, whose regulation is particularly simple, is the family of rRNA genes. Eukaryotic ribosomes contain one copy each of 5S, 5.8S, 18S, and 28S rRNA, and these are synthesized in equal numbers. The 5.8S, 18S, and 28S rRNA molecules are components of a single transcript that is cleaved to yield one copy of each molecule. The 5S rRNA is, however, part of another transcript. How the ratio of the 5S rRNA to the others is maintained is not known; possibly, the promoter strengths of the two transcripts are the same. However, this kind of gene organization is not particularly common, and related genes often reside on different chromosomes.

Of particular interest are the developmentally regulated gene families. Similar, but not identical, proteins having the same properties are made at different stages of development. A well-studied example is ***hemoglobin,*** a tetrameric protein containing two α subunits and two β subunits.

There are several different forms of both α and β subunits differing by only one or a few amino acids, and the forms that are present depend on the stage of development of an organism (Chapter 28). For example, the following temporal sequence shows the subunit types present in humans at various times after conception.

Subunit type	Embryonic (<8 wk)	Fetal (8–41 wk)	Adult (birth $\rightarrow$ henceforth)
α-like	$\zeta_2 \rightarrow \zeta_1$	α	α
β-like	ϵ	γ^G and γ^A	β and δ

Both the α- and β-like genes form distinct gene clusters. A property of each cluster is that the order of the genes is the order in which they are expressed in development.

26.4 Mechanisms of Gene Regulation in Eukaryotes

Gene expression in human cells is regulated primarily at the level of transcription as it is in prokaryotic cells. However, because transcription is more complex in eukaryotic cells, gene expression can be regulated at many different stages:

1. Initiation of transcription,
2. Processing the primary transcript (this includes capping, splicing out introns, addition of the polyA tail, and RNA editing in which specific nucleotides are changed),
3. Transport of the processed mRNA from the nucleus to the cytoplasm,
4. Translation of the mRNA (synthesis of protein), and
5. Posttranslational modification of the protein(s).

Each of these regulatory steps is crucial to the proper functioning of the cell and the organism.

Transcription Factors

The synthesis of mRNA molecules by RNA polymerase II (RNAPII) is a multistep process requiring the interaction of numerous proteins with DNA in the region of the promoter (Chapter 25). However, before the expression of a gene can be initiated, the ***transcription factor*** TFIID must attach to the promoter. This is followed by the attachment of other proteins, called ***general transcription factors (GTFs),*** to DNA in the region of the promoter. A description of the functions of some human GTFs is given in Table 26-3. The various GTFs facilitate attachment of RNAPII to the promoter at the correct nucleotide for initiation, destabilize the DNA at the promoter, and initiate transcription; together the GTFs and RNAPII are called the preinitiation complex. Once the first phosphodiester bond has formed, transcription has been initiated.

TABLE 26-3
The Role of General Transcription Factors in Initiation of Gene Expression in Human Cells

Factor	Subunits*	Function
TFIID	13	Recognizes TATA box; recruits TFIIB (composed of factors TBP and TAFs)
TFIIA	3	Stabilizes TFIID binding
TFIIB	1	Orients RNAPII to start site
TFIIE	2	Recruits TFIIH (helicase)
TFIIF	2	Destabilizes nonspecific RNAPII-DNA interactions
TFIIH	9	Promotes promoter melting by helicase activity

*RNAPII contains nine subunits.

The large number of GTFs required to initiate transcription does not completely solve the transcription problem. DNA is organized in chromatin and is wrapped around histone proteins and tightly packaged into nucleosomes (Chapter 24). These structures inhibit transcription as well as DNA replication. A distinct class of transcription factors has been identified that modifies chromatin structure so that transcription can occur. A complex of proteins termed RSF (***remodeling and spacing factor***) facilitates transcription initiation on chromatin templates *in vitro*. Another protein complex termed FACT (facilitates chromatin transcription) promotes elongation of RNA chains through nucleosomes. Together, RSF and FACT disassemble nucleosomes and permit transcription initiation to occur.

In addition to the GTEs that control the initiation of transcription at the TATA box and the protein factors that disassemble nucleosomes, other transcription factors are required to regulate the expression of particular genes or families of genes. Each transcription factor binds to DNA at a specific sequence, which ensures specificity (Table 26-4). The large number of transcription factors have certain structural motifs that facilitate their binding to DNA and, based on structural similarities, fall into four distinct groups (Figure 26-7).

1. ***Zinc finger:*** The structure of the zinc finger motif (discovered in TFIIA) consists of two cysteine residues and two histidine residues separated by 12 amino acids. The two cysteine residues are separated by two amino acids and the two histidine residues are separated by three amino acids. The cysteine and histidine residues are linked by a zinc ion and this

TABLE 26-4
Promoter and Enhancer Controlling Elements and Transcription Factors That Bind to Them

Consensus Sequence*	Response Element	Factor
TATAAAA	TATA Box	TFIID
GGCCAATCT	CAAT Box	CTG/NF1
GGGCGG	GC Box	SP1
ATTTGCAT	Octamer	Oct1, Oct2
CNNGAANNTCCNNG	HSE	Heat shock factor
TGGTACAAATGTTCT	GRE	Glucocortoid receptor
CAGGGACGTGACCGCA	TRE	Thyroid receptor
CCATATTAGG	SRE	Steroids

*Consensus sequence is for either a promoter or enhancer element. The letter N stands for any nucleotide (A,G, C,T).

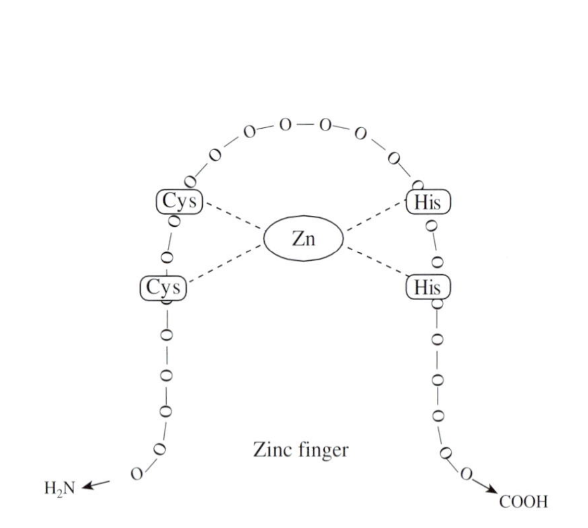

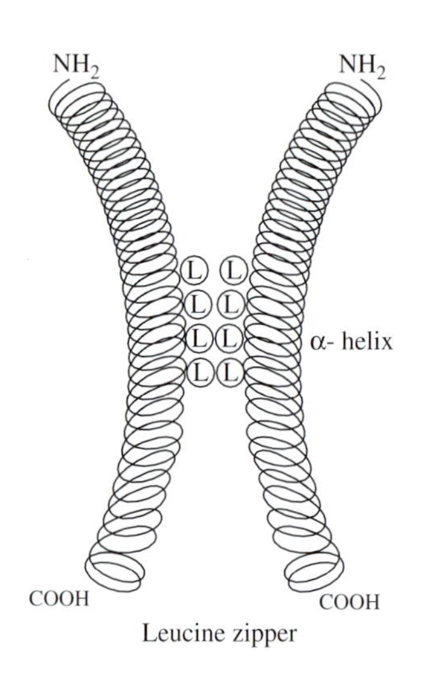

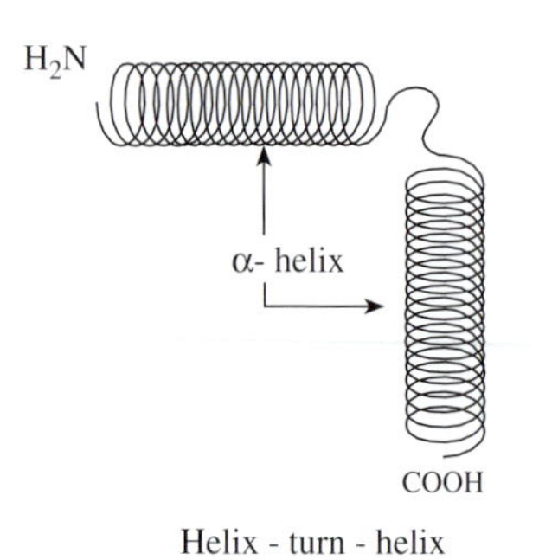

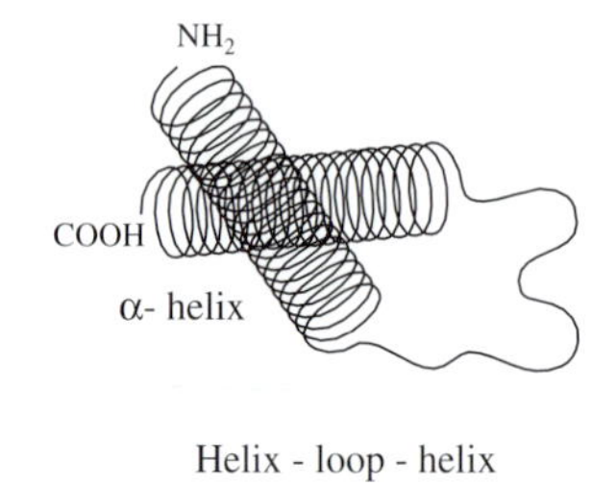

FIGURE 26-7
Structures of transcription factors that facilitate binding to DNA at specific sequences to regulate gene expression.

motif is repeated nine times in TFIIA protein. The 12 amino acids between the cysteine and histidine residues form a loop that can interact with the major groove in DNA; this loop is called the "zinc finger."

2. ***Leucine zipper:*** The structure of the leucine zipper (first discovered in C/EBP, CCAAT, and enhancer binding protein) consists of four leucine residues in an α-helical segment of the protein. Two polypeptides join to form a Y-shaped dimer whose arms can interact with the major groove of DNA. The stem of the Y-shaped structure is known as the "leucine zipper."
3. ***Helix-turn-helix (HTH):*** This structure (first discovered in the λ bacteriophage repressors, cI and cro) consists of two α-helical regions of a protein joined by an amino acid sequence that allows the α helices to turn. The two α helices form a dyad axis of symmetry, a structure that binds tightly to the major groove of DNA.
4. ***Helix-loop-helix (HLH):*** This structure is similar to the helix-turn-helix in that it consists of two α-helical segments joined by a long sequence of amino acids that can form a loop. This gives the two α-helical segments more flexibility so that the proteins can fit into the large groove of DNA at some distance from one another. This may facilitate looping of the DNA to make enhancer or promoter sites more accessible.

The serum albumin gene is one example of a gene that is regulated by several transcription factors. Although this gene is present in all tissues, it is only expressed in liver and spleen. The gene is activated by five different transcription factors that bind DNA in a region located between the CCAAT and TATA boxes (Figure 26-8). The transcription factor NF-1 actually binds to the right of the CCAAT box. Binding sites for transcription factors are determined by changing bases in the DNA one at a time and observing changes in the binding of transcription factors.

Steroid Receptors

Steroid hormones perform many functions in cells, one of which is to activate gene expression by binding to ***steroid receptors,*** proteins in the cytoplasm that, when activated, act as factors that initiate transcription. All steroid hormones are derived from cholesterol and, as a result, have similar chemical structures. Steroid hormones differ one from another primarily in hydroxylation of particular carbon atoms and by aromatization of the steroid A ring of the molecule. Once a steroid hormone binds to a steroid receptor protein, the complex undergoes a series of structural changes that result in the complex binding to DNA at a particular sequence called a ***steroid response element*** (SRE)

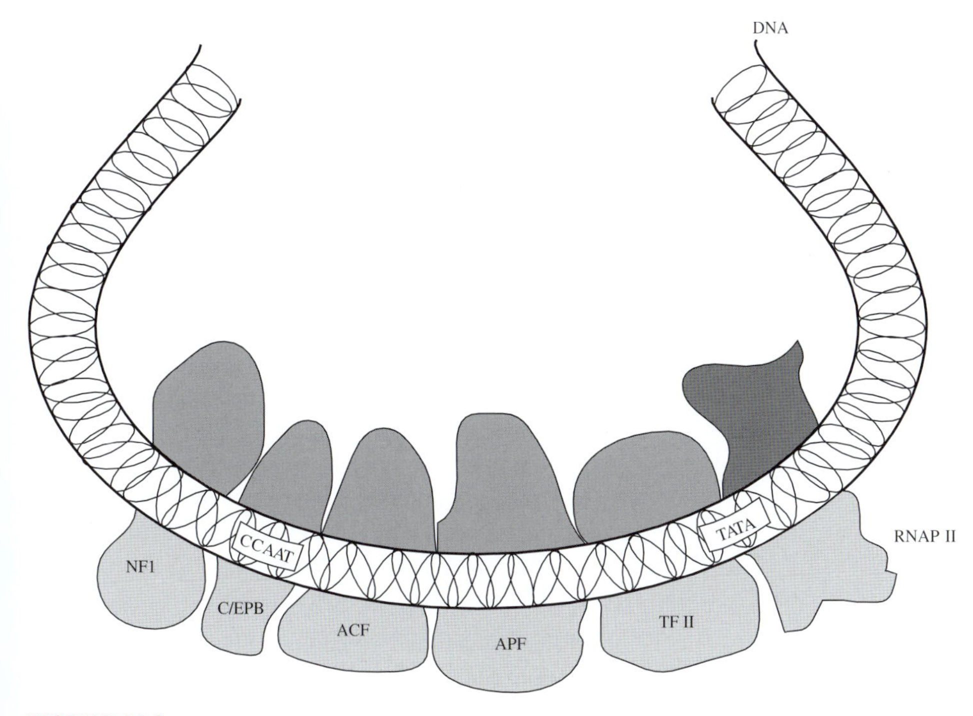

FIGURE 26-8
Expression of the human serum albumin gene is regulated by five transcription factors, four of which bind to the promoter region.

located at some distance upstream or downstream from the promoter.

Steroid receptor proteins are synthesized from a gene family that shows a high degree of homology in the DNA binding region. All steroid receptors belong to one of five classes: androgen receptor (AR), estrogen receptor (ER), glucocorticoid receptor (GR), mineralocorticoid receptor (MR), and progesterone receptor (PR). Also, all receptors contain a zinc finger motif which, if altered by mutation, destroys the steroid receptor's function.

G-Protein Diseases

Mutations that alter the functions of trimeric ***guanine nucleotide-binding proteins (G-proteins)*** are responsible for wide range of diseases (Table 26-5). While most of these diseases are quite rare, the identification of an altered G-protein in hypertension is estimated to affect as many as 15% of patients with hypertension. G-proteins relay signals from hormones, neurotransmitters, and other extracellular agents to different intracellular effectors that regulate enzymes and ion channels. G-proteins are responsible for signal transfer from more than 1000 cell receptors. Changes in G-proteins cause excessive or insufficient transmission of signals that are essential for proper cellular functioning.

G-proteins consist of three subunits: an α subunit that is loosely bound to a dimer consisting of a β subunit that is tightly bound to a γ subunit. Human chromosomes contain genes for 16α, 6β, and 12γ subunits. Thus, a large variety of Gα proteins define different G-protein trimers that regulate signaling pathways such as cAMP synthesis, cGMP breakdown, closing of Ca^{2+} channels, and opening of K^+ channels.

The activity of G-proteins is regulated by the binding and hydrolysis of guanosine triphosphate (GTP) by the Gα subunit. If guanosine diphosphate (GDP) is bound to the α subunit, it will associate with the $\beta\gamma$ subunits, but the trimer is inactive. When a cell receptor is activated, the trimer releases GDP and the α subunit is able to bind GTP. After binding GTP the α subunit dissociates from the $\beta\gamma$ subunit and from the receptor. Either the α–GTP complex or the $\beta\gamma$ subunits can then activate downstream effectors. Thus, defects in any of the numerous G-proteins may alter cellular biochemistry and cause disease.

TABLE 26-5
Diseases Caused by Mutations That Alter the Signal Activities of G Proteins

Diseases	G Proteins
Excess Signal	
Pituitary and thyroid adenomas	$G_3\alpha$
Adrenal and ovarian adenomas	$G_{12}\alpha$
McCune-Albright syndrome	$G_S\alpha$
Cholera	$G_2\alpha$
Deficient Signal	
Pseudohypoparathyroidism type Ia	$G_5\alpha$
Nightblindness	$G_t\alpha$
Deficient or Excessive Signal	
Pertussis	$G_i\alpha$
Essential hypertension	$\beta3$

Regulation of Transcription by Methylation

Most cells contain several enzymes that transfer a methyl group from S-adenosylmethionine (Chapter 17) to cytosine or adenine in DNA. These ***DNA methylases*** are both base-specific and sequence-specific. Another enzyme that methylates cytosine only in particular CpG sequences also is important for transcription. In a few cases, particularly in vertebrates, methylation of CpG sequences prevents transcription of some genes.

Evidence for a regulatory role of methylation is:

1. Certain genes are heavily methylated in cells in which the gene is not expressed and unmethylated in cells in which the gene is expressed.
2. *In vitro* methylation of the upstream site of a cloned γ-globulin gene prevents transcription. Methylation outside the upstream sequence does not inhibit transcription.
3. If the base analogue 5-azacytidine is added to cultures of growing cells, newly made DNA contains the analogue, and methylation of CpG sites containing the analogue fails to take place (5-azacytidine is also a general inhibitor of many methylases). Cells in such cultures gain the ability to make proteins whose synthesis is normally turned off (i.e., in cells growing in medium lacking the base analogue).
4. The housekeeping genes, which provide for general cell function and which are continuously transcribed, are rarely methylated in or near their initiation regions.

Undifferentiated and precursor cells often replicate. If methylation actually prevents gene expression in some types of cell, an inhibitory methylated site must be inherited as a methylated site in a daughter strand during DNA

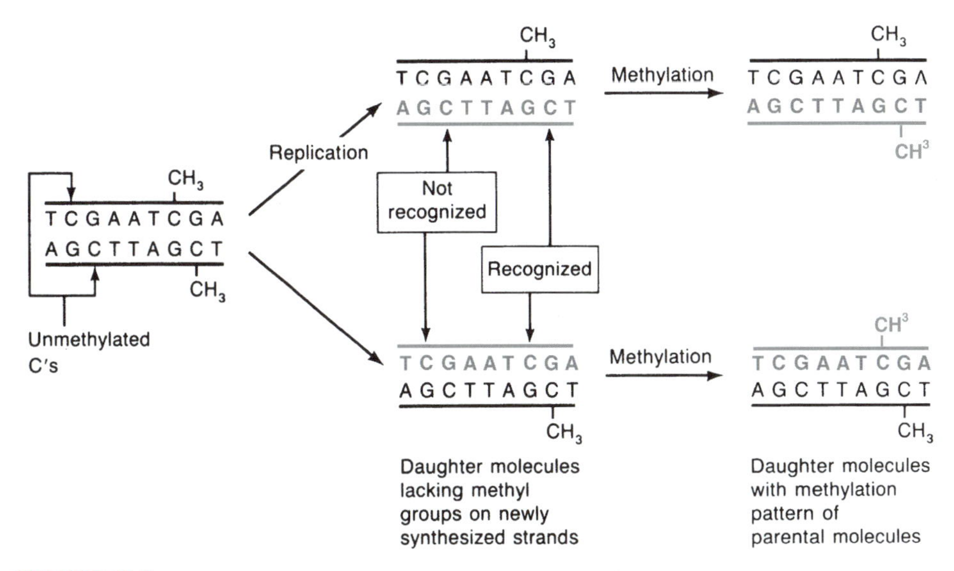

FIGURE 26-9
Mechanism by which the pattern of methylation in parental DNA is inherited in daughter molecules. The rule is that a C in a CpG sequence can be methylated only if the C in the complementary CpG sequence is already methylated. For clarity only, methyl groups have been drawn outside the sugar-phosphate strands.

replication. This requires methylation of a sequence complementary (rather than identical) to a methylated site. A property of certain DNA methylases shows how a methylated parental strand can direct methylation of the appropriate daughter sequence, i.e., these enzymes only methylate CpG (embedded in certain surrounding sequences) and only when the CpG in the opposite strand is already methylated (Figure 26-9). In this way, the methylation pattern of parental DNA strands is inherited by the daughter strands.

The sex chromosome composition of human males and females is XY and XX, respectively. If cells contain more than one X chromosome, all except one are inactivated. The cells of XX females contain a structure called a ***Barr body,*** which is a condensed, heterochromatic, transcriptionally inactive X chromosome. The cells of XXY males also contain a Barr body. XXY males suffer from ***Klinefelter's syndrome;*** they are usually mentally retarded, sterile, and suffer from physiological and developmental abnormalities. It is thought that extra X chromosomes must be prevented from gene expression to preserve the correct gene dosage in both males and females.

Evidence for inactivation of one X chromosome in XX females comes from the observation that females are mosaic for X-linked alleles that are heterozygous. For example, a woman who is heterozygous for a gene that controls production of sweat glands has patches of skin that perspire and patches that do not. Cells in the patches of skin that do not perspire express the mutant allele, and the wild type is silent in the inactive X chromosome. Cells in the patches of skin that do perspire express the wild type allele, and the mutant allele is silent. X chromosome inactivation occurs early in embryonic female development and the X chromosome that is inactivated is selected at random. Once selected, the same X chromosome is inactivated in subsequent cell divisions, which explains why patches of skin and other tissues differ in the expression of X-linked heterozygous genetic loci.

Inactivation of the X chromosome is thought to occur in three phases: initiation, spreading, and maintenance. Initiation involves the choice of chromosome to be inactivated and depends on a unique genetic locus on the X chromosome called the X-inactivation center (X*ic*). Once one X-inactivation center has been triggered in a cell, the other X chromosome is protected from inactivation by a "blocking signal." The inactivation signal spreads from the X*ic* locus, eventually to inactivating almost all of the genes on that X chromosome. However, at least 19 genes on inactive X chromosomes have been shown to have some transcriptional activity, so inactivation is not complete.

Specific maintenance mechanisms ensure that the inactive X chromosome is clonally transmitted and that it remains inactive in subsequent cell divisions. Methylation of CpG islands at the beginning of genes in the inactive X chromosome appears to play an important role in suppressing gene expression; the corresponding CpG islands in the active X chromosome are usually unmethylated.

Genomic Imprinting

Approximately two dozen autosomal genes in humans are inherited in a silent state from one parent and in a fully active state from the other parent. Such genes are said to be "imprinted" by the parents during gamete formation and the phenomenon is referred to as ***genomic imprinting,*** which does not change the nucleotide sequence of DNA. Rather, it is an epigenetic phenomenon in which the DNA at a particular locus is marked according to gender, and this determines whether or not the locus is expressed during embryonic development. Although regarded as a normal epigenetic phenomenon, two clinically different disorders are characterized by genomic imprinting.

Prader-Willi and ***Angelman syndromes*** are distinct disorders associated with multiple abnormalities and mental retardation. Both disorders are caused by mutations at the proximal end of chromosome 15 that silence one or more genes. In Prader-Willi syndrome, the maternally inherited chromosome is silent; in Angelman syndrome, the paternally inherited chromosome is silent. For both syndromes, a small deletion is usually responsible for the genomic imprinting that silences the relevant gene(s). Both syndromes can also result from having both copies of chromosome 15 derive from only one parent, a condition called ***uniparental disomy. Angelman syndrome*** is caused by mutation in a single gene whereas ***Prader-Willi syndrome*** is caused by mutations in more than one gene. Thus, Angelman syndrome can also be caused by mutations in the responsible gene itself, which is not the case for Prader-Willi syndrome. Diagnosis of both can be confirmed by analysis of DNA methylation in the respective genes since methylation is the mechanism used for imprinting.

Regulation of RNA Processing

Initiation of transcription ultimately leads to production of a primary transcript, which in higher eukaryotes is processed to form an mRNA. Alternative processing patterns can yield different mRNAs. One example comes from chicken skeletal muscle in which two forms of the muscle protein myosin, LC1 and LC3, are produced. The myosin gene has two different TATA sequences that yield two different primary transcripts. These two transcripts are processed differently to form mRNA molecules encoding distinct forms of the protein (Figure 26-10).

Another example of regulation of RNA processing is found in adenovirus, which has only a few promoters but which makes numerous primary transcripts and a large number of mRNAs. A fairly small amount of DNA is used efficiently because proteins are translated in all reading frames and different regions of the DNA are, by virtue of distinct RNA splicing patterns, used to form different proteins.

The amount of each mRNA is regulated with respect to the time after infection. RNAs 1 and 2 are both formed shortly after the primary transcript is made, although more of mRNA-2 is made. Later in the viral life cycle, mRNA-1 is not made and mRNA-2 is abundant. If cycloheximide, an inhibitor of protein synthesis, is added to the infected cells before the shift in splicing pattern takes place, the shift does not occur. This inhibition implies that a newly synthesized protein is a positive effector of the shift. Cycloheximide has a similar effect on other mRNA species derived from other primary transcripts at late times.

Adenovirus has a single promoter for all RNA made late in the cycle of infection. The primary transcripts terminate at five major polyadenylation sites. Each termination site influences the splicing pattern by allowing particular introns or intron termini to be present or not. The five sets are not used with equal frequency, with the result that most mRNAs encoding various genes are not the same. This is the primary mechanism for determining the relative amounts of the different structural proteins synthesized late in the adenovirus life cycle.

Another type of regulation of processing involves choice of different sites of polyadenylation. One example is the differential synthesis of the hormone ***calcitonin*** in different tissues; another is the synthesis of two forms of the heavy chain of immunoglobulins (Chapter 35). In both cases, the differential processing includes distinct patterns of intron excision (i.e., splicing), but they are necessitated by an earlier event in which differential poly(A) sites are selected from the primary transcript. That is, when the poly(A) site nearer the promoter is selected, a splice site used in the larger primary transcript is not present, so a different splice pattern results. Thus, slightly different proteins are synthesized.

The calcitonin gene consists of five exons and uses two alternative polyadenylation sites that respond to different signals in different tissues. In the thyroid calcitonin is produced by a signal that produces a pre-mRNA consisting of exons 1–4. The introns are then spliced out to give the mature mRNA. However, in neural tissue a different polyadenylation site is activated that produces a pre-mRNA consisting of exons 1–5 and the intervening introns. When this larger pre-mRNA is processed, the introns are spliced out but so is exon 4, producing a mature mRNA consisting of exons 1–3 and 5. This is translated into a growth factor called ***calcitonin-related gene peptide (CRGP).*** Mutations in the calcitonin gene can result in both adrenal and thyroid tumors,

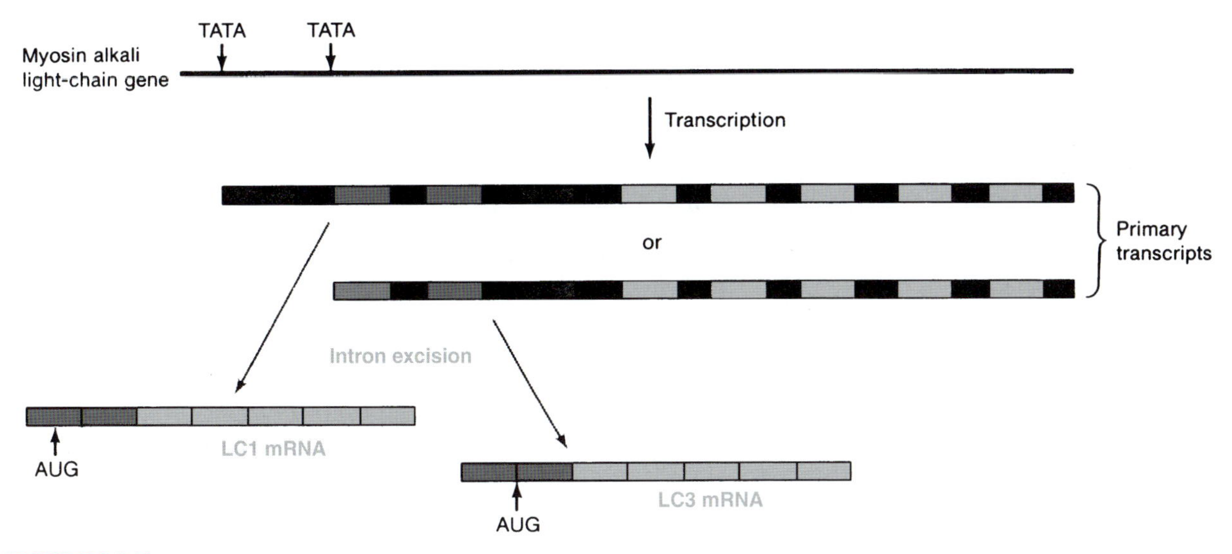

FIGURE 26-10
Chicken *LC1/LC3* gene. Two distinct TATA sequences lead to production of different primary transcripts, which contain the same coding sequences. Two modes of intron excision lead to the formation of distinct mRNA molecules that encode proteins having different amino terminal regions and the same carboxy terminal regions.

conditions that are classified as *multiple endocrine neoplasias*.

Alternative RNA Splicing and Editing

Most primary transcripts in eukaryotic cells derive from complete removal of all introns and complete joining of all exons. This results in only one species of mature mRNA being synthesized from each primary transcript. However, many eukaryotic genes can give rise to many different forms of mature mRNAs using different promoters and different polyadenylation sites (discussed above), as well as by ***alternative splicing.*** This form of regulation of RNA processing is especially useful because a single gene can be expressed differently at various developmental stages or in different tissues of the same organism. One example is the ***troponin T*** gene that synthesizes the fast skeletal muscle protein. This gene consists of 18 exons; however, only 11 are found in all mature mRNAs. Five of the exons can be included or excluded and two are mutually exclusive; if one is included, the other is excluded. Altogether, 64 different mature mRNAs can be produced by alternative splicing of the primary transcript of the troponin T gene.

RNA editing involves processing of RNA in the nucleus by enzymes that change a single nucleotide (insertion, deletion, or substitution). One example is ***apolipoprotein B*** (apo B) gene. In the liver this gene produces a 4536-amino-acid protein (apo B100), whereas in the small intestine the same gene produces a 2152-amino-acid protein (apo B48). The truncated protein is identical in amino acid sequence to the first 2152 amino acids in apo B100. This occurs because in cells of the small intestine one nucleotide (at position 6666) is edited by deamination of a cytosine residue. The conversion of a cytosine to uracil at this position produces a stop codon that terminates translation and produces the truncated protein. Thus, selective editing of mRNAs prior to translation in specific tissues is used to produce different proteins.

Regulation of Iron Utilization in Cells

Iron is an important constituent of human cells and regulates many biochemical functions (Chapter 29). Iron acts as an effector molecule in the translation of several mRNAs by binding to either 3′ or 5′ stem-loop structures, called ***iron-response the elements*** (IREs), that flank the coding sequences for several genes whose expression is regulated by iron. In particular, the inRNAs for the transferrin receptor, light and heavy chains of ferritin, an erythrolytic form of aminolevulinic acid synthetase, and a form of mitochondrial aconitase are regulated by IREs and an ***IRE-binding protein*** (IRE-BP). The IRE in the ferritin in RNA is in the 5′ flanking sequence. Translation of the mRNA is regulated by binding of IRE-BP whose activity, in turn, is regulated by the concentration of iron in the cell. In contrast, the IRE of transferrin mRNA is in the 3′ flanking sequence; binding of IRE-BP to this IRE regulates turnover of the mRNA. This example illustrates the variety of ways that gene expression can be regulated under varying physiological conditions in human cells.

Polyproteins

In prokaryotes, coordinated regulation of the synthesis of several gene products is accomplished by regulation of the synthesis of a single polycistronic mRNA molecule encoding all of the products. The analogue to this arrangement in eukaryotes is the synthesis of a ***polyprotein,*** a large polypeptide that is cleaved after translation to yield individual proteins. Each protein can be thought of as the product of a single gene. In such a system, the coding sequences of each gene in the polyprotein unit are not separated by stop and start codons but instead by specific amino acid sequences that are recognized as cleavage sites by particular protein-cutting enzymes. Polyproteins have been observed with up to eight cleavage sites; the cleavage sites are cut not simultaneously but rather in a specific order. Use of a polyprotein serves to maintain an equal molar ratio of the constituent proteins; moreover, delay in cutting at certain sites introduces a temporal sequence of production of individual proteins, a mechanism frequently used by animal viruses.

Some polyproteins are differentially cleaved in different tissues. An example is ***proopiomelanocortin,*** a polyprotein that is the source of several hormones synthesized in the pituitary gland. In the anterior lobe of the pituitary, the polyprotein is cleaved to release β-lipotropin and adrenocorticotropic hormone (ACTH). In the intermediate lobe, a different pattern of cleavage forms β-endorphin and α-melanotropin (Chapter 31).

Translational Regulation

Translational regulation refers to the number of times a finished mRNA molecule is translated. The three ways in which translation of a particular mRNA may be regulated are

1. By the lifetime of the mRNA,
2. By the probability of initiation of translation, and
3. By regulation of the rate of overall protein synthesis.

The silk gland of the silkworm *Bombyx mori* predominantly synthesizes a single type of protein, ***silk fibroin.*** Since the worm takes several days to construct its cocoon, it is the amount and not the rate of fibroin synthesis that must be great; hence, the silkworm can manage with a fibroin mRNA molecule that is very long lived. Silk production begins with chromatid amplification in which, over a period of several days, the single cell of which the gland is composed has a ploidy of 10^6. Each fibroin gene is transcribed from a strong promoter to yield about 10^4 fibroin mRNA molecules. An "average" eukaryotic mRNA molecule has a lifetime of 3 hours before it is degraded. However, fibroin mRNA survives for several days, during which each mRNA molecule is translated repeatedly to yield 10^5 fibroin molecules. Thus, the whole unicellular silk gland makes 10^{15} molecules or 300 mg of fibroin.

Production of a large amount of a single type of protein by means of a prolonged mRNA lifetime is common in highly differentiated cells. For example, cells of the chicken oviduct, which makes ovalbumin (for egg white), contain only a single copy of the ovalbumin gene per haploid set of chromosomes, but the cellular mRNA is long lived.

Translational and transcriptional control are sometimes combined. For example, ***insulin*** (which regulates the synthesis of a large number of substances) and ***prolactin*** (another hormone) are required together for production of casein (milk protein) in mammary tissue. Both hormones are needed to initiate transcription but prolactin in addition, increases the lifetime of casein mRNA.

The synthesis of some proteins is regulated by direct action of the protein on the mRNA. For instance, the concentration of one type of immunoglobulin is kept constant by self-inhibition of translation. This protein, like all immunoglobulins, consists of two H chains and two L chains. The tetramer binds specifically to H-chain mRNA and thereby inhibits initiation of translation.

Regulation of Protein Activity

Many enzymes contain several subunits and are regulated by a process known as ***allostery.*** A common arrangement in enzymes is that the binding sites for the molecule that is acted on (the substrate) and the inhibitor (which may be the product) are located on different subunits. If binding of the inhibitor prevents binding of the substrate, the information from a site on one subunit must somehow be transmitted to the other subunit. This can be accomplished by the following subunit interactions. Binding of the inhibitor molecule alters the shape of the subunit to which it is bound, resulting in changes in the reactive sites on other subunits. If the subunits remain in contact, all subunits adjoining the first will undergo a conformational change at their respective subunit interaction sites, altering, in turn, the substrate binding site of other subunits. Proteins capable of undergoing such conformational interactions are called ***allosteric proteins*** (Chapter 7).

In mammalian cells, ***cAMP*** is called a second messenger, because it regulates the activities of many proteins. Furthermore, certain hormones and cAMP work in concert to regulate enzymatic activities. Many hormones regulate metabolic processes, such as glucose metabolism and calcium utilization, through binding to specific receptors in the cell membranes of target cells. However, many hormones are not capable of penetrating the target cells, and instead the binding of a hormone to a membrane receptor induces intracellular synthesis of cAMP and other second messengers; these second messengers then cause the desired metabolic effects (Chapter 30).

26.5 Regulation of Cell Death: Apoptosis

Apoptosis (programmed cell death) is characterized by a complex series of biochemical changes that culminate in cell death without inflammation or swelling, which are signs of necrosis. Embryonic, fetal, and postnatal development involve cell death by apoptosis, which serves to eliminate excessive cell proliferation and migration. Apoptosis is initiated by a variety of external stimuli and molecular events such as oxidative stress, mitochondrial permeability transition, mitochondrial cytochrome c release, activation of caspase proteases, activation of endonucleases, transglutaminase activation, and poly(ADP-ribose) polymerase cleavage.

Necrosis is a form of passive cell death that is very distinct from apoptosis and usually occurs in a regional group of cells after a particular event, such as stroke, infarction, hemorrhage, or infection. In contrast to necrosis, apoptosis is an active process requiring RNA synthesis, protein synthesis, and new enzyme activities. Apoptosis usually involves isolated single cells that may undergo programmed cell death in a particular organ at different times. Certain features of a cell undergoing death by apoptosis distinguish it from one dying by necrosis. First, the cell shrinks, the plasma membrane and vesiculates change shape, and phosphatidylserine is redistributed on the cell surface. Specific nuclear changes also are diagnostic of a cell undergoing death by apoptosis. Nuclear fragmentation occurs from chromatin condensation and internucleosomal DNA breakdown. Analysis of the DNA shows a laddering fragment pattern that is the result of abnormally activated endonucleases. Although apoptosis is essential to normal development, it also appears to be the cause of pathogenic changes in several chronic neurological diseases including ***amyotrophic lateral sclerosis*** (ALS), ***Parkinson's disease, Huntington's disease,*** and ***spinocerebellar ataxias.*** Apoptosis also may play a key role in the development of Alzheimer's disease.

Regardless of the activating signal for apoptosis, dying cells follow the same series of events. ***Caspase*** proteinases play a central role in apoptosis; they are cysteine proteinases that are activated by apoptotic signals and degrade cellular proteins by cleavage after aspartate residues. The first caspase discovered (caspase-1) in mammalian cells was recognized as a "cell death" enzyme by its similarity to the cell death gene, *ced-3,* in the nematode *Caenorhabditis elegans.* Subsequent studies have identified almost a dozen mammalian caspases that function as apoptotic initiators (caspases 2, 8, 9, 10), apoptotic executioners (caspases 3, 6, 7), and cytokine processors (caspases 1, 4, 5). Activation of caspases 1 and 3 appears to play a key role in the pathogenesis of chronic neurological diseases resulting from progressive cell death. On the other hand, caspase inactivation appears to promote oncogenesis by allowing cell accumulation. Thus, caspases are prime targets for intervention in a variety of diseases that progress to death.

ALS was first described by the French neurologist Charçot in 1869. The disease involves rapid loss of motor neurons in the cortex, brain stem, and spinal cord; death usually follows 3-5 years after diagnosis. ALS is more commonly referred to as ***Lou Gehrig's disease*** because it struck down the New York Yankees immortal at the peak of his baseball career. ALS usually occurs in a sporadic form but in rare cases there is a familial factor; this provided the first clues to the molecular basis for ALS.

One form of ALS results from mutations in the *SOM* gene that codes for the cystolic enzyme copper/zinc ***superoxide dismutase*** (Cu,Zn-SOD). Mice lacking this enzyme do not develop ALS but mice that overexpress the mutant gene do develop symptoms characteristic of ALS and motor neuron death. It is now known that a mutant *SOD1* gene causes apoptotic motor neuron death by activating initiator caspases. Using drugs that inhibit caspase activity in ALS transgenic mice, the life span of the mice has been increased by about 70%. As more molecular details of caspase functions and apoptosis in cancer and neurological diseases are revealed, it eventually may be possible to develop drugs that ameliorate or suppress symptoms of these deadly diseases.

26.6 Regulation of Cell Proliferation: Oncogenes

In the course of differentiation, most adult cell types of higher eukaryotes lose the ability to divide. Only a small number of cell types, (e.g., cells of the intestinal mucosa, erythropoietic cells, and male germline cells) continue to do so. Many genes and regulatory elements determine whether a cell can or cannot divide. However, a class of genes has been discovered that, in certain circumstances, restores the ability of the cells to divide and, more significantly, to do so with little or no control. These ***oncogenes*** have been detected in many tumors and in RNA and DNA oncogenic viruses that cause cancers in infected animals.

The first oncogene discovered was *src*, a gene carried by ***Rous sarcoma virus*** that causes sarcomas in infected fowl.

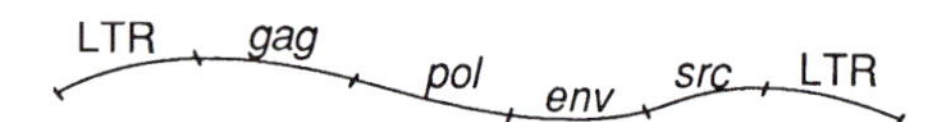

FIGURE 26-11
Genome of the Rous sarcoma virus (RSV). The single-stranded RNA is about 10 kb. Four proteins are encoded in this retrovirus: gag, a core protein; pol, reverse transcriptase; env, glycoprotein of the envelope; and src, a nonessential viral gene whose product is responsible for cellular transformation. LTR, Long terminal repeat sequences present at the ends of each viral RNA molecule.

This single-stranded oncogenic RNA virus infects chicken cells in which the RNA is copied by ***reverse transcriptase*** and converted to DNA (Figure 26-11). The DNA copy integrates into a chromosome, where it becomes a provirus. Integration of viral DNA is an essential step in the production of more virus particles. Expression of the *src* gene, which is nonessential to multiplication of the virus, causes tumor formation in infected fowl. Infected chicken fibroblast cells cultured *in vitro* become transformed and grow in an unregulated fashion that is visibly distinct from the growth of uninfected cells.

Study of DNA from avian tissue has indicated that the *src* gene is present in cells from normal tissue. However, the cellular *src* gene, denoted c-*src,* contains introns, whereas the viral gene, v-*src,* does not. This has led to the conclusion that at some time in the distant past, processed v-*src* mRNA was converted to DNA by reverse transcriptase, a virus-encoded enzyme that makes DNA from a single-stranded RNA template, and the virus picked up this DNA copy by genetic recombination. Similar oncogenes have been found in other viruses, and in each case a related cellular copy of the gene is present. The cellular counterpart is called a ***proto-oncogene.*** The oncogenes and proto-oncogenes are almost never identical because they have evolved separately. The product of the *src* gene is a tyrosine-specific protein kinase, an unusual enzyme because most protein kinases are active only against serine. This enzyme is capable of phosphorylating a large number of cellular proteins, causing either activation or inactivation.

The first human oncogene was discovered by testing DNA fragments isolated from human cancer cells and testing for their ability to transform a culture of mouse cells. This experiment utilized DNA from a line of human bladder cancer cells that was applied to a culture of National Institutes of Health 3T3 cells, an established mouse cell line. Some of the mouse cells became transformed, which suggested that an oncogene was being expressed in the bladder cancer cells. Only a single DNA fragment could induce the transformation; therefore, this fragment contained the human bladder cancer oncogene, which was later named *ras*. A search for a matching sequence in noncancerous cells (in analogy to the *src* study) uncovered a proto-oncogene that differed from *ras* by a single base pair in the fragment that was sequenced. That is, *ras* is a mutant form of a normal gene. *ras* has also been found in many other viruses and eukaryotes, including yeast. A list of most known oncogenes is given in Table 26-6.Their names are abbreviations or acronyms for the virus or tissue of origin.

Among the numerous oncogenes that have been identified in cancer cells, the *ras* genes are the most common. Expression of the oncogenic or mutant *ras* gene is found in 50% of human cancers. The protein produced by the *ras* genes is called p21 (M.W. 21,000) and contains 189 amino acid residues. The oncogenic *ras* gene proteins differ from the normal protein by a single amino acid substitution in most instances, and these changes may be responsible in some cases for converting a normal cellular gene product to a cancer-promoting one.

The normal *ras* p21 binds GTP and GDP; it also hydrolyzes GTP to GDP. The mutant *ras* p21 proteins show reduced or complete absence of GTPase activity. In this regard, normal p21 appears to be similar to the cellular G (GTP-binding) protein that activates adenylate cyclase and thereby mediates cellular activities.

Chromosome changes are the hallmark of many cancer cells. The most common type of change is ***translocation,*** a chromosome alteration in which two chromosomes have exchanged segments. For example, in ***Burkitt's lymphoma,*** fragments in chromosomes 8 and 14 are exchanged in 90% of affected individuals, between chromosomes 8 and 2 in 5%; and between chromosomes 8 and 22 in another 5%. In each case, a segment of chromosome 8 has been moved. The ***proto-oncogene c-myc*** is located on chromosome 8, and in each of the translocations, c-*myc* is relocated adjacent to a gene encoding an antibody. This relocation apparently places the proto-oncogene under control of the genes that regulate antibody synthesis, so that c-*myc* is made in large quantities.

Involvement of a different oncogenic protein has been demonstrated in ***chronic myelogenous leukemia (CML).*** This form of leukemia is associated with the so-called Philadelphia (Ph) chromosome and is characterized by a translocation from chromosome 9 to a new location on the short arm of chromosome 22. The translocation results in the production of a chimeric *abl* oncogenic protein. The chimeric *abl* oncogene has enhanced tyrosine protein kinase activity both *in vivo* and *in vitro* that is correlated with the development of leukemia. The translocated segment of DNA is inserted into chromosome 22 in a region consisting of 200 bp called the ***breakpoint cluster region (bcr).*** Consequently, in different CML patients the *bcr*/*abl* p21 proteins differ slightly in amino acid sequence, but all

TABLE 26-6
Some Oncogenes Involved in Human Cancers*

Oncogene	Protein**	Cancer**
K-ras	p21 GTPase	pancreas, lung, colorectal
N-ras	p21 GTPase	endometrial, other carcinomas
H-ras	p21 GTPase	bladder
erb-B	EGF receptor	gliomas, carcinomas
erb-B2	growth factor receptor	breast, ovarian, gastric
C-myc	transcription factor	Burkitt's lymphoma, SCLC
N-myc	transcription factor	neuroblastoma, SCLC
L-myc	transcription factor	SCLC
bcl-2	antiapoptosis protein	B-cell lymphoma
cycd-1	cyclin-D	B-cell lymphoma, carcinomas
bcr-abl	tryosine kinase	ALL (T cell), CML
cdk-4	cyclin-dependent kinase	sarcoma
β-cat	transcription factor	melanoma, colorectal
hst	growth factor	gastric
mdm-2	p53 binding protein	sarcoma
gl1	transcription factor	glioma, sarcoma
ttg	transcription factor	ALL (T cell)

*Each oncogene must be activated by a mutation. Most are activated by one or more somatic mutations. Germ line mutations that activate oncogenes lead to familial cancers.

** Abbreviations: EGF, epidermal growth factor; SCLC, small cell lung carcinoma; ALL, acute lymphocytic leukemia; CML, chronic myelogenous leukemia.

chimeric proteins are characterized by elevated protein kinase activity.

If oncogenes are altered forms (or aberrantly expressed forms) of normal genes that usually participate in growth regulation, the number of different oncogenes should be fairly small. Thus, we should expect that if a large number of tumor viruses and tumors are screened for oncogenes, the same ones should appear repeatedly. This is indeed the case. For example, the oncogene in human bladder carcinoma (*ras*) has been found in human lung and colon carcinomas and in rat mammary carcinoma. A similar sequence but with a different base change was found in a human neuroblastoma and a fibrosarcoma. This variant was called *N-ras.*

How oncogenes cause cellular transformation is unknown. However, when a tumor virus brings into a cell an oncogene whose sequence differs from the cellular proto-oncogene, the mutant gene product probably causes transformation because it is able to carry out some process that the proto-oncogene fails to do. That is, the proto-oncogene itself carries out a normal function or is silent; only a mutation or chromosomal rearrangement produces the cancerous cell.

Certain biological functions are not determined simply by the presence of a particular gene product but by its concentration. Thus, the conversion of a proto-oncogene to an active oncogene may occur simply by changes in concentration. The activity of a gene can be changed by altering the adjacent promoter or regulatory sequences (or both). Some oncogenes differ from the normal counterparts in that base changes exist only in the promoter and these changes presumably alter the rate of RNA synthesis.

Another important mechanism by which the expression of a proto-oncogene can be altered is by moving the gene to a new location, e.g., by relocating it next to a different promoter, separating it from an adjacent regulatory element, or placing it adjacent to an enhancer. Thus, when the viral DNA is inserted into the chromosome of an infected cell, the viral oncogene may be expressed at a much greater rate than the proto-oncogene, which remains at its normal location.

Whereas oncogenes are characterized by gain of function, another class of genes are characterized by loss of function; these are the ***tumor suppressor genes,*** which are often involved in familial (inherited) cancers (Table 26-7). The rare human cancer ***retinoblastoma*** provided the first significant insight into tumor suppressor genes. Retinoblastoma affects about 1 in 20,000 children who usually develop tumors in both eyes. Affected individuals have a small deletion in chromosome 13 that

TABLE 26-7
*Some Tumor Suppressor Genes Involved in Human Cancers**

Tumor Suppressor Gene	Protein	Cancer
Rb-1	E2F1 binding	retinoblastoma, osteosarcoma
p53	transcription factor	Li-Fraumeni syndrome, ~50% of all cancers
p16	cyclin-dependent, kinase inhibitor	breast, lung, bladder, pancreas, ~30% of all cancers
APC	regulates β-catenin	familial adenomatous polyposis
MSH1, MLH2, PMS1, PMS2	DNA mismatch repair	hereditary non-polyposis colorectal cancer
WT-1	transcription factor	Wilms' tumor
NF-1	p21ras-GTPase	melanoma, Neurofibromatosis type 1, neuroblastoma
VHL	protein stability regulator	renal (clear cell), von Hippel Lindau syndrome
BRCA-1	?DNA repair	ovarian, familial breast cancer
BRCA-2	DNA repair	familial breast cancer, pancreas
PTEN	tyrosine phosphastase	breast, prostate, thyroid, glioma
DPC4	TGF-β pathway	pancreas
E-CAD	transmembrane cell-cell adhesion molecule	gastric, breast
DCC	transmembrane receptor	colorectal, other carcinomas
MEN-1	undetermined	parathyroid, adrenal, pituitary

*Mutations in tumor suppressor genes (many of which code for transcription factors) alter an essential cellular activity that regulates growth. Mutations occur both in somatic cells and in germ line cell and are often the cause of hereditary cancers.

predisposes mutations to occur in the homologous chromosome by mitotic recombination. Thus, in the majority of cases of retinoblastoma, one mutation is transmitted in the germline and the other arises as the result of a somatic mutation.

It is estimated that 5–10% of breast cancers are due to germline mutations in the cancer susceptibility genes *BRCA-1* and *BRCA-2,* which are also tumor suppressor genes. Carriers of the *BRCA-1* mutation have an 80–90% lifetime risk of developing breast cancer; carriers of the *BRCA-2* mutation have a lesser risk. Ashkenazi Jews and the population of Iceland carry founder mutations since only one mutation, or a very few mutations, occurs repeatedly in the *BRCA-1* and *BRCA-2* genes among breast cancer patients in these populations. As a result, it is possible to genetically screen many individuals at high risk for harboring particular alleles of the **BRCA-1** and **BRCA-2 genes.** However, even with extensive counseling, learning that one is at high risk for developing breast cancer carries a heavy emotional burden. The only medical intervention offered at present for young, high-risk carriers of *BRCA-1* or *BRCA-2* genes is prophylactic mastectomy to avoid future development of breast cancer.

Tumorigenesis is a complex, multistep process involving acquisition of a number of genetic lesions. Whereas the mutations in oncogenes can cause unregulated growth, a deletion or a mutation in ***tumor suppressor genes*** (also called ***antioncogenes***) can also predispose normal cells to become cancer cells. Thus, induction of a malignant phenotype involves a mutation in oncogenes, tumor suppressor genes, and perhaps other genes. In human colon cancer, defects in both oncogene and tumor suppressor genes are known. The latter have been assigned to chromosomes 5, 17, and 18.

26.7 Retroviruses and AIDS

Since 1980 three types of infectious ***human T-cell lymphotrophic viruses*** (HTLVs) have been identified. HTLV-I is frequently associated with adult forms of ***T-cell leukemia-lymphoma,*** HTLV-II with ***hairy T-cell leukemia,*** and HTLV-III (now called human immunodeficiency virus; HIV) with ***acquired immunodeficiency syndrome (AIDS).*** Other related retroviruses have been isolated from human and primate populations. Molecular biologists have developed a wealth of information about the biology and genetics of retroviruses, including the complete sequencing of their genomes. Despite this enormous research effort it still

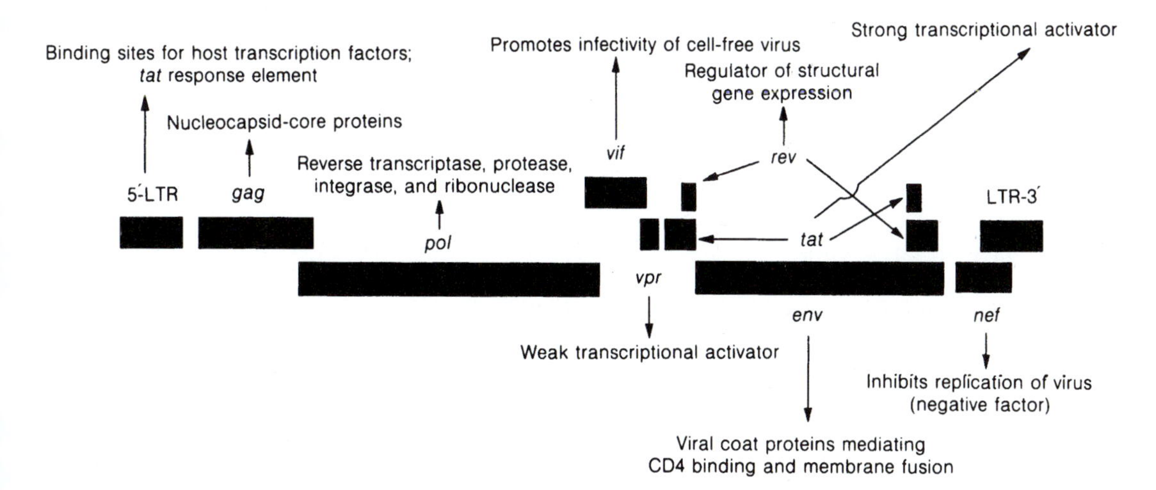

FIGURE 26-12
Genome of the human immunodeficiency virus (HIV, or the AIDS virus).

is not possible to prevent or cure infections by retroviruses.

The well-characterized retroviruses, HTLV-I, HTLV-II, and HIV, all exhibit a strong affinity for T cells; HIV specifically recognizes a surface antigen (CD4) on T cells of the immune system. The organization of the HIV genome resembles that of the oncogenic RNA viruses described above but it is considerably more complex. In addition to the *gag*, *pol*, and *env* genes and LTR sequences characteristic of all retroviruses, the HIV genome encodes some additional functions (Figure 26-12). At least four regulatory proteins have been provisionally identified encoded by the *vpr*, *tat*, *rev*, and *nef* genes. The envelope protein is synthesized as a single large polypeptide (gp 160), which is subsequently cleaved into two structural components. Most attempts at constructing an effective vaccine utilize part or all of the envelope proteins as antigens. Since these proteins do not elicit a strong antibody response by themselves, efforts are directed at coupling envelope antigens with more potent antigens to stimulate antibody response.

The human retroviruses represent a unique class of infectious agents, and the mechanisms whereby they ultimately result in a variety of diseases are unknown. In comparison to infections caused by most other viruses, retroviruses are weak antigens. This failure to evoke a strong immune system response presumably enables them to become established and remain active in cells that they infect. They rarely produce a strong viremia and consequently may remain quiescent in infected individuals for years before clinical symptoms develop. Other yet-to-be-identified extrinsic or intrinsic factors also may be necessary to activate the expression and proliferation of human retroviruses. At present, the only protection against retroviral infections is avoidance of exposure to these agents that are transmitted primarily through intercourse and exposure to contaminated blood.

Supplemental Readings and References

A. A. Antson, E. J. Dodson, G. Dodson, et al.: Structure of the trp RNA-binding attenuation protein, TRAP, bound to RNA. *Nature* **401,** 235 (1999).

C. Brenner and G. Kroemer: Mitochondria—the death signal integrators. *Science* **289,** 1150 (2000).

L. C. Brody and B. B. Biesecker: Breast cancer susceptibility genes: BRCA1 and BRCA2. *Medicine* **77,** 208 (1998).

S. B. Cassidy and S. Schwartz: Prader-Willi and Angelman syndromes: Disorders of genomic imprinting. *Medicine* **77,** 140 (1998).

G. M. T. Cheetham, D. Jeruzalini, and T. A. Steitz: Structural basis for initiation of transcription from an RNA polymerase-promoter complex. *Nature* **399,** 80 (1999).

D. Chen, H. Ma, N. Hong, et al.: Regulation of transcription by a protein methyltransferase. *Science* **284,** 2174 (1999).

P. R. Cook: The organization of replication and transcription. *Science* **284,** 1790 (1999).

Z. Farfel, H. R. Bourne, and T. Iiri: The expanding spectrum of G protein diseases. *New England Journal of Medicine* **340,** 1012 (1999).

X. Y. Li, A. Virbasivs, and M. R. Green: Enhancement of TBP binding by activators and general transcription factors. *Nature* **399,** 605 (1999).

D. A. Haber and E. R. Fearon: The promise of cancer genetics. *The Lancet* **351,** SII 1 (1998).

E. Heard, P. Clerc, and P. Avner: X-chromosome inactivation in mammals. *Annual Review of Genetics* **31,** 571 (1997).

S. John and J. L. Workman: The facts of chromatin transcription. *Science* **282,** 1836 (1998).

J. O. Kahn and B. D. Walker: Acute human immunodeficiency virus type I infection. *New England Journal of Medicine* **339,** 33 (1998).

L. S. Honig and R. N. Rosenberg: Apoptosis and neurologic disease. *American Journal of Medicine* **108,** 317 (2000).

P. W. Laird and R. Jaenisch: The role of DNA methylation in cancer genetics and epigenetics. *Annual Review of Genetics* **30,** 441 (1996).

G. Larola, R. Cuesta, G. Brewer, et al.: Control of mRNA decay by heat shock-ubiquitin-proteasome pathway. *Science* **284,** 499 (1999).

D. S. Latchman: Transcription-factor mutations and disease. *New England Journal of Medicine* **334,** 28 (1996).

M. Li, V. D. Ona, C. Guegan, et al.: Functional role of caspase-1 and caspase-3 in an ALS transgenic mouse model. *Science* **288,** 335 (2000).

A. J. Lopez: Alternative splicing of pre-mRNA: Developmental consequences and mechanisms. *Annual Review of Genetics* **32,** 279 (1998).

R. G. Roeder: The role of general initiation factors in transcription by RNA polymerase II. *Trends in Biochemical Sciences* **21,** 327 (1996).

C. H. Schein: From housekeeper to microsurgeon: The diagnostic and therapeutic potential of ribonucleases. *Nature Biotechnology* **15,** 529 (1997).

R. Singal and G. D. Ginder: DNA methylation. *Blood* **93,** 4059 (1999).

V. Skulachev: Mitochondria in the programmed death phenomena; a principle of biology: "It is better to die than to be wrong." IUBMB *LIFE* **49,** 365 (2000).

D. G. Tenen, R. Hromas, J. D. Licht, et al.: Transcription factors, normal myeloid development, and leukemia. *Blood* **90,** 489 (1997).

M-J. Tsai and B. W. O'Malley: Molecular mechanisms and action of steroid/thyroid receptor superfamily members. *Annual Review of Biochemistry* **63,** 451 (1994).

B. B. Wolf and D. R. Green: Suicidal tendencies: apoptotic cell death by caspase family proteinases. *Journal of Biological Chemistry* **274,** 20049 (1999).

J. L. Workman and R. E. Kingston: Alternation of nucleosome structure as a mechanism of transcriptional regulation. *Annual Review of Biochemistry* **67,** 275 (1998).

W. S. Yarnell and J. W. Roberts: Mechanism of intrinsic transcription termination and antitermination. *Science* **284,** 611 (1999).

CHAPTER 27

Nucleotide Metabolism

A nucleotide consists of a purine or pyrimidine base, a pentose (or deoxypentose), and a phosphate. The synthesis and degradation of nucleotides are discussed in this chapter. Structural features of nucleotides were discussed in Chapter 23 and how nucleotides are used in the synthesis of DNA and RNA in Chapters 24 and 25, respectively.

Table 27-1 gives the nomenclature of purine and pyrimidine nucleosides and nucleotides. Names of purine nucleosides end in *-osine,* whereas those of pyrimidine nucleosides end in *-idine*; guanine nucleoside is guanosine and should not be confused with guanidine, which is not a nucleic acid base; thymidine is a deoxyriboside.

Nucleotides are synthesized by two types of metabolic pathways: **de novo *synthesis*** and ***salvage pathways.*** The former refers to synthesis of purines and pyrimidines from precursor molecules; the latter refers to the conversion of preformed purines and pyrimidines—derived from dietary sources, the surrounding medium, or nucleotide catabolism—to nucleotides, usually by addition of ribose-5-phosphate to the base. *De novo* synthesis of purines is based on the metabolism of one-carbon compounds.

27.1 One-Carbon Metabolism

One-carbon moieties of different redox states are utilized in the biosynthesis of the purine nucleotides and thymidine monophosphate (also known as thymidylate), the metabolism of several amino acids (particularly serine and homocysteine), the initiation of protein biosynthesis in bacteria and mitochondria by formylation of methionine, and methylation of a variety of metabolites. These one-carbon reactions utilize coenzymes derived from ***folic acid,*** or ***folate.*** Folate is a vitamin for humans (and other animals because of their inability to synthesize it) (Chapter 38).

Folic acid is the common name for ***pteroylglutamic acid,*** a compound consisting of a heterobicylic pteridine, p-aminobenzoic acid (PABA), and glutamic acid (Figure 27-1). The combination of the first two produces pteroic acid. Humans lack the enzymes capable of synthesizing PABA or of linking pteroic acid to glutamate. The antimicrobial activity of sulfonamides is due to their competitive inhibition of the bacterial enzyme that incorporates PABA into dihydropteroic acid (Chapter 6).

Folates have a wide biological distribution; a rich dietary source is green leaves. They occur in nature largely as oligoglutamyl conjugates (e.g., in plants, predominantly pteroylheptaglutamate) in which the peptide linkages occur between the γ-carboxyl group of one glutamate and the α-amino group of the next (Figure 27-1). The mechanism of intestinal absorption of folate is not completely understood. The ingested folylpolyglutamates must be converted to folylmonoglutamate prior to absorption. The folylpolyglutamates are rapidly hydrolyzed in the intestines at neutral pH by the brush-border enzyme

TABLE 27-1
Nomenclature of Nucleosides and Nucleotides

Base	Nucleoside (Base-Sugar)*	Nucleotide† (Base-Sugar Phosphate)
Purines		
Adenine (6-aminopurine)	Adenosine	Adenosine monophosphate (AMP) or adenylic acid
	Deoxyadenosine	Deoxyadenosine monophosphate (dAMP)
Guanine (2-amino-6-oxypurine)	Guanosine	Guanosine monophosphate (GMP)
	Deoxyguanosine	Deoxyguanosine monophosphate (dGMP)
Hypoxanthine (6-oxypurine)	Inosine	Inosine monophosphate (IMP)
	Deoxyinosine	Deoxyinosine monophosphate (dIMP)
Xanthine (2,6-dioxypurine)	Xanthosine	Xanthosine monophosphate (XMP)
Pyrimidines		
Cytosine (2-oxy-4-aminopyrimidine)	Cytidine	Cytidine monophosphate (CMP)
	Deoxycytidine	Deoxycytidine monophosphate (dCMP)
Thymine (2,4-dioxy-5-methylpyrimidine)	Thymidine (thymine deoxyriboside)	Thymidine monophosphate (TMP) (thymine deoxyribotide)
Uracil (2,4-dioxypyrimidine)	Uridine	Uridine monophosphate (UMP)

*The sugar residue can be ribose or deoxyribose. If it is deoxyribose, it is identified as such; otherwise it is assumed to be ribose, with the exception of thymidine, which is deoxyriboside.

†A nucleotide is a nucleoside monophosphate; the monophosphates are sometimes named as acids.

pteroylpolyglutamate hydrolase (also called conjugase) to pteroylmonoglutamate (folic acid). If the folylpolyglutamates enter the intestinal epithelial cells intact, they may be converted to folylmonoglutamate within lysosomes by lysosomal hydrolase; however, the significance of this pathway appears to be minor. Folate transport in the intestine and the choroid plexus is mediated by a specific carrier, and the disorder ***hereditary folate malabsorption*** is associated with a defective folate carrier. Individuals affected with hereditary folate malabsorption exhibit early onset of failure to thrive, megaloblastic anemia, and severe mental retardation. Therapy requires the administration of large doses of oral and systemic folates. Folate is reduced and converted to N^5-methyltetrahydrofolate in intestines and secreted into the circulation.

In the plasma, about two-thirds of the folate is protein-bound. Tissue needs for folate are met by uptake from

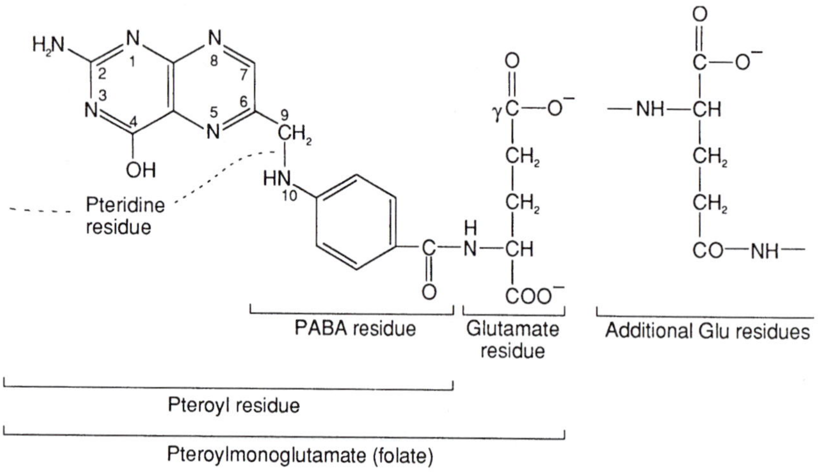

FIGURE 27-1
Structure of folic acid showing its components. The numbered part participates in one-carbon transfer reactions. In nature, folate occurs largely as polyglutamyl derivatives in which the glutamate residues are attached by isopeptide linkages via the γ-carboxyl group. The pteridine ring structure is also present in tetrahydrobiopterin, a coenzyme in the hydroxylation of phenylalanine, tyrosine, and tryptophan (Chapter 17).

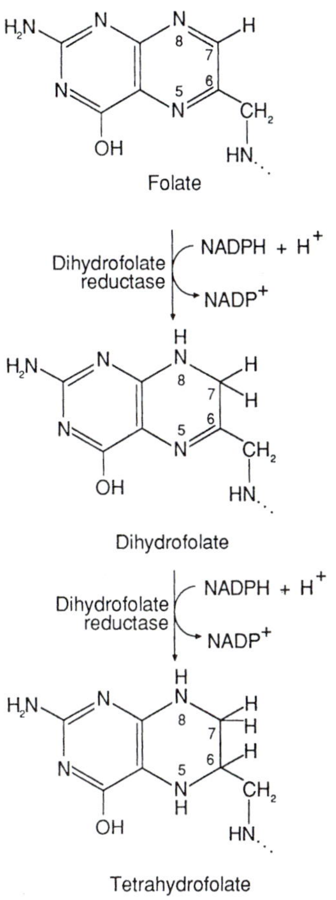

Folate

Dihydrofolate reductase NADPH + H^+ → $NADP^+$

Dihydrofolate

Dihydrofolate reductase NADPH + H^+ → $NADP^+$

Tetrahydrofolate

FIGURE 27-2
Reduction of folate to tetrahydrofolate.

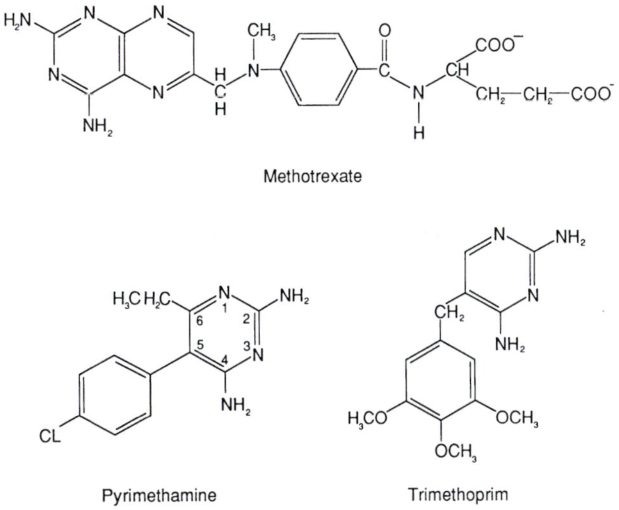

Methotrexate

Pyrimethamine

Trimethoprim

FIGURE 27-3
Inhibitors of dihydrofolate reductase. Methotrexate, a structural analogue of dihydrofolate, is effective against intact mammalian cells but ineffective against protozoa and some bacteria owing to permeability barriers. Trimethoprim and pyrimethamine (2,4-diaminopyrimidines) are effective against microorganisms. The former is antibacterial and antimalarial; the latter is primarily antimalarial.

plasma. Within tissue cells, N^5-methyltetrahydrofolate transfers its methyl group to homocysteine with the formation of methionine (Chapter 17). This reaction is catalyzed by homocysteine methyltransferase, a vitamin B_{12} coenzyme-dependent enzyme, and appears to be the major site of interdependence of these two vitamins.

In the tissues, tetrahydrofolate is converted to polyglutamyl forms by an ATP-dependent synthetase. In the liver, the major form is pteroyl pentaglutamate. Reduced polyglutamyl forms, each substituted with one of several one-carbon moieties, are the preferred coenzymes of folate-dependent enzymes. Reduction of folate (F) to tetrahydrofolate (FH_4) occurs in two steps: F is reduced to 7,8-dihydrofolate (FH_2), and FH_2 is reduced to 5,6,7,8-tetrahydrofolate (FH_4). Both of these reactions are catalyzed by a single NADPH-linked enzyme, dihydrofolate reductase (Figure 27-2).

Inhibitors of Dihydrofolate Reductase

Methotrexate (Figure 27-3), a structural analogue of FH_2, is a potent inhibitor of dihydrofolate reductase and is used in chemotherapy of neoplastic disease. Methotrexate is not effective against infections from bacteria and protozoa, since these organisms are impermeable to folate and its analogues. However, methotrexate inhibits dihydrofolate reductase of both bacterial and protozoal origin in cell-free preparations. Pyrimethamine is extremely effective against protozoan (e.g., malarial parasite) infections, is ineffective against bacterial infections., and is a mild inhibitor of the mammalian enzyme. Trimethoprim is an effective inhibitor of both bacterial and protozoal enzymes but has minimal inhibitory action against the mammalian enzyme. These selective enzyme inhibitors have been used in the treatment of bacterial and malarial infections.

Formation of One-Carbon Derivatives of Folate

There are several one-carbon derivatives of folate (of different redox states) that function as one-carbon carriers in different metabolic processes. In all of these reactions, the one-carbon moiety is carried in a covalent linkage to one or both of the nitrogen atoms at the 5- and 10-positions of the pteroic acid portion of tetrahydrofolate. Six known forms of carrier are shown in Figure 27-4. Folinic acid (N^5-formyl FH_4), also called ***leucovorin*** or ***citrovorum factor,*** is chemically stable and is used clinically to prevent or reverse the toxic effect of folate antimetabolites, such as methotrexate and pyrimethamine. The formation and interconversion of some metabolites of

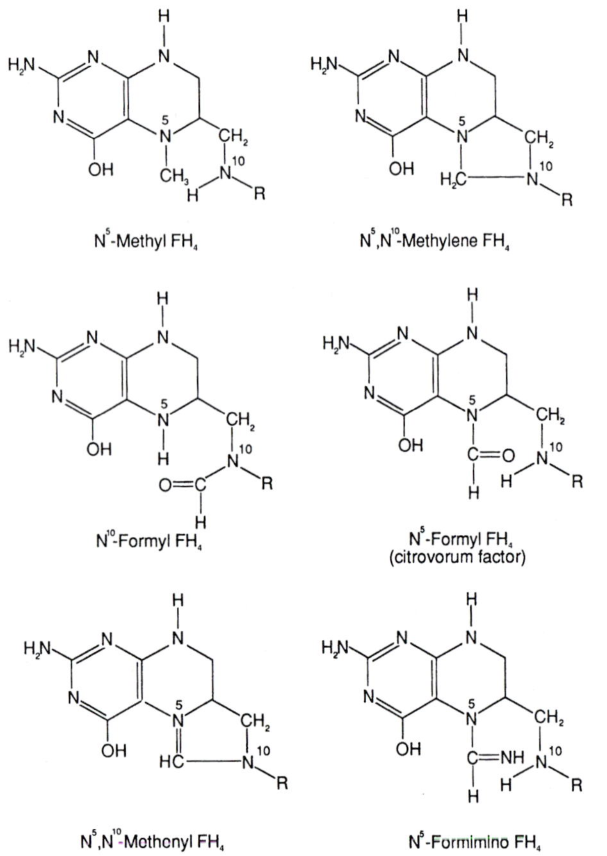

N^5-Methyl FH_4 N^5,N^{10}-Methylene FH_4

N^{10}-Formyl FH_4 N^5-Formyl FH_4 (citrovorum factor)

N^5,N^{10}-Methenyl FH_4 N^5-Formimino FH_4

FIGURE 27-4

The six one-carbon substituents of tetrahydrofolate. The oxidation state is the same in the one-carbon moiety of N^5-formyl, N^{10}-formyl, and N^5,N^{10}-methenyl FH_4. N^{10}-Formyl FH_4 is required for *de novo* synthesis of purine nucleotides, whereas N^5,N^{10}-methylene FH_4 is needed for formation of thymidilic acid.

FH_4 are shown below.

1. Serine is the principal source of one-carbon units.

$$FH_4 + \text{serine} \xrightleftharpoons[\text{pyridoxal phosphate}]{\text{serine hydroxymethyltransferase}} N^5,N^{10}\text{-methylene } FH_4 + \text{glycine} + H_2O$$

2. N^5,N^{10}-Methylene FH_4 plays a central role, since it can be reduced to N^5-methyl FH_4 or oxidized to N^5,N^{10}-methenyl FH_4.

Red.

$$N^5,N^{10}\text{-Methylene } FH_4 + NADH + H^+ \xrightleftharpoons{\text{reductase}} N^5\text{-methyl } FH_4 + NAD^+$$

Ox

$$N^5,N^{10}\text{-Methylene } FH_4 + NADP^+ \xrightleftharpoons{\text{dehydrogenase}} N^5,N^{10}\text{-methenyl } FH_4 + NADPH + H^+$$

where "Ox." is oxidation and "Red." is reduction.

3. Hydrolysis of N^5,N^{10}-methenyl FH_4 by a cyclohydrolase yields N^5-formyl FH_4 (folinic acid) and N^{10}-formyl FH_4. The oxidation state of the one-carbon moiety is the same in all three species.

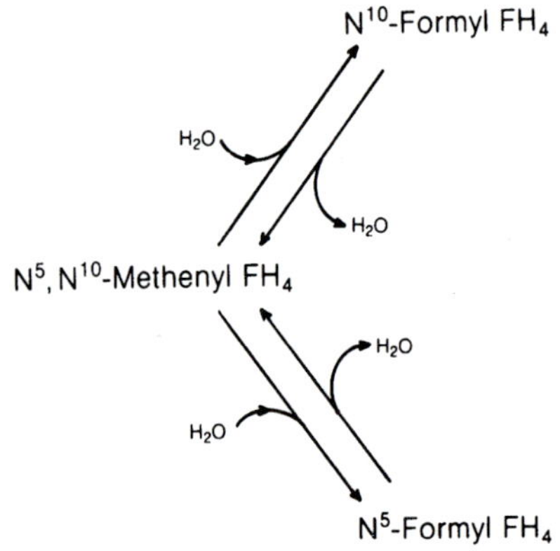

N^{10}-Formyl FH_4

H_2O H_2O

N^5,N^{10}-Methenyl FH_4

H_2O H_2O

N^5-Formyl FH_4

4. N^{10}-Formyl FH_4 can be formed directly from formate in an ATP-requiring reaction.

$$FH_4 + HCOO^- + ATP \xrightarrow[\text{synthetase}]{N^{10}\text{-formyl } FH_4} N^5\text{-formyl } FH_4 + ADP + P_i$$

5. The formimino group (–CH=NH) of formiminoglutamate (Figlu), a catabolite of histidine (Chapter 17), forms formimino FH_4.

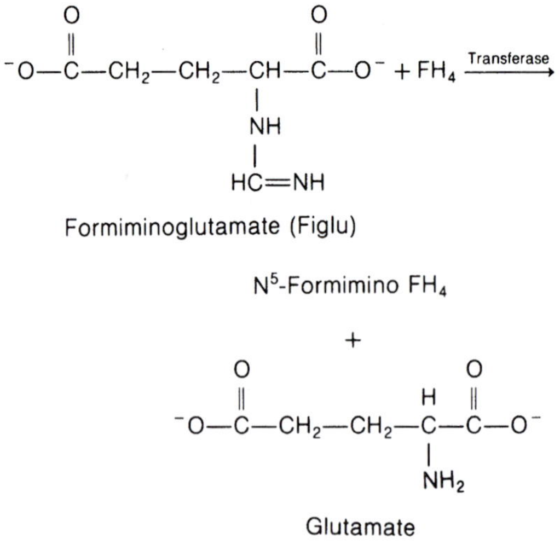

$$^-O-\overset{O}{\overset{\|}{C}}-CH_2-CH_2-\underset{\underset{HC=NH}{|}}{\underset{NH}{|}}{CH}-\overset{O}{\overset{\|}{C}}-O^- + FH_4 \xrightarrow{\text{Transferase}}$$

Formiminoglutamate (Figlu)

N^5-Formimino FH_4

+

$$^-O-\overset{O}{\overset{\|}{C}}-CH_2-CH_2-\underset{NH_2}{\underset{|}{\overset{H}{C}}}-\overset{O}{\overset{\|}{C}}-O^-$$

Glutamate

N^5-Formiminoglutamate undergoes deamination to N^5,N^{10}-methenyl FH_4. If a loading dose of histidine is given to a patient who is deficient in folic acid, urinary excretion of Figlu is increased. This ***Figlu excretion test*** is useful in the diagnosis of folate deficiency, which clinically is manifested as a megaloblastic anemia (Chapter 38).

The various catalytic roles of folate-mediated one-carbon transfer reactions in anabolic and catabolic

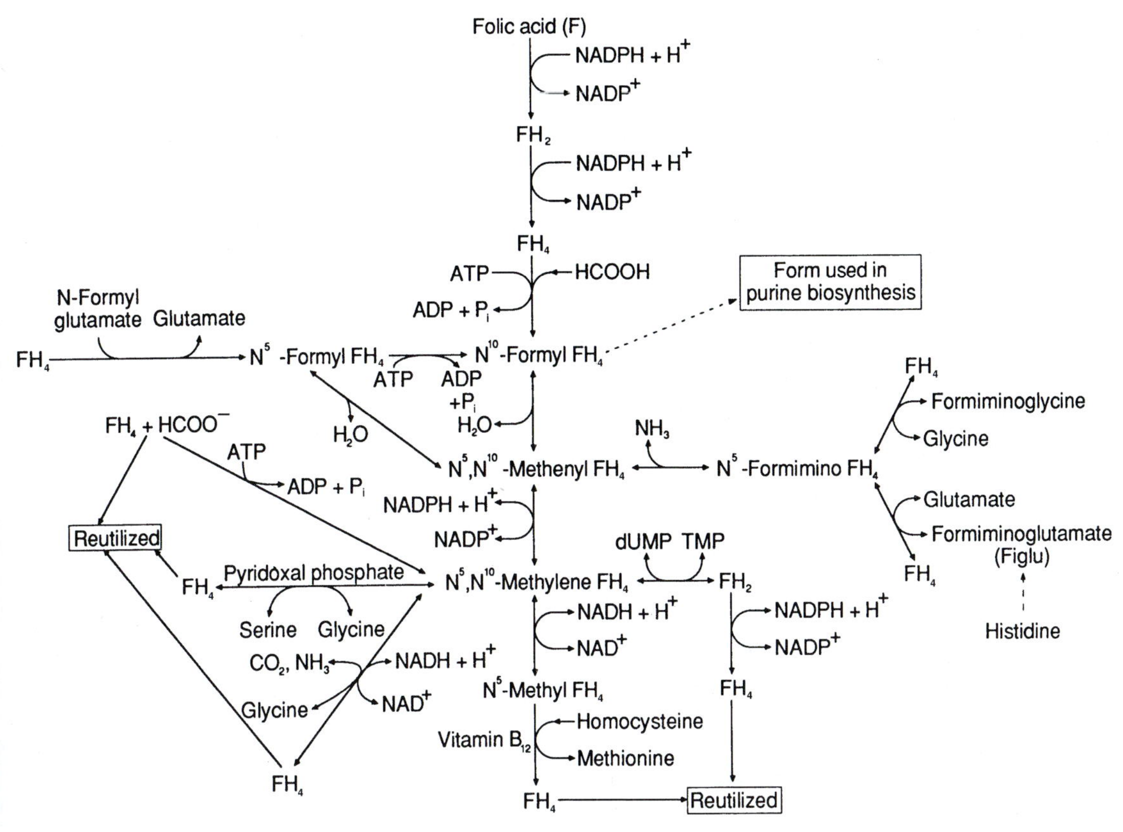

FIGURE 27-5
One-carbon transfer reactions involving folate-derived carriers. Dashed arrows indicate multiple-step reaction pathways; solid arrows represent direct, single-step reactions.

reactions are shown in Figure 27-5. Serine, a nonessential amino acid, is the major source of one-carbon fragments. Some key processes that utilize folate-mediated one-carbon transfer reactions are as follows:

1. Synthesis of purine nucleotides (N^{10}-formyl FH_4);
2. Methylation of deoxyuridylate to thymidylate (N^5,N^{10}-methylene FH_4) (an essential step in the biosynthesis of DNA; impairment of this reaction is responsible for the main clinical manifestations of folate deficiency);
3. Synthesis of formylmethionyl-tRNA (N^{10}-formyl FH_4) required for initiation of protein synthesis in prokaryotes (Chapter 25) and in mitochondria; and
4. Conversion of homocysteine to methionine (N^5-methyl FH_4) (Chapter 17).

Inherited disorders of folate transport and metabolism include defects in folate carrier (hereditary folate malabsorption, previously discussed), deficiency of N^5,N^{10}-methylene FH_4 reductase (Chapter 17), or functional deficiency of N^5-methyl FH_4 methyltransferase due to defects of vitamin B_{12} metabolism (Chapter 38) or formiminotransferase (previously discussed).

Several observational studies have shown that dietary supplementation of folate in women of childbearing age who might become pregnant can reduce the risk of fetal neural tube defects (e.g., spina bifida).

27.2 Formation of 5-Phosphoribosyl-1-Pyrophosphate

5-Phosphoribosyl-1-pyrophosphate (PRPP) is a key intermediate in nucleotide biosynthesis. It is required for *de novo* synthesis of purine and pyrimidine nucleotides and the salvage pathways, in which purines are converted to their respective nucleotides via transfer of ribose 1-phosphate group from PRPP to the base; that is,

$$\text{Purine} + \underset{\text{(PRPP)}}{\text{P-ribose-P-P}} \rightarrow \text{purine-ribose-P} + \text{PP}_i$$

PRPP is synthesized from ribose 5-phosphate by the following reaction:

$$\text{Ribose 5-phosphate} \xrightarrow[\text{PRPP synthetase}]{\text{ATP} \quad Mg^{2+}, P_i \quad \text{AMP}} \text{PRPP}$$

In this reaction, the pyrophosphate group of ATP is transferred to ribose 5-phosphate; the product PRPP is a high-energy compound. PRPP synthetase has an absolute requirement for inorganic phosphate (P_i), which functions as an allosteric activator. The enzyme is inhibited by many nucleotides, the end products of the pathway for which PRPP is an essential substrate. The gene for PRPP synthetase is located on the X chromosome. Mutations in this gene have given rise to PRPP synthetase variants with increased catalytic activity, which leads to overproduction of uric acid (discussed below, under "Gout"). The ribose 5-phosphate needed comes from three sources:

1. *The pentose phosphate pathway.* This pathway is probably the major source of ribose 5-phosphate in liver and in bone marrow.
2. The *uronic acid pathway* (Chapter 15). Xylulose 5-phosphate produced in this pathway is converted to ribose 5-phosphate as follows:

$$\text{D-Xylulose 5-phosphate} \xrightarrow{\text{epimerase}} \text{D-Ribulose 5-phosphate} \xrightarrow{\text{isomerase}} \text{D-ribose 5-phosphate}$$

   The contribution of this pathway to the supply of ribose 5-phosphate may be a minor one. The epimerase and isomerase are the same enzymes that participate in the pentose phosphate pathway.
3. *Glycolysis.* Ribose 5-phosphate can also be produced from intermediates of glycolysis (Figure 27-6). The enzymes involved are those of the nonoxidative phase of the pentose phosphate pathway, that occur in many tissues.

27.3 Biosynthesis of Purine Nucleotides

Purine nucleotides can be produced by two different pathways. The salvage pathway utilizes free purine bases and converts them to their respective ribonucleotides by appropriate phosphoribosyltransferases. The *de novo* pathway utilizes glutamine, glycine, aspartate, N^{10}-formyl FH_4, bicarbonate, and PRPP in the synthesis of inosinic acid (IMP), which is then converted to AMP and GMP.

De Novo Synthesis

The atoms of the purine ring originate from a wide variety of sources (Figure 27-7). C_2 and C_8 are derived from

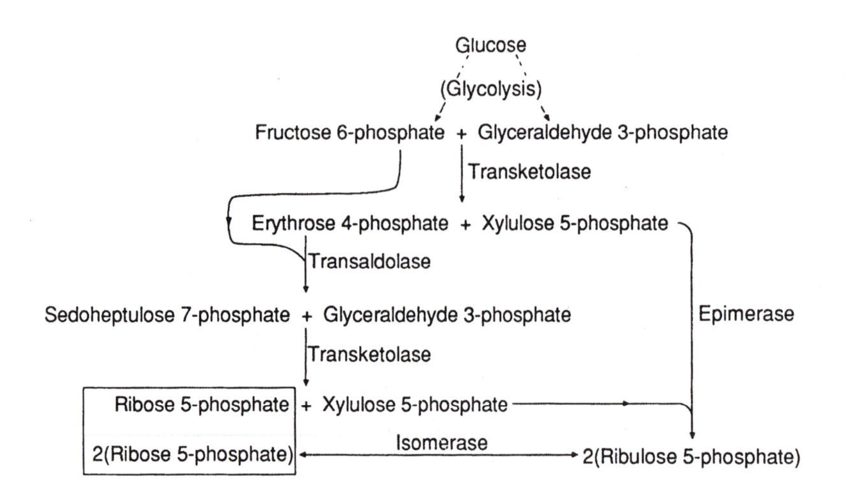

FIGURE 27-6
Synthesis of ribose 5-phosphate from intermediates of glycolysis by enzymes of the nonoxidative phase of the pentose phosphate pathway. (See Chapter 15.)

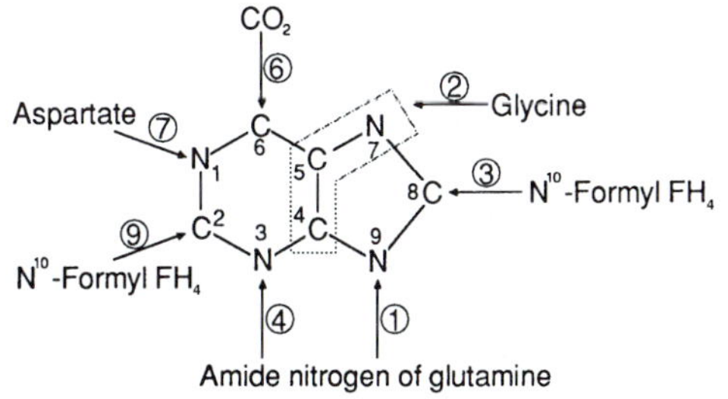

FIGURE 27-7
Sources of the atoms in *de novo* synthesis of the purine ring. The numbered reactions correspond to the reaction steps in Figure 27-8.

N^{10}-formyl FH_4. Glycine is incorporated entirely and provides C_4, C_5, and N_7. The α-amino group of aspartic acid provides N_1. The amide group of glutamine contributes N_3 and N_9. Carbon dioxide (or HCO_3^-) provides C_6. The stepwise synthesis of IMP is depicted in Figure 27-8. The following features of purine biosynthesis should be noted.

1. The glycosidic bond is formed when the first atom of the purine ring is incorporated (reaction 1, Figure 27-7). Free purines or purine nucleosides are not intermediates of this pathway. In pyrimidine biosynthesis, the pyrimidine ring is formed completely before addition of ribose 5-phosphate.
2. The biosynthesis is accomplished by several ATP-driven reactions. Ring formation (cyclization) is achieved by incorporating nucleophilic amino groups and electrophilic carbonyl groups at the appropriate positions.
3. The first purine nucleotide synthesized is inosinic acid (IMP), which is the precursor for the synthesis of

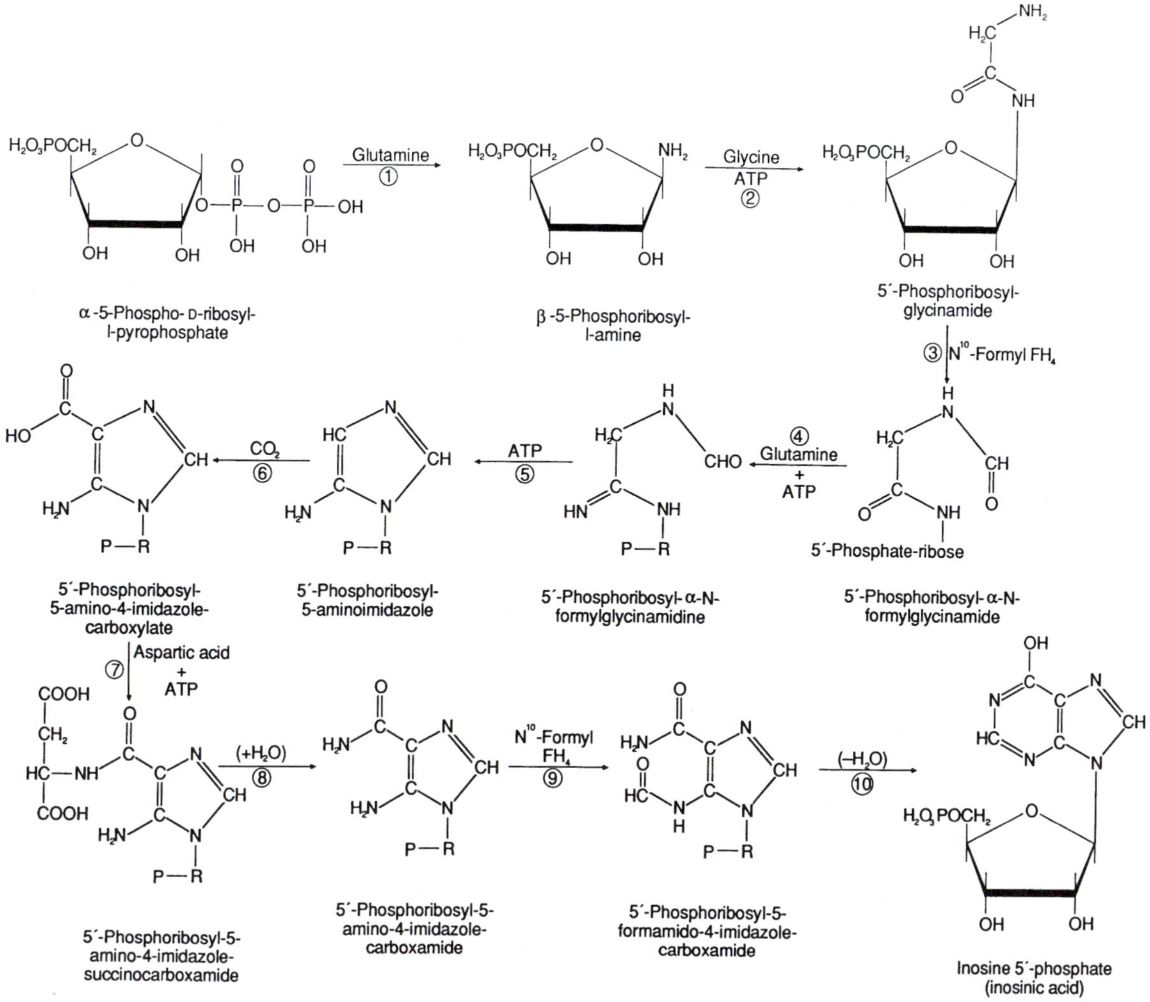

FIGURE 27-8
Biosynthesis of inosinic acid. The reactions are numbered. Enzymes: (1) amidophosphoribosyltransferase; (2) phosphoribosylglycinamide synthetase; (3) phosphoribosylglycinamide formyltransferase; (4) phosphoribosylformylglycinamidine synthetase; (5) phosphoribosylaminoimidazole synthetase; (6) phosphoribosylaminoimidazole carboxylase; (7) phosphoribosylaminoimidazolesuccinocarboxamide synthetase; (8) adenylosuccinate lyase; (9) phosphoribosylaminoimidazolecarboxamide formyltransferase; (10) IMP cyclohydrolase.

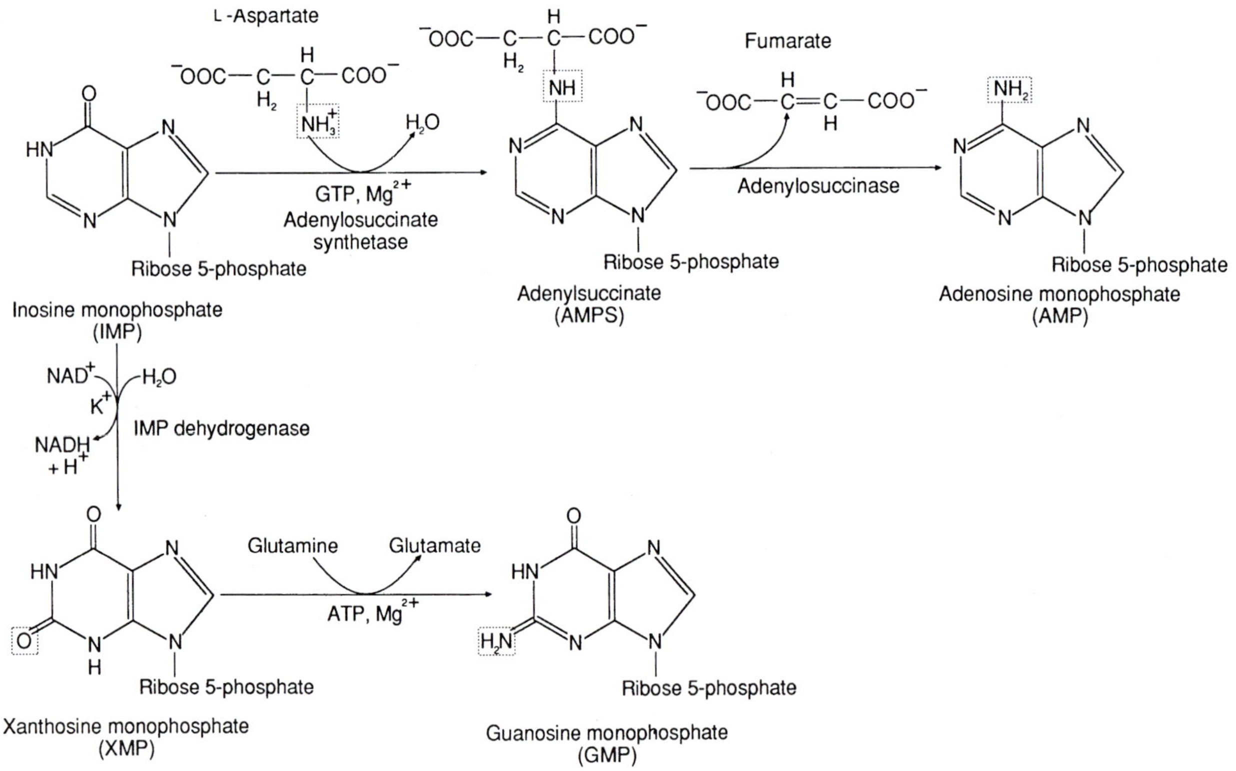

FIGURE 27-9
Biosynthesis of AMP and GMP from IMP.

adenylic acid (AMP) and guanylic acid (GMP) in two different pathways (Figure 27-9).

4. All the required enzymes occur in the cytosol. This is also true for enzymes of salvage pathways, nucleotide interconversion, and degradation.
5. *De novo* synthesis is particularly active in the liver and placenta. Nonhepatic tissues (e.g., bone marrow) depend on preformed purines that are synthesized in the liver and transported by red blood cells. They are very effective in salvaging the purines and exhibit little or no activity of xanthine oxidase, which oxidizes free purines.

Salvage Pathways

Reutilization of purine bases after conversion to their respective nucleotides constitutes salvage pathways. These pathways are particularly important in extrahepatic tissues. Purines arise from several sources: intermediary metabolism of nucleotides, degradation of polynucleotides, and dietary intake. Quantitatively, the first two sources are the more important. Salvage occurs mainly by the phosphoribosyltransferase reaction:

$$\text{Base} + \text{PRPP} \rightleftharpoons \text{base-ribose-phosphate} + \text{PP}_i$$

Human tissue contains two phosphoribosyltransferases (Figure 27-10). Adenine phosphoribosyltransferase (APRT) catalyzes the formation of AMP from adenine. Hypoxanthine guanine phosphoribosyltransferase (HPRT) catalyzes the formation of IMP and GMP from hypoxanthine and guanine, respectively. HPRT also catalyzes the conversion of other purines (6-thiopurine, 8-azaguanine, allopurinol) to their respective ribonucleotides. The K_{eq} of both phosphotransferases favors formation of ribonucleotides. During salvage, only two high-energy bonds are used, whereas in *de novo* synthesis of AMP or GMP, at least six high-energy bonds are expended. Deficiencies of these enzymes are discussed later.

A quantitatively less significant salvage pathway uses purine nucleoside phosphorylase and nucleoside kinase:

$$\text{Base} + \text{ribose 1-phosphate} \xrightleftharpoons{\text{nucleoside phosphoryiase}} \text{base-ribose} + \text{P}_i$$

$$\text{Base-ribose} + \text{ATP} \xrightarrow{\text{nucleoside kinase}} \text{base-ribose-phosphate} + \text{ADP}$$

The phosphorylase can catalyze the formation of inosine or deoxyinosine, and of guanosine or deoxyguanosine, but not adenosine or deoxyadenosine. However, the last two nucleosides can be converted to inosine and deoxyinosine by adenosine deaminase. The normal function of the phosphorylase appears to be the formation

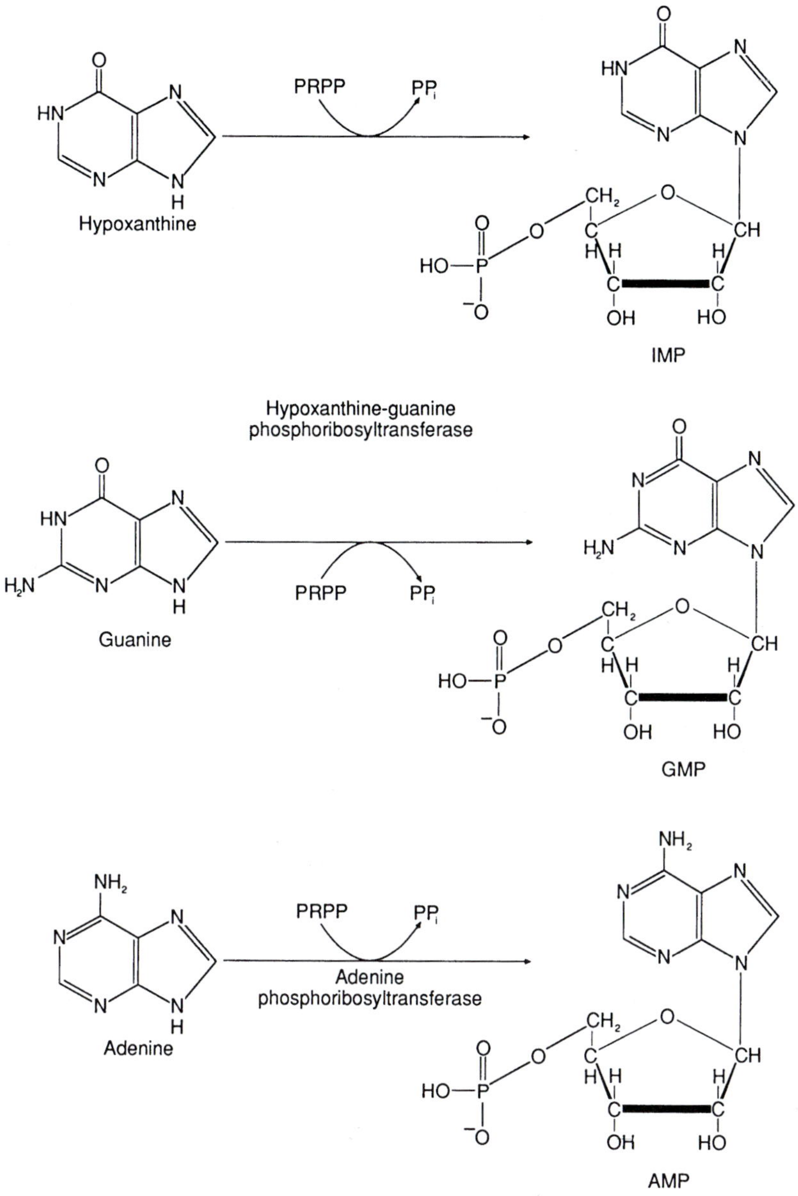

FIGURE 27-10
Salvage pathways of purine nucleotide synthesis. The preformed purines can be converted to mononucleotides in a single step, using PRPP.

of free hypoxanthine and guanine for conversion to uric acid. Deficiency of adenosine deaminase or purine nucleoside phosphorylase results in immunodeficiency diseas (Chapter 35).

In muscle, a unique nucleotide reutilization pathway, known as the *purine nucleotide cycle,* uses three enzymes: myoadenylate deaminase, adenylosuccinate synthetase, and adenylosuccinate lyase. In this cycle, AMP is converted to IMP with formation of NH_3, and IMP is then reconverted to AMP. Myoadenylate deaminase deficiency produces a relatively benign disorder of muscle (Chapter 21), which is characterized by muscle fatigue following exercise (see Myoadenylate Deaminase Deficiency).

Nucleoside kinases specific for inosine or guanosine have been described. Adenosine kinase is widely distributed in mammalian tissues.

Dietary Purines

Purines derived from food do not participate significantly in the salvage pathways described above; they are mostly

Nucleoproteins
↓ Proteolytic enzymes → Amino acids
Nucleic acids
↓ Nucleases and phosphodiesterases
Nucleotides
↓ Nucleotidases, phosphatases → Phosphate
Nucleosides
↓ Phosphate; Nucleoside phosphorylase → Ribose 1-phosphate
Free bases
↓ Xanthine oxidase
Uric acid

(Excreted in urine, or catabolized by bacteria in the large intestine)

FIGURE 27-11
Fate of dietary nucleoproteins. Only the predominant reactions are shown. Dietary purines are mostly converted to uric acid.

converted to uric acid (Figure 27-11). Dietary nucleic acids, present predominantly as nucleoproteins, are converted in the small intestine to nucleic acids and protein digestive products by the action of proteolytic enzymes (Chapter 12). The liberated nucleic acids are depolymerized to nucleotides by pancreatic nucleases and phosphodiesterases. The nucleotides are hydrolyzed to nucleosides by nucleotidases and phosphatases. The nucleosides are either absorbed or further cleaved to free bases by nucleoside phosphorylase. The free bases may be oxidized to uric acid by xanthine oxidase. Nucleoside phosphorylase and xanthine oxidase are very active in human small intestinal mucosa. Uric acid produced in the small intestine may be absorbed and excreted in urine, or it may be further catabolized by bacteria in the large intestine. In experiments in which normal or gouty subjects ingested ^{15}N-labeled nucleic acids, the labeled purines were converted to uric acid mainly by direct oxidative pathways without prior incorporation in tissue nucleic acids. Dietary purines (and pyrimidines; see below) do not serve significantly as precursors of cell nucleic acids in the body.

27.4 Conversion of Nucleoside Monophosphates to Diphosphates and Triphosphates

The triphosphates of nucleosides and deoxynucleosides are substrates for RNA polymerases and DNA polymerases, respectively. They are formed from the monophosphates in two stages. Conversion to diphosphates is catalyzed by kinases. These enzymes are base-specific but not sugar-specific. ATP is the usual source of phosphate; in some cases, other triphosphates or dATP may be used. Typical reactions are as follows:

$$\underset{\text{(dGMP)}}{\text{GMP}} + \text{ATP} \xrightarrow{\text{guanylate kinase}} \underset{\text{(dGDP)}}{\text{GDP}} + \text{ADP}$$

$$\text{AMP} + \text{ATP} \underset{}{\overset{\text{adenylate kinase}}{\rightleftharpoons}} \text{ADP} + \text{ADP}$$

The diphosphates are converted to the triphosphates by the ubiquitous enzyme nucleoside diphosphate kinase. Remarkably, the lack of base or sugar specificity applies to the phosphate acceptor and the phosphate donor. Hence, the general reaction is

$$\underset{\text{(acceptor)}}{\text{dXDP}} + \underset{\text{(donor)}}{\text{dYTP}} \xrightarrow{\text{nucleoside diphosphate kinase}} + \text{dXTP} + \text{dYDP}$$

Recall that conversion of ADP to ATP occurs mostly by mitochondrial oxidative phosphorylation coupled to electron transport (Chapter 14).

27.5 Formation of Purine Deoxyribonucleotides

Conversion of ribonucleotides to the deoxy forms occurs exclusively at the diphosphate level. Ribonucleoside diphosphate reductase (ribonucleotide reductase) catalyzes the reaction. This enzyme is found in all species and tissues. The immediate source of reducing equivalents is the enzyme (E) itself in which two sulfhydryl groups are oxidized to a disulfide. The general reaction is

$$\text{Ribonucleoside diphosphate} + \text{E(SH)}_2 \xrightarrow{\text{ribonucleoside diphosphate reductase}} \text{deoxyribonucleoside diphosphate} + \text{E(S-S)} + \text{H}_2\text{O}$$

In this reaction, the 2′-hydroxyl group of ribose is replaced by a hydrogen, with retention of configuration at the 3′-carbon atom:

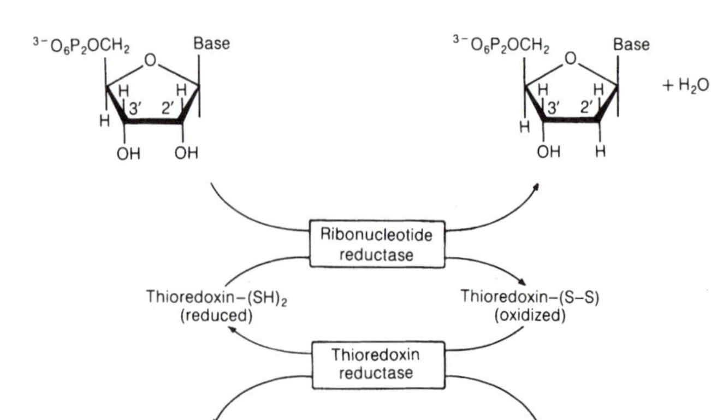

Regeneration of the ribonucleotide reductase is accomplished in *Escherichia coli* and in mammals by thioredoxin, a dithiol polypeptide (M.W. 12,000) coenzyme, which also plays a role in other protein disulfide reductase reactions. In thioredoxin, two cysteine residues in the sequence –Cys–Gly–Pro–Cys are converted to cystine. Reduced thioredoxin is regenerated by thioredoxin reductase, a flavoprotein enzyme that uses NADPH + H^+.

E. coli mutants unable to synthesize thioredoxin are still able to form deoxyribonucleotides. In these bacteria, a related substance, glutaredoxin, and two molecules of glutathione carry out the reduction. In *Lactobacillus*, the triphosphate is reduced and vitamin B_{12} is an essential coenzyme. Another example of this use of vitamin B_{12} is in *Euglena*, where the diphosphates are reduced. The mammalian system is nearly identical to that of *E. coli*.

Ribonucleotide reductase from *E. coli* consists of two subunits, B_1 and B_2, neither of which possesses catalytic function. B_1 is a dimer in which each monomer contains a substrate binding site and two types of allosteric effector binding sites. One type of effector site confers substrate specificity and the other is regulatory. B_1 contains a pair of sulfhydryl groups that are required for catalytic activity. B_2 also is a dimer, contains one nonheme Fe(III), and has an organic free- radical delocalized over the aromatic ring of a tyrosine residue in each of its polypeptide chains. The catalytic site is formed from the interaction of B_1 and B_2; a free-radical mechanism involves the tyrosyl residues, iron atom of B_2, and sulfhydryl groups of B_1. An antineoplastic agent, ***hydroxyurea,*** inhibits ribonucleotide reductase by inactivating the free radical.

Ribonucleotide reductase is regulated so as to ensure a balanced supply of deoxynucleotides for DNA synthesis. For example, if excess dATP is present, decreased synthesis of all the deoxyribonucleotides ensues, whereas ATP stimulates formation of dCDP and dUDP. Binding of TTP (also designated as dTTP) stimulates formation of dGDP and hence of dGTP; binding of dGTP stimulates the formation of dADP and hence of dATP. In this way, these nucleotide effectors, by binding to various regulatory sites, tend to equalize the concentrations of the four deoxyribonucleotides required for DNA synthesis.

27.6 Regulation of Purine Biosynthesis

Regulation of *de novo* purine biosynthesis is essential because it consumes a large amount of energy as well as of glycine, glutamine, N^{10}-formyl FH_4, and aspartate. Regulation occurs at the PRPP synthetase reaction, the amidophosphoribosyltransferase reaction, and the steps involved in the formation of AMP and GMP from IMP.

PRPP Synthetase Reaction

PRPP synthetase requires inorganic phosphate as an allosteric activator. Its activity depends on intracellular concentrations of several end products of pathways in which PRPP is substrate. These end products are purine and pyrimidine nucleotides (Figure 27-12).

Increased levels of intracellular PRPP enhance *de novo* purine biosynthesis. For example, in patients with HPRT deficiency, the fibroblasts show accelerated rates of purine formation. Several mutations of PRPP synthetase, which exhibit increased catalytic activity with increased production of PRPP, have been described in gouty subjects.

Amidophosphoribosyltransferase Reaction

This reaction is the first and uniquely committed reaction of the *de novo* pathway and the rate-determining step. Amidophosphoribosyltransferase is an allosteric enzyme and has an absolute requirement for a divalent cation. The enzyme is inhibited by AMP and GMP, which bind at different sites. The enzyme also is inhibited by pyrimidine nucleotides at relatively high concentrations.

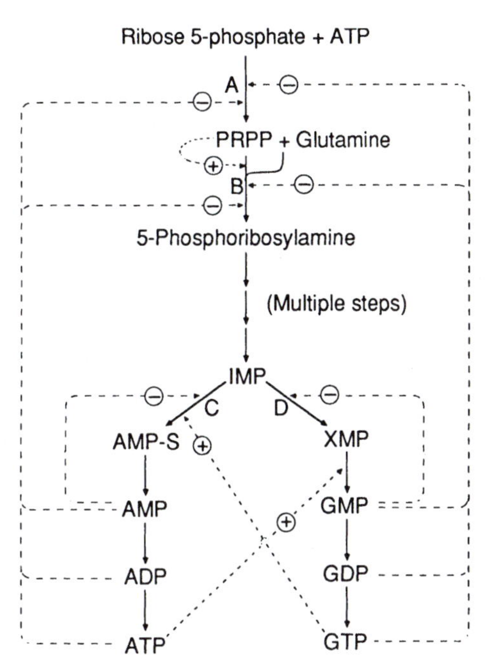

FIGURE 27-12
Feedback regulation of the *de novo* pathway of purine biosynthesis. Solid lines represent metabolic pathways, and broken lines represent sites of feedback regulation. ⊕, Stimulatory effect; ⊖, inhibitory effect. Regulatory enzymes: A, PRPP synthetase; B, amidophosphoribosyltransferase; C, adenylosuccinate synthetase; D, IMP dehydrogenase.

Inhibition by AMP and GMP, is competitive with respect to PRPP. The human placental enzyme exists in a small form (M.W. 133,000) and a large form (M.W. 270,000). The small form is catalytically active. Ribonucleotides convert the active form to the large form, whereas PRPP does the opposite. The regulatory actions of PRPP synthetase and amidophosphoribosyltransferase are coordinated. When there is a decrease in the intracellular concentration of adenine ribonucleotides, PRPP synthetase is activated; this results in increased synthesis of PRPP, which in turn converts the inactive form of amidophosphoribosyltransferase to the active form and increases production of purine nucleotides.

Regulation of Formation of AMP and GMP from IMP

In the formation of AMP and GMP from IMP, ATP is required for GTP synthesis, and GTP is needed to form ATP (Figure 27-9). In addition, adenylosuccinate synthetase is inhibited by AMP, and IMP dehydrogenase is inhibited by GMP (Figure 27-12).

27.7 Inhibitors of Purine Biosynthesis

A variety of inhibitors of purine biosynthesis function at different stages and are used as antimicrobial, anticancer, and immunosuppressive agents.

Inhibitors of Folate Biosynthesis

The ***sulfonamide*** drugs were the first effective antibacterial agents to be employed systemically in humans. These drugs resemble p-aminobenzoic acid in structure and inhibit utilization of that compound for the synthesis of folate in bacteria. Sulfonamides do not interfere with human metabolism.

Inhibitors of Formation of IMP

Folate analogues, such as methotrexate (Figure 27-3), are folate antagonists. They block production of FH_2 and FH_4 by dihydrofolate reductase and lead to diminished purine biosynthesis (inhibition of reactions 3 and 9 in Figure 27-8). Methotrexate also affects metabolism of amino acids and pyrimidine (inhibition of thymidylate synthesis) and inhibits DNA, RNA, and protein synthesis. It is effective in the treatment of breast cancer, cancer of the head and neck, choriocarcinoma, osteogenic sarcoma, and acute forms of leukemia. High doses of methotrexate can be tolerated provided that the patient also receives folinic acid (N^5-formyl FH_4), which decreases damage to bone marrow and prevents the development of leukopenia and thrombocytopenia. Methotrexate also produces mucositis, gastrointestinal symptoms, and liver damage by a direct toxic effect on hepatic cells. Chronic oral administration of methotrexate in psoriasis has been associated with an increased incidence of postnecrotic cirrhosis.

The "rescue" of normal, but not tumor, cells from methotrexate toxicity by folinic acid is partly explained by differences in membrane transport. For example, osteogenic sarcoma cells (which do not respond to conventional doses of methotrexate treatment) are not rescued by folinic acid administered after methotrexate, presumably owing to the absence of transport sites for folinic acid in the neoplastic cells. The therapeutic effects of administration of methotrexate and "rescue" with folinic acid are superior to those of methotrexate alone. Resistance to methotrexate can develop from increased activity of dihydrofolate reductase, synthesis of an enzyme having a lower affinity for the inhibitor, decreased transport of the drug into tumor cells, decreased degradation of the reductase, and genetic amplification of the gene for dihydrofolate reductase.

Azaserine (L-Serine diazoacetate), isolated from a species of *Streptomyces,* is a structural analogue of glutamine

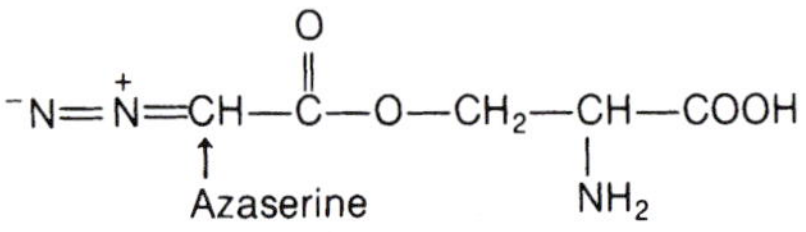

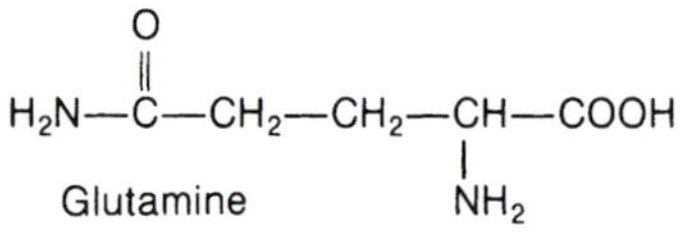

$$^{-}N{=}\overset{+}{N}{=}\underset{\uparrow}{C}H{-}\overset{O}{\overset{\|}{C}}{-}O{-}CH_2{-}\underset{NH_2}{\underset{|}{C}H}{-}COOH$$

Azaserine

$$H_2N{-}\overset{O}{\overset{\|}{C}}{-}CH_2{-}CH_2{-}\underset{NH_2}{\underset{|}{C}H}{-}COOH$$

Glutamine

that inhibits steps 1 and 4 of inosinic acid synthesis (Figure 27-8) and the conversion of XMP to GMP (Figure 27-9). A related substance, 6-diazo-5-oxo-L-norleucine (DON), is also a glutamine analogue. Both compounds function as alkylating agents and become attached to an essential sulfhydryl group (a nucleophilic group) of the enzyme to form an inactive thioether derivative. The linkage occurs via the electrophilic carbon atom (designated by the arrow) of azaserine and release of N_2.

Inhibitors of Formation of AMP and GMP

Hadacidin (N-formyl-N-hydroxyglycine) is an

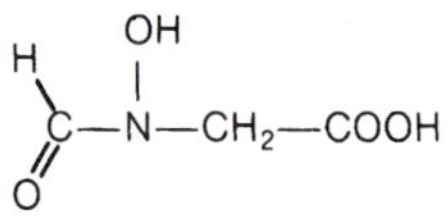

$$H{-}\underset{O}{\underset{\|}{C}}{-}\overset{OH}{\overset{|}{N}}{-}CH_2{-}COOH$$

analogue of aspartic acid isolated from fungi. By competing with aspartate, it inhibits adenylosuccinate synthetase and hence the synthesis of AMP, it is an experimental antineoplastic agent.

Mycophenolic acid and ***ribavarin monophosphate*** inhibit IMP dehydrogenase and hence GMP synthesis.

Inhibitors of Multiple Steps—Purine Analogues

6-Mercaptopurine is a structural analogue of hypoxanthine and is converted to thio-IMP in the salvage pathway.

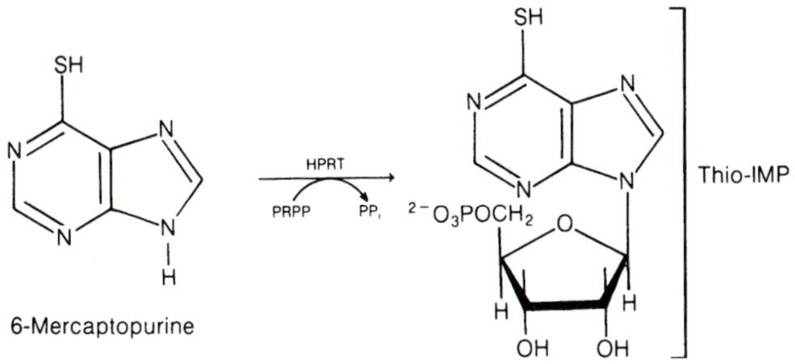

6-Mercaptopurine

HPRT

PRPP → PP_i

Thio-IMP

Thio-IMP prevents production of AMP and GMP by inhibiting the following reactions:

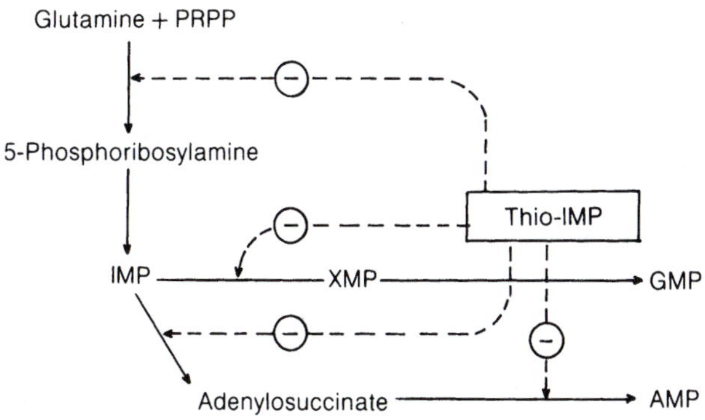

Glutamine + PRPP

5-Phosphoribosylamine

IMP → XMP → GMP

Adenylosuccinate → AMP

Thio-IMP

6-Mercaptopurine is used in the treatment of acute leukemias. However, resistant tumor cells develop rapidly, probably because of altered specificity or lack of phosphoribosyltransferases, so that thio-IMP (the active inhibitor) is not formed. In support of this mechanism, resistant cells respond to 6-methylmercaptopurine ribonucleoside, which is phosphorylated to the corresponding nucleotide. Other mechanisms may include altered cell permeability and increased rate of destruction of 6-mercaptopurine.

6-Mercaptopurine is partially metabolized to 6-thiouric acid (which lacks antitumor activity) by xanthine oxidase. Allopurinol, a xanthine oxidase inhibitor used in the treatment of gout, potentiates the action of 6-mercaptopurine by preventing its conversion to 6-thiouric acid. This effect is taken into consideration in treatment. Mercaptopurines are also inactivated by S-methylation carried out by thiopurine S-methyltransferase (TPMT), particularly in hematopoietic tissues which lack xanthine oxidase. Deficiency of TPMT due to polymorphisms causes profound toxicity with the regular therapeutic regime. This is another example of the use of pharmacogenomics.

Azathiopurine, a derivative of 6-mercaptopurine, functions as an antiproliferative agent. Its principal use is as an immunosuppressive agent. It presumably releases 6-mercaptopurine in the body by reacting with sulfhydryl compounds such as GSH.

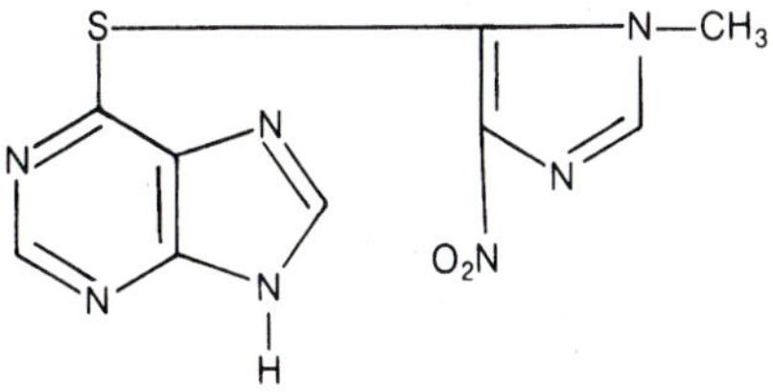

Azathiopurine

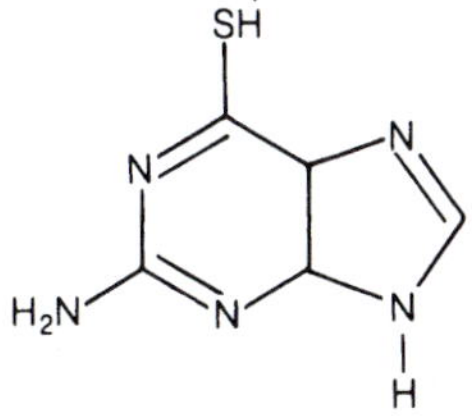

6-Thioguanine

6-Thioguanine is similar to 6-mercaptopurine in its action. The most active form is 6-thio-GMP, which inhibits guanylate kinase and, at higher concentrations, IMP dehydrogenase. Thio-IMP and thio-GMP also inhibit PRPP amidotransferase.

Vidarabine (adenine arabinoside), an analogue of adenosine, does not interfere with purine biosynthesis but affects DNA synthesis.

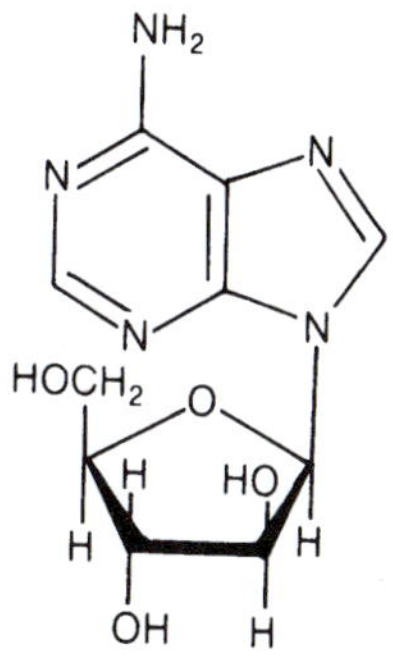

It is phosphorylated to adenine arabinoside triphosphate, which inhibits viral DNA polymerases, but not the

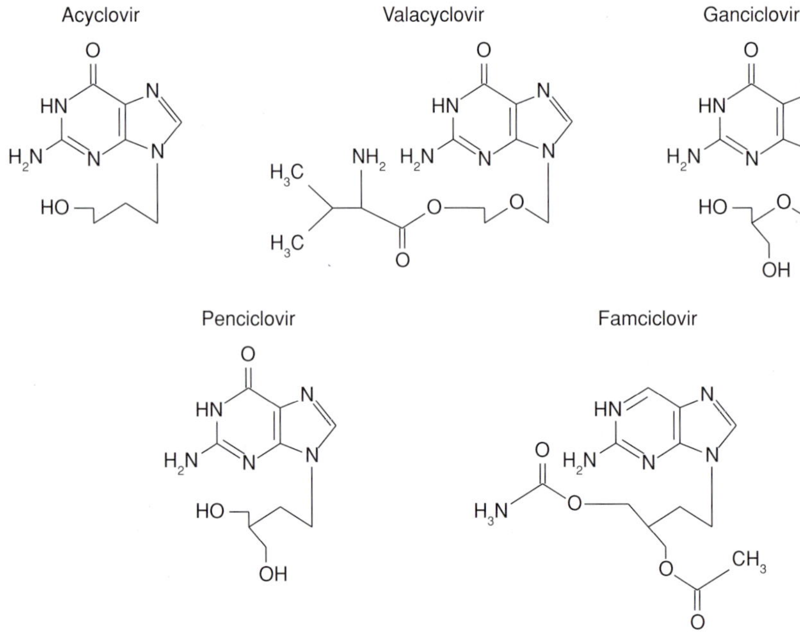

Acyclovir Valacyclovir Ganciclovir

Penciclovir Famciclovir

FIGURE 27-13
Structure of some of the purine nucleoside analogues used as antiviral agents.

mammalian enzymes by competition with dATP. Originally developed as an antileukemia drug, it has proved useful in the treatment of herpes virus infections. Several other purine nucleoside analogues are used as antiviral agents (Figure 27-13); these include acyclovir, valacyclovir, ganciclovir, penciclovir, and famiciclovir. After conversion to their respective triphosphate derivatives, these drugs inhibit viral DNA polymerase. Viruses inhibited by these drugs are herpes simplex, varicella zoster, cytamegalovirus, and hepatitis B. Antiviral agents that are not purine nucleoside analogues are amantadine analogues and α-interferon.

Inhibition of Conversion of Ribonucleoside Diphosphate to Deoxyribonucleoside Diphosphate

Hydroxyurea, an antineoplastic agent, acts by destroying an essential free radical in the active center of ribonucleotide reductase.

$$H_2N-\overset{\displaystyle O}{\overset{\|}{C}}-NHOH$$

An unexpected finding of hydroxyurea treatment is induction of hemoglobin F (HbF) production in red blood cells. Hydroxyurea therapy results in an increase in both content and the number of red blood cells that contain HbF (known as F cells). This property of hydroxyurea has been used in the treatment of ***sickle cell disease*** because of the antisickling effect of HbF (Chapter 28).

27.8 Catabolism of Purine Nucleotides

Degradation of purines and their nucleotides occurs during turnover of endogenous nucleic acids and degradation of ingested nucleic acids (Figure 27-11), during which most of the purines are converted to uric acid.

Degradation of purine nucleoside phosphates (AMP, IMP, GMP, and XMP) begins by hydrolysis by 5′-nucleotidase.(Figure 27-14) to produce adenosine, inosine, guanosine, and xanthosine, respectively, and phosphate. Adenosine is converted to inosine by adenosine deaminase (adenosine aminohydrolase). For inosine, guanosine, and xanthosine, the next step is catalyzed by purine nucleoside phosphorylase and involves a phosphorylation and a cleavage to produce ribose 1-phosphate and hypoxanthine, guanine, and xanthine, respectively. (The enzyme also acts on deoxyribonucleosides to release deoxyribose 1-phosphate.) The pentose sugars are metabolized further or excreted. Purine nucleoside phosphorylase functions in purine salvage, and deficiency of this enzyme results in decreased cell-mediated immunity. Hypoxanthine and guanine are converted to xanthine by xanthine oxidase and guanine aminohydrolase, respectively. Thus, all purine nucleosides produce xanthine. Xanthine oxidase is very active in the intestinal mucosa and the

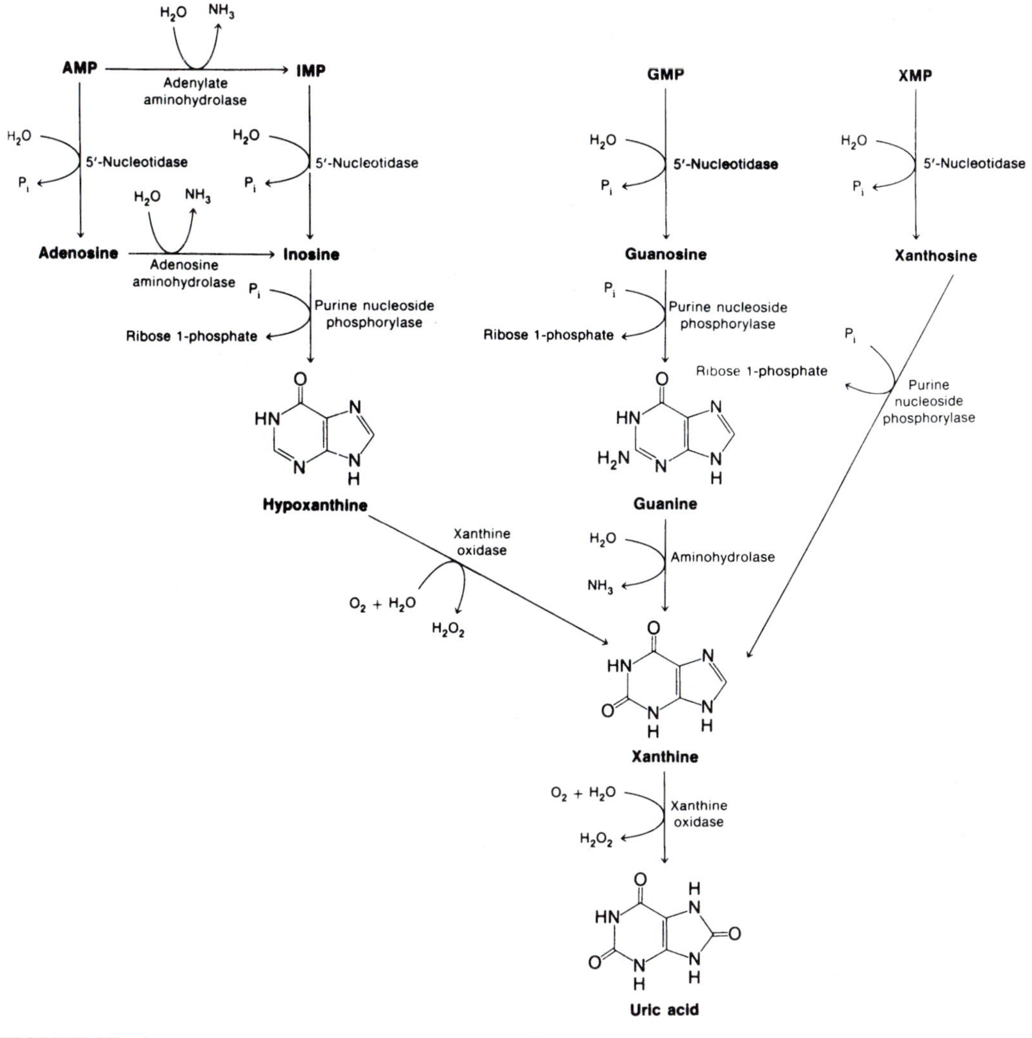

FIGURE 27-14
Pathways for the formation of uric acid from purine nucleotides. [Reproduced with permission from G. Zubay, *Biochemistry*, 2nd ed. New York: (Macmillan 1988). © 1988 by Macmillan Publishing Company.]

liver and is present in most tissues. The products of the xanthine oxidase reaction are uric acid and hydrogen peroxide. In primates (humans included) and birds, uric acid is excreted in the urine. In mammals other than primates, a liver enzyme, urate oxidase, converts uric acid to allantoin (Figure 27-15). In lower animals, allantoin is metabolized to allantoic acid (some fish), urea (most fish and amphibia), and ammonia (some marine invertebrates and crustaceans). In humans, uric acid, urea, and creatinine are the end products of nitrogen metabolism. All three are excreted in urine. The primary end product is urea (Chapter 17). Animals in which uric acid is the major end product excrete the rather insoluble uric acid as a semisolid mass, which allows for conservation of water. In humans, uric acid production and excretion is a balanced process. On a daily basis, about two-thirds of the uric acid produced is excreted by the kidneys and the remainder via intestinal bacterial degradation. In chronic renal failure and hyperuricemia, extrarenal uric acid disposal is enhanced. Renal excretion of uric acid is a multicomponent bidirectional process involving glomerular filtration (100%) and proximal tubular reabsorption (98–100%), followed by proximal tubular secretion (50%) and proximal tubular reabsorption (40–44%). Ultimately, the renal clearance of uric acid is only about 7–10% of creatinine clearance (Chapter 17).

Xanthine Oxidase Reaction

Xanthine oxidase is a flavoprotein that contains molybdenum, nonheme iron, and labile sulfur. The enzyme is present in two forms, one with dehydrogenase activity (xanthine dehydrogenase) and the other with oxidase activity. The former is converted to the latter by oxidation of

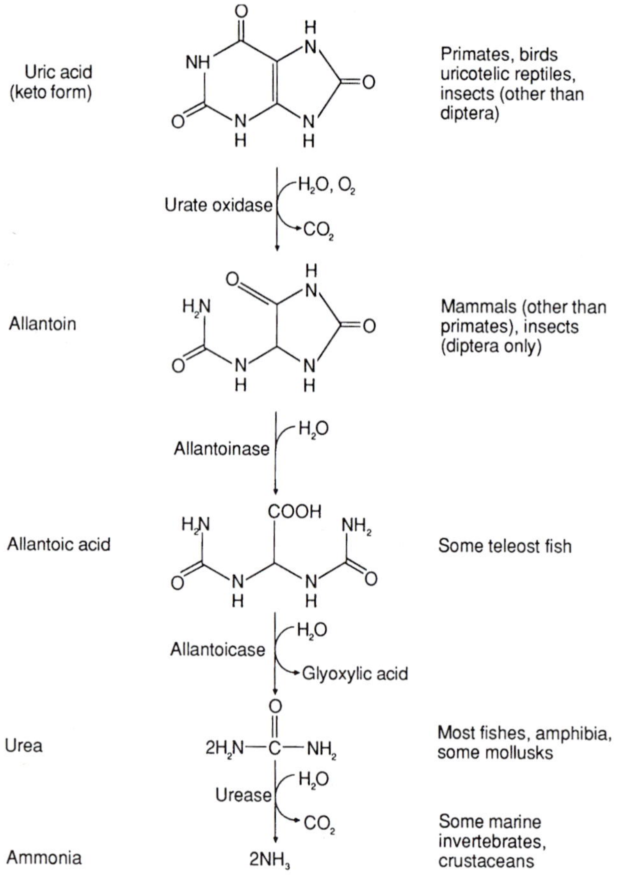

FIGURE 27-15
Metabolic degradation of uric acid in some animals.

thiol groups of the enzyme owing to the presence of high concentrations of oxygen. It is active on purines, aldehydes, and pteridines. The reactions catalyzed on purines are

$$\text{Hypoxanthine} + H_2O + O_2 \rightarrow \text{xanthine} + H_2O_2$$

and

$$\text{Xanthine} + H_2O + O_2 \rightarrow \text{uric acid} + H_2O_2$$

The reaction mechanism is incompletely understood. Molybdenum, an essential cofactor, is the initial acceptor of electrons during purine oxidation and undergoes reduction from Mo^{6+} to Mo^{4+}. Deficiency of molybdenum can result in ***xanthinuria.*** The electrons from molybdenum are passed successively to the iron-sulfur center, to FAD, and finally to oxygen. The oxygen incorporated into xanthine and uric acid originates in water. Xanthine oxidase also yields the superoxide radical, O_2^-, which is then converted to hydrogen peroxide by superoxide dismutase (Chapter 14). This may yield free radicals,

$$H_2O_2 + Fe^{2+} \rightarrow Fe^{3+} + OH^- + OH^{\cdot}$$

which can cause tissue destruction. Thus, the xanthine oxidase reaction can cause tissue injury during reperfusion of organs that have been deprived of oxygen (e.g., ischemic myocardium, transplanted organs). The xanthine oxidase reaction is further promoted in hypoxic organs by providing increased substrates owing to the enhanced adenine nucleotide breakdown that occurs during oxygen deprivation. Hydrogen peroxide is inactivated by catalase or by peroxidase. Xanthine oxidase is inhibited by allopurinol (a purine analogue), which is used in the treatment of gout.

Disorders of Purine Nucleotide Metabolism

Several disorders affect purine metabolism. They are gout and the syndromes associated with deficiency of HPRT, APRT, adenosine deaminase, nucleoside phosphorylase, myoadenylate deaminase, and xanthine oxidase.

Gout

Gout is a heterogeneous group of genetic and acquired diseases characterized by elevated levels of urates in blood (***hyperuricemia***) and of uric acid in urine (***uricuria***). Hyperuricemia in men is defined as serum urate concentration greater than 7 mg/dL (420 μmol/L) and in women 6 mg/dL (357 μmol/L,). In *gout,* hyperuricemia is a common biochemical occurrence; however, many hyperuricemic subjects may not develop clinical gout (asymptomatic hyperuricemia). All clinical symptoms of gout arise from the low solubility of urate in biological fluids. The maximum solubility of urate in plasma at 37°C is about 7 mg/dL. However, in peripheral structures and in the extremities, where the temperature is well below 37°C, the solubility of urate is decreased. When urate is present in supersaturated solutions, crystals of monosodium urate monohydrate form easily (Figure 27-16). Deposits of aggregated crystals, known as ***tophi,*** in and around joints of

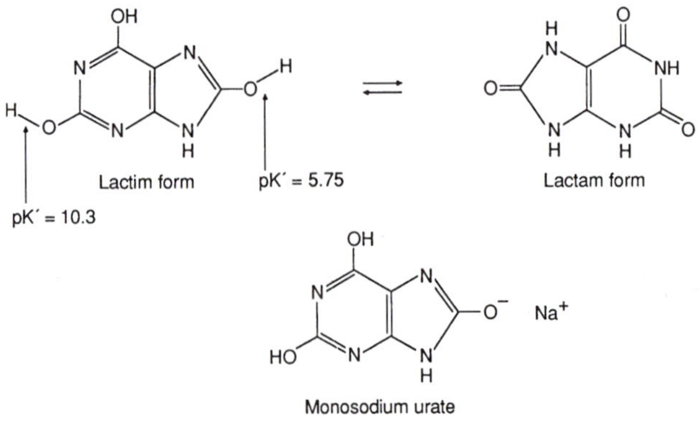

FIGURE 27-16
Structures of tautomeric forms of uric acid. In the lactim form, uric acid has two acidic hydrogens, with pK′ values of 5.75 and 10.3. At physiological pH (7.4), the predominant species of the lactim form is urate monoanion.

the extremities initiate an inflammatory foreign body reaction (acute arthritis) that involves leukocytes, complement, and other mediators (Chapter 35). This reaction causes severe pain, swelling, redness, and heat in the affected areas. Initial attacks are usually acute and frequently affect the metatarsophalangeal joint of one big toe. Tophi may also be present in subcutaneous tissues, cartilage, bone, and kidneys. Formation of urinary calculi (urolithiasis) of urate is common. Gout is potentially chronic and disfiguring.

Primary gout is a disorder of purine metabolism seen predominantly in men. The condition is multifactorial and involves genetic and nongenetic factors. Occurrence in women is uncommon; when it does occur, it is usually found in postmenopausal women. The blood urate concentration of normal men is ~1 mg/dL higher than that in women, but this difference disappears after the menopause. Thus, in women, the postmenopausal rise in serum urate levels may increase the risk of developing gout. Gout is very rare in children and adolescents.

Primary gout may be due to overproduction or underexcretion of uric acid or to a combination of both. Frequently, siblings and other close relatives of afflicted individuals have high levels of uric acid in blood but do not develop gout, indicating that hyperuricemia is not the only factor involved. Primary renal gout is due to underexcretion caused by a renal tubular defect in uric acid transport. Primary metabolic gout is due to overproduction of purines and uric acid. The prevalence of gout is high in some populations (e.g., 10% of adult male Maori of New Zealand). In Europe and the United States, the prevalence rate is 0.13–0.37%.

Secondary gout develops as a complication of hyperuricemia caused by another disorder (e.g., leukemia, chronic nephritis, polycythemia). This type of hyperuricemia usually is associated with abnormally rapid turnover of nucleic acids. The rare cases of gout in adolescents and children are usually of this type.

Overproduction of PRPP

The mechanism of the hyperuricemia in most individuals who have gout is unknown. Following is a discussion of biochemical lesions that lead to hyperuricemia and may eventually lead to gout. Enhanced PRPP synthesis results from X-chromosome-linked mutants of PRPP synthetase. Several variants show increased V_{max}, resistance to feedback inhibition, or a low K_m for ribose 5-phosphate.

PRPP levels can also be increased as a result of underutilization in purine salvage pathways. Thus, HPRT deficiency (partial or complete) causes hyperuricemia as an X-linked recessive trait. In situations in which ATP is consumed more rapidly than it is synthesized or in which ATP synthesis is impaired, ADP and AMP accumulate and eventually are converted to uric acid. Hyperuricemia can occur in hypoxic conditions (e.g., adult respiratory distress syndrome), glucose-6-phosphatase deficiency, and acute illness (e.g., severe hemorrhagic shock). Furthermore, hyperuricemia may serve as a marker for cellular energy crises. Ethanol ingestion causes hyperuricemia owing to increased degradation of ATP to AMP. The latter occurs in ethanol metabolism during the conversion of acetate to acetyl-CoA (Chapter 18). In all these instances of production of uric acid by xanthine oxidase, the formation of cytotoxic byproducts of the reaction, namely, hydrogen peroxide and superoxide radical, also is increased.

Patients with glucose-6-phosphatase deficiency (glycogen storage disease type 1; see Chapter 15) exhibit hyperuricemia from infancy and some develop gout later in life. The hyperuricemia is due to decreased excretion and increased production of uric acid. Because of their hypoglycemia, these patients develop marked hyperlactic acidemia, and lactate reduces the renal clearance of uric acid by suppressing its tubular secretion. Renal excretion of urate is complex, comprising glomerular filtration, tubular reabsorption, and tubular secretion. The urate in the urine is thought to arise almost entirely by tubular secretion. The increased production of uric acid in glucose-6-phosphatase deficiency has been attributed to enhanced ATP turnover and consequent adenine nucleotide depletion, with release of feedback inhibition of amidophosphoribosyltransferase, and to acceleration of *de novo* purine nucleotide synthesis. Diminished levels of ATP and P_i result from the presence of increased levels of phosphorylated glycolytic intermediates (Chapter 15). A similar mechanism has been proposed for the hyperuricemia that results when fructose is administered intravenously to humans.

Excessive production of organic acids leads to elevated levels of uric acid. Lactate, acetoacetate, and β-hydroxybutyrate (the latter two are known as ketone bodies) compete with uric acid for secretion by the kidney tubules. Lactic acidemia can occur in glucose-6-phosphatase deficiency and in alcohol ingestion. Ketonemia and ketonuria occur in untreated diabetes mellitus, starvation, glucose-6-phosphatase deficiency, etc. (Figure 27-17).

Treatment

A variety of drugs are used in the management of gout in three clinical situations:

1. To treat acute gout arthritis,
2. To prevent acute attacks, and
3. To lower serum urate concentrations.

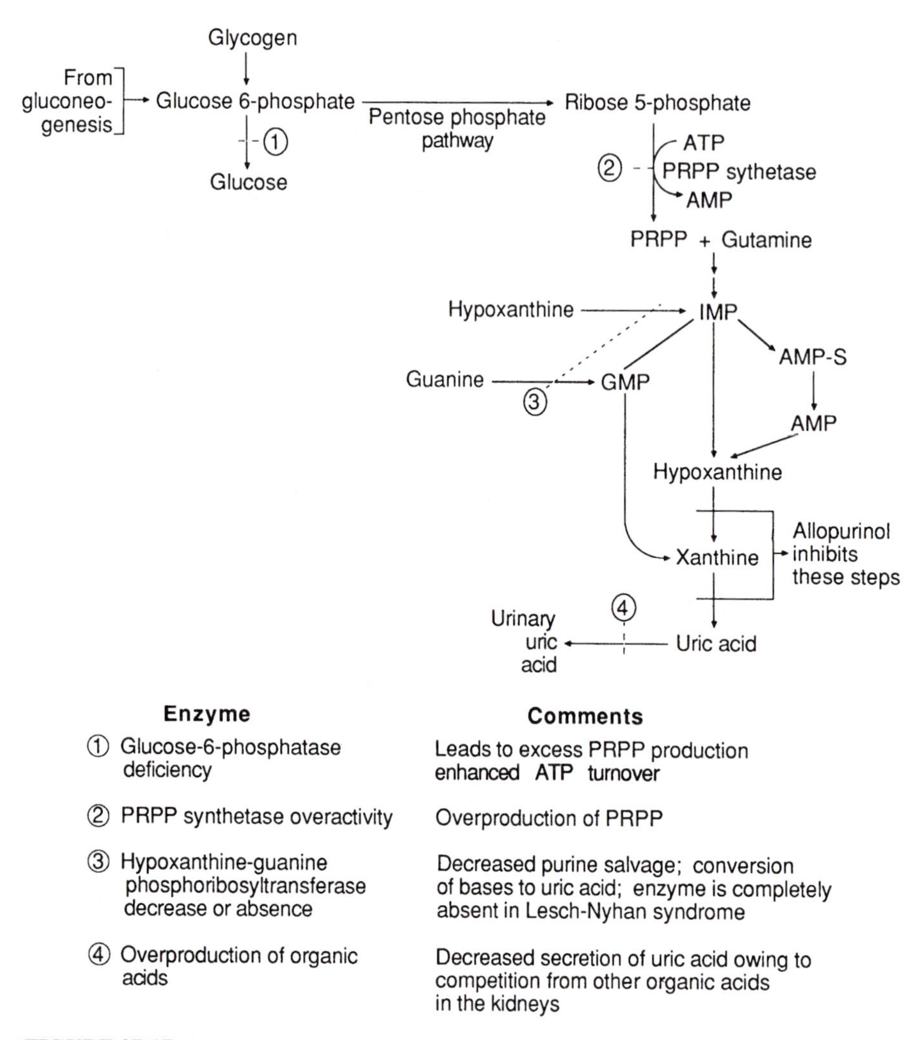

FIGURE 27-17
Composite diagram of the biochemical lesions that may lead to hyperuricemia.

Acute gouty attacks commonly affect the first metatarsal joint of the foot. In aspirated joint fluids, birefringent urate crystals can be seen in the polarized light microscope, which is used in the definitive diagnosis. The treatment of acute gout includes the administration of colchicine, nonsteroidal anti-inflammatory drugs (NSAIDs), corticosteroids, adrenocorticotropic hormone (ACTH), and analgesics. Colchicine and NSAIDs also can be used prophylactically to prevent acute attacks in patients with gout. Drugs used to lower serum urate concentrations include probenecid, sulfinpyrazone, and allopurinol.

Colchicine depolymerizes microtubules and structures (such as the mitotic spindle) consisting of microtubules. It is effective in decreasing pain and the frequency of attacks but its mechanism of action is obscure.

Allopurinol, an analogue of hypoxanthine, inhibits xanthine oxidase and reduces formation of xanthine and uric acid. It is converted by xanthine oxidase to alloxanthine, which binds tightly to the active site by chelation with Mo^{4+}. This greatly reduces reoxidation of Mo^{4+} to Mo^{6+} and affects catalytic activity (Figure 27-18). This type of inhibitor, in which a substrate analogue is converted to an inhibitor and not released from the active site, is known as a *suicide enzyme-inactivator* or a *mechanism-based inhibitor* (Chapter 6). Allopurinol also is converted to allopurinol ribonucleotide by HPRT. Uric acid production is decreased through depletion of PRPP. Allopurinol ribonucleotide also inhibits PRPP-amidotransferase allosterically. These effects are illustrated in Figure 27-19. Since xanthine is both a product and a substrate for xanthine oxidase, allopurinol therapy could be expected to cause accumulation of xanthine in the body. Since xanthine is only sparingly soluble in urine (but more soluble than uric acid), this could lead to urinary xanthine crystalluria, or stone formation. This complication has not been observed in patients who receive the drug for treatment of gout or uric acid stones, but it has occurred in a few patients with Lesch-Nyhan syndromeand lymphosarcoma.

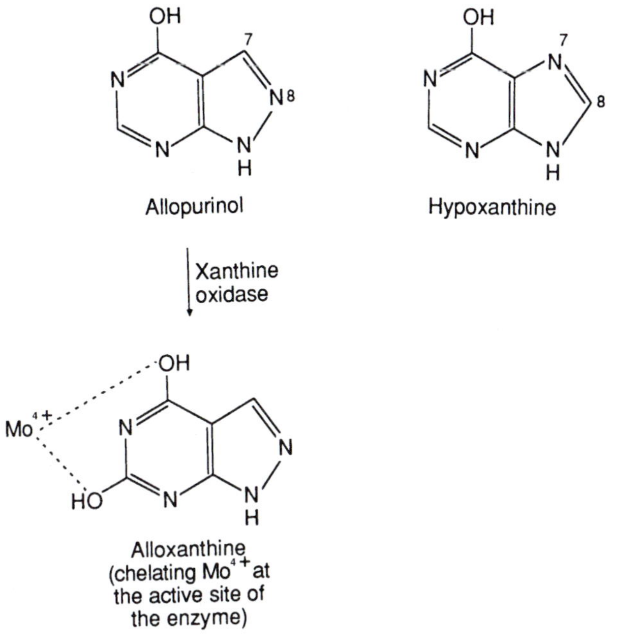

Allopurinol

Hypoxanthine

Xanthine oxidase

Alloxanthine (chelating Mo^{4+} at the active site of the enzyme)

FIGURE 27-18

Inhibition of xanthine oxidase by allopurinol, an analogue of hypoxanthine. N_7 and C_8 of hypoxanthine are reversed in allopurinol. Allopurinol, a suicide enzyme inactivator, is converted to alloxanthine, which binds tightly to the active site of the enzyme via Mo^{4+} and interrupts the reoxidation of Mo^{4+} to Mo^{6+} needed to initiate the next catalytic cycle.

Drugs that increase uric acid excretion in humans include ***probenecid,*** which is effective in the regulation of hyperuricemia and the resolution and prevention of tophi, and ***sulfinpyrazone,*** which has similar effects. Both agents are weak organic acids and probably act as competitive inhibitors of tubular reabsorption of uric acid.

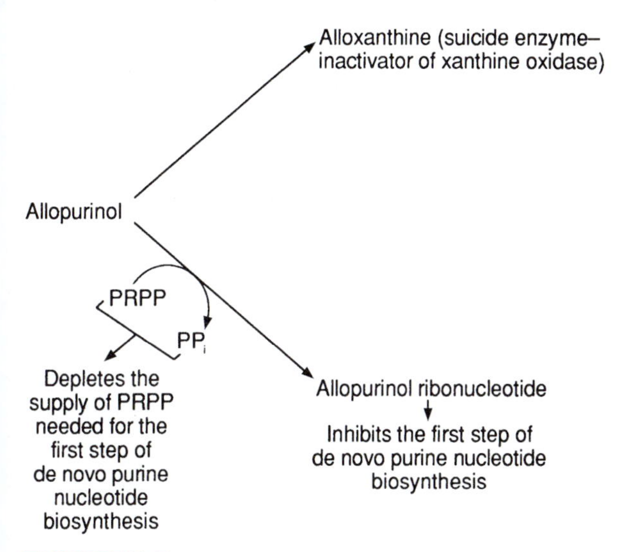

FIGURE 27-19

Multiple actions of allopurinol inhibit formation of uric acid.

Serum urate levels can be lowered by dietary and lifestyle changes. These include correction of obesity, avoidance of ethanol consumption, and avoidance of high-purine foods (e.g., organ meats). **Psuedogout** is a disorder caused by deposition of calcium pyrophosphate dihydrate usually in larger joints such as knee, wrist, and ankle.

Lesch-Nyhan Syndrome

Lesch-Nyhan syndrome is characterized by virtual absence of HPRT, excessive production of uric acid, and abnormalities of the central nervous system. These abnormalities include mental retardation, spasticity (increased muscle tension resulting in continuous increase of resistance to stretching), choreoathetosis (characterized by irregular, jerky, or explosive involuntary movements, and writhing or squirming, which may involve any extremity or the trunk), and a compulsive form of self-mutilation. The disorder associated with partial deficiency of HPRT also leads to hyperuricemia but lacks the devastating neurological and behavioral features characteristic of the Lesch-Nyhan syndrome. Both disorders are X-linked.

The hyperuricemia in Lesch-Nyhan patients is explained, at least in part, on the basis of intracellular accumulation of PRPP leading to increased purine nucleotide biosynthesis *de novo* and increased production of uric acid. Such patients do not usually develop gouty arthritis early in life but do exhibit uric acid crystalluria and stone formation.

In Lesch-Nyhan patients, all tissues are devoid of HPRT. The disorder thus can be detected by an assay for HPRT in erythrocytes and by cultured fibroblasts. The former test has been used in detection of the heterozygous state. HPRT is a 217-amino-acid cytosolic enzyme coded for by a single gene on the X chromosome.

Several mutations of the HPRT gene are known (Table 27-2). Mutations include major gene alterations and missense mutations causing either gouty arthritis or Lesch-Nyhan syndrome (Table 27-2). An example of a missense mutation causing Lesch-Nyhan syndrome is HPRT-Kinston, which has a substitution of asparagine for aspartic acid at position 194 (D194N). The mutant enzyme, although present in a concentration similar to that of the normal enzyme, is catalytically incompetent because of its very high K_m values for hypoxanthine and PRPP.

The mechanism by which deficiency of HPRT causes central nervous system disorders remains unknown. Lesch-Nyhan patients do not show anatomical abnormalities in the brain. In normal subjects, HPRT activity is high in the brain and, in particular, in the basal ganglia where *de novo* purine biosynthesis is low. This suggests the importance of the purine salvage pathway in this tissue. However, the relationship between HPRT deficiency and

TABLE 27-2

Examples of Missense Mutations That Alter Human Hypoxanthine-Guanine Phosphoribsoyl-Transferase (HPRT) and Cause Lesch-Nyhan Syndrome (LN) or Gout

Mutant HPRT	Nucleotide change	Amino acid change	Clinical presentation
Kinston	G→A	D194N	LN
Urangan	G→A	G16S	Gout
Mashad	A→T	D20V	Gout
Montreal	T→C	M57T	LN
Toowong	G→A	G58R	Gout
London	C→T	S110L	Gout
Ann Arbor	T→G	I132M	Gout
Tokyo	G→A	G140D	LN
Brisbane	C→T	T168I	Gout

neurological manifestations is not understood. The most significant abnormality identified in neurotransmitter systems is in the dopaminergic pathway (Chapter 32). There is no known treatment for the central nervous system dysfunction. Allopurinol has been used in the management of the hyperuricemia.

Adenine Phosphoribosyltransferase (APRT) Deficiency

APRT isolated from erythrocytes is a dimer with each subunit having a molecular weight of 19,481: the gene is located on chromosome 16. This autosomal recessive trait results in inability to salvage adenine, which accumulates and is oxidized to 2,8-dihydroxyadenine by xanthine oxidase (Figure 27-20). The main clinical abnormality is the excretion of 2,8-dihydroxyadenine as insoluble material (gravel) in the urine. These stones can be confused with those of urate on routine analysis. Thus, appropriate chemical analyses are required, particularly in the pediatric age group, to identify APRT-deficient individuals. The neurological disorders characteristic of HPRT deficiency are not found in APRT deficiency, indicating that APRT may not play a significant role in the overall regulation of purine metabolism in humans. APRT deficiency is treated with a low-purine diet and allopurinol.

Adenosine Deaminase (ADA) Deficiency and Purine Nucleoside Phosphorylase (PNP) Deficiency

ADA deficiency and PNP deficiency are two autosomal recessive traits that cause immune system dysfunction. The reactions catalyzed by ADA and PNP are shown in Figure 27-21. Both enzymes function in conversion of adenosine and deoxyadenosine to hypoxanthine. PNP is also involved in conversion of guanosine and deoxyguanosine to guanine. In ADA or PNP deficiency, the appropriate substrates accumulate along with other alternative

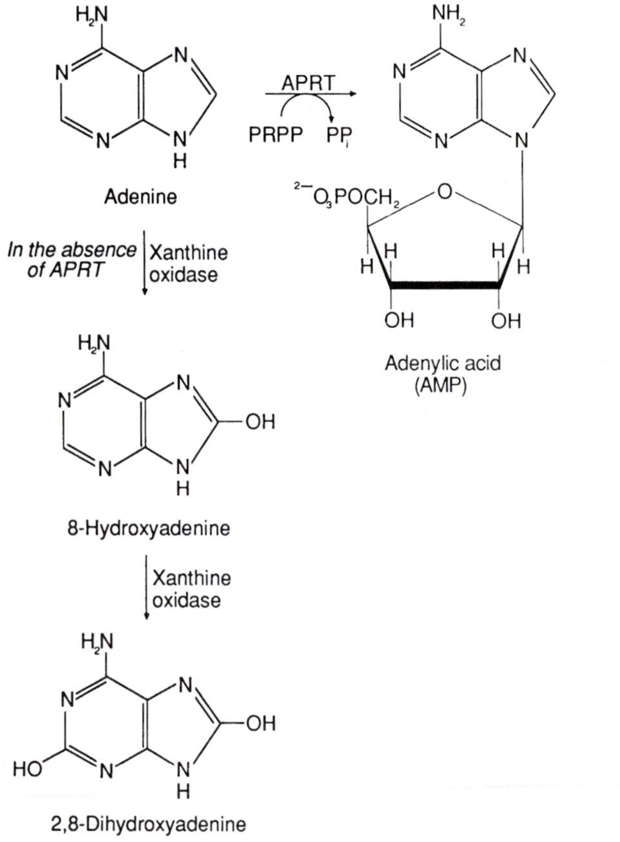

Adenine

APRT

PRPP PP_i

Adenylic acid (AMP)

In the absence of APRT

Xanthine oxidase

8-Hydroxyadenine

Xanthine oxidase

2,8-Dihydroxyadenine

FIGURE 27-20

Metabolic pathways for the conversion of adenine. Adenine is normally salvaged by conversion to AMP by adenine phosphoribosyltransferase (APRT). In the absence of APRT, adenine is oxidized to the highly insoluble product 2,8-dihydroxyadenine by xanthine oxidase.

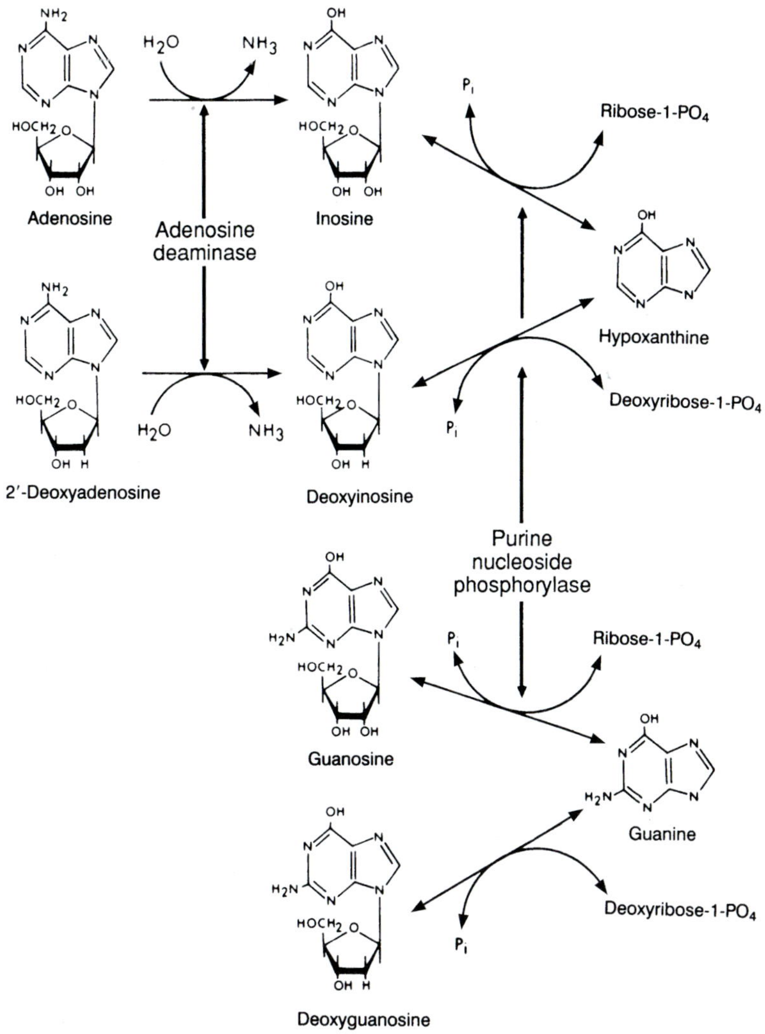

FIGURE 27-21
Reactions catalyzed by adenosine deaminase (ADA) and purine nucleoside phosphorylase (PNP). ADA and PNP participate in the purine catabolic pathway, and deficiency of either leads to immunodeficiency disease. [Slightly modified and reproduced, with permission, from N. M. Kredich and M. S. Hershfield, Immunodeficiency diseases caused by adenosine deaminase and purine nucleoside phosphorylase deficiency. In: *The Metabolic Basis of Inherited Disease,* 6th ed., C. S. Scriver, A. L. Beaudet, W. S. Sly, and D. Valle, Eds. New York: McGraw-Hill (1989).]

products to cause toxic effects on the cells of the immune system.

Patients with ADA deficiency lack both T- and B-lymphocyte-mediated functions, namely, cellular and humoral immunity, respectively, and exhibit a severe combined immunodeficiency (SCID) disorder. Other genetic defects can cause SCID (Chapter 35), but ADA deficiency is responsible for about one-third of patients who have SCID. PNP deficiency is associated only with T-lymphocyte dysfunction.

The exact mechanisms responsible for immune system dysfunction are not known. In ADA-deficient T lymphocytes, there is a 50- to 100-fold increase in the concentration of dATP; similarly, PNP-deficient T lymphocytes exhibit elevation in dGTP concentration. Both dATP and dGTP inhibit ribonucleotide reductase, cause a reduction of other deoxyribonucleotides (e.g., dCTP) required for DNA synthesis, and inhibit cell division. S-Adenosylhomocysteine (SAH) also accumulates and inhibits many methylases required for normal functioning of the cell (Chapter 17). Accumulation of SAH results from the suicide inactivation of SAH hydrolase by deoxyadenosine. Formation and degradation of SAH are shown in Figure 27-22. In the hydrolysis of SAH, the substrate undergoes a temporary oxidation by NAD^+, which is *tightly coupled* to the enzyme. This reaction eventually

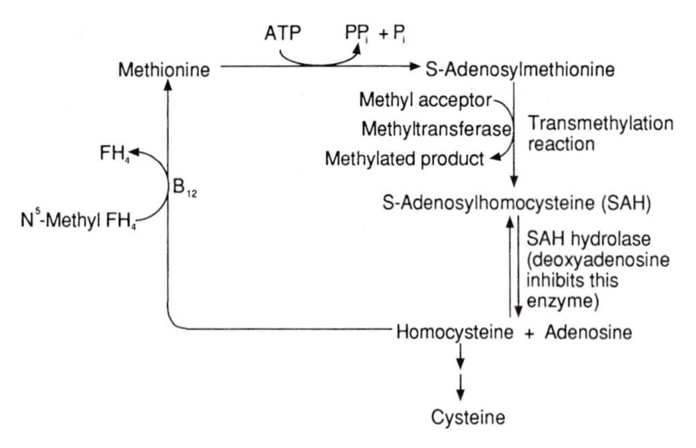

FIGURE 27-22
Formation and degradation of S-adenosylhomocysteine (SAH). In ADA deficiency, accumulation of deoxyadenosine, a suicide substrate, inhibits SAH hydrolase. This inhibition is accompanied by a higher steady-state level of SAH, which in turn inhibits several methyltransferase reactions. Normally, homocysteine is reconverted to methionine or cysteine.

leads to the hydrolysis of the substrate and to the reoxidation of NADH, (Figure 27-23). Deoxyadenosine binds to the enzyme and undergoes initial oxidation to a ketosugar intermediate, which is unstable and decomposes with elimination of adenine, leaving the enzyme in the reduced (NADH) state (Figure 27-23). The catalysis is stopped, and exogenous $NAD^+/NADH + H^+$ cannot affect the reaction. Thus, inactivation of SAH hydrolase by conversion of enzyme-bound NAD^+ to NADH is a suicide inactivation, in which no covalent modification of the enzyme is involved.

Immune system dysfunction in ADA deficiency has also been ascribed to the inhibition of pyrimidine nucleotide synthesis by adenosine, known as ***pyrimidine starvation.*** This may arise from inhibition of conversion of orotic acid to orotidine 5′-monophosphate or from inhibition of PRPP synthesis by excessive synthesis of adenine nucleotides.

ADA deficiency causes death from massive infection before the patient reaches the age of 2 years. Some children with ADA or PNP deficiency have benefited from periodic infusions of irradiated erythrocytes (which contain ADA and PNP). Irradiation of erythrocytes is necessary to inactivate any white blood cells that may be present and thereby to reduce the risk of graft-versus-host disease (Chapter 35). PNP deficiency usually causes hypouricemia and hypouricosuria and excretion of inosine, guanosine, deoxyinosine, and deoxyguanosine.

Another mode of enzyme replacement therapy is periodic administration of a polyethylene glycol-modified form of bovine intestinal ADA by intramuscular injection. The modified enzyme, which is prepared by conjugating polyethylene glycol with purified ADA (PEG-ADA), possesses a longer half-life as well as reduced immunogenicity. As a consequence of this treatment, the correction of two biochemical abnormalities, namely, accumulation of toxic phosphorylated metabolites and inhibition of SAH hydrolase, preceded clinical improvement (i.e., the absence of infection and resumption of weight gain).

Gene replacement therapy should be possible in both ADA and PNP deficiency, and such a trial was attempted with two patients having ADA deficiency. The patient's T cells were removed and a normal ADA gene was inserted into the T cells by means of a retroviral vector. The modified T cells were reintroduced into the patient's bloodstream. Patients were followed for clinical improvement while they continued to receive PEG-ADA treatment. No permanent cure was achieved.

Myoadenylate Deaminase Deficiency

Myoadenylate deaminase (or AMP deaminase) deficiency is a relatively benign muscle disorder characterized by fatigue and exercise-induced muscle aches. This disorder is presumably inherited as an autosomal recessive trait. The relationship between the exercise-induced skeletal muscle dysfunction and AMP deaminase deficiency is explained by an interruption of the purine nucleotide cycle.

The purine nucleotide cycle of muscle consists of the conversion of AMP $\rightarrow$ IMP $\rightarrow$ AMP and requires AMP deaminase, adenylosuccinate synthetase, and adenylosuccinate lyase (Figure 27-24). Flux through this cycle increases during exercise. Several mechanisms have been proposed to explain how the increase in flux is responsible for the maintenance of appropriate energy levels during exercise (Chapter 21).

1. During muscle contraction AMP deaminase activity increases. Nucleoside triphosphates are negative modulators, whereas nucleoside di- and monophosphates are positive modulators of the enzyme. The increased AMP deaminase activity prevents accumulation of AMP so that the adenylate kinase reaction favors the formation of ATP: ADP + ADP $\rightleftharpoons$ ATP + AMP.
2. The NH_3 produced in the AMP deaminase reaction and the decreased levels of ATP as a result of exercise stimulate phosphofructokinase to enhance the rate of glycolysis (Chapter 13).
3. The increased concentration of IMP may activate glycogen phosphorylase and further enhance glycolysis.
4. The production of fumarate (Figure 27-24; see also Chapter 13) may enhance the TCA cycle when demand for ATP production increases.
5. The formation of IMP may provide a means by which the intracellular purine nucleotide pool is maintained. AMP deaminase deficiency disrupts the purine

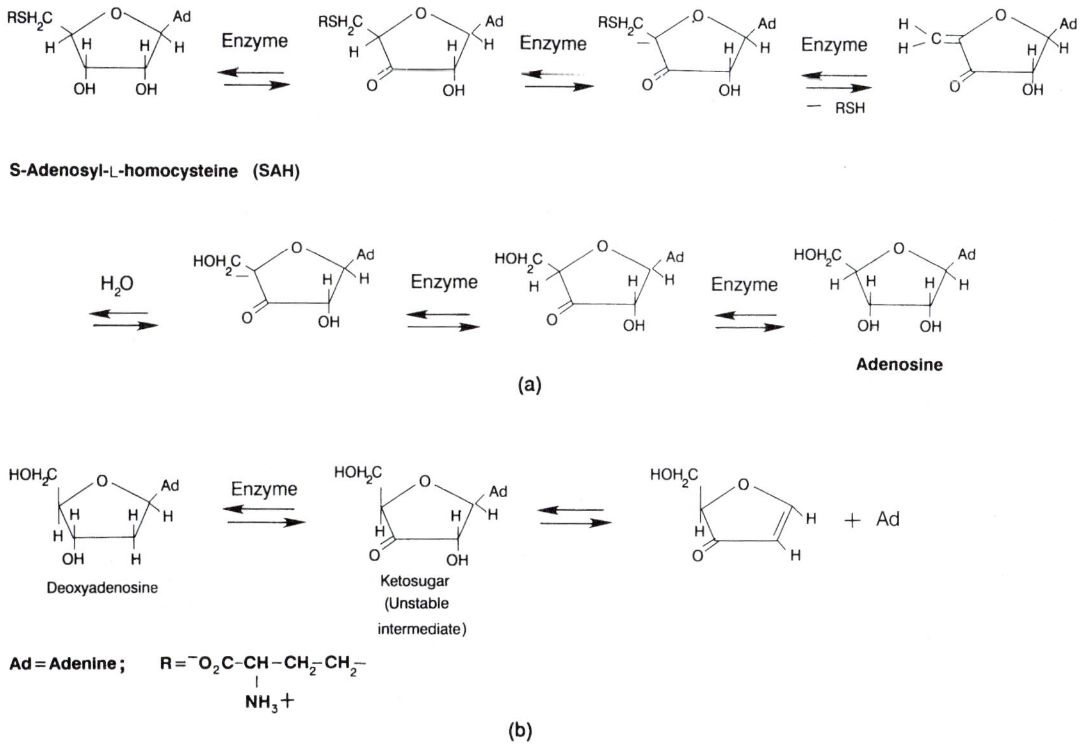

Enzyme
Enzyme
Enzyme
− RSH

S-Adenosyl-L-homocysteine (SAH)

H_2O
Enzyme
Enzyme

Adenosine

(a)

Enzyme

Deoxyadenosine

Ketosugar
(Unstable
intermediate)

+ Ad

Ad = Adenine; $R = {}^{-}O_2C\text{-}CH(NH_3^+)\text{-}CH_2\text{-}CH_2\text{-}$

(b)

FIGURE 27-23
Mechanism of action of S-adenosylhomocysteine hydrolase (enzyme) and its inhibition by deoxyadenosine. (a) The enzyme uses enzyme-bound NAD^+ to temporarily oxidize substrate and eventually hydrolyze it to adenosine and homocysteine. [Reproduced with permission from R. H. Abeles, Suicide enzyme inactivators. *Chem. Eng. News* **61**(38), 55 (September 19, 1983). ©1983 by the American Chemical Society.] (b) Deoxyadenosine, a suicide substrate, is also oxidized by the enzyme with the formation of a ketosugar, which undergoes decomposition, with the product dissociating from the enzyme and leaving the enzyme in the reduced state (NADH).

nucleotide cycle and leads to muscle dysfunction during exercise; however, the cycle operates minimally at rest so that AMP deaminase deficiency should not cause muscle dysfunction during rest. Type IIa (fast-twitch oxidative) and type I (slow-twitch oxidative) muscle fibers have greater oxidative capacity and are less dependent on the cycle than type IIb (fast-twitch glycolytic) fibers. Thus, gradual exercise programs that lead to production of a greater proportion of type IIa and type I fibers might improve exercise tolerance in AMP deaminase deficiency.

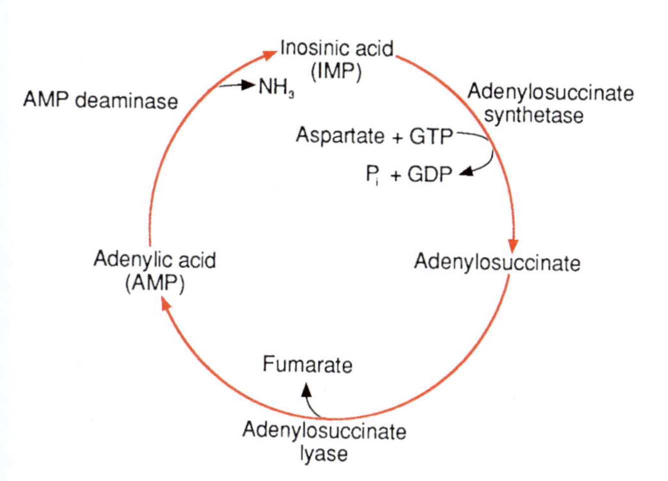

FIGURE 27-24
Purine nucleotide. The cycle plays an important role in energy production in skeletal muscle during exercise.

Xanthine Oxidase Deficiency

This autosomal recessive trait results in hypouricemia and in increased urinary excretion of hypoxanthine and xanthine. Patients frequently have xanthine stones.

27.9 Biosynthesis of Pyrimidine Nucleotides

Pyrimidine nucleotides, in common with purine nucleotides, are required for the synthesis of DNA and RNA. They also participate in intermediary metabolism. For example, pyrimidine nucleotides are involved in the biosynthesis of glycogen (Chapter 15) and of phospholipids

(Chapter 19). Biosynthesis of pyrimidine nucleotides can occur by a *de novo* pathway or by the reutilization of preformed pyrimidine bases or ribonucleosides (salvage pathway).

Salvage Pathways

Pyrimidines derived from dietary or endogenous sources are salvaged efficiently in mammalian systems. They are converted to nucleosides by nucleoside phosphorylases and then to nucleotides by appropriate kinases.

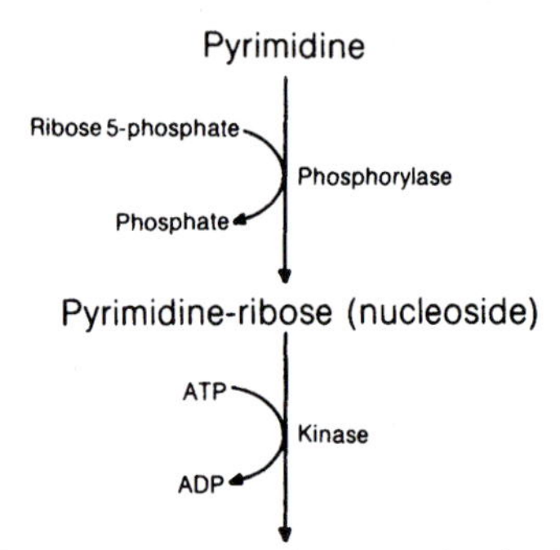

UMP also can by synthesized from uracil and PRPP by uracil phosphoribosyltransferase.

De Novo Synthesis

The biosynthesis of pyrimidine nucleotides may be conveniently considered in two stages: the formation of uridine monophosphate (UMP) and the conversion of UMP to other pyrimidine nucleotides.

Formation of UMP

The synthesis of UMP starts from glutamine, bicarbonate, and ATP, and requires six enzyme activities. The sources of the atoms of the pyrimidine ring are shown in Figure 27-25, and the pathway is shown in Figure 27-26.

1. The pyrimidine base is formed first and then the nucleotide by the addition of ribose 5-phosphate from PRPP. In contrast, in *de novo* purine nucleotide biosynthesis, ribose 5-phosphate is an integral part of the earliest precursor molecule.
2. In the biosynthesis of both pyrimidine and urea (or arginine) (Chapter 17), carbamoyl phosphate is the source of carbon and nitrogen atoms. In pyrimidine biosynthesis, carbamoyl phosphate serves as donor of the carbamoyl group to aspartate with the formation of carbamoyl aspartate. In urea synthesis, the carbamoyl moiety of carbamoyl phosphate is transferred to ornithine, giving rise to citrulline.

   In eukaryotic cells, two separate pools of carbamoyl phosphate are synthesized by different enzymes located at different sites. Carbamoyl phosphate synthetase I (CPS I) is located in the inner membrane of mitochondria in the liver and, to lesser extent, in the kidneys and small intestine. It supplies carbamoyl phosphate for the urea cycle. CPS I is specific for ammonia as nitrogen donor and requires N-acetylglutamate as activator. Carbamoyl phosphate synthetase II (CPS II) is present in the cytosol. It supplies carbamoyl phosphate for pyrimidine nucleotide biosynthesis and uses the amido group of glutamine as nitrogen donor. The presence of physically separated CPSs in eukaryotes probably reflects the need for independent regulation of pyrimidine biosynthesis and urea formation, despite the fact that both pathways require carbamoyl phosphate. In prokaryotes, one CPS serves both pathways.
3. In mammalian tissue, the six enzymes are encoded by three genes. One gene codes for a multifunctional polypeptide (Pyr 1–3) that is located in the cytosol and has carbamoyl phosphate synthetase II (Figure 27-27), aspartate transcarbamoylase, and dihydroorotase activity. Each subunit of Pyr 1–3 has a molecular weight of 200,000–220,000, and the native enzyme exists as multiples of three subunits. The second gene codes for dihydroorotate dehydrogenase which is located on the outer side of the inner mitochondrial membrane. Dihydroorotate, the product of Pyr 1–3, passes freely through the outer mitochondrial membrane and converted to orotate. Orotate readily diffuses to the cytosol for conversion to UMP. The third gene codes for another multifunctional polypeptide known as UMP synthase (Pyr 5,6). Pyr 5,6 (M.W. 55,000) contains orotate phosphoribosyltransferase and orotidylate (orotidine-5′-monophosphate) decarboxylase activity. Use of multifunctional polypeptides is very efficient, since the intermediates neither accumulate nor become consumed in side reactions. They are

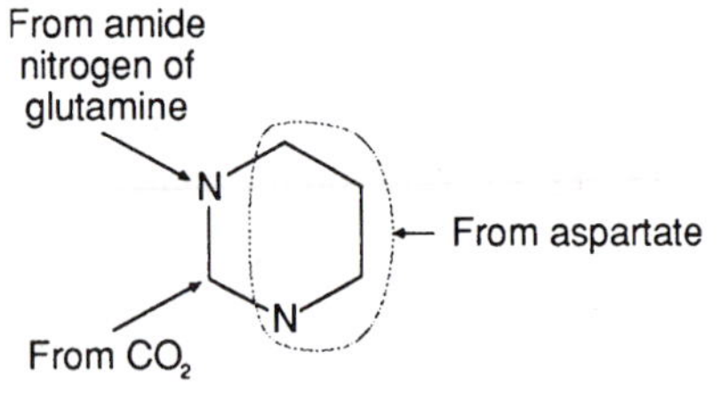

FIGURE 27-25
Sources of the pyrimidine ring atoms in *de novo* biosynthesis.

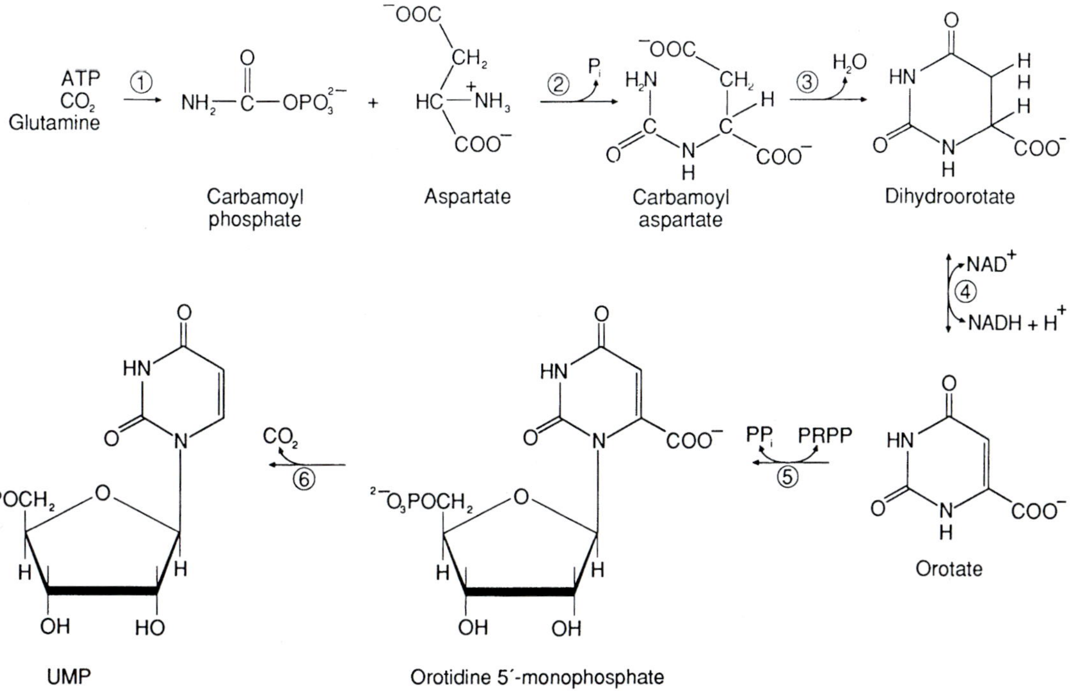

FIGURE 27-26
De novo pathway of uridine-5′-monophosphate (UMP) synthesis. Enzymes: (1) carbamoyl phosphate synthetase II; (2) aspartate transcarbamoylase; (3) dihydroorotase; (4) dihydroorotate dehydrogenase; (5) orotate phosphoribosyltransferase; (6) orotidine-5′-monophosphate decarboxylase (orotidylate decarboxylase).

rapidly channeled without dissociation from the polypeptide. Other pathways in eukaryotic cells, such as fatty acid synthesis, occur on multifunctional polypeptides.

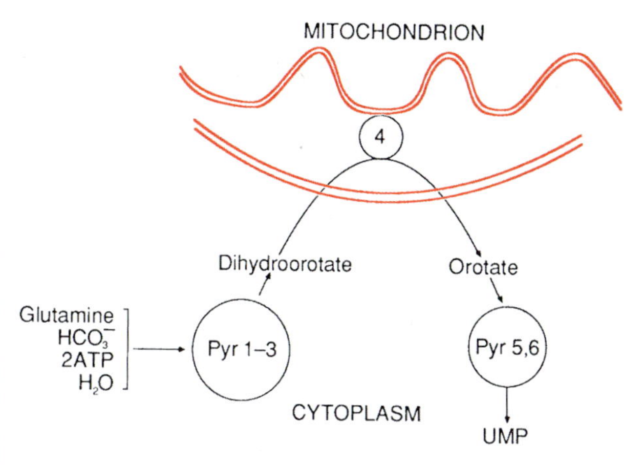

FIGURE 27-27
Schematic representation of the intracellular location of the six enzymes of UMP biosynthesis in animals. Pyr 1–3 = 1, Carbamoyl phosphate synthetase II; 2, aspartate transcarbamoylase; 3, dihydroorotase 4, dihydroorotate dehydrogenase; Pyr 5,6 = 5, orotate phosphoribosyltransferase; 6, orotidine-5′-monophosphate decarboxylase.

4. Conversion of dihydroorotate to orotate is catalyzed by dihydroorotate dehydrogenase, a metalloflavoprotein that contains nonheme-iron atoms and flavin adenine nucleotides (FMN and FAD). In this reaction, the electrons are probably transported via iron atoms and flavin nucleotides that are reoxidized by NAD^+.
5. Biosynthesis of purine and pyrimidine nucleotides requires carbon dioxide and the amide nitrogen of glutamine. Both use an amino acid "nucleus"—glycine in purine biosynthesis and aspartate in pyrimidine biosynthesis. Both use PRPP as the source of ribose 1-phosphate.
6. UMP is converted to UTP by uridylate kinase and nucleoside diphosphate kinase.

$$\text{UMP} + \text{ATP} \xrightleftharpoons{\text{uridylate kinase}} \text{UDP} + \text{ADP}$$

$$\text{UDP} + \text{ATP} \xrightleftharpoons{\text{nucleoside diphosphate kinase}} \text{UDP} + \text{ADP}$$

Formation of Other Pyrimidine Nucleotides

UMP is the parent compound in the synthesis of cytidine and deoxycytidine phosphates and thymidine nucleotides (which are deoxyribonucleotides).

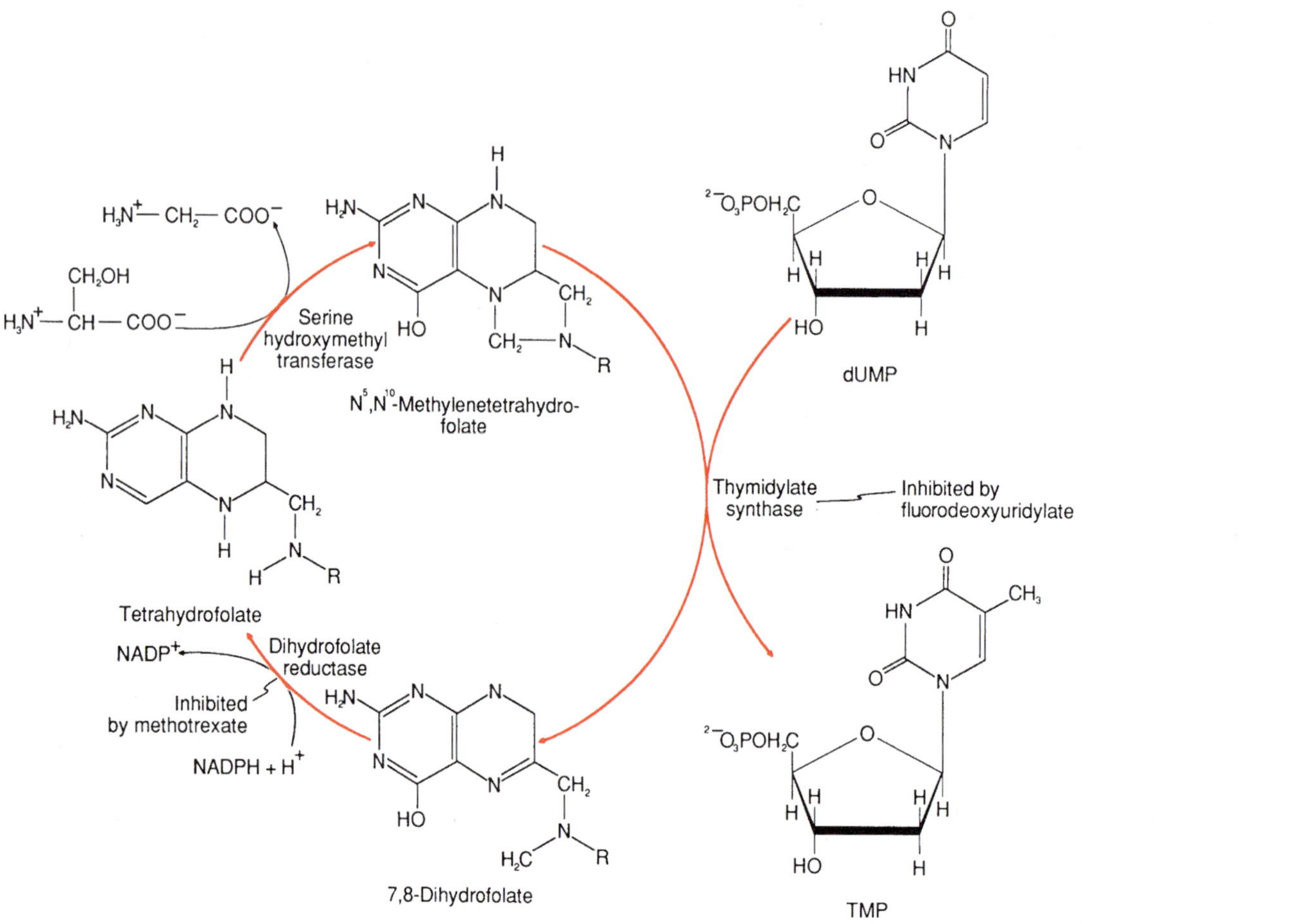

FIGURE 27-28
Synthesis of thymidylic acid (TMP). Fluorodeoxyuridylate inhibits conversion of dUMP to TMP, and methotrexate inhibits regeneration of the tetrahydrofolate coenzyme.

Synthesis of Cytidine Nucleotides

CTP is synthesized from UTP by transfer of the amide nitrogen of glutamine to C-4 of the pyrimidine ring of UTP. This reaction requires ATP as an energy source.

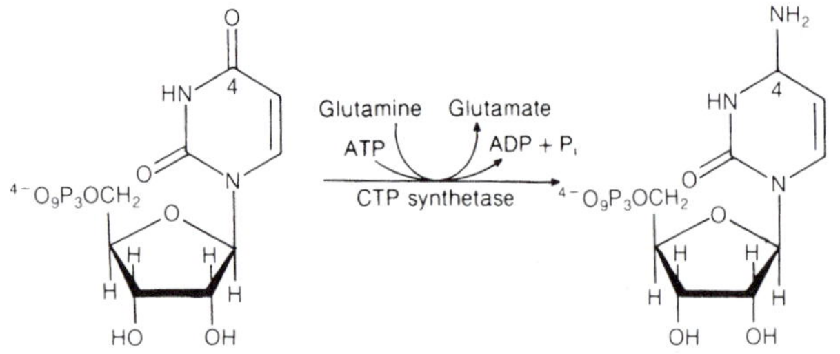

The deoxycytidine phosphates result from reduction of CDP to dCDP by a mechanism analogous to that described for the purine nucleotides. Then dCDP is converted to dCTP by nucleoside diphosphate kinase.

Synthesis of Thymidine Nucleotides

De novo synthesis of thymidilic acid (TMP) occurs exclusively by methylation of the C-5 of dUMP (Figure 27-28) by thymidylate synthase. The methylene group of N^5,N^{10}-methylene FH_4 is the source of the methyl group, and FH_4 is oxidized to FH_2. For sustained synthesis, FH_4 must be regenerated by dihydrofolate reductase. Recall that deoxynucleotides are formed at the diphosphate level by ribonucleotide reductase; thus, UDP is converted to dUDP, then to dUTP, dUMP is then generated mainly by dUTPase.

$$\text{dUTP} \xrightarrow{\text{dUTPase}} \text{dUMP} + \text{PPi}$$

The dUTPase reaction is very important because the DNA polymerases cannot distinguish dUTP from TTP and catalyze significant incorporation of dUMP into DNA when dUTP is present. Incorporation of the base U into DNA is not deleterious because all cells contain a uracil

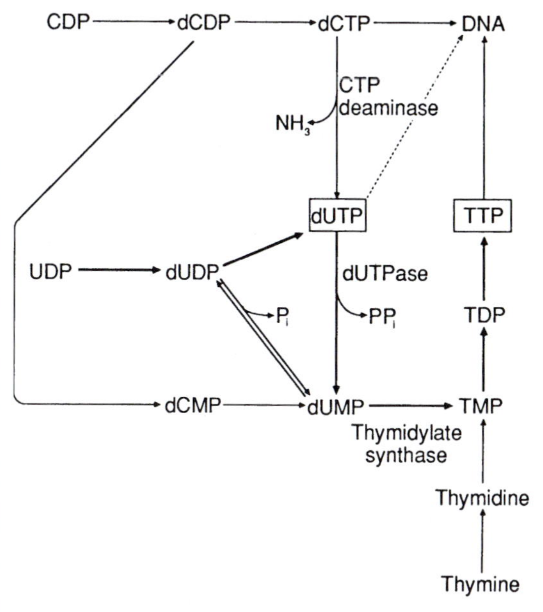

FIGURE 27-29
Pathway for TTP synthesis. The key intermediate is dUTP, which is converted to dUMP by dUTPase.

N-glycosylase that removes U from DNA. Figure 27-29 summarizes the pathway for TTP synthesis.

Because there is only one pathway for synthesis of TMP, it can be used to synthesize radioactively labeled DNA or to inhibit DNA synthesis selectively. In bacterial thymidylate synthase mutants (thy$^-$), DNA synthesis is not possible without added thymine or thymidine, both of which can be utilized by salvage pathways. Neither thymine nor thymidine engages in synthetic reactions other than production of TMP, so radioactive thymine or thymidine can serve as unique precursors for synthesis of radioactive DNA. The radioactivity usually is in the methyl group; since this group is not metabolized, the appearance of radioactivity in any other compound is avoided.

Thymidylate synthase is competitively inhibited by fluorodeoxyuridylate (FUDRP), with formation of a stable ternary complex with methylene FH_4. FUDRP is generated by salvage from exogenous 5-fluorouracil (FU) or fluorodeoxyuridine (FUDR). FUDR is a useful drug in cancer chemotherapy because it inhibits TMP formation in proliferating cells. Thymidine nucleotide deficiency can also be induced by competitive inhibitors of dihydrofolate reductase, e.g., aminopterin (4-aminofolate) and methotrexate (4-amino-10-methylfolate; see Figure 27-3).

Fluoropyrimidines are catabolized by enzymes that normally participate in the breakdown of endogeneous pyrimidines uracil and thymine (see page 643). The deficiency of the initial rate-limiting enzyme dihydropyrimidine dehydrogenase can cause adverse drug reactions in patients receiving standard chemotherapy. This is another use of pharmacogenomics.

Pyrimidine Analogues

This group of compounds (Figure 27-30) includes several drugs that are clinically useful in the treatment of neoplastic disease, psoriasis, and infections caused by fungi and DNA-containing viruses. The mechanism of action of ***5-fluorouracil,*** a halogenated pyrimidine, was mentioned above. An antifungal agent, ***flucytosine (5-fluorocytosine),*** acts through conversion to 5-fluorouracil by cytosine deaminase in the fungal cells. ***Idoxuridine (iododeoxyuridine),*** another halogenated pyrimidine derivative, is used in viral infections. It is a competitive inhibitor (via phosphorylated derivatives) of the incorporation of thymidylic acid into DNA.

Cytarabine (cytosine arabinoside) is an analogue of 2′-deoxycytidine. The 2′-hydroxyl group of the arabinose moiety is in a trans position with respect to the 3′-hydroxyl group (instead of in a cis position, as in the ribose) and causes steric hindrance to rotation of the base around the nucleoside bond (Chapter 23). Phosphorylated derivatives of cytarabine inhibit nucleic acid synthesis as well as being incorporated into nucleic acids.

6-Azauridine and its triacetyl derivative, azaribine, inhibit pyrimidine biosynthesis. Azaribine is better absorbed than 6-azauridine, but it is converted in the blood to 6-azauridine, which undergoes intracellular transformation to 6-azauridylic acid (6-AzUMP). 6-AzUMP, a competitive inhibitor of orotidylate decarboxylase, blocks formation of UMP (Figure 27-26), resulting in high rates of excretion of orotic acid and orotidine in the urine.

5-Azacytidine, after conversion to a phosphorylated derivative, arrests the DNA synthesis phase of the cell cycle. It also affects methylation of certain bases, which leads to hypomethylation (Chapter 26).

Azathymidine, 3′-azido-3′-deoxythymidine (AZT), after conversion to the corresponding 5′-triphosphate, inhibits viral reverse transcriptase. Hence, AZT is used in the treatment of AIDS (Chapter 25).

Regulation of *de Novo* Pyrimidine Biosynthesis

Regulation of the *de novo* pathway is complex. In prokaryotic cells, aspartate transcarbamoylase, an allosteric protein, is inhibited by the end products of pyrimidine

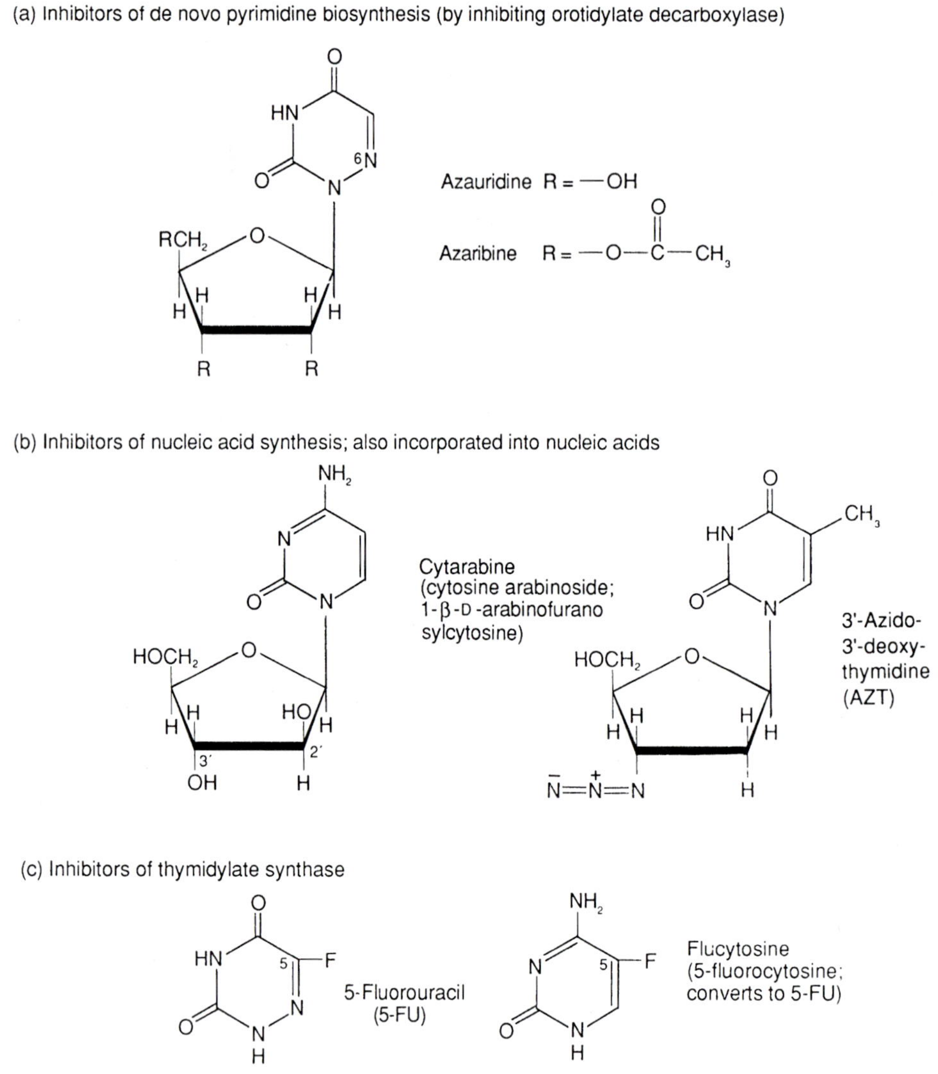

(a) Inhibitors of de novo pyrimidine biosynthesis (by inhibiting orotidylate decarboxylase)

Azauridine R = —OH

Azaribine R = —O—C(=O)—CH_3

(b) Inhibitors of nucleic acid synthesis; also incorporated into nucleic acids

Cytarabine (cytosine arabinoside; 1-β-D-arabinofuranosylcytosine)

3'-Azido-3'-deoxythymidine (AZT)

(c) Inhibitors of thymidylate synthase

5-Fluorouracil (5-FU)

Flucytosine (5-fluorocytosine; converts to 5-FU)

FIGURE 27-30
Pyrimidine analogues. All of these compounds require conversion to appropriate nucleotides before they become active.

biosynthesis, in particular by CTP (Chapter 7). However, mammalian aspartate transcarbamoylase is not regulated in this manner, and carbamoyl phosphate synthetase II (CPS II) appears to be a primary regulatory site. CPS II is inhibited by UTP and is activated by PRPP and ATP. Activation by ATP may be important in achieving a balanced synthesis of purine and pyrimidine nucleotides. PRPP is also an essential (and probably a rate-limiting) substrate for the orotate phosphoribosyltransferase reaction (reaction 5 in Figure 27-26) promoting *de novo* pyrimidine nucleotide synthesis by induction of Pyr 1–3. Another potential site of regulation is orotidine-5-phosphate decarboxylase, which is inhibited by UMP, CMP, allopurinol nucleotide, and oxypurinol.

27.10 Coordination of Purine and Pyrimidine Nucleotide Biosynthesis

A balanced synthesis of pyrimidine and purine nucleotides is essential for the biosynthesis of nucleic acids in growing cells. Several mechanisms exist for coordinating nucleotide synthesis.

1. PRPP affects purine and pyrimidine nucleotide biosynthesis. PRPP formation is activated by inorganic phosphate and inhibited by several end products of pathways that use PRPP. In the purine *de novo* pathway, PRPP activates amidophosphoribosyltransferase and is the rate-limiting substrate for the enzyme. In purine

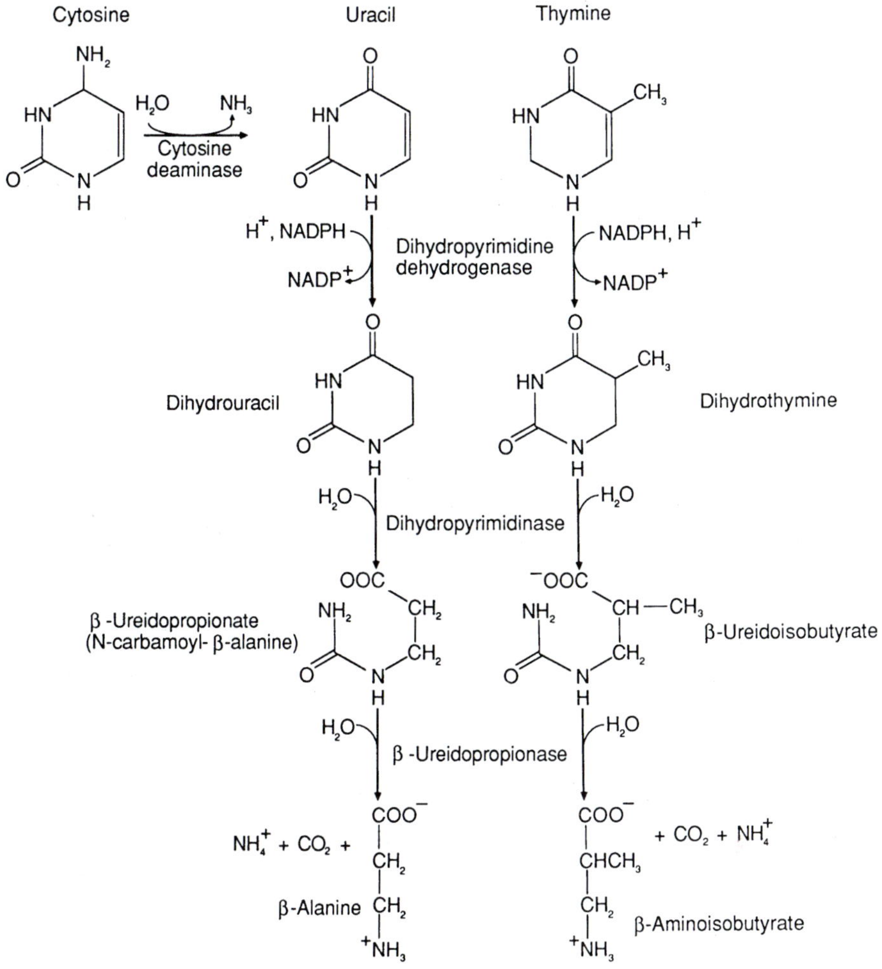

FIGURE 27-31
Pathways for pyrimidine catabolism. The major end product from cytosine and uracil is β-alanine, from thymine it is β-aminoisobutyrate.

salvage pathways, PRPP is the substrate for HPRT and APRT. In the pyrimidine pathway, PRPP activates CPS II, may induce Pyr 1–3, and is a rate-limiting substrate for orotate phosphoribosyltransferase. Thus, changes in the levels of PRPP bring about concordant changes, while enhanced utilization by one pathway might result in reciprocal changes in the other. This interrelationship between the synthesis of purine and pyrimidine nucleotides was seen in a patient who had a deficiency of PRPP synthetase and exhibited orotic aciduria and hypouricemia, consistent with decreased synthesis of purine and pyrimidine nucleotides.

2. Coordination of synthesis of purine and pyrimidine nucleotides is affected by activation of CPS II by ATP.
3. Cultured mammalian cells (e.g., human lymphocytes) fail to grow when exposed to adenosine and have increased pools of ADP, ATP, and GTP, decreased pools of UDP, UTP, and CTP, and decreased incorporation of [^{14}C] orotic acid. These effects are reversed by exogenous uridine. These results suggest that adenosine inhibits the conversion of orotic acid to orotidine-5′-monophosphate. Adenosine deaminase reduces the toxic effect of adenosine by converting it to inosine. Increased levels of purine nucleotides may also inhibit PRPP synthesis. This inhibition of formation of pyrimidine nucleotides in the presence of excess purine nucleosides and nucleotides has been termed "pyrimidine starvation."

27.11 Catabolism of Pyrimidine Nucleotides

Pyrimidine catabolism occurs mainly in the liver. In contrast to purine catabolism, pyrimidine catabolism yields highly soluble end products. Pyrimidine nucleotides are converted to nucleosides by 5′-nucleotidase.

Cytidine so formed is converted to uridine by cytidine aminohydrolase, while uridine and thymidine are converted to free bases by pyrimidine nucleoside phosphorylase.

In mammalian systems, catabolism of uracil and thymine proceeds in parallel steps, catalyzed by the same enzymes (Figure 27-31). The rate-determining step is reduction to a 5,6-dihydroderivative by dihydropyrimidine dehydrogenase. In the second step, dihydropyrimidinase hydrolyzes cleavage of the dihydropyrimidine rings to β-ureido compounds. In the third step, β-ureidopropionase hydrolyzes the β-ureido compounds to β-alanine or β-aminoisobutyrate (BAIB), with release of ammonia and carbon dioxide. Thus, the major end product of the catabolism of cytosine and uracil is β-alanine, whereas that of thymine is BAIB.

High concentrations of BAIB in urine follow excessive cell turnover or destruction (e.g., owing to leukemias or radiation therapy). High levels of BAIB excretion has been observed in some Asian families . Its significance is not known and the high excretors are otherwise normal. Degradation of β-alanine and BAIB or their reutilization in various biosynthetic pathways is possible.

27.12 Abnormalities of Pyrimidine Metabolism

Hereditary orotic aciduria is a rare autosomal recessive trait. In this disorder, both orotate phosphoribosyltransferase and orotidine-5′-phosphate decarboxylase activities (reactions 5 and 6 in Figure 27-26) are markedly deficient. Recall that these activities occur on the polypeptide Pyr 5,6.

Orotic aciduria is characterized by failure of normal growth and by the presence of hypochromic erythrocytes and megaloblastic bone marrow, none of which are improved by the usual hematinic agents (e.g., iron, pyridoxine, vitamin B_{12}, and folate). Leukopenia is also present. Treatment with uridine (2–4 g/d) results in marked improvement in the hematological abnormalities, in growth and development, and in decreased excretion of orotic acid. These patients are pyrimidine auxotrophs and require an exogenous source of pyrimidine just as all humans need vitamins, essential amino acids, and essential fatty acids.

Deficiency of folate or vitamin B_{12} can cause hematological changes similar to hereditary orotic aciduria. Folate is directly involved in thymidylic acid synthesis and indirectly involved in vitamin B_{12} synthesis. Orotic aciduria without the characteristic hematological abnormalities occurs in disorders of the urea cycle that lead to accumulation of carbamoyl phosphate in mitochondria (e.g., ornithine transcarbamoylase deficiency; see Chapter 17). The carbamoyl phosphate exits from the mitochondria and augments cytosolic pyrimidine biosynthesis. Treatment with allopurinol or 6-azauridine also produces orotic aciduria as a result of inhibition of orotidine-5′phosphate decarboxylase by their metabolic products.

Dihydropyrimidine dehydrogenase deficiency can cause unexpected toxic effects to 5-fluorouracil administration (see page 641).

Supplemental Readings and References

H. H. Balfour, Jr.: Antiviral drugs. *New England Journal of Medicine* **340,** 1255 (1999).

R. Curto, E. O. Voit, and M. Cascante: Analysis of abnormalities in purine metabolism leading to gout and to neurological dysfunctions in man. *Biochemical Journal* **329,** 477 (1998).

B. T. Emmerson: The management of gout. *New England Journal of Medicine* **334,** 445 (1996).

M. D. Harris, L. B. Siegel, and J. A. Alloway: Gout and hyperuricemia. *American Family Physician* **59,** 925 (1999).

S. Haste, E. De Clercq, and J. Balzarini: Role of antimetabolites of purine and pyrimidine nucleotide metabolism in tumor cell differentiation. *Biochemical Pharmacology* **58,** 539 (1999).

W. L. Nyhan and D. F. Wong: New approaches to understanding Lesch-Nyhan disease. *New England Journal of Medicine* **334,** 1602 (1996).

J. R. Pittman and M. H. Bross: Diagnosis and management of gout. *American Family Physician* **59,** 1799 (1999).

J. G. Puig, A. D. Michan, M. L. Jimenez, et al.: Clinical spectrum and uric acid metabolism. *Archives of Internal Medicine* **151,** 726 (1991).

D. Rush: Periconceptional folate and neural tube defect. *American Journal of Clinical Nutrition* **59**(Suppl), 511 S (1994).

A. Saven and L. Piro: Newer purine analogues for the treatment of hairy-cell leukemia. *New England Journal of Medicine* **330,** 691 (1994).

V. Serre, G. Hedeel, and X. Liu: Allosteric regulation and substrate channeling in multifunctional pyrimidine biosynthetic complexes: analysis of isolated domains and yeast-mammalian chemeric proteins. *Journal of Molecular Biology* **281,** 363 (1998).

CHAPTER 28

Hemoglobin

The easy availability of blood has resulted in many studies of its constituents. Probably the most extensively studied component is hemoglobin, the predominant protein in the red blood cell and the molecule responsible for transporting oxygen, carbon dioxide, and protons between the lungs and tissues. The study of hemoglobin has led to a detailed knowledge of how oxygen and carbon dioxide transport is accomplished and regulated and has provided insight into the functioning of other allosteric proteins (Chapter 7). Hemoglobin is the first allosteric protein for which molecular details of allosteric effector binding and the mechanism of allosteric action are known. Studies of the genes coding for the globin polypeptide chains have provided a better understanding of many anemias and of the regulation of expression of other eukaryotic genes. Correction of the genetic defects in sickle cell anemia and thalassemia by the introduction of new genetic information into bone marrow cells (gene therapy) is being actively explored.

28.1 Structure of Hemoglobins

Globin Chains

Mammalian hemoglobins are tetramers made up of two α-like subunits (usually α) and two non-α subunits (usually β, γ, or δ). These subunits differ in primary structure but have similar secondary and tertiary structures. However, the differences in tertiary structure among them are critical to the functional properties of each subunit. Each globin subunit has associated with it, by noncovalent interaction, an Fe^{2+}-porphyrin complex known as a ***heme group.*** Oxygen binding occurs at the heme iron. The predominant hemoglobin in adult erythrocytes is $\alpha_2\beta_2$, known as hemoglobin A_1 (HbA). The structural and functional characteristics of hemoglobin have been worked out almost entirely through studies of HbA and its naturally occurring variants.

Each tetramer has a molecular weight of about 64,500, and each α-like and β-like chain has a molecular weight of about 15,750 and 16,500, respectively. The subunits are situated at the corners of a tetrahedron (Figure 28-1). The structure changes slightly during the binding and release of oxygen. In a tetramer, dissimilar chains are more strongly joined than similar chains. In dilute solution, oxyhemoglobin dissociates as follows:

$$\begin{matrix}\alpha & \beta \\ \beta & \alpha\end{matrix} \longrightarrow 2\,(\alpha\beta) \longrightarrow 2\,\alpha + 2\,\beta$$

There is no evidence for the formation of $\alpha\alpha$ or $\beta\beta$, although in the absence of α-chains β_4 (HbH) forms. Association involves salt bridges, hydrogen bonds, van

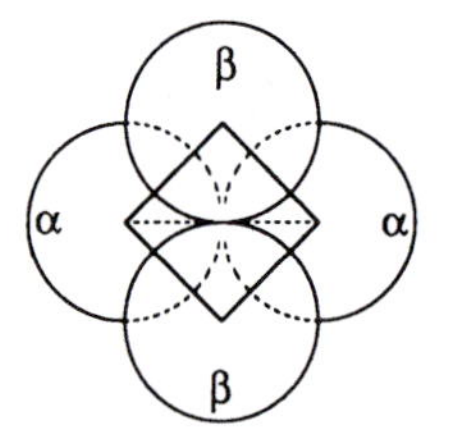

FIGURE 28-1
Arrangement of hemoglobin subunits. The four polypeptide subunits are located at the corners of a tetrahedron to give a roughly spherical structure.

der Waals bonds, and hydrophobic forces. About 60% of the contact between subunits does not change during reversible oxygen binding (packing contacts), while about 35% does (sliding contacts). $\alpha\alpha$ and $\beta\beta$ contacts form about 5% of the total intersubunit contacts, the remainder being $\alpha\beta$ contacts. In deoxyhemoglobin, there is no $\beta\beta$ contact. This information was derived principally from x-ray crystallographic studies of Perutz, Kendrew, and coworkers and clarifies how **2,3-bisphosphoglycerate** (BPG) or (**2,3-diphosphoglycerate, 2,3-DPG**) regulates oxygen transport (Note: 2,3-bisphosphoglycerate and 2,3-diphosphoglycerate are identical.)

If each subunit is labeled, the dissociation in dilute solution becomes clearer:

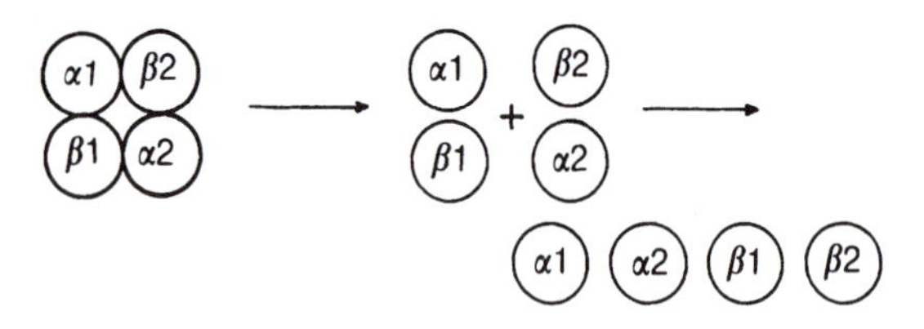

It also shows the asymmetry in the tetramer, since the $\alpha_1\beta_1$ contact differs from the $\alpha_1\beta_2$ contact, and $\alpha_2\beta_1$ differs from $\alpha_2\beta_2$. However, $\alpha_1\beta_1$ and $\alpha_2\beta_2$ contacts are similar, as are $\alpha_1\beta_2$ and $\alpha_2\beta_1$ contacts. During oxygenation and deoxygenation, the $\alpha_1\beta_1$ and $\alpha_2\beta_2$ contacts do not change, but the $\alpha_1\beta_2$ and $\alpha_2\beta_1$ contacts do. Thus, the dimers $\alpha_1\beta_1$ and $\alpha_2\beta_2$ function as a unit and can slide relative to each other. Stable $\alpha\beta$ dimers can be present as a result of a mutation that prevents the formation of the tetramer. An example in which hemoglobin dissociates into stable dimers in the liganded state is the variant hemoglobin Rothschild, which is due to a mutation affecting the β chain (β37 Trp $\rightarrow$ Arg).

Figures 28-2 and 28-3 are schematic diagrams of the secondary structures of α and β chains, respectively. Each chain contains helical and nonhelical segments and surrounds a heme group. Almost 80% of the amino acids exist in an α-helical conformation.

Heme Group

Heme consists of a porphyrin ring system with an Fe^{2+} fixed in the center through complexation to the nitrogens of four pyrrole rings. The porphyrin system is a nearly planar aromatic ring formed from four pyrrole rings linked by =CH– (methene) bridges. The pyrrole rings are substituted so that different porphyrins are distinguished by variations in their side chains (see also Chapters 14 and 29). The heme porphyrin is protoporphyrin IX. Iron has a coordination number of six, i.e., each atom of iron can bond with six electron pairs. In heme, two coordination positions of iron are not occupied by porphyrin nitrogens. When heme is associated with a globin chain, a histidyl nitrogen (from histidine F8, important in the allosteric mechanism) bonds with the fifth coordination position of iron, while the sixth position is open for combination with oxygen, water, carbon monoxide, or other ligands. Portions of the seven or eight helices of a globin chain form a hydrophobic crevice near the surface of the subunit. Heme lies in this crevice, between helices E and F. The low dielectric constant of the crevice prevents permanent oxidation of iron (from Fe^{2+} to Fe^{3+}) by oxygen and is responsible for the reversible binding of oxygen. When Fe^{2+} in solution combines with oxygen, it is oxidized to Fe^{3+}, which does not bind reversibly with oxygen. Evidence from electron paramagnetic resonance studies shows that, in oxyhemoglobin, the oxygen can be considered to oxidize the iron as long as the oxygen is bound. When the oxygen is released, Fe^{3+} is again reduced to Fe^{2+}. Reversible O_2 binding is a unique property of hemoglobin. If the Fe^{2+} in hemoglobin is permanently oxidized to the ferric state (as in methemoglobin), the Fe^{3+} binds tightly to a hydroxyl group, and oxygen will not bind.

Heme groups can be removed from hemoglobin by dialysis, indicating that they are not covalently attached. Not counting the proximal histidine [His β92 (F8) or α87 (F8)], about 60 amino acid residues come within 0.4 nm of one or more atoms of the heme group. This is approximately the maximal length of an effective hydrogen bond or hydrophobic interaction. Of these contacts, only one in the α subunit and two in the β subunit are polar, which emphasizes the highly hydrophobic environment of the hemes. The three polar interactions involve the carboxyl groups of the propionic acid side chains on heme.

28.2 Functional Aspects of Hemoglobin

Oxygen Transport

The primary function of hemoglobin is to transport oxygen from the lungs to the tissues. Hemoglobin forms a

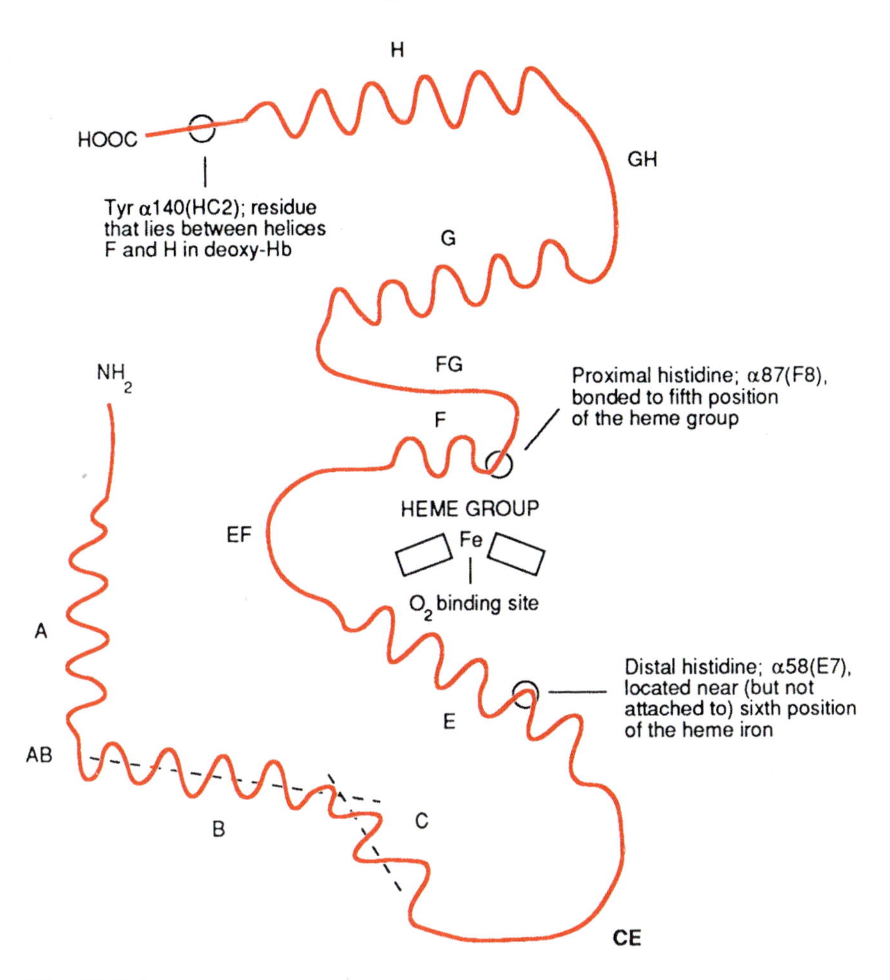

FIGURE 28-2
Secondary structure of the α chain of human hemoglobin. The helical regions (labeled A–H, after Kendrew), N and C termini, and the histidines located near the heme group are indicated. The axes of the B and C helices are indicated by dashed lines. Note that the α chain lacks helix D present in the β chain. The amino acid residues are numbered by two different methods: from the N terminus of the polypeptide chain and from the N-terminal amino acid residue of each helix. Nonhelical regions are designated by the letters of helices at each end of a region.

dissociable complex with oxygen.

$$\text{Deoxyhemoglobin} + 4O_2 \rightleftarrows \text{oxyhemoglobin}$$

This reaction goes to the right with an increase in oxygen pressure (as in the lungs) and to the left with a decrease in oxygen pressure (as in the tissues), in accordance with the law of mass action. Table 28-1 shows typical partial pressures of oxygen between the atmosphere and tissue mitochondria. Hemoglobin increases the solubility of oxygen in the blood 70-fold (from 0.3 to 20.3 mL of oxygen per milliliter of blood).

In Figure 28-4, the oxygen saturation percent for hemoglobin is plotted against oxygen pressure in the gas above the surface of the hemoglobin solution. The curves, known as binding or dissociation curves, are often characterized by their P_{50} value, the oxygen pressure at which the hemoglobin is 50% saturated with oxygen. For comparison, the dissociation curve for myoglobin is also shown. The hemoglobin curves are sigmoid (S-shaped), whereas the myoglobin curve is a rectangular hyperbola. This difference arises because hemoglobin is allosteric and shows cooperative oxygen binding kinetics (Chapter 7), whereas myoglobin is not allosteric. The binding of each molecule of oxygen to hemoglobin causes binding of additional oxygen molecules. This cooperativity occurs because hemoglobin is a tetramer. The cooperative binding of oxygen by hemoglobin is the basis for the regulation of oxygen and, indirectly, carbon dioxide levels in the body. The molecular understanding of this phenomenon is a major accomplishment of x-ray crystallographic studies of protein structure. Myoglobin is a heme protein found in high concentrations in red muscle, particularly cardiac muscle, where it functions as a storage site for oxygen. At the oxygen pressures found in

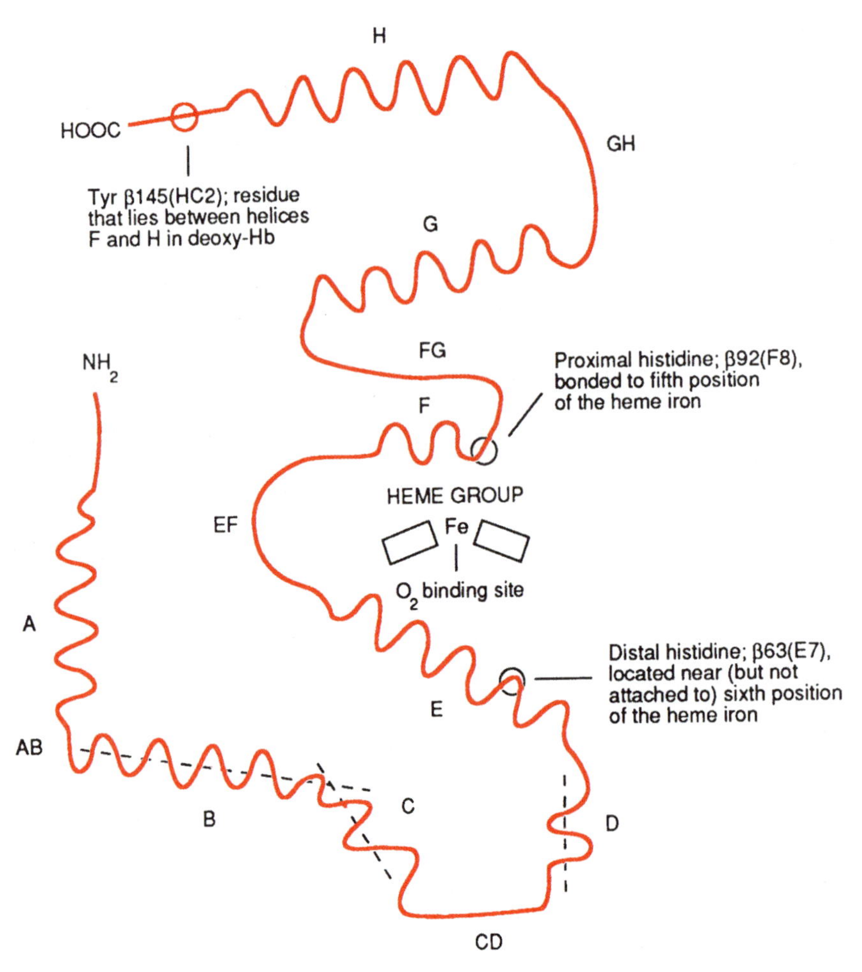

FIGURE 28-3
Secondary structure of the β chain of human hemoglobin. The helical regions (labeled A–H, after Kendrew), N and C termini, and the histidines located near the heme group are indicated. The axes of the B, C, and D helices are indicated by dashed lines.

tissues, the dissociation curve for myoglobin lies to the left of the hemoglobin curves; consequently, hemoglobin (Hb) can readily oxygenate myoglobin (Mb). In the equation

$$Hb(O_2)_4 + 4Mb \rightleftarrows Hb + 4MbO_2$$

the equilibrium lies far to the right.

In the hemoglobin curves, the change from A to B to C is termed a rightward shift. The farther to the right a curve is shifted, the larger is the P_{50} value and the lower is the oxygen affinity. In this example, the shift is due to an increase in P_{CO_2}, since pH, temperature, and 2,3-DPG concentration were held constant. However, the shift could also have been caused by an increase in temperature (as in strenuous exercise or fever), a decrease in pH (as in acidosis or exercise), an increase in 2,3-DPG concentration (see below), or a combination of these variables. For example, an increase in P_{CO_2} is usually accompanied by a decrease in pH. Substances that cause a rightward shift are called ***negative allosteric effectors*** or ***allosteric inhibitors***

TABLE 28-1
Partial Pressures of Oxygen

P_{O_2} (mm Hg)	Stage*
158	Inspired air
100	Alveolar air
95	Arterial blood
40	Peripheral capillary beds
<30 (est.)	Interstitial fluid
<10 (est.)	Inside cells in tissues
<2 (est.)	Inside mitochondria

*During transport from lungs to mitochondria in tissues, where O_2 is used as a terminal electron acceptor.

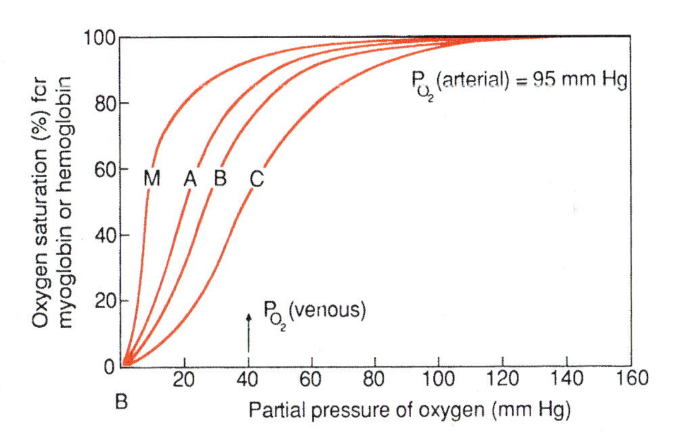

FIGURE 28-4
Oxygen dissociation curves for myoglobin and hemoglobin at several CO_2 pressures. The myoglobin plot (curve M) is very similar to the plot for a hemoglobin subunit (monomer). Curves A, B, and C are for hemoglobin at $P_{CO_2} = 20$, 40, and 80 mm Hg, respectively. P_{O_2} (arterial) and P_{O_2} (venous) indicate normal values for oxygen tensions. During exercise, P_{O_2} (venous) will be lower, around 20–25 mm Hg, because of the increased extraction of oxygen from blood by exercising muscle.

of hemoglobin. They decrease the stability of the oxy form (R-state) of hemoglobin or increase the stability of the deoxy form (T-state). ***Positive allosteric effectors,*** or ***allosteric activators,*** increase the affinity of hemoglobin for oxygen. They cause a leftward shift by increasing the stability of the R-state or by decreasing the stability of the T-state. Oxygen and carbon monoxide are positive allosteric effectors.

A rightward shift decreases the saturation of hemoglobin with oxygen at any particular P_{O_2}. If the blood is initially fully saturated (arterial blood, $P_{O_2} = 95$ mm Hg), a rightward shift would increase the amount of oxygen received by the tissues, as number of moles of oxygen per unit time. At $P_{O_2} = 40$ mm Hg, only about 15% of the oxygen is released from hemoglobin on curve A (85% saturation at 40 mm Hg), while 25% and 50% of the oxygen is released on curves B and C, respectively. On the other hand, if the arterial $P_{O_2} = 45$ mm Hg, a rightward shift would decrease both the initial and final percent saturations to roughly the same extent, and on a molar basis, the tissues would receive almost the same amount of oxygen regardless of the position of the curve.

The following examples illustrate this shift (refer to Figure 28-4). At a normal alveolar P_{O_2} of 95 mm Hg, the hemoglobin is fully saturated with oxygen. At this P_{O_2}, the degree of saturation is essentially independent of pH, temperature, and organic phosphate concentration; even the most right-shifted hemoglobin will be fully saturated with oxygen. This is the normal condition in arterial blood. Under abnormal conditions, such as an anatomical shunt or breathing at a high altitude, the arterial P_{O_2} might be only 45 mm Hg. In this case, the percent saturation in the lungs would depend heavily on the dissociation curve and on such factors as temperature, pH, and 2,3-DPG concentration in the erythrocyte. At P_{O_2} of 45 mm Hg, under the conditions for curve B, hemoglobin in the arterial blood would be about 82% saturated, whereas for curve C it would be only 62% saturated.

The amount of oxygen released by hemoglobin in the tissue capillary beds is almost always determined by the position of the oxygen dissociation curve. At tissue $P_{O_2} = 40$ mm Hg, hemoglobin will be 75% or 50% saturated with oxygen, depending on whether curve B or C describes the physiological state. Thus, for curve B at arterial $P_{O_2} = 95$ mm Hg, $100\% - 75\% = 25\%$ of the oxygen bound to hemoglobin in the lungs per unit time would be released to the tissues, whereas for curve C, the rate of delivery is $100\% - 50\% = 50\%$. If the arterial $P_{O_2} = 43$ mm Hg, the rates of delivery become $82\% - 75\% = 7\%$ (curve B) and $62\% - 50\% = 12\%$ (curve C). Thus, at low loading pressures, a rightward shift of the oxygenation curve makes much less difference in the amount of oxygen available to the tissues than it does at higher loading pressures (7% versus 25%), given that the unloading pressures are the same or nearly so. This consideration is important when evaluating the effect on oxygen delivery to the tissues of a shift in the oxygen dissociation curve.

However, this explanation is an oversimplification of actual events *in vivo*. As blood travels from the lungs to the tissues and back again, pH, P_{O_2}, temperature, and other factors vary continually. Consequently, the curve that describes the affinity of hemoglobin for oxygen differs from one moment to the next.

The need for careful evaluation of all the factors that affect tissue oxygenation is illustrated by the following experiment. As part of study of the adaptation of the respiratory system to changes in oxygen pressure, it was observed that erythrocytic 2,3-DPG concentration increased by 25–30% within 24 hours after going from 10 m to 4509 m above sea level, with a corresponding rightward shift of the oxygen dissociation curve. This finding was interpreted as evidence that 2,3-DPG plays a central role in the body's adaptation to anoxic anoxia caused by a change in altitude. However, the arterial P_{O_2} decreased from about 94 mm Hg at 10 m to about 45 mm Hg to 4509 m owing to a decrease in atmospheric oxygen tension from 149 to 83 mm Hg. Thus, the arterial P_{O_2} decreased to a point at which the supply of oxygen to the tissues was largely independent of the position of the curve. Shifting the curve by the change in 2,3-DPG has no significant effect on oxygen delivery. Once this fact was recognized, the principal adaptive factor was shown to be an increase in respiratory rate. Increases in hematocrit and mean corpuscular hemoglobin concentration were also observed.

A shift in the dissociation curve can be very important. In chronic conditions in which the supply of oxygen to the lungs is normal and arterial blood is saturated with oxygen but the ability of the blood to deliver oxygen to the tissues is impaired because of a low hemoglobin concentration (anemia) or a low cardiac output (cardiac insufficiency), the intraerythrocytic 2,3-DPG concentration increases. A rightward shift of the dissociation curve will not affect the percent saturation during loading in the lungs (since the upper part of the curve is quite flat), but it will decrease the percent saturation in the tissues, thereby providing a useful increase in moles of oxygen per unit time available for metabolism.

Respiration in a normal individual provides another example. Since P_{CO_2} is higher (and pH lower) in the tissues than in the lungs, oxygen unloading is facilitated at the tissues, and loading occurs more readily in the lungs. The decrease in oxygen saturation at constant P_{O_2} (i.e., decrease in oxygen affinity) with increasing P_{CO_2} or decreasing pH above pH 6.3 is known as the ***alkaline Bohr effect.*** An acid Bohr effect occurs below pH 6.3 and consists of an increase in oxygen affinity with decrease in pH; it is due to a chloride-induced proton uptake that is greater for oxyhemoglobin than for deoxyhemoglobin. It is not physiologically significant because a serum pH below $\sim$6.8 is not compatible with human life.

Mechanism of Oxygenation

Hemoglobin has two quaternary structures that correspond to the deoxygenated (deoxy; five-coordinate iron; T-state) and the oxygenated (oxy; six-coordinate iron; R-state) forms. X-ray crystallographic studies have demonstrated that the coordination number of the iron is the crucial difference between these forms. All six-coordinate hemoglobins (oxyhemoglobin, carbon monoxyhemoglobin, and methemoglobin) are in the R-state, whereas deoxyhemoglobin is in the T-state. Binding of one molecule of oxygen to deoxyhemoglobin causes a change in the tertiary structure of the binding subunits, which results in a change in quaternary structure (T-to-R transition) that enhances the binding of oxygen to the remaining subunits. Thus, up to four molecules of O_2 per molecule of Hb can be bound.

The "machinery" for this transition is composed of the globin peptide chain, amino acid side chains, and heme. In deoxyhemoglobin, the bond between the nitrogen of histidine F8 and the iron atom of heme is tilted at a slight angle from the perpendicular to the plane of the porphyrin ring. The porphyrin ring is domed "upward" toward the histidine, with the iron at the apex of the dome. Thus, the iron is in an unfavorable position for binding of a sixth ligand, such as oxygen. This is a primary cause of the weak binding of O_2 to T-state hemoglobin (Figure 28-5).

Deoxyhemoglobin is in an unstrained conformation, as expected for a stable molecular form, despite its designation as the T ("tense") state. It is stabilized by hydrogen bonds, salt bridges, and van der Waals contacts between amino acid side chains on the same subunit and on different side chains. Breaking of many of these bonds occurs during oxygen binding, destabilizing the deoxy structure and causing the release of the Bohr protons (see below). Oxyhemoglobin is also stabilized by several noncovalent bonds, formation of which aids in the T-to-R transition.

Upon binding of the first molecule of oxygen to completely deoxygenated hemoglobin, strain is introduced into the molecule as a result of competition between maintenance of the stable deoxy form and formation of a stable iron-O_2 bond. "Strain" refers to "long" bonds that have higher energy than do normal-length, minimal-energy bonds. When possible, the "stretched" bonds shorten to lower their energy and thereby produce movement within the molecule. In an effort to reduce the strain and return to an energetically more favorable state, the F helix slides across the face of the heme, causing the bond between histidine F8 and the iron atom to straighten toward the perpendicular (Figure 28-5b), and the heme rotates and sinks further into the heme pocket. These movements occur in both α and β subunits but are greater in the β subunits. The change in radius of the iron upon O_2 binding, originally thought to "trigger" the conformational changes, is now considered to be of minor importance in the T-to-R transition.

The movement of heme and surrounding amino acid residues accomplishes two things. First, it relieves the strain on the six-coordinated heme iron, increasing the strength of the Fe-O_2 bond (i.e., increasing oxygen affinity). Second, it strains other noncovalent bonds elsewhere in the subunit, causing some of them to break. This is the beginning of the quaternary structure transition from T-state to R-state. The energy needed for the conformation changes is provided by the energy of binding of oxygen to the heme iron and by the formation of new, noncovalent bonds typical of oxyhemoglobin. Under normal conditions of P_{CO_2} and P_{O_2} in the lungs, more energy is released by binding four oxygen molecules than is needed to break the noncovalent bonds in deoxyhemoglobin. The opposite situation prevails in extrapulmonary capillary beds.

The binding of an oxygen molecule in one subunit induces strain in another subunit and causes some noncovalent bonds to break. Thus, $Hb(O_2)$, $Hb(O_2)_2$, and $Hb(O_2)_3$ represent different transient structures. Which

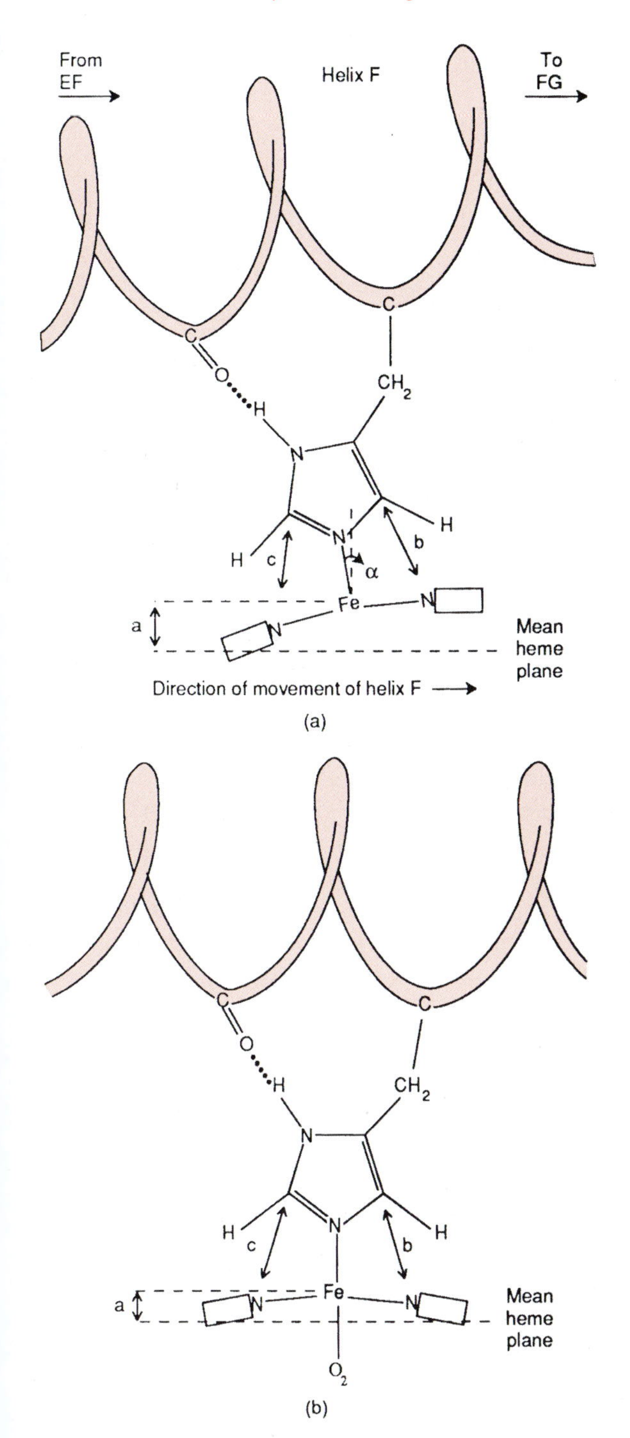

FIGURE 28-5
Schematic representation of the geometry of the unliganded (a) and the liganded (b) heme group, and the F helix in hemoglobin. The iron atom in the unliganded heme (a) is located out of the mean heme plane (distance a) and the iron and the porphyrin ring are slightly domed toward histidine (R). The histidine is tilted as indicated by the bond angle α (8–10 degrees) and the distance $b > c$. When O_2 binds to the iron (b), the iron atom is closer to the heme plane (distance a decreases), the N-Fe bond becomes more nearly perpendicular ($\alpha = 1$ degree and $b = c$), and the heme acquires a less domed conformation. These events pull helix F across the heme group, setting off a series of reactions leading to the T-to-R transition. (Modified and redrawn with permission from I. Geis and R. E. Dickerson.)

type of subunit (α or β) binds oxygen first is unclear, although a three- to fourfold difference in oxygen affinity exists between the subunits in the tetramer. It is assumed that oxygen binds first to the α subunit because its heme is more readily accessible.

The T-to-R transition occurs after the binding of two or three oxygen molecules, when the noncovalent interactions are too few to stabilize the deoxy form. The oxygen binding breaks the remaining noncovalent bonds that stabilize the deoxy structure, relieves the strain on the oxygenated subunits, and increases the affinity for oxygen of the unliganded subunits. The higher oxygen affinity of the oxy form and the formation of new noncovalent interactions account for the cooperative binding of oxygen.

The sum of the motions of individual side chains on binding and release of oxygen produces a relative reorientation of the four subunits (Figure 28-6). The motion can be considered as rotation of $\alpha\beta$ pairs, joined by $\alpha_1\beta_1$-type interactions, around the $\alpha\alpha$ axis. Movement occurs primarily at the $\alpha_1\beta_2$ interfaces. The effect is to move the β subunits closer to each other. Thus, a change in the tertiary structure of the subunits brings about the change in quaternary structure of the tetramer.

At pH 7.4, deoxyhemoglobin releases 0.7 mol of H^+ for each mole of oxygen it binds, the process being reversible. However, under physiological conditions, deoxyhemoglobin in whole blood releases only 0.31 mol of H^+ per mole of O_2 bound. This phenomenon is the alkaline Bohr effect (decrease in oxygen affinity of hemoglobin with decrease in pH or increase in P_{CO_2}); these H^+ ions are known as ***Bohr protons.*** An equivalent statement would be that oxyhemoglobin is a stronger acid than the deoxy form. In fact, the pK′ for oxyhemoglobin is 6.62, while that for deoxyhemoglobin is 8.18. This change in pK′ occurs because basic side chains involved in the salt bridges are broken during the T-to-R transition. Release of the protonated side chains decreases their pK′s, making it easier for the H^+ to separate. It is estimated that, at pH 7.4, 40% of the Bohr protons come from the C-terminal histidines on the β chains, 25% from the N-terminal amino groups on the α chain, and 35% from many amino acids, particularly

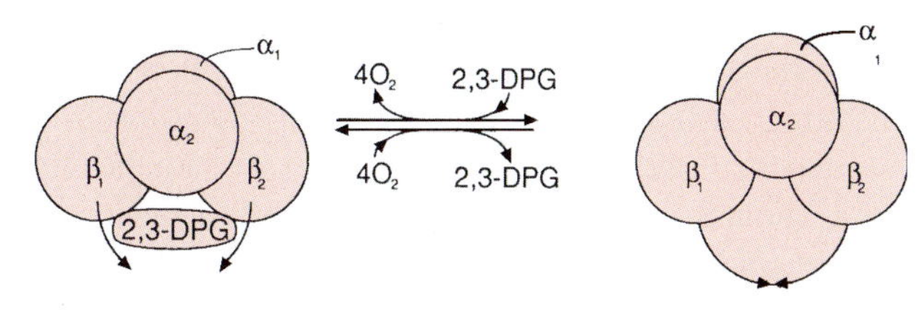

FIGURE 28-6
Relative motion of the β subunits of hemoglobin upon oxygenation and deoxygenation.

His α122, His β143, and Lys β144. The amount each of these groups contributes depends on pH and other conditions prevailing at the time of measurement.

A primary allosteric effector of hemoglobin in human erythrocytes, in addition to H^+ and CO_2, is the organic phosphate 2,3-DPG. Deoxyhemoglobin binds 1 mol of 2,3-DPG per mole of hemoglobin, whereas oxyhemoglobin does not. The reason for this differential binding is the location of the binding site (Figure 28-6). The negatively charged 2,3-DPG molecule fits between the β subunits, where it binds to positively charged residues (see below). In oxyhemoglobin, the β subunits are close together, the cleft between them being too small for entry of the 2,3-DPG molecule. Binding of 2,3-DPG inhibits this movement of the β chains and helps to stabilize the deoxy form. This binding is an equilibrium phenomenon, and the amount of 2,3-DPG bound depends on the concentrations of 2,3-DPG, O_2, CO_2, H^+, and hemoglobin. Chloride promotes H^+ binding to deoxyhemoglobin and acts as an allosteric effector, so its concentration must also be considered. All of these factors interact to determine the amount of oxygen bound to hemoglobin at any particular P_{O_2}.

Function, Metabolism, and Regulation of Organic Phosphates in Erythrocytes

2,3-DPG, ATP, inositol hexaphosphate (IHP), and other organic phosphates bind to hemoglobin and decrease its affinity for oxygen. IHP, the principal organic phosphate in avian erythrocytes, is the most negatively charged of these compounds and binds the most tightly; however, it is not found in human red cells.

2,3-DPG levels in the erythrocytes help regulate hemoglobin oxygenation (Figure 28-7). Unloading of oxygen at the P_{O_2} in tissue capillaries is increased by 2,3-DPG, and small changes in its concentration can have significant effects on oxygen release. At the pH prevailing in the erythrocyte, the net charge on the 2,3-DPG molecule is −5. The binding site between the two β chains of hemoglobin contains eight positively charged amino acid residue side chains contributed by the Val1, His2, Lys82, and His143 of each chain (Figure 28-8). These residues are highly conserved in mammalian hemoglobins, indicating their importance for normal hemoglobin function. In placental mammals, a fetus receives its oxygen by diffusion from the maternal circulation, across the placenta, into the fetal circulation. To ensure that the oxygen flow is adequate, the pressure gradient from mother to fetus is increased by increasing the affinity of fetal hemoglobin for oxygen. This decreases the partial pressure of oxygen in the fetal circulation, thereby increasing the rate of transplacental diffusion.

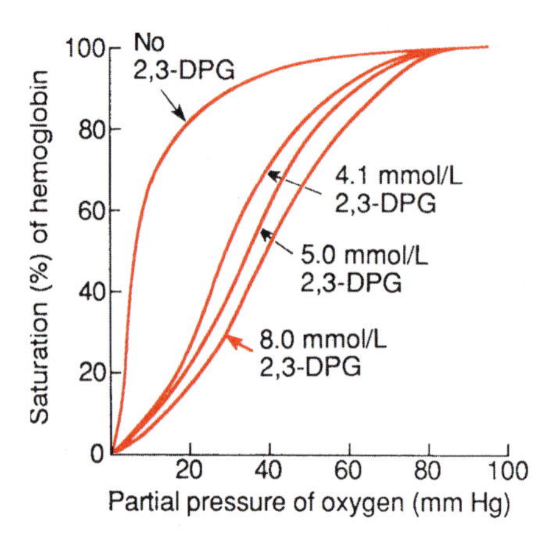

FIGURE 28-7
Effect of 2,3-bisphosphoglycerate (2,3-DPG) on the oxygen saturation curve of hemoglobin. Note that 2,3-DPG decreases the affinity of hemoglobin for oxygen.

Different species increase transplacental diffusion in different ways. In humans and other primates, adult hemoglobin contains two α and two β chains, while fetal hemoglobin has two α and two γ chains. Although the amino acid sequences of β and γ chains are similar, they differ in position 143, which is part of the 2,3-DPG binding site. In the γ chain, His143 is replaced by Ser, thus reducing the charge in the 2,3-DPG binding site from +8 to +6. Thus, 2,3-DPG binds less tightly to fetal hemoglobin, the oxygen affinity of which is thereby increased relative to that of HbA at the same concentration of 2,3-DPG. Thus, erythrocytes in fetal primates, despite having 2,3-DPG concentrations equal to those in adult erythrocytes, have higher oxygen affinity than maternal red blood cells, allowing for maternal to fetal oxygen transport.

In other mammals, including horse, dog, pig, and guinea pig, fetal hemoglobin is not structurally different from adult hemoglobin, and transplacental diffusion is facilitated by a reduced concentration of 2,3-DPG in fetal erythrocytes. In ruminants, hemoglobin does not bind 2,3-DPG because the β chains are too far apart. However, fetal hemoglobin in ruminants has a higher affinity for oxygen than does adult hemoglobin because of other structural differences. These three different solutions to the problem of the need for transfer of oxygen to the fetus are an example of ***convergent evolution.***

In humans, 2,3-DPG is the most abundant phosphate compound in the red cell. Its concentration is 5 mmol/L, approximately the same as that of hemoglobin tetramer. The concentration of ATP is also high, 1.3 mmol/L. Although ATP has roughly the same affinity for hemoglobin as does 2,3-DPG, it has little effect on oxygen affinity because it is mostly present as ATP-Mg^{2+}, which binds weakly to hemoglobin. 2,3-DPG is formed by rearrangement of 1,3-bisphosphoglycerate,

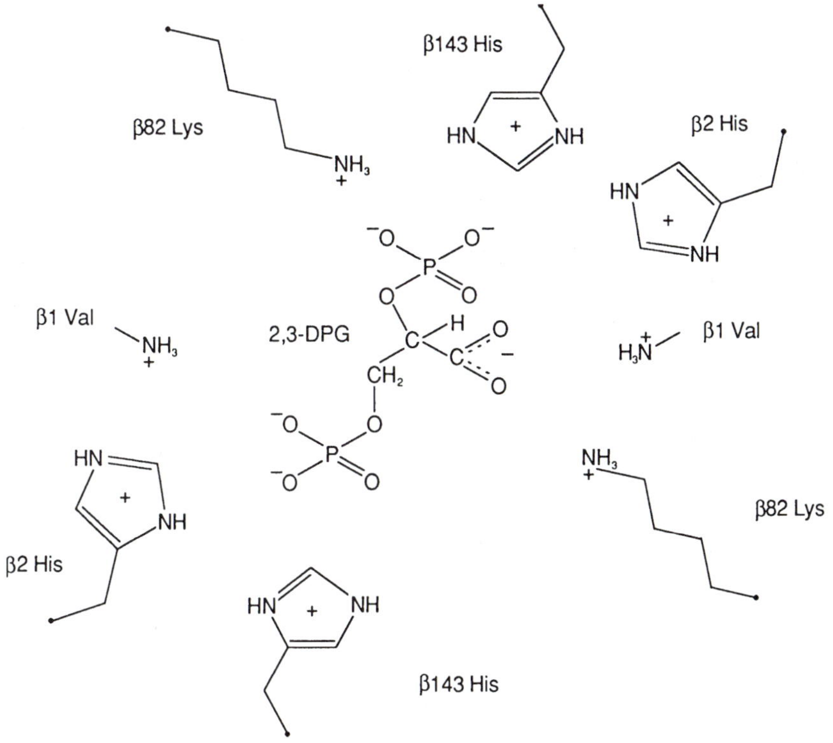

FIGURE 28-8
Schematic representation of the side chains of the β subunits of human hemoglobin that participate in binding to 2,3-DPG. The binding cavity is lined with eight positive charges (four from each β subunit) that react with five negative charges on 2,3-DPG. Fetal hemoglobin binds 2,3-DPG much less tightly than does maternal hemoglobin because its γ chains (the counterpart of β-chains) contain Ser at γ143 in place of His at β143.

an intermediate in glycolysis (Chapter 13). The rearrangement, catalyzed by bisphosphoglycerate mutase, requires 3-phosphoglycerate as cofactor and is allosterically stimulated by 2-phosphoglycerate (Figure 28-9). Inorganic phosphate appears to be a negative allosteric modifier. 2,3-DPG is also a cofactor for the phosphoglycerate mutase of glycolysis. Bisphosphoglycerate phosphatase converts 2,3-DPG to 3-phosphoglycerate. Identical electrophoretic and chromatographic patterns and copurification of the two activities suggest that the catalytic sites for bisphosphoglycerate mutase and phosphatase may reside in the same protein. This hypothesis is supported by the report of an individual with an extremely low intraerythrocytic concentration of 2,3-DPG whose red blood cells lacked both bisphosphoglycerate mutase and phosphatase activities. Trace amounts of 2,3-DPG were present to act as cofactor for phosphoglycerate mutase and permit glycolysis to proceed. The 2,3-DPG deficiency diminished oxygen delivery to the tissues and produced a mild erythrocytosis. There was no hemolysis, and the disorder was clinically silent. Patients who have pyruvate kinase deficiency have above-normal levels of 2,3-DPG, whereas those who have hexokinase deficiency have below-normal levels. Appropriate erythropoietic responses are seen in both types (see below). The concentration of 2,3-DPG in the red cell can be altered by 15–25% in less than 12 hours. The most probable mechanisms involved are summarized below.

1. The binding of 2,3-DPG to deoxyhemoglobin decreases the amount of free 2,3-DPG available for participation in other reactions and causes increased 2,3-DPG synthesis at the expense of 1,3-bisphosphoglycerate. A decrease in oxygen saturation of hemoglobin may act in the same way.
2. The intraerythrocytic pH affects 2,3-DPG concentration. A decrease in pH increases the amount of bound 2,3-DPG by increasing the concentration of deoxyhemoglobin, which acts as described in the preceding paragraph. An increase in pH stimulates glycolysis, which tends to increase the concentration of all glycolytic intermediates, including 2,3-DPG. A decrease in pH within the physiological range also decreases the activity of bisphosphoglycerate mutase and increases the activity of bisphosphoglycerate phosphatase.
3. As erythrocytes age *in vivo,* their oxygen affinity increases. The concentration of 2,3-DPG in young red

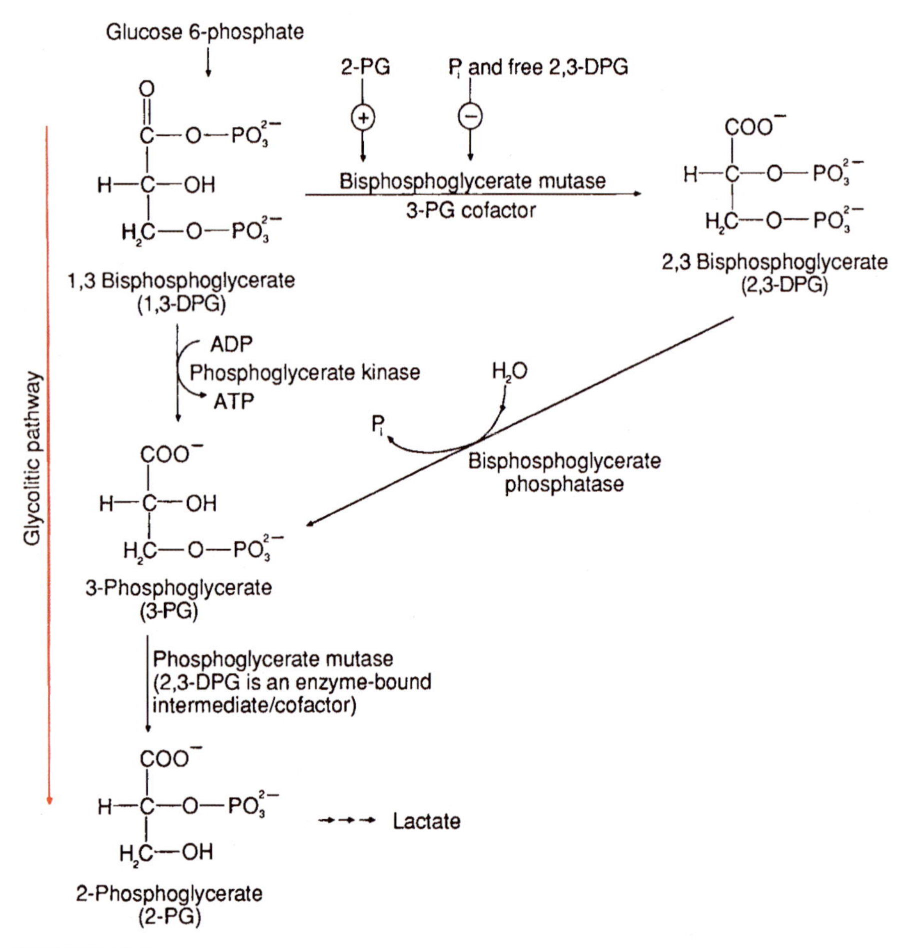

FIGURE 28-9
Formation of 2,3-bisphosphoglycerate (2,3-DPG) in erythrocytes. Formation of 2,3-DPG occurs as a shunt from the main pathway of glycolysis, and free energy is used that otherwise would have been employed in the formation of ATP. ⊕, Positive allosteric modifier; ⊖, negative allosteric modifier; P_i, inorganic phosphate.

cells is higher than that in old red cells. This may reflect a general change in activity of bisphosphoglycerate mutase and phosphatase. Since erythrocytes are unable to synthesize proteins, inactivated enzymes cannot be replaced.

4. There may also be genetic control over 2,3-DPG levels. The ATP concentration in erythrocytes is under hereditary control, and in hooded rats, levels of ATP and 2,3-DPG appear to be genetically influenced. However, this finding may be of no importance in producing rapid, short-term, adaptive changes.

These processes are final-step controls that directly influence 2,3-DPG concentrations. The primary stimuli that trigger these final steps include the following:

1. Decreased delivery of O_2 to tissues as a result of anemia, altitude, cardiac insufficiency, or pulmonary disease.
2. Thyroxine (which may directly stimulate bisphosphoglycerate mutase), androgens (which act partly by increasing erythropoiesis), and other hormones.
3. Polycythemia, which decreases the intraerythrocytic concentration of 2,3-DPG.

Whether the shift in the O_2-dissociation curve that accompanies changes in the concentration of 2,3-DPG is beneficial depends largely on the oxygen saturation of arterial blood. The 2,3-DPG concentration can vary widely among patients with the same disease. For example, in severe pulmonary disease, the increase in 2,3-DPG concentration ranges from 0% to 100%; in leukemia with depressed production of erythrocytes, elevations of 20–150% occur; in iron deficiency, increases range from 40% to 75%.

2,3-DPG and, to a lesser extent, ATP concentrations decrease rapidly in blood that is stored even for a few days in the acid-citrate-dextrose (ACD) medium used by many blood banks. As a result, oxygen affinity is increased and the ability of blood that is transfused to supply oxygen to the tissues is decreased. Volunteers who received such blood had an increase in oxygen affinity that did not return to normal for 6–24 hours. The therapeutic significance of these changes is not clear. The greatest effects should occur in patients who receive numerous transfusions over a period of about 6 hours, so that a significant fraction of their circulating erythrocytes has increased oxygen affinity.

Traditionally, red cell survival has been the main criterion of the quality of stored blood. However, cell survival does not necessarily correspond to maintenance of adequate organic phosphate levels. Studies on the composition of the storage medium needed to prevent this metabolic loss of organic phosphates show the following:

1. Citrate-phosphate-dextrose (CPD) medium is better than ACD for maintaining organic phosphate levels and for preventing a reduction in P_{50}, probably because of the higher pH of CPD. In 1971, 90% of blood banks in the United States used ACD and only 10% used CPD; by 1975, the reverse was true. The shelf life of cells is the same for both media (21 days), but cells stored in CPD function better physiologically when transfused.
2. Supplementation with inosine generates a supply of ribose 1-phosphate and provides a potential glycolytic substrate that can be metabolized to 2,3-DPG. Pyruvate and fructose, which help maintain a supply of oxidized NAD^+, potentiate the effect of inosine. This modification must be balanced against the possibility of hyperuricemia (Chapter 27) caused by transfusion of large amounts of inosine-containing blood.
3. Supplementation with dihydroxyacetone phosphate could provide a glycolytic substrate without the risk of hyperuricemia.

CO_2 Transport

In addition to carrying oxygen from the lungs to the tissues, hemoglobin helps move carbon dioxide from tissues back to the lungs, where it can be eliminated. This CO_2 movement is accomplished with little or no change in blood pH, owing to the buffering capacity of hemoglobin (Chapter 1). Figure 28-10 summarizes the pathways and intermediates involved. Of particular importance are the following:

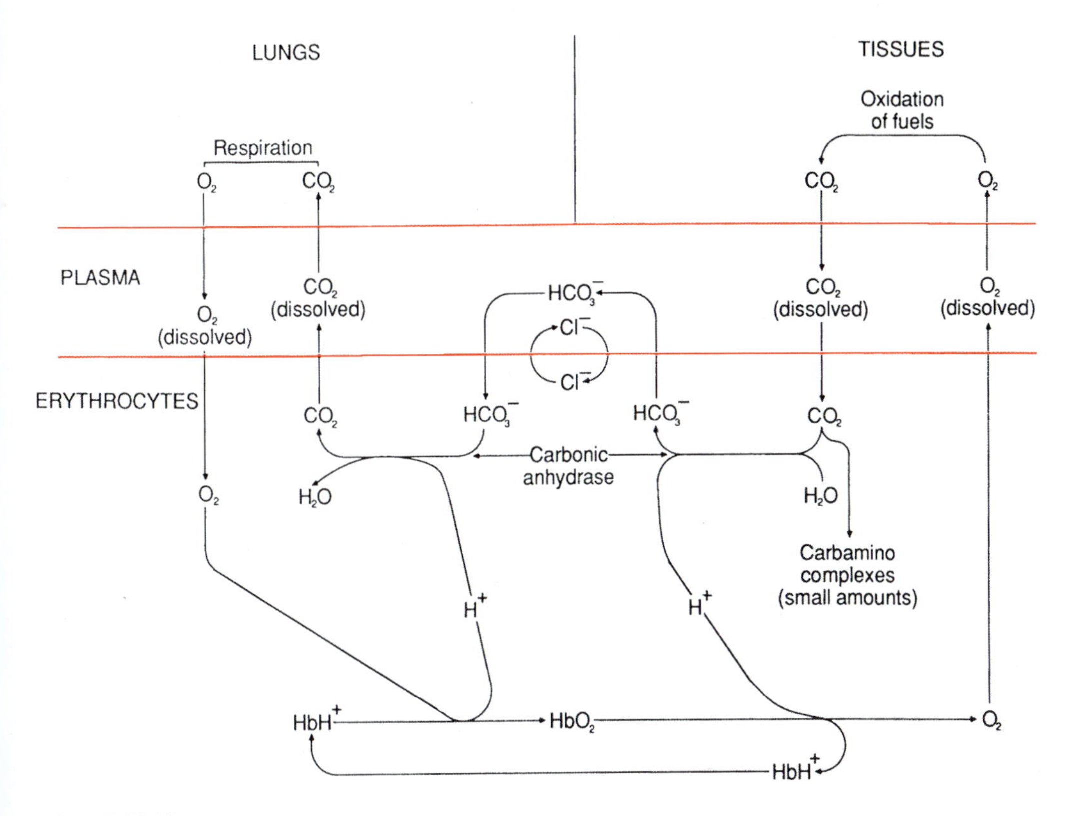

FIGURE 28-10
Interchange of carbon dioxide and oxygen between tissues and lungs. The reactions at right occur in plasma and erythrocytes in tissue capillaries during oxygen delivery to tissue cells; the reactions at left occur in the lungs during CO_2 release and O_2 uptake. The cyclic nature of these changes (lungs to capillaries to lungs) should be noted.

1. The exchange of chloride and bicarbonate across the erythrocyte membrane (the ***chloride shift***).
2. The role of H^+, and therefore CO_2 and carbonic anhydrase, in unloading oxygen from oxyhemoglobin.
3. The intermediary role of plasma as a carrier of CO_2 (as HCO_3^-).
4. The occurrence of oxygen in solution in plasma, in addition to that bound to hemoglobin; this satisfies the law of mass action (since oxygen binding to hemoglobin is an equilibrium reaction) and also moves oxygen from erythrocytes into tissues. The concentration of dissolved oxygen can be increased by breathing oxygen at pressures greater than atmospheric pressure. Hyperbaric oxygen therapy has been used to treat carbon monoxide poisoning, gas gangrene, decompression sickness, and other disorders in which hemoglobin cannot carry adequate oxygen to one or more tissues.
5. The occurrence of CO_2 chemically combined with hemoglobin, as carbamino compounds.

Nitric Oxide (NO) Binding to Hemoglobin

NO binds to hemoglobin at two sites. One is involved in scavenging NO through its binding to the ferrous iron of heme. The affinity of NO at the Fe^{2+}-heme site is about 8000 times greater than oxygen affinity at the same site. The binding of NO at the second site is a reversible process and occurs at the $\beta 93$ cysteine residue with the formation of S-nitrosothiol. This interaction of NO with $\beta 93$ cysteine is linked to the binding of oxygen to hemoglobin in the lungs and also to release of NO and O_2 into the tissues. It is thought that NO reduces regional vascular resistance. In the blood, NO is also transported by binding to plasma albumin; the exact physiological role of the transport of NO by hemoglobin or albumin is not understood.

Erythropoietin

Erythropoietin is a glycoprotein hormone that regulates red blood cell production in a feedback loop manner between kidney and bone marrow based on oxygen tension. It consists of 165 amino acids and has a molecular weight of 30,000–34,000; approximately 30% is accounted for by covalently linked carbohydrate. Erythropoietin is produced by the fetal liver and shortly after birth production switches from the liver to the kidney. In the fetus, erythropoietin functions in a paracrine-endocrine fashion because liver is the site of erythropoietin synthesis as well as erythropoiesis. The mechanism of this developmental switch is unclear. In the liver, erythropoietin synthesis occurs in the Ito cells and in a subset of hepatocytes. In the kidney, erythropoietin is synthesized in the peritubular interstitial fibroblast-like cells. Hypoxia is the primary physiological stimulus for the dramatic rise of erythropoietin levels, which can rise up to 1000-fold through an exponential increase in the number of erythropoietin-producing cells as well as in the rate of synthesis. Hypoxic states can occur due to loss of red blood cells, decreased ambient oxygen tension, presence of abnormal hemoglobin with increased oxygen affinity (discussed later), and other causes that limit oxygen delivery to the tissues (e.g., chronic obstructive lung disease). The expression of the erythropoietin gene is affected by a number of other physiological and pharmacological agents in addition to hypoxic states. Transition metal ions, Co^{2+}, Ni^{2+}, Mn^{2+}, and iron chelator desferrioxamine can stimulate erythropoietin gene expression. Carbon monoxide, nitric oxide, inflammatory cytokine tumor necrosis factor-α, and interleukin-1 can prevent expression of the gene. The latter is implicated as one of the factors causing anemia in chronic disease (*anemia of chronic disease*). Erythropoietin is the primary regulator of erythropoiesis and red blood cell mass. Erythropoietin action is mediated by its binding to plasma cell membrane receptors of erythroid progenitors and precursors. The sensitivity of erythroid progenitors to erythropoietin appears to be under developmental regulation. Colony-forming unit erythroid (CFU-E) and proerythroblasts have a peak amount of erythropoietin receptors. In peripheral reticulocytes, the receptors are undetectable. Erythropoietin's function is mediated through its receptor and includes proliferation, survival, and terminal differentiation of CFU-E cells.

The mechanism of oxygen sensing and signal transduction that leads to activation of the erythropoietin gene and erythropoietin synthesis in renal cells involves the following steps. The first step in the hypoxia-induced transcription of the erythropoietin gene consists of production of oxygen free radicals (O_2^-) by cytochrome b-like flavoheme NADPH oxidase in proportion to oxygen tension. Superoxide in the presence of iron is converted to other reactive oxygen species (e.g., $OH^\bullet$). During normal oxygen tension, superoxide and other reactive oxygen species oxidize hypoxia-inducible factor-α (HIF-α) and cause its destruction in the ubiquitin-proteosome pathway. However, at low oxygen tension, HIF-α is preserved and forms a heterodimer with constitutively expressed HIF-β. The dimer HIF-α/HIF-β is translocated into the nucleus where it interacts with hypoxia response elements to activate gene expression.

Recombinant human erythropoietin has been used in the correction of anemia of chronic renal failure. It has also been used in other disorders of anemia such as anemia of prematurity, anemia of inflammation, and anemia of malignancy.

Polycythemia is characterized by an increase in the number, and in the hemoglobin content, of circulating red cells. In patients who have chronic anoxia from impaired pulmonary ventilation or congenital or acquired heart disease, the increase in plasma erythropoietin leads to secondary polycythemia. Some renal cell carcinomas, hepatocarcinomas, and other tumors, which produce physiologically inappropriate amounts of erythropoietin, may also cause secondary polycythemia. Conversely, anemia can result from renal insufficiency and from chronic disorders that depress erythropoietin production. In *polycythemia vera* (primary polycythemia), which is a malignancy of erythrocyte stem cells of unknown cause, erythropoietin levels are normal or depressed.

28.3 Inherited Disorders of Hemoglobin Structure and Synthesis

Although hemoglobin disorders are extremely diverse, they can be generally classified into two somewhat overlapping groups.

1. Thalassemia syndromes are characterized by a decreased rate of synthesis of one or more of the globin peptides. Although there are exceptions, the globin chains produced in thalassemic states usually have normal amino acid sequence. In ***hereditary persistence of fetal hemoglobin (HPFH),*** there is a failure at birth to switch from synthesis of fetal hemoglobin (HbF) to adult hemoglobin (HbA), so that the individual continues to have high levels of HbF throughout life. Although HPFH does not cause any major hematological abnormalities and is compatible with normal life, it is grouped with the thalassemias because its molecular pathology is closely related to that of several of the ***β-thalassemias.***
2. Hemoglobinopathies are characterized by alterations in function or stability of the hemoglobin molecule arising from changes in the amino acid sequence of a globin chain. They include single-amino-acid substitutions (the largest group), insertion or deletion of one or more amino acids, drastic changes in amino acid sequence caused by frameshift mutations, combination of different pieces of two normal chains by unequal crossing over during meiosis to produce fusion hemoglobins, and increase or decrease in chain length from mutations that create or destroy stop codons. Frameshift mutants, chain termination mutants, and fusion polypeptides produce thalassemia-like syndromes. About 750 abnormal hemoglobins have been described, some of which cause no physiological abnormality; consequently, they are not true hemoglobinopathies.

Two systems of nomenclature are used for the abnormal hemoglobins. Initially, each hemoglobin, normal or abnormal, was assigned a different letter of the alphabet on the basis of its electrophoretic mobility, e.g., hemoglobins A (normal adult hemoglobin), C, D, E, F (normal fetal hemoglobin), G, H, J, M (methemoglobin), and S (sickle hemoglobin). It was soon realized that there were far more mutant hemoglobins than letters of the alphabet and that several hemoglobins might have the same electrophoretic mobility. Hemoglobins were therefore named for the geographic locations where they were discovered, e.g., Hb Cowtown, Hb Kankakee, and Hb Dakar. If a newly discovered hemoglobin has characteristics of a lettered hemoglobin, the letter and the new name are used together, e.g., Hbs J, J-Capetown, J-Rovigo, and J-Altgeld Gardens. Occasionally, abnormal hemoglobins discovered independently in two different places and given two different names are shown to have the same mutation. In such cases, both names are retained and used as synonyms. Examples include Hbs:Fort Gordon, Osler, and Nancy; and Hbs:Perth and Abraham Lincoln. Heterozygotes for an abnormal hemoglobin or a thalassemia are called carriers of the variant.

Normal Hemoglobins

Each molecule of human hemoglobin is a tetramer of two α-like (α or ζ) and two β-like (β, γ, δ, or ε) chains. The normal hemoglobins, in order of their appearance during development, are

1. Embryonic hemoglobins
   a) Gower I = $\zeta_2\varepsilon_2$
   b) Gower II = $\alpha_2\varepsilon_2$
   c) Portland = $\zeta_2\gamma_2$
2. Fetal hemoglobin. Hemoglobin F = $\alpha_2\gamma_2$; the γ may be Gγ or Aγ, depending on whether glycine or alanine is present at γ136. HbF, the major oxygen carrier in fetal life, accounts for less than 2% of normal adult hemoglobin and falls to this level by the age of 6–12 months.
3. Adult hemoglobins
   a) Hemoglobin A = $\alpha_2\beta_2$. This is the major adult type (HbA_1), composing 95% of the total hemoglobin.
   b) Hemoglobin A_2 = $\alpha_2\delta_2$. This type accounts for less than 3.5% of the total hemoglobin. Synthesis of δ chains is low at all times.

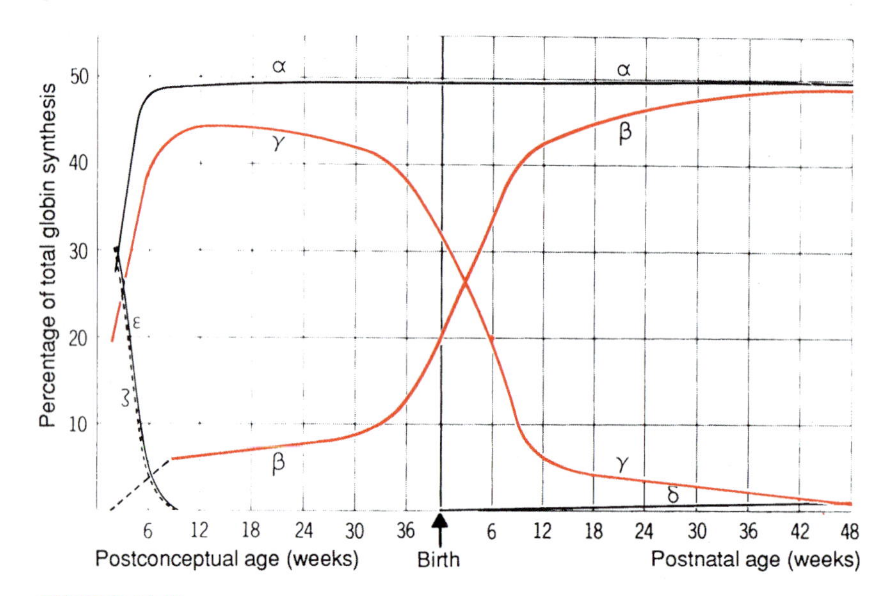

FIGURE 28-11
Changes in human globin chain concentration during development. [Reproduced with permission from W. G. Wood, Hemoglobin synthesis during human fetal development. *Br. Med. Bull.* **32,** 282 (1976).]

Embryonic hemoglobins normally are not found in the fetus after the first trimester of pregnancy because synthesis of ζ and ε chains ceases before the tenth week of gestation (Figure 28-11). HbF has a higher affinity for oxygen than does HbA, facilitating oxygen transfer from the maternal to the fetal circulation, because of the lower affinity of HbF for 2,3-DPG. Since β-chain synthesis remains low relative to γ-chain synthesis throughout intrauterine life, HbF predominates during that period. The mechanisms that control hemoglobin switching (change in expression of the globin genes during development) represent an active area of research.

Synthesis of the different hemoglobin chains is directed by different genes, at least one for each chain type. The chromosomal arrangement of these genes (Figure 28-12) was determined by restriction endonuclease mapping and by hybridization with probes (Southern blots) specific for the different globin genes (Chapter 23). The α-globin gene complex consists of two α genes and one ζ gene, separated by a pseudo-α_1 gene and a pseudo-ζ gene (previously designated the ζ_1 gene). Pseudogenes have extensive sequence homology with the corresponding normal gene but, because of mutation, do not produce functional polypeptides. Pseudogenes are detected by DNA-DNA hybridization mapping, making use of their sequence homology with the probe. Pseudogenes probably arose during evolution by gene duplication and subsequent mutations that inactivated the gene. The δ-globin gene, which is poorly expressed, may represent a duplication of the β-globin gene that is on its way to becoming a pseudogene. The

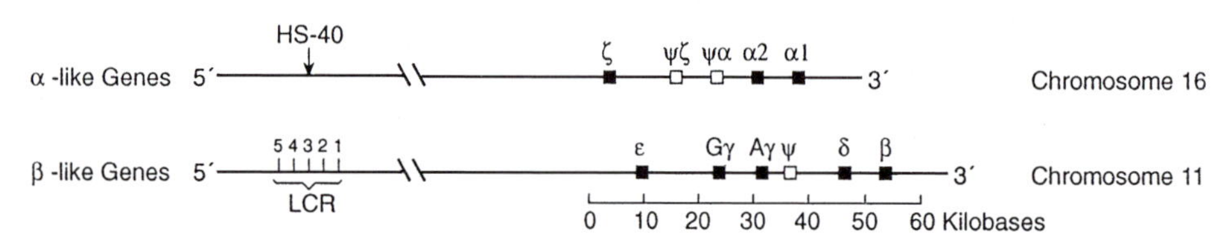

FIGURE 28-12
Arrangement of the α- and β-gene complex on human chromosomes 16 and 11 respectively. Functional genes are indicated by dark boxes; nonfunctional genes or pseudogenes (ψ), which are evolutionary remnants, are shown by open boxes. The α-gene complex consists of two pseudogenes, $\psi\zeta$ and $\psi\alpha$, and three functional genes, ζ, α_1, and α_2. The β-gene complex consists of at least one pseudogene ($\psi\beta$) and five functional genes, ε, Gγ, Aγ, δ, and β. They are arranged in the order in which the β-like genes are expressed during development. Both α- and β-like genes contain essential upstream sequences that can regulate gene expression. In chromosome 11, the locus control region (LCR) contains five DNase-hypersensitive sites (HS) that are essential for β-like gene expression. A similar region located in chromosome 16, designated HS-40, regulates α-like gene expression.

β-gene complex consists of ε, G_γ, A_γ, δ, β, and at least one pseudo-β gene. Clustering of the α-like and β-like genes is consistent with the hypothesis that members of each family arose by duplication of, and divergence from, a parent α or β gene. The order of the genes in each group, from 5′ to 3′, is the same as that of their expression during ontogeny. The order on the chromosome may be important for regulation of globin gene expression, as appears to be the case for the immunoglobulin heavy-chain genes (Chapter 35). The dynamics of globin gene expression and hemoglobin synthesis involve the participation of many genes. These include cis-acting promoter elements (e.g., TATA and CCAAT boxes, duplicated CACCC motifs), enhancer elements, and the presence of the genes that are absolutely required for expression. The gene cluster that is required for expression of β-globin-like genes is located in the locus control region (LCR) marked by five DNase-hypersensitive sites (HSs), and an analogous site in the α-globin gene cluster is known as HS-40 (Figure 28-12). Transcription of the β-gene cluster begins with the formation of the transcription complex on the promoter, which involves the interaction of the locus control gene with an individual hemoglobin gene along with trans-acting transcription factors and RNA polymerase II.

Although the α-like and β-like polypeptides contain 141 and 146 residues, respectively, the corresponding genes are considerably longer than required, since there are two introns within each globin gene. In contrast to the β-gene cluster located on chromosome 16, the α-globin genes are duplicated on chromosome 11 (Figure 28-12). Thus, in a diploid cell, four α genes are present. Introns are transcribed with exons, but they are removed ("spliced out") from the RNA before it leaves the nucleus. Although mutations in introns are generally under less "selection pressure" (have a lesser effect on fitness of the organism) than mutations in exons, mutations or deletions in introns may affect normal RNA splicing, causing thalassemia syndromes. The major steps involved in the synthesis of the β-globin polypeptide chain are shown in Figure 28-13, together with the sites in the β-globin gene that are involved in the synthesis.

The degree of relatedness of the globin genes can be estimated from the number of amino acid differences in the peptides or the number of base differences in the DNA.

Thalassemias

The thalassemias are a heterogeneous group of hypochromic, microcytic anemias caused by unbalanced synthesis of globin chains. In Southeast Asia, the Philippines, China, the Hawaiian Islands, and the Mediterranean countries, thalassemia syndromes are relatively common and constitute a significant public health problem. Worldwide, they are probably the most common hereditary diseases. The clinical severity of these disorders ranges from mild or totally asymptomatic forms to severe anemias, causing death *in utero*. Presentation depends on the specific globin genes, acquired conditions that modify the expression of these genes, race, and inheritance of genes for structural hemoglobin variants (most often HbS or HbC). Anemia and other characteristic hematological abnormalities are due to ineffective erythropoiesis and to precipitation of the excess free globin chains within the red cells. Ineffective erythropoiesis causes the appearance of hypochromic red cells, with a clear center and darker rim containing the hemoglobin (target cells). The precipitates form inclusions, called Heinz bodies, which are removed ("pitted out") by the spleen with consequent membrane damage. The resulting cells, called poikilocytes, have abnormal shapes and a shortened life span. Sequestration

Transcription Signals	Location (base pairs)
1. RNA polymerase promotor; this is the binding site for RNA polymerase II.	−50 to −1
2. 5′ End of the RNA transcript and base to which the 7-methyl guanylate-5′-diphosphate cap is attached.	1
3. Initiation site for translation of mature message into protein, i.e., the codon AUG for initiator methionyl tRNA. This amino acid is removed following translation.	51–53
4. Coding sequence for exon 1 (amino acids 1–30). In the mRNA, the 3′ end of this sequence (base 143) will be linked to the 5′ end of exon 2 (base 274).	54–143
5. Splice "donor" for intron 1. Cleavage of the bond between this and the preceding base marks the beginning of the RNA sequence to be excised.	144
6. Intron 1. Sequences of the recognition sites within the intron for splicing are not yet known for certain.	144–273
7. Splice "acceptor" for intron 1. Cleavage of the bond between this and the next base marks the end of the RNA sequence to be excised.	273
8. Coding sequence for exon 2 (amino acids 31–104). In the mRNA, bases 274 and 143 will be linked, as will bases 495 and 1346.	274–495
9. Splice "donor" for excision of intron 2.	496
10. Intron 2.	496–1345
11. Splice "acceptor" for excision of intron 2.	1345
12. Coding sequence for exon 3 (amino acids 105–146).	1346–1471
13. Translation termination codon read by the termination factors causing the ribosomes and nascent polypeptide chain to be released. For the β-globin gene, it is UAA.	1472-1474
14. The poly(A) tail is attached to the 3′end of this base. Transcription probably continues beyond this point, but the primary transcript is processed by removal of all RNA 3′ (downstream) from this base.	1606

FIGURE 28-13
Summary of regulation of transcription in human β-globin genes. Base pairs in DNA are numbered from the transcription initiation site. Bases downstream (in the direction of $5' \rightarrow 3'$) have positive numbers, and bases upstream have negative numbers. E = Exon; I = intron; A_n, represents the polyadenylate "tail" of length n. (Courtesy of Dr. R. Condit.)

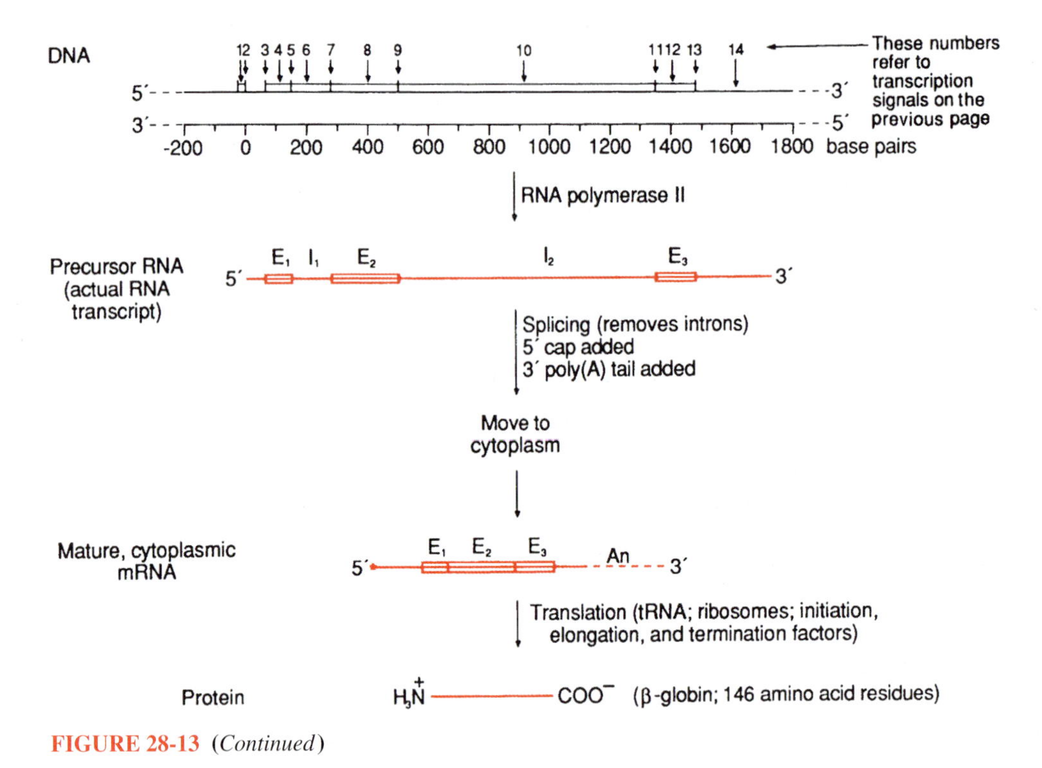

FIGURE 28-13 (*Continued*)

of abnormal cells by the spleen and the reticuloendothelial hyperplasia caused by hemolysis lead to splenomegaly.

The thalassemias are classified according to which chains are deficient. The most important types are the α- and β-thalassemias. In each type, globin chain synthesis may be reduced (α^+, β^+) or absent (α°, β°). By the use of probes to detect the presence or absence of specific globin genes, thalassemias due to deletion of globin genes have been distinguished from nondeletion varieties. δ-, δ-β-, and γ-δ-β-Thalassemias are extremely rare.

α-Thalassemias

In order of increasing severity, the four clinical entities are

1. the ***silent carrier state*** (α-thalassemia-2),
2. ***α-thalassemia trait*** (α-thalassemia-1),
3. ***HbH disease,*** and
4. ***hydrops fetalis with Hb Barts,*** or ***homozygous α-thalassemia.***

The first three are α^+-thalassemias, and the last is an α°-thalassemia. Since non-α-globin synthesis is not affected in α-thalassemia, the relative proportions of hemoglobins A, A_2, and F are unchanged, although all three are reduced in concentration.

These conditions are produced by deletion of one, two, three, or four copies of the α-globin gene; the more copies missing, the more severe the disease. Although deletion is probably the most common cause of the α-thalassemias (and is the *only* cause identified for α°-thalassemias), a number of nondeletion mutations are known. Most Asian cases of α-thalassemia trait result from heterozygosity for a double-deletion chromosome. The defective chromosomes and other data relevant to α-thalassemias are listed in Table 28-2. Since deletion mutations remove variable amounts of DNA on the 5′ and 3′ sides of the affected genes, single- and double-deletion chromosomes resulting from different mutations will not, in general, lack identical pieces of DNA. The deletion may extend to the ζ genes, in which case synthesis of embryonic hemoglobins is affected.

Decreased α-chain synthesis leads to formation of two abnormal hemoglobin tetramers. Hemoglobin Barts (γ_4) is found in the cord blood in association with all of the α-thalassemias. It arises from normal production of γ chains in association with a lack of α chains. After birth, if the infant survives, the γ chain is replaced by the β chain. If α-chain production is severely depressed, excess β chains form and self-associate (β_4). Neither Hb Barts nor HbH shows cooperativity, and their oxygen binding curves resemble that of myoglobin. Consequently, they bind oxygen very tightly, do not release it to the tissues, and are almost useless for oxygen transport.

In HbH disease, 5–30% of the hemoglobin in adults is HbH, which is detected as a rapidly moving hemoglobin during electrophoresis at pH 8.4. In some patients, a

TABLE 28-2
Characteristics of the α-Thalassemia Syndromes

Syndrome (or Phenotype)	Genotype	No. of α-Genes Present	Possible Parental Genotypes*	Frequency†	Clinical Severity	Hemoglobin Pattern
Hydrops fetalis with Hb Bart's	--/--	0	α-Thalassemia trait × α-Thalassemia trait	0.25	Lethal (death in utero)	Cord blood: mostly Hb Bart's (γ_4); small amounts of HbH (β_4) and Hb Portland
Hemoglobin H disease	--/-α	1	α-Thalassemia trait × silent carrier	0.25	Severe to moderate microcytosis; hemolysis	Cord blood: about 25% Hb Bart's; adult blood: 5–30% HbH.
α-Thalassemia trait (α-thalassemia 1)	(I) --/αα or (II) -α/-α	2	(I) α-Thalassemia trait × normal or (II) Silent carrier × Silent carrier	0.50 (I); 0.25 (II)	Mild microcytosis	Cord blood: about 5% Hb Bart's; adult blood: normal composition.
Silent carrier (α-thalassemia 2)	-α/αα	3	Silent carrier × normal	0.50	None	Cord blood: about 1–2% Hb Bart's; adult blood: normal composition.

*The parental genotypes are only representative examples. Matings of HbH genotypes with other genotypes are not shown.
†The probability that a child from the indicated mating will have this thalassemia syndrome.

slow-moving, minor band of Hb Constant Spring may also be detected. HbH is unstable and precipitates in older red cells and under oxidant stress (e.g., treatment of the patient with primaquine) as Heinz bodies, which are removed by the spleen. Following splenectomy, these inclusions can be seen in peripheral blood after staining with methylene blue. In erythrocytes from unsplenectomized patients, fine inclusions of precipitated HbH can be induced and visualized by staining for 1 hour with brilliant cresyl blue. Although other unstable hemoglobins also produce inclusion bodies, staining for 1 hour is usually not sufficient for their visualization. Patients with iron deficiency together with HbH disease may not show the presence of HbH.

Hemoglobin H has been described in several other disorders. In erythroleukemia (*Di Guglielmo's syndrome*), unbalanced globin chain synthesis may result from chromosomal aberrations associated with the leukemia. In a patient with acute myeloblastic leukemia, 50–65% of the hemoglobin was HbH. All four α-globin genes were intact, but α-chain synthesis was very low. Specific α-globin gene repressors may have been synthesized by the tumor cells.

Several nondeletion α^+-thalassemias have been described. Four mutations in the normal stop codon for the α-globin gene have been identified (Table 28-3). In these cells, the mutated codon specifies an amino acid, and translation continues beyond the normal point, extending the α chains by 31 additional amino acids. Synthesis is terminated by a stop codon downstream from the normal stop codon. Hb Constant Spring is relatively common in Southeast Asian populations. In heterozygotes, about 1% of the α-globin is Hb Constant Spring (25% is expected

TABLE 28-3
Hemoglobins Having an Extended α-Globin Chain Due to Mutation of the Chain Termination Codon

Hemoglobin	Codon Change at α-142
Hb Constant Spring	UAA→CAA (Gln)
Hb Icaria	UAA→AAA (Lys)
Hb Koya Dora	UAA→UCA (Ser)
Hb Seal Rock	UAA→GAA (Glu)

for the output of one α-globin gene), apparently because the mRNA is unstable. Because of the low rate of synthesis of Hb Constant Spring, when a chromosome carrying this mutation is present along with one lacking α-globin genes, HbH disease results. Extended α chains also result from frameshift mutations (Hb Wayne) and duplications (Hb Grady).

Other α-globin mutants have a mutation at the intron-exon boundary (mRNA splice boundary) that prevents RNA processing or a mutation in the recognition sequence needed for polyadenylation of the mRNA during RNA maturation. In other α^+-thalassemias with both α genes present but active only at a reduced level, the defect has not been identified.

Thalassemias of the β-Globin Gene Family

In thalassemias of the β-globin gene family, there is reduced synthesis of β chains, with or without reduced synthesis of γ or δ chains. An isolated decrease in γ- or δ-globin synthesis would probably be benign and likely to be detected only by chance. Hemoglobin Lepore is usually included with the β-thalassemias, since synthesis of normal β chains is reduced or absent.

The β-thalassemias are more important in terms of patient suffering and expense than the α-thalassemias. Because all of the normal hemoglobins of fetal and adult life require α-globin chains for normal function, homozygous α-thalassemia (hydrops fetalis) is usually fatal *in utero* by the third trimester of pregnancy. When ζ-chain synthesis is unusually prolonged and a living fetus is born, death invariably occurs soon after delivery. In the α-thalassemias, even a single α-locus seems sufficient to preclude serious morbidity. In contrast, the β-globin chain is not needed until after birth. Except for the rare instances in which γ-, δ-, and β-chain syntheses are all absent, β-thalassemic fetuses are delivered normally at term. In homozygous β-thalassemia, problems begin about 4–6 months postnatally, when γ-chain synthesis has declined and β-chain synthesis should have taken over. Thus, β-thalassemia is a crippling disease of childhood.

Characteristics of the more common β-thalassemia syndromes are summarized in Table 28-4. Not included are two different abnormal heterozygotes. Anemia results from precipitation of excess α-globin chains, premature red cell destruction in bone marrow and the circulation, and deficiency of functional hemoglobin tetramer. β-Thalassemia major, or Cooley's anemia, occurs when β-globin synthesis is markedly depressed or absent. The ineffective erythropoiesis causes massive erythroid proliferation, skeletal deformities, and extramedullary erythropoiesis. The usual treatment is frequent blood transfusion. However, this treatment leads to iron overload, usually by the early teenage years, and death due to iron deposition cardiomyopathy by the second decade. Treatment with chelating agents, particularly deferoxamine, has been of considerable value (Chapter 29). Transfusions also put these patients at risk for viral hepatitis and acquired immunodeficiency syndrome (AIDS). Immunization with the hepatitis B virus vaccine reduces the risk for hepatitis B infection. Two other therapeutic approaches in the treatment of severe β-thalassemia have been employed. These include bone marrow transplantation from HLA identical donors and augmentation of HbF synthesis by using hydroxyurea, butyric acid analogues or acylating agents (discussed later).

Thalassemia intermedia occurs in homozygotes for a δ-β deletion chromosome (δ°-β°-thalassemia). Because the production of HbF is greater than in β°-thalassemia, it is a milder clinical form of homozygous β-thalassemia. The increased γ-chain synthesis in thalassemia intermedia and the increased synthesis of γ- and δ-globins in β°- and β^+-thalassemias suggest that some site downstream from the γ and δ genes normally prevents their expression. Thalassemia intermedia also is seen in homozygous β^+-thalassemia if the reduction in β-globin synthesis is not severe. Under oxidative stress or following a febrile illness, transfusion may be needed. Because these patients are anemic, they have increased intestinal iron absorption and may develop secondary hemochromatosis in later life. β-Thalassemia trait is usually asymptomatic.

The molecular defects of β-thalassemias and related disorders are heterogeneous and nearly 200 mutations have been identified. Many of the mutations are single-nucleotide substitutions affecting critical loci in the expression of β-globin-like genes. For example mutations in the promoter region A$\rightarrow$G at position -29 or C$\rightarrow$T at position -88 cause defects in binding of transcription factors leading to decreased synthesis of β-globin mRNA. The same mutations however promote transcription of δ-globin gene, leading to increased synthesis of hemoglobin A_2. These mutations have been noted in African-Americans.

δ-β Thalassemias, Lepore Hemoglobins, and Hereditary Persistence of Fetal Hemoglobin (HPFH)

δ-β-Thalassemias, Lepore hemoglobin thalassemia, and HPFH are characterized by lack of β-globin synthesis and are caused by deletions in the β-gene family on chromosome 11 (Figure 28-14).

Hemoglobin Lepore contains abnormal δ-β fusion polypeptides. The N-terminal sequence in the non-β

TABLE 28-4
*Characteristics of the β-Thalassemia Syndromes**

		Hemoglobin Pattern			Molecular Defect	
Type	**Severity**	**A_1**	**A_2**	**F**	**mRNA**	**Genes**
Homozygous States						
β^+-Thalassemia	Cooley's anemia (thalassemia major)	–	+	+	β-mRNA markedly reduced	β-genes present
β°-Thalassemia	Cooley's anemia	0	+	+	Absent or nonfunctional β-mRNA	β-genes present
$\delta^\circ\beta^\circ$-Thalassemia	Milder disease (thalassemia intermedia)	0	0	100%	Absence of β- and δ-mRNA	Deleted β- and δ-genes
HB Lepore	Cooley's anemia	0	0 (and 25% Hb Lepore)	75%	Absence of normal β- and δ-mRNA	Normal β- and δ-genes absent, δβ-fusion gene present
Heterozygous States						
β^+-Thalassemia	Mild disease (β-thalassemia trait or β-thalassemia minor)	–	+ (usually not more than 6.5%)	= or slightly +	Deficiency of β-mRNA	β-genes present
β°-Thalassemia	Mild disease	–	+ (usually not more than 6.5%)	= or slightly +	Deficiency or nonfunctional β-mRNA	β-genes present
δβ-Thalassemia	Mild disease	–	–	5–20%	Deficiency of β- and δ-mRNA	Deletion of β- and δ-genes on one chromosome
Hb Lepore	Mild disease	–	– (and 5–15% Hb Lepore)	+	Deficiency of normal β- and δ-mRNA	δβ-fusion gene on one chromosome; other chromosome normal

*+ means that the fraction of the indicated hemoglobin in the erythrocyte is variably increased relative to that of a normal individual; – means decreased; = means normal fraction; and 0 means that the hemoglobin is absent.

chains is homologous with that of δ-globin, and the C-terminal sequence is homologous with that of β-globin. The fusion gene arose from nonhomologous crossing over between the δ gene on one chromosome and the β gene on the other. The products are a chromosome with a δ-β fusion gene but no normal β or δ loci, and a chromosome with normal β- and δ-globin genes and a β-δ fusion gene. Homozygotes for the δ-β (Lepore) chromosome will have a $\delta^\circ\beta^\circ$-thalassemia, while homozygotes for the β-δ (anti-Lepore) chromosome will be clinically normal, have normal β- and δ-chain synthesis, and produce a β-δ fusion chain. Three δ-β fusion genes (hemoglobins Lepore-Hollandia, Lepore-Boston, and Lepore-Baltimore) and three anti-Lepore genes (hemoglobins Miyada, P-Congo, and P-Nilotic), which differ in the crossover site, have been identified. Because of the high degree of homology between the δ and β genes (96%), it is difficult to pinpoint the exact crossover site. HPFH and δ-β-thalassemia are also thought to result from nonhomologous crossovers within the β-gene family.

In homozygotes for Hb Lepore and G_γ-A_γ-δ-β-thalassemia, the deletion does not extend beyond the 5′ end of the δ gene, and the G_γ and A_γ loci are only weakly expressed. In three deletions identified in homozygotes for HPFH, DNA upstream from the 5′ end of the δ locus is deleted together with some or all of the δ and β loci.

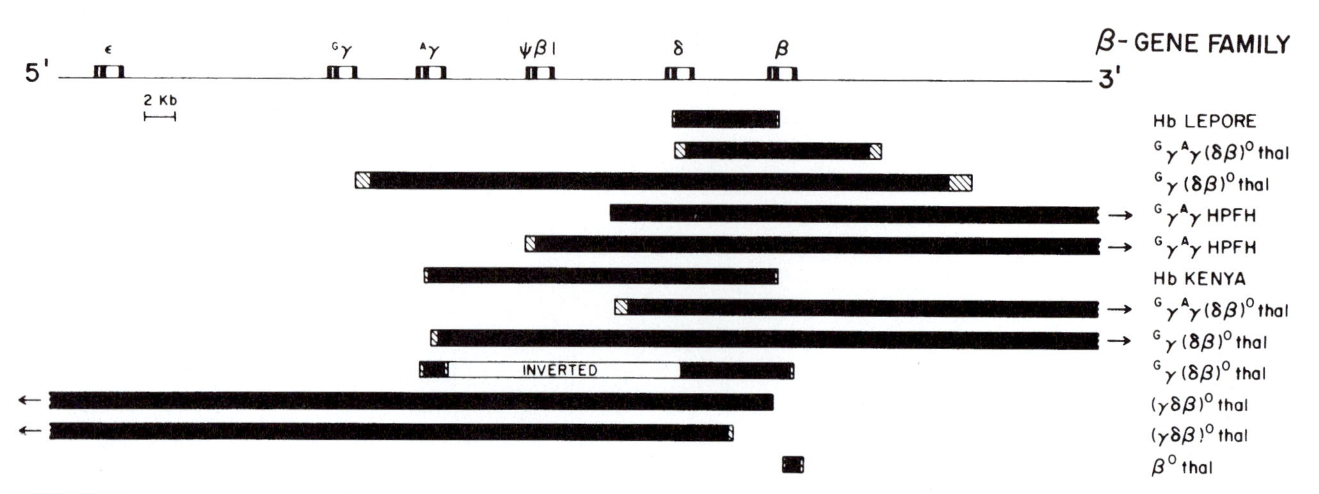

FIGURE 28-14
Chromosomal organization of the human β-globin gene family and deletions that cause thalassemia (thal). In the β-gene family, the exons are shown as dark boxes and the introns as open boxes. The various deletions are shown by horizontal bars under the β-gene family. Cross-hatched regions indicate uncertainty in the termination points of the deletion. Arrows indicate unmapped end points of the deletion. [Reproduced with permission from R. A. Spritz and B. G. Forget, The thalassemias: molecular mechanisms of human genetic disease. *Am. J. Hum. Genet.* **35,** 336 (1983). © 1983 by the University of Chicago Press.]

Presumably, sequences somewhere in that region repress expression of the γ genes; removal of these sequences allows unrestrained expression of the γ loci. If the deletion extends too close to the G_γ locus, however, expression of the G_γ gene is reduced, producing the more severe phenotype seen in G_γ-δ-β-thalassemia. Finally, in γ-β-thalassemia, the δ locus and both γ loci are deleted, while the β gene is normal; yet there is no synthesis of any of the β-like globins. The reason for the nonexpression of this normal β locus is unknown. Thus, these disorders hint at the presence of regulatory information within the β-gene complex. It may eventually be possible to activate the γ genes to produce HPFH-like states and reduce the morbidity associated with β-thalassemias and with mutations affecting the function or stability of the β-globin chain.

In normal adults, the small amount of HbF present is distributed nonuniformly among the erythrocytes. The distribution pattern was established by selective extraction of HbA from erythrocytes in a peripheral blood smear. The more poorly soluble HbF left behind stains pink with eosin. In most persons who have HPFH, all erythrocytes contain some HbF.

Hemoglobinopathies

The term *hemoglobinopathies* refers to hemoglobin disorders caused by normal synthesis of qualitatively abnormal globin chains. Transcription and translation of mutant genes usually proceed at a normal rate, but the products denature rapidly or function abnormally.

A selected list of mutant hemoglobins is given in Table 28-5. The most common mutations are the single-amino-acid substitutions caused by one nucleotide change in a globin gene. The table also lists several deletion mutations and one nonsense mutation. Globin chains that contain two mutations are extremely rare and appear to have resulted from a second mutation in a common mutant (usually HbS or HbC).

Molecular Pathology

The effect of a mutation on the structure and function of hemoglobin is determined by the effects of the amino acid change. Many residues in hemoglobin have been conserved during evolution, probably because they are essential for normal hemoglobin function. The most critical regions for normal structure and function are the heme contacts and $\alpha_1\beta_1$ and $\alpha_1\beta_2$ contacts. Substitutions near the heme group can cause weak or absent heme binding, resulting in loss of the heme group from the affected subunits (Hb Sydney, Hb Hammersmith). They can also stabilize the iron as Fe^{3+} and prevent reversible oxygen binding (as in hemoglobin M). The heme group is also less tightly bound in methemoglobin than normal hemoglobin.

Substitutions in amino acids involved in subunit contacts can have several effects. In Hb Philly, a Tyr → Phe change in the at $\alpha_1\beta_1$ interface eliminates a hydrogen bond needed to stabilize deoxyhemoglobin (T-state) and leads to dissociation into monomers, which precipitate. This dissociation decreases cooperativity and increases oxygen

TABLE 28-5

Molecular and Clinical Characteristics of Some Abnormal Hemoglobins

Hemoglobin	Position*	Amino Acid Change	DNA Codon[†] Alterations	Remarks[‡]
α-Globin Mutations				
Memphis	23(B4)	Glu→Gln	GAG→CAG	Reduces sickling when present in HbS homozygotes.
Torino	43(CEI)	Phe→Val	TTC→GTC	Decreases O_2 affinity; unstable; mild anemia; decreases heme binding.
Ann Arbor	80(F1)	Leu→Arg	CTG→CGG	Unstable.
Chesapeake	92(FG4)	Arg→Leu	CGG→CTG	Increases O_2 affinity; R state stabilized; $\alpha_1\beta_2$-contact mutant.
G-Georgia	95(G2)	Pro→Leu	CCG→CTG	Oxy form dissociates to dimers; decreases O_2 affinity; $\alpha_1\beta_2$-contact mutant.
Bibba	136(H19)	Leu→Pro	CTG→CCG	Dissociates.
β-Globin Mutations				
S	6(A3)	Glu→Val	GAG→GTG	See discussion of sickle cell anemia in this chapter
G-Makassar	6(A3)	Glu→Ala	GAG→GCG	Benign; does not interact with HbS.
C	6(A3)	Glu→Lys	GAG→AAG	Crystallizes in red blood cells; enhances sickling when heterozygous with HbS.
Leiden	6 or 7 (A3 or 4)	Glu→---	GAG→---	Deletion decreases O_2 affinity; mild hemolysis; unstable.
G-San Jose	7(A4)	Glu→Gly	GAG→GCG	Compare to HBS; benign.
Saki	14(A11)	Leu→Pro	CTG→CCG	Unstable; Pro disrupts α-helix.
Savannah	24(B6)	Gly→Val	GGT→GTT	Unstable.
E	26(B8)	Glu→Lys	GAG→AAG	Mild hemolytic anemia; increases subunit dissociation; normal O_2 affinity; $\alpha_1\beta_1$-contact mutant.
Genova	28(B10)	Leu→Pro	CTG→CCG	Increases O_2 affinity; unstable.
St. Louis	28(B10)	Leu→Gln	CTG→CAG	Iron readily oxidized; polar group in heme pocket; increases O_2 affinity; unstable.
Tacoma	30(B12)	Arg→Ser	AGG→AG(T,C)	Normal O_2 affinity; decreases Bohr effect; unstable.
Abraham Lincoln	32(B14)	Leu→Pro	CTG→CCG	Also called Perth; unstable.
Philly	35(CI)	Tyr→Phe	TAC→TTC	Loss of H bond needed to stabilize T state; high O_2 affinity; unstable.
Hammersmith	42(CD1)	Phe→Ser	TTT→TCT	Decreases O_2 affinity; unstable; poor heme binding severe anemia; inclusion bodies.
Louisville	42(CD1)	Phe→Leu	TTT→TT(A,G) or TTC	Decreases O_2 affinity; unstable; mild anemia; heme misoriented but not lost.
Tochigi	56–59 (D7–E3)	deletion	—	Shortening of E-helix; unstable.
Bristol	67(E11)	Val→Asp	GTG→GA(T,G)	Unstable; polar group in heme pocket; unusual codon needing 2 mutations.
Sydney	67(E11)	Val→Ala	GTG→GGG	Unstable; mild hemolysis; inclusion bodies; poor heme binding.
Seattle	70(E14)	Ala→Asp	GCC→GAC	Unstable; mild hemolysis; decreases O_2 affinity.
Shepherd's Bush	74(E18)	Gly→Asp	GGC→GAC	Unstable; increases O_2 affinity; negative charge in DPG site; decreases DPG binding.
Tours	87(F3)	Thr→---	ACA→---	Deletion shortens F-helix; unstable; increases O_2 affinity.
Gun Hill	91–95	deletion	—	Deletion shortens F-helix; unstable; increases O_2 affinity.
Köln	98(FG5)	Val→Met	GTG→ATG	Unstable; increases O_2 affinity; $\alpha_1\beta_2$-contact mutant.
Nottingham	98(FG5)	Val→Gly	GTG→GGG	Unstable; increases O_2 affinity; T-state destabilized.
Kempsey	99(G1)	Asp→Asn	GAT→AAT	Increases O_2 affinity; R-state stabilized.

(continued)

TABLE 28-5 *(Continued)*

Hemoglobin	Position*	Amino Acid Change	DNA Codon† Alterations	Remarks‡
Kansas	102(G4)	Asn→Thr	AAC→ACC	Very low O_2 affinity; cyanosis.
Beth Israel	102(G4)	Asn→Ser	AAC→AGC	
Yoshizuka	108(G10)	Asn→Asp	AAC→GAC	
Presbyterian	108(G10)	Asn→Lys	AAC→AA(A,G)	Unstable; increases O_2 affinity; $\alpha_1\beta_1$-contact mutant.
San Diego	109(G11)	Val→Met	GTG→ATG	
Peterborough	111(G13)	Val→Phe	GTC→TTC	
Indianapolis	112(G14)	Cys→Arg	TGT→CGT	Unstable.
D-Los Angeles	121(GH4)	Glu→Gln	GAA→CAA	Also called D-Punjab; benign; enhances sickling in heterozygotes with HbS.
O-Arab	121(GH4)	Glu→Lys	GAA→AAA	As for HbD-LosAngeles.
North Shore	134(H12)	Val→Glu	GTG→GAG	Unstable; mild hemolysis; normal O_2 affinity.
Syracuse	143(H21)	His→Pro	CAC→CCC	Increases O_2 affinity, decreases Bohr effect; loss of positive group at DPG binding site.
McKees Rocks	145(HC2)	Tyr→term.	TAT→TAA	Decreases DPG binding and Bohr effect; increases O_2 affinity; deletes C-terminal residues; eliminates C-terminal bonding.
Bethesda	1145(HC2)	Tyr→His	TAT→CAT	Increases O_2 affinity; decreases Bohr effect and DPG binding; disrupts C-terminal H bond, destabilizes T-state.
Cowtown	146(HC3)	His→Leu	CAC→CTC	
Mutations Causing Congenital Methemoglobinemia				
α-Chain:				
M-Boston	58(E7)	His→Tyr	CAC→TAC	
M-Iwate	87(F8)	His→Tyr	CAC→TAC	
β-Chain:				Only known in heterozygotes; all cause a benign cyanosis.
M-Freiburg	23(B5)	Val→---	GTT→----	
M-Saskatoon	63(E7)	His→Tyr	CAT→TAT	
M-Milwaukee	67(E11)	Val→Glu	GTG→GAG	
M-Hyde Park	92(F8)	His→Tyr	CAC→TAC	

*The first number is the residue position, with 1 being the N-terminal amino acid; Kendrew's helical notation is given in parentheses.

†Codon assignments are based on published sequences for human α- and β-globin genes. Sequences of the alpha-1 and alpha-2 globin genes do not differ from each other at any of the codons used above. In a few instances, the normal codon could have mutated to either of two or three codons; these ambiguities are indicated by: term = termination codon; --- means codon (and amino acid) deleted. The codons shown here would occur on the nontranscribed strand in DNA.

‡Benign means no observable symptoms.

affinity. In hemoglobins Yoshizuka, Presbyterian, and Peterborough, the $\alpha_1\beta_1$ interface also is affected, but they exhibit decreased oxygen affinity because the R-state (oxyhemoglobin) is more destabilized than the T-state.

Contacts at the $\alpha_1\beta_2$ interface stabilize the interaction of the two $\alpha\beta$ dimers through the packing contacts and allow subunit motion during the T⇄R transformation at the sliding contacts. Mutations in the β chains of Hb Kansas and Hb Beth Israel at this interface eliminate a hydrogen bond found only in the R-state, destabilizing the R-state and decreasing oxygen affinity and dissociation of the tetramer. The oxygen affinities of these variants are so low that they cause cyanosis, usually seen only in methemoglobinemia. Hb Chesapeake has an α-chain mutation that affects the $\alpha_1\beta_2$ interface, stabilizing the R-state and increasing oxygen affinity. Other $\alpha_1\beta_2$ interface mutants, such as HbG Georgia, dissociate into dimers upon oxygenation and reassociate to tetramers when oxygen is removed.

The deleterious effect of an amino acid substitution is due to replacement of a useful side chain by one that cannot perform the necessary function. The importance of a side chain may reside in its size, shape, or charge. A side chain that is too small (Hbs Torino, Hammersmith, and Sydney) or too large (Hbs Savannah and Peterborough) can be disruptive. Because 80% of the globin chain has a helical conformation, mutations that introduce proline residues are likely to disrupt this structure because proline cannot fit into an α helix (Chapter 4). The disruption can cause severe disease (Hbs Bibba, Abraham Lincoln, and Genova) or be benign (Hb Saki). If proline is introduced at one end of a helix, it does not disrupt it. However, in Hb Syracuse, new proline at the end of the H helix eliminates positive charge needed for 2,3-DPG binding.

Loss of a charged or polar group (Hb Syracuse) or introduction of a charged or polar group into the hydrophobic interior of the molecule (Hbs Ann Arbor, St. Louis, North Shore, and Indianapolis) usually distorts the structure and causes altered function. If the mutation occurs at a site where the side chain is not very important or where only the type of side chain is important, the mutation may well be benign. A number of mutations of this type have been detected during routine screening in many populations.

Unstable Hemoglobins

Many of the substitutions mentioned above adversely affect the stability of the tetrameric molecule and cause easy denaturation and precipitation within erythrocytes to form Heinz bodies. Deletion of one or more amino acids is also likely to cause instability, precipitation (e.g., Hbs Leiden, Tochigi, Tours, Gun Hill), and membrane damage. Consequently, these variants are associated with intravascular hemolysis, anemia, reticulocytosis, splenomegaly, and, in some patients, intermittent urobilinuria.

If the hemoglobin precipitates soon after synthesis, before the cells leave the bone marrow, a thalassemia-like syndrome may occur (Hbs Nottingham and Indianapolis). These mutations as well as Hbs Hammersmith, Abraham Lincoln, Bibba, and others produce a severe clinical disorder with massive hemolysis that is not improved by splenectomy. Other mutations (e.g., Hbs Torino, Ann Arbor, St. Louis, Koln, and Shepherd's Bush) in which precipitation occurs in the circulation cause severe hemolysis but are improved by splenectomy, which increases the erythrocyte lifetime. Although sometimes difficult, it is important to distinguish these two groups clinically when deciding whether splenectomy is indicated. Most unstable hemoglobins are associated with only mild hemolysis and occasional hemolytic crises, usually brought on by some stress such as a mild infection or treatment with sulfonamide or other oxidizing drugs (e.g., Hbs Philly, Sydney, North Shore, Leiden, Gun Hill, Seattle, and Louisville). Another group (e.g., Hbs Saki and Tacoma) is clinically benign, and the unstable hemoglobin is discovered incidentally by routine electrophoretic screening.

The difference in disease severity among the unstable hemoglobins may reside in the number and time of formation of Heinz bodies. Heinz body formation is believed to start when hemoglobin is oxidized to methemoglobin. The unstable methemoglobins are degraded to hemichromes, a process promoted by their tendency to dissociate (separate subunits form hemichromes more readily than tetramers). The tendency of the chains to form hemichromes decreases in the order $\alpha > \beta > \gamma$. Hemichromes precipitate and form Heinz bodies, which attach hydrophobically to the erythrocyte membrane, increasing membrane permeability and the rate of lysis. Loss of heme and oxidation of sulfhydryl groups do not appear to be necessary for precipitation.

The degree of morbidity associated with an unstable hemoglobin is also determined by the effect of the mutation on oxygen affinity. If the oxygen affinity is decreased, delivery of oxygen to the tissues will be higher than normal and compensate, in part, for the lower hemoglobin concentration. This is the case with Hbs Leiden, Seattle, Louisville, and Peterborough. In contrast, Hbs Köln, St. Louis, and Shepherd's Bush, which cause severe hemolytic anemia and Heinz body formation, have increased oxygen affinity. The effect of the anemia is exacerbated by the inability of erythrocytes to unload the oxygen they carry. Both groups have similarly low hemoglobin levels and rapid hemolytic rates. Thus, the phenotype caused by a mutation is the net result of all its effects on hemoglobin structure and function.

Secondary Polycythemia Syndromes

Mutations thatincrease oxygen affinity (reduce P_{50}) can decrease tissue oxygenation more than expected on the basis of hemoglobin concentration. At low oxygen tension, production of erythropoietin is stimulated, causing a secondary polycythemia that increases oxygen delivery to the tissues and partially compensates for the hemoglobin abnormality.

Any mutation that stabilizes the R (oxy) state (Hb Chesapeake) or destabilizes the T (deoxy) state (Hb Kempsey) can cause polycythemia. More than 28 such variants are known. Fourteen of these high-affinity hemoglobins have mutations at the $\alpha_1\beta_2$ interface. Because the C terminus of the β chain supplies a hydrogen bond that is important for stabilizing the T-state, its replacement (Hb Bethesda), deletion (Hb McKees Rocks), or alteration in its environment (Hbs Syracuse and Cowtown) markedly decreases P_{50}. Hb San Diego is unique in having high oxygen affinity due to a substitution in the $\alpha_1\beta_1$ interface. Mutations that decrease the stability of the hemoglobin and cause precipitation or dissociation can also increase oxygen affinity.

Congenital Methemoglobinemias and Cyanosis

Cyanosis is a bluishcoloration of the skin caused by the presence of more than 5 g of deoxyhemoglobin or methemoglobin per deciliter of blood in the subcutaneous capillary beds. It occurs in severe cardiopulmonary disease from poor oxygenation of normal hemoglobin or methemoglobin production. An extremely rare cause of cyanosis is the presence of a hemoglobin variant with very low oxygen affinity (e.g., Hb Kansas, Hb Beth Israel).

Two groups of genetic defects can cause methemoglobinemia.

1. Defects in oxidoreductase in erythrocytes (methemoglobin reductase, glutathione reductase, glucose-6-phosphate dehydrogenase; Chapter 15). Patients who have these defects usually are not cyanotic unless subjected to oxidative stress (e.g., treatment with primaquine or sulfonamide).
2. Defects in which substitution of an amino acid in the region of the heme pocket increases the stability of Fe^{3+} in the heme group and produces methemoglobin. Although heme oxidation occurs in several unstable hemoglobins, cyanosis is principally a manifestation of the Hbs M (Table 28-5). Since a single mutation can affect only one type of subunit (α or β), half of each tetramer is normal. Most of these hemoglobins undergo reduced T-to-R transitions but with more difficulty. In addition, since only heterozygotes, for the Hbs M have been identified, half the tetramers are normal, and the carriers have no functional impairment. Cyanosis is visible because only 1.5–2 g of methemoglobin are needed to produce the same degree of discoloration as 5 g of deoxyhemoglobin. Diagnosis of patients with Hbs M is important because of the seriousness of the alternative causes of cyanosis.

Hemoglobin S and Sickling Disorders

The most common, severe, and best studied hemoglobin mutation is HbS, which causes sickle cell anemia (or disease) in homozygous individuals and sickle cell trait in heterozygous individuals. The characteristic change in shape of erythrocytes from biconcave disk to curved and sickle-like occurs at low oxygen pressures. Carriers of sickle cell trait are asymptomatic, since their red cells do not sickle unless the P_{O_2} drops below about 25 mm Hg; this happens only under unusual circumstances, such as unassisted breathing at high altitudes, some severe forms of pneumonia, and occasionally during anesthesia.

The symptoms of sickle cell anemia are due to sequestration and destruction of the abnormal, sickled erythrocytes in the spleen and to the inability of many sickled cells to pass through capillaries. The normal life span of 120 days for red blood cells is decreased to 10–12 days. Painful vaso-occlusive crises can occur anywhere in the body and are the major complication in sickle cell anemia. These crises are caused by blockage of capillaries in the affected tissue by the deformed red cells, which causes hemostasis, anoxia, further sickling, and eventually infarction. Increased adhesion of sickled erythrocytes to capillary endothelial cells may also contribute to capillary obstruction. Crises occur only when the circulation slows or hypoxia is present because the normal circulatory rate does not keep cells deoxygenated for the approximately 15 seconds required for sickling to begin. In contrast to the splenomegaly usually seen in chronic hemolytic anemias, a small, fibrous spleen is usually seen in adults with sickle cell anemia. This autosplenectomy is due to repeated infarction of the spleen. The clinical complications of sickle cell disease are highly variable. The severe forms of the disease occur in homozygous SS disease and S/β°-thalassemia and a milder disease occurs in double heterozygous SC disease and S/β^{+}-thalassemia. Patients who have a concurrent α-thalassemia trait or high HbF levels have a mild course of the SS disease. The latter observation has been utilized in the pharmacological approaches that raise HbF levels (discussed later).

Some of the clinical consequences in SS disease include megaloblastic erythropoiesis, aplastic crisis, stroke, bone pain crisis, proneness to infection particularly by *Pneumococcus, Salmonella,* and *Haemophilus* due to hyposplenism and acute chest syndrome. Prophylactic use of penicillin and antipneumococcal and *Haemophilus* vaccines has aided in the management of life-threatening infectious complications of SS disease. Neonatal screening has been used in the identification of infants with sickle cell disease so that risk of infection can be modulated by appropriate immunizations and penicillin prophylaxis. The acute chest syndrome characterized by chest pain is due to clogged pulmonary capillaries; in a small number of studies, patients have been treated with inhaled nitric oxide, which dilates blood vessels with clinical improvement.

Sickle cell trait is present in about 8% of black Americans and to a much greater extent (as high as 45%) in some black African populations. The homozygous condition causes considerable morbidity and about 60,000–80,000 deaths per year among African children. HbS also occurs in some parts of India, the Arabian region, and occasionally in the Mediterranean area. The HbS mutation that occurs in eastern Saudi Arabia and India, known as the Asian haplotype, has different flanking DNA sequences surrounding the β-globin locus. Thus, the Asian haplotype may represent an independent occurrence of the HbS mutation that is distinct from its African counterpart. The deleterious gene likely has persisted in these populations because the HbS that increases resistance to malaria caused by *Plasmodium falciparum,* which was, until recently, endemic in those areas. A similar explanation has been advanced for the high frequencies of β-thalassemia and single α-locus genotypes in the same regions. The biological basis of resistance to malaria has been established in laboratory experiments. As schizonts of *P. falciparum* grow in erythrocytes, they lower the intracellular pH and generate hydrogen peroxide. The lower pH promotes sickling and the hydrogen peroxide damages cell membranes of thalassemic erythrocytes. In both cases, the erythrocyte membranes become more permeable to potassium ions; the resulting decrease in intracellular potassium kills the parasites.

The mutation in HbS replaces glutamic acid (a polar amino acid) by valine (a nonpolar residue) at position 6 of the β chains. The solubilities of oxy- and deoxy-HbA and oxy-HbS are similar, being about 50 times that of deoxy-HbS. Upon deoxygenation, HbS precipitates, forming long, rigid, polymeric strands that distort and stiffen the red cells. The very high concentration of hemoglobin in the erythrocyte (340 mg/mL), giving an average intermolecular distance of about 1 nm (10 Å), minimizes the time necessary for precipitations to occur. Dilution of HbS, as in sickle cell trait (HbS/HbA heterozygotes), reduces the concentration below the point at which sickling readily occurs. Similarly, the mildness of homozygous HbS disease in populations such as those of eastern Saudi Arabia and Orissa, India is attributed to an accompanying elevation of HbF, which reduces the effective concentration of HbS tetramers.

Other abnormal hemoglobins can interact with HbS and alter the course of the disease. The most common is HbC, which has a lysine in place of glutamic acid, also at $\beta 6$. The gene has a frequency second only to that of HbS in black Americans and in some black African populations. Persons heterozygotic for HbC are asymptomatic, but homozygotic individuals have a mild hemolytic anemia with splenomegaly. Because of its insolubility, crystals of HbC can sometimes be seen in peripheral blood smears from homozygous individuals. As a result of the coincidental distribution of the genes for HbS and HbC, heterozygotes for both hemoglobins are not uncommon. HbSC disease has a severity intermediate between that seen in persons homozygotic for HbS and those homozygotic for HbC. Unlike HbA, HbC copolymerizes with HbS. In contrast, replacement of the $\beta 6$ glutamic acid by alanine (HbG-Makassar) or deletion of $\beta 6$ (Hb Leiden) results in hemoglobin that neither precipitates nor interacts with HbS. Hb San Jose (Glu $\rightarrow$ Gly at residue $\beta 7$) is a harmless variant.

Individuals heterozygous for HbS and HbO Arab or HbD Los Angeles have a hemolytic anemia of a severity intermediate between that of HbSC disease and sickle cell anemia. Certain α-chain variants (e.g., Hb Memphis), when present in HbS homozygotes, can ameliorate the clinical course of the disease. The severity of HbS-β-thalassemia depends on whether the thalassemia is β° or β^{+} and, if it is β^{+}, on how much normal β chain is synthesized. Severity also depends on the HbF concentration.

A patient with heterozygosity for both HbS and Hb Quebec-chori exhibited clinical symptoms suggestive of sickle cell disease. Hb Quebec-chori, an electrophoretically silent variant at acid and alkaline pH (see Appendix VII) ($\beta 87$ Thr $\rightarrow$ Ile), polymerizes with HbS with the stabilization of the polymer under hypoxic conditions leading to sickling of red blood cells. Thus, Hb Quebec-chori provides an example of a hemoglobin that has the potential to polymerize with HbS and causing sickle cell disease in a sickle cell trait condition which is otherwise benign by itself.

Determination of the structure of crystalline HbS has shown that in the β subunits of oxy-HbA and oxy-HbS, a "hydrophobic pocket" between helices E and F is closed; this opens in the deoxy form. In HbA, the residues at the surfaces of the globin subunits are hydrophilic (polar) and do not interact with this pocket. In HbS, however, the $\beta 6$ valine is hydrophobic and fits into the hydrophobic pocket (formed by leucine and phenylalanine at $\beta 85$ and $\beta 88$) of an adjacent β chain to form a stable structure. Since each β subunit in deoxy-HbS has an "acceptor" hydrophobic pocket and a "donor" valine, linear aggregates form (Figure 28-15).

Understanding of the sickling process and of the structure of the HbS polymer provides a rational basis for ways of correcting the molecular defect. Thus, dilution of the HbS in the red cells, blockage of the interaction of the $\beta 6$ valine with the hydrophobic pocket, and decrease of

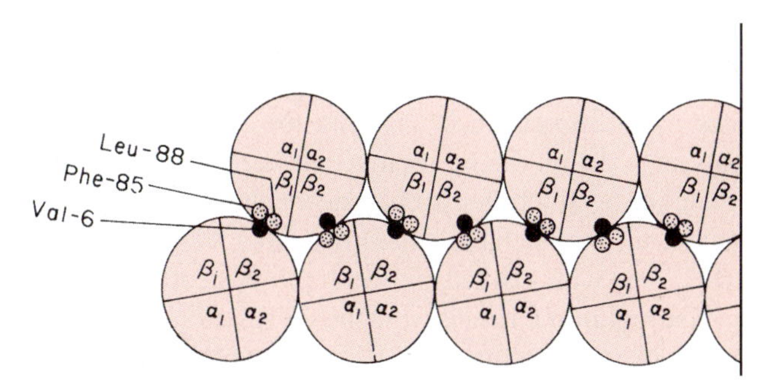

FIGURE 28-15
Structure of hemoglobin S (HbS) polymer. The valine at the $\beta 6$ position of the deoxy-HbS fits into the hydrophobic pocket formed by leucine and phenylalanine at $\beta 85$ and $\beta 88$ of an adjacent β chain. Since each β chain has an "acceptor" pocket and a "donor" valine, the HbS polymer has a double-stranded, half-staggered structure. [Reproduced with permission from S. Charache, Advances in the understanding of sickle cell anemia. *Hosp. Pract.* **21**(2), 173 (1986). J. E. Zupko, illustrator.]

the deoxy-HbS oxy-HbS ratio should reduce the likelihood of sickling and the severity of the disease. These approaches have been tried but so far have failed to ameliorate the disease.

Among Bedouin Arabs and some populations of central and southern India, high HbF levels reduce the severity of sickle cell disease by inhibiting the formation HbS polymers. This observation has led to therapeutic approaches to induce higher levels of HbF in patients with sickle cell disease. The therapeutic agents are hydroxyurea and short-chain fatty acid derivatives. Hydroxyurea is an antineoplastic agent that inhibits ribonucleotide reductase (Chapter 27). It is thought that in bone marrow, hydroxyurea selectively kills many precursor cells but spares erythroblasts that produce HbF. Hydroxyurea therapy also results in decreased circulating granulocytes, monocytes, and platelets. These changes, along with increased HbF, reduce vaso-occlusion due to a decreased propensity for sickling and adherence of red blood cells to endothelium. The long-term toxicity of hydroxyurea due to its myelosuppressive and teratogenic effects is not known. Induction of HbF by short-chain fatty acids such as butyrate was discovered in infants of diabetic mothers who had a delay in switching from HbF to HbA in association with elevated serum levels of amino-n-butyrate. This led to studies of HbF induction with several short-chain acids, including butyrate, all of which induce production of HbF. The mechanism by which butyrate and other short-chain fatty acids affect gene expression involves transcriptional activation of a γ-globin gene at the promoter site. One of the mechanisms of regulation of transcription, which occurs at the promoter site, is due to changes in histone acetylation and deacetylation (Chapter 26). Acetylation of specific histones by acetylases allows increased binding of transcription factors to target DNA that stimulates transcription. Deacetylation by histone deacetylases of histones prevents transcription. Butyrate is an inhibitor of deacetylase and it is thought that this action leads to induction of γ-globin gene expression. Since hydroxyurea acts by a different mechanism in the elevation of HbF, butyrate's action can be additive or synergistic. Sustained induction of HbF by pulse butyrate therapy in patients with sickle cell anemia has resulted in up to a 20% increase in HbF along with marked clinical improvement.

Hemoglobin E results from the substitution of glutamic acid with lysine at the 26th residue of the β chain; it has a high prevalence in Southeast Asia. Neither heterozygous nor homozygous states are associated with significant clinical abnormalities. The homozygous state exhibits a mild anemia, and both states show red blood cell indices resembling those of heterozygous thalassemic states, namely, hypochromic microcytosis. The latter has been attributed to activation of a cryptic splice site in exon 1 of β-globin mRNA caused by β^{E} mutation. This leads to abnormal splicing of β-globin mRNA, producing a less stable and ineffective mRNA. Coinheritance of the HbE gene with different forms of thalassemia genes is also known.

γ- and δ-Globin Mutants

The γ- and δ-chain mutants are difficult to study because of the small fraction of HbF and HbA_2 present in adult erythrocytes. Overt clinical symptoms associated with γ- and δ-chain variants are rare. By routine screening, 14 δ-globin variants have been discovered but are of no clinical consequence. Thirty-five mutant γ sequences (involving the G_γ or the A_γ chains) have been identified. They are all benign except for HbF Poole [G_γ130(H8) Tyr $\rightarrow$ Gly], an unstable hemoglobin that causes hemolytic disease in the newborn.

Screening and Prenatal Diagnosis

Screening programs are aimed at detection of the most common and most harmful hemoglobin genes and the identification of heterozygotes. Screening can provide a basis for genetic counseling and decrease the number of homozygous individuals conceived. Electrophoresis is used to differentiate among sickle cell trait, sickle cell anemia, and HbSC disease and to detect thalassemia, hereditary persistence of HbF, and HbC trait and HbC disease (see Appendix VII). Prenatal diagnosis of genetic disease often complements programs for detection of heterozygotes. If both parents are known to be carriers, a homozygous fetus can be detected early in pregnancy. The first prenatal tests measured the ability of fetal reticulocytes to direct the synthesis *in vitro* of normal or abnormal β chains. Fetal blood was obtained either by percutaneous aspiration from the placenta or transvaginally by direct aspiration from one of the fetal vessels on the surface of the placenta. The cells were incubated with radioactive amino acids, and the labeled globin chains were separated by chromatography and quantitated. This method detects normal and sickle cell anemia genotypes, absence of α- or β-globin synthesis (homozygous α°- or β°-thalassemia), and some types of β^{+}-thalassemia. DNA has been isolated directly from cultured fetal cells (obtained by amniocentesis and from trophoblast biopsy) and subjected to restriction endonuclease digestion. (Culturing of fetal cells is necessary, since their number in the amniotic fluid sample is small.) The restriction fragments are separated according to size by electrophoresis and reacted with a labeled DNA probe that is complementary to the gene being sought. The hybridization pattern is compared with

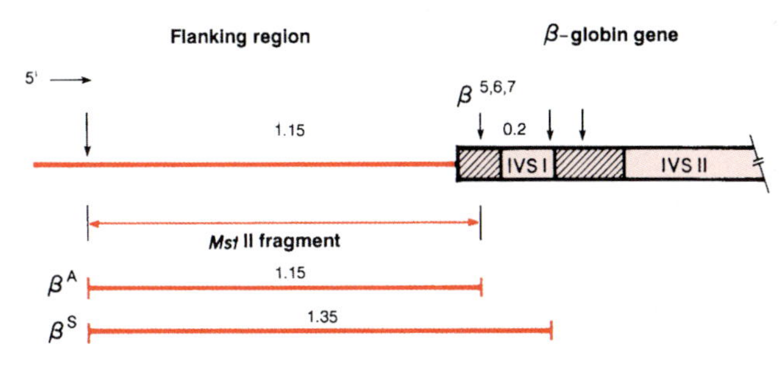

FIGURE 28-16
Diagram of the 5′ region of the human β-globin gene and its flanking region, showing the cleavage sites of the restriction endonuclease Mst II (vertical arrows). The fragments obtained by digestion of normal and sickle β-globin genes are shown in the lower part of the diagram. Shaded regions of the gene are the exons, and IVS indicates the introns (intervening sequences). [Reproduced with permission from J. C. Chang and Y. W. Kan, A sensitive prenatal test for sickle cell anemia. *N. Engl. J. Med.* **307,** 30 (1982).]

one from a normal individual (Chapter 23). Differences between the normal and some mutant globin genes at recognition sites for restriction endonucleases will cause differences in the cuts made and, hence, in the sizes of the fragments detected.

In the ideal situation, a restriction endonuclease recognizes a site in the DNA that includes the mutated base. Restriction enzymes Mnl I, Dde I, and Mst II recognize part of the coding sequence for amino acids 5, 6, and 7 of the normal β-globin gene. The mutation in the β-globin gene in HbS abolishes these recognition sites, converting two small pieces of DNA found in the digest of normal β-globin DNA into one larger fragment (Figure 28-16). Mst II is the most useful enzyme because it generates larger, more easily detected fragments. Loss of recognition sites due to deletions is seen in most α-thalassemias and occasionally in β-thalassemias, since most β-thalassemias are not caused by deletion. In gene deletion, multiple sites are lost. However, point mutations are more common and therefore of greater importance than deletions. The endonuclease that is most useful in the detection of each disorder must be carefully selected. Because β-thalassemias are caused by many different mutations with similar phenotypic expressions, it is not possible to find one or even a few restriction endonucleases that can detect all β-thalassemia genes. What has been accomplished so far is the detection of affected individuals among members of the same kindred or racial group in which a single type of β-thalassemia occurs.

Alternatively, sites adjacent to an abnormal gene may differ from those on normal chromosomes. Such linked polymorphisms have been found for the sickle cell gene and for several β-thalassemias. Because different individuals having a particular hemoglobin abnormality will not always have the same linked polymorphism, this method suffers from lack of sensitivity. Under proper conditions, short, synthetic oligonucleotide probes, complementary to the normal DNA sequence in the region that contains the mutation, hybridize with completely homologous sequences but not with sequences that differ by as little as one nucleotide. This method is potentially more powerful than restriction endonuclease mapping because an oligonucleotide probe can be synthesized for any mutation once its sequence is known. Furthermore, multiple mutations can be detected simultaneously by hybridizing the DNA with a mixture of different probes, making this technique useful for screening. This technique has been used successfully to detect the sickle cell gene and several β-thalassemias that could not be detected by restriction endonuclease mapping.

All these methods depend on a satisfactory sample being collected from the fetus. Studies of globin chain synthesis require an adequate volume of fetal blood. Collection methods result in 4–6% fetal death under optimal conditions. Furthermore, β-globin synthesis cannot be detected until about the middle of the second trimester. Use of amniotic fluid cells obtained by amniocentesis is less dangerous. However, an adequate volume of fluid cannot be obtained until the middle of the second trimester of pregnancy (14–20 weeks); if culturing of the cells is necessary, there is an additional delay of 1–2 weeks. Chorionic villus biopsy can be obtained at about 8–12 weeks of gestation. This allows more time for deliberation regarding termination or continuation of the pregnancy, thereby increasing the margin of safety for the mother, should termination be desired. This technique is no more dangerous than either of the other methods. Some other techniques for laboratory evaluation of hemoglobin disorders are presented in Appendix VII.

28.4 Derivatives of Hemoglobin

Carbon Monoxide-Hemoglobin

Carbon monoxide, an odorless gas, has an affinity for hemoglobin that is 210 times that of oxygen. Thus, in the equilibrium reaction

$$HbO_2 + CO \rightleftarrows HbCO + O_2 \qquad K_{eq} = 2.1 \times 10^2$$

the equilibrium lies far to the right. Like oxygen, CO binds to the sixth position of the heme iron. When CO and O_2 are present together in appreciable quantities, CO is bound preferentially and oxygen is excluded, effectively causing an anemic hypoxia. Even if some O_2 is bound to the hemoglobin, it cannot be released owing to the tight binding of CO and to its action as an allosteric activator, with the result that hemoglobin is trapped in the R-state.

Carbon monoxide poisoning may result from breathing automobile exhaust, poorly oxygenated coal fires in stoves and furnaces, or incomplete combustion of a carbon-containing compound. Breathing air containing 1% CO for 7 minutes can be fatal; automobile exhaust contains about 7% CO. Unconsciousness, a cherry-red discoloration of the nail beds and mucous membranes (due to large amounts of R-state carbon monoxide-hemoglobin), and spectrophotometric analysis of the blood are useful for clinical diagnosis.

Treatment of carbon monoxide poisoning consists of breathing a mixture of 95% O_2 and 5% CO_2, which will usually eliminate carbon monoxide from the body in 30–90 minutes, or of breathing hyperbaric oxygen.

The enzymatic oxidation of heme produces CO and biliverdin in equimolar amounts (Chapter 29). The CO is transported via the blood to the lungs, where it is released. Although no cases are known in which endogenous CO proved toxic, this carbon monoxide may contribute significantly to air pollution, particularly in crowded and enclosed areas.

Carbaminohemoglobin

Some of the CO_2 in the bloodstream is carried as carbamino compounds (Chapter 1). These compounds form spontaneously in a readily reversible reaction between CO_2 and the free α-amino groups in the N-terminal residues of the Hb chains.

Hemoglobin $+ 4CO_2 \rightleftharpoons$ Carbaminohemoglobin

Although Hb can carry as many moles of CO_2 as of O_2, the HCO_3^- system is a more important way of transporting CO_2. When the N-terminal amino groups are blocked by carbamylation with CNO^-, no marked degree of acidosis develops. The carbamino groups decrease the affinity of Hb for O_2. The P_{CO_2} thus can affect oxygen affinity independently of any effect it may have on pH and may influence the position of the oxygen dissociation curve under various conditions.

Methemoglobin

In the presence of oxygen, dissolved hemoglobin is slowly oxidized to methemoglobin, a derivative of hemoglobin in which the iron is present in the ferric (Fe^{3+}) state. In metHb, the ferric ion is bound tightly to a hydroxyl group or to some other anion. The heme porphyrin (protoporphyrin IX) that contains an Fe^{3+} ion is known as ***hemin.***

A small but constant amount of methemoglobin is produced in the body. It is reduced by specific enzymes (methemoglobin reductases) and NADH, which is generated in glycolysis. Reductases isolated from human red cells also use NADPH but to a lesser extent. Inability to reduce metHb produces ***methemoglobinemia*** and tissue anoxia.

Hereditary methemoglobinemia may arise from the following:

1. Deficiency of one or more of the metHb reducing enzymes (usually a recessive trait). MetHb values may range from 10% to 40% of the total Hb (normal = 0.5%). Treatment involves administration of an agent that will reduce the metHb (e.g., ascorbic acid, methylene blue).
2. Defects in the hemoglobin molecule that make it resistant to reduction by metHb reductases and ascorbate or methylene blue. These hemoglobins are collectively called the M hemoglobins (Table 28-5). HbM is inherited as a dominant trait, since homozygosity for an M β-chain hemoglobin would be lethal. In four types that involve an His → Tyr substitution, the phenolic hydroxyl group forms a stable complex with Fe^{3+}, making it resistant to reduction to Fe^{2+}. In HbM Milwaukee, a glutamic acid residue substitutes for a valine at position 67 near the distal histidine and forms a stable complex with Fe^{3+}. Although a brownish cyanosis is characteristic of blood containing HbM, in contrast to the purplish cyanosis caused by excess deoxyhemoglobin, diagnosis should be confirmed by the absorption spectrum of either the Hb or its CN^- derivative. The spectrum of the cyanide derivative is preferred, since it differs more from that of HbA. Some of the M hemoglobins cannot form CN^- derivatives, presumably because a Tyr or Glu blocks the sixth iron position. In these cases, the absorption spectrum of the methemoglobin itself must be examined. Because of the change in the environment of the heme due to the change in the globin structure, the spectra of M

hemoglobins differ among themselves and from that of normal metHb.

Theoretically, hereditary methemoglobinemia could occur if the hemoglobin was altered in that region where methemoglobin reductase binds in performing its function. In such a mutant Hb, the reductase would not bind. Acquired acute methemoglobinemia is a relatively common condition caused by a variety of drugs such as phenacetin, aniline, nitrophenol, aminophenol, sulfanilamide, and inorganic and organic nitrites and nitrates. The condition is less commonly produced by chlorates, ferricyanide, pyrogallol, sulfonal, and hydrogen peroxide. These compounds appear to catalyze the oxidation of Hb by oxygen. Symptoms include brownish cyanosis, headache, vertigo, and somnolence. Diagnosis is based on occurrence of brownish cyanosis and presence of excessive amounts of metHb (measured spectrophotometrically). Treatment usually consists of removal of the offending substance and administration of ascorbic acid (in mild cases) or methylene blue (in severe cases). These reducing agents function according to the reactions below (MB = methylene blue, oxidized; MBH_2 = methylene blue, reduced).

$$MBH_2 \rightleftarrows MB + H_{2(gas)}$$
$$2H_{2(gas)} + Hb(Fe^{3+})_4 \rightarrow Hb(Fe^{2+}) + 4H^+$$

(The release of H_2 by MBH_2 is spontaneous because methylene blue is autoxidizable.)

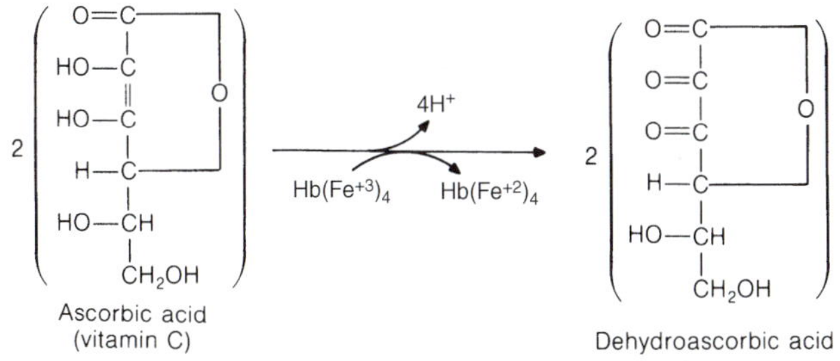

2 Ascorbic acid (vitamin C) [O=C–, HO–C, HO–C, H–C– (ring O), HO–CH, CH_2OH] + $Hb(Fe^{+3})_4$ → 2 Dehydroascorbic acid [O=C–, O=C, O=C, H–C– (ring O), HO–CH, CH_2OH] + $Hb(Fe^{+2})_4$ + $4H^+$

Ascorbic acid (vitamin C)

Dehydroascorbic acid

Sulfhemoglobin

Many drugs that cause methemoglobinemia also stimulate production of sulfhemoglobin, a greenish derivative. Methemoglobin and sulfhemoglobin may appear together in poisoning by phenacetin, acetanilid, or sulfanilamide. Dapsone (used to treat leprosy) and exposure to sulfur-containing compounds either occupationally or from air pollution can also produce sulfhemoglobinemia. The sulfur appears to attach covalently to the porphyrin ring rather than to the iron atom. It is unclear whether the sulfur is derived from hydrogen sulfide generated by anaerobic metabolism of intestinal bacteria, from glutathione, or from some other source.

Because of its color, patients with circulating sulfhemoglobin appear cyanotic like those with methemoglobinemia. However, unlike methemoglobinemia, sulfhemoglobinemia cyanosis becomes clinically apparent at extremely low concentrations and is usually not accompanied by cardiopulmonary pathology. This fact, together with the overlap in the list of causative agents, may lead to underdiagnosis of sulfhemoglobinemia. Both sulfHb and metHb show an absorption peak at about 620 nm that is not present in deoxyHbA (the most common cause of cyanosis). They can be discriminated by changes in this peak on addition of cyanide, carbon monoxide, or dithionite (a reducing agent). Isoelectric focusing of samples treated with the same chemicals is also useful for differential diagnosis.

Oxygen does not bind to a subunit having a sulfurated porphyrin ring. The presence of one or two such subunits, together with normal subunits in a hemoglobin tetramer, alters the cooperativity, favors the deoxy structure, and thereby decreases the oxygen affinity of the normal subunits. Thus, at low levels of sulfuration, the percentage of affected hemoglobin tetramers is greater than the percentage of sulfurated monomers. (This is also true of the methemoglobins). In contrast, binding of carbon monoxide to one or two subunits increases the oxygen affinity of the tetramer, since CO stabilizes the oxyhemoglobin conformation.

In a person who has HbA as the predominant hemoglobin, the presence of sulfhemoglobin is probably innocuous. Tissue oxygenation is relatively normal because the decrease in the number of binding sites for oxygen is offset by the lower binding affinity. However, since sulfuration stabilizes the deoxy form, sulfhemoglobin, in individuals who have sickle cell anemia, should increase the likelihood of sickling and thereby exacerbate the disease. No way is known to reverse formation of sulfhemoglobin. If the causative agent is removed, sulfhemoglobin will disappear from the circulation at the same rate as that of the erythrocytes that contain it.

Cyanmethemoglobin

Cyanide poisoning does not cause production of cyanohemoglobinemia or cyanosis. It does produce cytotoxic anoxia by poisoning cytochrome oxidase and other respiratory enzymes, thereby preventing utilization of O_2 by tissues. Cyanide poisoning is detected by the characteristic odor of HCN gas (odor of bitter almonds) on the breath and by laboratory tests (absorption spectra, tests for CN^-).

Treatment of cyanide poisoning consists of diverting the cyanide into the production of cyanmetHb. First, some of the normal hemoglobin is converted to methemoglobin by intravenous infusion of a solution of $NaNO_2$ or inhalation of amyl nitrite. Once metHb is formed, CN^- can replace OH^- at position 6 of the iron, since it has a higher affinity of Fe^{3+} than does OH^-. CyanmetHb is no more toxic than metHb and cells containing it can be eliminated by normal body processes. The cyanide bound to metHb is always in equilibrium with free CN^-, and this uncomplexed cyanide is converted to thiocyanate (SCN^-; nontoxic) by administration of thiosulfate (Chapter 6).

Glycated Hemoglobins

Several minor varieties of hemoglobin are produced by nonenzymatic posttranslational modification (e.g., glycation). Because these minor hemoglobins are present in such small amounts, none is pathological.

Both α- and ε-amino groups of hemoglobin form amino-1-deoxyfructose adducts upon reaction with glucose (Chapter 2). Other hexoses can give rise to similar adducts (e.g., galactose in galactosemia; Chapter 15). The major sites of *in vivo* glycation in order of prevalence are β-Val1, β-Lys66, α-Lys61, β-Lys17, and α-Val1. The adduct formed with the amino terminus of the β chains is known as HbA_{1C}, which makes up about 4–6% of the total hemoglobin in normal red blood cells. Its concentration is increased in uncontrolled diabetics who have hyperglycemia. Glycated hemoglobins are detected as a fast-moving hemoglobin in electrophoresis at alkaline pH. Because HbA_{1C} accumulates within the erythrocyte throughout the cell's normal life span, it is used as an indicator of the success of long-term blood glucose control in diabetics (Chapter 22). Also known are adducts of hemoglobin with glucose-6-phosphate and fructose-1,6-diphosphate, which probably compete with 2,3-DPG for binding to the β chains of deoxyhemoglobin, and a complex between hemoglobin and glutathione, particularly in older erythrocytes.

Supplemental Readings and References

M. O. Arcasoy and P. G. Gallagher: Molecular diagnosis of hemoglobinopathies and other red blood cell disorders. *Seminars in Hematology* **36,** 328 (1999).

G. F. Atweh, M. Sutton, L. Nassif, et al.: Sustained induction of fetal hemoglobin by pulse butyrate therapy in sickle cell disease. *Blood* **93,** 1790 (1999).

H. F. Bunn: Pathogenesis and treatment of sickle cell disease. *New England Journal of Medicine* **337,** 762 (1997).

H. F. Bunn: Induction of fetal hemoglobin in sickle cell disease. *Blood* **93,** 1787 (1999).

A. Cao, R. Galanello, and M. C. Rosatelli: Genotype-phenotype correlations in β-thalassemias. *Blood Reviews* **8,** 1 (1994).

M. Cazzola, F. Mercuriali, and C. Brugnara: Use of recombinant human erythropoietin outside the setting of uremia. *Blood* **89,** 4248 (1997).

D. H. K. Chui, R. Hardison, and C. Riemer: An electronic database of human hemoglobin variants on the World Wide Web. *Blood* **91,** 2643 (1998).

S. C. Davies and L. Oni: Management of patients with sickle cell disease. *British Medical Journal* **315,** 656 (1997).

B. L. Ebert and H. F. Bunn: Regulation of the erythropoietin gene. *Blood* **94,** 1864 (1999).

A. Ferster, C. Vermylen, G. Cornu, et al.: Hydroxyurea for treatment of severe sickle cell anemia: a pediatric clinical trial. *Blood* **88,** 1960 (1996).

J. R. Girman, Y-L. Chang, S. B. Hayward, et al.: Causes of unintentional deaths from carbon monoxide poisonings in California. *Western Journal of Medicine* **168,** 158 (1998).

F. Grosveld, E. De Boer, N. Dillon, et al.: Dynamics of globin gene expression and gene therapy vectors. *Annals of the New York Academy of Sciences* **550,** 18 (1998).

C. C. W. Hsia: Respiratory function of hemoglobin. *New England Journal of Medicine* **338,** 239 (1998).

E. Liakopoulou, C. A. Blau, L. Qiliang, et al.: Stimulation of fetal hemoglobin production by short chain fatty acids. *Blood* **86,** 3227 (1995).

J. A. Little, N. J. Dempsey, M. Tuchman, et al.: Metabolic persistence of fetal hemoglobin. *Blood* **85,** 1712 (1995).

J. M. Manning, A. Dumoulin, X. Lee, et al.: Normal and abnormal protein subunit interactions in hemoglobins. *Journal of Biological Chemistry* **273,** 19359 (1998).

P. G. McCaffrey, D. A. Newsome, E. Fibach, et al.: Induction of γ-globin by histone deacetylase inhibitors. *Blood* **90,** 2075 (1997).

R. T. Means Jr. and S. B. Krantz: Progress in understanding the pathogenesis of the anemia of chronic disease. *Blood* **80,** 1639 (1992).

M. Noor and E. Beutler: Acquired sulfhemoglobinemia. *Western Journal of Medicine* **169,** 386 (1998).

N. F. Olivieri: The β-Thalassemias. *New England Journal of Medicine* **341,** 99 (1999).

J. P. Scott, C. A. Hillery, E. R. Brown, et al.: Hydroxyurea therapy in children severely affected with sickle cell disease. *Journal of Pediatrics* **128,** 820 (1996).

G. Serjeant: Sickle-cell disease. *Lancet* **350,** 725 (1997).

M. H. Steinberg: Management of sickle cell disease. *New England Journal of Medicine* **340,** 1021 (1999).

I. A. Tabbara: Erythropoietin. *Archives of Internal Medicine* **153,** 298 (1993).

P. M. Tibbles and J. S. Edelsberg: Hyperbaric-oxygen therapy. *New England Journal of Medicine* **334,** 1642 (1996).

D. J. Weatherall and J. B. Clegg: Genetic disorders of hemoglobin. *Seminars in Hematology* **36,** 24 (1999).

H. E. Witkowska, B. H. Lubin, Y. Beuzard, et al.: Sickle cell disease in a patient with sickle cell trait and compound heterozygosity for hemoglobin S and hemoglobin Quebec-Chori. *New England Journal of Medicine* **325,** 1150 (1991).

CHAPTER 29

Metabolism of Iron and Heme

Heme, an iron-porphyrin complex, is the prosthetic group of many important proteins. The central role of hemoglobin and myoglobin in oxygen transport and storage was discussed in Chapter 28. Heme proteins or enzymes are involved in redox reactions (e.g., cytochromes) and participate in many oxidation reactions needed for synthesis of metabolically important compounds as well as for degradation and detoxification of waste products and environmental toxins.

Ionic forms of iron (referred to hereafter as iron) also participate in a variety of enzymatic reactions as nonheme irons, which are present as iron-sulfur clusters (e.g., mitochondrial electron transport). There are also both storage and transportable forms of iron that are bound to proteins. Under normal physiological conditions only trace amounts of free iron exist. In the body, if iron exceeds the sequestration capacity of the iron-binding proteins present in different physiological compartments, the free iron can cause tissue damage. Cellular injury is caused by reactive oxygen species that are produced from H_2O_2 in a reaction catalyzed by iron. Thus, iron homeostasis in the body is in a delicate balance. Either the deficiency or the excess results in abnormalities and presents as a common cause of human diseases.

29.1 Iron Metabolism

Total-body iron of a 70-kg adult is about 4.2–4.4 g. The distribution of iron in various body compartments is given in Table 29-1. The key players of iron metabolism include iron-responsive elements of appropriate mRNAs, iron regulatory proteins divalent metal transporter 1, major histocompatibility complex (MHC) class I-like protein designated as HFE protein, β_2-microglobulin, transferrin, transferrin receptor, and ferritin.

Absorption of Iron from the Diet

The dietary requirement for iron depends on the amount and composition of the food, the amount of iron lost from the body, and variations in physiological state such as growth, onset of menses, and pregnancy. The average North American diet contains about 6 mg of iron per 1000 calories and supplies about 10–15 mg/d. Of that ingested, 8–10% (1–1.5 mg/d) is absorbed. Thus, dietary factors that affect absorption are more important than the iron content of the diet and may be more important for correction of iron deficiency than addition of iron to the diet.

TABLE 29-1
Distribution of Iron in a 70-kg Adult[1]

Circulating erythrocytes	1800 mg[2]
Bone marrow (erythroid)	300 mg
Muscle myoglobin	300 mg
Heme and non-heme enzymes	180 mg
Liver parenchyma[3]	1000 mg
Reticuloendothelial macrophages[4]	600 mg
Plasma transferrin[5]	3 mg

[1] These are approximate values. Premenopausal women have lower iron stores due to periodic blood loss through menstruation. Iron balance in the body is maintained by intestinal absorption of 1–2 mg/day and by loss of 1–2 mg/day.
[2] 1 mg = 17.9 μmol
[3] Primarily storage forms of iron.
[4] Senescent red blood cells are catabolized by the macrophages, the salvaged iron is temporarily stored, and made available via transferrin for erythron and for hemoglobin synthesis.
[5] Transportable form of iron.

Iron in food exists mainly in the ferric (Fe^{3+}) state, complexed to proteins, amino acids, organic acids, or heme. It is absorbed in the ferrous state, reduction being accomplished in the gastrointestinal tract by ascorbate, succinate, and amino acids. Gastric acid potentiates iron absorption by aiding in formation of soluble and absorbable ferrous chelates. In achlorhydria or after partial gastrectomy, absorption is subnormal but is increased by administration of hydrochloric acid. Carbonates, tannates, phosphates, phytates, and oxalates may decrease iron absorption by formation of insoluble complexes, but their effect can be prevented by adequate dietary calcium, which complexes with them and makes them unavailable for reaction with iron. Absorption of heme iron is not affected by these agents. Heme is absorbed intact from food and more effectively than inorganic iron. Antacids, such as aluminum hydroxide and magnesium hydroxide, also decrease iron absorption.

In general, foods of animal origin provide more assimilable iron than foods of vegetable origin, since on a weight basis, vegetables contain less iron and more substances (e.g., phytates) that inhibit iron absorption. Foods that contain more than 5 mg of iron per 100 g include organ meats (liver, heart), wheat germ, brewer's yeast, oysters, and certain dried beans. Foods that contain 1–5 mg of iron per 100 g include muscle meats, fish, fowl, some fruits (prunes, raisins), most green vegetables, and most cereals. Foods that contain less than 1 mg of iron per 100 g include milk and milk products and most nongreen vegetables.

Ferrous iron is absorbed principally from the mature enterocytes lining the absorptive villi of the duodenum. The amount of iron absorption by these enterocytes is determined by the prior programming of the duodenal crypt cells based on iron requirements of the body as they undergo maturation. The regulation of intestinal iron absorption is critical because iron excretion from the body is a limiting physiological process (discussed later). The small intestine is also an excretory organ for iron, since that stored as ferritin in the epithelial cells is lost when they are shed and replaced every 3–5 days. Heme iron is transported intact into the mucosal cells and the iron removed for further processing.

The mechanism of entry of ferrous iron from the intestinal lumen into the enterocytes and its eventual transport into the portal blood is beginning to be understood. The first step is the programming of the duodenal undifferentiated deep-crypt cells with regard to sensing the iron requirements of the body. The programming of the crypt cells for capacity to absorb iron is thought to occur as follows. A protein (HFE) that spans the cell membrane like an HLA molecule associates with β_2-microglobulin like an HLA protein. The HFE protein is coded for by a gene (*HFE*) located on the short arm of chromosome 6 near the *MHC* gene loci. Mutations in the *HFE* gene are associated with an inherited disorder of excessive dietary iron absorption that is known as ***hereditary hemochromatosis*** (discussed later). HFE protein spans the cell membrane of the crypt enterocytes with its N-terminal domain projecting outside. Near the cell membrane a segment of HFE protein, like an HLA protein, associates with β_2-microglobulin (Figure 29-1) and stabilizes the HFE protein.

Plasma transferrin transports iron in the ferric state and is an indicator of iron stores in the body. Each molecule of transferrin binds with two ferric ions (diferric transferrin) and undergoes receptor-mediated endocytosis when bound to transferrin receptors (discussed later) with the aid of HFE protein β_2-microglobulin complex. In the cytosol, iron is released from the endosomes. The level of cytoplasmic iron regulates the translation of mRNA of a protein known as divalent metal transporter-1 (DMT1), which participates in iron entry into the enterocytes located on the villus tip (Figure 29-2). Regulation of DMT1 synthesis is coupled to cytoplasmic iron levels and involves the presence of a stem-loop hairpin structure in the 3′-untranslated region that resembles an iron regulatory element (IRE) and its interaction with an iron regulatory protein (IRP1). In the iron-deficient state, IRP binds to IRE and stabilizes the mRNA of DMT1. This stabilization of mRNA leads to increased production of DMT1 and its eventual localization on the cell surface. The transport of ferrous iron across the apical membrane of the villus enterocyte that is mediated by DMT1 occurs through a proton-coupled process. DMT1 also transports a number

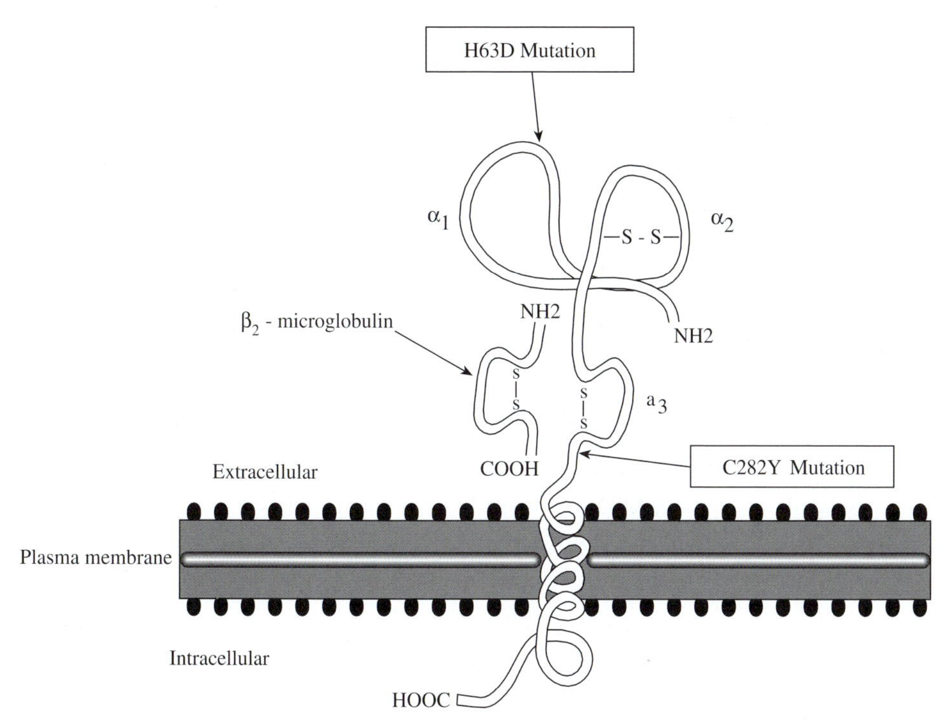

FIGURE 29-1
A diagrammatic representation of the transmembrane HFE protein. The extracellular portion of the HFE protein consists of three α domains. The β_2-microglobulin is noncovalently associated with the α_3 domain of HFE protein, stabilizing its structure. The extracellular missense mutations H63D and C282Y are identified in patients with hereditary hemochromatosis. HFE protein is involved in sensing circulating iron concentration and participates in the regulation of gene expression of products involved in iron absorption, transport, or storage. [Reproduced with permission from: A novel MHC class 1-like is mutated in patients with hereditary hemochromatosis. J. N.Feder, A. Gnirke, W. Thomas et al. *Nature Genetics* **13,** 399 (1996).]

of divalent metal ions, including Mn^{2+}, Cu^{2+}, Zn^{2+}, Cd^{2+}, and Pb^{2+}.

At normal levels of iron intake, absorption requires uptake from the intestinal lumen by the mucosa and transfer from the mucosa to the portal blood. Both events are inversely affected by the state of body iron stores. In iron deficiency states, nonferrous metals such as cobalt and manganese, which have an ionic radius similar to that of iron and form octahedral complexes with six-coordinate covalent bonds, also are absorbed at an increased rate. Oral administration of a large dose of iron reduces (or temporarily inhibits) the absorption of a second dose of iron by the absorptive enterocytes even in the presence of systemic iron deficiency. The mechanism of mucosal block, which resists acquiring additional iron by the enterocytes with high amounts of intracellular iron, is not yet understood. It probably involves set points established in the enterocytes for iron recently consumed in the diet (dietary regulator).

Iron absorption also is affected by erythropoiesis. When erythropoiesis is accelerated by bleeding, hemolysis, or hypoxia, iron absorption is increased. Conversely, diminished erythropoiesis due to starvation, blood transfusion, or return to sea level from a high altitude decreases iron absorption. How the size of body iron stores and the rate of erythropoiesis transmit information to the duodenum is not known. Feedback control seems to be weak or absent, since in iron-deficient subjects enhanced iron absorption continues long after hemoglobin is restored to normal levels. Furthermore, in chronic hemolytic anemia, iron absorption is increased, persists for prolonged periods, and leads to iron overload.

The need for dietary iron is ultimately determined by the rate of iron loss from the body and the amount required for maintenance and growth. Iron is tightly conserved once it is absorbed. Its excretion is minimal and unregulated, and facilitated by normal exfoliation from the surfaces of the body (dermal, intestinal, pulmonary, urinary), loss of blood by gastrointestinal bleeding, and loss in bile and sweat. Insignificant amounts are lost in urine, since iron in plasma is complexed with proteins that are too large to pass through the kidney glomerular membrane. Iron in feces is primarily unabsorbed dietary iron. Obligatory iron

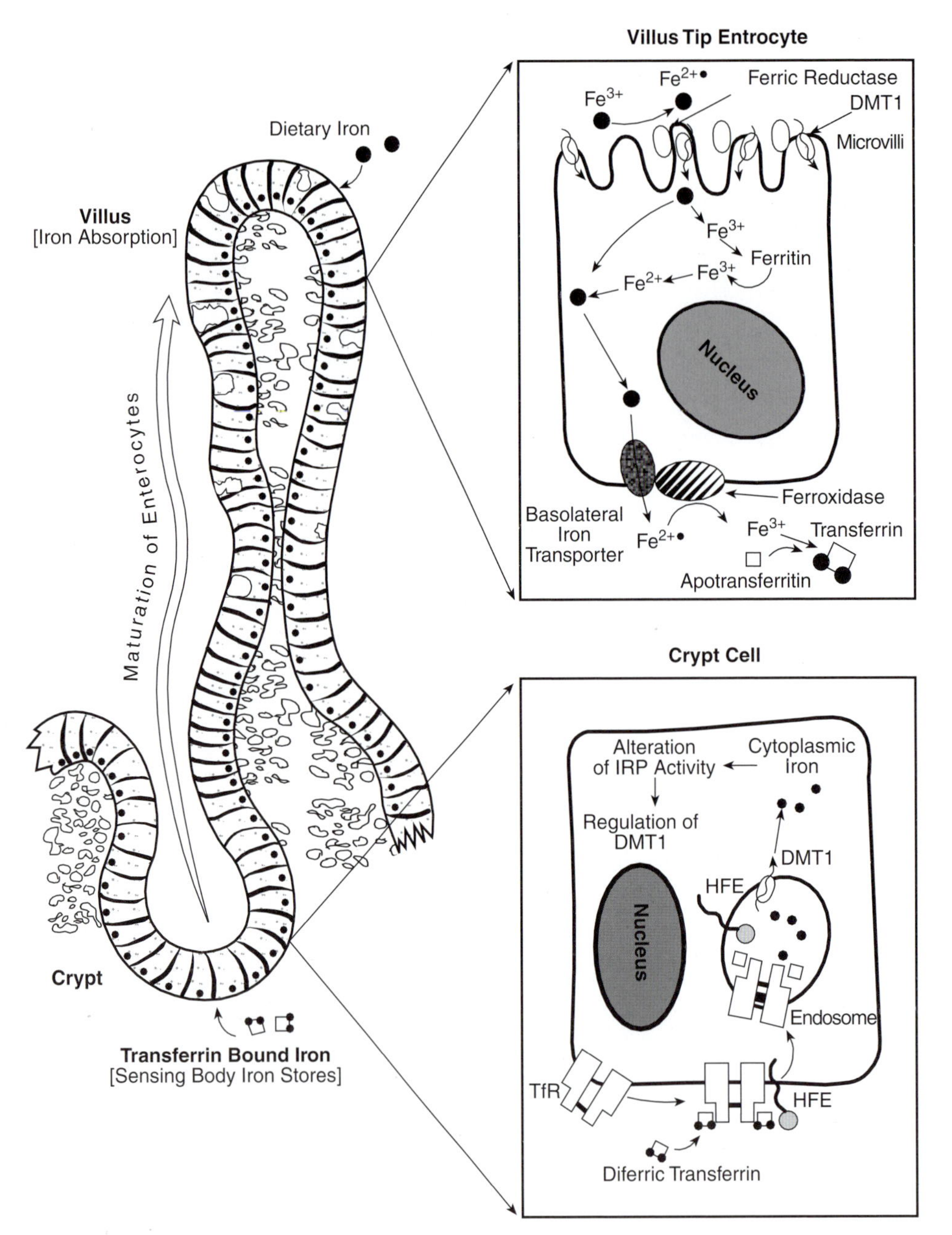

FIGURE 29-2

Intestinal absorption of dietary iron. Ferrous iron is absorbed by the duodenal villus tip enterocytes mediated by divalent metal transporter-1 (DMT1). Iron transport mediated by DMT1 of the apical surface and the basolateral transporter at the basolateral surface are coupled to ferric reductase and ferroxidase that change the iron oxidation state, respectively. The degree of iron entry is determined by the level of DMT1 and its level of expression is programmed in the crypt cells. The programming of the crypt cells is coupled to the body iron stores via transferrin-mediated and HFE protein-modulated iron transport. [Modified and reprinted with permission from B. R. Bacon, L. W. Powell, P. C. Adams, et al. Molecular medicine and hemochromatosis: at the cross roads. *Gastroenterology* **116,** 193 (1999).]

loss for a 70-kg man is 0.5–1 mg/d, an amount equal to that normally absorbed from the diet. During growth, menstruation, and pregnancy, the requirement reaches about 2–2.5 mg/d. Recommended daily iron intake for various groups is shown in Appendix IV.

The principal loss of iron in nonpregnant women during the reproducing years is through menstruation. In one study, the mean menstrual blood loss was 43.4 ± 2.3 mL. Since each milliliter of blood from a normal woman contains about 0.5 mg of iron, the amount lost every 27 days is about 20–23 mg. Increased menstrual flow (***menorrhagia***) significantly augments iron loss and leads to iron deficiency anemia (see below). In pregnancy use of supplemental iron is recommended. A newborn has about a 3- to 6-month supply of iron in its liver and may require iron-rich foods from the sixth month onward, since milk is poor in iron.

Plasma Iron Transport

Over 95% of plasma iron is in the Fe^{3+} state bound to the glycoprotein transferrin, a monomeric β_1-globulin (M.W. 80,000). Electrophoretic studies have revealed the existence of 21 genetic variants. In some, single-amino-acid substitutions account for variation in electrophoretic mobility. Transferrin is synthesized primarily in the liver and appears at the end of the first month of fetal development. Its half-life in humans is about 8 days. Desialylation may be a requirement for its removal from plasma by the liver, as it is for other plasma proteins (Chapter 10). In fact, asialotransferrin is more rapidly cleared from plasma than transferrin. It is not required for intestinal absorption of iron.

Each molecule of transferrin can bind two Fe^{3+} ions. The binding is extremely strong under physiological conditions, and the binding constants of the two sites are not significantly different. For each Fe^{3+} bound, one HCO_3^- ion is also bound and three H^+ ions are released from the protein.

Thus, diferric transferrin gains two net negative charges.

$$2Fe^{3+} + \text{apotransferrin} + 2HCO_3^- \rightleftharpoons [Fe_2\text{-transferrin-}(HCO_3)_2]^{2-} + 6H^+$$

The metal binding sites are located in N- and C-terminal domains. The protons released upon binding of each Fe^{3+} ion are probably derived from ionization of two tyrosyl residues and of a water molecule bound to Fe^{3+} ion.

The bulk of transferrin iron is delivered to immature erythroid cells for utilization in heme synthesis. Iron in excess of this requirement is stored as ferritin and hemosiderin. Unloading of iron to immature erythroid cells is by ***receptor-mediated endocytosis.*** The process begins in the clathrin-coated pits with the binding of diferric transferrin to specific plasma membrane transferrin receptors that are associated with the HFE protein complex. The next step is the internalization of the transferrin-transferrin receptor-HFE protein complex with formation of endosomes. The iron transporter DMT1 present in the cell membrane is also internalized into the endosomes. In the endosomes, a proton pump acidifies the complex to pH 5.4, and by altering conformation of proteins, iron is released from transferrin bound to transferrin receptor and HFE protein. This process of iron release from the complex is inhibited by HFE protein. Thus, dysfunctional HFE protein can cause excessive release of iron from the transferrin-transferrin receptor-HFE protein complex. In the acidified endosomes, DMT1 facilitates iron transport into the cytosol. Both apotransferrin (and a fraction of iron-bound transferrin) and transferrin receptor are returned to cell surfaces for reuse. In this type of receptor-mediated endocytosis of transferrin-transferrin receptor complex, the endosomes do not come into contact with lysosomes. The process is therefore unlike that of low-density lipoprotein receptor–mediated internalization (Chapter 20).

In the erythroid cells, most of the iron released from the endosomes is transported into mitochondria for heme synthesis (discussed later); in nonerythroid cells, the iron is stored predominantly as ferritin and to some extent as hemosiderin.

Storage of Iron

There are two storage forms of iron, ferritin and hemosiderin. Ferritin is the predominant storage form and contains diffusable, soluble, and mobile fractions of iron. Hemosiderin is aggregated deposits resulting from the breakdown of ferritin in secondary lysosomes and its level increases progressively with increasing levels of iron overload. Apoferritin is a protein shell consisting of 24 subunits of two types; a light (L) subunit (M.W. 19,000) and a heavy (H) subunit (M.W. 21,000). The H subunit has ferroxidase activity and the L subunit facilitates nucleation and mineralization of the core made up of hydrated ferric oxide phosphate complex.

Coordinate Regulation of Iron Uptake and Storage in Non-Erythroid Cells

Iron uptake is regulated by transferrin receptor and storage of iron as ferritin which occurs post-transcriptionally for these two proteins. The regulation maintains an optimal intracellular-transit-chelatable iron pool for normal functioning in the body. The regulatory process consists of an interaction between IREs and IRPs 1 and 2. One copy of each IRE has been identified in the 5′-untranslated region of H and L ferritin mRNAs and five copies in the 3′-untranslated region (UTR) of transferrin receptor mRNA. IRE sequences are highly conserved and have a stem-loop structure with a CAGUGN sequence at the tip of the loop. IRPs are RNA-binding proteins that bind to IREs and regulate the translation of the respective mRNAs.

During low levels of intracellular chelatable iron, iron storage declines due to inhibition of ferritin synthesis; cellular entry of iron increases due to enhanced transferrin receptor synthesis. An opposing set of events occurs

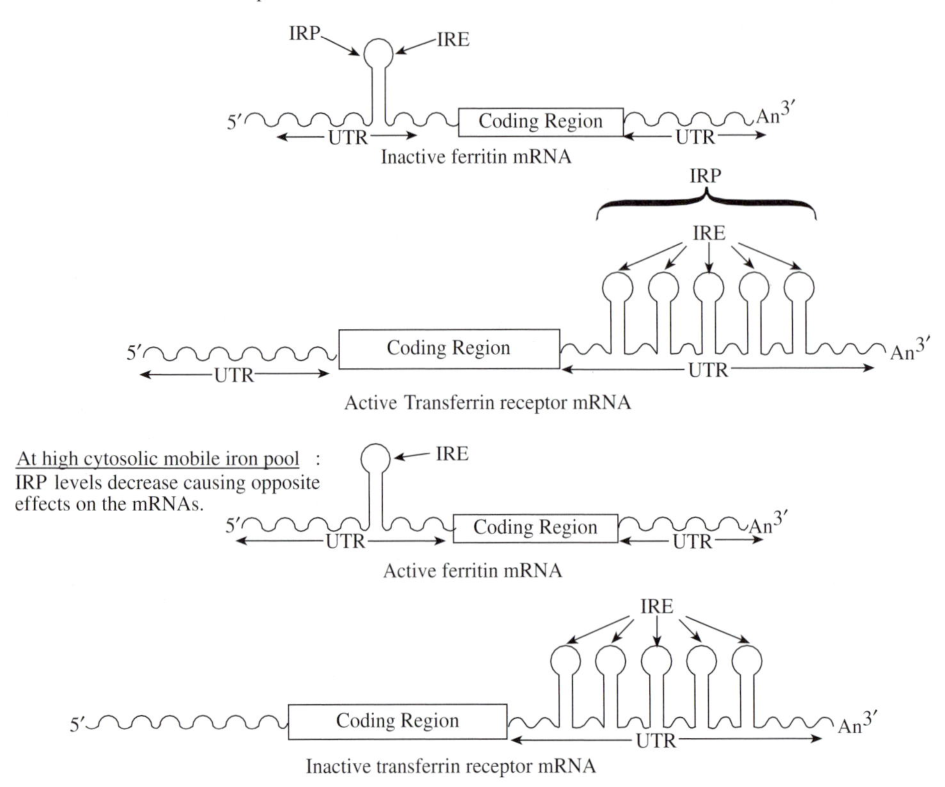

FIGURE 29-3

Coordinate translational regulation of ferritin mRNA and transferrin receptor mRNA in nonerythroid cells. Iron regulatory proteins (IRP) are RNA-binding proteins that bind to iron regulatory elements (IREs). IREs are hairpin structures with loops consisting of CAGUGN sequences and are located at the 5′-untranslated region (UTR) and 3′-UTR for ferritin mRNA and transferrin mRNA, respectively.

during intracellular chelatable iron excess or iron-replete states. A coordinated control occurs when IRP binds to IRE at the 5′-UTR of ferritin mRNAs inhibiting ferritin synthesis; simultaneously, the binding of IRP to IRE at the 3′-UTR of transferrin receptor mRNA stimulates transferrin receptor synthesis (Figure 29-3). Intracellular iron regulates the level of IRPs. During the expansion of the iron pool, IRPs are inactivated, leading to efficient translation of ferritin mRNA and rapid degradation of transferrin receptor mRNA. In iron-replete cells, IRP1 acquires iron by the formation of iron-sulfur clusters (4Fe-4S) that bind to IREs with low affinity. During iron deficiency states, IRP1 lacks a 4Fe-4S cluster and binds to IREs with high affinity. IRP1, when it possesses an iron-sulfur cluster, has aconitase activity, normally TCA cycle enzyme (Chapter 14). Mutations that change IREs can lead to constitutive ferritin biosynthesis. An autosomal dominant disorder of IRE leads to ***hyperferritinemia*** without any iron overload. Patients with this inherited disorder also exhibit early onset of cataract that may also be cotransmitted as an autosomal dominant trait. Other factors also regulate ferritin synthesis. For example, nitric oxide enhances binding of IRP to IRE and inhibits ferritin synthesis. One of the causes of anemia of chronic inflammatory diseases may be due to increased ferritin synthesis in the reticuloendothelial macrophages by inflammatory cytokines interleukins 1 and 6 (IL-1 and IL-6), which act by preventing efficient release of iron.

Measurement of serum ferritin levels has diagnostic utility. In iron deficiency anemia (discussed later), serum ferritin levels are low; in iron storage disease, the levels are high. However, serum ferritin levels can also be elevated under many other circumstances, including liver diseases and chronic inflammatory diseases.

Alterations of Plasma Transferrin Concentration

Plasma transferrin levels are commonly measured in the evaluation of disorders of iron metabolism (see below). It is customary to measure transferrin concentration indirectly from the maximum (or total) iron binding capacity (TIBC) of plasma (reference range for adults, 250–400 μg/dL). It can also be measured directly by immunological methods (reference range for adults, 220–400 mg/dL). ***Hypertransferrinemia*** (or increased TIBC) can occur with diminished body iron stores as in iron deficiency anemia or during pregnancy (because of enhanced mobilization of storage iron to supply maternal and fetal demands). Hypertransferrinemia of iron deficiency is corrected by oral iron supplementation, whereas that due to pregnancy is not. Exogenous administration of estrogens (e.g., oral contraceptives) also causes hypertransferrinemia.

Hypotransferrinemiacan result from protein malnutrition and accompanies hypoalbuminemia. Since transferrin has a much shorter half-life (8 days) than albumin (19 days), measurement of the transferrin level may be a more sensitive indicator of protein malnutrition than albumin measurement (see also chapter 17). Hypotransferrinemia also results from excessive renal loss of plasma proteins (e.g., in nephrotic syndrome).

Disorders of Iron Metabolism

Iron Deficiency Anemia

Iron deficiency anemia is the most prevalent nutritional disorder. Its cause may comprise many overlapping factors: dietary iron deficiency; absence of substances that favor iron absorption (ascorbate, amino acids, succinate); presence of compounds that limit iron absorption (phytates, oxalates, excess phosphates, tannates); lack of iron absorption due to gastrointestinal disorders (malabsorption syndrome, gastrectomy); loss of iron due to menstruation, pregnancy, parturition, lactation, chronic bleeding from the gastrointestinal tract peptic ulceration, hemorrhoids, cancer, colonic ulceration, or hookworm infestation or the genitourinary tract (uterine fibroids); enhanced demand for growth or new blood formation; deficiency of iron transport from mother to fetus; abnormalities in iron storage; deficiencies in release of iron from the reticuloendothelial system (infection, cancer); inhibition of incorporation of iron into hemoglobin (lead toxicity); and rare genetic conditions (transferrin deficiency, impaired cellular uptake of iron by erythroid precursors).

In the initial phase of depletion of the iron content of the body, the iron stores maintain normal levels of hemoglobin and of other iron proteins. With exhaustion of storage iron, hypochromic and microcytic anemia becomes manifest.

The clinical characteristics of iron deficiency anemia are nonspecific and include pallor, rapid exhaustion, muscular weakness, anorexia, lassitude, difficulty in concentrating, headache, palpitations, dyspnea on exertion, angina on effort, peculiar craving for unnatural foods (pica), ankle edema, and abnormalities involving all proliferating tissues, especially mucous membranes and the nails. The onset is insidious and may progress slowly over many months or years.

Physiological adjustments take place during the gradual progression of the disorder, so that even a severe hemoglobin deficiency may produce few symptoms. Iron deficiency may affect the proper development of the central nervous system. Early childhood iron deficiency anemia may lead to cognitive abnormalities.

Individuals who have ***congenital atransferrinemia*** lack apotransferrin and suffer from severe hypochromic anemia in the presence of excess iron stores in many body sites, susceptibility to infection (transferrin inhibits bacterial, viral, and fungal growth, probably by binding the iron required for growth of these organisms), and retardation of growth. This condition does not respond to administration of iron. Intravenous administration of transferrin normalizes the iron kinetics. A rare congenital defect in uptake of iron by red cell precursors has been reported that leads to severe hypochromic anemia with normal plasma iron and transferrin levels.

Microcytic anemia occurs frequently in thalassemia syndromes (Chapter 28), but these patients do not require iron supplementation unless they have concurrent iron deficiency as assessed by measurement of serum iron levels and TIBC. Serum iron concentration exhibits a morning peak and an evening nadir; this pattern is reversed in night-shift workers. The circadian variation is primarily due to differences in rate of release of iron by the reticuloendothelial system. Transferrin levels do not show circadian fluctuation. Iron deficiency anemia can also be assessed from the plasma ferritin concentration (which when decreased reflects depleted iron stores), red cell protoporphyrin concentration (increased because of lack of conversion to heme), and the number of sideroblasts in the bone marrow (which parallels iron stores). Sideroblasts are erythrocyte precursors (normoblasts) containing free ferritin-iron granules in the cytoplasm that stain blue with the Prussian blue reagent. There is a close correlation between plasma iron levels, TIBC, and the proportion of sideroblasts in bone marrow. In hemolytic anemias, pernicious anemia, and hemochromatosis, the serum iron level increases and sideroblast number reaches 70% (normal range, 30–50% of total cells). In iron deficiency, the sideroblasts are decreased in number or absent.

Before treatment is initiated, the cause of the negative iron balance must be established. Treatment should correct the underlying cause of anemia and improve the iron balance. In general, oral therapy with ferrous salts is satisfactory; however, sometimes parenteral therapy is preferred, e.g., in proven malabsorption problems, gastrointestinal disease and excessive blood loss, and for patients who cannot be relied on to take oral medication.

Iron-Storage Disorders

A type of iron storage disorder characterized by general increase in tissue iron levels without damage to parenchymal cells is known as ***hemosiderosis.*** Hemosiderin is a storage form of iron in which ferric hydroxide is present as micelles. It appears as insoluble granules that contain denatured aggregated ferritin, nonferritin proteins, lipids, heme, and other pigments. Hemosiderosis results when iron is present in excessive quantities in a diet that permits maximum iron absorption. For example, the African Bantu eat a diet high in corn (low in phosphate) cooked in iron pots, drink an indigenous beer brewed in iron pots, and suffer from ***Bantu siderosis.*** Hemosiderosis can progress to ***hemochromatosis*** with hepatic cirrhosis and diabetes mellitus. progressive deterioration in pancreatic, hepatic, gonadal, and cardiac function. Clinical manifestations include cirrhosis, diabetes mellitus, life-threatening arrhythmias, and intractable heart failure. Removal of excess iron produces clinical improvement, particularly of diabetes and congestive heart failure.

In iron storage diseases accompanied by normal erythropoiesis (e.g., hereditary hemochromatosis), removal of excessive iron is accomplished by repeated bloodletting (phlebotomy). Therapeutic phlebotomy of a unit of blood (which contains about 250 mg of iron) may be performed up to three times per week. When the iron stores become depleted, reaccumulation of iron is prevented by four to six phlebotomies per year. In asymptomatic patients, periodic determination of serum ferritin provides a measure of storage of iron.

In hemochromatosis secondary to refractory anemias (e.g., Cooley's anemia, sickle cell anemia), patients require repeated blood transfusions to survive childhood and adulthood. Therapy consists of administration of iron-chelating agents. Deferoxamine, an iron chelator isolated from *Streptomyces pilosis*, has the structure:

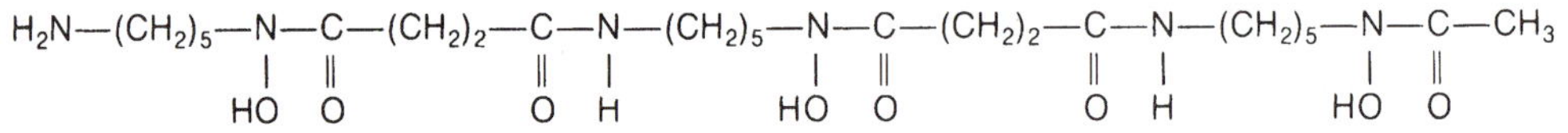

$$H_2N{-}(CH_2)_5{-}\underset{\displaystyle HO}{N}{-}\underset{\displaystyle O}{\overset{}{C}}{-}(CH_2)_2{-}\underset{\displaystyle O}{C}{-}\underset{\displaystyle H}{N}{-}(CH_2)_5{-}\underset{\displaystyle HO}{N}{-}\underset{\displaystyle O}{C}{-}(CH_2)_2{-}\underset{\displaystyle O}{C}{-}\underset{\displaystyle H}{N}{-}(CH_2)_5{-}\underset{\displaystyle HO}{N}{-}\underset{\displaystyle O}{C}{-}CH_3$$

Excessive accumulation of iron (chronic iron overload) can result from the following.

1. Defective erythropoiesis (dyserythropoiesis); impaired hemoglobin synthesis leading to lack of utilization and consequent accumulation of iron in mitochondria, e.g., from inhibition of ALA synthase activity by dietary vitamin B_6 deficiency; inhibition of heme synthesis by lead; impairment of pyridoxine metabolism in alcoholic patients; familial sideroblastic anemias; and **Cooley's anemia.**
2. Repeated blood transfusions, e.g., in Cooley's anemia or sickle cell disease.
3. Hereditary hemochromatosis, an autosomal recessive defect in which there is increased rate of absorption of iron in the presence of normal or enlarged iron stores and normal hematopoiesis (discussed later).
4. High dietary iron and substances that enhance its absorption (e.g., Bantu siderosis).
5. Hereditary atransferrinemia.

In all of these disorders, the gastrointestinal tract cannot limit absorption of iron to significant extent. Thus, the "mucosal block" responsible for keeping out unneeded iron on a daily basis is susceptible to disruption, perhaps at more than one point. Iron overload leads to

It contains six nitrogen atoms separated by fairly long, flexible stretches of methylene groups. Since each iron atom can bind six ligands, one molecule of deferoxamine is probably capable of occupying all six coordination sites and producing a 1 : 1 iron-deferoxamine complex.

For ferric iron, the K_{assoc} of deferoxamine is about 10^{30}, while the K_{assoc} for Ca^{2+} is about 10^2. Iron in hemoproteins is not affected by this agent, while the ferric iron of ferritin and hemosiderin is chelated in preference to that found in transferrin. Such selectivity makes the compound useful in treatment of iron storage problems and acute iron poisoning.

The deferoxamine-iron complex is excreted in urine. (Iron is not normally excreted by this route.)

Deferoxamine given orally complexes with dietary iron, making the drug and the iron unavailable for absorption. The preferred route of administration is by intramuscular injection. Irritation and pain at the site of administration and the need for daily injections make the treatment unpopular. In addition, even with coadministration of large amounts of ascorbic acid, the iron loss produced is far below that necessary to remove all of the iron accumulated during chronic transfusion therapy.

Slow, continuous intravenous infusion or continuous subcutaneous administration may be more effective in establishing negative iron balance and eliminating stored iron. A small, labile (chelatable) iron pool may be in slow equilibrium with a much larger (storage) pool. When deferoxamine is administered, the labile pool is rapidly emptied. Any deferoxamine that remains in the body or is administered thereafter finds no iron to bind. Thus, most of a single intramuscular dose is simply excreted unchanged. If the chelator is given as a continuous infusion, the labile pool is initially depleted and kept empty. As iron is released from storage sites, it is immediately chelated and removed. Removal of up to 180 mg of iron per day has been accomplished by this method, making it as effective as phlebotomy. Massive intravenous injections of deferoxamine have also been reported to produce excretion of large amounts of iron in iron-overloaded patients.

Hereditary Hemochromatosis

Hereditary hemochromatosis is a common inherited autosomal recessive disorder of excessive iron accumulation in parenchymal cells of liver, heart, pancreas, endocrine organs, skin, and joints. The term hemochromatosis is used when organ damage has occurred in the presence of impaired function. It occurs predominately in Caucasian populations; about 1 in 200–400 caucasians are at risk for developing clinical symptoms. Individuals with hereditary hemochromatosis absorb about 3–4 mg/d of iron as compared with a normal rate of about 1–2 mg/d. This excess iron, absorbed over several years, causes accumulation of as much as 20–40 g as compared to normal amounts of about 4 g. In untreated patients, progressive iron accumulation can cause organ damage resulting in hepatic dysfunction, diabetes, cardiomyopathy, hypogonadism with infertility, arthritis, and skin hyperpigmentation. Death can occur due to cirrhosis, diabetes, cardiomyopathy or hepatocellular carcinoma.

Hereditary hemochromatosis is associated with a gene on the short arm of chromosome 6 near the MHC gene complex. This gene is known as *HFE* and codes for HFE protein. The roles of HFE protein along with β_2-microglobulin in the regulation of intestinal iron absorption and iron sequestration in the form of ferritin and hemosiderin have been discussed previously. Gene knockout mice for either HFE protein or β_2-microglobulin develop an iron overload disorder similar to human hemochromatosis, thus substantiating the roles of HFE protein and β_2-microglobulin in iron homeostasis.

Two missense mutations (C282Y and H63D) in the *HFE* gene have been identified in hereditary hemochromatosis. The substitution of a tyrosyl residue for a cysteinyl residue at position 282 results in the loss of formation of a disulfide linkage necessary for the proper association with β_2-microglobulin (Figure 29-1). In the absence of this disulfide linkage, HFE protein fails to reach the normal membrane location and is rapidly degraded. Thus, C282Y is a loss of function (knockout) mutation. The H63D mutation has no effect on the HFE protein's association with β_2-microglobulin. However, this mutation may compromise the protein regulation of the interaction between transferrin and its receptor.

Population studies among Caucasians have shown that about 1 in 10 are heterozygous for the C282Y mutation. Homozygosity of C282Y has been observed in 85–90% of hereditary hemochromatosis patients, In about 4% of the hereditary hemochromatosis patients, heterozygosity of C282Y and H63D has been observed. There are still unanswered questions concerning hereditary hemochromatosis. For example, some C282Y homozygotes exhibit neither biochemical nor clinical evidence of iron accumulation. On the contrary, some hemochromatosis patients do not possess a C282Y mutation. Thus, other yet-to-be-identified genetic and environmental factors must play a role in the development of hemochromatosis.

Other forms of hemochromatosis include neonatal and juvenile types in which biochemical defects have not yet been identified. Patients with *aceruloplasminemia* resulting from mutations in the ceruloplasmin gene exhibit accumulation of iron in neural and glial cells in the brain, in hepatocytes, and in pancreatic islet cells. Ceruloplasmin, a copper-containing protein, has ferroxidase activity and participates in the release of iron from cells. Aceruloplasminemia associated with iron overload is a different disorder from that of ***Wilson's disease*** (hepatolenticular degeneration) in which biliary excretion of copper and incorporation of copper into ceruloplasmin are defective (Chapter 37). *Porphyria cutanea tarda,* a disorder of porphyrin biosynthesis (discussed later), usually is accompanied by iron accumulation. Thirty percent of patients with porphyria cutanea tarda are either homozygous or heterozygous for the C282Y mutation affecting the HFE protein.

Treatment of hereditary hemochromatosis is therapeutic phlebotomy (discussed earlier). This method is safe, effective, and life saving, and ideally should begin before symptoms develop. Serum ferritin levels are used as a surrogate marker for estimating total-body iron stores. Morphologic studies and quantitative determination of iron in liver tissue obtained by biopsy have been used in the assessment of early hereditary hemochromatosis and the degree of liver injury.

Finally, hereditary hemochromatosis is a treatable disease. Biochemical screening for the identification of the disease in the general population has been suggested. The biochemical tests include the measurement of serum levels of iron, transferrin saturation, and ferritin.

29.2 Heme Biosynthesis

The principal tissues involved in heme biosynthesis are the hematopoietic tissues and the liver. Biosynthesis requires participation of eight enzymes, of which four (the first and the last three) are mitochondrial and the rest are cytosolic (Figure 29-4). The reactions are irreversible. Glycine and succinate are the precursors of porphyrins.

Formation of δ-Aminolevulinic Acid

δ-Aminolevulinic acid (ALA) formation is catalyzed by mitochondrial ALA synthase, which condenses glycine and succinyl-CoA to ALA. The enzyme is located on the matrix side of the inner mitochondrial membrane. It is encoded by a nuclear gene and is synthesized in the cytosol on the free polyribosomes as a

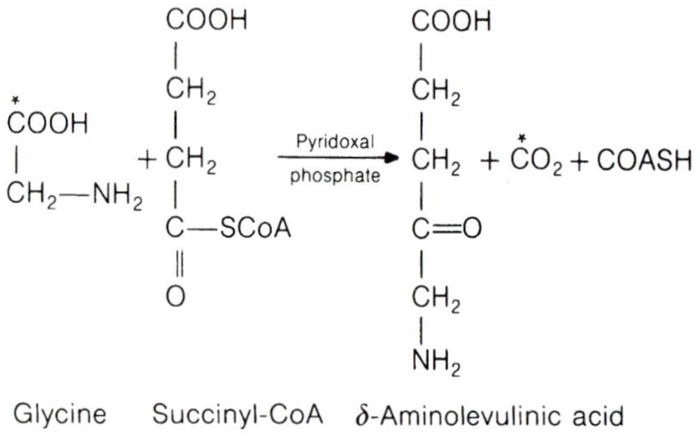

$$\underset{\text{Glycine}}{\overset{*}{C}OOH-CH_2-NH_2} + \underset{\text{Succinyl-CoA}}{HOOC-CH_2-CH_2-C(=O)-SCoA} \xrightarrow[\text{phosphate}]{\text{Pyridoxal}} \underset{\delta\text{-Aminolevulinic acid}}{HOOC-CH_2-CH_2-C(=O)-CH_2-NH_2} + \overset{*}{C}O_2 + COASH$$

precursor. The precursor protein is processed to active form during its translocation into mitochondria (Chapter 25). Pyridoxal phosphate is the required coenzyme.

The reaction mechanism consists of formation of a Schiff base by pyridoxal phosphate with a reactive amino group of the enzyme; entry of glycine and formation of an enzyme-pyridoxal phosphate-glycine-Schiff base complex; loss of a proton from the α carbon of glycine with the generation of a carbanion; condensation of the carbanion with succinyl-CoA to yield an enzyme-bound intermediate (α-amino-β-ketoadipic acid); decarboxylation of this intermediate to ALA; and liberation of the bound ALA by hydrolysis. ALA synthesis does not occur in mature erythrocytes.

In experimental animals, deficiency of pantothenic acid (needed for CoASH and, hence, succinyl-CoA synthesis), lack of vitamin B_6, or the presence of compounds that block the functioning of pyridoxal phosphate (e.g., isonicotinic acid hydrazide; Chapter 17) can prevent heme synthesis and cause anemia. Heme synthesis also requires a functional tricarboxylic acid cycle and an oxygen supply.

The primary regulatory step of heme synthesis in the liver is apparently that catalyzed by ALA synthase. The regulatory effects are multiple. The normal end product, heme, when in excess of need for production of heme proteins, is oxidized to ***hematin,*** which contains a hydroxyl group attached to the Fe^{3+} atom. Replacement of the hydroxyl group by a chloride ion produces ***hemin.*** Hemin and heme inhibit ALA synthase allosterically. Hemin also inhibits the transport of cytosolic ALA synthase precursor protein into mitochondria.

ALA synthase has a turnover rate of 70 minutes in adult rat liver and is inducible. Its induction is suppressed by hemin and increased by a variety of xenobiotics (e.g., environmental pollutants) and natural steroids. In erythropoietic tissues, where the largest amount of heme is synthesized, regulation of heme biosynthesis may also involve the process of cell differentiation and proliferation of the erythron, which occurs to meet change in requirements for the synthesis of heme. The differentiation and proliferation are initiated by erythropoietin.

Formation of Porphobilinogen

Two molecules of ALA are condensed by cytosolic zinc containing ALA dehydratase to

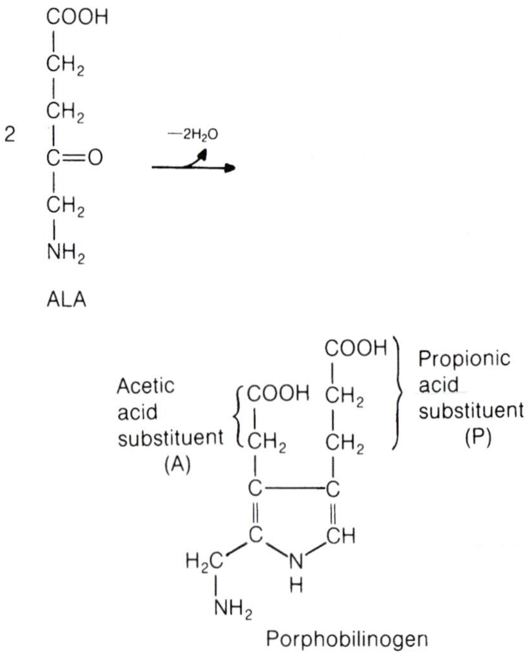

$$2\ \underset{\text{ALA}}{HOOC-CH_2-CH_2-C(=O)-CH_2-NH_2} \xrightarrow{-2H_2O} \text{Porphobilinogen}$$

Porphobilinogen: pyrrole ring bearing an acetic acid substituent (A), $-CH_2-COOH$; a propionic acid substituent (P), $-CH_2-CH_2-COOH$; and an aminomethyl group, $-CH_2-NH_2$ (H_2C-NH_2); ring N–H.

yield porphobilinogen (PBG). There are four zinc ions per

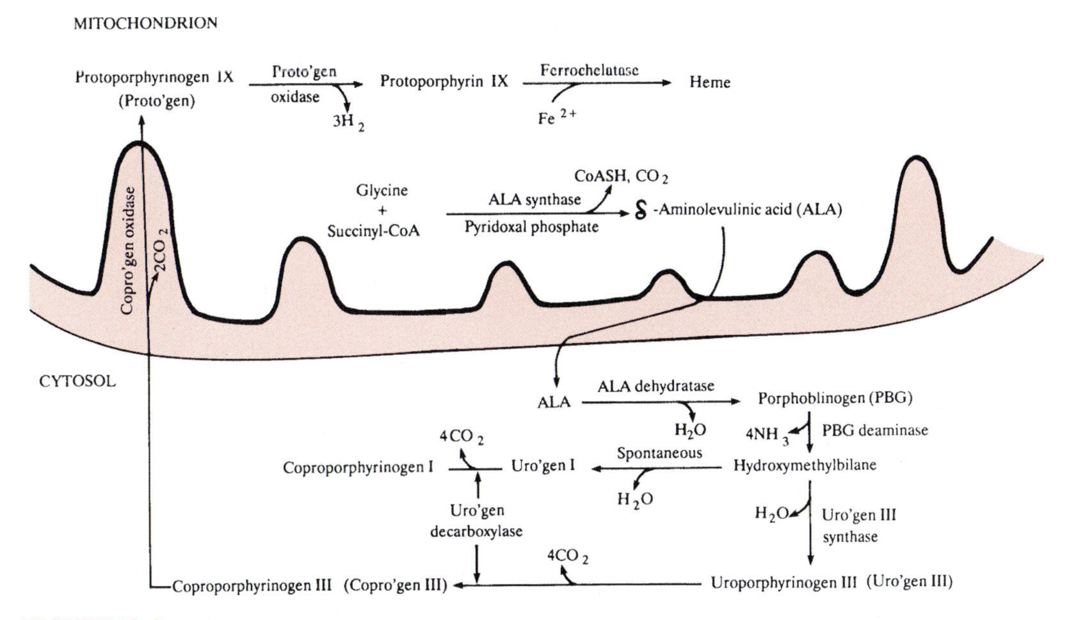

FIGURE 29-4
Biosynthetic pathway of heme. The pathway consists of eight irreversible reactions, four each in the mitochondrion and the cytosol. The primary site of regulation is the ALA synthase step.

octamer of the enzyme, and they are bound via the reduced thiol groups. Zinc is required for enzyme activity.

The reaction mechanism consists of Schiff base formation by the keto group of one molecule of ALA with the ε-amino group of a lysyl residue of the enzyme, followed by nucleophilic attack by the enzyme-ALA anion on the carbonyl group of a second ALA molecule with elimination of water. Then, a proton is transferred from the amino group of the second ALA molecule to the ε-amino group of the lysyl residue with formation of PBG. Lead is a potent inhibitor of ALA dehydratase, presumably by displacement of zinc by lead because the lead-inhibited enzyme can be reactivated by the addition of zinc. ALA dehydratase is inhibited competitively by succinyl acetone ($HOOC–CH_2–CH_2–CO–CH_2–CO–CH_3$), which occurs in urine and blood in hereditary tyrosinemia (Chapter 17). Genetic deficiency of ALA dehydratase is known to occur.

Formation of Uroporphyrinogen III

Uroporphyrinogen III formation occurs in the cytosol and requires the successive action of porphobilinogen deaminase (or methylbilane synthase) and uroporphyrinogen III synthase. Porphobilinogen deaminase catalyzes condensation of four porphobilinogen molecules in a symmetrical head-to-tail arrangement to form a straight-chain tetrapyrrole, hydroxymethylbilane. Uroporphyrinogen III synthase catalyzes the rearrangement of one of the pyrrole rings (ring D in Figure 29-5) to form an asymmetrical tetrapyrrole, followed by its cyclization to form uroporphyrinogen III. In the absence of uroporphyrinogen III synthase (e.g., in congenital erythropoietic porphyria), the hydroxymethylbilane cyclizes spontaneously to the uroporphyrinogen I isomer, which is not a precursor of heme (Figure 29-5).

Formation of Coproporphyrinogen III

Cytosolic uroporphyrinogen decarboxylase catalyzes successive decarboxylation of the four acetic groups to yield four methyl groups (Figure 29-6).

Formation of Protoporphyrinogen IX

Mitochondrial coproporphyrinogen oxidase is localized in the intermembrane space and is probably loosely bound to the outer surface of the inner membrane. It catalyzes the successive conversion of propionic acid groups of ring A and ring B to vinyl groups (Figure 29-7).

Formation of Protoporphyrin IX and Heme

Both of these steps occur in mitochondria (Figure 29-8). Porphyrinogen oxidase removes six hydrogen atoms (four from methane bridge carbons and two from pyrrole nitrogens) from protoporphyrinogen to yield protoporphyrin. The oxidase has an absolute requirement for oxygen. Protoporphyrinogen can also be oxidized nonenzymatically to protoporphyrin at physiological pH, temperature, and aerobic conditions. Protoporphyrin oxidase is bound to the

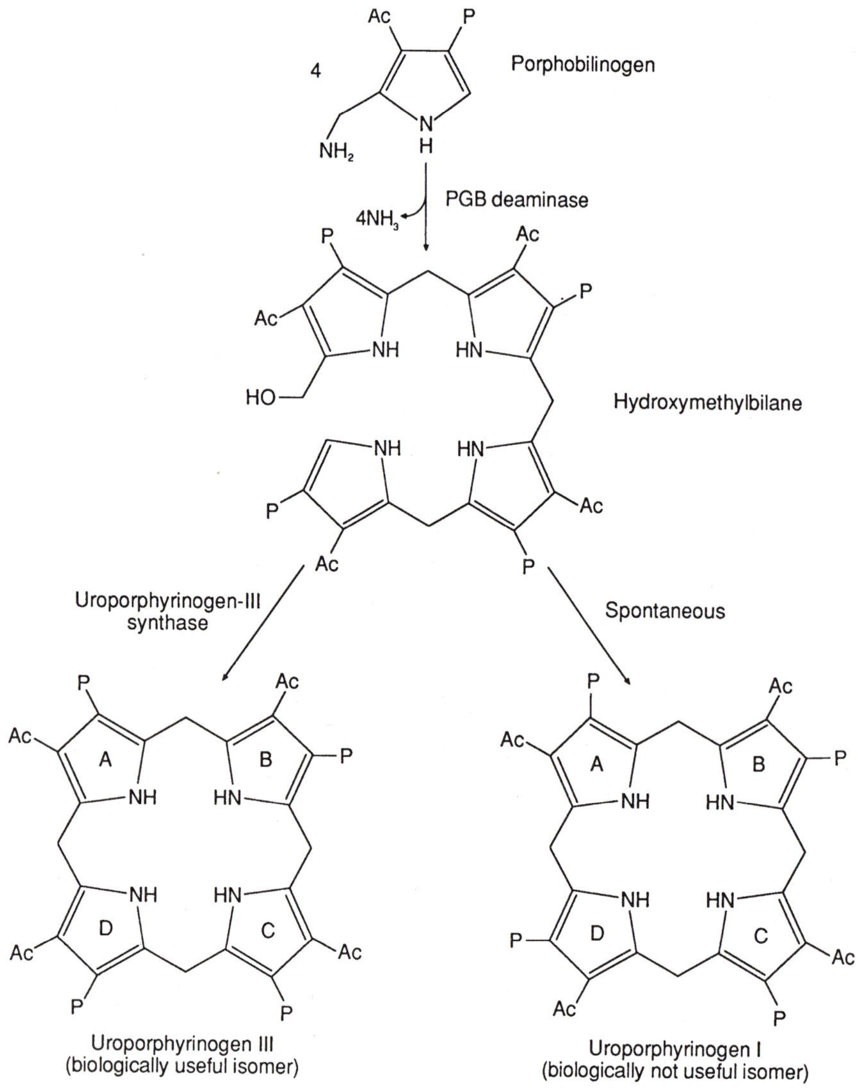

FIGURE 29-5
Synthesis of uroporphyrinogen I and III. The latter is the biologically useful isomer, and its formation requires the action of uroporphyrinogen-III synthase. Ac, $-CH_2COOH$; P, $-CH_2CH_2COOH$.

inner mitochondrial membrane, and its active site faces the cytosolic side of the membrane. Formation of heme is accomplished by ferrochelatase (or heme synthase), which incorporates Fe^{2+} into protoporphyrin and is inhibited by lead. Zinc can function as a substrate in the absence of iron.

Disorders of Heme Biosynthesis

The **porphyrias** are a group of disorders caused by abnormalities in heme biosynthesis. They are inherited and acquired disorders characterized by excessive accumulation and excretion of porphyrins or their precursors. Defects in any one of the eight enzymes involved in heme biosynthesis may cause inherited porphyrin-related disorders (Figure 29-9). Porphyrins have a deep red or purple color (Greek *porphyra* = purple). Porphyrins are excreted by different routes, depending on their water solubility. For example, uroporphyrin with its eight carboxylic group substituents is more water-soluble than the porphyrins derived from it and is eliminated in the urine, whereas protoporphyrin (which contains two carboxylic groups) is excreted exclusively in bile. Coproporphyrin has four carboxylic groups and is found in bile and urine.

These disorders are associated with acute or cutaneous manifestations (or both). In the acute state, the presentation may include abdominal pain, constipation, hypertension, tachycardia, and neuropsychiatric manifestations. Cutaneous problems consist of photosensitivity (itching, burning, redness, swelling, and scarring), hyperpigmentation, and sometimes hypertrichosis

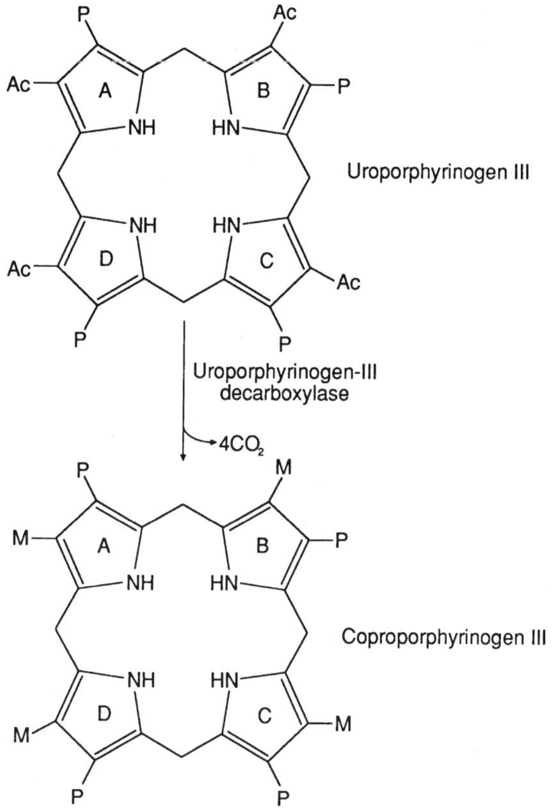

FIGURE 29-6
Formation of coproporphyrinogen III from uroporphyrinogen III. Acetic acid side chains (Ac) are decarboxylated to methyl groups (M), sequentially, starting clockwise from ring D. P, $-CH_2CH_2COOH$.

(an abnormally excessive growth of hair). Four porphyrias can manifest as acute disorders: δ-ALA dehydratase deficiency porphyria, acute intermittent porphyria, hereditary coproporphyria, and variegate porphyria.

Porphyria maybe classified as hepatic or erythropoietic. However, enzyme defects are sometimes common to both tissues. Porphyrias can be induced by alcohol, stress, infection, starvation, hormonal changes (e.g., menstruation), and certain drugs. These drugs presumably precipitate acute manifestations in susceptible subjects since they are inducers of cytochrome P-450 and increase the need for synthesis of heme as they deplete the mitochondrial pool of free heme. Major hepatic porphyrias include *acute intermittent porphyria, variegate porphyria, hereditary coproporphyria,* and *porphyria cutanea tarda.* The principal erythropoietic porphyrias are *hereditary erythropoietic porphyria* and *erythropoietic protoporphyria.*

Hepatic Porphyrias

Acute intermittent porphyria is associated with excessive urinary excretion of ALA and porphobilinogen. The lack of polymerization of porphobilinogen is due to deficiency of porphobilinogen deaminase in several cell types (e.g., hepatocytes, erythrocytes, fibroblasts, lymphocytes). Acute clinical manifestations include neuropsychiatric disorders and abdominal pain. The cause of these manifestations is not clear, but accumulation of porphyrin precursors (ALA and porphobilinogen) in pharmacological amounts has been implicated. Since afflicted subjects cannot make porphyrins to any great extent, they are not photosensitive. This disorder is inherited as an autosomal dominant trait.

Porphyria cutanea tarda is the most common form. It is inherited as an autosomal dominant trait and is due to deficiency of uroporphyrinogen III decarboxylase. Clinical

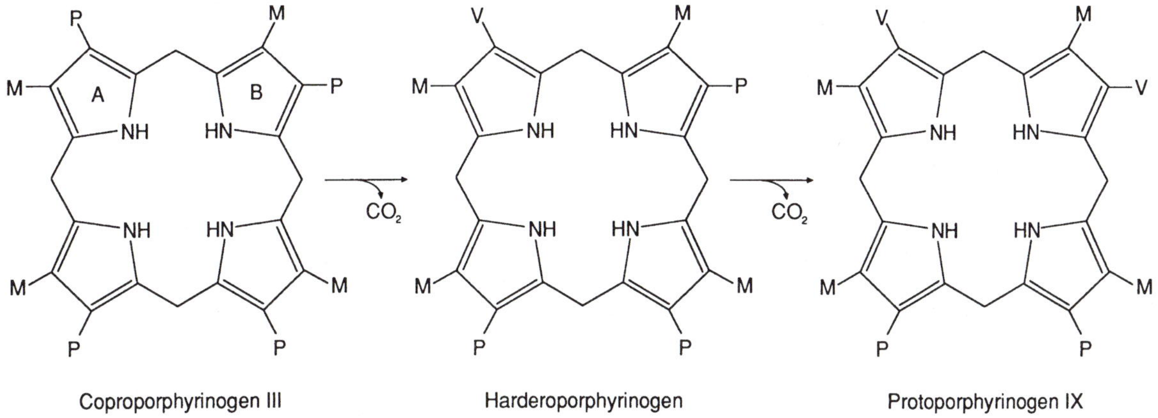

FIGURE 29-7
Formation of protoporphyrinogen IX from coproporphyrinogen III by coproporphyrinogen oxidase. Sequential oxidative decarboxylation of the propionic acid (P) sidechains of rings A and B produces vinyl (V) groups ($V = -CH{=}CH_2$). The reaction proceeds via the stereospecific loss of one hydrogen atom and decarboxylation of the propionic acid group. Molecular oxygen is the oxidant, and β-hydroxypropionate is a probable intermediate. M, CH_3.

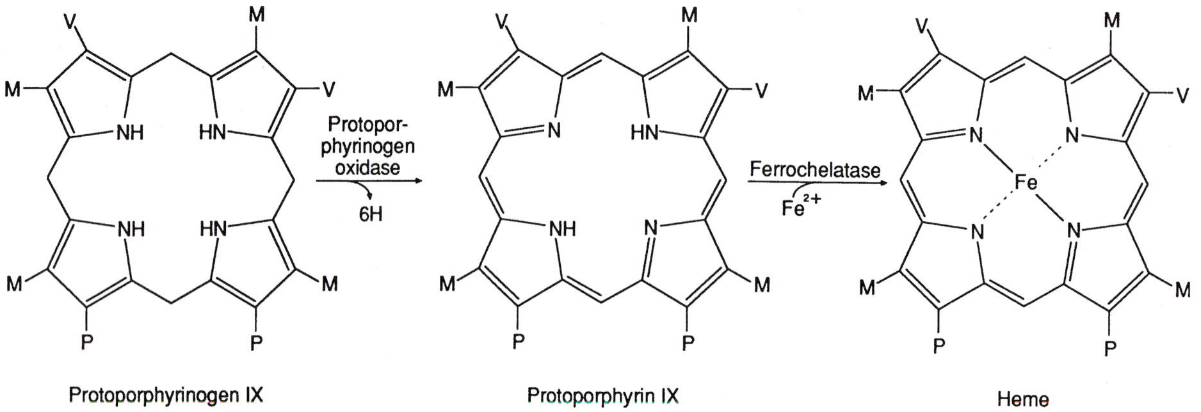

FIGURE 29-8
Formation of heme. In the reaction catalyzed by protoporphyrinogen oxidase, six hydrogens are removed and the primary electron acceptor is not known, but oxygen is required for enzyme activity. In the terminal step of heme synthesis, only Fe^{2+} is incorporated into protoporphyrin.

manifestations are mild to severe photosensitivity and liver disease. Most affected individuals have increased hepatic iron stores, which can be successfully decreased by phlebotomy. Acute episodes can be precipitated by overindulgence of alcohol or, less frequently, by therapy with estrogen.

Hereditary coproporphyria and ***variegate porphyria*** are inherited as autosomal dominant traits. They are caused by deficiency of coproporphyrinogen oxidase and protoporphyrinogen oxidase, respectively.

In most of these disorders, increased hepatic ALA synthase activity is due to decreased heme synthesis. There are also increased amounts of ALA and porphobilinogen in liver, plasma, and urine and specific metabolites produced before the metabolic block. ALA synthase is regulated by the heme by a feedback process and by gene repression. Hematin has been used to treat acute attacks with marked success. Although often inadequate, carbohydrate feeding has been associated with improvement in acute intermittent porphyria. This "glucose effect" may depend on repression of the gene for ALA synthase, but the mechanism is not known.

Erythropoietic Porphyrias

A defect in synthesis of type III isomers from hydroxymethylbilane, due to deficiency of uroporphyrinogen III

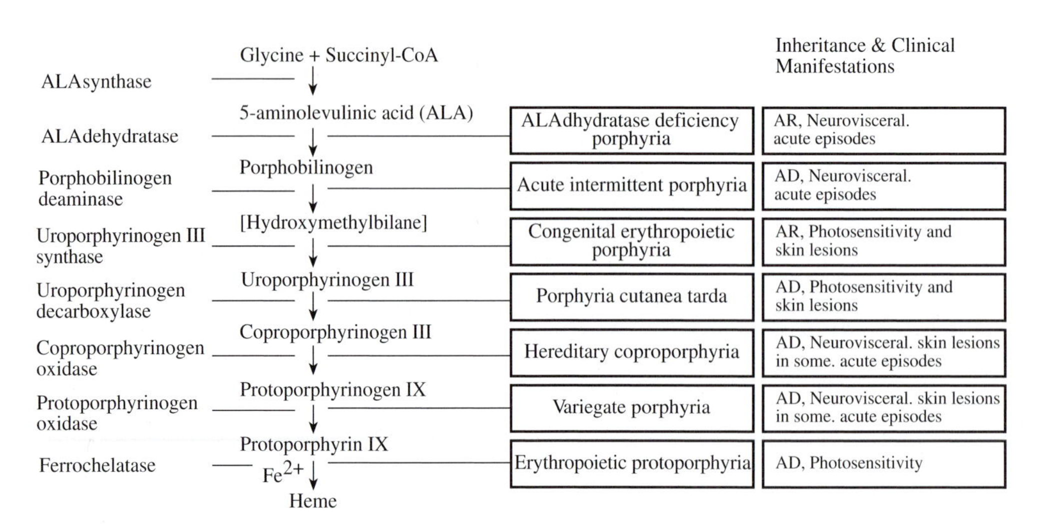

FIGURE 29-9
Heme biosynthetic pathway and the enzyme defects in various porphyrias. AD, autosomal dominant; AR, autosomal recessive.

synthase, produces ***congenital erythropoietic porphyria.*** Type I porphyrins (principally uroporphyrin I) are formed, accumulate in the tissues, and are excreted in the urine. The deficiency in production of the type III isomer further increases levels of the type I isomers by reducing the regulatory effect on ALA synthase. Excessive amounts of porphyrins in erythrocytes may produce hemolysis. A compensatory increase in hemoglobin formation can then exaggerate the already increased production of type I porphyrins. Their accumulation produces a pink to dark red color in teeth, bones, and urine. Red-brown teeth and urine are pathognomonic. Patients are sensitive to long-wave ultraviolet light and sunlight. The abnormality is transmitted as an autosomal recessive trait.

Erythropoietic protoporphyria results from deficiency of ferrochelatase in reticulocytes in bone marrow. It is transmitted as an autosomal dominant trait with variable penetrance and expressivity. In general, it is a benign disorder whose most prominent symptom is photosensitivity. Occasionally, it leads to liver disease. Reduced ferrochelatase activity results in accumulation of protoporphyrin in maturing reticulocytes and young erythrocytes. When a smear of these cells is exposed to fluorescent light, they exhibit red fluorescence. The protoporphyrin appears in the plasma, is picked up by the liver, and is excreted into the bile. Protoporphyrin accumulation in the liver can lead to severe liver disease. In contrast to individuals afflicted by other porphyrias, these patients have normal urinary porphyrin levels. High levels of protoporphyrin are found in erythrocytes, plasma, and feces. The photosensitivity may be caused by stimulation of protoporphyrin in dermal capillaries to an excited (triplet) state by visible light. This in turn converts molecular oxygen to singlet oxygen, which produces cell damage. Oral administration of β-carotene decreases the photosensitivity, possibly owing to a quenching effect on singlet oxygen and free-radical intermediates.

29.3 Heme Catabolism

When heme proteins are degraded in mammals, the polypeptides are hydrolyzed to amino acids while the heme groups are freed of their iron, which is salvaged, and are converted to bilirubin. After transport to the liver, bilirubin is coupled to glucuronic acid and the conjugated bilirubin is excreted into bile as the principal bile pigment. When increased production or decreased excretion of bilirubin causes its plasma concentration to exceed 0.1–1.0 mg/dL (2–17 μmol/L), it diffuses into tissues and produces jaundice. Although jaundice is relatively harmless unless due to extremely high concentrations of unconjugated bilirubin, it indicates the presence of a disease process that requires medical attention. The yellow coloration of jaundiced skin and sclerae has aroused much interest and has made bilirubin the subject of extensive research. Fractionation and quantitation of serum bilirubin are now widely used for diagnosis and prognosis of hepatobiliary disease.

Bilirubin is a waste product and has no known beneficial physiological function. However, both the conjugated and the unconjugated forms of bilirubin show antioxidative properties (e.g., inhibition of lipid peroxidation). The physiological role of the antioxidative property of bilirubin is not known.

Bilirubin is a yellow-orange pigment that in its unconjugated form is strongly lipophilic and cytotoxic. It is virtually insoluble in aqueous solutions below pH 8 but readily dissolves in lipids and organic solvents and diffuses freely across cell membranes. Bilirubin toxicity is normally prevented by tight binding to serum albumin. Only when the binding capacity of albumin is exceeded can a significant amount of unconjugated bilirubin enter cells and cause damage. Conjugated bilirubin is hydrophilic and does not readily cross cell membranes, even at high concentrations. Of the 250–300 mg (4,275–5,130 μmol) of bilirubin normally produced in 24 hours, about 70–80% is derived from hemoglobin. The remainder comes from several sources, including other heme proteins (e.g., cytochromes P-450 and b_5, catalase), ineffective hemopoiesis (erythrocytes that never leave the marrow), and "free" heme (heme never incorporated into protein) in the liver. Hemoglobin heme has a life span equal to that of the red cell (about 125 days), whereas heme from other sources (with the exception of myoglobin, which is also quite stable) turns over much more rapidly. Hepatic P-450 enzymes have half-lives of 1–2 days. When radioactively labeled glycine or ALA is injected and radioactivity in fecal bile pigments is monitored, two peaks are seen. The rapidly labeled bilirubin (early bilirubin) peak appears 3–5 days after injection and contains about 15–20% of the injected label. It is increased by drugs that induce hepatic P-450 oxygenases and in erythropoietic porphyria and anemias associated with ineffective erythropoiesis (lead poisoning, thalassemias, and some hemoglobinopathies). Thus, the bilirubin in the early peak is partly derived from these sources. The slowly labeled bilirubin (late bilirubin) peak appears at approximately 120 days, contains about 80–85% of the label, and is due to heme released from senescent erythrocytes.

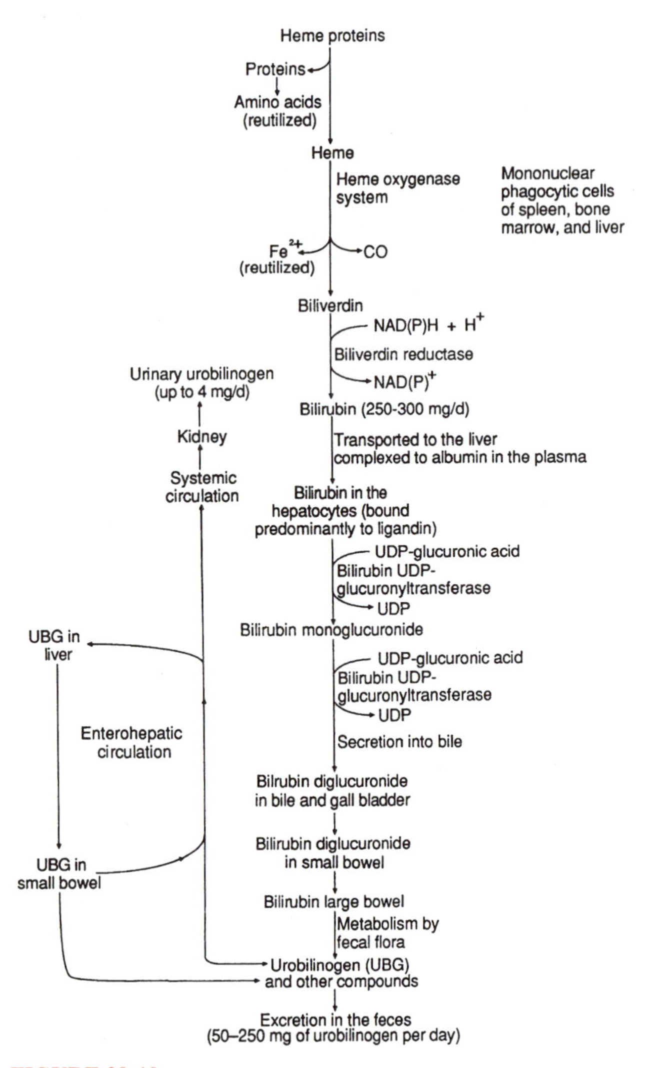

FIGURE 29-10
Catabolic pathway for the heme group from hemoproteins (predominantly hemoglobin).

Formation of Bilirubin

A summary of the pathway for bilirubin metabolism and excretion is shown in Figure 29-10. Release of heme from heme proteins and its conversion to bilirubin occur predominantly in the mononuclear phagocytes of liver, spleen, and bone marrow (previously known as the reticuloendothelial system), sites where sequestration of aging red cells occurs. Renal tubular epithelial cells, hepatocytes, and macrophages may also contribute to bilirubin formation under some conditions. Structures of the intermediates in the conversion of heme to bilirubin are shown in Figure 29-11. The initial step after the release of heme is its binding to heme oxygenase, a microsomal enzyme distinct from the microsomal P-450 oxygenases. Heme oxygenase catalyzes what appears to be the rate-limiting step in catabolism of heme. It is induced by heme and requires O_2 and NADPH for activity. The activity of the inducible isoenzyme form of heme oxygenase is highest in the spleen, which is involved in the sequestration of senescent erythrocytes. The constitutive form of heme oxygenase is mainly localized in the liver and brain. After binding, the α-methene carbon of heme is oxidized (hydroxylated) to α-hydroxyhemin, which undergoes autoxidation to biliverdin (a blue-green pigment) with consumption of O_2 and release of iron and carbon monoxide (derived from oxidation of the α-methene bridge). Since CO production in mammals occurs primarily by this pathway, measurement of expired CO has been used to estimate heme turnover. Values obtained exceed those derived from plasma bilirubin measurements by about 15%, probably because of bilirubin produced in the liver and excreted into the bile without entering the circulation. A potent competitive synthetic inhibitor of heme oxygenase is tin (Sn) protoporphyrin, which has a potential therapeutic use in treatment of neonatal jaundice (see below).

In nonmammalian vertebrates, biliverdin is the final metabolite in heme catabolism. Transport of biliverdin is much easier than that of bilirubin because biliverdin is water-soluble. Conversion of biliverdin to bilirubin may have evolved in mammals because, unlike biliverdin, bilirubin readily crosses the placenta. In this way, the fetus can eliminate heme catabolites via the mother's circulation. However, this explanation may not be complete, since the rabbit (a placental mammal) excretes biliverdin as the major bile pigment.

Biliverdin is reduced to bilirubin by NAD(P)H-dependent biliverdin reductase, a cytosolic enzyme that acts at the central methene bridge. Although both molecules have two propionic acid groups, the polarity of biliverdin is greater than that of bilirubin. Bilirubin can form six internal hydrogen bonds between the carboxylic groups, the two lactam carbonyl oxygens, and four pyrrolenone ring nitrogens, and thus prevents these groups from hydrogen-bonding with water (Figure 29-12). Biliverdin cannot form these hydrogen bonds because of the lack of free rotation imposed by the double bond at the central methene bridge. Esterification of the propionyl side chains of bilirubin with glucuronic acid disrupts the hydrogen bonds and increases its solubility and lability. "Activators," such as ethanol and methanol, used in the van den Bergh test to measure "indirect bilirubin," and phototherapy for neonatal jaundice also act by disrupting the hydrogen-bonded structure of unconjugated bilirubin.

Hemoglobin and heme released from intravascular hemolysis or blood extravasations (e.g., subcutaneous hematomas) are bound, respectively, by ***haptoglobin*** and

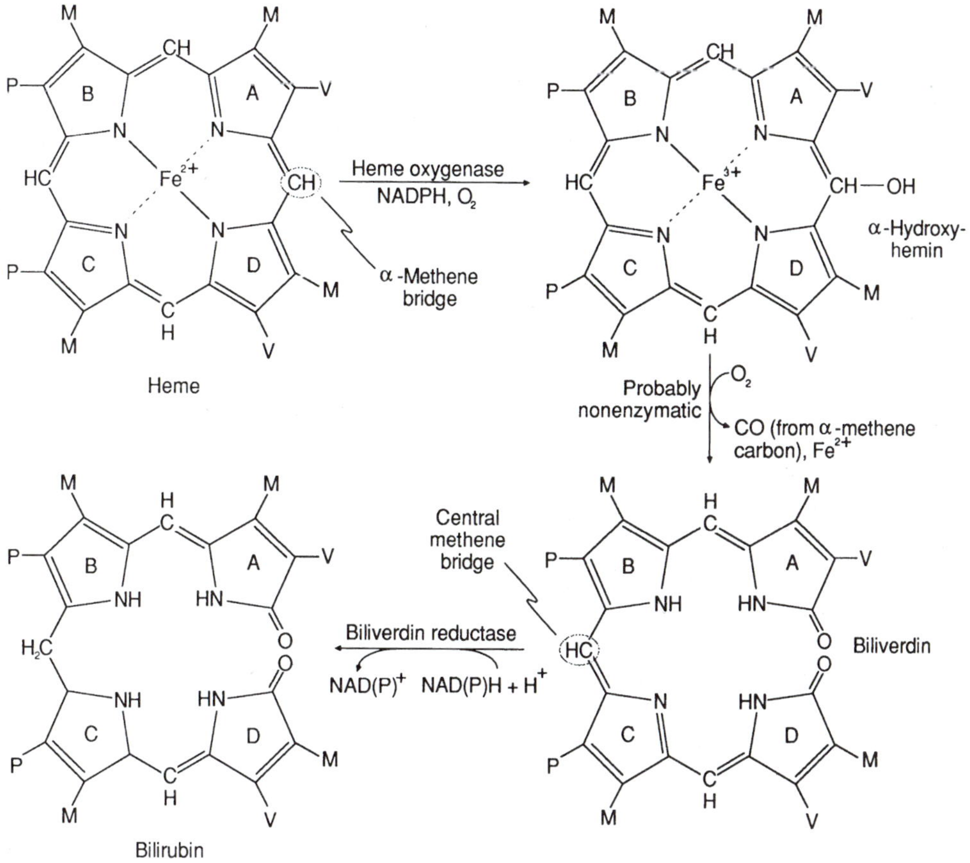

FIGURE 29-11
Conversion of heme to bilirubin in the monocytic phagocytic cells. Carbon monoxide and bilirubin are generated. Fe^{3+} released is conserved and reutilized. Biliverdin and bilirubin are lactams. P, Propionic acid; M, methyl; V, vinyl.

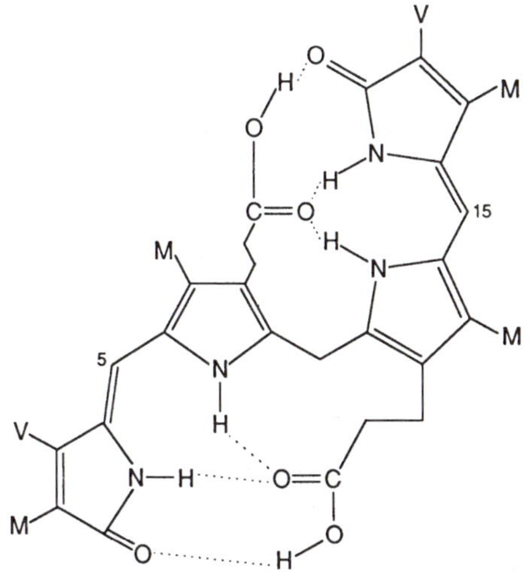

FIGURE 29-12
Conformation of bilirubin showing involuted hydrogen bonded-structure between NH/O and OH/O groups. Despite the presence of polar carboxyl groups, bilirubin is nonpolar and lipophilic. Disruption of hydrogen bonds by glucuronidation or by conversion of bilirubin to configurational or structural isomers yields water-soluble pigments.

hemopexin to form complexes that cannot be filtered by the kidney. This action prevents renal loss of the heme iron and protects the renal tubules from possible damage by precipitated hemoglobin. Haptoglobin-hemoglobin and hemopexin-heme complexes are processed in mononuclear phagocytic cells in a way similar to that for hemoglobin. Haptoglobin and hemopexin are glycoproteins synthesized in the liver. The former is an α_2-globulin and an acute-phase reactant (i.e., its synthesis and release into the circulation are augmented during an acute insult to the body); the latter is a β_1-globulin but not an acute-phase protein (see also Appendix VI).

Circulatory Transport of Bilirubin

Bilirubin formed in extrahepatic tissues is transported to the liver for excretion in bile. Since bilirubin is virtually insoluble in aqueous media, it is transported to the liver bound noncovalently to serum albumin. The bilirubin-albumin complex increases the amount of bilirubin carried per volume of plasma and minimizes diffusion of bilirubin into extrahepatic tissues, thereby preventing bilirubin toxicity. Because of formation of this complex, bilirubin does not normally appear in urine. Urinary bilirubin is almost invariably conjugated bilirubin (see below) and signifies the presence of a pathological process. An albumin molecule binds two molecules of bilirubin at one high-affinity site and at one to three secondary sites. Bilirubin conjugated with glucuronic acid also binds to albumin but with much lower affinity. Another form of bilirubin (probably conjugated), very tightly (probably covalently) bound to albumin, has been described. The mechanism of its formation is not known, although blockage of biliary flow associated with an intact hepatic conjugating system releases a chemically reactive form of bilirubin into the circulation.

If the capacity of albumin to bind bilirubin is exceeded because of increased amounts of unconjugated bilirubin or decreased concentration of albumin, bilirubin readily enters extrahepatic tissues. In neonates, this can cause ***kernicterus,*** a serious condition associated with permanent neurological damage (see below). Bilirubin can be displaced from binding to albumin by sulfonamides, salicylates (notably aspirin), and cholangiographic contrast media. Use of these substances in jaundiced newborn infants increases the risk of occurrence of kernicterus. Medium-chain fatty acids increase and short-chain fatty acids decrease bilirubin binding to albumin. Binding, at least to the primary site, is independent of pH. Estimation of reserve bilirubin binding capacity has been used to evaluate the risk of bilirubin toxicity in icteric patients.

Hepatic Uptake, Conjugation, and Secretion of Bilirubin

Hepatocytes take up bilirubin from the sinusoidal plasma and excrete it after conjugation with glucuronic acid across the canalicular membrane into the bile. The entry and exit steps and the transport of bilirubin within the cell are not completely understood. The following is a plausible interpretation of the available data.

Since binding of bilirubin to albumin is usually reversible, a small amount of free bilirubin is present in plasma in equilibrium with albumin-bound bilirubin. It is probably this free bilirubin that is taken up at a rate determined by its plasma concentration. As this free bilirubin concentration decreases, more bilirubin is released from albumin and becomes available for uptake. Alternatively, the albumin-bilirubin complex may bind to specific hepatocyte plasma membrane receptors, and thereby bilirubin is released to enter the cell. Both models are consistent with the finding that albumin does not accompany bilirubin into the hepatocyte.

The entry step seems to be carrier-mediated, is saturable, is reversible, and is competitively inhibited by sulfobromophthalein, indocyanine green, cholecystographic agents, and several drugs. Bile salts do not compete with bilirubin for hepatic uptake.

After it enters hepatocytes, bilirubin is transported to the smooth endoplasmic reticulum for glucuronidation bound to a protein. Two cytosolic proteins, ***Z protein*** (fatty acid–binding protein) and ***ligandin*** (Y protein), bind bilirubin and other organic anions. Ligandin, which constitutes about 2–5% of the total soluble protein in rat and human liver, has lower capacity but higher affinity for bilirubin than Z protein. Ligandin (M.W. 47,000) has two subunits, A and B, which appear to be identical except for a 30-amino-acid extension at the carboxyl terminus of the B subunit. Bilirubin is bound entirely to the A subunit (two molecules per A subunit). Ligandin also has glutathione S-transferase, glutathione peroxidase, and ketosteroid isomerase activities, which depend on both subunits. Glutathione S-transferases catalyze detoxification reactions for a number of substances. Binding of bilirubin and other organic anions to ligandin occurs at sites unrelated to its enzyme activities.

Under normal conditions, ligandin is probably the principal hepatic bilirubin-binding protein and may serve intracellularly the same protective and transport functions as albumin in plasma. It may also help limit reflux of bilirubin into plasma, since its affinity for bilirubin is at least five times greater than that of albumin. Z protein (M.W. 11,000) becomes important at high plasma bilirubin concentrations. The concentration of ligandin in the liver does not reach adult levels until several weeks after birth, whereas neonatal and adult levels of Z protein are the same. This lack of ligandin, together with low hepatic glucuronyltransferase activity, is the probable cause of transient, "physiological," nonhemolytic, ***neonatal jaundice.***

Glucuronidation of bilirubin in the endoplasmic reticulum by UDP–glucuronyltransferase produces an ester between the 1–hydroxyl group of glucuronic acid and the carboxyl group of a propionic acid side chain of bilirubin (Figure 29-13). In bile, about 85% of bilirubin is in the diglucuronide form and the remainder is in the monoglucuronide form. Glucuronidation increases the water solubility of several lipophilic substances. There appear to be many UDP-glucuronyltransferases in the endoplasmic reticulum, which differ in substrate specificity. (Biosynthesis of UDP-glucuronic acid was described in Chapter 15.)

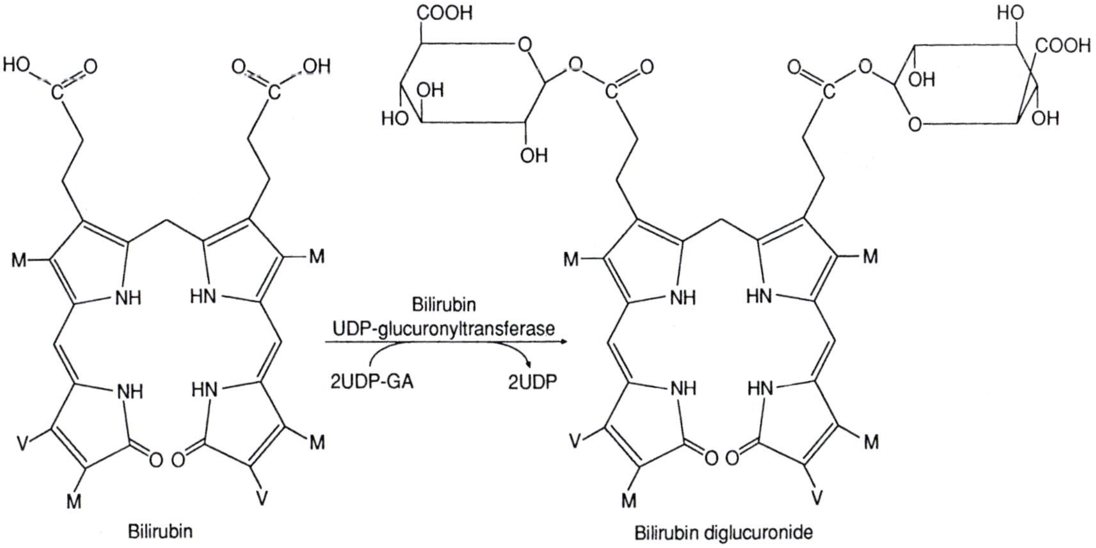

Bilirubin UDP-glucuronyltransferase

2UDP-GA 2UDP

Bilirubin

Bilirubin diglucuronide

FIGURE 29-13
Formation of bilirubin diglucuronide. Glucuronidation occurs in two steps via formation of monoglucuronide. Mono- and diglucuronides are more water-soluble and less lipophilic than bilirubin. Conversion of bilirubin to water-soluble products is obligatory for excretion of bilirubin from hepatocytes. M, Methyl; V, vinyl; UPD-GA, UDP-glucuronic acid.

Secretion across the canalicular membrane into bile appears to be the rate-limiting step in hepatic bilirubin metabolism. It is probably carrier-mediated, requires energy, is saturable, and is unaffected by bile salts. Bilirubin can be made water-soluble by conversion to its configurational isomers. These photobilirubins are formed when bilirubin is exposed to blue light of the 400- to 500-nm wavelength (Figure 29-14). Photobilirubins cannot form the intramolecular hydrogen bonds characteristic of the natural isomer of bilirubin (Figure 29-12). Thus, they are more polar and readily excreted in the bile without the requirement for glucuronidation. ***Lumirubin,*** a structural isomer of bilirubin, is formed by light-induced intramolecular cyclization of the vinyl side group of C-3. It contains a seven-membered ring, is stable, is polar, and is excreted without conjugation. These observations explain the mechanism of phototherapy commonly used for treatment of neonatal hyperbilirubinemia.

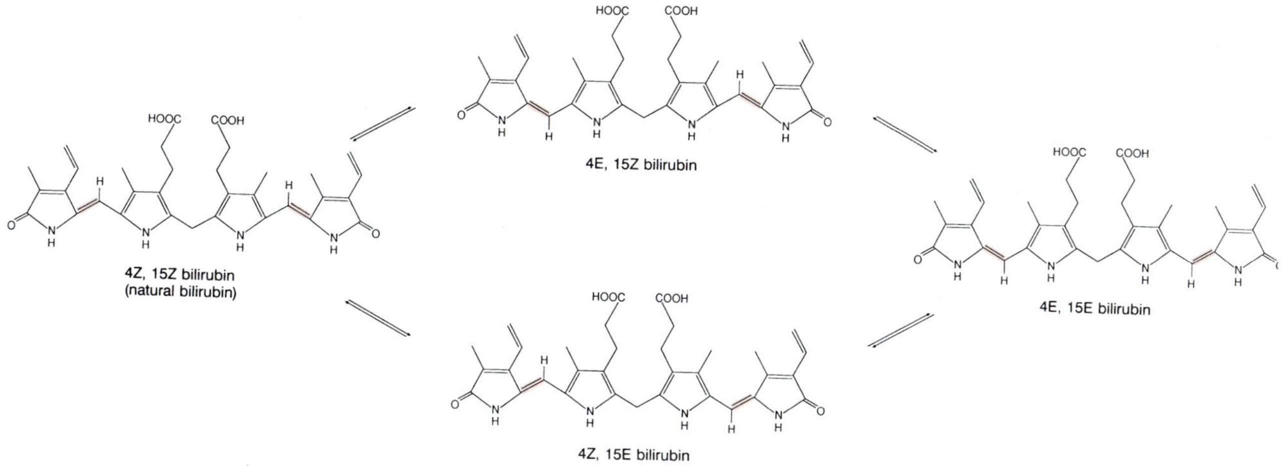

4Z, 15Z bilirubin (natural bilirubin)

4E, 15Z bilirubin

4Z, 15E bilirubin

4E, 15E bilirubin

FIGURE 29-14
Photoisomers of bilirubin. The presence of two methene bridges containing double bonds (colored areas) gives rise to configurational (geometrical) isomers of bilirubin. Each double bond can exist in the Z or E configuration. The naturally occurring, most stable, water-insoluble form is the Z, Z isomer. It undergoes photoisomerization to configurational isomers (Z, E; E, Z; and E, E), which are more polar owing to inability to form intramolecular hydrogen bonds and are excretable from the liver without glucuronidation. Some excretion of photoisomers in urine also occurs.

Bilirubin in the Intestinal Tract

Most bilirubin entering the intestine in bile is in the diglucuronide form, which is very poorly absorbed in the small and large intestines. In the lower small intestine and colon, bacteria remove glucuronic acid residues and reduce bilirubin to the colorless ***urobilinogen*** and ***stercobilinogen.*** Exposure to air oxidizes these to urobilin and stercobilin, respectively, (i.e., red-orange pigments that contribute to the normal color of stool and urine). Other degradation products of bilirubin are present in minor amounts in feces.

Urobilinogen is excreted mostly in the feces, but a small fraction is absorbed from the colon, enters the portal circulation, is removed by the liver, and is secreted into bile. That which is not removed from the portal blood by the liver enters the systemic circulation and is excreted by the kidneys. Urobilinogen excretion in urine normally amounts to 1–4 mg per 24 hours, as opposed to the 40–280 mg (67–470 μmol) excreted in feces.

Lack of urobilinogen in the urine and feces indicates biliary obstruction; stools are whitish ("clay-colored") owing to the absence of bile pigment. Urinary and fecal urobilinogen excretion increases in hemolytic anemia.

Disorders of Bilirubin Metabolism

The plasma of normal subjects contains 0.1–1 mg of bilirubin per deciliter (2–17 μmol/L), mostly in the unconjugated form. Unconjugated bilirubin is known as indirect-reacting bilirubin and conjugated bilirubin as direct-reacting bilirubin (see Table 29-2).

Jaundice occurs when plasma becomes supersaturated with bilirubin (>2–2.5 mg/dL) and the excess diffuses into the skin, sclera, and other tissues. The sclera is particularly affected because it is rich in elastin, which has a high affinity for bilirubin. Reddish yellow pigments, particularly carotene and lycopene, may give a yellowish tinge to the skin but they do not usually produce scleral coloration. Hyperbilirubinemia may result from elevation of unconjugated or conjugated bilirubin levels.

Unconjugated Hyperbilirubinemias

Unconjugated hyperbilirubinemias result from imbalance between the rates of production of pigment and of its uptake or conjugation in the liver. Because of the large reserve capacity of the liver for conjugation and excretion of bilirubin, increased production seldom elevates unconjugated serum bilirubin to more than 3–4 mg/dL. If a greater increase occurs, some degree of liver dysfunction probably also occurs. These disorders are usually due to decreased uptake of pigment by hepatocytes or to failure of these cells to store, transport, or conjugate bilirubin. Bilirubinuria does not accompany these disorders. Except in infancy or when pigment gallstones form, unconjugated hyperbilirubinemias are benign.

Gilbert's syndrome may be the most common cause of mild, persistent, nonhemolytic, unconjugated hyperbilirubinemia. Serum bilirubin concentration rarely exceeds 5 mg/dL and usually fluctuates between 1.3 and 3 mg/dL. Other liver function tests are normal. The syndrome is usually asymptomatic and is detected during routine laboratory testing or examination for other disease. Family studies suggest that Gilbert's syndrome is an autosomal dominant disorder. The unconjugated hyperbilirubinemia in Gilbert's syndrome is due to decreased UDP-glucuronyltransferase activity resulting from an insertion mutation found in the promoter region of the enzyme. The wild-type promoter $[TA]_6TAA$ is mutated to $[TA]_7TAA$. Mutations affecting the coding region of the enzyme, although rare, also occur.

In ***Crigler-Najjar syndrome type I*** activity of hepatic bilirubin UDP-glucuronyltransferase is undetectable and bilirubin conjugates are absent from the serum, bile, and urine, but biliary secretion of sulfobromophthalein and indocyanine green is normal. The disease is apparent shortly after birth, kernicterus develops, and death commonly occurs during the neonatal period. The effectiveness of phototherapy is often transient. The enzyme is not inducible by phenobarbital. This autosomal recessive defect occurs in all races. The Gunn strain of Wister rats has a similar genetic defect and has been used to study the syndrome.

Crigler-Najjar syndrome type II (Arias syndrome) is milder, usually benign, and caused by partial deficiency of bilirubin UDP-glucuronyltransferase. Jaundice may not appear until the second or third decade of life. The monoglucuronide is the predominant pigment in bile. Phenobarbital induces the enzyme. Dominant and recessive inheritance patterns have been described. An accurate diagnosis of type 1, as opposed to type 2 Crigler-Najjar syndrome, is essential since orthotopic liver transplantation is an important therapy for type 1 patients.

Conjugated Hyperbilirubinemias

Conjugated hyperbilirubinemias are due to intra- or extrahepatic reduction to bile flow (cholestasis) with spillage of conjugated bilirubin into the bloodstream, which may occur from injury to the endothelial cells lining bile ductules or from reverse pinocytosis, by the hepatocytes. Since the serum bilirubin is mostly the water-soluble glucuronide, bilirubinuria is usually present.

Abdominal tumors, gallstones, strictures, hepatitis, and cirrhosis can mechanically block the biliary canaliculi or ducts. If obstruction affects only intrahepatic bile flow, hyperbilirubinemia occurs when 50% or more of the liver is involved. Extrahepatic obstruction elevates serum bilirubin only if it increases the pressure in the canaliculi above the maximum secretion pressure of the hepatocytes (about 250 mm Hg). Nonmechanical cholestasis can be caused by bacterial infection, pregnancy, and sex steroids and other drugs, or it may be genetically determined.

In cholestasis, bile salts and bile pigments are retained and appear in the circulation, and steatorrhea and deficiencies of fat-soluble vitamins may occur. These deficiencies are often manifested as hypoprothrombinemia (from lack of vitamin K) and osteomalacia (from lack of vitamin D). The magnitude depends on the degree of obstruction. If blockage is complete, urinary urobilinogen will be absent and the stools will have a pale, clay-like color.

Familial diseases include ***Dubin-Johnson syndrome, Rotor's syndrome,*** and ***benign familial recurrent cholestasis.*** Serum bilirubin values in Dubin-Johnson syndrome are the same as those found in Rotor's syndrome (Table 29-2). Very little is known about benign familial recurrent cholestasis. All three disorders are uncommon or rare, and all are benign. The liver in Dubin-Johnson syndrome appears black on direct examination. The pigment is not identical to melanin but may be a catecholamine (perhaps epinephrine) polymer. The total urinary coproporphyrin excretion is normal but about 80% is the I isomer and the rest is the III isomer, the reverse of the normal ratio. This abnormality seems to be diagnostic for the syndrome, provided the history and physical examination are consistent with the diagnosis. Obligate heterozygotes have urinary coproporphyrin patterns intermediate between those of normal and affected individuals, indicating autosomal recessive inheritance. The hyperbilirubinemia, hepatic pigment, and urinary coproporphyrin abnormality may result from a defect in porphyrin biosynthesis. Studies of sulfobromophthalein clearance show normal storage but greatly decreased secretion. Investigation of a mutation in Corriedale sheep, which causes a similar condition, may help clarify this poorly understood disorder.

In Rotor's syndrome, patients lack the hepatic pigment but have urinary coproporphyrin levels 2.5–5 times greater than normal. Coproporphyrin type I is increased relative to the type III isomer but not as much as in Dubin-Johnson syndrome. Intrahepatic storage of sulfobromophthalein is

TABLE 29-2
Serum Bilirubin Levels in Normal and Some Abnormal Conditions

	Bilirubin Values in mg/dL of Serum*		
Condition	**Total**	**Unconjugated (indirect reacting)**	**Conjugated (direct reacting)**
Newborn	0.2–2.9		
1 day	0–6		
3 days	0.3–11		
7 days	0.1–9.9		
Normal (adult)	0.1–1	0.2–0.7	0.1–0.3
Hemolytic disorders (in adults)	2.2–3.4	2–3	0.2–0.4
Gilbert's disease (probably a heterogeneous group of diseases)	1.2–3 (rarely >5)	1.1–2.7 (rarely >5)	Normal
Glucuronyltransferase deficiency			
Type I: Crigler-Najjar syndrome—complete deficiency	15–48	15–48	Negligible
Type II: Crigler-Najjar syndrome—partial deficiency	6–22 (up to 40 with physiological stress)	6–22 (up to 40 with physiological stress)	Trace
Dubin-Johnson syndrome	2.5–20	<50% of total	>50% of total
Cholestasis (severe form)	10	19	
Cirrhosis (severe)	11	56	
Hepatitis (acute/severe)	10	1.5	8.5

*To convert mg/dL to μmol/L, multiply by 17.1.

markedly decreased, but the rate of secretion is only moderately depressed. Inheritance appears to be autosomal recessive.

Neonatal Hyperbilirubinemia

Normal neonates are frequently hyperbilirubinemic (Table 29-2). Birth interrupts normal placental elimination of pigment, and the "immature" liver of the neonate must take over. Normally serum bilirubin levels rise on the first day of life, reaching a maximum (rarely greater than 10 mg/dL) by the third or fourth day. This type is mostly unconjugated. If the placenta is functioning normally, jaundice will not be present at birth. If jaundice is present at birth, a cause other than hepatic immaturity must be sought.

The primary blocks to bilirubin metabolism are low activity of bilirubin glucuronyltransferase and low concentration of ligandin in the liver at birth. Secretion of conjugated bilirubin into the bile is also reduced.

Hepatic immaturity may be partly due to diversion *in utero* of blood from the liver by the ductus venosus. When this channel closes shortly after birth and normal hepatic blood flow is established, concentrations of a number of substances rise within the hepatocytes and may induce enzymes needed for their metabolism. Accumulation of bilirubin in plasma may play an important role in hastening the maturation. Although the liver normally matures within 1–2 weeks after birth, hypothyroidism can prolong this process for weeks or months.

The neonate is at risk for kernicterus if the serum unconjugated bilirubin level is higher than 17 mg/dL. Kernicterus is characterized by yellow staining of clusters of neuronal cell bodies in the basal ganglia, cerebellum, and brain stem, leading to motor and cognitive deficits or death. Immaturity and perhaps hypoxia make the blood-brain barrier permeable to bilirubin and contribute to the likelihood of kernicterus. The biochemical basis of bilirubin encephalopathy is due to many causes: inhibition of RNA and protein synthesis, carbohydrate metabolism (both cAMP-mediated and Ca^{2+}-activated), phospholipid-dependent protein kinases, enzymes involved in the electron transport system, and impaired nerve conduction.

A major complicating factor can be hemolytic anemia such as that of ***erythroblastosis fetalis*** caused by Rh incompatibility between mother and child. The hemolysis increases the rate of bilirubin formation, which soon overwhelms the liver and produces severe jaundice and kernicterus. Sickle cell anemia has a similar effect. Congenital absence of bilirubin UDP-glucuronyltransferase (Crigler-Najjar syndrome type 1) usually causes a kernicterus that is fatal shortly after birth. Inhibition of glucuronyltransferase by various drugs (e.g., novobiocin) or toxins can increase the severity of neonatal jaundice. "Breast milk jaundice" is due to the presence in breast milk of a substance (perhaps pregnane-$3\alpha,20\beta$-diol) that inhibits bilirubin glucuronyltransferase, although the resulting unconjugated hyperbilirubinemia is seldom serious enough to cause neurotoxicity or to require discontinuation of breast-feeding. Other risk factors for pathologic hyperbilirubinemia include Gilbert's syndrome (discussed earlier) and glucose-6-phosphate dehydrogenase deficiency (Chapter 15).

Conjugated hyperbilirubinemia is rare during the neonatal period. It can result from impaired hepatocellular function or extrahepatic obstruction. Hepatocellular defects can be caused by bacterial, viral, or parasitic infections, cystic fibrosis, α_1-antitrypsin deficiency, Dubin-Johnson and Rotor's syndromes, and other genetic disease. Extrahepatic obstruction can be congenital (biliary atresia) or acquired.

Treatment of neonatal jaundice is usually by phototherapy. A decrease in bilirubin production in the neonatal period can also be achieved by inhibiting the rate-limiting enzyme of bilirubin formation from heme, namely, the heme oxygenase. A potent competitive inhibitor of heme oxygenase is the synthetic heme analogue tin (Sn^{4+}) protoporphyrin. When administered parenterally, the tin protoporphyrin safely decreases bilirubin formation. Exchange transfusions also rapidly decrease plasma bilirubin levels.

Supplemental Readings and References

B. R. Bacon, J. K. Olynyk, E. M. Brunt, et al.: HFE genotype in patients with hemochromatosis and other liver disease. *Annals of Internal Medicine* **130,** 953 (1999).

N. C. Andrews: Disorders of iron metabolism. *New England Journal of Medicine* **341,** 1986 (1999).

J. D. Arnold, A. D. Mumford, J. O. Lindsay, et al.: Hyperferritinaemia in the absence of iron overload. *Gut* **41,** 408 (1997).

M. C. Augustine: Hyperbilirubinemia in the healthy term newborn. *Nurse Practitioner* **24,** 24 (1999).

B. R. Bacon, L. W. Powell, P. C. Adams, et al.: Molecular medicine and hemochromatosis: at the crossroads. *Gastroenterology* **116,** 193 (1999)

J. D. Bancroft, B. Kreamer, and G. R. Gourlev: Gilbert syndrome accelerates development of neonatal jaundice. *Journal of Pediatrics* **132,** 656 (1998).

T. H. Bothwell and A. P. MacPhall: Hereditary hemochromatosis: etiologic. pathologic. and clinical aspects. *Seminars in Hematology* **35,** 55 (1998).

S. S. Bottomley: Secondary iron overload disorders. *Seminars in Hematology* **35,** 77 (1998).

N. Chalasani, N. R. Chowdhury, J. R. Chowdhury, et al.: Kernicterus in an adult who is heterozygous for Crigler-Najjar syndrome and homozygous for Gilbert-type genetic defect. *Gastroenterology* **112,** 2099 (1997).

M. E. Conrad: Introduction: iron overloading disorders and iron regulation. *Seminars in Hematology* **35,** 1 (1998).

R. W. I. Cooke: New approach to prevention of kernicterus. *Lancet* **353,** 1814 (1999).

P. A. Dennery, D. S. Seidman, and D. K. Stevenson: Neonatal Hyperbilirubinemia. *New England Journal of Medicine* **344,** 581 (2001).

C. Q. Edwards, L. M. Griffen, R. S. Ajioka, et al.: Screening for hemochromatosis: phenotype versus genotype. *Seminars in Hematology* **35,** 72 (1998).

G. H. Elder, R. J. Hift, and P. N. Meissner: The acute porphyrias. *Lancet* **349,** 1613 (1997).

B. M. Gaston: Emissions testing for children. chapter 2: carbon monoxide. *Journal of Pediatrics* **135,** 537 (1999).

Y. Haimi-Cohen, P. Merlob, T. Marcus-Eidlits, et al.: Dubin-Johnson syndrome as a cause of neonatal jaundice: the importance of coporphyrins investigation. *Clinical Pediatrics* **37,** 511 (1998).

R. Kauppinen and P. Mustajoki: Prognosis of acute porphyria: occurrence of acute attacks, precipitating factors, and associated diseases. *Medicine* **71,** 1 (1992).

Y. Maruo, H. Sato, T. Yamano, et al.: Gilbert syndrome caused by a homozygous missense mutation (Tyr486Asp) of bilirubin UDP–glucuronyltransferase gene. *Journal of Pediatrics* **132,** 1045 (1998).

J. M. McCord: Iron: free radicals and oxidative injury. *Seminars in Hematology* **35,** 5 (1998).

G. Monaghan, A. McLellan, A. McGeehan, et al.: Gilbert's syndrome is a contributory factor in prolonged unconjugated hyperbilirubinemia of the newborn. *Journal of Pediatrics* **134,** 441 (1999).

E. Pollitt: Early iron deficiency anemia and later mental retardation. *American Journal of Clinical Nutrition* **69,** 4 (1999).

P. Ponka, C. Beaumont, and D. R. Richardson: Function and regulation of transferrin and ferritin. *Seminars in Hematology* **35,** 35 (1998).

R. D. Press: Hemochromatosis: a simple "genetic" trait. *Hospital Practice* **34,** 55 (1999).

Report of a Meeting of Physicians and Scientists at the Royal Free Hospital School of Medicine: Genetic hemochromatosis. *Lancet* **349,** 1688 (1997).

F. F. Rubaltelli, A. Novello, L. Zancan, et al.: Serum and bile bilirubin pigments in the differential diagnosis of Crigler-Najjar disease. *Pediatrics* **553,** (1994).

A. Tefferl, L. Solberg, and R. Ellefson: Porphyrias: clinical evaluation and interpretation of laboratory tests. *Mayo Clinic Proceedings* **69,** 289 (1994).

C. G. Uasuf, A. Jatakanon, A. James, et al.: Exhaled carbon monoxide in childhood asthma. *Journal of Pediatrics* **135,** 569 (1999).

J. N. Umbreit, M. E. Conrad, E. G. Moore, et al.: Iron absorption and cellular transport: the mobilferrin/paraferritin paradigm. *Seminars in Hematology* **35,** 13 (1998).

C. Utzel and M. E. Conrad: Absorption of heme iron. *Seminars in Hematology* **35,** 27 (1998).

CHAPTER 30

Endocrine Metabolism I: Introduction

Survival of multicellular organisms depends on integration and coordination of differentiated cell functions and the ability to react appropriately to internal and external influences that threaten to disrupt homeostatic conditions. These requirements are fulfilled by a form of intercellular communication in which chemical signals (messengers) released by one cell evoke a receptor-mediated response in another. There are two types of chemical messenger: neurotransmitters and hormones. Neurotransmitters convey signals from one neuron to another or from a neuron to an effector cell, travel very short distances to reach their target sites, and function within the specialized regions of synapses and junctions. Hormones are usually defined as messengers that are transported by the blood to distal target cells. Because they are released into the interstitial space and thence into blood, they are called "endocrine" (ductless; "secreted within") secretions to distinguish them from those which are released into the external environment ("exocrine" or ductal secretions).

This terminology and these definitions, though not always appropriate and possibly misleading, are used for want of better alternatives. The term "hormone" comes from the Greek *hormaein,* "to excite." This is a misnomer because some hormones have primarily inhibitory effects, and others may stimulate or inhibit, depending on the target cell type. The term *endocrine* is misleading because some hormones (steroids, prostaglandins, growth factors) are released via ducts to the external environment, where they affect reproductive activities or wound healing. The designation *pheromones* may apply to some of these hormones (steroids, prostaglandins) because they act on cells in another organism or individual. The usual definition of a hormone excludes those messengers which affect cells in the vicinity of their release and which do not rely on transport in the blood to reach their destination. For these, the term *parahormone* (paracrine secretion, local hormone, autacoid) is commonly used. However, if produced in sufficiently large amounts, parahormones diffuse into the vascular compartment and behave as authentic hormones (e.g., somatomedins). Conversely, hormones frequently assume paracrine functions. For example, androgens produced by Leydig cells of the testes diffuse and exert their effect on nearby Sertoli cells. Parahormones, therefore, should be regarded as authentic hormones that exert mainly local effects. Finally, neurotransmitters and hormones differ in respect to the systems that use them. Thus, the nervous and endocrine systems share some messengers (norepinephrine, dopamine, enkephalins, endorphins, cholecystokinin, bombesin, etc.) that exert receptor-mediated effects. The different events that are

Endocrine topics not discussed in Chapters 30–34 that are covered elsewhere in the text are as follows: gastrointestinal hormones, Chapter 12; eicosanoids, Chapter 18; pancreatic hormones, Chapter 22; parathyroid hormone and vitamin D, Chapter 37; renin-angiotensin system and antidiuretic hormone, Chapters 32 and 39. A list of expanded acronyms appears in Appendix VIII.

triggered by the same substance clearly manifest the different specialized cell functions that are affected.

Hormones and neurotransmitters may have evolved from the same, or similar, ancestral prototypes in unicellular organisms. Substances that resemble mammalian hormones in structure and biological activity serve as neurotransmitters in some invertebrates and lower vertebrates. Several peptide and steroid hormones long considered to be evolutionarily advanced mammalian forms have been found in certain prokaryotes and unicellular eukaryotes. Several mammalian brain-gut peptides occur in the skin of amphibians. Specific receptor sites for mammalian hormones are demonstrable at all stages of phyletic evolution. Thus, the structures of hormones and neurotransmitters and their functions as chemical messengers may have been highly conserved during evolution. The fact that many prototypic messengers in unicellular organisms predate the appearance of nerve cells implies that neurotransmitters may be a specialized form of hormone. Chemically, hormones are of four types:

1. Hormonal amine,
2. Peptide, protein, or glycoprotein,
3. Steroid, and
4. Eicosanoid.

Amine hormones and some agents of the second type have counterparts in the nervous system that function as neurotransmitters.

30.1 Hormonal Amines

Hormonal amines are derived from amino acids and, in most cases, represent simple modifications of the parent compound. Table 30-1 lists the important hormonal amines, parent amino acids, major sites of synthesis, and principal actions. All of these amines except the thyroid hormones (Chapter 33) are decarboxylated products that are synthesized both in and out of the nervous system. Within the nervous system, they are important neurotransmitters; outside the nervous system, the cells that produce

TABLE 30-1
Hormonal Amines

Hormone	Parent Amino Acid	Major Site of Synthesis	Neurotransmitter in Nervous System	Systemic Action
Dopamine	Tyr	Hypothalamus; adrenal medulla	Yes	Renal vasodilation (D_1) inhibition of anterior pituitary hormone release (D_2); inhibition of aldosterone synthesis (D_1).
Epinephrine	Tyr	Adrenal medulla	Minor	Cardiovascular and metabolic effects (via α- and β-adrenergic receptors).
Norepinephrine	Tyr	Peripheral nerve endings; adrenal medulla	Yes	Cardiovascular and metabolic effects (via α- and β-adrenergic receptors).
Histamine	His	Mast cells; basophils; regenerating cells	Yes	Systemic vasodilation (H_1, H_2); increased capillary permeability (H_1); increased gastric secretion (H_2); increased heart rate (H_2).
Serotonin	Trp	Mast cells; liver; sympathetic nerve endings; stored platelets	Yes	Vasodilation in skin and skeletal muscle, vasoconstriction elsewhere; positive inotropic and chronotropic effects.
3,5,3′-Triiodothyronine (T_3)[1]	Tyr	Peripheral conversion from thyroxine (T_4); thyroid	?	Many effects, e.g., regulation of metabolism.

[1] An amino acid.

them are modified postsynaptic neurons (e.g., adrenal medulla), blood-derived cells (e.g., basophils, mast cells), or APUD [1] cells (e.g., enterochromaffin cells). Regardless of where they are released, hormonal amines exert their effects through specific receptor sites located in various parts of the body. These systemic receptors (Table 30-1) and their subtypes probably have identical counterparts in the central and peripheral nervous systems. All of these amines except T_3 exert rapid systemic effects that usually involve smooth-muscle activity. Because the amines are hydrophilic, their receptors are located on the outer surface of target cells, and most if not all of their effects are mediated by intracellular mediators.

30.2 Peptide, Protein, and Glycoprotein Hormones

Cells that produce peptide, protein, or glycoprotein hormones are derived embryologically from the entoderm or ectoblast (progenitor of ectoderm and neuroectoderm). More than 40 hormones have been identified, containing from three to over 200 amino acid residues. Table 30-2 lists the important hormones and hormonal candidates, number of amino acid residues, and sites of synthesis.

All hormonal peptides and many hormonal proteins are synthesized as part of a larger molecule (***preprohormone***) that contains a leader sequence (signal peptide) at its amino terminal end. The leader sequence is removed as the nascent precursor enters the lumen of the endoplasmic reticulum, and the resultant prohormone undergoes posttranslational processing after being packaged into secretory granules by the Golgi complex. Posttranslational processing involves proteolytic cleavage at specific sites, usually paired basic amino acid residues (Lys-Arg, Arg-Lys, Arg-Arg, Lys-Lys), by endopeptidases within the secretory granule. In the synthesis of insulin, for example, removal of the 23-amino-acid leader sequence of preproinsulin results in proinsulin, which has two pairs of basic residues (Lys-Arg and Arg-Arg) that are cleaved to yield insulin and the C-peptide (Chapter 22). Likewise, the endogenous opiates arise from site-specific cleavages of their respective prohormones (Chapter 31). When the cell is stimulated to secrete, all major fragments of the prophormone, active and inactive, are released by calcium-dependent exocytosis.

[1] APUD (amine precursor uptake and decarboxylation) cells are derived from the neuroblast (stem cells that give rise to nerve cells and neural crest cells) or the entoderm. They have the ability to synthesize and release peptide hormones and, as their name implies, take up amine precursors (e.g., dopa) and decarboxylate them, producing hormonal amines (e.g., dopamine).

Although several peptide hormones have multiple anatomical sites of synthesis, there is usually only one important source of the circulating hormone. The presence of the same hormone in ancillary sites (Table 30-2) may indicate that it functions as a local hormone at those sites. For example, somatostatin produced by the hypothalamus is transported by blood to the anterior pituitary, where it inhibits the release of growth hormone (GH); somatostatin produced by the hypothalamus also functions as an inhibitory neurotransmitter when released into synapses in the central nervous system; somatostatin produced by pancreatic islet cells acts locally to inhibit the release of insulin and glucagon; and somatostatin produced by the gastrointestinal mucosa acts locally to inhibit the secretion of gastrin, secretin, and gastric inhibitory polypeptide (GIP). Other "brain-gut" peptides(Chapter 31) also exemplify this point, although they are not released into the general circulation in significant amounts.

Peptide hormones, because of their high information content,may have evolved from a limited number of ancestral molecules that functioned as intercellular messengers or as extracellular enzymes. Thus, some exhibit homology in their amino acid sequences, suggesting the same (or similar) evolutionary roots. Several families of peptides have been described, including the opiomelanocortin family (endorphin, adrenocorticotropic hormone, melanocyte-stimulating hormone), the somatomammotropin family (growth hormone, prolactin, human placental lactogen), the glycoprotein hormones (thyroid-stimulating hormone, luteinizing hormone, follicle-stimulating hormone, human chorionic gonadotropin), the insulin family (insulin, insulin-like growth factors, somatomedins, relaxin), and the secretin family (secretin, glucagon, glicentin, gastric inhibitory peptide, vasoactive intestinal peptide). Members in a family are related in structure and function. Thus, members of the secretin family stimulate secretory activity in target cells, while those of the insulin family promote cell growth. This preservation of structure and function can be traced to more primitive forms of life. For example, "α-mating factor," a primitive relative of mammalian gonadotropin-releasing hormone (GnRH) with which it shares 80% sequence homology, functions as a reproductive pheromone in yeast. Moreover, mammalian corticotropin-releasing hormone (CRH) is related to urotensin I of teleosts and to sauvagine of amphibians in structure and in the ability to cause hypotension, vasodilatation, and ACTH release when administered to mammals. Certain strains of bacteria synthesize substances indistinguishable from mammalian hCG and TSH and also synthesize serine proteases that are structurally homologous to the subunits of the glycoprotein hormones. Glycoprotein hormones

TABLE 30-2
Peptide, Protein, and Glycoprotein Hormones in Humans

Hormone	AA Residues	Released into Blood?[1]	Major Production Site[2]	Other Production Sites[3]
Thyrotropin Releasing Hormone (TRH)	3	Y	Hypoth	
Methionine Enkephalin (MENK)	5	N	CNS, Hypoth	
Leucine Enkephalin (LENK)	5	N	CNS, Hypoth	
Angiotensin II (AII)	8	Y	Blood Circulation	
Antidiuretic Hormone (ADH); Arginine Vasopressin (AVP)	9	Y	Hypoth	
Oxytocin (OT)	9	Y	Hypoth	
Gonadotropin Releasing Hormone (GnRH)	10	Y	Hypoth	
Somatostatin (SS)(SRIH)	14	Y	Hypoth	Islets (P)
Endothelin-1 (ET-1)	21	N	Blood vessels	
Atrial Natriuretic Peptide (ANP)	28	Y	Heart	
Pancreatic Glucagon	29	Y	Islets	
β-Endorphin (β-END)	31	Y	AP, Hypoth	
Calcitonin (CT)	32	Y	Thyroid C-cell	
Cholecystokinin (CCK)	33	Y	Intestinal mucosa	Hypoth (S)
Adrenocorticotropic Hormone (ACTH)(Corticotropin)	39	Y	AP	
Corticotropin Releasing Hormone (CRH)	41	Y	Hypoth	
Gastric Inhibitory Polypeptide (GIP)	43	Y	Intestinal mucosa	
GH-Releasing Hormone (GRH)	44	Y	Hypoth	
Insulin	51	Y	Islets	
Relaxin	57	Y	Ovary, Placenta	
Insulin-like Growth Factor I (IGF-I)	70	Y	Cartilage, liver, et al.	
Parathyroid Hormone (PTH)	84	Y	Parathyroid glands	
Growth Hormone (GH)(Somatotropin)	191	Y	AP	
Placental Lactogen (hPL) (Chorionic Somatomammotropin)	191	Y	Placenta	
Prolactin (PRL)	198	Y	AP	
Thyroid Stimulating Hormone (TSH)(Thyrotropin)	209	Y	AP	
Luteinizing Hormone (LH)	215	Y	AP	
Chorionic Gonadotropin (hCG)	231	Y	Placenta	
Follicle Stimulating Hormone (FSH)	236	Y	AP	

[1] Y = major route of dissemination to target cells is via blood transport; N = functions mainly as paracrine or autocrine hormone, although it may enter the blood stream.
[2] Hypoth= hypothalamus; CNS = central nervous system; AP = anterior pituitary (Adenohypophysis); Islets = Islets of Langerhans.
[3] Restricted to those sites that are physiologically important; P = paracrine function; S = released into blood.

therefore may have evolved from ancestral extracellular proteases.

30.3 Steroid Hormones

Four kinds of steroid hormones differ in structure and action; they are the androgens (C_{19}), the estrogens (C_{18}), the progestins (C_{21}), and the corticosteroids (C_{21}). All are synthesized from cholesterol and are produced by mesoderm-derived cells regulated by at least one peptide hormone. Structures of the steroid hormones are shown in Figure 30-1. The hormonal form of vitamin D_3 [1,25$(OH)_2$-D_3] is a sterol derived from an intermediate of cholesterol biosynthesis and is discussed in Chapter 37.

Synthesis begins with cholesterol, which in most instances is acquired from circulating low-density lipoprotein (Chapter 20); however, all steroidogenic cells are

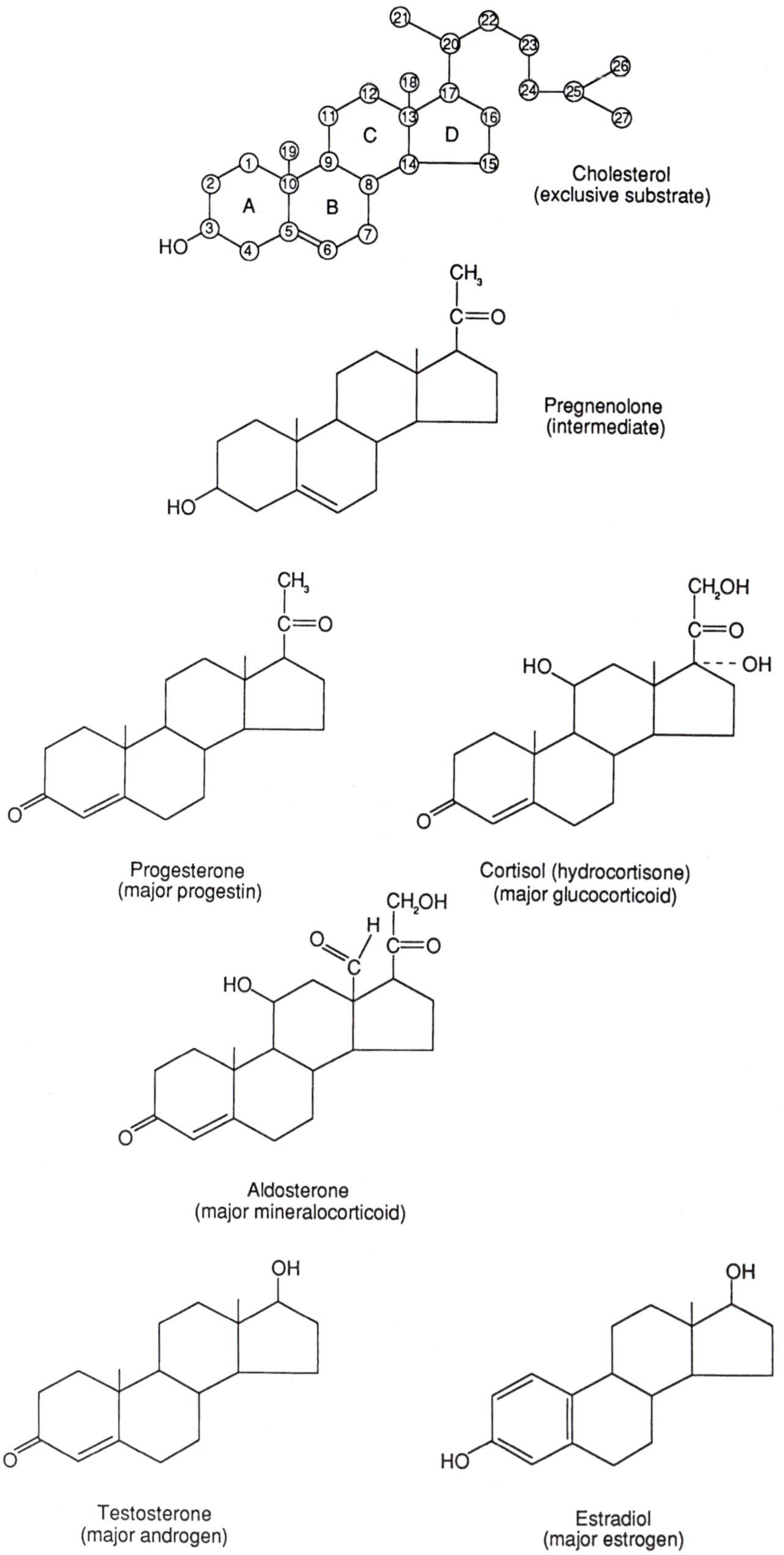
A
B
C
D
HO
Cholesterol
(exclusive substrate)
CH3
C=O
HO
Pregnenolone
(intermediate)
CH3
C=O
O
Progesterone
(major progestin)
CH2OH
C=O
HO
OH
O
Cortisol (hydrocortisone)
(major glucocorticoid)
CH2OH
H
O
C
C=O
HO
O
Aldosterone
(major mineralocorticoid)
OH
O
Testosterone
(major androgen)
OH
HO
Estradiol
(major estrogen)

FIGURE 30-1
Steroid structure and nomenclature.

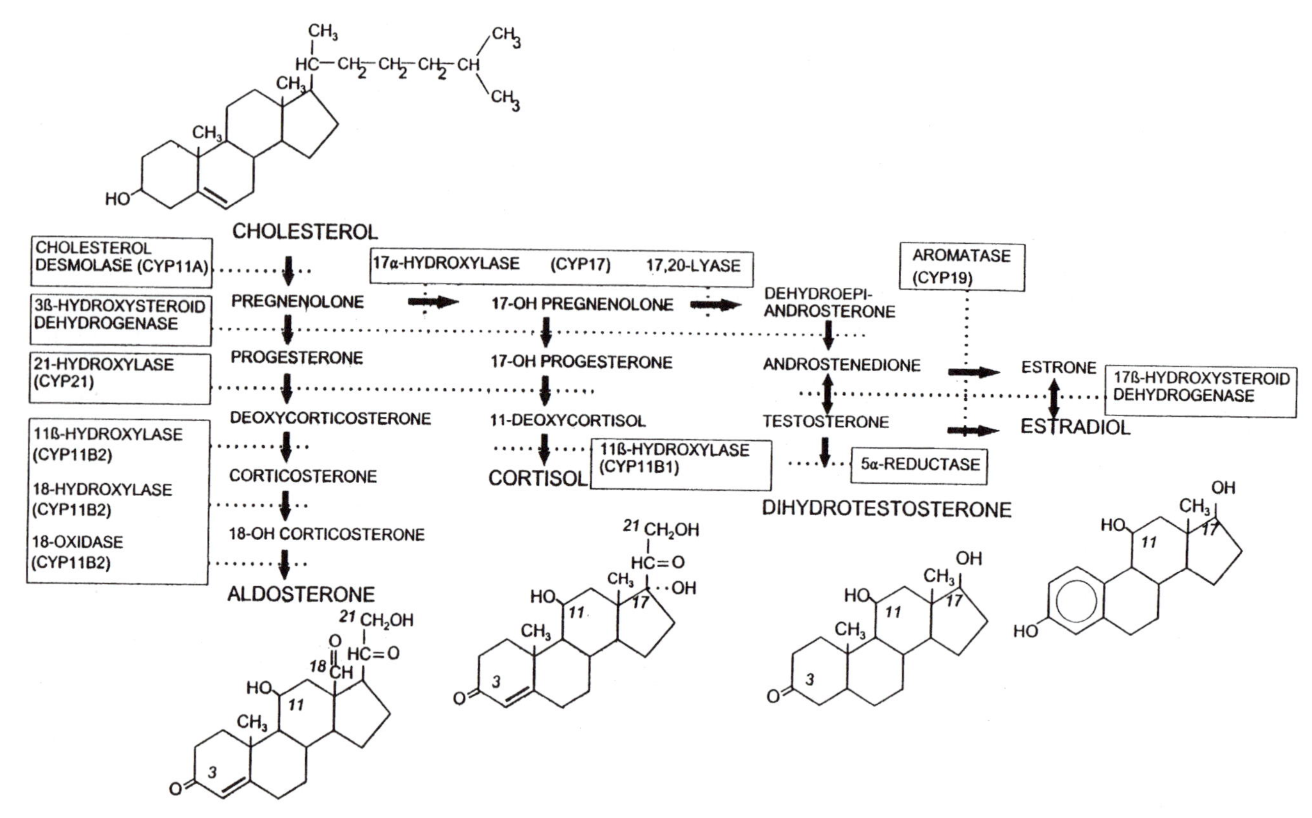

FIGURE 30-2

Biosynthesis of steroid hormones from cholesterol. The different pathways occur to varying extent in adrenals, gonads, and placenta. The systematic names for cytochrome P-450 enzymes, namely CYP followed by a number, are given in parentheses. CYP11B2 and CYP17 possess multiple enzyme activities. [Reproduced with permission from P. C. White and P. W. Speiser, Congenital adrenal hyperplasia due to 21-hydroxylase deficiency. *Endocrine Reviews* **21,** 245 (2000).]

capable of *de novo* cholesterol synthesis from acetate. Cholesterol either is stored as a fatty acid ester or is immediately utilized in steroidogenesis. The esterified cholesterol is tapped when the cell is stimulated by the activation of cholesterolesterase, a cytoplasmic enzyme that catalyzes the hydrolysis of cholesteryl esters to free cholesterol and fatty acid. The steroidogenic pathway begins in the mitochondria. The unesterified cholesterol is transported from the cytosol to the matrix side of the mitochondrial membrane; this process is mediated by a 37-kDa phosphoprotein known as stetoridogenic acute regulatory (StAR) protein. The importation of cholesterol from cytosol to mitochondria is the rate-limiting step in the steroidogenesis. Trophic stimuli such as ACTH in zona fasciculata and zona reticularis and elevated cytosolic Ca^{2+} in zona glomerulosa cause increased synthesis of the StAR protein within minutes. Other proteins in addition to StAR are also involved in cholesterol transfer across the mitochondrial membrane.

A number of enzymes in the steroidogenic pathway are cytochrome P-450 proteins that require molecular oxygen and NADPH (Chapter 14). These enzymes catalyze the clevage of carbon-carbon bonds and hydroxylation reactions. The P-450 enzymes are distinguished by specific notations. For example, CYP11A (also known as cholesterol desmolase, side chain clevage enzyme, P450scc) is a mitochiondrial enzyme and catalyzes the conversion of cholesterol (a C_{27} compound) to pregenolone (a C_{21}steroid). CYP21 refers to the hydroxylase that catalyzes the hydroxyation at carbon 21. The pathways of steroidogenesis are shown in Figure 30-2.

A true steroidogenic cell is equipped with the enzyme CYP11A, which catalyzes the initial step of steroidogenesis; the conversion of cholesterol to the steroid intermediate, pregnenolone. CYP11A catalyzes the hydroxylation of carbons 20 and 22 of cholesterol, followed by the cleavage of the bond between these two hydroxylated carbons. The resultant steroid, pregnenolone, is a 21-carbon derivative of cholesterol with a two-carbon remnant side chain. It retains the Δ^5 configuration and the 3β-hydroxyl group of cholesterol, and is lacking in biological activity; however, it is the universal intermediate in all steroid hormone synthesis. Because mitochondria have no enzyme that recognizes pregnenolone as a substrate, pregnenolone enters the smooth endoplasmic reticulum (SER) and is processed along whatever steroidogenic pathway is active in the given tissue. Because no single steroidogenic tissue normally expresses all of the steroidogenic enzymes, the presence or absence of enzymes in a given tissue will determine what sort of steroid hormone the tissue can produce from pregnenolone. Pregnenolone serves as a substrate for either of two enzymes; the SER:3β-hydroxysteroid dehydrogenase/isomerase (3βHSD) and 17α-hydroxylase/17,20-lyase (CYP17). In cells that contain both activities, pregnenolone is first processed by 3βHSD, followed by CYP17.

The Δ^4 Pathway: 3βHSD

3βHSD is a single, nonheme protein that catalyzes a two-step conversion of a Δ^5,3β-hydroxylated steroid to a Δ^4,3-ketosteroid in the SER of the adrenal cortex, ovary, and testes. When pregnenolone is the substrate, the product is progesterone, a biologically active steroid. In the corpus luteum and the placenta, which are lacking in progesterone-metabolizing enzymes, the nascent progesterone diffuses from the cells and enters the plasma. However, in other tissues progesterone serves as a substrate for at least one other enzyme.

Directional Flow Valve: CYP17

The presence or absence of the enzyme, CYP17, is a major determinant of the fate of pregnenolone (or progesterone). The presence of the enzyme allows access to either the cortisol or the androgenic (sex steroid) pathways but blocks off the aldosterone synthetic pathway. The absence of the enzyme allows for the formation of progesterone and aldosterone but precludes the formation of either cortisol or androgens. (See following subsections on *corticosteroid pathway* and *sex steroid pathway*.)

CYP17 is a single protein that catalyzes the 17α-hydroxylation of either pregnenolone or progesterone in the SER. In some tissues, such as the gonads, this enzyme catalyzes an additional reaction, involving lysis of the bond between carbons 17 and 20 of the 17α-hydroxylated product. This results in the formation of a 19-carbon androgenic steroid: dehydroepiandrosterone (DHEA) is formed from pregnenolone and Δ^4-androstenedione (Δ^4-A) is formed from progesterone. Thus, the lyase reaction initiates the sex steroid pathway. However, in other tissues that express the *CYP17* gene, such as the zona fasciculata of the adrenal cortex, the enzyme does not catalyze the 17, 20-lyase reaction to a significant extent, and the product is a 17α-hydroxylated pregnenolone or progesterone, which enters the cortisol synthetic pathway. In both tissues, the enzyme is identical, with the same 17α-hydroxylation activity; and it is not clear what confers the 17,20-lyase activity. In tissues that have CYP17 but are lacking in 3βHSD, such as the fetal zone and the zona reticularis (Chapter 32), the 17α-hydroxylation of pregnenolone favors the lyase reaction and the formation of dehydroepiandrosterone (DHEA), probably because no other options are available in the Δ^5 pathway.

The Corticosteroid Pathway Is Initiated by 21-Hydroxylase (CYP21)

The valve that allows entry of 17-OH progesterone into the corticosteroid pathway is CYP21, a SER enzyme that catalyzes the hydroxylation of carbon 21. In the absence of CYP21, neither aldosterone nor cortisol can be produced. In the presence of this enzyme, two corticosteroid pathways become available, each with a different substrate requirement. Normally, the gene encoding CYP21 is expressed only in the adrenal cortex, primarily in the outer two zones (glomerulosa and fasciculata). The kidney and other tissues are also capable of 21-hydroxylation of steroids; however, these tissues contain an enzyme that is not CYP21 and its activity is too low to interfere with the normal endocrine function of the adrenals.

The ***mineralocorticoid pathway*** is catalyzed by two enzymes and utilizes progesterone as the initial substrate. The first enzyme, CYP21, catalyzes the 21-hydroxylation of progesterone, which converts it to the mineralocorticoid, 11-deoxycorticosterone (DOC). This enzyme is present in the outer two zones of the adrenal cortex. The second enzyme is zone-specific in that each of the two outer zones of the cortex expresses a different gene for the enzyme. In the zona fasciculata, the *CYP11B1* gene expresses a mitochondrial enzyme that catalyzes the 11β-hydroxylation of DOC to form cortisol; in the zona glomerulosa, another gene, *CYP11B2,* expresses a mitochondrial enzyme (aldosterone synthase) that catalyzes the three-step conversion of DOC to aldosterone with the formation of the intermediates corticosterone and 18-hydroxycorticosterone. No other tissue in the body normally expresses either of these two genes. Therefore, production of DOC and corticosterone occurs only in the two outer zones of the adrenal cortex, and aldosterone is synthesized exclusively by the zona glomerulosa.

The ***glucocorticoid pathway*** is confined to the zona fasciculata of the adrenal cortex and it is catalyzed by the same CYP21 enzyme that is involved in the mineralocorticoid pathway in this zone along with CYP11B1. The principal difference between the two pathways in the zona fasciculata is the initial substrate, and this explains why the intermediates such as DOC and the products (corticosterone and cortisol) are different. Thus, whereas in the mineralocorticoid pathway the initial substrate is progesterone, in the glucocorticoid pathway it is 17α-hydroxyprogesterone. The only difference between the two substrates is the presence of a 17α-hydroxy group in the latter. Note that this same difference is seen between the two pathways at all three steps—substrate, intermediate, and product. The salient features here are:

1. The presence of 17α-hydroxy group confers maximal glucocorticoid activity to cortisol (Chapter 32); and
2. The zona fasciculata is the exclusive site of cortisol formation.

The Sex Steroid Pathway Is Initiated by 17α-Hydroxylase/17,20-Lyase (CYP17)

The valve that allows entry of 17-OH prognenolone into the "sex steroid" pathway is the 17,20-lyase component of CYP17, the SER enzyme that catalyzes the 17α-hydroxylation of steroids. As noted above, the lyase component of the enzyme is active in certain tissues for reasons that are not currently clear. Both Δ^4 and Δ^5 steroids are known to feed into the sex steroid pathways although the absence of 3βHSD precludes the former. It should be emphasized that the formation of androgenic steroids is necessary before estrogen formation is possible; therefore, the initial reaction in the sex steroid pathway invariably produces an androgenic steroid.

Synthesis of Androgens: 17,20-Lyase

In the adrenal cortex, the sites of sex steroid production (fetal zone and zona reticularis) are lacking in 3βHSD, and pregnenolone is processed by both components of CYP17 to become DHEA, as discussed above. Most of the DHEA is sulfated at the 3-hydroxyl group and the product, DHEAS, is secreted into the plasma. This occurs due to the paucity or absence of any other steroid-modifying enzymes in the tissue. In the Leydig cells of the testes and the theca interna cells of the ovaries, both components of CYP17 are active and will accept either pregnenolone and progesterone as a substrate; radioactive tracer studies indicate that the Δ^4 substrate (progesterone) is preferred. Progesterone processing by the enzyme results in the formation of androstenedione, which is secreted as such (by the theca interna) or converted to the biologically active androgen, testosterone (T) by a 17β-hydroxysteroid dehydrogenase (17βSDH)-catalyzed reaction (by the Leydig). Androstenedione and testosterone are 19-carbon steroids are quantitatively the major androgens produced by the gonads; however, the two biologically active androgens are testosterone and dihydrotestosterone (DHT). Androstenedione is inactive because its 17-keto group is not recognized by the androgen receptor. In the peripheral tissues, androstenedione can be converted to T because of the presence of 17βSDH and many androgen target tissues can convert either androstenedione or T into DHT because of the presence of 5α-reductase. CYP17 activity (both 17α-hydroxylase and 17,20-1yase activities) in the ovary and testis is inhibited by estrogen.

Synthesis of Estrogens: Aromatase (CYP19)

Formation of estrogen is catalyzed by the enzyme aromatase (CYP19) which is present in the SER of a few tissues (gonads, brain, placenta, and adipose). The enzyme is expressed by a single gene (*cyp19*) on chromosome 15; it catalyzes a multistep reaction in which a 19-carbon androgen undergoes hydroxylations at carbon 19 followed by the removal of carbon 19 and the aromatization of ring A. The overall aromatization reactions require an appropriate ring A as substrate (either testosterone or androst-4-ene-3,17-dione), 3 NADPH molecules, and 3 oxygen molecules. The regulation of CYP19 activity differs among tissues; FSH induces the enzyme in Sertoli and granulosa cells of the gonads, whereas glucocorticoids induce the enzyme in adipose tissue. The tissues that contain CYP19 are of two varieties; one that produces estrogen *de novo* from cholesterol (e.g., corpus luteum), and the other that lacks the ability to produce its own 19-carbon substrate and relies on a supply of androgens from other sources (e.g., Sertoli cells of testis, granulosa cells of ovary, placenta, adipose cells, hypothalamus, liver). CYP19 activity is inhibited by the nonaromatizable androgen DHT and is stimulated by the aromatizable androgen testosterone.

Steroidogenic enzymes are located mainly, but not exclusively, in the three tissues that are authentically steroidogenic (i.e., capable of converting cholesterol to pregnenolone): the adrenal cortex, the gonads, and, in pregnancy, the placenta. Steroidogenic enzymes are also present in a few tissues that are not capable of *de novo* steroid synthesis from cholesterol but are able to take up intermediates from plasma and convert them to active hormones; for example, fat cells and the hypothalamus contain CYP19, which enables them to convert circulating androgens into estrogens. The kidney, one of the few extra-adrenal sites with CYP21, converts progesterone to the mineralocorticoid DOC.

A "steroid-modifying" enzyme catalyzes a reaction that alters some structural feature of a steroid, resulting in the formation of an active hormone or an inactive metabolite. These enzymes are present in many tissues, although in most the result of steroid modification is inactivation; for example, the liver contains a large number of steroid-modifying enzymes, virtually all of which are involved in steroid inactivation. The kidney is another tissue that contains several steroid-modifying enzymes, most of which serve to inactivate steroids. On the other hand, at least three steroid-modifying enzymes are physiologically important, as demonstrated in the disorders resulting from a deficiency in any one of them:

5α-Reductase, an enzyme that is present in a few tissues, serves to convert testosterone and other androgens to DHT, a reduced form that is most active in those tissues. There are two types of the enzyme (type 1 and type 2), which are products of different genes. The two types of 5α-reductase do not differ in substrate specificity but differ in enzyme kinetics and tissue distribution. Tissues that express the type 1 enzyme appear to be those contributing to secondary sex characteristics (the hair follicles, sebaceous glands), while tissues that express the type 2 enzyme are structures of the reproductive system (external genital tissues of the fetus, prostate gland, and seminal vesicles). A deficiency of 5α-reductase type 2 enzyme results in a disorder of sexual differentiation in the male (Chapter 34). The synthesis of 5α-reductase activity in prostate and skin is regulated by androgens; activity in liver is regulated by thyroid hormone.

17β-Hydroxysteroid dehydrogenase (17βHSD) (also called 17-ketosteroid reductase; 17β-hydroxysteroid ketoreductase) catalyzes the reversible conversion of unconjugated 17-ketosteroids to 17-hydroxysteroids and supplies the physiologic mechanism for generating estradiol from estrone and testosterone from androstenedione, conversions that lead to the formation of biologically active hormones from inactive precursors. This enzyme is present in the testes and ovaries and its activity is critical for the gonads' ability to produce the biologically active form of androgen (testosterone, DHT) and estrogen (estradiol); a deficiency of this enzyme is known to cause disturbances in sexual differentiation. The gene encoding gonadal and placental 17βHSD is on chromosome 17. The same enzyme is also present in nonreproductive tissues such as liver, lung, erythrocytes, and platelets, but the gene encoding this enzyme is not subject to the same regulation as its gonadal counterpart. Accordingly, in ***pseudohermaphroditism*** due to hereditary 17βHSD deficiency, nongonadal 17βHSD activity is normal.

11β-Hydroxysteroid dehydrogenase (11βHSD) catalyzes the conversion of 11-hydroxysteroids into 11-ketosteroids, a physiologically important reaction that converts the biologically active cortisol to cortisone, an inactive metabolite. The presence of this enzyme in tissues that respond to aldosterone is critical in the hormonal regulation of blood volume (Chapter 32). It is not clear whether the reverse reaction, one that converts cortisone to cortisol, is catalyzed by a separate enzyme, 11-oxidoreductase. From the clinical standpoint, this is an important reaction because the effectiveness of cortisone treatment relies on it (Chapter 32). Evidence that 11-oxidoreductase and 11βHSD are different enzymes comes from the observation that patients with a deficiency of 11-oxidoreductase activity may have normal 11βHSD activity; however, there is evidence that the same enzyme catalyzes both reactions but that different isoforms of the enzyme exist. The activity of 11βHSD (cortisone formation) is inhibited by licorice and by carbenoxolone while the activity of 11-oxidoreductase is inhibited by carbenoxolone.

A number of drugs have an inhibitory effect on one or more steroidogenic/steroid modifying enzymes, and some of these are used clinically to inhibit the production of certain steroid hormones. Some of the important drugs used for this purpose are broad spectrum while others are hormone-specific; a few are listed below along with their principle therapeutic use:

- Aminoglutethimide—inhibits CYP11A and CYP19 (breast cancer, ***Cushing's syndrome***)
- Trilostane—inhibits 3βHSD (***Cushing's syndrome***)
- Ketoconazole—inhibits CYP17 enzymes (17α-hydroxylase and $C_{17,20}$-lyase), particularly those involved in testosterone synthesis. After 6 months of treatment in hirsute women, serum testosterone declines without a change in serum cortisol or DHEAS.
- Mitotane (o,p′-DDD)—inhibits adrenal CYP11A and CYP11B1, but not CYP11B2 (neoplasms of adrenal cortex)
- Metyrapone—inhibits CYP11B1 (hypercorticism due to adrenal neoplasms or ectopic ACTH production)
- Finasteride—inhibits 5α-reductase type 2 (prostatic hyperplasia)
- Testolactone—inhibits CYP19 (breast cancer)
- Fadrozole—inhibits CYP19 (breast cancer)
- Spironolactone—primarily an aldosterone antagonist but inhibits CYP17 in both ovary and testis. It is also an inhibitor of androgen binding to the AR (K^+-sparing diuretic)

Steroid-Binding Serum Proteins

Steroid hormones are hydrophobic molecules that convey information to diverse regions of the body by way of the bloodstream. Because of their very limited solubility in water, however, steroid hormones would precipitate

TABLE 30-3
Steroid-Binding Serum Proteins

	Albumin	CBG[1]	TeBG[2]
Site of production	Liver	Liver	Liver
Molecular weight	66,000	58,000	94,000
Normal concentration	550 μM	710 nM	25 nM in men 40 nM in women
Cortisol-binding capacity	>200 mg/L	250 μg/L	—
Testosterone-binding capacity	>400 mg/L	—	6 μg/L
Estradiol-binding capacity	>3,000 mg/L	—	2 μg/L
Steroid binding sites/molecule	>10	1	1
Affinity (Kd)	~2×10^{-4} M (cortisol) ~4×10^{-4} M (testosterone)	3×10^{-8} M (cortisol)	~2×10^{-9} M (testosterone)
Factors that increase concentration	Glucocorticoids Thyroid hormone	Increased estrogens Increased thyroid hormone	Increased thyroid hormone Increased estrogen (5–10×) Decreased androgens Stress Aging High carbohydrate diet Prolonged stress Cirrhosis of liver
Factors that decrease concentration	Hepatic disease Renal disease Protein malnutrition	Increased glucocorticoids Septic shock Pernicious anemia High carbohydrate diet	Increased androgens (2×) Obesity Hyperprolactinemia Increased GH Menopause Progestins Glucocorticoids

[1]Corticosteroid-binding globulin (also called transcortin).
[2]Testosterone–estradiol-binding globulin.

out of blood if it there were no means of rendering them water-soluble. Two processes normally operate to solubilize steroids in blood conjugation and protein binding. Steroids that are conjugated, either with sulfate or glucuronic acid, are unable to enter target tissues because of their water-soluble (hydrophilic) property; they are usually destined for excretion by the kidney. Although a few tissues (e.g., placenta) have sulfatases that can liberate a steroid from its sulfoconjugated state, most tissues are lacking in this ability and are therefore unresponsive to the high levels of conjugated steroids in the blood. The physiologically important means by which steroid hormones are solubilized for transport in blood is the use of certain serum proteins that solubilize steroids. The liver produces three serum proteins that serve to solubilize steroid hormones (Table 30-3): albumin, a nonspecific adsorber of hydrophobic substances that binds steroids with low affinity; corticosteroid-binding globulin (CBG, also called *transcortin*), which binds C_{21} steroids but has the greatest affinity for cortisol; and testosterone-estradiol-binding globulin (TeBG, also called *sex hormone-binding globulin,* SHBG), which binds steroids of the androstane (C_{19}) and estrane (C_{18}) series but has greater specificity for androgens than estrogens. Because of the high-affinity binding, steroids are not accessible for tissue uptake while bound to either CBG or TeBG; therefore, the greater the fraction of the steroid that is bound to these proteins, the longer the half-life of the steroid. Unlike that of albumin, the concentrations of CBG and TeBG in plasma are normally, near the concentration of the hormone they preferentially bind, and each protein has only one steroid binding site per molecule. The CBG concentration in plasma is

$\sim$0.71 μM, with a cortisol binding capacity of 250 μg/L (mean cortisol levels ($\simeq$100 μg/L); while the plasma TeBG concentration is $\sim$25 nM in men and $\sim$40 nM in women (mean testosterone levels in adult men is $\sim$22 nM). An increase in steroid hormone production will result in an increase in the fraction of the hormone bound by the protein until all of the sites on the protein are saturated (i.e., maximal binding capacity); any increase in hormone production beyond this will result in a pathological situation of hormone excess because the cells will be exposed to supraphysiological levels of albumin-bound and unbound hormone. The concentrations of CBG and TeBG in plasma have important influences on both the production and bioeffectiveness of steroid hormones; thus, factors that affect the plasma levels of these binding proteins will indirectly affect the physiology of the steroid hormones they bind.

Serum albumin is a quantitatively important binder of hydrophobic molecules that makes possible the transport of these otherwise insoluble molecules in blood; however, it binds very loosely, and this enables the molecules (ligands) to dissociate very rapidly. This means that the albumin-bound fraction serves primarily to solubilize the steroid in plasma and does not oppose tissue uptake of the steroid. For this reason, *the albumin-bound fraction should be regarded as being functionally "free."* Thus, plasma steroid hormones can be viewed as existing in only two fractions: those that are bound to specific binding protein and those that are not (albumin-bound + unbound). For the sake of simplicity, when dealing with specific steroid hormones, the fractions will be designated relative to the specific binding protein only; for example, when dealing with testosterone, the two plasma fractions will be designated "TeBG-bound" and "non-TeBG-bound." Likewise, plasma cortisol fractions will be referred to as "CBG-bound" and "non-CBG-bound."

Although the unbound form of steroid hormones is the "active" form, it is also the form that is more susceptible to rapid metabolism. The major site of steroid metabolism is the liver, which inactivates steroids by conjugation and by structural modification. Hepatic Δ^4-hydrogenase reduces the double bond at positions 4 and 5; 17-HSD oxidizes the hydroxyl group at position 17, thereby forming 17-ketosteroids; 3α-hydroxysteroid dehydrogenase reduces the 3-keto to a 3α-hydroxy group. The efficacy of this liver function explains why steroid hormones are ineffective when taken by mouth: they are transported via the hepatic portal system to the liver, where they are promptly metabolized. Minor modifications in the steroid structure, however, can render the hormone more resistant to liver inactivation and, hence, orally active. For example, the introduction of a double bond at position 1 of cortisol yields prednisolone, of a 17α-ethinyl group in estradiol gives ethinyl estradiol, and of a 17α-ethinyl group and removal of the 19-carbon in testosterone yields the orally active progestin, norethindrone.

Eicosanoids

The eicosanoids are discussed in Chapter 18.

30.4 Mechanism of Hormone Action

A target cell is equipped with specific receptors that enable it to recognize and bind a hormone selectively and preferentially. Usually there is one type of receptor for a hormone, but in some cases more than one type of receptor may exist. For example, α_1-, α_2-, β_1-, and β_2-adrenergic receptors are available for catecholamines. The number of receptors for a given hormone in a given cell varies from about 10,000 to more than 100,000.

Hormone recognition (and binding) by a receptor initiates a chain of intracellular events that ultimately lead to the effect of the hormone. Binding can be described as a lock-and-key interaction, with the hormone serving as the key and the receptor as the lock. The structural attributes of a hormone allow it to bind to its receptor and to unlock the expression of receptor function. The receptor, which frequently is a hormone-dependent regulatory protein, is functionally coupled to key enzyme systems in the cell, such that hormone binding initiates a receptor-mediated activation of enzymatic reactions, or it is functionally coupled to a region on chromatin, such that hormone binding initiates expression of one or more structural genes. Stated in another way, a hormone-responsive cell is programmed to carry out certain functions when a sensor (receptor) receives the appropriate signal (hormone).

In general, the number of receptors for a hormone determines how well the cell responds to that hormone. Several factors influence the number of receptors in a cell.

1. The genotype of the cell determines whether the cell is capable of receptor synthesis and, if so, how much and of what type. (Some endocrinopathies involve receptor deficiency or a defect in receptor function.)
2. The stage of cellular development.
3. Hormones themselves are important regulators of the number of receptors in a cell; this regulation may be homologous or heterologous.

Homologous regulation occurs when a hormone affects the number of its own receptors. "Downregulation" (decrease in receptor number) is seen when a target cell is exposed to chronically elevated levels of a hormone. Downregulation involves a decrease in receptor synthesis and is a means by which a cell protects itself against excessive hormonal stimulation. Insulin, the catecholamines, GnRH, endogenous opiates, and epidermal growth factor downregulate their receptors. "Upregulation" (increase in receptor number) also occurs. Prolactin, for example, upregulates its receptors in the mammary gland.

Heterologous regulation is more widespread and is a mechanism by which certain hormones influence the actions of other hormones. Some hormones diminish the production of receptors for another hormone thereby exerting an antagonistic effect. Growth hormone, for example, causes reduction in the number of insulin receptors. More prevalent, however, is augmentation of receptor number by a heterologous hormone. Estrogen, for example, increases the number of receptors for the progesterone, oxytocin, and LH, while thyroid hormone increases the number of β-adrenergic receptors in some tissues.

Hormone receptors can be classified into three types on the basis of their locations in the cell and the types of hormone they bind:

1. Nuclear receptors, which bind triiodothyronine (T_3) after it enters the cell;
2. Cytosolic receptors, which bind steroid hormones as they diffuse into the cell; and
3. Cell surface receptors, which detect water-soluble hormones that do not enter the cell (peptides, proteins, glycoproteins, catecholamines). The mechanism of action of each of these receptor types is different because each is associated with different postreceptor events in the cell.

30.5 Types of Hormone Receptors

Nuclear Receptors

Receptors for thyroid hormone (TR), 1,25-dihydroxyvitamin D (VDR), and retinoic acid (RAR) are called ***nuclear receptors*** because they are located in the nucleus already bound to DNA (nuclear chromatin) even in the absence of their respective hormone or ligand. These nuclear receptors closely resemble the steroid hormone receptors and belong to the same "superfamily" of DNA-binding proteins. The similarities among the members of this superfamily are striking, particularly in their DNA recognition domains and in the corresponding receptor recognition segment of DNA. The receptor molecule consists of three domains: a carboxy terminal that binds the ligand (hormone/vitamin), a central DNA binding domain (DBD), and an amino terminal that may function as a gene enhancer. The ligand binding domain has high specificity for the ligand, and is the trigger that initiates receptor-mediated regulation of gene expression by the ligand. The receptor is bound to DNA by way of two "zinc fingers" (Chapter 26) in the DBD, which associates with a specific nucleotide sequence in DNA called the "response element" for the ligand. The response element is usually located "upstream" (at the 5′ end) of a promoter for the gene that is regulated by the ligand, such that binding of the ligand to the receptor activates the response element (via the DBD), which then activates ("transactivates") transcription of the gene. Transactivation of gene transcription involves the binding of RNA polymerase II to the promoter region of the gene and construction of an RNA transcript (hnRNA) from the DNA gene sequence (Figure 30-3). The hnRNA is then spliced to yield mature mRNA (messenger RNA), which translocates to the cytoplasmic compartment and becomes associated with ribosomes. Ribosomal translation of the mRNA results in the synthesis of a nascent polypeptide, the primary sequence of which is encoded in the gene that was activated by the ligand.

Thyroid Hormone Receptors

Thyroid hormone receptors (TRs) are nonhistone proteins that function as transducers of the effects of thyroid hormone on gene expression. There are four isoforms of TR (TRα1, TRα2, TRβ1, TRβ2) that are products of two different genes, *c-erbA*α and *c-erbA*β. There is some tissue specificity in the distribution of these isoforms. Three of these isoforms (TRα1, TRα2, TRβ1) are found in almost all cells (although in different relative amounts); however, the TRβ2 isoform is found only in the brain. The TRα2 is unique because it does not bind thyroid hormone but it does bind to DNA. Its function is unclear. The other three TR isoforms are single proteins with three regions (domains): a carboxy terminal region that binds T_3, a central region that binds a specific region of DNA, and an amino terminal region that may function as a gene enhancer. The TR is synthesized on ribosomes and is actively transported through the nuclear membrane pores into the nucleus, where the DBD of the TR binds to a specific segment of DNA called the "thyroid hormone response element" (TRE). The TR binds to a TRE half-site as a monomer, a homodimer (α/α, α/β, β/β, etc.) or a heterodimer, in which a TR isoform dimerizes with another protein, e.g., the retinoid X receptor (RXR). In the absence of thyroid hormone, the TR association with the TRE does not result in any changes in gene expression. Thus, TRs exist in the nucleus, are bound to DNA, and await the arrival of thyroid hormone.

Thyroid hormones T_4 and T_3 arrive at their target cells by transport in plasma, for the most part bound to plasma proteins (Chapter 33). The very small fraction of T_4 and

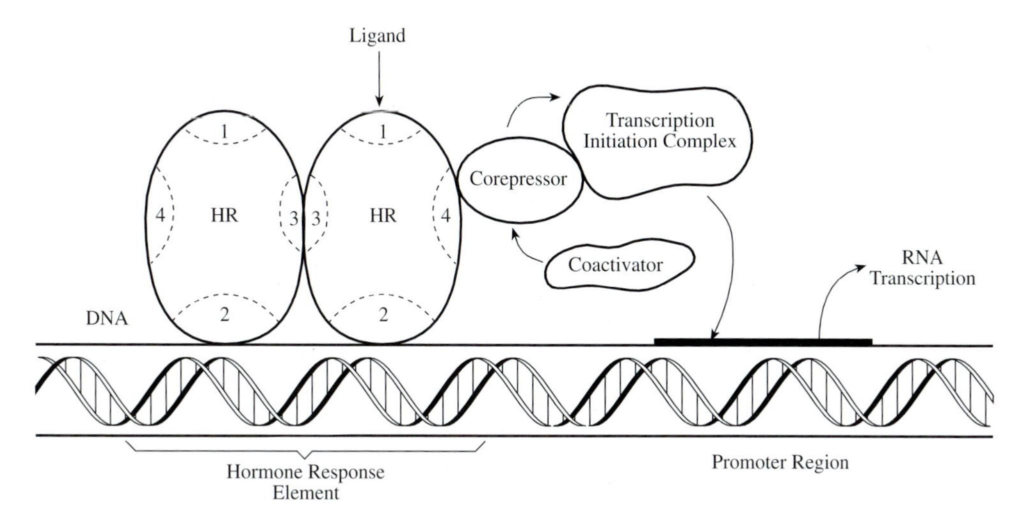

FIGURE 30-3
Thyroid hormone, 1,25-dihydroxyvitamin D, and retinoic acid receptor regulation of transcription. The hormone receptor (HR) is dimerized at site (3) and is bound to DNA at hormone response element site (2). Without the ligand, transcription is inactive due to the interaction of HR with corepressor at site 4. When the ligand (hormone) binds to HR, the bound corepressor dissociates leading to an interaction between the coactivator and HR. These regulatory changes result in increased transcription.

T_3 that is unbound at any given instant is accessible for cellular "uptake," i.e., the diffusional passage of the hydrophobic hormone through the lipid matrix of the cell membrane and into the cell. Once in the cell, the hormone associates loosely with cytoplasmic proteins that prevent its escape back into plasma. Microsomal 5′-deiodinase, which is present in thyroid target cells, catalyzes the removal of the 5′-iodine from T_4, converting it to T_3. This creates a concentration gradient that favors the movement of hormone into the nuclear compartment, in which TR (monomer or dimers) exists in association with the TRE sequence of DNA. Monomeric TR appears to bind to a TRE half-site, while dimeric (homodimer or heterodimer) TR binds to two half-sites. When activated by T_3 binding to TR, the TRE promotes transcription of the gene into hnRNA. After removal of the introns from hnRNA, the resultant mRNA is translated into a protein at the ribosomes. The protein, which may be regulatory or structural, is the final molecular expression of thyroid hormone action.

The outcome of T_3 binding to TR is not a stimulation of transcription in all cases. In the pituitary thyrotrophs, T_3 inhibits transcription of both the TSH-α and TSH-β genes by binding to TR-TRE complexes. The TR mediating this inhibitory effect appears to be a monomer that binds to a TRE half-site, in contrast to the usual dimeric TR that binds to two TRE half-sites and mediates the stimulatory effects of T_3 on transcription in other cell types.

In some instances, the actual cell response to thyroid hormone is a manifestation of the activity of the induced protein; therefore, the induced protein would first need to be activated (e.g., by phosphorylation) before the thyroid hormone effect on the cell can be seen.

Steroid Hormone Receptors

Steroid hormone receptors belong to a large family of DNA-binding proteins that include receptors for thyroid hormone, retinoic acid, and 1,25$(OH)_2$ vitamin D. There are specific receptors for glucocorticoids (GRs), mineralocorticoids (MRs), estrogens (ERs), androgens (ARs), and progestogens (PRs), all of which are coded for by different genes. Unlike the TR, there appears to be only one functional receptor protein for a given steroid that is encoded on a single gene. Like the TR, the steroid receptor is a single protein with three regions (domains): a carboxy terminal region that binds the steroid specifically, a central DBD, and an amino terminal region that may function as a gene enhancer. The DBD of the steroid receptor projects two zinc fingers that allows both recognition of and binding to the hormone response element (HRE) of DNA.

A generic depiction of the mechanism for steroid hormone activity at a target cell is shown in Figure 30-4. The first step consists of dissociation of the hormone from the plasma transport protein and entry into the cell by diffusion across the plasma membrane. In the second step the hormone binds with the receptors in the cytoplasm and nucleus. Receptors for glucocorticoid and aldosterone are found in the cytoplasm and receptors for estrogen and progesterone are found in the nucleus. Recall that receptors for

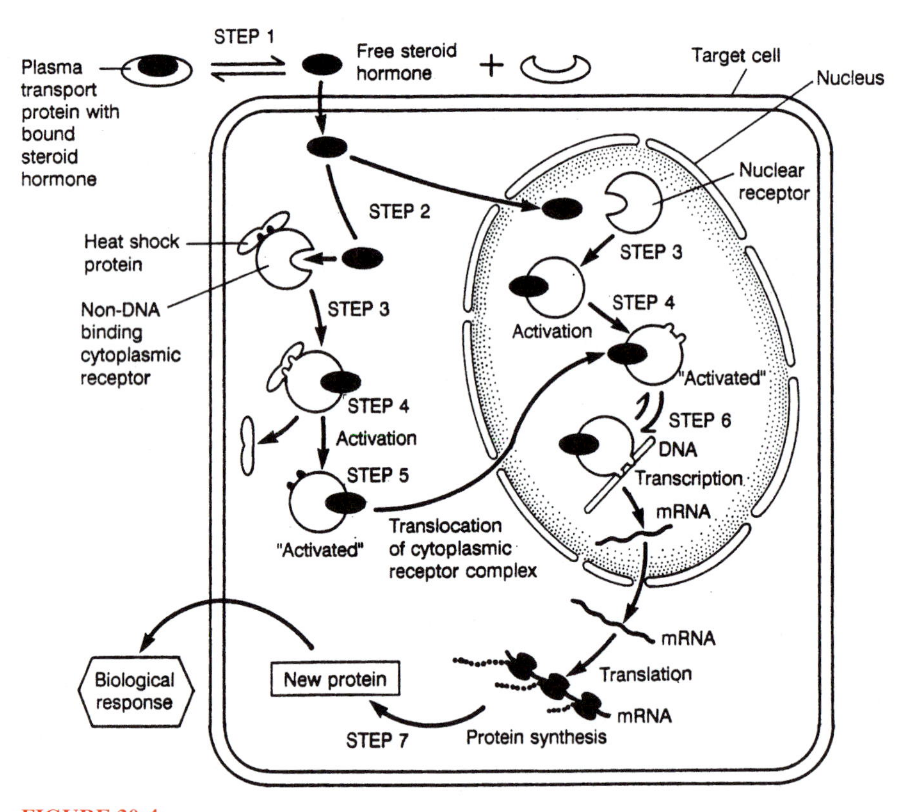

FIGURE 30-4
A generic depiction of the mechanism of steroid hormone action (see text for details). [Reprinted with permission from A. W. Norman and G. Liwack, *Hormones,* 2nd ed. San Diego; Academic Press, 1997, p. 40.]

thyroid hormone, retinoic acid, and 1,25$(OH)_2$ vitamin D are all found in the nucleus.

Activation of the receptor occurs in steps 3 and 4. The receptors exist in cells complexed with a dimer of heat-shock protein (HSP, 90-kDa), which conserves the functional status of the receptor molecule in the absence of the hormone. In addition, HSPs may function to prevent the unoccupied receptor from binding to DNA by masking the DNA binding domain. HSPs bound to the hormone binding domain of the receptor are displaced by the hormone for which the receptor is specific. Once formed, steroid-receptor complexes aggregate into homodimers, and in this form the receptor is said to be "activated." The activated cytoplasmic receptors are translocated via a nucleopore to the nucleus (step 5). In step 6, the activated receptor triggers (or inhibits) transcription of the hormone-regulated gene. In the event of mRNA production, the newly synthesized mRNA is translocated to the cytoplasm where it is translated (step 7). The biological effect of the hormone reflects the function of the protein (e.g., enzyme, transporter, hormone, receptor), and the duration of the effect is a function of the half-life of the mRNA (minutes to hours) and of the protein (hours to days).

The steroid receptor homodimer may also regulate gene expression by interacting with transcription factors that are associated with either the HRE or the promoter. By binding to certain transcription factors, the activated hormone receptor may

1. Stimulate gene expression by removing an inhibitory transcription factor;
2. Stimulate gene expression by facilitating the activation of the promoter by a transciption factor; or
3. Inhibit gene expression by removing a stimulatory transcription factor.

One important transcription factor with which steroid receptors interact is AP-1 (accessory/auxiliary/adapter protein-1), which promotes the expression of genes involved in cell proliferation. The growth-suppressing effect of glucocorticoids is probably mediated by the binding of the hormone-GR complex to one component of AP-1 (c-jun) which, in effect, removes the AP-1-mediated promotion of cell proliferation. Another transcriptional factor involved in steroid hormone action is the ***cAMP response element binding protein*** (CREB), which mediates the genomic effects of peptide hormones that utilize the cAMP second messenger system (see below). A number of steroid-regulated genes are subject to modulation by CREB and are thus subject to the regulatory influence of both steroid and peptide hormones.

Cell Surface Receptors

Many hormones, such as the hormonal amines and all peptidic hormones, are unable to penetrate the lipid matrix of the cell membrane, and thus depend on the presence of receptor sites at the surface of target cells. As listed in Table 30-4, there are several types of cell membrane receptors for these hormones, each of which is coupled to a distinct set of intracellular postreceptor pathways. The surface receptors all initiate postreceptor events that involve the phosphorylation of one or more intracellular proteins, some of which are enzymes whose activities depend on the state of phosphorylation. In two of these cases, an intracellular second messenger is utilized to implement the hormonal action and involves G-protein-coupled receptors. One is coupled to the adenylate cyclase-cAMP system and the other is associated with the phosphatidylinositol-Ca^{2+} pathway (IP_3 pathway).

G-Protein-Coupled Adenylate Cyclase-cAMP System

The effect on adenylate cyclase of ligand binding to specific receptors mediated by heterotrimeric G-proteins is shown in Figure 30-5. The G-proteins belong to a family of regulatory proteins each of which regulates a distinct set of signaling pathways. The cell membrane receptors coupled to intracellular G-proteins share a common seven transmembrane-spanning α-helical serpentine structure; however, they do not show a common overall sequence homology. The α-helical segments are linked with alternating intracellular and extracellular peptide loops with the N-terminal region located on the extracellular side and C-terminal region located on the intracellular side (Figure 30-6). Receptors coupled to G-proteins have been divided into three subfamilies and significant sequence homologies do exist within the families. the receptors belonging to the three families are described in Figure 30-6.

Extracellular signals detected by the membrane receptors are diverse and include hormones, growth factors, neurotransmitters, odorants, and light. Examples of G-proteins are G_s, G_i, products of *ras* oncogenes (Chapter 26), and transducin. Transducin is a constituent of the light-activated cGMP-phosphodiesterase system in the retina (Chapter 38). This system is similar to the adenylate cyclase system except that light is the ligand, rhodopsin is the membrane receptor, cGMP-phosphodiesterase is the effector, and the G-protein is the transducer. The two types of receptors, R_s and R_i, can be functionally linked to adenylate cyclase (Figure 30-5). Binding of an appropriate ligand to R_s stimulates the adenylate cyclase whereas binding to R_i inhibits the enzyme. The G-proteins are heterotrimeric consisting of an α subunit and a tightly coupled $\beta\gamma$ subunit. The α subunit is the unique protein in the trimer. It possesses sites for interaction with cell membrane receptors, $\beta\gamma$ subunit, intrinsic GTPase activity, and ADP ribosylation. In the unstimulated state, the G-protein is present in the heterotrimeric ($\alpha\beta\gamma$) form with GDP tightly bound to the α subunit. Upon activation of the receptor by the bound ligand, GDP is exchanged for GTP on the α subunit followed by dissociation of $G_{s\alpha GTP}$ from the $\beta\gamma$ subunit. The α subunits G_s and G_i are designated $G_{s\alpha}$ and $G_{i\alpha}$. The G-protein subunits undergo posttranslational covalent additions of lipids such as palmitoyl, farnesyl, and geranyl groups. These hydrophobic groups linked to G-protein subunits are necessary for proper anchoring in the cell membrane.

Figure 30-7 shows the cyclic functioning of the G_s protein. Binding of a hormone to R_s permits binding of GTP to $G_{s\alpha}$ to form a $G_{s\alpha \cdot GTP}$ complex and release of the G_β and G_γ subunits. The $G_{s\alpha \cdot GTP}$ complex activates adenylate cyclase which remains active as long as GTP is bound to $G_{s\alpha}$. Binding of GTP to $G_{s\alpha}$ also activates a GTPase intrinsic to $G_{s\alpha}$ that slowly hydrolyzes the bound GTP (GTP turnover is one per minute), eventually allowing reassociation and formation of the trimer and inactivation of adenylate cyclase. A similar series of events is initiated by binding of a ligand to R_i. In this instance, dissociation of the G_i trimer by formation of $G_{i\alpha \cdot GTP}$ inhibits adenylate cyclase and may allow the β,γ subunit to bind to $G_{s\alpha}$, preventing $G_{s\alpha}$ from activating adenylate cyclase. GTP hydrolysis by GTP-activated $G_{i\alpha}$-GTPase allows re-formation of the G_i trimer. GDP is displaced from $G_{i\alpha}$ by binding of another molecule of GTP with reactivation of $G_{i\alpha}$. The inhibitory effect occurs because the target cell contains much more G_i than G_s. $G_{i\alpha}$ may also interact with membrane calcium transport and the phosphatidylinositol pathway. The function of $\beta\gamma$ dimers is not completely understood but they may participate in the regulation of downstream effectors in the signaling pathway.

The G-protein-coupled receptor signaling pathway is regulated by several different mechanisms. Negative regulation of the signaling pathway can occur at both the receptor and G-protein levels. Receptor phosphorylation by G-protein-coupled receptor kinase and arrestin binding can lead to inability of the ligand to activate the signaling system (e.g., β-adrenergic receptor–G-protein transducing system). Negative regulation also occurs after the activation of G-protein, by promoting the GTPase activity of $G_{\delta\alpha}$. This is achieved by regulator of G-protein signaling

TABLE 30-4

Membrane-Bound Receptor Hormone Transduction Systems

Hormones acting via utilization of a cAMP system				
cAMP decrease	**cAMP increase**	**Hormones acting via a cGMP system**	**Hormones acting via IP_3 pathway**	**Hormones acting via ion channels**
Bradykinin A	ATH	Atrial natriuretic hormone	ACTH	Acetylcholine
Norepinephrine(α_2-adrenergic)	α-MSH		Angiotensin II/III	Nicotinic receptor (Na^+ channel)
Somatostatin	Calcitonin		Cholecystokinin	Nitric oxide (NO) (Ca^{2+} channel)
	FSH		EGF	
	Gastrin		Epinephrine	
	Glucagon		GnRH	
	GnRH		Histamine (HI)	
	Histamine (H_2)		NGF	
	LH		Norepinephrine (α_1 adrenergic)	
	Norepinephrine(β_1 adrenergic)		PGF	
	PGE_1		Thromboxanes	
	PTH		TRH	
	Secretin		TSH	
	TSH		Vasopressin	
	Vasopressin		VIP	
	VIP			

Reprinted with permission from A. W. Norman and G. Litwack, *Hormones* 2nd ed. Academic Press 1997, page 32.

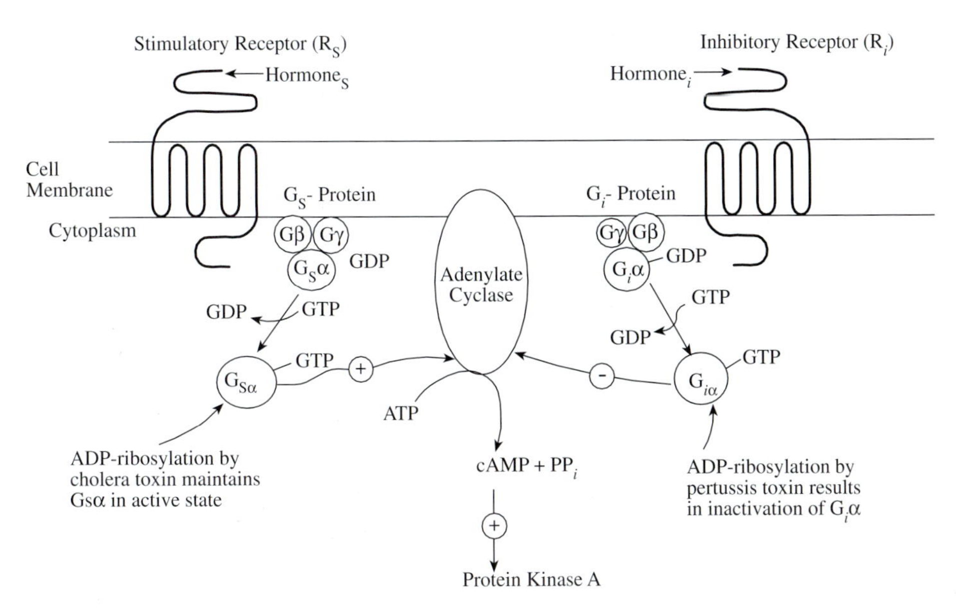

FIGURE 30-5
Dual control of adenylate cyclase activity by guanine nucleotide-binding proteins (G). Subscripts s and i denote stimulatory and inhibitory species, respectively. + and − indicate activation and inactivation, respectively.

(RGS) proteins, which bind to G_α and promote GTPase activity. This process of converting G_α from an active state to an inactive state is rapid. Transcriptional regulation and posttranslational modification of G-proteins provide other means for regulation of the G-protein-coupled signaling pathway.

Understanding of G-proteins has been aided by specific agents that modify G_s or G_i and by mutant mouse lymphoma cell lines deficient in G_s activity. Both $G_{s\alpha}$ and $G_{i\alpha}$ contain sites for NAD^+-dependent ADP ribosylation. Cholera toxin ADP-ribosylates a specific arginine side chain of $G_{s\alpha}$ and maintains it in a permanently active state, while islet-activating protein, one of the toxins in ***Bordetella pertussis,*** ADP-ribosylates a specific cysteine side chain of $G_{i\alpha}$, permanently blocking its inhibitory action because it maintains $G_{i\alpha}$ in the GDP form. Thus, both toxins stimulate adenylate cyclase activity and lead to excessive production of cAMP. Cholera toxin causes a diarrheal illness (Chapter 12), while toxins of *B. pertussis* cause whooping cough, a respiratory disease affecting ciliated bronchial epithelium.

Response of the adenylate cyclase system to a hormone is determined by the types and amounts of various constituent proteins. Cyclic AMP production is limited by the amount of adenylate cyclase present. When all the adenylate cyclase is fully stimulated, further hormone binding to R_s's cannot increase the rate of cAMP synthesis. In cells having many different R_s's (adipocytes have them for epinephrine, ACTH, TSH, glucagon, MSH, and vasopressin), maximal occupancy of the receptors may not stimulate cAMP production beyond what can be achieved by full occupancy of only a few of the receptor types. Therefore, the greatest stimulation that can be achieved by a combination of several hormones will not be simply the sum of the maximal effects of the same hormones given singly. A hormone's ability to stimulate cAMP production may depend on the cell type. For example, epinephrine causes large increases in cAMP concentration in muscle but has relatively little effect on liver. The opposite is true for glucagon (see Chapter 15). Within a particular cell type, destruction of one type of R_s does not alter the response of the cell to hormones that bind other stimulatory receptors.

In prokaryotic cells, cAMP binds to catabolite regulatory protein (CAP), which then binds to DNA and affects gene expression (Chapter 26). In eukaryotic cells, cAMP binds to cAMP-dependent protein kinase, which contains two regulatory (R) and two catalytic (C) subunits. Upon binding of cAMP, the catalytic subunits separate, become active,

$$\underset{\text{(inactive)}}{R_2C_2} + 4\text{cAMP} \rightleftharpoons 2(\text{R–2cAMP}) + \underset{\text{(active)}}{2\text{C}}$$

and catalyze ATP-dependent phosphorylation of serine and threonine residues of various cell proteins, often altering the activities of these proteins. Some, such as phosphorylase kinase, become activated, whereas others, such as glycogen synthase, are inactivated (Chapter 15). Cyclic AMP-dependent protein kinase remains active while intracellular cAMP concentration, controlled by the

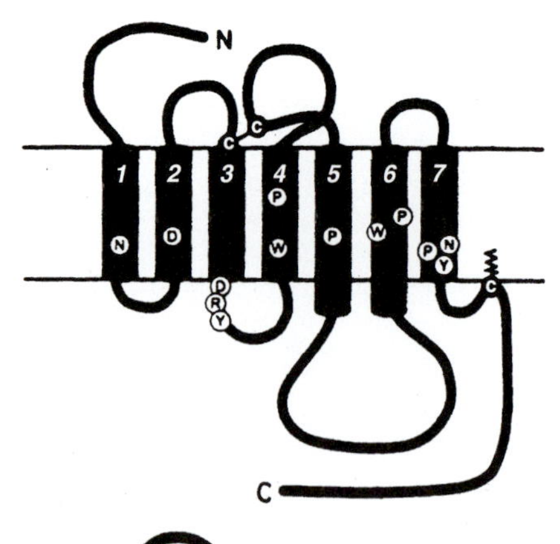

Family A. Rhodopsin/β2 adrenergic receptor-like

Biogenic amine receptors (adrenergic, serotonin, dopamine, muscarinic, histamine)

CCK, endothelin, tachykinin, neuropeptide Y, TRH, neurotensin, bombesin, and growth hormone secretagogues receptors plus vertebrate opsins

Invertebrate opsins and bradykinin receptors

Adenosine, cannabinoid, melanocortin, and olfactory receptors.

Chemokine, fMLP, C5A, GnRH, eicosanoid, leukotriene, FSH, LH, TSH, fMLP, galanin, nucleotide, opioid, oxytocin, vasopressin, somatostatin, and protease-activated receptors plus others.

Melatonin receptors and other non-classified

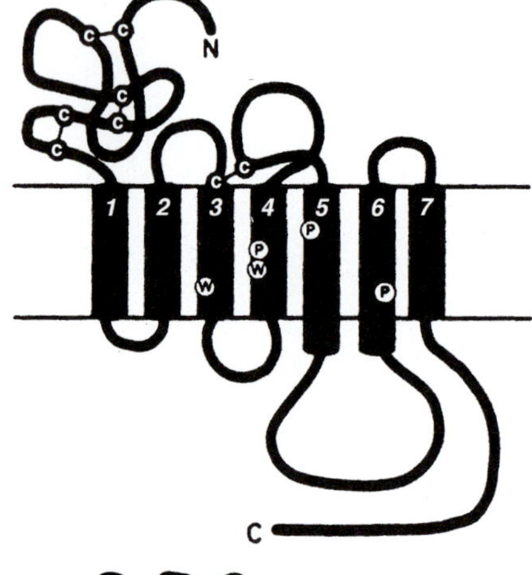

Family B. Glucagon/VIP/Calcitonin receptor-like

Calcitonin, CGRP and CRF receptors

PTH and PTHrP receptors

Glucagon, glucagon-like peptide, GIP, GHRH, PACAP, VIP, and secretin receptors

Latrotoxin

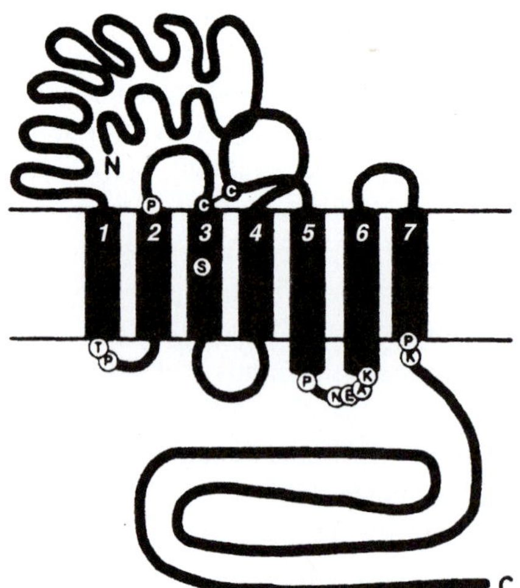

Family C. Metabotropic neurotransmitter/ Calcium receptors

Metabotropic glutamate receptors

Metabotropic GABA receptors

Calcium receptors

Vomeronasal pheromone receptors

Taste receptors

FIGURE 30-6

G-protein-coupled seven-transmembrane receptors (GPCR). These can be divided into three major subfamilies (1). A snake diagram for a prototypic member of each subfamily is shown. Family A receptor (*upper panel*) can be subdivided phylogenetically into six subgroups as indicated. Family A receptors are characterized by a series of highly conserved key residues (*black letter in white circles*). In most family A receptors a disulfide bridge connects the second (ECL2) and third extracellular loops (ECL3) (*white letters in black circles*). In addition, a majority of the receptors have a palmitoylated cysteine in the carboxy terminal tail causing formation of a putative fourth intracellular loop.

Family B receptors (*middle panel*) are characterized by a long amino terminus containing several cysteines that form a network of disulfide bridges. The B receptors contain, like the A receptors, a disulfide bridge connecting ECL2 and ECL3. However, the palmitoylation site is missing. Moreover, the conserved prolines are different from the conserved prolines in the A receptors and the DRY motif at the bottom of TM 3 is absent.

Family C receptors (*lower panel*) are characterized by a very long amino terminus (~600 amino acids). The amino terminal domain contains the ligand binding site. Except for two cysteines that form a putative disulfide bridge, the C receptors do not have any of the key features characterizing A and B receptors. Some highly conserved residues are indicated (*black letter in white circles*). A unique characteristic of the C receptors is a very short and highly conserved third intracellular loop. [Reproduced with permission from U. Gether, *Endocrine Reviews* **21,** 92 (2000).]

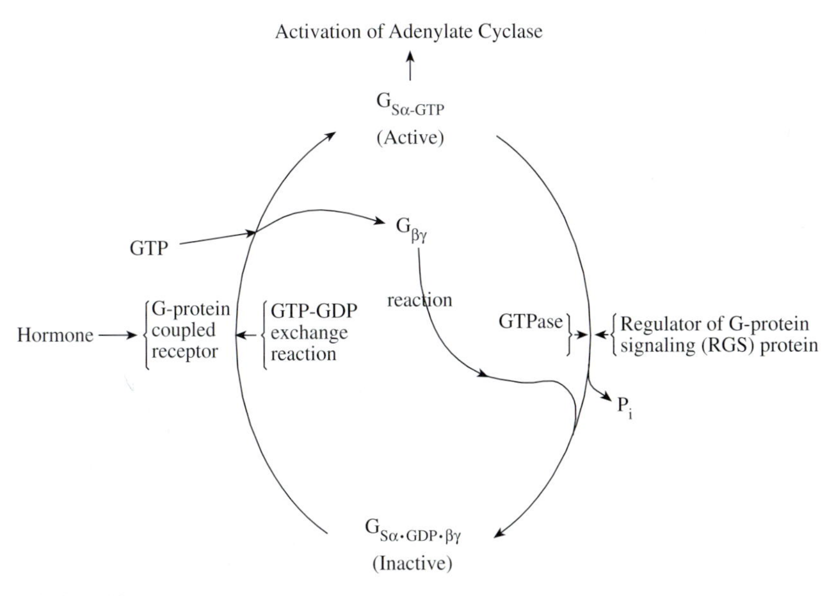

FIGURE 30-7
Activation and inactivation (or return to resting state) of signal transducer trimeric stimulatory G-protein (G_s). Hormone binding to membrane G-protein-coupled stimulatory receptor causes a GTP and GDP exchange reaction in $G_{s\alpha}$ and dissociation of $G_{\beta\gamma}$. $G_{s\alpha.GTP}$ activates adenylate cyclase and initiates the signal transduction pathway. The return of $G_{s\alpha.GTP}$ to the resting state is mediated by the intrinsic GTPase activity of the $G_{s\alpha}$ subunit and also through promotion of GTPase activity by the RGS protein. Activation of G_i protein occurs by analogous reactions of G_s when a hormone binds to its corresponding inhibitory receptor. This leads to the formation of $G_{i\alpha.GTP}$ which inhibits adenylate cyclase. Gain in function or loss in function can result from mutations in genes coding for receptor protein, $G_{s\alpha}$, $G_{i\alpha}$, or RGS protein.

relative rates of cAMP synthesis by adenylate cyclase and of degradation by phosphodiesterase, is elevated above basal levels. The reaction catalyzed by the phosphodiesterase is

$$3',5'\text{-cAMP} + H_2O \rightarrow 5'\text{-AMP}$$

As cAMP concentration decreases, reassociation of catalytic and regulatory subunits of the kinase occurs.

The cAMP system also derives some of its specificity from the proteins that are substrates for cAMP-dependent protein kinase. It is an indirect regulatory system, since cAMP modulates the activity of cAMP-dependent protein kinase and the kinase, in turn, affects the activities of a variety of metabolic enzymes or other proteins. This indirectness is the basis for amplification of the hormonal signal. Activation of adenylate cyclase by binding of a hormone molecule to a receptor causes formation of many molecules of cAMP and allows activation of many molecules of the protein kinase, each of which in turn can phosphorylate many enzymes or other proteins. This "amplification cascade" accounts in part for the extreme sensitivity of metabolic responses to small changes in hormone concentrations. Finally, the response to increased cAMP concentrations within a cell usually involves several metabolic pathways and a variety of enzymes or other proteins.

Abnormalities in Initiation of G-Protein Signal

Two-well studied bacterial toxins, cholera toxin and pertussis toxin, perturb normal functioning of G_α proteins by ADP ribosylation of specific amino acid residues (discussed earlier).

Germline and somatic mutations in the genes coding for G-proteins can cause endocrine disorders in several ways. Mutations can be either activating (gain of function) or inactivating (loss of function). Examples of gain-in-function mutations that result in the ligand-independent activator of $G_{s\alpha}$ include some adenomas of pituitary and thyroid, some adenomas of adrenals and ovary, and syndromes of hyperfunction of one or more endocrine glands. In most of these disorders, a point mutation alters an amino acid in $G_{s\alpha}$ leading to inhibition of GTP hydrolysis required for the termination of the activity of $G_{s\alpha}$. An example of autonomous hyperhormone secretion and cellular proliferation of several endocrine glands is *McCune-Albright syndrome*. In this disorder, two missense mutations result in the replacement of $Arg^{201} \rightarrow$ His or $Arg^{201} \rightarrow$ Cy in $G_{s\alpha}$ of affected endocrine glands. Some of the clinical manifestations of McCune-Albright syndrome include hyperthyroidism, polyostotic fibrous dysplasia, and hyperpigmented irregularly shaped skin lesions (known as ***café-au-lait lesions***). The cutaneous hyperpigmentation is caused by hyperfunctioning of MSH receptors coupled to $G_{s\alpha}$ in the melanocytes.

An example of decreased function due to impaired activation or loss of $G_{s\alpha}$ and resistance to hormone action is pseudohypoparathyroidism (***Albright's hereditary osteodystrophy***). Patients with this syndrome exhibit generalized resistance to the action of those hormones dependent on $G_{s\alpha}$ for their function despite elevated serum hormone levels. Several dysmorphic features are also characteristic of these patients.

A well-understood disorder of hypofunctioning $G_{s\alpha}$ is ***pseudohypoparathyroidism type Ia.*** This disorder exhibits a dominant inheritance pattern with one normal and one abnormal $G_{s\alpha}$ allele. In different families affected by pseudohypoparathyroidism Ia, two missense mutations have been identified in $G_{s\alpha}$: $Arg^{385} \rightarrow$ His and $Arg^{231} \rightarrow$ His. The former mutation prevents receptor-mediated G_s stimulation and the latter mutation prevents receptor-mediated binding of GTP to G_s. A paradoxical finding consisting of both gain and loss of function due to identical mutations in $G_{s\alpha}$ ($Ala^{366} \rightarrow$ Ser), in two unrelated boys with pseudohypoparathyroidism Ia. Both of these patients, while exhibiting hormone resistance in some tissues, showed autonomous production of testosterone by testicular Leydig cells due to gain of function in the receptors of luteinizing hormone. This results in precocious puberty (***testotoxicosis***). This particular mutation in the $G_{s\alpha}$ gene involves a gain in function property in $G_{s\alpha}$ enhancing the rate of release of GDP. However, the paradoxical effect is due to temperature stability. The mutant protein is unstable at 37°C, which causes pseudohypoparathyroidism; however, the mutant protein is stable at lower temperature and thus is active in testes, which are about 3°C–5°C cooler than the rest of the body.

G-protein abnormalities also may arise from mutations in genes encoding proteins that regulate G-protein signaling. Several RGSs have been identified. RGS proteins bind to G_α and accelerate the hydrolysis of GTP. The physiological responses of G-proteins regulated by RGSs are fast and include vagal slowing of the heart rate (G_i), retinal detection of photones (G_t), and contraction of vascular smooth muscle (G_Q).

G-Protein-Coupled Phosphatidylinositol-Ca^{2+} Pathway

The calcium system is more complex than the cAMP system and contains a variety of mechanisms for transducing the Ca^{2+} signal into changes in cellular function. It is also sensitive and responds to relatively small, transient changes in free Ca^{2+} concentration. This sensitivity is desirable from a regulatory point of view, but it also reflects the toxicity of even low concentrations of free Ca^{2+} (Chapter 37).

In a resting cell, the pool of cytosolic calcium is very small, and most of the intracellular Ca^{2+} is bound in mitochondria or the endoplasmic reticulum or is bound to the plasma membrane. The cytosolic calcium ion concentration is $\sim$0.1 μmol/L, in contrast to the high concentration ($\sim$ 1000 μmol/L) outside cells. This 10,000-fold gradient is maintained by the plasma membrane, which is relatively impermeable to calcium; by an ATP-dependent Ca^{2+} pump (which extrudes Ca^{2+} in exchange for H^+) in the plasma membrane; and by "pump-leak" systems in the endoplasmic reticulum and inner mitochondrial membranes. A change in the permeability of any of these membranes alters the calcium flux across the membrane and causes a change in cytosolic calcium concentration. This change is thought to act as the messenger in the calcium system. Plasma membranes regulate cytosolic calcium concentration by their role as a diffusion barrier and by containing the ATP-dependent Ca^{2+} pump and receptors for extracellular messengers. Activation of some of these receptors increases the Ca^{2+} permeability of the plasma membrane and thus the level of cytosolic calcium. In other cases, the activated receptor produces one or more second messengers that increase cytosolic Ca^{2+} concentration from the calcium pool in the endoplasmic reticulum. Calcium then becomes a "third messenger." The endoplasmic reticulum is the source of Ca^{2+} in the initial phase of cell activation by some hormones in many systems. Since the plasma membrane can bind Ca^{2+}, the rise in cytosolic calcium following binding of some extracellular messengers may be due to release from this intracellular calcium pool.

The inner mitochondrial membrane may function primarily as a calcium sink, taking up excess calcium in the cytosol that results from hormonal activation of the cell. At cytosolic Ca^{2+} concentrations greater than 0.6 μmol/L, the mitochondrial calcium pump is activated and stores calcium in the mitochondrial matrix as a nonionic, rapidly exchangeable, phosphate salt. At low cytosolic calcium concentrations, the inner mitochondrial membrane allows Ca^{2+} to "leak" into the cytosol. The capacity of the active influx pathway (the pump) is much greater than that of the passive efflux route (the leak). The mitochondrial pump-leak system may serve to fine-tune the cytosolic calcium concentration while the plasma membrane is the principal safeguard against entry of toxic amounts of calcium into the cell.

Mechanism of the Calcium Messenger System

Most extracellular messengers that cause a rise in cytosolic Ca^{2+} concentration also increase the turnover

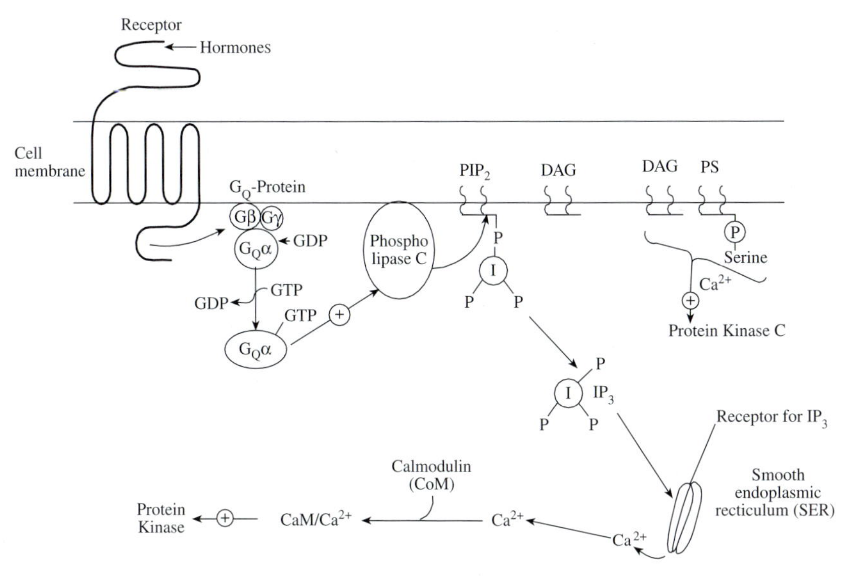

FIGURE 30-8

G_Q protein-coupled phosphatidylinositol-Ca^{2+} pathway. The binding of a hormone at a specific receptor site results in the activation of G-protein which, in turn, activates phospholipase C via $G_{Q\alpha \cdot GTP}$-protein. The action of phospholipase C on phosphatidylinositol 4,5-bisphosphate (PIP_2) yields inositol trisphosphate (IP_3) and diacylglycerol (DAG) which, along with phosphatidylserine (PS), activates protein kinase C. IP_3 binds to receptors on SER, releasing Ca^{2+} which, in turn, activates another set of protein kinases. +, Activation.

of phosphoinositides in the plasma membrane. The pathway is initiated by the binding of extracellular messengers to specific receptors located on the plasma membrane and activating G-proteins (e.g., G_Q). The activated $G_{Q\alpha \cdot GTP}$-protein, in turn, activates phospholipase C, which catalyzes hydrolysis of phosphatidylinositide 4,5-bisphosphate, initiating the intracellular second-messenger effects (Figure 30-8). The action of phospholipase C results in the formation of two products, inositol 1,4,5,-trisphosphate (IP_3) and diacylglycerol (DAG), which function as intracellular messengers (Figure 30-9). Most of the DAG released in response to stimulation of calcium-mobilizing receptors contains arachidonic acid esterified to C_2 of the glycerol. This arachidonate may be released from DAG by diacylglycerol lipase or from phosphatidic acid by phospholipase A_2, following phosphorylation of DAG. It can then serve as a substrate for the synthesis of eicosanoids (prostaglandins, prostacyclins, thromboxanes, and leukotrienes; Chapter 18), which also modulate a number of physiological functions. These reactions are part of lipid and phosphoinositide cycles (Figure 30-10). Many extracellular messengers in a variety of cells activate this "leak" pathway. They include acetylcholine (many tissues); epinephrine, angiotensin, and vasopressin (liver); angiotensin (adrenal cortex); glucose (pancreatic islets); photons (photoreceptor cells); and several different substances in neutrophils, leukocytes, and platelets. Many of the membrane receptors linked to phosphoinositide metabolism also activate guanylate cyclase, but the mechanism and function of this activation remain unknown.

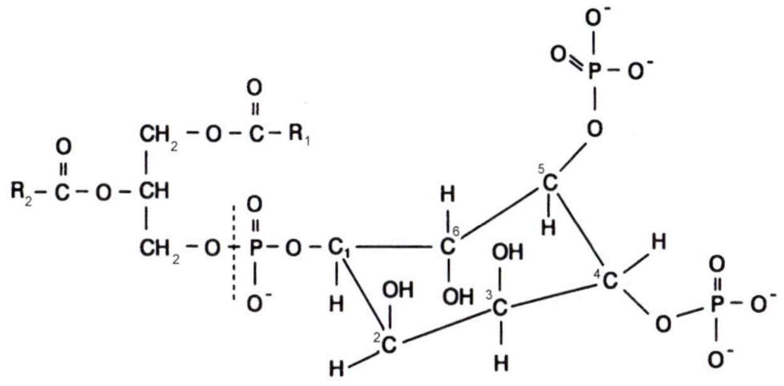

FIGURE 30-9

Structure of phosphatidylinositol 4,5-bisphosphate. Hydrolysis at the dotted line is catalyzed by phospholipase C and releases diacylglycerol and inositol 1,4,5-trisphosphate, which are the intracellular second messengers.

The starting point of these reactions is phosphatidylinositol 4,5-bisphosphate (PIP_2; Figure 30-10), which is synthesized from phosphatidylinositol (PI) in two steps catalyzed by specific kinases. The phosphates are removed by specific phosphomonoesterases setting up futile cycles. Stimulation of phospholipase C by binding of an agonist to its receptor hydrolyzes PIP_2, releasing IP_3 and DAG (usually 1-stearoyl-2-arachidonyl-*sn*-glycerol). IP_3 stimulates the release of Ca^{2+}, predominantly from the endoplasmic reticulum, and may increase plasma membrane permeability to Ca^{2+} in some cells. In most cells studied, the rise in level of cytosolic Ca^{2+} is preceded by an increase in that

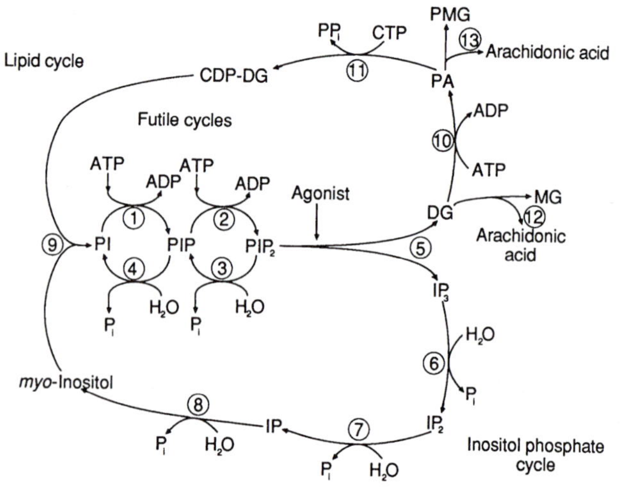

FIGURE 30-10
Inositol metabolism. An inositol phosphate cycle converts inositol 1,4,5-trisphosphate (IP_3) to free inositol, which can then combine with CDP-diacylglycerol (CDP–DAG) to re-form phosphatidylinositol (PI). Inositol 1,4-bisphosphate (IP_2) and inositol 1-phosphate (IP) are intermediates in this cycle. The CDP-DAG produced in the lipid cycle returns diacylglycerol for resynthesis of PI. It is uncertain whether arachidonic acid is released by hydrolysis of diacylglycerol, with release of a monoglyceride (MG), or of phosphatidic acid (PA), with release of a phosphomonoglyceride (PMG). The futile cycling of PI, phosphatidylinositol 4-phosphate (PIP), and phosphatidylinositol 4,5-bisphosphate (PIP_2) maintains a supply of PIP_2 to generate diacylglycerol and IP_3 in response to agonist binding. The enzymes involved are 1, phosphatidylinositol kinase; 2, PIP kinase; 3, PIP_2 phosphomonoesterase; 4, PIP phosphomonoesterase; 5, PIP_2 phosphodiesterase (phospholipase C); 6, IP_3 phosphatase; 7, IP_2 phosphatase; 8, IP phosphatase; 9, CDP-diacylglycerol inositol phosphatidate transferase (phosphatidylinositol synthase); 10, diacylglycerol kinase; 11, CTP-phosphatidate cytidyl transferase; 12, diacylglycerol lipase; and 13, phospholipase A_2. [Modified and redrawn with permission from M. J. Berridge and R. F. Irvine, *Inositol triphosphate, a novel second messenger in cellular signal transduction. Nature* **312,** 315 (1984). © 1984 by Macmillan Magazines Ltd.]

of IP_3, making IP_3 the second messenger and calcium the third messenger. In the endoplasmic reticulum, IP_3 seems to increase efflux of Ca^{2+}. Amplification occurs since an estimated 10–20 calcium ions are released from liver endoplasmic reticulum by each IP_3 molecule. IP_3 is rapidly hydrolyzed to free inositol, which is used for resynthesis of PI (Chapter 19). Removal of the 1-phosphate is blocked by lithium, with accumulation of inositol 1-phosphate and depletion of *myo*-inositol pools. The effect is most marked in the brain, where lithium also inhibits *de novo* synthesis of *myo*-inositol and plasma inositol is unable to cross the blood-brain barrier. Lithium carbonate, used to treat bipolar affective disorders, may have its therapeutic effect by inhibition of the phosphatidylinositol system.

In several types of cells, stimulation of plasma membrane receptors coupled to phosphoinositide turnover not only triggers a rapid mobilization of intracellular Ca^{2+} as discussed above, but also a prolonged mobilization. The latter is dependent extracellular Ca^{2+}, and another metabolite of inositol, namely, inositol tetrakisphosphate (IP_4),or Ins(1,3,4,5)P_4, has been implicated in the transport of extracellular Ca^{2+}. The molecular mechanism of this process is nuclear. IP_4, unlike IP_3, cannot mobilize Ca^{2+} from endoplasmic reticulum. IP_4 metabolism consists of its synthesis from IP_3 by a novel, specific ATP-dependent kinase and its conversion to inactive metabolite Ins(1,3,4)P_3 by a phosphatase. The regulation of kinase and phosphatase activity may be of importance in controlling the intracellular events. Diacylglycerol released by phospholipase C may increase protein phosphorylation. Protein kinase C (C-kinase) is a monomeric enzyme (M.W. 82,000) found free in the cytosol in most tissues except brain, kidney, and liver, where it is predominantly membrane-bound. Diacylglycerol increases both the binding of protein kinase C to the inner surface of the plasma membrane and the sensitivity of the kinase to activation by Ca^{2+} and phosphatidylserine. The effects of DAG related to protein phosphorylation are thought to be mediated by this kinase. Tumor-promoting phorbol esters such as 12-O-tetradecanoylphorbol-13-acetate (ingredient in croton oil that promotes skin tumor production by carcinogens) can directly activate protein kinase C, mimicking DAG and bypassing the receptors. When activated by phorbol esters, protein kinase C phosphorylates cytosolic myosin light-chain kinase (Chapter 21) and other proteins, particularly those related to secretion and proliferation. When activated by DAG, the kinase phosphorylates a number of proteins *in vivo,* but its physiological substrates are unknown. DAG is removed by phosphorylation to phosphatidic acid or by hydrolysis to monoacylglycerol and free fatty acid.

In addition to the effects of DAG and phorbol esters on protein kinase C, some evidence links malignant transformation by some oncogene products to phosphoinositide metabolism. Platelet-derived growth factor, the product of the *sis* oncogene, stimulates inositol lipid metabolism. Products of the *src* and *ras* oncogenes may be phosphoinositide kinases, which increase the concentration of PIP_3 and provide more substrate for phospholipase C. The product of the *ras* gene is a GTP-binding protein (G-protein) involved in receptor activation of PIP_2 hydrolysis. A point mutation activates the *ras* oncogene, and the resulting oncogene product binds GTP but lacks GTPase activity. This abnormal protein could activate PIP_2 hydrolysis independently of receptor occupancy.

Following these early events initiated by binding of a calcium-mediated hormone to its receptor, a transient rise in cytosolic calcium concentration and a more prolonged increase in DAG concentration occur. There is also a prolonged four- to fivefold increase in Ca^{2+} flux across the plasma membrane, which may maintain an elevated Ca^{2+} concentration at the cytoplasmic face of the membrane. This increase, together with the diacylglycerol, may continue to activate the membrane-bound protein kinase C. In this way, myosin light chains may remain phosphorylated and active during smooth muscle contraction (Chapter 21) despite a fall in myoplasmic Ca^{2+} to near-basal levels.

The rise in intracellular Ca^{2+} level also alters the activity of metabolic pathways by calcium-binding proteins. Of these proteins, calmodulin (M.W. 17,000), found in the cytoplasm of every eukaryotic cell, has received the most attention. Parvalbumin in skeletal muscle, troponin C in skeletal and cardiac muscle, and myosin light chain in smooth muscle serve similar functions. The affinity of these proteins for Ca^{2+} is high, as reflected by dissociation constants of 10^{-6}–10^{-8} M. Occupation of at least three of four equivalent Ca^{2+} binding sites on calmodulin induces a conformational change and, in most cases, increases its affinity for proteins having a calmodulin binding site. A major function of Ca^{2+}-calmodulin is regulation of the phosphorylation states of intracellular proteins. Systems affected include those involved in secretion (of insulin, for example), neurotransmitter release and neuronal function, cellular contractile processes (Chapter 21), and several protein kinases involved in regulation of glycogen and glucose metabolism.

A phosphatase activated by Ca^{2+} and calmodulin, known as calcineurin, functions as a negative modulator of some of the effects of Ca^{2+} related to protein phosphorylation. Calcineurin is a Ca^{2+}-calmodulin-dependent serine/threonine protein phosphotase and it participates in many Ca^{2+}-dependent signal transduction pathways. The immunosuppressant drugs cyclosporin A and FK506, after binding to their respective cytoplasmic binding proteins (cyclophilin and FK506-binding protien), inhibit the action of calcineurin. This blocks the translocation of a factor into the nucleus of activated T lymphocytes and prevents allograft rejection by blocking T cell cytokine gene transcription (Chapter 35).

Receptors for Insulin and Growth Factors

Insulin exerts its effects by altering the state of phosphorylation of certain intracellular enzymes by a mechanism that does not involve cAMP but that requires specific binding to surface receptors with tyrosine kinase activity. Insulin exerts acute (minutes), delayed-onset (hours), and long-term (days) effects entirely by way of a single receptor.

The insulin receptor is a transmembrane heterotetramer consisting of one pair each of two dissimilar subunits linked by disulfide bridges (Chapter 22). Both subunits are expressed entirely by a single gene, which is located on chromosome 19. The two subunits, α and β, are formed posttranslationally from a single proreceptor following glycosylation and proteolytic cleavage by an endopeptidase. The α subunit, which projects into the extracellular space, is a glycosylated 135-KDa protein that specifies the insulin binding site (one binding site per α subunit). The transmembrane β subunit is a glycosylated 95-KDa subunit that contains four domains, one of which has tyrosine kinase activity. The tyrosine kinase activity of the β subunit is suppressed by the α subunit in the absence of insulin; insulin binding to the α subunit removes the inhibition and allows the β subunit tyrosine kinase to phosphorylate itself (autophosphorylation of the β subunit). Autophosphorylation is followed by phosphorylation of one or more key intracellular proteins, one of which is insulin receptor substrate-1 (IRS-1) that initiates a cascade of intracellular phosphorylations mediated by a number of intracellular enzymes that are incompletely characterized, but which are critical in bringing about all of the known cellular effects of insulin (Chapter 22).

Autophosphorylation of the insulin receptor may also be required for inactivation of the insulin signal ("OFF" signal). Within minutes after autophosphorylation, the insulin-activated receptor complex undergoes lateral movement toward clathrin-coated pits, the site at which a number of activated insulin receptors congregate or "cluster." This clustering stimulates endocytosis of the congregation, a process referred to as "internalization" or "endocytosis," which is followed by fusion of the internalized insulin receptor membrane vesicle (endosome) with a lysosome, i.e., an endolysosome formation. This allows breakdown of both insulin and its receptor by lysosomal proteases. It has been suggested that insulin fragments resulting from this digestion process are mediators of some of the actions of insulin within the cell, but further evidence in support of this is needed.

In states of chronically elevated blood levels of insulin, there is a substantive decrease in the density of insulin receptors in insulin-dependent cells due to a decrease in the synthesis of insulin receptors. This phenomenon, which is referred to as "downregulation," represents a means by which a cell autoregulates its receptivity to the hormone, and is also detected by other hormones such as the catecholamines, endorphins, and GnRH. The precise mechanisms involved in downregulation are not known.

The insulin receptor is expressed in most cells and tissues examined to date, including the non-insulin-dependent tissues such as brain, kidney, and blood cells (both red and white). Although no specific cellular response to insulin has been described in these tissues, it is conceivable that insulin regulates one or more intracellular

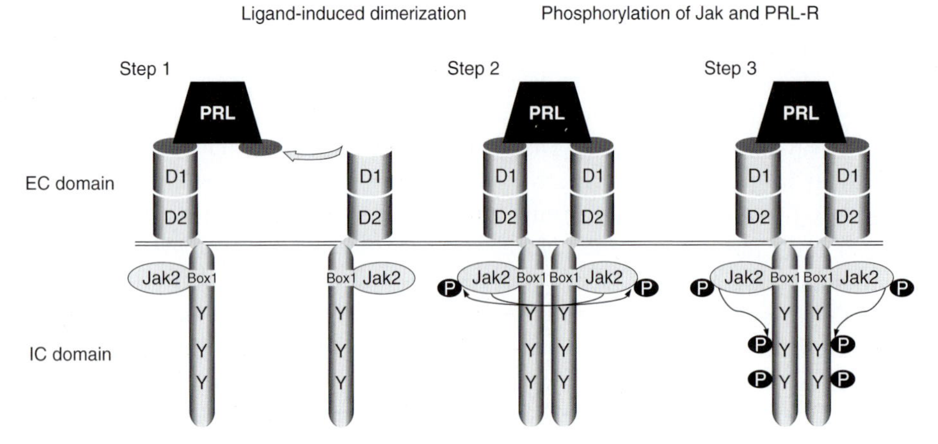

FIGURE 30-11
Mechanism of prolactin receptor activation. Activation of prolactin receptor consists of ligand-induced sequential receptor homodimerization driven by the two binding sites of prolactin. In the intracellular domain of the homodimer of the ligand-receptor complex, a tyrosine kinase [known as Janus kinase 2 (Jak-2)] is activated. Jak-2 kinase causes autophosphorylation and phosphorylation of the receptor. [Reproduced with permission from M. E. Freeman, B. Kanyicska, A. Lerant, and G. Nagy, *Physiological Reviews* **80,** 1530 (2000).]

activities that are unrelated to glucose uptake and utilization.

The IGF-I receptor is an $\alpha_2\beta_2$ heterotetramer that is structurally similar to the insulin receptor but is coded for by a different, single-copy gene located within bands q25-26 of chromosome 15. Like the insulin receptor, with which it shares 50–60% sequence homology, the IGF-I receptor has a cysteine-rich domain in the α subunit for ligand binding and a tyrosine kinase domain in the β subunit that catalyzes transphosphorylation of contralateral β-subunit sites. The receptor has highest affinity for IGF-I, a lesser affinity for IGF-II (about 2- to 15-fold less), and low affinity for insulin (about ~100- to 1000-fold less) due to the absence of insulin-binding determinants in the α subunit.

Receptors for Growth Hormone and Prolactin

Growth hormone (GH) and prolactin (PRL) belong to the "helix bundle peptide" (HBP) hormone family, which includes human placental lactogen (hPL), erythropoietin (EPO), and many interleukins and cytokines. All members of this family have similar mechanisms of action; they utilize membrane receptors that are straight-chain glycoproteins with extracellular, transmembrane and intracellular domains. The membrane receptors for this group of hormones when bound with their respective receptors undergo dimerization. Each dimerized receptor is bound to one molecule of hormone. The intracellular events of the ligand-mediated activation of the receptor consists of activation of a constitutively associated tyrosine kinase known as Janus kinase 2 (Jak-2). Jak-2 kinase transphosphorylates itself and phosphorylates the intracellular domains of the receptor. Figure 30-11 illustrates the mechanism of prolactin receptor activation. The signal transduction pathways for GH and PRL are discussed in Chapter 31.

30.6 Organization of the Endocrine System

The nervous and endocrine systems function in a coordinated manner to promote growth, homeostasis, and reproductive competence. The need to rapidly adjust physiological processes in response to impending disturbances in homeostasis is met by the nervous system, while the endocrine system effects the more prolonged, fine-tuned adjustments. Together, they effect appropriate organismic adaptation. Some interdependence also exists between the two systems. The nervous system, for example, relies heavily on a continuous supply of glucose, the circulating levels of which are under multiple hormonal control, while maturation of the nervous system is regulated by thyroid hormone, and maintenance of mental acuity in adults depends on the availability of both thyroid hormone and glucocorticoids. On the other hand, hormone production is generally dependent on the nervous system, albeit to varying extents. The synthesis of insulin, glucagon, calcitonin, parathyroid hormone, and aldosterone appears to require little or no neural regulation, whereas that of the hypothalamic peptides and adrenal medullary hormones is totally dependent on it. It may be significant that those hormones which are relatively independent of neural control serve

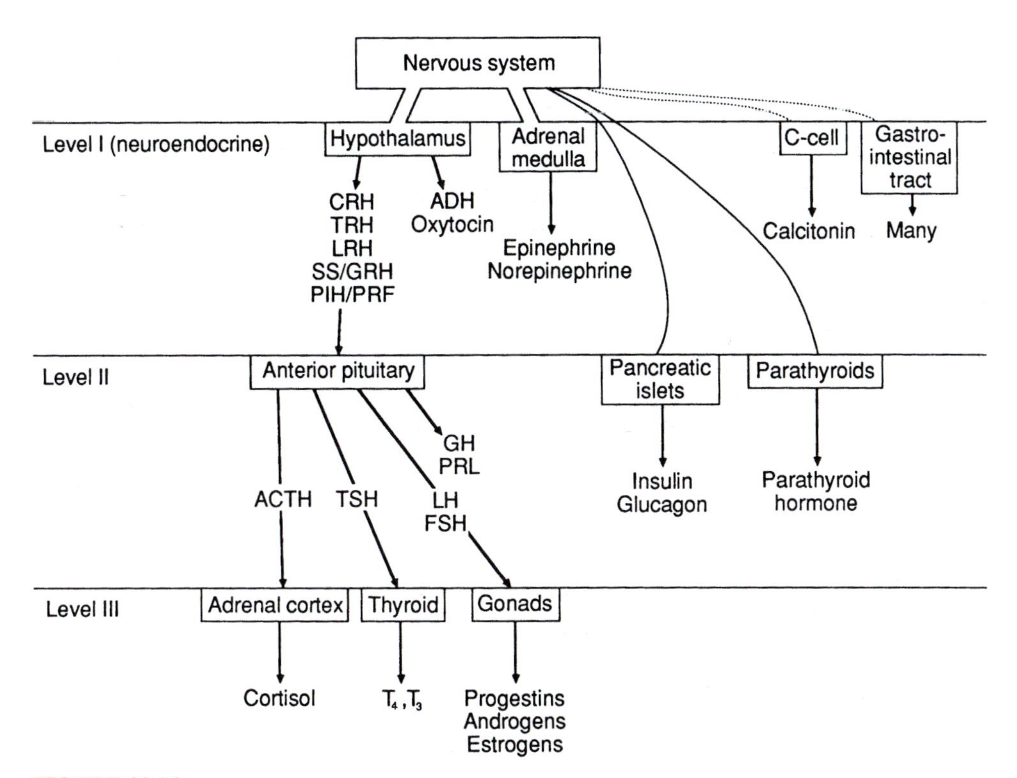

FIGURE 30-12

Organization of the endocrine system. The endocrine system can be classified into three levels. At level I are those endocrine tissues that are, or were embryologically derived from, nervous tissue; these include the hypothalamus, the adrenal medulla, the thyroid C-cell, and several of the gastrointestinal mucosal cells. Hormones produced by these tissues resemble neurotransmitters to the extent that their effects are rapid in onset and short term. At level II are those endocrine tissues that are directly or indirectly influenced by the nervous system; these include the anterior pituitary, the pancreatic islets, and the parathyroids. Hormones produced by these tissues, which are polypeptides, proteins, or glycoproteins, exert both short- and long-term effects. At level III are the adrenal cortex, the thyroid, and gonads; these endocrine tissues exhibit strong dependence on the anterior pituitary. Hormones produced by these tissues are small, hydrophobic molecules that ultimately affect gene expression in the target cells; their effects are thus slow in onset but long term. Although the level III tissues are not directly regulated by the central nervous system (CNS), they are, by virtue of their control by the anterior pituitary, indirectly regulated by the hypothalamus. Neural regulation by the CNS is critical; —, neural regulation by the CNS exists but is not critical; - - - -, little or no regulation by CNS; ⟶ , hormonal regulation.

to regulate the level of substances that profoundly affect nerve function (glucose, calcium, sodium, potassium).

The organization of the endocrine system can best be described in relation to the central nervous system. Three levels of endocrine tissues can be distinguished on the basis of their association with the central nervous system (Figure 30-12). The first level consists of those that are (or were) derived from nerve cells; these include the hypothalamus, adrenal medulla, thyroid C-cell, and gastrointestinal enterochromaffin cells. The hypothalamus and adrenal medulla still retain their neural connections and can therefore be regarded as endocrine extensions of the nervous system. The C-cell and the gut cells, however, are APUD cells and lack neural connections. These four tissues produce hormonal peptides or amines having, like neurotransmitters, rapid-onset, short-term effects.

The second level of endocrine tissues are the anterior pituitary, pancreatic islets, and parathyroids, which show varying degrees of dependence on neural regulation. Because it is highly dependent on the hypothalamus for control of hormone synthesis and release, the anterior pituitary is highly dependent on the nervous system. On the other hand, the pancreatic islets and parathyroids synthesize and release their hormones in the absence of neural signals; however, the rate of hormone release can be influenced by autonomic nerve stimulation. These three tissues produce polypeptide, protein, or glycoprotein hormones having both short- and long-term effects, often involving the regulation of cellular protein synthesis.

The third level of endocrine tissues includes the adrenal cortex, thyroid follicles, and gonads, all of which depend heavily on trophic stimulation by anterior pituitary hormones; therefore, although these peripheral endocrine tissues are anatomically and functionally separated from the central nervous system, they are indirectly dependent on the hypothalamus and, thus, on the nervous system. They produce low-molecular-weight hormones (steroids, iodinated amines) having long-term effects that involve regulation of gene expression.

The hierarchical organization of the endocrine system emphasizes the substantial influence of the nervous system on endocrine function; however, the influence of the central nervous system is not always obligatory. Thus, although C-cells and gut cells are of neuroblastic origin, they function well in the absence of neural input; similarly, the neural influence on hormone release from pancreatic islets and parathyroid glands is relatively unimportant. Nevertheless, the hypothalamus and adrenal medulla fail when their neural connections are severed, leading to atrophy of the anterior pituitary, adrenal cortex, thyroid, and gonads. However, two hormones are produced by these tissues in normal or elevated amounts following neural disconnection: prolactin from the anterior pituitary, and aldosterone from the adrenal cortex. Release of aldosterone is regulated by the renin-angiotensin system and by extracellular potassium levels, while that of prolactin normally is under tonic inhibition by the hypothalamus (Chapter 31).

A unidirectional scheme of endocrine control has been alluded to above in the discussion of three levels of endocrine tissues: the nervous system-hypothalamus controls the anterior pituitary, which in turn controls the peripheral endocrine tissues. Closer examination of how hormonal secretions are regulated reveals that, in many instances, a closed-loop feedback (servo-mechanism) circuit operates to ensure that the correct amount of hormone is released at any given time. In its simplest form, a circuit involves an endocrine cell that can monitor changes in the blood level of the substance it regulates. This sensing ability is coupled to a discriminator with an imprinted "set point" such that if the circulating level deviates significantly from the set point, the hormonal output is appropriately modified to counteract that deviation. This type of simple ***negative feedback*** regulation of hormone release—usually involving one endocrine tissue, one hormone, and one substance that is monitored—is the primary means by which secretion of insulin, glucagon, parathyroid hormone, and (to some extent) aldosterone is regulated (Figure 30-13).

Another way in which hormonal secretion is regulated involves a ***neuroendocrine reflex,*** which differs from a ***neural reflex*** in that the efferent neuron is replaced by a neuroendocrine cell (Table 30-5). As in a neural reflex arc, a sensory (afferent) neuron conveys, via synaptic connections, signals emitted from a sensory receptor to an efferent (neuroendocrine) cell, feedback which then affects the function of an effector organ or tissue. Release of oxytocin, ADH (vasopressin), and adrenal medullary hormones is regulated in this way (Table 30-5). No direct negative feedback comparable to that in the previous model operates; however, the efferent signal (hormone) counteracts, alleviates, or compensates for the stimulus that initiated the reflex. Several examples follow. By increasing water reabsorption, ADH promotes osmodilution and volume expansion, which counteract the dehydration that triggered distress signals from osmoreceptors and baroreceptors; by causing milk ejection, oxytocin helps appease the suckling infant. The adrenal medullary catecholamines produce

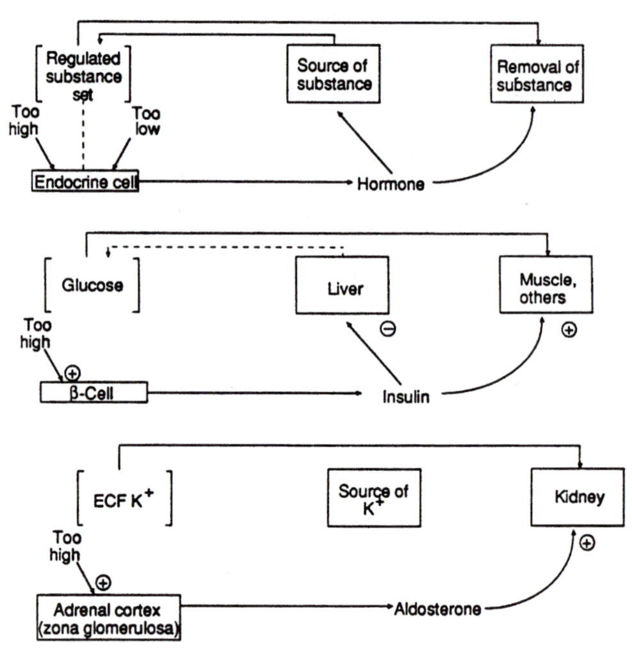

FIGURE 30-13
Simple negative feedback control of hormone release. (*Top*) A discriminator associated with an endocrine cell is capable of monitoring blood levels of a regulated substance. If the level of the substance deviates from the "set point," the endocrine cell adjusts its hormonal output to restore that level. The hormone may act either on the source of the substance or on its removal. (*Middle*) If the blood level of glucose rises above the set point imprinted in the β cell of the pancreatic islet, insulin secretion is increased. Insulin acts on the liver (source of glucose) to inhibit (⊖) further production of glucose, thereby diminishing the input of glucose into the blood glucose pool (broken line). Insulin also acts on several tissues (muscle, others) to promote (⊕) glucose utilization, thereby facilitating the removal of glucose from the blood glucose pool (solid line). (*Bottom*) Plasma K^+ concentration is monitored by the zona glomerulosa of the adrenal cortex, and an elevation will stimulate the release of aldosterone. Aldosterone acts only on the disposal of K^+, thereby serving to control the level of K^+ in blood. ECF, Extracellular fluid.

TABLE 30-5
Neuroendocrine Reflex

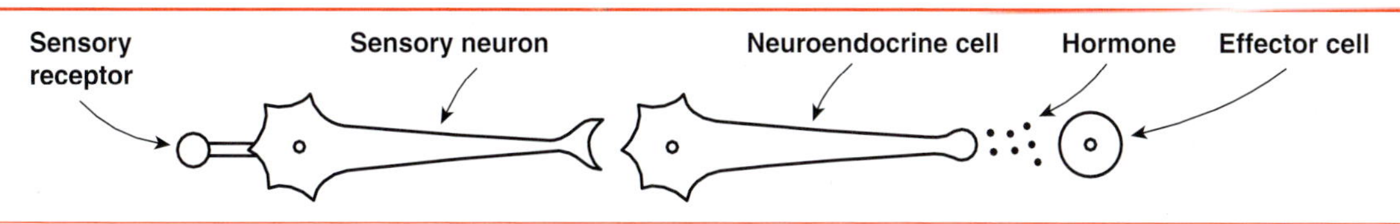

Sensory Receptor	Sensory Neuron	Neuroendocrine Cell	Hormone	Effector; Effects
Cardiovascular baroreceptors	Cranial nerves IX and X	Magnocellular neurons, hypothalamic PVN*, SON†	ADH	Kidney; water reabsorption
Osmoreceptors	? Osmoreceptor neuron	(As above)	(As above)	(As above)
Touch receptors in nipple	Ascending touch pathways	(As above)	Oxytocin	Mammary gland; milk ejection
Uterine stretch receptors	Unclear	(As above)	(As above)	Uterus; smooth muscle contraction
Various receptors of noxious stimuli	Many	Adrenal medulla	Epinephrine, norepinephrine	Cardiovascular augmentation and liberation of energy substrates

*PVN = paraventricular nucleus.
†SON = supraoptic nucleus.

cardiovascular and metabolic changes that enable the individual to cope with the stressful situation that triggered their release, while the adrenal medullary enkephalins probably serve to suppress the sensation of traumatic pain. The neuroendocrine reflex is usually polysynaptic, i.e., one or more interneurons connect the sensory neuron to the endocrine cell. Thus, the system is more amenable to modulation by other neurons or hormones, serving to fine-tune the system. For example, under conditions of normal extracellular fluid volume and osmolality (i.e., when ADH release is suppressed), the sensation of pain stimulates the release of ADH and promotes water retention. Pain-induced release of ADH may be due to neural discharge of endorphins or enkephalins in the central nervous system, which act on the hypothalamus. The peripheral hormones (level III, Figure 30-12) also influence the neuroendocrine reflex. Cortisol, for example, amplifies the adrenal medullary response to stress by increasing the excitability of interneurons in the reticular formation and by stimulating the synthesis of adrenal medullary phenylethanolamine N-methyltransferase (PNMT), which catalyzes the formation of epinephrine (Chapter 32). Estrogen is an important regulator of hypothalamic pro-oxyphysin synthesis (Chapter 31) and thus promotes the formation of oxytocin.

Some hormones are under complex ***feedback regulation*** involving both a neuroendocrine reflex and a negative feedback circuit. Invariably, the neuroendocrine reflex component involves the hypothalamus, which in one instance is also the site of positive feedback (Chapter 34). The "substance" that is monitored and exerts negative feedback may be a substrate (e.g., glucose, amino acid) or the hormone that is being regulated. Generally, two or more hormones are involved, which is not surprising, since most hypothalamic and anterior pituitary hormones subserve the function of regulating the release of another hormone (Figure 30-14). As examples, release of growth hormone is regulated by growth hormone-releasing hormone and somatostatin; that of prolactin by prolactin-releasing factor and prolactin-inhibiting factor (dopamine); that of thyroid hormone by thyrotropin-releasing hormone and thyroid-stimulating hormone; that of cortisol by corticotropin-releasing hormone and corticotropin; and that of the sex steroids by gonadotropin-releasing hormone, luteinizing hormone, and follicle-stimulating hormone.

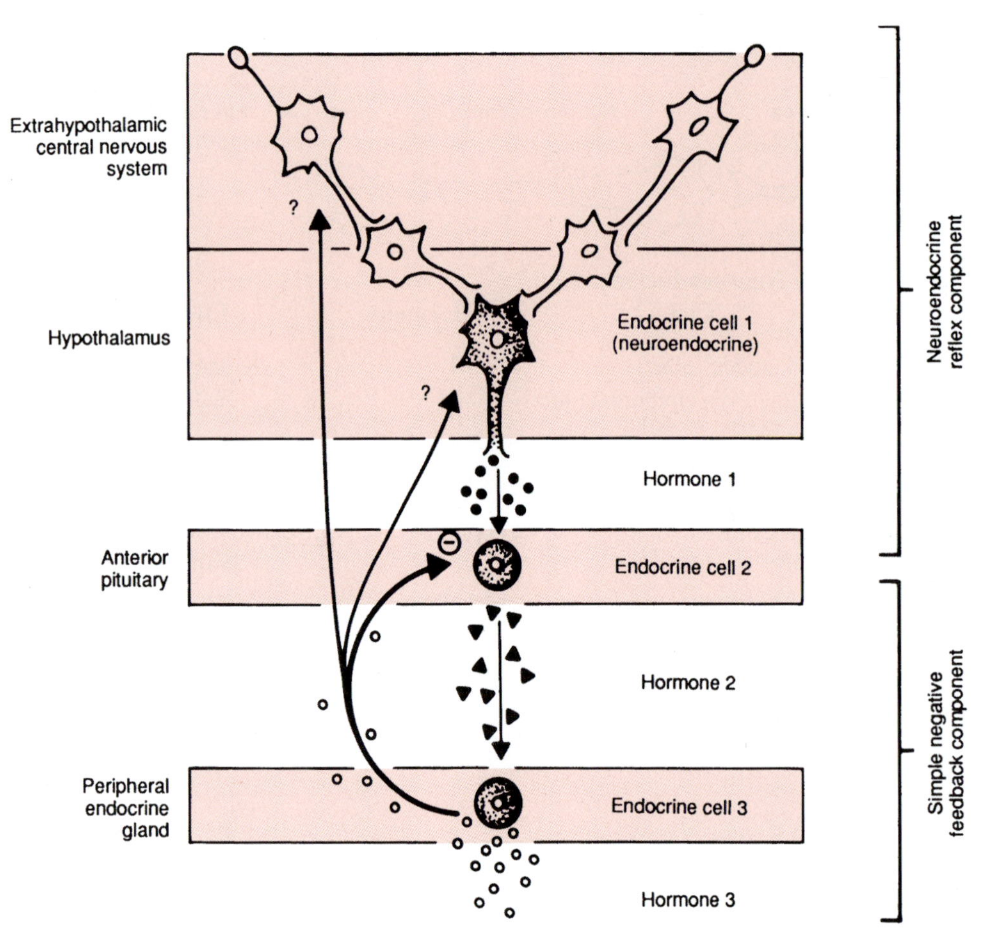

FIGURE 30-14
Complex feedback regulation of hormone secretion. Hormones of level III endocrine tissues are regulated by a complex system having both a neuroendocrine reflex component and a simple negative feedback component. Invariably, the former involves the extrahypothalamic central nervous system (sensory neurons), the hypothalamus (endocrine cell 1), and the anterior pituitary (endocrine cell 2), while the latter involves the anterior pituitary (endocrine cell 2) and a level III endocrine tissue (endocrine cell 3). Hormone 3 exerts negative feedback on endocrine cell 2 and hence on its own secretion. Hormone 3 may also feed back to endocrine cell 1 and to higher centers, although such cases of feedback may be positive ones.

Supplemental Readings and References

M. K. Bagchi, M. J. Tsai, B. W. O'Malley, and S. Y. Tsai: Analysis of the mechanism of steroid hormone receptor-dependent gene activation in cell-free systems. *Endocrine Reviews* **13,** 525 (1992).

D. M. Berman and A. G. Gilman: Mammalian RGS proteins: Barbarians at the gate. *Journal of Biological Chemistry* **273,** 1269 (1998).

D. L. Bodenner and R. W. Lash: Thyroid disease mediated by molecular defects in cell surface and nuclear receptors. *American Journal of Medicine* **105,** 524 (1998).

G. A. Brent: The molecular basis of thyroid hormone action. *New England Journal of Medicine* **331,** 847 (1994).

D. L. Cadena and G. N. Gill: Receptor Tyrosine kinases. *FASEB Journal* **6,** 2332 (1992).

M. A. Carson-Jurica, W. T. Schrader, and B. W. O'Malley: Steroid receptor family: structure and functions. *Endocrine Reviews* **11,** 201 (1990)

L. De Vries, B. Zheng, T. Fischer, et al.: The regulator of G protein signaling family. *Annual Review of pharmacology and Toxicology* **40,** 235 (2000).

Z. Farfel, H. R. Bourne, and T. Iiri: The expanding spectrum of G protein diseases. *New England Journal of Medicine* **340,** 1012 (1999).

M. E. Freeman, B. Kanyicska, A. Lerant, et al.: Prolactin: structure, function and regulation of secretion. *Physiological Reviews* **80,** 1523 (2000).

U. Gether: Uncovering molecular mechanisms involved in activation of G protein-coupled receptors. *Endocrine Reviews* **21,** 90 (2000).

C. K. Glass: Differential Recognition of target genes by nuclear receptor monomers, dimers and heterodimers. *Endocrine Reviews* **15,** 391 (1994).

M. A. Lazar: Thyroid hormone receptors: Multiple forms, multiple possibilities. *Endocrine Reviews* **14,** 184 (1993).

W. L. Miller: Molecular biology of steroid hormone synthesis. *Endocrine Reviews* **9,** 295 (1988).

M. Pfahl: Nuclear receptors/AP-1 interactions. *Endocrine Reviews* **14,** 651 (1993).

J. W. Putney, Jr. and G. S. Bird: The inositol phosphate-calcium signalling system in nonexcitable cells. *Endocrine Reviews* **14,** 610 (1993).

M. D. Ringel, W. F. Schwindinger, and M. A. Levine: Clinical implications of genetic defects in G proteins: The molecular basis of McCune-Albright

Syndrome and Albright Hereditary Osteodystrophy. *Medicine* **75,** 171 (1996).

W. Rosner: The functions of corticosteroid-binding globulin and sex hormone-binding globulin: Recent advances. *Endocrine Reviews* **11,** 80 (1990).

F. Rusnak and P. Mertz: Calcineurin: form and function. *Physiological Reviews* **80,** 1483 (2000).

E. R. Simpson, M. S. Mahendroo, G. D. Means, and others: Aromatase cytochrome P450, the enzyme responsible for estrogen biosynthesis. *Endocrine Reviews* **15,** 342 (1994).

M. Truss and M. Beato: Steroid hormone receptors: Interaction with DNA and transcription factors. *Endocrine Reviews* **14,** 459 (1993).

P. C. White and P. W. Speiser: Congenital adrenal hyperplasia due to 21-hydroxylase defeciency. *Endocrine Reviews* **21,** 245 (2000).

CHAPTER 31

Endocrine Metabolism II: Hypothalamus and Pituitary

31.1 Hypothalamus

The hypothalamus is a small region of the brain in the ventral aspect of the diencephalon. In the adult human, it is about 2.5 cm in length and weighs about 4 g. Ventromedially, it surrounds the third ventricle and is continuous with the infundibular stalk of the pituitary (hypophysis). This cone-shaped region of the hypothalamus, the ***median eminence,*** consists mainly of axonal fibers from hypothalamic neurons, which either terminate in the median eminence or continue down into the posterior lobe of the pituitary, and it is perfused by a capillary network (primary plexus) derived from the carotid arteries. Blood from the primary plexus is transported by portal vessels (hypophyseal portal vessels) to another capillary network (secondary plexus) in the anterior lobe of the pituitary (adenohypophysis) (Figure 31-1).

The hypothalamus contains a high density of nerve cell bodies clustered into nuclei or areas. Neurons in each of these nuclei tend to send their axons to the same regions in the form of tracts. These nuclei innervate the median eminence, other hypothalamic nuclei, the posterior pituitary, and various structures in the extrahypothalamic central nervous system. All of the hypothalamic neurons are presumably monoaminergic (i.e., they synthesize and release the neurotransmitter amines norepinephrine, serotonin, or dopamine); however, many are also peptidergic (i.e., they synthesize and release neuropeptides). At least 12 hypothalamic neuropeptides have been identified, and it is expected that many more will be discovered.

The hypothalamus is an important integrating area in the brain. It receives afferent signals from virtually all parts of the central nervous system (CNS) and sends efferent fibers to the median eminence, the posterior pituitary, and certain areas of the central nervous system. In the median eminence, a number of peptidergic fibers terminate in close proximity to the primary plexus, which transports their neuropeptide secretions via the portal blood flow to the anterior pituitary. Since these neuropeptides affect the function of the anterior pituitary cells, they are called "***hypophysiotropic.***" Two hypothalamic nuclei, the paraventricular nucleus (PVN) and the supraoptic nucleus (SON), consist of two populations of neurons that differ in size: the parvicellular (small-celled) neurons and the magnocellular (large-celled) neurons. The parvicellular neurons of the paraventricular nucleus produce a peptide (CRH; see below) that is transported by axoplasmic flow to the median eminence and then released into the hypophyseal portal blood. The magnocellular neurons of the paraventricular and supraoptic nuclei send their long axons directly into the posterior pituitary (neurohypophysis),

A list of expanded acronyms appears in Appendix VIII.

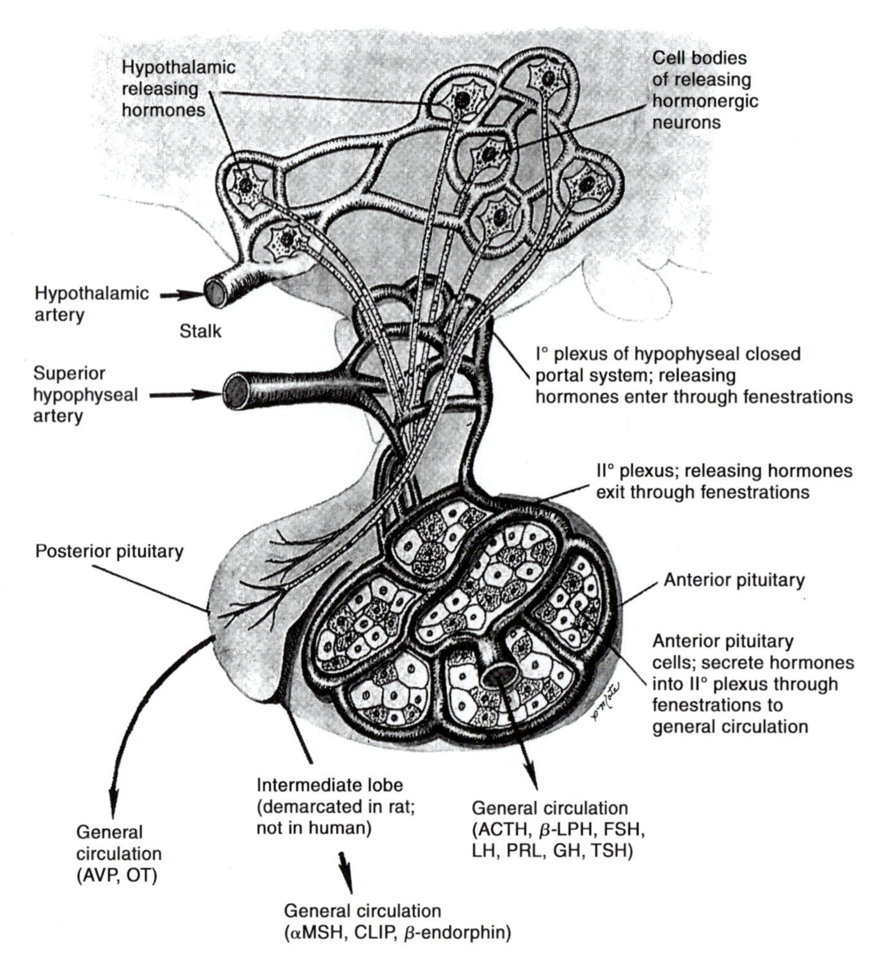

FIGURE 31-1
Hypothalamic-pituitary regulatory system. Hypothalamic releasing hormones and release-inhibiting hormones are synthesized in various neurons. These hypothalamic hormones are transported to the anterior pituitary via a portal venous system and the anterior pituitary target cells either release or inhibit the release of specific hormones into the general circulation. The posterior pituitary hormones are synthesized and packaged in cell bodies in the hypothalamus and then are transported to nerve endings in the posterior pituitary. Afterward they are released following appropriate stimuli. The explanation for the abbreviations appears in the text. [Slightly modified and reproduced with permission from A. W. Norman and G. Litwack, *Hormones,* 2nd ed. New York: Academic Press, 1997, p. 89.]

where their dilated nerve endings are closely positioned next to capillaries that drain the tissue. Peptides synthesized in these magnocellular neurons are transported to the neurohypophysis by axoplasmic flow, then released into the general circulation. These neuropeptides are called "***neurohypophyseal.***" Other hypothalamic neurons make synaptic contact with other neurons, and their neuropeptide products function as neurotransmitters or as neuromodulators. They are called "***neuroregulatory.***"

Hypophysiotropic Peptides (Table 31-1)

Thyrotropin-releasing hormone (TRH) is a tripeptide amide with the following structure:

Pyroglutamic acid — Histidine — Proline amide

All of its components are essential for activity. TRH is resistant to intestinal peptidase action and is active when taken orally. TRH is produced by the parvicellular neurons of the PVN and the periventricular nucleus, and is

TABLE 31-1

Hypophysiotropic Hormones of the Human Hypothalmus

Hormone	No. AA[2] (MW)	Gene (human)	Origin[1] (Prod. Site)	Functions	Mechanism of Action
Gonadotropin Releasing Hormone GnRH (LHRH)	10 (1,182)	8p21-q11.2	PON, OVLT	Stimulates LH and FSH secretions	$G_Q\alpha$-PLCβ-Ca^{2+}-PKC
Corticotropin Releasing Hormone (CRH)	41 (4,758)	8q13	PVN*parv*	Stimulates POMC synthesis and processing (⇑ ACTH secretion)	Gsα-AC activation
GH Releasing Hormone (GHRH)	44 (5,040)	20	TIDA (Arcuate, VMN)	Stimulates GH secretion	Gsα-AC activation
Somatostatin SS (SRIH)	14 (1,638)	3q28	PeriVN, PVN*parv*, PON	Inhibits GH secretion, also inhibits TSH secretion	Giα-AC inhibition
Thyrotropin Releasing Hormone (TRH)	3 (362)		PVN*parv*, PeriVN	Stimulates TSH secretion, also stimulates PRL release	$G_Q\alpha$-PLCβ-Ca^{2+}-PKC
Prolactin Inhibiting Hormone PIH (Dopamine)	na (153)	na	TIDA (Arcuate, VMN)	Inhibits PRL release, also indirectly inhibits LH, FSH release	Giα-AC inhibition (D_2)
Prolactin Releasing Factor (PRF)	unknown			Stimulates PRL release	

[1]PeriVN = periventricular nucleus; PVN*parv* = parvicellular paraventricular nucleus; PON = preoptic nucleus; OVLT = organum vasculosum of the lamina terminalis; VMN = ventromedial nucleus; TIDA = tuberoinfundibular dopaminergic system; [2]AA = amino acid residue; MW = molecular weight; na = not applicable.

released at the median eminence. TRH principally stimulates the synthesis and release of thyroid-stimulating hormone (TSH, thyrotropin) in the anterior pituitary, and also stimulates the release of prolactin (PRL). Both effects are mediated by membrane receptors coupled to the $G_{Q\alpha}$-phospholipase C-β-calcium-protein kinase C second-messenger system (Chapter 30).

Gonadotropin-releasing hormone (GnRH) or ***luteinizing hormone-releasing hormone*** (LHRH) is a decapeptide which, like TRH, has a pyroglutamic acid residue in its N terminus. The presence of histidine and tryptophan at positions 2 and 3, respectively, is required for biological activity, while the replacement of glycine at position 6 by D-Trp increases activity up to 100-fold. The potent synthetic GnRH agonist shown below lacks the glycinamide terminal and has instead an ethylamide residue and D-Trp at position 6:

$$\text{pGlu}^1\text{—His}^2\text{—Trp}^3\text{—Ser}^4\text{—Tyr}^5\text{—D-Trp}^6\text{—Leu}^7\text{—Arg}^8\text{—Pro}^9\text{—}\overset{\overset{\displaystyle O}{\|}}{C}\text{—NH—CH}_2\text{—CH}_3$$

GnRH-producing neurons are located in the preoptic area and in the organum vasculosum of the lamina terminalis but are known to have migrated to these areas from the olfactory placode during fetal development. Failure of these neurons to migrate into the hypothalamus during embryogenesis results in infertility secondary to GnRH deficiency (***Kallmann's syndrome***). GnRH stimulates the release of luteinizing hormone (LH) and follicle-stimulating hormone (FSH), the two gonadotropic hormones produced by the pituitary. GnRH also occurs in other regions of the brain and in the placenta. Its pivotal role in the regulation of gonadotropin release, and thus in reproductive competence, has led to the formulation of GnRH antagonists for contraceptive purposes. Paradoxically, however, synthetic GnRH agonists, when administered chronically in humans, cause a downregulation of GnRH receptors that results in suppression of gonadal steroid production. GnRH action on pituitary gonadotropes begins by occupancy of GnRH receptors by GnRH and involves intracellular mobilization of calcium ions through the action of G-protein, phospholipase C, and phosphatidylinositol pathways (Chapter 30). This suppression of gonadal function has been referred to as "medical castration."

Corticotropin-releasing hormone (CRH) is a 41-amino-acid polypeptide produced in the parvicellular neurons of the paraventricular nucleus and released into the hypophyseal portal system in the median eminence. Antidiuretic hormone (ADH, AVP) is also synthesized and released by the same neurons, but in humans, unlike the situation in rats, the amount of ADH released at the median eminence is too small to influence the function of the anterior pituitary. CRH stimulates the release of ACTH and β-endorphin by the anterior pituitary corticotrophs by stimulating the synthesis and posttranslational processing of their prohormone, pro-opiomelanocortin (POMC).

Growth hormone-releasing hormone (GHRH) is a 44-amino-acid polypeptide produced in the tuberoinfundibular dopaminergic system (TIDA) of the hypothalamus, which includes the arcuate and ventromedial nuclei. GHRH stimulates the synthesis and release of GH in anterior pituitary somatotrophs by a cAMP-mediated mechanism. The efficacy of GHRH in promoting GH release is strongly modulated by hypothalamic somatostatin.

Somatostatin(growth hormone release-inhibiting hormone) is a tetradecapeptide with an intrachain disulfide bridge. Somatostatin neurons are located in the parvicellular neurons of the PVN, the periventricular nuclei and the preoptic nuclei. Somatostatin inhibits the synthesis and release of GH from the somatotropes by neutralizing the effect of GHRH, an effect mediated by $G_{i\alpha}$ inhibition of adenylyl cyclase. Within the hypothalamus, somatostatin inhibits the release of GHRH by the same mechanism and therefore exerts an inhibitory effect at two levels of GH control. Somatostatin is also produced outside of the CNS, particularly in the D cells of the islets of Langerhans and in the gastrointestinal tract. Somatostatin produced at these extra hypothalamic sites does not normally influence GH secretion by the anterior pituitary because it is greatly diluted by systemic blood.

Prolactin-inhibiting hormone (PIH), also known as dopamine, is not a peptide (Chapter 17). It functions as a neurotransmitter in the CNS and as a precursor of norepinephrine and epinephrine in the adrenal medulla. In the hypothalamus, it originates in the TIDA and is released at the median eminence. Dopamine is a potent inhibitor of PRL release by the lactotropes (and mammosomatotropes) of the anterior pituitary, and this effect is mediated by D_2 receptors that are coupled to $G_{i\alpha}$ inhibition of adenylate cyclase. The lactotropes are unique in that they do not require stimulation by the hypothalamus to secrete PRL; in fact, blockage of the blood flow from the hypothalamus to the anterior pituitary results in elevated serum levels of PRL, due to withdrawal of dopamine. Thus, unlike somatostatin, the effectiveness of dopamine does not depend on the presence of a stimulating hormone (Chapter 34).

The existence ***prolactin-releasing factor*** (PRF), a yet-to-be-characterized stimulator of PRL release, remains controversial, particularly in view of the fact that TRH subserves this role. What is clear, however, is that the

TABLE 31-2
Neurohypophyseal Hormones

Hormone	Structure
Antidiuretic hormone (ADH) (arginine vasopressin)	$Cys^1–Tyr^2–Phe^3–Gln^4–Asn^5–Cys^6–Pro^7–Arg^8–Gly^9–NH_2$ (S–S bridge between Cys^1 and Cys^6)
Oxytocin	$Cys^1–Tyr^2–Ile^3–Gln^4–Asn^5–Cys^6–Pro^7–Leu^8–Gly^9–NH_2$ (S–S bridge between Cys^1 and Cys^6)

principal effect of the hypothalamus on PRL secretion is inhibitory. A stimulating hormone from the hypothalamus may be important in tempering the inhibition by dopamine.

Neurohypophyseal Peptides

The magnocellular neurons of the paraventricular and supraoptic nuclei synthesize antidiuretic hormone (ADH, vasopressin, or arginine vasopressin) and oxytocin, neuropeptides that are released into the general circulation in significant amounts (Table 31-2). These neuropeptides are synthesized in separate cells and transported by axoplasmic flow to nerve endings in the neurohypophysis. Cholinergic signals to the magnocellular neurons stimulate the release of ADH and oxytocin via nerve impulse propagation along the axon and Ca^{2+}-dependent exocytosis of secretory granules. Release mechanisms for ADH and oxytocin are under separate control.

ADH is a nonapeptide that contains an intrachain disulfide bridge. ADH is synthesized as part of a precursor glycoprotein (propressophysin), whose gene is located on chromosome 20. Propressophysin is packaged into secretory granules and transported to the neurohypophysis. During transit, propressophysin is split by proteases in the granule wall, yielding ADH and a "carrier" polypeptide called neurophysin II or "nicotine-sensitive neurophysin" (NSN), whose release is stimulated by nicotine. Upon neural stimulation, ADH and NSN are released in equimolar amounts. NSN has no known biological activity. In the collecting ducts of the kidney, ADH acts by a Gs-protein coupled cAMP-mediated mechanism to increase the permeability of ductal cells to water by mobilizing aquaporin proteins to the apical membrane which prevents diuresis (see also Chapter 39). At relatively high concentrations, ADH is a potent vasoconstrictor. Such concentrations are attained after massive hemorrhage, when constriction of blood vessels prevents further blood loss. Major stimuli for ADH release are an increase in plasma osmolality, monitored by hypothalamic osmoreceptors, and a decrease in blood volume, monitored by baroreceptors in the carotid sinus and in the left atrial wall. Afferent fibers to the vasomotor center in the hindbrain sense a fall in blood volume, and diminished noradrenergic signals from this center to the hypothalamic magnocellular neurons stimulate ADH release. Because norepinephrine inhibits ADH release, decrease of this inhibition constitutes a stimulation.

Oxytocin is a nonapeptide that differs from ADH at the residues in positions 3 and 8. It is synthesized in separate magnocellular neurons from a gene that codes for pro-oxyphysin, the precursor of oxytocin and its "carrier" (neurophysin I or estrogen-sensitive neurophysin). Neurons in the paraventricular and supraoptic nuclei synthesize pro-oxyphysin, which is packaged and processed during transit to the neurohypophysis. The principal action of oxytocin is ejection of milk from the lactating mammary gland ("milk let-down"), and it also participates in parturition (Chapter 34). Oxytocin is released by a neuroendocrine reflex mechanism and stimulates contraction of estrogen-conditioned smooth muscle cells. The mechanism of action of oxytocin does not involve cAMP but may involve regulation of increased intracellular Ca^{2+}. The oxytocin receptor belongs to a seven-membrane-spanning receptor family.

Neuroregulatory Peptides

Neuronal projections from the hypothalamus to other regions of the brain relay important output information that influence blood pressure, appetite, thirst, circadian rhythm, behavior, nociception (pain perception), and others factors. Although many of these neurons release neurotransmitter amines at synapses, some of them are known to release neurotransmitter peptides. These include, among others, peptides that closely resemble hormones formed in the gastrointestinal system as well as the endogenous opiates (Table 31-3).

Brain-Gut Peptides

Gut hormone-like neurotransmitters have been detected in the brain and are believed to function in a manner that

TABLE 31-3

Neuroregulatory Peptides of the Hypothalamus

Hypothalamic Peptide Neurotransmitter[1,2]	No. AA's	GI Counterpart	Functions in CNS Inferred or Postulated	Functions in Gut
Cholecystokininoctapeptide (CCK-8)*	8	CCK-33; "I" cell in duodenum and jejunum	Suppression of appetite Opiate antagonist	Gallbladder contraction; Pancreatic enzyme secretion
Gastrin*	17	Gastrin-17; "G" cells in stomach, duodenum	Appetite regulation?	Gastric acid secretion via histamine release from APUD cells in gastric mucosa (enterochromaffin-like (ECL) cells) (H_2 antagonist-sensitive)
Substance P*	11	Neural crest-derived cells, sensory C fibers	Nociception	Vasoconstriction (local effect) Mediates immune response to inflammation (?)
Vasoactive Intestinal Polypeptide (VIP)*	28	Enteric nerve endings	Probably vasomotor	Probably vasomotor
Gastrin-Releasing Peptide (GRP) (mammalian Bombesin)*	27	Stomach, duodenum	Thermoregulation Nociception	Gastrin release from G cells in stomach
Somatostatin*	14	Stomach, pancreatic islet, myenteric plexus	Suppression of GHRH release (?) Suppression of TRH release (?)	Suppresses release of glucagon, insulin, gastrin, CCK, secretin, VIP, GIP Suppression of acid production by parietal cells, pancreatic enzyme and bicarbonate release Suppression of proliferative growth in most cells
Neurotensin*	13	"N" cells of Ileum	Suppresses appetite	Not known
β-Endorphin	31	?	Suppresses avoidance/alarm behavior (μ_1) Suppresses behavioral reactiveness to pain, trauma, stresses (μ_1) Stimulates appetite (μ?)	?
Met-Enkephalin (MENK)*	5	Myenteric neurons ($A\delta$ and C fibers)	Suppresses nociception (δ)(μ_1) Substance P antagonism (δ)(μ_1) Depresses medulla respiratory center (μ_2)	Suppression of GI motility (μ_2) Stimulation of sphincter tone (μ_2)
Leu-Enkephalin (LENK)*	5	Myenteric neurons ($A\delta$ and C fibers)	Suppresses nociception (δ)(μ_1) Substance P antagonism (δ)(μ_1) Depresses medulla respiratory center (μ_2)	Suppression of GI motility (μ_2) Stimulation of sphincter tone (μ_2)
Dynorphin 1-8	8	?	Stimulates appetite (κ)? Stimulates ADH release (μ)(CNS) Inhibits ADH release (κ)(neurohypophysis)	?

[1] * = A "Brain-Gut" peptide; has a counterpart in the gastrointestinal tract.

[2] Many are also produced in other regions of the CNS. For example, the enkephalin pentapeptides are produced mainly in the periaqueductal gray (mesencephalon), the raphe nucleus (pons), and in the dorsal horn of the spinal cord, with minor amounts in the hypothalamus. In contrast, β-endorphin is produced mainly in the adenohypophysis and hypothalamus.

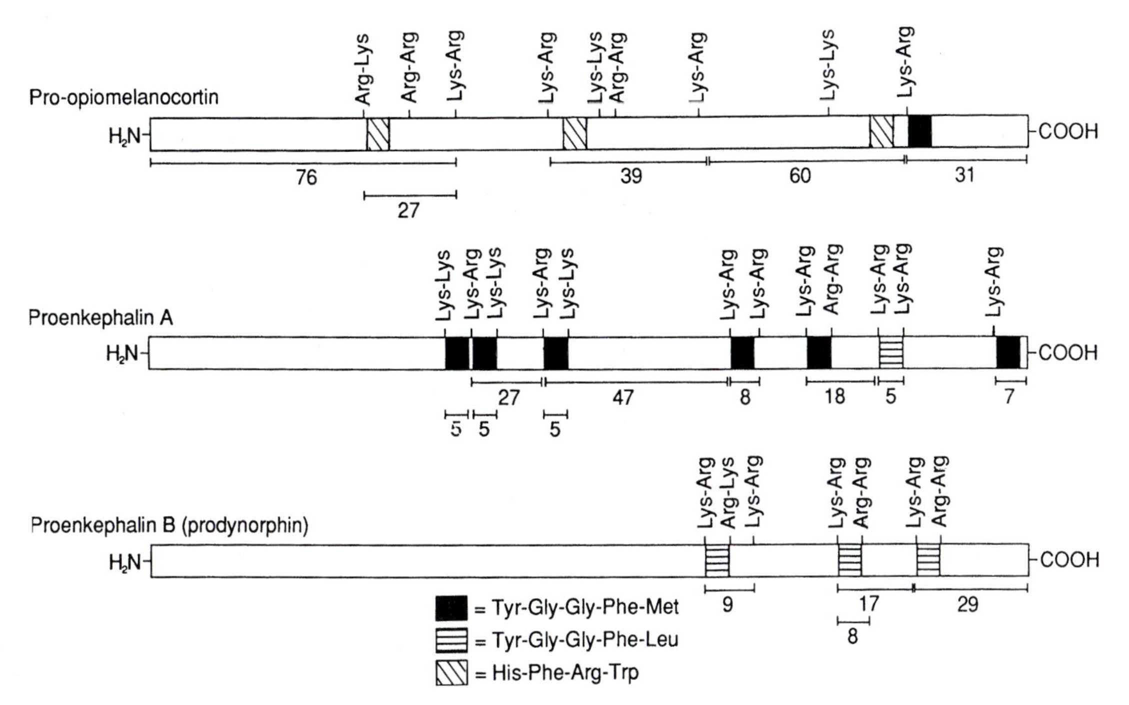

FIGURE 31-2

Dipeptide signals in the posttranslational processing of hormones. This figure presents schematic structures of three precursor molecules that contain one or more of three sequences characteristic of ■ met-enkephalin, ▤ leu-enkephalin, ▧ melanocyte-stimulating hormone. Bioactive peptides (like ACTH, endorphins, enkephalins, and dynorphins) are released by proteolytic enzymes during processing of the precursor molecules. Cleavage sites are predominantly dipeptide signals that contain Arg or Lys or both. The presence or absence of appropriate proteases determines the nature of the peptides generated from a given precursor. Numbers represent the lengths of peptides that can be generated from the precursors.

complements their counterparts in the gastrointestinal tract (see Chapter 12). For example, cholecystokinin (CCK) functions in the gut to promote digestion by acting on the gallbladder and exocrine pancreas. The gut-derived CCK is a large peptide (33 amino acid residues), the last 8 of which confer biological activity. In the brain, a smaller CCK with the same 8 carboxy terminal amino acids functions as a neurotransmitter for appetite suppression. Thus, in the brain and in the gut, CCK influences some facet of eating. Gastrin, a 17 residue hormone in the gut that stimulates gastric acid secretion and that has the same five carboxy terminal residues as CCK, has also been detected in the brain, where it is believed to have an effect on appetite.

Endogenous Opiates

Opioid neurotransmitters in the brain are peptides that modulate pain perception and/or the reaction to perceived pain; they include the enkephalins, endorphins, dynorphins, and neoendorphins. All exert their effects by binding to specific types of opiate receptors that are located in various parts of the CNS, but particularly in those regions that function in pain perception. Derivatives of opium, the extract of the poppy *Papaver somniferum,* exert their analgesic and psychological effects through these opiate receptors.

The endogenous opiates include β-endorphin, the enkephalins (met-enkephalin and leu-enkephalin), the dynorphins, and the neoendorphins. All are peptides, varying in size from 5 to 31 amino acid residues. All have in common an amino terminus consisting of either of two pentapeptide sequences: Try–Gly–Gly–Phe–Met (the met-enkephalin sequence) or Tyr–Gly–Gly–Phe–Leu (the leu-enkephalin sequence). The fundamental endogenous opioid peptides are the pentapeptides, met-enkephalin and leu-enkephalin which function as neurotransmitters in the CNS.

All of the known endogenous opioids are derived from three different prohormones: pro-opiomelanocortin (POMC), proenkephalin A, and proenkephalin B (prodynorphin) (Figure 31-2). The genes encoding these are expressed in the CNS and in other regions of the body (e.g., POMC is also produced in the adrenohypophysis and the gastrointestinal tract; proenkephalin A is also produced in the adrenal medulla). What is important is that sites outside of the CNS that produce the prohormones may not possess all of the posttranslational enzymes present in the brain and thus may be unable to generate the same opiate peptides. For example, proenkephalin A in the brain is processed to produce the pentapeptides met-enkephalin (MENK) and leu-enkephalin (LENK); in the adrenal medulla, however, larger fragments of the prohormone are produced. Another example is brain-derived POMC which is processed to form melanocyte-stimulating hormone (MSH) fragments, whereas adrenohypophyseal POMC is processed to form large fragments containing the MSH sequence (e.g., ACTH, γ-LPH).

At least three types of opiate receptors have been identified. The μ receptor exhibits greatest affinity for alkaloid morphine but also recognizes β-endorphin. There are two subtypes of the μ receptor: μ_1, which mediates the analgesic effects of opioids, and μ_2, which mediates the respiratory depression and gastrointestinal effects of opioids. The δ receptor exhibits greatest affinity for the pentapeptides MENK and LENK, while the κ receptor has high affinity for the dynorphins but does not recognize the enkephalins or endorphins. The opiate antagonists naloxone and naltrexone bind to μ and δ receptors, but not to the κ receptor. Although the endogenous opioid peptides were discovered more than 25 years ago, their physiological functions are still not completely understood. The biological importance of the endogenous opioids is primarily attributed to their documented analgesic effects. Opium and its derivatives (morphine, codeine, etc.) have long been used to reduce the perception of pain. They are known to amplify the existing analgesia (pain-modulating) system of the body, which the endogenous opioid peptides and their receptors normally operate to modulate the sensation of pain. The endogenous analgesia system consists of enkephalinergic neurons that inhibit the transmission of sensory pain signals at multiple levels of the ascending pain pathway. These neurons originate at three regions of the CNS: the periaqueductal gray area surrounding the aqueduct of Sylvius in the mesencephalon; the raphe magnus nucleus in the pons; and the dorsal horn of the spinal cord. Pain impulses transmitted by either Ad (acute) or C (slow) fibers can be blocked upon entry in the dorsal horn and as they ascend through the brain stem. Both μ_1 and δ opioid receptor sites are present throughout the analgesia system and mediate the inhibitory effects of enkephalins on pain signal transmission.

The cause of death in most cases of opioid overdosage is respiratory depression, i.e., an effect of opioids on the respiratory center of the brainstem that inhibits chemoreceptor responsiveness to CO_2. The medullary ventilatory control center is said to have a high density of μ_2 opioid receptors that mediate the depressing effect of opioids on respiration.

Opioids increase appetite and stimulate food intake, whereas opiate antagonists inhibit food intake in food-deprived animals. Although circulating levels of β-endorphin were found to be higher in obese than in lean women, there is no evidence that obesity is caused by an imbalance in endorphin production. However, a relationship between early-onset obesity and inherited defects in the *POMC* gene that leads to a lack of production of ACTH, MSH (α, β, γ), and β-endorphin has been observed in humans (discussed later).

The opioids probably do not play an important role in the regulation of neuroendocrine function because blockade of opioid receptor sites has either minor or no effect on the basal or stimulated release of most pituitary hormones. However, there is evidence that central opioids may tonically inhibit the rate of hypothalamic GnRH release after puberty in males.

31.2 Pituitary Gland (Hypophysis)

The pituitary is a small, bilobed gland connected to the base of the hypothalamus by the infundibular process (pituitary stalk). Embryologically it derives from Rathke's pouch (buccal epithelium) and the infundibulum (neuroectoderm). The former gives rise to the anterior lobe (anterior pituitary or adenohypophysis). The latter is a projection of the hypothalamus and gives rise to the posterior lobe (posterior pituitary or neurohypophysis). In many species, an "intermediate lobe" also exists, but in humans this is rudimentary and apparently nonfunctional. The pituitary in an average-sized adult weighs only about 0.5 g, 75% of which is anterior pituitary.

The anterior pituitary is not innervated by nerve fibers from the hypothalamus but is well vascularized by the portal blood that drains the median eminence. The portal blood flows primarily from the median eminence to the anterior pituitary; however, some vessels may transport blood in the opposite direction (retrograde flow). The posterior pituitary consists mainly of nerve endings of hypothalamic magnocellular neurons and contains no portal connections with the hypothalamus; its vascular connections are largely independent of those in the anterior pituitary. The hormones of the posterior pituitary (neurohypophyseal hormones) were discussed above.

The anterior pituitary has five endocrine cell types, each of which produces different hormones. In the human, at least seven are produced: growth hormone (GH), prolactin

TABLE 31-4
Somatomammotropin Family

Hormone	Site of Secretion	Chemistry	Location of Gene	Function	Regulation of Secretin
Growth Hormone (GH)	Anterior Pituitary (Somatotroph)	191 aa (22 kDa)	17q22-q24	Stimulates protein synthesis in many tissues; exerts protein-sparing effects; stimulates production of IGFs.	Primarily stimulated by GHRH and inhibited by somatostatin.
Prolactin (PRL)	Anterior Pituitary (Lactotroph)	198 aa (23 kDa)	6p22.2-q21.3	Stimulates growth and protein synthesis in breast; promotes milk secretion during lactation.	Primarily inhibited by PIH (dopamine); stimulated by PRF and TRH.
Placental Lactogen (hPL)	Placenta (Syncytiotrophoblast)	191 aa (22 kDa)	17q22-q24	Exerts protein-sparing effect in maternal tissues, promotes fetal growth by stimulating production of fetal IGF-1.	Progesterone-stimulated placental growth.

(PRL), luteinizing hormone (LH), follicle-stimulating hormone (FSH), thyroid-stimulating hormone (TSH), adrenocorticotropic hormone (ACTH, corticotropin), and β-endorphin. β-Lipotrophic hormone (β-LPH) is also secreted but serves mainly as a precursor of β-endorphin and is not regarded as a hormone. In species that possess a prominent intermediate lobe, the pituitary also secretes α- and β-melanocyte-stimulating hormones (MSH), which affect skin coloration.

Anterior pituitary hormones are classified into three families: the somatomammotropin family (GH and PRL), the glycoprotein hormones (LH, FSH, and TSH), and the opiomelanocortin family (ACTH, β-endorphin, and related peptides). These three families appear to have evolved from three separate ancestral polypeptides; homologous members of each family occur in other parts of the body. For example, human placental lactogen (hPL) is somatomammotropic, human chorionic gonadotrophin (hCG) is a glycoprotein hormone, and the brain and gut produce substances related to endorphins (enkephalins and dynorphin).

Somatomammotropin Family

Members of the somatomammotropin family are single-chain proteins with two or three intrachain disulfide bridges; their molecular weight is about 20,000, and they function mainly to promote protein synthesis (Table 31-4). The hypothalamus produces both a stimulating and an inhibiting hormone to regulate secretion of the adenohypophyseal hormone.

Growth Hormone (GH Somatotropin)

Human growth hormone consists of 191 amino acid residues (M.W. 22,000) and contains two disulfide bridges (Figure 31-3). It exhibits extensive sequence homology with prolactin (76%) and placental lactogen (94%) (Figure 31-4). The gene encoding GH is on chromosome 17 q22-q24, and its expression is regulated by a transcription factor (Pit-1/GHF-1) that is promoted by hormones that stimulate GH synthesis (e.g., GHRH, glucocorticoids, thyroid hormone). GH is the most abundant hormone in the human pituitary gland, averaging about 6 mg per adult gland. Its basal blood level averages 2 ng/mL in adults and 6 ng/mL in preadolescent and adolescent boys. GH is cleared from circulation with a half-life of about 25 minutes in lean adults, but is cleared more rapidly in obese subjects. GH is inactivated mainly by the liver but also by the kidney. About 40% of the hormone is bound to "GH-binding protein" (GHBP), a fragment of the GH receptor, which serves to prolong the half-life of GH 10-fold. Its primary structure is species-specific and antigenic (e.g., rat GH is antigenic

FIGURE 31-3
Structure of human growth hormone. [Reproduced with permission from R. K. Murray, D. K. Granner, P. A. Mayes, and V. W. Rodwell, *Harper's Biochemistry*, 21st ed. (Norwalk: Appleton & Lange, 1988). © 1988 Appleton & Lange.]

in humans, and vice versa). GH from nonprimate species does not promote growth in monkeys and humans; however, human GH promotes growth in subprimate species.

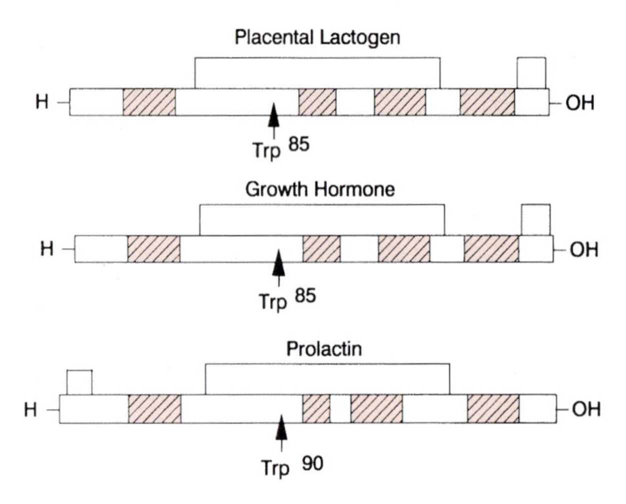

FIGURE 31-4
Structural homology in the somatomammotropin family. Diagram of basic structure of human chorionic somatomammotropin, human growth hormone, and ovine prolactin. Bars represent the peptide chains with the amino terminus at left and the carboxyl terminus at right. Shaded portions of each bar represent the areas of homology. Lines above the bars represent disulfide bridges. [Reproduced with permission from H. D. Niall, M. L. Hogan, et al.: Sequence of the pituitary and placental lactogenic and growth hormones: evolution from a primordial peptide by gene reduplication. *Proc. Natl. Acad. Sci. U.S.A.* **68,** 866 (1971).]

Actions of GH

The actions of GH are initiated by its binding to the membrane receptor of target cells. The GH and PRL receptors are single-membrane-bound proteins that belong to class I of the cytokine receptor superfamily (Chapter 30). Each of the receptor contains an extracellular, transmembrane, and intracellular domain. Receptor activation begins when one molecule of growth hormone (with its two binding sites) binds to the first receptor, followed by binding to the second receptor resulting in receptor dimerization. The hormone-receptor complex activates a tyrosine kinase known as Janus kinase 2 (Jak-2) that is associated with the proximal region of the intracellular domains of the dimerized receptor (see Figure 30-11). Two Jak-2 kinases transphosphorylate each other and also phosphorylate tyrosine residues of the receptor. Subsequent downstream single transduction pathways are complex and may involve activation of transcription proteins, mitogen-activated protein (MAP) kinase cascade, protein kinase C activation, and phosphatidylinositol and Ca^{2+} pathway activation.

GH promotes transport and incorporation of amino acids in skeletal muscle, cardiac muscle, adipose tissue, and liver and is responsible for the proportionate growth of visceral organs and lean body mass during puberty. Linear growth during childhood requires the presence of GH. GH acts directly on cartilage tissue to promote the endochondral growth that results in skeletal growth; however although GH has a direct effect on chondrocyte stem cells, the growth-promoting effect of GH is due to its stimulation of the chondrocytes to produce insulin-like growth factor I (IGF-I, discussed later), which then acts locally to stimulate cellular replication in the distal proliferative zone of the epiphyseal plate. Thus, the growth-promoting effect of GH can be abolished by blocking the IGF-I receptor and can be duplicated by exogenous IGF-I treatment in the absence of GH. The importance of GH is in its ability to stimulate IGF-I production within bone cartilage, an ability that is unique to GH. This explains why GH deficiency results in growth retardation despite the fact that bone cartilage has the ability to produce IGF-I. GH exerts a "protein-sparing" effect by mobilizing the body's energy substrates, such as glucose, free fatty acids, and ketone bodies, in the same tissues in which it stimulates protein synthesis. GH inhibits glucose uptake by skeletal muscle by inhibiting hexokinase activity and by desensitizing the tissue to the actions of insulin; the effect is to elevate the blood glucose level. GH promotes lipolysis in adipocytes, possibly by increasing the synthesis of hormone-sensitive lipase (HSL), and ketogenesis in the liver. In addition, GH increases the activity of hepatic glucose-6-phosphatase, increasing glucose secretion. These protein-sparing effects of GH are diabetogenic and explain how GH functions as an insulin antagonist (Chapter 22).

Regulation of GH Release

The protein-anabolic and protein-sparing actions of GH require the metabolic effects of insulin, glucagon, cortisol, and thyroid hormone in the unstressed individual. These actions depend on fine-tuned control of GH release, which is achieved mainly by substrate feedback to the hypothalamus (Figure 31-5).

A fall in blood glucose level stimulates GH release, whereas a rise inhibits GH release. GH release also occurs in response to certain amino acids, the most potent being L-arginine, but with a latency period of about 30 minutes. Circulating hormones exert their effects at the level of the hypothalamus or the pituitary. GH influences its own secretion by way of IGFs that exert a negative feedback effect at the median eminence. However, whether this feedback involves a decrease in GHRH, an increase in somatostatin, or both is not known. Other hormones promote the synthesis and release of GH at the level of the anterior pituitary.

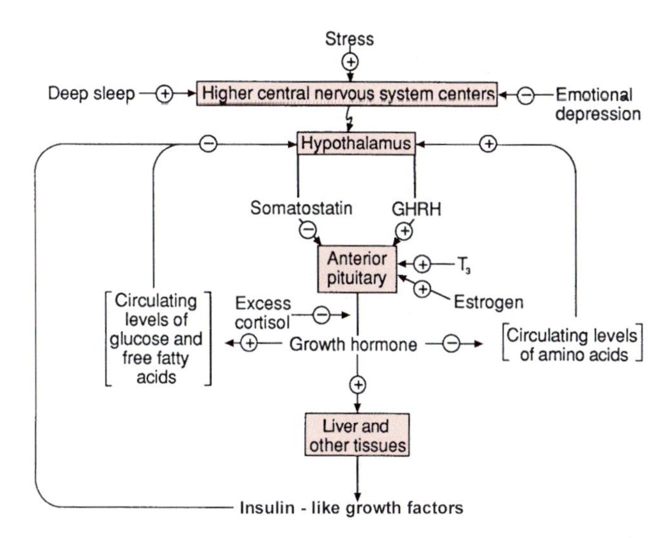

FIGURE 31-5
Regulation of growth hormone secretion in humans.

Estrogen promotes an increase in somatotroph numbers and GH mRNA levels, while androgens increase, and IGF-I decreases, the somatotroph response to GHRH. Glucocorticoids and thyroid hormone act in concert to stimulate GH gene expression; however, pharmacological concentrations of glucocorticoids strongly inhibit GH release in response to GHRH (Figure 31-5).

Superimposed on this fine regulation of GH secretion by substrates and hormones is the coarse regulation by higher brain centers that operate on an open-loop, substrate-independent basis. These influences are dramatic and result in the several-fold increase in GH levels during stress and during deep sleep. GH is one of several hormones released in response to stress (see below); it is also one of the few released during deep sleep (EEG stages III and IV) irrespective of the time of day. The amount of GH released during deep sleep is substantial, accounting for about 75% of the daily output of GH. In children who are unable to achieve deep sleep because of emotional disturbances, the absence of sleep-induced GH results in growth retardation.

Insulin-like Growth Factors (IGFs)

Insulin-like growth factors (IGFs) are GH-dependent polypeptide hormones that promote cell replication in most mesenchymally derived tissues and are responsible for the growth-promoting effects of GH. There are two IGFs, I and II, both of which are 7 kDa proteins that resemble proinsulin in structure. The IGFs exert insulin-like biological effects when tested in insulin bioassay systems *in vitro,* and account for the "nonsuppressible insulin-like activity" (NSILA) in plasma that had been described before the discovery of the IGFs. The similarities and

TABLE 31-5
Insulin and Insulin-like Growth Factors

	Insulin	IGF-I	IGF-II
MW	5,734	7,649	7,471
Site of production	β-cell, pancreatic islet	Many sites	Many sites
Major source of circulating form	β-cell, pancreatic islet	Liver	Liver
Regulation of secretion	Glucose, amino acids, catecholamines	GH, insulin, nutritional status	GH
GH-dependence	-0-	++++	+
Production rate	2 mg/day	10 mg/day	13 mg/day
Plasma levels (adults)	< 5 ng/ml	200 ng/ml	700 ng/ml
Characteristics in circulation	Pulsatile Variable	Nonpulsatile Steady, nonvariable	Nonpulsatile Steady, nonvariable
Binding proteins in serum	None	3 (types 1,3,4)	6 (types 1–6)
Half-life	10 min	12–15 h	15 h
Target tissues	Liver, muscles, adipose, skin, connective tissue, bone	Muscles, connective tissue, cartilage, bone	Muscles, connective tissue, cartilage, bone
Receptor type	Plasma membrane heterotetramer ($\alpha 2\beta 2$) Ins > IGF-I_{100}	Plasma membrane heterotetramer ($\alpha 2\beta 2$) IGF-I > II_{10} >> Ins_{1000}	IGF-I receptor; plasma membrane IGF-II/Man-6-P receptor (monomeric) IGF-II >> I_{500}
Regulation of receptor density	Ins inhib GH inhib	IGF-I inhib Starvation inhib	IGF-I inhib (IGF-IR); Starvation inhib (IGF-IR); Insulin, IGF-I stim (IGF-II R)

differences of IGFs with insulin and among themselves are summarized in Table 31-5.

Six serum proteins produced mainly in the liver that bind IGFs in the circulation have been identified; these are designated IGF-binding proteins (IGFBPs) 1 through 6. IGFBP-1 appears to retard target tissue uptake of IGF-I, while IGFBP-3 appears to enhance it. The latter accounts for most of the IGF found in blood. About 80–90% of the IGF-I in circulation is bound to IGFBP-3, along with an 88-kDa acid-labile subunit (ALS); this complex confers protection on the IGF and prolongs its half-life to 12–15 hours. About 5% of the IGF is unbound, and the remaining 5–15% is bound to IGFBP-1, 2, or 4 as smaller complexes.

Unlike insulin and other peptide hormones that are released from storage granules, IGFs are released as they are produced. Most tissues produce IGFs in small amounts that are sufficient for local (paracrine/autocrine) effects and do not contribute significantly to the circulating pool of IGFs. The highest concentration of IGF-I and II is found in circulation and is primarily derived from the liver. GH and insulin are positive regulators of IGF-I synthesis in the nonfasting state. During mild starvation when GH levels increase, there is a decline in hepatic production of IGF-I due, in part, to the decline in insulin. Refeeding, which causes insulin to rise and GH to fall, promotes IGF-I production only if the diet is adequate in caloric and protein content. Note that circulating IGF-I does not mediate the growth-promoting effect of GH but is an important feedback regulator of GH. The indispensability of IGF-I for linear growth and the fact that it mediates this effect of GH are firmly established. IGF-I, locally produced in endochondral bone in response to GH, promotes skeletal growth by stimulating clonal chondrocyte expansion of the distal proliferative zone of the epiphyseal plate.

Under physiological conditions, IGF-I is a relatively minor regulator of fuel homeostasis. It does not mediate the effects of GH on intermediary metabolism; in fact, many of the effects of IGF-I resemble those of insulin, not GH. IGF-I promotes production and actions of erythropoietin and is responsible for the increased packed volume of blood that results from elevated GH levels.

Disturbances in GH and IGF

Only rarely has a condition of IGF deficiency or excess been described that is not accompanied by a disturbance in

FIGURE 31-6
Structure of human prolactin. [Reproduced with permission from R. K. Murray, D. K. Granner, P. A. Mayes, and V. W. Rodwell, *Harper's Biochemistry*, 21st ed. Norwalk: Appleton & Lange, 1988. ©1988 Appleton & Lange.]

GH production or responsiveness. Thus, a GH deficiency results in IGF-I deficiency, whereas a GH excess results in IGF-I excess. GH deficiency in children results in reduced growth velocity and growth retardation (dwarfism) due to secondary IGF-I deficiency, whereas in adults there are no dramatic signs or symptoms. Adults who are deficient primarily in GH have decreased lean body mass, increased adiposity, and are at increased risk of cardiovascular disease.

GH hypersecretion in children causes increased growth rate and can result in ***gigantism.*** However, GH excess occurs infrequently in childhood, and most frequently in middle-aged adulthood, which leads to ***acromegaly*** (*acral* = extremities + *megas* = large), a condition in which the cartilaginous tissues proliferate, resulting in distorted overgrowth of the hands, feet, mandibles, nose, brow, and cheek bones. Because epiphyseal cartilage is absent in long bones, there is no gain in height. Acromegaly promotes insulin resistance and cardiovascular complications. Resistance to the effects of GH results in a condition similar to GH deficiency. The major cause of GH resistance is a genetic defect in the growth hormone receptor (GHR) and the resultant condition is known as ***Laron-type dwarfism.*** This is an autosomal recessive disorder characterized by normal to high levels of serum growth hormone and low levels of IGF-I. Analysis of the GHR gene in Laron-type dwarfism has shown point mutations, deletions, and splicing defects.

Growth hormone deficiency in children can be treated by administration of recombinant GH or, in some cases, growth hormone-releasing hormone. Potential targets for recombinant GH that are undergoing clinical investigation include children with idiopathic short stature, persons with wasting syndrome associated with human immunodeficiency virus infection, critically ill patients, and elderly individuals.

Excess production of GH due to somatotroph adenomas may be treated by surgical resection, irradiation, or in some cases with somatostatin analogues (e.g., octreotide) that suppress GH secretion, or by a combination of these above.

Prolactin

Human prolactin (PRL) contains 199 amino acid residues (M.W. 23,500) and three intramolecular disulfide bridges (Figure 31-6). In healthy adults the anterior pituitary releases very little PRL under nonstressed conditions, primarily because PRL release is under hypothalamic inhibition. This inhibition is exerted by dopamine or PIH. The hypothalamus also secretes PRF but the major regulator appears to be PIH.

TABLE 31-6
Opiomelanocortin Family

Hormone	Site of Secretion	Peptide Length	Location of Gene	Function	Regulation of Secretion
Adrenocorticotropic Hormone (ACTH) (Corticotropin)	Anterior Pituitary (Corticotroph)	39 aa	2p25	Stimulates steroidogenesis in all three zones of the adrenal cortex, thereby increasing secretion of cortisol, DHEAS, and aldosterone.	Stimulated by CRH and inhibited by cortisol.
β-Endorphin	Anterior Pituitary (Corticotroph)	31 aa	2p25	Possible role in stress analgesia.	Stimulated by CRH and inhibited by cortisol.

The biological actions of prolactin are initiated by its membrane receptor, followed by the associated intracellular signal transduction pathways. These actions are similar to growth hormone (discussed previously; also see Chapter 30). In humans, the function of PRL may be restricted to promotion of lactation, but there is some evidence that PRL suppresses gonadal function in females. There is no consensus regarding a physiological role for PRL in males. During late pregnancy, the maternal pituitary releases increasing amounts of PRL in response to rising levels of estrogen, a stimulator of PRL synthesis. Elevated levels of PRL stimulate milk production in the mammary gland (Chapter 34). After parturition, PRL promotes milk secretion via a neuroendocrine reflex that involves sensory receptors in the nipples (Chapter 34). In mammary tissue, it binds to alveolar cells and stimulates the synthesis of milk-specific proteins (casein, lactalbumin, and lactoglobulin) by increasing production of their respective mRNAs. Accordingly, there is a lag of several hours before this effect of PRL is seen. In the liver, PRL stimulates the synthesis of its own receptors. PRL receptors occur in the mammary gland, liver, gonads, uterus, prostate, adrenals, and kidney.

Disturbances in Prolactin

In women, ***hyperprolactinemia*** is often associated with amenorrhea, a condition that resembles the physiological situation during lactation (lactational amenorrhea). Excess PRL may inhibit menstrual cyclicity directly by a suppressive effect on the ovary or indirectly by decreasing the release of GnRH (Chapter 34). In men, hyperprolactinemia is not associated with altered testicular function but is often attended by diminished libido and impotence. This finding suggests that PRL may serve an important role in regulating certain behavior patterns. *Prolactinomas* are the most common hormone-secreting pituitary adenomas and, in addition to the above-mentioned clinical characteristics, the patient may exhibit visual field defects. Prolactinoma may be treated by surgery, radiotherapy, or pharmacotherapy. This last method consists of using a dopamine D_2 receptor agonist such as bromocriptine or cabergoline to suppress prolactin secretion.

The Opiomelanocortin Family

All members of this family are derived from a single prohormone, pro-opiomelanocortin (POMC). The prohormone molecule contains three special peptide sequences: enkephalin (opioid), melanocyte-stimulating hormone (MSH), and corticotropin (ACTH). These result from posttranslational cleavage of the prohormone but do not share the same biological actions (Table 31-6). In humans, the gene encoding POMC is normally expressed in the adenohypophysis, hypothalamus, and brain, as well as in other sites. It may also be expressed as a result of neoplastic transformation, notably in the lung. POMC is a large glycoprotein that contains three MSH sequences and one MENK sequence. Several paired basic amino acid residues exist along the molecule and are sites of potential cleavage by proteases during posttranslational processing (Figure 31-7). In humans, the important products of POMC that are secreted by the adenohypophysis are ACTH and β-endorphin, although other products (e.g., γ3-MSH, γ-LPH) may also be significant (see later).

After translation POMC is processed by proteases contained in tissue . In the adenohypophyseal corticotrophs there is no further processing of ACTH and β-endorphin , whereas in the intermediate lobe of other species, ACTH can be processed to form α-MSH, and γ-LPH can be processed to form β-MSH. The human hypothalamus

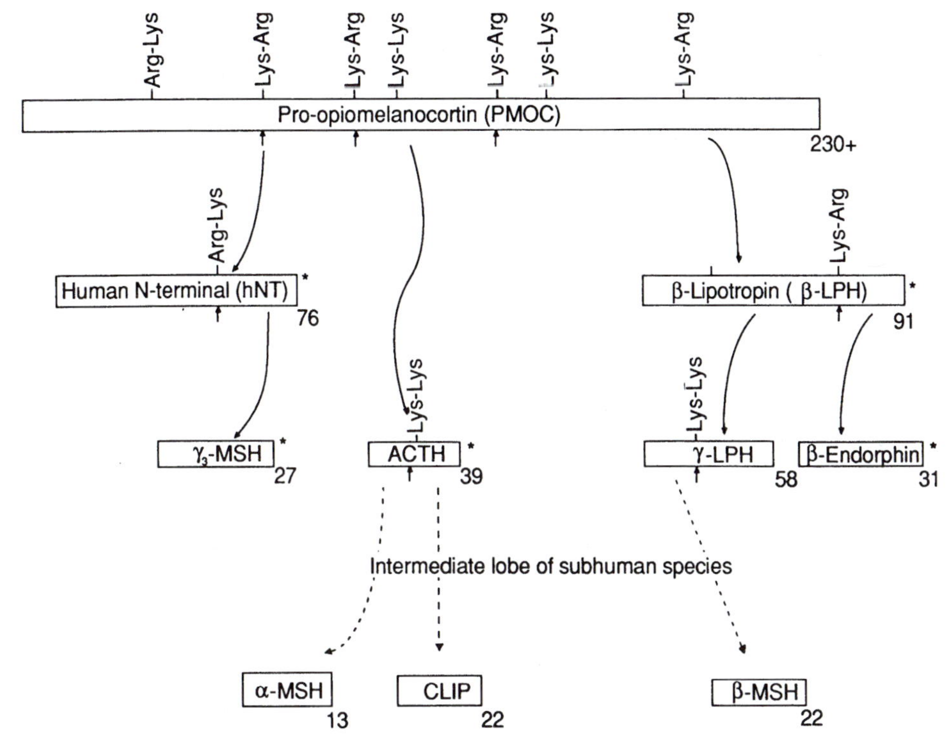

FIGURE 31-7
Posttranslational processing of pituitary pro-opiomelanocortin (pmoc). * , Released into blood.

and/or brain may be capable of forming α- and β-MSH from POMC, since these peptides appear to exert central effects.

Inherited defects in POMC production result in deficiency or a complete lack of secretion of ACTH, MSH (α-, β-, γ-), and β-endorphin. These patients exhibit adrenal insufficiency, red hair pigmentation, and early-onset obesity that has been related to deficiency of α-MSH production. Animal studies have shown that α-MSH regulates food intake by activation of melanocortin receptor-4 (Chapters 5 and 22). In mioe, a *POMC* gene defect also leads to adrenal deficiency, altered pigmentation, and obesity with loss of a significant portion of the patient's excess weight when he or she is treated with a stable α-MSH preparation. Other genetic causes of obesity include defects in genes encoding prohormone convertase-1 (i.e., inability to convert POMC to various peptides), leptin, and leptin receptor (Chapters 5 and 22).

ACTH (Table 31-6) is a polypeptide of 39 residues, the first 24 of which are required for corticotropic activity and do not vary among species. $ACTH_{1-24}$ has been synthesized and is used for diagnostic purposes. Because it contains the MSH sequence in residues 6–9 (His–Phe–Arg–Trp), ACTH has intrinsic melanocyte-stimulating activity; it can thus cause skin darkening if present in high concentrations. In normal adults, the pituitary contains about 0.25 mg of ACTH, and the basal level of ACTH in blood is about 50 pg/mL.

ACTH acts mainly on the cells of the zona fasciculata of the adrenal cortex to stimulate the synthesis and release of cortisol (Chapter 32). It also stimulates the secretion of adrenal androgens from the zona reticularis. Binding of ACTH to receptors activates formation of cAMP, which mediates cortisol formation and secretion and protein synthesis. Deficiency of ACTH leads to reduction in size and activity of adrenocortical cells in the inner two zones.

ACTH secretion is regulated so as to ensure constant, adequate levels of cortisol in blood. Corticotrophs in the pituitary are under tonic stimulation by CRH and are modulated by the negative feedback effect of blood cortisol, which inhibits POMC synthesis by repression of the *POMC* gene. Although cortisol also exerts an effect on the hypothalamus and related areas in the brain (limbic system, reticular formation), this activity may involve the regulation of emotion and behavior and not of CRH release. The regulation of the ACTH and cortisol is illustrated in Figure 31-8. Secretion of POMC in the intermediate lobe is not regulated by CRH and glucocorticoids, since the intermediate lobe is poorly vascularized and contains no glucocorticoid receptors. However, the intermediate lobe is rich in dopaminergic fibers, so that dopamine agonists (ergocryptine) decrease,

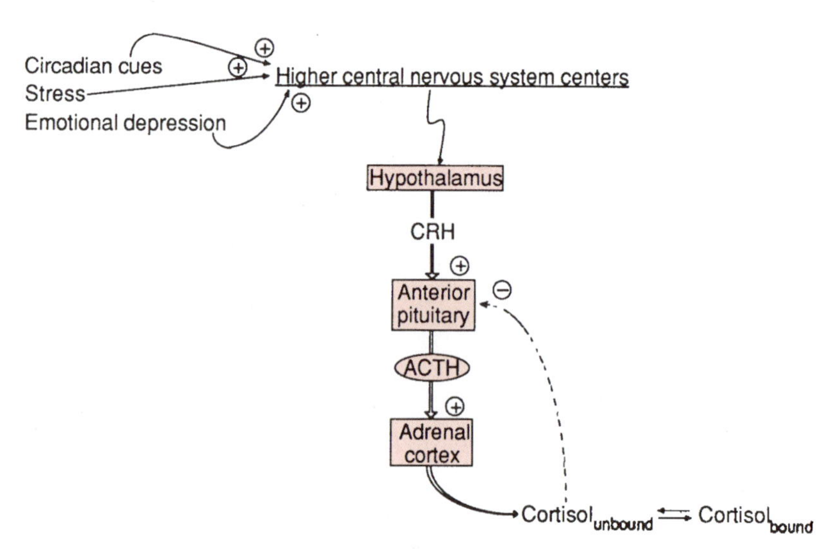

FIGURE 31-8
Regulation of ACTH secretion in humans.

and antagonists (haloperidol) increase, the synthesis and release of POMC-derived peptides.

β-Endorphin is a 31-amino-acid polypeptide released together with ACTH. When introduced into the third ventricle of the brain, it produces dramatic behavioral changes, but when injected systemically, it does not. Thus, the function of circulating β-endorphin remains unclear. The CNS and gastrointestinal effects of the hormone are probably produced by that secreted locally. Circulating β-endorphin may act in conjunction with enkephalins released by the adrenal medulla to produce stress analgesia.

Human N-terminal fragment (hNT) and **γ-melanocyte stimulating hormone** (γ-MSH) are released with ACTH and β-endorphin. γ-MSH, a fragment of hNT. (Figure 31-7), may be the putative aldosterone-stimulating hormone of the pituitary. hNT and γ-MSH may also be the major stimulators of adrenocortical proliferation, whereas ACTH is the regulator of adrenocortical steroidogenesis.

Glycoprotein Hormones

The glycoprotein hormones may have evolved from a single ancestral polypeptide (Table 31-7). In the anterior pituitary, the thyrotrophs produce thyrotropin, and the gonadotrophs produce LH and FSH. Each glycoprotein hormone is composed of two dissimilar subunits (α and β) that are glycosylated and noncovalently bound. Synthesis of β subunits is rate limiting. Glycosylation takes place posttranslationally on the endoplasmic reticulum membranes, at about the time when disulfide bridges form in each subunit. "Mature" subunits then dimerize and are packaged in secretory granules.

The α subunits of all pituitary glycoprotein hormones are identical but show minor differences from the α subunit of hCG. The β subunits differ in all and confer hormonal specificity. The association of α and β subunits is required for activity, but the subunits need not be homologous. Dissociation of the subunits of one hormone, followed by recombination with subunits of another hormone, results in hybrids that are biologically active and exhibit the biological activity characteristic of the β subunit. For example, the hybrids LH-α/FSH-β have FSH activity, while TSH-α/LH-β have LH activity.

The hallmark of glycoprotein hormones is the carbohydrate moiety, particularly in the content of their sialic acid component, which is proportionate to the half-lives of the circulating hormones. Their effects are mediated by intracellular cAMP.

Thyroid-stimulating hormone (TSH, thyrotropin) stimulates secretion of the thyroid hormones T_4 and T_3 (Chapter 33). It stimulates synthesis of thyroid hormone, synthesis of thyroglobulin, synthesis of RNA and protein, uptake and utilization of glucose, and synthesis of phospholipids. TSH also promotes thyroid cell growth (hypertrophy) and vascularization through the gland. All of these effects are mediated by intracellular cAMP. Release of TSH is stimulated by TRH and inhibited by circulating T_4. T_4 is converted to T_3 by 5′-deiodinase in the thyrotroph, and it is T_3 that inhibits the thyrotroph response to TRH. TRH and T_3 act directly on the thyrotrophs, but TRH elicits a prompt release of TSH (within minutes), presumably via cAMP, whereas T_3 represses the *TSH* gene, and its effect is evident after several hours. Release of TSH is also affected by circadian cues and psychogenic signals (see below). Unlike many subprimate species, the adult human

TABLE 31-7
Glycoprotein Family

Hormone	Site of Secretion	Peptide Length	Location of Gene	Carbohydrate (%)	Function	Regulation of Secretion
Thyroid-Stimulating Hormone (TSH) (Thyrotropin)	Anterior Pituitary (Thyrotroph)	TSHα = 89 aa TSHβ = 112 aa	6q21.1-23 1p22	16.2 (1% sialic acid)	Stimulates thyroid growth and function.	Stimulated by TRH and inhibited by thyroid hormone and somatostatin.
Luteinizing Hormone (LH)	Anterior Pituitary (Gonadotroph)	LHα=89 aa LHβ=112 aa	6q21.1-23 19q12.32	15.7 (1% sialic acid)	Stimulates steroidogenesis in gonads.	Stimulated by GnRH and inhibited by estrogen.
Follicle-Stimulating Hormone (FSH)	Anterior Pituitary (Gonadotroph)	FSHα = 89 aa FSHβ = 115 aa	6q21.1-23 11p13	18.2 (5% sialic acid)	Stimulates gametogenesis and estrogen formation in gonads.	Stimulated by GnRH and inhibited by estrogen.
Chorionic Gonadotropin (hCG)	Placenta (Syncytiotrophoblast)	hCGα = 89 aa hCGβ = 112 aa	6q21.1-23 19q13.32	31.0 (12% sialic acid)	Stimulates steroidogenesis in gonads and placenta.	Autocrine?

does not release TSH in response to cold exposure; only during infancy is TSH released in humans in response to cold. In the normal adult, the pituitary contains about 0.3 mg TSH, and the basal level of the hormone in blood is about 1 ng/mL (see also Chapter 33).

Luteinizing hormone (LH) and ***follicle stimulating hormone*** (FSH) are synthesized within the same gonadotrophs but are products of different genes. They differ in carbohydrate composition (and thus in clearance rates); although their β subunits have the same number of amino acid residues, they differ in amino acid sequences. In normal men and in women whose menstrual periods are regular, the pituitary contains about 0.08 mg LH and about 0.04 mg FSH. In postmenopausal women, the levels are about 2.5 times higher. The regulation of the release of LH and FSH and the functions of these hormones are discussed in Chapter 34.

Neural Regulation of Anterior Pituitary Function

Superimposed upon the fine-tuned regulation of anterior pituitary hormone release by the hypothalamic hypophysiotrophic hormones and by circulating substances is the coarse-tuned regulation by the CNS, which conveys the need for a large adjustment in hormone output. These CNS signals arise from different regions of the brain and influence one or more of the pituitary hormones via different hypothalamic nuclei. Some signals result from CNS programming in relation to circadian rhythms, some are psychogenic, and some are sensory inputs from exteroceptors (sight, sound, smell) and interoceptors (e.g., cardiovascular disturbances) and represent the afferent limbs of a neuroendocrine reflex. They affect pituitary hormone secretion and also, via the peripheral nerves, that of the adrenal medulla. As a rule, the release of a hormone in response to coarse tuning of the CNS is of high amplitude and stimulus-coupled, lasting only as long as the stimulus is applied. It generally overrides the fine-tuning, negative-feedback effect of peripheral hormones or blood substrates, unless these levels are abnormally elevated. The means by which the CNS influences the activity of the anterior pituitary are presented in Table 31-8. For purposes of comparison and to emphasize that the pituitary is not the only endocrine gland affected by neurogenic signals, the adrenal medulla and the neurohypophysis are included in the table.

"Stress" refers to any stimulus that can arouse the autonomic nervous system. It may be pleasurable or disturbing, or it may involve actual perception of pain or the anticipation of it. The nature of the stress appears to be important in the stress-induced release of ACTH (and cortisol), GH, and PRL, since these hormones are released in response to noxious stimuli (unpleasant or painful) that are relatively novel, unfamiliar, or unexpected. ACTH is released most readily (lowest threshold), while PRL is the least responsive. The intensity of the stress appears to be important

TABLE 31-8
Factors That Affect the Neuroendocrine System

	Changes in circulating levels of:					
	ACTH	PRL	GH	TSH	LH†	ADH
Stress	↑↑↑	↑	↑↑	↓	↓ testosterone	↑
Exercise	↑		↑↑		↑↑P ↓↓A	↑
Sleep		↑	↑↑↑	↓		↑
Insulin hypoglycemia	↑	↑	↑↑			↑
Mild starvation	↑		↑↑			
Arginine i.v.			↑↑			
Emotional depression	↑↑↑		↓↓	↓ response to TRH	↓	
L-DOPA		↓↓↓	↑↑			

† P = pubertal child; A = adult.
i.v. = intravenous.
↑ = increased; ↓ = decreased.

in the release of adrenal medullary hormones, since both unpleasant (e.g., painful) or pleasant (e.g., sexually arousing) stresses, if intense enough, will cause elevation of catecholamine release. Unlike that of ACTH, the epinephrine response to stress appears not to be diminished by adaptation.

Almost every process in the body is subject to daily rhythms. If the rhythm occurs once a day, it is called circadian; if it occurs once every 90–120 minutes, it is called ultradian; and if it occurs once every hour, it is called circhoral. Anterior pituitary hormones are released in low-amplitude ultradian pulses, which, in the case of LH and FSH, can become high-amplitude circhoral bursts under conditions of intense stimulation. These patterns of release reflect the pulsatile discharge of hypothalamic releasing factors and are components of a large, circadian rhythm. The daily peaks and troughs in circulating levels of ACTH and TSH are authentically circadian; they are coupled to the body temperature rhythm and oscillate in a predictable manner under normal conditions. In contrast, GH and PRL do not undergo regular circadian rhythmicity unless the individual maintains a regular sleep schedule, since circadian release of GH and PRL is sleep-induced, that of GH being stimulated by deep sleep (EEG stages 3 and 4) and that of PRL being stimulated by rapid eye movement (REM) sleep. For this reason, peak GH release occurs early during sleep (when deep sleep is prevalent), and PRL levels rise during the later periods of sleep (when REM sleep dominates). In adult humans, the level of LH shows no circadian variation, but during puberty the release of LH is sleep-induced. Table 31-8 summarizes the circadian regulation of pituitary hormone release.

Depressive states can alter significantly the basal rates of pituitary hormone secretion and their circadian rhythms. In fact, depression may be related to periods when hormonal rhythms are out of phase with other rhythms in the body. Circulating levels of sleep-inducible hormones (GH, PRL, LH) are lower in depressive states, whereas that of ACTH is elevated. The basal level of ACTH in a depressed individual is elevated to the extent that it flattens the circadian oscillations of the hormone. These changes resemble those seen during disruptive phase shifts (e.g., east-bound trip, altered work schedule, etc.) and emphasize the importance of CNS influence on the release of anterior pituitary hormones.

Supplemental Readings and References

A. Colao and G. Lombardi: Growth-hormone and prolactin excess. *Lancet* **352,** 1455 (1998).

N. De Roux, J. Young, M. Misrahi, et al.: A family with hypogonatropic hypogonadism and mutations in the gonatropin-releasing hormone receptor. *New England Journal of Medicine* **337,** 1597 (1997).

M. E. Freeman, B. Kanyicska, A. Lerant, and others: Prolactin: structure, function, and regulation of secretion. *Physiological Reviews* **80,** 1523 (2000).

S. E. Inzucchi and R. J. Robbins: Effects of GH on human bone biology. *Journal of Clinical Endocrinology and Metabolism* **79,** 691 (1994).

R. S. Jackson, J. W. M. Creemers, S. Ohagi, et al.: Obesity and impaired prohormone processing associated with mutations in the human prohormone convertase 1 gene. *Nature Genetics* **16,** 303 (1997).

P. A. Kelly, J. Djiane, M-C. Postel-Vinay, and M. Edery: The prolactin/growth hormone receptor family. *Endocrine Reviews* **12,** 235 (1991).

J. R. Kerrigan and A. D. Rogol: The impact of gonadal steroid hormone action on growth hormone secretion during childhood and adolescence. *Endocrine Reviews* **13,** 281 (1992).

H. Krude, H. Beiebermann, W. Luck, et al.: Severe early-onset obesity, adrenal insufficiency and red hair pigmentation caused by POMC mutations in humans. *Nature Genetics* **19,** 155 (1998).

S. W. J. Lamberts, W. W. de Herder, and A. J. van der Lely: Pituitary insufficiency. *Lancet* **352,** 127 (1998).

D. Le Roith: Insulin-like growth factors. *New England Journal of Medicine* **336,** 633 (1997).

C. B. Newman and D. L. Kleinberg: Adult growth hormone deficiency. *Endocrinologist* **8(3),** 178 (1998).

D. N. Orth: Corticotropin-releasing hormone in humans. *Endocrine Reviews* **13,** 164 (1992).

M. Schwanzel-Fukuda, K. L. Jorgenson, H. T. Bergen, et al.: Biology of normal luteinizing hormone releasing hormone neurons during and after their migration from olfactory placode. *Endocrine Reviews* **13,** 623 (1992).

I. Shimon and S. Melmed: Management of pituitary tumors. *Annals of Internal Medicine* **129,** 472 (1998).

N. A. Tritos and C. S. Mantzoros: Recombinant human growth hormone: old and novel uses. *American Journal of Medicine* **105,** 44 (1998).

N. C. Vamvakopoulos and G. P. Chrousos: Hormonal regulation of human CRH gene expression: implications for the stress response and immune/inflammatory reaction. *Endocrine Reviews* **15,** 409 (1994).

G. Vassart and J. E. Dumont: The thyrtropin receptor and the regulation of thyrocyte function and growth. *Endocrine Reviews* **13,** 596 (1992).

L. Yaswen, N. Diehl, M. B. Brennan, and U. Hochgeschwender: Obesity in the mouse model of proopiomelanocortin deficiency responds to peripheral melanocortin. *Nature Medicine* **5,** 1066 (1999).

CHAPTER 32

Endocrine Metabolism III: Adrenal Glands

The adrenal glands, a pair of well-vascularized glands positioned bilaterally above the cranial poles of the kidney, consist of two embryologically, histologically, and functionally distinct regions. The outer region of each (***adrenal cortex***) accounts for about 80% of the weight of the gland, is derived from the coelomic epithelium of the urogenital ridge (mesodermal), and produces steroid hormones. The inner core of each gland (***adrenal medulla***) is derived from the neural crest (neuroectodermal), represents a modified sympathetic ganglion that has assumed an endocrine function, and synthesizes and secretes catecholamines and enkephalins. Neural supply by preganglionic sympathetic fibers from the splanchnic nerve is mainly to the adrenal medulla. Blood is supplied mainly to the adrenal cortex via the inferior phrenic, celiac, and renal arteries and is drained from the adrenal medulla.

32.1 Adrenal Cortex

The adult adrenal cortex is divided into three zones histologically: the outermost ***zona glomerulosa,*** the middle ***zona fasciculata,*** and the inner ***zona reticularis.*** All three zones are exclusive steroid producers; together, they can produce all classes of steroid hormones. Under normal conditions the major products are 11-hydroxylated C_{21} steroids (corticosteroids) (Table 32-1). The two types of corticosteroids are the ***mineralocorticoids,*** which regulate sodium and potassium levels, and the ***glucocorticoids,*** which regulate carbohydrate metabolism. The major mineralocorticoid, aldosterone (4-pregnen-11β,21-diol-3,18,20-trione; see Figure 30-1), is produced exclusively by the zona glomerulosa, which contains the unique mitochondrial enzyme, aldosterone synthase (CYP11B2). The major glucocorticoid in humans and some other species is cortisol (or hydrocortisone) (4-pregnen-11β,17α,21-triol-3,20-dione; see Figure 30-1), which is produced exclusively by the zona fasciculata (major) and zona reticularis (minor). These two inner zones contain 17-α-hydroxylase (CYP17) but lack enzymes to convert corticosterone to aldosterone. Figure 32-1 reveals that regional differences in steroid production exist because of the enzyme distribution.

Aldosterone and cortisol are not the only adrenal steroids with mineralocorticoid and glucorticoid activities. Corticosterone (4-pregnen-11β,21-diol-3,20-dione), a glucocorticoid about one-fifth as active as cortisol, has little mineralocorticoid activity relative to aldosterone; 11-deoxycorticosterone (DOC) (4-pregnen-21-ol-3,20-dione) has about one-tenth the mineralocorticoid activity of aldosterone but is devoid of glucocorticoid activity (Table 32-2). Both are intermediates in the biosynthesis of aldosterone (Figures 30-2 and 32-1), but under normal conditions, circulating corticosterone and DOC

A list of expanded acronyms appears as Appendix VIII.

TABLE 32-1
Secretion of Hormones by the Human Adrenal Glands

Hormone	Major Site of Synthesis	Approximate Basal (Unstimulated) Concentration in Blood* (ng/dL)	Approximate Percent due to Adrenal Contribution
Cortisol	Cortex (zona fasciculata)	14,000	100
Corticosterone	Cortex (zona fasciculata)	500	100
Aldosterone	Cortex (zona glomerulosa)	6	100
11-Deoxycorticosterone (DOC)	Cortex (zona glomerulosa)	6	95–100
Dehydroepiandrosterone (DHEA)	Cortex (zona reticularis)	500	85
DHEA sulfate (DHEAS)	Cortex (zona reticularis)	200,000	100
Δ^4-Androstenedione	Cortex (zona reticularis)	135	30
Dopamine (DA)	Medulla	5	50
Norepinephrine (NE)	Medulla	19	20
Epinephrine (E)	Medulla	2	100

*Total concentration (includes protein-bound fraction).

originate mainly from the zona fasciculata. The sodium-retaining action of DOC is normally obscured by the pronounced effect of aldosterone; however, in certain disease states (e.g., CYP17 deficiency), the levels of DOC can result in hypokalemia and hypertension in the absence of aldosterone.

Synthesis of Corticosteroids (See Chapter 30)

Formation of active corticosteroids in each zone of the adrenal cortex involves a series of transformations of the cholesterol molecule (steroidogenesis). The enzymes for steroidogenesis are distributed in mitochondria and endoplasmic reticulum (Figure 32-2). Conversion of cholesterol to pregnenolone, the initial step, is regulated by ACTH and angiotensin II (see below). Pregnenolone in the zona glomerulosa is converted to aldosterone by transformations that involve formation of DOC (Figure 32-1). In the zona fasciculate pregnenolone is directed mainly at formation of cortisol and involves 17α-hydroxylation (Figures 30-2 and 32-1). In the zona reticularis, pregnenolone is

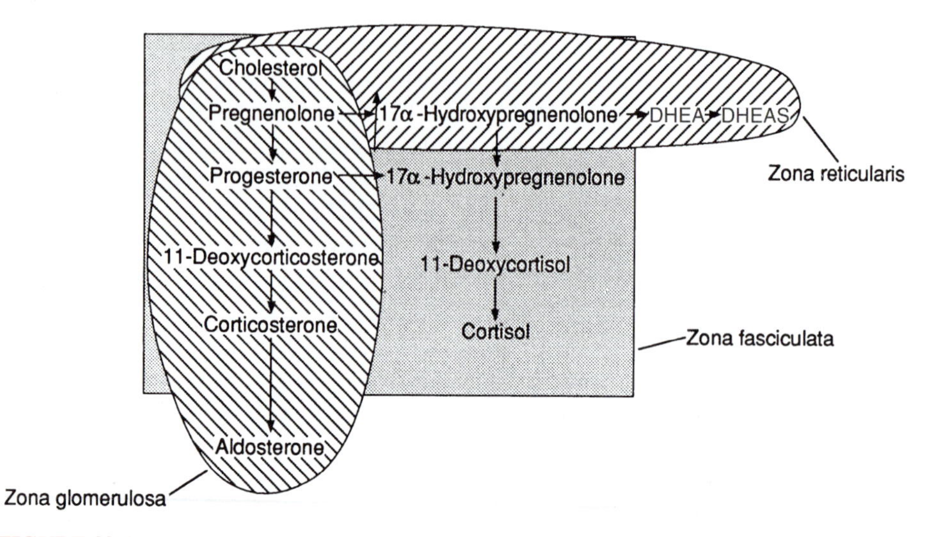

FIGURE 32-1
Diagrammatic representation of zone-specific steroid synthesis in the adrenal cortex. All three zones can convert cholesterol into pregnenolone, but because of the presence/absence of certain enzymes, the flow of pregnenolone is preferentially directed to the synthesis of the steroid for which the particular zone is specific. It has not been determined whether the zona fasciculata produces dehydroepiandrosterone (DHEA) and dehydroepiandrosterone sulphate (DHEAS) in physiologically significant amounts.

TABLE 32-2
Relative Activity of Natural and Synthetic Corticosteroids

Steroid	Structure	Relative Bioactivity	
		Glucocorticoid	Mineralocorticoid
Natural			
Cortisol	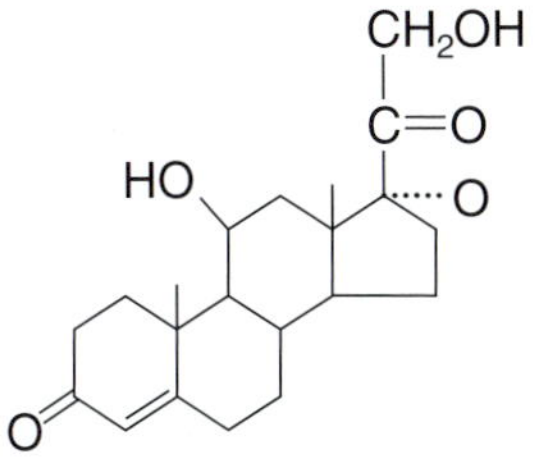	100	100
Corticosterone	See Figure 30–2	20	200
11-Deoxycorticosterone	See Figure 30–2	0	2000
Aldosterone	See Figure 30–2	10	40,000
Synthetic			
Prednisolone (Δ^1-cortisol)	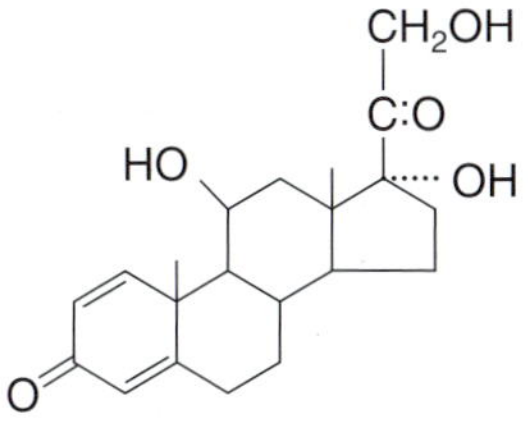	400	70
Methylprednisolone (6α-methylprednisolone)	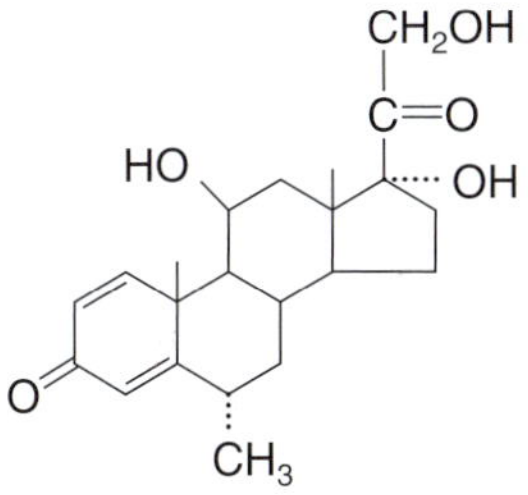	500	50
Dexamethasone (9α-fluoro-6α-methylprednisolone)	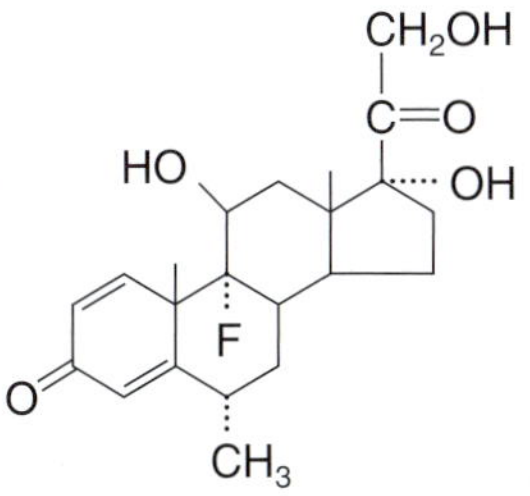	3000	0
Fludrocortisone (9α-fludrocortisol)	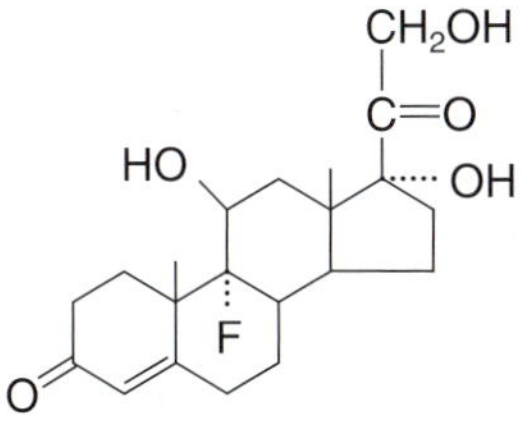	1000	40,000

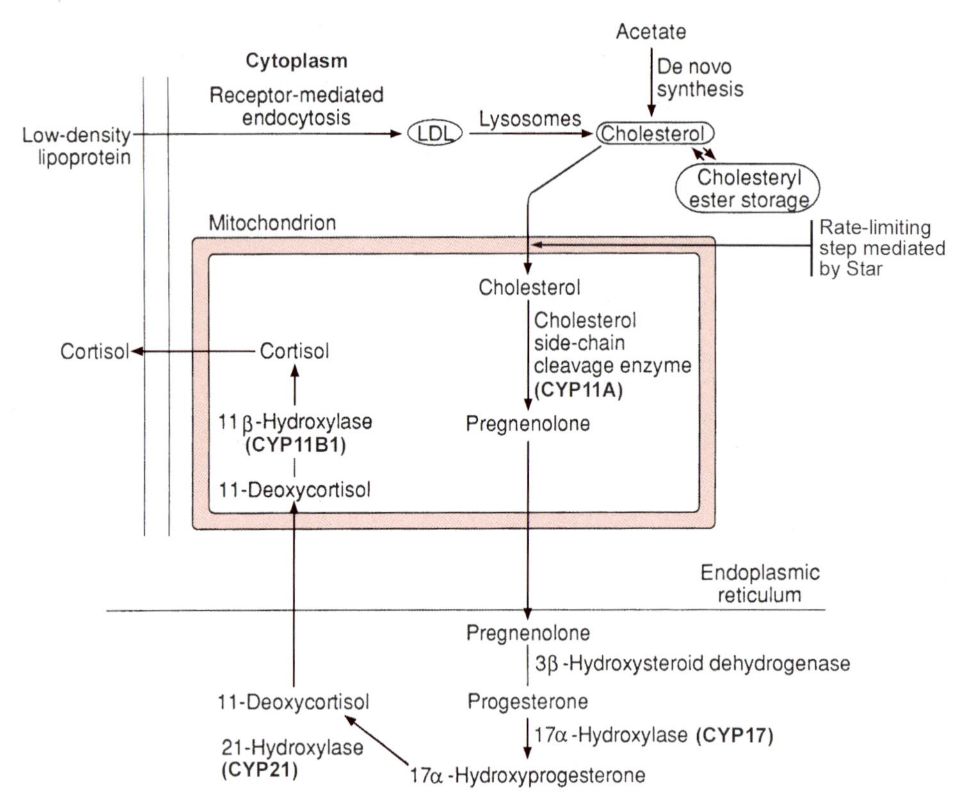

FIGURE 32-2

Cortisol biosynthesis by mitochondrial and endoplasmic reticulum enzymes of the adrenal cortex. The rate-limiting step in steroidogenesis is the importation of cholesterol from cytoplasm to mitochondria which is mediated by the steroidogenic acute regulatory protein (StAR).

utilized mainly for formation of the androgen dehydroepiandrosterone (DHEA) and its sulfate (DHEAS), which also requires 17α-hydroxylation (Figures 30-2 and 32-1). This hydroxylation is required for synthesis of cortisol and androgens (and estrogens as well) but not of aldosterone. Since the zona glomerulosa lacks 17α-hydroxylase, it cannot synthesize cortisol or androgens.

The adrenal cortex, the zona reticularis in particular, daily secretes substantial amounts of DHEA and DHEAS, equaling or exceeding the amount of cortisol. Table 32-1 shows approximate blood levels of the important corticosteroids. Most of the DHEA and all of the DHEAS come from the adrenals. Negligible amounts of testosterone, dihydrotestosterone (DHT), and estradiol are secreted by the cortex; however, DHEA and, to a lesser extent, DHEAS undergo conversion to estradiol in skeletal muscle and adipose tissue; they also can be converted to testosterone. The adrenal cortex accounts for about two-thirds of the urinary 17-ketosteroids, which are a measure of androgen production. This steroidogenic versatility makes the adrenal cortex an important factor in certain disease states (see below).

Regulation of Corticosteroid Secretion

The zona glomerulosa and zona fasciculata differ in their content of enzymes, major secretory products, and regulators of their functions. They can be regarded as separate endocrine tissues.

Regulation of Aldosterone Secretion

The major regulators of aldosterone secretion are the renin-angiotensin system and extracellular potassium ions (K^+). The former is sensitive to changes in intravascular volume and arterial pressure, while the latter is an aldosterone-regulated substance that feeds back to reduce aldosterone synthesis (simple negative feedback). Aldosterone secretion is also influenced (but not regulated) by ACTH and, directly and indirectly, by atrial natriuretic factor (ANF).

Potassium ions exert a direct, stimulatory effect on aldosterone secretion that is independent of the renin-angiotensin system. Small increases in serum potassium elicit a rise in serum aldosterone levels, whereas small decreases in serum potassium result in reduced levels of serum aldosterone. Aldosterone, by promoting the

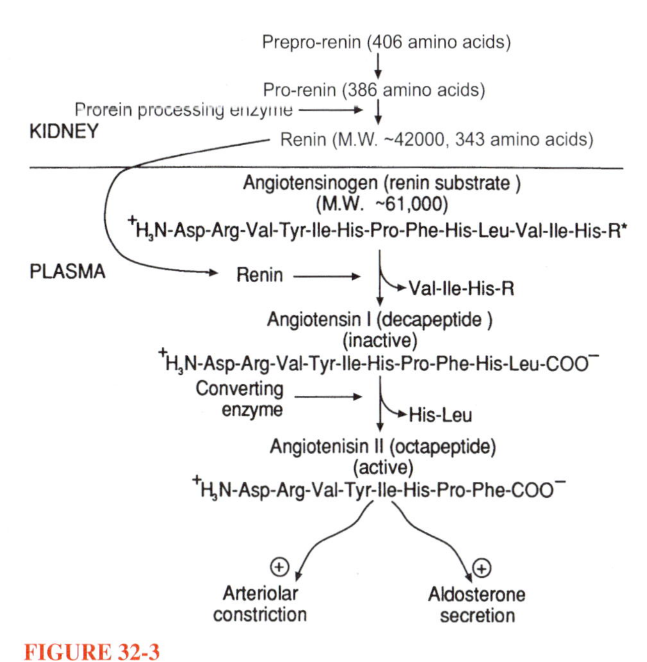

FIGURE 32-3
Renin-angiotensin system. R* is the remainder of the amino acid sequence.

excretion of potassium, lowers serum potassium levels, and thereby completes the negative feedback circuit. Potassium ions are believed to act by causing depolarization of the zona glomerulosa cells, allowing calcium ions to enter the cells through voltage-gated calcium channels. This increase in intracellular calcium ions fuels the calcium-dependent intracellular events triggered by angiotensin II (see below), and would explain why the angiotensin II effect on aldosterone secretion is augmented in hyperkalemic states and attenuated in hypokalemic states.

The ***renin-angiotensin system***consists of components derived from precursor molecules produced by the liver and kidney. The juxtaglomerular apparatus of the kidney is a specialized region of the afferent arteriole that releases ***renin,*** an aspartyl protease (about M.W. 42,000), in response to several stimuli (see below). Renin, derived from an inactive precursor (prorenin) by the action of prorenin processing enzyme (Figure 32-3), acts on the Leu-Val bond at the amino terminus of a circulating α_2-globulin called ***angiotensinogen*** or renin substrate secreted by the liver. The liberated amino terminal segment is an inactive decapeptide called angiotensin I. The physiologically important fate of this decapeptide is realized during passage through capillary beds in the lung and other tissues, where converting enzyme, a dipeptidyl carboxypeptidase present in endothelial cells, splits off the carboxy terminal dipeptide (His-Leu) to yield the highly active octapeptide ***angiotensin II*** (Figure 32-3). Angiotensin II exerts sodium-dependent arteriolar constriction, and it acts directly on the zona glomerulosa to stimulate the secretion of aldosterone. Both of these effects are mediated by membrane-bound receptors coupled via a G_s subunit to the phospholipase C-phosphatidylinositol 4,5-biphosphate (PLC-PIP_2) complex, which utilizes calcium ions and protein kinase C (PKC) as intracellular effectors. The effect of angiotensin II on aldosterone secretion involves an activation of both CYP11A and CYP11B2 (aldosterone synthase) activities in the zona glomerulosa and is influenced by dietary sodium intake; a low-sodium diet enhances the aldosterone response to angiotensin II, whereas a high-sodium diet tends to attenuate it.

Because renin release leads to formation of angiotensin II and secretion of aldosterone, and because, under normal conditions, the concentrations of angiotensinogen and converting enzyme are not rate limiting, any factor that influences the release of renin influences the secretion of aldosterone. Two important regulators of renin release are the mean renal arterial blood pressure and the extracellular fluid volume (blood volume). A decrease in the mean renal arterial blood pressure is sensed by baroreceptors in the juxtaglomerular apparatus, which responds by release of renin. A decrease in blood volume, such as that caused by hemorrhage or by standing from a reclining position, results in diminished venous return to the heart. Baroreceptors in the atrial walls of the heart signal this change via cranial nerves to the vasomotor center in the medulla oblongata, which then relays the information to the juxtaglomerular apparatus via β-adrenergic fibers, and renin output is stimulated. Aldosterone, which is secreted in response to renin release, promotes potassium excretion and extracellular fluid retention and thus inhibits further release of renin (Figure 32-4).

Regulation of Cortisol Secretion

The principal regulator of cortisol secretion is ACTH (see Chapter 31), the release of which is regulated by CRH and by circulating unbound cortisol. The balance of the effects of CRH and cortisol on the anterior pituitary maintains fairly constant ACTH stimulation of the adrenals and circulating levels of cortisol. This closed-loop feedback system is superseded by neural signals from higher regions of the brain in nonsteady-state conditions (e.g., circadian rhythm, stress).

ACTH exerts effects on the zona fasciculata by surface binding, calcium-dependent activation of membrane-bound adenylate cyclase and intracellular cAMP mediation. It increases activities of cholesterol esterase and cholesterol side-chain cleavage enzyme (CYP11A), thereby stimulating production of pregnenolone. Cholesterol side-chain cleavage enzyme is a cytochrome P-450 enzyme present in mitochondria. ACTH promotes the

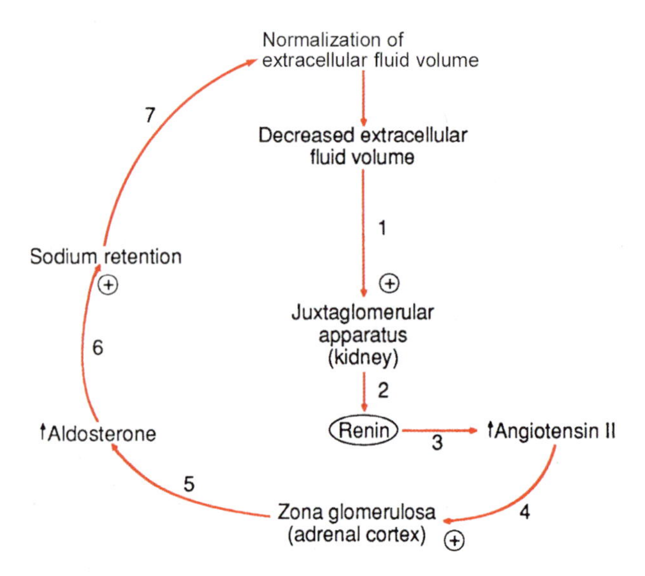

FIGURE 32-4

Regulation of renin release. Numbers 1–7 indicate the steps involved in the secretion of renin by the juxtaglomerular apparatus and its action. ⊕, Stimulatory, ⊖ inhibitory; ↑ increased production.

uptake of low-density lipoprotein (LDL) by increasing the number of LDL receptor sites. The adrenal cortex, although capable of *de novo* cholesterol synthesis, normally depends on LDL for its cholesterol needs. By increasing the size of the pregnenolone pool, ACTH promotes the flow of steroids through the major pathways of the zona fasciculata. It also stimulates DHEA and DHEAS secretion by the zona reticularis and aldosterone secretion from the zona glomerulosa. ACTH is not an important regulator of aldosterone secretion because there is no feedback communication between ACTH release and the zona glomerulosa.

Metabolism of Corticosteroids

Aldosterone and cortisol circulate in protein-bound states and have different fractional distributions (see Table 30-3). Because of the relatively large fraction of unbound (40%) and loosely bound (50%) forms, aldosterone is cleared from blood much faster than is cortisol (half-lives in blood: aldosterone, ~30 minutes; cortisol, ~10 hours). Both steroids are metabolized mainly in the liver. The major metabolite of aldosterone is tetrahydroaldosterone glucuronide and that of cortisol is tetrahydrocortisol glucuronide. These water-soluble conjugates are excreted by the kidney. Another metabolite of cortisol is cortisone (4-pregnen-17α,21-diol-3,11-20-trione), which has no intrinsic activity but can be reconverted to cortisol in certain peripheral tissues (e.g., lung, liver).

Synthetic Corticosteroids

Because of inactivation in the liver, the corticosteroids are ineffective when taken by mouth. Synthetic corticosteroids have been developed that are relatively resistant to hepatic inactivation and are active when taken orally (Table 32-2). Certain structural modifications enhance glucocorticoid activity while suppressing mineralocorticoid activity; other modifications enhance both activities. For example, the double bond at position 1 (Δ^1) of prednisolone increases glucocorticoid activity fourfold while decreasing mineralocorticoid activity by a one-third. If a 16-α-methyl group also is added to form methylprednisolone, glucocorticoid activity is further enhanced and mineralocorticoid activity is further reduced. Addition of a 9-α-fluoro group to produce dexamethasone gives one of the most potent glucocorticoids available, with negligible mineralocorticoid activity. If, however, the 9-α-fluoro group is introduced into cortisol to produce fludrocortisone, the result is a very potent mineralocorticoid with enhanced glucocorticoid activity. Thus, the 11-β-hydroxyl group of corticosteroids, which is protected from oxidation by a 9-α-fluoro group, is essential for glucocorticoid and mineralocorticoid activities. It also appears that steric hindrance of the A ring and the 20,21-α-ketol side chain enhances glucocorticoid activity while diminishing mineralocorticoid activity.

Biological Actions of Aldosterone

Mechanism of Action

Aldosterone exerts its effect by binding to type I corticosteroid receptors in the cytoplasm, translocating to the nucleus, and binding to an acceptor site in the chromatin, which results in gene activation and synthesis of a specific protein (see Chapter 30). Aldosterone induces the synthesis of aldosterone-induced protein (AIP), which is involved in transcellular Na^+,K^+-ATPase.

The type I corticosteroid receptor (mineralocorticoid receptor) binds cortisol and aldosterone with equal affinity. Because the circulating level of cortisol normally exceeds that of aldosterone by about 1000-fold (Table 32-1), activation of the receptor by aldosterone would probably not occur, were it not for the presence of a cortisol-inactivating enzyme in cells responsive to aldosterone. This enzyme, 11βHSD (Chapter 30), catalyzes the conversion of cortisol to cortisone, a metabolite that is not recognized by the receptor. Inhibition or absence of this enzyme leads to excessive aldosterone-like effects due to receptor activation by cortisol, a condition referred to as "apparent mineralocorticoid excess" (AME). AME can result from ingestion of large amounts of licorice, which contains glycyrrhetinic acid, an inhibitor of 11βHSD.

Physiological Effects of Aldosterone

The major effect is on the distal tubules of nephrons, where aldosterone promotes sodium retention and potassium excretion. Under the influence of aldosterone, sodium ions are actively transported out of the distal tubular cell into blood, and this transport is coupled to passive potassium flux in the opposite direction. Consequently, intracellular [Na^+] is diminished and intracellular [K^+] is elevated. This intracellular diminution of [Na^+] promotes the diffusion of sodium from the filtrate into the cell, and potassium diffuses into the filtrate. Aldosterone also stimulates sodium reabsorption from salivary fluid in the salivary gland and from luminal fluid in the intestines, but these sodium-conserving actions are of minor importance.

Biological Actions of Cortisol

Mechanism of Action

Cortisol and the synthetic glucocorticoids (GCs) bind to the type II corticosteroid receptor (glucocorticoid receptor, GR), and the resultant GC-GR homodimer activates glucocorticoid response elements (GRE) in chromatin, which ultimately leads to the genomic effects characteristic of GCs (Chapter 30). However, not all of the biological effects of GCs can be explained by this mechanism. In exerting its antimitogenic effects, GCs interfere with mitogen-stimulated AP-1 transactivation of late response genes by binding of c-jun by the GC-GR complex; thus, no induction of a GC-responsive gene is required for this effect.

Not all of the GC effects involve binding of GC to the GR. Elevated levels of cortisol and synthetic GCs are known to exert rapid-onset, nongenomic effects that are seen within minutes and do not involve the GR or a change in gene expression. Such nongenomic effects, which include the rapid suppression of ACTH release, the inhibition of exocytosis in inflammatory cell types, and the strong inhibition of GH release, are believed to involve the direct binding of extracellular GC-CBG complex to cell membrane receptors.

Physiological Effects of Cortisol

Normal circulating levels of cortisol, including the circadian early-morning rise and the moderate elevations after meals and minor stresses, help sustain basic physiological (vegetative) functions. Large amounts of cortisol released in response to major stresses enable the individual to withstand, or cope with, the metabolic, cardiovascular, and psychological demands of the situation. Cortisol (and other glucocorticoids) promotes the conservation of glucose as an energy source in several ways:

1. Cortisol induces and maintains the activity of all of the specifically gluconeogenic enzymes in the liver. By increasing hepatic formation of glucose, cortisol promotes its conversion to hepatic glycogen. Insulin reduces cortisol-stimulated gluconeogenesis but potentiates its effect on glycogenesis; glucagon, on the other hand, augments the gluconeogenic action of cortisol while inhibiting its effect on glycogen deposition.
2. Cortisol inhibits glucose utilization in peripheral tissues, such as skeletal muscle, adipose tissue, bone matrix, lymphoid tissue, and skin, by inhibiting glycolysis and promoting the use of fatty acids. This action is modulated by insulin and thyroid hormones but is potentiated by GH.
3. Cortisol promotes the liberation of fatty acids from adipose tissue by inducing and maintaining the synthesis of hormone-sensitive lipase (HSL), an effect supported by GH. The actual activity of HSL is controlled by those hormones that trigger its phosphorylation (glucagon, catecholamines) or dephosphorylation (insulin, PGE).
4. Cortisol, by inhibiting glucose utilization in peripheral tissues, exerts a mild antianabolic effect on these tissues; this effect diminishes their rate of amino acid incorporation, thus making available more amino acids for metabolism by these tissues, but mainly for hepatic protein synthesis and gluconeogenesis.

Cortisol has an important permissive influence on the cardiovascular system by conditioning many components of the system to respond maximally to regulatory signals. In the absence of GC, inadequacy of cardiovascular response can result in circulatory collapse and death, whereas in GC excess, increased cardiovascular responsiveness may result in hypertension. Cortisol is required for vascular smooth muscle to respond to the vasoconstrictor effect of catecholamines (norepinephrine), but the mechanism is incompletely understood. The hypotension in GC deficiency is normalized by intravenous cortisol treatment. Cortisol enhances the positive inotropic effect of catecholamines on the heart, possibly by promoting coupling of β-adrenergic receptors to adenylate cyclase. Elevated levels of cortisol increase both cardiac output and stroke volume, whereas cortisol deficiency has the opposite effect. Cortisol (GCs) increases the density of β_2-adrenergic receptors in vascular smooth muscle and promotes the vasodilating effect of epinephrine in some vascular beds.

Cortisol is an important modulator of the renin-angiotensin system. It stimulates the synthesis of angiotensinogen (renin substrate) by the liver, probably by prolonging the half-life of the angiotensinogen mRNA; it also stimulates the synthesis of angiotensin-converting enzyme (ACE) in vascular endothelial cells by promoting the expression of the *ACE* gene. By these actions, cortisol increases the magnitude of the renin-angiotensin response. Concurrently, however, cortisol promotes the expression of the atrial natriuretic peptide (*ANP*) gene in cardiac muscle cells and thus allows for better modulation of a renin-angiotensin overshoot.

Cortisol maintains the reactivity of the reticular activating system, the limbic system, and areas of the thalamus and hypothalamus to sensory "distress" signals and to endogenous opiates. Cortisol thus has an important role in behavioral and neuroendocrine responses to stress. The presence of glucocorticoids appears to be necessary for the stress analgesia that is attributed to endogenous opiates. Cortisol promotes tissue responsiveness to catecholamines and induces adrenal medullary phenylethanol amine-N-methyltransferase (PNMT), which converts norepinephrine to epinephrine (see below). Thus, even before it reaches the general circulation, cortisol promotes cardiovascular performance by promoting the formation of a cardiotropic hormone.

These central, metabolic, and cardiovascular effects of cortisol are accentuated when large amounts of cortisol are released in response to severe stress. Although the plasma cortisol concentration attained during stress often increases 10-fold, this hypercortisolism is relatively acute and ephemeral, and the effects do not resemble those seen after chronic excesses of cortisol or synthetic glucocorticoids.

Pharmacological Effects of Glucocorticoids

When the plasma glucocorticoid levels are chronically elevated, whether because of hyperactivity of the adrenal cortex or administration or consumption of synthetic glucocorticoids, some of the "physiological" effects become exaggerated, while other effects not normally seen appear. Protein catabolism is enhanced in skeletal muscle, skin, bone matrix, and lymphoid tissues by inhibition of protein synthesis and of cellular proliferation (DNA synthesis). Glucose utilization is severely inhibited and hepatic gluconeogenesis enhanced, which can lead to muscle weakness and atrophy, thinning and weakening of the skin, osteoporosis, diminished immunocompetence (from destruction of lymphocytes in lymphoid tissues), increased susceptibility to infections, and poor wound healing. These are classical features of *Cushing's syndrome* and *Cushing's disease,* in which the adrenal cortex secretes supraphysiological amounts of cortisol.

Chronic excess of glucocorticoids leads to elevated levels of glucose and free fatty acids in blood. Because these effects indicate inadequate counteraction by insulin, they are diabetogenic and are associated with an abnormal glucose tolerance test. Although glucocorticoid excess stimulates lipolysis, which leads to hyperlipemia, body fat is not depleted; in fact, a form of obesity ("central obesity") involving redistribution of body fat to the abdomen, upper back, and face appears to be characteristic of glucocorticoid toxicity.

Chronic glucocorticoid excess during the period of growth (e.g., peripubertal period) leads to suppression of cellular proliferation and of production of growth-promoting hormones and results in stunting. Proliferation of fibroblasts and other cell types required for longitudinal bone growth and for somatic growth in general is severely affected. The function of these cell types is also inhibited, and adequate formation of tissue matrices is not possible. Release of GH and formation of IGFs, 5′-deiodinase activity, and release of TSH and ACTH all are inhibited. Synthetic glucocorticoids can exert a long-lasting repression of POMC synthesis in the anterior pituitary. After cessation of chronic glucocorticoid treatment, ACTH levels do not return to normal before 2–3 months, and cortisol secretion resumes only after an additional 6 months. Thus, withdrawal of exogenous steroids results in a state of adrenocortical deficiency for 8–9 months, during which time the imposition of stressful stimuli may have undesirable consequences.

High local concentrations of glucocorticoids inhibit or diminish inflammatory and allergic reactions. Cell-mediated inflammation (of joints, bursae, etc.) and allergic reactions (IgE-induced) are caused by release of agents designed to combat infection (see Chapter 35). Thus, chemotaxis of neutrophils and other invasive cells to the affected area is followed by release of lysosomal enzymes (e.g., collagenase), histamine, prostaglandin E_2 (PGE_2), superoxide anion radicals, and other mediators of inflammation that cause tissue destruction and vascular permeability changes (see Chapters 15 and 18). Glucocorticoids counteract the inflammatory response by

1. Inhibiting phospholipase A_2 activity, thereby decreasing the synthesis of PGE_2 and of the potent chemotactic substance leukotriene B_4
2. Stabilizing membranes of the lysosomes and secretory granules, thereby inhibiting the release of their contents, and
3. Acting directly on the capillary endothelium to render it less permeable.

This endothelial effect and the permissive effect of glucocorticoids on catecholamine-induced vasoconstriction inhibit edema and swelling. Use of glucocorticoids to reduce inflammation due to bacterial infection should not be undertaken without concurrent use of antibiotics.

Adrenal Androgen, Dehydroepiandrosterone (DHEA)

The zona reticularis is the innermost layer of the cortex, about equal in size to the zona glomerulosa in the adult. Although it shares similarities with the zona fasciculata, both histologically and functionally the zona reticularis should be regarded as a distinct entity because it has certain features that are not found in the other zones. One distinct feature is the relatively late appearance and growth of this zone ("adrenarche") and its functional decline in late adulthood ("adrenopause"); another unique feature is that the zona reticularis is the exclusive site of DHEAS formation in the adult, owing to the presence of a steroid sulfotransferase that attaches a sulfate to DHEA. Although there is evidence that the zona fasciculata produces DHEA, no sulfate ester is formed at that site; and although the zona reticularis contains cortisol-producing cells from the zona fasciculata, the major steroid product of this zone is DHEAS. This sulfated steroid is the most abundant in circulation with basal plasma levels at least 10 times higher than peak levels of cortisol and about 100 times higher than that of DHEA. The levels are twice as high in men as in women. The circulating level of DHEAS increases at adrenarche, when the zona reticularis matures, and it declines at adrenopause, when the zona reticularis undergoes functional attenuation; in neither situation is there any significant change in the circulating levels of ACTH. However, the endocrine function of the zona reticularis is known to be strongly influenced by ACTH. In instances of ACTH deficiency, the level of DHEAS declines, and the administration of ACTH provokes a rise in plasma DHEAS levels. The physiological significance of DHEAS in the adult is unclear. This sulfated steroid is biologically inactive; however, cells that have steroid sulfatase are capable of converting the steroid to its unesterified form (DHEA), which can then be transformed to either estrone (EI) in the presence of P450arom (CYP19) or to DHT in the presence of 5α-reductase. The latter conversion pathway explains why virilization can occur in conditions of adrenal hyperactivity.

It has been suggested that the decline in DHEA and DHEAS production after adrenopause may be causally related to age-related health problems and that DHEA replacement may reduce the aging process. Evidence to date from animal and human studies that involved the administration of DHEA appears to support the idea that increasing or restoring the level of this steroid may be beneficial in promoting protein anabolism and immune function while decreasing body fat content, although the evidence is not conclusive. However, the notion that this steroid is a youth-promoting drug has led to successful marketing of this preparation as a food supplement. Although no known deleterious side effects of this steroid have been reported, there is evidence from *in vitro* studies that high levels may cause convulsions by inhibiting chloride currents in the CNS.

Disturbances in Adrenocortical Function

Deficiency

Primary adrenocortical insufficiency (Addison's disease) is a condition in which secretion of all adrenal steroids diminishes or ceases owing to deterioration of adrenocortical function. The adrenal cortex atrophies as a result of an infectious disease or autoimmune reaction. Circulating levels of aldosterone and cortisol decrease, whereas those of renin and ACTH increase. ACTH levels increase enough to produce darkening of the skin, a hallmark of Addison's disease. Aldosterone deficiency encourages potassium retention and sodium loss and leads to hyperkalemia, hypovolemia, and hypotension. Deficiency of cortisol renders the tissues more sensitive to insulin; hence, hypoglycemia may develop. In addition, cortisol deficiency diminishes responsiveness of tissues to catecholamines, particularly in vascular smooth muscles, which do not contract adequately in response to α-adrenergic stimulation. All of these changes can lead to circulatory collapse.

More specific adrenocortical deficiencies occur when only specific steroidogenic enzymes are affected. If there is deficient production of cortisol, which is the exclusive feedback suppressor of ACTH release, hypersectetion of ACTH with overstimulation of the adrenal cortex (hyperplasia) results. This condition is called ***congenital adrenal hyperplasia.*** The most common enzyme deficiency is that of 21-hydroxylase (CYP21), and neonatal diagnosis of this disorder is accomplished by measuring 17-hydroxyprogesterone. The incidence of 21-hydroxylase deficiency varies from 1:10,000 to 1:18,000 live births. The deficiency can be severe or mild, giving rise to a continuum of clinical symptoms. Deficiency of 21-hydroxylase causes a lack of production of both mineralocorticoids and glucocorticoids (Figure 30-2). Two steroid precursors prior to the enzyme deficiency, namely, progesterone and 17-hydroxyprogesterone, accumulate and are diverted to the synthesis of androgens. The decreased production of cortisol removes the negative feedback on ACTH production by the pituitary, resulting in excess production in ACTH and overstimulation of the adrenal glands. *In utero,* female fetuses exposed to high levels of androgen show masculinization (virilization) and ambiguous genitalia. In contrast, male newborns with 21-hydroxylase deficiency appear normal at birth.

In infants with classic congenital adrenal hyperplasia as a result of 21-hydroxylase deficiency, the mineralocorticoid deficiency causes a rapid evolving "salt wasting" crisis due to renal loss of Na^+ and accumulation of K^+. These changes result in hyponatremia, hyperkalemia, hyperreninemia, and hypovolemic shock. The electrolyte disturbances are treated with fluid and Na^+ replacement and administration of both mineralocorticoid and glucocorticoid. The administration of glucocorticoid also corrects hypoglycemia. Because this life-threatening disorder is treatable, newborn screening programs consisting of assaying 17α-hydroxyprogesterone from a filter paper-dried blood spot has been developed. In milder, attenuated forms of 21-hydroxylase deficiency, virilization may be the sole clinical manifestation. In young women, this condition may cause hirsutism and menstrual irregularities.

The 21-hydroxylase gene (CYP21) is located in the major histocompatibility complex (MHC)on chromosome 6p. The functional gene of 21-hydroxylase is present in a tandem arrangement with a pseudogene (CYP21P) and genes for complement proteins C4A and C4B. The pseudogene for 21-hydroxylase is nonfunctional as it contains several mutations. The importance of the pseudogene in causing the 21-hydroxylase deficiency is that it allows transfer of mutations to the functional gene. The transfer of mutations from pseudogene to functional gene can occur due to unequal crossovers and gene conversions. A variety of mutations in the pseudogene have been identified, which include premature 3′ splice site, deletions, and frameshift, nonsense, and missense mutations. *De novo* mutations of the functional gene (CYP21) that are not derived from the psuedogene have also been identified.

Prenatal diagnosis and fetal therapy of affected female fetuses may be desirable since corrective surgery of external genitalia is not always optimal. Oral administration of a synthetic glucocorticoid (dexamethasone, which crosses the placenta) to the mother during pregnancy can limit the virilization of a female fetus. Dexamethasone is used instead of cortisol because cortisol does not readily cross the placental barrier.

Deficiency of 17α-hydroxylase (CYP17), the enzyme required for formation of cortisol and androgens, also causes congenital adrenal hyperplasia. The genetic defect also affects the gonads, which consequently cannot synthesize androgens or estrogens. Thus, the anterior pituitary increases its output of ACTH, LH, and FSH.

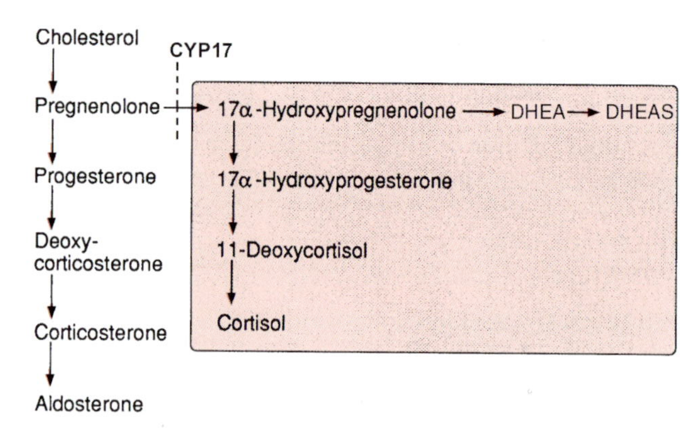

FIGURE 32-5
Steroidogenic pathways in 17α-hydroxylase (CYP17) deficiency. Synthesis of steroids within the boxed area is diminished as a result of deficiency in the enzyme. Production of those steroids will increase outside the boxed area.

Because 17α-hydroxylase is absent from the zona fasciculata, the flow of pregnenolone is directed at formation of DOC, corticosterone, and 18-hydroxycorticosterone, none of which is an effective suppressor of ACTH release (Figure 32-5). Levels of DOC rise sufficiently to produce sodium retention, hypokalemia, and hypervolemia, which suppress renin release and aldosterone secretion. Thus, low levels of cortisol and aldosterone are attended by a mineralocorticoid excess and sex steroid deficiency. Administration of a potent glucocorticoid (dexamethasone) corrects the DOC excess by suppressing ACTH release; this, in turn, restores blood volume, and the circulating levels of renin and aldosterone return to normal. However, neither cortisol nor sex steroids will be normalized because the enzyme that catalyzes their formation is lacking.

Deficiency of 11β-hydroxylase (CYP11B1) which is required for formation of cortisol, produces the second most common form of congenital adrenal hyperplasia. This condition also is characterized by excessive ACTH stimulation of the adrenal cortex and by elevated DOC secretion, which lead to mineralocorticoid excess in the face of low renin and aldosterone levels. Secretion of the adrenal androgens is elevated (Figure 32-6). The heightened stimulation of the adrenal cortex by ACTH promotes increased formation of pregnenolone and increased flow of pregnenolone into androgen synthesis. Although considerably less androgenic than testosterone, adrenal androgens (DHEA and DHEAS) exert virilizing effects at the concentrations that occur in CYP11B1 deficiency. In addition, these steroids are converted to testosterone and DHT by peripheral tissues. The effect of excessive adrenal androgen production is best manifested in prepubertal boys and in women since they do not normally exhibit

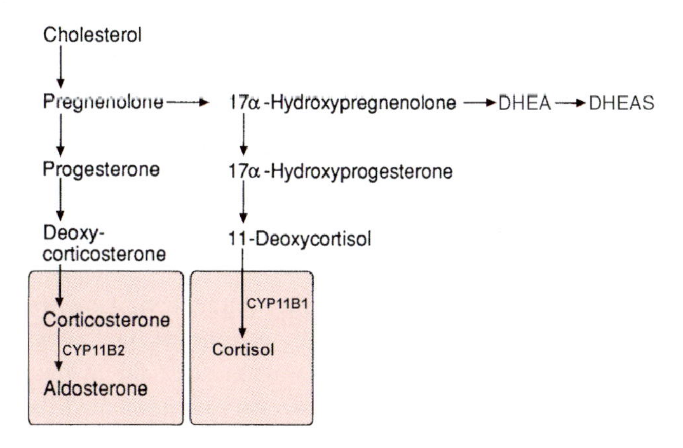

FIGURE 32-6
Steroidgenic pathways in 11β-hydroxylase CYPB1 and CYPB2 deficiency. A CYP11B1 defect causes a deficiency of cortisol and the disorder *congential adrenal hyperplasia* (CAH). A CYP11B2 defect causes an aldosterone deficiency. Synthesis of steroids within the boxed areas is decreased and steroids outside the boxed area increased, respectively, for each enzyme deficiency.

secondary sex characteristics that typify the adult male. Administration of dexamethasone suppresses ACTH release, restores electrolyte balance, and reduces the production of adrenal androgen toward normal. Deficiency of 11β-hydroxylase CYP11B2 (aldosterone synthase), which is required for the conversion of corticosterone, to aldosterone, causes aldosterone deficiency without causing congenital adrenal hyperplasia (Figure 32-6).

Deficiencies of other enzymes are rare. Deficiency of CYP11A (cholesterol desmolase) prevents all steroid biosynthesis, is incompatible with extrauterine life, and is the rarest adrenal cortex enzyme deficiency. 3β-Hydroxysteroid dehydrogenase deficiency impairs the synthesis of glucocorticoids, mineralocorticoids, and adrenal androgens and estrogens. Deficiency of 18-hydroxylase or 18-hydroxydehydrogenase decreases aldosterone production. 17-Hydroxysteroid dehydrogenase or 17,20-desmolase (or 17,20-lyase) deficiency causes androgen deficiency without affecting cortisol biosynthesis.

Steroid sulftase* (STS) *deficiency due to an X-linked inborn error of metabolism causes a skin disorder (***ichthyaosis***) in affected males. In heterozygous females, STS deficiency leads to decreased estrogen synthesis during the later stages of pregnancy due to inability to convert DHEAS to DHEA, which is then metabolized to estrogen in the maternal-fetal-placental unit. The diminished estrogen synthesis results in prolonged labor. STS deficiency also causes increased levels of cholesterol sulfate in the blood and skin. In the keratocytes, cholesterol sulfate inhibits cholesterol synthesis, which may be responsible for skin abnormalities.

Excess Inappropriately large amounts of aldosterone or cortisol result from a disturbance at the level of the adrenals (primary) or of the regulation of adrenal function (secondary).

Primary hyperaldosteronism (Conn's syndrome) usually results from an adrenocortical adenoma that produces aldosterone but is unresponsive to regulation by potassium or by the renin-angiotensin system. Aldosterone excess leads to excessive sodium retention and potassium wastage, hypokalemia, and hypervolemia. Muscle weakness or paralysis and hypertension are common symptoms. Because of hypernatremia and hypervolemia, renin levels are depressed. However, if the condition is chronic, blood volume increases to a point at which the glomerular filtration rate is increased, causing an aldosterone-insensitive urinary loss of sodium and water. This "escape" phenomenon, presumably not hormonally regulated, accounts for the rarity of severe hypertension in this syndrome. Conn's syndrome can be controlled by treatment with spironolactone, an aldosterone antagonist.

Secondary hyperaldosteronism is usually distinguished from the primary disorder because plasma renin levels are elevated. This disorder may be due to a defect in the juxtaglomerular apparatus (autonomous renin secretion) or to a dysfunction of a peripheral organ that affects vascular dynamics. Primary hyperaldosteronism leads to hypertension, whereas the secondary form may not. In the case of autonomous renin secretion, an elevated renin level leads to elevated angiotensin II and aldosterone levels, which in turn leads to an increase in plasma sodium concentration and to hypervolemia. Increased levels of angiotensin II, in the presence of elevated plasma sodium, promote vasoconstriction, which, in the face of hypervolemia, can lead to hypertension. Because the juxtaglomerular apparatus functions autonomously, it is not responsive to negative feedback by sodium, angiotensin II, blood pressure, or blood volume changes. An example of a peripheral organ disease affecting the vascular system is the combination of congestive heart failure, cirrhosis of the liver, and nephrosis, without any defect in the renin-angiotensin-aldosterone axis. The cardiac and hepatic failures decrease venous return to the heart, which, together with the low plasma albumin level (due to liver damage), results in hypovolemia. This lowers the mean renal arterial blood pressure, thereby stimulating the release of renin and aldosterone. Aldosterone increases sodium retention and blood volume. However, because plasma albumin levels are low, the hypervolemia does not increase blood pressure but instead causes the shunting of fluid into the interstitial spaces. The result is a condition of normotensive edema.

***Glucocorticoid-remediable aldosteronism* (*GRA*)** is due to mutations involving the 11β-hydroxylase gene

(*CYP11B1*) and the aldosterone synthase gene (*CYP11B2*), which are located in close proximity on chromosome 88 and are 95% homologous in nucleotide sequence. It is hypothesized that during meiosis, when the chromosomes duplicate, an unequal crossing over between the chromosomes occurs, resulting in a hybrid gene containing the regulatory sequence of the *CYP11B1* gene and the coding sequence of the *CYP11B2* gene. Since the 5′ portion of this hybrid gene contains the promoter region of the *CYP11B1* gene, it is under ACTH regulation. The 3′ region of the hybrid contains the coding region of the aldosterone synthase and therefore expresses an enzyme with aldosterone synthase activity. This enzyme converts corticosterone to aldosterone, but also converts cortisol to 18-hydroxycortisol and 18-oxocortisol. The important feature is that aldosterone synthesis is now under direct ACTH control, and this ultimately shuts down aldosterone production in the glomerulosa.

Primary hypercortisolism (Cushing's syndrome) is usually due to an autonomous adrenocortical tumor and is characterized by low levels of circulating ACTH. Virtually all of the effects that have been described for glucocorticoid excess occur in this syndrome. Skeletal muscle, skin, bone, and lymphoid tissue exhibit protein loss, which can lead to muscle weakness, fragility of the skin, osteoporosis, and diminished immunocompetence, respectively. Chronic excess of cortisol encourages lipolysis in adipose tissues and favors fat deposition in the face and trunk regions, resulting in the characteristic "moon face" and central obesity. Because cortisol is an insulin antagonist, some glucose intolerance may develop. Removal of the adrenal tumor corrects this syndrome, but it also creates adrenocortical insufficiency for which exogenous glucocorticoid treatment is required. In the case of a unilateral tumor, the other adrenal atrophies in Cushing's syndrome because ACTH secretion is chronically suppressed. Recovery of ACTH and cortisol secretion following removal of the tumor may take several months.

Overproduction of cortisol by hyperplastic adrenal cortices can result from oversecretion of ACTH (***secondary hypercortisolism***) due to a defect at the level of the pituitary or median eminence or to ectopic production of ACTH. The former condition is known as *Cushing's disease* and the latter as *ectopic ACTH syndrome*. In both conditions, the adrenal cortex functions normally and secretes cortisol in proportion to the level of ACTH.

Cushing's disease, which is more prevalent than ectopic ACTH syndrome, can be distinguished from the latter by its glucocorticoid suppressibility. Large doses of dexamethasone administered for three consecutive days suppress ACTH secretion. In ectopic ACTH syndrome, malignant transformation in some tissues (notably the lung) induces in that tissue the synthesis and release of ACTH. Because the transformed tissue is unresponsive to cortisol, ACTH and cortisol levels continue to rise and cannot be depressed by dexamethasone.

32.2 Adrenal Medulla

The adrenal medulla is a modified sympathetic ganglion that lacks postsynaptic axonal projections; as such, it can be regarded as the endocrine component of the sympathetic division of the autonomic nervous system. Like its sympathetic ganglia counterparts, the adrenal medulla receives numerous preganglionic cholinergic fibers from the spinal cord that transmit autonomic efferent signals from the brain stem and higher centers. Unlike its sympathetic ganglia counterparts, which release norepinephrine into synaptic junctions at target cells, the adrenal medulla releases mainly epinephrine, into the systemic circulation. Cells of the adrenal medulla are often referred to as "chromaffin cells" because they contain "chromaffin granules," electron-dense membrane-bound secretory vesicles with an affinity for chromic ions (hence the name "chromaffin"). Chromaffin granules contain catecholamines (~20%), various proteins (~35%), ATP (15%), lipids (~20%), calcium ions, ascorbic acid, and other substances; they are the adrenal medullary counterparts of secretory vesicles in ganglion cells.

Regulation of Release

The catecholamine content of mature chrornaffin granules has been estimated to consist of 80% epinephrine (E), 16% norepinephrine (NE) and 4% dopamine (DA), although the percentage of E depends on the rate of cortisol production (discussed later). Upon cholinergic stimulation and depolarization of the adrenal medullary cells, the intracellular Ca^{2+} ion concentration increases, promoting fusion of chromaffin granules with the plasma membrane. This leads to exocytosis of all soluble granule constituents, including the catecholamines, ATP, dopamine β-hydroxylase, calcium ions, and chromogranins.

Adrenal medullary cells have plasma membrane receptors for acetylcholine (ACh) of the neuronal nicotinic subtype (N_N). These receptors are cation channels that span the plasma membrane and are activated by ACh to rapidly increase Na^+ and K^+ permeabilities (Na^+ influx rate $\sim 5 \times 10^7$ ions/s), causing the cells to depolarize and release their catecholamines by exocytosis. The cholinergic stimulation of exocytosis is accompanied by an activation of tyrosine hydroxylase activity within the adrenal medullary cell, and this promotes biosynthesis of more catecholamines. This cholinergic activation of tyrosine hydroxylase involves phosphorylation of the enzyme, presumably by a Ca^{2+}-calmodulin-dependent protein kinase.

Synthesis of Epinephrine

Epinephrine (adrenaline) (Figure 32-7) is synthesized from tyrosine by conversion of tyrosine to 3,4-dihydroxyphenylalanine (dopa) by tyrosine-3-monooxygenase (tyrosine hydroxylase) in the cytosol. The mixed-function oxidase requires molecular oxygen and tetrahydrobiopterin, which is produced from dihydrobiopterin by NADPH-dependent dihydrofolate reductase. In the reaction, tetrahydrobiopterin is oxidized to dihydrobiopterin, which is reduced to the tetrahydro form by NADH-dependent dihydropteridine reductase. These reactions are similar to the hydroxylations of aromatic amino acids (phenylalanine and tryptophan), in which an obligatory biopterin electron donor system is used (Chapter 17).

The rate-controlling step of catecholamine synthesis is the tyrosine hydroxylase reaction, for which the catecholamines are allosteric inhibitors. The enzyme is activated by the cAMP-dependent protein kinase phosphorylating system. α-N-Methyl-p-tyrosine is an inhibitor of this enzyme and is used to block adrenergic activity in pheochromocytoma (see below).

Dopa is decarboxylated to 2-(3,4-dihydroxyphenyl) ethylamine (dopamine) by aromatic L-amino acid decarboxylase, a nonspecific cytosolic pyridoxal phosphate-dependent enzyme also involved in formation of other amines (e.g., 5-hydroxytryptamine).

$$\text{Dopa (HO, HO-C}_6\text{H}_3\text{-CH}_2\text{-CH(NH}_3^+\text{)-COO}^-) \xrightarrow[\text{Pyridoxal phosphate}]{\text{Aromatic L-amino acid decarboxylase}} \text{Dopamine (HO, HO-C}_6\text{H}_3\text{-CH}_2\text{-CH}_2\text{-NH}_3^+) + CO_2$$

Tyrosine can be decarboxylated to tyramine by aromatic L-amino acid decarboxylase of intestinal bacteria. Tyramine, which is present in large amounts in certain foods (e.g., aged cheeses, red wines), is converted by monoamine oxidase (MAO) to the aldehyde derivatives. However, individuals who are receiving MAO inhibitors for the treatment of depression can accumulate high levels of tyramine, causing release of norepinephrine from sympathetic nerve endings and of epinephrine from the adrenal medulla. This results in peripheral vasoconstriction and increased cardiac output, which lead to hypertensive crises that can cause headaches, palpitations, subdural hemorrhage, stroke, or myocardial infarction.

In dopaminergic neurons, dopamine is not metabolized further but is stored in presynaptic vesicles. In noradrenergic neurons and in the adrenal medulla, dopamine enters the secretory granules and is further hydroxylated to norepinephrine. This reaction is catalyzed by dopamine-β-monooxygenase, a copper protein that requires molecular oxygen and ascorbate or tetrahydrobiopterin.

$$\text{Dopamine (HO, HO-C}_6\text{H}_3\text{-CH}_2\text{-CH}_2\text{-NH}_3^+) \xrightarrow[\text{Ascorbate, } O_2 \;\to\; \text{Dehydroascorbate, } H_2O]{\text{Dopamine-}\beta\text{-monooxygenase}} \text{Norepinephrine (HO, HO-C}_6\text{H}_3\text{-CH(OH)-CH}_2\text{-NH}_3^+)$$

Norepinephrine diffuses into the cytosol, where it is converted to epinephrine by methylation of its amino group. This reaction, in which the methyl group is donated by S-adenosylmethionine, is catalyzed by S-adenosyl-L-methionine:phenylethanolamine-N-methyltransferase (PNMT). Epinephrine enters the granules and remains there until it is released.

$$\text{Norepinephrine (HO, HO-C}_6\text{H}_3\text{-CH(OH)-CH}_2\text{-NH}_3^+) \xrightarrow[\text{S-adenosylmethionine} \;\to\; \text{S-adenosylhomocysteine}]{\text{PNMT}} \text{Epinephrine (HO, HO-C}_6\text{H}_3\text{-CH(OH)-CH}_2\text{-NH-CH}_3)$$

Catecholamine neurotransmitters in the central nervous system are synthesized in that location itself because they cannot cross the blood-brain barrier. However, dopa readily crosses the blood-brain barrier, promoting the catecholamine synthesis. Thus, in disorders involving deficiency of catecholamine synthesis, administration of dopa may have beneficial effects. In ***Parkinson's disease,*** in which deficiency of dopamine synthesis affects nerve transmission in the substantia nigra of the upper brain stem, administration of dopa leads to some symptomatic relief. Parkinsonism is a chronic, progressive disorder characterized by involuntary tremor, decreased motor power and control, postural instability, and muscular rigidity.

Regulation of Catecholamine Secretion

Regulation of Synthesis

Dopamine (DA) and ***norepinephrine*** (NE) are allosteric inhibitors of tyrosine hydroxylase and regulate catecholamine synthesis when the adrenal medulla is quiescent (unstimulated). Continuous stimulation of the adrenal medulla (as during prolonged stress) promotes tyrosine hydroxylase activity primarily because the turnover of DA and NE is rapid. Tyrosine hydroxylase activity is also regulated by cAMP and by cholinergic nerve activity. The enzyme is active when phosphorylated (Chapter 30). Tonic cholinergic impulses maintain the activity of tyrosine hydroxylase, whereas

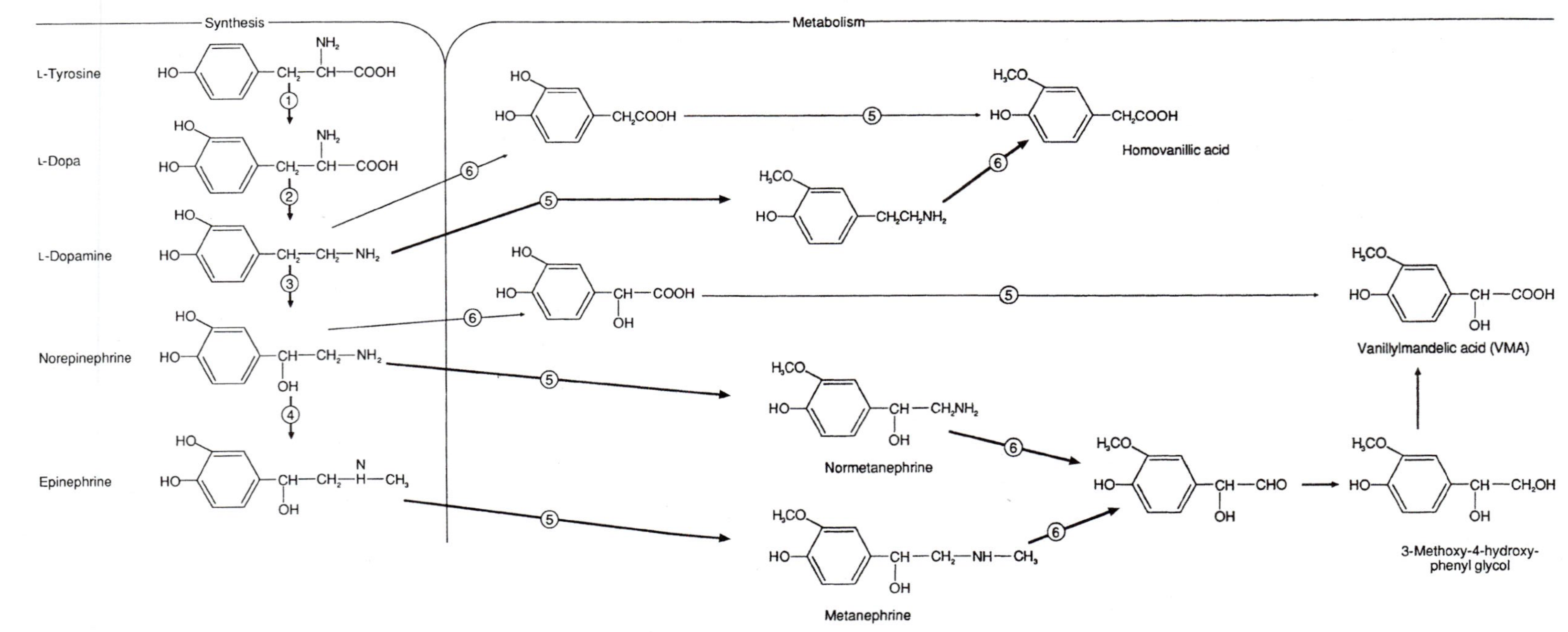

FIGURE 32-7

Synthesis and metabolism of catecholamines. Arrows indicate molecular conversions catalyzed by specific enzymes. Bold arrows indicate major (preferred) pathways. Enzymes: (1) tyrosine hydroxylase; (2) aromatic L-amino acid decarboxylase; (3) dopamine-$\beta\gamma$monooxygenase; (4) PNMT; (5) cateckel-o-methyltransferase; (6) monoamine oxidase.

chronic, intense cholinergic stimulation increases the activity of tyrosine hydroxylase and dopamine β-hydroxylase by increased protein synthesis.

Epinephrine production is catalyzed by PNMT, which is induced by glucocorticoids (cortisol). The venous drainage of the adrenal cortex, which contains very high concentrations of newly released cortisol, bathes the adrenal medulla before entering the general circulation.

Regulation of Release

The major regulator of catecholamine release from the adrenal medulla is cholinergic stimulation, which causes calcium-dependent exocytosis of the contents of the secretory granules. Exocytosis of the granular content releases epinephrine (E), NE, DA, dopamine β-hydroxylase, ATP, peptides, and chromaffin-specific proteins that are biologically inert. The amounts of DA and NE released are minor in comparison with that of E. Of the total catecholamine content in the granules, approximately 80% is E, 16% is NE, and the remainder is mostly DA.

Transport and Metabolism of Catecholamines

Under basal (unstimulated) conditions in the average adult, NE enters plasma at an average rate of ~5.8 ng/kg/min, almost entirely from adrenergic nerve endings that innervate peripheral tissues (liver, kidney, intestines, pancreas, skeletal muscle, heart, etc.), and with only negligible contribution by the adrenal medulla; E enters plasma at an average rate of ~3.6 ng/kg/min, entirely derived from the adrenal medulla. Under stimulated conditions, however, the degree to which plasma NE and E levels increase may differ, depending on the nature of the stimulus. For example, whereas physical exercise increases the level of both catecholamines similarly, acute hypoglycemia causes E levels to increase 5- to 10-fold without affecting the levels of NE significantly. As discussed below, the adrenal contribution to NE is physiologically important because it can elevate the plasma level of NE to what is required for a biological effect in peripheral tissues.

Circulating catecholamines have an estimated half-life of about one circulation time (20 seconds). They are rapidly taken up by various tissues (notably the liver), where they may exert their effects before they are inactivated. DA may be converted to NE after uptake by nerve endings, kidney, heart, and other tissues. Two enzymes responsible for inactivation of catecholamines are present in most tissues but are particularly abundant in the liver. ***Catechol-O-methyltransferase*** (COMT) is a cytosolic, Mg^{2+}-dependent enzyme that catalyzes methoxylation of catecholamines at the hydroxyl group at position 3. COMT utilizes S-adenosylmethionine as the methyl donor and usually initiates inactivation (Figure 32-7). ***Monoamine oxidase*** (MAO), a mitochondrial enzyme that oxidizes the amino side chain of catecholamines, acts generally (but not invariably) on methoxylated catecholamines (Figure 32-7). About 70% of the total output of urinary catecholamines is 3-methoxy-4-hydroxymandelic acid (also called vanillylmandelic acid, VMA) (Figure 32-7). Unmodified catecholamines represent 0.1–0.4% of the total.

Biological Actions of Catecholamines

Catecholamines exert their effects through specific receptors on the target cell surface. However, the effects elicited depend on the type or subtype of receptor with which they interact. There are three types of catecholamine receptor: dopamine, α-adrenergic, and β-adrenergic. Each of these consists of at least two or more subtypes, which differ with respect to ligand affinity, tissue distribution, postreceptor events, and drug antagonists (Table 32-3).

Mechanism of Action

Dopamine Receptors

Dopamine receptors (D_1 and D_2) have a greater affinity for DA than for NE or E (Table 32-3). Both subtypes exert effects by altering the activity of adenylate cyciase via a G-protein. D_1 receptors coupled to $G_{s\alpha}$ activate the enzyme and cause a rise in intracellular cyclic AMP. D_2 receptors are coupled to $G_{\alpha i}$, and inhibit the enzyme (Table 32-3). At very high concentrations, however, dopamine is capable of activating α-adrenergic receptors, which results in an effect resembling that of NE (vasoconstriction) instead of the characteristic vasodilatation elicited by low concentrations of DA.

Adrenergic Receptors

There are two types of adrenergic receptors, designated α and β. Two subtypes of α-adrenergic receptors (α_1, α_2) and three subtypes of β adrenergic receptors (β_1, β_2, β_3) have been identified according to their differing affinities or susceptibilities to synthetic ligands with agonist or antagonist biological activity.

Table 32-3 shows the five subtypes of adrenergic receptors, their relative affinities for E and NE, their subtype-specific agonists and antagonists, their tissue distribution, and the biological effects they mediate. Although both E and NE are capable of binding to and activating any of the adrenergic receptor subtypes (albeit to differing extents), the receptor subtypes may not be equally

TABLE 32-3

The Catecholamine Receptors

Receptor Subtype[1]	Relative Affinity[2]	Subtype-Specific Agonist	Subtype-Specific Antagonist	Post-receptor Events	Tissue	Physiologic Ligand[3]	Effect
D1	DA>>E=NE		Metoclopramide	Gs: AC stimulation Increased cAMP	Vascular smooth muscle (kidney, coronary, mesenteric)	DA	Relaxation (low levels)
D2	DA>>E=NE	Bromocriptine		Gi: AC inhibition Decreased cAMP	Lactotrope (anterior pituitary)	DA	Inhibition of prolactin release
$\alpha 1$[4]	E≥NE>>I	Methoxamine Phenylephrine	Phentolamine Phenoxybenzamine	Gp:PLC:IP_3:DAG Ca^{+2}-Protein Kinase	Vascular smooth muscle (skin)	NE	Constriction
					Vascular smooth muscle (arterioles)	NE	Constriction
					Vascular smooth muscle (abdominal viscera except liver)	NE	Constriction
					Intestinal smooth muscle	NE	Relaxation
					Intestinal sphincters	NE	Contraction
					Urinary smooth muscle (ureter, sphincter of bladder)	NE	Increased ureter tone, contraction of sphincter
					Iris (radial muscle)	NE	Contraction (mydriasis (pupil dilation))
					Skin (piloerector muscle)	NE	Contraction (piloerection)
					Skin (sweat gland) (palm of hands, sole of feet)[5]	NE	Secretion
					Hypothalamus	NE	Stimulation of ADH release
$\alpha 2$[6]	E≧NE>>I	Clonidine	Yohimbine	Gi:AC inhibition Decreased cAMP	β cell Islet of Langerhans	NE, E	Inhibition of insulin secretion
					Vascular smooth muscle (skin)	E	Constriction (vasoconstriction)
					Vascular smooth muscle (veins)	NE, E	Constriction
					Hypothalamus	NE	Decreased SRIH and/or increased GRH release (results in increased GH release from pituitary)
$\beta 1$[7]	I>E=NE	Dobutamine	Metoprolol Atenolol	Gs: AC stimulation Increased cAMP	Heart (SA node, AV node, His-Purkinjie system, ventricles)	NE, E	Increased firing rate; increased conduction velocity; increased contractility, increased heart rate, increased cardiac output
					Juxtaglomerular appartus of kidney	NE	Stimulation of renin release
					Adipose tissue (white)	E	Lipolysis
$\beta 2$	I>E>>NE	Terbutaline Salbutamol Soterenol	Butoxamine	Gs: AC stimulation Increased cAMP	Vascular smooth muscle (skeletal muscle)	E	Relaxation (vasodilation)
					Vascular smooth muscle (coronary)	E	Relaxation (vasodilation)
					Bronchiolar smooth muscle	E	Relaxation (bronchodilation)
					Skeletal muscle	E	Glycogenolysis; Increased contractility?
					α cells Islets of Langerhans	NE, E	Stimulation of glucagon release
					Intestinal smooth muscle		
					Genitourinary smooth muscle		
					Liver	E	Glycogenolysis; Gluconeogenesis
$\beta 3$	I=NE>E, Low affinity receptor	Fenoterol	none[8]	Gs: AC stimulation Increased cAMP	White adipose tissue (visceral > subcutaneous)	NE, E	Lipolysis
					Brown adipose tissue	NE (and E)	Thermogenesis

[1] $\alpha 1$ and $\alpha 2$ each have 3 subtypes, information about which can be found in: Bylund (1992) Subtypes of $\alpha 1$- and $\alpha 2$-adrenergic receptors. FASEB J 6:832–839.
[2] DA = dopamine; E = epinephrine; NE = norepinephrine; I = isoproterenol (a non-selective β-adrenergic agonist; stimulates all 3 subtypes of β-adrenoceptors).
[3] NE = norepinephrine as neurotransmitter released from sympathetic nerve endings at synaptic junctions, acting upon postjunctional adrenergic receptors; E = plasma-borne epinephrine that bind to extrajunctional adrenergic receptors that are not associated with nerve terminals.
[4] Localized entirely postsynaptically.
[5] Most of the sweat glands are innervated by *cholinergic* sympathetic fibers that release AChol at synaptic junctions and respond to sympathetic stimulation with an increase in secretion (increased sweating). The exceptions are the sweat glands in the palm of the hands and sole of the feet, which are innervated by adrenergic fibers that release NE at synaptic junctions. Emotional excitement activates these adrenergic nerves, and causes sweating in the hands and feet ("adrenergic sweating"). Emotional excitement usually has no effect on sweating in other regions because the cholinergic sympathetic nerves are mainly controlled by the thermoregulatory center of the hypothalamus, and are not affected by behavioral reactiveness. Note, however, that all sweat glands have $\alpha 1$ adrenoceptors, and are therefore capable of responding to plasma epinephrine (and NE?) with an increase in sweat secretion. This would result in a more generalized sweating.
[6] Located both pre- and postsynaptically.
[7] Primarily involved in neuronal functions.
[8] Metoprolol is a non-selective antagonist.

accessible to both circulating (E) and neurotransmitted (NE) catecholamines. In fact, thedistribution of the subtypes appcars to bc appropriate for the ligand that is prevalent in a given region. For example, the β_2-adrenergic receptors (E > NE) in bronchiolar smooth muscles are located in regions of the respiratory system that are poorly supplied by sympathetic nerve fibers but are accessible to agents that arrive by way of the blood supply. These receptors are most likely to respond to circulating E than to neural NE, and are responsive to inhaled synthetic adrenergic agonists (e.g., albuterol, salmeterol) that diffuse through the mucosal surface. On the other hand, the β_1-adrenergic receptors in the juxtaglomerular apparatus (JGA) of the kidney, the site of renin release that is richly supplied by sympathetic nerve fibers but is also well vascularized, tend to respond preferentially to neural NE rather than to plasma E, despite the fact that this subtype is known to have equal affinity for both catecholamines (Table 32-3). This suggests that the β_1 receptors of the JGA are localized in the synaptic clefts as postjunctional binding sites and are not accessible to plasma-derived E.

To facilitate the discussion on the effects of catecholamines (below), Table 32-3 includes a list of the "physiological ligand" (i.e., neural NE or plasma E) for the adrenergic receptor subtype in some of the tissues, based on their efferent sympathetic nerve supply. The different tissue distribution of adrenergic receptor subtypes indicates the genetic influence on the kinds of receptors; however, the density of the receptor subtypes also exhibits tissue variation. For example, although white adipose tissue contains both β_1 and β_3 receptors, the latter is in greater quantity in visceral adipose tissue.

The density of adrenergic receptors is strongly influenced by hormones. The adrenergic catecholamines exert an important regulation on their own receptors, causing downregulation when the concentrations are raised. All subtypes of adrenergic receptors except the β_3 subtype is subject to downregulation. Thyroid hormone increases the density of β-adrenergic receptors in some tissues (heart, fat, skeletal muscle), but decreases it in the liver. Glucocorticoids (GC) increase the density of β_2-adrenergic receptors in vascular and bronchiolar smooth muscle, and has been shown to be effective in reversing agonist-induced desensitization of β_2-adrenergic receptors in asthmatic patients.

Biological Effects of Epinephrine and Norepinephrine (Table 32-3)

Although NE is present at higher concentrations in plasma than E (Table 32-1), plasma NE is physiologically ineffective under most conditions. Plasma NE levels above 100 ng/dL are required for a systemic response, and such concentrations usually are produced only during very stressful situations (e.g., strenuous exercise or myocardial infarct). In contrast, E is capable of activating adrenergic receptors at threshold concentrations of 10–15 ng/dL, levels that are attained with relatively mild stimuli (e.g., cigarette smoking or hypoglycemia). E exerts important cardiovascular, pulmonary, renal, metabolic, endocrine, and thermogenic effects, most of which are supported or complemented by neural effects of NE that are exerted at the same time.

Cardiovascular, Pulmonary, and Renal Effects

E increases cardiac output by increasing both the strength and the rate of ventricular contractions (β_1). This effect is augmented by the constrictive effect of E (α_2) and NE (α_2) on the great veins, which increases venous return to the heart and leads to increased ventricular preload and increased stroke volume. The increased cardiac output is selectively directed to certain organs by the combined effects of circulating E and neurotransmitted NE. Thus, blood flow to the skin (subcutaneous) is reduced by the combined vasoconstrictive effects of E (α_2) and NE (α_1), while flow to the gastrointestinal tract, spleen, pancreas and kidney are reduced by sympathetic (NE) stimulated vasoconstriction (α_1). At the same time, E redirects the flow of blood through skeletal and cardiac muscles by promoting dilatation of the arterioles of the skeletal muscle (β_2) and the coronary arteries of the heart (β_2). Because the arteries in the brain, lungs, and liver are not directly affected by the adrenergic catecholamines, these tissues passively receive an increase in blood flow as a result of the augmented cardiac output and increased blood pressure secondary to increased peripheral resistance.

Pulmonary Respiratory System

E promotes air flow through the bronchioles by causing relaxation of their smooth muscle (bronchodilatation) (β_2), and thus allows for increased alveolar ventilation. The increased blood flow through the lungs (see above) and the increased alveolar ventilation ensure maximal oxygenation of blood.

Renal Urinary System

Autonomic discharge results in stimulation of ADH release due to a neurotransmitted NE effect (α_1) on the magnocellular neurons in the hypothalamic paraventricular nuclei. This promotes increased retention of water, which, in turn, promotes the effective blood volume. In addition, autonomic discharge causes the release of neurotransmitter

NE at the JGA of the kidney and this stimulates the release of renin (β_1), ultimately resulting in aldosterone-mediated sodium retention at the nephron. The overall result is preservation and promotion of vascular volume due to isotonic fluid retention, decreased urine flow, and decreased glomerular filtration rate, all of which are attributable to neurotransmitted NE.

Paradoxically, elevated levels of E and NE can cause a reduction in plasma volume that can lead to hemoconcentration and poor tissue perfusion; however, the mechanism of this effect is not known. This is said to be the cause of orthostatic hypotension in untreated patients with pheochromocytoma, a tumor of chromaffin tissue that produces excessive amounts of NE; most of these patients have reduced plasma volume but exhibit hypertension while in the reclining position.

Metabolic, Endocrine, and Thermogenic Effects

E promotes the production and release of glucose from the liver by two related mechanisms:

1. A direct stimulatory effect on hepatic glycogenolysis (β_2) mediated by cAMP-dependent phosphorylation of phosphorylase. This effect is dependent on prior storage of glycogen in the liver; therefore, both insulin and cortisol serve to condition the liver for this effect.
2. A direct activation of gluconeogenesis in the liver (β_2) via cAMP-dependent phosphorylation of the key gluconeogenic enzymes. This effect requires prior induction and maintenance of enzyme concentrations by cortisol.

These two glucose-generating actions of E on the liver are enhanced in an additive fashion by glucagon which, like E, depends on cAMP mediation and cortisol conditioning. The fact that neural NE simultaneously causes a fall in the insulin/glucagon ratio indicates that the sympathetic nervous system has a supportive role.

Neural NE and plasma E are both stimulators of adipocyte lipolysis, an effect that is mediated by β_1- and β_3-adrenergic receptors and involves cAMP-mediated phosphorylation of hormone-sensitive lipase (HSL). There are regional differences in β_3-adrenergic receptor density (visceral fat has more than subcutaneous fat), which determines that catecholamine is physiologically important. Note that β_3 receptors are low-affinity receptors that require high concentrations of catecholamines. They respond better to neural stimulation because of the higher local concentration of NE at the fat tissue site.

Cortisol, thyroid hormone, and GH support this effect of the catecholamines, presumably by promoting the synthesis of one or more components of the lipolytic pathway. This results in the release of free fatty acids (FFAs) and glycerol. Glycerol is taken up by the liver and is used for the production of glucose via gluconeogenesis (β_2). FFAs are taken up by cardiac muscle, skeletal muscle, kidney, and liver, and are oxidized for energy which obliges these tissues to consume less glucose. In the liver, where FFA uptake is related to circulating FFA levels, the oxidation of acyl-coA also is shunted to ketogenesis; ketone body production increases as a result (Chapter 18). E can promote FFA utilization and ketogenesis in the liver by inhibiting acetyl-CoA carboxylase activity (β_2); this results in unrestricted entry of FFA into the mitochondria for oxidation and subsequent ketogenesis. Ketone bodies are consumed by extrahepatic tissues that possess succinyl-CoA-acetoacetate-CoA transferase, also known as thiophorase (cardiac muscle, skeletal muscle, intestines, kidney, and the brain during starvation), and this also has a glucose-sparing effect.

E stimulates skeletal muscle glycogenolysis (β_2) and thus promotes glycolysis. The effects of E may depend on skeletal muscle activity because contraction alone is a stimulus of both glycogenolysis and glycolysis; the effects may also be influenced by the muscle fiber type (types I, IIa, or IIb). In resting muscle and presumably in type I (slow-twitch) fibers, there is preferential use of fatty acids and ketone bodies for energy, which also inhibits glycolysis; thus, the glycogenolytic and glycolytic effects of E may be minimal or overriden. In the contracting muscle and presumably in fast-twitch (types IIa and IIb) fibers, E may potentiate the active glycogenolytic and glycolytic pathways and promote the release of both lactate and alanine. Indirectly, E can inhibit skeletal muscle glycolysis by increasing the supply of FFAs and ketone bodies, which are glucose-sparing energy fuels for this tissue; however, the increased turnover of ATP in the contracting muscle probably allows the tissue to accommodate both glucose and fatty acids as fuels (Chapter 21).

A potentially harmful effect is the stimulation by E of potassium ion movement from plasma into skeletal muscle and liver (β_2), which can lead to *hypokalemia*. This effect is accompanied by a decrease in potassium ion excretion by the kidney. The ability of E to stimulate plasma membrane Na^+,K^+-ATPase activity in skeletal muscle (β_2) may partially explain this hypokalemic action, as well as the thermogenic effect of the hormone (see below).

The direct adrenergic innervation of the islets of Langerhans in the pancreas allows NE to stimulate the release of glucagon from the α cells (β_2) and to inhibit the release of insulin from the β cells (α_2), causing the insulin:glucagon (I:G) ratio to fall. Circulating E also exerts these effects and is probably important in maintaining the low I:G ratio following neural stimulation. Because the effects of a lowered I:G ratio are similar to those of NE and E on glycogenolysis, gluconeogenesis, lipolysis, and ketogenesis, the outcome is a great enhancement of the metabolic response to adrenergic discharge.

In the neonate, neural NE promotes "nonshivering thermogenesis," i.e., heat production via stimulation of lipolysis and fatty acid oxidation in brown adipose tissue (β_3) with the support of thyroid hormone. Brown adipose tissue contains a high density of mitochondria with an "uncoupling protein" that allows the cells to oxidize fatty acids and generate heat as a major product; NE released at nerve endings stimulates cAMP-mediated lipolysis in these cells and promotes thermogenesis by this β_3-mechanism (Chapter 14).

In the adult, nonshivering thermogenesis is probably a minor source of body heat production because brown adipose tissue is sparse and limited to only a few regions of the body. Adults depend on neurogenically induced shivering thermogenesis, which does not depend on adipose lipolysis. However, even in the resting state, catecholamines stimulate thermogenesis by promoting the thermogenic effect of thyroid hormone. Thus, when administered to adults, E increases resting body temperature and oxygen consumption (β), in part due to stimulation by E of skeletal muscle Na^+,K^+-ATPase activity (β_2).

Disturbances in Adrenal Medullary Function

Because there is considerable overlap in the functions of the sympathetic nervous system and the adrenal medulla, elimination of the adrenal medulla would be tolerated as long as the autonomic nervous system remains functionally intact. However, overproduction of adrenal medullary hormones would be disruptive. Such catecholamine excesses are seen in patients with tumors of the adrenal medullary chrornaffin cells and/or tumors of chromaffin tissue located outside of the adrenal gland. These tumors are called *pheochromocytomas,* and, although their incidence is rare, the pathophysiology should be understood for full appreciation of adrenergic catecholamine functions under normal conditions. One site at which pheochromocytomas often develop is the organ of Zuckerkandl, a collection of pheochromocytes at the bifurcation of the aorta. Familial predisposition to pheochromocytoma can occur due to activating mutations of the *RET* proto-oncogene that cause multiple endocrine neoplasia type II (Chapter 37) and mutations in the von Hippel-Lindau tumor suppressor gene. Pheochromocytoma is usually characterized by intermittent to permanent hypertension with potentially life-threatening consequences. The biochemical diagnosis in patients suspected of pheochromocytoma consists of measuring epinephrine, norepinephrine, and their metabolites—namely, metanephrine, normetanephrine, dihydroxyphenylglycol, and vanillylmandelic acid—in a 24-hour urine specimen. Measurement of plasma levels of total free metanephrine has a very high sensitivity for detecting pheochromocytoma. Tumors of the pheochromocytoma can be surgically removed.

Supplemental Readings and References

E.-E. Baulieu: Dehydroepiandrosterone (DHEA): a fountain of youth? *Journal of Clinical Endocrinology and Metabolism* **81,** 3147 (1996).

S. R. Bornstein, C. A. Stratakis, and G. P. Chrousos: Adrenocortical tumors: recent advances in basic concepts and clinical management. *Annals of Internal Medicine* **130,** 759 (1999).

P. R. Casson, S. A. Carson, and St. J. E. Buster: Replacement dehydroepiandrosterone in the elderly: rationale and prospects for the future. *Endocrinologist* **8,** 187 (1998).

K. Clément, C. Vaisse, B. S. J. Manning, et al.: genetic variations in the β_3-adrenergic receptor and an increased capacity to gain weight in patients with morbid obesity. *New England Journal of Medicine* **333,** 352 (1995).

A. Ganguly: Primary aldosteronism. *New England Journal of Medicine* **339,** 1828 (1998).

T. L. Goodfriend, M. E. Elliott, and K. J. Catt: Drug therapy: angiotensin receptors and their antagonists. *New England Journal of Medicine* **334,** 1649 (1996).

P. A. Insel: Adrenergic receptors—evolving concepts and clinical implications. *New England Journal of Medicine* **334,** 580 (1996).

M. New: The prismatic case of apparent mineralocorticoid excess. *Journal of Clinical Endocrinology and Metabolism* **79,** 1 (1994).

W. Oelkers: Current concepts: adrenal insufficiency. *New England Journal of Medicine* **335,** 1206 (1996).

D. N. Orth: Cushing's syndrome. *New England Journal of Medicine* **332,** 791 (1995).

T. M. Penning: Molecular endocrinology of hydroxysteroid dehydrogenases. *Endocrine Reviews* **18,** 281(1997).

K. Pacak, W. M. Linehan, G. Eisenhofer, et al.: Recent Advances in Genetics, Diagnosis, localization, and Treatment of Pheochromocytoma. *Annals of Internal Medicine* **134,** 315 (2000).

P. M. Stewart: Mineralocorticoid hypertension. *Lancet* **353,** 1341 (1999).

P. C. White: Disorders of aldosterone biosynthesis and action. *New England Journal of Medicine* **331,** 250 (1994).

P. C. White and P. W. Speiser: Congenital adrenal hyperlpasia due to 21-hydroxylase deficiency. *Endocrine Reviews* **21,** 245 (2000).

CHAPTER 33

Endocrine Metabolism IV: Thyroid Gland

The thyroid gland consists of two lobes connected by an isthmus and is positioned on the ventral surface of the trachea just below the larynx. It receives adrenergic fibers from the cervical ganglion and cholinergic fibers from the vagus and is profusely vascularized by the superior and inferior thyroid arteries. Histologically, thyroid tissue is composed of numerous follicles lined by a single layer of epithelial cells around a lumen filled with proteinaceous material called *colloid.*

The follicle cells, called ***thyrocytes,*** produce the thyroid hormones and are derived from the entodermal pharynx. Interspersed between follicles are specialized APUD cells derived from the neural crest, called *C*-cells or *parafollicular* cells. These cells produce calcitonin (CT), a polypeptide hormone discussed in Chapter 37.

33.1 Structure-Activity of Thyroid Hormones

The generic term "thyroid hormones" refers to the iodinated amino acid derivatives T_3 (3,3′,5-triiodo-L-thyronine) and T_4 (3,3′,5,5′-tetraiodo-L-thyronine), the only iodinated hormones produced endogenously. T_3 is the biologically active hormone and is, for the most part, produced from T_4 in extrathyroidal tissues. T_4 lacks significant bioactivity and is a hormone precursor; however, since thyroidal production of T_4 and its handling by the body are the pivotal determinants of thyroid hormone effects, the term "thyroid hormone" includes T_4.

The basic structure of T_3 and T_4 is that of thyronine, a substituted tyrosine [0-(4-hydroxyphenyl)tyrosine] (Figure 33-1). Iodine residues at positions 3 and 5 of the inner phenolic ring confer on the outer ring a preferred orientation approximately 120° to the plane of the inner ring (Figure 33-2). Hormonal activity is maximal when the following requirements are met:

1. Iodine is present at positions 3 and 5 of the inner ring (required for receptor binding). A natural analog of T_3, called reverse T_3 (rT_3, 3,3′,5′-triiodo-L-thyronine), is inactive. Substitution of iodine by bromine or a methyl group almost completely abolishes the activity.
2. A substituent is present at position 3′, but it need not be iodine. One of the most potent synthetic thyroid hormones has an isopropyl group at position 3′.
3. A hydroxyl group is present at position 4′. This group forms a hydrogen bond with the receptor protein.
4. A substituent is absent at position 5′. This permits full expression of the activity of the hormone. A substituent at this position would cause steric hindrance in binding with receptor and interfere with hydrogen bond formation. Thus, T_4 lacks significant activity whereas T_3 has about six times the activity of T_4.

A list of expanded acronyms appears as Appendix VIII.

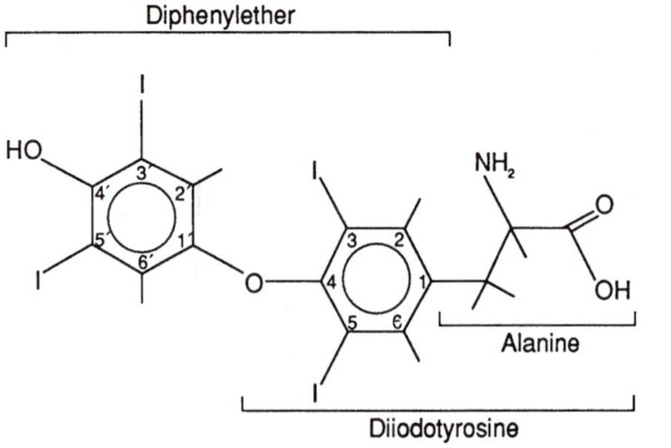

Diphenylether

HO

NH_2

O

OH

Alanine

Diiodotyrosine

FIGURE 33-1
Structure of 3,3′,5,5′-tetraiodo-L-thyronine (T_4). T_4 is a substituted tyrosine or a diphenylether derivative of alanine. Positions in the outer phenolic ring have prime designations, in contrast to those in the inner ring. T_4 is iodinated at positions 3 and 5 in both rings. [Modified and reproduced with permission from V. Cody, Thyroid hormone interactions: molecular conformation, protein binding, and hormone action. *Endocr. Rev.* **1,** 140 (1980). ©1980 by the Endocrine Society.]

5. The molecule must be the L form. D Isomers have minimal biological activity; for example, D-T_3 has about 7% the activity of L-T_3.

33.2 Thyroid Hormone Synthesis

Two substrates are required in the synthesis of thyroid hormones. The intrinsic substrate is ***thyroglobulin*** (Tg), a homodimeric glycoprotein (M.W. 669,000) synthesized in the rough endoplasmic reticulum of the thyrocytes and secreted into the follicular lumen by exocytosis. Tg contains 134 tyrosyl residues, only 25–30 of which undergo iodination and only four of which become a hormonogenic segment of the molecule.

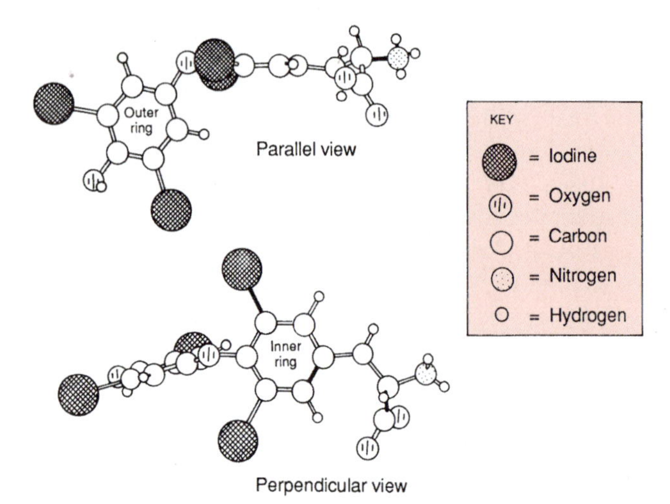

FIGURE 33-2
Skewed conformation of T_4. The molecule shown at the top is viewed in parallel to the plane of the inner phenolic ring, while that shown at the bottom is viewed perpendicularly to the plane of the inner ring. The ether linkage between the two rings is angled at about 120°. [Reproduced, with permission from V. Cody, Thyroid hormone interactions: molecular conformation, protein binding, and hormone action. *Endocr. Rev.* **1,** 140 (1980). © 1980 by the Endocrine Society.]

The extrinsic substrate is elemental ***iodine,*** present in food as inorganic iodide. Iodide is readily absorbed via the small intestine; it is almost entirely removed from the general circulation by the thyroid and kidney. Iodide that is taken up by salivary and gastric glands is secreted in salivary and gastric fluids and is returned to the plasma iodide pool by intestinal reabsorption. A small amount of iodide is taken up by the lactating mammary gland and appears in milk. The synthesis and release of thyroid hormone involves a number of steps (Figure 33-3).

1. ***Uptake of Iodide.*** Iodide (I^-) is actively taken up by the follicular cells against electrical and concentration gradients. The active uptake of I^- is mediated by the Na^+/I^- symporter, an intrinsic membrane protein with approximately 13 transmembrane segments. The Na^+ and I^- bound symporter releases both Na^+ and I^- on the cytoplasmic side. The empty symporter then returns to its original conformation exposing binding sites on the external surface of the cell. The internalized Na^+ is pumped out of the cell by the ATP-dependent Na^+, K^+-ATPase (oubain sensitive) to maintain an appropriate ion gradient. In the normal gland, the limiting step of thyroid hormone synthesis is uptake of I^-. Both I^- and thyroid-stimulating hormone (TSH) regulate the Na^+/I^- symporter function. TSH promotes I^- uptake whereas excess I^- decreases I^- uptake (discussed later). Anions of similar charge and ionic volume (e.g., ClO_4^-, BF^-, TcO_4^-, SCN^-) compete with I^- for transport in the follicular cell.
2. ***Activation and Organification of Iodide.*** Iodide that enters thyrocytes is "activated" by oxidation that is catalyzed by thyroperoxidase (TPO). Hydrogen peroxide is required and is supplied by an H_2O_2-generating system which may be an NADPH (NADH) oxidase system similar to that of leukocytes (Chapter 15). Thyroperoxidase is a glycosylated heme enzyme that is bound to the apical plasma membrane of the thyrocyte. Thyroperoxidase exists in two molecular forms (M.W. 105,000 and 110,000) and its catalytic domain faces the colloid space. In thyroid autoimmune disorders, one of the major microsomal antigenic components is thyroperoxidase.

In vitro, iodide peroxidase catalyzes the reaction $2H^+ + 2I^- + H_2O_2 \rightarrow I_2 + 2H_2O$. However, *in vivo*

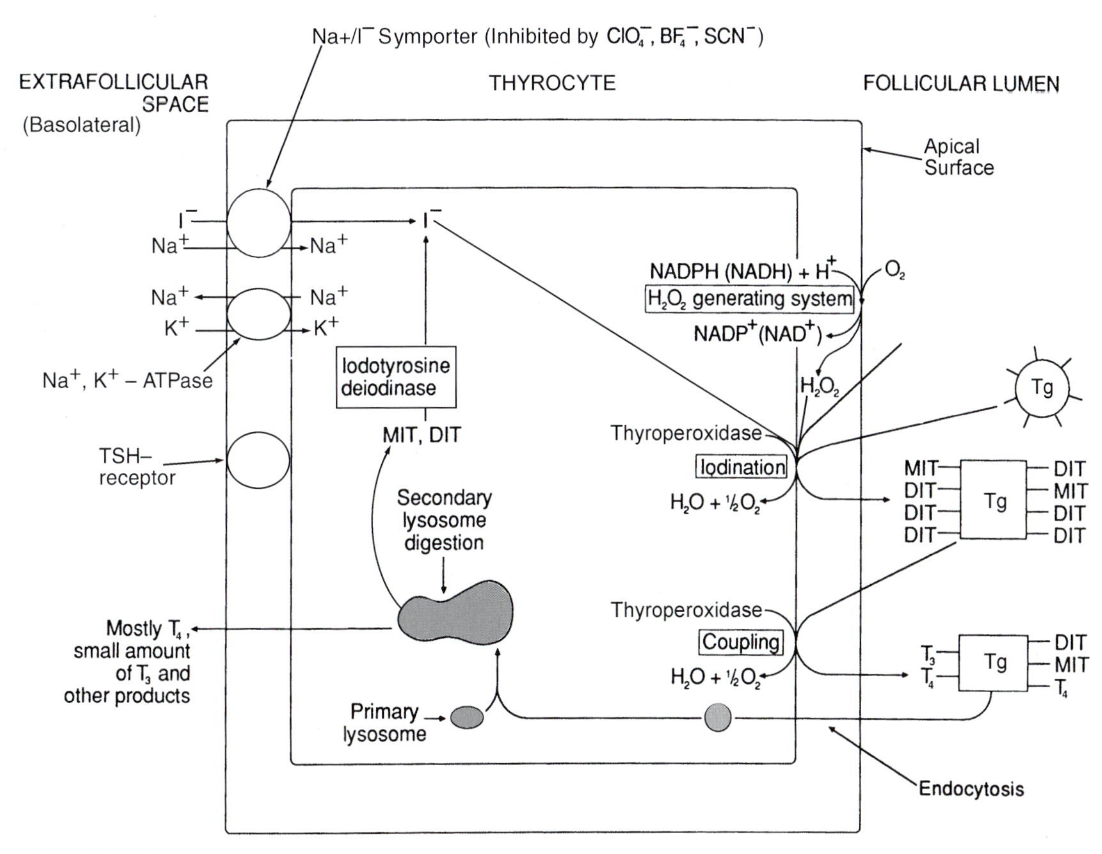

FIGURE 33-3
Synthesis of thyroid hormone in the thyrocyte. Tg, Thyroglobulin.

the enzyme probably forms an enzyme-bound iodinium ion (E-I^+) or a free radical of iodine. These activated derivatives iodinate the tyrosyl residues of Tg ("organification of iodide") by the action of thyroperoxidase and H_2O_2. Thionamides (e.g., thiourea, thiouracil, propylthiouracil, methimazole) inhibit this organification of iodide without affecting iodide uptake. The first position iodinated is position 3, which forms monoiodotyrosyl-Tg (MIT-Tg). The second is position 5, which forms diiodotyrosyl residues (DIT-Tg). Iodination is accompanied by structural changes (e.g., cystine formation) and includes a conformational change in the Tg molecule (illustrated diagrammatically in Figure 33-3). Activation and organification occur at the cell-colloid interface where thyroperoxidase activity is prevalent.

3. ***Coupling Reaction and Storage as Colloid.*** Iodide peroxidase or a "coupling enzyme" catalyzes the "coupling reaction" at the cell-colloid interface by intramolecular coupling of two iodotyrosyl residues with formation of an iodothyronyl residue. Coupling of DIT residues is favored; thus, formation of T_4 residues predominates. The coupling reaction requires that both substrates be iodinated and that one of the substrates be DIT-Tg. For this reason, no T_0, T_1, or T_2 residues are formed. Under normal conditions less than half of the iodotyrosyl residues in Tg undergo coupling such that of the total iodinated residues in Tg, 49% are MIT, 33% DIT, 16% T_4, 1% T_3, and a trace amount rT_3. The ratio of T_4 to T_3 is ≥ 10 on a typical American diet, but it rises or falls with dietary iodine content. The coupling reaction appears to be the most sensitive to inhibition by the thionamides (propylthiouracil, methimazole), being inhibited at doses that do not inhibit the organification reactions. As a final note, although proteins other than Tg are iodinated in nonthyroid tissues that take up iodine (e.g., salivary gland, mammary gland, intestinal mucosa), no coupling of the iodinated residues occurs. Thus, the requirements for the coupling reaction may be more stringent than previously believed.
4. ***Processing of TG and Release of Thyroid Hormone.*** Thyroglobulin, with its tetraiodothyronyl residues, is a pre-prohormone stored in the follicular lumen.

When follicle cells are stimulated by TSH or TSI (see below), their luminal border ingests colloid (and hence Tg) into endocytotic vesicles. These vesicles fuse with lysosomes to form phagolysosomes, which, during transit toward the basolateral surface of the cell, hydrolyze Tg and release T_4, DIT, MIT, and a small amount of T_3. These substances are released into the cytosol, and T_4 and T_3 diffuse into the blood. MIT and DIT do not enter the circulation in significant amounts because they are rapidly deiodinated by iodotyrosine deiodinase, an enzyme complex that contains ferredoxin, NADPH: ferredoxin reductase, and a deiodinase containing flavin mononucleotide (FMN). This reaction promotes recycling of iodide within the follicle cell. Some of the iodide, however, diffuses into plasma and constitutes the daily "iodide leak," estimated to be about 16 μg/d. By mechanisms that are not clear, a small amount of intact Tg also leaks out of the thyroid and can be found circulating in plasma in normal individuals. Tg leakage at a rate of 100 μg/d has been reported in euthyroid individuals, and a concentration of 15–25 μg Tg per liter of serum is considered to be normal. The route of Tg leakage is by way of the lympathic drainage and is increased when the gland is excessively stimulated. Increased entry of Tg into the circulation may result in an immune response because of the antigenic nature of this glycoprotein.

The release of thyroid hormone following endocytosis of colloid is inhibited by iodide and is known to be reduced with high dietary intake of iodine. This inhibition is due to the inverse relationship between the iodide content of Tg and the digestibility of iodo-Tg by lysosomal peptidase; that is, poorly iodinated Tg is more readily digested than richly iodinated Tg.

Regulation of Thyroid Hormone Synthesis

Thyroid-Stimulating Hormone (TSH)

TSH (also called thyrotropin), which is secreted by the pituitary, plays a central role in the regulation of growth and function of the thyroid gland. TSH receptors are functionally coupled to G-proteins; thus, the extracellular stimulus by TSH is transduced into intracellular signals mediated by a number of G-proteins. Activation of G_S-protein results in the stimulation of the adenylate cyclase-cAMP-protein phosphorylation cascade. Other G-proteins coupled to TSH-receptor activation include G_Q-protein, which mediates the phospholipase C-phosphatidylinositol 4,5-bisphosphate-Ca^{2+} signaling pathway (see Chapter 30 for a detailed discussion).

The TSH receptor is a glycoprotein that consists of three major domains: a long amino-terminal extracellular segment that confers binding specificity, a transmembrane segment, and a carboxy-terminal intracytoplasmic segment that mediates G-protein interactions (see Figure 30-6). TSH-receptor activation promotes iodide trapping, organification of iodide, and endocytosis and hydrolysis of colloid by a mechanism that involves cAMP.

Thyroid disorders consisting of either hypofunction or hyperfunction can result form constitutive inactivating or activating mutations, respectively, in the TSH receptor. Similarly, $G_{S\alpha}$-inactivating (loss of function) or activating (gain of function) mutations can result in *hypothyroidism* and *hyperthyroidism.* An example of the former is *Albright hereditary osteodystrophy* and the latter is *McCune-Albright syndrome.* Both syndromes are discussed in Chapter 30. All of the effects of TSH on the thyroid gland are exaggerated in hyperthyroidism due to excessive stimulation of the TSH receptor whether or not the ligand is TSH. For example, in *Grave's disease* hyperactivity of the thyroid gland is due to a thyroid-stimulating antibody that binds the TSH receptor and, in so doing, activates it. This *thyroid-stimulating immunoglobulin (TSI)* or *long-acting thyroid stimulator (LATS)* is an expression of an autoimmune disease that may be hereditary, although in some cases it may be the result of a normal immune response. For example, certain strains of bacteria contain a membrane protein which is homologous to the TSH receptor and which elicits an immune response in infected individuals. The antibodies generated resemble TSI. Thus, some cases of hyperthyroidism may be due to cross-recognition across phylogenetic lines (molecular mimicry).

Human chorionic gonadotropin (hCG), a placental hormone, has some TSH-like activity. hCG synthesis is initiated during the first week after fertilization and reaches its highest level near the end of the first trimester, after which levels decrease. The action of hCG on thyrocytes may help meet the increased requirement of thyroxine during pregnancy. The total thyroxine pool is increased in pregnancy due to elevated levels of thyroxine-binding globulin. However, the serum free T_4 level, which represents the biologically active form, remains the same compared to nonpregnant status. During the first trimester, the fetal requirement for thyroxine is met by maternal circulation until the fetal pituitary-thyroid axis becomes functional late in the first trimester. The maternal-fetal transport of thyroxine which occurs throughout pregnancy minimizes the fetal brain abnormalities in fetal hypothyroidism (discussed later). Maternal hypothyroidism (e.g., due to women consuming iodine-deficient diets) is associated with defects in the neurological functions of offspring.

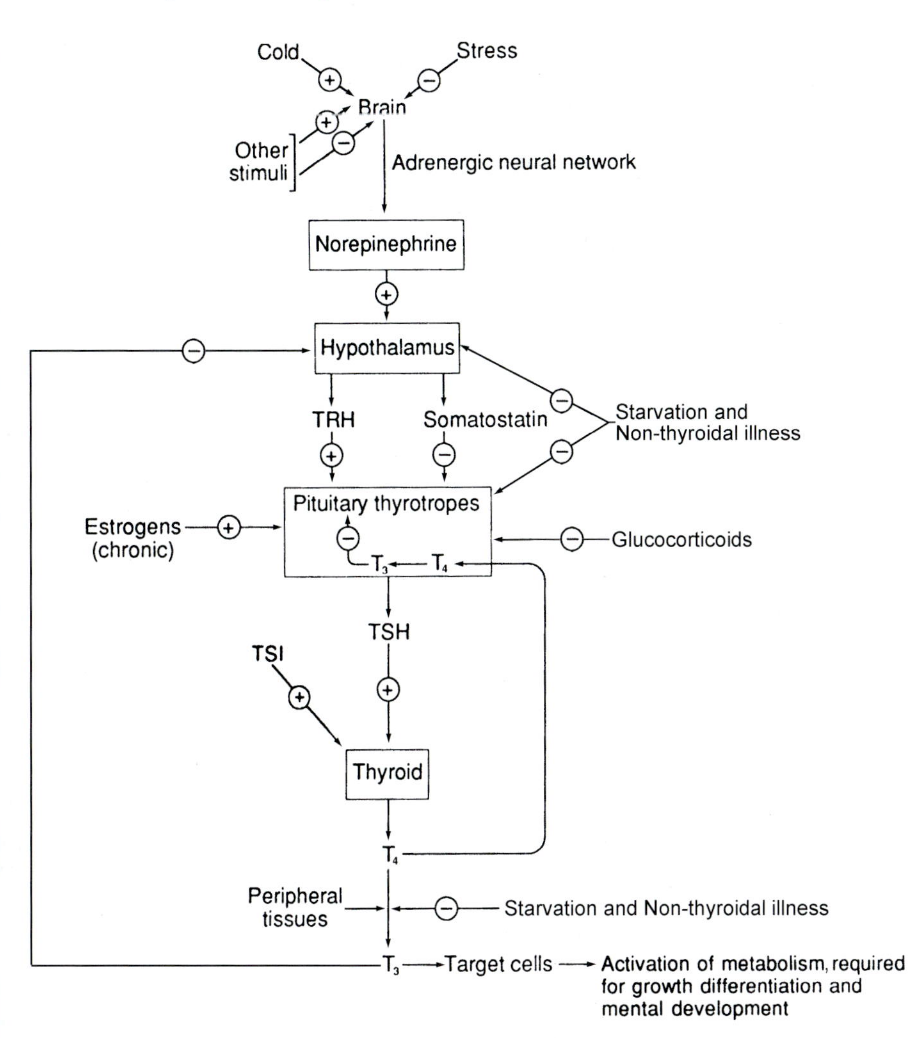

FIGURE 33-4
Schematic representation of the major steps in the regulation of thyroid hormone secretions and metabolism at five levels, namely, brain, hypothalamus, pituitary thyrotropes, thyroid, and peripheral tissues.

Release of TSH from the anterior pituitary is stimulated by hypothalamic TRH and inhibited by thyroid hormone (Chapter 31). These opposing signals to the anterior pituitary determine the magnitude of TSH secretion. Many other factors influence secretion of TSH and thyroid hormone (Figure 33-4). For example, starvation reduces conversion of T_4 to T_3 in peripheral tissues but not in the anterior pituitary; consequently, TSH secretion is not suppressed but circulating T_3 levels fall. Stress, somatostatin, estrogens, and cold exposure (in infants) also affect TSH release, but except for cold stress in infants, they are not regarded as important regulators of TSH and thyroid hormone secretion.

Iodine

The thyroid gland is capable of adjusting its synthetic activity to the supply of iodine. When the iodine supply is low, the thyroid makes maximal use of iodide; when the iodide supply is abundant, the thyroid defends itself against hormone overproduction by reducing iodide uptake. The thyroid gland autoregulates its iodine supply by an internal feedback mechanism that controls the intraglandular handling of iodide and the thyroid response to TSH. The exact nature of this feedback is not known, but the amount of the organified iodide pool is believed to have an inhibitory effect on the rate of iodide uptake, intraglandular T_4/T_3, the completeness of thyroglobulin hydrolysis, and the responsiveness to TSH.

This autoregulatory mechanism is not prominent when iodine intake is within the normal range but becomes physiologically significant when there is a moderate deficiency or excess of iodine. On an ordinary diet in which the iodine intake ranges between 300 μg and 1000 μg/d, T_4 production increases linearly with increased consumption of iodine. With lower iodine intake (e.g., 50–150 μg/d), T_4 production increases due to increased efficiency in iodide processing and to a greater response to TSH; in fact, although plasma TSH levels are within the normal range, a mild enlargement of the gland is not an uncommon consequence (e.g., nontoxic goiter). On the other hand, with a diet enriched in iodine (e.g., 1–2 mg/d), the increase in T_4 production as a function of iodine consumption is subnormal, due to negative feedback of the organified iodide pool on T_4 production and the thyroid response to TSH; as a result, plasma TSH levels are maintained within the normal range.

Abnormal thyroid function is seen at the extremes of iodine intake: at the lower end, less than 50 μg/d; at the upper end, 5 mg/d or more. At the lower end, frank hypothyroidism occurs because there is an insufficient supply of iodine to meet the T_4 production demands despite the efficiency with which the limited iodide is used. In this condition, the iodide deficiency creates an organified iodide pool consisting of MITs and a paucity of DITs, and, consequently, little or no coupling occurs because substrate requirements are not met. Products of Tg digestion are mainly MITs, small amounts of DITs, and possibly some T_3 and T_4. The decrease in thyroid hormone increases output of TSH which, in its futile attempt to increase T_4 production, creates glandular overgrowth due to thyrocyte hypertrophy and colloid accumulation ("hypothyroid goiter," "colloid goiter," or "endemic goiter"). At the upper end, further increase in iodine intake causes a shut-down of thyroid gland function and acutely reduces the thyroid response to TSH; consequently, there is a paradoxical decline in T_4 production as a function of iodine consumption accompanied by a dramatic reduction in vascular supply to the gland. This phenomenon is called the Wolff-Chaikoff effect and is observed only in thyroid glands that are stimulated (e.g., with TSH or in hyperthyroidism).

33.3 Transport and Metabolism of Thyroid Hormones

Thyroid hormones in blood are mainly bound to serum proteins. The major thyroid hormone–binding protein in blood is ***thyroxine-binding globulin*** (TBG), a glycoprotein (M.W. 63,000) of electrophoretic mobility between that of α_1- and α_2-globulins. TBG exhibits a high affinity for T_4 ($K_d = 50$ pM) and a moderate affinity for T_3 ($K_d = 500$ pM). Approximately three-fourths of circulating T_4 is firmly bound to TBG, while about one-half of circulating T_3 is TBG-bound (Table 33-1). *Thyroxine-binding prealbumin* (TBPA), more properly known as *transthyretin,* which functions in the transport of retinol, has about 1% of the affinity of TBG for T_4, but it accounts for about 20% of circulating T_4 and does not bind T_3. These differences in binding can be partially explained, since T_4 is almost completely ionized at physiological pH, whereas T_3 is not. The 4′-hydroxyl group of T_4 probably exists as a 4′-phenoxide ion, which optimizes binding to serum proteins. However, the 4′-hydroxyl group is required for receptor binding. Binding to albumin accounts for much of the circulating T_4 and T_3 that is not bound to TBG and TBPA. About 0.03% of T_4 and about 0.3% of T_3 is unbound, so that only a small fraction is accessible for target cell uptake. The albumin-bound hormone dissociates rapidly in the vicinity of target cells. The cellular uptake of T_4 and T_3 is mediated by carrier-mediated processes at stereospecific binding sites. At high-affinity sites the uptake is energy, temperature, and often Na^+-dependent. T_3 is metabolized about six times more

TABLE 33-1
Source, Transport, and Metabolism of the Thyroid Hormones

	T_4 3, 5, 3′, 5′-Tetraiodo-L-thyronine	T_3 3, 5, 3′-Triiodo-L-thyronine
Molecular weight	777	651
Production rate (µg/day)	90	35
% Produced by thyroid gland	100	20
Half-life	7 days	0.75 day
Total plasma concentration	51–142 nmol/L (4–11 µg/dl)	1.2–3.4 nmol/L (75–220 ng/dl)
% Free	0.02%	0.3%
% Bound to plasma proteins	99.98%	99.7%
% Bound to TBG	70–75	70–75
% Bound to albumin	5–10	20–25
% Bound to transthyretin	15–20	0–5

rapidly than T_4 and has a metabolic clearance rate (MCR) 25 times as high as that of T_4 (Table 33-1).

Mutations in a human serum albumin gene that substitute either histidine or proline for arginine at position 218 increase binding affinity for T_4. These mutations are autosomal dominant and occur with relatively high frequency. Carriers of these mutations have high levels of total serum T_4 but their free-T_4 and TSH concentrations are within the normal range. Individuals are euthyroid (normal thyroid) and their conditions are known as ***familial dysalbuminemic hyperthyroxinemia*** *(FDH)*.

Two reactions account for the metabolic fate of about 80% of the T_4 in plasma: about 40% is converted to T_3 via 5′-deiodination (activation), and another 40% of the T_4 is converted to rT_3 by 5′-deiodination (inactivation). These two reactions are catalyzed by three enzymes designated types I, II, and III iodothyronine deiodinases (Figure 33-5 and Table 33-2). Types I and II both catalyze the 5′-deiodination reaction but differ with respect to substrate specificity, tissue distribution, and regulation. Type III is a 5-deiodinase, which catalyzes the removal of iodine from position 5 of the inner ring. Type I is deiodinase selenocysteine–containing microsomal enzyme present in the liver, kidney, and thyroid, with specificity for thyronines bearing iodine at position 5′. It is responsible for catalyzing the formation of most of the circulating T_3 and is a high-capacity ($K_m \sim 1\ \mu M$) processor of T_4. A unique feature of the type I enzyme is that it is sensitive to inhibition by *propylthiouracil* (PTU), a known inhibitor of thyroid gland function that has little or no effect on the activity of the other deiodinases. It is also inhibited in conditions such as starvation, caloric restriction, and non-thyroid-related illnesses, which may deplete the cofactor supply and/or depress enzyme synthesis. The type II deiodinase is less abundant and has a higher specificity and a higher affinity for T_4 than type I. The type II enzyme is present in the pituitary, brain, brown fat, and skin, and is responsible for intracellular T_3 formation that leads to a cellular response; whereas the type I enzyme produces T_3 for export to other cells, and the type II enzyme is self-serving, producing its own supply of intracellular T_3. A unique feature of the type II enzyme is that it is sensitive to inhibition by T_4 by an unknown mechanism. The activity of the type II enzyme is increased in hypothyroidism and is depressed in hyperthyroidism, whereas these conditions have the opposite effect on the activity of the other two deiodinases. The type III deiodinase is mainly involved with the degradation of T_3 and the inactivation of

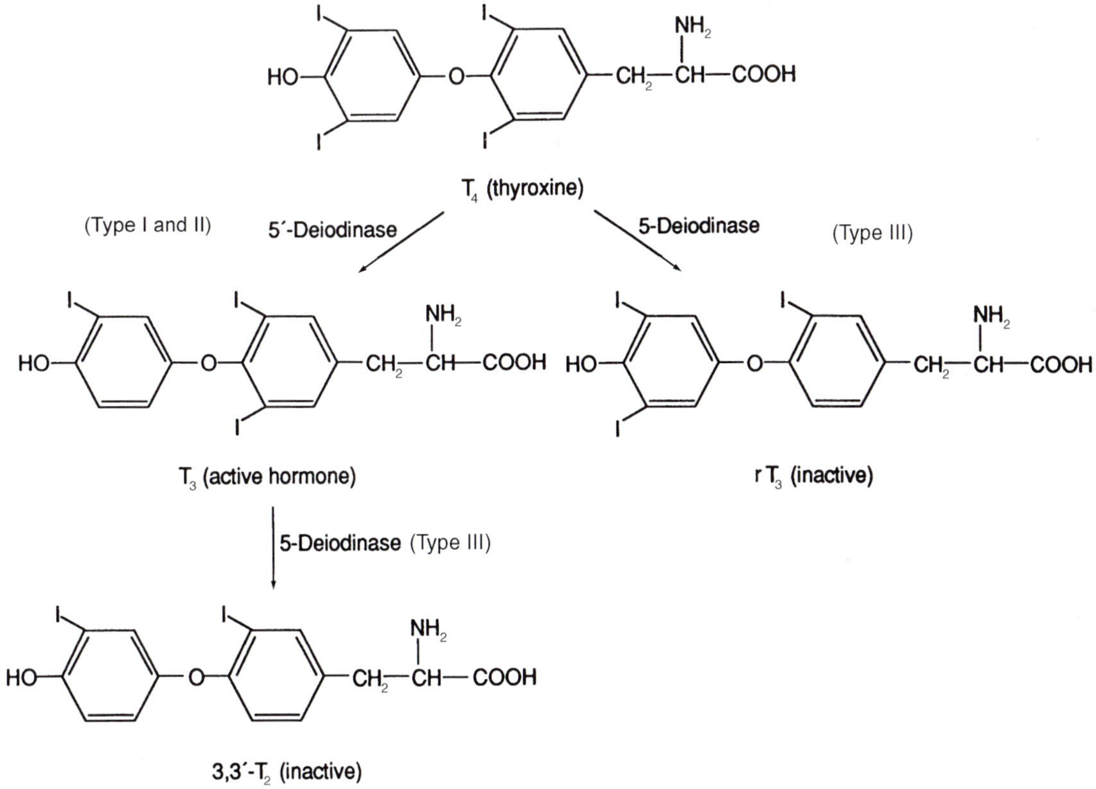

FIGURE 33-5
Activating and inactivating pathways of T_4.

TABLE 33-2
The Iodothyronine Deiodinases

	Type I	Type II	Type III
Reaction catalyzed	5′-deiodination (outer ring)	5′-deiodination (outer ring)	5-deiodination (inner ring)
Substrate preference	rT3 > T4 > T3	T4 > rT3	T3 > T4
Tissue distribution	Liver, kidney, thyroid, et al.	Brain, pituitary, skin, brown fat	Liver, kidney, et al.
Affinity (Km) for T4	Low (~1 μM)	High (~1 nM)	High (~40 nM)
Effect of:			
Iopanoic acid*	Decrease	Decrease	Decrease
Amiodarone**	Decrease	Decrease	Decrease
Fetal life	Decrease	Decrease	?
Propylthiouracil (PTU)	Decrease	None	None
Selenium deficiency	Decrease	None	
Starvation	Decrease	None	
Caloric restriction	Decrease	None	
Glucocorticoids‡	Increase	None	
Excess glucocorticoids	Decrease	None	
Nonthyroid illnesses	Decrease	None	
High carbohydrate diet	Increase	None	
Hypothyroidism	Decrease	Increase	Decrease
Hyperthyroidism	Increase	Decrease	Increase
Propranolol	Decrease	None?	?

*Iopanoic acid (3-amino-α-ethyl-2,4,6-*triiodo*-hydrocinnamic acid) is a radiopaque iodine compound used as a contrast medium in cholecystography.
**Amiodarone (2-butyl-3-benzofuranyl) [4- [2- (diethylamino) ethoxyl]-3,5-*diiodo*-phenyl] methanone) is a coronary vasodilator used in the control of ventricular arrhythmias, and in the management of angina pectoris.
‡The physiologic rise in cortisol levels at about the time of birth stimulates 5′-deiodinase and causes a rise in T3 and fall in rT3. The effect of physiologic levels of cortisol on 5′-deiodinase in adults is unclear.

T_4, both involving the removal of the inner ring 5-iodide. The type III enzyme is widely distributed among tissues and is responsible for the formation of almost all (95%) of the circulating rT_3. Its activity is not altered by most of the factors or conditions that inhibit the activity of types I or II deiodinases, and this explains why there is a reciprocal relationship between the plasma levels of T_3 and rT_3.

Not all metabolic processing of iodothyronines involves deiodination, although it is the major metabolic route for T_4, T_3, and rT_3, accounting for about 80% of the fate of each compound. The remaining 20% is conjugated in the liver with sulfate or glucuronide, and/or processed by oxidative deamination and decarboxylation to form thyroacetic acid derivatives with minimal biological activity (Figure 33-6).

Inorganic iodide liberated from deiodinations enter the extracellular compartment and is excreted in the urine (~488 μg/d) or stool (~12 μg/d via bile), or is reabsorbed by the thyroid gland (~108 μg/d) for resynthesis of thyroid hormones. At the average daily intake of ~500 μg inorganic iodide, this amounts to no net change in the iodide status, although the small amount of iodide that is lost through the skin and gastrointestinal tract would place the body in a negative iodide balance unless intake is adjusted accordingly.

33.4 Biological Actions of Thyroid Hormones

Mechanism of Action and Latency Period

All but a few of the thyroid effects that have been identified occur at the level of gene transcription, mediated by nuclear thyroid hormone receptors (Chapter 30). These effects have a longer latency period than for most steroids; some of the relatively early responses show a latency period of several hours.

Physiological Effects

Cardiovascular System

Thyroid hormone enhances cardiac contractility and exerts a positive chronotropic effect on the heart, increasing heart rate by a mechanism that may involve more than a potentiation of the β-adrenergic effect. In the

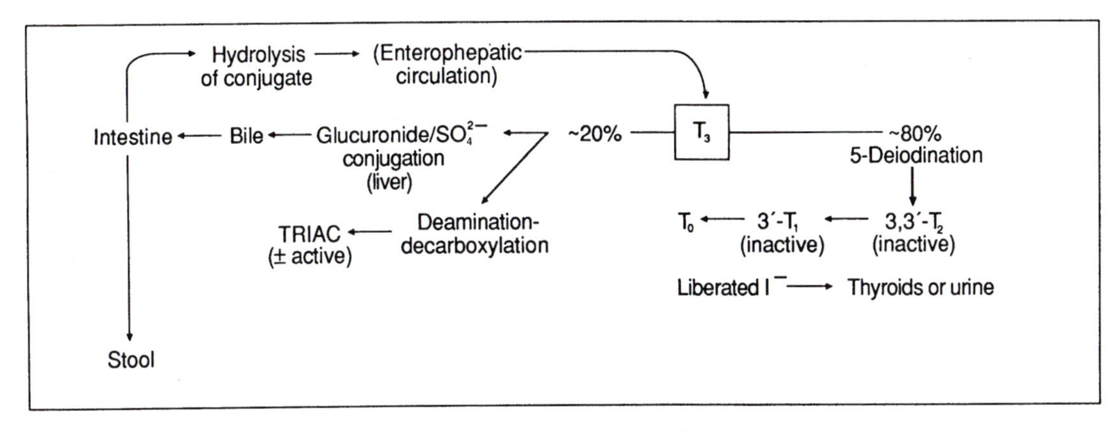

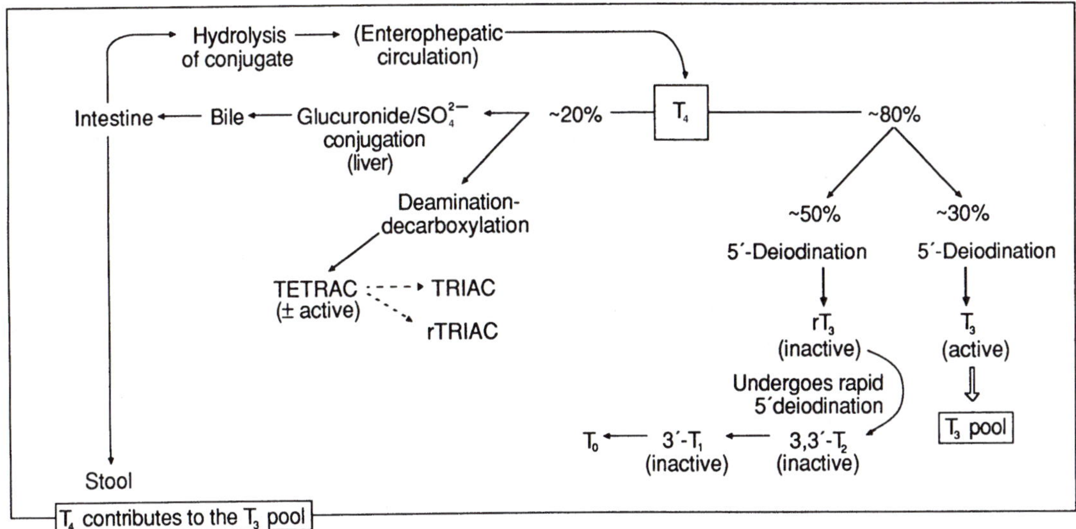

FIGURE 33-6
Metabolism of T_3 and T_4. T_0, Thyronine; 3,3′-T_2, 3,3′-diiodothyronine; 3′-T, 3′-iodothyronine; TETRAC, 3,3′,5,5′-tetraiodothyroacetic acid; TRIAC, 3,3′,5-triiodothyroacetic acid.

ventricular myocardium, thyroid hormone stimulates the synthesis of myosin heavy chain α (MHCα), while inhibiting the synthesis of MHCβ, by regulating the expression of the respective genes. The result is that myocardium with a higher MHCα content has higher calcium and actin-activated ATPase activities and increased velocity of muscle fiber shortening. Thus, the thyroid hormone–conditioned ventricle is capable of increased contractile performance.

Intermediary Metabolism

Thyroid hormone increases both lipolysis and lipogenesis, although lipogenesis is stimulated before lipolysis, due to early induction of malic enzyme (malate dehydrogenase), glucose-6–phosphate dehydrogenase, and fatty acid synthase. Thyroid hormone lowers serum cholesterol levels by maintaining low-density lipoprotein (LDL) receptor density, by maintaining hepatic lipase activity which catalyzes conversion of intermediate density lipoproteins (IDL) to LDL, and probably also by maintaining lipoprotein lipase activity which promotes triglyceride clearance. In liver, kidney, skeletal muscle, cardiac muscle, and adipose tissue, thyroid hormone stimulates Na^+,K^+–ATPase gene expression and promotes thermogenesis.

Growth and Maturation

Thyroid hormone stimulates production of IGF-I directly (liver) and indirectly (via increased growth hormone, GH). In adults with hypothyroidism, basal serum GH levels are normal, but the GH responses to provocative stimuli, nocturnal GH secretion, and serum IGF-I are all subnormal. In ***cretinism*** (infantile hypothyroidism), linear growth is severely retarded and the resultant dwarfism is characterized by a retention of the high upper to lower body ratio of infancy. The growth retardation in persons with cretinism is primarily due to delayed appearance of ossification centers in long bone, and secondarily to a deficiency in growth factors.

A deficiency of thyroid hormone during fetal development due to untreated or undertreated maternal hypothyroidism results in a neurological deficit in the offspring. In congenital hypothyroidism, a normal maternal thyroid can meet the fetal requirements for thyroid hormone. However, in the postnatal period these infants require a prompt thyroid hormone replacement therapy throughout their life, beginning in the first few weeks of life. If untreated, they inevitably develop growth and mental retardation. Thyroid hormone is essential for maturational development of the CNS and is required for the development of axonal projections and myelination. One of the important effects of thyroid hormone is to promote the synthesis of myelin basic protein.

Congenital hypothyroidism in Western populations occurs in about 1 in 4000 births. Congenital hypothyroidism is caused by embryogenic defects leading to thyroid agenesis (50–55%), dysgenesis (30–35%), errors in the synthesis of thyroid hormone (10–15%), and defects in the pituitary-hypothalamic axis (4%). Neonatal screening for congenital hypothyroidism is performed using blood collected on filter paper from newborns older than 12 hours. The laboratory test consists of measuring T_4, TSH, or both. Abnormal values must be reconfirmed and additional tests include other parameters (e.g., TBG).

Reproductive System

Thyroid hormone increases total plasma androgen levels by increasing the production of testosterone-binding globulin (TeBG) by the liver. Hyperthyroidism is associated with increased plasma TeBG levels, higher total testosterone levels, but normal free testosterone levels in adult men. There is a high incidence of gynecomastia in hyperthyroid men (40–83%) due to a higher circulating level of free estradiol in plasma. The elevated free-estradiol level may be explained by the lower affinity of estrogen for TeBG and also by increased conversion of androgen to estrogen in hyperthyroidism.

In prepubertal boys, but not in older boys or adult men, hypothyroidism is associated with a high incidence of increased testicular growth (macroorchidism) that is not accompanied by any change in testosterone levels. Elevated follicle-stimulating hormone levels are found in most cases, but there is no testicular maturation. The mechanism of this disorder is not known.

Pharmacological Effects

The effects of thyroid hormone may persist for up to a week because of its long half-life in the circulation, its long-acting effect on the genome of the target cell, and the slow rate of recovery of some target tissues (e.g., brain).

Thyroid hormone excess depletes body energy stores and imparts hypersensitivity of tissues to catecholamines. Protein synthesis is inhibited, protein catabolism is accelerated, and the antianabolic actions of cortisol are enhanced. During the period of growth, these factors would lead to growth suppression, muscle weakness, weight loss, and depletion of liver and muscle glycogen stores. Gluconeogenesis is fueled by elevated substrate levels (amino acids, glycerol, lactate, and pyruvate), and this contributes to the blood glucose pool; however, since peripheral glucose utilization is also increased, hyperglycemia is not severe. Thyroid excess also leads to depletion of triacylglycerol stores, mainly due to its action on catecholamine-induced lipolysis. Tissue consumption of free fatty acids and ketones is also increased. Thus, serum lipid levels fall, including the level of serum cholesterol. In essence, the storage forms of glucose, amino acids, and fatty acids are depleted, and these substrates are rapidly metabolized, resulting in the need for increased food intake.

Because thyroid hormone increases tissue responsiveness to catecholamines, many symptoms of thyroid hormone excess are those that characterize catecholamine excess. Effects of T_3 on cardiovascular hemodynamics consist of increased tissue thermogenesis, decreased systemic vascular resistance, decreased effective arterial filling volume, increased renal sodium reabsorption, increased blood volume, increased cardiac inotropy and chronotropy leading to increased cardiac output. Behavioral changes (such as nervousness, restlessness, short attention span, and emotional lability) are common, some of which require a longer period for decay than do the metabolic changes following normalization of thyroid hormone levels. Clearly, the central actions of thyroid hormone involve more than potentiation of adrenergic neurotransmission.

Thyroid disorders. Disturbances in thyroid metabolism can occur at any level of the hypothalams-pituitary-thyroid-peripheral tissue axis. Several of these disorders have been discussed previously. Hyperthyroidism is more prevalent in women than men. The three most common causes of hyperthyroidism are Graves' hyperthyroidism, toxic multinodular goiter, and toxic adenoma. The clinical features of hyperthyroidism include hyperkinesis, weight loss, cardiac anomalies (e.g., atrial fibrillation), fatigue, weakness, sweating, palpitations, and nervousness. The typical biochemical laboratory parameters are increased serum free T_4 and decreased serum TSH.

The most common cause of hypothyroidism is failure of the thyroid gland; this is known as primary hypothyroidism. In adults, the cause of primary hypothyroidism is often spontaneous autoimmune disease (e.g., Hashimoto's thyroiditis) or destructive therapy for hyperthyroid states (e.g., Graves' disease). In adults, hypothyroidism has an insidious onset with a broad range of symptoms. The typical biochemical laboratory parameters include the measurement of serum free T_4 and TSH, which are decreased and increased, respectively. In addition, measurement of serum autoantibodies for thyroglobulin and thyroperoxidase is also utilized.

Supplemental Readings and References

D. L. Bodenner and R. W. Lash: Thyroid disease mediated by molecular defects in cell surface and nuclear receptors. *American Journal of Medicine* **105,** 524 (1998).

S. C. Boyages: Clinical review 49: Iodine deficiency disorders. *Journal of Clinical Endocrinology and Metabolism* **77,** 587 (1993).

G. A. Brent: The molecular basis of thyroid hormone action. *New England Journal of Medicine* **331,** 847 (1994).

H. B. Burch and L. Wartofsky: Graves' ophthalmopathy: current concepts regarding pathogenesis and management. *Endocrine Review* **14,** 747 (1993).

G. N. Burrow: Thyroid function and hyperfunction during gestation. *Endocrine Review* **14,** 194 (1993).

G. N. Burrow, D. A. Fisher, and P. R. Larsen: Mechanisms of disease: maternal and fetal thyroid function. *New England Journal of Medicine* **331,** 1072 (1994).

I. J. Chopra: Clinical review 86: Euthyroid sick syndrome: is it a misnomer? *Journal of Clinical Endocrinology and Metabolism* **82,** 329 (1997).

D. Glinoer: The regulation of thyroid function in pregnancy: pathways of endocrine adaptation from physiology to pathology. *Endocrine Review* **18,** 404 (1997).

M. A. Lazar: Thyroid hormone receptors: multiple forms, multiple possibilities. *Endocrine Review* **14,** 184 (1993).

J. H. Lazarus: Hyperthyroidism. *Lancet* **349,** 339 (1997).

R. S. Lindsay and A. D. Toft: Hypothyroidism. *Lancet* **349,** 413 (1997).

S. M. McLachlan and B. Rapoport: The molecular biology of thyroid peroxidase: cloning, expression and role as autoantigen in autoimmune thyroid disease. *Endocrine Review* **13,** 192 (1992).

G. Medeiros-Neto, H. M. Targovnik, and G. Vassart: Defective thyroglobulin synthesis and secretion causing goiter and hypothyroidism. *Endocrine Review* **14,** 165 (1993).

A. J. Mixson, R. Parrilla, S. C. Ransom, et al.: Correlations of language abnormalities with localization of mutations in the B-thyroid hormone receptor in 13 kindreds with generalized resistance to thyroid hormone: identification of four new mutations. *Journal of Clinical Endocrinology and Metabolism* **75,** 1039 (1992).

J. C. Morris: The clinical expression of thyrotropin receptor mutations. *Endocrinologist* **8,** 195 (1998).

R. Paschke and M. Ludgate: Mechanism of disease: the thyrotropin receptor in thyroid disease. *New England Journal of Medicine* **337,** 1675 (1997).

S. P. Porterfield and C. E. Hendrich: The role of thyroid hormones in prenatal and neonatal neurological development—current perspectives. *Endocrine Review* **14,** 94 (1993).

S. Refetoff, R. E. Weiss, and S. J. Usala: The syndromes of resistance to thyroid hormone. *Endocrine Review* **14,** 348 (1993).

E. Roti, R. Minelli, E. Gardini, and L. E. Braverman: The use and misuse of thyroid hormones. *Endocrine Review* **14,** 401 (1993).

V. Singh and J. P. Catlett: Hematologic manifestations of thyroid disease. *Endocrinologist* **8,** 87 (1998).

M. I. Surks and R. Sievert: Drug therapy: drugs and thyroid function. *New England Journal of Medicine* **333,** 1688 (1995).

A. D. Toft: Thyroxine therapy. *New England Journal of Medicine* **331,** 174 (1994).

Y. Tomer and T. F. Davies: Infection, thyroid disease, and autoimmunity. *Endocrine Review* **14,** 107 (1993).

J. I. Totréns and H. B. Burch: Serum thyroglobulin measurement: utility in clinical practice. *Endocrinologist* **6,** 125 (1996).

G. Vassart and J. E. Dumont: The thyrotropin receptor and the regulation of thyrocyte function and growth. *Endocrine Review* **13,** 596 (1992).

A. De La Vieja, O. Dohan, O. Levy, et al.: Molecular analysis of the sodium/iodide symporter: impact on thyroid and extrathyroid pathophysiology. *Physiological Reviews* **80,** 1083 (2000).

A. P. Weetman and A. M. McGregor: Autoimmune thyroid disease: further developments in our understanding. *Endocrine Review* **15,** 788 (1994).

I. Klein and K. Ojamaa: Thyroid Hormone and the Cardiovascular System. *New England Journal of Medicine* **344,** 501 (2001).

CHAPTER 34

Endocrine Metabolism V: Reproductive System

The sex of an individual is the outcome of genetic determinants (genotypic sex) and of hormonal effects that confer structural features characteristic of a given sex (phenotypic sex). Both genetic and hormonal determinants normally operate only at two phases of life: during fetal development and at puberty. Genotypic sex is either homogametic (XX) or heterogametic (XY). In humans and other mammals, the homogametic (XX) pairing programs ovarian development and oocyte formation (oogenesis), while the heterogametic (XY) pairing leads to testicular development and spermatogenesis. In both genotypes, the embryonal gonads develop from the epithelium and stroma of the urogenital ridge, a thickening of the coelomic (ventromedial) aspect of the mesonephros that emerges at about the second or third week of pregnancy. Both the epithelium and stroma of the urogenital ridge are derived from the intermediate mesoderm of the embryo; however, invading this structure at about week 4 are primordial germ cells from the yolk sac, which take residence in association with the epithelial cells of the developing gonad and replicate. During the germ cell invasion, the epithelial cells of the gonads undergo proliferation and begin entering the stromal spaces as cord-like projections, called "primary sex cords." Until about week 6, the gonads are undifferentiated and uncommitted; that is, the structure has the potential to develop into either ovaries or testes.

A list of expanded acronyms appears as Appendix VIII.

The fetal gonads develop in parallel with the Wolffian (mesonephric) duct and the Müllerian (paramesonephric) duct, both of which are positioned bilaterally to the gonads within the urogenital ridge. The Wolffian duct begins forming from about week 4, starting at the mesonephros and terminating caudally at the cloaca. The Müllerian duct begins forming from about week 6 and merges caudally with the urogenital sinus. The Wolffian duct is the primordium of the male reproductive tract, whereas the Müllerian duct is the primordium of the female reproductive tract. Until about week 8, both ducts are present and undifferentiated. In the absence of a Y chromosome, a gene on the X chromosome directs the development of fetal gonads into ovaries. This critical sex-determining gene located on the X chromosome is known as the *DAX-1* gene. Subsequent to the action of this gene, the Müllerian ducts develop into a female genital tract and the Wolffian duct regresses.

In the presence of a Y-chromosome, the *SRY* gene (sex-determining region of the Y-chromosome) antagonizes the action of the *DAX-1* gene (the *SRY* gene was previously known as testis-determining factor, TDF). Mutations in the *SRY* gene can result in an XY female with gonadal dysgenesis. Testis differentiation may also require the participation of a gene located on chromosome 17 (*SOX9* gene). The precise interaction between *SOX9* and *SRY* in gonadal development is not yet understood; however, both genes code for DNA-binding proteins. Mutations in the *SOX9* gene can result in sex reversal in XY individuals with *camptomelic dysplasia*. The latter is a severe form of skeletal dysplasia associated with dysmorphic features and cardiac defects.

After the fetal gonads are converted into testes by *SRY*, testosterone and Müllerian-inhibiting substance (MIS) control subsequent sexual differentiation. MIS causes the regression of the Müllerian duct, which would otherwise develop into female genital tracts. Testosterone promotes the development of the Wolffian duct into male internal genitalia, including epididymis, vas deferens, and seminal vesicles.

Testosterone also causes differentiation of the urogenital sinus into the prostate gland and masculinizes the external genitalia; however, these tissues must first convert testosterone into dihydrotestosterone (DHT), a reaction catalyzed by cytoplasmic 5α-reductase type 2. The effects of testosterone on the Wolffian duct and urogenital sinus are dependent on the presence of androgen receptors and 5α-reductase, which are expressed by genes on the X chromosome and chromosome 2, respectively, and are present in both genotypic sexes. This explains why exposure of the fetus to androgens during the critical period may lead to masculinization of the internal and external genitalia of genotypic female fetuses. It also explains why lack of expression of either gene can lead to feminization in genotypic male fetuses.

The gonad has two interrelated functions: gametogenesis and hormone production. Gametogenesis depends on gonadal hormone production, which is influenced by gametogenesis. Both processes are controlled by luteinizing hormone (LH) and follical-stimulating hormone (FSH), which are released in response to hypothalmic gonadotropin-releasing hormone (GnRH) and to the levels of circulating gonadal hormones. In addition, gametogenesis is regulated by paracrine actions of the gonadal hormones.

The obvious difference between the sexes in gametogenesis is formation of ova (ootids) in the female and of sperm (spermatozoa) in the male. Spermatogenesis becomes operational from about the time of puberty and continues throughout life. The process takes about 74 days. Oogenesis occurs in three phases. The initial phase involves proliferation of the stem cells (oogonia) and occurs only in fetal life. The second phase involves the first maturational division (formation of the secondary oocyte) and occurs about the time of ovulation (see below). The third phase involves the second and final maturational division (formation of ova) and occurs at fertilization. Because the number of oogonia is determined before birth, the number of fertilizable ova that can be produced in a woman's lifetime is limited and is greatly reduced by normal degenerative processes (atresia), so that of the estimated seven million oogonia in the fetal ovaries, only about 300–500 ova ultimately develop. Unlike the continuous generation of sperm in the testes, the ovaries generally produce only one secondary oocyte every 22–28 days. The fertile lifetime in the average woman is about 35–40 years. Ovaries and testes produce identical steroid hormones, but the amounts and their patterns of secretion are different (Table 34-1). Although testosterone is present in both males and females, its level in the male is about 18 times that in the female; conversely, circulating levels of estradiol in the female are about 3–15 times those in the male. In either sex, the major sex steroid originates in the gonads, whereas in the opposite sex, this steroid is generated in sustantial amounts by the adrenal cortex or by peripheral conversion of another steroid. For example, almost all of the circulating estradiol in the female comes from the ovaries, whereas in the male, only about one-third comes from the tests, the rest being generated by peripheral conversion of androgenic precursors. Finally, the pattern of gonadal steroid secretion is different. In the male, the secretion of gonadal steroids is fairly constant, although minor fluctuations occur as a result of circadian rhythms. In the adult female, gonadal steroid secretion undergoes dramatic, cyclic changes at about monthly intervals. These cyclic changes, which are dictated by processes regulating oogenesis, are referred to as the menstrual cycle.

34.1 Testes

Regulation of Spermatogenesis: Sertoli-Neuroendocrine Axis

Sertoli cells are epithelial cells that line the seminiferous tubules of the testes. At their basal aspects, these cells form the basement membrane and tight junctions that make up the highly selective "blood–testes barrier," which normally prevents entry of immune cells into the lumen.

Their function is to provide nutritional and hormonal support for cells undergoing spermatogenic transformation (i.e., spermatocytes and spermatids). For this reason, they are often referred to as "nurse cells." Sertoli cells require FSH stimulation for their maintenance and specifically for production of ***androgen-binding protein*** (ABP). ABP, which has a high affinity for testosterone, maintains high levels of testosterone in the seminiferous tubules and thereby helps maintain spermatogenesis. FSH also stimulates aromatase activity in the Sertoli cell, thus stimulating conversion of testosterone into estradiol. This FSH-sensitive Sertoli cell activity accounts for most of the testicular output of estradiol.

FSH secretion by the anterior pituitary is held in check by negative feedback from the testes. This feedback is

TABLE 34-1

Circulating Levels and Origin of Sex Steroids

	Adult Men				Adult Women				
		Approximate Contribution (%) of					Approximate Contribution (%) of		
Steroid	Plasma Level (ng/dL)	Testes	Adrenal	Periphery	Cycle Phase*	Plasma Level (ng/dL)	Ovary	Adrenal	Periphery
Testosterone (T)	700	90	5	5 (A, DHEA)	—	40	60	10	30 (A, DHEA)
Dihydrotestosterone (DHT)	40	20	—	80 (T, A)	—	20	—	—	100 (A, T)
Androstenedione (A)	115	60	30	10 (DHEA, T)	—	160	60	30	10 (DHEA)
Dehydroepiandro-sterone (DHEA)	640	10	85	5 (DHEAS)	—	500	10	85	5 (DHEAS)
DHEA sulfate (DHEAS)	260,000	—	100	—	—	130,000	—	100	—
Estradiol (E_2)	3	30	—	70 (T, A)	EF	10			
					LF	50	95	—	5 (A, DHEA, E_1)
					ML	20			
Progesterone (P)	30	70	30	—	F	100	70	30	—
					L	1100	95	5	—

*EF = Early follicular phase; LF = late follicular phase; ML= midluteal phase; F = follicular phase; L = luteal phase.

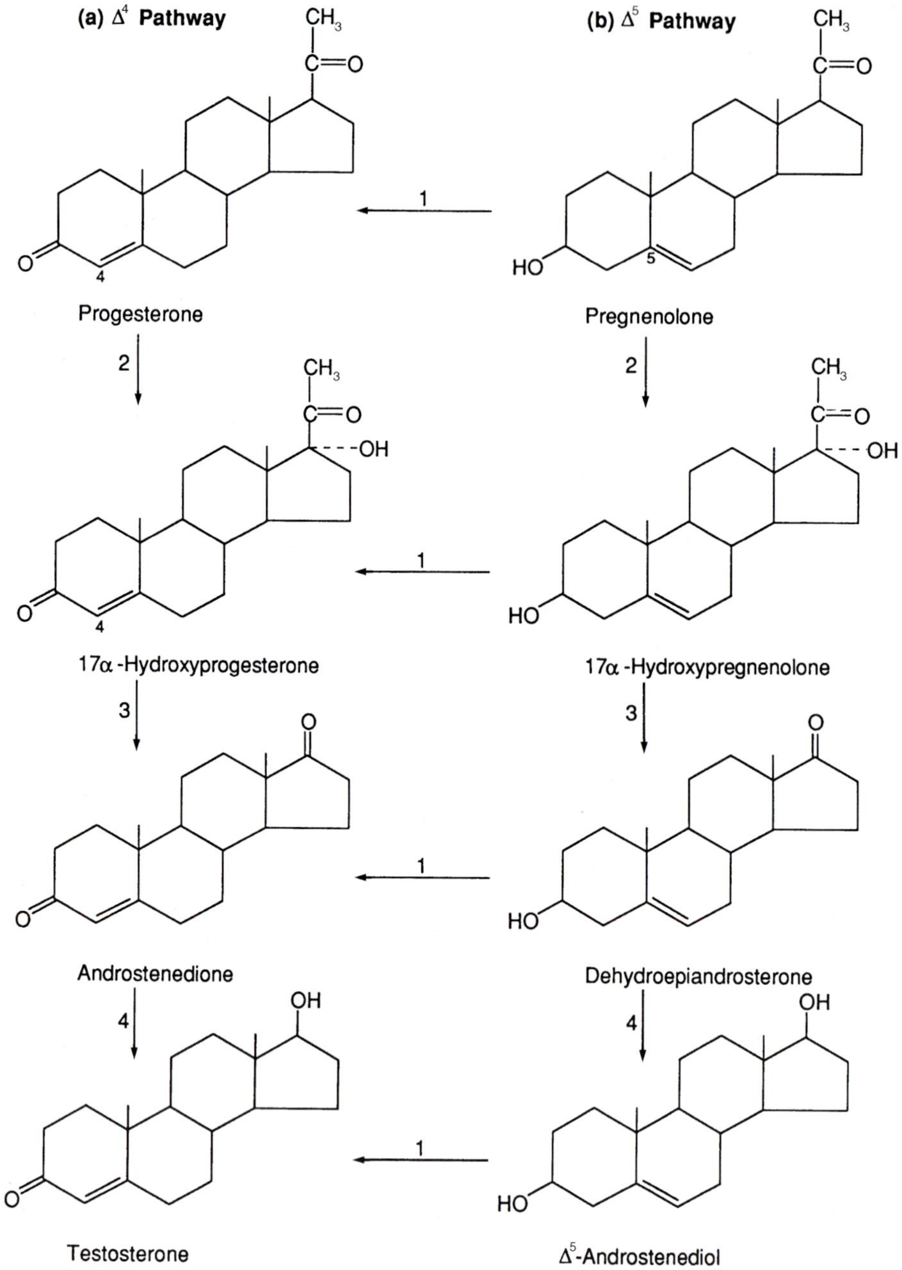

FIGURE 34-1
Testosterone biosynthetic pathways. (a) Δ^4 Pathway (preferred pathway in human testes). (b) Δ^5 Pathway (all enzymes located in the endoplasmic reticulum). Enzymes: 1,3β-hydroxysteroid dehydrogenase and Δ^5,Δ^4-isomerase; 2, 17α-hydroxylase (CYP 17); 3, $C_{17\text{-}20}$-lyase; 4, 17β-hydroxysteroid dehydrogenase.

achieved in part by a polypeptide produced by Sertoli cells called ***inhibin,*** which specifically inhibits FSH release. Presumably, the output of inhibin is proportional to the intensity of spermatogenesis and maintains spermatogenesis at a relatively constant rate. FSH release also is modulated by estrogens. Because the anterior pituitary lacks aromatase and therefore cannot convert testosterone to estradiol, it is possible that testicular estrogen exerts a negative feedback on FSH release.

Regulation of Testicular Steroidogenesis: Leydig-Neuroendocrine Axis

Leydig cells (interstitial cells) are situated in the vascularized compartments outside the seminiferous tubules in the testes. Leydig cells are steroidogenic cells, converting cholesterol to pregnenolone under regulation by LH. The major steroid product of the Leydig cells is testosterone, which accounts for most of the steroid output by the adult testes. The preferred pathway is the Δ^4 pathway, although detectable amounts of Δ^5 steroids (dehydroepiandrosterone, androstenediol) are secreted (Figure 34-1). LH acts at the cholesterol side chain cleavage step. LH release is held in check by the level of non-TeBG-bound testosterone in the circulation, although the principal inhibitor is DHT. Because the anterior pituitary contains 5α-reductase activity, it converts testosterone to DHT and responds to DHT by diminishing output of LH. LH release is also inhibited by testosterone via inhibition of GnRH release. In the preoptic hypothalamus, testosterone can be converted to DHT or estradiol, and may thereby inhibit the pulse frequency (DHT) or pulse height (estradiol) of GnRH released at the median eminence.

Secretion of testosterone is fairly constant and consistent in adult men, although higher levels occur in the morning and lower levels in late evening. This circadian rhythm is not accompanied by changes in the levels of LH; thus, it appears to be an intrinsic testicular rhythm. Testosterone levels decline by 10–15% between the ages of 30 and 70 years, accompanied by reduction in tissue responsiveness to androgenic stimulation.

Metabolism of Testosterone

Testosterone in plasma exists in two fractions: TeBG-bound (44%) and non-TeBG-bound (56%). The plasma level of TeBG in adult males is ~25 nM, which approximates the normal testosterone level in adult men (~22 nM); thus, an increase in testosterone production or a decrease in TeBG level will result in a higher level of testosterone available for tissue uptake.

In androgen target tissues that lack 5α-reductase, testosterone activates the androgen receptor (AR)–androgen response element (ARE) mechanism directly and induces the expression of androgen-dependent genes (Chapter 30). After dissociating from the AR, testosterone may be metabolized by the target cell or make its way to the liver for inactivation. The liver modifies three parts of the testosterone structure that are important for hormonal activity: oxidation of the 17β-OH group (to produce a 17-keto derivative), saturation of the Δ^4 in ring A (to produce an androstane derivative), and reduction of the 3-keto- group (to produce a 3α-hydroxylated derivative (Figure 34-2).

In androgen target tissues that contain 5α-reductase, testosterone is converted to DHT, which activates the AR–ARE sequence leading to an androgenic response. DHT undergoes inactivation when it dissociates from the AR; the major metabolites (except in the liver and kidney) are the reduced, hydroxylated metabolites, 3α-androstenediol and 3β-androstenediol, which are then conjugated to glucuronic acid for renal excretion (Figure 34-2). These and other polar compounds account for about 30–50% of the testosterone metabolites excreted daily.

In tissues that contain aromatase (CYP 19), testosterone can be converted to estradiol, which may exert an effect *in situ* if the tissue is an estrogen-responsive one, and/or may return to plasma for distribution to estrogen target tissues. Unlike the androgenic steroids, the only major site of estrogen inactivation appears to be the liver; therefore, estrogen theoretically can be recycled until being transported to the liver, which itself is an estrogen-responsive tissue. Testosterone-derived estradiol accounts for a small percentage (0.3%) of the testosterone metabolized.

Testosterone is known to exert some important biological effects in the liver and kidney, which are the major testosterone-inactivating tissues in the body. About 50% of the testosterone is removed from plasma with each passage through the liver. The major metabolites of testosterone in the liver are etiocholanolone, androsterone, and epiandrosterone, collectively referred to as the 17-ketosteroids (17KS), all of which are released as glucuronide conjugates for excretion by the kidney (Figure 34-2). This route accounts for 50–70% of the total testosterone metabolized daily.

An epimer of testosterone, epitestosterone (17α-hydroxylated testosterone), is produced by the testes and excreted as such in the urine in amounts approximately equal to that of testosterone(T:epiT ~1:1). Epitestosterone is biologically inactive, but it is not a metabolite, and is believed to be produced only by the gonads; thus, it is used as a gonadal steroid marker. In women, the ratio of T to epiT is also normally 1:1. Urinary T:epiT is useful in monitoring abuse of anabolic steroids by athletes because the ratio increases when any exogenous testosterone derivative is used.

Biological Effects of Androgens

The biological effects of androgenic hormones can be of two types: (a) reproductive (androgenic), i.e., promoting the primary and secondary sexual characteristics of a male; and (b) nonreproductive, which includes anabolic effects. Both types of effects are mediated by the same AR; therefore, testosterone and DHT are capable of exerting either types of effect; however, under physiologic conditions, a critical factor that determines what hormone is active in a given tissue is the presence or absence of 5α-reductase. As underscored above, pharmacological treatment with

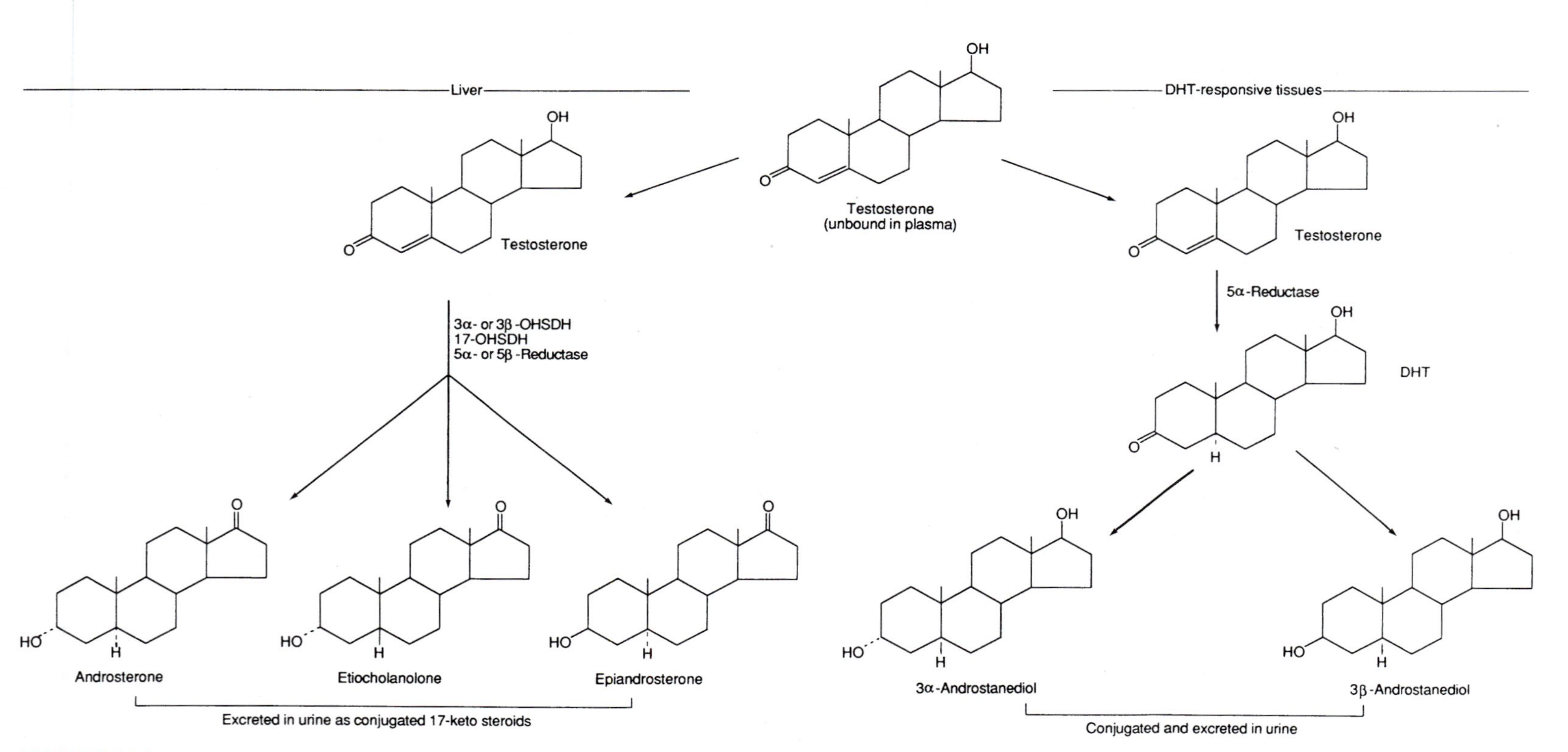

FIGURE 34-2
Metabolism of testosterone.

any steroid that can activate the AR will evoke both androgenic and anabolic responses.

Reproductive (Androgenic) Effects

Testes (Spermatogenesis) At puberty, locally produced testosterone promotes spermatogenesis in the seminiferous tubules, although it is not known with certainty whether or not the Sertoli cells contain 5α-reductase. As mentioned above, testosterone alone promotes the two meiotic (maturational) divisions if present in high concentrations within the tubule, but it is not capable of initiating spermatogenesis in the absence of FSH. Thus, in cases of drug-induced azoospermia, FSH (not androgens) is required for reinitiation of spermatogenesis.

Internal and External Genitalia In the fetus, testosterone causes differentiation of the Wolffian ducts into the male-type reproductive tract, and thereby rescues the tissue from pre-programmed destruction. The seminal vesicles, which differentiate from the Wolffian duct as a result of testosterone action, produce 5α-reductase-2 from week 13–14 and becomes DHT-responsive after that; thus, development and function (but not appearance) may be regulated by DHT from the second trimester through adulthood. On the other hand, the urogenital sinus and external genital primordium in the fetus contain 5α-reductase-2 before week 6 and require DHT for their differentiation into the male phenotype. These tissues do not respond to the very high levels of testosterone in fetal plasma during the critical period (weeks 8–12) unless they are able to convert it into DHT. This explains why ***hereditary 5α-reductase deficiency*** in genotypic male fetuses results in the development of female-type external genitalia, despite the normal production of testosterone by the fetal testes. At puberty, however, the derivative of the genital tubercle (clitoris and penis) becomes responsive to testosterone, which enhances the male-type morphology by promoting growth of the existing structure. On the other hand, the growth and initiation of function of the prostate (urogenital sinus derivative) from puberty through adulthood requires DHT, and testosterone alone does not substitute. The importance of DHT in prostate growth and function is underscored by the success with which ***benign prostatic hyperplasia*** (BPH) in adults can be controlled by blocking 5α-reductase activity.

Skin The hair follicles and sebaceous glands (collectively called pilosebaceous units) in the androgen-sensitive areas of the skin and sebaceous glands throughout the body contain 5α-reductase type 1 and respond to DHT. At puberty, DHT stimulates the appearance and growth of terminal hair in certain androgen-sensitive regions of the body (face, chest, upper pubic triangle, nostrils, external ear). Testicular testosterone in boys and adrenal androstenedione in girls stimulate the appearance and growth of terminal hair in the axillae and lower pubis, presumably after conversion to DHT in the hair follicles in these regions. Axillary and lower pubic hair growth requires less androgen than that in the face, chest, upper pubic area, nose, and ear; thus, hair in *mons pubis* is the most sensitive to androgen (i.e., responds to lower amounts), beard is least sensitive (i.e., requires higher amounts).

DHT stimulates the growth and secretory function (sebum production) of all sebaceous glands, and thus predisposes the skin to acne formation, This DHT-promoted effect can be reversed or attenuated by treatment with estrogens, which antagonize the actions of DHT on the sebaceous glands and cause a depression in glandular activity.

In adulthood, the changes at puberty are maintained by the same hormones, but in many men and few women varying degrees of hairline recession occur as a result of sex-linked genetic predisposition and a DHT effect. Hair on the frontal and vertex scalp is converted from terminal to vellus types in response to DHT (the opposite of what occurs in other androgen-sensitive areas), resulting in the "male pattern baldness"; in women, this is one of the signs of hirsutism. A reduction in circulating testosterone and/or inhibition of 5α-reductase activity (e.g., with finasteride) will prevent or retard the terminal-to-vellus transition.

Voice Pitch Testosterone promotes enlargement of the larynx and thickening of the vocal folds (vocal cords), causing a lowering of the voice pitch. The larynx and associated vocal fold mucosa do not contain 5α-reductase; thus, deepening of the voice occurs at puberty in boys due to the increased production of testosterone, and this is seen even in hereditary 5α-reductase deficiency. Once established at puberty, the laryngeal vocal structures do not require significant amounts of androgens for maintenance; in fact, bilateral testicular failure or loss in a previously normal adult male may not result in any change in voice pitch, possibly because of the adequacy of androgens from the adrenals.

Nonreproductive Effects

Skeletal Muscle Skeletal muscle has no 5α-reductase, and, although all skeletal muscles are believed to contain AR, some are known to contain a higher concentration than others. The most androgen-responsive skeletal muscle in man (presumably because of a higher concentration of AR) are those of the pectoral and shoulder regions. In other species (e.g., rodents), testosterone promotes skeletal muscle growth by stimulating both hypertrophy and mitosis of myofibrils and increases the fast-twitch isoform of myosin heavy chain. The extent to which skeletal muscle mass can be increased is limited by the concentration of androgen receptors in the muscle that can be activated. In normal adult males, the androgen receptors in most tissues are nearly saturated, and further increases in plasma androgens do not produce significant increases in tissue responsiveness. This explains why androgen treatment of boys, girls, and women results in increased skeletal muscle mass, whereas similar treatment of men does not; however, pharmacologic doses of potent synthetic androgens reportedly promoted skeletal muscle mass by a mechanism that is unclear. Conversion of the exogenous testosterone to estrogen may result in a stimulation of AR synthesis in muscle, since AR concentration in skeletal muscle is increased by estrogen and decreased (downregulated) by androgen.

Bone Androgens promote skeletal growth and maturation by a direct effect on bone tissue and by an indirect effect on growth hormone (GH) release. At puberty, the increasing levels of androgens stimulate the release of GH and result in accelerated endochondral growth of the epiphyses of long bones, which causes a doubling of height gain that is maximal at about mid- to late puberty. This peak height velocity (PHV) or pubertal growth spurt is dependent on the increased secretion of androgens (estrogen in girls) and GH at this time. By mechanisms that are not yet understood, androgens also increase bone mass and accelerate the ossification of the epiphyseal growth plate, and thereby reduce the rate of linear growth during late puberty. Ultimately, androgens bring about the fusion of the epiphysis with the diaphysis (i.e., complete endochondral ossification of the growth plate), resulting in the irreversible cessation of linear growth, at about age 19 in boys and age 17 in girls. Even after epiphyseal closure, androgens continue to promote an increase in cortical bone mass, a process that continues to age 30 years. Testicular failure (or any other cause of androgen deficiency through childhood) allows the epiphyseal growth plate to grow for several additional years, resulting in eunuchoidal proportions, as indicated by a reduction in the upper-to-lower body ratio due to longer legs, and an arm span that is greater than the height due to longer arms. In addition, the absence of pubertal androgen results in osteopenia due to inadequate cortical bone mass.

Blood Volume Testosterone acts on the proximal tubule of the nephron to promote the reabsorption of K^+, Na^+, and Cl^-, which, along with stimulated erythropoiesis (see below), contributes to the androgen-associated increase in blood volume. Athletes treated with synthetic androgens experience a weight gain that can largely be explained by an increase in blood volume, and men who undergo androgen treatment as a means of contraception also exhibit blood volume expansion and weight gain.

Erythropoiesis Androgens stimulate the production of erythropoietin (Chapter 28) by the kidney and, in part, cause an increase in hematocrit by this mechanism. This may explain why males have a higher hematocrit than females.

Adipose Tissue Androgens promote truncal-abdominal fat deposition and favor development of upper body obesity. In contrast to gluteofemoral (lower body) fat, upper body fat accumulation, particularly visceral fat, is characterized by an increase in fat cell size, increased lipoprotein lipase (LPL) activity, enhanced lipolysis, and reduced response to the antilipolytic effect of insulin. This explains why androgen-dominated states favor insulin resistance.

Liver Androgens cause a reduction in the plasma levels of testosterone-estradiol-binding globulin (TeBG), which results in an increased percentage of testosterone that is accessible for tissue uptake. This leads to a greater androgenic response by the tissues but also results in accelerated clearance of testosterone from circulation. Ultimately, the outcome of a reduced TeBG level will be a reduction in total plasma androgen concentration due to increased negative feedback suppression of LH release and a reduction of testosterone synthesis. Androgens also modify the production of other hepatic proteins such as fibrinogen (decreased), transferrin (decreased), haptoglobin (increased), α_1-antitrypsin (increased), and hepatic triglyceride lipase (increased). Synthetic steroids exhibiting enhanced anabolic activity and reduced androgenicity (anabolic steroids) have been developed (Figure 34-3). Athletes who use these steroids to increase their strength and durability often consume high doses, which can have adverse effects. The anabolic steroids exert their effects by binding to cytosolic androgen receptors, and the neuroendocrine system responds by reducing secretion of LH and FSH. This response leads to diminished Leydig and Sertoli cell function and to a reduction of endogenous testosterone production and of spermatogenesis. Testicular atrophy and reduced sperm counts are documented consequences of excessive anabolic steroid treatment. Most of the orally active androgenic steroids contain a 17α-alkyl group (17α-methyl or 17α-ethynyl) that makes the steroid resistant to liver inactivation. This explains why these steroids are orally active; however, long-term usage of these steroids is associated with hepatic disorders with varying severity (from abnormal liver function tests to jaundice to hepatic carcinoma). The probable cause of hepatic damage is the chronic demand on the liver to continue increasing its microsomal P450 redox activity, which is incapable of oxidizing 17α-alkyl-substituted steroids.

Structure–Function Relation The 17-hydroxyl group of testosterone and DHT appears to be important for androgenic activity, since testosterone enanthate and

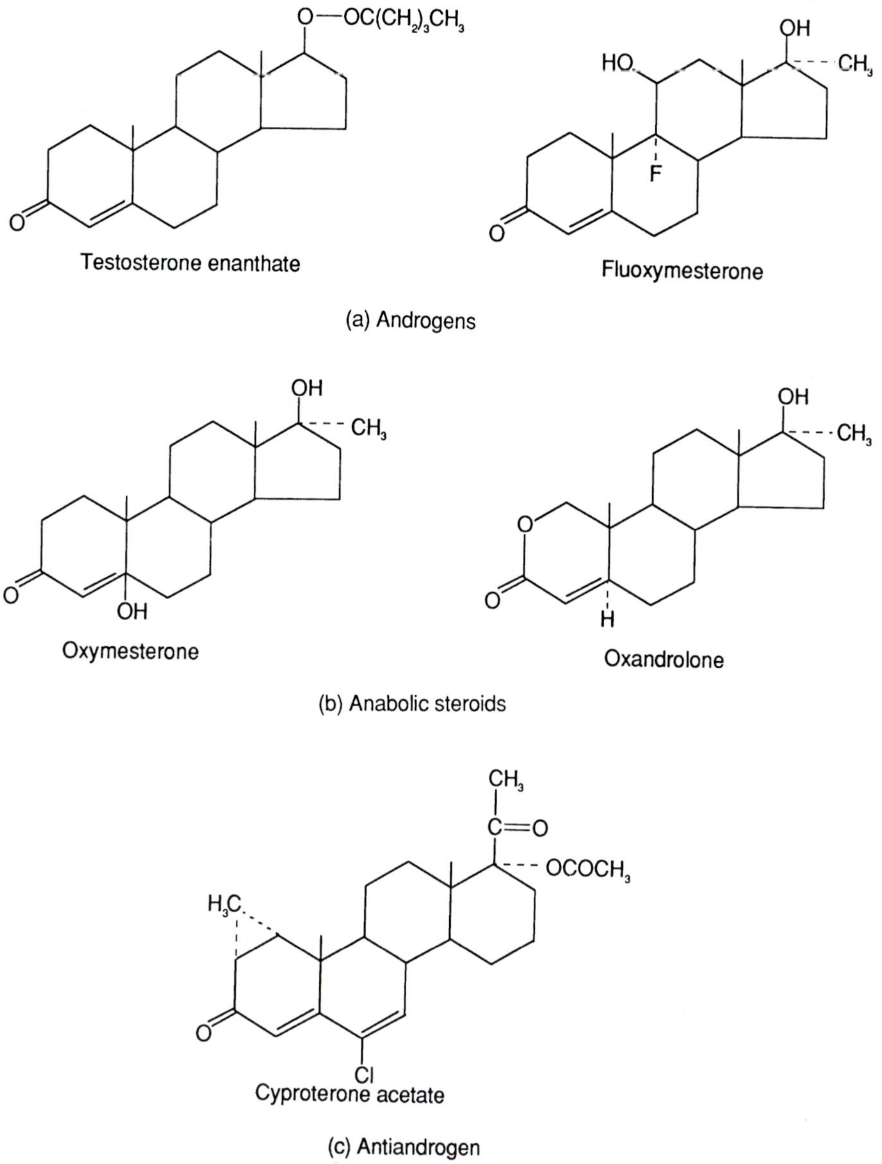

Testosterone enanthate

Fluoxymesterone

(a) Androgens

Oxymesterone

Oxandrolone

(b) Anabolic steroids

Cyproterone acetate

(c) Antiandrogen

FIGURE 34-3
Synthetic androgens, anabolic steroids, and an antiandrogen.

fluoxymesterone (Figure 34-3), which have protected 17-hydroxyl residues, are potent androgenic drugs. Substitutions in the A ring while protecting the 17-hydroxyl group decrease the androgenic potency of the steroid while enhancing the anabolic potency.

Pharmacological Inhibition of Spermatogenesis or Fertility (Male Contraception)

Attempts to devise a male contraceptive (antifertility) have involved the testing of a number of chemicals for their ability to interfere with spermatogenesis or spermatozoa viability. In order to be effective, a contraceptive would have to produce azoospermia (complete absence of spermatozoa in ejaculate) because pregnancies have resulted with sperm counts as low as 1×10^6 per mL or less.

Testosterone Synthetic derivatives of testosterone with long biological half-lives have been used to create azoospermia in most subjects after 42 weekly treatments. The mechanism involved is the suppression of both LH and FSH release by way of hypothalamic GnRH inhibition. Although exogenous testosterone treatment maintains the androgen supply to most tissues, the seminiferous tubule compartment is androgen-deficient because the Leydig cells are not functional in this LH-deficient condition. Thus, the requirement for high local concentration of testosterone within the seminiferous tubules is not met, and spermatogenesis ceases as a result.

Progestins A synthetic progestin, medroxyprogesterone acetate (Provera), has been used for male contraceptive trials alone or in combination with nandrolone (a synthetic androgen). When given once monthly for 2–3 months, the combined treatment caused azoospermia in about one-half of the subjects.

Synthetic GnRH Agonists Synthetic derivatives of GnRH with more biological activity than the native hormone have been used for the suppression of spermatogenesis after 2 or more weeks of treatment that downregulated GnRH receptors in the anterior pituitary. However, in most trials there was no uniform induction of azoospermia, probably because the reduction of LH and testosterone secretion was incomplete.

Gossypol Gossypol is a naphthaphenol present in crude cottonseed oil that causes infertility in those who consume it regularly, as was found to be the case during the 1950s. Oral consumption of this agent for one month caused a reduction in sperm count and immobilization of spermatozoa, the result of the drug on the seminiferous tubule and epididymis that affected both spermatogenesis and sperm maturation. Leydig cell function was not affected. Aside from symptomatic hypokalemia, the disadvantage of this contraceptive is that its effect is not completely reversible after cessation of treatment.

Androgen Antagonists Cyproterone acetate and flutamide have been shown to be ineffective in reducing fertility in men, primarily because their effects on the neuroendocrine system override any intratesticular effects they may exert. Thus, the increase in LH and FSH caused by these antiandrogens stimulates increased testosterone production by the Leydig cells leading to a compensatory increase in intratesticular androgen. Both spironolactone (a mineralocorticoid receptor antagonist) and cimetidine (an H_2 blocker) interfere with androgen effects in the testes, and may cause a reduction in sperm count in those who are treated with either drug on a chronic basis. However, the effect is mild and completely reversible.

34.2 Ovaries

Menstrual Cycle

The menstrual cycle consists of the ***follicular phase*** and the ***luteal phase,*** each lasting about 2 weeks. The follicular phase is a period of ovarian follicular growth (folliculogenesis) that results in ovulation (release of secondary oocyte). This phase is dominated by estrogen produced by the growing follicle itself. During the follicular phase, the uterine endometrium is stimulated by estrogen to proliferate and to synthesize cytosolic receptors for progesterone. Follicular production of estrogen depends on FSH and LH. The luteal phase is progesterone-dominated. The follicle that ruptures at ovulation becomes the corpus luteum, which produces progesterone and estradiol. During this period, the uterine endometrium becomes secretory under the influence of progesterone. About 5 days into the luteal phase, the endometrium is ready to accept a blastocyst for implantation; in the absence of fertilization, however, the corpus luteum degenerates after about 12 days, steroid production ends, and the endometrium deteriorates (menstruation). The first day of menstruation is the first day of the menstrual cycle.

Endocrine Control of Folliculogenesis

Two types of endocrine cells are associated with the ovarian follicle. One is the granulosa cell, which resides within the follicle and is encased by the basal lamina. Like the Sertoli cell in the testes, the granulosa cell is perfused by plasma transudate, not blood, and possesses FSH-sensitive aromatase activity; thus, FSH stimulates the formation of estrogen by the granulosa cell. Unlike Sertoli cells, however, granulosa cells proliferate in response to estrogen, and this proliferation is inhibited by androgens. The nonendocrine function of granulosa cells is to promote growth of oocytes by conditioning the follicular fluid. The other type of follicular cell with endocrine function is the theca interna cell. Such cells are positioned along the outer border of the basal lamina; thus, they are located outside the follicle and are perfused by blood. Theca interna cells, like Leydig cells of the testes, respond to LH with androgen production from cholesterol. Unlike Leydig cells, they produce mainly androstenedione, although some testosterone is also produced. Theca interna cells provide androgens for estrogen production by granulosa cells.

Hormonal Control of Follicle Growth

During the follicular phase, the ovarian follicle (preantral follicle) grows by pronounced proliferation of granulosa cells. During the second half of the follicular phase, the follicle accumulates fluid, which leads to formation of an antrum (antral follicle).

In the preantral follicle, LH stimulates the interna cells to produce androgens (mainly androstenedione), which diffuse through the basal lamina into the granulosa cell compartment. The granulosa cells, under the influence of FSH, aromatize the androgens into estrogens (mainly estradiol), which act directly on the granulosa cells to stimulate proliferation. This causes the follicle to grow and

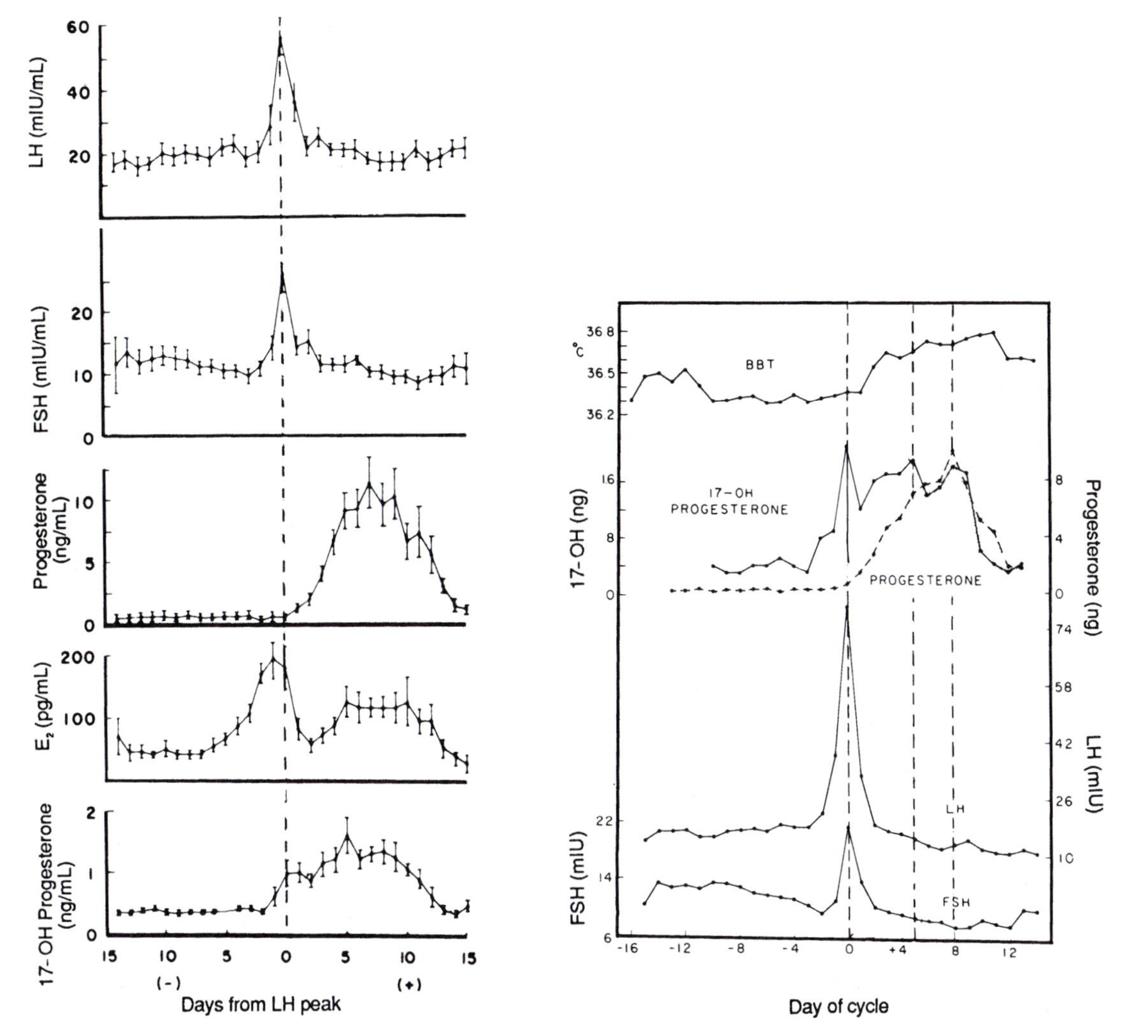

FIGURE 34-4
Hormonal and physiological changes during the menstrual cycle. BBT, Basal body temperature.

to accumulate fluid and leads to formation of an antrum. Although the intrafollicular concentration of estradiol is sufficient to stimulate proliferation of granulosa, it is not high enough to enter the general circulation. For this reason, the level of circulating estradiol does not increase during the first half of the follicular phase (Figure 34-4), and thus the release of LH and FSH is not inhibited but remains at a fairly constant level.

In the antral stage of follicular growth, production of estradiol by granulosa cells increases owing to acquisition of LH receptors by the granulosa cells. Induction of LH receptors is brought about by the combined effects of FSH and estradiol and enables the granulosa cells to begin producing estradiol from pregnenolone. Thus, the formation of estradiol by aromatization of androgens derived from the theca interna is augmented by estradiol synthesized *de novo*. The increased estradiol pool causes marked acceleration in follicle growth and spillage of estradiol into the general circulation. Consequently, a relatively rapid increase in circulating estradiol level is seen during the last 5 or 6 days of the follicular phase, which exerts an initial negative feedback on release of FSH (Figure 34-4). This rise in estradiol level is a critical cue for the neuroendocrine system because it indicates that the ovarian follicle is ready for ovulation. This message is in the form of an approximately threefold increase in estradiol level, which stimulates the release of a large surge of LH and FSH. In contrast to the negative feedback effect that estradiol normally exerts on the release of gonadotropins, the very high levels of estradiol presented over 2–3 days exert a positive feedback effect. Two mechanisms may operate in this positive feedback: High levels of estradiol for 2 or 3 days may increase the sensitivity of the pituitary to GnRH or increase the release of GnRH from the hypothalamus (or both).

The latter mechanism is believed to involve the formation of catecholestrogen (2-hydroxyestradiol) after uptake by the median eminence, which, because of its abundance, competes with hypothalamic norepinephrine for inactivation by catechol-O-methyltransferase (COMT). This inactivation results in a higher norepinephrine content in the median eminence, which favors GnRH release.

The large bolus of LH released (preovulatory LH surge) in response to positive feedback by estradiol induces ovulation in about 1 day, probably by stimulation of production of granulosa plasminogen activator, which leads to formation of plasmin, an enzyme that may be responsible for the digestion of the basal lamina and, consequently, for the rupture of the follicle (ovulation).

Hormonal Control of Luteal Function

After ovulation, the granulosa cells proliferate in response to the preovulatory LH surge, and the theca interna cells and perifollicular blood vessels invade the cavity of the collapsed follicle. Under the influence of LH, the granulosa and invasive theca cells differentiate into luteal cells, which are characterized by their high lipid content. These luteal cells are steroidogenic and produce large amounts of progesterone and moderate amounts of estradiol. The ruptured follicle thus becomes the corpus luteum. Morphogenesis of the corpus luteum is not complete until about 1 or 4 days after ovulation, and luteal production of progesterone and estradiol gradually increases to a maximum about 6 or 7 days after ovulation. Thus, there is an interval of about 3 days after ovulation during which levels of circulating estradiol are reduced (Figure 34-4), and this interval is required for proper transport of the ovum through the fallopian tube into the uterus. Exposure to high levels of estrogen during this interval would lead to expulsion of the ovum or to blockage of ovum transport. The rise in levels of progesterone and of estradiol during the first week of the luteal phase is required for the endometrium to become secretory in preparation for implantation and pregnancy. The corpus luteum has a life span of about 12 days; it can synthesize steroids autonomously without extraovarian hormonal stimulation. Although the corpus luteum has receptors for LH, release of LH (and FSH) during the luteal phase is strongly inhibited by the potent negative feedback effect of progesterone and estradiol. Thus, if fertilization and implantation do not occur, the corpus luteum degenerates (luteolysis), and its production of progesterone and estradiol rapidly declines (Figure 34-4). Withdrawal of progesterone and estradiol during luteolysis results in deterioration of the endometrium and its shedding (menstruation). If fertilization and implantation occur, secretion of chorionic gonadotropin (hCG) (Chapter 31) by the implanting blastocyst stimulates the corpus luteum to continue producing progesterone, and luteolysis is prevented.

The regularity of the menstrual cycle in women of reproductive age can be affected by anatomical defects of the uterus or vagina, or by functional or structural defects in the hypothalamic-pituitary-ovarian axis that affect hormonal secretions. Complete cessation of menses (for more than 6 months) is known as ***amenorrhea*** and a reduction in the frequency is known as ***oligomenorrhea.*** Physiologic states of amenorrhea include prepuberty, pregnancy, lactation, and postmenopause. A common pathologic cause of amenorrhea can result from a reduction in the secretion of GnRH from the hypothalamic neurons into the hypothalamic-hypophyseal portal system with a corresponding reduction in the secretion of FSH and LH from the pituitary. Decreased GnRH can occur due to weight loss, anorexia nervosa, excessive exercise, debilitating diseases, or psychological trauma.

Pregnancy

Human Chorionic Gonadotropin (hCG)

After fertilization of the ootid in the fallopian tube, the zygote develops into a blastocyst by the time it enters the uterine cavity (about 4 days after ovulation). By the seventh day after ovulation, the blastocyst is completely embedded in the endometrium, and its outer trophoblast has differentiated into an inner, mitotically active cytotrophoblast and an outer, nonmitotic syncytiotrophoblast. The syncytiotrophoblast begins producing hCG from about the eighth day after ovulation, and it continues producing increasing amounts of hCG until maximal levels are attained at about 6–8 weeks of pregnancy. hCG stimulates the corpus luteum to continue producing progesterone from about the eighth day after ovulation and thus protects the corpus luteum from luteolysis. As mentioned above, hCG promotes expression of the male genotype in the fetus by stimulating testosterone production by fetal Leydig cells from about the eighth week of pregnancy. Moreover, hCG may be the stimulus for the initial development of the fetal zone of the adrenal cortex, which appears at about the sixth week. Lastly, and perhaps most importantly, hCG stimulates steroidogenesis in syncytiotrophoblasts at 4–5 weeks of pregnancy by inducing CYP 11A (side chain cleavage enzyme), 3β-hydroxysteroid dehydrogenase (3β-HSDH/iso), and CYP 19 (aromatase). These enzymes are critical for maintenance of pregnancy because the placenta assumes the role of the ovaries as the major generator of progesterone and estrogen after the sixth week of pregnancy.

The implantation of the fertilized ovum at a site other than the endometrium is known as ***ectopic pregnancy.***

Patients with ectopic pregnancy frequently exhibit abdominal pain with amenorrhea. By far the most common cause is an impaired fallopian tubal function. The tubal pathology can result from pelvic infection, endometriosis, or previous surgery. The implanted embryo commonly dies in the early weeks of pregnancy. Tubal rupture and the associated bleeding require emergency care. Thus, early diagnosis is essential and consists of serial measurements of serum levels of hCG using antibodies specific for the β subunits of hCG (β-hCG) and serum progesterone, and pelvic ultrasonography. In ectopic pregnancy, the serum β-hCG levels are lower than those observed at the same gestational age in a normal pregnancy. A serum progesterone level of 25 ng/mL (79.5 nM/L) or higher excludes ectopic pregnancy with 97.5% sensitivity. The treatment for ectopic pregnancy consists of either surgery to remove the embryonic tissue or medical treatment with methrotrexate, a folic acid antagonist. Methotrexate inhibits both *de novo* purine and pyrimidine nucleotide biosynthesis (Chapter 27), blocking DNA synthesis and cell multiplication of the actively proliferating trophoblasts.

Placental Steroids

The absence of progesterone is incompatible with the gravid state. Progesterone acts in three ways to maintain pregnancy:

1. It maintains placental viability, thus ensuring adequate exchange of substances between maternal and fetal compartments;
2. It maintains perfusion of the decidua basalis (maternal placenta), presumably by inhibiting the formation of vasoconstrictive prostaglandins; and
3. It diminishes myometrial contractility, possibly by increasing the resting membrane potential or by inhibiting the formation of prostaglandins E_1 and $F_{2\alpha}$ (Chapter 18). The placenta is the major site of conversion of cholesterol to progesterone after the sixth week; however, it is not capable of synthesizing cholesterol from acetate. Moreover, the placenta lacks CYP 17 and CYP 21, which explains why the major steroid it produces is progesterone (Figure 34-5). Although the placenta cannot process cholesterol beyond progesterone, it contains some important steroid-modifying enzymes which enable it to produce substantial amounts of estrogen.

Small amounts of estrogen are required for maintenance of pregnancy because estrogen maintains tissue responsiveness to progesterone. The placenta forms estrogen from androgenic steroids. The major androgen used is

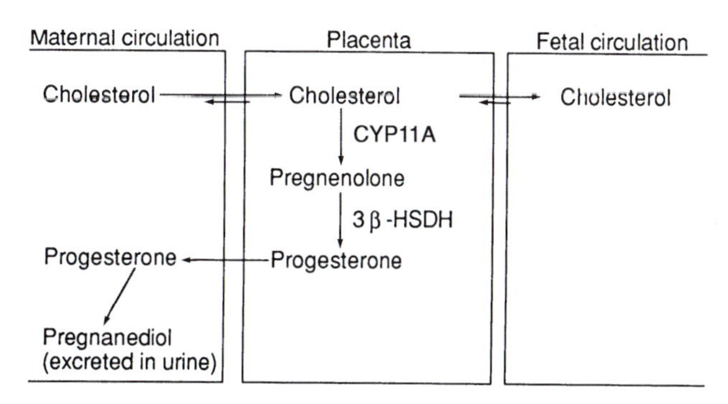

FIGURE 34-5
Placental synthesis of progesterone. The placenta contains cholesterol side-chain cleavage enzyme (CYP11A) and 3β-hydroxysteroid dehydrogenase (3β-HSDH), which enable it to form progesterone from cholesterol. However, it does not make cholesterol *de novo* and relies mainly on the maternal cholesterol pool for substrate. Progesterone exerts its effects on the uterus and placental vasculature and enters the maternal circulation. Maternal tissues reduce progesterone predominantly to pregnanediol, which is the major metabolite of progesterone in maternal urine.

dehydroepiandrosterone (DHEA), derived mainly from the fetal compartment in the form of DHEAS (DHEA sulfate). The source of DHEAS is the fetal zone of the fetal adrenal gland, which, produces mainly DHEAS because it lacks 3β-HSDH (Chapter 30). The placenta takes up DHEAS from the fetal circulation and, by the action of placental sulfatase, liberates DHEA, which is then aromatized into estrone (E_1) (Figure 34-6). Some E_1 is converted to E_2 by placental 17-HSDH, and both are released into the maternal circulation. A substantial fraction of DHEAS produced by the fetal zone is taken up by the fetal liver, which detoxifies it to 16β-hydroxy-DHEAS. The placenta also processes this metabolite through the aromatase system, and the product is estriol (E_3), a weak estrogen that is not produced in significant amounts in the nongravid state. The circulating level of E_3 is an important index of fetal–placental function because its production depends on normal functioning of the fetal adrenal, fetal liver, and the placenta. It is therefore used clinically as a diagnostic tool.

Human Placental Lactogen (hPL)

Human placental lactogen (hPL) is produced by the syncytiotrophoblast from about the time when production of hCG begins to diminish. Production of hPL is proportional to placental growth, and its level reflects placental well-being. hPL exerts GH-like effects in both fetal and maternal compartments. In the fetus, it promotes formation of insulin-like growth factor and growth factors believed to promote growth of most, if not all, fetal tissues. In the mother, hPL promotes nitrogen retention and utilization of free fatty acids, and it creates a state of mild insulin

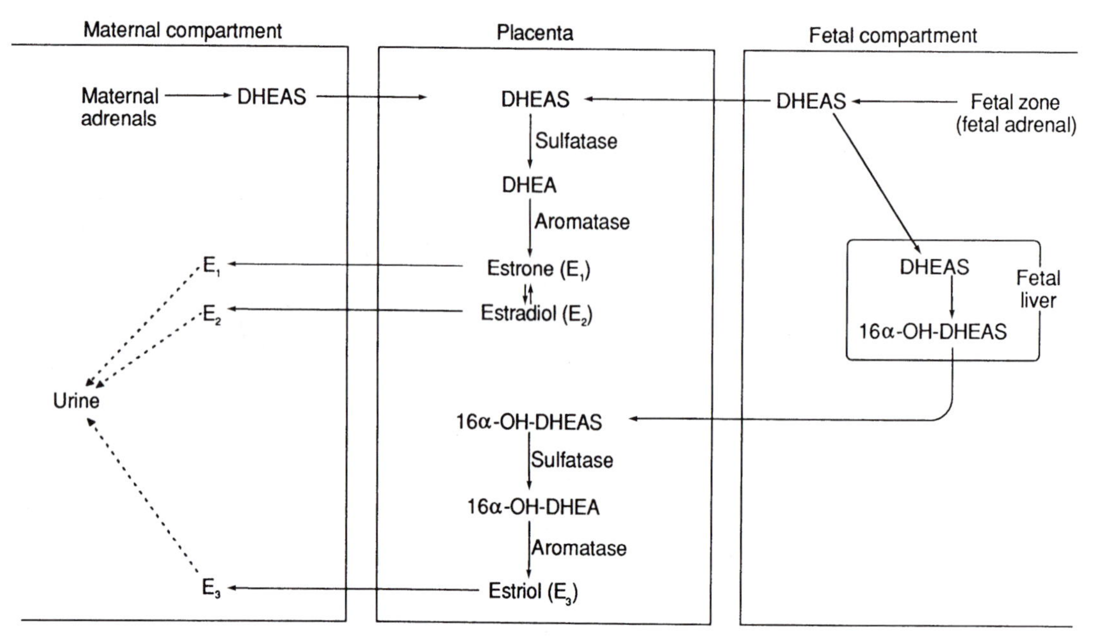

FIGURE 34-6
Placental synthesis of estrogens. The placenta lacks the key enzyme necessary for formation of estrogens from cholesterol (CYP 17) and relies on androgenic precursors from the maternal and fetal compartments. The major androgen used comes from the fetal zone of the fetal adrenal; this is DHEAS, which is also taken up and metabolized by fetal liver into 16α-hydroxy-DHEAS. The placenta converts DHEAS into estrone (E_1) and estradiol (E_2) and processes 16α-hydroxy-DHEAS into estriol (E_3). Estrogens enter the maternal circulation and appear in maternal urine as conjugated estrogens.

resistance that benefits the fetus because it increases the availability of maternal glucose for fetal consumption. In the mother, hPL also exerts a prolactin-like effect and, with estrogen and progesterone, promotes ductal and alveolar growth in the mammary gland during the third trimester of pregnancy. However, prolactin is believed to be more important than hPL in this effect.

Decidual Prolactin

The maternal placenta (decidua basalis) is believed to produce a hormone that is biologically and antigenically similar to pituitary prolactin. If released into the amniotic fluid during the third trimester of pregnancy, it may maintain amniotic fluid volume and osmolality.

Relaxin

The maternal placenta also produces relaxin, a hormone produced by the ovaries. Relaxin causes relaxation of the cervix during parturition.

Parturition

Precisely what initiates parturition in humans is not known. Unlike the situation in sheep, there is no induction of placental CYP 17 by fetal cortisol to bring about a reduction in progesterone production and, hence, parturition. During late gestation, rising levels of estrogen are thought to increase the synthesis of oxytocin in hypothalamic magnocellular neurons to induce oxytocin receptors in the myometrium and to increase myometrial contractility by lowering the membrane potential. During this period, relaxin from the decidua softens the cervix for impending delivery. At term an unidentified triggering factor in fetal urine stimulates uterine production of PGE_1 and $PGF_{2\alpha}$, which lead to myometrial contraction, mainly in the fundus, and encourages movement of the infant through the cervical canal. Dilation of the cervix by the infant stimulates the release of oxytocin by neuroendocrine reflex, further stimulating uterine contractions, which in turn causes more cervical stretching and a positive feedback to oxytocin release, until delivery is complete.

Lactation

During midpregnancy and late pregnancy, growth and differentiation of mammary glands are promoted by the combined effects of estrogen, progesterone, and prolactin (or hPL). In the presence of prolactin or hPL, estrogen stimulates branching of the mammary ducts and increases their growth. Estrogen also stimulates production of lactogenic receptors in mammary ductal cells and acts on the anterior pituitary to stimulate prolactin secretion. In the presence of prolactin or hPL, progesterone acts on the estrogen-conditioned mammary gland to promote differentiation of the terminal ductal buds into alveoli and, consequently, into lobules. Synthesis and secretion of milk occur in these lobules. Lactogenesis begins during the third trimester of pregnancy and involves synthesis of the milk-specific proteins casein, lactalbumin, and lactoglobulin. The primary regulator of lactogenesis is prolactin, although the participation of additional hormones is needed (i.e., insulin, cortisol, thyroid hormone). Lactation involves release of milk into the alveolar lumen. Lactation is inhibited by estrogen and, thus, is held in check by the high levels of estrogen during pregnancy; the withdrawal of estrogen after birth triggers lactation. When lactation is unwanted by the mother, a large amount of estrogen is administered at the time of labor. Lactation is maintained by prolactin, the release of which is triggered when the mother's nipple is stimulated (neuroendocrine reflex). Prolactin does not stimulate the actual release of milk from the nipple (milk let-down or milk ejection); this is brought about by oxytocin released in response to suckling. Oxytocin acts on myoepithelial cells, contractile cells that surround the alveoli and ducts, and causes them to contract and squeeze out the milk.

Mothers who nurse their infants have frequent prolactin surges in response to suckling. During the initial 6 months of lactation, the response amplitude of prolactin is high, LH and FSH levels are suppressed, and ovarian steroidogenesis is reduced. These effects are due to the antigonadotropic action of elevated levels of prolactin. They have been explained by reduced GnRH secretion by the hypothalamus. This condition, which is common during the first 6 months of lactation and usually does not persist beyond 9 months, is referred to as lactational amenorrhea. It resembles the gonadal quiescence of hyperprolactinemia.

Biological Effects of Estrogens

The biological effects of estrogen are many. Estrogen is the major determinant of female reproductive function, bone maintenance, and cardioprotection. Its effect on the brain includes reproductive behavior and function, learning, and memory. The mechanism of action of estrogen in the improvement of cognitive function is not understood, but it is thought that estrogen delays the onset of Alzheimer's disease. Estrogen is also a male hormone and has important actions in the male urogenital tract and skeleton. The key role of estrogen in the normal bone metabolism of both sexes is illustrated in two human syndromes, namely, defects in aromatase and estrogen receptor (ER). The latter is required for the conversion of testosterone to estrogen. Men and women with ***aromatase deficiency*** suffer from osteoporosis, incomplete epiphyseal fusion and continued linear growth in adulthood, and lack of sexual development. Treatment with estrogen in the aromatase- deficient patient increases bone mineral density and the epiphyseal closure. A mutation in the ER-α gene can cause severe osteoporosis with unclosed epiphyses and continued linear growth in males. An undesirable property of estrogen is its potential carcinogenicity. Estrogen exposure increases the risk of breast and endometrial cancer. Exposure to high-potency estrogens (e.g., diethylstilbestrol) during embryonic and neonatal development results in abnormalities of the reproductive tract and in increased incidence of reproductive tract tumors. The antiestrogenic action of selective estrogen response modifiers (e.g., tamoxifen) is used in the treatment of breast cancer (discussed later).

The diverse actions of estrogen and its differential effects in tissues are mediated by ER-α and ER-β. The ERs perdominantly found in the cell nucleus are transcription factors whose action is similar to that of other steroid hormones (Chapter 30). In the target cells, ER is in the inactive state bound to heat-shock proteins. Estrogen binding to ER releases the associated proteins and facilitates formation of ligand-bound receptors. Receptor dimers bind to the cognate DNA response elements. The overall complex, with the recruitment of coactivators and other transcription factors, becomes an active transcriptional complex capable of gene expression.

The dimerization of ER can involve either homo- or heterodimers—$(\text{ER-}\alpha)_2$, $(\text{ER-}\beta)_2$, or ER-α/ ER-β—thus expanding the physiological specificity and action of estrogen in the target tissue. Breast tumors that contain ER are treated with selective estrogen response modifiers (SERMs). Current assays for ER in breast tumors only identify ER-α, but ER-α- negative, ER-β- positive tumors may also respond to SERM therapy.

Ovary Estradiol and FSH stimulate proliferation of granulosa cells in recruited follicles and promote their secretory function. Estradiol, with FSH, transforms the basal subpopulation of granulosa cells (next to the basal lamina) into LH-responsive granulosa cells by inducing the synthesis of LH receptor sites. These intraovarian effects of estradiol require high levels of the hormone locally, which are normally achieved because of the autocrine action of these effects. Estradiol also exerts a paracrine effect in the ovary, inhibiting CYP17 (both 17α-hydroxylase and 17,20-lyase) activity in theca interna cells during their luteinization. In the corpus luteum, estradiol may be the cause of luteolysis in a nonfertile cycle, although the involvement of intraluteal $PGF_{2\alpha}$, is also implicated.

Endometrium Estrogen stimulates proliferation of the epithelial and stromal cells of the endometrium and is essential for its repair (regeneration) following menstruation. Estrogen induces the synthesis of progesterone receptors in the endometrium and thereby permits the endometrium to respond to progesterone during the luteal phase. This permissive effect of estrogen decays rapidly and requires the continued presence of low levels of estrogen for maintenance; this explains why the production of estradiol during the luteal phase and during pregnancy is important. In pathological states, the endometrial tissue may grow at ectopic sites of the body, and has been found in the fallopian tube, abdominal cavity, ovary, and other sites, including the lung. This condition (endometriosis) is dependent on estrogenic stimulation but is not caused by estrogen; treatment with either an estrogen antagonist or with an androgen is effective. The stimulatory effect of estrogen on endometrial growth also explains why estrogen treatment of postmenopausal women carries with it an increased risk of developing endometrial hyperplasia and cancer; however, when combined with progesterone, the risk is markedly reduced.

Myometrium Estrogen decreases the resting membrane potential of the myometrium and increases ciliary activity of the endometrial epithelium. Rising levels of estradiol prior to ovulation cause the smooth muscle of the fallopian tubes to become increasingly contractile at about the time of ovulation. This and the increased activity of cilia on the endometrial lining facilitate entry of the oocyte-cumulus complex into the lumen of the fallopian tube at the time of ovulation. In the pregnant uterus shortly before parturition, estradiol stimulates the formation of gap junctions and the production of myometrial receptors for oxytocin, PGF, and PGE, and thereby promotes the myometrial contractions that facilitate parturition.

Mammary Gland Estrogen, in the presence of prolactin (PRL), stimulates ductal proliferation and thereby increases the area occupied by the branching ductules. This effect also occurs in males if the level of estrogen is sufficiently high to overcome the inhibitory effect of androgens. Estrogen induces synthesis of progesterone receptors in the ductal epithelium and probably induces oxytocin receptor sites in the myoepithelial cells that surround the alveoli and ductules. The stimulatory effect of estrogen on mammary gland growth continues in the presence of progesterone, such that mitotic growth of the mammary gland peaks during the luteal phase, in contrast to the endometrium, in which mitotic growth peaks in the follicular phase and is inhibited by luteal progesterone. This explains why the risk of breast cancer is approximately the same ($\sim$32–41%) whether or not estrogen treatment is supplemented with progesterone. Estradiol inhibits the release of GnRH by reducing the amount of GnRH released per pulse (reduces pulse height) without affecting the pulse frequency. This results in a corresponding reduction in the pulse height of LH and FSH that reduces their plasma concentrations. Estradiol also directly inhibits FSH release by reducing the production of FSH-β subunits but has no inhibitory effect on either LH-β or the α subunit. In addition to its effect on GnRH and the pituitary gonadotropins, estrogen stimulates growth hormone (GH) synthesis and promotes an increase in the GH response to provocative stimuli. Estrogen decreases hypothalamic somatastatin content, increases somatotrope number, GH mRNA levels, and GH content.

Bone At puberty, estrogen promotes height gain and accounts for the peak height velocity in girls. As in pubertal boys, the pubertal growth spurt in girls leads to cessation of linear growth due to ossification of the epiphyseal plate of long bone. In the adult, estrogen has an overall protective effect on bone tissue that results from its action on osteoblasts and osteoclasts in both cancellous (trabecular) and cortical bone. This explains why estrogen replacement therapy begun at the time of menopause effectively prevents the loss of bone mass (osteoporosis) associated with the postmenopausal state, if begun prior to the onset of osteoporotic changes. The molecular mechanism of estrogen action on bone mass may partially occur through the regulation of the gene for transforming growth factor β (TGF-β). Of the number of specific effects estrogen has been found to have on bone, the two most important are as follows:

1. Estrogen inhibits osteoclast-mediated resorption of cancellous and cortical bone by a direct effect on the osteoclasts; and
2. Estrogen inhibits the bone-mobilizing effect of parathyroid hormone (PTH) by a direct effect on the osteoblasts (Chapter 37).

Liver Estrogen has an overall trophic effect on the liver and opposes some of the hepatic effects of androgens, while promoting the hepatotropic actions of glucocorticoids. Thus, estrogen increases circulating levels of TeBG and thereby reduces the percentage of bioactive androgen in plasma. Estrogen stimulates the hepatic production of corticosteroid-binding globulin and angiotensinogen, and provides support for glucocorticoids in an additive fashion. Estrogen promotes blood clotting by stimulating hepatic production of some of the clotting factors (VII–X), while inhibiting formation of antithrombin (Chapter 36). These and possibly other effects are believed to contribute to the increased incidence of ***thromboembolism*** and ***thrombophlebitis*** in women on estrogen treatment alone or in combination with progestin.

Skin Estrogen inhibits the stimulatory effect of androgen (DHT) on pilosebaceous activity, and thereby reduces the incidence of acne vulgaris. There is no evidence that estrogen can prevent the androgen-sensitive conversion of terminal to vellus hair in male pattern baldness.

Adipose Tissue Estrogenic states favor subcutaneous fat deposition in the gluteofemoral region (lower body) and promote lower body obesity. Current evidence suggests that a lower body fat pattern, or a low upper–lower body circumference ratio (referred to as "gynoid"), is associated with a lower incidence of coronary heart disease and may be due to the effect of estrogen on lipoprotein metabolism.

Cardiovascular System In women, estrogen protects against cardiovascular disease. The protective cardiovascular effects of estrogen include decreased serum LDL cholesterol and increased HDL cholesterol levels, vasodilatory effect, and antioxidation of LDL cholesterol (Chapter 20). Extensive clinical trials have shown that estrogen replacement therapy of postmenopausal women reduces the risk of heart disease.

Brain Estrogen mediates learning and memory functions and also enhances cognitive processes. The brain can produce significant levels of estrogen, and brain tissues possess ER-β.

Metabolism of Estrogen

The liver is the principal site of estrogen inactivation although target tissues are equipped with a means of controlling their estrogen microenvironment. The major route of estradiol metabolism is by way of estrone formation, which is essentially an inactivation process that converts an active estrogen (estradiol = 100% activity) into a less active one (estrone = 30% activity). Peripheral 17βHSD converts estradiol to estrone (15%) to a greater extent than the reverse reaction (5%). Estrone is metabolized by three independent pathways in the liver but is present in other tissues as well. The three pathways are as follows:

1. Estriol formation;
2. Catecholestrogen formation; and
3. Sulfotransferase reaction.

The estriol route is the most significant, judging from the fact that estriol is the major urinary estrogen in humans. As its name suggests, estriol contains three hydroxyl groups, two of which are identical to those of estradiol (3β, 17β), but the third is unique to estriol, a 16α-hydroxyl group that is attached to either estrone or estradiol. Estriol formation is increased in obesity and hypothyroidism, and is decreased in hyperthyroidism. Catecholestrogen formation is catalyzed by a 2-hydroxylase that is present in the liver, nerve cells, and other tissues, and both estrone and estradiol are converted to their respective 2-hydroxylated metabolites. As the name suggests, the A ring of the catecholestrogens has a catechol structure; this makes for their rapid processing by COMT, which is present in many tissues, including the erythrocytes. Thus, catecholestrogens are rapidly O-methylated at position 2 and excreted as such. Catecholestrogen formation is increased in hyperthyroidism and decreased in hypothyroidism. A potentially mutagenic and carcinogenic metabolite of estradiol is 4-hydroxyestradiol, which is also a catecholestrogen. The hydroxylation of estradiol at the 4-position is mediated by a specific hydroxylase (CYP1B1). The activated 4-hydroxyestradiol gives rise to reactive, free-radical semiquinone / quinone intermediates that can damage DNA. Finally, estrone is converted to estrone sulfate by sulfotransferase, present in the liver and other tissues such as the uterus. Estrone sulfate is the most abundant estrogen in plasma, with peak concentrations about 2–10 times higher than these of unconjugated estrogens, probably due to its long half-life in circulation. Estrone sulfate is biologically inactive and can be metabolized to estriol and catecholestrone, but it is usually excreted as such by the kidney.

Selective Estrogen Receptor Modulators (SERM)

As discussed earlier, estrogens have tissue selectivity in part, based on the type of the estrogen receptor and DNA-bound transcription complex present in the target cells. Drugs are being developed to mimic (agonist) or antagonize (antagonist) the effect of estrogen (Figure 34-7). An example of a synthetic estrogen is diethylstilbestrol. Tamoxifen and raloxifene are mixed agonists and antagonists of estrogen activity and are named SERMs (also known as designer estrogens). Tamoxifen antagonizes the action of estrogen in breast tissue and is used in the treatment of ER-positive breast cancer. However, tamoxifen is an estrogen agonist in the uterus and increases the risk for endometrial cancer. In bone and in the cardiovascular system, tamoxifen has beneficial effects similar to those of estrogen. Raloxifene is used in the treatment of osteoporosis as an antiresorptive agent in postmenopausal women but does not exhibit estrogen-related adverse effects on the endometrium. Raloxifene has a similar effect on the serum lipid profile compared to estrogen; however, neither tamoxifen or raloxifene increases the serum concentration of HDL-cholesterol. Many phytoestrogens contain nonsteroidal precursors of estrogenic substances (Figure 34-8), which may affect estrogen sensitive tissues if ingested in significant amounts. However, indications are that the intake must be very high for this to occur.

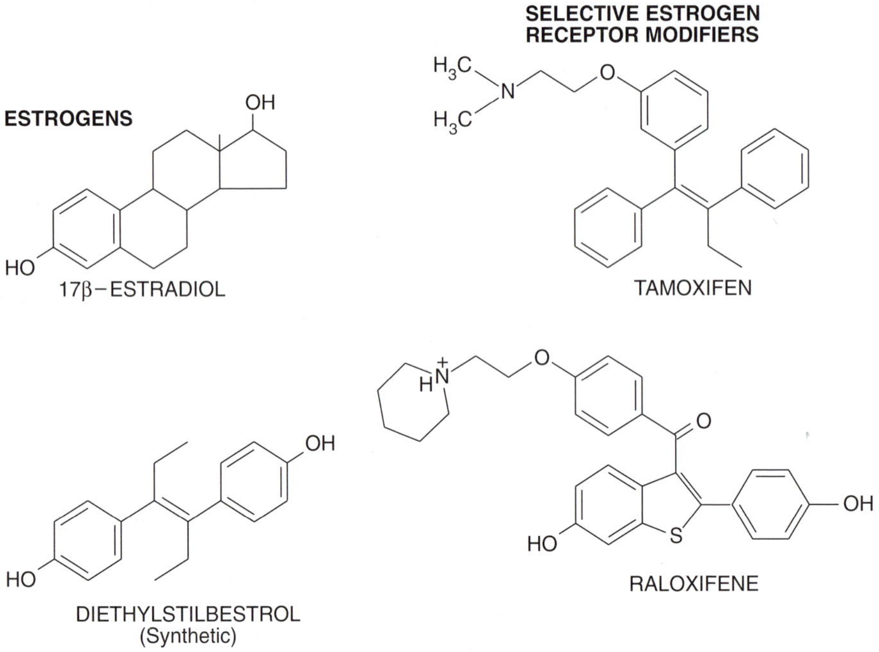

SELECTIVE ESTROGEN RECEPTOR MODIFIERS

ESTROGENS

17β-ESTRADIOL

TAMOXIFEN

RALOXIFENE

DIETHYLSTILBESTROL (Synthetic)

FIGURE 34-7
Structures of estrogens and selective estrogen receptor modifiers.

Biological Effects of Progesterone

The human progesterone receptor (PR) gene on chromosome 11 encodes a single PR protein that is homologous to the other steroid hormone receptors but that binds progesterone preferentially. Like the ER, PR is predominantly found in the nucleus; however, it shuttles between the nucleus and the cytoplasm. PR exists in a stable form bound to heat-shock proteins (hsp), but is activated when progesterone (P) displaces the hsp and binds to the hormone binding domain. The P–PR complex dimerizes and binds to two half-sites of progesterone response elements (PREs) in chromatin; this promotes transcription of a progesterone-sensitive gene. The PRE is identical to the glucocorticoid response element and androgen response element, but the significance of three hormones sharing the same response element in DNA is not known. PR production is stimulated by estrogen and in some tissues is downregulated by progesterone.

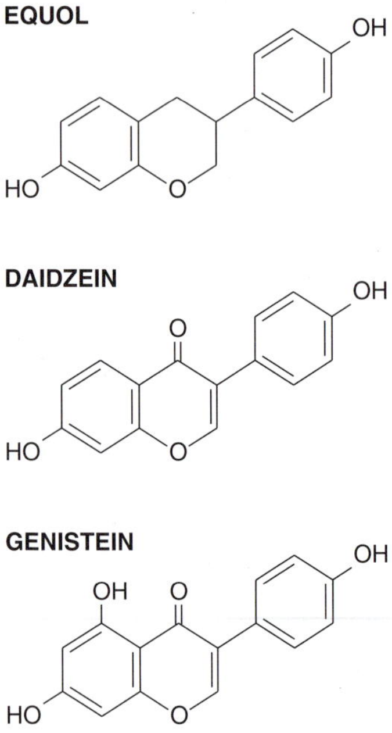

EQUOL

DAIDZEIN

GENISTEIN

FIGURE 34-8
Structures of phytoestrogens. These are diphenolic compounds found in plants, such as soybeans, sprouts of cloves, and alfalfa.

Endometrium Following adequate priming of the endometrium by estradiol, progesterone increases glycogen deposition in the glandular epithelium and stimulates secretory function of the glands from about day 4 after ovulation. Progesterone also stimulates fluid accumulation in the endometrial stroma, with maximal edema occurring at about the time of implantation (day 7). Progesterone promotes the decidual reaction in the stroma in response to the implanting blastocyst and stimulates the production of decidual prolactin (dPRL) by the decidual cells. In the absence of conception, progesterone causes predecidualization of the stroma from about day 9 in a process of terminal (irreversible) differentiation that leads to degeneration of the tissue when progesterone levels decline.

Progesterone is required for pregnancy to continue, in part because it keeps the spiral arteries patent and promotes maternal blood flow to the placenta. Progesterone is also involved in the pathogenesis of endometriosis, a condition in which endometrial tissue grows in ectopic sites (most commonly in the peritoneal cavity); treatment with danazol, a synthetic androgen that antagonizes progesterone, reduces growth of the tissue.

Myometrium Progesterone increases the resting membrane potential of the myometrium and thereby reduces its contractility. Progesterone antagonizes the effect of estrogen by inhibiting synthesis of ER and blocks $PGF_{2\alpha}$-induced contraction of the myometrium. Accordingly, the administration of an antiprogesterone causes a rise in myometrial ER and increased responsiveness to $PGF_{2\alpha}$-induced contractions. There is evidence that myometrial contractility is promoted by the formation of gap junctions between muscle cells; progesterone presumably inhibits gap junction formation and therefore inhibits myometrial contractions.

Mammary Gland In the presence of PRL, progesterone stimulates lobuloalveolar development in the breast of pubertal girls, but only after the tissue has been stimulated by estrogen. As discussed above, estrogen promotes the growth of the ducts and induces the synthesis of progesterone receptors in the tubular epithelium. During the luteal phase of the menstrual cycle and during pregnancy, progesterone (in the presence of estrogen and PRL) stimulates maximal proliferative growth of the breast mainly by alveolar (glandular) growth. In breast cancer cells, progesterone reduces the formation of estrogen from androgenic precursors but increases the production of some of the autocrine growth factors (e.g., TGF-α).

Central Nervous System In addition to its inhibitory effect on GnRH secretion, progesterone causes thermogenesis during the menstrual cycle and pregnancy. In a normal menstrual cycle, the oral basal temperature in most women increases by about 0.5°F around the time of ovulation and remains elevated until shortly before the onset of menses. In pregnant women, the oral basal temperature remains elevated to term. An anesthesizing and EEG-altering effect of progesterone administered to laboratory rodents and the increase in ventilatory drive (hyperventilation) stimulated by progesterone may also reflect CNS actions of the steroid.

Immunosuppression and Antiinflammatory Effects In addition to its ability to suppress myometrial and endometrial prostaglandin formation, progesterone suppresses T-lymphocyte proliferation, interleukin-8 synthesis, and increases prostaglandin dehydrogenase activity, all of which contribute to preventing maternal rejection of the implanting conceptus (an allograft).

Pharmacological Enhancement of Fertility

Female infertility due to inadequate gonadotropin stimulation of the ovary has been successfully treated by two approaches:

1. Increasing gonadotropin levels by supplying exogenous gonadotropins, and
2. Increasing gonadotropin levels by stimulating their release from the pituitary.

Exogenous Gonadotropin Treatment Women who are infertile because of a deficiency in gonadotropins are given exogenous FSH for 7–12 days to promote folliculogenesis followed by a single large dose of hCG to induce ovulation. This has resulted in ovulation in a majority of patients (~90%) and pregnancy in about half. However, about 17–20% of the patients have multiple births due to the secretion of more than one follicle.

Stimulation of Endogenous Gonadotropin Secretion In women who are infertile because of a neuroendocrine system that is overly sensitive to negative estrogen feedback, an antiestrogen is used to reduce the intensity of the negative feedback. By reducing the intensity of estrogen feedback, normal gonadotropin secretion should resume and lead to ovulation. Clomiphene (50 mg) is taken daily for 5 days to block the estrogen effect in the neuroendocrine system; in a majority of patients (~80%), ovulation occurs about a week later, and about 30–40% of the women conceive. The incidence of multiple birth by this method (~8%) is higher than in untreated patients (~1%) but lower than with gonadotropin treatment. In some laboratories that conduct *in vitro* fertilization, the two protocols (exogenous gonadotropins and clomiphene) are combined for a greater yield of secondary oocytes.

Pharmacological Prevention of Pregnancy (Female Contraception)

Currently, the most effective method to prevent conception is by oral steroid hormone treatment (Figure 34-9) to inhibit or modify the cyclic changes in endogenous reproductive hormones. Two effective types of oral contraceptives are used with proven success: one to prevent ovulation, the other to prevent implantation.

The most effective means of preventing ovulation is by the oral administration of ovulation-inducing gonadal steroid derivatives at dosages that will prevent cyclic changes in LH and FSH secretion. This method has proven to be more than 99% effective over the past 40 years and is currently used by more than 50 million women worldwide. The treatment protocol is of four different types, all of which involve the use of synthetic steroid

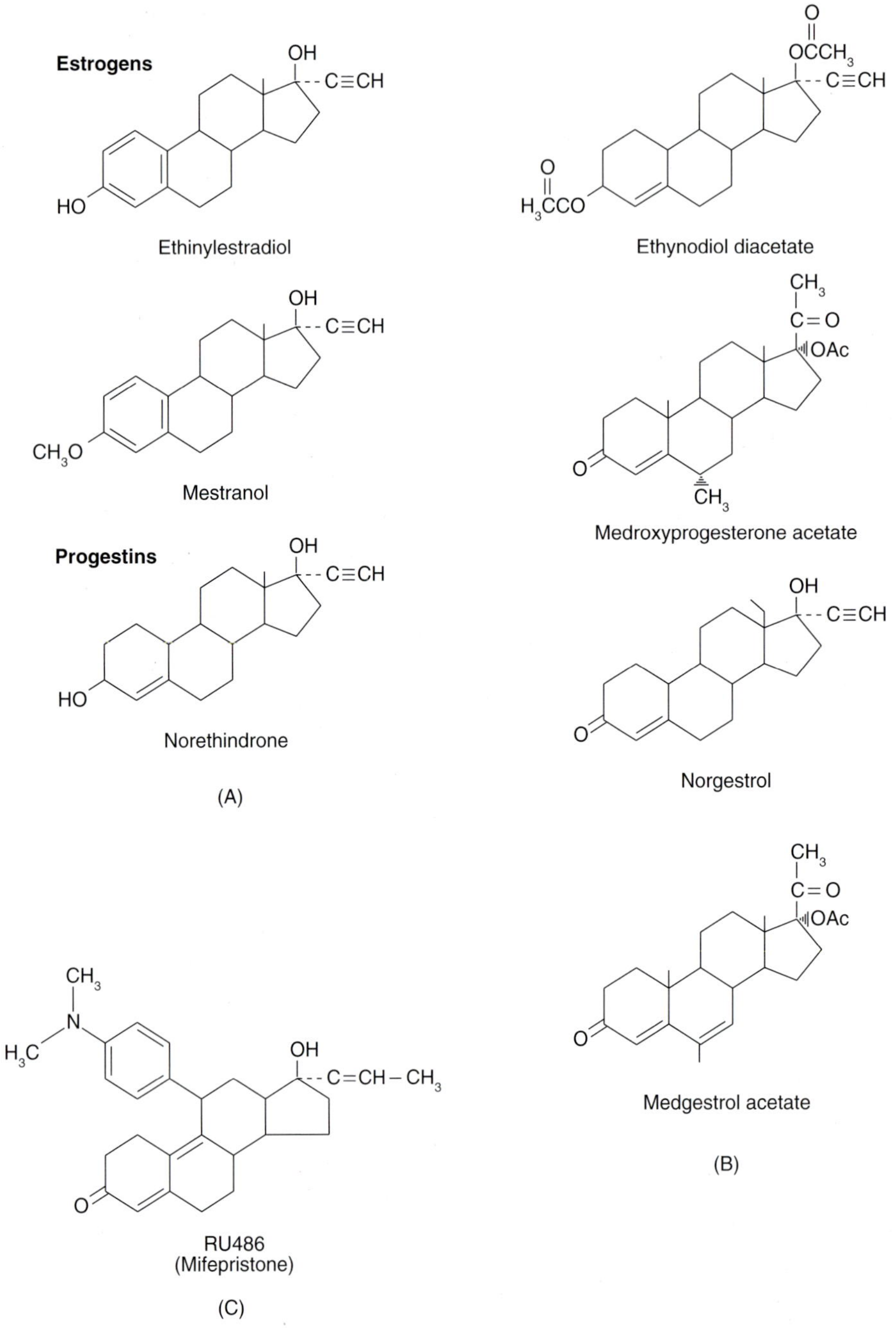
Estrogens
OH
C≡CH
HO
Ethinylestradiol
O
OCCH3
C≡CH
O
H3CCO
Ethynodiol diacetate
OH
C≡CH
CH3O
Mestranol
CH3
C=O
OAc
O
CH3
Medroxyprogesterone acetate
Progestins
OH
C≡CH
HO
Norethindrone
(A)
OH
C≡CH
O
Norgestrol
CH3
C=O
OAc
O
Medgestrol acetate
(B)
CH3
N
H3C
OH
C=CH−CH3
O
RU486
(Mifepristone)
(C)

FIGURE 34-9
Structures of steroids used as oral contraceptives. (A) Combination pills containing an estrogen + progestin. (B) The progestin-only pill. (C) RU-486 or mifepristone, which can be utilized as an abortifacient. [Reproduced with permission from S. W. Norman and G. Litwack, *Hormones,* 2nd ed. San Diego: Academic Press, 1997].

derivatives that have long biological half-lives. The first three protocols involve estrogen + progestin combinations taken for 3 weeks, followed by no treatment for 1 week.

1. Fixed combined dose of estrogen + progestin daily ("combination pill");
2. Biphasic type: fixed dose estrogen + two different doses of progestin;
3. Triphasic type: fixed dose estrogen + three different doses of progestin (stepwise increase at 1-week intervals); and
4. "Mini" pill, containing a progestin only; taken daily without interval of withdrawal.

Among these, combination pill is the most widely used and is probably the most effective at $> 99\%$, whereas the mini pill is the least effective at $\sim 95\%$ because it does not always prevent ovulation. The rationale for varying the dose of progestin was that by reducing the total amount of steroid, it would reduce the risk of myocardial infarction, hypertension, and stroke, all of which were subsequently shown not to be altered by varying the dose of progestin. The rationale for the progestin only protocol was to eliminate the risk of endometrial cancer and breast cancer associated with estrogen; however, some of the health benefits of estrogen (e.g., increased HDL, osteoprotection) are eliminated and some side effects (e.g., breakthrough bleeding) are introduced.

In women using the combination pill, the plasma levels of LH, FSH, estradiol, and progesterone were found to be greatly suppressed and unchanged throughout the 3 week period on the pill. Folliculogenesis did not occur during treatment but there was no evidence of increased atresia; when electing to bear a child, former pill users experienced normal menstrual cyclicity, pregnancy, and lactation. Long term observations indicated that pill users experienced menopause within the normal age range, and had no health problems that could be attributed to the long term use of the pill, although recent epidemiological studies suggest that combination pill users may be at increased risk of developing breast cancer prior to age 45.

Post-coital pharmacologic prevention of implantation can be achieved by oral administration of a synthetic estrogen (25 mg diethylstilbestrol) twice daily for 5 days. This "morning after pill" treatment stimulates fallopian tube contractions during the period of conceptus travel toward the uterus, such that the conceptus is propelled into the uterus prematurely and is resorbed, or is trapped within the fallopian tube because of spastic contraction of the isthmus. Although effective, the high dose estrogen produces nausea, vomiting, and menstrual problems. Alternatively, oral administration of a combination pill (50 μg ethinylestradiol + 0.5 mg norgestrol), two tablets within 72 h post-coitum and two tablets 12 h later, is also effective and does not produce undesirable side effects. This treatment does not involve an alteration in fallopian tube activity but the treatment may serve to convert the endometrium into a less receptive tissue for blastocyst implantation.

Another post-coital treatment is the use of a progesterone antagonist, mifepristone (RU-486), which apparently works by reversing the effects of progesterone on the endometrium, and thereby interferes with implantation. Administration of the drug two days after the midcycle LH surge prevents implantation by a direct effect on the endometrium, and has no effect on the circulating levels of gonadotropins or on luteal function.

Supplemental Readings and References

C. J. Bagatell and W. J. Bremner: Drug therapy: Androgens in men: Uses and abuses. *New England Journal of Medicine* **334,** 707 (1996).

S. Bhasin and W. J. Bremner: Clinical review 85: Emerging issues in androgen replacement therapy. *Journal of Clinical Endocrinology and Metabolism* **82,** 3(1997).

S. Bhasin, T. W. Storer, N. Berman, and others: The effects of supraphysiologic doses of testosterone on muscle size and strength in normal men. *New England Journal of Medicine* **335,** 1 (1996).

E. G. Biglieri: A prismatic case: 17α-Hydroxylase deficiency: 1963-1966. *Journal of Clinical Endocrinology and Metabolism* **82,** 48 (1997).

B. C. J. M. Fauser and A. M. Van Heusden: Manipulation of human ovarian function: physiological concepts and clinical consequences. *Endocrine Review* **18,** 71 (1997).

A. Gougeon: Regulation of ovarian follicular development in primates: facts and hypotheses. *Endocrine Review* **17,** 121(1996).

K. B. Horwitz: The molecular biology of RU486. Is there a role for antiprogestins in the treatment of breast cancer? *Endocrine Review* **13,** 146 (1992).

J. L. Jameson and A. N. Hollenberg: Regulation of chorionic gonadotropin (hCG) gene expression. *Endocrine Review* **14,** 203 (1993).

A. L. Kierszenbaum: Mammalian spermatogenesis *in vivo* and *in vitro*: a partnership of spermatogenic and somatic cell lineages. *Endocrine Review* **15,** 116 (1994).

C. S. Kovacs and H. M. Kronenberg: Maternal-fetal calcium and bone metabolism during pregnancy, puerperium, and lactation. *Endocrine Review* **18,** 832 (1997).

R. A. Wild: Metabolic and cardiovascular issues in women with androgen excess. *Endocrinologist* **6,** 120 (1996).

J. D. Wilson: Androgen abuse by athletes *Endocrine Review* **9,** 181 (1988).

J. D. Wilson, J. E. Griffin, and D. W. Russell: Steroid 5α-reductase 2 deficiency. *Endocrine Review* **14,** 577 (1993).

M. D. Pisarska, S. A. Carson, and J. E. Buster: Ectopic Pregnancy. *The Lancet* **351,** 1115 (1998).

D. T. Baird: Amenorrhoea. *The Lancet* **350,** 275 (1997).

F. Muscatelli, T. M. Strom, A. P. Walker, and others: Mutations in the DAXC-1 gene give rise to both X-linked adrenal hypoplasia congenita and hypogonadtropic hypogonadism. *Nature* **372,** 672 (1994).

A. Swain, V. Narvaez, P. Burgoyne, and others: Dax1 antagonizes Sry in mammalian sex determination. *Nature* **391,** 761 (1998).

H. E. Maclean, G. L. Warne, and J. D. Zajact: Intersex disorders: shedding light on male sexual differentiation beyond SRY. *Clinical Molecular Endocrinology* **46,** 101 (1997).

A. Glasier: Emergency postcoital contraception. *New England Journal of Medicine* **337,** 1058 (1997).

F. Muscatelli, T. M. Strom, A. P. Walker, et al.: Mutations in the DAXC-1 gene give rise to both X-linked adrenal hypoplasia congenita and hypogonado. tropic hypogonadism. *Nature* **372,** 672 (1994).

A. Swain, V. Narvaez, P. Burgoyne, et al.: Daxl antagonizes Sry in mammalian sex determination. *Nature* **391,** 761 (1998).

H. E. Maclean, G. L. Warne, and Zajact: Intersex disorders: shedding light on male sexual differentiation beyond SRY. *Clinical Molecular Endocrinology* **46,** 101 (1997).

G. El-Hajj Fuleihan: Tissue-specific estrogens—the promise for the future. *New England Journal of Medicine* **337,** 1686 (1997).

P. D. Delmas, N. H. Bjarnason, B. H. Mitlak, et al.: Effects of raloxifene on bone mineral density, serum cholesterol concentrations, and uterine endometrium in postmenopausal women. *New England Journal of Medicine* **337,** 1641 (1997).

W. Khovidhunkit and D. M. Shoback: Clinical effects of raloxifene hydrochloride in women. *Annals of Internal Medicine* **130,** 431 (1999).

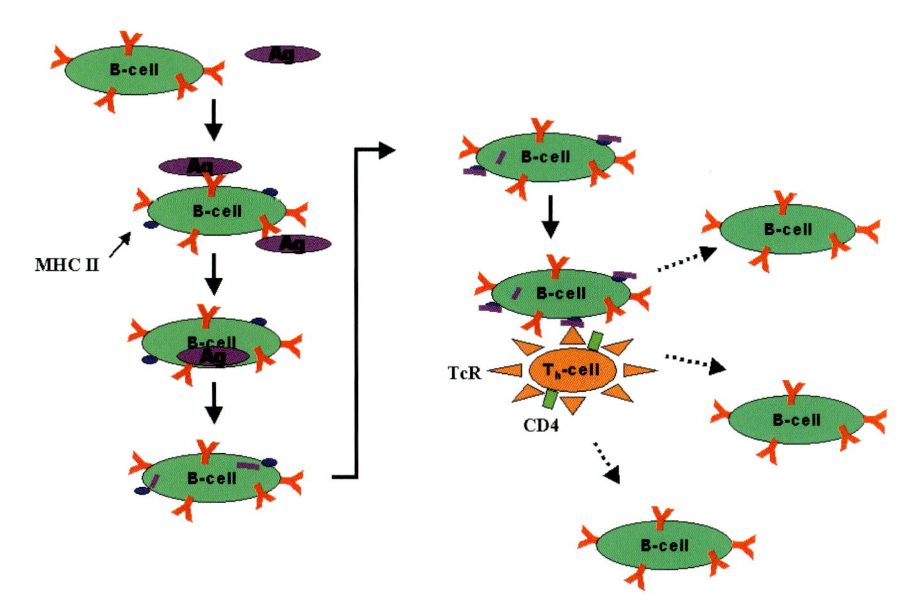

CHAPTER 35, FIGURE 1 Antigen recognition by B-cell surface receptors. Response by B cells begins with antigen binding to receptors on the B-cell surface (membrane-bound IgM and /or IgD molecules). Phagocytosis by the B cell transports the antigen into the cell where it is degraded to peptides by proteolytic enzymes. The antigen-derived peptides are next transported to the surface and bind to membrane proteins of the major histocompatibility complex (MHC). The peptides are presented by the MHC proteins (specifically MHC II) to T-helper (T_h) cells. The T_h cells bind to MHC II proteins via a specific T-cell receptor (TcR) and the peptides displayed on MHC II molecules. Interleukin-4 from the T_h cells (marked by the surface protein CD4) stimulate the B cells to proliferate and to synthesize and secrete antibodies.

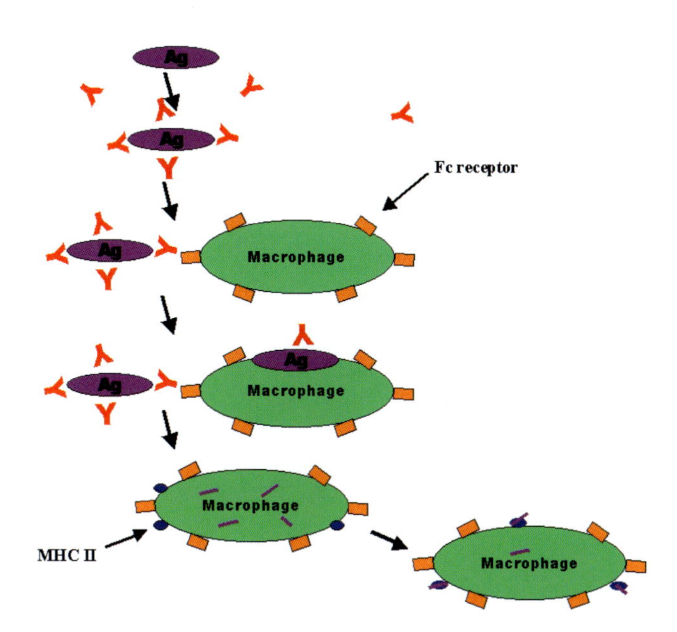

CHAPTER 35, FIGURE 2 Antigen-antibody complex formation and recognition by macrophages. Antigens that react directly with antibodies in the circulation and in the extravascular fluid are recognized and removed by macrophages. Recognition occurs with the Fc region of the antibody in the Ag-Ab complex. The antigen-antibody complex is bound to Fc receptors on the external surface of the macrophage. The antigen is taken into the cell by endocytosis and degraded proteolytically. The antigen-derived peptides are transported and bound to the surface MHC II proteins. The MHC II peptide complexes are presented to the antigen-specific T cells.

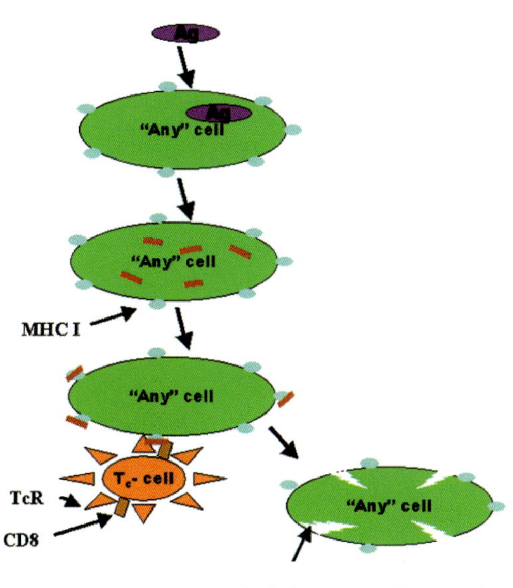

CHAPTER 35, FIGURE 3 Antigen uptake and processing by nonimmune system cells. Microbes and viruses can invade and infect any nucleated cell, not only those involved in the immune response. The invading organism is digested by lysosomal enzymes. Peptides encoded by infecting DNA or RNA from the microbe are synthesized in the cell and then transported to and displayed on MHC I molecules on the surface. The MHC I displayed peptides are recognized by a T-cell receptor (TcR) on the surface of cytotoxic T lymphocytes (CTLs). The infected cells are made permeable by the protein perforin produced by the CTLs and destroyed by apoptosis.

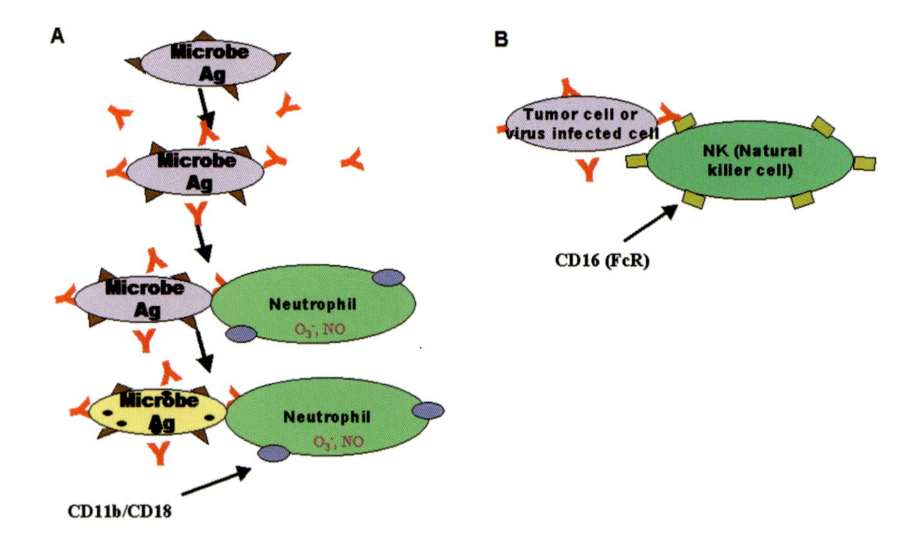

CHAPTER 35, FIGURE 4 Microbe destruction by neutrophils and antibody-dependent, cell-mediated cytotoxicity. (A) Neutrophils recognize and attack antibody-tagged microorgansims by a mechanism called antibody-dependent, cell-mediated cytotoxicity (ADCC). In this defense mechansim, microbes "tagged" by antibodies are destroyed by a respiratory burst of O_3^-, H_2O_2, and NO produced by the attacking neutrophils. (B) In a similar way, natural killer (NK) cells recognize antibody-tagged virus-infected cells and tumor cells.

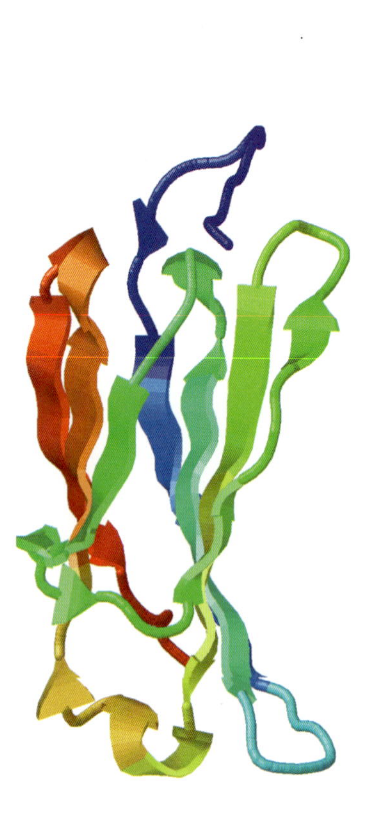

CHAPTER 35, FIGURE 5 Structural motif of the immunoglobulin family. Each domain of the immunoglobulin or other molecule of the immunoglobulin family is composed of a collection of β strands that form a globular structure. Colors indicate different β strands. There are seven strands in the constant region domains that form two β sheets; one sheet contains three antiparallel strands and the other four antiparallel strands. The orientation illustrates the two sheets that together create the β sandwich or immunoglobulin fold.

CHAPTER 35, FIGURE 6 Pairs of immunoglobulin motif structures (see Figure 35-7) connected by a "tether" polypeptide that creates a flexible fold. The disulfide bridges that connect the β sheets of the β sandwich are shown as yellow space-filling residues. Each IgG molecule contains six such folds arranged so that a heavy and light chain folds partially wrap around each other (Figure 35-7 B, C). The ribbon model shown is from an x-ray crystallographic structure of a Fab fragment. The figure is derived from the coordinates published in the Protein Data Bank file, 2FPB.

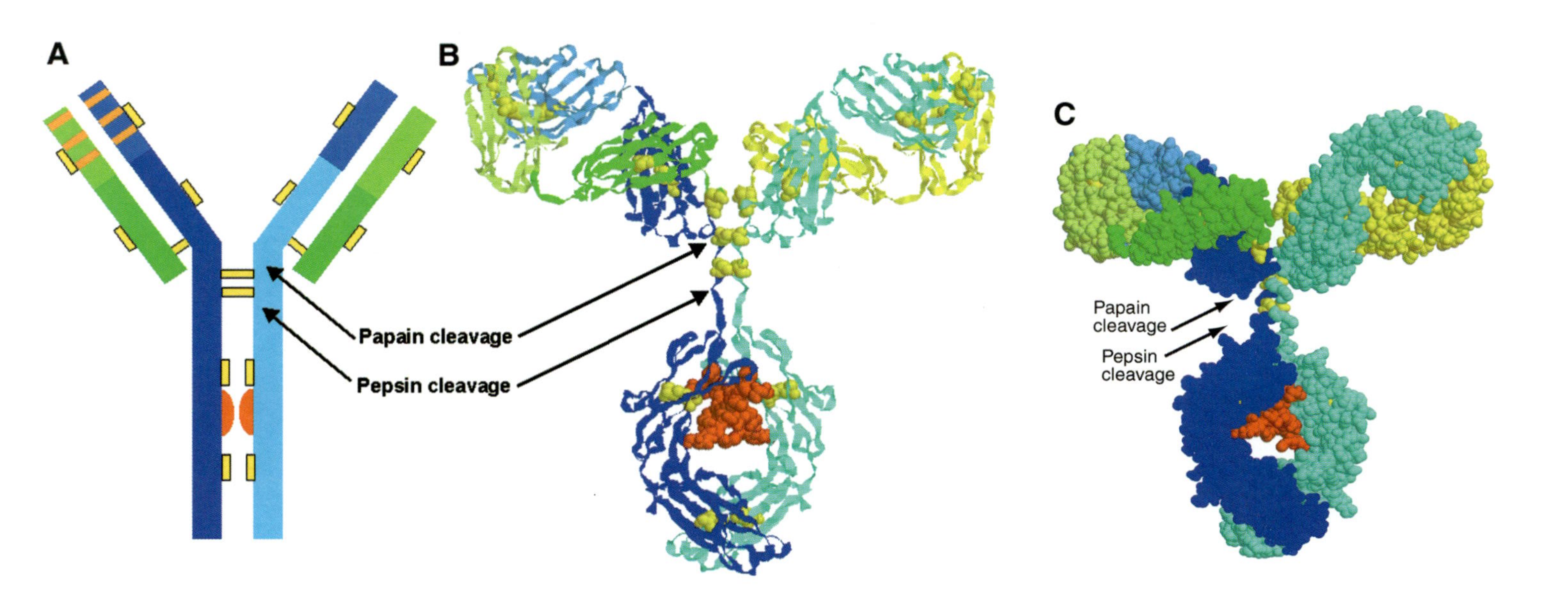

CHAPTER 35, FIGURE 7 Schematic diagram of an immunoglobulin. (A) The simplest immunoglobulin (Ig) molecule, IgG consists of four polypeptide chains: two heavy chains of ~50,000 Da and two light chains of ~25,000 Da. More complex Ig molecules are dimers or pentamers that consist of similar basic structures but may have slightly different heavy and/or light chains. Each of the two heavy chains (shown in blue) contains four immunoglobulin domains, a variable VH donain at the N terminus, and three "constant" domains, CH1, CH2, and CH3. Each of the light chains contains two domains, VL and CL. An immunoglobulin domain is 110–120 amino acids in length. The V domains of both chains contain three regions described as hypervariable regions (orange bands) within the VH and VL domains. Each domain contains one intrachain disulfide bridge. The CH1 and the CL1 domains are linked by a single, interchain disulfide bridge. Two disulfide bridges join the two heavy chains near the N-terminal residues of the CH2 domain. A single, Asn-linked, oligosaccharide chain is present within the CH2 domains of each of the heavy chains. A stretch of amino acids between the CH1 and CH2 domains forms a "hinge" region where the two heavy chains are connected by two disulfide bridges. The hinge region acts as a flexible "tether" between the CH1 and CH2 domains. NMR studies show this region to be very flexible, which permits the Ig molecule to assume various shapes, from the "Y" shown in the figure to a "T" with the "arms" perpendicular to the Fc region (see below). The antigen-binding residues are located at the upper "tips" of the Y-shaped Ig molecule. In the folded Ig molecule, the residues of the hypervariable regions are actually located at the tips, as can be seen in Figure 35-8. The portion of the IgG molecule above the dashed lines is the variable region; that portion below the dashed lines is the constant region. (B) Immunoglobulin secondary structure. The secondary structures of all immunoglobulin domains are essentially the same. Each domain contains seven antiparallel β sheets. The cystine residues (disulfide bonds) are space-filled in the figure. The open, accessible residues within the hinge are also evident in the structure. The figure is derived from the coordinates published by E.A. Padlan, *Mol. Immunol.* **31**:169, 1994. (C) Space-filling model for IgG. Chains are color coded the same throughout. The polypeptide regions comprising the hinge are readily accessible to proteolytic enzymes. Papain cleaves at the N-terminal side of the uppermost disulfide bridge in the hinge to create three fragments; two Fab fragments containing the VH and CH1 domains linked to the VL and CL domains by a single disulfide bond. The Fc fragment contains two CH2 and two CH3 domains of the heavy chains. The antigen-binding regions of each Fab retain their ability to bind to the epitope for which they are specific. The Fc fragment contains a regions that binds to the cellular Fc receptor. This receptor binding sire is sometimes referred to as the "biological activity" of the Ig molecule. Cleavage by pepsin produces a fragment $(Fab')_2$ and several fragments from the Fc region of the Ig chains. The figure is derived from the coordinates published by E. A. Padlan, *Mol. Immunol.* **31**:169, 1994.

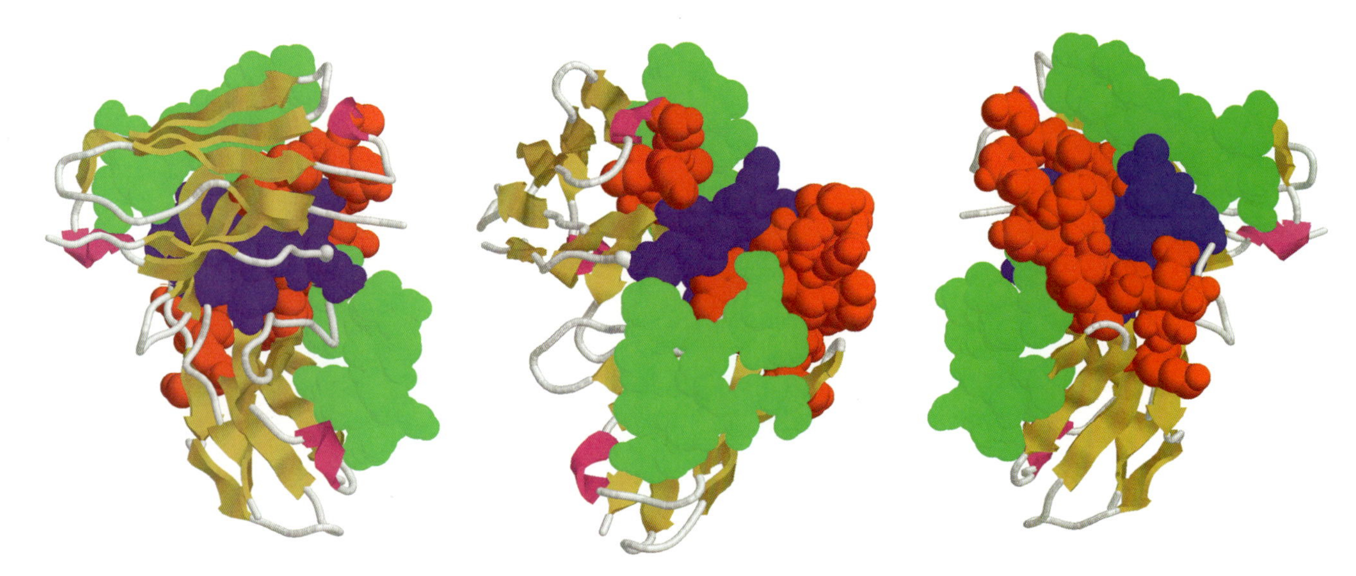

CHAPTER 35, FIGURE 8 Complementarity-determining regions. The actual locations of the complementarity-determining regions of the VH and VL chains are shown in this fragment. The orientation is approximately 90° clockwise relative to the orientations in Figure 35-7. The CDR residues are primarily found in the hairpin turns that connect the β sheets of the VH and VL domains. The crystallographic structure used for showing the CDR residues is derived from the structure of a Fab fragment interacting with the capsular polysaccharide from *H. influenzae B* published in the Protein Data Bank file 1HOU. The three orientations of the structure show (1) the backbone β sheets of the immunoglobulin domain; (2) a side view that shows how the CDR regions overlap; and (3) a view that shows the extensive area that the CDR regions provide for making contact with an epitope of the antigen. CDR1=red; CDR2=green; CDR3=blue.

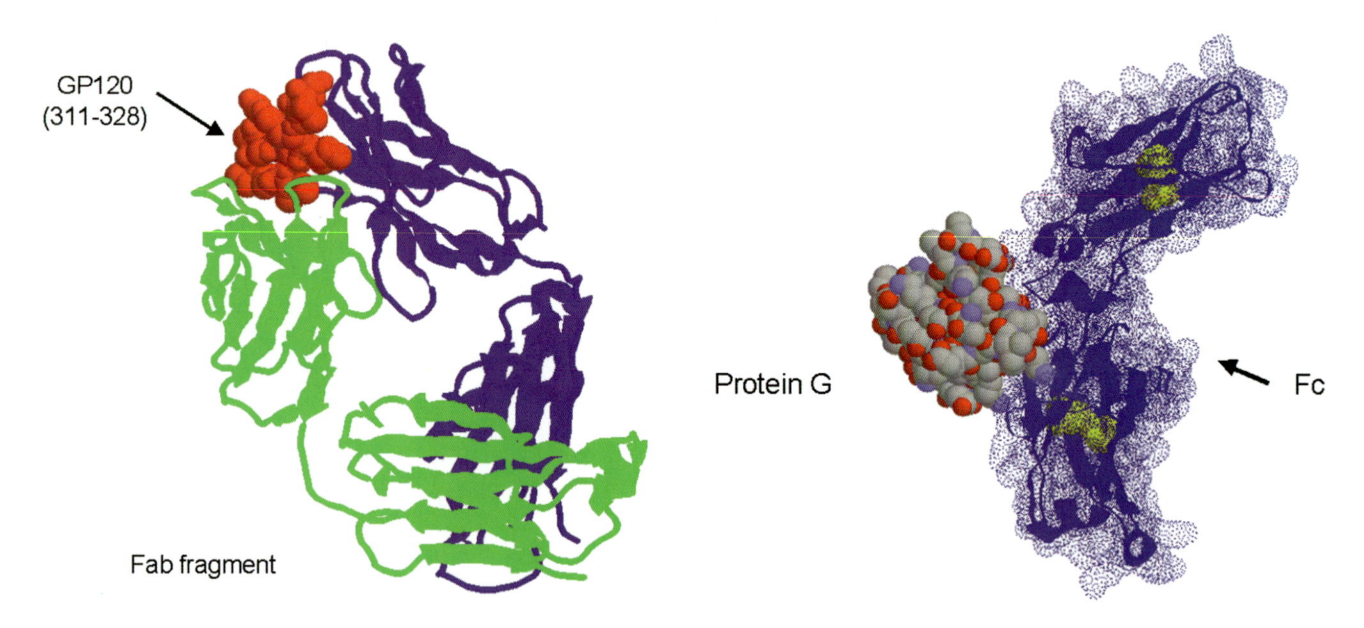

CHAPTER 35, FIGURE 9 HIV-1 Gp120 peptide bound to the antigen recognition site of an antibody Fab fragment. The binding of HIV-1 Gp120 residues 311–318 to the antigen binding site shows the actual contacts between antigen and antibody that underlie Ag/Ab recognition. The figure is derived from the coordinates published in the Protein Data Bank file 1GGI.

CHAPTER 35, FIGURE 10 Binding of protein G (*S. aureus*) to an Fc fragment. Protein G binding mimics the Fc receptor binding of an Ig in the Ig Fc region. Protein G is used for purification of antibodies and in their "capture" in immunoassays. The figure is derived from the coordinates published in the Protein Data Bank file 1FCC.

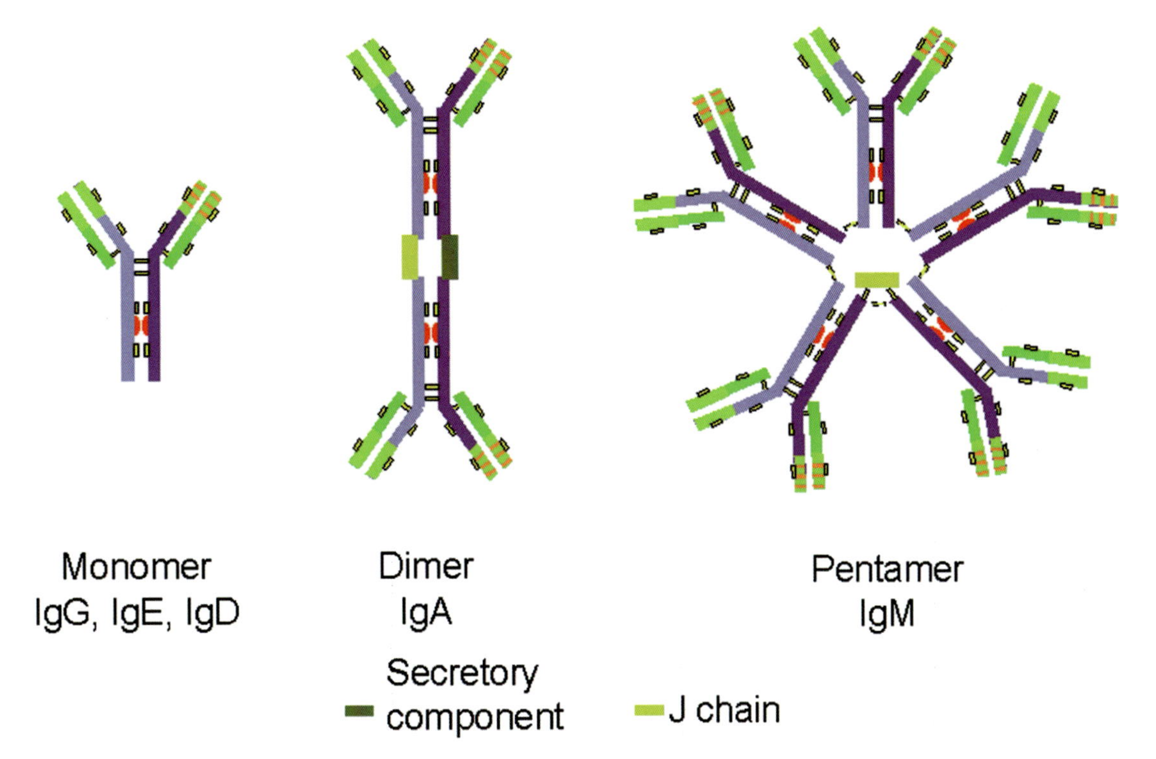

CHAPTER 35, FIGURE 11 Immunoglobulin classes. The five immunoglobulin classes are determined by the amino acid sequences of the constant regions of the heavy polypeptide chains (see Table 35-2).

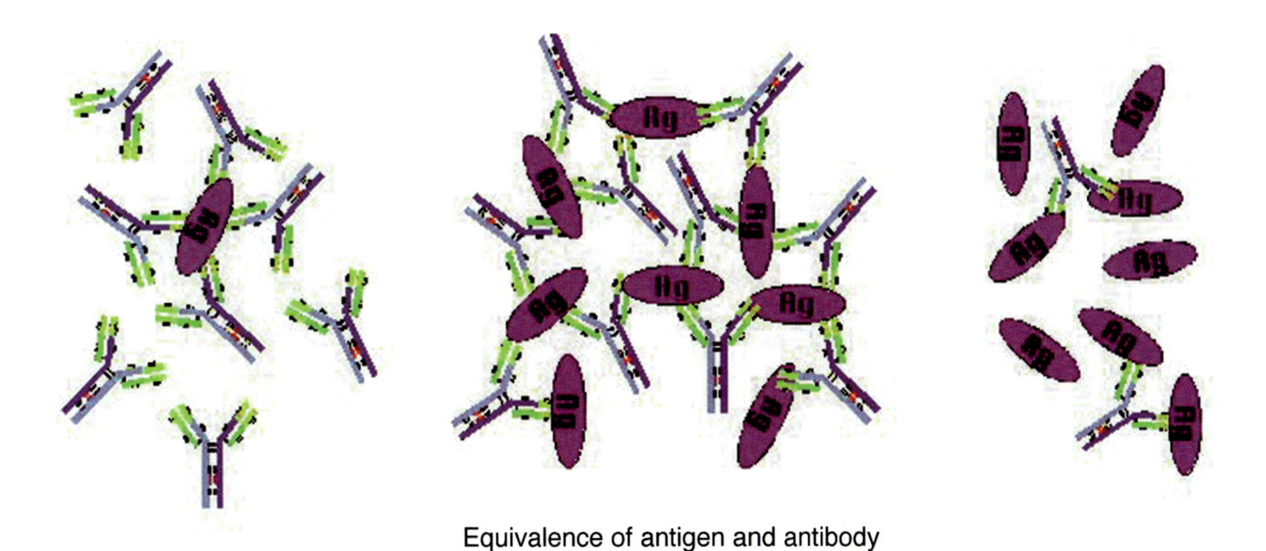

CHAPTER 35, FIGURE 12 Antigen-antibody complex formation. Optimal concentrations of antibody and antigen result in cross-linking and precipitation of the antigen-antibody complexes. Cross-linking is important in B-cell stimulation when antigens bind to the IgM B-cell receptors. If antibody is present in excess, then all of the epitopes on the antigen can be occupied by single antibody molecules (subject to steric exclusion) and the antigen-antibody complexes do not precipitate. If antigen is present in excess, then all antibody binding sites can be occupied by antigen molecules and no cross-linking occurs.

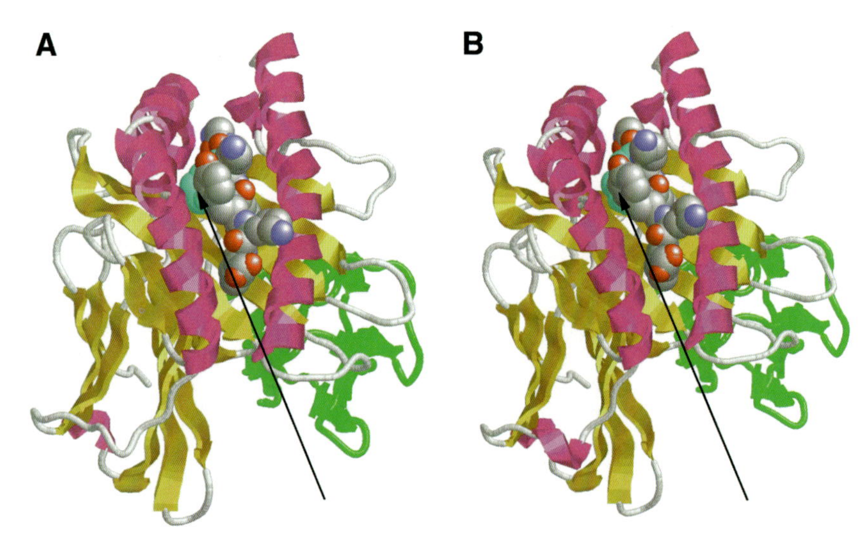

CHAPTER 35, FIGURE 13 Major histocompatibility complex (MHC) proteins class I. The MHC class I proteins of antigen-presenting cells comprise two polypeptide chains, α and β. The α-polypeptide (~360 residues) (helices red, β sheets dark yellow, and loops gray) has two extracellular domains shown here (α1 and α2) and a transmembrane and intracellular domain (α3, not shown). The α3 domain is structurally similar to the immunoglobulins. The non- MHC gene-related protein, β_2-microglobulin (99 residues), is shown in green on the right side of each structure. The other subunits of the complex that are involved in T-cell receptor recognition are not shown. The figures are derived from the coordinates published in the Protein Data Bank files 1AGD and 1AGB. (A) HIV-1 GAG peptide GGKKKYKL presented on the surface of the MHC class II protein. The peptide is shown with the residues "space-filled", the Lys residue, K is colored cyan and marked by the arrow. (B) HIV-1 GAG peptide GGRKKYKL presented on the surface of the MHC class II protein. The peptide is shown with the residues space-filled, the Arg residue, R is colored cyan and marked by the arrow. Note that the position of the α helix (part of the α1 domain) immediately adjacent to the Arg or Lys residues has undergone a change in position (conformation change) and has a small helical segment (lower left) as a result of this single-amino-acid substitution. Such sensitivity to small changes in antigen structure is critical to specificity.

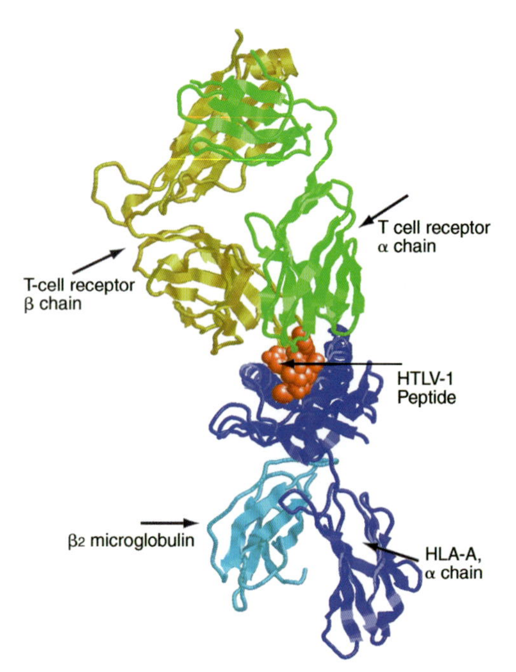

CHAPTER 35, FIGURE 14 Complex of MHC class I proteins, HTLV-1 peptide (antigen) and TcR. The MHC class I proteins bind and present peptides that are synthesized inside the infected cell to T-cell receptors (TcRs) specific for the bound peptides. Interaction and stimulation of T-cell proliferation occurs with CD8 positive, cytoxic Tc cells. The figure is derived from the coordinates published in the Protein Data Bank file 1BD2.

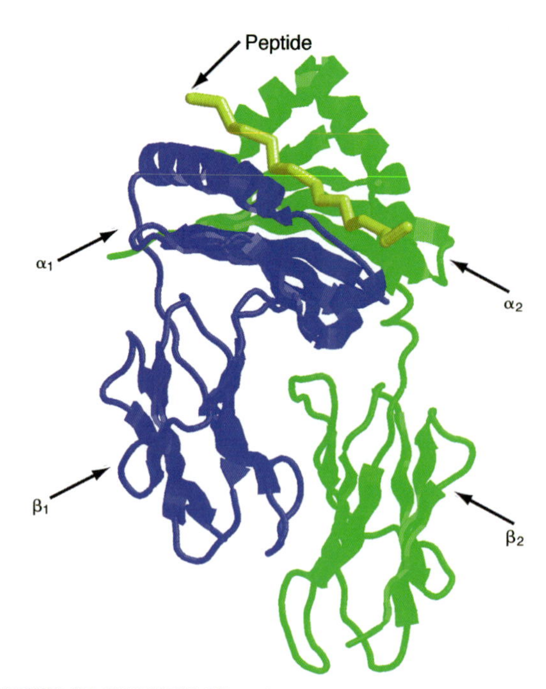

CHAPTER 35, FIGURE 15 Major histocompatibility complex (MHC) proteins class II. The MHC class II proteins bind and present and present peptides that are synthesized outside the infected cell, i.e., peptides that are derived from proteins of the infecting organisms, bound to their specific T-cell receptors (TcRs). Interaction and stimulation of T-cell proliferation occurs with $CD4^+$ positive T cells (Figure 35-16). The figure is derived from the coordinates published in the Protein Data Bank file 1A6A.

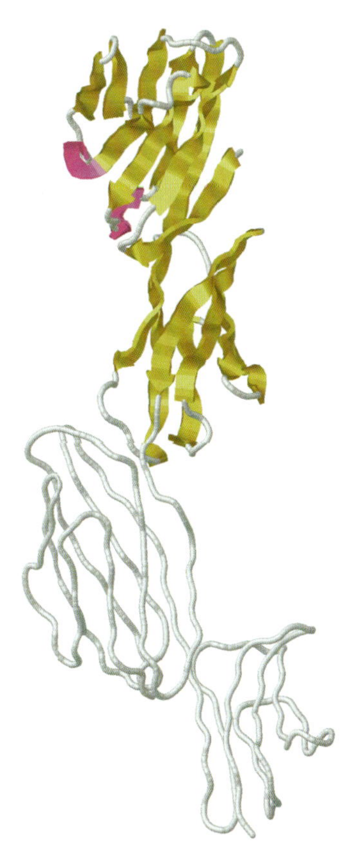

CHAPTER 35, FIGURE 16 CD4 (cluster differentiation 4) protein. CD4 on T cells is a component of the MHC class II, TcR complex. Shown are the extracellular domains that consist of two immunologlobulin-like folds. The β-sheet secondary structure is illustrated in one fold; only the α-carbon backbone is shown in the other. The figure is derived from the coordinates published in the Protein Data Bank file 1WIO.

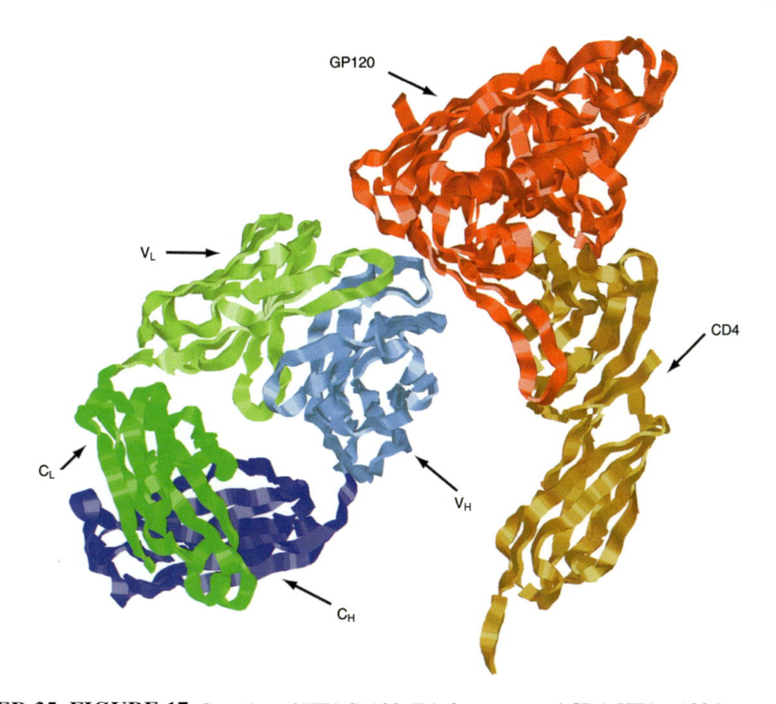

CHAPTER 35, FIGURE 17 Complex of HIV Gp120, Fab fragment, and CD4. HIV gp120 is an envelope protein of HIV-1 that is exposed on the outer surface of the virus. It is derived from gp160 by proteolysis that also produces another product, gp41. Only a small area of the surface of the gp120 and CD4 molecules is in contact. Interaction between CD4 and gp120 is dominated by residues Phe43 and Arg59 of CD4, and Gly370, Try427, and Asp368 of Gp120. The antibody fragment blocks the region of Gp120 that interacts with a chemokine receptor, CCR5. The figure is derived from the coordinates published in the Protein Data Bank file 1GC1.

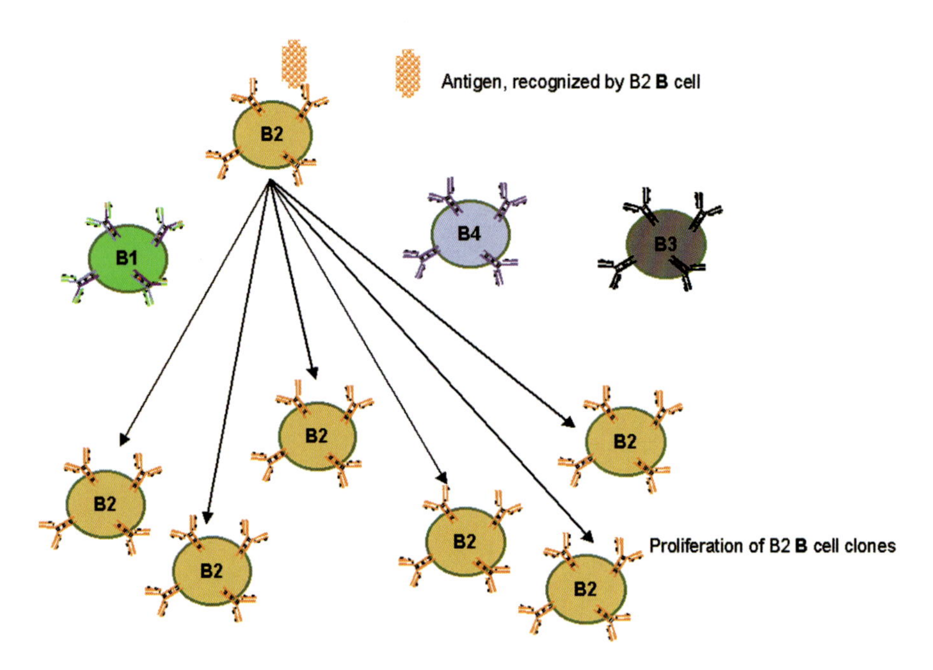

CHAPTER 35, FIGURE 18 B-cell clonal selection. Antigen binding to the surface Ig (monomeric IgM and IgD) receptors occurs only to the B-cell clone that recognizes a particular epitope on the antigen (Figure 35-1).

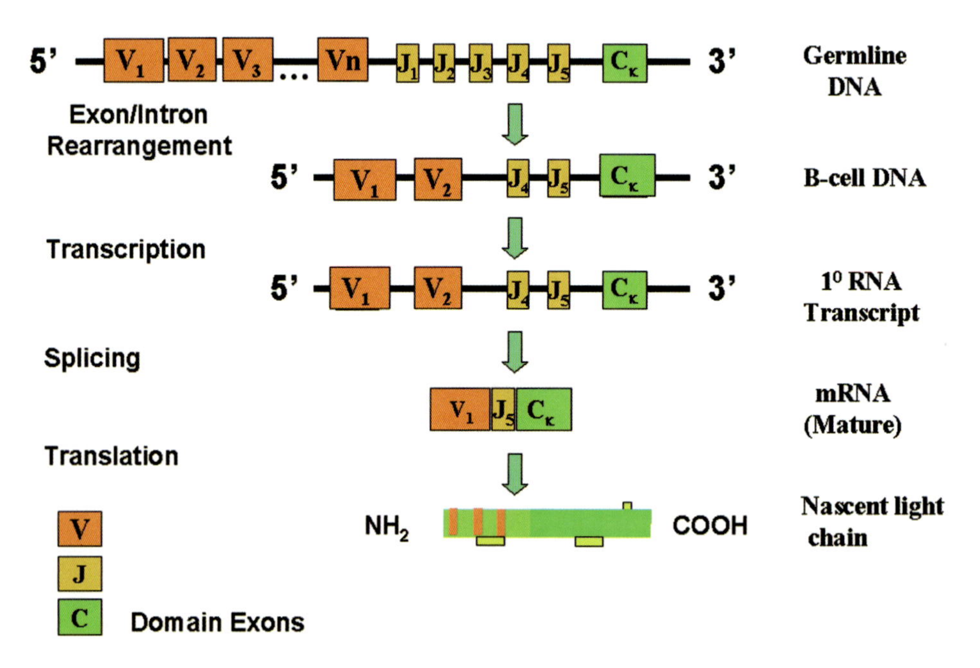

CHAPTER 35, FIGURE 19 Immunoglobulin light chain gene structure and gene processing. Germline immunoglobulin genes consist of exons selected during differentiation to generate the mature mRNA that is translated to give the immunoglobulin chain that recognizes a single specific epitope. Exons (>100) code for the first ~95 amino acid residues, the VL regions of the κ and λ chains. J (junction) exons code for the next ~15 amino acid residues. Separate exons exist for the CL regions of the κ and λ chains. Diversity in the CDR3 region of the VL chain arises from the various combinations of the 5 J exons that can occur with the ~100 VL exons. The excised DNA of the germline is degraded, leaving a specific B-cell DNA that "defines" the particular B-cell clone. The primary RNA transcript that is formed is spliced such that the primary transcript becomes a mature mRNA that is translated into a specific Ig light chain.

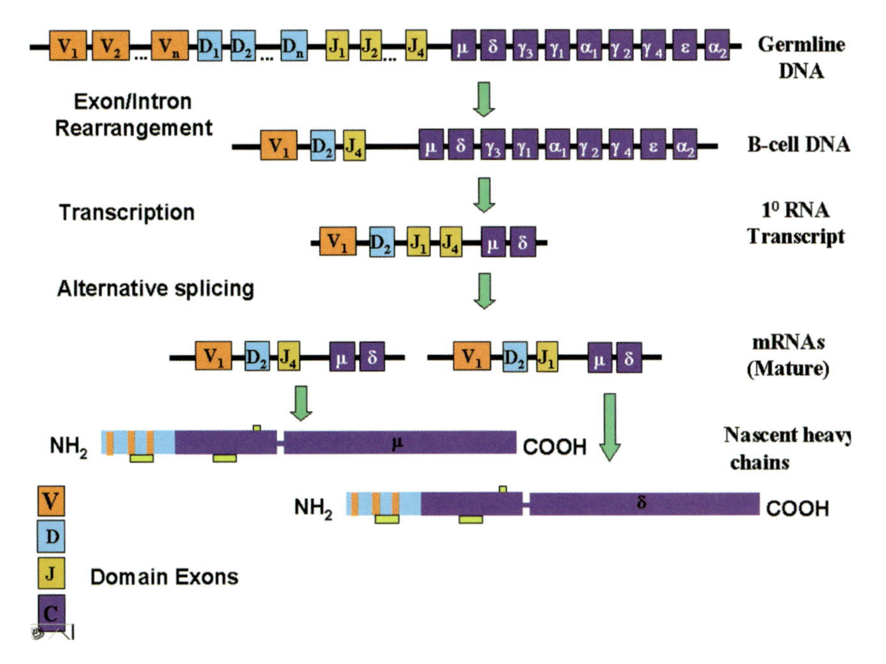

CHAPTER 35, FIGURE 20 Immunoglobulin heavy chain gene structure and gene processing. Exons of the heavy chain genes that encode the variable regions of the immunoglobulin molecule are shown in orange. Selection from these V exons during embryonic development produces the unique sequences of each B-cell clone. The germline genes for the immunoglobulin heavy chains contain an additional group of exons, the D (diversity, light blue) exons. Recombination between the V and D regions occurs more frequently than that between the V and J exons in the light chain exons. Introns between the V and D and between the D and J exons contain signal sequences that regulate the expression of the enzyme recombinase. This enzyme is responsible for the efficient recombination that gives rise to the epitope-specific B-cell clones with their individual Ig genes. The heavy chain genes contain exons that encode all of the isotype heavy chains. Class switching, i.e., the change in chain expression that occurs during antibody synthesis after B-cell activation, results from alternative splicing between the J exons and the exons for the various heavy chain isotypes.

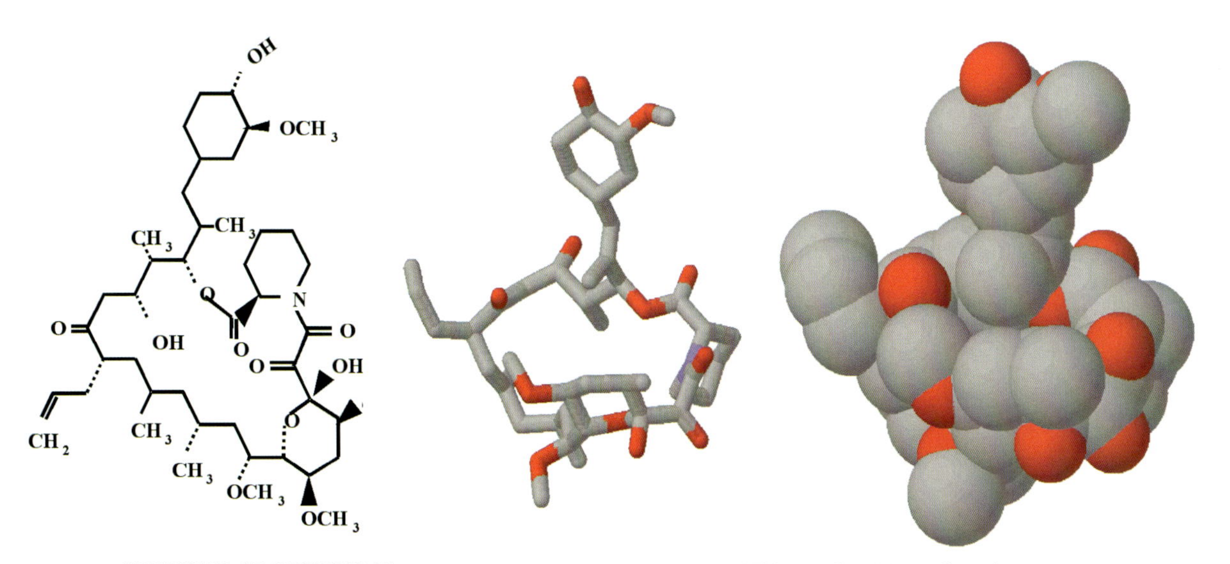

CHAPTER 35, FIGURE 21 Structures for the immunosuppresant, FK506 (tacrolimus). Tacrolimus is a macrolide isolated from *Streptomyces tsukubaensis.* Shown from left to right in approximately the same orientation are the simple line structure, a stick-structure, and a space-filling structure. The stick and space-filling structures are based on the 3-dimensional structure of the molecule bound to the immunophilin FKBP. Protein data bank designation 1FKF.

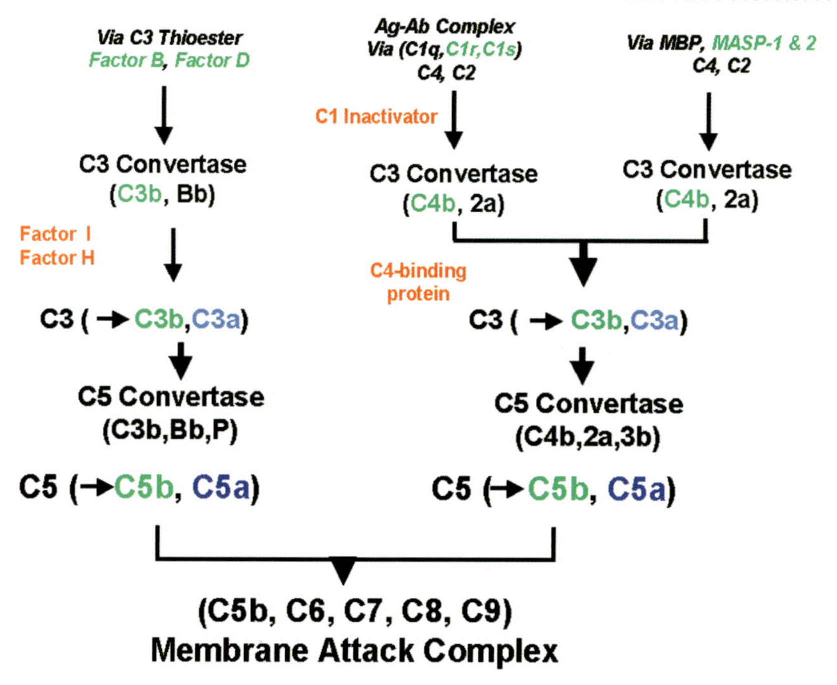

CHAPTER 35, FIGURE 22 Initiation of action by complement occurs via three pathways: classical, alternative (properdin), and lectin pathways.

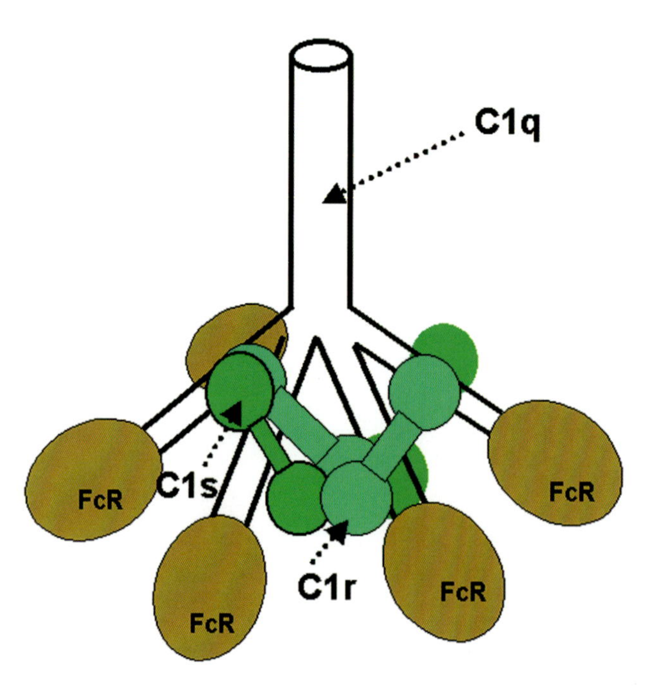

CHAPTER 35, FIGURE 23 A cartoon representation of complement protein C1. The C1 complex consists of one C1q, two C1r, and two C1s molecules. The formation of the complex requires the presence of Ca^{2+}. C1q contains six polypeptide chains that form a globular structure that binds to antigen-antibody complexes at one end of each chain.

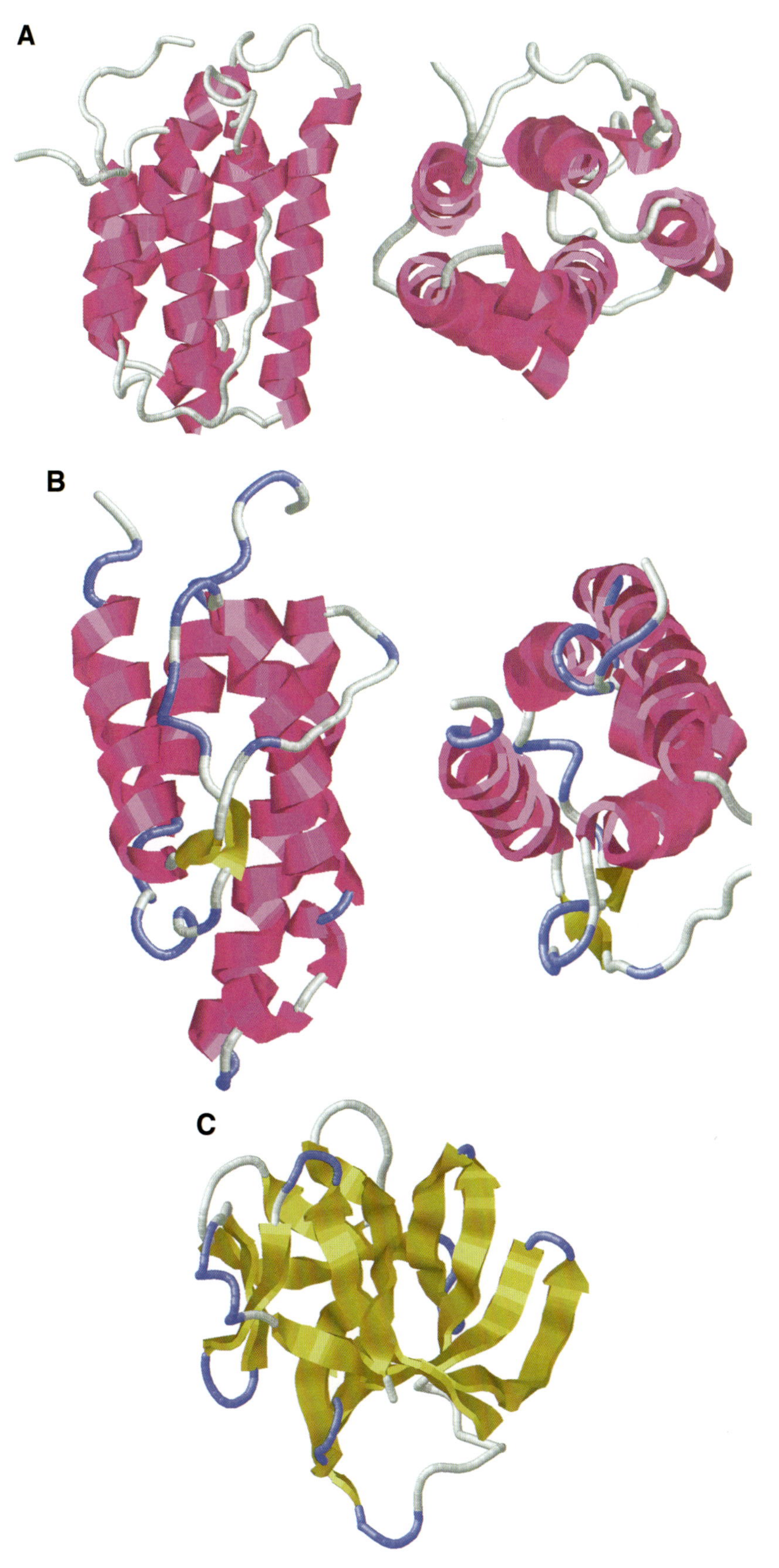

CHAPTER 35, FIGURE 24 Secondary structural features of cytokines. (A) IFN-α2A (shown with both end and side views of the 5 α helices). (B) IL-4, showing the α helices from both end and side views. This group of cytokines is characterized by the 4–6 distinct α helices that form the protein "backbone." (C) IL-1 showing the β-sheet structure that is characteristic of this group of cytokines.

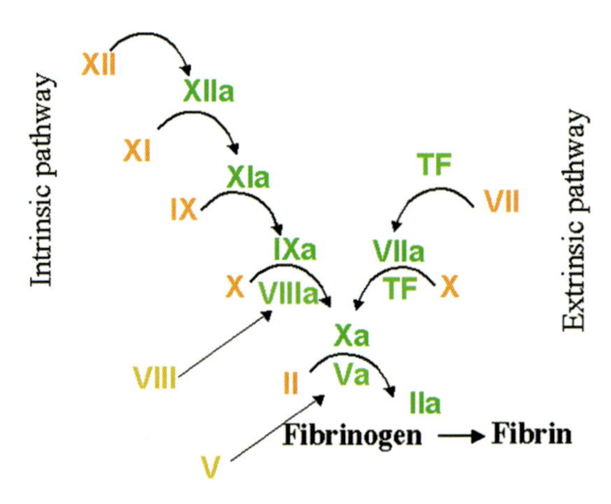

CHAPTER 36, FIGURE 2 Classical "cascade" or "waterfall" model for coagulation. This model represents the ordered sequence of *in vitro* transformations of precursor molecules from their inactive forms to their catalytically active forms. The publication of the "cascade" model for the coagulation system provided a milestone in rationalizing a confusing biological system. The figure has been corrected to place factors V and VIII outside the linear sequence that differs from the proposals made in 1964. Two components, TF (tissue factor) and factor VII, were designated as the "extrinsic pathway" components because the TF is provided by thromboplastin, a tissue-derived material extrinsic to the circulating blood. Four components—factors XII, XI, IX and VIII—were designated as "intrinsic pathway" components because they are always present in blood. Two other plasma proteins—prekallikrein and high-molecular-weight kininogen—were discovered later to be involved in the process by which factor XI becomes activated.

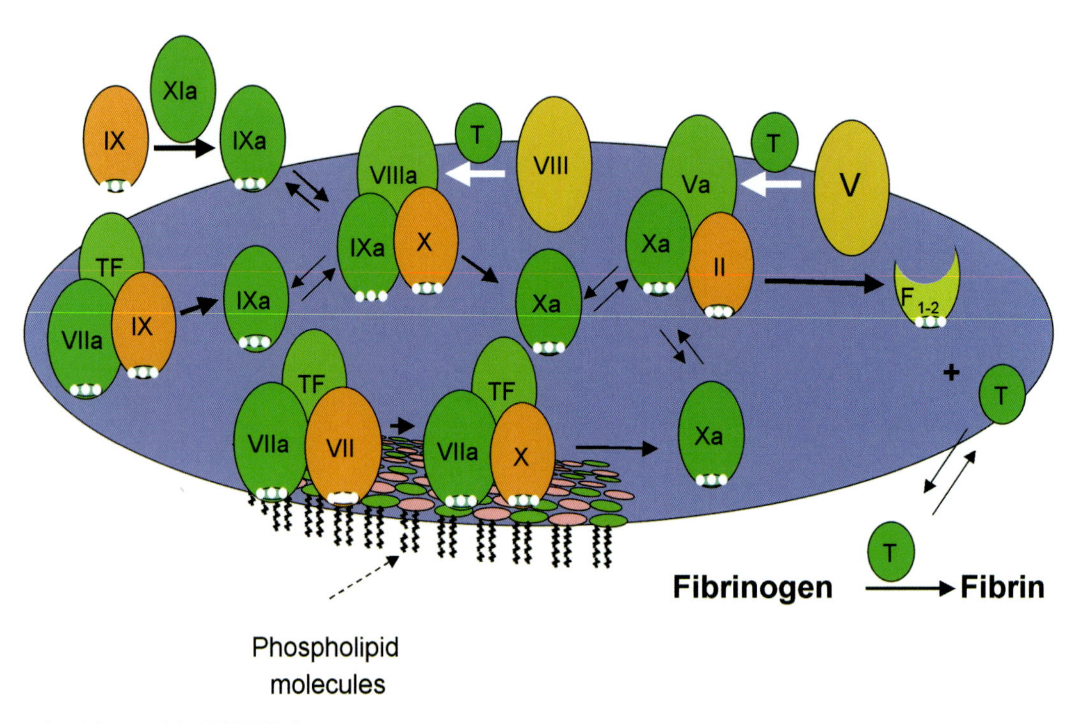

CHAPTER 36, FIGURE 3 Contemporary model for the procoagulant subsystem. This model recognizes that the transformations of proteinase precursors to active proteinases are organized on the surface to which the proteins bind. The binding of the freely circulating coagulation proteins to these surfaces localizes the reactions that make up the hemostatic response, as well as increases the rates of the reactions. Unlike the other coagulation factors, tissue factor does not circulate freely. It is an integral membrane protein that is exposed as a result of exposure of the subendothelium. In each stage of the sequence, a proteinase precursor (shown in orange), a proteinase from the preceding complex (shown in dark green), and an activated cofactor protein (shown in yellow-green) form a noncovalent complex on the membrane surface (shown as a blue "island"). Dissociation of the proteinase formed in one complex (the product proteinase) permits that proteinase to diffuse to the next complex and catalyze the transformation of a different precusor protein into an active proteinase. Thrombin is the only proteinase that does not bind to the surface of the membrane; it dissociates and diffuses so that it can convert the soluble fibrinogen into the insoluble fibrin clot. Vitamin K-dependent proteins are not shown, protein C and protein S. The *unactivated* cofactor proteins, tissue factor, and factors V and VIII, are shown in yellow.

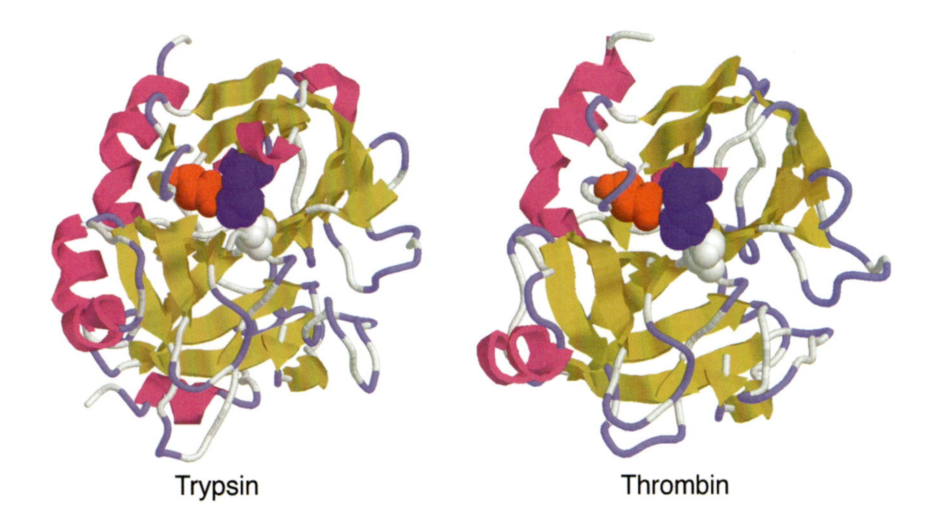

CHAPTER 36, FIGURE 4 Proteinase domains of serine proteinases trypsin and thrombin. The active site *amino acid residues* (*shown as space-filling structures*) are Ser^{195} (shown in white); His^{57} (shown in blue), and Asp^{102} (shown in red). The *specificity* pocket residue is Asp^{189}. Both trypsin, the reference serine proteinase, and thrombin are shown as cartoons with the secondary structure elements marked as defined in the coordinate files for the structures from the Protein Data Bank. The color scheme is that of R. Sayle, the creator of RasMol, the program used to produce these ribbon diagrams. Some of the functionally significant differences in thrombin are the residues in the helix above the catalytic triad and two ligand binding regions, exosites I and II (not shown).

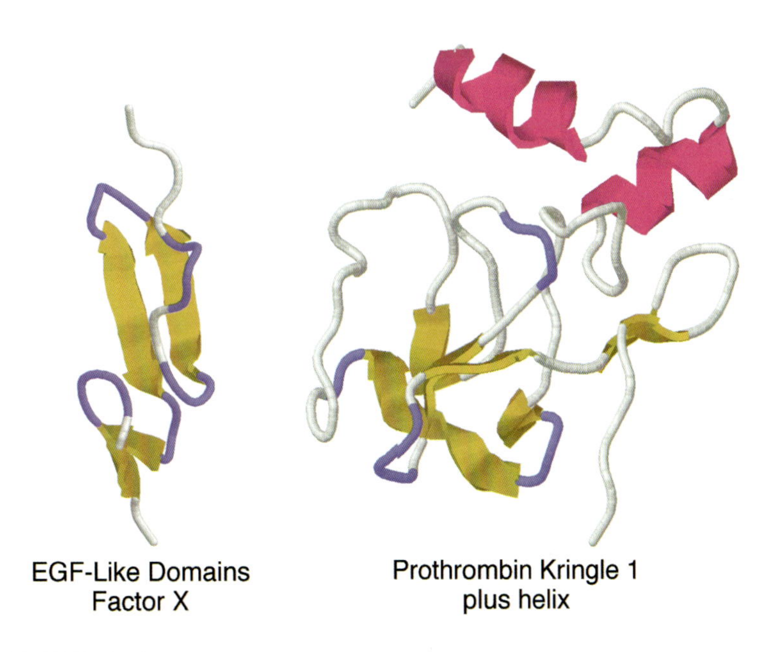

CHAPTER 36, FIGURE 5 Motif structures within coagulation proteins. Common motifs are found in the amino terminal regions of the proteinase precursor molecules. Shown are the kringle motifs and EGF-like motifs found in the vitamin K-dependent proteins and in plasminogen. Fibronectin (types I and II) motifs and "apple" motifs (named from their two-dimensional representations) are also present but not shown. Some epidermal growth factor–like domains contain β-hydroxylated Asp residues. The cartoon structures for the motifs are derived from three-dimensional structures determined by x-ray crystallography or by two-dimensional NMR spectroscopy.

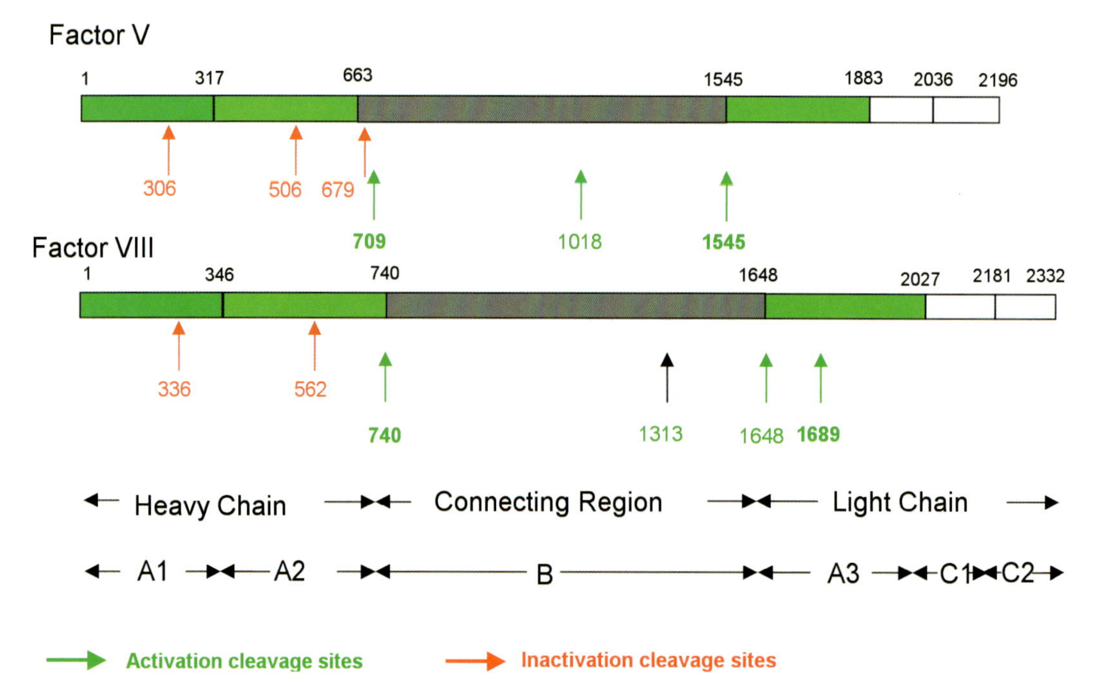

CHAPTER 36, FIGURE 6 Cofactor proteins, factor V and factor VIII. Factor V and factor VIII coagulant (not the von Willebrand factor carrier of factor VIII) contain six distinct structural domains. The two A domains, A_1 and A_2 at the N-terminal end of the polypeptide chain, are separated from the A_3 domain by a highly glycosylated B domain. The two C domains are at the C-terminal end of the molecule. The A domain sequences are homologous to the A domains of ceruloplasmin. Both factor Va and factor VIIIa act as catalysts in the activation of prothrombin and factor X, respectively. Activation sites are indicated by green arrows; inactivation sites by red arrows. In factor Va, complete inactivation requires cleavage of Arg^{306}.

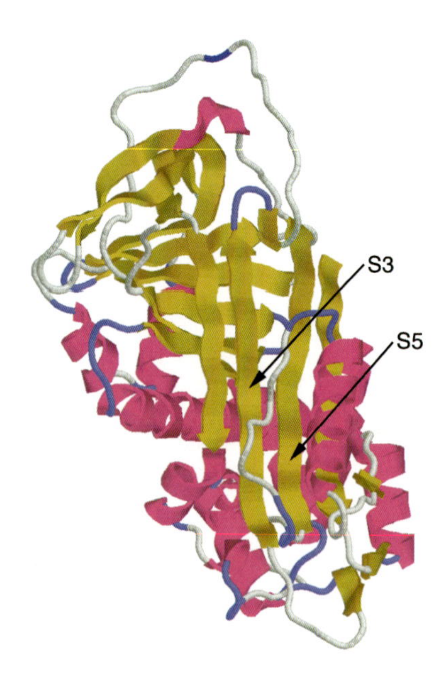

CHAPTER 36, FIGURE 7 A group of structurally similar protein inhibitors of the serine proteinases known as SERPINS (SERine Proteinase INhibitors). The structure shown is human antithrombin. The reference SERPIN, α_{-1}-proteinase inhibitor or α_{-1}-antitrypsin contains ~30% α helix (9 helices) and 40% sheet (5 β sheets). Other members of the SERPIN family contain both additional helices and β sheets. The reactive center loop of antithrombin, residues 378–396, contains the reactive site residues Arg^{393} and Ser^{394}. Upon reaction with the target proteinase or after cleavage by the target proteinase (a reaction that inactivates the inhibitor without inactivating the proteinase), the reactive center loop folds between the S3 and S5 sheets.

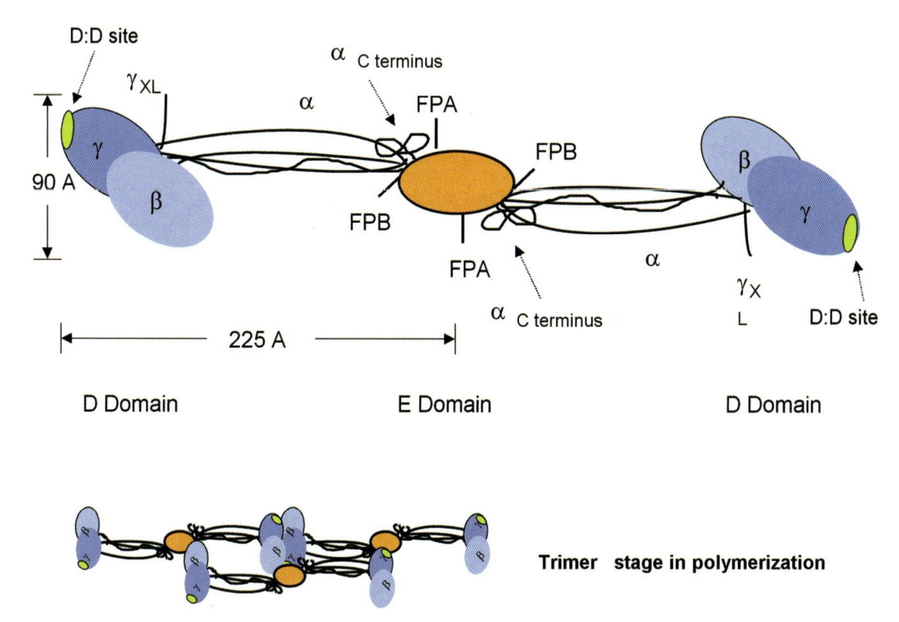

CHAPTER 36, FIGURE 8 Fibrinogen and polymerizing fibrin monomers. The fibrinogen molecule is a 340,000-Da protein consisting of three pairs of ploypeptide chains: two Aα chains, and two Bβ chains, and two γ chains. The A and B designations refer to the two A peptides and two B peptides cleaved from fibrinogen by thrombin to produce the self-polymerizing fibrin monomer (Fn_m), the building block of the fibrin blood clot. The polypeptide chains are linked together by disulfide bridges between Cy–SH residues of the chains. Only after chemical reduction of the disulfide bonds are the separate chains of the fibrinogen molecule discernible. Fibrinogen is frequently abbreviated as $(A\alpha, B\beta, \gamma)_2$. The 16-residue A (FPA) and 14-residue B (FPB) peptides are at the N-terminal ends of the A and the B chains, respectively. The central domain, also identified as the E domain, contains portions of all of the six chains of fibrinogen. The two terminal domains, the D domains, also contain polypeptide sequences from all six chains.

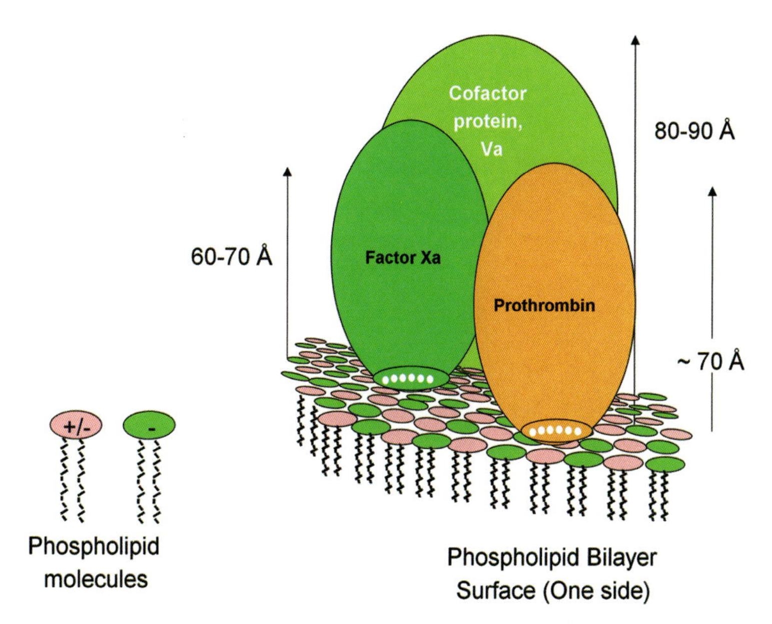

CHAPTER 36, FIGURE 9 Archetypal activation complex of coagulation: *prothrombinase*. Complex formation precedes rapid proteolysis of prothrombin. The presence of the specific domains and the binding sites associated with them are responsible for the formation of the complexes between the correct proteins. The A domains of factor Va provide the interacting sites for both factor Xa and prothrombin, the C domains the sites for binding to the surface. The Gla domains bind Ca^{2+} (white dots) and bind factor Xa and prothrombin to the surface. The locations of the various sites further orient the molecules optimally with respect to the surface and to each other to give the localized, large increases in rates of proteolysis. The conversion of the proenzyme form of a proteinase to its catalytically active form commonly requires proteolysis of a peptide bond that frees a Leu, Ile, or Val residue from the peptide linkage. This permits a conformation change within the active proteinase domain that is stabilized by the formation of a salt bridge between the newly freed residue and Asp^{102} (chymotrypsin numbering). Proteolysis, *per se*, is not the key element in "enabling" of the catalytic apparatus of the proteinase, but rather the conformation that is stabilized by the salt bridge. The dimensions indicated on the figure are measured distances between fluorescent labels on the proteins (active sites for factor Xa and meizothrombin) and a fluorophore in the membrane.

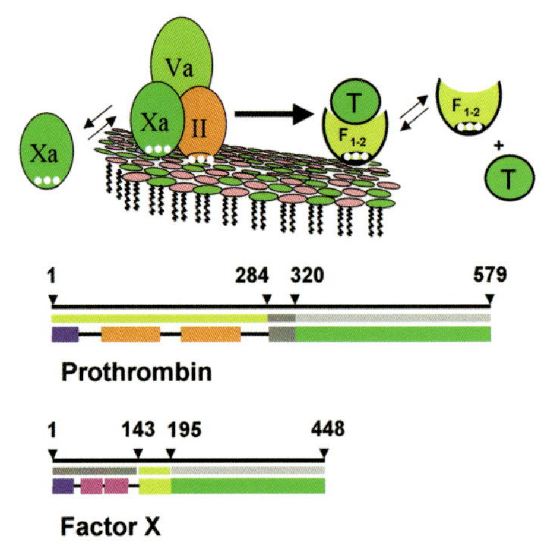

CHAPTER 36, FIGURE 10 Activation of prothrombin. Prothrombin is cleaved by factor Xa at two residues, Arg^{284} and Arg^{320}. The order of cleavage produces different products. If the first cleavage is at Arg^{284}, it produces two products: prothrombin fragment1-2 (residues 1–284, yellow bar) and prethrombin 2 (residues 285–579, gray bar), each of which represents approximately one-half of the prothrombin molecule. When Arg^{320} is cleaved first, the intermediate product meizothrombin is formed. Meizothrombin contains the entire sequence of the prothrombin molecule; the two halves are held together by a single disulfide bond (CyS^{283} to Cys^{439}). In the absence of factor Va, the rate constants for cleavage of the two bonds are essentially the same. In the presence of factor Va and phospholipids, Arg^{320} is cleaved first. Meizothrombin is detectable in reaction mixtures, but its rapid cleavage produces prothrombin fragment 1-2 and α-thrombin. Motifs and domains are color coded as follows: Gla domain (blue), kringle domains (orange), EGF-like domain (magenta), activation peptide (yellow), and proteinase domain (green). Light chains are indicated in dark gray, heavy chains in light gray. Regions connecting the motifs are black.

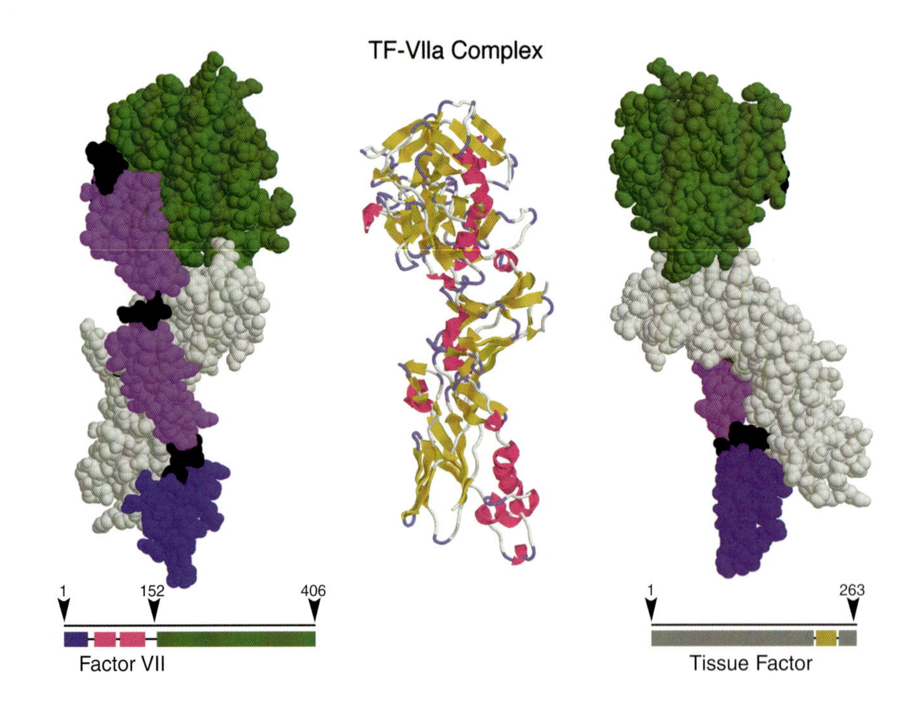

CHAPTER 36, FIGURE 11 Tissue factor–factor VIIa complex. The three-dimensional structure of the complex of factor VIIa and tissue factor (minus the transmembrane polypeptide domain of the tissue factor) in the absence of membrane surface. It is approximately 115Å in length and has a diameter of 40–50Å. Factor VIIa shows its four distinct domains: the Gla domain, two EGF-like domains, and the proteinase domain. Tissue factor contacts factor VIIa via the interface between the two "fibronectin type III-like" domains. All four domains of factor VIIa appear to be involved in the interaction between tissue factor and factor VIIa. The Gla domain of factor VIIa is folded very similarly to the Gla domain of prothrombin (Gla domain of prothrombin fragment 1). Activation of factor VII can be catalyzed by thrombin, factor Xa, factor VIIa and factor XIIa—all by cleavage at Arg^{152}–Ile^{153}. Secondary structures are shown in the center diagram; two views of the close interactions between TF and factor VIIa are shown in the two diagrams at each side.

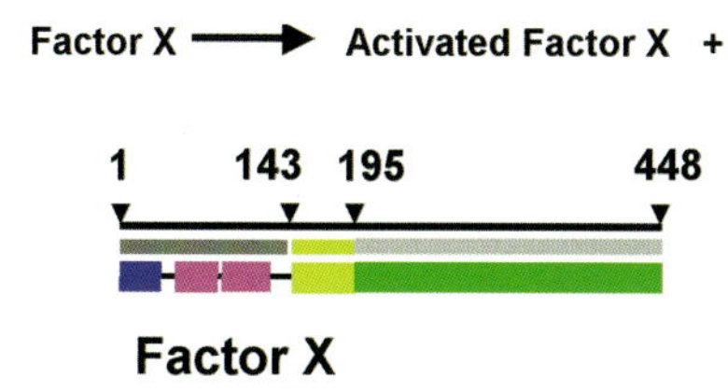

CHAPTER 36, FIGURE 12 Activation of factor X. Factor X is activated by factor VIIa through a single peptide bond cleavage, Arg^{51}–Ile^{52} in the heavy chain of the proenzyme. The activation peptide is highly glycosylated as are many of the activation peptides of the procoagulant proenzymes. Factor X is converted to factor Xa by factor IXa through cleavage of the same peptide bond that is cleaved by Factor VIIa. Motifs and domains are color coded as follows: Gla domain (blue), EGF-like domain (magenta), activation peptide (yellow), and proteinase domain (green). Light chains are indicated in dark gray, heavy chains in light gray. Regions connecting the motifs are black.

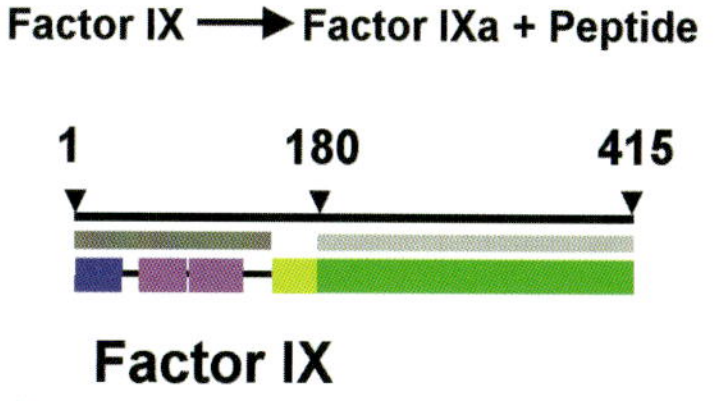

CHAPTER 36, FIGURE 13 Activation of factor IX. Factor IX is activated by cleavage of two peptide bonds, Arg^{145} and Arg^{180}-Val^{181}. The activation peptide that is released is composed of 35 amino acid residues and is highly glycosylated. The same peptide bonds are cleaved by factor VIIa and factor XIa. Glycosaminoglycans, particularly heparin, can increase the rate of activation as the result of the binding of both factor XIa and factor IX to the heparin molecule to form a ternary complex. Color coding is as described in Figure 36-12.

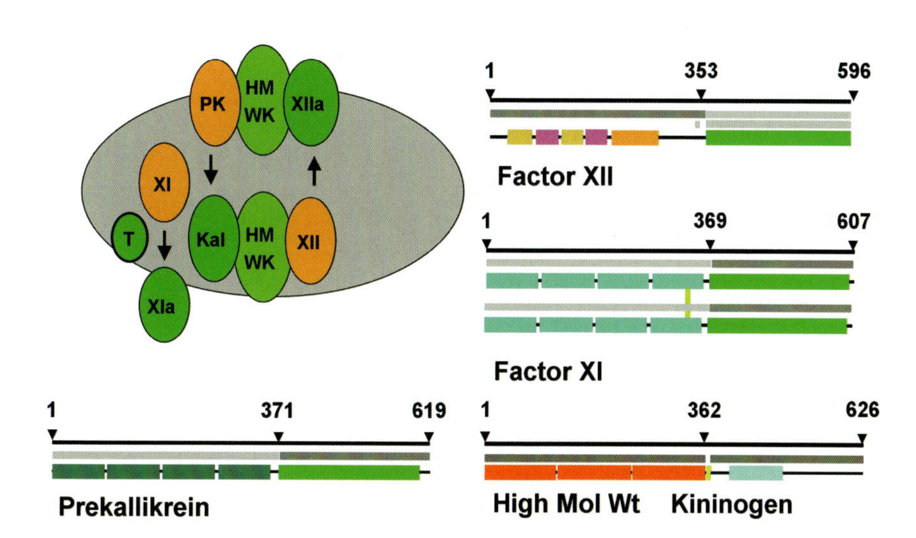

CHAPTER 36, FIGURE 14 Contact phase of *in vitro* coagulation. The contact phase reactions involve two proteinase precursors—factor XII and prekallikrein—and one cofactor protein—plasma high-molecular-weight kininogen (HAWK). The structures, in bar diagrams, show the presence of EGF-like fibronectin type I and type II, kringle, cystatin-like, and "apple" motifs. The cleavage site for proteolytic activation of factor XII is Arg^{353} and for prekallikrein Arg^{371}. The cofactor protein, HMWK, is cleaved by kallikrein at Arg^{362} and Arg^{371} to release bradykinin (yellow). HMWK contains a unique His-rich region that is associated with binding to the contact surface (shown in light blue). Factor XI is a dimer of two identical polypeptide chains that are linked by a single disulfide bridge. In the circulating blood, factor XIIa cleaves the two Arg^{369}-Ile^{370} peptide bonds in factor XI to form factor XIa. Other motifs and domains are color coded as follows: apple domains (cyan), EGF-like motifs (magenta), fibronectin motifs (yellow), kringle (orange), and proteinase domain (green). Light chains are indicated in dark gray, heavy chains in light gray. Regions connecting the motifs are black.

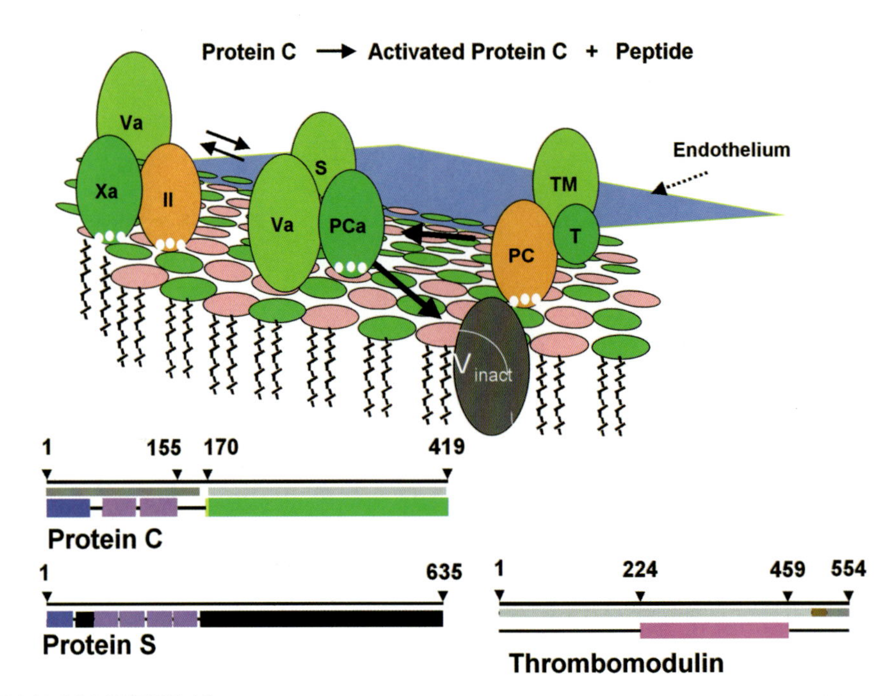

CHAPTER 36, FIGURE 15 Anticoagulant subsystem. Activation of protein C occurs adjacent to the injury site; inactivation occurs on the exposed surface at the injury site. Protein C is activated through cleavage by thrombin at Arg^{169}-Ile^{170}. Protein S is a 635-residue vitamin K-dependent protein that functions as a cofactor. There is a Gla domain and four EGF-like domains in the molecule, but no proteinase domain. Thrombomodulin is a 554-residue integral membrane protein. The extracellular region comprises residues 1–494, the transmembrane region residues 495–518, and the intracellular region residues 519–554. EGF-like structures are found from residue 224 to 459. Motifs and domains are color coded as follows: Gla domain (blue), EGF-like domain (magenta), activation peptide (yellow), and proteinase domain (green). Light chains are indicated in dark gray, heavy chains in light gray. Regions connecting the motifs are black.

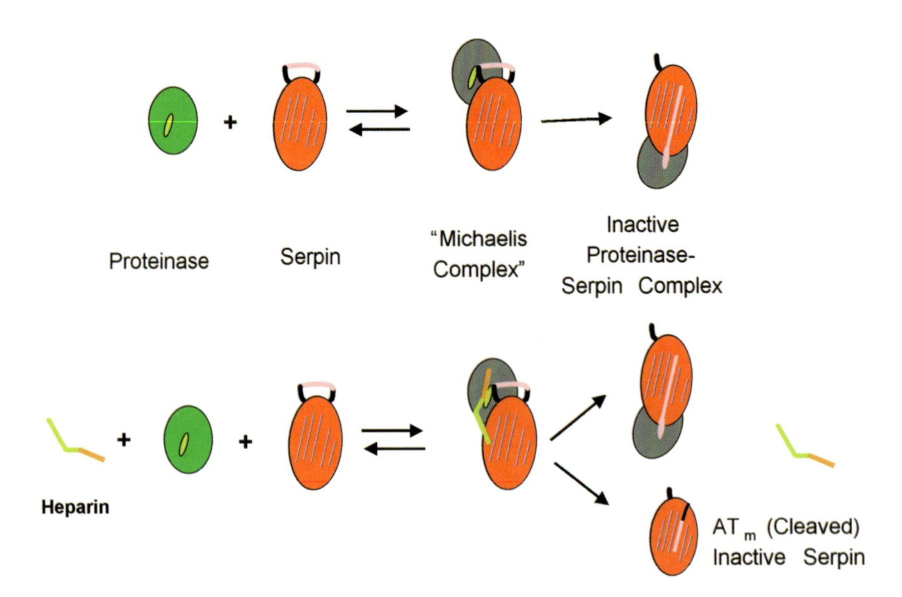

CHAPTER 36, FIGURE 16 Proteinase inactivation by SERPINS. Proteinase inactivation occurs by reaction between proteinase and inhibitor, e.g., antithrombin. The proteinase is a stoichiometric reactant in this instance but is not a catalyst. This reaction results in the formation of a covalent bond between the reactive site residue of the inhibitor (Arg^{393} in antithrombin) and the active site residue (Ser^{195} in the proteinase). This complex formation prevents the proteinase from hydrolyzing any other peptide bond. Proteinases, thrombin, factor Xa, factor IXa, and, less effectively, factor VIIa and factor XIa are all inactivated by the plasma protein inhibitor antithrombin (previously designated antithrombin III). The product AT_m is the cleaved form of antithrombin. It is formed in both the absence and presence of heparin, but more so in the presence. The proteinase is indicated is indicated by green, the inhibitor by red, and the inactivated proteinase by gray. Stripes on the inhibitor represent the helices (Figure 36-7).

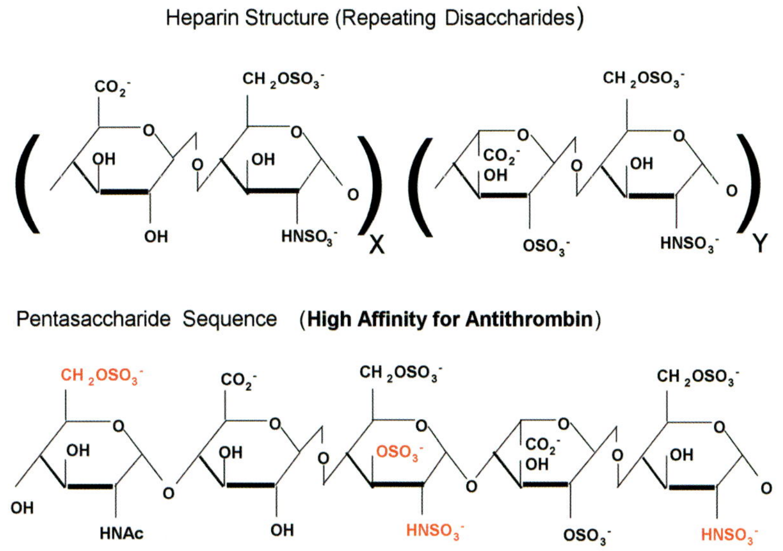

Heparin Structure (Repeating Disaccharides)

Pentasaccharide Sequence (**High Affinity for Antithrombin**)

CHAPTER 36, FIGURE 17 Structure of heparin. Heparin is a polymer of repeating disaccharide units that contain one uronic acid and one hexosamine residue. The uronic acid residues may be either glucuronic acid or iduronic acid, both of which are monosaccharide acids that differ in their stereochemistry. The hexosamine residue is glucosamine. Both the uronic and hexosamine residues can be modified by O- and N-sulfation and the glucosamine residue by N-acetylation. All heparins bind to antithrombin; however, heparin molecules that contain a unique pentasaccharide sequence bind with particularly high affinity (high-affinity heparins). Approximately 30% of the heparin molecules present in the commonly used therapeutic heparins have high affinity for antithrombin.

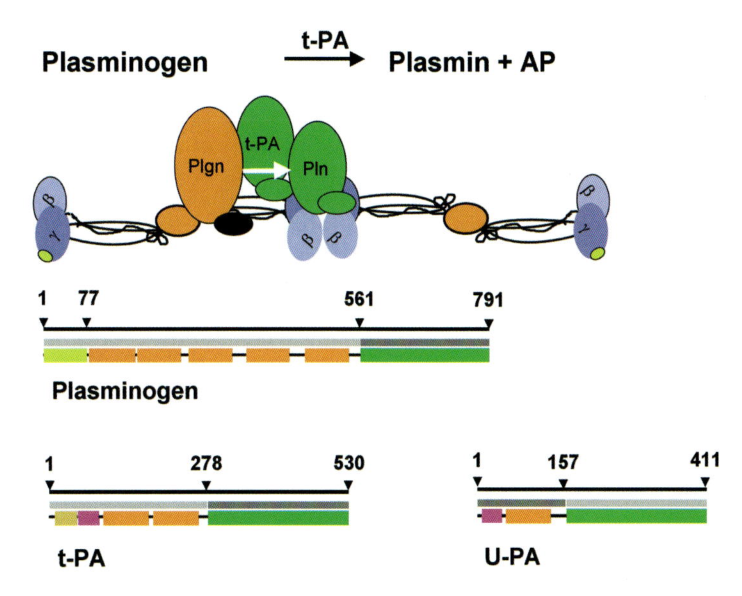

CHAPTER 36, FIGURE 18 Plasminogen activation. Plasminogen is converted to plasmin as the result of cleaving a single peptide bond, Arg^{561}-Val^{562}. The molecular mechanism of transforming the zymogen into its active form is the same as the activation of trypsinogen to trypsin. Two different proteinases are responsible for the physiological activation of plasminogen: tissue-type plasminogen activator (t-PA) and urinary plasminogen activator (u-PA). t-PA is the principle activator of plasminogen and is synthesized primarily in endothelial cells which are the principle source of t-PA in the circulating blood. t-PA is present in high concentrations in uterine tissue and in various malignant cells. u-PA is found in the kidney and may be particularly important in preventing the accumulation of fibrin in the glomeruli. In the circulation, more than one-half of the t-PA is in a complex with PAI-1. The activation of plasminogen is accompanied by the proteolytic action of plasmin on plasminogen at Arg^{68}-Met^{69} and at Lys^{77}-Lys^{78}. This reaction results in the formation of a derivative of plasminogen in which the first 77 amino acids at the N-terminal end of the molecule are removed. This derivative of plasminogen, *lys plasminogen*, is activated more rapidly than the native, *glu plasminogen*. Motifs and domains are color coded as follows: kringle domains (orange), EGF-like domain (magenta), activation peptide (yellow), and proteinase domain (green). Light chains are indicated in dark gray, heavy chains in light gray. Regions connecting the motifs are black.

CHAPTER 36, FIGURE 19 Products of fibrinolysis. The products of plasmin action are used to identify ongoing fibrinolysis and to distinguish fibrinolysis from fibrinogenolysis. The presence of larger fibrin fragments, which result from factor XIIIa–catalyzed cross-linking, is the basis for the distinction between fibrinogenolysis and fibrinolysis. Detection of D dimer (DD) is particularly useful. Some of the fibrin degradation products interfere with polymerization of fibrin. Digestion of fibrin. Digestion of fibrinogen by plasmin proceeds with initial cleavages that remove polypeptides of approximately 40,000 Da from the carboxy terminal regions of the A chains. The large fragment that is formed is designated fragment X, and it is heterogeneous. Fragment X has a molecular weight of ~300,000 (one A-chain peptide removed) or ~260,000 (two A-chain peptides removed). Further proteolysis by plasmin removes polypeptides of ~100,000 Da total mass from the disulfide bond–linked carboxy terminal regions of A, B, and γ chains. The larger product of this fibrogenolysis is designated fragment Y, which has a molecular weight of ~150,000 Da. The product that contains the cross-linked polypeptides from A, B, and γ chains is designated fragment D. A second round of proteolysis removes the remaining fragment D from fragment Y to form fragment E (M.W. ~33,000 Da) and a second fragment D. Fragment E is highly cross-linked and is also known as the N-terminal disulfide knot (DSK). In fribrinogen, an additional cleavage occurs early in the fibrinogenolytic process, i.e., cleavage of Arg^{42} in the Bβ chain to produce Bβ 1–42. This can be distinguished from the product that is produced by the cleavage of fibrin, β 15–42.

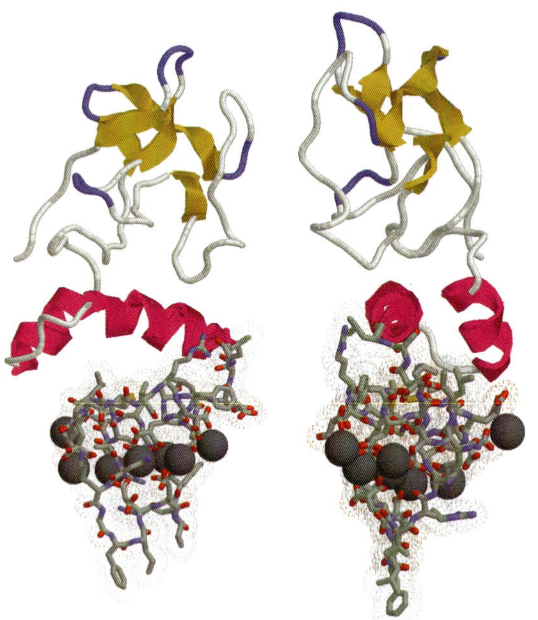

CHAPTER 36, FIGURE 21 Structure of a Gla domain of prothrombin. The Gla domains of the vitamin K-dependent proteins contain between 9 and 12 carboxyglutamic acid residues. Prothrombin contains 10 Gla residues: pairs of adjacent Gla residues at positions 6 and 7; 19 and 20; and 25 and 26 and four other Gla residues at positions 14, 16, 29, and 32. Factor VII also contains 10 Gla residues in the same positions as in prothrombin. Factor IX contains 12 Gla residues. Because factor IX has an additional amino acid between its amino terminal Tyr and the first Gla residue, the numbering of the residues is different from prothrombin and factor VII. However, the pattern is similar with pairs of adjacent Gla residues at positions 7 and 8; 20 and 21; and 26 and 27. The six other Gla residues are at positions 15, 17, 30, 33, 36, and 40. The last two Gla residues account for 12 Gla residues in factor IX. Factor X contains 11 Gla residues: pairs of adjacent Gla residues at positions 6 and 7; 19 and 20; and 25 and 26, with five other Gla residues at positions 14, 16, 29, 32, and 39. Protein C contains 9 Gla residues; pairs of adjacent Gla residues at positions 6 and 7; 19 and 20; and 25 and 26, and the other Gla residues at positions 14, 16, and 29. The cartoon structures show helices in magenta, β sheets in yellow, β bends in blue, and Ca^{2+} ions in gray. The space-filling dot surfaces for the residues of the Gla domains are standard CPK colors.

CHAPTER 36, FIGURE 22 Hepatic vitamin K metabolism. Oral anticoagulant drugs act indirectly on the process of glu carboxylation of the vitamin K dependent proteins. The vitamin K antagonists block the reduction of the reaction intermediate, vitamin K–epoxide, that results in the accumulation of vitamin K–epoxide and other nonfunctional forms of vitamin K. Without cycling of the vitamin K-related reaction intermediates in the cycle shown, a depletion of functional vitamin K occurs. Vitamin K antagonists do not block polypeptide synthesis and formation of non-carboxylated proteins occurs. These noncarboxylated proteins are still secreted from the liver and account for nearly normal levels of each of the vitamin K-dependent proteins that can be detected by immunoassay (sometimes called PIVKA, proteins induced in vitamin K absence).

CHAPTER 35

Molecular Immunology

The ability of vertebrate animals to resist and overcome disease caused by infectious agents and toxic proteins is provided by the immune system. The immune system comprises specialized tissues, cells, highly specific proteins (e.g., antibodies and receptors), and unique genetic mechanisms for creating the many different defenses that the organism requires for combating infectious diseases. Table 35-1 provides a glossary and cross-references to tables and figures that relate these components and their functions to the molecules of the immune system. The cells and molecules composing the immune system are designed to attack and destroy infectious and foreign substances, but do not under normal circumstances attack the host organism's tissues, cells, or molecules. This ability to discriminate between self and nonself is an essential property of the immune recognition system.

Immunity is obtained through reactions of cells and molecules that are components of two distinguishable immune response systems. The first, innate immunity, is the simpler of the two systems and remains essentially unchanged upon repeated exposure to an infectious agent or foreign substance. In contrast, the second system, acquired immunity, is enhanced upon repeated exposure to infectious agents. The "memory" capacity of acquired immunity enables the immune system to respond more rapidly and more extensively when an infectious agent or foreign substance is encountered a second time and is the basis for resistance to organisms previously encountered and explains the effectiveness of vaccination.

The term immunity, which is derived from the Greek *immunitas*, means exemption from service or duty to state; when extended to the organism, it means exemption from disease. Abbreviations: Ab, antibody; Ag, antigen; MAb, monoclonal antibody; TcR, T-cell receptor; T_h, T helper cell; T_c, cytotoxic T cell; CH, constant region, heavy chain; CL, constant region, light chain; VH, variable region, heavy chain; VL, variable region, light chain; CDR, complementarity-determining region; CD, cluster of differentiation; IL, interleukin; MAC, membrane attack complex C*X*, complement component (numbered); Fab, antibody fragment, single arm; F(ab)′, antibody fragment, two arms; Fc, antibody fragment, heavy chains; ADCC antibody-dependent, cell-mediated cytoxicity.

35.1 Innate Immunity

The first line of defense against infectious agents is provided by the innate immunity of the organism. This defense is the result of physical barriers such as skin, the cells that line the gastrointestinal and genitourinary tracts, and the mucus that these cells secrete.

Mucous secretions act to move infectious agents away from their sites of attachment and out of the body. Infections from bacteria are also opposed by the enzyme lysozyme, which is capable of degrading the carbohydrate structures of the bacterial cell walls. One particular type of antibody molecule, IgA (see below), is present in the mucous secretions and participates in this first line of defense against invading organisms.

TABLE 35-1
Components of the Immune System

Cellular (Leukocyte) Components	Principal Functions	Characteristic Receptors/Specific Recognition Molecules (CD's)	Related Figure(s) or Tables
Lymphoid Cells			
B-cells (B lymphocytes) Mature in bone marrow, reside in lymph nodes, spleen, alimentary and respiratory tract mucosa	Recognition of antigens, become antibody-forming plasma cells (synthesize Ig molecules), act as antigen presenting cells (APCs), some B cells defferentiate into memory cells for rapid recognition of previously encountered antigens	Antigen receptors are surface IgM (monomeric IgM), IgD. As APCs B-cells bind exogenous antigens and present processed antigen via MHC Class II to T cells. Express **CD3, CD19, CD20, CD21** (complement receptor), **CD32** (FCγReceptor II), **CD35** (complement receptor), **CD40** and **CD45** on surface	Figure 1 Figure 18
T-cells (T lymphocytes) Mature in thymus, reside in thymus	Regulation of T and B-cell immune response	Recognize antigens on APCs via **TcRs** (T-cell receptors) antigen (epitope) specific All T-cells express **CD3** on their surface	
Th (T helper cells)	T-cell subclass that release necessary cytokines for cell stimulation and proliferation	Recognize antigens (epitopes) on APCs via **TcR**s, specifically ***MHC Class II*** receptors	Figure 1
Th1	**Th1** - regulation of cellular immunity via IL2 and IFN-γ	Express **CD3** and **CD4**, on surface, **CD4** used to identify T-helper cells in cell sorting procedures. Also express **CD2, CD28** and **CD45R**	
Th2	**Th2** - B-cell stimulation via IL's 4, 5, 6 and 10		
T Suppressor cells	Suppress B and T cell responses to antigen stimulation	**TcR, CD3** Express ***CD8*** on surface	
CTLs (Cytotoxic T-lymphocytes)	Cytotoxic against tumor cells, and virus infected host cells. Destroy infected cells through perforin-dependent permeability changes. Produce TNFα and TNFβ. Promote apoptosis.	Possess **TcR, IL-2R**, (IL-2 receptor) recognize processed antigens (epitopes) via ***MHC Class I*** receptors on APCs. Express **CD3** and ***CD8***, (also **CD2, CD28**)	Figure 3
Natural killer cells (NK cells)	Granular lymphocytes that recognize and destroy virus-infected and cancer cells and use ADCC. Exclude destruction of cells that express MHC-I. Stimulated by IL-2, IFN-γ, TNF-α. Create pores that permit transfer of proteins which promote apoptosis of infected cells	**NO** Ig, **CD3** or **TcR**s Express **CD16 (FcγIIR),** i.e. surface receptors that bind via Fc portion of IgGs Express **CD56** on surface. **CD16** and **CD56** used for identification in cell sorting	Figure 4

(continued)

TABLE 35-1

Components of the Immune System (continued)

Cellular (Leukocyte) Components	Principal Functions	Characteristic Receptors/Specific APC Recognition	Related Figure(s) or Tables
Myeloid Cells			
Neutrophils Mature in bone marrow	First cells recruited to sites of inflammation, dominant phagocytic granulocyte, able to mediate ADCC (antibody-dependent cell-mediated cytotoxicity), major defense against pyogenic bacteria, use respiratory burst to kill cells	Adhere to foreign cells via **CD11b/CD18** No MHC proteins. Neutrophils are attracted to sites of complement activation by C5a, a chemotactic peptide from C5	Figure 4
Basophils	Blood "analog" of tissue mast cell, stimulated by complement C3	Possess Fc receptors that bind IgE causing degranulation and histamine release	
Mast cells Mature in bone marrow, resident in tissue	Tissue granulocytes that produce cytokines, release heparin, histamine	Fc receptors that bind IgE causing degranulation and histamine release	
Eosinophil	Attack parasites too large for phagocytosis, possess granules that contain toxic proteins, adhere to foreign organisms via C3b. Release protins and enzymes that damage foreign organism membranes, kill with active oxygen burst. Engage in IgE-mediated ADCC	High-affinity receptors for IgE, bind via Fc region of antibody	
Monocytes Mature in bone marrow, found in blood	Mononuclear phagocyte in blood, are precursors to tissue macrophage. Process protein antigens and present peptides to T-cells via MHC I and II. Engage in IgG-mediated ADCC	Express MHC Class II surface proteins, do not possess antigen-specific receptors	Figure 2
Macrophages Mature in bone marrow as monocytes, monocytes mature further to tissue resident macrophages	Tissue phagocytes (APCs) that processes antigens for presentation to T cells, possess MHC receptors, principal attack is on bacteria, viruses and protozoa that enter cells. Engage in IGG-mediated ADCC	Express MHC Class II surface proteins, do not possess antigen-specific receptors, but have receptors for C3b.	Figure 2
Dendritic cells			
Follicular	Antigen-antibody and Ag-Ab complexes with complement are localized in these cells. Involved in maintenance of B cell memory	*Possess* Fc receptors No MHC class II receptors	
Interdigitating	Found in skin (Langerhans cells) in T cell areas of lymph nodes and spleen. Recognize and endocytose foreign carbohydrate antigens that are not found in vertebrate animals. Act as APCs to CD 28-bearing T cells, stimulated by IFN α, TNF α.	Do not have Fc receptors *Possess* MHC class II receptors. Express **CD80** and **CD86**	

(continued)

TABLE 35-1
Components of the Immune System *(continued)*

Protein Components	Principal Functions	Structural Characteristics	Related Figure(s) or Tables
Immunoglobulins	Ig's confer humoral immunity, react with epitopes on antigens when bound to antigens initiate complement activation	Four chains, two heavy and two light chains, charateristic fold between β-sheet domains, "immunoglobulin" fold	Figure 7 Table 35-2
Cytokines	Proteins that can communicate with and stimulate immune and other cells to proliferate, synthesize specific proteins		Table 35-4
CD3 Complex CD3D(delta) CD3E(epsilon) CD3G(gamma) CD3H(eta) CD3Z(zeta)	T-cell surface glycoprotein found on all T-cells Forms a complex with TcR Mediates signal transduction across the cell membrane	Heterodimer of either: α/β or δ/γ chains	
CD4	T-cell surface glycoprotein, Interacts with **IL-16**, coreceptor with **TcR** in recognizing **MHC Class II** presented peptides, identifier of T-helper cells	Structure: Single polypeptide chain, 3 domains, 1) extracellular, (~370 res); 2) transmembrane, (~25 res) 3) intracellular (~38 res)	Figure 4 Figure 16 Figure 17
CD8	T-cell surface glycoprotein, Interacts with **MHC Class I** Identifier of cytotoxic or suppressor T-cells	Heterodimer of α/β chains, can form homodimers	Figure 3
Interleukins	See Table 35-4		Table 35-4
Perforin	Permeability induction for ADCC cell destruction		
B cell receptors	Specific to individual antigen (epitope)	IgM molecules anchored to B-cell membrane	Figure 1 Table 35-2
T cell receptors (TcR) (Many copies of each **TcR** on each cell, all directed for the same epitope (antigen))	T-cell surface glycoprotein, with **CD4** or **CD8** binds to MHC proteins on surface of antigen presenting cells	Structure: 2 $\alpha\beta$-polypeptide chains with immunoglobulin folds; $V_\alpha C_\alpha$ and $V_\beta C_\beta$ OR 2 $\delta\gamma$ polypeptide chains (less prevalent form)	Figures 1 and 3 Figure 14

(continued)

TABLE 35-1
Components of the Immune System (continued)

Protein Components	Principal Functions	Comments	Related Figure(s) or Tables
Complement	Plasma proteins that act sequentially to produce cell lysis, facilitate phagocytosis		Table 35-3
Classical pathway	Creates the proteolytic enzyme complex, C3 convertase (C4b2b)	Initiated by interaction with antigen-antibody complexes	Figure 21
Alternative pathway (Properdin pathway)	Creates the proteolytic enzyme complex, C3 convertase (C3bBb)	Initiated by reaction of the C3 thioester with nucleophilic molecules on cell surfaces	Figure 21
Lectin pathway	Recognizes mannose in oligosaccharide chains	Initiated by binding of mannose binding protein to foreign cell carbohydrate residues	Figure 21
C3 Convertase	Principal agent responsible for reactions that lead to formation of the membrane attack complex and lysis of the invading cells	NOTE: Two different C3 convertases are formed by the classical and alternative pathways	Figure 21
Opsonins	C3b, C5b, the mannose-binding protein, CRP (C-reactive protein) are examples of opsonins	Proteins that adhere to microbe surfaces making them "attractive" for attack by complement and phagocytes	
Antigen Presenting Proteins (MHC proteins)	Proteins that display peptide fragments of foreign (nonself) proteins on the surface of cells	Human leukocyte antigens *Class I*: HLA-A, HLA-B, HLA-C *Class II*: HLA-DP, HLA-DR, HLA-DQ	Figures 1, 2, and 3
Class I MHC HLA-A, HLA-B, HLA-C	Present on all nucleated cells, present peptides to **CD8** cytotoxic T-lymphocytes (CTLs, Tc), presents *endogenously* synthesized peptides (e.g. virus coded peptides)	Structure: α-polypeptide chain, 3 domains, α_1, α_2, α_3 and β_2 microglobulin (Not a MHC protein) High degree of sequence variability in $\alpha 1$ and $\alpha 2$ domains	Figure 3 Figures 13 and 14
Class II MHC HLA-DP, HLA-DR, HLA-DQ	Present on macrophages and B-cells, presents peptides from processed *exogenous* or vacuolar antigens for recognition by **CD4**, T_h-cells	Structure: $\alpha\beta$-polypeptide chain, 2 domains, α_1, β_1 and $\alpha_2\beta_2$ Variability in both α and β chains	Figures 2 and 3 Figure 15
β_2-microglobulin	Not MHC gene coded, but is a part of the MHC class I antigen processing complex		Figure 14 Figure 15

Cells and proteins that participate in the innate immune response recognize generic "marker" molecules (e.g., complex polysaccharides) on foreign microbes and other substances that are generally not found in the host. Attack on the "marked" invading microorganisms by phagocytes engulf and destroy the infectious organisms. These phagocytic cells kill infectious microbes by releasing reactive oxygen species and nitric oxide that are produced by the phagocyte and by releasing enzymes that initiate programmed cell death (apoptosis). Organisms that are too large for phagocytosis, such as large parasites or worms, are attacked by a specialized leukocyte, the eosinophil, which is capable of injecting toxic substances into the organism to kill it.

The phagocytic cells, i.e., neutrophils, monocytes, macrophages, and dendritic cells, are myeloid cells that actively participate in innate immunity and are one lineage of stem cell differentiation (Table 35-1). Mutations in myeloid cells are one of the causes of the *myelocytic leukemias.*

The proteins of the complement system provide a second line of defense for the infected host. *Complement* proteins are so named because they complement the action of the phagocytes and antibodies. However, antibodies are not involved in the action of complement in innate immune response. Complement proteins, which are present in the circulating blood, attach covalently to bacterial cells and undergo a series of proteolytic reactions that activate the complement system and enhance the innate immune response. Binding of the complement activation products C3b and C4b and another plasma protein, CRP (C-reactive protein, one of a group of proteins called acute phase reactants) to the surface of foreign organisms is a process called ***opsonization.*** This process labels the organism as nonself and promotes adhesion of phagocytes to the invading microbes, resulting in phagocytosis of the opsonized organisms. The terminal product of complement activation, the membrane attack complex (MAC), creates a pore in the organism, making the microbe permeable to water and other solutes and ultimately causing the microbe to lyse. The MAC is composed of components C5b, C6, C7, C8, and C9 (see Table 35-3). Fragments produced during complement activation also act as chemotactic and inflammatory agents and attract phagocytes to the sites of invasion and injury and to the opsonized microbes. These potent products of complement activation cause blood vessel dilation and facilitate the movement of phagocytic cells from the bloodstream into the surrounding tissues where the infectious organisms may be located. The innate immune response involves only one of three pathways by which complement is activated, i.e., the "alternative pathways of complement activation" (see complement).

35.2 Acquired or Adaptive Immunity

The general mechanisms responsible for innate immunity are inadequate to deal with all types of organisms, e.g., different strains of bacteria and viruses. These inadequacies are compensated for by the specific abilities of the acquired or adaptive immune system. Leukocytes, i.e., white blood cells (specifically B and T lymphocytes), are the agents of acquired immunity. B cells are so designated because they mature in the bone marrow; T cells because they mature in the thymus. B and T cells perform several biological processes that are uniquely responsive to repeated infection by foreign organisms. Adaptive immune responses that are responsible for these unique properties of the immune system occur in specialized regions (germinal centers) of the lymph nodes, the spleen, and mucosal lymphoid tissues, e.g., tonsils and adenoids.

Adaptive immunity provides a defence against some of the pathogens that avoid the innate immune system and can mount an attack against the evolving and ever changing characteristics of disease-causing organisms, e.g., different strains of bacteria and viruses, such as those that cause influenza.

Preexisting B-lymphocytes are present from birth and can recognize and bind antigens, more specifically ***epitopes*** (restricted portions of the antigen), on infectious agents or foreign macromolecules. The receptors on a B-cell surface that recognize these epitopes are antibody molecules that are synthesized by the B cell. After endocytosis and processing of the bound antigen by the B cell, the B cells are stimulated to divide and produce more B cells. This is a cooperative interaction between the B cells and specific T cells and signaling molecules called *cytokines* that result in cooperative interactions with T cells is B-cell proliferation and differentiation. Some of the stimulated and proliferating B cells undergo a change to a specialized variety of B cell, called a *plasma cell.* The plasma cells synthesize large amounts of a specific antibody that recognizes an epitope on the antigen to which the B cell was originally bound. The antibodies are secreted into the extracellular fluid where they provide a highly specific line of defense against current and future disease-causing agents. Other stimulated B cells differentiate to become very long-lived cells called *memory cells.*

The antibodies secreted by plasma cells are distributed to blood, lymph, and interstitial fluid where they are capable of binding to the infectious organism. Once these antibodies react with cell surface antigens on the infectious organism or foreign macromolecule, the cell or macromolecule has a "signal" that attracts circulating complement proteins. This antibody-based component of acquired immunity is commonly designated ***humoral***

immunity. The antigen–antibody complexes cause complement proteins to be activated so that they can attack and lyse invading microbes or signal phagocytic cells to destroy the invaders. Complement activation that occurs in response to antigen–antibody complexes occurs via the "classical pathway" of complement activation.

Acquired immunity is dependent on the presence of low levels of antibodies that recognise antigens of the infecting organism, phagocytic cells that respond to antibody tagged antigens, and memory B cells that are poised to produce large quantities of antibodies to the infecting agents. Memory B cells live for long periods and respond quickly with enhanced antibody production upon reinfection with organisms or foreign agents that have been previously encountered. Some T cells also possess memory and are similarly poised for rapid response should a reencounter with the same antigen occur. Immunological memory, involving both B cells and T cells, is the mechanism that makes vaccination effective and explains why an individual does not become ill a second time when exposed to the same pathogens, e.g., the pathogens that cause measles or whooping cough.

Infectious agents such as viruses and other microbes (e.g., mycobacteria), as well as parasites, enter the host cells or are engulfed by macrophages and become inaccessible to attack by B cells and antibodies. Lysosomal (endosomal) enzymes of the infected host cells degrade the components of the infecting organism. The resulting degradation products (e.g., peptides), bound to specialized host cell proteins, are transported to the external surface of the cell where they are presented to the extracellular milieu bound to specialized host cell membrane proteins. The proteins that bind these antigen degradation products are components of the ***major histocompatibility complex (MHC),*** or ***HLA system.*** Cells capable of presentation of peptides of the infectious organisms are described as "antigen-presenting cells" (APCs). The group of MHC proteins involved in presentation of antigen by macrophages and by B cells are members of "class II" MHC proteins. The antigen-presenting cells with digestion products from the invading agent now bound to their external surfaces attract $CD4^+$ T lymphocytes (***T-helper cells***) capable of recognizing the degraded pieces of the antigen from the infectious organism. The T-helper cells produce protein messengers or effector molecules (cytokines). Cytokines provided by the T cells "direct" the differentiation of B cells to form plasma cells and memory cells and also initiate mechanisms that result in inactivation of the virus. Cytokines also signal other specific T lymphocytes, thus preparing them to respond to the infectious organism(s). Recognition occurs between the MHC II-presented molecules on the APC and an antigen (epitope)-specific T-cell receptor (TcR). The TcR molecules are, like antibodies, specific for the antigen (e.g., epitopes on the peptides) and for the particular MHC protein. The recognition of molecules presented by MHC II proteins involves another T-cell membrane protein, CD4 (cluster differentiation protein 4). The immune response that involves B cells and MHC class II antigen presentation is schematically illustrated in Figure 35-1 and that which involves macrophages and MHC class II antigen presentation in shown in Figure 35-2.

Nucleated cells other than those specific to the immune system process infectious agents such as viruses somewhat differently from the MHC II mechanism. For example, viral DNA or RNA replication within the host cells results in the production of new virus, including viral proteins. These newly synthesized viral proteins are degraded to peptides within the infected cell and are bound to an MHC-related carrier protein. The bound peptides are transported to the endoplasmic reticulum and subsequently to the external surface of the cell membrane. Recognition of the degraded pieces of the infectious agent again occurs via proteins of the MHC system proteins; however, in this case it is with proteins of the "class I" MHC. The presentation of the viral protein–derived peptides on the surface of the cell labels the cell as infected and makes it a target for destruction.

A second group of T cells, $CD8^+$ ***cytotoxic T cells*** (frequently abbreviated CTLs or T_c cells), are similar to $CD4^+$ T cells in that each type of T cell clone interacts only with a particular peptide epitope presented by MHC proteins. Cytotoxic T cells bind to and kill the marked, infected cells via specific TcRs. These T cells undergo clonal proliferation to create many additional T cells that are specific for the particular epitopes that the MHC I protein presents and only that T-cell clone recognizes. Among theses T cells are some that persist for a long time and thus become "memory" T cells, poised to prevent later infection by the same agent. These T cells release the cytokine, interferon-γ, which prepares other cells to resist and destroy the invading virus. Cytotoxic T cells and their TcRs also act in conjunction with another membrane protein, CD8. Cellular immune response that involves almost any "ordinary" cell and MHC class I antigen presentation is shown schematically in Figure 35-3.

The cells that are the principal agents of adaptive or acquired immunity are from the lymphoid lineage of stem cells. Similar to the situation with myeloid cells, mutations in lymphoid precursor cells give rise to the ***lymphocytic leukemias.***

In summary, acquired immunity is a lymphocyte-dependent (B and T cell) process by which molecular properties of infectious agents are recognized, anti-infectious

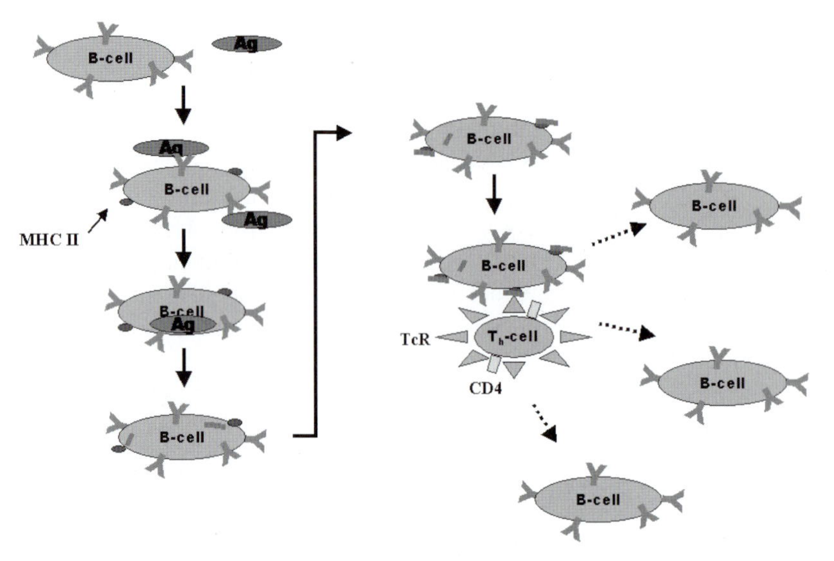

FIGURE 35-1
(Also see color figure.) Antigen recognition by B-cell surface receptors. Response by B cells begins with antigen binding to receptors on the B-cell surface (membrane-bound IgM and/or IgD molecules). Phagocytosis by the B cell transports the antigen into the cell where it is degraded to peptides by proteolytic enzymes. The antigen-derived peptides are next transported to the surface and bind to membrane proteins of the MHC. The peptides are presented by the MHC proteins (specifically MHC II) to T-helper (T_h)cells. The T_h cells bind to MHC II proteins via a specific T-cell receptor (TcR) and the peptides displayed on MHC II molecules. Interleukin-4 from the T_h cells (marked by the surface protein CD4) stimulate the B cells to proliferate and to synthesize and secrete antibodies.

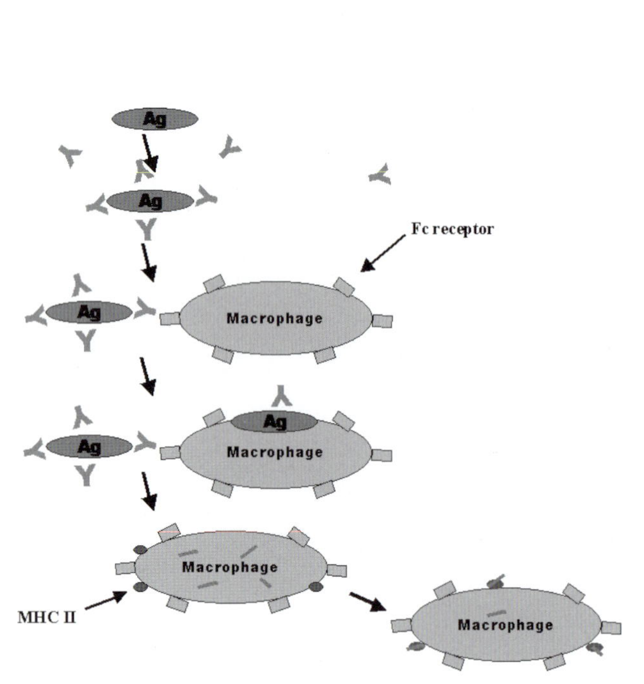

FIGURE 35-2
(Also see color figure.) Antigen-antibody complex formation and recognition by macrophages. Antigens that react directly with antibodies in the circulation and in the extravascular fluid are recognized and removed by macrophages. Recognition occurs with the Fc region of the antibody in the Ag-Ab complex. The antigen-antibody complex is bound to Fc receptors on the external surface of the macrophage. The antigen is taken into the cell by endocytosis and degraded proteolytically. The antigen-derived peptides are transported and bound to the surface MHC II proteins. The MHC II peptide complexes are presented to the antigen-specific T cells.

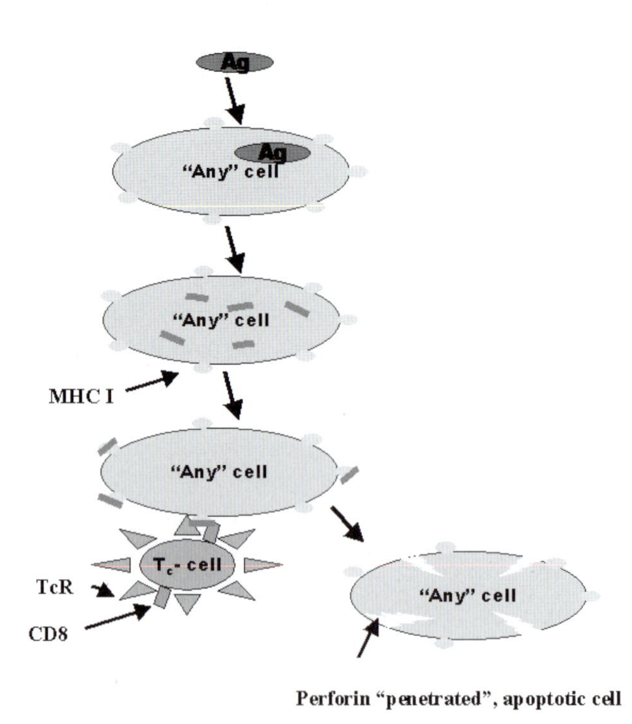

FIGURE 35-3
(Also see color figure.) Antigen uptake and processing by nonimmune system cells. Microbes and viruses can invade and infect any nucleated cell, not only those involved in the immune response. The invading organism is digested by lysosomal enzymes. Peptides encoded by infecting DNA or RNA from the microbe are synthesized in the cell and then transported to and displayed on MHC I molecules on the surface. The MHC I displayed peptides are recognized by a T-cell receptor (TcR) on the surface of cytotoxic T lymphocytes (CTLs). The infected cells are made permeable by the protein perforin and destroyed by apoptosis.

agent antibodies synthesized, and the infectious agents destroyed. Because the selection of B cells and T cells by an antigen results in proliferation of only clones that recognize a specific epitopes of the antigen, an amplified, specific response occurs. Persistence of some of the specific B and T lymphocytes provides memory of the first encounter with the infectious organism so that subsequent infections by the same agent are opposed by a faster and more effective immune response.

35.3 Antibody-Dependent Cell-Mediated Cytotoxicity

Another process involving antigen-antibody complexes exists for killing infected cells. In this process, circulating antibodies bind to epitopes of the viral proteins or glycoproteins exposed on the host cell membrane surface. These "antibody-tagged" cells are then recognized by granular lymphocytes called natural killer (NK) cells. Recognition of the "tagged," infected cells by NK cells occurs through a structurally similar region of all antibody molecules (the Fc region), not through the specific antigen-recognizing regions of the antibody. This mechanism is called antibody-dependent, cell-mediated cytotoxicity (ADCC). Immune response involving neutrophils and NK cells is schematically illustrated in Figure 35-4. (Interaction of a bacterial protein with the Fc region of an antibody is shown in Figure 35-10).

NK cells, monocytes, neutrophils, and macrophages are involved in IgG-mediated ADCC (see below); eosinophils are involved in IgE-mediated ADCC.

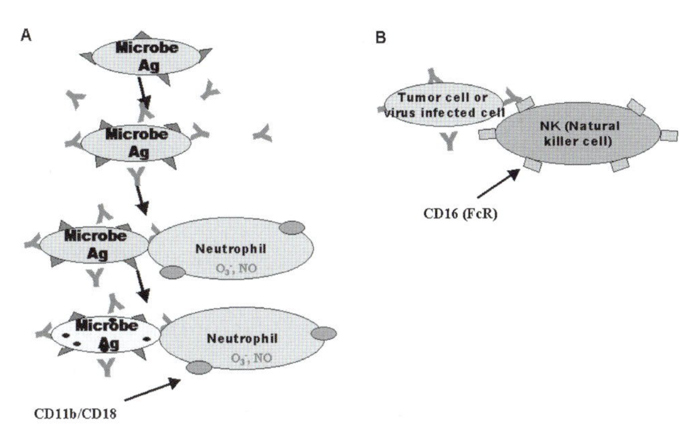

FIGURE 35-4
(Also see color figure.) Microbe destruction by neutrophils and antibody-dependent, cell-mediated cytotoxicity. (A) Neutrophils recognize and attack antibody-tagged microorganisms by a mechanism called antibody-dependent, cell-mediated cytotoxicity (ADCC). In this defense mechanism, microbes "tagged" by antibodies are destroyed by a respiratory burst of O_3^-, H_2O_2, and NO produced by the attacking neutrophils. (B) In a similar way, natural killer (NK) cells recognize antibody-tagged virus-infected cells and tumor cells.

35.4 Distinguishing Self from Nonself

Distinguishing self from nonself is critical to homeostatic immune system function. If self is "perceived" as foreign, autoimmunity and autoimmune disease are the consequence. The ability to recognize self constitutes one form of immune tolerance. Conversely, if pathogenic organisms or toxic substances are unrecognized or unopposed, the consequential infection and organ damage is debilitating and frequently fatal.

Tolerance or unresponsiveness to self-antigens is actively achieved. Exclusion of T cells that recognize epitopes on molecules that are "self" occurs in the thymus during T-cell differentiation and proliferation. Self-reactive T cells are "instructed" to die, a process generally described as negative selection. B cells that recognize epitopes on molecules that are "self" are distinguished from B cells that recognize foreign antigens by a mechanism generally described as positive selection. Positive selection occurs in lymphoid tissues. In this process, B cells stimulated by T-helper cells proliferate. B cells that are not stimulated to produce antibodies or are precluded from responding by the actions of T-suppressor cells die by apoptosis.

Apoptosis, programmed cell death, occurs by mechanisms that include proteolytic processes catalyzed by caspases (a group of enzymes that are responsible for the destruction of intracellular proteins). Failed apoptosis can be a cause of autoimmunity. Autoimmune diabetes mellitus, caused by destruction of β cells in islets of Langerhans in the pancreas, is an example of failed apoptosis. Similarly, chronic ***lymphocytic leukemia*** is an expression of B cells surviving beyond their normal, apoptosis-determined lifetimes. ***B-cell lymphomas*** are cancerous growths that are caused by delayed apoptosis, not by stimulation of cell division. One approach to the development of therapeutic drugs for leukemia includes designing molecules that regulate apoptosis.

Autoimmune disease can also arise from a foreign substance possessing an epitope very similar to an epitope that is found on a host molecule. This situation, termed "molecular mimicry," is the cause of post-streptococcal rheumatic fever because an epitope on the streptococcal M protein elicits formation antibody that reacts with an epitope on myosin.

35.5 Molecules and Chemical Processes of the Immune System

The highly coordinated and regulated responses of the cells of the immune system directly reflect properties of unique proteins of the immune system. Although new components and new functions for known components of the immune system continue to be discovered, the principal groups of molecules of immunity are immunoglobulins, T-cell receptors and coreceptors, cytokines, and the proteins of the complement system. Discrimination between self and nonself, which is the result of the elimination of B and T cells that recognize self-molecules, reflects the actions of substances that regulate programmed cell death. Recognition of epitopes on foreign substances and foreign cells by immunoglobulin molecules and T-cell receptors occurs because of complementarity between specialized "lock-like surface" regions of an immunoglobulin or T-cell receptor molecule and the "key-like" epitopes of an antigen molecule.

Cell proliferation in response to foreign antigens involves receptors, coreceptors, and intracellular processing of the antigens by proteolytic enzymes. The transport and binding of the processed antigens is directed by proteins that are encoded by genes of the major histocompatibility complex. Finally, T-cell receptor recognition of the processed, antigen-derived peptides results in the synthesis and release of cytokines. ***Cytokines*** are pleiotropic protein molecules with functions for each cytokine molecule determined both by the structure of the cytokine molecule and the cells and cell receptors with which the cytokine interacts. Cytokines act as the messengers through which communication and coordination among the cells of the immune system are obtained. Complement proteins act on the cell membrane of foreign organism by assembly of the MAC, which causes permeability changes that destroy the cell. In a highly coordinated manner, the cells and molecules of the immune system protect the organism from pathogenic organisms and bacterial toxins.

In the newborn infant, resistance to infectious agents is first achieved by maternal antibodies that cross the placenta during fetal development or antibodies from mammary gland secretions transferred to a nursing infant. Subsequent antibody-mediated protection is provided by antibodies synthesized by the B cells of the individual as described above. Following this natural sequence, the focus on molecules and how molecular structures account for biological function begins with antibody structure.

35.6 Immunoglobulin Structure and Function

Antibodies, i.e., immunoglobulin molecules, account for approximately 15–20% of the mass of proteins in human serum, the protein solution that remains after blood or plasma has clotted. The immunoglobulins can be divided into five classes—IgG, IgM, IgA, IgE, and IgD—based on their polypeptide chain sequences. The immunoglobulin classes are divided as follows: ~80% IgG, 5–10% IgM, ~15% IgA, and $<0.2\%$ for IgE and IgD combined.

FIGURE 35-5
(Also see color figure.) Structural motif of the immunoglobulin family. Each domain of the immunoglobulin or other molecule of the immunoglobulin family is composed of a collection of β strands that form a globular structure. There are seven strands in the constant-region domains that form two β sheets; one sheet contains three antiparallel strands and the other four antiparallel strands. The orientation illustrates the two sheets that together create the β sandwich or immunoglobulin fold.

The most common type of molecule of the immunoglobulin family is IgG (γ-immunoglobulin or γ-globulin). IgG will be used as the reference molecule for describing the other classes of Ig molecules. An IgG immunoglobulin molecule contains four polypeptide chains; two heavy polypeptide chains (H chains), ~50,000 Da each, and two light polypeptide chains (L chains), ~25,000 Da each. The difference between this "simple" immunoglobulin and more complex forms of immunoglobulin molecules, i.e., IgM and IgA, are described later (Figure 35-11). Because of a common motif in the immunoglobulins and some immune system cell membrane receptor molecules, the proteins with this motif are classified as belonging to the immunoglobulin superfamily.

Immunoglobulin superfamily members are chacterized by a common β-sheet structural motif. Each of the β sheets consists of several β strands. The β strands within each sheet are arranged antiparallel to each other, with each antiparallel strand connected to the next by a hairpin turn. The overall "up and down" arrangement of these strands is sometimes designated a Greek key topology based on the occurrence of a similar pattern on ancient pottery. Two β sheets are derived from 110- to 120-amino-acid residues of polypeptide chain and are folded in such a way that they are frequently described as a "β sandwich." This "β sandwich" forms the structural domain that is the building block of the immunoglobulin molecules in the Ig superfamily (Figure 35-5).This motif is also called the *immunoglobulin fold*.

In the four polypeptide chains of the immunoglobulin molecules(IgG, IgE, and IgD) there are 12 β-sandwich domains. Two domains are formed from each light chain and four domains are formed from each heavy chain. An extended amino acid segment connects adjacent β-sheet domains of each polypeptide chain that creates a flexible fold between the adjacent β-sandwich domains (Figure 35-6; see color plates). An IgG schematic structure, a cartoon secondary structure, and a space-filling tertiary structure based on x-ray crystallography of immunoglobulins are shown in Figure 35-7. The relationships between the heavy and light chains and the immunoglobulin β-sheet motifs of the immunoglobulin G molecule are illustrated in Figure 35-7B.

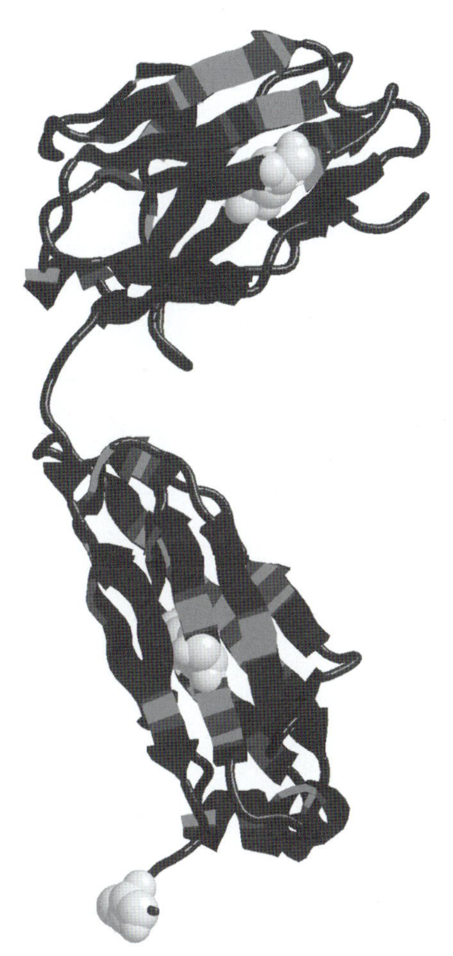

FIGURE 35-6
(Also see color figure.) Pairs of immunoglobulin motif structures (see Figure 35-7) connected by "tether" polypeptide that creates a flexible fold. The disulfide bridges that connect the β sheets of the β sandwich are shown as yellow space-filling residues. Each IgG molecule contains six such folds arranged so that heavy and light chain folds partially wrap around each other (Figure 35-7B,C). The ribbon model shown is from a x-ray crystallographic structure of a Fab fragment. The figure is derived from the coordinates published in the Protein Data Bank file, 2FPB.

The immunoglobulin molecule can also be divided into two regions that are both structurally and functionally distinct. These regions are designated the constant and variable regions (Figure 35-7A). The immunoglobulin heavy chains contain three regions with relatively constant amino acid sequences. These are designated CH1, CH2, and CH3. Each heavy chain also contains a region notable for its sequence variability: the "variable region" (VH). Within the variable region are three shorter segments in which the amino acid sequences are highly variable; these are designated *hypervariable regions* and are the basis for antigen recognition. The immunoglobulin light chains consist of only two structurally and functionally distinct regions, one "constant region" (CL) and one variable region (VL). As in the heavy chain, the variable region of the light chain contains three hypervariable regions that are involved in antigen recognition.

The two heavy and two light chains of the immunoglobulin molecule form a tertiary structure resembling the letter Y (Figure 35-7). Disulfide bonds within each of the variable and constant domains stabilize the immunoglobulin structures. The VL and CL domains of each light chain are linked to the VH and CH domains of the heavy chain via a single disulfide bridge. The region connecting the disulfide linked VH and CH1 to the CH2 and CH3 domains of the heavy chains forms a hinge in the immunoglobulin molecule. The two arms of the Y-shaped immunoglobulin molecule are flexible and when "open" can be as well described as a T shape. The immunoglobulin heavy chains are linked to each other via two disulfide bridges, both located within the hinge region of the immunoglobulin molecule.

Amino acid residues within the hinge region of the heavy chains are accessible to proteolytic enzymes. Cleavage within this region by the enzyme papain results in the formation of three products; one consists of the CH2 and

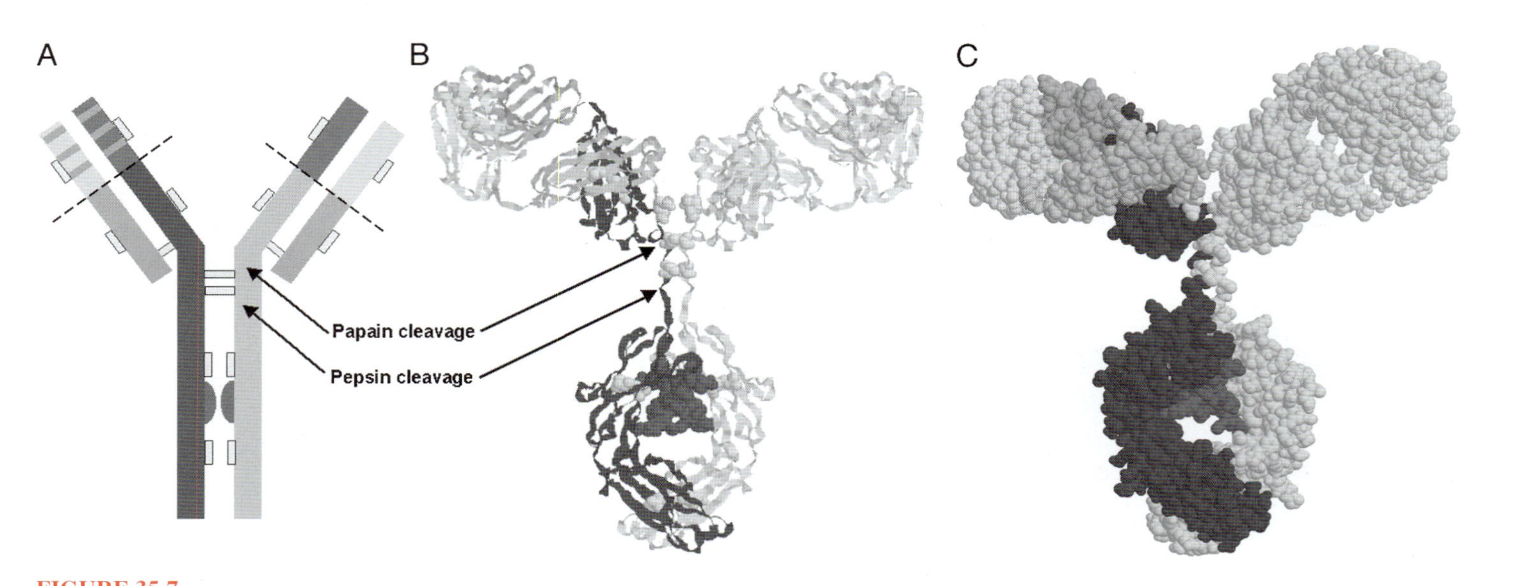

FIGURE 35-7
(Also see color figure.) Schematic diagram of an immunoglobulin. (A) The simplest immunoglobulin (Ig) molecule, IgG consists of four polypeptide chains: two heavy chains of ~50,000 Da and two light chains of ~25,000 Da. More complex Ig molecules are dimers or pentamers that consist of similar basic structures but may have slightly different heavy and/or light chains. Each of the two heavy chains (shown in blue) contains four immunoglobulin domains, a variable VH domain at the N terminus, and three "constant" domains, CH1, CH2, and CH3. Each of the light chains contains two domains, VL and CL. An immunoglobulin domain is 110–120 amino acids in length. The V domains of both chains contain three regions described as hypervariable regions (orange bands) within the VH and VL domains. Each domain contains one intrachain disulfide bridge. The CH1 and the CL1 domains are linked by a single, interchain disulfide bridge. Two disulfide bridges join the two heavy chains near the N-terminal residues of the CH2 domain. A single, Asn-linked, oligosaccharide chain is present within the CH2 domains of each of the heavy chains. A stretch of amino acids between the CH1 and CH2 domains forms a "hinge" region where the two heavy chains are connected by two disulfide bridges. The hinge region acts as a flexible tether between the CH1 and CH2 domains. NMR studies show this region to be very flexible, which permits the Ig molecule to assume various shapes, from the "Y" shown in the figure to a "T" with the "arms" perpendicular to the Fc region (see below). The antigen-binding residues are located at the upper "tips" of the Y-shaped Ig molecule. In the folded Ig molecule, the residues of the hypervariable regions are actually located at the tips, as can be seen in Figure 35-8. The portion of the IgG molecule above the dashed lines is the variable region; that portion below the dashed lines is the constant region. (B) Immunoglobulin secondary structure. The secondary structures of all immunoglobulin domains are essentially the same. Each domain contains seven antiparallel β sheets. The cystine residues (disulfide bonds) are space-filled in the figure. The open, accessible residues within the hinge are also evident in the structure. The figure is derived from the coordinates published by E.A. Padlan, *Mol. Immunol.* **31,** 169, 1994. (C) Space-filling model for IgG. Chains are color coded the same throughout. The polypeptide regions comprising the hinge are readily accessible to proteolytic enzymes. Papain cleaves at the N-terminal side of the uppermost disulfide bridge in the hinge to create three fragments; two Fab fragments containing the VH and CH1 domains linked to the VL and CL domains by a single disulfide bond. The Fc fragment contains two CH2 and two CH3 domains of the heavy chains.

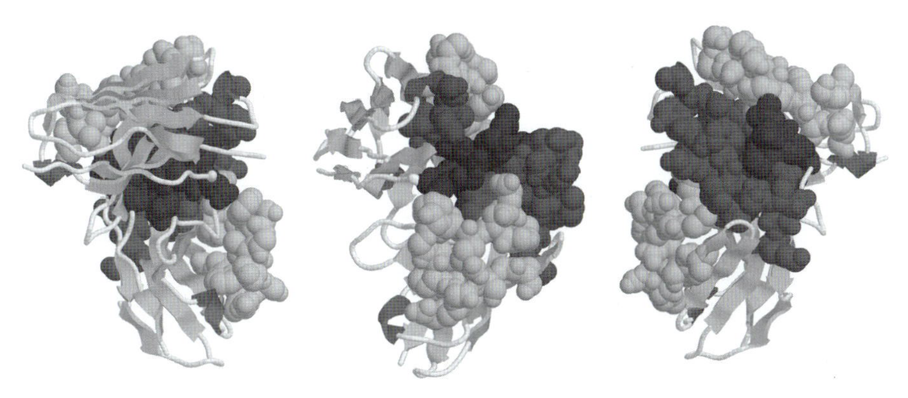

FIGURE 35-8
(Also see color figure.) Complementarity-determining regions. The actual locations of the complementarity-determining regions (CDRs) of the VH and VL chains are shown in this fragment. The orientation is approximately 90° clockwise relative to the orientations in Figure 35-7. The CDR residues are primarily found in the hairpin turns that connect the β sheets of the VH and VL domains. The cystallographic structure used for showing the CDR residues is derived from the structure of a Fab fragment interacting with the capsular polysaccharide from *H. influenzae B* published in the Protein Data Bank file 1HOU. The three orientations of the structure show (1) the backbone β sheets of the immunoglobulin domain; (2) a side view that shows how the CDR regions overlap; and (3) a view that shows the extensive area that the CDR regions provide for making contact with an epitope of the antigen.

CH3 domains of the two heavy chains (designated Fc, for fragment, crystallizable) and two products designated Fab. The Fab comprises the VL and CL domains of one light chain linked through a disulfide bridge to the VH and CH1 domains of a heavy chain. Two Fab fragments are formed from each immunoglobulin molecule. Because the Fab fragment contains the antigen binding sites, it has been a very useful tool for understanding antigen-antibody structural relationships. Proteolytic cleavage by the enzyme pepsin produces a fragment that contains the Fab regions of both heavy and light polypeptide chains. This fragment is designated $F(ab)'_2$. These fragments were particularly useful in the elucidation of the amino acid sequences and the crystal structures of immunoglobulin molecules. Because of the great flexibility of the hinge region, crystals of the complete IgG molecule are not suitable for x-ray crystallography. The models shown in Figure 35-7 are thus a composite of $F(ab)'_2$ and Fc structures.

The initial discovery and description of the constant, variable, and hypervariable domains and elucidation of the primary structures of the immunoglobulins was made possible because of the disease multiple myeloma. In multiple myeloma, a single B-cell clone (monoclone) synthesizes large quantities of structurally identical or monoclonal antibody molecules. Isolation of immunoglobulin from multiple myeloma patients who were producing structurally different immunoglobulins provided sufficient quantities of unique immunoglobulins for protein sequence analysis. These sequence comparisons led to identification of the hypervariable regions and their relationship to antibody specificity. This seminal accomplishment provided the basis for the now established relationships between the primary structures of the hypervariable regions and antibody epitope specificity.

Immunoglobulin molecules possess the ability to recognize structurally diverse epitopes because of the enormous variety of complementary tertiary structures that different amino acid sequences provide. Although the hypervariable regions are separated by intervening sections of several amino acids in the linear protein sequence, the folded polypeptide presents the hypervariable regions on one "face" of the immunoglobulin "arm" (Figure 35-8). Complementarity between antibody and antigen is responsible for the specificity and affinity of the antibody the epitope that is being recognised. Complementarity is determined by the amino acid sequences of the hypervariable regions; thus leading to these regions being now designated the complementarity-determining regions (CDRs).

Binding of small antigens, e.g., haptens, to antibodies occurs when hypervariable region amino acid residues form a pocket into which the hapten fits. Such binding is similar to that between enzymes and small-molecule substrates, i.e., a lock-and-key fit between antigen and

---

The antigen-binding regions of each Fab retain their ability to bind to the epitope for which they are specific. The Fc fragment contains a region that binds to the cellular Fc receptor. This receptor binding site is sometimes referred to as the "biological activity" of the Ig molecule. Cleavage by pepsin produces a fragment $(Fab')_2$ and several fragments from the Fc region of the Ig chains. The figure is derived from the coordinates published by E.A. Padlan, *Mol. Immunol.* **31,** 169, 1994.

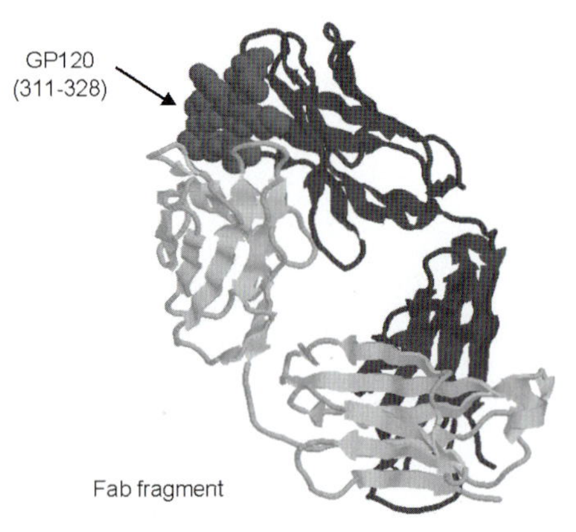

FIGURE 35-9
(Also see color figure.) HIV-1 Gp120 peptide bound to the antigen recognition site of an antibody Fab fragment. The binding of HIV-1 Gp120 residues 311–318 to the antigen binding site shows the actual contacts between antigen and antibody that underlie Ag/Ab recognition. The figure is derived from the coordinates published in the Protein Data Bank file 1GGI.

antibody. The affinity of the antibody for the antigen is determined by the close contact between the amino acid residues of the antibody and the antigen. The interactions may be electrostatic, hydrogen bonding, and apolar (van der Waals' interactions). Hypervariable region amino acid residue interactions with large protein antigens occurs via a relatively flat, large complementary surface. The specific CDR residues are commonly derived from several of the CDRs; thus, the interaction of the antibody with a macromolecule occurs by a discontinous and flexible extended antigen binding site.

A three-dimensional structure of an antigen-antibody complex in which a peptide from HIV-1 Gp120 is bound to a Fab fragment shows the complementarity between a relatively small peptide antigen and the antibody. A crystallographic determined structure is shown in Figure 35-9 for this antigen-antibody complex. Contacts between the HIV peptide and the CDRs of both heavy and light chain residues are clearly evident.

Although the binding of antigen to antibody is specific and occurs via the variable regions of the heavy and light chains, the binding of immunoglobulin molecules to cell surface receptors and to some microbes is nonspecific and occurs through binding sites within the constant regions of the heavy chains. This part of the antibody, the Fc region, marks the microbes for attack. Antibody that is bound to antigen to form antigen-antibody complexes binds to complement via determinants within the constant Fc regions of the antibody molecule. Antibodies that are bound to foreign organisms and foreign substance also signal to phagocytes via determinants that are found on the constant regions of the antibody. Figure 35-10 shows the structure of the *Streptococus* protein, protein G, and the Fc fragment of an antibody. Protein G and a similar protein, protein A (from *Staphylococcus aureus*), mask the constant region determinants noted above and thus protect the microorganisms from the immune system by preventing opsonization. In this way thus interfere with complement activation and block phagocytosis. This constant-region recognition of IgGs by protein G and protein A have made them useful tools for immobilizing antibodies for use in immunodiagnostic procedures and for protein purification.

Smaller differences in amino acid sequences and immunological properties also exist within the constant regions of both the heavy and light polypeptide chains that are unrelated to the hypervariable regions and antigen recognition. The amino acid sequences within the "constant" regions of the heavy chains are the basis for the immunoglobulin classes. The original classification of immunoglobulins was in terms of antibodies that distinguished the epitopes of the different polypeptide chain regions; thus, *isotype* is a synonym for *class*. The immunoglobulin class or isotype distinction is based on

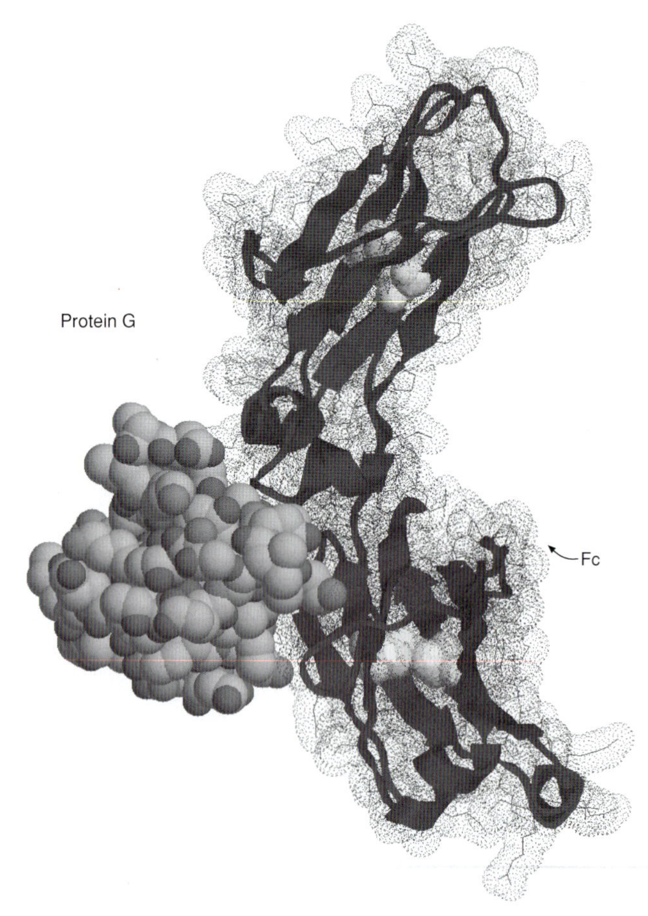

FIGURE 35-10
(Also see color figure.) Binding of protein G (*S. aureus*) to an Fc fragment. Protein G binding mimics the Fc receptor binding of an Ig in the Ig Fc region. Protein G is used for purification of antibodies and in their "capture" in immunoassays. The figure is derived from the coordinates published in the Protein Data Bank file 1FCC.

TABLE 35-2
Immunoglobulin Classes

	IgG	IgM	IgA	IgE	IgD
Molecular weight	150,000	950,000	160,000	190,000	180,000
Molecular form[1]	Monomer (1)	Pentamer (5)	Monomer (1), dimer	Monomer (1)	Monomer (1)
Sedimentation Coeffecient[2]	7S	19S	7S (monomer), 11S (dimer)	8S	7S
Heavy chain (subclasses)	$\gamma(\gamma 1, \gamma 2, \gamma 3)$	μ	$\alpha 1, \alpha 2$[3]	β	δ
Light chain	κ or λ	κ or λ	κ or λ	κ or λ	κ or λ
Additional chains		J	J, SC		
Valency (binding capacity)	2	5(10)[4]	2 (4, dimer)	2	2
Serum concentration, g/L)[5]	5.65–17.65	0.5–3	0.4 –3.5	$<3 \times 10^{-5}$	>0.01
% plasma immunoglobulin	80	7–10	7–15	0.2	0.2
Rate of synthesis g per 70 kg person per day	2.2	0.5	1.5	0.01	0.03
Plasma half-life	20	5	6	2	3
Carbohydrate (wt %)	3	~12	~7	~12	~13
Complement fixation					
Classical pathway	Y (IgG1–3)	Y	N	N	N
Alternative pathway	N	N	Y	N	N
Biological roles (selected)	Defends against microbes, crosses placenta, binds to receptor or surface of phagocytes	First antibody formed to microbial antigens, in monomeric form, B-cell receptor	Major immuno-globulin in secretions defends external surfaces[6]	Causes histamine release from mast cells, defense of external surfaces	Membrane-bound, regulates and triggers antibody synthesis

[1] Number of 4-chain units in class.
[2] Sedimentation coefficients are used in the older literature.
[3] $\alpha 2$ does not have the inter-chain disulfide bonds.
[4] Steric effects appear to restrict binding.
[5] Ranges are from Tietz, N.W., *Clinical Guide to Laboratory Tests,* W.B. Saunders, Philadelphia (1995).
[6] External surfaces include respiratory, digestive and genitourinary tract surfaces.

the heavy chain sequences. The molecules of the immunoglobulin classes are as follows: IgG has γ (gamma); IgM μ (mu); IgA α (alpha); IgE β (beta); and IgD δ (delta) heavy chains. Within the γ chain are three subclasses $\gamma 1$, $\gamma 2$, and $\gamma 3$, and within the α chain are two subclasses $\alpha 1$ and $\alpha 2$. Two classes of light chains also exist, κ (kappa) and λ (lambda), which also possess different epitopes. An individual immunoglobulin molecule contains either κ or λ light chains, but not both. Classes and subclasses are described more fully in Table 35-2. Schematic structures of the molecules of the immunoglobulin classes are shown in Figure 35-11.

The five classes of immunoglobulin molecules are functionally distinct, even when they recognize the same epitopes. IgM, which contains five "basic" four-chain immunoglobulin G–like structures, is decavalent, i.e., it possesses the capacity to bind 10 antigen molecules. Only two antigen molecules can bind to the two arms of an IgG molecule, i.e., the IgG is bivalent. As result, IgM molecules are more effective than IgG in binding antigen, although generally IgM molecules bind antigens more weakly than IgG molecules. IgM is the first immunoglobulin synthesized after B-cell activation and proliferation of plasma cells. IgM molecules, in a monomeric form, are the antigen receptors on B cells. B-cell membrane-bound IgD antibodies also are involved in the regulation of antibody synthesis. IgA molecules are present in the mucous secretions of the respiratory, gastrointestinal, and genitourinary tracts. IgE is the antibody formed in response to allergenic substances. It is responsible for the release of histamine from mast cells, the cause of the discomfort associated with allergies.

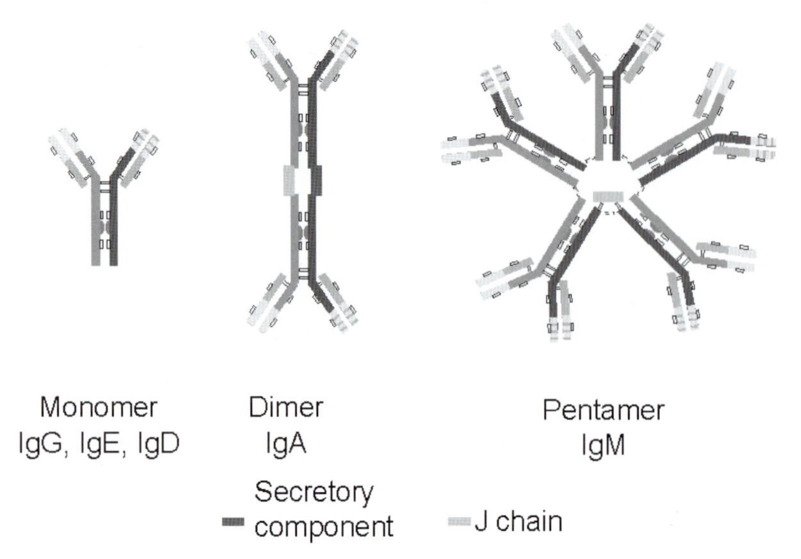

FIGURE 35-11

(Also see color figure; color coding is as the same as indicated in Fig. 35-7.) Immunoglobulin classes. The five immunoglobulin classes are determined by the amino acid sequences of the constant regions of the heavy polypeptide chains (see Table 35-2).

Antigen-Antibody Reactions

Under normal circumstaces an animal produces a mixture of antibodies because of there being several epitopes of an antigen. This is because antibodies that react with the different epitopes of the same antigen are produced by many different B-cell clones (see "B-Cell Clonal Selection and Proliferation"). Each B-cell clone synthesizes antibodies to a single epitope and thus the total antibody that is specific for the antigen will be a mixture of single-clone–derived antibodies. The total antibody is therefore polyclonal with the antibody molecules from each clone recognizing different epitopes on the same antigen molecule.

The two Fab "arms" of the reference immunoglobulin molecule, each with the ability to recognise antigen, make the antibody molecule bivalent or divalent. The bivalent immunoglobulin molecules are IgG, IgA, IgE, IgD. IgM molecules, as described above, are decavalent and can react with as many as 10 antigen molecules. One form of IgA can bind four antigen molecules. The consequence of bivalency and multivalency is that antibodies can react with antigens to form aggregates composed of alternating Ag-Ab-Ag-Ab molecules. This cross-linking results in the formation of large Ag-Ab aggregates. Such cross-linking is strictly true only for polyclonal antibodies, i.e., mixtures of monoclonal antibodies that contain antibodies to more than a single epitope on the antigen. If an antigen molecule possesses two copies of the epitope that the monoclonal Ab recognizes, then monoclonal antibodies can cross-link also.

The Ag-Ab aggregates may become sufficiently large so that solutions become visibly turbid. The complexes can precipitate from the solution. This phenomenon is called *immunoprecipitation* and is the basis of many antibody-based diagnostic tests. It is used to deplete plasma samples of a particular protein. Immuno-absorbed plasmas find widespread use in tests of specific coagulation factors. The formation of aggregates of red blood cells that occurs when cells of one blood type are mixed with plasma from an individual of a different blood type is another example of Ab-Ab cross-linking. Red blood cell aggregation and precipitation is the basis for blood typing and cross-matching procedures of immunohematology. The formation of bands in gels using immunoelectrophoresis or immunodiffusion procedures involves laboratory methods that exploit the property of bivalency (or multivalency) of immunoglobulins.

Because of the bi- and multivalency of antibodies, the formation of antigen-antibody aggregates is sensitive to the *relative* concentrations of antigen and antibody. Optimal precipitation occurs at a concentration called the equivalence point. At too high an antigen concentration, the binding of antigen molecules to antibody is saturated and the aggregates become simple complexes between each antibody and the two antigens. At too high an antibody concentration, all of the antibody binding sites on the antigen are occupied by individual antibodies and sites are not available to form cross-links. The complexes that contain either excess antigen or excess antibody are usually soluble and thus do not lead to identification of

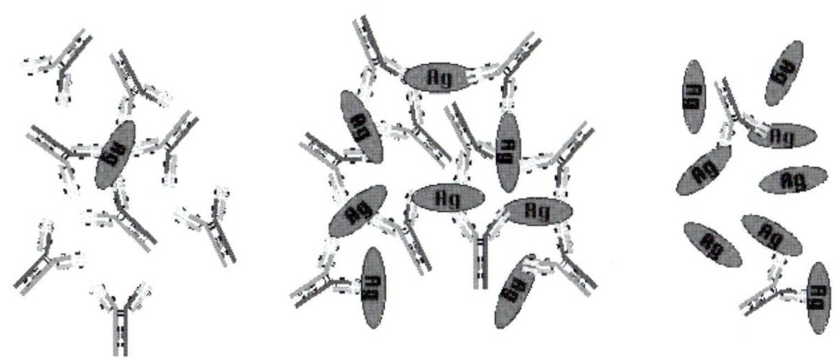

FIGURE 35-12
(Also see color figure.) Antigen-antibody complex formation. Optimal concentrations of antibody and antigen result in cross-linking and precipitation of the antigen-antibody complexes. Cross-linking is important in B-cell stimulation when antigens bind to the IgM B-cell receptors. If antibody is present in excess, then all of the epitopes on the antigen can be occupied by single antibody molecules (subject to steric exclusion) and the antigen-antibody complexes do not precipitate. If antigen is present in excess, then all antibody binding sites can be occupied by antigen molecules and no cross-linking occurs.

antigen-antibody complex by precipitate formation. These antigen-antibody complexes and the complexes at the equivalence point are illustrated in Figure 35-12.

The strength of monoclonal antibody binding to an antigen can be described by classical stoichiometric chemical equations for a bivalent acceptor (the antibody) and a monovalent donor (the epitope of the antigen). The strength of binding of an antigen to a monoclonal antibody is described by a single binding constant and is therefore called an affinity constant. Polyclonal antibodies may appear to be quantitatively more effective in forming immune complexes, i.e., seem to have higher affinity. However, binding strengths for polyclonal antibodies cannot be described in this way because they are complex mixtures of monoclonal antibodies that recognize different epitopes and have different affinities. To distinguish these two situations, the strength of polyclonal antibody binding is now called *avidity* to distinguish it from *affinity*. However, laboratory determinations of the effective strength of binding are performed in the same way.

Monoclonal Antibodies: Diagnostic Tools and Therapeutic Agents

Because each B-cell clone produces antibody to a single epitope on an antigen, an opportunity exists for producing chemically and functionally homogeneous antibody in substantial quantity. However, the short B-cell lifetime that is determined by normal apoptosis initially made this impractical. However, the discovery by Kohler and Milstein that B cells could be fused with "immortal" myeloma cells to create hybridomas that can be maintained almost indefinitely in tissue culture removed this limitation on monoclonal antibody production.

The now common *hybridoma method* begins with immunization of mice and collection of ascites (peritoneal white blood cells) that are synthesizing antibodies to epitopes on the immunogen administered to the mice. Fusion of the mouse cells with myeloma cells produces "immortal" hybridomas. Eliminating unfused myeloma cells from the B-cell, myeloma cell, and hybridoma cell mixture was also a challenge. The use of myeloma cells that are unable to use the purine salvage pathway for nucleotide synthesis, e.g., mutants missing active thymidine kinase and/or hypoxanthine-guanine phosphoribosyltransferase (HGPRT), provided the solution to this problem. The mutant myeloma cells die in a tissue culture medium that does not support their growth. However, hybridomas having the necessary enzymes from the B cells for the purine salvage pathway survive and divide. Unfused B cells die a "natural, apoptotic" death. Selective immortalization provides a way to produce monoclonal antibodies in large quantity.

The hybridoma cultures contain cells with genes that direct synthesis of antibody to the many epitopes of the immunogen; however, dilution of the hybridomas to form subcultures that contain single cells permits selection of clones. The single hybridoma cells grow, divide, and secrete monoclonal antibodies. Cultures from the single cells can be chosen that produce antibody to the desired epitope. Continued growth of the selected hybridomas in tissue culture results in production of monoclonal antibody specific for a single defined epitope.

Immunochemical methods have provided and continue to provide the most specific and sensitive analytical methods used in research and in monitoring the use of prescribed and illegal drugs. Monoclonal antibodies are used in highly sensitive and specific immunoassays. In many immunochemical assays monoclonal antibodies are preferred to polyclonal antibodies because of their homogeneity, single-epitope specificity, and consistent affinity.

Monoclonal antibodies have made possible a very useful, convenient, and highly specific type of immunoassay. This assay uses a monoclonal antibody to one epitope on the antigen to capture and bind the antigen. The captured antigen is then reacted with a second monoclonal (or polyclonal) antibody that is specific to the antigen and to which an enzyme has been covalently linked. After washing away the excess antibody and its linked enzyme, the amount of the antigen can be determined by measuring the activity of the enzyme that is bound to the second antibody. Such assays are designated "sandwich" enzyme-linked immunoassay (ELISA).

Immunoassays that detect viral antigens, e.g., various hepatitis viruses, lymphotropic viruses, and HIV, are used to exclude blood from donors who might be capable of transmitting disease because of a prior infection. The presence of antibodies to viral antigens can also determined by immunassay methods to ascertain if an individual might have been previously infected with the virus when the presence of the virus itself cannot be detected.

Monoclonal antibodies linked to therapeutic drugs can be used to target malignant cells for site-specific drug delivery. Monoclonal antibodies are also used to block cell receptors when interaction of the receptor with its natural ligand causes undesirable effects or disease. A monoclonal antibody called abciximab is specific for the platelet fibrinogen receptor (GpIIb/IIIa) and is administered after angioplasty and other cardiovascular surgical procedures to prevent platelet (hemostatic) plug formation at the surgical (injury) sites. However, mouse monoclonal antibodies are foreign molecules and are themselves immunogenic. To overcome this problem, monoclonal antibodies are "humanized" by replacing the mouse antibody genes for the constant region in the hybridoma with human antibody genes. Therapeutic antibodies such as abciximab have been "humanized" so that they can be used therapeutically with minimal adverse effects.

Recognition of Infected Cells by Cell Receptors

Recognition of the antigen by an antibody is the result of complementarity between the antigen and specific amino acid residues in the hypervariable regions of the immunoglobulin light and heavy chains. Some cell receptors (Figure 35-1) are immunoglobulin molecules, e.g., IgM, IgD, and IgA. Humoral immunity, antibody-dependent, cell-mediated cytotoxicity (Figure 35-4), macrophage attack on foreign cells via Fc receptor recognition of antibody-tagged cells (Figure 35-2), and complement-based cell membrane attack incorporate this antigen-antibody recognition mechanism for identifying and reacting to foreign substances and cells.

Acquired cell-mediated immunity involves complex interactions between antigen molecules and immune cells that recognize the antigens. Finely tuned, cell-based mechanisms are paramount in mounting responses to foreign organisms. T cells (lymphocytes) are the key players in this response. There are two principal types of T cells that are identified by the presence of either $CD4^+$ or $CD8^+$ coreceptor proteins on the T-cells surface. $CD4^+$ T cells are T-helper (T_h) cells. These cells recognize peptides from foreign proteins that are processed and presented on the infected cell membrane by MHC class II proteins. $CD8^+$ T cells recognize peptides from foreign proteins that are presented on the cell membrane by MHC class I molecules (see Figures 35-1 and 35-3).

Attack on infected cells by cytotoxic $CD8^+$ T cells (CTLs or T_c cells) requires that the peptides presented by the MHC I protein on the surface of the host cell be recognized as nonself. The antigen-presenting cell (APC) is thus marked as a target for destruction (Figure 35-3). The MHC I surface complex that culminates in destruction of virus-infected cell involves foreign peptides that have been synthesized by the infected cell and subsequently transported to the cell surface MHC I protein. Peptides from infecting bacteria may be synthesized by the bacteria themselves but are still presented by MHC I to mark the cell for destruction by the CTLs. The processing and transport of newly synthesized MHC I proteins and the foreign peptides through the endoplamic reticulum to the APC membrane involves several proteins. These additional proteins "usher" the MHC I proteins and the peptides until the MHC I proteins and peptides meet and then release them to the Golgi apparatus for final movement to the exterior of the APC. Arrival of an MHC I protein with its bound peptide is followed by binding of a CTL to the APC via a TcR specific for the forign peptide on the $CD8^+$ CTL. Action by the CTL is aided by the CD3 molecule, which acts as a cofactor to the TcR.

Discrimination between nonself-derived peptides, i.e., those from a pathogen, and self-derived peptides requires that the chemical interactions between the T cell via its TcR and the MHC protein with its bound peptides be extremely specific. This specificity is a property of the particular TcR of the individual T-cell clone.

The MHC I protein contains two polypeptide chains; α and β. The α chain contains two distinct domains, $\alpha 1$ and $\alpha 2$, that are on the outside surface of the antigen-presenting

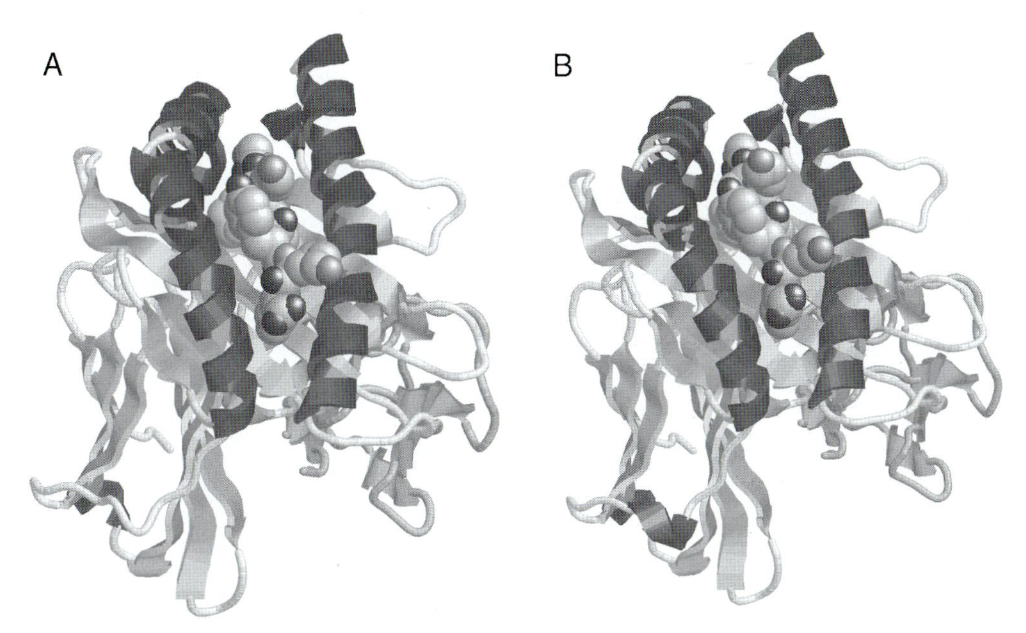

FIGURE 35-13
(Also see color figure.) Major histocompatibility complex (MHC) proteins class I. The MHC class I proteins of antigen-presenting cells comprise two polypeptide chains, α and β. The α-polypeptide (~360 residues) (helices red, β sheets dark yellow, and loops gray) has two extracellular domains shown here ($\alpha 1$ and $\beta 2$), and a transmembrane and intracellular domain ($\alpha 3$, not shown) The $\alpha 3$ domain is structurally similar to the immunoglobulins. The non-MHC gene-related protein, β_2-microglobulin (99 residues), is shown in green on the right side of each structure. The other subunits of the complex that are involved in T-cell receptor recognition are not shown. The figures are derived from the coordinates published in the Protein Data Bank files 1AGD and 1AGB. (A) HIV-1 GAG peptide GGKKKYKL presented on the surface of the MHC class II protein. The peptide is shown with the residues spacefilled, the Lys residue, K is colored cyan and marked by the arrow. (B) HIV-1 GAG peptide GGRKKYKL presented on the surface of the MHC class II protein. The peptide is shown with the residues spacefilled, the Arg residue, R is colored cyan and marked by an arrow. Note that the position of the α helix (part of the α_1 domain) immediately adjacent to the Arg or Lys residues has undergone a change in position (conformation change) and has a small helical segment (lower left) as a result of this single-amino-acid substitution. Such sensitivity to small changes in antigen structure is critical to specificity.

cell. A third domain, $\alpha 3$, crosses the membrane and is exposed on the inside of the cell. The second polypeptide is β-microglobulin, a protein that is functionally involved in antigen presentation by MHC I protein. This molecule is not encoded by genes within the MHC locus. Two MHC I protein structures with bound, virus-derived peptides have been determined by x-ray crystallography. The two MHC I proteins were identical, but the bound peptides differed in that an Arg residue had replaced a Lys residue in the 8-residue peptide from the HIV-1 GAG protein. The peptide that is bound to the MHC I complex "sits" in a groove formed by the $\alpha 1$ and $\alpha 2$ domains. The difference between these two peptides of one basic amino acid is sufficient to cause a movement of an α helix in the $\alpha 1$ domain of the MHC I α polypeptide. The dramatic change in conformation that is produced by a small change in antigen structure illustrates the almost absolute requirement for complementarity that underlies immunological specificity. Such stringent requirements enable the self/nonself distinction to be almost flawlessly in virtually all situations. The two MHC I protein structures are shown in Figure 35-13.

Action by the cytotoxic T cell requires interaction between the $CD8^+$ T cell and the MHC I antigen-presenting cell. The TcR consists of two polypeptides, α and β chains. The TcR is structurally related to immunoglobulin Fab fragments and contains both V and C domains. Recognition of the MHC I peptide complex by a TcR occurs through contact between the TcR α chains and MHC I α chains. A complex showing the MHC I protein and a peptide derived from the T-cell lymphotropic virus HTLV-I is given in Figure 35-14. The regions of contact between TcR molecules and the MHC I molecule vary and depend on the structure of the peptide antigen. The TcR on the cytotoxic T cell is associated with the CD3 complex, the transmembrane signaling protein.

Stimulation of B cells to proliferate and differentiate into memory and plasma cells depends on epitope binding to epitope-specific receptors on $CD4^+$ T-helper cells. Transmembrane signaling then requires the coreceptor protein CD3. Binding of the T-cell receptor to the antigen-presenting B cell occurs via the TcR that interacts with

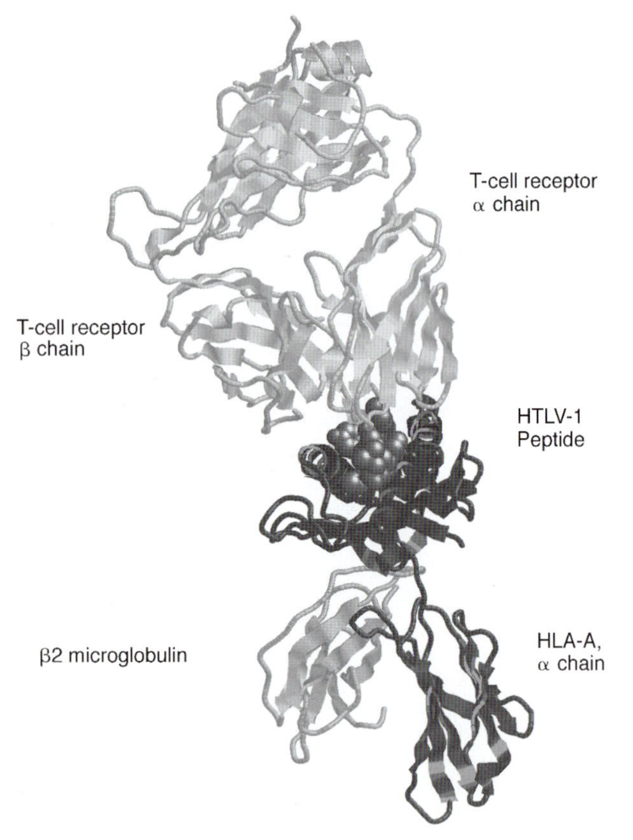

FIGURE 35-14
(Also see color figure.) Complex of MHC class I proteins, HTLV-1 peptide (antigen) and TcR. The MHC class I proteins bind and present peptides that are synthesized inside the infected cell to T-cell receptors (TcRs) specific for the bound peptides. Interaction and stimulation of T-cell proliferation occurs with CD8 positive, cytotoxic Tc cells. The figure is derived from the coordinates published in the Protein Data Bank file 1BD2.

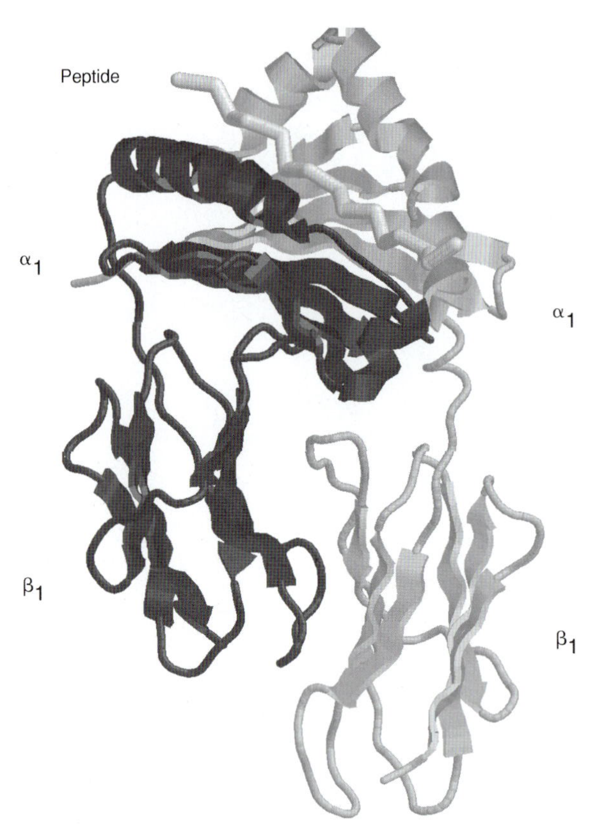

FIGURE 35-15
(Also see color figure.) Major histocompatibility complex (MHC) proteins class II. The MHC class II proteins bind and present peptides that are synthesized outside the infected cell, i.e., peptides that are derived from proteins of the infecting organisms, bound to their specific T-cell receptors (TcRs). Interaction and stimulation of T-cell proliferation occurs with $CD4^{+}$ T cells (Figure 35-16). The figure is derived from the coordinates published in the Protein Data Bank file 1A6A.

the MHC II protein thereby presenting the antigen to the T cell (Figure 35-1).

Presentation of foreign (nonself) peptides by the MHC II protein to T_h cells occurs in a similar manner to the MHC I proteins, requiring both the MHC II protein and T-cell surface proteins of the CD3 complex (see Table 35-1). The cell surface protein CD3 is involved in transmembrane signaling and initiation of cytokine (immune system messenger molecules) synthesis. Newly synthesized and recycled MHC II proteins are, like the MHC I proteins, "ushered" to the surface of the APC.

The MHC II antigen complex on the cell surface of also contains two polypeptide chains, α and β. The extracellular polypeptide of the two chains each consists of two domains, $\alpha 1$ and $\alpha 2$ and $\beta 1$ and $\beta 2$. Similar to the MHC I complex, the peptide lies in a groove between the α_1 and $\alpha 2$ domains. The structure of a human MHC II protein and the CLIP peptide is shown in Figure 35-15. As noted earlier, Recognition of the MHC II–peptide complex by the TcR on the T-helper cell occurs with the participation of two other T_h protein receptors: CD3 and CD4.

CD4 contains three domain structures: an extracellular domain, a transmembrane domain, and an intracellular domain. The extracellular domain, which itself contains two distinct immunoglobulin-like structures, interacts with the TcR on the T-helper cell and with the MHC II protein of the antigen-presenting cell. The transmembrane domain and the intracellular domain (not shown in the figure) are responsible for signaling to the inside of the cell to which binding has occurred. The structure of two immunoglobulin-like folds from a recombinant CD4 is shown in Figure 35-16.

The glycoprotein, gp120, part of the surface of the AIDS virus (HIV-1), is responsible for the binding of virus to the host cell. Although not an example of the region of the antigen molecule that typically interacts, the structure of gp120, an antibody Fab fragment, and CD4 is shown in Figure 35-17. Gp120 binds both to the CD4

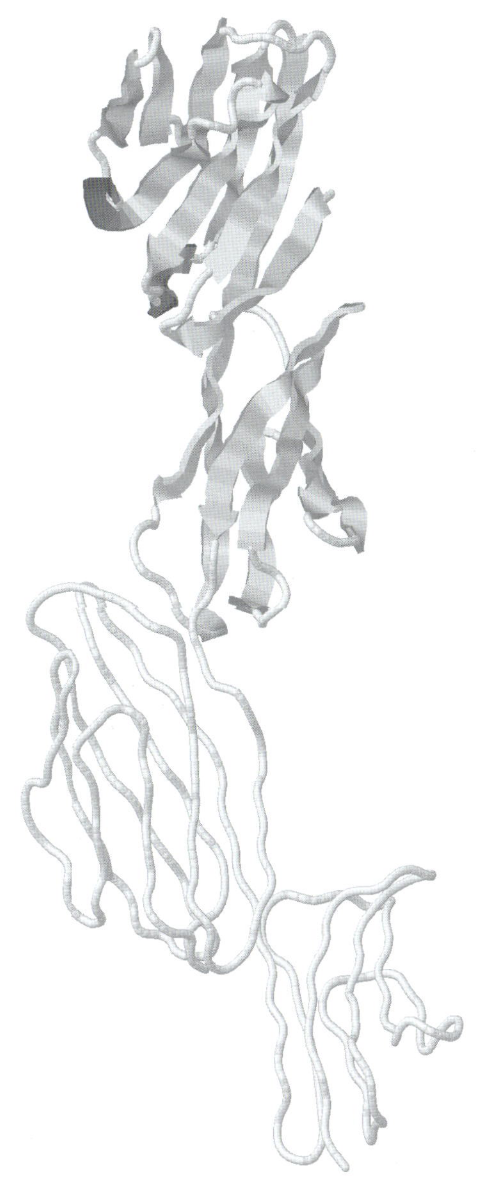

FIGURE 35-16
(Also see color figure.) CD4 (cluster differentiation 4) protein. CD4 on T cells is a component of the MHC class II, TcR complex. Shown are the extracellular domains that consist of two immunoglobulin-like folds. The β-sheet secondary structure is illustrated in one fold; only the α-carbon backbone is shown in the other. The figure is derived from the coordinates published in the Protein Data Bank file 1WIO.

receptor molecule and to another coreceptor (not shown), the chemokine receptor, CCR5. Only a few residues of gp120 and CD4 are in contact; two residues of CD4, Phe (43) and Arg (59), are particularly important for the interaction. The amino acid sequence of gp120 can vary extensively without affecting its binding to CD4, an observation that is explained by the limited contact that is required for the binding. Many HIV mutations occur on the gp120 surface that do not make contact with either CD4 or CCR5. However, three residues in gp120 appear to be conserved in primate lentiviruses (of which HIV is one), a Glu (370), a Trp (427), and an Asp (368). The residues of gp120 that interact with the CCR5 also are highly conserved. The high mutation rate of HIV is one of the problems for host cell recognition and in vaccine development. It also affects immunoassays that are used for detecting the HIV virus and/or antibodies to it.

It is evident from the examples describing immunoglobulin and TcR structures that only a few residues of each protein molecule are crucial for the recognition process. The remainder of the structure of the molecule, the receptor, or the antigen-presenting protein is involved in other functions, including localization by other cell surface proteins.

35.7 B-Cell Clonal Selection and Proliferation

The binding of an antigen, or more precisely the binding of a specific epitope on an antigen to an IgM or IgD antibody receptor on a B cell, is a prerequisite to proliferation of the B cell and the formation of plasma and memory cells. This restricted recognition of single epitopes on nonself antigens is one of the important self/nonself discrimination mechanisms. The interaction between the antigen and the IgM receptor, the internalization of the antigen by the B cell, and the proteolytic digestion of the antigen and presentation of the antigen-derived peptides is shown in Figure 35-1. After presentation of the peptides on the MHC II protein of the B cell, a T-helper cell recognizes the peptides through a specific recognition mechanism that is a property of the TcR molecule (Figure 35-15). After stimulation by the T_h cells through secreted interleukins (Table 35-4), proliferation of memory and plasma cells occurs. The clonal proliferation response of B cells to antigen binding is shown schematically in Figure 35-18.

Stimulation of B cells to proliferate, differentiate, and synthesize antibodies occurs through signals provided by molecules synthesized and secreted by T cells. These signaling molecules, the cytokines and particularly the interleukins, are described in Table 35-4.

The critical importance the $CD4^+$ T cells and their messenger molecules for B-cell proliferation is manifest in AIDS where HIV proteins promote death of $CD4^+$ T cells resulting in inadequate stimulation of B-cell proliferation. One of the early clues to this action of HIV was the change in the ratio of CD4 to CD8 cells in AIDS patients. The $CD4^+$ cell count and the CD4/CD8 ratio quickly became markers for the progression of the disease. The immunodeficiency that results from destruction of T_h cells

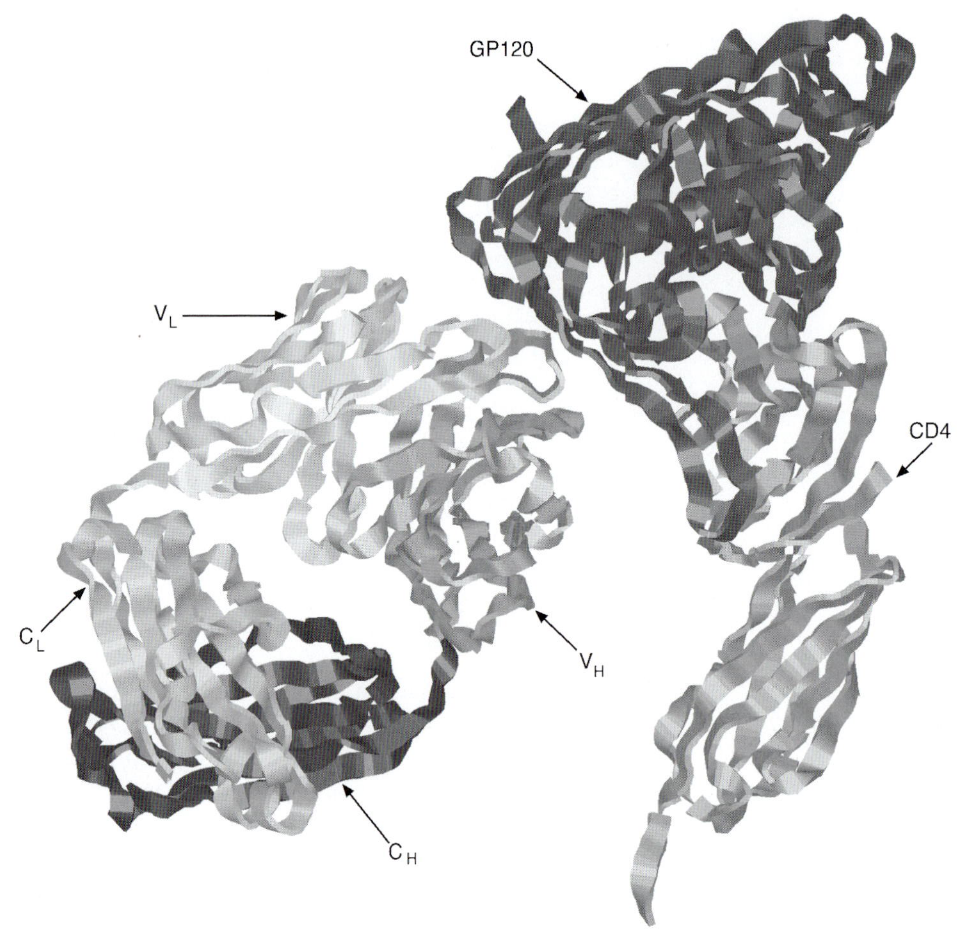

FIGURE 35-17
(Also see color figure.) Complex of HIV gp120, Fab fragment, and CD4. HIV gp120 is an envelope protein of HIV-1 that is exposed on the outer surface of the virus. It is derived from gp160 by proteolysis that also produces another product, gp41. Only a small area of the surface of the gp120 and CD4 molecules is in contact. Interaction between CD4 and gp120 is dominated by residues Phe43 and Arg59 of CD4, and Gly370, Try427, and Asp368 of Gp120. The antibody fragment blocks the region of gp120 that interacts with a chemokine receptor, CCR5. The figure is derived from the coordinates published in the Protein Data Bank file 1GC1.

permits infections by microorganisms that are not normally pathogenic, e.g., *Pneumocystis carinii,* to occur.

35.8 Antibody Diversity and Immunoglobulin Genes

The enormously large number of epitopes on antigens that must be recognized so that an organism can successfully defend itself from disease requires the existence of a comparable diversity of antibodies and antigen receptor molecules. Because each immunoglobulin molecule and T-cell receptor is produced by separate clones, mechanisms for generating such diversity are necessary. It is estimated that there are more than 100 trillion ($>10^{14}$) different B- and T-cell clones. The genes that code for the immune system molecules are capable of recognizing and interacting with the very large number of different epitopes. These are "created" during embryonic differentiation. The genetic mechanisms required to meet this need for diversity and adaptability are unique to the immune system.

There are three families of immunoglobulin genes: one for κ light chains, one for λ light chains, and one for the heavy chains (see Table 35-2). Within each of these families are germline genes that contain many different exons that code for the constant regions in chains of the immunoglobulin classes and for the variable regions that are responsible for epitope recognition. By combinatorial recombination of the different exons present in the germline genes, the genes are created that code for both immunoglobulins and the T-cell receptors that recognize

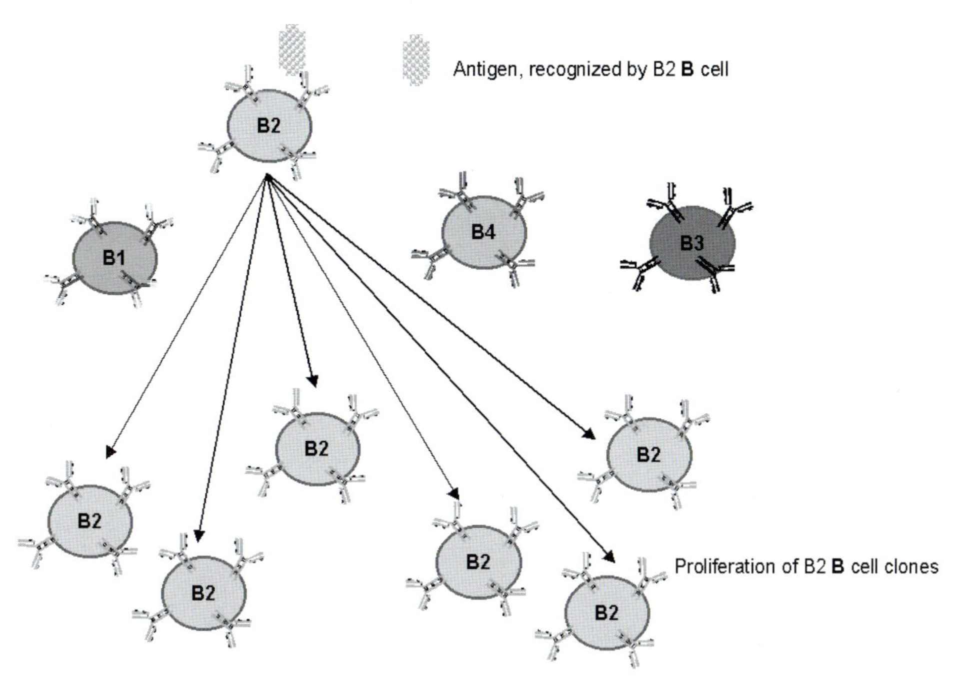

FIGURE 35-18
(Also see color figure.) B-cell clonal selection. Antigen binding to the surface Ig (monomeric IgM and IgD) receptors occurs only to the B-cell clone that recognizes a particular epitope on the antigen (Figure 35-1).

essentially all of the possible epitopes vertebrate animals encounter.

Germline immunoglobulin genes consist of exons that are selected and arranged during differentiation to generate a mature mRNA. The mRNA is translated to give an immunoglobulin chain that recognizes a single specific epitope. The V exons (~100) code for the first ~95 amino acid residues of the variable (VL) regions of the κ and λ chains. J (junction) exons code for the next ~15 amino acid residues. Separate genes exist for the constant (CL) regions of the κ and λ chains. Diversity in the CDR3 region of the VL chain arises from the various arrangements of the 5 J exons that can occur in combination with any one of the ~100 VL exons. The excised DNA of the germline that is not contained within the final-light chain gene is degraded leaving a single B-cell DNA structure that "defines" the particular B-cell clone. The primary RNA transcript that is formed is spliced such that the primary transcript forms a mature RNA that is translated to produce the single Ig light chain. The combinations of V, J, and C exons produce more than 3000 light chains having different amino acid sequences. Ig genes can also undergo an "editing" process during cell division to produce even more specific antibodies. The processing of the germline genes for immunoglobulin light chains is shown in Figure 35-19.

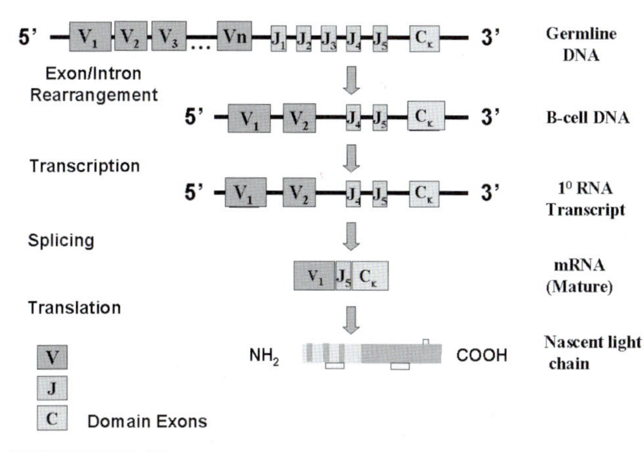

FIGURE 35-19
(Also see color figure.) Immunoglobulin light chain gene structure and gene processing. Germline immunoglobulin genes consist of exons selected during differentiation to generate the mature mRNA that is translated to give the immunoglobulin chain that recognizes a single specific epitope. Exons (>100) code for the first ~95 amino acid residues, the VL regions of the κ and λ chains. J (junction) exons code for the next ~15 amino acid residues. Separate exons exist for the CL regions of the κ and λ chains. Diversity in the CDR3 region of the VL chain arises from the various combinations of the 5 J exons that can occur with the ~100 VL exons. The excised DNA of the germline is degraded, leaving a specific B-cell DNA that "defines" the particular B-cell clone. The primary RNA transcript that is formed is spliced such that the primary transcript becomes a mature mRNA that is translated into a specific Ig light chain.

Heavy chain genes differ from the light chain genes in that they include an additional group of DNA segments; the D (diversity) exons and the genes for the different

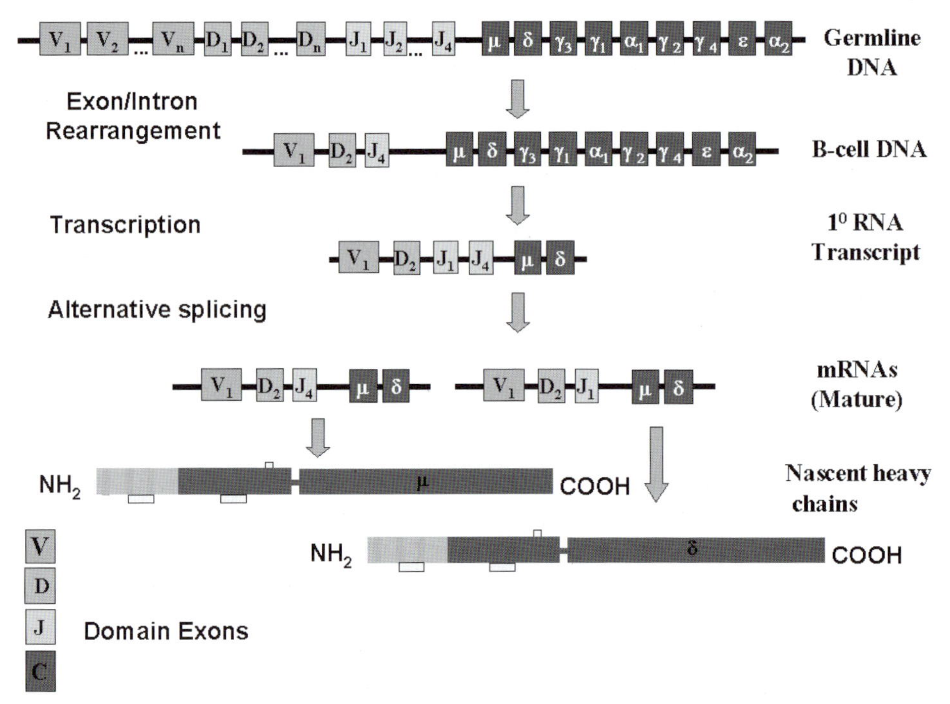

FIGURE 35-20
(Also see color figure.) Immunoglobulin heavy chain gene structure and gene processing. Exons of the heavy chain genes that encode the variable regions of the immunoglobulin molecule are shown in orange and labeled $V_1, V_2, \ldots V_n$. Selection from these V exons during embryonic development produces the unique sequences of each B-cell clone. The germline genes for the immunoglobulin heavy chains contain an additional group of exons, the D (diversity, light blue; labelled $D_1, D_2, \ldots D_n$) exons. Recombination between the V and D regions occurs more frequently than that between the V and J exons in the light chain exons. Introns between the V and D and between the D and J exons contain signal sequences that regulate the expression of the enzyme recombinase. This enzyme is responsible for the efficient recombination that gives rise to the epitope-specific B-cell clones with their individual Ig genes. The heavy chain genes contain exons that encode all of the isotype heavy chains. Class switching, i.e., the change in chain expression that occurs during antibody synthesis after B-cell activation, results from alternative splicing between the J exons and the exons for the various heavy chain isotypes.

heavy chain constant regions (Table 35-2). In a rearrangement process similar to that of the light chain genes, combinations of 1000 V exons, 10 D exons, and 4 J exons lead to more than 40,000 different polypeptide sequences. The constant-region exons, of which there are eight, create the different classes of antibodies. Recombination among the V, D, and J exons creates additional diversity. And, because there is "imprecision" in the regions joining these exons, even greater diversity is obtained. The processing of the germline genes that generate Ig heavy chains is shown in Figure 35-20.

The genes for T-cell receptors are also rearranged by mechanisms similar to the immunoglobulin genes. V, D, and J regions recombine to form the multiplicity of TcR variants that specifically recognize the peptides presented on the MHC proteins. The genes for the α and γ TcR proteins do not contain D regions. TcR genes also differ from Ig genes in that they do not commonly undergo changes during T-cell division.

The genes for Ig and TcR protein chains are located on different chromosomes. The genes for κ chains are located on chromosome 2 and for λ chains on chromosome 22. The genes for Ig heavy chains are on chromosome 14. Similarly, the TcR genes are distributed among different chromosomes. The genes for the α and δ chains are on chromosome 14; the genes for β and γ chains are on chromosome 7.

Class (Isotype) Switching

Immunoglobulin synthesis by B cells in response to an antigen initially results in the production of IgM. This is followed after a period of 1–2 weeks by the formation and appearance of IgG with the same epitope specificity as the original IgM. This "class switching" results from a change in the expression from μ (mu) to γ (gamma) heavy chain genes. Although the specificity of the antibody produced by a B-cell clone is determined during embryonic development, the class of antibody produced by the clone is influenced by T-cell communication with the B cell. This B-cell maturation is mediated by cytokines.

In addition to V gene recombination and class switching, Ig genes are alternatively spliced following transcription. Introns are spliced out of the RNA as expected, but the μ chain gives rise to both the pentameric, IgM molecules and the membrane-bound monomeric form of IgM. A region of the heavy chain gene that codes for constant regions of the polypeptide chain (the switch region) is the sequence within which alternative splicing occurs. IgD heavy chain variation also results from alternative splicing of RNA transcripts.

35.9 Major Histocompatibility Complex (MHC) Genes

The antigen-presenting proteins encoded by genes of the major histocompatibility complex have been described above. These proteins and the genes that code for them are highly polymorphic, but not because of the mechanisms that generate B- and T-cell diversity. The variability in MHC gene products is derived primarily from gene polymorphisms—different alleles carried by individuals in a population. Allelic exclusion does not occur with the MHC genes; thus, an individual has genes (haplotypes) contributed by both parents. Historically, the MHC I and II gene families were identified based on transplanted tissue compatibility and are alternatively known as HLA I and HLA II gene clusters.

The major histocompatibility genes (on chromosome 6) comprise a cluster that encode the class I and class II proteins (MHC I and MHC II) and some of the complement proteins (MHC III). A fourth group, the MHC IV genes, has been proposed. These additional genes encode at least three proteins associated with inflammation: tumor necrosis factor α (TNF-α) and the lymphotoxins TNF-β and TNF-γ. The histocompatibility genes are the most polymorphic genes known and thus they create a major challenge that is associated with donor/recipient matching for organ transplantation. In contrast to the immunoglobulin and T-cell receptor genes, the MHC genes do not undergo rearrangements; the polymorphisms are due to the different alleles (more than 100 for the MHC I heavy chain gene).

Transplant rejection is an example in which the distinction between self and nonself works to the disadvantage of the transplant recipient and provides the most formidable challenge facing transplantation medicine.The MHC and other cell surface proteins of the transplant are foreign to the receipt and thus are targets for antibodies, T cells, and phagocytes. This recognition of the transplanted cells and organ(s) as nonself initiates the immune response. Attack on the transplanted tissues then leads to destruction (rejection) of the organ. Complement activation also occurs, which both enhances the destructive process and produces additional biological effects that are caused by complement-derived products (see "Complement"). Immunosuppressive drugs, although they make transplantation possible, create risk to the recipient because they increase susceptibility to infection. Two immunosuppressive drugs, cyclosporin A and FK506 (tacrolimus), suppress the production of the interleukin (IL-2) and thus, because IL-2 is necessary for T-cell growth (see Table 35-4), suppress T-cell proliferation. This is accomplished through the binding of these drugs to a class of proteins called *immunophilins.* The immunophilin-bound cyclosporine A or FK506 inhibits calcineurin, a protein phosphatase that is involved in the T-cell receptor signaling pathway protein transcription. The structures of FK506 is shown in Figure 35-21. Humanized monoclonal antibodies have been developed that successfully reduce the rejection of transplanted organs. One such monoclonal antibody, OKT3, is directed to the T-cell antigen CD3. Binding to OKT3 renders the T cells incapable of functioning. Perhaps the most exciting discovery relevant to transplantation is that cytokines and other growth factors can direct host stem cells to produce tissues and organs that are immunologically compatible with the host.

MHC genes also determine the susceptibility of individuals to autoimmune diseases. Among the diseases clearly related to MHC genes are insulin-dependent diabetes, ***multiple sclerosis, systemic lupus, erythematosus, myasthenia gravis,*** and ***rheumatoid arthritis.*** Alleles of MHC genes are also associated with non–immune system diseases, e.g., ***hemochromatosis, narcolepsy,*** and ***dyslexia.***

35.10 Complement

The complement system, part of the innate immunity of animals, provides the third defense mechanism against infectious agents, the first and second being the physical barriers provided by skin and the mucous secretions. The complement system consists of more than 30 proteins that circulate as well as complement receptors and complement control proteins. There is no simple mnemonic for understanding the nomenclature of the complement system components. Some—but only some—of the complement proteins are designated by an uppercase letter C followed by a number, e.g., C1-C9. Complement proteins (C) that have been proteolytically cleaved are distinguished by lower case letter following the C-number, e.g., C3b, C3a, etc. Table 35-3 contains the list of complement proteins along with a description of some of the pertinent chemical, physical and physiological properties of each component.

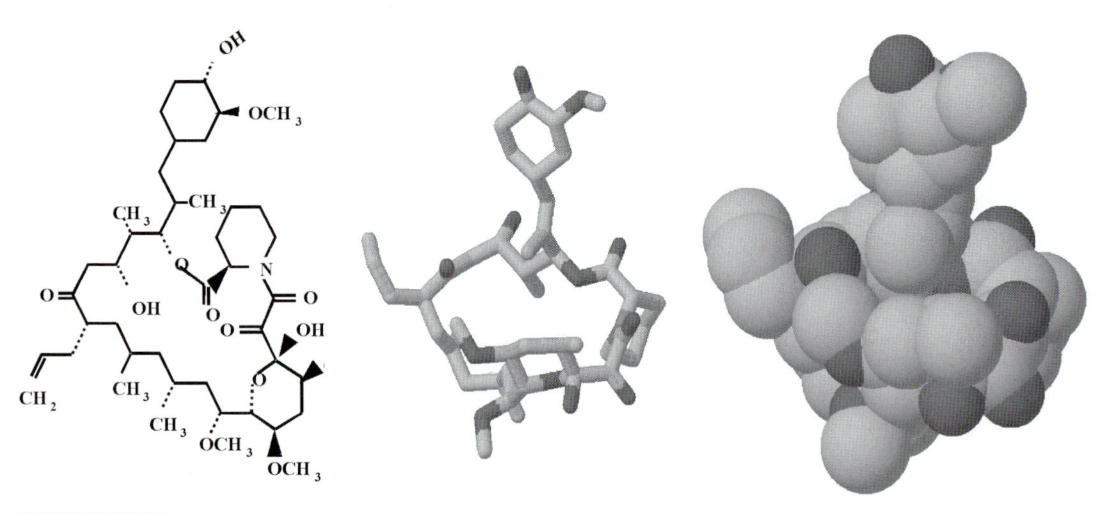

FIGURE 35-21
(Also see color figure.) Structures for the immunosuppressant, FK506 (tacrolimus). Tacrolimus is a macrolide isolated from *Streptomyces tsukubaensis*. Shown from left to right in approximately the same orientation are the simple line structure, a stick- structure, and a space-filling structure. The stick and space-filling structures are based on the 3-dimensional structure of the molecule bound to the immunophilin FKBP. Protein data bank designation 1FKF.

TABLE 35-3
Proteins of the Complement System: Alternative, Classical, Lectin, and Membrane Attack Pathways and Complement Regulating Proteins

ALTERNATIVE PATHWAY	Initiated by attack of nucleophilic cell surface molecules on thioester in C3 and/or spontaneous hydrolysis by H_2O of thioester in C3		
Protein/Subunit(s) Active Form	**Serum Conc. *(mg/L)***	**Molecular Weight (chains)**	**Description and Role of the Active Forms in These Pathways**
D	2	24,000	Cleaves B to Ba and Bb
B	200	93,000	Contains serine protease active site of C3bBb
Ba			Chemotactic peptide
Bb			Active site-containing chain of C3bBb, the C5 convertase
P (properdin)	20–30	~46,000	Stabilizer of C3bBb
C3 Convertase (C3b,Bb)			Alternative pathway C3 convertase
C3 (see classical pathway also)	1,110–1,710	185,000	Precursor to C3a and C3b
C3a			Anaphylatoxin, increases vascular permeability and causes mast cell degranulation
C3b			C3b contains a thioester bond that can attach C3b covalently to target cells. Forms complex with factor B, allowing D to cleave B. C3bBb forms complex with C5, allowing C3bBb to cleave C5. C3bBb stabilized by binding P to form C3BbP. Opsonin
C3d, C3d,g,			Fragments that interact with complement receptor, CR2
C3e, C3d-K			Involved in leukocytosis

(continued)

TABLE 35-3
Proteins of the Complement System (continued)

CLASSICAL PATHWAY	Initiated by IgG1, IgG2, IgG3 or IgM—antigen complexes		
Protein/Subunit(s) Active Form	**Serum Conc. *(mg/L)***	**Molecular Weight (chains)**	**Description and Role of the Active Forms in These Pathways**
C1 (multi-subunit)			Requires Ca^{2+} for subunit association
C1q	83–125	460,000 (6A x 26,500) (6B x 26,500) (6C x 24,000)	Shaped like a tulip with a "stalk" that contains collagen-like domains and "petals", globular domains. *Globular domains bind Fc regions of IgG and IgM antigen-antibody complexes, leading to activation of C1r and initiation of the classical pathway.*
C1r After activation becomes 2 chains	34–48	83,000 (α, ~56,000) (β ~27,000)	Proenzyme to serine protease; auto-activates and activates C1s. Activated C1r consists of two polypeptide chains, C1rα, C1rβ. Anaphylatoxin activity (C1rα), active site-containing chain (C1rβ)
C1s After activation becomes 2 chains	34	85,000 (α, ~56,000) (β ~27,000)	Proenzyme to serine protease; activates C2 and C4 to produce C4b and C2a which form the C4b2a (C3 convertase) that activates C3. Activated C1s consists of two chains, C1sα and C1sβ.
C4 (before activation)	350–600	204,000	
C4a			Amino terminal residues, anaphylatoxin (weak), increases vascular permeability and causes mast cell degranulation
C4b			C4b contains a thioester bond that can attach C4b can attach covalently to target cells. Forms a complex with C2, allowing C1s to cleave C2
C2	25	110,000	Contains serine protease active site of C4b2a, which can cleave C3 and C5
C2a			Component of C3 convertase
C3 Convertase (C4b2a)			Activates C3, i.e. is the classical pathway C3 convertase
C3 (See alternative pathway also)	1,110–1,710	185,000	Precursor to C3a and C3b
C3a			Anaphylatoxin, increases vascular permeability and causes mast cell degranulation
C3b			C3b can attach covalently to target cells via reaction with the C3 thioester. Opsonin

(continued)

TABLE 35-3
Proteins of the Complement System *(continued)*

LECTIN PATHWAY Intitiated by binding of MBP to mannose residues on foreign cell surface molecules

Protein/Subunit(s) Active Form	Serum Conc. (mg/L)	Molecular Weight (chains)	Description and Role of the Active Forms in These Pathways
MBP (mannan-binding protein)			Protein that recognizes mannose residues on oligosaccharides on cell surfaces
MASP-1	–		Homologous to C1r & C1s, cleaves C1, C2 and C4 directly
MASP-2	–		Homologous to C1r & C1s, cleaves C4 directly

MEMBRANE ATTACK PATHWAY Final, common pathway by which the cell is made permeable

Protein/Subunit(s) Active Form	Serum Conc. (mg/L)	Molecular Weight (chains)	Description and Role of the Active Forms of These Pathways
C5	55–113	196,000	The membrane attack complex (MAC) is formed by stepwise assembly of C5b through C9.
C5a	–	–	Anaphylatoxin (potent), has chemotactic activity, increases vascular permeability and causes mast cell degranulation
C5b			Part of the MAC
C6	179–239	125,000	Component of MAC, steps involving C5, C6 & C7 occur in solution
C7	27–74	120,000	Component of MAC
C8	49–106	150,000	Component of MAC on membrane surface
C9	33–95	66,000	Component of MAC on membrane surface

REGULATORY REACTIONS

Protein/Subunit(s) Active Form	Serum Conc. (mg/L)	Molecular Weight (chains)	Description and Role of the Active Form in These Pathways
C1 inhibitor	180	109,000	SERPIN, (**SER**ine **P**rotease **In**hibitor), Inactivates C1r, C1s by covalent bond formation
Anaphylatoxin inactivator (carboxypeptidase N)		310,000	Protease, Inactivates C3a and C5a anaphylatoxins by cleavage of X-Arg peptide bond
Factor I	25–50	88,000	Protease, Cleaves C3b that is bound to H or CR1 and C4b that is bound to C4BP
Factor H	405–717	15,000	Binding protein dissociates C3bBb
C4b binding protein (C4BP)	250	570,000	Binding protein dissociates C4b2a
Decay accelerating factor		70,000	Dissociates C3bBb and Binding protein C4b2a from host cells
Complement receptor (Type 1) (Cr1)		200,000	Binding protein when bound to C3b or C4b, allows them to be cleaved by Factor I
Vitronectin (S protein)	500	80,000	Binding protein Binds to MAC

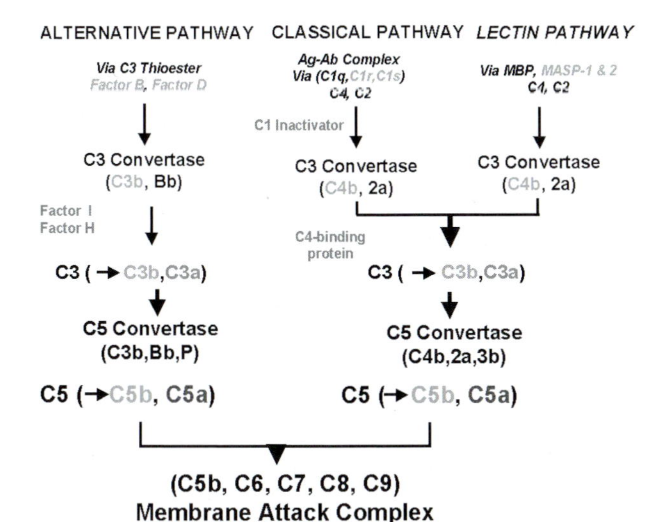

FIGURE 35-22
(Also see color figure.) Initiation of actron by complement occurs via three pathways: classical, alternative (properdin), and lectin pathways.

Action of the complement system has many consequences: lysis of bacterial cells, recruitment of phagocytic cells to the site(s) of invasion, opsonization (marking for phagocytic attack and removal) of foreign cells, and removal of immune complexes. Initiation of action by complement can occur via three distinct pathways: the "classical" pathway, the"alternative" or properdin pathway, and the lectin pathway (Figure 35-22).The physiological significance of the first two pathways has been established; however, the importance of the lectin pathway is not yet fully known. The lectin pathway has only been investigated *in vitro*.

The Alternative Pathway

The principal pathway of complement activation, which is part of the innate immunity of the organism, is the alternative pathway. The designation of this pathway as "alternative" reflects order of discovery and acceptance, not physiological importance.

Initiation of complement activation occurs during the "spontaneous" reaction of C3, the most prevalent complement protein with nucleophilic chemical groups on the surface of invading organisms. This reaction involves a relatively uncommon organic chemical in nature, a thioester that is found on C3, C4, and on one of the plasma protein inhibitors, α_2-macroglobulin. Reaction between the thiester and a cell surface nucleophile, i.e., an amino, hydroxyl, or thiol group, results in the covalent attachment of C3 to the cell surface. Spontaneous hydrolysis of the C3 thioester also produces a functionally active form of C3, C3(H_2O). C3(H_2O) forms a complex with factor B and makes factor B susceptible to proteolysis by factor D, a proteinase from adipocytes. Factor B is converted to Bb by factor D, becoming a proteolytic enzyme itself. The combined product of this reaction, C3b, Bb forms the enzyme C3 convertase (specifically the alternative pathway C3 convertase). C3 convertase proteolytically converts circulating C3 molecules to C3b and C3a. The C3 convertase has a short life time because the C3b and Bb dissociate rapidly. Factor H, another circulating complement protein, can bind to C3b and replaces Bb in the C3 convertase. C3b that is bound to factor H also becomes very susceptible to the complement proteinase factor I. Upon cleavage of C3b by factor I, C3b loses its ability to function in the C3 convertase complex. Proteolytically cleaved C3b (iC3b) still retains its ability to opsonize cells and thus mark them for attack by phagocytes. The product C3a is an anaphylatoxin, i.e., it causes histamine release from mast cells and increases vascular permeability through vasodilation. Increased permeability facilitates phagocyte access to pathogens in the tissues surrounding the blood vessels. C3a is also chemotactic for eosinophils suggesting an association with allergic responses and asthma.

P (properdin) can also bind to the C3 convertase. Its role is to stabilize the complex and hence it is considered a cofactor-activator in the alternative pathway. These components plus additional C3b molecules form the C5 convertase, an enzyme complex that peoteolytically converts C5 to C5a and CSb. Properdin stabilizes C3b and Bb in the complex and protects these proteins from proteolytic inactivation by factor I. Factor H competes for Bb in the C5 convertases, the same as it does in the C3 convertase. The alternative pathway has also been called the properdin pathway because of properdin's participation in alternative pathway C3 and C5 convertases. C5b is a component in the terminal complex of the complement activation process, the MAC. The MAC is composed of a self-assembled, noncovalent complex of C5b, C6, C7, C8, and C9. Together these components produce a pore-like structure that makes the membrane of the cell to which it is attached permeable and causes cell death. Under the electron microscope the MAC appears like an impact crater similar to those observed on the surface of the moon. C5a is also an anaphylatoxin like C3a, but it is more potent. C5a is also a chemokine and attracts phagocytic cells to the site of complement activation.

The Classical Pathway

The classical pathway of complement activation is closely linked to acquired immunity because it is initiated by antigen-antibody complexes binding to C1, the first component of this pathway. Antigen-antibody complexes that

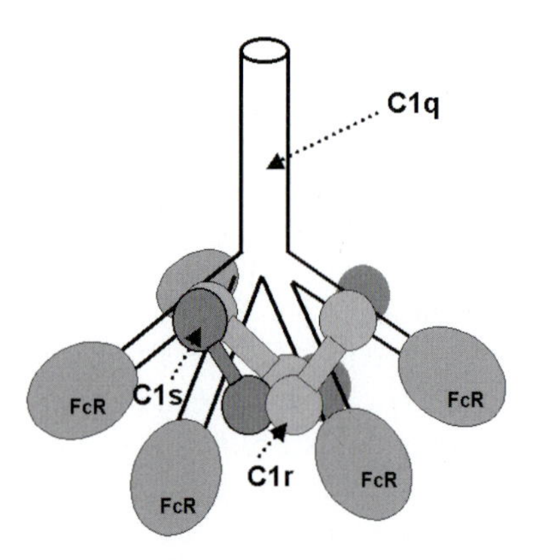

FIGURE 35-23
(Also see color figure.) A cartoon representation of complement protein C1. The C1 complex consists of one C1q, two C1r, and two C1s molecules. The formation of the complex requires the presence of Ca^{2+}. C1q contains six polypeptide chains that form a globular structure that binds to antigen-antibody complexes at one end of each chain.

bind to C1 can be those formed with epitopes on the surfaces of microbes, those formed on B cells by reaction with the surface IgM, or those formed with foreign proteins or other macromolecules. Immunoglobulins such as IgGl, IgG2, IgG3, and IgM are capable of initiating complement activation via C1; IgG4, IgA, and IgE do not activate complement. The activated C1 consists of three different polypeptides: C1q, C1r, and C1s. Within the complex are two C1r and two C1s molecules bound to a single C1q molecule. C1q is often described as a tulip-like structure because it contains a "stalk" with amino acid sequences that are collagen-like and six globular structures that form a tulip blossom at one end of the stalk. A cartoon structure for the C1q, $C1r_2$, $C1s_2$ is shown in Figure 35-23. Both C1r and C1s are precursors to proteolytic enzymes. Activation of C1r occurs as a result of the binding of the IgG or IgM antigen complexes to C1q. The globular domains of C1q bind to a site in the Fc region of the immunoglobulin molecule. This is followed by proteolysis of a single peptide bond in the C1r molecule. The activated C1r subsequently catalyzes proteolysis of a single peptide bond in C1s, creating activated C1s. Association of the components of C1 requires the presence of Ca^{2+}. *In vitro*, a distinction between the classical, alternative, and lectin pathways can be made by blocking formation of C1 by chelating Ca^{2+} with EDTA.

The next step in the sequence of classical pathway activation is cleavage of components C4 and C2. Two products these cleavages, C4b and C2a, form the "classical pathway" C3 convertase. Proteolysis of C3 leads to a second convertase. The second convertase is the C5 convertase and is composed of C4b, C2a, and C3b. This convertase cleaves C5 to form C5b and C5a.

As in the alternative pathway, C5b, in conjunction with C6, C7, C8, and C9, forms the MAC.

The Lectin Pathway

The lectin pathway has been described *in vitro*, but its physiological significance remains unclear. This pathway is initiated by binding of a lectin, the mannose binding protein (MBP or serum collectin) to mannose or N-acetyl glucosamine residues on the surface of the invading foreign cells. MBP is structurally similar to C1 and, as a consequence of its binding to the sugar residues, activates the first proteinase precursor of the lectin pathway, the MBP-associated serine proteinase (MASP-1). MASP-1 is homologous to C1r and C1s and can proteolytically activate MASP-2, the second proteinase precursor of the lectin pathway that is similar to Cls. MASP-1 can also activate C2 and C4, not directly but with the participation of MASP-2.

Activation of C2 and C4 produces the C3 convertase and, via the same reactions sequences of the "classical pathway," leads to formation of the MAC.

Formation of C3b C4b, by any of the pathways, results in "tagging" of the foreign cells by covalent linking of these molecules to the foreign cell surfaces via thioester bonds. This tagging or opsonization marks the foreign cells for phagocytosis by macrophages and the other phagocytic cells of the immune system. Alternatively, destruction of foreign cells is caused by the MAC. Antigen-antibody complexes are cleared from the blood by binding to complement receptors on erythrocytes. The red cell–bound Ag-Ab complexes are removed along with the erythrocytes by the liver.

Complement deficiencies are associated with several diseases. Alternative pathway deficiencies are rare, but when they exist more than one-half of factor D or properdin-deficient individuals suffer from ***Neisseria*** infections of which 75% are fatal. Individuals with deficiencies in the MAC components, e.g., C5, C6, C7, and C8, are also susceptible to infection with *Neisseria*. Deficiencies in C1, C4, and C2 are associated with ***systemic lupus erythematosus*** and ***glomerulonephritis***. ***Hereditary angioedema,*** a disease characterized by recurrent submucosal and subcutaneous edema, is caused by a deficiency in C1 inhibitor. Complexes and interactions similar to those of the complement system are also characteristic of the clotting system (Chapter 36).

35.11 Cytokines

Cytokines are a group of cell-derived proteins or peptide molecules that transmit signals between cells of the immune system. The consequence of cytokine binding to receptors on the target cells is modulation of processes occurring in the target cell. Cytokines can stimulate, inhibit, upregulate, or downregulate depending on the particular cytokines and cell types involved. Cytokines are similar to hormones in that they act at sites (cell receptors) different from their sites of synthesis. There are many types of cells that synthesize the various cytokines. Many of the cytokines are *pleiotropic*, i.e., they exhibit multiple biological activities.

Simple classification of individual cytokines is difficult because the function of a single cytokine can be different depending on the target cell. Subcategories of cytokines are defined, e.g., interleukins generally are cytokines produced by leukocytes, lymphokines are produced by lymphocytes, and monokines are produced by monocytes. However, there are exceptions to these generalizations. Other subcategories exist, e.g., interferons are historically related to antiviral activities and growth factors are predominantly related to cellular growth.

Based on function, cytokines are divided into four categories. The first category contains the cytokines that mediate natural immunity (IFN-α, IFN-β, or type I interferons), IL-1 and IL-6, and the chemokines of which more than 40 have been identified. The second category includes the cytokines that regulate lymphocyte activation, growth, and differentiation. Included in this group are IL-2, IL-4, and TGF-β. The third category includes cytokines that regulate immune-mediated inflammation. These are IFN-γ, TNF lymphotoxin, IL-5, IL-10, IL-12, and MIF (migration inhibition factor). The fourth group includes cytokines that stimulate hematopoiesis, e.g., IL-3, IL-7, and the colony-stimulating factors, GM-CSF, M-CSF, and G-CSF.

Cytokines use a common signal transduction mechanism to initiate the intracellular processes by which they achieve their diverse biological effects. The biological effects are obtained through changes in the expression of specific genes. Cytokine binding to a membrane receptor results in the activation of a tyrosine kinase (a Janus kinase of "Jak"). This in turn leads to intracellular phosphorylation of the receptor and prepares the receptor for binding of a transcription factor. The transcription factors are members of a group of signal transducer-activators of transcription (Stat proteins). The specificity in cytokine signaling is determined by a combination of receptor recognition and specific recognition of particular Stat molecules by different Jaks.

An annotated list of some of the cytokines involved in immune system modulation is given in Table 35-4. The rapid development of cytokine biology, however, guarantees that such tables are likely to be incomplete the moment that they are created and, more importantly, inaccurate in some respects. Internet sites exist that update cytokine information, e.g., the COPE (Cytokines Online Pathfinder Encyclopedia) site at http://www.copewithcytokines.de and the commercial site, http://www.rnsystems.com. Links to tables of synonyms, activity conversion units, and other useful information are available from the latter site.

As protein molecules, cytokines can be classified most simply based on their dominant secondary structures. Four groups are defined on this basis. In the first group, the structure is dominated by four α-helix bundles that are arranged in pairs. The interferons and most interleukins (IL-2 through IL-7, IL-9, IL-11, IL-13, and IL-15) are members of this group. Interaction of these cytokines with their receptors occurs via one of the helix pairs, the A–B pair. An example, IFN-α2A, is shown in Figure 35-23A. The second group contains molecules with short a helix segments, β-sheet motifs, and epidermal growth factor (EGF)–like motifs. Included within this group are the chemokines and members of the insulin-like cytokine subfamily. The third group of cytokines is dominated by long β-sheet motifs. IL1-α, IL1-β, TNF-α and TNF-β are members of this group. The last group is structurally heterogeneous and contains domains similar to EGF-like structures, immunoglobulin domains, and kringles and in one, a proteinase-like region. (See figures in Chapter 36 for examples of these motifs.) Examples of three-dimensional structures representing group 1 (IFN-α2A and IL-4) and 3 are shown (Figure 35-24a–c).

The products of cytokine genes, like immunoglobulin genes, can be varied by alternative splicing. Consequently, investigation of cytokines and their actions is complicated by this source of structural variation. Also, most cytokines are glycosylated and secreted into the circulation. Cytokine receptors on cells may indicate the state of activation of the cytokine-stimulated cells. Investigation of cytokine receptors is aided by flow cytometry using fluorescence-labeled antibodies bound to the receptors. Cytokine research is one of the most active areas of immunological research.

35.12 Vaccines

Immunological memory, i.e., B cells and T cells that are poised for rapid antibody formation and antigen destruction, provides the molecular basis for the successful use of vaccines. In vaccination, the stimulation of B- and T-cell clones leads to formation of antibodies to the foreign proteins. Differentiation of some of the B cells that recognize the antigen into memory cells provides the mechanisms for resisting subsequent infections.

Weakened or killed organisms, or organisms with epitopes common to those of pathogens, can stimulate an

TABLE 35-4

Human Cytokines (Interleukins, Interferons, Other Cytokines)

Name	Mol Wt kDA	Sources	Effects on Cells
Interleukins			
IL-1		Activated macrophages	Activation of T cells and macrophages, general inflammatory effect, promotes synthesis of acute-phase proteins in hepatocytes, fever, increased IL-2 release, differentiation, proliferation of antigen-stimulated cells
α	18		
β	17		
IL-2	18	Helper CD4 T cells (Th1)	Activation of NK cells, binds to activated T cells, increased IL-2 receptors, increased IL-2 release, promotes proliferation, growth and differentiation leading to antibody secretion
IL-3	15	Activated T cells, mast cells	Growth of multi-potential stem cells, myeloid, erythroid, mast cells
IL-4	14.9	CD4 T cells (Th2 cells), mast cells, eosinophils, basophils	Increases MHC Class II expression on macrophages, promotes mast cell proliferation, promotes hematopoietic precursors, isotype switching in IgG1 and IgE, T cell proliferation, activation of B cells from resting state, increased MHC Class II expression, proliferation in presence of Ag, increased IL-2R receptor expression
IL-5	26	CD4 (Th2) cells, mast cells, eosinophils	Eosinophil differentiation and proliferation, production of IgM and IgA, increased IL-2 receptors, increased IgA secretion
IL-6	17	IL-1 stimulated cells, CD4 (Th2) cells, monocytes, macrophages	Synthesis of acute-phase proteins by hepatocytes, synergistic with IL-3 on multipotential stem cells, growth and differentiation of T cells, differentiation to antibody secretion
IL-7	17	Stromal cells, thymus	Regulates early T cell progenitors, regulates early B cell progenitors, CD4 and CD8 cells
IL-8	8	Monocytes, macrophages	Promotes chemotaxis and granule release by neutrophils, basophils, T cells
IL-9	14	T, B cells, macrophages	Mitogenic for myeloid cells, supports antigen-independent growth of helper T cells
IL-10	18.6	CD4 T cells, macrophages	Inhibits interferon secretion, induces thymocyte proliferation in presence of other IL's, inhibits Th1 cytokine production
IL-11	19.1	Stromal cells	Induction of synthesis of acute phase proteins, enhances stimulation of hematopoietic cell proliferation by IL-3 and T-cell development of B cells
IL-12	75	Macrophages, dendritic cells (B cells, rarely)	Activates NK cells, stimulates production of IFN-α by Th1 cells
IL-13	10	1 cells	Inhibits mononuclear phagocyte inflammation, promotes growth and differentiation, similar to IL-4

(continued)

TABLE 35-4
Human Cytokines (Interleukins, Interferons, Other Cytokines) *(continued)*

Name	Mol Wt kDA	Sources	Effects on Cells
Interferons			
Interferon	19.2	Virally infected cells	Participates in the innate immune response, induces resistance to viral RNA synthesis
α	20		
β	16.7	Antigen-stimulated cells,	
γ		NK cells in response to IL-1 or TNF	Expression of MHC I & II, T cell differentiation, synthesis of IgG2, antagonizes IL-4 action
Other Cytokines			
TNFα	17	Macrophages, NK cells, B cells, T cells, mast cells	Activates monocytes, PMN cells, promotes inflammation
TNFβ lymphotoxin	18.6	Th1 cells, B cells,	Cytotoxic, promotes inflammation, can inhibit T cell activation
TGFβ	5.5	T cells, B cells, macrophages, mast cells	Immunosuppressive
GM-CSF	14	T cells, macrophages, B cells, mast cells	Promotes growth of granulocytes and monocytes

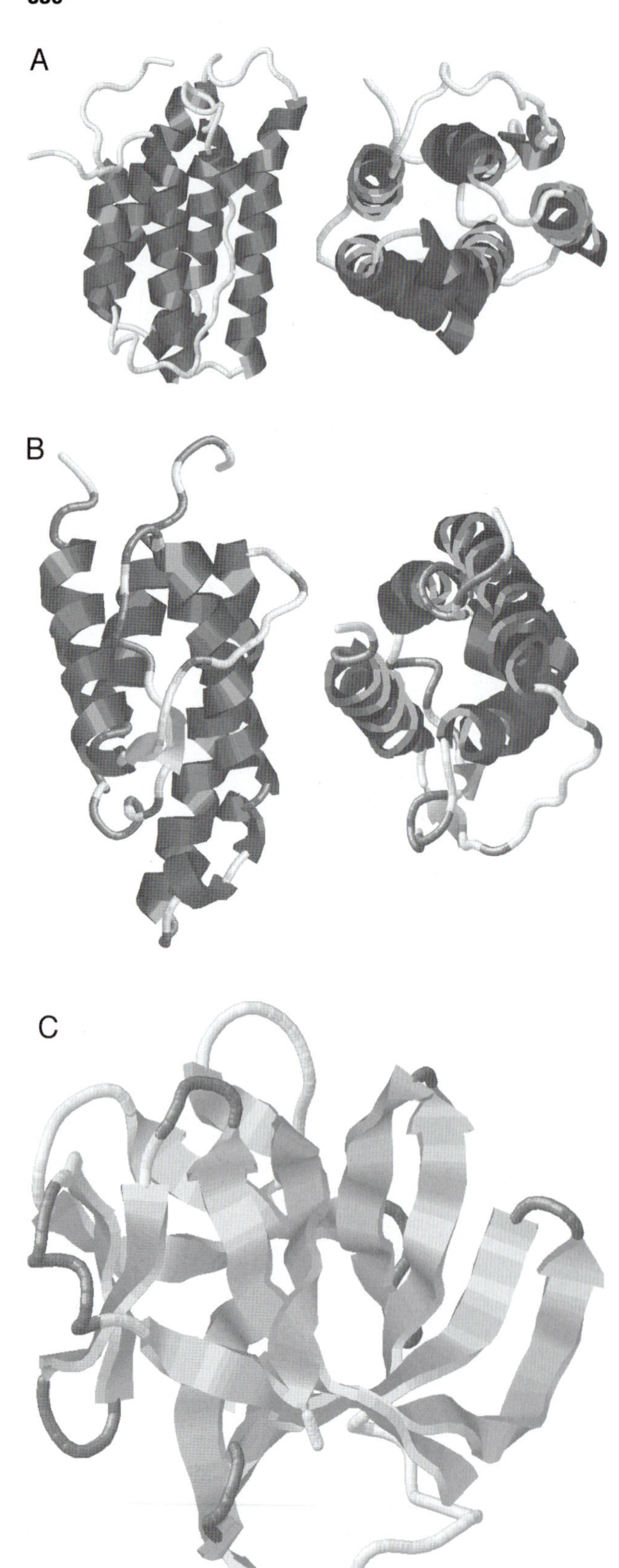

immune response without causing serious disease, e.g., cowpox virus stimulates formation of antibodies and memory cells that recognize epitopes common to cowpox and smallpox. Organisms grown under unfavorable conditions (attenuation) so as to become incapable of causing disease are also used as immunogens; measles, mumps, rubella, and polioviruses are examples of this approach to vaccine development. Vaccination must be credited with the tremendous success in preventing many communicable diseases, particularly those of childhood. Vaccination is not without risk, however. Contaminants introduced in the vaccine production, incompletely killed organisms, and virulence in immune-compromised individuals can produce serious illness and death in rare cases.

Vaccines against bacterial disease are commonly directed against the toxins produced by the bacteria. The toxins are purified and treated to prevent them from causing their debilitating effects on the host (toxoids). The toxoids retain their immunogenicity and thus their ability to cause antibody production, which provides immunity to the bacterial pathogen. A well known example is the diphtheria toxoid.

Proteins manufactured by recombinant DNA technology now enable production of immunogenic molecules that can be used to stimulate immune response to known pathogens. These immunogens can be modified so that they cannot cause disease. Chemically synthesized immunogens that present an epitope or epitopes of a pathogen are also used in the development of vaccines. As the knowledge of the chemical structures of pathogenic molecules increases, the ability to select suitable structures for vaccines becomes increasingly practical. Vaccines prepared from such molecules, e.g., a hepatitis B vaccine, permits immunization without risk of hepatitis to the recipient.

DNA vaccines may provide another safe method for vaccination. Instead of introducing a protein antigen into a subject, DNA containing a gene for a particular antigen is introduced into cells. The antigen is synthesized intracellularly and evokes an immune response (also discussed in Chapter 23).

Prevention of disease caused by infectious agents through vaccine development offers the best hope for the elimination of several of the most challenging contemporary public health crises. The continuing technical

FIGURE 35-24
(Also see color figure.) Secondary structural features of cytokines. (A) IFN-α2A (shown with both end and side views of the 5 α helices). (B) IL-4, showing the α helices from both end and side views. This group of cytokines is characterized by the 4–6 distinct α helices that form the protein "backbone." (C) IL-1 showing the β-sheet structure that is characteristic of this group of cytokines.

advances in molecular biology and in recombinant protein production methods can ultimately be expected to lead to the elimination of AIDS and hepatitis as well as other viral disease. Immune system recognition of transformed (cancerous) cells and discrimination between these and normal cells makes it possible to envision immunological targeting of transformed cells using humanized monoclonal antibodies and agents to selectively kill the cancer cells. Even more existing is the possibility for the development of vaccines that promote formation of antibodies to cancer cells and selective destruction of them by the immunological mechanisms described in this chapter. As our knowledge of immune response grows, the opportunities to prevent disease expand. Engineering molecules that participate in this system also provides opportunities for new therapeutic drugs to combat the disease that have not been prevented. Optimism regarding new immunotherapies is justified.

Acknowledgments

The images shown in this chapter were created using RasMol 2.6, Molecular Graphics Visualisation Tool by Roger Sayle, BioMolecular Structures Group, Glaxo Research & Development, Greenford, Middlesex, UK. The coordinates were obtained from the Protein Data Bank at Brookhaven National Laboratory as described in F. C. Bernstein, T. F. Koetzle, G. J. B. Williams, E. F. Meyer, Jr., M. D. Brice, J. R. Rodgers, O. Kennard, T. Shimanouchi, and M. Tasumi: "The Protein Data Bank: A Computer-Based Archival File for Macromolecular Structures." *Mol. Biol.* **112,** 535–542 (1977).

Supplemental Readings and References

P. J. Delves and I. M. Roitt: The immune system, Part I. *New England Journal of Medicine* **343,** 37 (2000).

P. J. Delves and I. M. Roitt: The immune system, Part II. *New England Journal of Medicine* **343,** 108 (2000).

J. R. Gruen and S. M. Weissman: Evolving view of the major histocompatibility complex. *Blood* **90,** 4252 (1997).

J. Kuby: *Immunology*, W. H. Freeman, San Francisco, 1997, p. 664.

P-H. Lambert and C-A. Siegrist: Vaccines and vaccination. *British Medical Journal* **315,** 1595 (1997).

A. D. Luster: Chemokines—chemotactic cytokines that mediate inflammation. *New England Journal of Medicine* **338,** 436 (1998).

J. J. Oppenheim and M. Feldman: *Cytokine Reference.* Academic Press, San Diego, 2000, p. 2000.

C. P. Price and D. J. Newman (eds.): *Principles and Practice of Immunoassay*, 2nd Ed. Stockton Press, New York, 1997, p. 667.

I. M. Roitt: *Essential Immunology,* Blackwell Scientific Publications, Oxford, 1997, p. 448.

B. J. Rollins: Chemokines. *Blood* **90,** 909 (1997).

P. Saikumar, Z. Dong, V. Mikhailov, et al.: Apoptosis: Definition, mechanisms, and relevance to disease. *American Journal of Medicine* **107,** 489 (1999).

M. Santos-Rosa, J. Bienvenu, and J. Whicher: Cytokines in *Tietz Textbook of Clinical Chemistry*, 3rd ed. C. A. Burtis and E. R. Ashwood (Eds); W.B. Saunders, Philadelphia, 1999, p. 541.

A. C. Ward, I. Touw, and A. Youshimura: The Jak-Stat pathway in normal and perturbed hematopoiesis. *Blood* **95,** 19 (2000).

K. Whaley and W. Schwaeble: Complement and complement deficiencies. *Seminars in Liver Disease* **17,** 297 (1997).

CHAPTER 36

Biochemistry of Hemostasis

Hemostasis, or hemostatic response, is the spontaneous arrest of blood loss from ruptured vessels. In human and other vertebrate animals, hemostatic response involves three distinct groups of components. Initially, in what is generally described as ***primary hemostasis,*** blood vessels contract because of the injury and platelets adhere to the subendothelium. The collagen fibers of smooth muscle cells and the glycosaminoglycans that form the extracellular matrix beneath the normal endothelium of the blood vessel become the site of the primary hemostatic response. Secondly, the platelets form a hemostatic plug that prevents extravasation without occluding the blood vessel. If vessel occlusion were to occur, it would prevent the delivery of essential nutrients to the adjacent tissues. Finally, the hemostatic response includes the actions of the blood coagulation system, a system that consists of nearly 20 plasma proteins. The sequence of reactions among these plasma proteins, collectively described as the blood clotting process, produces the enzyme thrombin. Thrombin then transforms circulating fibrinogen into the fibrin meshwork that mechanically reinforces the hemostatic plug. The reactions of the coagulation system are frequently designated as constituting ***secondary hemostasis.*** The characteristics of the hemostatic system that enable it to meet the requirements for normal hemostasis are as follows:

1. It responds rapidly to the injury.
2. It is localized to the site of injury.
3. It is spatially constrained so as not to occlude the ruptured blood vessel.
4. It is mechanically resistant to disruption by adjacent blood flow.
5. It creates a temporary structure.

The coagulation system consists of three subsystems. The "procoagulant subsystem" provides the rapid, localized response to the injury and, because of the enmeshing of the platelets by fibrin, a hemostatic plug that is spatially constrained and mechanically stable. The "anticoagulant subsystem" modulates two of the key reactions of the procoagulant system by inactivating the cofactor proteins that are components in these reactions. The "anticoagulant subsystem" thus also acts to ensure localized, spatially constrained response. The "fibrinolytic subsystem," by proteolytic digestion of the fibrin that reinforces the platelet (hemostatic) plug, is responsible for the temporary nature of the hemostatic plug. Digestion of the fibrin occurs after tissue repair has commenced and the risk of hemorrhage has been eliminated.

Primary and even tertiary structures are now known for many of the hemostatic system proteins. Gene structures have been identified for all of the known components of the hemostatic system. Databases listing mutations and their related hemorrhagic or thrombotic risks are growing rapidly. For many reactions, the molecular mechanisms responsible for the properties of the hemostatic

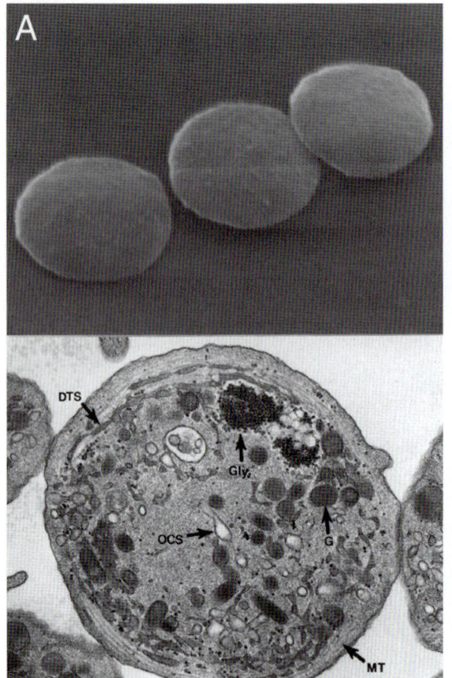

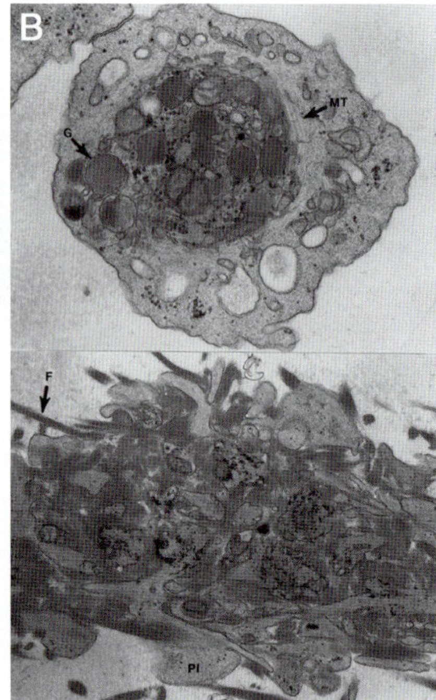

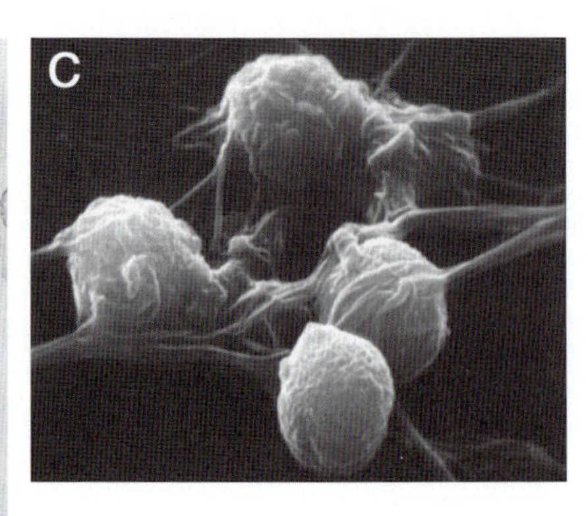

FIGURE 36-1
Platelet microanatomy. (A) (Upper) Normal, discoid resting (unstimulated) platelets. (Lower) Resting platelet cross-section showing the dense tubular system (DTS), glycogen storage granules (Gly), alpha granules (G), open cannicular system (OCS, "pores" that penetrate into the discoid platelet) and microtubules (MT). (B) (Upper) Platelet from sample activated by a low dose of thrombin. The cell has lost its discoid form and development internal transformation. Alpha granules (G) are mainly concentrated in the central area encircled by the coil of microtubules (MT) and microfilaments (not visible). Magnification is 33,000 times. Upon activation, platelets undergo dramatic changes; spreading on the surface of the exposed subendothelium, contracting via microtubule action to gather granules toward the center and releasing microparticles from the platelet membrane. (Lower) Platelets from a clot prepared for study in the electron microscope while under isometric tension. Fibrin (F) strands and transformed platelets (P) are closely associated like muscle cells and tendons in order to develop tension. Magnification is 15,000 times. Photograph is a gift from James G. White, MD, Regents' Professor, University of Minnesota. (C) Activated platelets are enmeshed by fibrin strands. Fibrin, shown in cross-section in the activated platelet photograph (C), is shown here as it consolidates the platelets into a stabilized hemostatic plug. Photograph was the gift of the late Marion I. Barnhart, Wayne State University, Detroit, Michigan.

response under physiological conditions have been identified. The underlying premise of this chapter is that understanding the chemical reactions of hemostasis and their relationships to the structures of the proteins is the basis for understanding normal hemostasis, hemorrhagic and thromboembolic diseases, and their relationships to the underlying genetic constitution of the individual.[1]

[1]The nomenclature for the hemostatic system components reflects the long history of the study of blood clotting. The initial identification of coagulation factors comes from patients suffering from hemorrhagic diseases. The designation of putative causes of disease as "factors" occurred long before any information about the molecular structures and functions of the components was discovered. The Roman numeral designations for the coagulation factors resulted from the action of a committee that was charged with providing single designations for factors that were clearly functionally the same but described by several different names. Because adoption of any of the prior names might imply priority of discovery, the numerals were assigned as a compromise. Albeit imperfect, the assignment of single designations for the coagulation factors was a milestone in advancement of our understanding of hemostasis.

36.1 Primary Hemostasis

Primary hemostasis is achieved rapidly as the result of vasoconstriction of the ruptured blood vessels and the adhesion of blood platelets to the exposed subendothelial surfaces. Recognition of subendothelial and basal membrane surface components by the platelets occurs through specific receptors on the external surface of the platelet membrane. Collagen, present in the subendothelial space, binds to the platelet receptor GpIa/IIa. The plasma proteins, von Willebrand factor, fibrinogen, and vitronectin also provide bridging molecules between the exposed endothelial surface and the platelet. Von Willebrand factor binds to the platelet via GpI; fibrinogen binds via the receptor GpIIb/IIIa. Another platelet protein, glycocalicin, binds the platelets via the receptor GpIb/IX. At virtually the same time that adhesion of the platelets to the newly exposed surface is occurring, the "procoagulant subsystem" (see below) is providing thrombin to activate the platelets. Thrombin activation of platelets occurs through a unique receptor mechanism that requires proteolytic action by thrombin followed by transmembrane signaling via a G-protein-coupled mechanism. Platelet activation causes the release of the contents of the granules present inside the platelet. The platelet-dense granules release substances that promote further platelet aggregation; the alpha granules release several coagulation proteins that supplement those present in the plasma.

Several hemorrhagic disease states are identified from defects in platelet membrane proteins. *Bernard-Soulier syndrome* is caused by a defect(s) in GpIb, causing impaired binding of von Willebrand factor in the plasma. The platelet receptor for fibrinogen and fibrin, GpIIb/IIIa, is defective in *Glanzman's thrombasthenia*. Figure 36-1C is a photomicrograph of fibrin-enmeshed activated platelets formed *in vitro*. This illustration is comparable to what is seen in a normal hemostatic plug. A scanning electron photomicrograph of resting platelets is shown in Figure 36-1A (top), and a transmission electron photomicrograph showing internal structures and fibrin strands in association with the platelet contractile proteins is shown in Figure 36-1A (bottom) and in Figure 36-1B. Figure 36-1C is a photomicrograph of fibrin-enmeshed activated platelets formed *in vitro*. This illustration is similar to what is seen in a normal hemostatic plug. In blood, a clot contains erythrocytes as well as platelets and fibrin. This type of clot is similar to that which forms outside the body from a skin-breaking injury and is similar to "red thrombus" that forms intravenously. Upon drying the external clot becomes a scab. An occlusive clot or venous thrombus altogether too frequently breaks up and is carried to the lung to create a pulmonary embolus.

36.2 Secondary Hemostasis

Secondary hemostasis refers to the reactions of the blood clotting factors that are circulating in the plasma *that do not require platelets*. It is this process that culminates in the transformation of blood from a readily flowing fluid to a gel, or, molecularly, the transformation of the soluble protein fibrinogen into the self-assembling, insoluble polymer, fibrin. The sequence of reactions that makeup the coagulation system (procoagulant reactions) was described as a "cascade" or "waterfall" in 1964. This description, an ordered sequence of transformations of precursor molecules from their inactive forms to their catalytically active forms, provided a context in which the hemorrhagic deficiencies that had been primarily identified from patients with bleeding disorders could be understood. This

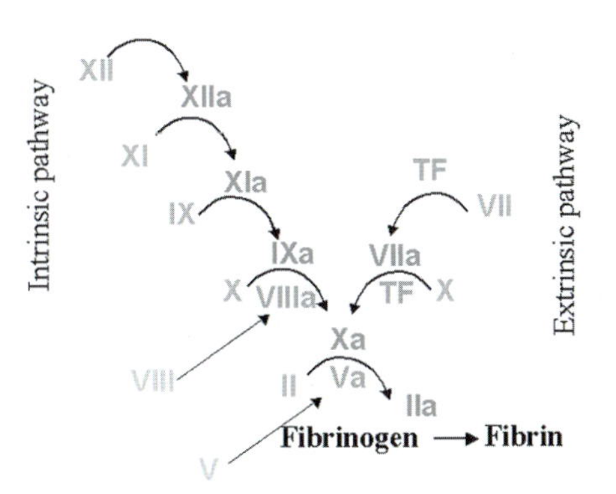

FIGURE 36-2
(Also see color figure.) Classical "cascade" or "waterfall" model for coagulation. This model represents the ordered sequence of *in vitro* transformations of precursor molecules from their inactive forms to their catalytically active forms. The publication of the "cascade" model for the coagulation system provided a milestone in rationalizing a confusing biological system. The figure has been corrected to place factors V and VIII outside the linear sequence that differs from the proposals made in 1964. Two components, TF (tissue factor) and factor VII, were designated as the "extrinsic pathway" components because the TF is provided by thromboplastin, a tissue-derived material extrinsic to the circulating blood. Four components—factors XII, XI, IX, and VIII—were designated as "intrinsic pathway" components because they are always present in blood. Two other plasma proteins—prekallikrein and high-molecular-weight kininogen—were discovered later to be involved in the process by which factor XI becomes activated.

conceptualization of secondary hemostasis (clotting cascade) is shown in Figure 36-2.

The "classical" clotting cascade depiction of the procoagulant subsystem indicates the order in which the various factors participate in the process and remains a convenient mnemonic device. However, this model does not represent additional properties of the procoagulant subsystem discovered since its presentation. Our current understanding of the procoagulant process is better represented by Figure 36-3. The occurrence of the reactions on the surface of membrane lipids, either from cells damaged by the injury that initiates hemostasis or from exogenous sources in laboratory tests, is now generally recognized. Figure 36-3 explicitly acknowledges formation of activation complexes on the cell surfaces and, by implication, the tremendous increases in the rates of reactions that occur in, and only in, the activation complexes. The names and descriptions of the components of the procoagulant, anticoagulant and fibrinolytic subsystems are given in Table 36-1.

Clot Dissolution—Fibrinolysis

The fibrinolytic subsystem, by proteolytic digestion of fibrin, eliminates the fibrin from the hemostatic plug. Similar to the other subsystems, the fibrinolytic subsystem

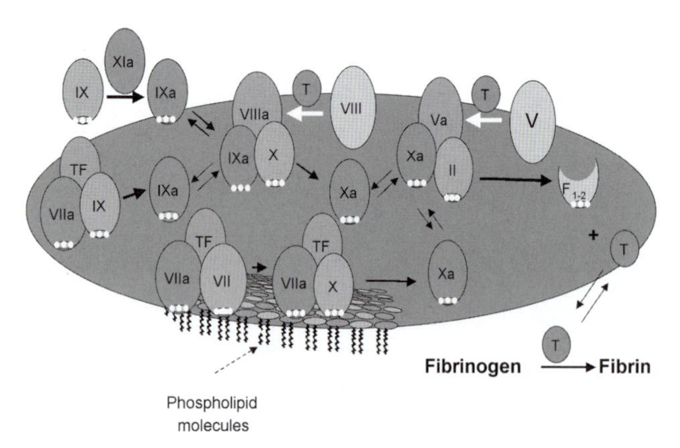

FIGURE 36-3
(Also see color figure.) Contemporary model for the procoagulant subsystem. This model recognizes that the transformations of proteinase precursors to active proteinases are organized on the surfaces to which the proteins bind. The binding of the freely circulating coagulation proteins to these surfaces localizes the reactions that makeup the hemostatic response, as well as increases the rates of the reactions. Unlike the other coagulation factors, tissue factor does not circulate freely. It is an integral membrane protein that is exposed as a result of exposure of the subendothelium. In the color version of this figure, each stage of the sequence a proteinase precursor (shown in orange), a proteinase from the preceding complex (shown in dark green), and an activated cofactor protein (shown in yellow-green) form a noncovalent complex on the membrane surface (shown as is distinguished a blue "island"). Dissociation of the proteinase formed in one complex (the product proteinase) permits that proteinase to diffuse to the next complex and catalyze the transformation of a different precursor protein into an active proteinase. Thrombin is the only proteinase that does not bind to the surface of the membrane; it dissociates and diffuses so that it can convert the soluble fibrinogen into the insoluble fibrin clot. Vitamin K-dependent proteins are prothrombin and factors VII, IX, and X. Two vitamin K-dependent proteins are not shown, protein C and protein S. The *unactivated* cofactor proteins, tissue factor, and factors V and VIII are shown in yellow.

components become functionally active as the result of an initiating event, the formation of the fibrin itself. Through association of the plasminogen, and plasminogen activators with the fibrin, the processes responsible for the formation of plasmin are localized to the fibrin. Fibrinolysis is responsible for the temporary *nature* of the fibrin clot.

Thrombosis: A Dark Side of Hemostatic System Function

Formation of a hemostatic plug or a red thrombosis (fibrin and entrapped red blood cells) at a site at which there is no injury produces occlusion. Death to the surrounding tissues can occur (infarction), or pieces of the thrombus may be "torn" away and moved to the lungs, producing pulmonary embolism. Such dysfunction of the hemostatic system is generally not attributable to a single cause. For example, heterozygous individuals with half of the normal amount of protein C do not necessarily present with thrombus. Rather, a combination of defects in more than one component appears to be required. Each defect can increase the risk of thrombosis at some time during an individual's lifetime. The cumulative increased risk reduces the age at which a thrombotic event may occur. Deficiencies or defects in the inhibitors of the procoagulant proteinases, an increased concentration of fibrinogen, antibodies to lipid-bound proteins, and hyperhomocysteinemia are examples of situations that create an increased risk of thrombosis. Hyperhomocysteinemia results from a decrease in methylene tetrahydrofolate reductase activity. This may be caused by a defect in this enzyme, by other enzymes in the pathway, and/or by a decrease in vitamin B_{12} (Chapter 17).

Estrogen therapy has been linked to changes in the concentrations of several components of the hemostatic system. However, there are no changes in components of the procoagulant, anticoagulant, or fibrinolytic system that can be clearly linked to an increase in risk of thromboembolic disease sometimes associated with estrogen therapy.

36.3 Functional Properties and Structures of the Hemostatic System Factors (Proteins)

Although there are approximately 20 plasma proteins or plasma clotting factors that participate in the reactions of blood clotting, these proteins can be divided into four categories based on their principal functions. Grouping in this way simplifies and facilitates our understanding of how and what these proteins do in the clotting process. The four categories are as follows:

1. Proteinase precursors,
2. Cofactor proteins,
3. Proteinase inhibitors, and
4. Other proteins not included in the first three categories. This fourth group contains one enzyme, a transglutaminase, one adhesion and carrier protein, and fibrinogen, the structural protein of the fibrin clot.

Proteinase Precursors

Proteinase precursors, also called zymogens or proenzymes, become catalytically active enzymes upon specific proteolytic cleavage. The precursor molecules circulate as inactive forms that do not exhibit enzymatic activity. The activation processes are irreversible. Extensive structural similarities are found among all proteinase precursors. The regions of the molecules that express proteinase activity after activation are found in the C-terminal one-half to one-third of each molecule. The portion of each precursor

TABLE 36-1

Components of the Hemostatic System

Common Name[1]	Factor Number[2]	Associated Disease(s) Hemorrhagic/ Thrombotic Risk	Disease Prevalence	Concentration in Plasma[3] mg/L	Concentration in Plasma[3] μM	Tissue Source	Half-life (hr)	PDB Codes for Proteins[4]	SwissProt Designations[4]
Plasma Coagulation Factors – Procoagulant Subsystem Components									
Fibrinogen	I	Afibrinogenemia, dysfibrinogenemias, increased thrombotic risk with elevated fibrinogen concentration	Uncommon	~2,500	~7.5	Liver	72–96	3FIB, 1FIC 1FID, 1FZA, 1FZD	FIBA_HUMAN FIBB_HUMAN FIBG_HUMAN
Prothrombin	II[5]	Dysprothrombinemias, defective thrombin function, increased risk with some polymorphisms	Extremely rare	100	1.5	Liver	~72	2PF1, 2PF2, 2HPP, 2HPQ, 1AOH, 1PPB	THRB_HUMAN
Thromboplastin	III[6]	Not applicable, (Exogenous material containing both tissue factor and phospholipids)	NA	NA		Homogenate of brain or lung in physiological NaCl ($\pm CaCl_2$)		NA	NA
Tissue Factor (TF)	III[6]	—	—	Membrane protein				1BOY, 1DAN, 2HFT	TF_HUMAN
Ca^{2+}	IV	—	—	3.3 mM (total) 1.5 mM (free)		—		Not applicable	NA
Proaccelerin Accelerator Globulin (Ac-G)	**V**	Factor V deficiency, (parahemophilia) Increased risk for thrombosis (Factor V Leiden)	Extremely rare	7–10	0.025	Liver, megakaryocytes, endothelial cells	11–15		FA5_HUMAN
Abandoned[7]	VI	F-VI is the activated form of F-V	—	—		—			NA
Proconvertin, Plasma thromboplastin component, (PTC)	VII	Factor VII deficiency, increased risk of thrombosis with elevated F-VII concentration	Rare; 2 per 1,000,000	0.5	0.01	Liver	4–5	1DAN	FA7_HUMAN
Antihemophilic factor (AHF), antihemophilic globulin (AHG)	**VIII**	Hemophilia A, "classical hemophilia"	100–200 per 1,000,000	0.001–0.002		Liver	8–12	1CFG, 1FAC	FA8_HUMAN
Christmas factor, antihemophilic factor B	IX	Hemophilia B, Christmas disease	20–30 per 1,000,000	5	0.1	Liver	12–24	1CFH, 1CFI, 1EDM, 1IXA, 1PFX	FA9_HUMAN

(continued)

TABLE 36-1
Components of the Hemostatic System (continued)

Common Name[1]	Factor Number[2]	Associated Disease(s) Hemorrhagic/ Thrombotic Risk	Disease Prevalence	Concentration in Plasma[3]		Tissue Source/Half-life (hr)		PDB Codes for Proteins[4]	SwissProt Designations[4]
				mg/L	μM				
Plasma Coagulation Factors – Procoagulant Subsystem Components (continued)									
Stuart-Prower factor (Autoprothrombin III)	X	Factor X deficiency	Very rare	8	0.15	Liver	24–50	1APO, 1CCF, 1FAX, 1HCG, 1KIG, 1WHE, 1WHF	FA10_HUMAN
Plasma thromboplastin antecedent (PTA)	**XI**	Factor XI deficiency	Rare, primarily in Ashkenazi Jews	4	0.025	Liver	40–80		FA11_HUMAN
Hageman factor	**XII**	None observed	NA	30	0.4	Liver	50–70		FA12_HUMAN
Fibrin stabilizing factor, Plasma transglutaminase Laki-Lorand factor, fibrinoligase	**XIII**	Spontaneous abortion, poor wound healing, excessive scar tissue formation	Very rare	30 11 (a subunit) 21 (b subunit)		Liver	100–200	1FIE, 1F13, 1GGT	F13A_HUMAN F13B_HUMAN
Prekallikrein Fletcher factor	None	None observed	—	50	0.6	Liver	35	1HIA, 2KAI	KAL_HUMAN
High Mol Wt Kininogen Fitzgerald factor, Flaujac factor, Williams factor	None	None observed	—	70	0.7	Liver	150		KNH_HUMAN
von Willebrand factor	None	von Willebrand's disease, many forms	1–125 per 1,000,000	5–10[8]		Endothelial cells, megakaryocytes/ ~12 (multimer) ~2 (propeptide)		1AO3, 1ATZ, 1AUQ, 1OAK	VWF_HUMAN

(continued)

TABLE 36-1

Components of the Hemostatic System (*continued*)

Common Name[1]	Associated Disease(s) Hemorrhagic/ Thrombotic Risk	Disease Prevalence	Concentration in Plasma[3] mg/L	μM	Tissue Source	Half-life (hr)	PDB Codes for Proteins[4]	SwissProt Designations[4]
Anticoagulant Subsystem Components								
Cofactor Protein Inactivation								
Protein C	Protein C deficiency, purpura fulminans	Unknown (~2–5% in patients with thromboembolic disease)	3–5	0.06	Liver		1PCU, 2PCT	PRTC_HUMAN
Protein S	Protein S deficiency	Unknown (~2–5% in patients with thromboembolic disease)	20–25	0.25	Liver, megakaryocytes, endothelial cells			PRTS_HUMAN
Thrombomodulin			NA		Endothelium		1ADX, 1EGT, 1FGD, 1FGE, 1TMR, 1ZAQ, 2ADX	TRBM_HUMAN
Thrombin	None							THRB_HUMAN
Protein C Inhibitor (PAI-3)			~4	0.07				IPSP_HUMAN
Proteinase Inactivation								
Antithrombin (antithrombin III)	Deficiency–homozygous–only with defective heparin binding Heterozygous, defective proteinase binding (low risk), defective heparin binding (any risk?)	1 per 5,000 to 1 per 350 (heparin-dependent)	125 (90% α-form, 10% β-form)	2.3	Liver	61–92	1ANT, 1ATH, 1ATT, 1BR8,	ANT3_HUMAN
Heparin Cofactor II (Leuserpin)	No clear link to risk for thrombosis	1 per 100,000 heterozygous	80	1.2	Liver	60		HEP2_HUMAN
Tissue Factor Pathway inhibitor (TFPI, LACI, EPI)	Decreased in massive DIC, no change in moderate DIC	Unknown	0.1	0.0025	Endothelium		1ADZ, 1TFX	TFPI_HUMAN
α-1 Proteinase Inhibitor (**α-1PI**) α-1 antitrypsin	None linked directly to hemostasis, primarily linked to emphysema	—	~1,300	23.5	Liver		1ATU, 1KCT, 2PSI	A1AT_HUMAN
α-2 Macroglobulin			2,500	3.5	Liver			A2MG_HUMAN

(*continued*)

TABLE 36-1
Components of the Hemostatic System (continued)

Common Name [1]	Associated Disease(s) Hemorrhagic/ Thrombotic Risk	Disease Prevalence	Concentration in Plasma [3] mg/L	Concentration in Plasma [3] μM	Tissue Source/Half-life (hr)	PDB Codes for Proteins [4]	SwissProt Designations [4]
Fibrinolytic System Components							
Proteinase Precursors	*Malignant tumors have high concentrations of fibrinolytic system components*						
Plasminogen			200	1.5–2	Liver 53	1CEA, 1CEB, 1HPJ, 1HPK, 1HTN, 5HPG	PLMN_HUMAN
Tissue plasminogen activator (**t-PA**)				80×10^{-6}	Endothelial cells ~ 5 *min*	1PK2, 1PK4, 1PKR, 1PMK 1PML, 1RTF, 1TPG, 1TPK, 1TPM, TPN, 2PK4	UROT_HUMAN
Urinary plasminogen activator (**scu-PA**)			0.04	150×10^{-6}	~7 *min*	1KDU, 1LMW, 1URK	UROK_HUMAN
Proteinase Inhibitors							
α-2 Antiplasmin Plasma Platelets (α-2 AP) α-2 plasmin inhibitor	Bleeding, usually hours after hemostasis (re-bleeding) due to premature digestion of the fibrin in the hemostatic plug	Extremely rare	70	1 0.09	Liver 62		A2AP_HUMAN
α-2 Macroglobulin (α-2 M)			2,500	3.5	Liver		A2MG_HUMAN
PAI-1 Plasma Platelets	Elevated in disease (acute phase reactant), elevation related to DVT, deficiency results in bleeding. Rises during pregnancy in 3rd trimester), *elevated* in pre-eclampsia	Septicemia –elevated PAI-1	0.02	0.0004 [9] 1.4	Endothelium, Liver ~7 min		PAI1_HUMAN

(continued)

TABLE 36-1

Components of the Hemostatic System *(continued)*

Common Name [1]	Associated Disease(s) Hemorrhagic/ Thrombotic Risk	Disease Prevalence	Concentration in Plasma [3] mg/L	μM	Tissue Source	Half-life (hr)	PDB Codes for Proteins [4]	SwissProt Designations [4]
Proteinase Inhibitors *(continued)*								
PAI-2	Rises during pregnancy in 2nd trimester), *decreased* in pre-eclampsia		<0.005	$<4 \times 10^{-3}$	Placenta	24		PAI2_HUMAN
PAI-3 (Protein C Inhibitor)	Decreases during pregnancy		~4	0.07			1PAI, 2PAI	IPSP_HUMAN

[1] Common names may be of primarily historical significance. The factor nomenclature was devised to reduce the number of synonyms for the same component which had made description of the process of coagulation unnecessarily awkward.

[2] Plasma coagulation factors are all designated by Roman numerals, platelet factors by Arabic numerals.

[3] Concentration is given for the common situation in which 9 volumes of blood collected are collected into one volume of anticoagulant solution. The anticoagulant dilutes the plasma, but does not affect the volume of the red blood cells. As a consequence of this, concentrations are dependent on the hematocrit, a consideration of clinical importance in polycythemic patients.

[4] Protein structure and gene databases can be found at many locations on the World Wide Web. A few are: http://www.ncbi.nlm.nih.gov/, http://www.gdb.org/, http://www.hgmp.mrc.ac.uk/, http://www.ebi.ac.uk/ebi_home.html, http://www.expasy.ch, http://www.biochem.ucl.ac.uk/bsm/pdbsum/, http://www.rcsb.org/.

[5] Color indicates that the factor is a vitamin K-dependent protein.

[6] Factor III is used as a synonym for thromboplastin, the mixture of tissue factor and phospholipid.

[7] Discovered to be the the activated form of Factor V after the numeral had been assigned.

[8] The molar concentration of vWF is difficult to express because of the many different polymeric forms that vary in mol wt from 500,000 to 20×10^6.

[9] Assumes a platelet volume of 1×10^{-14} L.

molecule found at the N-terminal end provides for recognition of cofactor proteins and membrane lipid surfaces.

Proteinase Domain Structures

The amino acid sequences of the proteinase domains are homologous to the pancreatic serine proteinases chymotrypsin and trypsin.[2] Primary specificity, i.e., recognition of the amino acid at which the peptide bond is hydrolyzed, is trypsin-like. All of the coagulation proteinases cleave peptide bonds at arginyl residues; however, plasmin shows a preference for lysyl residues. Peptide bond hydrolysis occurs in the proteinase active sites according to the mechanism established for chymotrypsin and other serine proteinases. The amino acid residues in the active site that are responsible for the hydrolytic reaction are Asp^{102}, His^{57} and Ser^{195}, respectively. These three residues form the charge relay system that facilitates the attack by water on the intermediate enzyme-substrate complex and the splitting of the peptide bond. Within the active site, His^{57}, acts as a general base in the hydrolytic process. Asp^{102}, the other residue in the charge relay with His^{57}, acts as an acid. The Ser residue forms an acyl enzyme (strictly a tetrahedral intermediate) with the arginyl peptide bond that is being cleaved. The attack of water on the acyl enzyme completes the hydrolysis. Specificity for basic amino acids is conferred by Asp^{189} (not shown), which interacts electrostatically with the guanidine group of the Arg side chain or amino group of the Lys side chain of the protein substrate.

In our "low-resolution" picture, the proteinase domains differ most obviously from the pancreatic proteinases in that they contain inserted sequences. These insertions are found primarily on the surfaces of the proteinase domains and are responsible for the high specificity that the coagulation proteinases have for their protein substrates, all of which are cleaved at Arg (or Lys) residues. Conventional secondary structure cartoons that illustrate the differences between trypsin and thrombin are shown in Figure 36-4.

Amino Terminal Domain Structures and Structural Motifs

The N-terminal domains of the hemostatic proteinase precursors and proteinases are constructed of several protein structural motifs (Figure 36-5). These motifs, in various combinations and by placement in different positions,

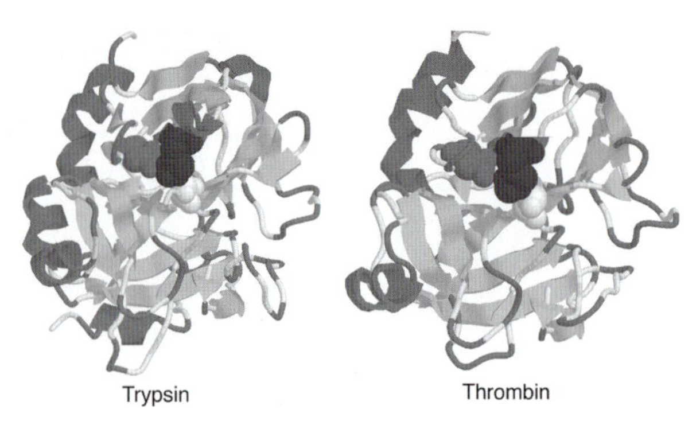

FIGURE 36-4

(Also see color figure.) Proteinase domains of serine proteinases trypsin and thrombin. In the color version of this figure, the active site *amino acid residues (shown as space-filling structures)* are Ser^{195} (shown in white); His^{57} (shown in blue), and Asp^{102} (shown in red). The *specificity* pocket residue is Asp^{189}. Both trypsin, the reference serine proteinase, and thrombin are shown as cartoons with the secondary structure elements marked as defined in the coordinate files for the structures from the Protein Data Bank. The color scheme is that of R. Sayle, the creator of RasMol, the program used to produce these ribbon diagrams. Some of the functionally significant differences in thrombin are the residues in the helix above the catalytic triad and two ligand binding regions, exosites I and II (not shown).

provide the building blocks of the non-proteinase-forming regions. With limited changes in their individual amino acid sequences, the motifs in the different proenzymes provide sites that are responsible for the formation of the particular activation complexes.

The four common motifs found within the amino terminal regions of the proteinase precursors and proteinases are "kringles," epidermal growth factor (EGF)–like motifs, fibronectin motifs, and "apple" motifs (Figure 36-5). The kringle is so named because in two dimensions, i.e., on paper prior to determination of its three-dimensional structure, it has the shape of a Danish pretzel. Similarly, apple motifs resemble drawings of apples.

Although the many components of the hemostatic system inevitably create a high level of complexity, the functional and structural similarities between and among the different molecules permit the simplification described above. Simplification can be obtained by representing many of the molecules as ellipses, the two-dimensional representations of the generally ellipsoid shapes of the molecules in solution (see Figure 36-9 below). However, the common motifs and similar proteinase domain structures can be presented even more simply in the form of bar diagrams, which are usually used in the descriptions of reactions that follow.

Cofactor Proteins

Cofactor proteins bind both the proteinase (enzyme) and the proenzyme (substrate). They bind to lipid membranes,

[2]Because of the initial identification of these residues in chymotrypsin, their numbering follows that of chymotrypsin. Actual numbers for amino acid residues are given in the figure legends or in the figures themselves.

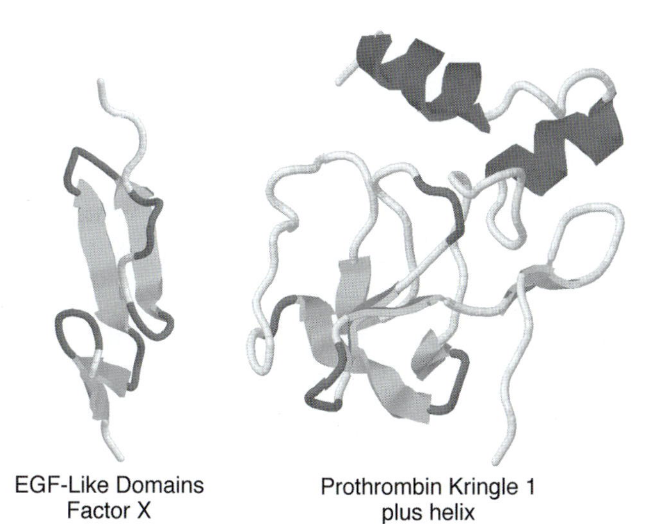

FIGURE 36-5
(Also see color figure.) Motif structures within coagulation proteins. Common motifs are found in the amino terminal regions of the proteinase precursor molecules. Shown are the kringle motifs and EGF-like motifs found in the vitamin K-dependent proteins and in plasminogen. Fibronectin (types I and II) motifs and "apple" motifs (named from their two-dimensional representations) are also present but not shown. Some epidermal growth factor–like domains contain β-hydroxylated Asp residues. The cartoon structures for the motifs are derived from three-dimensional structures determined by x-ray crystallography or by two-dimensional NMR spectroscopy.

either to the surfaces or, in the case of integral proteins (tissue factor and thrombomodulin), spanning the cell membrane. Cofactor proteins enhance the specificity of the reactions and increase the rate of activation of the proteinase precursor. Cofactor proteins are shown as bar diagrams in Figure 36-6.Cofactor protein activation occurs by proteolysis of the precursor forms by thrombin and factor Xa. Activation of the cofactor proteins exposes the binding sites for factor Xa and prothrombin (factor Va) and, by analogy, factor IXa and factor X (factor VIIIa). Sites of cleavage are indicated in the diagrams in the figure. Binding of factors V and Va and factors VIII and VIIIa to the phospholipid membrane is mediated by the C domains; binding to their respective proteinases and proteinase precursors is mediated by the A domains.

Proteinase Inhibitors

Proteinase inhibitors (proteins) that circulate in plasma inactivate the hemostatic system proteinases after they have "essentially completed" their proteolyses. With two exceptions, α_2-macroglobulin and tissue factor pathway inhibitor, all of the inhibitors of the procoagulant, anticoagulant, and fibrinolytic proteinases are **SERPINS**

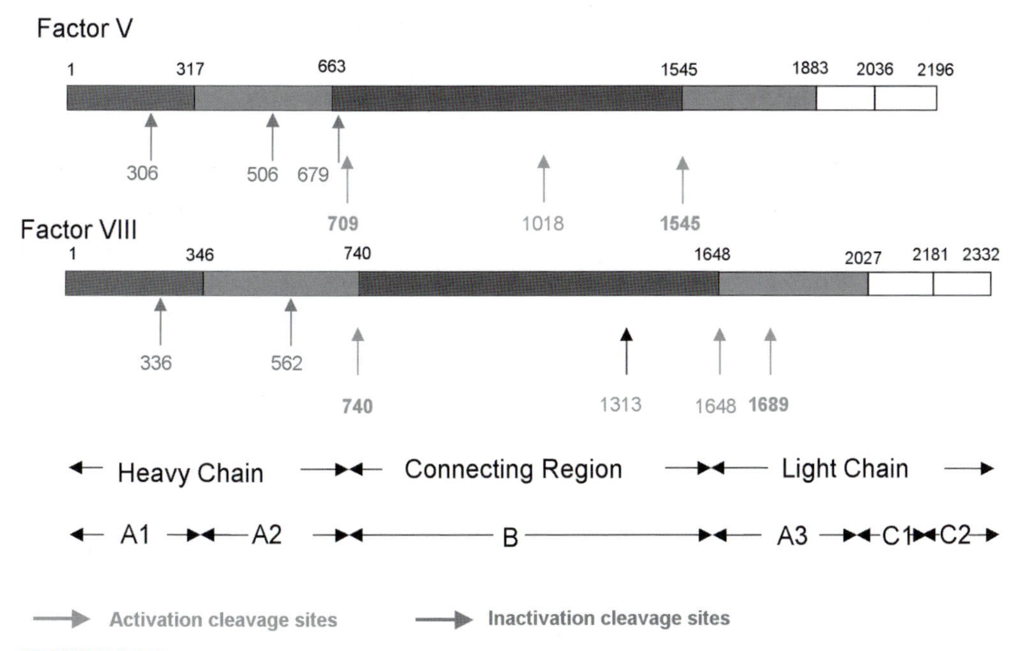

FIGURE 36-6
(Also see color figure.) Cofactor proteins, factor V and factor VIII. Factor V and factor VIII coagulant (not the von Willebrand factor carrier of factor VIII) contain six distinct structural domains. The two A domains A_1 and A_2 at the N-terminal end of the polypeptide chain are separated from the A_3 domain by a highly glycosylated B domain. The two C domains are at the C-terminal end of the molecule. The A domain sequences are homologous to the A domains of ceruloplasmin. Both factor Va and factor VIIIa act as catalysts in the activation of prothrombin and factor X, respectively. In the color version of this figure, activation sites are indicated by green arrows; inactivation sites by red arrows. In factor Va, complete inactivation requires cleavage of Arg^{306}.

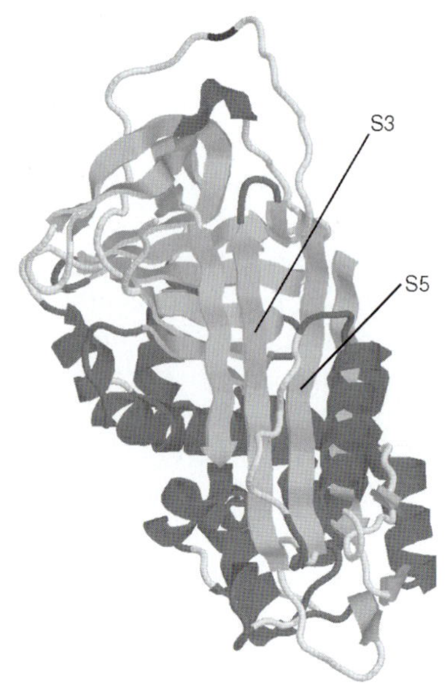

FIGURE 36-7
(Also see color figure.) A group of structurally similar protein inhibitors of the serine proteinases known as SERPINS (SERine Proteinase INhibitors). The structure shown is human antithrombin. The reference SERPIN, α_1-proteinase inhibitor or α_1-antitrypsin, contains ~30% α helix (9 helices) and 40% sheet (5 β sheets). Other members of the SERPIN family contain both additional helices and β sheets. The reactive center loop of antithrombin, residues 378–396, contains the reactive site residues Arg^{393} and Ser^{394}. Upon reaction with the target proteinase or after cleavage by the target proteinase (a reaction that inactivates the inhibitor without inactivating the proteinase), the reactive center loop folds between the S3 and S5 sheets.

(**SER**ine **P**roteinase **IN**hibitors) (Figure 36-7). There are three SERPINS that inactivate proteinases in the procoagulant subsystem, one in the anticoagulant subsystem, and three in the fibrinolytic subsystem. Regulation of the action of the proteinases requires a mechanism compatible with the fact that proteinase precursor activation is irreversible. This is achieved by the protein proteinase inhibitors acting as suicide substrates, i.e., the reaction between proteinase and inhibitor proceeds "half-way" when compared with the proteolytic reaction. The reaction between proteinase and inhibitor stops at the stage of acyl enzyme or tetrahedral intermediate. Water normally does reach the active site in the proteinase substrate complex to achieve hydrolysis, but the serpin prevents this, and thus the proteinase is "killed" by the inhibitor. All SERPINS react with their target proteinases irreversibly and form 1:1 stoichiometric complexes.

The reaction of the proteinase with the inhibitor occurs at a basic residue (Arg or Lys residue) in a loop that extends away from the globular inhibitor molecule. This basic residue, called the reactive site residue, is locked in the proteinase active site as the acyl enzyme. The reactive site and the complete amino acid sequences are similar in all SERPINS.

Other Proteins of the Hemostatic System

Proteins that do not fit into the previous functional categories are fibrinogen, factor XIII (the plasma protransglutaminase), and von Willebrand factor.

Fibrinogen Fibrinogen is a precursor to the spontaneously polymerizing protein fibrin. Fibrin is the substance of the gelatinous clot and, thus, fibrinogen and fibrin are best categorized as structural proteins. Fibrin strands, as noted above, provide the reinforcement necessary for an adequate hemostatic plug. Fibrin can also act as a "surface" for fibrinolytic system proteins. This latter function serves to localize fibrinolysis to the fibrin clot. The structure of fibrinogen, as deduced from electron microscopy and other techniques, is shown in schematic form in Figure 36-8.

Factor XIII (Plasma Protransglutaminase) Factor XIII, after activation to factor XIIIa, is a transglutaminase that catalyzes the formation of covalent cross-links in fibrin. In the absence of its action, the fibrin structure is unstable and physiologically inadequate. Factor XIII exists in two forms, as a tetramer in plasma, a_2b_2, and as a dimer, a_2, in the platelet granules. Activated factor XIII catalyzes the formation of "isopeptide," ε-(γ-glutamyl)lysine bonds between the ε-amino groups of Lys residues and γ-carboxamido groups of Gln residues of fibrin monomers in polymerized fibrin. The a subunits contain the active site residues that are directly involved in the transamidation reaction. The a_2b_2 subunits dissociate in the presence of Ca^{2+} to form a_2 and b_2 dimers. In plasma, which contains ~1.5 mM Ca^{2+}, free a_2 and free b_2 subunits are found in addition to the tetramer.

Von Willebrand Factor Von Willebrand factor (vWF) is a multisubunit protein that serves both to anchor the platelets to the subendothelium and as a carrier protein for factor VIII in the plasma. The circulating vWF is the largest protein of the hemostatic system. It is made of protomeric units which themselves are dimers of the 250,000-Da vWF polypeptide chain. The vWF gene product is a single polypeptide chain that associates to form the dimeric protomer. Disulfide bonds link the two vWF monomer chains in the protomer. The 500,000-Da dimeric protomers further associate to form the more than 20,000,000-Da vWF molecules present in the circulation. Recognition sites are present on the von Willebrand factor

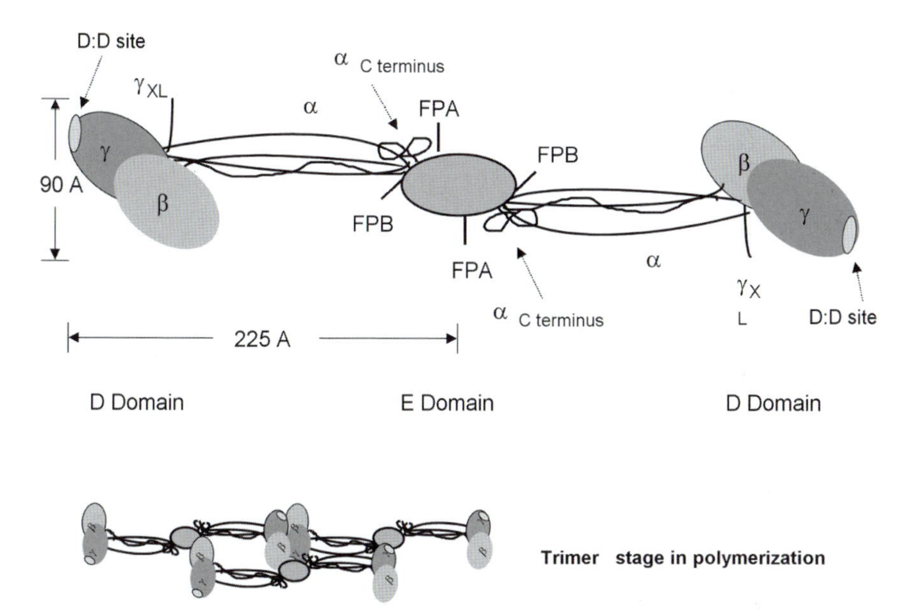

FIGURE 36-8
(Also see color figure.) Fibrinogen and polymerizing fibrin monomers. The fibrinogen molecule is a 340,000-Da protein consisting of three pairs of polypeptide chains: two Aα chains, two Bβ chains, and two γ chains. The A and B designations refer to the two A peptides and two B peptides cleaved from fibrinogen by thrombin to produce the self-polymerizing fibrin monomer (Fn_m), the building block of the fibrin blood clot. The polypeptide chains are linked together by disulfide bridges between Cy–SH residues of the chains. Only after chemical reduction of the disulfide bonds are the separate chains of the fibrinogen molecule discernible. Fibrinogen is frequently abbreviated as $(A\alpha, B\beta, \gamma)_2$. The 16-residue A (FPA) and 14-residue B (FPB) peptides are at the N-terminal ends of the A and the B chains, respectively. The central domain, also identified as the E domain, contains portions of all of the six chains of fibrinogen. The two terminal domains, the D domains, also contain polypeptide sequences from all six chains.

polypeptide chain that enable it to bind to the platelet receptors GpIb and GpIIb/IIIa and to collagen in the extracellular matrix.

Membrane Phospholipid Surfaces

In addition to these categories of proteins, the membranes of damaged cells and activated platelets are components of the hemostatic system and provide surfaces onto which the proteinases, proteinase precursors, and cofactor proteins bind. The formation of complexes of proteinase, proteinase precursor, and cofactor protein is spontaneous and is very likely triggered by the exposure of the membrane surfaces. The membrane surfaces that are capable of binding these clotting proteins contain phospholipids normally found only on the interior of the lipid bilayer cell membrane. The inner membrane phospholipids are phosphatidylserine and other negatively charged phospholipids. Platelets contain an enzyme that converts phosphatidylethanolamine to phosphatidylserine upon activation of the platelets by thrombin and a protein that promotes flip-flop of phospholipids to the outer surface. This coordinated binding of the coagulation proteins to the inner membrane phospholipids that are exposed as the result of cell membrane rupture serves to localize the coagulation reactions at the hemostatic plug.

36.4 Clotting

Proteolytic cleavage of fibrinogen to fibrin results in transformation of blood from a freely flowing fluid into the gel-like clot. In this final step of the procoagulant subsystem, thrombin cleaves four peptide bonds in fibrinogen. Two A fibrinopeptides are removed from the amino terminal ends of the Aα chains and two B fibrinopeptides are removed from the aminoterminal ends of the Bβ chains. The other product of fibrinogen proteolysis is designated *fibrin monomer* ($FnII_m$). The removal of the A and B peptides changes the net charge on the central domain of the fibrin monomer to net positive from net negative as it is in fibrinogen. This change fosters an attractive electrostatic interaction between the positive central domain and the net negative terminal domains of the $FnII_m$. Soluble fibrinogen is converted to fibrin monomers that, after polymerizing, form insoluble fibrin. This dramatic change occurs as the result of removal of less than 0.2% of the fibrinogen mass. Polymerization of $FnII_m$ occurs with both side-to-side and end-to-end association. Interaction between fibrin monomers is best visualized as occurring between specific sites that are structurally complementary to one another. Polymerization begins with the formation of a dimer. Two fibrin monomers lie side-to-side with each monomer overlapping one-half of the other. This produces a dimer that is approximately 1.5 times longer than the fibrin monomer or, because of the lack of contribution of the fibrinopeptides to the length of the fibrinogen, 1.5 times longer than fibrinogen (Figure 36-8). The third fibrin monomer is believed to lie adjacent to the "open half' of a monomer in the fibrin dimer, thus leading to a trimer with a length equal to approximately two times that of the fibrin monomer. Continued side-to-side and end-to-end association leads to the formation of the fibrin polymer. The regular association described above is not always maintained and thus "defects" cause formation of branches on the polymerizing fibrin. The fibrin monomer molecules are noncovalently associated in the growing fibrin polymer. Consequently, in the circulation where blood is flowing past the hemostatic plug and its fibrin mesh, dissociation of the fibrin monomers from the polymer occurs and the clot dissolves.

Fibrinogen is an acute phase reactant[3] and thus its concentration is substantially increased in several clinical situations. When the fibrinogen concentration is increased, the action of thrombin on fibrinogen is faster—the consequence of the greater extent of saturation of thrombin with fibrinogen.

Red blood cells are responsible for the color of the blood clot *in vitro* and also the clot that forms from blood released onto the skin. A clot formed from fibrin in the absence of entrapped red blood cells is a white, translucent gel. *In vitro,* fibrin forms a stringy mass that is easily wrapped around a glass rod.

Noncovalently associated fibrin is physiologically unsatisfactory because the dissociation of the fibrin results in recurrent bleeding. Fibrin monomer dissociation is prevented by formation of covalent cross-links between different $FnII_m$ molecules. The result of this covalent cross-linking is an insoluble fibrin and a stable hemostatic plug. These cross-links are formed by the action of factor XIIIa, plasma, and/or platelet transglutaminase (see below). Multiple cross-links are formed among a chains of several different fibrin monomers. This creates a molecular species designated α polymer (see "Fibrinolysis" below). Two adjacent γ chains from two different fibrin monomers are cross-linked also to form a species designated γ-γ dimer.

[3] Acute phase reactants are plasma proteins that undergo large increases in synthesis and thus concentration in plasma in response to acute inflammation caused by surgery, myocardial infarction, and infections.

36.5 The Procoagulant Subsystem of Coagulation

Activation Complexes

The proteolytic reactions of the hemostatic system are neither catalytically efficient nor localized when proteinase and proteinase precursor only are present. The rapid, localized proenzyme activation required for normal hemostatic response occurs only in a complex of proteinase, proteinase precursor, and cofactor protein assembled on the surface of a damaged cell membrane, or *in vitro,* on the surface of phospholipid bilayers. The catalytic efficiency of an enzyme-catalyzed reaction is expressed by the ratio of the kinetic constants k_c and K_m (k_c/K_m).[4] In the activation complexes, k_c/K_m values can be greater than 10^7 M^{-1} s^{-1}. With proteinase and proenzyme alone, the k_c/K_m values are only approximately 100 M^{-1} s^{-1} and thus the reactions are 10^5 times less efficient. Expressed in terms of the same amount of product formed in the two situations, a 10^5 increase represents the difference between requiring 1 minute and about 6 months to form the product!

It is helpful in the effort to understand activation complexes to consider complex formation, the reactions that occur in the complexes, and the "demise" of the complexes as proceeding in a sequence. First, a reversible, noncovalent association of proteinase, cofactor protein (strictly, activated cofactor protein), proteinase precursor, and membrane surface occurs to form the activation complex. This spontaneous association occurs as the result of complementary interaction sites on the protein molecules, e.g., the binding sites between proteinase and protein substrate, proteinase and cofactor protein, substrate and cofactor protein, and all three proteins with the membrane surface. Tissue factor normally exists as an integral membrane protein and is always associated with the membranes of cells in the vessel wall. The same processes are involved in the anticoagulant subsystem and, with a different surface, fibrin in the fibrinolytic system as well.

Second, irreversible proteolytic action in the complex converts the proteinase precursor in the complex into an active proteinase. This is followed by dissociation of the product proteinase from the complex in which it has been formed. Association of this proteinase with the next cofactor protein and the next protein substrate to form the next complex in the coagulation cascade then occurs.

[4] K_m is the Michaelis constantand k_c is the turnover rate constantin the classical mechanism for enzyme catalyzed reactions (See Chapter 6).

Third, as the proteinases complete their catalysis of precursor activation, they dissociate more completely from their respective complexes and are inactivated by the proteinase inhibitors present in the blood. Dissociation from the complexes is important; in the complexes and in the presence of substrates, the proteinase inhibitors are "blocked" from reacting with the proteinases because the substrate obstructs access to the active site of the proteinase.

Fourth, complete dissociation is promoted by the proteolytic inactivation of the activated cofactor proteins by activated protein C. Proteolytic inactivation of the activated cofactors ensures that reformation of the activation complex does not occur to any appreciable extent. Because the system is dynamic, all of these processes occur simultaneously. However, in the initial stages of the secondary hemostatic response, the first two "steps" predominate. As the hemostatic plug becomes consolidated by fibrin reinforcement, the third and fourth steps predominate.

The Prothrombin Activation Complex (Prothrombinase)

The first and most completely characterized of the activation complexes of the procoagulant subsystem is that of prothrombin activation. Because of this, the prothrombin activation complex will be described in greater detail than the other, similar activation complexes (Figure 36-9) The general principles described below apply to all of the complexes.

The prothrombin activation complex consists of factor Xa (the proteinase), prothrombin (the substrate), factor Va (the cofactor protein in an "activated" form), phospholipid bilayer surface, and Ca^{2+} ion. Ca^{2+} ion is involved in the binding of prothrombin and factor Xa to the phospholipid bilayer and cell membrane surfaces. Prothrombin and factor Xa associate with the membrane surface as the result of a Ca^{2+} ion–dependent conformation that provides a "face" on the molecule that interacts with the lipid molecules of the membrane bilayer. Ca^{2+} ion binding to these two proteins and the lipid-binding conformation are related to the presence of several (14–12) unique amino acid residues, γ-carboxyglutamate, in these proteins (see section "Vitamin K, Oral Anticoagulants, and Their Mechanisms of Action") The shapes of the proteins and their relative dimensions, and the half-thickness of the phospholipid bilayer are based on studies of the hydrodynamic properties of the proteins and on x-ray crystallographic structure data. Association of factor Xa and prothrombin with factor Va involves the EGF-like and kringle structures of the two proteins and regions (sites) on the A domains of factor Va.

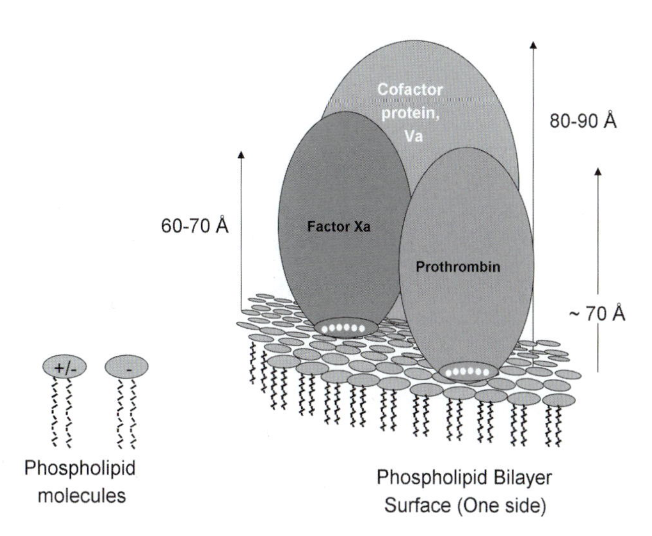

FIGURE 36-9
(Also see color figure.) Archetypal activation complex of coagulation: *prothrombinase*. Complex formation precedes rapid proteolysis of prothrombin. The presence of the specific domains and the binding sites associated with them are responsible for the formation of the complexes between the correct proteins. The A domains of factor Va provide the interacting sites for both factor Xa and prothrombin, the C domains the sites for binding to the surface. The Gla domains bind Ca^{2+} (white dots) and bind factor Xa and prothrombin to the surface. The locations of the various sites further orient the molecules optimally with respect to the surface and to each other to give the localized, large increases in rates of proteolysis. The conversion of the proenzyme form of a proteinase to its catalytically active form commonly requires proteolysis of a peptide bond that frees a Leu, Ile, or Val residue from the peptide linkage. This permits a conformation change within the active proteinase domain that is stabilized by the formation of a salt bridge between the newly freed residue and Asp^{102} (chymotrypsin numbering). Proteolysis, *per se*, is not the key element in "enabling" of the catalytic apparatus of the proteinase, but rather the conformation that is stabilized by the salt bridge. The dimensions indicated on the figure are measured distances between fluorescent labels on the proteins (active sites for factor Xa and meizothrombin) and a fluorophore in the membrane.

Activation of Prothrombin

The activation of prothrombin to thrombin and the conversion of fibrinogen to fibrin are frequently described as the final, common pathway of blood coagulation. This designation, the result of the convergence of the "intrinsic" and "extrinsic" pathways with the formation of factor Xa, is common in older literature. Prothrombin activation is shown in Figure 36-10. After proteolytic cleavage, fragment 1-2 retains the Gla domain and the two kringles of the prothrombin molecule. Although it may remain associated with the surface and ultimately contribute to displacement of other proteins, the amount of prothrombin converted to thrombin and fragment 1-2 makes this a minor role of the fragment.

The α-thrombin dissociates from the prothrombin fragment 1-2 and diffuses to its substrates: platelets (the platelet thrombin receptor), factor V, factor VIII,

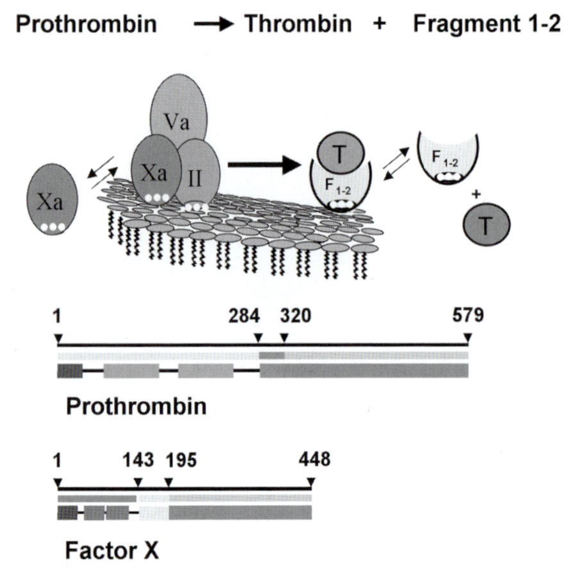

FIGURE 36-10
(Also see color figure.) Activation of prothrombin. Prothrombin is cleaved by factor Xa at two residues, Arg^{284} and Arg^{320}. The order of cleavage produces different products. If the first cleavage is at Arg^{284}, it produces two products: prothrombin fragment 1-2 (residues 1–284, yellow bar) and prethrombin 2 (residues 285–579, gray bar), each of which represents approximately one-half of the prothrombin molecule. When Arg^{320} is cleaved first, the intermediate product meizothrombin is formed. Meizothrombin contains the entire sequence of the prothrombin molecule; the two halves are held together by a single disulfide bond (CyS^{283} to Cys^{439}). In the absence of factor Va, the rate constants for cleavage of the two bonds are essentially the same. In the presence of factor Va and phospholipids, Arg^{320} is cleaved first. Meizothrombin is detectable in reaction mixtures, but its rapid cleavage produces prothrombin fragment 1-2 and α-thrombin. In the color version of this figure, motifs and domains are color coded as follows: Gla domain (blue), kringle domains (orange), EGF-like domain (magenta), activation peptide (yellow), and proteinase domain (green). Light chains are indicated in dark gray, heavy chains in light gray. Regions connecting the motifs are black.

fibrinogen, factor XIII, and protein C. It almost certainly acts on factor VII to generate factor VIIa. In contrast to the other vitamin K-dependent proteins, thrombin is not linked to the Gla domain-containing polypeptide and thus is free to diffuse away from the site of its formation. All of the other vitamin K-dependent proteins, both in their precursor and active proteinase forms, retain the Gla domain and are able to bind to the phospholipids of the cell membranes at the injury site. Prothrombin fragment 1-2 is measured as a marker of ongoing prothrombin activation and by inference, an indication of hypercoagulability (an imbalance of procoagulant subsystem activity over anticoagulant subsystem activity) when it is present in a higher than normal concentration.

Although there is an explosive increase in the reaction rate, only a small fraction ($\ll 1\%$) of the total prothrombin available ($\sim 1.5\ \mu M$) is converted to thrombin. The concentration of thrombin required to convert fibrinogen to fibrin that is sufficient to form a clot in 15 seconds is less than 1 nM. When the coagulation "cascade" is viewed as an amplifier, it must be noted that it is the rates of the reactions that are being amplified.[5]

An interesting mutation in the gene for prothrombin, a G-to-A transition in the 3′ untranslated region at nucleotide 20210, results in an elevated concentration of prothrombin in the circulation (>115% of normal). It is not known if the mutation causes the elevated prothrombin levels, but the defect is associated with a twofold increase in the risk of thrombosis.

36.6 The Extrinsic Pathway

Historically, the reactions of the coagulation system that are initiated *in vitro* by the addition of an exogenous substance such as tissue homogenates have been grouped under the title of "extrinsic pathway." The reaction pathway that is followed without tissue homogenate being added was designated the "intrinsic pathway" because it was assumed that all components were present in blood.[6] The two pathways are not independent, but the terms continue to be used both because of their historical roots and because the commonly performed *in vitro* coagulation tests can be related to them.

Initiation of the Procoagulant Subsystem and Activation of Factor VII

The first complex to form is almost certainly the "extrinsic" factor X activation complex. Factor VIIa binds to tissue factor, an integral membrane protein that is exposed upon injury of the blood vessel. The phospholipid bilayer surface is provided by the damaged cell membranes. *In vivo,* Ca^{2+} is always present in the blood.[7]

Two hypotheses exist for the origin of factor VIIa. In one, factor VII is considered to have some proteolytic activity prior to a proteolytic cleavage. In the other view, low concentrations of thrombin and/or factor Xa are considered present at all times in the circulating blood. Thrombin, factor Xa, and factor VIIa can activate factor VII to

[5] Amplification in a sequence of enzyme-catalyzed reactions suggests that more of the product is formed at each stage than in the previous stage. However, in the coagulation "cascade" the principal amplification is of reaction rate, not quantity of product formed.

[6] It is now known that the surface of the glass test tube provides a surface upon which proteins bind and initiate clot formation. See discussion of **contact system (phase)** following.

[7] *In vitro*, Ca^{2+} is added only because the anticoagulant solutions used for blood collection employ substances such as citrate ion, EDTA, or oxalate to bind Ca^{2+} to prevent activation complex formation. It is worth noting that factor XIIa can activate factor VII in the absence of Ca^{2+}.

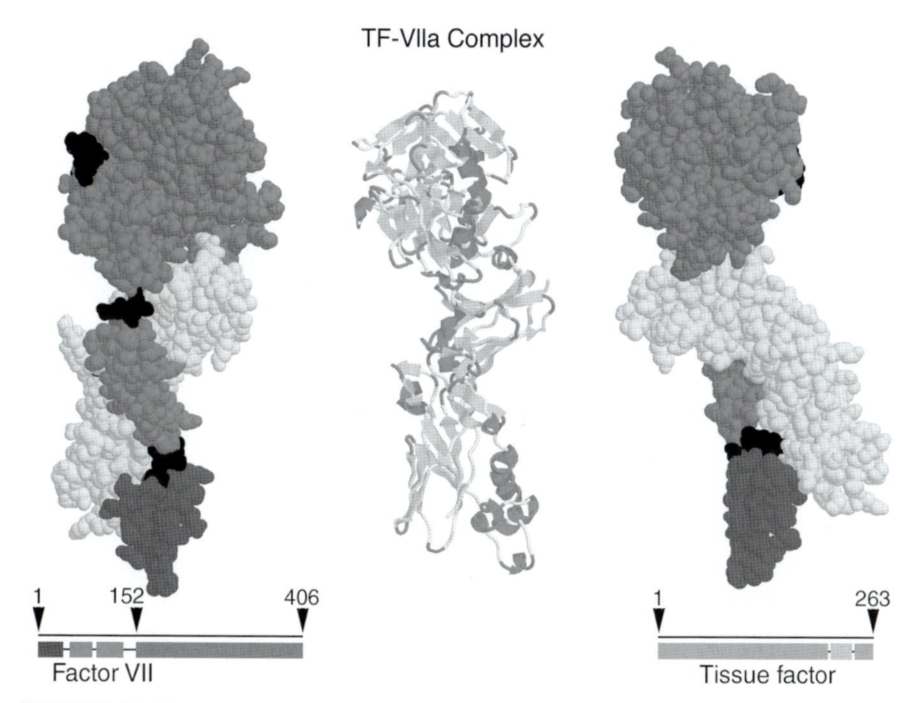

FIGURE 36-11
(Also see color figure.) Tissue factor–factor VIIa complex. The three-dimensional structure of the complex of factor VIIa and tissue factor (minus the transmembrane polypeptide domain of the tissue factor) in the absence of membrane surface. It is approximately 115 Å in length and has a diameter of 40–50 Å. Factor VIIa shows its four distinct domains: the Gla domain, two EGF-like domains, and the proteinase domain. Tissue factor contacts factor VIIa via the interface between the two "fibronectin type III-like" domains. All four domains of factor VIIa appear to be involved in the interaction between tissue factor and factor VIIa. The Gla domain of factor VIIa is folded very similarly to the Gla domain of prothrombin (Gla domain of prothrombin fragment 1). Activation of factor VII can be catalyzed by thrombin, factor Xa, factor VIIa, and factor XIIa—all by cleavage at Arg^{152}–Ile^{153}. Secondary structures are shown in the center diagram; two views of the close interactions between TF and factor VIIa are shown in the two diagrams at each side.

produce factor VIIa. Factor VIIa is detectable in extremely low concentration in the circulating blood of normal individuals. Low levels of factor VIIa can be measured in an assay that employs recombinant tissue factor that is missing the transmembrane and inner membrane domains. The trace amounts of factor VIIa that are present are probably sufficient to initiate the clotting process upon injury-related exposure of tissue factor and membrane lipids. The complex of tissue factor and factor VIIa is shown in Figure 36-11.

In vitro other activators of factor VII exist, e.g., factor XIIa, kallikrein, and factor XIa. The physiological significance, i.e., *in vivo* significance of these activators of factor VII is unknown, although the presence of factor XIIa may create risk of thrombosis. *In vitro,* factor VII becomes factor VIIa upon storage in the cold, a process that involves factor XIIa. Factor XIIa presumably is generated from factor XII through binding of the components of the *contact phase* of *in vitro* clotting to the containers in which the blood or plasma is being stored. Activation of factor VII by factor XIIa does not require the presence of Ca^{2+} ions, and therefore this activation occurs in anticoagulated blood or plasma. Higher concentrations of factor VIIa have been shown by epidemiological studies to increase the risk of thrombosis, whereas individuals with factor VII deficiency appear to have a lower thrombotic risk.

Activation of Factor X by Factor VII(VIIa)

Factor X is activated by a single peptide bond cleavage, Arg^{51}-Ile^{52} in the heavy chain of the proenzyme. The activation peptide is highly glycosylated, as are many of the activation peptides of the procoagulant proenzymes. The Ile^{52} residue subsequently forms the "salt bridge" with Asp^{194} (chymotrypsin numbering) in the heavy chain, thus enabling formation of the catalytic apparatus of factor Xa.

In the absence of tissue factor, factor VIIa cleaves the Arg^{51} peptide bond in factor X at a rate so low that it is not readily detectable *in vitro*. In the presence of tissue factor and membrane surface phospholipids, factor VIIa rapidly activates factor X. The estimates for the magnitude of the increase in activation rates in the presence of tissue factor

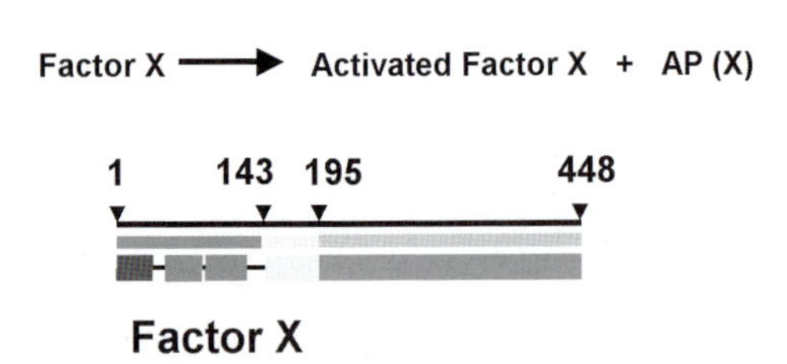

FIGURE 36-12
(Also see color figure.) Activation of factor X. Factor X is activated by factor VIIa through a single peptide bond cleavage, Arg^{51}-Ile^{52} in the heavy chain of the proenzyme. The activation peptide is highly glycosylated as are many of the activation peptides of the procoagulant proenzymes. Factor X is converted to factor Xa by factor IXa through cleavage of the same peptide bond that is cleaved by factor VIIa. In the color version of this figure, motifs and domains are color coded as follows: Gla domain (blue), EGF-like domain (magenta), activation peptide (yellow), and proteinase domain (green). Light chains are indicated in dark gray, heavy chains in light gray. Regions connecting the motifs are black.

and membrane surface are approximately 500,000 times. The activation of factor X by factor VIIa in the activation complex thus follows the same general behavior of the activation of prothrombin. A high rate of activation occurs only in the complex, partly the result of a decrease in K_m, and partly the result of an increase in k_c. The activation of factor X and the motifs present in the molecule are shown in Figure 36-12. A unique inhibitor unlike the SERPINS can modulate the activation of factor X by factor VIIa and tissue factor. The inhibitor, tissue factor pathway inhibitor (TFPI), or extrinsic pathway inhibitor (EPI), is related to the Kunitz-type pancreatic proteinase inhibitors.

36.7 The Intrinsic Pathway

Activation of Factor X

Factor IXa, factor VIIIa, and the phospholipid surface compose the "intrinsic pathway" activator of factor X. Factor VIII, as described above, is activated prior to its becoming functional in the activation of factor X by factor IXa. Although the similarities in the activation of factor V and factor VIII are extensive, factor VIII in the plasma differs from factor V in a unique way. Whereas factor V circulates as a single molecule, factor VIII is bound to the von Willebrand factor that acts as a carrier and as a stabilizer for factor VIII and may modulate its activity. Because of the binding of vWF to collagen in the subendothelium and to the platelet receptor GpIb, factor VIII is preferentially associated with the hemostatic plug at the injury site. This association provides an additional mechanism that is distinct from factor VIII binding to the membrane lipids. Some cases that appear to be factor VIII deficiency (classical hemophilia) can result from defective von Willebrand factor and can be treated by increasing the quantity of vWF rather than of factor VIII.

Factor X is converted to factor Xa by factor IXa by cleavage of the same Arg^{51}-Ile^{52} peptide bond that is cleaved by factor VIIa and with the release of the same activation peptide. The relative rates of activation are again 100,000–500,000 times greater in the complex than in solution, i.e., with the proteinase and protein substrate alone (Figure 36-12).

Activation of Factor IX by Factor VIIa and Tissue Factor

Factor IX can be activated by the complex containing factor VIIa, tissue factor, phospholipid bilayer surface, and Ca^{2+} ions. This "crossover" reaction for factor IX activation, although perhaps not strictly a reaction of the "intrinsic pathway," contributes about half of the factor IXa that is formed in situations in which tissue factor is present. Factor IX is activated as the result of the cleavage of two peptide bonds, Arg^{145} and Arg^{180}-Ile^{181} (Figure 36-13).

Activation of Factor IX by Factor XIa

In the classical reaction of the "intrinsic pathway," factor IXa is formed from factor IX by the action of factor XIa (the proteinase). The "surface" for this reaction is believed to be a glycosaminoglycan, a sugar polymer to which factor IX and factor XIa can bind. In this situation, the "surface" is actually a two-dimensional polymer that acts as a tether between the molecules and on which they can migrate toward or away from each other. As glycosaminoglycans are found on the surfaces of both endothelial and subendothelial cells, such a two-dimensional surface is a good candidate for promoting complex formation at this stage of the procoagulant pathway. The

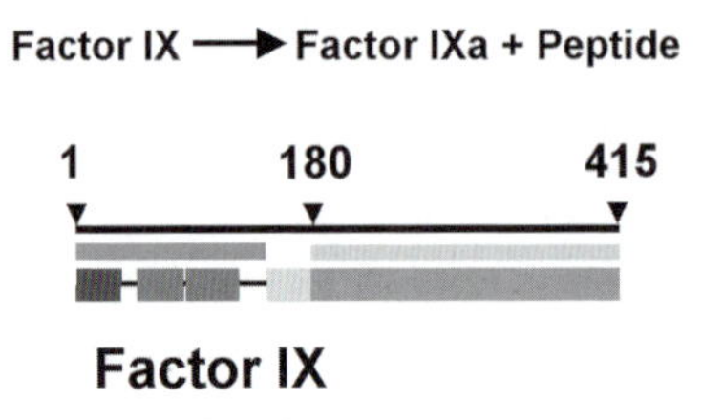

FIGURE 36-13
(Also see color figure.) Activation of factor IX. Factor IX is activated by cleavage of two peptide bonds, Arg^{145} and Arg^{180}-Val^{181} The activation peptide that is released is composed of 35 amino acid residues and is highly glycosylated. The same peptide bonds are cleaved by factor VIIa and factor XIa. Glycosaminoglycans, particularly heparin, can increase the rate of activation as the result of the binding of both factor XIa and factor IX to the heparin molecule to form a ternary complex. Color coding is as described in Figure 36-12.

glycosaminoglycan is acting as a catalyst of a procoagulant reaction. In contrast to the other stages of the procoagulant subsystem, there is no known cofactor protein involved in factor IX activation. The activation of factor IX by factor XIa is not dependent on the presence of Ca^{2+}, although Ca^{2+} is certainly present in blood.

The Contact Phase of the *in Vitro* Intrinsic Pathway of Coagulation

Figure 36-2 shows that two other reactions exist that are not shown in Figure 36-3. These reactions, part of the "contact phase" or "contact system," are only important for *in vitro* clotting. Individuals with deficiencies in the components of the "contact phase" do not suffer from bleeding.

The contact phase reactions are important for *in vitro* coagulation function testing using the activated partial thromboplastin time test. This test is so named because it was believed to involve nothing extrinsic to blood plasma. Historically, the partial thromboplastin was considered to be generated from plasma proteins alone and augmented by phospholipid, a platelet substitute. Because the surface of the glass test tube is a participant in the "contact reactions," "intrinsic" is no longer an accurate distinction between the two pathways.

The contact phase reactions involve two proteinase precursors—factor XII and prekallikrein—and one cofactor protein—plasma high-molecular-weight kininogen. Activation of prekallikrein to form kallikrein occurs as a consequence of proteolytic cleavage; however, activation of factor XII does not obligatorily require proteolysis. Factor XII can be converted from an enzymatically inactive molecule to an enzymatically active one through a conformation change that occurs when factor XII binds to the surface of the glass test tube in which the plasma is contained. In the commonly used laboratory test procedures, the surface of fumed silica (very fine silica powder), kaolin (an aluminosilicate clay), or ellagic acid (a dispersion of an aromatic compound as fine droplets in aqueous solution) is used (Figure 36-14).

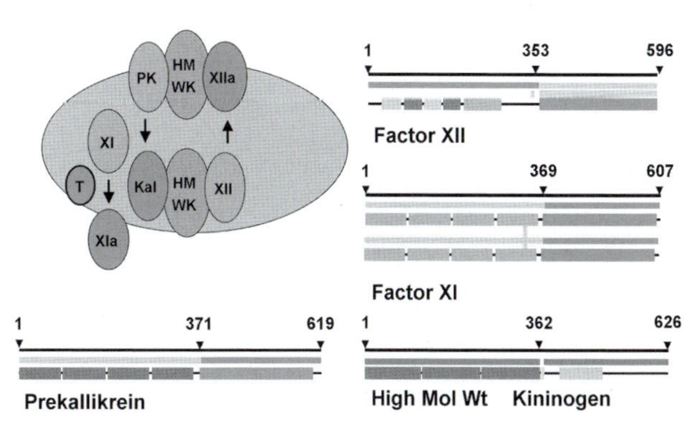

FIGURE 36-14
(Also see color figure.) Contact phase of *in vitro* coagulation. The contact phase reactions involve two proteinase precursors—factor XII and prekallikrein—and one cofactor protein—plasma high-molecular-weight kininogen (HAWK). The structures, in bar diagrams, show the presence of EGF-like fibronectin type I and type II, kringle, cystatin-like, and "apple" motifs. The cleavage site for proteolytic activation of factor XII is Arg^{353} and for prekallikrein Arg^{371}. The cofactor protein, HMWK, is cleaved by kallikrein at Arg^{362} and Arg^{371} to release bradykinin (yellow). HMWK contains a unique His-rich region that is associated with binding to the contact surface (shown in light blue). Factor XI is a dimer of two identical polypeptide chains that are linked by a single disulfide bridge. In the circulating blood, factor XIIa cleaves the two Arg^{369}-Ile^{370} peptide bonds in factor XI to form factor XIa. In the color version of this figure, other motifs and domains are color coded as follows: apple domains (cyan), EGF-like motifs (magenta), fibronectin motifs (yellow), kringle (orange), and proteinase domain (green). Light chains are indicated in dark gray, heavy chains in light gray. Regions connecting the motifs are black.

36.8 Anticoagulant Subsystem—Activation of Protein C and Inactivation of Factors Va and VIIIa

The very large increases in the rates of activation of prothrombin and factor X that occur in the presence of factors Va and VIIIa, respectively, make the hemostatic response both rapid and localized. However, if such rates were to continue unabated, the extension of the hemostatic plug into the blood vessel would occlude the vessel and result in ischemia and death to the adjacent cells and tissues. If the pathologically extended hemostatic plug is in the venous system, the separation of the occlusive plug (also designated clot, or red thrombus) can result in the clot being sent to the lungs with consequent pulmonary embolism. The proteinase activations of the procoagulant subsystem are opposed by the cofactor protein inactivations of the anticoagulant subsystem.

Protein C is activated by thrombin in the complex with thrombomodulin to produce activated protein C, the proteolytic inactivator of factors Va and VIIIa. The binding of thrombin to thrombomodulin changes thrombin from a procoagulant proteinase to an anticoagulant proteinase. Whereas the hemostatic reactions that prevent blood loss at the injury site are associated with the ruptured blood vessels, thrombomodulin is on the endothelium (Figure 36-15).

Activation of protein C by thrombin occurs adjacent to the injury site; inactivation of factors Va and VIIIa occurs on the exposed surface at the injury site. Inactivation of activated factors Va and VIIIa occurs by proteolysis of two peptide bonds, both in each of the heavy chains (domains A1 and A2, Figure 36-6) of factors Va and VIIIa. The consequence of these cleavages is that the cofactor proteins

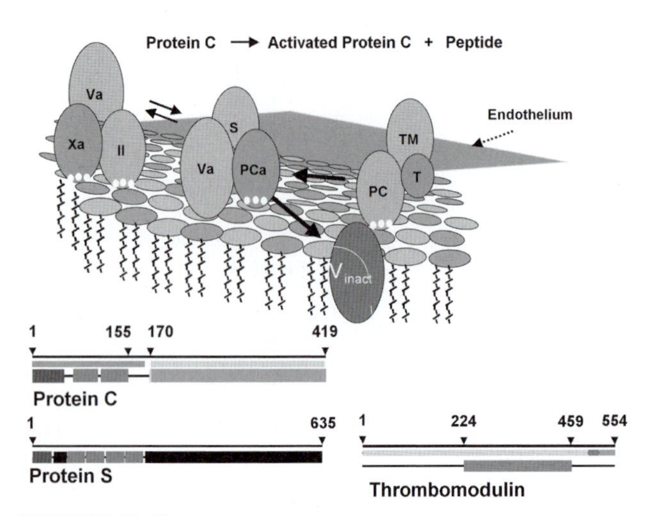

FIGURE 36-15

(Also see color figure.) Anticoagulant subsystem. Activation of protein C occurs adjacent to the injury site; inactivation occurs on the exposed surface at the injury site. Protein C is activated through cleavage by thrombin at Arg^{169}-Ile^{170}. Protein S is a 635-residue vitamin K-dependent protein that functions as a cofactor. There is a Gla domain and four EGF-like domains in the molecule, but no proteinase domain. Thrombomodulin is a 554-residue integral membrane protein. The extracellular region comprises residues 1–494, the transmembrane region residues 495–518, and the intracellular region residues 519–554. EGF-like structures are found from residue 224 to 459. In the color version of this figure, motifs and domains are color coded as follows: Gla domain (blue), EGF-like domain (magenta), activation peptide (yellow), and proteinase domain (green). Light chains are indicated in dark gray, heavy chains in light gray. Regions connecting the motifs are black.

become nonfunctional catalysts. The interaction of the activated cofactors with the proteinases in their respective complexes is eliminated because the interaction is a property of the factor Va and factor VIIIa heavy chains. The rates of proteolytic inactivation of factors Va and VIIIa are increased by participation of the cofactor protein, protein S. The increased rate of proteolysis with protein S (only about 500 times) is much less than observed with the cofactor proteins Va and VIIIa of the procoagulant subsystem (more than 10,000 times).

Mutations that result in amino acid substitution at the cleavage site in a protein substrate cause two changes in the proteolytic process. First, the rate of cleavage of the peptide bond is reduced because of the altered orientation of the substrate within the active site. Second, the altered substrate is very likely to be a competitive inhibitor of the proteinase responsible for the cleavage. Both of these effects are evident in the phenomenon called *activated protein C resistance*.

A mutation in the factor V gene, G to A at 1691, results in the replacement of the normal Arg residue at position 506 in the heavy chain of factor Va by a Gln residue. Individuals carrying this mutation, called factor V (Leiden) are at increased *risk* of venous thrombosis and venous thromboembolism. The inability of activated protein C to cleave Gln^{506} slows the cleavage of Arg^{306} in factor Va. A second mutation, factor V (Cambridge), that confers activated protein C resistance is at Arg^{306}. As a result, the cleavage essential for complete inactivation of factor Va does not occur.

The factor V (Leiden) mutation is very common in individuals of western European origin. The prevalence is approximately 5% in the Western Hemisphere in Caucasians; the mutation is almost completely absent from Africans and almost entirely absent from Asians.

Anticoagulant Subsystem—Proteinase Inhibitors

Proteinase inactivation occurs by a stoichiometric reaction between proteinase and inhibitor that results in the formation of a "covalent" ester bond between the reactive site residue of the inhibitor (Arg^{393} in antithrombin) and the active site residue (Ser^{195} in the proteinase). The proteinases thrombin, factor Xa, factor IXa, and, less effectively, factor VIIa and factor XIa are all inactivated by antithrombin (Figure 36-16). Other SERPINS can inactivate procoagulant proteinases, heparin cofactor II can inactivate thrombin, and α_1-proteinase inhibitor can inactivate factor Xa. An altered α_1-proteinase inhibitor (α_1-proteinase inhibitor

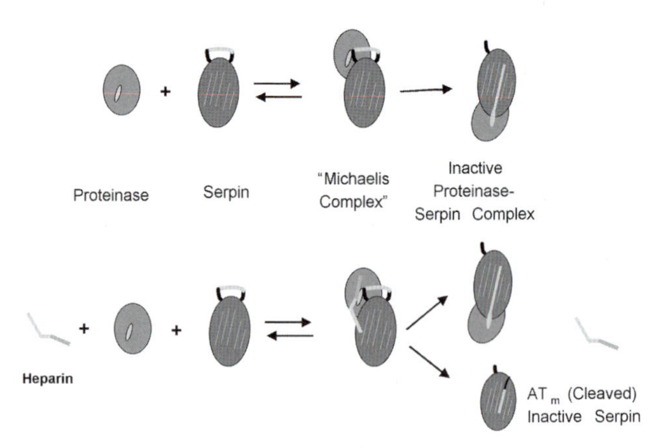

FIGURE 36-16

(Also see color figure.) Proteinase inactivation by SERPINS. Proteinase inactivation occurs by reaction between proteinase and inhibitor, e.g., antithrombin. The proteinase is a stoichiometric reactant in this instance but is not a catalyst. This reaction results in the formation of a covalent bond between the reactive site residue of the inhibitor (Arg^{393} in antithrombin) and the active site residue (Ser^{195} in the proteinase). This complex formation prevents the proteinase from hydrolyzing any other peptide bond. Proteinases, thrombin, factor Xa, factor IXa, and, less effectively, factor VIIa and factor XIa are all inactivated by the plasma protein inhibitor antithrombin (previously designated antithrombin III). The product AT_m is the cleaved form of antithrombin. It is formed in both the absence and presence of heparin, but more so in the presence. In the color version of this figure, the proteinase is indicated by green, the inhibitor by red, and the inactivated proteinase by gray. Stripes on the inhibitor represent the helices (Figure 36-7).

Pittsburgh), in which an Arg has replaced Ala at the reactive site, is a good inhibitor of thrombin.

Because inactivation of factors Va and VIIIa by activated protein C promotes the dissociation of the proteinases, cofactor protein inactivation complements the action of the proteinase inhibitors. This eliminates the "protection" that the proteinases have when bound to their cofactor proteins and substrates. Inactivation of proteinases by SERPINS occurs via a common mechanism that involves a Michaelis complex between the proteinase and the inhibitor (Figure 36-16). This mechanism applies to all serine proteinases of the hemostatic system, i.e., the procoagulant, anticoagulant, and fibrinolytic subsystem proteinases.

Mechanism of Action of Heparin as a Therapeutic Anticoagulant

The inactivation of procoagulant proteinases can be catalyzed by glycosaminoglycans, which are sulfated polysaccharide molecules found on the surface of the normal endothelial cells and in the basophilic granules of mast cells. The glycosaminoglycans act as catalysts, increasing the rates of inactivation of the proteinases as much as 100,000-fold. The therapeutic benefits of heparin in preventing and treating deep vein thrombosis are derived from its altering the balance between the procoagulant and anticoagulant subsystem reactions in favor of the anticoagulant subsystem.

Heparin is a polymer of repeating disaccharide "building blocks." All heparins bind to antithrombin; however, heparin molecules that contain a unique pentasaccharide sequence bind with particularly high affinity (designated *high-affinity heparins*). Heparins that contain this pentasaccharide sequence are also the most effective in catalyzing the inactivation of proteinases by antithrombin. This increased effectiveness of high-affinity heparins is due to a conformation change that they cause in the antithrombin molecule that makes the binding between antithrombin and heparin tighter.

The magnitude of the heparin-catalyzed increases in the rates of proteinase inactivation by antithrombin depends on the molecular weight of the high-affinity heparin molecules. Up to a molecular weight of about 20,000, the higher the molecular weight, the greater the increase in the rate of proteinase inactivation. Thrombin inactivation requires heparin molecules of sufficient molecular weight to permit the formation of a thrombin-heparin-antithrombin complex. Factor Xa inactivation is less sensitive to the molecular weight of the heparin molecules, and factor Xa is the only proteinase that can be efficiently inactivated by the very low-molecular-weight high-affinity pentasaccharide (Figure 36-17). Here, in contrast to the situation of factor XI activation, the glycosaminoglycan is acting as a catalyst of an anticoagulant (proteinase inactivation) process.

Inhibitors of the Contact Phase Proteinases

Several proteinase inhibitors inactivate the proteinases of the contact phase. Among the SERPINS are C-1 inactivator, α_1-proteinase inhibitor, and antithrombin. The target proteinases for these inhibitors are factor XIIa, kallikrein, and factor XIa. The molecular mechanisms are the same as those described for the procoagulant and fibrinolytic system proteinases.

Another inhibitor is present in plasma that can inactivate thrombin, plasmin, and, to a much lesser extent, the other proteinases as well. This inhibitor, α_2-macroglobulin, inhibits by a completely different mechanism from that of the SERPINS. It entraps the proteinases in a "cavity" that is created by the four subunits of the α_2-macroglobulin molecule. The active sites of the entrapped proteinases are sterically hindered from protein substrates but are accessible to low-molecular-weight chromogenic substrates that are used in some laboratory coagulation tests for heparin and antithrombin.

36.9 Fibrinolytic Subsystem

The fibrinolytic system removes the fibrin of the hemostatic plug and thus is responsible for the temporary existence of the fibrin clot. The proteolytic action of plasmin on fibrin and fibrinogen is extensive and more like the digestive proteolysis catalyzed by trypsin and chymotrypsin than the proteolysis involved in proteinase precursor activation. The fibrinolytic subsystem includes the reactions of plasminogen activation, plasmin inactivation, and fibrin digestion. As is common throughout the hemostatic system, irreversible activation reactions of the fibrinolytic system are opposed by irreversible proteinase inactivation.

Plasminogen Activation

Plasminogen is converted to plasmin as the result of the cleavage of a single peptide bond, Arg^{161}-Val^{562}. Two different proteinases are responsible for the physiological activation of plasminogen: tissue-type plasminogen activator (t-PA) and urinary plasminogen activator (u-PA). t-PA is the principal activator of plasminogen and is synthesized primarily in endothelial cells, the principal source of t-PA in the circulating blood.

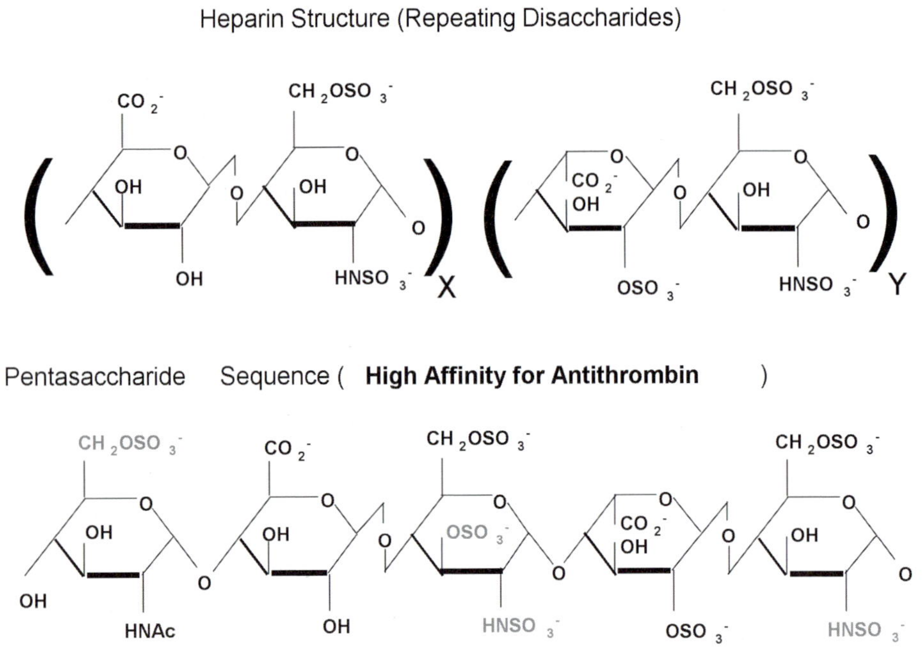

Heparin Structure (Repeating Disaccharides)

Pentasaccharide Sequence (**High Affinity for Antithrombin**)

FIGURE 36-17

(Also see color figure.) Structure of heparin. Heparin is a polymer of repeating disaccharide units that contain one uronic acid and one hexosamine residue. The uronic acid residues may be either glucuronic acid or iduronic acid, both of which are monosaccharide acids that differ in their stereochemistry. The hexosamine residue is glucosamine. Both the uronic and hexosamine residues can be modified by O- and N-sulfation and the glucosamine residue by N-acetylation. All heparins bind to antithrombin; however, heparin molecules that contain a unique pentasaccharide sequence bind with particularly high affinity (high-affinity heparins). Approximately 30% of the heparin molecules present in the commonly used therapeutic heparins have high affinity for antithrombin.

Activation of plasminogen occurs as a result of the binding of both plasminogen and t-PA to fibrin, which results in a nearly 400-fold decrease in the K_m for the activation reaction, from about 65 μM to 0.16 μM. In addition to this enhancement of ES binding, the common binding to the same "surface" localizes the activation reaction to fibrin. Binding to fibrin occurs via the lysyl residues in fibrin and, as a result of fibrinolysis, plasmin proteolysis of Lys-X peptide bonds increases the number of available sites to which the plasminogen and t-PA can bind. Kringles 1 and 4 of plasminogen and plasma bind to fibrin via their lysine binding sites. The activation of plasminogen by t-PA and u-PA is opposed by two protein inhibitors, plasminogen activator inhibitors 1 and 2 (PAI-1 and PAI-2; Figure 36-18). PAI-1 inactivates t-PA very rapidly, but less so when the t-PA is bound to fibrin. Higher than normal concentrations of PAI-1 are associated with the occurrence of thrombosis, indicative of the necessity for balance between tPA activation of plasminogen and t-PA inactivation by PAI-1.

Plasminogen can be activated rapidly by the formation of a "stoichiometric" complex between the plasminogen molecule and proteins from several strains of hemolytic *Streptococcus* and *Staphylococcus*. The "streptokinase-plasmin" complex (SK-plasmin) converts plasminogen to plasmin as well as digesting fibrin. In contrast to normal plasmin, SK-plasmin is not inhibited by α_2-antiplasmin. Streptokinase-plasmin(ogen) is used therapeutically in fibrinolytic therapy to remove thrombi. Another therapeutically useful product has been prepared from SK-plasmin. In this product, the active Ser of the plasmin is reacted with an acylating chemical to form an unstable acyl ester. Slow, spontaneous hydrolysis of the acylated SK-plasmin results in a more sustained level of SK-plasmin in the circulation during fibrinolytic therapy.

Plasmin is inactivated by α_2-antiplasmin in a very rapid reaction; k_2/K_d is approximately 2×10^7 M^{-1} s^{-1} (k_2/K_d is analogous to k_c/K_m, the best indicator of the efficiency of an enzyme-catalyzed reaction). Blocking of the kringle Lys binding sites reduces the rate of plasmin inactivation. This is most evident from the "protection" of plasmin that occurs when the plasmin is bound to the fibrin strands. This protection tips the balance between the proteolytic action of plasmin on its fibrin substrate and its inactivation by α_2-antiplasmin to proteolysis. The half-life (time for half of the plasmin to be inactivated by α_2-antiplasmin) is greater than 10 s for plasmin bound to fibrin, while it is approximately 0.1 s in the absence of fibrin. This is another

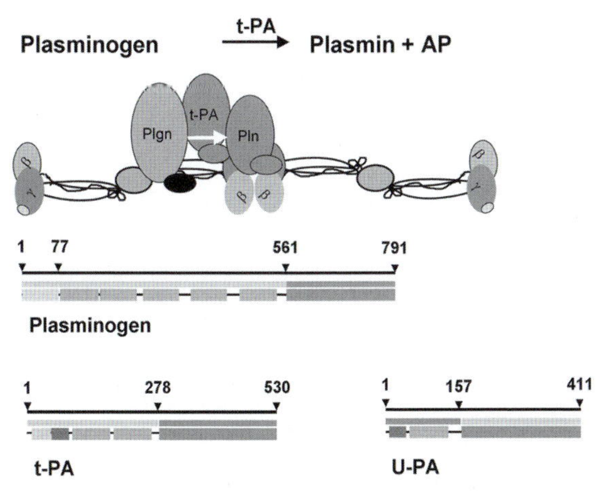

FIGURE 36-18
(Also see color figure.) Plasminogen activation. Plasminogen is converted to plasmin as the result of cleaving a single peptide bond, Arg^{561}-Val^{562}. The molecular mechanism of transforming the zymogen into its active form is the same as the activation of trypsinogen to trypsin. Two different proteinases are responsible for the physiological activation of plasminogen: tissue-type plasminogen activator (t-PA) and urinary plasminogen activator (u-PA). t-PA is the principal activator of plasminogen and is synthesized primarily in endothelial cells, which are the principal source of t-PA in the circulating blood. t-PA is present in high concentrations in uterine tissue and in various malignant cells. u-PA is found in the kidney and may be particularly important in preventing the accumulation of fibrin in the glomeruli. In the circulation, more than one-half of the t-PA is in a complex with PAI-1. The activation of plasminogen is accompanied by the proteolytic action of plasmin on plasminogen at Arg^{68}-Met^{69} and at Lys^{77}-Lys^{78}. This reaction results in the formation of a derivative of plasminogen in which the first 77 amino acids at the N-terminal end of the molecule are removed. This derivative of plasminogen, *lys plasminogen*, is activated more rapidly than the native, *glu plasminogen*. In the color version of this figure, motifs and domains are color coded as follows: kringle domains (orange), EGF-like domain (magenta), activation peptide (yellow), and proteinase domain (green). Light chains are indicated in dark gray, heavy chains in light gray. Regions connecting the motifs are black.

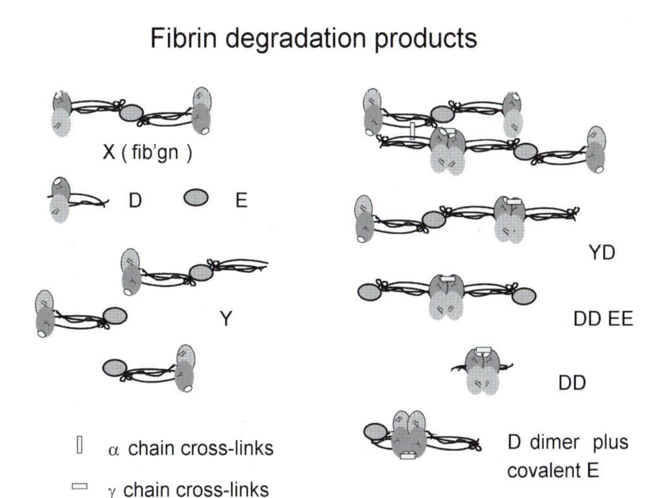

FIGURE 36-19
(Also see color figure.) Products of fibrinolysis. The products of plasmin action are used to identify ongoing fibrinolysis and to distinguish fibrinolysis from fibrinogenolysis. The presence of larger fibrin fragments, which result from factor XIIIa–catalyzed cross-linking, is the basis for the distinction between fibrinogenolysis and fibrinolysis. Detection of D dimer (DD) is particularly useful. Some of the fibrin degradation products interfere with polymerization of fibrin. Digestion of fibrinogen by plasmin proceeds with initial cleavages that remove polypeptides of approximately 40,000 Da from the carboxy terminal regions of the A chains. The large fragment that is formed is designated fragment X, and it is heterogeneous. Fragment X has a molecular weight of ~300,000 (one A-chain peptide removed) or ~260,000 (two A-chain peptides removed). Further proteolysis by plasmin removes polypeptides of ~100,000 Da total mass from the disulfide bond–linked carboxy terminal regions of A, B, and γ chains. The larger product of this fibrinogenolysis is designated fragment Y, which has a molecular weight of ~150,000 Da. The product that contains the cross-linked polypeptides from A, B, and γ chains is designated fragment D. A second round of proteolysis removes the remaining fragment D from fragment Y to form fragment E (M.W. ~33,000 Da) and a second fragment D. Fragment E is highly cross-linked and is also known as the N-terminal disulfide knot (DSK). In fibrinogen, an additional cleavage occurs early in the fibrinogenolytic process, i.e., cleavage of Arg^{42} in the $B\beta$ chain to produce $B\beta$ 1–42. This can be distinguished from the product that is produced by the cleavage of fibrin, β 15–42.

example of proteinase protection when the proteinase action is localized in a reaction complex.

Degradation of Fibrin (Fibrinolysis)

The principal substrate of plasmin is fibrin and except under pathological situations, e.g. disseminated intravascular coagulation (DIC), the concentrations of the products of plasmin action are usually less than 10 μg/mL (less than 0.5% of the circulating fibrinogen). Because of the opportunities for artifactual generation of FDPs (fibrin degradation products) during the collection of the blood sample and the common use of serum, the true concentrations of FDPs in normal plasma are probably less than 50 ng/mL. The opportunity for artifactual elevation of the FDPs is easily understood when the fibrin enhancement of plasminogen activation is recalled. The products of fibrin digestion, i.e., fibrin degradation products or FDPs, are distinguishable from fibrinogen degradation products. The cross-linking of fibrin by factor XIIIa produces D dimers and α polymers. The reactions of fibrin degradation and the products are shown in Figure 36-19.

Under pathological conditions, e.g., (DIC), fibrinogen can be digested by the fibrinolytic enzyme plasmin. This leads to bleeding, even though clotting is also occurring.

36.10 Vitamin K, Oral Anticoagulants, and Their Mechanisms of Action

Vitamin K acts in the liver as an enzyme cofactor in the reactions that add an additional carboxyl group to particular Glu residues in the vitamin K-related coagulation

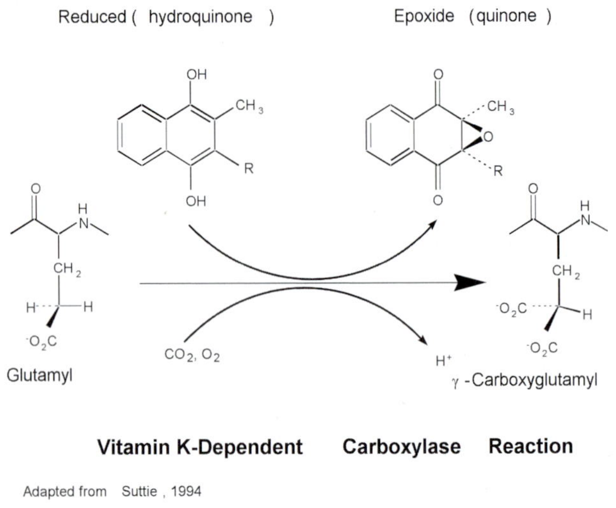

Reduced (hydroquinone) Epoxide (quinone)

OH CH3 R OH O CH3 O R O

Glutamyl CO_2, O_2 H^+ γ -Carboxyglutamyl

Vitamin K-Dependent Carboxylase Reaction

Adapted from Suttie , 1994

FIGURE 36-20
Vitamin K action in protein Glu carboxylation. The formation of Gla from Glu requires abstraction of a H ion from the γ position of Glu in the proteins that is followed by the addition of a carboxyl group from CO_2. The proteins to be carboxylated are marked by a propeptide that serves as a signal for the vitamin K-dependent carboxylase.

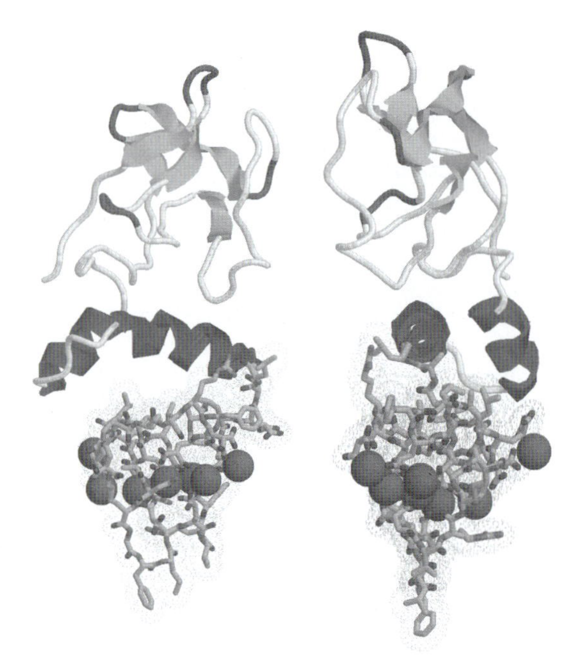

FIGURE 36-21
(Also see color figure.) Structure of a Gla domain of prothrombin. The Gla domains of the vitamin K-dependent proteins contain between 9 and 12 carboxyglutamic acid residues. Prothrombin contains 10 Gla residues: pairs of adjacent Gla residues at positions 6 and 7; 19 and 20; and 25 and 26, and four other Gla residues at positions 14, 16, 29, and 32. Factor VII also contains 10 Gla residues in the same positions as in prothrombin. Factor IX contains 12 Gla residues. Because factor IX has an additional amino acid between its amino terminal Tyr and the first Gla residue, the numbering of the residues is different from prothrombin and factor VII. However, the pattern is similar with pairs of adjacent Gla residues at positions 7 and 8; 20 and 21; and 26 and 27. The six other Gla residues are at positions 15, 17, 30, 33, 36, and 40. The last two Gla residues account for 12 Gla residues in factor IX. Factor X contains 11 Gla residues: pairs of adjacent Gla residues at positions 6 and 7; 19 and 20; and 25 and 26, with five other Gla residues at positions 14, 16, 29, 32, and 39. Protein C contains 9 Gla residues: pairs of adjacent Gla residues at positions 6 and 7; 19 and 20; and 25 and 26, and the other Gla residues at positions 14, 16, and 29. In the color versions of this figure, the cartoon structures show helices in magenta, β sheets in yellow, β bends in blue, and Ca^{2+} ions in gray. The space-filling dot surfaces for the residues of the Gla domains are standard CPK colors.

factors. Vitamin K is a substituted naphthoquinone. A methyl group at the 2-position and a prenyl (or phytyl) side chain of varying length is required for the vitamin to be active. The two primary sources of vitamin K in humans are green plants and intestinal flora. The designation "dietary factor K" derives from the initial description of the vitamin as the "koagulations vitamin." The active form of the vitamin is the reduced hydroquinone form, not the quinone form found in the diet. In the vitamin K-dependent carboxylase reaction, the vitamin is converted from the hydroquinone to an epoxide (Figure 36-20). During this reaction, an intermediate O^- anion is involved in the removal of a hydrogen from the γ-carbon of the Glu residues in the vitamin K-dependent protein. This stereospecific reaction requires O_2 and the hydroquinone for the formation of the Glu anion. Reaction of the Glu anion with CO_2 leads to the formation of the γ-carboxy Glu. A propeptide of 23 amino acids at the N-terminal end of vitamin K-dependent proteins provides the signal for carboxylation of the Glu residues in the "precursor" Gla domain of the protein on the ribosome. The propeptide is cleaved from the precursor molecule prior to secretion.

In the absence of vitamin K or in the presence of antagonists of vitamin K action, uncarboxylated vitamin K-dependent proteins are synthesized and secreted from the liver. The residues that would be Gla in the normal, posttranslationally modified proteins are ordinary Glu residues. These proteins do not bind Ca^{2+} normally and do not bind to the surface of the phospholipids in the cell membranes. The structure of the Gla domains with bound Ca^{2+} ions is shown in Figure 36-21.

The physiological expression of oral anticoagulant action is an increase in the time required for clotting in the prothrombin time assay. The slowing of all of the reactions that lead to the formation of thrombin is the direct result of the reduced concentrations of the vitamin K-related proteins in the reaction complexes on the membrane surface. The effects of oral anticoagulant blockage on the carboxylation reaction are common to all vitamin K-related proteins.

Action of Warfarin and Other Vitamin K Antagonists

Vitamin K antagonists were first identified by K. P. Link while he was investigating a hemorrhagic disease in cattle. The compound responsible for the hemorrhagic disease, a product of microbial metabolism (spoiling) of sweet clover, was identified as 3,3′-methylbis-4-(hydroxycoumarin). This compound was named dicoumarol. A synthetic, clinically useful compound with the ability to function like dicoumarol is warfarin [1-(4′-hydroxy-3′-coumarinyl)-1-phenyl-3-butanone]. This compound is sold in the United states under the name Coumadin.

The oral anticoagulant drugs act indirectly on the process of Glu carboxylation of the vitamin K-dependent proteins. The vitamin K antagonists prevent the reduction of the reaction intermediate, vitamin K epoxide. Blockage of the epoxide reductase reaction results in the accumulation of vitamin K epoxide and other nonfunctional forms of vitamin K. Without regeneration of the vitamin K-related reaction intermediates through the cycle shown in Figure 36-22, a depletion of functional vitamin K occurs.

Vitamin K antagonists are processed in the liver by the cytochrome P450 systems that metabolize many other drugs. Drug interactions commonly alter the effectiveness of oral anticoagulant drugs. This requires that oral anticoagulant therapy be monitored frequently during the early stages of therapy and at regular intervals after stable

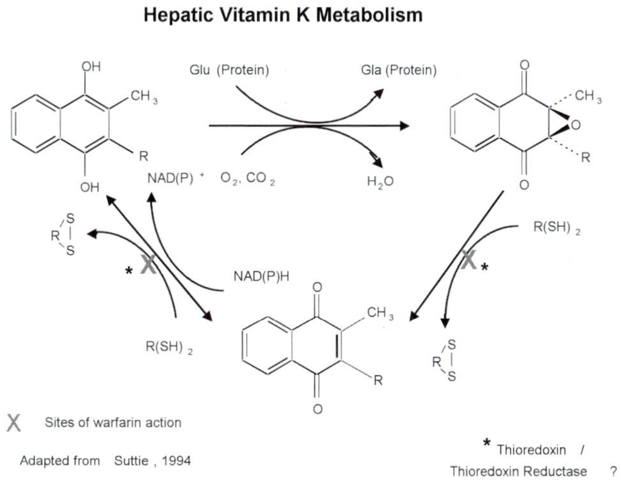

FIGURE 36-22
(Also see color figure.) Hepatic vitamin K metabolism. Oral anticoagulant drugs act indirectly on the process of glu carboxylation of the vitamin K dependent proteins. The vitamin K antagonists block the reduction of the reaction intermediate, vitamin K–epoxide, that results in the accumulation of vitamin K–epoxide and other nonfunctional forms of vitamin K. Without cycling of the vitamin K-related reaction intermediates in the cycle shown, a depletion of functional vitamin K occurs. Vitamin K antagonists do not block polypeptide synthesis and formation of noncarboxylated proteins occurs. These noncarboxylated proteins are still secreted from the liver and account for nearly normal levels of each of the vitamin K-dependent proteins that can be detected by immunoassay (sometimes called PIVKA, proteins induced in vitamin K absence).

anticoagulation (a predetermined level of effectiveness) is achieved to prevent both excessive hemorrhage and inadequately reduced risk of thrombosis.

36.11 Coagulation Factor Measurements

Contemporary methods of coagulation factor testing have their conceptual origins in the "classical theory of coagulation," circa 1905, and the long-standing recognition of inherited bleeding disorders in the royal families of Europe. A. J. Quick developed the first practical method for evaluating the functionality of the "classical" (extrinsic) pathway with the publication of the "prothrombin time" assay. This test was so named because the classical pathway recognized only fibrinogen and prothrombin, the postulated precursor of thrombin and thromboplastin. This test established a measurable relationship between the time required for clot formation and the presence or absence of the requisite components (activities) and was a milestone in the advancement of the study of blood coagulation.

The concept of qualitatively distinguishable components (factors) is related to the observation that one could "correct" the time for clotting of plasma from an individual with a bleeding disorder by mixing the patient's plasma with plasma from an individual without bleeding problems. Moreover, if plasmas from two individuals with bleeding disorders were mixed, the disorders could be classified as the same if there were no corrective effect, or different if there were a corrective effect. Variations of this approach, coupled with clinical presentation of the bleeding disorder, have provided the basis for the discovery of most of the procoagulant subsystem components.

Quantification was developed by measuring the relationship between the shortening of the clotting time of the plasma from the patient with the bleeding disorder (deficient plasma) and the amount of normal plasma added to the deficient plasma. This approach remains the basis for specific factor assays; however, the deficient plasmas are depleted of individual factors by immunoprecipitation and separation of the antibody-factor complex from the plasma.

Deficiencies

The term *coagulation factor deficiency* is frequently ambiguous. Previously, it implied a deficit in the functional activity of a component because the clotting time tests measured only the ability of the component to support normal coagulation. The reference for normal functionality is the response of dilutions of plasma from pooled plasma samples from individuals with no known bleeding or thrombotic tendencies. Consequently, both a decreased concentration of a factor (protein) and the presence of the factor in a functionally inactive form result in functional deficiency. Deficiencies in which the concentration of the normal factor is less than present in the pool are now categorized as type I. Deficiencies in which the normal concentration is present but some molecules are inactive or less than fully active are categorized as type II. Subcategories exist but are beyond the scope of this description. Frequencies of each of the coagulation factor deficiencies are given in Table 36-1. Deficiencies in procoagulant subsystem components that lead to hemorrhage are relatively rare. Deficiencies of anticoagulant subsystem components are found with much higher frequency in individuals presenting with venous thrombosis. Prospective, epidemiological studies indicate, however, that heterozygous individuals deficient in single components do not have a substantial additional risk of thrombosis. It is now generally believed that *thrombophilia* (a substantial risk of thrombosis) is a multigenic disorder.

Laboratory Assessment of Coagulation System Functions

Sample collection and sample handling are particularly important in coagulation factor testing. The venipuncture, or capillary puncture for some devices, is a blood vessel injury that elicits hemostatic response. When a tourniquet is employed to facilitate venipuncture, which is almost universally the case, the tourniquet should be in place for the shortest time possible. Plasminogen activator (t-PA) is released from the vessel in response to the compression and activates plasminogen. Badly collected samples from extensive tourniquet compression can be rendered unclottable because of the fibrinogenolysis that occurs from plasmin action. Normally, all abnormal coagulation system functional tests should be reconfirmed.

Prothrombin Time

The prothrombin time (PT), as proposed by Quick, is the most commonly performed coagulation function test. It is used to monitor oral anticoagulant therapy and also as a preoperative screening test to warn of possible bleeding risk in patients with a personal or family history of bleeding (Table 36-2). Measured clotting times are extremely dependent on the animal and tissue source and the quality of the thromboplastin used. Variability can be expected because of the assay's dependence on the number of tissue factor molecules and the quantity of membrane surface provided by the thromboplastin. To improve the comparability of prothrombin time measurements, thromboplastins are standardized by comparison with an international reference thromboplastin. Standardized thromboplastins are described by a number, the International Standardization Index (ISI), which relates the thromboplastin to the international standard. Although this has improved comparisons of oral anticoagulant therapy monitoring, it is still difficult to accurately compare the results of different laboratories and different thromboplastin manufacturers.

Further improvement in measurement of oral anticoagulation therapy has been achieved by relating the ratio of the prothrombin time of the patient to the mean value for "normal" individuals who are not on oral anticoagulant therapy. This ratio, called the International Normalized Ratio (INR), better relates the anticoagulation assays from different laboratories. Monitoring oral anticoagulant therapy is critical to maintaining a balance between the risk of thrombosis and the risk of bleeding.

Thrombin Time

Purified thrombin is added to plasma samples and the time for clotting is measured. This test, the thrombin time (TT), primarily reflects the concentration of fibrinogen. However, it also reflects the ability of the fibrinopeptides to be cleaved and the polymerization of fibrinogen. Separation of these three contributions requires a quantitative measurement of the concentration of fibrinogen that is not related to time for clot formation. The thrombin time can be prolonged if heparin is present in the patient's plasma sample because it will promote preferential inactivation of the added thrombin by antithrombin and reduce the amount of thrombin that can act on fibrinogen.

Activated Partial Thromboplastin Time

Intrinsic pathway components are measured in the activated partial thromboplastin time (APTT). The designation "activated" refers to the performance of the test in two stages. In the first stage, the citrate-anticoagulated plasma is preincubated with the surface activator (see "The Contact Phase of the *In Vitro* Intrinsic Pathway of Coagulation") to form activated factor XI. In the second stage, Ca^{2+} is added to initiate the activation of factor IX and the remaining reactions of the "intrinsic" pathway. The clotting time is determined from the time of addition of Ca^{2+}. The most common uses of the APTT are screening for possible factor VIII or factor IX deficiencies and for monitoring heparin therapy. Because the rate of activation of factor IX and the clotting time are very dependent on the

TABLE 36-2

Common Tests of Hemostatic Function

Test Name	Description	Component Measured Sensitivity	Common Use	Interferences/Complications
Plasma Procoagulants				
Prothrombin Time (PT)	Functional evaluation of the extrinsic and final common pathways. Time for clot formation is determined after addition of thromboplastin and Ca^{2+} to a plasma sample.	**Factor VII**, Factor X, Factor V, Prothrombin and **Fibrinogen** (afibrinogenemia, dysfibrinogenemias)	Monitoring oral anticoagulant therapy Preoperative screening Liver dysfunction (effects on synthesis and secretion of vitamin K-dependent factors)	Particularly dependent on the the thromboplastin (tissue factor and phospholipids) Vitamin K-deficiency arising from antibiotic therapy decreases γ-carboxy-glutamate formation Use for oral anticoagulant monitoring requires careful "standardization" by the INR method. Cannot compare results from different laboratories
Activated Partial Thromboplastin Time (aPTT)	Functional evaluation of the intrinsic and final common pathways. Activation of the contact phase is by addition of a surface reagent to a plasma sample. Preincubation in the absence of Ca^{2+} produces Factor XIa. Time for clot formation is determined after Ca^{2+} and phospholipid are added to the incubation mixture.	**Factor VIII, Factor IX,** Factor Va inactivation An abnormally short aPTT suggests resistance to activated Protein C.	Preoperative screening Diagnosis of bleeding disorders Heparin monitoring (poor), but widely used during surgery involving extracorporeal circulation Activation of Factor XI to XIa in the first stage is sensitive to Factor XII	Most common cause of prolongation is heparin Standardization is difficult. Cannot compare results from different laboratories. Can be affected by platelet fragments (microparticles) arising from poor sample acquisition and handling technique.

(continued)

TABLE 36-2
Common Tests of Hemostatic Function *(continued)*

Test Name	Description	Component Measured Sensitivity	Common Use	Interferences/Complications
Plasma Procoagulants *(continued)*				
Activated Clotting Time	A form of an aPTT performed using whole blood	Heparin	Heparin monitoring in the operating room	Insensitive to low mol wt heparins
Thrombin Time	Functional evaluation of fibrinogen concentration, polymerization	Fibrinogen, antithrombin/(*heparin*)	Fibrinogen 'concentration' and fibrin functionality (fibrinopeptide cleavage, fibrin polymerization)	Heparin causes artifactual prolongation because of enhanced thrombin inactivation by antithrombin
Specific Factor Assays Factor VIII Factor IX Other Factors	Evaluation of the functional concentration (activity) of individual factors, specificity created by using specific factor deficient plasmas. Commonly performed as an aPTT-like assay	Sensitivity is to the factor missing from the particular deficient plasma	Estimation of the activity level of a particular factor	Differences may be observed between deficient plasmas from individuals with hereditary deficiencies and immunodepleted plasmas
Primary Hemostasis, Cellular Components				
Platelet Count	Direct determination of the number of platelets per unit volume	Platelet number	Diagnosis of thrombocytopenia	Generally none, but insensitive to defective function
Bleeding Time	Measurement of vWF function, platelet function and indirect measure of platelet count and functionality	Platelet count, vWF function (multimers) and platelet function	General hemostatic function, without platelet count interpretation cannot be made	Interpretation requires an independently determined platelet count

(continued)

TABLE 36-2
Common Tests of Hemostatic Function (continued)

Test Name	Description	Component Measured Sensitivity	Common Use	Interferences/Complications
Plasma Coagulation Factors—Anticoagulant Subsystem Components				
Protein C	Determination of the activity of Protein C Determination of the amount of Protein C antigen	Other vitamin K-dependent proteins, especially in samples from patients receiving vitamin K antagonist oral anticoagulants. In some tests that use aPTT or PT based procedures dependence on Factor VIII and Factor V exists.	Determination of a possible cause of thrombosis, or thrombotic risk, performed only in specialized laboratories	Oral anticoagulant therapy. Different procedures are inconsistently biased by the decreased γ carboxylation of Protein C and the other vitamin K-dependent proteins
Protein S	Determination of the activity (ability) of Protein S to enhance the activity of activated Protein C	C4b binding protein can affect the assay, only free Protein S is measured	Determination of a possible cause of thrombosis, or identification of thrombotic risk, performed only in specialized laboratories	Oral anticoagulant therapy, Factor V^{R506Q} (Factor V Leiden), anti-phospholipid antibodies (lupus anticoagulants)
	Determination of the amount of Protein S antigen	C4b binding protein can affect the assay		
APC Resistance Assay	Determination of the ability of activated Protein C to inactivate Factor Va in patients in which the Protein C activity is normal	The most common assay is a modified aPTT. Elevated levels of Factor VIII bias the assay results, as do decreased levels of vitamin K-dependent factors.	Determination of a possible cause of thrombosis. May be used as a screen for women on oral contraceptives who are suggested to have other risk factor for thrombosis.	Oral anticoagulants, unless a dilution is made into Factor V-deficient plasma and a variation on a factor V specific assay is needed
Antithrombin	Determination of the activity (functional concentration) of antithrombin	Minimal other component sensitivity because of the use of exogenous thrombin or Factor Xa. Clinically antithrombin levels may be lowered by liver disease	Determination of the cause of thrombosis, risk of thrombosis in patients with history of thrombosis	Heparin cofactor II may interfere, bovine thrombin minimizes this interference

(continued)

TABLE 36-2
Common Tests of Hemostatic Function (continued)

Test Name	Description	Component Measured Sensitivity	Common Use	Interferences/Complications
Plasma Coagulation Factors—Anticoagulant Subsystem Components *(continued)*				
Heparin	Determination of the functional concentration of heparin	Specific heparin assays employ either exogenous thrombin or Factor Xa. Exogenous antithrombin may also be added to ensure that only heparin is measured	Monitoring "standard" heparin and low mol wt heparins. Routine heparin monitoring is commonly done using the aPTT, but the aPTT is insensitive to the new low mol wt heparins. Common OR procedure in surgeries employing extracorporeal circulation	Heparin binding to thrombin (if thrombin is used) can cause biased results. Thrombin-based assays are poorly sensitive to low mol wt heparins
Plasma Coagulation Factors—Fibrinolytic System Components				
Plasminogen/Plasmin	Determination of fibrinolysis or plasmin hydrolysis of a peptide chromogenic substrate	Plasmin inhibitors (antiplasmin)	Monitoring of fibrinolytic therapy, evidence of ongoing fibrinolysis, decreased levels of plasminogen	Incomplete activation of plasminogen by streptokinase, competing inactivation reaction with inhibitors
Fibrin D-Dimer	Immunoassay, formats are ELISA and latex agglutination	Fibrin	Determination of ongoing fibrinolysis and its extent, evidence for thrombosis	Fibrin degradation products that contain the D fragment
Plasminogen Activator(s)	Functional assay–PA's circulate in an active form and are released by blood pressure cuff compression, the activator's ability to convert plasminogen into plasmin is measured	Affected by plasminogen activator inhibitors (PAI's) Show substantial diurnal variation	Monitoring of fibrinolytic therapy, determination of possible cause for thrombosis or risk factor for thrombosis, may be useful in predicting ischemia	Inactivation by inhibitors, minimized by collection into pH 6 buffer (without pH 6, values are as low as 1/10 normal values) Prolonged cuff pressure
	Determination of the amount of tPA antigen by ELISA	Affected by plasminogen activator inhibitors (PAI's) Plasminogen (antibody specificity), glycosylation of tPA	Possible cause for thrombotic risk, performed only in specialized laboratories	Complicated by the presence of tPA and TPA-PAI 1 complexes

(continued)

TABLE 36-2
Common Tests of Hemostatic Function *(continued)*

Test Name	Description	Component Measured Sensitivity	Common Use	Interferences/Complications
Plasma Coagulation Factors—Fibrinolytic System Components *(continued)*				
Plasminogen Activator Inhibitor(s)	Determination of the amount of PAI antigen by ELISA	TPA reaction with PAI-1, Large variability of unestablished causes is observed among kits from different manufacturers	Determination of possible cause of increased bleeding tendency in the absence of procoagulant factor deficiency, performed only in specialized laboratories	Release of PAI-1 from platelets

factor XIa, the APTT is sensitive to deficiencies in factor XII, prekallikrein, high-molecular-weight kininogen, and factor XI.

Specific Factor Assays

Specific factor assays are variations on the APTT or PT tests. In the APTT and PT, dilutions of the patient's plasma are made into a deficient, or depleted, "substrate" plasma. The assays are then performed in the usual way. The clotting times are compared with those obtained from dilutions of pooled normal plasma, commonly 1:10, 1:20, 1:50, and 1:100. A graph of the logarithm of the clotting time (y axis) versus the logarithm of the concentration as percentage of normal (x axis) is used to determine the amount of the factor activity in the patient's plasma. The normal pooled plasma is conventionally assigned a value of 100% activity. Many variations exist for specific factor assays, e.g., the venom of *Vipera russellii* and phospholipids may be substituted for thromboplastin in a PT-like assay. An enzyme in Russell's viper venom rapidly and relatively specifically activates factor X. In conjunction with a factor X-deficient "substrate" plasma, this provides a specific factor X assay.

Immunoglobulin inhibitors to factor VIII and factor IX are found in patients with hemophilia A and B, respectively as a consequence of their treatment with factor VIII and factor IX preparations to arrest bleeding. Evidence for inhibitors can be seen from APTT assays using specific factor-deficient substrate plasmas. This evidence is obtained by making dilutions of the patient's plasma and observing that higher dilutions give higher levels of activity for the factor in the patient's plasma. This reflects the simultaneous dilution of the antibody inhibitor and the decreased inhibition at the higher dilution.

Assay of Heparin

Heparin therapy may be monitored by its increases in clotting time in the APTT, although this method for measuring heparin is difficult to standardize. Heparin is more specifically assayed by its effect on factor Xa inactivation by antithrombin. Such factor Xa-based heparin assays usually employ purified factor Xa as a reagent and factor X-deficient substrate plasma as the source of antithrombin. The prolongation of the clotting time that results from the heparin in the patient's plasma is compared with pooled normal plasma that is known to be free of heparin. Many variations of this heparin assay are available. Heparin assays can use thrombin rather than factor Xa, however, the low-molecular-weight heparins are not reliably measured in thrombin-based assays.

36.12 Case Studies

Fibrinogen

Testing on a plasma sample from a patient gave a PT of 12.9 s (normal 10.5–12.2 s) and an APTT of 38 s (normal 20–32 s). A TT was determined and the clotting time was 15 s (normal 10–12 s). The patient's plasma sample was incubated with an enzyme that digests heparin. After this treatment, the TT was 14.7 s. The concentration of fibrinogen was determined by a method that weighs the fibrin after clotting. The fibrinogen was 340 mg/dL, a value higher than the normal range. What might explain these results?

- *Fibrinogen proteolysis or fibrin monomer polymerization may be impaired as the result of a mutation that alters the fibrinogen molecule.*

Factor XIII

A 60-year-old women is referred to you because of bleeding that has been a problem since infancy. Childbirth (multiple) and surgery have required treatment with cryoprecipitate and other plasma protein fractions that contain fibrinogen, vonWillebrand factor, factor VIII, factor XIII, and many other plasma proteins. All ordinary coagulation functions tests were determined by you and found to be normal. Her platelet count was measured and found to be at the high end of the normal range. The addition of plasmin to the plasma clot resulted in almost immediate dissolution (lysis) in contrast to ~10 minutes for a control clot formed from pooled normal plasma. Plasmin was not detectable in a freshly collected plasma sample from the patient. After addition of Ca^{2+} and incubation for 1 h, the patient's plasma clot was mixed with an equal volume of 8 M urea and the clot dissolved. A plasma clot prepared in the same manner from normal plasma did not dissolve upon addition of 8 M urea. What is the likely cause of the patient's bleeding disorder and her laboratory findings?

- *Factor XIII deficiency. There is no expected effect on any of the clotting time measurements. The susceptibility to plasmin suggests an unstabilized clot. The solubility in urea is diagnostic for un-cross-linked fibrin.*

Activated Protein C Resistance

A 30-year-old man with recurrent thrombosis is referred to you for diagnosis. The PT and APTT results determined on his freshly drawn plasma sample are 10.1 s (normal 10–12 s) and 18 s (normal 20–32 s). Which components of the hemostatic system would you initially suspect as being responsible for the thrombosis and the test results? Measurements of antithrombin in the presence and absence of heparin are normal. What would you consider candidate factors to explain the thrombosis? If both protein C and protein S were normal, what would you suspect? If assays for the inactivation of factor V by activated protein C were normal, to which of the known factors in the anticoagulant subsystem would you ascribe his thrombosis?

- *The components are all mentioned in the case. Because of the current attention on APC resistance, it is almost certainly that this test would be performed first. Antithrombin is indicated as being measured first because of its simplicity compared to protein C and protein S measurements. If there is no APC resistance, thrombomodulin defects might be considered.*

Antibiotic-Induced Vitamin K Deficiency

A 4-year-old girl was brought to the "urgent care" department of your hospital after a minor automobile accident in which glass shards had caused some cuts. Continued bleeding from the cuts was noted from a pile of blood-spotted tissues collected during her 3-h wait in the emergency room! What questions would you ask her and/or her parents about the bleeding? If the parents indicated that she had been receiving high doses of antibiotics for an infection during the last few weeks, what coagulation test would you expect to be abnormal?

- *A history of bleeding by the girl, parents, or other relatives should be explored. If there was no bleeding, given the patient's age and any information about infectious diseases in the community, use of antibiotics might be indicated. The PT could be prolonged if the girl had used antibiotics. This is the result of the antibiotics killing the intestinal bacteria that produce vitamin K from dietary sources of phylloquinones.*

Hemophilia

A young African-American male patient is brought to you because of a hemarthrosis sustained after he twisted his knee while running. His PT is normal, but his APTT is 51 s (normal 20–32 s). Mixing equal volumes of the patient's plasma and normal plasma shortens the APTT to 25 s. Specific factor assays show a normal factor IX activity, but a factor VIII activity of ~25% of that in pooled normal plasma. Why was his PT normal? What else, prior to the factor IX and VIII assay results, might have been responsible for the prolonged APTT?

- *Factor VIII is not involved in the PT, i.e., "extrinsic," pathway. Before the normal factor IX result was obtained, factor XII deficiency might have causes a prolonged APTT. Because the patient was not under treatment for anything, heparin would be unlikely.*

Von Willebrand Disease

A 7-year-old boy is brought to you because of recurrent nose bleeds and bleeding after minor injuries associated with ordinary "horse play." There is no history of any type of bleeding disorder in the four brothers in the mother's family; she is the only girl. The PT and the APTT tests were performed and both were at the low end (shorter clotting times) of the normal range. A platelet count was done and found to be normal. What hemostatic system components might be responsible for the bleeding but not be detectable in the two tests for which results had been obtained? What test would you suggest to complete the diagnosis?

- *von Willebrand factor and factor XIII should be tested. Because testing of factor XIII is easier, it might be done first.*

Laboratory-Created Artifacts

A plasma sample was obtained from a female patient for routine coagulation testing, i.e., there was no history of bleeding. The PT was 11.9 s (normal 10.5–12.2 s). The APTT was 80 s (normal 20–32 s). Sample was sent to a reference laboratory specializing in coagulation factor testing. At the reference laboratory, the PT was 11.6 s and the APTT was 43 s. Because of this discrepancy, blood was drawn at the reference laboratory and a new plasma sample prepared. In a second set of tests performed at the reference laboratory, the PT was 10.9 s and the APTT was 94 s. What might account for the discrepancy between the results obtained on the original plasma sample in the two laboratories?

- *If the initial sample was not centrifuged appropriately to remove the platelets, platelet fragmentation and microparticle formation could have occurred during transport to the reference laboratory. The membrane fragments could artifactually shorten the APTT.*

Acknowledgments

The images shown in this chapter are created using RasMol 2.6, Molecular Graphics Visualisation Tool by Roger Sayle, BioMolecular Structures Group, Glaxo Research & Development, Greenford, Middlesex, UK. RasMol can be obtained at http://www.umass.edu/microbio/rasmol/. The coordinates were obtained from the Protein Data Bank at Brookhaven National Laboratory as described in F. C. Bernstein, T. F. Koetzle, G. J. B. Williams, E. F. Meyer, Jr., M. D. Brice, J. R. Rodgers, O. Kennard, T. Shimanouchi, and M. Tasumi, "The Protein Data Bank: A Computer-based Archival File for Macromolecular Structures", *J. Mol. Biol.,* **112,** 535–542 (1977). Gene information has been obtained from GenBank: Burks, C., Cassidy, M., Cinkosky, M. J., Cumella, K. E., Gilna, P., Hayden, J. E-D., Keen, G. M., Kelley, T. A., Kelly, M., Kristofferson, D., and Ryals, J. GenBank. Nucl. Acids Res. **19**(Suppl), 2221–2225 (1991). For the coordinates of the molecules for which structures are shown see http://www.rcsb.org/. The author thanks James G. White, M. D., Regents' Professor, University of Minnesota, Minneapolis, Minnesota for the photographs of platelets and Meir Rigbi, Professor Emeritus, Hebrew University of Jerusalem for his many helpful comments on the manuscript.

Supplemental Readings and References

R. D. Appleby and R. J. Olds: The inherited basis of venous thrombosis. *Pathology* **29,** 341 (1997).

D. W. Banner: The factor VIIa/tissue factor complex. *Thrombosis and Haemostasis* **78,** 512 (1997).

K. A. Bauer: New markers for *in vivo* coagulation. *Current Opinions in Hematology* **1,** 341 (1994).

W. Bode and M. T. Stubbs: Spatial structure of thrombin as a guide to its multiple sites of interaction. *Seminars in Thrombosis and Hemostasis* **19,** 321 (1993).

R. Carrell, D. Lomas, P. Stein, and J. Whisstock: Dysfunctional variants and the structural biology of the serpins. *Advances in Experimental Medicine and Biology* **425,** 207 (1997).

D. Collen and H. R. Lijnen: Molecular basis of fibrinolysis, as relevant for thrombolytic therapy. *Thrombosis and Haemostasis* **74,** 167 (1995).

R. A. Engh, R. Huber, W. Bode, and A. J. Schulze: Divining the serpin inhibition mechanism, a suicide substrate "spring?" *Trends in Biotechnology* **13,** 503 (1995).

T. Halkier: *Mechanisms in Blood Coagulation Fibrinolysis and the Complement System.* Cambridge University Press, Cambridge, 1991, pp. 466.

C. M. Jackson: Hemostasis and blood coagulation, *Encyclopedia of Human Biology*, R. Dulbecco, (Ed), Academic Press, Inc., San Diego, Vol. 2, Second Edition **23** (1997).

M. Kalafatis, J. O. Egan, V. K. M. van Cawthern, and K. G. Mann: The regulation of clotting factors. *Critical Reviews of Eukaryotic Gene Expression* **7,** 241 (1997).

H P. Kohler and P. J. Grant: Plasminogen-activator inhibitor type 1 and coronary artery disease. *New England Journal of Medicine* **342,** 1792 (2000).

D. A. Lane and P. J. Grant: Role of hemostatic gene polymorphisms in venous and arterial thrombotic disease. *Blood* **95,** 1517 (2000).

H. R. Lijnen and D. Collen: Mechanisms of physiological fibrinolysis. *Baillieres Clinical Haematology* **8,** 277–290 (1995).

J. P. Miletich: Thrombophilia as a multigenic disorder. *Seminars in Thrombosis and Hemostasis* **24** Supplement 1, 13 (1998).

M. W. Mosesson: Fibrinogen structure and fibrin clot assembly. *Seminars in Thrombosis and Hemostasis* **24,** 169 (1998).

Y. Nemerson and P. L. Giesen: Some thoughts about localization and expression of tissue factor. *Blood Coagulation and Fibrinolysis* **9**(Suppl. 1), S45 (1998).

S. T. Olson and I. Bjork: Regulation of thrombin activity by antithrombin and heparin. *Seminars in Thrombosis and Haemostasis* **20,** 373 (1994).

S. I. Rapaport and L. V. Rao: The tissue factor pathway, how it has become a "prima ballerina". *Thrombosis and Haemostasis* **74,** 7 (1995).

J. Rosing, H. C. Hemker, and G. Tans: Molecular biology and pathophysiology of APC resistance, current insights and clinical implications. *Seminars in Thrombosis and Haemostasis* **24,** 329 (1998).

Z. M. Ruggeri: Mechanisms initiating platelet thrombus formation. *Thrombosis and Haemostasis* **78,** 611 (1997) [published erratum appears in *Thrombosis and Haemostasis* **78,** 1304 (1997)] [see comments].

J. E. Sadler: Biochemistry and genetics of von Willebrand factor. *Annual Reviews of Biochemistry* **67,** 395 (1998).

A. H. Schmaier: Contact activation, a revision. *Thrombosis and Haemostasis* **78,** 101 (1997).

M. T. Stubbs and W. Bode: A player of many parts, the spotlight falls on thrombin's structure. *Thrombosis Research* **69,** 1 (1993).

M. T. Stubbs, M. Renatus, and W. Bode: An active zymogen, unraveling the mystery of tissue-type plasminogen activator. *Biological Chemistry* **379,** 95 (1998).

J. W. Suttie: Synthesis of vitamin K-dependent proteins. *FASEB Journal* **7,** (1993).

R. F. Zwaal, P. Comfurius, and E. M. Revers: Lipid-protein interactions in blood coagulation. *Biochimica et Biophysica Acta* **376,** 433 (1998).

A. L. Bloom, C. D. Forbes, D. P. Thomas, and E. G. D. Tuddenham (Eds): *Haemostasis and Thrombosis*, 3rd ed. Churchill Livingstone, New York, 1994, p. 1477.

R. Colman, J. Hirsch, V. Marder, and E. Salzman (Eds): *Hemostasis-and Thrombosis*, 3rd ed. J. B. Lippincott, Philadelphia, 1995, p. 1025.

J. Jespersen, R. M. Bertina, and F. Haverkate (Eds): *Laboratory Techniques in Thrombosis: A Manual,* 2nd ed. Kluwer Academic, Dordrecht, 1999, p. 308.

CHAPTER 37

Mineral Metabolism

The chemical elements—exclusive of the common elements carbon, hydrogen, nitrogen, oxygen, and sulfur—necessary for normal structure and function of the body are collectively known as ***minerals*** and their study as ***bioinorganic chemistry.*** The minerals can be classified as ***macrominerals*** and ***trace elements.*** Macrominerals are present in large amounts and include sodium, potassium, chloride, phosphate, calcium, and magnesium. Trace elements are needed in very small amounts. Improvement in the sensitivity of analytical methods has increased the number of known essential trace elements, and the list is likely to grow. Certain other elements have no known biological function and are toxic.

Iron, the central element in oxygen transport and utilization, is discussed in Chapter 29. Iodine, a constituent of thyroid hormones, is discussed in Chapter 33. Sodium, potassium, and chloride, which are important for maintaining proper osmolality and ionic strength and for generating the electrical membrane potential, are discussed in Chapter 39. Most of this chapter is devoted to the metabolism of calcium and phosphorus because of their importance in the skeleton and other body systems. Because of its chemical and biological relationship to calcium, magnesium is also covered. The trace elements are surveyed with emphasis on those for which a biochemical function is known.

37.1 Calcium and Phosphorus

Distribution and Function

Calcium is the fifth most abundant element on earth and the principal extracellular divalent cation in the human body. A healthy, 70-kg adult contains 1–1.25 kg of calcium (25–33 g/kg of fat-free tissue), while a 3.5-kg newborn contains about 25 g of calcium. About 95–99% of body calcium is in the skeleton as hydroxyapatite crystals. The remainder is in the extracellular fluid and is exchangeable with that in periosteal fluid, bone-forming surfaces, and soft tissues. Skeletal calcium is slowly exchangeable with extracellular fluid calcium, and the skeleton is thus a reservoir of calcium. The steady-state extracellular and periosteal fluid concentrations of calcium depend, in large part, on the balance between bone formation and bone resorption, which are regulated by hormones.

The plasma concentration of calcium is kept remarkably constant throughout life at about 8.8–10.3 mg/dL (2.20–2.58 mmol/L). The normal serum calcium concentration is maintained by the integrated actions of parathyroid hormone (PTH) vitamin D metabolites, calcitonin, and cytokines such as transforming growth factor β and interleukin-6. The principal target sites for these hormones in the regulation of both calcium and phosphorus homeostasis are the gastrointestinal tract, kidney, and bone. Based on physiological signals the gastrointestinal tract regulates absorption, the kidney regulates reabsorption and excretion, and bone regulates accretion and mobilization of calcium and phosphorus. Abnormal serum calcium concentration has deleterious physiological effects on diverse cellular processes involving muscular, neurological, gastrointestinal, and renal systems.

Three forms of calcium are in equilibrium in serum: nondiffusible calcium bound primarily to albumin; diffusible complexes of calcium with lactate, bicarbonate, phosphate, sulfate, citrate, and other anions; and diffusible ionized calcium (Ca^{2+}). Ionized calcium accounts for approximately half of total serum calcium, and nondiffusible and complexed calcium account for 45% and 5%, respectively. Ionized calcium is the physiological active form; its concentration is regulated by the parathyroid gland. A decrease in serum ionized calcium can cause **tetany** (involuntary muscle contraction) and related neurological symptoms, regardless of the total serum calcium concentration.

Although ionized and protein-bound calcium are in equilibrium, release from the protein-bound fraction is slow and changes in plasma protein (especially albumin) concentration result in parallel changes in total plasma calcium. A decrease in serum albumin of 1 g/dL results in a decrease of about 0.8 mg/dL in total serum calcium. The equilibrium among the three forms of serum calcium is affected by changes in blood pH. Thus, at pH 6.8 (acidosis), about 54% of serum calcium is in the ionized form, whereas at pH 7.8 (alkalosis), only 38% is ionized.

The most accurate and sensitive method of determining ionized calcium concentration is by an ion-specific electrode. The percentage of total ionized calcium can be estimated from the empirical equation:

$$\%\ \text{ionized calcium} = 100 - [8 \times \text{albumin (g/dL)} + 2 \times \text{globulin (g/dL)} + 3]$$

This formula reflects the fact that albumin has a greater affinity for calcium than do other serum proteins.

Besides being the principal component of teeth and skeleton, calcium is essential for blood coagulation (Chapter 36), muscle contraction (Chapter 21), secretion of digestive enzymes (Chapter 12), secretion and action of many hormones (Chapter 30), and other body systems.

Calcium is intimately involved in transfer of information and material across membranes. Hormone binding to a plasma membrane receptor or electrical depolarization of the plasma membrane (as in muscle contraction) causes calcium entry into the cell or release from binding sites on the inner surface of the membrane. Cytoplasmic calcium concentration in unstimulated cells is about 10^{-7} mol/L (compared to 10^{-2} mol/L in the extracellular fluid). Upon stimulation, the cytoplasmic ionic calcium content rises rapidly to about 10^{-6}–10^{-5} mol/L. If the intracellular concentration in resting cells is too high or the extracellular concentration is too low, cellular function is impaired. The intracellular concentration is maintained by a Ca^{2+}-ATPase located in the plasma membrane. Vitamin D and the other components of the calcium homeostatic system are responsible for keeping the extracellular calcium at the required concentration.

Intracellular ionized calcium acts as a second messenger, coupling the action of a hormone or electrical impulse (the first messenger) on the outside of the cell to intracellular events, such as hormone or protein secretion, protein kinase activity, or muscle contraction. The effect of Ca^{2+} on intracellular processes is often mediated by a small calcium-binding protein, such as troponin C in muscle (Chapter 21) or calmodulin in many other cells (Chapters 15 and 30). Synthesis of these calcium-binding proteins is not directly affected by vitamin D or any of its metabolites. Many stimuli that affect permeability to calcium also activate membrane-bound adenylate cyclase and increase the intracellular concentration of cAMP (Chapter 30).

About 80% of the phosphate in the body is combined with calcium as hydroxyapatite in the skeleton. The remainder is present in many organic compounds as phosphate esters and anhydrides, e.g., nucleic acids, nucleoside triphosphates (particularly ATP), membrane phospholipids, and sugar-phosphate metabolites. Phosphate is distributed fairly equally between extracellular and intracellular compartments. A typical male contains about 670 g of phosphorus and a female about 630 g, or an average of about 12 g/kg of fat-free tissue. Serum inorganic phosphorus concentration under fasting conditions in children is 4–6 mg/dL (1–1.5 mmol/L) and in adults 3–5 mg/dL (0.75–1.25 mmol/L). These values may vary considerably during the day, particularly following meals. Although total plasma concentrations of calcium and phosphate are at or near the solubility product for calcium phosphate, precipitation does not occur because 45% of the calcium and 10% of the inorganic phosphate are protein-bound. The remainder of the serum phosphate is present as HPO_4^{2-} and $H_2PO_4^-$, with the relative proportions dependent on pH (at pH 7.40, $[HPO_4^{2-}]/[H_2PO_4^-] \cong 4{:}1$). Inorganic phosphate is a substrate in oxidative phosphorylation (Chapter 14), in glycogen breakdown (Chapter 15), in formation of 1,3-bisphosphoglycerate from glyceraldehyde-3-phosphate (Chapter 13), in conversion of nucleosides to free base and sugar-phosphate (Chapter 27), and in several other reactions.

Bone Structure, Formation, and Turnover

The human body contains 206 bones whose size and shape are highly diverse; many contain joints at their ends that connect them to adjacent bones. The functions of bone include maintenance of external form, structural framework for attachment of muscles, weight-bearing support, and protection of internal organs. In addition, the interior medullary cavity of the bone is filled with soft, pulpy material known as bone marrow which houses the hematopoietic system. The cells involved in bone formation and resorption arise in the hematopoietic system. There are two basic types of bone: compact (or cortical) and cancellous (or trabecular). The outer layer is compact bone. It is dense, solid, and responsible for mechanical and protective functions. The interior part contains the cancellous bone, which has the appearance of sponge-like or honeycombed structures. Because of its large surface area, cancellous bone contains bone-forming cells and is the site of mineral-requiring bone formation.

The skeleton is the body's principal reservoir of calcium and phosphorus. Contrary to its appearance, bone is a dynamic tissue, and calcium and phosphate are continuously deposited and released. Bone is a modified connective tissue consisting of a cellular component, an organic matrix and an inorganic (mineral) phase. Its cells are osteoblasts, osteoclasts, osteocytes, and osteoprogenitor cells. The last are a type of mesenchymal cell that can differentiate into any of the other three types and to which the other types can revert.

During bone formation, osteoblasts secrete tropocollagen, mucopolysaccharides, sialoproteins, and lipids to form the organic matrix. When this matrix matures into an insoluble, fibrillar network (osteoid), mineralization begins with a nucleation step, followed by precipitation of calcium and phosphate from the surrounding interstitial fluid. The initial deposits are amorphous and have the composition of brushite ($CaHPO_4 \cdot 2H_2O$). This mineral changes to hydroxyapatite, a hard, crystalline compound of approximate composition $Ca_{10}(PO_4)_6(OH)_2$. Incorporation of fluoride ions into bones and teeth increases the ratio of crystalline to amorphous calcium phosphate, which increases the hardness of the mineral. The calcium phosphates are quite insoluble; their precipitation may be enhanced by an increase in the $Ca^{2+} \times PO_4^{3-}$ ion product, perhaps by the action of alkaline phosphatase on sugar phosphates and pyrophosphates in the bone matrix. Alkaline phosphatase may regulate bone mineralization by hydrolysis of pyrophosphate, which is a potent inhibitor of mineralization *in vitro*. This enzyme is localized in osteoblasts and its activity is increased in sera of patients afflicted by rickets, osteomalacia, and hyperparathyroidism, all of which are associated with increased osteoblastic activity.

As mineralization progresses, osteoblasts become surrounded by growing bone and differentiate into osteocytes, which reside in individual lacunae in the bone. These lacunae communicate with each other via canaliculi, exchanging substrates and metabolites. Osteocytes nourish the bone, which is a living, highly vascularized tissue. The bone cells account for 2–3% of mature bone volume. Bone mass is about 65% mineral and 35% organic matrix.

Bone turnover (remodeling) is a dynamic, continuous process. The adult skeleton is renewed about every 10 years. The remodeling process is tightly regulated and is coupled to resorption of old and defective bone and formation of new bone. The remodeling is accomplished by the formation of temporary anatomical structures known as basic multicellular units. In these units, multinucleated osteoclasts located in the front resorb the existing bone, and the osteoblasts coming from the rear carry out bone formation. The attachment of osteoclasts to bone surface occurs at specific target sites consisting of integrin receptors that recognize specific bone matrix proteins. Osteoclast attachment is mediated by mechanical stimuli or release of chemotactic substances from the damaged bone. Resorption of the bone requires hydrogen ions, lysosomal enzymes, and collagenase, which are secreted through the ruffled borders containing microvilli of osteoclasts. The low pH is responsible for the solubilization of the mineral component of the bone. The requisite hydrogen ions are derived from organic acids and H_2CO_3 formed locally by hydration of CO_2 by carbonic anhydrase present in the osteoclasts. The importance of carbonic anhydrase in the production of H^+ and bone resorption is evident in ***osteopetrosis*** (discussed later).

Osteoblasts recruited to the site of the erosion cavity carry out bone formation. During bone matrix synthesis, osteoblasts become lining cells or osteocytes, and some undergo apoptosis. Thus, in bone remodeling, regulators of apoptosis of osteoclasts and osteoblasts play a major role. For example, increased production of cytokines, namely interleukin-1, interleukin-6, and tumor necrosis factor that occurs due to estrogen deficiency, leads to more bone resorption than bone formation and causes ***osteoporosis*** (discussed later). Osteocytes compose more than 90% of bone cells. The osteocyte-canalicular system plays an important role in activating the bone remodeling process by functioning as a transducer that detects microfractures or other flaws in the bone structure. Osteocytes also undergo apoptosis with increasing age.

TABLE 37-1
Factors That Can Influence Calcium and Phosphate Homeostasis

Factor	Pathological Effects on Calcium and Phosphorous Metabolism*	Mechanism/Site of Action†
Vitamin D	Excess (vitamin D toxicosis) causes hypercalciuria and hypercalcemia, leading to urolithiasis and soft-tissue (especially renal) calcification; insufficiency causes rickets and osteoporosis.	As 1,25-$(OH)_2D$, increases bone resorption and intestinal and renal Ca^{2+} and phosphate absorption; other metabolites may have other activities, especially in bone.
Parathyroid hormone (PTH)	Primary hyperparathyroidism causes hypercalcemia, hypophosphatemia, and increased urinary cAMP; hypoparathyroidism causes hypocalcemia and hyperphosphatemia, often with soft-tissue calcification and tetany and convulsions.	Binds to cell-surface receptors and activates adenylate cyclase; increases bone mineralization and activity of renal 1α-hydroxylase; in kidney, reabsorption of Ca^{2+} increases and reabsorption of phosphate decreases.
Parathyroid hormone-related protein (PTHrP)	Predominant factor responsible for hypercalcemia of malignancy.	PTHrP functions through the activation of PTH receptor; it has paracrine functions, it regulates rate of cartilage differentiation and increase placental calcium transport, it may also have PTHrP specific receptors.
Calcitonin	Neither deficiency nor excess of calcitonin is known to have any pathological effects. Plasma calcitonin is increased in the medullary carcinoma of the thyroid.	Binds to cell-surface receptors in bone and kidney, increasing intracellular cAMP; may also function by activation of phospholipase C signal transduction pathway, decreases Ca^{2+} release from bone and stimulate Ca^{2+} and phosphate excretion in kidneys, generally antagonistic to PTH.
Magnesium	Hypermagnesemia decreases PTH secretion; mild hypomagnesemia increases PTH secretion; severe hypomagnesemia (<0.05mmol/L) inhibits PTH secretion, even with hypocalcemia.	Needed to maintain parathyroid responsiveness to serum Ca^{2+}.

(continued)

TABLE 37-1 *(continued)*

Factor	Pathological Effects on Calcium and Phosphorous Metabolism*	Mechanism/Site of Action†
Diet	Inadequate Ca^{2+} intake produces rickets or osteomalacia; hypophosphatemia rarely (if ever) results from dietary inadequacy; excess dietary Ca^{2+} causes hypercalciuria and risk of urolithiasis.	Ca^{2+} and inorganic phosphate are absorbed in the small bowel by active 1,25-$(OH)_2$D-requiring processes; Ca^{2+} decreases PTH secretion and may decrease 1α-hydroxylase activity; inorganic phosphate decreases 1α-hydroxylase activity.
Estrogens	Decreased estrogens are probably responsible for postmenopausal osteoporosis.	Estrogens increase renal 1α-hydroxylase activity. Promotes osteoblast differentiation.
Glucocorticoids‡	Osteoporosis occurs in Cushing's disease and in patients treated with glucocorticoids for immunosuppression. Glucocorticoids are used in the treatment of hypercalcemia.	Decrease intestinal Ca^{2+} absorption; may antagonize 1,25-$(OH)_2$D or PTH; may directly stimulate parathyroids; may have direct effects on intestine and bone independent of PTH and 1,25-$(OH)_2$D.
Growth hormone/insulin-like growth factors‡	Absence during bone growth causes slow growth and short stature (dwarfism).	Unknown; may not be directly related to calcium and vitamin D metabolism.
Insulin‡	Decreased bone mass in insulin-dependent diabetes; in rats, decreased insulin lowers intestinal Ca^{2+} absorption, intestinal CaBP, and serum 1,25$(OH)_2$D.	Probably increases conversion of 25-(OH)D to 1,25-$(OH)_2$D.
Prolactin‡	No pathological effects are known.	In chicks, prolactin administration increases renal 1α-hydroxylase activity and serum 1,25-$(OH)_2$D concentration.
Thyroid hormones‡	Bone abnormalities are seen in hyper- and hypothyroidism.	Unknown; may affect vitamin D metabolism.

*Effects observed in humans unless otherwise noted.
†Results are from animal studies; some of these mechanisms may also explain effects observed in humans.
‡These hormones have other major roles in the body, and their effects on calcium and phosphate homeostasis are generally of secondary importance.

Both osteoblasts and osteoclasts are derived from osteoprogenitor cells originating in the bone marrow. Osteoblast precursors are pluripotent mesenchymal stem cells and the osteoclast precursors are hematopoietic cells of the monocyte-macrophage lineage. The development of osteoblasts and osteoclasts is regulated by several growth factors and cytokines whose responsiveness in turn is modulated by systemic hormones.

Bone morphogenetic proteins (BMPs) are signaling molecules involved in the formation of bone and also govern other developmental functions. These proteins belong to the transforming growth factor β superfamily. BMPs regulate many steps in the formation of new bone, such as mobilization of progenitor cells and their differentiation and proliferation into chondrocytes and osteoblasts. Thus, BMPs have therapeutic implications in the enhancement of osteoblast differentiation and bone formation. For example, *in vitro* and rodent studies have shown that one particular BMP (BMP-2) promotes bone formation. HMG-CoA reductase inhibitors that decrease hepatic cholesterol biosynthesis also activate the promoter of the BMP-2 gene, thereby promoting new-bone formation. HMG-CoA reductase inhibitors (statins) cause decreased prenylation of proteins, such as GTP-binding proteins, and induction of osteoclast apoptosis. (For a discussion on mevalonate-cholesterol multifunctional pathway and HMG-CoA reductase inhibitors, see Chapters 19 and 20, respectively.) Population-based, case-controled epidemiologic studies have shown that the statin used by elderly individuals is associated with a decreased risk of bone fractures (e.g., hip fracture). Some bisphosphonates that are used clinically as potent antiresorptive agents may also affect prenylation of proteins by inhibiting enzymes in the mevalonate-cholesterol biosynthetic pathway more distal to the HMG-CoA reductase catalyzed step (e.g., farnesyl-pyrophosphate synthase).

The transformation of precursor cells to osteoclasts is stimulated by parathyroid hormone (PTH), thyroxine, growth hormone, and vitamin D metabolites, whereas calcitonin, estrogens, and glucocorticoids inhibit the formation of osteoclasts. Osteoblast formation is promoted by calcitonin, estrogen, growth hormone, inorganic phosphate, and mechanical stress, and is antagonized by PTH and vitamin D metabolites. The actions of some of these agents are mediated by cAMP.

Calcium and Phosphate Homeostasis

Four primary factors influencing calcium and phosphate homeostasis are diet, vitamin D and its metabolites, PTH, and calcitonin. Table 37-1 lists other hormones known to affect homeostasis of these elements by interacting with one or more of these factors.

Calcium and Phosphate in the Diet

Sufficient dietary calcium arid phosphate must be absorbed to support growth (including pregnancy) and replace mineral lost from the body. Phosphates are present in adequate quantities in a wide variety of foods, and it is very unlikely that hypophosphatemia results from inadequate dietary phosphorus. Phosphate is absorbed from the small intestine with Ca^{2+} as a counterion and by an independent process that requires vitamin D metabolites. Normal daily intake of phosphate is about 800–1500 mg. Phosphate is highly conserved by the body, and obligatory losses are minimal.

Calcium homeostasis is profoundly dependent on diet and intestinal absorption. ***Rickets*** and postmenopausal ***osteoporosis*** are related to inadequate intestinal absorption of calcium. In the adult, an unavoidable loss of about 300 mg of Ca^{2+} per day occurs in urine, feces, and sweat. Since only about 30–40% of dietary calcium is absorbed, the recommended daily allowance for adults is 800 mg of Ca^{2+} per day. Allowances for other age groups are shown in Table 37-2. These allowances take into account rates of skeletal calcium deposition of 80–150 mg/d between birth and 10 years of age, and of 200 mg/d (female) and 270 mg/d (male) during the pubertal growth period. With cessation of growth, accumulation of calcium normally ceases, although bone remodeling continues. During pregnancy and lactation, there is calcium deposition in the fetus and calcium loss in milk. About 80% of the 25 g of calcium present in a full-term fetus is deposited during the last trimester of pregnancy. Human breast milk contains 30 mg of Ca^{2+} per deciliter (7.5 mmol/L), and about 250 mg of calcium per day is lost in milk during lactation. Increased calcium intake is recommended for older women because

TABLE 37-2
Recommended Daily Allowances of Dietary Calcium

Group	Recommended Daily Allowance (RDA) (mg Ca^{2+})
Infants	360–540
Children	800
Teenagers (pubertal growth period)	1200
Adults	800
Pregnancy and lactation	1200
Women over 40 years old (both pre- and postmenopausal)	1200

of the occurrence of osteoporosis. Administration of vitamin D metabolites together with the calcium is indicated for older women.

Major dietary sources of calcium are milk and dairy products, such as cheese and yogurt. Cow's milk contains 120 mg of Ca^{2+} per deciliter (30 mmol/L) and is now usually supplemented with vitamin D. Sardines (and other small fish whose bones are consumed) and soybean products can provide significant amounts of calcium. Soybean curd, known as tofu and eaten widely in China and Japan, contains 128 mg of Ca^{2+} per 100 g. Dark green leafy vegetables, legumes, nuts, and whole-grain cereal products contribute to dietary calcium.

When the recommended daily allowance (RDA) for calcium is not met by the diet is (particularly in women), supplementation in the form of calcium salts is recommended. Calcium salts vary widely in calcium content; by weight, calcium gluconate has 9%, calcium lactate has 13%, and calcium carbonate has 40% calcium. Absorption of calcium from salts may vary; calcium carbonate is the most poorly absorbed. Bone meal and dolomite are not recommended sources of calcium, since they may contain lead, arsenic, mercury, and other toxic metals. A potential complication of excessive calcium intake is formation of urinary tract stones; this risk may be reduced by ample fluid intake.

Calcium is absorbed both actively and passively throughout the small intestine and, to a small extent, in the colon. The active transcellular transport occurs in the duodenum and the passive paracellular process takes place in the jejunum and ileum. The chemical gradient and the sojourn time of the food passing through the intestine determine the movement of calcium that occurs by a passive process. The absorption of calcium in the colon becomes nutritionally significant under conditions of small intestine resection.

The active, saturable calcium transport consists of three steps: uptake by the brush-border cell membrane, diffusion through the cytoplasm, and extrusion at the basolateral surface where calcium is transferred to the portal blood circulation. The first step of calcium uptake is not energy-dependent, the second step of transcellular calcium movement is thought to be a rate-limiting process, and third step of calcium extrusion from the enterocyte is an energy-dependent process. The energy-dependent process is mediated by Ca^{2+}-ATPase, which forms a transmembrane segment, and the enzyme undergoes phosphorylation-induced conformational changes. This transcellular active calcium transport requires 1α,25-dihydroxyvitamin D [1,25-$(OH)_2D$, calcitriol], which is responsible for inducing the synthesis of enterocyte calcium-binding proteins that promote absorption and transcellular movement as well as enhancing the number of Ca^{2+}-ATPase pumps. Increased physiological need during pregnancy, lactation, and growth enhances calcium absorption. The molecular mechanism by which this occurs is not understood.

Phosphate is also absorbed in the small intestine by an active process, with maximal absorption occurring in the middle of the jejunum. 1,25-$(OH)_2D$ also mediates phosphate absorption.

Intestinal calcium absorption is influenced by dietary factors. Lactose and other sugars increase water absorption, thereby enhancing passive calcium uptake. The effect of lactose is especially valuable because of its presence in milk, a major source of calcium. Lactose also increases absorption of other metal ions. This effect may contribute to the incidence of lead poisoning (plumbism) among young inner-city children exposed to high dietary levels of both lead and lactose.

Calcium absorption is reduced by high pH; complexing agents such as oxalate, phytate, free fatty acids, and phosphate; and shortened transit times. These factors are probably of clinical importance only when associated with vitamin D deficiency, marginal calcium intake, or malabsorption disorders. Absorption is also reduced by increased intake of protein, fat, and plant fiber; increasing age; stress; chronic alcoholism; immobilization (e.g., prolonged hospitalization); and drugs such as tetracycline, thyroid extract, diuretics, and aluminum-containing antacids.

As intestinal absorption of calcium increases, urinary calcium excretion also increases. When the latter exceeds 300 mg/d, formation of calcium phosphate or calcium oxalate stones (***urolithiasis***) may occur. Hypercalciuria may result from decreased reabsorption of calcium due to a renal tubular defect or from increased intestinal absorption of calcium. Hypercalciuria may be due to an intrinsic defect in the intestinal mucosa or secondary to increased synthesis of 1,25-$(OH)_2D$ in the kidney. Disordered regulation of 1,25-$(OH)_2D$ synthesis is relatively common in idiopathic hypercalciuria. Treatment usually includes reduction in dietary calcium. Increased vitamin D intake, hyperparathyroidism, and other disorders can also cause hypercalciuria and urolithiasis.

In some forms of steatorrhea, calcium, which normally binds to and precipitates oxalate in the intestine, binds instead to fatty acids producing increased oxalate absorption and hyperoxaluria. Even though urinary calcium is decreased under these conditions, the concentration of urinary oxalate may be elevated sufficiently to cause precipitation of calcium oxalate crystals. Stone formation can be exacerbated by a diet that contains foods rich in oxalate, such as rhubarb, citrus fruits, tea, and cola drinks.

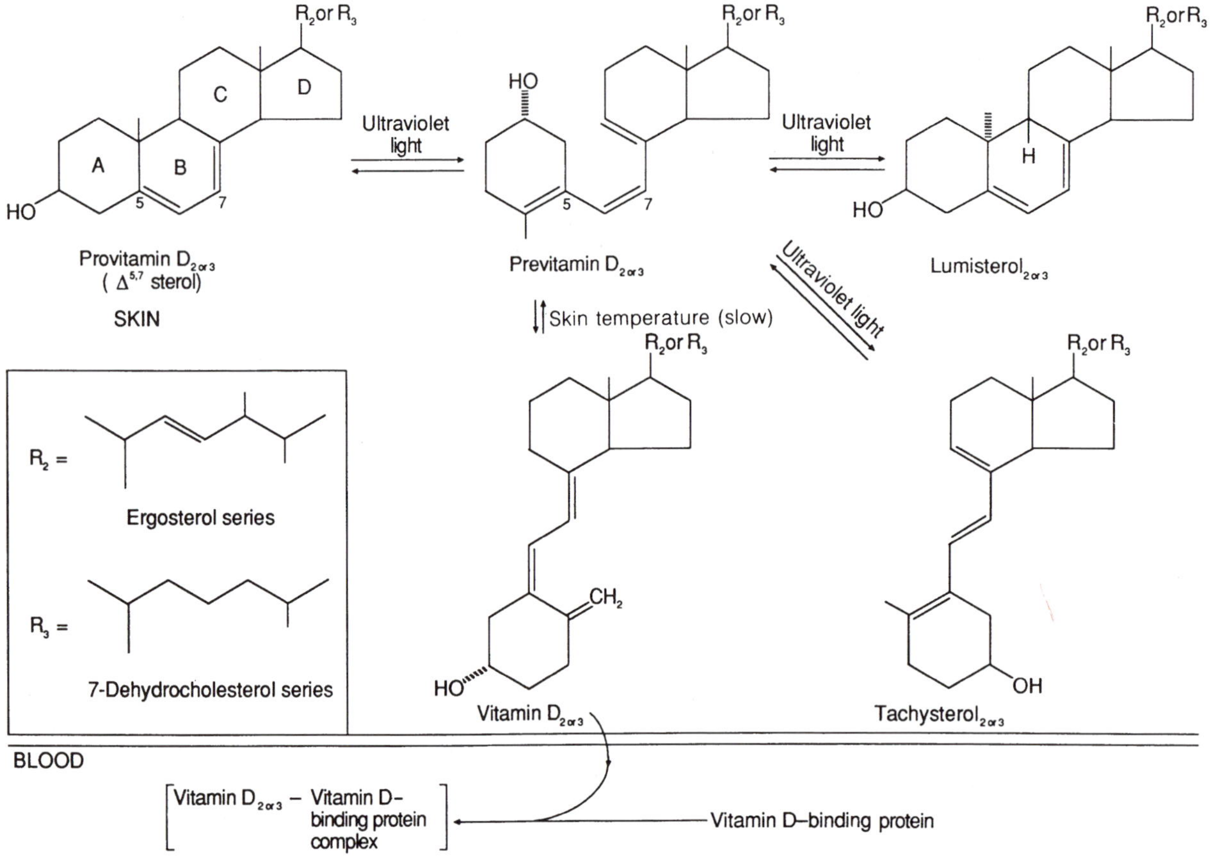

FIGURE 37-1

Conversion of provitamin D (ergosterol or 7-dehydrocholesterol) to vitamin D in the skin and its transport in the blood complexed with vitamin D-binding protein (also known as transcalciferin, G_C-protein, and group-specific component).

Increased oxalate absorption can be reduced by calcium administered with meals as a water-soluble salt.

Vitamin D Metabolism and Function

Although rickets was first described in the mid-1600s, it was not until the 1920s that deficiency of vitamin D was recognized as its cause. Despite its designation as a vitamin, dietary vitamin D is needed only if a person receives inadequate exposure to sunlight. Normally, vitamin D_3 is synthesized in the skin by irradiation of 7-dehydrocholesterol (Figure 37-1). Sufficient exposure to ultraviolet radiation can cure rickets.

The principal compound formed in the skin is ***cholecalciferol*** (vitamin D_3). The critical step requiring irradiation is the breaking of the 9,10 bond in the sterol B ring to form a secosterol. A secosterol occurs when one of the rings of the steroid skeleton cyclopentanoperhydrophenanthrene has undergone carbon-carbon breakage. Ring opening is accomplished by light of wavelength from 290 to 320 nm (ultraviolet-B radiation) with a maximal effect at 297 nm. 7-Dehydrocholesterol is present at high concentration in the stratum spinosum and stratum basale of the epidermis. Thermal isomerization of previtamin D_3 to vitamin D_3 occurs slowly, over 2–3 days. The vitamin D_3 formed diffuses gradually through the basal layers of the skin into the circulation. Excessive exposure to solar radiation causes conversion of previtamin D_3 to tachysterol$_3$ and lumisterol$_3$, which are biologically inactive. Melanin can reduce formation of previtamin D_3 by absorbing part of the solar radiation. This effect may partly explain the greater susceptibility of dark-skinned children to rickets. The concentration of 7-dehydrocholesterol in skin is reported to decrease with increasing age; if so, inadequate synthesis of vitamin D_3 may contribute to senile osteoporosis. The RDA for vitamin D is 400 IU.

Ergocalciferol (vitamin D_2) is formed by irradiation of ergosterol, a plant sterol common in the diet. It differs from cholecalciferol in the side chain attached to the D ring (Figure 37-1). Irradiation of ergosterol is an important commercial method for the synthesis of vitamin D_2, which is used for enriching cow's milk. The practice

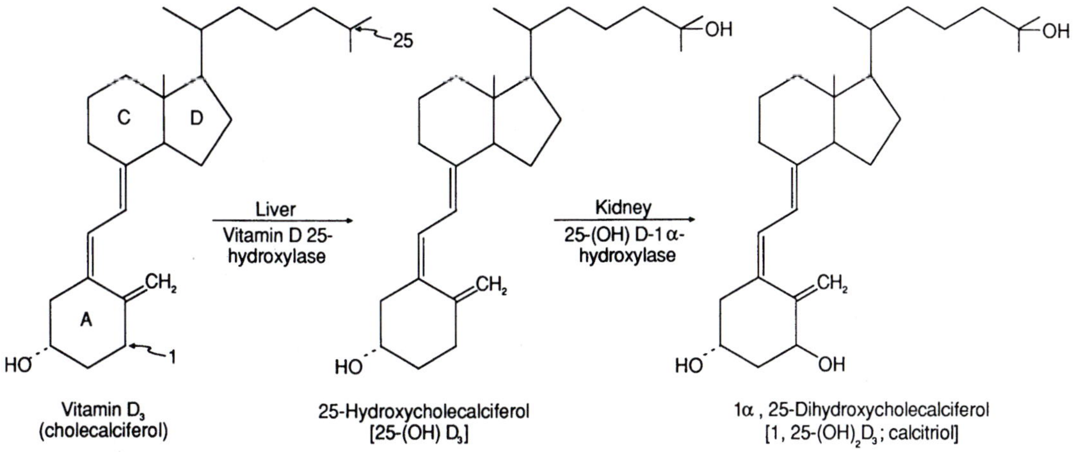

Liver
Vitamin D 25-hydroxylase

Kidney
25-(OH) D-1 α-hydroxylase

Vitamin D_3 (cholecalciferol)

25-Hydroxycholecalciferol [25-(OH) D_3]

1α, 25-Dihydroxycholecalciferol [1, 25-$(OH)_2D_3$; calcitriol]

FIGURE 37-2
Activation of vitamin D_3 by hydroxylation. Vitamin D_2 is metabolized to 1,25-$(OH)_2D_2$ (1α,25-dihydroxyergocalciferol), its active form, by the same enzymes.

of fortifying milk and milk products with vitamin D has nearly eliminated rickets as a major disease of infancy and childhood in industrialized countries. The metabolism and biological activity of vitamin D_2 in humans are identical to those of vitamin D_3. For this reason, the subscripts 2 and 3 are usually omitted from vitamin D and its metabolites except when it is important to distinguish D_2 from D_3. Several other sterols that differ from ergocalciferol and cholecalciferol only in the side chain can be converted to antirachitic compounds by ultraviolet irradiation.

The discovery and elucidation of modifications to vitamin D that render it biologically active represented a major advance in understanding how this vitamin functions and explained the lag between administration of vitamin D and its effect on serum calcium (Figure 37-2). Vitamin D and its metabolites are transported in the circulation bound mainly to ***vitamin D–binding protein*** (G_C-protein, transcalciferin; M.W. 56,000), an α-globulin. Serum normally contains 500–550 μg/mL (6×10^{-6} mol/L) of this protein. This concentration is unchanged in disorders of calcium homeostasis. Vitamin D-binding protein, albumin, and α-fetoprotein are all structural homologues. They are all synthesized in the liver, belong to the same multigene family, and probably arose from duplication of the same ancestral gene.

In the liver, vitamin D is hydroxylated to 25-hydroxyvitamin D [25-$(OH)_2D$], the principal circulating metabolite of vitamin D. In the kidney, hydroxylation at position 1 yields 1,25-$(OH)_2D$. This metabolite has the highest specific activity of the naturally occurring metabolites. 24,25-$(OH)_2D$ is also synthesized by renal mitochondria and in other tissues in relatively large amounts in animals with adequate intake of vitamin D, calcium, and phosphorus. These conditions are opposite to those that favor synthesis of 1,25-$(OH)_2D$. Consequently, the serum concentration of 24,25-$(OH)_2D$ varies inversely with that of 1,25-$(OH)_2D$. 24,25-$(OH)_2D$ may be of physiological importance, particularly in bone; patients with Paget's disease have normal plasma concentrations of 1,25$(OH)_2D$ but low concentrations of 24,25$(OH)_2D$. Furthermore, patients with chronic renal disease or osteomalacia that is resistant to 1,25-$(OH)_2D$ show clinical improvement in response to combined therapy with 1,25-$(OH)_2D$ and 24,25-$(OH)_2D$. Other metabolites of vitamin D have been characterized but have little or no activity in the usual assays.

There are two hepatic vitamin D 25-hydroxylases, the major one in the mitochondria and the other in the smooth endoplasmic reticulum (Figure 37-2). Both require NADPH and molecular oxygen. The microsomal enzyme appears to be a P-450 mixed-function oxidase. 25-Hydroxylase activity also occurs in intestine, kidney, and lung. 25-Hydroxylase is apparently regulated only by availability of its substrate, leading to a high plasma concentration of 25-(OH)D and a low concentration of vitamin D.

25-(OH)D-1α-hydroxylase is found in the inner mitochondrial membrane of the cells lining the proximal convoluted renal tubules. It is a mixed-function oxidase that requires molecular oxygen, a flavoprotein, a ferredoxin, and a cytochrome P-450 for activity. It is inhibited by carbon monoxide. Placental tissue contains 1α-hydroxylase activity, as do cultured bone cells and macrophages. This finding is of questionable importance under most circumstances, however, since 1,25-$(OH)_2D$ is not present in significant amounts in nonpregnant, nephrectomized animals, and bilateral renal failure or nephrectomy causes severe hypocalcemia in humans.

An exception seems to be the hypercalcemia that occurs in some patients with chronic granulomatous diseases, such as tuberculosis, sarcoidosis, and silicosis. The observed increase in circulating 1,25-$(OH)_2D$ is presumably due to the increased 25-(OH)D-1α-hydroxylase activity found in the inflammatory cells in the granulomas. The hydroxylase activity does not appear to be under the usual tight feedback control by serum calcium levels.

Renal osteodystrophy [due to decreased synthesis of 1,25-$(OH)_2D$ secondary to kidney failure] is treatable with synthetic 1,25-$(OH)_2D$ or 1α-(OH)D. These compounds are also useful in other renal disorders such as hypoparathyroidism and vitamin D-dependent rickets.

Deficiency of renal 1α-hydroxylase has been found in patients with hereditary vitamin D-dependent rickets type I, in which the serum concentration of 1,25-$(OH)_2D$ is low. Patients respond to treatment with 1,25-$(OH)_2D$.

Synthesis of 1,25-$(OH)_2D$ is the principal control point in vitamin D metabolism. Although the serum concentrations of vitamin D and 25-$(OH)D_3$ show seasonal and other types of variation, serum concentration of 1,25-$(OH)_2D$ remains constant owing to feedback control of its synthesis. Activity of renal 1α-hydroxylase is increased by PTH, hypocalcemia (both through PTH and directly), and hypophosphatemia. Calcitonin has no effect on the activity of the 1α-hydroxylase. The interaction of these factors in calcium and phosphate homeostasis is shown in Figure 37-3. Growth hormone, estrogens, androgens, prolactin, and insulin may also influence the activity of 1α-hydroxylase indirectly (Table 37-1). Thyroxine and glucocorticoids are necessary for bone mineralization and indirectly influence calcium homeostasis and synthesis of 1,25-$(OH)_2D$ (Table 37-1). Glucocorticoids appear to act as antagonists of 1,25-$(OH)_2D$, reducing intestinal calcium absorption independently of any effect on vitamin D metabolism. Because of the high plasma concentration of 25-(OH)D and the short half-life of circulating 1,25-$(OH)_2D$ (1–5 hours), small changes in activity of the 1α-hydroxylase rapidly change the plasma concentration of 1,25-$(OH)_2D$. Deviations from normal serum concentrations of calcium and phosphate are rapidly corrected by this mechanism.

The major target tissues for 1,25-$(OH)_2D$ are bone, intestine, and kidney (Figure 37-3). In the intestine, absorption of dietary calcium and phosphorus is increased. In bone, resorption is accelerated. In the kidney, 1,25-$(OH)_2D$ is localized in the nuclei of the distal convoluted

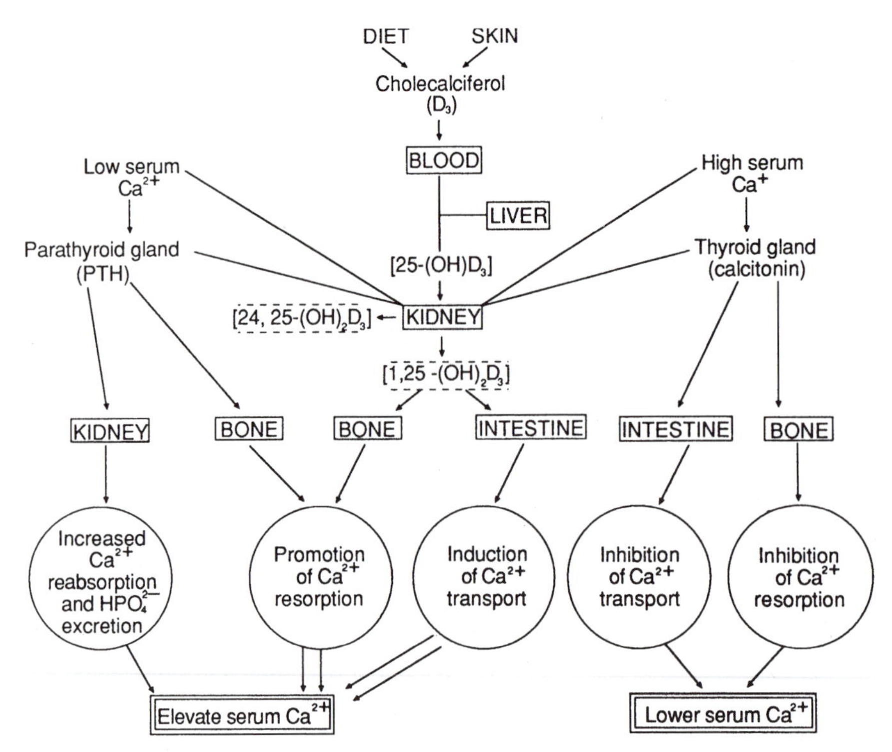

FIGURE 37-3
Interaction of PTH, calcitonin, and serum concentrations of calcium and phosphate in calcium and phosphate homeostasis. [Modified and reproduced, with permission, from A. W. Norman, *Vitamin D: The Calcium Homeostatic Hormone*. Academic Press, New York, 1979.]

tubule cells [recall that synthesis of 1,25-$(OH)_2D$ occurs in the proximal tubular cells], and reabsorption of calcium and phosphate is increased. In each case, the changes brought about by 1,25-$(OH)_2D$ act on the kidney to reduce production of 1,25-$(OH)_2D$, thereby forming a regulatory feedback loop.

Upon entry into target cells, 1,25-$(OH)_2D$ migrates into nuclei, where it binds with high affinity to the vitamin D receptor (VDR), which is a member of the nuclear receptor family. The binding of 1,25-$(OH)_2D$, which functions as a ligand to the nuclear receptor VDR, is analogous to the initiation of biochemical action by steroid-thyroid hormones (Chapters 30 and 33). The binding of the ligand to VDR brings about conformational changes so that the retinoid X receptor (RXR) can combine with it, forming a heterodimer complex. The ligand-VDR-RXR complex recruits additional coactivators and binds to vitamin D response elements (VDREs) in the promoter regions of target genes. The specific binding occurs via the two zinc finger modules of the DNA binding domain of the receptors. The complex interacts with the general transcription apparatus of 1,25-$(OH)_2D$-responsive genes. The target genes can be either upregulated or downregulated. An example of upregulation is calcium-binding proteins (CaBP), which facilitate intestinal calcium absorption. Inhibition of PTH gene expression in the parathyroid glands is an example of downregulation.

The vitamin D receptor is widely distributed not only in the target tissues but also in many other tissues such as thyroid, skin, adrenal, liver, breast, pancreas, muscle, prostate, lymphocytes, and in numerous malignant cells. Thus, the actions of vitamin D include regulation, proliferation, differentiation, and immunomodulation.

The importance of VDR in bone metabolism arises from studies of polymorphisms of the VDR gene. These show that normal allelic variations in the VDR gene may account for inherited variability in bone mass and density. Specific alleles of the VDR gene predict the density of femoral and vertebral bone in prepubertal American girls of Mexican descent. In the development of osteoporosis, a disorder determined by multiple genetic and environmental factors, variants of the VDR gene and its products may play a significant role. One of the important functions of the VDR protein is calcium absorption.

An autosomal recessive disease known as ***hereditary vitamin D-resistant rickets, type II*** is caused by defects in the gene for the vitamin D receptor which renders it nonfunctional. Thus, this knockout of VDR function illustrates its importance. Patients with this disease have high circulating levels of 1,25-$(OH)_2D$ (unlike type I disease which is due to 1α-hydroxylase deficiency, discussed earlier), severe rickets, and alopecia. The alopecia may be due to defective VDR in the hair follicles. Type II patients do not benefit from 1,25-$(OH)_2D$ therapy, and may require both intravenous calcium therapy and high-dose oral calcium administration.

In bone, a CaBP called osteocalcin contains 49 amino acids (M.W. 5500–6000). Its synthesis is stimulated by 1,25-$(OH)_2D$. Osteocalcin contains four residues of γ-carboxyglutamic acid, which require vitamin K for their synthesis and are important as binding sites for calcium (Chapter 36). Although vitamin K deficiency reduces the osteocalcin content of bone, it does not cause functional bone defects. For this reason, osteocalcin may function in calcium mobilization rather than deposition. Alternatively, as an effective inhibitor of hydroxyapatite formation, it may prevent overmineralization of bone. 1,25-$(OH)_2D$ increases γ-glutamyl carboxylase activity in the renal cortex. The relationship between vitamin D and vitamin K needs clarification.

Although only CaBP mRNA is known to increase in response to vitamin D, other vitamin D-dependent changes occur in the intestinal epithelium, including increases in activity of alkaline phosphatase, calcium ATPase, adenylate cyclase, and RNA polymerase. In response to vitamin D, several brush-border membrane proteins increase in concentration as does a calcium-binding complex. Non-cAMP-dependent phosphorylation of a brush-border membrane protein, increased synthesis and turnover of microvillar membrane phospholipids, and effects on mitochondria, Golgi membranes, and intracellular membrane vesicles are observed. Increased transport of Ca^{2+} across the basolateral membrane may be produced by 1,25-$(OH)_2D$.

1,25-$(OH)_2D$ increases reabsorption of phosphate in the kidney and intestinal absorption of phosphate. In the intestine, phosphate is absorbed as a counterion with Ca^{2+} and also by a calcium-independent route. Phosphate flux through both pathways is increased by 1,25-$(OH)_2D$ but more slowly than calcium transport. The calcium-independent pathway may involve alkaline phosphatase, the activity of which is increased by 1,25-$(OH)_2D$. In rat intestine *in vitro,* phosphate transport is greatest in the jejunum and least in the ileum, whereas calcium uptake is highest in the duodenum.

The major excretory route for vitamin D is bile. Vitamin D metabolites may undergo conjugation in the liver prior to secretion. An enterohepatic circulation of 25-(OH)D, 1,25-$(OH)_2D$, and other vitamin D metabolites has been demonstrated in humans.

In liver disease, decrease in hepatic 25-hydroxylase activity has little effect on serum 1,25-$(OH)_2D$ concentration. Even when serum 25-(OH)D concentration is reduced in primary biliary cirrhosis, the 1,25-$(OH)_2D$ level remains normal. Bone disease associated with hepatic failure may be due to malabsorption of vitamin D, interruption of enterohepatic circulation of vitamin D metabolites, insufficient exposure to sunlight, or the need for some vitamin D metabolite other than 1,25-$(OH)_2$D for normal bone mineralization.

At high doses, ***dihydrotachysterol*** (DHT), produced by chemical reduction of tachysterol (Figure 37-4), is more effective than vitamin D in mobilizing bone calcium and stimulating intestinal calcium absorption. DHT is activated by 25-hydroxylation in the liver by the same enzyme system as vitamin D (Figure 37-4). Unlike 25-$(OH)D_3$, 25-DHT does not require 1-hydroxylation for activity. The antirachitic activity of 25-(OH)DHT is apparently due to its A ring being inverted relative to that of vitamin D. Thus, its 3β-hydroxyl group is in approximately the same position as the 1α-hydroxyl group of 1,25-$(OH)_2$D. DHT is used to treat many disorders of vitamin D metabolism.

Vitamin D metabolites have been found in plants. *Enteque Seca,* a disease of cattle that graze on the plant *Solanum glaucophyllum (malacoxalon)* in South America, is characterized by calcinosis and soft-tissue calcification. Aqueous extracts of the leaves of this and other plants that have been used to treat uremic bone disease contain glycosides of 1,25-$(OH)_2$D.

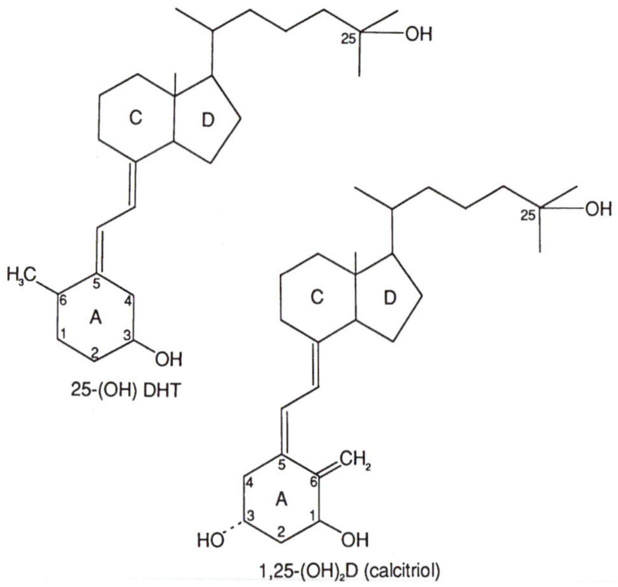

FIGURE 37-4
Structures of 25-hydroxydihydrotachysterols [25-$(OH)DHT_3$] and 1,25-$(OH)_2D_3$, the biologically active forms of dihydrotachysterol$_3$ and vitamin D_3, respectively. Note the similar spatial arrangement of the 3- and 25-hydroxyl groups in DHT and of the 1- and 25-hydroxyl groups in calcitriol.

Vitamin D toxicity (hypervitaminosis D) can be produced in humans by ingestion of 2000–5000 units of vitamin D per kilogram per day for several months. It is characterized by hypercalcemia and hypercalciuria. Because serum phosphate levels are usually normal, the $Ca^{2+} \times PO_4^-$ solubility product is exceeded. Ectopic calcification results and in the kidney leads to renal failure. Because excess vitamin D is stored in adipose tissue and is released gradually, hypercalcemia may persist after ingestion of vitamin D has stopped. The circulating concentration of 25-(OH)D is 1000 times normal, whereas that of 1,25-$(OH)_2$D is normal or low. These concentrations reflect the tight regulation of 1α-hydroxylase and the lack of regulation of 25-hydroxylase. The high concentration of 25-$(OH)_2$D and the observation that anephric patients can become intoxicated with vitamin D suggest that the hypercalcemia is caused by 25-(OH)D. In fact, 25-(OH)D is antirachitic but is much less potent than 1,25-$(OH)_2$D. *In vitro* binding studies using intestinal cytoplasmic binding proteins have shown their affinity for 1,25-$(OH)_2$D to be about 1000 times more than for 25-(OH)D. Transient hypercalcemia has been noted following oral therapeutic doses of 1,25-$(OH)_2$D, and acute vitamin D toxicity could presumably result from pharmacological doses of this compound.

Parathyroid Hormone (PTH). Humans usually have four parathyroid glands, located behind the thyroid. They were frequently removed during thyroidectomy, leading to hypocalcemia. The parenchyma of the glands is composed of chief cells and oxyphil cells. The chief cells are more numerous and are responsible for production of PTH. The oxyphil cells have no known function.

PTH is a polypeptide of 84 amino acids (M.W. 9,500). Full hormonal activity is present in the N-terminal 34-residue peptide [PTH(1–34)]. Removal of the first two residues eliminates biological activity, even though PTH(3–34) binds well to PTH receptors.

The mRNA for PTH codes for pre-proPTH, which contains 115 amino acids (M.W. 13,000). The N-terminal presequence of 25 residues is hydrophobic and is similar to leader sequences of other secreted proteins. It is removed within 1 minute of completion of synthesis of pre-proPTH, leaving proPTH. During transport of proPTH to the Golgi apparatus for packaging into secretory vesicles, the N-terminal hexapeptide is cleaved to form PTH. The time elapsed from formation of proPTH to conversion to PTH is about 15–20 minutes.

PTH can be secreted, sequestered in an intracellular storage pool, or degraded within the parathyroid gland. Secretion is thought to occur by exocytosis, although the number of secretory granules is inadequate to maintain the observed rate of sustained release of PTH. It appears that most proPTH is rapidly degraded at a rate that parallels changes in extracellular Ca^{2+} concentration.

Chronic hypocalcemia leads to hypertrophy and hyperplasia of the parathyroid glands, whereas chronic hypercalcemia has the opposite effect. DNA synthesis is probably involved.

At normal levels of calcium, a decrease in Mg^{2+} stimulates PTH secretion, while an increase in Mg^{2+} inhibits it. Although the effects of Mg^{2+} and Ca^{2+} are qualitatively similar, Ca^{2+} is two to four times as potent as Mg^{2+}. In humans, PTH secretion is suppressed by severe hypermagnesemia (three times normal), even when accompanied by hypocalcemia. Although moderate hypomagnesemia stimulates PTH output, extremely low levels of Mg^{2+} severely decrease the secretion and peripheral effectiveness of PTH, possibly because of the need for Mg^{2+} for adenylate cyclase activity. Epinephrine, other β-adrenergic agents, and exogenous cAMP stimulate PTH secretion, whereas 1,24 $(OH)_2D$ decreases PTH gene transcription and thus its synthesis and secretion. Target tissues for PTH are bone and kidney. PTH affects intestinal calcium absorption only indirectly, by regulating 1,25-$(OH)_2D$. The action of PTH on target cells is mediated by its binding to a distinct family of plasma membrane G-protein-coupled receptors, followed by stimulation of adenylate cyclase and cAMP production. Parathyroid hormone-related protein also binds to the same receptor in initiating its biological actions (discussed later).

In bone, PTH promotes resorption and new-bone formation. Bone dissolution predominates at high concentrations of PTH, while formation is more important at physiological levels. PTH also stimulates calcium release independently of resorption and more rapidly than it stimulates resorption. This pathway is probably most important in short-term regulation of the serum calcium level. Calcium released in this way may come from a pool of soluble calcium in the extracellular fluid of bone.

PTH promotes bone resorption by increasing the number of osteoclasts. PTH stimulates osteoclastic collagenase activity and release of lysosomal hydrolases. These enzymes degrade the organic matrix of bone and increase urinary excretion of hydroxyproline and hydroxylysine. Decrease in local pH from accumulation of organic acids may aid in this process. Although PTH causes cAMP accumulation in the skeleton and cAMP analogues increase bone resorption, PTH analogues that block the increase in cAMP have no effect *in vitro* on stimulation of bone resorption by PTH.

PTH increases activity of renal 25-(OH)D-1α-hydroxylase. Proximally in the renal tubule, it inhibits reabsorption of calcium, phosphate, sodium, and bicarbonate. The effect on sodium and bicarbonate is probably not physiologically important, since regulation of their excretion is accomplished in the distal tubule (Chapter 39), where PTH does not affect their concentrations. The proximal effects require higher concentrations of PTH than the distal effects.

Normally, 65–80% of filtered calcium and 85–90% of filtered phosphate are reabsorbed, mainly in the proximal tubule. The daily loss of 700–800 mg of phosphate is balanced by dietary intake. Fine-tuning of calcium excretion is accomplished by PTH in the distal convoluted tubules and collecting ducts. Phosphate excretion is regulated by PTH in the proximal tubules. Elevation of the PTH level increases reabsorption of calcium and decreases reabsorption of phosphate from the tubules. This phosphaturic action opposes the phosphate-sparing action of 1,25-$(OH)_2D$.

Urine concentrations of cAMP are normally 100 times higher than those in plasma or cytoplasm. Urinary cAMP is formed by action of PTH on the renal tubules. In hypocalcemia due to pseudohypoparathyroidism, reabsorption of Ca^{2+} from the collecting ducts is impaired because the cells do not respond to PTH. Serum levels of PTH are high, but the urine contains very little cAMP. The defect may be in the receptors for PTH or distal to the receptor. PTH may also affect tubular reabsorption by cAMP-independent mechanisms.

Thus, a complex relationship exists among serum Ca^{2+} and phosphate, PTH, and vitamin D and its metabolites. Release of PTH in response to low serum Ca^{2+} directly mobilizes calcium from bone and increases synthesis of 1,25-$(OH)_2D$, which in turn mobilizes skeletal Ca^{2+} and causes increased intestinal calcium absorption. These effects raise the serum Ca^{2+} level sufficiently to reduce PTH secretion. The effect of PTH on the kidneys occurs within minutes, whereas the effects of PTH on bone and (indirectly) on intestine take hours and days, respectively. An increase in serum phosphate acts in a way qualitatively similar to that of hypocalcemia to release PTH, increase excretion of phosphate in the proximal tubules, and decrease intestinal phosphate absorption. These events are mediated predominantly by the decrease in serum calcium that accompanies a rise in phosphate concentration. In addition, phosphate may inhibit 25-(OH)D-1α-hydroxylase.

Primary hyperparathyroidism results from hyperplasia, adenoma, or carcinoma of the parathyroid glands and from ectopic production of the hormone by squamous cell carcinoma of the lung or by adenocarcinoma of the kidney. In about 10% of hyperparathyroidism, hyperplasia or tumors of the parathyroid glands occur due to familial disorders known as multiple endocrine neoplasia (MEN). MEN syndromes consist of three subtypes (I, IIA, IIB) and are characterized by hyperplasia or tumors (or both) involving two or more endocrine glands in the same individual. MEN syndromes are inherited as an autosomal dominant trait; however, they may also arise sporadically. MEN I syndrome is characterized by tumors of the parathyroid, pancreatic islet cells, and anterior pituitary. The MEN I gene located on chromosome 11q13, which is thought to be a putative tumor suppressor gene, encodes for a novel 610-amino-acid protein. The precise role of MEN I protein in endocrine cell growth and regulation is not understood. In patients with MEN I syndrome, several inactivating mutations in the gene, namely, nonsense, missense, deletions, insertions, and donor-splice mutations, have been identified.

MEN II syndrome is characterized by medullary thyroid carcinoma, with or without pheochromocytoma and hyperparathyroidism. Unlike MEN I inactivating mutations that results in inactive putative tumor supressor protein, MEN II is caused by activating mutation of the RET (REarranged during Transfection) proto-oncogene, which codes for a membrane-bound receptor tyrosine kinase. The RET gene is developmentally regulated and the RET-encoded tyrosine kinase is expressed primarily in neural crest and urogenital precursor cell. Under normal conditions, glial-derived neurotrophic factor along with other cell surface components, by binding to tyrosine kinase, transmits growth and differentiation signals via a number of downstream signaling pathways. The intracellular signaling pathways involve adapter proteins containing SH2 domain, which interacts with amino acid residues surrounding phosphorylated tyrosine residues of tyrosine kinase. These interactions with adapter proteins stimulate RAS-mediated activation of the MAP kinase pathway, as well as other signaling pathways. Germline mutations of RET proto-oncogene can give rise to a constituitively activated tyrosine kinase (gain in function mutation), resulting in the transmission of unregulated growth and differentiation signals. In all these conditions, PTH secretion is not affected by serum calcium concentration. The signs and symptoms are due to hypercalcemia, hypophosphatemia, hypercalciuria, and hyperphosphaturia. Nephrocalcinosis, or urolithiasis progressing to nephrolithiasis, is the most common complication and can lead to renal failure. Bone involvement results in ***osteitis fibrosa cystica,*** in which normal bone is replaced by fibrous tissue. Serum intact PTH levels are used in the assessment of primary hyperparathyroidism. In addition measurement of PTH in venous blood draining individual parathyroid gland has been used for preoperative localization of malfunctioning glands.

Another example of a constitutively activated tyrosine kinase that results from a BCR-ABL fusion gene causes ***chronic myeloid leukemia.*** The fusion gene occurs due to reciprocal translocation beween the long arms of chromosome 9 and 22. An inhibitor (a phenylaminopyrimidine derivative) of the BCR-ABL tyrosine kinase has produced remission of clinical and hematological abnormalities.

Secondary hyperparathyroidism develops whenever hypocalcemia occurs. This condition frequently arises from chronic renal failure or intestinal malabsorption. In chronic renal failure, the hypocalcemia is secondary to hyperphosphatemia caused by inability of the diseased kidneys to excrete phosphate. The loss of renal tissue also decreases 1α-hydroxylase activity, which leads to decreased intestinal calcium absorption. This situation becomes clinically important when only about 25% of renal function remains. In long-term hemodialysis, hypocalcemia can be prevented by adequate amounts of ionized calcium in the dialysis fluid; decrease of phosphate intake; decrease of intestinal phosphate absorption; ingestion of aluminum hydroxide, calcium carbonate, or calcium lactate; and administration of 1α-(OH)D, 1,25-$(OH)_2D$, or dihydrotachysterol. However, chronic use of aluminum-containing agents can cause toxicity (see below). If secondary hyperparathyroidism is long standing, normalization of serum calcium level following successful renal transplantation may not result in immediate normalization of serum PTH concentration, although serum PTH usually returns to normal within several months.

Hypoparathyroidism is characterized by hypocalcemia, hyperphosphatemia and decrease in circulating PTH. Ectopic calcification is common. The most common form is due to inadvertent removal of or damage to the parathyroid glands during thyroid gland surgery or surgery to remove malignant tumors in the neck. The congenital absence of the parathyroids combined with thymic agenesis is known as ***diGeorge's syndrome.*** This condition is usually fatal by age 1–2 years because of hypocalcemia and immunodeficiency. ***Familial hypoparathyroidism*** may be an autoimmune disease, but its inheritance is complex and the presence of circulating antiparathyroid antibodies does not always correlate well with occurrence of the disease.

In ***pseudohypoparathyroidism,*** hypocalcemia, and hyperphosphatemia are accompanied by normal or high concentrations of PTH in plasma. In some forms, resistance to PTH is due to deficiency of a G-protein that couples PTH-receptor binding to cAMP synthesis (Chapter 30). ***Pseudopseudohypoparathyroidism*** is a variant that occurs in some relatives of patients with pseudohypoparathyroidism. These individuals have developmental and skeletal defects without clinical or metabolic evidence of hypoparathyroidism.

Parathyroid Hormone-Related Protein

A second member of the parathyroid hormone family, ***parathyroid hormone-related protein*** (PTHrP), is quite similar to PTH in amino acid sequence and protein structure. Like PTH, it activates the parathyroid hormone receptor causing increased bone resorption and renal tubular calcium reabsorption. Increased serum concentrations of parathyroid hormone-related protein are the predominant cause of hypercalcemia in cancer patients with solid tumors. This observation led to its discovery and to the elucidation of its many cellular functions in normal tissues. In contrast to PTH, which is expressed only in parathyroid glands, PTHrP is detected in many tissues in fetuses and adults; it is found in epithelia, mesenchymal tissues, endocrine glands, and the central nervous system. This protein is also the principal regulator of placental calcium transport to the fetus.

The functions of PTHrP have come primarily from studies in mice in which the parathyroid hormone-related gene has been deleted or altered in other ways. Loss of the protein is lethal in mice; without an active gene, mice die shortly after birth due to defects in cartilage development. The results of both loss of function (homozygous knockout of the genes) and gain-of-function experiments (targeted overexpression of the gene in cartilage) show that PTHrP accelerates the growth of cartilage cells and blocks their progression to a terminally differentiated state. It plays a key role along with other proteins (such as the hedgehog protein identified in *Drosophila*) in shaping the pattern of the trunk and limbs. Receptors for PTHrP are present on proliferating chondrocytes and help determine whether the cell (a) continues to proliferate or (b) terminally differentiates and prepares to mineralize the cartilaginous matrix. Patients with ***Jansen's disease*** have a constitutively activated PTHrP receptor gene in bone and kidney and suffer from hypercalcemia and delayed maturation of chondrocytes.

PTHrP is released during lactation and its levels increase about 10,000-fold over the levels found in the serum of nonlactating women. Receptors for PTHrP are present in the intestinal epithelium and the hormone may play a physiologic role in the development of the alimentary tract. However, infants raised on soy milk formulas that do not contain the hormone are healthy; thus, the hormone is not essential.

PTHrP is a polyhormone that is cleaved into three fragments by cells that secrete it. The N-terminal fragment consists of amino acids 1–36 and has a PTH-like activity in breast, skin, and cartilage. The mid-region fragment terminates at amino acid 94, 95, or 101, and is involved in placental calcium transport. The C-terminal fragment is called osteostatin and is involved in bone resorption.

Calcitonin. Parafollicular cells (C cells) scattered throughout the thyroid gland synthesize, store, and secrete calcitonin (thyrocalcitonin). These cells are derived from neural crest cells that fuse with the thyroid gland. In nonmammalian vertebrates, they remain together as discrete organs, i.e., ultimobranchial bodies. Continuous secretion of calcitonin occurs in eucalcemia, while hypercalcemia and hypocalcemia modulate this secretion. Gastrin, pentagastrin, cholecystokinin, and glucagon stimulate release of calcitonin, but there is no evidence that they are physiological regulators. The half-life of plasma calcitonin is about 10 minutes. It is increased in patients with renal failure, which suggests that, since little calcitonin is found in the urine, it is probably degraded in the kidney.

The calcitonin precursor contains about 135 amino acid residues (M.W. $\sim$15,000) from which the amino and carboxy terminal sequences are removed to yield the active peptide of 32-aminos-acid residues (M.W. 3590). Calcitonin from humans contains a disulfide bridge between amino acids 1 and 7 and a C-terminal proline amide (Figure 37-5). Salmon calcitonin, the type most frequently used clinically, is 25–100 times as potent as human calcitonin. Increased resistance to degradation and tighter binding to membrane receptors may contribute to the increased potency.

The principal target organ for calcitonin is bone, but renal excretion of calcium and phosphate is also directly affected. In bone and kidney, calcitonin activates adenylate cyclase by binding to a distinct class of G-protein-coupled receptors. Calcitonin may also exert effects on cytosolic Ca^{2+} and IP_3 levels by activation of the phospholipase

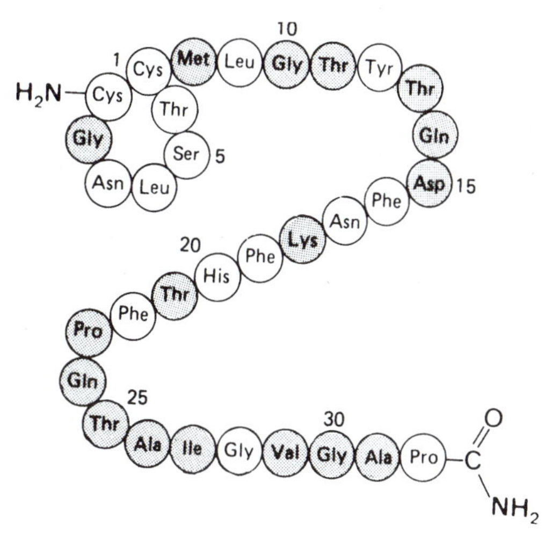

FIGURE 37-5

Amino acid sequence of human calcitonin. [Reproduced with permission from R. K. Murray, D. K. Granner, P. A. Mayes, and V. W. Rodwell, Norwalk *Harper's Biochemistry,* 21st ed. Appleton & Lange, Norwalk, 1988. 1988 © Appleton & Lange.]

C signal transduction pathway. Secretion of calcitonin is stimulated by hypercalcemia but the effect of the hormone on calcium transport appears to be secondary to increased phosphate uptake by target cells. The number and activity of osteoclasts are decreased, and urinary excretion of hydroxyproline is decreased, Calcitonin may also inhibit release of calcium from the extracellular fluid calcium pool, but it increases calcium and phosphate excretion by renal tubules. Some tubular cells respond to calcitonin, PTH, and vasopressin, while others respond only to one or two of these hormones. In general, the actions of calcitonin in kidney and in bone are antagonistic to those of PTH. Calcitonin decreases secretion of gastrin and of gastric acid, and inhibits bile flow.

The exact biological role of calcitonin remains elusive. Neither deficiency nor excess seems to produce pathological changes. Calcitonin may prevent hypercalcemia during childhood when calcium intake is high, or it may be important in adults during periods of hypercalcemia or high calcium intake.

Calcitonin is a useful marker for medullary carcinoma of the thyroid, which occurs both sporadically and as a dominantly inherited disease. In this type of tumor, the plasma concentration of calcitonin is 1–1000 mg/mL (normal concentration ranges from undetectable to 0.05 ng/mL). Also, urinary hydroxyproline excretion is decreased. Ectopic secretion of calcitonin also occurs from several types of pulmonary tumor in addition to other hormones.

Calcitonin is used in the treatment of ***Paget's disease*** (osteitis deformans), a chronic disorder characterized by increased bone remodeling, normocalcemia and normophosphatemia, frequent episodes of hypercalciuria leading to stone formation, and elevation of serum alkaline phosphatase and urinary hydroxyproline levels. The disease does not appear to be primarily a derangement of calcium metabolism. Calcitonin reduces the levels of serum alkaline phosphatase and urinary hydroxyproline, and may relieve other symptoms of the disease as well. Diphosphonates, especially etidronate disodium, also reduce bone resorption in this disease. Various cancers are accompanied by hypercalcemia and may respond to treatment with calcitonin.

Disorders of Calcium and Phosphorus Homeostasis

Several disorders have already been discussed; the following are additional disorders of calcium and phosphorus homeostasis. In ***rickets*** in children and ***osteomalacia*** in adults, there is failure of mineralization of osteoid, with consequent "softening" of bones. In rickets, there is defective mineralization of bone and of the cartilaginous matrix of the epiphyseal growth plates. In osteomalacia, since the epiphyseal plates are closed, only bone is affected. The most common cause is deficiency of vitamin D or, more rarely, deficiency of 1,25-$(OH)_2D$. Characteristic findings are eucalcemia or hypocalcemia, hypophosphatemia, hypocalciuria, and elevated serum alkaline phosphatase. Hypocalcemia or hypophosphatemia due to inadequate dietary intake or excess urinary loss of calcium or phosphate, respectively, causes identical lesions.

The most common metabolic bone disease is ***osteoporosis,*** which results from many environmental factors such as poor diet, smoking, alcohol consumption, and lack of exercise. However, recent evidence indicates that genetic variation may be the most important factor in the determination of bone mass, development of osteoporosis, and the risk of fracture. Osteoporosis can occur at any age. Poor diet and lack of exercise early in life contribute to the development of osteoporosis later in life, particularly in genetically susceptible individuals. Thus, the seeds of osteoporosis are sown early in life. Lack of intake of calcium and vitamin D in adequate quantities and an insufficient amount of weight bearing exercise (e.g., walking and running but not swimming) leads to failure in achieving peak bone mass in late adolescence and early adulthood. Under normal conditions, skeletal mass is a balance between bone formation, bone resorption, bone cell proliferation, and apoptosis. Only recently has it been recognized that the skeleton is a delicately balanced regenerating tissue, regulated as precisely as the destruction and synthesis of blood cells.

An estimated 75 million people are affected by osteoporosis to some degree in the United States, Europe, and Japan. Osteoporosis is a systematic skeletal disease characterized by bone mass and microarchitectural deterioration with a consequent increase in bone fragility and susceptibility to fracture. Operationally, osteoporosis can be defined as a certain level of bone mineral density. The definition of osteoporosis is somewhat arbitrary and is based on epidemiological data relating fracture incidence to bone mass. Uncertainty also is introduced due to variability in bone densitometry measurements. Other clinical measures to assess the skeleton include collagen cross-links (measure of bone resorption) and levels of bone-specific alkaline phosphatase and osteocalcin (bone formation). A list of biochemical markers of bone remodeling is provided in Table 37-3. Measurement of total serum alkaline phosphatase level and urinary hydroxyproline or calcium levels is of limited value.

Osteoporosis requires both increased resorption and one or more defects in bone formation. During adolescence, the rate of bone formation is higher than the rate of bone resorption; in older persons the rate of bone resorption

TABLE 37-3

Biochemical Markers of Bone Formation and Bone Resorption

Markers of Bone Formation (measured in serum)

1. Bone-specific alkaline phosphatase: Rich in osteoblasts
2. Osteocalcin: Major noncollagen protein of bone matrix
3. Amino- and carboxyl-terminal Procollagen 1 extension peptides: By-product of collagen biosynthesis

Markers of Bone Resorption

Serum marker

Bone specific acid phosphatase: A lysosomal enzyme of osteoclasts

Urine Markers: All of these are collagen breakdown products

1. N-Telopeptide and C-Telopeptide
2. Pyridinium cross-links (pyridinoline and deoxypyridinoline): Posttranslational modification of lysine and hydroxylysine residues of collagen
3. Hydroxlysine glycosides
4. Hydroxyproline

is greater than bone formation. Recent research indicates that peak bone mass, skeletal structure, and metabolic activity are determined by a large number of different genes interacting with environmental factors; thus, osteoporosis is a polygenic-multifactorial disease. Candidate genes that may play a role in the development of osteoporosis include vitamin D receptor, estrogen receptor, transforming growth factor β, interleukin-6, collagen type 1 genes, and collagenase. Gene knockout mice are used in assessing the contribution of various genes to bone formation and bone resorption. Knockout of the hemopoietic transcription factor PU-1 in mice results in the phenotype of osteoporosis; these mice are characterized by defective osteoclast formation and deficiency of macrophages. Knockout of the myeloid growth factor (M-CSF) gene produces a similar phenotype. However, knockout of the transcription factor c-*fos* gene causes osteopetrosis with normal macrophage synthesis while knockout of the transcription factor c-*src* allows formation of osteoclasts that are unable to resorb bone normally. Deficiencies of other molecules, such as cathepsin K, carbonic anhydrase II, and tartrate-resistant acid phosphatase, also impair the ability of mature osteoclasts to resorb bone. It is apparent that many factors influence development of osteoporosis.

Osteoporosis is a common diagnosis in postmenopausal women. Menopause results from the permanent cessation of ovarian function that usually precedes the final menses by several years and is diagnosed after 12 months of amenorrhea. Menopause and osteoporosis are causally related based on (a) the higher rates of osteoporotic fractures in postmenopausal women; (b) loss of bone mineral density in postmenopausal women; and (3) preservation of bone mineral density as a result of hormone replacement therapy. Most studies of hormone replacement therapy involving estrogen or estrogen plus progestagen showed increases in bone mineral density of 1–4% in postmenopausal women. Although the increase in bone density is small, significant reduction in fractures is reported for women using hormone replacement therapy.

Estrogen or other drugs are used to prevent osteoporosis in postmenopausal women. Bisphosphonates and calcitonin act as antiresorptive drugs. Actually, these drugs decrease the rates of both bone resorption and bone formation, but they affect the remodeling cycle so that there is a net increase in bone mineral density of 5–10%. Both etidronate and alendronate (bisphosphonates, Figure 37-6) are used in the treatment of osteoporosis in postmenopausal women and provide a treatment option if estrogen replacement therapy is contraindicated. It is thought that bisphosphonates are incorporated into bone matrix and incapacitate osteoclasts upon entry during resorption. Promotion of osteoclast apoptosis through inhibition of the mevalonate-cholesterol biosynthetic pathway, which leads to loss of G-protein prenylation, is also a possible mechanism of action. The use of HMG-CoA reductase inhibitors (statins), which also inhibits G-protein prenylation, decreases the risk of fractures in the elderly (discussed earlier). Statins, independent of their lipid-lowering activity, can improve endothelial function including the formation

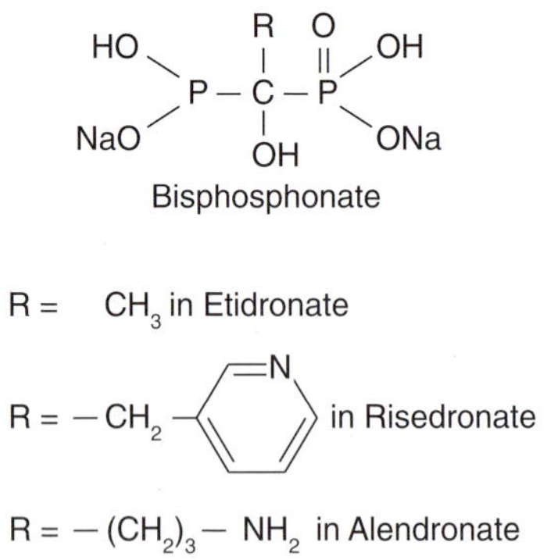

HO, NaO, P – C – P, R, O, OH, OH, ONa

Bisphosphonate

R = CH_3 in Etidronate

R = $-CH_2-$ (pyridinyl) in Risedronate

R = $-(CH_2)_3-NH_2$ in Alendronate

FIGURE 37-6
Structures of bisphosphonates. These antiresorptive therapeutic agents are characterized by a geminal bisphosphonate bond.

of new blood vessels (angiogenesis). The mechanism of action involves increased formation of nitric oxide and nitric oxide dependent pathways. These pathways are linked to the activation of *Akt*-protooncogene/protein kinase mediated cellular processes.

Calcitonin therapy results in decreased bone resorption. Osteoclasts have calcitonin receptors and calcitonin inhibits their activity. Sodium fluoride stimulates bone formation by unknown mechanisms. In women with osteoporosis, fluoride therapy produced an increased bone mineral density but no reduction in the rate of vertebral fractures. Other drugs known as selective estrogen receptor modulators (raloxifene, droloxifene, idoxifene, and levormeloxifene) may provide an alternative to estrogen replacement therapy (Chapter 34). Administration of low doses of PTH [or recombinant PTH(1–34)] does not affect serum calcium concentration, promotes bone formation, and increases mineral density. This anabolic action of PTH is probably mediated by decreasing osteoblast apoptosis.

Osteogenesis imperfecta, perhaps the most common hereditary disease of bone, is due to a defect in collagen formation (Chapter 25).

Osteopetrosis (marble bone disease) is marked by the formation of abnormally dense and condensed bone. It is a genetically, biochemically, and clinically heterogeneous disease. However, the underlying mechanism for the various types of osteopetrosis is a failure in bone resorption due to defects in osteoclasts. In some forms of osteopetrosis, providing normal osteoclastic precursors by bone marrow transplantation has yielded clinical improvement. Stimulation of osteoclast formation or activity with 1,25-$(OH)_2D$ or recombinant interferon-γ also has yielded modest clinical improvement.

One form of osteopetrosis is caused by a deficiency of one of the isoenzymes of carbonic anhydrase (carbonic anhydrase II). This is an autosomal recessive disorder and four different mutations in the structural gene of carbonic anhydrase II have been identified. As discussed earlier, the acidic environment necessary for resorption is generated by the action of carbonic anhydrase II. The patients with carbonic anhydrase II deficiency also exhibit renal tubular acidosis and cerebral calcification. The cause of renal tubular acidosis is due to failure of reclamation of bicarbonate from the glomerular filtrate because of the deficiency of carbonic anhydrase in the renal tubular cells (see Chapter 39 for the role of carbonic anhydrase in the reclamation of filtered bicarbonate in the kidney).

Three genes coding for isoenzymes of carbonic anhydrase I, II, and III, all belonging to the same gene family, are clustered on chromosome 8q22. These are all zinc metalloenzymes, soluble, monomeric, and have molecular weights of 29,000. Isoenzyme I is found primarily in erythrocytes (Chapters 1 and 39) and III in skeletal muscle. There are yet other isoenzymes coded for by different genes.

Hyperphosphatemia, often seen in renal failure, produces no specific signs or symptoms. In contrast, chronic ***hypophosphatemia*** can produce weakness, bone pain, congestive cardiomyopathy, dizziness, and hemolytic anemia.

Severe hypophosphatemia is potentially fatal and is usually due to hyperphosphaturia, shifting of phosphate from extracellular to intracellular fluid (as in electrolyte and pH imbalances), or diminished intestinal absorption of phosphate. Iatrogenic hypophosphatemia may occur in diabetic ketoacidosis. The immediate goals of therapy are to normalize glucose metabolism and to restore fluid and electrolyte balance. Since metabolism of glucose requires phosphate, rapid entry of glucose into insulin-dependent tissues as a result of insulin therapy shifts phosphate to the intracellular fluid. This effect can be exacerbated by intravenous rehydration with phosphate-poor fluids. Rehydration and hyperalimentation of alcoholics (who are often phosphate depleted due to poor diet, diarrhea, and vomiting) can lead to hypophosphatemia. Concurrent hypomagnesemia can increase phosphaturia.

37.2 Magnesium

Magnesium is the fourth most abundant cation in the body after sodium, potassium, and calcium, and is the second most abundant cation in intracellular fluid after K^+ Mg^{2+} is needed in many enzymatic reactions, particularly those in which $ATP^{4-} \cdot Mg^{2+}$ is a substrate. Magnesium binds to other nucleotide phosphates and to nucleic acids and is required for DNA replication, transcription, and translation. The DNA helix is stabilized by binding of histones and magnesium to the exposed phosphate groups. In green plants, chlorophyll—a magnesium-porphyrin complex similar to heme—is vital for photosynthesis.

A normal 70-kg adult contains about 25 g of magnesium (11–18 mmol/kg wet weight). Bone contains 60–65% (about 45 mmol/kg wet weight) of the body's Mg^{2+}, complexed with phosphate. Of the other tissues, liver and muscle contain the highest concentrations of Mg^{2+}, approximately 7–8 mmol/kg wet weight. Only 1% of total body magnesium is in extracellular fluid. The skeletal and extracellular fluid Mg^{2+} pools probably exchange freely with each other but not with the intracellular pool, which remains stable even when there are large fluctuations in the level of serum Mg^{2+}. Thus, the plasma concentration does not accurately reflect total body stores of magnesium.

Plasma contains 1.6–2.3 mg/dL (0.7–1.0 mmol/L) of magnesium, 20–25% of which is protein bound. As with Ca^{2+}, the unbound (ionized) portion of plasma magnesium is the biologically active fraction. Intracellular Mg^{2+} concentration is about 10 mmol/L, mostly bound to organelles. The intracellular concentration of free Mg^{2+} is 0.025–0.5 mmol/L.

The RDA for Mg^{2+} is 350 mg (14.5 mmol) for men and 300 mg (12.5 mmol) for women. During pregnancy and lactation, 450 mg (18.5 mmol/d) is needed to maintain positive Mg^{2+} balance. Leafy green vegetables are a good source because of the chlorophyll they contain. Whole grains, dried beans, peas, cocoa, nuts, soybeans, and some seafoods also are good sources of Mg^{2+}. Most dietary magnesium is absorbed in the small intestine, but small amounts are also absorbed in the colon. Although the fractional absorption of dietary magnesium is fairly constant and independent of intake, it ranges between 25% and 65% among individuals. Vitamin D and its metabolites are not important for its absorption. Calcium affects magnesium absorption, perhaps by competing with it for uptake sites. However, the occurrence of a rare genetic disease in which absorption of Mg^{2+} is selectively decreased suggests that transport for Mg^{2+} is independent of that for Ca^{2+}.

The kidney appears to be the main organ responsible for maintaining plasma Mg^{2+} concentration within normal limits. Most of the serum Mg^{2+} that is filtered at the glomerulus is reabsorbed, and only about 3–5% is excreted in the urine. Urinary excretion varies with plasma Mg^{2+} concentration. PTH enhances the tubular reabsorption of Mg^{2+}, whereas aldosterone decreases it.

Above, it was noted that hypermagnesemia suppresses PTH secretion whereas hypomagnesemia, if not too severe, stimulates PTH secretion or inhibits it at very low plasma Mg^{2+} concentrations. The suppressive effect of hypomagnesemia on PTH secretion occurs even in hypocalcemia. The hypocalcemia that accompanies marked hypomagnesemia can be corrected by magnesium repletion. Both hypomagnesemia and hypocalcemia cause tetany. At plasma Mg^{2+} concentrations of 20 mg/dL, anesthesia and paralysis of peripheral neuromuscular activity occur; they can be reversed by intravenous administration of calcium.

Hypermagnesemia occurs in acute or chronic renal failure, in hemodialysis, and in women receiving magnesium sulfate for treatment of preeclampsia. The clinical manifestations resemble the effects of curare. At serum Mg^{2+} levels of 2.5–5.0 mmol/L, cardiac conduction is affected, and at concentrations above 12.5 mmol/L, cardiac arrest occurs in diastole. Hypomagnesemia can occur in steatorrhea, alcoholism, diabetic ketoacidosis, and many other disorders. Tetany usually occurs at serum Mg^{2+} concentrations below 1 mmol/L.

37.3 Nonessential Trace Elements

The trace elements include all inorganic elements found in living systems at very low concentration (picograms to micrograms per gram of cells or wet tissue). Some trace elements are essential for normal health and development; others are toxic; and still others have not been shown to be either. Chemically similar elements may compete for intestinal or cellular uptake or for metal binding sites on metalloenzymes and proteins. Interactions may also be complementary or synergistic.

Many nonessential trace elements are found in the body. Depending on the local environment, at least 43 elements are normally incorporated into developing teeth; another 25 elements are seen less frequently. The rest, notably the heavy metals, have never been detected in teeth. Many trace elements, particularly the heavy metals, are considered when testing for metal poisoning. Many plants concentrate essential and nonessential elements from soil and water, including aluminum (several species of subtropical plants), selenium (many plants), strontium (mesquite beans), and lithium (wolfberries, used by Native Americans in the southwestern United States for jam). Ingestion of these plants can cause toxicity for the element involved.

Cadmium

Cadmium, a waste product from several industrial processes, is the cause of *itai-itai* ("ouch-ouch") disease, which is characterized by osteomalacia with painful fractures of affected, bones and by nephropathy with an excretory pattern similar to that seen in Fanconi's syndrome. In

TABLE 37-4

*Trace Elements Thought to Be Essential in Animals and Humans**

Element[†]	Function[††]	Deficiency Signs		Occurrence of Imbalances in Humans
		Animal	Human	
Arsenic	May be necessary for normal metabolism of an amino acid or protein that influences the urea cycle.	Impairment of growth and reproduction (several species); heart damage and decreased plasma triacylglycerol (goats); plasma amino acid changes.	Not known.	Not known.
Boron[‡]	May act in concert with magnesium in some aspect of the metabolism of calcium, phosphorous, or magnesium.	Depressed growth and abnormal bone development when magnesium or vitamin D nutrition is marginal.	Not known.	Not known.
Chromium	Potentiates insulin; may be involved in glucose transport into the cell.	Relative insulin resistance.	Relative insulin resistance; impaired glucose tolerance; elevated serum lipids.	Deficiency known in malnutrition, total parenteral nutrition.
Cobalt	Acts as part of vitamin B_{12}.	Anemia; growth retardation in ruminants.	Only known as vitamin B_{12} deficiency.	Inability to absorb B_{12} (pernicious anemia and other causes); low B_{12} intake from vegetarian diet.
Copper	Acts as cofactor for oxidative enzymes, together with iron; in elastin and collagen cross-linking; in superoxide dismutase.	Anemia; rupture of large vessels; disturbances of ossification.	Anemia; changes in ossification; possibly elevated serum cholesterol.	Deficiencies known in malnutrition, total parenteral nutrition.
Fluorine[§]	Influences structure of teeth, possibly of bone; has possible effect on growth; can stimulate adenylate cyclase.	Dental caries; possible growth depression; possibly depressed hematopoiesis and fertility.	Increased incidence of dental caries; possibly increased likelihood of osteoporosis.	Deficiency unknown; excess can cause osteofluorosis and mottled teeth.
Iodine[‖,¶]	Acts as constituent of thyroid hormones.	Goiter; depression of thyroid function.	Goiter; depression of thyroid function; cretinism.	Deficiencies are widespread; excess leads to thyrotoxicosis.
Iron[¶,#]	Acts in oxygen transport and electron transport; part of microsomal oxidases (p-450s).	Anemia; growth retardation.	Anemia.	Deficiencies widespread; excess causes hemochromatosis.
Lithium[‡]	Possibly regulates some endocrine function; has many similarities to Na^+ and K^+; disrupts cAMP-mediated processes.	Depressed fertility; depressed growth and longevity.	Not known.	Excess known secondary to use of lithium carbonate for treatment of mental disorders (low therapeutic index).
Manganese	Involved in glycosaminoglycan biosynthesis; as cofactor for glycosyl transferases; in mitochondrial superoxide dismutase.	Growth depression; bone deformities; pancreatic β-cell degeneration.	Not known.	Deficiency not known; toxicity by inhalation of dust.

Molybdenum	Is cofactor for xanthine, sulfide, and aldehyde oxidases.	Deficiency difficult to produce; growth depression.	Not known.	Excessive exposure in parts of the Soviet Union associated with goutlike syndrome.
Nickel	Interacts with iron absorption; may be a structural component or cofactor of a metalloenzyme.	Depressed growth, hematopoiesis, and reproduction; disturbed metabolism of iron, copper, and zinc; ultrastructural changes in liver.	Not known.	Not known.
Selenium**	Is glutathione peroxidase cofactor; in type I deiodinase; interacts with heavy metals.	Depending on species, muscle degeneration, pancreatic atrophy.	Endemic cardiomyopathy conditioned by selenium deficiency.	Deficiencies and excesses known in parts of China; deficiency due to total parenteral nutrition.
Silicon	Involved in calcification and as cross-linking agent contributing to the architecture and resilience of connective tissue.	Growth depression; structural abnormalities of the skull and long bones; decreased amounts of collagen and hexosamine in bone.	Not known.	Not known.
Vanadium	Perhaps involved in regulation of a phosphoryltransferase enzyme.	Growth depression; impairment of reproduction and lipid metabolism; none of these are reliably reproducible.	Not known.	Not known.
Zinc	Is cofactor for enzymes of energy metabolism, transcription, and translation.	Failure to eat; severe growth depression; sexual immaturity; skin lesions.	Growth depression; sexual immaturity; skin lesions; change of taste acuity; decreased immunocompetence.	Deficiencies known in Iran and Egypt; with total parenteral nutrition; due to genetic disease, traumatic stress.

*Modified and reproduced, with permission, from W. Mertz: The essential trace elements. *Science*, **213,** 1332 (1981); and F. H. Nielsen: Ultratrace elements in nutrition. *Annu. Rev. Nutr.*, **4,** 21 (1984).

†Bromine, cadmium, lead, and tin are not listed here because the evidence for their being essential is weak.

‡Probably essential, but further studies are needed to establish this.

§It is uncertain whether fluorine should be considered an essential element or a pharmacological agent. A fluoride concentration in drinking water of 0.9–1.0 mg/L is prophylactic for dental caries; consumption of water with fluoride concentration greater than 1.5 mg/L, during the years of tooth development, produces mottling.

||Discussed in Chapter 30.

¶Iron and iodine are not true "trace" elements because of their relatively high daily requirements and their high concentrations in the body.

#Discussed in Chapter 29.

**Interactions with vitamin E may also be important.

††Functions for arsenic, boron, fluorine, lithium, nickel, silicon, and vanadium are still partially or completely hypothetical. They are based on the symptoms of the deficiency diseases produced.

the 1950s, this disorder affected many residents of Japan's Jinzu River basin who consumed rice that had been irrigated with water contaminated from an upstream mine. Daily ingestion of cadmium was in excess of 300 μg, approximately six times the average intake in the United States and in parts of Europe. Many victims were multiparous women 40–70 years of age, a group at risk for osteomalacia. Their diet was deficient in calcium and vitamin D. Less severe examples of cadmium poisoning are known. Cadmium also disrupts the metabolism of iron, copper, and zinc in humans and in several species of farm animals.

Lead

Major sources of lead include fumes from combustion of leaded gasoline, flakes of lead-based paint, and alcoholic beverages brewed or distilled in lead-containing vessels. The first two sources have been implicated in lead poisoning of children from inner-city areas. Children are more susceptible to lead poisoning than adults; adults absorb about 10% of an oral dose of lead, while children 3–5 years of age absorb about 40% of the same dose. Lactose enhances intestinal absorption of lead (and calcium). Uptake of immunoglobulins from maternal milk in the intestines of very young children occurs by pinocytosis, which could provide a route for lead uptake. Lead inhibits renal 1α-hydroxylation of vitamin D metabolites, which may contribute to the hypocalcemia seen in children with high plasma concentration of lead (>60 μg/dL). Lead inhibits heme biosynthesis (Chapter 29). Severe lead poisoning can cause lead encephalopathy.

Aluminum

Aluminum accumulates in the bones of patients undergoing long-term hemodialysis. It is strongly implicated as the cause of dialysis encephalopathy ("dialysis dementia"), and it produces an osteodystrophy that responds poorly to vitamin D. Use of deionized water to prepare the dialysates has largely eradicated these problems. Deferoxamine, used to treat hemochromatosis (Chapter 29), increases aluminum excretion in these patients by forming a water-soluble complex with aluminum that can be removed by dialysis. Chronic use of aluminum-containing antacids to reduce intestinal phosphate absorption in uremic patients contributes to aluminum overload. An increase in serum aluminum concentration and in bone aluminum content, accompanied by severe osteomalacia, occurs in infants with azotemia who are given large doses of aluminum hydroxide to reduce phosphate absorption. Intestinal absorption of aluminum seems to be increased in hyperparathyroidism. The metal inhibits hexokinase, ALA-dehydratase, and isocitrate dehydrogenase and decreases Na^+, K^+-ATPase activity, Mg^{2+}-ATPase activity, and choline uptake into synaptosomes. *In vitro,* aluminum displaces magnesium from Mg^{2+}-ATP complexes, and it could thus antagonize virtually any phosphate-transferring reaction that uses Mg^{2+}-nucleotide triphosphate complexes.

Other Trace Elements

Chronic exposure to silver produces a bluish skin discoloration known as ***argyria.*** Gold poisoning has resulted from the use of gold salts for treatment of arthritis. Poisoning from antimony, arsenic, bismuth, mercury, and thallium is well known.

37.4 Metallothioneins

The ***metallothioneins*** are small proteins (M.W. ~7000), rich in sulfhydryl groups, that bind cadmium, zinc, and small amounts of copper, iron, mercury, and perhaps other heavy metals. Within a species, multiple metallothioneins occur that differ in amino acid composition. These proteins typically contain 20 cysteine residues but no disulfide bonds and no aromatic amino acids. They bind one atom of cadmium or zinc for each three sulfhydryl groups. Binding of copper to metallothionein is somewhat different, since a greater number of metal atoms are bound per protein molecule. Metallothioneins occur particularly in liver and kidney but also in the intestine and other tissues. Synthesis of the metallothioneins is increased by the metals that they bind, by dexamethasone, and by many other agents. Cadmium and zinc are potent inducers in the intestine and in cultured HeLa cells, while copper induces intestinal metallothionein only weakly. The metals and dexamethasone act by increasing transcription of metallothionein genes. Metal binding also increases the stability of the apoproteins; metals also cause greater accumulation of metallothionein than dexamethasone.

The physiological role of metallothioneins remains unknown, but it may be, in part, to protect cells from metal toxicity by binding metal ions. Metallothioneins may also be important for intestinal and renal absorption of metals and for metal storage and excretion.

37.5 Essential Trace Elements

Table 37-4 summarizes current knowledge of the essential trace elements. Bromine, cadmium, lead, and tin may also be essential, but the evidence is not convincing. Many of these elements are toxic when present in excess, and cause changes in membrane permeability or inhibition of vital enzyme processes such as protein synthesis, oxidative phosphorylation, and DNA replication. Chronic exposure to these elements may cause other, poorly understood, disorders. Copper and zinc are discussed below because they have been most thoroughly studied and their involvement in human disease is well documented. Iron and iodine are discussed in Chapters 29 and 33, respectively; cobalt is discussed in Chapter 38.

Copper

Copper is necessary, together with iron, for hematopoiesis, probably partly because it is needed for the synthesis of ferroxidase (ceruloplasmin). Many enzymes require copper for activity. Examples of some of the copper-enzymes and their functions are given in Table 37-5. Mitochondrial iron uptake may be blocked by deficiency of a cuproprotein, perhaps cytochrome oxidase. Several inherited diseases involving abnormalities in copper metabolism (Wilson's disease, Menkes' syndrome) or copper enzymes (X-linked cutis laxa, albinism) occur in human and in several animal species.

The oxidation state of copper in biological systems is +1 or +2. Copper(III) is found in inorganic systems and may occur as a reaction intermediate in galactose oxidase, laccase (a plant enzyme), and perhaps other enzymes. The coordination number of copper in these enzymes ranges from two to six and occasionally higher.

The average adult human contains 70–100 mg of copper. The highest concentrations (in decreasing order) are in liver, brain, heart, and kidney. Muscle contains about 50% of total body copper. Of the remainder, about one-fifth is in the liver (3–11 μg/g wet weight).

Human erythrocytes contain 1.0–1.4 μg of copper per milliliter, of which more than 60% is in superoxide dismutase. Normal serum contains of copper and 200–400 mg/L of ceruloplasmin. Whether copper is released from ceruloplasmin by endocytosis or by conformational change following binding of ceruloplasmin to a membrane receptor remains unknown.

Copper is absorbed from food in the upper small intestine. The absorption is primarily dependent on the quantity of the copper present in the diet. High intake of zinc diminishes copper absorption by inducing metallothionein formation in the mucosal cells. Metallothioneins, due to their high affinity for copper, bind it preferentially and the bound copper is lost during the sloughing of cells from the villi. Copper accumulation in patients with ***Wilson's disease*** can be reduced by giving oral zinc acetate, which decreases absorption (discussed later). Absorbed copper is transported to the portal blood where it is bound to albumin (and probably transcuprein), amino acids, and small peptides. Copper binds to albumin at the N-terminal tripeptide (Asp-Ala-His) site. The recently absorbed copper is taken up by the liver, which plays a central role in copper homeostasis.

Copper occurs in many foods; particularly good sources are liver, kidney, shellfish, nuts, raisins, and dried legumes. Copper deficiency due to diet is rare except in malnutrition and in children with chronic diarrhea. It occurs in total parenteral nutrition with fluids low in copper, particularly following intestinal resection and in patients who receive large amounts of zinc to improve wound healing or for management of sickle cell anemia. Copper is often removed from prepared foods to increase their shelf life.

TABLE 37-5
Examples of Copper-Containing Enzymes and Their Functions

Enzymes	Functional Significance
1. Cytochrome c oxidase	Terminal enzyme of mitochondrial electron transport; oxidative phosphorylation.
2. Superoxide dismutase	Inactivation of reactive oxygen species; antioxidant defense.
3. Ceruloplasmin	Ferroxidase; Iron metabolism.
4. Tyrosinase	Synthesis of melanin.
5. Dopamine β-monooxygenase	Synthesis of norepinephrine and epinephrine.
6. Lysyl oxidase (also called protein-lysine 6-oxidase)	Required for cross-linking of collagen and elastin; maturation of collagen.
7. Peptidylglycine mono-oxygenase	Required for removal for carboxy terminal residue and α-amidation; Processing and maturation of neuroendocrine and gastro-intestinal peptide hormones.
8. Amine oxidases	Deamination of primary amines.

Ascorbic acid, increased intestinal pH, sulfide ion, phytate (inositol hexaphosphate in plants), and some dietary proteins interfere with its intestinal absorption.

Copper sulfate causes ulceration of the oral and esophageal mucosa, and an acute dose of 7–8 g is usually fatal. Copper intoxication has occurred when copper salts were used to treat extensive skin burns or when copper-containing tubing or dialysis membranes were used for hemodialysis. Pulmonary fibrosis has been described in vineyard workers exposed for many years to fungicidal sprays (e.g., "Bordeaux mixture") containing copper sulfate. About 33–50 mg of copper per year ($\sim$100 μg/d) dissolves from copper-containing intrauterine contraceptive devices. Part of this copper is lost in menstrual flow, but part is rapidly absorbed. Whether this is harmful is not known but it seems unlikely. Penicillamine is the drug of choice for treatment of copper excess.

The RDA for copper increases with age, ranging from 0.5–0.7 mg in infants up to 6 months of age to 2–3 mg in adults. The average adult diet contains 2.5–5 mg/d. Healthy, full-term infants have a relatively high hepatic copper content. The lower copper content of premature and malnourished infants is exacerbated by feeding with cow's milk, which contains less copper than human milk.

About 40–70% of ingested copper is absorbed. Appearance in plasma of ingested copper with a peak occurring 1.5–2.5 hours after eating suggests that absorption begins in the stomach and proximal small intestine. Absorption may be inversely related to the metallothionein content of the mucosa. Although pinocytosis occurs in the intestines of human infants, their inability to absorb copper in Menkes' syndrome suggests that physiologically important absorption is by an active process, even at an early age. Absorption is enhanced by complex formation with L-amino acids and small peptides. Newly absorbed copper is taken up by the liver in a saturable transport process and secreted into plasma as ceruloplasmin (0.5–1.0 mg/d).

Copper is excreted in bile in an amount roughly equal to daily absorption ($\sim$1.7 mg/d). Toxic levels in liver occur in primary biliary cirrhosis, Indian childhood cirrhosis (a familial, probably genetic, disease, limited to Asia), and other liver diseases in which bile flow is disrupted. Toxicity is rare unless there is preexisting liver disease or ingestion of large amounts of copper salts. Biliary copper is in a form that cannot be readily reabsorbed. Some loss also occurs in urine and sweat. Renal copper reabsorption may be important for copper homeostasis and may be regulated by insulin or glucagon. Approximately 0.5 mg of copper is lost during each menstrual period. During lactation, copper loss averages 0.4 mg/d.

The biological functions of copper are mostly reflected in the proteins listed in Table 37-5. Like iron, copper is intimately involved in adaptation to an aerobic environment. The proteins contain copper bound directly to specific side chains. Copper is also essential in plants, lower eukaryotic animals, and prokaryotes.

Wilson's disease is an autosomal recessive disease of copper metabolism. It has a prevalence of 1 in 30,000 live births in most populations. The disease has a highly variable clinical presentation. It is characterized by impairment of biliary copper excretion, decreased incorporation of copper into ceruloplasmin, and accumulation of copper in the liver and, eventually, in the brain and other tissues. The biochemical findings include low serum ceruloplasmin, high urinary copper excretion, and high hepatic copper content. Some patients have normal serum ceruloplasmia levels, and heterozygous individuals do not consistently show reduced levels of this protein.

The genetic defect in Wilson's disease resides in the long arm of chromosome 13 and the gene codes for a copper-transporting ATPase. Thus far, more than 40 different mutations have been found. The defect in the ATPase results in reduced biliary excretion of copper from the hepatocytes. The end result may lead to hepatic failure. The Cu-ATPase belongs to a family of ubiquitous proteins that are involved in the translocation or movement of cations such as H^+, Na^+, K^+, Ca^{2+}, and other metal ions (discussed in the appropriate areas of the text). This family of ATPases is designated as P-type ATPase. The P designation stems from the fact that, during the catalytic cycle, an invariant aspartic acid residue undergoes phosphorylation by ATP and dephosphorylation by the phosphatase domain that results in a conformational change promoting cation transfer. Another common feature of P-type ATPase is that all possess an ATP binding domain at the carboxy terminus. Because intestinal copper absorption is unaffected, there is a net positive copper balance. Normal hepatic copper uptake ensures that copper accumulation occurs first in the liver. As the disease progresses, nonceruloplasmin serum copper increases, and copper deposits occur in various tissues, e.g., Descemet's membrane in the cornea (Kayser-Fleischer rings), basal ganglia (leading to lenticular degeneration), kidney, muscle, bone, and joints. Cultured fibroblasts from patients exhibit increased copper uptake. A disorder in Bedlington terriers that causes hepatic copper accumulation may serve as an animal model for Wilson's disease, although it differs from the human disorder in several ways.

Penicillamine, a chelating agent, solubilizes copper and other heavy metals and promotes their excretion in urine, analogous to the use of deferoxamine in the treatment of iron overload (Chapter 29). Long-term treatment with penicillamine increases the requirement for pyridoxine. Iron supplements should also be given, on a schedule that prevents their interference with penicillamine. Triethylenetetramine, another copper-chelating agent, may be of particular value in patients who are or who are becoming sensitive to penicillamine. As discussed earlier, since excessive zinc prevents copper absorption, oral intake of zinc acetate is used in the management of Wilson's disease.

Menkes' syndrome (Menkes' steely-hair syndrome) is a rare, X-linked recessive disorder in which infants have low levels of copper in serum and in most tissues except kidney and intestine, where the concentration is very high. They also have greatly reduced plasma ceruloplasmin levels. Hair of the affected infants has a characteristic color and texture (*pili torti,* "twisted hair"). It appears tangled and dull, has an ivory or grayish color, and is friable. Weakness and depigmentation of hair and defects in arterial walls (leading to aneurysms) are explained by loss of activity of copper-dependent enzymes (Table 37-5). Cerebral dysfunction may be due to a disturbance in energy metabolism or neurotransmitter synthesis secondary to decreased activity of cytochrome oxidase and dopamine β-hydroxylase.

The defect in copper metabolism is generalized, and the degenerative changes cannot be reversed by parenteral copper administration. However, parenteral copper administration corrects the serum ceruloplasmin and hepatic copper deficiencies. Other tissues take up copper administered parenterally, but activities of their copper enzymes are not normalized. The failure of postnatal treatment is due to the fact that many of the deleterious effects occur *in utero.*

The molecular defect in ***Menke's syndrome,*** like that in Wilson's disease, resides in a P-type ATPase. The gene for the enzyme is located on the X chromosome. Wilson's disease is characterized by defective biliary excretion; in Menkes' syndrome, the defect is a failure to transport copper to the fetus during development as well as failure to absorb copper from the gastrointestinal tract after birth.

Aceruloplasminemia is an autosomal recessive disorder characterized by progressive neurodegeneration and accumulation of iron in the affected parenchymal tissues. Iron accumulation in this disorder is consistent with ceruloplasmin's role as a ferroxidase in iron metabolism (discussed in Chapter 29).

Zinc

Zinc is a necessary nutrient in animals. The discovery that zinc deficiency is the cause of parakeratosis in pigs was the first demonstration of the practical importance of zinc in animal nutrition. More than 100 zinc enzymes are known, many of which are in the liver and include examples from all six classes of enzymes. In these enzymes, zinc may have a regulatory role or be required for structure or for catalytic activity. Zinc is not readily oxidized or reduced from its usual oxidation state of +2 and is not involved in redox reactions.

Table 37-6 lists some zinc-containing enzymes important for mammalian metabolism. Zinc has an important regulatory function in fructose-1,6-bisphosphatase and has structural catalytic and regulatory roles in DNA and RNA polymerases. Many of the transcription factors, namely, DNA-binding proteins, contain zinc finger motifs consisting of Zn^{2+} bound to four cysteine or two cysteine and two histidine residues in a coordinate covalent linkage (Chapter 26). Thymidine kinase is also a zinc enzyme. Zinc is present in gustin, a salivary protein secreted by the parotid glands that may be necessary for proper development of taste buds. A common sign of zinc deficiency is hypogeusesthesia. Gustin is structurally similar to nerve growth factor (NGF) isolated from male mouse submaxillary glands and other sources, which contains 1 mol of zinc per mole of protein. The metal prevents autocatalytic activation of NGF. Along with copper, zinc is required for activity of superoxide dismutase (Table 37-5).

In pregnant rats, zinc deficiency adversely affects fetal development and parturition and produces fetal abnormalities perhaps because of a requirement of zinc for gene expression. It is involved in synthesis, storage, and secretion of insulin by pancreatic islet β cells. Physiological stresses, such as trauma, infections, and low protein intake, cause a decrease in the plasma zinc level. Zinc can substitute for iron in protoporphyrin IX during heme synthesis, and quantitation of zinc protoporphyrin in erythrocytes is useful for evaluating iron deficiency anemia. Zinc interacts with plasma membranes of many cells, including erythrocytes (see below), and it may be vital for their structural integrity. The metal also has roles in wound healing, male fertility, bone formation, and cell-mediated immunity.

A healthy 70-kg man contains 1.4–2.3 g ($\sim$30 mmol) of zinc. The prostate gland is rich in zinc (about 100 μg Zn^{2+} per gram wet weight), as are prostatic secretions and semen. In rats, prostatic zinc concentration is decreased by castration and increased by administration of testosterone or gonadotropin. Testicular atrophy and failure of spermatogenesis occur in zinc-deficient rats. The skin contains approximately 20% of total body zinc. Dermatitis and poor wound healing are the most common symptoms of zinc deficiency in patients receiving total parenteral nutrition with zinc-deficient fluids. Zinc supplements do not promote wound healing when zinc stores are adequate. Bone has a high zinc content (>100 μg/g); however, bone zinc is not readily mobilized. Hair and nails also have high concentrations of the metal.

TABLE 37-6
Some Mammalian Zinc Proteins

Enzyme	Source	Molecular Weight	Zn Atoms/ Protein Molecule	Functions/Comments
Alcohol dehydrogenase	Human liver	87,000	4	Requires 2NAD$^+$; two zinc atoms used for catalysis; function of other two not known but may stabilize enzyme.
Alkaline phosphatase	Human placenta	125,000	~3	Unknown function; role of Zn^{2+} unknown but needed for activity.
Leucine aminopeptidase	Porcine kidney	300,000	6	N-terminal exopeptidase; Zn binds at two sites, one for catalysis, the other for regulation.
5′-AMP aminohydrolase	Rat muscle	290,000	2	—
Carbonic anhydrase B	Human erythrocyte	26,600	1	Catalyzes reversible hydration of CO_2; needed for CO_2 transport, buffering of ECF, gastric and renal acid secretion; Zn is catalytic.
Carboxypeptidase A	Bovine pancreas	34,500	1	C-terminal exopeptidase; involved in digestion of proteins; Zn functions in catalysis as a Lewis acid.
Glutamate dehydrogenase	Bovine liver	1,000,000	2–4	—
α_2-Macroglobulin	Human serum	840,000	3–8	—
Malate dehydrogenase	Bovine heart	40,000	1	—
Metallothionein	Human renal cortex	10,500	3	Usually also contains 4–5 Cd; may serve as source of Zn.
Rhodanese (sulfur transferase)	Bovine liver	37,000	2	Metabolizes SCN^-.

The zinc content of human blood is 8.8 μg/mL, of which 80–90% is within erythrocytes, mostly in carbonic anhydrase. Zinc binds to hemoglobin, increasing its oxygen affinity. By binding to erythrocyte membranes, it increases the flexibility of irreversibly sickled cells *in vitro* but plasma levels high enough to be therapeutic cannot be achieved. Leukocytes and platelets also contain an appreciable amount of zinc.

Plasma zinc levels vary with sex, age, time of day, geographic location, and time elapsed since the last meal prior to phlebotomy. The normal range for an adult is probably 70–95 μg/dL of plasma. Approximately 60% of zinc in plasma is bound to albumin, 30–40% to an α_2-macroglobulin of unknown function, and a small amount to transferrin (Chapter 29) and amino acids, particularly cysteine and histidine. Copper does not compete with zinc for binding sites on albumin. Zinc newly absorbed from the intestine is albumin-bound. This zinc is the fraction that changes most in disease states and that correlates best with total serum zinc. The zinc bound to amino acids accounts for urinary excretion of zinc, and it may be utilized by some tissues.

The RDA for zinc depends on age. For infants younger than 6 months, it is 3 mg; from 6 months to 1 year, 5 mg; for ages 1–10 years, 10 mg; and more than 10 years, 15 mg. For women, it increases during pregnancy (20 mg) and lactation (25 mg). A typical daily diet in the United States supplies 12–15 mg (about 200 μmol), of which 30–40% is absorbed. Dietary fiber binds and decreases the absorption of zinc. Human zinc deficiency was first reported in boys from Iran and Egypt who were small in stature, hypogonadal, and mildly anemic. The zinc content of their hair, plasma, and erythrocytes was below normal. Etiological factors included high dietary phytate content, loss of zinc through sweating, and geophagia (the practice of eating earth and clay, which bind zinc and other metals). All showed clinical improvement following oral supplementation with zinc.

The jejunum is the principal absorptive region, although zinc absorption occurs through out the small intestine. Uptake occurs both by an active transport process, the activity of which varies inversely with zinc status, and by a passive process. The increase in serum zinc following a single oral or parenteral bolus of zinc decreases absorption within about 6 hours. In contrast, with slow intravenous infusion, significant decrease in zinc absorption may not occur for 4 days. Intestinal metallothionein is induced by and binds zinc. It increases by about 6 hours following a bolus of zinc and may block absorption by sequestering the metal in the mucosal cells. The evidence is largely circumstantial. Uptake by the liver is by an active process.

Fecal zinc is derived from unabsorbed dietary zinc and zinc in desquamated epithelial cells and gastrointestinal secretions. Active intestinal secretion of zinc decreases in zinc-depleted animals and may be a regulated step in zinc homeostasis.

Excretion of zinc in urine (0.3–0.6 mg/d) is independent of dietary intake. It is increased in renal disease, in alcoholic cirrhosis, in hepatic porphyria, following major operations, in severely burned patients, and in response to treatment with ethylenediaminetetraacetic acid.

The normal concentration of zinc in sweat is 1.15 μg/mL, decreasing to about 0.6 μg/mL in zinc deficiency. In tropical climates, zinc loss from this source may reach 4 mg/d. Menstrual zinc loss is about 15 μg/d.

The zinc content of human milk is highest in colostrum. The average zinc concentration in human milk is approximately 1.5 μg/mL. Fresh cow's milk averages 4 μg/mL, but human milk is a better source of zinc for infant nutrition because of lower bioavailability of zinc in cow's milk.

Zinc is relatively nontoxic but, if present in the diet at high concentration and in readily available form, it can interfere with the absorption of calcium, copper, iron, and cadmium and can produce anemia. These elements, in turn, can decrease zinc absorption if present in sufficient quantities. Use of zinc salt lozenges has been claimed to reduce the duration of the common cold. However, many randomized, double-bind, placebo-controlled trials have failed to show any beneficial effects of zinc lozenges in reducing the duration of upper respiratory tract viral infections.

Acrodermatitis enteropathica (AE) is an autosomal recessive disorder characterized by defective absorption of zinc, thickened, ulcerated skin on the extremities and around body orifices, alopecia, diarrhea, growth failure, and abnormalities in immune function. The concentration of zinc in serum and hair is reduced. AE, and a similar disease of cattle known as adema disease or lethal trait A-46, respond to large amounts of oral zinc. Malabsorption of zinc occurs in both the human and animal diseases. Human milk seems to alleviate AE, presumably because it provides zinc in a form that can be readily absorbed.

Supplemental Readings and References

J. S. Adams and G. Lee: Gains in bone mineral density with resolution of vitamin D intoxication. *Annals of Internal Medicine* **127,** 203 (1997).

D. G. Barceloux: Manganese. *Clinical Toxicology* **37,** 293 91999).

D. G. Barceloux: Zinc. *Clinical Toxicology* **37,** 279 (1999).

D. Beshgetoor and M. Hambidge: Clinical conditions altering copper metabolism in humans. *Clinical Nutrition* **67,** 1017S (1998).

M. J. Bingham, T-J. Ong, K. H. Summer, et al.: Physiologic function of the Wilson disease gene product, ATP7B. *Clinical Nutrition* **67,** 982S (1998).

A. L. Boskey: Biomineralization: conflicts, challenges, and opportunities. *Journal of Cellular Biochemistry Supplements* **30,** 83 (1998).

F. Bronner: Calcium absorption—a paradigm for mineral absorption. *Journal of Nutrition* **128,** 917 (1998).

F. Bronner and D. Pansu: Nutritional aspects of calcium absorption. *Journal of Nutrition* **129,** 9 (1999).

D. A. Bushinsky and R. D. Monk: Calcium. *Lancet* **352,** 306 (1998).

S. Chan, B. Gerson, and S. Subramaniam: The role of copper molybdenum, selenium, and zinc in nutrition and health. *Toxicology* **18,** 673 (1998).

T. W. Clarkson: Mercury—an element of mystery. *New England Journal of Medicine* **323,** 1137 (1990).

A. Cordano: Clinical manifestations of nutritional copper deficiency in infants and children. *Clinical Nutrition* **67,** 1012S (1998).

H. F. DeLuca and C. Zierold: Mechanisms and functions of vitamin D. *Nutrition Reviews* **56,** S4 (1998).

J. S. Duthie, H. P. Solanki, M. Knshnamurthy, et al.: Milk-alkali syndrome with metastatic calcification. *American Journal of Medicine* **99,** 102 (1995).

R. Eastell: Treatment of postmenopausal osteoporosis. *New England Journal of Medicine* **338,** 736 (1998).

K. E. Ensrud, D. M. Black, L. Palermo, et al.: Treatment with alendronate prevents fractures in women at highest risk. *Archives of Internal Medicine* **157,** 2617 (1997).

E. Evron, S. Goland, J. von der Walde, et al.: Idiopathic calcitriol-induced hypercalcemia. *Archives of Internal Medicine* **157,** 2142 (1997).

D. Feldman: Vitamin D, parathyroid hormone, and calcium: a complex regulatory network. *American Journal of Medicine* **107,** 637 (1999).

P. Garnero, W. J. Shih, E. Gineyts, et al.: Comparison of new biochemical markers of bone turnover late postmenopausal osteoporotic women in response to alendronate treatment. *Journal of Clinical Endocrinology and Metabolism* **79,** 1693 (1994).

F. H. Gloneux, N. J. Bishop, H. Plotkin, et al.: Cyclic administration of pamidronate in children with severe osteogenesis imperfecta. *New England Journal of Medicine* **339,** 986 (1998).

G. A. Greendale, N. P. Lee, and E. R. Arriola: The Menopause. *Lancet* **353,** 571 (1999).

J. Guardiola, X. Xiol, R. Sallie, et al.: Influence of the vitamin D receptor gene polymorphism on bone loss in men after liver transplantation. *Annals of Internal Medicine* **131,** 752 (1999) .

J. G. Hall: A bone is not a bone is not a bone. *Journal of Pediatrics* **133,** 5 (1998).

J. R. Hansford and L. M. Mulligan: Multiple endocrine neoplasia type 2 and RET: from neoplasia to neurogenesis. *Journal of Medical Genetics* **37,** 817 (2000).

Z. L. Harris, L. W. J. Klomp, and J. D. Gitlin: Aceruloplasminemia: an inherited neurodegenerative disease with impairment of iron homeostasis. *Clinical Nutrition* **67,** 972S (1998).

R. P. Heaney: Bone mass, bone loss and osteoporosis prophylaxis. *Annals of Internal Medicine* **128,** 313 (1998).

J. L. Jackson, C. Peterson, and E. Lesho: A meta-analysis of zinc salts lozenges and the common cold. *Archives of Internal Medicine* **157,** 2373 (1997).

H. M. Kronenberg, B. Lanske, C. S. Kovacs, et al.: Functional analysis of the PTH/PTHrP network of ligands and receptors. *Recent Progress in Hormone Research* **53,** 283 (1998).

L. Kuritzky, G. P. N. Samraj, and D. M. Quillen: Improving management of type 2 diabetes mellitus: 6. Chromium. *Hospital Practice* **35,** 113 (2000).

M. C. Linder, L. Wooten, P. Cerveza, et al.: Copper transport. *Clinical Nutrition* **67,** 9655 (1998).

P. J. Malloy and D. Feldman: Vitamin D resistance. *American Journal of Medicine* **106,** 355 (1999).

R. Marcus: Exercise and the regulation of bone mass. *Archives of Internal Medicine* **149,** 2170 (1989).

E. Mariani, G. Ravaglia, A. Meneghetti, et al.: Natural immunity and bone remodelling hormones in the elderly. *Mechanisms of Ageing and Development* **102,** 279 (1998).

M. M. Maricic: Early prevention vs. late treatment for osteoporosis. *Archives of Internal Medicine* **157,** 2545 (1997).

J. C. Marini: Osteogenesis imperfecta—managing brittle bones. *New England Journal of Medicine* **339,** 986 (1998).

L. A. Martini: Magnesium supplementation and bone turnover. *Nutrition Reviews* **57,** 227.

M. McClung, B. Clemmesen, and A. Daifotis: Alendronate prevents postmenopausal bone loss in women without osteoporosis. *Annals of Internal Medicine* **128,** 253 (1998).

C. R. Meier, R. G. Schlienger, M. E. Kraenzlin, et al.: HMG-CoA reductase inhibitors and the risk of fractures. *Journal of the American Medical Association* **283,** 3205 (2000).

W. G. Meijer, E. Van Der Veer, and P. H. B. Willemse: Biochemical parameters of bone metabolism in bone metastases of solid tumors. *Oncology Reports* **5,** 5 (1998).

J. F. B. Mercer: Menkes syndrome and animal models. *Clinical Nutrition* **67,** 1022S (1998).

B. A. Michel, D. A. Bloch, and J. F. Fries: Weight-bearing exercise, overexercise, and lumbar bone density over age 50 years. *Archives of Internal Medicine* **149,** 2325 (1989).

W. L. Miller and A. A. Portale: Genetic disorders of vitamin D biosynthesis. *Pediatric Endocrinology* **28,** 825 (1999).

P. D. Miller, N. B. Watts, A. A. Licata, et al.: Cyclical etidronate in the treatment of postmenopausal osteoporosis: efficacy and safety after seven years. *American Journal of Medicine* **103,** 469 (1997).

G. Mundy, R. Garrett, and S. Harris: Stimulation of bone formation *in vitro* and in rodents by statins. *Science* **286,** 1946 (1999).

G. R. Mundy and T. A. Guise: Hypercalcemia of malignancy. *American Journal of Medicine* **103,** 134 (1997).

G. R. Mundy and T. A. Guise: Hormonal control of calcium homeostasis. *Clinical Chemistry* **45,** 1347 (1999).

K. M. Prestwood, C. C. Pilbeam, J. A. Burleson, et al.: The short term effects of conjugated estrogen on bone turnover in older women. *Journal of Clinical Endocrinology and Metabolism* **79,** 366 (1994).

L. G. Raisz: The osteoporosis revolution. *Annals of Internal Medicine* **126,** 459 (1997).

L. G. Raisz: Physiology and pathophysiology of bone remodeling. *Clinical Chemistry* **45,** 1353 (1999).

S. H. Ralston: Osteoporosis. *British Medical Journal* **315,** 469 (1997).

J. K. S. Rao, C. D. Katsetos, M. M. Herman, et al.: Experimental aluminum encephalomyelopathy. *Toxicology* **18,** 687 (1998).

J. Sainz, J. M. Van Tornout, M. L. Loro, et al.: Vitamin D-receptor gene polymorphisms and bone density in prepubertal American girls of Mexican descent. *New England Journal of Medicine* **337,** 77 (1997).

D. L. Schneider and E. L. Barrett-Connor: Urinary N-telopeptide levels discriminate normal osteopenic, and osteoporotic bone mineral density. *Archives of Internal Medicine* **157,** 1241 (1997).

E. Shane, D. Mancini, K. Aaronson, et al.: bone mass vitamin D deficiency, and hyperparathyroidism in congestive heart failure. *American Journal of Medicine* **103,** 197 (1997).

S. J. Silverberg, E. Shane, D. W. Dempster, and J. P. Bilezikian: The effects of vitamin D insufficiency in patients with primary hyperparathyroidism. *American Journal of Medicine* **107,** 561 (1999).

M. Sowers, D. Eyre, B. W. Hollis, et al.: Biochemical markers of bone turnover in lactating and nonlactating postpartum women. *Journal of Clinical Endocrinology and Metabolism* **80,** 2210 (1995).

P. Steindl, P. Ferenci, and H. P. Dienes: Wilson's disease in patients presenting with liver disease: a diagnostic challenge. *Gastroenterology* **113,** 212 (1997).

G. J. Strewler: The physiology of parathyroid hormone-related protein. *New England Journal of Medicine* **342,** 177 (2000).

R. Subramanian and R. Khardori: Severe hypophosphatemia: pathological implications, clinical presentations and treatment. *Medicine* **79,** 1 (2000).

T. D. Thacher, P. R. Fischer, J. M. Pettifor, et al.: A comparison of calcium, vitamin D, or both for nutritional rickets in Nigerian children. *New England Journal of Medicine* **341,** 563 (1999).

R. V. Thakker: Editorial: Multiple endocrine neoplasia-syndromes of the twentieth century. *Journal of Clinical Endocrinology and Metabolism* **83,** 2617 (1998).

J. R. Turnlund: Human whole-body copper metabolism. *Clinical Nutrition* **67,** 960S (1998).

R. Uauy, M. Olivares, and M. Gonzalez: Essentiality of copper in humans. *Clinical Nutrition* **67,** 952S (1998).

P. S. Wang, D. H. Solomon, H. Mogun, and J. Avorn: HMG-CoA reductase inhibitors and the risk of hip fractures in elderly patients. *Journal of the American Medical Association* **283,** 3205 (2000).

N. B. Watts: Clinical utility of biochemical markers of bone remodeling. *Clinical Chemistry* **45,** 1359 (1999).

R. S. Weinstein and S. C. Manolagas: Apoptosis and osteoporosis. *American Journal of Medicine* **108,** 153 (2000).

R. S. Weinstein and S. C. Manolagas: Bone cell apoptosis: implications for the pathogenesis and treatment of osteoporosis. *Physiology in Medicine* **108,** 153 (2000).

J. R. Weisinger and E. Bellorin-Font: Magnesium and phosphorous. *Lancet* **352,** 391 (1998).

P. Whittaker: Iron and zinc interactions in humans. *American Journal of Clinical Nutrition* **68,** 442S (1998).

G. Worley, S. J. Claerhout, and S. P. Combs: Hypophosphatemia in malnourished children during refeeding. *Clinical Pediatrics* **6,** 347 (1998).

Y. Kureishi, Z. Luo, I. Shiojima, et al: The HMG-CoA reductase inhibitor simvastatin activates the protein kinase Akt and promotes angiogenesis in normocholesterolemic. animals. *Nature Medicine* **6,** 1004 (2000).

CHAPTER 38

Vitamin Metabolism

The word *vitamin* is used to describe any of a heterogeneous group of organic molecules that are needed in small quantities for normal growth, reproduction, and homeostasis but that the human body is unable to synthesize in adequate amounts. The group includes the fat-soluble vitamins (A, D, E, and K) and the water-soluble vitamins (B-complex and C). Vitamins are generally needed in catalytic quantities and do not function as structural elements in the cell. Other organic compounds are not synthesized in the body but are required for maintenance of normal metabolism, such as essential fatty acids (sometimes inaccurately called "vitamin F"; Chapter 18) and essential amino acids (Chapter 2). These substances are needed in relatively large quantities, serve as nonregenerated substrates in metabolic reactions, and are used primarily as structural components in lipids and proteins, respectively. A number of other substances are essential food factors in various nonhuman species, e.g., biopterin, inositol, ubiquinone, lipoic acid, phosphatidylcholine, and paraaminobenzoic acid. They are often classified as "vitamin-like."

Vitamins discussed in other chapters include vitamin D (Chapter 37) and vitamin K (Chapter 36). All the B vitamins function as cofactors or precursors for cofactors in enzyme-catalyzed reactions and are discussed in appropriate chapters. Less well-defined actions are reviewed here.

General Considerations

Classification of vitamins into fat-soluble and water-soluble groups reflects the history of their discovery. This grouping is still useful, despite the lack of chemical relatedness within each class, because it mirrors other underlying similarities.

Vitamers are chemically similar substances that have a qualitatively similar vitamin activity. Thus, "vitamin D" refers to ergocalciferol (D_2) and cholecalciferol (D_3) and sometimes to their 25-hydroxy- and 1,25-dihydroxy derivatives (Chapter 37). Similarly, pyridoxine (pyridoxol), pyridoxal, and pyridoxamine are vitamin B_6 vitamers, riboflavin is the active form of vitamin B_2 and cobalamin is vitamin B_{12}. The members of a particular vitamin "family" are functionally interchangeable and protect against deficiency symptoms for that vitamin. A vitamin and its corresponding deficiency disease are related as follows:

1. The putative vitamin is a normal dietary constituent of healthy individuals.
2. The diet consumed by persons exhibiting symptoms of the putative deficiency disease is lacking in the vitamin.
3. Symptoms of such persons can be alleviated by addition of the vitamin to their diet.

TABLE 38-1
Genetic Disorders That Respond Favorably to Treatment with Pharmacological Doses of Vitamins*

Therapeutic Vitamin	Daily Dose	Disorder[†]	Biochemical/Biological Defect
Vitamin D (adult RDA = 200 IU = 50 μg of cholecalciferol)	>100,000 units	Familial X-linked hypophosphatemic rickets[‡]	Unknown
	>25,000 units	Vitamin D–dependent rickets, type I	Defect in 25-(OH)D-1-α-hydroxylase
Ascorbic acid (vitamin C; adult RDA = 60 mg)	4g	Ehlers-Danlos syndrome, type VI (ocular type)	Procollagen lysyl hydroxylase
Biotin (RDA has not been established)	10g	Multiple carboxylase deficiency[§]	Holocarboxylase synthetase, or defect in intestinal or cellular uptake, or intracellular transport
	10mg	Propionic acidemia; ketoacidosis	Propionyl-CoA carboxylase
Cobalamin (vitamin B_{12}; adult RDA = 3.0 μg)	<5μg	Megaloblastic anemia (juvenile pernicious anemia)	Decreased intestinal absorption due to abnormality of intrinsic factor or ileal mucosal cells, or absence of intrinsic factor
	>100 μg	Megaloblastic anemia	Transcobalamin II
	>250 μg	Methylmalonic aciduria; ketoacidosis	Adenosyl cobalamin formation
	>500 μg	Methylmalonic aciduria; homocystinuria	Formation of both adenosyl and methyl cobalamins
Folic acid (folacin; adult RDA = 400 μg)	>10 mg	Megaloblastic anemia and mental retardation[§]	Defect in intestinal and perhaps brain transport of folate
	>10 mg	Homocystinuria and hypomethioninemia	N^5,N^{10}-Methylenetetrahydrofolate reductase
	>5 mg	Formiminoglutamic aciduria and mental retardation	Probably formiminotransferase
Lipoic acid	3–10 mg	Lactic and pyruvic acidemia[‖]	Pyruvate carboxylase
Niacin (adult RDA = 13 mg = 13 niacin equiv.)	50–250 mg	Hartnup disease	Intestinal and renal transport of tryptophan (high protein diet needed in addition to niacin to relieve symptoms)
Pyridoxine	10–50 mg	Infantile convulsions	Perhaps glutamate decarboxylase
[vitamin B_6; adult RDA = 2.2 mg (men), 2.0 mg (women)]	>10 mg	B_6-responsive anemia[¶]	Unknown
	>25 mg	Cystathioninuria	γ-Cystathionase
	5–10 mg	Homocystinuria	Cystathionine synthetase
	> 100 mg	Primary hyperoxaluria, type I	Soluble glyoxalate, α-ketoglutarate carboligase
	5–10 mg	Xanthurenic aciduria[§]	Kynureninase
	18–30 mg	Gyrate atrophy of choroid and retina; hyperornithinemia	Ornithine: 2-oxoacid aminotransferase

(continued)

TABLE 38-1 (*continued*)

Therapeutic Vitamin	Daily Dose	Disorder[†]	Biochemical/Biological Defect
Thiamine (vitamin B_1; adult RDA = 1 mg/2000 kcal/d)	20 mg	Megaloblastic anemia	Unknown
	5–20 mg	Branched-chain ketoaciduria (maple syrup urine disease)	Branched chain α-ketoacid decarboxylase
	5–20 mg	Pyruvic acidemia	Pyruvate decarboxylase
	> 10 mg	Lactic and pyruvic acidemia; Leigh's encephalomyelopathy[‖]	Pyruvate decarboxylase (low K_m form)

*These disorders can be thought of as relative vitamin deficiency states. Not all patients will respond to vitamins owing to the existence of phenocopies. Data from L. E. Rosenberg: Vitamin-responsive inherited metabolic disorders. In *Advances in Human Genetics,* Vol. 6, H. Harris and K. Hirschhorn, Eds. (Plenum Press, 1976); and S. H. Mudd: Vitamin-responsive genetic abnormalities, In *Advances in Nutritional Research,* Vol. 4 (Plenum Press, 1982).
[†]All disorders have an autosomal recessive pattern of inheritance except as follows:
[‡]X-linked dominant inheritance;
[§]Pattern of inheritance unknown;
[‖]Autosomal, probably recessive, inheritance;
[¶]X-linked recessive inheritance.

4. The biochemical function of the vitamin should be related directly to the observed deficiency symptoms, although this is not always possible to demonstrate. The classic example is ***scurvy*** (scorbutus), which used to be common among sailors who ate no fresh fruit or vegetables during long sea voyages. Scurvy could be prevented by consumption of fresh citrus fruits, now known to be high in vitamin C (ascorbic acid). The deficiency arose because vitamin C was destroyed by the methods used to preserve food while at sea. Other examples of a single vitamin deficiency in human populations are rare, however, except where the dietary supply is adequate but utilization of the vitamin is impaired. Frequently, if one vitamin is lacking, others are as well, and the intake of protein, trace elements, and other nutrients is probably insufficient. This is particularly true of B vitamins.

In addition to nutritional inadequacy, vitamin deficiency may result from malabsorption, effects of pharmacological agents, and abnormalities of vitamin metabolism or utilization. Thus, in biliary obstruction or pancreatic disease, the fat-soluble vitamins are poorly absorbed despite adequate dietary intake because of steatorrhea. Absorption, transport, activation, and utilization of vitamins require the participation of enzymes or other proteins whose synthesis is under genetic control. Dysfunction or absence of one of these proteins can produce a disease that is clinically indistinguishable from one caused by dietary deficiency. In vitamin-dependent or vitamin-responsive disorders, use of pharmacological doses of the vitamin can sufficiently overcome the blockage for normal function to occur (Table 38-1; discussed at the end of the chapter).

Vitamin deficiency can result from treatment with certain drugs. Thus, destruction of intestinal microorganisms by antibiotic therapy can produce symptoms of vitamin K deficiency. Isoniazid, used to treat tuberculosis, is a competitive inhibitor of pyridoxal kinase, which is needed to produce pyridoxal phosphate. Isoniazid can produce symptoms of pyridoxine deficiency. To prevent this, pyridoxine is often incorporated into isoniazid tablets. Methotrexate and related folate antagonists act by competitively inhibiting dihydrofolate reductase (Chapter 27).

Two standards have been established to plan diets that contain adequate supplies of the vitamins and to aid in diagnosing vitamin deficiency diseases.

1. The ***minimum daily requirement*** (MDR) is the smallest amount of a substance needed by a person to prevent a deficiency syndrome. It is considered to represent the body's basic physiological requirement of the material. MDR values, which are established by the U.S. Food and Drug Administration (FDA), are not available for all vitamins.
2. The ***recommended daily allowance*** (RDA) is the amount of a compound needed daily to maintain good nutrition in most healthy people. RDA values are intended to serve as nutritional goals, not as dietary requirements. They are defined by the Food and Nutrition Board of the National Academy of Sciences

of the USA. RDA values have been established for mineral elements, energy, protein, electrolytes, and water. Those for the vitamins are given in Appendix IV.

Overdosage of vitamins A and D produces ***hypervitaminosis.*** In recognition of this, the FDA has ruled that vitamin A in doses greater than 10,000 IU and vitamin D in doses greater than 400 IU should be available on a prescription-only basis and that all products containing quantities of a vitamin in excess of 150% of its RDA be classified as a drug. Hypervitaminosis D was discussed in Chapter 37; vitamin A toxicosis is discussed below. The toxicity of high doses of vitamin B_6 is also covered later in the chapter.

Certain vitamins can be synthesized by humans in limited quantities. Niacin can be formed from tryptophan (Chapter 17). This pathway is not active enough to satisfy all the body's needs; however, in calculating the RDA for niacin, 60 mg of dietary tryptophan is considered equivalent to 1 mg of dietary niacin. In Hartnup's disease (see Table 38-1 and Chapter 17), a rare hereditary disorder in the transport of monoaminomonocarboxylic acids (e.g., tryptophan), a pellagra-like rash may appear, suggesting that over a long period of time dietary intake of niacin is insufficient for metabolic needs. This pattern also occurs in carcinoid syndrome in which much tryptophan is shunted into the synthesis of 5-hydroxytryptamine.

Vitamin D is synthesized in the skin, provided radiant energy is available for the conversion (Chapter 37):

$$\text{7-Dehydrocholesterol} \xrightarrow{\text{photons}} \text{cholecalciferol (vitamin } D_3) \quad (38.1)$$

This pathway is adequate to supply the body's need for vitamin D, provided exposure to sunlight is adequate. Vitamin D becomes a vitamin when environmental conditions prevent synthesis of an adequate supply by the skin.

Physiological age–related changes in the elderly can affect the nutritional status. Decreased active intestinal transport and atrophic gastritis impair the absorption of vitamins and other nutrients. Reduced exposure to sunlight can lead to decreased vitamin D synthesis. Many drugs may impair both appetite and absorption of nutrients. Some examples of unfavorable drug–nutrient interactions are drugs that inhibit stomach acid production (e.g., omeprazole); drugs that reduce vitamin B_{12} absorption; anticonvulsant drugs (e.g., barbiturates, phenytoin, primidone) that act by inducing hepatic microsomal enzymes which accelerate inactivation of vitamin D metabolites and aggravate osteoporosis(Chapter 37); interference with folate metabolism by antifolate drugs (methotrexate) used in the treatment of some neoplastic diseases; and vitamin B_6 metabolism affected by isoniazid, hydralazine, and D-penicillamine. Examples of negative impacts of vitamins on drug action are vitamin B_6-dependent action of peripheral conversion of L-3,4-dihydroxyphenylalanine (L-dopa) to L-dopamine that is mediated by aromatic L-amino acid decarboxylase and prevents L-dopa's transport across the blood-brain barrier; also ingestion of large amounts of vitamin K-rich foods or supplements, and action of warfarin on anticoagulation (Chapter 36). L-Dopa is the metabolic precursor of L-dopamine and is used in the treatment of Parkinson's disease (Chapter 32). Thus, L-dopa is administered along with a peripherally acting inhibitor of aromatic L-amino acid decarboxylase (e.g., carbidopa). Vitamins A (and retinoids), C, and E are antioxidants. Randomized trials of dietary supplementation of antioxidant vitamins to determine their favorable and/or unfavorable physiological effects on cardiovascular diseases, cancer, and other chronic diseases are in progress.

38.1 Fat-Soluble Vitamins

The fat-soluble vitamins share many properties despite their limited chemical similarity. They are absorbed into the intestinal lymphatics, along with other dietary lipids, after emulsification by bile salts. Lipid malabsorption accompanied by steatorrhea usually results in poor uptake of all the fat-soluble vitamins. Deficiency disease (except in the case of vitamin K) is difficult to produce in adults because large amounts of most fat-soluble vitamins are stored in the liver and in adipose tissue. The fat-soluble vitamins are assembled from isoprenoid units; this fact is apparent from examination of the structures of vitamins A, E, and K; cholesterol, the precursor of vitamin D, is derived from six isoprenoid units (Chapter 18). Specific biochemical functions for vitamins A, D, and K are known, but a role for vitamin E, other than as a relatively nonspecific antioxidant, remains elusive.

Vitamin A

Nutrition and Chemistry

The role of vitamin A in vision is fairly well understood, but the other functions of vitamin A are only beginning to be elucidated (e.g., mucus secretion, maintenance of the integrity of differentiated epithelia and of the immune system, growth, and reproduction). Loss of night vision (nyctalopia) is an early sign of vitamin A deficiency, and clinical features of well-developed deficiency include epidermal lesions, ocular changes, growth retardation, glandular degeneration, increased susceptibility to infection, and sterility.

Natural and synthetic compounds with vitamin A activity and inactive synthetic analogues of vitamin A are collectively termed ***retinoids.*** The most biologically active,

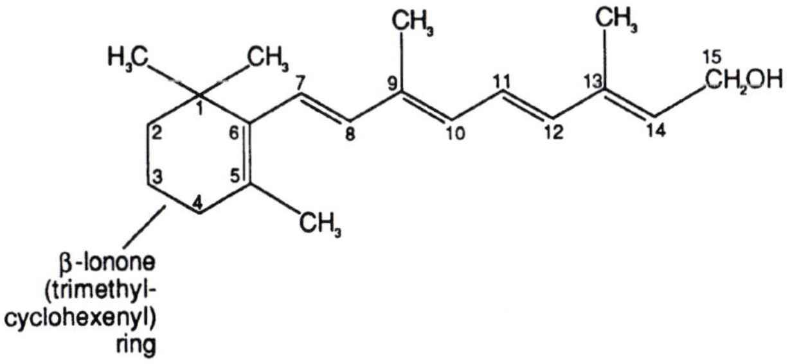

β-Ionone (trimethyl-cyclohexenyl) ring

FIGURE 38-1
Chemical structure of all-*trans* retinol (vitamin A_1), the most active form of vitamin A. Oxidation of C_{15} to an aldehyde or an acid produces, respectively, retinaldehyde (retinal) and retinoic acid. The cis-trans isomerization of the double bond between C_{11} and C_{12} occurs during functioning of retinaldehyde in vision.

naturally occurring retinoid is all-*trans* retinol, also called vitamin A (Figure 38-1). The β-ionone ring is required for biological activity. Vitamin A_1 exists free or as retinyl esters of fatty acids (primarily palmitic acid) in foods of animal origin, including eggs, butter, cod liver oil, and the livers of other vertebrates. In most of the dietary retinol, the four double bonds of the side chain are in the trans configuration. They are readily oxidized by atmospheric oxygen, inactivating the vitamin. They can be protected by antioxidants such as vitamin E.

Several provitamins are present in yellow and dark green leafy vegetables and fruits, such as carrots, mangoes, apricots, collard greens, and broccoli. They are collectively known as the carotenoid pigments or ***carotenes.*** The most widely occurring and biologically active is β-carotene (Figure 38-2). Other nutritionally important carotenoid pigments are cryptoxanthine, a yellow pigment found in corn, and α- and γ-carotenes. Other carotenoids, such as lycopene the red pigment of tomatoes, and xanthophyll, lack the β-ionone ring essential for vitamin A activity.

Absorption, Transport, and Metabolism

Retinyl esters are hydrolyzed in the intestinal lumen by pancreatic carboxylic ester hydrolase, which also hydrolyzes cholesteryl esters. In mucosal cells, ***retinol*** is reesterified, mostly with long-chain fatty acids, by acyl-CoA: retinol acyltransferase. The retinyl esters are incorporated into chylomicron particles and secreted into the lacteals. In humans and rats, 50% of the retinol is esterified with palmitic acid, 25% with stearic acid, and smaller amounts with linoleic and oleic acids. These esters are eventually taken up by the liver in chylomicron remnants. β-Carotene is cleaved in the intestinal mucosa to two molecules of retinaldehyde by β-carotene-15,15′-dioxygenase, a soluble enzyme. Other provitamins are also activated upon cleavage by this enzyme. Retinaldehyde is then reduced to retinol by retinaldehyde reductase, using either NADH or NADPH.

In the liver, retinyl esters are hydrolyzed and reesterified. More than 95% of hepatic retinol is present as esters of long-chain fatty acids, primarily palmitate. In an adult receiving the RDA of vitamin A, a year's supply or more may be stored in the liver.

More than 90% of the body's supply of vitamin A is stored in the liver. The hepatic parenchymal cells are involved in its uptake, storage, and metabolism. Retinyl esters are transferred to hepatic fat-storing cells (also called Ito cells or lipocytes) from the parenchymal cells. The capacity of these fat-storing cells may determine when vitamin A toxicosis becomes symptomatic. During the development of hepatic fibrosis (e.g., in alcoholic liver disease), vitamin A stores in Ito cells disappear and the cells differentiate to myofibroblasts. These cells appear to be the ones responsible for the increased collagen synthesis seen in fibrotic and cirrhotic livers.

Retinol is released from the liver and transported in plasma bound to ***retinol-binding protein*** (RBP), which is synthesized by hepatic parenchymal cells. Less than 5% circulates as retinyl esters. Retinol-binding protein from human plasma is a monomeric polypeptide (M.W. 21,000), which has a single binding site. Transfer of retinol into cells may be mediated by cell surface receptors that specifically recognize RBP. After binding and releasing its vitamin A, RBP appears to have decreased affinity for prealbumin (see below) and is rapidly filtered by the kidney and degraded or excreted.

The retinol-RBP complex circulates as a 1:1 complex with prealbumin (***transthyretin;*** M.W. 55,000), which

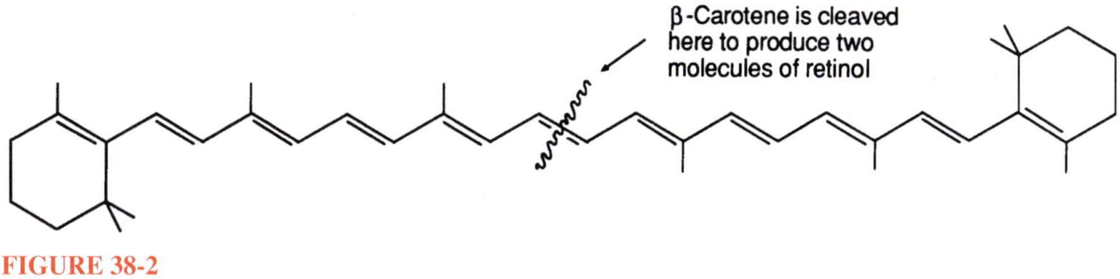

β-Carotene is cleaved here to produce two molecules of retinol

FIGURE 38-2
Structure of all-*trans* β-carotene.

also functions in transport of thyroid hormones. Interaction between transthyretin and RBP is quite specific (dissociation constant 10^{-6}–10^{-7} mol/L). Retinol-RBP-transthyretin complex (RTC) (M.W. 76,000) migrates electrophoretically as an α-globulin. Normal concentrations of RBP and transthyretin in human plasma are 40–50 and 200–300 μg/mL, respectively. Transthyretin, a glycoprotein with a high tryptophan content, is synthesized in the liver and migrates electrophoretically ahead of albumin. It has a half-life of 2 days and has a small pool size. These two properties make it a sensitive indicator of nutritional status (Chapter 17). Its plasma level decreases in protein-calorie malnutrition, liver disease, and acute inflammatory diseases. The serum concentration of retinol is held remarkably constant despite wide variation in dietary intake and hepatic stores of the vitamin by tight control over the concentration of RBP. In vitamin A-deficient rats, hepatic secretion of RBP is specifically blocked, and its level in serum falls whereas that in the liver increases. Administration of vitamin A causes a rapid proportional release of hepatic RBP. The rapidity indicates the release of pooled protein rather than new synthesis. Patients suffering from parenchymal liver disease or protein-calorie malnutrition have reduced amounts of serum RBP and vitamin A. Provision of an adequate diet to malnourished children increases plasma concentrations of RBP and retinol, even without vitamin A supplementation, suggesting the presence of stores of retinol that could not be mobilized on the inadequate diet. The plasma concentration of RBP increases in renal disease, indicating that the kidney is a major catabolic site for this protein. Complex formation with transthyretin reduces glomerular filtration and, consequently, renal catabolism of RBP. In rat hepatocyte cultures, glucocorticoids stimulate net synthesis of RBP.

Oxidation of retinol produces ***retinoic acid*** (***tretinoin***). The reaction is irreversible. Retinoic acid enters the portal blood, is transported bound to albumin, and is not stored to any great extent. The concentration of retinoic acid in plasma is normally 3–4 ng/mL. A biologically active metabolite, 5,6-epoxyretinoic acid, has been isolated from the intestinal mucosa of vitamin A-deficient rats following administration of ^{3}H-retinoic acid. Several tissues have specific ***cellular retinoic acid-binding proteins*** (CRABPs).

Function

Vitamin A and its derivatives retinal and retinoic acid have many essential biological roles in such processes as vision, regulation of cell proliferation and differentiation, and as morphogenetic agents during embryonic development and differentiation. Both natural and synthetic analogues (known as retinoids) possess varying degrees of biological activity ascribed to vitamin A. Other than the role of retinal in vision, retinoids exert their biological actions by binding to specific nuclear receptors, such as transcription factors that modulate gene expression. The retinoid receptors belong to two subfamilies: retinoic acid receptors (RARs) and retinoid X receptors (RXRs). Both the RARs and RXRs contain three isotypes, α, β, and γ encoded by separate genes. Thus, the retinoid receptor family includes RARα, RARβ, RARγ, RXRα, RXRβ, and RXRγ, which are members of the steroid and thyroid hormone superfamily of receptors (Chapter 30). All of these receptors possess at least two domains: a ligand binding domain and a DNA binding domain. The DNA binding domain of the receptor recognizes retinoic acid response elements within the target gene via the two zinc finger binding motifs.

The receptors form dimers of various combinations and permutations within the superfamily, and each dimer may perform a specific biological function by binding to specific DNA response elements. Activation of retinoid receptors can lead to inhibition of cell proliferation, induction of differentiation, and induction of apoptosis during normal cell development. The importance of RARs in cell proliferation and differentiation is illustrated by ***acute promyelocytic leukemia*** (PML), which is a subtype of acute myeloid leukemia. PML is characterized by abnormal hypergranular promyelocytes, a cytogenetic translocation, and a bleeding disorder secondary to consumptive coagulopathy and fibrinolysis. The bleeding problems are presumably initiated by procoagulant phospholipids present in the leukemic cells (Chapter 36).

The cytogenetic hallmark of PML consists of a balanced reciprocal translocation between the long arms of chromosomes 15 and 17, designated as t(15;17) (Figure 38-3). The translocation results in two recombinant chromosomes, one abnormally long 15q+ and one shortened 17q–. This translocation fuses the *PML* gene located on chromosome 15 with the RARα gene located on chromosome 17. The two chimeric genes formed are *PML-RARα* and RARα-PML. Of these, the *PML-RARα* gene is transcriptionally active and produces a PML-RARα protein. This protein, which is an oncoprotein, is responsible for the pathogenesis of PML by interfering with the normal functions of *PML* and *RARα* genes in differentiation of myeloid precursors. The normal *PML* gene product has growth suppressor activity. Treatment with all-*trans* retinoic acid has produced complete remissions in PML by promoting the conversion of leukemic blast cells into mature cells. This is because the PML-RARα oncoprotein maintains responsiveness to retinoic acid's biological actions. Retinoic acid

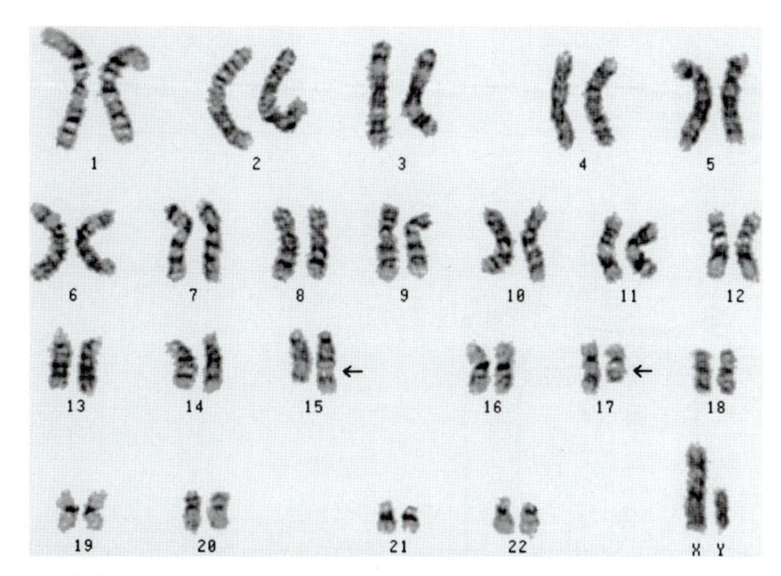

FIGURE 38-3
Karyotype of a promyelocytic leukemic cell, with balanced reciprocal translocation between chromosomes 15 and 17, t(15; 17). The rearranged chromosomes are identified by arrows. (Courtesy of Dr. Mark H. Bogart.)

therapy may also augment the degradation of the PML-RARα fusion protein. Transgenic mice with a PML-RARα fusion gene develop PML, which supports the pathogenic role of the fusion protein in human PML.

Retinyl phosphate, acting as a sugar carrier in a manner similar to that of dolichol phosphate (Chapter 16), may be necessary for normal glycosylation of a subgroup of essential glycoproteins. In vitamin A deficiency, synthesis of some glycoproteins is decreased. Retinyl phosphate and mannosyl phosphoryl retinol have been identified *in vivo* and have been synthesized *in vitro* using biological preparations. In rat liver, GDP-mannose: retinyl phosphate mannosyltransferase has its highest specific activity in the endoplasmic reticulum. However, retinoic acid can fully support growth in the whole animal even though it is not converted to retinol and does not participate in sugar transfer reactions.

More than 1500 retinoids have been synthesized. Two in particular, 13-*cis*-retinoic acid (isotretinoin) and etretinate (Figure 38-4), have generated considerable interest as agents for the treatment of dermatological disorders. 13-*cis* retinoic acid inhibits sebum production, is the drug of choice for treatment of severe cystic acne, and is useful in the treatment of other forms of acne. Etretinate is used in Europe for treatment of psoriasis and related disorders.

Hypo- and Hypervitaminosis A

Serious vitamin A deficiency is not a major public health problem in North America; however, where poverty and poor nutrition are common many persons suffer from vitamin A deficiency with active corneal involvement. This condition leads to permanent blindness in about half of cases, and many of those affected are preschool children. Night blindness occurs early in vitamin A deficiency. There is a reduction in rhodopsin concentration, followed by retinal degeneration and loss of photoreceptor cells. Degenerative changes may be due to instability of free opsin or may indicate an additional nutritive function for retinaldehyde (or retinoic acid) in retinal cells. The degenerative changes of retinitis pigmentosa, a relatively common, inherited cause of blindness, closely resemble those of vitamin A deficiency.

Ingestion of vitamin A in large excess of the RDA can cause toxicity. Daily intake of more than 7500 retinol equivalents (25,000 IU) is not recommended, and doses in excess of 3000 retinol equivalents (10,000 IU) should only be used with medical supervision. Acute toxicity after

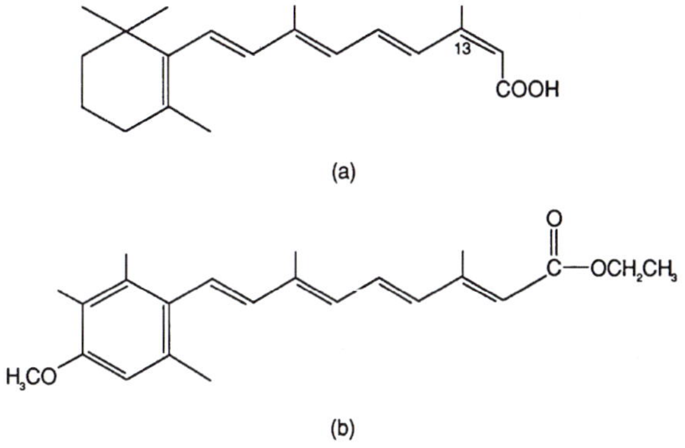

13
COOH
(a)

O
C—OCH_2CH_3
H_3CO
(b)

FIGURE 38-4
Structures of 13-*cis*-retinoic acid (isotretinoin) (a) and etretinate (b), synthetic retinoids used as pharmacological agents.

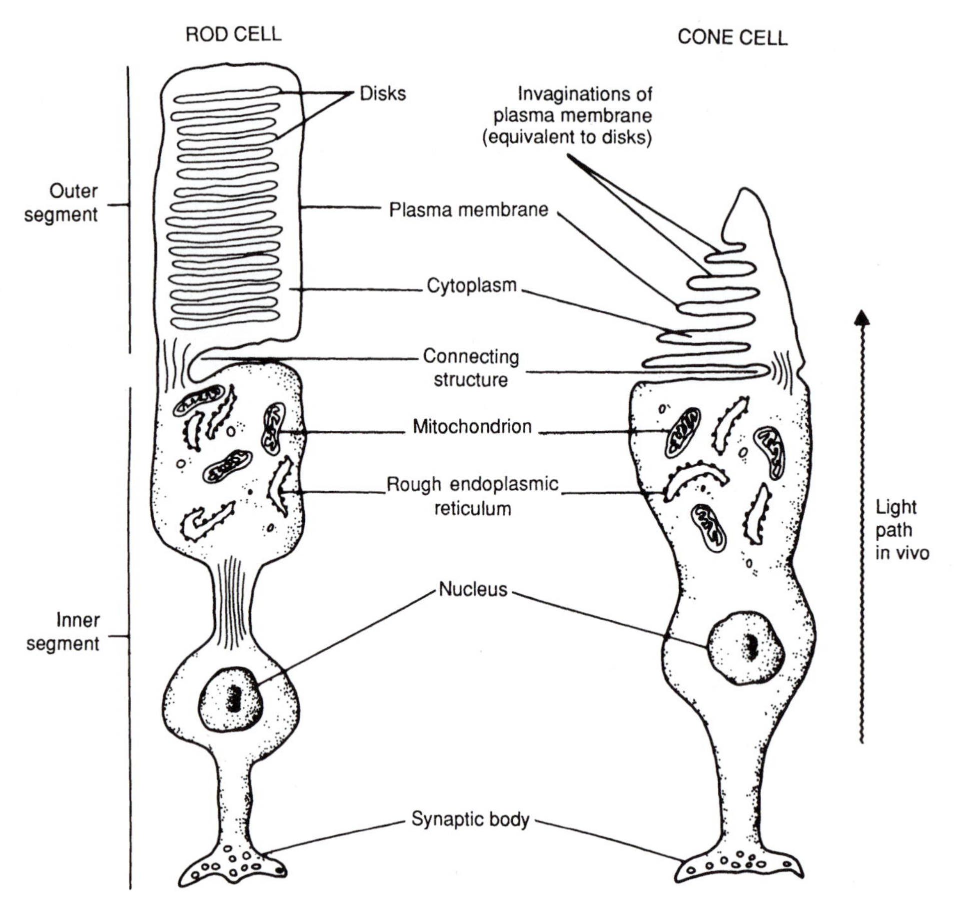

FIGURE 38-5
Schematic representations of mammalian retinal rod and cone cells. There are about 500–1500 disks in the outer segment of each rod cell. The deep membrane invaginations in the plasma membrane of the cone cell outer segment are thought to be functionally equivalent to the rod cell disks.

a single large dose may last several days. Chronic toxicity occurs in adults after daily doses of 7500–15,000 retinol equivalents (25,000–50,000 IU) taken for several months.

Ingestion of carotenes alone does not cause vitamin A toxicity, probably because of markedly decreased absorption at high doses and feedback inhibition of carotene conversion to retinaldehyde, but can cause harmless carotenemia with yellowing of the skin, particularly on the palmar and plantar surfaces. Carotenemia can cause falsely elevated values for the icterus index or for other direct reading methods for estimating serum bilirubin (Chapter 29).

In cases of toxicity, total serum vitamin A concentration is elevated two- to eightfold, mostly in the retinyl esters, which are associated with plasma lipoproteins, while the RBP level is normal or only slightly increased. Unesterified retinol is generally increased less than twofold.

Vitamin A is a surfactant, and its toxicity may be due to labilization and disruption of biological membranes. Excessive amounts of retinol also increase synthesis and release of lysosomal hydrolases. In *in vitro* systems, retinol bound to RBP is not toxic, and RBP alone protects against toxicity of previously added vitamin A. Thus, clinical toxicity occurs only when the binding capacity of RBP is exceeded and retinol and retinyl esters are present either free or in lipoproteins in the circulation.

Vision and Vitamin A

Photosensitivity in various organisms is based on photoisomerization of retinaldehyde. In humans, ***11-cis-retinaldehyde*** complexes with opsin to produce ***rhodopsin*** (visual purple). Absorption of a photon by the π-electron system of retinaldehyde changes it to the all-*trans* isomer; this reaction—the only light-catalyzed step in vision—is transduced to an action potential transmitted in the optic nerve. The human visual system functions over a range of light intensity. If properly dark adapted, the eye can detect a single photon. Vitamin A deficiency reduces the amount of rhodopsin in the retina, thereby increasing the minimum amount of illumination that can be detected (the visual threshold) and causing night blindness.

The retina consists of a thin layer of photoreceptor cells connected to the optic nerve via axons, synapses, and the bodies of several intermediate cells. Other structures of the eye provide the optical system for focusing light onto the retina. Visual pigments are present in the approximately 100 million rod and 5 million cone cells (Figure 38-5). The cones are less sensitive than the rods but are required for color vision. Three types of cone cells differ with respect to the wavelengths of light to which they are maximally sensitive. Comparison and integration of inputs from these three cell types by the visual cortex produce color vision. At low light levels, only the rods are active and color vision is lost. A higher pigment content and greater molar absorptivity of the pigment contribute to the higher sensitivity of rods.

Rhodopsin has a broad absorption spectrum with a maximum at 500 nm, making the rods most sensitive to green light. In the human retina, cone cell absorption maxima are 445 nm (blue), 535 nm (green), and 570 nm (red). Opsins in the three types of cone cell have different primary structures and are coded by three separate genes. Those for red- and green-sensitive opsins are on the X chromosome, while that for blue-sensitive opsin is on an autosome. Congenital color blindness, which affects about 9% of the male population, results from the absence of one or more cone cell types or from a decrease in amount of one of the pigments. Absence of red or green cones occurs in 2.5% of males. Absence of blue cones occurs in only 0.001% of males. Decrease in red cone pigment (protanomaly) or green cone pigment (deuteranomaly) occurs in 1.3% and 5.0%, respectively, of males. The decrease in pigment type presumably is due to mutation in a structural gene. Absence of two or three types of cone cell is extremely rare.

Light reception and energy transduction take place in the rod outer segments, while cellular metabolism occurs in the rod inner segment, which is rich in glycogen, mitochondria, and rough endoplasmic reticulum. The outer segment contains 500–1500 flattened membrane sacs (disks), each about 16 nm thick, which are electrically isolated from the surrounding plasma membrane and which contain rhodopsin as an integral, transmembrane protein. The disks are continually replaced from the outer segment nearest the nucleus. Phagocytosis of old disks by cells of the retinal pigment epithelium occurs at the top of the outer segment. In the rhesus monkey, the lifetime of a disk is 9–13 days. Disks are formed by the sealing off of invaginations of plasma membrane, and the composition of their contents differs from that of cytoplasm. Blindness due to hereditary defects in regeneration of disks has been observed in rats.

Cone cells are similar in many ways to rod cells (Figure 38-5). Because cone cells lack distinct disks, their visual pigments are located on deep invaginations of the plasma membrane that resemble disks and probably are formed by a similar mechanism. Disk membranes are composed primarily of phospholipid and rhodopsin at a molecular ratio of about 70:1. The phospholipids are 40% phosphatidylcholine, 40% phosphatidylethanolamine, and 13% phosphatidylserine. The membrane has high fluidity because of extensive unsaturation of the lipids, which gives rhodopsin considerable rotational and translational mobility. Average spacing of rhodopsin molecules is 5.6 nm.

Bovine rhodopsin (M.W. 39,000) has two asparagine-linked oligosaccharides with the structure (mannose)$_3$ (N-acetylglucosamine). Seven hydrophobic α-helical segments, each 21-28 amino acids long, are embedded in the disk membrane and are connected by hydrophilic sequences that are exposed to either side of the membrane. The retinaldehyde moiety is within the membrane, halfway down the first helical segment, attached by an aldimine linkage to the ε-amino group of lysine 53 residues from the carboxyl terminus by aldimine linkage (Figure 38-6).

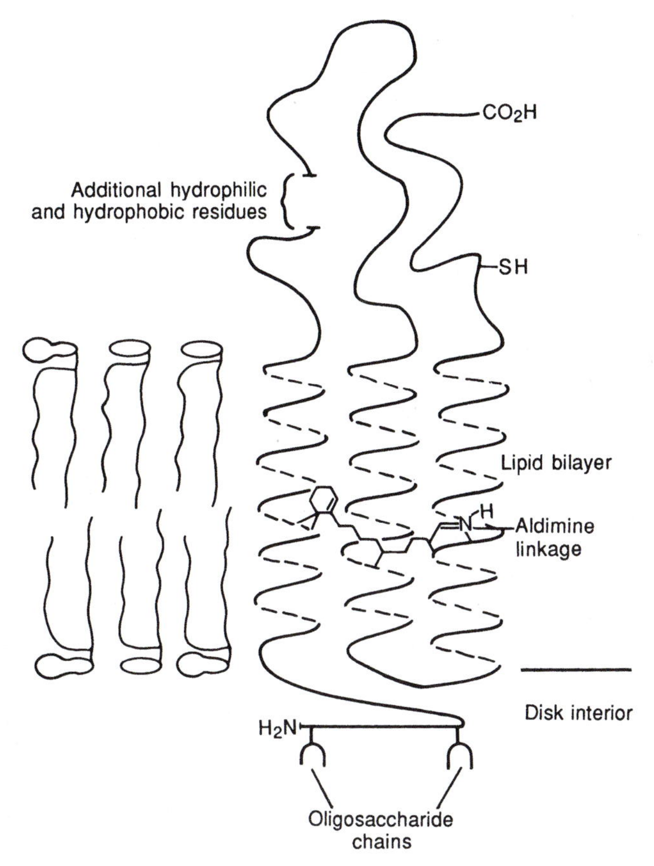

FIGURE 38-6
Diagrammatic representation of bovine rhodopsin embedded in the disk membrane. Only three of the seven α-helical segments are shown. The only point of attachment between retinaldehyde and opsin is the indicated aldimine bond. [Modified and reproduced, with permission, from D. F. O'Brien, The chemistry of vision. *Science* **218,** 961 (1982), © 1982 by the American Association for the Advancement of Science.]

In the dark, rhodopsin has a red color, which changes to pale yellow upon exposure to light. This bleaching takes place within a few milliseconds after absorption of a photon and occurs because all-*trans* retinaldehyde bound to opsin is unstable. At low temperatures, bleaching proceeds more slowly, and seven intermediates have been identified (Figure 38-7). The structures and order of appearance of these intermediates have yet to be clarified. The final

1 6 11 12
N—Lys (53′)-opsin
⊕ H

Rhodopsin
$h\nu$ ↓
Hypsorhodopsin
↓
Bathorhodopsin
↓
Lumirhodopsin
↓
Metarhodopsin I
↓
Metarhodopsin II
↓
Pararhodopsin
↓
N-Retinylideneopsin
H_2O ↓

11 12 O
all-*trans* -Retinaldehyde
+ H_2N-Lys (53′)-opsin

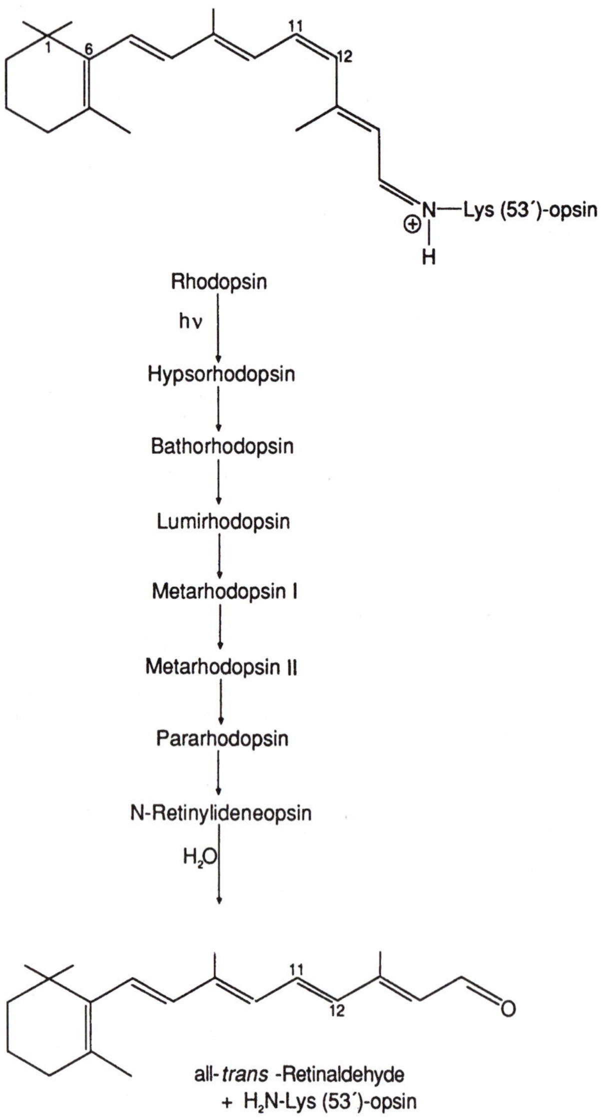

FIGURE 38-7
Bleaching of rhodopsin in bovine retina. Absorption of light by 11-*cis*-retinaldehyde initiates the configurational change to all-*trans* retinaldehyde, culminating in hydrolysis of the bond between retinaldehyde and Lys 53′ of opsin. The order of formation of the intermediates is tentative. The symbol $h\nu$ represents a photon of visible light. [Modified and reproduced, with permission, from C. D. B. Bridges, Retinoids in photosensitive systems. In: *The Retinoids,* Vol. 2, M. B. Sporn, A. B. Roberts, and D. S. Goodman, eds. Academic Press, San Diego, 1984).]

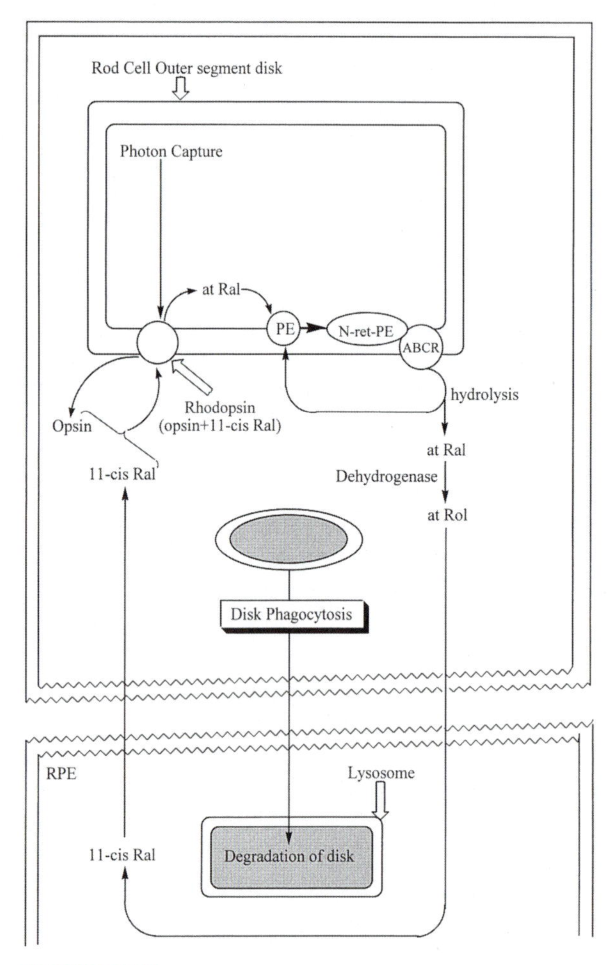

FIGURE 38-8
Regeneration of 11-*cis*-retinaldehyde (11-*cis*Ral). Following light exposure, rhodopsin (opsin + 11-*cis*Ral) breaks down to opsin and all-*trans* retinaldehyde (atRal) initiating a neural signal. The restoration of 11-*cis*Ral begins with the atRal combining with phosphatidylethanolamine (PE) to form the protonatal Schiff base N-retinylidine-PE complex. This complex is transported across the disk membrane by an ATP-binding cassette protein, exclusively found in the retinal outer segment (ABCR). Defects in ABCR cause early-onset maculopathy *(Stargardt's disease).* After reduction of atRal to all *trans*-retinol (atRol) by atRol dehydrogenase in the retinal outer segment, atRol is converted to 11-*cis*Ral in a series of reactions occuring in the retinal pigment epithelial cell (RPE). Also shown in the bottom is the phagocytosis of disk by RPE.

compound (perhaps N-retinylideneopsin) is unstable and hydrolyzes to all-*trans* retinaldehyde plus opsin.

The regeneration of rhodopsin following exposure to bright light requires the following steps. The all-*trans* retinaldehyde combines with phosphotidylethanolamine (PE) to form a protonated Schiff base N-retinylidene-PE. This complex is then exported out of the disk by an ATP-binding cassette (ABC) transporter exclusively located in the retinol disk membrane (ABCR). Outside the disk the all-*trans* retinaldehyde is reduced to all-*trans* retinol by a dehydrogenase and transported to retinal pigment epithelial (RPE) cells. In the RPE cells involving a series of reactions all-*trans* retinol is converted to 11-*cis* retinaldehyde which is transported to rod cells. In the rod cells, 11-*cis* retinaldehyde reassociates to form rhodopsin. The importance of ABCR becomes evident in early onset macular degeneration, due to mutations in the ABCR gene. This disorder is known as ***Stargardt's disease,*** an inherited autosomal recessive disease, which affects one person in 10,000. Age-related macular degeneration in susceptible individuals with risk factors (e.g., high lipid diet, smoking) may also be due in part to defects in ABCR gene.

ABC transporters are involved in the transport of a wide variety of molecules. For example, cystic fibrosis is caused by a defective ABC transporter for chloride ion (cystic fibrosis transmembrane regulator, Chapter 12) and in ***Tangier's disease,*** the abnormality of cholesterol efflux is due to defects in an ABC protein (Chapter 20).

In the dark, the cation (Na^+, Ca^{2+})–specific cGMP-gated channels located in the rod outer segment (ROS) are open, thus promoting the influx of Na^+ and Ca^{2+}. The steady state of cations is maintained by outward pumping of Ca^{2+} by the Na^+/Ca^{2+} exchanger and Na^+, K^+-ATPase pumps located in the inner segment. Exposure to light blocks cGMP-gated cation channels, causing the inside of the plasma membrane to become more negative and resulting in hyperpolarization. The signal of hyperpolarization is transmitted to the synaptic body and eventually to the brain. In the dark, because of the influx of Na^+ and Ca^{2+}, the rod cells are depolarized and neurotransmitters (possibly aspartate or glutamate) are released from the presynaptic membrane of the rod cell at relatively high rates. In the hyperpolarized state the neurotransmitter release is inhibited. Thus, the postsynaptic bipolar cells are either inhibited (hyperpolarized) or excited (depolarized).

Coupling of rhodopsin present in the disk membrane to cation-specific cGMP-gated channels of the plasma membrane involves signal amplification after photon absorption. The coupling mechanism (Figure 38-9) begins with the absorption of light by 11-*cis*-retinal-rhodopsin and its conversion to all-*trans* retinal and the photoactivated rhodopsin (R*). In the next step, activated rhodopsin initiates an amplification process that consists of activating transducin, a signal-coupling G-protein. The active form of transducin, $T_{\alpha\text{-GTP}}$, activates cGMP-phosphodiesterase (PDE). In the second amplification process, PDE hydrolyzes cGMP with the closure of cation channels. The opening and closing of these cation channels that is mediated by cGMP is similar to Ca^{2+} ion channel gating by cAMP in olfaction.

PDE belongs to a superfamily of enzymes that regulate cell function by maintaining cyclic nucleotide levels. There are cAMP-specific PDEs (PDE4, PDE7, and PDE8),

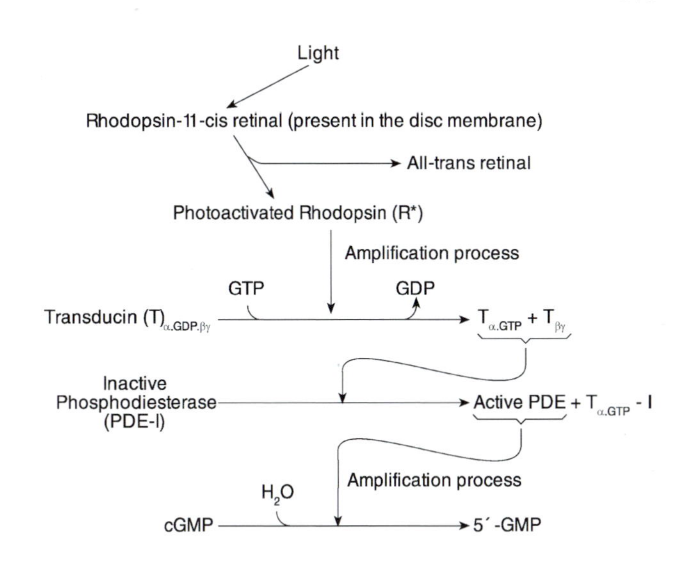

FIGURE 38-9
Mechanism for stimulus-response coupling of photon absorption to the closure of plasma membrane cation channels. I, Inhibitory subunit of phosphodiesterase.

cGMP-specific PDEs (PDE5, PDE6, and PDE9), and ones specific for both cAMP and cGMP PDEs (PDE1, PDE2, and PDE10). Light induced-activated transducin activates a cGMP-specific PDE, namely, PDE6. Another previously discussed cGMP-specific PDE is PDE5, which is abundant in vascular smooth muscle (Chapter 17). PDE5 is involved in the nitric oxide and cGMP signaling pathway. A specific inhibitor of PDE5, sildenafil, causes increased blood flow into the sinusoidal spaces of the corpus cavernosum and corpus spongiosum, resulting in erection of the penis (Chapter 17).

Photoactivation of rhodopsin produces a conformational change that allows it to interact with and activate transducin, resulting in the replacement of GDP by GTP at the α subunit. Metarhodopsin II (Figure 38-7) is the form that interacts with transducin at the disk membrane surface. Depending on the lifetime of the activated rhodopsin molecule, each molecule can activate up to 500 transducin molecules. The α-subunit GTP complex of transducin disinhibits PDE by removing its inhibitor subunit (I), causing it to hydrolyze cGMP to 5′-GMP. The decrease in cGMP concentration closes the plasma membrane cation channels, decreasing Na^+ and Ca^{2+} concentrations. Some cases of ***retinitis pigmentosa*** are caused by mutations in the transducin gene that prevent it from disinhibiting the PDE, leading to an accumulation of cGMP that is toxic to the retina. The transducin-GTP complex remains active as long as the GTP bound to G_α is not hydrolyzed.

The recovery and adaptation of the rod and the cone cells to the dark state begins shortly after illumination

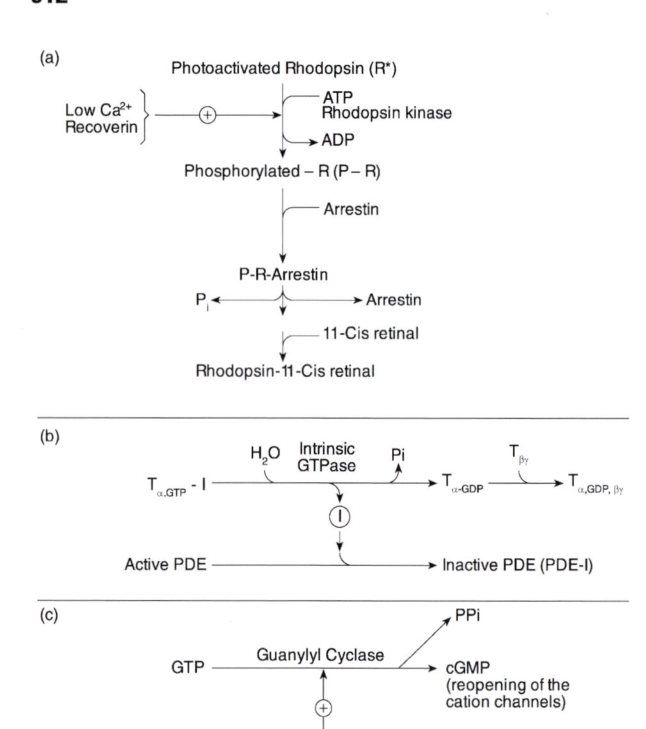

FIGURE 38-10
Recovery and the restoration to dark state in the rod and cone cells require three sets of reactions, a, b, and c. The light-induced lowering of cytosolic Ca^{2+} levels (see Figure 38-9) initiates the process of adaptation to dark (see text for details). P_i, Inorganic phosphate; ⊕, stimulation; I, inhibitory subunit of phosphodiesterase.

(Figure 38-10). The GTP bound to α subunit is hydrolyzed by the intrinsic GTPase activity of α subunit, followed by its reassociation with $T_{\beta\gamma}$ to form inactive $T_{\alpha.GDP.\beta\gamma}$. The formation of $T_{\alpha.GDP.\beta\gamma}$ results in the reversion of PDE to its inactive state. Photoactivated rhodopsin is inactivated by phosphorylation at multiple serine and threonine residues and is catalyzed by rhodopsin kinase and binding of an inhibitory protein known as arrestin. Low Ca^{2+} levels and recoverin stimulate rhodopsin kinase activity. Prestimulus cGMP levels are attained through the activation of guanylyl cyclase by low Ca^{2+} levels, which converts GTP to cGMP.

A disorder ***hyperornithinemia*** of the retina and choroid with gyrate atrophy and progressive degeneration is due to the deficiency of the enzyme ornithine-δ-aminotransferase (OAT). OAT deficiency is inherited as an autosomal recessive disorder and illustrates the metabolic importance of ornithine, a nonprotein amino acid (Chapter 17). Ornithine participates either as a substrate or a product of five enzymatic reactions. Two biochemical mechanisms have been proposed to explain the pathophysiology of gyrate atrophy of the choroid and the retina. One is that a high ornithine concentration causes reduced formation of creatine and phosphocreatine (an important energy source) by inhibiting the enzyme glycine transaminidase. The second is the inhibitory effect of ornithine on Δ'-pyrroline-5-carboxylate synthase preventing formation of proline (Chapter 17). Local proline synthesis in the retina is required because of the lack of entry of proline from plasma. Treatment of gyrate atrophy represents an attempt to reduce the concentration of ornithine by administering pharmacological doses of pyridoxal phosphate (vitamin B_6), which is a cofactor of OAT, restricting dietary intake of arginine, which is a precursor of ornithine, and augmenting renal loss of ornithine by inhibiting the renal dibasic amino acid transport using lysine or the nonmetabolizable α-aminoisobutyric acid. In some patients, therapeutic benefits also have been obtained by supplementing the decreased end products, creatine and proline.

Vitamin E

Nutrition and Chemistry

Vitamin E was crystallized and its structure determined in 1936. Eight plant vitamers are known (Figure 38-11), the most abundant and active being **α-tocopherol.**

Vitamin E vitamers are viscous, light yellow oils that are heat stable but readily degraded by oxygen or ultraviolet light. Principal sources are the vegetable oils,

Tocopherols (Tocols)

Tocotrienols

Tocopherol	Tocotrienol	R_1	R_2	R_3
Alpha-	Alpha-	$-CH_3$	$-CH_3$	$-CH_3$
Beta-	Beta-	$-CH_3$	$-H$	$-CH_3$
Gamma-	Gamma-	$-H$	$-CH_3$	$-CH_3$
Delta-	Delta-	$-H$	$-H$	$-CH_3$

FIGURE 38-11
Structures of naturally occurring plant compounds having vitamin E activity. The nucleus in each is 6-hydroxychroman. Attachment of a saturated 16-carbon chain produces the tocopherols; the tocotrienols have a 16-carbon unsaturated chain. Both groups are optically active. The tocotrienols have one chiral center at carbon 2, while the tocopherols have three, at carbons 2, 4′, and 8′. Tocol is a tocopherol in which R_1, R_2, and R_3 are all hydrogen atoms. The tocopherols can be viewed as methylated tocols.

particularly wheat germ and salad oils. Green vegetables, beef liver, butter, milk, and eggs contain appreciable amounts. Animals presumably obtain vitamin E from plants in their diet. Fish liver oils, although rich in vitamins A and D, are devoid of vitamin E.

The RDA depends on age and increases as the amount of polyunsaturated fatty acid (PUFA) in the diet increases. However, foods that are rich in PUFA are also rich in vitamin E.

Absorption, Transport, and Metabolism

Vitamin E is absorbed as free tocopherol, along with other fat-soluble vitamins and dietary lipids. Tocopheryl acetate, the form commonly used for dietary supplementation, is hydrolyzed before absorption. Uptake requires bile salts. A selective impairment of vitamin E absorption without malabsorption of other fat-soluble vitamins has been identified; it was corrected after a large oral intake of the vitamin. Patients with chronic fat malabsorption and abetalipoproteinemia (Chapter 20) may develop vitamin E deficiency.

About 75% of the absorbed vitamin E enters the lymphatics in chylomicrons, and the rest in other lipoproteins. In plasma, vitamin E is carried by lipoproteins and erythrocytes. In humans, vitamin E is present in greatest amounts in adipose tissue, liver, and muscle. Its principal excretory route is the feces, probably by way of bile.

Function

The function most consistent with symptoms of vitamin E deficiency in animals is that of a general, membrane-localized antioxidant, which protects cellular and subcellular membranes from attack by endogenous and exogenous free radicals. In membranes, vitamin E may be located near enzyme complexes that produce free radicals, such as NADPH-dependent oxidase systems. Selenium alleviates some symptoms of vitamin E deficiency, probably through its role as a cofactor for glutathione peroxidase, which reduces peroxides generated within cells, thereby preventing formation of free radicals. Vitamin E in the plasma membrane may act as a first line of defense against free radicals, while glutathione peroxidase may be a second line of defense. Most enzymes affected by vitamin E deficiency are membrane bound or are involved in the glutathione-peroxidase system.

Vitamin E may also possess antiatherogenic properties. In vitro, studies have shown that oxidized low-density lipoproteins (LDL) are proatherogenic (Chapter 20) and vitamin E retards LDL oxidation. Thus, it is thought vitamin E supplementation might reduce the morbidity and mortality from coronary artery disease. Nonantioxidant functions of vitamin E may involve several cellular signaling pathways (e.g., protein kinase C initiated pathways). The signaling pathways cause inhibition of proliferation of smooth muscle cells, platelet adhesion, and aggregation and function of adhesion molecules. Vitamin E also may attenuate the synthesis of leukotrienes and increase synthesis of prostacyclin by upregulating phospholipase A_2 and cyclooxygenase. All of these actions of vitamin E may contribute toward its protective properties against the development of atherosclerosis. In elucidating the various biological functions of vitamin E complex, the specific roles played by each of the four species of tocopherols and four of trienols are not understood. Some studies have suggested that tocotrienols may be superior to tocopherols in their cardiovascular related effects.

Hypo- and Hypervitaminosis E

Characteristic lesions of vitamin E deficiency in animals include necrotizing myopathy (inaccurately referred to as nutritional muscular dystrophy), exudative diathesis, nutritional encephalomalacia, irreversible degeneration of testicular tissue, fetal death and resorption, hepatic necrosis, and anemia. Several of these conditions are directly related to peroxidation of unsaturated lipids in the absence of vitamin E, and others can be prevented by synthetic antioxidants or vitamin E.

It is difficult to produce vitamin E deficiency in adult humans. Adult males who were depleted of vitamin E for 6 years showed no symptoms, although serum tocopherol concentrations became very low. However, their erythrocytes lysed more readily than normal when exposed to hydrogen peroxide or other oxidizing agents *in vitro*. This finding led to the use of low-plasma vitamin E and increased susceptibility of erythrocytes to oxidative hemolysis as criteria for vitamin E deficiency.

Premature infants and children with chronic cholestasis may develop spontaneous vitamin E deficiency. In premature infants, the deficiency manifests itself as increased red cell fragility and mild hemolytic anemia. It has been claimed, but not established, that these infants respond to administration of vitamin E. The anemia is not prevented by vitamin E, and only small improvements in red cell indices follow vitamin E treatment. A role has been claimed for vitamin E in prophylaxis of ***retrolental fibroplasia*** and ***bronchopulmonary dysplasia,*** two types of oxygen-induced tissue injury that occur in premature infants treated aggressively with oxygen.

Children with chronic cholestasis may exhibit a neuromuscular disorder that responds to treatment with vitamin E given parenterally or in large oral doses. Some patients show neurological signs despite normal serum levels of vitamin E, which may arise from the associated increase in levels of serum lipids, which contain vitamin E. In patients with cystic fibrosis, steatorrhea and fat malabsorption with subnormal plasma and tissue concentrations of vitamin E are common, but neuromuscular disorders such as those in chronic cholestasis do not occur.

Vitamin E deficiency occurs due to genetic defects in the formation of hepatic α-tocopherol transfer protein. This transport protein plays a central role in the liver and one of its functions is to facilitate incorporation of α-tocopherol into nascent very low density lipoproteins (VLDLs). Since there are no specific transport proteins for vitamin E in plasma, the delivery of vitamin E to the tissues is primarily mediated by VLDL-LDL transport mechanisms (Chapter 20). Thus, deficiency of hepatic α-tocopherol transport protein causes low plasma levels of vitamin E with impairment of delivery to the tissues. Patients with the transport protein deficiency exhibit peripheral neuropathy and ataxia. Early and vigorous vitamin E supplementation in patients with neurological symptoms and with low plasmal levels of vitamin E has yielded therapeutic benefits.

A pharmacological role for vitamin E may exist in claudication arising from peripheral vascular disease. Studies with small numbers of patients having cystic fibrosis, glucose-6-phosphate dehydrogenase deficiency, and sickle cell anemia conditions associated with decreased erythrocyte half-lives showed that many had chemical evidence of vitamin E deficiency. Administration of vitamin E supplements (400–800 IU/d) significantly increased red cell survival time. Claims that doses of vitamin E 10–20 times the RDA are beneficial for treatment of skin disorders, fibrocystic breast disease, sexual dysfunction, cancer, baldness, and other disorders have not been substantiated.

Self-medication with high doses of vitamin E appears to be relatively nontoxic. However, in patients receiving warfarin (Chapter 36), vitamin E in excess of 400 IU/d may further depress coagulability and produce coagulopathy.

38.2 Water-Soluble Vitamins

The B group vitamins and vitamin C serve as coenzymes or coenzyme precursors. The B complex includes thiamine, riboflavin, pyridoxine, niacin, pantothenic acid, biotin, folate, and cobalamin. Inositol, choline, and paraaminobenzoic acid, usually classified as vitamin-like substances in humans, are sometimes included with the B-complex vitamins. They will be discussed briefly at the end of the chapter. The B vitamins occur in protein-rich foods and in dark green, leafy vegetables. A deficiency of one B vitamin is usually accompanied by deficiencies of others in the group and of protein. Vitamin C is the antiscorbutic factor of citrus fruit and other fresh fruits and vegetables. Symptoms of deficiency of the water-soluble vitamins are similar and include disorders of the nervous system and of rapidly dividing tissues, such as the gastrointestinal epithelium, mucous membranes, skin, and cells of the hematopoietic system.

Thiamine (Vitamin B_1)

Nutrition and Chemistry

This vitamin, also called aneurin, is the antiberiberi factor. The active coenzyme form is thiamine pyrophosphate (TPP), or cocarboxylase. Thiamine triphosphate (TTP) may be an active form in the central nervous system. Of the thiamine in the body, 10% occurs as TTP, 80% as TPP, and 10% as TMP (thiamine monophosphate) (Figure 38-12).

The principal dietary sources include fish, lean meat (especially pork), milk, poultry, dried yeast, and whole-grain cereals. Bread, cereals, and flour-based products are frequently enriched with this vitamin. Thiamine is present in the outer layers of rice grains, from which it was first identified. Deficiency is common in Asian countries where polished rice is the principal dietary staple. The RDA depends on energy intake.

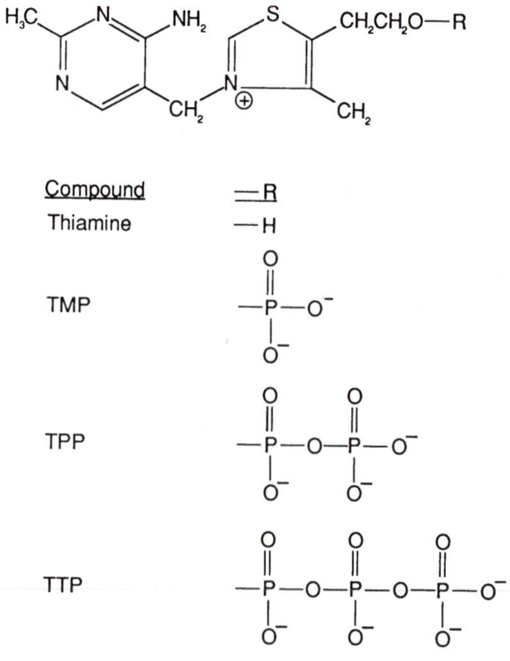

FIGURE 38-12
Structures of thiamine (vitamin B_1) and its phosphorylated metabolites. They consist of a six-membered pyrimidine ring and a five-membered thiazole ring, linked through a methylene group.

Absorption, Transport, and Metabolism

Thiamine is absorbed by a pathway that is saturable at concentrations of 0.5–1.0 μmol/L. Oral doses in excess of 10 mg do not significantly increase blood or urine concentrations of vitamin B_1. In the human, absorption occurs predominantly in the jejunum and ileum. Some ferns, shellfish, fish, and species of bacteria contain thiaminase, which cleaves the pyrimidine ring from the thiazole ring. This enzyme causes thiamine deficiency in cattle. In plasma, thiamine is transported bound to albumin and, to a small extent, other proteins. TPP is synthesized in the liver by thiamine pyrophosphokinase.

Function

In humans, TPP is a coenzyme for transketolation, an important reaction in the pentose-phosphate pathway, and for the oxidative decarboxylations catalyzed by pyruvate dehydrogenase, branched-chain α-ketoacid decarboxylase, and α-ketoglutarate dehydrogenase complexes. In lower organisms, TPP is also a cofactor for nonoxidative decarboxylations such as the conversion of pyruvate to acetaldehyde that occurs in yeast.

TPP and TTP occur in the central nervous system and play an important part in brain metabolism. Entry of thiamine into cerebrospinal fluid may occur via a saturable transport mechanism, perhaps located in the choroid plexus.

Hypo- and Hypervitaminosis

Major diseases caused by thiamine deficiency are the ***Wernicke-Korsakoff syndrome,*** commonly associated with chronic alcoholism, and ***beriberi,*** the classic thiamine deficiency syndrome. Both are related to diets high in carbohydrate and low in vitamins. The Wernicke-Korsakoff syndrome comprises *Wernicke's encephalopathy* and *Korsakoff's psychosis*. The encephalopathy presents with ataxia, confusion, and paralysis of the ocular muscles, which are relieved by administration of thiamine. Korsakoff's psychosis is characterized by amnesia and is only slightly responsive to thiamine. These disorders reflect different stages of the same pathological process. Some areas within the brain appear to be more susceptible than others to the effects of thiamine depletion. The nature of the biochemical lesion is unknown. Thiamine deficiency in chronic alcoholics probably has multiple causes, including poor diet, an inhibitory effect of alcohol and of any accompanying folate deficiency on intestinal transport of thiamine, and reduced metabolism and storage of thiamine by the liver due to alcoholic cirrhosis.

Beriberi occurs whenever thiamine intake is less than 0.4 mg/d for an extended period of time. It occurs where polished rice is a dietary staple, and, in Western society, in poor and elderly populations and alcoholics. Beriberi has wet, dry, and cardiac types, and an individual may have more than one type. "Wet" refers to pleural and peritoneal effusions and edema; "dry" refers to polyneuropathy without effusions. Cardiomyopathy is the principal feature of the cardiac type. An infantile form occurs in breast-fed infants, usually 2–5 months of age, nursing from thiamine-deficient mothers. The symptoms of beriberi remit completely upon thiamine supplementation. A subclinical deficiency of thiamine occurs in hospital patients and the elderly. Deficiency of thiamine and other vitamins may contribute to a generally reduced state of health in these populations.

Thiamine deficiency can be assessed by measuring blood levels. Increased blood levels of pyruvate and lactate suggest thiamine deficiency. Measurement of erythrocyte transketolase activity, which requires TPP as a coenzyme, confirms the deficiency.

Four inherited disorders responsive to treatment with pharmacological doses of thiamine are summarized in Table 38-1. However, at least for megaloblastic anemia and maple syrup urine disease, more nonresponsive than responsive cases are known. Since thiamine is promptly excreted in the urine, excessive intake is not associated with toxicity.

Riboflavin (Vitamin B_2)

This vitamin is the precursor of flavin mononucleotide (FMN) and flavin adenine dinucleotide (FAD), cofactors for several oxidoreductases that occur in all plants, animals, and bacteria (Figure 38-13).

The FAD-requiring enzymes in mammalian systems include the D- and L-amino acid oxidases, mono- and diamine oxidases, glucose oxidase, succinate dehydrogenase, α-glycerophosphate dehydrogenase, and glutathione reductase. FMN is a cofactor for renal L-amino acid oxidase, NADH reductase, and α-hydroxy acid oxidase. In succinate dehydrogenase, FAD is linked to a histidyl residue; in liver mitochondrial monoamine oxidase, to a cysteinyl residue. In other cases, the attachment is noncovalent but the dissociation constant is very low.

Use of oral contraceptives may increase the dietary requirement for riboflavin. Riboflavin status can be evaluated from the activity of erythrocyte glutathione reductase, an FAD-requiring enzyme, before and after addition of exogenous FAD. A low initial activity or a marked stimulation by FAD (or both) is indicative of ariboflavinosis.

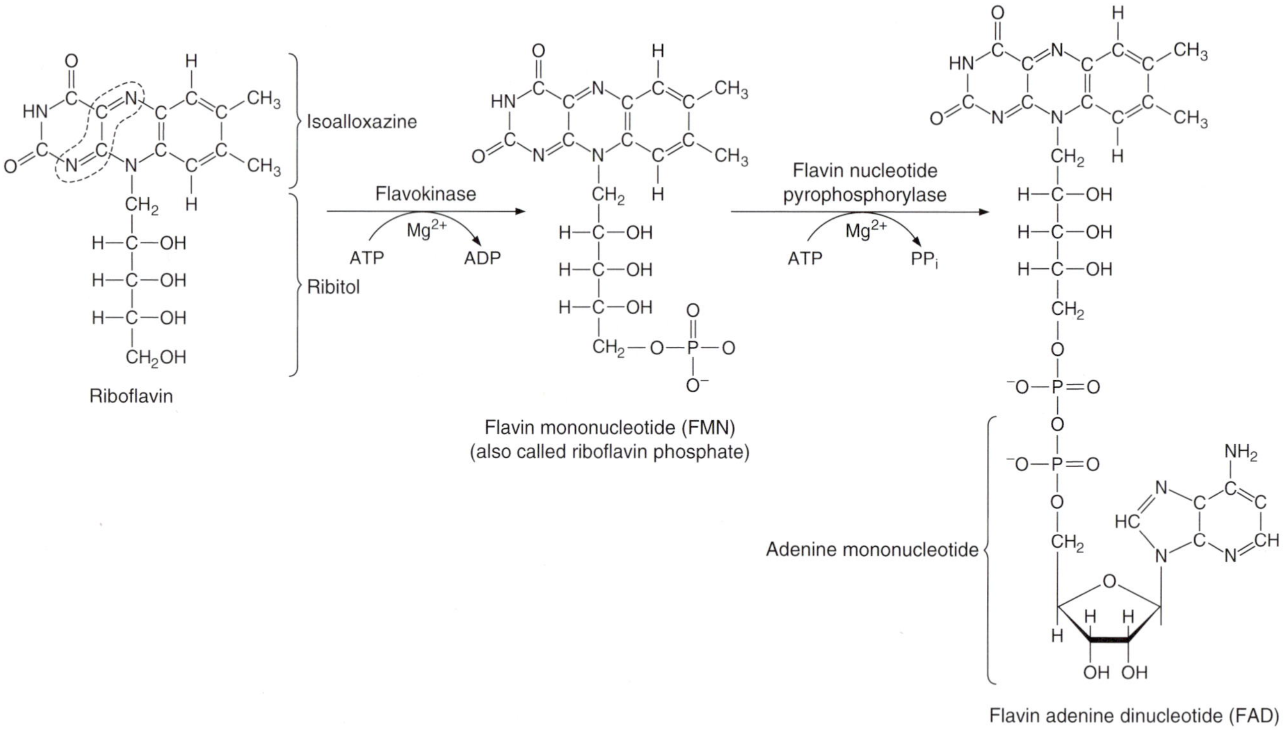

Riboflavin

Isoalloxazine

Ribitol

Flavokinase

Mg^{2+}

ATP ADP

Flavin mononucleotide (FMN)
(also called riboflavin phosphate)

Flavin nucleotide pyrophosphorylase

Mg^{2+}

ATP PP_i

Adenine mononucleotide

Flavin adenine dinucleotide (FAD)

FIGURE 38-13

Synthetic pathways connecting riboflavin (vitamin B_2) and its two cofactors, FMN and FAD. Riboflavin contains an isoalloxazine ring system joined to a ribityl group (a sugar alcohol) through a cyclic amine on the middle ring. Because of the isoalloxazine nucleus, the three compounds are yellow and exhibit a yellow-green fluorescence in aqueous solutions. FMN is not a true nucleotide because the bond between the flavin (isoalloxazine) ring and the ribityl side chain is not a glycosidic bond. Two nitrogen atoms and two carbon atoms, identified by the colored area in the isoalloxazine ring system, participate in the oxidation-reduction reactions of the flavin coenzymes.

Although only synthesized in plants, bacteria, and most yeasts, riboflavinis ubiquitous in plants and animals. Good dietary sources include liver, yeast, wheat germ, green leafy vegetables, whole milk, and eggs. Riboflavin is readily degraded when exposed to light, especially at elevated temperatures, and considerable decrease in its content in foods can occur during cooking accompanied by exposure to light. The greatest amounts of flavin nucleotides are in the liver, kidney, and heart.

Dietary riboflavin is present mostly as a phosphate, which is rapidly hydrolyzed before absorption in the duodenum.In humans, the rapid, saturable absorption of riboflavin following an oral dose suggests that it is transported by a carrier-mediated pathway located predominantly in duodenal enterocytes. The process may be sodium-dependent. Bile salts enhance absorption of riboflavin. Fecal riboflavin is derived from the intestinal mucosa and the intestinal flora. This is the predominant excretory route for the vitamin.

Signs of riboflavin deficiency include *cheilosis,* angular stomatitis,magenta tongue,and localized seborrheic dermatitis.Some of these conditions may be due to concurrent deficiency of other B-complex vitamins, since it is difficult to produce "pure" riboflavin deficiency in humans. No toxicity following large doses of riboflavin has been reported.

Pyridoxine (Vitamin B_6)

Vitamin B_6 was first recognized as an essential food factor in 1934; it was called pyridoxine when it was found to be a substituted pyridine. Three closely related compounds have vitamin B_6 activity (Figure 38-14).

With the exception of glycogen phosphorylase (Chapter 15), and kynureninase, all of the pyridoxine-requiring

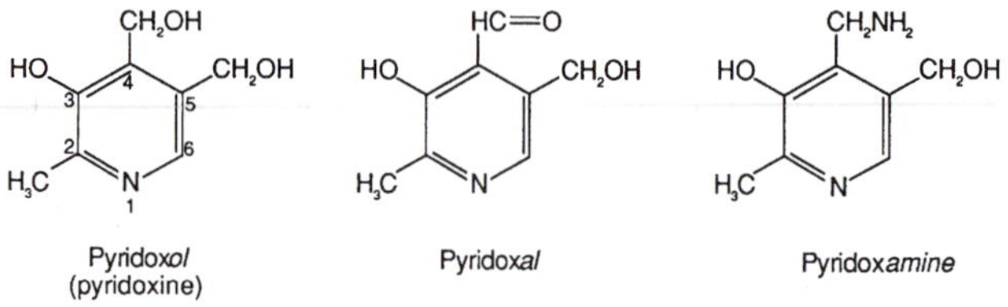

Pyridox*ol*
(pyridoxine)

Pyridox*al*

Pyridox*amine*

FIGURE 38-14

Structures of the three naturally occurring compounds having vitamin B_6 activity.

enzymes are involved in transamination or decarboxylation of amino acids (Chapter 17). Under physiological conditions, in the absence of apoenzymes, pyridoxal phosphate catalyzes similar reactions, but much more slowly (by a factor of as much as 10^6) and without substrate specificity.

Pyridoxine is synthesized by many plants and most bacteria but not by higher animals. Principal dietary sources are whole-grain cereals, wheat germ, yeast, meat, and egg yolk. Bioavailability of dietary pyridoxine is reduced by heat processing, possibly owing to reduction of Schiff base linkages between pyridoxal phosphate and ε-amino groups of lysine in proteins. Some infants fed a formula that was heat-sterilized during preparation and inadequately fortified with vitamin B_6 developed nervousness, irritability, and, in some cases, marked opisthotonos. Administration of pyridoxine relieved these symptoms. Rats and monkeys fed a vitamin B_6-deficient diet develop dermatitis and neuropathological changes. The neurological symptoms seen with pyridoxine deficiency in humans and animals may be due to decreased synthesis of γ-aminobutyric acid (GABA), a neurotransmitter. Glutamate decarboxylase, which catalyzes the formation of GABA from glutamic acid, requires pyridoxal phosphate as a cofactor.

A pyridoxine requirement greater than normal has been observed in patients with celiac disease, gastroenteritis, and Crohn's disease, presumably due to impaired intestinal absorption. Increased need for vitamin B_6 is observed in infections, uremia, severe burns, pregnancy, and lactation. Low circulating levels of pyridoxine occur in patients with active gastric ulcers, gastritis, gastric carcinoma, and benign gastric polyps. Isoniazid, used for treatment of tuberculosis, reacts with pyridoxal phosphate to form a hydrazone that is biologically inactive and also inhibits pyridoxal kinase and kynurenine transaminase. Deoxypyridoxine, a pyridoxine antagonist, produces symptoms of vitamin B_6 deficiency when administered to humans. Vitamin B_6 deficiency has been reported in rheumatoid arthritis, some malignancies, liver disease, alcoholism, diabetes mellitus, atherosclerosis, and some cases of hyperkinesis in children. At least seven inherited disorders responsive to pharmacological doses of pyridoxine have been described (Table 38-1).

Vitamin B_6 status can be evaluated by direct measurement of plasma pyridoxine or pyridoxal phosphate by microbiological, enzymatic, radioimmunological, or chemical methods. Measurement of urinary xanthurenic acid or other intermediates of the kynurenine pathway (Chapter 17) are used to assess indirectly the adequacy of vitamin B_6 for metabolic needs.

Vitamin B_6 is rapidly absorbed from the intestine by passive diffusion. Phosphorylated pyridoxine vitamers are hydrolyzed by intestinal membrane alkaline phosphatase before absorption.

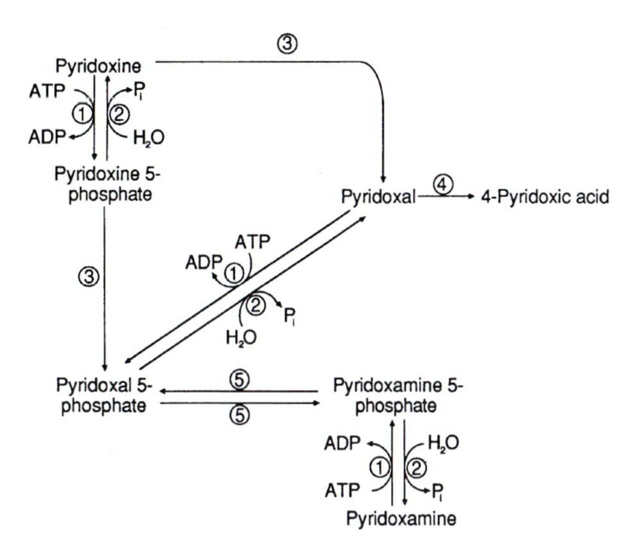

FIGURE 38-15
Metabolism of pyridoxine-related compounds in mammals. Enzymes: 1, pyridoxal kinase (present in all mammalian tissues); 2, nonspecific (probably alkaline) phosphatases; 3, pyridoxine oxidase (cofactor is FMN; O_2 is required; subject to product inhibition); 4, aldehyde oxidase or aldehyde dehydrogenase; 5, aminotransferase.

For most enzymes, the coenzyme form of pyridoxine is pyridoxal 5-phosphate. The transaminases can use pyridoxamine 5-phosphate because they interconvert pyridoxal and pyridoxamine during their activity. The three vitamers are readily converted to the active form (Figure 38-15).

The predominant circulating form of vitamin B_6 is pyridoxal phosphate. Absorbed pyridoxine is oxidized and phosphorylated in intestinal mucosal cells, liver, and erythrocytes. Pyridoxine enters hepatocytes and erythrocytes by passive diffusion and is mostly retained by phosphorylation. Pyridoxal phosphate is transported in the blood bound to albumin. The blood-brain barrier has limited permeability to pyridoxal.

Pyridoxic acid is the principal urinary excretory form of vitamin B_6 when physiological doses of the vitamin are given. Formation of pyridoxic acid is catalyzed by aldehyde oxidase or aldehyde dehydrogenase (Figure 38-15). The FAD-dependent aldehyde oxidase seems to occur only in liver, whereas the NAD^+-dependent dehydrogenase is present in many tissues. Under physiological conditions, the dehydrogenase is the more important enzyme. Pyridoxal, pyridoxamine, and pyridoxine are excreted in urine following pharmacological doses of vitamin B_6. Only 2% of an intravenous dose of pyridoxine appears in bile.

In rats and dogs, very high doses of pyridoxine cause neurotoxicity characterized by demyelination of dorsal nerve roots, ataxia, and muscle weakness. Sensory neuropathy with ataxia occurred in seven patients who received daily doses of 2–6 g of pyridoxine for 2–34 months. The central nervous system was spared, and all showed recovery following cessation of pyridoxine ingestion. Four of the patients had taken the vitamin to treat premenstrual edema, for which there is no clinical support. Others believed that "more is better." The toxicity may be caused by displacement of pyridoxine cofactors from pyridoxine-requiring enzymes by abnormal metabolites of vitamin B_6. These metabolites have little or no activity as enzyme cofactors, resulting in loss of the reactions catalyzed by the affected enzyme. The metabolites occur because of "activation" by high substrate (pyridoxine) concentration of minor pathways for B_6 metabolism.

Cobalamin (Vitamin B_{12})

Vitamin B_{12} is unusual among the vitamins because its two coenzyme forms contain an organometallic bond between cobalt and carbon—the only such bonds known in biological systems. The vitamin was first crystallized in 1948 as the cyano derivative (cyanocobalamin), the principal form of commercially available cobalamin. Anions such as hydroxyl, chloride, nitrite, and sulfate can replace the CN^- without affecting the activity. Hydroxocobalamin is the form usually found in the body. Animals and higher plants cannot synthesize cobalamin, although several animal tissues can concentrate it, making lean meat, liver, seafood, and milk important dietary sources. *De novo* synthesis of cobalamin is accomplished only by microorganisms, including some in the human colon. However, absorption of B_{12} occurs only in the ileum, making cobalamin in the colon of no nutritional value. Some species of bacteria that colonize the lower ileum synthesize B_{12} and may contribute to the cobalamin requirement of their host. Many species of bacteria cannot make vitamin B_{12}, and microbiological assays are based on stmulation of growth of some of these species by exogenous cobalamin.

Vitamin B_{12} is stable to temperatures up to 250°C (482°F) in acidic or neutral solutions. Dietary B_{12} deficiency is rare among meat eaters but not in strict vegetarians. The average total body content of vitamin B_{12} is about 2.5 mg, most of which is in the liver (1 μg of B_{12} per gram of hepatic tissue). There is extensive reutilization of cobalamin and an active enterohepatic circulation. The principal disease caused by vitamin B_{12} deficiency is *megaloblastic anemia*. Deficiency also causes neurological abnormalities that become irreversible if allowed to persist.

The structure of the cobalamin family of compounds is shown in Figure 38-16. The corrin ring system of four pyrrole rings linked by three methene bridges is similar to the porphyrin ring system. The 5,6-dimethylbenzimidazole ring is sometimes replaced by 5-hydroxybenzimidazole, adenine, or similar groups. Although the cobalt is shown in the +1 oxidation state (B_{12_s}), it is readily oxidized to +2 (B_{12_r}) or +3 (B_{12_a}). Dietary cobalamins generally contain Co^{3+} while the coenzyme forms have Co^{1+}. Reduction of B_{12_s} to B_{12_r}, and B_{12_r} to B_{12_a}, is catalyzed by specific NADH-dependent reductases (Figure 38-17). In the coenzyme form, the group attached to the sixth position of cobalt is 5′-deoxyadenosine (adenosylcobalamin) or a methyl group (methylcobalamin). In the liver and several other tissues, about 70% of the cobalamin is present as adenosylcobalamin in mitochondria, 1–3% as methylcobalamin in the cytoplasm, and the rest probably as hydroxocobalamin (Figure 38-17). Absorption, transport, and cellular uptake of vitamin B_{12} are summarized in Figure 38-18. Dietary cobalamins are released by gastric acid and bind to cobalophilins (R-proteins), derived primarily from saliva, and to intrinsic factor (IF), a glycoprotein secreted by gastric parietal cells. At pH 2, the relative affinities of cobalophilin and IF for cobalamin are about 50:1. In the duodenum, pancreatic proteases partially degrade cobalophilins, allowing more of the cobalamin to bind to

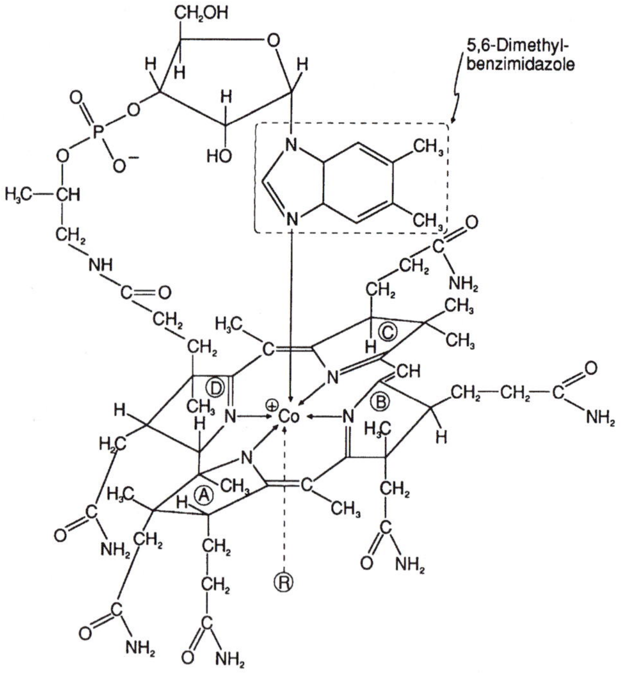

FIGURE 38-16
Structure of the cobalamin family of compounds. A through D are the four rings in the corrinoid ring system. The B ring is important for cobalamin binding to intrinsic factor. If R = –CN, the molecule is cyanocobalamin (vitamin B_{12}); if R = 5′-deoxyadenosine, the molecule is adenosylcobalamin; if R = $-CH_3$, the molecule is methylcobalamin. Arrows pointing toward the cobalt ion represent coordinate-covalent bonds.

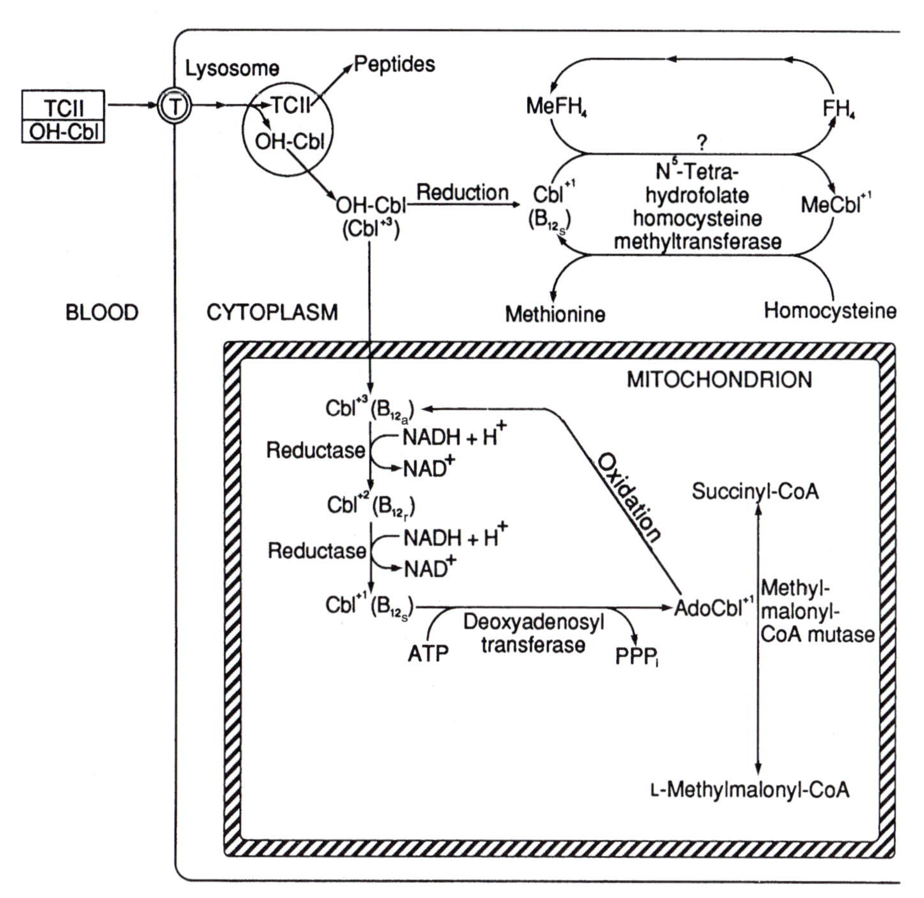

FIGURE 38-17
Metabolism of cobalamins in mammalian cells. Note the compartmentalization of the synthesis of the two coenzymes: adenosylcobalamin (AdoCbl) synthesis occurs in mitochondria, whereas that of methylcobalamin (MeCbl) occurs in cytoplasm. TCII, Transcobalamin II; OH-Cbl, hydroxocobalamin; T, transport protein for TCII-OHCbl complex; FH_4, $MeFH_4$, tetrahydrofolate and methyltetrahydrofolate, respectively; Cbl, a cobalamin in which the ligand occupying the sixth coordination position of the cobalt is not known. The numerical superscripts adjacent to some of the cobalamins indicate the oxidation state of the cobalt ion. Cobalamins in the +1, +2, and +3 oxidation states are also known as B_{12_s}, B_{12_r}, and B_{12_a}, respectively.

IF. Transfer to IF is aided by the bicarbonate of pancreatic juice. At duodenal pH, IF has a greater affinity than cobalophilin for B_{12}. Patients with pancreatic insufficiency may have vitamin B_{12} deficiency.

The IF-cobalamin complex binds to specific receptors on the luminal surface of ileal mucosal cells. Binding requires Ca^{2+} and occurs optimally at pH >6.6. IF may enter the mucosal cell with cobalamin, or it may bind to the receptor and release the vitamin into the cell interior. The IF-receptor complex may then remain on the cell membrane and transport other molecules of cobalamin into the cell.

Transfer of cobalamin to plasma is specifically to transcobalamin II (TCII). The TCII-B_{12} complex enters virtually all cells of the body via a specific cell surface receptor. The ligand-receptor complex is internalized and appears in lysosomes, where TCII is degraded, thus freeing the cobalamin. Passage of cobalamin from lysosomes into the cytoplasm may be mediated by a specific transport mechanism. Fibroblasts from a child with developmental delay and mild, B_{12}-responsive methylmalonic aciduria took up TCII-B_{12} normally and released B_{12} from the complex, but the free B_{12} accumulated within the lysosomes rather than being released into the cytoplasm. This finding suggests that the putative lysosomal B_{12} transport system was defective in this patient.

Thus, four major types of binding protein are needed for absorption of cobalamin and for its uptake by tissues.

1. ***Cobalophilins (R-proteins; R-binders).*** These proteins occur in plasma, gastric juice, saliva, granulocytes, and other tissues and body fluids. Most of the cobalophilin found in the stomach probably comes from saliva. Cobalophilins are glycoproteins (M.W. 56,000–66,000) that contain variable amounts

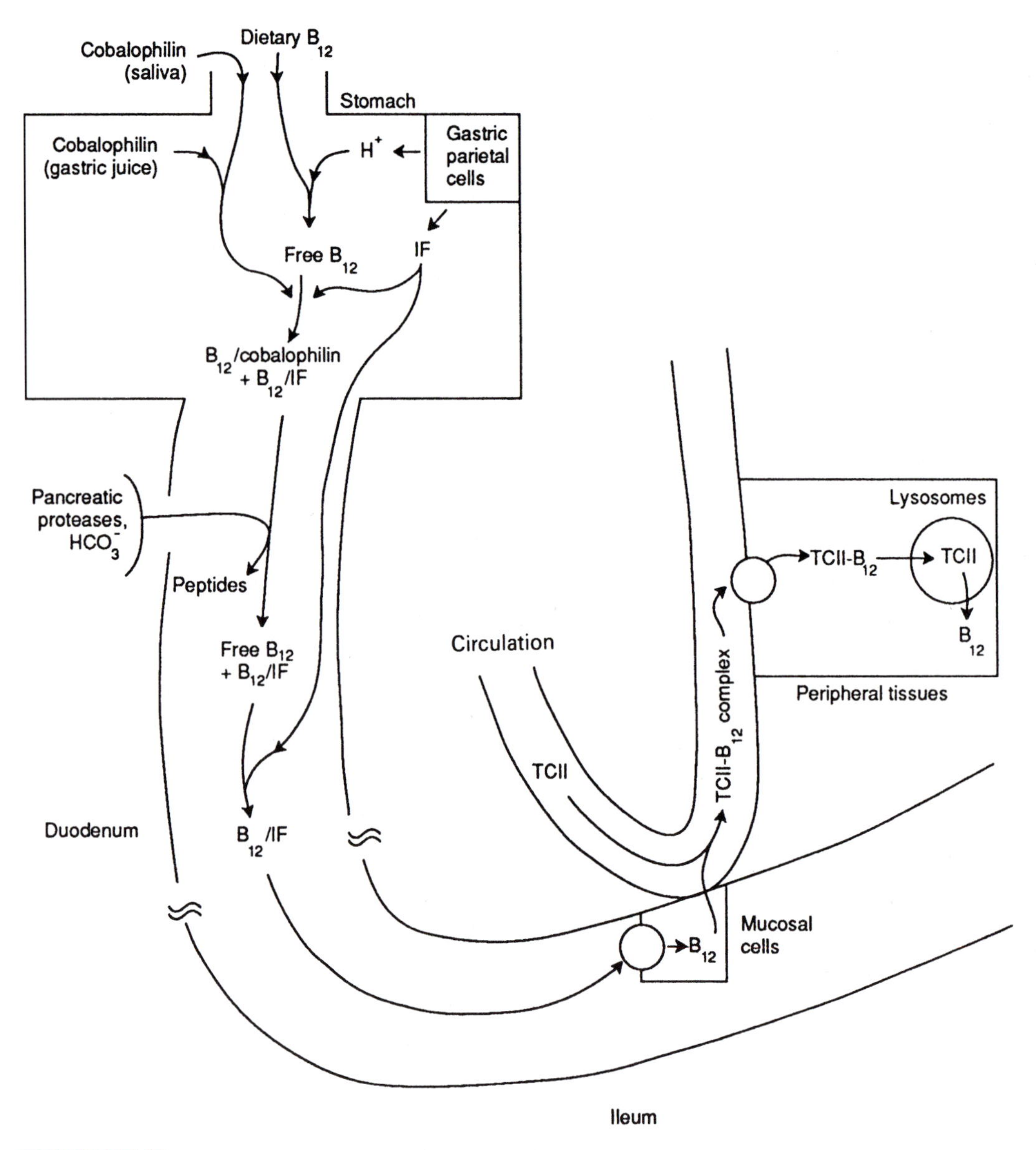

FIGURE 38-18
Absorption, transport, and cellular uptake of vitamin B_{12} in humans. IF, Intrinsic factor; TCII, transcobalamin II; circles in the membranes of the ileal mucosal cell and peripheral tissues represent transport molecules for IF/B_{12} and TCII/B_{12}, respectively.

of carbohydrate. Their isoelectric points range from 2.3 to 5.0 owing to the presence of 6–18 sialic acid residues per molecule. Because of the low pI values, these proteins have high anodic mobility during electrophoresis at alkaline pH; hence, their designation as "R" (for rapidly migrating) proteins.

Cobalophilins bind one cobalamin per molecule. They also bind corrinoids that have modifications in the nucleotide or corrinoid regions of the molecule and so may be important for excretion of inactive forms of the vitamin via bile.

The primary structures of several cobalophilins are similar. A single cobalophilin gene is suggested by the discovery of a family with congenital absence of cobalophilin from cerebrospinal fluid, gastric juice, granulocytes, saliva, and serum.

The serum cobalophilins, transcobalamin I and III (TCI and TCIII), are immunologically identical and may differ only in their carbohydrate content. TCI moves as an α_1-globulin in standard serum protein electrophoresis, while TCIII moves as an α_2-globulin owing to its lower sialic acid content. Plasma TCIII appears to be released from leukocytes during sample collection.

Normal plasma concentration of cobalamins is 200–1000 pg/mL, of which more than 90% is bound to TCI, mostly as methylcobalamin. A serum concentration of less than 100 pg/mL is usually diagnostic for vitamin B_{12} deficiency, but pronounced changes in concentrations of circulating cobalophilins could cause misleading changes in serum B_{12} concentration, which is usually measured by a competitive protein binding assay. If cobalophilins are used as the binding protein, falsely high levels of "vitamin B_{12} activity" may be obtained because of the presence of inactive cobalamins in the sera of some patients and the relative nonspecificity of cobalophilins toward corrinoids. This result is avoided by using microbiological assays or by, using gastric IF as the binding protein because it is more specific.

TCI has a high affinity for and is about 90% saturated with cobalamin, while TCIII binds less strongly and is largely unsaturated. However, TCII binds newly absorbed cobalamins, and the TCII-cobalamin complex appears to deliver cobalamin to peripheral tissues (Figure 38-18). TCII accounts for most of the unsaturated vitamin B_{12} binding capacity of serum. Therefore, cobalophilins (except TCII; see below) probably have a minimal role in cobalamin absorption, transport, and metabolism. Thus, members of the family with congenital cobalophilin deficiency showed no symptoms of vitamin B_{12} deficiency despite a very low serum concentration of the vitamin secondary to the absence of TCI.

2. ***Intrinsic Factor (IF).*** IF is a glycoprotein (M.W. ~45,000), approximately 15% of which is carbohydrate, including sialic acid. It binds one cobalamin per IF molecule. The B-corrinoid ring and the benzimidazole ring (Figure 38-15) are needed for optimal binding. It is synthesized in the gastric parietal cells, which also secrete HCl. The amount of IF normally secreted each day is sufficient to bind 40–80 μg of vitamin B_{12}. Secretion, and perhaps synthesis of IF is increased by vagal stimulation or by histamine, gastrin, or gastrin pentapeptide. In the small intestine, free IF is partially degraded by pepsin and chymotrypsin but is resistant to trypsin. The IF-B_{12} complex is resistant to all of these proteases.

   Pernicious anemia is a megaloblastic anemia caused by malabsorption of vitamin B_{12} secondary to inadequate secretion of normal IF. The word "pernicious" reflects the unremitting and usually fatal course of the disease prior to the discovery that it could be corrected by administration of vitamin B_{12}. Virtually all cases result from atrophic gastritis,which in turn may result from autoimmune attack of gastric parietal cells, leading to absence of IF and achlorhydria.Pernicious anemia may rarely be due to an inherited defect in structure or secretion of IF. Several cases of congenital absence of IF have been described. Anemia due to inborn errors usually manifests itself early in life, while the average age for appearance of atrophic gastritis is 60 years; the latter condition is rare, but not unknown, before the age of 30.

   The Schilling test differentiates pernicious anemia from other causes of megaloblastic anemia. Vitamin B_{12} containing radioactive cobalt is administered orally, and a large dose of unlabeled B_{12} is given intravenously (to saturate B_{12} binding sites in tissues). The radioactivity appearing in a 24-hour urine specimen is determined. A normal person excretes about one-third of the radioactive label in 24 hours, while less than 8% is found in the urine of a person with pernicious anemia (a positive Schilling test). The test is repeated several days later, this time including IF with the vitamin B_{12}. If the defect is lack of IF, the uptake (and excretion) of the radioactive label will be normal. Malabsorption of B_{12}-IF complex in the terminal ileum causes both parts of the Schilling test to be abnormal (see below). The megaloblastic anemia of folate deficiency is usually indistinguishable from that of vitamin B_{12} deficiency except by the Schilling test.

3. ***Transcobalamin II*** (TCM). TCII is the principal plasma-binding protein for cobalamins newly absorbed from the intestine. Its normal plasma concentration is 0.5–1.5 μg/mL. The protein (M.W. 38,000) is synthesized in hepatocytes, fibroblasts, macrophages, and perhaps enterocytes and other cells. It is probably not a glycoprotein, making it distinct from TCI and TCIII. It migrates with the β-globulins during serum protein electrophoresis and binds one cobalamin per protein molecule.

   TCII plays a major role in transport of cobalamins to tissues. A receptor for the TCII-B_{12} complex has been tentatively identified *in vitro* on HeLa cells,Ehrlich ascites cells,and human fibroblasts. The bound complex is transferred into the cell, where the cobalamin is released and the TCII is degraded in lysosomes (Figure 38-18). An inborn error of vitamin B_{12} metabolism has been attributed to a defect in vitamin B_{12} release from lysosomes. A congenital abnormality of TCII has been described that results in megaloblastic anemia (Table 38-1).

4. ***Intestinal Receptor for IF-B$_{12}$ Complex.*** Although this receptor is not, strictly speaking, a cobalamin-binding protein, it is essential for normal absorption of dietary cobalamin. It is present on the membrane of microvilli of ileal but not jejunal or duodenal cells, with the highest concentration in the distal 60-cm portion of the small intestine. The purified receptor is composed of two subunits (M.W. 90,000 and 140,000) and binds free IF and IF-B$_{12}$ complex, although free IF binds more slowly. Subsequent transport of cobalamin into enterocytes is accomplished by an active process.

   Patients have been described who have a vitamin B$_{12}$ deficiency that is not correctable by IF. In the Schilling test with or without added IF, these patients have a low urinary output of radioactivity. This finding could be due to a deficiency or defect in the ileal receptor protein. More commonly, it occurs when there is an acquired ileal defect such as a bacterial overgrowth (blind loop syndrome) or tropical sprue.

   The two reactions in mammalian systems in which cobalamins participate are conversion of L-methylmalonyl-CoA to succinyl-CoA by methylmalonyl-CoA mutase, and methylation of homocysteine to methionine by N^5-tetrahydrofolate homocysteine methyltransferase (Figure 38-17). The interrelationship of vitamin B$_{12}$ and folic acid derivatives are discussed below, under "Folic Acid."

   Inborn errors in the synthesis of adenosylcobalamin or of both adenosyl- and methylcobalamins have been described. They cause, respectively, methylmalonic aciduria alone or combined with homocystinuria (Table 38-1). These disorders respond to treatment with pharmacological doses of vitamin B$_{12}$. Methylmalonic aciduria that does not respond to vitamin B$_{12}$ is probably due to an abnormal methylmalonyl-CoA mutase.

   The cobalamins are excreted in urine, bile, and feces. Intact cobalamin molecules are the predominant forms in urine. Urinary excretion is highly variable, but it averages ~150 ng/d and is somewhat higher in smokers than nonsmokers. The principal excretory route is bile. Cobalamins destined for biliary excretion may be transported to the liver by TCI and TCIII. The liver is the only tissue that takes up the TCI-cobalamin complex.

   The importance and completeness of the enterohepatic circulation are demonstrated by strict vegetarians who consume no vitamin B$_{12}$. They require 20–30 years or more to develop vitamin B$_{12}$ deficiency, whereas deficiency symptoms occur within 2–10 years after ileal absorption is blocked by disease of the stomach, pancreas, or ileum or by ileal resection. Disruption of the absorptive pathway is the probable reason for vitamin B$_{12}$ deficiency in tropical sprue.

Nitrous oxide (N_2O) used as an anesthetic agent inactivates vitamin B$_{12}$. Subjects with marginal vitamin B$_{12}$ stores can develop vitamin B$_{12}$ deficiency within weeks after administration of the anesthetic. Major clinical manifestations of vitamin B$_{12}$ deficiency include hematologic abnormalities (e.g., megaloblastic macrocytic anemia, hypersegmentation of neutrophils) and neurological deficiency (e.g., sensory neuropathy). In addition, since dividing cells require vitamin B$_{12}$, all rapidly growing cells are affected. One visible sign of vitamin B$_{12}$ deficiency is sore mouth and tongue, and the latter may be bald and shiny.

In vitamin B$_{12}$ deficiency, neurological damage may sometimes occur in the absence of hematological abnormalities and may be mistaken for multiple sclerosis, diabetic neuropathy, or neuropsychiatric disorders. In some elderly subjects malfunctioning of gastric parietal cells leads to reduction in both intrinsic factor and H^+ production (achlorhydria) causing vitamin B$_{12}$ malabsorption. Two metabolites that accumulate due to reduced catalytic function of L-methylmalonyl-CoA mutase and methionine synthase as a result of vitamin B$_{12}$ deficiency are methylmalonic acid and homocysteine. In the clinical assessment of subtle vitamin B$_{12}$ deficiency in the presence of normal or low-normal serum levels of vitamin B$_{12}$, measurement of homocysteine and methylmalonic acid serum levels may prove valuable in detecting incipient vitamin B$_{12}$ deficiency. In folate deficiency, serum homocysteine levels are also increased, but methylmalonic acid levels remain in the normal range. In vitamin B$_{12}$-deficient subjects, folate therapy may normalize homocysteine levels along with some of the hematological abnormalities, but it will not correct the neurological manifestations. Thus, inappropriate folate therapy may mask vitamin B$_{12}$ deficiency (discussed in the next section).

Folic Acid (Pteroylglutamic Acid)

The common name *folic acid* is derived from Latin *folium*, "leaf," because this vitamin was originally isolated from dark green, leafy vegetables such as spinach. Folic acid metabolism was discussed in Chapter 27.

A close relationship exists between metabolism of the folates and of vitamin B$_{12}$. Deficiency of either produces megaloblastic anemia, and symptoms of vitamin B$_{12}$

deficiency anemia are reversed by large doses of folate. However, folate does not reverse the neurological abnormalities associated with vitamin B_{12} deficiency and may even exacerbate them. Correction of megaloblastosis by folate is not necessarily an indication of the absence of vitamin B_{12} deficiency; in fact, it could mask developing vitamin B_{12} deficiency until the advent of neurological damage.

The common point in metabolism of folate and B_{12} is transmethylation of homocysteine to methionine (Figure 38-17). In humans and some other species, this is the only route for conversion of N^5-methyl FH_4 to other folate derivatives. According to the ***methyl-folate trap hypothesis,*** inhibition of this reaction by cobalamin deficiency leads to accumulation of folate as N^5-methyl FH_4, deprives the cell of other folate cofactors, and leads to blockage of several enzymatic reactions. Administration of large amounts of folates may circumvent this inhibition and directly supply the folate derivatives. Cobalamin deficiency decreases the activity of the methyltransferase and brings about increased urinary excretion of formate, aminoimidazolecarboxamide, and formiminoglutamic acid (Figlu) in the histidine loading test (Chapter 17). Megaloblastosis occurs because of impaired DNA synthesis secondary to an inadequate supply of thymidylate. Synthesis of dTMP from dUMP requires N^5N^{10}-methylene FH_4. However, evidence for accumulation of N^5-methyl FH_4 is more equivocal. In patients with pernicious anemia, methionine decreases the excretion of Figlu, further depresses synthesis of dTMP, and increases the severity of the megaloblastosis. Methionine is a precursor of S-adenosylmethionine, an allosteric inhibitor of N^5,N^{10}-methylene FH_4 reductase. Several patients have been described who have abnormal cobalamin metabolism and decreased methyltransferase activity but no megaloblastosis or anemia. These observations, and the ineffectiveness of folates in relieving the neurological symptoms of vitamin B_{12} deficiency, suggest additional vitamin B_{12}-dependent functions in mammalian systems.

Homocysteine, an amino acid associated with coronary heart disease as a risk factor is elevated in folate deficiency (discussed earlier and also see Chapter 17). If the elevation of plasma homocysteine is due to folate deficency, supplementation of folate corrects the plasma homocysteine level and may decrease the morbidity and mortality from atherosclerotic disease which can lead to heart attack and stroke. Vitamin B_6 and Vitamin B_{12} deficencies can also cause elevated plasma homocysteine levels.

Folate supplementation of 400 μg/d, in addition to a healthy diet that includes foods rich in folate, has been recommended for women before and in the early weeks of pregnancy. Folate supplementation helps prevent the majority of birth defects of the brain and spinal cord, called neural tube defects (NTDs). Although the exact mechanism of folate in the prevention of NTDs is not understood, the role of folate in nucleic acid and amino acid metabolism in the raidly developing fetus is probably involved. Also, supplementation of dietary folate may correct metabolic defects in individuals who inherit defects in folate metabolism. A recent study indicates that as many as one in seven people carry a mutation affecting folate metabolism.

A screening test for NTDs has been developed using maternal serum to measure α-fetoprotein (AFP) levels at 15–20 weeks of gestation. AFP is synthesized in the fetal liver and yolk sac during development and is the major fetal serum protein (analgous to serum albumin in adults). AFP appears in amniotic fluid from fetal urination. Its amniotic fluid concentration parallels fetal serum levels except that the amniotic fluid level is about 150 times lower in concentration. Through placental transfer, AFP also appears in maternal serum. During the screening test period of 15–20 weeks, maternal serum AFP concentration increases by about 15% per gestational week. If NTDs are present (e.g., anencephaly, open spina bifida), fetal serum leaks into the fluid compartment raising the AFP amniotic fluid levels, as well as maternal serum levels. Thus, elevated levels of maternal serum AFP are used to identify women who may be carrying a fetus with a NTD. However, because of the overlap in AFP values between unaffected and affected pregnancies, the test is not diagnostic and requires confirmatory diagnostic procedures (e.g., acetylcholinesterase and AFP levels in the amniotic fluid, high-resolution ultrasonography).

Low levels of maternal serum AFP also may be informative in assessing other fetal abnormalities such as Down's syndrome. Some of the salient features of Down's syndrome include trisomy 21, malformations, dysmorphic features, and mental retardation. The risk of Down's syndrome increases with maternal age. At age 35, the risk at birth is 1 in 385 and at age 40 it is 1 in 105. In estimating the risk of Down's syndrome, two other biochemical serum parameters are measured: unconjugated estriol (which is decreased) and human chorionic gonadotropin (hCG, which is increased). The measurement of these three maternal serum markers is used in conjunction with maternal age, twin gestation, maternal insulin-dependent diabetes mellitus, maternal weight, ethnic derivation, and smoking to assess risk. Evaluating all of these parameters in comparison to values obtained with healthy women of comparable gestational age has increased the sensitivity and reduced false-positive rates in the assessment of Down's syndrome in the fetus. A four-marker maternal

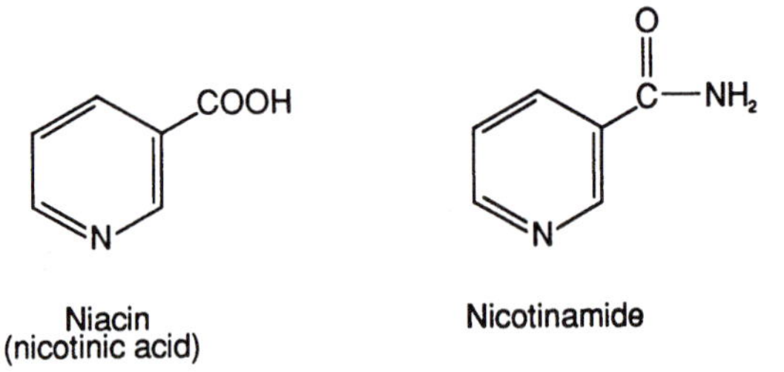

Niacin (nicotinic acid) — Nicotinamide

FIGURE 38-19
Structures of nicotinic acid and nicotinamide.

serum screening for Down's syndrome, which includes measuring separate subunits of hCG (free α-hCG and free β-hCG), unconjugated estriol, and AFP, has proven to be more effective than triple-marker assessment. The confirmatory test for Down's syndrome is karyotype analysis.

Serum AFP levels are elevated in hepatocellular carcinomas and malignancies involving ovaries and testes. Yolk sac tumors, which occur more frequently in the ovaries of young women, and girls, and in the testes of boys, raise serum levels of AFP. Serum hCG measurement is also helpful in the diagnosis and management of germ cell tumors. Thus, hCG and AFP are also used as tumor markers.

Niacin

Niacin (nicotinic acid; pyridine-3-carboxylic acid) and nicotinamide are precursors of NAD^+ and $NADP^+$ (Figure 38-19). Niacin occurs in meat, eggs, yeast, and whole-grain cereals in conjunction with other members of the vitamin B group. Little is known about absorption, transport, and excretion of niacin and its coenzyme forms. A limited amount of niacin can be synthesized in the body from tryptophan, but it is not adequate to meet metabolic needs.

Pellagra was originally thought to be due to inadequate dietary niacin and tryptophan. In some parts of the world, it is associated with consumption of diets high in maize (American corn), which, like other cereal grains, is relatively deficient in tryptophan and niacin. In addition, about 20% of the niacin in maize is protein-bound and not biologically available. Several bound forms of niacin have been characterized. They include niacinogens (peptides of M.W. 12,000–13,000) and niacytin, isolated from wheat (M.W. 2370).

Pellagra is currently thought to be due to imbalance of dietary amino acids and deficiency of niacin. The common variety of maize is rich in leucine, which inhibits synthesis of nicotinic acid mononucleotide and causes deficiency of NAD^+ and $NADP^+$. A strain of maize known as opaque 2 contains less leucine and does not cause pellagra unless excess leucine is added to the diet.

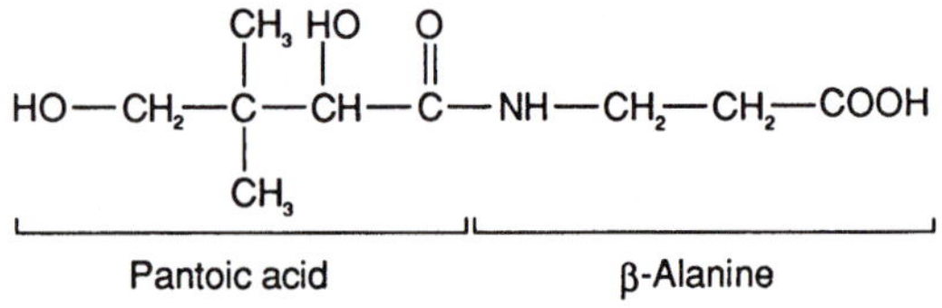

$$HO-CH_2-\underset{CH_3}{\overset{CH_3}{C}}-\overset{HO}{CH}-\overset{O}{\overset{\|}{C}}-NH-CH_2-CH_2-COOH$$

Pantoic acid — β-Alanine

FIGURE 38-20
Structure of pantothenic acid (pantoyl-β-alanine).

Pellagra-like symptoms can occur in ***Hartnup's disease*** and ***carcinoid syndrome.*** Hartnup's disease is an inherited disorder of amino acid transport (Chapter 17) in which niacin deficiency presumably develops because niacin intake is inadequate to supply metabolic needs when combined with the decreased absorption of dietary tryptophan. In carcinoid syndrome, up to 60% of available dietary tryptophan is diverted to formation of 5-hydroxytryptamine (serotonin) by what is normally a minor pathway.

Pantothenic Acid (Pantoyl-β-Alanine)

Pantothenic acid is ubiquitous in plant and animal tissues and especially abundant in foods rich in other B vitamins. No RDA has been established, but a daily intake of 5–10 mg is thought to be adequate for adults. Deficiency in humans is unknown.

Pantothenic acid is a precursor for the synthesis of coenzyme A (CoA, CoASH) and forms part of the "swinging sulfhydryl arm" of the fatty acid synthase complex (Chapter 19). Its structure is shown in Figure 38-20.

Biotin

Biotin is widely distributed in foods. Beef liver, yeast, peanuts, kidney, chocolate, and egg yolk are especially rich sources. The intestinal flora synthesizes biotin. Fecal excretion reflects this enteric synthesis. Total daily urinary and fecal excretion exceeds the dietary intake.

Biotin deficiency occurs when large amounts of raw egg white are consumed. Egg white contains ***avidin,*** a protein (M.W. 70,000) that binds biotin strongly and specifically, preventing its absorption from the intestine. Because of the tight binding and specificity of biotin, avidin-labeled probes have been used to detect proteins and nucleic acids to which biotin has been covalently attached ("biotinylated" molecules). Avidin is a homotetramer. Each subunit contains 128 amino acids and binds one molecule of biotin. The affinity of avidin for biotin is abolished by heat and other denaturants. Biotin deficiency can result from sterilization of the intestine by antibiotics and from administration of biotin analogues.

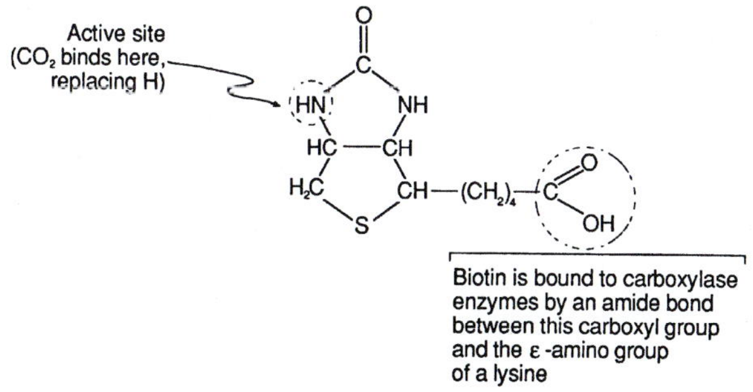

FIGURE 38-21
Structure of biotin.

The structure of biotin consists of fused imidazole and tetrahydrothiophene rings and a carboxyl-containing side chain (Figure 38-21). In oxybiotin, which can substitute for biotin in most species, the sulfur of the tetrahydrothiophene ring is replaced by oxygen, making it a tetrahydrofuran ring.

Biotin is a coenzyme for the carbon dioxide fixation reactions catalyzed by acetyl-CoA carboxylase (Chapter 19), propionyl-CoA carboxylase, pyruvate carboxylase, and β-methylcrotonyl-CoA carboxylase. Carboxylation reactions that do not require biotin are the addition of C_6 to the purine ring (Chapter 27), the formation of carbamoyl phosphate (Chapter 17), and the γ-carboxylation of glutamyl residues of several of the clotting factors, which requires vitamin K (Chapter 36).

Biotin is bound to an apoenzyme by an amide linkage to a lysyl ε-amino group (Figure 38-21). This binding occurs in two steps, catalyzed by holocarboxylase synthetase:

$$\text{Biotin} + \text{ATP} \rightarrow \text{biotinyl 5}'\text{-adenylate} + \text{PP}_i$$

$$\text{Biotinyl 5}'\text{-adenylate} + \text{apocarboxylase} \rightarrow \text{holocarboxylase} + \text{AMP}$$

Most dietary biotin is bound to protein, the amide linkage being broken prior to absorption. At least eight children have been described who have multiple carboxylase deficiency with low activities of several of the biotin-requiring carboxylases, i.e., multiple carboxylase deficiency (Table 38-1). Pharmacological doses of biotin restored the activities of the carboxylases in these patients, indicating that the defect was not in the apocarboxylases. Thus, the defect is presumably in the intestinal transport system, in holocarboxylase synthetase, or in some step in cellular uptake or intracellular transport of biotin.

Proteolysis of biotin-containing enzymes releases ε-biotinyllysine, or biocytin. Biotinidase cleaves biocytin and biotinylated peptides, resulting from degradation of endogenous carboxylases, to biotin and lysine. Thus, biotin is recycled. Deficiency of biotinidase may cause biotin deficiency, manifested clinically by neurological problems, cutaneous findings, and developmental delay. These defects can be corrected by pharmacological doses of biotin. Toxicity due to excessive consumption of biotin is not known.

Biotinidase deficiency is an autosomal recessive disorder with an estimated incidence of 1 in 72,000–126,000. Many newborn-screening programs of genetic diseases include testing for this enzyme. Prompt treatment with oral biotin administration of 5–20 mg/d in affected infants will prevent clinical consequences. If the treatment is delayed, neurological manifestations (e.g., hearing loss and optic atrophy) and developmental delay occur and may not revert to normal.

Ascorbic Acid (Vitamin C)

Humans and guinea pigs lack the enzyme that converts L-gulonolactone to 2-keto-L-gulonolactone required for biosynthesis of ascorbic acid (Chapter 15), L-ascorbic acid, and L-dehydroascorbic acid (Figure 38-22) are biologically equivalent in humans, presumably because of the ready reduction of dehydroascorbate to ascorbate in the body. Ascorbic acid (M.W. 176.1) is a six-carbon enediol lactone (ketolactone) having a configuration analogous to that of glucose. The enolic hydroxyl groups dissociate with $pK_1' = 4.17$ and $pK_2' = 11.57$. It is one of the strongest naturally occurring reducing agents known. Ascorbate is a specific electron donor for eight enzymes and also may participate in several nonenzymatic reactions as a reductant (Table 38-2). However, it should be emphasized that *in vivo* the role of ascorbate as a reductant in nonenzymatic reactions (based on its redox potential) is not established. Other reductants may participate or substitute for ascorbate in nonenzymatic reactions.

The RDA for ascorbate (Appendix IV) is the amount needed to cure or prevent scurvy while allowing adequate body reserves. However, rates of ascorbate synthesis in animals and the amounts needed to maintain serum

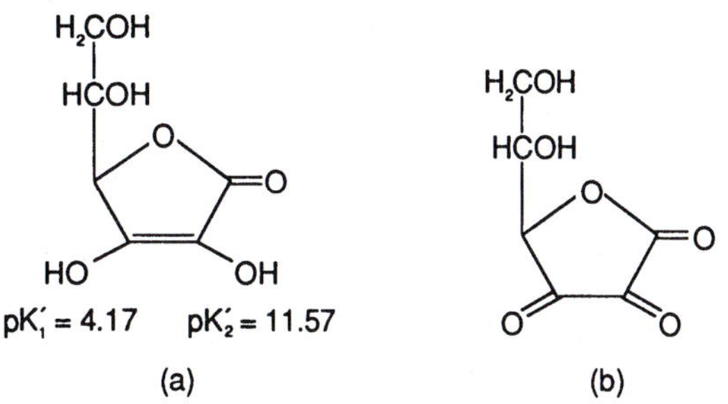

FIGURE 38-22
Structures of L-ascorbic acid (a) and L-dehydroascorbic acid (b).

TABLE 38-2
Metabolic Functions of Ascorbate

A: As specific electron donor for enzymes and their metabolic role.

Enzymes	*Metabolic Role*
Three collagen hydroxylases	Collagen biosynthesis
Two enzymes in carnitine biosynthesis	Carnitine is essential for mitochondrial fatty acid oxidation.
Dopamine β-monooxygenase	Necessary for the synthesis of norepinephrine and epinephrine
4-hydroxyphenylpyruvate dehydrogenase	Participates in tyrosine metabolism
Peptidyl-glycine α-monooxygenase	Required for amidation of peptide hormones

B: As potential chemical reductant and/or antioxidant in nonenzymatic reactions.

Reaction or Function	*Consequence*
Gastrointestinal iron absorption.	Increase
Oxidative DNA and/or protein damage.	Decrease
Low-density lipoprotein oxidation.	Decrease
Endothelial-dependent vasodilation.	Increase
Lipid peroxidation.	Decrease
Oxidants and nitrosamines in gastric juice.	Decrease
Extracellular oxidants from neutrophils.	Decrease

levels at values found in wild animals suggest that the RDA is low by as much as an order of magnitude. This discrepancy may arise from the use of a single criterion for ascorbate repletion—prevention of scurvy—which may not accurately reflect all the functions of ascorbate. For example, it is not known whether enzymes that require or are activated by ascorbate (see below) are fully active in persons maintained on the RDA. The biological functions requiring the greatest amount of ascorbate have to be defined. The RDA for ascorbate intake of 60 mg/d is currently under revision. Studies have suggested that an ascorbate intake of 100–200 mg/d may be needed to prevent adverse health effects due to vitamin C deficiency. To maintain optimal health, the dietary guidelines encourage consumption of five servings of fruits and vegetables per day.

Claims that consumption of massive doses of ascorbate (1–2 g/d or even higher quantities) can prevent or cure common cold, cancer, or other ailments have no basis in fact. Although ascorbate has low toxic effects, adverse effects at high dose, such as hyperoxalemia (oxalate is a catabolite of ascorbate) in patients undergoing dialysis, or hemolysis in subjects with glucose-6-phosphate dehydrogenase deficiency, have been reported.

Major dietary sources of vitamin C are fresh, frozen, and canned citrus fruits. Other fruits, leafygreen vegetables, and tomatoes are important contributors to ascorbate intake. Human milk contains 30–55 mg/L, depending on maternal intake of vitamin C. Exposure to copper, iron, and oxygen can destroy vitamin C by oxidation. The vitamin is heat-labile, so excessive cooking will degrade it. D-Ascorbate (isoascorbate or erythroascorbate), frequently used as a food preservative, has one-twentieth the biological activity of L-ascorbate.

Absorption of vitamin C from the small intestine is a carrier-mediated process that requires sodium at the luminal surface. Transport is most rapid in the ileum and resembles the sodium-dependent transport of sugars and amino acids, but the carrier is distinct for each class of compound. Some ascorbate may also enter by simple diffusion. With dietary intake less than 100 mg/d, efficiency of absorption is 80–90%. With intake equal to the RDA, plasma ascorbate is 0.7–1.2 mg/dL, and the ascorbate pool size is 1500 mg. Scurvy becomes evident when the pool is less than 300 mg, at which point plasma ascorbate is 0.13–0.24 mg/dL. Highest tissue concentrations of ascorbate are in the adrenal gland (cortex > medulla).

Most signs of scurvy can be related to inadequate or abnormal collagen synthesis. Ascorbate enhances prolyl and lysyl hydroxylase activities (Chapter 25). Collagen formed in scorbutic patents is low in hydroxyproline and poorly cross-linked, resulting in skin lesions, bone fractures, and rupture of capillaries and other blood vessels. The absolute amount of collagen made in scorbutic animals may also decrease independently of the hydroxylation defect. The anemia of scurvy may result from a defect in iron absorption or folate metabolism.

Ascorbate increases the activity of hydroxylases needed for the conversion of *p*-hydroxyphenylpyruvate to homogentisate (Chapter 17), synthesis of norepinephrine from dopamine (Chapter 32), and two reactions in carnitine synthesis (Chapter 18). It is not known whether decreased activity of these enzymes contributes to the clinical characteristics of scurvy. Although ascorbic acid is needed for maximal activity of these enzymes *in vivo* and *in vitro,* most show some activity when other reducing agents are used.

38.3 Vitamin-Responsive Inherited Metabolic Disorders

The vitamin requirements indicated by the RDA[1] represent the amounts needed for normal health by most individuals in the population. Following absorption,

[1] (See Table 38-1.)

transport, and metabolism, vitamins must interact with many biomolecules before they reach the location where they function. The active cellular form also must interact with one or more proteins in carrying out its biological function. This gives many potential sites for genetic disruption of vitamin metabolism and function. For example, because of mutation in an intestinal transport protein, a vitamin may be inadequately absorbed. Mutation of an enzyme may increase the K_m for a cofactor derived from a vitamin and result in relative deficiency of activity when the vitamin is present at normal concentration. Although such diseases cannot yet be cured, their clinical signs and symptoms can sometimes be relieved by administration of a very large oral or parenteral dose of the appropriate vitamin (Table 38-1). Megadose vitamin therapy in these patients has a basis in human biochemistry and differs from indiscriminate self-medication for diseases (real or imagined) that have no established relationship to the vitamins used.

Some inherited disorders respond to megadoses of a vitamin that is not directly related to the defective protein. Pyruvate carboxylase deficiency leads to encephalomyelopathy in which there is lactic and pyruvic acidemia. Pyruvate carboxylase requires biotin and converts pyruvate to oxaloacetate for the tricarboxylic acid cycle or for gluconeogenesis. A major metabolic pathway for pyruvate is oxidative decarboxylation to acetyl-CoA, catalyzed by the pyruvate dehydrogenase complex. Some patients have shown clinical improvement when treated with lipoic acid. Others have responded to thiamine (Table 38-1). Thiamine and lipoic acid participate in the pyruvate dehydrogenase complex. They presumably reduce the concentrations of pyruvate and lactate by increasing the flux through the pyruvate dehydrogenase complex. Other examples include the use of vitamin E to reduce hemolysis in some patients with deficiency of glutathione synthetase or glucose-6-phosphate dehydrogenase; folic acid, choline, or betaine in some cases of homocystinuria due to cystathionine β-synthase deficiency; and pyridoxine in some cases of primary hyperoxaluria due to deficiency of soluble glyoxalate-α-ketoglutarate carboligase. As in the case of pyruvate carboxylase deficiency, the effect is probably due to enhancement of alternative metabolic pathways that bypass the defective enzyme.

Not all patients who present the same clinical picture respond to vitamin therapy. Thus, if the structural gene for an apoenzyme or transport molecule is completely absent because of a gene deletion, no amount of vitamin or cofactor will correct the defect. If the mutation affects substrate rather than cofactor binding, the pathway is blocked just as effectively and cannot be relieved by increased concentration of cofactor. Thus, six mutations have been identified that cause methylmalonic aciduria, but not all respond to large doses of vitamin B_{12}. Similarly, only some patients with homocystinuria due to N^5,N^{10}-methylenetetrahydrofolate reductase deficiency respond to treatment with folic acid.

About two dozen inherited diseases respond to pharmacological doses of a vitamin (Table 38-1). Many have been mentioned elsewhere in this book in conjunction with the affected metabolic pathways. Although most are very rare, their study has contributed much to acknowledge of metabolism in the human body. Some are heterogeneous in symptoms and in responsiveness to therapy, suggesting genetic heterogeneity, as discussed above.

38.4 Vitamin-like Substances

Several compounds, e.g., vitamin D and niacin, are apparently required in the diet even though pathways for their synthesis occur in the body. Such a situation may arise if a pathway does not provide an adequate supply for the body's needs or if the material cannot be readily transported from the site of synthesis to the place of action. The compounds discussed below are essential dietary nutrients in one or more nonhuman species, but such a status in humans is not supported by evidence.

Choline (N,N,N-trimethyl-β-hydroxyethylamine) is an important constituent of phospholipids (lecithin is phosphatidylcholine) and of acetylcholine. It can be completely synthesized from serine (Chapter 19), but only in the form of phosphatidylserine and then only when the dietary supply of amino acids is adequate. Betaine (N,N,N-trimethylglycine) readily replaces dietary choline for all species. Choline is conserved by a salvage pathway. In the lung, this salvage route is the principal route for the synthesis of the phosphatidylcholine needed as a surfactant (see Chapter 19).

Inositol (hexahydroxycyclohexane) occurs in several isomeric forms. *Myo*-inositol (or *meso*-inositol) is an important constituent of phospholipids and is the only isomer with biological activity. Inositol hexaphosphate (phytic acid) is found in avian erythrocytes, where it binds to hemoglobin, thereby regulating the oxygen capacity of the blood. It is also important as an intracellular messenger in a number of pathways.

Lipoic acid (thioctic acid) functions as an intermediate in the oxidation-reduction reaction of the oxidative decarboxylation of certain ketoacids.

Ubiquinone (coenzyme Q) is involved in the mitochondrial electron transport. Although coenzyme Q supplements have been recommended as preventative measures for coronary artery disease and cancer, scientific studies do not confirm their effectiveness in preventing these diseases.

Supplemental Readings and References

A. C. Antony: The biological chemistry of folate receptors. *Blood* **79,** 2807 (1992).

L. M. Ausman: Criteria and recommendation for vitamin C intake. *Nutrition Reviews* **57,** 222 (1999).

R. J. Berry, Z. Li, J. D. Erickson, et al.: Prevention of neural-tube defects with folic acid in China. *New England journal of Medicine* **341,** 1485 (1999).

A. Bjørneboe, G-E. Bjørnboe, and C. A. Drevon: Absorption, transport and distribution of vitamin E. *Journal of Nutrition* **120,** 233 (1990).

J. Blanchard, T. N. Tozer, and M. Rowland: Pharmacokinetic perspectives on megadoses of ascorbic acid. *American Journal of Clinical Nutrition* **66,** 1165 (1997).

L. D. Botto, C. A. Moore, and M. J. Khoury: Neural-tube defects. *New England Journal of Medicine* **341,** 1509 (1999).

B. B. Bowman, D. B. McCormick, and I. H. Rosenberg: Epithelial transport of water-soluble vitamins. *Annu. Rev. Nutr.* **9,** 187 (1989).

R. Brigelius-Flohé and M. G. Traber: Vitamin E: function and metabolism. *FASEB Journal* **13,** 1145 (1999).

R. Carmel: Cobalamin, the stomach, and aging. *American Journal of Clinical Nutrition* **66,** 750 (1997).

A. C. Carr and B. Frei: Toward a new recommended dietary allowance for vitamin C based on antioxidant and health effects in humans. *American Journal of Clinical Nutrition* **69,** 1086 (1999).

Centers for Disease Control and Prevention: Knowledge and use of folic acid by women of childbearing age—United States. *Morbidity and Mortality Weekly Report* **46,** 721 (1997).

Centers for Disease Control and Prevention: Recommendations for the use of folic acid to reduce the number of cases of spina bifida and other neural tube defects. *Morbidity and Mortality Weekly Report* **41,** RR14 (1992).

P. Christian and K. P. West Jr.: Interactions between zinc and vitamin A: an update. *American Society for Clinical Nutrition* **68,** 435S (1998).

A. C. Chan: Vitamin E and atherosclerosis. *Journal of Nutrition* **128,** 1593 (1998).

Committee on Genetics: Newborn screening fact sheet. *Pediatrics* **98,** 473 (1996).

P. Di Mascio, M. E. Murphy, and H. Sies: Antioxidant defense systems: the role of carotenoids, tocopherols, and thiols. *Am. J. Clin. Nutr.* **53**(1 Suppl), 194S (1991).

M. N. Diaz, B. Frei, J. A. Vita, et al.: Antioxidants and atherosclerotic heart disease. *New England Journal of Medicine* **337,** 408 (1997).

A. T. Diplock: Safety of antioxidant vitamins and β-carotene. *American Journal of Clinical Nutrition* **62,** 1510S (1995).

J. W. Eikelboom, E. Lonn, J. Genest Jr., et al.: Homocyst(e)ine and cardiovascular disease: a critical review of the epidemiologic evidence. *Annals of Intemal Medicine* **131,** 363 (1999).

T. R. J. Evans and S. B. Kaye: Retinoids: present role and future potential. *British Journal of Cancer* **80,** 1 (1997).

P. Fenaux and L. Degos: Differentiation therapy for acute promyelocytic leukemia. *New England Journal of Medicine* **337,** 1076 (1997).

S. C. Guba, L. M. Fink, and V. Fonseca: Hyperhomocysteinemia. *American Journal of Clinical Pathology* **105,** 709 (1996).

J. Haller: The vitamin status and its adequacy in the elderly: an international overview. *Intemational Journal of Vitamin Nutrition Research* **69,** 160 (1999).

E. B. Healton, D. G. Savage, J. C. M. Brust, et al.: Neurologic aspects of cobalamin deficiency. *Medicine* **70,** 229 (1991).

A. Kamal-Eldin and L-A. Appelqvist: The chemistry and antioxidant properties of tocopherols and tocotrienols. *Lipids* **31,** 671 (1996).

H. Kasper: Vitamin absorption in the elderly. *International Journal of Vitamin Nutrition Research* **69,** 169 (1999).

M. Levine, S. C. Rumsey, R. Daruwala, et al.: Criteria and recommendations for vitamin C intake. *Journal of American Medical Association* **281,** 1415 (1999).

J. Lindenbaum, E. B. Healton, D. G. Savage, et al.: Neuropsychiatric disorders caused by cobalarnin deficiency in the absence of anemia or macrocytosis. *New England Journal of Medicine* **318,** 1720 (1988).

J. G. Locksmith and P. Duff: Preventing neural tube defects: the importance of periconceptional folic acid supplements. *Obstetrics and Gynecology* **91,** 1027 (1998).

J. C. Mattson: Acute promyelocytic leukemia: from morphology to molecular lesions. *Clinics in Laboratory Medicine* **20,** 83 (2000).

M. Meydani: Vitamin E. *Lancet* **345,** 170 (1995).

A. Molloy, et al.: Thermolabile variant of 5.10-methylenetetrahydrofolate reductase associated with low red-cell folate: implications for folate intake recommendations. *Lancet* **349,** 1591 (1997).

National Research Council: *Recommended Dietary Allowances,* 10th ed. National Academy of Sciences, Washington, D.C., 1989.

C. L. Rock, R. A. Jacob, and P. E. Bowen: Update on the biological characteristics of the antioxidant micronutrients: vitamin C, vitamin E, and the carotenoids. *Journal of the American Dietetic Association* **96,** 693 (1996).

K. J. Rothman, L. L. Moore, M. R. Singer, et al.: Teragenicity of high vitamin A intake. *New England Journal of Medicine* **333,** 1369 (1995).

K. Schümann: Interactions between drugs and vitamins at advanced age. *Intemational Journal of Vitamin Nutrition Research* **69,** 173 (1999).

R. H. Schwarz and R. B. Johnston: Folic acid supplementation—when and how. *Obstetrics and Gynecology* **88,** 886 (1996).

C. F. Snow: Laboratory diagnosis of vitamin B_{12} and folate deficiency. *Archives of Internal Medicine* **159,** 1289 (1999).

S. P. Stabler, R. H. Allen, D. G. Savage, et al.: Clinical spectrum and diagnosis of cobalamin deficiency. *Blood* **76,** 871 (1990).

S. P. Stabler, J. Lindenbaum, and R. H. Allen: Vitamin B_{12} deficiency in the elderly: current dilemmas. *American Journal of Clinical Nutrition* **66,** 741 (1997).

N. G. Stephens, A. Parsons, and P. M. Schofield: Randomised controlled trial of vitamin E in patients with coronary disease: Cambridge Heart Antioxidant Study (CHAOS). *Lancet* **347,** 781 (1996).

M. S. Tallman, J. W. Andersen, C. A. Schiffer, et al.: *New England Journal of Medicine* **337,** 1021 (1997).

B. Termanini, F. Gibril, V. Sutliff, et al.: Effect of long-term gastric acid suppressive therapy on serum vitamin B_{12} levels in patients with Zollinger-Ellison syndrome. *American Journal of Medicine* **104,** 422 (1998).

A. Theriault, J-T. Chao, Q. Wang, et al.: Tocotrienol: a review of its therapeutic potential. *Clinical Biochemistry* **32,** 309 (1999).

M. G. Traber and H. Sies: Vitamin E in humans: demand and delivery. *Annual Review of Nutrition* **16,** 321 (1996).

J. M. Upston, A. C. Terentis, and R. Stocker: Tocopherol-mediated peroxidation of lipoproteins: implications for vitamin E as a potential antiatherogenic supplement. *FASEB Journal* **13,** 977 (1999).

J. Virtamo, J. M. Rapola, S. Ripatti, et al.: Effect of vitamin E and beta carotene on the incidence of primary nonfatal myocardial infarction and fatal coronary heart disease. *Archives of Intemal Medicine* **158,** 668 (1998).

U-W. Weigand, S. Hartmann, and H. Hummler: Safety of vitamin A: recent results. *Intemational Journal of Vitamin Nutritional Resources* **68,** 411 (1998).

N. J. Wald, J. W. Densem, D. Smith, et al.: Four-marker serum screening for Down's syndrome. *Prenatal Diagnosis* **IL4,** 707 (1994).

CHAPTER 39

Water, Electrolytes, and Acid-Base Balance

39.1 Water Metabolism

Water is the most abundant body constituent; it is 45–60% of total body weight (Figure 39-1). In a lean person, it accounts for a larger fraction of the body mass than in a fat person. Since most biochemical reactions take place in an aqueous environment, control of water balance is an important requirement for homeostasis.

Although water permeates freely across cell membranes, other solutes are less mobile because of barriers imposed by membrane systems. These barriers give rise to fluid pools or compartments of different but rather constant composition (Figure 39-2).

Intracellular fluid makes up 30–40% of body weight, or about two-thirds of total body water. Potassium and magnesium are the predominant cations. The anions are mainly proteins and organic phosphates, with chloride and bicarbonate at low concentrations.

Extracellular fluid contains sodium as the predominant cation and accounts for 20–25% of body weight, or one-third of total body water. It makes up vascular, interstitial, transcellular, and dense connective tissue fluid pools. Vascular fluid is the circulating portion, is rich in protein, and does not readily cross endothelial membranes. Interstitial fluid surrounds cells and accounts for 18–20% of total body water. It exchanges with vascular fluid via the lymph system. Transcellular fluid is present in digestive juices, intraocular fluid, cerebrospinal fluid (CSF), and synovial (joint) fluid. These fluids are secretions of specialized cells. Their composition differs considerably from that of the rest of the extracellular fluid, with which they rapidly exchange contents under normal conditions. Dense connective tissue (bone, cartilage) fluid exchanges slowly with the rest of the extracellular fluid and accounts for 15% of total body water.

Movements of water are due mainly to osmosis and filtration. In osmosis, water moves to the area of highest solute concentration. Thus, active movement of salts into an area creates a concentration gradient down which water flows passively. In filtration, hydrostatie pressure in arterial blood moves water and nonprotein solutes through specialized membranes to produce an almost protein-free filtrate: This process occurs in formation of the renal glomerular filtrate. Filtration also accounts for movement of water from the vascular space into the interstitial compartment, which is opposed by the osmotic (oncotic) pressure of plasma proteins.

Cells move ions (especially Na^+ and K^+) against a concentration gradient by a "sodium pump" that actively transports sodium across the plasma membranes (Chapter 12).

The kidneys are the major organ to regulate extracellular fluid composition and volume. Three main processes occur in nephrons:

1. Formation of a virtually protein-free ultrafiltrate at the glomerulus;

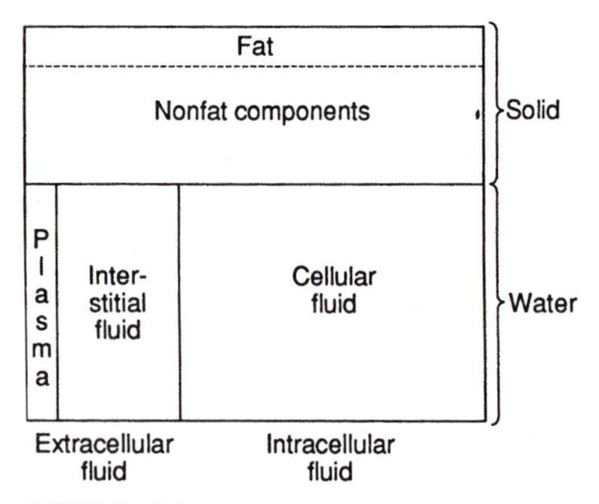

FIGURE 39-1
Proportional distribution of solids and water in a healthy adult.

2. Active reabsorption (principally in the proximal tubule) of solutes from the glomerular filtrate; and
3. Active excretion of substances such as hydrogen ions into the tubular lumen, usually in the distal portion of the tubule (Figure 39-3).

The normal ***glomerular filtration rate*** (GFR) is 100–120 mL/min; about 150 L of fluid passes through the renal tubules each day. Since the average daily urine volume is 1-1.5 L, 99% of the glomerular filtrate is reabsorbed. Approximately 80% of the water is reabsorbed in the proximal tubule, a consequence of active absorption of solutes. Reabsorption in the rest of the tubule varies according to the individual's water balance, in contrast to the *obligatory* reabsorption that occurs in the proximal tubule.

The facultative absorption of water depends on the establishment in the loop of Henle of an osmotic gradient

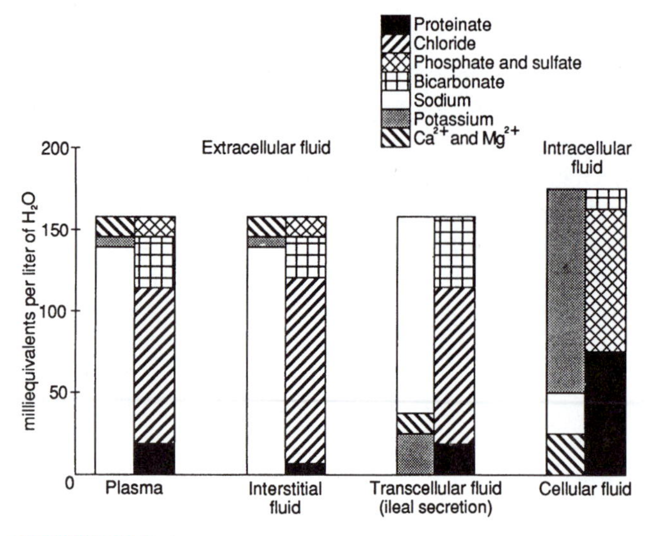

FIGURE 39-2
Composition of body fluids.

by the secretion of Na^+ from the ascending loop and their uptake by the descending loop. As a result, the proximal end of the loop is hyperosmotic (1200 mosm/kg) and the distal end hypoosmotic with respect to blood. The collecting ducts run through the hyperosmotic region. In the absence of antidiuretic hormone (ADH; see Chapter 31), the cells of the ducts are relatively impermeable to water. They become permeable to water in the presence of ADH, however, and the urine becomes hyperosmotic with respect to blood.

39.2 Homeostatic Controls

The composition and volume of extracellular fluid are regulated by complex hormonal and nervous mechanisms that interact to control its osmolality, volume, and pH.

The osmolality of extracellular fluid is due mainly to Na^+ and accompanying anions. It is kept within narrow limits (285–295 mosm/kg) by regulation of water intake (via a thirst center) and water excretion by the kidney through the action of ADH. The volume is kept relatively constant, provided the individual's weight remains constant to within ±1 kg. Volume receptors sense the effective circulating blood volume, which when decreased, stimulates the renin-angiotensin-aldosterone system and results in retention of Na^+ (Chapter 32). The increased Na^+ level leads to a rise in osmolality and secretion of ADH, with a resultant increase in water retention. Antagonistic systems exist that result in an increased Na^+ excretion. ***Atrial natriuretic peptide*** (ANP), also called atrial natriuetic factor or hormone, is released by the cardiocytes of the cardiac atria in response to mechanical stretch caused by the plasma volume expansion. ANF induces diuresis and natriuresis. These effects result from renal hemodynamic changes associated with increases in GFR and inhibition of Na^+ reabsorption from inner medullary collecting ducts. ANP is a 28-amino-acid peptide and has a single disulfide linkage (Figure 39-4). The precursor and the storage form of ANP in cardiocytes is a 126-amino-acid polypeptide. Some of the stimuli other than the blood volume which function as secretogosues of ANP include high blood pressure, elevated serum osmolality, increased heart rate, and elevated levels of plasma catecholamines. Activation of the ANP gene in cardiocytes by glucocorticoids leads to increased synthesis of ANP. ANP also regulates Na^+ and water homeostasis by different mechanisms that include inhibition of steps in the renin-angiotensin-aldosterone pathway and inhibition of ADH secretion from posterior pituitary cells.

The mechanism of action of ANP on target cells involves the formation of cGMP via the activation of plasma

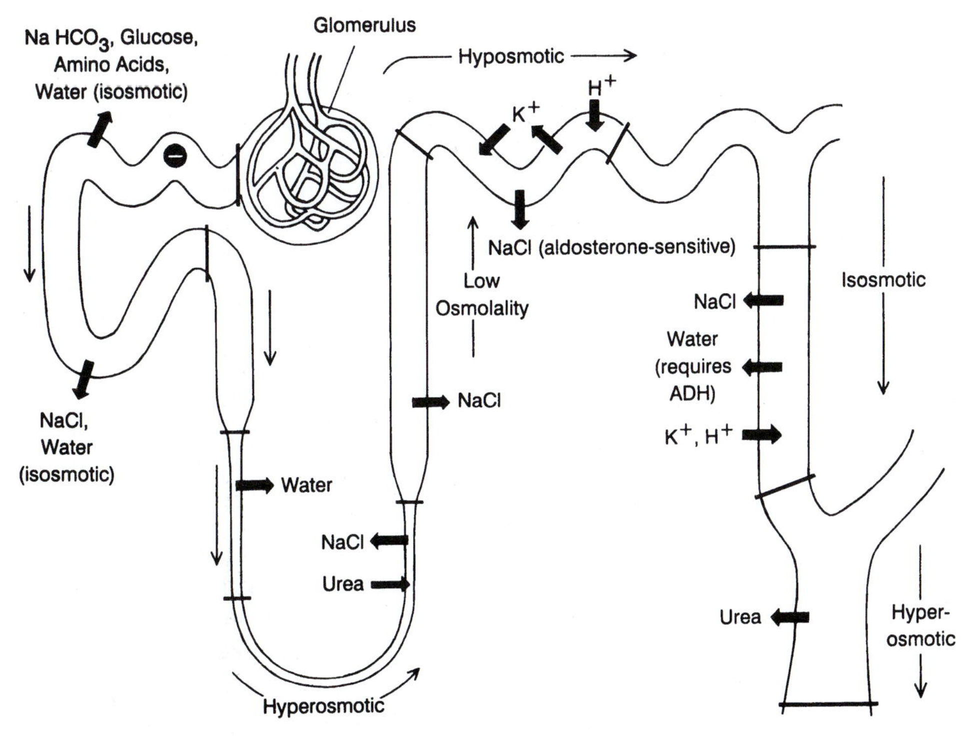

FIGURE 39-3
Principal transport processes in the renal nephron. ADH, Antidiuretic hormone. [Reproduced with permission from M. B. Burg, The nephron in transport of sodium, amino acids, and glucose. *Hosp. Pract.* **13**(10), 99 (1978). A. Iselin, illustrator.]

membrane receptor. The ANP receptor itself is a guanylyl cyclase with its ligand binding domain located in the extracellular space and its catalytic domain in the cytsolic domain. The receptor has only one membrane spanning domain. This ANP receptor–activated cGMP complex is unique and does not involve any G-proteins. A soluble cytosolic guanylyl cyclase that is activated by nitric oxide binding to the heme group of the enzyme causes vascular relaxation (Chapter 17). The intracellular formation of cGMP causes activation cGMP-dependent protein kinases which mediate the actions of ANP. An ANP of 32 amino acid residues (known as B type-ANP) found in the ventricles of the heart (and in the brain) is secreted in response to ventricular expansion and pressure. B type-ANP has similar physiologic function compared to atrial ANP.

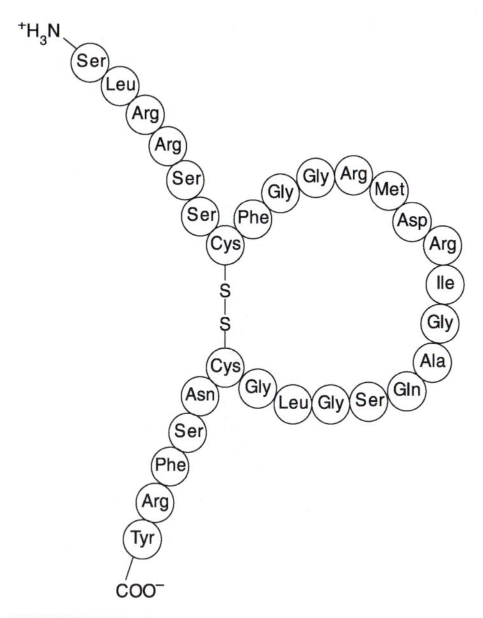

FIGURE 39-4
Amino acid sequence of atrial natriuretic peptide. It has a one-disulfide linkage.

The pH of extracellular fluid is kept within very narrow limits (7.35–7.45) by buffering mechanisms (see also Chapter 1), the lungs, and the kidneys. These three systems do not act independently. For example, in acute blood loss release of ADH and aldosterone restores the blood volume and renal regulation of the pH leads to shifts in K^+ and Na^+ levels.

39.3 Water and Osmolality Controls

Despite considerable variation in fluid intake, an individual maintains water balance and a constant composition of body fluids. The homeostatic regulation of water is

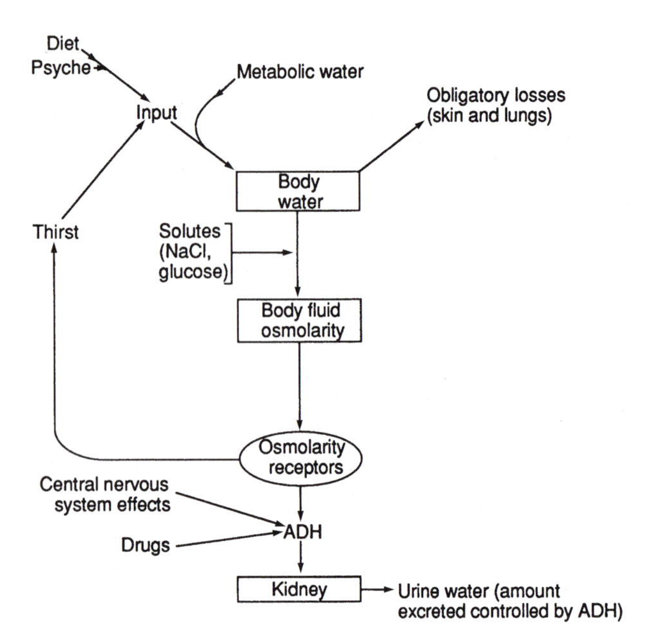

FIGURE 39-5
Regulation of osmolality in the body.

summarized in Figure 39-5. Body water is derived from 2–4 L of water consumed daily in food and drink and 300 mL of metabolic water formed daily by oxidation of lipids and carbohydrates. Water loss occurs by perspiration and expiration of air (~1 L/d), in feces (~200 mL) (Chapter 12), and in urine (1–2 L/d).

Water balance is regulated to maintain constant osmolality of body fluids. This osmolality is directly related to the number of particles present per unit weight of solvent. A solution that contains 1 mol of particles in 22.4 kg of water (22.4 L at 4°C) exerts an osmotic pressure of 1 atm and has an osmolality of 0.0446. Conversely, the osmotic pressure of an osmolal solution (1 mol of particles/kg of water) is 22.4 atm. In this sense, "number of particles" is roughly defined as the number of noninteracting molecular or ionic groups present. Since glucose does not readily dissociate, 1 mol dissolved 1 kg of water (a molal solution) produces 1 mol of "particles" and has an osmolality of 1. Sodium chloride dissociates completely in water to form two particles from each molecule of NaCl so that a molal solution of NaCl is a 2-osmolal solution. Similarly, a molal solution of Na_2SO_4 or $(NH_4)_2SO_4$ is a 3-osmolal solution. In practice, the milliosmole (mosm) is the unit used.

With aqueous solutions, osmolarity is sometimes used interchangeably with osmolality. Although this practice is not strictly correct (moles of particles per liter of solution versus moles of particles per kilogram of solvent), in water at temperatures of biological interest the error is fairly small unless solute concentrations are high (i.e., when an appreciable fraction of the solution is not water). Thus, with urine the approximation is acceptable, whereas with serum it is not because of the large amount of protein present. Although osmolarity is more readily measured, it is temperature dependent, unlike osmolality.

Osmolality is commonly measured by freezing point or vapor pressure depression. In terms of vapor pressure (P^{v}), the osmotic pressure (π) is defined as

$$\pi = P^{\text{v}}_{\text{pure solvent}} - P^{\text{v}}_{\text{solution}}$$

As defined above, osmolality $= \pi/22.4$, where π is measured in atmospheres. In one instrument, solution and solvent vapor pressures are measured by use of sensitive thermistors to detect the difference in temperature decrease caused by evaporation of solvent from a drop of pure solvent and from a drop of solution. Because the rate of evaporation (vapor pressure) of the solution is lower, the temperature change will be less and the vapor pressure difference can be calculated.

The freezing point of a solution is always lower than that of the solvent. The exact difference depends on the solvent and the osmolality of the solution. For water,

$$\text{Osmolality} = \frac{\Delta T}{1.86}$$

where ΔT is the freezing point depression in degrees Celsius. Instruments that measure the freezing point of a sample are used in clinical laboratories to determine serum and urine osmolality.

Since water passes freely through most biological membranes, all body fluids are in osmotic equilibrium so that the osmolality of plasma is representative of the osmolality of other body fluids.

The osmotic pressure of extracellular fluid is due primarily to its principal cation Na^+ and the anions Cl^- and HCO_3^-. Taking twice the Na^+ concentration gives a good estimate of serum osmolality. Thus, normal plasma contains 135–145 mEq of Na^+/L (3.1–3.3g/L) and normal plasma osmolality is about 270–290 mosm/kg (this corresponds to an osmotic pressure of 6.8–7.3 atm and a freezing point depression of 0.50–0.54°C). Glucose provides only 5–6 mosm/kg (or 0.1 atm to the osmotic pressure). Plasma protein contributes about 10.8 mosm/kg. Because of their size and general inability to pass through biological membranes, proteins are important determinants of fluid balance between intravascular and extravascular spaces. That portion of the osmotic pressure which is due to proteins is often referred to as the ***oncotic pressure.***

Since many molecules in plasma interact, the measured osmolality of a sample is an effective osmolality and is lower than the value calculated from the concentrations of all the ions and molecules it contains. A solution that has the same effective osmolality as plasma is said to be ***isotonic,*** e.g., 0.9% saline, 5% glucose, and Ringer's and Locke's solutions. If a solute can permeate a membrane freely, then a solution of that solute will behave like pure water with respect to the membrane. Thus, a solution of urea will cause red cells to swell and burst as does pure water because urea moves freely across erythrocyte membranes.

The osmolality of urine can differ markedly from that of plasma because of active concentrative processes in the renal tubules. The membranes of renal collecting ducts show varying degrees of water permeability and permit removal of certain solutes without simultaneous uptake of water.

Plasma osmolality can be calculated from the concentrations of plasma Na^+, glucose, and serum urea nitrogen:

$$\text{Osmolality} = 1.86(\text{Na}^+ \text{ mEq/L}) + \frac{\text{glucose (mg/dL)}}{18} + \frac{\text{serum urea nitrogen (mg/dL)}}{2.8}$$

The numerical denominators for glucose and urea nitrogen convert the concentrations to moles per liter. Such an estimated osmolality is usually 6–9 mosm less than the value determined by freezing point or vapor pressure measurements. If the latter value is much greater than the estimated value, molecules other than Na^+, glucose, and urea must account for the difference. Such "osmolal gaps" occur in individuals suffering from drug toxicity (alcohol, barbiturates, salicylates), acute poisoning due to unknown substances, and acidosis (keto-, lactic, or renal). Determination of osmolality is helpful in management of patients with fluid and electrolyte disorders, e.g., chronic renal disease, nonketotic diabetic coma, hypo- and hypernatremia, hyperglycemia, hyperlipemia, burns, sequelae to major surgery or severe trauma (particularly serious head injuries), hemodialysis, or diabetes insipidus. Changes of about 2% or more are detected by the hypothalamic osmoreceptors (Chapter 31) and elicit a sensation of thirst and production of hypertonic urine. Under conditions of fluid restriction, urine osmolality can reach 800–1200 mosm/kg (normal is 390–1090 mosm/kg), or three to four times the plasma levels. Decrease in plasma osmolarity (as in excessive water intake) produces urine with decreased osmolality. Water losses from skin and lungs are not subject to controls of this type.

ADH acts at the renal tubules and collecting ducts to raise cAMP levels. Urinary levels of cAMP are increased by ADH. Factors other than plasma hypertonicity may stimulate ADH secretion. Thus, in acute hemorrhage extracellular fluid volume drops abruptly, and ADH is secreted to increase the volume at the expense of a drop in osmolarity.

Inappropriate ADH secretion can occur in the presence of water overload and a decline in plasma Na^+ concentration and osmolality. Fear, pain, and certain hormone-secreting tumors can cause inappropriate ADH secretion. It leads to hyponatremia and water retention. Morphine and barbiturates increase, and ethanol decreases, secretion of ADH.

In ***diabetes insipidus*** due to defective ADH receptors or to diminished ADH secretion, renal tubules fail to recover the water from the glomerular filtrate. In cases of deficiency of ADH, hormone replacement by 8-lysine vasopressin or 1-deamino-8-D-arginine vasopressin (administered as a nasal spray or subcutaneously) is effective. In osmotic diuresis, e.g., in ***diabetes mellitus*** with severe glycosuria, the solute load increases the osmolality of the glomerular filtrate and impairs the ability of the kidney to concentrate the urine. Extracellular fluid volume in a normal adult is kept constant; body weight does not vary by more than a pound per day despite fluctuations in food and fluid intake. A decrease in extracellular fluid volume lowers the effective blood volume and compromises the circulatory system. An increase may lead to hypertension, edema, or both. Volume control centers on renal regulation of Na^+ balance. When the extracellular fluid volume decreases, less Na^+ is excreted; when it increases, more Na^+ is lost. Na^+ retention leads to expansion of extracellular fluid volume, since Na^+ is confined to this region and causes increased water retention. Renal Na^+ flux is controlled by the aldosterone-angiotensin-renin system (Chapter 32) and atrial natriuretic peptide (discussed earlier).

39.4 Electrolyte Balance

The major electrolytes are Na^+, K^+, Cl^-, and HCO_3^- (HCO_3^- is discussed below, under "Acid-Base Balance.")

Sodium

The average Na^+ content of the human body is 60 mEq/kg, of which 50% is in extracellular fluid, 40% is in bone, and 10% is intracellular. The chief dietary source of sodium is salt added in cooking. Excess sodium is largely excreted in the urine, although some is lost in perspiration. Gastrointestinal losses are small except in diarrhea.

Sodium balance is integrated with regulation of extracellular fluid volume. Depletional hyponatremia (sodium loss greater than water loss) may result from inadequate Na^+ intake, excessive fluid loss from vomiting or diarrhea, diuretic abuse, and adrenal insufficiency. Hyponatremia can decrease extracellular fluid volume, as occurs in congestive heart failure, uncontrolled diabetes, cirrhosis, nephrosis, and inappropriate ADH secretion.

Hypernatremia results from loss of hypoosmotic fluid (e.g., in burns, fevers, high environmental temperature, exercise, kidney disease, diabetes insipidus) or increased Na^+ intake (e.g., administration of hypertonic NaCl solutions, ingestion of $NaHCO_3$).

Potassium

The average K^+ content of the human body is 40 mEq/kg. K^+ occurs mainly in intracellular space. It is required for carbohydrate metabolism, and increased cellular uptake of K^+ occurs during glucose catabolism. K^+ is widely distributed in plant and animal foods, the human requirement being about 4 g/day. Insulin and catecholamines promote a shift of K^+ into the cells. Excess K^+ is excreted in the urine, a process regulated by aldosterone.

Plasma K^+ plays a role in the irritability of excitable tissue. A high concentration of plasma K^+ leads to electrocardiographic (ECG) abnormalities and possibly to cardiac arrhythmia, which may be due to lowering of the membrane potential. Low concentration of plasma K^+ increases the membrane potential, decreases irritability, and produces other ECG abnormalities and muscle paralysis.

Hyperkalemia may occur in renal disease and adrenal insufficiency owing to impairment of normal secretory mechanisms. Metabolic acidosis, in particular diabetic acidosis, and catabolism of cellular protein in starvation or fever cause K^+ release from cells. Treatment consists of correction of the acidosis and promotion of cellular uptake of K^+ by administration of insulin, which enhances glucose intake. In severe cases, ion exchange resins given orally bind K^+ in intestinal secretions.

Hypokalemia may occur from loss of gastrointestinal secretions (which contain significant amounts of K^+) and from excessive loss in the urine because of increased aldosterone secretion or diuretic therapy. Hypokalemia is usually associated with alkalosis.

Chloride

Chloride is the major extracellular anion. About 70% is in the extracellular fluid. The average $C1^-$ content of the human body is 35 mEq/kg. Chloride in food is almost completely absorbed. Plasma levels of Na^+ and $C1^-$ in general undergo parallel alterations. However, in metabolic alkalosis, chloride concentration increases.

39.5 Acid-Base Balance

Normal blood pH is 7.35–7.45 (corresponding to 35–45 nmol of H^+ per liter). Values below 6.80 (160 nmol of H^+ per liter) or above 7.70 (20 nmol of H^+per liter) are seldom compatible with life. A large amount of acid produced is a byproduct of metabolism. The lungs remove 14,000 mEq of CO_2 per day. From a diet that supplies 1–2 g of protein per kilogram per day, the kidneys remove 40–70 mEq of acid per day as sulfate (from oxidation of sulfur-containing amino acids), phosphate (from phospholipid, phosphoprotein, and nucleic acid catabolism), and organic acids (e.g., lactic, β-hydroxybutyric, and acetoacetic). These organic acids are produced by incomplete oxidation of carbohydrate and fats, and in some conditions (e.g., ketosis; see Chapter 18) considerable amounts may be produced.

The most important extracellular buffer is the carbonic acid-bicarbonate system:

$$CO_2 + H_2O \rightleftarrows H_2CO_3 \rightleftarrows HCO_3 + H^+$$

As discussed in Chapter 1, at a blood pH of 7.4, the ratio of $[HCO_3^-]$ to $[H_2CO_3]$ is 20:1 and the system's buffering capacity can neutralize a large amount of acid. The system is independently regulated by the kidneys, which control the plasma HCO_3^- level, and by the respiratory rate, which regulates the P_{CO_2}. Protein and phosphate buffer systems also operate in plasma and erythrocytes. Proteins are especially important buffers in the intracellular fluid. The hydroxyapatite of bone also acts as a buffer.

The medullary respiratory center senses and responds to the pH of blood perfusing it and is the source of pulmonary control. P_{CO_2} and perhaps P_{O_2} also influence the center, together with nervous impulses from higher centers of the brain. A decrease in pH results in an increased respiratory rate and deeper breathing with a consequent increase in the respiratory exchange of gases and lowering of P_{CO_2} which elevates the pH. Similarly, a decrease in respiratory rate leads to accumulation of CO_2, increase in P_{CO_2}, and decrease in pH. Pulmonary responses to fluctuations in blood pH are rapid, while renal compensatory mechanisms are relatively slower.

The kidneys actively secrete H^+ ions through three mechanisms:

1. Na^+/H^+ exchange,
2. Reclamation of bicarbonate, and
3. Production of ammonia and excretion of NH_4^+.

The proximal tubule is responsible for reclamation of most of the 4500 mEq of HCO_3^- filtered through the glomeruli each day. H^+ secreted into the tubules

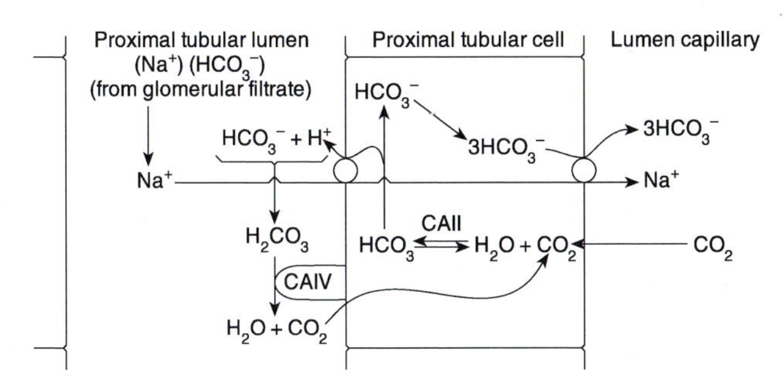

FIGURE 39-6
Reclamation of bicarbonate. The filtered Na^+ is reabsorbed by the proximal tubular cell in exchange for H^+. The filtered HCO_3^- is converted to H_2O and CO_2 catalyzed by the luminal carbonic anhydrase IV (CAIV). CO_2 diffuses in the tubular cell where it is hydrated to H_2CO_3 by carbonic anhydrase II and dissociated to H^+ and HCO_3^-. Three molecules of HCO_3 and one of Na^+ are transported to the peritubular capillary by the basolateral cotransporter.

in exchange for Na^+ from the tubular fluid (an energy-dependent process mediated by Na^+, H^+-ATPase) combines with HCO_3^- to form CO_2 and water. The CO_2 diffuses into the tubular cells, where it is rehydrated to H_2CO_3 by carbonic anhydrase and dissociates to bicarbonate and H^+. The HCO_3^- diffuses into the bloodstream, resulting in reclamation of bicarbonate (Figure 39-6).

The formation of H_2CO_3, from H^+ and HCO_3^- in the tubular lumen is catalyzed by the membrane-bound isoenzyme of carbonic anhydrase (CAIV). CAIV is located on the brush border lining the lumen of the proximal tubules of the kidney. The CO_2 diffuses into the tubular cells where it is rehydrated to H_2CO_3 by a different isoenzyme of carbonic anhydrase, namely, carbonic anhydrase II (CAII). HCO_3^- is transported from the cytosol of the proximal tubular cell to peritubular capillary blood by a basolateral cotransporter, which transports three molecules of HCO_3^- and one of Na^+. An inherited CAII deficiency is an autosomal recessive disorder causing renal tubular acidosis and osteopetrosis. The latter is due to lack of H^+ production required for bone remodeling processes (Chapter 37). The proximal tubular mechanism for reclamation of HCO_3^- becomes saturated at approximately 26 mEq HCO_3^-/L. At higher levels, HCO_3^- appears in the distal tubule and may be excreted in urine. Under conditions of elevated P_{CO_2}, H^+ secretion is more active, possibly owing to intracellular acidosis. In proximal renal tubular acidosis, HCO_3^- reabsorption is impaired and saturation occurs at a lower concentration (about 16–18 mEq/L), so that plasma $[HCO_3^-]$ is low while urine pH may be high because of the presence of HCO_3^-.

Na^+/H^+ exchange may also be coupled to formation of $H_2PO_4^-$ from HPO_4^{2-} in the lumen (Figure 39-7). This coupling is of particular importance in distal tubules and acidifies the urine to a maximum pH of about 4.4. The

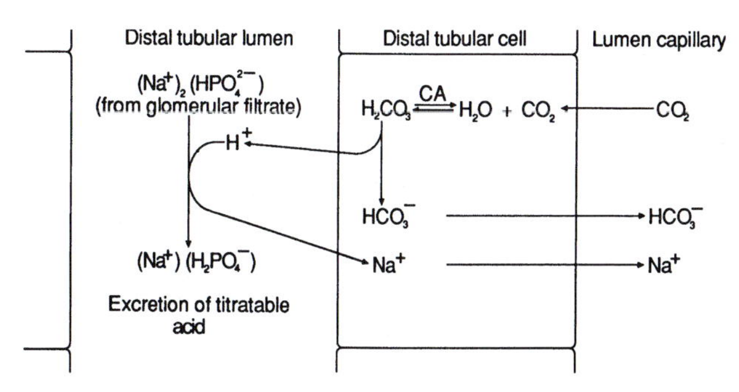

FIGURE 39-7
Hydrogen ion secretion and Na^+/H^+ exchange coupled to the conversion of HPO_4^{2-} to $H_2PO_4^-$ in the distal tubule lumen. CA, carbonic anhydrase.

proximal tubule cannot maintain an H^+ gradient of more than one pH unit between the lumen and the intracellular fluid. The $H_2PO_4^-$ present (termed the "titratable acidity") can be measured by titrating the urine to pH 7 (pK_2 for $H_3PO_4 = 6.2$). Normal values for titratable acidity are 16–60 mEq/24 h, depending on the phosphate load.

Generation of ammonia from glutamine and its excretion as NH_4^+ is an important mechanism for elimination of protons, particularly during severe metabolic acidosis, when it becomes a significant mode of nitrogen excretion. Most renal glutamine is derived from muscle (Chapters 17 and 22). Glutamine provides two molecules of NH_3:

$$\text{Glutamine} + H_2O \xrightarrow{\text{glutaminase}} \text{glutamate}^- + NH_3 + H^+ (\rightleftarrows NH_4^+)$$

and

$$\text{Glutamate}^- \xrightarrow[\text{NADH} \;\to\; \text{NADH} + H^+]{\text{glutamate dehydrogenase}}$$

$$\alpha\text{- ketoglutarate}^{2-} + NH_3 + H^+ \;(\rightleftarrows NH_4^+)$$

The ratio of $[NH_3]/[NH_4^+]$ depends on intracellular pH; however, the NH_3 readily diffuses into the tubular lumen and there forms an ammonium ion that is no longer able to pass freely through the membranes and remains "trapped" in the urine, where it is associated with the dominant counterion (Figure 39-8). Protons produced in these reactions are consumed when α-ketoglutarate is either completely oxidized or converted to glucose. Thus, they do not add to the existing "proton burden" due to severe acidosis (see also Chapters 13 and 15).

Disorders of Acid-Base Balance

These disorders are classified according to their cause and the direction of the pH change into respiratory acidosis, metabolic acidosis, respiratory alkalosis, or metabolic alkalosis. Any derangement of acid-base balance elicits

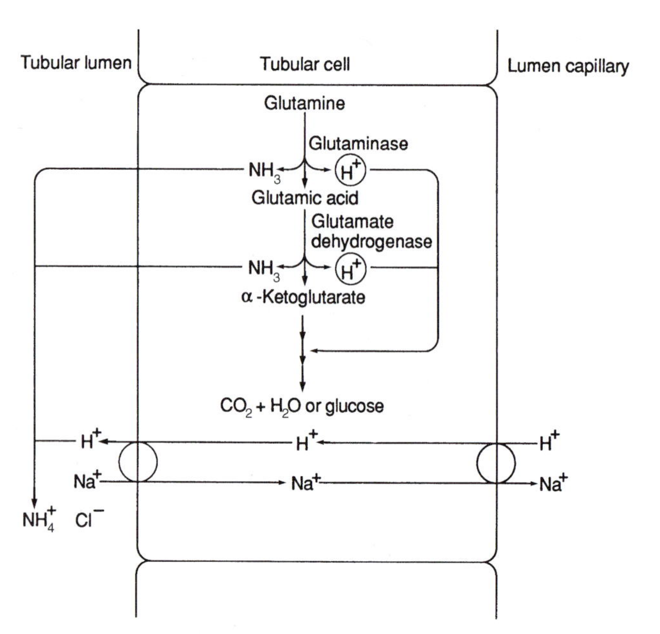

FIGURE 39-8
Formation of ammonia in the renal tubule cells from glutamine and secretion of ammonium ion in the urine.

compensatory changes in an attempt to restore homeostasis (Table 39-1). Acidosis due to respiratory failure leads to compensatory renal changes, which lead to increased reclamation of HCO_3^-.

In assessment of acid-base disorders, commonly measured electrolytes are serum Na^+, K^+, H^+ (as pH), Cl^-, and HCO_3^-. Other anions (e.g., sulfates, phosphates, proteins) and cations (e.g., calcium, magnesium, proteins) are not measured routinely but can be estimated indirectly, since (to maintain electrical neutrality) the sum of the cations must equal that of the anions. Serum Na^+ and K^+ content accounts for 95% of cations, and Cl^- and HCO_3^- for about 85% of anions. The concentration of phosphate, sulfate, and proteins can be calculated from the formula:

$$\text{Unmeasured anions} = [Na^+] + [K^+] - [Cl^-] - [HCO_3^-]$$

The unmeasured anion is commonly known as the ***anion gap,*** which is normally 12 ± 4 mEq/L. This value is useful in assessing the acid-base status of a patient and in diagnosing metabolic acidosis. Disorders that cause a high anion gap are metabolic acidosis, dehydration, therapy with sodium salts of strong acids, therapy with certain antibiotics (e.g., carbenicillin), and alkalosis. A decrease in the normal anion gap occurs in various plasma dilution states, hypercalcemia, hypermagnesemia, hypernatremia, hypoalbuminemia, disorders associated with hyperviscosity, some paraproteinemias, and bromide toxicity.

Respiratory acidosis is characterized by accumulation of CO_2, rise in P_{CO_2} (hypercapnia or hypercarbia), decrease in $[HCO_3^-]/[P_{CO_2}]$, and decrease in pH (see Henderson-Hasselbalch equation, Chapter 1). It may result from central depression of respiration (e.g., narcotic or barbiturate overdose, trauma, infection, cerebrovascular accident) or from pulmonary disease (e.g., asthma, obstructive lung disease, infection). Increased $[H^+]$ is in part buffered by cellular uptake of H^+ with corresponding loss of intracellular K^+. In acute hypercapnia, the primary compensatory mechanism is tissue buffering. In chronic hypercapnia, the kidneys respond to elevated plasma P_{CO_2} increasing the amount of HCO_3^- formed by carbonic anhydrase in the tubules and by excreting more H^+.

The primary goal of treatment is to remove the cause of the disturbed ventilation. Immediate intubation and assisted ventilation also aid in improving the gas exchange.

Metabolic acidosis with increased anion gap occurs in diabetic or alcoholic ketoacidosis; lactic acidosis from hypoxia, shock, severe anemia, alcoholism, cancer; toxicity from ingestion of salicylates, methanol, paraldehhyde, and ethylene glycol; and renal failure. Lactic acid acidosis caused by deprivation of tissue oxygenation, inhibition of gluconeogensis, and some drugs and toxins is due to accumulation of L-lactate, which is the end product of glycolysis (Chapter 13). Frequently, L-lactate (simply referred to as lactate), is the metabolite measured in assessing metabolic acidosis. However, D-lactate may be produced under certain clinical conditions, such as diminished colonic motility, short bowel syndrome, jejunoileal bypass, due to overgrowth of D-lactate producing gram-positive organisms (e.g., *Lactobacillus* species, *Streptococcus bovis*). Carbohydrate malabsorption and ingestion of large amounts of carbohydrate may also exasperate the development of D-lactate acidosis. In addition, an impairment of D-lactate metabolism may also contribute to D-lactic acidosis. D-Lactate is converted to pyruvate by D-2-hydroxy acid dehydrogenese, a mitochondrial enzyme present in liver, kidney, and other tissues. The clinical manifestations of D-lactic acidosis include episodes of encephalopathy and metabolic acidosis. Since serum D-lactate is not a normally measured clinical parameter, the critical indices of suspicion of D-lactic acidosis in the clinical conditions mentioned above include increased anion gap metabolic acidosis with negative tests for L-lactate and ketoacidosis. Metabolic acidosis with normal anion gap occurs in renal tubular acidosis, carbonic anhydrase inhibition, diarrhea, ammonium chloride administration, chronic pyelonephritis, and obstructive uropathy. In both groups, plasma HCO_3^- levels decrease and tissue buffering occurs by exchange of extracellular H^+ for intracellular K^+. Thus, plasma K^+ levels may increase.

Metabolic acidosis produces prompt stimulation of respiratory rate and decrease in P_{CO_2}. This effect cannot be sustained, however, because of tiring of respiratory

TABLE 39-1
*Classification and Characteristics of Simple Acid-Base Disorders**

	Primary Change	Compensatory Response	Expected Compensation
Metabolic acidosis	↓↓↓HCO_3^-	↓↓P_{CO_2}	$P_{CO_2} = 1.5\,(HCO_3^-) + 8 \pm 2$. P_{CO_2} falls by 1–1.3 mm Hg for each mEq/L fall in HCO_3^-. Last 2 digits of pH = P_{CO_2} (thus, if P_{CO_2} = 28, pH = 7.28). HCO_3^- + 15 = last 2 digits of pH (HCO_3^- = 15, pH = 7.30).
Alkalosis	↑↑↑HCO_3^-	↑↑P_{CO_2}	P_{CO_2} increases 6 mm Hg for each 10 mEq/L rise in HCO_3^-. HCO_3^- + 15 = last 2 digits of pH (HCO_3^- = 35, pH = 7.50).
Respiratory acidosis			
Acute	↑↑↑P_{CO_2}	↑HCO_3^-	HCO_3^- increases by 1 mEq/L for each 10 mm Hg rise in P_{CO_2}.
Chronic	↑↑↑P_{CO_2}	↑↑HCO_3^-	HCO_3^- increases by 3.5 mEq/L for each 10 mm Hg rise in P_{CO_2}.
Alkalosis			
Acute	↓↓↓P_{CO_2}	↓HCO_3^-	HCO_3^- falls by 2 mEq/L for each 10 mm Hg fall in P_{CO_2}.
Chronic	↓↓↓P_{CO_2}	↓↓HCO_3^-	HCO_3^- falls by 5 mEq/L for each 10 mm Hg fall in P_{CO_2}.

↑ = Increase; ↓ = decrease.
*Reproduced, with permission, from R. G. Narins and L. B. Gardner: Simple acid-base disturbances. *Med. Clin. North Am.* **65,** 321 (1981).

muscles. Renal compensation is slower but can be maintained for an extended period because of induction of glutaminase.

Treatment is by correction of the cause of the acidosis (e.g., insulin administration in diabetic ketoacidosis) and neutralization of the acid with $NaHCO_3$, sodium lactate, or TRIS [tris(hydroxymethyl)aminomethane] buffer. Problems that may occur following alkali replacement therapy include development of respiratory alkalosis, particularly if the low CO_2 tension persists, and further decline in the pH of CSF, which may decrease consciousness. The alkaline "overshoot" results from resumption of oxidation of organic anions (e.g., lactate, acetoacetate) with resultant production of bicarbonate from CO_2. Severe acidosis should be corrected slowly over several hours. Potassium replacement therapy frequently is needed because of the shift of intracellular K^+ to extracellular fluid and loss of K^+ in the urine.

Respiratory alkalosis occurs when the respiratory rate increases abnormally (hyperventilation), leading to decrease in P_{CO_2} and rise in blood pH. Hyperventilation occurs in hysteria, pulmonary irritation (pulmonary embolus), and head injury with damage to the respiratory center.

The increase in blood pH is buffered by plasma HCO_3^- and, to some extent, by exchange of plasma K^+ for intracellular H^+. Renal compensation seldom occurs because this type of alkalosis is usually transitory.

Metabolic alkalosis is characterized by elevated plasma HCO_3^- level. It may result from administration of excessive amounts of alkali (e.g., during $NaHCO_3$ treatment of peptic ulcer) or of acetate, citrate, lactate, and other substrates that are oxidized to HCO_3^-, and from vomiting, which causes loss of H^+ and Cl^-.

In excessive loss of extracellular K^+ from the kidneys, cellular K^+ diffuses out and is replaced by Na^+ and H^+ from the extracellular fluid. Since K^+ and H^+ are normally secreted by the distal tubule cells to balance Na^+ uptake during Na^+ reabsorption (see Figure 39-3), if extracellular K^+ is depleted, more H^+ is lost to permit reabsorption of the same amount of Na^+. Loss of H^+ by both routes causes hypokalemic alkalosis. Excessive amounts

TABLE 39-2
Important Causes of Mixed Acid-Base Disturbances*

Respiratory acidosis with metabolic acidosis	
Example:	Cardiopulmonary arrest
	Severe pulmonary edema
	Drug ingestion with central nervous system depression
Respiratory alkalosis and metabolic alkalosis	
Example:	Hepatic failure and diuretics
	Patients on ventilation given nasogastric suction
Respiratory alkalosis and metabolic acidosis	
Example:	Septic shock
	Renal failure with sepsis
	Salicylate overdose
Respiratory acidosis and metabolic alkalosis	
Example:	Chronic lung disease and diuretic use
Mixed acute and chronic respiratory acidosis	
Example:	Chronic lung disease and superimposed infection
Metabolic acidosis and metabolic alkalosis	
Example:	Renal failure and vomiting
	Vomiting and hypotension (lactic acidosis)

*Reproduced, with permission, from M. Bia and S. O. Thier: Mixed acid-based disturbances: A clinical approach. *Med. Clin. North Am.* **65,** 347 (1981).

of some diuretics and increased aldosterone production can cause the hypokalemia that initiates this type of alkalosis.

In compensation, the respiratory rate decreases, raising P_{CO_2} and lowering the pH of blood. This mechanism is limited because if the respiratory rate falls too low, P_{O_2} decreases to the point where respiration is again stimulated. Renal compensation involves decreased reabsorption of bicarbonate and formation of alkaline urine. Because the urinary bicarbonate is accompanied by Na^+ and K^+, if the alkalosis is accompanied by extracellular fluid depletion, renal compensation by this mechanism may not be possible.

Treatment consists of fluid and electrolyte replacement and NH_4Cl to counteract the alkalosis.

Acid-base disturbances frequently coexist with two or more simple disorders (Table 39-2). In these settings, blood pH is either severely depressed (e.g., a patient with metabolic acidosis and respiratory acidosis) or normal. Both plasma HCO_3^- and pH may be within normal limits when metabolic alkalosis and metabolic ketoacidosis coexist, as in a patient with diabetic ketoacidosis who is vomiting. In this situation, an elevated anion gap may be the initial abnormality that can be detected in the underlying mixed acid-base disturbance.

Supplemental Readings and References

H. J. Adrogue and N. E. Madias: Management of life-threatening acid-base disorders. *New England Journal of Medicine* **338,** 26 (1998).

D. G. Bichet: Nephrogenic diabetes insipidus. *American Journal of Medicine* **105,** 431 (1998).

S. L. Gluck: Acid-base. *Lancet* **352,** 474 (1998).

M. L. Halperin and K. S. Kamel: Potassium. *Lancet* **352,** 135 (1998).

V. L. Hood and R. L. Tannen: Protection of acid-base balance by pH regulation of acid production. *New England Journal of Medicine* **339,** 819 (1998).

S. Klahr and S. B. Miller: Acute oliguria. *New England Journal of Medicine* **338,** 671 (1998).

S. J. Scheinman, L. M. Guay-Woodford, R. V. Thakker, et al.: Genetic disorders of renal electrolyte transport. *New England Journal of Medicine* **340,** 1177 (1999).

J. Uribarri, M. S. Oh, and H. J. Carroll: D-Lactic acidosis. *Medicine* **77,** 73 (1998).

K. D. Wrenn, C. M. Slovis, G. E. Minion, et al.: The syndrome of alcoholic ketoacidosis. *American Journal of Medicine* **91,** 119 (1991).

H. J. Androgué and N. E. Madias: Hyponatremia *New England Journal of Medicine* **342,** 1581 (2000).

TABLE I-1
Median Heights and Weights and Recommended Energy Intake*

Category	Age (yr) or Condition	Weight (kg)	Weight (lb)	Height (cm)	Height (in)	REE† (kcal/day)	Average Energy Allowances (kcal)‡ Multiples of REE	Per kg	Per day§
Infants	0.0–0.5	6	13	60	24	320		108	650
	0.5–1.0	9	20	71	28	500		98	850
Children	1–3	13	29	90	35	740		102	1,300
	4–6	20	44	112	44	950		90	1,800
	7–10	28	62	132	52	1,130		70	2,000
Males	11–14	45	99	157	62	1,440	1.70	55	2,500
	15–18	66	145	176	69	1,760	1.67	45	3,000
	19–24	72	160	177	70	1,780	1.67	40	2,900
	25–50	79	174	176	70	1,800	1.60	37	2,900
	51+	77	170	173	68	1,530	1.50	30	2,300
Females	11–14	46	101	157	62	1,310	1.67	47	2,200
	15–18	55	120	163	64	1,370	1.60	40	2,200
	19–24	58	128	164	65	1,350	1.60	38	2,200
	25–50	63	138	163	64	1,380	1.55	36	2,200
	51+	65	143	160	63	1,280	1.50	30	1,900
Pregnant	1st trimester								+0
	2nd trimester								+300
	3rd trimester								+300
Lactating	1st 5 months								+500
	2nd 6 months								+500

*From: Recommended Dietary Allowances, 10th ed., Food and Nutrition Board, National Research Council—National Academy of Sciences, 1989.
†REE = Resting energy expenditure.
‡In the range of light to moderate activity, the coefficient of variation is ±20%.
§Figure is rounded.

TABLE II-1
Weights for Heights of Adults in the United States*[*,†]

	Weight, kg (lb)					
	Males, by Percentile			Females, by Percentile		
Height cm (in)	**15th**	**50th**	**85th**	**15th**	**50th**	**85th**
147 (58)				45 (99)	55 (122)	72 (159)
152 (60)				49 (107)	60 (132)	75 (164)
157 (62)	57 (125)	64 (142)	76 (168)	51 (112)	60 (132)	77 (170)
163 (64)	58 (129)	67 (148)	79 (174)	54 (118)	63 (139)	79 (175)
168 (66)	61 (134)	71 (158)	83 (183)	55 (122)	64 (141)	81 (179)
173 (68)	65 (143)	76 (167)	88 (195)	59 (130)	67 (148)	83 (184)
178 (70)	67 (149)	79 (173)	93 (206)	61 (133)	69 (152)	78 (171)
183 (72)	73 (161)	83 (183)	99 (218)			
188 (74)	77 (171)	88 (194)	99 (217)			
193 (76)	85 (187)	103 (227)	106 (234)			

*From: Recommended Dietary Allowances, 10th ed. Food and Nutrition Board, National Research Council—National Academy of Sciences, 1989.
†Unpublished data from NHANES II (1976–1980) provided by the National Center for Health Statistics. Values rounded to nearest whole number. Subjects were ages 18 to 74 years. Height determined without shoes. Weight includes clothing weight, ranging from an estimated 0.09 to 0.28 kg (0.20 to 0.62 lb).

TABLE III-1

Weight and Height of Males and Females up to 18 Years in the United States*

	Males, by Percentile						Females, by Percentile					
	Weight, kg (lb)			Height, cm (in)			Weight, kg (lb)			Height, cm (in)		
Age	**5th**	**50th**	**95th**	**5th**	**50th**	**95th**	**5th**	**50th**	**95th**	**5th**	**50th**	**95th**
Months												
1	3.16	4.29 (9.4)	5.38	50.4	54.6 (21.5)	58.6	2.97	3.98 (8.8)	4.92	49.2	53.5 (21.1)	56.9
3	4.43	5.98 (13.2)	7.37	56.7	61.1 (24.1)	65.4	4.18	5.40 (11.9)	6.74	55.4	59.5 (23.4)	63.4
6	6.20	7.85 (17.3)	9.46	63.4	67.8 (26.7)	72.3	5.79	7.21 (15.9)	8.73	61.8	65.9 (25.9)	70.2
9	7.52	9.18 (20.2)	10.93	68.0	72.3 (28.5)	77.1	7.00	8.56 (18.8)	10.17	66.1	70.4 (27.7)	75.0
12	8.43	10.15 (22.3)	11.99	71.7	76.1 (30.0)	81.2	7.84	9.53 (21.0)	11.24	69.8	74.3 (29.3)	79.1
18	9.59	11.47 (25.2)	13.44	77.5	82.4 (32.4)	88.1	8.92	10.82 (23.8)	12.76	76.0	80.9 (31.9)	86.1
Years												
2	10.49	12.34 (27.1)	15.50	82.5	86.8 (34.2)	94.4	9.95	11.80 (26.0)	14.15	81.6	86.8 (34.2)	93.6
3	12.05	14.62 (32.2)	17.77	89.0	89.9 (37.4)	102.0	11.61	14.10 (31.0)	17.22	88.3	94.1 (37.0)	100.6
4	13.64	16.69 (36.7)	20.27	95.8	102.9 (40.5)	109.9	13.11	15.96 (35.1)	19.91	95.0	101.6 (40.0)	108.3
5	15.27	18.67 (41.1)	23.09	102.0	109.9 (43.3)	117.0	14.55	17.66 (38.9)	22.62	101.1	108.4 (42.7)	115.6
6	16.93	20.69 (45.5)	26.34	107.7	116.1 (45.7)	123.5	16.05	19.52 (42.9)	25.75	106.6	114.6 (45.1)	122.7
7	18.64	22.85 (50.3)	30.12	113.0	121.7 (47.9)	129.7	17.71	21.84 (48.0)	29.68	111.8	120.6 (47.5)	129.5
8	20.40	25.30 (55.7)	34.51	118.1	127.0 (50.0)	135.7	19.62	24.84 (54.6)	34.71	116.9	126.4 (49.8)	136.2
9	22.25	28.13 (61.9)	39.58	122.9	132.2 (52.0)	141.8	21.82	28.46 (62.6)	40.64	122.1	132.2 (52.0)	142.9
10	24.33	31.44 (69.2)	45.27	127.7	137.5 (54.1)	148.1	24.36	32.55 (71.6)	47.17	127.5	138.3 (54.4)	149.5
11	26.80	35.30 (77.7)	51.47	132.6	143.3 (56.4)	154.9	27.24	36.95 (81.3)	54.00	133.5	144.8 (57.0)	156.2
12	29.85	39.78 (87.5)	58.09	137.6	149.7 (58.9)	162.3	30.52	41.53 (91.4)	60.81	139.8	151.5 (59.6)	162.7
13	33.64	44.95 (98.9)	65.02	142.9	156.5 (61.6)	169.8	34.14	46.10 (101.4)	67.30	145.2	157.1 (61.9)	168.1
14	38.22	50.77 (111.7)	72.13	148.8	163.1 (64.2)	176.7	37.76	50.28 (110.6)	73.08	148.7	160.4 (63.1)	171.3
15	43.11	56.71 (124.8)	79.12	155.2	169.0 (66.5)	181.9	40.99	53.68 (118.1)	77.78	150.5	161.8 (63.7)	172.8
16	47.74	62.10 (136.6)	85.62	161.1	173.5 (68.3)	185.4	43.41	55.89 (123.0)	80.99	151.6	162.4 (63.9)	173.3
17	51.50	66.31 (145.9)	91.31	164.9	176.2 (69.4)	187.3	44.74	56.69 (124.7)	82.46	152.7	163.1 (64.2)	173.5
18	53.97	68.88 (151.5)	95.76	165.7	176.8 (69.6)	187.6	45.26	56.62 (124.6)	82.47	153.6	163.7 (64.4)	173.6

*From: Recommended Dietary Allowances, 10th ed. Food and Nutrition Board, National Research Council—National Academy of Sciences, 1989. Source: Hamill et al., 1979.

APPENDIX IV

TABLE IV-1

Recommended Daily Dietary Allowances *†

Category	Age (yr) or Condition	Weight‡ (kg)	Weight‡ (lb)	Height‡ (cm)	Height‡ (in)	Protein (g)	Fat-Soluble Vitamins: Vitamin A (μg RE)§	Vitamin D (μg)‖	Vitamin E (mg α-TE)¶	Vitamin K (μg)#
Infants	0.0–0.5	6	13	60	24	13	375	7.5	3	5
	0.5–1.0	9	20	71	28	14	375	10	4	10
Children	1–3	13	29	90	35	16	400	10	6	15
	4–6	20	44	112	44	24	500	10	7	20
	7–10	28	62	132	52	28	700	10	7	30
Males	11–14	45	99	157	62	45	1,000	10	10	45
	15–18	66	145	176	69	59	1,000	10	10	65
	19–24	72	160	177	70	58	1,000	10	10	70
	25–50	79	174	176	70	63	1,000	5	10	80
	51+	77	170	173	68	63	1,000	5	10	80
Females	11–14	46	101	157	62	46	800	10	8	45
	15–18	55	120	163	64	44	800	10	8	55
	19–24	58	128	164	65	46	800	10	8	60
	25–50	63	138	163	64	50	800	5	8	65
	51+	65	143	160	63	50	800	5	8	65
Pregnant						60	800	10	10	65
Lactating	1st 6 months					65	1,300	10	12	65
	2nd 6 months					62	1,200	10	11	65

*From: Recommended Dietary Allowances, 10th ed. Food and Nutrition Board, National Research Council—National Academy of Sciences, 1989.

†The allowances, expressed as average daily intakes over time, are intended to provide for individual variations among most normal persons as they live in the United States under usual environmental stresses. Diets should be based on a variety of common foods in order to provide other nutrients for which human requirements have been less well defined.

‡Weights and heights of Reference Adults are actual medians for the U.S. population of the designated age, as reported by NHANES II. The median weights and heights of those under 19 years of age were taken from Hamill et al., 1979. The use of these figures does not imply that the height-to-weight ratios are ideal.

§Retinol equivalents. 1 retinol equivalent = 1 μg of retinal or 6 μg of β-carotene.

‖As cholecalciferol. 10 μg of cholecalciferol = 400 IU of vitamin D.

¶α-Tocophenol equivalents. 1 mg of d-α-tocopherol = 1 α-TE.

#1 NE (niacin equivalent) is equal to 1 mg of niacin or 60 mg of dietary tryptophan.

TABLE IV-2

Water-Soluble Vitamins							Minerals						
Vitamin C (mg)	Thiamine (mg)	Riboflavin (mg)	Niacin (mg NE)#	Vitamin B_6 (mg)	Folate (μg)	Vitamin B_{12} (μg)	Calcium (mg)	Phosphorus (mg)	Magnesium (mg)	Iron (mg)	Zinc (mg)	Iodine (μg)	Selenium (μg)
30	0.3	0.4	5	0.3	25	0.3	400	300	40	6	5	40	10
35	0.4	0.5	6	0.6	35	0.5	600	500	60	10	5	50	15
40	0.7	0.8	9	1.0	50	0.7	800	800	80	10	10	70	20
45	0.9	1.1	12	1.1	75	1.0	800	800	120	10	10	90	20
45	1.0	1.2	13	1.4	100	1.4	800	800	170	10	10	120	30
50	1.3	1.5	17	1.7	150	2.0	1,200	1,200	270	12	15	150	40
60	1.5	1.8	20	2.0	200	2.0	1,200	1,200	400	12	15	150	50
60	1.5	1.7	19	2.0	200	2.0	1,200	1,200	350	10	15	150	70
60	1.5	1.7	19	2.0	200	2.0	800	800	350	10	15	150	70
60	1.2	1.4	15	2.0	200	2.0	800	800	350	10	15	150	70
50	1.1	1.3	15	1.4	150	2.0	1,200	1,200	280	15	12	150	45
60	1.1	1.3	15	1.5	180	2.0	1,200	1,200	300	15	12	150	50
60	1.1	1.3	15	1.6	180	2.0	1,200	1,200	280	15	12	150	55
60	1.1	1.3	15	1.6	180	2.0	800	800	280	15	12	150	55
60	1.0	1.2	13	1.6	180	2.0	800	800	280	10	12	150	55
70	1.5	1.6	17	2.2	400	2.2	1,200	1,200	320	30	15	175	65
95	1.6	1.8	20	2.1	280	2.6	1,200	1,200	355	15	19	200	75
90	1.6	1.7	20	2.1	260	2.6	1,200	1,200	340	15	16	200	75

TABLE V-1
Classification of Selected Enzymes with Clinical Importance According to the Enzyme Commission (EC) and Their Common Names

EC Code	Systematic Name	Common Name (with Some Abbreviations)
1	Oxidoreductases	
1.1	Acting on CHOH groups	
1.1.1	With NAD^+ or $NADP^+$ as hydrogen acceptor	
1.1.1.1	alcohol:NAD^+ oxidoreductase	Alcohol dehydrogenase (AD)
1.1.1.14	L-iditol:NAD^+ oxidoreductase	Sorbitol dehydrogenase (SD or ID)
1.1.1.27	L-lactate:NAD^+ oxidoreductase	Lactate dehydrogenase (LD or LDH)
1.1.1.37	L-malate:NAD^+ oxidoreductase	Malate dehydrogenase (MD)
1.1.1.42	threo-D_s-isocitrate-NAD^+ oxidoreductase (decarboxylating)	Isocitrate dehydrogenase (ICD)
1.1.1.44	6-phospho-D-gluconate: $NADP^+$ 2-oxido-reductase (decarboxylating)	Phosphogluconate dehydrogenase
1.1.1.49	D-glucose 6-phosphate: $NADP^+$ 1-oxidoreductase	Glucose-6-phosphate dehydrogenase (GDP or G6PD)
1.2	Acting on aldehyde or keto groups	
1.2.1	With NAD^+ or $NADP^+$ as acceptor	
1.2.1.12	D-glyceraldehyde 3-phosphate:NAD^+ oxidoreductase (phosphorylating)	Glyceraldehydephosphate (triosephosphate) dehydrogenase
1.4	Acting on $CH{\cdot}NH_2$ groups	
1.4.1	With NAD^+ or $NADP^+$ as acceptor	
1.4.1.3	L-glutamate:NAD(P) dehydrogenase (deaminating)	Glutamate dehydrogenase
1.6	Acting on NADH or NADPH	
1.6.4	With a disulfide as acceptor	
1.6.4.2	NAD(P)H:glutathione oxidoreductase	Glutathione reductase
2	Transferases	
2.1	Transferring one-carbon groups	
2.1.3	Carboxyl- and carbamoyltransferases	
2.1.3.3	Carbamoylphosphate:L-ornithine carbamoyltransferase	Ornithine carbamoyltransferase (OCT)
2.3	Amino acid transferases	
2.3.2.2	γ-glutamyl-peptide:amino acid γ-glutamyl transferase	γ-Glutamyltransferase (GGT or γ-GT)
2.6	Transferring nitrogenous groups	
2.6.1	Aminotransferases (transaminases)	
2.6.1.1	L-aspartate:2-oxoglutarate aminotransferase	Aspartate aminotransferase (AST) (also known as glutamate oxaloacetate transaminase, GOT)
2.6.1.2	L-alanine:2-oxoglutarate aminotransferase	Alanine aminotransferase (ALT) (also known as glutamate alanine transaminase, GPT)
2.7	Transferring phosphorus-containing groups	
2.7.1	Phosphotransferases with alcohol group as acceptor	
2.7.1.1	ATP:D-hexose 6-phosphotransferase	Hexokinase (HK)
2.7.1.40	ATP:pyruvate 2-0-phosphotransferase	Pyruvate kinase (PK)
2.7.3	Phosphotransferases with nitrogenous group as acceptor	
2.7.3.2	ATP:creatine N-phosphotransferase	Creatine kinase (CK) (also known as creatine phosphokinase, CPK)
2.7.4	Phosphotransferases with phospho-group as acceptor	
2.7.4.3	ATP:AMP phosphotransferase	Adenylate kinase
2.7.5	Phosphotransferases catalyzing intramolecular transfers	

(continued)

TABLE V-1 *(continued)*

EC Code	Systematic Name	Common Name (with Some Abbreviations)
2.7.5.1	α-D-glucose 1,6-diphosphate:α-D-glucose 1-phosphate phosphotransferase	Phosphoglucomutase
2.7.7	Nucleotidyltransferases	
2.7.7.12	UDP-glucose:α-D-galactose 1-phosphate uridylyltransferase	Hexose-1-phosphate uridylyltransferase
3	Hydrolases	
3.1	Acting on esters	
3.1.1	Carboxylic ester hydrolases	
3.1.1.3	triacylglycerol acyl-hydrolase	Lipase
3.1.1.7	acetylcholine hydrolase	Acetylcholinesterase
3.1.1.8	acylcholine acyl-hydrolase	Cholinesterase
3.1.3	Phosphoric monoester hydrolases	
3.1.3.1	orthophosphoric monoester phosphohydrolase	Alkaline phosphatase (ALP)
3.1.3.2	orthophosphoric monoester phosphohydrolase	Acid phosphatase
3.1.3.5	5′-ribonucleotide phosphohydrolase	5′-Nucleotidase (5′-NT)
3.1.3.9	D-glucose 6-phosphate phosphohydrolase	Glucose-6-phosphatase
3.2	Acting on glycosyl compounds	
3.2.1	Glucoside hydrolases	
3.2.1.1	1,4-α-D-glucan glucanhydrolase	α-Amylase
3.2.1.31	β-D-glucuronide glycuronosohydrolase	β-Glucuronidase
3.4	Acting on peptide bonds	
3.4.1	α-Aminopeptide amino acid hydrolases	
3.4.11.1	α-Aminoacyl-peptide hydrolase	Leucine aminopeptidase
3.4.4	Peptide peptidohydrolases	
3.4.23.1	no systematic name	Chymotrypsin
3.4.21.4	no systematic name	Trypsin
3.4.21.1	no systematic name	Pepsin
3.5	Acting on C–N bonds, other than peptide bonds	
3.5.3	In linear amidines	
3.5.3.1	L-arginine amidinohydrolase	Arginase
4	Lyases	
4.1	C–C lyases	
4.1.1	Carboxy lyases	Pyruvate decarboxylase
4.1.2	Aldehyde lyases	Aldolase
4.2	C–O lyases	
4.2.1	Hydrolyases	Fumarate hydratase (= fumarase)
4.3	C–N lyases	Histidine-ammonia lyase (= histidase)
5	Isomerases	
5.1	Racemases and epimerases	
5.1.3	Acting on carbohydrates	Ribulose-5-phosphate epimerase
5.2	*cis-trans* isomerases	Maleylacetoacetate isomerase
5.3	Intramolecular oxidoreductases	
5.3.1	Interconverting aldose and ketoses	Glucosephosphate isomerase
5.4	Intramolecular transferases	Methylmalonyl-CoA mutase
6	Ligases	
6.1	Forming C–O bonds	
6.1.1	Amino acid–RNA ligases	Amino acid–activating enzymes
6.3	Forming C–N bonds	
6.3.1	Acid–ammonia ligases	Glutamine synthetase
6.3.2	Acid–amino acid ligases	Peptide synthetase, glutathione synthetase
6.4	Forming C–C bonds	
6.4.1	Carboxylases	Acetyl-CoA carboxylase

APPENDIX VI

VI.1 Serum Protein Electrophoresis and Its Diagnostic Significance

Serum contains more than 100 different proteins, each under separate genetic control. They are transport proteins for hormones, vitamins, lipids, metals, pigments, and drugs; enzymes; enzyme inhibitors (proteinase inhibitors); hormones; antibodies; clotting factors; complement components; and kinin precursors. Quantitation (by radial immunodiffusion, electroimmunoassay, nephelometric methods, enzyme-linked immunological methods, and radioimmunoassay) of the various constituents of serum is of value in diagnosing and following the course of certain diseases. Several of these proteins are discussed elsewhere in the text.

A simple and useful technique involves separation of serum proteins by an electric field at pH 8.6, using cellulose acetate as a support medium (electrophoresis). Agarose gel electrophoresis provides a higher resolution in the separation of proteins than cellulose acetate electrophoresis. The former gives about 12 protein bands and the latter provides the basic five-band pattern. Separation of these proteins is possible because each carries different charges and hence migrates at a differing rate when subjected to an electric potential. Serum is generally used instead of plasma because the fibrinogen found in plasma appears as a narrow band in the β region, which may be mistaken for the sign of monoclonal paraproteinemia (see below). The support medium, cellulose acetate, possesses several advantages over paper: the time required to separate the major proteins is short, albumin trailing is absent, and rapid quantitative determination of protein fractions by photoelectric scanning (after a suitable staining procedure) is possible. The factors that affect electrophoresis are ionic strength of the buffer, voltage, temperature, application width, and staining.

Figure VI-1 shows normal and some abnormal patterns of serum protein electrophoresis. The electrophoretic patterns obtained are *not* indicative of any one disease or class of disease. Furthermore, a characteristic pattern may be obscured or not found in a disease entity where normally such a pattern is expected. Serum electrophoretic patterns provide only a general impression of the disorder and require confirmation by other procedures. An alteration (depression or elevation) in a given fraction should be quantitated by more sensitive and specific methods.

Five major fractions seen in cellulose acetate serum protein electrophoresis are albumin, α_1-globulin, α_2-globulin, β-globulin, and γ-globulin. Adult reference ranges for these five fractions, expressed as grams per 100 mL, are 3.2–5.6 for albumin, 0.1–0.4 for α_1-globulin, 0.4–0.9 for α_2-globulin, 0.5–1.1 for β-globulin, and 0.5–1.6 for γ-globulin. With the exception of γ-globulin, adult values are attained by 3 months of age for all fractions. Cord blood

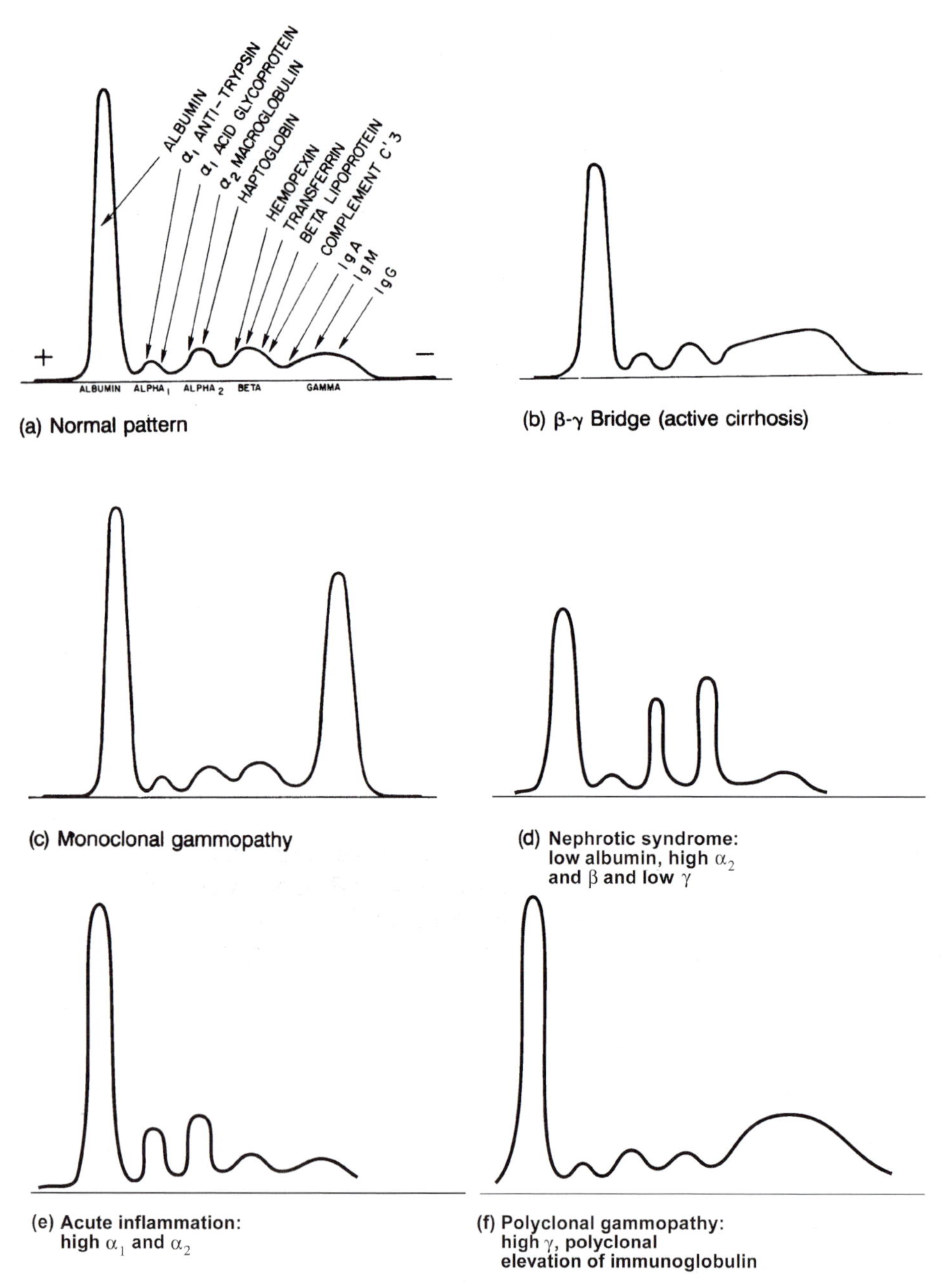

FIGURE VI-1
Serum protein electrophoretic patterns. [Pattern (a) is reproduced, with permission, from Helena Laboratories, Beaumont, Texas.]

γ-globulin is largely of maternal origin and undergoes catabolism, reaching its lowest value at about 3 months. Adult levels for γ-globulin are not reached until about 2–7 years. After age 40, the level of albumin gradually declines and the β-globulin fraction increases.

Albumin

Albumin is synthesized in the liver. Since albumin accounts for about 75% of the osmotic pressure in blood, it is responsible for the stabilization of blood volume and regulation of vascular fluid exchange. Therefore, hypoalbuminemia can give rise to edema. **Hyperalbuminemia** is uncommon, but many types of abnormalities lead to **hypoalbuminemia.** Some hypoalbuminemic conditions include nephrotic syndrome, proteinlosing enteropathies, cystic fibrosis (hypoalbuminemia with edema may be an early abnormality in infants with this disease), glomerulonephritis, cirrhosis, carcinomatosis, bacterial infections, viral hepatitis, congestive heart failure, rheumatoid arthritis, uncontrolled diabetes, intravenous feeding (the hypoalbuminemia is due to a deficiency of amino acids in portal blood), and dietary deficiency of proteins containing essential amino acids.

Hypoalbuminemia also occurs in the acute stress reaction. The sharp drop of albumin is essentially due to adrenocortical stimulation, which gives rise to enhanced catabolism of albumin and sodium retention. The latter is responsible for hemodilution and expansion of extracellular fluid. The α_1- and α_2-globulin fractions (see below) are also increased in the acute stress reaction. The immunoglobulins (γ-fraction) do not undergo significant alterations during the acute phase of a stress reaction but are elevated during the chronic phase. Some examples of situations in which acute stress patterns may be seen are acute infections in early stages, tissue necrosis (myocardium, renal, tumor), severe burns, rheumatoid disease of acute onset, surgery, and collagen disorders.

Analbuminemia is a rare autosomal recessive disorder. Affected individuals do not exhibit serious clinical symptoms, not even edema. The lack of clinical edema is presumably due to osmotic compensation by the mildly elevated globulins. Osteoporosis in analbuminemia has been corrected by the administeration of human serum albumin. Affected females exhibit minimal pretibial edema, mild anemia, normal liver function tests, absence of proteinuria, lowered blood pressure, elevated serum cholesterol levels, and lipodystrophy. Despite elevated plasma cholesterol levels, severe atherosclerosis was not present.

Bisalbuminemia is due to albumin polymorphism. Upon serum electrophoresis, a double albumin band is seen. These double peaks are due either to differences in electrical charge (11 electrophoretically distinct forms have been reported) or to albumin dimers. Both forms are expressed as autosomal dominant traits and apparently present no significant clinical abnormalities. However, acquired bisalbuminemia may be associated with diabetes mellitus, the nephrotic syndrome, hyperamylasemia, and penicillin therapy. In these instances, the bisalbuminemia disappears after correction of the underlying disorder.

α_1-Globulins

The proteins that migrate in this region are α_1-antitrypsin (which accounts for 70–90% of the α_1-peak), α_1-acid glycoprotein, or orosomucoid (which accounts for 10–20% of the α_1-peak), α_1-lipoprotein, prothrombin, transcortin, and thyroxine-binding globulin. In acute phase reactions, elevation of the α_1-peak is due to increases in α_1-antitrypsin and α_1-acid glycoprotein, the two principal constituents of the peak. α_1-Peak elevations are also seen in chronic inflammatory and degenerative diseases and are highly elevated in some cancers. When the α_1-peak is depressed, α_1-antitrypsin deficiency (Chapter 6) must be ruled out by direct measurements using sensitive quantitative methods. The presence of α_1-fetoprotein (AFP) in nonfetal serum is of clinical significance because of its close association with primary carcinoma of the liver. However, during pregnancy, AFP synthesized by the fetus gains access in small but measurable amounts to maternal circulation. Measurement of maternal serum AFP (MSAFP) levels during the second trimester is widely used to detect fetal abnormalities. In neural tube defects, MSAFP levels are elevated, and in Down's syndrome, they are decreased (Chapter 38). Routine serum electrophoresis seldom gives such sensitive information, and specific immunochemical methods should be employed to detect α_1-fetoprotein in the serum.

α_2-Globulin

The α_2 fraction includes haptoglobin, α_2-macroglobulin, α_2-microglobulin, ceruloplasmin, erythropoietin, and cholinesterase. Haptoglobin is nonspecifically increased in the acute stress reaction in the presence of inflammation, tissue necrosis, or destruction. A function of haptoglobin is to combine with hemoglobin to remove it from the circulation. Thus, during an episode of intravascular hemolysis, haptoglobin is depleted and may require a week or more to return to normal serum levels. Haptoglobin levels in these instances should be quantitated by immunochemical procedures. α_2-Macroglobulin functions as a protease inhibitor (Chapter 6). In the nephrotic syndrome, there is a characteristic elevation of both α_2 and β peaks with hypoalbuminemia (Figure VI-1d). The elevation in α_2 and β fractions is due to higher very-low-density lipoprotein (VLDL) and low-density lipoprotein (LDL) levels. Thus, characteristics of nephritic syndrome include proteinuria (>3.5 g per 24 hours), hypoalbuminemia (<3 g/dL), hyperlipidemia with elevated triglyceride and cholesterol levels, and lipiduria and generalized edema. The exact mechanism for the higher VLDL and LDL levels is not understood. In part, these elevations may be due to increased synthesis of VLDLs in the liver and then conversion to LDLs in the peripheral circulation (Chapter 20), as well as their decreased catabolism. It is thought that the decreased plasma oncotic pressure due to hypoalbuminemia may stimulate increased hepatic synthesis of VLDLs.

Increased glomerular permeability for protein due to derangement in capillary walls is an early event in the nephrotic syndrome that has several causes, as listed below with examples:

1. Primary glomerular disease: membranous glomerulonephritis, lipid nephrosis

2. Systemic disease: diabetes mellitus, amyloidosis, systemic lupus erythematosus
3. Neprotoxins and drugs: gold, mercury, penicillamine
4. Infections: poststreptococcal glomerulonephritis
5. Allergens, venoms, and vaccines
6. Miscellaneous: toxemia of pregnancy, malignant hypertension

The most common cause of primary nephrotic syndrome in adults is membranous nephropathy. It is characterized by deposition of immune complex in the subepithelial portion of glomerular capillary walls.

β-Globulins

Quantitatively significant β-globulins are composed of transferrin, β-lipoproteins, complement C3, and hemopexin. Transferrin, synthesized in the liver, has a half-life of about 8.8 days and is an iron transport protein in serum that accounts for about 60% of the β peak. It is increased in iron deficiency anemia and pregnancy (Chapter 29). Transferrin levels are decreased in metabolic (e.g., liver, kidney) and neoplastic diseases. Transferrin and transthyretin (Chapter 38) are both decreased in protein-calorie malnutrition. A pattern often observed in advanced cirrhosis is lack of separation of the β peak with the γ region, described as β-γ bridging (Figure VI-1). The pattern is presumably due to elevated levels of IgA. Hypoalbuminemia is also found. Complement levels (C3 and C4) are decreased (owing to their consumption) in diseases associated with the formation of immune complexes, e.g., glomerulonephritis (acute and membranoproliferative), systemic lupus erythematosus.

γ-Globulins

The five major classes of immunoglobulins in descending order of quantity are IgG, IgA, IgM, IgD, and IgE (see also Chapter 35). C-reactive protein migrates with the γ-globulins and is increased in trauma and acute inflammatory processes (discussed later). During an intense acute inflammatory process, its concentration may be highly elevated, giving rise to a sharp protein band that may be mistaken for monoclonal gammopathy. Variations in electrophoretic pattern due to γ-globulins can be categorized into the following three groups.

1. **Agammaglobulinemia and hypogammaglobulinemia:** May be primary or secondary. Secondary forms may be found in chronic lymphocytic leukemia, lymphosarcoma, multiple myeloma, the nephrotic syndrome, long-term steroid treatment, and occasionally overwhelming infection.
2. **Polyclonal gammopathy:** A diffuse polyclonal increase in γ-globulin, primarily in the IgG fraction (Figure VI-If). Recall that in chronic stress γ-globulins are elevated. The major causes of polyclonal gammopathy are listed below.
   a. **Chronic liver disease:** An increase in polyclonal IgG and IgA with a decrease in albumin levels is a characteristic finding regardless of the cause. The β-γ bridging observed in cirrhosis is presumably due to elevated levels of IgA that migrates to a position between the β and γ region, increasing the trough between these bands.
   b. **Sarcoidosis:** The electrophoretic pattern shows a stepwise descent of globulin fractions starting from the γ-globulins.
   c. **Autoimmune disease:** This group includes rheumatoid arthritis (increased IgA levels), systemic lupus erythematosus (increased IgG and IgM levels), and others.
   d. **Chronic infectious disease:** This group includes bronchiectasis, chronic pyelonephritis, malaria, chronic osteomyelitis, kala-azar, leprosy, and others.
3. **Monoclonal gammopathy:** Presence of narrow protein band in the β-γ region may indicate the

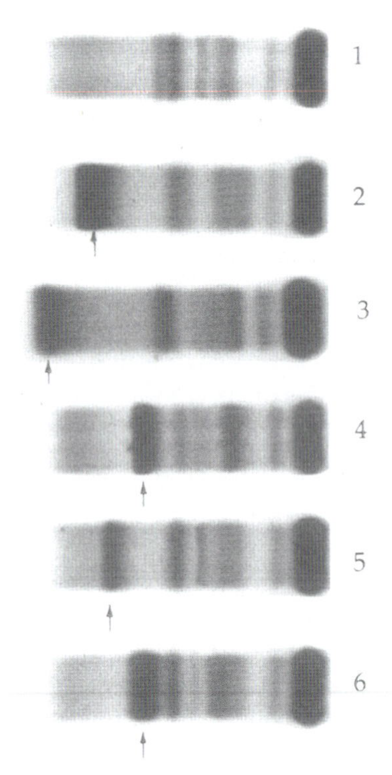

FIGURE VI-2
Serum electrophoretic protein patterns consisting of monoclonal immunoglobulins identified by arrows. Normal serum pattern is identified as 1.

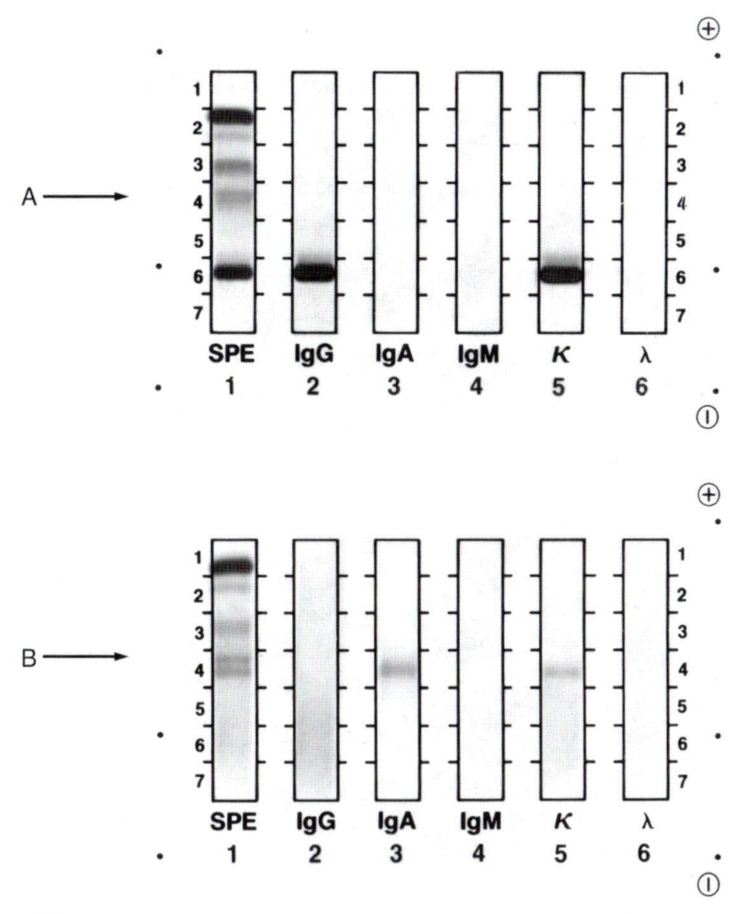

FIGURE VI-3
Serum immunofixation electrophoresis of two patients with monoclonal gammopathies. Pattern A represents IgG(κ) monoclonal gammopathy and pattern B represents IgA(κ) monoclonal gammopathy, as indicated by arrows. The procedure consists of serum protein electrophoresis (SPE) separation, reaction of each track with the exception of SPE with specific respective antiserum, followed by protein staining to make visible the respective bands.

presence of a homogeneous class of protein (Figures VI-1c and VI-2). Such a pattern should be further investigated using other laboratory and clinical findings to rule out multiple myeloma.

Means of investigation include quantitation of immunoglobins, immunofixation electrophoresis to identify the class and type of monoclonal immunoglobulin (Figure VI-3), bone marrow aspiration and biopsy to determine plasmacytosis, and radiological examination to identify osteolytic lesions. Multiple myeloma is a disorder of neoplastic proliferation of a single clone of plasma cells in the bone marrow that leads to overproduction of any one of the five immunoglobulins (IgG, IgA, IgM, IgD, or IgE) or light chains (κ or λ). Less commonly, biclonal or triclonal gammopathy also occurs. If the light chains are overproduced, they pass through the glomeruli and appear in the urine; this is known as *Bence Jones proteinuria*. In a given individual, Bence Jones proteinuria refers to either κ or λ light chains but not both. Persons over age forty may be affected and the initial findings may include spontaneous fractures, bone pain, anemia, or infections. Hypercalcemia may be found after bone destruction. Multiple myeloma is a progressive disorder and is usually fatal due to renal damage and/or recurrent infections. Cytotoxic measures (e.g., melphalan, a nitrogen mustard) are used in the treatment of multiple myeloma, and serum and/or urine monoclonal immunoglobulin or light chain levels serve in monitoring therapy.

Although the monoclonal spike is usually observed in the β-γ region, it is occasionally seen in the α_2 region. In pregnancy, a monoclonal spike in the β region may be due to elevated transferrin levels. A serum electrophoretic pattern of hypogammaglobulinemia requires further studies of serum and urine (e.g., immunoglobulin measurement and immunofixation electrophoresis) to rule out multiple myeloma. The presence of monoclonal immunoglobulin in the serum is a characteristic feature of a majority of multiple myeloma cases; however, 1–5% of patients do not show any detectable monoclonal protein in serum or urine. These disorders are known as nonsecretory myeloma. They fall into two classes: (a) producers of a paraprotein that remains in the cytosol of the malignant plasma cells due to secretory defects or (b) nonproducers of a paraprotein. Nonsecretory myelomas are characterized by immunoperoxidase staining methods using antibodies directed against light chain–specific immunoglobulins.

Monoclonal immunoglobulins in serum without significant clinical illness have also been found, and the incidence increases with age. As many as 3% of patients older than 70 show the presence of monoclonal immunoglobulin (benign monoclonal proteinemia, also known as monoclonal gammopathy of undermined significance). These individuals require periodic serum and urine studies. In some rare instances, monoclonal gammopathies of undetermined significance are associated with neuropathies and the monoclonal immunoglobulin possesses an antinerve activity. A lymphoproliferative disorder characterized by monoclonal lymphocytes that produce monoclonal IgM is known as *Waldenström's macroglobulinemia.* In this disorder, the elevated serum viscosity is palliated by plasmapheresis and cytotoxic treatment (e.g., chlorambucil and nucleoside analogs, respectively) due to the presence of large quantities of IgM and the tumor burden.

It should be noted that pseudomonoclonal bands appearing in the β-γ region may be misidentified as authentic monoclonal gammopathy. Pseudomonoclonal bands occur in those serum specimens due to hemolysis or to the presence of fibrinogen due to inappropriate blood coagulation techniques. Serum immunofixation studies should clarify these problems.

Secondary paraproteinemias may be seen in association with hematopoietic cancers (e.g., lymphomas and leukemias), other neoplasms (e.g., colon carcinoma), long-standing chronic urinary or biliary tract infection, rheumatoid factor related to IgM monoclonal protein, and amyloidosis.

Changes in Serum Proteins during Acute Phase of Disorders of Tissue Injury

The alterations in serum proteins (acute phase response) that occur within a few hours to a few days after tissue injury due to infection, trauma, burns, surgery, or infarction and inflammatory condition are divided into two categories. The first category includes those proteins that are increased by at least 25% (positive acute phase proteins), and the second category includes proteins that are decreased by at least 25% (negative acute phase proteins). During acute stress, macrophages and monocytes are activated and produce several intercellular signaling polypeptides, known as cytokines (Chapter 35). Some examples of the released cytokines are interleukin-6, interleukin-1β, tumor necrosis factor α, interferon-γ, and transforming growth factor β. One of the functions of these cytokines is to alter the synthesis of acute phase proteins in hepatocytes by regulating expression of acute phase protein genes by both transcriptional and posttranscriptional mechanisms. Glucocorticoids stimulate the action of cytokines by promoting the production of some acute phase proteins. The postulated function of the acute phase response is to protect the body from the injurious processes. One action is activation of the complement system to fight infection or antagonize the activity of proteolytic enzymes. Other actions initiate or sustain inflammation, whereas others reflect antiinflammatory and antioxidant properties.

Examples of positive and negative acute phase proteins are given in Table VI-1. C-reactive protein and serum amyloid A are elevated in serum by as much as 1000-fold from their basal values. Serum amyloid A is an apolipoprotein; it is synthesized in hepatocytes in response to inflammatory stimuli and associated with HDL. The function of serum amyloid A is not clear and it is not commonly measured as an acute phase reactant. However, serum C-reactive protein (so named because it reacts with pneumococcal C-polysaccharide) is measured. It binds with phosphocholine of pathogens, phospholipid constituents of damaged blood cells, and phagocytic cells, and it activates the complement system. All of the functions of C-reactive protein modulate inflammatory conditions of the body. These

TABLE VI-1
Examples of Positive and Negative Acute Phase Proteins

Positive

1. Several members of complement system (e.g., C3, C4)
2. Proteins of the coagulation and fibrinolytic systems (e.g., fibrinogen, plasminogen)
3. Antiproteases (e.g., α_1-antitrypsin)
4. Transport proteins (e.g., haptoglobin, ceruloplasmin, hemopexin)
5. Inflammatory response modulators and others (e.g., phospholipase A_2, C-reactive protein, amyloid A, fibronectin, ferritin)

Negative

Albumin
Transferrin
Transthyretin
Thyroxine-binding globulin

include induction of inflammatory cytokines, prevention of adhesion of neutrophils to endothelial cells by inhibiting the surface expression of L-selectin, and inhibition of superoxide production by neutrophils. Measurement of serum C-reactive protein is useful in differentiating an acute inflammatory condition from a noninflammatory one, as well as in the assessment of the severity of inflammation and its prognosis. Another widely used test in the assessment of acute phase disorders is erythrocyte sedimentation rate (ESR). ESR determines the rate at which erythrocytes fall through the plasma to the bottom (sediment) of the test tube, and it depends largely on the plasma concentration of fibrinogen, an acute phase reactant (Table VI-1). Thus, ESR is decreased during acute phase disorder. Compared to ESR measurement, serum C-reactive protein measurement has several advantages. ESR changes occur relatively slowly and increase with age, whereas C-reactive protein concentrations change rapidly and levels do not change with age. Since inflammation may play a role in cardiovascular disorders, measurement of a serum inflammation marker, such as C-reactive protein, is useful as a predictor of subsequent coronary events (Chapter 20). In one prospective study, healthy subjects with higher baseline serum C-reactive protein levels had increased risk of myocardial infarction and ischemic stroke. The use of aspirin, an antiinflammatory agent (Chapter 18), also was associated with reduction in the risk of myocardial infarction. Another inflammatory marker, lipoprotein-associated phospholipase A_2, is an independent predictor of risk for abnormal cardiovascular events.

Acute phase phenomena also include a variety of metabolic and neuroendocrine changes. For example, fever is a neuroendocrine change that occurs as an acute phase response.

Supplemental Readings and References

G. C. Blobe, W. P. Schiemann, and H. F. Lodish: Role of transforming growth factor β in human disease. *New England Journal of Medicine* **342,** 1350 (2000).

C. Gabay and I. Kushner: Acute-phase proteins and other systemic responses to inflammation. *New England Journal of Medicine* **340,** 448 (1999).

J-M. Halimi, J Ribstein, G. Du Cailar, et al.: Nephrotic-range proteinuria in patients with renovascular disease. *American Journal of Medicine* **108,** 120 (2000).

E. Kallee: Bennhold's analbuminemia: a follow up study of the first two cases (1953–1992). *Journal of Laboratory and Clinical Medicine* **127,** 470 (1996).

R. A. Kyle and M. A. Gertz: Monoclonal gammopathies and related disorders. *Hematology/Oncology Clinics of North America* **13** (1999). This complete volume is dedicated to topics on monoclonal gammopathies.

D. A. Morrow and P. M. Ridker: High sensitivity c-reactive protein (hs-CRP): a novel risk marker in cardiovascular disease. *Preventive Cardiology* **1,** 13 (1999).

C. J. Packard, D. S. J. O'Reilly, M. Caslake, et al.: Lipoprotein-associated phospholipase A_2 as an independent predictor of coronary heart disease. *New England Journal of Medicine* **343,** 1148 (2000).

D. J. Rader: Inflammatory markers of coronary risk. *New England Journal of Medicine* **343,** 1179 (2000).

P. M. Ridker and P. Haughie: Prospective studies of c-reactive protein as a risk factor for cardiovascular disease. *Journal of Investigative Medicine* **46,** 391 (1998).

P. M. Ridker, C. H. Hennekens, J. E. Buring, et al.: C-reactive protein and other markers of inflammation in the prediction of cardiovascular disease in women. *New England Journal of Medicine* **342,** 836 (2000).

APPENDIX **VII**

VII.1 Laboratory Evaluation of Hemoglobin Disorders

The differential diagnosis of hemoglobinopathies and some thalassemias is made in the laboratory. Some of the laboratory methods are described below.

Electrophoretic Procedures

Most of the *common* hemoglobin mutants can be identified by electrophoretically separating the hemoglobins on cellulose acetate at pH 8.4–8.6 and on citrate agar at pH 6.0–6.2. At pH 8.4, the hemoglobins are negatively charged and migrate toward the anode (positive electrode) in an electric field at a rate dependent on their net charge (see Figure VII-1 and Table VII-1). Separation of the hemoglobins with citrate agar as the support medium is based on net charge and adsorption of the hemoglobin to the supporting gel. The degree of adsorption may depend on hemoglobin solubility and may be due to an impurity in the agar rather than to the agar itself. Patterns obtained on citrate agar with various hemoglobins are shown in Figures VII-2 through VII-4.

The value of using these two electrophoretic systems can be seen in the following examples.

(1) Hbs S, D, and G migrate together about halfway between Hbs A and A_2 on cellulose acetate electrophoresis. With citrate agar gels, Hbs D and G migrate cathodally with HbA, while HbS continues to migrate anodally separating it from the other hemoglobins.
(2) Similarly, HbC comigrates with Hbs A_2, E, O, and C–Harlem on cellulose acetate but is readily separated from them on citrate agar.
(3) On cellulose acetate at pH 8.4, HbC–Harlem (two mutations, each changing the net charge on the tetramer by +2) migrates with HbC (one mutation, change of +4). These two hemoglobins are well separated from each other on citrate agar.

Two modifications of classical hemoglobin electrophoresis are also useful for studying these proteins: isoelectric focusing and electrophoresis of individual globin chains, with or without the heme. These techniques can be used to identify some hemoglobins that are not separable by electrophoresis of the complete hemoglobin tetramers.

Nonelectrophoretic Procedures

It is apparent from the above discussion that many hemoglobins with different amino acid substitutions demonstrate identical electrophoretic mobilities. Methods that rely on differences other than net charge are needed to establish the identity of these hemoglobins. Definitive

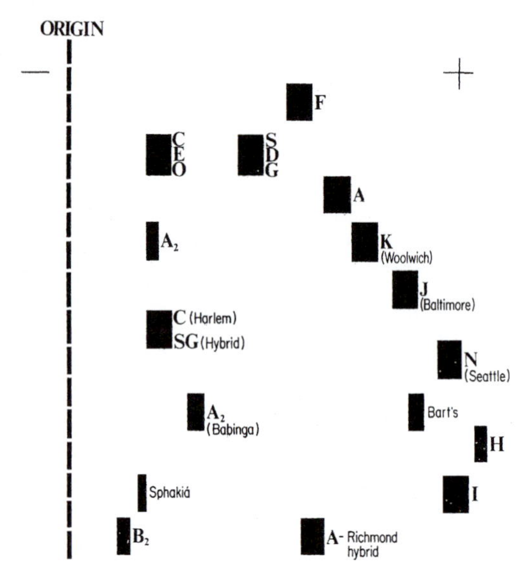

FIGURE VII-1
Relative mobilities of some hemoglobins on cellulose acetate (pH 8.4).

molecular biological techniques employed in the evaluation of hemoglobin disorders are discussed in Chapters 23 and 28.

HbS can be differentiated from most other hemoglobins by its **insolubility upon deoxygenation.** The test is carried out in the presence of a reducing agent (sodium dithionite or sodium metabisulfite, which consumes dissolved oxygen) in an appropriate buffer. When HbS is added, the solution becomes turbid owing to precipitation of deoxy-HbS. The turbidity can be quantitated spectrophotometrically and related to the amount of HbS added. By using the same reagents with intact erythrocytes, HbS can be precipitated within the cells, causing them to sickle. The extent of sickling can be assessed by microscopic examination. Because HbC–Harlem and Hb–Bart's behave like HbS in the solubility test, they may give false-positive results in tests for HbS.

HbF is identified and quantitated by its **resistance to denaturation by strong alkali.** The hemolysate is mixed with an alkaline buffer (pH 12.8) and incubated to denature the nonfetal hemoglobins. The denatured hemoglobin is

TABLE VII-1
Amino Acid Substitutions and Net Charge Alterations in Hemoglobins Whose Relative Mobilities Are Shown in Figure VII-1

Hemoglobin	Amino Acid Change*	Charge Alteration in Tetramer*
C	Glu→Lys	+4
E	Glu→Lys	+4
O–Arab	Glu→Lys	+4
C–Harlem	{Glu⁶→Val = Asp→Asn}	+4
S	Glu→Val	+2
G–San Jose	Glu→Gly	+2
D–Los Angeles	Glu→Gln	+2
Sydney	Val→Ala	0
I–Philadelphia	Lys→Glu	−4

*As compared to HbA. Hemoglobins A_2, F, Bart's and H have multiple amino acid differences from HbA, and their electrophoretic mobilities depend on the net effect of all these differences on their net charge at pH 8.4–8.6.

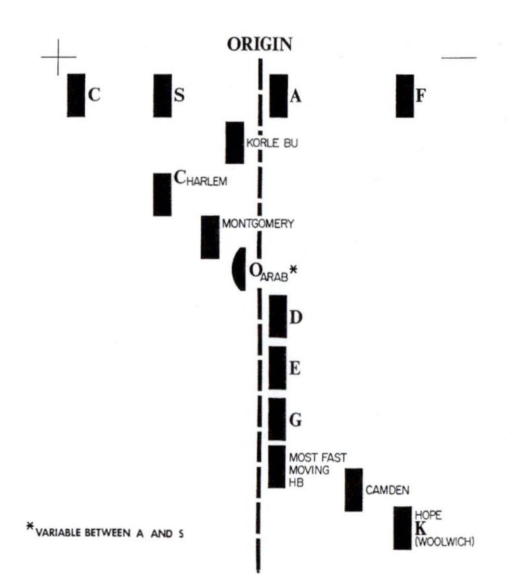

FIGURE VII-2
Relative mobilities of some hemoglobins on citrate agar (pH 6.0). The differences between this gel and cellulose acetate are the differential interactions of the hemoglobins with the gel and the changes in net charge caused by the lower pH.

Alpha Chain Mutants

Hemoglobin	Structure	Mobility in Citrate Agar pH 6.0 (+ C S A F −)
J Paris	α 12 Ala → Asp	
J Oxford	α 15 Gly → Asp	
I	α 16 Lys → Glu	
Fort Worth	α 27 Glu → Gly	
Hasharon (Sealy)	α 47 Asp → His	
Montgomery	α 48 Leu → Arg	
Russ	α 51 Gly → Arg	
Shimonoseki	α 54 Gln → Arg	
G-Philadelphia	α 68 Asn → Lys	
Winnipeg	α 75 Asp → Tyr	
Inkster	α 85 Asp → Val	
Broussais	α 90 Lys → Asn	
Titusville	α 94 Asp → Asn	
G-Georgia	α 95 Pro → Leu	
Rampa	α 95 Pro → Ser	

FIGURE VII-3
Mobilities of α-chain mutants on citrate agar electrophoresis relative to hemoglobins C, S, A, and F. (Courtesy of Dr. Rose Schnieder. Reproduced with permission from Helena Laboratories, Beaumont, Texas.)

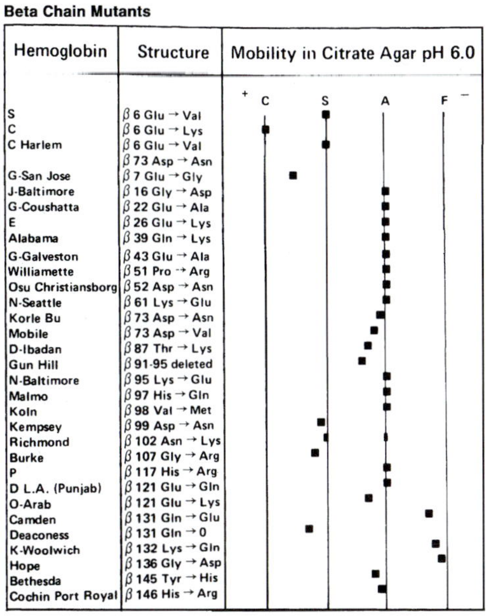

Beta Chain Mutants

Hemoglobin	Structure	Mobility in Citrate Agar pH 6.0
		+ C S A F −
S	β 6 Glu → Val	
C	β 6 Glu → Lys	
C Harlem	β 6 Glu → Val β 73 Asp → Asn	
G-San Jose	β 7 Glu → Gly	
J-Baltimore	β 16 Gly → Asp	
G-Coushatta	β 22 Glu → Ala	
E	β 26 Glu → Lys	
Alabama	β 39 Gln → Lys	
G-Galveston	β 43 Glu → Ala	
Williamette	β 51 Pro → Arg	
Osu Christiansborg	β 52 Asp → Asn	
N-Seattle	β 61 Lys → Glu	
Korle Bu	β 73 Asp → Asn	
Mobile	β 73 Asp → Val	
D-Ibadan	β 87 Thr → Lys	
Gun Hill	β 91-95 deleted	
N-Baltimore	β 95 Lys → Glu	
Malmo	β 97 His → Gln	
Koln	β 98 Val → Met	
Kempsey	β 99 Asp → Asn	
Richmond	β 102 Asn → Lys	
Burke	β 107 Gly → Arg	
P	β 117 His → Arg	
D L.A. (Punjab)	β 121 Glu → Gln	
O-Arab	β 121 Glu → Lys	
Camden	β 131 Gln → Glu	
Deaconess	β 131 Gln → 0	
K-Woolwich	β 132 Lys → Gln	
Hope	β 136 Gly → Asp	
Bethesda	β 145 Tyr → His	
Cochin Port Royal	β 146 His → Arg	

FIGURE VII-4
Mobilities of β-chain mutants on citrate agar electrophoresis relative to hemoglobins C, S, A, and F. (Courtesy of Dr. Rose Schnieder. Reproduced with permission from Helena Laboratories, Beaumont, Texas.)

then precipitated by addition of ammonium sulfate, and the undenatured HbF in the clear supernatant is measured by spectrophotometry.

HbA_2 can be separated from most other common hemoglobins by **anion exchange chromatography** using diethylaminoethylcellulose. Separation results from differences in the interactions of the charged groups of the various hemoglobins with the positively charged groups on the anion exchange resin. Following separation, the HbA_2 can be quantitated by spectrophotometry. Although several other hemoglobins (including C, E, O, and D) are eluted with HbA_2, it is unlikely that one of these hemoglobins would be present in a person being tested for elevated HbA_2. HbA_2 measurement is used in the diagnosis of β-thalassemia trait, in which the HbA_2 is elevated to about twice the normal level (the upper normal limit is about 3–3.5%).

HbH and many unstable hemoglobins spontaneously precipitate within the red cells, forming **Heinz bodies,** which can be detected in splenectomized patients by staining with methylene blue. Alternatively, **precipitation** of these hemoglobins can be induced and the precipitates visualized by incubation of the erythrocytes with a redox dye such as brilliant cresyl blue (Chapter 28).

Many abnormal hemoglobins are also much more readily **denatured by heat** than normal hemoglobins. Heating an unstable hemoglobin for 30 minutes at 60°C usually causes complete denaturation, whereas HbA is hardly precipitated at all under the same conditions. The sulfhydryl groups of the mutant globins are generally more exposed than those in normal hemoglobin, making them **more reactive toward parachloromercuribenzoate** (PCMB). Treatment with PCMB for several hours precipitates many mutant hemoglobins, whereas it only causes dissociation of HbA.

A sensitive method for identifying changes in the primary structure of hemoglobin is **peptide mapping** or **fingerprinting** (Chapter 3). High-performance liquid chromatographic techniques have been used as a sensitive method to detect various hemoglobins, including fetal hemoglobins γ^A and γ^G.

Molecular biology techniques, such as Southern blot analysis and polymerase chain reaction (PCR), have been used in the diagnosis of hemoglobinopathies (see the following discussion).

Thalassemia Syndromes Thalassemia syndromes (see also Chapter 28) are a group of heterogeneous inherited disorders. They are relatively common in persons of Mediterranean, African, and Southeast Asian ancestry and are characterized by an unbalanced and defective rate of (absent or reduced) synthesis of one or more globin chains of hemoglobin. Defects in α-globin chain synthesis are designated as α-thalassemia and defects in β-globin chain synthesis are designated as β-thalassemia. The clinical manifestations in thalassemia syndromes occur due to inadequate hemoglobin synthesis and the accumulation of unused globin subunits due to unbalanced synthesis. The former leads to hypochromic microcytic anemia and the latter leads to inaffective erythropoiesis and hemolytic anemia. The clinical severity of these disorders depends on the nature of mutation. They range from asymptomatic mild hypochromic microcytosis (α-thalassemia silent trait) to early childhood mortality (homozygous β-thalassemia) to *in utero* death (Hb–Bart's hydrops fetalis).

β-Thalassemia: The β and β-like genes are located in chromosome 11 and their organization is discussed in Chapter 28. In homozygous β-thalassemia syndrome the β-globin chain synthesis is either absent (β^0) or severely reduced (β^+). The molecular defects are numerous and include gene deletion, as well as defects in transcription, processing, transport, and translation of mRNA. The clinical symptoms manifest after 6 months of age, when the switch of synthesis from HbF to HbA occurs. Numerous clinical problems follow and require transfusion therapy with its consequent complications (e.g., iron overload; see Chapter 29). Heterozygous β-thalassemia (β-thalassemia trait) is accompanied by mild hypochromic microcytic anemia without any significant clinical abnormalities. β-Thalassemia trait can be identified by elevated

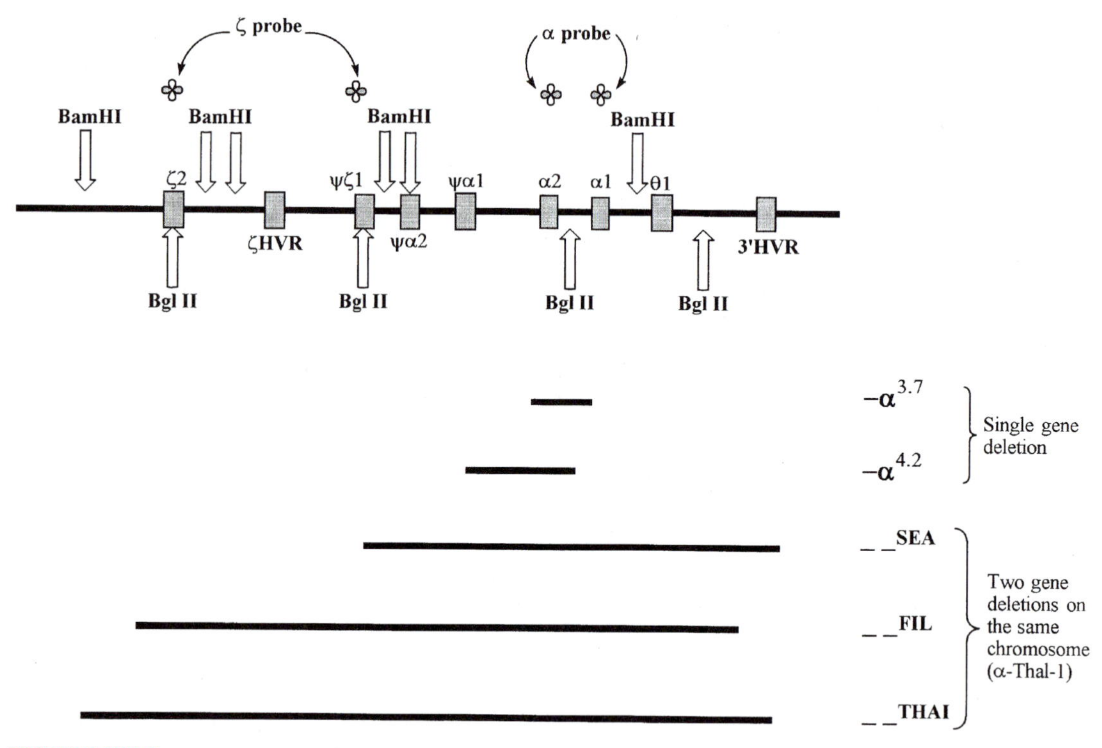

FIGURE VII-5
Restriction enzyme map of α-globin gene cluster and some of the deletions (indicated as dark lines).

HbA_2 (>3.5%) and, in some mutations, by HbF and normal serum iron studies.

α-Thalassemia: The α-globin gene cluster is found in chromosome 16 and each chromosome contains two α genes (α_2 and α_1). Thus each individual has four copies of α gene. The genes are present in the order: 5′-ζ2-$\psi\zeta$1-$\psi\alpha$2-$\psi\alpha$1-α_2-α_1-θ1-3′ (Figure VII-5). The ψ genes (pseudogenes) and θ1 gene are nonfunctional. The ψ genes are thought to be relics of the past evolutionary gene mutations. The θ1 gene does not yield a viable globin protein product and its function is not known. α-Thalassemia is probably the most common genetic defect, and its prevalence correlates with geographic areas of malaria incidence. The clinical manifestations of α-thalassemia depends on the number of genes affected. Gene deletions are the most common defect, and point mutations account only for about 5% of α-thalassemia.

α-Thalassemia silent carrier ($\alpha\alpha/\alpha$-): This genotype is caused by a single α_2-globin gene defect, primarily due to $-\alpha^{3.7}$ or $-\alpha^{4.2}$ deletion. The superscript indicates the length of deletion in kilobases of DNA. Both deletions are caused by unequal crossover points occurring at different locations. The reciprocal product of unequal crossover yields a triplicated α-globin gene. Individuals who have inherited single α-gene defect do not exhibit any hematological or clinical manifestations and the relative proportion of hemoglobins A, A_2, and F are normal.

α-Thalassemia-2 (α-/α-): Inheritance of two single-gene defects gives rise to clinically significant microcytosis, and the relative proportions of hemoglobins A, A_2, and F are normal.

α-Thalassemia-1 (--/$\alpha\alpha$): Two-gene deletions on the same chromosome are characterized by microcytic anemia. Different deletions have been observed. Southeast Asian (--^{SEA}) and Filipino (--^{Fil}) are examples of two-gene deletions occurring in the designated geographic areas. Deletion of a regulatory element (HS-40, see Chapter 28) also results in thalassemia syndrome.

Hemoglobin H disease: Compound heterozygosity for α-thalassemia-1 and α-thalassemia-2 deletions give rise to hemoglobin H (HbH) disease. These individuals exhibit hemolytic, microcytic anemia, which is not life threatening. HbH(β_4) inclusions are identified by supravital staining with brillian cresyl blue (discussed earlier) and as fast-moving hemoglobin in cellulose acetate, alkaline pH 8.4 electrophoresis technique (Figure VII-1).

Hemoglobin Bart's hydrops fetails: This occurs due to inheritance of two α-thalassemia-1 alleles and the offspring does not have any functional α genes. This condition is incompatible with life and results in stillborn or critically ill fetus with hydrops fetalis. Hemoglobin

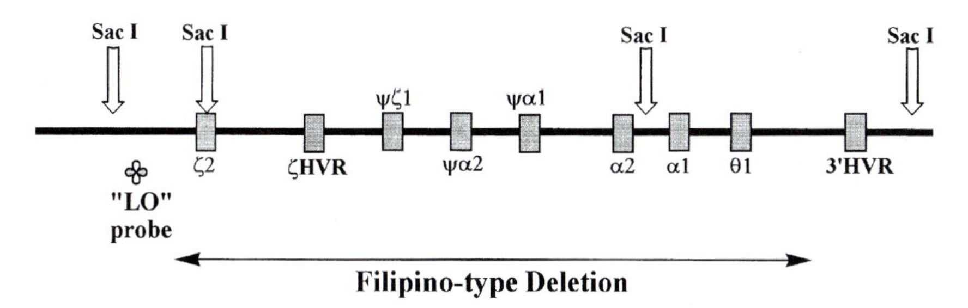

FIGURE VII-6
Restriction enzyme map of Filipino-type deletion in α-globin gene cluster.

Bart's is γ-tetramer and does not function as a normal hemoglobin.

Thalassemia syndromes also occur along with other mutations in the α-globin gene. One example is hemoglobin constant spring which is a result of mutation in the stop codon and gives rise to an abnormally large α-globin chain (Chapter 28). In rare cases, α-thalassemia is associated with mental retardation. An X-linked form of mental retardation, known as ATR-X syndrome, is a result of mutations in XH2. gene. The XH2 gene is a member of a subgroup of the family of genes that codes for proteins within many diverse functions (e.g., DNA helicase, DNA repair enzymes, putative global transcriptional regulatory proteins). It is thought that XH2 mutations result in down-regulation of α-globin gene expression.

Laboratory Diagnosis of α-Thalassemia Syndromes: DNA-based analysis is utilized for definitive testing in prospective parents for assessing the risk of transmission of α-thalassemia-1 genes and in prenatal diagnosis to identify Hb-Bart's hydrops fetalis. The methods may include Southern blot analysis and PCR methodology using appropriate α-globin gene cluster probes and primers, respectively. In Southern blot analysis, DNA obtained from leukocytes or fetal cells is subjected to digestion with specific restriction endonucleases, and the resulting fragments are separated by electrophoresis. Blotting is performed by alkaline transfer to nylon membranes and exposed to labeled (e.g., radioactive, fluorescent) DNA probes. The genotypes are established by using previously defined restriction fragment length patterns. Figure VII-5 shows the map of α-globin gene locus with restriction sites useful for diagnosis. In the Filipino α-thalassemia-1 ($\alpha\alpha$/--^{Fil}), due to large deletion involving the entire α-globin gene locus, digestion with restriction endonuclease, Bam H1, and Bg1II is not useful in identifying the deletion. Thus, Filipino α-thalassemia-1 is identified by using a probe ("LO Probe") derived from sequences of about 4 kb 5′ to the ζ-globin gene and digestion with the resctriction endonuclease SacI or SstI (Figure VII-6).

Supplemental Readings and References

B. J. Bain, R. J. Amos, D. Bareford, et al.: The laboratory diagnosis of haemoglobinopathies. *British Journal of Haematology* **101,** 783 (1998).

D. K. Bowden, M. A. Vickers, and D. R. Higgs: A PCR-based strategy to detect the common severe determinates of α-thalassaemia. *British Journal of Haematology* **81,** 104 (1992).

D. H. K. Chui, R. Hardison, C. Riemer, et al.: An electronic database of human hemoglobin variants on the world wide web. *Blood* **91,** 2643 (1998).

N. Fischel-Ghodsian, M. A. Vickers, M. Seip, et al.: Characterization of two deletions that remove the entire human ζ-α globin gene complex (--^{Thai} and --^{Fil}). *British Journal of Haematology* **70,** 238 (1988).

R. J. Gibbons, D. J. Picketts, L. Villard, et al.: Mutations in a putative global transcriptional regulator cause X-linked mental retardation with α-thalassemia (ATR-X syndrome). *Cell* **80,** 845 (1995).

C. H. Joiner: Universal newborn screening for hemoglobinopathies. *Journal of Pediatrics* **136,** 145 (2000).

K. J. Skogerboe, S. F. West, C. Smith, et al.: Screening for α-thalassemia. *Archives of Pathology and Laboratory Medicine* **116,** 1012 (1992).

APPENDIX VIII

TABLE VIII-1
Acronyms and Abbreviations Found in This Text

ABP	Androgen-binding protein
ACAT	Acyl-CoA:cholesterol acyltransferase
ACTH	Adrenocorticotropic hormone, corticotropin
ADH	Alcohol dehydrogenase
ADH	Antidiuretic hormone, vasopressin
ADP	Adenosine diphosphate
AFP	α_1-Fetoprotein
AIP	Aldosterone-induced protein
Ala	Alanine
ALA	Aminolevulinic acid
AMP	Adenosine monophosphate
Arg	Arginine
Asn	Asparagine
Asp	Aspartic acid
AST	Aspartate aminotransferase
ATP	Adenosine triphosphate
BDGF	Bone-derived growth factor
BMP	Bone morphogenic protein
BMR	Basal metabolic rate
BPG	Bisphosphoglycerate
CaBP	Calcium-binding protein
cAMP	3′,5′-Cyclic adenosine monophosphate, cyclic AMP
CBG	Corticosteroid-binding globulin
CCK	Cholecystokinin
cDNA	Complementary DNA
CDP	Cytidine diphosphate
cGMP	3′,5′-Cyclic guanosine monophosphate, cyclic GMP
CLIP	Corticotropin-like intermediate lobe polypeptide
CMP	Cytidine monophosphate
CoASH	Free (uncombined) coenzyme A
COMT	Catechol-O-methyltransferase
CPK	Creatine phosphokinase
CRBP	Cellular retinol-binding protein
CRH	Corticotropin-releasing hormone
CT	Calcitonin
CTP	Cytidine triphosphate
Cys	Cysteine
D	Dopamine receptor
D-	Geometric isomer of L form of chemical compound
DA	Dopamine
DCC	Dicyclohexylcarbodiimide
DES	Diethylstilbestrol
DHEA	Dehydroepiandrosterone
DHEAS	Dehydroepiandrosterone sulfate
DHT	Dihydrotestosterone
DIT	Diiodotyrosine
DNA	Deoxyribonucleic acid
DOC	Deoxycorticosterone
Dopa	3,4-Dihydroxyphenylalanine
DPF	Diisopropylphosphofluoridate
DPG	Diphosphoglycerate (bisphosphoglycerate)
DPN^+	Diphosphopyridine nucleotide (oxidized) (now replaced by NAD)
DPNH	Diphosphopyridine nucleotide (reduced) (now replaced by NADH)
dsRNA	Double-stranded RNA
E	Epinephrine
EC	Enzyme Commission number (International Union of Biochemistry) system
EDTA	Ethylenediaminetetraacetic acid
EFA	Essential fatty acid
EGF	Epidermal growth factor
ELISA	Enzyme-linked immunosorbent assay
emf	Electromotive force
Enz	Enzyme (also E)
FAD	Flavin adenine dinucleotide (oxidized)
$FADH_2$	Flavin adenine dinucleotide (reduced)
FFA	Free fatty acid
FH_4	Tetrahydrofolic acid
Figlu	Formiminoglutamatic acid
FMN	Flavin mononucleotide (oxidized)
$FMNH_2$	Flavin mononucleotide (reduced)
FSH	Follicle-stimulating hormone
FST	Foam stability test
GABA	γ-Aminobutyric acid
GDP	Guanosine diphosphate
GFR	Glomerular filtration rate
GH	Growth hormone
GIP	Gastric inhibitory polypeptide
Gla	γ-Carboxyglutamic acid
Gln	Glutamine
Glu	Glutamic acid
Gly	Glycine
GMP	Guanosine monophosphate
GnRH	Gonadotropin-releasing hormone
GRH	Growth hormone-releasing hormone
GTP	Guanosine triphosphate
HAL	Hepatic acylglycerol lipase
Hb	Hemoglobin
hCG	Human chorionic gonadotropin
HDL	High-density lipoprotein
5-HETE	Monohydroxyeicosatetraenoic acid
5-HIAA	5-Hydroxyindoleacetic acid
His	Histidine
HLA	Major histocompatibility (MHC) locus
HMG-CoA	β-Hydroxy-β-methylglutaryl-CoA
HMM	Heavy meromyosin
HMP	Hexose monophosphate

(continued)

TABLE VIII-1 *(continued)*

hnRNA	Heterogeneous nuclear RNA
hNT	Human N-terminal fragment
5-HPETE	5-Hydroperoxyeicosatetraenoic acid
hPL	Human placental lactogen
HSL	Hormone-sensitive lipase
I	Isoproterenol
IDL	Intermediate-density lipoprotein
IF	Intrinsic factor
IFN	Interferon
Ig	Immunoglobulin
IGF	Insulin-like growth factor
IL	Interleukin
Ile	Isoleucine
IMP	Inosine monophosphate
ITP	Inosine triphosphate
kat	Katal
kcal	Kilocalorie (calorie)
kJ	Kilojoule
K_m	Michaelis constant
L-	Geometric isomer of D form of chemical compound
LATS	Long-acting thyroid stimulator
LCAT	Lecithin-cholesterol acyltransferase
LDH	Lactate dehydrogenase
LDL	Low-density lipoprotein
Leu	Leucine
LH	Luteinizing hormone
LHRH	Luteinizing hormone-releasing hormone
LMM	Light meromyosin
LPH	Lipotropic hormone
LPL	Lipoprotein lipase
LT	Leukotriene
Lys	Lysine
M	Molar
MAC	Membrane attack complex
MAO	Monoamine oxidase
Mb	Myoglobin
MCR	Metabolic clearance rate
MDR	Minimum daily requirement
Met	Methionine
MIF	MSH release-inhibiting factor
MIF	Müllerian inhibiting factor
MIT	Monoiodotyrosine
MJ	Megajoule (1000 kJ)
MLCK	Myosin light chain kinase
MLCP	Myosin light-chain phosphatase
mol	Mole(s)
MRF	MSH-releasing factor
MRI	Magnetic resonance image
mRNA	Messenger RNA
MSAFP	Maternal serum AFP
MSH	Melanocyte-stimulating hormone
M.W.	Molecular weight
NAD^+	Nicotinamide adenine dinucleotide (oxidized); same as DPN
NADH	Nicotinamide adenine dinucleotide (reduced); same as DPNH
$NADP^+$	Nicotinamide adenine dinucleotide phosphate (oxidized); same as TPN
NADPH	Nicotinamide adenine dinucleotide phosphate (reduced); same as TPNH
NE	Norepinephrine
NGF	Nerve growth factor
NMR	Nuclear magnetic resonance
NSN	Nicotine-sensitive neurophysin
OHSDH	Hydroxysteroid dehydrogenase
PABA	ρ-Aminobenzoic acid
PAF	Platelet-activating factor
PAPS	Phosphoadenosine-phosphosulfate
PBPs	Penicillin-binding protein
PCR	Polymerase chain reaction
PDGF	Platelet-derived growth factor
PE	Phenylephrine
PEPCK	Phosphoenolpyruvate carboxykinase
PG	Prostaglandin
PGE	Prostaglandin E
PGI	Prostacyclin
Phe	Phenylalanine
P_i	Inorganic phosphate
pI	Isoelectric point
PIF	Prolactin-inhibiting factor
PIH	Prolactin-inhibiting hormone
PITC	Phenylisothiocyanate (Edman's reagent)
PKU	Phenylketonuria
PNMT	Phenylethanolamin-N-methyltransferase
POMC	Pro-opiomelanocortin
PP	Pancreatic polypeptide
PRF	Prolactin-releasing factor
PRL	Prolactin
Pro	Proline
PRPP	5-Phosphoribosyl-1-pyrophosphate
PTH	Parathyroid hormone
PUFA	Polyunsaturated fatty acid
RBP	Retinol-binding protein
RDA	Recommended daily (dietary) allowance
rDNA	Recombinant DNA
RFLP	Restriction fragment length polymorphism
RNA	Ribonucleic acid
RQ	Respiratory quotient

(continued)

TABLE VIII-1 *(continued)*

rRNA	Ribosomal RNA
SAM	S-Adenosylmethionine
SC	Secretory component
SDA	Specific dynamic action
SDS	Sodium dodecyl sulfate
Ser	Serine
SH	Sulfhydryl
SMP	Submitochondrial particle
snRNA	Small nuclear RNA
SRS-A	Slow-reacting substance of anaphylaxis
T^3	Triiodothyronine, thyroid hormone
T^4	Tetraiodothyronine, thyroxine
TBG	Thyroxine-binding globulin
t-BOC	Tertiary butyloxycarbonyl
TBPA	Thyroxine-binding prealbumin
TCA	Tricarboxylic acid
TCII	Transcobalamin
TeBG	Testosterone-estrogen-binding globulin, sex hormone-binding globulin
TeBP	Testosterone-estradiol-binding globulin
TETRAC	3,3′,5,5′-Tetraiodothyroacetic acid
TG	Thyroglobulin
Thr	Threonine
TMP	Thiamine monophosphate
TMP	Thymidine monophosphate
TPA	Tissue plasminogen activator
TPN^+	Triphosphopyridine nucleotide (oxidized) (now replaced by $NADP^+$)
TPNH	Triphosphopyridine nucleotide (reduced) (now replaced by NADPH)
TPP	Thiamine pyrophosphate
TRH	Thyrotropin-releasing hormone
TRIAC	3,3′,5-Triiodothyroacetic acid
TRIS	Tris(hydroxymethyl)aminomethane
tRNA	Transfer RNA
Trp	Tryptophan
TSH	Thyroid-stimulating hormone
TSI	Thyroid-stimulating immunoglobulin
TTP	Thymidine triphosphate
TX	Thromboxane
Tyr	Tyrosine
U	International Unit(s)
UDP	Urine diphosphate
UMP	Uridine monophosphate
UTP	Uridine triphosphate
V_{max}	Maximal velocity
Val	Valine
VIP	Vasoactive intestinal polypeptide
VLDL	Very low density lipoprotein
VMA	3-Methoxy-4-hydroxymandelic acid, vanillylmandelic acid
XMP	Xanthosine monophosphate

TABLE IX-1
Reference Intervals for Clinical Laboratory Parameters*

Component	Present Reference Interval (Examples)	Present Unit	Conversion Factor	SI Reference Interval	SI Unit Symbol	Significant Digits	Suggested Minimum Increment
Clinical Hematology							
Erythrocyte count (B)							
female	3.5–5.0	10^6/mm^3	1	3.5–5.0	10^{12}/L	X.X	0.1 10^{12}/L
male	4.3–5.9	10^6/mm^3	1	4.3–5.9	10^{12}/L	X.X	0.1 10^{12}/L
Erythrocyte count (Sf)	0	mm^{-3}	1	0	10^6/L	XX	1 10^6/L
Erythrocyte sedimentation rate (ESR) (BErc)							
female	0–30	mm/h	1	0–30	mm/h	XX	1 mm/h
male	0–20	mm/h	1	0–20	mm/h	XX	1 mm/h
Hamatocrit (BErcs) volume fraction							
female	33–43	%	0.01	0.33–0.43	1	0.XX	0.01
male	39–49	%	0.01	0.39–0.49	1	0.XX	0.01
Hemoglobin (B)							
mass concentration							
female	12.0–15.0	g/dL	10	120–150	g/L	XXX	1 g/L
male	13.6–17.2	g/dL	10	136–172	g/L	XXX	1 g/L
substance conc. Hb (Fe)							
female	12.0–15.0	g/dL	0.6206	7.45–9.30	mmol/L	XX.XX	0.05 mmol/L
male	13.6–17.2	g/dL	0.6206	8.45–10.65	mmol/L	XX.XX	0.05 mmol/L
Leukocyte count (B)	3200–9800	mm^{-3}	0.001	3.2–9.8	10^9/L	XX.X	0.1 10^9/L
number fraction ("differential")		%	0.01		1	0.XX	0.01
Leukocyte count (Sf)	0–5	mm^{-3}	1	0–5	10^6/L	XX	1 10^6/L
Mean corpuscular hemoglobin (MCH) (BErc)							
mass	27–33	pg	1	27–33	pg	XX	1 pg
amount of substance Hb (Fe)	27–33	pg	0.06206	1.70–2.05	fmol	X.XX	0.05 fmol
Mean corpuscular hemoglobin concentration (MCHC) (BErc)							
mass concentration	33–37	g/dL	10	330–370	g/L	XX0	10 g/L
substance conc. Hb (Fe)	33–37	g/dL	0.6206	20–23	mmol/L	XX	1 mmol/L
Mean corpuscular volume (MCV) (BErc)							
erythrocyte volume	76–100	μm^3	1	76–100	fL	XXX	1 fL
Platelet count (B)	130–400	10^3/mm^3	1	130–400	10^9/L	XXX	5 10^9/L
Reticulocyte count (B)—adults	10000–75000	mm^{-3}	0.001	10–75	10^9/L	XX	1 10^9/L
number fraction	1–24	0/00 (number per 1000 erythrocytes)	0.001	0.001–0.024	1	0.XXX	0.001
	0.1–2.4	%	0.01	0.001–0.024	1	0.XXX	0.001
Clinical Chemistry							
Acetaminophen (P)—toxic	>5.0	mg/dL	66.16	>330	μmol/L	XX0	10 μmol/L
Acetoacetate (S)	0.3–3.0	mg/dL	97.95	30–300	μmol/L	XX0	10 μmol/L
Acetone (B, S)	0	mg/dL	172.2	0	μmol/L	XX0	10 μmol/L
Acid phosphatase (S)	0–5.5	U/L	16.67	0–90	nkat/L	XX	2nkat/L
Adrenocorticotropin (ACTH) (P)	20–100	pg/ml	0.2202	4–22	pmol/L	XX	1 pmol/L
Alanine aminotransferase (ALT) (S)	0–35	U/L	0.01667	0–0.58	μkat/L	X.XX	0.02 μkat/L
Albumin (S)	4.0–6.0	g/dl	10.00	40–60	g/L	XX	1 g/L
Aldolase (S)	0–6	U/L	16.67	0–100	nkat/L	XX0	20 nkat/L
Aldosterone (S)							
normal salt diet	8.1–15.5	ng/dL	27.74	220–430	pmol/L	XX0	10 pmol/L
restricted salt diet	20.8–44.4	ng/dL	27.74	580–1240	pmol/L	XX0	10 pmol/L
Aldosterone (U)—sodium excretion							
= 25 mmol/d	18–85	μg/24 h	2.774	50–235	nmol/d	XXX	5 nmol/d
= 75–125 mmol/d	5–26	μg/24 h	2.774	15–70	nmol/d	XXX	5 nmol/d
= 200 mmol/d	1.5–12.5	μg/24 h	2.774	5–35	nmol/d	XXX	5 nmol/d
Alkaline phosphatase (S)	30–120	U/L	0.01667	0.5–2.0	μkat/L	X.X	0.1 μkat/L
Alpha$_1$-antitrypsin (S)	150–350	mg/dL	0.01	1.5–3.5	g/L	X.X	0.1 g/L
Alpha-fetoprotein (S)	0–20	ng/mL	1.00	0–20	μg/L	XX	1 μg/L

TABLE IX-1 *(continued)*

Component	Present Reference Interval (Examples)	Present Unit	Conversion Factor	SI Reference Interval	SI Unit Symbol	Significant Digits	Suggested Minimum Increment
Alpha-fetoprotein (Amf)	Depends on gestation	mg/dL	10	Depends on gestation	mg/L	XX	1 mg/L
Alpha$_2$-macroglobulin (S)	145–410	mg/dL	0.01	1.5–4.1	g/L	X.X	0.1 g/L
Aluminum (S)	0–15	µg/L	37.06	0–560	nmol/L	XX0	10 nmol/L
Amino acid fractionation (P)							
Alanine	2.2–4.5	mg/dL	112.2	245–500	µmol/L	XXX	5 µmol/L
Alpha-aminobutyric acid	0.1–0.2	mg/dL	96.97	10–2	µmol/L	XXX	5 µmol/L
Arginine	0.5–2.5	mg/dL	57.40	30–145	µmol/L	XXX	5 µmol/L
Asparagine	0.5–0.6	mg/dL	75.69	35–45	µmol/L	XXX	5 µmol/L
Aspartic acid	0.0–0.3	mg/dL	75.13	0–20	µmol/L	XXX	5 µmol/L
Citrulline	0.2–1.0	mg/dL	57.08	15–55	µmol/L	XXX	5 µmol/L
Cystine	0.2–2.2	mg/dL	41.61	10–90	µmol/L	XXX	5 µmol/L
Glutamic acid	0.2–2.8	mg/dL	67.97	15–190	µmol/L	XXX	5 µmol/L
Glutamine	6.1–10.2	mg/dL	68.42	420–700	µmol/L	XXX	5 µmol/L
Glycine	0.9–4.2	mg/dL	133.2	120–560	µmol/L	XXX	5 µmol/L
Histidine	0.5–1.7	mg/dL	64.45	30–110	µmol/L	XXX	5 µmol/L
Hydroxyproline	0–trace	mg/dL	76.26	0–trace	µmol/L	XXX	5 µmol/L
Isoleucine	0.5–1.3	mg/dL	76.24	40–100	µmol/L	XXX	5 µmol/L
Leucine	1.0–2.3	mg/dL	76.24	75–175	µmol/L	XXX	5 µmol/L
Lysine	1.2–3.5	mg/dL	68.40	80–240	µmol/L	XXX	5 µmol/L
Methionine	0.1–0.6	mg/dL	67.02	5–40	µmol/L	XXX	5 µmol/L
Ornithine	0.4–1.4	mg/dL	75.67	30–400	µmol/L	XXX	5 µmol/L
Phenylalanine	0.6–1.5	mg/dL	60.54	35–90	µmol/L	XXX	5 µmol/L
Proline	1.2–3.9	mg/dL	86.86	105–340	µmol/L	XXX	5 µmol/L
Serine	0.8–1.8	mg/dL	95.16	75–170	µmol/L	XXX	5 µmol/L
Taurine	0.3–2.1	mg/dL	79.91	25–170	µmol/L	XXX	5 µmol/L
Threonine	0.9–2.5	mg/dL	83.95	75–210	µmol/L	XXX	5 µmol/L
Tryptophan	0.5–2.5	mg/dL	48.97	25–125	µmol/L	XXX	5 µmol/L
Tyrosine	0.4–1.6	mg/dL	55.19	20–90	µmol/L	XXX	5 µmol/L
Valine	1.7–3.7	mg/dL	85.36	145–315	µmol/L	XXX	5 µmol/L
Amino acid nitrogen (P)	4.0–6.0	mg/dL	0.7139	2.9–4.3	mmol/L	X.X	0.1 mmol/L
Amino acid nitrogen (U)	50–200	mg/24 h	0.07139	3.6–14.3	mmol/d	X.X	0.1 mmol/d
Delta-aminolevulinate [as levulinic acid] (U)	1.0–7.0	mg/24 h	7.626	8–53	µmol/d	XX	1 µmol/d
Amitriptyline (P, S)—therapeutic	50–200	ng/mL	3.605	180–720	nmol/L	XX0	10 nmol/L
Ammonia (vP)							
as ammonia (NH_3)	10–80	µg/dL	0.5872	5–50	µmol/L	XXX	5 µmol/L
as ammonium ion (NH_4+)	10–85	µg/dL	0.5543	5–50	µmol/L	XXX	5 µmol/L
as nitrogen (N)	10–65	µg/dL	0.7139	5–50	µmol/L	XXX	5 µmol/L
Amylase (S)	0–130	U/L	0.01667	0–2.17	µkat/L	XXX	0.01 µkat/L
Androstenedione (S)							
male >18 years	0.2–3.0	µg/L	3.492	0.5–10.5	nmol/L	XX.X	0.5 nmol/L
female >18 years	0.8–3.0	µg/L	3.492	3.0–10.5	nmol/L	XX.X	0.5 nmol/L
Angiotensin converting enzyme (S)	<40	nmol/mL/min	16.67	<670	nkat/L	XX0	10 nkat/L
Arsenic (H) (as As)	<1	µg/g (ppm)	13.35	<13	nmol/g	XX.X	0.5 nmol/g
Arsenic (U) (as As)	0–5	µg/24 h	13.35	0–67	nmol/d	XX	1 nmol/d
(as As_2O_3)	<25	µg/dL	0.05055	<1.3	µmol/L	XX.X	0.1 µmol/L
Ascorbate (P) (as ascorbic acid)	0.6–2.0	mg/dL	56.78	30–110	µmol/L	X0	10 µmol/L
Aspartate aminotransferase (AST) (S)	0–35	U/L	0.01667	0–0.58	µkat/L	0.XX	0.01 µkat/L
Barbiturate (S)—overdose total expressed as:							
phenobarbital	Depends on composition of mixture. Usually not known.	mg/dL	43.06		µmol/L	XX	5 µmol/L
sodium phenobarbital		mg/dL	39.34		µmol/L	XX	5 µmol/L
barbitone		mg/dL	54.29		µmol/L	XX	5 µmol/L
Barbiturate (S)—therapeutic							
see phenobarbital							
see pentobarbital							
see thiopental							

TABLE IX-1 *(continued)*

Component	Present Reference Interval (Examples)	Present Unit	Conversion Factor	SI Reference Interval	SI Unit Symbol	Significant Digits	Suggested Minimum Increment
Bile acids, total (S) (as chenodeoxycholic acid)	Trace–3.3	μg/mL	2.547	Trace–8.4	μmol/L	X.X	0.2 μmol/L
cholic acid	Trace–1.0	μg/mL	2.448	Trace–2.4	μmol/L	X.X	0.2 μmol/L
chenodeoxycholic acid	Trace–1.3	μg/mL	2.547	Trace–3.4	μmol/L	X.X	0.2 μmol/L
deoxycholic acid	Trace–1.0	μg/mL	2.547	Trace–2.6	μmol/L	X.X	0.2 μmol/L
lithocholic acid	Trace	μg/mL	2.656	Trace	μmol/L	X.X	0.2 μmol/L
Bile acids (Df) (after cholecystokinin stimulation)							
total as chenodeoxycholic acid	14.0–58.0	mg/mL	2.547	35.0–148.0	mmol/L	XX.X	0.2 mmol/L
cholic acid	2.4–33.0	mg/mL	2.448	6.8–81.0	mmol/L	XX.X	0.2 mmol/L
chenodeoxycholic acid	4.0–24.0	mg/mL	2.547	10.0–61.4	mmol/L	XX.X	0.2 mmol/L
deoxycholic/acid	0.8–6.9	mg/mL	2.547	2.0–18.0	mmol/L	XX.X	0.2 mmol/L
lithocholic acid	0.3–0.8	mg/mL	2.656	0.8–2.0	mmol/L	XX.X	0.2 mmol/L
Bilirubin, total (S)	0.1–1.0	mg/dL	17.10	2–18	μmol/L	XX	2 μmol/L
Bilirubin, conjugated (S)	0–0.2	mg/dL	17.10	0–4	μmol/L	XX	2 μmol/L
Bromide (S)—toxic							
as bromide ion	>120	mg/dL	0.1252	>15	mmol/L	XX	1 mmol/L
as sodium bromide	>150	mg/dL	0.09719	>15	mmol/L	XX	1 mmol/L
	>15	mEq/L	1.00	>15	mmol/L	XX	1 mmol/L
Cadmium (S)	<3	μg/dL	0.08897	<0.3	μmol/L	X.X	0.1 μmol/L
Calcitonin (S)	<100	pg/mL	1.00	<100	ng/L	XXX	10 ng/L
Calcium (S)							
male	8.8–10.3	mg/dL	0.2495	2.20–2.58	mmol/L	X.XX	0.02 mmol/L
female <50 y	8.8–10.0	mg/dL	0.2495	2.20–2.50	mmol/L	X.XX	0.02 mmol/L
female >50 y	8.8–10.2	mg/dL	0.2495	2.20–2.56	mmol/L	X.XX	0.02 mmol/L
	4.4–5.1	mEq/L	0.5000	2.20–2.56	mmol/L	X.XX	0.02 mmol/L
Calcium ion (S)	2.00–2.30	mEq/L	0.5000	1.00–1.15	mmol/L	X.XX	0.01 mmol/L
	4.00–4.60	mg/dL	0.2495	1.00–1.15	mmol/L	X.XX	0.01 mmol/L
Calcium (U), normal diet	<250	mg/24 h	0.02495	<6.2	mmol/d	X.X	0.1 mmol/d
Carbamazepine (P)—therapeutic	4.0–10.0	mg/L	4.233	17–42	μmol/L	XX	1 μmol/L
Carbon dioxide content (B, P, S) (bicarbonate + CO_2)	22–28	mEq/L	1.00	22–28	mmol/L	XX	1 mmol/L
Carbon monoxide (B) (proportion of Hb that is COHb)	<15	%	0.01	<0.15	1	0.XX	0.01
Beta-carotenes (S)	50–250	μg/dL	0.01863	0.9–4.6	μmol/L	X.X	0.1 μmol/L
Catecholamines, total (U) (as norepinephrine)	<120	μg/24 h	5.911	<675	nmol/d	XX0	10 mg/d
Ceruloplasmin (S)	20–35	mg/dL	10.0	200–350	mg/L	XX0	10 mg/L
Chlordiazepoxide (P)—							
therapeutic	0.5–5.0	mg/L	3.336	2–17	μmol/L	XX	1 μmol/L
toxic	>10.0	mg/L	3.336	>33	μmol/L	XX	1 μmol/L
Chloride (S)	95–105	mEq/L	1.00	95–105	mmol/L	XXX	1mmol/L
Chlorimipramine (P) (includes desmethyl metabolite)	50–400	ng/mL	3.176	150–1270	nmol/L	XX0	10 nmol/L
Chlorpromazine (P)	50–300	ng/mL	3.136	150–950	nmol/L	XX0	10 nmol/L
Chlorpropamide (P)—therapeutic	75–250	mg/L	3.613	270–900	μmol/L	XX0	10 μmol/L
Cholestanol (P) (as a fraction of total cholesterol)	1–3	%	0.01	0.01–0.03	1	0.XX	0.01
Cholesterol (P)							
<29 years 30–39 years 40–49 years >50 years	<200	mg/dL	0.02586	<5.20	mmol/L	X.XX	0.05 mmol/L
Cholesterol esters (P) (as a fraction of total cholesterol)	60–75	%	0.01	0.60–0.75	1	0.XX	0.01
Cholinesterase (S)	620–1370	U/L	0.01667	10.3–22.8	μkat/L	XX.X	0.1 μkat/L
Chorionic gonadotrophin (P) (beta-hCG)	0 if not pregnant	mIU/mL	1.00	0 if not pregnant	IU/L	XX	1 IU/L
Citrate (B) (as citric acid)	1.2–3.0	mg/dL	52.05	60–160	μmol/L	XXX	5 μmol/L
Complement, C3 (S)	70–160	mg/dL	0.01	0.7–1.6	g/L	X.X	0.1 g/L
Complement, C4 (S)	20–40	mg/dL	0.01	0.2–0.4	g/L	X.X	0.1 g/L
Copper (S)	70–140	μg/dL	0.1574	11.0–22.0	μmol/L	XX.X	0.2 μmol/L
Copper (U)	<40	μg/24 h	0.01574	<0.6	μmol/d	X.X	0.2 mmol/d

TABLE IX-1 *(continued)*

Component	Present Reference Interval (Examples)	Present Unit	Conversion Factor	SI Reference Interval	SI Unit Symbol	Significant Digits	Suggested Minimum Increment
Coproporphyrins (U)	<200	μg/24 h	1.527	<300	nmol/d	XX0	10 nmol/d
Corticol (S)							
0800 h	4–19	μg/dL	27.59	110–520	nmol/L	XX0	10 nmol/L
1600 h	2–15	μg/dL	27.59	50–410	nmol/L	XX0	10 nmol/L
2400 h	5	μg/dL	27.59	140	nmol/L	XX0	10 nmol/L
Cortisol, free (U)	10–110	μg/24 h	2.759	30–300	nmol/d	XX0	10 nmol/d
Creatine (S)							
male	0.17–0.50	mg/dL	76.25	10–40	μmol/L	X0	10 μmol/L
female	0.35–0.93	mg/dL	76.25	30–70	μmol/L	X0	10 μmol/L
Creatine (U)							
male	0–40	mg/24 h	7.625	0–300	μmol/d	XX0	10 μmol/d
female	0–80	mg/24 h	7.625	0–600	μmol/d	XX0	10 μmol/d
Creatine kinase (CK) (S)	0–130	U/L	0.01667	0–2.16	μkat/L	X.XX	0.01 μkat/L
Creatine kinase isoenzymes (S)							
MB fraction	>5 in myocardial infarction	%	0.01	>0.05	1	0.XX	0.01
Creatinine (S)	0.6–1.2	mg/dL	88.40	50–110	μmol/L	XX0	10 μmol/L
Creatinine (U)	Variable	g/24 h	8.840	Variable	mmol/d	XX.X	0.1 mmol/d
Creatinine clearance (S, U)†	75–125	mL/min	0.01667	1.24–2.08	mL/s	X.XX	0.02 mL/s
Cyanide (B)—lethal	>0.10	mg/dL	384.3	>40	μmol/L	XXX	5 μmol/L
Cyanocobalamin (S) (vitamin B_{12})	200–1000	pg/mL	0.7378	150–750	pmol/L	XX0	10 pmol/L
Cyclic AMP (S)	2.6–6.6	μg/L	3.038	8–20	nmol/L	XXX	1nmol/L
Cyclic AMP (U)							
total urinary	2.9–5.6	μmol/g creat.	113.1	330–630	nmol/mmol creat.	XX0	10 nmol/mmol creatinine
renal tubular	<2.5	μmol/g creat.	113.1	<280	nmol/mmol creat.	XX0	10 nmol/mmol creatinine
Cyclic GMP (S)	0.6–3.5	μg/L	2.897	1.7–10.1	nmol/L	XX.X	0.1 nmol/L
Cyclic GMP (U)	0.3–1.8	μmol/g creat.	113.1	30–200	nmol/mmol creat.	XX0	10 nmol/mmol creatinine
Cystine (U)	10–100	mg/24 h	4.161	40–420	μmol/d	XX0	10 μmol/d
Dehydroepiandrosterone (P, S) (DHEA)							
1–4 y	0.2–0.4	μg/L	3.467	0.6–1.4	nmol/L	XX.X	0.2 nmol/L
4–8 y	0.1–1.9	μg/L	3.467	0.4–6.6	nmol/L	XX.X	0.2 nmol/L
8–10 y	0.2–2.9	μg/L	3.467	0.6–10.0	nmol/L	XX.X	0.2 nmol/L
10–12 y	0.5–9.2	μg/L	3.467	1.8–31.8	nmol/L	XX.X	0.2 nmol/L
12–14 y	0.9–20.0	μg/L	3.467	3.2–69.4	nmol/L	XX.X	0.2 nmol/L
14–16 y	2.5–20.0	g/L	3.467	8.6–69.4	nmol/L	XX.X	0.2 nmol/L
premenopausal female	2.0–15.0	g/L	3.467	7.0–52.0	nmol/L	XX.X	0.2 nmol/L
male	0.8–10.0	g/L	3.467	2.8–34.6	nmol/L	XX.X	0.2 nmol/L
Dehydroepiandrosterone (see steroid fractionation) (U)	—	—	—	—	—	—	—
Dehydroepiandrosterone sulfate (DHEA-S) (P, S)							
newborn	1670–3640	ng/mL	0.002714	4.5–9.9	μmol/L	XX.X	0.1 μmol/L
prepubertal children	100–600	ng/mL	0.002714	0.3–1.6	μmol/L	XX.X	0.1 μmol/L
male	2000–3350	ng/mL	0.002714	5.4–9.1	μmol/L	XX.X	0.1 μmol/L
female (premenopausal)	820–3380	ng/mL	0.002714	2.2–9.2	μmol/L	XX.X	0.1 μmol/L
female (post-menopausal)	110–610	ng/mL	0.002714	0.3–1.7	μmol/L	XX.X	0.1 μmol/L
pregnancy (term)	230–1,170	ng/mL	0.002714	0.6–3.2	μmol/L	XX.X	0.1 μmol/L
11-Deoxycortisol (S)	0–2	μg/dL	28.86	0–60	nmol/L	XX0	10 nmol/L
Desipramine (P)—therapeutic	50–200	ng/mL	3.754	170–700	nmol/L	XX0	10 nmol/L
Diazepam (P)—therapeutic	0.10–0.25	mg/L	3512	350–900	nmol/L	XX0	10 nmol/L
toxic	>1.0	mg/L	3512	>3510	nmol/L	XX0	10 nmol/L
Dicoumarol (P)—therapeutic	8–30	mg/L	2.974	25–90	μmol/L	XX	5 μmol/L
Digoxin (P)—therapeutic	0.5–2.2	ng/mL	1.281	0.6–2.8	nmol/L	X.X	0.1 nmol/L
	0.5–2.2	μg/L	1.281	0.6–2.8	nmol/L	X.X	0.1 nmol/L
toxic	>2.5	ng/mL	1.281	>3.2	nmol/L	X.X	0.1 nmol/L
Dimethadione (P)—therapeutic	<1.00	g/L	7.745	<7.7	mmol/L	X.X	0.1 mmol/L

TABLE IX-1 *(continued)*

Component	Present Reference Interval (Examples)	Present Unit	Conversion Factor	SI Reference Interval	SI Unit Symbol	Significant Digits	Suggested Minimum Increment
Disopyramide (P)—therapeutic	2.0–6.0	mg/L	2.946	6–18	μmol/L	XX	1μmol/L
Doxepin(P)—therapeutic	50–200	ng/mL	3.579	180–720	nmol/L	XX0	10nmol/L
Electrophoresis, protein (S)							
Albumin	60–65	%	0.01	0.60–0.65	1	0.XX	0.01
Alpha$_1$-globulin	1.7–5.0	%	0.01	0.02–0.05	1	0.XX	0.01
Alpha$_2$-globulin	6.7–12.5	%	0.01	0.07–0.13	1	0.XX	0.01
Beta-globulin	8.3–16.3	%	0.01	0.08–0.16	1	0.XX	0.01
Gamma-globulin	10.7–20.0	%	0.01	0.11–0.20	1	0.XX	0.01
Albumin	3.6–5.2	g/dL	10.0	36–52	g/L	XX	1 g/L
Alpha$_1$-globulin	0.1–0.4	g/dL	10.0	1–4	g/L	XX	1 g/L
Alpha$_2$-globulin	0.4–1.0	g/dL	10.0	4–10	g/L	XX	1 g/L
Beta-globulin	0.5–1.2	g/dL	10.0	5–12	g/L	XX	1 g/L
Gamma-globulin	0.6–1.6	g/dL	10.0	6–16	g/L	XX	1 g/L
Epinephrine (P)	31–95 (at rest for 15 min)	pg/mL	5.458	170–520	pmol/L	XX0	10 pmol/L
Epinephrine (U)	<10	μg/24 h	5.458	<55	nmol/d	XX	5 nmol/d
Estradiol (S)—male >18 y	15–40	pg/mL	3.671	55–150	pmol/L	XXX	1 pmol/L
Estriol (U) (nonpregnant)							
onset of menstruation	4–25	μg/24 h	3.468	15–85	nmol/d	XXX	5nmol/d
ovulation peak	28–99	μg/24 h	3.468	95–345	nmol/d	XXX	5 nmol/d
luteal peak	22–105	μg/24 h	3.468	75–365	nmol/d	XXX	5nmol/d
menopausal women	1.4–19.6	μg/24/h	3.468	5–70	nmol/d	XXX	5 nmol/d
male	5–18	μg/24 h	3.468	15–60	nmol/d	XXX	5 nmol/d
Estrogens (S) (as estradiol)							
female	20–300	pg/mL	3.671	70–1100	pmol/L	XXX0	10 pmol/L
peak production	200–800	pg/mL	3.671	750–2900	pmol/L	XXX0	10 pmol/L
male	<50	pg/mL	3.671	<180	pmol/L	XX0	10 pmol/L
Estrogens, placental (U) (as estriol)	Depends on period of gestation	mg/24 h	3.468	Depends on period of gestation	μmol/d	XXX	1 μmol/d
Estrogen receptors (T)							
negative	0–3	fmol estradiol bound/mg cytosol protein	1.00	0–3	fmol estradiol/mg cytosol protein	XXX	1 fmol/mg protein
doubtful	4–10	fmol estradiol bound/mg cytosol protein	1.00	4–10	fmol estradiol/mg cytosol protein	XXX	1 fmol/mg protein
positive	>10	fmol estradiol bound/mg cytosol protein	1.00	>10	fmol estradiol/mg cytosol protein	XXX	1 fmol/mg protein
Estrone (P, S)							
female 1–10 days of cycle	43–180	pg/mL	3.699	160–665	pmol/L	XXX	5 pmol/L
female 11–20 days of cycle	75–196	pg/mL	3.699	275–725	pmol/L	XXX	5 pmol/L
female 20–39 days of cycle	131–201	pg/mL	3.699	485–745	pmol/L	XXX	5 pmol/L
male	29–75	pg/mL	3.699	105–275	pmol/L	XXX	5 pmol/L
Estrone (U)—female	2–25	μg/24 h	3.699	5–90	nmol/d	XXX	5 nmol/d
Ethanol (P)							
legal limit (driving)	<80	mg/dL	0.2171	<17	mmol/L	XX	1 mmol/L
toxic	>100	mg/dL	0.2171	>22	mmol/L	XX	1 mmol/L
Ethchlorvynol (P)—toxic	>40	mg/L	6.915	>280	μmol/L	XX0	10 μmol/L
Ethosuximide (P)—therapeutic	40–100	mg/L	7.084	280–780	μmol/L	XX0	10 μmol/L
Ethylene glycol (P)—toxic	>30	mg/dL	0.1611	>5	mmol/L	XX	1 mmol/L
Fat (F) (as stearic acid)	2.0–6.0	g/24 h	3.515	7–21	mmol/L	XX	1 mmol/L
Fatty acids, nonesterified (P)	8–20	mg/dL	10.00	80–200	mg/L	XX0	10 mg/L
Ferritin (S)	18–300	ng/mL	1.00	18–300	μg/L	XX0	10 μg/L
Fibrinogen (P)	200–400	mg/dL	0.01	2.0–4.0	g/L	X.X	0.1 g/L
Fluoride (U)	<1.0	mg/24 h	52.63	<50	μmol/d	XX0	10 μmol/d
Folate (S) (as pteroylglutamic acid)	2–10	ng/mL	2.266	4–22	nmol/L	XX	2 nmol/L
		μg/dL	22.66		nmol/L		

TABLE IX-1 *(continued)*

Component	Present Reference Interval (Examples)	Present Unit	Conversion Factor	SI Reference Interval	SI Unit Symbol	Significant Digits	Suggested Minimum Increment
Folate (Erc)	140–960	ng/mL	2.266	550–2200	nmol/L	XX0	10 nmol/L
Follicle stimulating hormone (FSH) (P)							
female	2.0–15.0	mIU/mL	1.00	2–15	IU/L	XX	1 IU/L
peak production	20–50	mIU/mL	1.00	20–50	IU/L	XX	1 IU/L
male	1.0–10.0	mIU/mL	1.00	1–10	IU/L	XX	1 IU/L
Follicle stimulating hormone (FSH) (U)							
follicular phase	2–15	IU/24 h	1.00	2–15	IU/d	XXX	1 IU/d
midcycle	8–40	IU/24 h	1.00	8–40	IU/d	XXX	1 IU/d
luteal phase	2–10	IU/24 h	1.00	2–10	IU/d	XXX	1 IU/d
menopausal women	35–100	IU/24 h	1.00	35–100	IU/d	XXX	1 IU/d
male	2–15	IU/24 h	1.00	2–15	IU/d	XXX	1 IU/d
Fructose (P)	<10	mg/dL	0.05551	<0.6	mmol/L	X.XX	0.1 mmol/L
Galactose (P), children	<20	mg/dL	0.05551	<1.1	mmol/L	X.XX	0.1 mmol/L
Gases (Ab)							
pO_2	75–105	mmHg (= Torr)	0.1333	10.0–14.0	kPa	XX.X	0.1 kPa
pCO_2	33–44	mmHg (= Torr)	0.1333	4.4–5.9	kPa	X.X	0.1 kPa
Gamma-glutamyltransferase (GGT) (S)	0–30	U/L	0.01667	0–0.50	µkat/L	X.XX	0.01 µkat/L
Gastrin (S)	0–180	pg/mL	1	0–180	ng/L	XX0	10 ng/L
Globulins (S) (see immunoglobulins)							
Glucagon (S)	50–100	pg/mL	1	50–100	ng/L	XX0	10 ng/L
Glucose (P)—fasting	70–110	mg/dL	0.05551	3.9–6.1	mmol/L	XX.X	0.1 mmol/L
Glucose (Sf)	50–80	mg/dL	0.05551	2.8–4.4	mmol/L	XX.X	0.1 mmol/L
Glutethimide (P)							
therapeutic	<10	mg/L	4.603	<46	µmol/L	XX	1 µmol/L
toxic	>20	mg/L	4.603	>92	µmol/L	XX	1 µmol/L
Glycerol, free (S)	<1.5	mg/dL	0.1086	<0.16	mmol/L	X.XX	0.01 mmol/L
Gold (S)—therapeutic	300–800	µg/dL	0.05077	15.0–40.0	µmol/L	XX.X	0.1 µmol/L
Gold (U)	<500	µg/24 h	0.005077	<2.5	µmol/d	X.X	0.1 µmol/d
Growth hormone (P, S)							
male (fasting)	0.0–5.0	ng/mL	1.00	0.0–5.0	µg/L	XX.X	0.5 µg/L
female (fasting)	0.0–10.0	ng/mL	1.00	0.0–10.0	µg/L	XX.X	0.5 µg/L
Haptoglobin (S)	50–220	mg/dL	0.01	0.50–2.20	g/L	X.XX	0.01 g/L
Hemoglobin (B)							
male	14.0–18.0	g/dL	10.0	140–180	g/L	XXX	1 g/L
female	11.5–15.5	g/dL	10.0	115–155	g/L	XXX	1 g/L
Homogentisate (U) (as homogentistic acid)	0	mg/24 h	5.947	0	µmol/d	XX	5 µmol/d
Homovanillate (U) (as homovanillic acid)	<8	mg/24 h	5.489	<45	µmol/d	XX	5 µmol/d
Beta-hydroxybutyrate (S) (as beta-hydroxybutyric acid)	<1.0	mg/dL	96.05	<100	µmol/L	XX0	10 µmol/L
5-Hydroxyindoleacetate (U) as 5-hydroxyindole acetic acid; 5 HIAA)	2–8	mg/24 h	5.230	10–40	µmol/d	XXX	5 µmol/d
17-Alpha-hydroxyprogesterone (S, P)							
children	0.2–1.4	µg/L	3.026	0.5–4.5	nmol/L	XX.X	0.5 nmol/L
male	0.5–2.5	µg/L	3.026	1.5–7.5	nmol/L	XX.X	0.5 nmol/L
female	0.3–4.2	µg/L	3.026	1.0–13.0	nmol/L	XX.X	0.5 nmol/L
female, postmenopausal	0.3–1.7	µg/L	3.026	1.0–5.0	nmol/L	XX.X	0.5 nmol/L
Hydroxyproline (U)							
1 wk–1 y	55–220	mg/24 h/m^2	7.626	420–1680	µmol/(d·m^2)	XX0	10 µmol/(d·m^2)
1–13 y	25–80	mg/24 h/m^2	7.626	190–610	µmol/(d·m^2)	XX0	10 µmol/(d·m^2)
22-65 y	6–22	mg/24 h/m^2	7.626	40–170	µmol/(d·m^2)	XX0	10 µmol/(d·m^2)
>65 y	5–17	mg/24 h/m^2	7.626	40–130	µmol/(d·m^2)	XX0	10 µmol/(d·m^2)
Immunoglobulins (S)							
IgG	500–1200	mg/dL	0.01	5.00–12.00	g/L	XX.XX	0.01 g/L
IgA	50–350	mg/dL	0.01	0.50–3.50	g/L	XX.XX	0.01 g/L
Ig M	30–230	mg/dL	0.01	0.30–2.30	g/L	XX.XX	0.01 g/L

TABLE IX-1 (*continued*)

Component	Present Reference Interval (Examples)	Present Unit	Conversion Factor	SI Reference Interval	SI Unit Symbol	Significant Digits	Suggested Minimum Increment
Ig D	<6	mg/dL	10	<60	mg/L	XX0	10 mg/L
IgE							
0–3 y	0.5–10	IU/mL	2.4	1–24	μg/L	XX	1 μg/L
3–80 y	5–100	IU/mL	2.4	12–240	μg/L	XX	1 μg/L
Imipramine (P)—therapeutic	50–200	ng/mL	3.566	180–710	nmol/L	XX0	10 nmol/L
Insulin (P, S)	5–20	μU/mL	7.175	35–145	pmol/L	XXX	5 pmol/L
	5–20	mU/L	7.175	35–145	pmol/L	XXX	5 pmol/L
	0.20–0.84	μg/mL	172.2	35–145	pmol/L	XXX	5 pmol/L
Iron (S)							
male	80–180	μg/dL	0.1791	14–32	μmol/L	XX	1 μmol/L
female	60–160	μg/dL	0.1791	11–29	μmol/L	XX	1 μmol/L
Iron binding capacity (S)	250–460	μg/dL	0.1791	45–82	μmol/L	XX	1 μmol/L
Isoniazid (P)—							
therapeutic	<2.0	mg/L	7.291	<15	μmol/L	XX	1 μmol/L
toxic	<3.0	mg/L	7.291	>22	μmol/L	XX	1 μmol/L
Isopropanol (P)	0	mg/dL	0.1664	0	mmol/L	XX	1 mmol/L
Lactate (P) (as lactic acid)	0.5–2.0	mEq/L	1.00	0.5–2.0	mmol/L	X.X	0.1 mmol/L
	5–20	mg/dL	0.1110	0.5–2.0	mmol/L	X.X	0.1 mmol/L
Lactate dehydrogenase (S)	50–150	U/L	0.01667	0.82–2.66	μkat/L	X.XX	0.02 μkat/L
Lactate dehydrogenase isoenzymes (S)							
LD1	15–40	%	0.01	0.15–0.40	1	0.XX	0.01
LD2	20–45	%	0.01	0.20–0.45	1	0.XX	0.01
LD3	15–30	%	0.01	0.15–0.30	1	0.XX	0.01
LD4	5–20	%	0.01	0.05–0.20	1	0.XX	0.01
LD5	5–20	%	0.01	0.05–0.20	1	0.XX	0.01
LD1	10–60	U/L	0.01667	0.16–1.00	μkat/L	X.XX	0.02 μkat/L
LD2	20–70	U/L	0.01667	0.32–1.16	μkat/L	X.XX	0.02 μkat/L
LD3	10–45	U/L	0.01667	0.22–0.76	μkat/L	X.XX	0.02 μkat/L
LD4	5–30	U/L	0.01667	0.08–0.50	μkat/L	X.XX	0.02 μkat/L
LD5	5–30	U/L	0.01667	0.02–0.50	μkat/L	X.XX	0.02 μkat/L
Lead (B)—toxic	>60	μg/dL	0.04826	>2.90	μmol/L	X.XX	0.05 μmol/L
		mg/dL	48.26		μmol/L	X.XX	0.05 μmol/L
Lead (U)—toxic	>80	μg/24 h	0.004826	>0.40	μmol/d	X.XX	0.05 μmol/d
Lidocaine (P) (Xylocaine)	1.0–5.0	mg/L	4.267	4.5–21.5	μmol/L	X.X	0.5 μmol/L
Lipase (S)	0–160	U/L	0.01667	0–2.66	μkat/L	X.XX	0.02 μkat/L
Lipids, total (P)	400–850	mg/dL	0.01	4.0–8.5	g/L	X.XX	0.1 g/L
Lipoproteins (P)							
low-density (LDL)— as cholesterol	<100	mg/dL	0.02586	<2.60	mmol/L	X.XX	0.05 mmol/L
high-density (HDL)— as cholesterol							
Optimal	40	mg/dL	0.02586	1.03	mmol/L	X.XX	0.05 mmol/L
Cardioprotective	>60	mg/dL	0.02586	1.55	mmol/L	X.XX	0.05 mmol/L
Lithium ion (S)—therapeutic	0.50–1.50	mEq/L	1.00	0.50–1.50	mmol/L	X.XX	0.05 mmol/L
		μg/dL	0.001441		mmol/L	X.XX	0.05 mmol/L
		mg/dL	1.441		mmol/L	X.XX	0.05 mmol/L
Luteinizing hormone (S)							
male	3–25	mIU/mL	1.00	3–25	IU/L	XXX	1 IU/L
female	2–20	mIU/mL	1.00	2–20	IU/L	XXX	1 IU/L
peak production	30–140	mIU/mL	1.00	30–140	IU/L	XXX	1 IU/L
Lysozyme (S) (muramidase)	1–15	μg/mL	1.00	1–15	mg/L	XXX	1 mg/L
Lysozyme (U) (muramidase)	<2	μg/mL	1.00	<2	mg/L	XX	1 mg/L
Magnesium (S)	1.8–3.0	mg/dL	0.4114	0.80–1.20	mmol/L	X.XX	0.02 mmol/L
	1.6–2.4	mEq/L	0.5000	0.80–1.20	mmol/L	X.XX	0.02 mmol/L
Maprotiline (P)—therapeutic	50–200	ng/mL	3.605	180–720	nmo/L	XX0	10 nmol/L
Meprobamate (P)—							
therapeutic	<20	mg/L	4.582	<90	μmol/L	XX0	10 μmol/L
toxic	>40	mg/L	4.582	>180	μmol/L	XX0	10 μmol/L
Mercury (B)—							
normal	<1.0	μg/dL	49.85	<50	nmol/L	XX0	10 nmol/L
chronic exposure	>20	μg/dL	0.04985	>1.00	μmol/L	X.XX	0.01 μmol/L

TABLE IX-1 *(continued)*

Component	Present Reference Interval (Examples)	Present Unit	Conversion Factor	SI Reference Interval	SI Unit Symbol	Significant Digits	Suggested Minimum Increment
Mercury (U)—							
normal	<30	μg/24 h	4.985	<150	nmol/d	XX0	10 nmol/d
exposure—							
organic	>45	μg/24 h	4.985	>220	nmol/d	XX0	10 nmol/d
inorganic	>450	μg/24 h	0.004985	>2.20	μmol/d	X.XX	0.01 μmol/d
Metanephrines (U) (as normetanephrine)	0–2.0	mg/24 h	5.458	0–11.0	μmol/d	XX.X	0.5 μmol/d
Methanol (P)	0	mg/dL	0.3121	0	mmol/L	XX	1 mmol/L
Methaqualone (P)—							
therapeutic	<10	mg/L	3.995	<40	μmol/L	XX0	10 μmol/L
toxic	>30	mg/L	3.995	>120	μmol/L	XX0	10 μmol/L
Methotrexate (S)—toxic	>2.3	mg/L	2.200	>5.0	μmol/L	X.X	0.1 μmol/L
Methsuximide (P) (as desmethylsuximide)—therapeutic	10–40	mg/L	5.285	50–210	μmol/L	XX0	10 μmol/L
Methyprylon (P)—							
therapeutic	<10	mg/L	5.457	<50	μmol/L	XX0	10 μmol/L
toxic	>40	mg/L	5.457	>220	μmol/L	XX0	10 μmol/L
$Beta_2$-microglobulin (S)—<50 y	0.80–2.40	mg/L	84.75	68–204	nmol/L	XXX	2 nmol/L
$Beta_2$-microglobulin (U)—<50 y	<140	μg/24 h	0.08475	<12	nmol/d	XXX	2 nmol/L
Nitrogen, total (U)	Diet dependent	g/24 h	71.38	Diet dependent	mmol/d	XX0	10 mmol/d
Norepinephrine (P)	215–475 (at rest for 15 minutes)	pg/mL	0.005911	1.27–2.81	nmol/L	X.XX	0.01 nmol/L
Norepinephrine (U)	<100	μg/24 h	5.911	<590	nmol/d	XX0	10 nmol/d
Noretriptyline (P)—therapeutic	25–200	ng/mL	3.797	90–760	nmol/L	XX0	10 nmol/L
Osmolality (P)	280–300	mOsm/kg	1.00	280–300	mmol/kg	XXX	1 mmol/kg
Osmolality (U)	50–1200	mOsm/kg	1.00	50–1200	mmol/kg	XXX	1 mmol/kg
Oxalate (U) (as anhydrous oxalic acid)	10–40	mg/24 h	11.11	110–440	μmol/d	XX0	10 μmol/d
Palmitic acid (Amf)	Depends on gestation	mmol/L	1000	Depends on gestation	μmol/L	XXX	5 μmol/L
Phentobarbital (P)	20–40	mg/L	4.419	90–170	μmol/L	XX	5 μmol/L
Phenobarbital (P)—therapeutic	2–5	mg/dL	43.06	85–215	μmol/L	XXX	5 μmol/L
Phensuximide (P)	4–8	mg/L	5.285	20–40	μmol/L	XX	5 μmol/L
Phenylbutazone (P)—therapeutic	<100	mg/L	3.243	<320	μmol/L	XX0	10 μmol/L
Phenytoin (P)							
Therapeutic	10–20	mg/L	3.964	40–80	μmol/L	XX	5 μmol/L
Toxic	>30	mg/L	3.964	>120	μmol/L	XX	5 μmol/L
Phosphate (S) (as phosporus, inorganic)	2.5–5.0	mg/dL	0.3229	0.80–1.60	mmol/L	X.XX	0.05 mmol/L
Phosphate (U) (as phosphorus, inorganic)	Diet dependent	g/24 h	32.29	Diet dependent	mmol/d	XXX	1 mmol/d
Phospholipid phosphorus, total (P)	5–12	mg/dL	0.3229	1.60–3.90	mmol/L	X.XX	0.05 mmol/L
Phospholipid phosphorus, total (Erc)	1.2–12	mg/dL	0.3229	0.40–3.90	mmol/L	X.XX	0.05 mmol/L
Phospholipids (P)—substance fraction of total phospholipid							
Phosphatidyl choline	65–70	% of total	0.01	0.65–0.70	1	0.XX	0.01
Phsphatidyl ethanolamine	4–5	% of total	0.01	0.04–0.05	1	0.XX	0.01
Sphingomyelin	15–20	% of total	0.01	0.15–0.20	1	0.XX	0.01
Lysophosphatidyl choline	3–5	% of total	0.01	0.03–0.05	1	0.XX	0.01
Phospholipids (Erc)—substance fraction of total phospholipid							
Phosphatidyl choline	28–33	% of total	0.01	0.28–0.33	1	0.XX	0.01
Phosphatidyl ethanolamine	24–31	% of total	0.01	0.24–0.31	1	0.XX	0.01
Sphingomyelin	22–29	% of total	0.01	0.22–0.29	1	0.XX	0.01
Phosphatidyl serine + phosphatidyl inositol	12–20	% of total	0.01	0.12–0.20	1	0.XX	0.01

TABLE IX-1 *(continued)*

Component	Present Reference Interval (Examples)	Present Unit	Conversion Factor	SI Reference Interval	SI Unit Symbol	Significant Digits	Suggested Minimum Increment
Lysolphosphatidyl choline	1–2	% of total	0.01	0.01–0.02	1	0.XX	0.01
Phytanic acid (P)	Trace–0.3	mg/dL	32.00	<10	μmol/L	XX	5 μmol/L
(Human) placental lactogen (S) (HPL)	>4.0 After 30 wk gestation	μg/mL	46.30	>180	nmol/L	XX0	10 nmol/L
Phorphobilinogen (U)	0.0–2.0	mg/24 h	4.420	0–9.0	μmol/d	X.X	0.5 μmol/d
Porphyrins							
coproporphyrin (U)	45–180	μg/24 h	1.527	68–276	nmol/d	XXX	2 nmol/d
protoporphyrin (Erc)	15–50	μg/dL	0.0177	0.28–0.90	μmol/L	X.XX	0.02 μmol/L
uroporphyrin (U)	5–20	μg/24 h	1.204	6–24	nmol/d	XX	2 nmol/d
Uroporphyrinogen synthetase (Erc)	22–42	mmol/mL/h	0.2778	6.0–11.8	mmol/(L·s)	X.X	0.2 mmol/(L·s)
Potassium ion (S)	3.5–5.0	mEq/L	1.00	3.5–5.0	mmol/L	X.X	0.1 mmol/L
		mg/dL	0.2558		mmol/L	X.X	0.1 mmol/L
Potassium ion (U) (diet dependent)	25–100	mEq/24 h	1.00	25–100	mmol/d	XX	1 mmol/d
Pregnanediol (U)							
normal	1.0–6.0	mg/24 h	3.120	3.0–18.5	μmol/d	XX.X	0.5 μmol/d
pregnancy	Depends on gestation						
Pregnanetriol (U)	0.5–2.0	mg/24 h	2.972	1.5–6.0	μmol/d	XX.X	0.5 μmol/d
Primidone (P)—							
therapeutic	6.0–10.0	mg/L	4.582	25–46	μmol/L	XX	1 μmol/L
toxic	>10.0	mg/L	4.582	>46	μmol/L	XX	1 μmol/L
Procainamide (P)—							
therapeutic	4.0–8.0	mg/L	4.249	17–34	μmol/L	XX	1 μmol/L
toxic	>12.0	mg/L	4.249	>50	μmol/L	XX	1 μmol/L
N-acetylprocainamide (P)—therapeutic	4.0–8.0	mg/L	3.606	14–29	μmol/L	XX	1 μmol/L
Progesterone (P)							
follicular phase	<2	ng/mL	3.180	<6	nmol/L	XX	2 nmol/L
luteal phase	2–20	ng/mL	3.180	6–64	nmol/L	XX	2 nmol/L
Progesterone receptors (T)							
negative	0–3	fmol progesterone bound/mg cytosol protein	1.00	0–3	fmol progesterone bound/mg protein	XX	1 fmol/mg protein
doubtful	4–10	fmol progesterone bound/mg cytosol protein	1.00	4–10	fmol progesterone bound/mg cytosol protein	XX	1 fmol/mg protein
positive	>10	fmol progesterone bound/mg cytosol protein	1.00	>10	fmol progesterone bound/mg cytosol protein	XX	1 fmol/mg protein
Prolactin (P)	<20	ng/mL	1.00	<20	μg/L	XX	1 μg/L
Propoxyphene (P)—toxic	>2.0	mg/L	2.946	>5.9	μmol/L	X.X	0.1 μmol/L
Propranolol (P) (Inderal)—therapeutic	50–200	ng/mL	3.856	190–770	nmol/L	XX0	10 nmol/L
Protein, total (S)	6.0–8.0	g/dL	10.0	60–80	g/L	XX	1 g/L
Protein, total (Sf)	<40	mg/dL	0.01	<0.40	g/L	X.XX	0.01 g/L
Protein, total (U)	<150	mg/24 h	0.001	<0.15	g/d	X.XX	0.01 g/d
Protryptyline (P)	100–300	ng/mL	3.797	380–1140	nmol/L	XX0	10 nmol/L
Pyruvate (B) (as pyruvic acid)	0.30–0.90	mg/dL	113.6	35–100	μmol/L	XXX	1 μmol/L
Quinidine (P)—							
therapeutic	1.5–3.0	mg/L	3.082	4.6–9.2	μmol/L	X.X	0.1 μmol/L
toxic	>6.0	mg/L	3.082	>18.5	μmol/L	X.X	0.1 μmol/L
Renin (P)							
normal sodium diet	1.1–4.1	ng/mL/h	0.2778	0.30–1.14	ng/(L·s)	X.XX	0.02 ng/(L·s)
restricted sodium diet	6.2–12.4	ng/mL/h	0.2778	1.72–3.44	ng/(L·s)	X.XX	0.02 ng/(L·s)

TABLE IX-1 *(continued)*

Component	Present Reference Interval (Examples)	Present Unit	Conversion Factor	SI Reference Interval	SI Unit Symbol	Significant Digits	Suggested Minimum Increment
Salicylate (S) (salicylic acid)—toxic	>20	mg/dL	0.07240	>1.45	mmol/L	X.XX	0.05 mmol/L
Serotonin (B) (5-hydroxy-tryptamine)	8–21	μg/dL	0.05675	0.45–1.20	μmol/L	X.XX	0.05 μmol/L
Sodium ion (S)	135–147	mEq/L	1.00	135–147	mmol/L	XXX	1 mmol/L
Sodium ion (U)	Diet dependent	mEq/24 h	1.00	Diet dependent	mmol/d	XXX	1 mmol/d
Steriods							
17-hydroxycortico-steroids (U) (as corticol)							
female	2.0–8.0	mg/24 h	2.759	5–25	μmol/d	XX	1 μmol/d
male	3.0–10.0	mg/24 h	2.759	10–30	μmol/d	XX	1 μmol/d
17-ketogenic steroids (U) (as dehydroepian-drosterone)							
female	7.0–12.0	mg/24 h	3.467	25–40	μmol/d	XX	1 μmol/d
male	9.0–17.0	mg/24 h	3.467	30–60	μmol/d	XX	1 μmol/d
17-Ketosteroids (U) (as dehydroepian-drosterone)							
female	6.0–17.0	mg/24 h	3.467	20–60	μmol/d	XX	1 μmol/d
male	6.0–20.0	mg/24 h	3.467	20–70	μmol/d	XX	1 μmol/d
Ketosteroid fractions (U)							
androsterone							
female	0.5–3.0	mg/24 h	3.443	1–10	μmol/d	XX	1 μmol/d
male	2.0–5.0	mg/24 h	3.443	7–17	μmol/d	XX	1 μmol/d
dehydroepiandrosterone							
female	0.2–1.8	mg/24 h	3.467	1–6	μmol/d	XX	1 μmol/d
male	0.2–2.0	mg/24 h	3.467	1–7	μmol/d	XX	1 μmol/d
etiocholanolone							
female	0.8–4.0	mg/24 h	3.443	2–14	μmol/d	XX	1 μmol/d
male	1.4–5.0	mg/24 h	3.443	4–17	μmol/d	XX	1 μmol/d
Sulfonamides (B) (as sulfanilamidea)—therapeutic	10.0–15.0	mg/dL	58.07	580–870	μmol/L	XX0	10 μmol/L
Testosterone (P)							
female	0.6	ng/mL	3.467	2.0	nmol/L	XX.X	0.5 nmol/L
male	4.6–8.0	ng/mL	3.467	14.0–28.0	nmol/L	XX.X	0.5 nmol/L
Theophylline (P)—therapeutic	10.0–20.0	mg/L	5.550	55–110	μmol/L	XX	1 μmol/L
Thiocyanoate (P) (nitroprusside toxicity)	10.0	mg/dL	0.1722	1.7	mmol/L	X.XX	0.1 mmol/L
Thiopental (P)	Individual	mg/L	4.126	Individual	μmol/L	XX	5 μmol/L
Thyroid stimulating hormone (TSH) (S)	2–11	μU/mL	1.00	2–11	mU/L	XX	1 mU/L
Thyroxine (T_4) (S)	4.0–11.0	μg/dL	12.87	51–142	nmol/L	XXX	1 nmol/L
Thyroxine binding globulin (TBG) (S) (as thyroxine)	12.0–28.0	μg/dL	12.87	150–360	nmol/L	XX0	1 nmol/L
Thyroxine, free (S)	0.8–2.8	ng/dL	12.87	10–36	pmol/L	XX	1 pmol/L
Triiodothyronine T_3) (S)	75–220	ng/dL	0.01536	1.2–3.4	nmol/L	X.X	0.1 nmol/L
T_3 uptake (S)	25–35	%	0.01	0.25–0.35	1	0.XX	0.01
Tolbutamide (P)—therapeutic	50–120	mg/L	3.699	180–450	μmol/L	XX0	10 μmol/L
Transferrin (S)	170–370	mg/dL	0.01	1.70–3.70	g/L	X.XX	0.01 g/L
Triglycerides (P) (as triolein)	<160	mg/dL	0.01129	<1.80	mmol/L	X.XX	0.02 mmol/L
Trimethadione (P)—therapeutic	<50	mg/L	6.986	<350	μmol/L	XX0	10 μmol/L
Trimipramine (P)—therapeutic	50–200	ng/mL	3.397	170–680	nmol/L	XX0	10 nmol/L
Urate (S) (as uric acid)	2.0–7.0	mg/dL	59.48	120–420	μmol/L	XX0	10 μmol/L
Urate (U) (as uric acid)	Diet dependent	g/24 h	5.948	Diet dependent	mmol/d	XX	1 mmol/d
Urea nitrogen (S)	8–18	mg/dL	0.3570	3.0–6.5	mmol/L urea	X.X	0.5 mmol/L

TABLE IX-1 *(continued)*

Component	Present Reference Interval (Examples)	Present Unit	Conversion Factor	SI Reference Interval	SI Unit Symbol	Significant Digits	Suggested Minimum Increment
Urea nitrogen (U)	12.0–20.0 Diet dependent	g/24 h	35.70	450–700	mmol/d urea	XX0	10 mol/d
Urobilinogen (U)	0.0–4.0	mg/24 h	1.693	0.0–6.8	μmol/d	X.X	0.1 μmol/d
Valproic acid (P)—therapeutic	50–100	mg/L	6.934	350–700	μmol/L	XX0	10 μmol/L
Vanillylmandelic acid (VMA) (U)‡	<6.8	mg/24 h	5.046	<35	μmol/d	XX	1 μmol/d
Vitamin A (retinol) (P, S)	10–50	μg/dL	0.03491	0.35–1.75	μmol/L	X.XX	0.05 μmol/L
Vitamin B_1 (thiamine hydrochloride) (U)	60–500	μg/24 h	0.002965	0.18–1.48	μmol/d	X.XX	0.01 μmol/d
Vitamin B_2 (riboflavin) (S)	2.6–3.7	μg/dL	26.57	70–100	nmol/L	XXX	5 nmol/L
Vitamin B_6 (pyridoxal) (B)	20–90	ng/mL	5.982	120–540	nmol/L	XXX	5 nmol/L
Vitamin B_{12} (cyanocobalamin) (P, S)	200–1000	pg/mL	0.7378	150–750	pmol/L	XX0	10 pmol/L
		ng/dL	7.378		pmol/L		
Vitamin C (see ascorbate) (B, P, S)	—	—	—	—	—	—	—
Vitamin D_3 (cholecalciferol) (P)	24–40	μg/mL	2.599	60–105	nmol/L	XXX	5 nmol/L
25 OH-cholecalciferol	18–36	ng/mL	2.496	45–90	nmol/L	XXX	5 mmol/L
Vitamin E (alpha-tocopherol) (P, S)	0.78–1.25	mg/dL	23.22	18–29	μmol/L	XX	1 μmol/L
Warfarin (P)—therapeutic	1.0–3.0	mg/L	3.243	3.3–9.8	μmol/L	XX.X	0.1 μmol/L
Xanthine (U)—hypoxanthine	5–30	mg/24 h	6.574	30–200	μmol/d	XX0	10 μmol/d
		mg/24 h	7.347		μmol/d	XX0	10 μmol/d
D-xylose (B) (25-g dose)	30–40 (30–60 min)	mg/dL	0.06661	2.0–2.7 (30–60 min)	mmol/L	X.X	0.1 mmol/L
D-xylose excretion (U) (25-g dose)	21–31 (excreted in 5 h)	%	0.01	0.21–0.31 (excreted in 5 h)	1	0.XX	0.01
Zinc (S)	75–120	μg/dL	0.1530	11.5–18.5	μmol/L	XX.X	0.1 μmol/L
Zinc (U)	150–1200	μg/24 h	0.01530	2.3–18.3	μmol/d	XX.X	0.1 μmol/d

*Source: *American Journal of Clinical Pathology* **87,** 140 (1987).

†Creatinine clearance (corrected for body surface area) = $\left(\frac{\mu\text{mol/L (urine creatinine)}}{\mu\text{mol/L (serum creatinine)}} \times \text{mL/s} \times \frac{1.73}{A}\right)$ (where A is the body surface area in square metres [m^2])

‡This is a misnomer, but because of its popularity the name VMA has been retained in this publication. In many publications it is being referred to as 4-hydroxy-3-methoxy mandelic acid.

INDEX

A

B

C

D

F

G

H

I

J

K

M

N

O

P

S

T

W

X

Y

Z